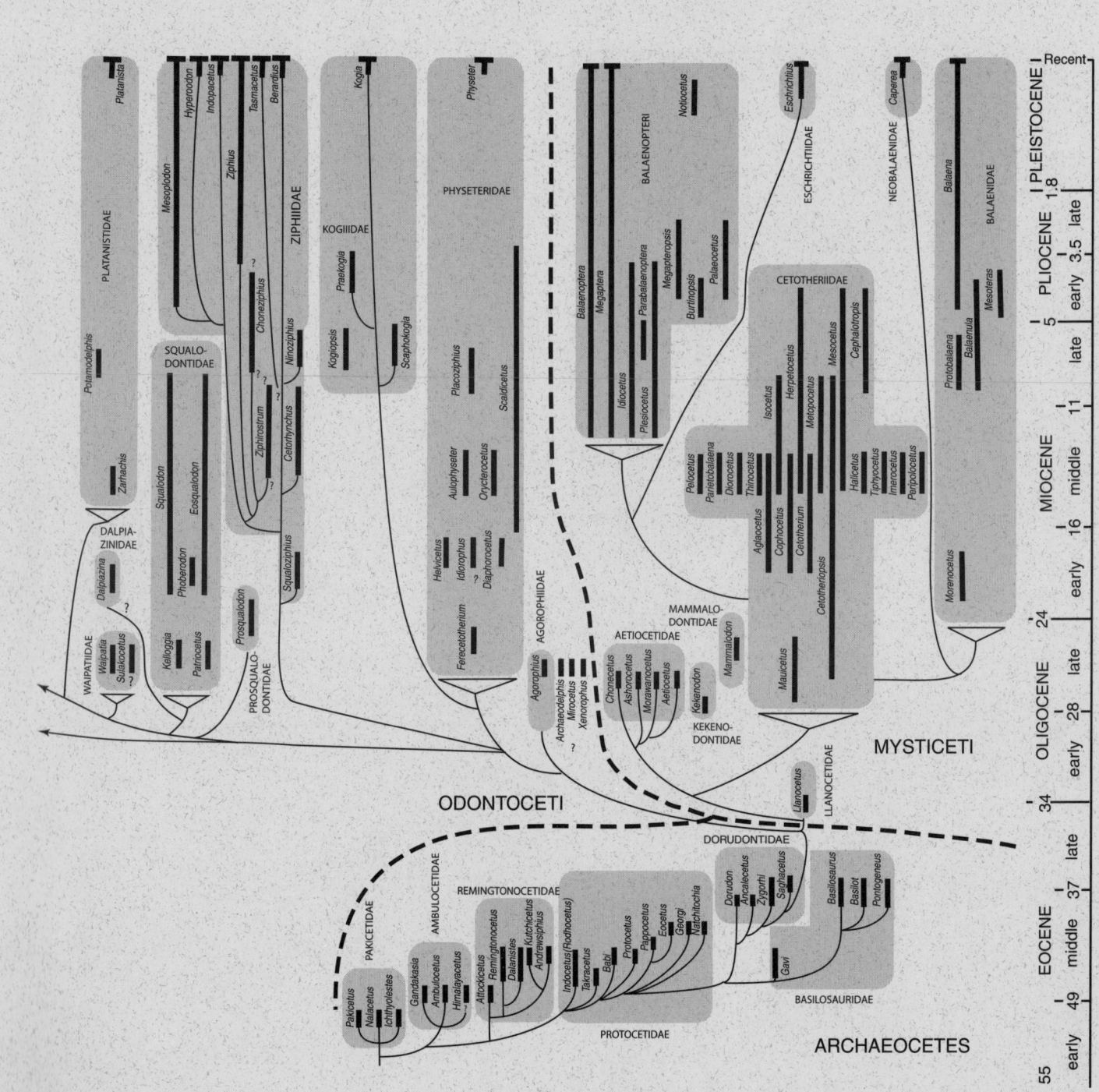

Encyclopedia of

MARINE

MAMMALS

Encyclopedia of
MARINE
MAMMALS

Editors

William F. Perrin
Southwest Fisheries Science Center, NOAA
La Jolla, California

Bernd Würsig
Texas A&M University
Galveston, Texas

J. G. M. Thewissen
Northeastern Ohio Universities College of Medicine
Rootstown, Ohio

ACADEMIC PRESS
A Division of Harcourt, Inc.

San Diego   San Francisco   New York   Boston   London   Sydney   Tokyo

Academic Press
A Division of Harcourt, Inc.
525 B Street, Suite 1900, San Diego, California 92101-4495, USA
http://www.academicpress.com

Academic Press
Harcourt Place, 32 Jamestown Road, London NW1 7BY, UK
http://www.academicpress.com

Library of Congress Catalog Card Number: 2001087128

International Standard Book Number: 0-12-551340-2

PRINTED IN THE UNITED STATES OF AMERICA
01 02 03 04 05 06 MB 9 8 7 6 5 4 3 2 1

This work is dedicated to Kenneth S. Norris,
great naturalist and first president of the
Society for Marine Mammalogy.
His passing in 1998 left a gap in all our lives.

CONTENTS

T

U

V

W

CONTENTS BY SUBJECT AREA

Human Effects and Interactions

Organisms and Faunas

RESEARCH METHODOLOGY

CONTRIBUTORS

Alejandro Acevedo-Gutiérrez
University of California, Santa Cruz
Santa Cruz, California, USA
Group Behavior

Peter J. Adam
University of California, Los Angeles
Los Angeles, California, USA
Pelvic Anatomy

Alex Aguilar
University of Barcelona
Barcelona, Spain
Fin Whale (Balaenoptera physalus)
Pollution and Marine Mammals

Masao Amano
Otsuchi Marine Research Center, University of Tokyo
Otsuchi, Japan
Finless Porpoise (Neophocaena phocaenoides)

Paul K. Anderson
University of Calgary
Calgary, Alberta, Canada
Steller's Sea Cow (Hydrodamalis gigas)

George A. Antonelis
National Marine Fisheries Service
Honolulu, Hawaii, USA
Rookeries

Frederick I. Archer II
Southwest Fisheries Science Center
La Jolla, California, USA
Striped Dolphin (Stenella coeruleoalba)

Peter W. Arnold
Museum of Tropical Queensland
Queensland, Townsville, Australia
Irrawaddy Dolphin (Orcaella brevirostris)

John P. Y. Arnould
Macquarie University
Sydney, Australia
Southern Fur Seals (Arctocephalus spp.)

Shannon Atkinson
University of Alaska
Seward, Alaska, USA
Male Reproductive Systems

Whitlow W. L. Au
University of Hawaii
Kaneohe, Hawaii, USA
Echolocation

Deborah A. Austin
Dalhousie University
Halifax, Nova Scotia, Canada
Pinniped Ecology

F. Javier Aznar
University of Valencia, Spain
Valencia, Spain
Parasites

Robin W. Baird
Dalhousie University
Halifax, Nova Scotia, Canada
False Killer Whale (Pseudorca crassidens)
Risso's Dolphin (Grampus griseus)

C. Scott Baker
University of Auckland
Auckland, New Zealand
Forensic Genetics
Whaling, Modern

Juan A. Balbuena
University of Valencia, Spain
Valencia, Spain
Parasites

Lisa T. Ballance
Southwest Fisheries Science Center
La Jolla, California, USA
Cetacean Ecology

John L. Bannister
Western Australian Museum

Perth, Australia
Baleen Whales (Mysticetes)

Jay Barlow
Southwest Fisheries Science Center
La Jolla, California, USA
Management
Population Status and Trends

Lawrence G. Barnes
Natural History Museum of Los Angeles County
Los Angeles, California, USA
Cetacea, Overview
Delphinoids, Evolution of the Modern Families

Nélio Barros
Mote Marine Laboratory
Sarasota, Florida, USA
Diet

Carrie A. Beck
Dalhousie University
Halifax, Nova Scotia, Canada
Pinniped Ecology

Marc Bekoff
University of Colorado, Boulder
Boulder, Colorado, USA
Ethics and Marine Mammals

John L. Bengtson
National Marine Mammal Laboratory
Seattle, Washington, USA
Crabeater Seal (Lobodon carcinophaga)

Annalisa Berta
San Diego State University
San Diego, California, USA
Pinnipedia, Overview
Pinniped Evolution
Systematics, Overview

Martine Bérubé
University of California, Berkeley
Berkeley, California, USA
Hybridism

Carol A. Beuchat
University of Arizona, Tucson
Tucson, Arizona, USA
Kidney, Structure and Function

Arne Bjørge
Institute of Marine Research
Bergen, Norway
Harbor Porpoise (Phocoena phocoena)

James L. Bodkin
U.S. Geological Survey

Santa Cruz, California, USA
Otters

Meghan E. Bolen
Florida Marine Research Institute
St. Petersburg, Florida, USA
Anatomical Dissection: Thorax and Abdomen

Daryl J. Boness
National Zoological Park, Smithsonian Institution
Washington, DC, USA
Estrus and Estrous Behavior
Sea Lions, Overview

W. Don Bowen
Department of Fisheries and Oceans Canada
Dartmouth, Nova Scotia, Canada
Pinniped Ecology

Ian L. Boyd
University of St. Andrews
St. Andrews, Scotland, U.K.
Antarctic Marine Mammals
Pinniped Life History

Jeffrey M. Breiwick
National Marine Mammal Laboratory
Seattle, Washington, USA
Stock Assessment

Robert L. Brownell, Jr.
Southwest Fisheries Science Center
La Jolla, California, USA
Illegal and Pirate Whaling
*Minke Whales (Balaenoptera acutorostrata and
 B. bonaerensis)*

Emily A. Buchholtz
Wellesley College
Wellesley, Massachusetts, USA
Convergent Evolution

Stephen T. Buckland
University of St. Andrews
St. Andrews, Scotland, U.K.
Abundance Estimation

John J. Burns
Fairbanks, Alaska, USA
Arctic Marine Mammals
*Harbor Seal and Spotted Seal (Phoca vitulina and
 P. largha)*

Douglas S. Butterworth
University of Cape Town
Cape Town, South Africa
Competition with Fisheries

Claudio Campagna
Centro Nacional Patagónico
Puerto Madryn, Argentina
Aggressive Behavior, Intraspecific
Infanticide and Abuse of Young

Humberto Luis Cappozzo
Museo Argentino de Ciencias Naturales
Buenos Aires, Argentina
South American Sea Lion (Otaria flavescens)

Michael Castellini
University of Alaska, Fairbanks
Fairbanks, Alaska, USA
Thermoregulation

Susan J. Chivers
Southwest Fisheries Science Center
La Jolla, California, USA
Cetacean Life History

Frank Cipriano
San Francisco State University
San Francisco, California, USA
Atlantic White-Sided Dolphin (Lagenorhynchus acutus)
Evolutionary Biology, Overview

Phillip J. Clapham
Northeast Fisheries Science Center
Woods Hole, Massachusetts
Humpback Whale (Megaptera novaeangliae)
Whaling, Modern

Malcolm R. Clarke
Cornwall, U.K.
Diet

Rochelle Constantine
University of Auckland
Auckland, New Zealand
Folklore and Legends

Peter Corkeron
James Cook University
Townsville, Queensland, Australia
Captivity

Daniel P. Costa
University of California, Santa Cruz
Santa Cruz, California, USA
Energetics
Osmoregulation
Pinniped Physiology

Daniel F. Cowan
University of Texas Medical Branch
Galveston, Texas, USA
Pathology

Enrique A. Crespo
Centro Nacional Patagónico
Puerto Madryn, Argentina
Franciscana (Pontoporia blainvillei)
South American Aquatic Mammals

Daniel E. Crocker
Sonoma State University
Rohnert Park, California, USA
Pinniped Physiology

Donald A. Croll
University of California, Santa Cruz
Santa Cruz, California, USA
Filter Feeding

Jim Darling
West Coast Whale Research Foundation
Tofino, British Columbia, Canada
Song

Stephen M. Dawson
University of Otago
Dunedin, New Zealand
Cephalorhynchus Dolphins (Cephalorhynchus spp.)

Susan D. Dawson
University of Prince Edward Island
Charlottetown, Prince Edward Island, Canada
Kentriodontidae

Douglas P. DeMaster
National Marine Fisheries Service
Seattle, Washington, USA
Endangered Species and Populations

Lawrence M. Dill
Simon Fraser University
Burnaby, British Columbia, Canada
Feeding Strategies and Tactics

Andrew E. Dizon
Southwest Fisheries Science Center
La Jolla, California, USA
Genetics for Management

M. Louella L. Dolar
Tropical Marine Research
San Diego, California, USA
Fraser's Dolphin (Lagenodelphis hosei)

Daryl P. Domning
Howard University
Washington, DC, USA
Desmostylia
Sirenian Evolution
Steller's Sea Cow (Hydrodamalis gigas)

Meghan A. Donahue
Southwest Fisheries Science Center
La Jolla, California, USA
Pygmy Killer Whale (Feresa attenuata)

Gregory P. Donovan
International Whaling Commission
Cambridge, U.K.
International Whaling Commission (IWC)

Alton C. Dooley, Jr.
Virginia Museum of Natural History
Martinsville, Virginia, USA
Baleen Whales, Archaic

Etienne Douaze
Singapore
Communication

Kathleen A. Dudzinski
Dolphin Communication Project
Oxnard, California, USA
Communication

Deborah A. Duffield
Portland State University
Portland, Oregon, USA
Extinctions, Specific

Richard Ellis
American Museum of Natural History
New York, New York, USA
Whaling, Early and Aboriginal
Whaling, Traditional

Robert Elsner
University of Alaska, Fairbanks
Fairbanks, Alaska, USA
Cetacean Physiology, Overview

Sergio Escorza-Treviño
Southwest Fisheries Science Center
La Jolla, California, USA
North Pacific Marine Mammals

James A. Estes
U.S. Geological Survey
Santa Cruz, California, USA
Otters

Peter G. H. Evans
University of Oxford
Oxford, U.K.
Habitat Pressures

Michael A. Fedak
University of St. Andrews
St. Andrews, Scotland, U.K.
Reproductive Behavior

Gennadiy Fedoseev
Marine Mammals Council
Usman, Russia
Ribbon Seal (Histriophoca fasciata)

Mercedes Fernández
University of Valencia
University of Valencia, Spain
Parasites

Dagmar Fertl
Minerals Management Service, USDOI
New Orleans, Louisiana, USA
Albinism
Barnacles
Fisheries, Interference with
Remoras

Paul C. Fiedler
Southwest Fisheries Science Center
La Jolla, California, USA
Ocean Environment

Frank E. Fish
West Chester University
West Chester, Pennsylvania, USA
Speed
Streamlining

Warren Fitch
University of Calgary
Calgary, Alberta, Canada
Lutrinae

Paulo A. C. Flores
PUCRS (Pontifícia Universidade Católico do Rio
    Grande do Sol), Porto Alegre, Brazil
Tucuxi (Sotalia fluviatilis)

Jaume Forcada
Southwest Fisheries Science Center
La Jolla, California, USA
Distribution
*Monk Seals (Monachus monachus, M. tropicalis, and
    M. schauinslandi)*

John K. B. Ford
Fisheries and Oceans Canada
Nanaimo, British Columbia, Canada
Dialects
Killer Whale (Orcinus orca)

R. Ewan Fordyce
University of Otago
Dunedin, New Zealand
Cetacean Evolution
Fossil Record
Fossil Sites
Neoceti

Paul H. Forestell
Southampton College, Long Island University
Southampton, New York, USA
Popular Culture and Literature

Karin A. Forney
Southwest Fisheries Science Center
La Jolla, California, USA
Surveys

Charles W. Fowler
National Marine Mammal Laboratory
Seattle, Washington, USA
Sustainability

Stuart M. Frank
Kendall Whaling Museum
Sharon, Massachusetts, USA
Scrimshaw

Adam S. Frankel
Marine Acoustics, Inc.
Arlington, Virginia, USA
Sound Production

Nicholas C. Fraser
Virginia Museum of Natural History
Martinsville, Virginia, USA
Baleen Whales, Archaic

Nicholas J. Gales
Australian Antarctic Division
Tasmania, Australia
New Zealand Sea Lion (Phocarctos hookeri)

John Gatesy
University of Wyoming, Laramie
Laramie, Wyoming, USA
Hippopotamus

Roger L. Gentry
National Marine Fisheries Service
Silver Spring, Maryland, USA
Eared Seals (Otariidae)
Northern Fur Seal (Callorhinus ursinus)

Joseph R. Geraci
National Aquarium in Baltimore
Baltimore, Maryland, USA
Health
Stranding

Timothy Gerrodette
Southwest Fisheries Science Center
La Jolla, California, USA
Tuna–Dolphin Issue

William G. Gilmartin
Hawaii Wildlife Fund
Volcano, Hawaii, USA
Monk Seals (Monachus monachus, M. tropicalis, and
*   M. schauinslandi)*

R. Natalie P. Goodall
Centro Austral de Investigaciones Científicas
Tierra del Fuego, Argentina
Hourglass Dolphin (Lagenorhynchus cruciger)
Peale's Dolphin (Lagenorhynchus australis)
Spectacled Porpoise (Phocoena dioptrica)

Shannon Gowans
Dalhousie University
Halifax, Nova Scotia, Canada
Bottlenose Whales (Hyperoodon ampullatus and
*   H. planifrons)*

Ailsa Hall
University of St. Andrews
St. Andrews, Scotland, U.K.
Gray Seal (Halichoerus grypus)

Mike O. Hammill
Department of Fisheries and Oceans Canada
Mont Joli, Quebec, Canada
Earless Seals (Phocidae)

John Harwood
University of St. Andrews
St. Andrews, Scotland, U.K.
Mass Die-Offs

Carolyn B. Heath
Fullerton College
Fullerton, California, USA
California, Galapagos, and Japanese Sea Lions
*   (Zalophus californianus, Z. wollebaeki, and*
*   Z. japonicus)*

M. P. Heide-Jørgensen
Greenland Institute of Natural Resources
Nuuk, Greenland
Narwhal (Monodon monoceros)

Ronald E. Heinrich
Ohio University, Athens
Athens, Ohio, USA
Carnivora
Mustelidae
Ursidae

Michael R. Heithaus
Simon Fraser University
Burnaby, British Columbia, Canada
Feeding Strategies and Tactics

Louis M. Herman
University of Hawaii
Honolulu, Hawaii, USA
Language Learning

Roger Hewitt
Southwest Fisheries Science Center
La Jolla, California, USA
Krill

John E. Heyning
Natural History Museum of Los Angeles County
Los Angeles, California, USA
Cuvier's Beaked Whale (Ziphius cavirostris)
Museums and Collections
River Dolphins, Relationships

Mark A. Hindell
University of Tasmania
Hobart, Australia
Breeding Sites
Elephant Seals (Mirounga angustirostris and
*   M. leonina)*

A. Rus Hoelzel
University of Durham
Durham, U.K.
Molecular Ecology

Aleta A. Hohn
National Marine Fisheries Service
Beaufort, North Carolina, USA
Age Estimation

Sascha K. Hooker
University of St. Andrews
St. Andrews, Scotland, U.K.
Toothed Whales, Overview

Joseph Horwood
Centre for Environment, Fisheries and Aquaculture
    Science
Lowestoft, U.K.
Sei Whale (Balaenoptera borealis)

Erich Hoyt
North Berwick, Scotland, U.K.
Whale Watching

Sara J. Iverson
Dalhousie University
Halifax, Nova Scotia, Canada
Blubber

Nathalie Jaquet
Texas A&M University
Galveston, Texas, USA
Lobtailing

Armando Jaramillo-Legorreta
Instituto Nacional de Ecología
Ensenada, Baja California, Mexico
Vaquita (Phocoena sinus)

Thomas A. Jefferson
Southwest Fisheries Science Center
La Jolla, California, USA
Clymene Dolphin (Stenella clymene)
Dall's Porpoise (Phocoenoides dalli)
Rough-Toothed Dolphin (Steno bredanensis)

Anne M. Jensen
Ukpeagvik Iñupiat Corporation Science Division
Barrow, Alaska, USA
Inuit and Marine Mammals

Mary Lou Jones
Southeast Fisheries Science Center
Miami, Florida, USA
Gray Whale (Eschrichtius robustus)

Ronald A. Kastelein
Harderwijk Marine Mammal Park
Harderwijk, The Netherlands
Walrus (Odobenus rosmarus)

Toshio Kasuya
Teikyo University of Science and Technology
Uenohara, Japan
Giant Beaked Whales (Berardius bairdii and
*   B. arnuxii)*
Japanese Whaling

Hidehiro Kato
National Research Institute of Far Seas Fisheries
Shimizu, Japan
Bryde's Whales (Balaenoptera edeni and B. brydei)

Akito Kawamura
Kyoto, Japan
Plankton

Catherine M. Kemper
South Australian Museum
Adelaide, Australia
Pygmy Right Whale (Caperea marginata)

Robert D. Kenney
University of Rhode Island, Narragansett
Narragansett, Rhode Island, USA
North Atlantic, North Pacific, and Southern Right
*   Whales (Eubalaena glacialis, E. japonica, and*
*   E. australis)*

Carl C. Kinze
Zoological Museum, University of Copenhagen
Copenhagen, Denmark
White-Beaked Dolphin (Lagenorhynchus albirostris)

Gerald L. Kooyman
Scripps Institution of Oceanography
La Jolla, California, USA
Diving Physiology

Kit M. Kovacs
Norwegian Polar Institute
Tromsø, Norway
Bearded Seal (Erignathus barbatus)
Hooded Seal (Cystophora cristata)

Jeffrey T. Laitman
Mt. Sinai School of Medicine
New York, New York, USA
Prenatal Development in Cetaceans

André M. Landry, Jr.
Texas A&M University
Galveston, Texas, USA
Remoras

David M. Lavigne
International Fund for Animal Welfare
Guelph, Ontario, Canada
Harp Seal (Pagophilus groenlandicus)

Rick LeDuc
Southwest Fisheries Science Center
La Jolla, California, USA
Biogeography
Delphinids, Overview
Speciation

Gina M. Lento
University of Auckland
Auckland, New Zealand
Forensic Genetics

Jon Lien
Memorial University
St. John's, Newfoundland, Canada
Entrapment and Entanglement

John K. Ling
Clare, South Australia
Australian Sea Lion (Neophoca cinerea)

Jessica D. Lipsky
Southwest Fisheries Science Center
La Jolla, California, USA
Krill
*Right Whale Dolphins (Lissodelphis borealis and
  L. peronii)*

Thomas R. Loughlin
National Marine Mammal Laboratory
Seattle, Washington, USA
Steller's Sea Lion (Eumetopias jubatus)

Valerie J. Lounsbury
National Aquarium in Baltimore
Baltimore, Maryland, USA
Health

Mary C. Maas
Northeastern Ohio Universities College of Medicine,
  Rootstown, Ohio, USA
Bones and Teeth, Histology of

Stephen A. MacLean
Texas A&M University
Galveston, Texas, USA
Inuit and Marine Mammals

Janet Mann
Georgetown University
Washington, DC, USA
Parental Behavior

Lori Marino
Emory University
Atlanta, Georgia, USA
Brain Size Evolution

Helene Marsh
James Cook University
Townsville, Queensland, Australia
Dugong (Dugong dugon)

Christopher D. Marshall
University of Washington
Seattle, Washington, USA
Morphology, Functional

Alla M. Mass
Russian Academy of Sciences
Moscow, Russia
Vision

Donald F. McAlpine
New Brunswick Museum
Saint John, New Brunswick, Canada
*Pygmy and Dwarf Sperm Whales (Kogia breviceps and
  K. sima)*

Guram A. Mchedlidze
Georgian Academy of Sciences
Tsibilisi, Georgia
Sperm Whales, Evolution

William A. McLellan
University of North Carolina
Wilmington, North Carolina, USA
Skull Anatomy

James G. Mead
National Museum of Natural History, Smithsonian
    Institution
Washington, DC, USA
Beaked Whales, Overview
Gastrointestinal Tract
Shepherd's Beaked Whale (Tasmacetus shepherdi)

Sarah L. Mesnick
Southwest Fisheries Science Center
La Jolla, California, USA
Mating Systems
Sexual Dimorphism

Edward H. Miller
Memorial University
St. John's, Newfoundland, Canada
Baculum
Territorial Behavior

Nobuyuki Miyazaki
Otsuchi Marine Research Center, University of Tokyo
Otsuchi, Japan
*Ringed, Caspian, and Baikal Seals (Pusa hispida,
    P. caspica, and P. sibirica)*
Teeth

Christian de Muizon
National Museum of Natural History
Paris, France
Odobenocetops
River Dolphins, Evolutionary History

Simon Northridge
University of St. Andrews
St. Andrews, Scotland, U.K.
Fishing Industry, Effects of
Incidental Catches

Gregory M. O'Corry–Crowe
Southwest Fisheries Science Center
La Jolla, California, USA
Beluga Whale (Delphinapterus leucas)

Daniel K. Odell
SeaWorld, Inc.
Orlando, Florida, USA
Captive Breeding
Marine Parks and Zoos
Sirenian Life History

Helmut H. A. Oelschläger
Johann Wolfgang Goethe University, Frankfurt am
    Main, Germany
Brain

Jutta S. Oelschläger
Johann Wolfgang Goethe University, Frankfurt am
    Main, Germany
Brain

Maureen A. O'Leary
State University of New York, Stony Brook
Stony Brook, New York, USA
Mesonychia

Paula A. Olson
Southwest Fisheries Science Center
La Jolla, California, USA
*Pilot Whales (Globicephala melas and
    G. macrorhynchus)*

D. Ann Pabst
University of North Carolina
Wilmington, North Carolina, USA
Skull Anatomy

Debra L. Palka
Northeast Fisheries Science Center
Woods Hole, Massachusetts, USA
North Atlantic Marine Mammals

Per J. Palsbøll
University of California, Berkeley
Berkeley, California, USA
Genetics, Overview

William F. Perrin
Southwest Fisheries Science Center
La Jolla, California, USA
Atlantic Spotted Dolphin (Stenella frontalis)
Coloration
*Common Dolphins (Delphinus delphis, D. capensis,
    and D. tropicalis)*
Geographic Variation
*Minke Whales (Balaenoptera acutorostrata and
    B. bonaerensis)*
Pantropical Spotted Dolphin (Stenella attenuata)
Species
Spinner Dolphin (Stenella longirostris)
Stranding

Wayne L. Perryman
Southwest Fisheries Science Center
La Jolla, California, USA
Melon-Headed Whale (Peponocephala electra)
Pygmy Killer Whale (Feresa attenuata)

Carl J. Pfeiffer
Virginia Polytechnic Institute and State University
Blacksburg, Virginia, USA
Whale Lice

Robert L. Pitman
Southwest Fisheries Science Center
La Jolla, California, USA
Indo-Pacific Beaked Whale (Indopacetus pacificus)
Mesoplodont Whales (Mesoplodon spp.)

Éva E. Plagányi
University of Cape Town
Cape Town, South Africa
Competition with Fisheries

Paddy P. Pomeroy
University of St. Andrews
St. Andrews, Scotland, U.K.
Reproductive Behavior

Paul J. Ponganis
Scripps Institution of Oceanography
La Jolla, California, USA
Circulatory System

James A. Powell
Florida Marine Research Institute
St. Petersburg, Florida, USA
Manatees (Trichechus manatus, T. senegalensis, and
*   T. inunguis)*

J. Antonio Raga
University of Valencia
Valencia, Spain
Parasites

Katherine Ralls
National Zoological Park, Smithsonian Institution
Washington, DC, USA
Mating Systems
Sexual Dimorphism

Andrew J. Read
Duke University Marine Laboratory
Beaufort, North Carolina, USA
Porpoises, Overview
Telemetry

Randall R. Reeves
Okapi Wildlife Associates
Hudson, Quebec, Canada
Conservation Efforts
Hunting of Marine Mammals
Population Status and Trends
River Dolphins

Joy S. Reidenberg
Mt. Sinai School of Medicine
New York, New York, USA
Prenatal Development in Cetaceans

Peter J. H. Reijnders
Alterra, Marine and Coastal Zone Research
Den Burg, The Netherlands
Pollution and Marine Mammals

Stephen B. Reilly
Southwest Fisheries Science Center
La Jolla, California, USA
Pilot Whales (Globicephala melas and
*   G. macrorhynchus)*

Julio C. Reyes
Areas Costeras y Recursos Marinos (ACOREMA)
Pisco, Peru
Burmeister's Porpoise (Phocoena spinipinnis)

John E. Reynolds III
Eckerd College, St. Petersburg, Florida,
    and U.S. Marine Mammal Commission,
    Bethesda, Maryland, USA
Anatomical Dissection: Thorax and Abdomen
Endangered Species and Populations
Manatees (Trichechus manatus, T. senegalensis, and
*   T. inunguis)*
Skeletal Anatomy

Dale W. Rice
National Marine Mammal Laboratory
Seattle, Washington, USA
Ambergris
Baleen
Classification
Spermaceti

W. John Richardson
LGL, Ltd.
King City, Ontario, Canada
Noise, Effects of

Todd R. Robeck
SeaWorld, Inc.
San Antonio, Texas, USA
Captive Breeding

Tracey L. Rogers
University of Sydney
Sydney, Australia
Leopard Seal (Hydrurga leptonyx)

Lorenzo Rojas-Bracho
Instituto Nacional de Ecología
Ensenada, Baja California, Mexico
Vaquita (Phocoena sinus)

Sentiel A. Rommel
Florida Marine Research Institute
St. Petersburg, Florida, USA
Anatomical Dissection: Thorax and Abdomen
Skeletal Anatomy
Skull Anatomy

Patricia E. Rosel
National Marine Fisheries Service
Charleston, South Carolina, USA
Albinism

Graham J. B. Ross
Australian Biological Resources Study
Canberra, Australia
*Humpback Dolphins (Sousa chinensis, S. plumbea, and
    S. teuszi)*

Peter Rudolph
National Museum of Natural History
Leiden, The Netherlands
Indo-West Pacific Marine Mammals

David J. Rugh
National Marine Mammal Laboratory
Seattle, Washington, USA
Bowhead Whale (Balaena mysticetus)

Laela S. Sayigh
University of North Carolina
Wilmington, North Carolina, USA
Signature Whistles

Michael D. Scott
InterAmerican Tropical Tuna Commission
La Jolla, California, USA
*Bottlenose Dolphins (Tursiops truncatus and
    T. aduncus)*

Richard Sears
Mingan Island Cetacean Study, Inc.
Longue Pointe de Mingan, Quebec, Canada
Blue Whale (Balaenoptera musculus)

Glenn W. Sheehan
Barrow Arctic Science Consortium
Barrow, Alaska, USA
Inuit and Marine Mammals

Kim E. W. Shelden
National Marine Mammal Laboratory
Seattle, Washington, USA
Bowhead Whale (Balaena mysticetus)

Gregory K. Silber
National Marine Fisheries Service
Silver Spring, Maryland, USA
Bioluminescence
Endangered Species and Populations

Vera M. F. da Silva
Instituto Nacional de Pesquisas da Amazônia (INPA)
Manaus, Brazil
Amazon River Dolphin (Inia geoffrensis)

Chris Smeenk
National Museum of Natural History
Leiden, The Netherlands
Indo-West Pacific Marine Mammals

Brian D. Smith
Aquatic Biodiversity Associates
Eureka, California, USA
*Susu and Bhulan (Platanista gangetica gangetica and
    P. g. minor)*

Rachel Smolker
University of Vermont
Burlington, Vermont, USA
Tool Use

David J. St. Aubin
Mystic Aquarium
Mystic, Connecticut, USA
Endocrine Systems

S. Jonathan Stern
Florida State University
Tallahassee, Florida, USA
Migration and Movement Patterns

Barbara E. Stewart
Sila Consultants
St. Norbert, Manitoba, Canada
Female Reproductive Systems

Brent S. Stewart
Hubbs-SeaWorld Research Institute
San Diego, California, USA
Diving Behavior
Hair and Fur

Robert E. A. Stewart
Department of Fisheries and Oceans Canada
Winnipeg, Manitoba, Canada
Female Reproductive Systems

Ian Stirling
Canadian Wildlife Service
Edmonton, Alberta, Canada
Polar Bear (Ursus maritimus)

James L. Sumich
Grossmont College
El Cajon, California, USA
Blowing

Alexander Ya. Supin
Russian Academy of Sciences

Moscow, Russia
Vision

Steven L. Swartz
Southeast Fisheries Science Center
Miami, Florida, USA
Gray Whale (Eschrichtius robustus)

Pascal Tassy
National Museum of Natural History
Paris, France
Paenungulates
Tethytheria

Barbara L. Taylor
Southwest Fisheries Science Center
La Jolla, California, USA
Conservation Biology

Michael A. Taylor
National Museums of Scotland
Edinburgh, Scotland, U.K.
Origins of Marine Mammals

Bernie R. Tershy
University of California, Santa Cruz
Santa Cruz, California, USA
Filter Feeding

Jessica M. Theodor
University of California, Los Angeles
Los Angeles, California, USA
Artiodactyla

J. G. M. Thewissen
Northeastern Ohio Universities College of Medicine
Rootstown, Ohio, USA
Archaeocetes, Archaic
Hearing
Mammalia
Musculature
Paleontology
Perissodactyla

Jeanette A. Thomas
Western Illinois University
Moline, Illinois, USA
Communication
Ross Seal (Ommatophoca rossii)
Weddell Seal (Leptonychotes weddellii)

Krystal A. Tolley
Institute of Marine Research
Bergen, Norway
Harbor Porpoise (Phocoena phocoena)

Fritz Trillmich
University of Bielefeld

Bielefeld, Germany
Sociobiology

Andrew W. Trites
University of British Columbia
Vancouver, British Columbia, Canada
Predator–Prey Relationships

Ted Turner
Behavior International
Aurora, Ohio, USA
Training

Peter L. Tyack
Woods Hole Oceanographic Institution
Woods Hole, Massachusetts, USA
Behavior, Overview
Mimicry

Mark D. Uhen
Cranbrook Institute of Science
Bloomfield Hills, Michigan, USA
Basilosaurids
Dental Morphology (Cetacean), Evolution of

Koen Van Waerebeek
Peruvian Centre for Cetacean Research (CEPEC)
Lima, Peru
Pacific White-Sided Dolphin and Dusky Dolphin
(Lagenorhynchus obliquidens and L. obscurus)

Paul R. Wade
National Marine Mammal Laboratory
Seattle, Washington, USA
Population Dynamics

Michael M. Walker
University of Auckland
Auckland, New Zealand
Biomagnetism

John Y. Wang
FormosaCetus Research and Conservation Group
Thornhill, Ontario, Canada
Stock Identity

Gordon T. Waring
Northeast Fisheries Science Center
Woods Hole, Massachusetts, USA
North Atlantic Marine Mammals

Douglas Wartzok
Florida International University
Miami, Florida, USA
Breathing

Mason T. Weinrich
Whale Center of New England
Gloucester, Massachusetts, USA
Callosities

David W. Weller
Southwest Fisheries Science Center
La Jolla, California, USA
Predation on Marine Mammals

Randall S. Wells
Chicago Zoological Society
Chicago, Illinois, USA
*Bottlenose Dolphins (Tursiops truncatus and
  T. aduncus)*
Identification Methods

Hal Whitehead
Dalhousie University
Halifax, Nova Scotia, Canada
Breaching
Culture in Whales and Dolphins
Sperm Whale (Physeter macrocephalus)

Ellen M. Williams
Northeastern Ohio Universities College of Medicine
Rootstown, Ohio, USA
Geological Time Scale

Terrie M. Williams
University of California, Santa Cruz
Santa Cruz, California, USA
Swimming

Ben Wilson
University of St. Andrews
St. Andrews, Scotland, U.K.
Reproductive Behavior

Loran Wlodarski
SeaWorld Florida
Orlando, Florida, USA
Marine Parks and Zoos

Bernd Würsig
Texas A&M University
Galveston, Texas, USA
Bow-Riding
Courtship Behavior
Ecology, Overview
Fluking
History of Marine Mammal Research
Intelligence and Cognition
Leaping Behavior
Noise, Effects of
*Pacific White-Sided Dolphin and Dusky Dolphin
  (Lagenorhynchus obliquidens and L. obscurus)*
Playful Behavior

André R. Wyss
University of California, Santa Barbara
Santa Barbara, California, USA
Locomotion, Terrestrial

A. V. Yablokov
Center for Russian Environmental Policy
Moscow, Russia
Illegal and Pirate Whaling

Pamela K. Yochem
Hubbs-SeaWorld Research Institute
San Diego, California, USA
Hair and Fur

Anne E. York
National Marine Mammal Laboratory
Seattle, Washington, USA
Abundance Estimation
Stock Assessment

Zhou Kaiya
Nanjing Normal University
Nanjing, China
Baiji (Lipotes vexillifer)

PREFACE

Marine mammals are awe inspiring, whether one is confronted with the underwater dash of a sea lion, a breaching humpback, or simply the sheer size of a beached sperm whale. It is no surprise that we are fascinated and intrigued by these creatures. Such fascination and curiosity brought us, the editors, to the study of marine mammals at the beginning of our careers, and they keep us excited now. To share the excitement and feed the curiosity of others, scientists or laypersons, we here attempt to summarize the field of marine mammalogy, in a very broad sense, including aspects of history and culture. This was the first reason to compile this encyclopedia.

The science of marine mammals goes back at least to Aristotle, who observed in 400 BC that dolphins gave birth to live young which were nursed with the mother's milk. Observations on the biology of marine mammals expanded throughout the Middle Ages, usually mixing freely with imagination and superstition. Konrad Gesner's *Historia Animalium* (1551), for instance, is a pictorial guide to the animals known in his time. Next to rhinos and seals, it also depicts the unicorn, the fabled mix of a horse and a narwhal. Interest greatly increased with the advent of hunting marine mammals on a large scale. Herman Melville's *Moby Dick* (1851) chronicles 19th century Western whaling and displays a curious mix of accurate natural history observations on whales with stubborn misconceptions (such as "whales are fish"). The great whaler/naturalist Charles Scammon accurately described the behavior and aspects of natural history of many species, albeit of necessity from his view behind gun and harpoon.

From these roots, marine mammal science has grown exponentially, especially since the Second World War. Unlike in earlier days, most contemporary research on marine mammals is carried out by observing living animals. Modern marine mammal studies combine aspects of mammalogy, ethology, ecology, animal conservation, molecular biology, oceanography, evolutionary biology, geology, and—in effect—all major branches of the physical and biological sciences, as well as some of the social sciences. This enormous breadth unfortunately necessitates that most marine mammalogists specialize, concentrating on one or a few aspects of marine mammal science and limiting the number of species that they study. Therein lies the second reason for compiling this encyclopedia: we aim to present a summary of the entire field for the scientist who needs information from an unfamiliar subfield.

As editors, we constrained what authors wrote as little as possible, applauding diversity and keeping to minimal guidelines. We consider modern marine mammals to include the mammalian order Cetacea (including whales, dolphins, and porpoises), the order Sirenia (dugongs and manatees), and many members of the order Carnivora: the polar bear, the sea otter and marine otter, and the pinnipeds (true seals, sea lions, fur seals, and walruses). We asked the authors to follow Rice (1998) for the species-level taxonomy and nomenclature of the modern marine mammals (with certain exceptions, as noted in the Marine Mammal Species list), as his work is an excellent, generally accepted listing of diversity.

There is some overlap among the articles. This is not an accident. As in every scientific field, different workers in marine mammalogy have different perspectives on many technical issues and disagree strongly on some of them. We urge the reader to use the cross-indexing to peruse different accounts relating to the same question; on some matters the jury is still very much out, and the range of views is interesting and important.

Ours is an encyclopedia, an alphabetically arranged compilation of articles that are independent and multi-authored, the only such work on marine mammals. However, some other recent books form excellent complements to our work. For example, *Handbook of Marine Mammals* (S. H. Ridgway and R. J. Harrison, Academic Press, 1985–1999) is a series of compendia presenting descriptions of the marine mammal species. *Biology of Marine Mammals* (J. E. Reynolds III and S. A. Rommel, Smithsonian Institution Press, 1999) presents an overview of marine mammals based on a number of long review chapters. *Marine Mammals: Evolutionary Biology* (A. Berta and J. L. Sumich, Academic Press, 1999) presents a current review of the evolutionary aspects of marine mammal science in a textbook format. There are many other authored and edited books, monographs, and research papers, often on more specific topics or particular species. These are listed here in the bibliographies that follow each entry, and the interested reader is encouraged to make use of university libraries, major research libraries (such as in the Smithsonian Institute in Washington, DC, for example), and World Wide Web search engines to find out how to obtain specific reference works. In our modern computer-accessible information era, it is hardly ever appropriate to use the excuse "I cannot find the reference," and we hope that this encyclopedia serves as a text to help point the way.

We hesitated before agreeing to edit this encyclopedia. Marine mammalogy is an exceptionally broad field, ranging across many taxa and across disciplines from molecular genetics and microstructure to whaling history and ethics. We three are all cetologists: we study the evolution and biology of whales, dolphins, and porpoises, and we personally know relatively little about seals, sea cows, or whaling. But we rub shoulders with those who do know much about these things, in our laboratories and universities, in advisory bodies, and at conferences, so we were considered to be in a good position to elicit and edit articles from our colleagues. The project has been fatiguing and sometimes exasperating but elevating nonetheless. We have learned a lot along the way. We owe a great deal to many people. First we thank our editors at Academic Press: Chuck Crumly (the Encyclopedia was his concept and owes its existence to his drive), Gail Rice, and Chris Morris, who all put up bravely with our editing and publishing amateurism and endless missteps and interventions. A very large number of colleagues acted as anonymous peer reviewers for the articles. But the most credit must go to the authors, who gave so freely of their time and expertise. The Encyclopedia is appropriately an international project: articles were authored by scientists in Argentina, Australia, Brazil, Canada, China, Denmark, France, Georgia, Germany, Japan, Mexico, The Netherlands, New Zealand, Norway, Peru, Russia, South Africa, Spain, the United Kingdom, and the United States. The difficulties of such wide participation were eased by the Internet.

We and the authors have engaged in our tasks as a labor of love of our field. We hope that you find not only information in these pages, but also a sense of the excitement of the known and the mystery of the yet-to-be-explored. If this work so affects you, it will have been successful. We also hope that it will help stimulate our growing cadres of young colleagues, naturalists, conservationists, and citizens of earth to contribute to the efforts to save and protect these marvelous creatures of the seas.

W. F. Perrin
B. Würsig
J. G. M. Thewissen

GUIDE TO THE ENCYCLOPEDIA

The *Encyclopedia of Marine Mammals* is a complete source of information on the subject of marine mammals, contained within a single volume. Each article in the Encyclopedia provides an overview of the selected topic to inform a broad spectrum of readers, from researchers to students to the interested general public.

In order that you, the reader, will derive the maximum benefit from the *Encyclopedia of Marine Mammals*, we have provided this Guide. It explains how the book is organized and how the information within its pages can be located.

SUBJECT AREAS

The *Encyclopedia of Marine Mammals* presents 283 separate articles on the entire range of marine mammal study. Articles in the Encyclopedia fall within seven general subject areas, as follows:

- Anatomy and Physiology
- Behavior and Life History
- Ecology and Population Biology
- Evolution and Systematics
- Human Effects and Interactions
- Organisms and Faunas
- Research Methodology

ORGANIZATION

The *Encyclopedia of Marine Mammals* is organized to provide the maximum ease of use for its readers. All of the articles are arranged in a single alphabetical sequence by title. An alphabetical Table of Contents for the articles can be found beginning on p. vii of this introductory section.

As a reader of the Encyclopedia, you can use the alphabetical Table of Contents by itself to locate a topic. Or you can first identify the topic in the Contents by Subject Area and then go to the alphabetical Table to find the page location.

So that they can be more easily identified, article titles begin with the key word or phrase indicating the topic, with any descriptive terms following this. For example, "Noise, Effects of" is the title assigned to this article, rather than "Effects of Noise" because the specific term *Noise* is the key word.

ARTICLE FORMAT

Each article in the Encyclopedia begins with an introductory paragraph that defines the topic being discussed and summarizes the content of the article. For example, the article "Baculum" begins as follows:

> The baculum (os penis) is a bone in the penis of Insectivora, Chiroptera, Primates, Rodentia, and Carnivora; it is absent from Cetacea and Sirenia. The corresponding element in females is the os clitoridis. In marine mammals the baculum and os clitoridis have been studied mainly in pinnipeds.

Major headings highlight important subtopics that are discussed in the article. For example, the article "Intelligence and Cognition" includes the topics "Brain Size and Characteristics," "Learning," and "Behavioral Complexity in Nature."

CROSS-REFERENCES

The *Encyclopedia of Marine Mammals* has an extensive system of cross-referencing. References to other articles appear in three forms: as marginal headings within the A–Z article sequences; as designations within the running text of an article; and as indications of related topics at the end of an article.

As an example of the first type of reference cited above, the following marginal entry appears in the A–Z article list between the entries "Antarctic Marine Mammals" and "Archaeocetes, Archaic":

Aquariums
SEE *Marine Parks and Zoos*

This reference indicates that the topic of Aquariums is discussed elsewhere, under the article title "Marine Parks and Zoos."

Top: Aggregations of beluga whales (Delphinapterus leucas) *interacting and rubbing on the substrate of a shallow estuary during the summer molt. (Photo by Flip Nicklin.) See article BELUGA WHALE. Bottom: Cephalorhynchus dolphins* (Cephalorhynchus *spp.) are among the smallest dolphins. The four dolphins of this genus are found only in Southern Hemisphere waters. (Photo by Steve Dawson.) See article CEPHALORHYNCHUS DOLPHINS.*

Top: Recently born ringed seal (Pusa hispida) pup, showing lanugo (fetal covering of fur). See article COLORATION. Bottom: Crabeater seal (Lobodon carcinophaga); most crabeater seals possess long, raking scars on their torsos resulting from attacks in their first year of life by leopard seals. (Photo by John L. Bengtson.) See article CRABEATER SEAL.

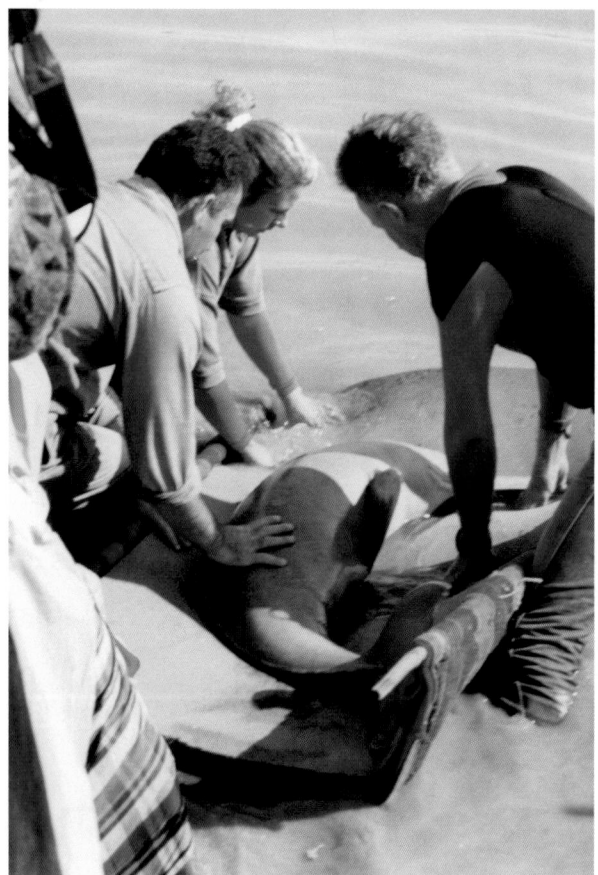

Top: A team from the Center for Coastal Studies in Provincetown, Massachusetts, attempts to cut away fishing line wrapped around the tail of a right whale in Cape Cod Bay. Specially designed knives are deployed at the ends of long poles. The men are wearing helmets as a precaution while undertaking this dangerous work. (Photo courtesy of Center for Coastal Studies.) Bottom left: An Indus river dolphin that had been marooned in an irrigation canal is released back into the main river channel upstream of Sukkur, Sind Province, after being rescued by a team of conservationists from Sind Wildlife Department and WWF-Pakistan. (Photo courtesy of World Wide Fund for Nature-Pakistan.) See article CONSERVATION EFFORTS. Bottom right: When a marine mammal strands, the public effort to provide assistance is typically immediate and highly committed. (Paul H. Forestell/Pacific Whale Foundation.) See article POPULAR CULTURE AND LITERATURE.

Top: California sea lions often move into a purse seine net to start feeding before the net is pursed, and continue to feed until all the fish are pumped aboard the fishing vessel. (Photo by Jon Stern.) See article FISHERIES, INTERFERENCE WITH. Bottom: A harbor porpoise entangled in a cod gillnet in the North Sea, one of several thousand dying this way each year in European gillnet fisheries. (Photo by Nigel Godden/Sea Mammal Research Unit.) See article INCIDENTAL CATCHES.

Top: The flukes of the gray whale (Eschrichtius robustus) are over 3 m wide, frequently bear scars from the teeth of killer whales, and are often lifted before a deep dive. Bottom: The gray whale's narrow head is usually covered with patches of barnacles. Inset: Dense clusters of barnacles surrounded by whale lice develop shortly after birth. Barnacles leave white scars on the whale's skin that slowly re-pigment over time. See article GRAY WHALE.

Top: Gray whales breach frequently while migrating and at their winter breeding grounds. One animal was observed to breach 40 consecutive times. Bottom: A "friendly" gray whale cow and calf allow whale watchers to pet them (note the tip of the female's lower jaw in the foreground). (Gray whale photos © Mary Lou Jones and Steven L. Swartz.) See article GRAY WHALE.

Top: Killer whale (Orcinus orca) attacking a gray whale calf. (Photo by Sue Flood/BBC Natural History Unit on Marine Mammals.) See article PREDATION ON MARINE MAMMALS. Bottom: Communal hunting pattern allows dolphins to combine pursuing efforts. (Photo © Alejandro Acevedo-Gutiérrez.) See article GROUP BEHAVIOR.

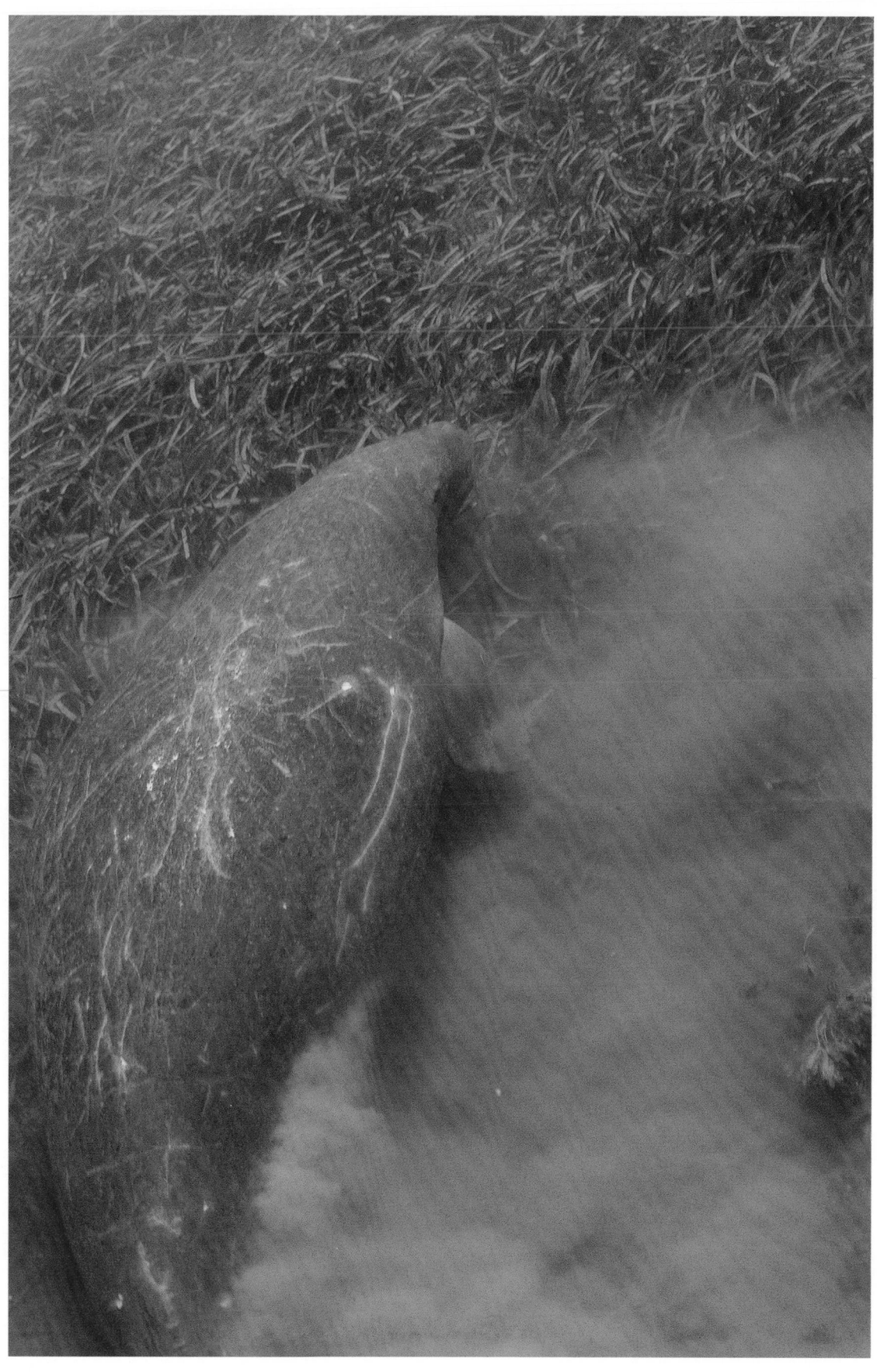

Dugongs (Dugong dugon) preferentially feed on below-ground portions of seagrasses, creating a cloud of sediment during their foraging activity. (Photo by Michael R. Heithaus.) See article FEEDING STRATEGIES AND TACTICS.

Top: A female sea otter (Enhydra lutris) *with her pup. See article OTTERS. Bottom: The polar bear,* Ursus maritimus, *is aptly named, as members of the species are often observed miles from the nearest land on polar pack ice and swimming between ice floes, where they hunt ringed seals and sometimes bearded seals. (Photo by François Gohier.) See article POLAR BEAR.*

Top: California sea lions (Zalophus californianus) are highly polygynous and form dense aggregations on rookeries commonly found along the shoreline at breeding sites on the California Channel Islands. Bottom left: Adult male California sea lions compete for territories at San Miguel Island, California. Bottom right: A much larger and darker adult male attempts to block an adult female from leaving his territory at San Miguel Island. (Photos courtesy of George Antonelis, NMFS.) See article ROOKERIES.

Top: Female New Zealand fur seal (Arctocephalus forsteri). *(Photo by John Arnould.) See article SOUTHERN FUR SEALS. Bottom: A weaned hooded seal* (Cystophora cristata) *pup 4 days old, reported as weighing 44 kg. See article EARLESS SEALS.*

A false killer whale (Pseudorca crassidens) *displays the typical fusiform body that aids the cetacean in locomotion. See article* STREAMLINING.

Whales and dolphins have become emblematic of the diversity and health of marine ecosystems and therefore have found their way onto the stamps of nations wishing to celebrate the sea and promote conservation. Examples are the many marine-mammal stamps and souvenir sheets issued during the UNESCO International Year of the Ocean in 1998, by the United Nations and a large number of its member countries; some are shown here. Series of colorful stamps showing wildlife are popular with topical collectors in general and provide important revenue for small, less-developed countries and subnational entities; an example is the whale and dolphin series issued in 2000 by Jersey (below), actually part of the UK but authorized to issue its own stamps. The current trend is toward attractive souvenir sheets showing species in their natural environments (for example, the Irish souvenir sheet on facing page, top). Whaling history has also become a popular topic (see souvenir sheet from Tristan da Cunha). Collecting (and using) these stamps contributes to public awareness of marine mammals, highlighting their diversity, beauty, grace, and conservation problems.

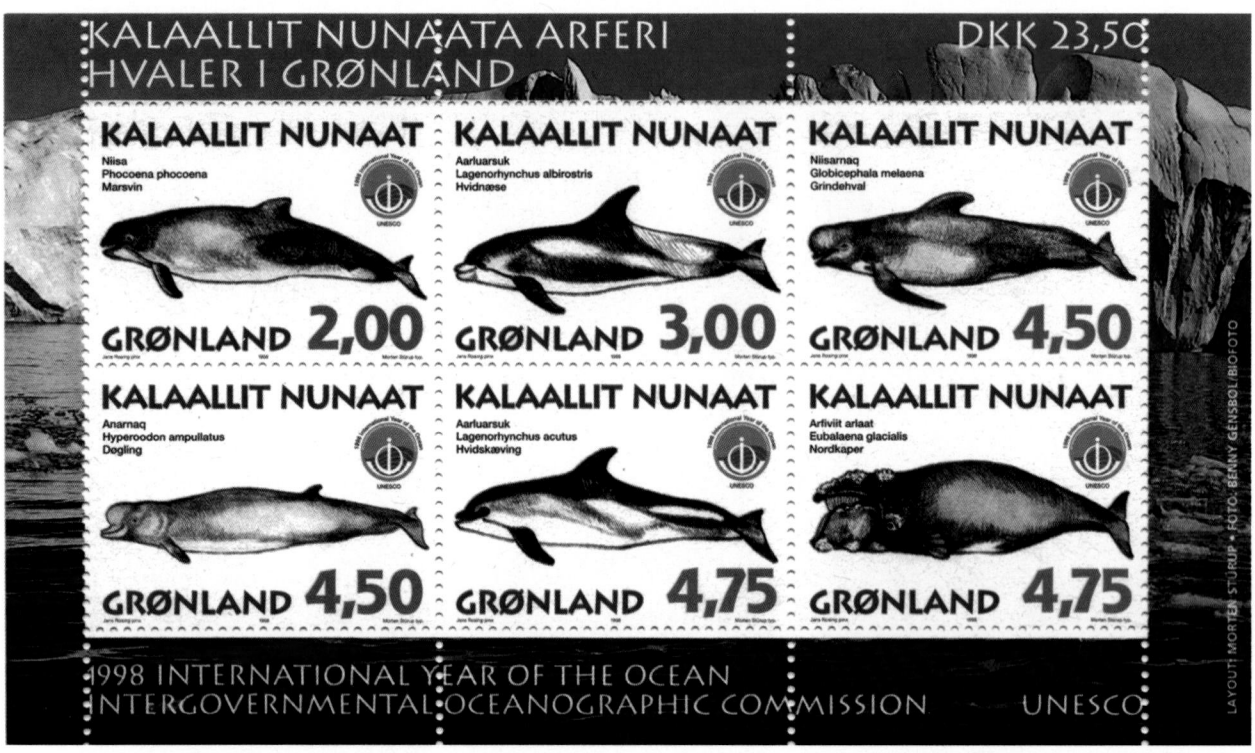

Abundance Estimation

ivations of estimators and variances: whether an object
is independent of whether any other object is detected.
imates are robust to the assumption of independence,
st variance estimates are obtained by taking the line to
mpling unit, either by bootstrapping lines or by calcu-
weighted sample variance of encounter rates by line.
o not need to assume that animals are randomly dis-
throughout the survey area, provided that lines are
andomly with respect to the animals. This ensures that
n the surveyed strip are uniformly distributed with dis-
om the line.

imation Perpendicular distances x are measured from
to each detected animal. (We will consider the case that
occur in groups later.) In practice, for shipboard surveys,
n distances r and detection angles θ are usually recorded,
ich perpendicular distances are calculated as $x = r\sin(\theta)$.
. Suppose there are k lines of lengths $l_1, \ldots, l_k$ (with
, and n animals are detected, at perpendicular distances
x_n. Suppose that animals further than some distance w
e line are not recorded. Then the surveyed area is $a =$
thin which n animals are detected. However, not all an-
ithin the surveyed area are detected. Let the effective
th of the strip be $\mu < w$ (so that the proportion of ani-
thin the surveyed strip that are detected is μ/w). Then
density (number of animals per unit area) is estimated by

$$\hat{D} = \frac{n}{2\hat{\mu}L} \qquad (1)$$

nce is estimated as $\hat{N} = A\hat{D}$, where A is the size of the
rea. We therefore need an estimate $\hat{\mu}$ of μ. The software
e provides comprehensive options for these analyses.

ariance and Interval Estimation The variance of $\hat{D}$
approximated using the delta method, assuming no cor-
between n and $\hat{\mu}$:

$$\hat{V}(\hat{D}) = \hat{D}^2 \left[\frac{\hat{V}(n)}{n^2} + \frac{\hat{V}(\hat{\mu})}{\hat{\mu}^2} \right]$$

ariance of n is generally estimated from the sample
e in encounter rates, n_j/l_j, weighted by line lengths l_j.
$\hat{\mu}$ is estimated by maximum likelihood, its variance is esti-
from the information matrix.

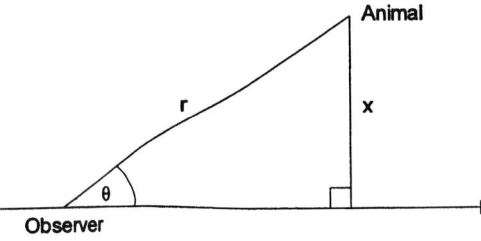

e **1** *The observer records an animal at detection dis-
r and detection angle θ, from which the perpendicular
e is calculated as $x = r\sin(\theta)$.*

If we assume that $\hat{D}$ is log-normally distributed, approximate 95% confidence limits are given by $(\hat{D}/C, \hat{D}C)$ where

$$C = \exp\{1.96[\hat{V}(\log_e \hat{D})]^{0.5}\}$$

with $\hat{V}(\log_e \hat{D}) = \log_e [1 + \hat{V}(\hat{D})/\hat{D}^2]$.

Often, bootstrap variance and interval estimation are preferred. Resamples are usually generated by sampling with replacement from the lines so that independence between the
lines is assumed, but independence between detections on the
same line is not.

5. Estimation When Animals Occur in Clusters Animals
often occur in groups, which we term "clusters." If one animal
in a cluster is detected, it is assumed that the whole cluster is
detected, and the position of the cluster is recorded. Equation
(1) then gives an estimate of the density of clusters. To obtain
the estimated density of individuals, we must multiply by an estimate of mean cluster size in the population, $E(s)$:

$$\hat{D} = \frac{n\hat{E}(s)}{2\hat{\mu}L} \; .$$

The corresponding variance estimate is

$$\hat{V}(\hat{D}) = \hat{D}^2 \left[\frac{\hat{V}(n)}{n^2} + \frac{\hat{V}(\hat{\mu})}{\hat{\mu}^2} + \frac{\hat{V}\{\hat{E}(s)\}}{\{\hat{E}(s)\}^2} \right].$$

Because the probability of detection is often a function of cluster size, the sample of cluster sizes exhibits size bias. In the absence of size bias, we can take $\hat{E}(s) = \bar{s}$, the mean size of detected clusters. Several methods exist for estimating $E(s)$ in the
presence of size bias (Buckland *et al.*, 1993).

The just-described methods assume that once a cluster of
animals is detected, it is possible to record the size of that cluster accurately. For shipboard and aerial surveys, this often dictates that at least part of the survey is conducted in "closing
mode." After detection, search effort ceases, and the vessel
closes with the detected cluster, to allow more accurate estimation of cluster size. This strategy also eases the difficulties of
species identification. If "passing mode" is adopted, then underestimation of the size of more distant clusters might be anticipated. Regression methods for correcting size bias also correct for this bias, provided that the sizes of clusters on or close
to the transect line are estimated without bias. Where cluster
size estimation is problematic, observer training is usually necessary to ensure that bias is not large.

6. Estimation When Detection on the Line Is Not Certain
The standard line transect method assumes that animals on the
line are certain to be detected. Double-platform methods in
which observers search simultaneously from two platforms are
therefore becoming commonplace. This allows extension of the
standard methods to the case that animals on the line are not
certain to be detected and also, given appropriate field methods,
allows adjustment for responsive movement of animals prior to
detection. Several researchers have made advances in developing methodology for analyzing such data, notably Borchers *et al.*

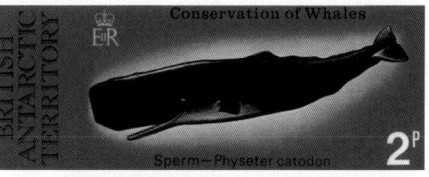

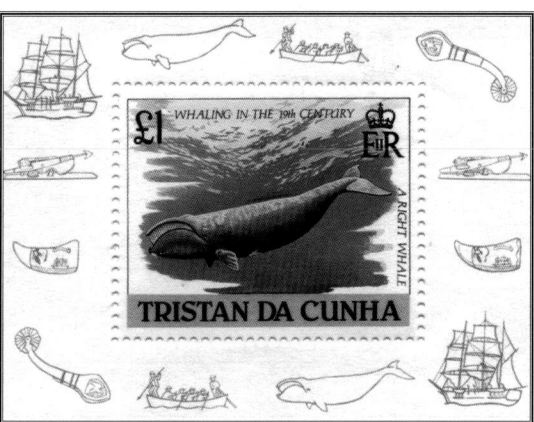

Top: An adult male Pacific walrus (Odobenus rosmarus). Note the large diameter of its tusks, the bulging eye, which is turned to look backward, and the prounounced tubercles of the skin on the neck. The last feature is seen only in adult males. Bottom: A large group of male Pacific walruses on Round Island, Bristol Bay, Alaska. Note how closely the animals are packed together. (Photos by Steve Rice, U.S. Fish and Wildlife Service.) See article WALRUS.

Abundance Estimation

Stephen T. Buckland
*University of St. Andrews,
Scotland, United Kingdom*

Anne E. York
*National Marine Mammal Laboratory,
Seattle, Washington*

Abundance estimation covers the range of techniques by which the size of a population of marine mammals can be estimated. Such population size estimates are often referred to as "absolute" abundance estimates. When it is difficult to estimate absolute abundance with an acceptably low bias, relative abundance indices are often used instead. These are indices that are believed to be proportional to population size, apart from stochastic variation, allowing trends in the population in space and/or time to be assessed. The main techniques for abundance estimation (relative or absolute) are distance sampling, mark-recapture, migration counts, and colony counts.

I. Distance Sampling

Distance sampling (Buckland *et al.*, 1993) is the most widely used technique for estimating the abundance of animal populations. Because it is particularly suited to populations of animals that are readily detectable (at least at close quarters) and sparsely distributed over a large area, it is unsurprising that the methods are widely used by marine mammalogists. Indeed, most of the innovative developments in distance sampling in the last two decades have arisen through applications of the methodology to cetaceans. Distance sampling has also been applied on several populations of ice seals.

The two primary methods of distance sampling are line transect sampling and point transect sampling. The latter method has never to our knowledge been applied to marine mammal populations, and we therefore concentrate mostly on line transect sampling. Another distance sampling method is cue counting, which was developed specifically for [...] of large whales, and the theory for which is close [...] that for point transects. Distance sampling data [...] lyzed using software Distance (Thomas *et al.*, 199[...]

A. Line Transect Sampling

1. Survey Design In line transect sampling, t[...] sign comprises a set of straight lines, spanning [...] area for which an abundance estimate is require[...] mammal surveys, the lines are covered by a tea[...] on a ship or boat or by one or more observers [...] The methodology requires that lines are rand[...] the study area or that a grid of systematical[...] equally spaced) lines is randomly located in th[...] efficiency is improved if lines are placed perpe[...] sity contours, a common design for inshore s[...] a series of parallel lines, randomly spaced or [...] tematically spaced, as far as possible perpendi[...] line. The study area is often divided into geo[...] strata, allowing different orientations of the g[...] ferent strata and allowing effort to be great[...] strata. For shipboard surveys especially, syst[...] signs are often used because there is then n[...] ship time in traversing off-effort (i.e., not se[...] mammals) from one line to the next. The sh[...] tinuously searching for marine mammals du[...]

2. Assumptions The following three [...] hold:

1. Animals on or very close to the line [...] tected (but see later).
2. Animals are detected before they res[...] of the observer, and nonresponsive m[...] tive to the speed of the observer.
3. Distances are measured accurately (f[...] data), or objects are correctly allocat[...] (for grouped data).

Bias from nonresponsive movement [...] provided that the average speed of the [...] half of the speed of the observer. A four[...]

(1998). Their method uses Horvitz–Thompson-type estimators and also yields a unified approach to the analysis of line transect data in which animal density is estimated in a single step, which contrasts with the conventional strategy of independently estimating the three parameters: encounter rate, effective strip width, and mean cluster size. Double-platform surveys are used widely in cetacean surveys and have also been used for estimating the abundance of POLAR BEARS (*Ursus maritimus*).

7. Current Areas of Research Active research topics likely to become available shortly in software Distance include spatial modeling of line transect data; integrated analysis of double-platform data; general methods for incorporating covariates in analyses; automated survey design algorithms; and adaptive line transect sampling.

B. Strip Transect Sampling

Strip transect sampling is a special case of line transect sampling in which it is assumed that all animals out to the truncation distance w are detected. This simplifies analysis, and distances of detected animals from the line need not be measured, except to ensure that they are within distance w of the line. However, the method is seldom efficient for marine mammals; if the strip is narrow enough to ensure that all animals out to w are detected, then many animals are detected beyond w, and these observations must be ignored. Abundance of sirenian populations has traditionally been obtained by strip transect methods.

C. Cue Counting

In cue counting, observers on a ship or in an aircraft cover a sector ahead of their observation platform and record all cues detected within the sector and the distances of the cues from the platform. In principle, the method can be used for any marine mammal, but in practice, it has been used primarily for large whales, for which the cue is the blow. The same design considerations apply as for line transect sampling, although the analysis is essentially the same as for point transect sampling.

If cues are well defined (as blows of large whales are), then cue counting has the advantage over line transect sampling that the recording unit is the individual cue. Observers need not identify whether different cues are from different animals or how many animals are in a cluster. It also does not matter if some whales stay down so long that they will be undetectable even if they are on the transect line, provided that all cues within the recording sector and very close to the observation platform are detected. Another advantage is that the method requires observer-to-animal distances, which are easier to estimate than perpendicular distances of animals from the line. The main disadvantage is that the method yields estimates of cue density per unit time, which can only be converted into whale density by estimating the cue rate from additional costly surveys. The estimated cue rate is prone to bias, both because animals may behave differently when a survey ship is close by and because it is easier to monitor animals that cue frequently, thus biasing the cue rate upward. Additionally, if animals cue more frequently when a ship is bearing down on them, an excess of short distances will be observed in the distance data, biasing the estimation of cue density.

The number of cues per unit area per unit time is estimated by

$$\hat{D}_c = \frac{2n}{\phi \hat{\rho}^2 T},$$

where n is the number of cues detected in time T, ϕ is the angle of the sector in radians, and $\hat{\rho}$ is the estimated effective radius of detection. Estimated density is then

$$\hat{D} = \frac{\hat{D}_c}{\hat{\lambda}},$$

where $\hat{\lambda}$ is the estimated number of cues per animal per unit time (the cue rate). As before, abundance is estimated as $\hat{N} = A\hat{D}$, where A is the size of the study area.

Because cues may be from the same whale, or the same pod of whales, they cannot be assumed independent. However, this is not a problem given the robust variance estimation methods provided by software Distance.

If cues immediately ahead of the vessel might be missed, double-platform methods similar to those for line transect sampling may be used. This has the advantage over those analyses in that it is easier to identify whether a single cue is seen from both platforms, e.g., by recording exact times of cues, than to identify whether a single animal or animal cluster is seen by both platforms, as the two platforms may see different cues from the same animal.

II. Mark-Recapture

Mark-recapture tends to be more labor-intensive and more sensitive to failures of assumptions than distance sampling. However, it is applicable to some species that are not amenable to distance sampling methods and can yield estimates of survival and recruitment rates, which distance sampling cannot do. Mark-recapture methods can be useful for populations that aggregate at some location each year, whereas distance sampling methods are more effective on dispersed populations. They should therefore be seen as different tools for different purposes. Among marine mammals, mark-recapture has been used most often to estimate the abundance of pinnipeds, usually for the estimation of young of the year. Polar bears have also been the subject of mark-recapture studies. Perhaps the most comprehensive software currently available for analyzing mark-recapture data is MARK (White, 1999).

A. Estimation from a Tagged Subset of Animals

1. The Petersen Estimator In its simplest form, mark-recapture consists of marking a sample of M animals from a population of unknown size N, returning the animals to the population and then removing, capturing, or observing a sample of n animals. Suppose that, of these n animals, m were marked. We assume that the proportion of marked animals in the second sample is a valid estimate of the proportion of marked animals in the population, giving the following "Petersen" estimate of population size: $\hat{N} = nM/m$.

2. Chapman's Modified Estimator   Inference for this model is complicated by the fact that the variance of $\hat{N}$ is infinite unless $n + M > N$, in which case m cannot be zero. Chapman's estimator $\hat{N}_c = (n + 1)(M + 1)/(m + 1) - 1$ has a lower bias (and no bias for $n + M > N$), provided that the assumptions of the estimator are satisfied. It has variance

$$\hat{V}(\hat{N}_c) = \frac{(n + 1)(M + 1)(n - m)(M - m)}{(m + 1)^2(m + 2)}.$$

Many variations on this theme have been developed, including extensions to multiple samples, extensions to open populations, "single release" methods, and "single recapture" methods (particularly suited for when the mark is recovered from a dead animal). We make no attempt to cover these methods here, but refer the reader to Seber (1982).

3. Assumptions   The assumptions required for $\hat{N}$ to be a reasonable estimate of population size are that the population of interest is closed over the survey period and that animals are marked and resighted or recaptured at random. In using the ordinary Petersen estimate, it is also assumed that marks are not lost during the survey period and that marking does not affect the probability of resighting or recapturing the animal. Methods have been developed to circumvent these assumptions and the literature for this topic is rich (e.g., Otis *et al.*, 1978; Pollock, 1990). For most wildlife populations, probabilities of recapture or resighting tend to vary among animals for a variety of reasons. This heterogeneity can be problematic to model and can lead to a large bias in abundance estimates, so that the design of a mark-recapture survey should be carefully addressed to minimize heterogeneity.

4. Estimation of Pinniped Numbers by Mark-Recapture   Mark-recapture techniques have been successfully used to estimate the abundance of young of the year for several species of fur seals. Chapman and Johnson (1968) described the first successful application of this technique for the population of northern fur seals (*Callorhinus ursinus*) on the Pribilof Islands. They marked seals by shearing some hair from their heads and later went back to the colony and counted numbers of marked and unmarked animals within groups of animals. They calculated a Petersen estimate of abundance, which they verified on small colonies where direct counts of young of the year could be made. Resighting was replicated on each colony and several procedures for estimating the variance of the total population size were investigated. These included (1) an empirical estimate calculated as the variance of the mean of replicated estimates for each colony, and the variance of the total calculated by summing the individual colony variances, and (2) a variance for each replicated colony estimate assuming the hypergeometric distribution, with the variance of the mean count for each colony estimated from the variances of the individual counts. They also discussed the use of interpenetrating subsamples to estimate the variance. This procedure is similar in flavor and intent to the bootstrap procedure.

5. Mark-Recovery Methods   Before the development of line transect methods for estimating the size of populations of large baleen whales, mark-recovery studies were carried out in which "Discovery" marks were fired into whales, a proportion of which would later be recovered by whalers. Disadvantages of this approach included a requirement for very large sample sizes to ensure an adequate number of recaptures; a long delay before sufficient data accumulated to allow abundance to be estimated; and strong sensitivity of abundance estimates to failures of assumptions. The methods were largely unsuccessful. For a review of the mark-recapture models that are potentially relevant to such data, and of the numerous sources of potential bias in the abundance estimates, see Buckland and Duff (1989). Before the development of mark-recapture or mark-resight techniques for northern fur seal pups, there were many attempts to estimate the population size by tagging pups at birth and recovering the tags in a commercial harvest. This application failed for similar reasons that the use of Discovery tags failed to properly estimate the size of cetacean populations.

B. Use of Natural Markings

Studies that use natural markings to identify individual animals in a population have become widespread in recent years. These usually rely on photo identification of individuals. A significant milestone in the use of such methods was Hammond and colleagues (1990), which comprises an edited collection of papers from a workshop on the topic. While the technique is undoubtedly of great value, it is important to be aware of its limitations.

Natural markings data can be very effective for estimating survival rates of marine mammals. Abundance estimation is more problematic, as this involves extrapolation from the identified subpopulation. If a high proportion of the population (>80%) can be identified, then abundance estimates are likely to have small bias, especially as there is a tendency to underestimate population size. It is possible to achieve such high rates, e.g., for small coastal populations of bottlenose dolphins (*Tursiops truncatus*) and pinniped colonies, provided individuals have distinct markings. The method is then useful because it allows enumeration of almost the whole population without fear of double counting individuals or of seriously underestimating population size. When smaller proportions of animals are identified, estimates of population size can be badly compromised for a variety of reasons. Severe heterogeneity in the "capture" probabilities is common, e.g., because some natural markings are identified much more readily than others or because some individuals are more approachable than others. It is notoriously difficult to model such mark-recapture data reliably. Another problem is that the population being estimated is not always well defined, with some animals from elsewhere temporarily entering the population and others temporarily absent. A severe problem for large populations, in which only a small proportion can be identified, is that false positives in the matching procedure, even if they occur only rarely, can lead to a substantial underestimation of population size. Genetic fingerprinting, if feasible, can reduce this problem substantially.

Natural markings studies are invaluable for estimating survival and birth rates, for MIGRATION routes, and for detailed studies, including abundance estimation, of a small population. However, they are rarely a cost-effective or reliable method for estimating the size of large populations of marine mammals. Given the current research interest in this topic, especially for estimating abundance of feeding or breeding aggregations of humpback (*Megaptera novaeangliae*), blue (*Balaenoptera musculus*), and right whales (*Eubalaena* spp.), further advances can be anticipated that will widen the applicability of the methods.

III. Migration Counts

Many populations of large whales conveniently file past coastal watch points on migration, allowing observers to count a large proportion of the population. This count can then be corrected for animals passing outside watch periods to estimate population size. In practice, further corrections are needed, e.g., to adjust for pods that pass undetected during watch periods, for biased estimation of pod size, for different rates of passage between day and night, and for a component of the population that fails to pass the watch point. Despite the need for various correction terms, migration count data yield very precise estimates of abundance with low bias, provided that the more significant correction factors are estimated reliably. This is unsurprising given that typically 30–40% of the population might be seen by the observers, a much higher fraction than is normal in a distance sampling survey.

The methods usually used for modeling migration counts were developed for the analysis of surveys of the California gray whale (*Eschrichtius robustus*). To estimate numbers of undetected pods passing during watch periods, two count stations operate independently, and these double-count data are modeled using logistic regression. A polynomial model is used to estimate the rate of passage as it varies through the season, from which numbers of whales passing outside watch periods are estimated. A Bayesian approach is used for analyzing similar data on bowhead whales (*Balaena mysticetus*).

IV. Colony Counts

Many populations of pinnipeds gather for breeding and pupping at certain times of the year. Researchers often make counts of these populations from cliffs above the colonies, from planes flying overhead, or sometimes from ships passing the colony. Often photographs are taken of the colonies. These are brought back to the laboratory for analysis and form a permanent record of the population. In most pinniped populations, it cannot be assumed that all the animals are on shore at any given time, although in fur seal populations, there is a time window in which almost all of the young of the year and breeding males are present, and in certain phocid populations all the young of the year and breeding females are present. Thus, colony counts alone cannot be used to determine absolute abundance of the population size, except for certain classes of animals, and this depends on the reproductive patterns of the population of interest, which must be taken into account when the survey is designed. Serial colony counts can be used to determine the rate of increase of the population if the same proportion of animals is present each year at the colony when the counts are made. This assumption is most likely valid for young of the year. For other segments of the population, this assumption fails if the timing of reproduction changes or if conditions at sea change so that animals need to spend a different amount of time at sea feeding and consequently a different amount of time at the colony.

The size of the harbor seal (*Phoca vitulina*) population in the state of Washington is estimated by combining colony counts made during aerial surveys and mark-recapture to account for animals not present during the aerial surveys. Transponder tags with unique frequencies are attached to animals before the surveys. During the flyovers, animals on shore are counted and radio searches are made to determine the proportion of animals that is ashore. The total population is estimated as $\hat{N}_{tot} = \overline{N}/\hat{p}$ where $\overline{N}$ is the average count of animals on shore and $\hat{p}$ is the estimated fraction of marked animals on shore. The variance of $\hat{N}_{tot}$ is estimated as

$$\hat{V}(\hat{N}_{tot}) = \frac{\hat{V}(\overline{N})}{\hat{p}^2} + \overline{N}^2 \hat{V}\left(\frac{1}{\hat{p}}\right) - \hat{V}(\overline{N})\hat{V}\left(\frac{1}{\hat{p}}\right),$$

where $\hat{V}(\overline{N}) = s^2/n$, with s^2 equal to the sample variance of the n counts, and $\hat{V}(1/\hat{p}) = \hat{V}(\hat{p})/\hat{p}^4 = (1 - \hat{p})/M\hat{p}^3$, with M equal to the number of marked animals.

A corrected count method is also used to estimate numbers of southern elephant seal (*Mirounga leonina*) and fur seal pups (*Arctocephalus gazella*) on South Georgia. In those surveys, counts of adult females are made from shore or ship along the whole coastline during the pupping season. The counts made at any particular site are then used to estimate the total production for that site based on the adult female haul-out curves and pregnancy/pupping rate estimates from sites that are monitored regularly (twice daily in the case of fur seals) through the breeding season. Similarly, the abundance of northern fur seal pups on the Pribilof Islands is sometimes estimated from mark-recapture estimates on sample colonies coupled with counts of breeding males on all colonies (York and Kozloff, 1987). The ratio of pups to breeding males, estimated on the sampled rookeries, is multiplied by the total count of breeding males on all colonies.

The sizes of colonies of pinnipeds can also be determined using estimates of the area of all colonies coupled with estimates of the density of animals on those colonies. Although this method is often used to estimate the sizes of bird colonies, it has only been used occasionally to estimate pinniped population sizes. The estimates of the areas of the colonies were made from maps of the colonies. Counts or corrected counts, or mark-recapture estimates of the population of interest, are determined on a subsample of colonies. It is assumed that the density of animals in the sampled colonies is representative of the density on all colonies and that the total population is estimated by multiplying the total area by the estimated density. Researchers attempted to use this method to estimate the size of the Pribilof northern fur seal population in the late 1940s. At that time, it was thought that the variability of the estimates was too large and efforts were begun to design mark-recapture studies.

Counts, or more often corrected counts, are also sometimes attempted on other marine mammals. For example, because sea otters (*Enhydra lutris*) are difficult to survey by other means, they tend to be counted from a boat. Such counts typically underestimate population size, sometimes substantially so (Udevitz *et al.*, 1995).

See Also the Following Articles

References

Borchers, D. L., Zucchini, W., and Fewster, R. M. (1998). Mark-recapture models for line transect surveys. *Biometrics* **54**, 1207–1220.

Buckland, S. T., Anderson, D. R., Burnham, K. P., and Laake, J. L. (1993). "Distance Sampling: Estimating Abundance of Biological Populations." Chapman and Hall, London.

Buckland, S. T., and Duff, E. I. (1989). Analysis of the Southern Hemisphere minke whale mark-recovery data. *In* "The Comprehensive Assessment of Whale Stocks: The Early Years" (G. P. Donovan, ed.), pp. 121–143. International Whaling Commission, Cambridge.

Chapman, D. G., and Johnson, A. M. (1968). Estimation of fur seal populations by randomized sampling. *Trans. Am. Fish. Soc.* **97**, 264–270.

Hammond, P. S., Mizroch, S. A., and Donovan, G. P. (eds.) (1990). "Individual Recognition of Cetaceans: Use of Photo-Identification and Other Techniques to Estimate Population Parameters." International Whaling Commission, Cambridge.

Otis, D. L., Burnham, K. P., White, G. C., and Anderson, D. R. (1978). Statistical inference from capture data on closed animal populations. *Wildlife Monogr.* **62**, 1–135.

Pollock, K. H. (1990). Modelling capture, recapture and removal statistics for estimation of demographic parameters for fish and wildlife populations: Past, present and future. *In* "Proceedings of the American Statistical Association Sesquicentennial," pp. 26–50.

Seber, G. A. F. (1982). "The Estimation of Animal Abundance and Related Parameters," 2nd Ed. Macmillan, New York.

Thomas, L., Laake, J. L., Derry, J. F., Buckland, S. T., Borchers, D. L., Anderson, D. R., Burnham, K. P., Strindberg, S., Hedley, S. L., Marques, F. F. C., Pollard, J. H., and Fewster, R. M. (1998). Distance 3.5. Research Unit for Wildlife Population Assessment, University of St Andrews, St Andrews, UK. Available from http://www.ruwpa.st-and.ac.uk/distance/.

Udevitz, M. S., Bodkin, J. L., and Costa, D. P. (1995). Detection of sea otters in boat-based surveys of Prince William Sound, Alaska. *Mar. Mammal Sci.* **11**, 59–71.

White, G. C. (1999). Program MARK. Department of Fishery and Wildlife Biology, Colorado State University. Available from http://www.cnr.colostate.edu/~gwhite/mark/mark.htm.

York, A. E., and Kozloff, P. (1987). On the estimation of numbers of northern fur seal, *Callorhinus ursinus*, pups born on St. Paul Island, 1980–86. *Fish. Bull.* **85**, 367–375.

Age Estimation

ALETA A. HOHN
National Marine Fisheries Service,
Beaufort, North Carolina

Age estimation is a tool for obtaining a numerical value of age for animals for which actual age is not known. Currently, age is estimated primarily from counts of growth layers deposited in several persistent tissues, primarily TEETH, less often bone, and in some cases from other layered structures or from chemical signals. Growth layers in the persistent structures are similar in concept to growth rings in trees. Until use of growth layers became a feasible means of age estimation, relative measures of age, such as tooth wear, pelage or skin color, or fusion of cranial sutures, allowed individuals to be placed in age groups; these techniques largely have been replaced with methods that allow for estimation of absolute age by counting growth layers. Marine mammalogists pioneered age estimation from counting growth layers in teeth; this discovery was followed by widespread use for terrestrial mammals as well. Much of the development of this field has focused on how to ensure that age estimates are accurate and precise. That focus has been directed toward verifying the amount of time represented by a growth layer (i.e., calibration or validation), developing increasingly better ways to prepare samples for optimizing counts, and standardizing methods to ensure that growth layer counts are consistent among studies.

Age is fundamental to interpreting and understanding many aspects of the biology of marine mammals. The traditional and most obvious use of age is for estimating parameters used in population dynamics models. Age-specific estimates of fecundity or mortality can be used in these models to project population growth, for example. Estimates of age at sexual maturation are used in absolute terms in population models, whereas changes in this parameter have been interpreted to reflect changes in population abundance or resource availability and, therefore, indicate a density-compensatory response. Population age structure would also be a useful parameter, although it is rarely known. It is possible, however, to determine the age structure of individuals removed from a population intentionally, such as through directed fisheries, or incidentally, such as bycatch. This information then can be used to refine estimates of the impact of fisheries on those populations.

The need for accurate and precise estimates of age does not end with traditional population modeling. Of late, there has been increasing concern about the effects of contaminants on the health of marine mammals. Because many of these contaminants bioaccumulate, interpretation of the measured levels of organic or inorganic compounds must be taken as a function of the age, and reproductive condition, of the individual. Furthermore, because indices of health such as blood parameters change naturally with age, understanding the effects of contaminants or other factors on the health of individuals also requires knowing their age. With the recent epizootic events involving morbilliviruses (Tautenberger *et al.*, 1996), the ages of individuals infected as well as those with titers indicating previous infections become important in understanding the epidemiology of these outbreaks.

I. Growth Layer Terminology

In the context of age estimation, the term growth layer is ambiguous. That is because annual increments, as a rule, comprise more than the minimum two growth layers, e.g., a broad layer and a fine layer, needed to differentiate one annual increment from the next. Other layers are usually present; these layers are often referred to as accessory or incremental layers. These are all growth layers. It has been suggested that the existence of an annual layering pattern is controlled endogenously whereas individual growth layers represent events that have a systemic effect on the animal and therefore influence the deposition of the collagen matrix or mineral in teeth or bone. Events that have been suggested include lunar cycles, maturation, pregnancy, lactation, weaning, and feeding bouts. In essence, the annual increments themselves, as well as any layers formed within the annual increments, are recording structures that reflect the physiology of an individual at the time of deposition (Klevezal, 1996). Interpreting these structures is an interesting pursuit itself and can serve as another tool for elucidating life history events for individuals. In the context of age estimation and identifying annual increments, however, they can cause errors and confusion and have resulted in semantic controversies with regard to the term "growth layer."

To help remedy confusion in terminology, a more descriptive phrase, growth layer group (GLG), was coined at a workshop held in 1977 on estimating age in toothed whales and sirenians (Perrin and Myrick, 1980), predominantly in reference to dentine. Its use has expanded, however, to other marine mammal species and to cement as well as dentine. A GLG is a group of layers that occur with cyclical and predictable repetition. Strictly speaking, a GLG is a generic term and does not automatically imply deposition that occurred over a 1-year period. It needs to be defined for each species and each use. For practical purposes, however, a GLG generally is defined by authors to represent 1-year's deposition, i.e., "annual" is implied. The term "annual layer" is equivalent to "annual GLG."

II. Calibration of Annual Layers

Verification that annual layers exist within the complement of visible layers derives from validation or calibration studies. Notably, the first confirmation of annual layers in pinniped teeth occurred soon after teeth were examined for the possibility of age estimation; Scheffer (1950) found external layers (ridges) in the canines of northern fur seals, *Callorhinus ursinus*, that corresponded to the known age of seals branded as pups and recovered up to 8 years later. Further studies to validate annual layering patterns and to show that patterns are consistent among individuals and species have involved three approaches: (1) examining teeth or bone from animals of known age or known history; (2) examining teeth or bone marked with tetracycline; and (3) comparing growth layers in teeth that have been removed at known intervals (multiple extractions).

For cetaceans, animals of known age or with a known history most often were captive for all or much of their lives. In the latter situation, support for annual layers then hinges on counts of the number of presumed annual layers corresponding to the known age or to the known approximate age of the animal given the length of time it spent in captivity and other data, such as its body length when removed from the wild. Initial encouragement that growth layers in dolphin teeth were annual was from three captive common bottlenose dolphins, *Tursiops truncatus* (Sergeant, 1959). Teeth obtained from free-ranging *Tursiops* of known age and known history were significant for confirming and identifying annual layering patterns and determining that annual layers in free-ranging bottlenose

dolphins were similar to those in their captive conspecifics (Hohn *et al.*, 1989). Within pinnipeds, sirenians, and sea otters (*Enhydra lutris*), numerous studies of free-ranging tagged or individually identified animals have compared the number of growth layers in tooth sections to known ages (e.g., Bowen *et al.* 1983; Arnbom *et al.*, 1992). In many of these studies, as in cetacean studies, the actual age of individuals is greater than the "known age" because animals were captured or tagged some time after their birth. Thus, the number of growth layers counted is compared to that minimum age plus an additional number of years estimated as a function of the size of the animal at the time it was first tagged or identified. The most recent and rigorous studies counted growth layers without knowledge of the known ages of specimens in the sample, which eliminates a bias in counting. What is notable about all of these studies is that the authors concluded that they were able to identify annual growth layers (annual GLGs) that correspond to known ages or known approximate ages of the individuals in their samples at least up to some minimum age.

True calibration of growth layer deposition over extended periods of time relative to the life span of an animal has not been attempted. To do so would require direct marking of layers, such as through administration of tetracycline, preferably at the same time each year and ideally on the animal's birthday. Tetracycline binds permanently to actively growing mineralized tissue and fluoresces when a bone or tooth is viewed under ultraviolet light, hence its ability to serve as a marker. Two tetracycline treatments or one treatment followed by extraction have been used to unambiguously identify growth layer deposition over the period of time between the marks, providing a limited calibration; annual layers were determined from this method for several dolphin species, most extensively for spinner dolphins, *Stenella longirostris* (Myrick *et al.*, 1984) and common bottlenose dolphins (Myrick and Cornell, 1990). Alternatively, multiple extractions of teeth from an individual allow for calibration but with much restricted sampling opportunities. This method has been used with free-ranging bottlenose dolphins where two teeth were extracted and growth layer deposition between extractions compared (Hohn *et al.*, 1989). Limited opportunities exist for extensive direct calibration, although captive animals could be used for such studies as could free-ranging populations where individuals are resighted each year and could be caught, administered tetracycline, and released.

III. Tissues Commonly Used to Obtain Absolute Age Estimates

Given the importance of obtaining age estimates, various tissues and methods have been investigated for elucidating growth layers (Klevezal, 1996; McCann, 1993; Perrin and Myrick, 1980). The most commonly used tissue has been teeth, as for terrestrial mammals (Klevezal and Kleinenberg, 1967). Fortunately, odontocetes (Figs. 1–3), pinnipeds, sea otters (Fig. 4), and polar bears (*Ursus maritimus*) have teeth that are suitable for use in estimating age. In contrast, because teeth cannot be used for baleen whales and manatees, other tissues or methods have been investigated. As alternatives, incremental layers have been found in bone, baleen, and ear plugs. Teeth

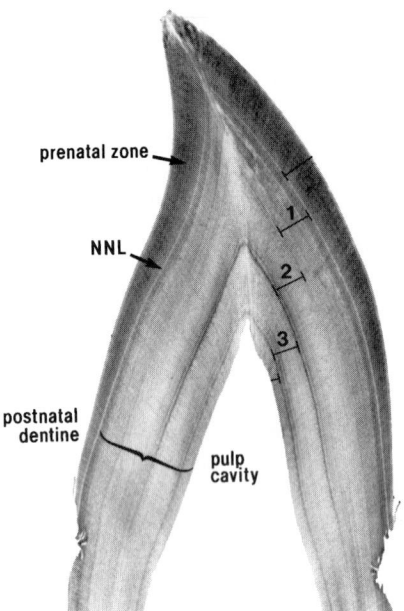

Figure 1 *Decalcified and stained midlongitudinal section in the buccal–lingual plane from a free-ranging bottlenose dolphin known to be 3 years of age. This view shows only the upper half of the section. The neonatal line (NNL) represents the time when the animal was born and, therefore, is age "0" for the purpose of estimating age. Dentine external to the neonatal line was deposited before birth and is known as prenatal dentine, whereas the neonatal line and dentine internal to it is postnatal dentine. A thin layer of enamel covered the prenatal dentine but was removed when the tooth was decalcified. The first three complete presumed annual growth layers or GLGs are marked in the sequence they were deposited. Teeth from young dolphins have very little cement and none can be seen in this photograph. Photograph from Hohn* et al. *(1989).*

have several advantages over these other tissues. The normal process of remodeling (resorption and reconstruction) in bone results in resorption of all but the most recent growth layers. For young animals, the number of bone layers may accurately reflect age; otherwise, the number of layers will be less than the age of the animal. The most useful bones are those that show negative allometry, i.e., growing more slowly than the skeleton as a whole (Klevezal, 1996). Growth layers have also been identified in baleen. Unfortunately, baleen abrades fairly quickly during normal use, and relatively few growth layers accumulate. Ear plugs are restricted to just a few species of whales and are challenging to collect.

In the normal course of events, teeth do not remodel, and growth layers continue to be deposited throughout the life of the individual. Teeth are easy to collect, store, and section and have become the preferred means of age estimation for most species with teeth. Within a tooth, two tissues have been used for aging: dentine and cement. New dentine is deposited on the internal surface, i.e., from the pulp cavity side, so that layers deposited when the animal was youngest are found on the

outer edges of the tooth or at the crown (Figs. 1–3). Cement or cementum wraps around the outer dentine and functions in anchoring the tooth to its alveolus. In contrast to dentine, new cemental layers are deposited on the external surface (Fig. 3). In most species of cetaceans, the cemental layer is very thin and the resulting growth layers so fine that they can be difficult to differentiate. As a result, dentine is used primarily for estimating age. Notable exceptions include the franciscana, *Pontoporia blainvillei,* and the beaked whales, family Ziphiidae, where dentine is useful only for the first few years and then cement, which is extensive, must be used. In addition, for sperm whales (*Physeter macrocephalus*) and the beluga whale (*Delphinapterus leucas*), both cement and dentine are well developed and can be used. Because cetaceans have homodont dentition (the teeth are all the same), each tooth contains the same layering pattern except for the underdeveloped teeth found most anteriorly and posteriorly in the tooth rows.

For pinnipeds, sea otters, and polar bears cement is used most frequently for age estimation (Bodkin *et al.,* 1997; Garlich-Miller *et al.,* 1993) (Fig. 4), similar to most terrestrial mammals. For many species, dentine can give accurate age estimates for young animals, but the pulp cavity either becomes

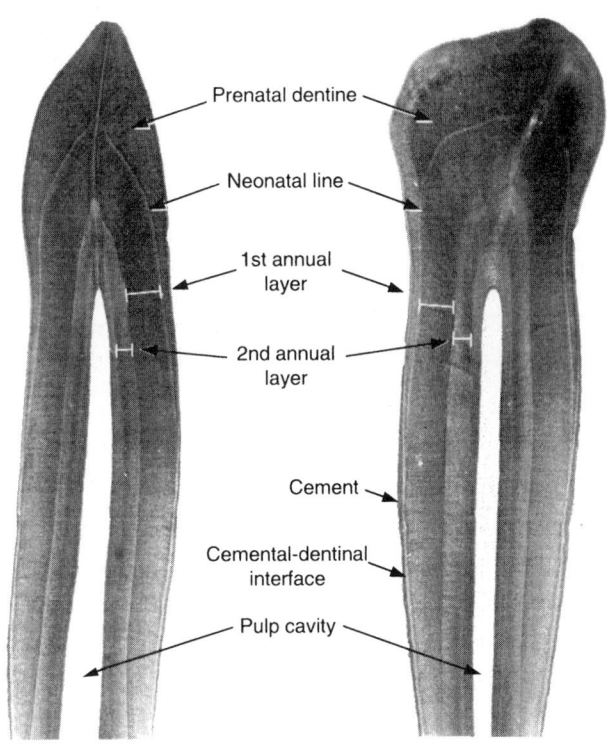

Figure 2 *Decalcified and stained midlongitudinal sections of teeth from a harbor porpoise (*Phocoena phocoena*). Porpoise teeth are spatulate. When sectioned along the buccal–lingual plane, they appear similar to dolphin tooth sections; when sectioned sagittally, the spatulate shape is apparent. The results are comparable in both orientations for this group. A narrow layer of cement occurs external to the dentine in the part of the tooth that was below the gum line.*

occluded or the dentine deposited is too irregular to resolve additional growth layers. Notable exceptions occur in some of the phocids, such as the ringed seal (*Pusa hispida*), Caspian seal (*Pusa caspica*), and the harbor seal (*Phoca vitulina*), where more than 15–20 dentinal layers can be found (Stewart *et al.*, 1996). For these species, which have heterodont dentition, canines are best for counting dentinal layers whereas postcanines are better for counting cement.

Although dentine and cement do not remodel like bone, teeth do wear down. When this occurs, it generally is not a problem for age estimation for species whose teeth show limited growth, i.e., do not continue to grow from the root but reach a maximum length when the animal is still relatively young. That is because an important marker for accurate age estimates is the neonatal line, which is deposited at birth and represents time zero for the purposes of counting growth layers (Figs. 1 and 2). As long as the neonatal line is visible, it is possible to obtain a complete count of growth layers. Initially, the neonatal line extends below the gum line. In species for which tooth growth is limited, even when the tooth wears down above the gum line, the neonatal line remains visible in the remaining tooth that was below the gum. In species with continuously growth teeth, such as the walrus (*Odobenus rosmarus*) (including mandibular teeth), bearded seal (*Erignathus barbatus*), narwhal (*Monodon monoceros*), members of the sperm whale family (Fig. 3), and the dugongs (*Dugong dugon*), wear continues as the tooth grows up from the root and eventually the neonatal line is worn away. When this occurs, the count of growth layers of dentine or cement is only a minimum. In some species, such as the beluga whale (*Delphinapterus leucas*), tooth wear is not equal and the best estimates of age are made from the least worn tooth.

Manatees (*Trichechus* spp.) present an unusual case for age estimation. In the related dugong, tusks (incisors) and other teeth provide a means for aging using techniques similar to those used for teeth from other species. Manatees lack tusks. Furthermore, they have an indeterminate number of molars that are constantly lost and replaced throughout life. Therefore, except in young animals, the number of growth layers in a tooth will reflect the age of the tooth but not the age of the individual manatee. As an alternative, it has been demonstrated in manatees that growth layers in tympano-periotic (auditory) bones are annual (Fig. 5) and that resorption occurs at a much slower rate than in other bones, meeting the requirement of a bone with negative allometry. More than 20 annual layers were found in many specimens and 59 found in a single animal (Marmontel *et al.*, 1996).

Baleen whales also present a special case for age estimation because they lack teeth. The rorquals (family Balaenopteridae) have ear plugs that are deposited in an annual layering pattern (Fig. 6) throughout life that are considered accurate for obtaining age estimates. These structures are more difficult to collect and are more fragile than teeth or bone. An advantage of ear plugs is that they do not resorb or wear. Other methods of aging have been investigated for balaenopterids, as well as other species of baleen whales. As in manatees, layers occur in the tympanic bullae (auditory) bone in bowhead (*Balaena mysticetus*), gray (*Eschrichtius robustus*), and common minke (*Balaenoptera acutorostrata*) whales (Christensen, 1995), often

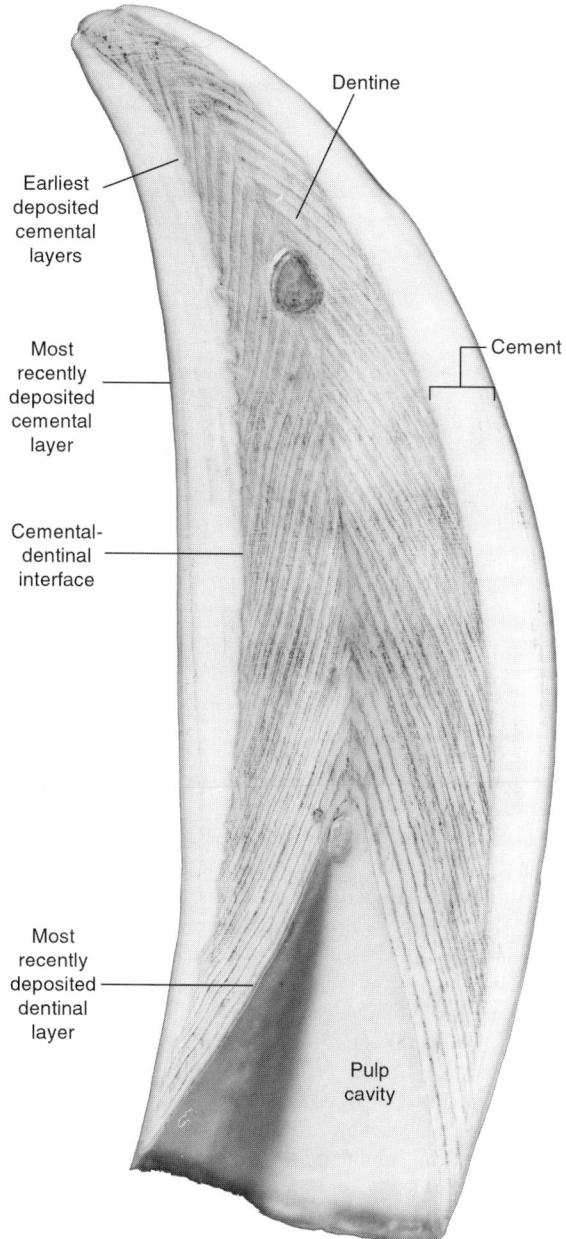

Figure 3 *Mandibular tooth from a sperm whale* (Physeter macrocephalus) *cut midlongitudinally in the buccal–lingual plane, etched in acid, and then rubbed with pencil to highlight growth layers in the dentine. Sperm whales have thick cement from which age can be estimated. In contrast to cement in the dolphin or porpoise tooth, here cement covers all of the dentine. In sperm whales and other species with continuously growing teeth, the tooth adds layers at the bottom (root end) and pushes upward. The cement was deposited when the dentine was still below the gum line. Erupted teeth wear continuously, and in older animals the earliest deposited layers are no longer present. The circular structures are pulp stones that form at the edge of the pulp cavity as globular masses of secondary dentine. Pulp stones are common in some species. Photograph courtesy of the International Whaling Commission.*

Labels on figure: Dentine; Earliest deposited cemental layers; Most recently deposited cemental layer; Cemental-dentinal interface; Cement; Most recently deposited dentinal layer; Pulp cavity

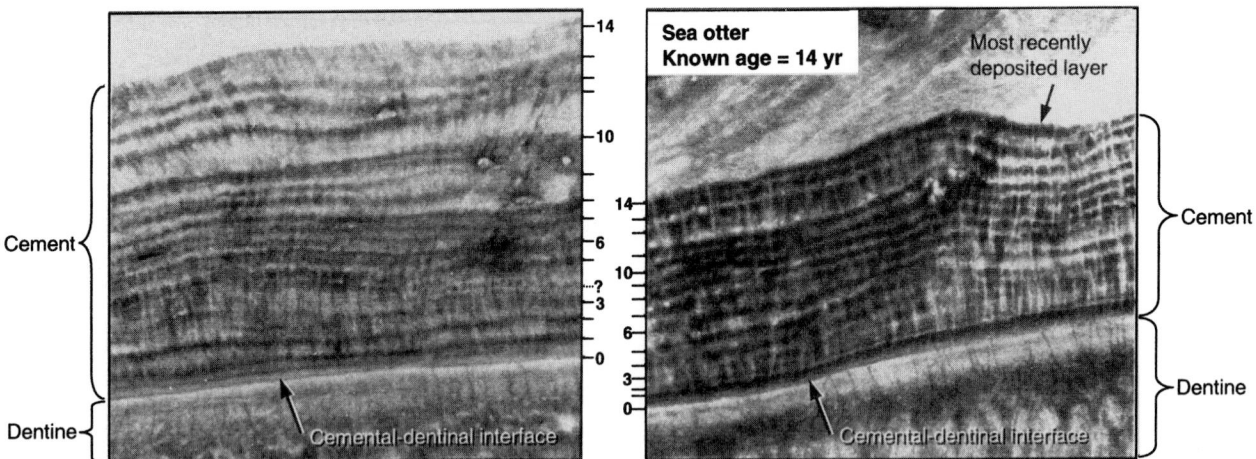

Figure 4 *Growth layer deposition in cement of a known-age sea otter (14 years). Images are from the same tooth section at different locations. In one location (right image), 14 well-defined, presumably annual layers are visible. These layers are exceptionally clear. In another location (left image), growth layers split and merge: on the right side there appear to be fewer layers, whereas on the left side there appear to be more layers. Presumed annuli are marked on the two images, with the marks on the left image before more subjective and a particularly uncertain layer marked with a dashed line. Counts begin at the interface where the dentine meets the cement, which represents time zero for counting growth layers. Positive identification of annual layers is made by carefully following layers along the tooth to watch for splitting and merging. Photographs of decalcified and stained thin sections courtesy of James Bodkin, USGS, and Gary Matson, Matson's laboratory. 250×.*

with no concomitant layers in other cranial or skeletal bones (Klevezal, 1996). Use of tympanic bullae is challenging because extensive effort is required to determine where exactly within the bullae the least amount of resorption and greatest resolution of growth layers will occur. When this region is located the maximal number of layers will be found. Otherwise, ages will be underestimated. Chemical signals, specifically amino acid racemization, have been used for dolphins and small and large species of whales (Bada *et al.* 1980), most recently fin (*B. physalus*) and bowhead whales (George *et al.*, 1999). Age is estimated as a function of the proportion of D and L isomers of aspartic acid in the lens of the eye.

IV. Collection and Preparation of Tissue for Age Estimation

When the primary tissue to be examined is dentine, especially for old animals, it is critical that a full midlongitudinal section be obtained. Otherwise, the very fine layers deposited in old animals will be missed. In toothed whales and dolphins (the odontocetes), the possibility of obtaining this midlongitudinal section is increased greatly if a tooth that is straight in the buccal–lingual plane (check to tongue) is used. Generally, the largest and straightest teeth occur near the center of the tooth row, and generally teeth are sectioned in the buccal–lingual plane. In some species, sections in the anterior–posterior plane are comparable (Fig. 2). It has become convention for studies on small odontocetes to use teeth from the center of the left ramus when possible (Perrin and Myrick, 1980). When using specimens from MUSEUM COLLECTIONS, often the teeth will have fallen out of the alveoli and so the straightest, largest (in that priority) teeth will be optimal. For studies using cemental

layers, postcanines or molars generally are the preferred tissue. In terrestrial mammals, some differences in counts of cemental growth layers among tooth sections from the same individuals have demonstrated that the thickness of the cement influences the deposition pattern, either because the cement is so narrow that layers are not readily distinguishable or because the cement is so thick that other incremental layers are apparent and may appear as annual layers (Klevezal, 1996). Differences in cemental thickness can occur both within a molar and between molars (Fig. 4). Ideally, a full investigation of the best site for sectioning can be made to select the optimal tooth and location within that tooth. When that selection has been made, midlongitudinally sections are more likely than cross sections to show all of the cemental layers, although cross sections are commonly used (Klevezal, 1996). As noted earlier , there is also variability in compact bone thickness in tympanic bullae, resulting in variability in number of growth layers visible; an investigation of the optimal site for sectioning is required. The bone is then cross-sectioned at that site. Ear plugs are sectioned centrally along the long axis of the plug.

Because growth layers are integral to bone and tooth structure, growth layer counts are not sensitive to most of the common ways of storing bones and teeth: cleaned of soft tissue and stored dry, such as in museum bone collections, or in alcohol, formalin, or glycerin. It has been suggested that long-term storage in formalin will affect growth layer counts if formalin degrades to formic acid (Perrin and Myrick, 1980). Some teeth will crack at the tip when stored dry, making sectioning a bit more difficult but not affecting the growth layers. Earplugs are stored in 5–10% buffered formalin (Lockyer, 1984). For amino acid racemization, eye lenses must be collected fresh and frozen immediately (George *et al.*, 1999)

Figure 5 *Growth layer deposition in the tympano-periotic bone of a Caribbean manatee* (Trichechus manatus) *that was maintained in captivity for 9 years. Eleven to 12 growth layers can be seen. These layers are primarily on the outer surface of the bone. Even at this age, the bone tissue is being resorbed and is beginning the remodeling process. Photograph of decalcified and stained thin section courtesy of Miriam Marmontel and the USGS Sirenia Project.*

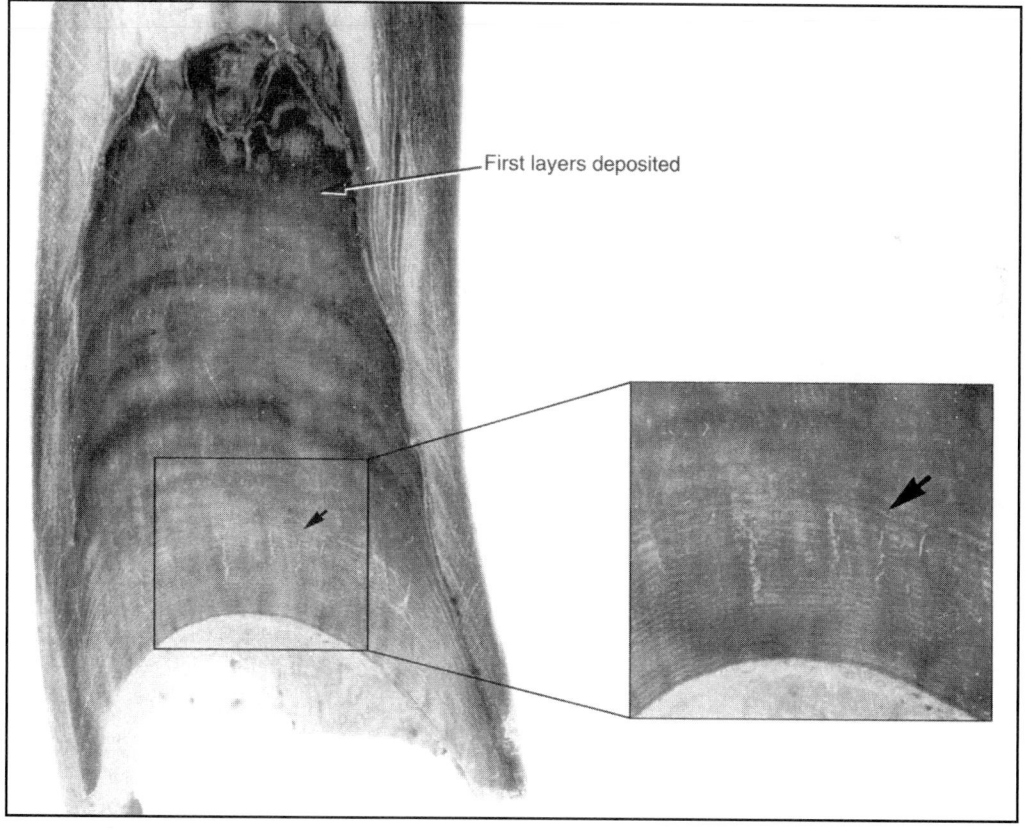

Figure 6 *Ear plug from a fin whale cut midlongitudinally to expose the growth layers. The arrow denotes a significant and abrupt change in growth layer characteristics that coincides with the transition of the animal from sexually immature to sexually mature. It is called the transition phase. Photograph from Lockyer (1984) and provided courtesy of C. Lockyer and the International Whaling Commission.*

Many creative methods have been tried to obtain the best resolution of growth layers (Perrin and Myrick, 1980). Two of these methods have persisted and become the most widely used: untreated sections (i.e, not decalcified and stained) and decalcified and stained thin sections. The former method generally involves using a low-speed saw with diamond blades to cut a section ranging from 50 to 200 μm thick, depending on the species and tissue, and counting layers directly from that section. The section may be mounted permanently on a microscope slide. This method was initially the most prevalent one for age estimation from teeth, a little less so for bone, and continues to be widely used because it is fast, easy, and requires little specialized equipment. The increasingly used alternative, decalcified and stained thin sections, requires additional preparation. For this method, whole teeth or thick sections from teeth or bone are decalcified in acid, sectioned on a microtome from 6 to 25 μm, depending on the species, tissue, and microtome used, and then stained in hematoxylin and sometimes counterstained with eosin, two routine histological stains. Sections are mounted on a microscope slide. It has become increasingly evident that using the easier method produces inaccurate results for both bone and tooth sections (Stewart *et al.*, 1996; Hohn and Fernandez, 1999). Stained thin sections allow for a much better resolution of growth layers in dentine, cement, and bone to the extent that many layers not apparent in untreated sections are visible and countable in stained sections. This difference is especially apparent in older animals where growth layers become increasingly thinner; staining is required to separate adjacent fine layers. As a result, many estimates of age using untreated sections are underestimates.

V. Consistency and Repeatability of Age Estimates

Because annual layers are not the only growth layers present, the interpretation of annual layers is often subjective. Misinterpretation of annual layers or differences in interpretation between investigators or studies lead to errors. Is one population but not another actually responding differently to exploitation or is an apparent difference simply caused by differences in age estimation? Is a population failing to recover because a growth model is incorrect or because the parameters used in that model were incorrect due to misinterpretation of annual layers?

Accuracy and precision are, in large part, influenced by the species being examined. For some species, growth layers are well defined and easily identified, whereas for other species growth layers are inherently indistinct. Annual layers in polar bear cement are notably difficult to interpret, at least during the first few years of life. Different areas in the cement have more or less distinct annual layers and accessory layers. Furthermore, different populations within a species may have different growth layer characteristics. For example, harbor porpoise from the Bay of Fundy have very distinctive growth layers, whereas those from California do not. Within studies it is common to conduct multiple readings of sections by one or more researchers to test for differences between readers or between readings for an individual reader. Measures of precision can be incorporated in models or can be used to evaluate the reliability of ages estimated for a sample.

Consistency and repeatability of age estimates can be increased if the tooth or bone sections are well prepared. Preparing these sections is a multistep process, and at each step the potential for error exists. If the end product is not well done, then the age estimate may be inaccurate or imprecise. For counts using dentine, a large source of error is using a section that is not midlongitudinal. For all sections, the incorrect stain or degree of staining (light or dark) and under- or overdecalcification also affect the final product in ways that prevent optimal resolution of all growth layers.

Even when sections are perfect, the subjective nature of counting growth layers still results in different age estimates. Descriptive models of the appearance, size, and complexity of annual layers have been developed to increase consistency, particularly between investigators. These models include photographs with the growth layers interpreted to be annual clearly marked (Hohn, 1990). Such photographs are equally valuable in individual studies to allow other investigators to determine whether the age estimates were obtained using comparable annual layering patterns. To date, such descriptive models have been prepared for bottlenose dolphins (Hohn *et al.*, 1989) and the franciscana (Pinedo and Hohn, 2000). Development of models for other species would increase the accuracy and precision of age estimates. Such models are particularly important and valuable when known-age specimens are available.

See Also the Following Articles

Bones and Teeth, Histology of ■ Cetacean Life History ■ Dental Morphology, Evolution of ■ Pinniped Life History ■ Sirenian Life History

References

Arnbom, T. A., Lunn, N. J., Boyd, I. L., and Barton, T. (1992). Aging live antarctic fur seals and southern elephant seals. *Mar. Mamm. Sci.* **8**, 37–43.

Bada, J. L., Brown, S., and Masters, P. M. (1980). Age determination of marine mammals based on aspartic acid racemization in the teeth and lens nucleus. *In* "Age Determination of Toothed Whales and Sirenians" (W. F. Perrin and A. C. Myrick, Jr., eds.), pp. 113–118. International Whaling Commission, Cambridge.

Bodkin, J. L., Ames, J. A., Jameson, R. J., Johnson, A. M., and Matson, G. M. (1997). Estimating age of sea otters with cementum layers in the first premolar. *J. Wildl. Manage.* **61**, 967–973.

Bowen, W. D., Sergeant, D. E., and Øritsland, T. (1983). Validation of age estimation in the harp seal, *Phoca groenlandica*, using dentinal annuli. *Can. J. Fish. Aqu. Sci.* **40**, 1430–1441.

Christensen, I. (1995). Interpretation of growth layers in the periosteal zone of tympanic bulla from minke whales *Balaenoptera acutorostrata*. *In* "Whales, Seals, Fish and Man" (A. S. Blix, L. Walløe, and Ø. Ulltang, eds.), pp. 413–423. Elsevier Science, New York.

Garlich-Miller, J. L., Stewart, R. E. A., Stewart, B., and Hiltz, E. A. (1993). Comparison of mandibular with cemental growth-layer counts for ageing Atlantic walrus (*Odobenus rosmarus rosmarus*). *Can. J. Zool.* **71**, 163–176.

George, J. C., Bada, J., Zeh, J., Scott, L., Brown, S. E., O'Hara, T., and Suydam, R. (1999). Age and growth estimates of bowhead whales (*Balaena mysticetus*) via aspartic acid racemization. *Can. J. Zool.* **77**, 571–580.

Hohn, A. A. (1990). Reading between the lines: Analysis of age estimation in dolphins. *In* "The Bottlenose Dolphin" (S. Leatherwood and R. R. Reeves, eds.), pp. 575–585. Academic Press, New York.

Hohn, A. A., and Fernandez, F. (1999). Biases in dolphin age structure due to age estimation technique. *Mar. Mamm. Sci.* **15**, 1124–1132.

Hohn, A. A., Scott, M. D., Wells, R. S., Sweeney, J. C., and Irvine, A. B. (1989). Growth layers in teeth from known-age, free-ranging bottlenose dolphins. *Mar. Mamm. Sci.* **5**, 315–342.

Klevezal, G. A. (1996). "Recording Structures of Mammals." A. A. Balkema, Rotterdam.

Klevezal, G. A., and Kleinenberg, S. E. (1967). "Age Determination of Mammals by Layered Structures of Teeth and Bone." Translated from Russian by Israel Progr. Sci. Transl. Jerusalem.

Lockyer, C. H. (1984). Age determination by means of the earplug in baleen whales. *Rep. Int. Whal. Comm.* **34**, 692–696.

Marmontel, M., O'Shea, T. J., Kochman, H. I., and Humphrey, S. R. (1996). Age determination in manatees using growth-layer-group counts in bone. *Mar. Mamm. Sci.* **12**, 54–88.

McCann, T. S. (1993). Age determination. *In* "Antartic Seals: Research Methods and Techniques" (R. M. Laws, ed.), pp. 199–226. Cambridge Univ. Press, Cambridge.

Myrick, Jr., A. C., and Cornell, L. H. (1990). Calibrating dental layers in captive bottlenose dolphins from serial tetracycline labels and tooth extractions. *In* "The Bottlenose Dolphin" (S. Leatherwood and R. R. Reeves, eds.), pp. 587–608. Academic Press, New York.

Myrick, Jr., A. C., Shallenberger, E. W., Kang, I., and MacKay, D. B. (1984). Calibration of dental layers in seven captive Hawaiian spinner dolphins, *Stenella longirostris,* based on tetracycline labeling. *Fish. Bull.* **82**, 207–225.

Perrin, W. F., and Myrick, A. C., Jr. (eds.) (1980). "Age Determination of Toothed Whales and Sirenians." International Whaling Commission, Cambridge.

Pinedo, M. C., and Hohn, A. A. (2000). Growth layer patterns in teeth from the franciscana, *Pontoporia blainvillei:* Developing a model for precision in age estimation. *Mar. Mamm. Sci.* **16**, 1–27.

Scheffer, V. B. (1950). Growth layers on the teeth of Pinnipedia as an indication of age. *Science* **112**, 309–311.

Sergeant, D. E. (1959). Age determination in odontocete whales from dentinal growth layers. *Norsk Hvalf.-Tid.* **48**, 273–288.

Stewart, R. E. A., Stewart, B., Stirling, I., and Street, E. (1996). Counts of growth layer groups in cementum and dentine in ringed seals *(Phoca hispida)*. *Mar. Mamm. Sci.* **12**, 383–401.

Tautenberger, J., Tsai, M., Krafft, A., Lichy, J., Reid, A., Schulman, F., and Lipscomb, T. (1996). Two morbilliviruses implicated in bottlenose dolphin epizootics. *Emerg. Infect. Dis.* **2**, 213–216.

Aggressive Behavior (Intraspecific)

CLAUDIO CAMPAGNA

*Centro Nacional Patagónico,
Puerto Madryn, Argentina*

The heterogeneous phenomenon considered as intraspecific aggressive or agonistic behavior represents a conglomerate of social responses, including male disputes over territorial boundaries, female fights to protect an offspring, female harassment and forced copulations, and infant abuse and killing. Agonistic encounters:

1. Mediate competition for limited resources economically defendable and valuable to the fitness of an individual (Bartholomew, 1970). Finite resources that can be monopolized would lead to social conflict between individuals of different sexes and generations and of the same sex and similar age class and status. Most often, agonistic confrontation (at least the most conspicuous interactions) involves sexually mature males.

2. Are more common in some social contexts, such as breeding on land in a polygynous mating systems, in which competition for resources is typically solved via aggressive disputes. Size and strength (but also agility) correlate positively with winning a contest through exerting dominance over individuals subdued by the costs of rebellion.

3. Have a broad range of costs for actors and recipients, from simple rejection after a ritualized threat display to injury or even death after an overt physical encounter.

The form and frequency of agonistic behavior partially reflect the sophistication of a social system. Aquatic mammals vary widely in the complexity of their societies, thus in the manifestation of agonistic behaviors. The most openly competitive societies characterize the otariids, the walrus (*Odobenus rosmarus*), and phocids that live in crowded conditions (e.g., elephant seals, *Mirounga* spp., and gray seals, *Halichoerus grypus*), a fertile ground for aggressive social interactions (Riedman, 1990). Conversely, polar bears (*Ursus maritimus*), all the mysticetes and river dolphins, and some other phocids (e.g., Ross and leopard seals, *Ommatophoca rosii* and *Hydrurga leptonyx*) generally occur in smaller social groups, except for periods during reproduction in which breeding males engage in scramble competition over receptive females (Berta and Sumich, 1999). The most complex social systems in the aquatic mammals would characterize some of the odontocete cetaceans, such as killer whales, *Orcinus orca*, pilot whales, *Globicephala* spp., bottlenose dolphins, *Tursiops* spp., or sperm whales, *Physeter macrocephalus* (Connor *et al.*, 1998). These species live in stable social units and show coordinated, cooperative behaviors. The long-term shared history among individuals of the group would have ritualized many of the overt aggressive responses typical of the polygynus pinnipeds.

I. Male–Male Competition for Mates

Competition over limited resources to maximize reproductive success would be the most common origin of agonistic encounters. It is likely that in all aquatic mammals, males compete for access to reproductive females, by either direct or indirect monopolization, through achieving the best place for reproduction or the highest status in a dominance rank. Defense systems can set the stage for the evolution of sexually selected traits, such as dimorphism in size and in special morphological structures (e.g., tusks, manes, elongated snouts).

The behavioral manifestations of conflict directed to the intimidation of rivals is often referred to as agonistic display or agonistic social signaling. Behavioral displays include vocal signals, facial expressions, and stereotyped postures and movements, such as static open-mouth threats, open-mouth sparring, foreflipper raise or waving, and oblique staring. Overt fighting is commonly a last-resort solution to conflict.

A. Pinnipeds

The form of male agonistic encounters and their outcome has been described in detail for several pinnipeds. Within the highly polygynous otariids and phocids there are examples of resource-defense (territorial) and female-defense polygynous systems (Riedman, 1990). Both types of polygyny may occur in the same species, such as in the South American sea lion, *Otaria flavescens,* as a function of different ecological conditions (Campagna and Le Boeuf, 1988).

The establishment and defense of a territory involve vocal displays, stereotyped postures and movements, and fights. During territorial displays, male contenders may rush toward each other with the mouth open or vocalizing, weave the head from side to side, puff out the chest, or perform the "oblique stare" posture at one another, but physical contact is usually avoided. Much of the fighting between otariid males takes place early in the breeding season, when territorial boundaries are being established. When physical contact occurs, it typically lasts a few seconds but may be violent, particularly in the largest sea lions. Fights involve chest-to-chest pushing, vigorous biting of the neck and face, lunging, and slashing at the opponent's flippers, chest, and hindquarters.

In female-defense polygyny, females cannot be sequestered or attracted to a particular place. Males then compete to achieve a position among the females in the breeding colony and move with the shifting population of females [see Boness and James (1979) for gray seals and Campagna and Le Boeuf (1988) for South American sea lions]. Association to a particular group of females is loose and may change even during the same day due to female redistribution related to the physical environment (high temperatures, variable space due to tidal movements) or to social behaviors (e.g., group raids of ousted males into the colony; Campagna *et al.,* 1988).

In phocids such as elephant seals, males aggressively establish a dominance hierarchy rather than a resource or female defense system. Only the highest ranking individuals have undisturbed reproductive access to females (Le Boeuf and Laws, 1994). During the establishment of hierarchies, males attempt to intimidate each other with vocal displays. If none of the contestants retreats, then a chest-to-chest fight takes place. Fights in elephant seals are violent confrontations and may last half an hour. Each bull throws his weight against the other and slashes at his opponent's face, neck, and back with long canines. Most fights end with multiple lacerations and bloody wounds or even a broken canine tooth; even death may occur on rare occasions.

Vocal threats are a common component of agonistic encounters. Pinnipeds vocalize both in air and underwater (Riedman, 1990). Harp (*Pagophilus groenlandicus*), ringed (*Pusa hispida*), Weddell (*Leptonychotes weddellii*), and bearded (*Erignathus barbatus*) seals and the walrus have a rich underwater vocal repertoire. Males maintain underwater territories and vocalizations seem to be part of territorial displays. Vocal displays can be of a repetitive nature and are then termed songs. Otariid males, particularly among the fur seals (*Arctocephalus* spp.), have a rich variety of airborne threat vocalizations associated with boundary display postures. The California sea lion, *Zalophus californianus,* vocalizes both in air and underwater (several phocids also produce airborne and underwater sounds).

The strong airborne calls or barks of *Zalophus* occur during breeding and nonbreeding seasons and may serve to advertise dominance. In elephant seals, airborne threat displays consist of loud and directional pulsed sounds that tend to precede fights.

B. Cetaceans

There is comparatively little description of agonistic encounters in the rest of the aquatic mammals. Agonistic behaviors to establish dominance relationships were described among dolphins in captivity. Observations of free-ranging cetaceans described a range of behaviors interpreted as agonistic, such as LOBTAILING, tail and flipper slaps to the body of other individuals, open-mouth postures, jaw claps, forceful exhalations, chases, body charges and leaps and body slaps, and vocal threat displays (Wells *et al.,* 1999). Escalated agonistic displays involve striking with flukes, biting, and jousting with tusks, the latter in narwhals (*Monodon monoceros*). Humpback whale (*Megaptera novaeangliae*) males fight vigorously in surface-active groups and receive not only scrapes and scratches but also deep gouges and bloody wounds as a result (Tyack and Whitehead, 1983).

The scar pattern of some odontocetes has been interpreted as the consequence of tooth marks and violent interactions. Several odontocetes have conspicuous scars. In Risso's dolphins, *Grampus griseus,* narwhals and several of the beaked whales, most of the body is covered with scars. At least for the narwhal, scars have been associated with intraspecific agonistic encounters (see later). Scrape marks are also common in baleen whales. It has been suggested for the southern right whales, *Eubalaena australis,* that males may use the thorny callosities during scramble competition over females.

Agonistic contests in cetaceans also involve vocal displays. Males of the humpback whale escort receptive females and vigorously rebuff other males by threatening displays such as thrashing of the flukes. The underwater songs of humpback whales in Hawaiian breeding grounds are performed by males and likely serve as communication signals in the context of male competition.

An example of male–male competition involves Australian bottlenose dolphins, *Tursiops aduncus* (Connor *et al.,* 1998). Males of this population form stable alliances of two or more individuals that cooperate to obtain and control reproductive females. Two alliances occasionally combine efforts to sequester or defend females from another alliance. Alliances in dolphins and group raids in sea lions (see later) represent special cases in which competition involves the participation of several individuals simultaneously.

C. Other Aquatic Mammals

Sea otters (*Enhydra lutris*), polar bears, and sirenians tend to be more solitary or live in low-density societies with little interaction among individuals (Berta and Sumich, 1999). Male sea otters are polygynous, establish breeding territories, and mate in the water. Females live in low-density areas chosen in relation to the distribution and abundance of food.

During the mating season, polar bear males rove to locate receptive females that are dispersed and solitary. Males access one female at a time. Competition involving physical interactions has been observed rarely but is indicated by broken teeth and scarring on the head and neck.

Manatees (*Trichechus* spp.) form mating groups in which several males compete for access to a receptive female by pushing and shoving each other. Physical competition for females also occurs in dugongs (*Dugong dugon*) with some males obtaining scars probably made by the tusks of other males.

II. Tusks as Special Structures for Aggression?

Two species of marine mammals have extraordinarily developed tusks: the walrus and the narwhal. The two upper canines in both male and female walrus are extraordinarily elongated (Riedman, 1990). The massive tusks of a male can weigh up to 10 pounds and be almost 1 m long. Both sexes use tusks in squabbles, to threaten one another, and, perhaps, to establish dominance. Males may force their way to selected places in crowded colonies by pushing and jabbing other walruses with their tusks.

The tusks of narwhals are even more exceptional morphological traits. As a general rule, the left canine in males extends anteriorly into a spiraled tusk to a length that may exceed 2.5 m. Some males have two tusks and a few females also develop one or even two shorter and less robust tusks. It has been suggested that narwhal tusks may be used to disturb or pierce prey, to open breathing holes in the ice, as defense weapons against predators, or as organs of sexual display. Although tusks may be used in more than one context, evidence shows that they serve in aggressive encounters (Silverman and Dunbar, 1980; Gerson and Hickie, 1985). Evidence includes direct observations of males crossing tusks and striking them against one another, scar patterns (with adult males having more and larger scars on the head after attaining sexual maturity), significantly higher incidences of broken tusks in mature males compared to immature individuals or females, and imbedded splinters and tusk tips found in the head of males. Tusks are also used to spear individuals of other species or, apparently, at times even female narwhals.

III. Sexual Selection and Special Morphological Traits

Pronounced SEXUAL DIMORPHISM in the direction of males being heavier and larger than females is common in all otariids, the walrus, and some phocids (e.g., elephant and gray seals). This kind of dimorphism often indicates direct physical confrontation among reproductive males involving pushing or strength contests. Dimorphism is not apparent, is slight, or is even reversed in most other phocids. A lack of or even reversed dimorphism is often accompanied by the defense of aquatic territories, aquatic mating, and serial monogamy. Females in these species are usually dispersed and breeding occurs over a protracted period. Social and ecological conditions do not favor frequent direct physical confrontations, but competition does occur, and may for more agile rather than larger individuals.

Among other aquatic mammals, males are much larger than females in some odontocetes, such as killer and sperm whales, whereas dimorphism is reversed in all the mysticetes. Mysticetes may have promiscuous MATING SYSTEMS in which competition for insemination takes place at the level of males displacing each other from the vicinity of a female and of sperm cells displacing or diluting sperm of other males. Gray (*Eschrichtius robustus*), right (*Eubalaena* spp.), and bowhead (*Balaena mysticetus*) whales have larger testes than expected based on their body weight, suggesting selection for sperm competition.

In addition to dimorphism in body size, males of some species evolved special secondary sexual features that may function in the context of competition for mates. Examples include the enlarged snouts of male elephant seals and gray seals and the inflatable nasal cavity of hooded seals (*Cystophora cristata*). Hooded seal males can blow a red, balloon-like sac from one nostril that is similar in shape to the long proboscis of elephant seals. These organs have visual or acoustic effects and may allow other males and females to judge the quality of a contender or a sexual partner. The developed neck and mane of sea lions with long and thick guard hairs also has visual effects and serves as a shield that protects internal organs from bites.

IV. Avoiding Fights

Competition for resources by direct aggression is a costly experience in species capable of inflicting serious injuries that could lower future fitness of the contestants. Thus, contenders with low chances of success should avoid physical confrontations. Theory predicts that the assessment of the fighting ability of competitors and of resource value prior to an escalation of violence may allow differential adaptive responses on the basis of the perceived asymmetries (Maynard Smith and Parker, 1974). Once a territory or social hierarchy is established, disputes tend to be asymmetric contests in which territory owners or high-ranking males almost always win. Threat displays may then serve as indicators of a quality and motivational state of a contender. Individual variation in vocal displays may help territorial males to recognize one another and to forgo direct competition if each knows its respective status.

In female-defense systems, the proportion of sexually receptive females accessible to a male is variable in space and time. Thus, the level of asymmetry can vary within the same day of a breeding season. This social context would favor behaviors that are unusual in strict territorial or hierarchical systems, such as group raids in South American sea lions.

V. Group Raids and Other Forms of Male Harassment of Reproductive Females

In the South American sea lion, losers in male–male competitions at times raid breeding colonies in groups of dozens of individuals (Campagna *et al.*, 1988). Raiders abduct females from the harems of established males and attempt to mate with them. A male seizes a female in his jaws and hurls her into the air to a spot where he can hold his ground against other males while aggressively keeping her in place. In the process, females are often wounded and can be killed. Perhaps group raids represent a primitive stage of a male alliance or coalition.

Violent behavior toward females is relatively common in pinnipeds. Harassed females are injured and sometimes killed by males during mating attempts. Le Boeuf and Mesnick (1990) suggested some social conditions that can increase mortality risks to a female during mating: (a) marked male sexual dimorphism, (b) males outnumbering females, (c) use of force or potentially dangerous weapons in mating, and (d) monopolization of mating by a few individuals through direct or indirect control of resources (space, females, food, etc.) with forcible exclusion of the majority of the competitors. All of these traits are common in the most polygynous mating systems.

The majority of female deaths during the breeding season of elephant seals, the most sexually dimorphic of all the pinnipeds, occurs by traumatic injuries inflicted by males during mating attempts as the females depart the harems for the sea at the end of lactation. Male South American sea lions and elephant seals are three to five times heavier than females, have large canines, and often bite the neck of the female when copulating. Breeding colonies early and late in the season have a high number of males that intercept departing females and attempt to mate with them. Mating injuries inflicted by males to females have also been reported for several other species [e.g., gray seals, Boness *et al.* (1995), Hawaiian monk seals (*Monachus schauinslandi*), Hiruki *et al.* (1993)]. Male aggression toward females may be a selective force in shaping female behavior, female choice, maternal performance, and reproductive synchrony (Boness *et al.,* 1995).

VI. Female Agonistic Behavior

In polygynous pinnipeds, females are aggressive toward one another and rarely tolerate neighbors close by, which helps to regulate density of a site. A common context of female agonistic encounters is that of protection of a pup in a crowded breeding colony. Alien pups are often bitten by females. Aggressive mothers react rapidly and intensively to the threat to their pup by a neighbor, which enhances chances of pup survival by decreasing the risks of mother–pup separation and pup injury (Christenson and Le Boeuf, 1978).

At times, females threaten transient males when the latter approach or protest vocally when males mount them. As a result, a harassing male will then be more likely challenged by another male who hears the female vocalizing. These challenges generally interrupt a male's approach or mount, and hence a potential copulation. By resisting male copulatory attempts, females increase their likelihood of mating with a dominant individual, which may be viewed as an indirect form of mate choice.

VII. Abuse and Killing of Young

Infanticide is the killing by conspecifics of young still dependent on their mothers. Infant abuse implies injury of a young either via active violent behaviors or via passive neglect. Violent abuse of pups by males (most often young individuals but also adults) occurs in several pinniped species, particularly in sea lions and elephant seals. The killing of young is most often the by-product of abuse, although it may also occur as a directed behavior. In addition to pinnipeds, infanticide has been described in polar bears and is inferred in at least one odontocete, the common bottlenose dolphin.

See Also the Following Articles

Infanticide and Abuse of Young ■ Parental Behavior ■ Territorial Behavior

References

Bartholomew, G. A. (1970). A model for the evolution of pinniped polygyny. *Evolution* **24,** 546–559.

Berta, A., and Sumich, J. L. (1999). "Marine Mammals: Evolutionary Biology." Academic Press, San Diego.

Boness, D. J., and James, H. (1979). Reproductive behavior of the grey seal (*Halichoerus grypus*) on Sable Island, Nova Scotia. *J. Zool. (Lond.)* **188,** 477–500.

Boness, D. J., Bowen, W. D., and Iverson, S. J. (1995). Does male harassment of females contribute to reproductive synchrony in the grey seal by affecting maternal performance? *Behav. Ecol. Sociobiol.* **36,** 1–10.

Campagna, C., and Le Boeuf, B. J. (1988). Thermoregulatory behavior in the southern sea lion and its affect on the mating system. *Behaviour* **107,** 72–90.

Campagna, C., Le Boeuf, B. J., and Cappozzo, H. L. (1988). Group raids in southern sea lions. *Behaviour* **105,** 224–249.

Christenson, T. E., and Le Boeuf, B. J. (1978). Aggression in the female northern elephant seal, *Mirounga angustirostris. Behaviour* **64,** 158–172.

Connor, R. C., Mann, J., Tyack, P. L., and Whitehead, H. (1998). Social evolution in toothed whales. *Trends Ecol. Evol.* **13**(6), 228–232.

Gerson, H. B., and Hickie, J. P. (1985). Head scarring on male narwhals (*Monodon monoceros*): Evidence for aggressive tusk use. *Can. J. Zool.* **63**(9), 2083–2087.

Hiruki, L. M., Gilmartin, W. G., Becker, B. L., and Stirling, I. (1993). Wounding in Hawaiian monk seals (*Monachus schauinslandi*). *Can. J. Zool.* **71,** 458–468.

Le Boeuf, B. J., and Laws, R. M. (1994). "Elephant Seals." Univ. of California Press, Berkeley.

Maynard Smith, J., and Parker, G. A. (1974). The logic of asymmetric contests. *Anim. Behav.* **24,** 159–175.

Riedman, M. (1990). "The Pinnipeds: Seals, Sea Lions and Walruses." California Univ. Press.

Silverman, H. B., and Dunbar, M. J. (1980). Aggressive tusk use by the narwhal (*Monodon monoceros* L.). *Nature* **284,** 57–58.

Tyack, P. L., and Whitehead, H. (1983). Male competition in large groups of wintering lumpback whales. *Behaviour* **83,** 132–154.

Wells, R. S., Boness, D. J., and Rathbun, G. B. (1999). Behavior. *In* "Biology of Marine Mammals" (J. E. Reynolds III and S. A. Rommel, eds.). Smithsonian Institution Press.

Albinism

DAGMAR FERTL
Minerals Management Service, U.S. Department of the Interior, New Orleans, Louisiana

PATRICIA E. ROSEL
National Marine Fisheries Service, Charleston, South Carolina

Albinism refers to a group of inherited conditions resulting in little or no pigment (hypopigmentation) in the eyes alone or in the eyes, skin, and hair. In humans, all types of albinism exhibit abnormalities in the optic system, including

Figure 1 *Anomalously white humpback whale sighted off Australia. Photo by Paul Forestell, Pacific Whale Foundation.*

misrouting of the optic fibers between the retina and the brain, and incomplete development of the fovea, the area of the retina where the sharpest vision is located. Thus, these characteristics can provide useful diagnostic criteria for identifying albinism. Inheritance of an altered copy of a gene that does not function correctly is the cause of most types of albinism. Albinos have white or light skin and hair, and often pink eyes, although the eye color can vary from dull gray to brown. The "pink" eyes are due to the reflection from choroid capillaries behind the retina. Albinism is differentiated from piebaldism (body pigmentation missing in only some areas) and leucism (dark-eyed anomalously white animals). Pigmentation patterns should not be the only criterion used to define albinism, as some mutant phenotypes (pseudo-albinism) may be due to the action of genes at other loci.

I. Pigmentation

Mammalian color is almost entirely dependent on presence (or absence) of the pigment melanin in the skin, hair, and eyes. Melanin is produced through a stepwise biochemical pathway in which the amino acid tyrosine is converted to melanin. The enzyme tyrosinase plays a critical role in this pathway, and al-

terations or mutations in the tyrosinase gene can result in a defective enzyme that is unable to produce melanin, or does so at a reduced rate. Mutations in five other genes have also been identified in different types of albinism in humans.

II. Problems Associated with Albinism

Humans with albinism often are photophobic and have other vision impairments, such as extreme far-sightedness, near-sightedness, and astigmatism. There are unpublished reports of apparent vision problems for albino seals, when they are on shore. Costs of this aberrant pigmentation for marine mammals may include reduced heat absorption in colder waters, increased conspicuousness to predators, and impaired visual communication.

III. Albinism and Marine Mammals

Anomalously white individuals have been reported for 20 cetacean species (Fertl *et al.*, 1999) (Fig. 1); they have also been reported for pinnipeds (e.g., Rodriguez and Bastida, 1993). No reports are known of anomalously white sea otters (*Enhydra lutris*) or sirenians. Anomalously white individuals are often

Figure 2 *An albino killer whale ("Chimo") postmortem diagnosed with Chédiak–Higashi syndrome. Photo by Peter Thomas.*

presumed to be true albinos. Some of those individuals match the description of true albinism [e.g., there are well-documented reports of albino sperm whales (*Physeter macrocephalus*) and bottlenose dolphins (*Tursiops truncatus*)], but many do not. "Chimo," an anomalously white killer (*Orcinus orca*) captured for display in Canada, was diagnosed postmortem with Chédiak-Higashi Syndrome, (Fig. 2), a type of albinism. This inherited disorder is characterized by diluted pigmentation patterns that appear pale gray, white blood cell abnormalities, and a shortened life span. Whales and dolphins also may appear white if extensively scarred, or covered with a fungus, such as Lobo's disease.

See Also the Following Articles

Coloration ■ Hair and Fur ■ Vision

References

Alhaidari, Z., Olivry, T., and Ortonne, J.-P. (1999). Melanocytogenesis and melanogenesis: Genetic regulation and comparative clinical diseases. *Vet. Dermatol.* **10**, 3–16.

Fertl, D., Pusser, L. T., and Long, J. J. (1999). First record of an albino bottlenose dolphin (*Tursiops truncatus*) in the Gulf of Mexico, with a review of anomalously white cetaceans. *Mar. Mamm. Sci.* **15**, 227–234.

Hain, J. H. W., and Leatherwood, S. (1982). Two sightings of white pilot whales, *Globicephala melaena*, and summarized records of anomalously white cetaceans. *J. Mammal.* **63**, 338–343.

Oetting, W. S., and King, R. A. (1994). Molecular basis of oculocutaneous albinism. *J. Invest. Dermatol.* **103**, 131S–136S.

Oetting, W. S., and King, R. A. (1999). Molecular basis of albinism: Mutations and polymorphisms of pigmentation genes associated with albinism. *Hum. Mutat.* **13**, 99–115.

Rodriguez, D. H., and Bastida, R. O. (1993). The southern sea lion, *Otaria byronia* or *Otaria flavescens? Mar. Mamm. Sci.* **9**, 372–381.

Searle, A. G. (1968). "Comparative Genetics of Coat Colour in Mammals." Logos Press and Academic Press, London.

Taylor, R. F., and Farrell, R. K. (1973). Light and electron microscopy of peripheral blood neutrophils in a killer whale affected with Chediak-Higashi syndrome. *Fed. Proc.* **32**, 822.

Amazon River Dolphin
Inia geoffrensis

VERA M. F. DA SILVA
Instituto Nacional de Pesquisas da Amazônia, Manaus, Brazil

I. Genus and Species: Common Names and Taxonomy

The Amazon River dolphin, *Inia geoffrensis*, is known by different names throughout its distribution: boto in Brazil; bufeo and bufeo colorado in Colombia, Ecuador, and Peru; and tonina and delfin rosado in Venezuela. It is also known in English as pink dolphin, although the Brazilian name "boto" is considered the international common name.

The boto belongs to the superfamily Platanistoidea. The genus *Inia* is monospecific, with three currently recognized subspecies: *Inia geoffrensis geoffrensis, I. g. boliviensis,* and *I. G. humoldtiana.*

II. Distribution, Abundance, and Density

The boto has an extraordinarily wide distribution, occurring almost everywhere it can physically reach without venturing into marine waters. It occurs in six countries of South America—Bolivia, Brazil, Colombia, Ecuador, Peru, and Venezuela—in a total area of about 7 million km^2 (Fig. 1). It can be found along the entire Amazon River and its principal tributaries, smaller rivers and lakes, from the delta near Belém to its headwaters in the Ucayali and Marañon Rivers in Peru. Its principal limits are impassable falls such as those of the upper Xingú and Tapajós Rivers, and the Teotônio falls in the upper Madeira River in the southern part of the Amazon basin. The boto is also found throughout the Orinoco river basin, with the exception of the Caroni River and upper Caura River above Para falls in Venezuela. An isolated population occurs above Teotônio and Abuña falls in the upper Madeira River and in the Beni/Mamoré basin of Bolivia.

The boto is the most common river dolphin. Its current distribution and abundance apparently do not differ from in the past, although relative abundance and density are highly seasonal and appear to vary among rivers. During the dry season the dolphins are concentrated in the main channels of the rivers, whereas during the flooded season they dispense into the flooded forest (igapó) and river floodplains (várzea).

No quantitative estimation of the relative abundance of the boto between rivers or basins exists. Differences in survey methodology used by different authors and lack of effort make the comparison between the results of the different surveys available in the literature very difficult. The only long-distance surveys of the species were carried out on the Solimões-Amazon River, from Manaus to Santo Antônio do Içá-Tabatinga over a total of ca. 1200 km. The number of sightings per unit effort gave an average number of 332 ± 55 botos per survey ($n = 9$), and the estimated density was of 0.08–0.33 botos/km in the main river and 0.49–0.98 botos/km in the smaller channels. Another boat survey along ca. 120 km of the Amazon River bordering Colombia, Peru, and Brazil carried out by Vidal and collaborators (1997) estimated 345 (CV = 0.12) botos in the study area with a density per square kilometer of 4.8 in tributaries, 2.7 around islands, and 2.0 along the main banks. These figures suggest that the boto shows the highest densities among any cetacean.

III. External Characteristics

The boto (Fig. 2) is the largest of the river dolphins, with a maximum recorded body length of 255 cm and mass of 185 kg for males and 215 cm and ca. 150 kg for females. The body is corpulent and heavy but extremely flexible. Nonfused cervical vertebrae allow the movement of the head in all directions. The flukes are broad and triangular; the dorsal fin is long, low, and keel-shaped, extending from the midbody to the strong laterally flattened caudal peduncle. The flippers are large, broad, and paddle-like and are capable of circular movements. Although most of these characteristics restrict speed during swim-

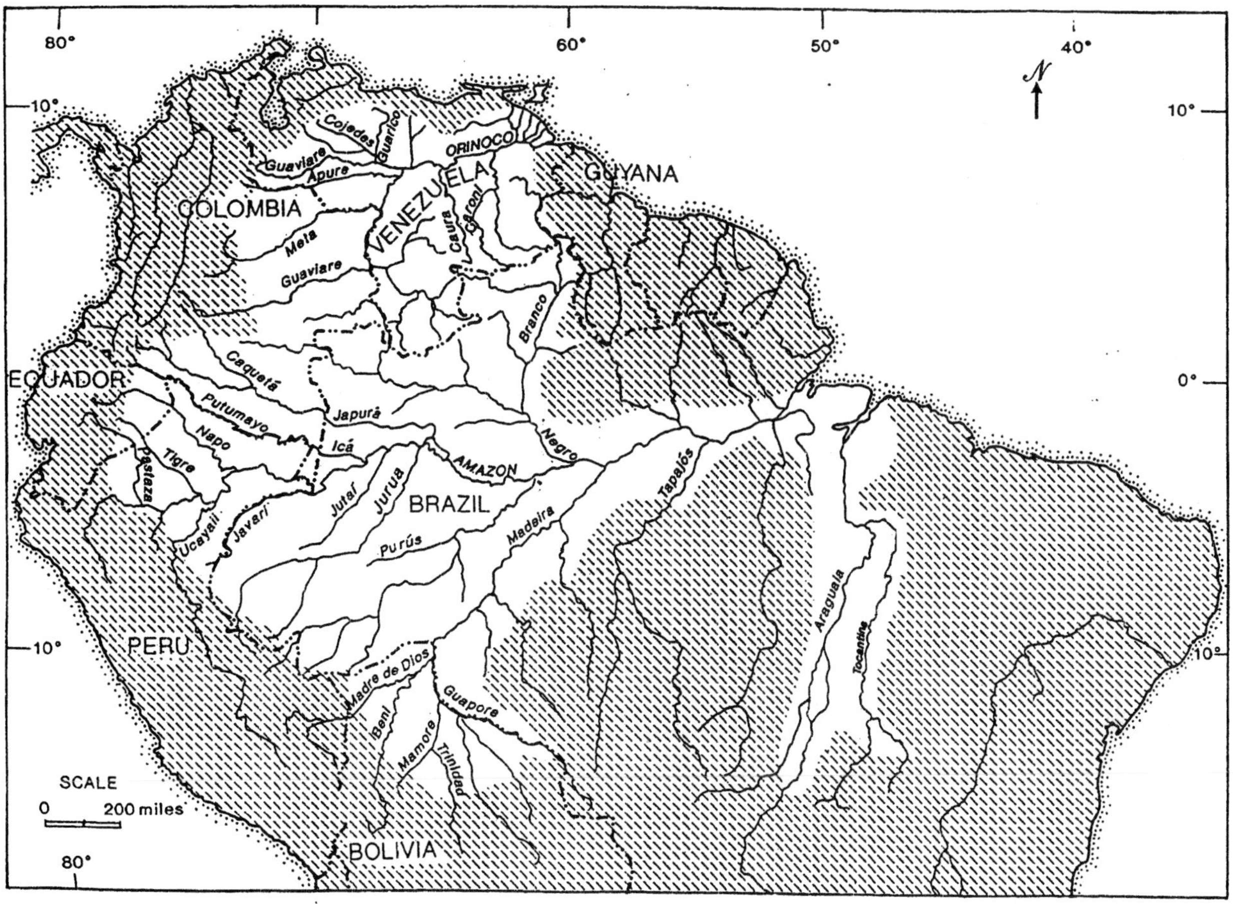

Figure 1 *Map showing the general distribution of the boto* (Inia geoffrensis) *in South America.*

ming, they allow this dolphin to maneuver between trees and submerged vegetation to search for food in the flooded forest. The rostrum and mandible are prominent, long, and robust. Short bristles on the top of the rostrum persist during adulthood. The melon is small and flaccid, but the shape can be altered by muscular control. The small, round eyes are functional and the vision is good, both under and above water.

Body color varies with age. Fetuses, neonates, and young animals are dark gray. Juveniles and subadults are uniform medium gray to pinkish, and older botos are completely pink or blotched pink. When adult botos are dark on the dorsum, the flanks and underside are pinkish. One albino was captured and maintained in captivity for more than 1 year in an aquarium in Germany.

IV. Behavior and Life History

The boto is at times solitary and is not often seen in cohesive groups of more than three individuals; most groups of two are mother–calf pairs. Loose aggregations may be seen at the mouth or in bends of rivers and canals due to the large concentrations

Figure 2 *Often characterized by pink body color, Amazon River dolphins* (Inia geoffrensis) *are the largest of the platanistoid dolphins. Pieter A. Folkens/Higher Porpoise DG.*

of fish or for purposes of courtship and mating. The boto is known to react protectively to injured or captured individuals.

The boto is a slow swimmer with a normal speed of 1.5–3.2 km/hr, but bursts of >14–22 km/hr have been recorded. The boto is capable of strong swimming for some length of time. When surfacing, the melon, tip of the rostrum, and long dorsal keel are out of the water simultaneously in a very conspicuous way. The boto does a high-arching roll in which these parts appear sequentially thrust well out of the water. The tail is rarely raised out of the water prior to a dive. Botos also wave a flipper, show the head or tail above the surface, lob-tail, and rarely jump clear of the water.

Studies in captivity indicated that botos are less timid and show less social contact, aggressive behavior, play, and aerial behavior than bottlenose dolphins (*Tursiops truncatus*). However, botos in captivity may not show their true range of behaviors. The boto is very curious and playful, rarely showing fear of strange objects. Wild botos grasp fishermen's paddles, rub against canoes, pull grass under water, throw sticks, and play with logs, clay, turtles, and fish. Several observers have reportedly seen botos in a stationary position, often upside down with the eyes closed.

The boto is active day and night. The greatest fishing activity occurs at 0600–0900 and 1500–1600 hrs. It feeds on over 43 species of fish belonging to 19 families. Stomach content analysis has revealed up to 11 fish species in one animal. The mean size of consumed fish is 20 cm (range 5–80 cm), with larger fish torn to pieces. In captivity, food sharing has been recorded. Daily consumption is about 2.5% of the body weight. The boto's diet is unique among cetaceans in that its heterodont dentition allows it to tackle and crush armored prey.

Males attain sexual maturity much later than females at about 200 cm in length. In females, sexual maturity occurs at around 5 years of age at body lengths between 160 and 175 cm. Reproductive events are seasonal. Gestation time has been estimated at about 11 months, and the calving season is apparently long, with most births occurring at the peak of the river's flood level. Length at birth is about 80 cm. Lactation lasts more than 1 year and the birth interval is 2 to 3 years. Studies of marking and recaptures carried out by da Silva and Martin in Central Brazil have shown that some individuals are resident in a particular area during the entire year.

V. Human Effects and Interactions

The boto is part of the FOLKLORE and culture of Amazonian people, and several legends and myths are commonly known throughout its distribution. Because of these legends, often giving the boto supernatural powers, the boto was protected and respected in the past, although body parts of incidentally captured animals have been used by local people for medical purposes and as love charms. With increased use of nylon gill nets, machine-made lampara seines, and other new fishing techniques, the incidental catching of botos has become more common. With greater demand for fish due to rapid increases in human populations, the boto's food sources are being reduced. Other threats to the species are the construction of hydroelectric dams on major tributaries affecting the abundance and

presence of some species of fish. Dams separate and isolate populations and may reduce the gene pool, and thereby increase chances of extinction. Analysis of milk from botos from the upper Amazon River (Letícia) and Central Amazon (Manaus) revealed that chemical pollution of the river systems by pesticides and mercury poses serious threats to the species.

VI. Status and Conservation

Inia geoffrensis is listed in Appendix II of the Convention on International Trade in Endangered Species of Wild Fauna and Flora (CITES) and is classified by the IUCN as Vulnerable because of serious threats throughout its range.

See Also the Following Articles

Endangered Species and Populations ▪ Folklore and Legends ▪ River Dolphins

References

Best, R. C., and da Silva, V. M. F. (1984). Preliminary analysis of reproductive parameters of the Boutu, *Inia geoffrensis,* and the tucuxi, *Sotalia fluviatilis,* in the Amazon River System. Reports of the International Whaling Comission, Special Issue 6.

Best, R. C., and da Silva, V. M. F. (1989). Amazon River dolphin, Boto–*Inia geoffrensis* (de Blainville, 1817). *In* "Handbook of Marine Mammals." (S. H. Ridgway and R. Harrison, eds.), 1st Ed. Academic Press, London.

da Silva, V. M. F., and Martin, R. A. (2000). A study of the boto, or Amazon River dolphin *Inia geoffrensis* in the Mamirauá Reserve, Brazil. IUCN, Occasional Papers SSC.

Layne, J. N. (1958). Observations on freshwater dolphins in the upper Amazon. *J. Mammal.* **39,** 1–22.

Martin, R. A., and da Silva, V. M. F. (1998). Tracking aquatic vertebrates in dense tropical forest using VHF telemetry. *MTS J.* **32**(1), 83–88.

Vidal, O., Barlow, J., Hurtado, L. A., Torre, J., Cendón, P., and Ojeda, Z. (1997). Distribution and abundance of the Amazon River dolphin (*Inia geoffrensis*) and the tucuxi (*Sotalia fluviatilis*) in the upper Amazon River. *Mar. Mamm. Sci.* **13**(3), 427–445.

Ambergris

DALE W. RICE
*National Marine Mammal Laboratory,
Seattle, Washington*

A mbergris is a substance that forms only in the intestines of the sperm whale (*Physeter macrocephalus*). The word comes from the Old French *ambre gris* or "gray amber," as opposed to *ambre jaune,* "yellow amber," which refers to the true, resinous amber. Most ambergris is found in the large intestine, but smaller pieces have been found in the small intestine, and it may be that it initially forms in the small intestine and subsequently passes into the large intestine. Probably most lumps of ambergris are eventually voided during defecation, unless they grow too large to pass through the anus. Ambergris occurs in only 1 to 5% of whales of both sexes and all ages, but the circumstances that induce its production remain unknown.

Ambergris forms as concretions that usually weigh 0.1 to 10.0 kg, but rarely much bigger pieces have been recovered; the largest on record, weighing 420 kg., was removed from a 14.9-m bull sperm whale killed in the Southern Ocean on December 21, 1953 (Clarke, 1954). Such huge masses greatly distend the whale's large intestine. Most pieces of ambergris are in the form of an irregular roundish lump, somewhat resembling a potato. Their specific gravity is 0.73 to 0.95. In consistency they are solid and friable, similar to nearly dry clay. Internally they usually show no laminations, but when broken apart they tend to fracture along concentric cleavage surfaces. In color they are pale yellowish to light gray on the inside, whereas the outer surface is dark brown with a varnished appearance. The chitinous beaks of cephalopods are almost invariably found imbedded in the lumps. Fresh ambergris has the highly distinctive pungent odor of sperm whale feces, but aged pieces have an almost pleasant musty or even musky smell.

Chemically, ambergris is a nonvolatile solid consisting mainly of a mixture of waxy, unsaturated, high molecular weight alcohols. The principal component is an ester of ambrein ($C_{23}H_{39}OH$), which gives it its peculiar properties and odor (Gilmore, 1951). One analysis gave the following chemical composition: ambrein, 25–45%; epicoprosterol, 30–40%; coprosterol, 1–5%; coprostanone, 3–4%; cholesterol, 0.1%; pristane, 2–4%; ketone, 3–4%; free acids, 5–8%; and residues insoluble in ether, 10–16% (Berzin, 1971; this analysis was mistranslated in the 1972 English edition of Berzin's book).

Contrary to the prevalent notion, ambergris is hardly ever found on beaches; most is recovered directly from whale carcasses. Through the years many people have brought me malodorous globs that they picked up on the seashore in hopes that it was ambergris; none of it ever was. If a suspected specimen of ambergris fits the physical description, the simplest way to confirm its identity is to heat a wire or needle in a flame and thrust it into the sample to a depth of about a centimeter; if the substance is really ambergris it will instantly melt into an opaque fluid the color of dark chocolate. When the needle is withdrawn, the ambergris will leave a tacky residue on it.

Ambergris was known throughout the Moslem world as early as the 9th century. There it was highly valued as an incense, an aphrodisiac, a laxative, a spice, an ingredient in candles and cosmetics, and as a medication for treating a diversity of ailments. Its reputation soon spread around the globe. In those days, ambergris was picked up on beaches or found floating on the sea, and its origin remained a complete mystery, thus giving rise to many fanciful and hotly debated theories. In 1574 the Flemish botanist Carolus Clusius was the first author to deduce from the inclusions of squid beaks in ambergris that it was the product of the digestive tract of whales. It was not until after the commencement of the American sperm whale fishery in 1712 that it became generally recognized that ambergris was produced solely by the sperm whale (Beale, 1839; Dannenfeldt, 1982). In the ensuing years, ambergris was prized mainly as a fixative for fragrances in perfumes. In the 20th century, synthetic chemicals replaced it so it no longer has much value.

See Also the Following Articles

Gastrointestinal Tract ■ Sperm Whale

References

Beale, T. (1839). "The Natural History of the Sperm Whale," pp. 1–393. John Van Voorst, London [Reprinted 1973 by The Holland Press, London].

Berzin, A. A. (1971). "Kashalot," p. 368. Izdatel'stvo "Pishchevaya Promyshlennost," Moscow.

Clarke, R. (1954). A great haul of ambergris. *Norsk Hvalfangst-Tidende* **43**(8), 286–289.

Dannenfeldt, K. H. (1982). Ambergris: The search for its origin. *Isis* **73**(268), 382–397.

Gilmore, R. M. (1951). The whaling industry: Whales, dolphins, and porpoises. *In* "Marine Products of Commerce" (D. K. Tressler and J. M. Lemon, eds.), pp. 680–715. Reinhold, New York.

Anatomical Dissection: Thorax and Abdomen

JOHN E. REYNOLDS III
Eckerd College, St. Petersburg, Florida

SENTIEL A. ROMMEL AND MEGHAN E. BOLEN
Florida Marine Research Institute, St. Petersburg

The general organization of the postcranial soft tissues does not vary appreciably among mammals. Factors that may influence the relative proportions or positions of organs and organ systems include phylogeny and adaptations to a particular environment or trophic level.

The structure and function of specific postcranial organs or organ systems are described in other articles of this encyclopedia. This article provides a "road map" that orients a prosector to the organs and organ systems of marine mammals. For comparative purposes, we focus on the California sea lion (*Zalophus californianus*), Florida manatee (*Trichechus manatus latirostris*), harbor seal (*Phoca vitulina*), and common bottlenose dolphin (*Tursiops truncatus*). Our descriptions are at the gross anatomical level.

To recognize variations on a theme, one must first recognize the theme. Although there is no "typical" mammal, we shall use our own species and the domestic dog as the norms against which to make comparisons. To appreciate human and dog anatomy, we suggest Hollinshead and Rosse (1985) and Evans (1993), respectively. Anatomy of internal organs of domestic mammals is covered by Schummer *et al.* (1979). For discussions of the anatomy of various types of marine mammals, consult Fraser (1952), Green (1972), Herbert (1987), Howell (1930), King (1983), Murie (1872, 1874), Pabst *et al.* (1999), von Schulte (1916), Slijper (1962), and St. Pierre (1974).

Wherever possible, anatomical terms follow the Nomina Anatomica Veterinaria as illustrated by Schaller (1992).

I. Mammalian Postcranial Landmarks

Marine mammals are generally dissected either ventrally or laterally, but some large, stranded animals must be examined in whatever position they are found. For consistency, we provide figures that describe anatomy in terms of a lateral view, and we discuss organs and organ systems in the order in which they are revealed during necropsy. Although this approach may take some getting used to if one is accustomed simply to the ventral approach, the lateral orientation approximates the living condition more closely.

A. The Diaphragm

The diaphragm of most marine mammals is generally similar in orientation to that of the diaphragm in both the human and the dog. It lies in a transverse plane and provides a musculotendinous sheet to separate the heart and its major vessels, the lungs and their associated vessels and airways, the thyroid, thymus, and a variety of lymph nodes (all located cranial to the diaphragm) from the major organs of the digestive, excretory, and urogenital systems (all typically caudal to the diaphragm). The diaphragm is generally confluent with the transverse septum (a connective tissue separator between the heart and liver) and, thus, attaches medially at its ventral extremity to the sternum.

Although the diaphragm separates the heart and lungs from the other organs of the body, the diaphragm is traversed by nerves and other structures, such as the aorta (crossing in a dorsal and medial position), the vena cava (crossing more ventrally than the aorta, and often slightly right of the midline, although appearing to approximate the center of the liver), and the esophagus (crossing slightly right of the midline, at roughly a midhorizontal level). This approximately transverse orientation exists in most marine mammals, although the orientation of the diaphragm may be more or less diagonal, with the ventral portion being more cranial than the dorsal portion (Fig. 1A).

The West Indian manatee's diaphragm differs from this general pattern of orientation and attachment. The diaphragm and the transverse septum are separate, with the septum occupying approximately the "typical" position of the diaphragm and the diaphragm itself occupying a horizontal plane extending virtually the entire length of the body cavity (Fig. 1B). This apparently unique orientation contributes to buoyancy control (Rommel and Reynolds, 2000). Additionally, there are two separate hemidiaphragms in the manatee (Figs. 2B and 2C). The central tendons attach firmly to the ventral aspects of the thoracic vertebrae, producing two isolated pleural cavities. The position of the manatee

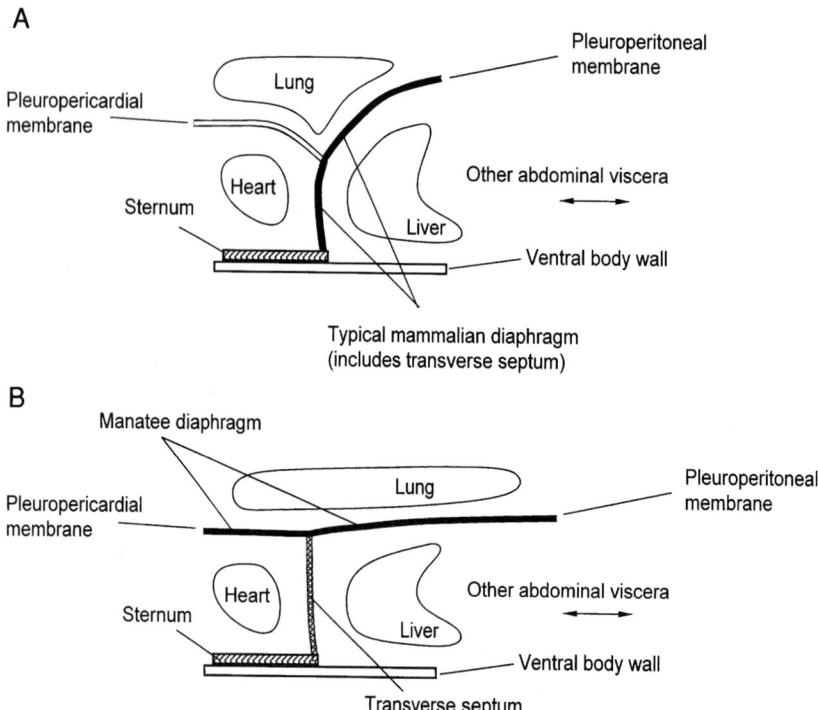

Figure 1 *Schematic arrangements of mammalian diaphragms. Modified after Rommel and Reynolds (2000). (A) The typical mammalian diaphragm extends ventrally from the dorsal midline to attach to the sternum. The typical diaphragm is a separator between the heart and lungs in the front and the liver and other abdominal organs in the back. (B) The manatee diaphragm extends dorsally to the heart and does not touch the sternum. There is a mechanical barrier between the heart and the liver and other abdominal organs, but it is a relatively weak barrier called the transverse septum.*

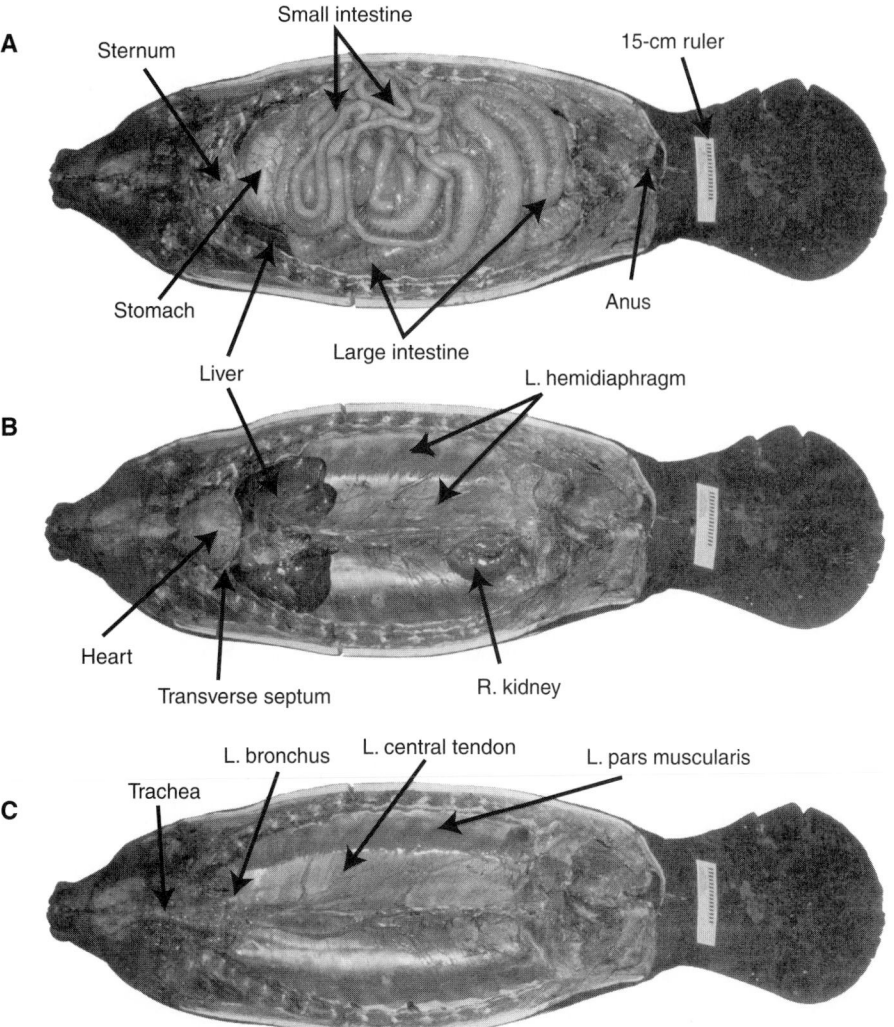

Figure 2 *Ventral views of the Florida manatee. Modified after Rommel and Reynolds (2000). The ruler is 15 cm long. (A) After removal of the ventral skin, fat, and musculature, the small and large intestines are exposed; the large intestine (with contents) may account for 10% of the total body weight and can measure 20 m long. Portions of the stomach and ventral margins of the liver are visible caudal to the sternum. (B) Removal of the GI tract reveals the heart, transverse septum, liver, hemidiaphragms, and right kidney (the left kidney was removed to expose that portion of the hemidiaphragm). (C) The two central tendons of the hemidiaphragms attach medially to the ventral aspects of the vertebral column. The diaphragm muscles attach laterally to the ribs. The lungs are flattened, elongate structures dorsal to the hemidiaphragms; when fully inflated, the lungs extend almost the entire length of the region dorsal to the hemidiaphragms. Note the junctions of the central tendon and the pars muscularis of each hemidiaphragm; this approximates the lateral margin of each lung.*

diaphragm stands in contrast with the curved, oblique diaphragms (DIA, Fig. 3) of the sea lion, seal, and dolphin.

B. Regions and Structures Cranial to the Diaphragm

The region cranial to the diaphragm is typically compartmentalized into three sections: (1) the pericardium (containing the heart), (2) the pleural cavities (containing the lungs), and (3) the mediastinum (Figs. 3 and 4).

The pericardium is a fluid-filled sac surrounding the heart (HAR, Fig. 3); in manatees, it often contains more fluid than is found in the pericardia of the typical mammal or in those of other marine mammals. The heart occupies a ventral position in the thorax (immediately dorsal to the sternum), making it easy

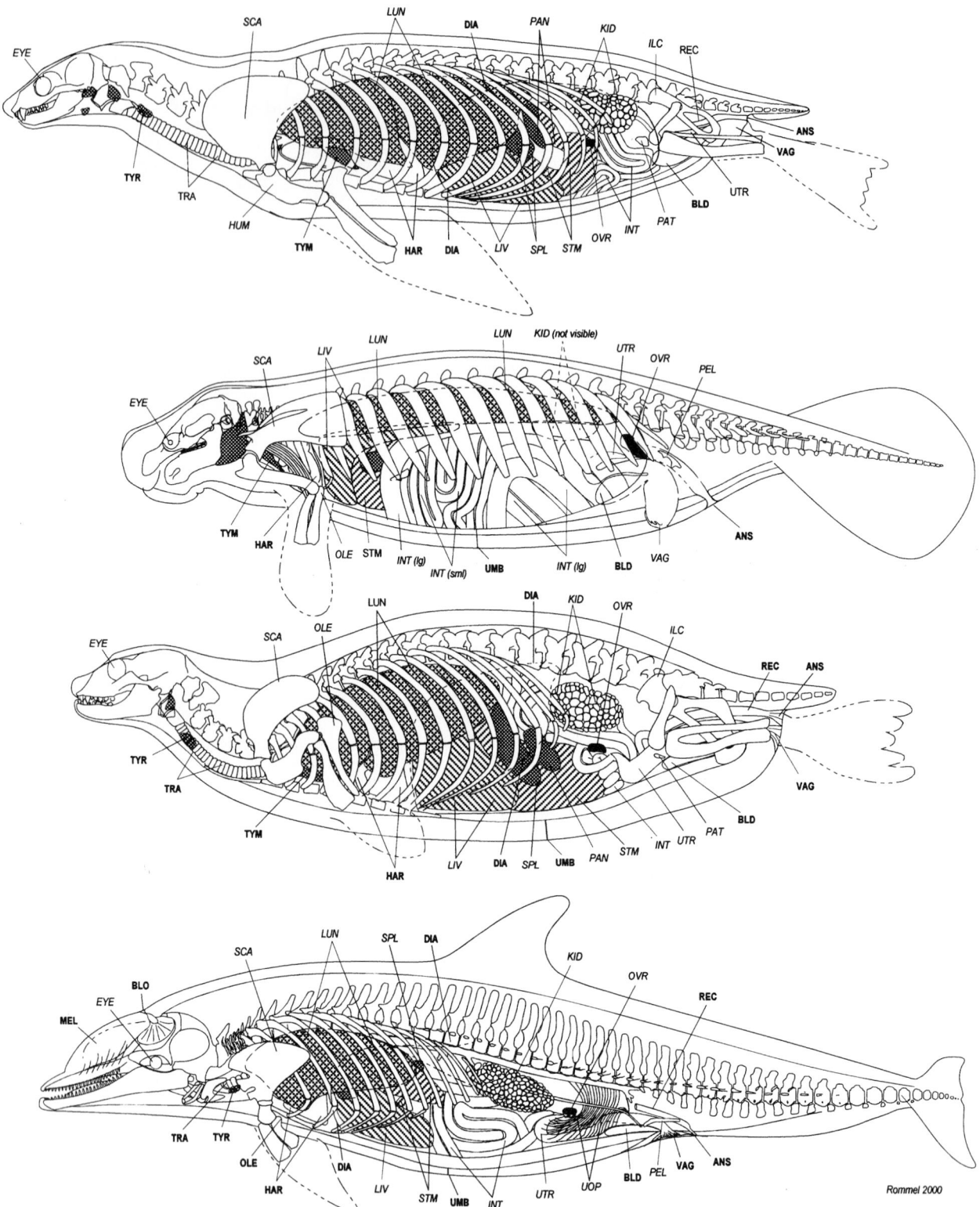

Rommel 2000

to see when the overlying muscles, ribs, and sternum are removed. The heart lies immediately cranial to the central portion of the diaphragm (or just the transverse septum in the manatee). Some lungs may embrace the caudal aspect of the heart, separating the heart from the diaphragm. As do the hearts of all other mammals, marine mammal hearts have four chambers, separate routes for pulmonary and systemic circulation, and the usual arrangements of great vessels (venae cavae, aorta, coronary arteries, pulmonary vessels). Cardiac fat is commonly found in manatees but is typically absent in pinnipeds and cetaceans.

The pleural cavities and lungs of mammals are generally found dorsally and laterally to the heart and are separated along the midline by the heart and mediastinum (see later). In the manatee, the lungs are unusual in that they extend virtually the length of the body cavity and remain dorsal to the heart (Rommel and Reynolds, 2000). Lungs of some marine mammals (cetaceans and sirenians) tend to be unlobed. The size of the lungs of marine mammals varies according to each species' diving proficiency. Marine mammals that make deep and prolonged dives (e.g., elephant seals, *Mirounga* spp) tend to have smaller lungs than expected (based on allometric relationships), whereas shallow divers (e.g., sea otters, *Enhydra lutris*) tend to have larger than expected lungs.

The mediastinum is typically considered to be the area between the lungs, excluding the heart and pericardium. The mediastinum contains the major vessels leading to and emanating from the heart, nerves (e.g., the phrenic nerve to the diaphragm), and lymph nodes. The thymus, which is larger in younger individuals, is found on the cranial aspect of the pericardium (sometimes extending caudally to embrace almost the entire heart) and may extend into the neck in some species. The thyroid gland is located in the cranial part of the mediastinum along either side of the distal part of the trachea, cranial to its bifurcation into the bronchi (in sea lions, but not in other marine mammals, the bifurcation is cranial to the thoracic inlet).[1] In most marine mammals, the mediastinum is generally not remarkable; in the manatee, however, the unusual placement of the lungs and the unique diaphragm change how one must define the mediastinum (Rommel and Reynolds, 2000).

One additional structure, located on the cranial aspect of the diaphragm in seals and sea lions, is an atypical mammalian muscular feature associated with the heart. This is the caval sphincter (CAS, Fig. 3), which can regulate the flow of oxygenated[2] blood in the large venous hepatic sinus to the heart during dives (Elsner, 1969).

C. Structures Caudal to the Diaphragm

Easy to find landmarks caudal to the diaphragm include a massive liver and the various components of the gastrointestinal (GI) tract. The urogenital organs are generally found only after removal of the GI tract (note that the exception is the uterus of a pregnant female).

1. The Liver   Typically, the liver is located immediately caudal to the diaphragm. It is a large, brownish, multilobed organ positioned so that most of its volume/mass is to the right of the midline of the body. Although marine mammal livers are generally similar to the livers of other mammals, in manatees, the organ is displaced somewhat to the left and dorsal relative to its location in most other mammals. The size, color, and "sharpness" of the liver margins can be used to assess the nutritive state and health of individual animals. Bile may be stored in a gallbladder (often greenish in color) located ventrally between the lobes of the liver, although some species (e.g., cetaceans, horses, and rats) lack a gallbladder. Bile enters the duodenum to facilitate the chemical digestion of fats.

---

[1]The thoracic inlet is the cranial opening of the thoracic cavity and is bounded by the vertebral and sternal ribs and sternum.

[2]Diving mammals with abundant arteriovenous anastomoses (shunts between arteries and veins before capillary beds) can have high blood pressure and highly oxygenated blood in their veins. One such venous reservoir of oxygenated venous blood is the hepatic sinus of seals (King, 1983).

Figure 3 *Left lateral illustrations of the superficial internal structures and "anatomical landmarks" of the California sea lion, Florida manatee, harbor seal, and common bottlenose dolphin with the skeleton (minus the distal appendicular elements) superimposed for reference. Our view is a left lateral view, focused on relatively superficial internal structures (labeled in bold) visible from that perspective; the other important bony or soft "landmarks" are not necessarily visible from a left lateral view but they are useful for orientation and are labeled in italics. Skeletal elements are included for reference, but not all are labeled. Each drawing is scaled so that there are equivalent distances between the shoulder and the hip; thus, the thoracic and abdominal cavities are roughly equal in length. The shoulder joints are aligned. The left kidney (not visible from this vantage in the manatee) is illustrated. The relative sizes of the lungs represent partial inflation—full inflation would extend margins to distal tips of ribs (except in the manatee). The following abbreviations are used as labels (structures on the midline are in **bold**, those off-midline are not): **ANS,** anus; **BLD,** urinary bladder; **BLO,** blowhole of dolphin; **DIA,** diaphragm, midline extent (except manatee); EYE, eye (note small size in manatee); **HAR,** heart; ILC, iliac crest of the pelvis; INT, intestines, note the large diameter of the large intestines in the manatee; KID, left kidney (not visible from this vantage in the manatee); LIV, liver; LUN, lung (note that in this illustration, the lung extends under the scapula except in the seal); **MEL,** melon, dolphin only; OLE, olecranon; OVR, left ovary; PAN, pancreas (in this view visible only in seal and sea lion); PAT, patella; PEL, pelvic vestige; **REC,** rectum; SCA, scapula; SPL, spleen; STM, stomach; **TRA,** trachea (not visible in this view of the manatee); **TYM,** thymus gland; **TYR,** thyroid gland; **UMB,** umbilical scar; UOP, uterovarian plexus in dolphins; UTR, uterine horn; **VAG,** vagina. Copyright S. A. Rommel.*

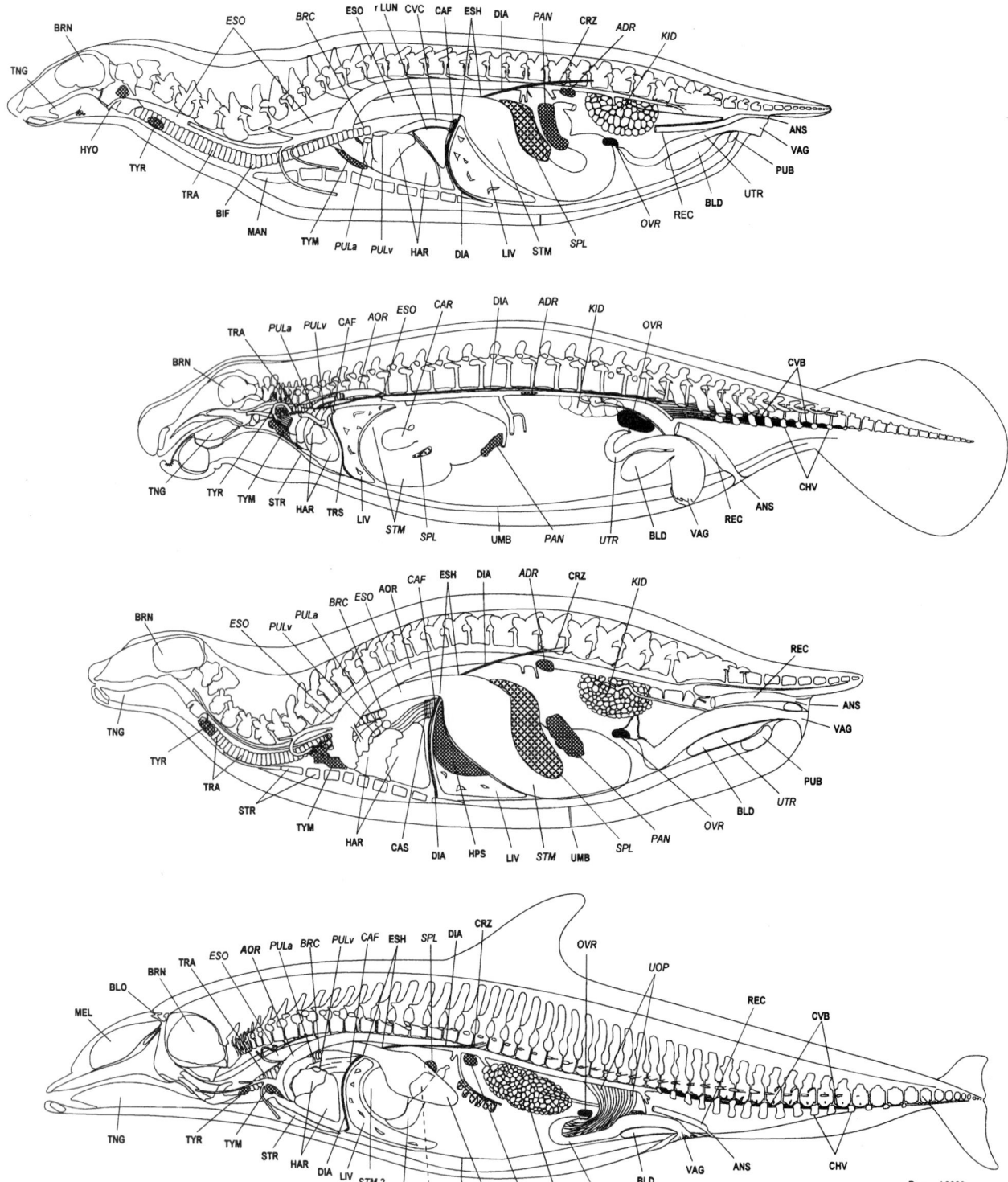

Rommel 2000

2. The GI Tract Most of the volume of the cavity caudal to the diaphragm (the abdominal cavity) is occupied by the various components of the GI tract: the stomach, the small intestine (duodenum, jejunum, and ileum), and the large intestine (cecum, colon, and rectum). The proportions and functions of these components reflect the feeding habits and trophic levels of the different marine mammals. Therefore, the gastrointestinal tracts of marine mammals vary considerably.

Food and water travel from the mouth, through a muscular pharynx, and into the esophagus. As noted earlier, the latter pierces the diaphragm to join the stomach, which is typically a single, distensible sac. The distal end of the stomach (the pylorus) is marked by a strong sphincter before it connects with the small intestine (duodenal ampulla in cetaceans). The separation between jejunum and ilium of the small intestine is difficult to distinguish grossly, although the two sections are different microscopically. The junction of the small and large intestines is often (but not in cetaceans) marked by the presence of a cecum (homologous to the human appendix). In manatees, the midgut cecum has two blind pouches called cecal horns. In some marine mammals, the large intestine, as its name implies, has a larger diameter than the small intestine.

The gastrointestinal tracts of pinnipeds and other marine mammal carnivores follow the general patterns outlined earlier, although the intestines can be remarkably long in some species. Cetaceans, however, have some unique specializations (Gaskin, 1978). Cetaceans can have two or three stomachs (usually three), depending on the species being examined. The multiple stomachs of cetaceans function in much the same way as the single stomach found in most other mammals. The first stomach of cetaceans, called the forestomach (essentially an enlargement of the esophagus), is muscular and very distensible, and it acts much like a bird crop, i.e., as a receiving chamber. The second or glandular stomach is the primary site of chemical breakdown among the stomach compartments; it contains the same types of enzymes and hydrochloric acid that characterize a "typical" stomach. Finally, the "U-shaped" third or pyloric stomach ends in a strong sphincteric muscle that regulates the flow of digesta into the duodenum (the duodenal ampulla is sometimes mistakenly called a fourth stomach) of the small intestine. The cetacean duodenum is expanded into a sac-like ampulla. The only other remarkable feature at the gross level is the lack of a cecum, which makes it difficult to tell where the small intestine ends and the large intestine begins. The intestines of some cetaceans may be extremely long (especially in the sperm whale, *Physeter macrocephalus*; Slijper, 1962), but they are not especially long in many other marine mammal species.[3]

Among marine mammals, sirenians have the most remarkably developed gastrointestinal tract. Sirenians are herbivores and hindgut digesters (similar to horses and elephants) so the large intestine (specifically the colon) is extremely enlarged, enabling it to act as a fermentation vat (see Marsh *et al.*, 1977; Reynolds and Rommel, 1996). In horses, the cecum is the region of the large intestine that is enlarged, but in sirenians, the cecum is relatively small and has two "horns." The sirenian stomach is single chambered and has a prominent accessory secretory gland (the cardiac gland) extending from the greater curvature. The duodenum is capacious and has two obvious diverticulae projecting from it. The GI tract and its contents can account for more than 20% of a manatee's weight.

The length and mass of the gastrointestinal tract are impressive and create three-dimensional relationships that can be complex. Simplifying the organization is the fact that tough sheets of connective tissue called mesenteries suspend the organs from the dorsal part of the abdominal cavity and shorter bands of connective tissue (ligaments)[4] hold organs close to one another in predictable arrangements (e.g., the proximal spleen is always found

---

[3]Assessing the length of intestines is fraught with potential bias because it is extremely difficult not to stretch the intestines to unnatural lengths after they are freed from the mesenteries and straightened. Linear measurements of gastrointestinal tract are, therefore, highly subjective.

[4]Ligament has several meanings in anatomy: a musculoskeletal element (e.g., the anterior [cranial] cruciate ligament), a vestige of a fetal artery or vein (e.g., the round ligament of the bladder), the margin of a fold in a mesentery (e.g., broad ligament), and a serosal fold between organs (e.g., the gastrolienal ligament).

Figure 4 *A view slightly to the left of the midsagittal plane illustrates the circulation, body cavities, and selected organs of the California sea lion, Florida manatee, harbor seal, and common bottlenose dolphin, with the skeleton for reference. The left lung is removed. Note that the diaphragm separates the heart and lungs from the liver and other abdominal organs. Each drawing is scaled so that there are equivalent distances between the shoulder and the hip; thus, the thoracic and abdominal cavities are roughly equal in length. The shoulder joints are aligned. Note that the manatee's diaphragm is unique and that the distribution of organs and the separation of thoracic structures from abdominal structures require special consideration in these beasts. The following abbreviations are used as labels (structures on the midline are in* **bold**, *those off midline are not): ADR, adrenal gland;* **ANS,** *anus;* **AOR,** *aorta;* **BLD,** *urinary bladder;* **BLO,** *blowhole; BRC, bronchus;* **BRN,** *brain;* **CAF,** *caval foramen; CAR, cardiac gland, in manatee only;* **CAS,** *caval sphincter, surrounding the vena cava in the seal and sea lion;* **CHV,** *chevron bones; CRZ, crus (plural crura) of the diaphragm;* **CVB,** *caudal vascular bundle, in manatee and dolphin;* **DIA,** *diaphragm, cut at midline, extends from crura dorsally to sternum ventrally (except in manatees); ESH, esophageal hiatus;* **ESO,** *esophagus (to the left of the midline cranially, on the midline caudally);* **HAR,** *heart; HPS, hepatic sinus within liver, in seals only; KID, right kidney;* **LIV,** *liver, cut at midline;* **LUN,** *lung, right lung between heart and diaphragm; MEL, melon, dolphin only; PAN, pancreas;* **PUB,** *pubic symphysis (seals and sea lions only); PULa, pulmonary artery, cut at hilus of lung; PULv, pulmonary vein, cut at hilus of lung;* **REC,** *rectum, straight part of terminal colon; SPL, spleen; STM1, forestomach; STM2, main stomach (STM in noncetaceans); STM3, pyloric stomach;* **STR,** *sternum, sternabrae;* **TNG,** *tongue;* **TRA,** *trachea;* **TRS,** *transverse septum;* **TYM,** *thymus gland;* **TYR,** *thyroid gland;* **UMB,** *umbilicus; UOP, right uterovarian vascular plexus in dolphin; UTR, uterus;* **VAG,** *vagina. Copyright S. A. Rommel.*

along the greater curvature of the stomach and is connected to the stomach by the gastrolienal, or gastrosplenic, ligament). Also suspended in the mesenteries are numerous lymph nodes and fat.

Accessory organs of digestion include the salivary glands (small in most marine mammals but very large in the manatee), pancreas, and liver (where bile is produced and then stored in the gall bladder). The pancreas is sometimes a little difficult to locate because it can be a rather diffuse organ and it decomposes rapidly; however, a clue to its location is its proximity to the initial part of the duodenum, into which pancreatic enzymes flow. Another organ that is structurally, but not functionally, associated with the GI tract is the spleen, which is suspended by a ligament, generally from the greater curvature of the stomach (the first stomach in cetaceans) on the left side of the body. The spleen may be a single organ accompanied by accessory spleens in some species. The spleen is bluish in color and varies considerably in size among species; in manatees and cetaceans it is relatively small, but is more massive in some deep-diving pinnipeds (Zapol *et al.,* 1979) and acts as a storage region for red blood cells.

3. Urogenital Anatomy The kidneys lie in a retroperitoneal position, typically against the musculature of the back (epaxial muscles) at or near the dorsal midline attachment of the diaphragm (crura). In the manatee, the unusual placement of the diaphragm means that the kidneys lie against the diaphragm, not against the epaxial muscles. All mammals have metanephric kidneys (i.e., containing cortex, medulla, calyces). In many marine mammals, the kidneys are specialized as reniculate (multilobed) kidneys, where each lobe (renule) has all the components of a complete metanephric kidney. Why marine mammals have reniculate kidneys is uncertain, but the fact that some large terrestrial mammals also have reniculate kidneys has led to speculation that they are an adaptation associated simply with large body size (Vardy and Bryden, 1981).

The renal arteries of cetaceans enter the cranial poles of the kidneys, whereas in other marine mammals, they enter the hilus (typical of most mammals). Additionally, in manatees, there are accessory arteries on the surface of the kidney. The kidneys are drained by separate ureters, which carry urine to a medially and relatively ventrally positioned urinary bladder. The urinary bladder lies on the floor of the caudal abdominal cavity and, when distended, may extend as far forward as the umbilicus in some species. The pelvic landmarks are less prominent in fully aquatic mammals. In the manatee, the bladder can be obscured by abdominal fat.

Pabst *et al.* (1999) noted that the reproductive organs tend to reflect phylogeny more than adaptations to a particular niche. If one were to examine the ventral side of cetaceans and sirenians before removing the skin and other layers, one would discover that positions of male and female genital openings are different, permitting rather easy determination of sex without dissection. In all marine mammals, the female urogenital opening is more caudal than the opening for the penis in males. One way to approach dissection of the reproductive tracts is to follow structures into the abdomen from their external openings.

The position and general form of the female reproductive tract in marine mammals are generally similar to those of the female reproductive tracts in terrestrial mammals. The vagina opens cra-

nial to the anus and leads to the uterus, which is bicornuate in marine mammal species. The body of the uterus is found on the midline and is located dorsally to the urinary bladder (the ventral aspect of the uterus rests against the bladder). Although the body of the uterus lies along the midline, it has bilaterally paired, relatively large diameter projections called uterine horns (cornua), which extend laterally. The relatively small-diameter oviducts conduct eggs from the ovaries to the uterine horns where implantation of the fertilized egg and subsequent placental development occur. The dimensions of the uterine horns vary with reproductive history and age. Often the fetus may expand the pregnant horn to the point that it fills a substantial portion of the abdominal cavity. The horns terminate abruptly, narrowing and extending as uterine tubes (fallopian tubes) to paired ovaries. The uterus and the uterine horns are held in place in the abdominal cavity by the broad ligaments. Uterine and ovarian scarring may provide information about the reproductive history of the individual.

The ovaries of mature females may have one or more white or yellow-brown scars, called corpora albicantia and corpora lutea, respectively. Although ovaries are usually solid organs, in sirenians they are relatively diffuse.

Mammary glands are ventral, medial, and relatively caudal in most marine mammals, but they are axillary in sirenians. Many marine mammals have a single pair of nipples, sea lions and polar bears, *Ursus maritimus,* (DeMaster and Sterling, 1981), have two pairs of nipples, and cetaceans have mammary slits (note that some male cetaceans have distinct mammary slits).

The male reproductive tracts of marine mammals have the same fundamental components as the tracts in "typical" mammals, but positional relationships are significantly different. This difference is due to the testicond (ascrotal) position of the testes in most marine mammal species [sea otters are scrotal (J. Bodkin personal communication); polar bears are seasonally scrotal (I. Stirling personal communication); sea lion testes become scrotal when temperatures are elevated]. The testes of some marine mammals are intraabdominal, but in phocids, for example, they lie outside the abdomen, partially covered by the oblique muscles and BLUBBER. The position of marine mammal testes creates certain thermal problems because spermatozoa do not survive well at body (core) temperatures; in some species, these problems are solved by the circulatory adaptations mentioned later.

The penis of marine mammals is retractable and it normally lies within the body wall. The general structure of the penis relates to phylogeny (see Pabst *et al., 1999*).

4. Adrenal Glands The term "suprarenal gland" is often used interchangeably with "adrenal gland." Although the suprarenals often lie immediately atop or very close to the kidneys of terrestrial mammals, adrenals of marine mammals may lie several centimeters cranial to the kidneys, along either side of the median. Adrenal glands can be confused with lymph nodes, but if one slices the organ in half, an adrenal gland is easy to distinguish grossly by its distinct cortex and medulla.

5. Circulatory Structures Basically, blood vessels are named for the regions they feed or drain. Thus, the fully aquatic marine mammals (cetaceans and sirenians) lack femoral arteries that supply the pelvic appendage. However, most organs in marine mammals are similar to those of terrestrial mammals so their blood supply is also similar. Therefore, readers who want to learn details of typical circulatory anatomy should consult one of the anatomy references cited earlier. The thoracic aorta leaves the heart and lies ventral to the vertebral column, giving off segmental arteries to the vertebrae and epaxial muscles (and in the case of cetaceans and manatees to the thoracic retia). The aorta continues through the aortic hiatus of the diaphragm (between the crura) and into the abdomen as the abdominal aorta and lumbar aorta, which give off several paired (e.g., renal, gonadal) and unpaired (e.g., celiac, mesenteric) arteries. The caudal aorta follows the ventral aspect of the tail vertebrae. In the permanently aquatic marine mammals, there are robust ventral chevron bones that form a canal in which the caudal aorta, its branches, and some veins are protected.

Some of the diving mammals (e.g., seals, cetaceans, and sirenians) have few or no valves in their veins (Rommel *et al.*, 1995); this adaptation simplifies blood collection.[5] Other exceptions to the general pattern of mammalian circulation are associated with thermoregulation and diving. Countercurrent heat exchangers abound, and extensive arteriovenous anastomoses exist to permit two general objectives to be fulfilled: (1) regulating loss of heat to the external environment, while keeping core temperatures high; and (2) permitting cool blood to reach specific organs (e.g., testes, uteri, spinal cord) that cannot sustain exposure to high body temperatures (see reviews by Rommel *et al.*, 1998; Pabst *et al.*, 1999).

In mammals, several paths for supplying blood to the brain exist; via the internal carotid, the external carotid, and the vertebral/basilar arteries. Some species use only one, others use two, and manatees use all three pathways. In cetaceans, the path for supplying blood to the brain is unique. The blood destined for the brain first enters the thoracic rete, a plexus of convoluted, small diameter arteries in the dorsal thorax. Blood leaves the thoracic rete and enters the spinal rete where it surrounds the spinal cord and enters the base of the skull (McFarland *et al.*, 1979). There are two working explanations for this convoluted path of blood to the brain: (1) the elasticity of the retial system allows mechanical damping of the blood pulse pressure wave (McFarland *et al.*, 1979) and (2) the juxtaposition of the thoracic retia to the dorsal aspect of the lungs may provide thermal control of the blood entering the spinal retia. Combined with cooled blood in the epidural veins, the spinal retia may provide some temperature control of the central nervous system (Rommel *et al.* 1993, paper presented at the Tenth Biennial Conference on the Biology of Marine Mammals).

II. Overview

Marine mammal postcranial soft tissue anatomy is, in many regards, similar to that of "typical" mammals. However, the relative proportions of and, to some extent, the positions of organs may be somewhat different from the norm.

---

[5]The near absence of valves in the veins of seals and dolphins allows two-way flow to occur, increasing the blood available when venipuncture is used; in contrast, sea lions have numerous valves in their hind flipper veins.

We close with a reminder about orientation: namely the orientation of the prosector relative to the orientation of the specimen and the orientation of the specimen to the orientation of that animal when it was alive. The position of animals during necropsy may be belly-up, obviously not the usual position of the living animals. Thus, gravitational forces make the positional relationships we may observe during necropsy somewhat artificial; we assess "dead anatomy" rather than "living anatomy." We suggest that people examining marine mammal postcranial anatomy bear this fact in mind and try to constantly picture how the structures being observed during necropsy might be arranged in a free-ranging animal. The more the latter perspective can be maintained, the easier it will be to envision dynamic relationships among organs and systems and to relate function (physiology) to structure (anatomy).

Acknowledgments

We thank Ian Stirling and Jim Bodkin for providing information on the polar bear and sea otter, respectively. We thank Derek Fagone, Judy Leiby, Tom Pitchford, and James Quinn at the Florida Marine Research Institute for reviewing the manuscript. We thank Llyn French for help with Fig. 2. Anatomical illustrations were created with FastCAD (Evolution Computing, Tempe, AZ).

See Also the Following Articles

Female Reproductive Systems ■ Male Reproductive Systems ■ Musculature ■ Pelvic Anatomy ■ Skeletal Anatomy ■ Skull Anatomy

References

DeMaster, D. P., and Stirling, I. (1981). *Ursus maritimus. Mammal. Spec.* **145**, 1–7.

Elsner, R. W. (1969). Cardiovascular adjustments to diving. *In* "The Biology of Marine Mammals" (H. T. Andersen, ed.), pp. 117–145. Academic Press, New York.

Evans, H. E. (1993). "Miller's Anatomy of the Dog," 3rd Ed. Saunders, Philadelphia.

Fraser, F. C. (1952). "Handbook of R. H. Burne's Cetacean Dissections." Trustees of the British Museum, London.

Gaskin, D. E. (1978). Form and function of the digestive tract and associated organs in cetacea, with a consideration of metabolic rates and specific energy budgets. *Oceanogr. Mar. Biol. Annu. Rev.* **16**, 313–345.

Green, R. F. (1972). Observations on the anatomy of some cetaceans and pinnipeds. *In* "Mammals of the Sea, Biology and Medicine." (S. H. Ridgway, ed.), pp. 247–297. Thomas, Springfield, IL.

Herbert, D. (1987). "The Topographic Anatomy of the Sea Otter *Enhydra lutris*. Unpublished MS Thesis. Johns Hopkins University, Baltimore, MD.

Hollinshead, W. H., and Rosse, C. (1985). "Textbook of Anatomy." Harper & Row, Philadelphia.

Howell, A. B. (1930). "Aquatic Mammals: Their Adaptations to Life in the Water." Thomas, Springfield, IL.

King, J. E. (1983). "Seals of the World," 2nd Ed. Comstock, Ithaca, NY.

Marsh, H., Heinsohn, G. E., and Spain, A. V. (1977). The stomach and duodenal diverticulae of the dugong (*Dugong dugon*). *In* "Functional Anatomy of Marine Mammals" (R. J. Harrison, ed.), Vol. 3, pp. 271–295. Academic Press, London.

McFarland, W. L., Jacobs, M. S., and Morgane, P. J. (1979). Blood supply to the brain of the dolphin, *Tursiops truncatus*, with comparative observations on special aspects of the cerebrovascular supply of other vertebrates. *Neurosci. Biobehav. Rev.* (Suppl. 1) **3**, 1–93.

Murie, J. (1872). On the form and structure of the manatee. *Trans. Zool. Soc. Lond.* **8**, 127–202.

Murie, J. (1874). Researches upon the anatomy of the Pinnipedia. 3. Descriptive anatomy of the sealion (*Otaria jubata*). *Trans. Zool. Soc. Lond.* **8**, 501–582.

Pabst, D. A., Rommel, S. A., and McLellan, W. A. (1999). The functional morphology of marine mammals. *In* "Biology of Marine Mammals" (J. E. Reynolds III and S. A. Rommel, eds.), pp. 15–72. Smithsonian Institution Press, Washington, DC.

Reynolds, J. E., III, and Rommel, S. A. (1996). Structure and function of the gastrointestinal tract of the Florida manatee, *Trichechus manatus*. *Anat. Rec.* **245**, 539–558.

Rommel, S. A., Early, G. A., Matasa, K. A., Pabst, D. A., and McLellan, W. A. (1995). Venous structures associated with thermoregulation of phocid seal reproductive organs. *Anat. Rec.* **243**, 390–402.

Rommel, S. A., Pabst, D. A., and McLellan, W. A. (1998). Reproductive thermoregulation in marine mammals. *Am. Sci.* **86**, 440–448.

Rommel, S. A., and Reynolds, J. E., III. (2000). Diaphragm structure and function in the Florida manatee (*Trichechus manatus latirostris*). *Anat. Rec.* **259**, 41–51.

Schaller, O. (1992). "Illustrated Veterinary Anatomical Nomenclature." Ferdinand Enke Verlag, Stuttgart.

Schummer, A., Nickel, R., and Sack, W. O. (1979). The viscera of the domestic mammal. "The Anatomy of the Domestic Animals" (R. Nickel, A. Schummer, and E. Seiferle, eds.), 2nd Ed., Vol. 2. Verlag Paul Parey, Berlin.

Slijper, E. J. (1962). "Whales." Hutchinson & Co., London.

St. Pierre, H. (1974). The topographical splanchnology and the superficial vascular system of the harp seal *Pagophilus groenlandicus* (Erxleben 1777). *In* "Functional Anatomy of Marine Mammals" (R. J. Harrison, ed.), pp. 161–195. Academic Press, London.

Vardy, P. H., and Bryden, M. M. (1981). The kidney of *Leptonychotes weddelli* (Pinnipedia: Phocidae) with some observations on the kidneys of two other southern phocid seals. *J. Morphol.* **167**, 13–34.

von Schulte, H. (1916). Anatomy of a fetus of *Balaenoptera borealis*. *Mem. Am. Mus. Nat. Hist.* **1**(VI), 389–502 + plates XLIII–XLII.

Zapol, W. M., Liggins, G. C., Schneider, R. C., Qvist, J., Snider, M. T., Creasy, R. K., and Hochachka, P. W. (1979). Regional blood flow during simulated diving in the conscious Weddell seal. *J. Appl. Physiol.* **47**(5), 986–973.

Antarctic Marine Mammals

IAN L. BOYD

*University of St. Andrews,
Scotland, United Kingdom*

The Southern Ocean is the oceanic region surrounding the continent of Antarctica. Its southern boundary is defined by the narrow coastal continental shelf of Antarctica itself. To the north the boundary is defined by an oceanic frontal feature known as the Antarctic convergence or southern polar frontal zone. This zone marks the boundary between cold southern polar waters and northern temperate waters. The ocean temperature can change by as much as 10°C across the front, which may be only a few miles across. The polar front is an important physical feature that determines marine mammal distributions. It defines the normal southern extent of the distributions of most tropical and temperate marine mammals (Fig. 1).

A second feature that is important to marine mammals in the Antarctic is the annual sea ice. The seasonal changes in sea ice cover can lead to up to 50% of the Southern Ocean being covered in ice during late winter, but by late summer this can have contracted to 10% of the winter maximum. These large seasonal fluctuations in the sea ice have profound implications for the ecology of the Southern Ocean, including that of marine mammals. Many marine mammals, including most cetaceans, migrate north across the polar front in winter.

I. Antarctic Species

This section deals with true Antarctic species defined as those species whose populations rely on the Southern Ocean as a habitat, i.e., critical to a part of their life history, either through the provision of habitat for breeding or through the provision of the major source of food. Species that inhabit the sub-Antarctic, which is generally seen as including the islands that circle Antarctica in the region of the polar front or the polar frontal zone itself, are not included.

The Southern Ocean accounts for about 10% of the world's oceans but it probably supports >50% of the world's marine mammal biomass, including six species of pinnipeds, eight species of baleen whales, and at least seven species of odontocete whales. Therefore, in terms of the diversity of species, the Antarctic is host to only one-fifth of the world's pinniped and a little less than one-fifth of the world's cetacean species. This low diversity may be attributed partly to the lack of land masses to cause isolation and speciation and also because, although large in its total area, the Southern Ocean does not have the diversity of habitats and prey species seen in other ocean basins.

Among the pinnipeds, there is one species from the family Otariidae (eared seals, which include fur seals and sea lions) and there are five species from the family Phocidae (earless or "true" seals), but all of these come from a single subfamily, the Monachinae (see Table I). This list is as notable as much by its absences as it is for those that are present. For example, there is no representative of the phocid subfamily Phocinae, which contains a diverse collection of species of Northern Hemisphere seals. There are also no representatives of the subfamily Otariinae, which includes all of the sea lions, and there is only one representative of the diverse Southern Hemisphere subfamily Arctocephalinae, which includes the southern fur seals.

Where pinnipeds are concerned, historically it would appear that there have been only two or three species immigrating into the Antarctic. The main immigration was of an ancestral phocid, possibly related to the nearly extinct tropical phocids of today known as monk seals, which gave rise to the four most closely related Antarctic phocids: the crabeater, Weddell, Ross, and leopard seals. At some later date it is likely that elephant seals arrived. Although they extend their distribution into south temperate latitudes, as much as 90% of the world population relies on the Southern Ocean as a critical habitat.

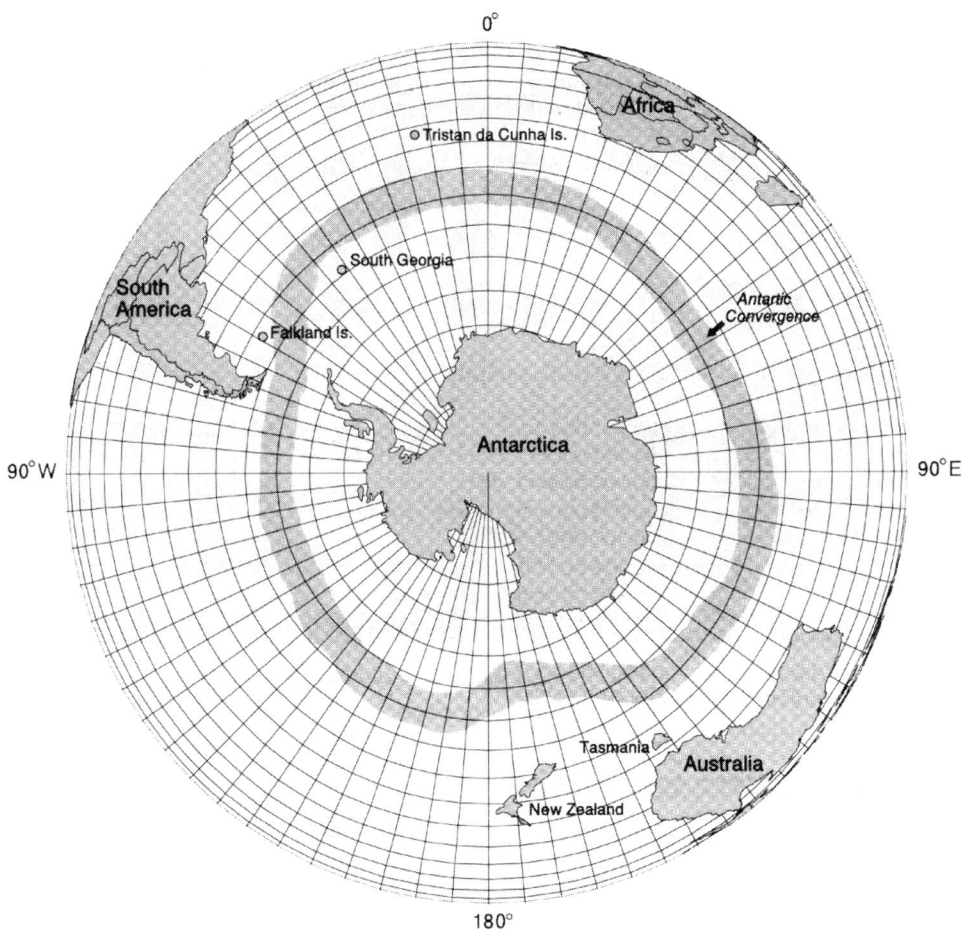

Figure 1 *Waters of warmer northern seas meet the perpetually colder waters around Antarctica to create the Antarctic Convergence. This confluence of relatively warm and frigidly cold waters occurs in some of the roughest seas known.*

These were likely to have been followed, or perhaps even preceded, by Antarctic fur seals. The taxonomic status of Southern Hemisphere fur seals, a group within which eight species are currently recognized, is uncertain and it seems probable that many of these are not true species but are instead subspecies. Therefore, the Antarctic fur seal may simply be an Antarctic race or subspecies of the southern fur seal.

Among cetaceans, there are only three Antarctic species within the highly diverse family Delphinidae, which includes all of the dolphins and porpoises. These three are the hourglass dolphin, long-finned pilot whale, and killer whale. The beaked whales are represented by only three species, but because these species are very difficult to identify in the field, it is possible that among the very large number of these individuals that are found in the Southern Ocean, several other species could be present.

II. Ecology of Antarctic Marine Mammals

The presence of a large biomass of marine mammals in the Antarctic is probably a result of the unusual food chain structure of the Southern Ocean. The marine mammals of the Antarctic

with large numbers, such as crabeater seals and Antarctic fur seals, rely on krill as their main food source (see Section IV). This is in contrast to marine mammal communities elsewhere that rely mainly on a fish-based diet. Energy enters the food chain through photosynthesis and carbon sequestration by phytoplankton. The relative efficiency with which this energy is passed up the food chain to predators with a krill- or fish-based diet is illustrated in Fig. 2. The efficiency of energy transfer at each step in the food chain can be as low as only a few percent. The fewer steps there are between phytoplankton and marine mammals, the more efficiently will energy be transferred to marine mammals. In the Antarctic, there is on average one less step than there is in other oceanic ecosystems, which has led to the very large biomasses of marine mammals found in the Southern Ocean.

III. Distribution and Abundance

Antarctic marine mammals can be divided ecologically among those associated with fast ice, pack ice, or found in the open ocean. Weddell seals are most associated with fast ice, Ross seals with open water or pack ice. Leopard seals are

molting grounds on sub-Antarctic islands to shallow regions along the coast of Antarctica. Most of these types of preferences for different locations are assumed to reflect the distribution of food so that marine mammals migrate to the areas of greatest food abundance.

The crabeater seal is probably the most abundant seal in the world, with a population of somewhere between 7 and 14 million. There are considerably fewer Weddell seals and leopard seals (Table I). Ross seals are rarely seen and the total number is very uncertain, but it is probably the least abundant Antarctic pinniped. The Antarctic fur seal population is >3 million and is increasing at about 10% each year. In contrast, the southern elephant seal population within the Antarctic appears to have been relatively stable since the early-1960s, even though the number of elephant seals breeding at sites outside the Antarctic has declined steadily over the same period. The elephant seal population at South Georgia is estimated at 470,000, which probably represents 58% of the world population of the species.

In general, whale populations are in a highly depleted state (Table I). Blue whales are numbered in the hundreds for the whole of the Antarctic, and the sighting of a blue whale is a rare event. The number of fin whales appears to be increasing, as are humpback whales and southern right whales.

Within the Antarctic, there are no significant threats to pinniped species. However, some cetacean populations have been depleted to such a high degree that several are endangered. In particular, blue whales are so rare in the Antarctic that they are possibly close to extinction from the area (Fig. 4). Similarly, severely depleted southern right whale and humpback whale populations have very specific migratory routes between summering grounds in the Antarctic and winter grounds in temperate and tropical regions, which make them more vulnerable to threats such as disturbance, habitat loss, and reduced genetic diversity.

IV. Diet

Among seals, there is a progression of dietary specialization from those that mainly eat krill to those that mainly eat fish (Fig. 5). The leopard seal has seabirds and other seals as a major component of its diet, and it is probable that some individuals specialize in feeding on other seals or penguins instead of krill, fish, or squid. Among whales, dietary specializations are divided along taxonomic lines between odontocetes that mainly eat squid and mysticetes that forage primarily on zooplankton.

The crabeater seal is one of the most ecologically specialized of all seals because it feeds almost entirely on Antarctic krill that it gathers from the underside of ice floes where the krill themselves feed on the single-celled algae that grow within the brine channels in the ice. Antarctic fur seals also feed on krill to the north of the Antarctic pack ice edge, and many of the other Antarctic seals rely, to varying degrees, on krill as a source of food. Antarctic krill probably sustains more than half of the world's biomass of seals and also sustains a substantial proportion of the biomass of the world's seabirds and whales.

Although the dentition of crabeater seals is modified to help strain krill from the water, the feeding apparatus of the baleen whales is the most highly modified for a DIET of plankton. Krill

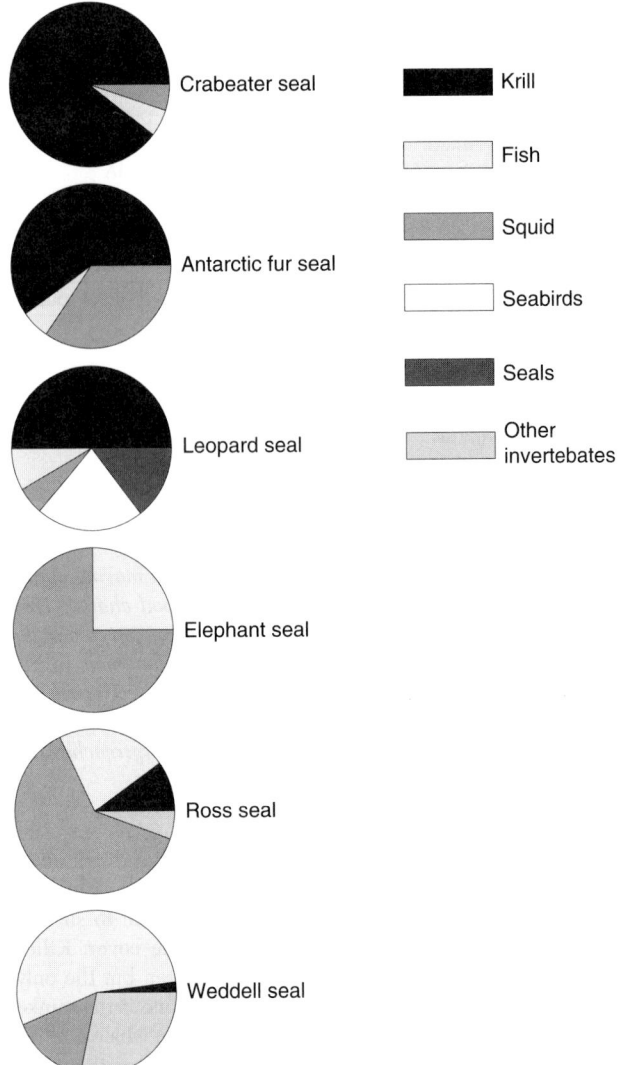

Figure 5 *Pie charts showing the composition of diets of Antarctic seals. The charts are arranged with those species that depend most on krill at the top and those that depend most on fish at the bottom; Between these are those species that have squid as a major component of the diet.*

is the major component in the diet of most of the Antarctic baleen whales, although copepods may also be strained from the water, especially by right whales. The Antarctic KRILL, *Euphausia superba*, often occurs in dense swarms in the open ocean, and the baleen whales have probably evolved to exploit these dense patches of food. Baleen whales eat 30–50 million tons of krill in the Antarctic each year and seals probably eat a similar or slightly lower total amount as whales. Consumption of squid by beaked whales and sperm whales is estimated to be about 14 million tons each year. Killer whales prey on fish and squid but also hunt seals and penguins. Pods of killer whales have been observed tipping over ice floes to push crabeater seals into the water in an effort to catch them.

V. Exploitation

Throughout the 19th and early 20th centuries, the Antarctic was viewed as an almost limitless source of marine mammals to be hunted for skins, oil, and other products that found expanding markets in Europe and North America. However, industrialization of whale and seal HUNTING brought both greater efficiency and the inevitability that the resources would be exhausted, much to the detriment of the ecology of the Antarctic and its populations of marine mammals.

There were three phases of exploitation: exploratory sealing (late 18th and early 19th centuries), preindustrial sealing and whaling (19th century), and industrial whaling (20th century). There are very few records of the exploratory sealing and the preindustrial era. During the exploratory era, exploitation was mainly targeted at fur seals to supply skins for the Chinese market, where they were turned into felt to supply the European market. By about 1830, fur seals in the Antarctic and elsewhere in the Southern Hemisphere had been all but extinguished. In 1825, James Weddell, himself the captain of a sealing vessel, noted that "the number of skins brought from off Georgia cannot be estimated at fewer than 1,200,000." He was referring to South Georgia, where >95% of the current world population of Antarctic fur seals resides. This species was considered to be extinct until the early 1920s when whalers saw several individuals at South Georgia. Since then, the numbers have increased rapidly and the population is conservatively estimated to now be on the order of 3 million. The preindustrial era was mainly targeted at whaling and the larger seals, particularly elephant seals, for their oil. This activity was mainly undertaken from sailing vessels. The introduction of steam power to the Antarctic was largely responsible for the transition to industrial whaling.

Industrial whaling began in the early years of the 20th century. This industry operated for more than 60 years and in that time it removed about 71 million tons of whale biomass involving 1.4 million individual whales from the Antarctic; about 10% of these were taken at South Georgia. Antarctic fur seals feed on krill (Fig. 5), and may have benefitted by the reduction in numbers of krill-feeding baleen whales and therefore had less competition for their food.

The industry was selective about which species of whales it targeted. The largest and most profitable were selected first, followed by progressively smaller species (Fig. 6). Eventually, the industry became unprofitable because only minke whales were left to exploit and these were too small to be profitable.

VI. Conservation Measures

Concerns about the effects of industrial whaling on the populations of whales began early in the industrial era. By the early 1920s, the "Discovery Investigations" had been established to determine whale populations mainly around South Georgia. These were funded by a levy on the industry, but they were free from control of the industry. They are one of the first examples of the fledgling field of ecology being used to solve a wildlife management problem. Even though the "Discovery Investigations" made ground-breaking scientific progress and were influential in the introduction of some conservation mea-

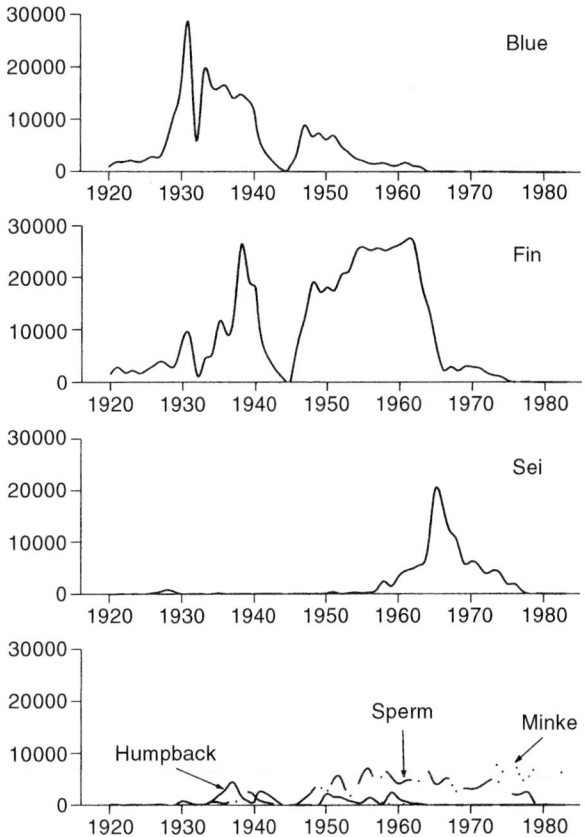

Figure 6 *Changes in the number of each species of whale caught in the industrial whale harvest in the Southern Ocean.*

sures, they came too late to influence the power of the industry and the fate of the populations of whales in the Southern Ocean.

The story of overexploitation of a marine resource in the Southern Ocean repeated itself in the 1960s and 1970s when industrial fisheries targeted fin fish populations and reduced them to uneconomic levels. This stimulated a renewed effort to ensure that there was proactive conservation of marine living resources in the Southern Ocean. The result was the Convention for the Conservation of Antarctic Marine Living Resources (CCAMLR) and the Convention for the Conservation of Antarctic Seals (CCAS), which came into effect in 1982 and 1978, respectively. One of the unique features of the CCAMLR convention is that it accepts that exploitation has effects on components of the ecosystem far beyond those that are being targeted for exploitation. This means that any proposals for the exploitation of living resources in the Antarctic must consider the effects that such exploitation is likely to have on marine mammals, whether or not they are the target species. Therefore, even though marine mammals enjoy legal protection in the Antarctic from unregulated exploitation under the environmental protocol within the Antarctic Treaty, they are also protected from other activities within the Southern Ocean ecosystem. Only time will tell if this is sufficient to ensure their long-term survival.

References

Brown, S. G., and Lockyer, C. H. (1984). Whales. *In* "Antarctic Ecology"
(R. M. Laws, ed.), Vol. 2, pp. 717–782. Academic Press, London.
Kasamatsu, F., and Joyce, G. G. (1995). Current status of odontocetes
in the Antarctic. *Antarct. Sci.* **7**, 365–379.
Laws, R. M. (1984). Seals. *In* "Antarctic Ecology" (R. M. Laws, ed.),
Vol. 2, pp. 621–716. Academic Press, London.
Laws, R. M. (ed.) (1993). "Antarctic Seals: Research Methods and
Techniques." Cambridge Univ. Press, Cambridge.

Aquariums

SEE *Marine Parks and Zoos*

Archaeocetes, Archaic

J. G. M. THEWISSEN

*Northeastern Ohio Universities College of Medicine,
Rootstown*

Archaeocetes is the common name for a group of primitive whales that lived in the Eocene Period (approximately 55–34 million years ago). Archaeocetes are important because they represent the earliest radiation of
cetaceans and because they include the ancestors of the two
modern suborders of cetaceans (Mysticeti and Odontoceti), Archaeocetes are also the main source of information about the
great morphological changes that were associated with the acquisition of aquatic features in cetaceans; archaeocetes document the initial amphibious stages in cetacean evolution [for a
semipopular account, see Zimmer (1998)]. Six families of
cetaceans are commonly included in archaeocetes (Thewissen,
1998; Fig. 1): Pakicetidae, Ambulocetidae, Remingtonocetidae,
Protocetidae, Basilosauridae, and Dorudontidae. Basilosaurids
and dorudontids are discussed separately, and the remaining
four families are treated here. Williams (1998) discussed the
taxonomy of these archaeocetes.

I. Pakicetidae

Pakicetidae are only known from the early-to-middle
Eocene and lived approximately 50 million years ago in India
and Pakistan. Dozens of fossils of pakicetids are known, but
none consist of complete skeletons. Known skeletal elements
of pakicetids include mainly SKULLS, teeth, and jaw fragments.
The smallest pakicetids were as small as a fox, with the largest
as large as a wolf. The dentition of pakicetids varied greatly: the
smaller species had TEETH that resemble those of modern fish
eaters, and teeth of the largest pakicetids resemble those of
hyenas in some respects. Pakicetids may have been predators
or carrion feeders. The nasal opening of pakicetids was near
the front of the head, and the eyes faced dorsally, similar to
crocodiles (Fig. 2). Pakicetids had small BRAINS flanked by
enormous chewing muscles. The skull and dentition of pakicetids do not resemble those of modern whales and dolphins,
but the ear of pakicetids clearly shows that they were cetaceans:
there is a sigmoid process, an involucrum, and ear ossicles that

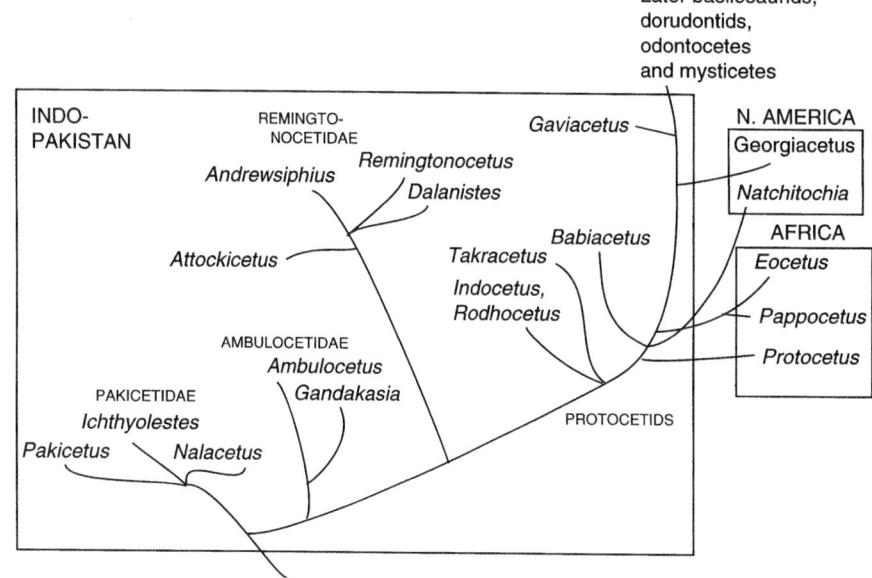

Figure 1 *Phylogeny of early and middle Eocene cetaceans, indicating on which continent each of the families occurred. Protocetids is considered a paraphyletic group, and
Gaviacetus is not considered a protocetid.*

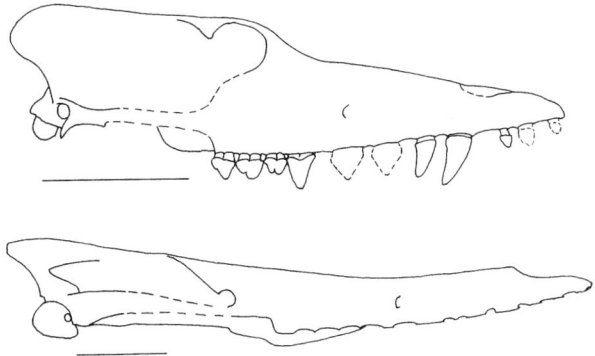

Figure 2 *Reconstructed skulls of* Pakicetus *and* Remingtonocetus. *No teeth are indicated for* Remingtonocetus, *as their
morphology is poorly known. Scale bars are 10 cm.*

are pachyostotic and rotated (Thewissen and Hussain, 1993).
Pakicetid fossils are only found in freshwater deposits and most
are known from deposits that represent shallow ephemeral
streams in an arid climate, which makes it unlikely that pakicetids were good swimmers. Three genera are included in
Pakicetidae: *Pakicetus, Ichthyolestes,* and *Nalacetus.*

II. Ambulocetidae

Ambulocetids are known from middle Eocene rocks in northern India and Pakistan. There are fewer than 10 described am-
bulocetid fossils, but one of these consists of a nearly complete
skeleton of a single individual of *Ambulocetus natans* (Thewissen *et al.,* 1994, 1996; Fig. 3). This skeleton is a prime source of
information about early cetacean biology. *Ambulocetus* was similar in size to a large male sea lion. It had a large head, with a
long snout and eyes that were dorsal on the skull, but faced laterally. The teeth are robust and strongly worn. Skull and vertebrae indicate that the muscles of the head and neck were strong,
indicating that *Ambulocetus* was a powerful animal. The shape
of the lower jaw of *Ambulocetus,* unlike that of the pakicetids,
shows that there was an unusual soft tissue connection between
the back of the jaw and the middle ear. In modern odontocetes,
this connection consists of a large fat pad that functions as part
of the sound-receiving system. This connection is small in *Ambulocetus* and was probably not as important functionally as it is
in modern cetaceans. It does show that hearing adaptation arose
early in cetacean phylogeny. *Ambulocetus* had a back that was
strong and muscular, and the tail was long and lacked flukes. The
hindlimbs were relatively short, but the feet were long, and there

were four toes. The long paddle-shaped feet indicate that it
swam like a modern otter, by swinging its hindlimbs through the
water and creating additional propulsive force with its tail
(Thewissen and Fish, 1997). The forelimbs were short, with five
fingers that each terminated in a short hoof. The hands were
much shorter than the feet. The skeleton of *Ambulocetus* indicates that it was probably slow on land. *Ambulocetus* was probably an ambush hunter, attacking prey in or near shallow water.
This method of hunting is used by modern crocodiles.

Ambulocetus is only known from nearshore marine environments, including estuaries or bays. Geochemical analyses of am-
bulocetid bones indicate that it drank a mixture of fresh and seawater and that different individuals may have inhabited different
microenvironments (Roe *et al.,* 1998). Genera included in Ambulocetidae are *Ambulocetus, Gandakasia,* and *Himalayacetus.*

III. Remingtonocetidae

Remingtonocetids are only known from India and Pakistan,
from sediments approximately 46 to 43 million years old (middle
Eocene; Bajpai and Thewissen, 1998). Dozens of remingtonocetid
fossils have been described, but most of these document only the
morphology of skull and lower jaw (Fig. 2). Dental and postcranial
remains are scarce. The smallest remingtonocetids may have been
as small as *Pakicetus,* and the largest may have been close in size to
Ambulocetus. All early cetaceans had long snouts, but those of remingtonocetids are proportionally even longer than those of other archaeocetes. Skull shape varied between different remingtonocetid
genera and possibly reflected different dietary specializations. In
Andrewsiphius the snout is very narrow and high, and the chewing
muscles are weak, suggesting that it may have eaten small, slippery
fish. In *Remingtonocetus,* the snout is rounded and robust, and the
chewing muscles are large, as would be expected in an animal that
attacks larger, struggling prey. No remingtonocetid displays the robust masticatory morphology of *Ambulocetus.* The nasal opening of
remingtonocetids is near the front of the skull, similar to pakicetids.
The eyes are small, unlike ambulocetids and protocetids. The ear of
remingtonocetids is larger than that of pakicetids and ambulocetids,
and the connection between the lower jaw and the ear is larger than
in ambulocetids. The ears are also set far part, possibly to increase
directional hearing. These features are consistent with an increased
emphasis on underwater hearing in remingtonocetids.

The postcranial morphology of remingtonocetids is only
known from fragmentary specimens. These indicate that the
neck was long and mobile and that the hindlimbs were large
and that remingtonocetids were certainly able to support their
body weight with their limbs, similar to ambulocetids (Gingerich *et al.,* 1995).

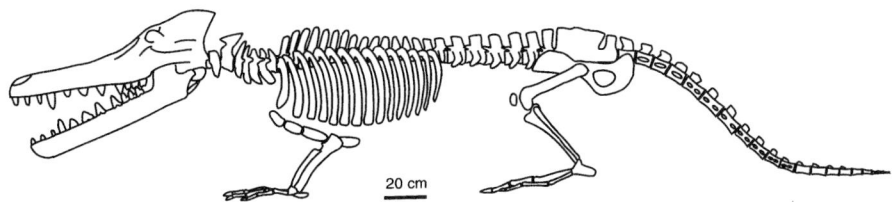

20 cm

Figure 3 *Skeleton of* Ambulocetus. *Scale bar is 20 cm.*

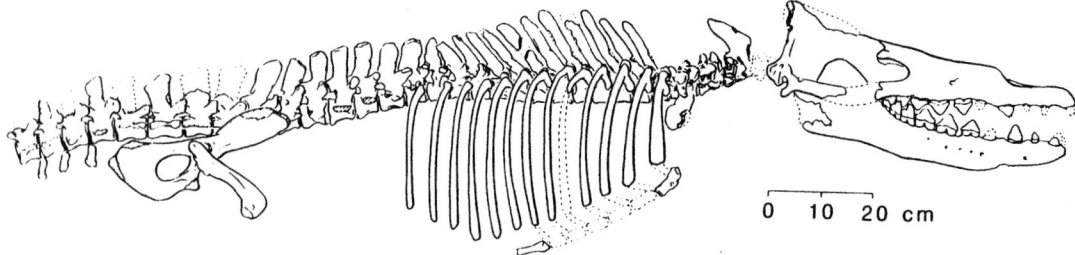

Figure 4 *Skeleton of the protocetid* Rodhocetus. *Reprinted by permission from* Nature 368, 845 (1994). © *Macmillan Magazines Ltd.*

The most primitive and oldest remingtonocetid (*Attockicetus*) is found in the same deposits as *Ambulocetus*. The other remingtonocetids are known from marine, nearshore deposits and may have lived in bays and saltwater swamps. Stable isotope geochemistry indicates that remingtonocetids ingested seawater (Roe *et al.*, 1998). Remingtonocetid genera include *Remingtonocetus, Andrewsiphius, Attockicetus,* and *Dalanistes*.

IV. Protocetidae

Protocetids are found in middle Eocene rocks in Indo-Pakistan, Africa, Europe, and North America. They are the oldest whales to disperse across the oceans, although they probably only inhabited the warm seas near the tropics. Many protocetid genera are known, and several of these include several partial skeletons (Fig. 4) (Hulbert *et al.*, 1998). Protocetids are diverse; their average size was somewhat smaller than *Ambulocetus*.

Protocetids had long snouts, large eyes, and their nasal opening was farther caudally than in earlier archaeocetes. This suggests that protocetids could breathe while holding much of their head horizontally, similar to modern cetaceans. It foreshadows the origin of the cetacean blowhole. The teeth of protocetids vary between genera, but there is trend toward the reduction of occlusal basins on the teeth. Basins on the teeth of primitive mammals, including pakicetids and ambulocetids, are an essential part of chewing morphology. Modern cetaceans do not chew and lack basins, and protocetids show an early stage in the reduction of masticatory function of the teeth. It is likely that protocetids were active hunters of marine animals, possibly similar to modern pinnipeds. Protocetid locomotor morphology was varied. In general, the tail is well developed and was probably involved in creating propulsive forces (Buchholtz, 1998). The hindlimbs are reduced, and in some species the innominate is not connected by bone to the vertebral column, suggesting that the hindlimb did not support the body weight. There are no fossils that document all of protocetid hindlimb morphology, but some preserved elements suggest that the hindlimbs were short. Indo-Pakistani protocetids inhabited the same environments as the remingtonocetids, and protocetids from other continents are known from shallow marine environments. Known genera of protocetids are *Protocetus, Babiacetus, Eocetus, Georgiacetus, Indocetus, Natchitochia, Pappocetus, Rodhocetus,* and *Takracetus*.

These archaeocete families document that the Eocene cetacean evolution is characterized by increasing aquatic adaptations, starting at amphibious early whales (pakicetids, ambulocetids) to more marine protocetids. Late Eocene whales (Dorudontids, basilosaurids) were probably obligate marine mammals.

See Also the Following Articles

Basilosaurids ■ Cetacea, Overview ■ Cetacean Evolution ■ Morphology, Functional ■ Paleontology

References

Bajpai, S., and Thewissen, J. G. M. (1998). Middle Eocene cetaceans from the Harudi and Subathu Formations of India. *In* "The Emergence of Whales, Evolutionary Patterns in the Origin of Cetacea" (J. G. M. Thewissen, ed.), pp. 213–233. Plenum Press, New York.

Buchholtz, E. A. (1998). Implications of vertebral morphology for locomotor evolution in early Cetacea. *In* "The Emergence of Whales, Evolutionary Patterns in the Origin of Cetacea" (J. G. M. Thewissen, ed.), pp. 325–352. Plenum Press, New York.

Gingerich, P. D., Arif, M., and Clyde, W. C. (1995). New archaeocetes (Mammalia, Cetacea) from the middle Eocene Domanda Formation of the Sulaiman Range, Punjab (Pakistan). *Contrib. Mus. Paleontol. Univ. Mich.* **30,** 291–330.

Gingerich, P. D., Raza, S. M., Arif, M., Anwar, M., and Zhou, X. (1994). New whale from the Eocene of Pakistan and the origin of cetacean swimming. *Nature* **368,** 844–847.

Hulbert, R. C. (1998). Postcranial osteology of the North American middle Eocene protocetid Georgiacetus. *In* "The Emergence of Whales, Evolutionary Patterns in the Origin of Cetacea" (J. G. M. Thewissen, ed.), pp. 235–268. Plenum Press, New York.

Roe, L. J., Thewissen, J. G. M., Quade, J., O'Neil, J. R., Bajpai, S., Sahni, A., and Hussain, S. T. (1998). Isotopic approaches to understanding the terrestrial to marine transition of the earliest cetaceans. *In* "The Emergence of Whales, Evolutionary Patterns in the Origin of Cetacea" (J. G. M. Thewissen, ed.), pp. 399–421. Plenum Press, New York.

Thewissen, J. G. M. (1998). Cetacean origins: Evolutionary turmoil during the invasion of the oceans. *In* "The Emergence of Whales, Evolutionary Patterns in the Origin of Cetacea" (J. G. M. Thewissen, ed.), pp. 451–464. Plenum Press, New York.

Thewissen, J. G. M., and Fish, F. E. (1997). Locomotor evolution in the earliest cetaceans: Functional model, modern analogues, and paleontological evidence. *Paleobiology* **23,** 482–490.

Thewissen, J. G. M., and Hussain, S. T. (1993). Origin of underwater hearing in whales. *Nature* **361,** 444–445.

Thewissen, J. G. M., Hussain, S. T., and Arif, M. (1994). Fossil evidence for the origin of aquatic locomotion in archaeocete whales. *Science* **263,** 210–212.

Thewissen, J. G. M., Madar, S. I., and Hussain, S. T. (1996). *Ambulocetus natans,* an Eocene cetacean (Mammalia) from Pakistan. *Courier Forschungs-Institut Senckenberg* **190,** 1–86.

Williams, E. M. (1998). Synopsis of the earliest cetaceans: Pakicetidae, Ambulocetidae, Remingtonocetidae, and Protocetidae. *In* "The Emergence of Whales, Evolutionary Patterns in the Origin of Cetacea" (J. G. M. Thewissen, ed.), pp. 1–28. Plenum Press, New York.

Zimmer, C. (1998). "At the Water's Edge: Macroevolution and the Transformation of Life." Free Press, New York.

Arctic Marine Mammals

JOHN J. BURNS
Fairbanks, Alaska

There is a popular tendency to speak in rather nebulous terms about arctic marine mammals without defining the Arctic, the role and diversity of sea ice as a major component in high-latitude ecosystems, or the diversity of marine mammals adapted to live in various ice-dominated habitats. There are, in fact, few truly arctic marine mammals. This introductory discussion is about those that occur in ice-covered seas, at least during winter and spring, and in most cases give birth when ice is present. They include one ursid, eight pinnipeds, and three cetaceans. Only three species have a continuous circumpolar distribution. A few species (or stocks thereof) maintain a mostly year-round association with sea ice whereas most do not. The different marine mammals show various degrees of adaptation to ice. There is a continuum of ice-influenced habitats, and of mammalian adaptations to those habitats.

I. Northern Ice-Covered Marine Environments

Traditionally the Arctic is viewed as an ill-defined region around the North Pole that is further subdivided into the high arctic and the low arctic. We are here concerned with much broader, although still poorly defined, areas within which ice-associated bears, pinnipeds, and cetaceans occur. Some freshwater seals are included. It is useful to think in terms of regional climate, oceanography, annual ice dynamics, and life history strategies. For most marine environments the definitions advanced by Dunbar (1953) are particularly useful. The arctic seas are those in which unmixed polar water from the upper layers of the Arctic Ocean occurs in the upper 200–300 m. A large portion of this zone is ice covered throughout the year. The maritime subarctic includes those seas contiguous with the Arctic Ocean in which the upper water layers are of mixed polar and nonpolar origin. There are, however, some noncontiguous subarctic seas (no water of polar origin) adjacent to terrestrial ecosystems that lie in the subarctic zone. Examples include the Okhotsk Sea, the northern part of the Sea of Japan (Tartar Strait), Lake Baikal in Siberia, and Cook Inlet in Alaska (Fig. 1). In the subarctic there is a complete annual ice cycle, from formation in autumn to disappearance in summer. Finally, there are areas in the temperate zone where unique climate conditions produce a winter ice cover of relatively short duration.

Such areas include the Baltic Sea, the northern Yellow Sea, and the western Sea of Japan.

In late summer the average annual minimum extent of sea ice is 5.2 million km^2, restricted mainly to the Arctic Ocean. The average maximum extent in late winter–early spring is 11.7 million km^2, including all of the subarctic seas (or parts thereof), and parts of some in the temperate zone. Most species of the so-called arctic marine mammals are associated with the seasonal ice during the breeding period. They cope with the annual expansion and contraction of the ice cover in a variety of different species-specific ways. Clearly there are many kinds of ice-dominated habitats formed in response to factors such as regional climate, weather, latitude, currents, tides, winds, land masses, proximity of open seas, and others.

II. Sea Ice Habitats

Sea ice in the Arctic and subarctic occurs in more complex forms than ice in the Antarctic. This is because of the central location of the Arctic Ocean with its perennial drifting ice, its partially landlocked nature, and the complexity of the subarctic seas encircling it. The annual expansion and contraction of the ice cover provides conditions ranging from the thick and relatively stable multiyear ice of the high latitudes to the transient and highly labile southern pack ice margins that border the open sea. Marine mammals must have regular access to air above the ice, as well as to their food in the ocean below it. During the breeding season the ice on which pinnipeds haul out must be thick enough and persist long enough for completion of the critical stages of birth, nurture of their young, and, in many cases, completion of the annual molt. Additionally, by virtue of location, behavior, reproductive strategies, and/or physical capabilities, they must be able to avoid excessive predation on dependent and often nonaquatic young. All of the marine mammals must also cope with the great reduction or complete absence of ice during the open water seasons.

There are many different features of the varied types of ice cover that provide marine mammals access to air and allow the pinnipeds to haul out. There are also some features, characteristics, or types of ice that exclude most marine mammals. Important ice features or types include stable land-fast ice (excludes most marine mammals); annually recurring persistent polynyas (irregular shaped areas of open water surrounded by ice); recurrent stress and strain cracks, coastal, and offshore lead systems (long linear openings); zones of convergence and compaction (as against windward shores or in constrictions such as narrow straits); zones of divergence (where boundary constraints are eased); the generally labile pack ice of the more southerly seas; and the margins of front zones of broken ice, the characteristics of which are strongly influenced by the open sea. Ice margins are particularly productive in that ice-edge blooms of phytoplankton and the associated consumers extend many tens of kilometers away from them.

III. Role of Sea Ice

There are great differences in how marine mammals exploit ice-dominated environments (Fig. 2). Many are a function of evolutionary constraints imposed on the different linages of

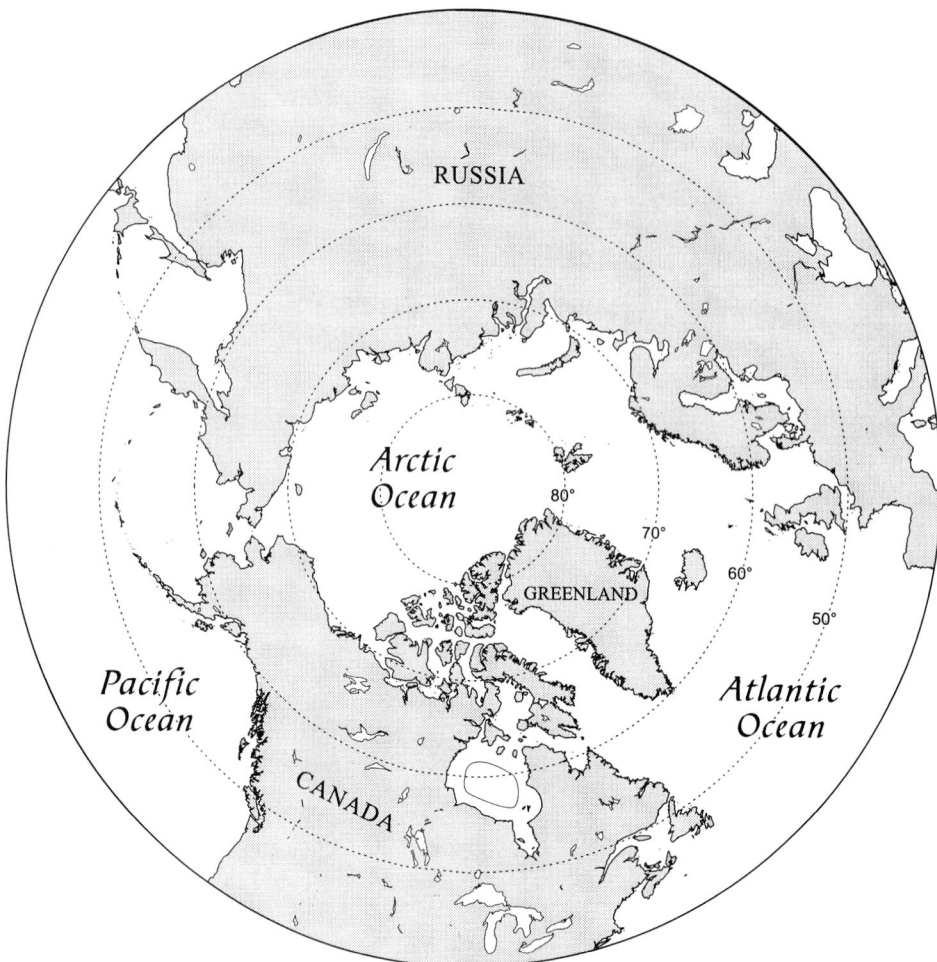

Figure 1 *Map of the Arctic Ocean.*

mammals. Polar bears (*Ursus maritimus*) are the most recent arrivals in the high-latitude northern seas, having evolved directly from brown bears (*U. arctos*). They utilize relatively stable ice as a sort of *terra infirma* on which to roam, hunt, den, and rest. Like their contemporary terrestrial cousins, they are generally not faced with the problem of ice being a major barrier through which they must surface to breathe. Cetaceans are at the other extreme. They live their entire lives in the water and have limited (though differing) abilities to make breathing holes through ice and are therefore constrained to exist where natural openings or thin ice are present. Pinnipeds spend most of their time in the water, but they must haul out to bear their young. Most of them also haul out on ice to suckle their young, to molt, and to rest.

For cetaceans, obvious benefits are protection from predators, access to ice-associated prey without competition from other animals, and a less turbulent winter environment shielded from perpetual and often storm force winds.

Pinnipeds have flourished in ice-dominated seas both in terms of the number of different species and in the number of individuals. All are obliged to haul out either on land or on ice

for at least part of the year. As noted by Fay (1974), ice has several special advantages over land, including *isolation* from many predators and other disturbing terrestrial animals; vastly increased *space* away from seashores; a *variety* of different habitats that accommodate more species than does land; easy access to their *food supply*, especially for those that are benthic feeders or that utilize concentrations of prey associated with ice fronts and polynyas; passive *transportation* to new feeding areas and during migrations; *sanitation* resulting from the ability to avoid or reduce crowding and to haul out on clean ice; and *shelter* among pressure ridges or in snow drifts.

IV. Ice-Breeding Marine Mammals

Ice-breeding marine mammals in the Northern Hemisphere include eight pinnipeds: gray (*Halichoerus grypus*) (some populations), harp (*Pagophilus groenlandicus*), hooded (*Cystophora cristata*), bearded (*Erignathus barbatus*), ringed (*Pusa hispida*), spotted (*Phora largha*), and ribbon seals (*Histriophora fasciata*) as well as the walrus (*Odobenus rosmarus*); three cetaceans: narwhal (*Monodon monoceros*), beluga (or bulukha)

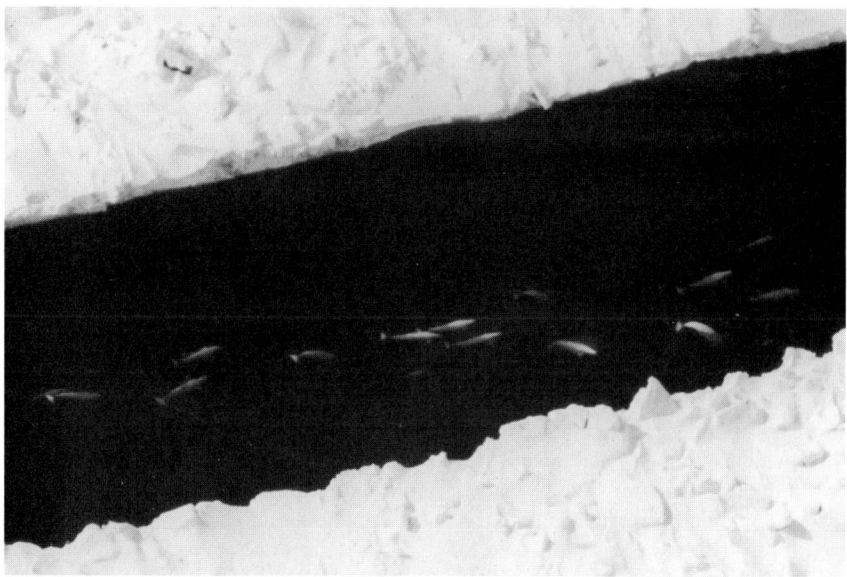

Figure 2 *Distribution of sea ice influences the distribution of many marine mammals. In the winter, thickening ice can threaten the survival of individuals if they become trapped in areas away from open sea. In the spring and summer, pack ice fragments provide avenues of transit for whales, such as the beluga whales pictured. Photo by S. Leatherwood.*

(*Delphinapterus leucas*), and bowhead whale (*Balaena mysticetus*); and one fissiped: the polar bear.

A. Pinnipeds

A common theme in the ecology of ice-breeding pinnipeds is that of an obligatory, or nearly obligatory, association with ice during the breeding season, which occurs during or shortly after the period of maximal ice extent and relative stability. Seal pups become independent during the spring onset of ice disintegration and retreat. Most species also molt on the ice, after which they disperse to a variety of habitats during the open water season, a few continuing to remain with the diminishing cover. They resume their increasing association with ice during autumn, as it again forms and expands. They haul out on the ice in all seasons during which it is present, although with highly variable frequency depending on species and weather. The maximum number of species and the greatest total number of seals are associated with ice when it is most extensive, and vice versa.

The lanugo of most seals born on ice or in snow lairs, and remaining in one place for long periods of time, is primarily an adaptation for maintaining body heat. Such pups tend to be small, have little insulating BLUBBER, and have a relatively large surface area to body mass ratio at birth. White-coated pups presumably also benefit from the cryptic coloration it provides during the period before they are weaned and begin to enter the water. Prenatal molting occurs in those ice-breeding pinnipeds that are relatively large at birth and can enter the water within hours or days. Detailed discussions of northern ice-breeding seals are presented in the following species accounts, although general comments are noted below.

Gray seals are usually not included in the category ice-associated marine mammals. However, some populations breed on the ice. Gray seals largely inhabit the temperate zone in the North Atlantic region. Their distribution is coastal, often in association with harbor seals (*Phoca vitulina*). There are three populations: those in the Baltic Sea, the eastern North Atlantic, and the western North Atlantic. There is a very wide range in timing of the breeding season. In the eastern Atlantic, pups are born on shore during late autumn to early winter. In both the Baltic and the western Atlantic, however, pups are born during mid- to late winter on ice near shore, or on shore when ice is absent. At birth, gray seals weigh about 15 kg. In all populations almost all pups are born with a silky, whitish coat of lanugo that is retained during the nursing period. They remain on ice or land until after weaning. The late pupping season of the marginally ice-associated breeding populations is thought to be an adaptation to that environment. Gray seals move extensively, although they are not considered to be migratory. None are associated with sea ice during late spring through autumn.

Spotted seals (or larga seals) occur in continental shelf waters of the Pacific region that are seasonally ice covered. During winter and spring they mainly inhabit the temperate/subarctic boundary areas, occurring in the southern ice front (mainly) of the Bering and Okhotsk seas or in the very loose pack ice of the northern Yellow Sea and Sea of Japan. The birth season is from January through April, depending on latitude. All populations give birth and nurture their pups on the ice. Newborn pups weigh about 10 kg and have a dense, whitish, wooly lanugo, which is shed toward the end of the month-long nursing period. Seals older than pups usually haul out on the ice to molt, although they also use land when the ice disappears early. As the seasonal ice disintegrates and recedes, all spotted seals disperse, moving to the ice free coastal zone where they use haulouts on land. The seasonal dispersal can be extensive: in the Okhotsk Sea to its entire perimeter and from the central Bering Sea to most of its perimeter, as well as northward into the northern Chukchi and Beaufort seas. Therefore, some spotted seals reside in the higher latitudes of the subarctic zone during the open water season. They range widely over the continental shelves. There is a close association with sea ice during autumn through spring.

Ribbon seals are animals of the temperate and temperate/subarctic boundary zones in the North Pacific region. Breeding populations are in the Bering and Okhotsk seas and Tartar Strait. During the open water season they live a completely pelagic existence in the cold temperate waters along and beyond the continental shelves, often far from the locations of their winter habitat. The breeding cycle is similar to that of the spotted seal, and the two occur in relative close proximity to each other during late winter and spring. At the time of pupping and molting, ribbon seals utilize ice of the inner ice front where floes are larger, thicker, more deformed, and more snow covered than in the adjacent ice margin favored by spotted seals. They are noted for hauling out on very clean ice. They pup in late March and April. At birth the pups weigh about 10.5 kg and have a coat of dense, white lanugo. During the nursing period the pups remain on the ice and gradually shed their lanugo. They remain on the ice for some time after they are weaned. In the opinion of this writer the preference for heavier ice of the inner front, which persists longer than that of the spring ice margin, is because it permits all age classes of these otherwise pelagic seals to haul out until the molt is completed. Ribbon seals do not come ashore unless debilitated. They appear to be the pinniped analogue of the Dall's porpoise (*Phocoenoides dalli*) during the pelagic phase of their annual cycle (June through late autumn), dispersing near the shelf breaks and the deeper waters beyond. They have the morphological and physiological attributes of a seal that can dive to great depths and remain submerged for a long time. Relatively few move north of their breeding range, except during years of minimal spring ice cover.

Harp seals occur in the North Atlantic region. There are three breeding populations: those of the White Sea, the Greenland Sea, and the Gulf of St. Lawrence. They are a gregarious and highly migratory species that lives primarily in the subarctic zone during winter and spring and is broadly distributed in the open sea from the coastal zone to near the ice margin during the open water season. The birth period extends from late January to early April, depending on the region. During the pupping season they form large aggregations in which pups are born in close proximity to each other (often closer than 2.5 m). They prefer large ice fields within the ice front, usually at some distance from the pack ice margins. Here the floes are extensively deformed and ridged, providing shelter to the otherwise exposed pups. At birth the pups weigh about 11.8 kg and have a coat of dense white lanugo. The nursing period lasts from 10 to 12 days and they fast, remaining on the ice floes, for some time after weaning. MATING, which occurs after pups are weaned, is followed by the molt. As with the ribbon seal (which is also pelagic after the molt) it seems that the preference of harp seals for the thicker and more stable ice of the inner front zone is because it provides the selective advantage of persisting until the molt is completed. Harp seals make one of the longest

annual migrations of any pinniped, with some traveling more than 3000 miles from wintering to summering areas. Part of the spring migration is passive as the seals drift on the receding ice.

Hooded seals are a high subarctic, strongly migratory, deep water species that occurs in the North Atlantic region and whelps in four different areas: near Jan Mayen, in Davis Strait, off the Labrador coast, and in the Gulf of St. Lawrence. Shifts to heavier ice in the more northerly whelping areas reportedly occur during periods of warmer climate and diminished ice (drift ice pulsations). Pups are mainly born on thick heavily ridged ice floes well within the subarctic pack during late March and early April. At birth the pups weigh about 22 kg (relatively large) and are comparatively precocious. Their lanugo is shed *in utero* and their birth coat (the blue-back stage) does not resemble the pelage of adults. The nursing period is amazingly brief, averaging 4 days, during which the mothers remain on the ice with their pups. Pups enter the water shortly after weaning, although they spend considerable time on the ice during the postweaning fast. Mating occurs after lactation, and molting after mating. They migrate, both passively on the ice and by swimming, and disperse widely in the open sea (to the Grand Banks), near high-latitude shores, and along the edge of the summer pack ice. Extralimital occurrences are common, even to the North Pacific region.

Bearded seals are primarily benthic feeders that have a circumpolar distribution in arctic and subarctic seas. They have evolved in the face of heavy predation pressure by polar bears. Their range broadly overlaps that of all the other ice-breeding pinnipeds. They are the least selective of the seals with respect to ice type, provided that it generally overlies water less than about 200 m deep. Bearded seals are usually solitary and occur from the southern ice margins and fronts (few) to the heavy drifting pack around the rim of the arctic basin, although infrequently in landfast and multiyear ice. Within the heavier pack ice they occur mainly in association with those features that produce open water or thin ice (polynyas, persistent leads, flaw zones, etc.). They are capable of breaking holes in thin ice (≤10 cm) and can make or at least maintain breathing holes in thicker ice with their stout foreclaws. The large pups (about 34 kg) are usually born on the edges of small detached, first year floes very close to the water. The lanugo is shed *in utero*. The pups can swim from birth if necessary, and usually do so, at least in order to move away from the afterbirth. Beyond that they remain on the ice for a day or so. Nursing, which is usually on the ice, lasts 12 to 18 days, during which time the pups spend a considerable amount of time in the water and begin independent feeding prior to the end of the nursing period. Mating occurs after pups are weaned. The main period of molt is during May and June, and the greatest numbers of all age classes haul out on the ice during that time. However, molting seals are encountered throughout the year. In some areas, such as the Bering and Chukchi seas, the adults and most juveniles migrate to maintain a year-round association with ice. They haul out on it throughout the year, although infrequently during winter. In areas where ice disappears during summer (i.e., the Okhotsk Sea) or recedes beyond the continental shelf, they occur in the open sea, in nearshore areas, in bays and estuaries, and sometimes haul out on land.

Ringed seals have a circumpolar distribution that includes the arctic and subarctic seas. They have evolved in the face of heavy predation pressure, primarily by arctic foxes, which take pups, and polar bears, which take all age classes. Unique species and subspecies of the genus *Pusa* also occur as landlocked populations in Eurasia and include the seals of lakes Baikal, Ladoga, and Saimaa, as well as the Caspian Sea. Ringed seals are the most numerous and widely distributed of the northern ice-associated pinnipeds. During winter to early summer they utilize all ice habitats from the drifting ice margins and fronts (relatively few) to thick stable shore-fast and multiyear ice. Their range extends farther north and includes areas of heavier ice cover than that of any other marine mammal except the polar bear. They occur from shallow coastal waters to the deeps of the Arctic Basin. During winter through late spring the adults tend to be solitary and territorial and are most abundant in moderate to heavy pack and shore-fast ice. Ringed seals can make and maintain holes through the ice and crawl out to construct snow lairs above them. In regions where conditions permit, they migrate and maintain a year-round association with ice. In some regions where the pack ice completely or mostly disappears during summer (i.e., the Okhotsk Sea, Baffin Bay, Lake Baikal) they move to nearshore areas and sometimes haul out on land.

Pups are born during late March through April, in snow lairs or cavities in pressure ridges. The pups are small, averaging about 4 kg at birth, and have a thick woolly lanugo, which is usually shed by the end of the nursing period. Lactation lasts 4 to 6 weeks. Pups mostly remain in the birth lair for the first several days, but are soon capable of entering the water and periodically returning to a lair. Mating occurs after the nursing period and is followed by the molt. The peak period of molt in nonpups is during May and June, when the seals haul out above collapsed (melted) lairs, at enlarged breathing holes, or next to natural openings in the ice. Ringed seals are extremely wary when hauled out. During the open water season, depending on the region, they occur in the much reduced pack ice and in open water over a broad area. In some regions they haul out on land.

Walruses are the largest and most gregarious of the ice-breeding northern pinnipeds. They have a discontinuous although nearly circumpolar distribution around the perimeter of the Arctic Ocean and the contiguous subarctic seas. They are benthic feeders mainly restricted to foraging in waters less than 110 m deep. In all areas their distribution is limited by water depth and in some (i.e., the Laptev and Kara seas) it is further constrained by the severity of ice conditions. In most regions, walruses haul out on ice in preference to land. However, during the open water season, they (mainly males) use land haulouts near the wintering grounds and, in more northerly areas, most come ashore to rest when ice drifts beyond shallow water, as occurs frequently in the Chukchi Sea. During autumn, walruses that migrate southward ahead of the advancing ice also come ashore to rest. All populations are associated with seasonal pack ice during winter to spring/early summer. They mainly use moderately thick floes well into the winter/spring ice cover. The combined requirements for floes low enough to haul out on, but thick enough to support these large animals (usually herds of them) and that are also over shallow productive continental shelves, make walruses particularly dependent on regions within which persistent natural openings are present. They make (batter) holes through ice as thick as 22 cm, using the head, and sometimes maintain them with the aid of their tusks.

Calves are born mainly in early May, which for the Bering Sea population is during the northward spring migration of females, calves, subadults, and some adult males. Walruses shed their lanugo *in utero*. Calves are born on the ice. They weigh about 60 kg and enter the water from birth, although they haul out frequently. Cows with young calves often form large nursery herds and migrate passively on the drifting ice, as well as by swimming. The nursing period lasts more than a year. Walruses haul out in all months of the year.

B. Cetaceans

Ice-associated cetaceans include two odontocetes (toothed whales), the beluga and the narwhal, and one mysticete (baleen) whale, the bowhead. None have a completely circumpolar distribution. Morphological adaptations to ice seem minimal and include the lack of a dorsal fin in all three and the high "armored" promontory atop, which the blowhole of the bowhead is situated. In winter, all three species occur in drifting ice where there are persistent natural openings or where the ice cover is thin. Polynyas, shear zones, and leads are important features for them in regions of heavy pack ice.

The narwhal is a North Atlantic species of the high subarctic and low arctic, which, in winter, consistently occurs in regions of heavy drifting ice over deep water or shelf edges. Adult males have a unique, long unicorn-like tusk that has an unknown function but is presumably used in male sexual display. The largest population is that in Davis Strait and Baffin Bay. Seasonal movements of narwhals are directly tied to the advance and retreat of ice. During summer they move to high-latitude, ice-free coastal and nearshore areas, which are often penetrated by deep fjords. Calves are born during the summer, reportedly during July and August, and are nursed for more than a year. This whales' preference for heavy pack ice during winter and spring makes them particularly vulnerable to entrapment during periods of rapid ice formation or when the pack becomes tightly compressed. Most episodes of entrapment are probably brief, although prolonged confinement and rapid ice formation sometimes result in death either by drowning or by polar bears, for which entrapped whales are easy and plentiful prey. Confined whales are also harvested by Inuit hunters whenever they are found.

Beluga whales have a nearly circumpolar distribution that extends from roughly 40°N (the Gulf of St. Lawrence and the northern Sea of Japan) into the summer multiyear pack of the Arctic Ocean. During winter they are most abundant near the southern ice margins and fronts and as far into the seasonal pack as conditions permit. Again, polynyas, flaw zones, persistent leads, and other features that permit belugas to surface for air are important in the more northerly regions. Belugas often make holes through thin (to about 10 cm) newly formed ice by pushing it up with their head and back. They also surface in openings made by bowheads, with which they often associate during spring migration.

DISTRIBUTION during the open water season is quite variable depending on region. In most cases these whales move into the coastal zone in May to July or early August, where they enter lagoons and estuaries to feed, bear calves in warmer water, and molt. They frequently ascend rivers to feed on seasonally abundant fishes. Telemetry studies have shown that belugas in the Beaufort Sea and the Canadian high arctic spend slightly less than 2 weeks in lagoons, and spend most of their summer feeding in offshore waters (unlike belukhas farther south). Some males from the eastern Chukchi and Beaufort Sea stocks are now known to penetrate much farther into the pack ice of the Arctic Ocean during summer than was previously supposed (to beyond 80°N). Other belugas range widely throughout Amundsen Gulf and the Beaufort and northern Chukchi seas during summer and early autumn.

The Bering Sea population includes multiple stocks, some of which migrate north through the disintegrating ice cover in spring and use both ice-free coastal waters and the summer pack of the Arctic Ocean. Most belugas leave the coastal zone by September, although some remain or revisit areas where food is abundant. This habit has resulted in some large and fatal entrapments. Smaller entrapments at sea are not uncommon. All move with the advancing ice in autumn, either migrating southward with it or moving into it as it forms and expands.

Bowhead whales occur in subarctic waters during winter and spring and, depending on the population, in productive marginal arctic waters during the open water season. These large whales are highly specialized zooplankton feeders and seek areas of high prey abundance. Bowheads may be the slowest growing and latest maturing mammal on earth. Females are thought to become sexually mature between their late teens to midtwenties (later than humans or elephants). They may live to be well over 100 years old.

The range of bowheads includes the North Atlantic region (three stocks) and North Pacific region (two stocks), with extensive gaps between the two. During winter through early spring they occur from the southern margins of the ice pack to as far into it as persistent natural opening in the ice permit. Large polynya systems are of great importance during winter and spring. In the Pacific sector the Okhotsk Sea stock remains there after the ice has completely disappeared. Most whales of the other stocks migrate northward during spring and southward during autumn. Most whales of the Bering Sea stock maintain a loose association with the summer ice margin, mainly feeding in the open waters south of it. The northward migration begins in late March or early April when they move from the Bering Sea into the eastern Chukchi, and then the Beaufort seas through heavy ice in a very long corridor cleaved by a linear system of stress cracks, polynyas, shore leads, and flaw zones. Some migrate into the western Chukchi Sea. Beluga whales commonly migrate with bowheads. Bowheads can stay submerged for long periods and push up through relatively thick ice. These abilities allow them to reside and travel in waters where natural openings in the ice are continually forming and refreezing. Calves are born mainly during April to early June, during the spring migration.

C. Fissipeds

Two fissipeds roam the high-latitude ice-covered seas: the polar bear and the arctic fox (*Vulpes lagopus*). The latter, which rarely enters the water and pups in dens on shore, is not usually considered to be a marine mammal.

Polar bears have a circumpolar distribution in the Arctic and contiguous high subarctic. They are not "marine" in the sense that whales or seals are, but occupy a marine environment in which ice is the substrate on which they live. They prey on other marine mammals, particularly the ringed seal. Depending on the region, they remain with the ice and hunt year round or, where it completely disappears, they come ashore and usually fast. Exceptions to the latter are some islands (i.e., Wrangel and Herald) where they feed on beachcast carcasses or hunt animals that haulout on shore, particularly walruses. On the ice, the availability (access) of prey seems to be a more important factor affecting the distribution of bears then is maximum prey abundance. It is difficult for bears to catch marine mammals, except pups, when there are unlimited escape routes and places to surface in a very labile ice cover. For example, few polar bears range south of the northern Bering Sea during winter, even though the majority of other marine mammals (except ringed seals) are south of there. Also, polar bears are not present in the Okhotsk Sea.

Pregnant females make and enter snow dens in early November. These maternity dens can be on the heavy pack ice, on shore-fast ice (relatively few), or on land. The altricial cubs are born in late December or early January, during the arctic winter, and do not emerge with their mothers until late March or early April. Sows that bore their cubs on shore go immediately back to the sea ice. Ringed seal pups, born in lairs beneath the snow starting in late March, are important prey for sows with cubs.

V. Possible Effects of Climate Change

It is now well recognized that we are in a phase of accelerated global warming and that the multiyear and seasonal ice cover is being affected. The seasonal ice cover is becoming generally less extensive and thinner, and it is forming later and disintegrating earlier than at any time in recorded history. In the Arctic Ocean the ice cover is also thinning. Similar changes have occurred in the past. In addition to a somewhat diminished ice cover, warming conditions also produce rising sea levels, increased ocean circulation, and increased nutrient flow into the northern seas. These changes are likely to have varying effects on the different species of ice-breeding marine mammals. For some species, e.g., the walrus, ringed seal, and polar bear, ameliorating conditions might be positive as they would result in more favorable habitat over a broader area then at present. For others, especially those now dependent on limited ice habitat, the changes are likely to have negative impacts. Spotted seals in the Yellow Sea and the Sea of Japan are likely to be affected negatively. At a minimum, global warming will likely result in geographic shifts of the seasonal centers of abundance of all ice-associated marine mammals.

See Also the Following Articles

Antarctic Marine Mammals ■ Cetacean Ecology ■ Pinniped Ecology

References

Burns, J. J. (1970). Remarks on the distribution and natural history of pagophilic pinnipeds in the Bering and Chukchi seas. *J. Mammal.* **51,** 445–454.

Burns, J. J. (1981). Ice as marine mammal habitat in the Bering Sea. *In* "The Eastern Bering Sea Shelf: Oceanography and Resources" (D. W. Hood and J. A. Calder, eds.), pp. 781–797. University of Washington Press, Seattle, WA.

Central Intelligence Agency. (1978). "Polar Regions Atlas." Publication GC 78-10040, Washington, DC.

Dunbar, M. J. (1953). Arctic and subarctic marine ecology: Immediate problems. *Arctic* **6,** 76–90.

Fay, F. H. (1974). The role of ice in the ecology of marine mammals of the Bering Sea. *In* "Oceanography of the Bering Sea" (D. W. Hood and E. J. Kelley, eds.), pp. 383–399. University of Alaska, Fairbanks, AK.

George, J. C., Bada, J., Zeh, J., Scott, L., Brown, S. E., O'Hara, T., and Suydam, R. (1999). Age and growth estimates of bowhead whales (*Balaena mysticetus*) via aspartic acid racemization. *Can. J. Zool.* **77,** 571–580.

National Research Council. (1996). "The Bering Sea Ecosystem." National Academy Press, Washington, D.C.

Niebauer, H. J., and Alexander, V. (1989). Current perspectives on the role of ice margins and polynyas in high latitude ecosystems. *In* "Proceedings of the Sixth Conference of the Comité Arctique International" (L. Rey and V. Alexander, eds.), pp. 121–124. E. J. Brill, New York.

Richard, P. R., Heide-Jørgensen, M. P., and St. Aubin, D. J. (1998). Fall movements of belugas (*Delphinapterus leucas*) with satellite-linked transmitters in Lancaster Sound, Jones Sound, and northern Baffin Bay. *Arctic* **51,** 5–16.

Ridgway, S. H., and Harrison, R. J. (eds.) (1981). "Handbook of Marine Mammals," Vol. 2. Academic Press, New York.

Stirling, I., and Cleator, H. (eds.) (1981). "Polynyas in the Canadian Arctic." Canadian Wildlife Service Occasional Paper 45, Ottawa, Ontario.

Stringer, W. J., and Groves, J. E. (1991). Location and areal extent of polynyas in the Bering and Chukchi seas. *Arctic* **44**(Suppl. 1), 164–171.

Vibe, C. (1967). Arctic mammals in relation to climate fluctuations. *Meddelelser on Grønland* 1–227.

Wadhams, P. (1990). Evidence for thinning of the arctic ice cover north of Greenland. *Nature* **345,** 795–797.

Weller, G., and Lange, M. (1999). "Impacts of Global Change on the Arctic Regions." Center for Global Change and Arctic System Research. University of Alaska, Fairbanks, AK.

Artiodactyla

JESSICA M. THEODOR
University of California, Los Angeles

Artiodactyls are the closest living relatives of the Cetacea (whales and dolphins), although whales may be more closely related to an extinct group of mammals, the mesonychids. It seems odd to think of the whales being closely related to terrestrial ungulates, but biologists have long appreciated that artiodactyls and whales share several unusual morphologies despite their vastly divergent ways of life. One such feature is that lumps of tissue, called umbilical pearls, are found along the umbilical cord between the fetus and the mother. Male whales and artiodactyls share distinctive features of the penis, which is composed primarily of fibroelastic tissue, unlike the penis of other mammals, largely made up of spongy tissue. Because the penis is fibroelastic and not spongy, the mechanism of erection of the penis differs from most mammals, utilizing specialized retractor muscles, which are found in both artiodactyls and cetaceans.

I. Diagnostic Characters

Artiodactyls are even-toed ungulates because they have two or four toes on each foot. They have elongated legs and stand on the tips of their toes, which bear hooves instead of nails. Artiodactyls are characterized by paraxonic foot symmetry, in which the main axis of weight bearing of the foot passes between the third and the fourth digits. The ankle of artiodactyls is uniquely specialized. The astragalus, the main bone transferring force from the leg to the foot, has a deeply grooved, pulley-shaped surface on both the tibial and foot ends, whereas other mammals have only the tibial pulley. The double-pulley joint strongly limits lateral rotation of the foot on the hind leg, permitting only fore and aft movements of the leg.

II. Taxonomy

Artiodactyla includes nine families today: the Suidae (pigs), Tayassuidae (peccaries), Hippopotamidae (hippos), Camelidae (camels and llamas), Tragulidae (mouse deer), Giraffidae (giraffe and okapi), Antilocapridae (pronghorn antelope), Cervidae (deer), and Bovidae (cows, sheep, goats, and antelope). Of the 211 extant artiodactyl species, close to 65% are in the Bovidae. The families Hippopotamidae, Tragulidae, Giraffidae, and Antilocapridae are very low in diversity, with one or two species each.

III. Distribution and Ecology

Living artiodactyls are native to North America, Eurasia, Africa, and South America, and they have been introduced to other areas. Artiodactyls occupy a wide range of habitats, from deserts to taiga, and show a wide range of dietary specialization, social structure, locomotor modes, and habitat preferences. However, the majority of artiodactyls are terrestrial, and only the hippo is semiaquatic. There is a considerable range of body sizes, from 0.7 to 3000 kg. Most artiodactyls are herbivores, except the pigs and peccaries, which also eat small prey and eggs. The pigs, peccaries, hippos, and mouse deer are nocturnal, whereas most other species are diurnal. Most artiodactyls bear precocial young in small litters, although pigs have large litters of relatively helpless young who are reared in nests.

IV. Notable Anatomy

The most notable anatomical system of artiodactyls is foregut rumination, which allows ruminant artiodactyls (bovids, cervids, antilocaprids, giraffids, and tragulids) the ability to break down cellulose, a plant material ordinarily indigestible to vertebrates.

This is accomplished by symbiotic bacteria housed in an elaborate system of chambers within the stomach (part of the foregut). Ruminants break down the plant material by swallowing the food, which is fermented by the bacteria in the first stomach chamber, the rumen. The cud, or fermented food, is later regurgitated, chewed, and swallowed again, passing into the rest of the digestive tract. Bacteria break down the cellulose from the food into components that can be digested by the ruminant. All ruminants have some ability to use foregut rumination, and it is also found to a more limited degree in the camels.

V. Phylogeny

Artiodactyla is a diverse group today, but it also contains a number of extinct types. This diversity has made it difficult to understand the interrelationships within this group. Artiodactyls have been thought to be monophyletic based on their morphology, but recent molecular studies indicate that whales may be closely related to hippos, which implies that whales actually belong within Artiodactyla.

Most morphologists place the living families into three major groups: the Suiformes (pigs, peccaries, and hippos), the Tylopoda (camels, protoceratids, and oromerycids), and the Ruminantia (bovids, cervids, antilocaprids, giraffids, and tragulids). Tylopoda and Ruminantia are usually considered to be more closely related to one another than either is to the Suiformes, based on their dental and gut morphology (see Fig. 1 for summary). However, molecular evidence linking whales to hippos

also suggests that the Ruminantia may be more closely related to Suiformes and whales than to the Tylopoda. This hypothesis conflicts strongly with morphological data from living and extinct artiodactyls and whales. Further investigation may help clarify the relationships between whales and artiodactyls, but it is unlikely that molecular analyses of the extinct forms will be feasible.

VI. Fossil Record

The oldest artiodactyls are known from the earliest Eocene of Europe, North America, and Pakistan, about 56 million years ago, around the same time as the earliest fossil whales. The oldest fossil artiodactyls, usually referred to the genus *Diacodexis*, were rabbit-sized omnivores, but during the middle Eocene many other new groups evolved (see Fig. 1). The majority of these lineages became extinct, including some which were enormously diverse and successful for long periods of time, such as the oreodonts, an endemic North American group of sheep-sized herbivores. A number of these fossil groups are usually described as pig-like, meaning that they have long trunks and short legs, with low-crowned teeth suitable for an omnivorous diet. These groups include the entelodonts, which were very large animals, with skulls as long as 1 m in length and with large bony bumps on the facial region. The anthracotheres are another of these groups, and they are usually thought to be related to the hippos. Other groups were obviously herbivores, including the oreodonts, and seem to be related either to the

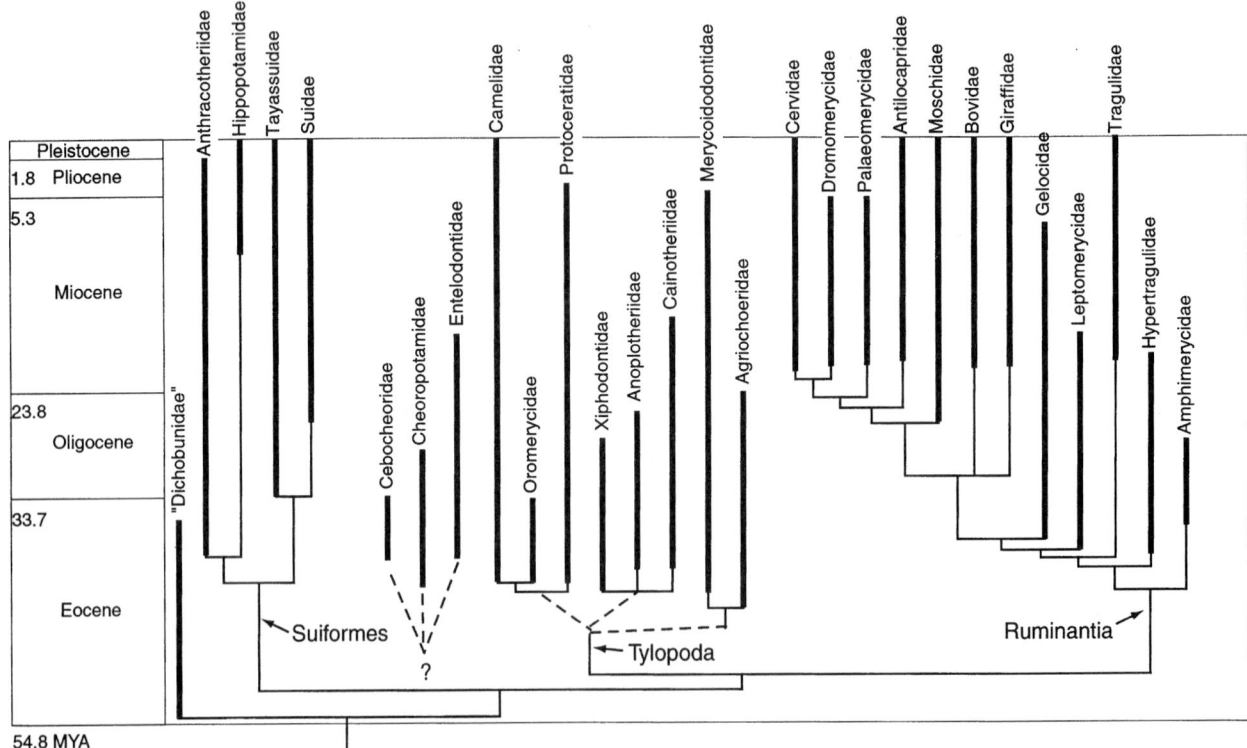

Figure 1 *Phylogeny and fossil occurrences of major lineages of Artiodactyla during the Cenozoic era. Epochs and radiometric ages are marked on the left.*

Tylopoda (oromerycids, protoceratids, and xiphodontids) or to the Ruminantia (hypertragulids, leptomerycids, gelocids, dromomerycids, and paleomerycids). A number of these groups had very complex antlers or horns, especially the extinct protoceratids, some of whom had large slingshot-shaped horns on the snout. Many extinct members of the Dromomerycidae, Antilocapridae, and Giraffidae had much more elaborate cranial ornamentation than any species alive today.

Another successful group was the Camelidae, which became very diverse during the Miocene in North America. The camels and llamas emigrated out of North America and were the only members of the family to survive the extinctions of large mammals in the Pleistocene. The bovids underwent a rapid radiation during the Miocene and Pliocene. The cervids diversified during the Pleistocene, and both groups are diverse today. It is especially important to realize that the modern geographic distribution of the order Artiodactyla is very different from what it was at different times in the past and has been heavily influenced by human activity.

See Also the Following Articles

Cetacean Evolution ▪ Fossil Record ▪ Hippopotamus ▪ Perissodactyla

References

Carroll, R. L. (1988). Ungulates, edentates and whales. *In* "Vertebrate Paleontology and Evolution," pp. 501–568. Freeman, New York.

Clutton-Brock, T. H., Guiness, F. E., and Albon, S. D. (1982). "Red Deer: Behavior and Ecology of Two Sexes." Univ. of Chicago Press, Chicago.

Geist, V. (1971). "Mountain Sheep: A Study in Behavior and Evolution." Univ. of Chicago Press, Chicago.

Janis, C. M. (1989). A climatic explanation for patterns of evolutionary diversity in ungulate mammals. *Palaeontology* **32**(3), 463–481.

Janis, C. M., Scott, K. M., and Jacobs, L. (eds.) (1998). Evolution of Tertiary Mammals of North America. "Terrestrial Carnivores, Ungulates, and Ungulate-like Mammals," Vol. 1, pp. 337–510. Cambridge Univ. Press, Cambridge.

Nowak, R. M. (1991). Artiodactyla. *In* "Walker's Mammals of the World," 5th Ed., Vol. II, pp. 1334–1499. Johns Hopkins University Press, Baltimore, MD.

Slijper, E. J. (1936). Die Cetaceen vergleichend-anatomisch und systematisch. *Capita Zool* **7**, 1–590.

Atlantic Spotted Dolphin
Stenella frontalis

WILLIAM F. PERRIN
*Southwest Fisheries Science Center,
La Jolla, California*

This sturdy spotted dolphin (Fig. 1) is found only in the Atlantic and is commonly seen around the "100-fathom curve" along the southeastern and Gulf U.S. coasts, in the Caribbean, and off West Africa.

I. Characters and Taxonomic Relationships

The Atlantic spotted dolphin is not always spotted. A large heavy-bodied form found along the coast on both sides of the Atlantic (formerly called *Stenella plagiodon* along the U.S. coast) may be so heavily spotted as to appear white from a distance, but a smaller more gracile form occurring in the Gulf Stream and out into the central North Atlantic can be lightly spotted or entirely unspotted as an adult (Perrin *et al.*, 1987; Viallelle, 1997). A constant diagnostic external feature of *S. frontalis* is a spinal blaze sweeping up into the dorsal cape; this distinguishes it from the very similar pantropical spotted dolphin, *S. attenuata*, also found in the tropical Atlantic. In addition, the peduncle does not exhibit the dorsoventral division

Figure 1 *Atlantic spotted dolphin in the Gulf of Mexico. Photo by R. L. Pitman, from Perrin* et al. *(1994a).*

into darker upper and lighter lower halves present in *S. atten-uata*. The calf of the heavily spotted form is born unspotted, with a three-part color pattern of dark dorsal cape, medium-gray lateral field, and white ventral field. Spots first appear at 2–6 years and increase in size and density up to 16 years (Herzing, 1997).

The beak is of medium length (intermediate between those of *Tursiops truncatus* and *S. attenuata*) and sharply demarcated from the melon. The dorsal fin is tall and falcate. Measured adults range from 166 to 229 cm in body length (*n* = 106) and weigh up to 143 kg (*n* = 37) (Perrin *et al.*, 1994a; Nieri *et al.*, 1999). Weight at length is greater than for *S. attenuata* (Perrin *et al.*, 1987).

As in *S. attenuata*, *T. truncatus*, and *T. aduncus*, the skull is characterized by a long rostrum, distal fusion of maxillae and premaxillae in adults, convergent premaxillae, large rounded temporal fossae, and arcuate mandibular rami. Tooth counts are 32–42 in the upper jaw (*n* = 115) and 30–40 in the lower (*n* = 107) vs 35–48 (*n* = 315) and 34–47 (*n* = 315) (Perrin *et al.*, 1994a; Perrin and Hohn, 1994; Nieri *et al.*, 1999). This species and *S. attenuata* overlap in all skull measurements as well as in tooth counts (Perrin *et al.*, 1987). Both species vary greatly geographically. Some specimens of the two species can be identified only with multivariate analysis. However, vertebral counts for the two species do not overlap [67–72 (*n* = 52) in *S. frontalis* vs 74–84 (*n* = 75) in *S. attenuata*].

Taxonomy of the spotted dolphins was long confused, with specimens of this species and the pantropical spotted dolphin (*S. attenuata*) classified or identified under various permutations of the nominal species *S. attenuata*, *S. frontalis*, *S. plagiodon*, *S. froenatus*, *S. pernettyi*, and *S. dubia* (e.g., see Hershkovitz, 1966). A revision (Perrin *et al.*, 1987) recognized one pantropical species (*S. attenuata*) and a second species endemic to the tropical Atlantic (*S. frontalis*), both highly variable geographically in size, tooth size, and color pattern. While the skull of the Atlantic spotted dolphin shows close affinities with that of the pantropical spotted dolphin, the two species did not emerge as sister taxa in a cladistic phylogenetic analysis based on cytochrome *b* mtDNA sequences (LeDuc *et al.*, 1999). *S. frontalis* was imbedded in a strongly supported polytomic clade with *S. coeruleoalba* and *S. clymene* (sister taxa), *Tursiops aduncus*, and *Delphinus* spp. *T. truncatus* was a sister taxon to this clade, with the resulting higher clade imbedded in the five-part polytomic delphinine clade with *S. attenuata*, *S. longirostris*, *Sousa chinensis*, and *Lagenodelphis hosei*. Despite a high degree of cranial similarity, this wide phylogenetic separation suggests that the similarity represents either convergence (homoplasy) or retention of primitive character states (plesiomorphy). The interspecific relationships in color pattern may accord better with the molecular phylogeny; e.g., the pattern of head stripes in *S. frontalis* is closer to those of *T. truncatus* and *T. aduncus* than that of *S. attenuata* (Perrin, 1997). In any case, the existing genus-level taxonomy of the group badly needs revision; *Stenella* is presently polyphyletic and *Tursiops* paraphyletic (LeDuc *et al.*, 1999). A cladistic analysis of morphology (not yet attempted) is in order.

II. Distribution and Ecology

This species is endemic to the tropical and warm-temperate Atlantic; it is not known to occur in the Pacific or Indian Oceans. The range extends from about 50°N to about 25°S (Jefferson *et al.*, 1993). In the western Atlantic, the large heavily spotted form inhabits shallow, gently sloping waters of the continental shelf and the continental-shelf break, usually within or near the 200-m curve but occasionally coming close to shore in pursuit of prey (Perrin *et al.*, 1994a; Davis *et al.*, 1998; Würsig *et al.*, 2000). It is usually replaced in nearshore waters by the coastal form of the bottlenose dolphin, *Tursiops truncatus*. Shallow water (6–12 m) over sand flats is utilized as habitat in the Bahamas (Herzing, 1997). A wide variety of prey items has been recorded, including small-to-large epipelagic and mesopelagic fishes and squids and benthic invertebrates; diet may differ between coastal and Gulf Stream forms. Sharks are the only known predators, but it is probably also preyed on by killer whales and other small-toothed whales.

III. Behavior and Life History

Dives to 40–60 m and lasting up to 6 min have been recorded, but most time is spent at less than 10 m (Davis *et al.*, 1996). Behavior of the Atlantic spotted dolphin has been studied extensively in the Bahamas (e.g., Herzing, 1997; Herzing and Johnson, 1997), where it associates closely with bottlenose dolphins during foraging and traveling. Schools may be segregated by age and sex and fluctuate in size and composition, consisting of up to 100 individuals (Perrin *et al.*, 1994a).

Little is known of the life history of this species. Age at sexual maturation is estimated at 8–15 years in females (Herzing, 1997). First parturition is associated with the mottled phase of spotting development. The average calving interval is about 3 years, with a range of 1–5 years. Nursing has been observed to last up to 5 years. Average first-year natural mortality in a study in the Bahamas was 24%.

IV. Interactions with Humans

This species does not do well in CAPTIVITY; most captive animals have died within a year or less, many refusing to eat (Perrin *et al.*, 1994a). It is killed incidentally in fisheries in Brazil, the Caribbean, the western North Atlantic, and Mauretania, mostly in small numbers (Perrin *et al.*, 1994b; Nieri *et al.*, 1999).

See Also the Following Articles

References

Davis, R. W., Worthy, G. A. J., Würsig, B., and Lynn, S. K. (1996). Diving behavior and at-sea movements of an Atlantic spotted dolphin in the Gulf of Mexico. *Mar. Mamm. Sci.* **12**, 569–581.
Davis, R. W., Fargion, G. S., May, N., Leming, T. D., Baumgartner, M., Evans, W. D., Hansen, L. J., and Mullin, K. (1998). Physical habi-

tat of cetaceans along the continental slope in the north-central and western Gulf of Mexico. *Mar. Mamm. Sci.* **14**, 490–507.

Hershkovitz, P. (1966). Catalog of living whales. *U. S. Nat. Mus. Bull.* **246**, 259.

Herzing, D. L. (1997). The life history of free-ranging Atlantic spotted dolphins (*Stenella frontalis*): Age classes, color phases, and female reproduction. *Mar. Mamm. Sci.* **13**, 576–595.

Herzing, D. L., and Johnson, C. M. (1997). Interspecific interactions between Atlantic spotted dolphins (*Stenella frontalis*) and bottlenose dolphins (*Tursiops truncatus*) in the Bahamas, 1985–1995. *Aquat. Mamm.* **23**, 85–99.

Jefferson, T. A., Leatherwood, S., and Webber, M. A. (1993). "FAO Species Identification Guide: Marine Mammals of the World," p. 320. FAO, Rome.

LeDuc, R. G., Perrin, W. F., and Dizon, A. E. (1999). Phylogenetic relationships among the delphinid cetaceans based on full cytochrome *b* sequences. *Mar. Mamm. Sci.* **15**, 619–648.

Nieri, M., Grau, E., Lamarche, B., and Aguilar, A. (1999). Mass mortality of Atlantic spotted dolphins (*Stenella frontalis*) caused by a fishing interaction in Mauretania. *Mar. Mamm. Sci.* **15**, 847–854.

Perrin, W. F. (1997). Development and homologies of head stripes in the delphinid cetaceans. *Mar. Mamm. Sci.* **13**, 1–43.

Perrin, W. F., Caldwell, D. K., and Caldwell, M. C. (1994a). Atlantic spotted dolphin *Stenella frontalis* (G. Cuvier, 1829). *In* "Handbook of Marine Mammals" (S. H. Ridgway and R. Harrison, eds.), Vol. 5, pp. 173–190. Academic Press, London.

Perrin, W. F., Donovan, G. P., and Barlow, J. (eds.) (1994b). "Gillnets and Cetaceans." *Rep. Int. Whal. Commn Spec. Iss.* **15**.

Perrin, W. F., and Hohn, A. A. (1994). Pantropical spotted dolphin *Stenella attenuata. In* "Handbook of Marine Mammals" (S. H. Ridgway and R. Harrison, eds.), Vol. 5, pp. 71–98.

Perrin, W. F., Mitchell, E. D., Mead, J. G., Caldwell, D. K., Caldwell, M. C., van Bree, P. J. H., and Dawbin, W. H. (1987). Revision of the spotted dolphins, *Stenella* spp. *Mar. Mamm. Sci.* **3**, 99–170.

Vialelle, S. (1997). "Dauphins et Baleines des Açores." Espaço Talassa, Lajes do Pico, Azores, Portugal.

Würsig, B., Jefferson, T. A., and Schmidly, D. J. (2000). "The Marine Mammals of the Gulf of Mexico." Texas A&M Univ., College Station, TX.

Atlantic White-Sided Dolphin
Lagenorhynchus acutus

FRANK CIPRIANO
San Francisco State University, California

I. Distribution

Inhabitants of the cold-temperate North Atlantic, Atlantic white-sided dolphins are usually encountered in waters over the continental shelf and slope, extending into deeper oceanic waters and occasionally into coastal areas (Fig. 1). The southern limit of this species in the western Atlantic is Cape

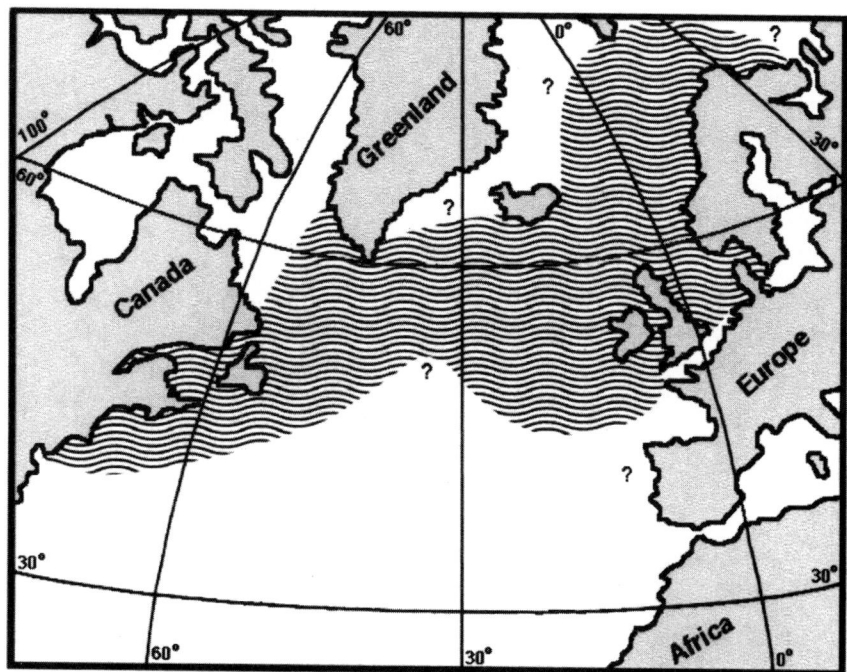

Figure 1 *Known distribution limits of the Atlantic white-sided dolphin. The patterned area indicates areas of regular occurrence and question marks indicate uncertainty about occurrence in particular areas.*

Cod and the submarine canyons south of Georges Bank, and Brittany in the east. Groups of Atlantic white sides are often seen by fishermen and deep-water sailors off the coasts of New England, Nova Scotia, and Newfoundland, the western British Isles, northern Europe, and in the Norwegian Sea. There are no records of this species from the inner Baltic Sea, although some sightings and strandings are known from the straits between Denmark, Norway, and western Sweden. The northern distribution limits are poorly known but extend at least to southern Greenland, southern Iceland, and the south coast of Svalbard Island.

II. Appearance and Life History

Atlantic white-sided dolphins are robust and powerful, impressively patterned, and more colorful than most dolphins. A narrow, bright white patch on the side extends back from below the dorsal fin and continues toward the flukes as a yellow-brown blaze above a thin dark stripe (Fig. 2). The back and dorsal fin are black or very dark gray, as are the flippers and flukes, whereas the belly and lower jaw are white, and the sides of the body a lighter gray. A black eye ring extends in a thin line to the upper jaw, and a very thin stripe extends backward from the eye ring to the external ear. A faint gray stripe may connect the leading edge of the flipper with the rear margin of the lower jaw. The beak is short and grades smoothly into the "melon" (forehead). The upper jaw contains 29–40 and the lower jaw 31–38 small, conical teeth.

Male Atlantic white-sided dolphins are known to reach a maximum body length about 270 cm and a weight of 230 kg, whereas adult females reach a maximum size about 20 cm shorter and 50 kg lighter. This is smaller than that well-known oceanarium inhabitant, the bottlenose dolphin, *Tursiops truncatus* (around 380 cm/270 kg maximum), and a bit longer and a lot heavier than the short-beaked common dolphin, *Delphinus delphis* (around 230 cm/75 kg). Females reach sexual maturity at 200–220 cm, at ages from 6 to 12 years. Males reach sexual maturity at lengths of 215–230 cm, corresponding to ages of 7–11 years. Maximum ages recorded were 22 and 27 years for males and females, respectively. At birth, Atlantic white-sided dolphins are around 120 cm long, after an approximately 11-month gestation period, and weigh about 25 kg. In the western Atlantic, the calving season peaks in midsummer, whereas in the eastern Atlantic the calving season may extend several months longer. The lactation period lasts around 18 months, and some stranded individuals were observed to be both pregnant and lactating, suggesting that some individuals may breed annually.

III. Group Size, Behavior, and Ecology

The number of Atlantic white-sided dolphins observed in a group ranges from a few individuals to several hundred, and mean group size appears to vary with location. In Newfoundland inshore waters, 50–60 dolphins in a group are typical; in inshore waters of the British Isles and near Iceland, groups usually contain less than 10 individuals; off the New England coast, group size ranges from a few to around 500, but the usual group size is around 40. Some segregation by sex and age has been suggested from mass stranding records—larger juveniles are absent from some mass-stranded groups that contain many calves, adult males, and pregnant females.

Analysis of the stomach contents of mass-stranded, incidentally entangled, and drive-caught dolphins is used to assess their diet, as diagnostic "hard parts" (crustacean shells, fish ear bones, and squid beaks) accumulate in stomach chambers. A general indication of the importance of particular prey items can be inferred from the percentage contribution of each type, although the number of meals represented by such traces is usually unknown and there may be bias in the retention of different types. Major prey species of Atlantic white-sided dolphins include herring, small mackeral, gadid fishes (codfish and their relatives), smelts and hake, sand lances, and several types of squid. Different prey species may predominate at different times of year, representing seasonal movements of prey, or in different areas, indicating prey and habitat variability in the environment. For example, different species of squid are eaten by these dolphins on opposite sides of the Atlantic, whereas in spring and autumn, sand lance and dolphin distributions in the Gulf of Maine appear to mirror each other. Atlantic white-sided dolphins are probably not deep divers—the maximum dive time recorded from a tagged dolphin was 4 min and most dives were less than a minute in duration.

IV. Abundance and Human Impacts

Censusing oceanic dolphins is a difficult task, requiring extensive aerial surveys or long observation tracks from survey ships (or both) and then extrapolation of the densities observed to immense ocean areas. Given the wide distribution of At-

Figure 2   *Color pattern of the Atlantic white-sided dolphin. Pieter A. Folkens/Higher Porpoise DG.*

lantic white-sided dolphins across the northern reaches of the Atlantic, rather wide confidence limits on abundance estimates are to be expected. Estimates for the western Atlantic total about 40,000 animals, and for the entire Atlantic a few hundred thousand. This species is not currently hunted on a large scale anywhere throughout its range, although historically many were killed in drive fisheries in Norway and Newfoundland, and smaller numbers taken off Greenland and the Faroe Islands. Incidental mortality has been documented in many areas and these dolphins may be particularly susceptible to ENTANGLE-MENT in midwater trawl nets, including some recent large catches in pelagic trawl nets in the Atlantic Frontier off Ireland.

V. Evolutionary Relationships

Molecular analysis has been used to examine the evolutionary relationships of *Lagenorhynchus acutus* and other dolphins, including the five other currently recognized members of the genus *Lagenorhynchus*. Although formal taxonomic revision awaits a more comprehensive review of morphological and molecular characters, molecular evidence suggests that *L. acutus* is not closely related to other putative members of *Lagenorhynchus* and should be recognized in a separate genus.

See Also the Following Articles

Abundance Estimation ▪ Molecular Ecology ▪ North Atlantic Marine Mammals ▪ Pacific White-Sided Dolphin and Dusky Dolphin

References

Berrow, S. D., and Rogan, E. (1997). Review of cetaceans stranded on the Irish coast, 1901-95. *Mamm. Rev.* **27,** 51–76.

Geraci, J. R., Testaverde, S. A., St. Aubin, D. J., and Loop, T. H. (1978). "A Mass Stranding of the Atlantic White-Sided Dolphin, *Lagenorhynchus acutus:* A Study into Pathobiology and Life History." Contract report to U.S. Marine Mammal commission. Available from National Technical Information Service, Springfield, Virginia, PB-289 361.

Leatherwood, S., and Reeves, R. R. (1983). "Sierra Club Handbook of Whales and Dolphins." Sierra Club Books, San Francisco.

LeDuc, R. G., Perrin, W. F., and Dizon, A. E. (1999). Phylogenetic relationships among the delphinid cetaceans based on full cytochrome *b* sequences. *Mar. Mamm. Sci.* **15,** 619–648.

Nelson, D. L., and Lien, J. (1994). Behavior patterns of two captive Atlantic white-sided dolphins, *Lagenorhynchus acutus. Aqu. Mamm.* **20,** 1–10.

Palka, D., Read, A., and Potter, C. (1997). Summary of knowledge of white-sided dolphins (*Lagenorhynchus acutus*) from U.S. and Canadian Atlantic waters. *Rep. Int. Whal. Commn.* **47,** 729–734.

Perrin, W. F., and Reilly, S. B. (1984). Reproductive parameters of dolphins and small whales of the family Delphinidae. *Rep. Int. Whal. Commn.,* Spec. Issue No. 6, pp. 97–133.

Reeves, R. R., Smeenk, C., Brownell, R. L., Jr., and Kinze, C. C. (1999). Atlantic white-sided dolphin *Lagenorhynchus acutus* (Gray, 1828). *In* "Handbook of Marine Mammals" (S. H. Ridgway and R. Harrison, eds.), Vol. 6, pp. 31–56. Academic Press, San Diego.

Sergeant, D. E., St. Aubin, D. J., and Geraci, J. R. (1980). Life history and northwest Atlantic status of the Atlantic white-sided dolphin, *Lagenorhynchus acutus. Cetology* **37,** 1–12.

Australian Sea Lion
Neophoca cinerea

JOHN K. LING
Clare, South Australia

The indigenous Australian sea lion (Fig. 1) is one of the world's rarest and most unusual seals: rare in terms of very small numbers and unusual in its having a sesquiennial reproductive cycle. It is also a temperate species, inhabiting waters around much of the southern part of the island continent between latitudes 28 and 38°S. Its current breeding range extends from Houtman Abrolhos (29°S,114°E) in Western Australia to The Pages Islands (36°S,138°E), just east of Kangaroo Island, South Australia, with stragglers reaching central New South Wales on the east coast (Fig. 2).

I. Color and Size

At birth the pups are a dark chocolate brown to charcoal gray in color, which changes to the smoky gray (hence the specific name *cinerea*) and cream adult color after the postnatal molt. Females retain this coloration throughout life, but males gradually develop a brownish-black coat with increasing age. Males of breeding age have a cream patch on the back of the head and nape of the neck. This species has flattened guard hairs but no underfur—the pelage apparently being adapted to a temperate environment. It also has a relatively thin layer of blubber beneath the skin, about 2 cm thick. Pups measure 62 to 68 cm in length (nose–tail) and weigh 6.4 to 7.9 kg at birth, males tending to be heavier than females. Adult females range in length from 132 to 181 cm and weigh between 61 and 105 kg (for a pregnant specimen); males measure up to 200 cm in length and attain weights well in excess of 200 kg.

II. Distribution and Abundance

There are only 9300 to 11,700 Australian sea lions occupying their wide geographic range in more than 50 scattered colonies, of which only 5 produce more than 100 pups in a breeding season. Three-quarters of the population resides in South Australia and a quarter occurs in Western Australia, where more than half the breeding colonies are located, all of which are small. The largest breeding colonies are on Purdie Island (32°S,133°E), Dangerous Reef (35°S,135°E), Seal Bay (36°S,137°E) on Kangaroo Island and the two islands of The Pages. Australian sea lions once ranged as far as the eastern end of Bass Strait, but today only stragglers occur there and beyond. They were hunted during the sealing era from the late 18th to the mid-19th century for their skins and oil, when only a few thousand skins are reported to have been harvested. It is not possible to estimate the number killed for oil, because "seal oil" included fur seal oil as well in the cargoes. However, there may not have been many sea lions to be taken anyway compared with the large fur seal populations, which are increasing today after having been almost exterminated by early sealers. However, because the first complete census over the sea lion's en-

Figure 1   Neophoca cinerea, *adult male (note white "cap" on head), three adult females, and juvenile (circa 18 months) resting on center female at Seal Bay, Kangaroo Island, South Australia. Photo by John K. Ling.*

tire breeding range has been carried out only recently, it cannot be determined at this stage whether numbers are increasing or decreasing. Future surveys of all breeding colonies will need to be undertaken, as counts of live and dead pups provide the most accurate estimates of the size of the total population.

III. Life History

The life history of *Neophoca* is unique in a number of aspects: the approximately 17.5-month reproductive cycle, a protracted (i.e., 5-month) pupping season, and prolonged gestation and lactation until the next pup is born. This life history is likely to be an adaptation to the sea lion's environment, which is characterized as a low-energy, stable milieu associated with the eastward flowing Leeuwin Current.

Pups are born a few days after the females have moved to their BREEDING SITES, to which they are known to return for successive birthings. Viable twins have never been observed, but two aborted fetuses, believed to be twins and estimated to be about 3 months postimplantation, were found on Kangaroo Island in 1985.

MATING takes place 7 to 10 days postpartum; there is a 3- to 4-month-free blastocyst (embryonic diapause) stage, followed by a gestation period up to 14 months in duration. The pup is suckled for the next 15–18 months and during this time it learns to forage for food that it will consume in later life. The milk is low in energy (fat) compared with other pinnipeds and its quality may vary according to the foraging success of the mother and stage of lactation. After they are weaned, sea lions feed on cephalapods, crustaceans, and fish. It is not known how far offshore they forage, but diving appears to begin as soon as females leave the rookery and continues to mean depths of 92 m for up to 8 min per dive over the continental shelf. This is a longer duration than dives of most other otariids (eared seals). Prey is generally seized in the mouth and shaken violently at the surface to remove cuttle bones, skin, or skeletons before swallowing. Depending on size, experimental markers take from 5 to 48 hr to pass through the alimentary tract. Australian sea lions are infected by the usual array of external and internal parasites: lice and mites, and acanthocephala, nematodes, cestodes, and trematodes. Dissections often reveal very heavy infestations.

While the pelage is unlikely to be involved in thermoregulation, the flattened hairs overlap each other and provide a smooth but flexible outer surface that reduces turbulence when swimming. Periodic renewal of the hair coat ensures that it functions efficiently in whatever role it has. The timing of pelage renewal or molt is variable. Immature sea lions molt during the breeding season, females begin their molt about 4 months after parturition, and adult males do not start their molt until about 9 months after breeding occurs.

The Australian sea lion's unusual life history enables it to survive as a very small population scattered over a wide, nutrient-poor longitudinal range. The longer than normal (pinniped) gestation and lactation periods allow the female to nurture a developing fetus and growing pup while having to forage for scarce food resources. Normal growth rates can also be achieved despite the low-energy content of the milk. At the same time, the long maternal association confers many learning and protective advantages on the young sea lion. The protracted, asynchronous pupping season spread out over a wide geographic area again means that food resources can be better shared and there is not a sudden influx of newly independent sea lions, such as occurs with more highly synchronized species that occupy nutrient-rich, higher latitudes.

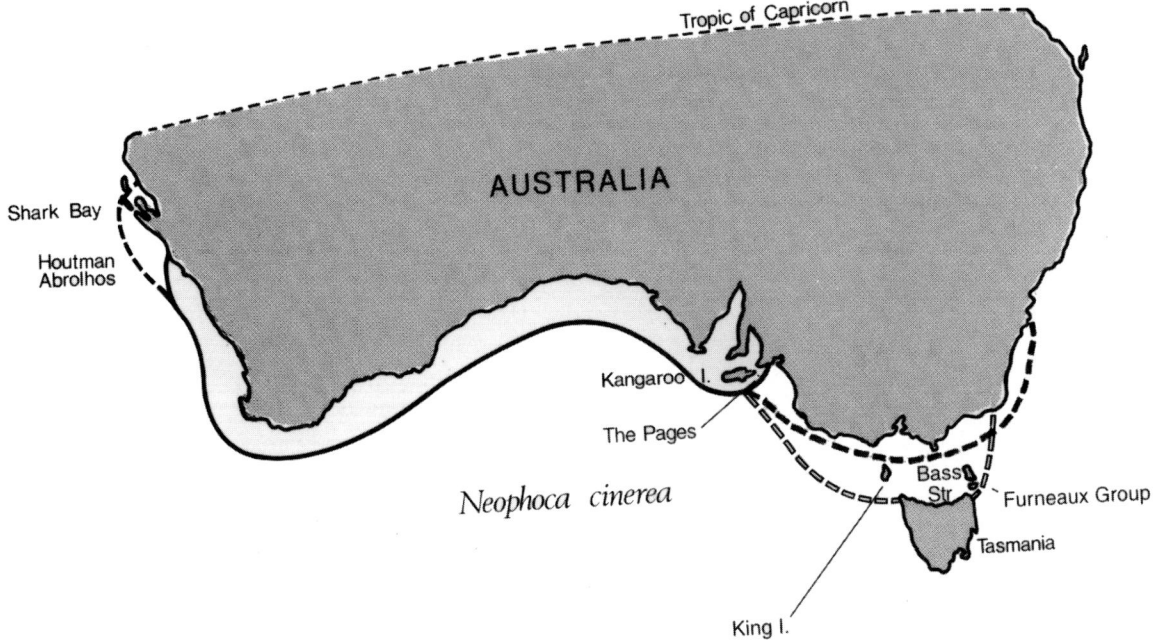

Figure 2 *Present and past distribution of* Neophoca cinerea. *Unbroken solid line depicts current known breeding range, broken solid line depicts extent of seasonal stragglers, and broken open line depicts extent of former breeding range. Reproduced with permission from the Royal Zoological Society of South Australia.*

IV. Behavior

The protracted pupping season, during which mating is effected, ensures that there is a high turnover of territories and a breakdown of any harem system. In contrast to many otariids in which dominant males control small to large numbers of females, *Neophoca* practices what is known as sequential polygyny, which still allows males access to several females in a season, but one at a time. Nevertheless, aggressive encounters do take place between rival breeding males and are a significant cause of mortality among young pups that are unfortunate enough to be attacked or trampled by rampaging bulls. Their lumbering gait resembles something of an ungainly gallop a little above a fast human walking pace and punctuated by frequent rests.

Females are most solicitous of their young, and several tourist visitors to Seal Bay and other breeding colonies have received nasty bites when they approached too close to a cow with her pup. When returning from a foraging trip, a female will call from the sea with a soft "moo" and wait for her pup's answering call, which resembles a lamb's bleat. Site fidelity is very strong, so little searching for each other is necessary. Once the two are reunited, recognition is confirmed by smelling.

Sea lions are powerful and skillful swimmers, using their large front flippers to propel them rapidly through the water. They are also excellent surfers and can often be seen riding the waves right into the shallows or "porpoising" along wave crests and troughs further out to sea.

Large males tend to lie apart from other sea lions, but females and immature animals often lie close together wriggling, squirming, and scratching constantly. On hot days when the sun temperature may exceed 45°C they will occasionally go into the sea and return a short while later to allow evaporative cooling to take effect. Sea lions may also venture some distance inland to lie under bushes or up steep slopes to find a shelter: they are quite agile on land.

V. Conservation and Management

The Australian sea lion colony at Seal Bay on Kangaroo Island, South Australia, is internationally famous because of its proximity to a large city (Adelaide) and the public being able to view the animals at close quarters. In view of the near-threatened (IUCN) status of this species and its importance to the tourist industry on Kangaroo Island (no other colonies are so easily accessible), the South Australian government has embarked on an intensive management strategy.

The whole Seal Bay area has been designated a conservation park. Public access is limited to the main beach and only in the company of authorized personnel, but there are also viewing platforms overlooking the beach and other restricted areas. The main pupping sites in coves to the west of Seal Bay have been declared prohibited areas. Regular classified censuses are conducted to monitor the status of the Seal Bay colony and enhance its chances of survival and value to tourism.

However, many other Australian sea lion colonies appear to be suffering very high pup mortality (30–40%) and decreasing pup production. Only with a widespread and cooperative research and management effort will the species be perhaps more secure.

See Also the Following Articles

Eared Seals ■ Reproductive Behavior ■ Rookeries

References

Dennis, T. E., and Shaughnessy, P. D. (1996). Status of the Australian sea lion, *Neophoca cinerea*, in the Great Australian Bight. *Wildl. Res.* **23**, 741–754.

Gales, N. J., Cheal, A. J., Pobar, G. J., and Williamson, P. (1992). Breeding biology and movements of Australian sea lions, *Neophoca cinerea*, off the west coast of Western Australia. *Wildl. Res.* **23**, 405–416.

Gales, N. J., and Costa, D. P. (1997). The Australian sea lion: A review of an unusual life history. *In* "Marine Mammal Research in the Southern Hemisphere" (M. Hindell and C. Kemper, eds.), Vol. I, pp. 78–87. Surrey Beatty & Sons, Chipping Norton.

Gales, N. J., Shaughnessy, P. D., and Dennis, T. E. (1994). Distribution, abundance and breeding cycle of the Australian sea lion *Neophoca cinerea* (Mammalia: Pinnipedia). *J. Zool. Lond.* **234**, 353–370.

Gales, N. J., Williamson, P., Higgins, L. V., Blackberry, M. A., and James, I. (1997). Evidence for a prolonged post-implantation period in the Australian sea lion (*Neophoca cinerea*). *J. Reprod. Fertil.* **111**, 159–163.

Higgins, L. V. (1990). "Reproductive Behavior and Maternal Investment of Australian Sea Lions." Ph.D. Thesis, University of California, Santa Cruz.

Higgins, L. V. (1993). The nonannual, nonseasonal breeding cycle of the Australian sea lion, *Neophoca cinerea. J. Mammal.* **74**, 270–274.

Ling, J. K. (1992). *Neophoca cinerea*. Mammalian Species No. 392. *Am. Soc. Mammalog.* **392**, 1–7.

Ling, J. K., and Walker, G. E. (1978). An 18-month breeding cycle in the Australian sea lion? *Search* **9**, 464–465.

Robinson, A. C., and Dennis, T. E. (1988). The status and management of seal populations in South Australia. *In* "Marine Mammals of Australasia Field Biology and Captive Management" (M. L. Augee, ed.), pp. 87–110. Royal Zoological Society of New South Wales, Sydney.

Shaughnessy, P. D. (1999). "The Action Plan for Australian Seals." Environment Australia, Canberra.

Walker, G. E., and Ling, J. K. (1981). Australian sea lion *Neophoca cinerea* (Peron 1816). *In* "Handbook of Marine Mammals" (S. H. Ridgway and R. J. Harrison, eds.), Vol. I, pp. 99–118. Academic Press, New York.

Baculum

EDWARD H. MILLER
Memorial University, St. Johns,
Newfoundland, Canada

The baculum (os penis) is a bone in the penis of Insectivora, Chiroptera, Primates, Rodentia, and Carnivora; it is absent from Cetacea and Sirenia. The corresponding element in females is the os clitoridis. In marine mammals the baculum and os clitoridis have been studied mainly in pinnipeds.

The baculum develops in the proximal part of the penis in association with the fibrous septum between the paired corpora cavernosa penis or in their fibrous noncavernous portion; in the dog, centers of ossification on left and right sides fuse early in development. The developing baculum grows distally above the urethra and thickens. The bacular base becomes firmly attached to the corpora cavernosa and the fibrous tunica albuginea which surrounds them. In rodents, the bacular shaft is true bone and includes hemopoietic and fatty tissue in the enlarged basal portion.

During development, bacula may become progressively straighter (southern elephant seal, *Mirounga leonina;* northern fur seal, *Callorhinus ursinus*) or curved (harbor seal, *Phoca vitulina;* hooded seal, *Cystophora cristata*). Pinniped bacula grow throughout life in mass and thickness (particularly at the basal end) but not in length. Growth is rapid around puberty and may exhibit a secondary spurt at social maturity (e.g., Cape fur seal, *Arctocephalus p. pusillus*). In Otariidae, the bacular apex, shaft, and base differ from one another in their growth patterns.

The urethral groove is shallow to absent in bacula of marine mammals (Fig. 2; lower drawings in Figs. 3A–3D). Typically the baculum is slightly arched dorsally (Figs. 1, 2, and 3). The apex is relatively elaborate in Mustelidae and Otariidae (Fig. 4) and is simple in Ursidae and Phocidae. In adults of some Antarctic seals a prominent crest develops on the anterior dorsal surface, however (upper drawing in Fig. 3D; Fig. 3E). In some species

the bacular apex has a prominent cartilaginous cap (e.g., hooded seal). Bacula are variable in size, shape, cross section, and specific structural features within species (even among animals of the same age). For example, a dorsal keel may be present or absent in adult southern elephant seals; processes on the shaft near the apex may be present or absent in adult California sea lions (*Zalophus californianus*); and bacula may be bilaterally asymmetrical or slightly twisted (Fig. 3D; Fig. 3E1).

Bacula of Carnivora are fairly large. In pinnipeds and indeed among all mammals, walruses (*Odobenus rosmarus*) have the largest baculum both absolutely (to 62.4 cm in length and 1040 g in mass) and relatively (18% of standard body length). Bacular length in a large polar bear (*Ursus maritimus*) was 17.3 cm. Interspecific differences in bacular size in mammals have been linked to aquatic copulation, copulatory duration or pattern, mating strategy, and risk of fracture. Fractures possibly result from accidents (e.g., falls in walruses), sudden movements during intromission (e.g., in aquatically copulating species such as Caspian seal, *Pusa caspica*), or aggressive social interactions (e.g., fights between adult male sea otters, *Enhydra lutris*). Healed fractured bacula have been observed in several species.

Bacula likely serve several functions, including (1) a mechanical aid to intromission (especially in the absence of full erection) or maintenance of intromission in aquatic copulations and (2) to initiate or engage neural or endocrinological responses in females. Bacular form and diversity reflect multiple functions and hence have multiple adaptive explanations.

Bacular characteristics have been informative in phylogenetic studies of Otariidae and MUSTELIDAE, suggesting needed taxonomic revisions and revealing interesting developmental and evolutionary trends.

Bacular size can aid in estimating age, and bacula grow quickly at puberty so are a good indicator of when it occurs. Bacular growth may be affected by pollutants, so bacular size and form may also be informative in studies on pollution biology.

See Also the Following Articles

Age Estimation ■ Carnivora ■ Male Reproductive Systems

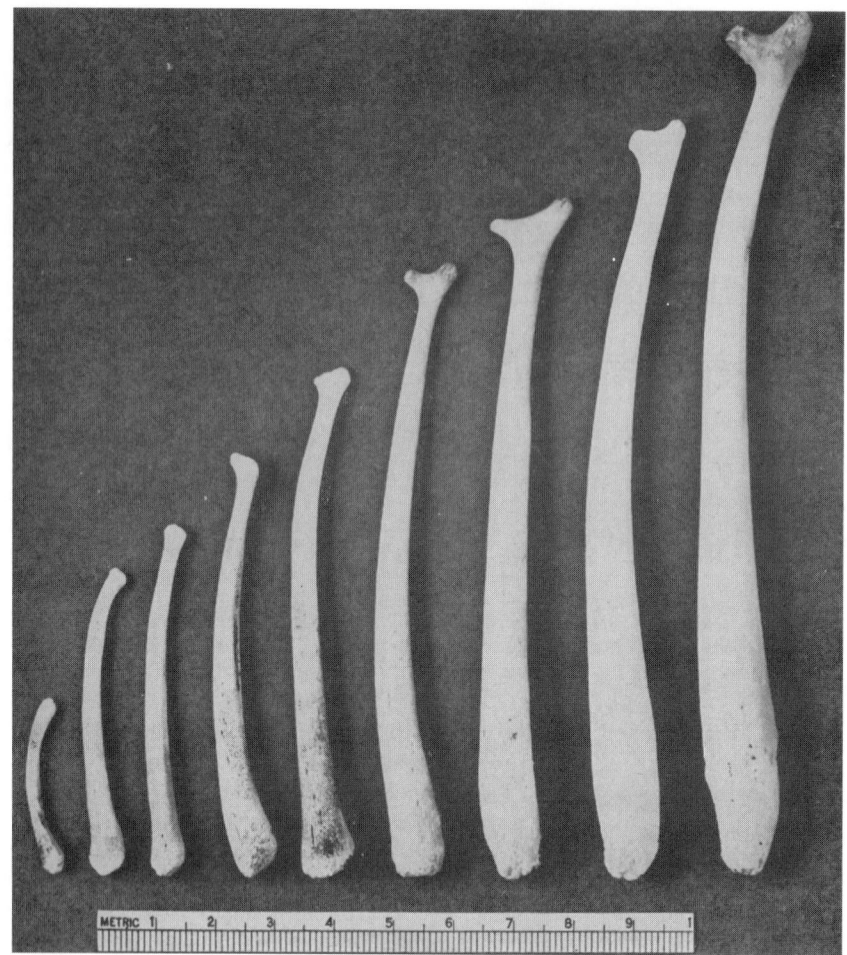

Figure 1   *Developmental changes in bacular size and shape, illustrated by representative bacula from northern fur seals ranging in age from newborn (left) to 8 years (right). From Scheffer (1950).*

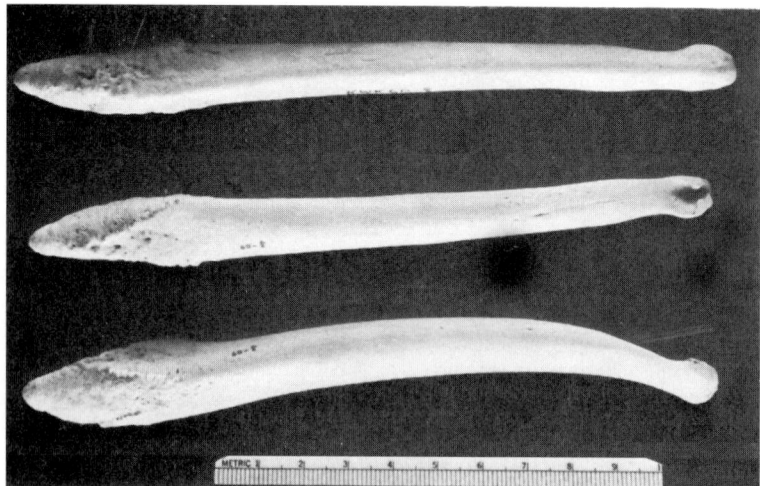

Figure 2   *The baculum of an adult sea otter showing general form and elaborate apex (center); apex is to the right. Top, dorsal view; center, ventral view; bottom, right lateral view. From Kenyon (North Am. Fauna 68, 1–352; 1969).*

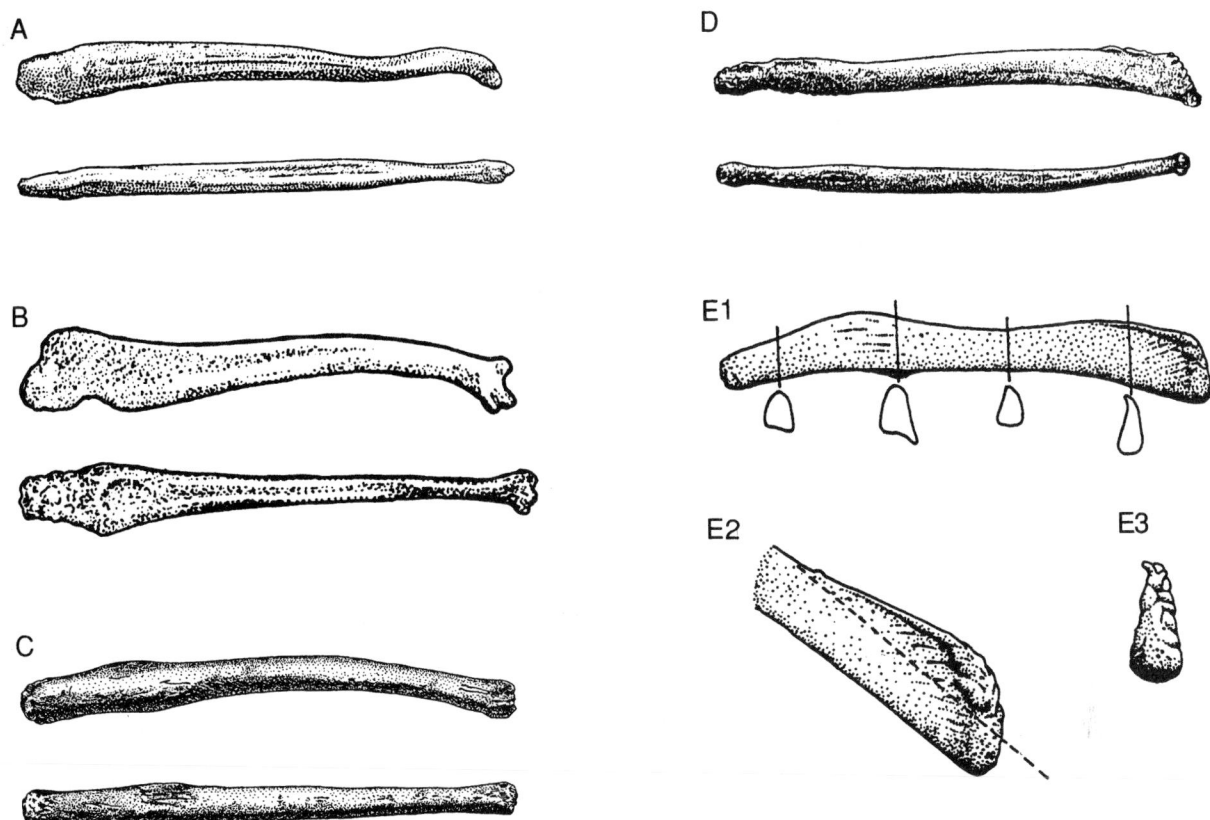

Figure 3 *Bacula of adult (A) polar bear, (B) sub-Antarctic fur seal* (Arctocephalus gazella), *(C) Mediterranean monk seal* (Monachus monachus), *(D) crabeater seal* (Lobodon carcinophaga), *and (E) Weddell seal* (Leptonychotes weddellii). *Bacula in A–D are shown in right lateral (upper) and ventral (lower) views and are not shown to the same scale. E1: baculum in right lateral view; note cross-sectional shapes at the indicated points. E2: oblique view (right side) of the bacular apex (same baculum); dashed line emphasizes how much growth occurs in the crest (above the line) following sexual maturity. E3: apical view (dorsal surface above; same baculum). A from Didier* (Mammalia 15, 11–22; 1951); *B from Didier* (Mammalia 16, 228–239; 1952); *C from van Bree* (Mammalia 58, 498–499; 1994); *D from Didier* (Mammalia 17, 21–26; 1953); *and E from Morejohn* (J. Mammal. 82, 877–881; 2001).

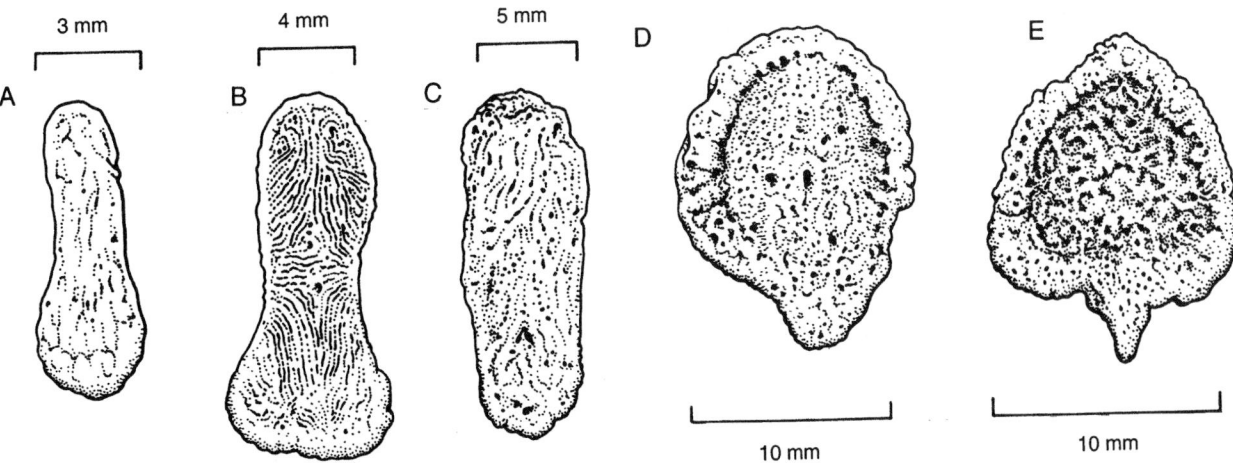

Figure 4 *Structural diversity of the bacular apex in Otariidae (dorsal surface up), illustrated by adult specimens of (A) unknown species of southern fur seal* (Arctocephalus), *(B) northern fur seal, (C) California sea lion. (D) Australian sea lion* (Neophoca cinerea), *and (E) Hooker's sea lion* (Phocarctos hookeri). *From Morejohn (1975).*

References

Dixson, A. F. (1995). Baculum length and copulatory behaviour in carnivores and pinnipeds (Grand Order Ferae). *J. Zool Lond.* **235**, 67–76.

Eberhard, W. G. (1985). "Sexual Selection and Animal Genitalia." Harvard Univ. Press, Cambridge, MA.

Miller, E. H., Pitcher, K. W., and Loughlin, T. R. (2000). Bacular size, growth, and allometry in the largest extant otariid, the Steller sea lion (*Eumetopias jubatus*). *J. Mammal.* **81**, 134–144.

Miller, E. H., Stewart, A. R. J., and Stenson, G. B. (1998). Bacular and testicular growth, allometry, and variation in the harp seal (*Pagophilus groenlandicus*). *J. Mammal.* **79**, 502–513.

Mohr, E. (1963). Os penis und os clitoridis der Pinnipedia. *Zeits. Säugetierk.* **28**, 19–37.

Morejohn, G. V. (1975). A phylogeny of otariid seals based on morphology of the baculum. *Rapports et Proces verbaux des Reunions, Conseil International pour l'Exploration de la Mer* **169**, 49–56.

Oosthuizen, W. H., and Miller, E. H. (2000). Bacular and testicular growth and allometry in the Cape fur seal *Arctocephalus p. pusillus* (Otariidae). *Mar. Mamm. Sci.* **16**, 124–140.

Patterson, B. D., and Thaeler, C. S., Jr. (1982). The mammalian baculum: Hypotheses on the nature of bacular variability. *J. Mammal.* **63**, 1–15.

Scheffer, B. B. (1950). Growth of the testes and baculum in the fur seal, *Callorhinus ursinus.* *J. Mamma.* **31**, 384–394.

Baiji
Lipotes vexillifer

ZHOU KAIYA
Nanjing Normal University, China

The baiji, or Chinese river dolphin (*Lipotes vexillifer*), which inhabits the middle and lower reaches of the Yangtze River in China, is the rarest and most endangered cetacean in the world. Together with the vaquita in the eastern central Pacific, it is one of the two cetaceans classified as critically endangered in the 2000 IUCN red list of threatened species. In China, it is known by the name baiji, meaning white dolphin. The name dates from classical times and is to be found in the ancient dictionary, "Erh Ya," published as long ago as 200 B.C. At that time, the Yangtze flowed clean and largely unencumbered from its source to the rich, sprawling Yangtze Delta, in the western Yellow Sea and East China Sea. The baiji has lived generation after generation and multiplied in number in the desirable environment of the Yangtze River for thousands of years. However, the population has declined in recent decades because of habitat degradation, accidental capture, and collisions with vessels. Recent information estimates a total population of only a few dozen animals.

I. Diagnostic Characters

The baiji is a graceful animal with a very long, narrow and slightly upturned beak. It can be easily identified by the rounded melon, longitudinally oval blowhole, very small eyes, low triangular dorsal fin, and broad rounded flippers (Fig. 1). The color is generally bluish gray or gray above and white or ashy-white below. Females are larger than males. The maximum recorded length for females is 253 cm and for males 229 cm (Zhou, 1989). Significant differences between the sexes in external proportions have been demonstrated in nine characters, and skull size is also sexually dimorphic (Gao and Zhou, 1992).

The Yangtze River is turbid. Visibility from the surface downward is about 25–35 cm in April and 12 cm in August. A corresponding regression has taken place in the eye of the baiji. The eyes are much smaller and placed much higher than those of other dolphins. However, they are functional, and objects on the surface or near the surface directly in front of the eye can be distinguished (Zhou, 1989).

II. Taxonomy

The species was classified previously as either in the family Platanistidae or Iniidae (Brownell and Herald, 1972). Zhou *et al.* (1978) established the new family Lipotidae for it based

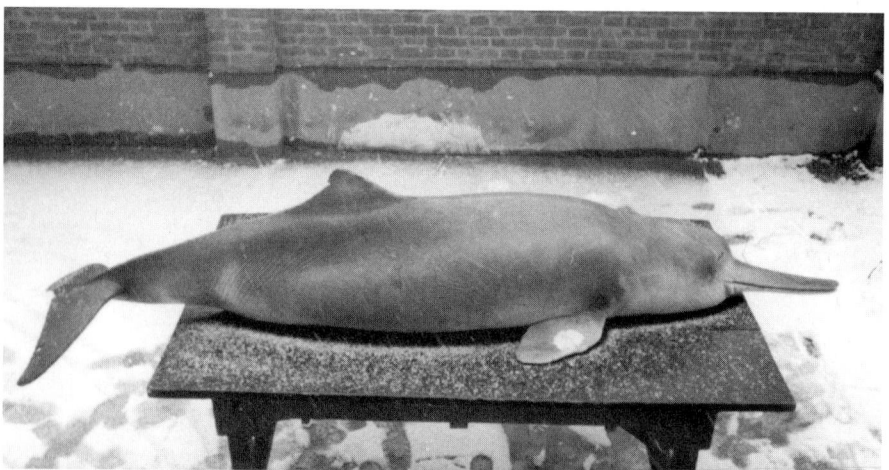

Figure 1   *Carcass of a 2.45 m adult female baiji with a notch on dorsal fin; found drifting down river near Jiangyin on January 15, 1996. From Zhou et al. (1998).*

on osteological studies and anatomy of the stomach. Barnes (1985) recognized a fossil, *Parapontoporia*, as morphologically intermediate between *Pontoporia* and *Lipotes* and placed *Lipotes* and the fossil taxon each in a subfamily of Pontoporiidae. Conversely, de Muizon (1988) placed the fossil taxon with *Lipotes* in Lipotidae, and Rice (1998) following de Muizon ranked Lipotidae as a family. Studies of mitochondrial cytochrome *b* gene sequences indicate that the sequence difference between *Lipotes* and *Pontoporia* is similar to or greater than those among other odontocete families. Molecular evidence also supports ranking Lipotidae as a family rather than as a subfamily of Pontoporiidae (Yang and Zhou, 1999). The only fossil placed close to *Lipotes* is *Prolipotes*, based on a fragment of mandible with teeth of Tertiary age from China (Zhou *et al.*, 1984). This indicates that the baiji is a relict species.

III. Distribution

Baiji are found mainly in the mainstream of the middle and lower reaches of the Yangtze River, although individuals might occasionally enter some tributary lakes during intense flooding. Dongting Lake was originally incorrectly reported to be the only habitat. The presence of this dolphin in the Yangtze River

is noted in documents going back about 2000 years. In the 1940s, the uppermost records in the Yangtze River were at Huanglingmiao and Liantuo in the Three Gorges area, approximately 50 km upstream of the Gezhouba Dam near Yichang (Zhou *et al.*, 1977). Baiji could be found up to Yichang in the 1970s, which is about 1700 km up from the mouth of the river. However, the range was no further upstream than Zhicheng in the 1980s, and than Shashi in the 1990s; the latter is approximately 150 km downstream of the dam site. In the lower part of the river, specimens of the baiji were obtained in the Yangtze estuary, off the eastern end of Chongming Island, Shanghai, in the 1950s and 1960s. The range has been no further downstream than Liuhe since the 1970s (Fig. 2). Results of recent surveys of almost all the species' previous range, Shanghai to Yichang, suggest that the population is very small and is still declining. Baiji were also seen in the Fuchun River during the great flood of 1955 but disappeared from that river after the construction of a hydropower station in 1957 (Zhou *et al.*, 1977).

IV. Ecology

Baiji are generally found in eddy countercurrents below meanders and channel convergences. They live in small groups.

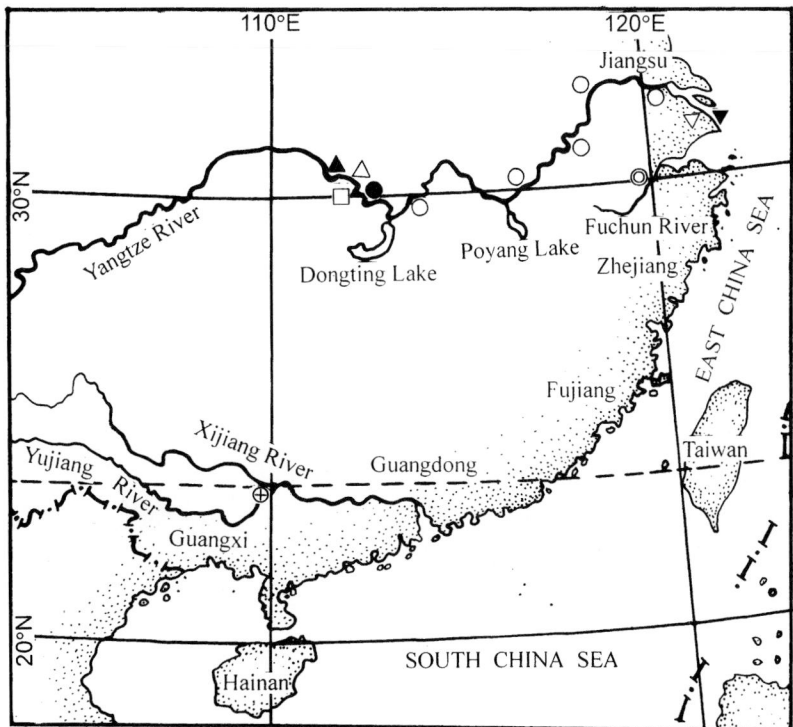

Figure 2 *Map of China showing distribution of baiji,* Lipotes vexillifer, *and the fossil* Prolipotes yujiangensis. *Baiji,* L. vexillifer: ▲, *Huanglingmiao and Liantuo, the uppermost records in the 1940s;* △, *Yichang, the uppermost records in the 1970s;* □, *Zicheng, the uppermost records in the 1980s;* ●, *Shashi, the upstream limit of distribution in the 1990s;* ○, *distribution in the 1990s;* ▼, *Yangtze estuary, the lowest records in the 1950s and 1960s;* ▽, *Liuhe, the downstream limit of distribution in the 1990s; and* ◎, *Tonglu and Fuyang, records in the 1950s in the Fuchun River.* P. yujiangensis: ⊕, *Guiping, locality where the fossil was found.*

Figure 3 *Free-ranging baiji in the Yangtze River near Tongling. From Zhou and Zhang (1991).*

growth layer groups (GLGs), which is the age at sexual maturation for males. After sexual maturation, males grow more slowly than females. Females attain sexual maturation at about 6 years. They continue to grow until about age 8. The smallest mature male and female described were 180 and 200 cm long, respectively. The oldest animal that was age-determined was a 242 cm female of 24 years of age, and a 21-year-old male was estimated to be about 214 cm in body length (Gao and Zhou, 1992). The baiji probably breeds and gives birth in the first half of the year. The peak calving season appears to be February to April.

V. Threats

The Yangtze is suffering massive habitat degradation. The banks of the Yangtze have been modified extensively to prevent destructive flooding of agricultural areas, thus reducing the floodplain area. Construction of dams and other barriers along the river and its tributaries has led to changes in fish abundance and distribution. Wastewater volume discharged into the Yangtze is about 15.6 billion cubic meters per year. Approximately 80% of the wastewaters are discharged directly into the environment without treatment.

The threats against the baiji include river traffic, fishing gear (Fig. 4), reduction of fishery resources, and water pollution (Chen and Hua, 1989; Zhou et al., 1998). Unfortunately, the major threats are continuing and appear to be rapidly becoming more serious. Although it is one of the nationally protected animals in China, the baiji population is reported to have declined from about 400 animals in 1979–1980, to 200 or so in 1986, to about 100 in 1989–1991, and to a few dozen in 1998. The species is facing a very high risk of EXTINCTION.

In the 1980s, the most common group size was 3 to 4 animals; the largest group observed was about 16 individuals in the middle reaches of the Yangtze River. These larger groups were probably temporary aggregations of several groups. The baiji usually surfaces without causing white water and breathes in a smooth manner (Fig. 3). It has a sequence of several short breathing intervals (10–30 sec) alternating with a longer one, the longest one up to 200 sec (Zhou et al., 1994). Photographic identifications and sighting records show that baiji groups make both local and long-range movements. The largest recorded range of a recognizable baiji was 200+ km from the initial sighting location (Zhou et al., 1998).

The baiji appears to take any available species of freshwater fish, the only selection criterion being size. The fish should not be so large that it cannot go down the throat. Sometimes, dead fish are seen floating on the Yangtze with patches of scales torn off. They are believed to have been prey of baiji. At times, a baiji may try a number of times to swallow a large fish, but in vain and finally letting go (Zhou and Zhang, 1991).

Body length at birth is estimated to be 91.5 cm for both males and females. The sexes have about the same growth rate until they are about 4 years old based on estimates of dentinal

See Also the Following Articles

Endangered Species and Populations ▪ River Dolphins, Relationships ▪ Vaquita

Figure 4 *Baiji taken from the Yangtze River near Nanjing in 1982, with 103 hook scars and 5 ulcers on the skin.*

References

Barnes, L. G. (1985). Fossil Pontoporiid dolphins (Mammalia: Cetacea) from the Pacific coast of North America. *Nat. Hist. Mus. Los Angeles County Contrib. Sci.* **363**, 1–34.

Brownell, J. R., Jr., and Herald, E. S. (1972). *Lipotes vexillifer. Mammal. Spec.* **10**, 1–4.

Chen, P., and Hua, Y. (1989). Distribution, population size and protection of *Lipotes vexillifer*. In "Biology and Conservation of the River Dolphins" (W. F. Perrin, R. L. Brownell, Jr., K. Zhou, and J. Liu, eds.), pp. 81–85. Occasional Papers of the IUCN Species Survival Commission, No. 3, Gland, Switzerland.

Gao, A., and Zhou, K. (1992). Sexual dimorphism in the baiji, *Lipotes vexillifer. Can. J. Zool.* **70**, 1484–1493.

Muizon, C. de (1988). Les relations phylogénétiques des Delphinida (Cetacea, Mammalia). *Ann. Paléontol. (Vert. Invert.)* **74**, 157–227.

Rice, D. W. (1998). "Marine Mammals of the World, Systematics and Distribution." Special Publications of the Society for Marine Mammalogy, No. 4, Lawrence, KS.

Yang, G., and Zhou, K. (1999). A study on the molecular phylogeny of river dolphins. *Acta Theriolog. Sin.* **19**, 1–9 [Chinese with English abstract].

Zhou, K. (1989). Review of studies of structure and function of the baiji, *Lipotes vexillifer. In* "Biology and Conservation of the River Dolphins" (W. F. Perrin, R. L. Brownell, Jr., K. Zhou, and J. Liu, eds.), pp. 99–113. Occasional Papers of the IUCN Species Survival Commission, No. 3, Gland, Switzerland.

Zhou, K., Ellis, S., Leatherwood, S., Bruford, M., and Seal, U. (eds.) (1994). "Baiji (*Lipotes vexillifer*) Population and Habitat Viability Assessment Report." IUCN/SSC Conservation Breeding Specialist Group, Apple Valley, MN.

Zhou, K., Qian, W., and Li, Y. (1977). Studies on the distribution of baiji, *Lipotes vexillifer. Acta Zool. Sin.* **23**, 72–79 [Chinese with English abstract].

Zhou, K., Qian, W., and Li, Y. (1978). Recent advances in the study of the baiji, *Lipotes vexillifer. J. Nanjing Norm. Coll. (Nat. Sci.)* **1978**(1), 8–13 [Chinese with English abstract].

Zhou, K., Sun, J., Gao, A., and Würsig, B. (1998). Baiji (*Lipotes vexillifer*) in the lower Yangtze River: Movements, numbers, threats and conservation needs. *Aqu. Mamm.* **24**, 123–132.

Zhou, K., and Zhang, X. (1991). "Baiji, the Yangtze River Dolphin and Other Endangered Animals of China." Stone Wall Press, Washington, DC.

Zhou, K., Zhou, M., and Zhao, Z. (1984). First discovery of a Tertiary Platanistoid fossil from Asia. *Sci. Rep. Whales Res. Instit.* **35**, 173–181.

Baleen

Dale W. Rice
National Marine Mammal Laboratory, Seattle, Washington

The term baleen (also called whalebone) is a mass noun that refers collectively to the series of thin keratinous plates ("baleen plates") that make up the filtering apparatus in the mouth of a baleen whale. The word derives from the classical Latin *Balaena* and ultimately from the Greek φάλλαινα [phallaina], "whale."

Baleen plates are suspended from the whale's palate and are arranged in a row down each side of the mouth, extending from the tip of the rostrum back to the esophageal orifice. The left and right sides are separated by a prominent longitudinal ridge down the midline of the palate, but in the rorquals the two sides are continuous around the tip of the palate. Depending on the species, each "side" of baleen may contain anywhere betweeen 140 and 430 plates. The plates are transversely oriented and are spaced 1 or 2 cm apart, leaving a narrow gap or slot between adjacent plates. The plates are roughly triangular, with their horizontal basal edges embedded in the palate, their near-vertical labial edges facing outward, and their oblique, fringed lingual edges facing the inside of the mouth. Each plate is slightly curved, with its convex side facing forward, so that its labial edge is directed slightly backward; when the whale is swimming forward, this arrangement helps direct the flow of water through the interplate gaps from the mouth cavity to the exterior side of the baleen row. The sizes of the plates are smoothly graded, with the longest ones half to two-thirds of the way back from the tip of the rostrum and only rudimentary ones at the anterior and posterior ends of the row (Williamson, 1973; Pivorunas 1976, 1979).

Each baleen plate is made up of a middle layer, the *medulla*, which is sandwiched between the thin, smooth outer layers or *cortex*. The medulla consists of a mass of fine, hollow, hair-like keratinous tubules that run parallel to the labial side of the plate and terminate along the lingual side; the tubules are embedded in and cemented together by a horny matrix.

Evoluntarily the plates were presumably derived by an elaboration of the transverse ridges present on the palates of many terrestrial mammals. In whale fetuses the baleen first appears as a series of crosswise ridges along each side of the palate. The palatial tissue of baleen whales is arranged in three layers. The basal layer, several centimeters thick, is the *corium*. This is overlain by a thin *epithelial layer* only a few millimeters thick. The outermost epidermal layer, several centimeters thick, is simply called the *gum tissue*. The corium gives rise to, and is continuous with, the medulla of each baleen plate, whereas the adjacent epithelial layer is deflected downward to produce the cortical layers of each plate. The dense, rubbery gum tissue does not contribute to the formation of the plates, but simply fills the spaces between their bases, where it provides them a firm support. As each plate grows downward, its cortical layers become cornified sooner than the medulla does. This leaves the first few centimeters of the base of the plate with a layer of soft, highly vascular, corial tissue sandwiched between the keratinous outer layers; this soft layer is often called the pulp, by analogy with the pulp in mammalian teeth (van Utrecht, 1965).

Throughout the life of the whale its baleen plates grow continuously at their base and wear away along their lingual margin. The cortex and the matrix of the medulla erode away first, freeing the ends of the fibrous tubules for a distance of about 10–20 cm. The freed tubules form a hairy fringe along the entire lingual side of the plate. The fringes of each plate lie back across the lingual edges of the plates immediately behind them, with the whole forming a dense hairy mat that covers the internal apertures to the gaps between the plates. This mat effectively filters out the food organisms while allowing the water to flow out of the whale's mouth through the gaps.

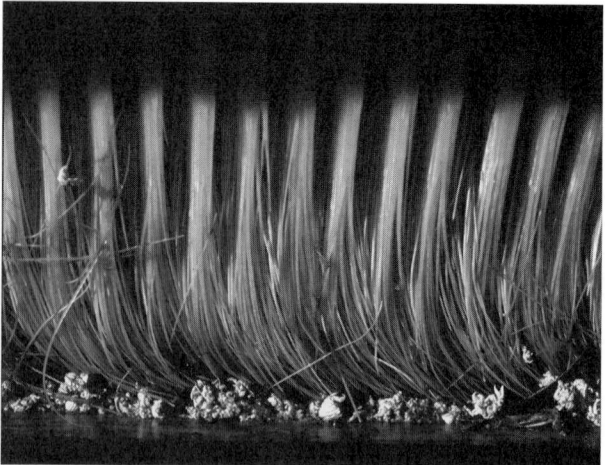

Figure 1 *Baleen in whales such as the gray whale, Es-chrichtius robustus, are used to filter seawater for food. Notice the arrangement of these keratinous plates in rows. Photo by François Gohier.*

Like human fingernails, the thickness of the baleen plates varies with the nutritional state of the whale. Alternating periods of summer gorging and winter fasting leave a regular series of visible growth zones on the surfaces of the plates. These zones have been used to infer the ages of whales, but because of the constant wear, it is rare for more than five or six zones to remain in a plate (Ruud, 1945). A claim that evidence of individual ovulations could be detected in the growth patterns of baleen plates was never confirmed (van Utrecht-Cock, 1965).

The number of baleen plates per side and their maximum size, shape, color, and other physical attributes are diagnostic for each species of whale. The right whales (family Balaenidae) with their narrow, highly arched rostrum have 250 to 390 narrow and extremely long plates, about 0.15–0.25 m wide and up to 2.50 m long in the black right whales (*Eubalaena* spp.) and 4.00 m in the bowhead whale (*B. mysticetus*); they are black with a fine whitish fringe. The pygmy right whale (*Caperea marginata;* family Neobalaenidae) has about 230 narrow, short plates up to 0.70 m long and 0.12 m wide; they are white with a black labial margin. The gray whale (*Eschrichtius robustus;* family Eschrichtiidae) has 140 thick but narrow and short plates, up to 0.10 m wide and 0.50 m long; they are white with a coarse white fringe that resembles excelsior (Fig. 1). The rorquals (family Balaenopteridae) with their wide, flat rostrum have 270 to 430 plates with a basal width 50 to 95% of their length, which varies from about 0.20 m in small minke whales to 1.00 m in huge blue whales. Each species of rorqual has a different color pattern on its baleen plates: humpback (*Megaptera novaeangliae*)—black with dirty-gray fringe; common minke (*Balaenoptera acutorostrata*)—white, sometimes with a narrow black stripe along labial margin; Antarctic minke (*B. bonaerensis*)—white with a wide black stripe along the labial margin; Bryde's (*B. edeni*)—black with light gray fringe; sei (*B. borealis*)—black with fine, silky, white fringe; fin (*B. physalus*)—gray and white longitudinal bands, with fringe the

same colors; and blue (*B. musculus*)—solid black with black fringe. All of the species of *Balaenoptera,* except the blue whale, usually have at least a few all-white baleen plates at the tip of the rostrum, mostly on the right side; this asymmetry is most prominent in the fin whale.

In the 19th century, the long baleen plates of the bowhead and right whales were much in demand for uses where a tough but limber material was needed so they were the most valuable product of the whale fishery. Landings of whalebone at United States ports reached their highest in 1853, with 5,652,300 pounds worth $1,950,000. The last year that any baleen reached the commercial market was 1930. Much of it was made into umbrella ribs, corset busks, and hoops for skirts. The fibrous fringes were used for brooms and brushes (Stevenson, 1907).

See Also the Following Articles

Dental Morphology, Evolution of ■ Filter Feeding ■ Prenatal Development in Cetaceans ■ Teeth

References

Pivorunas, A. (1976). A mathematical consideration on the function of baleen plates and their fringes. *Sci. Rep. Whales Res. Inst.* **28**, 37–55.

Pivorunas, A. (1979). The feeding mechanisms of baleen whales. *Am. Sci.* **67**, 432–440.

Ruud, J. T. (1945). Further studies on the structure of the baleen plates and their application to age determination. *Hvalrådets Skrifter* **29**, 1–69.

Stevenson, C. H. (1907). "Whalebone: Its Production and Utilization," pp. 1–12, Bureau of Fisheries Document 626.

van Utrecht, W. L. (1965). On the growth of the baleen plate of the fin whale and the blue whale. *Bijdr. Dierk.* **35**, 1–38.

van Utrecht-Cock, W. L. (1965). Age determination and reproduction of female fin whales, *Balaenoptera physalus* (Linnaeus, 1758) with special regard to baleen plates and ovaries. *Bijdr. Dierk.* **35**, 39–100.

Williamson, G. R. (1973). Counting and measuring baleen and ventral grooves of whales. *Sci. Rep. Whales Res. Inst.* **25**, 279–292.

Baleen Whales
Mysticetes

JOHN L. BANNISTER
Western Australian Museum, Perth

I. Diagnostic Characters and Taxonomy

Baleen or whalebone whales (Mysticeti) comprise one of the two recent (nonfossil) cetacean suborders. They differ from the other suborder (toothed whales, Odontoceti), particularly in their lack of functional teeth. Instead they feed on relatively very small marine organisms by means of a highly specialized filter-feeding apparatus made up of BALEEN plates ("whalebone") attached to the gum of the upper jaw.

Other differences from toothed whales include the baleen whales' paired blowhole, symmetrical skull, and absence of ribs articulating with the sternum.

Baleen whales are generally huge (Fig. 1). In the blue whale they include the largest known animal, growing to more than 30 m long and weighing more than 170 tons. Like all other cetaceans, baleen whales are totally aquatic. Like most of the toothed whales, they are all marine. Many undertake very long migrations, and some are fast swimming. A few species come close to the coast at some part of their life cycle and may be seen from shore; however, much of their lives is spent remote from land in the deep oceans. Baleen whale females grow slightly larger than the males. Animals of the same species tend to be larger in the Southern than in the Northern Hemisphere.

Within the mysticetes are four families: right whales (Balaenidae, balaenids), pygmy right whales (Neobalaenidae, neobalaenids), gray whales (Eschrichtiidae, eschrichtiids), and "rorquals" (Balaenopteridae, balaenopterids). Within the sub-order, 13 species are now generally recognized (Table I).

Right whales are distinguished from the other three families by their long and narrow baleen plates and arched upper jaw. Other balaenid features include, externally, a disproportionately large head (*ca.* one-third of the body length), long thin rostrum, and huge bowed lower lips; they lack multiple ventral grooves.

TABLE I
Mysticetes (Baleen Whales)

Family	Genus	Species	Subspecies	Common name	Maximum length (m)	Generalized distribution
Balaenidae				Right whales		
	Balaena	B. mysticetus		Bowhead whale	19.8	Circumpolar in the Arctic
	Eubalaena	E. glacialis		North Atlantic right whale	17.0	Temperate–Arctic
		E. australis		Southern right whale	17.0	Temperate–Antarctic
		E. japonica		North Pacific right whale	17.0	Temperate N. Pacific
Neobalaenidae				Pygmy right whales		
	Neobalaena	Caperea marginata		Pygmy right whale	6.4	Temperate, Southern Hemisphere only
Eschrichtiidae				Gray whales		
	Eschrichtius	E. robustus		Gray whale	14.1	North Pacific–Arctic
Balaenopteridae				Rorquals		
	Megaptera	M. novaeangliae		Humpback whale	16.0	Worldwide
	Balaenoptera	B. acutorostrata		Common minke whale		Worldwide
			B. a. acutorostrata	N. Atlantic minke whale	9.2	Temperate–Arctic
			B. a. scammoni	N. Pacific minke whale	?	Temperate–Arctic
			B. a. subsp.	Dwarf minke whale	?	Temperate–sub-Antarctic, Southern Hemisphere only
		B. bonaerensis		Antarctic minke whale	10.7	Temperate–Antarctic
		B. edeni		Bryde's whale	14.0	Circumglobal, tropical–subtropical
		B. borealis		Sei whale	17.7	Worldwide, largely temperate
		B. physalus		Fin whale	26.8	Worldwide
		B. musculus		Blue whale		Worldwide
			B. m. musculus	Blue whale	26.0	N. Atlantic, N. Pacific
			B. m. indica	Great Indian rorqual	?	N. Indian Ocean
			B. m. brevicauda	Pygmy blue whale	24.4	Southern Hemisphere, temperate–sub-Antarctic
			B. m. intermedia	"True" blue whale	30.5	Southern Hemisphere, temperate–Antarctic

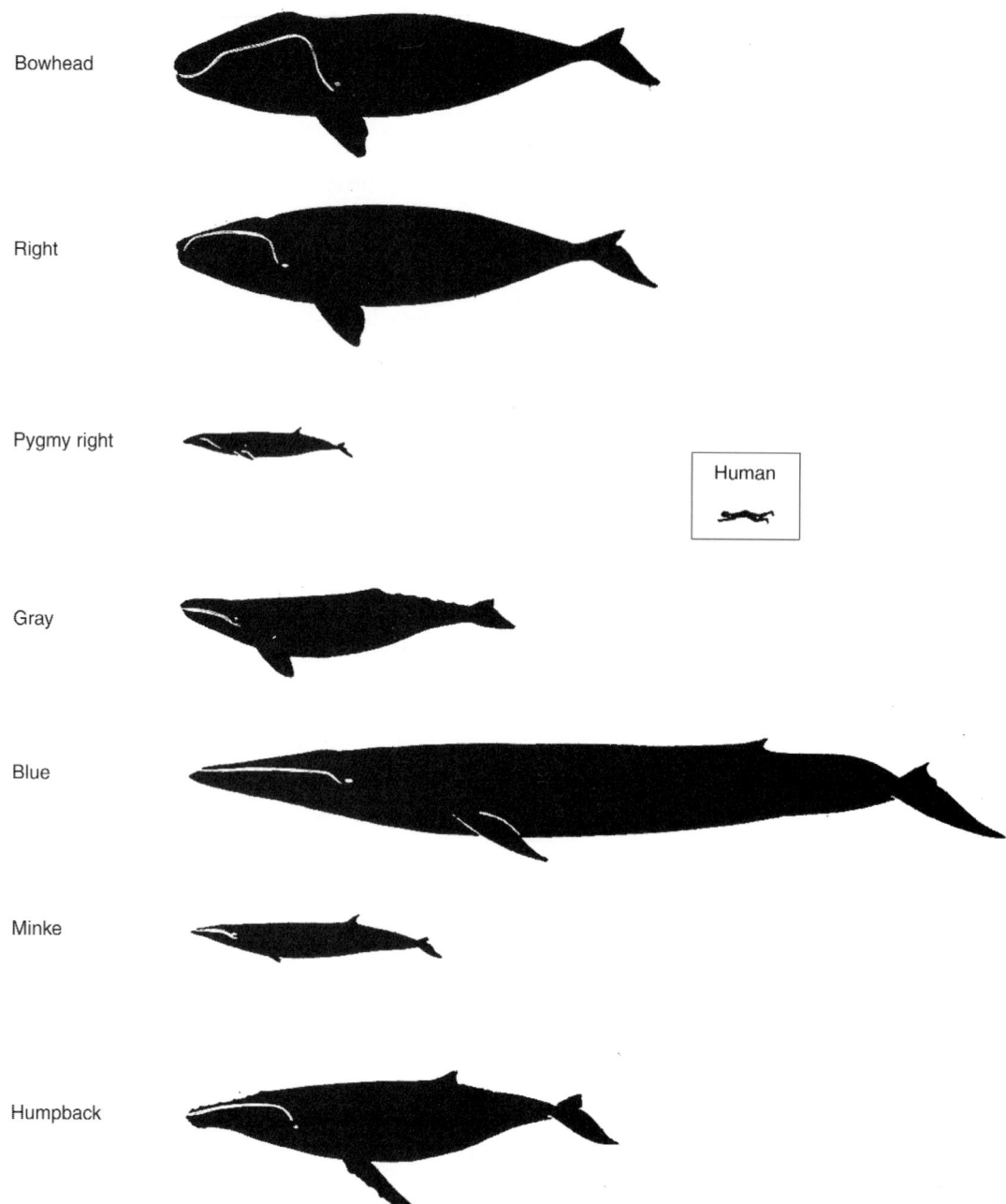

Bowhead

Right

Pygmy right

Human

Gray

Blue

Minke

Humpback

Figure 1 *Lateral profiles of representative baleen whales, with a human figure, to scale.*

Internally, there is no coronoid process on the lower jaw and cervical vertebrae are fused together. Within the family are two distinct groups: the bowhead (*Balaena mysticetus*) of northern polar waters (formerly known as the "Greenland" right whale) and the three "black" right whales (*Eubalaena* spp.) of more temperate seas (so called to distinguish them from the "Greenland" right whale). All balaenids are robust.

Pygmy right whales (*Caperea marginata*) have some features of both right whales and balaenopterids. The head is short (*ca* one-quarter of the body length), although with an arched upper jaw and bowed lower lips, and there is a dorsal fin. The relatively long and narrow baleen plates are yellowish-white, with a dark outer border, quite different from the all-black balaenid baleen plates. Internally, pygmy right whales have numerous broadened and flattened ribs.

Gray whales (*Eschrichtius robustus*) are also somewhat intermediate in appearance between right whales and balaenopterids. They have short narrow heads, a slightly arched rostrum, and between two and five deep creases on the throat instead of the balaenopterid ventral grooves. There is no dorsal fin, but a series of 6 to 12 small "knuckles" along the tail stock. The yellowish-white baleen plates are relatively small.

Balaenopterids comprise the six whales of the genus *Balaenoptera* (blue, *B. musculus;* fin, *B. physalus;* sei, *B. borealis;* Bryde's, *B. edeni;* common minke, *B. acutorostrata* and Antarctic minke, *B. bonaerensis*) and the humpback whale (*Megaptera novaeangliae*). All have relatively short heads, less than a quarter of the body length. In comparison with right whales, the baleen plates are short and wide. Numerous ventral grooves are present, and there is a dorsal fin, sometimes rather small. Internally, the upper jaw is relatively long and unarched, the mandibles are bowed outward, and a coronoid process is present; cervical vertebrae are generally free. All seven balaenopterids are often known as "rorquals" (said to come from the Norse "whale with pleats in its throat"). Strictly speaking, the term should probably be applied to the six *Balaenoptera* species, recognizing the rather different humpback in its separate genus, but many authors now use it for all seven balaenopterids.

Baleen whales are sometimes called "great whales." Despite their generally huge size, some of the species are relatively small, and it seems preferable to restrict the term to the larger mysticetes (blue, fin, sei, Bryde's, humpback) together with the largest odontocete (the sperm whale, *Physeter macrocephalus*).

In a recent review of the systematics and distribution of the world's marine mammals, Rice (1998) has drawn attention to the derivation of the Latin word Mysticeti and clarified the status of a variant, Mystacoceti. He describes the former as coming from Aristotle's original Greek *mustoketos*, meaning "the mouse, the whale so-called" or "the mouse-whale" (said to be an ironic reference to the animals' generally vast size). Mystacoceti means "moustache whales," and although used occasionally in the past (and more obviously appropriate for whales with baleen in their mouths) it has been superseded by Mysticeti.

The 13 species in Table I differ somewhat from those listed by Rice. Some authors disagree with his use of the genus *Balaena* for *Eubalaena* and his preference for the single species *glacialis* rather than the three species, *Eubalaena glacialis, E. japonica,* and *E. australis* (the North Atlantic, North Pacific and southern right whales). While acknowledging the need for further investigation, they refer to present-day biologists' usage, and genetic information, in preferring a separation between Northern and Southern Hemisphere animals and in recognizing two species in the northern hemisphere; *Eubalaena* is, however, the only mysticete genus where one or more separate species is recognized in each hemisphere. Rice also distinguishes between two species of Bryde's whale: *Balaenoptera edeni* and *B. brydei.* The taxonomic status of these "inshore, smaller" and "offshore, larger" forms has yet to be determined and here they are subsumed within *B. edeni.* In the case of the blue whale, Rice's inclusion of a northern Indian Ocean form (*B. m. indica,* referred to by Rice as "the great Indian rorqual") has been followed. Similarly, his listing of three subspecies of minke whale, including the Southern Hemisphere dwarf minke, which has yet to be formally described, has been retained. However, other sub-

species, e.g., two sei whales and two fin whales, have not been included.

II. Distribution and Ecology
A. Habitat

In addition to the subspecies listed in Section I, many stocks or populations have been recognized, some mainly for management purposes, based on more or less valid biological grounds. Some significant examples include:

1. Bowhead Whales In addition to the currently most abundant population (the Bering-Chukchi-Beaufort Seas stock), four others are recognized: Baffin Bay/Davis Strait, Hudson Bay, Spitzbergen, and Okhotsk Sea.

2. Right Whales In the North Atlantic species, two populations are currently recognized, western and eastern, with calving grounds off the southeastern United States and northwestern Africa. The latter may now represent only a relict population(s). In the North Pacific species, the current view is that there well may once have been two or more stocks, based on feeding ground information: at least one now centered in summer on the Sea of Okhotsk and another, although possibly not now a functioning unit, summering in the Gulf of Alaska.

In the southern right whale, there are several populations, defined by currently occupied calving grounds, but these cover only a proportion of the many areas known from historical whaling records to have once been occupied by right whales. Up-to-date information is available on presumed discrete populations off eastern South America, South Africa, southern Australia, and sub-Antarctic New Zealand.

3. Gray Whales A western North Atlantic population may have persisted until the 17th or 18th centuries, but is now extinct. The species now survives only in the North Pacific, where, in addition to a flourishing "Californian" stock, wintering on the coast of Baja California, and summering in the Bering Sea, animals are now being reported from a remnant western stock, summering in the northern Okhotsk Sea.

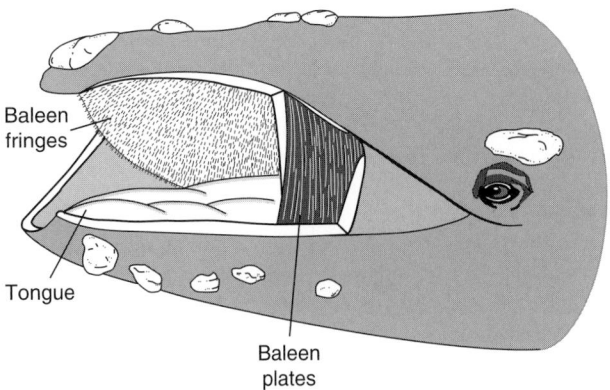

Figure 2 *Head of a right whale showing the arrangement of the filter-feeding apparatus. From Bonner (1980).*

4. Humpback Whales   In the North Atlantic, two major populations are recognized: one based on animals wintering in the West Indies and the other, now possibly only a relict population, wintering around the Cape Verde Islands. In the North Pacific, three discrete wintering grounds have been recorded: around the Bonin, Mariana, and Marshall Islands in the west; around the Hawaiian Islands in the center; and off Mexico in the east.

In the Southern Hemisphere, seven populations have been postulated. Six are well defined, based on calving (wintering) grounds on either side of each continent (one off eastern Australia is closely related to animals wintering off Fiji and Tonga), and a possible seventh in the central Pacific. In the northwest Indian Ocean, there seems to be a separate population where animals have been reported present throughout the year.

Baleen whales thus occupy a wide variety of habitats, from open oceans to continental shelves and coastal waters, from the coldest waters of the Arctic and Antarctic, through waters of both hemispheres and into the tropics.

Most specialized is the bowhead, *Balaena*, restricted to the harsh cold and shallow seas of the Arctic and sub-Arctic. The black right whales (*Eubalaena*) are more oceanic and prefer generally temperate waters, but come very close to coasts in winter to give birth, particularly in the Southern Hemisphere. Once believed not to penetrate much further south than the Antarctic convergence (*ca* 50–55°S), there have been recent records in the Antarctic proper, south of 60°S. Whether this is a new phenomenon is unclear: a report by Sir James Clark Ross of many "common black" (i.e., right) whales in the Ross Sea (eastern Antarctic) at 63°S in December 1840 was discounted when their presence there later that century could not be confirmed. It has been suggested that the currently greatly reduced population of the western North Atlantic right whale, now wintering off the southeastern United States and summering in coastal waters north to the Bay of Fundy (*ca* 45°N), may represent the peripheral remnant of a more widely distributed stock, formerly summering north to Labrador and southern Greenland, i.e., to at least 60°N.

The pygmy right whale (*Caperea*) is restricted to Southern Hemisphere temperate waters, between about 30 and 52°S; it can be found coastally in winter in some areas, and all-year round in others.

Gray whales (*Eschrichtius*) are the most obviously coastal baleen whales. The long coastal migration of the "Californian" stock, from Mexico to Alaska, supports a major whale-watching industry from December to April. In spring the animals migrate through the Bering Strait into the more open waters of the Bering Sea, but still favoring more shallow waters.

Among the balaenopterids, fin and sei whales are probably the most oceanic, with the former penetrating into colder waters than the latter in summer. Blue whales can be found closer inshore, but are often associated with deep coastal canyons, e.g., off central and southern California. The Southern Hemisphere pygmy blue whale (subspecies *B. m. brevicauda*) has been regarded as restricted to more temperate waters than the "true" blue whale (*B. m. intermedia*), not often being found much beyond 55°S. The most coastal balaenopterid is the humpback (*Megaptera*), with long migrations between temper-

ate/tropical breeding grounds and cold water feeding grounds. In the Southern Hemisphere, much of its journey occurs along the east and west coasts of the three continents. In the Northern Hemisphere, humpbacks are rather more oceanic, but still coastal at some stage in their migration: in the North Pacific they can be found wintering off the Hawaiian Islands and summering off Alaska, and in the western North Atlantic they winter in the Caribbean and summer between New England, the west coast of Greenland, and Iceland.

Minke whales are wide ranging, from polar to tropical waters in both hemispheres. In the Southern Hemisphere the Antarctic species can, with blue whales, be found closest to the ice edge in summer. Elsewhere they can often occur near shore, in bays and inlets. Their migrations are less well defined and predictable than other migratory balaenopterids; in some regions they are present year-round.

The most localized balaenopterid is Bryde's whale. It is the only species restricted entirely to tropical/warm temperate waters and probably does not undertake long migrations. The two forms—inshore and offshore, in several areas—can differ in their movements. Off South Africa, for example, the inshore form is thought to be present throughout the year, whereas the offshore form appears and disappears seasonally, presumably in association with movements of its food, shoaling fish.

B. Food and Feeding

Although they include the largest living animals, baleen whales feed mainly on very small organisms and are strictly carnivorous, feeding on zooplankton or small fish. In "filter feeding"—sieving the sea—baleen whales are quite different from toothed whales, where the prey is captured individually.

FILTER FEEDING has been described as requiring, in addition to a supply of food in the water, three basic features: a flow of water to bring prey near the mouth; a filter to collect the food but allow water to pass through; and a means of removing the filtered food and conveying it to the stomach for digestion. Baleen whales meet those requirements by (a) seeking out areas where their food concentrates, (b) either swimming open-mouthed through food or gulping it in, (c) possessing a highly efficient filter formed by the baleen plates, and (d) forcing the water containing the food out through the baleen plates and then transferring the trapped food back to the gullet and hence to the stomach. In the latter the tongue is presumed to be involved; in balaenopterids the process is aided by the distensible throat.

While all baleen whales possess a filter based on baleen plates, two rather different systems—essentially "skimming" and "gulping"—have evolved to filter a large volume of water containing food. Each relies on a series of triangular baleen plates, borne transversely on each upper jaw. The inner, longer border (hypotenuse) of each plate bears a fringe of fine hairs, forming a kind of filtering "doormat." Quite unrelated to teeth (which appear as early rudiments in the gums of fetal baleen whales), baleen is closest in structure to mammalian hair and human fingernails. In the right whales, filtration is achieved with very long and narrow plates in the very large mouth, itself carried in the very large head. The plates, up to 4 m long in bowheads and 2.7 m in other right whales, are accommodated in the mouth by an arched upper jaw and are enclosed in massively enlarged and upwardly bowed lower lips. There is a gap between the rows at the front of the mouth, and the whole arrangement allows the whale to scoop up a great quantity of water while swimming slowly forward. In balaenopterids, with their much smaller heads, the baleen plates are shorter and broader and the rows are continuous at the front. Taking in a large volume of water and food is usually achieved by swimming through a food swarm and gulping, while simultaneously enlarging the capacity of the mouth greatly by extending the ventral grooves and depressing the tongue. The two systems allow, on the one hand, the relatively slow-swimming balaenids to concentrate their rather sparse slow-swimming food over a period, and on the other, the faster-swimming balaenopterids to take in large amounts of their highly concentrated fast-swimming prey over a shorter time.

Typically, baleen whales feed on zooplankton, mainly euphausiids or copepods, swarming in polar or subpolar regions in summer. That is particularly so the Southern Hemisphere,

where the summer distributions of several balaenopterids depends on the presence of *Euphausia superba* (known to whalers by the Norwegian word "krill") in huge concentrations in the Antarctic. In the Northern Hemisphere, with a more variable availability of food, balaenopterids are more catholic in their feeding. Humpbacks and fin whales, for example, feeding almost exclusively on KRILL in the south, commonly take various species of schooling fish in the north.

The variety of organisms taken by the various species in different regions is listed in Table II. While most feeding occurs in colder waters, baleen whales may feed opportunistically elsewhere. All baleen whales but one, the gray whale, feed generally within 100 m of the surface and, consequently, unlike many toothed whales, do not dive very deep or for long periods. Gray whales feed primarily on bottom-living organisms, almost exclusively amphipods, in shallow waters.

The baleen plate structure, particularly the inner fringing hairs, to some extent mirrors the food organisms taken or (in

TABLE II
Baleen Whale Food Items

			Food items	
Species	*Subspecies*	*Common name*	*Northern Hemisphere*	*Southern Hemisphere*
B. mysticetus		Bowhead whale	Mainly calanoid copepods; euphausiids; occasional mysids, amphipods, isopods, small fish	
E. glacialis		North Atlantic right whale	Calanoid copepods; euphausiids	
E. australis		Southern right whale		Copepods; postlarval *Munida gregaria; Euphausia superba*
Caperea marginata		Pygmy right whale		Calanoid copepods
E. robustus		Gray whale	Gammarid amphipods; occasional polychaetes	
M. novaeangliae		Humpback whale	Schooling fish; euphausiids	*E. superba* (Antarctic); euphausiids, postlarval *M. gregaria*, occasional fish (ex-Antarctic)
B. acutorostrata	*B. a. acutorostrata*	N. Atlantic minke	Schooling fish; euphausiids	
	B. a. scammoni	N. Pacific minke	Euphausiids; copepods; schooling fish	
	B. a. subsp.	Dwarf minke		? Euphausiids, schooling fish
B. bonaerensis		Antarctic minke		*E. superba*
B. edeni		Bryde's whale	Pelagic crustaceans, including euphausiids	Schooling fish; euphausiids
B. borealis		Sei whale	Schooling fish	Copepods, including *Calanus; E. superba*
B. physalus		Fin whale	Schooling fish; squid; euphausiids; copepods	*E. superba* (Antarctic); other euphausiids (ex-Antarctic)
B. musculus	*B. m. musculus*	Blue whale	Euphausiids	
	B. m. indica	Great Indian rorqual	?Euphausiids; copepods	
	B. m. intermedia	"True blue"		*E. superba* (Antarctic); other euphausiids (ex-Antarctic)
	B. m. brevicauda	Pygmy blue		Euphausiids, mainly *E. vallentini*

the case of *E. superba*) different size classes. Thus there is some correlation between decreasing size of prey and fineness of baleen by species, viz. gray, blue, fin, humpback, minke, sei, and right whales. Where food stocks are very dense, e.g., around sub-Antarctic South Georgia, fin, blue, and sei whales may all overlap in their feeding on *E. superba*.

Baleen whale food consumption per day has been calculated as some 1.5–2.0% of body weight, averaged over the year. Given that feeding occurs mainly over about 4 months in the summer in the larger species, the food intake during the feeding season has been calculated at some 4% of body weight per day, *ca.* 400 kg per day for a large blue whale. To survive the enormous drain of pregnancy and lactation, it has been calculated that a pregnant female baleen whale needs to increase its body weight by up to 65%. The ability to achieve such an increase in only a few months' feeding indicates the great efficiency of the baleen whales' feeding system.

C. Predators and Parasites

Apart from humans, the most notable baleen whale predator is the killer whale (*Orcinus orca*). Minke whales have been identified as a major diet item of some killer whales in the Antarctic. Killer whale attacks have been reported on blue, sei, bowhead, and gray whales, although their frequency and success are unknown. Humpbacks often have killer whale tooth marks on their bodies and tail flukes. Humpback and right whale calves in warm coastal waters are susceptible to attack by sharks. There are anecdotal reports of calving ground attacks on humpbacks by false killer whales (*Pseudorca crassidens*).

A form of harassment, only recently described, occurs on right whales on calving grounds off Peninsula Valdes, Argentina. Kelp gulls have developed the habit of feeding on skin and blubber gouged from adult southern right whales' backs as they lie at the surface. Large white lesions can result. The whales react adversely to such gull-induced disturbance and calf development may be affected.

External parasites, particularly "WHALE LICE" (cyamid crustaceans) and BARNACLES (both acorn and stalked) are common on the slower-swimming more coastal baleen whales, such as gray, humpback, and right whales. In the latter, aggregations of light-colored cyamids on warty head callosities have facilitated research using callosity-pattern photographs for individual identification. External parasites are much less common on the faster swimming species, although whale lice have been reported on minke whales (in and around the ventral grooves and umbilicus); the highly modified copepod *Penella* occurs particularly on fin and sei whales in warmer waters. The commensal copepod *Balaenophilus unisetus* often infests baleen plates in such waters, especially on sei and pygmy blue whales.

A variety of internal PARASITES have been recorded, although some baleen whales seem less prone to infection than others. They appear, for example, to be less common in blue whales, but prevalent in sei whales. Records include stomach worms (*Anisakis* sp), cestodes, kidney nematodes, liver flukes, and acanthocephalans ("thorny-headed" worms) of the small intestine.

The cold water diatom *Cocconeis ceticola* often forms a brownish-yellow film on the skin of blue and other baleen whales in the Antarctic. Because the film takes about a month to develop, its extent can be used to judge the length of time an animal has been there. Its presence led to an early common name for the blue whale: "sulfur bottom."

For many years the origin of small scoop-shaped bites on baleen whale bodies in warmer waters remained a mystery until they were found to be caused by the small "cookie-cutter" shark, *Isistius brasiliensis*. Some species are highly prone to such attacks. In Southern Hemisphere sei whales the overlapping healing scars can impart a galvanized-iron sheen to the body.

III. Life History

A. Behavior

1. Sound Production   Unlike toothed whales, baleen whales are not generally believed to use sound for echolocation, although bowheads, for example, are thought to use sound reflected from the undersides of ice floes to navigate through ice fields. However, sound production for communication, for display, establishment of territory, or other behavior, is well developed in the suborder. Blue whales produce the loudest sustained sounds of any living animal. At up to nearly 190 decibels, their long (half-minute or more), very low-frequency (<20 Hz) moans may carry for hundreds of kilometers or more in special conditions. Fin whales produce similarly low (20 Hz) pulsed sounds. Minke whales also produce a variety of loud sounds. Right whales produce long low moans; bowhead sounds, recorded on migration past hydrophone arrays in nearshore leads, have been used in conjunction with sightings to estimate population size off northern Alaska. Southern right whales, at least, seem to use sound to communicate with their calves.

Humpbacks produce the longest, most complex sound sequences in "songs," described as an array of moans, groans, roars, and sighs to high-pitched squeaks and chirps, lasting 10 or more minutes before repetition, sometimes over hours. It seems that only the adult males sing, generally only in or close to the breeding season. In any one breeding season, all the males sing the same SONG, changing slightly over successive seasons. Different populations have different songs; so much so, for example, that those off western Australia have a distinctly different song—less complex, less "chirpy"—than that heard on breeding grounds separated by the Australian continent, off the east coast. "Songs" may also be heard in migrating humpbacks, but less so on the cold water feeding grounds, where if they occur at all, they appear generally only as "snatches" or isolated segments.

2. Swimming and Migration   With their streamlined bodies, rorquals include the fastest-swimming baleen whales. Sei whales have been recorded at around 35 knots (more than 60 km/hr) in short bursts; minke and fin whales are also known as fast swimmers, the latter up to 20 knots (37 km/hr). Blue whales are among the most powerful swimmers, able to sustain speeds of over 15 knots (28 km/hr) for several hours. On migration, humpbacks and gray whales average about 4 knots (8–9 km/hr) and bowheads only about 2.7 knots (5 km/hr). Migration speeds for southern right whales are not known, but medium range coastal movements off southern Australia indicate 1.5–2.3 knots (2.7–4.2 km/hr) over 24 hrs for cow/calf pairs.

Baleen whales undertake some of the longest MIGRATIONS known. Gray whales cover 10,000 nautical miles (18,000 km) on the round trip between the Baja California breeding grounds and the Alaskan feeding grounds, among the longest migrations of any mammal. Southern Hemisphere humpbacks may cover as much as 50° of latitude either way between breeding and feeding grounds, a round trip of some 6000 nautical miles (11,000 km). Not all baleen whale migrations are so well marked. The biannual movements of Bering Sea bowheads are governed by the seasonal advance and retreat of sea ice, which varies from year to year. Although Southern Hemisphere blue and fin whales all feed extensively in the Antarctic in summer, the locations of their calving grounds are not known. Sei whale migrations are relatively diffuse and can vary from year to year in response to changing environmental conditions. By comparison, Bryde's whales hardly migrate at all, presumably being able to satisfy both reproductive and nutritional needs in tropical/warm temperate waters. Even among such migratory animals as humpbacks, it may be that not all animals migrate every year; studies off eastern Australia indicate that a proportion of adult females may not return to the calving grounds each year, and individuals have even been reported in summer farther north. However, Southern Hemisphere migrating humpbacks show segregation in the migrating stream: immatures and females accompanied by yearling calves are in the van of the northward migration, followed by adult males and nonpregnant mature females; pregnant females bring up the rear. A similar pattern occurs on the southward journey, with cow/calf pairs traveling last. Very similar segregation is recorded among migrating gray whales.

Baleen whale migrations have generally been regarded as taking place in response to the need to feed in colder waters and reproduce in warmer waters. Explanations for such long-range movements have included direct benefits to the calf (better able to survive in calm, warm waters), evolutionary "tradition" (a leftover from times when continents were closer together), and the possible ability of some species to supplement their food supply from plankton encountered on migration or on the calving grounds. Corkeron and Connor (1999) have rejected these explanations, suggesting that there may be a major advantage to migrating pregnant female baleen whales in reducing the risk of killer whale predation on newborn calves in low latitudes. They cite in its favor the greater abundance of killer whales in higher latitudes, that their major prey (pinniped seals) is more abundant there, and that killer whales do not seem to follow the migrating animals.

3. Social Activity Large aggregations of baleen whales are generally uncommon. Even on migration, in those species where well-defined migration paths are followed (e.g., gray whales and humpbacks), individual migrating groups are generally small, numbering only a few individuals. It has been stated that predation is a main factor in the formation of large groups of cetaceans, e.g., open ocean dolphins. Given the large size of most adult baleen whales, predation pressure is low and group size is correspondingly small.

Blue whales are usually solitary or in small groups of two to three. Fin whales can be single or in pairs; on feeding grounds they may form larger groupings, up to 100 or more. Similarly,

sei whales can be found in large feeding concentrations, but in groups of up to only about six elsewhere. The same is true for minke whales, found in concentrations on the feeding grounds, but singly or in groups of two or three elsewhere. Social behavior has been studied most intensively in coastal humpbacks, e.g., on calving grounds. Male humpbacks compete for access to females by singing and fighting. The songs seem to act as a kind of courtship display. Males congregate near a single adult female, fighting for position. Such aggression can involve lunging at each other with ventral grooves extended, hitting with the tail flukes, raising the head while swimming, fluke and flipper slapping, and releasing streams of bubbles from the blowhole. As a result of such encounters, individuals can be left with raw and bleeding wounds caused by the sharp barnacles. Among southern right whales, similar "interactive" groups are often observed on the coastal calving grounds in winter, involving a tight group with up to seven males pursuing an adult female, but not generally resulting in wounded animals. As for humpbacks, it is not yet certain whether such behavior results in successful mating, although, at least in such right whale groups, intromission is often observed.

Feeding balaenopterids have often been reported as circling on their sides through swarms of plankton or fish. It has been suggested that gray whales feed on their right sides, as those baleen plates are more worn down, presumably through contact with the seabed. The most remarkable behavior, however, is reported from humpbacks. In the Southern Hemisphere, on swarms of krill, they may feed in the same "gulping" way as other balaenopterids. In the Northern Hemisphere, two methods are commonly reported: "lunging" and "bubble netting." In the former, individuals emerge almost vertically at the surface with their mouths partly open, closing them to force the enclosed water out through the baleen. In the latter, an animal circles below the food swarm; as it swims upward, it exhales a series of bubbles, forming a "net" encircling the prey. It then swims upward through the prey with its mouth open, as in lunging.

B. Growth and Reproduction

Young baleen whales, particularly the fetus and the calf, grow at an extraordinary rate. In the largest species, the blue whale, fetal weight increases at a rate of some 100 kg/day toward the end of pregnancy. The calf's weight increases at a rate of about 80 kg/day during suckling. During that 7-month period of dependence on the cow's milk, the blue whale calf will have increased its weight by some 17 tons and increased in length from around 7 to 17 meters. Blue whales attain sexual maturity at between 5 and 10 years, at a length of around 22 meters, and live for possibly 80–90 years. Adult female blue whales give birth every 2–3 years, with pregnancy lasting some 10–11 months.

Other balaenopterids follow the same general pattern (Fig. 3). Mating takes place in warm waters in winter, with birth following some 11 months later. A 7- to 11-month lactation period may be followed by a year "resting," or almost immediately by another pregnancy. Most adults are able to reproduce from between 5 and 10 years of age and reach maximum growth after 15 or more years. The smallest balanopterid, the common minke whale, is born after a pregnancy of some 10 months, at

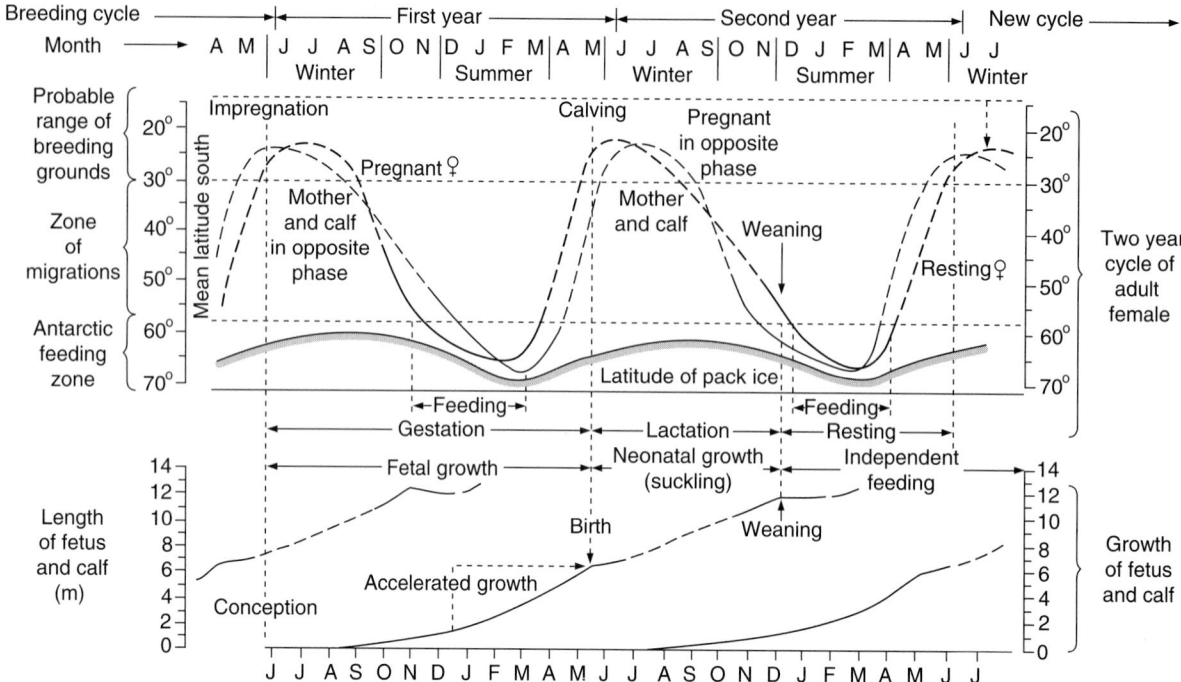

Figure 3 *"Typical" life cycle of a southern baleen whale. As modified by Bonner (1980), from Mackintosh (1965).*

a length of just under 3 meters. Weaning occurs at just under 6 meters, after 3–6 months. The adult female can become pregnant again immediately following birth, but the resulting short calving interval is generally uncommon in baleen whales: 2–3 years is the norm, although female humpbacks can achieve a similar birth rate, enabling their stocks to recover rapidly after depletion (see Section IV).

Right whales follow a similar general pattern, but there are some differences. In right whales, gestation lasts about 11 months and weaning for about another year. Females are able to reproduce successfully from about 8 years (there are records of successful first pregnancies from 6 years), but the calving interval is usually a relatively regular 3 years. For bowheads, it has been reported, rather surprisingly, that while growth is very rapid during the first year of life (from ca. 4.5 meters), it may be followed by a period of several years with little or no growth. Sexual maturity occurs at 13–14 meters; at the reduced growth rate, that would not be reached until 17–20 years. Similarly, evidence shows considerable longevity in this species: stone harpoon heads found in harvested whales and last known to be used off Alaska early this century suggest that individual animals can be at least 100 years old.

IV. Population Status

For centuries, baleen whales have borne the brunt of human greed, for products and profit. Only the sperm whale, largest of the toothed whales, has rivaled them as a whaling target. Black right whales (*Eubalaena*) were taken in the Bay of Biscay from the 12th century, with the fishery extending across the North Atlantic by the 16th century. Attention then shifted to the Greenland whale (*Balaena*) near Spitzbergen and later off southern and western Greenland. Both species' numbers were reduced to only small remnants, and in several areas (e.g., Spitzbergen and Greenland for *Balaena* and the northeast Atlantic and the North Pacific for *Eubalaena*) the stocks were virtually exterminated. That destruction was undertaken using the "old" whaling method, with open boats and hand harpoons, on the "right" species—"right" because they were relatively easy to approach, floated when dead, and provided huge quantities of products [oil for lighting, lubrication, and soap and baleen ("whalebone") for articles combining flexibility with strength, such as corset stays, umbrella spokes, and fishing rods].

Development of the harpoon gun and steam catcher, from 1864, increased the rate of catching greatly, but also allowed attention to turn to the largest baleen whales, the blue and fin whales, whose size, speed, and tendency to sink when dead had prevented capture by the old methods. From its beginning in the North Atlantic, then, by the end of the century, in the North Pacific, "modern" whaling's next and most intensive phase moved south, first in 1904 at South Georgia in the South Atlantic, just within the Antarctic zone. Initially on humpbacks [up to 12,000 were taken in one year (1912), leading to very rapid stock decline] and then on blue and fin whales, southern whaling based on such land stations—in the Antarctic in summer and the tropics in winter—was overtaken from the late 1920s by the great development of pelagic whaling using floating factory ships. Huge annual Southern Hemisphere catches resulted—a maximum of over 40,000 in 1931—averaging around 30,000 animals per year in the later 1930s and again after World War II until 1965. Whereas blue whales had been the preferred target in the 1930s, their great reduction in numbers led to a shift in attention progressively over the years to fin whales, to sei whales in the 1960s, and finally to minke. With depletion of stocks and more stringent conservation measures (killing of humpbacks, blue, and fin whales was banned from the mid-1960s, even though some illegal catching continued until the early 1970s or even later), catches fell to between 10,000 and 15,000 per year in the 1970s. The "old" whaling story had virtually repeated itself—enormous reductions through overfishing of one species or stock leading to exploitation of other species and stocks until, apart from minke whales, only remnants were left. Since 1989, a moratorium on all commercial whaling has eliminated that pressure, with the exception of limited whaling carried our under exemption for scientific research, and, since 1993, limited commercial catching of minke whales in the eastern North Atlantic. Some "aboriginal" whaling has also continued in the Northern Hemisphere, on bowheads off northern Alaska, on gray whales in the Bering Sea, on fin and minke whales off Greenland, and on humpbacks in the Caribbean.

Despite the great scale of the kill in "old" and "modern" whaling, no whale species has become extinct through whaling, although a number of individual stocks have been reduced greatly; at least one, the North Atlantic gray whale, has disappeared within the past 200–300 years. In its most recent (1996) "Red List" of threatened animals (Table 3), the World Conservation Union (IUCN) includes no baleen whale species or stocks as either **extinct (EX)** or **critically threatened (CR)** (the latter within the **threatened** category). Within the threatened category, eight taxa—four species, one subspecies, and three stocks—are listed as **endangered (EN);** four taxa—one species and three stocks—are **vulnerable (VU).** Six taxa—two

TABLE III
IUCN Red List Categories for Baleen Whales (1996)

Species	Subspecies	Common name	EX	CR	EN	VU	LR	DD
					Category Threatened			
B. mysticetus		Bowhead whale			°a	°b	°(cd)^c	
E. glacialis°		North Atlantic right whale			°d			
E. australis		Southern right whale					°(cd)	
Caperea marginata^e		Pygmy right whale						
E. robustus		Gray whale			°f		°(cd)^g	
M. novaeangliae		Humpback whale				°		
B. acutorostrata	*B. a. acutorostrata*	N. Atlantic minke					°(nt)	
	B. a. scammoni	N. Pacific minke						
	B. a. subsp.	Dwarf minke						
B. bonaerensis		Antarctic minke					°(cd)	
B. edeni	Bryde's whale							°
B. borealis		Sei whale			°			
B. physalus		Fin whale			°			
B. musculus	*B. m. musculus*	Blue whale				°h	°(cd)^i	
	B. m. indica	Great Indian rorqual						
	B. m. intermedia	"True" blue			°			
	B. m. brevicauda	Pygmy blue						°

°Includes *E. japonica*
^a Okhotsk Sea bowhead whale, Spitzbergen bowhead whale
^b Hudson Bay bowhead whale, Baffin Bay/Davis Strait bowhead whale
^c Bering-Beaufort-Chuckchi Seas bowhead whale
^d North Atlantic and North Pacific northern right whales
^e Pygmy right whale removed from 1996 Red List
^f Northwest Pacific gray whale
^g Northeast Pacific Gray whale
^h North Atlantic blue whale
^i North Pacific blue whale

species, one subspecies, and three stocks—are listed as at **lower risk (LR),** and two taxa—one species and one subspecies—as **data deficient (DD).**

Those species under greatest current threat **(EN)** are the North Atlantic and North Pacific right, the sei, and fin whales, together with the "true" blue subspecies, two of the five bowhead stocks (Okhotsk Sea, Spitzbergen), and the northwest Pacific gray. Next most threatened **(VU)** are the humpback, two bowhead stocks (Hudson Bay, Baffin Bay/Davis Strait), and the North Atlantic blue. At lower risk **(LR)** are the southern right and Antarctic minke, one bowhead stock (Bering-Chukchi-Beaufort Seas), the North Atlantic minke, northeast Pacific gray, and North Pacific blue; all but one are further qualified as **conservation dependent (cd,** not vulnerable because of specific conservation efforts). The exception is the North Atlantic minke, listed as **near threatened (nt,** not conservation dependent but almost qualifying as vulnerable). The two taxa for which insufficient information is currently available **(DD)** are Bryde's whale and the pygmy blue.

The INTERNATIONAL WHALING COMMISSION's Scientific Committee, responsible for the assessments of such stocks' current status, has reported encouraging recent reversals of stock decline for some stocks of some species. One, the northeast Pacific gray whale, has recovered under protection from commercial whaling (but with aboriginal catches up to some 150 per year) to at or near its "original" (prewhaling) state (ca. 26,000 animals). Similarly, the northwest Atlantic humpback and several Southern Hemisphere humpback populations have been showing marked increases. The latest estimate of the North Atlantic stock, some 10,600 animals in 1993 (*cf.* 5500 in 1986), must reflect some population growth in the intervening period, whereas two Southern Hemisphere stocks (off eastern and western Australia) have been increasing steadily, at 10% or more per year since the early 1980s. Indeed, in all areas where surveys have been undertaken recently on Southern Hemisphere humpback populations they have been shown to be undergoing some recovery. Three southern right whale stocks (off eastern South America, South Africa, and southern Australia) have been increasing since the late 1970s at around 7–8% per year, although at some 3000, 3000, and 1200 animals, respectively, all are still well below their "original" stock size. Even the "true" blue whale, whose future has been of considerable concern, with estimates for the late 1980s at fewer than 500 animals for the whole Antarctic, has shown recent encouraging signs. Based on a series of Antarctic sightings cruises, mainly for minke whales but including other large whales, the most recent calculations (admittedly using only small absolute numbers sighted) show that the population must have been increasing since the last estimate (1991). As yet the analyses do not permit a firm conclusion on the number now present, although it seems likely to be more than 1000.

The one species or stock for which there is now very great concern is the North Atlantic right whale. At very low absolute abundance (only some 300 animals), not recovering despite protection from whaling in the 1930s, now even decreasing through a reduced survival rate and an increase in calving interval, and subject to increasing removals from ship strikes and fishing gear entanglement, the only way to ensure the species'

survival is to reduce such anthropogenic mortality (ship strike and entanglement) to zero. While research on mortality reduction measures should be pursued, immediate management action is urgently needed.

It has been calculated that the great reduction of baleen whales by whaling, for the Antarctic to around one-third of original numbers and one-sixth in biomass, must have left a large surplus of food—some 150 million tons per year—available for other consumers, such as seals, penguins, and fish. (In a different way, early whaling in the North Atlantic, particularly on right whales, is believed to have influenced the spread of one sea bird—the fulmar—by providing food in the form of discarded whale carcasses.) In response to an increase in available food, there may well have been increases in growth rates, earlier ages at maturity, and higher rates of pregnancy in some baleen whale species. However, the evidence is equivocal, as it is for competition between individual whale species. For some, e.g., right whales and sei, it has been suggested that an increase in one (right whales) could be inhibited by competition with another (sei whales). In the North Pacific, both sei and right whales can feed on the same prey—copepods—and sei whales can at times be "skimming" feeders, like right whales. However, evidence that they actually compete on the same prey, in the same area, at the same time, and even on the same prey patch is lacking. Similarly, there has been much debate and speculation on whether the recovery of the Southern Hemisphere "true" blue whale has been inhibited by an apparent increase in Antarctic minke whales. In that case, there may in fact be very little direct competition for food where the common prey is not limited in abundance (as in the Antarctic) and is available in large patches. The well-authenticated increases in the substantial annual rates for several stocks of Southern Hemisphere humpbacks and right whales and the possibility of at least a limited increase in numbers for the "true" blue whale suggest that such competition is unlikely, at least where, as in the Antarctic, food supplies are abundant.

See Also the Following Articles

Cetacean Life History ▪ Hunting of Marine Mammals ▪ Migration and Movement Patterns ▪ Toothed Whales

References

Bonner, W. N. (1980). "Whales." Blandford Press, Poole, Dorset.

Corkeron, P. J., and Connor, R. C. (1999). Why do baleen whales migrate? *Mar. Mamm. Sci.* **15**(4), 1228–1245.

Harrison, R., and Bryden, M. M. (eds.) (1988). *"Whales, Dolphins and Porpoises."* Intercontinental, Hong Kong.

IUCN (1996). "1996 Red List of Threatened Animals." IUCN, Gland, Switzerland.

Laws, R. M. (1977). Seals and whales of the Southern Ocean. *Phil. Trans. R. Soc. Lond. B* **279;** 81–96.

Leatherwood, S., and Reeves, R. R. (1983). "The Sierra Club Handbook of Whales and Dolphins." Sierra Club Books, San Francisco.

Rice, D. W. (1998). "Marine Mammals of the World: Systematics and Distribution." Special Publication Number 4, the Society for Marine Mammalogy. Lawrence, KS.

Ridgway, S. H., and Harrison, R. (1985). "Handbook of Marine Mammals," Vol. 3.

Baleen Whales, Archaic

NICHOLAS C. FRASER AND ALTON C. DOOLEY, JR.
Virginia Museum of Natural History, Martinsville

The modern mysticetes (baleen whales) are typically characterized by the development of specialized epithelial tissue in the roof of the mouth that becomes keratinized: the so-called baleen. The baleen is used to filter organisms from water or sediment, ranging in size from microscopic copepods to crustaceans and fish as much as 5 cm in length. There is evidence for baleen in whales dating from the Oligocene, but some archaic mysticetes retained functional teeth all the way to the late Oligocene. Modern baleen whales do not have functional teeth, although incipient teeth are found in embryos. Archaic baleen whales are considered to be those mysticetes, both toothed and baleen-bearing, that do not belong to extant families. They comprise four named families: Mammalodontidae, Llanocetidae, Aetiocetidae, and Cetotheriidae, with over 20 named genera.

Modification of the rostrum is a characteristic of even the earliest mysticetes. The rostral elements extend backward to the orbits and are expanded laterally, while at the same time the occipital region extends forward. The telescoping of the skull also occurs in odontocetes, but there is an asymmetry in these forms, and the rostral elements tend to overide the frontals.

The remainder of the skeleton of mysticetes is greatly simplified. The mandibular symphysis is lost, with the two dentaries held together by a ligament. The sternum is greatly reduced. The hindlimbs are almost totally absent, the only remnant being a small splint of bone thought to be homologous with the pelvis, which no longer articulates with the vertebral column. There are trends toward compression, and ultimately fusion, in the cervical vertebrae and reduction of rib–vertebrae articulations.

I. Origins and Affinities

Although typically considered as a monophyletic group distinct from the odontocetes, certain molecular studies have indicated that mysticetes are the sister group to physeterids (sperm whales) and nested within a paraphyletic Odontoceti. This relationship was first suggested on the basis of mitochondrial ribosomal DNA sequences. However, other molecular studies (DNA–DNA hybridization, karotypes, and cytochrome DNA sequences) support the more traditional, morphology-based classification. Further support for odontocete and mysticete monophyly comes from studies of the inner ear of a number of Miocene and Pliocene "cetotheres" that shows that they did not have the high-frequency hearing that permits echolocation, and which is so typical of odontocetes.

Recent discoveries indicate that baleen whales evolved directly from archaeocete ancestors. The Llanocetidae from the late Eocene or early Oligocene of the South Pacific are heterodont and may have used their denticulate and palmate cheek teeth for filter feeding. Other primitive mysticetes from the late Oligocene of South Carolina include two heterodont genera but with skull development similar to later baleen whales. Homodont toothed mysticetes are seen in the Aetiocetidae (Fig. 1), also from the late Oligocene. This group was quite diverse, with eight species in three genera described to date. The aetiocetids retain marginal teeth in both upper and lower jaws, yet exhibit the loose mandibular symphysis, which is a characteristic feature of all later mysticetes. Another late Oligocene form is *Mammalodon*, which has an unusually short rostrum unlike any other mysticete.

Even though these toothed mysticetes are too young to be actual ancestors to the cetotheres, which also dated from the early Oligocene, they may indicate the pattern of mysticete evolution. The earliest fossils of baleen-bearing whales come from the early Oligocene of New Zealand. While baleen does not fossilize, the pattern of skull vascularization in these earliest members is consistent with those of living mysticetes.

II. Diversification of Mysticetes

The earliest records of both odontocetes and mysticetes are in early Oligocene sediments of New Zealand. While in the Northern Hemisphere cetaceans are commonly considered to have had limited diversity during Oligocene times, in the Austral realm both odontocetes and mysticetes radiated, while at the same time the archaeocetes declined. These radiations have been attributed to a series of major geographical and climatic changes. First, during the latest Eocene and earliest Oligocene major continental glaciations developed, probably together with sea ice, which resulted in marked temperature drops in the oceans. At the same time with the opening of the Drake Passage between Antarctica and South America the Circum-Antarctic Current was established. Coupled with the initiation of the psychosphere this brought about increases in

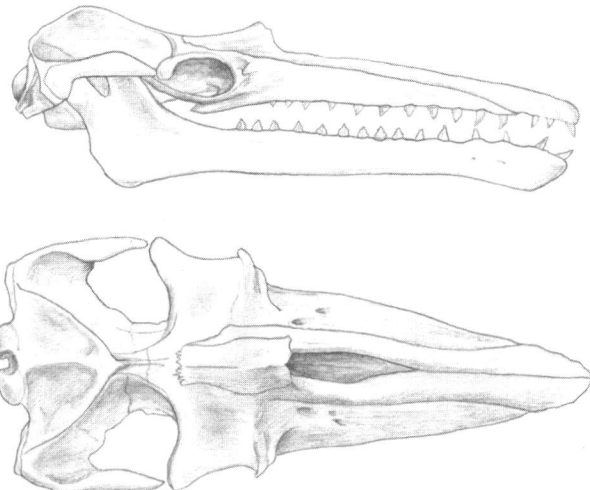

Figure 1 Aetiocetus polydentatus, *an aetiocetid from the late Oligocene Morawan Formation of Japan. Top, skull and mandible, left lateral view. Bottom, skull, dorsal view. Length of skull approximately 55 cm. After Barnes* et al. *(1994).*

local upwelling and areas of high productivity offshore. This may have led to the development of a FILTER-FEEDING lifestyle in the mysticetes and the eventual appearance of baleen.

Thus by the middle Miocene the mysticetes had peaked in diversity, with some 15 known genera compared to only 5 today. Moreover, the modern whales are largely characterized on the basis of external features. Clearly this is not possible with fossil forms, and it might be argued that solely on the basis of skeletal morphology that certain modern whales species would not have been separated as fossils. On this basis, and assuming that fossil forms also exhibited similar external variation, then the diversity of the Miocene cetotheres may have been even greater than is usually considered.

Most early mysticetes (particularly those from the late Oligocene to early Pliocene) are included in a single family, the Cetotheriidae (Fig. 2). However, this family has been acknowledged to be a "catch-all" assemblage of different taxa that really reflects our poor understanding of the early evolution of baleen whales. Cetothere species are mostly differentiated on the basis of skull proportions and, in some cases, differences in the ear region. No clearly derived shared characters have been recognized in any cetotheres that would permit a better understanding of the relationships within this family.

The cetothere grade whales still exhibit strong vertebral–rib articulations and do not have fusion of the cervical vertebrae. The cervicals tend to be quite compressed, and the hindlimbs are apparently absent. The skull is relatively smaller in the ce-

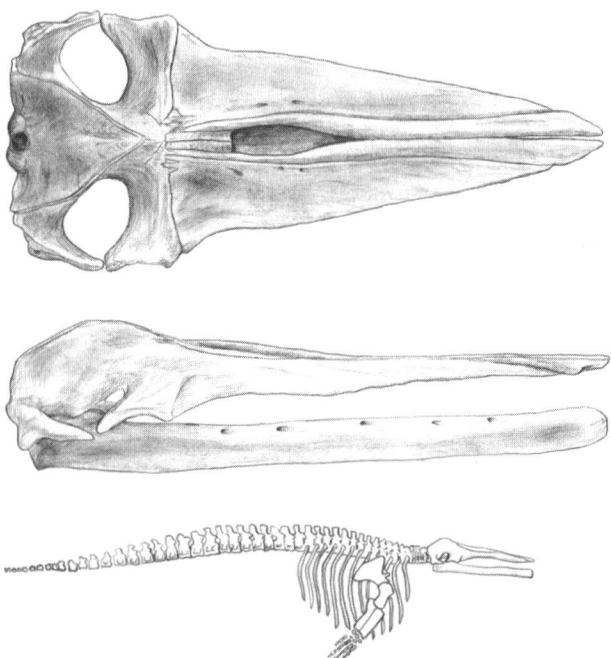

Figure 2 Parietobalaena palmeri, *a cetothere from the middle Miocene Calvert Formation of Maryland and Virginia. (Top) Skull, dorsal view. (Middle) Skull and mandible, lateral view. (Bottom) Skeletal reconstruction. Length of skull approximately 120 cm. A and B after Kellogg (1965).*

totheres than in the modern mysticetes, and the rostrum tends to be broad and tapering, making it more similar in outline to that of the Balaenopteridae and Eschrichtiidae than the Balaenidae. Cetotheres are also quite small. Body lengths in the Miocene range from about 3 m up to about 11 m (although only one specimen greater than 10 m is known). In modern mysticetes, adult body lengths are from 6 to 30 m, but only three species are normally less than 12 m.

Cetotheres have a worldwide distribution with a number of well-established genera, usually based at least on partial skulls. From eastern North America are *Pelocetus, Parietobalaena, Agalocetus, Diorocetus, Metopocetus, "Mesocetus," Thinocetus, Halicetus,* and *Cephalotropis.* From western North America are *Parietobalaena, Agalocetus, Herpetocetus, Tiphyocetus, Peripolocetus,* and *Cophocetus.* South American genera include *Morenocetus, Agalocetus,* and *"Plesiocetus."* European genera include *Mesocetus, Herpetocetus, Isocetus, Cetotheriopsis,* and *Cetotherium. Imerocetus* and *Cetotherium* are known from the Tethyan realm. From New Zealand is *Mauicetus* and from Japan is *Herpetocetus.* In addition, there are several unnamed specimens from most of these areas, some of which may represent new genera. There are also many nominal genera based on fragmentary remains.

III. Emergence of Modern Mysticete Fauna

The first members of some modern groups appeared in the middle part of the Miocene. These faunal changes may be correlative with worldwide environmental change and physical changes in the oceans. Among these changes were the establishment of a West Antarctic ice sheet, closure of the western opening of the Mediterranean Sea, and closure of the Indo-Pacific Seaway, all of which may have contributed to cooling, steeper temperature gradients and more complex ocean current patterns. In turn this would have resulted in a variety of pelagic habitats and increased partitioning of food resources. Thus in the same way that major changes in ocean currents resulted in the radiation of the "archaic" mysticetes (cetotheres) during the early and middle Oligocene, another major shift in ocean currents and sea temperatures could have initiated the radiation of modern mysticetes and brought about the demise of the archaic forms.

It is generally thought that all four modern mysticete families have separate origins among the cetotheres, but this is still speculative. The modern baleen whales [Balaenopteridae (rorquals and humpback whale, *Megaptera novaeangliae*), Balaenidae (right whales), Eschrichtiidae (gray whale, *Eschrichtius robustus*), and Neobalaenidae (pygmy right whale, *Caperea marginata*) all have have derived characters not presently known from any cetothere. All the modern families have weak-to-absent rib–vertebrae articulations and skulls which represent at least 25% of the body length. The balaenopterids have maxillary spurs that extend posteriorly along the top of the skull, and the supraorbital portions of the frontals are depressed well below the top of the skull. *Eschrichtius* has enlarged nasals and posterior ends of the premaxillae, a moderately arched rostrum, and posterior maxillary processes similar to balaenopterids, and there is no coronoid process on the mandible. The balaenids are perhaps the most divergent mysticetes, with an extremely arched and narrow rostrum, an anteroposteriorly compressed cranium, divergent zygomatic processes, the supraoccipital extending to a point anterior to the orbit, and fused cervical vertebrae. *Caperea* (Neobalaenidae) has a rostrum less arched and broader than in balaenids, although the supraoccipital extends anterior to the orbit. The balaenids and balaenopterids have fossil records extending to the Miocene, but it is not clear how they are related to the cetotheres. *Caperea* is thought to be close to the balaenids, but it does not have a fossil record. *Eschrichtius* is known only from the Quaternary. The body proportions of *Eschrichtius* are more similar to the cetotheres than to other modern mysticetes, and they still have strong rib articulations and a relatively long neck, making this genus more primitive than any of the other modern mysticetes.

Research programs on Oligocene and Miocene mysticetes are very active and new finds are continually being made. For example, additional finds from the Oligocene of New Zealand, Japan, and South Carolina will help shed further light on the origin of baleen whales from the archaeocetes, and exciting new discoveries in Virginia promise to elucidate separate lineages within the cetothere grade. One new Miocene cetothere appears to be more derived than other contemporary cetotheres in having compressed cervical vertebrae and exhibiting reduction of the capitulum. Furthermore, characters of the periotic indicate affinities with modern balaenopterids.

See Also the Following Articles

Cetacean Evolution ■ Neoceti ■ Systematics, Overview

References

Barnes, L. G., Kimura, M., Furusawa, H., and Sawamura, H. (1994). Classification and distribution of Oligocene Aetiocetidae (Mammalia; Cetacea; Mysticeti). *Island Arc* **3**, 392–431.

Fordyce, R. E. (1980). Whale evolution and Oligocene southern ocean environments. *Palaeogeogr. Palaeoclimatol. Palaeoecol.* **31**, 319–336.

Fordyce, R. E. (1992). Cetacean evolution and Eocene/Oligocene environments. *In* "Eocene-Oligocene Climatic and Biotic Evolution," (D. Prothero and W. Berggren, eds.), pp. 368–381. Princeton Univ. Press, Princeton, NJ.

Fordyce, R. E., and Barnes, L. G. (1994). The evolutionary history of whales and dolphins. *Annu. Rev. Earth Planet. Sci.* **22**, 419–455.

Geisler, J. H., and Luo, Z. (1996). The petrosal and inner ear of *Herpetocetus* sp. (Mammalia: Cetacea) and their implications for the phylogeny and hearing of archaic mysticetes. *J. Paleontol.* **70**, 1045–1066.

Kellogg, R. (1965). Fossil marine mammals from the Miocene Calvert formation of Maryland and Virginia, Pt. 1: A new whalebone whale from the Miocene Calvert Formation. *Bull. U.S. Nat. Mus.* **247**, 1–45.

Kellogg, R. (1969). Cetothere skeletons from the Miocene Choptank Formation of Maryland and Virginia. *Bull. U.S. Nat. Mus.* **297**, 1–40.

Mitchell, E. D. (1989). A new cetacean from the Late Eocene La Meseta Formation, Seymour Island, Antarctic Peninsula. *Can. J. Fish. Aqu. Sci.* **46**, 2219–2235.

Barnacles

DAGMAR FERTL
Minerals Management Service, U.S. Department of the Interior, New Orleans, Louisiana

Barnacle is the common name for over 900 marine species of the subclass Cirripedia. Barnacles are unique among crustaceans in being sessile; they attach to a variety of inanimate and animate objects. Barnacles live in polar regions of the world, as well as tropical and temperate waters. The principal superorder is Thoracica, which is composed of stalked (order Lepadomorpha) and sessile (order Balanomorpha) barnacles. At least 20 barnacle species have some association with marine mammal species; those barnacles also belong to Thoracica. Barnacles attaching to marine mammals are often referred to as ectoparasites. In actuality, the barnacles do not parasitize the whale, but use their host as a substrate from which they feed on plankton in passing water. Epizoic is a more appropriate term describing this lifestyle. Barnacles attached to marine mammals have been described as an example of phoresis bordering on commensalism.

I. Life History

Barnacles have been described as "nothing more than a little shrimp-like animal, standing on its head in a limestone house and kicking food into its mouth." The barnacle's life cycle includes six free-swimming planktonic stages (nauplius) progressing by molts into a cypris (final larval stage), which searches for a place to settle. After settling, the cypris secretes cement from glands located in the base of the first antenna to anchor itself and metamorphoses to grow and molt. After attachment, the barnacle then begins the secretion of calcareous plates that become its home. Through an opening in the plates, six pairs of feathery, leg-like appendages (cirri) can emerge and spread out to sweep through the water, like a net, to entrap planktonic organisms. Most barnacles are hemaphroidites, i.e., each individual possesses reproductive structures of both sexes. The breeding season of barnacles that cling to whales is synchronous with that of the whales' breeding season.

II. Sessile Barnacles

The balanomorphs or sessile barnacles are stalkless; the barnacle "shell" attaches directly to a surface. Marine mammals have interactions with *Balanus*, *Cetopirus*, *Chelonibia*, *Coronula*, *Cryptolepas*, *Platylepas*, *Tubicinella*, and *Xenobalanus*. Three of these, *Xenobalanus*, *Platylepas*, and *Tubicinella* superficially resemble stalked barnacles, but are actually aberrant sessile barnacles and can be considered pseudo-stalked. Because of their superficial resemblance to acorns of oak trees, coronuline (*Coronula* sp.) barnacles are called acorn barnacles.

III. Stalked Barnacles

The stalked, pedunculate, or goose barnacle is considered to be the more primitive. This barnacle is mounted on a muscular, flexible stalk (peduncle) that is attached to a firm base, and the upper portion of the animal is the major part of the body, bearing shell plates. *Conchoderma*, *Lepas*, and *Pollicipes* are all commensals of marine mammals.

IV. Barnacles and Marine Mammals

Barnacles appear to settle in greatest numbers on slower-moving cetaceans, as evidenced by the great number of barnacles found on gray whales (*Eschrichtius robustus*) and other mysticetes, compared to delphinids. Bottlenose (*Tursiops truncatus*) and striped (*Stenella coeruleoalba*) dolphins involved in mass mortality events on the U.S. Atlantic coast and in the Mediterranean, respectively, had an abundance of barnacles that might have been due to the reduced movement of sick dolphins and/or impaired immune functioning of the skin.

Cryptolepas rhachianecti is consider to be host-specific to gray whales but has been found on a killer whale (*Orcinus orca*) that stranded in southern California and on belugas (*Delphinapterus leucas*) housed in San Diego Bay. *Xenobalanus* is almost always found on the trailing edges of the dorsal and pectoral fins and on the tail flukes of cetaceans (Fig. 1). *Platylepas* may be mistaken for *Xenobalanus*; there is one record of *Platylepas hexastylos* (found associated with *Xenobalanus*) from a bottlenose dolphin. *Tubicinella major* has been found buried cryptically among callosities of southern right whales (*Eubalaena australis*). A *Tubicinella* sp. has been collected from the flank of a stranded northern bottlenose whale (*Hyperoodon ampullatus*). *Coronula* is a large barnacle that attaches almost exclusively to the skin of mysticetes and is rarely found on odontocetes. *C. reginae* and *C. diadema* are common epizoites of humpback whales (*Megaptera novaeangliae*). Acorn barnacles can be found attached to flukes, flippers, ventral grooves, genital slit, and head of humpback whales (Fig. 2). One humpback was reported to have as much as 450 kg of acorn barnacles attached to it. Humpback males scrape each other with their barnacle-encrusted flippers on the breeding grounds. The two *Conchoderma* species (*C. auritum* and *C. virgatum*) are the only true stalked barnacles recorded from cetaceans, with the exception of one record of *Pollicipes polymerus* that was found associated with those two species on a humpback whale. The stalked barnacle *Conchoderma* requires a hard surface for attachment. *Conchoderma auritum* is identified by its rabbit-eared appendages. *C. auritum* may be found at a site where teeth are exposed and unprotected, such as on erupted teeth of adult male beaked and bottlenose whales, or because of a malformation (including bone injury) in the jaw (Fig. 3). *Conchoderma* is less commonly found on BALEEN plates and has been once recorded from the penis of a stranded sperm whale (*Physeter macrocephalus*). *C. auritum* is also often found attached to *Coronula* (most commonly to *C. diadema* and much more rarely to the barnacle *Pollicipes*). *C. virgatum* is never attached directly to a cetacean, but is epizoic on other barnacles, usually *C. auritum*, as well as the parasitic copepod *Pennella*, which may attach to cetaceans. *Cetopirus complanatus* closely resembles *C. reginae*; there are only two well-documented records from southern right whales. *Lepas* usually occurs on floating objects, yet *L. pectinata* and *L. hilli* have been found between the TEETH of some Mediterranean striped dolphins.

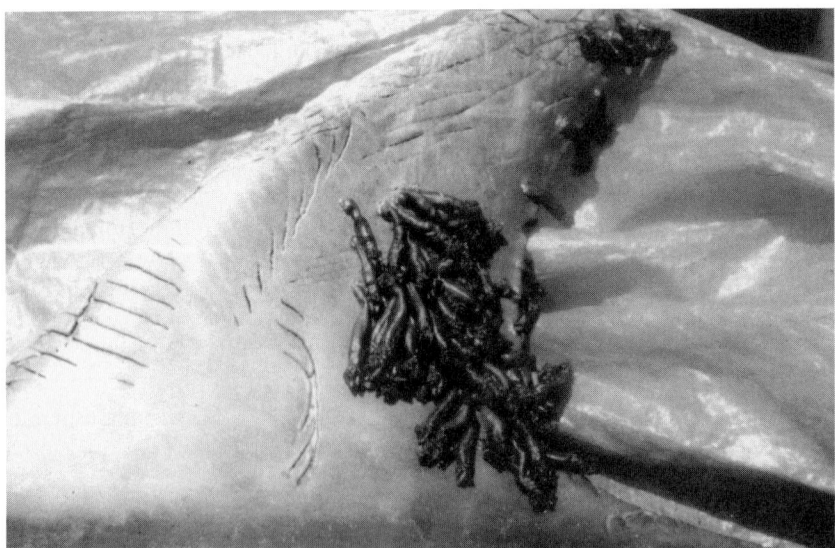

Figure 1 Xenobalanus *attached to the dorsal fin of a bottlenose dolphin. Courtesy of V. Thayer and K. Rittmaster, NC Maritime Museum.*

Figure 2 *Humpback whale with acorn barnacles* (Coronula) *and* Conchoderma au-ritum *attached to the barnacles. Also visible are white-rim scars from acorn barnacles that have dropped off. (Courtesy of Yuki Ogino, off California, 1999.)*

Figure 3 Conchoderma *barnacles attached to the teeth/jaw of a pantropical spotted dolphin* (Stenella attenuata). *(Courtesy of W. F. Perrin.)*

There are few published records of barnacles on pinnipeds. Three species of *Lepas* (*L. pacifica*, *L. australis*, and *L. hilli*) and *Conchoderma auritum* have been recorded from the dorsal body surface of various seal species. Manatees may acquire barnacles when in salt water; these barnacles die in fresh water and drop off, leaving scars. The common barnacle found embedded in the skin of West Indian (*Trichechus manatus*) and West African (*T. senegalensis*) manatees is *Chelonibia manati*. *Platylepas hexastylos* has been found on dugongs (*Dugong dugon*) and West Indian manatees. *Balanus amphitrite*, *B. eburneus*, *B. reticulatus*, *B. trigonus*, and *B. improvisus* attach to *Chelenibia* on manatees, but not directly to the skin. When population of preferred prey are reduced, sea otters (*Enhydra lutris*) in California and Alaska will eat *Balanus nubulis*.

See Also the Following Articles

Callosites ■ Health ■ Parasites ■ Plankton

References

Anderson, D. T. (1994). "Barnacles: Structure, Function, Development and Evolution." Chapman & Hall, New York.

Aznar, F. J., Balbuena, J. A., and Raga, J. A. (1994). Are epizoites biological indicators of a western Mediterranean striped dolphin die-off? *Dis. Aquat. Organ.* **18**, 159–163.

Best, P. B. (1971). Stalked barnacles *Conchoderma auritum* on an elephant seal: Occurrence of elephant seals on South African coast. *Zool. Afr.* **6**, 181–185.

Clarke, R. (1966). The stalked barnacle *Conchoderma*, ectoparasitic on whales. *Norsk Hvalfangsttidende* **55**, 153–168.

Newman, W. A., and Ross, A. (1976). Revision of the balanomorph barnacles; including a catalog of the species. Memoirs of the San Diego Society of Natural History, No. 9.

Raga, J. A. (1994). Parasitismus bei den Cetacea. *In* "Handbuch der Säugetiere Europas Band 6: Meeressäuger, Teil 1A: Wale and Delphine 1 (D. Robinuea, R. Duguy, and M. Klima, eds.), pp. 133-179. AULA-Verlag, Wiesbaden.

Ridgway, S. H., Lindner, E., Mahoney, K. A., and Newman, W. A. (1997). Gray whale barnacles *Cryptolepas rhachianecti* infest white whales, *Delphinapterus leucas*, housed in San Diego Bay. *Bull. Mar. Sci.* **61**, 377–385.

Ruppert, E. E., and Barnes, R. D. (1994). "Invertebrate Zoology," 6th Ed. Saunders, New York.

Scarff, J. E. (1986). Occurrence of the barnacles *Coronula diadema, C. reginae* and *Cetopirus complanatus* (Cirripedia) on right whales. *Sci. Rep. Whal. Res. Instit.* **37**, 129–153.

Basilosaurids

MARK D. UHEN
Cranbrook Institute of Science, Bloomfield Hills, Michigan

Basilosaurids are a paraphyletic group of archaeocete cetaceans known from the late middle to early late Eocene of all continents except Antarctica and South America. The family includes 11 species in 8 genera in 2 subfamilies, although some authors elevate the subfamilies to familial rank. They range in size from around 4 m (*Saghacetus osiris*) to around 16 m (*Basilosaurus cetoides*). Basilosaurids are probably the earliest fully aquatic cetaceans (Uhen, 1998) and are thought to have given rise to modern cetaceans (Barnes *et al.*, 1985; Uhen, 1998).

I. Basilosaurid Characteristics

Like all archaeocetes, basilosaurids lack telescoping of the skull like that seen in modern mysticetes or like that seen in modern odontocetes (see Fig. 1; Miller, 1923). In addition, basilosaurids are diphyodont, lack polydonty, and retain a heterodont dentition in which incisors, canines, premolars, and molars are easy to distinguish based on their morphologies (Kellogg, 1936; Uhen, 1998).

Basilosaurids also share a number of characteristics that distinguish them from other archaeocetes. All basilosaurids lack upper third molars and the upper molars lack protocones, trigon basins, and lingual third roots. Also, the cheek teeth of basilosaurids have well-developed accessory denticles (Fig. 1). The hindlimbs of basilosaurids are greatly reduced (see Fig. 2; Gingerich *et al.*, 1990; Uhen and Gingerich, 2000) and lack a bony connection to the vertebral column. Basilosaurids also lack sacral vertebrae, although vertebrae that are likely to be homologs of sacral vertebrae are identifiable (Kellogg, 1936; Uhen, 1998).

Other characteristics may be found only in basilosaurids (within archaeocetes) but are currently not known from other archaeocetes. For instance, basilosaurid forelimbs had broad, fan-shaped scapulae with the distal humerus, radius, and ulna flattened into a single plane (Fig. 2). In addition, the elbow joint motion was restricted to the same plane, and pronation and supination of the forelimb were not possible based on the articular surfaces of the distal humerus, proximal radius, and proximal ulna. Because forelimbs are unknown in protocetids (Thewissen *et al.*, 1994), it is unclear whether these features are found only in basilosaurids or whether they are characteristic of a larger group.

Some of the characteristics of basilosaurids can be seen in some protocetid archaeocetes, such as *Georgiacetus*. Although the innominate of *Georgiacetus* is large, it does not appear to have been connected to the vertebral column (Hulbert, 1998). None of the vertebrae are fused into a sacrum, yielding a condition similar to that seen in basilosaurids. In addition, the cheek teeth of *Georgiacetus* have small accessory denticles, somewhat different from those in basilosaurids, but certainly larger than any of the serrations seen in other nonbasilosaurid archaeocetes.

II. Taxonomy

Taxonomy for the family Basilosauridae is after Uhen (1998). References therein and in Kellogg (1936) are the basis for the discussion of each taxon. Basilosaurinae and Dorudontinae are included here in the single family Basilosauridae following Miller (1923) because a single character state (elongate trunk vertebrae) distinguishes basilosaurines from dorudontines

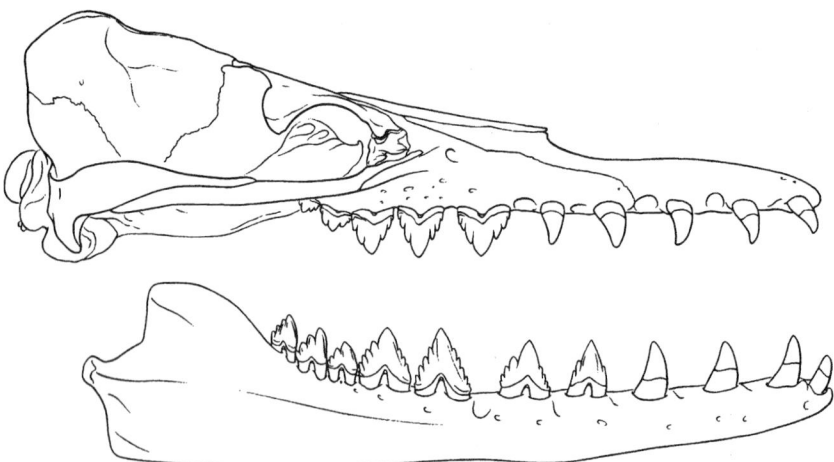

Figure 1 *Skull and lower jaw of* Dorudon atrox, *lateral view. This drawing is a composite drawn from specimens of* D. atrox *at the University of Michigan Museum of Paleontology by Bonnie Miljour.*

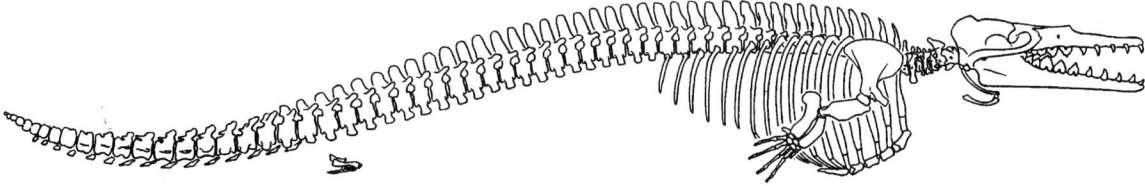

Figure 2 *Skeleton of* Dorudon atrox, *lateral view. This drawing is a composite drawn from specimens of* D. atrox *at the University of Michigan Museum of Paleontology by Bonnie Miljour.*

(Uhen, 1998). The names "zeuglodonts" and "zeuglodons" are often used colloquially to refer to basilosaurines or archaeocetes in general. These common names are derived from the generic name *"Zeuglodon"* (discussed later), and their usage should be avoided in formal works.

A. Basilosauridae Cope 1868

1. Basilosaurinae Cope 1868 Basilosaurines are basilosaurids with elongate posterior thoracic, lumbar, and anterior caudal vertebrae. All known basilosaurids are also considerably larger than all known dorudontines, with the exception of *Pontogeneus* (see later).

A. *BASILOSAURUS* HARLAN 1834 *Basilosaurus* was the first archaeocete whale named. The name was first coined in 1834 by Richard Harlan, who mistook the large vertebrae for those of a large marine reptile, thus the misnomer of "king lizard" for a cetacean. This mistake was pointed out by Richard Owen in 1842, when he attempted to rename the animal *Zeuglodon cetoides*. *Zeuglodon* is considered a junior subjective synonym of *Basilosaurus*, but it has been applied to so many archaeocete whales as to have become a common name for archaeocete or basilosaurid. *Basilosaurus* contains three species from the late middle and early late Eocene: *B. cetoides* is known from the southeastern United States, *B. isis* is known from Egypt and Jordan, and *B. hussaini* is known from Pakistan (Uhen, 1998).

B. *BASILOTERUS* GINGERICH ET AL. 1997 *Basiloterus* can be distinguished from *Basilosaurus* based on its antero-posteriorly long neural arch and more anteriorly-projecting metapophyses (Gingerich *et al.*, 1997). *Basiloterus* contains a single species, *B. drazindai*, based on a single lumbar vertebra from the late middle Eocene of Pakistan.

2. Dorudontinae Miller 1923

A. *DORUDON* GIBBES 1845 The genus *Dorudon* was erected in 1845 by Gibbes for a specimen of a small archaeocete that he dubbed *D. serratus*. This specimen is of a juvenile individual with deciduous teeth, making it difficult to compare to other specimens of adult individuals. Nonetheless, the number of species in *Dorudon* grew when Kellogg (1936) removed a number of species from the genus *Zeuglodon* and placed them in *Dorudon*. Subsequently, many of these species were synonymized and/or placed in other genera. Only *D. serratus* from the late Eocene of the southeastern United States and *D. atrox* (formerly *Prozeuglodon atrox*) from the late Eocene of Egypt remain (Uhen, 1998).

B. *ZYGORHIZA* TRUE 1908 The genus *Zygorhiza* was erected in 1908 by True for specimens of a small archaeocete from North America that he felt were different from *Dorudon*

serratus. Some of these specimens had been part of Koch's *Hydrarchos* and had been called by many different names (Kellogg, 1936). *Zygorhiza kochii* can be distinguished from all other dorudontines by the presence of well-developed cuspules on the cingula of the upper premolars. *Zygorhiza* currently includes *Z. kochii* from the late Eocene of the southeastern United States and *Zygorhiza* sp. from New Zealand, as European specimens assigned to *Zygorhiza* are identified more appropriately as Dorudontinae incertae sedis (Uhen, 1998).

C. *PONTOGENEUS* LEIDY 1852 The name *Pontogeneus* was first used by Leidy for a single cervical vertebrae of a large basilosaurid archaeocete from Louisiana, which he dubbed *Pontogeneus priscus*. Some of the vertebrae of Koch's *Hydrarchos* were large in size, but not elongate like those of *Basilosaurus*, and were given the name *Zeuglodon brachyspondylus*. Kellogg (1936) suggested that these two taxa were the same and used Leidy's generic name and the specific epithet *brachyspondylus* for the new combination. Currently, the concept of *P. brachyspondylus* is a bit fuzzy, but basilosaurids of large size that lack elongate vertebrae are found in the late Eocene of the southeastern United States and Egypt to which this name could be applied (Uhen, 1998).

D. *SAGHACETUS* GINGERICH 1992 The generic name *Saghacetus* was coined in 1992 to subsume the former species *Dorudon osiris*, *Dorudon zitteli*, *Dorudon sensitivius*, and *Dorudon elliotsmithii* within a single species, *Saghacetus osiris*. *S. osiris* can be distinguished from other dorudontines based on its small size and its slightly elongate lumbar and anterior caudal vertebrate. *S. osiris* is known only from the late Eocene of Egypt.

E. *ANCALECETUS* GINGERICH AND UHEN 1996 *Ancalecetus* includes one species, *A. simonsi*, which is similar to *Dorudon atrox*, but has greatly modified forelimbs that were highly restricted in their range of motion. *A. simonsi* is known from the late Eocene of Egypt (Uhen, 1998).

F. *CHRYSOCETUS* *Chrysocetus* includes one species, *C. healyorum*, which differs from all other dorudontines in the smoothness of the tooth enamel, height of the premolar crowns, and eruption of its adult teeth in a skeletally juvenile state. *Chrysocetus* is also the only dorudontine for which the innominate is known. *Chrysocetus* is known from the late Eocene of South Carolina (Uhen and Gingerich, 2000).

B. Questionable Basilosaurids

Excluded from this list is the genus *Gaviacetus*, which was referred to the Basilosauridae by Bajpai and Thewissen (1998). The identification of *Gaviacetus* as a basilosaurid was based on the likely absence of upper third molars in both the type specimens of *Gaviacetus razai* and the type specimen of *Gaviacetus sahnii*

Figure 3 *Cladogram of basilosaurids, selected nonbasilosaurid archaeocetes, mysticetes, and odontocetes.*

(Bajpai and Thewissen, 1998). In addition, Bajpai and Thewissen (1998) referred some elongate vertebrae to *G. sahnii*, further supporting their placement of *Graviacetus* in the Basilosauridae. Since no specimen of *Gaviacetus* clearly shows that the upper third molar is absent or that any of the cheek teeth have accessory denticles and since reference of postcrania to unassociated cranial material has proven problematic in the past, the author prefers to leave *Gaviacetus* in the Protocetidae as it was originally described until it can be shown clearly to have basilosaurid synapomorphies.

Species that may not be basilosaurids are *Basilosaurus hussaini* and *Basiloterus drazindai*. These species (as well as the genus *Basiloterus*) are based solely on one and two vertebrae, respectively. These vertebrae are thought to represent basilosaurines because they are elongate, like the vertebrae of *Basilosaurus*. While this feature is a distinguishing characteristic of Basilosaurinae within Basilosauridae, it is clear that vertebral elongation is not restricted to basilosaurids. *Eocetus*, a protocetid from Egypt and North America, also has elongate vertebrae, although they are not as elongate as those of *Basilosaurus*. It is possible that *B. hussaini* and *B. drazindai* are also protocetids. Once cranial or dental material associated with vertebrae is found, it will be obvious whether these taxa should be retained in Basilosauridae.

III. Phylogenetic Relationships

The phylogenetic relationships among basilosaurids and their relationships to other archaeocetes, mysticetes, and odontocetes are shown in Fig. 3. Many of the character state transformations that occur between basilosaurids and protocetids are associated with the adoption of a fully aquatic existence, such as presence of pterygoid air sinuses, extreme reduction of the hindlimb, loss of the sacrum, increase in the number of trunk vertebrae, and the presence of dorsoventrally flattened posterior caudal vertebrae (Uhen, 1998). Others, such as the loss of M^3, loss of lingual roots on the upper molars, and the

development of accessory denticles on the cheek teeth, have to do with changes in feeding that are not as easy to interpret.

Within Basilosauridae, basilosaurines are united by the presence of elongate trunk vertebrae, which dorudontines lack. *Pontogeneus* may be the sister taxon to Basilosaurinae based on its large size. Each genus of dorudontine is distinguishable from other genera based on the presence of autapomorphies, but it is difficult to confidently link any of the genera based on any clear synapomorphies. The result is a polytomous relationship among the genera or an imbalanced tree with Mysticeti + Odontoceti nested well within Dorudontinae. *Chrysocetus* is preferred as the sister taxon to Mysticeti + Odontoceti based on the interpretation of it and early mysticetes and odontocetes as monophyodont. It is hoped that some of the relationships among basilosaurids will become more secure as more of the anatomy of more of the species becomes known.

See Also the Following Articles

Archaeocetes, Archaic ▪ Dental Morphology, Evolution of ▪ Origins of Marine Mammals

References

Bajpai, S., and Thewissen, J. G. M. (1998). Middle Eocene cetaceans from the Harudi and Subathu Formations of India. *In* "The Emergence of Whales" (J. G. M. Thewissen, ed.), pp. 213–234. Plenum Press, New York.

Barnes, L. G., Domning, D. P., and Ray, C. E. (1985). Status of studies on fossil marine mammals. *Mar. Mamm. Sci.* **1**, 15–53.

Gingerich, P. D., Smith, B. H., and Simons, E. L. (1990). Hind limbs of Eocene *Basilosaurus:* Evidence of feet in whales. *Science* **229**, 154–157.

Hulbert, R. C. J. (1998). Postcranial osteology of the North American Middle Eocene protocetid *Georgiacetus*. *In* "The Emergence of Whales" (J. G. M. Thewissen, ed.), pp. 235–268. Plenum Press, New York.

Kellogg, R. (1936). A review of the Archaeoceti. *Car. Inst. Wash.* **482**, 1–366.

Uhen, M. D. (1998). Middle to Late Eocene Basilosaurines and Dorudontines. *In* "The Emergence of Whales" (J. G. M. Thewissen, ed.), pp. 29–61. Plenum Press, New York.

Uhen, M. D., and Gingerich, P. D. (2000). New genus of dorudontine archaeocete (Cetacea) from the middle-to-late Eocene of South Carolina. *Marine Mammal Science* **17**, 1–34.

Beaching

SEE *Stranding*

Beaked Whales, Overview

Ziphiidae

JAMES G. MEAD

National Museum of National History, Smithsonian Institution, Washington, DC

Beaked whales belong to the odontocete family Ziphiidae. They are medium-sized cetaceans, with adults ranging from 3 to 13 m. They are characterized by a reduced dentition, elongate rostrum, accentuated cranial vertex, and enlarged pterygoid sinuses. There are currently 20 recognized species in five genera. They are all pelagic, living in the open oceans and feeding on deep water squid and fish.

I. Classification

Family Ziphiidae
  Subfamily Ziphiinae

Berardius arnuxii	Arnoux's beaked whale
Berardius bairdii	Baird's beaked whale
Tasmacetus shepherdi	Shepherd's beaked whale
Ziphius cavirostris	Cuvier's beaked whale

  Subfamily Hyperoodontinae

Hyperoodon ampullatus	Northern bottlenose whale
Hyperoodon planifrons	Southern bottlenose whale
Indopacetus pacificus	Longman's beaked whale
Mesoplodon bahamondi	Bahamonde's beaked whale
Mesoplodon bidens	Sowerby's beaked whale
Mesoplodon bowdoini	Andrews' beaked whale
Mesoplodon carlhubbsi	Hubbs' beaked whale
Mesoplodon densirostris	Blainville's beaked whale
Mesoplodon europaeus	Gervais' beaked whale
Mesoplodon ginkgodens	Ginkgo-toothed beaked whale
Mesoplodon grayi	Gray's beaked whale
Mesoplodon hectori	Hector's beaked whale
Mesoplodon layardii	Strap-toothed whale
Mesoplodon mirus	True's beaked whale
Mesoplodon peruvianus	Pygmy beaked whale
Mesoplodon stejnegeri	Stejneger's beaked whale

II. Common Names

The common name of the family, beaked whales, refers to their pronounced rostrum or beak. The rostrum of beaked whales is, admittedly, relatively shorter than in most dolphins but relatively longer than most "whales." The origin of the English term "beaked whale" is a translation of the Norwegian *nebhval*, which means "whale with a beak." It is an extremely old Norwegian word, the origin of which is long lost to history.

Most beaked whales are encountered rarely enough that they do not have "common names" but rather "vernacular names." There seems to be a tendency toward the feeling that the Latin names that are used in science are somehow confusing and scientists will have to coin vernacular names in the language that is used by the people.

The only beaked whales that were seen on a regular basis by fishermen (and whalers) were the northern bottlenose whale (*Hyperoodon ampullatus*) and Baird's beaked whale (*Berardius bairdii*). The English name "bottlenose whale" was actually in common use, as were the Norwegian name *nebhval* or *naebhval* and the Danish and German name *dögling* or their derivatives in other northern European languages.

The name *tsuchi-kujira* or just *tsuchi* is the Japanese common name for Baird's beaked whale (*Berardius bairdii*).

III. Diagnostic Characters and Comments on Taxonomy

Living beaked whales are characterized externally by a pronounced rostrum (beak), which blends into a high forehead (or melon) without a break (Fig. 1); a pair of throat grooves; relatively small flippers with short digits and relatively long arm bones; a small triangular dorsal fin that is placed far back on the body; and lack of fluke notches. Internally they have a reduction in teeth; fusion of the bones of the rostrum and development of extremely dense rostral elements in males; expansion of the pterygoid air sinus and elimination of its lateral bony wall; and elevation of the bones associated with the nose into a bony protuberance called the vertex (Fig. 2).

It seems that the concept of the beaked whales as a separate group of cetaceans became common in the 1860s and 1870s, as Gray used the family Ziphiidae in his "Catalogue of Seals and Whales in the British Museum" (1866) as did Van Beneden and Gervais in their epic "Ostéographie des Cétacés" (1868–1879).

IV. Notable Anatomy, Physiology, and Life History

Several similarities between beaked whales and sperm whales became evident early. Partly these were due to the retention of ancestral characters and partly due to similarities in ecology. Both groups of whales feed at considerable depth and are specialized to feed on squid.

Ziphiids in general have reduced their teeth to the point that teeth in the upper jaw are vestigial or absent and teeth in the lower jaw are reduced to one or two pairs that usually erupt only in adult males. The only exception to this is Shepherd's beaked whale (*Tasmacetus shepherdi*), which has a full dentition in both jaws.

A. Externally

The pronounced rostrum results from an anterior extension of the rostral and palatal elements of the skull, the maxilla,

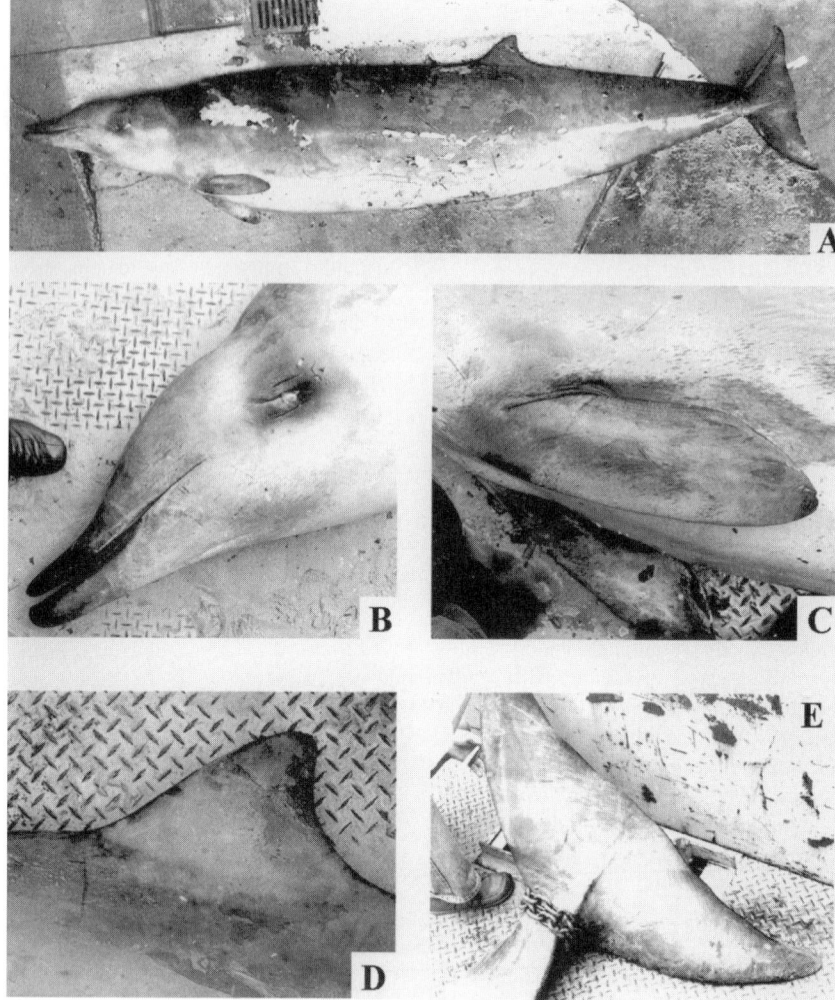

Figure 1 *Details of the external morphology of an adult male* Mesoplodon mirus *(USNM 504612). (A) Lateral view of the whole animal, (B) lateral view of head, (C) lateral view of flipper, (D) lateral view of dorsal fin, (E) oblique ventral view of flukes.*

premaxilla and vomer, coupled with a lateral compression to form a beak. Normally these bones are moderately extended in cetaceans to form a pincer-like beak, and, in fact the relative length of the rostra of some of the other toothed whales, like the river dolphins, exceeds that of beaked whales.

Beaked whales have a high forehead, which sets off the long rostrum. This forehead is composed of the soft tissue that forms the facial apparatus and the elevated cranium on which it rests. This soft tissue is responsible for sealing the nasal passages against water and modifying the emitted sound. The blowhole is crescent shaped with the horns pointing anteriorly, except in the genus *Berardius* where they point posteriorly.

The forehead merges with the rostrum without a break or groove that is characteristic of most other toothed whales. There is only one dolphin (rough-toothed dolphin, *Steno bredanensis*) that shares this character (Fig. 1B).

Beaked whales have a pair of throat grooves, which are in the shape of a "v" with its apex pointing forward. The anterior end of the throat grooves lies posterior to the symphysis of the lower jaws and anterior to the jaw joint (i.e., in the throat region). Throat grooves are present in gray whales (*Eschrichtius robustus*) and sperm whales (*Physeter macrocephalus*) but absent in all other species. They are not to be confused with ventral grooves, which stretch from the tip of the jaw back to the umbilicus in rorquals (Balaenopteridae).

Beaked whales have relatively small flippers, which are unspecialized. They consist of a relatively large forearm (radius and ulna) portion followed by a short phalangeal (finger) portion (Fig. 1C). This is also true of porpoises (Phocoenidae) and rorquals and appears to be a primitive cetacean character. Dolphins (Delphinidae) have an entirely different flipper shape, which tends to be falcate (hook shaped) and is the result of a lengthening of the phalangeal portion.

The dorsal fin is small and triangular instead of falcate and is situated on the posterior third of the body. The dorsal fin is usually placed over the anus at the junction of the abdomen and tail.

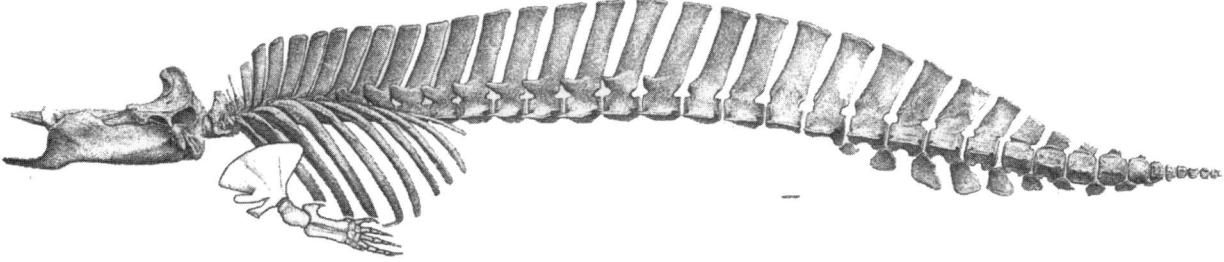

Figure 2 *Skeleton of an adult male* Mesoplodon densirostris *in the Australian Museum, Sydney (after Van Beneden and Gervais, 1868–1879). Forelimb and pelvic rudiment are from an adult male of the same species in the American Museum of Natural History (after Raven, 1942).*

The position of the dorsal fin in beaked whales correlates with a relatively long thorax and abdomen and short tail (Fig. 1D).

Beaked whales normally do not have fluke notches, and the trailing edge of the flukes is unbroken. Embryologically, the fluke notch is formed when the trailing edge of the flukes moves back beyond the end of the caudal vertebrae. Caudal vertebrae anchor the midline of the trailing edge, resulting in a notch in other whales (Fig. 1E).

B. Internally

The reduction in TEETH has proceeded to the point where all functional teeth are lost in females and immature males, and dentition is only represented by a single pair of teeth in the lower jaw of males. Females and immature males have a pair of vestigial teeth. The dentition seems to be only useful in male aggression. The two exceptions to this are the whales of the genus *Berardius,* which have two pairs of mandibular teeth, and *Tasmacetus,* which has a normal odontocete dentition in both the upper and the lower jaws. In *Tasmacetus,* the apical pair of mandibular teeth is enlarged, which strengthens the hypothesis that the single pair of teeth in all other ziphiids represents the apical pair. A row of vestigial teeth is sometimes present in the gums of both the upper and the lower jaws of some beaked whales, particularly *Mesoplodon grayi* and *Ziphius.*

Rostral fusion in some males takes place with increasing age. Because age estimation in beaked whales has been difficult, we are not sure of the age in which this process is active. The mesorostral canal is filled in by dorsal expansion of the vomer and the individual rostral elements fuse together. This is accompanied by an increase in density of the rostrum. The density of the core of the rostrum has been measured at 2.4 g/cc in a male of *Mesoplodon carlhubbsi* and 2.6 g/cc in a male *M. densirostris.*

The pterygoid air sinus has become enlarged in the ziphiids. It still is confined to the pterygoid bone but has become relatively larger and lost its lateral wall. The anterior sinus is not developed in ziphiids.

The vertex of the SKULL has been expanded both laterally and vertically beyond the state in all other odontocetes. The vertex is composed of the posterodorsal ends of the maxilla and premaxilla, the nasals, and the medial ends of the frontals. The dorsal tip of the vertex has expanded laterally and anteriorly like a mushroom. This region is deeply involved with sound production and modification.

V. Fossil Record

Ziphiids first appeared in the fossil record in the early Miocene (de Muizon, 1991). These early ziphiids had long rostra, full dentitions with the first mandibular tooth often hypertrophied, an elevated synvertex with a premaxillary crest, strong development of the pterygoid sinus with a reduction of the lateral wall of the pterygoid and an increase in the hamular process, and an auditory region of the skull that had minimal fenestration.

By the middle Miocene ziphiids were abundant. This was a period of maximum diversity of the entire order Cetacea and certainly was for the ziphiids. However, with all of this diversity, the origin of the modern genera is still in doubt.

There are about 14 genera of fossils currently recognized as ziphiids. Of these 14 genera there are at least 28 species that are based solely on rostral fragments. Critical work has demonstrated that 2 genera are based on nondiagnostic fragments and have been regarded as *nomina dubia.* With further study, particularly of the genera that are based on rostral fragments, there is bound to be a lot of demonstrated synonymy.

Muizon (1991) classified the Ziphiidae into three subfamilies: the Hyperoodontinae, which contains *Hyperoodon* and *Mesoplodon;* and Ziphiinae, which contains *Ziphius, Berardius, Tasmacetus,* and the fossil genera *Choneziphius, Ziphirostrum, Cetorhynchus,* and *Ninoziphius;* and the Squaloziphiinae, which currently contains only *Squaloziphius.* Figure 3 shows a cladogram of Ziphiidae.

VI. Interactions with Humans

Because of their pelagic habits and general lack of concentrated populations, ziphiids have not had much contact with humans. The only fisheries that had ziphiids as a target species were the bottlenose whale fishery in the North Atlantic and the *Berardius* fishery in the North Pacific.

The bottlenose whale was hunted from the middle of the 19th century by Norwegian and British whalers. Because the catches of the bottlenose whale were part of a multispecies small whale fishery, where catches of one species may serve to subsidize catches of another when the population of the second species has fallen to such a point that fishing of it would not be economical, the population was overexploited and became protected by the INTERNATIONAL WHALING COMMISSION in the late 1970s.

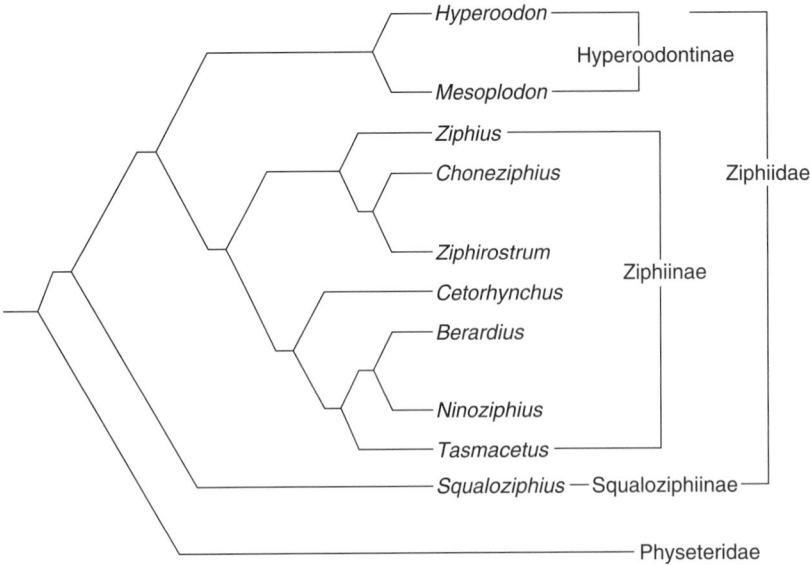

Figure 3 *Cladogram of Ziphiidae (after de Muizon, 1991). Indopacetus is included in Mesoplodon.*

Berardius is hunted primarily by the Japanese, who have fished it out of shore stations on the northeast coast of Japan since at least the 17th century. It was taken incidentally by other nations in the process of whaling for other species. Because the Japanese market was local to the whaling stations and would consume other species of ziphiids, they formerly sometimes took *Ziphius cavirostris* and the occasional *Mesoplodon*.

In the Southern Hemisphere, whalers rarely took the southern forms of *Berardius* (*B. arnuxii*) and *Hyperoodon* (*H. planifrons*). There were no fisheries based on them as the target species.

Ziphiids are moderately large, difficult to find and catch, and have habits (deep diving) that do not suit them to captivity. The occasional live strandings are sometimes maintained in CAPTIVITY in hopes of rehabilitating them and learning something of their behavior. The rehabilitation attempts have never been successful and the animals always died. One *Mesoplodon* calf that stranded in California in 1989 lived 25 days in an aquarium.

See Also the Following Articles

Giant Beaked Whales ■ Mesoplodont Whales ■ Skull Anatomy

References

Cuvier, G. (1829). Le règne animal distribué d'après son organisation pour servir de base a l'histoire naturelle des animaux et d'introduction a l'anatomie comparée. Nouvelle édition. Tome I, Paris, Déterville.

Gray, J. E. (1866). "Catalogue of Seals and Whales in the British Museum," 2nd Ed. British Museum, London.

Lacépède, B. G. E. (1804). "Histoire Naturelle des Cétacés." Plassan, Paris.

Moore, J. C. (1968). Relationships among the living genera of beaked whales with classifications, diagnoses and keys. *Field. Zool.* **53**(4), 209–298.

Muizon, C. de (1991). A new Ziphiidae (Cetacea) from the early Miocene of Washington state (USA) and phylogenetic analysis of the major groups of odontocetes. *Bull. Musée Natl. Hist. Nat. Paris* **12**(3–4), 279–326.

Ridgway, S. H., and Harrison, R. (1989). "Handbook of Marine Mammals," Vol. 4. Academic Press, San Diego.

True, F. W. (1910). An account of the beaked whales of the family Ziphiidae in the collection of the United States National Museum, with remarks on some specimens in other American museums. *U.S. Natl. Mus. Bull.* **73**, v + 89.

Van Beneden, P. J., and Gervais, P. (1868–1979). Ostéographie des cétacés vivants et fossiles, comprenant la description et l'iconographie du squelette et du système dentaire de ces animaux, ainsi que des documents relatifs a leur histoire naturelle. Paris, A. Bertrand.

Bearded Seal
Erignathus barbatus

KIT M. KOVACS
Norwegian Polar Institute, Tromsø

I. Description and Distribution

The bearded seal is a large, arctic phocid. Adults are 2–2.5 m long and are gray-brown (Fig. 1). Weight varies markedly on an annual cycle, but an average would be 250–300 kg. Females, which are somewhat larger than males,

Figure 1 *Bearded seal pup, 2 days old (top). Adult male bearded seal with rust-colored face (middle). An adult bearded seal in a typical habitat (bottom).*

can weigh in excess of 425 kg in the spring. Bearded seals often have irregular light-colored patches and can, in some geographic regions, also have rust-red faces and fore flippers. The rust color comes from contact with ferrous compounds in the bottom sediments that adhere to the surface of the hair and then oxidize when brought to the surface. Pups are approximately 1.3 m long at birth and weigh an average of 33 kg. They are born partially molted but usually still bear a significant quantity of fuzzy gray-blue lanugo in combination with a second darker-gray smooth coat. Their faces have white cheek patches and white eyebrow spots, which give them a "bandit" or "teddy-bear" look (Fig. 1). Yearlings look very similar to pups, but the facial patterns are less distinct and they often have small, distinct dark spots on their bellies. Similar to adults, young animals often have patches of white here and there. Bearded seals have several distinctive physical features. Their

body shape is very rectangular. Their heads appear to be small compared to their body size. They have very square-shaped fore flippers that bear very strong claws; Inuit people in the Canadian Arctic refer to this seal as "square flippers" because of the shape of their front flippers. They also have extremely elaborate, smooth, facial whiskers that tend to curl when dry; this trait gives them their other common name: bearded seal. Females have four mammary glands (a characteristic shared with the monk seals, *Monachus* spp.).

Bearded seals have a patchy distribution throughout much of the Arctic and sub-Arctic (Fig. 2). Their preferred habitat is drifting pack ice in areas over shallow water shelves. They are often found in coastal areas. Some populations are thought to be resident throughout the year, whereas others follow the retraction of the pack ice northward during the summer and the advance southward once again in the late fall and winter. They can maintain holes in relatively thin ice but avoid heavy-ice areas. During winter they concentrate in areas that contain polynyas, in areas where leads in the ice tend to be a regular feature, or along the outside of pack-ice areas. Juvenile animals wander quite broadly and can be found far south of the normal adult range. A neonate equipped with a satellite tag in Svalbard traveled south to Jan Mayen and then almost to the Greenland coast within the month following weaning, when the tag unfortunately ceased to transmit.

II. Behavior and Ecology

Bearded seals are largely solitary, although it is not unusual to see them hauled out together in small groups along leads or at holes in the spring or early summer. It is quite unusual to see a bearded seal on land, as they prefer to haul out on moving ice. They are rarely more than a body length from the water, and almost always face toward the water. However, they are not wary in a general sense; in some areas such as Svalbard, Norway, they are very tame and can be approached by humans to within meters by boat without reaction.

The time of breeding appears to vary somewhat geographically, with peaks occurring sometime between late March and mid-May, depending on the locality. Females give birth in a solitary fashion, on small drifting floes in areas of shallow water. The lanugo that is shed prior to birth is passed in the form of small, compact disks along with the afterbirth. Pups are born with a thin layer of subcutaneous BLUBBER, which is thought to be an adaptation to entering the water shortly after birth. Bearded seal pups swim with their mothers within hours following their births. This precocial entry into the sea is likely a mechanism to avoid polar bear PREDATION. Neonatal swimming skills develop quickly in this species, and pups can dive to depths > 90 m and remain submerged for periods in excess of 5 min when they are only a few weeks old. They spend approximately half of their time in the water during the nursing period, which lasts a total of 18–24 days and commence foraging on solid food while still accompanied by their mother. Female bearded seals spend little time on the surface with their pups, beyond that which is necessary for nursing. Most of the time they attend the pups from the water next to

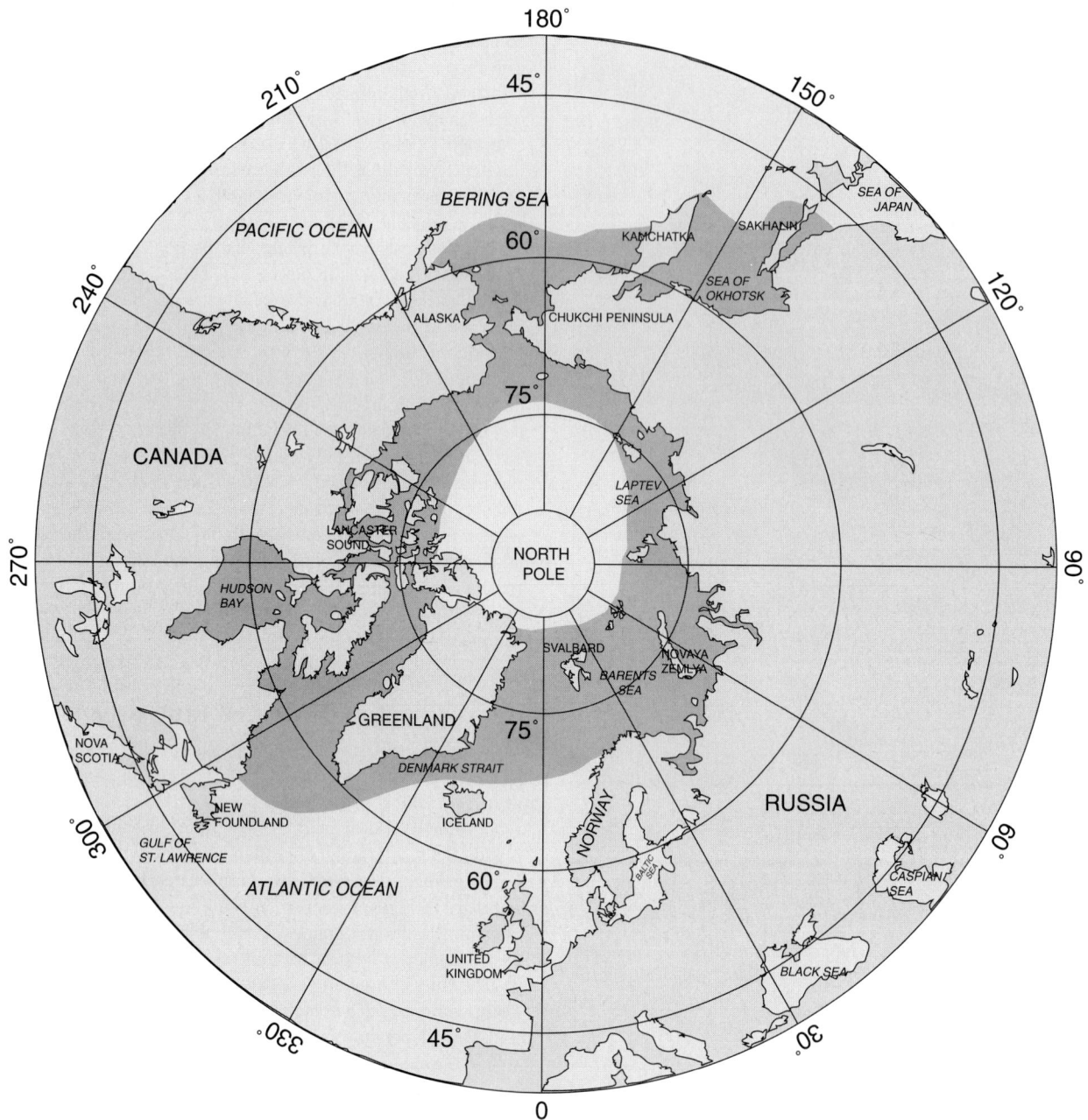

Figure 2 *Map showing the circumpolar, Arctic distribution of bearded seals (dark shaded area).*

the floe on which the pup is resting. Females do leave their pups unattended for periods to forage during the lactation period. Mother–pup pairs tend to remain in an area for some days at a time, but can also move tens of kilometers from one day to the next. Pups grow quickly during the nursing period, gaining about 3.3 kg per day while drinking > 7.5 liters of milk per day. The fat content of the milk is quite stable through lactation, at about 50%. Bearded seals pups have leaner bodies at the time of weaning than less active phocid pups but still have significant blubber stores and a body composition that is

about one-third fat. Pups are about 110 kg when they are weaned. Weaning does not appear to be as abrupt as it is in most phocid species.

MATING takes place around the time that females leave their offspring. Male bearded seals sing underwater to attract females and they also fight with other males during the breeding season. Their songs are a series of complex, downwardly spiraling trills that can be heard over tens of kilometers in quiet weather. The eerie yet melodious song is well known throughout Arctic regions; it is a sign of spring. Little is known regarding the specifics of mating behavior of this species because pairing takes place in the water.

Bearded seals seem to have quite a diffuse molting period and seem to shed hair almost constantly. However, there does appear to be a concentrated period of hair loss and renewal during the early summer. During this period bearded seals like to remain hauled out. Relatively large concentrations of animals can occur at this time because of the limited amount of ice available in some areas of the Arctic by this time in June.

Female bearded seals reach sexual maturity when they are about 5 years old, whereas males are normally 6 or 7. Bearded seals live 20–25 years.

Bearded seals tend to be shallow divers. They are for the most part benthic feeders. They likely use their whiskers to help them find food in soft substrates. Most diving is to depths less than 100 m, although adults have been recorded to dive to almost 300 m. Pups a few months old hold the current depth record for the species, which is > 488 m. Most dives are less than 10 min in duration, although they can dive for up to 20–25 min. They feed on polar cod (*Boreogadus saida*), sculpins (*Cottidae*), rough dabs (*Hippoglossoides platessoides*), and eelblennies (*Lumpenus medius*). Bearded seals also eat a variety of invertebrates, including spider crab (*Hyas araneus*), shrimps, a variety of molluscs, cephalopods, polychaete worms, and amphipods.

Polar bears (*Ursus maritimus*) and walruses (*Odobenus rosmarus*) are bearded seal predators. Killer whales (*Orcinus orca*) and Greenland sharks (*Somniosus microcephalus*) may also take bearded seals, particularly pups and juveniles.

III. Exploitation and Conservation Status

Bearded seals are an important subsistence resource in the Arctic. A few thousand animals are harvested annually for use as food, dog food, and for their thick leather, which is important for various traditional articles of clothing in Alaska, the Canadian Arctic, and in Greenland. Russia is the only country that has had a commercial scale harvest of bearded seals. Soviet ships took catches that exceeded 10,000 animals in some years during the 1950s and 1960s. Quotas were introduced to limit the harvests of the declining populations in the Okhotsk and Bering seas, and the catch dropped to a few thousand seals annually through the 1970s and 1980s. This hunt provided food for people and dogs and also fur-farm animal feed. Sinking losses are a serious problem when hunting bearded seals. During much of the year they sink when shot in open water or too close to ice edges; sinking loss is estimated to be 50%.

The global population of bearded seals cannot be accurately assessed. However, it is likely that this species numbers in the hundreds of thousands in the Arctic. Regional populations that are subjected to harvests can easily be depleted because densities are naturally low and most populations are quite sedentary.

See Also the Following Articles

References

Burns, J. J. (1967). The Pacific bearded seal. Alaska Dept. Fish and Game, Juneau, Alaska.

Burns, J. J. (1981). Bearded seal *Erignathus barbatus* Erxleben, 1977. *In* "Handbook of Marine Mammals" (S. H. Ridgway and R. J. Harrison, eds.), Vol. A, pp. 145–170. Academic Press, London.

Kelly, B. P. (1988). Bearded seal. *In* "Selected Marine Mammals of Alaska: Species Accounts with Research and Management Recommendations" (J. W. Lentfer, ed.), pp. 77–94. U.S. Marine Mammal Commission, Washington, DC.

Lydersen, C., and Kovacs, K. M. (1999). Behaviour and energetics of ice-breeding, North Atlantic phocid seals during the lactation period. *Mar. Ecol. Pro. Ser.* **187,** 265–281.

Behavior, Overview

PETER L. TYACK
Woods Hole Oceanographic Institution, Massachusetts

Marine mammalogists often divide behavioral research into categories defined by mode of study: "acoustics" is studied by recording underwater sounds with a hydrophone, "behavior" is often informally defined as that which can be seen by an observer watching animals, and "diving" is often studied by attaching tags to animals. This method-oriented view of behavior may be convenient for sorting different research traditions, but it obscures the integrated whole of behavior as it has been shaped by evolution. Each method yields its own view of behavior, but no one view alone can provide a complete picture.

Most behavioral ecologists divide behavior along functional lines, i.e., what is the problem the behavior has evolved to solve (Alcock, 1998)? The following is a short list of such problems:

- Foraging behavior: how to find, select, and process prey
- Predator avoidance or defense: the flip side of foraging from the prey's point of view
- Dispersal and migration
- Competition and agonistic behavior
- Sexual behavior: how to find, court, and choose mates
- Parental behavior
- Social behavior and social relationships

This functional taxonomy of behavior is mirrored by Bradbury and Vehrencamp's (1998) functional analysis of animal communication. A receiver can often be viewed as paying attention to a signal to answer a question related to one of these behavioral problems. When the receiver detects one signal out of a larger signal set, the signal can potentially help the receiver to reduce uncertainty about the correct answer. Bradbury and Vehrencamp (1998) suggest that a receiver's questions can be divided into three categories: sender identity, sender location, and behavioral context. Depending on the problem, the receiver may be interested in different levels of recognition of the signaler: species, group, sex, age, or individual. Receivers almost always need to know something about the location of the signaler: how far away is it? Is it within the receiver's territory? Is it approaching or moving away? The behavioral contexts of animal communication bear a striking resemblance to the functional behavioral problems listed previously: conflict resolution, territory defense, sexual interactions, parent–offspring interactions, social integration, and environmental contexts such as those related to prey and predators. This article discusses marine mammal examples for each of these basic problems in the behavioral ecology of all animal species, with special emphasis on how the marine environment may affect adaptations of marine mammals.

I. Foraging Behavior: How to Find, Select, and Process Prey

The earliest studies of foraging in marine mammals focused on the stomach contents of dead animals in order to define what kinds of organisms were in the DIET of marine mammals. The best that observers could do in early field studies of living marine mammals was to identify behavior associated with feeding, where feeding was linked to observation of prey at the surface or chases, and so on. However, these observations do not do justice to the complex process by which animals find, select, and handle their prey. Increased efforts in foraging theory to identify the kinds of decisions faced by a foraging individual have focused attention on a more detailed view of foraging, and new techniques such as time-depth recorders have improved our ability to collect the required data. This section discusses the various phases of foraging.

Marine mammals use every sensory modality available to find and select their prey. The optimal senses for solving a particular foraging problem depend on the setting. For example, vision is an excellent distance sense in air, but has a limited range underwater. Even though polar bears (*Ursus maritimus*) are classed as marine mammals, they often hunt their prey in air and may use vision in air to search for their pinniped prey. Many seals and dolphins chase fish prey close enough to the surface to search for images of fish reflected from sunlight. Biologists have used video recorders attached to seals to capture images of fish as seals hunt. In many coastal areas, seals can see fish at ranges of 10 m or so. Deep-diving seals such as elephant seals (*Mirounga* spp.) have eyes specially adapted to the wavelengths and low light levels of the deep sea. Many deep-sea organisms have light-producing organs, and researchers have speculated that marine mammals may use vision to find bioluminescent organisms in the dark.

As terrestrial mammals, we humans are accustomed to thinking of vision as the best distance sense, but sound carries much better underwater than light. Some marine mammals have developed sophisticated adaptations to use sound for finding prey. Perhaps the best-known example is the sonar of dolphins. Dolphins and most toothed whales have an auditory system that is specialized for high frequencies, and they can produce a directional beam of intense high-frequency pulses of sound. Most toothed whales echolocate by producing a click and then listening for echoes from surrounding targets. When they are in a search mode, they may produce a slow series of clicks, lis-

tening for echoes. Most cetaceans wait until they hear an echo to produce another click; in search mode, this means that the interval between clicks usually is greater than the maximum expected range to a target. Once they detect an echo, they may adjust their click rate to match the expected range to the target. If they click soon after they hear an echo, then the interval between clicks can give an indication of target range. Sound travels underwater at about 1500 m/sec. If an animal waits for 1 sec for the click to travel out to a target and then return, the target could be as far as 750 m away. Dolphins can detect a target as small as a 2.5-cm sphere at a range of >100 m. Once the animal detects the target, it then knows when to expect the echo, and it can set a click rate for that target range. Dolphins may take several tens of milliseconds to process information from a returning echo, and the click rate during the target-tracking mode usually equals the round trip travel time plus this processing time. When foraging, many toothed whales produce series of these clicks. For example, foraging narwhals (*Monodon monoceros*) make a regular series of slow clicks as they dive, but these slow clicks are interrupted by rapid and accelerating bursts of pulses. One hypothesis for this behavior is that slow clicks represent a search phase of the sonar and that when a narwhal detects a potential target, it switches to a target-tracking mode. As the narwhal closes on the target, the round trip travel time for the echo would decrease, which might explain why the narwhal produces an accelerating series of clicks. As with echolocating bats, these rapid accelerating click trains probably indicate detection and pursuit of prey.

While echolocation has been well studied with captive dolphins and artificial targets, we actually know very little about how odontocetes use sound in foraging. When wild dolphins (*Tursiops truncatus*) from inshore waters near Sarasota, Florida, are feeding, they produce echolocation clicks at very low rates. Echolocation has been studied in wild dolphins either with an acoustic recording tag or by following them with a small boat and using a towed hydrophone to listen to the sounds produced by the dolphins. Dolphins in Sarasota had an overall average click rate of 0.39 click trains/min while foraging and a rate of 0.10 click trains/min while not foraging. Clicks trains lasted about 2 sec and occurred every 200–450 sec on average. Dolphins had different modes of foraging, and different senses were used in the different feeding modes. Dolphins had the highest click rates and appeared to rely more upon echolocation when they were feeding on fish hiding in seagrass. In contrast, when dolphins were feeding in clear water over sand, they seldom clicked and appeared to rely primarily on VISION.

Dolphins and toothed whales are hunters who chase down individual prey items. Many species feed on highly mobile prey such as schooling fish. When a dolphin charges into a fish school, the fish usually disperse, and this can make it less efficient to find and chase down the remaining fish. Dolphins in many settings are reported to coordinate their feeding so that some dolphins keep the fish in a tight school as other dolphins feed. Toothed whales that disperse to feed on deep prey also usually feed in groups. While members of a group of whales such as sperm whales (*Physeter macrocephalus*) often disperse at depth, they usually converge as they surface. This pattern may help them find the densest concentrations of deep prey. When sperm whales dive to forage, they produce click sounds at rates of 0.5–2 clicks/sec, and they may also produce fast and accelerating series of clicks while feeding at depth. Most marine mammalogists have assumed that when sperm whales feed in the dark cold waters hundreds of meters below the sea surface, they use these clicks to detect their squid prey. This is a plausible hypothesis, but because we have no data on how sperm whales detect, pursue, and capture their prey, this is an area that requires further research.

Baleen whales have evolved to capture entire patches of prey in one mouthful. Balaenid whales, such as right (*Eubalaena* spp.) and bowhead whales (*Balaena mysticetus*), are specialized to feed on calanoid crustaceans. When balaenid whales feed, they swim through the prey patch with an open mouth. Because their baleen is very long and their head has a large cross-sectional area, they catch their prey by engulfing them in the water that flows into the mouth and out through the BALEEN. The basic problem faced by a feeding balaenid whale is to find a dense enough patch of prey to pay for the time and additional expense of swimming in the open-mouth foraging mode. Their prey move slowly enough that the feeding of balaenid whales is more like grazing then hunting. At times, balaenids coordinate their feeding by swimming in a staggered or "u shape" of up to 13 whales side by side. It is believed that such coordinated feeding keeps prey from escaping to the side, and may therefore more effectively "herd" prey towards each whale mouth.

Balaenopterid whales also feed on crustaceans, but their euphausiid prey are faster and more evasive than calanoid crustaceans. Balaenopterids also may capture schools of baitfish such as capelin, anchovy, sand lance, or even herring. These prey are more mobile than crustaceans, and balaenopterids have evolved a feeding mode that allows them to trap mobile prey. Balaenopterids have accordion-like pleats in the lower jaw, which can expand rapidly. When a balaenopterid feeds, it lunges while opening its mouth, forcing hundreds of gallons of prey and water into the mouth as the pleats expand. The whale then quickly closes its mouth, trapping the prey. The pleats then slowly contract, forcing the water through the baleen and leaving the prey behind. As with toothed whales, when balaenopterids such as humpback whales (*Megaptera novaeangliae*) feed on the most mobile schooling prey, such as herring, they may feed in coordinated groups. Perhaps the most striking reports concern a group of half a dozen or more female humpback whales who associated together each summer for several years in Southeast Alaska. Each individual played a specific role in prey capture, and their movements appeared to be coordinated with a regular series of vocalizations.

Marine mammals have evolved several different ways to feed on benthic prey that hides submerged in the sediment on the seafloor. Some bottlenose dolphins have been observed to echolocate on small sand dabs buried in the sand. The mustache of the walrus (*Odobenus rosmarus*) is exquisitely sensitive to touch, and walruses use the vibrissae in their mustache to detect prey in the sediment. Trained walruses have demonstrated remarkable abilities to use this mustache to determine the shape of objects, and presumably wild walruses use this ability to identify their favored prey within the sediment. Gray whales (*Eschrichtius robustus*) feed on benthic organisms by sucking mud and prey into the side of their mouth and then straining out the prey with their baleen. Gray whales make distinctive pits, measuring about 1 m wide by up to 3 m long and about one-half meter deep, in the seafloor when they feed in this way and these pits are big enough to be detected by sonar on surface ships.

Most marine mammals just swallow their prey whole, but some species face problems in handling their prey. Dolphins feeding on prey such as catfish with sharp spines may need to learn how to snap off the head and spines before they eat the rest of the fish. False killer whales (*Pseudorca crassidens*) measuring perhaps 3 m in length can capture a large mahi-mahi nearly half their size. It has been reported that one false killer whale may hold the fish while others rip off flesh. The most impressive prey handling among marine mammals involves the sea otter (*Enhydra lutris*), which feeds on shellfish such as abalone (*Haliotis* spp.) and sea urchins. Because these prey are too large or too strong for the otter to break the shell by biting it, most otters use a stone as a tool to open the shells. The sea otter will dive to get a shell, often carrying its stone tool in the axilla. When the otter surfaces, it lies stomach up with the stone on its stomach. It then uses the stone either as a hammer or as an anvil, smashing the shell open on the stone and then eating the flesh. It must be awkward to carry this stone while diving to feed, and some otters have simplified this by using a bottle for the same task. When the otter dives, it can leave the bottle floating on the surface and can then use it when it surfaces with shellfish.

II. Avoiding Predators and Defense from Predators

Many marine mammals are top predators and historically may not have faced a heavy predation pressure. The primary predators for pelagic marine mammals over evolutionary time are the killer whale (*Orcinus orca*) and sharks. However, humans have in the last few centuries been extremely effective predators of marine mammals, driving some species such as the Steller's sea cow (*Hydrodamalis gigas*) to extinction. Seals on ice face the risk of predation from polar bears, and seals hauled out on beaches, especially pups, are at risk from other pinnipeds and terrestrial predators such as foxes. Seals on land are less mobile than at sea and appear to be at a higher risk of predation, for they will usually respond to the approach of a terrestrial predator by entering the water.

The great whales such as baleen and sperm whales are so large that their main predator is the killer whale. Killer whales live in stable groups. They attack large whales in groups and appear to coordinate their attacks in much the same way as a pride of lions or African wild dogs will attack a herd of ungulates, isolate an individual, and then hold it down to kill it. Baleen and sperm whales use their flukes as a weapon during such an attack, lashing them sideways through the water. After they have been detected, most small odontocetes appear to rely on speed to escape killer whales, whereas some pinnipeds may hide from them, either on land or on the seafloor. Many cetaceans fall silent when a killer whale is detected nearby.

Early whalers had a predator's view of their marine mammal prey, and these observations make up an unusual body of data on predator defense in some species. While baleen whales often travel in groups, there is little sign of social defense from predators. The young calves of baleen whales may be more vulnerable to predation than adults, but females with young are less likely to be seen in groups, suggesting that they may disperse to reduce the chance a predator will detect them. Sperm whales, however, appear to have a well-developed social defense from dangerous predators such as killer whales and human whalers. Sperm whale calves are born into groups of about 10 adult females with young. Because newborn calves cannot dive deep enough to follow their mothers for their 40- to 50-min foraging dives, they remain nearer the surface when the adults feed. The adults desynchronize their dives, however, so that there is less time that young calves are unattended by an adult. If a predator attacks the group, calves or wounded animals in the group will be surrounded by the rest of the adults. Most adults will face in toward the animal needing protection and will lash their tails facing outside. This must have been a formidable defense against killer whales, but was less successful with human whalers. Whalers knew how predictable this behavior was and would often intentionally injure one animal and leave it. They then could slowly kill each adult attending the injured animal, knowing that adults would be unlikely to abandon an injured group member.

The first step in lowering the risk of PREDATION is to avoid detection. Some seals alter their foraging behavior apparently to avoid visual detection by predators. Female Galápagos fur seals (*Arctocephalus galapagoensis*) are less likely to make their normal foraging trips when the moon is full. It has been suggested that this avoids the risk that a predator will see them in the moonlight. Because most small cetaceans have little chance to defend themselves from killer whales, they must emphasize strategies to avoid detection by these predators. Dusky dolphins, *Lagenorhynchus obscurus*, mill in the surf zone as killer whales pass by offshore. They will even hide in tidal lagoons and, at times, become stranded in these lagoons until the next tide. Baleen whales also have strategies to avoid detection by killer whales. For example, gray whales that were exposed to experimental playback of the sounds of killer whales fled into shallow water. The surf zone and kelp beds may be particularly good places to hide from an echolocating predator because they absorb and reflect sound, making echolocation more difficult. There is some evidence that killer whales may even have evolved countermeasures to these predator-avoidance strategies. There are two sympatric populations of killer whales in the inshore waters of the Pacific Northwest. One population, called residents, feeds primarily on fish; the other population, called transients, feeds primarily on marine mammals. When residents feed on salmon, a fish with poor hearing, the killer whales make regular series of loud clicks. When transients feed on acoustically sensitive marine mammals, such as porpoises, dolphins, or seals, they have a much more stealthy pattern of echolocation. They produce fewer clicks, and those clicks that are produced are fainter and are produced with an erratic timing. These features appear to be designed to make marine mammal prey less likely to detect and avoid the killer whale.

Cetaceans are also subject to parasitism from animals that bite tissue without causing serious injury. In the tropics, many dolphins are subject to attack from the cookie cutter shark (*Icistius brasiliensis*), which takes bites of skin and blubber about 3 to 5 cm in diameter. Right whales (*Eubalaena aus-

tralis) in coastal bays in Argentina are subject to attack from seagulls, which peck chunks of skin and blubber from the back of a whale floating at the surface. While this can evoke a strong behavioral reaction from the whale, right whales do not seem to have an effective defense from this attack, which may be made worse by the growth of seagull populations in areas with human settlements.

Evidence in several cetacean species shows that when an animal has been injured, other members of the group may support them for hours or days. Because marine mammals must breathe air, if they cannot surface on their own, they are at great risk of drowning. This caregiving behavior may cost the caregivers, but at potential benefit to the incapacitated member of the group.

III. Migration and Orientation

Most marine mammals are excellent swimmers, and many species make annual MIGRATIONS of thousands of kilometers. Most baleen whales have an annual migratory cycle that affects many aspects of their life. These whales are adapted to take advantage of a burst of productivity in polar waters during the summer. Baleen whales store enough energy reserves during their intensive summer feeding season to last for most of the year, and this annual feast/fast cycle helps to select for large size. A humpback whale that is born in the winter in a tropical breeding ground near 20° of latitude will typically migrate in the spring to summer feeding grounds in polar waters near 40–60° of latitude. Humpbacks have traditional FEEDING grounds and an individual will often visit specific banks or inshore feeding areas of scales of tens of kilometers. Dolphins on both coasts of the United States also show annual migrations of 1000 km or more. Off the east coast of the United States, harbor porpoises (*Phocoena phocoena*) and bottlenose dolphins tend to move north in summer and south in winter. It is not known whether the colder temperatures in the north during winter are more important for this seasonal migration than are seasonal changes in prey distribution. Some pinnipeds also have annual migrations of thousands of kilometers. For example, northern elephant seals (*Mirounga angustirostris*) that breed and calve near San Francisco may migrate as far as the Aleutian Islands to feed. Both males and females swim north after the breeding season, following the California Current. Some male elephant seals feed along ocean fronts on the boundary of the Alaska Stream. Very little is known about how marine mammals orient and navigate during migration, and even less is known about how they find oceanographic features such as fronts.

Other species have more limited home ranges over the year. The home ranges of sea otters may be limited to 1–17 km of coastline. Sea otters show strong fidelity to their home range. When 139 sea otters were flown 200 km to an offshore island as part of a reintroduction program, most of the otters left and at least 31 managed to return to the area where they had been captured. Bottlenose dolphins in the inshore waters of Sarasota, Florida, tend to be sighted within a home range of 125 km². "Resident" killer whales in the inshore waters of Puget Sound have seasonal ranges limited to an area several tens of kilometers by about 100 km. Even nonmigratory species can be highly mobile. For example, resident killer whales will often swim 100 km or more in a day. Bottlenose dolphins (*Tursiops truncatus*) and sea otters may suddenly leave their home ranges and swim 100 km away from the normal range.

IV. Competition and Agonistic Behavior

When animals are competing for the same resource, they may fight for access. Among animals that exploit a specific substrate, this competition may be for territory. This kind of territorial defense has been well described for many pinnipeds during the breeding season. Female pinnipeds haul out onto beaches to give birth, and many species mate on land as well. This concentration of females creates a valuable resource for males. Males in many species will defend an area of beach from other males and may attempt to monopolize opportunities to mate with females there. For animals that live in the open ocean, resources are not likely to be as tied to a particular location, but rather will move. Animals in this setting are more likely to defend a particular resource at one time than to defend a patch of real estate. For example, a male humpback whale will not defend a specific location during the breeding time, but a male escorting a female will fight other males to limit their access to the female.

This pattern of males competing for access to females is common among mammals and leads to behavioral and morphological adaptations. Males in these species are often larger than females. Some behaviors appear to function to increase the apparent size of a male, and may function as visual displays. For example, male humpback whales competing for access to females may lunge with their jaws open, expanding the pleated area under the lower jaw with water. Several observers have suggested this may function to increase the apparent size of a competitor. Males may have larger weapons such as teeth or tusks than females. This is particularly striking in beaked whales. In most beaked whale species, the teeth may not erupt at all in females, whereas one or two pairs of teeth erupt in the lower jaw of males at about the time of sexual maturity. Males have scarring patterns, suggesting that these "battle teeth" are used in fights. Males may also have protection such as areas of toughened skin. Male elephant seals, for example, often strike one another on the chest, and this area has thickened and hardened skin.

Fighting often involves a gradually escalating series of threats and responses. This kind of escalated display has been hypothesized for bottlenose dolphins in captivity. The earliest stages of a threat may involve one dolphin directing pulsed sounds toward another. The threat may escalate if the dolphin produces an open-mouth threat display while emitting distinctive bursts of pulses. The longer in duration or louder in sound intensity the pulses are, the stronger the threat may be. As another step in escalation, the animal may accentuate this display with abrupt vertical head movements. One of the most intense threat displays in dolphins is called the jaw clap. A dolphin starts the jaw clap display with an open mouth. The jaw clap consists of an abrupt closure of the gaping jaw, accompanied by an intense pulsed sound. Many of the agonistic visual displays used by bottlenose dolphins are related to movements used to inflict injury. For example, the open-mouth display looks like the first step in preparing to bite.

Some animals live in situations where they interact repeatedly with the same individuals over and over again. In this setting, animals may develop a predictable hierarchy of who wins and loses in agonistic interactions. Male elephant seals establish a dominance hierarchy on the breeding beaches. The pace of competition is highest before the females appear on the beach. When males are competing using territory or dominance for access to females, they often sort out their competitive relations before the peak of the mating season. Dominance relations have also been studied in captive bottlenose dolphins (*Tursiops truncatus*). The most obvious competitive behaviors are violent fights in which each opponent responds to aggression with an aggressive response. This is not as useful for determining winners or losers as observation of more subtle submissive behaviors. A fight in which each opponent produces aggressive behaviors with no submission does not have an obvious winner or loser, but an animal can be identified as a loser of an interaction if it responds to a neutral or aggressive behavior with a submissive one. Systematic observations of winners and losers in dyadic agonistic interactions reveal that adult males are dominant over adult females. The rate of agonistic interactions is higher in males than in females. The low rate of female agonism means that dominance is rarely contested among females, and female dominance can be stable over years. Two male dolphins in a pool reversed dominance status several times over the years of study. Male dominance relations were characterized by periods of relatively low agonism interspersed with periods of high rates of agonism when one male challenged the other. Little is known about dominance relations among wild cetaceans, but because individuals in many species interact repeatedly with the same conspecifics and can recognize different individuals, dominance relations are likely to be important.

V. Courtship and Sexual Behavior

Charles Darwin made a distinction between features selected to improve chances of mating and features selected for survival. He called selection for mating sexual selection to discriminate it from natural selection. Darwin defined two kinds of sexual selection: intersexual and intrasexual. Intersexual selection can increase the likelihood that an animal will be chosen by a potential mate; intrasexual selection can increase the likelihood that an animal will outcompete a conspecific of the same sex for fertilization of a member of the opposite sex. More recent reviews have included a third mode of sexual selection where a male may attempt to limit the choice of a female by coercing her to mate with him and not to mate with other males.

Differences between male and female mammals alter the costs and benefits of different elements of reproduction. Female mammals all gestate the young internally and are specialized to provide nutrition to the young after birth. In many species, and most marine mammal species, the female provides most of the parental care. REPRODUCTION in most female mammals is limited by the amount of energy and nutrition they can acquire for pregnancy and lactation. Male mammals usually provide much less parental care to their young. This means that reproduction in most male mammals is limited by the number of females they can mate with. This situation often leads to a polygynous mating system in which there is high variability in the mating success of different males, with some males mating with many different females and other males mating with none. Males in polygynous species often fight other males for access to females. This often leads them to have weapons and to be larger than females; the intensity of polygyny is sometimes estimated by assessing the difference in size of males vs females. Some of the most extreme cases of sexual dimorphism occur in marine mammals. Adult sperm whales may be up to 16 m long, whereas adult females seldom grow beyond 12 m.

Most traditional discussions of mating systems emphasize male strategies. For example, polygyny occurs where one male mates with more than one female; the number of males a female mates with is not included in the definition. While the variance of reproductive success is higher in males than in females for most marine mammal species, female reproductive strategies can influence male strategies and impact other areas of social behavior. Areas in which female reproductive strategies vary include the following:

How seasonal and synchronized is estrus?
Do females have one (monoestrous) or more (polyestrous) estrous cycles per year?
Do females ovulate spontaneously or do they require the presence of a male to ovulate?
How many males are available during estrus?
Can the female select a mate?
If a female mates with more than one male, can she influence which male fertilizes the egg?

There are different patterns for the reproductive strategies of males and females in different polygynous MATING SYSTEMS. This article describes five different categories of male strategy that are used in the literature. The **resource defense strategy** is adopted by males who defend a resource used by females around the time of mating. In this case females do not select a mate but rather select an area for breeding and mate with the male defending this area. The **female defense strategy** is used by males who stay with a female and prevent other males from mating with her while she is receptive. The **sequential defense strategy** differs from the female defense strategy in that a male will defend a female through mating, but then leave in search of other mating opportunities. The distinction between these two male strategies depends in part on whether the female is mono- or polyestrous and on the degree of synchronization of different females. The strategy called by the name **"scramble competition"** occurs when a male searches for a receptive female, mates with her, and the moves on to search for another female without preventing access for other males. The last three models lie on a continuum of male strategies between pure guarding and pure roving. Models suggest that males should rove between groups of females if the duration of estrus is greater than the time it takes males to swim from group to group. At any one time, a male's decision to leave or stay with a group probably includes other factors, such as his assessment of what other males are doing. **Lekking strategy** occurs when males aggregate in an area with no resources needed by the female and produce displays to attract the female. In leks, males provide no parental care and females select a male for mating.

Some of these male strategies preempt the ability of a female to select a male for mating. In the resource defense model, the female does not select a particular male, but rather will select a particular place with the resource she needs. She will then be most likely to mate with the male who happens to be defending this location. When a female can and does choose a male for mating, she may select a mate based on several different bases. A female may select a male for inherent qualities based on indicators such as size, age, or an advertisement display. She may assess competition between males and select one based on this performance. In some species, males may compete for access to a particular location, and a female can select a good competitor by mating with a male in such a preferred spot. In some species, a female may mate with several males and allow competition between their sperm to determine which male fertilizes the egg. The males in this system would be likely to devote more resources to sperm production, sperm swimming speed, and so on than species that compete by fighting. Evidence shows that sperm competition may play a role in some cetaceans. Odontocete cetaceans have larger ratios of testis:body weight than most mammals. This contrast is also seen among mysticetes. Balaenid whales form mating groups with multiple males, but there is little sign of fighting between the males. Male right whales have testes weighing more than 900 kg; their testes weigh more than six times what would be predicted for a typical mammal of their body size. In contrast, humpback whale males, which fight for access to females, have testes weighing under 2 kg.

Marine mammals are highly mobile, and in the open ocean it seems unlikely that males could defend a resource in a way that would preempt the ability of a female to use the resource yet mate with another male. Resource defense is much easier to envisage on land. While pinnipeds spend much of their life at sea, they haul out on a solid substrate (land or ice) to give birth. Some of these species, such as elephant seals, also mate on land. Females have specific requirements for a place to give birth, and they often return to traditional areas. This scarce resource creates an opportunity for males to defend these sites in order to increase their chances of mating with the females who are selecting the site. In most otariid seals, males appear to employ resource defense strategies for mating. In many of these species, males will arrive before the females and will fight to establish territories that they defend from other males. In some phocid species that mate at sea, males may establish and defend territories just off the beach where females give birth. Genetic analyses of paternity, however, show that the fathers of some pups are not among the TERRITORIAL males. This suggests that some males have alternate mating strategies.

There are marine mammal species in which males may adopt a strategy of attempting to preempt female choice by guarding a receptive female and preventing her from mating with other males. Northern elephant seal males arrive at breeding beaches before females and compete for dominance status and for position on the breeding beach. A dominant male can guard a group of females and prevent access for other males. If an alpha male can maintain his status, he can prevent access to a group of females for the entire breeding season. This pattern of guarding a group of females is less likely for cetaceans, which are highly mobile. Most male cetaceans would take a shorter time to swim between groups than the duration of female estrus, thus favoring a roving strategy. There is some evidence for sequential female defense in bottlenose dolphins. In field studies of Indian Ocean bottlenose dolphins (*Tursiops aduncus*) in Shark Bay Western Australia and bottlenose dolphins (*Tursiops truncatus*) in Sarasota, Florida, groups of two or three adult male bottlenose dolphins may form consortships with an adult female. A coalition of males may start such a consortship by chasing and herding a female away from the group in which they initially find her. Some of these consortships appear to be attempts by the males to limit choice of mate by the female, who may try to escape from the males. Males in these alliances may form consortships with several different females during a breeding season.

Many pinnipeds and some baleen whales produce reproductive advertisement displays that may play a role in mediating male–male competitive interactions and may also used for female choice of a mate. Male humpback whales sing long complex songs during the winter breeding season. Singing males are usually alone and they usually stop singing when joined by another whale. Aggressive behavior is often seen when a male joins a singer; when a female joins, apparent sexual behavior has been observed. Male humpbacks do not seem to be able to defend any resource needed by females on the breeding grounds, so this mating system has been described as a kind of floating lek. Vocal reproductive advertisement displays have also been reported for bowhead whales and many species of seal, including polar ice-breeding seals and harbor seals (*Phoca vitulina*). Most of the phocid seals known to produce songs mate at sea. These seals breed in conditions that foster the development of leks. Females gather to breed on isolated sites, but they mate after they have weaned their pups so there are few resources males could defend. Females are so mobile that it would be difficult for males to prevent them from gaining access to other males. The females are already concentrated in hot spots around the places where they give birth. This creates an ideal setting for males to cluster near the females, producing advertisement displays to attract females for mating. Some of the songs of whales, of ice-loving seals and the bell-like sounds of the walrus (*Odobenus rosmarus*) stand as testimony to the power of sexual selection to fashion complex and fascinating advertisement signals.

VI. Parental Behavior

All mammals have some parental care when the female lactates and suckles the young. The mothering role of the female is critical to mammalian life, and female parental care impacts many aspects of social behavior. There is enormous variability in parental care among marine mammals. Some phocid seals give birth to their young on unstable ice floes, where they cannot count on a stable refuge for the young. The hooded seal (*Cystophora cristata*) has responded to this situation by an intense 4-day period of lactation when the young pup doubles in weight. While female phocid seals do not generally leave their young pup to feed, otariid females will leave their young in order to feed at sea and then they return to suckle the pup. This pattern leads to a large difference in duration of lactation, from 4 days to 2 months in phocids and from 4 months to 2 years in

otariids. Most pinnipeds have yearly breeding seasons, so the longest periods of lactation are limited to about 12 months. However, tropical Galapagos sea lions at times nurse their young for >12 months and produce young at greater intervals than that. Phocoenid porpoises and some baleen whale species also have a strong annual breeding cycle. Some baleen whales, such as the blue whale (*Balaenoptera musculus*), wean their young after about 7 months so that the young can start taking solid food during the summer feeding season. All porpoises and baleen whales wean the young within a year. The toothed whales stand at the other extreme of having very prolonged periods of parental care when the young are dependent. Bottlenose dolphins only 3 m or so in length often suckle the young for 3–5 years, which is remarkably long considering that the 30-m blue whale can wean the young in 7 months.

The longest periods of PARENTAL CARE known among marine mammals involve sperm whales and short-finned pilot whales (*Globicephala macrorhynchus*). In both species, mothers appear to suckle some calves for up to 13–15 years. The young may start to take some solid food by the first few years of life, but this suckling indicates a remarkably long period of dependency for the young. Adult female pilot whales typically start having young by 8–10 years, but by the time they are near 30, many cease to reproduce. The ovaries become nonfunctional in these nonreproductive females, showing changes similar to those of human females after menopause. Female pilot whales may live into their 50s, suggesting that they may have a life expectancy 15–20 years after becoming nonreproductive. Most students of life history believe that life history evolves to maximize lifetime reproductive success. If this has influenced the life history of pilot whales, it suggests that females switch their reproductive effort from having new offspring to parental care of their existing young. The 15- to 20-year duration of this period suggests that 15–20 years of parental care are required for the young to succeed or that these older females are caring for other kin, perhaps in a grandparental role.

VII. Social Behavior and Social Relationships

Not only do marine mammals show a broad range in the duration of the maternal bond, but there is great diversity in the duration of social bonds in general and especially in the importance of individual-specific social relationships. Resident killer whales have the most stable social groups known among mammals: no dispersal of either sex has been described. The only way group composition changes among the resident killer whales of the Pacific Northwest is for an animal to die or for a new animal to be born. The best-known vocalizations from killer whales are group-distinctive repertoires of stereotyped pulsed calls. In contrast, bottlenose dolphins have very fluid social groups. In their fission–fusion society, group composition changes on a minute-by-minute and hour-by-hour basis. However, some individuals may have strong social bonds and be sighted together for years at a time. As was just discussed in the section on parental care, bottlenose dolphin calves suckle for 3–5 years. Adult male bottlenose dolphins may also form coalitions with one to two other unrelated males. Members of a coalition tend to be sighted together 70–100% of the time, and alliances may last for over a decade. It is thought that males form alliances to improve their chances of mating with females, but lone males are also successful breeders. Males within a coalition often have highly coordinated displays, both when feeding and when escorting a female. Each bottlenose dolphin produces an individually distinctive whistle vocalization called a signature whistle, which is probably used for individual recognition.

In sperm whales, males have different life history patterns than females. Calves are born into matrilineal groups of females and young. Each matrilineal unit numbers about 10 animals, but often two units associate for days at a time. Males may leave their natal groups when 5–10 years of age, and they then will join bachelor groups. As males grow, they move to higher latitudes and associate in smaller groups of males. As the males approach social and sexual maturity at 20–25 years of age, they are increasingly likely to temporarily associate with a female group during the breeding season, when they may mate with females. The social relationships of males thus change over their lifetime, and adult males appear to have only temporary associations. Young females may stay with their natal group or may leave, but once they reach sexual maturity at 8–10 years of age, they will tend to associate with the same adult females for decades at a time. Because the matrilineal groups often join with other groups but segregate into the original groups, the females must recognize group members over periods of decades. Sperm whales make rhythmic patterns of click sounds called codas. Early reports reported individually distinctive codas, but there are also shared codas that vary with the geographic region. Most of the variation in codas involves differences between groups, and it has been suggested that codas may reaffirm social bonds when a group joins after dispersing to forage. Female sperm whales must have stable social relationships with specific other individuals. These family groups appear to be the basic social unit of sperm whales, with a primary function of vigilance against predators and social defense of calves.

Baleen whales may feed in groups as do sperm whales, but female baleen whales appear to differ from sperm whales in the importance of group care of young. On the feeding grounds, baleen whales of all sexes are often seen in groups of varying sizes. For humpback whales, the size of the feeding group correlates with the horizontal extent of the prey patch. However, during the breeding season, when a female humpback has a calf, she is extremely unlikely to associate with another adult female. When one or more adults escort a female during the breeding season, the escorts are almost always males. In baleen whales there is much less evidence for long-term social bonds than among most toothed whales. Odontocetes with little evidence for stable bonds are species such as the harbor porpoise and delphinids of the genus *Cephalorhynchus*, which also appear to have fluid groupings with few social bonds more stable than the mother–calf bond, which lasts less than 1 year in the porpoise. However, future research may find social bonds that have not yet been described.

There appears to be a correlation between the social relations of marine mammals and their communication patterns. Baleen whales and pinnipeds with large apparently anonymous breeding aggregations use reproductive advertisement displays to mediate male–male and male–female interactions on the breeding grounds. Killer whales with highly stable groups produce group-specific repertoires of stereotyped calls. Seals and dolphins with strong individual-specific bonds use a variety of different vocalizations for individual recognition, but no such recognition signals are known for porpoises or *Cephalorhynchus*. Sperm whales appear to use deceptively simple clicks to produce a diverse set of signals consistent with their diverse social groupings.

VIII. Conclusions

Marine mammals face the same basic problems that have been identified by behavioral ecologists for all animals. However, marine mammals live in an environment that differs in many important ways from the terrestrial environment. Studies since the 1980s have provided ever-growing opportunities for fascinating comparisons between marine mammals and their terrestrial relatives and between the diverse taxa that live in the sea.

See Also the Following Articles

Communication ■ Feeding Strategies and Tactics ■ Group Behavior ■ Migration and Movement Patterns ■ Predator–Prey Relationships ■ Sexual Dimorphism ■ Territorial Behavior

References

Alcock, J. (1998). "Animal Behavior: An Evolutionary Approach." 6th Ed. Sinauer Associates, Sunderland, MA.

Bradbury, J. W., and Vehrencamp, S. L. (1998). "Principles of Animal Communication." Sinauer Associates, Sunderland, MA.

Mann, J., Connor, R., Tyack, P. L., and Whitehead, H. (2000). "Cetacean Societies: Field Studies of Whales and Dolphins." Univ. of Chicago Press, Chicago.

Pryor, K., and Norris, K. S. (1991). "Dolphin Societies: Discoveries and Puzzles." Univ. of California Press, Berkeley, CA.

Reynolds III, J. E., and Rommel, S. A. (1999). "Biology of Marine Mammals." Smithsonian Press, Washington, DC.

Beluga Whale
Delphinapterus leucas

GREGORY M. O'CORRY-CROWE
Southwest Fisheries Science Center, La Jolla, California

The beluga or white whale inhabits the cold waters of the Arctic and sub-Arctic (Fig. 1). Its name, a derivation of the Russian "beloye," meaning "white," appropriately enough captures its most distinctive feature, the pure white color of adults (Fig. 2). The evolutionary history and ecology of

Figure 1 *Worldwide distribution of the beluga whale. The northernmost extent of its known range is off Alaska and northwest Canada and off Ellesmere Island, West Greenland, and Svalbard (above 80°N). The southern limit of distribution is in the St. Lawrence River in eastern Canada (47° to 49°N).*

belugas are inextricably linked to the extreme seasonal contrasts of the north and the dynamic nature of the sea ice. As well as adaptation to the cold, life in this region has necessitated the evolution of discrete calving and possibly mating seasons, annual migrations, and a unique feature distinguishing it from most other cetaceans, an annual molt.

I. Taxonomy and Evolutionary History

The beluga whale is a member of the Monodontidae, the taxonomic family it shares with the narwhal, *Monodon monoceros.* The Irrawaddy dolphin, *Orcaella brevirostris,* was at one point considered by some to also be a member of this family. Although superficially similar to the beluga, recent genetic evidence strongly supports its position as a member of the family Delphinidae (Lint *et al.,* 1990; LeDuc *et al.,* 1999). The earliest fossil record of the monodontids is of an extinct beluga *Denebola brachycephala* from late Miocene deposits in Baja California, Mexico, and indicate that this family once occupied temperate ecozones (Barnes, 1984). Fossils of *D. leucas* found in Pleistocene clays in northeastern North America reflect successive range expansions and contractions of this species associated with glacial maxima and minima.

II. Description and Life History

The beluga whale is a medium-sized toothed whale, 3.5–5.5 m in length and weighing up to 1500 kg. Males are up to 25% longer than females and have a more robust build. As their genus name ("apterus"–without a fin) implies, they lack a dorsal fin and are unusual among cetaceans in having unfused cervical vertebrae, allowing lateral flexibility of the head and neck. They possess a maximum of 40 homodont teeth, which become worn with age, and may live to 40 years of age, possibly longer. Neonates are about 1.6 m in length and are born a gray-cream color that quickly turns to a dark brown or blue-gray. They become progressively lighter as they grow, changing to gray, light gray, and finally becoming the distinctive pure white by age 7 in females and 9 in males (Fig. 3).

Belugas are supremely adapted to life in cold waters. They possess a thick insulating layer of BLUBBER up to 15 cm thick beneath their skin, and their head, tail, and flippers are relatively small. The absence of a dorsal fin is believed by some to be an adaptation to life in the ice or perhaps as a means to reduce heat loss. In its place, belugas possess a prominent dorsal ridge that is used to break through thin sea ice.

Females become sexually mature at age 5, males at age 8, although males may not become socially mature until some

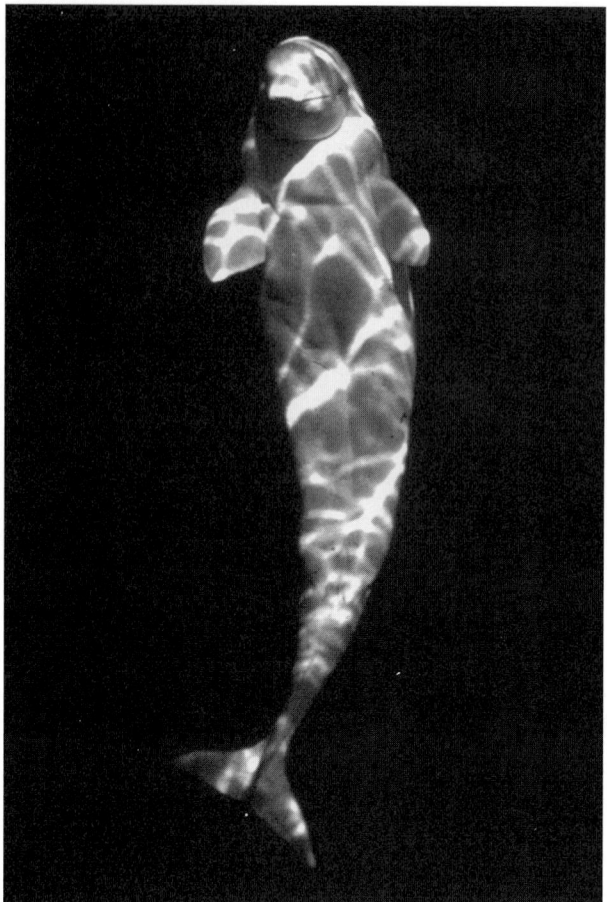

Figure 2 *Beluga whale,* Delphinapterus leucas. *Flip Nicklin/Minden Pictures.*

Figure 3 *Beluga whales concentrating near the coast during the brief summer. Note the dark to light gray color of younger animals compared to the white of adults. Flip Nicklin/Minden Pictures.*

eas, however, such as Cunningham Inlet on Somerset Island in the Canadian High Arctic, belugas may engage in more demonstrative behaviors, including spy hopping, tail waving, and tail slapping (Fig. 4).

Recent findings from studies using satellite-linked transmitters attached to free-swimming whales have confirmed that belugas are capable of covering thousands of kilometers in just a few months, in open water and heavy pack ice alike, while swimming at a steady rate of 2.5–3.3 km/h (Suydam *et al.,* 2001). Sensors on these transmitters have also recorded belugas regularly diving to depths of 300–600 m to the sea floor. In

time later. Gestation is 14–14.5 months with a single calf born in late spring–early summer prior to, or coincident upon, entry into warm coastal waters. Mothers produce milk of high caloric content and nurse their young for up to 2 years; the entire reproductive interval averaging 3 years. Little is known about the mating behavior or mating season of beluga whales. Mating is believed to primarily occur in late winter–early spring, a period when most belugas are still on their wintering grounds or on spring migration. Mating behavior, however, has been observed at other times of the year and the question of whether they have delayed implantation is unresolved.

III. Behavior

In contrast to the frozen smile of the oceanic dolphins, the ability of belugas to alter the shape of their mouth and melon enables them to make an impressive array of facial expressions. The lateral flexibility of the head and neck further enhances visual signaling and enable beluga whales to maneuver in very shallow waters (1–3 m deep) in pursuit of prey, to evade predators, and generally exploit a habitat rarely used by other cetaceans (see later).

Belugas typically swim in a slow rolling pattern and are rarely given to aerial displays. In nearshore concentration ar-

Figure 4 *Aggregation of beluga whales interacting and rubbing on the substrate of a shallow estuary during the summer molt. Flip Nicklin/Minden Pictures.*

the deep waters beyond the continental shelf, belugas may dive in excess of 1000 m, where the pressure is 100 times that at the surface, and remain submerged for up to 25 min (Richard *et al.*, 1997; Martin *et al.*, 1998)!

In areas of open water beluga whales may divide their days into regular feeding and resting bouts. Belugas appear to predominantly hunt individually, even when within a group, but have also been observed to hunt cooperatively. A typical hunting sequence begins with slow directed movement combined with passive acoustic localization (search mode) followed by short bursts of speed and rapid changes of direction using ECHOLOCATION for orientation and capture of prey (hunt mode) (Bel'kovich and Sh'ekotov, 1990).

IV. Acoustics–Vocalizations

The beluga possesses one of the most diverse vocal repertoires of any marine mammal and has long been called the "sea canary" by mariners awed by its myriad sounds reverberating through the hulls of ships. Communicative and emotive calls are broadly divided into whistles and pulsed calls and are typically made at frequencies from 0.1 to 12 kHz. As many as 50 call types have been recognized: "groans," "whistles," "buzzes," "trills," and "roars" to name but a few. Although some geographic variation is apparent, efforts to determine whether there are substantial regional differences or dialects have been hampered by differences among bioacousticians in the categorization of vocalizations. Belugas are capable of producing individually distinctive calls to maintain contact between close kin and can conduct individual exchanges of acoustic signals, or dialogues, over some distance (Bel'kovich and Sh'ekotov, 1990).

The echolocation system of the beluga whale is well adapted to the icy waters of the Arctic. Their ability to project and receive signals off the surface and to detect targets in high levels of ambient noise and backscatter enable belugas to navigate through heavy pack ice, locate areas of ice-free water, and possibly even find air pockets under the ice (Turl, 1990).

V. Ecology

As the sea ice recedes in spring, belugas enter their summering grounds, often forming dense concentrations at discrete coastal locations, including river estuaries, shallow inlets, and bays (Fig. 3). Several explanations have been proposed as to why belugas return to these traditional summering areas. In some regions, sheltered coastal waters are warmer, which may aid in the care of neonates. The occupation of estuarine waters also coincides with the period of seasonal molt. Belugas have been observed to actively rub their body surface on nearshore substrates (Smith *et al.*, 1992; Fig. 4), and the relatively warm, low-salinity coastal waters may provide conditions that facilitate molting of dead skin and epidermal regrowth (St. Aubin *et al.*, 1990; Smith *et al.*, 1994). Belugas feed on a wide variety of both invertebrate and vertebrate benthic and pelagic prey. In some parts of their range it is clear that belugas are feeding in nearshore waters on seasonally abundant anadromous and coastal fish such as salmon, *Oncorhynchus* spp., herring, *Clupea harengus*, capelin, *Mallotus villosus*, smelt, *Osmerus mordax*, and saffron cod, *Eleginus gracilis* (Kleinenberg *et al.*, 1964; Sea-

man *et al.*, 1982). The relative importance of these factors in determining coastal distribution patterns may vary among regions, depending on environmental and biological characteristics (Frost and Lowry, 1990). It is clear, however, that belugas exhibit some degree of dependence on specific coastal areas.

In many areas of the Arctic, belugas soon leave these coastal areas to range widely off shore. Satellite tracking has recorded belugas moving up to 1100 km from shore and penetrating 700 km into the dense polar cap where ice coverage exceeds 90% (Suydam *et al.*, 2001). How these animals find breathing holes in this environment is still a mystery. Analysis of dive profiles suggests that beluga whales may combine the use of sound at depth to find cracks in the ice ceiling overhead. Diving data also indicate that belugas are probably feeding on deepwater benthic prey as well as ice-associated species closer to the surface (Martin *et al.*, 1998).

Little is known about the DISTRIBUTION, ECOLOGY, or BEHAVIOR of beluga whales in winter. In most regions, belugas are believed to migrate in the direction of the advancing polar ice front. However, in some areas, belugas may remain behind this front and overwinter in polynyas and ice leads. In the eastern Canadian Arctic some belugas overwinter in the North Water, a large area of open water in northern Baffin Bay (Finley and Renaud, 1980), whereas in the White, Barents, and Kara Seas, belugas occur year-round, remaining in polynyas in the deeper water during winter (Kleinenberg *et al.*, 1964).

Killer whales, *Orcinus orca*, polar bears, *Ursus maritimus*, and humans prey upon beluga whales. Belugas sometimes become entrapped in the ice where large numbers may perish or be hunted intensively by humans.

VI. Social Organization

Belugas are sometimes seen singly, but more commonly occur in groups of 2–10 that may aggregate at times to form herds of several hundred to more than a thousand animals. Single animals are always large adults, whereas in mixed herds adult males may form separate pods of 6–20 individuals. Adult females form tight associations with newborns and sometimes a larger juvenile, presumably an older calf. These "triads" may join similar groupings to form large nursery groups. At certain times of the year, age and sex segregation may be more dramatic than at others with males migrating ahead of or feeding apart from females, young, and immatures. In general, group structure appears to be fluid, with individuals readily forming and breaking brief associations with other whales. Apart from cow–calf pairs, there appear to be few stable associations. However, considering the diverse vocal repertoire of beluga whales, including individual signature calls, their wide array of facial expressions, and the variety of interactive behaviors observed, as well as the numerous accounts of cooperative behavior, this species appears capable of forming complex societies where group members may not always be in close physical proximity to each other.

VII. Population Structure

Variation in body size across the species range has been taken as evidence of separate populations. The nonuniform pattern of distribution and predictable return of belugas to specific coastal areas further suggests population structure and has led to the treatment of these summering groups as separate management stocks. Resightings of marked or tagged individuals, as well as differences in contaminant signatures and limited evidence of geographic variation in vocal repertoire, add support to the independent identification of a number of these stocks. Although all are valid to varying degrees, many of these methods used for stock identification have limitations due to incomplete knowledge on year-round distribution, movement patterns, breeding strategies, and social organization. They provide little or no information on rates of individual or genetic exchange, and although phenotypic differences are highly suggestive, they may not provide evidence of evolutionary uniqueness.

A number of recent molecular genetic studies examined variation within the mitochondrial genome and confirmed that whales tend to return to their natal areas year after year and that dispersal among many separate summering concentrations is limited, even in cases where there are few geographic barriers (O'Corry-Crowe *et al.*, 1997; Brown Gladden *et al.*, 1997). These molecular findings reveal that knowledge of migration routes and destinations appears to be passed from mother to offspring, generation after generation. Such cultural inheritance of information leads to the evolution of discrete subpopulations, among which there is little dispersal. It is possible that many of these subpopulations may overwinter in a common area and that a certain amount of interbreeding may occur at this time. Regardless of such potential gene flow, in situations where management is concerned with the degree of demographic connectivity among areas, demonstrating that few animals disperse among subpopulations is sufficient evidence to designate them as separate management stocks.

VIII. Interactions with Humans

Because of their predictable MIGRATION routes and return to coastal areas, beluga whales have long been an important and reliable resource for many coastal peoples throughout the Arctic and sub-Arctic. A prudent subsistence tradition continues in many areas. However, primarily because of past commercial harvest excesses, current levels of subsistence taken from a number of populations may not be sustainable. Increasing human activity in the beluga's environment brings with it the threat of habitat destruction, disturbance, and pollution. In areas where there are large commercial fishing operations, belugas, particularly neonates, may be incidentally caught in gill nets. In a number of regions of the Arctic, beluga whales exhibit strong avoidance reactions to ship traffic, whereas in some coastal locations they appear to have developed a certain tolerance to boat traffic. The potential impacts of an emerging whale watching industry in more populated areas are as yet unknown. In some areas, belugas may also be victims of industrial pollution. A high incidence of various pathologies have been found in beluga whales in the St. Lawrence River in Canada and have been linked to high levels of heavy metals and organohalogens found in these whales. Some of these toxins may act by suppressing normal immune response, and there is concern that contaminants are adversely affecting

population growth (Béland, 1996). Finally, there is concern over the possible downstream effects of hydroelectric dams on estuarine habitats and the environmental and health risks associated with oil and gas development and mining in the Arctic.

Beluga whales were one of the first cetaceans to be held in captivity when in 1861 a whale caught in the St. Lawrence River went on display at Barnum's Museum in New York. Today, beluga whales are one of the more common and popular marine mammals in oceanaria across North America, Europe, and Japan. The majority of these animals are wild caught, but successful breeding programs at a number of facilities are increasing the number of belugas born in CAPTIVITY. Although the majority of beluga whales in captivity educate and entertain the public, a number of whales have been put to work by the navies of the United States and former Soviet Union.

See Also the Following Articles

Arctic Marine Mammals ■ Genetics for Management ■ Narwhal ■ Sound Production ■ Toothed Whales, Overview

References

Barnes, L. G. (1984). "Fossil Odontocetes (Mammalia: Cetacea) from the Almejas Formation, Isla Cedros, Mexico. PaleoBios, Museum of Paleontology, University of California.

Béland, P. (1996). The beluga whales of the St. Lawrence River. *Sci. Am.* May, 74–81.

Bel'kovitch, V. M., and Sh'ekotov, M. N. (1990). "The Belukha Whale: Natural Behaviour and Bioacoustics. USSR Academy of Science, Moscow. [Translated by Woods Hole Oceanographic Institution, 1993.]

Brown Gladden, J. G., Ferguson, M. M., and Clayton, J. W. (1997). Matriarchal genetic population structure of North American beluga whales *Delphinapterus leucas* (Cetacea: Monodontidae). *Mol. Ecol.* **6**, 1033–1046.

Finley, K. J., and Renaud, W. E. (1980). Marine mammals inhabiting the Baffin Bay North Water in winter. *Arctic* **33**, 724–738.

Frost, K. J., and Lowry, L. F. (1990). Distribution, abundance, and movements of beluga whales, *Delphinapterus leucas,* in coastal waters of western Alaska. *In* "Advances in Research on the Beluga Whale, *Delphinapterus leucas*" (T. G. Smith, D. J. St. Aubin, and J. R. Geraci, eds.), pp. 39–57. Can. Bull. Fish. Aquat. Sci. Vol. 224.

Kleinenberg, S. E., Yablokov, A. V., Bel'kovich, B. M., and Tarasevich, M. N. (1964). "Beluga (*Delphinapterus leucas*): Investigation of the Species." Academy of Sciences of the USSR, Moscow. [Translated by Israel Program for Scientific Translations, 1969.]

LeDuc, R. G., Perrin, W. F., and Dizon, A. E. (1999). Phylogenetic relationships among the delphinid cetaceans based on full cytochrome *b* sequences. *Mar. Mamm. Sci.* **15**, 619–648.

Lint, D. W., Clayton, J. W., Lillie, W. R., and Postma, L. (1990). Evolution and systematics of the beluga whale, *Delphinapterus leucas,* and other odontocetes: A molecular approach. *In* "Advances in Research on the Beluga Whale, *Delphinapterus leucas*" (T. G. Smith, D. J. St. Aubin, and J. R. Geraci, eds.), pp. 7–22. Can. Bull. Fish. Aquat. Sci. Vol. 224.

Martin, A. (1996). "Beluga Whales." Voyager Press, Stillwater.

Martin, A. R., and Smith, T. G. (1992). Deep diving in wild, free-ranging beluga whales, *Delphinapterus leucas. Can. J. Fish. Aquat. Sci.* **49**, 462–466.

Martin, A. R., Smith, T. G., and Cox, O. P. (1998). Dive form and function in belugas *Delphinapterus leucas* of the eastern Canadian High Arctic. *Polar Biol.* **20**, 218–228.

O'Corry-Crowe, G. M., Suydam, R. S., Rosenberg, A., Frost, K. J., and Dizon, A. E. (1997). Phylogeography, population structure and dispersal patterns of the beluga whale *Delphinapterus leucas* in the western Nearctic revealed by mitochondrial DNA. *Mol. Ecol.* **6,** 955–970.

Richard, P. R., Martin, A. R., and Orr, J. R. (1997). Study of summer and fall movements and dive behaviour of Beaufort Sea belugas, using satellite telemetry: 1992–1995. *Environ. Stud. Res. Funds* No. 134, 38p.

Seaman, G. A., Lowry, L. F., and Frost, K. J. (1982). Foods of belukha whales (*Delphinapterus leucas*) in western Alaska. *Cetology* **44,** 1–19.

Smith, T. G., and Martin, A. R. (1994). Distribution and movements of beluga, *Delphinapterus leucas,* in the Canadian high arctic. *Can. J. Fish. Aquat. Sci.* **51,** 1653–1666.

Smith, T. G., St. Aubin, D. J., and Hammill, M. O. (1992). Rubbing behaviour of belugas, *Delphinapterus leucas,* in a High Arctic estuary. *Can. J. Zool.* **70,** 2405–2409.

St. Aubin, D. J., Smith, T. G., and Geraci, J. R. (1990). Seasonal epidermal molt in beluga whales, *Delphinapterus leucas. Can. J. Zool.* **68,** 359–367.

Suydam, R. S., Lowry, L. F., Frost, K. J., O'Corry-Crowe, G. M., and Pikok D., Jr., (2001). Satellite tracking of eastern Chukchi Sea beluga whales in the Arctic Ocean. *Arctic*

Turl, C. W. (1990). Echolocation abilities of the beluga, *Delphinapterus leucas:* A review and comparison with the bottlenose dolphin, *Tursiops truncatus. In* "Advances in Research on the Beluga Whale, *Delphinapterus leucas*" (T. G. Smith, D. J. St. Aubin, and J. R. Geraci, eds.), pp. 119–128. Can. Bull. Fish. Aquat. Sci. Vol. 224.

Biogeography

RICK LEDUC

Southwest Fisheries Science Center,
La Jolla, California

Biogeography is the study of the patterns of geographic distribution of organisms and the factors that determine those patterns. Although marine mammals are very mobile and there is an apparent lack of physical barriers in the world ocean, only the killer whale (*Orcinus orca*), the sperm whale (*Physeter macrocephalus*), and perhaps some of the balaenopterids could arguably be considered to have cosmopolitan distributions. Other species have restricted distributions (e.g., coastal South America, Indo-West Pacific), reflecting their ecological requirements and their geographic centers of origin. Because related species tend to have similar ecological requirements and dispersal abilities, the distribution of higher taxa can also show distinct tendencies and restrictions, which reflect the cumulative distributions of their included species. For example, while delphinids, river dolphins, and sirenians have their highest diversity in tropical latitudes, the vast majority of pinniped, ziphiid, and phocoenid species occurs in temperate and polar regions. From a geographic perspective, specific regions can thus be characterized as centers of diversity for these higher taxa, and past global changes in the environment will have influenced their evolutionary history. For example, cooling of the world climates during the Tertiary may

have contributed to the radiation of cold water-adapted pinnipeds and mysticetes.

I. Types of Distributions

At the species level, distribution patterns can be described at different spatial scales. Broadly speaking, individual species are usually limited to certain latitudinal zones, such as tropical, temperate, or polar regions. These descriptions can be refined further into subtropical, cold temperate, and so on and incorporated into patterns of ocean basin or hemisphere endemism. For example, the Clymene dolphin (*Stenella clymene*) occurs only in the tropical Atlantic, the Steller sea lion (*Eumetopias jubatus*) in the cold temperate North Pacific, and the dugong (*Dugong dugon*) in the tropical Indo-West Pacific. On even smaller scales, species may be associated with specific physical features, such as nearshore coastal areas (e.g., the hump-backed dolphins, *Sousa* spp.) or the continental slope (e.g., Baird's beaked whale, *Berardius bairdii*), or with oceanographic features, such as specific water masses or even bodies of freshwater (e.g., the baiji, *Lipotes vexillifer*, and the Baikal seal, *Pusa sibirica*).

A few species, notably some of the baleen whales, are highly migratory, summering at high latitudes and spending the winter breeding season at lower latitudes. Some of the migrating rorqual species occupy (at least seasonally) a wide range of latitudes in both hemispheres, although the movements of the Northern and Southern Hemisphere populations are seasonally offset so that they do not normally co-occur in the tropics. At the other end of the spectrum, there are some species (e.g., the vaquita, *Phocoena sinus,* and the Hawaiian monk seal, *Monachus schauinslandi*) that have highly restricted ranges. If a formerly wide-ranging species is now limited to a small area, its distribution is considered relict.

There are distributions that are described as pan-tropical (or pan-tropical/temperate), exhibited by some delphinids (many species), ziphiids (e.g., Blaineville's beaked whale, *Mesoplodon densirostris*), kogiids (e.g., the pygmy sperm whale, *Kogia breviceps*), and balaenopterids (e.g., Bryde's whale, *Balaenoptera edeni*). A few species and species pairs occur at higher latitudes in both hemispheres but are absent from tropical waters, the so-called antitropical species and species pairs. These are seen in the families Delphinidae (e.g., the right whale dolphins, *Lissodelphis* spp.), Ziphiidae (e.g., the bottlenose whales, *Hyperoodon* spp.), Phocoenidae (e.g., *Phocoena sinus* and Burmeister's porpoise, *P. spinipinnis*), Phocidae (e.g., the elephant seals, *Mirounga* spp.), and Otariidae (e.g., the Guadalupe fur seal, *Arctocephalus townsendi*, and Juan Fernandez fur seal, *A. philippii*).

II. Ecology and History
Determine Distribution

Beyond these descriptive aspects of biogeography are the factors that determine a given species' distribution. Generally, distributions are determined by the ecology and the history of the species. In some cases, distribution is limited because a species may not be adapted for living in certain environments.

For example, tropical delphinids may not range into higher latitudes due to limitations on their abilities to thermoregulate in colder water or find food in different habitats. Tied into this is competition, either from closely related species or from ecologically similar species, which may exclude a species from a particular region in which it could otherwise survive. In the case of South American manatees, it is reasonable to surmise that competition places at least one boundary on species' ranges. Throughout most of its range, the Caribbean manatee (*Trichechus manatus*) occurs in both coastal and riverine habitats. However, it does not range into the Amazon River, where the exclusively freshwater Amazonian manatee (*T. inunguis*) occurs, although it occupies the coastal areas on either side of the river mouth. Here, the two species are parapatric, and competitive exclusion is likely at work.

The role that history plays in biogeographic patterns should not be overlooked, but it is not always evident from contemporary distributions. The dispersal abilities of organisms may partly explain why species occur in some areas and not in others. For example, the lack of otariids in the North Atlantic is probably not due to the lack of suitable habitat, but rather lies in the inability of any North Pacific or South Atlantic species to get there. Of course, one could also tie this into their ecological requirements, in that dispersal to the North Atlantic would be more likely if North Pacific species ranged far enough north for animals to disperse via the Arctic Ocean across northern North America or Eurasia. For some species that have widely separated allopatric populations (e.g., Commerson's dolphin, *Cephalorhynchus commersonii*), dispersal from one region to the other is a likely explanation for their distribution. In other cases, vicariance events can explain allopatric distributions. For example, the two subspecies of the Indian river dolphin (*Platanista gangetica*) occur in different river systems: the Indus and Ganges-Brahmaputra River. Although the two forms are not presently in contact, these rivers were all part of a single system until the late Pliocene and probably had sporadic connections through stream capture even until historical times. Therefore, the geographic separation of the populations is from a rather recent vicariance event.

Large-scale changes in the environment can have dramatic influences on species' distributions. For example, in times of global cooling, cold boundary currents in the ocean basins may have extended further toward the equator. This, in turn, could have enabled temperate species to disperse across the equator to similar habitats in a different hemisphere, giving rise to the antitropical species. Among the antitropical species and species pairs, some tendencies in their distributions are apparent. For example, for only one group (*Eubalaena* spp.) does the northern counterpart occur in both the North Atlantic and the North Pacific. Although the long-finned pilot whale, *Globicephala melas*, has only been recorded from the North Atlantic and the Southern Hemisphere, 1000+-year-old skulls of this species have been unearthed in Japan. For the rest of the eight or so recognized antitropical species and species pairs or trios, all except *Hyperoodon* have their northern members limited to the North Pacific. Perhaps the oceanographic and climatic conditions that allow transequatorial dispersal for temperate species occur more frequently or become more developed in the Pa-

cific basin than in the Atlantic. These comparisons do not include the latitudinal migrant species, such as many of the species of balaenopterids. For these, their seasonal occurrence at low latitudes greatly facilitates transequatorial dispersal and would not likely require any significant change in oceanographic or climatic conditions.

Beyond the consideration of the underlying mechanisms of a single species' distribution, it is possible to make inferences about the origins of entire ecological communities. Vicariance biogeographers look for congruence between the phylogenetic relationships among species and their geographical distributions. Species distributions can be superimposed on phylogenetic trees to create what are called area cladograms (Fig. 1). If the area cladograms of several unrelated but geographically similar higher taxa are congruent, it is good evidence that a specific sequence of vicariance events operated on all of those taxa as SPECIATION mechanisms. Furthermore, it may allow the researcher to make inferences about the centers of origin for the higher taxa being considered.

Finally, one should try to incorporate the fossil and geologic record when inferring historical mechanisms in biogeography, especially among distantly related taxa. A case in point can be seen in the river dolphins. Among the river dolphins, the Amazon *Inia geoffrensis* and the franciscana *Pontoporia blainvillei* appear to be the closest (albeit very distant) relatives among the extant species, with the former occupying several South American rivers that flow into the Atlantic and the latter occurring along the Atlantic coast of South America. However, the closest (albeit even more distant) living relative of this pair is probably the baiji (*Lipotes vexillifer*), which is only found in the Yangtze River in China. Considering their freshwater and nearshore habits, it is not obvious from their contemporary distributions how they came to occupy areas a world apart. However, the fossil record has yielded intermediate species from various localities across the North Pacific. Furthermore, the geologic record shows that in the late Pliocene, the major river system in northern South America flowed westward, into what is now the Gulf of Guayaquil in Ecuador. It is thought that the ancestors of *Inia* entered this system from the Pacific. With the uplift of the Andes, much of the river reversed direction and flowed eastward, eventually becoming rivers such as the Amazon and Orinoco. With the North Pacific intermediate forms dying off and the *Inia* lineage splitting to give rise to the ancestors of *Pontoporia* along the coast, one can see how the present-day species distributions came to be. It should be kept in mind, however, that determining the distributions of fossil taxa is notoriously difficult, especially for offshore species that are poorly represented in accessible deposits.

III. Taxonomic Patterns

As mentioned earlier, species within higher taxa share characteristics of their distributions to some degree. It is therefore possible to characterize the distributions of the different groups of marine mammals. The sirenians are primarily a tropical group, with mostly allopatric species occurring in warm coastal waters and some rivers of the Indo-West Pacific and both sides of the Atlantic. The trichechids are represented by two species

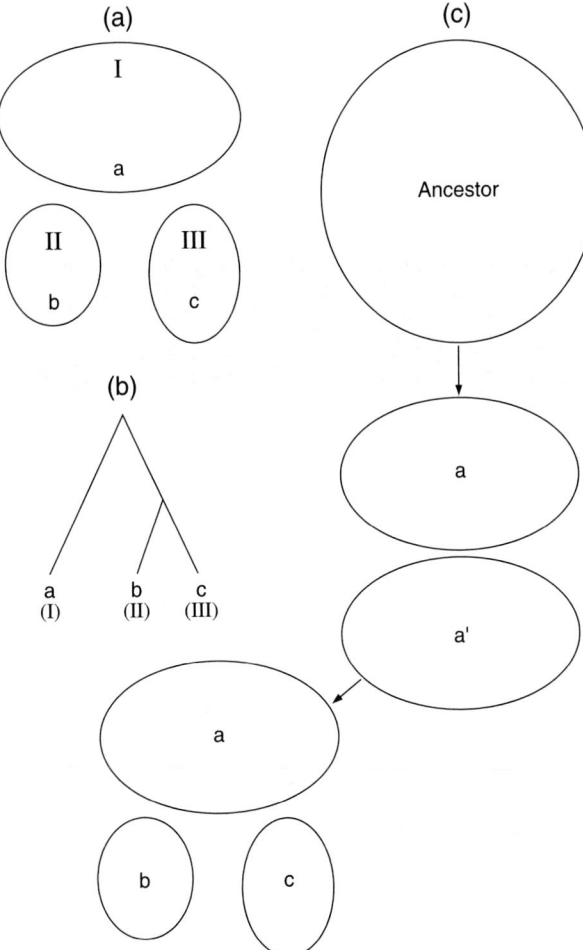

Figure 1 *In vicariance biogeography, speciation patterns are determined by vicariance events. The analysis attempts to reconstruct the sequence of vicariance events using the pattern of evolutionary relationships within a group of related species with allopatric distributions. (a) Species "a," "b," and "c" occupy ranges I, II, and III, respectively. (b) If a phylogenetic analysis determines that "b" and "c" are sister species to the exclusion of "a," this pattern of relationships is applied to their respective geographic ranges in an area cladogram. (c) Under this scenario, the range of the ancestral species is first divided by a vicariance event into a northern and a southern half. Populations in these two areas speciate into species "a" and "a'." Species "a'" is the inferred immediate common ancestor to "b" and "c." A later vicariance event divides the range of "a'" into eastern and western halves, giving rise to species "b" and "c." If unrelated species groups occupying these areas show congruent area cladograms, the support for this sequence of vicariance events is strengthened.*

in the new world (*Trichechus manatus* and *T. inunguis*) and a single congener, the African manatee (*T. senegalensis*), in western Africa, indicating the occurrence of a past trans-Atlantic dispersal event within that lineage. The family Dugongidae, formerly more diverse and widespread, now has only one ex-

tant species, the dugong (*Dugong dugon*), that occurs in the Indian and west Pacific Oceans. One recently extinct species of dugongid, Steller's sea cow (*Hydrodamalis gigas*), had a restricted range in the Commander Islands of the North Pacific, an anomalously cool habitat for a sirenian.

The majority of phocid species inhabit cold temperate and polar regions. Although no species occurs in both Northern and Southern Hemispheres, there are numerous species that are circumpolar either in the Arctic (e.g., the bearded seal, *Erignathus barbatus*) or in the Antarctic (e.g., the crabeater seal, *Lobodon carcinophagus*). In fact, all of the southern phocid species have very broad distributions; their range expansions were probably assisted by the oceanic currents that traverse all longitudes in the Southern Ocean. In the Northern Hemisphere, however, the habitats and ocean currents are much more fragmented by the continental land masses. In addition to the circumpolar species, there are northern species that have more restricted ranges, either endemic to a single ocean basin (e.g., the gray seal, *Halichoerus grypus*), or limited to landlocked bodies of water (e.g., the Caspian seal, *Pusa caspica*). In contrast to the rest of the family, the three recent (two extant) species of monk seals (*Monachus* spp.) inhabit(ed) warmer waters of the Mediterranean and eastern Atlantic, Caribbean, and Hawaii. The spread of monk seals to Hawaii must have occurred prior to the rising of the Isthmus of Panama, which has separated the Caribbean and Pacific basins for the past 3 million years.

As a group, the otariids are similar to the phocids in their distribution, although they are less well represented at very high latitudes (near the pack ice) and do not occur in the North Atlantic at all. Also, individual species tend to have more restricted ranges that are widely allopatric from their congeners. For example, the fur seal genus *Arctocephalus* is very widespread in the Southern Hemisphere, represented by six species (with an additional species endemic to the Galapagos Islands and another to the eastern North Pacific). However, there are only a handful of localities where more than one species occurs together; for the most part, the species are allopatric. It appears then that the dispersal abilities of fur seals have allowed them to colonize many areas in the Southern Hemisphere but have not prevented the resulting disjunct populations from speciating. Odobenids are represented by a single circumpolar Arctic species, the walrus (*Odobenus rosmarus*).

Cetacean species exhibit a wide range of distribution patterns. The family Balaenidae includes one circumpolar Arctic species (*B. mysticetus*) and a trio of very closely related species distributed in the North Atlantic, North Pacific, and Southern Hemisphere (*Eubalaena* spp.); until recently, these were thought to comprise a single species. The gray whale (*Eschrichtius robustus*) and the various species of balaenopterids are mostly latitudinal migrants in both hemispheres, although two species (the Bryde's whales, *Balaenoptera brydei* and *B. edeni*) are restricted to tropical and warm temperate waters, and some primarily migratory species include isolated populations that may be nonmigratory (e.g., the humpback whale, *Megaptera novaeangliae*, in the northern Indian Ocean). In addition to the widespread species of minke whale (*Balaenoptera acutorostrata*), the Southern Hemisphere also contains an endemic species of minke whale (*B. bonaerensis*). Similarly, the Southern Hemisphere is also home to two distinct forms (considered subspecies at present) of blue whale (*B. musculus*). In both of these cases, it is not known if the two southern forms represent divergent lineages that arose within the southern ocean or if they were the result of independent dispersal events across the equator.

Sperm whales are virtually cosmopolitan, and the kogiids (*Kogia sima* and *K. breviceps*) are worldwide in tropical and warm temperate waters. Beaked whales show a variety of distribution patterns, including pan-tropical species (e.g., *Mesoplodon densirostris*), antitropical species pairs (*Berardius* spp.), and ocean basin endemics (e.g., Sowerby's beaked whale, *M. bidens*). Some (e.g., the pygmy beaked whale, *M. peruvianus*) are only known from a few strandings within limited geographic areas. For most species of sperm whales and beaked whales, so little is known about their habits and ecological needs that it is difficult to hypothesize about the mechanisms that have led to their present distributions.

Three of the four species of river dolphins (*Inia geoffrensis*, *Lipotes vexillifer*, and *Platanista gangetica*) are strictly freshwater in specific tropical river systems, with the fourth species (*Pontoporia blainvillei*) having a restricted marine coastal range. The two species of monodontids (the narwhal, *Monodon monoceros*, and the beluga, *Delphinapterus leucas*) are circumpolar in the north and are among the few resident polar cetaceans, although fossil species of this family occurred as far south as San Diego, California.

Apart from a single Indo-West Pacific coastal species that also ranges into freshwater (the finless porpoise, *Neophocaena phocaenoides*), the phocoenids are strictly marine and cold temperate to warm temperate in distribution, some with very restricted ranges (e.g., *Phocoena sinus*). Only one phocoenid, the harbor porpoise (*P. phocoena*), has invaded the North Atlantic, becoming very widespread in both oceans of the Northern Hemisphere and even establishing isolated populations in the Black Sea and off West Africa.

The most speciose family of marine mammals, the delphinids, shows a wide variety of distributions, from pan-tropical species (e.g., the pantropical spotted dolphin, *Stenella attenuata*) to ocean basin endemics (e.g., the white-beaked dolphin, *Lagenorhynchus albirostris*) to species with wide-ranging but disjunct populations (e.g., the long-beaked common dolphin, *Delphinus capensis*). Many delphinids are pelagic, although some inhabit coastal waters (e.g., *Cephalorhynchus* spp.) and some even invade freshwater (e.g., the tucuxi, *Sotalia fluviatilis*). Only one, *Orcinus orca*, seems to regularly range to the pack ice in the far north and south. For the many pan-tropical/warm temperate species, the continental land masses effectively separate the populations inhabiting the Indian and Pacific Oceans from those inhabiting the Atlantic Ocean, raising the question of how they came to inhabit all the ocean basins. It has been hypothesized that during warm climatic periods, warm water extended far enough south to allow interchange and range expansion around the Cape of Good Hope. This would enable some species to become pan-tropical in their distribution, and the subsequent retreat of the warm weather and isolation of populations could provide a speciation mechanism for the es-

tablishment of the tropical species endemic to the Atlantic Ocean (the Atlantic spotted dolphin, *S. frontalis*, and *S. clymene*).

IV. Conclusion

Why do species live where they do? Answering such a simple question requires the examination of clues from the past as well as the present. Biogeography involves such diverse disciplines as geology, paleontology, ecology, physiology, behavior, and systematics. For marine mammals, studying biogeographic patterns presents real challenges. There is a paucity of information about past distributions and habitats, gaps in our knowledge of contemporary and recent distributions, uncertainties about evolutionary relationships, and a tremendous amount to learn about the basic ecology and physiology of many marine mammals.

See Also the Following Articles

Cetacean Evolution ■ Distribution ■ Ecology, Overview ■ Fossil Record ■ Geological Time Scale ■ Speciation

References

Au, D. W. K., and Perryman, W. L. (1985). Dolphin habitats in the eastern tropical Pacific. *Fish. Bull.* **83,** 623–643.

Brown, J. H., and Gibson, A. C. (1983). "Biogeography." Mosby, St. Louis, MO.

Cox, C. B., and Moore, P. D. (1985). "Biogeography," 4th Ed. Blackwell Scientific, Cambridge, MA.

Davies, J. L. (1963). The antitropical factor in cetacean speciation. *Evolution* **17,** 107–116.

Myers, A. A., and Giller, P. S. (eds.) (1988). "Analytical Biogeography." Chapman and Hall, London.

Nelson, G., and Rosen, D. E. (eds.) (1981). "Vicariance Biogeography: A Critique." Columbia Univ. Press, New York.

Rice, D. W. (1998). "Marine Mammals of the World: Systematics and Distribution." Soc. Mar. Mamm. Spec. Pub. 4. Society for Marine Mammalogy, Lawrence, KS.

Wiley, E. O. (1988). Vicariance biogeography. *Annu. Rev. Ecol. System.* **19,** 513–542.

Bioluminescence

GREGORY K. SILBER
National Marine Fisheries Service,
Silver Spring, Maryland

I. Description

Bioluminescence is the ability of certain organisms to generate visible light as a result of a chemical reaction. Although different animals use different chemicals, these are all variants of two chemicals: the substrate is called luciferin and the enzyme is luciferase. The chemical production of light is very efficient and does not generate heat like all other light sources.

Bioluminescence has a number of important biological and ecological functions. Light emission is used to avoid or confuse predators (e.g., shrimp and squid), attract prey (e.g., angler and flashlight fishes), and help attract mates (e.g., lanternfish and sea fireflies). Many organisms use bioluminescence for predator avoidance using a mechanism called counterillumination, in which the color and intensity of sunlight or moonlight entering the water column is matched to reduce the animal's silhouette as it appears from below.

II. Occurrence of Bioluminescence

Although rare on land, bioluminescence is extremely widespread in the marine environment and it would be difficult to find a location where it does not exist. It occurs in all oceans, all latitudes, and at all depths. Bioluminescent systems are believed to have evolved independently at least 30 different times.

In the sea, the photic zone occurs, in ideal conditions, only tens to hundreds of meters into the water column. Therefore, light that occurs below the photic zone is generated by the organisms themselves. In fact, most bioluminescence occurs in the very deep ocean, and more than 90% of marine organisms from 300 to 3000 feet glow—a tribute to the importance of bioluminescence to a wide variety of organisms that inhabit waters where sunlight does not penetrate. However, a number of invertebrate species that occur within the photic zone also exhibit bioluminescence.

Among the thousands of bioluminescent marine organisms, the majority are planktonic. There are also enormous numbers of other species, including lanternfish (Myctophidae), squid, decapod shrimp (e.g., the oplophorids and sergestids), krill, copepods, ostracods, amphipods, and gelatinous zooplankton. The benttooth bristlemouth (*Cylcothone* spp.) bioluminesces and may be one of the most abundant vertebrates on earth. These creatures are vastly important components of marine ecosystems and help form the basis of many marine food webs.

Near the surface, bioluminescence is often caused by dinoflagellates, single-celled algae. The bioluminescent flash this alga produces when physically agitated is probably related to predator avoidance. The flash may cause flight in copepod predators because the flash may have attracted the attention of something larger that might eat the copepod. Frequently found in vast groups, these organisms glow in the wake of swimming fish or dolphins (see later) or passing ships.

III. Bioluminescence and Marine Mammals

There is no known example of marine mammals emitting true bioluminescence. Nonetheless, it is undoubtedly a common part of the marine mammal ecosystem and it may be used by some species to locate prey. Much of the information about marine mammals and bioluminescence is anecdotal.

As noted earlier, many squid species bioluminesce. Inasmuch as squid are the basis of the diet for many marine mammals, bioluminescing prey may be an important cue for locating prey. Sperm whales (*Physeter macrocephalus*) dive to great depths and are known to feed on a variety of species, including many deepwater species and giant squid. Thus, sperm whales may be aided in locating prey that are bioluminescent, although the species' echolocation capabilities are almost certainly the primary means of locating its prey. A number of deep-diving marine mammals [e.g., elephant seals (*Mirounga* spp.), beaked whales, and other odontocete cetaceans] probably key on bioluminescing organisms when searching for prey. This raises the question, however, of bioluminesce being maladaptive if it increases the chances that an organism will fall victim to predation. The social and ecological significance of bioluminescence must outweigh the cost of announcing an organism's presence to a predator.

Some copepod species are among those that bioluminesce, including some that are preyed upon by skim-feeding large whale species, such as right whales (*Eubalaena* spp.). Some scientists believe that this attribute is an important visual cue to prey location for foraging right whales.

Vast swarms of dinoflagellates that bioluminesce when agitated are well known to all mariners and are occasionally seen by divers and beachgoers. These organisms bioluminesce when stirred by the mechanical action of a boat hull passing through them or by wave action. The same phenomenon has been seen when dolphins, riding on the pressure wave in front of a boat, for example, swim through the dinoflagellate swarms.

Vision, and therefore light, is vital to prey capture in many marine mammal species. While not true bioluminescence, cetaceans may use reflected ambient light in the capture of prey. For example, the inside surfaces of the mouth and the lower jaw of some sperm whales often have light-colored pigment. Some scientists believe that the reflective surface may be attractive to some prey, e.g., squid, and that prey actually move toward the reflective surface. This reflectivity may be enhanced by bioluminescence occurring on the white lower jaws of sperm whales who have been feeding on bioluminescent squid, a suggestion made because sperm whale jaws on whaling ships have been seen to "glow." Also, several authors have suggested that the asymmetric (i.e., present on the left side, but absent on the right) coloration patterns on the jaw and head of fin whales (*Balaenoptera physalus*) may serve as a light-colored backdrop by which small schooling fish are corralled. These are intriguing hypotheses that have not been tested. Of course, epidermal pigmentation patterns also play important social signaling roles for marine mammals that live in groups.

See Also the Following Articles

Coloration ■ Diet ■ Plankton ■ Vision

References

Gaskin, D. E. (1967). Luminescence in a squid *Moroteuthis* sp. (probably *ingens* Smith), and a possible feeding mechanism in the sperm whale *Physeter catodon* L. *Tuatara* **15,** 86–88.

Herring, P. J. (1990). Bioluminescent communication in the sea. *In* "Light and Life in the Sea" (P. J. Herring, A. K. Campbell, M. Whitfield, and L. Maddock, eds.), pp. 245–264. Univ. Press, Cambridge.

Herring, P. J., Campbell, A. K., Whitfield, M., and Maddock, L. (eds.) (1990). "Light and Life in the Sea." Univ. Press, Cambridge.

Nealson, K. H. (ed.) (1981). "Bioluminescence: Current Perspectives." Burgess, Minneapolis.

Biomagnetism

Michael M. Walker
University of Auckland, New Zealand

Biomagnetism is the study of the magnetic mineral, magnetite, in living organisms (Kirschvink *et al.*, 1985) and is extended here to include the hypothesis that magnetite is used for magnetic field detection. The earth's magnetic field may well be the only reliable source of navigational information in the open ocean through which cetacean and pinniped mammals often travel large distances (tens to thousands of kilometers). There has naturally been widespread interest in the hypothesis that marine mammals navigate magnetically, but there is as yet no direct experimental evidence for a magnetic sense in any cetacean or pinniped species. Studies of any kind have only been done on cetaceans, which are difficult to study experimentally because of their large size. In recent years, however, much has been learned from experimental studies of other vertebrates and it is reasonable to guess that the findings for other vertebrates will also largely apply to marine mammals.

I. Association of Mass-Stranding Sites with Magnetic Field Features

The locations where whales mass strand have been correlated with variations in the intensity of the earth's magnetic field. Klinowska (1985) proposed the hypothesis that otherwise healthy whales that strand themselves alive have made a serious navigational mistake and that examination of the circumstances surrounding stranding sites should provide clues as to the sensory system(s) responsible for the error. Klinowska (1985), Kirschvink *et al.*, (1986), and Kirschvink (1990) found that whales tended to strand at magnetic intensity "valleys," or minima, along the coastlines of Great Britain and the United States. Although similar studies in other areas could not extend these results (e.g., Brabyn and McLean, 1992), strandings have thus provided the first tantalizing evidence that whales use the earth's magnetic field for navigation.

II. Response by Migrating Fin Whales to the Earth's Magnetic Field

An approach derived from that used to investigate strandings has also shown that fin whales (*Balaenoptera physalus*) apparently respond to variations in the intensity of the earth's magnetic field during migration. Sighting positions off the eastern coast of the United States for fin whales, as well as for sharks, turtles, and other whales, were recorded by the Cetacean and Turtle Assessment Program (CETAP) during the late 1970s and early 1908s. During the spring and autumn migration seasons, sighting positions for fin whales were preferentially associated with areas of low magnetic intensity and in-

tensity gradient (Walker *et al.*, 1992), much like the intensity valleys or minima with which strandings were associated. Exclusion of whales that were engaged in feeding behavior when sighted increased the strength of the association with areas of low magnetic intensity and intensity gradient. These results suggested that the fin whales responded to the earth's magnetic field during active migration but not necessarily at other times.

III. Experimental Studies of the Magnetic Sense

Although the just-described findings are consistent with the existence of a magnetic sense in whales, whether whales or pinnipeds actually have a magnetic sense will require experimental demonstration of at least behavioral responses to magnetic fields. Large size and constraints on experimentation have severely restricted laboratory studies on the magnetic sense of cetaceans. Thus Bauer *et al.* (1985) unsuccessfully tested Atlantic bottlenose dolphins (*Tursiops truncatus*) for sensitivity to magnetic fields in a series of conditioning experiments. It is probably premature, however, to conclude that cetaceans do not have a magnetic sense as only a single species has been tested under a limited range of experimental conditions.

The search for evidence of a magnetic field detection mechanism in whales is complicated again by their large size and the fact that the sensory cells responsible for detecting magnetic fields may be relatively small in number and need not be organized into a sense organ. Searches for the detector cells in whales have concentrated on locating deposits of the magnetic mineral, magnetite, in the heads of four odontocete and one mysticete species. The crystals of magnetite that are most suitable for use in magnetic field detection are extremely small (<100 nm) and searching for them in the bodies of whales truly is a search for a needle in a haystack. Bauer *et al.* (1985) did find, however, that magnetic material is consistently located in the dura mater, the tissue that surrounds the brain. At least some of this magnetic material is consistent with the very small magnetite particles that will be most useful for magnetic field detection. What has not been possible so far is the unique identification of magnetite in the dura mater tissue and demonstration that it is associated with magnetic field detection in any way.

IV. Evidence from Studies of Other Vertebrates

In the last decade, a good deal has been learned about detection and use of the earth's magnetic field by other vertebrates, much of which is also likely to apply to marine mammals. Turtles, fish, and birds are all known to respond to both the intensity and the direction of magnetic fields in laboratory or field experiments, and a variety of magnetic effects have been demonstrated on the navigation of homing pigeons (e.g., Lohmann and Lohmann, 1996; Semm and Beason, 1990; Walker *et al.*, 1997; Wiltschko and Wiltschko, 1995). In addition, candidate magnetite-based magnetoreceptor cells have been identified in the rainbow trout (Walker *et al.*, 1997). A magnetic sense thus appears to be widespread among homing and migratory vertebrates, and it seems likely from the evidence available so far that marine mammals also have a magnetic sense that they use to navigate.

V. Outlook

Given the difficulties encountered so far, it is worth considering how to make better progress toward the understanding of biomagnetism in marine mammals. What is required in the first instance is the careful selection of subject animals and robust experimental techniques to demonstrate and characterize the magnetic sense. Although research has in the past focused on cetaceans, the results that have been obtained with other vertebrates suggest that pinnipeds are also likely to have a magnetic sense. The opportunity therefore exists for investigation of an important vertebrate group that behaves well in experimental situations and that may in fact be easier to study than the cetaceans. If biomagnetism is used by marine mammals as a part of a system of navigation, it may be used more regularly by open-ocean migrators such as pelagic whales and dolphins than by nearshore bottlenose and other dolphins.

See Also the Following Articles

Brain ▪ Echolocation ▪ Migration and Movement Patterns ▪ Stranding

References

Bauer, G. B., Fuller, M., Perry, A., Dunn, J. R., and Zoeger, J. (1985). Magnetoreception and biomineralization of magnetite in cetaceans. *In* "Magnetite Biomineralization and Magnetoreception in Living Organisms: A New Biomagnetism" (J. L. Kirschvink, D. S. Jones, and B. J. MacFadden, eds.), pp. 489–507. Plenum Press, New York.

Brabyn, M. W., and McLean, I. G. (1992). Oceanography and coastal topography of herd-stranding sites for whales in New Zealand. *J. Mamm.* **73**, 469–476.

Kirschvink, J. L. (1990). Geomagnetic sensitivity in cetaceans: An update with live stranding records in the United States. *In* "Sensory Abilities of Cetaceans" (J. A. Thomas and R. A. Kastelein, eds.), pp. 639–649. Plenum Press, New York.

Kirschvink, J. L., Dizon, A. E., and Westphal, J. A. (1986). Evidence from strandings for geomagnetic sensitivity in cetaceans. *J. Exp. Biol.* **120**, 1–24.

Kirschvink, J. L., Jones, D. S., and MacFadden, B. J. (eds.) (1985). "Magnetite Biomineralization and Magnetoreception in Living Organisms: A New Biomagnetism." Plenum Press, New York.

Klinowska, M. (1985). Cetacean live stranding sites relate to geomagnetic topography. *Aquat. Mamm.* **1**, 27–32.

Lohmann, K. J., and Lohmann, C. M. F. (1996). Detection of magnetic field intensity by sea turtles. *Nature* **380**, 59–61.

Semm, P., and Beason, R. C. (1990). Responses to small magnetic field variations by the trigeminal system of the bobolink. *Brain Res. Bull.* **25**, 735–740.

Walker, M. M., Diebel, C. E., Haugh, C. V., Pankhurst, P. M., Montgomery, J. C., and Green, C. R. (1997). Structure and function of the vertebrate magnetic sense. *Nature* **390**, 371–376.

Walker, M. M., Kirschvink, J. L., Ahmed, G., and Dizon, A. E. (1992). Evidence that fin whales respond to the geomagnetic field during migration. *J. Exp. Biol.* **171**, 67–68.

Wiltschko, R., and Wiltschko, W. (1995). "Magnetic Orientation in Animals." Springer-Verlag, Berlin.

Blowing

JAMES L. SUMICH
Grossmont College, El Cajon, California

A blow is one of the most visible behaviors of whales when they are observed at the sea surface. Whale blows represent the rapid emptying, or expiration, of whales' lungs through their blowholes in preparation for the next inspiration. The visibility of a blow is due to a mixture of vapor and sea water entrained into the exhaled column of air at the sea surface (Fig. 1). When a blow occurs below the sea surface, as it sometimes does, such as bubble blasts in gray whales (*Eschrichtius robustus*) or bubble trains in humpback whales (*Megaptera novaeangliae*), it may be intended as a signaling function to other nearby whales.

A complete BREATHING cycle typically consists of a very rapid expiration (the blow) immediately followed by a slightly longer and much less obvious inspiration and then an extended yet variable

Figure 1 *Towering blow of a blue whale. Courtesy of Phil Colla.*

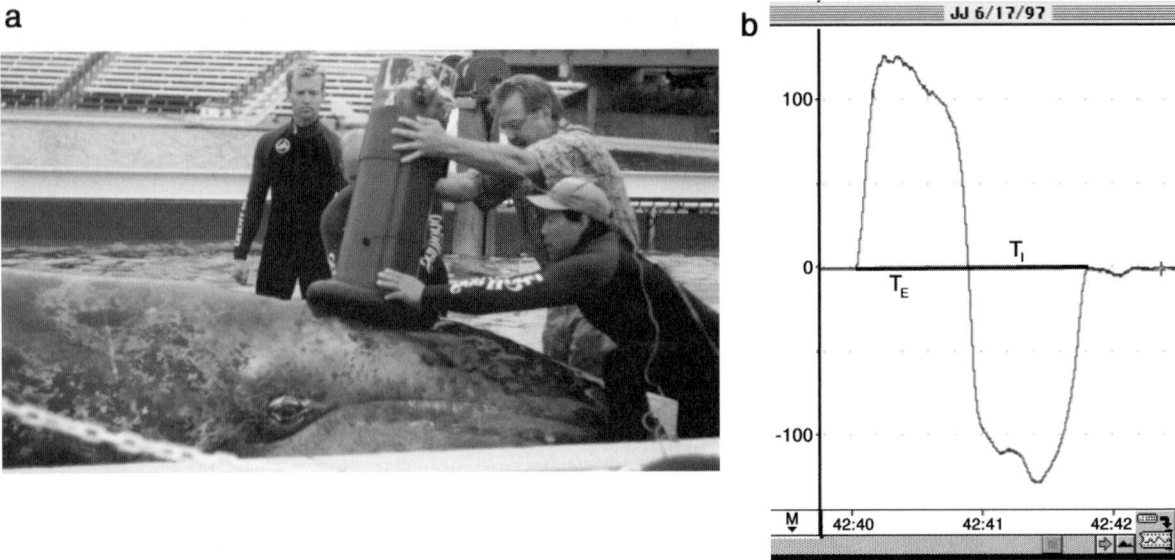

Figure 2   *(a) Measuring ventilatory flow rates of a young captive gray whale with a pneumotachograph. (b) An example of a single expiratory/inspiratory event (vertical scale = flow rate, 1/sec; horizontal scale = time, min, sec).*

period of breath holding, or apnea. The rapid expiration of a typical whale blow (from as little as 0.3 sec in small dolphins and porpoises to 1–2 sec in large baleen whales) provides more time to complete the next inspiration as the blowhole of a swimming animal breaks through the sea surface and results in little delay before submerging again (Kooyman and Cornell, 1981).

The rapidity of the blow is accomplished by maintaining high flow rates throughout almost the entire expiration, which is in strong contrast to humans and other land mammals. The high expiratory flow rates of cetaceans are enhanced by very flexible chest walls and by cartilage reinforcement of the smallest terminal air passages of the lungs to prevent them from collapsing until the lungs are almost completely emptied.

Rates of air flow during a blow are measured with a pneumotachograph (Fig. 2), a device that measures instantaneous flow rates through the duration of the blow, which can be integrated over time to determine the volume of expired air for each blow. The largest lung volume measured in this manner was nearly 200 liters for a 10-month-old gray whale, and lung volumes approaching 5000 liters are estimated for very large blue whales (*Balaenoptera musculus*).

Blow patterns of whales vary, depending on their behaviors. In small dolphins and porpoises SWIMMING at low SPEEDS, blowhole exposure during a blow is minimal and gradually changes to porpoising above the sea surface at higher speeds. Porpoising behavior may serve to conserve energy because it is the most efficient way to breathe at high swimming speeds (Fish and Hui, 1991). When migrating, larger whales typically surface to blow several times in rapid succession and then make an extended dive of several minutes duration (Fig. 3). In this

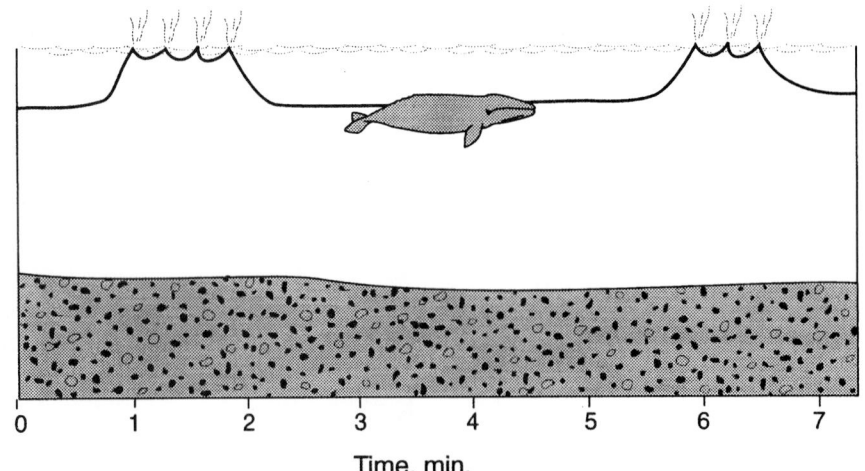

Figure 3   *Typical blowing and diving pattern of a migrating gray whale.*

manner, they reduce the amount of time they spend in the zone of high fractional drag near the sea surface. Thus, blowing in these mammals represents their best compromise in satisfying their oxygen demands while simultaneously avoiding large additional costs of energy for locomotion.

See Also the Following Articles

Breathing ■ Communication ■ Diving Physiology

References

Fish, F. E., and Hui, C. A. (1991). Dolphin swimming: A review. *Mamm. Rev.* **21,** 181–195.

Kooyman, G. L., and Cornell, L. H. (1981). Flow properties of expiration and inspiration in a trained bottle-nosed porpoise. *Physiol. Zool.* **54,** 55–61.

Blubber

SARA J. IVERSON
*Dalhousie University, Halifax,
Nova Scotia, Canada*

Blubber, a dense vascularized layer of fat beneath the skin, is one of the most well-known and universal characteristics of marine mammals. Although it is not strictly present in polar bears (*Ursus maritimus*) or sea otters (*Enhydra lutris*), all cetaceans, sirenians, and pinnipeds have blubber and it may comprise up to 50% of the body mass of some species at certain life stages. Blubber has long been recognized as the primary and most important site of fat, and thus also energy, storage in marine mammals. However, blubber also has a number of other important functions. The blubber layer serves as an insulator in mammals living in often cold marine environments and is thus central to their entire process of thermoregulation. Blubber also affects buoyancy and functions as a body streamliner and elastic spring for efficient hydrodynamic locomotion. Although blubber is a dynamic tissue, which can reflect both nutritional state and life history stage of individuals, the tissue itself has likely evolved to best suit the lifestyles, stresses, and constraints of specific groups and even individual species of marine mammals. Hence, the study of blubber can provide unique insights not only into phylogenetic relationships and environmental adaptations, but also into aspects of individual feeding habits, foraging ecology, species DISTRIBUTION and demography that are otherwise difficult to study.

I. Structure of Blubber

A. Tissue Characteristics

Blubber is a specialized subcutaneous layer of fat found only in marine mammals and is different from other types of adipose tissue in that it is anatomically and biochemically adapted to serve as an efficient and adjustable thermal insulator. The blubber layer is almost continuous across the body of marine mammals, lying over (but not tightly fixed to) the underlying musculature but absent on appendages. Although nearly continuous, the thickness, structure, and biochemical composition of the blubber can vary greatly over the body of an individual in some species, and these differences are likely associated with localized differences in function. Consistent with its role as an insulator, there also usually exists a thermal, as well as biochemical, gradient through the depth of the blubber layer. The outer layer (nearest the skin) is usually cooler than the inner layer (nearest the muscle or body core) and activities of individual enzymes in each of these locations may be adapted to function at the different respective temperatures. The polar bear also deposits huge quantities of fat subcutaneously, which likely provides some degree of insulation. However, this superficial adipose tissue does not appear to be a specific anatomical adaptation for that purpose as it does not differ in structure from the superficial fat depots of other large terrestrial carnivores (Pond *et al.,* 1992).

Blubber, like other adipose tissue, is composed of numerous fat cells called adipocytes. Adipocytes develop prior to filling with fat and are composed, like other cells, of mostly protein and water. Once developed, adipocytes can alternately fill and empty with lipid and thus can change greatly in size. Mature adipocytes are generally large and spherical and packed densely into adipose tissue. The cells are surrounded and held in place by a mesh of structural collagen fibers. While most other types of adipose tissue contain small to moderate amounts of collagen, blubber is distinct in being greatly enriched in collagen and elastic fibers. This gives blubber a firm, tough, and fibrous character from which it derives much of its mechanical and functional properties. The histological structure of the blubber in pinnipeds is relatively uniform throughout its depth. However, in some cetaceans, there is a distinct stratification of the tissue into an inner, middle, and outer layer based on the size, shape, and metabolic characteristics of adipocytes, as well as on the lipid and collagen content of the tissue. Blubber also contains numerous blood vessels and specialized shunts called arterio-venous anastomoses (AVAs), which allow larger and swifter blood flow than would be possible through capillaries alone and are important to the thermoregulation process. The blubber of manatees is unusual in that a layer of muscle is imbedded in the middle of the ventral blubber layer; however, a possible functional significance for this arrangement is not known.

B. Variation in Thickness and Proximate Composition

The thickness of the blubber layer varies among species. In general, because body volume increases more rapidly than body surface area, larger species tend to have greater maximum blubber thickness. Thus, the depth of the blubber layer can be commonly 7–10 cm in adult pinnipeds, 20–30 cm in fin whales (*Balaenoptera physalus*), and up to 50 cm in the bowhead whale (*Balaena mysticetus*). In contrast, in one of the smallest odontocetes, the harbor porpoise (*Phocoena phocoena*), blubber depth generally reaches only 2.5–3.0 cm.

Beyond general species characteristics, the amount, depth, and chemical composition of the blubber also vary with age, nutrition, and reproductive status. The adipose tissue of many newborn mammals is empty of lipid, filling quickly after birth during the lactation period. Proliferation of fat depots in immature mammals is due to an increase in both adipocyte numbers and size. However, in adults, changes in the size of fat depots are primarily due to filling or emptying of adipocytes. The same appears largely true in the case of blubber. Although neonates of large baleen whales are born with a blubber layer that is several centimeters thick, most pinniped neonates are born with very little blubber, at less than 3 mm in depth in some otariids and accounting for less than 5–6% of body mass in most phocids. Most newborn pinnipeds rely instead primarily on fur and delayed entry into the water. For instance, in newborn harp seals (*Pagophilus groenlandicus*), blubber represents less than 6% of body mass and contains only 20% lipid (Worthy and Lavigne, 1983). This changes rapidly during the 12-day lactation period, such that the blubber of a newly weaned harp seal can comprise up to 50% of body mass and contains greater than 90% lipid, representing abundant and replete fat cells. In contrast, during reduced food intake or fasting, lipid is mobilized rapidly from adipocytes and hence undernourished marine mammals are characterized by greatly reduced blubber thickness and lipid content. Likewise, during annual events associated with fasting, such as lactation or molting in some species, blubber is also reduced in depth and lipid content with fat mobilization. For instance, the sternal blubber of female harbor seals (*Phoca vitulina*) changes during the 24-day lactation period from 3.8 to 1.4 cm in depth and from 92.3% lipid, 2.2% protein, and 5.5% water to 76.9% lipid, 5.9% protein, and 17.2% water (Bowen *et al.*, 1992); i.e., the increases in protein and water content reflect the larger proportion of "emptier" fat cells.

C. Lipids in Blubber

Depot lipid in animals is stored predominantly as triacylglycerols, which consist of three fatty acids esterified (i.e., linked by an ester bond) to a glycerol (three-carbon alcohol) molecule. The synthesis, storage, and catabolism of fatty acids are the components of lipid energy metabolism. Fatty acids in the marine food web are exceptionally complex and are characterized by high levels of long-chain polyunsaturated fatty acids (PUFA). During digestion of triacylglycerols by mammals, fatty acids are released from the glycerol backbone but not degraded and they are carried in the bloodstream and taken up by tissues the same way. These fatty acids are then either used for energy or stored as triacylglycerols in adipose tissue. Thus, fatty acids travel up the food chain intact, and because the kinds of fatty acids that can be biosynthesized or modified in mammals are quite limited, most fatty acids found in marine mammal blubber arise from the dietary intake of fish and other prey lipids. Hence, marine mammal blubber lipid is usually characterized by high levels of long-chain PUFA as well as unique fatty acids produced at lower trophic levels of the marine ecosystem.

Marine mammals, like other mammals, can also synthesize some of their own fatty acids from sources such as dietary amino acids consumed in excess of body needs (glucose would be another source but is scarce in diets of marine mammals). These synthesized fatty acids are usually restricted to those with 16 or 18 carbon atoms and usually, at most, one double bond (i.e., 16:0, 16:1n-7, 18:0, and 18:1n-9). Although these fatty acids are also common in all prey items of marine mammals, some are undoubtedly deposited in marine mammal blubber from biosynthesis.

The are several exceptions to the general characteristics of marine mammal blubber lipids described earlier. In addition to the usual fatty acids that are synthesized by all mammals, one very unusual fatty acid, isovaleric acid, is also found in the blubber of some species of toothed whales (odontocetes), which can arise only from biosynthesis. Isovaleric acid is unusual in that it is both very short (5 carbons) and branched. When present, it is found in highest concentrations in the outermost layer (nearest the skin) of blubber (Koopman *et al.*, 1996). Additionally, besides the most common form of storage lipid, triacylglycerols, some marine mammals (primarily some of the odontocetes) store some or all of their fatty acids in blubber as wax esters. A wax ester is a single fatty acid esterified to a long chain (22–34 carbon) alcohol. In general, wax esters are firm, stable, and resistant to degradation. This is why sperm whale (*Physeter macrocephalus*) oil was popular as an illuminant in the last century.

II. Role of Blubber in Temperature Regulation: Heat Conservation and Dissipation

As a whole body envelope of insulation, blubber is central to thermoregulation in marine mammals. Marine mammals, like all mammals, are homeothermic endotherms and hence need to maintain a stable body core temperature of about 37°C in cooler (usually <25°C) and often much colder (−1–5°C) fluid environments. Additionally, heat is always lost far more rapidly to water than to air because the thermal conductivity of water is 25 times that of air. There are several ways marine mammals have dealt with this problem. One is to increase body size, which decreases the surface-to-volume ratio and thus provides less surface area per unit volume over which to lose heat. Even the smallest marine mammals are considered large mammals, being one to two orders of magnitude larger than small terrestrial mammals such as rodents and insectivores. Additionally, and perhaps more importantly, large body size generally allows for thicker insulation (be it fur or blubber), which further decreases heat conductance.

Although fur is a far more effective insulator than blubber in air (and is used as the sole means of insulation by sea otters), fur acts by trapping pockets of air (a poor thermal conductor) among hairs, which then forms the effective insulative layer. Thus potentially when fur is wetted, but more significantly when diving under pressure (as most marine mammals do), fur gets compressed, expelling the air layer and thus losing its insulative properties. In contrast, blubber does not compress with depth and it is a good insulator because, like air, it has a lower thermal conductivity than water.

Adipose tissue is also less metabolically active than other tissues and thus requires less perfusion by blood, which would otherwise tend to cause heat loss at the body surface. Nevertheless, because blubber is vascularized, circulatory adjustments allow both heat conservation and dissipation as necessary. An important means of regulating heat transfer in marine mammals is by restricting and diverting blood flow through the blood vessels and AVAs in blubber. Restricting blood flow to the blubber's surface (i.e., skin) conserves body heat and allows blubber to act as an effective insulator against cold. Conversely, increasing blood flow into the blubber allows sometimes massive redistribution and dumping of body heat in cases of either very warm water or air or during intense activity (e.g., Heath and Ridgway 1999).

The effectiveness of blubber as an insulative layer depends on its thickness, lipid content, and lipid composition. As an insulation layer increases in thickness, the lower critical temperature of an animal decreases and thus the animal can accommodate a broader ambient temperature range without having to increase its metabolism for heat production (i.e., to remain thermoneutral). As mentioned previously, many marine mammals, especially those of larger body size, possess a thick blubber layer, allowing them to remain thermoneutral at most of the temperatures of the world's oceans and, for some pinnipeds, even at air temperatures of −10 to −20°C on polar ice. However, smaller species are limited in the depth of blubber they can carry and also have relatively more surface area over which to lose heat. Hence most of the smallest cetaceans do not occur at high latitudes. Less insulation increases the lower critical temperature of an animal and requires increased metabolism for heat production. The harbor porpoise is the smallest cetacean species to inhabit temperate waters of the Northern Hemisphere. Although its blubber depth is only several centimeters thick, it is generally twice the thickness and contains more lipid than does a similarly sized dolphin inhabiting tropical waters. These properties appear to confer up to four times greater insulative capacity (Worthy and Edwards, 1990).

Depletion of lipid from blubber stores will decrease the insulative capacity of the tissue and may seriously compromise an individual, especially if nutritionally stressed. Thus, small species such as the harbor porpoise must feed nearly continually to maintain metabolism and to preserve their blubber's thickness and insulative capacities. In contrast, large whales can fast and mobilize blubber reserves for weeks or months, yet can remain thermoneutral due to a low surface-to-volume ratio as well as the maintenance of a still relatively thick blubber layer. Especially in cetaceans, the thickness, structure, and insulative properties of the blubber may vary across different regions of the body and thus the function of the blubber as an insulator may also vary regionally.

Variation in the lipid composition of blubber may also confer differing insulative capacities. As stated previously, the blubber of marine mammals is composed of large amounts of unsaturated fatty acids. Unsaturated fatty acids have lower melting points than saturated fatty acids. Thus even when the temperature of the outermost layer of the blubber and skin is near that of cold ambient temperatures, blubber tissue can remain fluid and an effective insulator if the melting point of its fatty acids is low. Saturated and monounsaturated fatty acids abundant in marine mammal blubber (e.g., 16:0, 16:1, 18:0, 18:1) have melting points of 13–70°C. However, nutritionally important polyunsaturated fatty acids are usually plentiful in marine mammal diets and thus in blubber, conferring an overall melting point in blubber lipid of less than −15°C. Additionally, in some small coldwater cetaceans such as the harbor porpoise, Dall's porpoise (*Phocoenoides dalli*), the bottlenose dolphin (*Tursiops truncatus*), and the beluga (*Delphinapterus leucas*), high concentrations of the very unusual branched short-chain isovaleric acid are biosynthesized and deposited in blubber (see earlier discussion). Isovaleric acid has an extremely low melting point of −37.6°C, which clearly provides fluidity to especially the outer blubber layer of these animals (e.g., Koopman *et al.*, 1996). Although the exact physiological function of isovaleric acid is not understood, its presence may contribute to the superior insulative properties observed previously in harbor porpoise blubber. In contrast to most other marine mammals, while manatees (*Trichechus* spp.) can also store large amounts of blubber, they generally do not tolerate temperatures below 20°C. As plant eaters, manatees must synthesize the majority of their blubber fatty acids, which would thus be restricted in their degree of unsaturation. However, little is known about the effectiveness of manatee blubber as an insulator in cold temperatures or the role that lipid composition might play in this ability.

III. Role of Blubber in Energy Storage and Water Balance

Blubber, as a rich energy store, is important in the lives of marine mammals because of the critical role that stored lipid plays in their ecology, reproduction, and survival. Perhaps surprisingly, even though marine mammals obviously live in the environment within which they also forage, reproduction and especially lactation are often spatially and temporally separated from their feeding grounds. For instance, the greatest areas of feeding activity for the large baleen whales are in polar regions during the high primary productivity of summers. However, they migrate in winter to warm tropical waters of low food availability to give birth and nurse their young. In phocid and otariid pinnipeds, parturition and lactation occur on land or ice and thus these activities are also separated from the feeding environment of the lactating female.

In all female mammals, lactation represents the greatest energetic cost of reproduction, requiring large amounts of nutrient transfer and elevated maternal maintenance costs. Hence, lactation is usually associated with increased maternal food consumption. However, because large energy reserves can be stored in blubber in the form of lipid, baleen whales and large phocid seals are the only mammals (besides holarctic bears) that can complete much or all of lactation without feeding. Again, because a smaller body size constrains the size of blubber stores, the smaller phocids and otariids are able to fast for only portions of lactation. All species of marine mammals produce high fat milks (usually 30–60% fat) to maximize the efficiency of fat transfer from maternal blubber into milk and the subsequent efficiency of neonatal fattening and growth. In species that fast throughout lactation, females switch almost completely to a fat-based metabolism. For instance, during a 16-day lactation period, a gray seal (*Halichoerus grypus*) female draws 97% of the energy required for her own metabolic needs and 90% of the milk energy supplied to her pup solely from her blubber stores. Furthermore, the extent to which she can both maintain lactation and produce a fat pup depends on the size of the blubber layer she starts out with (Mellish *et al.*, 1999). Fasting female polar bears, which begin the first 3–4 months of lactation in winter dens, use their extensive subcutaneous adipose tissue in a similar manner.

Blubber deposition is equally critical to the suckling neonate, both for thermoregulation and to act as an energy reservoir. For example, most newly weaned phocid pups rely on blubber deposited during the suckling period to survive their own subsequent fast of several weeks or months after their mothers have departed the breeding grounds. The energy supplied from blubber is critical to survival of the young while they learn how to forage on their own.

Adult males of many marine mammal species also fast or greatly reduce food intake during the breeding season and during their annual molting period. During these times they rely on stored lipid in blubber as their energy source. Sirenians also use blubber during fasting. For instance, in the Amazon, manatees (*Trichechus inunguis*) face dry seasons of up to 6 months at a time, where low waters restrict them to the deep water areas of larger lakes where the aquatic plants they feed upon are unavailable. Hence food intake during these periods is nil.

Finally, besides being an important fat and energy source for marine mammals, blubber is a critical source of water that is essential to maintaining water balance during fasting. Each kilogram of lipid that is mobilized from blubber and oxidized for energy use by an animal generates a net production of 1.07 kg of metabolic water. In fact, oxidation of blubber yields enough water such that individuals usually do not require an additional external source. This is true even of lactating females that are exporting large quantities of water in milk daily. For instance, a gray seal female exports about 23 kg of water in milk over a 16-day lactation period while fasting and has no external access to water during this time (Iverson *et al.*, 1993). Thus, in most species, blubber functions to maintain both water balance and energy metabolism during periods of fasting.

IV. Role of Blubber in Locomotion

Several forces act on animals swimming in fluids, and blubber plays a significant role in the way marine mammals deal with these forces. The predominant restrictive force is drag, but the vertical forces of gravity and buoyancy also exist. Drag is the force that resists the movement of a body through a medium and is much greater in seawater than in air due to seawater's higher density and greater viscosity. The single most effective way to reduce both drag and the power required for forward motion through a fluid is to have a smooth streamlined shape. Although all marine mammals tend to be somewhat streamlined in body shape as defined by their musculoskeletal system, blubber provides their form with a smooth sculpted contour. Blubber thickness is often distributed across an animal in a nonuniform manner that ensures this. For instance, the blubber over the hind end of a seal may be thicker than would be necessary for insulative purposes. The blubber layer here

instead serves to taper the animal more gradually than would be dictated by the musculoskeleton. In fact, another very important means by which to reduce drag on a body is to be spindle shaped, i.e., to have a gradually tapering tail end. This acts to reduce the wake left by the animal moving through the water and hence further reduces the forces of drag. Again, blubber creates this effect in cetaceans by a thickening and sculpting of the tailstock (Pabst *et al.*, 1999). This locomotor function may actually constrain the way in which animals utilize their blubber as energy reserves. In large baleen whales as well as the smallest harbor porpoise, blubber may be greatest in depth and fat content, even during nutritional stress, in the posterior dorsal and tail areas of the body (Lockyer, 1987; Koopman, 1998), as blubber in these areas serves important locomotory functions by both streamlining and possibly acting as a biomechanical spring, capable of temporarily storing and releasing elastic strain energy (Pabst, 1996).

Finally, blubber also plays a role in the buoyancy of marine mammals. Buoyancy is the force that acts on a body submerged in water where, if the mass of the body is greater or less than the volume of water it displaces, it will experience either a net downward or net upward force, respectively. In most marine mammals (except the sea otter), buoyancy will be determined primarily by the ratio of its adipose tissue to lean body tissue and body mass. Fat-filled adipose tissue is less dense than seawater, whereas lean tissue is more dense. Thus, the degree to which marine mammals store blubber will affect their buoyancy and thus the energy expended in moving or maintaining position in water. Although some newly weaned phocid pups may be positively buoyant at greater than 43% adipose tissue, most adult marine mammals will not be positively buoyant and thus are not likely to require any counteracting of this force when at the bottom of dives or when feeding at the benthos. However, changes in blubber stores will clearly affect the degree to which they are negatively buoyant. Only two studies of pinnipeds have directly addressed the degree to which buoyancy affects diving in marine mammals. Both studies found that seals descended faster during diving when they were more negatively buoyant than when they were less negatively buoyant, providing evidence that seals adjust their diving behavior in relation to seasonal changes in buoyancy (Webb *et al.*, 1998; Beck *et al.*, 2000).

V. Insights from the Study of Blubber

Marine mammals are widely distributed in tropical, temperate, and cold oceans of the world and show a diversity of distributional patterns and apparent physiological adaptations. However, our understanding of these patterns, as well as of the foraging ecology of most marine mammals, is hindered by the difficulties in observing free-ranging animals that spend most or all of their lives at sea. Blubber is clearly of central importance to the structure and function of marine mammals. Due to the fact that the composition and amount of blubber carried by an individual can change rapidly and yet blubber has evolved to serve complex functions, its study can provide unique insights into the lives of marine mammals as well as the ecosystems in which they live.

The ultrastructure, thickness, and proximate composition of blubber can provide insights into the feeding status of individuals as well as the functional significance of the blubber itself. As stated previously, the proximate composition, especially lipid content, of blubber changes radically in response to feeding and fasting behavior and thus, along with other nutritional indices, may be used to indicate nutritionally stressed versus robust individuals. Because many marine mammals go through predictable annual periods of fasting and fattening, the proximate composition of blubber can also be used to indicate the life cycle stage of an individual. In some cetaceans, the characteristics of blubber differ greatly across sites of the body and thus study of these properties can provide insight into the functions of blubber. For instance, the structure and composition of blubber at specific sites suggest that in some areas on the body (e.g., the thoracic-abdominal area) blubber may play a more important role in insulation and energy storage, whereas at other sites (e.g., the thick ridge posterior to the dorsal fin or at the caudal peduncle) blubber may serve more important roles in maintaining hydrodynamic shape and other locomotory functions (Koopman, 1998; Pabst et al., 1999). Thus, the study of how blubber at these various sites is utilized during times of fat mobilization may provide further insight into adaptations of blubber structure. For example, the finding that blubber in the area of the caudal peduncle is rarely used and always thicker than needed for insulation, even during severe nutritional stress, lends support to the hypothesis that it may be more important in that region for structural support and locomotory functions than as an insulator or energy provider.

Blubber can also provide insight into adaptation and phylogenetic relationships. For instance, the characteristic of storing blubber lipid primarily as wax esters appears to be confined to a group of the odontocetes (i.e., beaked whales and the sperm whale). The species in which blubber consists primarily of wax esters, although all closely related, are also all deep divers. Hence the study of their blubber may provide insight into phylogenetic patterns as well as roles that wax esters may play in deep-diving animals. The presence of isovaleric acid is likewise confined to a fairly restricted group of animals, which also may be under special thermal constraints (see earlier discussion). Thus the study of isovaleric acid in blubber may provide clues to its function and potential value in insulation. Additionally, in at least one species, the presence of isovaleric in the outer layer of blubber increases in direct proportion with age, suggesting the possibility of using its level in blubber to estimate ages of unknown individuals in the same population (Koopman et al., 1996).

Finally, the fatty acids in blubber can provide powerful insights into the foraging ECOLOGY and diets of both individuals and populations of marine mammals. As stated previously, fatty acids in the marine ecosystem are complex and diverse, fatty acids often travel up the food chain intact, and there are narrow limitations on their biosynthesis in marine mammals. Hence the fatty acids of marine mammal blubber arise in large part from dietary intake and therefore can be used to study aspects of diet and foraging ecology (Iverson, 1993). Given the dynamic nature of lipid mobilization and deposition in marine mammal blubber, fatty acids can provide insight into diets over both time and space. Studies on wild and captive animals demonstrate that

there is direct deposition of dietary fatty acids in both marine mammals and their prey and that the influence of dietary fatty acid intake on blubber composition can be both substantial and predictable. In seals, ingested fatty acids can be deposited directly into adipose tissue, such that blubber composition may mirror that of diet when a seal is rapidly fattening on a high-fat diet (Iverson et al., 1995), or blubber may reflect an integration of diet over a period of weeks or months even when not fattening (Kirsch et al., 2000). Blubber fatty acids have been used to detect differences in diets both among and within populations of various pinnipeds, including the distinction of species and populations occupying freshwater vs marine habitats. For instance, in harbor seals, blubber fatty acids have been used to assess when fasting lactating females resumed feeding (Smith et al., 1997) and to elucidate the use of spatial scales of foraging in Alaska populations across both large (400–800 km) and small geographical (10–50 km) scales (Iverson et al., 1997).

A blubber biopsy (100–500 mg) from a free-ranging animal provides relatively noninvasive information about diet that is not dependent on prey with hard parts, nor limited to only the last meal, as are analyses of fecal or stomach contents. This is accomplished most easily in pinnipeds where, using a medical biopsy punch, one can safely obtain a complete sample through the full depth of 5–10 cm. However, in cetaceans, blubber is generally much thicker and layering of fatty acids in the blubber is more pronounced, with dietary fatty acids being most reflected in the inner and middle layers nearest the deep body core. Thus, less work has been done on live animals in these species.

In conclusion, blubber plays a number of major roles in the lives of marine mammals. Blubber can also be a powerful tool in trying to understand adaptive solutions of species living in marine environments, as well as insights into their ecology and behavior.

See Also the Following Articles

Energetics ■ Hair and Fur ■ Streamlining ■ Thermoregulation

References

Beck, C. A., Bowen, W. D., and Iverson, S. J. (2000). Seasonal changes in buoyancy and diving behaviour of adult grey seals. *J Exp. Biol.* **203**, 2323–2330.

Bowen, W. D., Oftedal, O. T., and Boness, D. J. (1992). Mass and energy transfer during lactation in a small phocid, the harbor seal (*Phoca vitulina*). *Physiol. Zool.* **65**, 844–866.

Heath, M. E., and Ridgway, S. H. (1999). How dolphins use their blubber to avoid heat stress during encounters with warm water. *Am. J. Physiol.* **276**, R1188–R1194.

Iverson, S. J. (1993). Milk secretion in marine mammals in relation to foraging: Can milk fatty acids predict diet? *Symp. Zool. Soc. of Lond.* **66**, 263–291.

Iverson, S. J., Bowen, W. D., Boness, D. J., and Oftedal, O. T. (1993). The effect of maternal size and milk output on pup growth in grey seals (*Halichoerus grypus*). *Physiol. Zool.* **66**, 61–88.

Iverson, S. J., Frost, K. J., and Lowry, L. L. (1997). Fatty acid signatures reveal fine scale structure of foraging distribution of harbor seals and their prey in Prince William Sound, Alaska. *Mar. Ecol. Progr. Ser.* **151**, 255–271.

Iverson, S. J., Oftedal, O. T., Bowen, W. D., Boness, D. J., and Sampugna, J. (1995). Prenatal and postnatal transfer of fatty acids from mother to pup in the hooded seal. *J. Comp. Physiol.* **165**, 1–12.

Kirsch, P. E., Iverson, S. J., and Bowen, W. D. (2000). Effect of a low-fat diet on body composition and blubber fatty acids of captive juvenile harp seals (*Phoca groenlandica*). *Physiol. Biochem. Zool.,* **73**, 45–59.

Koopman, H. N. (1998). Topographical distribution of the blubber of harbor porpoises (*Phocoena phocoena*). *J. Mammal.* **79**, 260–270.

Koopman, H. N., Iverson, S. J., and Gaskin, D. E. (1996). Stratification and age-related differences in blubber fatty acids of the male harbour porpoise (*Phocoena phocoena*). *J. Comp. Physiol.* **165**, 628–639.

Lockyer, C. (1987). Evaluation of the role of fat reserves in relation to the ecology of North Atlantic fin and sei whales. *In* "Approaches to Marine Mammal Energetics" (A. C. Huntley, D. P. Costa, G. A. J. Worthy, and M. A. Castellini, eds.), pp. 184–203. Society for Marine Mammalogy Special Publication No. 1. Allen Press, Lawrence, KS.

Mellish, J. E., Iverson, S. J., and Bowen, W. D. (1999). Individual variation in maternal energy allocation and milk production in grey seals and consequences for pup growth and weaning characteristics. *Physiol. Biochem. Zool.* **67**, 677–690.

Pabst, D. A. (1996). Springs in swimming animals. *Am. Zool.* **36**, 723–735.

Pabst, D. A., Rommel, S. A., and McLellan, W. A. (1999). The functional morphology of marine mammals. *In* "Biology of Marine Mammals" (J. E. Reynolds and S. A. Rommel, eds.), pp. 15–72. Smithsonian Institution Press, Washington, DC.

Pond, C. M. (1998). "The Fats of Life." Cambridge Univ. Press, Cambridge.

Pond, C. M., Mattacks, C. A., Colby, R. H., and Ramsay, M. A. (1992). The anatomy, chemical composition, and metabolism of adipose tissue in wild polar bears (*Ursus maritimus*). *Can. J. Zool.* **70**, 326–341.

Smith, S., Iverson, S. J., and Bowen, W. D. (1997). Fatty acid signatures and classification trees: New tools for investigating the foraging ecology of seals. *Can. J. Fish. Aqu. Sci.* **54**, 1377–1386.

Webb, P. M., Crocker, D. E., Blackwell, S. B., Costa, D. P., and Le Boeuf, B. J. (1998). Effects of buoyancy on the diving behavior of northern elephant seals. *J. Exp. Biol.* **201**, 2349–2358.

Worthy, G. A. J., and Edwards, E. F. (1990). Morphometric and biochemical factors affecting heat loss in a small temperate cetacean (*Phocoena phocoena*) and a small tropical cetacean (*Stenella attenuata*). *Physiol. Zool.* **63**, 432–442.

Worthy, G. A. J., and Lavigne, D. M. (1983). Changes in energy stores during postnatal development of the harp seal, *Phoca groenlandica*. *J. Mammal.* **64**, 89–96.

Blue Whale

Balaenoptera musculus

RICHARD SEARS

Mingan Island Cetacean Study, Inc.,
Longue Pointe de Mingan, Quebec, Canada

The blue whale, *Balaenoptera musculus* (Linnaeus, 1758), is a baleen whale belonging to the family Balaenopteridae, which includes the group of cetaceans known as rorquals (Fig. 1). Common names are blue whale, sulfurbottom, Sibbald's rorqual, great blue whale, and great northern rorqual. The largest animal known to have existed on Earth, it is found worldwide, ranging into all oceans. Because of its great size and the commercial value of the products it yielded, the blue whale was hunted relentlessly beginning in the late 1800s. The greatest number of blue whales was taken from the early 1900s until the late 1930s, with the peak being in the 1930–1931 season when nearly 30,000 were killed. The height of blue whale whaling coincided with the advent of explosive harpoons, steam power vessels, and the construction of factory ships, which could process whale carcasses at sea. The blue whale was severely depleted by whaling, particularly in the Southern Hemisphere, where during the first half of the 20th century 325,000–360,000 were killed in Antarctic waters alone. A further 11,000 were taken in the North Atlantic, primarily in Icelandic waters and 9500 in the North Pacific. This unbridled hunt for blue whales, which lasted until its worldwide protection in 1966, brought the blue whale to the brink of extinction and it is still an endangered species today.

I. Diagnostic Characters

On average, Southern Hemisphere blue whales are larger than Northern Hemisphere animals, with the largest recorded at lengths of 31.7–32.6 m (104–107 feet) and 33.6 m (110 feet) for individuals caught off the South Shetlands and South Georgia. The largest blue whale recorded for the Northern Hemisphere was a 28.1-m (92 feet) female reported in whaling statistics from catches in Davis Strait. In the North Pacific female blue whales of 26.8 m (88 feet) and 27.1 m (89 feet) have been recorded. A 190-ton female was reported taken off South Georgia in 1947; however, body weights of adults generally range from 80 to 150 tons. For maximum size descriptions, female measurements are used because female baleen whales are larger than males.

Blue whales are observed most commonly alone or in pairs; however, concentrations of 50 or more can be found spread out in areas of high productivity. Blue whales project a tall (up to 10–12 m) spout, denser and broader than that of the fin whale, (*B. physalus*) which in calm conditions can help distinguish between the two species. When surfacing, the blue whale raises its massive shoulder and blowhole region out of the water more than other rorquals. The prominent fleshy ridge just forward of the blowhole, known as the "splash guard," is strikingly large in this species.

When seen from above, blue whales have a tapered elongated shape, with a huge broad, relatively flat, U-shaped head, adorned by a prominent ridge from the splash guard to the tip of the upper jaw or rostrum and massive mandibles (Fig. 2). The baleen is black, half as broad as its maximum 1-m length and 270–395 plates can be found on each side of the upper jaw (Yochem and Leatherwood, 1985). There are 60–88 throat grooves or ventral pleats running longitudinally parallel from the tip of the lower jaw to the navel, which enable the throat or ventral pouch to distend when feeding.

The dorsal fin, proportionally smaller than in other balaenopterids, yet varied in shape, ranging from a small nubbin to triangular and falcate, is positioned far back on the body.

Blue whales have long blunt-pointed flippers, slate gray, with a thin white border dorsally and white ventrally, reaching

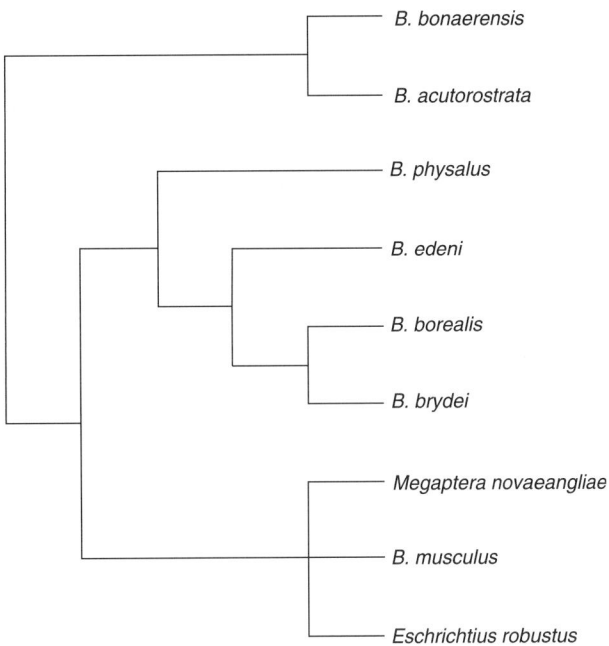

Figure 1 *Phylogenetic relationships among the species of the genera* Balaenoptera, Megaptera, *and* Eschrichtius *based on several DNA sequencing studies.*

up to 15% of their body length. In the field, particularly on bright days, blue whales generally appear much paler in coloration than all species of large whale except for the gray (*Eschrichtius robustus*), with which it should not be confused due to a great difference in size.

The characteristic mottled pigmentation of blue whales is a blend of light and dark shades of gray displayed in patches of varying sizes and densities (Fig. 3). The pigmentation is slate blue on overcast days to silvery or turquoise blue on bright sunny days depending on the clarify of the water. The mottling is found along the body dorsoventrally, occasionally on the flip-pers, but not on the head and tail flukes. Two prominent pigmentation configurations are found in blue whales, one where a darker, dominant background is mottled with sparser pale patches of pigmentation, while in the other there is a predominantly pale background mottled with sparser dark patches. Blue whale pigmentation can vary, however, from very sparse mottling, where the individual appears uniformly pale or dark, to densely mottled individuals, where the pigmentation is a highly contrasted variegation of spots unique to each whale. Distinct chevrons curving down and angled back from the apex on both sides of the back behind the blowholes and either pale

Figure 2 *Blue whale, aerial view. Richard Sears, MICS Photo.*

Figure 3 *Blue whale showing the characteristic mottled pigmentation of the species. Drawing by Daniel Grenier.*

or dark in tone can be found on certain blue whales. Blue whales are individually identified from the distinctive mottled pigmentation found on their back and sides.

A yellow-green to brown cast, caused by the presence of a diatom (*Cocconeis ceticola*) film, can be seen covering all or part of the body of blue whales found in cold waters. The yellowish, diatom-induced tint is the reason the "sulphur-bottom" moniker was once used for blue whales.

Although not noted for raising their flukes when diving, approximately 18% of blue whales observed in the western North Atlantic and northeast Pacific do so. This is an individual characteristic, and if the individual is relaxed it will generally raise its flukes high up in the air on each sounding dive. When disturbed, blue whales that raise their flukes when diving will often not raise their tails as high out of the water or not at all and dive more quickly from the surface. The tail flukes, sometimes striated ventrally, are predominantly gray above and below; however, some individuals do have white patches of pigmentation on the ventral surface that are used for individual identification. The trailing margin of the tail is either straight or curves very slightly from each tip to the median notch.

II. Distribution and Migration

Three subspecies have been designated: what has been considered the largest, *B. musculus intermedia,* found in Antarctic waters; *B. musculus musculus* in the Northern Hemisphere; and *B. musculus brevicauda,* from the subantartic zone of the southern Indian Ocean and south western Pacific Ocean, also colloquially known as the "pygmy" blue whale. Although the latter designation is now generally accepted, its validity still remains in question. Despite having being reduced greatly due to whaling, the blue whale remains a cosmopolitan species separated into populations from the North Atlantic, North Pacific and Southern Hemisphere (Fig. 4). In the North Atlantic, eastern and western subdivisions are recognized. Photo-identification work from eastern Canadian waters indicates that blue whales from the St. Lawrence, Newfoundland, Nova Scotia, New England, and Greenland all belong to the same stock, whereas blue whales photo-identified off Iceland and the Azores appear to be part of a separate population. The best known population in the North Atlantic is that found in the St. Lawrence from April to January, where 350 individuals have been catalogued photographically. Apart from the Icelandic and Azores sightings, few blue whales have been reported from eastern North Atlantic waters recently. North Atlantic blue whale abundance probably ranges from 600 to 1500 at this time, although more extensive photo-identification and shipboard surveys are needed for more reliable estimates.

In the North Atlantic, blue whales reach as far north as Davis Strait and Baffin Bay in the west, while to the east they travel as far north as Jan Mayen Island and Spitzbergen during summer months. Blue whales sighted recently in winter and spring off the Azores and Canary Islands could be migrating north along the mid-Atlantic ridge to Iceland, where they are seen from May to September. Other blue whales probably migrate along the European coast, far offshore and out around Ireland north to either Iceland or Norway. It is not clear where blue whales winter in the North Atlantic. Some have been observed in the St. Lawrence as late as February; however, acoustic studies have revealed that they are spread out across the North Atlantic basin, south as far as Bermuda and Florida, with concentrations south of Iceland, off Newfoundland and Nova Scotia. The southernmost observations on the eastern side of the North Atlantic are in the waters between the coast of Africa and the Cape Verde Islands (Yochem and Leatherwood, 1985).

In the North Pacific, where as many as five subpopulations were thought to exist, acoustic analysis of blue whale vocalizations now indicates there are no more than two. The best known is that from the eastern North Pacific where blue whales can be found as far north as Alaska, but are regularly observed from California in summer, south to Mexican and Costa Rican waters in winter. An abundance estimate of approximately 2000–3000 animals has been determined for this population, which has been studied extensively over a good portion of its range. From late fall to spring, blue whales can be found in the Gulf of California, Mexico, and south to offshore waters of central America. By April and May they migrate north along the West coast of North America, where a large proportion is found in California waters. From there some reach Canadian waters, and other groups may disperse north to the Gulf of Alaska or west toward the Aleutian Islands.

Few blue whales have been reported recently from the western North Pacific, including the Aleutian Islands, Kamchatka/Kurils/northern Japan, and southern Japan. In the western North Pacific, blue whales are thought to migrate to Kamchatka or the Kuril Islands and probably further northeast.

Blue whales are also found in the northern Indian Ocean; however, it is not clear whether these form a distinct population.

In the Southern Hemisphere, where the blue whale was historically most abundant, it is very rare today, with abundance estimates ranging from 710 to 1255, although reliable population assessment is poor due to limited sighting data. Al-

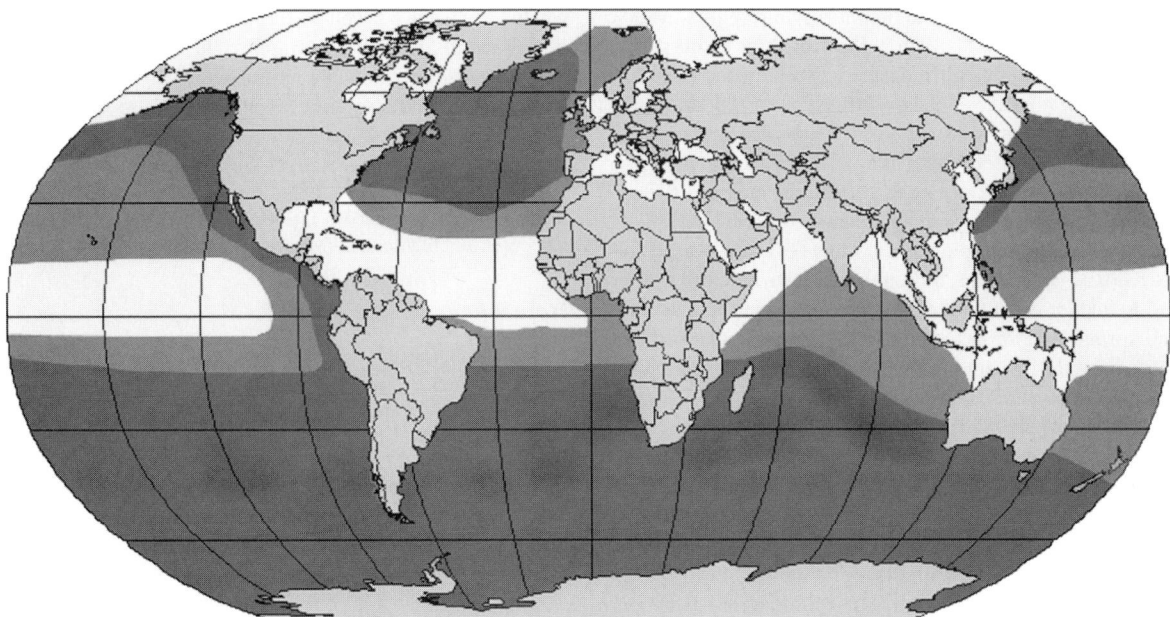

Figure 4 *Global distribution of blue whales. Darker gray indicates higher densities.*

though the general population structure in the Southern Ocean is not well understood, evidence shows discrete feeding stocks. Consistent with these feeding areas, the International Whaling Commission has assigned six stock areas for blue whales in the Southern Hemisphere.

III. Natural History and Ecology

A. Feeding

Food availability probably dictates blue whale DISTRIBUTION for most of the year. Although blue whales can be found in coastal waters of the St. Lawrence, Gulf of California, Mexico, and California, they are found predominantly offshore. Blue whales appear to feed almost exclusively on euphausiids (KRILL) worldwide in areas of cold current upwellings. When blue whales locate suitably high concentrations of euphausiids, they feed by lunging mouth wide open and gulping large mouthfuls of prey and water. The mouth is then almost completely closed and the water is expelled by muscular action of the distended ventral pouch and tongue through the still exposed baleen plates. Once the water is expelled, they swallow their prey.

When blue whales feed just a few meters below the surface, they often surface slowly, belly first, exposing the throat grooves of the ventral pouch, roll to breathe, and evacuate the water before diving to take their next mouthful. If the prey is close to the surface, blue whales lunge vigorously on their sides or lunge up vertically by projecting their cavernous lower jaws 4–6 m up through the surface. Although surface feeding has often been observed during the day, it is more usual for blue whales to dive to at least 100 m into layers of euphausiid concentrations during daylight hours and rise to feed near the surface in the evening, following the ascent of their prey in the water column. In the North Atlantic, blue whales feed on the krill species *Meganyctiphanes norvegica*, *Thysanoessa raschii*, *T. inermis*, and *T. longicaudata;* in the North Pacific, *Euphausia pacifica*, *T. inermis*, *T. longipes*, *T. spinifera*, and *Nyctiphanes symplex.* In Antarctic waters they prey on *E. superba*, *E. crystallorophias*, and *E. vallentini.*

When foraging or feeding at depth, blue whales will generally dive for 8–15 min; 20-min dives are not uncommon. The longest dive recorded was of 36 min; however, dives of more than 30 min are rare.

Blue whales generally swim at 3–6 km when feeding. When traveling, they can attain speeds of 5–30 km and when chased by boats, predators, or interacting with other blue whales, they can reach upward of 35 km.

B. Vocalization

Blue whales vocalize regularly throughout the year with peaks from midsummer into winter months. The majority of blue whale vocalizations are low frequency or infrasonic sounds of 17 to 20 Hz, lower than humans can detect. Blue whale sounds, at 188 decibels (louder than a jumbo jet at full power a few meters away), are one of the loudest and lowest made by any animal.

These calls can be heard easily for hundreds of kilometers, thousands of kilometers under optimal oceanographic conditions and may cover whole ocean basins. The low frequencies produced by blue whales are ideal for communication between individuals of a widely dispersed and nomadic species through water without much loss of information.

IV. Reproduction and Longevity

Blue whales reach sexual maturity at 5–15 years of age; however, 8–10 years appear to be more usual for both sexes. Length at sexual maturity in females from the Northern Hemisphere is

21–23 m and is 23–24 m in the Southern Hemisphere. Males reach sexual maturity at 20–21 m in the Northern Hemisphere and at 22 m in the Southern Hemisphere (Yochem and Leatherwood, 1985). Mating takes place starting in late fall and continues throughout the winter. Female blue whales give birth every 2–3 years in winter after a 10- to 12-month gestation period. The calves, which weight 2–3 tons and measure 6–7 m long at birth, are weaned when approximately 16 m long at 6–8 months. No specific breeding ground has been discovered for blue whales in any ocean, although mothers and calves are sighted regularly in the Gulf of California, Mexico, in late winter and spring. A portion of the northeast Pacific Ocean blue whale stock could be using this region as a breeding ground.

Little is known of their mating behavior; however, female/male blue whale pairings have been noted with regularity in the St. Lawrence from summer into fall, some lasting for as long as 3 weeks. When a female/male pair is approached by a third blue whale, even a fin whale, vigorous surface displays, where all three animals can be seen racing high out of the water, almost breaching, porpoising forward causing an explosive splash of a bow wave, can result. Such interactions between blue whales usually last for 5–15 min.

Blue whales are thought to live for at least 80–90 years and probably longer. What is certain, however, after extensive photo-identification fieldwork on known individuals in the St. Lawrence and northeast Pacific, is that they live for at least 31 years.

V. Mortality

Documentation of natural mortality in blue whales is rare. The blue whale's principal predator is the killer whale *Orcinus orca*, but there is little evidence of attacks on blue whales in the North Atlantic or Southern Hemisphere. However, in the Gulf of California, Mexico, 25% of the blue whales photo-identified carry rake-like killer whale teeth scars on their tails, indicating that attacks occur with some regularity but are probably rarely successful.

In the St. Lawrence, ice entrapment, where animals have been crushed, stranded, or suffocated by current and wind-driven ice floes in the late winter/early spring, has been reported.

While reports of blue whales approaching vessels are rare, at least 25% of the blue whales photo identified in the St. Lawrence carry scars that can be attributed to collisions with shipping. This type of scarring has been reported for a few northeast Pacific blue whales as well. Ship strikes in heavy shipping areas, such as the St. Lawrence and California coast, may have an impact on populations, but data are not available on this point.

Few entanglements in fishing gear have been reported, and it is thought that the size of the blue whale enables it to tear through fishing gear relatively unscathed.

Persistent contaminants accumulated over time, such as PCBs commonly found in blue whales from eastern Canadian waters, may have an impact on reproduction and limit the recovery of certain populations.

It has been shown that blue whales react strongly to approaching vessels. The degree of reaction depends on the whale's behavior, as well as the distance, speed, and direction of the vessel at the time of approach. The increasing anthropogenic noise probably has an impact on blue whales and their habitat and could also limit recovery of this species.

See Also the Following Articles

Baleen Whales ■ Cetacean Life History ■ Fluking ■ Noise, Effects of ■ Pollution and Marine Mammals

References

Barlow, J., and Calambokidis, J. (1995). Abundance of blue and humpback whales in California: A comparison of mark-recapture and line-transect estimates. *In* "Abstracts Eleventh Biennial Conference on the Biology of Marine Mammals, Orlando, Florida, 14-18 December 1995," p. 8. Society for Marine Mammalogy, Lawrence, KS.

Calambokidis, J., Steiger, G. H., Cubbage, J. C., Balcomb, K. C., Ewald, C., Kruse, S., Wells, R., and Sears, R. (1990). Sightings and movements of blue whales off central California 1986-88 from photo-identification of individuals. *Rep. Whal. Comm.* **12,** 343–348.

Clark, C. W., and Charif, R. A. (1998). Acoustic monitoring of large whales to the West of Britain and Ireland using bottom-mounted hydrophone arrays. October 1996–September 1997, JNCC Report, No. 281.

Sears, R., Williamson, J. M., Wenzel, F., Bérubé, M., Gendron, D., and Jones, P. W. (1991). "The Photographic Identification of the Blue Whale (*Balaenoptera musculus*) in the Gulf of St. Lawrence, Canada," pp. 335–342. IWC Rep. Sc/A88/ID23.

Yochem, P. K., and Leatherwood, S. (1985). Blue whale (*Balaenoptera musculus*) (Linnaeus, 1758). *In* "Handbook of Marine Mammals" (S. H. Ridgway and R. Harrison, eds.), Vol. 3, pp. 193–240. Academic Press, London.

Bones and Teeth, Histology of

MARY C. MAAS
Northeastern Ohio Universities College of Medicine, Rootstown

The bones and teeth of marine mammals, like those of other vertebrates, consist of both organic and mineral components. Because the mineral component (mostly calcium phosphate) predominates, the constituents of bones (bone and calcified cartilage) and teeth (cementum, dentine, and enamel) are referred to as "hard tissues." Each of these hard tissues is distinguished both by its composition and by its microscopic structure. Many of the histological features of marine mammal teeth and bones are typical for mammals, and vertebrates, in general, but others are unique or unusual. Some of these may have evolved in conjunction with their shifts to marine habitats.

I. Bone

A. Bone Structure and Composition

Bone consists of highly calcified, intercellular bone matrix, and three types of cells: osteocytes, osteoblasts, and osteoclasts. The outer surface of bone is covered by periosteum, which is bound to bone by bundles of collagen fibers known as Sharpey's fibers, and the inner bone surface is lined with endosteum (Fig. 1). Periosteum is thicker than endosteum, but both consist of fibrous connective tissue lined with osteoprogenitor cells from which osteoblasts are derived. Osteoblasts are the cells that synthesize bone matrix proteins and are active in bone matrix mineralization. Bone matrix (also known as osteoid) consists of about 33% organic matter (mostly type I collagen) and 67% inorganic matter (calcium phosphate, mostly hydroxyapatite crystals). Osteoblasts occur as a simple, epithelial-like layer at the developing bone surface. As the bone matrix mineralizes, some osteoblasts become trapped in small spaces within the matrix (lacunae). These trapped osteoblasts become osteocytes, the cells responsible for maintenance of the bony matrix. Each lacuna holds only a single osteocyte but is connected with adjacent lacunae by microscopic canaliculi, which house cytoplasmic processes of the osteocytes. Osteoclasts are large, multinucleated cells that occur in shallow erosional depressions (Howship's lacunae) on the resorbing bone surface and secret enzymes that promote the local digestion of collagen and dissolution of mineral crystals.

Bone is commonly classified according to its gross appearance as cancellous bone (bone with numerous, macroscopic interconnecting cavities, or trabeculae, also known as spongy or trabecular bone) or compact bone (dense lamellar bone without trabeculae), but both types have the same basic histological structure. In a typical mammalian long bone the diaphysis (shaft) is composed predominantly of compact bone, with cancellous bone confined to the inner surface around a central, medullary cavity (Fig. 1a), whereas the epiphyses (articular ends) consist mostly of cancellous bone overlain by a thin, smooth layer of compact bone. In short bones a core of cancellous bone is completely surrounded by compact bone, and in the flat bones of the skull, inner and outer plates of compact bone are separated by the diploë, a layer of cancellous bone.

Bone also can be classified histologically, as woven (primary) bone and lamellar (secondary) bone. Woven bone, or primary bone, has an irregular structure and is usually replaced in adults by the more highly mineralized lamellar bone. Lamellar bone is organized into thin layers (lamellae), usually 3–7 μm thick, which contain parallel collagen fiber bundles. Lacunae containing osteocytes are located between lamellae. There are three types of lamellae: concentric, interstitial, and circumferential (Fig. 1b). Concentric lamellae are arranged in circular layers around a long axis, the haversian canal, which is a vascular channel containing blood vessels, nerves, and connective tissue. Adjacent vertical channels are connected by more horizontally oriented vascular

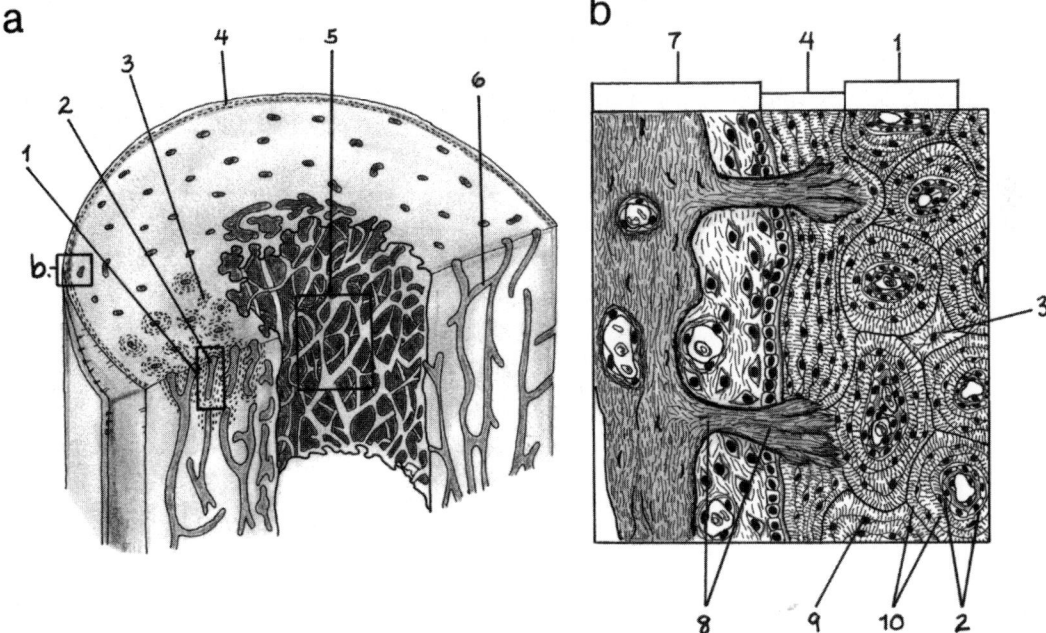

Figure 1 *(a) Schematic model of the wall of a mammalian long-bone diaphysis consisting of an outer layer of compact bone and an inner layer of cancellous bone, surrounding a central medullary cavity. Periosteum covers the outer bone surface, and endosteum covers the inner bone surface. (b) Enlarged diagram of periosteum and compact bone in (a) 1, osteon; 2, haversian canal; 3, interstitial lamellae; 4, outer circumferential lamellae; 5, cancellous bone; 6, Volkman's canal; 7, periosteum; 8, Sharpey's fibers; 9, lacuna; 10, concentric lamellae. Adapted from Ten Cate (1989).*

channels (Volkmann's canals). The entire complex, consisting of several layers of concentric lamellae around a vascular channel, is known as an osteon or haversian system. Interstitial lamellae, which appear as irregularly shaped areas between adjacent osteons, consist of lamellae that are remnants of osteons destroyed during bone remodeling. Circumferential lamellae are arranged parallel to each other and comprise the outer circumferential lamellae laid down next to the periosteum and the inner circumferential lamellae laid down next to the endosteum.

B. Bone Formation, Growth, and Remodeling

Osteogenesis (bone formation) of membrane bone (intramembranous bone, dermal bone) occurs directly by the mineralization of matrix formed by osteoblasts within condensations of mesenchyme. Osteogenesis of endochondral bone (cartilage bone) occurs indirectly by the deposition and mineralization of bone matrix on a preexisting cartilage matrix. Bone continues to grow through remodeling, as old bone is resorbed through activities stimulated by osteoclasts, and new bone is laid down through the activities of osteoblasts. Influences on bone remodeling include strain and stress imposed by movement and muscle action, hormones, and growth factors. For example, the parathyroid hormone is linked to osteoclast proliferation and activity, and calcitonin, a hormone synthesized within the thyroid, has an inhibitory effect on osteoclast activity (Marks and Popoff, 1988). Modifications of hormonal controls on bone growth and remodeling, in specific parts of the skeleton, are probably responsible for specializations in bone density patterns in some marine mammals (see later). In addition to this typical osteoblastic bone formation, osteogenesis can occur through the direct transformation of cartilage or fibrous tissue into bone (metaplastic bone). The metaplastic transformation of chondrocytes into osteoblasts may account for the formation of osseous globules found in the endosteal endochondral bone of some archaeocetes (Buffrénil et al., 1990). Osseous globules are pseudolamellar bone deposited in empty lacunae that once housed chondrocytes, but their mode of origin is controversial.

Because bone growth occurs throughout life, periodic growth marks in skeletal tissue, particularly periodical deposition of periosteal bone layers, are potentially useful in mammalian age determination. Techniques of skeletal tissue age determination involve the counting of growth layer groups. Growth layer groups are sets of incremental growth lines defined by at least one change in mineral density, such as between more stained and less stained layers or dark and light layers. However, the dynamic nature of bone growth and remodeling limits the accuracy of bone growth layer group counts.

C. Marine Mammal Bone

Marine mammals show two very different trends in bone architecture and histology, reduced bone density and increased bone density, both of which are associated with their aquatic habits (Wall, 1983). Deep-diving marine mammals, especially Recent cetaceans, have bones that are less dense than homologous elements in terrestrial mammals. They efficiently overcome buoyancy at depth by the active mechanism of lung collapse, whereas at the surface their lighter bones enhance buoyancy, allowing them to float with relatively little expenditure of energy. A pattern of reduced bone density has been documented thoroughly in small to medium-sized odontocetes, some of the large-bodied cetaceans, and some phocids (notably the elephant seals, Mirounga spp.) and is characterized by the replacement of cortical bone (the compact bone surrounding medullary cavities) with cancellous bone, which also fills the medullary cavities. This condition is apparently caused by an imbalance between bone resorption and redeposition beginning early in ontogeny and is probably under hormonal control. An increase in cancellous bone in these mammals does not appear to be pathological—the microscopic architecture of cancellous bone in cetacean limbs is significantly more organized than that of typical osteoporotic bone.

In contrast, shallow-diving marine mammals, such as sirenians, overcome buoyancy while diving in large part by the static mechanism of increased bone density. Their bones are much denser than typical mammal bones. This is achieved in different ways: by osteosclerosis, by pachyostosis, or by a combination of both conditions (pachyosteosclerosis) (Domning and Buffrénil, 1991). Sirenians show pronounced pachyosteosclerosis, especially in the thoracic and occipital regions. Similarly, walruses and some seals have unusually dense limb bones.

Ongoing research in the bone histology of extinct marine mammals indicates that both increased bone density and reduced bone density have evolved independently several times in different groups of marine mammals. The earliest sirenians show pronounced pachyosteosclerosis. Likewise, in contrast to Recent cetaceans, some bones of extinct Eocene archaeocetes are osteosclerotic. In Basilosaurus (Basilosauridae), the osteosclerosis of ribs is very pronounced, with the total replacement of medullary trabecular bone by compact bone (Buffrénil et al., 1990). Similarly, hyperostosis of the periosteal cortex and infilling of the medullary cavity with cancellous bone occur in ribs and vertebrae of Eocetus (Protocetidae) (Uhen, 1999) and in ribs of Zygorhiza (Durodontidae) (Buffrénil et al., 1990). In contrast, long bones of some other durodontids show a reduced thickness of periosteal compact bone, as in modern cetaceans (Madar, 1998).

Bone of one marine mammal, the toothed whale Mesoplodon densirostris, exhibits unique histological features. The rostral bone of this odontocete, which is among the densest bone known among tetrapods, is characterized by hypermineralized secondary osteons. These osteons have unusually well-aligned parallel and platy hydroxyapatite crystals and a tubular network of unusually thin collagen fibrils, and thus differ markedly from the structure of haversian systems of typical mammalian lamellae bone (Zylberberg et al., 1998).

The periodic deposition of periosteal bone layers has been used in studies of age determination in mammals, although limited in use by the fact that mammalian bone undergoes remodeling throughout life (Klevezal, 1996). Bone growth layer groups have been studied in a variety of marine mammals, including sirenians, pinnipeds, and odontocetes.

II. Cementum
A. Cementum Structure and Composition

Teeth of marine mammals, like all mammals, consist of a crown, which extends above the gums, and one or more roots, which extend below the gum line and hold the teeth in bony sockets (alveoli). The roots are covered by cementum (also known as cement), which sometimes extends to cover part of the crown, overlapping the cervical enamel. Cementum, along with the periodontal ligament, comprises the periodontium, the attachment apparatus of teeth. Cementum is similar in composition to bone. Its mineral component (65% by wet weight) consists of crystals of an impure form of hydroxyapatite similar in shape and size to those of bone. Its organic component (up to 20% of the total tissue) includes cementocytes (cementum cells), ground substance containing proteoglycans, intrinsic collagen fibers, and extrinsic collagen fibers (Sharpey's fibers). Intrinsic fibers and ground substance are the primary constituents of cementum. Intrinsic fibers, like collagen fibers of lamellar bone, are small, on the order of 1–2 μm in diameter. The extrinsic Sharpey's fibers are much larger, typically 3–12 μm in diameter. Intrinsic fibers, ground substance, and cementocytes are derived from cementoblasts, but the extrinsic fibers are derived from fibroblasts of the periodontal ligament.

Cementum is classified according to the relative proportions of the different components, although the different types are gradational. Thus, cementum can be classified as cellular or acellular depending on the relative proportions of cementocytes and ground substance, or it can be classified according to its fiber composition (Fig. 2). Extrinsic fiber cement occurs close to the alveolar bone and is dominated by well-mineralized Sharpey's fibers contained within a highly mineralized acellular ground substance. Mixed fiber cement contains intrinsic collagen fibers as well as Sharpey's fibers and ground substance and may contain cementocytes. Intrinsic fiber cement, which contains only intrinsic collagen fibers, ground substance, and cementocytes, occurs close to roots. In cellular mixed fiber cement and intrinsic fiber cement, cementocytes are contained in lacunae of variable shape that form within the mineralizing ground substance.

Incremental growth layers known as cementing lines or resting lines are sometimes a prominent histological feature of both cellular and acellular cementum. Cementing lines, like the incremental growth layers found in bone, dentine, and enamel, are distinct layers that parallel the developing surface. Due to periodic variation in mineralization during development, they contrast with adjacent layers. Cementum growth layer groups, like those of bone and dentine, can be defined empirically by at least one change in mineral density, such as between translucent and opaque layers, dark and light layers, ridge and groove, or more stained and less stained layers. Empirical studies have shown that cementum growth layer groups record the periodicity of tissue formation and thus are useful in age determination.

B. Marine Mammal Cementum

Cementum in marine mammals is, for the most part, structurally similar to that of other mammals. Cementum growth layer groups are used in conjunction with dentine and bone growth layer groups to estimate age in marine mammals, although their relative clarity varies among species. In some

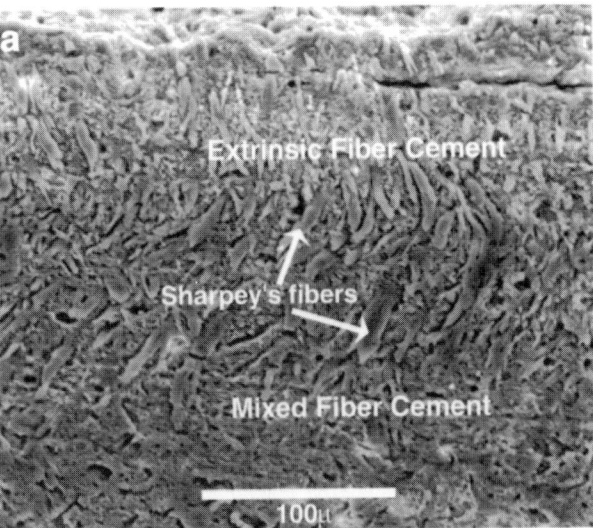

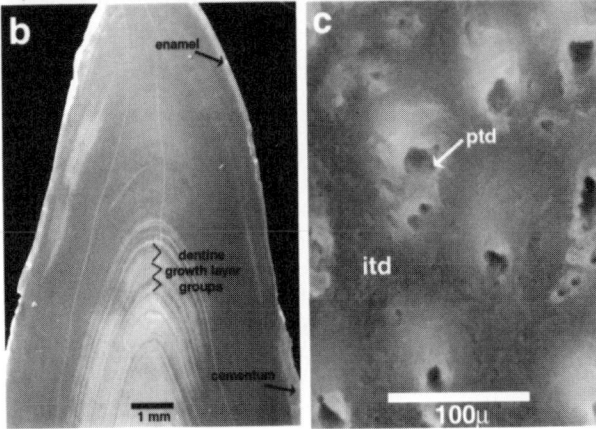

Figure 2 *Scanning electron micrographs of an isolated tooth of an unidentified delphinoid cetacean (Yorktown Formation, Pliocene, from the Lee Creek Mines, North Carolina). The specimen has been sectioned longitudinally, polished, and etched with dilute HCl. (a) High magnification view of cementum, which grades from extrinsic fiber cement on the outer periphery (top) to mixed fiber cement closer to the cementum–dentine junction (bottom). Classification of cementum depends on the proportion of Sharpey's fibers contained within the matrix. (b) Thin layers of enamel and cementum lie peripheral to dentine of the crown and root. Dentine growth layer groups appear as pairs of dark/light bands. (c) High magnification view of dentine. The walls of cross-sectioned dentine tubules contain hypermineralized peritubular dentine (ptd). Less mineralized intertubular dentine (itd) occurs between tubules.*

species, cementum formation continues beyond that of dentine, which is an advantage in age determination. In ziphiid whales, where the cementum typically extends over most of the crown and may comprise the bulk of the tooth, cementum growth layer groups are distinguishable without magnification. Ziphiids also have been reported to have an unusual, possibly vascular cementum (Boyde, 1980).

III. Dentine

A. Dentine Structure and Composition

Dentine comprises the bulk of the volume of teeth of most mammals. In the crown, dentine is covered by enamel, whereas in the root it is covered by cementum. Circumpulpal dentine surrounds the pulp cavity, which contains connective tissue, nerves, and blood vessels. Circumpulpal dentine is distinguishable histologically from a thin outer layer known as mantle dentine (in the tooth crown) or hyaline dentine (in the root). Dentine tubules, which radiate out from the pulp to the outer dentine surface, are distinctive features of dentine. They are narrow (1–4 μm diameter) tubular structures that form during dentine development around odontoblast (dentine-forming cells) cell processes. In adult teeth, tubules contain mostly fluid and amorphous cell debris.

The organic component of dentine consists mainly of very small (on the order of 50 nm in diameter) collagen fibrils. The collagen fibrils in circumpulpal dentine are laid down parallel to the developing dentine surface and perpendicular to the dentine tubules, but the mantle dentine contains some large (more than one micron in diameter) collagen fiber bundles known as von Korff fibers. Von Korff fibers are oriented parallel to tubules.

Dentine is 75% mineral (hydroxyapatite). Most of the small (2–3 nm in thickness and probably 20–100 nm in length) hydroxyapatite crystals are aligned parallel to each other and to the small collagen fibrils, but others are oriented radially and form spherical or semispherical structures known as calcospherites. Calcospherites are difficult to distinguish histologically because they typically fuse together. Areas where mineralization is incomplete and calcospherites have not fused are called interglobular dentine. Most mineralization of dentine takes place along the developing dentine front, but dentine deposited in tubule walls (peritubular dentine) undergoes further mineralization (Fig. 2c). In some cases, tubules become occluded by mineralization, forming sclerotic dentine. Denticles (smooth-surfaced, spherical mineralized bodies with a laminar structure) sometimes form by the mineralization of collagen fibers within the pulp cavity. These denticles may become attached to the inner surface of the dentine or become embedded in it during continued dentine formation.

B. Marine Mammal Dentine

The dentine of most marine mammals is structurally similar to that of other mammals, but there are some exceptions. In some, notably the narwhal (*Monodon monoceros*) and sperm whale (*Physeter macrocephalus*), the large von Korff fibers are not restricted to the mantle dentine but extend throughout the thickness of dentine, where they are located in the walls of dentine tubules. Denticles have been reported in some odontocetes and sclerotic dentine is found in some marine mammals, especially in seals.

Marine mammal dentine is characterized by prominent incremental growth layers (Fig. 2b) that lie at angles to dentine tubules and vary in their intensity, both within and among individuals. The finest scale layers are the incremental von Ebner lines, which probably reflect diurnal variation in matrix fiber arrangement. Von Ebner lines appear as alternating dark and light lines in ground sections under polarized light. Other larger-scale incremental growth layers reflect changes in density due to differences in mineralization. These include the neonatal line, a very prominent growth layer that marks physiological disturbance associated with birth, and other less distinct and consistent growth layers whose physiological bases are uncertain. In some seals, the growth layer groups are accentuated by layers of interglobular dentine. Whatever their origins, there is a regular repetition to growth layer groups that seems to reflect annual or semiannual growth cycles, and counting of dentinal growth layer groups is a primary basis of age determination in pinnipeds, sirenians, and odontocetes.

IV. Enamel

A. Enamel Structure and Composition

Enamel covers the tooth crown in most mammals. It is the most highly mineralized tissue in the body, consisting almost entirely (95% by weight) of highly structured arrangements of hydroxyapatite crystallites. The remaining fraction consists of water and two classes of proteins unique to enamel: enamelins, which predominate in mature (fully mineralized) enamel, and amelogenins, which predominate in developing enamel. The histological structure of enamel reflects the organization of crystallites into units of increasing scale, two of which are enamel prisms and enamel types. This structural organization is determined during enamel development. Unlike bone, cement, and dentine, enamel does not remodel after its initial deposition.

Enamel matrix is secreted by ameloblasts. The activity of these enamel-secreting cells commences at the enamel–dentine junction (EDJ) and continues as ameloblasts retreat outward, away from the EDJ. Mineral crystals precipitate and grow within the enamel matrix left by the retreating ameloblasts. The orientation and arrangement of the crystallites, and thus the structure of the mature enamel, depend on the shape of the secretory end of the ameloblast. The simplest enamel structure is formed by ameloblasts with flat secretory surfaces. In most mammal teeth, however, the bulk of the enamel is laid down by ameloblasts whose secretory ends form protrusions, called Tomes processes, surrounded by flattened areas called ameloblast shoulders. Because enamel crystallites grow perpendicular to the differently oriented secretory surfaces of the Tomes process and the ameloblast shoulder, there is a regular pattern of discontinuities in crystallites' orientations. These discontinuities define the boundaries of enamel prisms and interprismatic enamel.

Enamel prisms are cylindrical bundles of largely parallel hydroxyapatite crystals extending outward from the EDJ toward the outer tooth surface. The prism boundaries are defined by differences in orientations between prismatic crystallites and those of the adjacent enamel that fills the spaces between prisms. This enamel is called interprismatic enamel. It is compositionally identical to enamel prisms, but differs in crystallite orientation. The submicroscopic gap produced by the change in crystallite orientations at the prism–interprismatic boundary is known as the prism sheath (Fig. 3). Prism sheaths contain

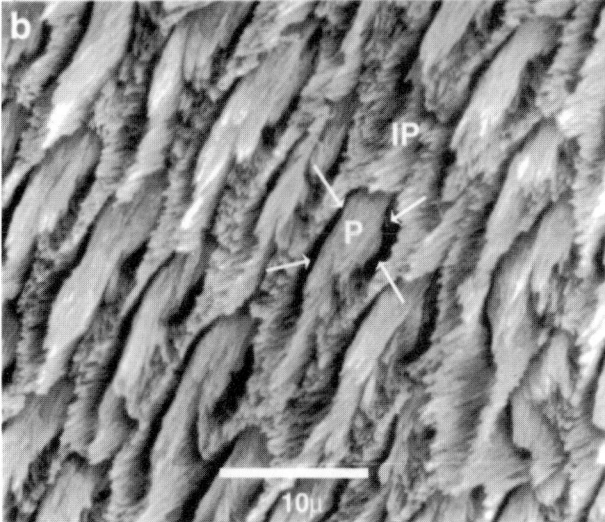

Figure 3 *(a) Scanning electron micrograph of fractured enamel near tip of tooth (unidentified Pliocene odontocete, Lee Creek Mine, North Carolina). The naturally fractured surface (at top) shows that the prisms take a straight course from the enamel–dentine junction (edj) to the outer surface, as is typical in radial enamel. (b) High magnification view showing enamel prisms (P) sectioned oblique to their long axes. Prism crystallites are parallel to each other, but not to crystallites in adjacent interprismatic enamel (IP). Arrows indicate the position of the prism sheath, which has been artificially enlarged in this acid-etched specimen.*

slightly greater concentrations of water and protein than the surrounding enamel, and thus are less dense. This allows prism patterns (the cross-sectional shapes and packing arrangement of prisms and interprismatic enamel) to be distinguished in ground sections or in acid-etched scanning electron microscope preparations. Prisms may have closed, circular cross sections or open, arc-shaped cross sections. Prism patterns have

been used to distinguish among some mammalian groups, but there is considerable variation within individuals, and considerable parallelism among different groups.

Enamel types describe the organization of enamel at a scale greater than individual crystallites or prisms. Common enamel types include parallel crystallite enamel, radial enamel, and decussating enamel. Parallel crystallite enamel, a type of nonprismatic enamel, is a volume of enamel in which hydroxyapatite crystallites are parallel to each other with no discontinuities in orientation and lacking larger-scale structural features, other than incremental lines. Radial enamel refers to a volume of prismatic enamel where prism long axes are parallel to one another and directed radially outward from the EDJ. Decussating enamel is a volume of enamel characterized by layers of parallel prisms, one or more prisms in thickness, whose long axes alternate in orientation with prisms in adjacent layers. Decussating enamel, also known as Hunter–Schreger bands (HSB), includes undulating HSB, where layers of similarly oriented prisms have a gently undulating course from the EDJ to the surface, and zigzag HSB, where the layers undulate with a pronounced vertical amplitude. Differences in enamel types have a phylogenetic component, but also have different mechanical properties that can be important functionally—parallel crystallite enamel may be harder than prismatic enamel, but prismatic enamel, especially decussating enamel, is more resistant to cracks induced by chewing stress. Zigzag enamel is thought to be especially resistant to cracking. Most mammal teeth are composed of more than one enamel type.

Cross-striations, a record of the daily incremental deposition of enamel, are sometimes evident in both prismatic and nonprismatic enamel. In the scanning electron microscope, cross-striations appear as alternating constrictions and varicosities along the length of the prism, suggesting that they reflect variations in the rate of enamel secretory activity. More prominent incremental lines, known as brown striae of Retzius, also transect prisms or crystallites. They are oriented parallel to the developing enamel surface and probably reflect regular interruptions in growth, although their causes and periodicity are not clear.

B. Marine Mammal Enamel

Although the crowns of most marine mammal teeth are covered with enamel, there is considerable variation in its structural complexity among and within orders. Likewise, prism patterns vary among and without orders, although there is no compelling evidence that prism patterns are diagnostic of particular marine mammal groups.

Most extant cetaceans have thin, structurally simple enamel. In some the enamel consists of a thin layer of radial prismatic enamel with or without an outer layer of nonprismatic parallel crystallite enamel, and in many species the tooth enamel consists entirely of nonprismatic parallel crystallite enamel. In contrast, the most primitive cetacean, the fossil *Pakicetus*, had relatively thick enamel with a more complex structure consisting of parallel crystallite enamel, radial enamel, and a thick inner layer of undulating Hunter–Schreger bands. Later archaeocetes show the same arrangement of enamel types, but almost all more derived odontocetes have much less complex enamel.

This has led some workers to conclude that the enamel of most extant cetaceans is evolutionarily degenerate. Only two extant odontocetes, the Indian river dolphin *Platanista gangetica* and Amazon dolphin *Inia geoffrensis* have well-developed, undulating HSB. It is unclear whether these were acquired independently in response to functional demands of their diet or a primitive retention from archaeocete ancestors.

Extant sirenians (*Dugong dugon* and *Trichechus* spp.) are reported to have radial enamel with variably circular and arc-shaped prism cross sections. Similar enamel has been reported for some fossil sirenians, and it is likely that this is primitive for the group. Pinniped enamel has not been described in detail, but enamel of some species appears to be more complex than that of sirenians. *Phoca vitulina* has undulating HSB, and walrus (*Odobenus rosmarus*) enamel shows a transition from undulating HSB to zigzag HSB near cusp tips.

Enamel incremental lines generally are not used in the age determination of marine mammals. The thin enamel of many species makes resolution of these lines difficult and, more importantly, enamel only records the period of tooth development during which enamel is laid down, which, is most cases, is before birth.

See Also the Following Articles

Age Estimation ▪ Dental Morphology, Evolution of ▪ Skeletal Anatomy ▪ Teeth

References

Boyde, A. (1980). Histological studies of dental tissues of odontocetes. *Rep. Intl. Whal. Comm. Spec. Iss.* **3**, 65–87.

Buffrénil, V.d., Ricqlés, A.d., Ray, C. E., and Domning, D. P. (1990). Bone histology of the ribs of the archaeocetes (Mammalia: Cetacea). *J. Vertebr. Paleontol.* **10**, 455–466.

Domning, D. P., and Buffrénil, V.d. (1991). Hydrostasis in the Sirenia: Quantitative data and functional interpretations. *Mar. Mamm. Sci.* **7**, 331–368.

Klevezal, G. (1996). "Recording Structures of Mammals, Determination of Age and Reconstruction of Life History." Translated by M. V. Mina and A. V. Oreshkin, A. A. Balkema, Rotterdam.

Koenigswald, W.v., and Sander, P. M. (1997). "Tooth Enamel Microstructure." A. A. Balkema, Rotterdam.

Maas, M. C., and Thewissen, J. G. M. (1995). Enamel microstructure of *Pakicetus* (Mammalia: Archaeoceti). *J. Paleontol.* **69**, 1154–1162.

Madar, S. I. (1998). Structural adaptations of early archaeocete long bones. *In* "The Emergence of Whales" (J. G. M. Thewissen, ed.), pp. 353–378. Plenum Press, New York.

Marks, S. C., and Popoff, S. N. (1988). Bone cell biology: The regulation of development, structure, and function in the skeleton. *Am. J. Anat.* **183**, 1–44.

Perrin, W. F., and Myrick, A. C., Jr. (eds.) (1980). Age determination of toothed whales and sirenians. Growth of odontocetes and sirenians: Problems in age determination. *Rep. Intl. Whal. Comm. Spec. Iss.* **3**.

Sahni, A., and Koenigswald, W.v. (1997). The enamel structure of some fossil and recent whales from the Indian subcontinent. *In* "Tooth Enamel Microstructure" (W.v. Koenigswald and P. M. Sander, eds.), pp. 177–191. Balkema, Rotterdam.

Ten Cate, A. R. (1989). "Oral Histology: Development, Structure, and Function." 3rd Ed. Mosby, St. Louis.

Uhen, M. D. (1999). New species of protocetid archaeocete whale, *Eocetus wardii* (Mammalia: Cetacea) from the middle Eocene of North Carolina. *J. Paleontol.* **73**, 512–528.

Wall, W. P. (1983). The correlation between high limb-bone density and aquatic habits in recent mammals. *J. Paleontol.* **57**, 197–207.

Zylberberg, L., Traub, W., Buffrénil, V.d., Allizard, F., Arad, T., and Weiner, S. (1998). Rostrum of a toothed whale: Ultrastructural study of a very dense bone. *Bone* **23**, 241–247.

Bottlenose Dolphins
Tursiops truncatus and *T. aduncus*

RANDALL S. WELLS
Chicago Zoological Society, Illinois

MICHAEL D. SCOTT
*InterAmerican Tropical Tuna Commission,
La Jolla, California*

I. Genus and Species

B ottlenose dolphins are arguably the best known of all cetaceans. They figured prominently in the legends of the ancient Greeks and Romans and were described in the writings of Aristotle, Oppian, and Pliny the Elder. Several books for scientific and public audiences have focused on these species (e.g., Caldwell and Caldwell, 1972; Leatherwood and Reeves, 1990; Reynolds *et al.*, 2000), and a number of comprehensive review articles have been produced as well (Tomilin, 1957; Leatherwood and Reeves, 1982; Shane *et al.*, 1986; Wells and Scott, 1999). The name *Tursiops* can be translated as "dolphin-like," deriving from the Latin *Tursio* ("dolphin") and the Greek suffix *-ops* ("appearance"). Two species of *Tursiops*, *T. truncatus*, the "common bottlenose dolphin," and *T. aduncus*, the "Indian Ocean bottlenose dolphin," are currently recognized (Rice, 1998), pending revisions based on recent genetic information.

Bottlenose dolphins are cosmopolitan in distribution and demonstrate a great deal of GEOGRAPHICAL VARIATION in morphology. *T. truncatus* is found in most of the world's warm temperate to tropical seas, in coastal as well as offshore waters; *T. aduncus* is limited to the coastal waters of the Indian Ocean and Western Pacific Ocean, from eastern Africa to Taiwan, southeastward to Australia (Fig. 1). They are recognizable by their generalized appearance—a medium-sized, robust body, a moderately falcate dorsal fin, and dark coloration, with a sharp demarcation between the melon and the short rostrum (Figs. 2 and 3). Adult lengths range from under 2 m to about 3.8 m, varying by geographic location (Cockcroft and Ross, 1990; Mead and Potter, 1990; Wells and Scott, 1999). Body size appears to vary inversely with water temperature in many parts of the world (but not the eastern Pacific). Bottlenose dolphins are colored light gray to black dorsally and laterally, with a light belly (Fig. 3). A light blaze

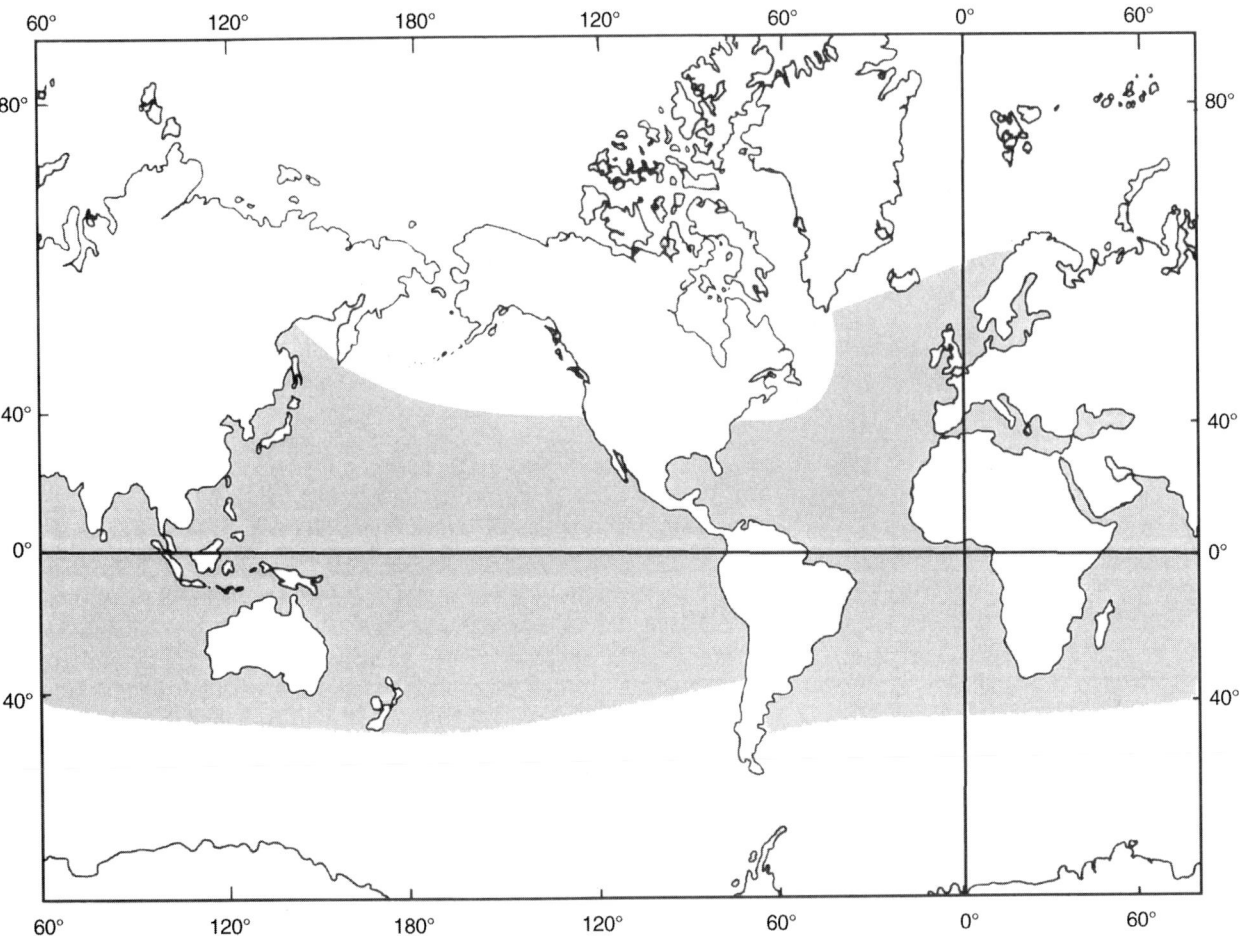

Figure 1 *Shading indicates the species range of the common bottlenose dolphin,* Tursiops truncatus. *It is not possible to be explicit regarding the range of the Indian Ocean bottlenose dolphin,* Tursiops aduncus, *because of uncertainties regarding the taxonomic status of bottlenose dolphins in the Indian Ocean, but the species generally inhabits the coastal waters of the Indian and Western Pacific Oceans, along the entire eastern coast of Africa, through the Red Sea and Persian Gulf, eastward as far as Taiwan, and southeastward to the coastal waters of Australia.*

or brush marking is sometimes observed on their sides. A distinct cape may be visible or may be obscured when the color pattern is very dark. *T. aduncus* tends to be smaller than *T. truncatus,* has a proportionately longer rostrum, and develops ventral spotting at about the time of sexual maturity (Ross and Cockcroft, 1990).

Variation in size, COLORATION, and cranial characteristics associated with feeding have led to descriptions of at least 20 nominal species of *Tursiops* (Hershkovitz, 1966, Rice, 1998). Recognition of the polymorphic nature of *Tursiops* and the existence of clinical variation had led to general agreement for many years that *Tursiops* was a single-species genus. However, recent genetic, morphologic, and physiologic studies suggest that revision of the genus may be necessary to acknowledge significant differences between forms from different oceans, as well as differences between forms in inshore vs offshore habitats within ocean basins (Hersh and Duffield, 1990; LeDuc *et al.,* 1999; Mead and Potter, 1995; Rice, 1998). Inshore bottlenose dolphins in the Atlantic and some other regions tend to

be smaller, lighter in color, have proportionately larger flippers, and differ in hematologic and mitochondrial DNA features from offshore forms (Hersh and Duffield, 1990; LeDuc *et al.,* 1999); however, eastern Pacific offshore bottlenose dolphins are smaller and darker than inshore forms. The taxonomic status of *Tursiops* is made even more confusing by observations of hybridization with several other odontocete species.

Recent genetic evidence suggests that *T. aduncus* is more closely related to pelagic *Stenella* and *Delphinus* species, in particular *S. frontalis,* than to *T. truncatus* (LeDuc *et al.,* 1999). This interpretation has more than just taxonomic implications. It could change the previous focus on differences in morphology and social behavior that were thought to reflect the polymorphic nature and behavioral plasticity of a single species to a renewed exploration of the many similarities perhaps brought about by evolutionary convergence in adaptations to a coastal/estuarine environment or, alternatively, a retention of primitive characteristics (plesiomorphy).

Figure 2 *Lateral view of an adult male common bottlenose dolphin. Photo by R. S. Wells.*

II. Distribution and Habitat

Common bottlenose dolphins are found in the temperate and tropical marine waters of the world (Fig. 1). In the North Pacific, they are commonly found as far north as the southern Okhotsk Sea, the Kuril Islands, and central California. In the North Atlantic, they are seen inshore during summer months off New England, offshore as far north as Nova Scotia, and have been recorded off Norway and the Lofoten Islands. Bottlenose dolphins occur as far south as Tierra del Fuego, South

Figure 3 *Ventral view of a common bottlenose dolphin. Photo by R. S. Wells.*

Africa, Australia, and New Zealand. Limits to the species' range appear to be temperature related, either directly or indirectly, through distribution of prey. Off the coasts of North America they tend to inhabit waters with surface temperatures ranging from about 10 to 32°C. At the northern limit of the species' range in the western North Atlantic, they are seasonally migratory, with a more southerly DISTRIBUTION in the winter.

Indian Ocean bottlenose dolphins are found in the coastal waters of eastern Africa from South Africa northward to the Red Sea and the Persian Gulf, and eastward to Taiwan and Australia (Rice, 1998), where they typically occur year-round.

Tursiops inhabits most warm temperate and tropical shorelines, adapting to a variety of marine and estuarine habitats, even ranging into rivers. Both species of bottlenose dolphins tend to be primarily coastal, but *T. truncatus* is also found in pelagic waters, near oceanic islands, and over the continental shelf, especially along the shelf break. In the Indian Ocean, *T. truncatus* tends to inhabit offshore waters, whereas *T. aduncus* is the more common coastal species.

III. Food and Feeding

The DIETS of common bottlenose dolphins have been described from many regions. A large variety of fish and/or squid forms most of the diets, although bottlenose dolphins seem to show a consistent preference for sciaenids, scombrids, and mugilids. Most fish prey are bottom dwellers, but some surface dwellers or pelagic fish are also represented in the diets. Differences in diets have been found where both inshore and offshore *Tursiops* ecotypes have been identified. In some cases, bottlenose dolphin groups feed in different areas depending on sex and size, with lactating females and their calves frequenting and feeding in the nearshore zone, adolescents feeding slightly farther offshore, and resting females and adult males feeding farther still.

Indian Ocean bottlenose dolphins also consume a variety of fish and squid (Cockcroft and Ross, 1990). Their prey includes fish with preferences for reefs or sandy bottoms and some inshore schooling fish.

IV. Predation

Sharks are probably the most important predators of bottlenose dolphins, although killer whales may also occasionally prey on them as well. Mutual tolerance during encounters between sharks and dolphins is probably typical, but as many as half of all bottlenose dolphins bear shark-bite scars as evidence of occasional encounters, depending on the region. Populations of Indian Ocean bottlenose dolphins in Australian waters are apparently subject to more frequent shark attacks than most populations of common bottlenose dolphins (Wood *et al.*, 1970; Corkeron *et al.*, 1987; Wells *et al.*, 1987; Cockcroft *et al.*, 1989a; Mead and Potter, 1990). In at least some areas, *Tursiops* appears to be a relatively minor and occasional part of the diets of sharks. Most wounds and scars from sharks tend to be found on the posterior and ventral regions of the dolphins, suggesting that the dolphins were ambushed from behind and below; some attacks may have been something other than a PREDATION attempt (e.g., sharks defending a territory). The primary shark predators of both *T. truncatus* and *T. aduncus* are the bull shark (*Carcharhinus leucas*), tiger shark (*Galeocerdo cuvier*), great white shark (*Carcharodon carcharias*), and dusky shark (*Carcharhinus obscurus*) (Wood *et al.*, 1970; Corkeron *et al.*, 1987; Connor *et al.*, 1999). Observations of captive dolphins suggest that they may recognize certain species of sharks as potential threats.

Anecdotal accounts describe common bottlenose dolphins attacking sharks by butting them with their rostra or by striking them with their flukes (summarized by Wood *et al.*, 1970). Defense may explain the apparently high survival rate indicated by the shark-bite scars on living dolphins. The relatively infrequent occurrence of shark-bite scars on young dolphins indicates either that the calves are well protected by their mothers or that attacks on young dolphins are generally fatal.

Stingrays have also been implicated in the deaths of common bottlenose dolphins (Wells and Scott, 1999). The dolphins were wounded externally or internally from ingestion of small rays, and deaths resulted from physical trauma, infection, or toxicosis.

V. Ranging Patterns

Coastal common bottlenose dolphins exhibit a full spectrum of movements, including seasonal MIGRATIONS, year-around home ranges, periodic residency, and a combination of occasional long-range movements and repeated local residency (Shane *et al.*, 1986; Wells and Scott, 1999). Much less is known about the ranging patterns of pelagic and Indian Ocean bottlenose dolphins. In some places, coastal dolphins living at the high-latitude or cold-water extremes of the species' range may migrate seasonally, as is the case along the Atlantic coast of the United States. Long-term residency has been reported from many parts of the world and may take the form of a relatively permanent home range or repeated occurrence in a given area over many years. For example, the year-round residents of several dolphin communities along Florida's west coast have maintained relatively stable, slightly overlapping home ranges during more than 30 years of observations, and through at least four generations; seasonal changes in habitat use may occur within the ranges (Wells and Scott, 1999). Home range bounds are often demarcated by physiographic features such as passes or abrupt changes in water depth. Some common bottlenose dolphins may use seasonal home ranges joined by a traveling range. At Shark Bay, Western Australia, a group of Indian Ocean bottlenose dolphins has returned to the same beach for more than 20 years to interact with humans (Connor *et al.*, 1999).

Longer-distance movements have been reported for some coastal common bottlenose dolphins, including range shifts of several hundred kilometers in an apparent response to environmental changes such as an El Niño warm-water event and a 600-km roundtrip for several identifiable dolphins in Argentina. Average daily movements of 33–89 km, monitored through travel distances of as much as 4200 km, have been reported for bottlenose dolphins in offshore waters (Wells *et al.*, 1999).

VI. Social Behavior

Common bottlenose dolphins are typically found in groups of 2–15 individuals, although groups of more than 1000 have been reported. In general, bottlenose dolphins in bays and estuaries tend to form smaller groups then those in offshore waters, but the trend does not continue linearly with increasing distance from shore (Wells *et al.*, 1999). Group composition tends to be dynamic, with sex, age, reproductive condition, familial relationships, and affiliation histories apparently being the most important determining factors. Subgroupings may be stable or repeated over periods of years. Basic social units include nursery groups, mixed-sex groups of juveniles, and adult males as individuals or strongly bonded pairs or trios.

Indian Ocean bottlenose dolphins are also found in groups of variable size. Numbers observed off South Africa range from 3 to 1000, around a mean of 140, whereas off Australia, groups average about 10 individuals in Moreton Bay, and average about 5, ranging up to 22, in Shark Bay (Connor *et al.*, 1999; Corkeron, 1990).

Dominance hierarchies have been observed for both common and Indian Ocean bottlenose dolphins in captivity, with large adult males dominating all other pool mates, females forming a less-rigid hierarchy, and with the largest females dominant over smaller animals. Aggressive behaviors, including contact and posturing, are used to establish and maintain hierarchies. In Shark Bay, Australia, coalitions of male *T. aduncus* fight with other male coalitions and aggressively herd females (Connor *et al.*, 1999).

VII. Activity Patterns

Common bottlenose dolphins in the wild appear to be active both during the day and night, interspersing bouts of feeding, traveling, socializing, and idling or resting (Shane *et al.*, 1986; Wells *et al.*, 1999). The duration and frequency of activities are influenced by such environmental factors as season, habitat, time of day, and tidal state and by physiological factors such as reproductive seasonality. Bottlenose dolphins feed in a large variety of ways, primarily as individuals, but cooperative herding of schools of prey fish also occurs. Individual prey capture involves behaviors as diverse as high-speed chases with a pin-wheeling capture at the surface, "fish whacking" in which a fleeing fish is struck with the dolphin's flukes and often knocked clear of the water, pushing fish onto shore and then partially beaching to capture them, and herding and perhaps disorienting fish with percussive leaps and tail lobs referred to as "kerplunking." Many of these same activities and behaviors have been observed for *T. aduncus* as well.

VIII. Life History

More detailed life history information is available for *T. truncatus* than for *T. aduncus*. Analyses of dentinal and cemental growth layer groups in teeth (Hohn *et al.*, 1989) have shown that female common bottlenose dolphins can live to more than 50 years, and some males have reached 40–45 years of age (Wells and Scott, 1999). Calves achieve most of their growth during the period of suckling, i.e., the first 1.5–2 years of life. Females typically reach sexual and physical maturity before males, leading to sexual dimorphism in some regions. Age at sexual maturity varies by region, but females usually reach sexual maturity at 5–13 years. Sexual maturity for males tends to occur at 9–14 years, often many years before they reach physical maturity (late teens) and achieve breeding status. Most breeding males in captivity are at least 20 years old (Wells *et al.*, 1999).

Indian Ocean bottlenose dolphins often develop ventral spotting at sexual maturity, when males are >235 cm in length and females are >230 cm (Ross and Cockcroft, 1990). Sexual maturity may be attained at an older age for *T. aduncus* than for *T. truncatus*, with females producing their first calf at age 12 or older (Connor *et al.*, 1999).

IX. Reproduction

Although births have been reported from all seasons, calving tends to be diffusely seasonal for both *T. truncatus* and *T. aduncus*, with peaks during warmer months (Connor *et al.*, 1999). Hormonal monitoring of captive common bottlenose dolphins indicates that females are spontaneous sporadic ovulators, ovulating repeatedly during a given season, whereas males may be active throughout the year with a prolonged elevation of testosterone concentrations over the months that different females may be ovulating. The reproductive life span for *T. truncatus* is prolonged; females up to 48 years of age have successfully given birth and raised young (Wells and Scott, 1999). Calves are born after a gestation period of about 1 year and range in length from about 84 to 140 cm depending on the geographic region. Calving intervals of 3–6 years are common for *T. truncatus*, whereas 4- to 6-year intervals are more common for *T. aduncus* (Connor *et al.*, 1999).

Lactation is the primary source of nutrition for the first year of life and may continue for several more years. Solid food has been found along with milk in the stomachs of calves as young as 4 months old. Maternal investment for free-ranging *T. truncatus* calves typically extends for about 3–6 years, with separation often coinciding with the birth of the next calf. *T. aduncus* calves in Shark Bay, Australia, typically attempt to nurse for 3–5 years (Connor *et al.*, 1999). Simultaneously pregnant and lactating females have been noted on occasion for both species.

X. Sound Production

Most of what is known about bottlenose dolphin SOUND PRODUCTION has resulted from detailed studies of *T. truncatus*, but it is reasonable to assume that *T. aduncus* acoustic patterns are comparable. Bottlenose dolphins produce these categories of sounds: whistles, echolocation clicks, and burst-pulse sounds. Dolphins produce a large variety of whistles, including largely stereotypic "signature whistles" that are individually specific, hypothesized to be used to communicate identity, location, and emotional state (Caldwell *et al.*, 1990). Once the signature whistle develops in neonates, it remains stable for at least many years, and probably for life. The signature whistles of many male calves are similar to the whistles of their mothers, whereas those of female calves are not. Dolphin echolocation involves the production of "clicks," with peak frequencies of about

40–130 kHz (Au, 1993). Echolocation is hypothesized to be used in navigation, foraging, and predator detection, among other possible functions. Burst pulses ("squawks") tend to be produced during social interactions.

XI. Fossil Record

Although no conclusive fossil evidence of the origin of *Tursiops* exists, FOSSIL RECORDS extend back several million years (Barnes, 1990). The geographic distribution of fossils falls within the range of modern animals. Anatomic features suggest that *Tursiops* evolved from some ancestral group of extinct fossil Delphininae, perhaps related to the subfamily Stenoninae, which might have evolved from the Kentriodontidae.

XII. Human Interactions

Common bottlenose dolphins take advantage of human activities in order to facilitate prey capture in a variety of ways. In Mauritania and Brazil, dolphins regularly drive schools of mullet toward fishermen wading with nets in shallow water, and in many parts of the world dolphins collect discarded fish from behind shrimp trawls and small purse seines or steal fish from various types of fishing gears.

A. Live Maintenance

The first *Tursiops* were publicly displayed at the Brighton Aquarium in 1883, at the New York Aquarium in 1914, and have been a regular attraction at Marineland of Florida since 1938. Common bottlenose dolphins continue to be the most common dolphins maintained in CAPTIVITY throughout the world. According to a May 2000 National Marine Fisheries Service inventory, 35 U.S. facilities held 392 common bottlenose dolphins. In addition, several hundred bottlenose dolphins, mostly *T. truncatus* but also including some *T. aduncus*, were held in at least 16 other countries. Within the United States, approximately 70% of the dolphins are held primarily for public display, whereas the remainder are used primarily for research or military purposes. Improved facilities and increased knowledge about the requirements for the care of dolphins have led to increasing success in the long-term maintenance of the animals, to the point where birth and survivorship rates at the better facilities approach and, possibly in a few cases, surpass those of wild populations (Wells and Scott, 1999).

B. Directed Fisheries

The largest of the historical FISHERIES for common bottlenose dolphins involved several countries surrounding the Black Sea, where dolphins were caught for oil, meat, and leather. Because of declines in dolphin populations, these countries have since outlawed the fishery. Directed takes still occur in other parts of the world, such as Peru, Sri Lanka, and Japan, for human consumption, to reduce the perceived competition with commercial fisheries, or for bait. Live-capture fisheries for dolphins for public display have existed for more than 100 years. More than 1500 *Tursiops* were removed from the waters of the United States, Mexico, and the Bahamas by 1980 for display, research, or military applications in many parts of the world (Leatherwood and Reeves, 1982). Although no bottlenose dolphins have been collected in U.S. waters since 1989, some small-scale, live-capture fisheries for *T. truncatus* continue in other countries such as Mexico, Cuba, and Japan, and some *T. aduncus* have been captured in recent years in Taiwan.

C. Incidental Fisheries

Incidental catches of small numbers of *T. truncatus* have been reported for several fisheries, including purse-seine fisheries for tunas, sardines, and anchovetas (Wells and Scott, 1999). In some cases, dolphins have been killed by fishermen to prevent damage to their fishing gear or stealing of the catch or bait by the dolphins (Leatherwood and Reeves, 1982). A large incidental take of *T. aduncus* has apparently occurred in the Taiwanese gill-net fishery off Australia, with an annual mortality perhaps exceeding 2000 animals (Northridge, 1991). Large numbers of Indian Ocean bottlenose dolphins have also been killed in nets set for sharks off the swimming beaches of South Africa and Australia. In the United States, entanglement in or ingestion of recreational fishing gear has resulted in dolphin mortality.

D. Habitat Alteration, Pollution, and Vessel Impacts

Although the impact of habitat alteration and POLLUTION on dolphins has not been studied systematically, anecdotal accounts suggest that human-caused degradation may have led to declines in some dolphin populations (Wells and Scott, 1999). Extremely high concentrations of chlorinated hydrocarbon residues have been found in the tissues of *Tursiops* in many parts of the world, with males accumulating higher concentrations than females with age (O'Shea, 1999). Cockcroft *et al.* (1989b) reported that first-born calves of South African bottlenose dolphins (identified by the authors as *T. truncatus*) received 80% of their mother's body burden of contaminant residues (polychlorinated biphenyls and dieldrin), perhaps leading to increased neonatal mortality, but also reducing levels of contaminants in the mothers. The accumulation of contaminants in tissues of males has reached levels that theoretically could impair testosterone production, and thus reduce reproductive ability. Preliminary findings suggest that even relatively low levels of PCBs and DDT metabolites can result in a decline in bottlenose dolphin immune system function. Responses to other human use of dolphin habitat through boating, dolphin feeding, and swimming with dolphins are receiving much current research attention, but it is clear that common bottlenose dolphins suffer mortality and serious injury from collisions with boats and that their behaviors change in the presence of vessels.

See Also the Following Articles

Convergent Evolution ■ Folklore and Legends ■ Incidental Catches ■ Signature Whistles

References

Au, W. L. (1993). "Sonar of Dolphins." Springer-Verlag, New York.

Barnes, L. G. (1990). The fossil record and evolutionary relationships of the genus *Tursiops*. In "The Bottlenose Dolphin" (S. Leatherwood and R. R. Reeves, eds.), pp. 3–26. Academic Press, San Diego.

Caldwell, D. K., and Caldwell, M. C. (1972). "The World of the Bottlenosed Dolphin." Lippincott, Philadelphia.

Caldwell, M. C., Caldwell, D. K., and Tyack, P. L. (1990). Review of the signature-whistle hypothesis for the Atlantic bottlenose dolphin. In "The Bottlenose Dolphin" (S. Leatherwood and R. R. Reeves, eds.), pp. 199–234. Academic Press, San Diego.

Cockcroft, V. G., Cliff, G., and Ross, G. J. B. (1989a). Shark predation on Indian Ocean bottlenose dolphins *Tursiops truncatus* off Natal, South Africa. *S. Afr. J. Zool.* **24**(4), 305–309.

Cockcroft, V. G., De Kock, A. C., Lord, D. A., and Ross, G. J. B. (1989b). Organochlorines in bottlenose dolphins *Tursiops truncatus* from the east coast of South Africa. *S. Afr. J. Mar. Sci.* **8**, 207–217.

Cockcroft, V. G., and Ross, G. J. B. (1990). Age, growth, and reproduction of bottlenose dolphins *Tursiops truncatus* from the east coast of southern Africa. *Fish. Bull.* **88**(2), 289–302.

Connor, R. C., Wells, R. S., Mann, J., and Read, A. J. (1999). The bottlenose dolphin, *Tursiops* spp: Social relationships in a fission-fusion society. In "Cetacean Societies: Field Studies of Dolphins and Whales" (J. Mann, R. C. Connor, P. L. Tyack, and H. Whitehead, eds.), pp. 91–126. Univ. of Chicago Press, Chicago.

Corkeron, P. J. (1990). Aspects of the behavioral ecology of inshore dolphins *Tursiops truncatus* and *Sousa chinensis* in Moreton Bay, Australia. In "The Bottlenose Dolphin" (S. Leatherwood and R. R. Reeves, eds.), pp. 285–293. Academic Press, San Diego.

Corkeron, P. J., Morris, R. J., and Bryden, M. M. (1987). Interactions between bottlenose dolphins and sharks in Moreton Bay. *Aqu. Mamm.* **13**(3), 109–113.

Hohn, A. A., Scott, M. D., Wells, R. S., Sweeney, J. C., and Irvine, A. B. (1989). Growth layers in teeth from known-age, free-ranging bottlenose dolphins. *Mar. Mamm. Sci.* **5**(4), 315–342.

Leatherwood, S., and Reeves, R. R. (1982). Bottlenose dolphin (*Tursiops truncatus*) and other toothed cetaceans. In "Wild Mammals of North America: Biology, Management, Economics" (J. A. Chapman and G. A. Feldhamer, eds.), pp. 369–414. Johns Hopkins Univ. Press, Baltimore.

Leatherwood, S., and Reeves, R. R. (1990). "The Bottlenose Dolphin." Academic Press, San Diego.

LeDuc, R. G., Perrin, W. F., and Dizon, A. E. (1999). Phylogenetic relationships among the delphinids cetaceans based on full cytochrome *b* sequences. *Mar. Mamm. Sci.* **15**, 619–648.

Mead, J. G., and Potter, C. W. (1990). Natural history of bottlenose dolphins along the central Atlantic coast of the United States. In "The Bottlenose Dolphin" (S. Leatherwood and R. R. Reeves, eds.), pp. 165–195. Academic Press, San Diego.

Mead, J. G., and Potter, C. W. (1995). Recognizing two populations of the bottlenose dolphin (*Tursiops truncatus*) off the Atlantic coast of North America: Morphological and ecological considerations. *Int. Biol. Res. Inst. Rep.* **5**, 31–43.

Northridge, S. P. (1991). An updated world review of interactions between marine mammals and fisheries. *FAO Tech Paper* **251**(Suppl. 1), 1–58.

O'Shea, T. J. (1999). Environmental contaminants and marine mammals. In "Biology of Marine Mammals" (J. E. Reynolds, III and S. A. Rommel, eds.), pp. 485–563. Smithsonian Institution Press, Washington, DC.

Reynolds, J. E., III, Wells, R. S., and Eide, S. D. (2000). "Biology and Conservation of the Bottlenose Dolphin." Univ. of Florida Press, Gainesville, FL.

Rice, D. W. (1998). "Marine Mammals of the World: Systematics and Distribution." Special Publication No. 4, Society for Marine Mammalogy, Allen Press, Lawrence, KS.

Ross, G. J. B., and Cockcroft, V. G. (1990). Comments on Australian bottlenose dolphins and the taxonomic status of *Tursiops aduncus*

(Ehrenberg, 1832). In "The Bottlenose Dolphin" (S. Leatherwood and R. R. Reeves, eds.), pp. 101–128. Academic Press, San Diego.

Shane, S. H., Wells, R. S., and Würsig, B. (1986). Ecology, behavior and social organization of the bottlenose dolphin: A review. *Mar. Mamm. Sci.* **2**(1), 34–63.

Tomilin, A. G. (1957). "Zveri SSSR i prilezhashchikh stran. IX: Kitoobraznye" ["Mammals of the U.S.S.R. and Adjacent Countries. Vol. IX: Cetacea"] (V. G. Heptner, ed.). Nauk U.S.S.R., Moscow. English translation, 1967, Israel Program for Scientific Translations, Jerusalem.

Wells, R. S., Boness, D. J., and Rathbun, G. B. (1999). Behavior. *In* "Biology of Marine Mammals" (J. E. Reynolds, III and S. A. Rommel, eds.), pp. 324–422. Smithsonian Institution Press, Washington, DC.

Wells, R. S., and Scott, M. D. (1999). Bottlenose dolphin *Tursiops truncatus* (Montagu, 1821). *In* "Handbook of Marine Mammals" (S. H. Ridgway and R. Harrison, eds.), Vol. 6, the Second Book of Dolphins and Porpoises, pp. 137–182. Academic Press, San Diego, CA.

Wells, R. S., Scott, M. D., and Irvine, A. B. (1987). The social structure of free-ranging bottlenose dolphins. In "Current Mammalogy" (H. H. Genoways, ed.), Vol. 1, pp. 247–305. Plenum Press, New York.

Wood, F. G., Jr., Caldwell, D. K., and Caldwell, M. C. (1970). Behavioral interaction between porpoises and sharks. *Invest. Cetacea* **2**, 264–277.

Bottlenose Whales
Hyperoodon ampullatus and *H. planifrons*

SHANNON GOWANS
Dalhousie University, Halifax, Nova Scotia, Canada

Bottlenose whales are relatively large beaked whales (6–9 m), found in deep waters of the North Atlantic and southern ocean. They are excellent divers, capable of diving for over an hour, and routinely dive deeper than 800 m. Their primary prey is deep-water squid of the genus *Gonatus*, although some fish and benthic organisms are also consumed. Northern bottlenose whales are the best-studied beaked whale, with data from whaling records and a long-term study on live animals concentrated in a submarine canyon known as the Gully off Nova Scotia, Canada. Relatively little is known about the southern bottlenose whale. Individuals are found in small groups, ranging from 1 to 20. In the Gully population, males form long-term bonds, whereas females live in a loose network of associates. Northern bottlenose whales were heavily whaled from the 1850s to the 1970s and numbers are believed to be reduced throughout their range. Only a few southern bottlenose whales were harvested.

I. Description and Diagnostic Characteristics

Bottlenose or bottle-nosed whales are large, robust-beaked whales distinguished by their large bulbous forehead and short dolphin-like beak (Fig. 1). They are chocolate brown to yellow in color, being lighter on the flanks and belly. This COLORATION is believed to be caused by a thin diatom layer. Newborns are gray in color with dark eye patches and a light-colored forehead.

Figure 1 *Spyhop of a northern bottlenose whale showing prominent beak and V-shaped throat grooves.*

The maxillary crests of males become larger and heavier with age, leading to a change in the shape of the forehead, with mature males having a flat, squared-off forehead whereas female/immature males have a smooth-rounded forehead. The dense bone in the males forehead may be used for male–male competition, as males head butt one another. Males possess a single pair of conical TEETH at the tip of the lower jaw; however, these teeth are rarely visible in live animals.

II. Distribution and Range

Northern bottlenose whales are found in cold and temperate waters of the North Atlantic, from the ice edge to the Azores, almost always in waters deeper than 500 m. They concentrate in submarine canyons, the shelf edge, and other areas of high relief. A resident year-round population is found in the Gully, a large submarine canyon off the coast of Nova Scotia, Canada. Whales found in different areas have different length distributions, indicating that there may be geographic isolation between the different whaling grounds. Preliminary analysis of mitochondrial DNA suggests that there may be reproductive isolation between bottlenose whales off Labrador and in the Gully.

Southern bottlenose whales are found throughout the southern hemisphere, from the ice edge to 30°S. There are no known areas of concentration, although relatively little effort has been made to identify these animals. Recent molecular work on southern bottlenose whales indicates that there may be more than one species (Dalebout *et al.*, 1998). Many sightings of a large beaked whale in the tropical Pacific have been identified as a bottlenose whale and may represent a more tropical DISTRIBUTION of *H. planifrons*, or a third undescribed *Hyperoodon* species. Pitman *et al.* (1999) suggested that the tropical bottlenose whale is actually Longman's beaked whale *Indopacetus pacificus*, known hitherto only from skeletal remains.

III. Ecology

Bottlenose whales are deep divers feeding predominantly on squid of the genus *Gonatus*, although other species of squid are eaten. Adult *Gonatus* are primarily benthic, although juveniles may inhabit the water column. Fish (including herring and redfish) and benthic invertebrates such as starfish and sea cucumbers are occasionally consumed. Time-depth recorders on two northern bottlenose whales in the Gully indicated that these whales were routinely diving to or near the sea floor, over 1400 m below the surface.

Evidence from whaling suggests that northern bottlenose whales migrate north in spring and south in the fall; however, evidence for this is weak and this migration may actually represent a migration of whaling vessels. Stomach contents of stranded animals indicate that both northern and southern bottlenose whales travel over long distances (ca. 1000 km), although it is not known if these movements are routine. Individuals routinely return to the Gully after spending time outside the canyon, and bottlenose whales are found in the Gully year round.

IV. Social Organization

Both northern and southern bottlenose whales are typically found in small groups (one to four individuals), although groups of up to 20 have been observed. Nothing is known about the social organization of southern bottlenose whales, and only the Gully population of northern bottlenose whales has been studied. In the Gully, individuals live in fision–fussion groups and most associations are brief (on the order of minutes to a few days). Females form a loose network of associates with most members of the community. However, mature males form long-term companionships with other mature males and these associations last for years. The function of these associations is unknown, but they may be linked to mating and may be similar to male coalitions in bottlenose dolphins.

V. Interactions with Humans

Northern bottlenose whales are often described as curious, as they will often approach boats. Whalers exploited this behavior to find groups of bottlenose whales, and as healthy whales would often remain near wounded individuals, the entire group was often captured.

The commercial hunt for northern bottlenose whales began in the 1850s and extended until the 1970s. Over 80,000 whales were captured during this period, and many more were harpooned but not recovered. Preexploitation numbers are estimated at 40–50,000 whales, although this number is at best a rough guess. There is no current estimate for the size of the North Atlantic population, but it is unlikely that it has fully recovered from whaling. Only 130 individuals reside in the Gully currently and this population is likely still recovering from the whaling catch of approximately 60 animals taken from the area in the 1960s.

The study in the Gully represents the first long-term study of live beaked whales. Crews from several documentary films and magazines have visited the Gully, as it is one of the few places where beaked whales can be observed routinely.

See Also the Following Articles

Beaked Whales, Overview ■ Hunting of Marine Mammals ■ North Atlantic Marine Mammals

References

Benjaminsen, T., and Christensen, I. (1979). The natural history of the bottlenose whale, *Hyperoodon ampullatus* (Forster). *In* "Behavior of Marine Animals" (H. E. Winn and B. L. Olla, eds.), Vol. 3. Plenum Press, New York.

Dalebout, M., van Heldon, A., van Waerebeek, K., and Baker, C. S. (1998). Molecular genetic identification of southern hemisphere beaked whales (Cetacea: Ziphiidae). *Mol. Ecol.* **6**, 687–692.

Gowans, S. (1999). "Social Organization and Population Structure of Northern Bottlenose Whales in the Gully." Ph.D. Dissertation, Dalhousie University, Halifax.

Hooker, S. K. (1999). "Resource and Habitat Use of Northern Bottlenose Whales in the Gully: Ecology, Diving, and Ranging Behaviour." Ph.D. Dissertation, Dalhousie University, Halifax.

Mead, J. G. (1989). Bottlenose whales *Hyperoodon ampullatus* (Forster, 1770) and *Hyperoodon planifrons* (Flowers, 1882). *In* "Handbook of Marine Mammals" (S. H. Ridgway and R. Harrison, eds.), Vol. 4. Academic Press, London.

Pitman, R. L., Palacios, D. M., Brennan, P. L., Brennan, B. J., Balcomb, K. C., and Miyashita, T. (1999). Sightings and possible identity of a bottlenose whale in the tropical Indo-Pacific: *Indopacetus pacificus?* *Mar. Mamm. Sci.* **15**, 531–549.

Reeves, R. R., Mitchell, E., and Whitehead, H. (1993). Status of the northern bottlenose whale, *Hyperoodon ampullatus. Can. Field-Nat.* **107**, 490–508.

Bowhead Whale
Balaena mysticetus

DAVID J. RUGH AND KIM E. W. SHELDEN
*National Marine Mammal Laboratory,
Seattle, Washington*

Bowhead whales (*Balaena mysticetus*), sometimes called Arctic right whales, Greenland right whales, great polar whales, or ahvik, are the only members of the family Balaenidae (suborder Mysticeti, order Cetacea) that live most of the year associated with sea ice in northern latitudes. Bowheads have never been seen in the Southern Hemisphere.

I. Description

Bowheads are readily identifiable by their large size, rotund shape, lack of a dorsal fin, dark color, white chins, triangular head (in profile), and neck (an indentation between the head and back). They are predominantly black, but most have characteristic white patterns on their chins, undersides, around their tail stocks, and/or on their flukes (Fig. 1). These patterns distinguish them from the similar-appearing right whales (*Eubalaena* spp.) and are unique to each individual. The white patterns around the tail and on the flukes increase with age. In addition, most bowheads accumulate distinctive, permanent marks on their backs, perhaps resulting from contact with sea ice. The bowed appearance of the mouth gives them their name.

These huge marine mammals are among the largest animals on earth, weighing as much as 75–100 tons. Males grow to

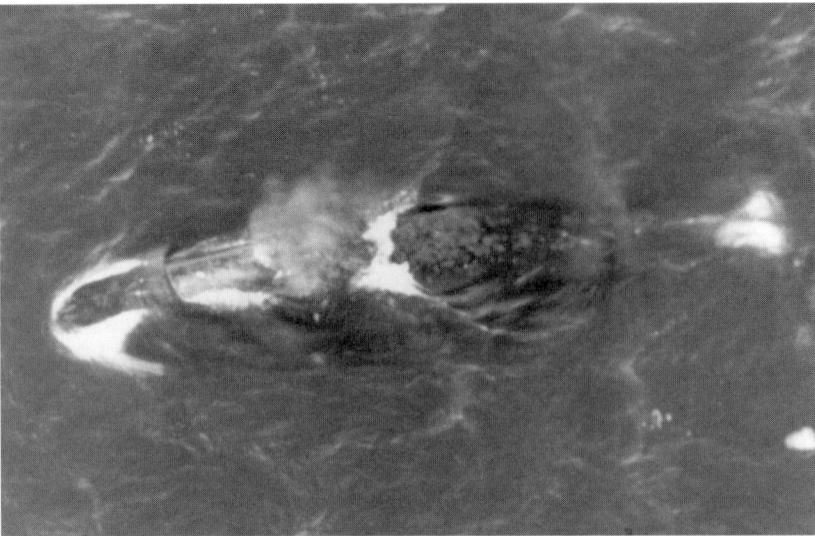

Figure 1 *Bowhead whales are large, black cetaceans with various amounts of white on their chins (the arched, bright white area on the left), tail stocks (the paired white spots on the far right), and ventral surfaces (out of sight in aerial photographs). Note that the whale's left eye is visible deep underwater directly below the blow (exhalation vapor). This adult bowhead was photographed in March at the start of its long migration from the Bering Sea to the Canadian Beaufort Sea for the summer. Photo by David Rugh.*

14–17 m in length and females 16–18 m, perhaps as much as 20 m. Their flukes are 2–6 m across. The heads of these whales constitute over a third of the bulk of their bodies, and their baleen may be as long as 4 m (no other whale has baleen longer than 2.8 m) with 230 to 360 plates on each side of the mouth, making their capacious mouths quite possibly the largest of any animal ever. To insulate them from the icy water, bowheads are wrapped in blubber 5.5 to 28 cm thick covered by an epidermis up to 2.5 cm thick. This combination of blubber and skin is the thickest of any whale species.

Bowheads are well adapted to the risky occupation of being air-breathing mammals in seas often covered with thick ice: they can withstand breaking through ice as much as 60 cm thick, and their diving abilities are exceptional—possibly exceeding an hour—which is critical to finding breathing holes when swimming under sea ice. The very low and very loud calls that bowheads produce may help them find mates or assist in following each other while navigating through sea ice. The only other whales commonly found as far north as bowheads are belugas (*Delphinapterus leucas*) and narwhals (*Monodon monoceros*), which are toothed whales with some of the same characteristics seen in bowheads: smooth backs and relatively thick blubber.

II. Breeding

Bowheads probably mate in later winter or early spring, but sexual activity may occur in any season. MATING groups consist of a male–female pair or several males and a female. Acoustics probably play a vital role in reproduction as bowheads are vocally active during the mating season and can hear each other 5–10 km away. Breaching (leaping completely out of the water) and fluke slapping (where the tails smash down on the water surface) may also play a role in attracting a mate or asserting dominance. Dominance is sometimes expressed through physical contact or sperm competition. However, there is a possibility that cooperation occurs among males during mating, making it more likely that at least one of them is able to inseminate a female. Over a year after mating (13–14 months), calves are born, usually during the spring migration between April and June. Calves are about 4 m long at birth. Females have calves 3 to 4 years apart. The following spring, the young whales, now 6 to 8 m long, are weaned from their mothers. After this, growth is slow compared to other baleen whales. At roughly 15 years of age, when 12 to 14 m long, females become sexually mature, and males become sexually active when 12 to 13 m long. Bowhead whales may live longer than other mammals; ancient harpoon points collected in whales recently indicate that the whales may have lived for more than a century.

III. Feeding

Bowheads feed throughout the water column, sometimes on the surface (called "skimming") and sometimes at or near the seafloor (as evidenced by mud smeared across their heads and backs). A bowhead's huge mouth can engulf large volumes of water, including prey, and, as the tongue rises, the water is pushed out, trapping prey on the inside fringed surfaces of the baleen, which serves as a filter all the way around the mouth. The massive tongue (as much as 5 m long and 3 m wide) then sweeps the food off the baleen into a very narrow digestive tract. As many as 60 species of animals have been found in bowhead stomachs, but their preferred prey are copepods (11 species) and euphausiids (2 species), as well as mysids and gammarid amphipods. Sometimes as many as a dozen bowheads will feed together in an echelon formation, similar to a line of migrating geese. Perhaps this coordinated effort helps the whales entrap their prey.

The only predators of bowhead whales, other than humans, are killer whales (*Orcinus orca*). Killer whale scars were found on approximately 4 to 8% of the whales taken by Alaskan Eskimos. In part, the bowheads' close association with sea ice may be a way of seeking refuge from killer whales.

IV. Distribution and Abundance

Bowhead whales may have once been a single panmictic (randomly interbreeding) population that emerged in the Northern Hemisphere during the Pliocene (roughly 8 million years ago), according to fossil records. Genetic mixing between stocks in different areas was possible during the relatively warm interglacial periods (such as in A.D. 1000–1200) when reduction in sea ice meant whales could move between the Atlantic and Pacific Basins. Bowheads could have moved freely between the Beaufort Sea and Hudson Bay until the "Little Ice Age" between 1400 and 1850. Today's temperatures are cool enough to keep ice across most of the east–west passages of the Arctic, isolating these whale stocks.

While ice may have contributed to this isolation, commercial whaling had a more profound effect. The bowhead's large size, long baleen, thick BLUBBER, slow speed, and gentle disposition have made them such a valuable commodity that whalers went to great lengths to harvest them. Commercial whalers from the 17th to 19th centuries were so efficient that they eliminated stock after stock of these whales. In fact, even a century after commercial whaling ceased, all bowhead stocks are still considered endangered.

Currently there are five stocks of bowheads defined as geographically distinct segments of the species' total population: the Bering Sea (or Bering-Chukchi-Beaufort stock around Alaska), Okhotsk Sea (eastern Russia), Davis Strait (northeastern Canada), Hudson Bay (perhaps a part of the Davis Strait stock), and Spitsbergen (North Atlantic). The largest remnant stock, the Bering Sea stock, consists of approximately 8000 whales that migrate from the Bering Sea in the winter through the Chukchi Sea to the Beaufort Sea in the summer. This is the only stock that appears to be recovering from commercial whaling, growing at an annual rate of 3%. Originally there may have been 10,000 to 23,000 whales in this stock. Currently, Native Alaskans harvest approximately 40 whales per year through quotas set by the International Whaling Commission (IWC). The Chukotka Natives of Siberia have been allotted 5 bowheads per year from the Alaska quota. Independent of the IWC quota, the Canadian government has allowed a limited hunt of bowheads in the Bering Sea stock as well as from the stocks in Hudson Bay and Davis Strait.

There is very little known about other stocks of bowheads. Available evidence indicates that most stocks are very small: only about 300–400 currently live in the Okhotsk Sea (originally more than 3000); approximately 350 (originally 11,700) are in the Davis Strait stock; roughly 270 (originally about 580) live in Hudson Bay; and the Spitsbergen stock numbers "only in the tens" where there may have been as many as 24,000.

See Also the Following Articles

Baleen ■ Beluga Whale ■ Breaching ■ Filter Feeding ■ Narwhal

References

Bockstoce, J. R. (1986). "Whales, Ice and Men: The History of Whaling in the Western Arctic." Univ. of Washington Press, Seattle, WA.

Braham, H. W., Marquette, W. M., Bray, T. W., and Leatherwood, J. S. (1980). The bowhead whale: Whaling and biological research. *Mar. Fish. Rev.* **42**(9-10), 1–96.

Burns, J. J., Montague, J. J., and Cowles, C. J. (eds.) (1993). "The Bowhead Whale." Spec. Publ. No. 2, The Society for Marine Mammalogy, Lawrence, KS.

George, J., Philo, L., Hazard, K., Withrow, D., Carroll, G., and Suydam, R. (1994). Frequency of killer whale (*Orcinus orca*) attacks and ship collisions based on scarring on bowhead whales (*Balaena mysticetus*) of the Bering-Chukchi-Beaufort Seas stock. *Arctic* **47**(3), 247–255.

McCartney, A. P. (ed.) (1995). "Hunting the Largest Animals: Native Whaling in the Western Arctic and Subarctic." The Canadian Circumpolar Institute, Studies in Whaling No. 3, Occasional Publication No. 36.

Moore, S. E., George, J. C., Coyle, K. O., and Weingartner, T. J. (1995). Bowhead whales along the Chukotka coast in autumn. *Arctic* **48**(2), 155–160.

Nerini, M., Braham, H., Marquette, W., and Rugh, D. (1984). Life history of the bowhead whale, *Balaena mysticetus* (Mammalia: Cetacea). *J. Zool. (Lond.)* **204**(4), 443–468.

Nicklin, F. (1995). Bowhead whales: Leviathans of icy seas. *Natl. Geograph. Mag.* **188**(2), 114–129.

Rice, D. W. (1998). "Marine Mammals of the World, Systematics and Distribution." Spec. Publ. No. 4, Society for Marine Mammalogy, Lawrence, KS.

Richardson, W. J., Finley, K. J., Miller, G. W., Davis, R. A., and Koski, W. R. (1995). Feeding, social and migration behavior of bowhead whales, *Balaena mysticetus*, in Baffin Bay vs. the Beaufort Sea: Regions with different amount of human activity. *Mar. Mamm. Sci.* **11**(1), 1–45.

Shelden, K. E. W., and Rugh, D. J. (1995). The bowhead whale, *Balaena mysticetus*: Its historic and current status. *Mar. Fish. Rev.* **57**(3-4), 1–20.

Vladimirov, V. L. (1994). Recent distribution and abundance levels of whales in Russian far-eastern seas. *Russ. J. Mar. Biol.* **20**(1), 1–9.

Würsig, B. (1988). The behavior of baleen whales. *Sci. Am.* **258**(4), 102–107.

Bow-Riding

BERND WÜRSIG
Texas A&M University, Galveston

One of the most fascinating BEHAVIORS of dolphins is when they ride the bow pressure waves of boats. Dolphins probably have been bow-riding ever since swift vessels plied the seas, propelled by oar, sail, or very recently in the history of seafaring, motor. The Greeks wrote of bow-riding in the eastern Mediterranean and Aegean Seas by what were most likely bottlenose (*Tursiops truncatus*), common (*Delphinus delphis*), and striped dolphins (*Stenella coeruleoalba*).

Bow-riding consists of dolphins, porpoises, and other smaller toothed whales (and occasionally sea lions and fur seals) positioning themselves in such a manner as to be lifted up and pushed forward by the circulating water generated to form a bow pressure wave of an advancing vessel (Lang, 1966; Hertel, 1969). Dolphins are exquisitely good at bow-riding, able to fine-tune their body posture and position so as to be propelled along entirely by the pressure wave, often with no tail (or fluke) beats needed. Bow-riders at the periphery of the pressure wave do need to beat their flukes, and so do bow-riders of a slowly moving vessel or one with a very sharp cutting instead of pushing bow.

There is often quite a bit of jostling for position at the bow, as dominant animals of a group edge others to a less favorable position, or as one is displaced from the bow by another one approaching (Fig. 1). It is great fun for a person to lean over the bow of a vessel and watch these interanimal antics, as well as the fine-tuning of positioning, effected by slight body turns and almost imperceptible movements of the flippers. Bow-riding dolphins also tend to emit what sounds to the human listener like a cacophony of underwater whistles and "screams," sounds implicated in high levels of social activity (Brownlee and Norris, 1994). Bow-riding is probably the dolphin behavior most noted, and most enjoyed, by seafaring people the world over.

Of course, riding the bow also makes these animals susceptible to being lanced or harpooned in areas where they are taken by humans. Where this occurs nearshore and in apparent smaller populations, dolphins become shy of the bow (Norris, 1974), but on the high seas or in deeper water, probably in larger populations, dolphins often still ride the bow after tens to hundreds of years of (generally small-scale) human hunting.

While many species of dolphins, porpoises, and small toothed whale ride the bow, some do not; and in some species, certain populations do not. Bottlenose dolphins are well-known bow-riders the world over, but even they do not ride in some

areas (even where they are not hunted) or on some types of vessels. For example, off the shores of Texas in the Gulf of Mexico, they generally do not approach any vessel smaller than about 20 m long to bow-ride, apparently finding the smaller bows not worth their while. Instead, they "hitch a ride" on the oil tankers and freighters that ply in and out of major harbors, at times bow-riding for 20 or more kilometers at a stretch. Dolphins ride underwater, and must leave their position to breathe, leaping forward and at an angle to the surface before falling back toward the advancing bow in a welter of foam (Fig. 1). Dolphins also ride the stern waves (or wakes) of boats, which present a different hydrodynamic challenge than bow-riding; and in some areas, dolphins that do not approach the bow will nevertheless ride in the influence of a large (or fast small) vessel's wake.

Most oceanic dolphins ride bow waves, with notable exceptions in areas of intensive hunting, such as by tuna vessels of the eastern Tropical Pacific, where vessels chase dolphins in order to net the tuna often affiliated with a dolphin school (Perrin, 1968). However, riding the bow is also "mood dependent"; dusky dolphins (*Lagenorhynchus obscurus*), for example, will not approach vessels when they have not fed for 2 or more days. These same dolphins will race toward a boat from several kilometers during and after social/sexual activities that take place immediately after bouts of feeding on schooling anchovy (Würsig and Würsig, 1980).

Why do dolphins bow-ride? It has been proposed that it is a mechanism to efficiently travel from one place to another. However, this is unlikely, for one often sees bow-riding dolphins after some time heading back to whence they picked up the vessel. Instead, it is more likely that riding the bow is done for enjoyment, for the sport of it; in other words, play. This is of great interest to behaviorists, for there are not too many nondomesticated adult mammals that habitually engage in activities just for the fun of them, although the list is growing with detailed observations in nature.

Bow-riding was certainly not "invented" by dolphins as a sport when human-made vessels first came on the scene. Instead, it appears to have been adapted from other wave-riding forms. Dolphins ride on the lee slopes of large oceanic waves and on the curling waves (or surf) that are formed as oceanic waves touch near-shore bottom (these two "rides" are hydrodynamically quite different; Hertel, 1969). Yes, dolphins "body surf" much as do humans, but dolphins are generally much better surfers than humans. Dolphins also ride the bow waves of surging whales, such as of the larger of the baleen whales, and sperm whales (*Physeter macrocephalus*). Dolphins even "entice" whales to surge ahead by rapidly crossing back and forth a whale's eyes and snout. The whale surges forward in response (and apparent annoyance), often blowing forcefully during the surge. An abrupt bow wave is formed, and the previously heckling dolphins are all lined up in that wave, apparently enjoying its momentary pressure effect. This activity can go on with one whale for 20 min or more, until the whale tires, the bow wave becomes less distinct, and the dolphins abandon it to try with another whale or to go about other activities. They have had their fun, and we are left to wonder what is going on in that large BRAIN during these bouts of quite obvious play.

Figure 1   *Common dolphins on the bow of a vessel off Panama.*

See Also the Following Articles

Group Behavior ■ Playful Behavior

References

Brownlee, S. M., and Norris, K. S. (1994). The acoustic domain. *In* "The Hawaiian Spinner Dolphin" (K. S. Norris, B. Würsig, R. S. Wells, and M. Würsig, eds.), pp. 161–185, University of California Press, Berkeley, CA.

Hertel, H. (1969). Hydrodynamics of swimming and wave-riding dolphins. *In* "The Biology of Marine Mammals" (H. T. Anderson, ed.), pp. 31–63. Academic Press, New York.

Lang, T. G. (1966). Hydrodynamic analysis of cetacean performance. *In* "Whales, Dolphins, and Porpoises" (K. S. Norris, ed.), pp. 410–434, University of California Press, Berkeley, CA.

Norris, K. S. (1974). "The Porpoise Watcher." Norton Press, New York.

Perrin, W. F. (1968). The porpoise and the tuna. *Sea Front.* **14,** 166–174.

Würsig, B., and Würsig, M. (1980). Behavior and ecology of the dusky dolphin *Lagenorhynchus obscurus*, in the south Atlantic. *Fish. Bull.* **77,** 871–890.

Brain

**HELMUT H. A. OELSCHLÄGER AND
JUTTA S. OELSCHLÄGER**
*Johann Wolfgang Goethe University,
Frankfurt am Main, Germany*

It takes a complicated brain to deal with a complicated environment.   —Christopher Wills, *The Sciences,* 1993

Adaptation to the aquatic environment is a multiconvergent phenomenon seen in a number of mammals. Most fascinating is the degree of specialization in different groups that comprise the semiaquatic insectivores, otters, and pinnipeds, and the fully aquatic sirenians and cetaceans. Both the body shape and the morphology of the sensory organs and brain intimate which selective pressures may have led to exclusively aquatic life. There are some obstacles, however, in understanding brain evolution in aquatic mammals. First, we are only marginally familiar with brain morphology of a very few species, and here mainly the bottlenose dolphin (*Tursiops truncatus*; Figs. 2–3). Second, the brain itself does not fossilize; only the outer shape can be studied in natural endocasts, and these are biased by covering blood vessels, meninges, and geological artifacts. Thus, the tracing of brain evolution in fossils is difficult and should be supplemented by phylogenetic reconstruction on the basis of extant relatives. This is particularly true for the cetaceans and the lack of adequate data from their closest relatives, the ungulates (hoofed animals; Fig. 1) and the more distantly related paenungulates [hyraxes or conies, sirenians or sea cows, and pro-

boscideans or elephants; for a survey see Berta and Sumich (1999)]. Third, although the comparative consideration of analogous developmental trends (primates) may be useful for the understanding of brain evolution in highly encephalized aquatic mammals, the paucity of data often leads to an overestimation of these analogies. Particularly the large size of the cetacean brain sometimes has led to an anthropocentric approach to these mammals, thereby resulting in inadequate questions and wrong answers. This article first focuses on the cetacean, particularly the odontocete brain, and then discusses its specifics together with those of convergent adaptive trends seen in semiaquatic and other aquatic mammals so as to provide an idea of what may have happened during the evolution of these mammals.

Among the most fascinating characteristics of cetaceans are their exceptionally large brain, both in absolute and relative terms, and their extremely convoluted neocortex. Whereas dolphins usually have a brain mass of about 200–2000 g, the maximal size is attained in killer whales (*Orcinus orca*) and sperm whales (*Physeter macrocephalus*) with nearly 10,000 g. Basically, cetacean brains show the typical mammalian bauplan and are as complicated morphologically as those of other mammalian groups. To some extent they parallel the simian and human brains. In this respect, however, it has to be kept in mind that cetaceans have been subject to profound modifications in brain morphology and physiology during 50 million years of separate evolution in the aquatic environment. Moreover, it is still very difficult to correlate the results of behavioral and physiological research on dolphins with the existing neuroanatomical data. Because invasive experimentation is not possible in cetaceans, the functional significance of such data can only be elucidated via comparison with other aquatic or terrestrial mammals.

Most studies during the last decades have focused on the morphology and physiology of the adult odontocete brain and its functional systems [for reviews, see Jansen and Jansen (1969), Pilleri and Gihr (1970), Morgane and Jacobs (1980), Glezer *et al.* (1988), and Ridgway (1986, 1990)]. Concerning the development of the cetacean brain, very few recent papers have been dedicated to the striped dolphin (*Stenella coeruleoalba*; Kamiya and Pirlot 1974), harbor porpoise (*Phocoena phocoena*; Buhl and Oelschläger 1986), pantropical spotted dolphin (*Stenella attenuata*; Wanke 1990), narwhal (*Monodon monoceros*; Holzmann (1991), and sperm whale (Oelschläger and Kemp 1998). Some studies of the morphology and ultrastructure of the cetacean cortex are by Supin *et al.* (1978) and Manger *et al.* (1998). Other relevant publications are those by Bauchot and Stephan (1966) on semiaquatic insectivores; Schwerdtfeger *et al.* (1984), Schulmeyer (1992), and Marino (1998) on the encephalization of toothed whales and the quantitative composition of their brain; and Pirlot and Kamiya (1985), Reep *et al.* (1989), and Reep and O'Shea (1990) on the brain of manatees.

I. Morphology of the Cetacean Brain
A. General Appearance of the Brain

Whereas its development in the embryonal and early fetal period is similar to that of other mammals, the brain of adult whales and dolphins is rather spherical in comparison with that

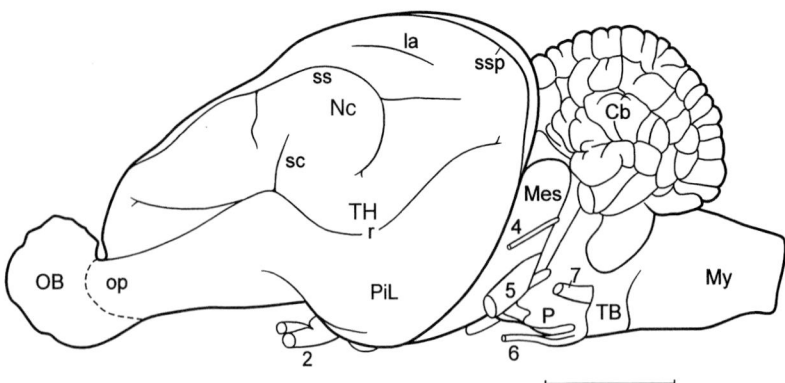

Figure 1 *Lateral aspect of the brain of a generalized land mammal, the mouse deer* (Hyemoschus aquaticus), *as a representative of the hoofed animals. Here the telencephalic hemisphere (TH) is rather flat, the neocortex (Nc) is moderately folded, and the olfactory bulb (OB) and the olfactory cortex in the piriform lobe (PiL) are large. In more advanced mammals such as cetaceans (see later), the hemispheres are much larger, the neocortex is much more extended, i.e., extensively folded, and it strongly dominates the olfactory cortex, which then is only found on the rostral basal surface of the telencephalic hemisphere. Cb, cerebellum; la, lateral sulcus; Mes, mesencephalon; My, myelencephalon; op, olfactory peduncle; P, pons; r, rhinal sulcus; sc, Sylvian cleft; ss, suprasylvian sulcus; ssp, splenial sulcus; TB, trapezoid body; 2, optic nerve; 4, trochlear nerve; 5, trigeminal nerve; 6, abducent nerve; 7, facial nerve. Scale: 1 cm. Modified after L. Sigmund* (1981; Vest Cs. Spolec. Zool. 45, 144–156).

of generalized land mammals (Fig. 1) and is somehow reminiscent of a boxing glove (Fig. 2a). In correlation with the so-called "telescoping" of the skull along the beak–fluke axis, both the cranial vault and the brain are short but wide (Figs. 2–7), and more so in the toothed whales (odontocetes) than in baleen whales (mysticetes). In the bottlenose dolphin (Fig. 2), the hemispheres are rounded and high, and the anterior profile is rather steep. In ventral aspect, the contour of the odontocete forebrain is more trapezoidal, whereas in mysticetes it is more trilobar, with the area of the insula being visible as an indentation between orbital and temporal lobes (Figs. 3, 4, and 7). In comparison with hoofed animals (Fig. 1), the telencephalic hemisphere seems to be rotated rostralward and ventralward leading to a subvertical position of the corpus callosum (Figs. 2b and 6). In some odontocetes (bottlenose dolphin, sperm whale), the posterior myelencephalon and the anterior spinal cord arch

around the cerebellum. Via an S-shaped transition, the spinal cord then continues straight along the body axis, thus accounting for the shortening of the cetacean neck region.

The ventricular system reflects the foreshortening of the brain in the tight coiling of the lateral ventricles, the shortness of the fronto-orbital region (anterior horn), the lack of an occipital pole of the hemisphere (no posterior horn), and the large size of the midbrain (cerebral aqueduct). Only rarely is a posterior horn of the lateral ventricle reported.

B. Telencephalon

1. Cortex In comparison with generalized tetrapod mammals (Fig. 1), the surface of the telencephalic hemispheres is extremely convoluted, particularly in toothed whales (Figs. 2–7). Gyrification in baleen whales is less extreme because of the greater width of their cortical layers. The neocortex accounts for

Figure 2 *Bottlenose dolphin brain: (a) lateral [after Langworthy (1932), modified after Morgane and Jacobs (1972) and Pilleri and Gihr (1970)] and (b) another specimen, mediosagittal aspect (after Morgane and coworkers). Cortex and structures containing nuclei are labeled with capital letters, fiber tracts (white matter) and sulci with small letters. Arrow pointing into sylvian cleft; a, interthalamic adhesion; ac, anterior commissure; An, anterior lobule; aq, cerebral aqueduct; c, "calcarine" cleft; cc, corpus callosum; Ch, cerebellar hemisphere; crs, cruciate sulcus; e, elliptic nucleus; E, epithalamus; en, entolateral sulcus; es, ectosylvian sulcus; ES, ectosylvian gyrus; f, fornix; H, hypothalamus; Hy, hypophysis; IC, inferior colliculus; IO, inferior olive; L, limbic lobe; la, lateral sulcus; La, lateral gyrus; Li, lingual lobule; Met, metencephalon; My, myelencephalon; oc, optic chiasm; OL, olfactory lobe; OrL, orbital lobe; Ov, oval lobule; P, pons; pc, posterior commissure; PC, perisylvian cortex; PL, perilimbic lobe; SC, superior colliculus; ss, suprasylvian sulcus; SS, suprasylvian gyrus; ssp, suprasplenial (limbic) sulcus; T, thalamus; TB, trapezoid body; Ve, vermis; 2, optic nerve; 5, trigeminal nerve; 7, facial nerve; 8, vestibulocochlear nerve; 10, vagus nerve; III, third ventricle. Scale: 1 cm.*

a

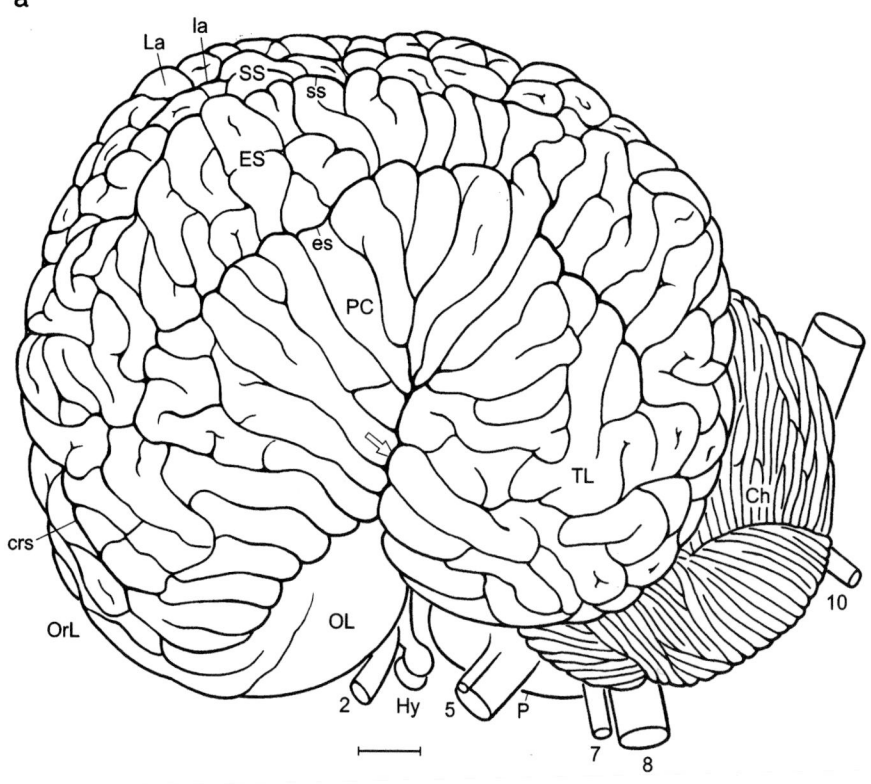

b

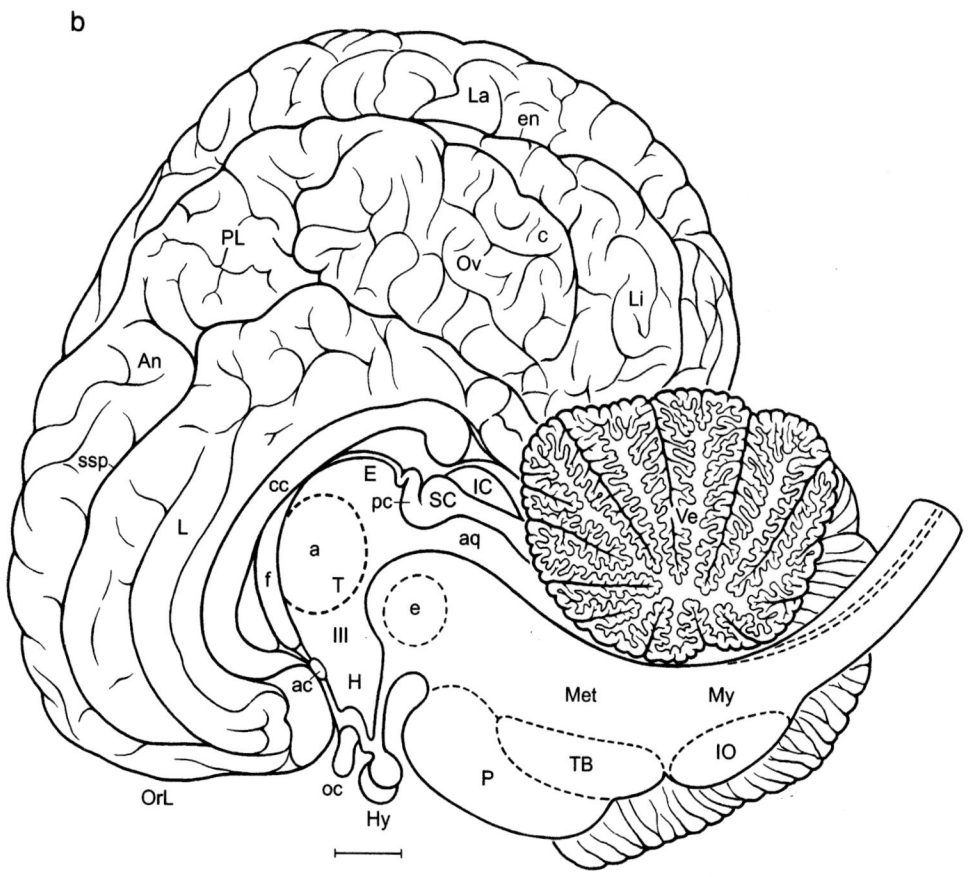

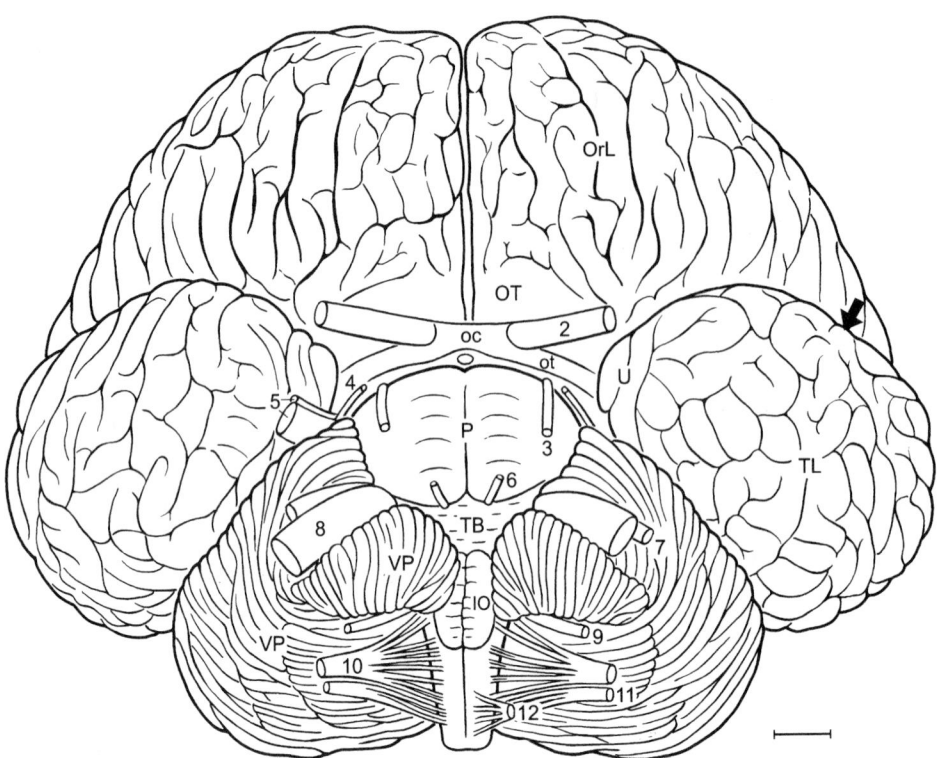

Figure 3 *Bottlenose dolphin brain in basal aspect. Arrow pointing into sylvian cleft. ot, optic tract; OT, olfactory tubercle; TL, temporal lobe; U, uncus; VP, ventral paraflocculus; 2-12, cranial nerves; 3, oculomotor nerve; 4, trochlear nerve; 6, abducens nerve; 9, glossopharyngeus nerve; 11, accessory nerve; 12, hypoglossus nerve. Scale: 1 cm. After Langworthy (1932), modified after Pilleri and Gihr (1970) and Morgane and Jacobs (1972).*

the large size of the telencephalon and thus the large size of the brain [percentage of the neocortex: 63% in the franciscana (La Plata dolphin or *Pontoporia blainvillei*); 87% in the sperm whale.

As in higher primates, the cortex of the cetacean olfactory and limbic systems (allocortex) is restricted to the rostral base of the hemisphere (paleocortex; olfactory system) or is located at the inferior horn of the lateral ventricle in the temporal lobe (archicortex: hippocampus as part of the limbic system). The archicortex in cetaceans, above all in toothed whales (Fig. 5), is much smaller than in terrestrial mammals. This correlates well with the small size of other components of the limbic system, e.g., the fornix as the main fiber tract of the hippocampus and the mammillary body as a main relay within the limbic system, whereas the cortical fields above the corpus callosum ("limbic lobe") and the entorhinal cortex on the temporal lobe are well developed. As in primates, the cortex of the cetacean limbic lobe presumably does not have an immediate relationship to olfaction. In adult baleen whales, the nose is small but

obviously functional, and the same holds true for the components of the rhinencephalon. In adult toothed whales, these components are very much reduced: there is no olfactory part of the nose, olfactory bulb or tract, and the central parts of the olfactory system are small.

A. SURFACE CONFIGURATIONS. The fissural or gyral pattern of the cetacean cortex, which has been discussed in many papers in the past, bears general resemblance to that of carnivores and ungulates (Figs. 1–7 and 12). On the convex lateral surface and the vertex of the hemisphere, the main fissures (ectosylvian, suprasylvian, lateral sulcus) run at different distances around the Sylvian cleft. Thus, the ectosylvian gyrus is bordered by the ectosylvian and suprasylvian sulci, the suprasylvian gyrus by the suprasylvian and lateral sulci, and the lateral gyrus by the lateral and entolateral (paralimbic) sulci. As in other high-encephalized mammals, the insular area (Figs. 4 and 5) is covered by so-called "opercula" of the neighboring neocortex, which meet at the lateral hemispheral fossa (Sylvian cleft) and are com-

Figure 4 *Common dolphin,* Delphinus delphis *(horizontal sections): (a) through middle of brain and (b) through basal brain. Arrowheads: locations for measurement of length and width of the brain, ventricular spaces black. Am, amygdaloid body; C, caudate nucleus; cp, cerebellar peduncles; fi, fimbria; GP, globus pallidus; J, insula; ic, internal capsule; Pu, putamen; sc, sylvian cleft; SCh, spinal cord; ssp, suprasplenial (limbic) sulcus; I–IV, ventricles. Scale: 1 cm. Modified after Pilleri et al. (1980).*

a

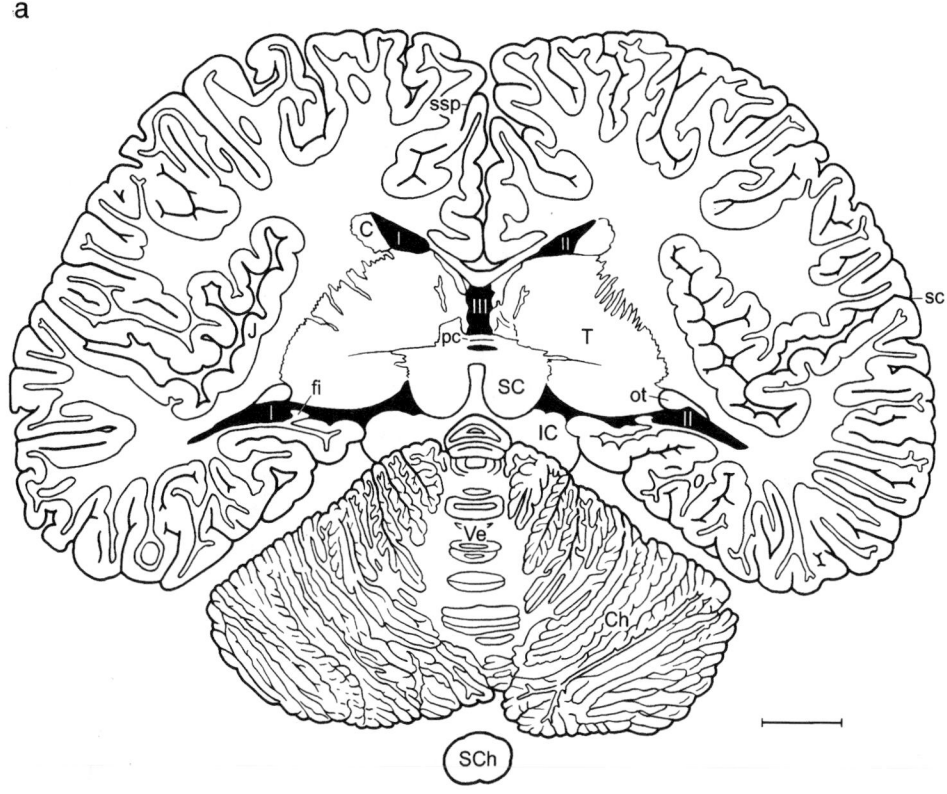

b

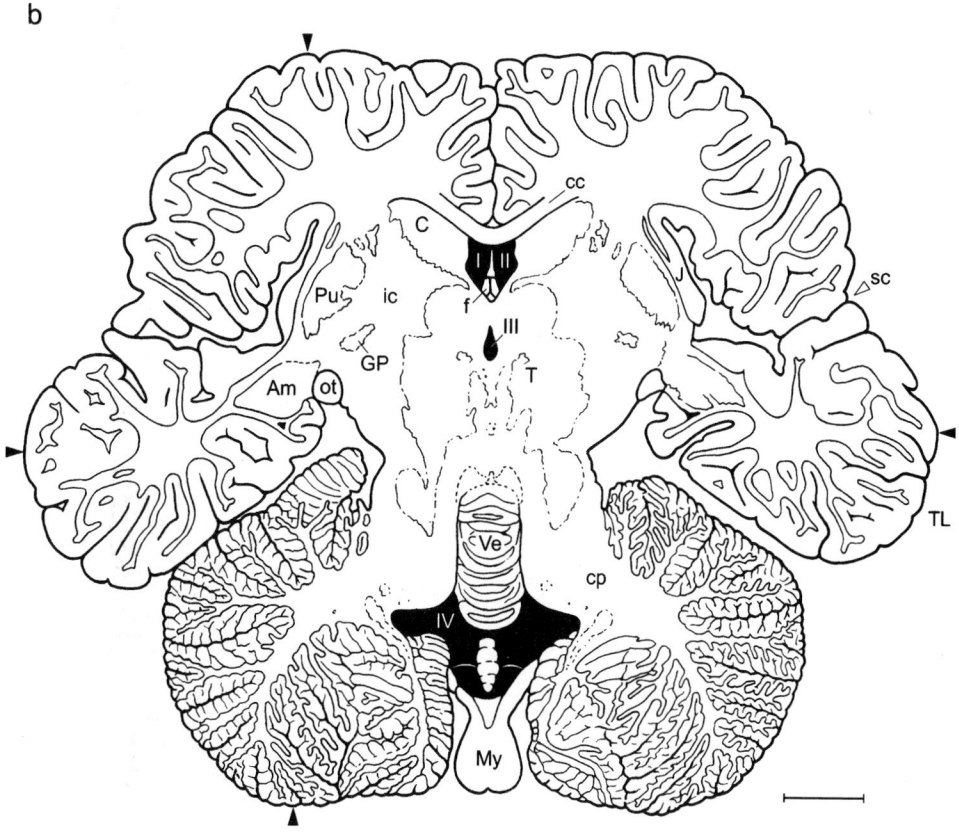

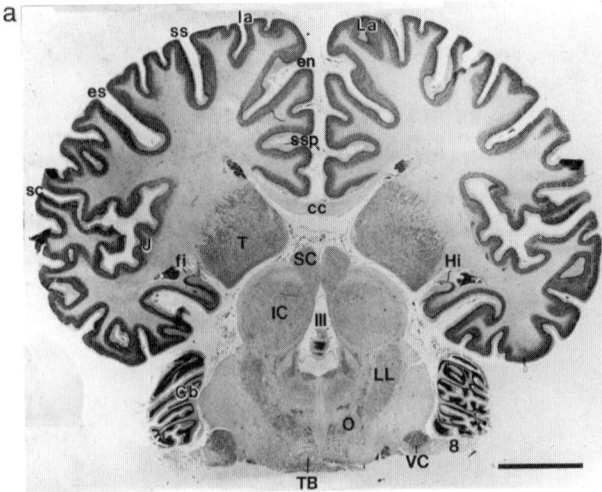

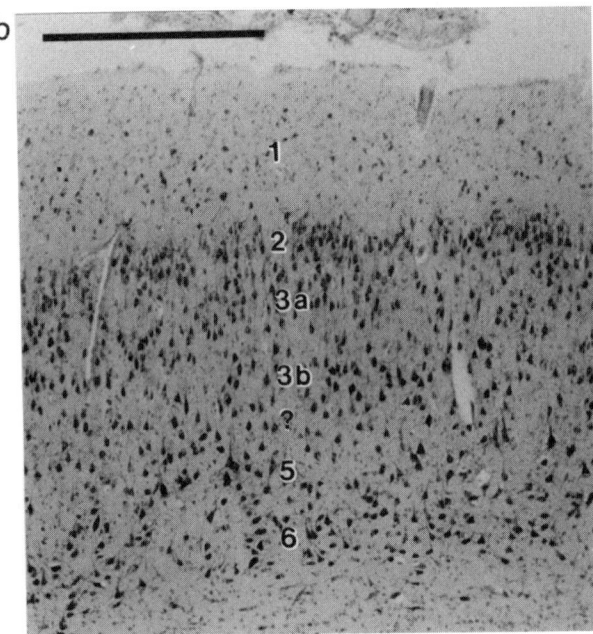

Figure 5 *La Plata dolphin: transverse section of 20 μm thickness through adult brain. Cresyl violet stain. (a) Overview and (b) cortex sample from the lateral gyrus. Cb, cerebellum; Hi, hippocampus; LL, lateral lemniscus nuclei; O, superior olive; VC, ventral cochlear nucleus; 8, vestibulocochlear nerve. Numbers 1–6 in insert, layers I–VI of the neocortex. Scale: (a) 1 cm and (b) 500 μm.*

bined in the term perisylvian cortex (Fig. 2a). The medial cortex of the hemisphere (Fig. 2b) is subdivided by the suprasplenial sulcus or limbic cleft. The cruciate sulcus (Figs. 2a and 12) separates an anterior (medial) motor cortical field from a posterior (lateral) somatosensory field and is therefore a candidate for homologization with the ansate sulcus in hoofed animals as well as the central sulcus in primates. It is doubtful whether the "cal-

carine sulcus," which originates from the entolateral paralimbic cleft and encircles the oval lobule (Fig. 2b), is the homologue of the primate calcarine fissure that houses the primary visual field. Electrophysiological mapping experiments in dolphins have detected visually responding cortical fields in a more anterior and lateral position on the vertex of the hemisphere.

In one of the most plesiomorphic whales (Susu or Ganges river dolphin; *Platanista gangetica*), gyrification is still relatively simple. In the dorsal aspect, the main fissures are straight, smooth, and similar to the mesonychid *Synoplotherium*, a fossil terrestrial relative of the cetaceans (Geisler and Luo in Thewissen, 1998). Brain length in the latter still exceeded brain width, the formation of a temporal lobe had only just begun, and the olfactory system was well developed. Archaic fossil cetacean (archaeocete) brains are difficult to interpret morphologically because they apparently had large retia mirabilia on the surface of the brain as is seen in living baleen whales, which largely conceal the posterior (cerebellar) part. Their telencephalic hemispheres were obviously small and showed no signs of gyrification. As a result, the pattern of gyrification in cetaceans is not necessarily homologous in detail to that of their potential terrestrial relatives, the extant hoofed animals (Fig. 1).

B. LOCALIZATION OF CORTICAL AREAS. Electrophysiological cortical mapping experiments in the bottlenose dolphin located the *motor neocortical field* in the frontal (orbital) lobe rostral to the paralimbic lobe (Fig. 12). The motor cortex is characterized by the presence of giant pyramidal neurons and gives rise to the pyramidal tract. The *somatosensory field* is situated rostral to the visual and auditory fields. The position of the *visual fields* is somewhat more complicated. Although different in many aspects from other mammalian visual cortices, those of the dolphin are apparently well developed and highly differentiated. Whereas all authors place the visual cortex in the lateral gyrus, some distinguish a primary visual field near the medial border of the suprasylvian gyrus from a secondary field in the lateral gyrus. Other authors find an additional visual area in the medially adjacent part of the paralimbic lobe (Fig. 12). On the basis of histological analysis, a visual field has also been reported for the area along the supposed "calcarine" sulcus that separates the oval and lingual sublobules of the paralimbic lobe (Fig. 2b). The large *primary auditory field* lies on the vertex of the hemisphere in the suprasylvian gyrus and lateral to the visual field(s), and the *secondary auditory field* lies more laterally in the medial part of the ectosylvian gyrus.

Viewed as a whole, the topography of the motor and sensory projection fields of dolphins differs from that in other mammals. However, the primary cortical fields in cetaceans have retained the sequence found in plesiomorphic terrestrial mammals. Therefore, it seems as if the hemisphere would have been expanded to such a degree in a caudal and ventral direction (huge temporal lobe) that the auditory cortex extends as a belt along the vertex of the hemisphere and reaches further caudally in the dolphin than the visual field, which now is located more in the center of the hemisphere. In contrast to carnivores (cat, dog, seal), however, where all of the auditory cortex is located in the ectosylvian gyrus, this holds only for the secondary auditory field in toothed whales. The lateral surface of the whale hemisphere may be interpreted as a large "association cortex"

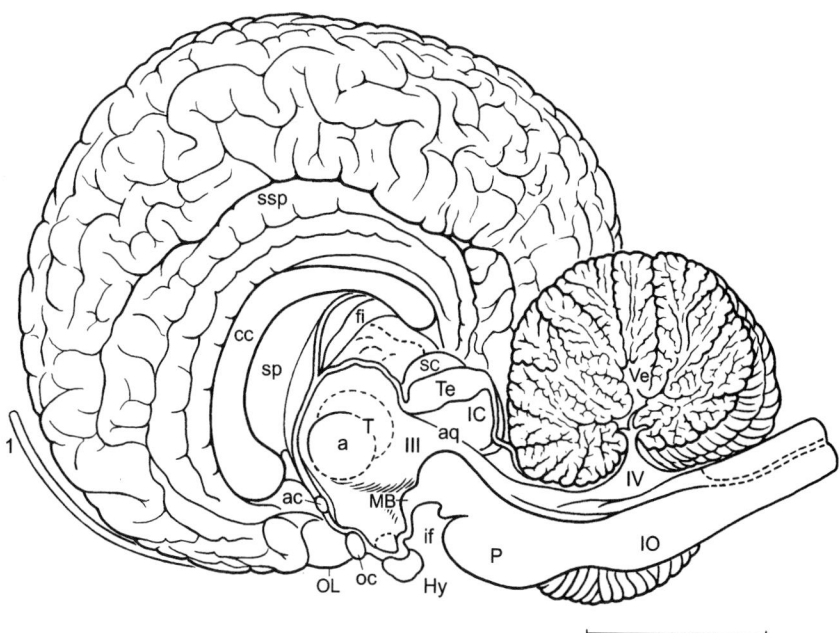

Figure 6 *Brain of humpback whale* (Megaptera novaeangliae) *in mediosagittal section. if, interpeduncular fossa; sp, septum pellucidum; Te, tectum; 1, olfactory nerve. Scale: 5 cm. After Breathnach (1955, 1960).*

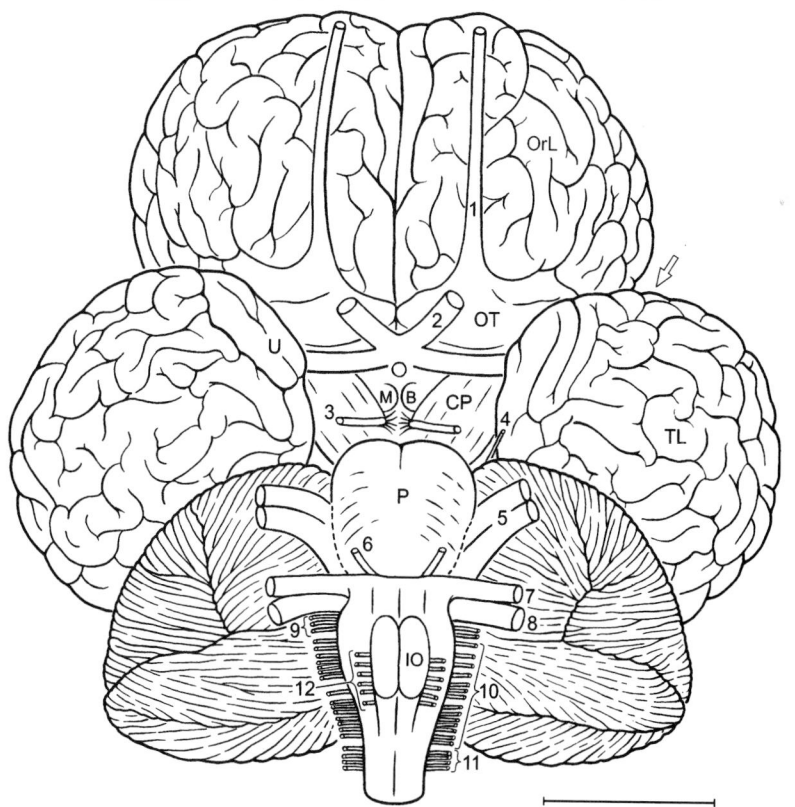

Figure 7 *Brain of humpback whale in basal aspect. CP, cerebral peduncle; MB, mammillary bodies. Scale: 5 cm. After Haug (1970), modified after Breathnach (1955), Pilleri (1966), and Pilleri and Gihr (1970).*

connecting the auditory fields with other sensory and motor modalities.

2. Commissures The size of the commissural systems is specific for cetaceans and shows correlations with cortical and nuclear structures throughout the forebrain (Figs. 2b, 4–6, and 8a).

The *anterior commissure*, which links neocortical and paleocortical areas of both temporal lobes, obviously is weak due to the strong reduction of the olfactory system. The *corpus callosum* as the main link between the neocortical fields of both hemispheres is rather thin in dolphins relative to total brain mass and in comparison to the situation in other mammals. Tarpley and Ridgway (1994) have shown that in cetaceans the cross-sectional area of the corpus callosum (defined by its area in the midsagittal plane) related to brain mass generally decreases in larger-brained toothed whales (and pinnipeds), thereby suggesting that increases in brain weight are not necessarily accompanied by increases in callosal linkage between the telencephalic hemispheres. Thus, the corpora callosa of a killer whale and a human brain show the same cross-sectional area, with the killer whale brain being some five times heavier than that of the human. Furthermore, this relative reduction of the corpus callosum in larger species obviously has not been compensated by an enlargement of other commissural tracts. In conclusion, the interhemispheric connectivity seems to correlate inversely with brain weight insofar as larger brains possess a lower neocortical neuronal density with the result that relatively fewer fibers constitute the corpus callosum. An additional explanation for the regression of the corpus callosum in larger brains could be a smaller percentage of cortical neurons establishing interhemispheric connections and thus a certain independence on the part of both hemispheres. In electroencephalographic experiments, sleeping bottlenose dolphins have been reported to show signs of wakefulness (low voltage, fast activity waveforms) in one hemisphere and sleep (high voltage, slow wave) in the opposite hemisphere. The *posterior commissure* (Fig. 8a) is very well developed. Provided that its connections to other brain structures are similar to those in other mammals, the considerable size of the posterior commissure in cetaceans may suggest massive projections from the somatosensory relay nuclei of the brain stem and the cerebellum to the contralateral pretectum and thalamus.

3. Basal Ganglia All components of the basal ganglia as known from other mammals are present in cetaceans (corpus striatum, globus pallidus, claustrum, amygdaloid complex) and, for the most part, show the usual topographic relationships to each other and to neighboring structures (Fig. 4). Moreover, their histological organization corresponds well with that in other mammals. The caudate nucleus as the main part of the large cor-

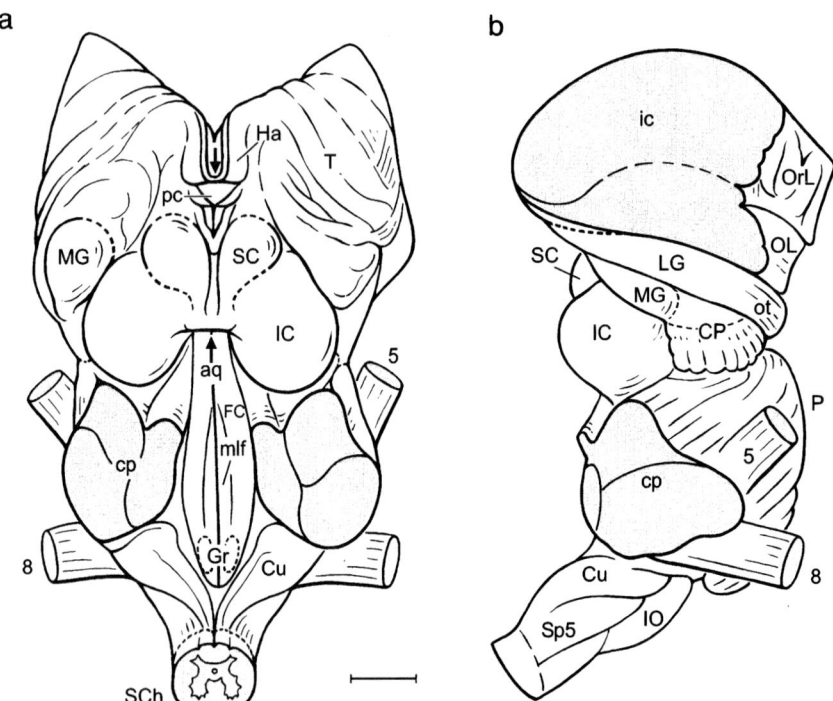

Figure 8 *Brain stem of the bottlenose dolphin: (a) dorsal and (b) lateral aspects. cp, cerebellar peduncle; CP, cerebral peduncle; Cu, cuneate nucleus; FC, facial colliculus; Gr, gracile nuclei; Ha, habenula; LG, lateral geniculate body; MG, medial geniculate body; mlf, medial longitudinal fascicle; pc, posterior commissure; Sp5, spinal nucleus of trigeminal nerve. Arrows pointing into cerebral aqueduct (aq). Scale: 1 cm. After Langworthy (1931), modified after Pilleri and Gihr (1970), and Morgane and Jacobs (1972).*

pus striatum, which bulges distinctly in the area of the "olfactory tubercle" (olfactory lobe), together with the putamen, is largely separated from the latter by a well-developed internal capsule.

Reports regarding the size of the basal ganglia in cetaceans are contradictory. Schwerdtfeger *et al.* (1984) have shown that in the franciscana (La Plata dolphin), the corpus striatum, one of the most important centers for locomotion, is large and attains a size index between that of prosimians and anthropoids. Moreover, as in primates, a size correlation between the striatum and the neocortex seems to be valid for the dolphins as well. The structure of the amygdaloid complex resembles that of other mammals very closely. The size of the amygdala as a whole seems to have been affected only slightly by the reduction of the paleocortex in odontocetes, giving the impression that this nucleus, as in primates, is largely independent of the olfactory system. Its relative size in the franciscana seems to be larger than that in primates, presumably on account of its interconnections with the hypertrophied auditory system and the temporal lobe. However, the corticomedial group of the amygdaloid nuclei, which functionally depends largely on the olfactory system, occupies the same proportion of the entire amygdaloid complex in the harbor porpoise as in the macrosmatic sheep. The lateral amygdaloid nucleus may bear some relation to auditory function, as this nucleus is extremely well developed both in whales and in bats.

C. Diencephalon

The relative size of the diencephalon in plesiomorphic dolphins (franciscana; Ganges river dolphin) is approximately the same as that in simian monkeys, including the human. There are no reliable data for advanced, i.e., pelagial, dolphins. The predominant structure is the thalamus (Figs. 2b, 4a, 5, and 6). The shape of the diencephalon in mediosagittal aspect is often rather wedge-like in adult cetaceans, with the hypothalamus bending slightly caudalward and tapering in the direction of the hypophysis, particularly in larger toothed whales. Whereas in late embryos and early fetuses the floor of the hypothalamus is rather long, it is foreshortened rapidly in later stages during the telescoping process, especially in larger toothed whales. Thus, the transverse interpeduncular fossa between the optic chiasm and pons appears slit-like in adult toothed whales, whereas it is much wider in baleen whales (Fig. 2b, 3, 6, and 7).

1. Epithalamus The habenular complex is large and the habenular commissure well developed. The pineal organ is reduced or even lacking in cetaceans. A pineal rudiment is present between the habenular and posterior commissures in embryos and early fetuses of dolphins and the sperm whale, but not in the early fetal narwhal (Holzmann, 1991). In adult whales and dolphins, many observers found the pineal organ to be lacking; rudiments, however, have been found in the humpback whale (*Megaptera novaeangliae*) and fin whale (*Balaenoptera physalus*).

2. Thalamus Basically, the organization of the large thalamus in cetaceans corresponds well with that in a variety of terrestrial mammals, among them ungulates and primates. There are four groups of nuclei in the dorsal thalamus that constitute about 92% of the thalamus in the bottlenose dolphin: anterior, medial, ventral, and lateral. (1) The anterior group of nuclei, which is related to the cortex of the highly developed cetacean limbic lobe, is well developed, but constitutes only a small part of the total dorsal thalamus. The anteroventral nucleus, which projects to the anterior limbic cortex, dominates this group. In contrast, the mammillary body and the interconnecting mammillothalamic tract are comparatively small and thin. (2) In the medial group of the thalamus, the mediodorsal nucleus is remarkably large and merits special interest because of various connections with olfactory and limbic structures as well as a presumed phylogenetic size correlation with the frontal (orbital) lobe of the mammalian telencephalic hemisphere. (3) The ventral group consists mainly of somatosensory nuclei and constitutes a large part of the dorsal thalamus. In mammals generally, its ventral posterior nucleus (VPN) receives afferents via the medial lemniscus and the spinothalamic and trigeminothalamic tracts and dispatches a main projection to the somatosensory cortex. In the bottlenose dolphin, the ventral posterior nucleus is relatively small and projects to the neocortex anterior to the suprasylvian auditory area (Fig. 12). Compared to its lateral subnucleus, where the body region is represented, the medial subnucleus of the VPN with the head representation is relatively large. In dolphins, the limited somatosensory representation of the body is also reflected in the spinal cord (see later). (4) As in higher primates, the lateral group of thalamic nuclei in cetaceans is dominated by the massive pulvinar, the largest single complex in the thalamus of the bottlenose dolphin. The pulvinar more or less blends into both the strongly protruding medial geniculate nucleus (MG; auditory) and the large lateral geniculate nucleus (LG; visual). The main projection of the inferior pulvinar targets the suprasylvian gyrus, and that of the medial pulvinar the ectosylvian gyrus, whereas the lateral pulvinar projects to the border of the lateral and suprasylvian gyri. The MG is impressively large in cetaceans (Fig. 8) and reflects the outstanding development of the auditory system in these animals. Ventral portions of the MG project to the primary auditory area of the suprasylvian gyrus (Fig. 12) and dorsal portions to the "secondary" auditory area in the ectosylvian gyrus, as well as to the temporal operculum (perisylvian cortex). In the bottlenose dolphin, the LG is surprisingly well developed, although less so than the MG. The LG projects to the visually excitable part of the lateral gyrus (Fig. 12), but does not show the laminar organization usually associated with biretinal projection. This may be related to the fact that the fibers in the optic nerve show a complete or almost complete decussation in cetaceans.

3. Hypothalamus The basal part of the diencephalon exhibits an organization similar to that encountered in other mammals. Anterior, tuberal, and posterior hypothalamic nuclei are evident but not particularly prominent. Paraventricular and supraoptic nuclei are obvious because of their large hyperchromatic cells, with the latter nucleus being especially well formed. As in other mammals the supraoptic commissure is well developed and well organized. The small size of the mammillary bodies, which in the postnatal animal do not protrude at the brain surface, correlates with the weak development of the hippocampus, postcommissural fornix, and mammillothalamic tract.

4. Hypophysis   The pituitary gland in both toothed and baleen whales is rather wide (adenohypophysis), a pars intermedia and residual lumen are lacking. The neurohypophysis, which is separated from the adenohypophysis by a meningeal septum, is slender and consists of a pars nervosa, an infundibular stalk, and a median eminence. The adenohypophysis is composed of chromophobe, basophilic, and eosinophilic cells. The hypothalamohypophysial tract was reported to be wide in the bottlenose dolphin.

D. Brain Stem and Cerebellum

The percentage of the dolphin midbrain in the total brain volume is relatively low and ranges between that of prosimian and simian primates. Nevertheless, the size index, which is related to body size and the regression line of basal insectivores (shrews, hedgehogs, tenrecs), shows a remarkable increase of this structure even in plesiomorphic "river dolphins" (franciscana; Ganges river dolphin). This may be attributable to the growth of auditory system components. The cerebellum and pons are well developed, and the myelencephalon (medulla oblongata) is very large in comparison with that of other mammals. This may be due to the considerable growth of cranial nerve nuclei and their connectivity, particularly those of the trigeminal, auditory, and motor systems.

1. Selected Nuclei   The cetacean brain stem comprises nuclei known from other mammals; some of them, however, are rather exotic and their functional properties are virtually unknown. Perhaps because they have changed considerably in appearance and/or location during evolution, a few nuclei have been described as unique to cetaceans so that they cannot be homologized easily with potentially corresponding nuclei in other mammals. Many of the features, however, presented in the literature on the cetacean brain stem could be confirmed in the work of Holzmann (1991) on the fetal narwhal (Fig. 11).

The *oculomotor nucleus* is the largest eye muscle nucleus, a fact that correlates well with the diameter of the oculomotor nerve. In comparison, the trochlear and abducent nuclei and nerves are rather small and thin in most cetaceans. The *trigeminal nuclei* (motor, principal, spinal nucleus) are very well developed, reflecting the large relative size of the cetacean head and the diameter of the trigeminal nerve, which is maximal or submaximal among the cranial nerves. Within cetaceans, the motor nucleus is reported to be larger in mysticetes but subdivided much better in odontocetes. Also the sensory principal nucleus is reported to be larger in baleen than in toothed whales, with its dorsal part giving rise to the well-developed trigeminothalamic tract (Wallenberg). The *facial nucleus* is very large in cetaceans and often bulges at the ventral surface of the medulla (tuberculum faciale). The nucleus is divided into a number of cell groups that can be differentiated from each other cytologically. These cell groups are believed to be responsible for specific muscles or muscular systems, e.g., the dorsal group for muscles of the upper respiratory tract around the blowhole (epicranial complex; Cranford *et al.*, 1996), which are involved in the generation and emission of sonar signals in toothed whales. In comparison with other mammals, the *ambiguus nucleus* is large in cetaceans, which recalls to mind the situation in bats (mouse-eared bat; *Myotis myotis*). This nucleus, which is larger in mysticetes than in odontocetes, innervates the muscles of the pharynx, larynx, and the striated muscles of the esophagus via the glossopharyngeus–vagus–accessorius nerve complex. Comparable to other mammals, it should be involved in cetaceans in respiration, food processing, and sound production in the larynx. The *nucleus of the accessory nerve*, as part of the anterior horn in the spinal cord, is moderately well developed. This may be related to the extreme foreshortening of the cervical region, restrictions in head and shoulder girdle motion, and the transformation of the forelimb into a steering device (flipper). The *hypoglossal nucleus*, a derivative of the ventral horn of the spinal cord, is well developed, although the flexibility of the tongue is reported to be restricted in most cetaceans. In large baleen whales, the tongue may attain the body mass of a full-size elephant.

Nuclei related to or belonging to the extrapyramidal motor system are located in the rostral mesencephalon and in the formatio reticularis throughout the rhombencephalon. The *elliptic nucleus*, which is situated within the central gray either rostral or dorsal to the oculomotor nuclear complex, is very conspicuous and in the past was thought to be unique for the Cetacea until a similar nucleus was found in the elephant. Moreover, it was unclear whether the *nucleus of Darkschewitsch* is integrated into the elliptic nucleus or is even equivalent to this nucleus. Today, the latter opinion is the generally accepted one: In cetaceans, the elliptic nucleus projects via the medial tegmental tract to the rostral medial accessory inferior olive and correlates with a hypertrophy in cerebellar structures (see later). The *red nucleus* in cetaceans is little known. In contrast to ungulates, which possess a large rubrospinal tract and lack a spinal pyramidal tract, cetaceans have both weak rubrospinal and corticospinal tracts (see later). *Pontine nuclei* are exceptionally well developed in cetaceans, giving rise to the bilaterally ascending large brachia pontis, and the transverse pontine bundles are very numerous. The size and the caudal extent of the pons are directly related to the size of the neocortex.

The cetacean *inferior olive* is characterized by an extraordinary development of the medial accessory olive, particularly its rostral portion; in comparison, the principal olive and the dorsal accessory olive appear small. In the two cetacean suborders, there are only minor differences in the relative development of the subnuclei, and both inferior olives join each other in the midline. The rostral part of the medial accessory olive receives massive input from the elliptic nucleus via the medial tegmental tract, and its pronounced development in cetaceans seems to be related to the immense size of the posterior interposed nucleus and of the paraflocculus. In terrestrial mammals, the medial accessory olive is part of a fiber system involved in directional hearing.

2. Cerebellum   The cetacean cerebellum is very large (Figs. 2–4, 6, 7, 9, and 10), its size obviously being linked phylogenetically with that of the neocortex. In older studies, the relative mass of the cerebellum in baleen whales with respect to total brain mass (average: 20%) was reported to represent a maximal development within the mammalia as a whole. A recent study has shown that in relation to body mass the cerebellum of baleen whales is not as voluminous as in larger dolphins such as the killer whale. Concomitantly, it became obvious that the large proportion of the cerebellum in the total brain volume of baleen whales is attributable to the relatively small size of the forebrain. Indeed, in double-logarithmic regressions, baleen whales rank a little higher than sperm whales, beaked whales, and "river" dolphins but distinctly below the delphinid cetaceans. With respect to the regression line in basal insectivores (Stephan), the cerebellum of the plesiomorphic La Plata dolphin ranks higher than the averages of prosimian and simian monkeys but lower than humans. In a group of delphinid species, indices of the total brain mass and cerebellum mass relative to body mass exceeded other groups (sperm whales, river dolphins, baleen whales) by up to three times. Within cetaceans, the cerebellum of the baleen whales is much better understood due to ontogenetic histological studies by Jansen (1950). Only minor structural differences exist, however, between the cerebella of toothed and baleen whales: Thus, the mysticete cerebellum is more rounded and slightly hourglass shaped in the dorsal aspect, whereas the odontocete cerebellum is somewhat more flattened dorsoventrally as a consequence of the stronger telescoping of the brain and resultant overlapping of the cerebellum by the cerebral hemispheres.

The cerebellum consists of two large hemispheres and a comparatively narrow vermis in between (Figs. 10 and 12). Two transverse fissures separate three cerebellar lobes: the primary fissure separates the small anterior (rostral third) from the large posterior lobe (caudal two-thirds), and the posterolateral fissure the posterior lobe from the small flocculonodular lobe. These size relations between the lobes are characteristic for cetaceans. In midsagittal section (Fig. 10), the conventional subdivision of the vermis into nine lobules of the mammalian cerebellum is obvious. In cetacean cerebellar hemispheres the small size of the anterior lobe may be explained by electrophysiological findings indicating that the hemispheral parts of this lobe comprise the cortical representation of the fore- and hindlimbs (cf. Jansen, 1950) that are highly modified or even have vanished in these animals. The caudally adjacent ansiform lobule, which also receives input from the extremities, is similarly small. However, the representation of the head in the simple lobule of the posterior lobe is rather large, and the considerable size of the paramedian lobule (body representation) has been related to the enormous significance of the tail in cetaceans. The paraflocculus, situated between the parafloccular and posterolateral fissures, is exceptionally large, particularly the ventral parafloccular lobule (Fig. 3). The latter comprises about half of the surface of the cerebellar hemisphere. In mammals, generally speaking, the paraflocculus usually receives climbing fibers from the rostral part of the medial accessory inferior olive. In cetaceans, both structures are exceptionally large, which strongly indicates a functional relationship between the paraflocculus on the one hand and trunk and tail on the other (Jansen, 1950). The flocculonodular lobe as the principal terminus of primary and secondary vestibulocerebellar and visceromotor fibers ("vestibulocerebellum") is very small in cetaceans, particularly the floccular component (Fig. 10). In mammals generally, the latter is responsible for the regulation of compensatory vestibulo-ocular and optokinetic movements as well as compensatory movements of the neck in the so-called "smooth pursuit" movements of the eyes, particularly in

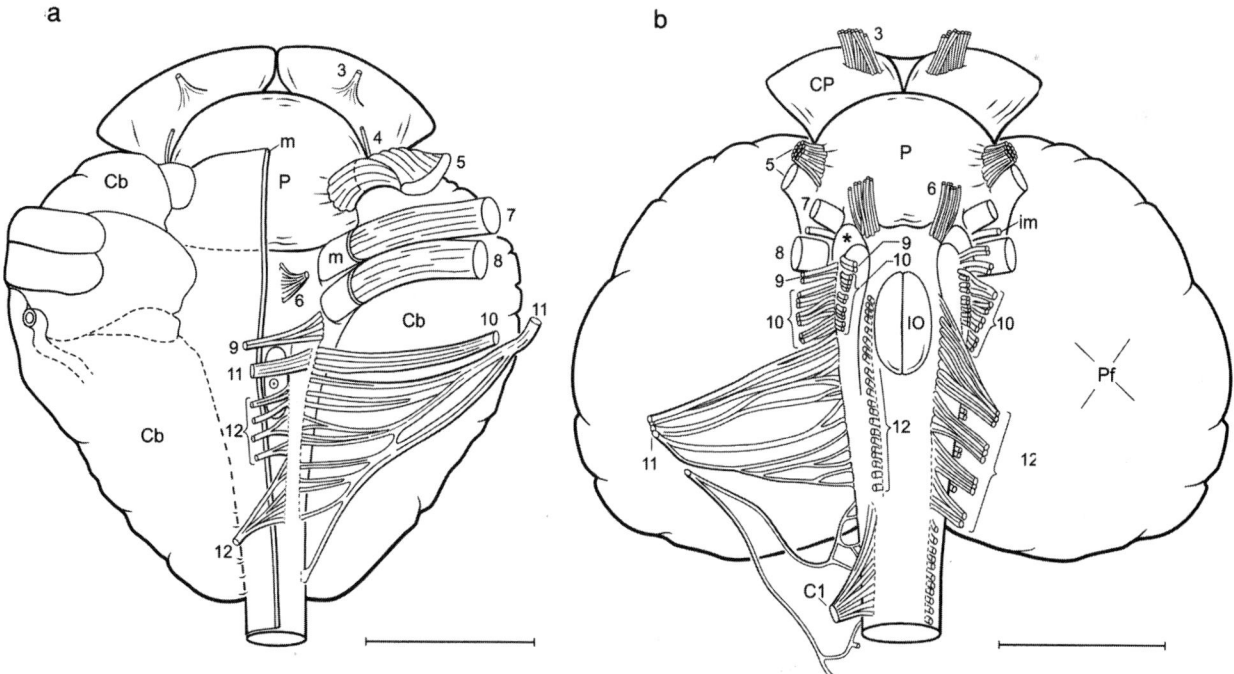

Figure 9 *Cetacean brain stems in basal aspect. (a) Sperm whale, (b) Fin whale [after Jansen (1953), modified after Pilleri and Gihr (1970)]. Asterisk, facial tubercle; dotted circle, inferior olive; m, cut meninx. C1, motor root of first cervical spinal nerve; im, intermedius nerve; pf, paraflocculus. Scale: 5 cm.*

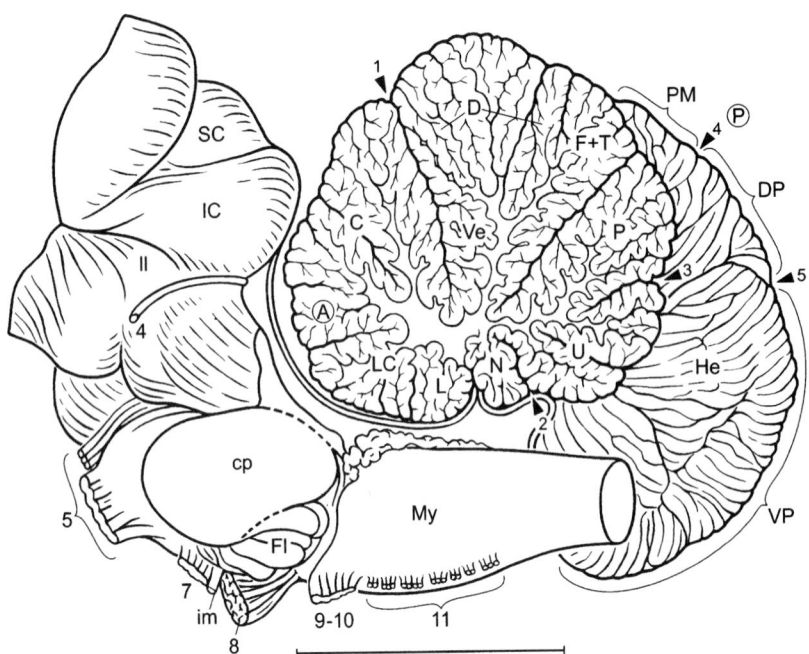

Figure 10 *Brain stem of the fin whale (mediosagittal aspect): 1, primary fissure; 2, posterolateral fissure; 3, secondary fissure; 4, paraflocular fissure; 5, intra-paraflocular fissure; (A), anterior lobe of cerebellum; C, culmen, D, declive; DP, dorsal paraflocculus; Fl, flocculus; F+T, folium and tuber; He, cerebellar hemisphere; L, lingula; LC, lobus centralis; ll, lateral lemniscus; N, nodulus; (P), posterior lobe of cerebellum; P, pyramis; PM, paramedian lobule; U, uvula. Modified after Jansen (1953). Scale: 1 cm.*

carnivorous animals that lure and hunt. Dolphins, however, which have a restricted neck mobility and can use their visual system only during daylight, may have to rely on their auditory system instead to follow their prey effectively.

In the ontogenesis of the cetacean cerebellar cortex, which is three layered, the fundamental mammalian pattern of transverse and longitudinal zones is discernible. These longitudinal zones obviously are related topographically to the development of the cerebellar nuclei (anterior, medial, and posterior interposed nuclei, lateral cerebellar nucleus). In cetaceans, the lateral intermediate cortical zone (C2 zone) is developed enormously, occupying about three-fourths of the cerebellar surface (paraflocculus) and correlating with the huge posterior interposed nucleus.

3. Main Fiber Systems   Medial lemniscus: In cetaceans, the afferent spinal system (proprioceptive sensitivity) is moderately

developed in accordance with the reduction of the hindlimbs and pelvic girdle. In these animals, the dorsal funiculi (gracile and cuneate fascicles) are strikingly small; they are thought to convey input predominantly from the flippers and the tail (sense of position). Nevertheless, cutaneous sensitivity in the trunk was reported to be high (Slijper, 1962). The medial lemniscus is weak in the caudal medulla, but becomes considerably stronger at more rostral medullary levels, presumably due to the input of afferent systems of the head (auditory, trigeminal systems).

Trigeminothalamic tract (Wallenberg): The dorsal part of the principal sensory trigeminal nucleus gives rise to the ipsilateral (uncrossed) trigeminothalamic tract. The latter terminates in the medial part of the ventral posterior thalamic nucleus as the main somatosensory thalamic nucleus. As in ungulates and the elephant where it is extremely well developed, the tract is thought to be responsible for intra- and peri-

Figure 11 *Brain stem of fetal narwhal (Monodon monoceros) with major nuclei (from Holzmann, 1991, modified). The physiological quality of the nuclei is indicated by different textures. (a) Mediosagittal aspect, (b) detail of (a) with auditory nuclei, and (c) synopsis of all nuclei in basal aspect. Mes, mesencephalon; 1, elliptic nucleus; 2, interstitial nucleus; 3, nucleus of oculomotor nerve; 4, nucleus of trochlear nerve; 5, interpeduncular nucleus; 6, nucleus of abducens nerve; 7, pontine nuclei; 8, motor nucleus of trigeminal; 9, principal nucleus of trigeminal; 10, spinal nucleus of trigeminal; 11, nucleus nervi facialis; 12, nucleus ambiguus; 13, nuclei of solitary tract; 14, dorsal nucleus of vagus; 15, nucleus of hypoglossal nerve; 16, spinal nucleus of accessory nerve; 17, cuneate nucleus; 18, gracile nucleus; 19, medial accessory inferior olivary nucleus; 20, principal and dorsal accessory inferior olivary nuclei; 21, nuclei of trapezoid body; 22, superior olivary complex; 23, nuclei of lateral lemniscus; 24, ventral cochlear nucleus; 25, pontobulbar nucleus. Scale: 1 mm.*

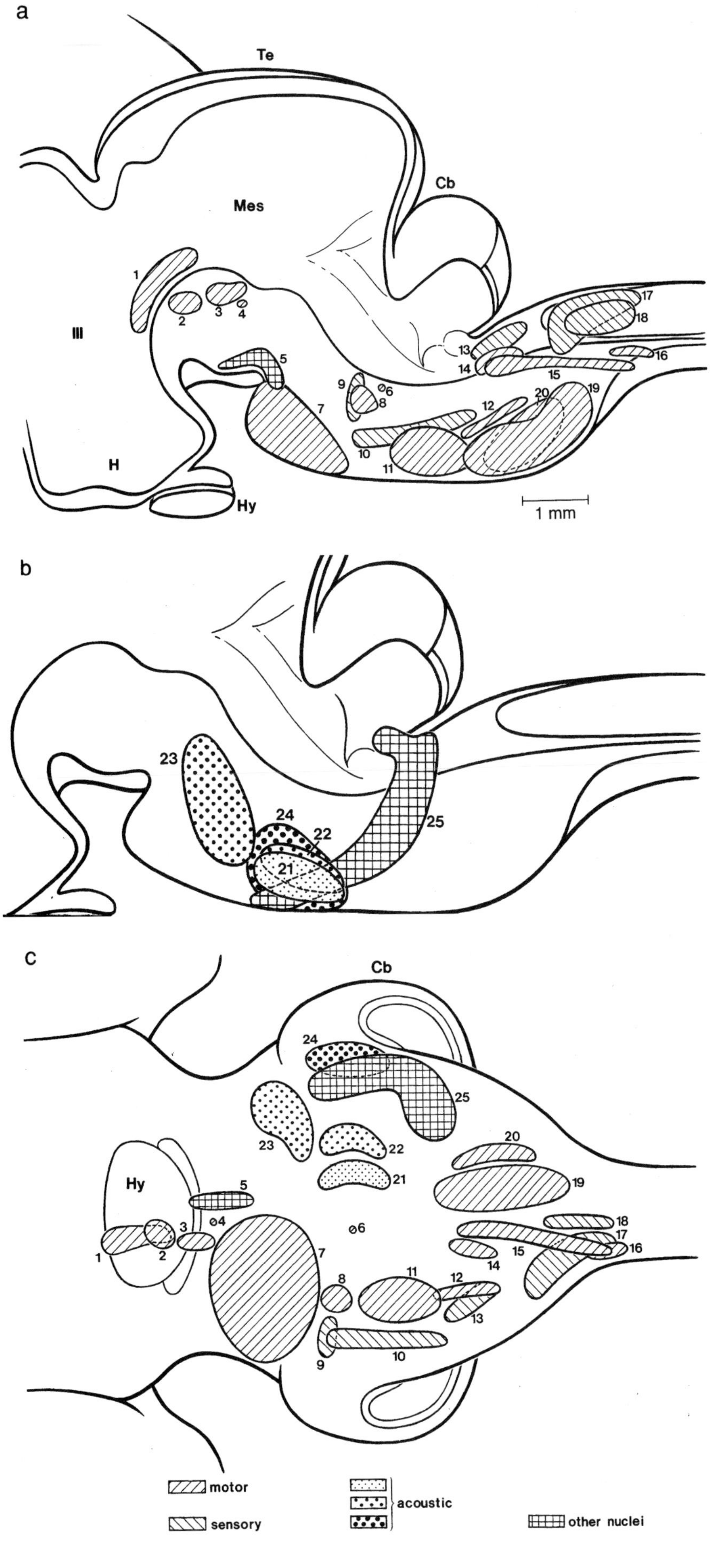

a

Te

Mes

Cb

III

1

2

3

4

5

9

8

6

7

13

14

10

11

12

20

15

19

16

18

17

H

Hy

1 mm

b

23

24

22

21

25

c

Cb

24

25

23

22

21

20

19

Hy

5

4

6

3

2

1

7

8

11

12

15

14

18

17

16

13

9

10

motor

acoustic

sensory

other nuclei

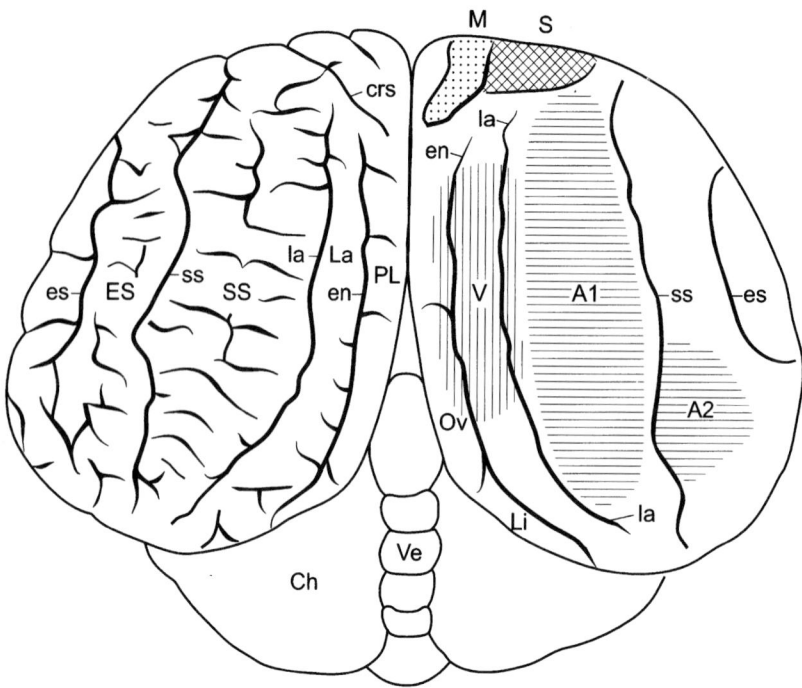

Figure 12 *Neocortical motor and sensory fields in the bottlenose dolphin. A1, A2, auditory fields; crs, cruciate sulcus; M, motor field; PL, paralimbic lobe; S, somatosensory field; V, visual field. After Morgane* et al. *(1986).*

oral sensitivity in cetaceans innervating, e.g., tactile bodies on the lips of fin whales and sei whales (*Balaenoptera borealis*), and the epicranial complex (Cranford *et al.*, 1996) in toothed whales.

Medial tegmental tract: The elliptic nucleus (Figs. 2b and 11), which almost "replaces" the red nucleus in Cetacea, is developed enormously in whales and elephants and gives rise to the strong medial tegmental tract that proceeds to the rostral part of the huge medial accessory inferior olive. This nucleus also receives afferents from the spinal cord (spino-olivary tract) and projects to the lateral intermediate (C2) zone in the huge paraflocculus. The paraflocculus, which was shown to be the main target of auditory pontocerebellar projections in the rat and was estimated to receive three-fifths of the pontocerebellar fibers in the blue whale (*Balaenoptera musculus*), has a massive projection to the posterior interposed nucleus of the cerebellum. From the latter nucleus, ascending fibers run to the elliptic nucleus and other nuclei at the diencephalic/mesencephalic border, which, in turn, project to the inferior olive. Like the medial accessory olive, the paraflocculus also has been associated with mass movements of the trunk-and-tail region, the only region where the axial skeleton possesses a reasonable range of motion. The medial tegmental tract thus seems to be part of a recurrent circuit (elliptic nucleus–inferior olive–paraflocculus–posterior interposed nucleus-elliptic nucleus), which combines auditory input with locomotor activity and illustrates the dominant position of hearing among sensory systems in whales.

Pyramidal tract: The tract originates in the neocortical motor area rostral and medial to the "cruciate" sulcus (Figs. 2a and 12) and runs through the internal capsule. At mesencephalic

levels, the localization of the pyramidal tract is difficult and it is very small at high medullary levels. Typical macroscopical "pyramids" as seen in terrestrial mammals are not present in cetaceans. Here, the pyramidal tracts are weak and situated lateral to the inferior olivary complex, with both inferior olives joining each other midsagittally (Figs. 3, 7, and 9). Obviously, the extremely well-developed rostral medial accessory olive, which occupies the ventromedial area in the medulla, has pushed the pyramids lateralward. Caudal to the inferior olives, the pyramidal tract disappears in baleen whales. In toothed whales, both pyramids merge with each other; their crossing (decussation) is described by most authors as indistinct and the pyramidal tract as small and hardly visible. Thus, it is likely that pyramidal (corticospinal) fibers do not descend more than a few (cervical) segments in the spinal cord. This pattern resembles much that is seen in hoofed animals and the elephant. In terrestrial mammals, an inverse relationship exists between the development of the pyramidal tract and of the rubrospinal tract. Perissodactyls and artiodactyls have small pyramidal tracts, whereas their rubrospinal tracts are large. The opposite is seen in primates whose small rubrospinal tract is coexistent with a large spinal pyramid. Such an inverse relationship is not encountered in Cetacea: Here, both the rubrospinal and corticospinal tracts are small, obviously a speciality in cetaceans.

II. The Cetacean Spinal Cord

Within the nervous system, the spinal cord (Figs. 4a and 8) is responsible for the innervation of the trunk and tail, and the pectoral and pelvic girdles together with their appendages. Thus, the spinal cord mediates between the locomotory apparatus of the body and the brain by transmitting sensory vs motor information.

All cetaceans are characterized by (1) the subtotal reduction of the pelvic girdle and hindlimb and the transformation of the pectoral girdle and forelimb into a steering device and (2) the extraordinary development of the axial musculature, thus contributing to the spindle shape of the body required for efficient locomotion by the trunk-and-tail complex.

With respect to the sensory innervation of the body in whales and dolphins, we do not know anything about the innervation of joints and tendons. The skin of the body is hairless. Nevertheless, the assumption of poor sensory innervation of the skin in cetaceans has been contradicted by Slijper (1962), who observed reactions of stranded and captive dolphins to the gentlest touch.

In correlation with the large amount of axial musculature, the motor ventral roots of the cetacean spinal nerves appear to be larger than the dorsal ones, particularly in the posterior part of the system (cauda equina). There are up to 97 pairs of spinal nerves (e.g., 44 in the harbor porpoise: 8 cervical, 11 thoracic, and 25 lumbocaudal). On each side of the body, along the vertebral column, there are two plexus (cervical plexus: C_1–C_3; brachial plexus: C_4–Th_1). The peripheral nerves supplying the tail can be very long (up to 15 m in large baleen whales).

Other remarkable features of the cetacean spinal cord are its relative length and shape. Whereas in the harbor porpoise the relation of body length to spinal cord length is 3.75:1, baleen whales have values from 4:1 (fin whale) to 6.7:1 (right whale; *Eubalaena* sp.). A cervical enlargement (intumescence) for the innervation of the pectoral girdle and flipper has been described in different odontocetes and in the fin whale. The presence of a caudal intumescence, however, is controversial.

The internal structure of the spinal cord reflects peculiarities of the cetacean motor and sensory systems: long and slender anterior horns, occupying half of the gray matter, the posterior horns modest to stunted in appearance (Fig. 8a). A short and pointed lateral horn is discernible at thoracolumbar levels. The gelatinous substance, in general, is remarkably poorly developed in cetaceans, with only traces found in some spinal segments of the harbor porpoise and the fin whale. In mammals, this crescentic mass of small interneurons is involved intimately in the connectivity of cutaneous afferents of the dorsal spinal root with delicate fibers of the spinothalamic tract (pain and temperature). In cetaceans, the lack of gelatinous substance may correlate with the loss and modification of the posterior and anterior limb, respectively.

Whereas cell density and numbers of neurons in the cetacean spinal cord are not known, the dimensions of the cells in the anterior horn have been investigated in baleen whales. Despite the large difference in body mass, the sizes of the perikarya in the cervical and thoracic spinal cord are similar in the whale and human, but in the lumbocaudal area the whale has much larger perikarya with a volume presumably several times larger than in comparable human perikarya.

III. Functional Systems

A. Chemoreceptor Systems

Adult baleen whales still have a small, but functional, nose probably equipped with olfactory mucosa and an olfactory nerve, and they possess an olfactory bulb, slender olfactory peduncle, and an olfactory tubercle (Figs. 3 and 7), a situation resembling that of the human. Presumably as a consequence of the adaptation to sonar orientation, prey detection, and communication (high-energy sonar clicks in the upper respiratory tract), toothed whales have reorganized their nasal system to such a degree that (at most) a short olfactory peduncle is found in the adult animal very rarely [sperm whale; northern bottlenosed whale (*Hyperoodon ampullatus*)]. In general, toothed whale embryos display the anlage of an olfactory bulb, but the latter is reduced in early fetal stages. Interestingly, however, a "residue" of the bulb persists in the large ganglion of the terminal nerve (Buhl and Oelschläger, 1986) that is responsible for the establishment of the hypothalamo–hypophyseal–gonadal axis in prenatal mammals, and thus for sexual behavior and reproduction in adults. In dolphins, the terminal ganglia contain the highest number of neurons found within the mammalia (Ridgway *et al.*, 1987). These facts argue both for the nonolfactory nature of the terminal nerve and its possible implications in the control of blood flow in the nose and basal forebrain, in maintenance of the epithelial lining of the upper respiratory tract, and in the sensory control of sonar signal emission as well. A vomeronasal organ (Jacobson's organ) and nerve are absent in cetaceans and sirenians.

Despite the strong general reduction of the olfactory system in toothed whales, the olfactory tubercle seems to be well developed even in comparison with that in baleen whales, which possess a small but presumably functional olfactory system. In many cases, the tubercle may not be separated clearly from the neighboring diagonal band and prepiriform cortex, a configuration often called the "olfactory lobe" (Fig. 2a). In reality, it is the large corpus striatum (basal ganglia) that protrudes here at the basal surface of the brain as an "olfactory" tubercle and is covered by an incomplete layer of thin paleocortex analogous to the situation in humans. The remaining paleocortex (diagonal band, piriform cortex) is moderately developed. The amygdala as a whole, however, is rather large in cetaceans for other reasons.

Dolphins are clearly sensitive to chemical stimuli (both natural and artificial compounds), and some cetaceans were reported to have functional taste buds. With the olfactory part of the nose disappearing during prenatal development, dolphins still may resort to their taste buds and the trigeminal innervation of the oral cavity for chemoreception.

B. Visual System

In most cetaceans, the visual system is reported to be fairly well developed (Figs. 2–4, 6–8, 10, and 12). In the adult harbor porpoise, the optic nerve contains 81,700 axons, 147,000–390,000 axons in the bottlenose dolphin, 193,000–250,000 axons in the domestic cat, and 1,200,000 in the human. The Amazon river dolphin (*Inia geoffrensis*) shows a rather low axon count (15,500), and the optic nerve of the

Ganges river dolphin, whose eye lacks a lens and may be capable of serving as a light receptor only, contains a few hundred axons. Large whales have moderate numbers of axons. In baleen whales, the optic nerve contains approximately the same number of axons (252,000–347,000) as in the bottlenose dolphin, whereas the sperm whale, with roughly the same body weight and relatively smaller eyes, has only 172,000 axons.

Bottlenose dolphins have a thick retina composed of rods and cones. The fovea centralis is band shaped, and neurons in the ganglionic layer are very large (up to 150 μm in diameter) with thick dendrites and myelinated axons of up to 9 μm thick. Cetaceans have laterally placed eyes and, if at all, one small binocular visual field rostrally and ventrally as well as another one dorsally and slightly caudally. The optic fibers show a complete or almost complete decussation, and the lateral geniculate body is not laminated. The superior colliculus is large in most cetaceans, as is the case in many land mammals, including carnivores, ungulates, and primates. Dolphins show a definite stratification of the superior colliculi, where layers typical of terrestrial mammals are recognizable. In some baleen whale species, the superior and inferior colliculi are approximately the same size (length and width), whereas in others [Southern right whale (*Eubalaena australis*); blue whale (*Balaenoptera musculus*)] the inferior colliculi have double the surface of the superior colliculi. In toothed whales, this relation may range from 2:1 up to 7:1 (maximum: susu).

Electrophysiological mapping studies in dolphins have located the visual cortex not in the dorsocaudal or occipital part of the hemisphere but in a more central position near the midline (Fig. 12). Although well developed, the physiological importance of the cetacean visual system theoretically should be inferior to that of the auditory system because of functional restrictions in murky water and in darkness.

C. Auditory System

The auditory system in whales and dolphins basically corresponds to that in terrestrial mammals; however, the various components have been adapted morphologically and physiologically to the specific conditions of hearing under water (Figs. 2, 3, 4a, and 5–12).

In general, auditory structures are smaller in baleen whales than in toothed whales (compare Figs. 3 and 7–10). According to the impressive diameter of the vestibulocochlear nerve and the number of axons in the cochlear nerve, the cetacean ventral cochlear nucleus (VCN) and other auditory centers are very large (Figs. 5, 11, and 12). The secondary auditory fiber tracts (trapezoid body, acoustic striae) are well developed. In the La Plata dolphin as well as in the harbor porpoise and the common dolphin (*Delphinus delphis*), the absolute volume of the cochlear nuclei is 6–10 times larger than in the cat and the human; in the beluga (*Delphinapterus leucas*) it is 32 times larger, and in the fin whale it is 20 times larger. Numbers of cochlear neurons range between 583,000 and 1,650,000, i.e., 6–17 times that of the human. The ratio of primary cochlear nerve fibers to secondary cochlear neurons is between 1:5 and 1:8 (human: 1:4). The volume of other auditory nuclei is either very large in comparison or very low, even if we take into account the different body mass of the animals. In the La Plata

dolphin, multiples 17× the volume of the cat's nuclei are found in the nucleus of the trapezoid body and 39× in the intermediate nucleus of the lateral lemniscus. A comparison between volumes of auditory nuclei in the common dolphin and the human yields similar results. There are multiples for the dolphin VCN (16×), lateral superior olive (150×), single nuclei of the lateral lemniscus (up to 200×), and for superior auditory centers such as the inferior colliculus (12×) and laminated medial geniculate nucleus (7×). The cetacean VCN is composed of five subunits consisting of specific neuron populations that are also found in terrestrial mammals. This has been proved by cytological investigations in the franciscana (Schulmeyer, 1992) involving the morphology of the axon terminations. The nucleus of the trapezoid body is well developed: A great mass of fibers passes from the VCN into the trapezoid body, to the ipsilateral and contralateral nuclei of the lateral lemniscus, and the inferior colliculus. Interestingly, the dorsal cochlear nucleus (DCN) could not be found in many toothed and baleen whales, perhaps because of its reduction to the point of insignificance. Obviously, this nucleus is engaged in the assessment and/or elimination of "auditory artifacts" caused by positional changes of the head and pinnae toward a sound source. In terrestrial mammals (including bats) with "normal" external ears and good movability of the head and pinnae, the DCN is well developed and even laminated.

The existence of the medial superior olive (MSO) is discussed in the literature on toothed whales. In some species, the two subnuclei of the superior olive cannot be distinguished, and whether the medial nucleus is very small or even lacking in other species, as has been reported for bats, is not clear at present. In mammals, the MSO is believed to be engaged in the processing of low frequencies and the lateral superior olive (LSO) in the processing of high frequencies and ultrasound. Therefore, it has to be expected that the LSO is larger in toothed whales and the MSO in baleen whales, reflecting the actual neurophysiological adaptation (audiogram) of these animals. The components of the midbrain tectum differ in size in baleen whales and toothed whales. Whereas in mysticetes the superior colliculi (vision) may be larger or smaller than the inferior colliculi (audition), odontocetes always have very large inferior colliculi (Figs. 2b, 4a, 5, 6, 8, and 10). The striking dominance of the sense of hearing in adult toothed whales has been emphasized by a number of investigators, among them Jansen and Jansen (1969), Morgane and Jacobs (1972), Schulmeyer (1992), and Oelschläger and Kemp (1998). Indeed, the auditory system has stamped the morphology, size, and connectivity of the whole odontocete brain due to the necessity of processing vast amounts of acoustic information and on account of the high propagation velocity of sound in water (Ridgway, 1986, 1990) and the need of background noise discrimination. In the adult sperm whale and other toothed whales, the auditory system seems to be the major source of information. Even when the animals dive to considerable depths and the visual system progressively loses its importance for orientation and the detection of prey, the auditory system may still be functional (navigation by auditory input). Toothed whales scan their surroundings with a sonar system (clicks) and are probably able to integrate the acoustic input together with visual information into two- or even three-dimensional ephemeral images within their extended neocortical auditory projection fields (Supin *et al.*, 1978). In addition, communication between individuals by means of acoustic signals clearly is very important for whales and dolphins, especially during hunting activity and/or when vision is reduced. Pelagic dolphins, in particular, tend to live in larger groups. In the open sea, their natural environment is largely represented by the distributional pattern of their kin, which changes more or less continually.

D. Vestibular System

The four major vestibular nuclei usually found in mammals are also present in cetaceans. The lateral vestibular nucleus of Deiters in cetaceans is most conspicuous and its dimensions fairly large, but the number of neurons does not seem to be exceptional. The other nuclei are rather small. The lateral vestibular nucleus does not receive projections via the vestibular nerve, and its connections to the lateral B-zone of vermis, dentate nucleus, fastigial nucleus, and the oculomotor complex are very similar to those of cerebellar nuclei. In mammals generally, the other vestibular nuclei receive input from the semicircular canals, but these are minute in cetaceans. All of the vestibular nuclei project via the medial vestibulospinal tract and medial longitudinal fasciculus to motor neurons of the neck musculature and to the nuclei of the ocular muscles, and they are involved in reflex arcs harmonizing the position and precision movements of the head with those of the eyes in the pursuit of prey.

E. Limbic System

In odontocetes, the various cortical and subcortical components of the limbic system show different degrees of development (Figs. 2b and 3–7). In adult toothed whales, including the sperm whale, the hippocampus and mammillary body are unusually small and interconnected by a relatively thin fornix, whereas the anterior thalamic nuclei and the habenulae are better developed as in other mammals. The amygdaloid complex is large in toothed whales (Schwerdtfeger *et al.*, 1984). In contrast to the nucleus of the olfactory tract, which is totally dependent on olfactory input, the amygdala (taken as a whole) is well developed in microsmatic species (baleen whales, human) and even in anosmatic species (toothed whales). This indicates that, apart from olfactory stimuli, input to the amygdala arises from other sources, among them the auditory system. Furthermore, the remarkable development of the cortical limbic lobe (periarchicortex) in the dolphin again points to the largely nonolfactory character of this system.

F. Characteristics of Major Peripheral Cranial Nerves

The eye muscle nerves are thin in small cetaceans, particularly in river dolphins with their reduced eyes where they may even be completely lacking (Pilleri and Gihr, 1970). In marine dolphins, the oculomotor nerve comprises the highest number of axons (about one-third that in the human), followed by either the trochlear or abducent nerves. In large whales (especially baleen whales), the numbers of axons in the eye muscle nerves are distinctly higher and correspond to the situation in humans.

The trigeminal nerve is the thickest cranial nerve in baleen whales and sometimes in the sperm whale as well, whereas in all of the other toothed whales the vestibulocochlear nerve has the maximal diameter (Figs. 2a, 3, and 7–9). The diameters of the axons tend to be thin in the trigeminal nerve, whereas cochlear axons rank among the thickest axons known. In the smaller toothed whales investigated, the trigeminal nerve contains 82,000–156,000 axons (human: 140,000). The fact that trigeminal nerves in sperm and baleen whales have about the same large number of axons (490,000 vs 370,000–500,000) may be explained by the extreme size of the forehead region innervated in these animals.

The vestibulocochlear complex is the largest cranial nerve in toothed whales, and occasionally in the sperm whale as well, and it is the second largest in baleen whales (Figs. 2a, 3, 5a, and 7–10). In the sperm whale, it contains 215,000 axons, whereas in baleen whales, axon numbers vary from 154,000 to 179,000. Smaller toothed whales have 84,000 (harbor porpoise) to 171,000 axons (beluga), with the latter thus ranking near the much larger baleen whales (human: 50,000). The ratio of auditory and vestibular axons within the eighth nerve is disputed in the literature. Whereas some authors found that cochlear fibers in the bottlenose dolphin (total: 116,500 fibers) comprise about 60% of all the vestibulocochlear fibers (a percentage known from the human), other authors reported a similar total axon number (113,000) but a much higher fraction of cochlear axons (97%) as opposed to a very low number of vestibular axons. In comparison with humans (19,000 vestibular axons), dolphins possess only one-fifth in vestibular axons, a fact that correlates well with their small semicircular canals, with their low number of vestibular ganglion cells, and with the observation that the diameter of the vestibular nerve is barely one-tenth that of the cochlear nerve.

The variation in thickness of the facial nerve and in the diameter of its fibers throughout the cetaceans again sheds some light on biological correlations. Thus, the facial nerve in the sperm whale is nearly as thick as the vestibulocochlear nerve, containing about three times as many axons as in large baleen whales and three to eight times more axons than in smaller toothed whales. Baleen whales rival sperm whales in body size, and in all of these giants the head may attain one-third of the total length and mass of the body. In contrast, the absolute and relative size of the head in other toothed whales is comparatively small. Accordingly, it may be speculated that the prominent thickness of the facial nerve and the large number of its motor axons in the sperm whale are attributable to the extreme size of the forehead, which has to be regarded as an oversized "sound machine" (Cranford). Here, the forehead is characterized by the unique amount of acoustic fat tissues and massive blowhole musculature (innervated by the facial nerve) that may help stabilize the giant fat bodies and adjust their shape (acoustic lenses) during the emission of sonar signals.

IV. Neocortex

A. Layering and Cell Morphology

For a biological interpretation of the cetacean brain, we must review some existing data about the neocortex, e.g., brain/body mass relationship, absolute and relative cortex mass, and the structure of the neocortex, i.e., the layering, neuron density, synapse number and density, as well as density of gliocytes. The latter cells nourish, isolate, protect, and possibly help stimulate and regenerate the neurons and therefore are as essential for neocortical activity as the neurons themselves.

No other brain surpasses the cetacean brain in richness and complexity of neocortex gyrification. Moreover, toothed whales exhibit encephalization indices similar to those of higher primates. The cetacean neocortex shows an extremely tight folding and has a maximal surface within mammalia. At the same time, dolphins have cortical widths known from ungulates (Hummel, 1975), and even large whales do not seem to attain the average cortical thickness of higher primates (human).

In principle, the cetacean neocortex is six layered and similar to that in other mammals (Fig. 5b). Regional differentiation, however, is less obvious, and cortex lamination is not well expressed, particularly due to the widespread absence of a distinct layer IV in adult toothed whales and the moderate granularization of the cortex in general. In whales, layer one (molecular layer) can comprise up to about one-third of the total cortical width, whereas layer II is thin but rich in small pyramidal neurons that stain intensively with cresyl violet (Nissl stain). This second layer corresponds to the outer granular layer in the human. Its pyramidal cells that border on the molecular layer are mostly "extraverted" in whales, i.e., they show a clear predominance of apical (subpial) dendrites over basal dendrites. Layer III is thick and characterized by a variety of pyramidal cells with comparatively smaller neurons populating the outer half of the layer, whereas larger cells occupy the inner half. In view of the special role played by layer IV (corresponding to the inner granular layer in the human) as a major input layer for thalamocortical specific afferents in advanced land mammals, the precise cytoarchitectonic definition of this layer in cetaceans and its neuronal composition are crucial. It was postulated that in eutherian mammals layer IV appears and develops in proportion to the progressive displacement of specific thalamic afferents from the first (molecular) layer to midlevels of the cortex. In evolutionary advanced primates and ungulates, layer IV of the sensory neocortex is relatively wide and consists mainly of small granular, i.e., nonpyramidal cells mostly of the spiny stellate type. The fact that a lower degree of granularization also exists in insectivores and insectivorous bats and that layer IV is often not discernible here led to the assumption that the structure of the cetacean neocortex is "primitive" and allegedly shows a considerable number of other plesiomorphic features. These are the lack of definite boundaries, gradual morphological transitions between functionally separate neocortical areas, a poor lamination pattern, a thick layer I, extraverted pyramidal neurons in layer II, a dense band of large pyramidal neurons referred to as layer III$_c$/V, overall weak granularization with a predominance among nonpyramidal neurons on the part of large isodendritic stellate cells, a well-developed layer VI, and a lack of giant pyramidal cells. In some cortical areas of the fin whale brain, however, a narrow zone where nerve cells are sparse or lacking seems to mark the site of layer IV, and the corresponding external stria of Baillarger was found in a dolphin. Very young postnatal bottlenose dolphins also show a "remnant" of this layer. The innermost part of the cetacean cortex, presumably corresponding to layers V and VI in terrestrial mammals, usually displays no clear stratification and fades out into the white matter.

According to Deacon (1990), the common ancestor of all cetaceans very likely possessed layer IV granule cells as do the hyraxes or conies, plesiomorphic members of the paenungulate order (animals with precursors of genuine hooves), which includes the sirenians and elephants. In the highly encephalized elephants, layer IV was found to be lacking in all cortical areas, and their neocortex architecture is strikingly similar to that of the fin whale. The loss of granule cells in layer IV in cetaceans was therefore regarded a rare derived trait that may be correlated with the shift of the orphaned terminations of the specific thalamic afferents (originally terminating in layer IV) back to layers I and II, thus allowing more neurons in layer II to persist.

However, based on Golgi studies, there is little reason for thinking that the neocortex in the dolphin exhibits less variety than in other mammals, and a fair proportion of true stellate and other nonpyramidal cells exists in the dolphin cortex scattered throughout layers III and V, which may combine functions of afferent, efferent, and associative layers. Moreover, the distribution of neuronal size classes is rather similar to that occurring in other mammals.

Comparative immunocytochemistry (GABA, calcium-binding proteins) in the primary visual and auditory cortex of dolphins has revealed that the overall quantitative characteristics of the neurons are similar to those in other mammalian orders, being closer to insectivores and rodents in some features and to bats and primates in other characteristics.

B. Neuron Density

The density of cells is subject to considerable variation throughout the cetacean neocortex not only from area to area but also within the individual layers, although the relatively small and densely packed neurons in layer II account for at least 50% of all the neurons in a given cortical block. Tower (1954) measured the neuronal density in unspecified cortical areas of two fin whales, arriving at a mean figure of 6800 cells/mm^3. He also reported a similar neuronal density (6900/mm^3) in an elephant brain of about the same size. In comparison with approximately 18,000/mm^3 in the human and more than 100,000/mm^3 in rat and mouse, Tower concluded that there was a direct relationship between cortical neuronal density and brain mass. Other cell counts, however, reported an average of 57,000 neurons/mm^3 for the adult human. Cell counts in a series of toothed whales again revealed some decrease in density with increasing body size and absolute brain mass (harbor porpoise, 13,200; bottlenose dolphin, 13,000–44,200; beluga, 12,300; humpback whale, 8300/mm^3). Total counts of neurons beneath 1 mm^2 of cortical surface do not take into consideration the different width of the neocortical plate in various mammals and result in 28,500 cells for the adult bottlenose dolphin and 46,400 in an animal of 18 days. By comparison, the figure of 147,000 was given for all neocortical areas studied in a series of mammals ranging from mouse to human, except for the primate visual cortex, which harbors about twice as many neurons. Cell counts in single cortical sampling columns from pial surface to white matter with a cross-sectional area of $25 \times 30 \mu m$ resulted in 19 cells for the elephant, 21 for the bottlenose dolphin (Güntürkün, personal communication), and 110 in the nonstriate cortex of other mammals. Although Rockel's assumption may be restricted to primates, carnivores, and some rodents, dolphins obviously have a much smaller number of neurons in a given volume sample of cortex than would be expected from the size of their brains.

The mean size of neuronal perikarya has no definitive relation to brain size. Whereas primates tend to have a similar mean neuronal size and size distribution, other mammals, including cetaceans, show a tendency to increase neuronal size with increasing brain weight. Dolphins have moderate neuron sizes more or less similar to those found in ungulates (sheep, cow, horse) and elephants. Volume measurements reported a maximal neuronal volume of over $20,000 \mu m^3$ in the harbor porpoise, which is about double the maximal volume found in ungulates and small to medium sized for the human.

C. Synapses

The synaptic parameters of the visual neocortex show many of the qualitative and quantitative features of the general mammalian bauplan. For example, the quantitative relationship between the number of synapses with contacts to different components of cortical neurons such as perikaryon, dendritic shafts, and dendritic spines in the dolphin cortex does not differ significantly from these parameters in most other mammals. The same holds true for the distribution of other synaptic parameters throughout the cortical layers such as the area and form factor of individual synaptic boutons, the length of active zones (densities) of synaptic membranes, and several parameters relating to synaptic vesicles.

In their neocortex, dolphins show a mixture of conservative and advanced features at the cortical architectonic level as well as at neuronal and synaptic levels. When looking at the cortex as a whole, the aforementioned so-called conservative features occur mostly in superficial layers I and II, whereas deeper layers III–VI are characterized as more or less equivalent to those in advanced terrestrial mammals with the exception of layer IV. To date, adequate comparative data on ungulate groups or other aquatic mammals are not available. Moreover, cetaceans have a much lower neuronal density but, at the same time, a much higher synaptic density per volume unit and per neuron than terrestrial mammals. The majority of all synapses (70%) is found in cortical layers I and II: The latter seem to receive the brunt of cortical input as opposed to layer IV in many terrestrial mammals. Thus, in cetaceans, layer II seems to be the main relay element conveying information from the subcortical and intracortical afferents via layer I to the other cortical layers. All of these data, however, are difficult to interpret as to their significance both for function and/or physiology and for the evolution of the cetacean neocortex.

With respect to the total number of synapses in their cortices (0.87 vs 1.3×10^{14}), the dolphin and human resemble each other much more closely than other mammalian species. This appears to reflect primarily the generally large volume of neocortices in both dolphin and human brains as well as the maximal number of synapses per neuron compensating for minimal neuronal density in the dolphin.

D. Glia

Only a few data are available on gliocytes in the cetacean neocortex. In the bottlenose dolphin, glial density was found to vary in different cortical areas from 28,000 to 93,200 cells/mm^3, values rather similar to those in the human (average: 40,000–100,000 cells/mm^3). The number of gliocytes per number of neurons (glia/neuron ratio) is species-specific, i.e., varies among the mammalian groups and during ontogenesis due to changes in neuron density. This basically implies that larger brains have higher glia/neuron ratios. Thus, the ratio rises from small rodents (mouse, rabbit: 0.35) via ungulate species (pig, cow, and horse: 1.1), the human (1.68–1.78) and bottlenose dolphin (2–3.1), to large whales (fin whale: 4.54–5.85). Consequently, the glia/neuron ratio within each species increases from birth to maturity, thereby signaling the importance of glia for growing neurons and, thus, for neocortical function. Another implication is the necessity of comparing only mature specimens. Accordingly, in species with similar neocortex volumes, the glia/neuron ratio may be interpreted both ways, i.e., in favor of neurons or glia as factors of equal importance.

In summary, the morphology of the dolphin neocortex seems to be equivalent to that of advanced terrestrial mammals, but its specific features are not yet understood. With respect to ontogenesis and evolution, the allometric process of thinning out the number of neuroblasts and/or neurons via apoptosis and via the generation of increasing amounts of glia and neuropil for their connectivity seems to proceed faster in cetaceans than in other mammals. At the same time, the cortical plate seems to spread out more than in other mammals, leading to an extremely extended and convoluted neocortex with a minimal neuron density but maximal synaptic density.

V. Characteristics of Semiaquatic and Aquatic Mammals

A. General Remarks

The brain can be regarded as the true center of the body responsible for the maintenance of physiological conditions (homeostasis) and survival. By means of the cranial and spinal nerves, the brain collects all of the sensory information available, thereby evaluating, synthesizing, and using it for optimal behavior in manifold aspects: orientation, FEEDING, defense, COMMUNICATION, and reproduction. Thus, the brain not only represents all parts of the body but, at the same time, mirrors the situation of the animal within its ecological niche. On account of these tight correlations, evolutionary changes in the environment and in the biology of the species must show in the morphology of the nerves and brain. Rapid transgressions of mammalian groups into totally different habitats are related to strong "selection pressure," i.e., adaptational processes unfold rather quickly and may lead to profound changes in brain morphology and function, reflecting changes in the body's periphery. One obvious principle is seen in the increase or decrease in size of individual brain structures, which may result in qualitative changes. Comparative analysis of the brain, therefore, can help unveil evolutionary strategies of the species by looking for qualitative and quantitative correlations between various brain structures and body mass and between those structures belonging to the same functional system.

B. Insectivores

The most obvious changes in the quantitation of the brain in all semiaquatic insectivores are as follows: a marked reduction of the primary and secondary olfactory centers, a slight to moderate size increase of the neocortex and myelencephalon, and a moderate to marked growth of the diencephalon (dorsal and ventral thalamic nuclei, geniculate nuclei), as well as other structures related to the trigeminal, auditory, or motor systems. There is a general but weak size increase of the limbic system, with the hippocampus showing the least change among the components of this system in comparison to terrestrial relatives. The trigeminal complex is thought to replace the olfactory system as the main sensory system in the search for food (vibrissae), whereas the accessory olfactory (vomeronasal) system seems to be unaffected by the animal's adaptation to water.

C. Carnivores

Other examples of CONVERGENT EVOLUTION are single mustelids (otters) and the pinnipeds. In comparison with their terrestrial relatives and with respect to basal insectivores, otters are distinctly more encephalized, and the same seems to be valid for pinnipeds (seals) and their potential terrestrial relatives. In contrast, sirenians have been reported to have low encephalization indices in comparison with herbivorous terrestrial "relatives," such as the hyraxes, a fact that may be related to the comparatively high body mass of sea cows and the arboricolous mode of life in hyraxes.

1. Eurasian River Otter (Lutra lutra)   Here, the adaptations to the aquatic medium are only moderate: The olfactory system is markedly smaller than in terrestrial carnivores but better developed than in the seal. The hippocampus and amygdala are large. The visual system is well developed (thick optic nerve, lateral geniculate nucleus layered). The same holds true for the auditory system, but the acoustic tubercle (dorsal cochlear nucleus) is small. The trigeminal system is well developed (vibrissae), but the facial nerve and nucleus are of moderate size. As in the seal, the basal ganglia are moderately developed, but the globus pallidus is large. In the cerebellum of the otter, the vermis is considerably smaller with respect to the hemispheres than in terrestrial mammals but larger than in pinnipeds. The inferior olive in the otter is organized similarly to that in terrestrial mammals, but it is larger than in these animals and smaller than in the seal and porpoise. The pyramids are prominent thick bundles. The gracile and cuneate nuclei (proprioceptive body sensitivity) are also well developed in the otter.

2. Harbor Seal (Phoca vitulina)   Pinnipeds spend much of their time on land. They have very good vision and their hearing is acute. The pineal organ is present. Both pairs of limbs are present although more or less modified, thereby indicating high somatic sensitivity in general. Apart from important differences, seals share many features with terrestrial carnivores, e.g., large telencephalic hemispheres with a well-developed corpus callosum. The hemispheres are covered with highly convoluted neocortex and nearly conceal the large cerebellum in dorsal aspect. The pons is flat but extended (Fig. 13). The olfactory bulb and tract are small and thin, and the anterior commissure is small. Hippocampus and amygdala are large, and the habenula is well developed. The visual system is well developed (thick optic nerve, large and layered lateral geniculate body, very large superior colliculus). The myelencephalon is wide, with the auditory system generally being well developed (thick eighth nerve, large ventral cochlear nucleus, trapezoid body, lateral lemniscus, medial geniculate nucleus and inferior colliculus), but the tuberculum acusticum (dorsal cochlear nucleus) is very small as in whales (no pinna). The trigeminal system is very pronounced (vibrissae), with the trigeminal being much thicker than the vestibulocochlear nerve, whereas the facial nerve and nucleus are of normal size (Fig. 13). The inferior olive is larger than in land carnivores but very similar to theirs in structure.

Among the fiber systems, the cerebral peduncles are very large and the pyramids thick, whereas the gracile and cuneate fascicles are smaller than in terrestrial carnivores but thicker than in cetaceans.

D. Sirenians: Dugongs and Manatees

In comparison with plesiomorphic dolphins such as the susu and franciscana, sea cows have small brains (Fig. 14). At about the same brain mass (220 g), the body of the young dugong (*Dugong dugon*) is about 4.5 times as heavy as that of the franciscana. The relative size of the telencephalic hemisphere (with respect to the total brain) is similar to that in the dolphin. With respect to basal insectivores, however, the size index of the cortex in the young susu is 3.5 times higher than in the dugong. In sirenians, gyrification of the cortex is only rudimentary. Also, there is little underlying white matter surrounding the very large ventricles, and the size of the corpus callosum is moderate. In the frontal region, cortical thickness (gray matter) is averaged at an exceptional 4.0 mm, and histological analysis suggests a potential motor cortical field at the rostroventral tip of the hemisphere as well as extended potential somatosensory fields on the adjacent lateral (convex) surface. Presumably, the primary auditory field and visual fields are located on the posterior lateral surface of the hemisphere and in the dorsal occipital region as in many other mammals. The cell number in the neocortex is reported to be sparse, but the subdivision into cortical fields and the lamination are more distinct than in cetaceans. The molecular layer (I) is of normal width. Layer II, which is most conspicuous, contains densely packed small pyramidal neurons with long apical dendrites. In the potential visual cortex, layer IV is composed of small almost granular pyramidal cells. Layer V is most pronounced in the potential motor field, and layer VI contains unique neuron clusters (Rindenkerne) in the potential somatosensory, visual, and auditory cortices. Sea cows have small olfactory bulbs and thin, ribbon-shaped olfactory tracts with small olfactory tubercles (Figs. 14a and 14b). Nevertheless, the olfactory system as a whole is distinctly larger in the sea cow than in the dolphin, but the vomeronasal organ

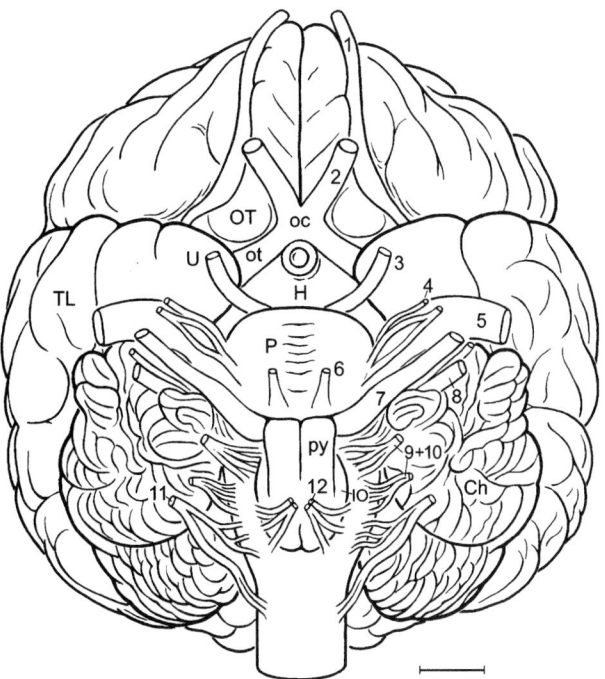

Figure 13 *Harbor seal brain in basal aspect. py, pyramidal tract. Scale: 1 cm.*

is also lacking. The anterior commissure is well developed. The hippocampus and fornix are reported to be even smaller than in plesiomorphic dolphins, and the mammillary bodies are small and not visible superficially. The visual system seems to be moderately developed (relatively thin optic nerve, nuclei of ocular muscle nerves highly reduced in size, small superior colliculus). A pineal gland is lacking. The brain stem of the dugong is small in comparison with that of the dolphin: The diencephalon is moderately developed (small visual thalamic nuclei) and the size increase of the mesencephalon with respect to body mass is very low. The myelencephalon is also rather small, probably due to the moderate size of the auditory and motor systems, and despite the large trigeminal nuclei and very thick nerve supplying the rostral part of the head (particularly the extremely important vibrissae). The long ascending and descending fiber systems are moderately developed. Whereas most components of the auditory system, such as the cochlear nerve, ventral cochlear nucleus, medial geniculate nucleus, and inferior colliculus, are more or less well developed (inferior colliculus distinctly larger than superior colliculus), the dorsal cochlear nucleus is very small. The facial nerve and nucleus are much better developed (motor activity and control of the muzzle in seaweed grazing). With reference to basal insectivores and in comparison with dolphins, the cerebellum is rather small in the sea cow. This may be partly attributable to the sluggish swimming behavior of this animal, which can hardly evade attacks by humans. The vermis, however, is relatively large as in aquatic carnivores (otter, seal). In the sea cow, the paraflocculus is also relatively large but much smaller than in cetaceans. Whereas the pyramids are thin,

the decussation in the cervical spinal cord is distinct, but the pyramidal tract cannot be followed further caudalward. The gracile and cuneate nuclei (proprioceptive body sensitivity) are fairly well developed. The inferior olivary complex in sirenians is characterized as being small, but it bulges at the basal surface of the myelencephalon, with its medial (accessory) and lateral (principal) nuclei being of similar size. As a whole, the inferior olive very much resembles that of ungulates, but is distinct from that of the elephant.

E. Cetaceans

Cetaceans presumably originated from an extinct basal ungulate group (Condylarthra, Mesonychidae) that also gave rise to the living artiodactyls (Thewissen, 1998), and close relationships with single extant ungulate groups have been discussed as well. Although generally very sparse in the fossil record, skeletal parts can give a reasonable impression of the evolutionary changes from terrestrial ancestors to living cetaceans. Unfortunately, this does not hold true for the brain, which only is known from endocasts. The latter formed in the cranial vaults of fossilizing precetacean or cetacean mammals and can provide only limited information about the external morphology of the actual brain of those animals. Furthermore, the understanding of cetacean brains has been hampered in yet another respect: Comparative considerations led to the idea that, similar to primates, cetaceans could be derived phylogenetically directly from insectivores, which represent some kind of "ancestral group" for many extant mammals. Nevertheless, although the cetacean neocortex exhibits some features that seem to re-

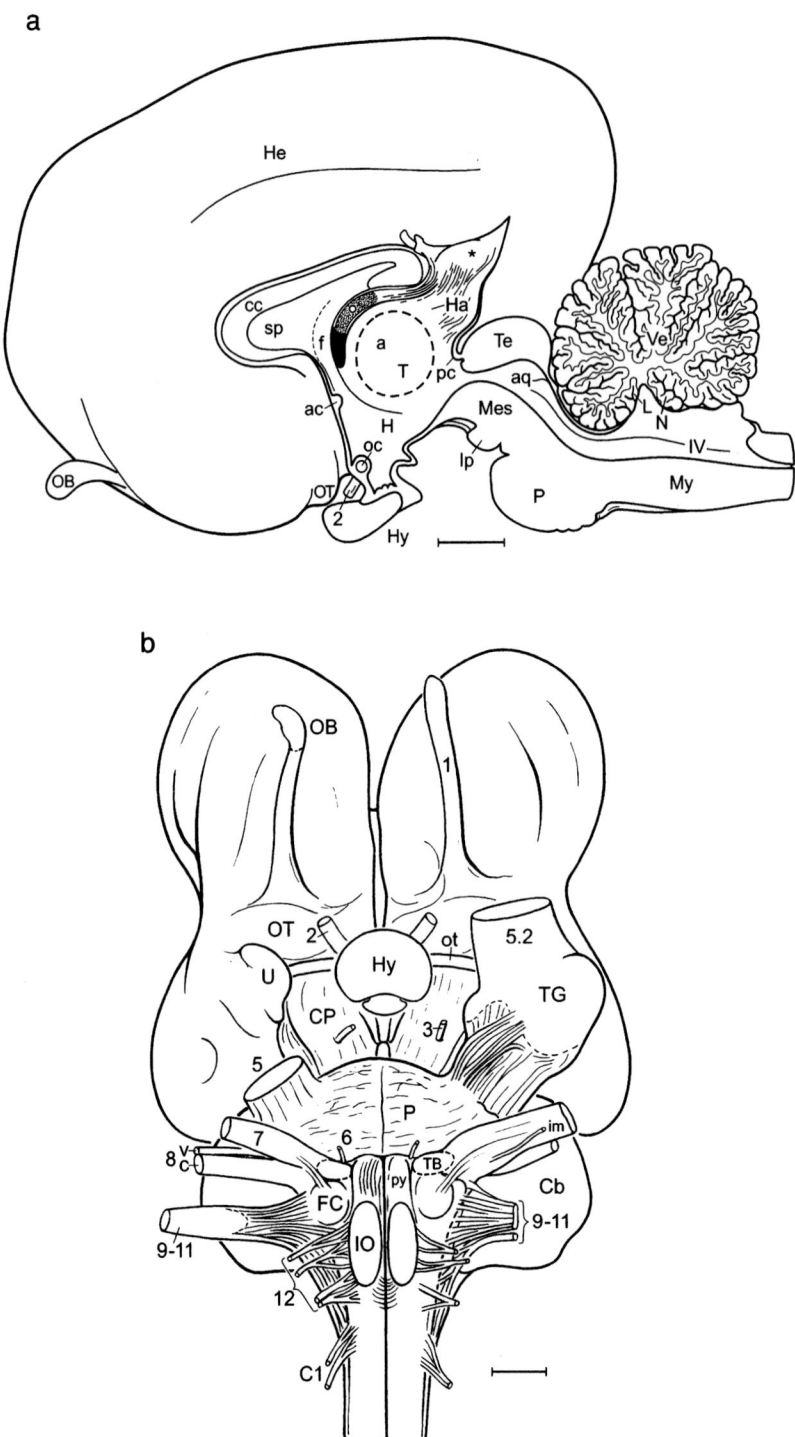

Figure 14 *Sea cow (dugong) brain: (a) mediosagittal aspect and (b) basal aspect. After Dexler (1913), modified. Asterisk, dorsal evagination of diencephalic roof; circle, choroid plexus and interventricular foramen; FC, facial tubercle; lp, interpeduncular nucleus; L, lingula; OB, olfactory bulb; TG, trigeminal ganglion; 5.2, maxillary branch of trigeminal nerve. Scale: 1 cm.*

call the situation in insectivorous brains, a more promising approach would be to compare cetaceans with plesiomorphic paenungulates such as the hyracoids (hyraxes or conies), proboscids (elephants), generalized artiodactyls such as the tragulids (Fig. 1), and also so-called "advanced" (derived) hoofed animals. Such a reconstruction of the hypothetical ancestors' brains on the basis of "closely" related extant mammalian groups would be of considerable value for understanding cetacean evolution in general.

In the franciscana (*Pontoporia blainvillei*), one of the least specialized living TOOTHED WHALES (brain mass of the adult: 220 g; body mass: 35 kg), the encephalization index (EI: 5.7) ranges above the level of prosimians (lemurs, *Tarsius*) and in the lower echelon of simians (monkeys, apes, human), whereas the EI level of marine dolphins (13.5) is above that of simian primates but clearly below that of humans. From what we know today, the franciscana brain shows all the features typical of toothed whales and mostly of whales, in general: (1) a large telencephalon and, at the same time, total loss of the rostral olfactory and the whole accessory olfactory system, and the reduction of the remaining paleocortex. The archicortex and some other components of the limbic system are very small, but the limbic cortex seems to be well developed. The neocortex is very much maximally extended and convoluted. As in primates, the neocortex is by far the largest brain structure in river dolphins and other cetaceans. Its size index is higher in river dolphins than the average of prosimians but clearly lower than that of simians. Even higher values for the neocortex can be expected for marine dolphins. Giant cetaceans are difficult to interpret because their brains, although approaching 10 kg in total mass, are dwarfed by their huge bodies. In addition to the neocortex, the striatum is one of the most progressive telencephalic structures and indicates a high functional capacity of the motor systems. The amygdala, another component of the basal ganglia and belonging to the limbic system, is considerable in size, although its corticomedial component largely consists of paleocortex. (2) The thalamus is large due to the exceptional volume of different nuclei (medial, dorsal, ventral, medial geniculate, and pulvinar). (3) The large volume of the midbrain is significant and is due to the extreme size of some components of the auditory system. Thus, for example, the inferior colliculus is much larger than the superior colliculus. (4) Another major character is the large size of the cerebellum and particularly of the paraflocculus and posterior interposed nucleus as well as associated structures (elliptic nucleus, pons, inferior olive and accessory fiber tracts). (5) Finally, the medulla oblongata is, comparatively speaking, also very large due to the outstanding development of most of the auditory nuclei as well as those of the trigeminal and facial nerves.

These five adaptational trends seen in the dolphin brain are more or less paralleled in semiaquatic insectivores (Bauchot and Stephan, 1966) and other aquatic mammals (otter, pinnipeds) where growth of single brain regions surpasses the reduction in other regions such that total brain size increases. Sirenians, however, seem to be rather different in many respects. Nevertheless, in some ways and to some degree, these different groups of mammals have been regarded as "stages" of convergent adaptation to the aquatic habitat.

F. Comparative Aspects

According to the present state of our knowledge, aquatic mammals differ in the quantity and thickness of the *neocortex* and in the degree of gyrification of the telencephalic hemispheres. Whereas in pinnipeds and terrestrial carnivores the primary neocortical fields coincide in location, their position is rather different in dolphins in comparison with other mammals. The changes and differences in neocortical structure and composition found in aquatic mammals are not very well understood and cannot be explained to date functionally.

The general appearance of the neocortex in otters and pinnipeds corresponds fairly well to that of their terrestrial relatives. Sirenians show a small neocortex volume but a reasonable layering and cell density. In contrast, the neocortex in cetaceans is voluminous and extended but rather uniform, layer IV is weakly developed, and neuron density seems to be lowest within the Mammalia. Whether in sirenians the neuron clusters (Rindenkerne) in layer VI correlate with their vibrissae (bristles) and are equivalent to the "barrel fields" in rodents remains unsubstantiated to date.

The *olfactory system* of aquatic mammals seems to exhibit a clear relationship to their lifestyle. Whereas in aquatic carnivores the olfactory bulbs and tracts are clearly reduced in size, sea cows come even close to the baleen whales. Toothed whales only very rarely have vestigial olfactory tracts and are characterized by residues of a secondary olfactory cortex. Obviously, the olfactory part of the nose inherited from their terrestrial ancestors could not be adapted successfully to the requirements of aquatic life and therefore was reduced in extant aquatic mammals or even lost in cetaceans. As another example of such an adjustment, Jacobson's organ (vomeronasal organ) is lacking in sirenians and cetaceans.

As in the cetaceans, parts of the *limbic system* (particularly the hippocampus) are reduced in some other aquatic mammals. Whereas in aquatic carnivores the hippocampus seems to be as large as in terrestrial mammals, sea cows and cetaceans have a minimal amount of archicortex. Other parts of the limbic system (e.g., fornix, mammillary body, mammillothalamic tract) are much reduced in toothed whales, but the anterior thalamic nuclei and the limbic lobe of Morgane are well developed. Here, we have a rather mysterious situation which, so far, defies explanation.

In the cetacean species investigated, the *anterior commissure* is generally very thin, obviously correlating with the reduction of its olfactory component. In sirenians and pinnipeds, the anterior commissure is neither as strongly reduced as in cetaceans (with respect to the whole brain) nor as thick as in the otter, and it is thin in comparison with that in land mammals. The *corpus callosum* is short and thin in dolphins but seems to be thicker in baleen whales. Its size (cross-sectional mediosagittal area) was shown to be inversely related to the mass of the brain (Tarpley and Ridgway, 1994). This potentially moderate interhemispheric connectivity may be related to the fact that neuronal density is exceptionally low in cetaceans and may indicate some interhemispheric independence as seen in the example of unihemispheral sleep in bottlenose dolphins. In sirenians, the corpus callosum is small in comparison to that in terrestrial mammals (ungulates) but it is well developed in aquatic carnivores (pinnipeds).

The *basal ganglia* of dolphins are well developed, particularly the nucleus basalis of Meynert, whose size seems to correlate positively with the neocortex (temporal lobe) and is maximally developed in cetaceans and primates. The caudate nucleus and putamen unite in a thick fundus striati, which encroaches on the surface of the basal forebrain in the area of the olfactory lobe (olfactory tubercle, diagonal band). In comparison with the porpoise, the putamen of the seal and otter is much smaller, but the pallidum is well developed. In the sea cow, the basal ganglia are small compared with those of other aquatic mammals.

The *visual system* shows a very different degree of development: Otters and seals have large eyes and central visual structures, whereas in cetaceans the eyes and optic nerves are of medium size and in sea cows small. It is not clear whether there is an inverse relationship in mammals between eyesight and the degree of adaptation to aquatic life.

The *auditory system* reaches its maximal development in the Cetacea. Whereas all aquatic and semiaquatic mammals exhibit the same bauplan, pinnipeds possess well developed but much smaller auditory structures, and in Sirenia most auditory components are developed only moderately. The dorsal cochlear nucleus, which is believed to be engaged in the integration of positional effects of the head and pinnae in the hearing process, is reported to be of very small size in aquatic mammals generally. This might correlate with the fact that in these animals the external ear is reduced in size or even lacking.

Most aquatic mammals are characterized by a well-developed *cerebellum*. The highest relative cerebellar mass is found in cetaceans (dolphins) and it is lower in pinnipeds and in the otter, whereas the good-natured and restful herbivorous sirenians have a comparatively small cerebellum. Within the cerebellum, there was obviously an allometric shift in the size relationship between the vermis and hemispheres during the adaptation to aquatic life, thus resulting in a small vermis but large cerebellar hemispheres in the fully adapted cetaceans. In aquatic mammals, the paraflocculus is very large or even the dominant part of the cerebellum. In the porpoise, the paraflocculus is connected to a specific cerebellar nucleus (posterior interposed nucleus), which may have evolved as an extension of the dentate nucleus and is very large. The paraflocculus is less developed in sirenians than in cetaceans, and even less so in the seal and otter. The latter approach the situation found in terrestrial carnivores, paenungulates, and ungulates, where the paraflocculus is already well developed, especially in the amphibious hippopotamus.

In cetaceans, the size of the *pons* and its nuclei seems to correlate with the size of the cerebellum and neocortex and with the massive projections to and from the huge elliptic nucleus as well as with the size of the rostral part of the medial accessory inferior olive.

The *inferior olives* of semiaquatic and aquatic mammals have been characterized by morphological criteria. As is the case for most components of the cetacean brain, experimental data are not available here. Cetaceans show an extreme thickening of their medial (accessory) olive, more accurately the rostral part of

this subnucleus, which bulges at the base of the myelencephalon and seems to be a secondary acquisition. The caudal part of the cetacean medial accessory olive resembles in shape more that in sirenians (dugong) and in ungulates than that in terrestrial carnivores. In the harbor seal, where the medial accessory olive is the largest of the subnuclei, it is larger than its counterpart in terrestrial carnivores but by far smaller than that in cetaceans. A similar situation is found in the otter, in which the olivary complex, however, is somewhat smaller than in the seal.

The *pyramids* appear to be somewhat better developed in sirenians than in the harbor porpoise; their crossing is very distinct. In comparison, pinnipeds and the otter have thick pyramidal cords, possibly as a heritage of their terrestrial carnivore ancestors. In contrast, ungulates (artiodactyls and perissodactyls) tend to have moderately developed pyramids. With their very weak spinal pyramidal tract, cetaceans seem to show an "overlapping" of phylogenetic heritage *and* morphological as well as physiological adaptations to aquatic life, i.e., a reduction of the peripheral locomotor apparatus (hindlimbs and pelvic girdle) and a strong modification of the anterior limb and girdle (flipper, shoulder girdle).

VI. Strategies in Aquatic Adaptation

In the semiaquatic and most aquatic mammals investigated (insectivores, otters, pinnipeds, river dolphins, marine dolphins), the brain taken as a whole and a number of its components show a convergent moderate to strong increase in size with respect to potential terrestrial relatives within their groups. There is considerable variability, however, in the different brain structures of the species concerned. Whereas in semiaquatic insectivores and the sirenians the total brain and telencephalon are moderately enlarged for the body in comparison with basal insectivores, the otters, pinnipeds, and particularly dolphins are increasingly more encephalized. Otters and pinnipeds have larger brains than their potential terrestrial relatives, but nothing is known about their quantitative composition.

In some groups of semiaquatic and aquatic mammals (single species of insectivores, otter, pinnipeds), the limbic system is well developed. In cetaceans and sirenians, however, the situation is more complicated. In cetaceans, the archicortex (hippocampus) is very small, as are the fornix and mammillary body, but not the amygdala. Moreover, the limbic lobe as a transitional area between archicortex and adjacent neocortical areas is large. It might be concluded, therefore, that the limbic system in cetaceans differs markedly from that in terrestrial mammals. In sirenians, the hippocampus was reported to be even smaller than in cetaceans and its size is obviously minimal within the Mammalia. It is unclear whether the strong reduction of the hippocampus, which in mammals is primarily concerned with attentiveness and vigilance, and serves as an intermediate store for all kinds of information and their "labeling" (short-term memory), correlates with the strong reduction (baleen whales) and even loss of the rostral part of the olfactory system (toothed whales). If we subscribe to the principle of a central position for the hippocampus in informational processing of mammals, we are unable to explain the coincidence in size between two groups otherwise so disparate in brain volume, physiology, and behavior, and so different from other mammals. The small size of the hippocampus, however, could explain why dolphins have difficulties learning new activities that are not already part of their normal lives. In this respect, the extremely large surface area and the considerable volume of the dolphin cortex, which deviates much in architecture from the pattern seen in "progressive" terrestrial mammals, indicates that dolphins may possess an "intelligence" very different from our own as the sum of their adaptations to the aquatic environment during approximately 50 million years of separate evolution.

In conclusion, it seems that, in most cases, the adaptation of mammals to aquatic life, including the necessity of three-dimensional locomotion and diving in a turbid or dark environment, causes (1) an increase in the size of the brain due to the growth of special functional systems such as the auditory and motor systems, (2) an increase in neocortex volume, and (3) the strong reduction or even loss of the rostral olfactory system and its possible compensation or replacement by the trigeminal system. Interestingly, toothed whales and sirenians as the most advanced groups in terms of aquatic adaptation differ in a number of major characteristics. With respect to the basal insectivores (Stephan), even plesiomorphic odontocetes are much more encephalized than sirenians, and most of their brain components and functional systems show a distinctly stronger growth (earlier discussion). The profound differences in the morphology and quantitative composition of the brain in dolphins and sirenians may be related to the contrasting behavior in these animals. Whereas dolphins are extremely agile, swift animals with an intense social life, hunting different types of highly mobile prey with a powerful sonar detection system, sirenians have no sonar and are sluggish plant eaters with a thick integument and limited diving capabilities.

Concerning neocortex morphology, cetaceans and sirenians differ with respect to its thickness, extension, arealization, degree of layering, and presumably neuronal density. Taken as a whole, it seems that in contrast to the situation in the sirenians the extended neocortex of dolphins is responsible for the processing of large amounts of acoustic data needed for orientation, feeding, and social communication. Because of the extreme propagation velocity of sound in water, the time intervals left over for meaningful decisions are very short for swift swimmers and active hunters. A large number of modules in the extended cetacean neocortex may facilitate the quick scanning and subsequent reconstruction of ephemeral two- or three-dimensional images of their "complicated" environment.

Acknowledgments

The authors cordially thank Professor Dr. Jürgen Winckler (Dr. Senckenbergische Anatomie, University of Frankfurt am Main) for the enduring and generous support of our work. We are indebted to Dr. Thomas Holzmann (Frankfurt am Main) for permission to publish a figure from his dissertation. Special thanks are due to Mrs. Inge Szász-Jacobi for her excellent work in the preparation of the originals of the figures. We are grateful to Dipl.Biol. Michaela Haas-Rioth, Professor Dr. Onur Güntürkün (Ruhr-University Bochum), Professor Dr. Gerhard Hummel (University of Gießen), PD Dr. Walter K. Schwerdtfeger (German Ministry of Health, Bonn), and PD Dr. Helmut Wicht (Frankfurt am Main) who kindly read the manuscript and helped with discussions of the content and many valuable suggestions. Thanks are also due to Jenny Narraway (Hubrecht Laboratory, Utrecht, The Netherlands) for the loan of the microslide series of the fetal narwhal and to Kelly Del Tredici, Ph.D. (Frankfurt am Main) for carefully editing the English manuscript.

See Also the Following Articles

Circulatory System ■ Hearing ■ Intelligence and Cognition ■ Mammalia ■ Prenatal Development in Cetaceans ■ Skull Anatomy ■ Vision

References

Bauchot, R., and Stephen, H. (1966). Donneés nouvelles sur encéphalisation des insectivores et des prosimiens. *Mammalia* **30**, 160–196.

Berta, A., and Sumich, J. L. (1999). "Marine Mammals: Evolutionary Biology." Academic Press, London.

Breathnach, A. S. (1955). The surface structures of the brain of the Humpback Whale. *J. Anat.* **89**, 343.

Breathnach, A. S. (1960). The Cetacean Central Nervous System. *Biol. Reviews* **35**, 532 pp.

Buhl, E. H., and Oelschläger, H. A. (1986). Ontogenetic development of the N. terminalis in toothed whales: Evidence for its nonolfactory nature. *Anat. Embryol.* **173**, 285–295.

Cranford, T. W., Amundin, M., and Norris, K. S. (1996). Functional morphology and homology in the odontocete nasal complex. *J. Morphol.* **228**, 223–285.

Deacon, T. W. (1990). Rethinking mammalian brain evolution. *Am. Zool.* **30**, 629–705.

Dexler, H. (1913). Des Hirn von Halicore dugong. *Morphol. Jahrb.* **45**, 97.

Glezer, I. I., Jacobs, M. S., and Morgane, P. J. (1988). Implications of the "initial brain" concept for the brain evolution in cetacea. *Behav. Brain Sci.* **11**, 75–116.

Holzmann T. (1991). "Morphologie and microscopische Anatomie des Gehirns beim fetalen Narwal, Monodon monoceros." Thesis, Faculty of Human Medicine, University of Frankfurt.

Hummel, G. (1975). Lichtmikroskopische, elektronenmikroskopische und enzymhistochemische Untersuchungen an der Großhirnrinde von Rind, Schaf und Ziege. *J. Hirnforsch.* **16**, 245–285.

Jansen, J. (1950). The morphogenesis of the cetacean cerebellum. *J. Comp. Neurol.* **93**, 341.

Jansen, J. (1953). Cetacean Brain. *Hvalrådets Skriftern* **37**.

Jansen, J., and Jansen, J. K. S. (1969). The nervous system of Cetacea. *In* "The Biology of Marine Mammals." (H. T. Anderson, ed.), pp. 175–252. Academic Press, London.

Kamiya, T., and Pirlot, P. (1974). Brain morphogenesis in *Stenella coeruleoalba*. *Scientific Rep. Whales Research Institute, Tokoyo* **26**, 245–253.

Langworthy, O. R. (1931). Factors determining the differentiation of the cerebral cortex in sea living mammals (Cetacea). A study of the brain of the porpoise *Tursiops truncatus*. *Brain* **54**, 225.

Langworthy, O. R. (1932). A description of the central nervous system of the porpose (*Tursiops truncatus*). *Journ. Compar. Neurol.* **54**, 437.

Manger, P., Sum, M., Szymanski, M., Ridgway, S., and Krubitzer, L. (1998). Modular subdivisions of dolphin insular cortex: Does evolutionary history repeat itself? *J. Cogn. Neurosci.* **10**, 153–166.

Marino, L. (1998). A comparison of encephalization between odontocete cetaceans and anthropoid primates. *Brain Behav. Evol.* **51**, 230–238.

Morgane, P. J., and Jacobs, M. S. (1972). Comparative anatomy of the cetacean nervous system. *In* "Functional Anatomy of Marine Mammals" (R. J. Harrison, ed.), pp. 117–244. Academic Press, London.

Morgane, P. J., Jacobs, M. S., and Glezer, I. I. (1986). Ultrastructural features of visual cortex of the dolphin. *Society for Neuroscience abstracts*, **12**, 105.

Nieuwenhuys, R., ten Donkelaar, H. J., and Nicholson, C. (eds.) (1998). "The Central Nervous System of Vertebrates," Vol. 3. Springer, Berlin.

Oelschläger, H. H. A., and Kemp, B. (1998). Ontogenesis of the sperm whale brain. *J. Comp. Neurol.* **399**, 210–228.

Pilleri, G. (1966). Morphologie des Gehirnes des Buckelwals, Megaptera novaeangliae Borowski (Cetacea, Mysticeti, Balaenopteridae). *J. Hirnforschung* **8**, 437–491.

Pilleri, G. and Gihr, M. (1970). The central nervous system of the Mysticete and Odontocete whale. *Investigations on Cetacea* (Berne) **2**, 87–135.

Pilleri, G., Chen, P., and Shao, Z. (1980). "Concise Macroscopical Atlas of the Brain of the Common Dolphin (*Delphinus delphis* LINNAEUS, 1758)." Brain Anatomy Institute, University of Berne, Waldau–Berne, Switzerland.

Pirlot, P. and Kamiya, T. (1985). Qualitative and quantitative brain morphology in the sirenian Dugong dugon, Erxleben. *Z. zool. Syst. Evol. Forsch.* **23**, 147–155.

Reep, R. L. and O'Shea, T. J. (1990). Regional brain morphometry and lissencephaly in the Sirenia. *Brain, Behavior and Evolution* **35**, 185–194.

Reep, R. L., Johnson, J. I., Switcher, R. C., and Welker, W. I. (1989). Manatee cerebral cortex: cytoarchitecture of the frontal region in Trichechus manatus latirostris. *Brain, Behavior, and Evolution* **34**, 365–386.

Ridgway, S. H. (1990). The central nervous system of the bottlenose dolphin. *In* "The Bottlenose Dolphin" (S. Leatherwood and R. R. Reeves, eds.), pp. 69–97. Academic Press, New York.

Ridgway, S. H. (1986). The central nervous system of the bottlenose dolphin. Pp. 31–60. *In* "Dolphin cognition and behavior: A comparative approach" (R. J. Shusterman, J. A. Thomas, and F. G. Wood, eds.). Lawrence Erlbaum Associates, Hillsdale, New Jersey.

Ridgway, S. H., Demski, L. S., Bullock, T. H., and Schwanzel-Fuduka, M. (1987). The terminal nerve in odontocete cetaceans. *Ann. of the New York Acad. Sciences* **519**, 201–212.

Schulmeyer, F. J. (1992). "Vergleichend morphologische Untersuchungen am Hirnstamm der Delphine mit besondere Berücksichtigung des akustischen Systems." Med. Diss. Univ. Frankfurt.

Schwerdtfeger, W. K., Oelschläger, H. A., and Stephan, H. (1984). Quantitative neuroanatomy of the brain of the La Plata dolphin, *Pontoporia blainvillei. Anat. Embryol.* **170**, 11–19.

Slijper, E. J. (1962). "Whales." 475 pp. Basic Books Publ. Co.

Stephan, H., and Andy, O. J. (1969). Quantitative comparative neuroanatomy of primates: An attempt at a phylogenetic interpretation. *Ann. N. Y. Acad. Sci.* **167**, 370–387.

Supin, A. Y., Mukhametov, L. M., Ladygina, T. F., Popov, V. V., Mass, A. M., and Poljakova, I. G. (1978). "Electrophysiological studies of the dolphin's brain" (in Russian). Moscow, Izdatel'sto Nauka.

Tarpley, R. J. and Ridgway, S. H. (1994). Corpus callosum size in delphinid cetaceans. *Brain Behav. Evol.* **44**, 156–165.

Thewissen, J. G. M. (ed.) (1998). "The Emergence of Whales: Evolutionary Patterns in the Origin of Cetacea." Plenum Press, New York.

Tower, D. B. (1954). Structural and functional organization of mammalian cerebral cortex: The correlation of neurone density with brain size. Cortical neurone density in the fin-whale (*Balaenoptera physalus* L.) with a note on the cortical neurone density in the Indian elephant. *J. Compar. Neurol.* **101**, 19–51.

Wanke, T. (1990). "Morphogenese des Gehirns beim Schlanken Delphin Stenella attenuata (Gray, 1846)." Med. Diss. Univ. Frankfurt.

Brain Size Evolution

Lori Marino
Emory University, Atlanta, Georgia

The study of brain size among three major living groups of marine mammals, cetaceans, pinnipeds, and sirenians, is a study in contrasts. Phylogenetic and ecological factors have shaped the course of brain evolution in each group in distinct ways. The resulting diversity of brains provides an illustration of the different successful paths that were taken in the evolution of marine mammals.

I. The Meaning of Encephalization

Brain size evolution is embodied in the concept of encephalization, which was originally put forth as an encephalization quotient (EQ) by psychologist Harry Jerison. Jerison widely applied this measure to comparisons across different species. EQ is a measure of observed brain size relative to expected brain size derived from a regression of brain weight on body weight for a sample of species. EQ values of one, less than one, and greater than one indicate a relative brain size that is average, below average, and above average, respectively. For example, a species with an EQ of 2.0 possesses a brain twice as large as expected for an animal of its body size. The EQ values reported here are based on a large sample of living mammalian species from Jerison (1973).

In addition to the whole brain, changes in the size of various brain components have also occurred throughout marine mammal evolution and contribute to changes in the overall brain size. Some of these changes are measurable, but undoubtedly many changes in the relative size of structures occurred in all marine mammals that are not apparent from the fossil record.

II. Accessing the Fossil Record

Studies of brain size evolution in fossil marine mammals have depended on measuring the volume of the endocranial cavity in fossil specimens. Because the specific gravity of brain tissue is nearly the value of water, volumetric data have been typically converted to units of weight. In addition to the general problem of finding intact fossil crania, early studies were hampered by difficulties accessing the sediment-filled endocranium. Researchers have taken advantage of the fortuitous occurrence of intact natural endocasts but these are rare. Also, earlier studies often resulted in overestimates of brain mass from endocranial volume because they did not take into account that total endocranial volume is partly composed of non-neural, e.g., vascular, components. Cetaceans, for instance, possess a massive endocranial system of blood vessels, called the rete mirabile, that surrounds the brain and can sometimes account for nearly 20% of total endocranial volume. Most recent studies take these vascular structures into account when estimating brain size from endocranial volume. Computed tomography (CT) has proven to be an important tool in the study of fossil endocranial features because it is nondestructive and enables more accurate, precise, and reliable measurement of endocranial features than traditional methods.

III. Brain Sizes in Fossils and Modern Species

A. Brain Size Evolution in Archaeocetes

The fossil record of early cetacean brain evolution includes the transition from the immediate land ancestor of cetaceans to the extinct aquatic forms known as archaeocetes. The best candidates for the cetacean land ancestor are the Paleocene and early Eocene Mesonychia. With EQs from 0.18 to 0.51, the mesonychians were not particularly encephalized compared with modern mammals. The range of EQs estimated for early, middle, and late archaeocetes is from 0.25 to 0.49 (Table I and Fig. 1). Therefore, archaeocetes appear to have experienced little increase in encephalization above their land precursors.

B. Brain Size Evolution in Odontocetes

The suborder Odontoceti appeared during the early Oligocene and radiated rather dramatically in that epoch and into the Miocene. Despite the existence of a fair number of specimens, there is surprisingly little systematic data on brain size in fossil odontocetes. Those data that do exist suggest that by the early-mid Miocene at least several odontocete species possessed encephalization levels substantially above that of archaeocetes and within the midrange of living species (Table I and Fig. 1). These data imply that some important changes in brain size occurred during the Oligocene after the turnover from archaeocetes to early odontocetes. Unfortunately, there are no brain size data on odontocetes during the Oligocene. Data on brain size in odontocetes during the Pliocene and Pleistocene are likewise lacking.

The EQs of living odontocetes are generally on a par with nonhuman primates, but some species have achieved a level of encephalization second only to modern humans (EQ ~ 7.0) and equal to or above that of the recent hominid ancestor *Homo habilis* (EQ ~ 4.4). Therefore, a number of odontocetes species are significantly more encephalized than other mammals, including nonhuman primates. There is, however, a range of encephalization levels within the odontocetes. The sperm whale (*Physeter macrocephalus*), with an EQ of 0.58, is an example of an odontocete species subject to disproportionate body enlargement for which the measure of EQ is not particularly meaningful. The Delphinidae, however, are the family that contains several species with exceptionally high EQs above 4.0. These include the bottlenose dolphin (*Tursiops truncatus*), the tucuxi (*Sotalia fluviatilis*), the Pacific white-sided dolphin (*Lagenorhynchus obliquidens*), and the shortbeaked common dolphin (*Delphinus delphis*).

Odontocete brain evolution was also characterized by increased foreshortening and widening of the brain, which coincided with telescoping of the skull. There was a trend toward increased relative size of auditory processing regions, such as the acoustic cranial nerve and inferior colliculus. In living odontocetes this is evident in the larger relative size of the inferior colliculus to the analogous midbrain visual processing area, the superior colliculus. In addition, structures associated with the processing of olfactory information regressed. Furthermore, the cerebral cortex of odontocetes (and cetaceans in general) has achieved an extremely high level of gyrification. Although surface morphology is not always discernible from fossil endocasts, it is generally thought that this was not a feature of archaeocete brains.

C. Brain Size Evolution in Mysticetes

The suborder Mysticeti appeared and diversified in the Oligocene and consisted of primitive toothed taxa in addition to the earliest baleen-bearing whales. Extant groups appeared in the mid-late Miocene. There are two problems associated with examining brain size evolution in mysticetes. First, data on fossil and living mysticete brain size are scarce. This is partly due to the difficulties associated with extracting and measuring such large brains. Second, mysticete brains tend to be smaller than expected relative to body size despite their large absolute size. This is partly due to the enormous body masses achieved by mysticetes. As in the sperm whale, mysticete bodies are greatly enlarged in ways that do not necessarily require a concomitant increase in neural tissue. EQs of living mysticetes are therefore unrepresentative of actual brain enlargement, with all values falling substantially below 1.0 (Table I and Fig. 1). For this reason, although encephalization has probably occurred throughout mysticete evolution, EQ is not an appropriate measure of it in this group, particularly in comparison with terrestrial mammals. In fact, to the extent that disproportionate increases in body size have played a role in body enlargement in any fully aquatic species, EQ will be underestimated relative to terrestrial mammals.

Many of the changes in morphology and size of brain components that occurred in odontocetes also characterize mysticete brain evolution, but to a lesser extent. For instance, unlike in odontocetes, olfactory tracts have remained in some mysticete species and the hypertrophy of the auditory processing regions is not as extreme as in odontocetes.

D. Brain Size Evolution in Sirenia

Sirenian brain evolution has been markedly conservative regarding relative brain size. Fossil endocasts of early Eocene sirenians (among the earliest) were small in relation to the skull and already very similar to modern forms. Sirenian encephalization levels are among the lowest of modern mammals. According to the same formula used to derive cetacean encephalization quotients in cetaceans, the Florida manatee (*Trichechus manatus*) possesses an EQ of about 0.35 and the dugong (*Dugong dugon*) about 0.5. The EQ of the extinct Steller's sea cow (*Hydrodamalis gigas*) was approximately 0.25 (Table I). Body size enlargement explains some of the reason for these low EQs, but, given that cetaceans of the same body size possess higher EQs, not all.

Unlike odontocetes, sirenians do possess olfactory bulbs. Perhaps the most striking contrast, however, is the fact that cetacean and sirenian brains anchor the two ends of the spectrum of cortical gyrification. Whereas the cetacean cerebral cortex is thin and highly convoluted, the sirenian cortex is unusually thick and almost lissencephalic (smooth). Interestingly, despite these differences, the relative volume of the cerebral

TABLE I
Estimates of Brain and Body Weight and EQ for Some Fossil and Living Marine Mammal Species

Species	Estimated brain weight (g)	Estimated body weight (g)	EQ[a]
Order Cetacea			
Suborder Odontoceti[b]			
Family Ziphiidae			
Mesoplodon mirus	2355	929,500	1.97
M. europaeus	2149	732,500	2.11
M. densirostris	1463	767,000	1.39
Ziphius cavirostris	2004	2,273,000	0.92
Family Kogiidae			
Kogia breviceps	1012	305,000	1.78
K. simus	622	168.500	1.63
Family Physeteridae			
Physeter macrocephalus	8028	35,833,330	0.58
Family Monodontidae			
Delphinapterus leucas	2083	636,000	2.24
Monodon monoceros	2997	1,578,330	1.76
Family Lipotidae			
Lipotes vexillifer	510	82,000	2.17
Family Iniidae			
Inia geoffrensis	632	90,830	2.51
Family Platanistidae			
Platanista gangetica	295	59,630	1.55
Family Pontoporiidae			
Pontoporia blainvillei	221	34,890	1.67
Family Phocoenidae			
Phocoena phocoena	540	51,193	3.15
Phocoenoides dalli	866	86,830	3.54
Family Delphinidae			
Tursiops truncatus	1824	209,530	4.14
Lagenorhynchus obliquidens	1148	91,050	4.55
Delphinus delphis	815	60,170	4.26
Grampus griseus	2387	328,000	4.01
Globicephala melas	2893	943,200	2.39
Stenella longirostris	660	66,200	3.24
Orcinus orca	5059	1,955,450	2.57
Sotalia fluviatilis	688	42,240	4.56
Suborder Mysticeti[b]			
Family Eschrichtiidae			
Eschrichtius glaucus	4305	14,329,000	0.58
Family Balaenopteridae			
Balaenoptera physalus	7085	38,421,500	0.49
B. musculus	3636	50,904,000	0.21
Megaptera novaeangliae	6411	39,295,000	0.44
Extinct species[c]			
Family Protocetidae			
Rodhocetus kasrani	290	590,000	0.25
Family Remingtonocetidae			
Dalanistes ahmedi	400	750,000	0.29
Family Basilosauridae			
Basilosaurus isis	2520	6,480,000	0.37
Saghacetus osiris	388	350,000	0.49
Dorudon atrox	976	2,700,000	0.40
Zygorhiza kochii	745	3,351,000	0.26
Family Squalodontidae			
Prosqualodon davidi	750	880,000	0.65
Family Physeteridae			
Aulophyseter morricei	2500	1,100,000	1.90

(continues)

TABLE I *(Continued)*

Species	Estimated brain weight (g)	Estimated body weight (g)	EQ[a]
Family Argyrocetus			
Argyrocetus sp.	650	72,000	3.01
Family Eurhinodelphidae			
Schizodelphis sulcatus	368	260,000	0.72
Order Carnivora			
Family Phocidae			
Phoca vitulina	250	30,000	2.08
Pusa hispida	253	39,570	1.75
Leptonychotes weddellii	520	400,000	0.76
Order Sirenia			
Family Trichechidae			
Trichechus manatus	364	756,000	0.35
Family Dugongidae			
Dugong dugon	266	281,000	0.50
Hydrodamalis gigas	1158	7,102,500	0.25

[a]Based on a reference group of modern mammals from Jerison (1973).
[b]For living species, estimated brain and body weights are averaged across several specimens in most cases.
[c]Body weight estimates for fossil specimens are often general estimates of adult species-specific values and are not necessarily from the same specimen(s) for which brain weight estimates are obtained.

cortex in both sirenians and cetaceans is on a par with nonhuman primates.

E. Brain Size Evolution in Pinnipedia

Pinnipeds diverged from terrestrial carnivores during the early Miocene. This is a relatively more recent date than cetaceans and sirenians diverged from their land ancestors. The pinniped brain, therefore, still resembles that of terrestrial carnivores. Living pinnipeds possess EQs that hover around the average for terrestrial mammals. For instance, the ringed seal (*Pusa hispida*) possesses an EQ of 1.75, the harbor seal (*Phoca vitulina*) 2.08, and the Weddell seal (*Leptonychotes*

weddellii) 0.76 (Table I). These values are fairly representative of pinniped EQ in general. Pinniped olfactory structures are reduced but not to the same degree as in cetaceans. The cerebral cortex is highly convoluted (and more so than most terrestrial carnivores) but lies somewhere in between the extreme degrees of gyrification and thickness found in cetaceans and sirenians. The pinniped brain is somewhat more spherical in shape than in terrestrial carnivores but did not undergo the dramatic change in overall morphology exhibited in cetaceans.

IV. Conclusions

Much more information is needed before we can obtain a complete picture of patterns of brain size evolution in marine mammals. However, what does seem clear is that the different marine mammal groups evolved along distinct paths that led to a great variety of levels of encephalization in modern species. For instance, among odontocetes there was a substantial increase in encephalization in the Oligocene lineages, which has led to the existence of a number of dolphin and porpoise species with relative brain sizes challenging only the hominid mammalian line. The relationship between mass and organization must be explored further, as well as the phylogenetic and ecological factors that led to the differential development of brain size among the various marine mammal groups.

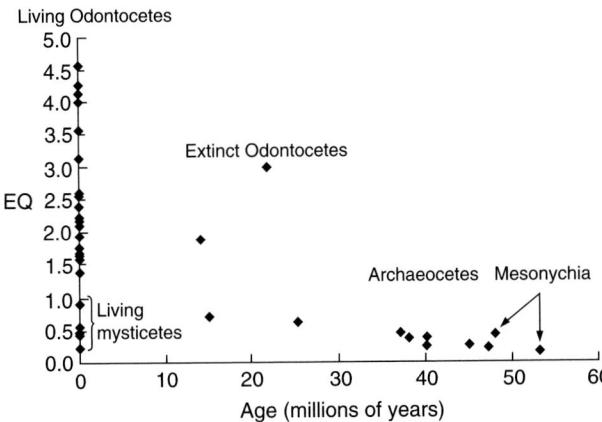

Figure 1 *Pattern of change in encephalization over geological time in Mesonychia, archaeocetes, and extinct odontocetes compared with living odontocetes and mysticetes. Encephalization is plotted as EQ where the reference group is a large sample of living mammals.*

See Also the Following Articles

Archaeocetes, Archaic ■ Brain ■ Cetacean Evolution ■ Mesonychia ■ Sirenian Evolution ■ Skull Anatomy

References

Breathnach, A. S. (1955). Observations on endocranial casts of recent and fossil cetaceans. *J. Anat.* **89,** 533–546.

Gingerich, P. D. (1998). Paleobiological perspectives on Mesonychia, Archaeoceti, and the origin of whales. *In* "The Emergence of Whales" (J. G. M. Thewissen, ed.), pp. 423–449. Plenum Press, New York.

Jerison, H. J. (1973). "Evolution of the Brain and Intelligence." Academic Press, New York.

Marino, L. (1998). A comparison of encephalization between odontocete cetaceans and anthropoid primates. *Brain Behav. Evol.* **51**(4), 230–238.

Ridgway, S. H., and Wood, F. H. (1988). Dolphin brain evolution. *Behav. Brain Sci.* **11**, 99–100.

Breaching

HAL WHITEHEAD
Dalhousie University, Halifax, Nova Scotia, Canada

Breaches are leaps from the water. The breach of a large whale is almost certainly the most powerful action performed by any animal; that of a dolphin rising many body lengths above the surface is one of the most breathtaking, but breaching is not an immediately functional activity for an aquatic animal. Observers of cetaceans have long pondered the breach, and science has made some inroads into our ignorance of breaching, but important questions remain open, including, "Why breach?"

I. What Is a Breach?

A breach may be defined as an intentional jump from the water in which at least 40% of the animal's body emerges. It is distinct from "lunging" (often the result of feeding at the surface) when less than 40% emerges and from "porpoising" when small cetaceans (and other marine mammals) reduce drag while swimming fast close to the surface by making horizontal leaps. In breaching, emergence from the water seems an intent rather than a result of another activity. While breaching is used for all cetaceans, the term leaping is more common for dolphins and porpoises.

When breaching, sperm whales (*Physeter macrocephalus*) and dolphins approach the surface vertically from depth. Other animals swimming in water less than a few body lengths deep, e.g., humpback and right whales (*Megaptera novaeangliae* and *Eubalaena* spp.), make a horizontal approach to the breach, gaining speed until, at the last moment, they raise their heads and flukes, pivoting on their flippers so converting horizontal momentum into vertical motion, and thus rising through the surface. To make a full breach, a humpback whale must break the surface at about 15 knots (~8 m/sec), close to its maximum speed. It is likely that some of the most spectacular breaches of other species also represent the full power of the animal.

After breaking the surface, whales and dolphins have many styles of breaches. In the classic breach of a large whale (such as a humpback or right) the animal emerges from the water at about 20–30° from the vertical, twisting so as to land on its back or side, having shown about 90% of its body above the surface at peak emergence (Fig. 1). However, about 20% of the breaches of sperm and humpback whales are "belly flops," with the animal landing ventrally (Fig. 2), and smaller cetaceans make what are sometimes called "clean-entry leaps" when, after leaping, the animal returns to the water smoothly, beak first (Fig. 3). Young whales and dolphins will frequently completely clear the water with their breaches (Fig. 3), sometimes by several body lengths. Some dolphin breaches, such as the spinning breaches, which give the spinner dolphin (*Stenella longirostris*) its name, are remarkably elaborate. Breaching animals produce large splashes upon reentry into the water, which can be visible at many kilometers.

Breaches are often performed in bouts. Extreme is 130 breaches in 75 min, probably all performed by the same humpback whale on Silver Bank, West Indies. As such sequences progress, animals tend to show less and less of their bodies, visibly appearing fatigued.

II. Which Whales and Dolphins Breach?

Quantitative breaching rates are only available for a few species and are usually not comparable. However, it is clear that there are substantial differences between species in the rates of this and other forms of aerial activity. Frequent breachers include the humpback, right, and sperm whales, as well as virtually all offshore dolphins. In contrast, balaenopterids (blue, *Balaenoptera musculus*; fin, *B. physalus*; sei, *B. borealis*; Bryde's, *B. edeni*; minke, *B. acutorostrata, B. bonaerensis*), most beaked whales (except the northern bottlenose, *Hyperoodon ampullatus*), porpoises, and river dolphins breach much more rarely.

Breaching rates, then, are not related to size and, at least in the large whales, are inversely related to speed—stouter, slower animals tend to breach more. Instead, interspecifically the best correlate of breaching rate is sociality. Animals found in larger groups, and for whom social structure seems more important, breach more frequently.

Figure 1 *Sperm whale breaching off the Galápagos Islands. Photograph courtesy of H. Whitehead laboratory.*

Most baleen whales have pronounced seasonal cycles, feeding in winter at high latitudes and breeding in winter nearer the equator. Humpbacks in the western North Atlantic breach about seven times more frequently on their winter breeding grounds in the West Indies than when feeding off Newfoundland.

In a number of species (including humpback, right, and sperm whales) breaching is observed more frequently when groups are merging or splitting. Male humpback whales may breach when they stop singing on the Hawaiian breeding grounds. Breaching is often observed together with lobtailing and/or flippering, with different animals performing different activities at the same time, or one animal switching between the different activities. Studies of humpback and right whales have also shown that breaches by one animal seem to trigger breaches by neighbors. On Silver Bank, breaching humpbacks form clusters about 10 km across. One of the most interesting, and in some ways unexpected, findings that has emerged from several studies of different species is that breaching rates of large whales increase with wind speed. There is currently no generally accepted explanation for this widespread pattern.

Figure 2 *Humpback whale belly flopping on Silver Bank, West Indies. Photograph courtesy of H. Whitehead laboratory.*

III. When Do Whales Breach?

The circumstances in which breaches occur can provide important clues as to their purpose or function. In some species, different segments of the population breach more frequently than others. For instance, calves of many species breach more frequently than adults. In sperm whales, the gregarious females breach more often than the much larger, and more solitary, males.

IV. When Do Dolphins Breach?

Variations in the breaching rates of dolphins seem less consistent across species. For instance, spinner dolphins off Hawaii have maximum breaching rates before a foraging hunt; in dusky dolphins (*Lagenorhynchus obscurus*) off Argentina and Peale's dolphins (*Lagenorhynchus australis*) off Chile, rates increase during hunts, peaking afterward; bottlenose dolphins (*Tursiops* spp.) off South Africa also show peak breaching rates after the hunt, but bottlenose dolphins around Coco Island, Costa Rica, breach at roughly the same frequency in all behavioral contexts.

One of the most detailed studies of breaching in dolphins is that of K. S. Norris and coworkers on the very aerobatic spinner dolphins off Hawaii. In this species, breaching was closely related to diurnal activity patterns, being mostly seen from fast-moving active schools in the afternoon.

V. Why Breach?

Few animal activities can have been ascribed so many, and so varied, functions as the breaches of whales and dolphins. Some suggested functions can now be discarded, including breaches resulting from animals being chased by swordfish (*Xiphius gladius*) or being accidental when animals travel too fast near the surface. However, the available scientific results are consistent with quite a range of other potential reasons for breaching.

A number of authors have suggested that the breach may help cetaceans feed by scaring, stunning, herding, or trapping fish or other prey. Although many, probably the majority, of breaches of most species occur in nonfeeding circumstances, the closely related activity of lobtailing is known to sometimes assist feeding, so a direct benefit to food capture cannot be ruled out as a function for some breaches, perhaps especially the "clean-entry leaps" of dolphins.

Similarly, we cannot completely rule out some rather prosaic potential benefits of breaching such as stretching, looking

Figure 3 *Hector's dolphins* (Cephalorhynchus hectori) *breaching off South Island, New Zealand. Photograph courtesy of Lars Bejder.*

above the water, or inhaling water-free air in rough weather. However, these too are unlikely to be important functions of most breaches.

One proposed benefit that is more consistent with at least some of the evidence is ectoparasite removal. Among the baleen whales, the more heavily infested species tend to be the most frequent breachers, and in Hawaiian spinner dolphins, 44% of breachers had remoras attached. However, spinners without remoras also breached, and much of the circumstantial evidence points to a very different function—communication.

As noted earlier, the more social species of cetacean breach more and, in most studies, breaching occurs most often when socially important activities are occurring, such as during changes in group composition or group activity. As sociality is based on communication between members of the same species, the strong inference from these results, then, is that breaching is a form of communication.

However, there is a paradox in that while breaches are excellent at conveying information to visually based human observers above the surface, they are far less prominent for the potential or actual social companions of the breacher. Other whales and dolphins cannot generally see the breacher's body arcing above the surface, and while the reentry makes a noticeable underwater sound, it seems to be less loud than the natural vocalizations of the animals. The paradox may be resolved by considering the theory of "honest signaling": signals are especially useful in animal communication if they convey some important attribute of the signaler that cannot be faked. The distinctive underwater sound or bubble pattern produced by a full breach, while not especially prominent in its own right, is an honest signal of the physical abilities of the breacher (which often seems to leave the surface at close to its maximum speed) and its desire to communicate this by using a significant amount of energy (about 1% of a humpback whale's estimated daily resting metabolic expenditure per breach).

Thus the breach may be a useful signal to nearby potential or actual social companions. What might it be signaling? Suggestions for large whales include aggression, "extreme annoyance" (perhaps with a nearby vessel), an "act of defiance," COURTSHIP, or a display of strength by males. Some scientists have suggested that a breach may be used to add emphasis to some other signal, perhaps a vocalization or visual display. By showing the extent of its physical prowess, and expending a significant amount of energy, the whale accentuates the importance of a companion signal. For dolphins, breaches have mainly been considered signals concerning schooling. For instance, it has been suggested that breaches may be used to define the deployment of a school, to recruit dolphins to a cooperative feeding event, or as social facilitators that reaffirm social bonds.

Finally, there is play. This is probably the most commonly attributed function of breaching by the general public, but it is also seriously considered by scientists who recognize play as a valid, but hard to define, type of BEHAVIOR. Biological definitions of play usually focus on its lack of immediate biological function, resulting in play becoming a "garbage can" into which activities without an obvious function, such as breaching, get placed in a haphazard fashion. It is likely that many breaches described by human observers as "playful" actually function as important signals. However, it is also likely that some breaches, especially those by young animals, provide no immediate benefit, instead, like other forms of play, helping equip the breacher with abilities that may be beneficial in later life.

See Also the Following Articles

Communication ▪ Leaping Behavior ▪ Lobtailing ▪ Playful Behavior ▪ Speed

References

Acevedo-Gutiérrez, A. (1999). Aerial behavior is not a social facilitator in bottlenose dolphins hunting in small groups. *J. Mammal.* **80**, 768–776.

Norris, K. S., Würsig, B., and Wells, R. S. (1994). Aerial behavior. *In* "The Hawaiian Spinner Dolphin" (K. S. Norris, B. Würsig, R. S. Wells, and M. Würsign, eds.), pp. 103–121. Univ. of California Press, Berkeley, CA.

Pryor, K. (1986). Non-acoustic communicative behavior of the great whales: Origins, comparisons, and implications for management. *Rep. Intl. Whaling Commission* (special issue) **8**, 89–96.

Waters, S., and Whitehead, H. (1990). Aerial behaviour in sperm whales, *Physeter macrocephalus. Can. J. Zool.* **68**, 2076–2082.

Whitehead, H. (1985). Humpback whale breaching. *Invest. Cetacea* **17**, 117–155.

Whitehead, H. (1985). Why whales leap. *Sci. Am.* **252**, 84–93.

Breathing

Douglas Wartzok

Florida International University, Miami

Oxygen is the final electron receptor in the metabolism of marine mammals as it is in all other mammals. Marine mammals obtain oxygen from the air they breathe at the surface. In contrast, most feeding, mating, and other activities essential to survival occur beneath the surface. Thus most marine mammals minimize the time they are at the surface and have evolved to load oxygen quickly and use it efficiently. For marine mammals, the breath-holding portion of the breathing cycle is significantly extended compared to the oxygen intake portion.

Because a distinguishing feature of marine mammals is their breath-holding ability, this characteristic has received much more attention than their breathing. However, a number of aspects of marine mammal breathing have necessarily been modified from terrestrial mammals in order to accommodate their submerged lifestyles. As with many modifications from terrestrial mammals, those related to breathing show the greatest difference in those species that dive the deepest and the longest.

Cetaceans, sea lions, and manatees usually take only one breath per surfacing. Manatees return to a normal, shallow dive after a single breath. The deeper-diving cetaceans take a series of breaths, each in a subsequent surfacing, before another dive. Seals remain at the surface for a series of breaths after a dive.

I. Lung Oxygen Stores

Every inspiration that fills the lungs with air brings in four times as much nitrogen as oxygen. Because nitrogen is neither bound to a carrier in the blood nor metabolized in the tissues, the partial pressure of nitrogen in the blood will equilibrate with that in the lungs. If gas exchange is allowed to take place during a dive, the resulting higher partial pressure of nitrogen in the blood and tissues will result in the formation of nitrogen gas bubbles when the external pressure is reduced as the animal comes to the surface. Thus deep-diving marine mammals limit the exchange of gas from lungs to blood during dives.

Most phocids have been observed to exhale before diving. Weddell seal (*Leptonychotes weddellii*) pups, which are observed to dive after an inspiration, are an occasional exception. At this developmental stage they are shallow divers. California sea lions (*Zalophus californianus*) may dive after a partial inspiration, but then vent air during descent. Dolphins and porpoises making shallow dives routinely dive on inspiration. These breathing behaviors correlate well both with the proportion of total oxygen stores in the lung at the start of a dive and with lung size in proportion to the body size of various marine mammals. Phocids dive with 7% of total oxygen stores in the lung, fur seals with 13%, and delphinids with 22% (Fig. 1). The proportionate size of the lung in phocids and manatees is about the same as in terrestrial mammals such as the horse and human (Fig. 2), whereas the lung size is greater than expected in delphinids and smaller than expected in whales. An outlier in these considerations is the sea otter (*Enhydra lutris*) whose lung is close to three times the expected size for its body mass and accounts for 75% of oxygen stores. The sea otter is not a deep-diving marine mammal, and the relatively large lung in the sea otter may be primarily used for buoyancy when the animal is resting at the surface.

II. Tidal Volume

The tidal volume of marine mammals is a larger proportion of the total lung capacity (TLC) than it is of terrestrial mammals. In a typical terrestrial mammal the volume of air inhaled and exhaled in one breath is in the range of 10 to 15% of TLC. In marine mammals, tidal volume is typically greater than 75% of TLC. The maximum tidal volume or vital capacity (VC) in terrestrial mammals is no more than 75% of TLC, whereas in marine mammals the VC can exceed 90% of TLC. Several factors contribute to the large tidal volume in marine mammals. Marine mammal lungs contain more elastic tissue than those of terrestrial mammals. The ribs contain more cartilage and are thus more compliant than those of terrestrial mammals. The lung is also more compliant. Marine mammal lungs can collapse and reinflate repeatedly, whereas in terrestrial mammals, lung collapse is a serious situation that requires intervention to reinflate. Although both terrestrial mammals and marine mammals inspire actively and expire passively, the features noted earlier allow a much greater elastic recoil of the lungs, chest cavity, and diaphragm, and thus a greater tidal volume in proportion to TLC.

The terminal portions of the airways in all marine mammals are supported and reinforced by cartilage or muscle. One purpose of this reinforcement is to provide a less collapsible region into which alveolar gases can be forced during a dive to prevent gas exchange with blood at high pressures. This prevents increased nitrogen tensions in the blood and tissues as noted previously. A second purpose of the reinforcement is to keep the terminal airways open even at high flow rates of gases in and out of the lung during a breath and to allow high expiratory flow rates even as the lung volume decreases. Figure 3 shows the flow volume profile comparison during exhalation between a

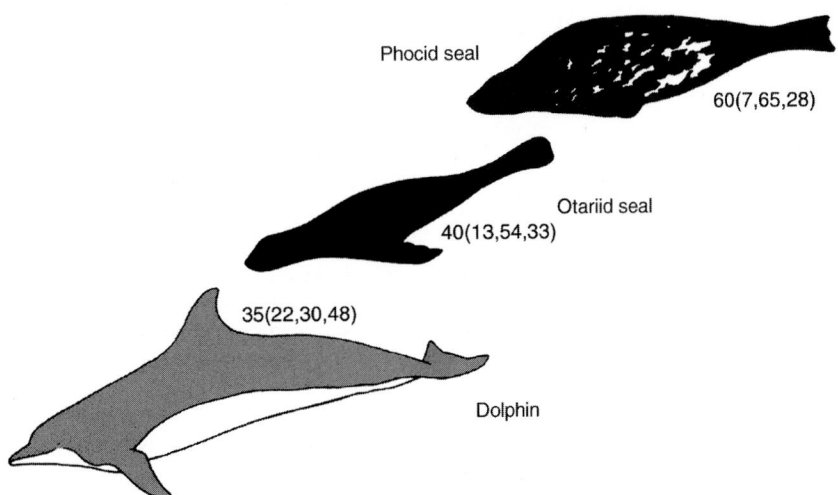

Figure 1 *Generalized total oxygen store of major taxa of marine mammals expressed in ml O_2 kg^{-1}. Numbers in parentheses are the percentage of total oxygen store found in the lungs, blood, and muscle, respectively. Modified, with permission, from Kooyman, G. L., "Diverse Divers: Physiology and Behavior." © 1989 Springer-Verlag.*

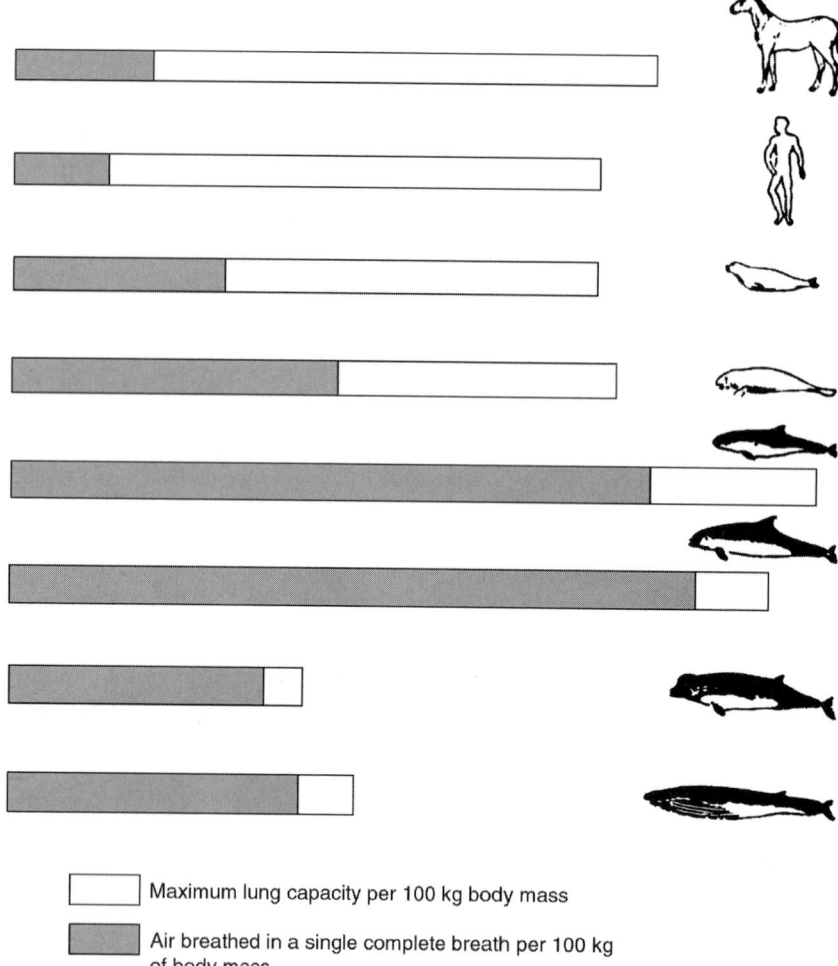

Maximum lung capacity per 100 kg body mass

Air breathed in a single complete breath per 100 kg of body mass

Figure 2 *The maximum amount of air the lungs can hold, and the amount of air breathed in and out with each breath, calculated per 100 kg of body mass for a horse, a human, a seal, a manatee (Trichechus* sp.*), a harbor porpoise (Phocoena phocoena), a bottlenose dolphin, a bottlenose whale (Hyperoodon* sp.*), and a fin whale (Balaenoptera physalus). Modified, with permission, from Slijper, E. J. "Whales and Dolphins." © 1976, University of Michigan.*

harbor porpoise and a human. There are two striking differences. First, the flow rates are much higher in terms of VC/sec. Second, the flow rates remain very high; even down to a small fraction of the VC. These two factors together allow very rapid exhalation of the full VC. Inspiration takes somewhat longer.

The bottlenose dolphin (*Tursiops truncatus*) completes an exhalation and inhalation cycle in approximately a third of a second. With a tidal volume of 10 liters, flow rates through the air passages can be as high as 70 liters/sec. In gray whale (*Eschrichtius robustus*) calves the duration of expiration and inhalation is closer to half a second, but the tidal volume can be as great of 62 liters, and the maximum flow rate as great as 202 liters/sec. Gas flows through the external nares at speeds up to 44 m/sec during inspiration and 200 m/sec during expiration.

Cetaceans usually initiate expiration prior to the blowholes breaking the surface. The explosive nature of the expiration creates the small droplets that make the blow visible and clears the upper respiratory passages and the area around the blowholes of any residual water. Most of the time the blowholes are above the surface is used for inspiration.

The large tidal volume allows for more oxygen loading and greater carbon dioxide unloading during a single breath at the surface. Even in a resting state, the carbon dioxide content of expired air in seals is twice as great as it is in humans. After extended breath holds, alveolar oxygen levels can be as low as 1.5%. The oxygen and carbon dioxide content of expired air after surfacing can provide indirect evidence of physiological adjustments to diving. In bottlenose dolphins, the oxygen content in the first

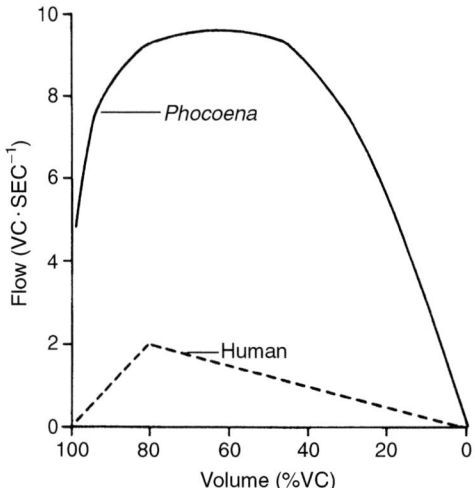

Figure 3 *Comparison of the flow–volume curves of a human (dashed line) and a harbor porpoise (solid line). Note that in the human, after the volume falls below about 80% of vital capacity, the flow rate declines steadily, but this is not the case in the porpoise. Modified, with permission, from Kooyman, G. L., and Sinnett, E. E., Mechanical properties of the harbor porpoise lung,* Phocoena phocoena, *Respiration Physiology 36, 287–300, © 1979, Elsevier Science.*

breath after a dive to 200 m is greater than it is in the first breath after an equivalent amount of swimming at 20 m. The interpretation is that the collapse of lungs in the deeper dive prevented the exchange of oxygen with the blood during the dive. For the same reason, the content of carbon dioxide in the first breath is always less after a dive to depth than after a dive near the surface. In gray seals (*Halichoerus grypus*), the end tidal partial pressure of oxygen in the first exhalation after surfacing is similar to that in the last breath before submergence, again indicating that the lungs were collapsed at depth and there was no gas exchange.

III. Hyperventilation

Marine mammals often hyperventilate both before and after a dive. Hyperventilation before a dive leads to increased oxygen tensions and reduced carbon dioxide tensions. Hyperventilation arises through an increase in the frequency of breathing and the tidal volume. Although both can initially increase during hyperventilation, lung dynamics requires an eventual reciprocal relationship between tidal volume and frequency of breathing. In Weddell seals, hyperventilation of five to six times resting is accomplished by increasing the tidal volume 1.5 to 2 times and the respiratory frequency 3 times. If the respiratory frequency rises above 25 breaths/min, the tidal volume, as a proportion of the TLC, is close to that of terrestrial mammals. Harp seals with respiratory rates of 27 breaths/min and harbor seals (*Phoca vitulina*) with rates of 35 breaths/min have tidal volumes between 20 and 30% of TLC. Because marine mammals normally have such large tidal volumes, they have much less ability to increase total ventilation in response

to exercise than terrestrial animals. A human can increase respiration by 4 times and can increase tidal volume by greater than 4 times for an overall increase in ventilation of more than 16 times resting compared to the 5 to 6 times resting maximum for a Weddell seal.

Marine mammals often exhibit an increase in heart rate on approach to the surface. It has been suggested that this anticipatory tachycardia coincides with the restored perfusion of peripheral tissues so the oxygen levels in the blood drop even lower as the carbon dioxide levels rise. These changes in blood gases increase the gradient between partial pressures in the blood and the lungs and lead to more rapid oxygen uptake and carbon dioxide exhausting during the first breaths at the surface. When gray seals show no anticipatory tachycardia, they do not achieve the maximum rate of oxygen uptake during the first few breaths.

The increased heart rate and breathing on surfacing lead to a rapid restoration of oxygen tensions. In fact, in Weddell seals the blood oxygen partial pressures in the postdive recovery period routinely end up exceeding the resting values. It appears that the purging of carbon dioxide is more critical for determining readiness to dive again than is the replenishing of oxygen. In Weddell seals the partial pressure of carbon dioxide falls to resting levels within a couple minutes after aerobic dives, but may take up to 10 min to reach resting levels after long dives, which rely on anaerobic metabolism. The fact that hyperventilation can continue for an hour after the partial pressures of oxygen and carbon dioxide return to resting levels indicates that prolonged hyperventilation is driven more by lactate-induced changes in the acid–base balance in the blood.

IV. Ventilation Control

Seals show a more fully developed mammalian reflexive response of increased ventilation to decreases in inspired oxygen concentration or increases in inspired carbon dioxide concentration at an earlier age than terrestrial mammals. However, adult seals show little ventilatory response to decreased oxygen concentration. Instead the adult seals respond by increasing the proportion of time they are at the surface relative to the total time between dives. Apparently the adult seals are able to substitute behavioral control of diving patterns for reflexive control of ventilation.

Breathing hyperoxic gas increases the dive time of Weddell seals, has no effect on the dive time of manatees, and shortens the dive time of hooded seals. A suggested explanation for the latter surprising finding is that the seal breathing a hyperoxic gas mixture dove before it had exchanged all the carbon dioxide and the increased carbon dioxide tensions resulted in the shortened dive time.

Both the nostrils of pinnipeds and the blowhole of cetaceans are normally closed when the controlling muscles are relaxed. The closure in pinnipeds is maintained by muscle tone and pressure of the moustacial pad. Contraction of the nasal and moustacial pad muscles results in a movement of the pad and opening of the nostrils. In cetaceans, muscles must contract to open the blowhole and to move the nasal plug so that it is not blocking the airway.

Pinnipeds on land often show a breathing pattern similar to that during diving with breathing periods (eupnea) being shorter than breath hold periods (apnea). The ratio between apnea and eupnea while on land is greatest in those species that normally dive for the longest periods. The periods of apnea also tend to be longer when the animals are asleep than when they are awake. Even the longest bouts of sleep apnea appear to be aerobic: plasma lactate and glucose remain stable even though oxygen tensions drop to very low levels, carbon dioxide tensions increase, and respiratory acidosis occurs. In elephant seal (*Mirounga* spp.) pups, awake apnea does not exceed about 5 min whereas sleep apnea can be as long as 14 min. The pups show a parallel increase in mean sleep apnea and mean dive duration during their first year of life.

Pups of several species have been observed to have a higher breathing rate than adults. Weddell seal pups take 16 breaths/min compared to the 8 breaths/min rate of adults. Pups of Australian (*Neophoca cinerea*) and New Zealand (*Phocarctos hookeri*) sea lions take 13 breaths/min whereas adults of these species typically breathe 3 to 5 times per minute.

On land and in the water different stages of sleep in pinnipeds are associated with different breathing patterns in different species. During rapid eye movement sleep, breathing is regular and at rates up to 16/min in gray seals, irregular in harp seal pups, and absent in elephant seal pups. Elephant seal pups sleeping in shallow water rise to the surface to breathe without showing brain wave patterns associated with wakefulness. In contrast, Caspian seals (*Pusa caspica*) sleeping below the surface awake prior to surfacing and breathing.

Some species of delphinids show unihemispheric brain waves associated with sleep. Thus there is one cerebral hemisphere that is always awake to control surfacing and breathing patterns. Northern fur seals (*Callorhinus ursinus*) sleeping in water also sometimes show unihemispheric sleep patterns with one cerebral hemisphere awake to control surfacing and breathing. In contrast to delphinids, no pinniped has shown exclusively unihemispheric sleep brain waves.

V. Oxygen Loading and Dive Time

Several authors have modeled the diving behavior of marine mammals based on oxygen loading curves at the surface compared with energy expenditure while below the surface. There have been various models based on what the animal may be attempting to maximize, be it time in a deep prey patch, gross energy intake during a dive, net energy intake, energetic efficiency, etc. All the models conclude that there should be some relationship between the duration of a dive and the time breathing at the surface, either predive or postdive. Although some species do show such relationships over certain dive time intervals, not all species show the expected patterns. The time Weddell seals spend at the surface is independent of preceding dive duration up to the aerobic dive limit. Beyond that time, the surface time increases exponentially with the preceding dive time. However, gray seals show a direct proportionality between dive time and surface time up to dive times of 7 min. Surface time is independent of dive time for dive times greater than 7 min. Surface time of sperm whales (*Physeter macrocephalus*) shows a slight trend to decrease

with increasing dive time, but is basically independent of dive time. Elephant seals can maintain, over periods greater than 24 hr, a pattern of long, deep dives followed by surface intervals of 3 min or less. Some of these differences are attributable to species variation among groups with different diving strategies and oxygen loading needs. Additional explanations of the breakdown of models relating surface oxygen uptake to underwater duration and activity include lowered metabolic rates underwater, passive gliding descents, and animals not maximizing any of the foraging-related parameters. For example, ringed seals (*Pusa hispida*) appear to be constrained by a risk aversive strategy. Diving under shore fast ice, ringed seals gain access to air only at a few breathing holes. If a seal finds a breathing hole occupied by another seal or detects a polar bear above the hole, it will need oxygen reserves to locate an alternate breathing site.

VI. Water Conservation during Breathing

Because marine mammals obtain most of their water requirements from their prey and through metabolic production of water, conservation of water is an important adaptation in marine mammals. Renal adaptations for water conservation are discussed elsewhere, but water can also be lost through ventilation. Both pinnipeds and cetaceans exhale air that is not saturated with water vapor. In bottlenose dolphins the respiratory water loss is reduced by 70% over what it would be compared to a terrestrial mammal breathing dry air. Countercurrent heat exchange and induced turbulence in the nares and nasal sac system allow for the extraction of a majority of the water vapor in the air coming from the lungs. In seals, the bones in the anterior part of the nasal cavity (turbinates) create a very dense mesh through which the expired air must pass. Moist air flowing over this large surface area gives up much of this water before being exhaled.

VII. Breathing Patterns in Response to Disturbance

Changes in breathing patterns have been used extensively as indicators of disturbance of marine mammals in response to human activity. In many cases, a statistically significant change has been observed in the interbreath interval, total number of breaths during a surfacing, or proportion of time spent at the surface. Although these changes may be statistically significant, it is questionable whether they are biologically significant for an individual animal or indicative of any long-term consequences for the population.

See Also the Following Articles

Anatomical Dissection: Thorax and Abdomen ∎ Blowing ∎ Brain ∎ Circulatory System ∎ Diving Physiology

References

Butler, P. J., and Jones, D. R. (1997). Physiology of diving of birds and mammals. *Physiol. Rev.* **77**, 837–899.

Kooyman, G. L. (1973). Respiratory adaptations in marine mammals. *Am. Zool.* **13**, 457–468.

Kooyman, G. L. (1989). "Diverse Divers: Physiology and Behavior." Springer-Verlag, Berlin.

Kooyman, G. L., and Sinnett, E. E. (1979). Mechanical properties of the harbor porpoise lung, *Phocoena phocoena. Resp. Physiol.* **36,** 287–300.

Ridgway, S. H. (1972). Homeostasis in the aquatic environment. *In* "Mammals of the Sea: Biology and Medicine." (S. H. Ridgway, ed.), pp. 590–747. Charles C. Thomas, Springfield, IL.

Slijper, E. J. (1976). "Whales and Dolphins." Univ. of Michigan, Ann Arbor.

Breeding Behavior

SEE *Mating Systems; Reproductive Behaviors*

Breeding Sites

MARK A. HINDELL
University of Tasmania, Hobart, Australia

Giving birth to young and the subsequent nursing of those young present unique problems to marine mammals. The strategies that marine mammals employ to deal with these problems can be divided into two groups: (a) those animals that need to leave the water to breed (seals and polar bears, *Ursus maritimus*) and those that remain at sea to breed (cetaceans, sirenians, and sea otters, *Enhydra lutris*). Whichever strategy is used, a crucial component of the reproductive process is the site used for breeding (here, the term breeding is restricted to parturition and suckling and does not include mating). Breeding sites used by marine mammals are quite diverse, both in terms of geography and physical characteristics, ranging from polar to equatorial regions and from sandy beaches to deep ocean basins. Which sites are used for breeding by a particular species is determined by a complicated mixture of factors, including evolutionary history, requirements of the young, requirements of the adults, biological characteristics (such as the proximity of prey), and physical characteristics (such as water temperature or beach substrate).

I. Species That Leave the Sea to Breed

All species of seals and polar bears leave the sea in order to give birth, and most also need to be ashore to suckle their young. Some species (predominantly the phocids) remain ashore for the entire lactation period, whereas in others (the otariids), the females regularly return to sea to forage during lactation. Of the phocids, only harbor seals (*Phoca vitulina*) and walruses (*Odobenus rosmarus*) have been reported to suckle young at sea. The sites used by these animals for breeding activities can be divided into two groups: those that breed on land and those that breed on ice.

A. Marine Mammals That Breed on Land

Pregnant polar bears leave the arctic pack ice with the onset of spring to breed on land. The deteriorating summer pack ice is too unstable a substrate for the bears, which have a rela-

tively long period of cub dependence. Females excavate caves in the side of riverbanks or hillsides, thereby getting close to the permafrost and providing an environment with a stable temperature. As summer progresses and snowfall increases, these caves get snowed in and the females need to maintain a cavity with sufficient wall thickness to provide insulation, but also thin enough to allow the passage of air.

About twenty species of seals breed on land. This includes all of the species of otariids (fur seals and sea lions) and six of the phocids (true seals) (Table I). None of these species use any kind of shelter and all give birth and suckle on the beach exposed to wind, rain (or snow), and waves. Land-breeding seals often occur in very large aggregations during the breeding season, which may offer some protection for young seals from the weather, but this also puts them at risk of being damaged in fights between adult seals. Land-based breeding sites occur at all but the most extreme polar latitudes.

Some species utilize both land- and ice-breeding sites. Gray seals (*Halichoerus grypus*) breed on beaches in northern Europe and America, but on pack ice in Northern Canada. Although primarily a land-breeding species, harbor seals also breed in pack ice in northern Canada.

Most land-breeding seals use islands as their principal breeding sites, with only a few species utilizing mainland beaches. This is likely to be an attempt to avoid large, mainland preda-

TABLE I
Land-Breeding Marine Mammals, Including the Primary Geographic Type of Breeding Site

Species	Geographic type[a]
Otariids	
Antarctic fur seal, *Arctocephalus gazella*	Island
Galápagos fur seal, *A. galapagoensis*	Island
Guadalupe fur seal, *A. townsendi*	Island
Juan Fernández fur seal, *A. philippii*	Island
New Zealand fur seal, *A. forsteri*	Island/mainland
South African/Australian fur seal, *A. pusillus*	Island/mainland
South American fur seal, *A. australis*	Island/mainland
Sub-antarctic fur seal, *A. tropicalis*	Island
Northern fur seal, *Callorhinus ursinus*	Island
Australian sea lion, *Neophoca cinerea*	Island/mainland
California sea lion, *Zalophus californianus*	Island/mainland
Galápagos sea lion, *Z. wollebaeki*	Island
Japanese sea lion, *Z. japonicus*	Island/mainland
New Zealand sea lion, *Phocarctos hookeri*	Island/mainland
Southern sea lion, *Otaria flavescens*	Island/mainland
Steller sea lion, *Eumetopias jubatus*	Island
Phocids	
Hawaiian monk seal, *Monachus schauinslandi*	Islands
Mediterranean monk seal, *M. monachus*	Island/mainland
Northern elephant seal, *Mirounga augustirostris*	Island/mainland
Southern elephant seal, *M. leonina*	Island/mainland
Harbor seal, *Phoca vitulina*	Island/mainland/ice
Gray seal, *Halichoerus grypus*	Island/mainland/ice

[a]From Reidman (1990).

tors, including humans. Generally, species have quite specific requirements of their island-breeding sites, and so suitable island-breeding sites are often limited. Consequently, those sites that are suitable tend to hold very large numbers of seals, which provides an ideal condition for the evolution of polygeny.

Different species have different substrate requirements. Most species use gradually sloping sandy beaches, such as those used by northern elephant seals. Fur seals, which are generally more agile than phocids, can also breed on rocky substrates, but again this is quite species specific. For example, at Madquarie Island, where both Antarctic (*Arctocephalus gazella*) and sub-Antarctic (*A. tropicalis*) fur seals breed, *A. gazella* use open beaches, whereas *A. tropicalis* breed on nearby rocky headlands. The reasons for these different preferences are unknown, but may arise from resource partitioning.

Proximity to a food source may be another factor that determines the location of breeding sites for some land-breeding seals. For example, female otariids need to return to sea to replenish their energy reserves regularly during the lactation period. As their pups are fasting during these trips to sea, faster growth and heavier weaning masses can be achieved if the foraging trips are kept as short as possible. Breeding sites are often located close to the continental shelf break or other oceanographic features that tend to have enhanced primary productivity. In cases where this is not possible, such as the sub-Antarctic fur seal breeding site on Amsterdam Island, female foraging trips are much longer than in other species (or populations) and the pups have correspondingly lower growth rates.

Land-breeding phocid seals generally have no requirement for feeding during lactation, and therefore the breeding sites can be located considerable distances from the foraging sites. Southern elephant seals (*Mirounga leonina*), for example, tend to feed in high-latitude waters, but breed thousands of kilometers away on sub-Antarctic islands. In this case the primary requirement of a breeding site is suitable beach structure and perhaps a moderate climate to help the pup in its early life.

B. Marine Mammals That Breed on Ice

Thirteen species of pinnipeds breed on ice, either floating pack ice or fast ice attached to land (Table II). Ice-breeding seals tend to be monogamous, and this is likely to be a consequence of the breeding habitat. Unlike suitable beaches for land-breeding seals, ice is not a limited resource, and females can haul out anywhere to breed, so aggregations of females tend not to occur. This limits the opportunities that males have to monopolize access to several females, and the best strategy for them is to find a female and remain with her until estrus. Weddell seals (*Leptonychotes weddellii*) are an exception to this rule, as they breed on fast ice, with limited access to open water. Female Weddell seals therefore tend to aggregate around tide cracks and other sources of permanent open water.

Most ice-breeding seals give birth and suckle their young on the ice. Exceptions to this rule are ringed seals (*Pusa hispida*) and Baikal seals (*P. sibirica*), which can use ice lairs under ice ridges. These lairs afford some protection from the elements and perhaps from predators, although polar bears are adept at locating and breaking into these lairs.

TABLE II
Ice-Breeding Marine Mammals, Including the Primary Ice Type

Species	Ice type[a]
Obodenids	
Walrus, *Odobenus rosmarus*	Pack ice
Phocids	
Weddell seal, *Leptonychotes weddellii*	Fast ice
Ross seal, *Ommatophora rossii*	Pack ice
Crabeater seal, *Lobodon carcinophaga*	Pack ice
Leopard seal, *Hydrurga leptonyx*	Pack ice
Hooded seal, *Cystophora cristata*	Pack ice
Harp seal, *Pagophilus groenlandicus*	Pack ice
Ribbon seal, *Histriophora fasciata*	Pack ice
Baikal seal, *Pusa sibirica*	Fast ice
Caspian seal, *P. caspica*	Pack ice
Ringed seal, *P. hispida*	Fast ice
Harbor seal, *Phoca vitulina*	Pack ice/land
Larga seal, *Phoca largha*	Pack ice
Gray seal, *Halichoerus grypus*	Pack ice/land
Bearded seal, *Erignathus barbatus*	Pack ice

[a]From Reidman (1990).

Although pack ice is the breeding substrate, many species have preferred geographic regions to which they move for the breeding season. Harp seals (*Pagophilus groenlandicus*), for example, migrate to the southerly edge of the pack ice and occupy only a small part of their overall range during the breeding season. The biggest areas are around Newfoundland and off western Greenland. Hooded seals (*Cystophora cristata*) show a similar pattern of migration, moving from widespread northerly feeding areas to more proscribed breeding areas, which appear to be associated with the continental shelf at the southern extent of the summer pack ice. This may provide the adults with access to the shelf for feeding immediately after weaning their pups.

II. Species That Stay at Sea to Breed

Three groups of marine mammals are sufficiently adapted to a marine existence that do not need to leave the water to give birth or suckle their young. These are the cetaceans, the sirenians, and the sea otters. By giving birth at sea, these animals are no longer constrained to use what is essentially a limiting resource: land. Nonetheless, many species still have quite specific requirements of their breeding sites and migrate thousands of kilometers to reach their breeding grounds. Because such profound separation of feeding and breeding sites incurs large energetic costs, the specific characteristics of these areas must be of considerable importance.

Dugongs (*Dugong dugon*) and manatees (*Trichechus* spp.) have no specialized breeding site requirements, although newborns may be kept close to shore or in protected bays or inlets. Young are born and remain with their mother while she forages on sea grass beds. Likewise, sea otters tend to give birth at sea (often among kelp beds) and the cubs remain with their mothers. Neither group appears to migrate to specific regions for breeding and remain within their foraging areas.

Cetaceans are the only group of ocean-breeding marine mammals in which some species have clear separation of feeding areas and breeding areas. Within the cetaceans, breeding sites can be loosely categorized as (i) coastal, (ii) open ocean, or (iii) nonspecific. Species with nonspecific-breeding sites are those that show no evidence of requiring different environments for breeding. This group contains most of the odontocetes and several of the mysticetes. Many of the smaller odontocete species do make seasonal migrations, but these are not clearly linked to breeding and seem more related to changes in prey distribution.

Several species of beaked whales appear to have year-round high-latitude distributions and do not migrate for breeding, giving birth and suckling their young in polar waters. This behavior is also seen in two mysticete species; the common minke (*Balaenoptera acutorostrata*) and the bowhead (*Balaena mysticetus*) whales. Bowhead whales are never far from pack ice, including during the breeding season.

A. Cetaceans with Coastal Breeding Sites

Humpback whales (*Megaptera novaeangliae*) have several recognized breeding grounds and all are associated with coastal regions. The Southern Hemisphere populations use either the coast of the major southern continents (South Africa, Australia, or South America) or smaller oceanic islands such as Tonga and Fiji, with the preferred sites generally north of 30°. The Northern Hemisphere humpbacks move to the Caribbean, Hawaii, or Cape Verde Islands.

Gray whales (*Eschrichtus robustus*) move south from arctic feeding grounds to breed in Baja California and, as with humpbacks, the breeding sites are close inshore. In fact, they are so inshore that they are close enough to be seen from land and form the basis of a tourist industry.

Of the inshore breeding cetaceans, the right whales (*Eubalaena* spp.) are least migratory. Neither the southern (*E. australis*) or the northern right whale species (*E. glacialis* and *E. japonica*) make long migrations to their breeding sites, but all three species favor coastal sites and sheltered bays for giving birth. Individual southern right whales show strong breeding site fidelity and use the same bays on several consecutive breeding events.

Only one odontocete species, the beluga (*Delphinapterus leucas*), seems to have an inshore migration, and often breeds in shallow inshore waters. Why these species seek out inshore waters in not really known. Aside from the thermal advantages common to all migrating whales (see later), the specific advantages associated with inshore breeding are likely to be related to environmental conditions and predator avoidance.

B. Cetaceans with Oceanic Breeding Sites

Less is known about the characteristics of the breeding sites for these species due to their lack of coastal aggregations, so much of what is known comes from early tagging studies and whaling records.

Fin whale (*Balaenoptera physalus*) breeding sites are widely spread over oceanic waters in temperate and subtropical waters. Some fin whales move toward land and even form loose aggregations, but they do not cross the continental shelf, remaining in deep water. The little information available for blue whales suggests that they use similar breeding sites to fin whales. This may also be the case for sei whales (*B. borealis*), although they may not penetrate tropical waters. Brydes whales (*B. edeni*) do not appear to migrate to breed, presumably this is a reflection of this species' largely tropical habitat all year round.

There is ongoing debate regarding the reasons for the use of temperate or equatorial breeding sites by mysticete whales. Thermal and energetic advantages to newborn calves is one possibility, but if so, why do some species such as the minke and bowhead whales, as well as many smaller odontocetes, remain in high-latitude waters to breed? An alternative view is that the abundant and predictable food supply of food at high latitudes declines over the dark winter months, but only species with a large body size (and lower mass specific metabolic rates) are able to make the long migrations that allow them to take advantage of warmer waters.

See Also the Following Articles

Habitat Pressures ■ Mating Systems ■ Migration and Movement Patterns ■ Parental Behavior ■ Predation on Marine Mammals

References

Riedman, M. (1990). "The Pinnipeds: Seals, Sealions and Walruses." University of California Press, Berkeley.

Reynolds, J. E., and Rommel, S. A. (1999). "The Biology of Marine Mammals." Smithsonian Institute Press.

Bryde's Whales
Balaenoptera edeni and *B. brydei*

HIDEHIRO KATO
National Research Institute of Far Seas Fisheries, Shimizu, Japan

B ryde's whales are the least known of the large baleen whales. We are not even sure how many species are represented in this complex of temperate and tropical whales.

I. Characters and Taxonomic Relationships

Bryde's whales were long confused with sei whales (*Balaenoptera borealis*) because of morphological similarities; this confusion lasted into the 1970s. Bryde's whales were first described by Anderson (1878) based on examination of a stranded balaenopterid on Thaybyoo Creek beach, Gulf of Marataban,

Burma. He gave it the scientific name *Balaenoptera edeni* in honor of the British high commissioner in Burma, Sir Ashley Eden. Olsen (1913) found another new species among "sei whales" caught in Durban, South Africa, and gave it the scientific name *B. brydei* in honor of his sponsor Johan Bryde, a pioneer in South African whaling from the traditional whaling port of Sandefjord. Junge (1950) concluded that the two names were synonymous based on examination of a skeleton collected in Pulu Sugi, Singapore. Further studies by Omura (1959) and Best (1960) supported Junge's view, and their conclusion had been generally accepted until recently, with *B. edeni* having priority as the scientific name. (The common name remained "Bryde's whale" probably due to its wide popularity.) However, today it is not clear how many species of Bryde's whale-like baleen whales there are or what their scientific names should be (Rice, 1998). Wada and Numachi (1991) found that a small form occurring off the Solomon Islands and Java did not accord with other Bryde's whales in allozymes. These results suggesting the existence of two species were supported by mtDNA analyses reported by Yoshida and Kato (1999). Rice (1998) proposed provisional recognition of the existence of two species, with the nomenclature unsettled. The problem is that it is not yet known to which of the two species the holotype specimen of *B. edeni* (in Calcutta) belongs. It is not distinctive morphologically (at the upper end of the size range for the small form, but still physically immature) and comes from an area of overlap of the two species. If GENETIC analysis determines that it belongs to the small species, then the small species will bear the name *B. edeni*. If it proves to be of the large species, then that species will bear the name, *B. brydei*, will fall into synonymy, and a new name (and holotype specimen) will be needed for the small species.

Bryde's whales are medium-sized balaenopterids. Females are larger than males throughout life, by about 2 feet (0.5–0.6 m) at full maturity. It is believed they reach 15.5 m, but most are much smaller. As demonstrated by Best (1977) for South African Bryde's whales (Table I) animals from coastal stocks or stocks inhabiting rather areas are generally smaller than those from migratory pelagic stocks. Southern Hemisphere animals are also larger than Northern Hemisphere animals. In the South Africa and western North Pacific stocks, body length increases rapidly until 4–5 years, reaches about 90% of asymptotic length for both sexes at about 10 years, and ceases to increase at about 20–25 years (Best, 1977). The length–weight relationship is given by the equation (Ohsumi, 1980):

$$W = 0.0126 \, L^{2.76},$$

where W is body weight in metric tons and L is body length in meters.

If mean lengths at physical maturity in the western North Pacific stock are substituted into the equation, weight estimates are 15.0 (at 13.0 m) and 16.6 (at 13.5 m) tons for males and females, respectively.

Bryde's whales closely resemble sei whales but have a number of distinctive characteristics (Fig. 1). Body color is principally dark smoky gray above and white below, but the dark area extends down to include the throat grooves and the flippers. The boundary between dark and light areas is diffuse. The rostrum is V shaped as in other balaenopterids (Fig. 1). The head occupies about 24–26% of the body. The dorsal fin is extremely falcate with a tapering tip and is located at 25.2–26.6% of body length anterior to the flukes. The flukes are broad (23.5–24.4% of body length), with rather straight posterior margins.

TABLE I
Distinguishing Characteristics of the Two Forms of Bryde's Whales off Donkergat[a]

Characteristic	Inshore form	Offshore form
Distribution	<20 miles off the coast	>50 miles off the coast
External appearance	Very few oval scars, several with scrapes under tail	Numerous oval scars, no scrapes under tail
Baleen shape (length: breadth quotient)	2.22–2.43	1.83–2.24
Food	Anchovies, pilchards, maasbankers	Euphausiids, myctophids, *Lestidium*
Size at sexual maturity	Males 39–40 ft. Females 41 ft.	Males 42–43 ft. Females 41–42 ft.
Breeding season	Unrestricted	Principally autumn
Mean number of ovulations per reproductive cycle	3.75	1.00
Calculated ovulation rate	2.35	0.42
Size at physcial maturity (= L∞)	Males 43 ft. Females 45 ft.	Males 45 ft. Females 47–47 ft.

[a]After Best (1977).

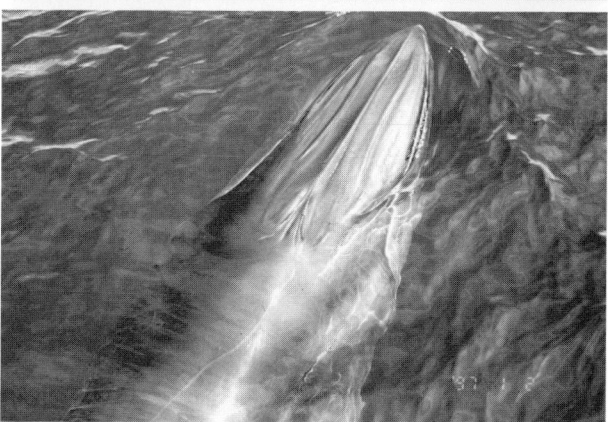

Figure 1 *(Top) Bryde's whale in western North Pacific in summer 1993 (photograph by Tomio Miyashita); (Bottom) head region has a lateral ridge on the rostrum of an animal off South Africa (photograph by Keiko Sekiguchi, 1997).*

Pelagic Bryde's whales, such as those in the western North Pacific stock or the offshore form off South Africa, bear large numbers of oval pit-like scars from bites by the tropical cookie cutter shark (*Isistius* sp.), evidence of migration to tropical waters. This scarring is usually most extensive on the lateral side of the peduncle, leading to an appearance like that of galvanized iron. Such scarring is rare on animals of the coastal form off South Africa; this may also be true for coastal animals in neritic waters off Kochi, southwest Japan, indicating the whales do not migrate to tropical waters. Best (1977) further noted that the coastal form has thin scratches on the undersurface of the tail and ventral keel and suggested that these scratches are due to accidental touching of the sea bottom in shallow waters.

The throat grooves extend to or beyond the navel, whereas those of the sei whale do not reach the navel. The number of grooves between the flippers is 54–56.

The most distinctive external character allowing the discrimination of Bryde's whales from other baleen whales is the presence of three prominent ridges on the rostrum (Fig. 1). The ridges run from just behind the tip of the snout to anterior to the blowholes and are composed of one central ridge and two lateral subridges.

Bryde's whales have 285–350 dark slate-gray baleen plates on each side of the mouth. They are much broader and have coarser bristles than those of sei whales. The longest plate may reach 40 cm in length above the gum. Best (1960, 1977) reported a clear difference in the proportion of length to breadth of the plates in South Africa, with those of the inshore form being more slender than those of the offshore form. Kawamura (1978) found animals in the South Pacific to have finer bristles than those in the western North Pacific stock, probably reflecting a difference in their feeding habits.

The SKULL occupies about 24.1–25.8% of total body length. It is relatively broad, short, and flat for a balaenopterid skull (Fig. 2). The rostrum is also relatively short and pointed. Its sides are nearly parallel posteriorly but slightly convexly curved anteriorly. The curved and robust mandible is also conspicuous among balaenopterids. The vertebrae formula is C 7 + D 13 + L 13 + Ca 21 − 22 = 54–55 (Omura *et al.*, 1981), for a total slightly lower total count than in sei whales (56–57). Cervical vertebrae are unfused, and thoracic (dorsal) vertebrae have short neural processes. The ribs are relatively thin and broad and usually number 13–14. The first rib has a double head, a characteristic shared with the sei whale. The phalangeal formula is I (6), II (5), IV (5), V (3).

II. Distribution, Stocks, and Ecology

Although there is a general pattern of MIGRATION toward the equator in winter and to higher latitudes in summer, Bryde's whales are seen throughout tropical and warm temperate waters of 16.3°C or warmer year round. Their occurrence has been reported from all tropical and temperate waters in the North and South Pacific, Indian Ocean, and South and North Atlantic between 40°N and 40°S (Fig. 3).

In the western North Pacific, Bryde's whales occur in temperate and tropical waters off the Pacific coasts of Japan, Taiwan, and the Philippines to 150°W, with the northern limit corresponding approximately to the southern margin of the sub-Arctic boundary at about 40°N and the southern limit at about 2°S. This is the western North Pacific stock, with ABUNDANCE ESTIMATED at about 24,000 (CV = 0.20; IWC, 1997). Bryde's whales inhabit the east China Sea; this stock extends to the coastal waters of the Pacific side of southwest Japan but is restricted to the west of the Kuroshio warm current (Kato *et al.*, 1996). In the Philippines, Bryde's whales are significantly smaller than those of the western North Pacific stock. They are distributed from there to the Gulf of Thailand, Burma, New Guinea, and the Solomon Islands, all likely "small or dwarf forms" considered to constitute a different species. Bryde's whales also occur in eastern tropical Pacific waters mainly west of 150°W between 20°N and 10°S;

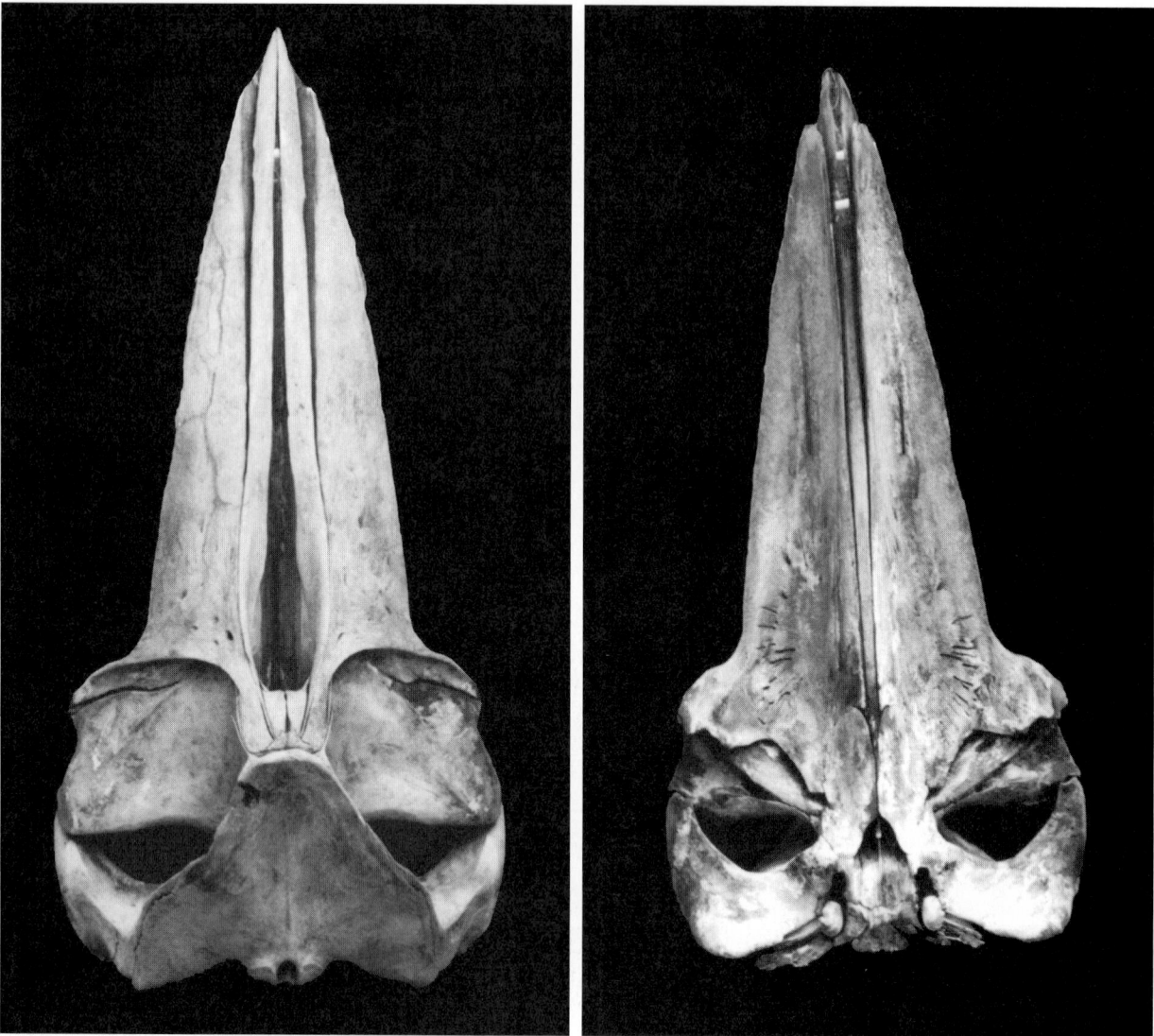

Figure 2   *Dorsal and ventral views of a skull of Bryde's whale caught in the western North Pacific and landed at Bonin Island in July 1983. Photograph by Hidehiro Kato.*

abundance is estimated at 13,000 (CV = 0.202; Wade and Gerrodette, 1993). Bryde's whales also inhabit the Gulf of California throughout the year; these are assumed to constitute an independent stock. They are also widely seen in the south, occurring continuously from the east coast of Australia to 120°W in a zone between the equator and about 30°S, including northern New Zealand. In the eastern South Pacific, the species is distributed in coastal waters off South America from the equator to 37°S. Occurrence seems to be related somewhat to seasonal upwelling events. Little is known about stock structure in these areas, but it is expected that the coastal stock(s) is genetically different from those of the western South Pacific.

Bryde's whales are common throughout the Indian Ocean, from waters north of 40°S such as the Bay of Bengal and the Arabian Sea. Geographical concentrations occur south of Java to the west coast of Australia (east of 90°E), in the central In-

dian Ocean (65–90°E), off Madagascar (35–65°E), and off South Africa (east of 35°E). Stock structure in these areas has not been fully examined other than to confirm the existence of inshore and offshore forms off South Africa (see later). However, it would not be realistic to assume that one homogeneous stock is distributed over the Indian Ocean.

Two allopatric forms of Bryde's whale known as the inshore and offshore forms are found off the west coast of South Africa (Best, 1977). The inshore form is restricted to within 20 miles from the coast and is seen there throughout the year. The offshore form occurs in waters over 50 miles from the coast and migrates north to the equator in winter. Accurate estimates of population abundance are not available for either stocks. Little is known about distribution and stock structure in the rest of the Atlantic, especially in the Northern Hemisphere. However, Bryde's whales have been sighted or stranded from Morocco

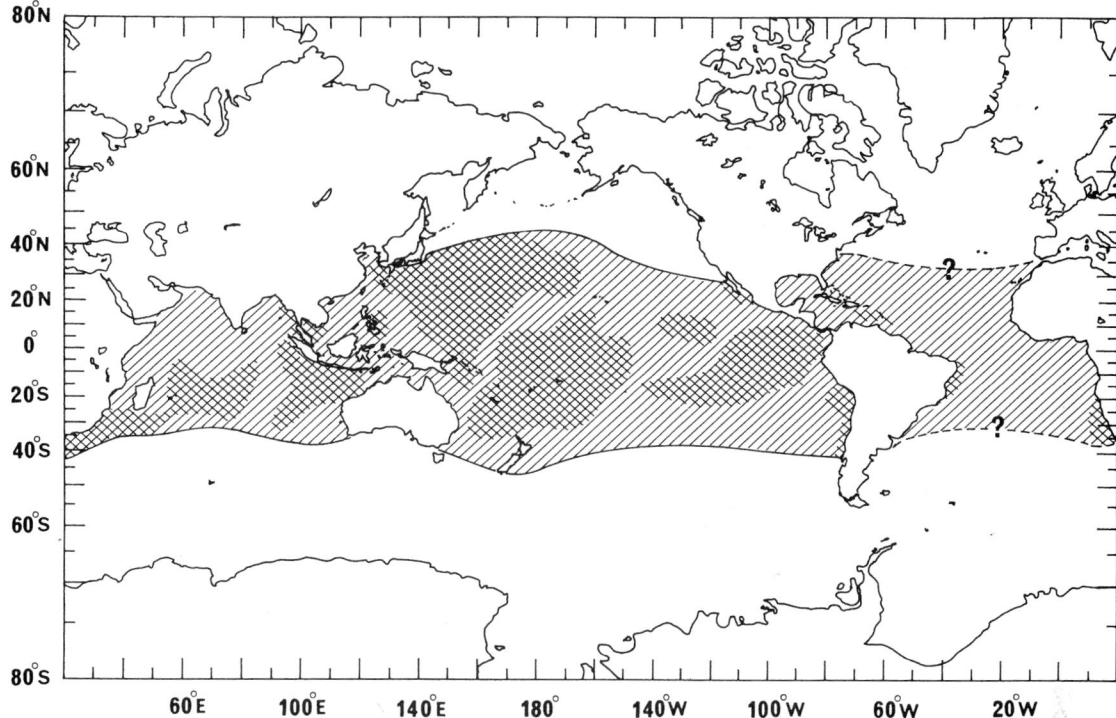

Figure 3 *Worldwide distribution of Bryde's whales based on published or available information. Dense hatch represents areas in which higher densities are expected.*

south to the Cape of Good Hope in the east and from Virginia south to Brazil in the west, including the Gulf of Mexico, the Caribbean, and off Venezuela.

Bryde's whales feed mainly on pelagic schooling fishes such as pilchard, anchovy, sardine, mackerel, herring, and others. However, they also feed on small crustaceans such as euphausids and copepods and on cephalopods and pelagic red crabs (Pleuroncodes) (Nemoto and Kawamura, 1977; Best, 1960, 1977; Kawamura, 1980; Ohsumi, 1977a). They are considered opportunistic feeders, unlike sei whales, which concentrated on copepods. Off South Africa, they tend to be dependent on euphausiids in pelagic waters and feed on schooling fishes in coastal waters; thus feeding habits may be characteristic of stocks (Best, 1977). Balaenopterids consume about 4% of body weight daily at the peak of the feeding season; this corresponds to approximately 600–660 kg per day for a Bryde's whale.

Bryde's whales are sometime seen within high-density patches of bonito (*Sarda*) in pelagic waters in the North Pacific. This may be a result of two predators chasing the same prey. Similarly, off the coasts of Kochi and Kasasa, southwest Japan, it is very common to see Bryde's whales feeding in patches of sardines or juvenile tuna, especially in summer. They have also been observed to utilize bubble net foraging with slow circle swimming under the surface.

III. Behavior and Life History

Bryde's whales do not gather into large groups. They are usually seen singly or in groups of 2–3 in the North Pacific, with a maximum group size of 12. They usually surface steeply like other balaenopterids. The blow is 3–4 m high. The dorsal fin is seen after the blow and then sometimes the dorsal keel. They seldom fluke up before DIVING. It is generally believed that they usually move at 2–7 km/hr, but can swim as fast as 20–25 km/hr and dive up to 300 m. Bryde's whales breach more often than other balaenopterids (Fig. 4). Bryde's whales produce powerful low-frequency moaning sounds averaging 0.4 sec in duration (range, 0.2–1.5 sec) with most of the energy concentrated at 124–250 Hz and a frequency modulation of as much as 15 Hz (Cummings, 1985).

While Bryde's whales have a life history similar to that of other balaenopterids, there are species-specific aspects due to their remaining in tropical and temperate waters throughout the year. As for many other migratory large cetaceans, little is known about the breeding grounds, even for inshore or coastal stocks, although it is generally believed that they must be somewhere in lower latitudes from the migratory stocks. In waters off Kochi (east China Sea stock), females accompanied by small calves sometimes appear in early spring, but there is no evidence that they give birth there. In pelagic stocks, peaks of both conception and calving are in winter, although these are much more diffuse than in other migratory balaenopterids. Best (1977) reported that breeding is not seasonally restricted for inshore animals off South Africa; this may be true for other local stocks.

Gestation lasts for about 11 months. Length at birth is about 4 m. The sex ratio at birth is not different from parity (Best, 1977). Lactation lasts about 6 months and calves wean at about 7 m in body length. Males attain sexual maturity at 11–11.4 m

Figure 4 *Breaching of a Bryde's whale off Kochi, Japan. Photograph by Akinobu Mochizuki.*

and females at 11.6–11.8 m in the western North Pacific stock. Taking into account bias due to operations and regulation of whaling, Best (1977) found length at sexual maturity for the inshore form to be less than for the offshore form off South Africa, by 1 foot in females and 3 feet in males. The mean age at sexual maturity is slightly less than 7 years. Based on the annual ovulation rate (0.42–0.46; Best, 1977; Ohsumi, 1977b) for pelagic Bryde's whale stocks, the calving interval is about 2 years. Inshore waters off South Africa are very frequent ovulators (1.88 per year), but this does indicate a higher pregnancy rate but rather probably results from the extended breeding season. In summary, Bryde's whales have a 2-year reproductive cycle composed of 11–12 months gestation, 6 months of lactation, and 6 months resting.

IV. Conservation Status

Bryde's whales were not harvested commercially or substantially until recent times; their value became relatively important in the late 1970s with the shift of whaling to the smaller species. However, commercial harvest of this species has been prohibited by a moratorium imposed by the INTERNATIONAL WHALING COMMISSION (IWC) in 1987.

Because Bryde's whales had been mainly exploited after substantial improvement of IWC stock management procedures adopted in 1975 (the MNP or new management procedure), stocks have kept relatively stable. A reliable estimate of the population trend for North Pacific Bryde's whales has been available for the western North Pacific stock from a comprehensive

assessment conducted by the IWC in 1995 and 1996. According to the assessment (IWC, 1997), the population has been increasing since 1987, and the current population level (mature females in 1996 relative 1911) ranges from 56.7 to 81.4%.

See Also the Following Articles

Baleen Whales ■ Population Status and Trends ■ Sei Whale ■ Species

References

Anderson, J. (1878). "Anatomical and Zoological Researchers: Comprising an Account of the Zoological Results of the Two Expeditions to Western Yunnan in 1868 and 1875." B. Quaritch, London.

Best, P. B. (1960). Further information on Bryde's whale (*Balaenoptera edeni* Anderson) from Saldanha Bay, South Africa. *Nor. Hvalfangst-Tid.* **49,** 201–215.

Best, P. B. (1977). Two allopatric forms of Bryde's whale off South Africa. *Rep. Int. Whal. Commn. (Spec. Issue I),* 10–38.

Cummings, W. C. (1985). Bryde's whale *Balaenoptera edeni* Anderson, 1878. *In* "Handbook of Marine Mammals" (S. H. Ridgway and R. Harrison, eds.), Vol. 3, pp. 137–154. Academic Press, London.

IWC (International Whaling Commission) (1997). Report of the subcommittee on North Pacific Bryde's whales, annex G, report of the scientific committee. *Rep. Int. Whal. Commn.* **47,** 163–168.

Junge, G. C. A. (1950). On a specimen of the rare fin whale, *Balaenoptera edeni* Anderson, stranded on Pulu Sugi near Singapore. *Zool. Verhandelingen* **9,** 1–26.

Kato, H., Shinohara, E., Kishiro, T., and Noji, S. (1996). Distribution of Bryde's whales off Kochi, Southwest Japan, from the 1994/95 sighting survey. *Rep. Int. Whal. Commn.* **46,** 429–436.

Kawamura, A. (1978). On the baleen filter area in the South Pacific Bryde's whales. *Sci. Rep. Whales Res. Inst.* **30,** 291–300.

Kawamura, A. (1980). Food habits of the Bryde's whales taken in the South Pacific and Indian Oceans. *Sci. Rep. Whales Res. Inst.* **32,** 1–23.

Nemoto, T., and Kawamura, A. (1977). Characteristics of food habits and distribution of baleen whales with special reference to the abundance of North Pacific sei and Bryde's whales. *Rep. Int. Whal. Commn (Spec. Issue I),* 80–87.

Ohsumi, S. (1977a). Bryde's whales in the Pelagic whaling ground of the North Pacific. *Rep. Int. Whal. Commn. (Spec. Issue I),* 140–150.

Ohsumi, S. (1977b). Further assessment of population of Bryde's whales in the North Pacific. *Rep. Int. Whal. Commn.* **27,** 156–160.

Ohsumi, S. (1980). Population study of the Bryde's whale in the Southern Hemisphere under scientific permit in the three seasons, 1976/77–1978/79. *Rep. Int. Whal. Commn.* **30,** 319–331.

Olsen, O. (1913). On the external characteristics and biology of Bryde's whale (*Balaenoptera brydei*) a new rorqual from the coast of South Africa. *Proc. Zool. Soc. Lond.* 1913, 1073–1090.

Omura, H. (1959). Bryde's whale from the coast of Japan. *Sci. Rep. Whales Res. Inst.* **14,** 1–33.

Omura, H., Kasuya, T., Kato, H., and Wada, S. (1981). Osteological study of the Bryde's whale from the central South Pacific and eastern Indian Ocean. *Sci. Rep. Whales Res. Inst.* **33,** 1–26.

Rice, D. W. (1998). "Marine Mammals of the World, Systematic and Distribution." Society for Marine Mammalogy Special Publication 4, Society for Marine Mammalogy, Lawrence, KS.

Wada, S., and Numachi, K. (1991). Allozyme analyses of genetic differentiation among the populations and species of the Balaenoptera. *Rep. Int. Whal. Commn. (Spec. Issue 13)* 125–154.

Wade, P. R., and Gerrodette, T. (1993). Estimates of cetacean abundance and distribution in the eastern tropical Pacific. *Rep. Int. Whal. Commn.* **43**, 477–493.

Yoshida, H., and Kato, H. (1999). Phylogenetic relationships of Bryde's whales in the western North Pacific and adjacent waters inferred from mitochondrial DNA sequences. *Mar. Mamm. Sci.* **15**, 1269–1286.

Buoyancy

SEE *Diving Physiology*

Burmeister's Porpoise
Phocoena spinipinnis

JULIO C. REYES
Areas Costeras y Recursos Marinos, Pisco, Peru

Described by Hermann Burmeister in 1865, this is one of the five species of the family Phocoenidae, a group of cetaceans whose members share the presence of small, spade-shaped TEETH. The Spanish name for this porpoise, "marsopa espinosa" or "spiny porpoise," refers to the series of tubercles present in the dorsal fin. It was long considered a rare species, but research in the last decade has shed light on the natural history and conservation problems of this cetacean.

I. Distribution

Burmeister's porpoise is restricted to South American waters, from Santa Catarina, in southern Brazil, around the Cape Horn north to Bahía de Paita, in northern Peru. As more studies focus on this species, its range appears to be continuous. The existence of Pacific and Atlantic stocks has been proposed, but no studies have yet addressed this matter.

II. External Form

In the Atlantic part of its range, Burmeister's porpoises may grow up to 200 cm; off the Pacific coast, the maximum size recorded corresponds to a female 183 cm long from Peruvian waters. The males, however, are significantly larger than females. The body is robust, with a small, blunt head, and proportionately large flippers (Fig. 1). A conspicuous feature of the species is the dorsal fin, which is triangular in shape and canted backward in an unusual fashion for a cetacean. Moreover, the leading edge of the dorsal fin has a series of small tubercles that become sharper as the animal grows (Fig. 2).

COLORATION varies from dark gray to brownish gray on the back and sides, and a light gray ventral region. Surrounding the eye, a dark patch can often be noted. A dark gray stripe runs from the chin to the base of the flipper. This "flipper stripe" is asymmetrical, being markedly wider on the left side (Fig. 3). A

Figure 1 *Side view of a Burmeister's porpoise caught in Peruvian waters.*

Figure 2 *The dorsal fin of a Burmeister's porpoise.*

Figure 3 *Close-up of a Burmeister's porpoise head, showing the eye patch and flipper stripe.*

pair of stripes is also present on the abdominal region. The pattern of these ventral stripes is sexually dimorphic.

III. Habitat and Ecology

Although it is considered a coastal, inshore species, Burmeister's porpoise has been observed up to 50 km from shore, at a depth of 60 m. These movements to offshore waters appear to respond to migration of the prey species. Some animals have been recorded upstream in the Valdivia River, in southern Chile.

Burmeister's porpoise feeds on a variety of both pelagic and demersal fish, including anchovies, hake, jack mackerel, sardine, drums, and, less frequently, squids and shrimps.

At sea this porpoise is difficult to observe. It swims with a gentle roll on the surface, with little of its body exposed. This behavior may change to very fast swimming when approached by a boat, although not porpoising or performing bow riding. The group size usually ranges from 2 to 8 animals, but aggregations of 22 and up to 70 porpoises have been reported.

IV. Reproduction

Females reach sexual maturity when they are 154.8 cm long, shorter than the length of attainment of sexual maturity for males, estimated to be 159.9 cm. There does not appear to be seasonality in the reproductive cycle of the male. Pregnancy may last 11–12 months, and the length at birth is approximately 86 cm. Some pregnant females may be simultaneously lactating, indicating that annual reproduction may occur. Based on the counting of dentine layers in the teeth, one porpoise 174 cm long was estimated to be 8 years old.

V. Interaction with Fisheries

The coastal habitat of Burmeister's porpoise makes it vulnerable to fishing activity. Through its range, the species become incidentally entangled in fishing nets. In Uruguay, several porpoises became entangled in shark nets. In the southern part of Argentina, several fisheries account for the capture of a small number of porpoises. In southern Chile, entanglements and possible harpooning of animals have been reported. From central Chile to northern Peru the species becomes incidentally entangled in nets set for a variety of small pelagic and demersal fishes, sharks, and rays. Only in Peru do captures of Burmeister's porpoises reach an annual figure of 2000 animals.

See Also the Following Articles

Harbor Porpoise ■ Porpoises, Overview ■ Spectacled Porpoise ■ Vaquita

References

Brownell, R. L., and Praderi, R. (1984). *Phocoena spinipinnis. Mamm. Species* **217**, 1–4.

Goodall, R. N. P., Norris, K. S., Harris, G., Oporto, J. A., and Castello, H. P. (1995). Notes on the biology of Burmeister's porpoise, *Phocoena spinipinnis*, off southern South America. *Rep. Int. Whal. Commn. (Spec Issue 16)*, 318–347.

Jefferson, T. A., Leatherwood, S., and Webber, M. A. (1993). FAO Series Identification Guide. Marine Mammals of the World. FAO, Rome.

Reyes, J. C., and Van Waerebeek, K. (1995). Aspects of the biology of Burmeister's porpoise from Peru. *Rep. Int. Whal. Commn. (Spec Issue 16)*, 349–364.

Simoes-Lopes, P. C., and Ximenez, A. (1989). *Phocoena spinipinnis* Burmeister 1865, na costa sul do Brasil (Cetacea-Phocoenidae). *Biotemas* **2**, 83–89.

Würsig, M., Würsig, B., and Mermoz, J. F. (1977). Desplazamientos, comportamiento general y un varamiento de la marsopa espinosa, *Phocoena spinipinnis*, en el Golfo San José (Chubut, Argentina). *Physis* **36**, 71–79.

California, Galapagos, and Japanese Sea Lions

Zalophus californianus, Z. wollebaeki, and *Z. japonicus*

CAROLYN B. HEATH
Fullerton College, California

The California, Galapagos, and Japanese sea lions are closely related species that together comprise the genus *Zalophus*. They occupy (or occupied, in the case of the presumably extinct Japanese sea lion) widely separated regions of the Pacific, including the temperate western and eastern North Pacific and the tropical Galapagos Archipelago (Fig. 1). The varying environmental conditions experienced throughout this range, along with the corresponding variations in behavior, provide fascinating insights into their behavioral ecology.

I. Taxonomy and Distribution

The California, Galapagos, and Japanese sea lions are now regarded as separate species: *Zalophus californianus* (Lesson, 1828), *Z. wollebaeki* (Sivertsen, 1953), and *Z. japonicus* (Peters, 1866), respectively. Previously they were typically considered to be geographically isolated subspecies (*Z. californianus californianus, Z. c. wollebaeki,* and *Z. c. japonicus*), but recent discoveries of substantial morphological and behavioral differences among them led to their reclassification. Further studies are needed to delineate the differences among species and to identify any subspecies or stocks. Differences in mitochondrial DNA indicate that the Pacific and Gulf of California populations of California sea lions have long been genetically isolated. However, no differences have been found in the cranial morphology of the two populations, and subadult males appear to migrate between colonies in the southern Gulf of California and those along the Pacific Coast of Mexico. The degree of interchange between Mexican and U.S. populations is not known.

The main breeding areas of California sea lions include the Channel Islands in southern California and Mexican islands off the Pacific coast of Baja and in the Gulf of California. Rarely, births may occur as far north as the Farallon Islands off central California. Outside the breeding season, animals (mostly males) are common as far north as Vancouver Island. Vagrants occur as far north as Prince William Sound, Alaska, and as far south as Chiapas, Mexico. Females and immatures may disperse from the breeding islands to forage, but apparently do not migrate as extensively as males.

Galapagos sea lions breed on all the islands of the archipelago. In 1986 a small rookery was established outside of the Galapogas on Isla de La Plata off the coast of Ecuador. Vagrants have been reported along the mainland coast of Ecuador and on Islas del Coco (Costa Rica) and Gorgona (Columbia).

Although historical and archaeological records are incomplete, Japanese sea lions appear to have lived in coastal areas from Kyushu to southern Kamchatka. Their range was likely centered along the west and east coasts of Honshu, off Shikoku and Kyushu, in the Seto Inland Sea, and on islands in the Sea of Japan and the Izu region. Known rookeries include Takeshima and Ullung-do in the southern Sea of Japan, the northwest and also central-eastern coasts of Honshu, and four islands in the Izu region. Vagrants have been noted to southwestern Sakhalin, the Kuril Islands, southern Kamchatka, and the east coast of South Korea.

II. Population Trends and Exploitation

The annual return of pinnipeds to predictable breeding areas makes them particularly vulnerable to exploitation. Subsistence HUNTING of California sea lions probably occurred for several thousand years without much of an effect on the population. However, commercial harvesting during the 1800s and early 1900s in southern California and Mexico reduced the population to only about 1500 animals by the 1920s. The harvest, which at various times was for hides, blubber, meat, predator control, or the whiskers and BACULA sold as aphrodisiacs, probably focused on adult males. A floundering market coupled with protective legislation allowed the population to start increasing by the 1940s, although some killing and live collecting continued until the 1970s. The 1972 U.S. Marine Mammal Protection Act and similar legislation in Canada and Mexico

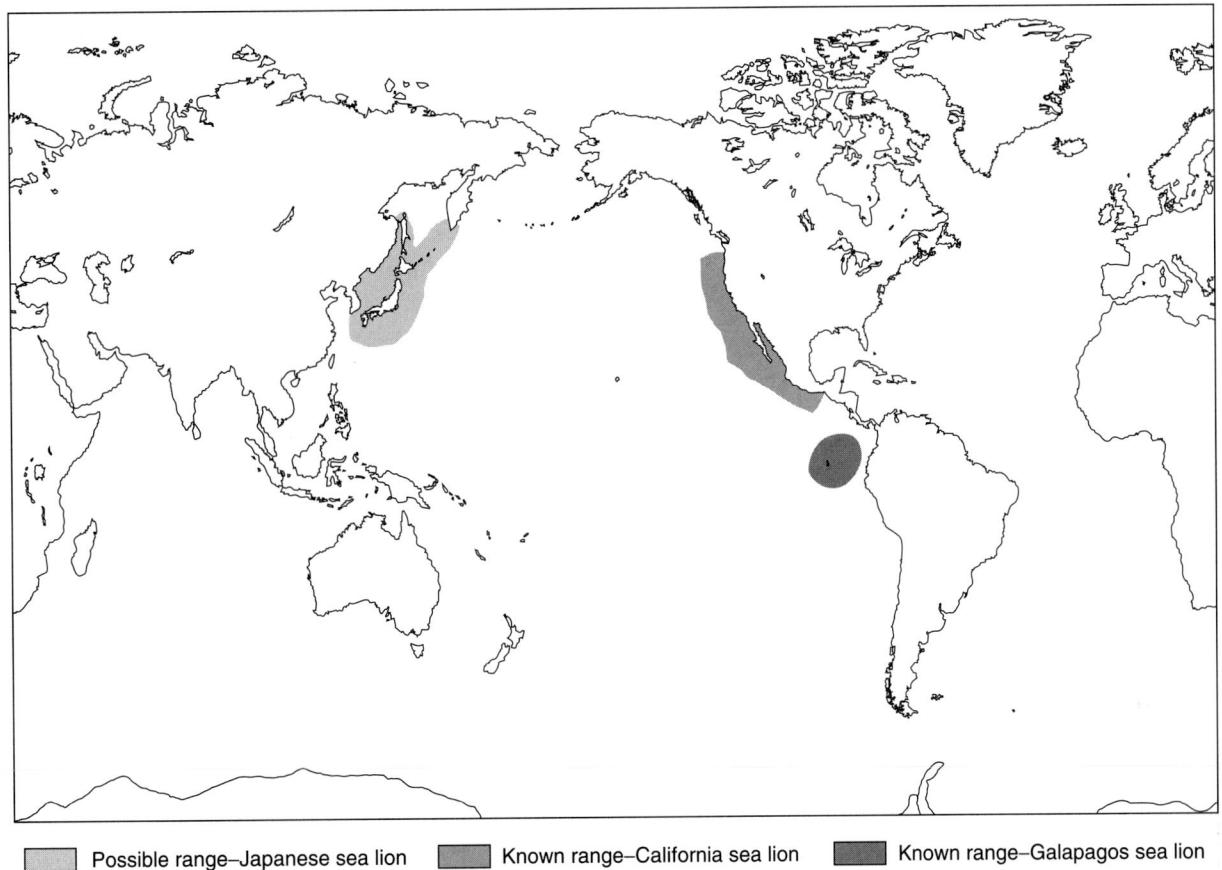

Possible range–Japanese sea lion   Known range–California sea lion   Known range–Galapagos sea lion

Figure 1 *Ranges of California, Galapagos, and Japanese sea lions.*

greatly facilitated the population's recovery. The U.S. population of California sea lions is currently estimated to be 167,000–188,000, with about 34,000 pups produced annually. Overall, the population is increasing at a rate of roughly 5% per year; however, periods of growth are frequently reversed during strong El Niño events. A 1992 census of the Mexican Pacific coast recorded about 10,000 California sea lions, compared to nearly 28,000 counted a decade earlier (Maravilla and Lowry, 1996). Allowing for uncensused animals at sea, the current Mexican Pacific population may be roughly 13,000 to 22,000. Clearly the population fluctuates in this region; its relationship to changes in the U.S. population is not known. A 1992 census of the Gulf of California population recorded about 23,000 animals; correction factors yielded a population estimate of about 31,000 (Aurioles and Zavala, 1994). Thus, the total California sea lion population is about 211,000–241,000.

The Galapagos sea lion has been largely spared the commercial exploitation that depleted the California sea lion population. However, the number of Galapagos sea lions apparently does fluctuate frequently due to El Niño events and epidemics of seal pox. The population was estimated at 20,000–50,000 in 1963. Following the 1997–1998 El Niño, which was accompanied by up to 90% pup mortality and 45% overall mortality, the population was estimated at only about 14,000 individuals (Salazar, 1999).

Between 30,000 and 50,000 Japanese sea lions may have existed in the mid-1800s. Heavy, unregulated hunting depleted the population such that by the 1950s 50 to 60 animals on Take-shima were the only ones reported. The IUCN lists the species as extinct; some scientists believe this status needs to be confirmed by surveys in remote regions.

III. Habitat and Environment

California and Galapagos sea lions breed on sandy beaches and rocky areas on remote islands. Because females must forage at sea during lactation, breeding areas are restricted to regions of high marine productivity. California sea lion rookeries along the Pacific coast are in a very productive upwelling zone, and productivity in the Gulf of California is also very high due to tide- and wind-generated upwelling. The waters of the continental margin adjacent to areas once used by breeding Japanese sea lions are quite productive. While low productivity generally excludes otariids from breeding in the tropics, the islands in the Galapagos Archipelago are bathed in nutrient-rich upwelling currents. This creates an isolated pocket of high productivity, which supports the Galapagos sea lions. The importance of high productivity can be seen in the devastation that occurs during El Niño events, when a plume of warm, nutrient-poor water emanating from the equatorial Pacific decreases the availability of the sea lions' prey. These frequent, but unpredictable events are most severe in the eastern tropical Pacific where the Galapagos sea lion lives, with lesser impacts at the higher latitudes occupied by California and Japanese sea lions. Sea lions living in the Gulf of California, Mexico, may be largely protected from the effects of El Niños; strong tidal mixing there should be able to resupply nutrients to surface waters during an El Niño event.

The breeding habitat occupied by *Zalophus* ranges from temperate to tropical regions. As a result, breeding animals are often subjected to high temperatures while on land. The affects of these high temperatures and of El Niño events are described in Section VIII.

IV. Description

California sea lions are highly sexually dimorphic: the weight and length for adult males is about 350 kg and 2.4 m compared to about 100 kg and 1.8 m for adult females. Males in the Gulf of California appear to be smaller than their Pacific counterparts. Newborn California sea lion pups weigh 6–9 kg for females and males, respectively. They are dark brown to black until they molt to a tawny brown color at 4–6 months (Fig. 2). Females remain this color throughout adulthood, whereas male coats typically darken as they age. Adult males usually are a dark brown, but can range from light brown to black. All individuals appear darker when wet. It is difficult to distinguish females from young males until the latter begin developing the broad chest, dark color, and sagittal crest of adult males (Fig. 3). The sagittal crest, which is unique to *Zalophus*, is usually topped by white fur and is quite conspicuous. Galapagos sea lions are smaller than California sea lions and appear to be much less sexually dimorphic. Females weigh about 80 kg. No weights are available for males. Young California and Galapagos sea lions are especially playful, spending much time chasing and mock fighting with each other, playing with objects such as kelp

Figure 3   *An adult male California sea lion on his territory. The broad neck and chest are indicative of an adult male; the fat deposits they contain will sustain him while fasting for the duration of his tenure. Note the pups playing in protected shallow waters. Photo by C. B. Heath.*

and feathers, and bodysurfing. All age classes are fairly vocal; territorial males are exceedingly so. Little is known about the physical characteristics of the Japanese sea lion.

V. Life History and Annual Cycle

In southern California, where California sea lions have been studied extensively, animals can be found on the breeding islands year-round. The number ashore increases rapidly in May with the onset of the breeding season. At this time adult males begin fighting for territories along the shorelines of the rookeries. Most are unsuccessful and retreat to sea or to "bachelor" beaches nearby. Those that establish territories maintain their boundaries with ritualized displays and frequent barking. Territorial males fast throughout their tenure, surviving on fat accumulated during the off-season. Tenure lasts from 1 to 45 days and may end when residents are displaced by another male or when their fat reserves are depleted. Some males maintain territories for multiple breeding seasons. Throughout May and June females give birth to a single pup a few days after coming ashore. Vocal and olfactory imprinting follows birth and is used by mothers and pups to reunite after separations (Fig. 4). Mothers spend the first week postpartum with their pup and then begin alternating feeding trips at sea with suckling bouts on land. The feeding trip length is largely determined by the distance to the foraging grounds and the availability of prey. It averages 2–3 days, but varies with location and year. Stays on land average 1–2 days. This pattern continues until the pup is weaned. Most pups are weaned by 10–12 months of age. The number that continue to suckle as yearlings or even 2 year–olds varies among years (Francis and Heath, 1991). Females mate

Figure 2   *An adult female California sea lion suckling her pup after returning from a feeding trip at sea. Photo by C. B. Heath.*

Figure 4 *An adult female California sea lion vocalizes to her newborn pup. Mothers and pups imprint on each others' calls and smell at birth, which helps them recognize each other and reunite after separations. Note remnants of amniotic sack on pup. Photo by C. B. Heath.*

occur from June to March, but the peak pupping period varies among rookeries and years. The interval between birth and mating has been estimated at about 3 weeks. Territorial males sometimes go to sea to feed, thus often losing their territory to another male, but sometimes reclaiming it upon their return. Migration within the archipelago is minimal. Normal pup mortality to age 6 months is about 5%. Most pups are weaned within 1 year, but may continue to suckle as yearlings or 2 year olds if their mother does not have a new pup.

As with all other parameters, we have virtually no information regarding the life history of Japanese sea lions.

VI. Social Relationships

California sea lions are highly intelligent and adaptable animals, capable of learning a simple sign language in captivity (Schusterman and Krieger, 1986). Both California and Galapagos sea lions can recognize individuals in the wild through scent, sound, and probably sight. They are also gregarious animals, with much opportunity for social interaction during the breeding season, at least. These traits would seem to dispose them to sustained relationship with other individuals; however, the only obvious social unit is between mothers and their offspring. Perhaps longer term studies of permanently marked individuals will reveal other types of relationships.

VII. Diet

California sea lions eat a wide variety of prey, which is determined to some degree by its relative availability. The most common prey items in southern California are northern

about 27 days after giving birth, an unusually long interval for otariids. Not all females breed every year. At many rookeries, females form "milling" groups as a prelude to mating (Fig. 5). In these groups of 2–20 females, the females often mount each other and the territorial male. Eventually 1–2 of the females mate and the milling activity ends (Heath, 1989). While ashore, both males and females make regular movements to the water to cool off (Peterson and Bartholomew, 1967).

After the breeding season ends in August, most adult and subadult males leave the southern California rookeries and migrate north, where they feed throughout the fall and winter. Females and juveniles appear to disperse to feed in the general vicinity of the breeding islands. Pup mortality is roughly 15–20% for the first 6 months of life. In the southern Gulf of California, at least, it then increases to about 40% for the next 6 months as pups venture into the water (Aurioles and Sinsel, 1988). Sexual maturity occurs at about 4–5 years of age, although males are not large enough to hold breeding territories for several more years. Longevity is estimated at 15–24 years. Sources of mortality include starvation, infection, sharks, killer whales, toxic phytoplankton blooms, entanglement, shooting, and disease.

California sea lions living in the Gulf of California, Mexico, experience a similar annual cycle with some notable differences. The pupping season lasts 1–2 weeks longer for at least some rookeries in the northern half of the Gulf (Morales, 1990), and the interval between birth and mating appears more variable than in the U.S. population (Heath, 1989). About 40% of the adult males in this region remain around the breeding islands year-round (Zavala, 1990), and the age of weaning appears to be both older and more variable than in southern California (Heath *et al.*, 1996).

The breeding season is very protracted in the Galapagos sea lion, and territorial males are observed most of the year. Births

Figure 5 *A "milling" group of California sea lions. These groups form in some areas as a prelude to mating. Females often mount each other or the territorial male while in these groups. Note the male's saggital crest, which is unique to Zalophus. Photo by C. B. Heath.*

anchovy, Pacific whiting, rockfish, cephalopods, jack and Pacific mackerel, and blacksmith (Lowry *et al.*, 1986). The first three species are also important in the Mexican Pacific and Gulf populations, as are midshipmen. Myctophids, sardines, cutlassfish, alopus, cusk eels, and bass are frequently prey in various areas of the Gulf (Sanchez, 1992). Diet varies greatly among years, seasons, locations, and probably individuals. El Niño events cause shifts in the DIET, and species otherwise rarely consumed, such as the pelagic red crab, may become more common in the diet. Feeding can occur at any hour of the day. Dives typically last for about 2 min, but can be as long as 10 min. Dive depth averages 26–98 m, but can be well over 200 m (Feldkamp *et al.*, 1991). The staple of the Galapagos sea lion diet is sardines. During El Niño events, however, partial shifts to green-eyes (1982–1983) and myctophids (1997–1998) have occurred (Trillmich and Dellinger, 1991; Salazar, 1999). Galapagos sea lions forage within a few kilometers of the coast, feeding during the daytime on a near-daily basis. Dive depth averages 37 m, but can reach 186 m. There is no information on what the Japanese sea lions ate.

VIII. Behavioral Ecology

Life is a compromise, and pinnipeds have evolved many adaptations in response to the sometimes conflicting pressures of breeding on land and feeding at sea. Certain environmental conditions can increase the costs or benefits of some of these adaptations and bring about compensatory changes in behavior. Two such conditions—high air temperatures on ROOKERIES and decreased prey availability during El Niño events—play particularly important roles in shaping the breeding and foraging behavior of California and Galapagos sea lions.

A. Temperature and Its Effects on Breeding Behavior

California sea lion females are very particular about male behavior. Boisterous, overly attentive, or aggressive males are typically abandoned and left to sit alone on their territories. Any interference with female movements is simply not tolerated. Should a male attempt to block a female's path, she needs only to extend her neck out and up, and sway it side to side as she walks. This long-neck display signals males that she requires free passage; the rare male that does not respond to this will likely be subjected to a display of jerky hopping and flipper slapping, which will dissuade him from interfering further. Males seem to have little option but to acquiesce if they are to be successful at breeding.

Why might this situation exist, especially in such apparent contrast to some otariid species where females may be herded, threatened, and even injured by territorial males? The explanation appears to partly lie in the animals' thermoregulatory needs. The particularly warm climate in which *Zalophus* breed increases the cost of moving between marine and terrestrial habitats. The blubber, fur, and large size, which insulate against cold ocean waters, can lead to overheating while ashore, thus making necessary regular access to wet substrate or water for cooling. The breeding fat of adult males intensifies their thermal stress, thereby limiting the distribution and number of ter-

ritories. Successful males have territories containing access to water; others must abandon their territories (and any hopes of mating) during the heat of the day. Breeding females must also have access to cooler shoreline areas on a near-daily basis, and they regularly travel through several males' territories while moving from resting to cooling areas. Furthering this pattern of female movements is the unusually long interval (about 21–27 days) between birth and MATING. During that interval females must also make regular feeding trips to sea to replenish their milk supplies. A male that prevented these premating thermoregulatory and feeding excursions by herding the females would be left with, at best, only severely stressed females, which are not likely to mate. Thus the combination of warm breeding areas and delayed mating together foster a system in which males may control each other, but not the females. Rather, when it is time to mate, many females leave the male's territory in which they have given birth and mate with another male that they may have encountered in their movements about the rookery. Females show a surprising degree of unity in their selection of mates. As a result, many males holding territories during the breeding season never or rarely mate, while a few males mate with many females (Heath, 1989). This dramatically increases the degree of polygyny in California sea lions: males make the first cut by excluding many of their gender from the rookeries and females make another cut via their selection of mates.

The influence of temperature on behavior can also be seen in the high percentage of copulations that occur in contact with the water. This percentage increases in hotter regions, as does the amount of time females spend cooling off at the shoreline. In the Gulf of California, where air temperatures are very high, nearly all territorial defense and breeding activities are restricted to the water (Garcia, 1992; Heath, 1989).

Galapagos sea lions show a similar response to high air temperatures with their great reliance on shoreline areas for cooling. However, observations of their breeding behavior are inadequate to compare to California sea lions. Environmental conditions in at least part of the former range of the Japanese sea lion are similar to those of its congeners. However, barring the discovery of historical records, we will never know if their behavioral responses were the same as those of the California and Galapagos sea lions.

B. Responses to El Niño

The environmental changes that occur during El Niño events also elicit behavioral responses by *Zalophus*. However, unlike the fairly constant heat stress experienced on rookeries, the environmental stresses associated with El Niño are unpredictable and only occur every few years. In addition, because the degree of stress varies among events and locations, the sea lions' responses must be somewhat flexible. El Niños cause a reduction in prey availability for *Zalophus* throughout much of its range. The potential consequences of this reduction are mitigated somewhat by adaptations that have evolved over the sea lions' long history of coexistence with El Niños. However, the severe impacts of some El Niño events demonstrate the limits of these adaptations. The 1982–1983 El Niño was a particularly strong one, and much is known about its effects on the California sea lions that breed in southern California. Some non-breeding sea lions in this region responded to local prey depletion by migrating north to more productive areas. Many immatures and some adult females left their normal winter foraging areas and migrated to central California (Huber, 1991). Emigration was thus apparently an option for some individuals to reduce the effects of El Niño. Territorial males in southern California showed no measurable affects from this event, most likely due to their preseason foraging farther north. Adult females, however, appeared to be more tied to the general vicinity of the BREEDING SITES, where prey reduction was more pronounced. The increase in spontaneous abortions during the 1982–1983 winter indicates that some of these females were unable to find adequate prey (Francis and Heath, 1991). Females that did manage to produce full-term pups then faced the greater challenge of nourishing them. Feeding during lactation makes females quite vulnerable to localized decreases in food availability. In southern California they attempted to compensate for decreased prey by increasing their foraging effort while at sea, partially shifting their prey, and by slightly prolonging their feeding trips (Lowry *et al.*, 1986; Ono *et al.*, 1987). These efforts, however, were inadequate to compensate for the strength of the 1982–1983 El Niño. Females apparently made less milk: pups suckled less, grew more slowly, and weighed less at age 2 months (Ono *et al.*, 1987). Pup mortality increased, and pup production decreased by 30–71% at various islands. Fewer of the male pups were weaned by age 1 year, and more of them stayed on their birth island and suckled into their second year. Fewer females mated during the El Niño summer, presumably a sign that they were undernourished. As a result, pup production was still low in the following year. Because pup production took several years to return to pre-El Niño levels, it is possible that there may have been some mortality of breeding females and juveniles associated with this event. In Mexico, pup production on at least one Pacific island decreased by 50% during the 1982–1983 El Niño, while effects appeared to be very weak in the Gulf (Aurioles and Le Bouef, 1991). An even stronger El Niño occurred in 1997–1998. While not as widely monitored for its effects on California sea lions, they appear to have been even greater.

The reduction of prey during El Niño events is particularly strong in the eastern tropical Pacific. Because Galapagos sea lions are isolated from alternative feeding areas by vast expanses of unproductive tropical waters, emigration to better feeding areas is not an option for them. Mortality has thus been very high for this species during El Niño events. Between 80 and 95% of the pups born in 1982 did not survive their first year of life. Pup production at various rookeries in 1983 was between 3 and 65% of normal years. Adult female mortality was estimated at 20%, and territorial male mortality was particularly severe (Trillmich and Dellinger, 1991). During the 1997–1998 El Niño, pup mortality was close to 90%, and mortality for the overall population was about 45% (Salazar, 1999).

Oceanic conditions also change in Japan during El Niño events, but what effects this may have had on the Japanese sea lions is not known.

The oceanographic counterpart to El Niños are Las Niñas, periods of generally cooler ocean temperatures and greater productivity. Little is known of their effects on pinnipeds or any role they might play in the recovery from El Niño events.

IX. Interactions with Humans

Like all marine mammals, California and Galapagos sea lions spend a good portion of their lives in remote areas or underwater, hidden from our view. However, the California sea lion is one of the most familiar marine mammals. This is due in part to their being the most commonly used "seal" performer in animal park shows and also to their habituation to human presence in some areas, especially where there is a comfortable dock or buoy to be acquired by this tolerance. These activities are indicative of their intelligence and flexible nature, which itself can sometimes lead to less positive interactions with humans.

A. Fisheries Interactions

Enviable hunter or lowly thief? The answer to this question is largely a matter of perspective. Certainly, California sea lions are highly skilled at catching their prey of fish and squid, and their growing population consumes many tons of them yearly. As with many predators, they are also flexible and opportunistic in their search for food, as their diet and thus foraging patterns vary with age, location, and environmental changes caused by things such as El Niño and commercial fishing. While this flexibility is partly responsible for the recovery of this species, it at times also brings them into direct COMPETITION WITH HUMAN FISHERIES. Healthy populations of fish, sea lions, and humans have coexisted throughout much of our history; however, the demand for marine resources generated by a rapidly increasing human population, coupled with its increasingly efficient exploitation of those resources, has heightened concerns about competition between humans and other marine predators. This, combined with the highly visible actions of individual sea lions that have learned to exploit the easy take from fishing lines and nets, has led some to view California sea lions as marine pests rather than an integral part of a healthy ecosystem.

B. Habitat Overlap

Another form of competition occurs when sea lions make themselves at home on docks and other man-made resting areas. While this can be quite inconvenient, in areas such as Pier 39 in San Francisco, the situation has been converted into a popular tourist attraction.

C. Pollution

The growing arsenal of toxic chemicals and waste that makes its way into marine mammals' habitats and prey has generated much concern. This is particularly relevant for the Channel Islands population of California sea lions; their proximity to the major metropolitan areas of southern California exposes them to a great deal of urban and industrial runoff, waste, and debris. Because they are high-level predators, sea lions are vulnerable to compounds such as organochlorines (e.g., DDT and PCBs) that become increasingly concentrated as they move up the food chain. Laboratory studies of such compounds have revealed that they can suppress pinniped immune systems, rendering them more vulnerable to disease. However, establishing such clear cause-and-effect relationships in wild populations exposed to organochlorines is more difficult due to confounding factors. California sea lions were found to have elevated levels of organochlorines associated with increased stillbirths and premature pupping, but the level of contribution of disease to this problem could not be determined. Although the specific links among chemicals, immune system responses, and disease or mortality are incompletely understood, enough indications of problems exist to warrant caution and further research.

D. Entanglement

Entanglement with marine debris is a problem found in all populations of California and Galapagos sea lions. Materials such as packing bands and discarded fishing line or nets can become caught on the animals' necks or flippers, leading to injury, infection, reduced feeding efficiency, or death. Sea lions are also killed incidentally in some fisheries.

The southern California population of California sea lions is currently thriving and is thus apparently quite able to recover from its interactions with humans. The extinction of the Japanese sea lion, however, reminds us that there can be limits to this recovery. This is especially true for smaller populations, such as those of the Galapagos sea lions or the Mexican population of California sea lions. If harmful human activities were to increase substantially or happened to coincide with natural stresses such as epidemics or El Niño events, recovery might not be so rapid or complete.

See Also the Following Articles

Extinctions, Specific ■ Habitat Pressures ■ Pinniped Ecology ■ Territorial Behavior

References

Aurioles G. D., and Le Boeuf, B. J. (1991). Effects of the El Niño 1982–83 on California sea lions in Mexico. In "Pinnipeds and El Niño: Responses to Environmental Stress" (F. Trillmich and K. A. Ono, eds.), Ecological Studies 88, pp. 112–118. Springer-Verlag, Berlin.

Aurioles G. D., and Sinsel, F. (1988). Mortality of California sea lion pups at los Islotes, Baja California Sur, Mexico. J. Mammal. **69**, 180–183.

Aurioles G. D., and Zavala G. A. (1994). Ecological factors that determine distribution and abundance of the California sea lion Zalophus californianus in the Gulf of California. Ciencias Marinas **20**(4), 535–553.

Feldkamp, S. D., DeLong, R. L., and Antonelis, G. A. (1991). Effects of El Niño 1983 on the foraging patterns of California sea lions (Zalophus californianus) near San Miguel Island, California. In "Pinnipeds and El Niño: Responses to Environmental Stress" (F. Trillmich and K. A. Ono, eds.), Ecological Studies 88, pp. 146–155. Springer-Verlag, Berlin.

Francis, J. M., and Heath, C. B. (1991). Population abundance, pup mortality, and copulation frequency in the California sea lion in relation to the 1983 El Niño on San Nicolas Island. In "Pinnipeds and El Niño: Responses to Environmental Stress" (F. Trillmich and K. A. Ono, eds.), Ecological Studies 88, pp. 119–128. Springer-Verlag, Berlin.

Garcia, R. M. C. (1992). Conducta territorial del lobo marino Zalophus californianus en la lobera Los Cantiles, Isla Angel de la Guarda, Golfo de California, Mexico. Tesis de Licenciatura, Facultad de Ciencias, UNAM, Mexico, D. F.

Heath, C. B. (1989). "The Behavioral Ecology of the California Sea Lion." Ph.D. dissertation, University of California, Santa Cruz, CA.

Heath, C. B., Adams, M., and Garcia, M. (1996). Geographic variation in the duration of maternal care in the California sea lion. In "Abstracts, Symposium on Otariids." Smithsonian Insitution, Washington, DC.

Huber, H. R. (1991). Changes in the distribution of California sea lions north of the breeding rookeries during the 1982–83 El Niño. In "Pinnipeds and El Niño: Responses to Environmental Stress" (F. Trillmich and K. A. Ono, eds.), Ecological Studies 88, pp. 129–137. Springer-Verlag, Berlin.

Lowry, M. S., Oliver, C. W., and Wexler, J. B. (1986). The food habits of California sea lions at San Clemente Island, California; April 1983 through September 1985. NOAA Admin. Rept. LJ-86-33, La Jolla, CA.

Maravilla, C. O., and Lowry, M. (1996). Censos de pinnipedos en islas de la costa occidental de la peninsula de Baja California, Mexico (Julio/Agosts, 1992). INP. SEMARNAP, Ciencia Pesquera No. 13. Mexico.

Morales, V. J. B. (1990). Parametros reproductivos del lobo marino en la Isla Angel de la Guarda, Golfo de California, Mexico. Tesis de Maestria, Fac. de Ciencias, UNAM, Mexico, D. F.

Ono, K. A., Boness, D. J., and Oftedal, O. T. (1987). The effect of a natural environmental disturbance on maternal investment and pup behavior in the California sea lion. Behav. Ecol. Sociobio. **21**, 109–118.

Peterson, R. S., and Bartholomew, G. A. (1967). "Natural History and Behavior of the California Sea Lion." Amer. Soc. Mammal. Special Publ. No. 1. Allen Press, Lawrence, KS.

Salazar, S. K. (1999). Dieta, tamano poblacional y interaccion con desechos costeros del lobo marino Zalophus californianus wollebaeki en las Islas Galapagos. Disertacion de Licenciatura, Pontificia Universidad Catolica del Ecuador.

Sanchez, A. M. (1992). Contribucion al conocimiento de los habitos alimentarios del lobo marino Zalophus californianus en las Islas Angel de la Guarda y Granito, Golfo de California, Mexico. Tesis Profesional, UNAM, Mexico, D.F.

Schusterman, R. J., and Krieger, K. (1986). Artificial language comprehension and size transposition by a California sea lion (Zalophus californianus). J. Comp. Physiol. **100**, 348–355.

Trillmich, F., and T. Dellinger (1991). The effects of El Niño on Galapagos pinnipeds. In "Pinnipeds and El Niño: Responses to Environmental Stress" (F. Trillmich and K. A. Ono, eds.), Ecological Studies 88, pp. 66–74. Springer-Verlag, Berlin.

Zavala, G. A. (1990). La poblacion del lobo marino comun Zalophus californianus californianus (Lesson, 1828) en las islas del Golfo de California, Mexico. Tesis profesional, Fac. de Ciencias, UNAM.

Callosities

MASON T. WEINRICH
Whale Center of New England, Gloucester, Massachusetts

Perhaps no external feature on any baleen whale is as distinctive as the hardened patches of skin, called callosities (pronounced cal-OS-it-ies), found on the head of North Atlantic, North Pacific, and southern right whales (Eubalaena glacialis, E. japonica, and E. australis, respectively; Fig. 1). These features are characteristic of the genus and are immedi-

Figure 1 *The head of a northern right whale* (Balaena glacialis) *showing the most prominent callosity (usually called the bonnet).*

ately notable and visible upon sighting the whale. Old whalers called the most visible callosity, on the tip of the rostrum, the "bonnet"; that name has stuck through the present day.

The term callosity gained acceptance in the first part of the 20th century. The word extends from the term "callus," which refers to a variety of thickened tissues in many species. A variety of terms have been used in the past, primarily based on the function assumed by the author. Speculation included the possibility that the callosity was an "excrescence" (a commonly used term in the late 19th century) from BARNACLES found on the head, abrasions from rubbing the head, or that they were irritations from WHALE LICE. A number of whalers and scientists have noted the coincident occurrence between hair clusters on the whale and callosities, and Payne *et al.* (1983) noted that callosities also occur in the same locations as facial hair in humans, e.g., above the eyes, between the nostrils (blowholes) and upper lip, and on the skin covering the mandible. In fact, there are several locations where callosities are found and hairs are not present, and vice versa (e.g., small callosities are often also found immediately posterior to the nostrils). However, large callosities have at least a scattering of hairs over their surface, and smaller ones often have a single hair near the center of the callosity.

Callosities are a naturally occurring physical feature of the whale. They have been reported to be visible on both late-term (2.5 m) fetuses and newborn calves, although in some cases they may develop shortly after birth. In calves, the callosities are smooth and gray in color, but quickly acquire a pitted, jagged texture. It is thought that this may come from whale lice (*Cyamid* spp.), which, through eating a portion of the skin which comprises the callosity, create an area of lowered laminar flow in which they could adhere more easily to the whale. Certainly whale lice are widely present over the surface area of callosity tissue. To date, *Cyamus ovalus* and *C. gracilis* have been found on both northern and southern right whales. In addition, *C. erraticus* has also been reported from southern right whales. While the callosities maintain their gray color throughout life, they often appear white, yellow, or orange because of the coloration of the whale lice living there.

The function of callosities remains unknown. Male right whales have, on average, a greater portion of the surface area of their head covered by callosity tissue than females. These may be used by males in mating competition, and observations have been reported where males in mating groups deliberately ran the dorsal side of their heads along the backs of other males, with the recipient of the scrape reacting by "twisting and writhing" (Payne and Dorsey, 1983). Given the sensitivity of cetacean skin, it would be likely that contact from the callosity of another animal would be painful. While use in competition may account for the greater amount of callosity tissue in males, it does not explain why callosities are also present in females.

In the past 30 years callosities have received increased attention from cetologists photo-identifying individual right whales, as the shape and area of the callosities vary between individuals. In the southern right whale, individual identification is facilitated by a configuration where the bonnet covers only the front portion of the rostrum, and there are several additional "rostral islands" between the bonnet and the blowholes. This is referred to as a "broken" callosity. Researchers can then use the shape of the bonnet in addition to the number, location, and shape of the rostral islands to identify individuals. In the North Atlantic right whale, however, the bonnet can cover the entire area between the tip of the rostrum and the blowholes, referred to as a "continuous" callosity (found rarely, but occasionally, in southern right whales). Identification is further confounded because whale lice on and around the callosities are mobile, significantly masking the true edge of the callosity. Callosity edges in the same animal can then appear different in several sightings of the same individual. By using additional distinctive features, including the three-dimensional configuration of the callosities, additional scars or marks, and crenulations on the lower lip, North Atlantic researchers have still been able to reliably identify each individual. Photographic catalogs of identified right whales, primarily of their callosity patterns, have been published for the North Atlantic (by the New England Aquarium, Boston, MA) and Peninsula Valdez, Argentina (by the Whale Conservation Institute, Lincoln, MA).

Additional collections of photographs of individual right whales based on callosities and other natural markings exist in various institutions around the world.

See Also the Following Articles

Identification Methods ■ North Atlantic, North Pacific, and Southern Right Whales ■ Whale Lice

References

Kraus, S. D., Moore, K. E., Price, C. A., Crone, M. J., Watkins, W. A., Winn, H. E., and Prescott, J. E. (1986). The use of photographs to identify individual northern right whales (*Eubalaena glacialis*). *Rep. Int. Whal. Comm.* **10**, 145–151.

Payne, R., Brazier, O., Dorsey, E. M., Perkins, J. S., Rowntree, V. J., and Titus, A. (1983). External features in southern right whales (*Eubalaena australis*) and their use in identifying individuals. *In* "Communication and Behavior of Whales" (R. Payne, ed.), pp. 371–445. Westview Press, Boulder, CO.

Payne, R., and Dorsey, E. M. (1983). Sexual dimorphism and aggressive use of callosities in right whales (*Eubalaena australis*). *In* "Communication and Behavior of Whales" (R. Payne, ed.), pp. 295–329. Westview Press, Boulder, CO.

Rowntree, V. (1983). Cyamids: The louse that moored. *Whalewatcher* (*J. Am. Cetacean Soc.*) **17**(4), 14–17.

Captive Breeding

Daniel K. Odell

SeaWorld, Inc., Orlando, Florida

Todd R. Robeck

SeaWorld Inc., San Antonio, Texas

I. Marine Mammals in Captivity

Animals have been held in CAPTIVITY in one form or another for hundreds, if not thousands, of years. Private collections turned into "public" collections. A private animal collection at Schloss Schönbrunn, Vienna, Austria, was opened to the public in 1765 and is considered to be one of the first modern zoos.

The first marine mammals to be held in captivity may have been polar bears (*Ursus maritimus*) and various pinnipeds. Reeves and Mead (1999) provide an excellent overview of marine mammals in captivity. Harbor porpoises (*Phocoena phocoena*) may have been held as early as the 1400s, polar bears since about 1060, and walruses (*Odobenus rosmarus*) since 1608. As with terrestrial animals, marine mammals were held in private collections.

However, most pinnipeds and some sirenians were not held in captivity until the late 1800s and early 1900s. Being considerably more difficult to capture, transport, and maintain, cetaceans, with few exceptions, have only been held in captivity since the mid-1900s. According to Reeves and Mead (1999), 4

species of sirenians, 33 pinnipeds, 51 cetaceans, the polar bear, and the sea otter (*Enhydra lutris*) have been held in captivity. Of these, 2 species of sirenians, 15 cetaceans, 22 pinnipeds, the polar bear, and the sea otter have reproduced in captivity. Among these, however, only a few species, such as the polar bear, California sea lion (*Zalophus californianus*), harbor seal (*Phoca vitulina*), bottlenose dolphin (*Tursiops truncatus*), and killer whale (*Orcinus orca*), have enough numbers and have been reproductively managed with enough production to be considered part of a successful captive breeding program.

Successful captive breeding of any marine mammal requires more than just holding the animals in a pool. Not only does it, as a bare minimum, require sexually mature males and females, but also adequate housing, nutrition, and consideration of the animals' social needs. It becomes obvious when analyzing the history of successful births and survivorship of these species in captivity that early animal managers had little thought or, in some cases, knowledge of the requirements necessary for the development of successful breeding programs. Contrast past records of breeding and survivorship with recent trends beginning in the mid-1980s where captive breeding successes have equaled or, in some cases, surpassed the best scientific estimates of wild population breeding and survivorship and one can see just how far the captive marine mammal community has evolved. Detailed censuses of captive marine mammals in North America (Andrews *et al.*, 1997; Asper *et al.*, 1990) have shown the increasing numbers of captive-bred marine mammals, particularly California sea lions, harbor seals, and bottlenose dolphins. In 1995, 70% of the California sea lions, 56% of the harbor seals, and 43% of the bottlenose dolphins on display in North American facilities were captive-born. This compares with 3, 4, and 6%, respectively, in 1975.

II. Why Breed Marine Mammals in Captivity?

A. Legal Necessity

In the "early days," when animals were just being displayed as "curiosities," it was easier to collect replacements when animals died. In most countries today this is simply not possible. Despite only a few species of marine mammals being threatened or endangered, they and their habitats are protected by a myriad of national and international laws and regulations. While a number of countries are considering regulation of the minimum conditions (e.g., pool size and volume, water quality, food quality and handling, medical care) under which marine mammals may be held in zoos, marine parks, or research facilities, apparently only the United States has such regulations in place. Even though the natural habitat of most marine mammals cannot be duplicated in captivity, the trend is toward larger, more complex, habitat-oriented displays and exhibits. Together, the various laws and regulations have reduced the collection of wild marine mammals and have eliminated smaller facilities that did not have the financial resources to adequately provide for their animals as required by law.

For obvious reasons, the just-mentioned laws and regulations favor captive breeding programs. A successful captive breeding program eliminates costly field expeditions and animal transports. Wild-caught animals are often of uncertain health status and may require routine treatments such as de-worming.

B. Maintaining/Enlarging a Captive Population

A successful captive breeding program can provide animals for other institutions with adequate holding facilities but without the financial resources to maintain a breeding colony or (if even possible) to collect wild animals. Captive-born animals have a known medical history and, to some extent, are imprinted on their keepers. Captive breeding programs have, out of necessity, reduced the impact on wild populations of marine mammals. Professional organizations such as the American Zoo and Aquarium Association (AZA) have established studbooks and other programs to assist with the management of captive animal breeding colonies. For example, studbooks track individual animals from birth to death and their reproductive histories. Computer programs are used to pair animals to optimize genetic diversity. In the United States, formal studbooks exist for the beluga whale (*Delphinapterus leucas*), common bottlenose dolphin, West Indian manatee (*Trichechus manatus latirostris*), gray seal (*Halichoerus grypus*), harbor seal, northern fur seal (*Callorhinus ursinus*), and polar bear and several others are under development. Similar studbooks are in place on other continents. The AZA also hosts a Marine Mammal Taxon Advisory Group whose function is to promote managed captive breeding of marine mammals.

C. A Breeding Program as a Scientific Resource

A successful captive breeding program is a unique scientific resource in that it allows one to document, in great detail, various aspects of reproductive behavior, reproductive physiology, and the subsequent birth, growth, and development of the off-spring. This is particularly valuable for cetaceans, which are typically difficult to study in great detail in their natural habitat due to various environmental factors and the fact that they spend most of their lives underwater. It is, however, most important to recognize that studies on captive marine mammals, even in the best breeding colonies, do not and cannot replace field studies. Both types of studies (i.e., laboratory and field) are necessary to fully describe the biology of a species.

Routine components of a proper animal husbandry program include regular physical exams, collection of blood, urine, and fecal samples, body measurements, and body weights. These samples and data are virtually impossible to collect from wild marine mammals. Consider, for example, what it would take to get a daily urine sample from a wild bottlenose dolphin or killer whale! Captive animals are easily conditioned to provide urine samples and to station for blood sampling, body measurements, and so on. Figures 1, 2, and 3 illustrate the kinds of observations that can be made and the kinds of data that can be gathered.

D. A Breeding Program as a Conservation Resource

A successful captive breeding program may provide the physical and human resources necessary to save some species of marine mammals in imminent danger of extinction (see Ralls

Figure 1   *Birth of a killer whale at SeaWorld Florida.*

and Meadows, 2000). These resources may allow us to maintain a viable gene pool until the habitat can be restored or other reasons for endangerment are eliminated. While such an undertaking is certainly honorable and the right thing to do, the magnitude of the job should not be underestimated and there are, at the present time, obvious limits based in good measure on the sheer size of the animals. For example, the population of right whales (*Eubalaena glacialis*) in the western North Atlantic Ocean is about 350 animals and may be decreasing due to human activities. These animals may reach lengths of 18 m and weights on the order of 20 tons. No facility in existence today (or likely to be in the foreseeable future) could hold a breeding group of right whales. However, threatened or endangered marine mammals such as the Hawaiian and Mediterranean monk seals (*Monachus schauinslandi* and *M. monachus*, respectively) and the river dolphins (families Platanistidae, Iniidae, Pontoporiidae, and Lipotidae) could be maintained in viable captive breeding colonies.

III. Assisted Reproductive Technologies

The domestic animal industry (cattle, hogs, horses) long ago realized that it is much more efficient to move genetic material (e.g., semen) to different facilities than it is to move the whole animal. Methodologies for semen collection, preservation, and transportation, along with methods for artificial insemination (AI), induction of ovulation, and even embryo transplantation, were developed and are in widespread use worldwide today. Some of the techniques have been applied successfully to endangered animals. It seems a logical next step in the captive breeding of marine mammals to apply some of these techniques.

The development of AI, the most common assisted reproductive technology (ART), for commercial use began in the 1950s. The successful application of AI and other ART to domestic species and humans was in part because of their accessibility for research into their reproductive mechanism or reproductive physiology. Of critical importance for the successful application of ART in any species is the determination of how reproductive hormones and behavior relate to ovulation (Sorensen, 1979).

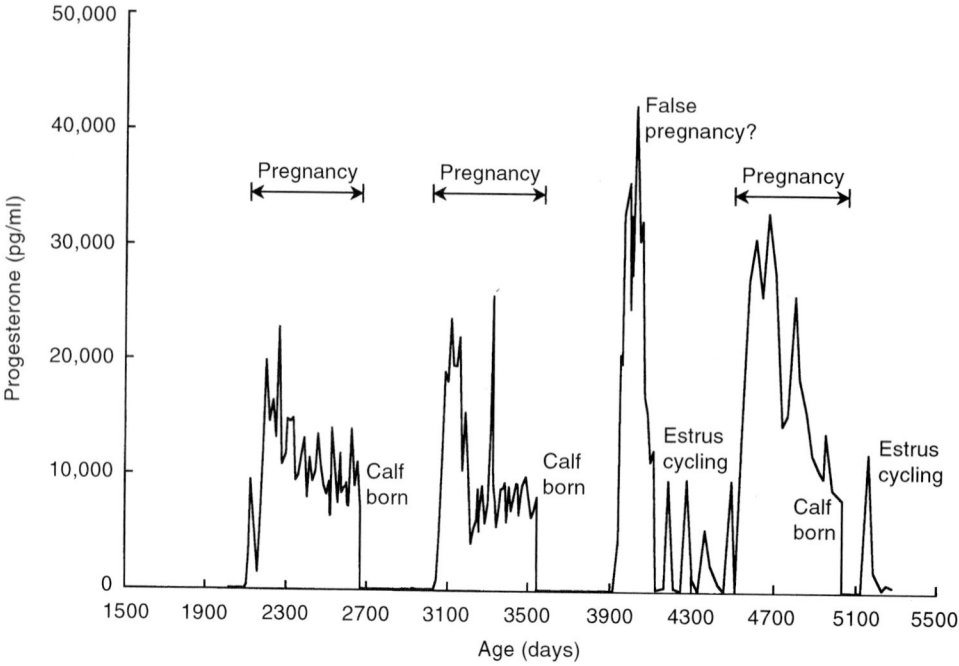

Figure 2 *Serum progesterone of killer whale from birth through sexual maturity and calving and showing estrous cycling and pregnancies. Progesterone levels were "zero" until after 1900 days of age.*

Once ART were developed in domestic species, it was naively believed that they could easily be transferred to exotic species. However, relatively little success was realized with exotic animals until the late 1970s when an endocrinological breakthrough occurred. This breakthrough was the ability to analyze reproductive hormones in urine. This technology was successfully used to characterize the endocrinological events in a wide range of exotic species, including for the first time marine mammals, or the killer whale (Walker *et al.*, 1988). The killer whale has proven to be easily trained to provide urine samples for endocrinologic evaluation and, as such, has provided detailed insight into its reproductive function. Despite this early success, no further attempts at applying ART to this species have occurred until recently.

The most common marine animal in captivity, the bottlenose dolphin, has been the subject of a few intensive investigations into its reproductive function. These investigations include the development of semen cryopreservation techniques and attempts at inducing or regulating ovulation (Robeck *et al.*, 1994). Despite the information obtained, detailed information concerning endocrinological events as they relate to ovulation has not been obtained and, as a result, AI was never successfully developed.

What then are the needs for the development and application of ART to marine mammals? The most obvious ART that should be developed and the one that is most likely to have an immediate impact on the genetic management of captive marine mammals is artificial insemination. Once developed, AI would provide an immediate mechanism for marine mammal managers to increase the genetic fitness of their respective populations without having to rely on the transportation of animals between facilities. Wild animal population studies have

shown that dolphins develop strong social ties to other animals and that these bonds can be maintained for the life of the animals. To what extent these bonds are important for the health of these animals can only be speculated, but it seems prudent for managers who have groups of compatible animals to hesitate changing the social structure by either removing or introducing more conspecifics. Further, bringing animals from other locations may expose the new population of animals to bacterial or viral organisms, which they have no natural resistance against.

What steps are required to develop AI? As stated earlier, one must first be able to monitor, by visualization (surgically or through sonography) or endocrine methods, ovarian activity. Endocrine markers can predict ovulation only after they have been related to actual visualization of the ovarian activity. The ability to consistently find and observe ovarian activity transabdominally with ultrasonography was first demonstrated in the bottlenose dolphin by Brook (1997). This simple technique has since been successfully utilized to consistently locate ovaries in many other captive marine mammals, including the killer whale, the Pacific white-sided dolphin (*Lagenorhynchus obliquidens*), and the false killer whale (*Pseudorca crassidens*) (Fig. 4). This ability then opens the door, with frequent body fluid sampling (most likely urine) for the identification of an effective endocrine marker of ovulation. Once ovulation can be predicted, then the ability to collect semen is an obvious requirement. Currently, only a few marine mammal species have been successfully collected on a regular basis. Early on during the process of trying to collect semen from these animals, electroejaculation was utilized with limited success. However, stress caused to the animals during collection and inconsistent results

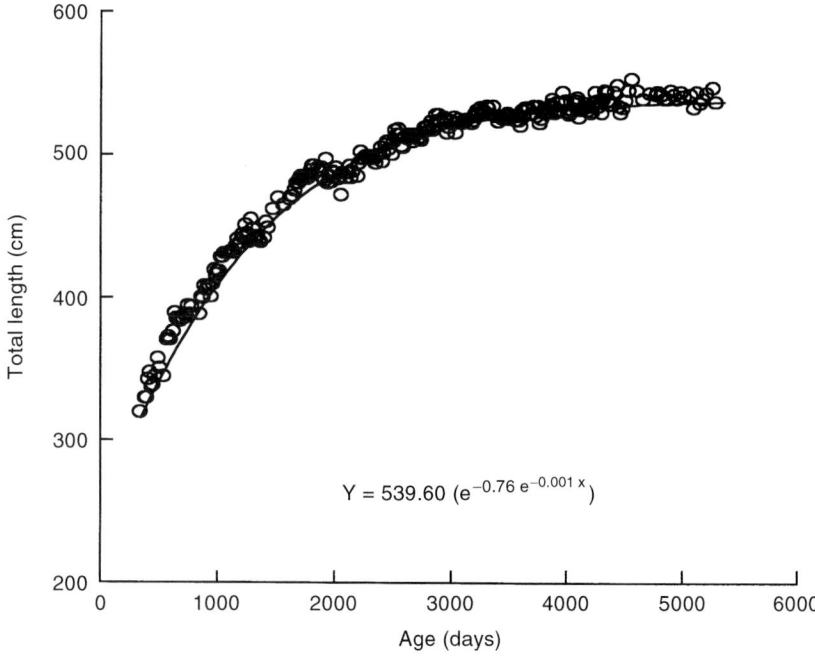

Figure 3 *Killer whale growth (length) curve as an example of data that can be gathered from captive-bred animals.*

$$Y = 539.60 \, (e^{-0.76 \, e^{-0.001 \, x}})$$

made this method unsuitable. Also, it was determined that bottlenose dolphins, animals that seem to have consistently elevated libidos, were relatively easily trained to provide semen via manual stimulation. Since the success with bottlenose dolphins, a killer whale and two Pacific white-sided dolphins have been trained to provide semen on a regular basis. Obviously, for AI to be widespread, all genetically valuable males should be trained to provide semen.

Once the semen has been collected, it must be stored temporarily for immediate use or permanently by cryopreservation.

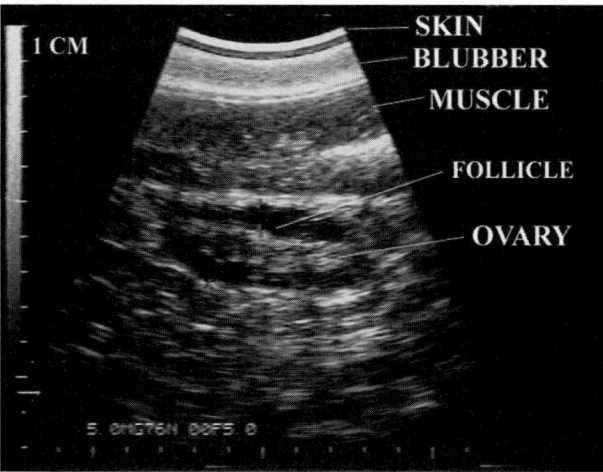

Figure 4 *Sonographic image of an ovary of a Pacific white-sided dolphin showing a follicle.*

Semen cryopreservation has been successfully accomplished by different methods (straws, pellets, or cryovials) in all of the species for which semen has been obtained. Although cryopreserved semen provides a long-term supply of genetic material from a particular animal and is the only method currently available to store semen for long (greater than a few days) periods of time, when used with AI it results in reduced conception rates. The reduced conception is believed to be the result of damage that occurs to the semen during the freezing process. As a general rule, cryopreserved semen must be deposited in the uterus and at close proximity to ovulation. Fresh cooled semen, having better motility and viability, can be placed with less accuracy, generally in the cervix, and with a greater time interval prior to ovulation. Thus, when attempting to develop AI, first attempts are generally made using fresh cooled semen.

As must be obvious from this discussion, a thorough understanding of female reproductive tract anatomy is essential to developing effective catheterization techniques for semen deposition. Despite the close phylogenetic relationship between captive delphinids, significant variation exists in their reproductive tract morphology. For example, bottlenose dolphins, being the best documented of captive species, have a pseudocervical vaginal fold or flap that must be traversed before encountering the true cervix. The space formed by the fornix surrounding the true cervix and the distance from the pseudocervical flap to the distal opening or external os of the cervix has been identified as the spermathecal recess. Deposition of fresh semen into this space (as opposed to vaginal placement) appears mandatory for a successful procedure. However, successful use of frozen semen for AI will probably require navigation through the pseudo and true cervix for uterine deposition. Knowledge of the anatomy of the bottlenose dolphin would do little to help placement of semen into the closely related member of the delphinid family, the Pacific white-sided dolphin. This species has a series of three annular folds that present anatomical barriers to the placement of semen. Finally, another member of the delphinid family, the killer whale, has a completely different arrangement consisting of two longitudinal ridged cervices in series.

Once the semen has been deposited and ovulation has occurred, how do you determine if you have had success? With the application of ultrasound for pregnancy detection, more and more marine mammal practitioners have become aware of the fact that all species of captive delphinids can exhibit variable periods of false pregnancies. These periods of elevated progesterone are currently endocrinologically indistinguishable from pregnancy. Thus, pregnancy can only be confirmed with the use of ultrasound.

Once AI has been developed successfully, other ART may begin to have effects in captive and potentially wild animal populations. One of the most intriguing developments in ART is the continued improvement of semen sexing techniques. Currently, semen sexing has been used to influence the sex ratio by close to 90% in either direction. However, the semen is damaged in the process, resulting in a significant reduction in viability. This translates into a requirement for interuterine or oviduct semen deposition. Unless or until the technique can be improved, semen deposition might be possible with laparoscopic techniques currently and tenuously being applied to cetaceans.

IV. Challenges for the Future

If self-sustaining populations are to be developed, then all reproductively capable animals must contribute to the gene pool. Producing viable offspring is the primary goal of any captive breeding program, but it must be managed to avoid overpopulation of the facility and to prevent or minimize inbreeding. This may require separate facilities for adult males, preparturient females, and females with new offspring. Breeding may be regulated simply by separating the sexes or by neutering (physically or chemically) both males and females.

Currently, many populations of marine mammal are kept in small groups or same-sex groups. Therefore, a large portion of reproductively mature animals within these groups are not reproducing. These animals are functionally excluded from contributing to the collective captive gene pool. Therefore, valuable financial resources and pool space are being used for a minority of the available genetic lines. This inefficient use of animal resources must be corrected before long-term population stability can be achieved. Judicious use of contraception and development of AI would help managers reach maximum utilization levels. However, if the maximum utilization of genetic resources does not result in a predictable stable population, then genetic infusions from wild stocks will be necessary. ART may provide another answer to this future dilemma if semen and possible oocytes can be collected from wild animals that are incidentally or purposely killed in fisheries or other commercial activities. These gametes can be placed into suitable captive animal recipients for reproductive purposes. As an example, researchers in Japan have created embryos by IVF (*in vitro* fertilization) from gametes collected postmortem from fin whales. Unfortunately, suitable captive recipients for this species do not exist. As an alternative to the ART solution or until ART were perfected, mangers could "borrow" adult males from wild populations for 1–2 years for breeding purposes and then return them. This, of course, involves considerable expense and there is no guarantee that any given animal would breed successfully.

See Also the Following Articles

Conservation Biology ■ Endangered Species and Populations ■ Hybridism ■ Reproductive Behavior

References

Andrews, B., Duffield, D. A., and McBain, J. F. (1997). Marine mammal management: Aiming at the year 2000. *IBI Rep.* **7**, 125–130.

Asper, E. D., Duffield, D. A., Dimeo-Ediger, N., and Shell, D. (1990). Marine mammals in zoos, aquaria and marine zoological parks in North America: 1990 census report. *Int. Zoo Yearbook* **29**, 179–187.

Brook, F. (1997). The use of diagnostic ultrasound in assessment of the reproductive status of the bottlenose dolphin, *Tursiops aduncus*, in captivity and applications in management of a controlled breeding programme. The Hong Kong Polytechnic University, Kowloon, Hong Kong.

Ralls, K., and Meadows, R. (2000). Captive breeding and reintroduction. *In* "Encyclopedia of Biodiversity." Academic Press, New York.

Reynolds, J. E., III and Rommel, S. A. (eds.) (1999). "Biology of Marine Mammals." Smithsonian Institution Press, Washington, DC.

Robeck, T., Curry, B.E., McBain, J.F., and Kraemer, D.C. (1994). Reproductive biology of the bottlenose dolphin (*Tursiops truncatus*) and the potential application of advanced reproductive technologies. *J. Zoo Wild. Med.* **25**, 321–336.

Sorensen, A. M. (1979). "Animal Reproduction: Principles and Practices." McGraw-Hill, New York.

Twiss, J. R., Jr., and Reeves, R. R. (eds.) (1999). "Conservation and Management of Marine Mammals." Smithsonian Institution Press, Washington, DC.

Walker, L. A., Cornell, L., Dahl, K. D., Czekala, N. M., Dargen, C. M., Joseph, B. E., Hsueh, A. J. W., and Lasley, B. L. (1988). Urinary concentrations of ovarian steroid hormone metabolites and bioactive follicle-stimulating hormone in killer whales (*Orcinus orca*) during ovarian cycles and pregnancy. *Biol. Reprod.* **39**, 1013–1020.

Captivity

PETER CORKERON

James Cook University, Townsville, Queensland, Australia

I. The Debate

The debate over the ethics of marine mammals in captivity is essentially about cetaceans (whales, dolphins, and porpoises) because other marine mammals such as seals and sea lions do not inspire the same passion as whales and dolphins. The allure of cetaceans is so powerful that they are now a symbol of the animal liberation movement. This ongoing debate over whether cetaceans should be kept in captivity is relatively recent in contrast to the history of human/marine mammal interactions. For centuries, human interest in marine mammals was based on the commercial value of killing seals and whales for oil, meat, and hides. The larger animals represented greater profit, so small marine mammals such as dolphins were mostly considered to be pests to fishermen. Occasional reports of marine mammals being kept in captivity as curiosities are scattered throughout history: polar bears were kept by Scandinavian rulers prior to the Middle Ages; a killer whale (*Orcinus orca*) that had been live stranded was kept and used for sport by Roman guards during the first century A.D.; and seals were kept in menageries by the 18th century. In the mid-1800s, P.T. Barnum displayed belugas (*Delphinapterus leucas*) and bottlenose dolphins (*Tursiops truncatus*) in his New York museum for a short time, and in the late 1800s, the Brighton Aquarium in England displayed harbor porpoises (*Phocoena phocoena*) for several months. A new era of modern marine mammal exhibits began in the late 1930s, when Marine Studios at Marineland opened in Florida.

At first, marine mammal facilities were quite popular. During the 1950s, 1960s, and 1970s, the number of aquaria and zoological parks displaying marine mammals increased rapidly to meet public demand, especially in Europe, North America, and Australasia. Simultaneously, technology and methods for the capture, transport, and maintenance of marine mammals improved with increasing knowledge and experience. Scientists took advantage of the availability of the animals at these facilities to conduct groundbreaking research on dolphin acoustics, human/dolphin communication, dolphin brain function, hearing and echolocation, and behavior. Soon the public became more familiar with dolphins through shows at aquaria and from a popular TV show called "Flipper" in which a dolphin was portrayed as an intelligent family pet. With this heightened awareness, the public began to understand that dolphins were not large fish, but intelligent and friendly marine mammals. However, the image transformation inspired by animals at marine mammal facilities would soon become a public relations nightmare for the industry that first made these animals popular.

II. The Impact

The boom in aquaria and oceanaria experienced through the 1970s came to a near halt during the mid-1980s due to the growing debate over keeping cetaceans in human care. Pressure and sometimes inaccurate information from the animal rights lobby forced the closure of some existing facilities and prevented some new facilities from being opened. While the 1990s saw a decline in the number of facilities keeping cetaceans in Australia and some parts of Europe, there was an increase in the establishment of new marine mammal facilities in other parts of the world such as Asia and South America where there was little to no animal activist movement.

Compared with most terrestrial mammals, marine mammals are expensive and difficult to maintain in captivity. They require a good deal of logistical support, such as high-quality food sources, specialized medical care, large enclosures, and expensive water-quality maintenance systems. Cetaceans and sirenians (manatees and dugongs), being wholly aquatic, present greater logistic difficulties than other marine mammals. A few species predominate at these facilities because they have shown greater success in human care. Of these animals, those most often used in public performances include bottlenose dolphins, belugas or killer whales, and California sea lions (*Zalophus californianus*), whereas phocids such as harbor seals (*Phoca vitulina*), polar bears (*Ursus maritimus*), sea otters (*Enhydra lutris*), and sirenians are more typically maintained in exhibits. While a few facilities, especially those in developing countries, have fallen short of being able to maintain the highest standards of operation, most modern facilities successfully meet these challenges. However, there is word that many oceanaria are being planned to be built outside the control of effective regulations in the coming years—more than 13 in China alone. There is concern that many capture and holding facilities that are effectively unregulated and pretty much undercover are being built in the Philippines, Indonesia, and other places to supply the resulting demand for animals. It is on these types of facilities that we should focus.

Even if for only short periods of time, almost every species of marine mammal other than most of the great whales have been maintained at a marine mammal facility at some point. Animals are kept captive for different reasons: display in zoos and aquaria; military work; scientific research; and temporary maintenance for rehabilitation for injured or sick animals, although these categories are not mutually exclusive. Size and temperament are generally the limiting factors in keeping some marine mammal species for long periods of time. Most are kept in zoos or aquaria; some live in open ocean enclosures. However, many commercial facilities with cetaceans are not traditional zoological gardens but marine parks where there tends to be more emphasis on performances by animals. (Temporary restraint from a few minutes to hours for research purposes is not considered as captivity here.)

III. Regulations for Collection, Care, and Maintenance
A. International Regulations

In general, regulations dealing with marine mammals in human care cover collection, care and maintenance of animals, and movements of animals between countries. The extent to which existing laws are administered and enforced varies internationally. The INTERNATIONAL WHALING COMMISSION (IWC) is the central international instrument for the protection of whales; however, its effectiveness has been hotly debated. Both the pro-whaling and the anti-whaling factions try to load the IWC to influence the outcome of voting, and qualifying animal activist groups are allowed to attend and lobby at annual IWC meetings.

The major instrument regulating the international trade in captive cetaceans is the Convention on International Trade in Endangered Species of Wild Fauna and Flora (CITES). Most cetaceans are listed in CITES Appendix II, which provides a means of regulating and monitoring trade for species not threatened with extinction but which are vulnerable to overuse. This listing allows for international trade with properly issued permits. The remaining species are listed in Appendix I of CITES indicating that they are either currently threatened with extinction or may be affected by trade. The smaller species/subspecies include the baiji or Chinese river dolphin (*Lipotes vexillifer*), Ganges river dolphin (*Platanista gangetica gangetica*), Indus River dolphin (*Platanista g. minor*), tucuxi (*Sotalia fluviatilis*), Indo-Pacific humpbacked dolphin (*Sousa chinensis*), Atlantic humpback dolphin (*Sousa teuszii*), finless porpoise (*Neophocaena phocaenoides*), and the vaquita (*Phocoena sinus*). With regard to marine animals, the CITES convention includes the requirement that suitable housing and care are available and, for Appendix I-listed animals, that they are not to be used for "primarily commercial purposes." Most nations keeping marine mammals in captivity are signatories to CITES. International agreements can regulate capture, but most trade in wild-caught animals now comes from a few nations with few or no regulations on capture, or with regulations that are ignored. International pressure on some nations (e.g., Iceland in 1989) has resulted in the closure of their capture industry.

In the United States, Congress passed unprecedented regulatory legislation in 1972 to bring under its protection all marine mammals within the borders of its jurisdiction. The legislation was called the Marine Mammal Protection Act (MMPA). Its intent was to protect marine mammals from human actions (predominantly fishing) that lead to extinction. However, the MMPA specifically authorized the collection of animals from the wild for scientific research and public display and education. Depending on the species involved, the collection of marine mammals is governed by a permit process administered by either the National Marine Fisheries Service or the Fish and Wildlife Service. The standards for the maintenance of marine mammals in research or public display facilities are established and monitored by the Animal and Plant Health Inspection Service under the U.S. Animal Welfare Act, and all marine mammal-related activities are monitored by the presidentially appointed Marine Mammal Commission. While collection is still permitted in the United States, there have not been any bottlenose dolphins collected for U.S. facilities since 1989 due to a self-imposed moratorium observed by the members of the Alliance of Marine Mammal Parks and Aquariums on the collection of bottlenose dolphins (*Tursiops truncatus*) from the Gulf of Mexico. In addition, there is a changing trend regarding holding and breeding bottlenose dolphins in North America. In 1976, 94% of the population of bottlenose dolphins were wild caught while 6% were captive born. From 1989 to 1996, no wild-caught animals were added to the captive population; all additions were a result of successful breeding programs. In 1996, 44% of bottlenose dolphins in North American facilities were captive born.

Several countries have legislated regulations or guidelines to govern the collection and keeping of marine mammals since 1972. New Zealand passed a MMPA in 1978, and in 1980, Australia passed the Whale Protection Act of Australia. In Australia, the state of Victoria banned the issuance of permits for keeping cetaceans for display or collecting them for export. Here, legislation does not absolutely preclude issuing a permit for the capture of free-ranging animals; however, general government policy, the legal requirement for public comment on

an application for capture, and the need for signed Ministerial approval for capture permits mean that it is highly unlikely that permits will be issued. Canada developed guidelines that forbid the capture of killer whales and gives priority to Canadian institutions in considering permits for the capture of belugas. Legislation can interact with government policy and public opinion to affect the capture industry. For example, guidelines established in some countries such as the United Kingdom do not specifically prohibit the collection and display of marine mammals, but they effectively force closure of some facilities by making it almost impossible for facilities to meet building codes and specifications. This is perhaps epitomized by British cetacean display facilities, the last of which closed in 1991. Movements of cetaceans into and within the European Union are regulated under EU wildlife trade regulation, established to fulfill EU member nations responsibilities under CITES. Trade in animals listed under Annex A of this regulation (including all cetaceans) is permitted for "research or education aimed at the preservation or conservation of the species," breeding for conservation, and biomedical research (the latter is not relevant for cetaceans), but not simply for commercial use. Other places such as the Republic of South Africa still permit the capture and display of cetaceans.

In Japan, multispecies drive fisheries that combine capturing animals alive for aquaria and dead for food have, in some drives, exceeded their allowed quota of animals. United States legislation requires that captures be conducted humanely, effectively denying animals from the Japanese fishery to institutions in the United States.

Regional conservation agreements have also been developed. Several nations that are signatories to the Agreement on the Conservation of Small Cetaceans of the Baltic and North Seas (ASCOBANS) are mandated to take part in conservation and research measures. Such measures include preventing the release of potentially harmful substances into the environment, developing fishing gear to reduce bycatch, and reducing the impact of other potentially harmful human activities. They are obliged to prohibit intentional killing of small cetaceans and to release immediately any healthy small cetaceans incidentally caught.

B. Care and Maintenance

The first published accounts of the behavior of captive cetaceans were provided by Charles Townsend in the early 1900s when he was the director of the New York Aquarium. His observations of a group of bottlenose dolphins described their social behavior and some sensory capabilities. Townsend understood the importance of developing health care and water treatment regimes to promote long-term survival. Requirements for the care and maintenance of marine mammals can vary dramatically between countries and between jurisdictions within a country. These requirements include regulations regarding pool dimensions and construction materials; food quality and feeding schedules; water quality; air quality; veterinary care; and educational message. There are countries (e.g., the United States) where the agency responsible for overseeing capture and international transport is different from that responsible for care and maintenance. United States regulations require that facilities importing animals from the United States must meet U.S standards of care and maintenance. These standards should become the *de facto* standards anywhere importing animals from the United States.

A core aspect of the argument against maintaining captive cetaceans is that it is impossible to provide an adequate environment for cetaceans in captivity. The basic reasons put forward for this are that pools can never be of adequate size; that regardless of size, pool construction is inappropriate; and that it is impossible to keep animals in suitable social groups. Even the larger commercial facilities, for example, one that includes a complex of four linked (sand-bottomed, rock-lined) pools of 30,000,000 liters, with a maximum depth of 7.5 m, holding 12 bottlenose dolphins, are considered inadequate to some. Facilities differ greatly in their resources, and so the quality of their environment.

IV. Issues

A. Experiencing Captive Marine Mammals

Zoos and aquariums in North America alone are visited by over a hundred million people each year. A Roper Poll titled "Public Attitudes Toward Aquariums, Animal Theme Parks, and Zoos" taken in the United States in 1992 and again in 1995 revealed that more Americans visit zoos and aquariums in the course of a year than the total of those that attend all professional sporting events. The assumption is that people, having experienced living marine mammals in close proximity, will be more likely to develop (or enhance) their marine conservation ethic. It is clear that public support for marine mammal conservation increased substantially in the latter half of the 20th century. In many parts of the country, images of dolphins are stenciled near storm drains with a reminder that "I live downstream." The extent to which commercial aquaria contributed to this change is questioned by some of those calling for closures, as is the extent to which this remains true today. The public view of killer whales changed radically at the same time as they appeared in captivity, supporting the "captive animals as ambassadors" argument. Unfortunately, there is little current research (by either side) on the extent to which visiting aquaria affects peoples' conservation ethic.

Some anticaptivity proponents suggest that it would be better if people ventured out to the ocean to see cetaceans through commercial whale watching. Whether most people who visit aquaria would go whale watching if the aquaria did not exist is unknown. However, whale watching has become an industry in itself. In 1989, the whale-watching industry off the Atlantic seaboard of the United States alone brought in 23 million dollars. Along with the increasing popularity of these trips, there is also an increasing number of reports of inappropriate behavior of whale watchers and a growing need for regulation. At marine mammal facilities, it is much easier to monitor and guard against inappropriate human behavior than it is out in the open ocean. Additionally, the environmental impact of an extra several million people a year going whale watching has not been estimated, but could be substantial. Most opponents of captivity are in favor of whale watching. However, if viewing captive cetaceans enhances an inappropriate worldview (as opponents of captive cetaceans contend), can large-scale commercial whale watching do the same? The argument that equal or greater conservation benefit could be gained through multimedia presentations remains untested.

Most of this discourse is set in a Western context. Public awareness of marine mammals in most other nations is less developed than it is in the West. It may be that zoos and aquaria can make a significant contribution in these countries.

B. Scientific Value of Captive Marine Mammals

The value of studies conducted with captive marine mammals has also come under scrutiny. However, before field studies of living cetaceans burgeoned (after the late 1970s), captive animals were the major means by which scientists collected data on biology and behavior. Some phenomenon such as echolocation may still have been unknown if not for captive animals. Still today, the echolocation capabilities of most dolphin species remain unknown, and controlled experiments with captive animals are the only way by which we are likely to determine them. Because there are so many other questions that have not or cannot be answered using wild marine mammals, captive animals are still the primary source of data for several fields such as comparative psychology, cognition, physiology, acoustics, toxicology, immunology, reproduction, and medicine. The remarkable sensing abilities of dolphins and their perceived intelligence are, paradoxically, two of the arguments used against their maintenance in captivity.

Most of the opposition to the scientific value of captive marine mammals appears to disregard the value of experiments in biological research generally and of the capacity for "pure" research" to alter peoples' outlook on conservation issues. Interplay remains between work on captive and free-ranging animals. Recent experiments with captive dolphins have provided new insights into the role of bottlenose dolphins' whistle repertoires, allowing development of better focused studies of free-ranging animals. The role of animal language experiments generally in providing direct conservation benefit through redefining our view of humanity's place in the natural world is (like the conservation value of captive facilities or whale watching) difficult to estimate, but may be substantial.

More recently, open ocean training has made it possible to use captive animals to conduct some studies in the wild. Incorporating both wild and captive marine mammals is also gaining favor in recent years. In a report to the Marine Mammal Commission, international experts advised the use of "model [marine mammal] species that have been well studied and are readily available in captivity . . . [to] provide basic insight about the variability in and relationships between contaminants and basic health and physiological processes, biomarkers, reproduction, and survival. Such insight could prove critical to interpretation of contaminant impacts on wild marine mammal populations" (MMC, 1999). Ken Norris, a pioneer in the field of marine mammalogy—both captive and wild—contended that "Both kinds of study are crucial, really" (Norris, 1984).

C. Captive Breeding for Conservation

While some populations of wild marine mammals have been reduced or depleted as a result of high mortality, populations of marine mammals in zoos, aquariums, and research facilities have increased as a result of successful breeding programs. Since the first birth of a bottlenose dolphin in an exhibit facility at Marine Studios in 1947, breeding programs have become more and more successful, increasing the number of animals at these facilities and providing animals to other marine mammal facilities. By 1996, 44% of bottlenose dolphins on exhibit in North America were a result of successful breeding programs. The success of bottlenose dolphin breeding programs is probably due in large part to the fact that more of these animals are available for breeding programs and more is known about the physiology and husbandry of this species than any other cetacean. Although there is still a paucity of data from free-ranging animals, it has also been shown that once acclimated, bottlenose dolphins in well-maintained facilities have survival rates comparable to those in the only free-ranging population that has been studied over a long time.

Breeding animals in zoological facilities has a role in the conservation of some endangered species. It has been suggested that such *ex situ* conservation supports the existence of aquaria. However, the development of such programs should not be an excuse to ignore our responsibility to implement conservation strategies to protect wild populations and their habitat.

There are major difficulties confronting those wishing to develop *ex situ* breeding programs for marine mammals, but the advances in reproductive technologies and the establishment of population management protocols will help overcome those challenges. After all, the successful 1999 birth of a giant panda at the San Diego Zoo in California was a result of artificial insemination. Even though the marine mammal species most commonly maintained in captivity are not endangered, knowledge and technologies developed with these animals can be applied to those that are at greater risk. While permits for the capture or importation of such animals may be difficult to obtain in some countries, demonstrating success with captive breeding programs may help overcome these difficulties.

Opponents to oceanariums worry that taking animals from the wild will threaten wild populations. To prevent this, no captures should be allowed unless stock structure is well understood and the takes can be shown to be sustainable. While in the 1990s, only seven cetaceans, three Pacific white-sided dolphins (*Lagenorhynchus obliquidens*) and four belugas, were collected from North American waters for public display and research in U.S. facilities, marine mammals in the wild are dying in large numbers. Gaskin (1982) warns, "If there is one lesson which modern ecological science has for us, it is that if one protects the habitat, then animal populations can show remarkable resilience in the face of other external pressures, even those such as intensive predation by man."

D. Rehabilitation

At times, free-ranging marine mammals that are ill, injured, or have suffered some misadventure require rehabilitation. In most cases, these are species that are not at risk of extinction, and so the issue is one of animal welfare, not conservation. An exception to this is the efforts made to rehabilitate Florida manatees (*Trichechus manatus latirostris*) where the rehabilitation and subsequent return to the wild of each individual manatee have demonstrable conservation value. Whether or not a species is endangered, with each rescue comes the possibility of acquiring new knowledge that will benefit all marine mammal species whether they be endangered or otherwise. Generally, zoos and aquaria provide needed support to rescue operations whether they be expertise, facilities, medicine, or food, and their role in rehabilitation has been rather significant.

E. Release

Perhaps due to the decrease in collecting animals from the wild, attention has now focused more on releasing those already in facilities. Several attempts have been made to release captive dolphins back to the wild. Following the closure of "Atlantis" in Perth, Western Australia, nine bottlenose dolphins were released in 1992. Animals were radiotracked after a gradual release back to the waters from which some had been caught 11 years previously (three were captive born). After a few weeks, three animals in very poor condition were returned to a sea pen, but the fate of most was unknown due to the failure of the radio tags. The animals that were recovered died of poisoning in late 1999.

The release program for Keiko, the killer whale that starred in the movie "Free Willy," encapsulates some of the issues regarding captivity. Caught as a calf by the Icelandic capture industry, he was imported to Canada and then to Mexico where he was held in an inadequate pool with no other members of his own species. Activists organized his importation to another captive facility in the United States where his condition improved before he was moved to Iceland for a planned release. At this writing, he is in a sea pen and is being trained to catch prey prior to his release. However, the example of the "Atlantis" release demonstrated that it is insufficient to teach cetaceans to catch prey—they need to be able to locate and capture their prey using less energy to do so than the prey provides. Although there are plans for Keiko to be fitted with a satellite tag, recovering him from oceanic waters off Iceland will be logistically challenging if there are indications that his health is deteriorating after release.

Just as the conservation value of zoos and aquaria should be questioned, so too should the value of release projects. Do the perceived benefits of using an individual animal as an ambassador to send a message about captivity outweigh the costs (e.g., funding for conservation projects foregone, greenhouse gas emissions used in the release) and possible risks of the project?

Too little money is available already for needy conservation programs, particularly for projects in developing countries. Clearly, people are more likely to be stirred into supporting a grandiose cause such as the Keiko release rather than more mundane low-profile projects with no immediate emotive attraction. Paul Watson, founder of the Sea Shepherd Conservation Society, stated: "Free Keiko, free Lolita, free Corky, free Hondo. These are wonderful and appealing ideals, but not all captive cetaceans can or should be freed. Not all facilities holding marine animals are the enemy. And the huge sums raised to free a few individuals could be more positively directed toward ending the slaughter of hundreds of thousands of nameless whales, dolphins, and seals on the world's oceans. Never in the history of the animal protection movement have so many given so much for so few and so many given so little for such large numbers" (from the June 1995 edition of Animal People).

F. Funding

Just as some zoos and aquaria contribute significant funds and resources to rescue and rehabilitation programs, some agencies keeping captive marine mammals also support research programs. For example, most of the research into cetaceans' acoustic faculties has been funded through the U.S. Office of Naval Research (ONR). There have been suggestions that the significant contribution made by the ONR to funding cetacean research affects the capacity of scientists to comment openly on issues relating to the U.S. Navy, although this is disputed. Some captive facilities and groups opposed to captivity either employ full-time research staff or provide funding for research projects. The relative funding provided by the two groups varies dramatically between countries, and there are places where one or both of the groups contribute significantly to research efforts.

V. Conclusion

The ongoing debate continues over whether marine mammals, particularly whales and dolphins, should be kept in captivity. Basically the debate between pro- and anticaptivity groups boils down to one thing: does the benefit achieved by holding animals in captivity outweigh the costs involved—both to the individual animal and to the population from which the animal came if all parties view "benefit" as contributing to the conservation of free-ranging marine mammals. Both sides believe their positions to be valid and rational. Neither side can afford to lose sight of the need to identify what is, in fact, better for the animal as opposed to what makes the participants in the debate feel satisfied.

In this debate, scientific data are often ignored because the debate has become based more on personal philosophy than on science, or the value of the science is debated. If nothing else, the debate brings attention to marine mammals, not only to those in facilities, but to those in the wild as well. The passing of the MMPA in the United States effectively curbed the mortality of dolphins in tuna nets from over 400,000 dolphins before the MMPA to less than 4000 in 1995. This is most likely a reflection of a new-found feeling of stewardship for marine mammals evoked by both marine parks housing them and activists trying to release them. However, the ongoing debate may be hurting them as well. Energy and money spent on a debate that may never be resolved may be more helpful to the animals if spent on improving the quality of the oceans. Klamer *et al.* (1991) suggested "If the increase in ocean PCB concentration continues, it may ultimately result in the extinction of fish-eating marine mammals." Ocean warming and other anthropogenic changes in the oceans are also potential threats along with the undermining of legislation such as the Endangered Species Act. Reynolds and colleagues (2000) ask whether the natural world today is a healthier place to live than the captive environment. They state that there is evidence that humans have damaged coastal and marine environments so much that the well-being of some wild dolphins may be in greater jeopardy than collection animals.

See Also the Following Articles

Conservation Efforts ■ Ethics and Marine Mammals ■ Marine Parks and Zoos ■ Whale Watching

References

Andrews, B. (2000). MMTAG: Bottlenose dolphin management past, present and future. *In* "Report from the Bottlenose Dolphin Reproduction Workshop" (D. Duffield and T. Robeck, eds.), pp. 7–15. June 3–6 1999, San Diego, CA.

DeMaster, D. P., and Dervenak, J. K. (1988). Survivorship patterns in three species of captive cetaceans. *Mar. Mamm. Sci.* **4**, 297–311.

Gaskin, D. E. (1982). "The Ecology of Whales and Dolphins," p. 393. Heinemann Educational Books, London.

Lacy, R. C. (2000). Management of limited animal populations. *In* "Report from the Bottlenose Dolphin Reproduction Workshop" (D. Duffield and T. Robeck, eds.), pp. 75–93. June 3–6, 1999, San Diego, CA.

Marine Mammal Commission (1999). "Marine Mammals and Persistent Ocean Contaminants: Proceeding of the Marine Mammal Commission Workshop," Keystone, CO.

Reeves, R. R., and Mead, J. M. (1999). Marine mammals in captivity. *In* "Conservation and Management of Marine Mammals" (J. R. Twiss and R. R. Reeves, eds.), pp. 412–436. Smithsonian Institute Press, Washington, DC.

Reynolds III, J. E., Wells, R. S., and Eide, S. D. (2000). "The Bottlenose Dolphin: Biology and Conservation." University Press of Florida.

Roper Organization Poll (1995). "Public Attitudes Toward Zoos, Aquariums and Animal Theme Parks. Roper Starch Worldwide, Inc.

Rose, G. (1996). International law and the status of cetaceans. Pages 23–53 *In* "The Conservation of Whales and Dolphins" (M. P. Simmonds and J. D. Hutchinson, eds.) pp. 23–53. Wiley, New York.

Samuels, A., and Tyak, P. (2000). Flukeprints: A history of studying cetacean societies. *In* "Cetacean Societies: Field Studies of Dolphins and Whales" (J. Mann, R. D. Connor, P. L. Tyack, and H. Whitehead, eds.), pp. 9–44. The University of Chicago Press, Chicago, IL.

Waples, K. A., and Stagoll, C. S. (1997). Ethical issues in the release of animals from captivity. *BioScience* **47**, 115–121.

Carnivora

RONALD E. HEINRICH
Ohio University, Athens

Living members of the mammalian order Carnivora exhibit an incredibly diverse array of ecological and behavioral adaptations. Among the 11 families and 275 or so extant species are animals that range in body size from 35 g (the least weasel) to 3700 kg (elephant seals) and that exhibit locomotor specializations for terrestrial speed (e.g., cheetah), arboreality (e.g., kinkajous), marine environments (e.g., pinnipeds), and almost everything in between. The origin of this diversity is traced to the miacoids, a paraphyletic assemblage of small and medium-bodied carnivorans that appear in the fossil record 65 million years ago. Miacoids possess carnassial teeth, the key diagnostic character of the order, involving the transformation of the last upper premolar and first lower molar into blade-like structures (Fig. 1 top), that enable carnivorans to effectively shear through the skin and flesh of prey. Carnassials are retained by nearly all members of the order, although they are greatly modified in some taxa such as pinnipeds where they resemble the premolars positioned in front of them (Fig. 1 bottom), an adaptation for grasping fish and other aquatic prey.

I. Diversity and Origin of Basal Carnivorans

Miacoid fossils are restricted to the northern hemisphere with the most abundant materials, primarily fragmentary jaws and isolated teeth, coming from localities in western North America, western Europe, and China. Nearly all of the proposed miacoid taxa are referred to one of two extinct families—Viverravidae and Miacidae—the common ancestor of which probably lived in the latest Cretaceous. Viverravids appear several million years earlier than miacids but are dentally more derived than miacids in having only two molar teeth rather than the primitive carnivoran condition of three. Viverravids ranged in size from as little as 150 g to about 10 kg and their skeletal anatomy suggests that they were predominantly terrestrial, although there is little doubt that at least the smallest members of the family could climb well. While similar in body size (estimates range from 1 to 10 kg), miacids were more arboreal than viverravids, possessing relatively larger and more powerful muscles that would have aided in climbing. The miacid fossil record extends to about 38 million years ago (mya) in North America, almost 10 million years beyond the last known occurrence of viverravids.

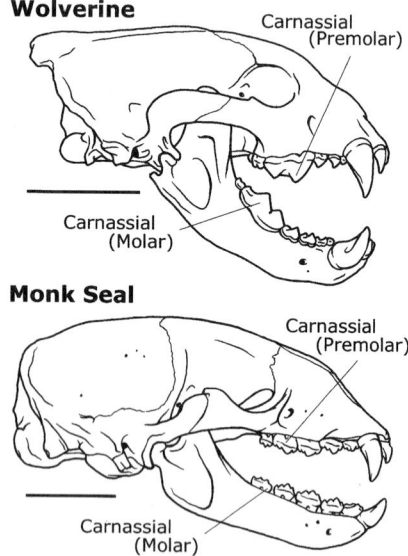

Wolverine

Carnassial (Premolar)

Carnassial (Molar)

Monk Seal

Carnassial (Premolar)

Carnassial (Molar)

Figure 1 *Skulls of a wolverine* (Gulo gulo) *and a Caribbean monk seal* (Monachus tropicalis) *showing the typically enlarged and blade-like carnassial teeth of the wolverine in comparison to the highly derived and premolar-like carnassial teeth of the seal. Scale bar = 5 cm.*

The origin of miacoids and their relationship to other mammalian orders remains poorly understood. Among fossil taxa the extinct order Creodonta has long been considered to be the most likely candidate as sister taxon to Carnivora, but recent efforts by several workers have found little support for this hypothesis. Likewise, higher level molecular and biochemical phylogenies suggesting sister group relationships with such disparate mammals as lagomorphs (rabbits), pangolins (scaly anteaters), primates, chiropterans (bats), and artiodactyls + cetaceans (whales) have few supporters. At present the most likely candidate for sister group affiliations appears to be Lipotyphla (Wyss and Flynn, 1993), an apparently monophyletic group of insectivores that includes hedgehogs, tenrecs, shrews, and moles. None of the known fossil lipotyphlans ("palaeoryctids" and "leptictids" among others), however, appear to be any more closely related to Carnivora than creodonts.

II. Phylogenetic Relationships of Feliform and Caniform Carnivorans

Based largely on his studies of basicranial anatomy, W. H. Flower outlined a tripartite division of extant terrestrial carnivorans that has been retained with little modification since 1869. Morphologic, biochemical, and molecular analyses provide nearly unanimous support for the monophyly of these three groups (Aeluroidea, Cynoidea, and Arctoidea) and for the hypothesis that Cynoidea and Arctoidea are sister taxa (Fig. 2A). In addition, there is a growing consensus that the clade Cynoidea + Arctoidea is derived from Miacidae and that aeluroids are descendant from viverravids, phylogenetic relationships that have been formalized taxonomically using suborder designations Caniformia and Feliformia, respectively (Flynn, 1998). Familial level relationships within Caniformia and Feliformia have proven more difficult to resolve.

Included within Feliformia are the families Felidae (cats), Hyaenidae (hyaenas), Herpestidae (mongooses), Viverridae (civets and genets), and Nimravidae, an extinct family of cat-like carnivorans. Appearing in the late Eocene, nimravids bore striking similarities to modern felids in several dental, cranial, and postcranial characteristics and to sabre-toothed cats in particular because of the large, dagger-like canines possessed by many of the included taxa. While basicranial evidence clearly demonstrates the uniqueness of these Holarctic carnivorans, the phylogenetic position of nimravids is less clear and the family has been considered to be basal caniforms, the sister taxon to Caniformia + Feliformia, and, what is probably the best supported hypothesis, the sister group to living aeluroids (Fig. 2B).

Aeluroidea is primarily an Old World radiation with the earliest members represented by a diverse array of small-bodied (generally <5 kg) closely related forms from late Eocene–early Oligocene fossil localities in France and Mongolia. The basicranial morphology exhibited by these carnivorans is plesiomorphic for the group, and it was not until about 32 mya that carnivorans belonging to the extant families Felidae and Viverridae make their first appearance in the FOSSIL RECORD, with the earliest herpestids and hyaenids showing up closer to 18 mya (Hunt and Tedford, 1993). Felids are the only member of this Old World radiation to successfully colonize the New

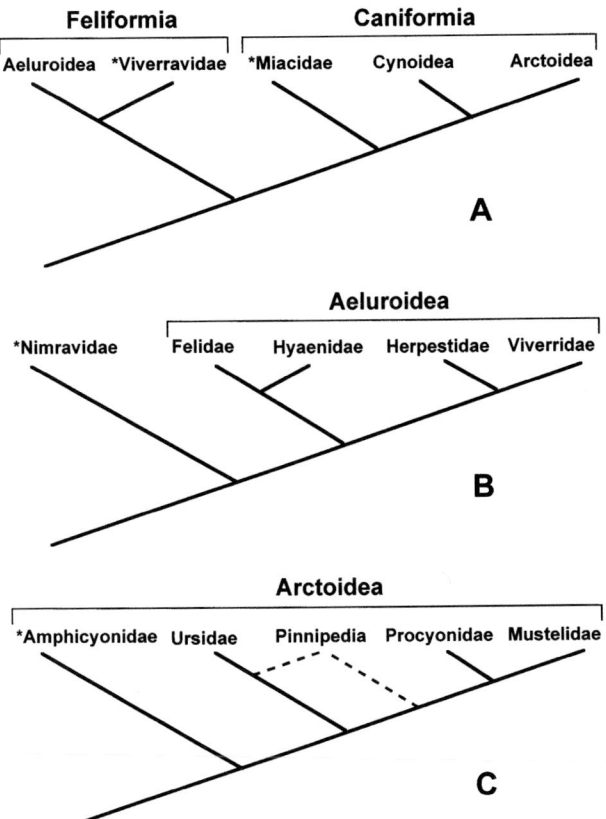

Figure 2 *Cladograms depicting several of the more strongly supported phylogenetic hypotheses of higher taxon relationships within Carnivora (A–C) and two hypotheses regarding pinniped relationships among arctoid carnivorans (C). See text for a discussion of alternative hypotheses. Asterisks denote fossil taxa.*

miacids through the earliest canids such as *Hesperocyon*, a strictly North American taxon that appeared in the fossil record about 42 mya. The skeleton of hesperocyonids possesses several characteristics indicative of the increased cursoriality common to modern canids but retains several miacid features, suggesting that they had not completely abandoned arboreal habitats.

Arctoidea includes the smallest and largest of the terrestrial carnivorans as well as all marine-adapted members of the order. Its origins are traced to the late Eocene–early Oligocene and its distribution is Holarctic with fossil material coming primarily from western North America, France, and Mongolia. Many of these earliest forms are attributed to the extinct family Amphicyonidae (the bear–dogs), which have been variably considered basal members of the arctoid clade (Fig. 2C), the sister taxon to Ursidae, and the sister taxon to Cynoidea + Arctoidea. Amphicyonids appear almost simultaneously in the fossil records of North America and Europe, are taxonomically and morphologically diverse, attain an extremely large body size (>200 kg), and remain Holarctic in their distribution until disappearing between 7 and 9 mya (Hunt, 1996).

A number of recent molecular and morphological studies have addressed phylogenetic relationships among the extant arctoid families, and the result has been reduced support for what had been a nearly consensus view, that Arctoidea included two clades: Mustelidae + Procyonidae and Ursidae + a monophyletic Pinnipedia (Fig. 2C). All subsequent analyses have upheld pinniped monophyly (relationships among the three included families are not as clear), but several molecular studies have suggested that pinnipeds may be the sister taxon to the clade Procyonidae + Mustelidae rather than Ursidae (Fig. 2C). In contrast, other molecular and morphological studies indicate that Procyonidae may be the sister group to Ursidae or to a clade Ursidae + Pinnipedia. One of the keys to resolving arctoid relationships will be in deciphering phylogenetic relationships of *Ailurus fulgens* (the lesser panda), which has been variously included within Procyonidae, as the sister taxon to the clade Mustelidae + Procyonidae, and as the most immediate sister taxon to the clade Ursidae + Pinnipedia. As the number of species-level phylogenies increases in coming years, it seems likely that the resolution of Carnivora phylogeny may become even more problematic and more interesting as key taxa such as *Ailurus* and *Nandinia* are identified and analyzed.

See Also the Following Articles

Dental Morphology, Evolution of ■ Mustelidae ■ Systematics ■ Ursidae

References

Bininda-Emonds, O. R. P., Gittleman, J. L., and Purvis, A. (1999). Building large trees by combining phylogenetic information: A complete phylogeny of the extant Carnivora (Mammalia). *Biol. Rev.* **74**, 173–175.

Flower, W. H. (1869). On the value of the characters of the base of the cranium in the classification of the order Carnivora. *Proc. Zool. Soc. Lond* **1869**, 4–37.

Flynn, J. J. (1996). Carnivoran phylogeny and rates of evolution: Morphological, taxic and molecular. *In* "Carnivore Behavior, Ecology, and Evolution" (J. L. Gittleman, ed.), Vol. 2, pp. 541–581.

World, although one hyaenid lineage does appear in North America and Mexico about 3.5 mya. Attempts to resolve interfamilial relationships among these four families have resulted in at least some support for nearly every possible phylogenetic pairing. A recent analysis (Binida-Emonds *et al.*, 1999) combining nearly all known data sets bearing on the question of aeluroid relationships does corroborate some earlier studies in identifying sister taxon relationships between Felidae and Hyaenidae and between Herpestidae and Viverridae (Fig. 2B), as well as argue for a split between these two clades about 38 mya. There is also growing evidence, both morphologic and molecular, that *Nandinia binotata* (palm civet) is the sister taxon to all living aeluroids rather than a member of Viverridae as traditionally assumed (Flynn, 1996).

Caniformia includes the families Canidae (dogs), Ursidae (bears), Procyonidae (raccoons and their relatives), Mustelidae (skunks, weasels, badgers, and otters), the extinct Amphicyonidae, and the pinnipeds Odobenidae (walruses), Otariidae (fur seals and sea lions), and Phocidae (seals). There is little contention that canids (i.e., Cynoidea) are the sister group to the remaining caniforms (i.e., Arctoidea) and there is good documentation of the transition from middle and late Eocene

Flynn, J. J. (1998). Early Cenozoic Carnivora ("Miacoidea"). *In* "Evolution of Tertiary Mammals of North America" (C. M. Janis, K. M. Scott, and L. L. Jacobs, eds.), Vol. 1, pp. 142–151. Cambridge Univ. Press, Cambridge.

Hunt, R. M., Jr. (1996). Biogeography of the order Carnivora. *In* "Carnivore Behavior, Ecology, and Evolution" (J. L. Gittleman, ed.), Vol. 2, pp. 485–541.

Hunt, R. M., Jr., and Tedford, R. H. (1993). Phylogenetic relationships within the aeluroid Carnivora and implications of their temporal and geographic distribution. *In* "Mammal Phylogeny: Placentals" (F. S. Szalay, M. J. Novacek, and M. C. McKenna, eds.), pp. 53–73. Springer-Verlag, New York.

Wyss, A. R., and Flynn J. J. (1993). A phylogenetic analysis and definition of the Carnivora. *In* "Mammal Phylogeny: Placentals" (F. S. Szalay, M. J. Novacek, and M. C. McKenna, eds.), pp. 53–73. Springer-Verlag, New York.

Carrying Capacity

SEE *Population Dynamics; Sustainability*

Cephalorhynchus Dolphins
C. heavisidii, C. eutropia, C. hectori, and C. Commersonii

STEPHEN M. DAWSON
University of Otago, Dunedin, New Zealand

The four dolphins of the genus *Cephalorhynchus* are small, coastal species found only in Southern Hemisphere waters (Fig. 1). Haviside's dolphin (*C. heavisidii*) occurs off the west coast of South Africa and Namibia. Hector's dolphin (*C. hectori*) is found solely off New Zealand. The Chilean dolphin (*C. eutropia*) is found in the coastal waterways of Chile and along the exposed west coast. The remaining species, Commerson's dolphin (*C. commersonii*), has the strangest distribution. Its principal stronghold is in the inshore waters of Argentina and in the Strait of Magellan, but it also occurs at the Falkland Islands and has an isolated population 8500 km away at the Kerguelen Islands, in the Indian Ocean. The Kerguelen Commerson's dolphins are larger, retain the darker juvenile coloration into adulthood, and have some skeletal and genetic differences. It appears that this population was founded by only a very few individuals, perhaps as recently as 10,000 years ago.

Cephalorhynchus dolphins are among the smallest dolphins. Indeed, in length, Hector's dolphins from New Zealand's South Island (maximum 145 cm; ca. 50 kg) and South American Commerson's dolphins (maximum 146 cm; ca. 45 kg) are the smallest of all dolphins (franciscanas are longer, but weigh less). Both Hector's and Commerson's dolphins have isolated populations in which individuals grow larger. Commerson's dolphins at Kerguelen grow to 174 cm (86 kg), and the few North Island Hector's dolphins reach at least 152 cm (65 kg). In both these species females are 5–10% larger than males. Far fewer Havi-

Figure 1 *The four* Cephalorhynchus *species, top to bottom: Haviside's, Chilean, Hector's, and Commerson's dolphins. Pieter A. Folkens/Higher Porpoise DG.*

side's and Chilean dolphins have been measured so it is not clear whether females are also larger in these species. Heaviside's dolphins reach about 174 cm (ca. 75 kg), and Chilean dolphins reach at least 167 cm (ca. 63 kg).

All four species share similar morphology. They are all small, blunt-headed (hence the frequent mistake of calling them porpoises), chunky dolphins with rounded, almost paddle-shaped flippers. The most characteristic feature of the genus is the dorsal fin, which is proportionately large and with a rounded, convex trailing edge (like a Mickey Mouse ear; Hector's, Commerson's, and Chilean dolphin) or triangular (Haviside's dolphins). In color pattern, Chilean dolphins and Hector's dolphins are most similar.

I. Distribution and Abundance

Of the four species, only Hector's dolphin has been surveyed throughout its range. It is most common along the east and west coasts of the South Island between 41° 30′ and 44° 30′S, with hot spots of abundance at Banks Peninsula and between Karamea and Makawhio Point. These South Island east coast and west coast populations are genetically distinct, as is the very small population occurring on the west coast of the North Island between 36° 30′ and 38° 20′S. An apparently isolated population exists in Te Wae Wae Bay, on the Southland coast. Recent surveys suggest South Island populations number ca. 1900 (east coast) and ca. 5400 (west coast). The species is rarely seen more than 8 km from shore. We have not seen it in water more than 75m deep.

South American Commerson's dolphins are found off the east coast between Rio Negro (40°S) and Cape Horn (55° 15′S) down into the Drake Passage (61° 50′S) and also at the Falkland Islands. They are common between Peninsula Valdez and Tierra del Fuego (42° to 54°S) and very common in the Straits of Magellan, where an aerial survey in 1984 suggested a population of around 3200. At Kerguelen, Commerson's dolphins are restricted to the immediate vicinity of the islands, where they are seen most frequently on the eastern side in the Golfe du Morbihan.

Haviside's dolphins are found on the west coast of Southern Africa from 17° 09′S) on the Namibian coast to around Cape Town (34° 13′S). Sightings are clustered between Walvis Bay and Cape Town. There are no estimates of abundance.

Chilean dolphins have been seen over a very wide latitudinal range, from Valparaíso (33° S) to near Cape Horn (55° 15′S), on both open and sheltered coasts. They are seen regularly in only a few places, however, including Valdivia, Golfo de Arauco, and near Chiloé. The total population appears to be very small (low thousands at most). Suggestions that the species is becoming very rare (Leatherwood, personal communication) are worrying and impossible to refute without dedicated survey work.

All *Cephalorhynchus* species favor waters of less than 200 m deep and are often seen in the surf zone, especially in summer.

II. Reproduction and Population Biology

Reproduction has been studied thoroughly only in Hector's dolphins and Commerson's dolphins, but the limited information available for Haviside's and Chilean dolphins suggests they are similar. Females bear their first calf between 6 and 9 years, in spring to late summer. Males reach sexual maturity between 5 and 9 years. MATING and calving occur in spring to late summer, and the gestation period is around 10–11 months. The maximum ages so far found in this genus are 20 (Hector's dolphin) and 18 (Commerson's dolphin).

Studies of individually identified Hector's dolphins suggest that mature females calve every 2–4 years. Late maturity, a relatively short life, and long intervals between calves inevitably result in a low population growth rate. Modeling studies have shown that Hector's dolphin populations, in the absence of human impact, would be expected to grow by about 2% per year.

III. Diet

All four species feed on a wide variety of coastal prey, focusing on benthic and small pelagic schooling fish and squid. The South American species supplement their fish/squid diet with crustaceans (mysids, euphausids, and *Munida* spp.) and, strangely, algae. Haviside's dolphins are apparently alone in preying on octopus to any significant degree. The DIET of Hector's dolphins is more varied on the South Island east coast, where eight species of fish and squid make up 80% of the diet by mass, than it is on the corresponding west coast (where four species make up 80%).

IV. Movement

Individual dolphins in this genus appear to be seasonally resident in a local area. Long-running studies on Hector's dolphins have shown that individuals usually range over about 30 km of coastline. In Haviside's, Commerson's, and Chilean dolphins, at least some individuals are found in local areas year-round, although inshore abundance is lower in winter. Individual Hector's dolphins have been resighted in the same general area year-round, but they tend to spread out, form smaller groups, and range further from shore in winter. There is no evidence for large-scale MIGRATION. Despite wide-ranging surveys over more than a decade, no two sightings of the same individual Hector's dolphin are more than 106 km apart.

V. Behavior

Small group sizes are characteristic of this genus. Although occasional sightings of more than 50 have been made, most sightings are of between 2 and 10 individuals. In high-density areas, there may be several such groups nearby. In Hector's dolphin, these often coalesce to form a large, temporary group of 25 or so. Large groups often show boisterous behavior, such as chases, leaps, and lobtailing. Sexual displays and copulation are also more common after groups have joined. Such large groups are usually short-lived; after 10–30 min they split into small groups again. Frequently groups lose, gain, or swap members in this process. Associations among adult Hector's dolphins are weak. It is likely that the other species are similar.

All members of the genus except Chilean dolphins are strongly attracted to boats and readily ride the bow wave. Chilean dolphins have been heavily hunted and their wariness of boats has probably been acquired. All species are thoroughly at home in turbulent water close to shore and are frequently seen surfing.

VI. Acoustic Behavior

Only Hector's dolphin has been recorded comprehensively in the wild with equipment that could capture the full range of its sounds. Almost all the sounds are short [ca. 140 μsec (roughly 1/7000 sec)], high-frequency (ca. 125 kHz) narrow-band, ultrasonic clicks (Fig. 2) that are used in "trains" of a few dozen to several thousand. Clicks may contain one, two, or three main pulses and appear to have a role in COMMUNICATION as well as sonar. Complex clicks are used disproportionately in social contexts, and a particular type of click is used preferentially in feeding. Click rates can be exceptionally high (maximum 1149 clicks/sec). Such high click rates generate an audible tone of the same frequency as the click rate (Fig. 3). These signals, which sound like a "cry" or "squeal," are strongly linked to

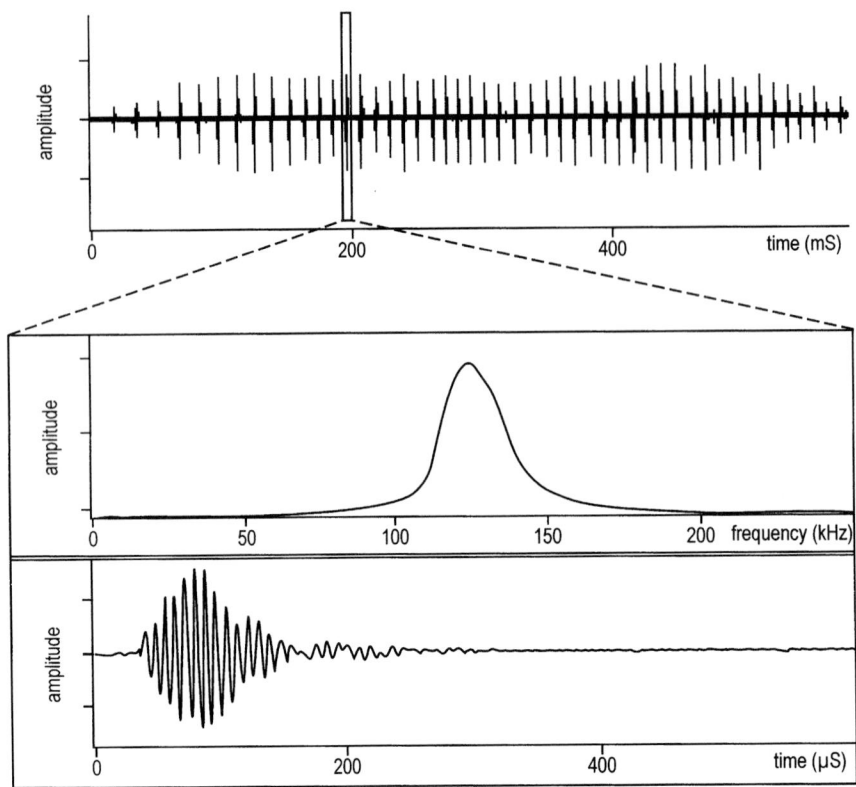

Figure 2 *A "train" of high-frequency Hector's dolphin clicks, with a "zoomed" section showing the waveform and spectrum shown for an example click.*

aerial behaviors and apparently indicate excitement. "Cry"-type sounds, and the lower emphases of occasional broadband clicks, are Hector's dolphins' only vocalizations that are audible to humans. Very similar click and "cry"-type sounds have been described from Commerson's dolphins.

Haviside's dolphin and Chilean dolphin both make the "cry" sounds just described. Neither has yet been recorded with equipment capable of recording ultrasonic frequencies. Given the close similarity of the sounds of Hector's and Commerson's dolphins, it is likely that the sounds of Haviside's and Chilean dolphins will be similar also.

VII. Threats and Conservation

All four species have, at some stage, been hunted for food or bait. The South American species have fared worst: they have been hunted extensively for bait for crab pots and are reported to be no longer common in areas frequented by crab fishers. As many as 1300–1500 Chilean dolphins may have been harpooned each year near the Western Strait of Magellan in the late 1970s and early 1980s. Such hunting is now illegal, but fishers in the area are poor, and the crab processing companies supply insufficient bait. Enforcement of this law in such a remote and convoluted area is practically impossible.

All four *Cephalorhynchus* species suffer some degree of incidental catch in fishing gear. In South America these animals are used as bait. Hence fishers have no motive to avoid areas where captures occur and may favor them. Numbers taken currently are unknown for any of the four species. It is clear, however, that gill nets pose a larger problem than any other fishing method. In the mid-1980s an average of 57 Hector's dolphins were caught each year in gill nets in the Canterbury region of the east coast of New Zealand's South Island. Given the low reproductive rate of this species, it was clear that this level of catch was unsustainable. An 1170-km² sanctuary, in which commercial gill netting is illegal and amateur gill netting is restricted to particular times and places, was established in 1989. The problem has not been entirely solved. Analysis of observed catches in the Canterbury region in 1997/1998 suggests that 18 Hector's dolphins were caught in gill nets to the north and south of the current sanctuary.

There appear to be fewer threats to Haviside's dolphin than to the others, and it seems that Commerson's dolphin, despite the impacts on it, is probably the most numerous member of this genus. The Chilean dolphin and Hector's dolphin need urgent consideration. Currently there is insufficient information to judge the conservation status of Chilean dolphins. Population surveys are needed before we can say even whether the species numbers hundreds (as the most dire of the warnings imply) or thousands. In a small population of slow-breeding animals, even a very low level of directed or incidental catch can be enough to continue the decline.

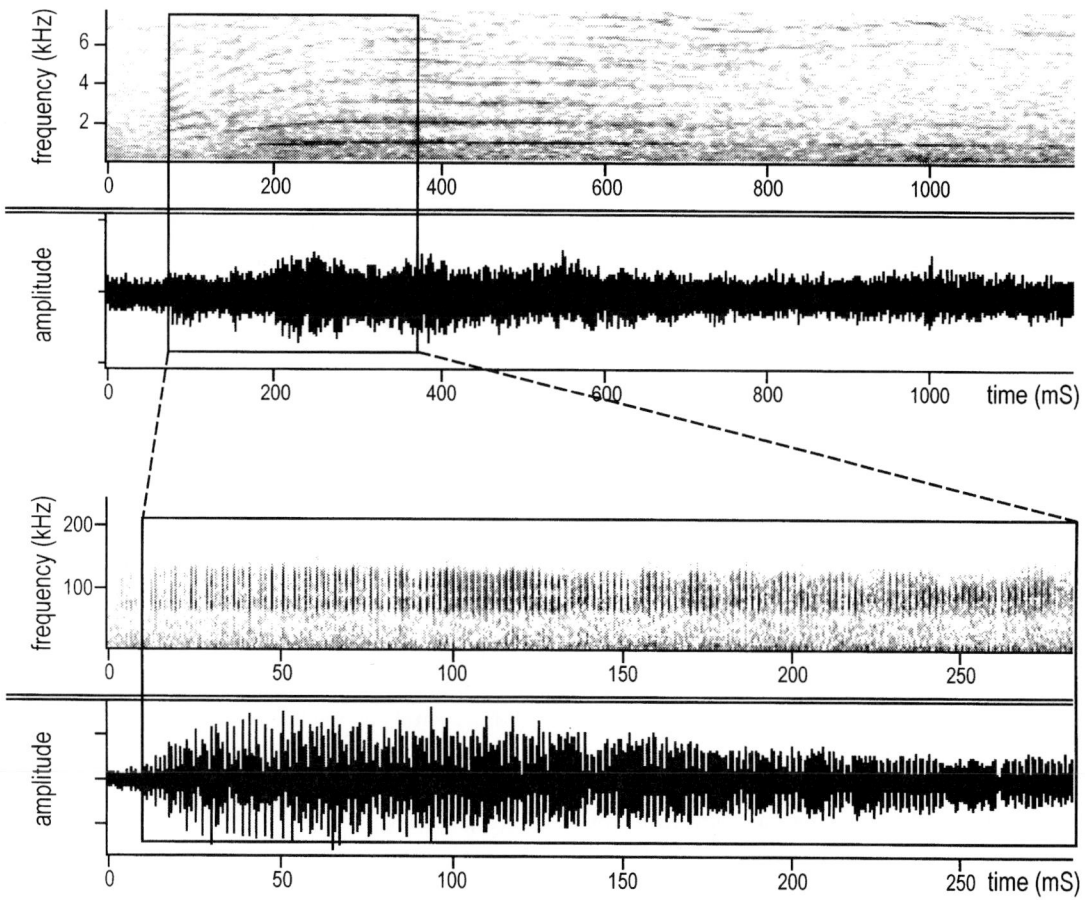

Figure 3 *Audible "cry" sound from Hector's dolphin. The "zoomed" section shows that the "cry" sound is made up of high-frequency clicks emitted at a very high rate (ca. 1000 clicks/sec)*

Hector's dolphin has at least received some direct conservation action, via the Banks Peninsula Marine Mammal Sanctuary. The latest modeling suggests that this population is still declining. Hector's dolphins off the North Island west coast are genetically as distinct from those off the South Island east coast as the South American Commerson's dolphins are from those 8500 km away at Kerguelen. Surveys suggest that the North Island west coast population may number fewer than 100 animals. Gill netting effort along this coast is high by New Zealand standards. Population models show that this population is declining rapidly. Unless effective conservation action is taken very quickly, the outlook for North Island Hector's dolphins is imminent extinction.

See Also the Following Articles

Franciscana ▪ Group Behavior ▪ Incidental Catches

References

Brownell, R. L., and Donovan, G. P. (eds) (1988). "Biology of the Genus *Cephalorhynchus*." Report of the International Whaling Commission (Special Issue) 9.

Dawson, S. M. (1991). Clicks and communication: The behavioural and social contexts of Hector's dolphin vocalisations. *Ethology* **88**(4), 265–276.

Dawson, S. M., and Slooten, E. (1993). Conservation of Hector's dolphins: The case and process which led to establishment of the Banks Peninsula Marine Mammal Sanctuary. *Aqua. Conserv.* **3**, 207–221.

Kamminga, C., and Wiersma, H. (1982). Investigations on cetacean sonar. V. The true nature of the sonar sound of *Cephalorhynchus commersonii. Aquat. Mamm.* **9**(3), 95–104.

Martien, K. K., Taylor, B. L., Slooten, E., and Dawson, S. M. (1999). A sensitivity analysis to guide research and management for Hector's dolphin. *Biol. Conserv.* **90**, 183–191.

Pichler, F. Baker, C. S., Dawson, S. M., and Slooten, E. (1998). Mitochondrial differences between east and west coast populations of Hector's dolphin. *Conserv. Biol.* **12**(3), 1–8.

Ridgway, S. H., and Harrison, R. (eds.) (1994). "Handbook of Marine Mammals," Vol. 5. Academic Press, London.

Slooten, E. (1991). Age, growth and reproduction in Hector's dolphins. *Can. J. Zoo.* **69**, 1689–1700.

Slooten, E., Fletcher, D. and Taylor, B. L. (2000). Accounting for uncertainty in risk assessment: Case study of Hector's dolphin mortality due to gillnet entanglement. *Conservation Biology* **14**, 1264–1270.

Watkins, W. A., Schevill, W. E., and Best, P. B. (1977). Underwater sounds of *Cephalorhynchus heavisidii* (Mammalia: Cetacea). *J. Mammol.* **58**(3), 316–320.

Cetacea, Overview

LAWRENCE G. BARNES
*Natural History Museum of Los Angeles County,
California*

The most highly aquatically adapted group of marine mammals are the mammalian order Cetacea, which are commonly called whales, dolphins, and porpoises. Living cetaceans are a diverse group of mammals whose specialized anatomy and behaviors mask their origin from terrestrial mammals. Ranging from under 5 feet to over 90 feet in length, cetaceans are ecologically diverse; they live in rivers and oceans and in tropical and polar latitudes. About 78 species live today, classified in 13 or 14 families, and in the suborders Mysticeti (baleen whales) and Odontoceti (toothed whales, dolphins, and porpoises). Hundreds of fossil species are known, going back to the Middle Eocene, about 55 million years (Ma) ago, including members classified in an extinct third suborder, the Archaeoceti.

I. Suborder Archaeoceti: Archaic Cetaceans

The oldest Cetacea are the toothed, extinct suborder Archaeoceti, a paraphyletic group. Archaeocetes are clearly related to living Cetacea; they show typical cetacean features such as an elongate upper jaw with bony nostrils set back from the tip, a broad shelf of bone above the eye, lack the incisive foramina, incisors aligned anteroposteriorly, a dense outer ear bone (tympanic bulla), and an enlarged mandibular canal on the inner side of the lower jaw; later archaeocetes have an expanded basicranial air sinus, like living cetaceans. Archaeocetes lack the "telescoped" SKULL, in which bones of the snout and cranium slide over one another, as in the later odontocetes and mysticetes.

Sinuous eel-like bodies have been depicted by authors and artists, especially for the later archaeocetes, but this would be contradictory to the vertical tail movements and caudal flukes typical of other cetaceans. A quite whale-like body shape can be reconstructed around the skeletal outlines of even the longest of the basilosaurine archaeocetes. Stomach contents of *Basilosaurus cetoides* suggest that it was an ambush predator on bony fishes and small sharks in shallow water. There is fossil evidence that most, if not all, archaeocetes had external hindlimbs. Early archaeocetes have well-developed pelves, sacral vertebrae, and hindlimb bones. The smaller yet still-functional hind limbs of *Basilosaurus isis* may have aided locomotion in shallow waters. Archaeoceti also have a primitively mobile elbow joint, with a long humerus and less streamlined foreflipper than in extant Cetacea. The more primitive forelimbs might indicate that they were semiamphibious-like pinnipeds.

Most older archaeocete fossils are from India and Pakistan, in sediments deposited around the ancient Tethys Seaway, which extended between India and Africa to the south and Eurasia to the north. Since 1980, new fossils have greatly expanded our knowledge of protocetids and other early archaeocetes.

A. Family Pakicetidae: Amphibious Early Cetaceans

Nalacetus is mainly known from isolated teeth, which show that it is the most primitive cetacean. *Pakicetus* was a small archaeocete with eyes on top of its head. Oxygen isotope ratios in its tooth enamel indicate that it drank only fresh water.

B. Family Ambulocetidae: Walking Whales

Ambulocetus had feet that were much longer than its hands and a long tail that probably was without flukes.

C. Family Remingtonocetidae: Gavial-Convergent Cetaceans

Remingtonocetus and *Andrewsiphius* had a long and narrow skull, very tiny eyes, and a long symphysis (junction) between the two lower jaws.

D. Family Protocetidae: First Pelagic Cetaceans

Indocetus had large molars that have a complex, primitive morphology. *Rodhocetus* was approximately 3 m long and had nostrils near the front of the snout. It retained a large pelvis and small hind legs. *Protocetus* had a skull that lacks the complex tooth crowns and expanded basicranial air sinuses of later archaeocetes. It had a small pelvis and flexible posterior part of the trunk.

E. Family Basilosauridae: Zeuglodonts

Basilosaurus was a large whale (approximately 20 m). It had cheek teeth with many small denticles (saw-tooth like) and had enlarged air sinuses around the ears. *Basilosaurus* also had large and elongate vertebral centra and short hindlimbs.

Sometimes included in the basilosaurids, but sometimes treated as a separate family are the Dorudontinae, including *Dorudon, Saghacetus,* and *Zygorhiza.*

II. Suborder Mysticeti: Baleen Whales

Mysticetes are small to very large filter-feeding whales with a loose mandibular symphysis and a type of jaw articulation that allows the mandible to open widely for bulk feeding. All mysticetes also have a smooth ventral surface of the maxilla (the palate), which in the baleen-bearing species has grooves that hold the blood vessels that nourish the baleen plates, an infraorbital plate of the maxilla ventral to the eye, and a laterally projecting antorbital process of the maxilla. Mysticeti are a natural clade that originated from the Archaeoceti. The oldest mysticete is Late Eocene in age and the group survives today, having been represented throughout most of its history by at least five or six contemporaneous family-level groups.

Just a few years ago, it would have seemed paradoxical to need to differentiate between baleen-bearing mysticetes and tooth-bearing mysticetes, the latter seemingly a misnomer. However, this need was brought about largely by the discovery of the Aetiocetidae and other Late Eocene and Oligocene whales that are transitional between the toothed Archaeoceti and the later baleen-bearing mysticetes. There is as much recognized diversity now at the family level among both named, and known but undescribed, fossil tooth-bearing mysticetes as there is among the fossil and living baleen-bearing mysticetes.

The primitive, extinct tooth-bearing mysticetes had heterodont dentitions, differentiated into incisors, canines, premolars, and molars, and these animals are classified in the families Llanocetidae, Kekenodontidae, Mammalodontidae, and Aetiocetidae, and in possibly two other undescribed families. Some of these families had skull characters that were very much like those of archaeocetes, thus confirming the origin of baleen whales from them.

More highly evolved toothed mysticetes, such as the Aetiocetidae, had skulls and jaws that were similar to the later baleen-bearing mysticetes, e.g., the cetotheriids, so they represent a transitional morphology. In many ways, the "intermediate" structure of aetiocetids helps us understand how baleen-bearing mysticetes might have evolved from archaeocetes. A possibly unanswerable question is which toothed mysticete taxon was the first to have rudimentary baleen plates between its teeth. It is most parsimonious to assume that baleen originated only once in cetacean evolution.

General evolutionary trends among the later Mysticeti include the acquisition of large body size, shortening of the intertemporal region as anterior and posterior parts of the skull "telescope" over and under each other, reduction of the coronoid process and mandibular foramen, lateral bowing of the mandibles, and shortening of the neck. Mysticetes are virtually cosmopolitan in their distribution, being found both in the past and now from polar to equatorial latitudes, with fossils known from all ocean basins. The great increase in size among baleen-bearing mysticete whales was a phenomenon of Miocene and later time, with the largest known species being among the living species.

A. Family Llanocetidae: Archaic-Toothed Mysticetes

Llanocetus had a large triangular braincase and its brain was shaped much as in later mysticetes. It had a deep mandible with widely spaced, large, multicusped teeth.

B. Family Kekenodontidae: Toothed Mysticetes

Kekenodon had long-rooted, multicusped cheek teeth, with large accessory cusps.

C. Family Mammalodontidae: Southern Ocean-Toothed Mysticetes

Mammalodon was a toothed mysticete with a strangely abbreviated and blunt rostrum. The posterior end of its mandible was deeper and more heavily built than that of the more highly derived Aetiocetidae. It had more than the primitive eutherian number of teeth per jaw (11). The teeth crowded each other, were implanted somewhat obliquely in the mandible, and had small accessory cusps.

D. Family Aetiocetidae: Small-Toothed Mysticetes

Aetiocetidae had heterodont denitions that were differentiated as incisors, canines, premolars, and molars, with small cuspules on comparatively small molar and premolar crowns. *Chonecetus* had a small braincase and relatively narrow rostrum. *Morawanocetus* had large molars, a wide and dorsoventrally flattened braincase, large tympanic bullae, and small, an-

teroposteriorly compressed cervical vertebrae. *Aetiocetus* was of relatively large size, with wide flat rostrum, anteroposteriorly foreshortened intertemporal region, and large zygomatic process of the squamosal. Some Aetiocetinae had increased numbers of teeth (as do embryonic modern baleen whales) and smaller teeth, possibly indicating the transition to incipient baleen plates.

E. Family Cetotheriidae: Primitive Baleen-Bearing Mysticetes

Cetotheriidae were flat-snouted, edentulous, baleen-bearing mysticetes. Their palates had characteristic grooves marking the courses of blood vessels that nourished the baleen plates, and fossil baleen has even been found with some specimens. The dentaries were elongate and relatively slender and had nutrient foramina along the dorsal border marking the previous locations of dental alveoli. Cetotheres had relatively small heads compared to their body size, similar in this regard to Recent rorquals of the family Balaenopteridae.

Herpetocetus and *Nannocetus* are characterized by their dentary-squamosal articulation: the zygomatic process of the squamosal is elongate anteroposteriorly and deepened dorsoventrally, and the angular part of the dentary projects posteriorly ventral to the postglenoid process and the ear region.

F. Family Balaenopteridae: Rorquals and Relatives

Balaenopterids have a deep transverse sulcus between the rostral portion of the maxilla and the supraorbital process of the frontal. They also have an elevated cranial vertex (relative to the supraorbital process of the frontal), a tapered and somewhat twisted (in derived taxa) horizontal ramus of the dentary, and a large, spheroid mandibular condyle. They also lack any transverse exposure of the parietals at the apex of the skull posterior to the nasal bones.

G. Family Eschrichtiidae: Gray Whales

Eschrichtius robustus is characterized by a relatively narrow and dorsally arched rostrum with large nares that are located at the apex of the curve, two tuberosities on the braincase near the nuchal crest, and a massive mandible with a low coronoid process and a large rounded mandibular condyle.

H. Family Balaenidae: Right Whales, Bowheads, and Relatives

Morenocetus, a fossil, probably had an upper jaw that was narrow and arched. Living balaenids have extremely long baleen plates, similar to those of the extant *Balaena mysticetus*.

I. Family Neobalaenidae: Pygmy Right Whales

Extant *Caperea marginata* is very small, has expanded ribs, an extremely large occipital shield of the braincase, a short and tapered rostrum, and tapered and twisted mandible.

III. Suborder Odontoceti: Toothed Cetaceans

Odontocetes are echolocating toothed cetaceans that have extensions of the basicranial sinuses, around their ears, into other parts of the skull. The large pterygoid sinus extends anteriorly around the sides of the nasal passages and, in derived taxa, in front of them. The Odontoceti are known as far back as the Eocene–Oligocene boundary, approximately 35 Ma.

Odontoceti are commonly called "toothed whales," but this is a bit of a misnomer because all of the Archaeoceti, some of the primitive fossil Mysticeti, and all fetal baleen-bearing mysticetes are toothed, whereas some Odontoceti had undergone dramatic tooth loss. Therefore, "presence of teeth" is not diagnostic of the Odontoceti. The earliest odontocetes have a primitive dental formula, differentiated into incisors, canines, premolars, and molars. Later odontocetes have all teeth alike, homodonty, and some taxa have additional teeth (polydonty) whereas others have reduced numbers of teeth.

Similarly, Odontoceti are commonly thought of as having asymmetrical crania, but this is not true for all Odontoceti, especially many of the fossil species, and, again, is not diagnostic of the group. Cranial asymmetry, a derived character state, is related to the development of complex echolocation abilities involving the production of both high- and low-frequency sound. Different odontocete species have either symmetrical or asymmetrical crania, and asymmetry is a derived character state that apparently has been independently acquired within different lineages at different times. In most odontocetes having asymmetrical crania, the skew is to the left side, but the skulls of members of the extinct kentriodontid delphinoid subfamily Pithanodelphinae are skewed to the right side.

The type of cranial bone telescoping of odontocetes differs from that of the mysticetes; the maxilla extends posteriorly *over* the supraorbital process of the frontal and there is no laterally projecting antorbital process of the maxilla. The pterygoid sinus fossa in Odontoceti extends from the ear region anteriorly around the nasal passage to invade the posterior part of the palatine bone. In the earliest Odontoceti, the nasal bones primitively were elongate, flat, and overhung the posterior part of the narial opening, as in the derived Mysticeti. In later Odontoceti, the nasal bones are reduced, being knob-like and retracted posterior to the nasal opening. All Odontoceti have a premaxillary foramen, a branch of the infraorbital foramen system, located in each premaxilla anterior to the narial opening. Around or anterior to the narial opening is a widened and flattened part of the premaxilla where the premaxillary sac, a diverticulum of the dorsal nasal passage, lies. This flat part of the premaxilla is called the premaxillary sac fossa or the spiracular plate.

Odontocetes have the ability to echolocate and to selectively take individual prey items. This separates them from the Mysticeti, which cannot echolocate and are bulk feeders. During echolocation, high-frequency sound, in the form of clicks, is produced by the movement of air, recycled within the diverticula, sacs, and valves of the nasal passages. The sound is focused by the melon on the face, which is an acoustic lens, and projected into the environment. Sound is reflected off of objects and animals in the water and is returned to the odontocete. The external acoustic meatus is closed; sound is instead transmitted to the ears via the side of the face, through the thin posterior part of the dentary, through a fat body, and thence to the ear region. Directional hearing is possible because the ear bones are isolated in fat bodies, and the sound arrives at each ear at a different time. Even the earliest odontocetes had spiracular plates and enlarged peribullary sinuses, and this implies that they could actively echolocate, at least to some extent.

Throughout their history, odontocetes have been more diverse taxonomically than mysticetes, with several different families existing simultaneously. Different families of odontocetes are diagnosed particularly on details of sutures around the bony nostrils and the base of the rostrum, the basicranial sinuses, and the ear region. Trends among odontocetes include the evolution of paedomorphism (adults showing a spectrum of juvenile-like features), repeated convergent evolution of very long rostra, increasing facial asymmetry, development of more complex basicranial sinuses, and shortening of the neck.

A. Family Agorophiidae: Primitive Toothed Whales

Agorophius was a medium-sized odontocete with heterodont dentition, wide mesorostral groove, elongate nasal bones, very large temporal fossae bordered by large cranial crests for the attachment of very massive jaw musculature, and a relatively elongate and narrow intertemporal constriction.

B. Family Physeteridae: Giant Sperm Whales

In physeterids the periotic is distinctive, having a large posterior process, a round internal acoustic meatus, and a part of the outer whorl of the bulla (called the "accessory ossicle") that is attached to the anterior process of the periotic. The teeth of all physeterids are homodont. *Ferecetotherium* had both upper and lower functional teeth. *Diaphorocetus* had a skull with the distinctive physeterid type of supracranial basin.

Scaldicetus had a relatively long, slender rostrum, large zygomatic arches, large tympanic bullae, prominent occipital condyles, and a forward-sloping occipital shield that is deeply emarginated laterally by large temporal fossae. It appears to have had 12 teeth on each side of the mandible. The living sperm whale, *Physeter macrocephalus*, lacks upper teeth.

C. Family Kogiidae: Pygmy and Dwarf Sperm Whales

Kogiids and physeterids share highly asymmetrical crania, supracranial basin, derived lachrimal-jugal structure, and a large mastoid process. Kogiids differ from physeterids by having lesser development of the melon and thus a less concave facial skeleton. In addition, the kogiid skull has a blunt and squared zygomatic process of the squamosal and an anteroposteriorly aligned bony crest or eminence in the middle of the facial basin of the skull.

Extant *Kogia* have skulls that are largely composed of spongy and inflated bone, with very short and tapered rostra, no upper teeth, an elevated cup or basin on the sagittal crest in the supracranial basin, and the orbit placed posteriorly within the temporal fossa.

The latest Miocene fossil pygmy sperm whale, *Praekogia*, differed from the Recent species of *Kogia* by having dense rather than spongy (=oil-filled) cranial bone, cranial crests that are not so elevated, and the orbit situated anterior to the end of the zygomatic process of the squamosal (the typical cetacean condition), not posteriorly within the temporal fossa.

Scaphokogia had an elongate and narrow skull, a long, parallel-sided, thick, dense snout, and nearly circular supracranial basin.

D. Family Ziphiidae: Beaked Whales

Ziphiidae have reduced the number and size of teeth. Male modern ziphiids generally retain one or two prominent teeth in the lower jaw, but primitive species showed rows of small dolphin-like teeth.

E. Family Squalodontidae: Shark Toothed Dolphins

Squalodontidae were medium-sized to large, dolphin-like cetaceans with long rostra and a relatively primitive, heterodont dentition composed of incisiform anterior teeth and premolars and molars commonly with triangular crowns and two roots. Their skulls were fully telescoped, the narial openings being located between the eyes, and the premaxillae and maxillae extending posteriorly to reach the cranial vertex. In *Patriocetus* and *Kelloggia*, the skull had a narrow intertemporal region. In *Squalodon* and related genera, the skull had a more disk-like facial region.

F. Family Eurhinodelphinidae

Eurhinodelphinidae (also called Rhabdosteidae) were medium to large sized and have fully telescoped skulls with extremely long, narrow rostra bearing many small, homodont teeth with single roots and conical crowns. The vertebrae are elongate and the neck is flexible. The lower jaw was shorter than the upper jaw. The front part of the upper jaw that overhung carried no teeth.

Argyrocetus and *Rhabdosteus* had narrow facial regions, but *Eurhinodelphis* had a wide, rounded facial region. *Macrodelphinus* attained the size of a living killer whale.

G. Family Waipatiidae

Waipatia was a dolphin-like cetacean with heterodont teeth and a relatively wide rostrum. The cranial vertex was slightly asymmetrical.

H. Family Squalodelphinidae

Squalodelphinids were medium-sized cetaceans with stout rostra bearing homodont denitions composed of conical-crowned teeth that somewhat resembled those of modern dolphins, but which had somewhat rugose enamel caps. The cranial vertex was asymmetrical and there was a moderate-sized elevated crest aligned anteroposteriorly over each orbit.

I. Family Dalpiaziniidae

Dalpiazinia had a nasal region and cranial vertex that were symmetrical, the rostrum was relatively short, and the dentition was homodont.

J. Family Acrodelphinidae

Acrodelphinids (also called Eoplatanistidae) were medium-sized cetaceans with relatively large vertebrae and pectoral limbs, and on the skull there was a relatively low anteroposteriorly aligned crest on top of the supraorbital process of the frontal. The pterygoid bone was continuous with the alisphenoid bone to form a thin, vertically oriented sheet of bone within the back of the orbit (similar to platanistids, a so-called "reduplicated pterygoid"). The lower jaw was the same length as the rostrum and had an elongate groove on each side, mirroring a similar groove on each side of the rostrum. Such grooves are also present in some Pontoporiidae.

K. Family Platanistidae: Indian River Dolphin (*Platanista gangetica*) and Relatives

Platanistidae have a generalized vertebral column, but a highly modified foreflipper. The skull has tiny eyes with reduplicated pterygoids. The narial region and cranial vertex show strong left-skew asymmetry. The jugal is thick, and the zygomatic process of the squamosal is dorsoventrally expanded and medially excavated. The maxillary crests are large and form a parabolic surface on the facial region, "accessory ossicles" surround bulla and periotic, and the rostrum is long and transversely compressed with elongate anterior teeth and short back teeth.

L. Family Iniidae: Amazon River Dolphin and Relatives

Inia geoffrensis has a delphinid-like basicranium and ear region, but its combination of elevated maxillary crests, premaxillary eminences, asymmetrically left-skewed narial region, elevated cranial vertex, and elongate zygomatic processes of the squamosals differentiate it from other families of Odontoceti.

M. Family Pontoporiidae: Franciscana and Relatives

Pontistes, Pliopontos, and the living *Pontoporia blainvillei* have symmetrical crania and long rostra. *Brachydelphis* was extremely short snouted, but its cranium was otherwise similar in construction to those of the Pontoporiinae and was symmetrical around the narial region. *Parapontoporia* (also classified as a lipotid) was extremely long snouted and had 77 to 88 teeth on each side of the slender rostrum and mandible. Fossil species are known from 8 to 3 Ma ago in the North Pacific.

N. Family Lipotidae: Yangtze River Dolphin

Lipotes vexillifer shares important derived characters of the cranial vertex and facial region with *Parapontoporia*. Its postcranial morphology is relatively primitive, but its skull has several unique derived characters, including crests on the supraorbital process of the maxilla over the orbit, a stout rostrum, a left-skewed highly asymmetrical nasal region, and a highly elevated asymmetrical cranial vertex.

O. Family Kentriodontidae: Primitive Delphinoid Dolphins

Kentriodontids were small to medium-sized fossil dolphins. The cranial vertices of all kentriodontids, with the exception of *Pithanodelphis* and *Atocetus*, were symmetrical, the skulls were fully telescoped, and their dentitions were more or less homodont. *Kampholophus* was relatively large, with a large skull that bore prominent cranial crests, and large teeth with vestiges of heterodonty—the posterior teeth had accessory cups. *Kentriodon* was similar in body form to the small living dolphins. *Pithanodelphis* and *Atocetus* displayed cranial asymmetry: the cranial vertex was skewed to the right side. This skew was in the opposite direction of that present in all other odontocetes with an asymmetrical skull. These two genera were also characterized by having exceptionally large and elevated nasal bones. *Lophocetus* and *Hadrodelphis* had large and elevated nasal bones, which were constricted medially between the posterior ends of the maxillae.

P. Family Albireonidae

Albireo was a *Tursiops*-sized dolphin that had a large, massive skull with an up-turned rostrum containing many large, conical-crowned teeth, a large brain, and an extensive basicranial air sinus system. Its flippers resembled those of *Platanista gangetica* and the stout body form that of *Lagenorhynchus* spp. It had numerous thoracic and lumbar vertebrae and these bear long processes.

Q. Family Odobenocetopsidae: Walrus-like Whales

Odobenocetops had asymmetrical, posteroventrally directed tusks, the right much longer than the other, similar to those of a narwhal, and a blunt snout. There were no other upper teeth. The facial region retained the odontocete nasal morphology, but lacked the bony indications of nasal sacs and air sinuses. The optic nerve tracts were exceptionally large for a cetacean.

R. Family Monodontidae: Belugas, Narwhals, and Relatives

Delphinapterids have a relatively narrow and slightly downturned snout like modern beluga (*Delphinapterus leucas*). *Delphinapterus* and *Monodon monoceros* have similar skulls, but the latter bears a large tusk. *Denebola* was a strange fossil monodontid of temperate latitudes that had a broad skull and a snout that was wide, flat, and blunt.

S. Family Delphinidae: True Dolphins

Delphinidae are characterized by skull osteology, especially details of the left-skewed asymmetrical cranial vertex and narial region, the basicranium, and the periotic. Delphinids range from small to large species with long or short rostra, and narrow or broad rostra, and are the most diverse living family of Cetacea.

T. Family Phocoenidae: Porpoises

Phocoenids have a raised eminence on each premaxilla anterior to the narial opening, a lobe of the pterygoid air sinus extending dorsal to the orbit between the frontal and maxillary bones, and anteriorly retracted posterior premaxillary terminations. The short rostra, rounded crania, and spatulate teeth of living phocoenids are paedomorphic characters that only appear in the more recent members of the group, known from the Late Pliocene and more recently. Late Miocene phocoenids had longer snouts, more prominent cranial crests, and conical tooth crowns. *Piscolithax* was a large phocoenid, approximately the size of extant *Tursiops truncatus*, and had relatively large pectoral flippers.

See Also the Following Articles

References

Barnes, L. G. (1977). Outline of eastern North Pacific fossil cetacean assemblages. *Syst. Zool.* **25**(4), 321–343 [for 1976].

Barnes, L. G., Inuzuka, N., and Hasegawa, Y. (eds.) (1995). Thematic issue: Evolution and biogeography of fossil marine vertebrates in the Pacific realm. *Island Arc* **3**(4), 392–431 [for 1994].

Fordyce, R. E., and Barnes, L. G. (1994). The evolutionary history of whales and dolphins. *In* "Annual Review of Earth and Planetary Sciences" (G. W. Wetherill, ed.), Vol. 22, pp. 419–455. Annual Reviews, Inc., Palo Alto, CA.

Kellogg, A. R. (1936). A review of the Archaeoceti. *Carnegie Inst. Wash. Publ.* **482**, 1–366.

Rice, D. W. (1998). "Marine Mammals of the World: Systematics and Distribution." The Society for Marine Mammalogy, Special Publication 4, i–ix, 1–231.

Thewissen, J. G. M. (ed.) (1998). "The Emergence of Whales: Evolutionary Patterns in the Origin of Cetacea." Plenum Press, New York.

Cetacean Ecology

LISA T. BALLANCE
Southwest Fisheries Science Center, La Jolla, California

Ecology is the study of the natural environment and of the relationships of organisms to each other and to their surroundings. From its natural history beginnings in the late 1800s, the field of ecology has blossomed into a broad discipline, encompassing empirical and theoretical research in fields as diverse as mathematics, conservation, physiology, geography, and behavior. The study of cetacean ecology is very much in its infancy, however. Until fairly recently, most of what was known about cetacean ecology was largely composed of anecdotes and observations handed down by early whalers (Herman Melville's "Moby Dick" provides for classic examples). Studying whales or dolphins in their natural environment is a formidable challenge. This is due to the logistical constraints of attempting to study highly mobile, oceanic animals that spend nearly all of their lives underwater, as well as the political and legal constraints of working on protected species, which include most cetaceans.

Cetaceans are a diverse group that includes some 83 species. They range in size from less than 1 m long for a newborn vaquita (*Phocoena sinus*) to 33 m in an adult blue whale (*Balaenoptera musculus*); they occupy water whose temperatures range from -2 to over 30°C; they exhibit a diverse array of life history strategies. Consider the sperm whale (*Physeter macrocephalus*), which can remain beneath the water for over an hour and dive to depths of several thousand meters; the Indian river dolphin (*Platanista gangetica*), which inhabits fresh water so turbid it is functionally blind; beaked whales of the genus *Mesoplodon*, which are so pelagic and so elusive that new species are still being described; the gray whale (*Eschrichtius robustus*), which annually migrates some 15,000 to 20,000 km between breeding and feeding areas; and the bowhead whale (*Balaena mysticetus*), which uses its rostrum to break ice in the Arctic.

One of the challenges of ecology is to search for pattern within diversity. Despite their diversity of form, behavior, and habitat, all cetaceans have some key features in common that underscore the fact that they are secondary marine forms, derived from terrestrial ancestors. That they are all air-breathing, live-bearing homeotherms provides a unifying theme. This article provides an overview of cetacean ecology with the ultimate goal of identifying some unifying principles in the ways that cetaceans interact with each other and with their environment.

I. Habitat

A. Where Do Cetaceans Live?

On a global scale, cetaceans have invaded a large proportion of the ocean's habitats. They inhabit coastal waters up to and including the surf zone (gray whale; some populations of bottlenose dolphins, *Tursiops truncatus*; harbor porpoise, *Phocoena phocoena*; Commerson's dolphin, *Cephalorhynchus commersonii*), neritic waters over continental shelves (long-beaked common dolphin, *Delphinus capensis*; *Lagenorhynchus* spp.; *Cephalorhynchus* spp.; *Phocoena* spp.), and the most oceanic of systems (sperm whale; Fraser's dolphin, *Lagenodelphis hosei*; beaked whales). They are found in tropical waters (pantropical spotted dolphin, *Stenella attenuata*), temperate seas (Risso's dolphin, *Grampus griseus*), and polar oceans, up to and within pack ice (beluga, *Delphinapterus leucas*; bowhead whale). They utilize much of the water column, some being confined to relatively shallow depths (most dolphins and baleen whales), and others diving to thousands of meters (sperm whale, many beaked whales). They have also invaded several of the world's major river systems (Ganges, Indus, Amazon/Orinoco, Yangtze).

Cetaceans in different habitats might be expected to show differential development of adaptations that reflect selective pressures of the environments in which they function. For example, species in polar seas must conserve heat and so bowhead whales have relatively large bodies, thick BLUBBER layers, and short appendages. Deep-diving species (sperm and beaked whales) must conserve oxygen and might be expected to have large blood volumes, a high hematocrit, and a well-developed diving response. Species that forage in low-light conditions (night feeders, deep divers, species living in turbid rivers) should have well-developed ECHOLOCATION abilities relative to those that function in habitats with greater light levels and better visibility. To date, these types of comparative studies have been rare and are a promising line of future investigation.

The geographic range of cetaceans runs from cosmopolitan to extremely local. For example, the killer whale (*Orcinus orca*) can be found throughout the world's oceans and, with the exception of humans, is the most wide-ranging animal on earth. At the other extreme is the vaquita, a tiny porpoise that occupies a few hundred square kilometers in the northern Gulf of California. Why some species are habitat generalists and others specialists remains largely a mystery.

On a smaller spatial scale, we are beginning to identify those features that correlate with centers of distribution for some species and so possibly identify what may be called critical habitat. For example, some species associate with ice edges (beluga), some with continental shelf edges or seamounts (beaked whales), and some with shorelines (gray whale, bottlenose dolphin, harbor porpoise). For oceanic species, habitat preferences are often defined by less obvious features: physical and chemical characteristics of the water itself, which define water masses and current boundaries. So, for example, some species associate with cold-water currents (Heaviside's *Cephalorhynchus heavisidii*, Commerson's, Peale's *Lagenorhynchus australis* dolphins). Blue whales in the eastern Pacific are found in relatively cool, upwelling-modified waters with high primary and secondary productivity. In the eastern tropical Pacific, pantropical spotted and spinner (*Stenella longirostris*) dolphins segregate from common dolphins according to thermocline depth and strength, sigma-t (a measure of seawater density computed from surface temperature and salinity), and surface water chlorophyll content. These differences are statistically significant, and these species-specific distribution patterns track oceanographic variation on a seasonal and interannual basis.

We know very little about why these species-specific preferences exist. Most often, habitat preferences are suggested to relate to prey abundance or availability, which in turn are determined by physical oceanographic patterns. In fact, a number of studies have linked general distribution and movement patterns of cetacean species (humpback, *Megaptera novaengliae*; fin, *Balaenoptera physalus*; long-finned pilot whales, *Globicephala melaena*; Atlantic white-sided, *Lagenorhynchus acutus*; bottlenose and common dolphins, *Delphinus* spp.) with those of their prey. In a very few instances, physical features have been directly linked with the processes that cause them to aggregate prey or increase productivity, and indirectly with cetacean distribution. This is a productive area for future investigation.

B. How Do Cetaceans Use Their Habitat?

Home range can be defined as an area of regular use that typically provides for all of an animal's requirements. The size of these areas ranges from 125 km² for bottlenose dolphins along the west coast of Florida, to 600 km² for minke whales (*Balaenoptera acutorostrata*) off the west coast of north America, to thousands of kilometers for many baleen whales, which annually migrate between the tropics and polar seas (and can be thought to occupy two home range areas, separated by large distances). Particularly for odontocetes, very little is known about home range size, but there are indications that it can be quite large. Bottlenose dolphins along the coast of Mexico and California regularly traverse 900 km. Pelagic spinner and pantropical spotted dolphins have been tracked over 500 and 1800 linear km, respectively, in a single year.

Within these home ranges, many cetaceans have well-defined habitat use patterns. This is perhaps best exemplified in species that migrate. These individuals move from high latitudes in the winter to low latitudes in the summer, distances spanning thousands of kilometers in some cases. They feed in their high-latitude habitat, where waters high in nutrients combined with seasonal sun produce a strong bloom in productivity and superabundant food. They mate and calve in their low-latitude habitat, where little feeding occurs because prey are scarce.

Differential habitat use patterns have also been identified for some non-migratory cetaceans. For example, island populations of spinner dolphins use sheltered bays to rest during the daytime and move to deep waters offshore during the night to feed. Coastal bottlenose dolphins feed in some areas of their range preferentially, sometimes with a seasonal shift, presumably in response to prey movement. Most knowledge of habitat use patterns comes from coastal or neritic species; little is known about how or if pelagic species differentially use their habitat.

II. Food, Feeding, and Foraging

One of the most striking differences between marine and terrestrial ecosystems has to do with the form of primary producers. Whereas terrestrial ecosystems tend to be dominated by large macroscopic plants that are long-lived and provide substantial resources to other organisms in the form of food and physical structure, macroscopic plants are almost completely absent in marine ecosystems. In contrast, marine primary producers are dominated by microscopic, short-lived plants. This has profound implications. Marine herbivores are dominated by small, often microscopic animals themselves, whereas the majority of large marine animals are carnivores, including all cetaceans. Because the vast majority of oceanic habitat is pelagic, i.e., without any benthos, primary producers, herbivores, and consumers tend to be patchy in space and time. Thus the form of marine primary producers affects community composition, resource distribution, and so food, feeding, and foraging of cetaceans.

A. What Do Cetaceans Eat?

Most of what is known about the food of cetaceans comes from data collected from dead animals, through directed fisheries, incidental mortality, or strandings. Prey of cetaceans fall into four general categories. The first prey type consists of small individuals that school at relatively shallow depths (surface to several hundred meters). These are primarily planktonic crustaceans (euphausiids, copepods, amphipods) and small fish [e.g., herring (*Clupea*), sardine (*Sardinops*), anchovy (Engraulidae), sandlance (Ammodytidae)]. They tend to occur in temperate or polar seas or in those tropical latitudes that are associated with unusually high productivity. They generally occupy low trophic levels, have small body sizes, and occur in dense aggregations. Accordingly, the cetaceans feeding on them capture multiple individuals simultaneously, have large body sizes, and have evolved filtering mechanisms (baleen) to strain prey items from the water. All mysticetes feed on this prey type.

The second prey type is composed of larger organisms that also school at relatively shallow depths (surface to several hundred meters) or migrate up to shallow depths during the night. This includes many pelagic fishes [e.g., hake (*Merluccius*), pollock (*Pollachius, Theragra*), myctophids (Myctophidae)] and schooling squids (*Loligo, Dosidicus*), which occur throughout the world's oceans. Because these prey are larger, they generally occupy higher trophic levels and are captured individually. Their cetacean predators typically have smaller body sizes. They in-clude all of the large-schooling dolphins (e.g., dusky, *Lagenorhynchus obscurus*; common, striped, *Stenella coeruleoalba*; spotted dolphins) and some small-schooling or solitary species (e.g., bottlenose, Commerson's, Indian river dolphin). These cetaceans tend to have a high tooth count, pointed TEETH, and pointed snouts, all adaptations for pursuing fast, individual prey.

The third prey type is composed of large, solitary squid (e.g., *Gonatus*). These are most often found in deep waters throughout the world's oceans. Because of their size and solitary habits, they are captured individually. Cetacean predators of these prey include the sperm whales (*Physeter, Kogia* spp.), all of the beaked whales (Ziphiidae), and pilot whales (*Globicephala* spp.). They are deep divers and tend to have reduced dentition, rounded heads, and well-developed melons, the latter perhaps indicative of the importance of echolocation for prey detection in the dark depths.

The final prey type includes species at high trophic levels that are themselves top predators. These include predatory fishes [e.g., tunas (Scombridae), sharks, salmonids], marine birds, pinnipeds, and cetaceans, including the largest of whales [rorquals (*Balaenoptera* spp.) and sperm whales]. Few cetaceans are able to take these prey items. They include the killer whale and, possibly, false killer whale (*Pseudorca crassidens*), pygmy killer whale (*Feresa attenuata*), and pilot whales. Two distinct forms of killer whales occur in waters off the west coast of North America: those that take fish and those that take mammals and birds. There is some indication that these two forms are found in Antarctic waters, and perhaps throughout the world's oceans.

B. How Do Cetaceans Capture Prey?

Cetaceans have two main types of feeding apparatus: baleen and teeth. Baleen is used for straining prey items from the water or, in the case of the benthic-feeding gray whale, from the sediment. Teeth are used for catching individual prey items. Species with a high tooth count use them to grasp individual prey; those with a low tooth count tend to be suction feeders (see later).

We know the most about prey capture strategies for cetaceans that feed on small prey that school at relatively shallow depths (the mysticetes). This is because it is relatively easy to observe these animals feeding in the wild. Mysticetes have BALEEN plates suspended from the roof of their mouths, which they use to strain prey items from the water. The number of baleen plates, their length, and the density of baleen fibers per plate vary between species and are correlated with prey size. Right whales (Balaenidae) and sei whales (*Balaenoptera borealis*) have the greatest number of plates with the finest filtering strands and feed mainly on tiny copepods. Blue whales and most other rorquals have an intermediate number of plates with coarser filtering strands and feed on larger prey items such as euphausiids and small fishes. Gray whales have the fewest number of plates with the coarsest strands and are largely bottom feeders, sifting benthic infauna from muddy substrate.

In addition to specializing on different prey sizes, baleen whales have specialized feeding methods that also correlate with the morphology of their baleen. "Skimmers," the right whales, swim slowly with their mouths open through dense clouds of slow-swimming copepods. "Gulpers," including most rorquals, lunge into dense schools of euphausiids or fishes with their mouths open, closing them rapidly to trap their prey. All rorquals have throat grooves that run along the ventral surface of the mouth and throat, which allow the buccal cavity to expand during a lunge, taking in huge quantities of water and, with this, prey (Fig. 1). A variation of this type of feeding is used by humpback whales when they form "bubble nets": streams of bubbles emitted from the blowhole as the whale swims in a circular pattern toward the surface. The bubbles form an ascending curtain, which concentrates prey inside. Most of these cetaceans are solitary feeders but they regularly aggregate in areas of high-prey density and, when prey are extremely dense, will feed cooperatively at times, through bubble-net feeding or in staggered echelon formations (Fig. 2).

Cetaceans that feed on larger fish and squid that school at relatively shallow depths capture individual prey items and swallow them whole. High speed is important, as is vision. Typically these predators forage cooperatively, herding prey into tight aggregations and capturing them in turn. Acoustic signaling is presumably important for the coordination of schooling activities. Some cetaceans in this group feed as individuals, particularly those found in coastal areas. They show a wide range of prey capture behaviors, including slapping fish with their flukes and deliberately stranding themselves on the beach in pursuit of fishes.

Cetaceans taking large, solitary squid feed at depth, in partial to full darkness. For this reason, not much is known about how they capture prey. They probably do not feed cooperatively because their prey do not school and because most of these cetaceans occur in small schools and are slow swimming. Most have reduced dentition, and evidence indicates that they are suction feeders, using the gular muscles and tongue in a piston-like action to suck prey into their mouths. How they are able to get close enough to their prey to suck them in remains a mystery. One intriguing idea is that they are able to partially stun prey with echolocation bursts. To date, this hypothesis remains largely untested.

Figure 2 *A pair of lunge-feeding fin whales* (Balaenoptera physalus) *in echelon formation. The animals are moving left with right sides to the surface. Baleen can be seen attached to the upper jaws and the throat pleats are expanded.*

Cetaceans that prey on top predators show a wide range of prey capture methods: hunting as individuals when prey are small and cooperatively in groups when prey are large. For example, killer whales may take pinnipeds by beaching themselves intentionally to grab adults and pups from ROOKERIES but hunt cooperatively to take dolphins and large whales. Cooperative behaviors include prey encirclement and capture, division of labor during an attack, and sharing of prey.

C. How Do Cetaceans Locate Prey?

Most cetaceans are visual predators, at least in part. For odontocetes, echolocation is equally important in locating and targeting prey, more so than vision in some species. Although only confirmed for a handful of captive species, all odontocetes are assumed to be able to echolocate and to use this sense extensively when foraging. At present, there is no evidence that mysticetes have the ability to echolocate, although they do produce low-frequency sounds that travel long distances (hundreds of kilometers). The long wavelengths of these pulses cannot resolve fine features and are transparent to most schooling prey, so it is doubtful that they could be used to locate and target prey patches.

The effective range of VISION and echolocation is a function of water clarity and the specific echolocation abilities of a species, but both are probably limited to distances on the order of hundreds of meters to a few kilometers. On a larger spatial scale, patchiness and variability in space and time are characteristic of most marine ecosystems and little is known about how cetaceans locate prey in such environments. Presumably, many species simply travel large distances in a continuous search. Here, schooling may increase the chances of encountering a patch (the more eyes and ears, the better), and dolphin schools have been observed moving through the water in wide line-abreast formations, apparently searching for prey.

However, there are circumstances under which prey occur predictably in space and time, and it is likely that cetaceans search for and exploit these opportunities. For example, oceanographic

Figure 1 *A lunge-feeding fin whale* (Balaenoptera physalus) *with throat grooves expanded to allow water into the buccal cavity. Note the animal's eye (top, center).*

features (e.g., boundaries between currents, eddies, and water masses) increase prey abundance or availability by enhancing primary production, by passively carrying planktonic organisms, and by maintaining property gradients (e.g., fronts) to which prey actively respond. Topographic features (e.g., islands, seamounts) are also sites of prey aggregation. Therefore, a good foraging strategy is simply to locate these physical features, and many species of cetaceans (right, blue, fin, humpback, sperm, killer whales, spinner, Risso's, common, Atlantic spotted *Stenella frontalis* dolphins) have been found to associate with them.

Many species of cetaceans locate and associate with predictable point sources of prey. For example, killer whales aggregate around pinniped rookeries when young seals and sea lions are weaning. Rough-toothed dolphins (*Steno bredanensis*) associate with flotsam in the oceanic tropics, which serves to aggregate communities of animals at a wide range of trophic levels. A wide variety of cetaceans associate with fishing operations to take their discards or their target species.

And there are times when prey are more accessible than others. The pelagic community of fishes and invertebrates, which live at depth during the day but migrate to the surface at night, provides an opportunity for cetaceans to predictably locate prey near the surface, and some dolphins (spotted, spinner, dusky, common) are known to feed on organisms in this community at night.

III. Cetacean Predators

By far the most important predator of cetaceans is the killer whale (Fig. 3). Its pack-hunting behavior allows it to take everything from the fastest dolphins and porpoises to the largest whales, including blue and sperm whales. Other predators known to occasionally prey on smaller or weakened individuals include large sharks, and possibly false killer, pygmy killer, and pilot whales. Polar bears (*Ursus maritimus*) take cetaceans along the ice edge.

The ecological significance of this predation pressure in the lives of whales and dolphins is difficult to assess, but it may be significant. Individual large whales often show signs of killer

Figure 3 *Killer whale* (Orcinus orca) *preying on a Dall's porpoise* (Phocoenoides dalli) *off the coast of Oregon. Photo by Karin Forney.*

whale tooth rake marks on their flippers, fins, and flukes, and up to one-third of the bottlenose dolphins off eastern Australia bear shark bite scars, suggesting that they regularly encounter predators. It has been hypothesized that large whales may undergo their annual migrations in order to reach calving grounds in areas of lower killer whale densities (i.e., the tropics). Aggregative behavior is a common defensive strategy among prey species and it is possible that schooling evolved in dolphins primarily as a defense mechanism against predators (see later). These kinds of behavioral adaptations have cascading effects influencing not only distribution and abundance, but also social structure, timing and mode of reproduction, foraging strategies, and speciation patterns. Predation therefore could play a major role in shaping cetacean communities and life history strategies.

IV. Schooling

Like many other animals, cetaceans form aggregations for two main reasons: feeding and protection. Feeding can bring animals together as passive aggregations in areas of high resource abundance. Alternatively, animals may actively seek others to take advantage of benefits provided by other school members. Schools also serve to protect members from predation, by providing cover for individual members, by confusing predators with synchronized movements of many individuals, by reducing the probability of predation on any one individual, by increasing the chance of detection of a predator, and by providing for coordinated defense. Although occurring in large groups also increases the potential for social interactions, including reproduction, this may only be a secondary benefit of schooling.

The majority of cetaceans occur in schools, although there are some species that regularly occur solitarily or in very small groups of pairs or trios (many mysticetes, large male sperm whales, most beaked whales, *Kogia* spp., and river dolphins). Most schooling species have characteristic school sizes (although they can vary somewhat area to area). For example, rough-toothed dolphins typically occur in groups of 10–20, pilot whales occur in schools of dozens, and some oceanic dolphins (*Stenella* spp., *Delphinus* spp.) regularly occur in groups of hundreds or thousands (Fig. 4).

School size correlates with feeding habits: species that form large schools are almost all shallow-diving species that feed mainly on schooling prey, whereas those that occur in school sizes of 25 or fewer tend to be (a) deep-diving species that feed mainly on larger squids or (b) coastal species feeding on dispersed prey. School size also correlates with predation pressure; large cetaceans, presumably subject to lower predation pressure than small species, occur only in small groups, whereas small cetaceans, subject to higher predation pressure, occur in schools whose size correlates with the openness of habitat: the more open, the larger the school size. School size should correlate with resource availability and will affect reproductive strategies, although the nature of these relationships remains largely unexplored.

Although most schools are monospecific, several species regularly occur in mixed-species schools. Some of these associations appear to be opportunistic: bottlenose dolphins, for ex-

Figure 4 *A school of long-beaked common dolphins* (Delphinus capensis) *in the Gulf of California. These schools typically include thousands of individuals.*

ample, have been recorded to occur with over 20 different species of whales and dolphins. Other associations appear to be more prescribed: spotted and spinner dolphins regularly occur together in mixed schools. Risso's, Pacific white-sided (*Lagenorhynchus obliquidens*), and northern right whale (*Lissodelphis borealis*) dolphins are commonly found in association. The nature of these interactions (e.g., why these species-specific associations occur, how these species avoid competition) is unknown.

V. Communities and Coexistence

Studies of communities typically focus on identifying member species and their interactions and then address mechanisms for their coexistence. These kinds of studies comprise a large part of the ecological knowledge for many terrestrial species. In contrast, almost nothing is known about this aspect of the ecology of cetaceans.

We do know that there are regularly occurring species assemblages (Fig. 5). For example, pantropical spotted and spinner dolphins are frequently found in mixed-species schools in association with yellowfin tuna (*Thunnus albacares*) and are accompanied by large and speciose flocks of seabirds; this association is particularly prevalent in the eastern tropical Pacific, as opposed to other tropical oceans. We know that there are variations in typical co-occurrence patterns. In the Gulf of Mexico, for example, five species of *Stenella* coexist in a relatively small area, more *Stenella* species than any other tropical ocean. The nature of the interactions between species in these assemblages, why they associate, and the reasons for variations in community membership patterns are almost completely unknown.

Coexisting species, particularly those that are closely related or have similar ecological roles, potentially compete for resources. An often cited example is the Southern Ocean, where the relative abundances of cetaceans, pinnipeds, and seabirds, all krill consumers, have been reported to have changed between pre- and postwhaling years. One plausible explanation is competitive release: the decrease in biomass of cetacean predators

released a huge prey base of krill to pinnipeds and seabirds, both of which were able to increase in abundance. However, available data on prey biomass, predator consumption, and population status are largely lacking, so the purported changes are in question. In fact, there is little hard evidence to indicate the degree, or existence, of competition between ecologically similar cetaceans.

Ecological theory states that stable communities of coexisting species must differ in resource utilization in some way: prey species or size specialization, differential habitat use, or diel pattern. Such niche partitioning is fairly clear for cetaceans on a broad scale. For example, there are species that feed on fish and those that feed on squid. There are species feeding in shallow water and those that feed at depth. Some cetaceans feed at night and others during the day.

On a smaller scale, one of the best known examples of niche partitioning is for baleen whales. In this group, there is a fair degree of prey specialization that presumably allows for niche partitioning in areas of sympatry. Blue whales feed almost entirely on euphausiids; fin whales and humpbacks feed mainly on fishes but take euphausiids when they are abundant; and right whales and sei whales feed mainly on copepods. Odontocetes provide additional possible examples. Bottlenose, short-beaked common (*Delphinus delphis*), pantropical spotted dolphins, and harbor porpoises exhibit diet specialization among age, sex, and reproductive class, although this diet specialization could be due to differing energy requirements. Aside from these examples, very little is known about how or if cetaceans partition resources.

Ultimately, in order to understand community structure, the mechanisms by which species partition resources, not merely the presence of differences in resource use, are of principal interest. The question then becomes, given that there are differences, what mechanisms can explain them? Community ecologists have identified interference and exploitative competition, mutualism, morphological or physiological factors, and habitat structure as potential mechanisms for maintaining resource utilization differences. This is an area that remains

Figure 5 *A feeding assemblage of long-beaked common dolphins* (Delphinus capensis), *skipjack* (Euthynnus affinis), *brown pelicans* (Pelecanus occidentalis), *Heerman's gulls* (Larus heermanni), *and brown boobies* (Sula leucogaster) *chasing schooling fishes in the Gulf of California.*

almost completely unexplored for cetaceans and the communities in which they are found.

VI. The Role of Cetaceans in Marine Ecosystems

What role do cetaceans play in marine ecosystems and what is their significance? Most cetaceans are apex predators. As such, they take tons of prey from the ecosystem. (Some estimate that cetacean consumption equals or exceeds that of fisheries in the Georges Bank, the continental shelves of the northeastern United States, the northwestern Mediterranean, and the Southern Ocean ecosystems.) In so doing, it seems likely that cetaceans affect the life history strategies and population biology of their prey, as well as organisms at other trophic levels that interact in various ways with these prey. Little is known about the details of these dynamics, although this may be the most significant way in which cetaceans impact marine ecosystems.

More specific effects have been documented. For example, benthic feeders such as gray whales alter habitat by regularly turning over substrate (between 9 and 27% of the benthos in the northern Bering Sea) and therefore significantly affect the species composition of benthic communities. Feeding cetaceans provide feeding opportunities for seabirds by driving prey to the surface, sometimes injuring or disorienting it; in one study, up to 87% of all feeding individuals from four seabird species in the Bering Sea associated with gray whale mud plumes. Large whales dying at sea may sink to the bottom and provide rare but superabundant food and habitat for deep-water species. There is evidence that mollusc communities may have specialized on these resources for the past 35 million years, and some speculate that whale carcasses may have been instrumental in the dispersal of hydrothermal-vent faunas. Feces of some cetaceans, particularly large whales in areas of low productivity, may play a significant role in nutrient cycling. Cetaceans are host to a variety or commensal or parasitic species; in some cases (Cyamid WHALE LICE), these species are completely dependent on cetaceans through all life stages.

VII. Concluding Remarks

The field of cetacean ecology is very much in its infancy. Technological advances and heightened interest will undoubtedly bring greater insight to this discipline in the near future. However, any attempts to make ecological sense of cetaceans as marine organisms and to interpret their distribution patterns, foraging ecology, community structure, and role in ecosystems must take into account the fact that many cetaceans today exist as remnant populations that have been reduced drastically through anthropogenic effects: commercial exploitation, incidental mortality, and habitat destruction. This means that cetacean ecologists must also add conservation biology to the list of disciplines that will likely affect our search for ecological patterns.

See Also the Following Articles

Biogeography ■ Competition with Fisheries ■ Feeding Strategies and Tactics ■ Ocean Environment ■ Predator–Prey Relationships

References

Berta, A., and Sumich, J. L. (1999). "Marine Mammals: Evolutionary Biology." Academic Press, San Diego.

Evans, P. G. H. (1982). Associations between seabirds and cetaceans: A review. *Mamm. Rev.* **12**, 187–206.

Evans, P. G. H. (1987). "The Natural History of Whales and Dolphins." Facts on File, New York.

Felleman, F. L., Heimlich-Boran, J. R., and Osborne, R. W. (1991). The feeding ecology of killer whales (*Orcinus orca*) in the Pacific northwest. *In* "Dolphin Societies: Discoveries and Puzzles" (K. Pryor and K. S. Norris, eds.), pp. 113–147. University of California Press, Berkeley.

Gaskin, D. E. (1982). "The Ecology of Whales and Dolphins." Heinemann Educational Books, London.

Katona, S., and Whitehead, H. (1988). Are Cetacea ecologically important? *Oceanog. Mar. Bio. Annu. Rev.* **26**, 553–568.

Mangel, M., and Hofman, R. J. (1999). Ecosystems: Patterns, processes, and paradigms. *In* "Conservation and Management of Marine Mammals" (J. R. Twiss, Jr., and R. R. Reeves, eds.), pp. 87–98. Smithsonian Institution Press, Washington, DC.

Norris, K. S., and Dohl, T. P. (1980). The structure and functions of cetacean schools. *In* "Cetacean Behavior" L. M. Herman, ed.), pp. 211–261. Wiley, New York.

Norris, K. S., and Møhl, B. (1982). Can odontocetes debilitate prey with sound? *Am. Nat.* **122**, 85–104.

Norris, K. S., Würsig, B., Wells, R. S., and Würsig, M. (1994). "The Hawaiian Spinner Dolphin." University of California Press, Berkeley.

Reynolds, J. E., III., and Rommel, S. A. (1999). "Biology of Marine Mammals." Smithsonian Institution Press, Washington/London.

Wells, R. S., Irvine, A. B., and Scott, M. D. (1980). The social ecology of inshore odontocetes. *In* "Cetacean Behavior" (L. M. Herman, ed.), pp. 263–317. Wiley, New York.

Würsig, B., Cipriano, F., and Würsig, M. (1991). Dolphin movement patterns: Information from radio and theodolite tracking studies. *In* "Dolphin Societies: Discoveries and Puzzles" (K. Pryor and K. S. Norris, eds.), pp. 79–111. University of California Press, Berkeley.

Cetacean Evolution

R. EWAN FORDYCE
University of Otago, Dunedin, New Zealand

I. Patterns of Evolution

Fossils show that cetaceans arose from terrestrial ancestors more than 50 million (M) years ago. They have evolved to become the dominant group of marine mammals in terms of both taxonomic and ecological diversity and geographic range. Fossils provide the only direct evidence of cetacean evolution and extinction, revealing many changing patterns over time, but rarely details of rates, modes, and mechanisms of ancestor-to-descendant transitions. Structural, genetic, and ecological patterns in living species also elucidate how Cetacea have evolved. At the species level, evolutionary processes include natural selection, sexual selection, coevolution, founder effects, vicariance, and hybridization. At larger scales, the distribution and abundance of global food resources have also played major roles in evolution, for there are strong links between changes in ocean structure and ocean ecosystems on one hand and structural/ecological patterns among Cetacea on the other.

II. Paleoecology and Evolution

Ecology encompasses the sum of interactions that species have with their environment, biological, physical, and chemical, whereas paleoecology explores such relationships for the past. At the scale of geological time, evolution and paleoecology are inextricably linked, for evolution occurs through natural selection and adaptation to the environment. An appreciation of paleoecology thus helps understand the history of Cetacea. Several ecological factors have been considered repeatedly in accounts of cetacean evolution (Fig. 1), including feeding and predator/prey interactions, migration, thermal adaptations, and habitat shifts.

For any ancient species, paleoecology can be inferred several ways. A common approach, termed "taxonomic uniformitarianism," assumes that a fossil species had similar ecological strategies to its closest living relatives. Thus, fossil sperm whales

presumably were deep divers, and fossil mysticetes were filter feeders. Fossils show that all cetacean lineages have evolved through a long-term change in structure. If such evolution is caused by ongoing adaptation to changing environments, then even within a lineage, ecological strategies will change. For these reasons, taxonomic uniformitarianism must be supported by other methods.

In groups that lack close living relatives, ancient ecologies may be inferred by analogy. The bizarre extinct Peruvian *Odobenocetops* species, or walrus whales, have no structural equivalents among living Cetacea, but their jaw form was remarkably similar to that of the unrelated living walrus, *Odobenus rosmarus*. A comparable style of suction FEEDING on molluscs is inferred. This approach to paleoecology can be expanded by studying functional morphology, which includes reconstructing soft tissues onto fossil bones using known conservative patterns of muscle, nerve, and vessel to bone. Finally, geological evidence (sediment patterns) and geochemical signals (isotopes) can identify the ancient environments in which cetaceans lived, sometimes indicating dramatic or novel ancient ecologies.

Changes in structure and diversity of fossils reveal major functional, ecological, and taxonomic shifts in cetacean history. In particular, these involve an Eocene radiation of archaeocetes and an Oligocene radiation of Neoceti. Ecological shifts in the

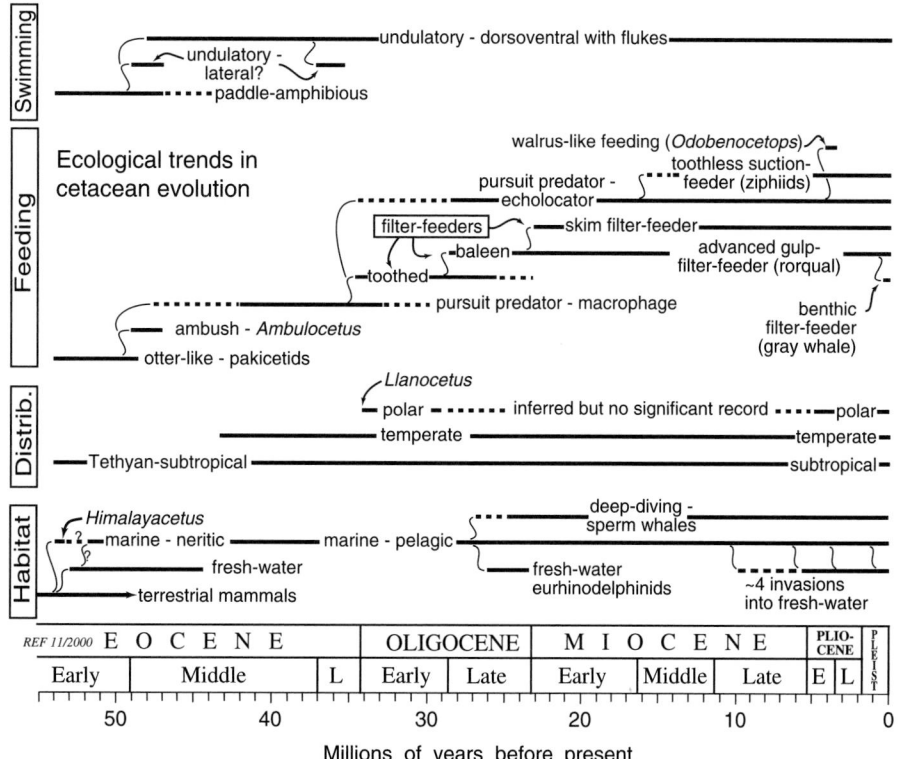

Figure 1 *Summary of inferred ecological strategies important in cetacean evolutionary history. Paleoecology is inferred mainly using taxonomic uniformitarianism (see text). Only selected taxa are shown.*

later Miocene radiation of some odontocetes and mysticetes are less clear.

III. Major Phases of Cetacean Evolution

The fossil record reveals three major radiations in cetacean history (Fig. 2). First, cetaceans diversified in near-tropical shallow waters of the Tethys seaway between India and Asia about 45–53 million years before the present, (Ma), spreading into midtemperate waters by 40 Ma. Presumably this early phase of evolution was influenced by local marine productivity. Up to 6 species in three families of Archaeoceti (stem or basal Cetacea) were present at some times, and >35 species have been reported for the interval 35–53 Ma. This initial radiation of archaeocetes involved shifts from riverine and nearshore marine settings to fully oceanic habits, linked with changes in, for example, locomotory and hearing mechanisms. In terms of structural variation (and, by inference, ecology), these early cetaceans were comparable with living mysticetes and did not show the wide disparity seen among living odontocetes. Archaeocete diversity dropped late in Eocene times, foreshadowing the rise of Neoceti.

The second major radiation, involving the Neoceti or crown group Cetacea, occurred early in the Oligocene. ECHOLOCATING odontocetes and FILTER-FEEDING mysticetes arose from Late Eocene (~35–36 Ma) archaeocetes, which neither echolocated nor filter-fed, and diversified rapidly in about 5 M years. (No certain archaeocetes are known from the Oligocene.) Concurrent events included the final breakup of Gondwanaland, opening of the Southern Ocean, cooling and increased tropics-to-polar temperature gradients, and changes in ocean ecosystems and productivity. Probably, cetaceans radiated in direct response to new ecological opportunities in rapidly changing oceans. By the Late Oligocene interval (23.5–~29 Ma), there were 13+ families and probably >50 species, with some species ranging into polar and fresh waters.

Third, in about the middle of the Miocene (12–15 Ma), seven extant families of mysticete and odontocete appeared. Balaenopterid whales radiated to become the most speciose mysticetes, concurrent with a decline in archaic grades of mysticete ("cetotheres"). Delphinoids—dolphins and relatives—diversified dramatically, particularly the family Delphinidae, but also the Phocoenidae and Monodontidae. Two extinct delphinoid families (Odobenocetopsidae, Albireonidae) appeared briefly. Some archaic cetacean groups, including the Eurhinodelphinidae and Squalodontidae, went extinct at about this time, and the superfamily Platanistoidea (which includes the Squalodontidae) declined almost to extinction. Evolutionary patterns during the Quaternary ice ages (<2.5 Ma) are not clear because of a patchy fossil record. It has been suggested that many north–south species pairs evolved late in Quaternary times, but molecular studies point to older evolutionary events (see later).

IV. Processes: Mechanisms of Evolution

Darwin's original evolutionary mechanism, natural selection, is the process that leads to the adaptation or "fine-tuning"

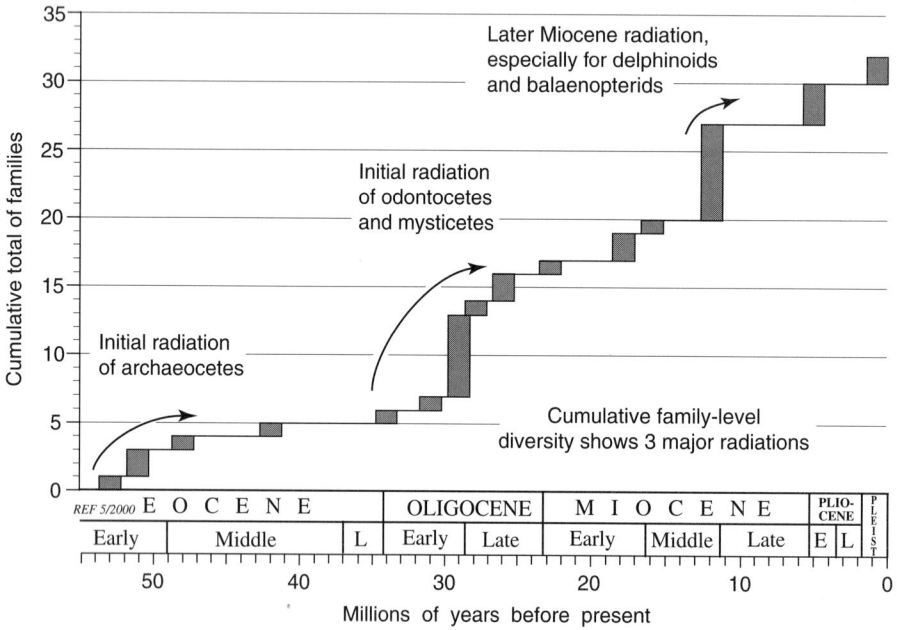

Figure 2 *Cumulative graph of changes in family level diversity over time based on the cetacean fossil record. Three major radiations are apparent. Most of the earliest records for families are based on individual specimens with likely age errors of ±1 million years. With ongoing refinement of dating and the breakup of family level grades, more families are likely to be added in the future, some families of crown group Cetacea will be pushed more firmly back into the Oligocene, and dating errors will be reduced.*

of organisms to the environment through beneficial structure or behavior, resulting in differential reproductive success. Natural selection alone does not, however, explain the origins of novel structures. It is inescapable that the process occurs, although it is not the only one significant in evolution.

Intra- or interspecific competition for limited resources, especially food, has long been viewed as fundamental in evolution. Many cetologists have suggested or implied that competition for food is important among species of baleen whales and between baleen whales and other plankton-eating vertebrates, with implications for the origins of the species involved. However, there is little direct evidence for this. The structure of the feeding complexes of baleen whales, mode (behavior) of feeding, and geographic distribution indicate that niche overlap and thus competition are limited between species. Perhaps previous competition among ancestral species led to taxonomic and ecological divergence seen today. Among fossils, platanistoid dolphins declined in diversity later in the Miocene, as the delphinoids diversified about 12 Ma; this pattern of extinction in one group and radiation in another could indicate that delphinoids outcompeted platanistoids. However, differences in skull structure, especially involving the jaws, imply different ecological requirements, and thus limited competition, between the two groups. Perhaps changing oceanic circulation and climate at about this time caused the extinction of platanistoids and allowed the radiation of delphinoids.

Darwin identified sexual selection as a mechanism in evolution. Here, one sex chooses a specific member of the other sex as a mating partner. Sexual selection might explain the origin of marked SEXUAL DIMORPHISM, particularly in structures involved in display, and has been linked with polygamous mating systems. Dimorphic and possibly sexually selected examples among male Cetacea include the large dorsal fin in *Orcinus orca,* conspicuous mandibular teeth in many ziphiids, prominent foreheads in species of *Hyperoodon* and *Globicephala,* the genital hump in some species of *Stenella,* and the prominent tusks in *Monodon monoceros* and the extinct *Odobenocetops leptodon.* Perhaps the marked size dimorphism in some species also results from sexual selection; examples include male *Physeter macrocephalus*—with its grossly enlarged forehead—and female *Balaenoptera* species.

Hybridization is potentially important in the origin of new species, and cetacean hybrids are known, for example, in *Balaenoptera,* between *Phocoenoides* and *Phocoena,* and between several genera of Delphinidae. However, no convincing cases for evolution by hybridization have been identified in Cetacea.

Coevolution, especially involving mimicry, predator/prey, and host/parasite interactions, has an acknowledged role as an evolutionary process. For cetaceans, mimicry, which graphically illustrates natural selection, is seen in pygmy and dwarf sperm whales, *Kogia* spp.; these have a remarkably shark-like form complete with underslung jaw and pigmented false gill slit. Presumably, to look like a predator will lessen the chance of being preyed upon. Predation has other roles in evolution. The presence or absence of predators (such as the killer whale, *Orcinus orca*) has been used to explain species-specific DISTRIBUTION patterns, including the distinctive and presumably ancient migration patterns of mysticetes from polar to warmer waters for breeding.

Convergent evolution occurs when species show similarity not inherited from a common ancestor. Mimicry is a form of convergence. Among Cetacea, the "river dolphins," long discussed as if they form a real group, encompass species in four different families and in two superfamilies. However, *Platanista* is convergent in its riverine habits, small body size, and long rostrum with the three delphinoid genera *Inia, Pontoporia,* and *Lipotes.* In the past, the now-extinct Eurhinodelphinidae included some riverine or lacustrine species, also with long rostra. For living species, the southern delphinid *Cephalorhynchus hectori* is similar in body form and some aspects of ecology and behavior to the unrelated porpoise *Phocoena phocoena* (Phocoenidae). Unrelated groups have convergently reduced and lost teeth, as seen in Physeteridae, Ziphiidae, and some Delphinidae (e.g., *Grampus*).

In biology, there is great interest in the developmental mechanisms leading to rapid evolutionary change in structure. A widely discussed mechanism is heterochrony, which involves a change in timing or rate of development of structures relative to the equivalent processes in an ancestor. Thus, some features might arise when juvenile structures persist into adult stages (paedomorphosis), whereas others arise when structures develop "beyond" that of the ancestral adult stage (peramorphosis). Either process could generate new structures. Apparent paedomorphic features in cetaceans include the shortened intertemporal region and longer vomer and mesorostral groove in Neoceti, the rounded cranium and persistent interparietal bone on the skull of many Delphinidae and Phocoenidae, and the down-turned rostrum and relatively symmetrical SKULL in Phocoenidae. Possible peramorphic features include extra body parts (e.g., increase in number of vertebrae, as in Dall's porpoise, *Phocoenoides dalli,* or phalanges) generated through a delayed halt in development.

V. Geography and Evolution

Darwin realized that speciation is often related to geography. The range of a species may be split by a change in physical habitat (namely, split by a vicariant event) or can be expanded by dispersal beyond normal limits. Populations that become geographically isolated through such events can diverge and, via allopatric speciation, may become new species. Most discussion of such ideas focuses on terrestrial habitats, but there are clear marine parallels. During 50+ M years of cetacean history, geographic changes have included the closure or opening of some straits and ocean basins, dramatic swings in continental shelf habitat area through sea level fluctuations, and major shifts in current systems, upwellings, and latitudinal water masses. Oceanic temperature regimes changed in parallel. This physical evolution of the oceans probably influenced cetacean evolution at many levels. The distributions of modern and fossil Cetacea indicate an important role for geography in evolution (Fig. 3).

Some living species have obvious northern and southern populations or closely related north–south species pairs, but do not occur in the tropics (Fig. 3). Such bipolar or antitropical distributions probably arose allopatrically when populations became isolated either side of the tropics through

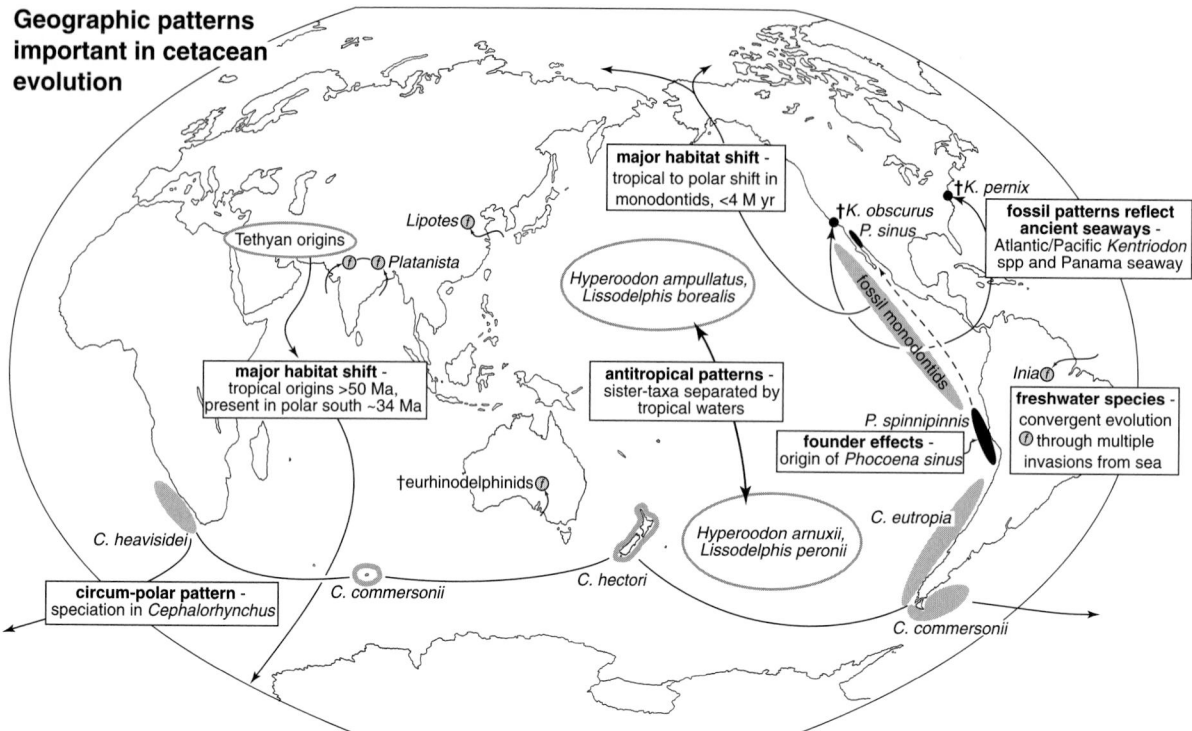

Figure 3 *Distribution patterns illustrating the possible role of geography in cetacean evolution. Patterns indicated include antitropical and circum-Antarctic distributions, allopatric species pairs between oceans, founder effects, convergent origins for various freshwater dolphins, habitat expansion in general for Cetacea (tropical origins, followed later by spread as far as the poles), and geologically recent changes in habitat. Only selected taxa are shown. Fossils are marked with a dagger.*

changing sea temperatures or current regimes, sometimes leading to speciation. Antitropical distributions are marked in populations of rorquals (*Balaenoptera*), species of right whales (*Eubalaena*), some beaked whales (Ziphiidae; species of *Berardius*, *Hyperoodon*, and *Mesoplodon*), porpoises (Phocoenidae), and dolphins (Delphinidae, e.g., *Lissodelphis* species).

Among delphinids, molecular studies reveal that six antitropical species of *Lagenorhynchus* split at varying times, not just during the geologically recent (<2.5 Ma) ice ages, and that these species represent two or more different lineages. At least one genus appears to have speciated around an ocean: four species in the small delphinid genus *Cephalorhynchus* occur around the circum-Antarctic Southern Ocean (Fig. 3). There is some fossil evidence for allopatric species pairs. Closely related species of the small archaic dolphin *Kentriodon* that occur in Miocene strata in California and Maryland perhaps evolved from an ancestor that ranged through the central American seaway before the uplift of Panama.

For porpoises, the endangered vaquita (*Phocoena sinus*), from the Gulf of Mexico, perhaps originated when a few of Burmeister's porpoise, *Phocoena spinipinnis*, crossed the equator only tens of thousands of years ago (Fig. 3). The vaquita, with its limited distribution and low genetic variability, illustrates founder effects; a new population, for example, in an isolated region on the limits of a normal range, may be established by only a few original founders, which are not genetically representative of the original population.

Sea level fluctuations have changed the extent of continental shelf habitat dramatically, especially during cycles of glaciation and cooling over the last 2.5 M years. At peak glaciation, the sea level was >100 m lower than at present, leading to fragmentation or loss of shelf habitat for long intervals, along with colder conditions. This sort of habitat change could lead to extinctions. However, if species' durations typically exceed 100,000 years, as is very likely, then most living cetaceans have survived several of these fluctuations. The record of Late Pliocene and Early Pleistocene Cetacea includes a few species from shallow-water habitats (the dolphin *Parapontoporia*, and cetothere *Nannocetus*), which are now extinct, perhaps as a result of Pleistocene climate changes and/or habitat loss.

Some cetacean groups show significant change in range, even in geologically recent times. The beluga and narwhal (family Monodontidae) are now restricted to cold north polar waters, but during warmer Pliocene times (~4 Ma), closely related extinct monodontids occurred far south in temperate to subtropical waters. Further evidence for changing geographic ranges comes from archaeocetes, which, for about 10 M years after their origin, were subtropical to tropical. In contrast, the maximum diversity for Cetacea is now in temperate waters.

Allopatric speciation results from lineage splitting, or cladogenesis. New species could also evolve by anagenesis, which involves transformation from an ancestral to descendant species without lineage splitting. Anagenetic change would produce a fossil record with several species in unbranched succession and might be expected in lineages containing geographically wide-ranged species. (Species with limited ranges seem more likely to experience allopatric speciation.) Among living Cetacea, the sperm whale *Physeter macrocephalus* and the killer whale represent genera with a single abundant and widely distributed species with no immediate relatives (sister species), and anagenesis might be considered in explaining their history.

Freshwater settings were important in the early transition of cetaceans from land to sea in the Tethys (Fig. 1). Early archaeocetes were amphibious, with well-developed hindlimbs, and they include species from strata deposited in freshwaters. Freshwater habits for some archaeocetes are indicated by oxygen isotopes from TEETH. Fossil Neoceti also show several reinvasions of freshwaters. For example, one unnamed Late Oligocene species of Eurhinodelphinidae occurs in freshwater sediments of central Australia. The dolphin shows no obvious structural adaptations to a riverine or lacustrine habitat, but it is represented by several specimens of slightly differing geological age, indicating a significant long-term occupation of freshwaters. A single large ziphiid fossil is known from Miocene freshwater sediments of Kenya. All the living "river dolphins" arose from marine ancestors, some well known from fossils, which invaded freshwaters in Asia and South America in up to four separate events. Mysticetes seem never to have occupied river systems.

VI. Life History Strategies

In terms of life history traits, cetaceans appear to be *K* strategists. That is, compared with many other mammals, cetaceans are large animals that have slow reproductive rates, produce a single offspring, show significant PARENTAL CARE of young, have long reproductive lives, and have relatively low mortality rates. This rather inflexible reproductive strategy has been linked, in an evolutionary sense, to the nutritional requirements of both the young and parents and thus to food availability. Physical disturbance of the environment and biological effects, such as predation, might be less important in cetacean evolution. The FOSSIL RECORD of cetaceans, which shows major evolutionary change linked to oceanic change, supports the idea that food resources have been fundamental in cetacean history.

VII. Taxonomic Duration

How long do species and genera persist? The fossil record shows that few, if any, species have ranges longer than one geological stage (a stage is a time unit of variable length, mostly 4–5 M years). For extinct Cetacea, the close-spaced succession of Eocene archaeocetes points to species' durations of 1–2 M years. Living species can be traced back into the Pleistocene but not reliably longer, i.e., <2.5 Ma. Molecular studies provide estimated dates for separation of lineages of living species, including 1.9–3 and 3.8–9.6 M years for species of *Lagenorhynchus* and >5 M years for *Balaenoptera musculus* and *B. physalus*. Dates for lineage separation, however, are not

necessarily the same as dates for species' durations. For groups beyond the species' level, some living genera have reliable fossil records back to the Early Pliocene or Late Miocene (~5 Ma), including *Balaenoptera*, *Megaptera*, *Balaena*, *Eubalaena*, *Delphinus* and *Mesoplodon*. Generally, though, generic ranges are uncertain because of inconsistent taxonomic concepts among paleontologists.

VIII. Rates of Evolution

Little is known about rates of evolution in Cetacea. The fossil record identifies phases of rapid radiation, and presumably rapid evolution, in the Early Oligocene and late Middle to early Late Miocene. Unlike the situation for some terrestrial mammals, the cetacean fossil record is not dense enough to reveal quantifiable change in structure over time. For living species, no speciation event has been dated reliably enough to clearly reveal evolutionary rate, but some species (vaquita, *Phocoena sinus*; gray whale, *Eschrichtius robustus*) probably evolved over tens, rather than hundreds, of thousands of years.

IX. Diversity and Disparity

Diversity is the number of species within a taxon such as a genus or family (effectively, an index of taxonomic richness), whereas disparity indicates the variation in structure or basic design within a taxon. Diversity is easy to assess, particularly now that advances in the philosophy and practice of classification have produced a widely accepted species level classification of living cetaceans, but the study of disparity is still developing.

A comparison of the two living clades Odontoceti and Mysticeti reveals quite different patterns of diversity (Fig. 4) and disparity. Mysticetes include 12 species in three or four families. The Balaenopteridae is most speciose, with 8 species in two genera (*Balaenoptera*, 7; *Megaptera*, 1). Broadly speaking, species of *Balaenoptera* vary in size, distribution, and behavior, but species boundaries may be blurred. Species are distinguished mainly on aspects of the feeding apparatus (baleen size and spacing, size and shape of the upper jaw), and skeletal differences are rather minor. Thus, disparity appears low. The gray whale, *Eschrichtius robustus*, is structurally quite different (disparate) from other mysticetes and is placed in its own monotypic family. The gray whale probably arose geologically recently (~0.5 Ma?) among the balaenopterids and might be classified with that family despite its different form. In this case, disparity does not indicate taxonomic distance.

Odontocetes include 71–72 species in 10 families. With its 36 species, Delphinidae is the most diverse family of cetaceans, and disparity seems much higher than, for example, within Balaenopteridae. Among delphinids, there is a greater variation in body size and proportion, skull form, proportions of the feeding apparatus and teeth, and distribution of air sinuses in the skull base. Among beaked whales (Ziphiidae; 20 species), the genus *Mesoplodon* has 13–14 rather similar species in which only adult males are separated easily. Disparity here appears low and awaits explanation in terms of evolutionary ecology within the genus. Among other odontocetes, 4 species of small long-beaked "river dolphins" each represent a single family (Iniidae, Pontoporiidae, Lipotidae, and Platanistidae). Although

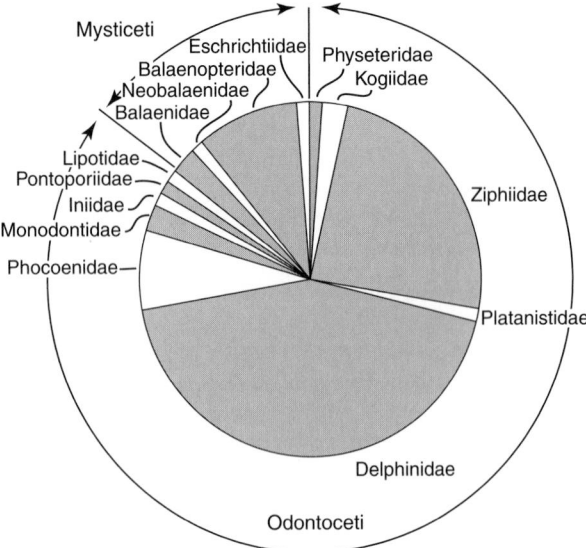

Figure 4 *Patterns of species' diversity at the family level among living Cetacea. Traditionally, families are clusters of species and genera with a similar body plan; thus, families usefully reveal disparity at high levels within the Cetacea. Mysticetes are less speciose and less disparate than odontocetes; delphinids and then ziphiids are the most speciose and most disparate of the odontocetes. Species' counts are derived from Rice (1998).*

these "river dolphins" are superficially similar externally, they are highly disparate in terms of skull form; this disparity supports the separation into 4 families. Of note, Iniidae, Pontoporiidae, and Lipotidae are related closely to delphinoids, whereas *Platanista gangetica* is a highly disparate species in its own family (Platanistidae), in turn in the ancient and once diverse superfamily Platanistoidea, which originated about 30 M years ago. For *Platanista*, Platanistidae, and Platanistoidea, diversity is low (1 living species of *Platanista*, with no fossil record) and disparity is high.

As a group, odontocetes are clearly quite disparate. Huge variation is seen in the skull, involving the shape, size, and sometimes basic construction of the feeding apparatus and teeth, facial region, including bony origin for nasofacial muscles implicated in echolocation, nasal passages, acoustic system, and air sinuses in the skull base. Disparity might be viewed as caused by reduced constraints on body form, allowing specialization in many different directions. This explanation for disparity, however, does not explain the ultimate origin of structural diversity.

X. Extinction

EXTINCTION, the disappearance of lineages, is a fundamental complement to evolution. Extinction obviously occurs when the number of individuals and geographic range drops to nil. The fossil record reveals that extinction is inevitable and, for lineages, usually terminal; only a few species go extinct by

evolving into descendants. Among different styles of extinction, there is no evidence that cetaceans have been involved in mass extinction comparable to that affecting dinosaurs. Taxonomic extinction (the disappearance of a real group or taxon) has occurred, as shown by the fossil record of such well-defined clades as the odontocetes Eurhinodelphinidae and Squalodontidae, and the *Cetotherium* group mysticetes. As noted earlier, environmental change might explain the demise of such groups through loss of habitat or food or of intolerance of cold. Most extinction has probably involved the piecemeal extinction of single species, but the cetacean fossil record is too patchy to expect pattern or cause to be clear. Species susceptible to extinction are those in low-diversity clades (e.g., one or two species in a genus, with no close relatives) occurring in geographically limited physical settings that are unstable over geological time; for Cetacea, this means particularly the "river dolphins." Conversely, widely distributed oceanic species would seem resistant to extinction by other than human agencies.

Patterns of extinction beg a question that ecologists might consider unthinkable: are there vacant modern niches that formerly were occupied by Cetacea? For example, Platanistidae lived in shallow marine settings until about 10 Ma and now occur only in freshwaters, whereas species of Squalodontidae and Eurhinodelphinidae were widely distributed before their demise in the later Miocene. Judging from the functional complexes seen in the latter fossils, there are no modern equivalents to these groups. Functional studies are needed to see if some extant taxa may be the ecological equivalents of the extinct forms.

See Also the Following Articles

Biogeography ■ Cetacea, Overview ■ Convergent Evolution ■ Geographic Variation ■ Geological Time Scale ■ Speciation

References

Clapham, P. J., and Brownell, R. L. (1996). Potential for interspecific competition in baleen whales. *Rep. Int. Whal. Comm.* **46,** 361–367.

Corkeron, P. J., and Connor, R. C. (1999). Why do baleen whales migrate? *Mar. Mamm. Sci.* **15**(4), 1228–1245.

Endler, J. A. (1986). "Natural Selection in the Wild." Princeton Univ. Press, Princeton.

Estes, J. A. (1979). Exploitation of marine mammals: r-selection of K-strategists? *J. Fish. Res. Board Can.* **36**(8), 1009–1017.

Fordyce, R. E., and Barnes, L. G. (1994). The evolutionary history of whales and dolphins. *Annu. Rev. Earth. Plane. Sci.* **22,** 419–455.

Gatesy, J. (1998). Molecular evidence for the phylogenetic affinities of Cetacea. *In* "The Emergence of Whales," (J. G. M. Thewissen, ed.), pp. 63–111. Plenum, New York.

Reynolds, J. E., and Rommel, S. A. (eds.) (1999). "Biology of Marine Mammals." Smithsonian Institution, Washington, DC.

Rice, D. W. (1998). "Marine Mammals of the World: Systematics and Distribution." Society for Marine Mammalogy, Lawrence, KS.

Rosel, P. E., and Rojas-Bracho, L. (1999). Mitochondrial dna variation in the critically endangered vaquita *Phocoena sinus* Norris and Macfarland, 1958. *Mar. Mamm. Sci.* **15**(4), 990–1003.

Thewissen, J. G. M. (ed.) (1998). "The Emergence of Whales: Evolutionary Patterns in the Origin of Cetacea." Plenum, New York.

Cetacean Life History

SUSAN J. CHIVERS
Southwest Fisheries Science Center,
La Jolla, California

A species' life-history strategy is defined by parameters that describe how individuals allocate resources to growth, reproduction, and survival. The allocation of resources presumably results from natural selection maximizing the reproductive fitness of individuals within a species. Biologists studying life-history strategies collect data to answer questions about how long individuals of a species live, the ages at which they become sexually mature and first reproduce, and where and when they travel to find sufficient food to survive (Fig. 1). In search of answers, these biologists may be found in a laboratory estimating an individual animal's age, at a computer modeling growth rates, or at sea observing animals in their natural habitat.

Among cetacean species, the life-history strategies are diverse and differ markedly between the two cetacean suborders: baleen whales (suborder Mysticeti) and toothed whales (suborder Odontoceti). This diversity demonstrates the range of successful strategies that have evolved and enable cetaceans to live in a completely aquatic environment as well as the influence of their phylogeny on adapting them to a particular niche. Reviewing the strategies of species within each suborder reveals that the baleen whales share more similar life-history characteristics. All species are large and long-lived, and all of

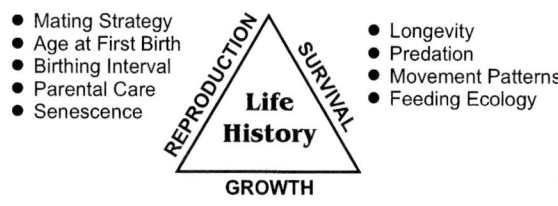

• Mating Strategy
• Age at First Birth
• Birthing Interval
• Parental Care
• Senescence

REPRODUCTION SURVIVAL

Life History

GROWTH

• Longevity
• Predation
• Movement Patterns
• Feeding Ecology

• Rates / Patterns of Growth
• Maximum Size or Length
• Size-at-Birth
• Morphology

Figure 1 *Studies of cetacean life history integrate data describing an individual animal's allocation of resources to growth, reproduction, and survival. Compiling data from many individuals allows the parameters listed next to each category to be estimated, which in turn describes the growth, reproduction, and survival strategies of a species. Life-history data may be collected by observing individual animals in directed photo identification, tagging, marking, or telemetry studies or by necropsying animals to collect teeth for aging, body length measurements for quantifying growth rates, gonads for determining reproductive condition, and skin for estimating individual relatedness or determining "local" adaptations using molecular genetic techniques.*

the baleen whales, except the bowhead whale (*Balaena mysticetus*) and Bryde's whale (*Balaenoptera edeni*), make long-range annual migrations between breeding grounds in tropical waters and feeding grounds in temperate or polar waters. However, the life-history patterns observed among odontocetes are more varied. These species range in size from the small, relatively short-lived (<24 years) harbor porpoise (*Phocoena phocoena*) to the large, relatively long-lived (>70 years) sperm whale (*Physeter macrocephalus*) and occupy diverse habitats, ranging from pelagic and coastal ocean waters to estuarine and fresh waters.

The life-history strategies for relatively few cetacean species are known in detail. Most of the biological data available for baleen whales were collected during whaling operations, whereas odontocetes have been studied from animals incidentally taken during fishery operations, taken in directed fisheries, found stranded on the beach, or observed in the wild or in captivity. Among the most well-known cetacean life histories are those of the humpback whale (*Megaptera novaeangliae*), fin whale (*B. physalus*), common bottlenose dolphin (*Tursiops truncatus*), and killer whale (*Orcinus orca*). Three of these species—the humpback whale, common bottlenose dolphin, and killer whale—have each been the subject of long-term studies, which have provided unique data about the natural variability of a species' life-history strategy based on the observed demographics of individual animals. Our knowledge about the life-history strategies of cetacean species is still incomplete, particularly for rarely encountered species, but our knowledge is expanding rapidly as more specimens are collected and new techniques are developed.

I. Characteristics of Cetacean Life Histories

Although a diversity of life-history strategies is exhibited by cetacean species, there are a few common characteristics that are likely the necessary adaptations for life in a completely aquatic environment. All species give birth to single, large, and precocial young. The presence of multiple fetuses or multiple births has been documented only rarely, and there are no known cases of successfully reared multiple offspring. Gestation times are approximately a year. Among the baleen whales, the estimates for gestation range from 10 to 12 months, and among the odontocetes, estimates range from 10 months for the harbor porpoise (Gaskin *et al.*, 1984) to 14 or 15 months for the sperm whale (Best *et al.*, 1984) and 17 months for the killer whale (Olesiuk *et al.*, 1990; Baird, 2000). Most of the small delphinids (e.g., *Stenella* spp.) have gestation periods of 11 to 12 months (Perrin and Reilly, 1984). The length of the gestation period in part balances the cost of producing a large neonate. Additionally, all cetaceans are relatively long-lived. Among odontocetes, estimates of longevity range from approximately 2 decades for the harbor porpoise to up to 7 decades for the sperm whale, and for the baleen whales, estimates of longevity range from 6 decades for the common minke whale (*B. acutorostrata*) up to 9 or 10 decades for the fin whale. Additional generalizations about life-history strategies are presented in Section III.

II. Methods of Studying Life Histories

Longitudinal and cross-sectional studies of cetacean species have provided data necessary for understanding their life-history strategies. Longitudinal studies are rare but valuable because they provide unique data on the variability of individual demographics. Three species—the humpback whale (Clapham, 1996), the common bottlenose dolphin (Wells and Scott, 1990), and the killer whale (Olesiuk *et al.*, 1990)—have been the subject of ongoing studies that originated during the 1970s. These studies are possible because individuals are relatively accessible and easily distinguishable in the field by natural markings. These studies have quantified individual variability in reproduction and survival through time and have provided unique insights into the species' life-history strategy by incorporating observations of the species' social behavior and ecology. However, most of our knowledge about cetacean life-history strategies is the result of cross-sectional studies. In these studies, data are collected from individual animals sampled primarily from directed or incidental takes. The primary advantage of these studies is that a complete suite of morphological and biological data can be collected, which allows explicit determination of reproductive and physical maturity as well as an estimate of age. Estimates of age are made from the layering patterns evident in the ear plugs or ear bones of baleen whales (Lockyer, 1984; Christensen, 1995) and in the teeth of odontocetes (Perrin and Myrick, 1980). Accurate determination of reproductive maturity in both sexes requires examination of the gonads. In females, the presence of one corpus or more in the ovaries indicates sexual maturity, and in males, the presence of spermatazoa and large seminiferous tubules in histologically prepared testes tissue indicates sexual maturity (Perrin and Reilly, 1984). Physical maturity is determined in both sexes by examining the vertebral column for evidence of fusion. That is, when the vertebral epiphyses are fused with the centrum, an animal is considered physically mature.

Life-history studies are designed to collect data on body size, age and reproductive and physical maturity from many individuals in order to estimate parameters that characterize a species' allocation of resources to growth, reproduction, and survival. Estimated parameters may include age-specific growth and pregnancy rates, the average age at attainment of sexual maturity, calving interval, and longevity. Age is the primary independent variable for all studies because age explicitly demonstrates the trade-off in resource allocation to growth and reproduction during an individual animal's life. The expected pattern of resource allocation from birth through attainment of sexual maturity is primarily for growth and then for reproduction once sexual maturity is attained. Also, the probability of an individual surviving to the next age class increases with increasing age after weaning until sexual maturity is attained and then remains high throughout the individual's reproductive years. Data on age-specific growth and reproductive rates, combined with estimates of age-specific survival rates, are essential to comparing and contrasting the life-history strategies of different species.

III. Cetacean Life-History Patterns

Neonates of all cetacean species are relatively large when compared to those of other mammal species. In fact, neonate size ranges from approximately 29% of the female's asymptotic total body length in most of the baleen whales to between 40 and 48% of the female's length in odontocetes. The large size of neonates, combined with their ability to swim and grow rapidly immediately after birth, increases their probability of survival. The lactation period for the baleen whales lasts only about 6 months, and the young grow rapidly during that period because the fat content of the milk is high. However, the calves of odontocetes grow more slowly, and the lactation period lasts approximately a year or more. The difference in calf growth rates between the two suborders of cetaceans is probably due to the transfer of energy to the young through the milk. Oftedal (1997) estimated that the energy output through milk ranges from 0.40 to 1.06 $MJ/kg^{0.75}$ for mysticetes and from 0.09 to 0.17 $MJ/kg^{0.75}$ for odontocetes. For species with a lactation period of more than a year, the additional investment likely further increases the calf's probability of survival by facilitating the learning of social behaviors [e.g., common bottlenose dolphin, short-finned pilot whale (*Globicephala macrorhynchus*)].

Patterns of growth differ between the sexes of many cetacean species, resulting in some degree of sexual dimorphism. Both males and females have high growth rates while suckling, but growth slows after weaning and again after reaching sexual maturity. However, the sex that grows largest tends to grow for a longer period of time and may have higher growth rates after weaning. Among baleen whales, females attain lengths that are generally 5% larger than males. Similarly, among odontocetes, females are slightly larger than males in the porpoises and river dolphins. However, for other odontocetes, males are larger than females. Sexual dimorphism is most marked in sperm whales, in which males are 60% larger than females. Among the smaller delphinids, such as the common bottlenose dolphin, pantropical spotted dolphin (*Stenella attenuata*), and common dolphins (*Delphinus* spp.), males are approximately 2 to 10% larger than females.

The breeding cycle for all cetacean species has three parts: a gestation period, a lactation period, and a resting, or anestrous, period. This cycle is 2 years or more for most cetacean species. Exceptions are the minke whale and harbor porpoise, which can breed annually. The breeding cycle of blue (*B. musculus*), Bryde's, humpback, sei (*B. borealis*), and gray (*Eschrichtius robustus*) whales includes an 11-month gestation period, a 6- to 7-month lactation period, and a 6- to 7-month resting, or anestrous, period for a minimum of a 2-year cycle, while the breeding cycle for the bowhead and right whales (*Eubalaena* spp.) is 3 to 4 years starting with a 10- to 12-month gestation period. Furthermore, the breeding season of baleen whales is synchronized with their migration cycle. These species travel long distances to breed in tropical waters. Exceptions are the Bryde's whale, and the pygmy Bryde's whale, which spend all year in tropical waters and do not breed synchronously (Lockyer, 1984). Several hypotheses have been proposed for the adaptive significance of the large-scale migrations of baleen whales. Although the phenomenon remains unexplained, hypotheses of increased survival rates for neonates in tropical waters by reducing thermoregulatory demands or the risk of predation by killer whales have been proposed (Corkeron and Connor, 1999).

Similar to other life-history characteristics, the breeding cycle for odontocetes is more variable than that of mysticetes. Porpoises have the shortest breeding cycle, which is approximately 1 year and includes a 10-month gestation. In fact, annual breeding among porpoises has been well documented for the harbor porpoise. The smaller delphinid species seem to have 2- to 3-year calving intervals, which include an 11- to 12-month gestation and a 1- to 2-year lactation period. However, larger odontocetes, such as the killer whale, short-finned and long-finned (*G. melas*) pilot whales, and sperm whale, have calving intervals of >3 years, which includes a 12- to 17-month gestation period and a 2- to 3-year, or longer, lactation period. Breeding synchrony also varies among odontocetes. Species inhabiting temperate waters, such as the harbor porpoise, have been found to have more synchronous breeding seasons than species inhabiting tropical waters. For example, studies of the pantropical spotted dolphin (Fig. 2) and the striped dolphin (*S. coeruleoalba*), which inhabit tropical waters in the Pacific Ocean, found that young are born throughout the year, although most births occur during the spring and fall (Perrin and Reilly, 1984).

Age at attainment of sexual maturity is delayed in all cetacean species as would be expected for large, long-lived mammals. However, the range of ages is quite broad and reflects the unique set of adaptations that characterize the life-history strategy of each species. The range in age of sexual maturity among baleen whales is from approximately 4 years for the bowhead and humpback whales to approximately 10 years for fin and sei whales (Lockyer, 1984). Among odontocetes, the range in age at attainment of sexual maturity is about the same as that observed for baleen whales and seems to be correlated to a degree with longevity and body size. The youngest age at attainment of sexual maturity is 3 years for the harbor porpoise, which is among the smallest odontocetes and is estimated to live approximately two decades (<24 years). However, many of the larger odontocetes reach sexual maturity at ages of 10 years or more and live for four or more decades (Perrin and Reilly, 1984).

Reproductive success varies throughout the life of female cetaceans. Initially, reproductive success is relatively low, peaks several years after the age at attainment of sexual maturity, and then declines as the female ages. This phenomenon is also characteristic of large terrestrial mammals and is probably due in part to a trade-off in costs between reproduction and growth that must occur because physical maturity is attained several years after sexual maturity and to learning to care for young. Evidence for low reproductive success among newly matured females has been documented in the common bottlenose dolphin and the fin whale. Lower reproductive rates for older females have also been documented in the common bottlenose dolphin as longer interbirth intervals for older females that include a 3- to 8-year lactation period. Postreproductive females with senescent ovaries have been identified in only a few odontocetes, including the short-finned pilot whale (Marsh and Kasuya, 1986) and the pantropical spotted dolphin (Myrick *et al.*, 1985), but senescence has not yet been identified in any of the baleen whales. The adaptive significance of senescence is not yet understood but likely contributes to increased reproductive success. For example, several species that exhibit senescence also have fairly complex social structures (e.g., sperm whale, short-finned pilot whale), and the role of postreproductive females in their societies may be associated with increased survival rates of the young by these females participating in the care of young that are not their own.

IV. Characteristics of Male Life Histories

The life-history characteristics of males are less well known than those of females, primarily because this knowledge is less critical to understanding a species' reproductive potential and population dynamics. In this sense, females are the limiting sex. However, knowledge about the life-history strategies of males provides a more complete picture of a species population dynamics and provides information about the species' breeding strategy and social structure.

One of the major differences between the life-history strategies of male and female cetaceans is the age at attainment of sexual maturity. In species with the greatest degree of sexual dimorphism, the difference in age at attainment of sexual maturity for males and females is greatest. This difference reflects the additional time required to grow to about 85% of their asymptotic length, which is the approximate size at which all mammals become sexually mature. For example, sperm whale males reach sexual maturity at a much later age than females. The estimated age at attainment of sexual maturity for the female sperm whale is from 7 to 13 years and for males is approximately 20 years (Rice, 1989; Best *et al.*, 1984). The difference is similar in the killer whale and the short- and long-finned pilot whales (Baird, 2000; Lockyer, 1993). However, the smaller delphinid species that show less sexual dimorphism

Figure 2 *Mother and calf pantropical spotted dolphin* (Stenella attenuata) *in the eastern tropical Pacific. Photograph by R. L. Pitman.*

reach sexual maturity at more similar ages. In fact, the difference in age between the sexes is about 3 years for the common bottlenose dolphin, pantropical spotted dolphin, and spinner dolphin (*S. longirostris*), with males reaching sexual maturity at the older age (Perrin and Reilly, 1984).

Sexual dimorphism has been used as a predictor of cetacean mating systems. For example among odontocetes, the degree of sexual dimorphism exhibited by the sperm whale, short-finned pilot whale and killer whale has been hypothesized to indicate male-male competition in a polygynous mating system. The presence of scars inflicted by other males provides evidence of male-male competition in sperm whales. However, fairly recent data that the short-finned pilot whale and killer whale have limited male dispersal from natal pods and likely breed promiscuously suggests that sexual dimorphism may have evolved for reasons other than mating. One interpretation of these data is that the presence of large males in their natal pod enhances their reproductive fitness by improving, for example, the foraging efficiency of the pod (Wells *et al.*, 1999).

V. Life-History Parameters and Demography

Knowledge of a species' life-history strategy provides the foundation for understanding the species' demography because their life-history characteristics reflect the species' adaptations to a particular niche, which is bounded by constraints of the environment as well as their morphology and physiology. Whereas life-history studies focus on individual variability in traits that express these adaptations, each study can usually only focus on a particular group of animals within the species. The comparison of studies made on different groups of animals within a species' range, however, reveals variability in life-history parameters. For example, pantropical spotted dolphins north and south of the equator have different breeding seasons, and the estimates of asymptotic length for animals in the western Pacific are 4 to 7 cm longer than those from the eastern Pacific (Perrin and Reilly, 1984). Similar examples exist for other cetacean species. There are also examples in the literature of cetacean populations responding to changes in the availability of resources through time. This is called density dependence. For example, changes in the age at attainment of sexual maturity for fin, sei, and minke whales through time have been reported and are presumed to be a response to increased per capita resource availability following reductions in population abundance that resulted from commercial whaling (Lockyer, 1984). Similarly for the striped dolphin and the spinner dolphin, changes in the age of sexual maturity and pregnancy rates have been reported and explained as responses consistent with increased resource availability that resulted from decreased population abundance (Perrin and Reilly, 1984). In addition to understanding a species' life-history strategy and its inherent variability, recognition of these types of population-level responses is important to consider when developing conservation and management plans.

Estimates of age-specific reproductive rates and survival rates are critical to quantifying a species' demography. However, for nearly all cetacean species, age-specific survival rates are unknown and are likely to remain so. Because demographic studies must include age-specific survival rates, unique solutions have been sought to allow the estimation of survival rates based on imperfect knowledge (Barlow and Boveng, 1991). Longitudinal studies like those of the common bottlenose dolphin and the humpback whale provide the only source of data to estimate survival rates, and these data are generally used as a guide for estimating survival rates for other species with similar, but less well-known, life histories.

VI. Life-History Studies and the Future

Several new technologies are being actively applied to studies of cetacean species and contribute to our knowledge about the adaptive significance of their life-history strategies. Specifically, the expansion of molecular genetic techniques and the development of satellite and VHF (Very High Frequency) tracking technology allow more detailed data collection on individual animals. For example, the application of molecular genetic markers as tags for individuals has been demonstrated successfully with the humpback whale data set (Palsbøll, 1999). Application of this technique to cetacean species whose individuals cannot be recognized readily by natural marks may facilitate life-history studies for those species. Additionally, the results of molecular genetic studies on several cetaceans, including the beluga whale (*Delphinapterus leucas*) and Dall's porpoise (*Phocoenoides dalli*), have confirmed hypotheses of male-biased dispersal (O'Corry-Crowe *et al.*, 1999; Escorza-Treviño and Dizon, 2000). Although this is not a surprising result because male-biased dispersal is common among large terrestrial mammals, molecular genetics provided the tool to examine large enough data sets to ask this question for cetacean species. In addition to the expansion of molecular genetic analy-

Figure 3 *A pantropical spotted dolphin* (Stenella attenuata) *wearing a radio tag and time-depth recorder to study diving behavior. Photograph by M. D. Scott.*

ses to studying cetaceans, the development of satellite and VHF tracking technology is continuing. There have been notable successes in the use of satellite tags to study beluga whales in the Arctic (Martin *et al.*, 1998) and blue whales in the North Pacific (Mate *et al.*, 1999). Broader application of this technology to study more individuals of more species has been limited in part by problems with tag attachment, but this technology, together with the expansion of molecular genetic techniques, will provide us opportunities for new insights into the life-history strategies of more cetacean species (Fig. 3).

See Also the Following Articles

Age Estimation ■ Female Reproductive Systems ■ Pinniped Life History ■ Population Dynamics ■ Sexual Dimorphism

References

Baird, R. W. (2000). The killer whale: Foraging specializations and group hunting. *In* "Cetacean Societies: Field Studies of Dolphins and Whales" (J. Mann, R. C. Connor, P. L. Tyack, and H. Whitehead, eds.), pp. 127–153. University of Chicago Press, Chicago.

Barlow, J., and Boveng, P. (1991). Modeling age-specific mortality for marine mammal populations. *Mar. Mamm. Sci.* **7,** 50–65.

Best, P. B., Canham, P. A. S. and MacLeod, N. (1984). Patterns of reproduction in sperm whales, *Physeter macrocephalus. Rep. Int. Whal. Comm. Spec. Issue* **6,** 51–79.

Christensen, I. (1995). Interpretation of growth layers in the periosteal zone of *tympanic bulla* from minke whales *Balaenoptera acutorostrata. In* "Whales, Seals, Fish and Man" (A. S. Blix, L. Walløe, and Ø. Ulltang, eds.), pp. 413–423. Elsevier Science, Amsterdam.

Clapham, P. J. (1996). The social and reproductive biology of humpback whales: An ecological perspective. *Mamm. Rev.* **26,** 27–49.

Corkeron, P. J. and Connor, R. C. (1999). Why do baleen whales migrate? *Mar. Mamm. Sci.* **15,** 1228–1245.

Escorza-Treviño, S., and Dizon, A. E. (2000). Phylogeography, intraspecific structure, and sex-biased dispersal of Dall's porpoise, *Phocoenoides dalli,* revealed by mitochondrial and microsatellite DNA analyses. *Mol. Ecol.* **9,** 1049–1060.

Gaskin, D. E., Smith, G. J. D., Watson, A. P., Yasui, W. Y., and Yurick, D. B. (1984). Reproduction in the porpoises (Phocoenidae): Implications for management. *Rep. Int. Whal. Com. Spec. Issue* **6,** 135–148.

Lockyer, C. (1984). Review of baleen whale (*Mysticeti*) reproduction and implications for management. *Rep. Int. Whal. Comm. Spec. Issue* **6,** 27–50.

Lockyer, C. (1993). A report on patterns of deposition of dentine and cement in teeth of pilot whales, genus *Globicephala. Rep. Int. Whal. Comm. Spec. Issue* **14,** 137–161.

Marsh, H., and Kasuya, T. (1986). Evidence for reproductive senescence in female cetaceans. *Rep. Int. Whal. Comm. Spec. Issue* **8,** 57–74.

Martin, A. R., Smith, T. G., and Cox, O. P. (1998). Dive form and function in belugas *Delphinapterus leucas* of the eastern Canadian High Arctic. *Pol. Biol.* **20,** 218–228.

Mate, B. R., Lagerquist, B. A., and Calambokidis, J. (1999). Movements of North Pacific blue whales during the feeding season off southern California and their southern fall migration. *Mar. Mamm. Sci.* **15,** 1246–1257.

Myrick, Jr., A. C., Hohn, A. A., Barlow, J., and Sloan, P. A. (1985). Reproductive biology of female spotted dolphins, *Stenella attenuata,* from the eastern tropical Pacific. *Fish. Bull.* **84,** 247–259.

O'Corry-Crowe, G. M., Suydam, R. S., Rosenberg, A., Frost, K. J., and Dizon, A. E. (1997). Phylogeography, population structure and dispersal patterns of the beluga whale *Delphinapterus leucas* in the western Nearactic revealed by mitochondrial DNA. *Mol. Ecol.* **6,** 955–970.

Oftedal, O. T. (1997). Lactation in whales and dolphins: Evidence of divergence between baleen- and toothed-species. *J. Mamm. Gland Biol. Neoplasia* **2,** 205–230.

Olesiuk, P. F., Bigg, M. A., and Ellis, G. M. (1990). Life history and population dynamics of resident killer whales (*Orcinus orca*) in the coastal waters of British Columbia and Washington State. *Rep. Int. Whal. Comm. Spec. Issue* **12,** 209–243.

Palsbøll, P. J. (1999). Genetic tagging: Contemporary molecular ecology. *Biol. J. Linn. Soc.* **68,** 3–22.

Perrin, W. F., and Myrick, Jr., A. C., (eds.) (1980). Age determination of toothed whales and sirenians. *Rep. Int. Whal. Comm. Spec. Issue* **3.**

Perrin, W. F., and Reilly, S. B. (1984). Reproductive parameters of dolphins and small whales of the family Delphinidae. *Rep. Int. Whal. Comm. Spec. Issue* **6,** 97–133.

Rice, D. W. (1984). Sperm whale. *Physeter macrocephalus* Linnaeus, 1758. *In* "Handbook of Marine Mammals" (S. H. Ridgway and R. Harrison, eds.), Vol. 4, pp. 177–233. Academic Press, London.

Wells, R. S., Bonnes, D. J., and Rathbun, G. B. (1999). Behavior. *In* "Biology of Marine Mammals" (J. E. Reynolds III and S. A. Rommel, eds.) pp. 324–422. Smithsonian Institution Press, Washington.

Wells, R. S., and Scott, M. D. (1990). Estimating bottlenose dolphin population parameters from individual identification and capture-release techniques. *Rep. Int. Whal. Comm. Spec. Issue* **12,** 407–415.

Cetacean Physiology, Overview

ROBERT ELSNER
University of Alaska, Fairbanks

Though whales are the strangest of all mammals—the farthest out from the mainstream of mammalian life—they have no structures fundamentally new but only familiar ones reworked.
—V. B. Scheffer, "The Year of the Whale," 1969.

Dolphins, porpoises, and whales occupy a special place in human interactions with the world of nature. Our fascination with them has its origins in a distant past. However, despite that long association, they remain the least accessible and least known of the mammals. Important pioneer investigations have been successfully applied to small captive cetaceans, but the obvious constraints of sheer magnitude have prevented all but the most rudimentary physiological studies of great whales.

The obligate cetacean aquatic lifestyle imposes environmental constraints with physiological consequences vastly different from those of terrestrial mammals in their atmospheric environments. However, most, if not all, of their adaptations are extensions of or variations on the general mammalian theme. The physical properties of water dominate these considerations:

water is 1000 times more dense than air, its viscosity is 60 times greater, heat is transferred through it at a rate about 25 times that of air, light penetration is much reduced, and the salt concentrations in sea water are substantially higher than those of the body fluids. Cetacean lungs, characteristically mammalian, are designed for aerial respiration. Without looking further, one may well suspect intuitively that living in water can have big disadvantages, imposing severe problems on respiration, foraging in long and deep dives, temperature regulation, vision, and navigation.

In contrast, what advantages are there in aquatic residence for a mammal that evolved originally from land-dwelling ancestors? Virtual weightlessness, for one; generally abundant and easily obtained food sources, for another. Whatever potential problems in the transition to the sea have been well dealt with by evolutionary natural selection. Cetacean adaptations range over the wide gamut of physiology and affect the responses of species from small porpoises to the blue whale (*Balaenoptera musculus*), the largest mammal ever to have lived, a range in body mass of more than 2000-fold. These adaptations for swimming, diving, maintaining temperature and salt equilibrium, navigating, and foraging endow the cetaceans with extraordinary facility in the sea. We have only the most rudimentary knowledge of the physiological mechanisms underlying these adaptations. Expressions of behavior are intimately related to these questions, and many physiological reactions are shaped by behavioral aspects of cetacean life (Wells *et al.*, 1999).

The long history of human interactions with cetaceans has been marked with wonder and mystery (Matthews, 1978). We admire the exploits of dolphins, and we have engaged in titanic battles for the harvest of great whales. Human exploitation of whales dates at least from the 12th century. They have been major suppliers of oil, food, and BALEEN—the "plastics" of previous ages. In addition to active hunting, whale carcasses that washed ashore in prehistoric times were doubtless used for whatever food value could be salvaged. Consumption, rather than study, has been the major preoccupation with this animal resource. However, incidental examinations of anatomy aroused curiosity of how these massive bodies operated.

Captive smaller cetaceans have been studied to good advantage (Irving *et al.*, 1941; Kanwisher and Ridgway, 1983; Williams *et al.*, 1993), but the massive size of whales severely limits their experimental manipulation. Even simple observations in their natural habitats challenge our ingenuity. Speculations about relations between structure and function have dominated our approach to understanding how whales work. With a few exceptions, we still engage in free deduction from morphology for our best guesses. The practice of extrapolation from anatomy to mechanism, functional morphology, has a long history. It works sometimes, but its indulgence must always be regarded with care because it can lead to misconceptions and error. Nothing beats controlled experimentation, in the field or laboratory, for the elucidation of physiological mechanisms and the regulatory processes that govern them.

Experimental physiological investigations of how the living cetaceans function are gradually becoming a reality, and they offer an especially promising research frontier in which field and laboratory studies will be advantageously combined. Accord-

ingly, with the prospect of a new era in cetacean physiology in the offing, it can be useful to review briefly what we know and to consider the prospects for improved investigative efforts.

I. Diving

Clearly, the ability to dive, whether by brief and shallow breath holding or in long and deep submergence, is essential for productive existence in the oceans. We have some information about the dive durations and depths of which cetaceans are capable, but with the exception of relatively few studies on small cetaceans, we know only very little about the mechanisms that sustain them. Most of what we do know about marine mammal DIVING PHYSIOLOGY is derived from studies of seals, especially the phocid, or earless, seals [e.g., harbor (*Phoca vitulina*), elephant (*Mirounga* spp.), Weddell (*Leptonychotes weddellii*) and gray (*Halichoerus grypus*) seals], but there is no obvious rationale for simple extrapolation from that knowledge to diving cetaceans. However, the similarities among diving responses of many mammals and birds in resisting the effects of diving suggest that the cetacean responses are variations on that central theme (Kooyman, 1989; Elsner, 1999).

Cessation of respiration and increased pressure are conditions of submergence. It is reasonable to suspect that the cardiovascular system plays an important role in dives because of the observed lowering of heart rate (bradycardia) while diving that has been observed in several studies of captive small cetaceans. Protection against the problems of high pressure experienced during deep dives are poorly understood, but the structure of thorax and lung suggests some adaptive reactions. Protective lung collapse that prevents absorption of lung gases in deep dives has been demonstrated in captive dolphins and in excised whale lungs. However, these are but the beginnings of what research will be required for the comprehensive understanding of reactions to such dives.

The consequence of breath holding during submergence is an unrelenting depletion of the oxygen stores that are primarily sequestered in muscle myoglobin and blood hemoglobin, both of which are augmented in most cetacean (and pinniped) species when compared with oxygen stores of terrestrial mammals. Decreased oxygen availability (hypoxia) is accompanied by other important consequences of breath holding: increasing levels of carbon dioxide (hypercapnia) and steady accumulation of the acid by-products of anaerobic metabolism (acidosis). This combination—hypoxia, hypercapnia, and acidosis—is defined as asphyxia, and it is this steadily progressing condition that ultimately limits the duration of the dive.

II. Temperature Regulation

Of all the world's marine mammals, only sirenians (manatees and dugongs) and cetaceans remain immersed in water throughout their lives, and they venture onto land only at their peril. The sirenian requirement for warm water contrasts with the more thermally robust nature of cetaceans and requires restriction to warm coastal tropic seas, but cetacean species are found in all of the world's oceans and in several tropical freshwater rivers. Some species are regularly exposed to the high heat transfer characteristics of cold polar water, and this condition raises questions about how they maintain body temperature. Other questions concern the ability of great whales to avoid overheating while exercising.

The thermal characteristics of the aquatic medium raise questions about the maintenance of body temperature. The high conductivity and convective heat transfer of water require marine mammals to have effective insulation, in cetaceans consisting mostly of subcutaneous BLUBBER. Body size becomes an important consideration here because heat loss is determined in part by the relationship of body mass and surface area. The geometry of great whales and their blubber thickness reveal that they have more insulation than they need for purposes of maintaining body temperature (Kanwisher and Ridgway, 1983; Hokannen, 1990). The enormous variations in size among cetaceans, and related surface/volume variations, imply that a simple extrapolation of thermal relationships from the small to the large species is invalid. The countercurrent circulatory structures in the tail and flippers indicate a variable heat regulatory function (Scholander and Schevill, 1955), but we lack evidence of how they operate in the living animal.

III. Metabolism

The enormous adult sizes of some whales and their apparently phenomenal growth rates are dependent on the vast quantities of small zooplankton animals that make up their diet, or a large part thereof. Food consumption is enormous, of course, as has been acknowledged in casual observations of the vast stomach contents revealed on the flensing decks where whale carcasses have been processed. However, actual metabolic determinations are limited almost exclusively to small toothed whales from the few reported observations on captive animals maintained in oceanaria. Rare indeed are determinations of food intake and growth in baleen whales. Two juvenile gray whales (*Eschrichtius robustus*) have been maintained in captivity for 1 year and released. Some of their metabolic and rather extraordinary growth characteristics have been described (Wahrenbrock *et al.*, 1974).

The metabolic reactions of cetaceans vary, just as they do in other mammals, depending on body size and surface area. These, in turn, will determine food requirements. Some controversy exists about the issue of possible metabolic consequences of lifelong immersion in water and whether metabolic rates of small cetaceans are elevated to maintain body temperature. Subcutaneous blubber represents a substantial metabolic storage, and whale migrations are supported by this resource and whatever relatively modest feeding may occur along the way. The detailed consequences, however, of the associated several month's fasting are unknown. The present state of knowledge of marine mammal metabolism and energy budgets has been well reviewed by Costa and Williams (1999).

IV. Migration and Navigation

Some cetaceans perform prodigious MIGRATIONS from regions of polar food abundance toward the more moderate climates where birth, nurture of newborn animals, and breeding occur. The energetic costs of such travel and its navigation techniques are poorly understood, but even more pertinent are

questions relating to the reasons for migrations, especially in view of controversies derived from speculations and calculations relating to feeding, thermal insulation, and heat production (Hokkanen, 1990; Wells *et al.*, 1999).

Some whales travel by the season between polar and temperate or tropic regions with the result that whales of one hemisphere migrate poleward, where prey is generally more abundant, whereas those of the opposite hemisphere congregate in warmer waters. Apparently little or no food is consumed during these excursions. The possible reasons for these regular long migrations have been the source of extended speculations. It is probable that the energetic storage represented by massive quantities of subcutaneous blubber serves another function than what is required for thermal insulation. Calculations suggest that this fat reserve can provide enough metabolic energy to last several months, even including the requirement for maternal nurture of newborn calves (Kanwisher and Ridgway, 1983).

V. Osmotic Considerations

Consumption of marine food resources, especially zooplankton and other invertebrates, is accompanied by obligatory salt intake. What little we know of the composition of cetacean blood indicates that it conforms approximately to the concentrations of terrestrial mammal blood, i.e., the osmotic pressure of sea water is about three times that of whale blood. Obligatory sea water intake accompanying food consumption requires that excess salts be excreted by kidney concentration of urine. Human kidneys cannot handle such salt loads, but cetaceans have a better capability for excreting salt and for conserving water. Incidentally, even that facility is far exceeded by desert rodents that are confronted with more extreme requirements for water conservation.

VI. Sensory Systems

Vision and olfactory senses are modified and reduced in some cetaceans, although just how compromised these abilities are is poorly understood. Sound production is common among them, and hearing is acute. Some species are especially vocal, and there is good evidence that echolocation is highly developed in at least some of these. Training techniques applied to captive animals have been especially useful in support of these studies (Wartzok and Ketten, 1999), but many questions remain concerning how the senses are employed in navigation, for aging, and social interactions.

Other topics excite our interests. For example, how long do whales live? Evidence indicates the extraordinary old age of some bowhead whales (*Balaena mysticetus*), even exceeding 150 years (George *et al.*, 1999). Reproductive physiology and newborn responses are other fields fertile for new research (Boyd *et al.*, 1999). Digestive processes are little known. Physiological responses interact with behavioral reactions and their combined study yields important insights that can clarify environmental influences on animals. Mammalian responses are governed by neural and humoral mechanisms and their interactions, and our understanding of cetacean physiology will be only superficial without knowledge of these regulatory mechanisms.

Healthy populations of marine mammals are indicators of healthy marine ecosystems, as revealed by their dependence on prey abundance, minimal water pollution, and lack of disturbance. However, the ecological physiology upon which that state of health depends is not well defined. Some cetacean populations, northern right whale (*Lissodelphis borealis*) and Chinese river dolphin (*Lipotes vexillifer*), for example, are critically endangered. Their survival will depend on many factors, not the least of which will be our understanding and appreciation for optimum physiological adaptations to their habitats.

VII. Research Prospects

Research on whatever can be recovered from dead whales during processing has historically depended on the good will of commercial whale-harvesting companies. However controversial that activity may be, it is undeniable that new knowledge of cetacean physiology has been forthcoming from that source. Investigations of this kind have been limited in the past, with the exception of a few like that of Scholander (1940), but important new work is being done on whale metabolism and digestion (Folkow and Blix, 1992; Olsen *et al.*, 2000). Other resources and occasional access to captive animals for benign observations are being utilized through the cooperation of public display oceanaria. Contemporary techniques of cell biology and genetics can be applied usefully in studies related to the expression of some physiological characteristics as revealed by tissue sampling from wild animals. Such sampling has been accomplished successfully at sea and by the cooperation of both commercial and Native cetacean harvests (Haldiman and Tarpley, 1993).

As for future research prospects, we are justified in believing that marine mammal science is on the threshold of new prospects for understanding whale physiology. Among the many prospective challenges for the new millenium, one stands out as especially attractive: the extension of the techniques and understandings that work so well in other, more manageable species to the great whales. Indeed, the possibilities for new discoveries regarding the physiology of whales are not beyond the realm of dreams; they are practical and realistic. New knowledge continues to be steadily derived at an ever-increasing rate from attachment of instrumentation and tracking operations that reveal much about the activities, lifestyles, diving, feeding, and migratory habits of whales. These await further application of ingenuity for recording more and detailed variables relating to the adaptions of free-ranging whales.

However successful and ingenious the instrumentation of free-swimming marine mammals may be, those techniques are insufficient for detailed and controlled investigations of cell chemistry, neurophysiology, energetics, and other topics necessary for the full understanding of cetacean physiology. We know from experience with the captive cetaceans that management of very large animals in CAPTIVITY is limited only by imagination and funding. In addition to long-term healthy captivity for study and conventional public display, one can foresee the more restricted capture for days or weeks of research and later release of whales at sea. These procedures will require exploration of what is practical, but such an approach is not beyond

the realm of possibility and would open a new era of research into how these animals function.

See Also the Following Articles

Breathing ▪ Brain ▪ Circulatory System ▪ Diving Physiology ▪ Health ▪ Hearing ▪ Energetics ▪ Osmoregulation ▪ Swimming ▪ Thermoregulation ▪ Vision

References

Boyd, I. L., Lockyer, C., and Marsh, H. D. (1999). Reproduction in marine mammals. *In* "Biology of Marine Mammals" (J. E Reynolds III and S. A. Rommel, eds.), pp. 218–286. Smithsonian Press, Washington, DC.

Costa, D. P., and Williams, T. M. (1999). Marine mammal energetics. *In* "Biology of Marine Mammals" (J. E. Reynolds III and S. A. Rommel, eds.), pp. 176–217. Smithsonian Institution Press, Washington, DC.

Elsner, R. (1999). Living in water: Solutions to physiological problems. *In* "Biology of Marine Mammals" (J. E. Reynolds III and S. A. Rommel, eds.), pp. 73–116. Smithsonian Institution Press, Washington, DC.

Folkow, L. P., and Blix, A. S. (1992). Metabolic rates of minke whales (*Balaenoptera acutorostrata*) in cold water. *Acta Physiol. Scand.* **146,** 141–150.

George, J. C., Bada, J., Zeh, J., Scott, L., Brown, S. E., O'Hara, T., and Suydam, R. (1999). Age and growth estimates of bowhead whales (*Balaena mysticetus*) via aspartic acid racemization. *Can. J. Zool.* **77,** 571–580.

Haldiman, J. T., and Tarpley, R. J. (1993). Anatomy and physiology. *In* "The Bowhead Whale" (J. J. Burns, J. J. Montague, and C. J. Cowles, eds.), pp. 71–156. Special Publ. No. 2, Soc. Mar. Mammalogy.

Hokkanen, J. E. I. (1990). Temperature regulation of marine mammals. *J. Theoret. Biol.* **145,** 465–485.

Irving, L., Scholander, P. F., and Grinnell, S. W. (1941). The respiration of the porpoise, *Tursiops truncatus. J. Cell. Comp. Physiol.* **17,** 145–168.

Kanwisher, J. W., and Ridgway, S. (1983). The physiological ecology of whales and porpoises. *Sci. Am.* **248**(6), 111–120.

Kooyman, G.L. (1989). "Diverse Divers, Physiology and Behavior." Springer-Verlag, Berlin.

Matthews, L. H. (1978). "The Natural History of the Whale." Columbia Univ. Press., New York.

Olsen, M. A., Blix, A. S., Aagnes, T. H., Sørmo, W., and Mathiesen, S. D. (2000). Chitinolytic bacteria in the minke whale forestomach. *Can. J. Microbiol.* **46,** 85–94.

Scholander, P. F. (1940). Experimental investigations on the respiratory function in diving mammals and birds. Hvalrådets Skrifter, Det Norske Videnskaps Akademi I Oslo **22,** 1–131.

Scholander, P. F., and Schevill, W. E. (1955). Counter-current vascular heat exchange in the fins of whales. *J. Appl. Physiol.* **8,** 279–282.

Wahrenbrock, E. A., Maruschak, G. F., Elsner, R., and Kenney, D. W. (1974). Respiratory function and metabolism of two baleen whale calves. *Mar. Fish. Rev.* **36,** 3–9.

Wartzok, D., and Ketten, D. R. (1999). Marine mammal sensory systems. *In* "Biology of Marine Mammals" (J. E. Reynolds III and S. A Rommel, eds.), pp. 117–175. Smithsonian Institution Press, Washington, DC.

Wells, R. S., Boness, D. J., and Rathbun, G. B. (1999). Behavior. *In* "Biology of Marine Mammals" (J. E. Reynolds III and S. A. Rommel, eds.), pp. 324–422, Smithsonian Institution Press, Washington, DC.

Williams, T. M., Friedl, W. A., and Haun, J. E. (1993). The physiology of bottlenose dolphins (*Tursiops truncatus*): Heart rate, metabolic rate and plasma lactate concentration during exercise. *J. Exp. Biol.* **179,** 31–46.

Circulatory System

PAUL J. PONGANIS
Scripps Institution of Oceanography,
La Jolla, California

lthough the circulatory systems of marine mammals follow the general mammalian plan, they are most notable for features associated with the diving response, thermoregulation, and large body mass. This article emphasizes anatomical and functional aspects of the circulatory system in these animals. The cardiovascular reflexes and adjustments that occur during diving are reviewed in other articles. Specific features of the circulatory system vary with orders, families, and species. These adaptations include large blood volumes, large capacitance structures (spleens and venous sinusses), venous sphincter muscles, vascular adaptations for thermoregulation, aortic windkessels, and vascular retia. The heart, arterial and venous systems, and blood volume are discussed first. Then specific structural adaptations of the circulation in various groups are considered.

I. General Anatomy

A. Heart

The basic structure and size of hearts in pinnipeds and cetaceans are typical of mammals. The four-chambered heart, with right ventricular outflow to the lungs and left ventricular output to the systemic circulation, weighs 0.5–1% of body mass in most pinnipeds and small cetaceans. In the great whales, relative heart mass is smaller, about 0.3–0.5% of body mass. Chamber size, stroke volume, and resting cardiac output and heart rate (where measured) are also in the general mammalian range and in agreement with mammalian allometric equations. Both the foramen ovale and the ductus arteriosus are closed in adult seals and cetaceans as in other mammals. Therefore, utilization of an intermittent fetal circulatory pathway does not appear to be a mechanism to bypass a potential increase in pulmonary vascular resistance during diving.

B. Arterial/Venous Systems

General aspects of the arterial and venous systems in marine mammals are remarkable for several features. First, dense sympathetic nerve innervation of proximal as well as distal arteries in seals may represent a mechanism by which the intense sympathetic vasoconstriction of the dive response can be maintained independent of local tissue metabolite-induced vasodilatation in the periphery. Angiography during forced submersions of seals supports this model. Second, venous capacitance is highly developed, especially in phocid seals and whales. This includes a large hepatic sinus and inferior vena cava, the latter of which in seals has been estimated to be capable of storing a fifth of the seal's blood volume. Presumably, this large venous capacitance is related to the large blood volume of seals.

In some species, the spleen appears to be a significant storage organ for red blood cells. Increased splenic volumes in several pinniped species, and extensive sympathetic nerve innervation and smooth muscle development in the splenic capsule, are consistent with this storage role. Fluctuations in hematocrit between resting and diving states, or anesthetized and stressed states, also support such a role for the spleen in seals. It has been estimated that 30% of the blood volume can be stored in the spleen in Weddell seals (*Leptonychotes weddellii*).

Another feature of the venous system, again well developed in both seals and whales, is the extradural venous system. This system, located within the vertebral canal and above the spinal cord, receives blood flow from the brain, back, and pelvic regions. It is linked with both the inferior and the superior vena cava via paravertebral communicating veins. The direction and magnitude of flow within the extradural vein is complex, varies with the respiratory cycle, and bears further investigation. In seals and cetaceans, the extradural vein is the primary venous drainage of the brain; the internal jugular vein is poorly developed or absent. The function of such a prominent vertebral venous system in these animals is unclear. In humans, it has been noted that extradural vein flow may participate in brain temperature regulation, and that, in the upright posture, the vertebral veins, kept open by attachment to the bony walls of the vertebral canal, are the primary cerebral venous drainage as blood flow decreases in the jugular veins due to venous collapse in the upright posture.

C. Blood Volume

Blood volumes are elevated and contribute to increased blood oxygen stores in marine mammals. On a mass-specific basis, most measurements indicate that blood volumes are two to three times the 70-ml kg^{-1} human value and that blood volume is greater in more active and in longer-diving species. The largest blood volumes (200–260 ml kg^{-1}) have been found in some of the best divers, including elephant seals (*Mirounga* spp.), Weddell seals, and sperm whale (*Physeter macrocephalus*).

II. Structural Adaptations

A. Vascular Thermoregulatory Adaptations

The parallel pattern of counterflowing arteries and veins, characteristic of counter current exchange units, is present in the flukes and flippers of cetaceans. Such arrangements, characteristic of blood vessel patterns in the limbs of many animals, are considered to conserve body heat by transferring heat from warm, outgoing arterial blood to cool venous blood returning from the limb. A superficial venous system, which does not return in conjunction with outgoing arteries, also occurs in the skin. These veins, which have well-developed muscular walls, have been considered to represent a route by which heat could be dissipated to the environment during periods of thermal stress.

Another structural adaptation observed in pinnipeds is the presence of numerous arteriovenous (a-v) anastomoses in the skin. These structures are connections between distal arteries and veins through which blood bypasses tissue capillary networks, and instead shunts directly from the arterial to venous system. The a-v anastomoses are distributed uniformly over the body surface of phocid seals, but, in otariids, are found in greater densities in the flippers. It is presumed that flow through these vessels allows heat exchange at the skin surface.

More recently, countercurrent anatomy has been observed in the reproductive organs of dolphins and pinnipeds. It has been proposed that the return of blood from the skin via vascular anastomoses allows relatively cool venous blood to prevent overheating of these organs. Temperature patterns along the length of the colon in the dolphin have been consistent with this hypothesis.

B. Aortic Bulbs/Windkessels

In pinnipeds, again particularly in phocid seals, the aortic root (ascending aorta) is dilated, forming the so-called aortic bulb. The bulb can accommodate the stroke volume ejected by the heart, and it is more distensible than the distal aorta. It has been proposed that the aortic bulb acts as a windkessel: the bulb expands and accepts blood ejected from the heart; then, gradual contraction of the bulb due to elastic fibers within its wall contributes to the maintenance of blood flow, especially to the brain and heart during diastole (relaxation phase of the cardiac cycle).

The ascending aorta, aortic arch, and proximal carotid arteries of whales are also very compliant and have also been hypothesized to act as a windkessel and preserve blood flow during diastole. This is especially important in whales because long diastoles accompany slow heart rates. Low heart rates can occur in whales due to the slow heart rates found in animals of large body mass and due to the bradycardias (decreased heart rates) that occur during diving.

The maintenance of blood flow and pressure during a long diastole is of course critical to the brain, but also to the heart. This is because coronary perfusion occurs during diastole when the heart is relaxed. Myocardial flow is dependent on the diastolic blood pressure as the driving pressure. Thus, species that are either large or have more profound diving responses are likely to have some form of an aortic windkessel.

A compliant ascending aorta may also contribute to a reduction in the impedance that the left ventricle must pump against during the peripheral vasoconstriction of the diving response. This reduction in afterload will decrease the work and oxygen consumption of the heart, which is of course beneficial to a diver with a limited oxygen supply.

C. Vascular Retia

The retia mirabilia (wonderful nets) of cetaceans have long been noted by anatomists. These plexuses of anastomosing arteries and veins occur along the vertebrae and base of the skull and are especially prominent in the thorax. The vascular retia are well developed in dolphins, in fact, more so than in large whales; they are also found in sirenians.

The thoracic rete is supplied by vessels from the aorta, which anastomose to form a complex, spongiform structure beneath the dorsal thoracic wall. This vascular tissue extends around the vertebrae into the vertebral canal and forms the primary arterial blood supply to the brain in cetaceans. The carotid arteries are vestigial or absent. The spinal meningeal artery in dolphins extends from the rete to the brain. Although

mean blood pressure in the spinal meningeal artery of dolphins is equal to the aortic pressure, it is notable that the pressure is nonpulsatile; there is no systolic peak or diastolic trough. Thus, the cetacean brain appears to receive nonpulsatile blood flow. The significance of such a flow pattern, as well as the function of the retia, is unknown.

Large venous retia have also been reported in the abdomens of whales. Hypotheses about the role of the retia have included windkessel functions, intrathoracic vascular engorgement to prevent "lung squeeze" during diving, thermoregulation, and modification of composition of the blood.

D. Inferior Vena Caval Sphincter

In most pinnipeds, the inferior vena cava is associated with a striated muscle sphincter at the level of the diaphragm. Again, this is most well developed in phocid seals. The sphincter is innervated by the right phrenic nerve and is located cranial to the large hepatic sinus and inferior vena cava. Relaxation/contraction of the sphincter has been observed angiographically during forced submersions, and it is assumed that this is a mechanism to regulate venous return to the heart during diving bradycardias. Vena caval sphincters are also described in whales; they presumably regulate blood return from the large venous capacitance vessels in the abdomen (inferior vena cava and venous rete).

Another venous structure, especially developed in phocid seals, is the pericardial venous plexus. This extensive venous network drains the pericardium (membrane surrounding the heart) and empties into the inferior vena cava just cranial to the vena caval sphincter. It is especially developed in the better diving seals such as elephant seals. Its significance is unknown, although it has been reported to be associated with brown fat and has been hypothesized to function in thermoregulation.

See Also the Following Articles

Anatomical Dissection ■ Endocrine Systems ■ Diving Physiology ■ Health ■ Morphology, Functional ■ Thermoregulation

References

Blix, A. S., Grav, H. J., and Ronald, K. (1975). Brown adipose tissue and the significance of the venous plexuses in pinnipeds. *Acta Physiol. Scand.* **94,** 133–135.

Bryden, M. M., and Molyneux, G. S. (1978). Arteriovenous anastomoses in the skin of seals. II. The California sea lion (*Zalopohus californianus*) and the northern fur seal (*Callorhinus ursinus*) (Pinnipedia: Otariidae). *Anat. Rec.* **191,** 253–260.

Butler, P. W., and Jones, D. R. (1997). The physiology of diving of birds and mammals. *Physiol. Rev.* **77,** 837–899.

Elsner, R. W., and Gooden, B. (1983). "Diving and Asphyxia: A Comparative Study of Animals and Man." Cambridge Univ. Press, Cambridge.

Gauer, O. H., and Thron, H. L. (1965). Postural changes in circulation. *In* "Handbook of Physiology: Circulation" (W. F. Hamilton and P. Dow, eds.), Vol. III, pp. 2409–2439. American Physiological Society, Washington, DC.

Harrison, R. J., and Tomlinson, J. D. W. (1956). Observations on the venous system in certain Pinnipedia and Cetacea. *Proc. Zool. Soc. Lond.* **126,** 205–233.

Kooyman, G. L., and Ponganis, P. J. (1998). The physiological basis of diving to depth: Birds and mammals. *Annu. Rev. Physiol.* **60**, 19–32.

McFarland, W. L., Jacobs, M. S., and Morgane, P. J. (1979). Blood supply to the brain of the dolphin, *Tursiops truncatus*, with comparative observations on special aspects of the cerebrovascular supply of other vertebrates. *Neurosci. Biobehav. Res.* **3**, 193.

Rhode, E. A., Elsner, R., Peterson, T. M., Campbell, K. B., and Spangler, W. (1986). Pressure-volume characteristics of aortas of harbor and Weddell seals. *Am. J. Physiol.* **251**, R174–R180.

Rommell, S. A., Early, G. A., Matassa, K. A., Pabst, D. A., and McLellan, W. A. (1995). Venous structures associated with thermoregulation of phocid seal reproductive organs. *Anat. Rec.* **243**, 390–402.

Ronald, K., McCarter, R., and Selley, L. J. (1977). Venous circulation of the harp seal (*Pagophilus groenlandicus*). *In* "Functional Anatomy of Marine Mammals" (R. J. Harrison ed.), pp. 235–270. Academic Press, New York.

Scholander, P. F., and Schevill, W. E. (1955). Counter-current vascular heat exchange in the fins of whales. *J. Appl. Physiol.* **8**, 279–282.

Shadwick, R. E., and Gosline, J. M. (1994). Arterial mechanics in the fin whale suggest a unique hemodynamic design. *Am. J. Physiol.* **267**, R805–R818.

Slijper, E. J. (1962). "Whales." Hutchinson and Co., London.

Zenker, W., and Kubik, S. (1996). Brain cooling in humans: Anatomical considerations. *Anat. Embryol.* **193**, 1–13.

Classification

DALE W. RICE
*National Marine Mammal Laboratory,
Seattle, Washington*

Four clades of placental mammals (class Mammalia: cohort Placentalia) independently evolved adaptations for life in the oceans. These are the still-living pinnipeds (sea lions, walruses, and seals), cetaceans (whales, dolphins, and porpoises), and sirenians (manatees and dugongs) and the extinct desmostylians. The pinnipeds are amphibious animals capable of terrestrial locomotion and must haul out on shore to give birth. The cetaceans and sirenians (except for a few primitive Eocene species) are totally aquatic, having lost their hindlimbs and evolved huge muscular tails with terminal flukes for swimming. The extinct desmostylians were quadrupedal amphibious creatures. The pinnipeds and cetaceans are carnivorous, and the sirenians and desmostylians are herbivorous. Although primarily oceanic, several members of each of the three living groups have secondarily invaded freshwater habitats. The systematics of all of these sea mammals, living and fossil, is a flourishing field of research; many details are currently contested, and several paraphyletic groupings await resolution, so changes in the prevailing classification (Table I, McKenna and Bell, 1997) may be anticipated (Fordyce and Barnes, 1994; Rice, 1998; Berta and Sumich, 1999).

I. Pinnipeds

The pinnipeds were long classified as order Pinnipedia, separate from but closely related to the terrestrial carnivores of the

TABLE I
Classification and Geologic Ranges of Living and Fossil Families of Marine Mammals[a]

Order Carnivora (in part)
  Suborder Caniformia (in part)
    Superfamily Phocoidea
      †Family Enaliarctidae.[c] L. Olig.-M. Mioc.
      Family Otariidae (fur seals and sea lions). M. Mioc.-Rec.
      Family Odobenidae (walruses). L. Mioc.-Rec.
      †Family Desmatophocidae.[c] E.-M. Mioc.
      Family Phocidae (true seals). L. Mioc.-Rec.

Order Cetacea [or Order Cete: Suborder Cetacea]
  †Suborder Archaeoceti[c] [or Infraorder Archaeoceti]
    †Family Pakicetidae. E. Eoc.
    †Family Ambulocetidae. E.-M. Eoc.
    †Family Remingtonocetidae. M. Eoc.
    †Family Protocetidae.[c] M. Eoc.
    †Family Dorudontidae[c] (zeuglodonts). M.-L. Eoc.
    †Family Basilosauridae (zeuglodonts). M.-L. Eoc.
  Suborder Mysticeti [or Infraorder Autoceta: Parvorder Mysticeti]
    †Family Llanocetidae. L. Eoc.
    †Family Aetiocetidae. L. Olig.
    †Family Mammalodontidae. L. Olig.
    †Family Kekenodontidae. L. Olig.
    †Family Cetotheriidae.[c] L. Olig-E. Plioc.
    Family Balaenidae (right whales). E. Mioc.-Rec.
    Family Neobalaenidae (pygmy right whales). Rec.
    Family Eschrichtiidae (gray whales). L. Pleist.-Rec.
    Family Balaenopteridae (rorquals). L. Mioc.-Rec.
  Suborder Odontoceti [or Infraorder Autoceta: Parvorder Odontoceti]
    †Superfamily *incertae sedis*
      †Family Agorophiidae. L. Olig.
    Superfamily Physeteroidea
      Family Physeteridae (sperm whales). L. Olig.-Rec.
      Family Kogiidae (pygmy sperm whales). L. Mioc.-Rec.
    Superfamily Ziphioidea
      Family Ziphiidae (beaked whales). M. Mioc.-Rec.
    Superfamily Platanistoidea
      Family Platanistidae (Indian river dolphins). M. Mioc.-Rec.
      †Family Waipatiidae. L. Olig.
      †Family Squalodelphinidae. E. Mioc.
      †Family Dalpiazinidae. E. Mioc.
      †Family Squalodontidae (shark-toothed dolphins). L. Olig.-M. Mio
    †Superfamily Eurhinodelphinoidea
      †Family Eurhinodelphinidae (long-snouted dolphins). E.-L. Mioc.
      †Family Eoplatanistidae. E. Mioc.
    Superfamily *incertae sedis*
      Family Iniidae (Amazon river dolphins). L. Mioc.-Rec.
      Family Lipotidae (Chinese river dolphins). L. Mioc.-Rec.
      Family Pontoporiidae (La Plata dolphins). L. Mioc.-Rec.
    Superfamily Delphinoidea
      †Family Kentriodontidae.[c] L. Olig.-L. Mioc.
      †Family Albireonidae. L. Mio.-E. Plio.
      Family Monodontidae (belugas and narwhals). Mioc.-Rec.
      †Family Odobenocetopsidae.[b] E. Plio.
      Family Delphinidae (dolphins). M. Mioc.-Rec.
      Family Phocoenidae (porpoises). L. Mioc.-Rec.

(continues)

NOTE: See Species List.
*ᵃ*From McKenna and Bell, 1997. Extinct taxa are marked with a dagger, and taxa that appear to be paraphyletic are marked with an asterisk. Abbreviations: E, early, M, middle, L, late; Eoc., Eocene; Olig., Oligocene; Mico., Miocene; Plioc., Pliocene; Pleist, Pleistocene; Rec., Recent.
*ᵇ*Allocation of the family Odobenocetopsidae to the order Cetacea is disputed.

order Carnivora. In recent years, cladistic analyses of both morphological and and molecular data have clearly shown them to be members of the suborder Caniformia of the order Carnivora.

Two strongly differentiated groups of living pinnipeds were long recognized: the eared seals, or sea lions and fur seals (family Otariidae), and the earless, or true, seals (family Phocidae); the walruses (family Odobenidae) were usually associated with the former group. Some taxonomists maintained that pinnipeds are a diphyletic assemblage and that the eared seals shared a common ancestry with the bears (family Ursidae), whereas the true seals were most closely related to the weasel group (family Mustelidae) or, more specifically, the otters (subfamily Lutrinae). Those authorities allocated the pinnipeds to two superfamilies: Phocoidea for the family Phocidae and Otarioidea, which included the extinct families Enaliarctidae and Desmatophocidae, and the extant Otariidae and Odobenidae. The Enaliarctidae included several late Oligocene and early Miocene genera that were postulated to have given rise to the other three families of otarioids in the Miocene (Repenning and Tedford, 1977).

With the advent of cladistic methods, a different picture emerges. All of the molecular analyses and most of the morphological analyses have supported the hypothesis that all pinnipeds shared a common ancestry. However, the position of the pinnipeds within the suborder Caniformia is still disputed. Although most investigators favor a sister group relationship to the Ursidae, a few studies (Bininda-Emonds and Russell, 1996) do support a closer affinity to the Mustelidae. One analysis placed the genus *Kolponomos* as the sister taxon of the pinnipeds. *Kolponomos*, an amphibious bear-like creature that lived along the coasts of Washington and Oregon during the Miocene, is currently listed in the family Amphicynodontidae, a paraphyletic group from which the Ursidae descended. With the present preponderance of evidence, the most appropriate classification is that of McKenna and Bell (1997), who rank the pinnipeds as the superfamily Phocoidea and the bear-like terrestrial carnivores as the superfamily Ursoidea, both under the parvorder Ursida, and place the mustelids in parvorder Mustelida; those authors also included all the genera of amphicynodonts as unallocated stem groups of the Phocoidea.

The cladistic studies have also reopened the question of interfamilial relationships of pinnipeds. A total evidence analysis

indicated the following "phyletic sequence": Desmatophocidae–Phocidae–Odobenidae–Otariidae. The traditional pairing of the Odobenidae with the Otariidae was likewise supported by molecular and morphological analyses of the living taxa (Árnason *et al.*, 1995; Bininda-Emonds and Russell, 1996). However, a comprehensive morphological analysis (Berta and Wyss, 1994) affirmed the paraphyletic nature of the Enaliarctidae and arranged the other families in the following phyletic sequence: Otariidae–Odobenidae–Desmatophocidae (paraphyletic)–Phocidae. Berta and Wyss (1994) proposed a new classification, but did not assign formal Linnean ranks to all of their suprageneric taxa; their taxon Pinnipedimorpha equals McKenna and Bell's Phocoidea minus the amphicynodontids.

Mention must be made of three other fossil genera of "otter-like seals" or "seal-like otters." These are *Potamotherium* with two species from freshwater Oligocene and Miocene deposits in Europe and North America, *Semantor macrurus* from freshwater Miocene or Pliocene deposits of Kazakhstan, and *Necromites nestoris* from a marine Pliocene stratum of Azerbaijan. The latter two are known only by the hinder halves of their skeletons. All three have been proposed as primitive pinnipeds or phocids, but their phylogenetic placement remains totally problematic.

II. Cetaceans

Whales differ so much from other placental mammals that their evolutionary relationships long remained conjectural. The marked anatomical dissimilarities between baleen whales and toothed whales led a few earlier cetologists to question the monophyly of the Cetacea, but none of them ever proposed an explicit hypothesis of diphyly (Rice, 1984). However, many recent studies, both morphological and molecular, overwhelmingly confirm the monophyletic origin of cetaceans and indicate that their closest living relatives are the even-toed ungulates (order Artiodactyla). A comprehensive cladistic analysis of morphological data from fossil and living taxa showed the cetaceans (order Cetacea) and the extinct mesonychids (order Acreodi) as monophyletic sister groups, which together constitute the sister group to the monophyletic Artiodactyla (O'Leary and Geisler, 1999). The mesonychids were cursorial, wolf-like creatures whose feet had five toes that bore hoof-like claws; they lived throughout the Holarctic from the early Paleocene to the early Oligocene.

During the past two decades a remarkable series of Eocene fossil cetaceans have been unearthed, mostly near the shores of the ancient Tethys Sea in Pakistan, India, and Egypt (Thewissen, 1998). These finds document the origin of the early cetaceans from the terrestrial mesonychids and their rapid evolutionary transition from amphibious quadrupeds to fully aquatic forms during the interval from 54 to 34 million years ago (Ypresian through Priabonian). The phylogenetic relationships among these primitive Eocene cetaceans have yet to be fully resolved. For the interim they are allocated to six families, some of which are paraphyletic; the entire assemblage is traditionally included in the paraphyletic suborder Archaeoceti. The close relationship between mesonychids and cetaceans led McKenna and Bell (1997) to include both in order Cete and to reduce Acreodi and Cetacea to subordinal rank; they further divided the Cetacea into two infraorders, Archaeoceti and Autoceta, with the latter including the Mysticeti (baleen whales) and Odontoceti (toothed whales) as parvorders.

In several cladistic analyses of molecular data, the Cetacea have appeared as the sister taxon to various subclades within the order Artiodactyla—most often the hippos (family Hippopotamidae) (Gatesy *et al.*, 1999; Nikaido *et al.*, 1999; Shimamura *et al.*, 1999). The strongest support for these hypotheses was claimed to come from short interspersed elements in the genomes, but their potential for resolving this problem has been questioned (Buchanan *et al.*, 1999). The veracity of these conclusions is further suspect because of the unavailability of molecular data from mesonychids, archeocetes, and any of the other extinct clades of ungulates. As noted earlier, the morphological analysis of living and fossil taxa by O'Leary and Geisler (1999) showed the Artiodactyla as well as the Cetacea as monophyletic clades, but when they repeated their analysis without the fossil species, the Cetacea appeared nested within a paraphyletic Artiodactyla, as they had in the molecular studies. Another critical assessment of morphological and molecular data from living taxa likewise corroborated the monophyly of the Artiodactyla (Luckett and Hong, 1998). The rich FOSSIL RECORD of the artiodactyls has provided no evidence whatever that would support the derivation of cetaceans from hippos or any other subclade within the Artiodactyla. Some authors now group the living orders Artiodactyla and Cetacea under a supraordinal taxon Cetartiodactyla, which with the addition of several extinct taxa becomes approximately equivalent to mirorder Eparctocyona of McKenna and Bell (1997).

The more advanced post-Eocene cetaceans are postulated to have descended from an archeocete, most plausibly a member of the family Dorudontidae. The monophyly of each of the two modern suborders, Mysticeti and Odontoceti, is strongly corroborated by a suite of complex morphological synapomorphies. Some years ago much publicity was given to a contradictory idea, based on molecular data, that sperm whales (superfamily Physeteroidea) are phylogenetically closer to baleen whales than they are to the other toothed whales. In addition to the overwhelming morphological evidence, subsequent molecular studies have corroborated the conventional hypothesis of odontocete monophyly (Luckett and Hong, 1998; Messenger and McGuire, 1998; Gatesy *et al.*, 1999).

The oldest fossil cetacean allocated to the suborder Mysticeti is *Llanocetus denticrenatus* (family Llanocetidae) from the end of the Eocene of the Antarctic Peninsula. It and several Oligocene genera of the families Aetiocetidae, Mammalodontidae, and Kekenodontidae all possessed teeth rather than baleen. The first toothless, baleen-bearing cetaceans appeared in the late Oligocene; they are assigned to the family Cetotheriidae, a paraphyletic assemblage from which the four living families descended.

The earliest members of the suborder Odontoceti appeared in the late Oligocene. During that epoch lived a number of unusual genera whose phylogenetic relationships remain unresolved. One distinctive superfamily, the Eurhinodelphinoidea, or long-snouted dolphins, diversified and then died out during the Miocene. All of the living odontocetes other than the peculiar river dolphins clearly fall into three superfamilies—Physeteroidea (sperm whales), Ziphioidea (beaked whales), and Delphinoidea (dolphins, porpoises, etc.)—all of which first appeared in the late Oligocene (Muizon, 1988, 1991). Beaked whales were long thought to be closely related to sperm whales, but a majority of recent cladistic analyses suggest that they are closer to the delphinoids. Studies of river dolphins have come up with no consistent phylogenetic pattern, other than an emerging consensus that the Platanistidae are only distantly related to the others, and are closer to the family Squalodontidae, or sharktoothed porpoises, that lived during the Miocene. The Iniidae, Lipotidae, and Pontoporiidae appear to constitute one or more branches from the ancestral lineage of the Delphinoidea.

Other putative cetaceans are *Odobenocetops peruvianus* and *O. leptodon*, bizarre, tuskbearing sea mammals unearthed from Pliocene deposits on the coast of Peru; they have been allocated to a new family (Odobenocetopsidae) and are thought to be related to the narwhal (family Monodontidae), but their cetacean affinities are disputed.

III. Sirenians and Desmostylians

Because of their superficially whale-like physique, many 19th century naturalists classified the sirenians as the "herbivorous cetacea." Modern studies have revealed that the Sirenia, along with the extinct Desmostylia, are marine members of a supraordinal group called the Tethytheria, which also embraces elephants (order Proboscidea) and several other extinct groups. The earliest sirenians had four limbs and appeared capable of terrestrial locomotion. The early to middle Eocene *Prorastomus* (family Prorastomidae) probably swam only with its hindlimbs, but the middle Eocene *Protosiren* (family Protosirenidae) probably used its well-developed tail as well. The latter genus thus foreshadowed the still-living Dugongidae and Trichechidae, fully aquatic forms that lost their hindlimbs and swim by means of caudal flukes (Domning, 1996). The affinities of the Oligocene and Miocene desmostylians long remained problematic because they were known only from skulls, but many authors classified them as a suborder of the Sirenia. The discovery of complete skeletons finally showed them to be quadrupedal hippopotamus-like animals, sufficiently different from sirenians to be ranked as a separate order. The most recent cladistic analysis places the Desmostylia closer to the Proboscidea than to the Sirenia. If true, this would imply that sirenians and desmostylians acquired their aquatic adaptations independently. McKenna and Bell (1997) demoted the Tethytheria to a suborder of their new order Uranotheria, under which they ranked Sirenia as one infraorder, and also included the Desmostylia and Proboscidea as parvorders under the infraorder Behemota.

IV. Other Marine Species

Several species of mammals that belong to otherwise terrestrial groups have become facultative or obligate members of the marine ecosystem (Rice, 1998). The polar bear (*Ursus maritimus*; family Ursidae) and the arctic fox (*Vulpes lagopus*; family Canidae) range widely over the north polar pack ice. Among the 10 or so species of otters (family Mustelidae: subfamily Lutrinae), the sea otter (*Enhydra lutris*) of the North Pacific

and the marine otter (*Lutra felina*) of western South America
are strictly marine, and local populations of at least six other
species feed in coastal marine waters. Two species of bats (order Chiroptera) also catch fish in coastal waters; the greater
bulldog bat (*Noctilio leporinus*; family Noctilionidae) of the
neotropics and the fishing bat (*Myotis vivesi*; family Vespertil-
ionidae) of the Gulf of California. Finally, the most unexpected
marine mammal was a large ground sloth (*Thalassocnus natans*;
order Xenarthra; family Megalonychidae) discovered in
Pliocene deposits on the coast of Peru, where it evidently
grazed on algae or sea grasses.

See Also the Following Articles

Cetacean Evolution ▪ Desmostylia ▪ Pinniped Evolution ▪
Sirenian Evolution ▪ Systematics

References

Árnason, Ú., Bodin, K., Gullberg, A., Ledje, C., and Mouchaty, S.
(1995). A molecular view of pinniped relationships with particular
emphasis on the true seals. *J. Mol. Evol.* **40,** 78–85.

Berta, A., and Sumich, J. L. (1999). "Marine Mammals: Evolutionary
Biology." Academic Press, San Diego.

Berta, A., and Wyss, A. R. (1994). Pinniped phylogeny. *Proc. San Diego
Soc. Nat. Hist.* **29,** 33–56.

Bininda-Emonds, O. R. P., and Russell, A. P. (1996). A morphological perspective on the phylogenetic relationships of the extant phocid seals
(Mammalia: Carnivora: Phocidae). *Bonner Zool. Monogr.* **41,** 1–256.

Buchanan, F., Crawford, A., Strobeck, C., Palsbøll, P., and Plante, Y.
(1999). Evolutionary application of MIRs and SINEs. *Anim. Genet.*
30, 47–50.

Domning, D. P. (1996). Bibliography and index of the Sirenia and
Desmostylia. *Smithson. Contrib. Paleobiol.* **80,** 1–611.

Fordyce, R. E., and Barnes, L. G. (1994). The evolutionary history of
whales and dolphins. *Annu. Rev. Earth Planet. Sci.* **22,** 419–455.

Gatesy, J., Milinkovitch, M., Waddell, V., and Stanhope, M. (1999). Stability of cladistic relationships between Cetacea and higher-level artiodactyl taxa. *Syst. Biol.* **48**(1), 6–20.

Luckett, W. P., and Hong, N. (1998). Phylogenetic relationships between
the orders Artiodactyla and Cetacea: A combined assessment of morphological and molecular evidence. *J. Mammal. Evol.* **5**(2), 127–182.

McKenna, M. C., and Bell, S. K. (1997). "Classification of Mammals
above the Species Level." Columbia Univ. Press, New York.

Messenger, S. L., and McGuire, J. A. (1998). Morphology, molecules,
and the phylogenetics of cetaceans. *Syst. Biol.* **47**(1), 90–124.

Muizon, C. de (1988). Les relations phylogénétiques des Delphinida
(Cetacea, Mammalia). *Ann. Paléontol. (Vertebr. Invertebr.)* **74**(4),
157–227.

Muizon, C. de (1991). A new Ziphiidae (Cetacea) from the Early
Miocene of Washington State (USA) and phylogenetic analysis of
the major groups of odontocetes. *Bull. Mus. Natl. Hist. Nat. Paris*
(*4ᵉ sér.*) **12,** sect. C (3–4), 279–326.

Nikaido, M., Rooney, A. P., and Okada, N. (1999). Phylogenetic relationships among cetartiodactyls based on insertions of short and
long interspersed elements: Hippopotamuses are the closest extant
relatives of whales. *Proc. Natl. Acad. Sci. USA* **96,** 10261–10266.

O'Leary, M. A., and Geisler, J. H. (1999). The position of Cetacea
within Mammalia: Phylogenetic analysis of morphological data from
extinct and extant taxa. *Syst. Biol.* **48**(3), 455–490.

Repenning, C. A., and Tedford, R. H. (1977). Otarioid seals of the
Neogene. *Geol. Surv. Prof. Paper* **992,** 1–93.

Rice, D. W. (1984). Cetaceans. *In* "Orders and Families of Recent
Mammals of the World" (S. Anderson and J. K. Jones, Jr., eds.), pp.
447–490. Wiley, New York.

Rice, D. W. (1998). "Marine Mammals of the World: Systematics and
Distribution." Soc. Mar. Mammal. Spec. Publ. No. 4, pp. 1–231.

Shimamura, M., Abe, H., Nikaido, M., Ohshima, K., and Okada, N.
(1999). Genealogy of families of SINEs in cetaceans and artiodactyls: the presence of a huge superfamily of tRNA^{Glu}-derived
families of SINEs. *Mol. Biol. Evol.* **16**(8), 1046–1060.

Thewissen, J. G. M. (ed.) (1998). "The Emergence of Whales: Evolutionary Patterns in the Origin of Cetacea." Plenum Press, New York.

Clymene Dolphin
Stenella clymene

THOMAS A. JEFFERSON
*Southwest Fisheries Science Center,
La Jolla, California*

B ecause the Clymene dolphin is one of the most recently
recognized species of dolphins (Perrin *et al.*, 1981), very
few papers have been published on this species (Perrin
and Mead, 1994). Its restricted range, limited to the tropical
and warm temperate waters of the Atlantic Ocean, has not
been well studied cetologically. These two facts make the Clymene dolphin one of the least-known delphinids.

I. Characters and Taxonomic Relationships

The Clymene dolphin is a small, but rather stocky, dolphin
with a moderately long beak, separated from the melon by a
distinct crease (Fig. 1). The dorsal fin is tall and nearly trian-

Figure 1 *Clymene dolphins ride the bow wave of a research
vessel offshore the Gulf of Mexico. Most of the species' distinctive characteristics are visible.*

gular to slightly falcate, and the flippers and flukes are typical of dolphins of the genera *Stenella* and *Delphinus*.

The COLOR pattern is tripartite, with a white belly, light gray flanks, and dark gray cape. The cape dips below the dorsal fin, somewhat lower than in the spinner dolphin. There is a dark gray line running down the length of the top of the beak, and often a dark, indistinct band between the white belly and gray sides. The most distinctive feature is a black "moustache" marking of variable extent on the top of the beak. With the exception of the moustache, most of the species' external characters are very similar to those of the spinner dolphin. This is one of the reasons why the Clymene dolphin was not fully recognized as a distinct species until 1981 (Perrin *et al.*, 1981).

These small dolphins probably do not reach much over 2.0 m in length, with males somewhat larger and heavier than females (Jefferson *et al.*, 1995; Jefferson, 1996). Adult-sized females have been between 171 and 190 cm, and males between 176 and 197 cm (Jefferson, 1996). The maximum weight known is about 80 kg, but they may get somewhat heavier than this.

The skull of this species is very similar to that of *Stenella longirostris* and *S. coeruleoalba* (especially the latter) (Fig. 2) It can be distinguished by its small size (<415 mm) combined with a short, broad rostrum.

Taxonomically, *S. clymene* has been considered to be most closely related to *S. longirostris* and *S. coeruleoalba* (Perrin *et al.*, 1981). Genetic studies by LeDuc *et al.* (1999) indicate that its cytochrome *b* sequence is actually closer to *S. coeruleoalba*.

II. Distribution and Ecology

The Clymene dolphin is found only in the Atlantic Ocean, in tropical to warm temperate waters. The range is not well documented, especially in South Atlantic and mid-Atlantic waters. Most sightings have been in deep, offshore waters.

There is very little known about the ecology of the species. It apparently feeds mostly on mesopelagic fishes and squids, including some species that are vertical migrators. Many dolphins bear bite marks and scars from cookie cutter sharks on their bodies.

III. Behavior and Life History

Schools of this species are often moderately large, although most appear to consist of less than a few hundred individuals (Perrin and Mead, 1994). Schools may be segregated by age and sex class, as evidenced by several mass stranded herds that were composed largely of individuals of one or the other sex (Jefferson *et al.*, 1995).

Clymene dolphins are active bow riders, often approaching ships from many kilometers away for a free ride. They are also often aerially active and they do spin (Fig. 3), although apparently not as frequently or as elaborately as the spinner dolphin.

There have been no studies of the life history of this species based on large samples of specimens. Most of what we know is based on scant information. Both males and females appear to reach sexual maturity by the length of 180 cm (Jefferson, 1996). Nothing is known of other life history parameters, but they are thought to be broadly similar to those of other members of the genus.

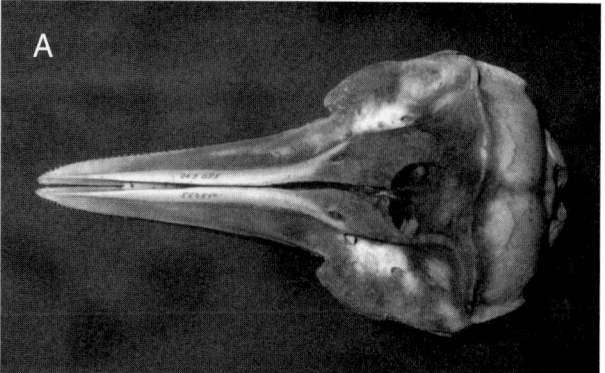

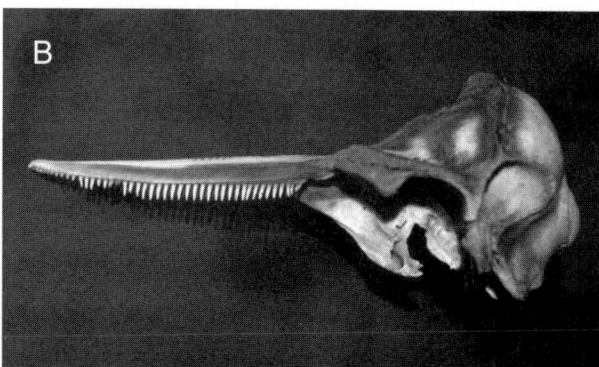

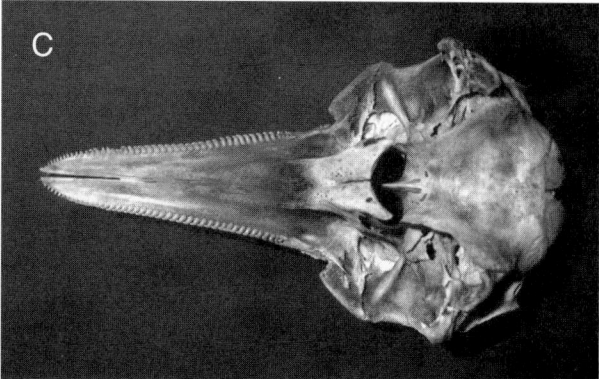

Figure 2 *Dorsal (A), lateral (B), and ventral (C) views of the skull of a Clymene dolphin from the Gulf of Mexico.*

IV. Interactions with Humans

No major conservation problems are known for this species, but it is likely that some undocumented problems exist. Some dolphins are known to be killed in directed fisheries in the Caribbean and incidentally in nets throughout most parts of the range (in particular, West Africa). There has been essentially no work on environmental contaminants in this species. Clymene dolphins have not been held captive, except for some animals that were held temporarily after stranding alive.

Figure 3 *Clymene dolphins are active and acrobatic animals, often leaping and spinning. Photo by Barbara E. Curry.*

See Also the Following Articles

Spinner Dolphin ■ Striped Dolphin

References

Jefferson, T. A. (1996). Morphology of the Clymene dolphin (*Stenella clymene*) in the northern Gulf of Mexico. *Aquat. Mamm.* **22,** 35–43.

Jefferson, T. A., Odell, D. K., and Prunier, K. T. (1995). Notes on the biology of the Clymene dolphin (*Stenella clymene*) in the northern Gulf of Mexico. *Mar. Mamm. Sci.* **11,** 564–573.

LeDuc, R. G., Perrin, W. F., and Dizon, A. E. (1999). Phylogenetic relationships among the delphinid cetaceans based on full cytochrome *b* sequences. *Mar. Mamm. Sci.* **15,** 619–648.

Perrin, W. F., Mitchell, E. D., Mead, J. G., Caldwell, D. K., and van Bree, P. J. H. (1981). *Stenella clymene*, a rediscovered tropical dolphin of the Atlantic. *J. Mamm.* **62,** 583–598.

Perrin, W. F., and Mead, J. G. (1994). Clymene dolphin *Stenella clymene* (Gray, 1846). *In* "Handbook of Marine Mammals" (S. H. Ridgway and R. Harrison, eds.), Vol. 5, pp. 161–171. Academic Press, San Diego.

Cognition

SEE *Intelligence and Cognition*

Coloration

WILLIAM F. PERRIN
*Southwest Fisheries Science Center,
La Jolla, California*

Marine mammals are not as colorful as birds or fishes or reptiles, but many have striking and distinctive coloration patterns that are useful in their taxonomy, presumably have function and adaptive value, and can vary in-

dividually and with age, sex, geographic region, and even time of the year.

I. Terminology

A number of schemes have been proposed for naming the elements of color patterns in cetaceans; the usage here follows Perrin (1973, 1997) and Perrin *et al.* (1991). In delphinids and phocoenids (Fig. 1), the *bridle* is composed of the *blowhole stripe* running from the blowhole to the apex of melon and the *eye stripe* from the eye to the apex of melon. Both stripes may have complex internal structure. An *eye spot* may be visible, and there may also be a small *ear stripe* or *spot*. The *eye-to-anus stripe* runs from the eye to the anal/genital region and may have *accessory stripes*. The *flipper stripe* runs forward from the base of the flipper variously to the eye (e.g., in spinner dolphin, *Stenella longirostris*), corner of the mouth (e.g., pantropical spotted dolphin, *S. attenuata*), or forward along the rostrum to join the *lip mark* ventrolaterally (common dolphins, *Delphinus* spp.).

The overall color pattern in at least some delphinids can be analyzed in terms of interacting independent components (Fig. 2). A basic *cape* is covered with a *dorsal overlay* of varying extent and intensity and may not be visible except in fetal or anomalously pigmented specimens. A crisscross of the boundaries of these two elements in *Delphinus* spp. yields a complex four-part pattern of a dark-gray *dorsal field* (cape and overlay combined), buff or yellowish *thoracic patch* (cape alone), light-gray *flank patch* (overlay alone), and white *ventral field* (outside both cape and overlay). In some anomalous individuals, the overlay may be absent, yielding a simplified pattern of cape only [e.g., in *Delphinus delphis* (Perrin *et al.*, 1995) and *Stenella longirostris* (Perrin, 1973)]. Spotting appears to be yet another independent component that develops with maturation in some species.

In pinnipeds, coloration can be a property of different *pelages* or pelage elements, through a range from white to silver, gray or bluish gray, brown and black. The *lanugo* is a fetal pelage that develops and can be lost before birth, although in many species it is shed a few days or weeks after birth. Juveniles may undergo additional *molts* and changes of color. The coarse *guard hairs* can differ in color from the hairs of the *undercoat*. Many seals are simply uniformly colored or countershaded, but some have bold patterns, such as the harp seal (*Pagophilus groenlandicus*) and ribbon seal (*Histriophoca fasciata*), and others are spotted.

II. Development

Coloration typically changes between birth and adulthood. In some cases appearance changes radically, whereas in others the change is more in contrast and distinctness of pattern elements. Only a few examples are discussed here.

The spotted dolphins, *Stenella attenuata* and *S. frontalis*, are unspotted at birth (Fig. 3). Small dark spots appear in large juveniles in the gular region and spread over the ventral surface, enlarging as maturity approaches. Light spots appear on the back and spread in a similar fashion, although not in an even distribution over the back. In *S. attenuata* the dark ventral spots fuse and lighten to yield a light-gray, faintly dappled

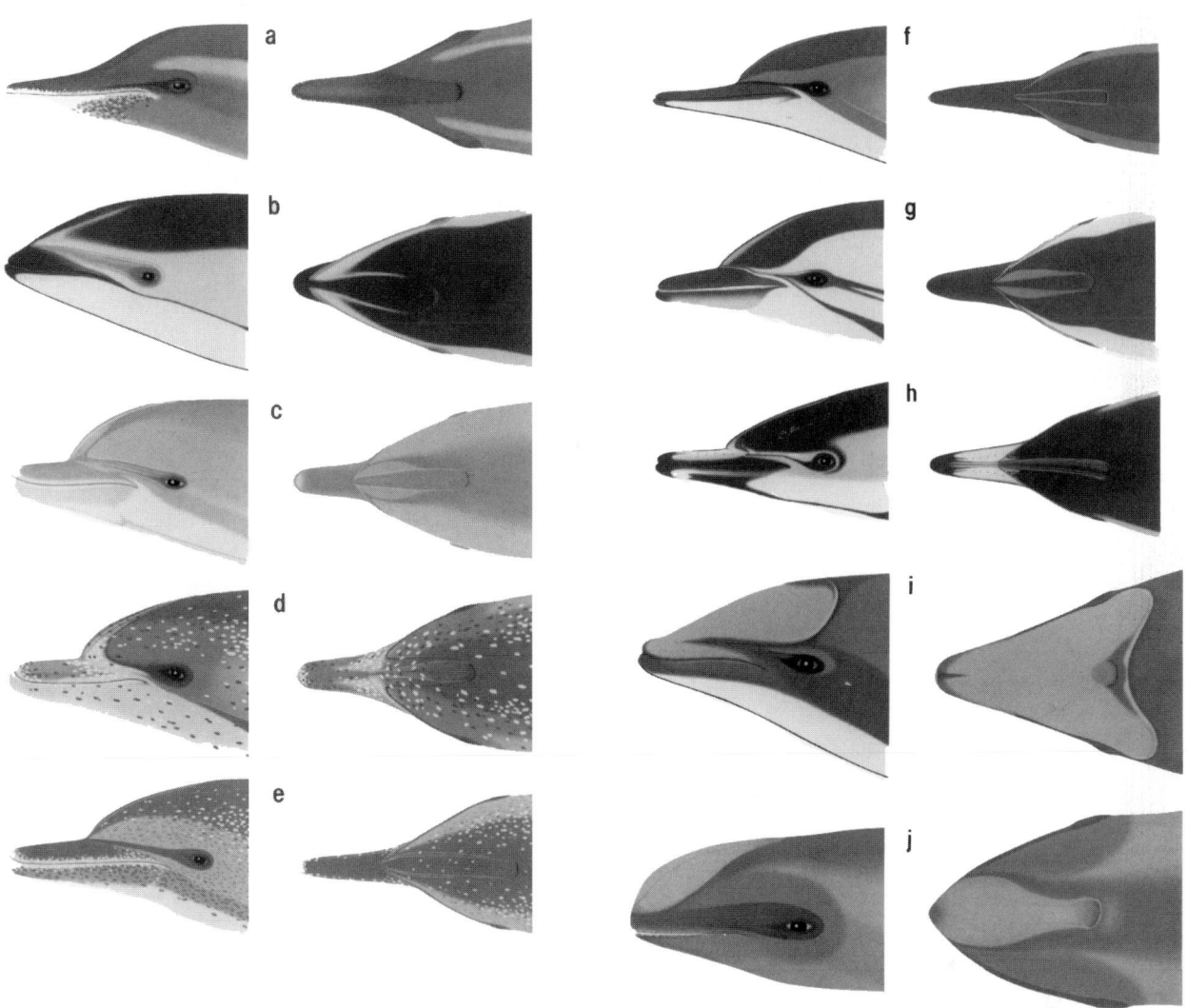

Figure 1 *Typical appearance of bridle in 10 delphinid species: (a)* Steno bredanensis, *(b)* Lagenorhynchus obliquidens, *(c)* Tursiops truncatus, *(d)* Stenella frontalis, *(e)* S. attenuata, *(f)* S. longirostris, *(g)* S. coeruleoalba, *(h)* Delphinus delphis, *(i)* Cephalorhynchus eutropia, *and (j)* Peponocephala electra. *From Perrin (1997).*

ventral surface. In *S. frontalis*, both ventral and dorsal spots persist into maturity.

The beluga, *Delphinapterus leucas*, is dark gray at birth but lightens as it grows; adults are white. A similar trend is seen in Asian populations of the Indopacific humpbacked dolphin *Sousa chinensis*. The reverse of this trend is seen in many other cetaceans, such as pilot whales and beaked whales; calves are lighter at birth and darken with age, although neonates of the bottlenose dolphin and the finless porpoise *Neophocaena phocaenoides* in some regions are darker than juveniles and adults.

The development of coloration tends to take opposite courses in different groups of pinnipeds (Bonner, 1990). Otariids are born dark and become lighter as juveniles, most darkening again as adults. However, most northern hemisphere phocids (the phocinines), including *Phoca largha, Pusa* spp., *Pagophilus groenlandicus, Histriophoca fasciata,* and *Halichoerus grypus*, are born with a white or yellowish lanugo

(Fig. 4), which is shed at 2–5 weeks. This molt exposes either the adult pattern of spots or other marks or a juvenile countershaded coloration that later changes to the adult patterned state. The harbor seal, *Phoca vitulina*, and the hooded seal, *Cystophora cristata*, are unusual in that the lanugo is molted before birth (always in the hooded seal and usually in the harbor seal). The bearded seal, *Erignathus barbatus*, is born with a grayish-brown lanugo (with white muzzle and white blotches), and in the monk seals (*Monachus* spp.) and elephant seals (*Mirounga* spp.) the lanugo is black. In the southern phocids, the Weddell seal *Leptonychotes weddellii*, Ross seal *Ommatophoca rossii*, crabeater seal *Lobodon carcinophaga*, and leopard seal *Hydrurga leptonyx*, pups are born with pale-gray to brownish-gray coats; in the leopard seal the birth coat resembles the adult state in color and pattern. The walrus (*Odobenus rosmarus*) molts the lanugo *in utero* and has sparse whitish, yellowish, or silver-gray coat at birth. The lanugo, whether light

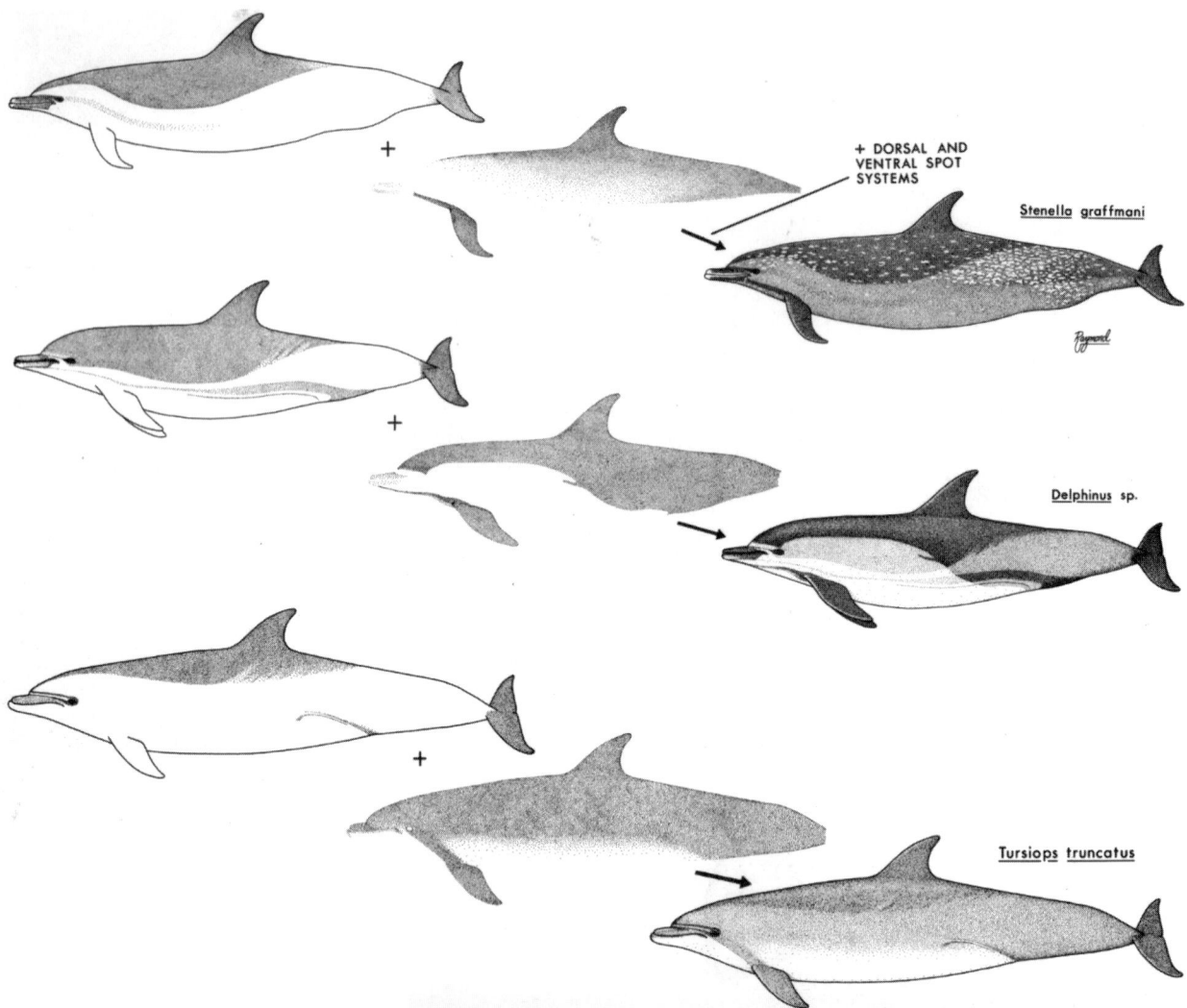

+ DORSAL AND
VENTRAL SPOT
SYSTEMS

Stenella graffmani

Delphinus sp.

Tursiops truncatus

Figure 2   *Component analyses of color patterns of* Stenella attenuata, Delphinus delphis, *and* Tursiops truncatus *(from left to right): basic cape plus dorsal overlay yields a complex color pattern. From Perrin (1973).*

or dark, is usually thick and wooly, and it has been suggested that it functions in heat conservation until a BLUBBER layer accumulates. Another suggested function, at least for the white lanugo, is camouflage against predators on the ice, although the southern ice-breeding monachines do not have white coats at birth. It has been posited that intrauterine loss of the lanugo in the harbor seal is a secondary adaptation to breeding on land since its descent from an ice-breeding ancestor, implying a camouflage function. Why some birth coats are light and others dark is still a matter for speculation.

III. Sexual Dimorphism

In delphinid cetaceans, sexually dimorphic color pattern elements are typically associated with the genital region. For example, a black tear-drop-shaped patch surrounding the genital slit in Commerson's dolphin *Cephalorhynchus commersonii*

has its apex directed posteriorly in adult males and anteriorly (sometimes with a posterior invagination around the genital slit) in females (Fig. 5). A lateral stripe extending from the eye to the genital region in Fraser's dolphin *Lagenodelphis hosei* is broader and darker in adult males than in females (Jefferson *et al.*, 1997). Adult male beaked whales (Ziphiidae) of many species tend to develop white areas on the head; in some species the entire head becomes white, whereas in others the white area may be confined to the front or top of the head or to the rostrum (Ridgway and Harrison, 1989).

Ventral coloration is dimorphic in the Mediterranean monk seal *Monachus monachus* from birth onward; as in some dolphins, the posterior boundary of a mark around the genital opening is arcuate in females and straight in males (Badosa and Grau, 1998). Adult males are also much darker than females.

In other phocids, the color pattern is usually more distinct and with darker elements in males than in females, e.g., in the

Figure 3 *Development of color pattern in two spotted dolphins:* Stenella frontalis *(top three) and* S. attenuata *(bottom two). From Perrin* et al. *(1987).*

Figure 4 *Recently born ringed seal pup, showing lanugo. From Ridgway and Harrison (1981).*

ribbon seal and harp seal. Some adult male gray seals are almost black, whereas females tend to be lighter colored. However, some species, e.g., the bearded seal and Weddell seal, are not noticeably dimorphic in coloration (Ridgway and Harrison, 1981). For walruses, the pattern seen in many phocids is reversed; old adult males tend to be lighter colored than females. The pattern is variable in the otariids; in some, e.g., the Cali-

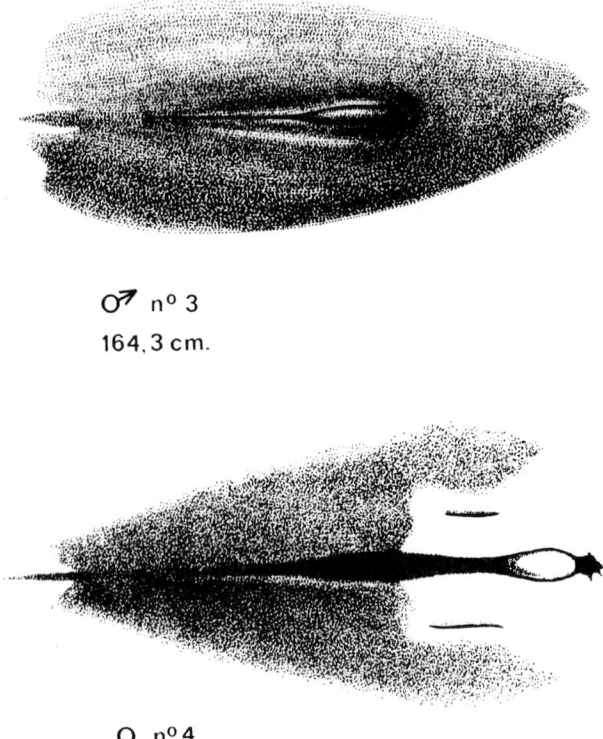

Figure 5 *Sexual dimorphism in the form of a genital patch in* Cephalorhynchus commersonii. *From Robineau (1984).*

fornia sea lion *Zalophus californianus* and northern fur seal *Callorhinus ursinus*, the adult male is darker than the female, whereas in others, e.g., the Steller sea lion *Eumetopias jubatus*, there is no apparent dimorphism.

IV. Geographic and Individual Variation

Color patterns vary among adults in marine mammal species, both individually and geographically. Biologists use individual variations in coloration and other natural marks as "tags" in studies of abundance, movements, and life history. "Tag and recapture" studies using natural marks can estimate population size; this has been applied to a number of cetacean species, including killer whales (*Orcinus orca*), minke whales (*Balaenoptera acutorostrata*), fin whales (*B. musculus*), blue whales (*B. physalus*), humpback whales (*Megaptera novaeangliae*), right whales (*Eubalaena* spp.), bowhead whales (*Balaena mysticetus*), bottlenose dolphins, and others (Hammond *et al.*, 1990). Long-range movements of migratory whales have been documented using natural marks, clarifying migratory cycles and stock structure and affiliations. Longitudinal studies of individuals and groups over generational time have been vital in arriving at estimates of such important life history parameters as age at first reproduction, calving interval, and survivorship, and these have been made possible by the use of natural marks to keep track of individuals (for the Atlantic spotted dolphin *Stenella frontalis* in the Caribbean, the killer whale in the Pacific Northwest, and a variety of other small cetaceans). Examples of coloration features that have been or could be used as natural marks include shape of the dorsal saddle and postocular spot in the killer whale (Fig. 6), details of the eye and blowhole stripes in dolphins and porpoises, color and patterning of the underside of the flukes in the humpback whale, and patterns of spots, blotches, and white areas on the body in right, gray (*Eschrichtius robustus*), bowhead, blue, and minke whales. Some naturally imposed marks, such as scars from infraspecific fighting and from bites by predators or cookie cutter sharks (*Isistius* sp.), have also proven useful, although factors such as fading and acquisition of new marks during a study must be taken into consideration. For some drably colored marine mammals, e.g., manatees (*Trichechus* spp.), scars (including those from collisions with boats and fishing gear) are the only marks available for use.

The use of natural marks in individually identifying pinnipeds from their patterns of spots and blotches has been complicated by the difficulty of photographing seals from a standard angle or in a standard posture. This problem has been approached by the development of computer-aided matching of images using a three-dimensional model to correct for orientation and posture (Hammond *et al.*, 1990).

As for other morphological characters, coloration tends to vary geographically most in those features that vary individually most within a population. For example, in spinner dolphins in the eastern Pacific, the degree to which the dorsal overlay obscures the underlying cape is highly variable; in some animals the cape is prominent whereas in others it is invisible (Perrin *et al.*, 1991). It is in this feature of the color pattern that spinner dolphins vary most from region to region around the world. Similarly, adult spotted dolphins of both species (*Stenella*

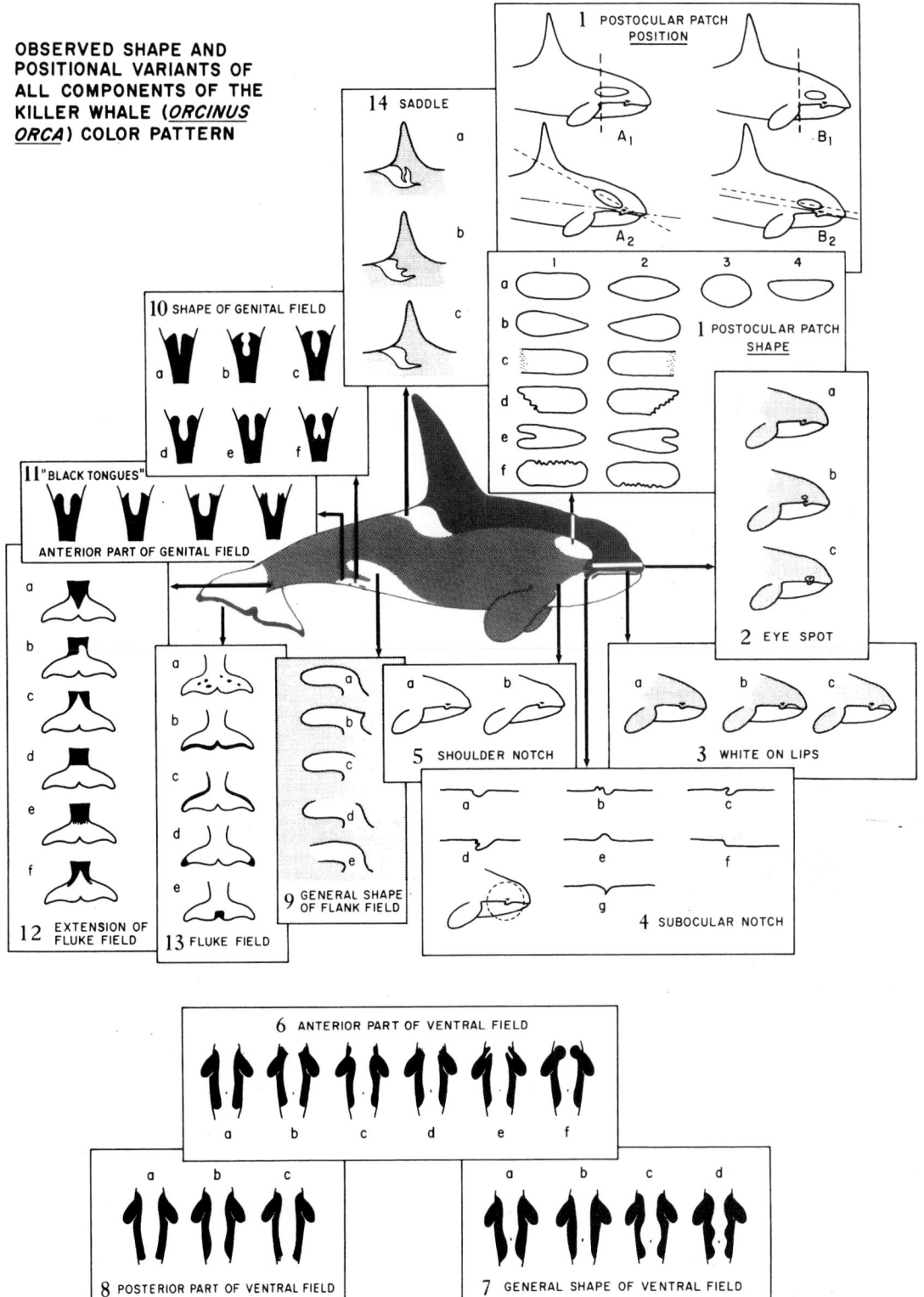

Figure 6 *Variation (individual and geographic) in components of the color pattern of* Orcinus orca. *From* Evans et al. *(1982).*

attenuata and *S. frontalis*) vary individually in the degree of spotting, and average spotting varies geographically as well; the offshore Gulf Stream form of *S. frontalis* is all but unspotted (Perrin *et al.,* 1987). The *truei* and *dalli* color morphs of Dall's porpoise (Fig. 7), originally described as different species, typify regional populations but include the range of coloration in a single population (Ridgway and Harrison, 1999). Baleen whales vary geographically in coloration of the baleen and in details of color pattern of the body and appendages, which also vary greatly within populations. Spotting is highly variable among individual harbor seals, and differences can be found between populations even on different islands in an archipelago (Hammond *et al.,* 1990). Humpback whales of the southern hemisphere tend to have more white pigmentation in their bodies than their northern hemisphere counterparts.

V. Genetics

Little is known of the genetic basis of coloration in marine mammals. Albinism is possibly an autosomal dominant trait as in humans. Analysis of populational data for the North Atlantic right whale, *Eubalaena glacialis,* suggests that the presence of white ventral skin patches is an autosomal recessive trait (Schaeff and Hamilton, 1999); it is not evident from data that the trait is subject to selection pressure.

VI. Microanatomy

Pigmentation in cetaceans is limited to the occurrence of melanin in the epidermis. The distribution of melanin in the skin of *Delphinus delphis* (Gwinn and Perrin, 1975) is de-scribed here as an example. The pigment is usually concentrated around the bases of the dermal papillae and extends in bands from their apices. Portions of the skin that appear white show very small amounts of diffuse pigment (particles unresolvable at 1250 × magnification) and small granules (less than 5 μm in diameter). The buff color characteristic of the thoracic patch is associated with an equal prominence of diffuse and small granules or higher density of diffuse pigment. The gray flank patch has some small granules but mostly large granules (>5 μm). The black regions of the back and flukes have the highest density of large granules. It may be hypothesized that the type of melanin that produces the buff color is of a composition that does not allow further polymerization but favors its combination with a protein instead of aggregation into granules. In gray and black areas of the color pattern, the aggregation of particles proceeds, and melanocytes containing them migrate toward the surface of the epidermis until diffuse pigment is largely replaced by granular pigment.

VII. Function and Evolution

An early analysis of coloration in cetaceans (Yablokov, 1963) proposed several functions: acquisition of prey, protection from predators, and communication with conspecifics. Cetacean patterns were divided into three types: (1) uniform or finely spotted, adapted to planktonic feeding or feeding in murky water or great depths where vision is not important; (2) strongly spotted, striped, or patterned, for intraspecific recognition; and (3) countershaded, as camouflage against predators in animals foraging near the surface. However, spots would seem also to

Figure 7 *Two color morphs of* Phocoenoides dalli: dalli *type (top) and* truei *type (bottom). From Jefferson (1988).*

be useful in camouflage, as the surface of the sea appears dappled when seen from below. Notably, spotted dolphins are strongly countershaded at birth and only begin to develop spots at about the age when they begin to forage on their own; in this case the likely function of the spots is camouflage against prey. The pattern of development is the reverse of that in many species of deer, which are spotted at birth and lose their spots as they become self-sufficient; here the function is camouflage against predators. The stripes and marks on the typical delphinid head (eye stripe, blowhole stripe, flipper stripe, eye spot, and lip mark) may also serve as camouflage against prey, obscuring the eye and mouth as in many terrestrial mammals and other tetrapods, including reptiles, amphibians, birds, and fishes. Many cetaceans have prominent white color pattern elements or patches, often bordered with dark pigmentation; these may function in species recognition or serve to signal the positions of school mates in low-light conditions. Similar functions have been suggested for the bold color patterns seen in some phocid pinnipeds, together with possible uses in signaling gender and age. The patterns may also serve in disruptive coloration for camouflage against predator or prey.

The fin whale is unique among marine mammals in being asymmetrically patterned. The left anterior third of the body and the baleen on the left side are dark, but the lower jaw and the anterior baleen on the right side are white. The white area has been proposed to function to maintain countershading when the whale rolls on its side during feeding or to startle prey during prey herding.

A simple countershaded pattern has been proposed to be the most primitive and generalized for the delphinid cetaceans (Mitchell, 1975) because it is a pattern shared by many taxa of marine organisms inhabiting near-surface waters and used for concealment through counterlighting. The crisscross pattern of *Delphinus* spp. is perhaps the most derived (as the most complex), with a possible function of obscuring the presence of a small calf swimming side by side with the mother.

The color pattern can evolve through paedomorphosis, the retention of fetal or juvenile characteristics into adulthood. The dark "hoods" on the heads of some dolphins and small-toothed whales may have evolved this way. The system of stripes on the head (eye stripe and blowhole stripe) in delphinid and phocoenid cetaceans develops from a single mark across the back of the head behind the blowhole in small fetuses (Perrin, 1997). A progressive forward invagination of the mark on each side of the head creates the two stripes, which are initially broad and then narrow to varying degrees. The stripes vary among species in width and definition, but in some species, e.g., *Cephalorhynchus* spp. and some of the globicephalinine species such as *Peponocephala electra*, a fetal condition is retained and the "blowhole stripe" effectively covers the entire top of the head (Fig. 1). It is interesting to note that the *Cephalorhynchus* species are also paedomorphic in their osteology, convergent on the phocoenids in body and skull size and shape.

Paedomorphosis may account for geographic variation in the expression of color pattern elements in some species. For example, the cape is visible in the fetus of the killer whale but is usually not expressed in the postnatal animal. However, the cape is distinctly visible in all whales in an Antarctic population.

Similarly, the adult of the Kerguelen Islands form of Commerson's dolphin *Cephalorhynchus commersonii* retains a grayish portion of the color pattern seen only in calves in the South American population (Robineau, 1986).

VIII. Coloration as a Taxonomic Character

Color pattern features are useful in taxonomy. In some cases, what were thought to be color variants of a single species have proved to be distinct species [e.g., the common dolphins *Delphinus delphis* and *D. capensis* (Heyning and Perrin, 1994) and the minke whales *Balaenoptera acutorostrata* and *B. bonaerensis* (Best, 1985)]. However, in other cases, species defined on the basis of color pattern differences have been subsequently lumped [e.g., the striped dolphin, *Stenella coeruleoalba*, and its nominal synonyms (Fraser and Noble, 1970) and the *dalli* and *truei* forms of Dall's porpoise, *Phocoenoides dalli* (Ridgway and Harrison, 1999)], demonstrating that the same caution must be used in employing color pattern characters as in the use of any other morphological characters. The contributions of ontogenetic, individual, and geographic variation must be delineated before species-level differences and higher-level taxonomic relationships emerge.

One use of coloration in higher-level taxonomy has been in the cladistic analysis of relationships among the delphininoid cetaceans (Fig. 8); the delphinids and the phocoenids are sister taxa in sharing the derived *bridle*, or system of head stripes, consistent with phylogenetic relationships in the group based on other morphological characters (Heyning, 1989).

IX. Ephemeral and Anomalous Variation

Real and apparent changes in coloration can be ephemeral or environmentally induced. For example, some dolphins (and old walruses) are at some times pink and other times white. The Amazon river dolphin *Inia geoffrensis* is also known as *"bufeo colorado"* because of its pink color when seen in its natural habitat; in captivity in a temperate-latitude aquarium it is white. Some adult individuals of the Indo-Pacific humpback dolphin, *Sousa chinensis*, in the waters of Hong Kong have been observed to be bright pink whereas some other tropical dolphins sometimes exhibit pink bellies. This fact is even incorporated in the scientific name of one dolphin, the dwarf spinner dolphin of Southeast Asia, *Stenella longirostris roseiventris* (Perrin *et al.*,

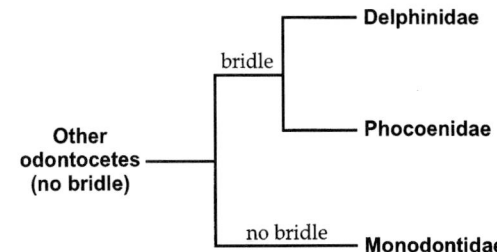

Figure 8 *Cladistic relationships among the living families of delphinoid cetaceans based on the presence or absence of the bridle. From Perrin (1997).*

1999). The pink color is due to dilation of subcutaneous blood vessels, presumably for purposes of thermoregulation; dolphins everywhere must dispose of excess heat generated by metabolism, which is a greater problem in warm tropical water. Even the Arctic walrus, *Odobenus rosmarus*, must "dump" heat, and palely pigmented old males may appear "rosy" during relatively warm weather. In cetaceans, the pink coloration is visible only in animals or parts of animals that are normally white (lacking melanin), although light-gray dolphins, such as immature individuals of *Sousa chinensis,* may sometimes appear purplish because of the subcutaneous suffusion with blood. When a "pink" walrus enters the water, the skin becomes ischemic (deprived of blood) and the animal appears white.

Other pinnipeds change color when wet; e.g., the brown California sea lion becomes almost black. Seasonally molting pinnipeds change color, becoming drabber as the molt approaches. Reddish pelage in harbor seals can be the result of deposition of iron oxide precipitates on the hair shaft. Even conditions such as water color, cloud cover, and angle of sun can cause cetaceans to apparently change color in the water. The striped dolphin, also known as the blue-and-white dolphin, can appear brown and white in turbid water or under overcast skies. A sojourn in high latitudes can lead to an accumulation of a coat of yellowish diatoms in the blue whale *Balaenoptera musculus,* leading to one of its other common names, "sulfurbottom." In one minke whale taken in the Antarctic, the areas normally white were pinkish and the baleen, blubber, and connective tissue were orange; the cause of the "carotenoid" coloration was not apparent (Kato, 1979) but may have been metabolic. Cetaceans in captivity in shallow tanks can become darker when exposed to the sun. The intensity of coloration has been reported to vary seasonally in common dolphins in the Black Sea. In some cetacean species, apparent stripes or spots can be scars. Adult male beaked whales inflict long parallel rakes on each other with their teeth (the behavior also occurs in species in which the teeth erupt in females). Most cetaceans that frequent tropical waters bear oval or stellate scars from the bites of cookie cutter sharks; these can be so numerous as to give the animal an overall spotted appearance.

Anomalous conditions such as albinism, melanism, and piebald coloration have been recorded for many cetacean and pinniped species. In one anomalous color pattern variety that has been seen in the short-beaked common dolphin in widely separated parts of the world, the dorsal overlay is missing, eliminating the typical crisscross pattern (Perrin *et al.,* 1995).

Coloration in cetaceans changes with death. Subtle elements of color pattern disappear quickly in a dead stranded animal as the skin dries, and a dolphin with a complex color pattern can turn solid black when long dead or frozen (careful thawing in water can sometimes bring back some of the pattern). The outer layers of the skin can be abraded away during stranding, which has resulted in more than one report of a stranded "white whale". The accounts of coloration in some original species descriptions of cetaceans were based on long-dead specimens and thus are defective (e.g., Heaviside's dolphin, *Cephalorhynchus heavisidii,* is not uniformly black as originally described, but boldly patterned). Much remains to be learned about the color pattern of the living animal for many species.

See Also the Following Articles

Albinism ■ Geographic Variation ■ Sexual Dimorphism ■ Species

References

Badosa, E., and Grau, E. (1998). Individual variation and sexual dimorphism of coloration in Mediterranean monk seal pups (*Monachus monachus*). *Mar. Mamm. Sci.* **14,** 390–393.

Best, P. B. (1985). External characters of southern minke whales and the existence of a diminutive form. *Sci. Rep. Whales Res. Inst. Tokyo* **36,** 1–33.

Bonner, W. N. (1990). "The Natural History of Seals". Facts On File, New York.

Evans, W. E., Yablokov, A. V., and Bowles, A. E. (1982). Geographic variation in the color pattern of killer whales (*Orcinus orca*). *Rep. Int. Whal. Commn,* **32,** 687–694.

Fraser, F. C., and Noble, B. A. (1970). Variation of pigmentation pattern in Meyen's dolphin, *Stenella coeruleoalba* (Meyen). *Invest. Cetacea* **2,** 147–163, pl. 1–7.

Gwinn, S., and Perrin, W. F. (1975). Distribution of melanin in the color pattern of *Delphinus delphis* (Cetacea: Delphinidae). *Fish. Bull. U.S.* **73,** 439–444.

Hammond, P. S., Mizroch, S. A., and Donovan, G. P. (eds.) (1990). Individual recognition of Cetaceans: Use of photo-identification and other techniques to estimate population parameters. *Rep. Int. Whal. Commn,* (special issue) **12.**

Heyning, J. E. (1989). Comparative facial anatomy of beaked whales (Ziphiidae) and a systematic revision among the family of extant Odontoceti. *Nat. Hist. Mus. Los Angeles County Contr. Sci.* **405,** 1–64.

Heyning, J. E., and Perrin, W. F. (1994). Evidence for two species of common dolphins (genus *Delphinus*) from the eastern North Pacific. *Nat. Hist. Mus. Los Angeles County Contr. Sci.* **442,** 1–35.

Jefferson, T. A. (1988). *Phocoenoides dalli. Mamm. Species* **319,** 1–7.

Jefferson, T. A., Pitman, R. L., Leatherwood, S., and Dolar, M. L. L. (1997). Developmental and sexual variation in the external appearance of Fraser's dolphins (*Lagenodelphis hosei*). *Aquat. Mamm.* **23,** 145–153.

Kato, H. (1979). Carotenoid colored minke whale from the Antarctic. *Rep. Whales Res. Inst. Tokyo* **31,** 97–99, pl. 1.

Mitchell, E. (1970). Pigmentation pattern evolution in delphinid cetaceans: An essay in adaptive coloration. *Can. J. Zool.* **48,** 717–740.

Perrin, W. F. (1973). Color pattern of spinner porpoises (*Stenella* cf. *S. longirostris*) of the eastern tropical Pacific and Hawaii, with comments on delphinid pigmentation. *Fish. Bull. U.S.* **70,** 983–1003.

Perrin, W. F. (1997). Development and homologies of head stripes in the delphinoid cetaceans. *Mar. Mamm. Sci.* **13,** 1–43.

Perrin, W. F., Akin, P.A., and Kashiwada, J. V. (1991). Geographic variation in external morphology of the spinner dolphin *Stenella longirostris* in the eastern Pacific and implications for conservation. *Fish. Bull. U.S.* **89,** 411–428.

Perrin, W. F., Armstrong, W. A., Baker, A. N., Barlow, J., Benson, S. R., Collet, A. S., Cotton, J. M., Everhart, D. M., Farley, T. D., Mellon, R. M., Miller, S. K., Philbrick, V., Quan, J. L., and Lira Rodriguez, H. R. (1995). An anomalously pigmented form of the short-beaked common dolphin (*Delphinus delphis*) from the southwestern Pacific, eastern Pacific, and eastern Atlantic. *Mar. Mamm. Sci.* **11,** 241–247.

Perrin, W. F., Mitchell, E. D., Mead, J. G., Caldwell, D. K., Caldwell, M. C., van Bree, P. J. H., and Dawbin, W. H. (1987). Revision of the spotted dolphins, *Stenella* spp. *Mar. Mamm. Sci.* **3,** 99–170.

Ridgway, S. H., and Harrison, R. J. (eds.) (1981–1999). "Handbook of Marine Mammals", Vols, 1–6. Academic Press, San Diego.

Robineau, D. (1984). Morphologie externe et pigmentation du dauphin de Commerson, *Cephalorhynchus commersonii* (Lacépède, 1804), en particulier celui des îles Kerguelen. *Can. J. Zool.* **62,** 2465–2475.

Schaeff, C. M., and Hamilton, P. K. (1999). Genetic basis and evolutionary significance of ventral skin color markings in North Atlantic right whales (*Eubalaena glacialis*). *Mar. Mamm. Sci.* **15**, 701–711.

Yablokov, A. V. (1963). Types of color of the Cetacea. *Byul. Morsk. Obshch. Ispytat. Pri. (Ot. Biol.)* **68(6),** 27–41. [In Russian]

Common Dolphins

Delphinus delphis, D. capensis, and *D. tropicalis*

WILLIAM F. PERRIN
*Southwest Fisheries Science Center,
La Jolla California.*

The short-beaked common dolphin (*Delphinus delphis*) is the most numerous dolphin in offshore warm-temperate waters in the Atlantic and Pacific, often coming from a distance to join a boat and ride the bow wave. The very closely related long-beaked common dolphin (*D. capensis*) is seen less often in most regions and is difficult to distinguish from its congener at sea. The very-long-beaked endemic Indian Ocean common dolphin (*D. tropicalis*) is poorly known and of uncertain taxonomic status.

I. Characters and Taxonomic Relationships

All the common dolphins are slender and have a long beak sharply demarcated from the melon (Fig. 1). The dorsal fin is high and moderately falcate, although in some areas it may be erectly triangular. The three species are distinguished from other dolphins by a unique crisscross color pattern formed by interaction of the dorsal overlay and cape, resulting in distortions of the usual delphinine lateral and ventral fields. The lower margin of the dorsal overlay passes high anteriorly and dips to cross the ventral margin of the low-riding cape, yielding a four-part pattern of dark gray to black uppermost portion or spinal field (cape under dorsal overlay), buff to pale yellow anterior portion or thoracic patch (undiluted cape), light to medium gray posterior portion or flank patch (undiluted dorsal overlay/lateral field), and white abdominal field. A dark flipper-to-anus stripe may parallel the lower margin of the cape and extend to the genital region. A dark flipper stripe runs forward to join the lip patch on the underside of the beak. In the short-beaked species, the color pattern is more crisp, the thoracic patch is more yellowish, the subcape stripe tends to be anteriorly narrow and faint, and the flipper stripe tends to be narrow and pass low below the corner of the gape. In the long-beaked species, the pattern is more muted, the spinal field may be grayish, the thoracic patch tends to be pale buff, the flipper-to-anus stripe tends to be broad anteriorly and may be pronounced and contiguous with the flipper stripe, and the flipper stripe tends to wander toward the corner of the gape before passing ventrally to join the lip patch mark.

The Indian Ocean common dolphin is similar to the long-beaked species in coloration. In the short-beaked species, the dorsal fin and flippers may be all white or have white centers. These relative features may not be evident in juveniles.

Full species status for the three forms has not been widely recognized until recently (Evans, 1994; Heyning and Perrin, 1994; Rice, 1998), and data on size and weight for the three species have been conmingled in the literature for some parts of the world. Data exist for well-documented series of adults of two species from California waters. In the short-beaked species, males were 172–201 cm long (*n* = 28) and females were 164–193 cm (*n* = 37) *vs* 202–235 cm (*n* = 15) and 193–224 cm (*n* = 10), respectively, for the long-beaked form. The short-beaked species ranged to about 200 kg and the long-beaked species to about 235 kg. However, this pattern of differential size may not hold globally; a geographic form of the short-beaked species in the eastern tropical Pacific ranges in length to 235 cm, as large as the long-beaked species in California waters.

The SKULLS of *Delphinus* spp. are different from those of all the other delphinines (*Stenella* spp., *Tursiops* spp., *Sousa* spp., and *Lagenodelphis hosei*) in the combination of long narrow rostrum and deep palatal grooves. They are similar to the skulls of *Stenella longirostris, S. clymene, S. coeruleoalba,* and *L. hosei* in having a strongly dorsoventrally flattened rostrum with distally splayed teeth, about 40–60 teeth in each row, relatively small temporal fossae, and sigmoid mandibular rami. The two sympatric species differ in proportional length of the rostrum in adults: the ratio of rostral length to zygomatic width in *D. delphis* is 1.21–1.47 and in *D. capensis* is 1.52–1.77 (Heyning and Perrin, 1994). Upper/lower tooth counts in California were 42–54/41–53 (*n* = 49/47) and 47–60/47–57 (*n* = 53/51), respectively. Vertebral counts were 74–80 (*n* = 80) in *D. delphis* and 77–80 (*n* = 25) in *D. capensis*. The osteology of *D. tropicalis* has not been well described, but the rostral/zygomatic ratio of the holotype specimen is 2.06 and the tooth counts 65–65/57–58, both well beyond the upper range of *D. capensis* for California. Six *Delphinus* skulls from the Arabian Peninsula ranged continuously from 1.72 to 1.94 in rostral/zygomatic ratio. Three (with the lowest ratios) had been identified as *D. capensis* and the others as *D. tropicalis*; the range overlaps that of *D. capensis* in California (Smeenk *et al.,* 1996). It has been suggested that *D. tropicalis* may be a very long-beaked form at the end of a cline of *D. capensis* extending into the Indian Ocean rather than a separate third species of *Delphinus* (Heyning and Perrin, 1994; Smeenk *et al.,* 1996; Rice, 1998).

The common dolphins are members of the delphinid subfamily Delphininae *sensu stricto* (LeDuc *et al.,* 1999). In a cladistic phylogenetic analysis based on cytochrome *b* mtDNA, they share a strongly supported polytomic clade with *Stenella coeruleoalba* and *S. clymene* (sister taxa), *S. frontalis,* and *Tursiops aduncus. D. tropicalis* is basal in the *Delphinus* clade, with California *D. delphis,* Black Sea *D. delphis,* and *D. capensis* in a subsidiary three-part polytomy, making *D. delphis* seemingly paraphyletic. However, apparent paraphyly among terminal taxa can result from lineage sorting in newly evolved species (LeDuc and Perrin, 1999); reciprocal monophyly has not yet had a chance to evolve in such cases. As noted earlier, *Delphinus* shares several skull features with *S. longirostris,*

Figure 1 *Short-beaked (top) and long-beaked (bottom) common dolphins,* Delphinus delphis *and* D. capensis, *respectively. From Heyning and Perrin (1994).*

S. coeruleoalba, S. clymene, and *L. hosei,* but that grouping is not supported by the molecular results to date.

II. Distribution and Ecology

Common dolphins occur in warm-temperate and tropical waters worldwide from about 40–60°N to about 50°S (Jefferson *et al.,* 1993), but because the three species were considered to be forms of one until recently, many DISTRIBUTION records and much documentation of range have not been identified to species (Rice, 1998). Based on records with diagnostic characters (Fig. 2; Heyning and Perrin, 1994), *D. delphis* occurs from southern Norway to West Africa in the eastern Atlantic (including the Mediterranean and Black Seas), from Newfoundland to Florida in the western Atlantic, from southern Canada to Chile along the

coast and pelagically in the eastern Pacific; in the central North Pacific (but not in Hawaii); from central Japan to Taiwan and around New Caledonia, New Zealand, and Tasmania in the western pacific; and is possibly absent from the South Atlantic and Indian Oceans. *D. capensis* occurs disjunctly in warm-temperate and tropical coastal waters in West Africa, from Venezuela to Argentina in the western Atlantic, from southern California to central Mexico, and in Peru in the eastern Pacific, around Korea, southern Japan and Taiwan in the western Pacific, and in waters of Madagascar, South Africa, and possibly Oman in the Indian Ocean. The identification of the Oman specimens is in some doubt (Smeenk *et al.,* 1996). *D. tropicalis* is known only from the northern Indian Ocean and Southeast Asia.

D. delphis and *D. capensis* are narrowly sympatric in some nearshore waters; schools of the two species may be seen in the

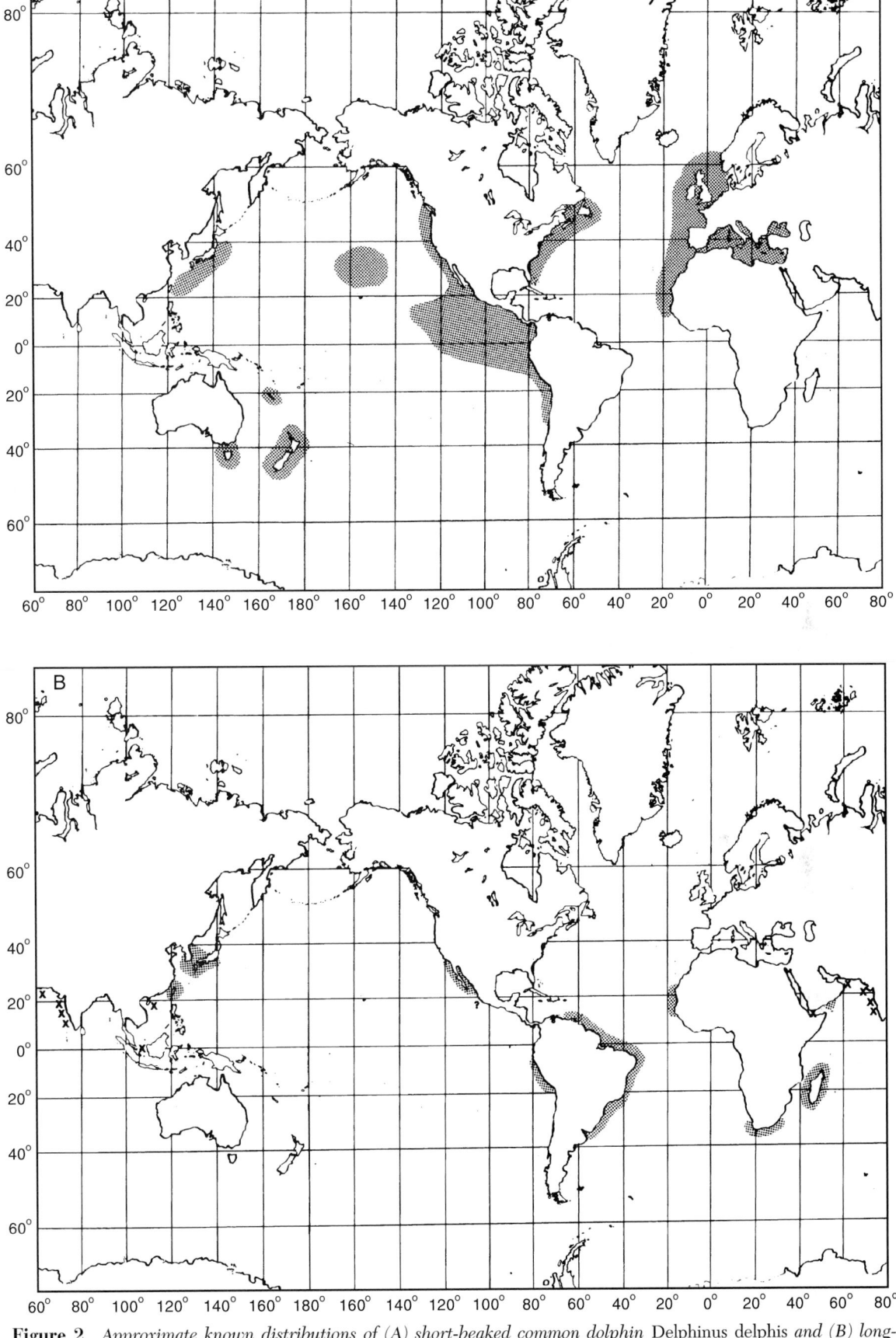

Figure 2 *Approximate known distributions of (A) short-beaked common dolphin* Delphinus delphis *and (B) long-beaked common dolphin* D. capensis. *Records of nominal species* D. tropicalis *represented by Xs. From Heyning and Perrin (1994).*

same general area on the same day. However, *D. capensis* seems to prefer shallower and warmer water and generally occurs closer to the coast. In the eastern North Pacific, substantial seasonal and interannual changes in abundance of *D. delphis* suggest migrations that vary with oceanographic conditions; the movements may be north–south and/or inshore–offshore (Forney and Barlow, 1998). In the tropical eastern Pacific, *D. delphis* occupies primarily upwelling-modified habitats with less tropical characteristics than surrounding water masses (Reilly and Fiedler, 1990); separate northern, central, and southern stocks associated with different upwelling areas are recognized in the management of incidental mortality in tuna fisheries (Perrin *et al.*, 1985).

As would be expected for a species occupying a wide range of habitats, common dolphins feed on a variety of prey, including small mesopelagic fishes and squids found in the deep scattering layer and epipelagic schooling species such as small scombroids, clupeoids, and market squids (Evans, 1994; Ohizumi *et al.*, 1998). Foraging dives to 200 m have been recorded. Diet varies with season as well as with region (Evans, 1994). FEEDING habits of the three species have not been compared.

III. Behavior and Life History

Schools of hundreds or thousands are thought to be composed of smaller subunits of about 20–30 perhaps closely related individuals (Evans, 1994). Differences between schools in cranial measurements and color pattern suggest that schools have temporal integrity at some level. There may be segregation in schools by age and sex. Association with schools of pilot whales (*Globicephala* spp.) and other dolphins (*Lagenorhynchus* spp.) have been observed, as has "bow riding" on mysticete whales, possibly the origin of bow riding on vessels. Common typical aerial behavior includes "pitch poling," in which the dolphin leaps high vertically and falls lengthwise back into the water to create a large splash. In some areas in the eastern tropical Pacific, schools of *D. delphis* "carry" yellowfin tuna and are chased and captured by tuna fishermen using purse seines.

Gestation in *D. delphis* in the Black Sea is 10–11 months (Perrin and Reilly, 1984), with length at birth at 80–90 cm. The calving interval varies from about 1 (Black Sea) to about 3 years (eastern Pacific). Weaning occurs at 5–6 months in the Black Sea, but possibly later in other areas. Estimates of age at sexual maturation vary with region from 3 years (Black Sea) to 7–12 years (eastern Pacific) for males and 2–4 and 6–7 years, respectively, for females. Part of the range of variation may be a density-dependent effect due to exploitation. Maximum estimated age is 22 years (Black Sea).

IV. Interactions with Humans

Short-beaked common dolphins die in large numbers in tuna purse seines in the eastern tropical Pacific (Evans, 1994) and, with long-beaked and Indian Ocean common dolphins, in gill net fisheries around the world (Perrin *et al.*, 1994); the impact of these kills on the populations is for the most part unknown.

Common dolphins have been kept successfully in captivity for up to 8 years in eastern European oceanaria and reproduced successfully in captive facilities in the United States, New Zealand, and Russia (Evans, 1994). *D. capensis* has hybridized with *T. truncatus* in captivity, producing viable offspring of unknown fertility (T. Goff, unpublished data).

See Also the Following Articles

Coloration ▪ Delphinids ▪ Hybridism ▪ Incidental Catches ▪ Tuna-Dolphin Issue

References

Evans, W. E. (1994). Common dolphin, white-bellied porpoise *Delphinus delphis* Linnaeus, 1758. *In* "Handbook of Marine Mammals" (S. H. Ridgway and R. Harrison, eds), Vol. 5, pp. 191–224.

Forney, K. A., and Barlow, J. (1998). Seasonal patterns in the abundance and distribution of California cetaceans, 1991–1992. *Mar. Mamm. Sci.* **14**, 460–489.

Heyning, J. E., and Perrin, W. F. (1994). Evidence for two species of common dolphins (genus *Delphinus*) from the eastern North Pacific. *Nat. His. Mus. Los Angeles County Contrib. Sci.* **442**, 35.

Jefferson, T. A., Leatherwood, S., and Webber, M. A. (1993). "FAO Identification Guide: Marine Mammals of the World," p. 320. FAO, Rome

LeDuc, R. G., Perrin, W. F., and Dizon, A. E. (1999). Phylogenetic relationships among the delphinid cetaceans based on full cytochrome *b* sequences. *Mar. Mamm. Sci.* **15**, 619–648.

Ohizumi, H., Yoshioka, M., Mori, K., and Miyazaki, N. (1998). Stomach contents of common dolphins (*Delphinus delphis*) in the pelagic western North Pacific. *Mar. Mamm. Sci.* **14**, 835–844.

Perrin, W. F., Donovan, G. P., and Barlow, J. (Eds). (1994). Gillnets and Cetaceans. *Rep. Int. Whal. Commn. Spec. Iss.* **15**, 629.

Perrin, W. F., and Reilly, S. B. (1984). Reproductive parameters of dolphins and small whales of the family Delphinidae. *Rep. Int. Whal. Commn. Spec. Iss.* **6**, 97–133.

Perrin, W. F., Scott, M. E., Walker, G. J., and Cass, V. L. (1985). Review of geographical stocks of tropical dolphins (*Stenella* spp. and *Delphinus delphis*) in the eastern Pacific. *NOAA Tech. Rep. NMFS* **28**, 28.

Reilly, S. B. (1990). Seasonal changes in distribution and habitat differences among dolphins in the eastern tropical Pacific. *Mar. Ecol. Prog. Ser.* **66**, 1–11.

Rice, D. W. (1998). Marine mammals of the world. *Soc. Mar. Mamm. Spec. Pub.* **4**, 231.

Smeenk, C., Addink, M. J., van den Berg, A. B., Bosman, C. A. W., and Cadée, G. C. (1996). Sightings of *Delphinus* cf. *tropicalis* van Bree, 1971 in the Red Sea. *Bonn. Zool. Beitr.* **46**, 389–398.

Communication

KATHLEEN M. DUDZINSKI
Dolphin Communication Project, Oxnard, California

JEANETTE A. THOMAS
Western Illinois University, Moline

ETIENNE DOUAZE
Greenleaf Walk, Singapore

Social behavior is organized into patterns of coordinated activities among individuals of a species. This coordination is achieved through the exchange of information about conspecifics, other species, and the environment. Communication is the glue that holds animal societies together. To understand communication in a given species, it is important to view the mode of the signal (*i.e.,* visual, audible, tactile, gustatory, or olfactory), medium in which the signal is transmitted (air and/or water), mechanisms of signal production (anatomical and/or physiological), and functions of signal (*e.g.,* aggression/submission, mate attraction, parental care, territorial defense).

This article is a brief overview of communication in marine mammals. Even with more than 40 years of focused studies on the social lives of marine mammals, relatively little is understood about the majority of species within marine mammal groups—cetaceans, pinnipeds, sirenians, mustelids (sea otters, *Enhydra lutris*) and ursids (polar bears, *Ursus maritimus*). Behavioral characteristics and social relationships are adapted to each species' unique ecology. Marine mammals are either amphibious or totally aquatic. Each life mode imposes different restraints on signal exchange and communication. A paucity of studies on communication exists for many marine mammals, especially polar bears, sea otters, dugongs (*Dugong dugon*), and manatees (*Trichechus* spp.). The majority of research on communication has been conducted on pinnipeds (such as Weddell seals, *Leptonychotes weddellii*; and California sea lions, *Zalophus californianus*) and cetaceans (particularly bottlenose dolphins, *Tursiops* spp.; killer whales, *Orcinus orca*; and humpback whales, *Megaptera novaeangliae*). Thus, our discussion highlights and compares species predominantly represented in the literature.

I. Definition

A clear definition of communication is needed to facilitate consistency among studies and to avoid ambiguities in methods. Bradbury and Vehrencamp (1998) provided this definition: "communication involves the provision of information (via a signal) by a sender to a receiver, and subsequent use of this information by the receiver in deciding how to respond." Determining how communication contributes to the maintenance and coordination of behavior among individuals of a group is pivotal to understanding social behavior.

The signal is the vehicle by which the sender and receiver exchange information or communicate. Both the sender and the receiver rely on the accurate interpretation of signals to meet common group challenges such as reproduction, predator defense, foraging, and parental care. Communication promotes group cohesiveness (Vauclair, 1996). Signals are mechanisms or "tools" specialized over time to be informative, salient to interactions among individuals, and adapted for optimum transmission in their environment(s). Active communication signals provide intentional information to a conspecific that benefits the sender. However, communication can also be passive and information transfer unintentional. For example, a killer whale is alerted to the presence of a sea lion jumping in the water by a splashing sound. This inadvertent noise by the sea lion cues the killer whale to the presence of a tasty morsel and members of a killer whale pod coordinate a hunt through various acoustic and visual signals. The acoustic cue does not benefit the sender at all and there was no intentional communication from the prey to the predator. Predators are adapted especially to detect and

exploit passive cues, whereas prey are adapted to minimize the sending of passive cues. In active communication, an efficient signal from the sender matches the most adaptive response by the receiver. For example, sound exchange among killer whales probably enhanced group foraging on the aforementioned sea lion. It is beneficial to both the sending and the receiving killer whales to communicate effectively.

Signals can be subtle, such as a unique color pattern or a slight change in body orientation, or signals can be more dramatic, as exemplified by a 193-dB trill made by a male Weddell seal patrolling its underwater territory or a lob-tailing dolphin warning an intruder. Detailed examination of signaling behavior in an individual or group provides the genesis for understanding information exchanged and interpreting the biological basis for the intent and meaning of a signal and the resulting response.

Signals are adapted for optimal transmission in an environment, to convey appropriate information for a given context, and often are accentuated by other behaviors. Recipients use diverse sensory abilities to decipher the meaning of a signal. A sender's signal can be modified based on location, spatial relationship with the receiver, past experience, age, sex, reproductive status, environmental factors, presence of other conspecifics, and other referents. Likewise, the interpretation of signals can depend on similar factors about the receiver. Therefore, the meaning a signal conveys has a contextual basis. For example, the same signal given by a sender in one context could result in a different meaning to the receiver in another context, i.e., a jaw pop from a dolphin might be used to warn an intruder in one situation or reprimand a youngster in another circumstance. The message is the information the sender encodes in the signal, whereas the meaning is the significance attached to the signal by the receiver and the potential effects on the receiver's subsequent actions (Bradbury and Vehrencamp, 1998).

Sensory channels or modes in mammals include chemical (*i.e.,* taste and olfaction), mechanical (*i.e.,* tactile and acoustic), photic (visual), and electromagnetic (Herman, 1980; Reynolds and Rommel, 1999). These modes are used only if both the sender and the receiver are capable of accessing the pathway, detecting the signal, and interpreting the message. While most terrestrial mammals evolved signals in each of these sensory channels (Hauser, 1997), marine mammals have not, primarily because of limitations of the aquatic environment. In marine mammals, communication is achieved primarily by acoustic, tactile, visual, and gustatory sensing (Reynolds and Rommel, 1999). Marine mammals, and fully aquatic groups in particular, place less reliance on olfaction. Messages often have cross-modal components that are partly conveyed in one mode and refined in another mode. For example, a threat vocalization often is accompanied by a visually distinct threatening body posture.

II. Communication in Air and Water

Transmission of signals in water varies depending on water temperature, salinity, clarity, depth, ice cover, ambient noise, local topography, vegetation, and surface and bottom reflections (Anderson, 1969; Herman, 1980). In general, sound in water travels 4.5 times faster and with less attenuation than in air, whereas the distance light penetrates is affected more by particulate matter suspended in the water column and depth. Thus, in water, sound provides a more affective method of communication over long distances, whereas visual, postural, or tactile signals are for short-range communication. In contrast, communication in air can include olfaction and long-distance visual displays. In air, sound attenuates more rapidly, is masked by wind and ambient noise, and is used only over shorter ranges for communication.

Because mysticetes and odontocetes are totally marine, signal exchange in air is limited to only a few aerial displays, such as breaches, lob tails, leaps, and spins. Olfaction is useless and the predominant signal modes are acoustic and tactile.

Because pinnipeds are amphibious, their sensory systems evolved as a compromise between the need to signal both in air and in water. While hauled out, many pinnipeds produce social sounds with an open mouth. These sounds often are concurrent with bared teeth or head-up postures that convey additional information. In contrast, social interactions under water occur with the mouth closed and often are associated with sounds, swimming postures, or a change in swim speeds or direction that accentuates the message. In contrast to water, where scent marks would diffuse quickly, land provides a "permanent" site for scent marking; thus olfaction likely plays a more important role in communication in pinniped colonies on land.

The communication environment of amphibious sea otters is most similar to their terrestrial mustelid cousins and pinnipeds. Sounds are most likely used in airborne communication, but there are no published records of underwater sounds in sea otters. The churning intertidal zone in which sea otters forage is not a good environment for acoustic signaling.

Although totally aquatic, sirenians, unlike cetaceans, do not dive deeply or echolocate. They do communicate at the water surface, using tactile and visual exchanges, but aerial displays, such as leaps, have not been observed and are likely not used. Manatees and dugongs use visual communication at short ranges, but rely heavily on audition at ranges beyond their vision. The dugong has a relatively elaborate sound repertoire that apparently is more complex than that of manatees.

Polar bears have similar sensory abilities to other ursids. Polar bears spend the majority of their time on land, only rarely entering the water to swim or forage. Although there are no published accounts of underwater sounds from polar bears, it is doubtful these predators have a large repertoire underwater. As in other marine mammals, olfaction and vision are of limited use in water for polar bears. Communication is primarily through airborne visual and olfactory signals. Communication among bears is limited to a short mating period, aggression between territorial males, and mother/cub interactions.

III. Context of Signal

A behavior characterizing a specific message for adults could have an entirely different meaning when exhibited by a juvenile. A good example is the "S" posture reported for many cetaceans (*e.g.*, Herman, 1980; Norris *et al.*, 1994). This posture has the head "up," anterior ventral surface (*i.e.*, chest) "down," peduncle "up," and flukes "down." This posture was observed in old and young Atlantic spotted (*Stenella frontalis*) and bottlenose (*Tursiops truncatus*) dolphins (Fig. 1). During play, juveniles perform a modified "S" with oblique angles of approach and much rubbing contact. During adult aggression, the posturing dolphin, usually a male, performs the "S" maneuver toward conspecifics accompanied by direct approaches, a bubble cloud, jaw clap, and often hits. Another context of the "S" posture is that of males during courtship indicating a willingness to copulate. It is possible that some behaviors (e.g., jaw claps, hits) are not specific to particular expressions of aggression (or play, etc.) unless modified by or coupled with a specific posture or approach.

PLAY often is characterized by mock fighting behaviors that are modified with subtle cues reminding conspecifics that the intent of ongoing activities is not truly aggressive. Play occurs in young during the formation of long-term social attachments and while young learn the meaning and proper use of the signals within their social structure. Because pinniped pups do not manipulate objects with their fore flippers, are usually kept separate from other pups by their mothers, and develop rapidly (in comparison to cetaceans), evidence of play is limited. Occasionally, Weddell seal pups chew on or manipulate a chunk of ice with their fore flippers in a playful manner. Under water, these pups spend a great deal of time learning to adjust their buoyancy. Initially, their fat bodies float at the surface, the pups struggle to go deeper by dramatic undulations of their hind flippers, and as soon as flipper motion stops the pup bobs back to the surface like a floating cork. The pups repeat the diving/pop-up sequence many times before they perfect the seemingly effortless swimming motion of older seals.

A signal can have different meanings to a male and female of the same species. For example, the repeated thumping sound, head-up posture, and display of the large proboscis in a male northern elephant seal (*Mirounga angustirostris*) warn an intruding male to stay away from a harem master's territory and attract females as potential mates. At the same time, this display warns nearby pups to move or be trampled.

Figure 1 *Two Atlantic spotted dolphins engaged in play behavior. The individual in the lower position is in the classic "S" posture. Photograph by K. M. Dudzinski.*

IV. Chemical Communication

Chemoreception is common among terrestrial mammals, but little is understood about how marine mammals sense chemical signals in the water (Reynolds and Rommel, 1999). The olfactory sense and anatomy are ill-suited for communication in water, and this sensation declines with greater adaptation to an aquatic lifestyle (Reynolds and Rommel, 1999). Jansen and Jansen (in Anderson, 1969) found that adult odontocetes lack olfactory nerves, bulbs, and tracts; the same are reduced greatly in adult mysticetes. The possibility that scents are pheromonal in nature and function has not been examined in marine mammals, although anecdotal accounts exist. Yablokov (1961) suggested that belugas (*Delphinapterus leucas*) release a "pheromone" when alarmed. Belugas react to blood in the water by either quickly escaping or becoming unusually excited. Trails of both feces and urine deposited by schools of dolphins could contain sexual pheromones. At times, spinner dolphins (*Stenella longirostris*) appear to swim deliberately through dispersing excrements deposited by schoolmates (Norris *et al.*, 1994).

Most studied marine mammals have the ability to taste, although with somewhat different receptive qualities than terrestrial mammals (Reynolds and Rommel, 1999). Taste buds have been documented behaviorally and physiologically for both cetaceans and sirenians (*e.g.*, Herman, 1980; Schusterman *et al.*, 1986; Reynolds and Rommel, 1999). Friedl *et al.* (1990) experimentally tested the taste sensitivities of bottlenose dolphins (*T. truncatus*) and found they can discriminate sour, sweet, bitter, and salty solutions. They were least sensitive to different salt concentrations, which seems adaptive given they live in a marine environment.

The olfactory anatomy of pinnipeds is variably reduced, more for phocids or odobenids than for otariids (Reynolds and Rommel, 1999). Pinnipeds employ scents to exchange or gain information about colony members. For example, male northern fur seals (*Callorhinus ursinus*) sniff the hindquarters of females to assess their estrous state (Reynolds and Rommel, 1999). The largest glands of pinnipeds are around the vibrissae and could play a role in mother/pup recognition. Mothers and pups maintain a great deal of nose-to-nose contact and use odor cues for recognition in air. Little work has been conducted on pinniped taste sensations. Friedl *et al.* (1990) examined taste abilities in California sea lions (*Zalophus californianus*), and Kuznetsov (1990) demonstrated gustatory abilities in Steller sea lions (*Eumetopias jubatus*). Both studies found similar taste abilities with some sensitivity to acidic, basic, and salty solutions, but not to sweet.

Sirenians have a rudimentary olfactory system (Reynolds and Rommel, 1999) and likely rely, to a limited degree, on chemicals for signal exchange among conspecifics. However, because aquatic plants are known to have different smells, manatees and dugongs could use this sense in foraging. No information is available on taste abilities in sirenians.

Unlike other mustelids, known for their musky smell, the sea otter has no scent glands. This is likely a result of the aquatic environment in which scent marking would have limited usefulness. Kenyon (1975) reported that sea otters in water commonly surface and sniff the air, and male sea otters smell the genital area of an estrous female during precopulatory behavior. It is assumed that sea otters and polar bears have taste abilities similar to their terrestrial counterparts, although the exact extent of this sense is unknown for these species.

Polar bears have a keen sense of smell, especially useful in foraging. While little is known about how olfaction is used for communication among polar bears, patterns are likely similar to those observed in other ursids and used by males to find potential mates (Ovsyanikov, 1996; Stirling, 1999). No studies are available on taste abilities of polar bears.

V. Visual Communication

Behavioral displays are well documented for many marine mammals with visual detection and acuity levels being good both above and under water for all species studied (Reynolds and Rommel, 1999). Under water, VISION is limited by light levels, the concentration and type of organic matter, and depth (for a thorough discussion of light in the ocean and visual adaptations by marine mammal species for visual detection and acuity above and below the water, see Reynolds and Rommel, 1999). Visual cues can be simple displays, such as body postures or coloration patterns, or they can be elaborate sequences of behaviors that indicate a context, species, age, sex, or reproductive condition. Several pinnipeds incorporate body coloration or postures into visual displays. For example, a territorial Weddell seal patrols the water underneath a crack in the fast ice, using loud trill sounds accompanied by an S-shaped posture that thrusts the chest forward and the hind flippers downward. Movements and postures often are highlighted in species with conspicuous color patterns. In clear water, visual signals provide cetaceans and other marine mammals a close-range alternative to acoustic signaling; however, displays could inadvertently alert predators or prey.

SEXUAL DIMORPHISM is a distinct visual cue and prevalent in many marine mammals. Females tend to be larger in mysticetes, whereas most odontocete males are larger. In mysticetes, size differences are likely due to the half-year fast that females endure each breeding/calving season. In odontocetes, the size discrepancy between genders is likely related to social signals: either direct competition among conspecifics or as an indication of fitness to potential mates. In many pinnipeds and the polar bear, males are two to three times larger than the females. Sea otter males are slightly larger than females, while there is no indication that adult male and female manatees differ in size. Secondary sexual characters often represent a passive visual display that could function to serve in mate selection (for an overview, see Thomas and Kastelein, 1990). Examples include the postanal keel in male spinner dolphins, tusk in male narwhals (*Monodon monoceros*), large dorsal fin in male killer whales, a bulbous, pronounced nose and larger chest in male elephant seals, and white fur heads in male sea otters (Fig. 2).

The variety of patterns and shades of black and white colors on cetacean bodies have been discussed elsewhere (see Thomas and Kastelein, 1990); however, use of COLORATION differences as communication signals is rarely addressed. Color

patterns can signal intention, species, individual identification, age, and reproductive status. Recurrent types of body marks and color patterns in marine mammals include spots, streaks, longitudinal stripes, saddle patches, capes, blazes, chevrons, and eye patches. Generally, color patterns in cetaceans take three forms: uniform, countershaded, and disruptive (Fig. 3). Uniform and countershaded color patterns likely result from predator pressures to be inconspicuous, with little communicative value. At close range, disruptive markings likely facilitate intraspecific signaling and thus schooling and other coordinated behavior. Dusky dolphins (*Lagenorhynchus obscurus*) likely use their longitudinal striping and white flank patches to corral fish during coordinated feeding bouts, but may also employ these same markings to communicate with conspecifics about the ongoing activity (in Thomas and Kastelein, 1990). Vivid eye patches in killer whales might intimidate prey. Atlantic spotted dolphins possess a similar increase in spot pigmentation with age. Belugas exhibit different colors between juveniles (gray) and adults (white).

In pinnipeds, otarids and walrus have no color markings used in communication. Walrus do change color from gray to brownish red as they warm up while hauled out. In contrast, phocids have a variety of color patterns, including stripes, streaks, spots, rings, and bands (Fig. 4). The ribbon seal (*Histriophoca fasciata*) and the harp seal (*Pagophilus groenlandicus*) have bold black and white or piebald markings that are distinguishable from a distance. Color patterns can be used in camouflage, as well as in signaling. Harp seals pups possess a pure white coat for blending into the arctic ice, whereas juveniles show black spots. The number and prominence of spots increase as harp seals age. Although observational and experimental evidence is lacking to support a communicative value for developmental changes in pigmentation or patterning in marine mammals, the possibility that color pattern differences coupled with age and behavioral differences might convey information to conspecifics cannot be discounted.

To enhance the communicative function, color patterns and morphology often are modified by postures or behaviors. Postures signal intent and demeanor of a sender and provide insight to the meaning concluded by a recipient. Terrestrial social mammals communicate much information through subtle and overt kinesic and nonlinguistic expressions, including changes in facial expressions, irregularities in respiration, and overall carriage of the body. Evolution has streamlined the cetacean body for swimming, but limited communication via facial or overt kinesic expressions. Still, subtle shifts in posture or slight body movements could be communicative between individuals or groups. For example, a slight shift in placement of a pectoral fin might signify a change in swimming direction or other shift in position among dolphins in a school. Similarly, aggression and threats in dolphins appear to be expressed through a direct vertical or horizontal approach that often is coupled with jaw claps, head shakes, body hits or slams, and emission of a bubble cloud. Submissive behavior is associated with an oblique angle of approach, mutual rubbing, or tactile contact. Often the ventral or lateral sides of the body are displayed to conspecifics.

Similarly, some pinnipeds enhance morphological characters with behaviors. The inflated red nasal sac of the hooded seal (*Crystophora cristata*) and the proboscis of the male elephant seal are used to defend territories from competing males and to attract females. Thus, the nasal sac has evolved as a signal demonstrating male fitness and intent.

VI. Tactual Communication

Visual displays are useful for close-range communication among marine mammals and, because of close proximity, visual displays readily become tactile signals. Extensive touching and rubbing occur in both captive and free-ranging animals during play, sexual, maternal, and social contexts using the nose or rostrum, flippers, pectoral fins, dorsal fin, flukes, abdomen, and the entire body. Tactile contacts often are observed during aggressive interactions, but are characterized by more overt actions, such as biting, raking, ramming, and butting. Obviously, tactile contact is limited to short distances for signal exchange. Tactile signals can be modified to increase the information content—who, where, and how animals touch, as well as the intensity of a touch, factor into the signal content.

A. Tactile Anatomy

Adult odontocetes do not have vibrissae. Some newborn odontocetes have hairs on the rostrum that could have a sensory function, but information is not available. Similarly, some adult mysticetes, bowhead (*Balaena mysticetus*) and right whales (*Eubalaena* spp.), for example, have a few hairs on their rostrum that could serve a tactile function. Cetaceans have well-developed skin sensitivity (Reynolds and Rommel, 1999). Ridgway used galvonic responses to test tactile sensitivity in different body regions of bottlenose dolphins and found the most sensitive areas were the insertion of the pectoral fin, gape of the mouth, tip of the rostrum, and abdomen.

Pinnipeds, sirenians, sea otters, and polar bears possess vibrissae, which function as a mechanical sensor for touch or pressure (Fig. 5). Pinnipeds and sea otters have mystachial and supercilliary vibrissae; the number varies by species. Vibrissae are heavily innervated and pinnipeds are capable of waving them forward to inspect an object. Pinniped vibrissae detect slight pressure changes similar in magnitude to the lateral line system of fish.

The manatee is unique among marine mammals in its ability to use its vibrissae or bristles in a prehensile manner. Manatees have bilobed, prehensile lips that are highly vascularized to support the numerous mystachial bristles. These bristles are likely

Figure 2 *Examples of passive visual displays in marine mammals. (A) The inflatable red nose and nasal sac of hooded seal adult males, (B) the sexual dimorphism in sea lions and fur seals, and (C) the sexual dimorphism in killer whale dorsal fins. Sketches by Jennifer Schmid-Webster.*

Figure 3 *Examples of different coloration patterns observed among dolphin species. From top to bottom, the color patterns include uniform, disruptive, and countershaded. Sketches by Jennifer Schmid-Webster.*

modified vibrissae and used in tactile exploration of the environment, vegetation, or conspecifics (Reynolds and Rommel, 1999). Manatees rub on each other, as well as inanimate objects, and females with calves maintain much body contact. The dugong has the most developed sensory hairs of any marine mammal; they are present over the entire dugong body surface, being most dense on the muzzle (Reynolds and Rommel, 1999).

Sea otters, like other carnivores, have mystachial (eight rows, 62 each side), superciliary (four each side), and nasal (three each side) vibrissae. Kenyon (1975) reported that mystachial vibrissae play the most important role in tactile sensing

while hauled out or during underwater foraging. Mystachial vibrissae can be voluntarily moved forward to investigate objects and could be used for tactile signal exchange. Sea otters have unique front paws: heavily padded, coarse-textured, webbed digits with retractable claws. The exaggerated structure of these paws might aid in tactile sensing of prey, in grooming, or in conspecific communication (Fig. 5).

As with other ursids, polar bear mothers probably groom their cubs and remain in close physical contact, especially in the maternal den. Polar bears have large paws with bare pads that function as snowshoes to disperse their weight evenly to

Figure 4 *Examples of color patterns in three seals: (top) harbor, (middle) harp, and (bottom) ribbon. Sketches by Jennifer Schmid-Webster.*

avoid breaking through thin ice. These large feet help with propulsion, and footpads are covered with small, soft papillae that increase friction between the foot and ice. Bare body regions are most likely to have a communication function, i.e., the muzzle, nose, ears, footpads, and insides of the thighs. Still, whether these anatomical traits function in tactile signal exchange is unknown.

B. Responsiveness to Touch

An inclination for tactile responsiveness has been noted in wild and captive studies of all cetaceans. Among mysticetes, the "friendly" gray whales (*Eschrichtius robustus*) of San Igna-

cio Lagoon are noted for approaching and rubbing under small boats and tolerance for petting by tourists. In the wild, both Atlantic spotted and bottlenose dolphins rub body parts into the sand or along rocky edges. Belugas rub molting skin off their bodies at a few distinct pebbly beaches and under the ice surface in the Arctic. All odontocetes in captivity seek and are receptive to gentle body contact. Mild tactile stimulation (*i.e.,* rubbing of gums, flippers, or dorsal fin) serves as an effective re-enforcer in training situations using bottlenose dolphins or the Amazon River dolphin (*Inia geoffrensis*). Trainers suggest that tactile stimulation is reinforcing and perhaps rubbing by another dolphin might also be rewarding.

Figure 5 *Composite drawing depicting specific characters useful to tactition in pinnipeds, fissipeds, and sirenians. Vibrissae differ in the extent of beading per unit between phocids and otariids. Sirenian bilobed lips, vibrissae, and a portion of their flipper are shown (center). The walrus tusk and vibrissae are presented in the lower right. The sea otter paw vibrissae and fur are shown at the lower left. Sketches by Jennifer Schmid-Webster.*

Pinnipeds vary in the degree of gregarious behavior and thus tolerance for tactile stimulation by conspecifics. Leopard seals (*Hydrurga leptonyx*) are solitary predators and rarely seen in close proximity. In contrast, Weddell seals congregate in breeding colonies, but each mother/pup pair maintains an individual space. The more polygynous pinnipeds, such as walrus (*Odobenus rosmarus*) and California sea lions, often crowd onto beaches, piling on top of each other, with little regard for personal space. Regardless of adult spacing, mothers and pups maintain close tactile communication. Young pinniped pups often crawl over their mothers and sleep touching their mother. There is, however, no maternal grooming of the young in pinnipeds.

In Crystal River, Florida, some manatees seek physical contact with divers, whereas others avoid divers. Florida manatees sometimes "body surf" on currents generated below dams when floodgates are partially opened. This surfing can last for up to an hour with manatees repeatedly riding the currents in parallel formation. Often, nuzzling and vocalizations accompany manatee body surfing (Reynolds and Odell, 1991). When not eating (nearly 8 hr per day), manatees curiously investigate objects, socialize by mouthing and rubbing against each other, and play together (Reynolds and Odell, 1991). Thus, touch is likely an important information component in manatee life.

Mustelids [sea and marine otters (*Lutra felina*)] possess thick layers of fur for warmth and protection and thus groom-ing is part of their social structure, as with many social terrestrial mammals. Unlike other marine mammals, sea otters groom their pups by licking their fur. They are the only marine mammal to have the ability to hold and manipulate their young for grooming. Mother sea otters spend a great deal of time grooming their pup and remove feces by licking the fur in the pup's anal region. Sea otter grooming behavior is probably at least partially hygienic in function; however, in other mammal species (e.g., primates, canids), grooming behavior signals affection, appeasement, or reconciliation.

More data are required to better understand how polar bears use touch in signal exchange. As in other ursids, mother polar bears likely have close tactile contact with their young for nursing and grooming in the den. Adult males are seen in intense fights grasping each other with "bear holds," nose-to-nose open mouth threats, growls, and biting.

C. Sexual Contacts

Head or body rubbing, pectoral fin rubs, and other touching contribute to simple affiliation, greetings, or represent courtship activities. Head rubbing was reported in a number of mysticete and odontocete species including, but not limited to, southern right whales (*Eubalaena australis*), gray whales, humpback whales, killer whales, bottlenose dolphins, Atlantic spotted dolphins, and pilot whales (*Globicephala* spp.). Lengthy and frequent contact with pectoral fins, flukes, dorsal fin, or body during social, play, or precopulatory behavior is common among cetaceans (Fig. 6; Pryor, 1990).

In mysticetes, such as the right whales, sexual activity occurs in groups of one or more females and several males. As many as three males surround a single female and two of them push her into position, enabling the third male to align himself for mating. Researchers believe this is an example of cooperation among males, but this hypothesis has not been substantiated. This behavior certainly requires good communication among the whales.

Copulation often is preceded by gentle mouthing at the genital area for many delphinids. In mature dolphins, precopulatory play becomes progressively more violent, with participants DIVING forcefully after each other and at other times ramming melons with such force that loud reverberations result. This behavior apparently signals sexual stimulation and receptivity because it almost always ends in copulation. Most dolphins also seek self-stimulation: captive dolphins expose various body parts to tactile stimulation, including opening the mouth under a water hose or rubbing against a brush, wall, or the tank floor for long periods of time. Both male and female dolphins masturbate by rubbing their genitals against objects in a tank, even when available mates are present (Herman, 1980). In a study that played back sounds from an oil-drilling platform, the dominant male beluga responded by rubbing his genitals against the vibrating transducer. In harem-forming seals and sea lions, several strategies are employed during mating, with varying degrees of signal exchange. Subadult males often attempt *not* to communicate to harem masters while sneaking to mate with females. Females, however, may choose not to mate with the potentially less desirable subadult male by signaling to the harem master vocally, behaviorally, or posturally.

Some pinnipeds mate on land (e.g., elephant seal and California sea lion) whereas others mate in the water (e.g., Weddell and leopard seals, walrus). Mating in the two media requires different signal exchange. In all pinniped species, mating is accomplished by the male mounting the female from the rear and clasping her abdomen with his fore flippers.

Once a sea otter male has found an estrous female, the two actively roll and splash about together at the surface. They nuzzle and fondle each other with forepaws (Kenyon, 1975). This behavior lasts for up to 1 hr. During this time, they make their way to a suitable haul-out site, which acts as the center of their courtship activity, but mating occurs in water. Sea otter females often exhibit bloody noses that attest to this species' rather aggressive mating habits. Mating can last up to 14 min with the male mounting her back, biting and grabbing her nose, and grasping her abdomen. It is possible that some of these actions could function as signals to induce estrous in the female.

D. Affiliative Contacts

Affiliative contact is assumed to strengthen bonds between individuals with high association indices but also could indicate to competitors that individuals exchanging contact have a strong bond, at least for the time being. Affiliative contact is observed most commonly between mother/calf dyads for all marine mammal species.

Nose touching or nuzzling often is observed between mothers and offspring in sea otters, pinnipeds, and sirenians. However, nuzzling is not only exchanged between mothers and offspring, but between older individuals during social exchanges.

In pinnipeds, the degree of affiliative contact varies with the social system. For example, the male monogamous crabeater seal (*Lobodon carcinophaga*) maintains a season-long bond with a female, remaining paired on their own ice floe until mating. Gregarious walrus and elephant seals pile on top of each other while hauled out. Weddell seal mothers and pups maintain discrete pairs while hauled out, often chasing away intruders. Hawaiian monk seals haul out similarly in well-spaced mother/pup pairs, but the mothers tolerate an occasional wandering pup and sometimes even nurse it.

Sea otter mothers maintain close tactile contact with their pups. Mothers float on their backs and support their pup on their abdomen for nursing or for the pup to rest. If disturbed, the mother grasps the pup with her front paws and swims away or dives with the pup. Wild sea otters form resting, floating groups called rafts. Sea otters rest in close proximity, often bumping into each other as they sleep. In California, these rafts are formed in kelp beds, with the sea otters rolling in the kelp to serve as an anchor. In Alaska, there is no kelp so sea otters group together in calm water areas. The rafting behavior likely provides "safety in numbers" for resting sea otters who could become prey of killer whales. Rafting behavior could also help animals remain close to shore in rough seas. Surely there is communication among raft members to coordinate the group resting behavior.

Sirenians maintain close tactile contact between mother and calf. Manatee calves swim parallel and directly behind the mother's flipper. It is possible that mother/calf communication is most effective in this formation or that there is some hydrodynamic advantage for the calf. Neonate dugongs swim in near contact to their mothers and occasionally reach out with a flipper to touch mother, possibly for reassurance of her presence or to maintain their bond (Reynolds and Rommel, 1999).

Figure 6 *Two subadult Atlantic spotted dolphins engaged in precopulatory petting exchanges. The female is above the male and is rubbing the male's melon with her left pectoral fin. Photograph by K. M. Dudzinski.*

Because mammary glands are pectoral in sirenians, mothers embrace calves close to them. Manatees grasp conspecifics in a hug-like embrace (at times, captive manatees "hug" their trainers). So far as is known, all sirenians are polyestrous. A single female manatee has been seen pursued by males on widely separate dates. In manatees, the mating herd can remain together for up to 1 month, with males vying for the closest position to the female. Dugong mating has several phases. The "follow phase" with groups of animals moving into a tight cluster around an estrous female; the "fighting phase" consisting of violent activity, including splashing, tail thrashing, rolls, and body lunges; and the "mounting phase" in which one male mounts the female from beneath with several others clinging to them. It is not clear if more than one male mates at that time. Dugong mating is much more violent than in manatees.

Polar bears are solitary except for a single female and her young. During the breeding season, females are polyandrous with males competing for access to a female. Touch in polar bears is likely important during interactions, but little is reported for more than play-fights or aggressive contacts. A nose-to-nose greeting is sometimes offered as a begging signal by one bear that approaches another with a food source.

E. Aggressive Contact

Exchanges of aggressive behavior (*i.e.*, fights) occur among cetaceans over objects, space, proximity to conspecifics, food, or mates. Mysticetes show obvious wounds from intraspecific aggression. Behaviors such as breaching and lob tailing often are used during aggressive exchanges or to indicate irritation or anger; however, in other situations, these actions could possess a noncommunicative function (e.g., parasite removal). Similarly, many odontocete species show intraspecific rake marks from the teeth of their peers. Male long-beaked common dolphins (*Delphinus capensis*) are extensively scarred on trunks and fins during the mating season and males have more teeth missing than females. Risso's dolphins (*Grampus griseus*) represent the extreme case with their adult coloration being light gray or white from the multitude of rake marks inflicted by competitive cohorts. Fighting dolphins slash and bite with their teeth, slash and ram with their jaws, and strike with their flukes and rostra. Encounters sometimes result in injuries and permanent scars.

Large males are considered the most aggressive and direct behaviors at peers and immature males most frequently, although females are not immune to aggressive attacks from males. In the wild, vertically organized swimming arrangements are seen among groups of pantropical spotted dolphins (*Stenella attenuata*), Pacific white-sided dolphins (*Lagenorhynchus obliquidens*), and bottlenose dolphins. This regime could be a dominance arrangement expressed as highest priority given to top dolphins who have the quickest access to the surface. For sperm whales (*Physeter macrocephalus*), male–male competitions, in apparent dominance contests, include extensive tooth-raking that sometimes escalates into violent head ramming and locking of the jaws by two animals while twisting and biting.

Mother cetaceans punish their young with less violent, yet still effective methods. Calves or juveniles can be held against the sea or tank floor, bitten, or even held out of the water by mothers or other adults—behavior that is likely disciplinary given the participants and context.

Male Stellar sea lions (*Eumetopias jubatus*), elephant seals, walrus, and Weddell seals all bear scars and often sport bloody chests that result from aggressive encounters with other males during the height of the mating season. Male elephant seals compete and exchange aggressive bellows, blows, bites, and lunges when competing for access to females. Unfortunately, Hawaiian monk seal (*Monachus schauinslandi*) females sometimes die from aggressive mating attempts by a male, even a group of males. Presently, the adult sex ratio is skewed toward more males than females in several Hawaiian atolls.

In contrast, manatees are slow moving and seemingly docile, not known for interanimal aggression.

A sea otter in a dangerous situation reverts to typical mustelid aggressive behavior, including biting and aggressive sounds.

Among polar bears, hissing and snorting signify aggression, and an attacking bear will charge forward with head down, ears back, and teeth bared. Aggressive contacts can include growling, hitting, and biting. A 3:1 adult male to female breeding ratio produces intense competition for mating rights (Stirling, 1999). Broken canine teeth and other scars often result from these battles.

F. Touching without Contact

1. Echolocation   Because dolphin skin is exceptionally sensitive to touch, some researchers suggest that sound might be used as an acoustic probe or means of physical contact among individuals. Human divers report the sensation of being echolocated on by dolphin as "feeling like a zipper was opened along their chest or as a dotted line going along their chest." Atlantic spotted dolphins in the Bahamas and spinner dolphins in Hawaii have been observed to produce a "genital buzz"; this behavior includes one dolphin inspecting the genital region of a second while producing a pulsed buzzing sound (Fig. 7). No physical contact between dolphins is observed concurrent with genital buzzing—"touch" occurs via sound.

Figure 7 *A genital buzz from a juvenile female Atlantic spotted dolphin to a second juvenile female of the same species. The context of this action was play. Photograph by K. M. Dudzinski.*

2. Bow Riding Bow riding is a behavior in members of the cetacean family Delphinidae. Groups of dolphins often approach vessels of various sizes to surf or swim inside the wake produced by the bow as it moves through the water. These bow-riding bouts can last for a few minutes or hours. Good communication among individuals is required for such intricate swimming. Dolphins are seen vying for the position closest to the bow. Perhaps shifts in dolphin movements by the group are sensed by pressure changes in the spaces among individuals. It is possible that dolphins detect subtle movements in the water column that signal a change in direction to the group. This ability also would be useful to dolphins traveling in turbid environments where subtle posture shifts could not be detected visually.

VII. Auditory Communication

Marine mammals use both vocal and nonvocal sounds for acoustic communication because sound travels faster in water than in air and acoustics are a more reliable means of information exchange among individuals separated by meters to kilometers.

A. Nonvocal Communication

1. Cetacea Nonvocal communication can include noise from various body parts projected into the air, striking the water surface as well as the percussive sounds of jaw claps, teeth nashing, or bubble emissions. Breaches, leaps, tail slapping, and chin slapping produce sounds under water that likely carry a communicative message. Humpback, right, and gray whales and Pacific white-sided, Hawaiian spinner, and bottlenose dolphins are known to leap vigorously into the air, called breaching. A breach produces airborne and underwater sounds upon reentry that carry for several kilometers. Breaching could be a spacing mechanism or help whales remain in acoustic contact. Breaching often indicates general excitement or arousal deriving from any of several causes, including sexual stimulation, location of food, or a response to injury or irritation. Calves and their mothers also breach on occasion and sometimes in unison. Clearly, there can be many immediate causes of breaching (*e.g.*, parasite removal), and further study is needed to clarify and understand the multiple contexts in which breaching occurs. Dusky dolphins are well known for the three leap types they produce in association with three stages of cooperative feeding: head-first reentry leaps, noisy leaps, and social, acrobatic leaps. The latter two create sounds that function to signal peers or, as for the noisy leaps, could act as a sound "barrier" to disorient prey and keep them tightly schooled. Upon water reentry after a spin, a breach, a back slap, or a head or tail slap, spinner dolphins generate omnidirectional noise that propagates over short to intermediate distances. Spinner dolphins' aerial behavior seems designed to produce noise, as many leaps are common at night, when visual contact is limited, or in daytime occur in fully alert, but dispersed schools. Visual signals during leaps likely convey position information to schoolmates and could facilitate aerial inspection for feeding sites or for detecting environmental features. In many cetaceans, ventral chevrons or white underflukes are displayed above water during leaps or breaches. These color patterns vary among individuals, and leaps certainly draw attention to the whale's identity.

Most observers agree that tail slaps convey threat or accompany frustration in addition to establishing contact. Tail slapping, which produces extensive, low-frequency underwater and aerial sound, often occurs dozens of times in succession. It is likely that tail slaps among mysticetes and odontocetes have an agonistic component. Pectoral fin slapping is observed predominantly in humpback whales, although other baleen whales exhibit this behavior. The exact communicative nature of pectoral fin slapping is not clear, although again it may signal irritation, maintain individual spacing, or invite play or socializing.

The percussive sound of a jaw clap with the accompanying posture of an approaching animal is often considered a threat or warning signal. Altogether, the social functions of nonvocal auditory signals seem mainly limited to affiliation, recruitment, and expressions of excitement, annoyance, or aggression.

2. Pinnipedia Pinnipeds do not use nonvocal communication as much as cetaceans, *i.e.*, fewer tail or flipper slaps or breaches. The most common example of nonvocal communication in pinnipeds is teeth chattering, which provides both an acoustic and a visual aggressive sign. Weddell seals, for example, teeth chatter both under water and on land to seals intruding their space. Weddell, ringed, (*Pusa hispida*), and Baikal (*P. sibirica*) seals repeatedly slap their sides with fore flippers while hauled out. This sound seems to warn intruders and is quite loud, especially from a wet seal. Many pinnipeds also produce a loud snort or hiss as they exhale, especially after a long dive. This exhalation could have communication significance because it is more forceful under some situations.

3. Sirenia Manatees are known to slap the water surface with their tail as a form of communication. Manatees sometimes play "follow-the-leader" (Reynolds and Odell, 1991), in which animals swim single file and perform behaviors synchronized with the lead animal. This behavior requires close communication, perhaps using many sensory channels. Still, more observations are required to better understand nonvocal signal exchange among sirenians.

4. Mustelidae Little is known about nonvocal auditory communication in sea otters.

5. Ursidae Little is known about nonvocal auditory communication in polar bears.

B. Sound Communication

1. Mysticete Cetaceans Generally, baleen whale sounds are very different from those of odontocetes, with a wide range of types and quantity of phonation across mysticete species. Social functions proposed for mysticete sounds include long-range contacts, assembly calls, sexual advertisement, greeting, spacing, threat, and individual identification; however, only rarely has a specific sound been associated with a given behavioral event. It is probable that sounds produced by mysticetes serve to synchronize biological or behavioral activities in listeners that promote subsequent feeding or breeding. More data are required to describe sound communication more completely in baleen whales. Known and examined baleen whale sounds seem to fall into three basic categories: low-frequency moans, short thumps or knocks, and chirps and whistles.

A. MYSTICETE LOW-FREQUENCY MOANS   Low-frequency moans range from 1 to 30 sec, with fundamental frequencies between 20 and 200 Hz. These sounds can be either pure tones, as in the second-long 20-Hz sounds of fin whales (*Balaenoptera physalus*), or more complex tones with a strong harmonic structure. Theoretically, these low-frequency, long-wavelength sounds are ideal for long-range communication. A 20-Hz moan from a fin whale has a wavelength of almost 75 m, which means that it passes unimpeded over most obstacles, only bouncing off something large, like a seamount with a diameter greater than 75 m. These sounds could travel hundreds, even thousands, of kilometers to reach conspecifics for signal exchange. Payne predicted that theoretically the low-frequency, high-amplitude signals of mysticetes could travel from pole to pole if it were not for interfering oceanic bottom topography.

Low-frequency sounds (20 Hz) of blue whales (*Balaenoptera musculus*) are recorded across ocean basins at distances of several hundred kilometers. Blue whales are the largest creatures to inhabit the earth; they traverse large expanses in relatively short time. It is no wonder then that the social structure of these animals reflects a scale that we are only beginning to comprehend.

B. MYSTICETE SHORT THUMPS OR KNOCKS   Short thumps or knocks are under 200 Hz, but <1 sec long, and are currently known to be produced by right, bowhead, gray, fin, and common minke whales (*Balaenoptera acutorostrata*). In the 1970s,

Clark recorded and studied southern right whale sounds in relation to behavior and found that their sounds were not random but were related to social context and activity. Blow and body sounds (e.g., flipper slaps) were documented as well as vocal emissions. Resting whales were least soniferous, whereas mildly social groups produced the most varied suite of sounds, including high, hybrid, and pulsive calls, body and flipper slaps, and forceful blows. Clark's work showed that "up calls" functioned as a request for contact between whales: lone swimming whales often produced up calls that were returned by other whales in the vicinity prior to joining (Clark, 1983, Figure 8).

C. MYSTICETE CHIRPS AND WHISTLES   Mysticete chirps and whistles are always >1 kHz, but change frequency rapidly and are less than 0.10 sec in duration. These pure tones involve harmonics and seem to be produced by most baleen whales.

Cynthia D'Vincent, Fred Sharpe, and colleagues documented underwater sounds from humpbacks on the Alaskan feeding grounds. During the summer feeding season, humpbacks often forage in a coordinated manner. Several individuals emit bubbles around fish schools and surface in a ring, with each whale always in the same position. Concurrent with these feeding bouts are several sounds that are individually distinct (Fig. 9). Any coordinated, even cooperative, behavior requires communication among individuals. If specific roles are played within a certain activity, then individuals need a method of recognizing each other. Humpback feeding sounds exhibit characters distinct to each individual and thus probably help coordinate their respective positions or roles at each stage of feeding.

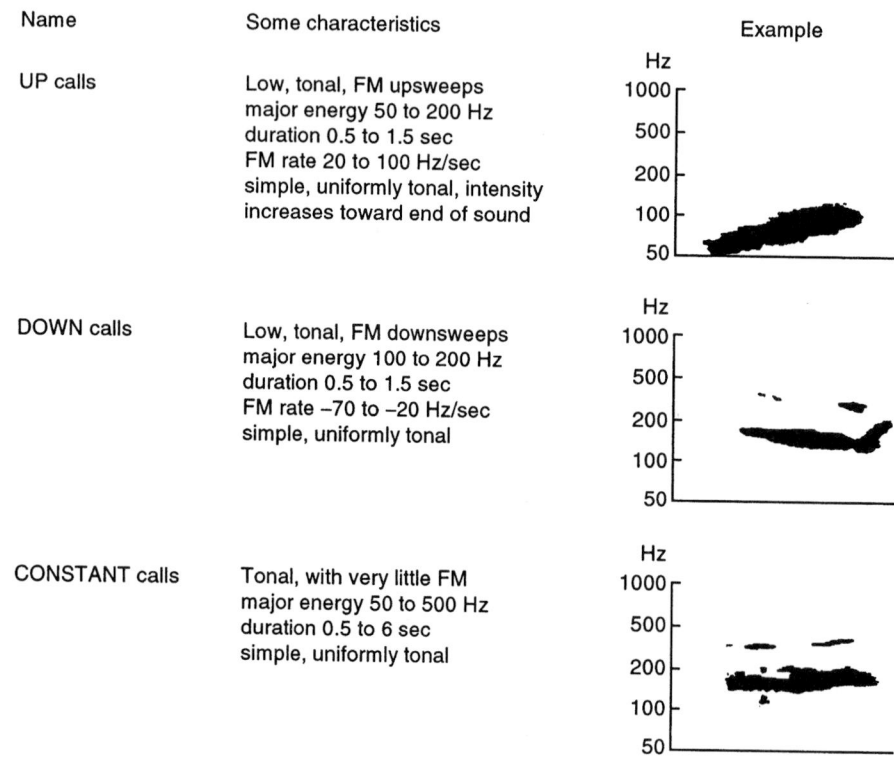

Figure 8   *Example of southern right whale "up calls" and two other calls for comparison. Adapted from Clark (1983).*

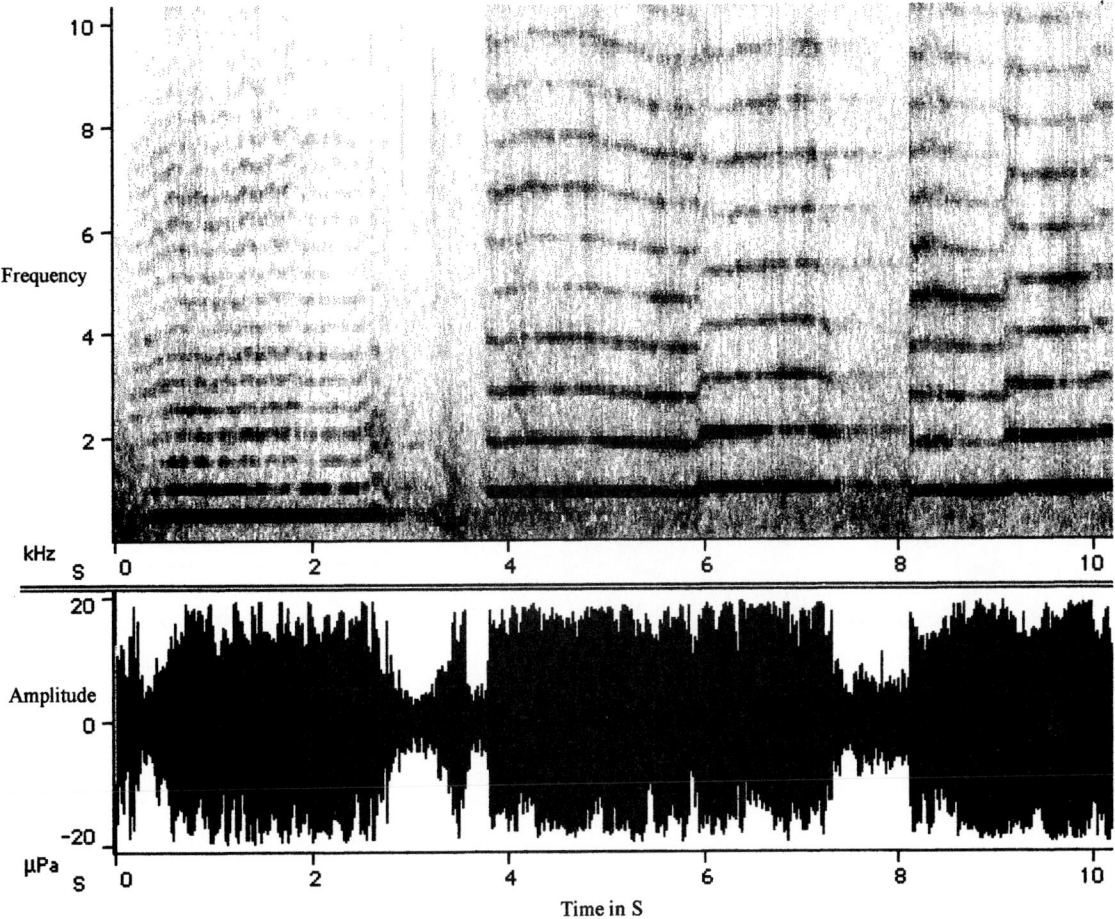

Figure 9 *Example of individually identifiable humpback whale feeding calls. Y axis is frequency in kHz on sonogram and amplitude on waveform. X axis is time in seconds. Sound example and figure courtesy of F. Sharpe.*

Sounds that are variable through time and do not exhibit a continuous or consistent pattern have been lumped as "social sounds." In 1986, Greg Silber determined that social sounds from humpback whales were produced from groups of three or more individuals and were usually linked with intraspecific competition among males. Silber suggested that social sounds in this context reflected agitation or acoustic threats. Still, detailed information on specific functions of social sounds is lacking.

D. MYSTICETE SONG   Humpback whale songs are probably the most recognized and well known of mysticete vocalizations. Only males sing while solitary and all sing the same song during each breeding season. While males do not sing outside of the breeding season, the song remains relatively constant from the end of one breeding season to the start of the next. The song could advertise each male's fitness as a mate and control male spacing when advertising to females. For whatever the specific purpose, humpback songs represent an evolved signal used by males to communicate information about their internal (e.g., reproductive condition or fitness) and external (e.g., location, proximity) state to conspecifics, likely both females and other males.

Humpback whales also produce social sounds that are correlated with group size and surface activity on the breeding grounds. Larger groups with much surface activity produce more sounds than smaller groups. Researchers observed that singers cease their song to join noisy social groups. It is possible that songs represent a strategy of lone males in finding a mate, whereas social sounds indicate more direct competition between males for a female.

2. Odontocete Cetaceans   Odontocete sounds can be divided broadly into two signal types: pulsed and narrow-band tonal sounds. Some pulsed sounds (clicks) are implicated in echolocation and can be of broad spectral composition as in the bottlenose dolphin or of narrow-band composition as in the narwhal. Other burst pulsed sounds, described in the literature as barks, squawks, squeaks, blats, and moans, have social functions (Fig. 10). Narrow-band tonal sounds are continuous signals called whistles. Limitations in audio equipment had suggested that whistles, frequency-modulated (FM) pure tones, were limited to the human mid- to upper sonic range of frequency (5–15 kHz) and of 0.5–2.0 sec in duration (Fig. 11).

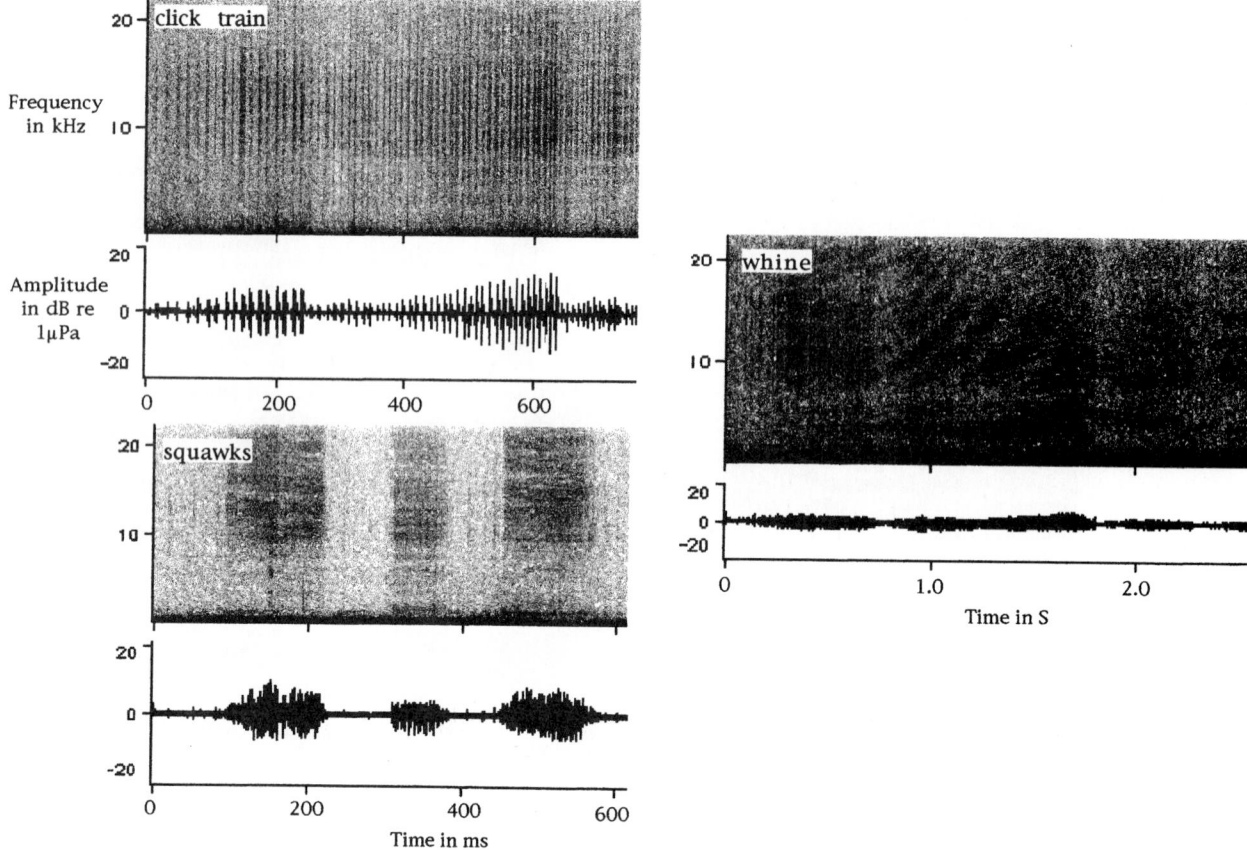

Figure 10 *Examples of dolphin click trains, squawk sounds, and a whine. All are pulsed or amplitude modulated in structure and vary primarily in repetition rate. Click trains function for investigation of objects or peers, whereas the latter two examples communicate social information and potentially referential details to conspecifics. Sounds courtesy of K. M. Dudzinski.*

Improvements in technology yielding a more complete bandwidth recording of dolphin sounds indicate that dolphins produce FM pure tones across a broad frequency range, from 5 to at least 85 kHz. Other FM tonal sounds include screams and chirps.

Research on sound communication in bottlenose dolphins and other delphinids has centered on whistle sounds for pragmatic reasons. The sonic range of whistles is recorded and analyzed easily. Also, whistles are produced by the most common captive species, the bottlenose dolphin, and appear to have no function other than communication. Because the number of nonwhistling species, such as the harbor porpoise (*Phocoena phocoena*) and Commerson's dolphin (*Cephalorhynchus commersonii*), is relatively large, it is premature to regard whistles as the principal means for sound communication among odontocetes.

A. ODONTOCETE PULSED SOUNDS   All recorded toothed cetaceans produce pulsed sounds. These sounds can be used for echolocation or communication (Herman, 1980; Au, 1993). They can be divided into two subclasses: pulse trains and burst-pulse sounds. Pulse trains, also called click trains, are sequences of acoustic pulses repeated over time. Individual pulses are about 50 µsec, with varying peak frequencies of 5–150kHz.

The repetition rate of pulses within a click train can vary from 1–2 to several hundreds per second. Click trains are thought to function mainly for echolocation.

Burst-pulse sounds can be defined as high repetition rate pulse trains where the interpulse interval is <5 msec which are similar in shape to echolocation pulses, but with about a 6- to 10-dB lower amplitude. Because of the high repetition rate in burst pulses, these sounds are no longer perceived as discrete sequences of sounds by the human ear but are heard as a continuous sound. Their peak frequencies vary with species from 20 kHz in killer whales to above 100 kHz in Commerson's dolphins.

Burst-pulse sounds are proposed as functioning primarily for communication. The directional characteristics of many of the pulsed sounds, the relative ease with which they can be localized, their variability, and possibly the power with which they can be produced enhance their potential value as communication signals. Indeed, in situations described as alarm, fright, or distress, broad-band high-intensity squeaks have been heard from bottlenose dolphins and harbor porpoises. River dolphins, killer whales, and Commerson's dolphins, as well as Physeteridae and Phocoenidae, which do not whistle, most likely communicate via pulsed sounds.

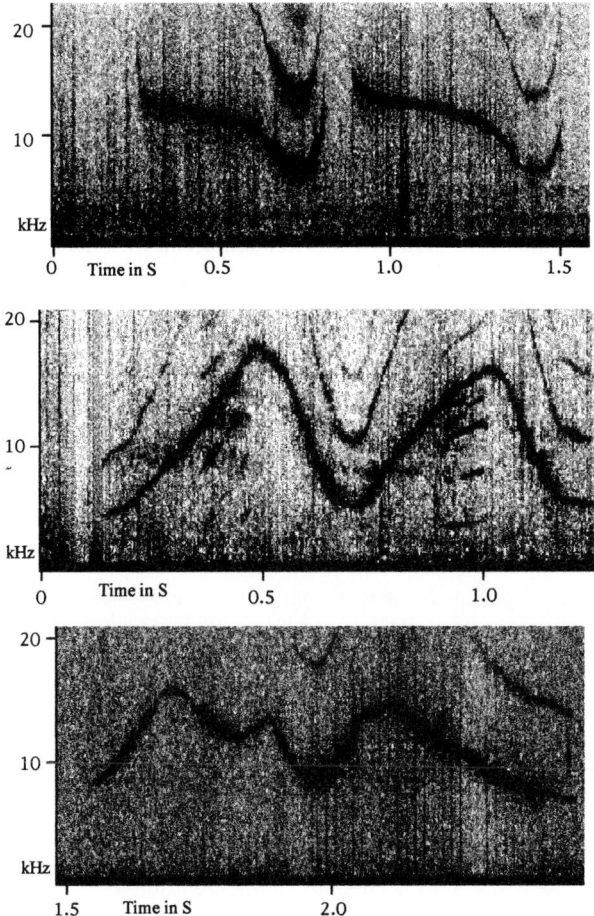

Figure 11 *Examples of narrow-band tonal sounds (i.e., frequency-modulated or whistles) from Atlantic spotted dolphins recorded in the Bahamas. Y axis is frequency in kHz and X axis is time in seconds. Sounds courtesy of K. M. Dudzinski.*

It is possible that the timing of clicks contains social significance. The timing of sperm whale clicks (codas) is thought to communicate the identity of the phonating whale and perhaps other social information.

Another possibility for information exchange via pulsed sounds would be through passive communication, i.e., "eavesdropping." Depending on their position relative to another clicking dolphin, one individual might listen to the emissions and echoes of others rather than actively transmitting signals. Xitco and Roitblat (1996) showed that a bottlenose dolphin trained to listen to echolocation clicks of an actively echolocating dolphin also could discriminate targets. This provides evidence that dolphins are capable of detecting and interpreting the echoes of another's sonar. Definitive evidence is still lacking for free-ranging odontocetes. Their results raise the interesting question about the proper "etiquette" of echolocation and how the task of clicking is organized within a group of cetaceans.

B. ODONTOCETE NARROW-BAND TONAL SOUNDS   Narrow-band tonal sounds, *i.e.*, whistles, are produced over a range of 5–20 kHz, are frequency modulated, and can last from mil-

liseconds to a few seconds. These sounds sometimes have a rich harmonic content that extends into the ultrasonic range of frequencies up to 70–80 kHz for some dolphin species. Whistles vary greatly in contour from simple up-or-down sweeps to frequency-modulated warbles to U loops and inverted U loops. Whistles often grade from one type to another. Whistles are thought to function only for communication, but are not produced by all odontocetes. In at least two odontocete species, false killer whales (*Pseudorca crassidens*) and belugas, animals shift whistle frequency ranges upward to avoid low-frequency ambient noise.

Why some odontocete species whistle and others do not is a very intriguing question. If the whistle appears only under certain ecological or social conditions, these should be found through careful comparisons of whistlers and nonwhistlers. Although a low degree of gregariousness is a feature characterizing some nonwhistling species, this idea is not consistent with findings concerning Hector's dolphins (*Cephalorhynchus hectori*) and killer whales, which live in groups of up to 10 members that occasionally congregate in large numbers but still do not produce whistles.

Of relatively low frequency, whistles travel longer distances in water than pulsed sounds. Although less directional than pulsed sounds, whistles probably are localized easily by cetaceans. Bottlenose dolphins and probably other whistling species can produce whistles and clicks simultaneously. Given these attributes, whistles provide a potential vehicle for maintaining acoustic communication and coordination during food search by echolocation. Also, whistles possess little overlap with the major portion of the echolocation frequency spectrum, minimizing potential masking effects. If whistles have species, regional, or individual specificity, this would at least allow for the identification of schoolmates or familiar associates or aid in the assembly of dispersed animals and in the coordination, spacing, and movements of individuals in rapidly swimming, communally foraging herds.

Assigning particular sounds to specific individual dolphins in the wild presently is difficult, if not impossible. A wealth of data has been gathered on dolphin vocal behavior as related to general behavioral activity. Hawaiian spinner dolphins, bottlenose dolphins, beluga whales, and pilot whales produce a variety of sounds that vary in type and rate with behavioral activities. For these species, the highest rate and variety of sounds (*e.g.*, whistles, screams, barks) were recorded during social activities. In contrast, when resting, acoustic behavior was almost nonexistent. These results seem intuitively correct, i.e., signal exchange would be higher when dolphins were engaged in more interactive behaviors with conspecifics.

While data on sound rates and occurrences are valuable, it is not an examination of communication by the strictest definition. Studies of animal communication require the ability to identify individuals and the signals used. It is especially important to identify the emitter and receiver when examining the intricacies of subtle interactions among social animals. Until recently, cetacean researchers were not able to identify individual soniferous dolphins within free-ranging groups.

Many different situations can elicit whistling. Whistling could appear as a simple phono-reaction in response to hearing another animal's whistle. Mimicry of whistles or of artificial sounds has been documented in bottlenose dolphins and belugas, revealing the plasticity of the sound production system. Studies by Peter Tyack and Louis Herman proved dolphins to be skilled mimics of tank-mate whistles and artificial frequency tones. A correlation between whistling and feeding was noted among wild and captive delphinids. For example, pods of false killer whales produce more whistles while feeding than during traveling. Dolphins accidentally captured in tuna seine nets whistle intensely. Captive bottlenose dolphins newly introduced into a tank or temporarily separated from a familiar tank mate whistle nearly continuously.

There have been several attempts to inventory the whistle repertoire of wild and captive delphinids. This is often difficult because the whistle contours are so variable. The size of a repertoire, including both whistles and pulsed sounds, is probably limited to fewer than 40 discrete types. However, it is possible that whistles are graded rather than discrete signals. In a graded system, several basic types of signals transition to one another through a series of intermediate forms. Still, strong evidence is lacking for many odontocetes. Although sophisticated devices can faithfully describe the frequency versus time function (contour) of a whistle sound, it is left to human judgment to decide when one contour is significantly different from another. Standardized analysis of whistles by taking 20 measurements along the contour of each whistle was conducted by Brenda McCowan and Diana Reiss, and Vincent Janik reviewed four methods for analyzing whistles contours with respect to the categorization of patterns into separate classes. Human observation is routinely used to classify whistle contours because human observers were more likely than computers to identify signatures used while an animal was isolated. Discrepancies in methodologies led Janik to stress the importance of external validation of categories when categorizing behavior patterns, including animal sounds.

In 1965, the Caldwells presented the idea of signature whistles from observations indicating that each dolphin in a captive group tended to produce whistles that were individually distinctive, stereotyped in certain acoustic features, and therefore called "signature" whistles. In the late 1980s and early 1990s, Tyack and colleagues proposed the hypothesis that dolphins use "signature whistles" to refer to each other and themselves (Leatherwood and Reeves, 1990).

3. *Pinnipedia* Pinnipeds typically produce frequency-modulated and pulsed sounds. Except for male walruses, pinnipeds do not whistle. The number of vocalizations produced by pinnipeds is correlated with their mating system and whether mating occurs underwater or on land. In general, otarids are much more vocal on land, often obtaining high densities in highly soniferous colonies. Phocids tend to be more vocal underwater, especially in the true seals that mate underwater, *e.g.*, polar pinnipeds.

Polar pinnipeds are much more vocal underwater than temperate or tropical pinnipeds. Early polar explorers reported hearing "eerie, ghost-like sounds from underneath the water." Because of polar bear predation, Arctic pinnipeds, are essentially silent while hauled out. In contrast, Antarctic pinnipeds

are vocal when they haul out. Comparing the vocal repertoire size and MATING SYSTEM of three species of Antarctic phocids, we find distinctive differences. The Weddell seal congregates in colonies up to 100 mothers with pups, whereas males establish underwater territories beneath the fast ice that are vigorously patrolled and defended with an elaborate repertoire of 34 sounds. Mating in Weddell seals is polygamous; males mate with as many females as will enter their territory. Presumably, the 2-month period that males defend their underwater territory assists females, hauled out on the ice above, in mate selection. The polygynous, but solitary, leopard seal has an intermediate number of underwater vocalizations, apparently used to establish short-term underwater territories in the pack ice and attract females to mate. The crabeater seal is seasonally monogamous: a male hauls out on an ice floe with a female, guards her and her pup against attacks from predators, and then conveniently is available for mating when the pup weans. It is unlikely that this male is the pup's father, and the pair bond is well established for the season. This pinniped has a single monotonous call.

In all pinnipeds, mothers and pups exchange vocalizations that are important in pup recognition and reuniting the mother with her pup after she returns from a foraging bout. Recognition of one's pup is especially important in some otarids mothers that go to sea to forage for up to 7 days before returning to nurse their pup. In many pinnipeds, the vocal repertoire of the mother and pup is unique and distinct from their sounds during other social activities. This repertoire occurs mostly while hauled out, but is also used by mothers coaxing their pups into the water or to haul out.

The majority of documented pinniped sounds are within the range of human hearing. Only one study on a captive leopard seal has examined ultrasonic frequencies, and underwater sounds were found up to 164 kHz. More research is needed in this area. Some species are nearly silent, whereas others have large repertoires that vary by season, sex, age, and whether the animal is in the air or water. Pinniped calls have been described as grunts, rasps, rattles, growls, creaky doors, warbles, trills, chirps, chugs, clicks, and whistles. Clicks are produced, but experimental attempts to demonstrate echolocation have not been successful. These studies, however, were on California sea lions and harbor seals (*Phoca vitulina*) and some researchers suggest that ECHOLOCATION, if present in pinnipeds, would more likely occur in the polar pinnipeds who live in ice-covered waters and total darkness during the polar winters.

Phocid calls are primarily between 100 Hz and 15 kHz, with peak spectra <5 kHz. Typical source levels of underwater sounds are 130 dB re 1 μPa, but are as high as 193 dB re 1 μPa in a territorial Weddell seal (Fig. 12). Elephant seals are reported to produce infrasonic to seismic vibrations while vocalizing in air. One of the most elaborate repertoire of sounds is from the Weddell seal, who has a separate repertoire of sounds for communicating while hauled out from its sounds for underwater contexts. At one colony in the Antarctic, the Weddell seal had 34 types of underwater sounds (Thomas and Kuechle, 1982), including trills, chugs, chirps, guttural glugs, clicks, and mews. Eleven types of trills are used exclusively by males for territorial advertisement and defense and could be used in a

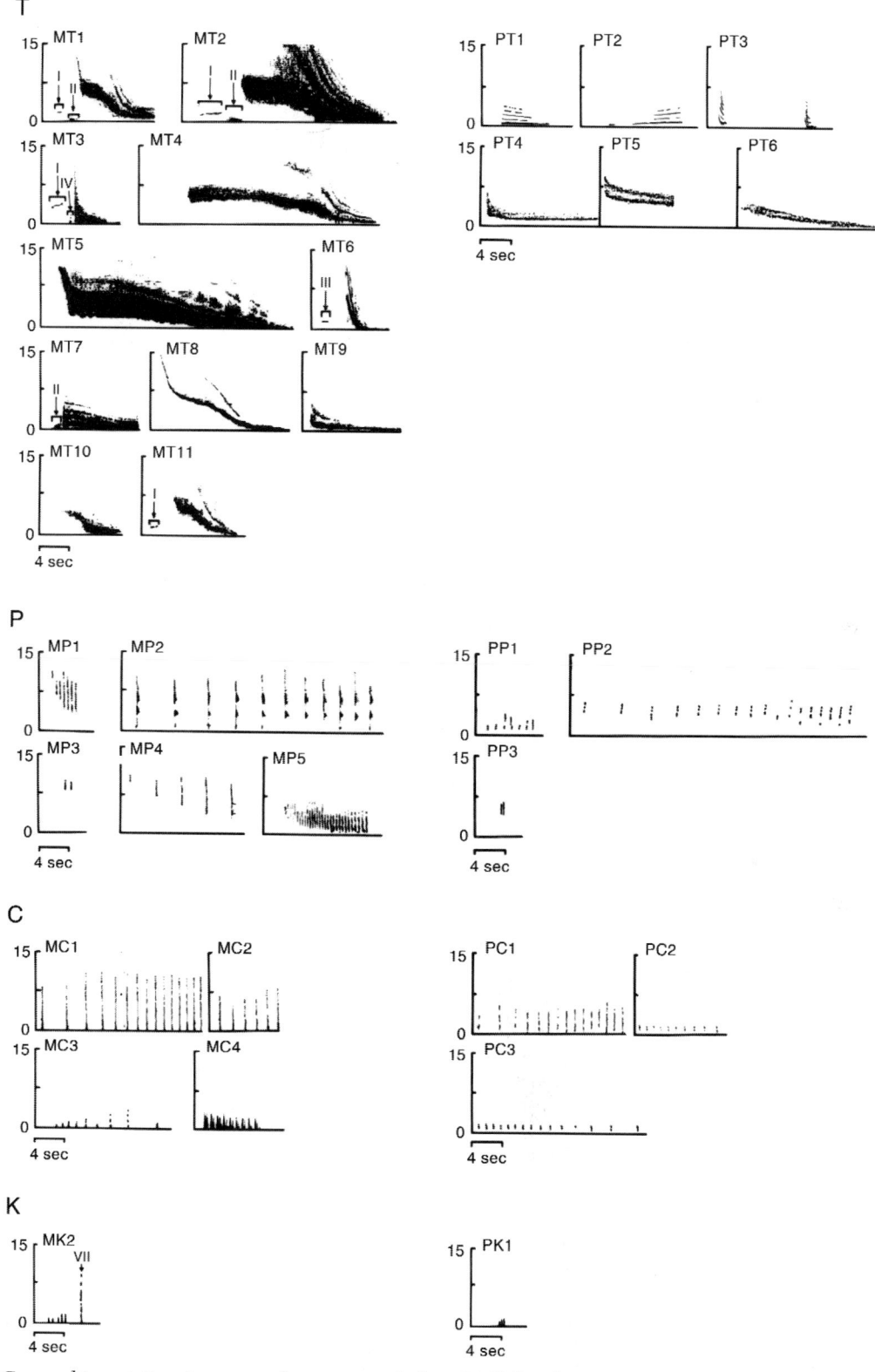

Figure 12 *Geographic variations in some underwater sounds from Weddell seals at McMurdo Sound and Palmer Peninsula, Antarctica. Sonograms have upper frequency scale 15 kHz, time designated by scale bar. Each sound has a location, call, and type designation, e.g., MT1 is trill type 1 from McMurdo Sound and PC1 is chirp type 1 from Palmer Peninsula. T = trill, C = chug, P = chirp, K = knock. Missing sonograms under a location indicate no corresponding call type at that location. Sounds courtesy of J. A. Thomas.*

graded context to convey the degree of warning, *i.e.*, shortest, quietest trills are just a reminder, but long, loud trills are an emphatic warning to an intruder. This species also uses prefixes and suffixes with main call types, seeming to warn or emphasize a message. The trills are as long as 75 sec (Fig. 12). The repertoire on the opposite side of the Antarctic shows geographic variations, including some unique usage of "mirror-image" calls, *i.e.*, an upsweep followed by the mirror-image, downsweep. Male pups as young as 2 months try to perfect the long, loud trills, using comical, voice-cracking sweeps reminiscent of adolescent adults. The leopard seal, the largest pinniped predator, produces a surprisingly musical repertoire of sounds, but primarily during mating. This seal also exhibits geographic variations (Fig. 13) in sounds around the Antarctic.

Otariid airborne sounds range from 1 to 4 kHz, with harmonics up to 6 kHz. Barks in water are slightly louder than in air and both center around 1.5 kHz. Individual California sea lion sounds have unique variations suggesting signal components for identity. Odobenid sounds are low in frequency, 500 Hz, with a peak of 2 kHz. Underwater, walruses have a unique bell-like sound, but also produce clicks and whistles. Recent studies indicate that territorial male walruses have their own distinctive sound patterns.

4. Sirenia Sounds of sirenians are low in amplitude and probably only propagate short distances. From field observations of mother/calf manatee pairs, it appears that vocalizations play a key role in keeping the mother and calf together. Some researchers even describe this vocal exchange as dueting, where the mother and calf exchange chirps (see Reynolds and Odell, 1991). Another example of communication in manatees again included a mother/calf pair on opposite sides of a flood control gate. For nearly 3 hr, the mother placed her head in the narrow opening and vocalized to the calf until the gate opened enough for the calf to swim through. Although most evidence is anecdotal, sirenians (at least manatees) seem to use sounds to communicate with conspecifics.

Dugongs are highly social, occurring in groups up to several hundred animals. Sound and vision probably play the most important roles in communication. Vocalizations of dugongs are low frequency, ranging from 1 to 8 kHz and seem to be especially important in maintaining the mother/calf bond. Studies in the clear waters of Shark Bay suggest that dugong males establish territories to attract estrous females. Reynolds and Odell (1991) suggested that low-frequency vocalizations play a role in mate attraction in dugongs.

Manatee and dugong sounds are described as chirps, whistles, squeals, barks, trills, squeaks, and frog-like calls. West Indian manatee sounds range from 0.6 to 5.0 kHz, whereas Amazonian manatees have higher sounds near 10 kHz, with distress calls having harmonic structure up to 35 kHz (Reynolds and Rommel, 1999). Dugongs produce calls between 0.5 and 18 kHz with maximum energy between 1 and 8 kHz (Reynolds and Rommel, 1999).

5. Mustelidae In sea otters, social interactions, pup care, and mating occur at the water surface; still, little is known about sea otter vocal behavior. No underwater sounds have been reported for sea otters. The intertidal zone is a noisy, churning environment that would not be a good environment for exchange of sounds and

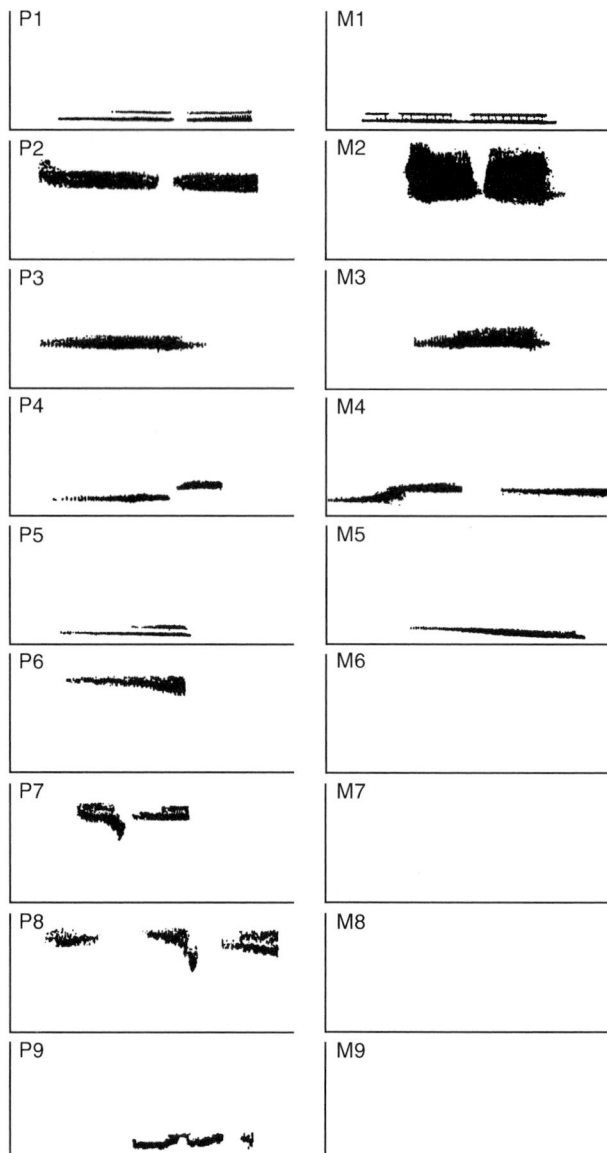

Figure 13 *Geographic variations in underwater sounds from leopard seals. Sonograms have an upper frequency scale of 5 kHz and time frame is 10 sec. P1–P9 are from Palmer Peninsula. M1–M5 are from McMurdo Sound, Antarctica. M6–M9 represent no corresponding sound types with those at Palmer Peninsula. Sounds courtesy of J. A. Thomas.*

would make recording difficult. In addition, sea otters forage singly and probably do not need to communicate while foraging. Kenyon (1975) provided the only detailed summary of sea otter sounds and these were based on sounds heard in air. He described their sounds as (1) baby cry—a sharp, high-pitched "waah-waah" sound used by pups in distress situations or when wanting to attract mothers attention; (2) scream—given by adults in distress or when a female has lost her young (it is an "ear-splitting" version of the pup cry detectable 0.5 miles away in the wild); (3) whistle or whine ("whee-whee")—a high-pitched sound resembling a human whistle (denotes frustration or mild distress, given by captive sea otters when feeding is delayed, detectable 200 m away in the wild); (4) coo ("ku-ku-ku")—produced by females before and after mating or while eating a "particularly pleasing food," detectable only 34 m away; (5) snarl or growl—originating deep in the throat produced by a newly captured sea otter, audible only a few meters away; (6) hiss—short, explosive cat-like hiss used in startle situations; (7) grunt—soft groaning sound produced while eating; (8) bark—a staccato bark trailing off into a whistle, indicative of frustration produced by a young male; and (9) cough, sneeze, and yawn as in other mammals.

6. Ursidae Polar bears have a variety of sounds used in different contexts. Growls serve as a warning to others bears and defense of a food source. Hissing, snorting, loud roars, and groans or grunts are aggressive sounds. Chuffing was documented as a response to stress, whereas mother bears will produce soft cuff sounds or low growls when scolding cubs (Ovsyanikov, 1996).

VIII. Acoustic Signal Exchange: Language or Not?

In the 1950s and 1960s, Lilly tried unsuccessfully to find behavioral correlates with sounds emitted by the captive bottlenose dolphin and then concentrated on research about vocal mimicry in dolphins. While Lilly's early work was groundbreaking in its approach to observing and studying the communication of another species, his later work and conclusions are considered controversial at best. He believed and wrote that dolphins could mimic human sounds and inferred that dolphins had their own natural language and were amazingly intelligent beings worthy of our worship. More recently, Sigurdson showed it to be extremely difficult to train bottlenose dolphins to acoustically mimic even the simplest sounds for communication with humans.

There is little evidence that dolphins possess a language of sounds, as humans. However, laboratory experiments showed that bottlenose dolphins can understand the meaning of sounds organized with simple syntactic rules to represent objects, actions, and modifiers. Using match-to-sample methodology, Herman *et al.* (1984) trained bottlenose dolphins to respond to coded instructions. For example, the trainer could present three cues for "*fetch* the *surfboard* to the *person*." One dolphin was taught using acoustic cues, while a different individual learned a vocabulary of gestures by the trainer. Herman *et al.* showed that dolphins could understand three dozen words: nouns and actions; generalize known words into concepts, regardless of color, size, and shape; decode five- to six-word sentences, such as "pipe bottom hoop surface place in"; understand novel sentences; and interpret both gestures and acoustic signals according to a word-order grammar.

In short, Herman *et al.* showed that dolphins understand the syntax and word order of trained commands. However, Herman has yet to explore the capacity of dolphins for logical reasoning, problem solving, or the ability to answer acoustic commands with their own sounds. Nonetheless, the capacity for sound mimicry, condition-ability of vocalizations, advanced rule-learning capabilities, and other demonstrated intellectual traits suggest that dolphins, like some of the greater apes, could be capable of learning an artificial language in a laboratory setting.

Along a similar track, Schusterman trained California sea lions to conduct many of the same "language-like" tasks based on hand cues from a trainer. California sea lions adapted quickly and accurately to the tasks. Interestingly, Pepperberg trained an African gray parrot to answer questions in mimicked human speech. The quality of mimicry and apparent understanding of this bird is truly remarkable, thus questioning the use of "language-like abilities" as an indication of superior cognition by dolphins or sea lions.

Whether marine mammals use syntax, grammar, or language in the wild remains to be documented. However, whales, dolphins, seals, sea lions, sea otters, and manatees do use several different signals to exchange information among conspecifics about activities and relationships. Signals include visual displays and cues, sound emissions, postures, and natural markings to indicate age, sex, and individual identification.

IX. Conclusions

Animals live in an ever-changing world. Reactions and responses to environmental and social variables must be flexible and adaptive for survival and reproduction. Examination of signaling behavior and subsequent receiver responses provides a window into nonhuman minds, as well as to the social complexity of other species. It can be assumed that evolutionary processes are at work on signals to keep them informative and useful to individuals. Ecological factors, coupled with social relationships and interactions, provide the principal force in the evolution of communication systems (Hauser, 1997). Foraging, mating, and parental strategies are examples of components that influence signaling behavior. In marine mammals, coastal or oceanic species living in relatively clear water may be more likely to use visual signals (e.g., postures, coloration patterns) than species inhabiting riverine or turbid environments. Similarly, amphibious species require a suite of signals useful both in air and underwater. Differential communication is also evidenced in the foraging methods of several delphinid species. Communal foragers have more complex signals compared with more solitary hunters. Frequent interactions with conspecifics necessitate a higher rate of information exchange than for solitary species. Observing and examining the social and ecological differences among individuals and groups will help elucidate the mechanisms underlying the use and evolution of different signals to exchange information among individuals, i.e., communicate.

See Also the Following Articles

Aggressive Behavior ■ Courtship Behavior ■ Hearing ■ Language Learning ■ Signature Whistles ■ Song ■ Sound Production ■ Swimming

References

Anderson, H. T. (ed.) (1969) "The Biology of Marine Mammals." Pergamon Press, New York.

Au, W. W. L. (1993). "The Sonar of Dolphins". Springer-Verlag, New York.

Bradbury, J. W., and Vehrencamp, S. L. (1998). "Principles of Animal Communication". Sinauer, Sunderland, MA.

Clark, C. W. (1983) Acoustic communication and behavior of the southern right whale (*Eubalaena australis*). *In* "Communication and Behavior in Whales" (R. Payne, ed.) pp. 163–198. Westview Press, Boulder, CO.

Hauser, M. D. (1997). "The Evolution of Communication". MIT Press, Massachusetts.

Herman, L. M. (1980). "Cetacean Behavior: Mechanisms and Functions". Wiley, Inc, New York.

Herman, L. M., Richards, D. G., and Wolz, J. P. (1984). Comprehension of sentences by bottlenosed dolphins. *Cognition* **16**, 129–219.

Leatherwood, S., and Reeves, R. R. (1990). "The Bottlenose Dolphin". Academic Press, New York.

Kenyon, K. (1975). "The Sea Otter in the Eastern Pacific Ocean". Dover Publications, New York.

Norris, K. S., Würsig, B., Wells, R. S., and Würsig, M. (1994). "The Hawaiian Spinner Dolphin". Univ. of California Press, Berkeley.

Ovsyanikov, N. (1996). "Polar Bear: Living with the White Bear". Voyageur Press, Stillwater, MN.

Reynolds, J. E., III, and Odell, D. K. (1991). "Manatees and Dugongs". Facts on File, New York.

Reynolds, J. E., and Rommell, R. (1999). "The Biology of Marine Mammals". Smithsonian Institution Press, Washington, DC.

Schusterman, R. J., Thomas, J. A., and Wood, F. G. (eds.) (1986). "Dolphin Cognition and Behavior: A Comparative Approach". Lawrence Erlbaum Associates.

Stirling, I. (1999). "Polar Bears". Univ. of Michigan Press, Ann Arbor, MI.

Thomas, J. A., and Kastelein, R. A. (eds.) (1990). "Sensory Abilities of Cetaceans: Laboratory and Field Evidence." NATO Life Science Series, Vol. 196. Plenum Press, New York.

Vauclair, J. (1996). "Animal Cognition: An Introduction to Modern Comparative Psychology". Harvard Univ. Press, Cambridge.

Xitco, M. J., Jr., and Roitblat, H. L. (1996). Object recognition through eavesdropping: Passive echolocation in bottlenose dolphins. *Anim. Learn. Behav.* **24**. 355–365.

Competition with Fisheries

ÉVA E. PLAGÁNYI AND DOUGLAS S. BUTTERWORTH
University of Cape Town, South Africa

From an ecological perspective, competition is a situation where the simultaneous presence of the two competitors is mutually disadvantageous. This article focuses on biological interactions and, specifically, the direct competition for food and fishery resources between marine mammals and fisheries, in contrast to operational interactions in which marine mammals damage or become entangled in fishing gear with negative consequences for both the fishery and the animals. These two forms of conflict are sometimes difficult to separate because, for example, animals may damage fishing gear in the process of removing fish therefrom. A third oft-cited marine mammal–fishery interaction concerns facilitation of the spread of PARASITES to commercial fish species, but this is not of direct relevance to the current topic.

Competitive interactions between marine mammal populations and fisheries can be either "direct" or "indirect." In the former case the two groups share a common prey species, whereas in the latter case, for example, a marine mammal may prey on a species that is also an important component in the diet of a commerical fish species.

Perceived conflicts between marine mammals and humans in pursuit of common sources of food have come increasingly to the fore in recent years. Escalating pressures on shared resources are expected in the future because of both increasing marine mammal populations and an increasing human population. Reductions in direct takes in response to recognition that several populations of marine mammals were heavily overexploited in the 19th and earlier part of the 20th century, as well as a widespread change in people's perceptions of whether marine mammals should still be regarded as resources available for harvest, have meant that several marine mammal populations are currently on the increase, sometimes by as much as 5–10% per annum. From the human population perspective, the Food and Agriculture Organization of the United Nations (FAO) has estimated that 950 million people worldwide currently rely on fish and shellfish for more than one-third of their animal protein. Based on the past trend in annual harvests from marine capture fisheries, FAO have predicted that, by as soon as the year 2010, supply may be unable to meet the global demand.

Commerical fisheries and marine mammals frequently target the same fish species so that, faced with shortages in marine food production in the future, it is likely that the possible impacts of growing marine mammal populations on the sustainable harvest of commercial fisheries will be vigorously questioned. Concerns about the consequences for fisheries of an increasing marine mammal population have already been expressed in southern Africa, for example, where in 1990, Cape fur seals (*Arctocephalus pusillus pusillus*) were estimated to consume some two million tons of food a year. Considering that this amount was about the same as the annual human catch of fish in the region and that the fur seal population was anticipated to increase further, the reasons for concerns and potential for conflict are obvious. A second example concerns the Pacific Ocean, where marine mammals are estimated to consume about 150 million tons of food per annum, which is some three times the current annual fish harvest by humans.

This article first presents a brief summary of some specific examples that address the question of whether marine mammal populations have negatively impacted the potential yields from fisheries through competition. Examples of perceived competitive interactions are included because, in most cases, the evidence is inconclusive. Second, some examples pertinent to the reverse—whether fisheries negatively impact marine mammals—are summarized.

I. Detrimental Effects of Marine Mammals on Fisheries

A. Pinnipeds (Seals, Sea Lions, and Walruses)

In the early 1990s, catastrophic collapses occurred in the cod (*Gadus morhua*) fisheries on the east coast of Canada. Although several hypotheses have been posited to explain this, the most likely cause was overfishing. Harp seal (*Pagophilus groenlandicus*) populations off Newfoundland and Labrador have been increasing at an estimated rate of 5% per annum since the mid-1980s and are known to consume a substantial tonnage of juvenile cod. The socioeconomic implications of the collapse of the cod fishery were huge, with some 40,000 fishermen rendered out of work, and there is an obvious temptation to argue a causal relationship between the failure of these cod populations to recover as rapidly as expected after their protection and the increase in harp seal abundance. Although the results of at least one ecosystem modeling study support the hypothesis that the recovery of these cod populations is being retarded to some extent by the increased biomass of harp seals, ecosystem models generally have poor predictive reliability, largely because of data limitations.

Demonstrating that either a fishery or a marine mammal will be affected adversely as a result of an increase in the removals by one party of a limited resource is not simple. Inferences based on assumptions of a linear relationship between predator and prey abundance are often incorrect because of the complex nonlinear interactions in an ecosystem. For example, off the west coast of South Africa, seals consume almost as much hake as is taken by the commercial fishery (Fig. 1). However, the commercially valuable hake consists of two species, a shallow-water (*Merluccius capensis*) and a deep-water species (*M. paradoxus*), with the larger of the shallow-water species eating the smaller individuals of the deep-water species. The results of multispecies models suggest that the net effect of a seal cull would be less hake overall because fewer seals would mean more shallow-water hake, and hence more predation on small deep-water hake. This study by Andre Punt and colleagues highlights the complexity of predation, food–fish, and

fishery interactions and hence the difficulties of demonstrating conclusively that marine mammals are in direct competition with humans for food fish, as may superficially appear to be the case.

B. Whales

Numerous multispecies modeling studies have been employed to investigate the direct and indirect effects of minke whales (*Balaenoptera acutorostrata*) on cod, herring (*Clupea harengus*), and capelin (*Mallotus villosus*) fisheries in the Greater Barents Sea. Minke whales are abundant in this region and prey on all three species, prompting the question of whether fishermen could expect greater catches if the populations of these marine mammals were reduced. The indications of these studies are that there is competition between whales and fishers in this region and that the fisheries are likely to respond linearly to changes in whale abundance. They estimate that each minke whale reduces the potential annual catches of both cod and herring by some 5 tons. Similarly, studies off Iceland suggest that the piscivorous minke, humpback (*Megaptera novaeagliae*), and fin whales (*B. physalus*) may be having a considerable impact on the region's cod stock. The cod fishery is of key importance to the Icelandic economy, and the rebuilding of the cod population and catches are recognized as an important economic consideration. It is therefore not surprising that arguments have been put forward that there is a need to reduce whale populations to permit commercial fisheries to increase.

Whereas marine mammals are thought to exert relatively minor influences on systems such as the North Sea and Baltic Sea, they have been identified as potentially serious competitors off, for example, the northeastern United States, a region that includes important fishery areas such as the Gulf of Maine and Georges Bank. The latter region exemplifies the conflicts

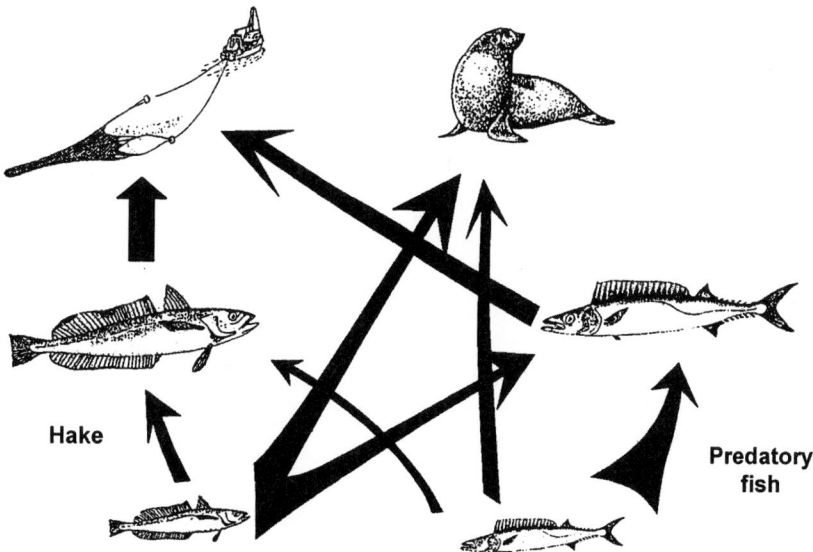

Figure 1 *Schematic showing the complexities of predation, food–fish, and fishery interactions as summarized in a minimal realistic model of Cape fur seals and Cape hake interactions off South Africa. Reproduced from Punt (1994), with permission.*

that can arise between fishery management plans tasked with rebuilding prey populations and prescriptions, by the U.S. Marine Mammal Protection Act in this case, to facilitate an increase in the abundance of marine mammal predators.

C. Small Cetaceans

In many areas of the world, coastal fishermen consider dolphins as serious competitors, although retaliation by the fishermen usually has only minor effects on the populations or on their perceived damage to fisheries. Short-beaked common dolphins (*Delphinus delphis*) in the Mediterranean have often been perceived as a threat to the purse seine and trawl fisheries operating in these waters and, as a result, have been deliberately caught in direct retaliation. Declines in this population have been attributed to both direct and incidental catches by these fisheries.

The largest hunt designed to reduce the perceived level of competition with fisheries took place in the Black Sea from 1870. In the mid-1900s, tens of thousands of dolphins and porpoises were killed every year as a result of fishing industry claims of competition. Other examples of cetacean kills due to perceived competition include the bombing of belugas (*Delphinapterus leucas*) from the air by the Quebec government in the 1920s and 1930s, the commissioning by the Icelandic government in 1956 of a United States naval vessel to kill killer whales (*Orcinus orca*), and the use of explosives and firearms by Alaskan fishers in the mid-1980s to eliminate local killer whales.

D. Sea Otters

Sea otters prey on a variety of marine invertebrates such as urchins and abalone. Off southern California, southern sea otters (*Enhydra lutris nereis*) have been labeled by some as responsible for the decline of the abalone fishery, but there is little direct evidence to support this notion. The commercial abalone fishery in California was closed in late 1997, and factors such as commercial fishing, poaching, disease, and changing environmental conditions are all thought to have contributed to the decline of this commercially valuable shellfish. Although the southern sea otter population is not increasing, perceptions of the level of competition between otters and commercial abalone fishers have increased because of the recent southward movements of otters, which has increased the overlap between otter fishing grounds and abalone fishing areas.

II. Detrimental Effects of Fisheries on Marine Mammal Populations

A. Pinnipeds

The western Alaska population of Steller sea lions (*Eumetopias jubatus*) has shown a continuous decline since the 1970s and was listed in 1997 as an endangered species under the U.S. Endangered Species Act. Several groups have argued that this decline is due in part to the large fishery harvest of walleye pollock (*Theragra chalcogramme*), which is both a key source of food for sea lions and currently the most important U.S. commercial fishery. Measures to reduce the perceived competition between sea lions and fisheries for groundfish stocks include the establishment of "buffer" (no-trawl) areas to include important locations where the sea lions breed, feed, and rest, as well as specifying a pollock harvest which is distributed more evenly over the remaining areas and spread throughout the year. However, the results of modeling studies indicate that the observed sea lion population decline cannot be explained solely through trophic interactions, and rather is more likely linked to shifts in environmental conditions, which lead to changes in the favored complex of species. Moreover, studies such as that by Andrew Trites and colleagues highlight the difficulties of predicting the direction and magnitude of a change in an ecosystem arising from a reduction in predation or fishing pressure. They posit that, paradoxically, Steller sea lion and northern fur seal (*Callorhinus ursinus*) populations might realize greater benefits if adult pollock and large flatfish were fished more heavily. This competitive release effect may result because, for example, pollock are cannibalistic and hence decreased adult pollock abundance as a result of heavier fishing may result in increased numbers of juvenile pollock available to marine mammals.

B. Whales

Competition effects are difficult to quantify, but it has been proposed by Whitehead and Carscadden (1985) that the collapse of the eastern Canadian capelin fishery in the 1970s had a negative effect on fin whales. They suggested that a shortage of capelin might have allowed humpback whales to outcompete fin whales because the latter rely principally on capelin as a prey source. If competitive predation between a marine mammal and a fishery occurs, this implies that the marine mammal population is limited by food availability and hence it should be possible to demonstrate a response of some vital population parameters to a change in food availability. Recent probable population increases of several krill-eating marine mammals, such as Antarctic minke whales (*Balaenoptera bonaerensis*), crabeater seals (*Lobodon carcinophaga*), and fur seals (*Arctocephalus gazella*), have been attributed by some investigators to a likely large increase in the availability of KRILL (*Euphausia superba*) in Antarctic waters. Following the substantial reduction through overexploitation of large whale populations during the early 20th century, some 150 million tons of "surplus" krill is argued to have become available annually to other predators. This "krill surplus" theory enunciated by Laws has yet to be universally accepted, and scientists are addressing questions such as whether the mean age at maturity of minke whales and crabeater seals has dropped in recent years as a response that might result from an increase in food availability.

Modeling studies by Andrew Trites and colleagues suggest that marine mammal populations can be reduced quickly through reductions in prey abundance, but show a generally slow recovery when abundant food becomes available.

C. Small Cetaceans

Dolphin populations that have localized coastal distributions, such as the Indo-Pacific humpbacked dolphins (*Sousa chinensis*) off KwaZulu-Natal, South Africa, may be vulnerable to commercial fishery expansions because of increased competition with fishermen for limited food resources.

D. Sea Otters

Recent declines in northern sea otter (*Enhydra lutris kenyoni*) populations in Alaska have been linked indirectly to competition with fisheries. As outlined earlier, fishing is argued to be one of the factors contributing to the decline of pinniped populations (harbor seals *Phoca vitulina* and Steller sea lions) in some of the Aleutian Islands. Killer whales preferentially feed on pinnipeds in this region, but as a result of the decline in pinnipeds, they have switched to sea otters as prey. The work of Estes and others argues that reduced populations of fish prey that provide high caloric and nutritive value to pinnipeds may impact not only directly on pinniped populations, but also indirectly on killer whale and sea otter populations.

E. Sirenians (Dugongs and Manatees)

Although direct kills and incidental capture in fishery gear are problems, these mammals feed mostly on vascular aquatic plants so that there is no direct competition with humans for a shared food resource.

III. Assessing the Competitive Effects

Whereas commercial fishermen in many parts of the world perceive marine mammals as serious competitors for a scarce resource, other organizations argue that marine mammals are being used simply as scapegoats for failed fisheries management policies. Scientific evidence is therefore increasingly being sought to settle these disputes, but it is becoming increasingly appreciated that the scientific methodologies required to address them are complex, time-consuming, and beset with difficulties.

Initial attempts to quantify the impact of consumption by marine mammals on fish catches used a simple approach. They took account of the fact that, particularly for pinnipeds, the sizes of fish eaten tend to be smaller than are taken by commercial fisheries. Thus 1 ton of a commercially desired fish species eaten by seals does not translate into exactly 1 ton less in the allowable catch for fishers. This is because although a fish eaten by a seal would have grown larger by the time it became vulnerable to fishing, it might also have died before reaching that size as a result of other sources of natural mortality.

It is now acknowledged that such computations, which essentially treated marine mammals as the equivalent of another fishing fleet, are likely to be inadequate because of oversimplification. Three complicating factors need to be addressed in performing more realistic computations, while still accepting that both data and computing power limitations necessarily restrict the degree of complexity that is viable to incorporate in multispecies models. The first concerns how many of the large number of interacting species in any ecosystem need to be considered. Second, do age-structure effects need to be taken into account? This can become important when, for example, one species that predates on the small juveniles of a second finds itself the prey of the larger adults of that same species. Thus, whiting *Merlangius merlangus* feeds extensively on the youngest (0+ and 1+) age classes of the commercially valuable cod, in turn a major predator on the smaller individuals of whiting. Fi-

nally, the customary modeling assumption that species interactions occur homogeneously over space may well be sufficiently flawed to invalidate results. Moreover, the distribution of seal breeding and resting sites does not necessarily reflect their feeding distributions. Modern animal tagging technology has demonstrated, for example, that grey seals (*Halichoerus grypus*) and southern elephant seals (*Mirounga leonina*) may travel hundreds of kilometers to a preferred feeding site.

More recent attempts to quantify marine mammal–fishery interactions have taken account of at least one of these complicating factors, although studies have yet to reach the stage where all three are considered simultaneously. However, with the development of generalized multispecies modeling tools such as Ecosim and Ecospace by Carl Walters and colleagues, groundwork is being laid to provide a more reliable basis for scientific evaluation of these competitive effects.

IV. Considering the Influence of Fish Harvesting on the Ecosystem

The adoption of the Convention for the Conservation of Antarctic Marine Living Resources (CCAMLR) was a watershed in international fishing agreements in that it was the first to acknowledge the importance of maintaining the ecological relationships among harvested, dependent, and related populations of marine resources. Krill is the primary food source of a number of marine mammal species, and concern has been expressed that the rapidly expanding Antarctic krill fishery might negatively impact or retard the recovery of previously overexploited populations, such as the large baleen whales of the Southern Hemisphere. In response to this CCAMLR mandate, modeling procedures have been developed to assess the impact of Antarctic KRILL harvesting on krill predator populations and to explore means of incorporating the needs of these predators into the models that are used in recommending annual krill catch levels. For example, Antarctic fur seals from Bird Island, South Georgia, are another example of a previously overexploited population that is now growing rapidly such that its needs for krill as a food source may in due course conflict with the objectives of krill harvesters. These competition effects have thus far been addressed by Robin Thomson and colleagues who calculate, within a modeling procedure, the level of krill fishing intensity that would reduce krill availability, and hence the population of a predator to a particular level. In general, initiatives such as these pursued under CCAMLR recognize the need to balance the needs of predators with the socioeconomic pressures underlying fishery harvests and represent a realistic step forward in resolving some of the management quandaries resulting from competition for limited marine resources.

V. Food Web Competition

Andrew Trites and colleagues assessed the competition between fisheries and marine mammals for prey and primary production in the Pacific Ocean. Although the 84 species of marine mammals inhabiting the Pacific consume about three times as much food as humans harvest, most of the species consumed are not of current commercial interest. The greatest overlaps occur with pinnipeds (60%) and dolphins and porpoises (50%). The least overlap is with baleen whales and beaked whales. The observed dietary overlap between the prey items of marine mammals and fisheries is less than expected because specialized feeding habits mean, for example, that some of the targeted prey either are not desired for human consumption or are not currently viable for commercial harvest. Trites and others argue that while direct competition between fisheries and marine mammals for prey appears limited, indirect competition for primary production may be cause for concern. Such so-called food web competition may occur if there is overlap between the trophic flows supporting the two groups (see Fig. 2). Evidence in support of food web competition between marine mammals and fisheries is provided by a negative correlation between estimates of primary production required to support fisheries catches and to support the number of marine mammals estimated in the different FAO areas of the Pacific Ocean.

VI. Summary

Despite a persistent notion worldwide that there is a mass for mass equivalence in the prey of marine mammals and the yields available to fishers, the balance of the evidence collected thus far indicates that this is not the case. Furthermore, the complexity of ecosystems could well be such that the response

to a marine mammal cull could, for example, be highly diffused through the food web, involving many other species. In some cases, competition effects are reduced because, for example, one of the putative competitors in fact reduces the abundance of a predatory fish species, in turn affecting the abundance of the target prey species. It is worth noting that although marine mammals are the most obvious scapegoat of fishers because of their visibility, there is typically greater competitive overlap of the feeding "niches" of fish predators with those of fishermen.

Because of the difficulties of providing definitive scientific advice on such questions, scientists often equivocate. It is currently virtually impossible to wholly substantiate claims that predation by marine mammals is adversely impacting a fishery or vice versa. In the absence of definitive answers, fisheries managers are increasingly applying the "precautionary principle," which requires that "where there are threats of serious or irreversible damage, lack of full scientific certainty shall not be used as a reason for postponing cost-effective measures to prevent environmental degradation," but this has been argued both ways in this context: either that marine mammal culls should not take place in the absence of clear evidence that they will benefit fisheries or alternatively that marine mammals should be culled in the absence of clear evidence that their consumption of fish will not possibly damage fisheries.

As more and better information on marine mammal DIETS becomes increasingly available, one of the key uncertainties in resolving questions as to the degree of competitive overlap be-

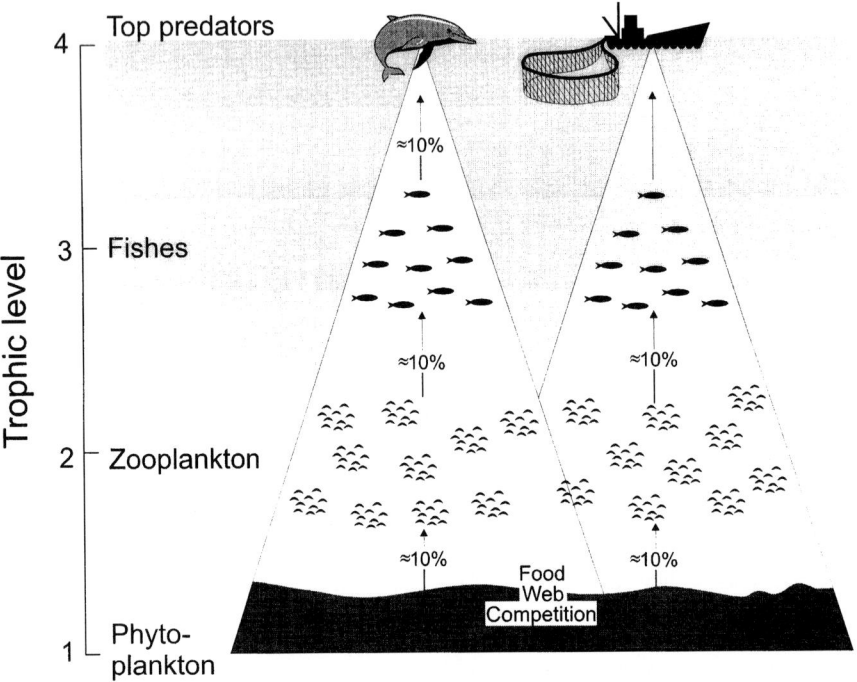

Figure 2 *Schematic example of indirect competition for food by marine mammals and fisheries. The representation shows how top predators, such as marine mammals, may be affected by fisheries because of limits on the primary productivity available to support the two groups. Thus even though the mammals' prey and species taken by fisheries may not overlap, so-called food web competition occurs at the base of the food pyramids. Reproduced from Trites et al. (1997), with permission.*

tween marine mammals and fisheries relates to limited understanding at present of the feeding strategies of marine mammals. There is a need to quantify not only spatial and temporal variability in diet, but also the conditions under which predators switch to alternative prey species as the abundances of the various species change. It is also important to bear in mind that some marine mammals that are highly specialized feeders (or, conversely, highly specialized fishers) are most vulnerable to competition effects because they cannot easily change their diet in response to overfishing of a vital food source.

See Also the Following Articles

Ecology, Overview ■ Fisheries, Interference with ■ Sustainability

References

Beddington, J. R., Beverton, R. J. H., and Lavigne, D. M. (1985). "Marine Mammals and Fisheries." Allen & Unwin, London.

Butterworth, D. S. (1999). Do increasing marine mammal populations threaten national fisheries? *In* "Issues Related to Indigenous Whaling" (M. R. Freeman, ed.), chapter 7. World Council of Whalers.

Estes, J. A., Tinker, M. T., Williams, T. M., and Doak, D. F. (1998). Killer whale predation on sea otters linking oceanic and nearshore ecosystems. *Science* **282,** 473–476.

Laws, R. M. (1977). The significance of vertebrates in the Antarctic marine ecosystem. *In* "Adaptations within Antarctic ecosystems; Third Symposium on Antarctic Biology" (G. A. Llano, ed.). Scientific Committee for Antarctic Research.

Northridge, S. P. (1991). An updated world review of interactions between marine mammals and fisheries. FAO Fisheries Technical Paper 251, Suppl. 1.

Punt, A. E. (1994). Data analysis and modelling of the seal-hake biological interaction off the South African West Coast. Report submitted to the Sea Fisheries Research Institute, South Africa.

Punt, A. E., and Butterworth, D. S. (1995). The effects of future consumption by the Cape fur seal on catches and catch rates of the Cape hakes. 4. Modeling the biological interaction between Cape fur seals *Arctocephalus pusillus pusillus* and the Cape hakes *Merluccius capensis* and *M. paradoxus. S. Afr. J. Mar. Sci.* **16,** 255–285.

Schweder, T., Hagen, G. S., and Hatlebakk, E. (2000). Direct and indirect effects of minke whale abundance on cod and herring fisheries: A scenario experiment for the Greater Barents Sea. *NAMMCO* Scientific Publications **2,** 120–132.

Thomson, R. B., Butterworth, D. S., Boyd, I. L. and Croxall, J. P. (2000). Modeling the consequences of Antarctic krill harvesting on Antarctic fur seals. *Ecol. Appl.* **10**(6), 1806–1819.

Trites, A. W., Christensen, V., and Pauly, D. (1997). Competition between fisheries and marine mammals for prey and primary production in the Pacific Ocean. *J. Northw. Atl. Fish. Sci.* **22,** 173–187.

Trites, A. W., Livingston, P. A., Mackinson, S., Vasconcellos M. C., Springer, A. M., and Pauly, D. (1999). Ecosystem change and the decline of marine mammals in the Eastern Bering Sea. Testing the ecosystem shift and commercial whaling hypotheses. Fisheries Centre Research Reports, Vol. 7.

Walters, C., Pauly, D., and Christensen, V. (1999). Ecospace: Prediction of mesoscale spatial patterns in trophic relationships of exploited ecosystems, with emphasis on the impacts of marine protected areas. *Ecosystems* **2**; 539–554.

Whitehead, H., and Carscadden, J. E. (1985). Predicting inshore whale abundance: Whales and capelin off the Newfoundland coast. *Can J. Fish. Aquat. Sci.* **42**(5), 976–981.

Yodzis, P. (2000). Diffuse effects in food webs. *Ecology* **81**(1), 261–266.

Conservation Biology

Barbara L. Taylor
*Southwest Fisheries Science Center,
La Jolla, California*

To many, "save the whales" is nearly synonymous with "conservation of nature." Whales have become flagship species for wildlife conservation because many of us have lived through the near EXTINCTION of many species of whales at the hands of our fellow humans. It is amazing that humans were able to come so close to exterminating many marine mammals from the huge oceans that cover our planet. What makes marine mammals so vulnerable? Or the flip side of that same coin, why must we go to special lengths to conserve marine mammals?

I. Causes of Vulnerability

The field of conservation biology has arisen to meet the special needs of conserving biodiversity. Prior to the 1980s natural resource management was focused primarily on managing species that were of economic or recreational importance and hence the emphasis then on "fish and game." The tools used to manage these generally more abundant species were not always appropriate for the more rare species that were increasingly facing extinction. Conservation biology has not only developed many important tools to assess risk and prescribe the needed conservation actions, but in the process has developed a new management philosophy. Because many marine mammals were nearly extirpated, many of the tools and management models were developed in this field.

A. Understanding Risk through the History of Exploitation

When species come close to extinction, scientists analyze what risks the species face in order to take management actions to reduce those risks. Marine mammals share a number of these risk factors that make them particularly vulnerable to human impact. Before considering the biological components of marine mammal vulnerability, it is important to note that their exploitation has an economic basis: marine mammals are and have been a very profitable natural resource to exploit. Although marine mammals have always been valued for their meat by native peoples such as the Eskimo, it was the ability to use modern weaponry and transport the goods to distant markets that set the scene for the first wave of decimation of marine mammals. The attributes that make marine mammals vulnerable are easily explained in the context of their exploitation through history.

The first species to be lost were those that were especially vulnerable because they were naturally rare and had limited distributions. Steller's sea cow (*Hydrodamalis gigas*), for example, lasted only a few years after being discovered in the Aleutians. However, even relatively abundant species could be rapidly overexploited if they congregated in easily accessible areas. The more these species congregated, the more vulnerable they were. Thus, most land-breeding species of pinnipeds were quickly driven to near extinction. Although records of the extent of the decimation of populations are poor, we know that some populations, such as northern elephant seals (*Mirounga angustirostris*) and sea otters (*Enhydra lutris*) in California, were reduced to very low size because at some point they were thought to be extinct.

What made the great whales vulnerable? Right whales (*Eubalaena* spp.) are slow moving and congregate in predictable areas to feed. Bowhead whales (*Balaena mysticetes*) are similarly slow moving and were also found in areas in sufficient density and predictability to be rapidly overharvested. Among bowhead whales the populations that fared the best occupied the least accessible habitat: heavy ice pack that served as a refuge from ships. Pacific gray whales (*Eschrichtius robustus*) were easily exploited because they concentrated in breeding lagoons and migrated in a dense, easily accessible coastal corridor to get from their Alaskan feeding grounds to their Mexican breeding grounds. Sperm whales (*Physeter macrocephalus*) were valued for their oil, which was said to have lit the lamps of the Western world. Again, sperm whales swam slowly enough that row boats deployed from sailing ships could be used to hunt them. Also, the blows of the whales could be spotted for miles from the tall crows nests of ships. They were also found in concentrations in predictable areas, although they were taken opportunistically from most of the world's oceans. Sperm whale products were sufficiently valuable that thousands of ships hunted them even when the average trip lengthened from one to several years as populations declined. The Yankee whaling era did eventually end because petroleum products became less expensive than whale products.

Technological advances allowed whale products to be sufficiently profitable in the 1920s: fast catcher boats, lethal explosive harpoons, and factory ships that allowed rapid processing anywhere in the world. Now the large fast-moving baleen whales, like blue whales (*Balaenoptera musculus*), could be exploited for their vast quantities of meat and BLUBBER. The past overexploitation of other whale species led to the formation of the International Whaling Commission to attempt to manage the harvest. The failure of this organization is legendary (Gambell, 1999) as species after species was successively severely overharvested. In the decade of the highest catch (1956–1965), official numbers killed were 631,980, and this is known to be an underestimate because the Soviets were guilty of vast underreporting in this era.

Technology had now reached the point that even species that lived in remote areas and sometimes in relatively low densities could be located and harvested with remarkable efficiency. For example, sperm whales were hunted by a fleet of catcher boats that ranged in front of the mother factory ship spaced apart roughly equally. These catcher boats methodically searched the sea and converged once they found their prey. World War II antisubmarine sonar technology was used to follow whales under water so as to be poised for the kill when the whales surfaced to breathe. The average time to kill all whales of legal length in the group (which often numbered about 20) was only 1 hr. Perhaps more amazing was that individual whales,

which were at least 30 feet long, were completely butchered aboard the factory ship and sent below to be made into various products in only 20 min! Blue whales in the Antarctic were reduced from an estimated 350,000 to approximately 460 (Gambell, 1999), and the vast majority of this whaling (predominantly by the British fleet) occurred over a period of only 10 years (1930–1940).

B. The Biology of Vulnerability

Why were these species so vulnerable to rapid overexploitation? The risk factor that all marine mammals share that makes them vulnerable to such rapid overexploitation is that they reproduce relatively slowly. Most fish mature in a few years and can produce thousands of eggs. Their life history strategy allows for many individuals to be consumed as prey. Marine mammals, to the contrary, are the predators. They have not evolved to withstand high rates of predation. At most, marine mammals produce one offspring per year. Most pinnipeds and some porpoises can mature in 3 to 4 years, have offspring annually, and only live up to about 20 years old. Some pinnipeds have been documented to grow at rates of up to 12% per year (Wade, 1998). Thus, 6–10% per year could be harvested and allow the population to persist. Of course, commercial sealers of the last century were harvesting at rates much greater than 12%/year and thus drove populations to commercial or actual extinction.

Most cetaceans have a much lower population growth rate. Perhaps the slowest growing species is the predator of predators: the killer whale. Killer whales (*Orcinus orca*) not only mature late at about age 12, but they invest large amounts of parental care into their offspring and thus have calves on average only once every 5 years. Recent research estimated that a population that was harvested for purposes of public display in the 1970s is only just now recovering to the numbers found then. This population was growing at only 3% per year (Brault and Caswell, 1993). Many of the great whale populations are now recovering at rates from 7 to 12% per year. Because marine mammal population size is relatively small compared with the fishes of the sea, a harvest that is substantial from an economic point of view is usually unsustainable from a biological point of view. Consider, for example, the Pacific gray whale, which was recently removed from the U.S. Endangered Species List. The recovery of this species is one of the great success stories of marine mammal management and some think that numbers are now approaching historical numbers. Still, there are only 22,000 individuals in this species, which would be considered to be a very small town of humans virtually anywhere on the planet. To put this in perspective, the average catch in the 1970s, which was the most recent period of commercial whaling, was over 27,000 whales/year. During the growth recovery phase for gray whales the population grew at about 3% per year (Buckland *et al.*, 1993). Allowing half of that growth to be harvested would result in 330 whales/year, which is a very small number compared to common commercial catches.

Another reason that marine mammals are vulnerable to extinction is that individual animals seem most likely to return to the same areas to feed and to breed. Great progress has been made in the past two decades in understanding how marine mammals organize themselves spatially. The ability to photo-graphically identify individuals over many years has revealed that despite the ability of marine mammals to move thousands of miles, many (if not most) individuals come back to the same area each summer to feed and migrate to the same BREEDING grounds (called site fidelity). Data from satellite tags and molecular genetics are adding focus to this same picture: population structure is the rule not the exception. Finding this structure makes us rethink the history of whaling and the more recent history of recovery. Are populations in some areas, like New Zealand that once supported shore-based whaling of humpback whales, failing to recover because all the whales that once went there were killed? If site fidelity is so strong, does this explain why whalers were able to expend about the same amount of effort to catch a whale even though the total abundance of whales was declining drastically? Whalers were simply able to sequentially harvest different feeding grounds like a combine harvesting grain. Such a theory is certainly compatible with many historical whaling records that show the harvest shifting to different places each year.

II. New Risks from Old Vulnerabilities

Because of various protective laws, a current ban on the commercial harvest of whales in the IWC, and the restriction in international trade because of current listings in the Conventional on International Trade in Endangered Species (CITES), commercial harvest of marine mammals on a global scale is not currently the risk factor it once was. However, all the risk factors that made marine mammals susceptible to large-scale commercial harvest (slow population growth rate, concentrations in accessible areas, ease of sighting and capture) make them vulnerable to other risks today. The species most vulnerable to extinction today are a few of the great whales that have failed to recover, the two species of remaining monk seals (*Monachus monachus* and *M. schauinslandi*), and a list of small cetaceans, including many river dolphins. Thus, the species most vulnerable have shifted from the seals and sea lions to the great whales and finally to the small cetaceans.

Once again, the ultimate cause of marine mammals' vulnerability is their life history strategy that leads to slow population growth rate, but in addition to the old risks such as overharvest, a new series of risks are posed by humans that are taking many small cetaceans to the brink of extinction. Recently added risks to marine mammals are entanglement in fishing gear (bycatch), POLLUTION, habitat destruction, and ship/boat injuries. Given all the risks posed to marine mammals, regulation of human activities will continue to be needed to conserve these species. Strong national laws and international treaties have played an important role in conservation and many species are making strong recoveries. Many species, however, fall outside of the purview of such protection and it is notable that many of these are the small cetaceans that are now in serious decline.

III. New Conservation Practices

As human abundance increases there will be increased use of marine natural resources putting humans and marine mam-

mals in competition for food. No one knows the indirect effect of how current fishing practices affect the level of food available to marine mammals. Reduction of pollock has been implicated as one of the causes for the decline in Steller sea lions (*Eumatopias jubatus*) (NMFS, 1992), which resulted in this species being listed as endangered under the U.S. Endangered Species Act. Scientifically it is very difficult to prove either that reduction in prey led to the decline or that a major oceanographic shift led to a reduced capacity of the environment to maintain large numbers of Steller sea lions. Past management required proof of harm, i.e., the burden of proof was that scientists prove human exploitation results in harm to marine mammals. Because these management practices failed, there has been a recent trend to reverse this burden of proof such that the users of natural resources should prove they will not cause harm. Until such proof is provided the use should not be allowed. This is called the "precautionary principle."

This principle is part of the philosophical shift in natural resource management. Experience has shown that better management can result from harvesting more conservatively as our ignorance (reflected in measures of uncertainty) increases. Past management was a failure in both economic and biological terms so there is agreement between those who wish to exploit natural resources and those that wish to preserve them that the philosophical move to incorporate uncertainty into management models is generally good. However, the level of conservatism in the models is a matter of choice, and beyond a certain point conservative management practices are economically costly. For example, allowing no capture of dolphins during tuna fishing would provide the greatest safety to the dolphins and would please animal rights activists, but there is no doubt that such management would be costly to the fishing industry. There is also no doubt that the dolphin populations could sustain some level of bycatch and remain healthy.

The role of conservation biology is to prioritize actions through a scientific analysis of risks and to guide management actions by coupling the policy decisions together with the analysis of data. Conservation biology is developing the new tools needed to create management models that successfully implement public policy. For example, consider the Marine Mammal Protection Act objective to maintain marine mammals as functioning elements of their ecosystem. This seemingly straightforward objective required developing the following tools: a model that incorporated our uncertainty about abundance, kill and population growth rates to determine the number of animals that can be safely killed by humans, defining the unit of conservation that will maintain ecosystem functionality, which in turn required developing new techniques to estimate how animals disperse between different areas, and developing decision analysis techniques that are based on how well the management model performs with respect to management objectives.

Although conservation biology can develop suitable tools, a more fundamental question about the conservation of marine mammals remains: just how precautionary do we want to be? What would we, as the world community of humans, like to see conserved with respect to marine mammals? One objective is to at a minimum conserve the biodiversity of marine mammals, i.e., assure the continuance of all species. Given current species definitions, this would allow the extirpation of populations of marine mammals in all ocean basins except one. For example, it would be acceptable to allow blue whales to go locally extinct everywhere except in one area, say the population that summers off California. It is likely that those living around the Atlantic ocean basins would find this unacceptable. There are also questions about species definitions: is there only one species of blue whale, if there are subspecies, should they be conserved, and so on. Another objective might be persistence of all species in all ocean basins or river systems. This definition would allow the extirpation of local populations, such as the humpback whales (*Megaptera novaeangliae*) off the east coast of North America. Such a loss would certainly be unacceptable to people in that area that not only enjoy the animals but enjoy a whale-watching industry of economic importance to local communities. Another objective would be to maintain all populations or even to maintain marine mammals as healthy, functioning elements of their ecosystems (an objective of the U.S. Marine Mammal Protection Act).

Thus, conservation is linked to societal values. One of the greatest difficulties in conserving marine mammals as human populations and demands on natural resources continue to grow will be finding acceptable global conservation values that must underlie the international will to regulate our actions that either directly or indirectly affect the lives of marine mammals. Global conservation of the species at highest risk of extinction will take both international will and international financial support to provide direct aid to failing populations and indirect aid through economic alternatives for people directly affected by conservation of an international resource.

See Also the Following Articles

Endangered Species and Populations ▪ International Whaling Commission ▪ Management

References

Brault, S., and Caswell, H. (1993). Pod-specific demography of killer whales (*Orcinus orca*). *Ecology* **74**(5), 1444–1454.

Buckland, S. T., Breiwick, J. M., Cattanach, K. L., and Laake, J. L. (1993). Estimated population size of the California gray whale. *Mar. Mamm. Sci.* **9**, 235–249.

Gambell, R. (1999). The International Whaling Commission and the contemporary whaling debate. *In* "Conservation and Management of Marine Mammals" (J. R. Twiss, Jr. and R. R. Reeves, eds.). Smithsonian Institution, Washington, DC.

National Marine Fisheries Service (1992). Recovery plan for the Steller sea lion (*Eumatopias jubatus*). Prepared by the Steller Sea Lion Recovery Team for the National Marine Fisheries Service, Silver Springs, Maryland.

Soulé, M. E. (1986). "Conservation Biology: The Science of Scarcity and Diversity." Sinauer Associates, Sunderland, MA.

Soulé, M. E. (1987). "Viable Populations for Conservation." Cambridge Univ. Press, Cambridge.

Soulé, M. E., and Wilcox, B. A. (1980). "Conservation Biology: An Evolutionary-Ecological Approach." Sinauer Associates, Sunderland, MA.

Twiss, J. R., Jr., and Reeves, R. R. (1999). "Conservation and Management of Marine Mammals." Smithsonian Institution, Washington, DC.

Wade, P. R. (1998). Calculating limits to the human-caused mortality of cetaceans and pinnipeds. *Mar. Mamm. Sci.* **14**, 1–37.

Conservation Efforts

RANDALL R. REEVES
Okapi Wildlife Associates, Hudson, Quebec, Canada

Efforts to conserve marine mammals began early in the 20th century. The impetus for these efforts came from the recognition that populations of several highly valued species—fur seals and the sea otter (*Enhydra lutris*)—had been nearly extirpated by hunting. In most instances, self-regulation through market feedback had been the only thing that prevented extinctions. In other words, as the animal populations became reduced by overkill, it became increasingly difficult to hunt them profitably, so the hunting effort declined. This mechanism was clearly inadequate to protect the stocks of whales because modern whaling was a multispecies enterprise. As right whales (Balaenidae) and blue whales (*Balaenoptera musculus*) became scarce, the fleets simply redirected their attention to humpback, fin, and sei whales (*Megaptera novaeangliae, Balaenoptera physalus,* and *B. borealis*, respectively), but any right or blue whale encountered would still be killed. By the late 1920s and 1930s, the whaling industry had begun to place limits on oil production and had given full protection to the depleted right whales and gray whales (*Eschrichtius robustus*). Eventually, international agreements emerged to manage the industry on terms more favorable to conservation. It was not until the 1970s, however, that the multispecies problem was addressed properly. Indeed, it is fair to say that efforts to conserve marine mammals were few and far between until the late 1960s and early 1970s.

A discussion of marine mammal conservation can be organized in a number of ways according to different types of threat (e.g., directed hunting, bycatch in fisheries, chemical pollution), on a species or population basis, by geographical region, or chronologically (see Whitehead *et al.*, 2000). The first part of this article is organized according to levels of governance. Conservation efforts have been and should be made at many different levels, from that of global international agreements all the way "down" to the scale of actions by local communities and individual citizens. Thus, some of the efforts to conserve marine mammals at the international, regional, national, and local levels are reviewed, and this review is followed by a discussion of some of the principal threats and the ways these are being addressed. Next is a brief overview of the zoogeography of marine mammal conservation, which considers regional differences in the seriousness of threats and in the ways these threats are being addressed. Finally, an attempt is made to identify the most seriously threatened marine mammal species and populations.

I. What Is "Conservation"?

This question may seem trivial at first glance, but it is crucial to define terminology rigorously in order to avoid misunderstanding. Here, "conservation" is defined as the preservation of wild populations so that they can continue to replicate themselves, in a natural context, for an indefinite (but long) time into the future. This necessarily means that not only the animals themselves, but also the environments ("ecosystems") that sustain them and even the biotic communities of which they are a part, must be preserved. It should go without saying that maintenance of a few individuals in zoo-like conditions, or the preservation of frozen DNA, are far from what constitutes conservation.

The unit of conservation has traditionally been the species, classically defined as a group of interbreeding natural populations that are isolated reproductively from other such groups (Mayr and Ashlock, 1991). In practice, conservation biologists generally agree that it is insufficient to be concerned only with preserving species. They argue that it is also important to preserve the natural variety within species, including genetic and behavioral variants. One way of achieving this more ambitious objective is by ensuring the survival of all local or geographical populations ("stocks") of a species. There is a substantial and growing body of literature on the "stock" concept as it applies to marine mammals (e.g., Dizon *et al.*, 1992, 1997).

The term "conservation" has a long history and is often cast in three different perspectives: biocentric, economic, and ecologic. *Biocentric* conservation emphasizes the intrinsic value of all life forms and is rooted in religious or philosophical beliefs that place humans on the same plane as other organisms. Although the concept of "animal rights" shares similar roots, it differs from biocentric conservation in that it focuses on the importance of individuals rather than on populations or genomes. Concerns about animal welfare and humane treatment also focus on individuals and are not always central to conservation, as defined here. *Economic* conservation regards wild animal populations as resources to be used for human benefit. A central tenet is sustainability. That is, killing or other forms of extractive, or consumptive, use are allowed and perhaps even encouraged, but only on the condition that such use does not compromise the ability of a wild population to regenerate itself. Finally, *ecologic* conservation places a premium on the maintenance of natural systems and processes. Individuals, populations, and species derive importance from their functional relationships with the communities of which they are a part.

II. International Conservation Efforts

Organized conservation efforts are made at the international level mainly by intergovernmental organizations that are established under the terms of treaties or conventions (Table I). A few nongovernmental organizations (NGOs) also operate on a global basis. Some of these, such as the World Conserva-

tion Union (IUCN), International Fund for Animal Welfare (IFAW), and World Wide Fund for Nature (WWF), address a wide range of environmental issues, while a few, such as the Whale and Dolphin Conservation Society (WDCS), are concerned only with the conservation and protection of marine mammals (Lavigne *et al.*, 1999).

The scale of any particular effort depends, first and foremost, on the geographical distribution of the organism(s) being conserved or the threat(s) being managed. There are relatively few international conservation instruments that focus solely on marine mammals. The best known of these is unquestionably the INTERNATIONAL WHALING COMMISSION (IWC), established under the International Convention for the Regulation of Whaling (ICRW) signed in Washington, DC, in 1946 (Gambell, 1999). A global conservation body was clearly necessary to manage the exploitation of the great whales, customarily defined as the baleen whales (Fig. 1) plus the sperm whale (*Physeter macrocephalus*). Most of these animals migrate over long distances and have been hunted on a truly worldwide scale.

Although the IWC's authority as the body responsible for managing whaling worldwide has been challenged in recent years, it is reconfirmed in Agenda 21, an agreement emanating from the United Nations Conference on Environment and Development in Rio de Janeiro in 1992. There is much controversy about the IWC's scope of responsibility. Some member states (e.g., Japan, Mexico, Denmark, and Russia) have traditionally taken the position that "small cetaceans," meaning all toothed species except the sperm whale and the "bottlenose" whales (defined in the IWC schedule as the northern and southern bottlenose whales *Hyperoodon ampullatus* and *H. planifrons*, respectively, Arnoux's beaked whale, *Berardius arnuxii*, and Baird's beaked whale, *B. bairdii*), are not covered by the ICRW and that their exploitation and conservation are national, or at most regional, concerns. This interpretation of the IWC's competence overlooks the fact that many populations of small cetaceans move seasonally across national borders or onto the high seas. It also ignores the close biological relationships among the cetaceans, which means that they face common threats (e.g., bycatch in fisheries, bioaccumulation of pollutants) and are similarly vulnerable to overexploitation (Fig. 2). In the absence of IWC oversight, various bilateral and multilateral instruments have been developed to manage takes of small cetaceans (see Section III) and national programs of full protection or managed exploitation are typical (see Section IV).

Any international agreement is effective only if the parties ensure compliance and enforcement. Typically, sovereign states are unwilling to accede to a convention without assurance of being able to opt out of provisions with which they disagree. Under the ICRW, for example, member countries have 90 days to consider their options before any amendment to the schedule of whaling regulations comes into effect. Once an objection has been lodged, the regulation is no longer binding on the objecting country. It is on this basis that Norway has continued commercial whaling for North Atlantic minke whales (*Balaenoptera acutorostrata*) in defiance of the IWC's global moratorium established in 1986. Japan used another "loophole" to continue whaling. The ICRW allows contracting governments to grant

TABLE I
Current International Conservation Conventions and Institutions

Name of entity	Year of initiation	Location of secretariat or HQ	Primary mandate or responsibility in relation to marine mammals	Comments on effectiveness
International Convention for the Regulation of Whaling (ICRW); International Whaling Commission (IWC)	Signed 1946, entered into force 1948; IWC established 1951	Cambridge, UK	Conservation of whale stocks (officially concerned only with baleen, sperm, and bottlenose whales)	Very strong scientific component; controversial but highly effective in 1970–1980; suffered loss of credibility and authority in 1990s
Convention on International Trade in Endangered Species of Wild Fauna and Flora (CITES)	Signed 1973, entered into force 1975	Geneva, Switzerland	Regulation and monitoring of international trade in products from species and populations classified as threatened	Highly politicized and rancorous, but continued through 1990s to be largely effective
World Conservation Union (IUCN)	Established 1948	Gland, Switzerland (with country offices)	Maintains red list of threatened species, sponsors specialist groups (e.g., cetacean, seal, sirenian, polar bear, otter), provides advice to CITES and IWC	Specialist groups provide scientific expertise, promote and coordinate conservation research
World Wide Fund for Nature (WWF)	Established as World Wildlife Fund in 1961	Gland, Switzerland (with many national affiliates)	Lobbies for conservation, supports conservation research, and participates in international conservation fora	Influences policies of IWC and CITES, many national affiliates conduct local or regional marine mammal research and conservation programs (e.g., Philippines, USA, Canada, Hong Kong, Malaysia)
TRAFFIC Network (trade monitoring program of IUCN and WWF)	1976	Cambridge, UK (with regional or national offices)	Monitoring international trade in wildlife, works in close cooperation with CITES Secretariat	Important role in documentation of trade, with emphasis on threatened species
Convention on the Conservation of Migratory Species of Wild Animals (CMS; or Bonn Convention)	Signed 1979, entered into force 1983	Bonn, Germany	Conservation of "entire populations or any geographically separate part of the population of any species or lower taxon . . . , a significant proportion of whose members cyclically and predictably cross one or more national boundaries"	Has recognized cetaceans, but not pinnipeds or sirenians, as highly migratory species; see Table II for relevant regional agreements
Convention for the Conservation of Antarctic Seals	1972, entered into force 1978	None, but scientific advice is provided by Scientific Committee on Antarctic Research's Group of Specialists on Seals, based in Cambridge, UK	Conservation of Antarctic seals, regulation of sealing, facilitation of scientific research on seals	First international conservation agreement to be established *prior to* the initiation of exploitation
Convention on the Conservation of Antarctic Marine Living Resources (CCAMLR)	1980, entered into force 1982	Hobart, Tasmania, Australia	Facilitation of recovery of depleted whale stocks; prevention of further irreversible human-caused changes in Antarctic ecosystem	Krill monitoring program, ecosystem focus, strong scientific base
United Nations General Assembly Drift-net Resolution 46/215	1991, took effect end of 1992	None	Elimination of large-scale (longer than 2.5 km), high-seas driftnet fishing (and thus elimination of the large associated bycatch of marine mammals)	More than 1000 vessels were withdrawn from this type of fishing, but drift netting continues inside national 200-nautical-mile exclusive economic zones (and probably to some extent illegally in international waters)

Figure 1 *(Top) A fin whale* (Balaenoptera physalus) *is butchered at a whaling station in Iceland on July 11, 1988 and (Bottom) a young Icelander poses with baleen. Iceland used a scientific rationale to justify continued whaling operations for a few years after the International Whaling Commission's moratorium took effect in 1986. Later, in 1992, Iceland withdrew from the commission. Along with Norway and Japan, Iceland is a strong advocate of a resumption of whaling and the reopening of international trade in whale products. Photographs by Steve Leatherwood.*

Figure 2 *Sri Lankans begin the butchering of a Risso's dolphin* (Grampus griseus) *in August 1985. A diverse array of dolphins and whales are killed in Sri Lanka, partly as a bycatch of net fisheries and partly by direct harpooning. There is no information on stock identity or population abundance for any of the species taken along the coasts of Sri Lanka, but estimates of the total annual kill during the 1980s were in the tens of thousands. Photograph by Steve Leatherwood.*

special permits to take whales for scientific research. Although the IWC's scientific committee reviews and comments on permit proposals, its advice is nonbinding. Thus, Japan has continued to kill 300–400 Antarctic minke whales (*B. bonaerensis*) each year and, since 1994, an additional 100 North Pacific minke whales (*B. acutorostrata*), under research permits issued unilaterally by the national government.

The United Nations Convention on the Law of the Sea (UNCLOS) was ratified in 1982. Rather than strengthening efforts to conserve marine mammals, however, this framework convention has tended to provide states with a rationale for opting out of agreements such as the ICRW. Under the convention, the idea of countries having exclusive sovereign rights to manage resources within 200 nautical miles of their coastlines became firmly entrenched. This has been interpreted to mean that the hunting of coastal stocks of marine mammals should not be subject to international oversight. Also, although Article 65 of the convention calls explicitly for member states to "work through the appropriate international organizations" for the conservation, management, and study of cetaceans, it leaves considerable latitude to governments in deciding what this means. Canada, for example, withdrew from the IWC in 1982, arguing that a bilateral commission with Greenland suffices as an "appropriate international organization" for the management of beluga (*Delphinapterus leucas*) and narwhal (*Monodon monoceros*) hunting (see Section III) and that the obligation of "working through" an appropriate international body for managing bowhead whale (*Balaena mysticetus*) hunting could be discharged simply by sending an occasional representative to meetings of the IWC Scientific Committee as well as an "observer" delegation to meetings of the commission.

The Antarctic, an area that hosts several endemic seal species and provides seasonal feeding grounds for numerous migratory whale populations, is a global commons. As such, it requires its own international regime of protection and conservation (Kimball, 1999). The Antarctic Treaty system consists of four separate instruments: the initial framework treaty signed in 1959 (entered into force in 1961), the seals convention of 1972, the marine ecosystem-oriented Convention on the Conservation of Antarctic Marine Living Resources (CCAMLR) of 1980, and a 1988 convention on mineral resources. As a whole, this system is nearly comprehensive, particularly taking into account the overlapping responsibilities of the IWC and other instruments such as the 1972 Convention on the Prevention of Marine Pollution by Dumping of Wastes and Other Matter (the London Dumping Convention) and the 1973–1978 Convention for the Prevention of Pollution from Ships (the MARPOL Convention). It should provide an adequate legal basis for protecting ANTARCTIC MARINE MAMMAL populations. What it cannot do, of course, is reverse the devastation of the southern stocks of baleen whales caused by the whaling industry. Nor can it protect the seal and whale populations from the ongoing (and possibly worsening) effects of climate change and ozone depletion (see Section VIII).

Another treaty that is relevant to marine mammal conservation is the Convention on International Trade in Endangered Species of Wild Fauna and Flora (CITES). This convention has been in force since 1975 and has provided a high-profile, controversial forum in which many of the most searing conservation issues of the day have been addressed. The economic stakes are often high, and this is certainly the case with trade in whale meat and blubber, which are in great demand in Japan. CITES has essentially no relevance for exploitative regimes in which the products are for domestic consumption only or for those instances in which animals are captured alive and placed in institutions within the country of origin. It becomes relevant only when the animals or their products or derivatives cross international borders. Under CITES, species and geographical populations are subject to listing in one of three appendices. Appendix I species or populations are threatened with extinction, and trade in products from them needs to be strictly regulated. Those in Appendix II are not considered to be in immediate danger of extinction but may become so unless trade is strictly monitored. The third appendix includes species or populations that are subject to national regulation and for which

multilateral cooperation is necessary to avoid overexploitation. The goals of monitoring and regulation are achieved through a system of permits and certificates for export or import, issued by national governmental authorities. As of 2000, when the 11th meeting of the conference of the parties took place in Nairobi, Kenya, all of the commercially valuable baleen whales were listed in CITES Appendix I, as were some of the odontocetes, including the sperm whale. Also, a resolution adopted in 1979, calling for CITES member states to honor IWC restrictions on whaling and the trade of whale products, remained in force.

The combined restrictions on trade under CITES and the IWC have not eliminated concern about the role of international commerce in undermining conservation efforts. The meat and blubber from Japanese "scientific whaling" are legally sold in Japan's large domestic market. Also, the blubber from minke whales killed off Norway is not all consumed domestically, and there is strong interest in exporting the surplus to Japan. Forensic techniques developed by geneticists have been used to monitor the species and populations represented in stores and restaurants in Japan, Hong Kong, and the Republic of Korea. A sliver of meat or blubber is sufficient to identify the source to species and in some cases to ocean basin or stock (Dizon *et al.*, 2000). Forensic analyses have revealed that products from "protected" species or populations are sometimes sold. However, it has usually been impossible to prove that these came from animals that were killed illegaly rather than from strandings, bycatch, or stockpiles acquired prior to the IWC moratorium. In a few instances, illegal shipments of whale meat have been intercepted by authorities in Norway, Japan, and Russia before reaching their destination.

An agreement by the U.S. and Japan governments in November 1991 included the following statement: "Data collected in 1990 indicated that over 41 million non-target fish, sharks, sea birds, marine mammals and sea turtles were killed in the Japanese squid driftnet fishery alone." The high-seas squid drift net fisheries in the North Pacific were killing approximately 15,000–30,000 northern right whale dolphins (*Lissodelphis borealis*), 11,000 Pacific white-sided dolphins (*Lagenorhynchus obliquidens*), and 6000 Dall's porpoises (*Phocoenoides dalli*) annually in the late 1980s. Drift nets set for salmon, tuna, and billfish were taking thousands more dolphins, porpoises, whales, and pinnipeds each year. The threats to populations of pelagic cetaceans, pinnipeds, seabirds, and many other organisms posed by large-scale, high-seas drift gill nets were judged to be sufficiently severe and global in scope to necessitate action by the United Nations. This took the form of a resolution, passed by the UN General Assembly in 1991, calling on all member nations to enact a moratorium on such drift netting by the end of 1992 (Northridge and Hofman, 1999). The global ban was a valuable step and undoubtedly helped to avert catastrophic declines in some marine animal populations. What the UN decree could not affect, however, was the use of these nets inside the 200 nautical mile limit of coastal states. Consequently, pinnipeds and cetaceans continue to be killed in large numbers by drift gill nets deployed in coastal waters.

III. Regional and Bilateral Conservation Efforts

In cases involving species or populations with well-defined distributions that cross several national boundaries, multilateral regional bodies have sometimes been established to monitor and manage exploitation (Table II). Included among these are some "international" instruments that are in fact regional because their scope is defined by the limited geographical ranges of the animals involved. For example, membership in the Inter-American Tropical Tuna Commission (IATTC) has been geographically diverse since its founding in 1949 under a treaty between the United States and Costa Rica. Only states with an interest in fishing for tuna in the eastern tropical Pacific Ocean have joined the commission, and over the years this has included France, Japan, Vanuatu, the United States, and Mexico in addition to a number of Central and South American countries. Thus, while the commission is international in the sense of having a geographically diverse membership, its purview is distinctly regional. Similarly, the International Agreement on the Conservation of Polar Bears and their Habitat involves only the northern circumpolar countries where polar bears regularly occur, and it is therefore treated here as a regional agreement.

The 1911 Treaty for the Preservation and Protection of Fur Seals (often referred to as the North Pacific Fur Seal Convention) involved four countries: Great Britain (signing on behalf of Canada), the United States, Russia, and Japan. This was essentially an agreement among the states involved in the exploitation of northern fur seals (*Callorhinus ursinus*), which are endemic to the North Pacific Ocean. Pelagic sealing was banned, and as part of the agreement, Japan and Canada were allocated a portion of the profits from the controlled killing (mainly of "surplus" male seals) on the Pribilof (USA) and Commander (Russia) islands. This treaty lapsed in 1941, when Japan withdrew, and was replaced in 1957 by the Interim Convention on the Conservation of North Pacific Fur Seals. The North Pacific fur seal is frequently cited as a conservation success story. Elimination of pelagic sealing, in combination with regulations limiting the kill at the breeding rookeries, allowed the seal population to make a strong recovery from its depleted state in 1910. The population reached about 2 million in the 1950s but dipped below a million by the early 1980s. Unfortunately, the 1957 convention lapsed in 1984 and has not been replaced. The fur seal population in U.S. waters stood at about a million in the late 1990s, and since 1988 it has been designated as "depleted" under the Marine Mammal Protection Act.

The polar bear (*Ursus maritimus*) treaty mentioned earlier is often cited as an example of an effective international agreement. Discussions among the range states—Canada, the United States, Denmark (on behalf of Greenland), Norway, and the Soviet Union (now Russia)—began in the mid-1960s, when the future of the polar bear was of great concern because of overhunting and habitat deterioration (Lyster, 1985). The three main objectives of the agreement, which was signed in 1973 and took effect in 1976, were to ensure that appropriate restrictions were placed on HUNTING, that polar bear habitat was preserved, and that needed research was conducted in a coor-

dinated fashion. The Polar Bear Specialist Group of the IUCN's Species Survival Commission has served as a *de facto* scientific committee, meeting every few years to share information, discuss research needs, and assess the state of polar bear conservation. At its 12th working meeting in 1997, the group concluded that the total population of wild polar bears was between about 22,000 and 27,000, more than half of them in Canada. It also stressed the need for better protection of female bears and their denning habitat and expressed concern about the high levels of PCBs found in Svalbard's bears.

The multinational hunt for harp and hooded seals (*Pagophilus groenlandicus* and *Cystophora cristata,* respectively) in the northern North Atlantic proceeded without meaningful regulation until the late 1950s, when Norway and the Soviet Union established a bilateral commission to set quotas for commercial takes of harp and hooded seals as well as walruses (*Odobenus rosmarus*) in the northeastern Atlantic. The reach of this agreement has been interpreted as including large areas of the Greenland and Barents seas, Denmark Strait, and waters near the island of Jan Mayen. A similar bilateral agreement pertaining to the northwestern Atlantic was signed by Canada and Norway in 1971.

A succession of regional bodies have become involved in monitoring the North Atlantic seal hunt and assessing the harp and hooded seal populations in order to provide management advice. Starting in the 1960s, a Sealing Panel of the International Commission for Northwest Atlantic Fisheries (later the Northwest Atlantic Fisheries Organization, or NAFO) recommended overall quotas and other conservation measures (e.g., opening and closing dates for sealing from ships) related to the hunting of the western populations. In recent years, scientific advice on harp and hooded seal stocks has been developed by a working group convened jointly by the International Council for the Exploration of the Sea (ICES) and NAFO. This group's advice is presented to the North Atlantic Marine Mammal Commission (NAMMCO), which in turn offers management advice to its members. Canada decides unilaterally on sealing quotas in the western North Atlantic, whereas Norway and Russia continue to allocate quotas in the West Ice (Jan Mayen) and East Ice (White Sea) on a bilateral basis.

NAMMCO is a regional body established in 1992 by several countries who had become frustrated by the IWC's unwillingness to allow the resumption of commercial whaling. Its membership presently consists only of Iceland, Norway, Greenland, and the Faroe Islands, the latter two belonging to the Kingdom of Denmark but with "home rule" governments. Considerable effort has been made to entice Canada and Russia to join, and Japan regularly sends an official observer to meetings of the NAMMCO Council. Although Iceland withdrew from the IWC in 1992, Norway and Denmark have maintained their IWC membership, thus trying to balance their involvement in both the IWC and NAMMCO. In its early years, NAMMCO focused much of its attention on species for which there was little or no direct conflict with the IWC, notably harp and hooded seals, ringed seals (*Pusa hispida*; Fig. 3), gray seals (*Halichoerus grypus*), walruses, long-finned pilot whales (*Globicephala melas*), and northern bottlenose whales (*Hyperoodon ampullatus*).

TABLE II
Regional or Bilateral Conservation Agreements Currently in Effect

Name of entity	Year of initiation	Location of secretariat or HQ	Mandate or objectives	Comments on effectiveness
Inter-American Tropical Tuna Commission (IATTC)	1949	La Jolla, California	Initially to document and manage tropical tuna fisheries; since then, expanded to include documentation, mitigation, and regulation of dolphin mortality incidental to fishing operations in eastern tropical Pacific	Operates programs to place observers aboard tuna vessels, reduce dolphin mortality through diagnosis and solution of gear problems, and training for captains and crews; provides mechanism for linking tuna industry with government agencies and environmental NGOs
North Atlantic Marine Mammal Commission (NAMMCO)	1992	Tromsø, Norway	Sustainable use and management of marine mammals in the North Atlantic Ocean	Emphasis on ecological interactions (e.g., rationales for culling marine mammals to protect fish stocks), hunting rights of coastal communities, forum for scientific information exchange
Canada/Greenland Joint Commission on Narwhal and Beluga	1989	None (Ottawa, Canada; Nuuk, Greenland)	Cooperative research and management related to "shared" stocks of narwhals and white whales	Forum for bilateral studies and sharing of information, with management measures left to national authorities and local "comanagement" bodies
Agreement on the Conservation of Small Cetaceans of the Baltic and North Seas (ASCOBANS)	Signed 1991 (concluded under CMS, the Bonn Convention; see Table I), entered into force 1994	Bonn, Germany	Cooperation to achieve and maintain a "favourable conservation status" for small cetaceans in the region	Most effort in the 1990s was directed at estimating abundance and incidental takes of harbor porpoises (and white-beaked dolphins) and seeking ways to reduce bycatch
Agreement on the Conservation of Cetaceans of the Black Sea, Mediterranean Sea and Contiguous Atlantic Area (ACCOBAMS)	1996 (concluded under CMS, the Bonn Convention; see Table I), not yet in force	Monaco	Cooperation to achieve and maintain a "favourable conservation status" for cetaceans in the region, including the complete prohibition of deliberate taking and establishment of a network of "specially protected areas to conserve cetaceans"	In its early days at the time of this writing
International Agreement on the Conservation of Polar Bears and their Habitat	Signed 1973, entered into force 1976	None; follows rotating chairmanship of IUCN Polar Bear Specialist Group	To prevent polar bear populations from becoming endangered because of hunting or other human activities	Provides a framework for communication and cooperation among circumpolar countries, emphasis on research and monitoring; signatory states are supposed to "enact and enforce such legislation and other measures as may be necessary to give effect to the Agreement"
U.S.–Russia Agreement on Cooperation in the Field of Environmental Protection	1972	None (Washington and Moscow)	Marine Mammal Project, under Area V of the agreement, provides for information exchange, coordination of research activities, and joint or cooperative research	Annual scientific meetings, with focus formerly on Bering and Chukchi sea regions, recently broadened to consider, e.g., Caspian seals, bycatch in Japanese salmon drift nets operating within Russian EEZ

Figure 3 *A Greenlandic hunter with a dead ringed seal tied onto his sled on May 31, 1988. Ringed seals are a staple of many maritime communities in the Arctic. They provide food for people and dogs, and their skins are either used to make clothing or sold for cash. Photograph by Steve Leatherwood.*

However, it remains uncertain whether, and to what extent, NAMMCO will eventually infringe upon the IWC's authority to manage commercial whaling.

An international treaty, the Convention on the Conservation of Migratory Species of Wild Animals (or Bonn Convention), provides a mechanism for regional conservation agreements. Two dealing explicitly with cetaceans have been negotiated thus far. The Agreement on the Conservation of Small Cetaceans in the Baltic and North Seas (ASCOBANS) entered into force in 1994, with a membership that included Belgium, Denmark, Germany, the Netherlands, Poland, Sweden, and the United Kingdom. In its early years, ASCOBANS addressed a wide range of issues related to harbor porpoises (*Phocoena phocoena*), including assessments of abundance and bycatch, pollution, and effects of noise disturbance. A major achievement was the instigation, planning, and completion of a multinational abundance survey of cetaceans throughout the Baltic and North Seas. ASCOBANS has a close working relationship with the IWC and provides the kind of international oversight for small cetaceans, at least in this one region, that the IWC is prevented from offering (see Section II). The other instrument, the Agreement on the Conservation of Cetaceans of the Black Sea, Mediterranean Sea and Contiguous Atlantic Area (ACCOBAMS), was concluded in 1996 and entered into force in 2001. One provision of this agreement is that parties must prohibit the deliberate killing of cetaceans in their national waters.

Canada and Greenland have adopted bilateralism as a preferred approach to manage the hunting of "transboundary stocks" of white whales and narwhals. In 1989, the two governments signed a memorandum of understanding that recognized the importance of hunting to the Inuit and called for "the

rational management, conservation and optimum utilization of living resources of the sea" as reflected in the UN Convention on the Law of the Sea (see Section II). The Joint Commission on Conservation and Management of Narwhal and Beluga meets annually, as does its Scientific Working Group. In addition to management advice directed at government agencies and, in Canada, the Nunavut Wildlife Management Board (a comanagement body established under an aboriginal land-claim agreement), the commission's Scientific Working Group plans and undertakes collaborative research on "shared" stocks of narwhals and belugas.

IV. National Conservation Efforts

In the United States, the Marine Mammal Protection Act was passed by Congress in 1972, and this act has been the cornerstone of a massive domestic commitment to conservation (Baur *et al.*, 1999). Although the MMPA is not the only such law in the world (many countries confer full legal protection to marine mammals), it is undoubtedly the most sweeping of its kind. At the time of the act's passage, a preeminent concern was the annual slaughter of more than 100,000 young harp and hooded seals on the spring pack ice off Newfoundland and in the Gulf of St. Lawrence. Public outrage at film footage of helpless, unweaned seal pups being clubbed to death was probably the most influential single factor in forcing Congress to act. In addition, however, there was growing concern about the deplorable condition of the world's stocks of large whales, especially the blue whale (McVay, 1966). Moreover, controversy swirled around the killing of pelagic dolphins by the American tuna fleet in the eastern tropical Pacific (Gosliner, 1999). The estimated annual kill from 1960 through 1971 had been more than 370,000 dolphins (nearly 4.5 million total during that period), and environmentalists were understandably outraged.

The resulting legislation was both comprehensive and innovative. An immediate embargo was placed on the importation of marine mammal products, with only a few specified exceptions. Deliberate taking was banned, although Alaskan Eskimos and other aboriginal people were allowed to continue hunting marine mammals for food, skins, and other products as long as the main purpose was to meet basic community needs. A strong emphasis in the act was placed on research, and science was accorded a prominent role in influencing how decisions were to be made. Existing federal departments were given the responsibility of implementing the new law, with the Department of Commerce to manage cetaceans and most pinnipeds and the Department of the Interior to manage polar bears, sea otters, sirenians, and walruses. In addition, however, an entirely new and independent federal agency, the Marine Mammal Commission, was established to oversee the process of implementation.

The goals of management, as set forth in the Marine Mammal Protection Act, were to achieve and maintain "optimum sustainable populations" of marine mammals and to reduce incidental mortality from fishing operations (including tuna seining) to "insignificant levels approaching zero." An optimum sustainable population has been defined operationally as having a lower bound at the maximum net productivity level and an upper bound at the unexploited population (carrying capacity) level. By defining population status in terms of productivity, the act placed the emphasis on the health and stability of ecosystems rather than on economic yield, as had traditionally been the case. The ambitious and lofty goals of the Marine Mammal Protection Act have been pursued over the past quarter century with what appears to be an undiminished national commitment to conservation.

The European Union, functioning as though a single state, has used a more selective approach to achieve certain objectives related to marine mammal conservation. In 1983, the European Community (as it was then called) established a controversial ban on the importation of products from seal pups. The explicit goal was to stop the clubbing of young white-coated harp seals and blue-backed hooded seals, a concern more directly related to animal welfare than to conservation. In combination with closure of the U.S. import market from 1972, the European ban effectively destroyed the profitability of sealskin production in North America and Greenland, with serious unintended economic and social consequences in Eskimo communities where the hunting of ringed seals was a major source of cash income. Canada was forced to stop the commercial hunt for unweaned harp and hooded seal pups, essentially bringing the large-scale, ship-based sealing industry to a halt. For more than a decade, the populations of harp and hooded seals were allowed to recover. Since the mid-1990s, however, Canada's commercial sealing industry has been reinvigorated with government subsidies and aggressive product marketing, particularly focused on the export of seal penises to China and other Asian countries, so the kill of harp and hooded seals has returned to levels not seen since the early 1970s (Lavigne *et al.*, 1999).

The European Community also effectively banned the importation of whale products in 1982–1983 by declaring that all cetaceans would be treated as though they were in Appendix I of CITES (no commercial trade allowed). Greenland, with its special relationship to Denmark, was exempted from the ban, meaning that narwhal tusks could be imported to EU countries under the normal provisions of CITES Appendix II. Thus, although the EU measure caused a steep decline in the value of narwhal ivory from Canada (Fig. 4), it had comparatively little effect on the market for tusks from Greenland, which was traditionally centered in Europe. Having lost access to the American market and much of the European market for narwhal ivory, entrepreneurs in Canada found new buyers in Asia, a pattern similar to that seen with seal products.

National conservation efforts are often influenced by international legal instruments, and such influence can be for either good or ill. The U.S. government, in implementing the Marine Mammal Protection Act, has had to take account of what are sometimes conflicting commitments under international agreements. For example, the United States has always belonged to the Inter-American Tropical Tuna Commission (IATTC), whose primary goal is to maximize tuna catches. A sometimes uneasy alliance has been forged over the years between the IATTC and the U.S. National Marine Fisheries Service, the agency responsible for pursuing the "zero mortality rate" goal for dolphins mandated by the Marine Mammal Protection Act. Although substantial progress had been made toward that goal by the late 1980s, animal protection groups continued to mount

Figure 4   *The long, spiraled tusk of the male narwhal has commercial value and is traded in the global marketplace. Here, Inuk hunters on northern Baffin Island, Canada, have dragged a carcass onto a gravel beach (A) and are proceeding to remove the tusk from the animal's upper jaw (B). In the 1970s, many tusks, like the one shown (C), were sold to a non-Inuk who exported them to Europe (mainly the United Kingdom). August 1975. Photographs by Randall Reeves.*

legal challenges, insisting that the procedure of setting purse seines around dolphins in order to catch tuna should cease altogether. Their efforts led to the "dolphin-safe" labeling of canned tuna and to embargoes on U.S. imports of tuna from countries continuing to "fish on dolphins." The IATTC took the position that by redirecting all fishing effort away from "dolphin sets" and toward "school sets" and "log sets" (neither of which involve dolphin encirclement), the bycatch of other species (e.g., billfish and turtles) and the proportion of undersized tuna in the catch would both increase. In general, the Fisheries Service has tended to assign a higher priority to dolphin protection than has the IATTC, and the working relationship has been strained in recent years because of this and other differences. To make matters worse, Mexico mounted a challenge to the tuna embargoes under the General Agreements on Tariffs and Trade (GATT), insisting that they were unwarranted and unacceptable impediments to free trade. The dispute-resolution panel ruled that the embargoes were indeed inconsistent with GATT provisions, and the United States consequently has had to seek a delicate balance between its commitment to marine mammal protection and its support for the principle of free trade (Gosliner, 1999).

In a more positive vein, the IWC has managed to influence the conservation of small cetaceans in Japan, despite Japan's refusal to recognize the commission's authority to impose measures related to the conservation of dolphins, porpoises, and smaller species of toothed whales (Perrin, 1999). A variety of small cetaceans have been hunted in Japanese coastal waters for many decades (Fig. 5). The Subcommittee on Small Cetaceans, a standing subcommittee of the IWC's Scientific Committee, meets annually to review the status of species, particular threats, and technical approaches to eliminating or managing the threats. In their reviews of stocks, the subcommittee has repeatedly found evidence of overexploitation by the Japanese coastal cetacean fisheries. As a result, the government of Japan has been forced, through international pressure from governments and nongovernmental organizations, to implement research programs and management measures that almost certainly would not have been contemplated in the absence of this outside scrutiny.

The most glaring example is the striped dolphin (*Stenella coeruleoalba*). A drive hunt for striped dolphins, in which entire schools are herded toward shore and killed *en masse*, has taken place annually on Japan's Izu Peninsula for more than a century. Catches of as many as 22,000 animals occurred in some years during the 1940s and 1950s. By the 1980s, when the hunters introduced a voluntary catch limit of 5000 dolphins, the annual average catch had declined to less than 3000 per year, presumably because the population was seriously depleted. Finally, in 1989, government quotas were imposed. Although ideally the hunt should have stopped entirely to allow the dolphin population to recover, the issue of overexploitation

Figure 5 *Japanese fishermen have hunted small cetaceans for many decades, often driving hundreds of animals toward shore where they are killed en masse. Rough-toothed dolphins* (Steno bredanensis), *shown here, are rarely taken in drive hunts, and even more rarely is a photographer on hand to record the carnage. Photograph by Rusty White, courtesy of Hubbs Marine Research Institute.*

of striped dolphins has at least helped demonstrate the need for stronger measures by the national government in Japan to prevent other similar disasters.

V. Local and Individual Conservation Efforts

Top-down approaches to managing resource use have often failed. The cost of policing human actions is likely to be unacceptably burdensome when local people attribute no legitimacy to the management regime. It is generally agreed that the greater the local or community involvement, the more likely it is that conservation efforts will succeed in the long run (Mangel *et al.*, 1996). Marine mammal hunting communities in the Arctic and in Australia [where dugongs (*Dugong dugon*) are the principal prey species] have been at the forefront of developing cooperative management ("comanagement") agreements. Ideally, such agreements recognize the interests and rights of local people while the broader concerns for conservation from a national or international perspective are represented by central governmental authorities.

The most prominent example is the Alaska Eskimo Whaling Commission (AEWC), which was established in 1977 by whalers in northern Alaska in reaction to the IWC's controversial decision to ban bowhead whaling. After several years of difficult negotiations, marked by threats, lawsuits, and even a grand jury investigation into violations of the agreed bowhead quota in 1980, a cooperative agreement was reached between the AEWC and the National Oceanic and Atmospheric Administration (NOAA), the federal body directly responsible by law for implementing IWC decisions within the United States. Under this agreement, the AEWC has assumed responsibility for managing the hunt, monitoring compliance with the quota and other regulations, and reporting each year's results. Quotas on the allowed number of strikes and landings of bowheads are still negotiated through the IWC.

Singling out the contributions of individuals to the cause of conservation is a questionable proposition. How does one recognize, measure, or compare the value of the conscientious daily efforts of bureaucrats, scientists, writers, educators, fishermen, engineers, veterinarians, lobbyists, lawyers, political activists, and others? Several individuals are mentioned here, but only with the caution that their work, while exemplary, is not necessarily exceptional.

As discussed further later (under Section VII), the rescue and rehabilitation of injured or otherwise incapacitated individual marine mammals can be of questionable conservation value. Nevertheless, in some circumstances, especially when an endangered species is involved, intervention is certainly appropriate. Jon Lien, a professor and researcher at Memorial University in Newfoundland, began working with fishermen in the late 1970s to devise ways of extricating whales, especially humpback whales, from fishing gear. The problem of entanglement was a concern of conservationists because, at the time, humpbacks in the North Atlantic were considered to be endangered (their status has improved since then). It was of concern to fishermen because of the economic losses associated with damaged gear and lost fishing time, as well as the personal danger involved when dealing with these large animals at sea. Lien gained the confidence of the Newfoundland fishermen and developed a highly successful program for assisting in the safe release of entangled or ENTRAPPED whales. Subsequently, Charles Mayo, David Mattila, and their associates at the Center for Coastal Studies on Cape Cod began rescuing whales from fishing gear on the U.S. coast, with an emphasis on critically endangered right whales (Fig. 6). Disentanglement teams are now integral to right whale (*Eubalaena glacialis*) recovery efforts in the eastern United States and southeastern Canada, thanks to the pioneering efforts of these three scientists. A similar, highly successful program centered on rescuing harbor porpoises, minke whales, and occasionally right whales trapped in herring weirs has been in operation in Canada's Bay of Fundy for more than a decade.

One of the greatest obstacles to conservation can be the difficulty of defining a threat and providing incontrovertible evidence of its effects. Marine debris pollution provides a clear example (Laist *et al.*, 1999). Although it is widely accepted today that marine debris, such as derelict fishing gear and plastic packaging material, is a menace to wildlife, the problem's seriousness was not recognized until the early 1980s. Charles Fowler, a scientist with the National Marine Fisheries Service in Seattle, was engaged in research to determine the cause of the North Pacific fur seal population's continuing decline (see Section III). Despite bureaucratic resistance and the skepticism of scientific colleagues, Fowler pressed ahead with the task of marshaling data to test the hypothesis that entanglement in debris was a major cause of juvenile mortality in fur seals. His painstaking compilation of evidence, together with mathematical models, finally convinced others that at least this one marine

Figure 6 *A team from the Center for Coastal Studies in Provincetown, Massachusetts, attempts to cut away fishing line wrapped round the tail of a right whale in Cape Cod Bay on July 24, 1998. Specially designed knives are deployed at the ends of long poles. Note that the men are wearing helmets as a precaution while undertaking this extremely dangerous work. Photograph courtesy of Center for Coastal Studies, Provincetown, Massachusetts.*

mammal species was being affected by debris pollution at the population level. Fowler's work provided the impetus for a chain of events, from beach clean-up campaigns to the signing of international treaties, intended to reduce the ocean's burden of debris and therefore lessen the risks to seals and cetaceans, to say nothing of seabirds, turtles, and other marine wildlife.

One final example of an individual's ability to change the course of conservation policy again relates to dolphin mortality in the eastern tropical Pacific tuna fishery. By the late 1980s, many conservationists had forgotten about this issue, assured that dolphin mortality had been reduced substantially as a result of changes in fishing techniques and the imposition of annual quotas on the number of dolphins from each species that could be killed before fishing would have to cease. In 1987–1988, however, Sam LaBudde, who described himself as an "itinerant biologist," spent 5 months aboard a Panama-registered tuna boat. Although he had signed on as an ordinary seaman and cook, he carried a video camera and clandestinely recorded grisly footage of dolphins being killed. When the scenes were aired on national television, it galvanized public support within the United States for strong measures to be taken against the non-American tuna fleet. While LaBudde's actions can be viewed as either heroic or deceitful, depending on one's point of view, there is no doubting his courage or his influence on the course of conservation.

VI. Protected Areas

The designation of specially protected areas (e.g., reserves, sanctuaries, parks) is a tool increasingly used to achieve conservation goals. Of the many such areas around the world (Table III), relatively few exist for the explicit purpose of benefitting marine mammals. Mexico declared Scammon's Lagoon (Laguna Ojo de Liebre) a refuge for gray whales in 1971, and San Ignacio Lagoon was given similar status in 1979. Also, Mexico established a Biosphere Reserve in the upper Gulf of California in 1993 mainly to protect the highly endangered vaquita (*Phocoena sinus*) and the totoaba, an endangered sea bass (*Totoaba macdonaldi*). New Zealand created the Banks Peninsula Marine Mammal Sanctuary in 1988 to protect Hector's dolphins (*Cephalorhynchus hectori*), like the vaquita an endemic coastal species, from entanglement in gill nets. In 1999, the parliament of the North German state of Schleswig-Holstein established a sanctuary for small cetaceans off the islands of Sylt and Amrum in the North Sea, intended to protect harbor porpoises from the dangers associated with gillnet fishing, jet skiing, and high-speed motor boating. In the United States, the Hawaiian Islands Humpback Whale Sanctuary was declared in 1993, and several other marine sanctuaries were established in large part because of public interest in the marine mammals that use them for feeding, breeding, or both (e.g., the California Channel Islands, Gulf of the Farallones, Monterey Bay, and Stellwagen Bank sanctuaries).

All too often, protected areas are created in response to a public outcry, but without an accompanying ongoing commitment to establish and enforce meaningful restrictions on human activities within them. The marine sanctuary program in the United States, for example, has failed to meet the public's

TABLE III
Protected or Managed Areas Intended, at Least in Part, to Benefit Marine Mammals

Name of area	Location or country	Marine mammals affected	Year established	Comments
International				
IWC Indian Ocean Sanctuary	Indian Ocean, 20–130°E and north from 55°S	Mainly baleen whales and sperm whale	1979	Commercial whaling banned
IWC Southern Ocean Sanctuary	Circumpolar south of Antarctic Convergence	Mainly baleen whales and sperm whale	1994	Commercial whaling banned
Oceania				
Point Labatt Conservation Park	South Australia mainland	Australian sea lion	1973	Protects haul-out beach as well as waters to 1 nautical mile offshore
Seal Bay Conservation Park	Kangaroo Island, South Australia	Australian sea lion (New Zealand fur seal also present)	1971 (having existed in other forms since 1954), expanded 1974	Intensively managed to promote tourism and maintain the sea lion population
Banks Peninsula Marine Mammal Sanctuary	East coast of South Island, New Zealand	Hector's dolphin	1988	Prohibits commercial gill netting, limits amateur gill netting
Great Barrier Reef Marine Park	Off Cape York, Queensland, Australia	Dugong, humpback whale, Indo-Pacific hump-backed dolphin, dwarf minke whale	1975	Intensively and complexly managed, with dugong as a major focus of conservation efforts
Great Australian Bight Marine National Park	Approximately 300 km of coastline in South Australia, including waters to three nautical miles offshore	Southern right whale, Australian sea lion	1996	Includes exclusion zone in Head of the Bight where right whales congregate in winter (with their calves), and prohibited areas, each 1 nautical mile radius, around sea lion breeding colonies
Maza Wildlife Management Area	Western Province, Papua New Guinea	Dugong	1978	Intended to protect dugong habitat and manage continued hunting; effectiveness uncertain
Europe				
Northeast Greenland National Park	972,000 sq. km of land in NE Greenland	Walrus, polar bear	1974, expanded 1988	Prevents commercial hunting; restricts mineral development
Wadden Sea National Park of Schleswig-Holstein	German Wadden Sea, east side of North Sea	Harbor porpoise	1999, expanded to include coastal waters of isles of Sylt and Amrum	Bans high-speed water craft and industrial fishing in porpoise "nursery" area
International Sanctuary for Mediterranean Cetaceans	Ligurian Sea, Northwestern Mediterranean Basin	Fin and sperm whales, various smaller odontocetes	1999, by agreement of Italy, France, and Monaco	In initial stages; whale watching regulated, high-speed offshore competitions to be regulated or prohibited, monitoring of cetacean populations
National Marine Park of Alonnissos Northern Sporades	Greece	Mediterranean monk seal	1986	Awareness and enforcement have been effective in reducing monk seal mortality
Nature Reserve of Las Desertas Islands	Madeira Natural Park, Madeira	Mediterranean monk seal	1990	Protection has been effective, and the very small seal population has increased
Valaam Archipelago Nature Reserve	Lake Ladoga, Karelian Republic of Russia	Ladoga ringed seal	1992	Further restrictions needed to reduce disturbance of hauled-out seals by campers
Franz Josef Land State Nature Reserve	NE Russia	Arctic seals and whales, polar bear	1994	Nominal protection from disturbance

(continues)

TABLE III (*Continued*)

Name of area	Location or country	Marine mammals affected	Year established	Comments
Asia				
Sha Chau and Lung Kwu Chau Marine Park	Hong Kong	Indo-Pacific hump-backed dolphin (locally called Chinese white dolphin)	1996	Prohibits high-speed watercraft and some kinds of fishing
National Chambal Sanctuary	Chambal River, India	Ganges river dolphin	1978	Primary purpose: protection of gharial (*Gavialus gangeticus*); effectively protects small dolphin population (tens of animals)
Vikramshila Gangetic Dolphin Sanctuary	50-km segment of Ganges River, Bihar, India	Ganges river dolphin	1991	Nominal protection to dolphins, regular surveys
Sind Dolphin Reserve	Indus River between Sukkur and Guddu barrages, Pakistan	Indus river dolphin	1974	Nominal protection to dolphins, regular surveys
Africa				
Saloum Delta National Park and Biosphere Reserve	Central Senegal	West African manatee, Atlantic hump-backed dolphin, bottlenose dolphin	1976?; Biosphere Reserve 1980	Poaching of manatees and bycatch of dolphins and manatees in fisheries occurs, but these and other harmful activities would be worse without the protection provided by park staff
Niumi National Park	Gambia	Atlantic hump-backed dolphin, probably West African manatee	1986	Same as for Saloum
Kiang West National Park	Gambia	Bottlenose dolphin, probably West African manatee	1987	Same as for Saloum; some commercial dolphin watching is facilitated
National Park of Banc d'Arguin	Mauritania; ca. 12,000 km², including coastal waters mainly <5 m deep	Mediterranean monk seal, Atlantic hump-backed dolphin, bottlenose dolphin	1976	Artisanal fishing is allowed within the park and probably involves substantial bycatch of marine mammals, but exclusion of industrial fishing and other general protection given to wildlife and fish probably benefits marine mammals to some extent
More than 20 designated protected areas	South Africa coast and nearshore waters	Right whale, various small odontocetes, Cape fur seal	1970s to present	Some explicit protection from disturbance by whale watching and research vessels
North America				
21 manatee sanctuaries	Throughout Florida	West Indian manatee	Florida Manatee Sanctuary Act (1978)	Restrictions on boat speed primarily
Blue Spring State Park	Upper St. Johns River, NE Florida	West Indian manatee		Very effective protection and public education program
Bahía de Chetumal manatee refuge	Mexico	West Indian manatee		
Stellwagen Bank National Marine Sanctuary	Off Cape Cod, MA	Humpback whale and other cetaceans	1993	Fishing activity unrestricted; prohibits sand and gravel mining, dumping or discharging of harmful substances
Saguenay-St. Lawrence Marine Park	Lower St. Lawrence River, Quebec, Canada	White whale (beluga), baleen whales	1998	Zoning for multiple use, provides protection from "harassment" by tourist vessels
Wapusk National Park	West coast of Hudson Bay, south and east of Churchill, Manitoba, Canada	Polar bear	1996	Protects maternity dens and resting areas, facilitates management of bear-watching tourism and of human–bear conflicts

(continues)

TABLE III (*Continued*)

Name of area	Location or country	Marine mammals affected	Year established	Comments
Gulf of the Farallones National Marine Sanctuary	Off San Francisco, CA	Various baleen and toothed cetaceans, otariids, and phocids	1981	Prohibits exploration for oil and gas, waste dumping and discharging; no special restrictions on fisheries
Channel Islands National Marine Sanctuary	Southwest of Santa Barbara, southern California	Various baleen and toothed cetaceans, otariids, and phocids	1980	Prohibits exploration for oil and gas, waste dumping and discharging; no special restrictions on fisheries
Monterey Bay National Marine Sanctuary	350 miles of coastline and adjacent waters, northcentral California	Sea otter, pinnipeds, cetaceans	1992	Prohibits exploration for oil and gas, waste dumping and discharging; no special restrictions on fisheries
Glacier Bay National Park and Preserve	Alaska	Humpback whale	1980	Vessel traffic regulated to prevent whale disturbance
El Vizcaino Biosphere Reserve (includes lagoon whale refuges)	Pacific coast of Baja California Sur, Mexico	Gray whale	Whale sanctuaries designated 1971, 1979, 1980, Biosphere Reserve 1988, UNESCO World Heritage Site 1993	Restrictions on human activity including tourism, fishing; Laguna Ojo de Liebre was world's first whale refuge
Upper Gulf of California and Colorado River Delta Biosphere Reserve	Upper quarter of Gulf of California, Mexico	Vaquita	1993	Nominal protection to vaquitas but fails to address main problem: bycatch in artisanal gillnet fisheries
South America				
Galápagos National Park; Galápagos Islands Marine Resources Reserve and Whale Sanctuary	Galápagos Islands, Ecuador	Galápagos fur seal, Galápagos sea lion, coastal cetaceans	National Park 1979, Whale Sanctuary 1990	Full protection for all wildlife
Fernando de Noronha National Marine Park	Archipelago off equatorial Brazil	Spinner dolphin	1988	Nominal protection to dolphins as well as support for dolphin-watching tourism
Lobos Island Ecological Reserve	Near border of Santa Catarina and Rio Grande do Sul, Brazil	South American sea lion and fur seal	1983	Protected haul-out site
Environmental Protection Area of Anhatomirim/ Bay of Dolphins	Florianópolis, Santa Catarina, Brazil	Tucuxi (marine type)	1992	Nominal protection from harassment
Golfo San José Marine Park (Whale Sanctuary)	Peninsula Valdés, Argentina	Southern right whale	1974	
Provincial Wildlife Reserve of North Point (Punta Norte)	Peninsula Valdés, Argentina	Southern elephant seal	1967	Protects breeding haul-out beach
Provincial Wildlife Reserve of Punta Pirámides	Peninsula Valdés, Argentina	Southern sea lion	1974	Protects breeding haul-out beach
Pacaya-Samiria National Reserve	Peruvian Amazon	Boto, tucuxi, Amazonian manatee	1972, enlarged 1982	Hunting and fishing by local people allowed, nominal protection to animals, some restrictions on commercial exploitation and industrial activity, some dolphin-watching tourism
Mamirauá Sustainable Development Reserve	Western Brazilian Amazon	Boto, tucuxi, Amazonian manatee	1990, upgraded 1996	Strong research focus, hunting and fishing by local people allowed

high expectations, largely because no serious attempt has been made to regulate fishing within the sanctuaries. To the program's credit, though, the dumping of wastes and the exploration for oil and gas have been strictly regulated, and this may be seen as having conservation value for marine mammals and other organisms. So-called "paper parks" and "paper reserves" are probably counterproductive for conservation. They provide false assurance that space and resources have been set aside for wildlife, thereby relieving the public pressure for meaningful conservation action.

Table III does not include the many small sites in the Antarctic and on the sub-Antarctic islands that are designated as specially protected areas or sites of special scientific interest under the Agreed Measures for the Conservation of Antarctic Fauna and Flora (1964), or as nature reserves under national legislative instruments, many of which protect vital haul-out habitat for seals in particular. The Svalbard (Spitsbergen) archipelago in the northeastern North Atlantic was the site of some of the worst excesses of early whaling and walrus killing, yet since 1973 about half of the land area has been declared to be inside nature reserves and national parks, and the Svalbard population of walruses is expanding rapidly.

VII. Strategies to Enhance Individual Survival and Reproduction

It is not often that human intervention can be effective in improving the chances for individual marine mammals to survive and reproduce. Organized programs for rescuing marine mammals that strand (come ashore) alive or that are injured and debilitated do manage to release some animals after rehabilitation. However, the success rate is low, and the conservation value of such programs is questionable. Many strandings represent "natural" mortality. Thus, while intervention may be justified as a humane gesture intended to improve the welfare of the stranded individual animals, it can also be argued that natural processes should be allowed to proceed without human interference. Only in a few special cases does rescue, rehabilitation, and release seem to be making a clear, direct difference in improving the health of a marine mammal population.

Most of the rivers in southern Asia inhabited by river dolphins are partitioned by irrigation dams (called barrages). When dolphins on the upstream side of such dams get too close to the intake structures of adjacent canals, they run the risk of becoming marooned in the canals, unable to return to the safety of the main river channel. In recent years, wildlife officers and conservationists have occasionally attempted to locate and rescue these ill-fated dolphins. In January 2000, at least two Indus dolphins, or bhulans (*Platanista gangetica minor*), trapped in a canal near Sukkur, Pakistan, were successfully caught and returned to the river (Fig. 7). In the same month, a Ganges dolphin, or susu (*Platanista gangetica gangetica*), that had become stranded in a small pool of water along the Damodar River in West Bengal, India, was rescued and safely transported to the Hooghly River for release. Pakistan's WWF office is working to develop a systematic procedure for notification and response, along with a protocol for rescuing river dolphins that enter irrigation canals.

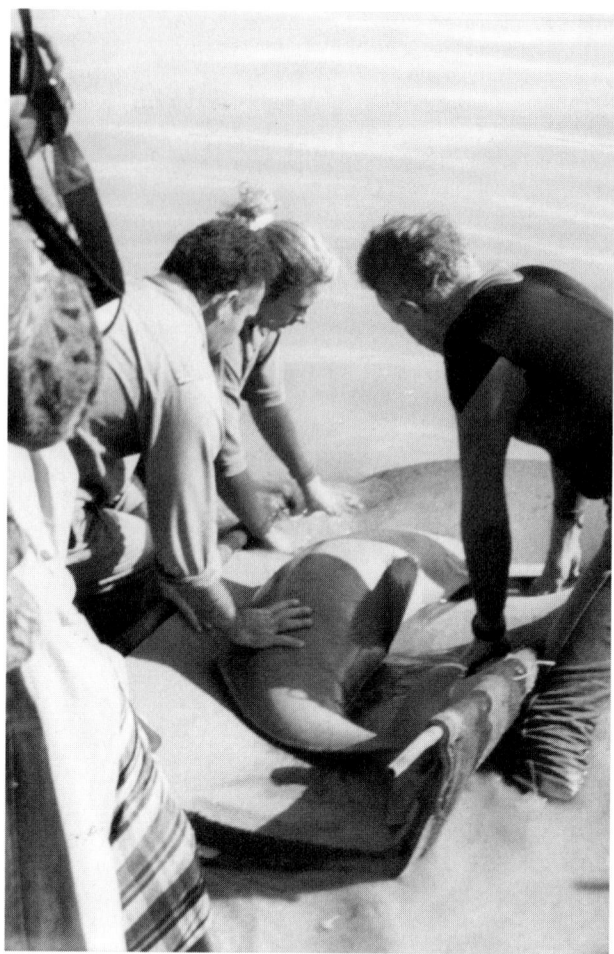

Figure 7 *An Indus river dolphin that had been marooned in an irrigation canal is released back into the main river channel upstream of Sukkur, Sind province, after being rescued by a team of conservationists from Sind Wildlife Department and WWF-Pakistan, January 2000. Photograph courtesy of World Wide Fund for Nature–Pakistan.*

In Florida, several facilities that display captive marine mammals have been collaborating for many years with the U.S. Fish and Wildlife Service to rehabilitate injured or orphaned manatees (*Trichechus manatus*). The animals are cared for and either maintained permanently in educational exhibits or, if judged healthy enough, released back into the wild. A single facility (Sea World in Orlando) was reported to have responded to 160 requests for assistance with distressed manatees from 1976 to 1995. More than half of the animals brought into captivity died, but nearly 60 individuals were eventually returned to the wild. Because virtually all of the injuries to manatees in Florida waterways are caused by human activity (mainly boating), the mortality being "offset" by the rehabilitation program is almost entirely nonnatural (Fig. 8).

Rescue and rehabilitation programs can contribute to conservation in less direct ways, too. For example, as John Reynolds (1999) has pointed out, "Educating people about manatee con-

Figure 8 *This badly injured manatee was rescued at Fernandina Beach, Florida, on November 10, 1997. She and her dependent calf were taken into captivity for medical treatment and rehabilitation. Although the mother died a little more than 5 months later, the calf survived and was released back into the wild in 1999. A record number of manatees (82) died in Florida from watercraft injuries in 1999, and more than 55 had been documented through the end of May in 2000. Photograph by Robert K. Bonde, U.S. Geological Survey–Sirenia Project.*

servation as they watch recuperating animals in a zoo setting can make a strong impression that may do more to encourage actual conservation than reading an article or watching a documentary about manatees." The whale and river dolphin rescue efforts mentioned earlier also serve to heighten awareness, educate people about conservation issues, and inspire actions to prevent further entanglement and entrapment. Reynolds also points out that manatees in captivity have allowed scientists to study their species' reproduction, osmoregulatory capabilities, and sensory abilities. Knowing more about manatee biology and physiology is important for conservation.

Finally, rescue and rehabilitation programs offer opportunities to instrument and monitor animals after release. This can lead to new discoveries about the animals and allow researchers to test new study methods. There are many examples, but one in particular stands out. In 1997 an adult male bottlenose dolphin (*Tursiops truncatus*), nicknamed "Gulliver," stranded in Florida. He was treated for a variety of ailments and, after about 4 months in captivity, released far offshore bearing a satellite-linked transmitter (Fig. 9). Gulliver's travels were impressive. After a week moving northward along the continental shelf, he headed southeast, swimming against the North Equatorial Current. He traversed waters more than 5000 m deep and reached an area northeast of the Virgin Islands before his transmitter stopped working, having covered 4200 km in 47 days. This study showed that bottlenose dolphins can be extremely mobile and that previous assumptions about the distributional limits of pelagic stocks would need to be reconsidered.

Another example of human intervention to enhance survival comes from the northwestern Hawaiian Islands, where biologists from the U.S. National Marine Fisheries Service have captured and translocated endangered monk seals (*Monachus schauinslandi*). In one program on Laysan Island, they caught a number of adult males that had been seen participating in "mobbing," or collective attacks on adult females and juvenile seals. The males were moved by ship to Johnston Island, some 600 miles south of Laysan, and released in the hope that they would survive but not return to carry on their destructive behavior toward other monk seals. In another program, called "Headstart," female pups at Kure Atoll have been collected after weaning and kept in a fenced beach enclosure for several months. The watered portion of the enclosure is kept well stocked with fish taken from nearby reefs, and the young seals have a chance to learn to forage in safety from large sharks, adult male monk seals, and hazardous fishing gear—all potential causes of mortality. The idea is that by the time they are released, they will have survived a critical stage in the life cycle and be ready for independence.

CAPTIVE BREEDING, with the intention of using captive-born young to reestablish a species in its former range or to supplement and reinvigorate a depleted wild population, is often discussed as a conservation strategy. Only one serious attempt has been made to restock a wild population of marine mammals with animals that were conceived, born, and reared in captivity. A number of captive-born harbor seals were released into the Dutch Wadden Sea, where their species is depleted severely (although harbor seals are not considered globally threatened). The released seals were monitored with TELEMETRY devices, and early results suggest that they have survived and adapted reasonably well.

Although captive breeding programs have been discussed in relation to Yangtze River dolphins, or baiji (*Lipotes vexillifer*),

Figure 9 *"Gulliver," an adult male bottlenose dolphin that had been rescued and rehabilitated after stranding, is released from a Coast Guard inflatable off Cape Canaveral, Florida, on May 20, 1997. The satellite-linked transmitter on Gulliver's dorsal fin allowed him to be tracked for 47 days while he traveled to an area northeast of the Virgin Islands, causing scientists to rethink hypotheses concerning offshore populations of this species. Photograph courtesy of Randy Wells, Mote Marine Laboratory.*

and Mediterranean monk seals (*Monachus monachus*), both gravely endangered, none of these programs have come to fruition. A much-publicized "seminatural reserve" was established for river dolphins in a Yangtze River oxbow, but this facility was stocked primarily with finless porpoises (*Neophocaena phocaenoides*) rather than dolphins. The single female baiji introduced to the reserve became entangled in fishing gear and died. No attempt was made to place this female with the lone male baiji in captivity, so there was no prospect of captive breeding. Efforts to capture additional baiji for captivity or for stocking the seminatural reserve have failed.

Translocation efforts played a role in the sea otter's reoccupation of parts of its original range. More than 700 otters were taken from high-density areas in Alaska during the 1960s and early 1970s and released at unoccupied sites in British Columbia, Washington, and Oregon. Populations are now well established in the first two of these three areas. A controversial at-

tempt was also made during the 1980s to establish a new population of sea otters in the California Channel Islands in order to reduce the risk that an oil spill would destroy the mainland population. More than 135 otters were captured and translocated to San Nicolas Island, but by the late 1990s only a few (less than 20) remained, and the program was abandoned.

VIII. Reduction of Environmental Pollution (Chemical and Acoustic)

The role of POLLUTION in impairing the productivity and survival of marine mammals was first realized in the 1970s, when the rate of reproductive failure (premature births, still births, and abortions) in California sea lions (*Zalophus californianus*) was found to be positively correlated with elevated tissue levels of DDT (O'Shea *et al.,* 1999). Also during the 1970s, studies of seals in the Baltic and North seas provided suggestive evidence that organochlorine pollutants pose serious risks to the health and reproductive potential of marine mammals. The production and use of DDT, PCBs, and some other dangerous persistent organochlorine chemicals began to be restricted in North America and western Europe in the 1970s, and there has been a general trend toward further restrictions since then. Unfortunately, however, the problem is far from solved. For example, India continued to produce 4000 metric tons of DDT at least as recently as the mid-1990s, and at least some of the former Soviet states have continued to manufacture and use PCBs. Moreover, the persistent nature of these chemicals means that they continue to be present in the environment, either temporarily sequestered in sediments or recycling in food webs, and therefore marine mammals continue to be vulnerable to their effects. While it must be acknowledged that the principal motivation for banning the release of harmful substances into the environment has had less to do with protecting marine mammals than with protecting human health (and birds, in the case of DDT), there is no doubt that reports of high levels of contaminants in marine mammals have contributed to public concern.

Acoustic pollution is thought to be especially damaging to cetaceans, as they depend heavily on sound for information about their environment, for foraging, and for communication. Noise associated with the exploration, development, and transport of hydrocarbon resources has been a particular source of concern, and many millions of dollars have been invested in studies of effects. In some instances, notably those involving seismic and drilling noise in the Arctic, steps have been taken to minimize the exposure of whales and seals to high-energy sounds. Government agencies and companies have conducted monitoring programs to determine when marine mammals are present in an area so that operations can be suspended or moved as necessary. In a similar vein, the sites and timing of military exercises off the east and west coasts of North America have, in a few instances, been planned with the safety of marine mammals and other marine wildlife as a primary consideration.

Human-induced changes in global climate have been the subject of intense debate in recent years, and it is fair to assume that marine mammals will be among the organisms affected by such changes. The effects are perhaps most obvious for ice-associated species: the phocid seals in the Arctic and Antarctic and the walrus and polar bear in the Arctic. These animals use sea ice as a platform for resting, giving birth, or, in the case of polar bears, hunting. As the extent and thickness of pack ice decrease consequent to global warming, these species are likely to lose critical habitat. Once again, as in the case of toxic chemical pollution, the primary motivation for taking steps to reduce emissions of greenhouse gases and ozone-depleting substances has been concern about human welfare rather than a desire to conserve marine mammals.

IX. Reduction of Conflicts with Fisheries

Fishery policies are the key to many of the most pressing marine mammal conservation problems. While there are examples of effective action to reduce marine mammal mortality in fishing gear, such as the UN ban on high-seas drift netting, the seasonal or permanent closure of certain areas to gill netting, and the development and implementation of deterrence programs using pingers and similar devices, the sad truth is that many critical situations simply continue to deteriorate. For example, although some legal limits have been placed on gill netting and commercial fishing in a portion of the upper Gulf of California, there has been little effective enforcement, and vaquitas remain in jeopardy. In China's Yangtze River, it is illegal to fish with electricity and explosives, yet there is almost no enforcement, and river dolphins and porpoises continue to be killed and injured as unintended victims.

X. Reduction of Disturbance and Direct Harm from Vessel Traffic

Manatees living in Florida's motorboat- and barge-infested waterways are frequently struck and injured, if not killed outright, by watercraft. About 50 Florida manatees are killed by boats each year, and many more are injured and harassed by vessel traffic. Although this problem had long been recognized, it was not until the late 1970s and early 1980s that serious efforts were made to reduce the risk of collisions and disturbance. More than 20 areas have been designated as protection zones for manatees, where vessel speed is regulated and signs warn visitors of the need to exercise caution. In some key manatee congregation areas, all waterborne human activity, including diving, boating, and swimming, is prohibited.

The other marine mammal species that is clearly threatened by ship strikes is the right whale. Where many thousands of North Atlantic right whales were present in the past, all that remains is a small population of about 300 or so centered along the east coast of Canada and the United States. Unfortunately, the whales regularly congregate in deep basins to feed, and these same deep waters are used by large cargo ships, tankers, and military vessels (Katona and Kraus, 1999). Several right whales are killed by ship collisions each year (Fig. 10). This mortality, combined with that caused by entanglement in fishing gear, is considered sufficient to have stalled population recovery and contributed to a recent downturn. Efforts have been made in both Canada and the United States to monitor the seasonal distribution and movements of individual right whales so that vessel captains can be notified to watch out for them. In Canada, consideration is being given to changing ship traffic lanes to reduce the risks to right whales. In the United States, a mandatory ship reporting system was put in place in 1999, requiring all vessels larger than 300 gross tons to report their location, speed, and port of destination when traversing designated right whale critical habitat. In return, they are provided with information on the presence of right whales in their vicinity.

It remains to be seen whether manatees and right whales will be able to withstand the effects of human activities in the nearshore and inshore waters we share with them. Thus far, our own species' recreational and commercial use of the

Figure 10 *This young right whale died after being struck by a ship off the coast of Florida, near Jacksonville, on January 10, 1993. Although right whales have been protected from whaling since the 1930s, their populations in the Northern Hemisphere have not recovered. One reason is the mortality caused by ship strikes and entanglement in fishing gear. Photograph by Robert K. Bonde, U.S. Geological Survey–Sirenia Project.*

marine environment has been regarded as sacrosanct, and the few gestures made to accommodate the needs of these other species have had to overcome strenuous resistance from boaters, the shipping industry, military authorities, and others.

XI. Giving Economic Value to Living Wild Marine Mammals

In the 1950s, a few nature enthusiasts in southern California began venturing into nearshore waters to watch gray whales. At the time, scientists were just beginning to document the remarkable recovery of these animals—a result of the protection from whaling afforded by the IWC and, in recent years, Mexico's protection of the breeding lagoons in Baja California. Interest in watching whales grew steadily, and by the mid-1970s, conservationists were suggesting that the economic value of this "nonconsumptive" form of use might eventually rival that of whaling. The 1980s and 1990s saw the rapid proliferation of tour enterprises for observing whales and dolphins. Even in the whaling countries of Norway, Iceland, and Japan, whale watching has become a popular and at least locally remunerative form of recreation. In eastern Canada, helicopter tours to the pack-ice pupping grounds of harp seals have been encouraged by animal-welfare groups as a way of demonstrating that seals can generate income without having to be killed.

XII. Zoogeography of Marine Mammal Conservation

Threats to marine mammal species and populations are relatively well understood and are being addressed aggressively in the United States, western Europe, South Africa, Australia, and New Zealand. That is not to say that the problems will all be solved and that threatened species and populations in these areas are safe. In fact, right whales, northern sea lions (*Eumetopias jubatus*), and California sea otters in U.S. waters, Mediterranean monk seals and some local populations of bottlenose dolphins in Europe, dugongs in parts of Australia, and Hector's dolphins in New Zealand are still in trouble. Rather, the point is that elsewhere in the world, by comparison, marine mammal populations are slipping away even before there has been a chance to document their distribution and abundance, or to elucidate their ecological roles.

Table IV lists 20 of the world's most threatened marine mammal taxa. It is important to caution that the list is arbitrary and potentially misleading. For example, it is dominated by taxa that have been relatively well studied, to the exclusion of ones that have been less closely studied. Some seriously threatened populations of finless porpoises, Irrawaddy dolphins (*Orcaella brevirostris*), and hump-backed dolphins (*Sousa* spp.) probably represent valid subspecies, or even species, and should be included, but they are simply not well enough known. Also, scientists have been concerned for many years about the large numbers of La Plata dolphins, or franciscanas (*Pontoporia blainvillei*), being killed accidentally in artisanal gill nets along the east coast of South America (Fig. 11). Until very recently, however, there was no estimate of franciscana abundance, and the information about mortality was not sufficiently precise or

accurate to support a proper assessment. Without a strong scientific underpinning to demonstrate the seriousness of the problem, conservationists are unlikely to be able to force governments to act to reduce dolphin mortality. Similarly, manatees in West Africa (*T. senegalensis*) and South America (*T. inunguis*) are hunted, trapped, and caught in fishing gear, but no one has the vaguest idea how many there are or the extent to which their distribution has already been reduced by overexploitation and habitat deterioration. For these and many other situations, there has been little or no active conservation.

Another difficulty is that by limiting the list in Table IV to recognized species and subspecies, many geographical populations are left out. For example, several populations of baleen whales (e.g., eastern North Pacific gray whales; Okhotsk and Spitsbergen/Barents Sea bowheads) and toothed whales (e.g., Cook Inlet and Ungava Bay belugas, North Island Hector's dolphins) are known to number only in the tens or low hundreds. Their disappearance might be likened to the world losing an entire branch of artistic expression, such as painting or dance. Art would still exist, but there would be a noticeable void, and our lives would be less rich.

Finally, assessing the extent to which different groups of animals are threatened or endangered is extremely complex. It involves far more than simply considering current abundance, or even trends in abundance. In 1996, IUCN attempted, for the first time, to evaluate the conservation status of the world's mammal species and subspecies according to standardized quantitative criteria. This process led to the listing of all four extant species of sirenians (the three manatees and the dugong) as vulnerable, meaning that they face a high risk of extinction in the medium-term future. Of the pinnipeds, one species (the Mediterranean monk seal) is listed as critically endangered, meaning that it faces an extremely high risk of extinction in the immediate future, and two (Hawaiian monk seal and northern sea lion) as endangered, meaning that they face a very high risk of extinction in the near future. Six pinniped species are listed as vulnerable and one as lower risk (near threatened), the latter meaning that the taxon is close to qualifying as vulnerable. Of the cetaceans, 2 species (baiji and vaquita) are considered critically endangered, 6 endangered, 6 vulnerable, 1 lower risk (near threatened), and 14 lower risk (conservation dependent). The last of these designations (which has also been applied to the polar bear) means that the species is the focus of a continuing taxon-specific or habitat-specific conservation program, which, if stopped, would result in the species qualifying for one of the threatened categories (i.e., vulnerable, endangered, or critically endangered) within 5 years. Yet another category, data deficient, was applied to 38 cetacean species. It signifies that there is inadequate information to assess the risk of extinction and acknowledges the possibility that, with further research, one of the threatened categories will be found to be appropriate for the species. In addition to the species listings, 5 pinniped and 12 cetacean subspecies or geographical populations are listed as either endangered or vulnerable.

Endemism is a feature that seems to be associated with vulnerability. Many of the species and subspecies in Table IV are on the list because they occur in only one place, and the smaller that place, the more vulnerable the taxon. The vaquita is limited to the upper portion of the Gulf of California, the baiji to

TABLE IV
Twenty of the World's Most Threatened Marine Mammal Taxa, Including Recently Extirpated Ones[a]

Taxon	Range states	Approx. abundance	Main threats
1. Caribbean monk seal, *Monachus tropicalis*	Mexico, USA, Bahamas, Jamaica, Cuba, Haiti, Dominican Republic, Guadeloupe, and other Caribbean states	Probably extinct	Deliberate killing, loss of habitat due to development
2. Japanese sea lion, *Zalophus japonicus*	Japan, Korea, Russia	Probably extinct	Deliberate killing
3. Baiji or Yangtze river dolphin, *Lipotes vexillifer*	China	Tens	Fishery bycatch, loss of suitable habitat and resources, possibly vessel strikes and disturbance
4. Saimaa ringed seal, *Pusa hispida saimensis*	Finland	200	Fishery bycatch, chemical contamination
5. Mediterranean monk seal, *Monachus monachus*	Turkey, Greece, Italy, Mauritania, Morocco, Western Sahara, Libya, Madeira (Portugal), Tunisia	250–500	Fishery bycatch, shooting by fishermen, loss of suitable pupping and pup-rearing habitat
6. Ungava harbor seal, *Phoca vitulina mellonae*	Canada	120–600	Hunting, changes to habitat caused by hydroelectric development
7. Vaquita, *Phocoena sinus*	Mexico	About 600 (200–1000)	Fishery bycatch
8. Bhulan or Indus river dolphin, *Platanista gangetica minor*	Pakistan	1000	Fishery bycatch, loss of suitable habitat (including habitat fragmentation), accidental movement into canals and other unsafe areas
9. North Atlantic and North Pacific right whales, *Eubalaena glacialis* and *E. japonica*	Canada, USA, Iceland, Norway, UK, Spain, Portugal, France (No. Atlantic population); Russia, Japan, Korea, Canada, USA, Mexico (No. Pacific population)	300–350 in North Atlantic; possibly < 1000 in North Pacific	Ship strikes, fishery bycatch, possibly effects of small population size (depletion from past over-hunting)
10. Antarctic blue whale, *Balaenoptera musculus intermedia*	Australia, New Zealand, Madagascar, South Africa, Namibia, Angola, Gabon, Argentina, Brazil, Chile, Peru	Hundreds	Uncertain, possibly effects of small population size (depletion from past over-hunting)
11. Hawaiian monk seal, *Monachus schauinslandi*	USA (mainly Hawaiian archipelago)	1000–1500	Fishery bycatch, disturbance on pupping beaches
12. California sea otter, *Enhydra lutris nereis*	USA	About 2000	Fishery bycatch
13. Red Sea dugong, *Dugong dugon hemprichii*	Egypt, Saudi Arabia, Yemen, Eritrea, Sudan	Probably low thousands at most, possibly only hundreds	Fishery bycatch, hunting?
14. Susu or Ganges river dolphin, *Platanista gangetica gangetica*	India, Bangladesh, Nepal	Possibly low thousands	Fishery bycatch, deliberate hunting, loss of suitable habitat (including habitat fragmentation), accidental movement into canals and other unsafe areas, chemical contamination
15. Yangtze River finless porpoise, *Neophocaena phocaenoides asiaeorientalis*	China	High hundreds or low thousands	Fishery bycatch, loss of suitable habitat and resources, possibly vessel strikes and disturbance
16. Florida manatee, *Trichechus manatus latirostris*	USA, Bahamas (occasionally)	2500–3000	Vessel collisions, fishery bycatch, exposure to toxic organisms (possibly related to human activities)

(continues)

TABLE IV *(Continued)*

Taxon	Range states	Approx. abundance	Main threats
17. Hector's dolphin, *Cephalorhynchus hectori*	New Zealand	Low thousands	Fishery bycatch
18. Antillean manatee, *Trichechus manatus manatus*	Caribbean and Atlantic mainland coastal states from Mexico to Brazil, Cuba, Puerto Rico, Trinidad, Dominican Republic, and other Caribbean island states	Unknown	Fishery bycatch, deliberate hunting and trapping
19. Ladoga ringed seal, *Pusa hispida ladogensis*	Russia	About 5000	Fishery bycatch, disturbance at haul-out sites
20. Caspian seal, *Pusa caspica*	Russia, Kazakhstan, Turkmenistan, Iran, Azerbaijan	Possibly tens of thousands but rapidly declining	Hunting, fishery bycatch, chemical contamination, loss of suitable habitat and resources, disease

*a*Taxonomy follows Rice, 1998, except for right whale. See Species List.

the mainstem of the Yangtze River, the bhulan to the mainstem of the Indus River, and the Saimaa, Ungava, and Ladoga ringed seals to single networks of freshwater lakes. Scale obviously matters in considering the effects of endemism. A species or population that ranges throughout, or on both sides of, an ocean basin would seem intrinsically less vulnerable than one limited to a single stretch of coastline or a single river or large lake.

Figure 11 *Franciscanas killed accidentally in gill nets near San Bernardo, Buenos Aires, Argentina. For decades, scientists have expressed concern about the scale of this mortality, but very little progress has been made toward assessing the impact on dolphin populations or reducing the bycatch rate. The franciscana is one of several marine mammal species with a restricted distribution (coastal waters of eastern South America between approx. 18°30'S and 41°10'S) that experiences substantial incidental mortality but for which information on abundance and productivity is far from adequate. Photograph courtesy of Pablo Bordino.*

The challenges that lie ahead are truly endless. As the world economy becomes more integrated and as the human appetite for consuming our planet's resources expands, marine mammals will inevitably experience new threats, even as long-standing threats persist. Not only are we in danger of losing numerous populations, some species, and a few genera (e.g., *Monachus, Eumetopias*), several entire families of cetaceans are already at risk, ranging from the "river dolphins" (Lipotidae, Platanistidae, and Pontoporiidae) to the gray and right whales (Eschrichtiidae and Balaenidae). However impressive the array of conservation efforts may seem, on paper, it is far from adequate. Only with a genuine, broad-scale change in how we value the remnants of the world's natural variety and abundance, and thus in how we use and care for the Earth's precious resources, can we hope to head off a cascade of marine mammal EXTINCTIONS in the coming decades.

See Also the Following Articles

Captive Breeding ▪ Competition with Fisheries ▪ Conservation Ecology ▪ Distribution ▪ Fishing Industry, Effects of ▪ Illegal and Pirate Whaling ▪ Pollution and Marine Mammals

References

Baur, D. C., Bean, M. J., and Gosliner, M. L. (1999). The laws governing marine mammal conservation in the United States. *In* "Conservation and Management of Marine Mammals" (J. R. Twiss, Jr., and R. R. Reeves, eds.), pp. 48–86. Smithsonian Institution Press, Washington, DC.

Dizon, A., Baker, C., Cipriano, F., Lento, G., Palsbøll, P., and Reeves, R. (2000). "Molecular Genetic Identification of Whales, Dolphins, and Porpoises: Proceedings of a Workshop on the Forensic Use of Molecular Techniques to Identify Wildlife Products in the Marketplace, La Jolla, California, 14–16 June 1999." U.S. Department of Commerce, NOAA Technical Memorandum NOAA-TM-NMFS SWFSC-286.

Dizon, A. E., Chivers, S. J., and Perrin, W. F. (eds.) (1997). "Molecular Genetics of Marine Mammals." Society for Marine Mammalogy, Special Publication No. 3.

Dizon, A. E., Lockyer, C., Perrin, W. F., DeMaster, D. P., and Sisson, J. (1992). Rethinking the stock concept: a phylogeographic approach. *Conser. Biol.* **6,** 24–36.

Gambell, R. (1999). The International Whaling Commission and the contemporary whaling debate. *In* "Conservation and Management of Marine Mammals" (J. R. Twiss, Jr., and R. R. Reeves, eds.), pp. 179–198. Smithsonian Institution Press, Washington, DC.

Gosliner, M. L. (1999). The tuna-dolphin controversy. *In* "Conservation and Management of Marine Mammals" (J. R. Twiss, Jr., and R. R. Reeves, eds.), pp. 120–155. Smithsonian Institution Press, Washington, DC.

Katona, S. K., and Kraus, S. D. (1999). Efforts to conserve the North Atlantic right whale. *In* "Conservation and Management of Marine Mammals" (J. R. Twiss, Jr., and R. R. Reeves, eds.), pp. 311–331. Smithsonian Institution Press, Washington, DC.

Kimball, L. A. (1999). The Antarctic Treaty system. *In* "Conservation and Management of Marine Mammals" (J. R. Twiss, Jr., and R. R. Reeves, eds.), pp. 199–223. Smithsonian Institution Press, Washington, DC.

Laist, D. W., Coe, J. J., and O'Hara, K. J. (1999). Marine debris pollution. *In* "Conservation and Management of Marine Mammals" (J. R. Twiss, Jr., and R. R. Reeves, eds.), pp. 342–366. Smithsonian Institution Press, Washington, DC.

Lavigne, D. M., Scheffer, V. B., and Kellert, S. R. (1999). The evolution of North American attitudes toward marine mammals. *In* "Conservation and Management of Marine Mammals" (J. R. Twiss, Jr., and R. R. Reeves, eds.), pp. 10–47. Smithsonian Institution Press, Washington, DC.

Lyster, S. (1985). "International Wildlife Law: An Analysis of International Treaties Concerned with the Conservation of Wildlife." Grotius Publications Limited, Cambridge, UK.

Mangel, M., Talbot, L. M., Meffe, G. K., Agardy, M. T., *et al.* (1996). Principles for the conservation of wild living resources. *Ecol. App.* **6,** 338–362.

Mayr, E., and Ashlock, P. D. (1991). "Principles of Systematic Zoology," 2nd ed. McGraw-Hill, New York.

McVay, S. (1966). The last of the great whales. *Sci. Am.* **215**(2), 13–21.

Northridge, S. N., and Hofman, R. J. (1999) Marine mammal interactions with fisheries. *In* "Conservation and Management of Marine Mammals" (J. R. Twiss, Jr., and R. R. Reeves, eds.), pp. 99–119. Smithsonian Institution Press, Washington, DC.

O'Shea, T. J., Reeves, R. R., and Long, A. K. (1999). "Marine Mammals and Persistent Ocean Contaminants: Proceedings of the Marine Mammal Commission Workshop, Keystone, Colorado, 12–15 October 1998." Marine Mammal Commission, Bethesda, MD.

Perrin, W. F. (1999). Selected examples of small cetaceans at risk. *In* "Conservation and Management of Marine Mammals" (J. R. Twiss, Jr., and R. R. Reeves, eds.), pp. 296–310. Smithsonian Institution Press, Washington, DC.

Reynolds, J. E., III. (1999). Efforts to conserve the manatees. *In* "Conservation and Management of Marine Mammals" (J. R. Twiss, Jr., and R. R. Reeves, eds.), pp. 267–295. Smithsonian Institution Press, Washington, DC.

Rice, D. W. (1998). "Marine Mammals of the World: Systematics and Distribution." Society for Marine Mammalogy, Special Publication No. 4.

Whitehead, H., Reeves, R. R., and Tyack, P. L. (2000). Science and the conservation, protection, and management of wild cetaceans. *In* "Cetacean Societies: Field Studies of Dolphins and Whales" (J. Mann, R. C. Connor, P. L. Tyack, and H. Whitehead, eds.), pp. 308–332. University of Chicago Press, Chicago.

Convergent Evolution

EMILY A. BUCHHOLTZ
Wellesley College, Massachusetts

Convergent evolution is the independent development of the same characters or group of characters in species whose common ancestor does not possess them. The resulting similarity in appearance is termed a homoplasy. Convergent evolution is typically associated with adaptation to similar environments. Because the constraints of the aquatic environment are extreme, they provide very powerful selection pressures on animals, such as mammals, that secondarily enter the water from terrestrial environments. Often these animals have independently evolved similar functional solutions in response to the same selective pressures and therefore show convergence in anatomy and/or behavior. Seacows, desmostylians, whales, pinnipeds, hippopotamuses, and otters almost certainly represent separate invasions of aquatic environments by mammals. Some authors believe that pinnipeds, which include sea lions, seals, and walruses, represent not a single but as many as three different invasions of the water.

In contrast to the air, water is very dense, has a high thermal conductance and capacity, and has a very low oxygen concentration. Even marginal marine species must evolve new modes of aquatic locomotion; most exploit aquatic plants or animals as food. Animals that spend most or all of their lives in water often have layers of BLUBBER to conserve heat, higher than average rates of basal metabolism, and a wide range of physiological adaptations that ensure high levels of oxygen uptake, storage, and transport. Because chemicals, light, and sound travel differently in water than in air, chemosensation, vision, and audition are usually modified to allow efficient defense, orientation, and COMMUNICATION.

Ample evidence exists that the highly restrictive marine environment has repeatedly selected for similar characters that enhance life in the water across a wide range of mammalian taxa. Nevertheless, anatomical and/or functional similarity need not imply convergence if the taxa involved are in fact more closely related to each other than they are to another terrestrial lineage. In such a case, the similarities of the two groups could be ascribed to common inheritance. This point, made by Wyss (1989), was used to argue for a single, monophyletic origin of all pinnipeds. To the extent that similar anatomical and physiological traits for dealing with life in the water occur in mammals with different terrestrial ancestors, they are the homoplastic results of convergent evolution. This article presents examples of convergence among marine mammals in locomotor pattern and body shape, sense organ choice and design, and circulatory physiology.

I. Convergence in Locomotor Pattern and Body Shape

All marine mammals had terrestrial ancestors with paraxial appendicular locomotion, and it is probable that transitional, semiaquatic taxa employed limb-based, paddling locomotor patterns to at least some extent. Fish (1996; Fig. 1) contrasted such drag-based propulsion with the lift-based propulsion of fully aquatic mammals. Lift-based propulsion may employ fore-limbs, hindlimbs, or a caudal fin as the lift-generating organ. Despite different swimming proficiencies and speeds, sirenians and cetaceans provide a striking example of the convergent evolution of dorsoventrally oscillating, lift-generating caudal flukes. The swimming style of the (reptilian) Jurassic ichthyosaurs was at least superficially convergent on the same pattern, although oscillation of the caudal fin was lateral instead of vertical.

Many marine mammals, especially those specialized for speed, have smooth, fusiform bodies that enhance movement through the dense aquatic medium. A smooth body reduces viscous drag produced by friction between water and the animal's body, whereas a streamlined, teardrop shape reduces surface area and inertial drag from the pressure differences caused by movement through the water. In test objects, minimum drag is achieved when body length is approximately four to five times body width, and many fast swimmers have evolved these approximate proportions convergently. When the functional constraints of feeding allow it, hydrodynamic shape is often further enhanced by reduction of neck length. The degree of aquatic specialization of both living and fossil mammals can be estimated by measuring the relative length of cervical vertebrae (Fig. 2). Other common convergent changes in body form include the loss of external ear pinnae and the enclosure of external genitalia in body pouches to reduce drag.

II. Convergence in Choice and Design of Sense Organs

Transition from land to water is typically accompanied by the modification of the sense organs for efficient use in the new environment. In very broad terms, sight plays a relatively smaller role, and hearing a larger role, in the lives of marine mammals than in those of terrestrial mammals. Nevertheless, most marine mammals see very well, especially in clear water and at short distances. Light is absorbed by water and scattered by suspended material. As a result, vision is restricted at distance, at depth, and in turbid nearshore environments. At least one species of river dolphin (*Platanista gangetica*) is nearly blind. Modification of the terrestrial eye for sight in lighted aquatic environments requires compensation for the lack of corneal refraction. In subaerial eyes, the cornea is a fixed refractor that depends on the different refractive indices of air and corneal tissue to bend light toward the retina. The near identity of the corneal refractive index with that of water makes it virtually useless in aquatic mammals.

Sound travels several times faster in water than in air, making hearing an optimal sense for aquatic communication and

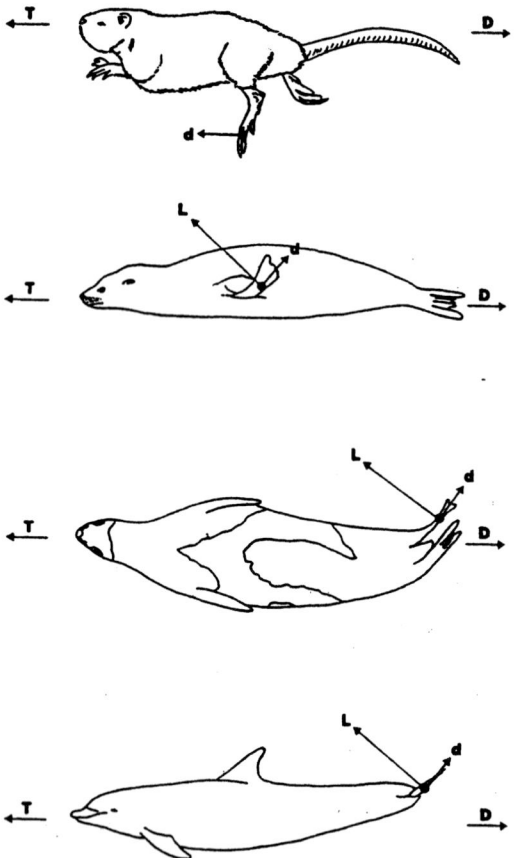

Figure 1 *Propulsive modes in marine mammals, from Fish (1996). From top, paddling (drag-based) in the muskrat (Ondatra zibethicus), pectoral oscillation (lift-based) in the California sea lion (Zalophus californianus), pelvic oscillation (lift-based) in the harp seal (Pagophilus groenlandicus), and caudal oscillation (lift-based) in the common bottlenose dolphin (Tursiops truncatus). Drag (d) and lift (L) generated by the propulsive organ contribute to total thrust force (T) and drag force (D). Reprinted with permission of the* American Zoologist.

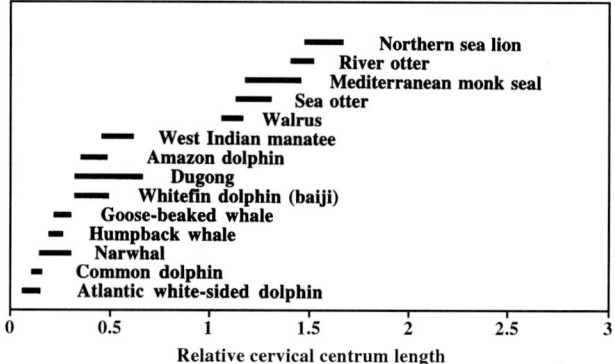

Figure 2 *Convergent reduction of centrum length relative to centrum height of cervical vertebrae in marine mammals (top to bottom: Eumetopias jubatus, Lutra sp., Monachus monachus, Enhydra lutris, Odobenus rosmarus, Trichechus manatus, Inia geoffrensis, Dugong dugon, Lipotes vexillifer, Ziphius cavirostris, Megaptera novaeangliae, Delphinus sp., Lagenorhynchus acutus). A short relative length enhances hydrodynamic shape and ease of movement through the water.*

navigation. Long wavelength sound that bends around obstacles is used for long-distance communication in both whales and pinnipeds. The "songs" generated by giant mysticetes may travel hundreds or even thousands of kilometers. Navigation via short wavelength sound is highly developed in odontocete cetaceans. Although short wavelength clicks or chirps are produced by many sea lions, seals, and walruses (*Odobenus rosmarus*) and by the Caribbean manatee (*Trichechus manatus*) (Norris, 1969), their function in either communication or navigation has not been reliably documented.

Adaptation of the terrestrial ear for aquatic hearing presents several significant challenges. These include the increased pressure at depth on the ear drum or tympanum, the impedance mismatch produced by the entry of sound through the air-based middle ear, and the ease with which sound from the environment enters the inner ear directly through the body. Reduction in the size of the tympanum and increases in both the density and the size of the ear ossicles have occurred convergently in whales and pinnipeds. In some cases, sound reaching the middle ear is supplemented or supplanted by sound traveling via the jaw. Both whales and pinnipeds have also responded to the disruption of directional information caused by the transmission of sound via many osseous routes by isolating and insulating the inner ear from the rest of the skull. In whales, this insulation involves not only soft tissue separation, but also the very dense tympanic bone.

III. Convergence in Circulatory Physiology

All marine mammals retain lungs as their dominant organs of gas exchange and are dependent on the air at the water's surface for their oxygen supplies. Despite this restriction, many are capable of dives that persist long after oxygen in the lungs is depleted. They may tolerate such dives by decreasing the heart rate (bradycardia), by utilizing oxygen stored in muscle or blood that has unusually high levels of myoglobin or hemoglobin, and by switching from aerobic to anaerobic metabolism. Additionally, blood may be directed away from the muscles, kidneys, and digestive tract, but its supply maintained or diverted to vital organs that constantly require oxygen, such as the brain and the heart. Hedrick and Duffield (1991) documented unusually high concentrations of red blood cells in serum (hematocrits) in those whales (belugas, *Delphinapterus leucas*) and seals (northern elephant seals, *Mirounga angustirostris*) that employ deep, long-duration dives. Interestingly, high hematocrits are also associated with increased blood viscosity and a reduced oxygen transport capacity, which may preclude fast sustainable swimming in these species. The extensive phylogenetic distance between whales and seals dictates that these similarities must have evolved separately in each group, making them a convincing example of convergent evolution.

Convergent evolution of animals whose aquatic habits demand long periods without breathing extends to the molecular level as well. Joysey (1988) studied the sequence of amino acids in the myoglobin molecule as a means of determining phylogenetic relationships. He found that his phylogenetic reconstructions were in some cases complicated by the occurrence of amino acids that could be associated with similar physiological requirements. When compared with those of terrestrial mammals, the myoglobins of phocids, otariids, penguins, and cetaceans all include additional arginines on surficial sites. Their presence changes the electrical charge of the protein, allowing high-density concentration of myoglobin in the cell.

See Also the Following Articles

Circulatory System ■ Habitat Pressures ■ Morphology, Functional

References

Fish, F. E. (1996) Transitions from drag-based to lift-based propulsion in mammalian swimming. *Am. Zool.* **36**(6), 628–641.

Hedrick, M. S., and Duffield, D. A. (1991). Haematological and rheological characteristics of blood in seven marine mammal species: Physiological implications for diving behaviour. *J. Zool. (Lond.)* **225**, 273–283.

Joysey, K. A. (1988). The use of amino acid sequences in phylogenetic analysis. *In* "Molecular Evolution and the Fossil Record, Short Courses in Paleontology" (B. Runnegar and J. W. Schopf, eds.), pp. 34–74. University of Tennessee, Knoxville.

Norris, K. S. (1969). The echolocation of marine mammals. *In* "The Biology of Marine Mammals" (H. T. Anderson, ed.), pp. 391–423. Academic Press, New York.

Wyss, A. R. (1989). Flippers and pinniped phylogeny: Has the problem of convergence been overrated? *Mr. Mamm. Sci.* **5**(4), 343–360.

Copulation
SEE *Mating Systems*

Courtship Behavior

BERND WÜRSIG
Texas A&M University, Galveston

Courtship behavior is defined as "the act of wooing a member of the opposite sex" and is thus a limited part of overall mating or reproductive behavior. Courtship is particularly well developed in monogamous animals or those where direct partner choice is important. Marine birds represent especially good examples. Many marine mammals do not so much as woo members of the opposite sex, but instead males appear to force themselves onto females in estrus, often after having worked out dominance/subservience relationships among themselves. Our discussion here will therefore be a blend of pure courtship behavior and other social interactions that must be a prelude for successful mating. The mere recognition of the opposite sex of the same species, e.g., by sound and sight, is considered a part of courtship in this broader definition (Krebs and Davies, 1993).

I. Pinnipeds

Pinnipeds tend to have polygynous mating systems, with one male attempting to inseminate numerous females. Extreme polygyny is especially well developed on solid land or ice and less so on moving pack or loose ice (Stirling, 1983). Males work out dominance relationships among themselves, usually by aggressive interactions and behavioral bluffing. In elephant seals (*Mirounga* spp.), only a few of the very largest and most aggressive males attain a status of dominance and father by far the most offspring on a section of beach. Nevertheless, the female has a choice in who mates with her. When a male mounts her, she protests by squawking loudly. This action attracts other males, and the mounting male may be challenged by one or more males who consider themselves as or more dominant (LeBoeuf and Laws, 1994).

While male displays of dominance are believed to function overwhelmingly in male–male advertisements, a part of intrasexual selection, it is nevertheless possible that male cues may predispose females toward particular males and away from others. This is perhaps most likely in species that use underwater sound for displays, such as the long trills of bearded seals (*Erignathus barbatus*) or the knocking and whistling sounds of walruses (*Odobenus rosmarus*). Such a potential choice by females of males aggregated on a breeding ground may be an example of lekking behavior, but there is considerable debate among pinniped researchers on this topic (Wells *et al.*, 1999).

Those species where single males occur with single females for some time probably consist of males mating with that female and staying with her during her period of estrus. This is termed "female defense," but does not imply true monogamy, as it is likely that the male will leave and search for another female in estrus after some time (also termed serial monogamy). Ringed seals (*Pusa hispida*) on loose ice appear to be serially monogamous. Nevertheless, casual observations indicate that there is much social interaction and sniffing by these males and females when they first meet, and one wonders whether future more detailed studies might find a much more developed courtship than is presently known. After all, males and females in this species are of similar size and strength; the male is therefore unlikely to be able to force himself on the female, and it need not be automatic that a female accepts any male who happens to come by and show an interest in her (Riedman, 1990).

II. Sea Otters

Sea otter (*Enhydra lutris*) males actively defend a near-shore area from other males, but females are free to come into or leave this apparent underwater territory, also called a *maritory*. When a female comes into estrus, she tends to associate with only one male for several days. The male mates with her repeatedly and grips her nose, which becomes heavily bloodied and scarred during mating, with his teeth and claws. This behavior may help to stimulate ovulation, as neck biting does in some other carnivores. It is unclear whether it is simply the goodness of the space that attracts a female to a particular male or whether there is courtship behavior beyond grabbing and biting involved.

The marine otter (*Lutra felina*) of the Chilean Pacific coast probably has a similar maritorial system.

III. Sirenians

The West Indian manatee (*Trichechus manatus*) is the best known of the three extant manatees and one dugong that form the tropical, small, taxonomic order Sirenia (one cold water dugong, Steller's sea cow, *Hydrodamalis gigas*, was extirpated in the 1700s). These manatees aggregate in herds of one female and from 2 to 22 pushing, shoving males. It is believed that the female will therefore mate with more than one male, and it is possible that a form of sperm competition as well as the obvious physical competition for access to the female has evolved. Males also aggregate in herds without females present and with much apparent homosexual mating. Such interactions may be important in practicing the gaining of access to females, what has been termed "scramble competition polygyny." It is also possible, and not mutually exclusive with the idea of practicing for competition, that males are working out dominance hierarchies in the all-male groups. Females move into shallow water to avoid amorous males, and it is possible that females exercise considerable choice of who gets to mate with them by making it impossible for some males to gain access. Manatees also use objects in their environment as traditional "rubbing posts" and rub especially their anal area, which has glandular secretions, on these objects. It is not known whether females in estrus might advertise themselves by chemical smell or taste or whether other communication may take place at rubbing areas (Reynolds and Odell, 1991).

Mating in dugongs (*Dugong dugon*) has not been well described. It appears that at times a single male stays with a single female, whereas at other times, mating aggregations of at least one female and several males occur. Adult male dugongs have small but sharp and hard tusks and can do considerable damage with them to each other and to females. However, it is unknown whether this secondary sexual characteristic functions only as an intrasexual display or as a male to female, or intersexual, display as well. As in manatees, it is likely that females have the capability of at least some choice of mating partner(s), and we therefore assume that some element of "impressing a female," or courtship behavior, may have evolved (Reynolds and Odell, 1991; Wells *et al.*, 1999).

IV. Polar Bears

Polar bears (*Ursus maritimus*) tend to be solitary animals, but males will stay with a female, "guarding" her from other males during a period of estrus. We can assume that these bears, as others, have greeting ceremonies and other interactions during a period of courtship, but details are lacking. Polar bear males have been seen sniffing the tracks of another bear, and it is likely that they—as other bears and many carnivores—find a female in estrus by her smell. It is also likely that multiple copulations are needed to induce ovulation in the female (Stirling and Guravich, 1988).

V. Cetaceans

Whales, dolphins, and porpoises have a diversity of precopulatory behaviors, but how much of these can be described as courtship is unclear. Cetaceans use sound during almost all social interactions, but vision and touch are well developed and appear of great importance in precopulatory displays. Nevertheless, to date, almost all observations have been from above the surface, and to properly describe courtship behaviors, dedicated focal individual and focal group studies underwater will have to proceed.

A. Baleen Whales

Humpback whales (*Megaptera novaeangliae*) are arguably the best-studied large whale, yet no researcher has reported seeing humpbacks mate. Humpback males on the mating/calving grounds produce series of sounds ("moans, screams, rumbles, etc.") that are repeated in exact fashion after some time. These series are therefore properly called song and may last from about 10 to 30 min per song. Interestingly, every male of an extended area (such as around the Hawaiian Islands, in the Caribbean, or off Japan) sings pretty much the same song. The song evolves during the singing season and is therefore different from the beginning to the end of season. It is presently unknown whether males space themselves underwater by song, whether they attract females by some subtle yet undescribed individual variations in each song, or both. If the latter, the males might be "courting" females by song, although evidence of only males joining each other after singing hints at a male–male aggression function, not courtship. If males space each other by song, perhaps the loudest individual has males farthest from him and may therefore be more eagerly selected by females. However, this is conjecture; it is rather interesting (and somewhat exciting) that there are huge whales out there for which we really still know very little. Humpbacks also aggregate in boisterous surface-active groups of several males and generally only one female. While the males appear to battle for access to the female, it is unknown whether the female mates with a male by his choice or whether she has an active say in who gets to mate with her. A third, potentially alternative, mating strategy appears to be a male who "shadows" a female for hours to several days in the possible hope that she comes into estrus or is in estrus. This association is called a cow-escort group (Wells *et al.*, 1999).

Even less is known about other whales of the taxonomic family Balaenopteridae, such as blue (*Balaenoptera musculus*) and fin (*B. physalus*) whales. These two species produce low-frequency many-second-long moans that travel for tens to hundreds of kilometers. The whales likely communicate with these infrasounds, but whether some aspect of infrasound advertises sexual readiness is not known.

The right whales (*Eubalaena* spp), bowhead whale (*Balaena mysticetus*), and gray whale (*Eschrichtius robustus*) all aggregate into surface active groups of several males and one or a few females. However, these groups are much less aggressive-seeming than those of humpbacks, and one has the impression that males are vying for position near females in a more fluid nonbattling manner. It may be that females incite such aggregations by sound or other displays and that females then have the ability to "choose" the male or males that are most successful in interactions with others. There is some evidence that females at times appear to make it easier for one male than another to gain access to her, but more studies need to be done. At any rate, actual copulations have often been witnessed at the water's surface for the three species mentioned here, and it is therefore known that both males and females have multiple mating partners over the course of hours or days. The social system is therefore a multimate or promiscuous one, but again the cues used to gauge successful courtship behavior are unknown. Bowhead whales during spring migration, when most mating occurs, also sing, but not as long and complicated songs as humpbacks have. Also, it is unknown whether only males sing. Right, bowhead, and gray whale males have larger testes than predicted by body size, which may allow them to "compete" at the sperm level simply by the axiom that the most sperm in the most females is most likely to lead to most successful pregnancies (Brownell and Ralls, 1986).

B. Toothed Whales

The taxonomic group of odontocete cetaceans encompasses a large variety of animals, from the relatively closed societies of killer (*Orcinus orca*) and pilot (*Globicephala* spp.) whales to the large and fluid associations of spinner (*Stenella longirostris*) and pan-tropical spotted dolphins (*S. attenuata*). Still, all toothed whales are social creatures, and all have long-term mother–calf bonds of association. In general, we can guess that those animals where males are morphologically different from females (sexually dimorphic) tend toward polygyny, with the male somehow advertising to males or females to gain greater access than others to females in estrus. Examples would be killer whale males' larger body size and relatively huge erect dorsal fin; sperm whale (*Physeter macrocephalus*) males' huge head and larger size, erupted lower jaw TEETH or "tusks" of male beaked whales; and the enormous narwhal (*Monodon monoceros*) male tusk, not usually present in females.

Sperm whale females stay together in their own matriarchal pods, probably for life, whereas maturing males leave. Therefore, mothers and their female offspring are long-term companions. Mature, large, usually single or "lone" males reapproach these matriarchal pods during the mating season. At that time, some different sounds (from the usual staccato-like clicks that make a pod of sperm whales sound like fat sizzling in a frying pan) are heard, and it is possible that males find the matriarchal group by the calls of females in estrus. It is also possible that the male calls themselves initiate estrus in sexually ready females and/or that the calls are used for inter-male displays.

Species that tend to be generally sexually monomorphic, such as almost all of the smaller dolphins, appear to be multimate or promiscuous. Nevertheless, as one observes these dolphins underwater—and here there is a bit more known as several species have been kept in captivity or have been studied in clear tropical waters—one recognizes that some animals appear to be more preferred partners than others, and that a "pecking" order of who gets to mate with whom may exist. There is much flipper to flipper, tail to belly, rostrum to genital slit, and so on caressing by both males and females, and it is clear that copulation can be initiated by either sex. In these animals, true courtship behavior can be seen, mainly in the form of such caressing but also in interactions by sound. Several dolphin species have larger-than-predicted testes, and it is likely that while both males and females have multiple mating partners, sperm competition exists for these animals as for several species of baleen whales.

A particularly aggressive MATING STRATEGY by males has been clearly documented for Indian-ocean bottlenose dolphins (*Tursiops aduncus*) in Shark Bay, Western Australia. There, several males band together to form a coalition that kidnaps a female from her group, and males repeatedly copulate with her, at times for days. Coalitions attempt to steal females from each other as well. Possible similar situations have been seen elsewhere, but it is unknown whether this behavior, on the margin of "courtship," is a generalized alternative mating strategy (perhaps by subadult males with less chance to mate with females on their own?) or represents cases of an aberrant behavior (Connor and Peterson, 1994).

See Also the Following Articles

Aggressive Behavior, Intraspecific ■ Estrus and Estrous Behavior ■ Reproductive Behavior

References

Connor, R. C., and Peterson, D. M. (1994). "The Lives of Whales and Dolphins." Henry Holt and Co., New York.

Brownell, R. L., Jr., and Ralls, K. (1986). Potential for sperm competition in baleen whales. *Rep. Int. Whal. Comm.* (special issue) **8**, 97–112.

Krebs, J. R., and Davies, N. B. (1993). "An Introduction to Behavioural Ecology." Blackwell Scientific, London.

LeBoeuf, B. J., and Laws, R. M. (1994). "Elephant Seals: Population Ecology, Behavior, and Physiology." Univ. of California Press, Berkeley, CA.

Reynolds, J. E., III and Odell, D. K. (1991). "Manatees and Dugongs." Facts on File, New York, NY.

Riedman, M. (1990). "The Pinnipeds: Seals, Sea Lions, and Walruses." Univ. of California Press, Berkeley, CA.

Stirling, I. (1990). The evolution of mating systems in pinnipeds. *In* "Recent Advances in the Study of Behavior" (J.F. Eisenberg and D.G. Kleiman, eds.), pp. 489–527. American Society of Mammalogists, Special publ.No.7, Provo, UT.

Stirling, I., and Guravich, D. (1988). "Polar Bears." Univ. of Michigan Press, Ann Arbor, MI.

Wells, R. S., Boness, D. J., and Rathbun, G. B. (1999). Behavior. *In* "Biology of Marine Mammals" (J.E. Reynolds III and S.A. Rommel, eds.), pp. 324–422. Smithsonian Institution Press, Washington, DC.

Crabeater Seal
Lobodon carcinophaga

JOHN L. BENGTSON
*National Marine Mammal Laboratory,
Seattle, Washington*

The crabeater seal may be the most abundant pinniped in the world, existing in the millions around Antarctica. The scientific name is derived from Greek and means "lobed tooth" (*Lobodon*) "crab eater" (*carcinophaga*).

I. Diagnostic Characters and Comments on Taxonomy

Adult crabeater seals are generally about 205 to 240 cm long, with some older individuals reaching up to 260 cm in length. Females may attain a slightly larger size than males. During the summer molting period, adults typically exhibit weights in a range of about 180 to 225 kg. Pups weigh about 35 kg at birth, but can grow to more than 100 kg by the time they are weaned.

The pelts of crabeater seals usually have medium brown to silver hair over most of their body, although darker COLORATION and spotting are not uncommon on the front and rear flippers and flanks (Fig. 1). Because hair fades in color throughout the year, recently molted seals may appear darker than those about to begin their molt, whose pelts can appear silvery white. The body form is relatively slender compared to other phocids, and crabeater seals' faces have a somewhat pointed snout. Crabeater seals have a high incidence of obvious scarring on their bodies, mostly caused by leopard seal (*Hydrurga leptonyx*) attacks (Fig. 2). Adults typically also have small scars from bites around their front and rear flippers (both sexes) and around their lower jaws and throat (mostly males) from intraspecific interactions during the breeding season.

Crabeater seals are highly mobile on ice and, when disturbed, often raise their heads and arch their backs. They can move surprisingly quickly over ice and snow, and on a cold day (when not subject to overheating) they may be capable of outrunning a fit human.

II. Distribution/Range Map

Crabeater seals have a circumpolar Antarctic distribution, spending the entire year in the pack ice zone as it seasonally advances and retreats. Occasionally, crabeater seals are found along the southern fringes of South America, Australia, New Zealand, and Africa, but such sightings or strandings are rare. Genetic analyses suggest that the circumpolar crabeater seal population is panmictic; there are no known subspecies.

Figure 1 *Crabeater seal head and shoulders illustrating spotting around front flippers. Photo by J. L. Bengtson.*

Figure 2 *Most crabeater seals possess long, raking scars on their torsos resulting from attacks in their first year of life by leopard seals. Photo by J. L. Bengtson.*

III. Ecology

There is presently no reliable estimate of the total abundance of crabeater seals. Estimates have ranged from 2 to 75 million individuals, although many scientists currently consider a population estimate in the range of 10–15 million may be reasonable. The observed densities of crabeater seals censused in 1983 were lower than densities observed in the late 1960s and early 1970s (4.3 versus 11.4 seals per nm^2 in the Weddell Sea and 1.9 versus 4.9 seals per nm^2 in the Pacific Ocean Sector, respectively). However, it is unclear whether these differences in densities reflected a change in population abundance or a shift in distribution within the sea ice zone. An international research initiative, the Antarctic Pack Ice Seals (APIS) Program, was undertaken in the late 1990s to refine estimates of the abundance and distribution of crabeater seals.

In their first year, crabeater seals experience a surprisingly high mortality rate that may be as high as 80%, which is perhaps double that which might normally be expected. For the approximately 20% of crabeater seals that survive past their first birthday, as many as 78% exhibit large, raking scars on their bodies resulting from previous attacks by leopard seals, suggesting that leopard seals may have a significant negative impact on crabeater seal populations. Most attacks by leopard seals on crabeater seals occur in the crabeater seals' first year; fresh wounds, indicating a recent attack, are rarely seen on crabeater seals that are older than 1 year.

Studies of crabeater seal diet have shown that these seals depend almost exclusively on Antarctic krill (*Euphausia superba*). Most investigators have reported that krill comprise over 95% of the crabeater diet, with the remainder being made up of small quantities of fish and squid. As specialist krill predators, crabeater seals do not seem to switch their prey seasonally.

IV. Behavior

Crabeater seals migrate over large distances in association with the annual advance and retreat of the pack ice. Although they can be found anywhere within the pack ice zone, it is typical to find higher densities of crabeater seals in the marginal ice zone as well as in pack ice that is present near or over the continental shelf. In a peculiar behavioral twist, crabeater seals likely hold the record for any pinniped wandering inland from the coast. Carcasses have been found up to 113 km from open water and as high as 1100 m above sea level. Seals that wander inland become lost and die, may eventually become mummified in the cold, dry Antarctic air, and can remain in this "freeze-dried" state for many decades or centuries.

Similar to other Antarctic pack ice seals, crabeater seals exhibit a daily haulout pattern that generally involves hauling out on ice floes during the middle of the day. However, only about 80% of crabeater seals haul out simultaneously on the ice, even during the height of the molting period in January and February. Haulout patterns also vary markedly among seasons, with as few as 40% of seals hauling out at the peak of daily haulout during winter months.

During daily foraging periods, which normally occur during the night, crabeater seals dive nearly continuously for periods of up to 16 hr. In one study, a single crabeater seal continued diving for 44 hr without interruption. Foraging dives made during crepuscular periods are often deeper than those made during the darkest hours, suggesting that the seals may prefer dark conditions when catching their principal prey, Antarctic krill.

V. Anatomy, Physiology, and Life History

Crabeater seals have finely divided, lobed TEETH, presumably an adaptation to their specialized diet on krill. The multiple cusps of upper and lower postcanine teeth interlock to form a sieve that can be used to filter crustaceans from seawater. A bony protrusion on the lower jaw behind the most posterior postcanine tooth fills the gap in this sieve so that prey cannot escape at the rear of the mouth.

During the breeding season, crabeater seals form "family groups," consisting of a female, her pup, and an attendant male who guards the female from other males until she completes lactation. Pups are born in September and October, with a light brown lanugo that is molted at about 2 weeks of age. Following weaning, the attendant male and the female form a "mated pair" and remain together for an estimated 1 to 2 weeks or until copulation. Females without pups also form mated pairs as they come into estrus. Crabeater seals can live up to 40 years, but adults dying at about 20–25 years is more typical.

VI. Interactions with Humans

Crabeater seals experienced an incident of mass mortality in 1955 in the vicinity of an Antarctic base where sledge dogs were active. Up to 97% of mixed-age aggregations died during that event. It was speculated that a viral infection may have been associated with the die-off, and circumstantial evidence suggests that it may have been caused by a distemper-like virus. Blood samples taken from crabeater seals in the late 1980s confirmed that populations of crabeater seals along the Antarctic Peninsula had antibodies similar to those related to canine distemper and phocine distemper, viruses that were responsible for major epizootic die-offs of harbor seals in the Northern Hemisphere in the late 1980s.

Crabeater seals were harvested commercially twice during the past century: in 1964/1965 by Norway and in 1986/1987 by the former Soviet Union. In both cases, the sealing ventures were judged to be economically unsuccessful. However, the concern generated by the earlier harvest was sufficient to mobilize an international effort to prevent potential overexploitation of the seals. This concern resulted in the Convention for the Conservation of Antarctic Seals, which came into effect in 1978 and provides international oversight for the CONSERVATION and MANAGEMENT of crabeater seals throughout their range.

See Also the Following Articles

Antarctic Marine Mammals ■ Krill ■ Leopard Seal

References

Bengtson, J. L., Boveng, P., Franzén, U., Have, P., Heide-Jøgensen, M. P., and Härkönen, T. J. (1991). Antibodies to canine distemper virus in Antarctic seals. *Mar. Mamm. Sci.* **7,** 85–87.

Bengtson, J. L., Hill, R. D., and Hill, S. E. (1993). Using satellite telemetry to study the ecology and behavior of Antarctic seals. *Kor. J. Pol. Res.* **4,** 109–115.

Bengtson, J. L., and Laws, R. M. (1985). Trends in crabeater seal age at maturity: An insight into Antarctic marine interactions. *In* "Antarctic Nutrient Cycles and Food Webs" (W. R. Siegfried, P. R. Condy, and R. M. Laws, eds.), pp. 670–675. Springer-Verlag, Berlin.

Bengtson, J. L., and Siniff, D. B. (1981). Reproductive aspects of female crabeater seals (*Lobodon carcinophagus*) along the Antarctic Peninsula. *Can. J. Zool.* **59,** 92–102.

Bengtson, J. L., and Stewart, B. S. (1992). Diving and haulout behavior of crabeater seals in the Weddell Sea, Antarctica, during March 1986. *Pol. Biol.* **12,** 635–644.

Boveng, P. L., and Bengtson, J. L. (1997). Crabeater seal cohort variation: Demographic signal or statistical noise? *In* "Antarctic Communities: Species, Structure, and Survival" (B. Battaglia, J. Valencia, and D. W. H. Walton, eds.), pp. 241–247. Cambridge Univ. Press, Cambridge.

Erickson, A. W., and Hanson, M. B. (1990). Continental estimates and population trends of Antarctic ice seals. *In* "Antarctic Ecosystems: Ecological Change and Conservation" (K. R. Kerry and G. Hempel, eds.), pp. 253–264. Springer-Verlag, Berlin.

Kooyman, G. L. (1981). Crabeater seal, *Lobodon carcinophagus* (Hombron and Jacquinot, 1842). *In* "Handbook of Marine Mammals" (S. H. Ridgeway and R. J. Harrison, eds.), pp. 221–235. Academic Press, London.

Laws, R. M. (1958). Growth rates and ages of crabeater seals, *Lobodon carcinophagus,* Jacquinot and Pucheron. *Proc. Zool. Soc. Lond.* **130,** 275–288.

Laws, R. M. (1977). The significance of vertebrates in the Antarctic marine ecosystem. *In* "Adaptations within Antarctic Ecosystems" (G. A. Llano, ed.), pp. 411–438. Smithsonian Institution, Washington.

Laws, R. M. (1984). Seals. *In* "Antarctic Ecology" (R. M. Laws, ed.), Vol. 2, pp. 621–715. Academic Press, London.

Øritsland, T. (1977). Food consumption of seals in the Antarctic pack ice. *In* "Adaptations within Antarctic Ecosystems" (G. A. Llano, ed.), pp. 749–768. Smithsonian Institution, Washington.

Siniff, D. B., and Bengtson, J. L. (1977). Observations and hypotheses concerning the interactions among crabeater seals, leopard seals, and killer whales. *J. Mammal.* **58,** 414–416.

Siniff, D. B., Stirling, I., Bengtson, J. L., and Reichle, R. A. (1979). Social and reproductive behavior of crabeater seals (*Lobodon carcinophagus*) during the austral spring. *Can. J. Zool.* **57,** 2243–2255.

Siniff, D. B., and Stone, S. (1985). The role of the leopard seal in the tropho-dynamics of the Antarctic marine ecosystem. *In* "Antarctic Nutrient Cycles and Food Webs" (W. R. Siegfried, P. R. Condy, and R. M. Laws, eds.), pp. 561–565. Springer-Verlag, Berlin.

Cranial Anatomy

SEE *Skull Anatomy*

Culture in Whales and Dolphins

HAL WHITEHEAD
Dalhousie University, Halifax, Nova Scotia, Canada

Evidence is growing that culture is an important determinant of the behavior of whales and dolphins. Among the many definitions of culture, one that is commonly used by evolutionary biologists and is useful when studying the phenomenon in whales and dolphins, is behavioral variation between sets of animals maintained and transmitted by social learning. There are two principal approaches to the study of nonhuman culture. Because some scientists will only ascribe culture to a behavioral pattern if it can be proved to be transmitted between animals by imitation or teaching, they investigate transmission mechanisms experimentally. Others, who use a broader definition of culture encompassing any form of social learning (not just imitation or teaching), look for patterns of behavioral variation in wild populations that cannot be explained by either genetic factors or environmental differences plus individual learning. This has been called the "ethnographic" approach to the study of culture.

The common bottlenose dolphin (*Tursiops truncatus*) has been shown experimentally to posses sophisticated social learning abilities, including vocal and motor imitation, but these have not been closely tied to observed patterns of behavior in the wild. Although social learning of other cetacean species has not been studied experimentally, there is observational evidence for imitation and teaching in some other whales and dolphins, especially killer whales (*Orcinus orca*).

Taking the second, ethnographic, approach, there is good evidence for cultural transmission in several cetacean species. Most notably, the complex and stable vocal (call dialects) and behavioral (foraging patterns and techniques) cultures of sympatric groups of killer whales have no known parallel outside humans and represent an independent evolution of cultural faculties. Although evidence is less firm, sperm whales (*Physeter macrocephalus*) also seem to have important group-based cultures, which include distinctive dialects. Perhaps most remarkable of all cetacean cultures is the song of male humpback whales (*Megaptera novaeangliae*). All males on any breeding ground sing nearly the same song, but it evolves over periods of months and years. This evolution is usually gradual, but over a 2-year period, the males off eastern Australia unanimously adopted the radically different western Australian song, which they had heard from a few itinerant males.

Several factors may be implicated in the apparent importance of cultural transmission of behavior among cetaceans. Long lives, prolonged parental care, and substantial cognitive abilities are often associated with the evolution of cultural faculties, and these are generally characteristic of cetaceans, as well as other cultural animals, such as primates and some birds. The wide movements of cetaceans and the greater variability of the marine biotic environment relative to that on land, as well as the stable matrilineal social groups of some species, are potentially important factors in the evolution of some of the more unusual aspects of cetacean culture.

Culture can affect the evolution of other aspects of the lives of animals. There have been a number of suggestions for gene–culture coevolution in cetaceans, and culture may be implicated in some of their unusual behavioral and life history traits. For instance, it has been proposed that the separation between "resident," fish-feeding and "transient," mammal-feeding forms of killer whales (which now show morphological and genetic differences) was originally driven by culture. Another suggestion is that cultural selection may have caused the low diversity of mitochondrial genes found in matrilineal whales, as these genes and beneficial cultural traits may have been inherited together by daughters from their mothers. Culture may also be implicated in mass strandings, as well as in the pronounced menopause shown by killer and pilot whale females (which is known only from humans among noncetaceans).

The focused study of culture in whales and dolphins is just beginning. Despite denials from those who demand experimental proof of imitation or teaching before attributing culture, there are strong indications that, in common with humans and chimpanzees, much of the behavioral repertoire of many cetaceans is learned socially and constitutes culture. Culture may also be an important attribute of other marine mammals, with the foraging techniques of sea otters (*Enhydra lutris*) perhaps forming the clearest example.

See Also the Following Articles

Communication ■ Intelligence and Cognition ■ Mimicry ■ Song

References

Boran, J. R., and Heimlich, S. L. (1999). Social learning in cetaceans: Hunting, hearing and hierarchies. *Symp. Zool. Soc. Lond.* **73**, 282–307.

Boyd, R., and Richerson, P. (1985). "Culture and the Evolutionary Process." Chicago Univ. Press, Chicago.

Deecke, V. B., Ford, J. K. B., and Spong, P. (2000). Dialect change in resident killer whales: Implications for vocal learning and cultural transmission. *Anim. Behav.* **40**, 629–638.

Ford, J. K. B. (1991). Vocal traditions among resident killer whales (*Orcinus orca*) in coastal waters of British Columbia. *Can. J. Zoo.* **69**, 1454–1483.

Noad, M. J., Cato, D. H., Bryden, M. M., Jenner, M.-N., and Jenner, K. C. S. (2000). Cultural revolution in whale songs. *Nature* **408**, 537.

Rendell, L., and Whitehead, H. (2001). Culture in whales and dolphins. *Behav. Brain Sci.* **24**, 309–382.

Whitehead, H. (1998). Cultural selection and genetic diversity in matrilineal whales. *Science* **282**, 1708–1711.

Cuvier's Beaked Whale
Ziphius cavirostris

JOHN E. HEYNING
*Natural History Museum of
Los Angeles County, California*

The original description of *Ziphius cavirostris* is based on a partial cranium collected near the village of Fos, France, in 1804. In his species description, Cuvier mistakenly identified the specimen as a fossil because he thought the skull was "petrified." The specimen actually represents part of a skull, including a densely ossified rostrum with a well-developed prenarial basin of an adult male. This basin, or cavity, on the top of the skull just anterior to the bony nares led to the trivial name of *cavirostris*. The densely ossified rostrum is found in adult males of Cuvier's beaked whale and mesoplodont beaked whales (Fig. 1). The function of this rock-hard snout is unknown, but it has been postulated to either reinforce the skull when males fight or serve as a sound conduit. The most common English names for *Z. cavirostris* are Cuvier's beaked whale and the goose-beaked whale, both of which are in wide usage.

The general body shape of *Z. cavirostris* is similar to other beaked whales with a rather robust, cigar-shaped body. The falcate dorsal fin is relatively small and set approximately two-thirds of the body length posterior to the rostrum. The flippers are also relatively small, narrow, and can be tucked into a slight depression or pocket along the body wall. This flipper pocket is also found in other ziphiids and is assumed to allow the flippers to be held tight against the body while swimming. As with other ziphiids, the flukes are proportionally large and, as a rule, lack the distinctive medial notch found in all other cetaceans. The head is rather blunt in profile with a small, poorly defined rostrum that grades into a gently sloping melon region. There is no significant difference in total length between sexes for *Z. cavirostris*, with an average adult size of 6.1 m. There are several reports of specimens that exceed 7.0 m in length, but virtually all of these appear to be either estimates of lengths or based on misidentified animals. The largest accurately measured specimen of *Z. cavirostris* is an adult male that measured 6.93 m from the Falkland Islands. There is one pair of throat grooves that converge, but do not meet anteriorly. Beaked whales feed primarily by suction and these grooves allow the throat region to expand as they slurp in their prey.

The pigmentation pattern for male *Z. cavirostris* is a dark slate gray over most of its body with a distinctively white head. This white coloration continues slightly posterior along the dorsum. This appears to be the pattern found in most mature males (Fig. 2). Adult females tend to vary in general color from a dark gray to a reddish-brown, with a slight lightening of the skin on the head. This is not as dramatic a contrast as in males and does not appear to extend posteriorly on the dorsal aspect of the body. Usually there are some distinctive patterns of dark pigment on the head of adult females. The eye is typically dark and there is a highly variable pair of dark crescents surrounding the

Figure 1 *As males of* Ziphius cavirostris *become sexually mature, they begin to reabsorb bone in front of the nasal passages, creating over time a distinct cavity or prenarial basin seen easily in this dorsal view (right). Adult females and immature males lack this basin (left).*

eye, one anteriorly and one posteriorly. Newborns are dark black or bluish-black above and lighter below. This pigmentation pattern is very similar to that found on young beaked whales of the genus *Mesoplodon* and may represent the primitive pigmentation pattern for the calves of many ziphiids. Light oval patches and linear marks are quite common on the skin of *Z. cavirostris*, which can give an animal a mottled appearance. The oval patches on ziphiids have been attributed to scars left by lampreys or cookie-cutter sharks of the genus *Isistius*. Linear marks have been attributed to scars resulting from the teeth of males raking along the skin during intraspecific fighting. The only erupted teeth are the apical pair in adult males, and linear scars are most prevalent in mature males.

As with most uncommon cetaceans, the distribution of *Z. cavirostris* is known primarily from strandings. This type of information may be somewhat biased, especially with regard to the

Figure 2 *This dead, stranded* Z. cavirostris *shows the whitish head characteristic of adult males. Photo by W. Perrin, National Marine Fisheries Service.*

abundance for a particular area. Stranding records indicate that *Z. cavirostris* is the most cosmopolitan of the beaked whales and is distributed in all oceans and most seas except in the high polar waters. Strandings of *Z. cavirostris* are the most numerous of all beaked whales, indicating that they are probably not as rare as originally thought. Observations of live animals in the field reveal that the blow of *Z. cavirostris* is low, diffuse, and directed forward, making sightings more difficult, and there is some evidence that they avoid vessels by diving. These two facts may explain why these whales are rarely seen at sea. Single animals are frequently observed with pods ranging in size up to seven animals. There are several records of mass strandings of *Z. cavirostris*.

Although most general accounts of *Z. cavirostris* list squid as the primary prey item, very few actual stomach contents have been analyzed, and care must be invoked in any interpretation. Stomach contents from *Z. cavirostris* caught off Japan varied consistently with a predominance of squid from animals taken in waters slightly under 1000 m in depth, but fish are the most abundant prey item found in animals in deeper waters. This evidence has been interpreted to suggest that *Z. cavirostris* is somewhat opportunistic in its feeding habits. Most of the prey items listed are open ocean, mesopelagic, or deep-water benthic organisms, concurring with the idea that *Z. cavirostris* is an offshore, deep-diving species.

Ectoparasites that have been reported include the barnacles *Xenobalanus* sp. from the flukes and dorsal fin and *Conchoderma* sp. on the erupted apical teeth. The following internal parasites have been reported from *Z. cavirostris*; Nematoda, *Anisakis* sp., *Crassicauda boopis,* and *Crassicauda crassicauda*; and Cestoda, *Phyllobothrium* sp. In the past, there have been few small cetacean fisheries that have taken *Ziphius*. In the Japanese *Berardius bairdii* fishery, *Z. cavirostris* have been taken on an opportunistic basis with catches of varying from 3 to 35 animals taken yearly. Although the *Berardius* fishery still continues, there been no takes of *Z. cavirostris* in recent years. It is probable that killer whales occasionally predate on *Z. cavirostris*.

See Also the Following Articles

Barnacles ▪ Beaked Whales, Overview ▪ Mesoplodont Whales

Reference

Heyning, J. E. (1989). Cuvier's beaked whale, *Ziphius cavirostris. In* "Handbook of Marine Mammals" (S. H. Ridgway and R. Harrison, eds.), Vol. 4, pp. 289–308. Academic Press, London.

D

Dall's Porpoise
Phocoenoides dalli

THOMAS A. JEFFERSON
*Southwest Fisheries Science Center,
La Jolla, California*

Dall's porpoise is one of the better-known species in the family Phocoenidae. Largely because many individuals have been killed in fisheries recently, much has been learned of the biology of this species throughout its range in the North Pacific Ocean (Houck and Jefferson, 1999).

I. Characters and Taxonomic Relationships

Typical of the porpoise family, Dall's porpoise has a stocky body with a short, wide-based, triangular dorsal fin (Fig. 1). The dorsal fin is slightly falcate at the tip, but the entire fin may

Figure 1 *A Dall's porpoise swims just below the surface on a calm day in Monterey Bay, California. Because of their normally high-speed swimming, clear views such as this are somewhat uncommon.*

be canted forward in large adult males. The tail stock is deepened, especially in adult males. There is an extremely short, poorly defined beak. The flippers and flukes are small, and the fluke blades may also be canted forward.

The color pattern is diagnostic. "Dall's" are largely dark gray to black with a large, ventrally continuous white patch that extends up about halfway on each flank. In addition, there is light gray to white frosting or trim on the upper part of the dorsal fin and on the trailing edge of the flukes. Some other light patches may exist, particularly around the base of the tail stock.

There are two major color morphs: one with a flank patch that extends forward to about the level of the dorsal fin (*dalli* type) and the other with a flank patch extending to about the level of the flippers (*truei* type). These forms were variously called separate species and subspecies in the past, but most recent work suggests that they are in fact color variants, with little or no other phenotypic difference. However, genetic analyses have confirmed that they do appear to form separate populations (Escorza, 1998).

Dall's porpoises reach maximum known lengths and weights of about 239 cm and 200 kg. Males grow longer and heavier than females, and adult males have secondary sexual characteristics (mentioned earlier). There is a great deal of geographical variation; size, shape, and coloration differences have been documented among different areas of the species' range.

The SKULL of the Dall's porpoise is larger than that of most other phocoenids and may reach 340 mm. The rostrum is wide at the base and relatively short (Fig. 2). There are prominent "maxillary shields" that make an angle of about 130° with the rostrum axis. The TEETH are extremely small, the smallest of any species of cetacean. They often do not rise above the level of the gums and are considered by many to be rudimentary.

Recent studies of mtDNA suggest that the previous classification of Dall's porpoise and the spectacled porpoise (*Phocoena dioptrica*) in the same subfamily was erroneous (Rosel *et al.*, 1995). These two species do not appear to be closely related, and their similarities may be the result of CONVERGENT EVOLUTION.

II. Distribution and Ecology

This species is found only in the North Pacific Ocean and adjacent seas (Bering Sea, Okhotsk Sea, and Sea of Japan),

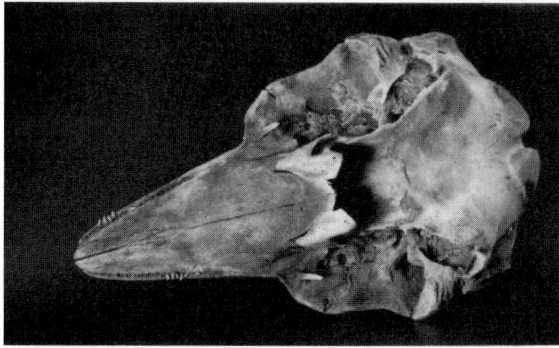

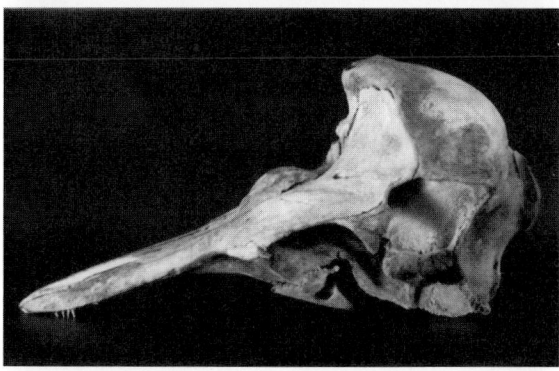

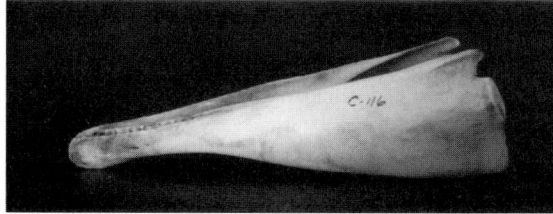

Figure 2 *Dorsal, ventral, and lateral views of the cranium, and a lateral view of the mandible, of a Dall's porpoise from central California.*

from about 32°–35°N (southern California and southern Japan) in the south to about 63°N (central Bering Sea) in the north. Up to 10 different populations and stocks are recognized, based on studies of morphology, genetics, and ecological parameters (see Amano and Miyazaki, 1992; Escorza, 1998). Sex-biased dispersal and migration patterns have been elucidated from molecular genetic analyses (Escorza, 1998).

This is an oceanic species that is found in deep offshore waters but also in deeper nearshore and inshore waters along the west coast of North America. There are seasonal inshore–offshore and north–south movements in both the eastern and western Pacific (Forney and Barlow, 1998), but in most areas these are very poorly defined.

Prey species include a wide variety of fishes and cephalopods. The most common prey items include small schooling fishes (such as herring, anchovies, mackerels, and sauries), mesopelagic fishes (such as myctophids and deep-sea smelts), and squids. KRILL, decapods, and shrimps have been found in some stomachs, but these are not considered to be common prey items. Amano and Miyazaki (1992) found that the skulls of porpoises grew to larger sizes in areas with higher productivity, suggesting that food availability affects growth.

Dall's porpoises in some areas appear to feed preferentially at night on vertically migrating fish and squid associated with the deep scattering layer. They are thought to be probably deep divers and capable of feeding at great depths.

III. Behavior and Life History

Small groups are most common, although large aggregations of several hundreds to about 1000 have been reported on occasion. Groups of over 20–30 porpoises are rather uncommon. Very little is known of the group structure of this species except that group composition is probably quite fluid.

These are very fast-swimming and active porpoises. They are often seen moving very quickly, slicing along the surface, creating a sloppy, V-shaped splash. These are called roostertail splashes (Fig. 3). Dall's porpoises are willing and capable bow riders and will converge on the bow of a fast-moving boat from all around. They have even been seen to "snout ride" on bow waves pushed forward by the heads of large whales.

Growth and reproductive parameters have been estimated for several populations in the central and western Pacific. Length at birth is about 100 cm. Estimates of length and age

Figure 3 *Most often when Dall's porpoises are seen at sea, they are swimming very fast and "roostertailing," as this porpoise is doing while riding the bow wave of a vessel in Southeast Alaska.*

at sexual maturity range from about 172 to 187 cm and 4 to 7 years for females and from 175 to 196 cm and 3.5 to 8 years for males (Houck and Jefferson, 1999; Ferrero and Walker, 1999). Gestation lasts about 10–12 months, and the length of lactation is not well known (but is most likely shorter than 1 year). The calving season is in the summer for all populations that have been studied to date, although some calves may be born outside the main season.

An intergeneric hybrid between a Dall's porpoise and a harbor porpoise (*Phocoena phocoena*) has been described (Baird *et al.*, 1998). Free-ranging animals resembling the confirmed hybrid are regularly observed around Vancouver Island, British Columbia, suggesting that such hybridization events may not be all that rare.

IV. Interactions with Humans

A number of threats to Dall's porpoise populations have been identified, including environmental POLLUTION and habitat alteration. However, the most serious threats are the various fisheries kills of this species. These include directed kills in Japanese waters and INCIDENTAL CATCHES in various fisheries (most prominently oceanic drift net fisheries) throughout the range. The most heavily impacted populations have been those in the central and western North Pacific.

Small numbers of Dall's porpoises have been kept in captivity in oceanaria and research institutes in the United States and Japan, but most individuals have not survived long. They are not currently sought after for captive display.

See Also the Following Articles

Bow-Riding ■ Geographic Variation ■ North Pacific Marine Mammals ■ Porpoises, Overview

References

Amano, M., and Miyazaki, N. (1992). Geographic variation and sexual dimorphism in the skull of Dall's porpoise, *Phocoenoides dalli. Mar. Mamm. Sci.* **8,** 240–261.

Baird, R. W., Willis, P. M., Guenther, T. J., Wilson, P. J., and White, B. N. (1998). An intergeneric hybrid in the family Phocoenoidae. *Can. J. Zool.* **76,** 198–204.

Escorza Trevino, S. (1998). "Molecular Ecology and Evolution of Dall's Porpoise, *Phocoenoides dalli* (True, 1885)." Ph.D. thesis, University of California, San Diego.

Ferrero, R. C., and Walker, W. A. (1999). Age, growth, and reproductive patterns of Dall's porpoise (*Phocoenoides dalli*) in the central North Pacific Ocean. *Mar. Mamm. Sci.* **15,** 273–313.

Forney, K. A., and Barlow, J. (1998). Seasonal patterns in the abundance and distribution of California cetaceans, 1991–1992. *Mar. Mamm. Sci.* **14,** 460–489.

Houck, W. J., and Jefferson, T. A. (1999). Dall's porpoise *Phocoenoides dalli* (True, 1885). *In* "Handbook of Marine Mammals" (S. H. Ridgway and R. Harrison, eds.), Vol. 6, pp. 443–472. Academic Press, San Diego.

Rosel, P. E., Haygood, M. G., and Perrin, W. F. (1995). Phylogenetic relationships among the true porpoises (Cetacea: Phocoenidae). *Mol. Phylogen. Evol.* **4,** 463–474.

Delphinids, Overview

RICK LeDUC
*Southwest Fisheries Science Center,
La Jolla, California*

The family Delphinidae is one of three extant (with Phocoenidae and Monodontidae) families in the cetacean superfamily Delphinoidea (which also includes three extinct families, Kentriodontidae, Odobenocetopsidae, and Albireonidae).

The popularity of oceanic dolphins (family Delphinidae) is arguably among the highest for wild animals, both for the general public and for scientists conducting research. However, despite this, they are very poorly understood compared to most terrestrial mammals of similar size. For most species of delphinids, basic aspects of their evolution, physiology, ecology, behavior, and population structure are virtually unknown. For the biologist, this presents real challenges and opportunities for dolphin research, not just to understand individual species, but also toward a better understanding of how they fit into marine ecosystems.

I. Taxonomic Overview

Delphinids likely arose in the mid- to late Miocene (11–12 mya) from kentriodontid-like ancestors and quickly radiated into many different morphological and ecological types. This early radiation produced precursors of many modern forms; many of the early delphinid fossils can be assigned to extant genera, particularly *Tursiops*. Today the Delphinidae is the most speciose family of marine mammals, with 34–36 recognized extant species arranged into 17–19 genera. At present, there is much uncertainty about the evolutionary relationships among the species of delphinids. Of the many recent classifications that have been proposed, two are depicted here. One represents a more traditional view of dolphin taxonomy (Table Ia) and the other a recent provisional classification based on molecular phylogenetic analyses (Table Ib). It should be mentioned that the latter classification is considered tentative and is based on analysis of a single gene. A more highly resolved representation of systematic relationships among delphinid species is not currently available. Also, there will no doubt be additional revisions proposed in the future, especially involving the apparently paraphyletic genera *Stenella* and *Tursiops*. In part, this changing nature of delphinid taxonomy is due to the new molecular and analytical tools currently available to researchers, but it also reflects the uncertainties about evolutionary relationships that have long been recognized by morphological systematists but have yet to be addressed.

II. Morphology

Dolphins have the typical morphological characteristics of toothed whales, such as spindle-shaped bodies, single external blowholes, telescoping of the skull such that the maxillary bones overlap the frontals in the supraorbital region, left-skewed cra-

---

TABLE I
Two Recent Classifications of the Family Delphinidae

---

a. A classification of the family Delphinidae from Perrin (1989) reflecting a traditional view of species interrelationships

Family Delphinidae
  Subfamily Steninae
    Steno bredanensis
    Sousa chinensis
    S. teuszii
    Sotalia fluviatilis
  Subfamily Delphininae
    Lagenorhynchus albirostris
    L. acutus
    L. obscurus
    L. obliquidens
    L. cruciger
    L. australis
    Grampus griseus
    Tursiops truncatus
    Stenella frontalis
    S. attenuata
    S. longirostris
    S. clymene
    S. coeruleoalba
    Delphinus delphis
    Lagenodelphis hosei
  Subfamily Lissodelphinae
    Lissodelphis borealis
    L. peronii
  Subfamily Cephalorhynchinae
    Cephalorhynchus commersonii
    C. eutropia
    C. heavisidii
    C. hectori
  Subfamily Globicephalinae
    Peponocephala electra
    Feresa attenuata
    Pseudorca crassidens
    Orcinus orca
    Globicephala melas
    G. macrorhynchus
  Subfamily Orcaellinae
    Orcaella brevirostris

b. Revised classification of the family Delphinidae based on molecular systematic analysis; adapted from LeDuc *et al.* (1999).

Family Delphinidae
  Subfamily Stenoninae
    Steno bredanensis
    Sotalia fluviatilis
  Subfamily Delphininae
    Sousa chinensis
    Stenella clymene
    S. coeruleoalba
    S. frontalis
    S. attenuata
    S. longirostris
    Delphinus delphis
    D. capensis
    Tursiops truncatus
    T. aduncus
    Lagenodelphis hosei
  Subfamily Lissodelphininae
    Lissodelphis borealis
    L. peronii
    Cephalorhynchus heavisidii
    C. hectori
    C. eutropia
    C. commersonii
    Sagmatias obscurus
    S. obliquidens
    S. cruciger
    S. australis
  Subfamily Globicephalinae
    Feresa attenuata
    Peponocephala electra
    Globicephala melas
    G. macrorhynchus
    Pseudorca crassidens
    Grampus griseus
  Subfamily Orcininae
    Orcinus orca
    Orcaella brevirostris
  Incertae sedis
    Lagenorhynchus albirostris
    Leucopleurus acutus

---

nial asymmetry, polydonty, and homodonty (in most). In some species, there has evolved a secondary reduction in the number of teeth, often seen as an adaptation for feeding on squid. The evolution of the delphinoid lineage saw the development of elaborate systems of pterygoid sinuses and better-isolated earbones, probably increasing their ability to echolocate and perhaps giving them an advantage over some of the other odontocete groups of the Miocene. The ancestors of the delphinids, the kentriodontids, were small dolphins with short to medium length rostra and, unlike most modern delphinoids, had symmetrical cranial vertices. The development in the Delphinidae of asymmetry in the cranial vertex and in the premaxillary bones suggests a further refinement of their echolocation capabilities and may partly explain their evolutionary success. The most noticeable difference between delphinids and their closest relatives, the phocoenids (true porpoises), is that the latter have spade-shaped teeth, whereas delphinids have conical or peg-like TEETH, as do most other odontocetes (toothed whales). They also differ from phocoenids in the shape of the facial region of the SKULL, including having a more distinct vertex. Within the Delphinidae, the most obvious variation among species relates to the feeding apparatus: the development of the rostrum, jaws, and teeth. There is a broad spectrum of rostrum lengths and widths, tooth counts, and tooth sizes, reflecting the range of ecological niches occupied by the different species. For example, total tooth counts range from less than 10 in Risso's dolphin (*Grampus griseus*) (which has no teeth in the upper jaw) to 250 in the spinner dolphin (*S. longirostris*).

Delphinids also show wide variation in their external morphology. Only a few species (e.g., killer whale, *Orcinus orca;* pilot whales, *Globicephala* spp.) are dramatically sexually dimorphic, although many others may have more subtle dimorphism in body size and shape, coloration, and dorsal fin shape. In size they are small to medium cetaceans, with adults ranging from less than 1.5 m (*S. longirostris roseiventris* from the Gulf of Thailand, some species of *Cephalorhynchus*) to over 9 m (killer whale). Beak length varies widely, from very long on some (e.g., Indian Ocean common dolphin, *Delphinus tropicalis*) to very short on others (e.g., white-beaked dolphin, *Lagenorhynchus albirostris*). The external beak is completely absent in a number of delphinid genera, particularly in *Orcinus, Globicephala, Feresa, Pseudorca,* and *Peponocephala*. Those species that lack a beak often have heads that are rounded or even bulbous in profile. In most delphinid species, the dorsal fin is pointed and falcate, although it is triangular in some subspecies of *S. longirostris* and in male *O. orca*, round in Hector's dolphin (*C. hectori*), and even forward canted in males of the eastern spinner dolphin (*S. longirostris orientalis*). The dorsal fin is completely missing in the right whale dolphins (*Lissodelphis* spp.), and some hump-backed dolphins (*Sousa* spp.) have a pronounced hump at the base of the dorsal fin. The color patterns of delphinids are similarly varied, from bold black and white patterns (e.g., *O. orca* and some of the *Cephalorhynchus* spp.) to complex patterns of black, white, and gray (e.g., *Delphinus* spp. and *Stenella* spp.) to rather simple patterns of black (e.g., *Globicephala* spp.) or gray (e.g., *Sousa* spp.). The complex color patterns exhibited by some delphinids are composites of various elements, including stripes, capes, overlays, spots, and blazes. Regardless of their overall color and pattern, all dolphins tend to have a countershaded aspect, where the ventral surface is lighter than the dorsum. This countershading reduces an animal's visibility in the marine environment, where the ambient light comes from above. Besides being a mechanism for species recognition, the more complex patterns may also play a role in camouflage against waves at the surface or against the dappling of light penetrating the water.

III. Distribution and Habitat

As a group, the family Delphinidae reaches its highest diversity in tropical and warm temperate latitudes. There are numerous species with pantropical distributions and others that occur in tropical waters but are limited to one or two ocean basins. For example, the Atlantic spotted dolphin (*Stenella frontalis*) and the Clymene dolphin (*S. clymene*) are limited to the lower latitudes of the Atlantic Ocean, whereas the Indo-Pacific bottlenose dolphin (*Tursiops aduncus*) and the Irrawaddy dolphin (*Orcaella brevirostris*) only occur in the Indian and west Pacific Oceans. In colder areas one can find species in the genera *Globicephala, Cephalorhynchus, Lissodelphis,* and

Lagenorhynchus (including *Sagmatias* and *Leucopleurus* of Table Ib). Interestingly, new genetic evidence suggests that the majority of these cold-temperate species appear to be closely related (subfamily Lissodelphininae in the classification of Table Ib). Only one recognized species, the long-finned pilot whale (*G. melas*), has an antitropical distribution, although some antitropical species pairs [e.g., *Lissodelphis* spp., and the dusky and Pacific white-sided dolphins *Lagenorhynchus* (or *Sagmatias*) *obscurus/obliquidens*] have been hypothesized as being single species with antitropical populations. Only one species, *O. orca,* ranges into high latitudes near the polar ice. Indeed, occurring also into mid- and low latitudes, this species is probably the most cosmopolitan of all the cetaceans.

Within this broad range of geographic distributions, delphinids occupy an equally diverse array of habitats. Many species occur far offshore in deep water, where the specifics of their ecological requirements are poorly known. In fact, in tropical seas, only the cetacean fauna of the eastern tropical Pacific has been studied extensively and systematically, and some differences in the species composition have been observed in different water masses. Here, areas with a stable mixed layer and a shallow thermocline are frequented by *Stenella attenuata, S. longirostris,* and the rough-toothed dolphin (*Steno bredanensis*), whereas areas with more variable conditions and some amount of upwelling contain species such as the short-finned pilot whale (*G. macrorhynchus*), short-beaked common dolphin (*Delphinus delphis*), striped dolphin (*Stenella coeruleoalba*), and melon-headed whale (*Peponocephala electra*). In any ocean, some of the offshore species may also range closer to the coast (e.g., *D. delphis*) or even have populations or sister species that are restricted to the coastal waters or the nearshore habitat [e.g., the long-beaked common dolphin (*D. capensis*), coastal populations/species of *Tursiops* spp]. In a few cases, coastal populations may ascend a short distance up rivers, but only two species [the tucuxi (*Sotalia fluviatilis*) of eastern South America and *Orcaella brevirostris* of the Indo-West Pacific] regularly occur far upstream. Both of these riverine species also have marine coastal populations.

IV. Social Organization and Behavior

All dolphin species are social to some degree. However, characteristic group sizes for the different species range from small pods of just a few individuals to large schools numbering in the thousands. Due to the difficulty of observing dolphins in the wild, especially those occupying offshore habitats, very little is known about the BEHAVIOR and social organization of most species. The populations that have been studied over longer time scales (e.g., *Orcinus orca* in the northeast Pacific and *T. truncatus* in the western Atlantic) are those that form relatively stable and small social groups within a short distance of the coast and whose movements do not regularly take them out of their study areas. Using photos to identify individuals via fin markings and color patterns, and supplementing observations with genetic data, associations of individuals and their genealogical relatedness have been recorded and monitored over generations. In these populations, some long-term associations and patterns have been noted. For example, it appears that bonds between individuals in a pod and/or between individuals and a particular area (philopatry) tend to be stronger for females than males; the society can even be considered matrilineal for *O. orca*. A similar pattern was inferred from molecular data on *G. melas* in the North Atlantic. However, one must be cautious in extrapolating these social patterns to other delphinid species. For example, some species that occur on the high seas are found in schools that number in the thousands, and these associations appear much more fluid in their composition. In fact, social patterns observed for inshore groups of species like *T. truncatus* and *O. orca* may not even reflect the organization of offshore populations of the same species. What little is known about the social organization of offshore dolphins comes from direct observations of school sizes and life history data collected from mass strandings and fishery kills. In the few pelagic species studied (primarily *Stenella attenuata* and *S. longirostris*), there is evidence for promiscuity, strong mother–calf bonds, and some segregation by age and sex both within and between schools.

In addition to their intraspecific social organization, most dolphin species are seen at least occasionally in the company of other species. One famous association is between *S. attenuata* and *S. longirostris* in the eastern tropical Pacific, an aggregation that also includes yellowfin tuna (*Thunnus albacares*) and numerous species of seabirds. Other associations are frequently observed, such as *P. electra* associated with Fraser's dolphin (*Lagenodelphis hosei*) or *Tursiops truncatus* with *G. macrorhynchus*. A few species, such as the northern right whale dolphin (*Lissodelphis borealis*), have been observed with a wide variety of marine mammals, including mysticetes and pinnipeds.

Most of the information on the behavior and cognition of individual dolphins comes from studies of captive animals, primarily of *T. truncatus*. Apart from humans, dolphins have the highest ratio of BRAIN SIZE to body mass of any animal. Their intelligence and behavioral versatility are legendary and are still being explored. Using controlled experiments, dolphins have been shown to be capable of understanding complex commands, including the incorporation of abstract concepts and variations of syntax, and of devising novel behaviors of their own volition. Although behavioral activities in the wild are more difficult to interpret, there are well-documented observations of cooperative hunting (e.g., *O. orca*) and play behavior (e.g., *T. truncatus*). In fact, *T. truncatus* has been observed surfing in many coastal areas, and many dolphin species are avid riders of the bow wakes of ships and large whales. Some species regularly perform aerial maneuvers such as high leaps and flips and, in the case of *S. longirostris*, spins. At the present time, the function of many of these behaviors can only be guessed at. Dolphins also have an array of vocalizations, such as clicks, whistles, and squeals, that are used in part for their well-developed echolocation and in part for communication. However, only those species kept in captivity have been studied extensively. For the rest, due to the logistic difficulties of collecting data on fast-swimming dolphins in the pelagic realm, the vocalizations of only a few species have even been recorded, and those vocalizations are difficult to understand in terms of functionality.

V. Feeding

Ecologically, dolphins have also radiated dramatically. There are species that forage on fish, squids, and/or other invertebrates. A few species (e.g. *O. orca*, false killer whale, *Pseudorca crassidens*) even take mammalian prey, including large whales and pinnipeds. While some have fairly specific diets, a few species have rather broad tastes. In those species that eat a wide variety of prey items, one type of food (e.g., fish vs squid) may predominate. For some species (e.g., *S. coeruleoalba*), the preferred food type varies among populations, whereas in others (e.g., *S. attenuata*), it may even vary among individuals within a population, depending on their sex and life history stage. Perhaps the most dramatic segregation of foraging strategies is seen in the populations of *O. orca* in the eastern North Pacific, where two distinct groups exist sympatrically: one specializing on mammals and the other on fish. Individual species of delphinids usually forage in a particular part of the water column, specializing in epipelagic prey (e.g., *S. attenuata*), mesopelagic prey (e.g., *Lissodelphis* spp.), or even benthic prey (e.g., some species of *Cephalorhynchus*). The few measurements of swimming speeds (often anecdotal) suggest that some may exceed 20 knots in short bursts, although prolonged cruising speeds are generally on the order of 5–9 knots (9–17 km/hr). Direct data on diving depths for wild delphinids are practically nonexistent. However, *S. longirostris*, a relatively small species, is thought to dive to at least 200–300 m, based on an analysis of prey items. Greater depths are no doubt possible for larger species (e.g., *Globicephala* spp.) that feed mainly on larger squid. In fact, a trained pilot whale (*Globicephala* sp.) is thought to have reached 610 m in an experimental situation. At the other extreme, certain populations of two dolphin species have been known to intentionally beach themselves in pursuit of prey. In one case, groups of *T. truncatus* drive and pursue schooling fish onto the beach, and in the other, *O. orca* beach themselves to take pinnipeds hauled out near the water's edge. The behavioral adaptability of dolphins may be best illustrated by those species who incorporate human activities into their foraging, such as the *Sousa chinensis* that feed on trawl discards or the *O. orca* that raid long lines for their catch. Along the Gulf of Mexico coast of the United States, there are even some *T. truncatus* that feed within shrimp trawl nets, many of whom are "caught," released, and netted again.

VI. Life History

As with many aspects of dolphin biology, there is much that is unknown about the reproductive biology of most species, although many of the parameters seem to be correlated with body size. Estimated ages at sexual maturity range from about 6 (*D. delphis*) to 16 (*O. orca*) years. Like most mammals, when the age at sexual maturity differs between the sexes, it is usually the females that reach maturity at a younger age. Like other cetaceans, dolphins bear single young. Gestation periods are rarely well documented and are thought to range from about 9 to 16 months, although most have a period of less than a year (16 months was an estimate for *G. melas*). Most species appear to show at least some seasonality to their breeding, although this varies in degree. Estimates of calving intervals similarly vary, even among populations within species or among studies, ranging from just over a year for many species to approximately 8 years in one study of *O. orca* from the eastern North Atlantic. Reliable estimates of longevity are quite rare, but range from around 20 years for smaller species up to about 60 years for females of the larger species. In the large species, which tend to be more sexually dimorphic, females may live 15–20 years longer than the males. The causes of natural mortality, when they can be ascertained, are usually attributed to parasites and pathogens or to predation by killer whales or sharks.

VII. Abundance and Conservation

Some delphinid species are no doubt the most numerous of cetaceans, occurring in schools of thousands in large portions of the world ocean. Reliable global estimates for most species do not exist, but in the eastern tropical Pacific alone, some species (*Stenella attenuata, D. delphis*) number in the low millions. At the other end of the spectrum, species with restricted ranges (e.g., *C. hectori*) may have total population numbers of only a few thousand. Despite (and in some cases because of) their general abundance, dolphins face numerous anthropogenic threats. They are still hunted in some parts of the world by harpoon, drive fisheries, or nets. Two well-known examples are the drive fisheries for *S. coeruleoalba* in Japan and for *G. melas* in the Faroe Islands. There are many other dolphin fisheries still in operation, mostly in developing countries, often despite the protection of assorted laws and treaties. Usually the meat from intentional takes is for human consumption, although in some areas it is used as bait in crab pots or shark long lines. In addition to the mortalities from directed fisheries, many dolphins are also taken incidentally in the course of other fishing operations. The dolphin mortality from the eastern tropical Pacific tuna purse seine fishery has been well studied and greatly reduced in recent years. However, many dolphins of a variety of species are still caught in coastal and offshore gill nets all over the world. It is difficult to accurately assess the severity of these threats to dolphin populations given that the extent of the mortality can only be roughly guessed at in most cases and the sizes of the affected populations are largely unknown. Nevertheless, mortality from fisheries bycatch may present major threats to some dolphin populations.

Compounding these direct kills are indirect effects from human activities. In some areas, large-scale fishing operations may adversely affect dolphin populations, either by direct competition for prey or by alteration of a region's ecology. Pollution also undoubtedly takes its toll on dolphin health, particularly in coastal areas, either by direct poisoning effects or by making the animals more susceptible to pathogens and parasites. Subtle effects on fitness such as decreases in reproductive capacity or shortened life spans are almost impossible to detect, but large-scale mortalities are difficult to ignore and are not at all unusual. There have been mass STRANDINGS in recent years along the Mediterranean coast, in the southeastern United States, and in the Gulf of California in Mexico. These large-scale dieoffs often involve multiple species (although one species usually predominates) and may occur over periods of several months. It is difficult to determine if such mass mortalities are increasing in frequency in recent years because historical events may not have been documented adequately. It is also difficult to assess their impact on dolphin populations. Nevertheless,

they are a cause for concern if for no other reason than as indicators of the declining health of marine ecosystems.

See Also the Following Articles

Cetacea, Overview ■ Coloration ■ Echolocation ■ Kentriodontidae

References

Au, D. W. K., and Perryman, W. L. (1985). Dolphin habitats in the eastern tropical Pacific. *Fish. Bull.* **83**, 623–643.

Davies, J. L. (1963). The antitropical factor in cetacean speciation. *Evolution* **17**, 107–116.

Leatherwood, S., and Reeves, R. R. (eds.) (1990). "The Bottlenose Dolphin." Academic Press, San Diego.

Mann, J., Connor, R. C., Tyack, P. L., and Whitehead, H. (eds.) (2000). "Cetacean Societies: Field Studies of Dolphins and Whales." Uni. of Chicago Press, Chicago.

Mitchell, E. (1970). Pigmentation pattern evolution in delphinid cetaceans: An essay in adaptive coloration. *Can. J. Zool.* **48**, 717–740.

Perrin, W. F. (1972). Color patterns of spinner porpoises (*Stenella* cf. *S. longirostris*) of the eastern tropical Pacific and Hawaii, with comments on delphinid pigmentation. *Fish. Bull. U.S.* **70**, 983–1003.

Perrin, W. F. (1989). "Dolphins, Porpoises, and Whales: An Action Plan for the Conservation of Biological Diversity: 1988–1992." 2nd Ed. IUCN, Gland, Switzerland.

Perrin, W. F., Donovan, G. P., and Barlow, J. (eds.) (1994). "Gillnets and Cetaceans. Rep. Int. Whal. Commn.," Spec. Issue No. 15. International Whaling Commission, Cambridge.

Perrin, W. F., and Reilly, S. B. (1984). Reproductive parameters of dolphins and small whales of the family Delphinidae. *In* "Reproduction in Whales, Dolphins and Porpoises, Rep. Int. Whal. Commn." (W. F. Perrin, R. L. Brownell, and D. P. Demaster, eds.), Spec. Issue No. 6, pp. 97–133. International Whaling Commission, Cambridge.

Rice, D. W. (1998). "Marine Mammals of the World: Systematics and Distribution." Soc. Mar. Mamm. Spec. Pub. 4. Society for Marine Mammalogy, Lawrence, KS.

Ridgway, S. H., and Harrison, R. (eds.) (1994). "Handbook of Marine Mammals," Vol. 5. Academic Press, London.

Ridgway, S. H., and Harrison, R. (eds.) (1999). "Handbook of Marine Mammals", Vol. 6. Academic Press, London.

Delphinoids, Evolution of the Modern Families

LAWRENCE G. BARNES
Natural History Museum of Los Angeles County, California

The odontocete families usually included in Delphinoidea are Kentriodontidae, Odobenocetopsidae, Albireonidae, Delphinidae, Phocoenidae, and Monodontidae. Among these, only the last three have Recent (=living) representatives and are discussed in this article. Odobenocetopsidae and Albireonidae are small families, consisting of a few known fossil species. The fossil kentriodontids are diverse and are the basal delphinoid family.

I. Family Monodontidae: Belugas, Narwhals, and Relatives

The family Monodontidae includes two living Arctic species: the living beluga (*Delphinapterus leucas*), in the subfamily Delphinapterinae, and the tusked narwhal (*Monodon monocerus*), in the subfamily Monodontinae. The SKULLS of these two cetaceans are similar, with the tusk of the narwhal differentiating them superficially. The fossil record of Monodontidae extends back to the Late Miocene (circa 11 or 12 Mya). Monodontids appear most closely related to the porpoises of the family Phocoenidae and to the dolphins of the family Delphinidae, but detailed study of their interrelationships has yet to be made.

Fossil representatives of Delphinapterinae, animals having a relatively narrow and slightly downturned snout like that of the modern beluga, *Delphinapterus,* have been collected from deposits bordering the western North Atlantic and from California and Japan.

Fossils of latest Miocene and Pliocene members of the extinct, aberrant, broad-headed monodontid *Denebola* are known from coastal sites in central and southern California and various localities in Baja California. The skull of this genus is broad and the snout is wide, flat, and blunt, causing it to mimic the skull of the pilot whales (family Delphinidae, genus *Globicephala*). Phylogenetic analysis is needed that would elucidate the relationships of this group to the other monodontids. We do not yet understand the evolutionary origin of the bizarre spiral-tusked narwhal of the Arctic because no fossils have been found.

The fossil record of the Monodontidae demonstrates that the group was much more numerous, diverse, and widespread in the past than it is now and that it was characteristic of the temperate and subtropical latitudes of the North Atlantic and the North Pacific oceans. No monodontid has been found in the Southern Hemisphere. The fossil distribution is seemingly paradoxical to the present strictly Arctic distribution, and the living monodontid species are relics. Past members of the family may have occupied the ecologic niches of several species of modern dolphins. The sole surviving species have relict distributions in a possibly marginal habitat.

II. Family Delphinidae: Dolphins

The modern, mostly pelagic dolphins of the family Delphinidae are the most diverse living family of Cetacea and are nearly cosmopolitan in the world's oceans. The family includes a taxonomically and ecologically diverse range of small to large species with long or short rostra, and narrow or broad rostra. Rostral shape and numbers and size of TEETH are related to diet, with the various species ranging from generalists, to predators, to squid eaters. Common names reflect the animal's sizes: some of the large "dolphins" are called pilot whales (*Globicephala* spp.) and killer whales (*Orcinus orca*), and smaller species are called dolphins, such as *Delphinus* spp. All Delphinidae are unified by skull osteology, especially details of the left-skewed asymmetrical cranial vertex and narial region, the basicranium, and the periotic.

Although the delphinids are the most diverse group of living cetaceans, they are surprisingly rare in the fossil record. The oldest documented delphinid is Late Miocene (about 10 to 12 million years ago) in age from southern California, and delphinids are not very common in subsequent latest Miocene and Pliocene deposits. All pre-Late Miocene supposed fossil delphinids are now known to be members of other odontocete families.

The present abundance of delphinids is apparently the result of explosive evolution that probably occurred in the later part of the Pliocene, resulting in the replacement of the earlier more diverse kentriodontids, phocoenids, and monodontids. The rather abrupt appearance of Delphinidae, in relative abundance, in Pliocene time is a notable phenomenon in cetacean evolution.

Morphology of the earliest fossil delphinid indicates that the family probably arose from within the primitive, extinct family Kentriodontidae, but no definitive study has been made that would prove this.

III. Family Phocoenidae: Porpoises

The true porpoises of the family Phocoenidae have a fossil record dating back to the Late Miocene, or about 11 or 12 million years ago. All pre-Pliocene fossil records of this family from the Atlantic realm are suspect. Phocoenids have a raised eminence on each premaxilla anterior to the nasal openings, a lobe of the pterygoid air sinus extending dorsal to the orbit between the frontal and maxillary bones, and anteriorly retracted posterior premaxillary terminations. The short rostra, rounded crania (paedomorphic characters) and spatulate teeth of living phocoenids only appear in the more recent members of the group, known from the Late Pliocene and more recently. Late Miocene phocoenids have longer snouts, more prominent cranial crests, and conical tooth crowns, all of which are primitive characters, which the early phocoenids share with kentriodontids or delphinids. Phocoenids were much more diverse in the past than now, and in rocks dating from 3 to 12 million years ago around the North Pacific margin their fossils are more diverse and more abundant than those of delphinids.

The oldest apparent phocoenid is represented by a single periotic of Middle Miocene age from Baja California Sur, Mexico. In morphology it closely matches the holotypic periotic of the most primitive named phocoenid, *Piscolithax tedfordi*, also from Baja California, which is considerably younger (latest Miocene, circa 9 Mya). However, *P. tedfordi*, while being the most primitive named phocoenid is for its time a relict taxon, because the somewhat more derived *Salumiphocaena stocktoni* from southern California is about 10–12 million years old. *Piscolithax* was a large phocoenid, approximately the size of extant *Tursiops truncatus*, and had relatively large and wide pectoral flippers.

Piscolithax and other relatively primitive genera of Phocoenidae, some with unusually long rostra, have been discovered in Late Miocene and Pliocene deposits in coastal Peru. There is no published fossil record that would elucidate the origins of the living genera *Neophocaena, Phocoena,* and *Phocoenoides.*

See Also the Following Articles

Cetacean Evolution ■ Classification ■ Delphinids ■ Kentriodontidae ■ Odobenocetops

References

Barnes, L. G. (1977). Outline of eastern North Pacific fossil cetacean assemblages. *Syst. Zool.* **25**, 321–343.

Barnes, L. G., Inuzuka, N., and Hasegawa, Y. (eds.) (1995). Thematic issue: Evolution and biogeography of fossil marine vertebrates in the Pacific realm. *Island Arc* **3**(4), 243–548.

Fordyce, R. E., and Barnes, L. G. (1994). The evolutionary history of whales and dolphins. *In* "1994 Annual Review of Earth and Planetary Sciences" (G. W. Wetherill, ed.), Vol. 22, pp. 419–455. Annual Reviews, Inc., Palo Alto, CA.

LeDuc, R. G., Perrin, W. F., and Dizon, A. E. (1999). Phylogenetic relationships among the delphinoid cetaceans based on full cytochrome *b* sequences. *Mar. Mamm. Sci.* **15**, 619–648.

Rice, D. W. (1998). Marine mammals of the world. *Soc. Mar. Mamm. Spec. Pub.* **4**, 1–231.

Rosel, P. E., Haygood, M. G., and Perrin, W. F. (1995). Phylogenetic relationships among the true porpoises (Cetacea: Phocoenidae). *Mol. Phyl. Evol.* **4**, 463–474.

Demography

SEE *Population Dynamics*

Dental Morphology (Cetacean), Evolution of

MARK D. UHEN
Cranbrook Institute of Science, Bloomfield Hills, Michigan

Modern cetaceans exhibit some of the most highly derived dentitions in all of Mammalia. Living odontocetes exhibit a wide variety of conditions from some of the highly polydont delphinids to the complete lack of teeth in some monodontids. Living mysticetes completely lack teeth as adults. Add to this even more divergent conditions in fossil mysticetes, odontocetes, and their early forebears, the archaeocetes, and the result is a broad range of conditions in which is embedded the story of how modern cetacean dentitions arose from the more typically mammalian teeth.

Acreodi (Mesonychia + Triisodontidae) is currently thought to be the sister taxon to Cetacea based on paleontological evidence. Some molecular studies have indicated that Cetacea is nested within ARTIODACTYLA, possibly with Hippopotamidae as its sister taxon. Acreodi are diphyodont and heterodont, with a plesiomorphic dental formula of 3.1.4.3/3.1.4.3. Their incisors are single rooted and in an arc across the front of the jaw. Mesonychians and cetaceans all have reduced trigonids and talonid basins, and lack an M_3 hypoconulid.

I. Archaeocetes

Archaeocetes are diphyodont and heterodont with their incisors in line with the cheek teeth and separated by diastemata. The anterior TEETH are conical and single rooted, and the premolar series forms a morphological gradient from the anterior teeth to the posterior premolars along with an increase in size such that P_3 or P_4 is usually the largest upper or lower tooth. Upper and lower P_1 may be single or double rooted, depending on the taxon, and the other premolars are double rooted (but see Albright, 1996). The premolars are generally equilaterally triangular in lateral view. The lower molars have steep mesial edges and more gradually sloping distal edges. The lower molars lack trigonid basins and retain hypoconids but have lost the talonid basin. The lower molars also have a vertical groove on their mesial margins. The tooth row is closed in this region such that there are no diastemata between the teeth, and the next anterior tooth rests in the groove of the distally adjacent molar. Archaeocetes often exhibit vertical wear facets on the buccal surfaces of the lower molars that resulted from contact with the lingual surfaces of the upper cheek teeth during chewing (O'Leary and Uhen, 1999).

A. Nonbasilosaurid Archaeocetes

Nonbasilosaurid archaeocetes have a dental formula of 3.1.4.3/3.1.4.3 that they share with mesonychians and other early mammals and that distinguishes them from the later basilosaurids. Nonbasilosaurid archaeocetes have a paracone (mesial) and metacone (distal) on the buccal margins of their molars, and a protocone projecting lingually. Pakicetids have single-rooted P_1s, which distinguish them from ambulocetids. Ambulocetids have a protocone lobe on the upper molars that is smaller than that in pakicetids and larger than that in remingtonocetids and protocetids (Thewissen *et al.*, 1996). Remingtonocetids have labiolingually narrow incisors and premolars; upper molars large and narrow, sometimes lacking a lingual third root, and crenulations on the cutting edges of the teeth, which distinguishes them from other nonbasilosaurid archaeocetes (Kumar and Sahni, 1986). Protocetids have robust teeth with upper molars having three roots and the cheek teeth lacking accessory denticles, (Fig. 1, but see *Georgiacetus* later).

B. Basilosauridae

Basilosaurids have a number of synapomorphies that distinguish them from other archaeocetes. Upper and lower first premolars are replaced in dorudontine basilosaurids, but it is unclear how broadly this character is distributed within the archaeocetes (Uhen, 2000). Basilosaurids lack M^3, giving them a dental formula of 3.1.4.2/3.1.4.3. The cheek teeth of basilosaurids have accessory denticles along their mesial and distal margins, flanking an apical cusp. These apical cusps are homologous with the paracone on the uppers and paraconid on the lowers. It is unclear whether any (or all or none) of the accessory denticles are homologous with more primitive mammalian cusps. There are usually no accessory denticles on the mesial margins of the lower molars (for exceptions, see Uhen, 1996). While accessory denticles of the size and number seen in basilosaurids are unknown in earlier cetaceans, *Georgiacetus* possesses small incipient accessory denticles on some of the cheek teeth. Also, accessory denticles similar to those of basilosaurids can be seen on dP_4 of *Pappoce-*

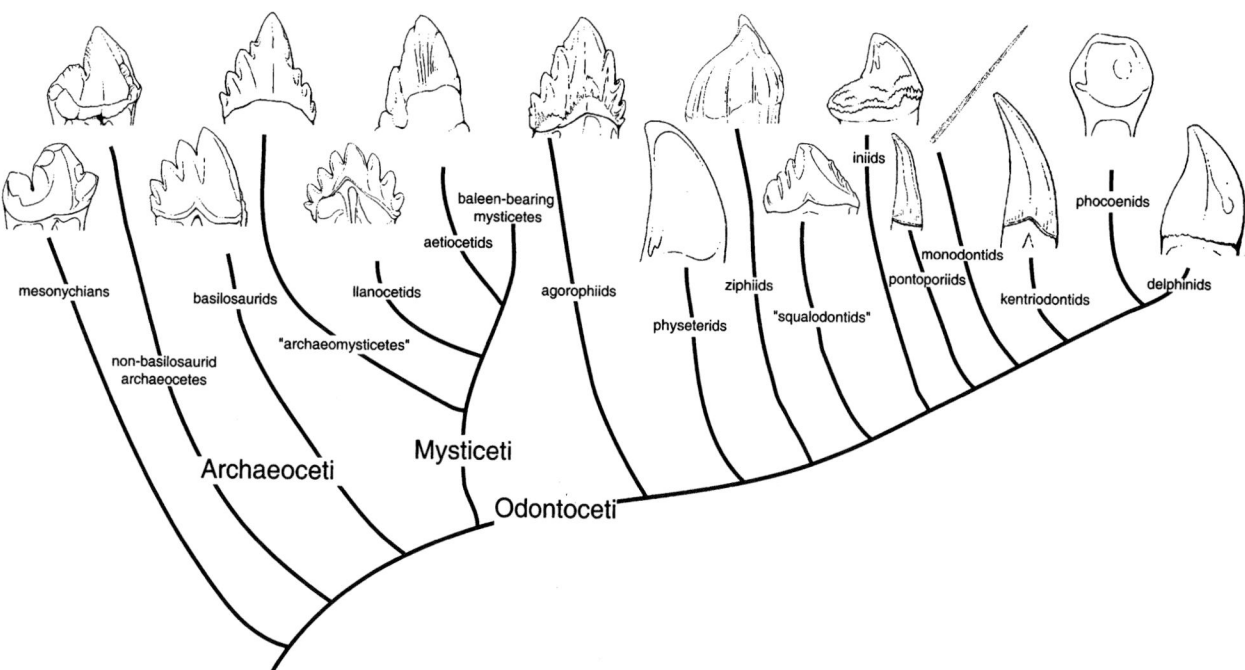

Figure 1 *Phylogenetic relationships of cetaceans with examples of dental morphology. For those cetaceans with heterodont teeth, lower molars (posterior cheek teeth) are shown.*

tus lugardi. The lingual posterior root found on the upper posterior cheek teeth of earlier cetaceans is fused to the buccal posterior root on these teeth in basilosaurids, as shown by the presence of a groove in the root and a lingual expansion of enamel on the crown, which lacks a trigon basin and protocone.

II. Mysticeti + Odontoceti

The fossil record of archaeocetes ends in the Late Eocene. Mysticeti and Odontoceti can be distinguished from archaeocetes based on the presence of some kind of telescoping (Barnes, 1984), but telescoping can hardly be used as a synapomorphy of Mysticeti + Odontoceti, as telescoping is manifested in different ways in each group and is almost certainly not homologous between them. Monophyodonty may serve as a synapomorphy for Mysticeti + Odontoceti. Even though modern mysticetes lack teeth as adults, they develop embryonic tooth buds that are later resorbed (Karlsen, 1962). In addition, early fossil mysticetes have well-developed teeth as adults (Emlong, 1966; Mitchell, 1989; Barnes and Sanders, 1996a,b). Even a late-occurring archaeocete has been described that may be monophyodont (Uhen and Gingerich, 2000). Early mysticetes and odontocetes are similar to archaeocetes in that they are heterodont, with a morphological gradient from conical, single-rooted anterior teeth to triangular (or rounded) multirooted cheek teeth with accessory denticles.

A. Odontoceti

Modern odontocetes can be generally characterized as being polydont and monophyodont, with single-rooted teeth that grow throughout life, but there are numerous counterexamples to these generalizations in both modern and fossil cetaceans (discussed later). The earliest odontocetes show only limited polydonty, with

Agorophius having 8 and *Xenorophus* with 10 teeth (Fordyce, 1982 and references therein). Polydonty could have originated by intercalation of deciduous and permanent teeth and could have been further increased in later odontocetes by the addition of supernumerary teeth, but it would be difficult to explain why no odontocete is known that exceeds three teeth in the premaxilla under this scenario (Fordyce, 1982). The recent description of early mysticetes with one extra molar in each quadrant (Barnes and Sanders, 1996a,b) may indicate that polydonty arose prior to the split of Mysticeti and Odontoceti by the terminal addition of teeth, although many other toothed mysticetes appear not to possess polydonty at all (Emlong, 1966; Fordyce, 1982).

All modern and fossil odontocetes are thought to be monophyodont (Fordyce, 1982). It is unclear whether the teeth of modern odontocetes are homologous with the deciduous or adult teeth (or both) of archaeocetes. A newly described archaeocete that appears to possess adult teeth in a skeletally juvenile individual suggests that the teeth of all Mysticeti + Odontoceti may be adult teeth, but Karlsen (1962) suggested that the embryonic tooth buds of mysticetes were homologous with the deciduous dentition of more primitive mammals.

Early odontocetes Agorophiidae (including *Xenorophus*), Simocetidae, Waipatiidae, and Squalodontidae are all considerably heterodont. Their anterior teeth are conical with long, single roots. These teeth grade into teeth with a more triangular shape with multiple accessory denticles on the anterior and posterior edges of the teeth and two roots. The most posterior lower cheek teeth may lack accessory denticles on their anterior edges. These differences in tooth form are reminiscent of the differences along the tooth rows of archaeocetes, but the teeth are smaller compared to the size of the skull.

Physeteroidea includes two genera, *Physeter* and *Kogia*. Modern physeteroids usually lack upper teeth (or have tiny upper teeth that do not erupt), but some fossil representatives (such as *Scaldicetus*) have similarly sized teeth in the lower and upper jaws (Hirota and Barnes, 1994).

Ziphiidae are thought to have secondarily reduced the dentition from a primitively polydont condition. One modern species, *Tasmacetus shepherdi*, and additional fossil species are polydont. In some other modern species, the females usually lack erupted teeth, and males have one tooth on each side of the lower jaw and none in the upper jaw. The teeth of male ziphiids take on different forms in different species.

Iniidae have rugose heterodont teeth that are expanded lingually in the posterior portion of the jaw. Monodontidae have very reduced dentitions. Male narwhals (*Monodon monoceros*) have a single, long, spiral tusk, usually on the left side with a small unerupted tooth on the right. Females usually have two unerupted teeth. Both male and female belugas (*Delphinapterus leucas*) generally have nine teeth in each upper and eight in each lower row. One related fossil, *Odobenocetops,* has a walrus-like skull with one large tusk-like tooth and another shorter tusk on the opposite side (de Muizon, 1993). Members of the extinct Kentriodontidae have small conical teeth that are generally homodont, a feature shared with one of their likely descendants, the Delphinidae, but not with another, the Phocoenidae, which have numerous small spatulate teeth or (in one species) peglike vestigial teeth that may be embedded in gum tissue.

B. Mysticeti

Despite the lack of teeth in modern mysticetes, they are thought to have evolved from toothed ancestors in part because they develop tooth buds as embryos that are later resorbed (Karlsen, 1962), and many tooth-bearing early mysticetes are known from the Oligocene epoch (Fordyce and Barnes, 1994). All post-Oligocene mysticetes are baleen-bearing and lack teeth (Fordyce and Barnes, 1994).

One group of toothed mysticetes includes species with archaeocete-like teeth (Fordyce, 1989; Barnes and Sanders, 1996a,b). This group could be delimited to include *Llanocetus denticrenatus* from the late Eocene of Antarctica (Mitchell, 1989). Mitchell (1989) placed *Llanocetus* in the family Llanocetidae, which he included in a new infraorder Crenataceti from which he excluded all other toothed mysticetes known at the time. *Kekenodon onamata* and *Phococetus vasconum* may also be members of this group (Fordyce, 1992). The teeth of this group are similar in size to those of archaeocetes, and the cheek teeth are similarly double rooted and have accessory denticles. Some species have an extra molar in each quadrant over the count of basilosaurids, yielding three upper and four lower molars (Barnes and Sanders, 1996a,b). These teeth may have been used to filter food from seawater, as the teeth generally lack wear resulting from tooth/food contact (Mitchell, 1989).

The second group of toothed mysticetes includes species that all have tiny teeth that are conical anteriorly and have tiny denticles posteriorly, but are all single rooted. All of these species have been placed in the Aetiocetidae. Some aetiocetids have the plesiomorphic number of teeth for archaeocetes (11), whereas others are polydont. Aetiocetids are not thought to have had baleen and may have used their small teeth for filter

feeding. The small, toothed mysticete *Mammalodon* from the Late Oligocene or early Miocene of Australia may be related to the aetiocetids, but it is difficult to compare in part because it is known only from the type specimen, and its teeth are highly worn (Pritchard, 1939; Fordyce and Barnes, 1994).

The earliest baleen-bearing mysticetes that lack teeth are placed in the Cetotheriidae, but this taxon has long been recognized as a wastebasket taxon in need of serious revision. The earliest cetotheres are from the Late Oligocene (contemporaneous with toothed mysticetes) and have worldwide distribution. Cetotheres may be ancestral to balaenopterids, but their relationship to other modern mysticete families is unclear.

See Also the Following Articles

Archaeocetes, Archaic ■ Baleen Whales, Archaic ■ Filter Feeding ■ Toothed Whales

References

Albright, L. B. (1996). A protocetid cetacean from the Eocene of South Carolina. *J. Paleontol.* **70**, 519–522.

Barnes, L. G. (1984). Whales, dolphins and porpoises: Origin and evolution of the Cetacea. Mammals: notes for a short course, University of Tennessee Department of Geological Sciences Studies in Geology **8**, 139–154.

Barnes, L. G., and Sanders, A. E. (1996a). The transition from Archaeoceti to Mysticeti: Late Oligocene toothed mysticetes from South Carolina, U.S.A. *J. Vertebr. Paleontol.* **16**, 21A.

Barnes, L. G., and Sanders, A. E. (1996b). The transition from archaeocetes to mysticetes: Late Oligocene toothed mysticetes from near Charleston, South Carolina. Sixth North American Paleontological Convention Abstracts of Papers. The Paleontological Society Special Publication 8:24.

Emlong, D. R. (1966). A new archaic cetacean from the Oligocene of northwest Oregon. *Bull. Oregon Univ. Mus. Nat. Hist.* **3**, 1–51.

Fordyce, R. E. (1989). Problematic early Oligocene toothed whale (Cetacea, ?Mysticeti) from Waikari, north Canterbury, New Zealand. *New Zeal. J. Geol. Geophy.* **32**, 395–400.

Fordyce, R. E. (1982). Dental anomaly in a fossil squalodont dolphin from New Zealand, and the evolution of polydonty in whales. *New Zealand Journal of Zoology* **9**, 419–426.

Fordyce, R. E. (1992). Cetacean evolution and Eocene/Oligocene environments. In "Eocene-Oligocene Climatic and Biotic Evolution" (D. R. Prothero and W. A. Berggren, eds.), pp. 368–381. Princeton Univ. Press, Princeton, NJ.

Fordyce, R. E. and L. G. Barnes (1994). The evolutionary history of whales and dolphins. *Annual Review of Earth and Planetary Sciences* **22**, 419–455.

Gingerich, P. D., and Russell, D. E. (1990). Dentition of early Eocene *Pakicetus* (Mammalia, Cetacea). *Contrib. Mus. Paleontol. Univ. Mich.* **28**, 1–20.

Hirota, K., and Barnes, L. G. (1994). A new species of middle Miocene sperm whale of the genus *Scaldicetus* (Cetacea; Physeteridae) from Shiga-mura, Japan. *Island Arc* **3**, 453–472.

Karlsen, K. (1962). Development of tooth germs and adjacent structures in the whalebone whale [*Balaenoptera physalus* (L.)]. *Hvalrådets Skrifter* **45**, 1–56.

Kumar, K., and Sahni, A. (1986). *Remingtonocetus harudiensis*, new combination, a Middle Eocene archaeocete (Mammalia, Cetacea) from western Kutch, India. *J. Vertebr. Paleontol.* **6**, 326–349.

Mitchell, E. D. (1989). A new cetacean from the late Eocene La Meseta Formation, Seymour Island, Antarctic Peninsula. *Can. J. Fish. Aqua. Sci.* **46**, 2219–2235.

Muizon, C. de (1993). Walrus-like feeding adaptation in a new cetacean from the Pliocene of Peru. *Nature* **365,** 745–748.

O'Leary, M. A., and Uhen, M. D. (1999). The time of origin of whales and the role of behavioral changes in the terrestrial-aquatic transition. *Paleobiology* **25.**

Pritchard, B. G. (1939). On the discovery of a fossil whale in the older tertiaries of Torquay, Victoria. *Vict. Nat.* **55,** 151–159.

Thewissen, J. G. M., Madar, S. I., and Hussain, S. T. (1996). *Ambulocetus natans,* an Eocene cetacean (Mammalia) from Pakistan. *Courier Forschungsinstitut Senckenberg* **191,** 1–86.

Uhen, M. D. (1996). *Dorudon atrox* (Mammalia, Cetacea): Form, Function, and Phylogenetic Relationships of an Archaeocete from the Late Middle Eocene of Egypt. Doctoral Dissertation. University of Michigan, Ann Arbor. 608 pp.

Uhen, M. D. (1998). Middle to Late Eocene Basilosaurines and Dorudontines. *In* "The Emergence of Whales." (J. G. M. Thewissen, ed.), pp. 29–61. Plenum Press, New York.

Uhen, M. D. (2000). Replacement of deciduous first premolars and dental eruption in archaeocete whales. *J. Mamm.* **81,** 123–133.

Desmostylia

DARYL P. DOMNING
Howard University, Washington, DC

The Desmostylia are the only completely extinct order of marine mammals. They were HIPPOPOTAMUS-like amphibious herbivores (Fig. 1) that were confined to the North Pacific Ocean and are known only as fossils of Oligocene and Miocene age. [For comprehensive references to the published literature on desmostylians, see Domning (1996).]

I. Desmostylian Relationships, Origins, and Distribution

Desmostylians have undergone more of a taxonomic odyssey than perhaps any other mammals. Although they were placed initially (and afterwards, most commonly) among either the Sirenia or the Proboscidea, various authors later assigned them to the Monotremata, Marsupialia, Multituberculata, or an order of their own. The latter view prevailed only in 1953. Today they are classified in the supraordinal group Tethytheria, together with the living Sirenia and Proboscidea; the Proboscidea are probably their sister group. The teeth of the most primitive desmostylians closely resemble those of anthracobunids, early tethytheres, which are sometimes considered true proboscideans.

Because the most primitive known desmostylians, as well as other early tethytheres, are found in Asia, this continent, bordering the eastern part of the ancient Tethys Sea, is likely to have been the area where this order arose, probably during the Paleocene or Eocene. However, desmostylians do not appear in the fossil record until the Late Oligocene, less than 30 million years ago. By that time they had already spread to the North American shore of the Pacific; and from then to their extinction some 20 million years later, they inhabited the North Pacific littoral from Japan to Baja California.

II. Anatomy and Mode of Life

The desmostylian skull (Fig. 2) features a more or less long, narrow, and little-deflected rostrum; dorsally protruding orbits; a stout paroccipital process; an external auditory meatus nearly enclosed ventrally by contact of the posttympanic and postglenoid

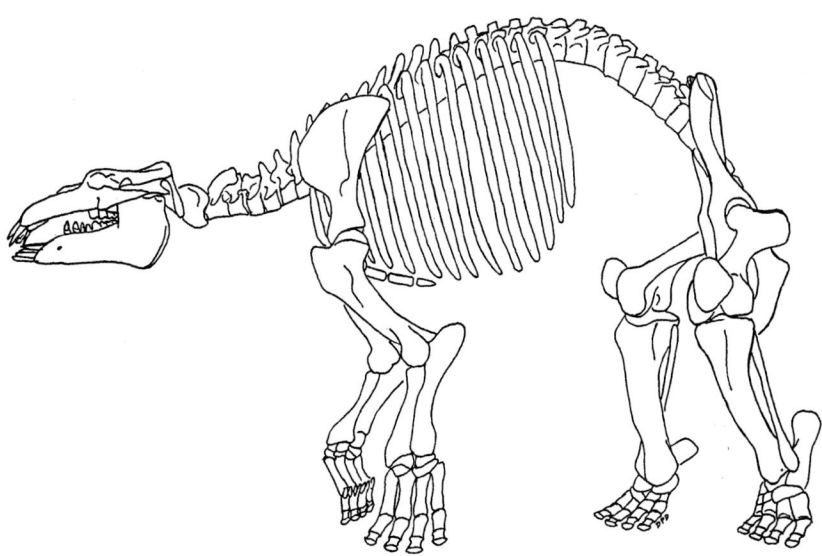

Figure 1 *Skeleton of* Paleoparadoxia tabatai, *a Miocene desmostylian, in terrestrial pose. Total length is about 2.2 m. Note hyperextension and anterolateral direction of front toes, anterior direction of hind toes, and strong abduction of knees. From Domning (2001), reproduced with permission of the Smithsonian Institution Press.*

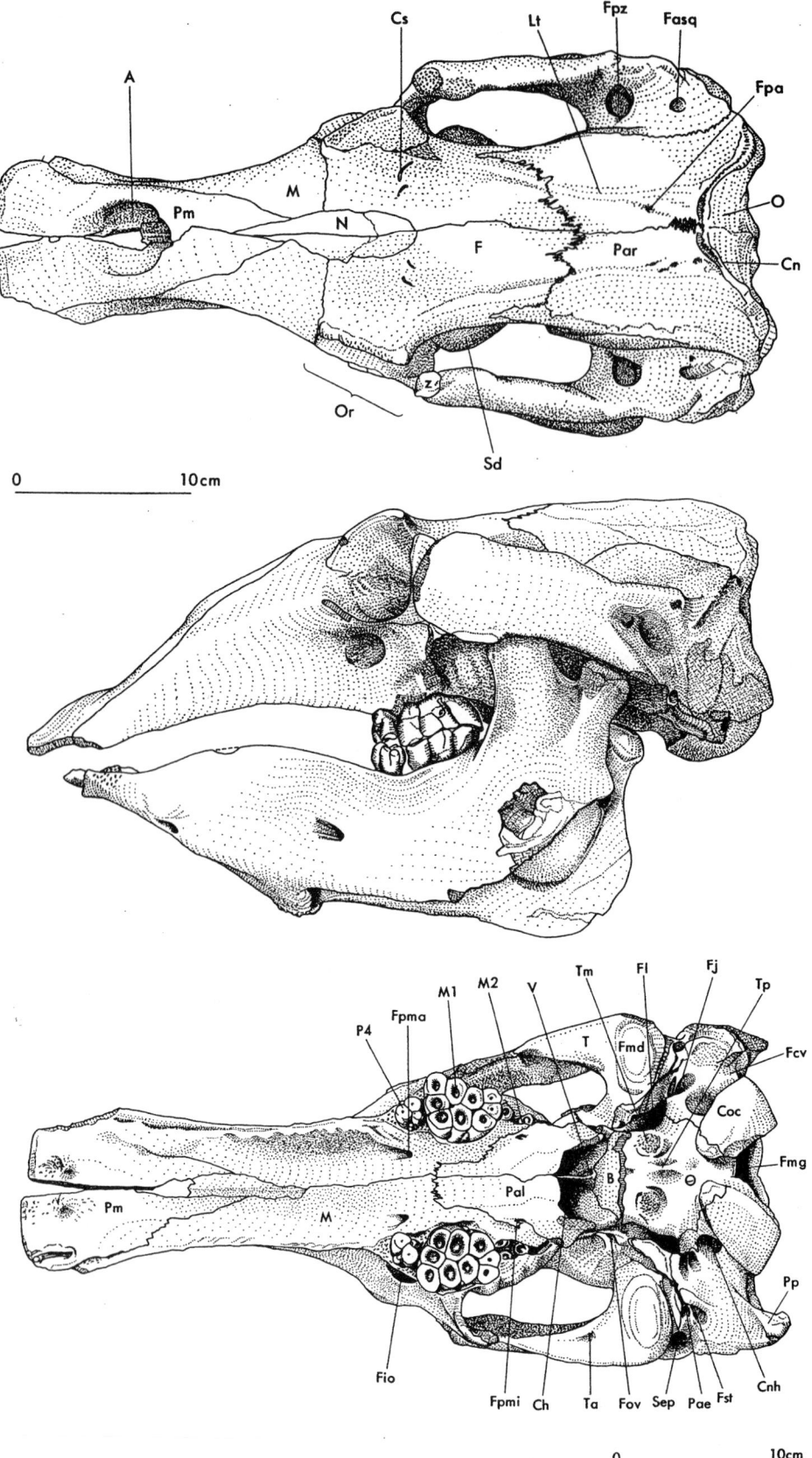

processes; and a large epitympanic sinus opening into the temporal fossa. The dental formula is primitively 3.1.4.3, with procumbent, transversely aligned incisors and canines; the lower canine is especially enlarged. The fourth lower deciduous premolar is primitively trilobed. The body (Fig. 1) is stout and compact, with a relatively short neck, a deep thorax, a broad sternum with paired plate-like sternebrae, a strongly arched lumbar spine, and a very reduced tail. The limbs are robust, with considerable torsion in the tibia and an ankle joint that is oblique to the tibial shaft (conditions also seen to varying degrees in many noncursorial land mammals). The metacarpals are longer than the metatarsals. The bones show some osteosclerosis (increased density) but no pachyostosis (increased volume).

Controversy over desmostylians' structure and posture has even surpassed the disagreements over their classification. Once complete skeletons were discovered, showing that they possessed four stout limbs and were capable of some sort of locomotion on land, paleontologists and artists created a startling variety of reconstructions, including ones resembling hippopotami, tapirs, sea lions, crocodiles, and creatures unlike anything else (see Inuzuka, 1982). The interpretations presented most recently and defended in the most detail have portrayed them as hauling out in the manner of sea lions (Repenning), as walking with the limbs sprawled in a "herpetiform" stance (Inuzuka), or as keeping the legs under the body in a more conventional land-mammal fashion, with resemblances to ground sloths (Domning, 2001; Fig. 1).

In contrast, their aquatic behavior has occasioned little argument. Their style of swimming is generally agreed to have been like that of polar bears; i.e., alternate paddling with the forelimbs while the hindlimbs were used for steering. Because desmostylian fossils are found exclusively in marine deposits, there has also never been any doubt expressed that they were strictly marine mammals, despite the lack of clearly aquatic specializations in their skeletons.

The peculiar, heavily enameled tooth structure of the more highly derived desmostylians has occasionally led to suggestions that they fed on molluscs or other shelled prey. However, these teeth tend to be high crowned, like those of grazing ungulates, rather than low, broad, and pavement-like as seen in animals that do crush shellfish. Moreover, the TEETH of earlier desmostylians closely resemble those of undoubted herbivores. Hence a diet of marine plants for all desmostylians is now generally conceded.

III. Diversity

Only about six genera and fewer than a dozen species are currently recognized in this order (Table I). They could be grouped into at least two families, but conflicting schemes of family-level classification have not been completely reconciled. All these taxa, so far as is known, had broadly similar postcranial skeletons, and they are distinguished mainly by details of SKULL and dentition. The cladistic relationships of the desmostylian genera are generally agreed to be as follows: (*Behemotops* (*Paleoparadoxia* (*Cornwallius* (*Desmostylus*, *Kronokotherium*, and *Vanderhoofius*)))). The relationships among the latter three genera have yet to be clarified.

The most primitive form named so far is the Late Oligocene *Behemotops*, an animal nearly the size of a Nile hippopotamus and with low-crowned, anthracobunid-like teeth. The Miocene *Paleoparadoxia* (Fig. 1) is similar, but has a more retracted nasal opening, fewer premolars, and a long postcanine diastema. *Cornwallius*, another Late Oligocene genus, shows a tendency for the molar cusps to become more columnar. This trend continued in the Miocene with the genera *Kronokotherium*, *Vanderhoofius*, and *Desmostylus*; the latter name, meaning "bundle of columns," expresses the appearance of a molar in these most highly derived and characteristic members of the group (Fig. 2). The procumbent incisors and canines are variously reduced, sometimes leaving only the large lower canines as digging organs, and the adult cheek dentition comprises only molars.

The diversity of desmostylians was constrained by their limited geographic range and dietary opportunities. Apparently they were adapted to cooler climates than the tropical sirenians, but they seem to have succumbed to competition from these more fully aquatic herbivores as soon as hydrodamaline sirenians evolved the ability to spread into the cool home waters of the desmostylians.

TABLE I
Genera of Desmostylia and Their Temporal Range

Genus	Range
Behemotops	Late Oligocene
Paleoparadoxia	Early to Late Miocene
Cornwallius	Late Oligocene
Desmostylus	Early to Late Miocene
Kronokotherium	?Early to ?Middle Miocene
Vanderhoofius	Middle Miocene

Figure 2 *Skull and mandible of* Desmostylus hesperus, *a Miocene desmostylian, in dorsal, lateral, and ventral views (immature specimen). The left mandibular angle is broken and the large right dental capsule is visible. Note the columnar cusps of the molars. A, nasal aperture; B, basisphenoid; Ch, choana; Cn, nuchal crest; Cnh, hypoglossal canal; Coc, occipital condyle; Cs, supraorbital canal; F, frontal; Fasq, anterior squamosal foramen; Fio, infraorbital foramen; Fj, jugular foramen; Fl, foramen lacerum; Fmd, mandibular fossa; Fmg, foramen magnum; Fov, foramen ovale; Fpa, parietal foramen; Fpma, greater palatine foramen; Fpmi, lesser palatine foramen; Fpz, postzygomatic foramen; Fst, stylomastoid foramen; Lt, temporal crest; M, maxilla; M1, M2, upper first and second molars; N, nasal; O, occipital; Or, orbit; P4, upper fourth premolar; Pae, external auditory meatus; Pal, palatine; Par, parietal; Pp, paroccipital process; Sd, dental capsule; Sep, epitympanic sinus; T, squamosal (temporal); Ta, articular tubercle; Tm, muscular tubercle; Tp, pharyngeal tubercle; V, vomer; Z, zygomatic (jugal). From Inuzuka (1988), reproduced with permission of the Geological Survey of Japan.*

See Also the Following Articles

Sirenian Evolution ■ Tethytheria

References

Domning, D. (1996). Bibliography and index of the Sirenia and Desmostylia. *Smith. Contr. Paleobiol.* **80**, 1–611.

Domning, D. (2001). The terrestrial posture of desmostylians. *Smith. Contr. Paleobiol.* (in press).

Domning, D., Ray, C., and McKenna, M. (1986). Two new Oligocene desmostylians and a discussion of tethytherian systematics. *Smith. Contr. Paleobiol.* **59**, 1–56.

Inuzuka, N. (1982). "Atlas of Reconstructed Desmostylians." *Abstracts, 36th Annual Meeting, Association for the Geological Collaboration in Japan, Saitama,* 44–61. [In Japanese]

Inuzuka, N. (1988). The skeleton of *Desmostylus* from Utanobori, Hokkaido. I. Cranium. *Bull. Geol. Surv. Japan* **39**, 139–190. [In Japanese]

Inuzuka, N., Domning, D., and Ray, C. (1995). Summary of taxa and morphological adaptations of the Desmostylia. *Island Arc* **3**, 522–537.

Ray, C., Domning, D., and McKenna, M. (1994). A new specimen of *Behemotops proteus* (Order Desmostylia) from the marine Oligocene of Washington. *Proc. San Diego Soc. Nat. Hist.* **29**, 205–222.

Dialects

John K. B. Ford
Fisheries and Oceans Canada,
Nanaimo, British Columbia

Consistent variations in the underwater vocalizations of marine mammals can exist among individuals, groups, or populations. Differences in vocal patterns between geographically isolated populations have been described in several species of pinnipeds and cetaceans. Good examples are regional variations in the vocal repertoire of the Weddell seal (*Leptonychotes weddellii*) (Thomas and Stirling, 1983) or the major differences in the songs of humpback whales (*Megaptera novaeangliae*), in different ocean areas (Payne and Guinee, 1983). These types of variations have occasionally been referred to as dialects, but are more appropriately termed geographic variations (Conner, 1982). Such differences likely result from long-term geographic and, therefore, reproductive isolation, leading to genetically determined distinctiveness, although the potential for cultural variation cannot be ruled out. Pinnipeds and cetaceans are among the few mammalian orders in which vocal mimicry and learning have been documented (Janik and Slater, 1997). Transmission of vocal patterns across generations may thus depend more on cultural than on genetic mechanisms. Copying errors and other forms of cultural mutation and drift may be responsible for at least a portion of the vocal differences between distant populations. Geographic variation in vocal patterns most likely represents epiphenomena, or by-products, of social and genetic isolation.

Vocal learning is most probably the primary mechanism involved in the evolution of dialects among local, interacting populations or groups of marine mammals. Dialects are well known in birds, but appear to be quite rare among marine mammals or, for that matter, mammals generally. They have only been described in two marine mammal species: killer whales (*Orcinus orca*) and sperm whales (*Physeter macrocephalus*). In British Columbia, matrilineal kinship groups, or pods, of "resident" killer whales have repertoires of 7–17 call types that vary among pods (Ford, 1989, 1991). All pods have distinctive features in their call repertoires, and thus each has a unique dialect. Certain pods share a portion of their call repertoires with others, and these are considered to belong to the same acoustic clan, each of which is acoustically distinct. Pods belonging to different clans have overlapping ranges and interact frequently, despite having very different call repertoires. New pods form by gradual fission of older, larger pods, along maternal lines. This process appears to be accompanied by divergence of common dialects, thus dialects reflect the historical matrilineal genealogy of pods within clans. Evidence indicates that minor though significant repertoire divergence exists among closely related matrilines within pods (Deecke *et al.*, 2000; Miller and Bain, 2000). Dialects likely serve as acoustical "badges" that help maintain cohesion and integrity of matrilineal groups. Genetic studies by Barrett-Lennard (2000) suggest that dialects in British Columbian killer whales also serve as a mechanism promoting outbreeding. Group-specific dialects have also been found in Norwegian killer whales (Strager, 1995) and are to be expected in other populations of the species that are strongly matrilineally organized.

The other cetacean species with group-specific dialects, the sperm whale, tends also to live in matrilineal groups, although these lack the long-term stability seen in "resident" killer whales. Sperm whale groups were found to have repertoires that consistently varied in the proportional usage of different coda types and classes (Weilgart and Whitehead, 1997). No coda type was unique to any particular group. As in killer whales, groups of sperm whales with distinct dialects interact regularly. Geographic variations in coda repertoires were also noted in different oceans and in different areas within oceans, but such variations were weaker than those observed in group-specific dialects within local regions.

See Also the Following Articles

Communication ■ Culture in Whales and Dolphins ■ Language Learning ■ Song

References

Barrett-Lennard, L. G. (2000). "Population Structure and Mating Systems of Northeastern Pacific Killer Whales." Ph.D. Dissertation, University of British Columbia, Vancouver, BC.

Conner, D. A. (1982). Dialects versus geographic variation in mammalian vocalizations. *Anim. Behav.* **30**, 297–298.

Deecke, V. B., Ford, J. K. B., and Spoug, P. (2000). Dialect change in resident killer whales (*Orcinus orca*): implications for vocal learning and cultural transmission. *Anim. Behav.* **60**, 617–628.

Ford, J. K. B. (1989). Acoustic behaviour of resident killer whales (*Orcinus orca*) in British Columbia. *Can. J. Zool.* **67**, 727–745.

Ford, J. K. B. (1991). Vocal traditions among resident killer whales (*Orcinus orca*) in coastal waters of British Columbia. *Can. J. Zool.* **69**, 1454–1483.

Janik, V. M., and Slater, P. J. B. (1997). Vocal learning in mammals. *Adv. Study Behav.* **26,** 59–99.

Miller, P. J. O., and Bain, D. E. (2000). Within-pod variation in the sound production of a pod of killer whales, *Orcinus orca. Anim. Behav.* **60,** 617–628.

Payne, R., and Guinee, L. N. (1983). Humpback whale, *Megaptera novaeangliae,* songs as an indicator of "stocks." *In* Communication and Behavior of Whales (R. Payne, ed.), pp. 333–358.

Strager, H. (1995). Pod-specific call repertoires and compound calls of killer whales, *Orcinus orca* Linnaeus, 1758, in waters of northern Norway. *Can. J. Zool.* **73,** 1037–1047.

Thomas, J. A., and Stirling, I. (1983). Geographic variation in the underwater vocalizations of Weddell seals (*Leptonychotes weddelli*), from Palmer Peninsula and McMurdo Sound, Antarctica. *Can. J. Zool.* **61,** 2203–2212.

Weilgart, L., and Whitehead, H. (1997). Group-specific dialects and geographical variation in coda repertoire in South Pacific sperm whales. *Behav. Ecol. Sociobiol.* **40,** 277–285.

Diet

NÉLIO B. BARROS
*Mote Marine Laboratory,
Sarasota, Florida*

MALCOLM R. CLARKE
Cornwall, United Kingdom

The ancestors of present-day marine mammals moved into water, possibly to escape competition on land for food resources, to escape predation, or to take advantage of relatively abundant food supplies in the seas. Most likely, it was a combination of factors. Eventually, the development of echolocatory capabilities in odontocetes (toothed whales) and physiological adaptations for deep and prolonged dives allowed for the exploration of deep waters in search for food. Some groups of cetaceans, such as beaked whales, were able to evolve the ability to dive to great depths and take advantage of food resources unavailable to other predators. From a presumed terrestrial insectivore diet, marine mammals switched largely to fish, squid, and shrimp as main prey (in addition to other crustaceans, mollusks, and zooplankton organisms).

I. Methods of Study

A variety of different methods have been used to gain insight into what marine mammals eat. Their aquatic lifestyle usually limits direct observations of feeding, except in shallow waters in geographical areas where water visibility is good. The following methods have been used to draw inferences on marine mammal diet.

A. Direct Observations of Feeding

This method is limited to what can be observed above the surface, from a vessel, from a vantage point on land, or from the air. Although much can be learned from these observations,

especially when made systematically in a particular area, subsurface feeding behavior is generally not observed, and the picture obtained on feeding is incomplete, at best. Prey that present aerial behavior (e.g., mullet, flying fish) tend to be overrepresented, whereas bottom-dwelling species (e.g., toadfish, flatfish, octopus) are underrepresented.

B. Traditional Methods

The traditional method to study marine mammal food habits has been the analysis of food remains present in vomit or scat from living animals, and the stomachs and intestines of stranded animals. This method relies on the finding and identification of structures representing a typical meal, e.g., fish bones and the jaws of cephalopods, often referred to as "beaks" due to their resemblance to beaks of parrots. Fish ear stones (or otoliths), in particular, are diagnostic structures in the identification of prey because their size and shape vary considerably from species to species (Fig. 1). Fish otoliths are calcareous structures (their primary composition is calcium carbonate) and are more resistant to digestion than bones. In life, they are housed in capsules inside the fish's skull, and their main function in a fish is to provide information on balance and sound reception. Each species of fish has three pairs of otoliths (the sagitta, lapillus, and asteriscus). With a few exceptions (e.g., catfish), the largest pair of otoliths (the sagitta) is also the one most distinctive and recognizable, and the one used in species identification. Similarly, cephalopod beaks possess morphological features that vary and can be used in the identification of species (Fig. 2). Although the upper beaks have some taxonomic value, the lower beaks are generally the ones used in prey identification. Beaks are composed of chitin, a similar material to our fingernails and to mammal horns and are not dissolved by digestive processes. Otolith and cephalopod beak lengths correlate well with the length and weight of the animal from which they came; relationships with their weights are described by a power curve. These allow reconstruction of the original meal by weight. Estimates can then be made of consumption of particular prey species by single mammals and, sometimes, their populations. The relationships relating otolith/beak length to animal length are best described by straight lines, which indicate the target lengths of prey species.

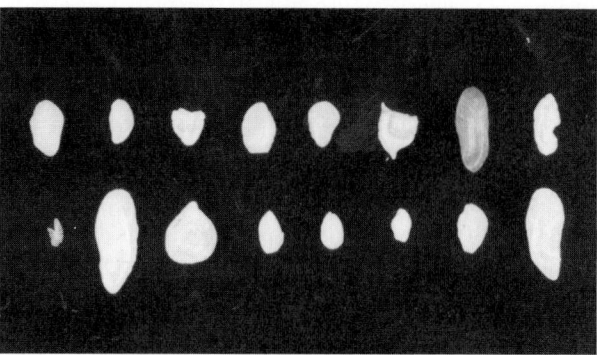

Figure 1 *Fish ear stones (otoliths) of different sizes and shapes retrieved from stomachs of bottlenose dolphins stranded in Florida.*

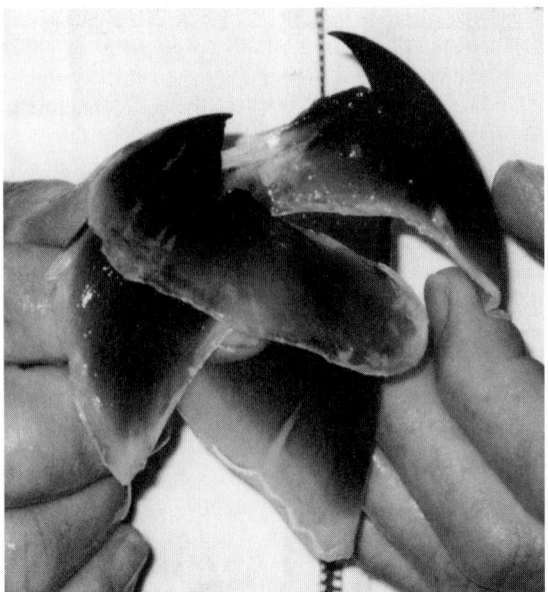

Figure 2 *Cephalopod lower (left) and upper (right) jaws or "beaks" from a giant squid* (Architeuthis).

The advantages of this method are (1) knowledge of prey composition and size classes allows for understanding spatial and temporal distribution of predators; (2) studies of predator–prey dynamics are possible; (3) prey species may be very poorly sampled by humans using other methods, and diets can give considerable information on the species in an area available to predators; (4) changes in diet during growth and over time can be monitored; (5) analysis requires low cost or little equipment; and (6) samples can be collected from carcasses in an advanced stage of decomposition.

The disadvantages of this method are (1) prey with no hard parts (e.g., invertebrates) will be underrepresented; (2) different digestion rates of prey can make calculations of reconstructed meal sizes complicated (fish otoliths can last for about a day in the gastrointestinal tracts of marine mammals, whereas squid beaks may accumulate for several days); (3) there can be potential bias in using feeding data gathered from stranded (possibly sick) animals, as they may not be representative of the population at large; and (5) a comprehensive reference collection of hard structures (fish otoliths and cephalopod beaks) of the most common prey in a particular area is of great advantage in species identification, but is not always available.

C. Use of Novel Tools (New Technology) to Understand Feeding Ecology

1. Stable Isotopes The principle of this method is that ratios of heavier vs lighter isotopes of particular elements (carbon, nitrogen, oxygen, sulfur) in tissues of predators can be traced to those of their prey as they are assimilated through the diet. The following are advantages of this method: (a) it is ideal to detect shifts in diet; (b) different tissues of the predator yield information reflecting the feeding history relative to the last days, months, or the entire life of the animal; these data can be

used to gain insights into the distribution, movements, and migratory habits of the animal; (c) isotopic ratios of carbon reflect those of the primary producers in the area; isotopic ratios of nitrogen are indicative of the trophic level occupied by the organism, which are helpful in understanding habitat utilization and trophic relationships; and (d) historical reconstruction of values through time can yield intra- and interannual variability in feeding. The disadvantages of this method are that a reference database for the isotopic signature of prey is needed and the cost of the equipment used in the analyses is high.

2. Fatty Acids The fatty composition of a prey is species specific and, as these compounds are assimilated through the diet and accumulated in fatty tissues of predators (e.g., blubber of marine mammals), they can be used as tracers of diet. Fatty acids analysis can be useful in (a) reconstructing diets in time and space; (b) population studies of various marine mammal species using different feeding grounds; (c) studies of energetic transfers between mother and their offsprings; and (d) the application of the technique to free-ranging animals, which can be done by the relative noninvasive collection of tissue (e.g., through biopsy darts). Similarly to stable isotopes, this method requires a reference database for the chemical signature of the various prey species, and the cost of the equipment is high. In addition, there is stratification of fat in the outer and inner blubber layer of marine mammals, and incomplete sampling of the BLUBBER layer may yield misleading results of dietary information. Additional variability may be associated with what part of the body the sample is taken, making interstudy comparisons difficult.

3. Molecular Identification of Prey This method involves the genetic identification of material from scats, stomach contents, and gut bacteria, which must be separated prior to analysis. The disadvantages of this method are that a reference database for the genetic signature of prey is needed and it has yet to be widely applied to marine mammal dietary studies.

4. Video-Taping Studies (Using "Critter Cams") of Animals Feeding at Depth This method has the following advantages: (a) it documents the actual feeding behavior of the predator, and the identity of the prey species can be verified by the images; (b) prey behavior during detection and capture can be documented; and (c) different feeding strategies can also be observed (e.g., cooperative feeding). Among the disadvantages: (a) only captures of a few species can be observed; (b) ambient light must be relatively bright and the water must be clear; (c) there is difficulty in the attachment and recovery of the equipment (video camera); and (d) there are high costs associated with the equipment and its operation.

In summary, although identifying and measuring items in vomit, scats, and stomach contents have many disadvantages, it provides more information at considerably less cost than other methods and cannot be replaced effectively by any other method at present.

II. Diets in General

Marine mammals, all together, eat a great variety of animals from minute crustaceans, less than 1 mm long, to giant squids, over 15 m in length. These live in a wide range of habitats, from the shallow shelf seas and estuaries to over 2000 m deep in the deepest oceans, from near the water surface to the ocean bottom. The animals consumed vary in texture from soft-skinned and gelatinous octopods to hard-scaled, muscular fish and vary in mobility from sedentary clams to jet-propelled squids. The three species of manatees (*Trichechus* spp.) and the dugong (*Dugong dugon*) of the order Sirenia rely on sea grasses and river plants; they are grazers in the true sense.

All this variety in food organisms has led to many specializations in structure. Most obviously, the mouth has developed a great variety of tooth numbers, sizes, and shapes or, for those eating very small organisms, a special filter made of horny plates, frayed on one edge, called "baleen." Various species dive to greater depths, so permitting an extension of their feeding grounds from the continental shelves, down the continental slopes, to depths exceeding 2000 m. Similarly, thickening of their fatty blubber layer has permitted further extension of their feeding grounds into the cold waters of the Antarctic and Arctic. The diet of any species reflects its adaptations; its mouth adaptations make it possible for it to catch certain types and sizes of prey. The actual species in the diet depends on its own and the prey's depth and geographic distribution.

Another property of the food is quality. Crustaceans, fish, and cephalopods vary in protein, fat, and mineral constituents and proportions. Marine mammal species can sometimes shift between these three groups during a year, as supply fluctuates or during their own migrations, but this may well drastically affect their physical condition and health. Even within one of these major groups the protein content, for instance, may vary greatly so that change to another species of prey as the main food item might affect the predator markedly. This is likely to be important when the mammal shifts from a diet of shellfish, which are very muscular, to deep-living oceanic fish, which are generally lower in protein content. Similarly, shelf-living cephalopods are mainly soft bodied, have weak muscles, and are low in protein so that twice as much has to be eaten. The high acidity and presence of ammonium salts in high concentration in some soft-bodied squids also probably require special adaptations of digestive processes.

The food of rarer marine mammals, and species that have not been caught for their oil, is not well known. Knowledge depends on information from occasional strandings, and stomach contents are often difficult to collect because of the size of the carcass. Debris from the stomachs only occasionally includes complete, readily identifiable, prey animals. Usually, information on diet has to be obtained from hard pieces, mainly fish otoliths and cephalopod beaks.

A. Cetaceans

1. Baleen Whales (Mysticeti) These possess baleen plates, and all but the gray whales collect swarming animals by skimming through the water or by gulping. They therefore primarily eat shoaling plankton or small nekton together with a few larger animals, such as fish and squids, caught with the shoals.

A. RIGHT WHALES (BALAENIDAE AND NEOBALAENIDAE) The five species have very long baleen plates hanging from the roof of the mouth, whose finely frayed inner edges can trap very small plankton. They mainly eat small crustaceans ranging from minute copepods less than 1 mm long, favored by the bowhead whale (*Balaena mysticetus*) of the northern seas and the pygmy right whale (*Caperea marginata*), to small euphasiid crustaceans called "krill" as much as 25 mm long, eaten by the southern right whale (*Eubalaena australis*). The bowhead is also known to eat a small molluscan called a pteropod.

B. GRAY WHALE (*ESCHRICHTIUS ROBUSTUS*) This species has very tough baleen plates that become worn, particularly at the right side, by rubbing on the sea floor from which it principally sucks, by piston action of its tongue, bottom amphipod crustaceans but also mollusks and bristle worms.

C. RORQUALS (BALAENOPTERIDAE) The eight species have shorter baleen plates than right whales and generally favor larger prey than copepods. Blue whales (*Balaenoptera musculus*) eat midwater crustaceans, mainly KRILL, in the Antarctic and other euphausiid species in the North Pacific and North Atlantic. The fin whale (*B. physalus*) eats krill in the Antarctic but in the North Pacific it broadens its diet to include fish such as clupeids, muscular squids, and a copepod. It eats the fish capelin in the North Atlantic. Humpback whales (*Megaptera novaeangliae*) eat mainly krill or "lobster krill" in the Southern Hemisphere, but mainly anchovies and cod in the Northern Hemisphere. Assorted squid are also eaten by humpback whales. The sei (*B. borealis*) whale eats 20 species of densely shoaling midwater crustaceans, including krill and copepods, in addition to anchovy, cod, and assorted oceanic squids. Minke whales (*B. acutorostrata* and *B. bonaerensis*) eat assorted crustaceans in the Arctic and Antarctic seas, and also fish, including anchovy, in the North Pacific and herring in the North Atlantic. They also take assorted midwater squid in the south tropical seas and appear to rely more on fish than other baleen whales. Bryde's whales (*B. brydei* and *B. edeni*) eat crustaceans, including krill, but also various fish, including mullet and anchovy in the Southern Hemisphere and anchovy in the North Pacific.

2. Toothed Whales (Odontoceti) Within this group, comprising seven families and 52 species, teeth are developed and the main prey items are fish and cephalopods. The cetacean species that live on the continental shelf eat muscular fish, such as herring, pilchards, whiting, and soles; muscular cephalopods, such as inshore squid, cuttlefish, and octopods, are occasionally taken as well. Odontocetes living in the deep ocean eat mainly lantern fishes and soft-bodied, often gelatinous, squids. Around oceanic islands, both lantern fishes and more muscular species, such as horse mackerel and trumpeter fish, are often taken.

A. DOLPHINS (DELPHINIDAE) In 45% of the species in this family, cephalopods comprise over 75% and fish less than 25% of the diet. In 24% of the species, cephalopods comprise 50–75% and fish 25–50%. Depending on the species of dolphin, the diet can be muscular species living on the continental shelf, as for many populations of the short-beaked common dolphin (*Delphinus delphis*) and the common bottlenose dolphin (*Tursiops truncatus*); or as in spinner (*Stenella longirostris*) and spotted dolphins (*S. attenuata* and *S. frontalis*) the diet may include many soft-bodied oceanic species. However, on the whole, dolphins favor muscular squid rather than soft-bodied ones, even in oceanic waters; this also applies to pilot whales (*Globicephala* spp.). A few species include other prey groups in their diet, e.g., Commerson's dolphins (*Cephalorhynchus commersonii*) eat some krill and killer whales (*Orcinus orca*) also prey on seals and other cetaceans.

B. PORPOISES (PHOCOENIDAE) In half of the species in this family, cephalopods comprise over 75% and fish less than 25% of the diet. In the other half of the species, cephalopods comprise 50–75% and fish 25–50%. Being inshore cetaceans, food consists of common (often economically important to humans) species: muscular inshore squid, cuttlefish, and octopus, as well as fish such as herrings, whitings, and bottom-living soles.

C. BEAKED WHALES (ZIPHIIDAE) In over half the species of beaked whales, more than half of the food is cephalopod and the rest is fish. These whales are deep divers and at least one species favors soft-bodied squids. The number of teeth is very reduced in this family and, in one species, the two teeth in the lower jaw grow over the upper jaw and limit it to a narrow gape. It is remarkable that this does not seem to inhibit capture of oceanic squids.

D. NARWHAL AND BELUGA (MONODONTIDAE) The diet of the narwhal (*Monodon monoceros*) includes fish such as Greenland halibut and polar cod, muscular squid, and shrimp. The suggestion that the long tooth of male narwhals is regularly used to stir prey from the mud is unlikely to be true. Belugas (*Delphinapterus leucas*) feed on fish such as capelin and sand lance, as well as larger species such as cod and flounder. Sand- and bottom-living worms show that they probably feed on the bottom as well as in midwater.

E. SPERM WHALES (PHYSETERIDAE AND KOGIIDAE) All three species feed mainly on squids, although a few fish are taken, including large sharks. Sperm whales (*Physeter macrocephalus*) eat mainly deep-living oceanic squids and most of these are soft-bodied or gelatinous, luminous, and weak swimmers. Contrary to common belief, the average weight of their prey is not great, varying from 0.5 kg off South Africa to 7 kg in the Antarctic, although some large sperm whales can eat squids over 15 m in length. Pygmy and dwarf sperm whales (*Kogia breviceps* and *K. sima*) eat some of the same species as their larger relative but, because they spend some time on continental shelves, they also include muscular, shelf-living squids and octopods.

F. RIVER DOLPHINS (INIIDAE, PONTOPORIIDAE, LIPOTIDAE, PLATANISTIDAE) This group includes four species of dolphins, three of which inhabit the freshwater systems of major rivers in South America, China, and the Indian subcontinent. The fourth species has a marine distribution and is found in coastal waters of the Atlantic coast of South America. The riverine species feed on a variety of freshwater fish (including sharks) and prawns, and occasionally also prey on other groups, such as freshwater turtles. As cephalopods are strictly marine in distribution, they are not part of the diet of riverine dolphins. The marine species in this group (the "franciscana," *Pontoporia blainvillei*) eats mainly bottom-dwelling fish, coastal species of cephalopods, and several species of shrimp.

B. Pinnipeds (Seals, Sea Lions, Walruses)

All but 2 of the 36 species of pinnipeds (seals and sea lions) probably include both fish and cephalopods in their diet. The exceptions inhabit freshwater systems where cephalopods do not occur. Most pinniped species inhabit coastal regions or seas close to oceanic islands, which partly influences their choice of diet.

1. Fur Seals and Sea Lions (Otariidae) Of the 16 species, 10 take benthic cephalopods and 11 eat midwater squids. At least 14 eat muscular cephalopods and at least 3 of these eat oily squids while only 1 eats soft-bodied squids. Three species consume all these on the continental shelf while 8 eat them in oceanic waters.

A. NORTHERN FUR SEAL (*CALLORHINUS URSINUS*) On the continental shelf this species eats primarily small shoaling fish and muscular squids. In offshore waters, they eat mainly muscular oceanic squids.

B. GUADALUPE FUR SEAL (*ARCTOCEPHALUS TOWNSENDI*) They eat oceanic cephalopods and lantern fish.

C. JUAN FERNANDEZ FUR SEAL (*A. PHILIPPII*) They apparently eat muscular, oceanic squids.

D. GALAPAGOS FUR SEAL (*A. GALAPAGOENSIS*) This species mainly eats muscular, oceanic squids.

E. CAPE AND AUSTRALIAN FUR SEALS (*A. PUSILLUS*) They eat both shelf and oceanic fish and cephalopods, as well as both midwater and bottom species. They favor muscular rather than soft-bodied cephalopods.

F. NEW ZEALAND FUR SEAL (*A. FORSTERI*) This species eats midwater fish and muscular cephalopods, but also takes penguins.

G. ANTARCTIC FUR SEAL (*A. GAZELLA*) They eat mainly krill in the Antarctic, but further north take oceanic fish, as well as muscular and soft-bodied squids.

H. SUB-ANTARCTIC FUR SEAL (*A. TROPICALIS*) They eat oceanic squids, both muscular and soft-bodied.

I. STELLER SEA LION (*EUMETOPIAS JUBATUS*) They eat muscular, bottom octopods as well as oily oceanic squids and polar cod.

J. CALIFORNIA, GALAPAGOS AND JAPANESE SEA LIONS (*ZALOPHUS CALIFORNIANUS, Z. WOLLEBAEKI, AND Z. JAPONICUS*) These species eat shelf-living, muscular squids and octopods, as well as muscular, oceanic squids.

K. SOUTH AMERICAN SEA LION (*OTARIA FLAVESCENS*) They eat mainly on bottom and midwater shelf fish but also take some shelf cephalopods.

L. AUSTRALIAN SEA LION (*NEOPHOCA CINEREA*) This species eats shelf octopods and cuttlefish.

M. NEW ZEALAND SEA LION (*PHOCARCTOS HOOKERI*) This species eats shelf fish, cephalopods, and crustaceans.

2. Earless (true) Seals (Phocidae) This group eats a variety of fish, cephalopods, and crustaceans of both inshore and oceanic species, depending on their locality. Of the 19 species, at least 15 eat muscular cephalopods, 5 eat oily species and 4 eat soft-bodied squids. Some slight deviations from this pattern are given.

A. HARP SEAL (*PAGOPHILUS GROENLANDICUS*) This species eats mainly fish and crustaceans, especially amphipods and euphausiids, although both bottom and midwater oceanic cephalopods are also taken.

B. BEARDED SEAL (*ERIGNATHUS BARBATUS*) They eat mainly bottom shelf invertebrates, such as clams, and also fish, in addition to a few species of octopods.

C. GRAY SEAL (*HALICHOERUS GRYPUS*) Gray seals eat schooling fish, squids, octopods, and occasionally sea birds.

D. CRABEATER SEAL (*LOBODON CARCINOPHAGA*) They eat krill almost exclusively.

E. ROSS SEAL (*OMMATOPHOCA ROSSII*) Ross seals eat oceanic species of fish and squids of both muscular and soft-bodied species.

F. LEOPARD SEAL (*HYDRURGA LEPTONYX*) This species eats krill, fish, soft-bodied squids, and occasionally mammals.

G. WEDDELL SEAL (*LEPTONYCHOTES WEDDELLII*) They eat mainly cephalopods, including muscular and soft-bodied species, and bottom octopods.

H. ELEPHANT SEALS (*MIROUNGA ANGUSTIROSTRIS* AND *M. LEONINA*) These species eat oceanic species, including muscular, soft-bodied, and oily species and, seasonally, shelf squids.

3. Walrus (Odobenus rosmarus) These are benthic feeders in shallow Arctic seas at depths less than 100 m. Their main food is clams but they also eat a small quantity of bottom octopods and have been known to attack other seals.

C. Sea Otter (*Enhydra lutris*)

These eat bottom invertebrates on the continental shelf, usually very close to shore, including clams, sea urchins, and other invertebrates.

D. Polar Bear (*Ursus maritimus*)

This species eats mainly harp seals but also feeds on other seals, young beluga and narwhal, young walrus or sick animals, and fish such as Arctic char. Polar bears also feed on terrestrial species of mammals and birds, on carcasses of bowhead and gray whales, and occasionally on humans. Polar bear males are known to kill and eat cubs of their own kind, possibly in order to incite the female to come into estrus again rapidly.

E. Sirenians

The manatees and dugong feed on tropical grasses and roots and rhizomes in nearshore areas in saline environments and on water hyacinths, water lilies, and other vegetation in rivers and lakes. The extinct Steller's sea cow (*Hydrodamalis gigas*) was a cold-adapted species, last found off the Kamchatcka Peninsula in far east Russia; it fed on cold water kelp.

See Also the Following Articles

Baleen ▪ Feeding Strategies and Tactics ▪ Filter Feeding ▪ Predator–Prey Relationships

References

Bowen, W. D., and Siniff, D. B. (1999). Distribution, population biology, and feeding ecology of marine mammals. *In* "Biology of Marine Mammals" (J. E. Reynolds, III and S. A. Rommell, eds.), pp. 345–384. Smithsonian Institution Press, Washington, DC.

Clarke, M. R. (1986). Cephalopods in the diet of odontocetes. *In* "Research on Dolphins" (M.M. Bryden and R. Harrison, eds.), pp. 281–321. Clarendon Press, Oxford.

Clarke, M. R. (1986). "A Handbook for the Identification of Cephalopod Beaks." Clarendon Press, Oxford.

Clarke, M. R. (1996). Cephalopods as prey. III. Cetaceans. *Phil. Trans. R. Soc. Lon. B.* **351**, 1053–1065.

Härkönen, T. (1986). "Guide to the Bony Fishes of the Northeast Atlantic." Danbiu ApS., Hellerup, Denmark.

Iverson, S. J., Frost, K. J., and Lowry, L. F. (1997). Fatty acid signatures reveal fine scale structure of foraging distribution of harbor seals and their prey in Prince William Sound, Alaska. *Mar. Ecol. Prog. Ser.* **151**, 255–271.

Kawamura, A. (1980). A review of the food of balaenopterid whales. *Sci. Rep. Whales Res. Ins.* **32**, 155–197.

Klages, N. T. W. (1996). Cephalopods as prey. II. Seals. *Phil. Trans. R. Soc. Lon. B* **351**, 1045–1052.

Latja, K., and Michener, R. H. (1994). "Stable Isotopes in Ecology and Environmental Science." Blackwell Scientific Publications, Oxford.

Read, A. J. (1998). Possible application of new technologies to marine mammal research. *Mar. Mamm. Comm. Contr. Rep.* T30919695, 1–36.

Walker, J. L., and Macko, S. A. (1999). Dietary studies of marine mammals using stable carbon and nitrogen isotopic ratios of teeth. *Mar. Mamm. Sci.* **15**, 314–334

Distribution

JAUME FORCADA
*Southwest Fisheries Science Center,
La Jolla, California*

The majority of marine mammals are found in marine environments; some are found in rivers and estuaries and a few in freshwater lakes. Amphibious marine mammals (the pinnipeds, sea otters, and polar bears) also occur on land and ice. Distribution is the part of ecology that deals with how they use different geographic ranges in space and time. The complexity of distribution patterns of individuals or aggregations of individuals depends on how the different factors affect the species' habitats, biological requirements, interactions with other organisms, and the patchiness of the environment.

I. Distribution Patterns and Preferences

Marine mammals are found in almost all the different marine environments, and their distribution varies according to the physical, chemical, and biological characteristics of the water masses. The effects of oceanographic phenomena, wind-induced movements (e.g., water currents, local divergence, and upwelling areas and water fronts) and the topography, can be used to characterize distribution. In the case of pinnipeds, the breeding and molting habitats on land or ice also characterize their distribution. In the polar bear, breeding and cub-rearing habitats are also relevant.

In freshwater environments, marine mammals are found in rivers and lakes. Examples of riverine species are the river dolphins (Platanistidae, Iniidae, Lipotidae, and Pontoporiidae) and the manatees. A few Phocidae live in freshwater inland lakes: the Saimaa seal (*Pusa hispida saimensis*) in Finland; the Caspian seal (*Pusa caspica*) in the Caspian Sea; the Baikal seal (*Pusa sibirica*) in lake Baikal; and the Ungava (common) seal

(*Phoca vitulina mellonae*) in freshwater lakes of the Hudson Strait.

In marine environments, distribution can be generally described as coastal (in estuarine or near shore waters), neritic (in waters on the continental shelf), or oceanic (in waters beyond the continental slope, in the open seas or oceans). Examples of marine mammals that reside primarily in coastal waters are populations of bottlenose dolphins (*Tursiops* spp.), sea otters (*Enhydra lutris*), and dugongs (*Dugong dugon*). Primarily neritic species include gray whales (*Eschrichtius robustus*), harbor porpoise (*Phocoena phocoena*), and California sea lions (*Zalophus californianus*). Primarily oceanic species include the sperm (*Physeter macrocephalus*) and beaked whales (family Ziphiidae). These generalizations should be used with caution, as many species occur in multiple habitats. Some species shift from one habitat to another seasonally, such as the switch from neritic feeding grounds to coastal migratory routes and breeding grounds by gray whales. Some species have populations that reside in a variety of habitats, such as the bottlenose dolphin, which occurs in coastal, neritic, oceanic, and, occasionally, riverine habitats.

Marine mammal distribution can also be classified according to general geographic areas. These are characterized by latitudinal bands and average water temperatures. Thus, marine mammals have tropical and/or subtropical, temperate, Antarctic, or Arctic distributions. Some species can be strictly included in just one of these categories, such as exclusively Arctic species: bowhead whales (*Balaena mysticetus*), polar bear (*Ursus maritimus*), narwhal (*Monodon monoceros*), and beluga (*Delphinapterus leucas*), but, again, other species often have multiple classifications. A clear example are the baleen whales that migrate from cold high latitudes to tropical low latitudes. Some species, such as the killer whale (*Orcinus orca*), are found in all the marine waters of the world, from the Equator to the Arctic and Antarctic. Finally, similar and closely related species may occupy different latitudinal (hemispheres, ocean basins) or longitudinal (different oceans and seas) ranges. Examples of pairs of similar species that occur in different hemispheres are the northern (*Hyperoodon rostratus*) and southern (*Hyperoodon planifrons*) bottlenose whales and the northern (*Mirounga angoustirostris*) and southern (*Mirounga leonina*) elephant seals. An example of very similar cetacean species with different distribution preferences within the same ocean basin are the long- and short-finned pilot whales (*Globicephala melas* and *G. macrorhynchus*).

Detailed data on distribution are provided in the species account of this encyclopedia and therefore only overall patterns by taxa are given in this section to avoid redundancy. Additional detailed description of marine mammal distribution can be found in Gaskin (1982) for cetaceans, in Riedman and Estes (1988) for the sea otter (*Enhydra lutris*), in Riedman (1990) for pinnipeds, in Reynolds and Odell (1991) for sirenians, and in Wiig *et al.* (1995) for the polar bear.

A. Cetaceans

Cetaceans live permanently in aquatic environments. They can be found in all the oceans and most of the seas of the world, and distribution patterns vary between and within fam-ilies. The Balaenidae, the Balaenopteridae, the gray whale, the sperm whale, and the killer whale are found in polar, temperate, and tropical waters. They are found in the Northern and Southern Hemispheres, except gray and bowhead whales, which are only found in the Northern Hemisphere. As noted earlier, other strictly northern and also Arctic species are the narwhal and beluga. The pygmy right whale (*Caperea marginata*) is only found in the Southern Hemisphere. Most delphinids live in tropical and temperate waters of both hemispheres. More tropical Delphinidae are *Stenella attenuata*, *S. longirostris*, *S. frontalis*, *Steno bredanensis*, *Sotalia fluviatilis*, *Globicephala macrorhynchus*, *Pseudorca crassidens*, *Peponocephala electra*, and *Feresa attenuata*. Other tropical odontocetes are the pygmy (*Kogia breviceps*) and dwarf (*K. sima*) sperm whales, Irrawaddy dolphin (*Orcaella brevirostris*), and many Ziphiidae. Most Phocoenidae live in temperate or subtropical waters with some species exclusively in the Northern Hemisphere (*P. phocoena*, *P. sinus*, and *Phocoenoides dalli*) and some exclusively in the Southern Hemisphere (*P. spinipinnis* and *P. dioptrica*). All Delphinidae of the genus *Cephalorhynchus* live in temperate waters of the Southern Hemisphere. Of the river dolphins, the Amazon River dolphin, *Inia geoffrensis*, lives in the large lakes and tributaries of the Amazon and Orinoco basins. The franciscana (*Pontoporia blainvillei*) lives in the coastal central Atlantic waters of South America, but is commonly found in the mouth of the rivers and ocean waters surrounding estuaries. Similarly, the tucuxi is distributed in both fresh and marine waters. The two subspecies of the family Platanistidae (*Platanista gangetica gangetica* and *P. g. minor*) live in the major rivers of India and Pakistan, the Indus and Ganges, and the baiji (*Lipotes vexillifer*) lives in the Yangtze and formerly lived in some of the lakes along this extremely large inland river system.

B. Pinnipeds

Pinnipeds are amphibious mammals and spend most of their life in aquatic environments. However, they must return to land or ice for breeding (giving birth and rearing their offspring) and molt after breeding. Other possible reasons for hauling out are resting, THERMOREGULATION, and escape from predators. Some common characteristics of nonaquatic habitats are space availability, isolation from predators, and proximity to food supply. Pinnipeds with tropical and temperate distributions find these conditions in isolated ROOKERIES or beaches of remote places, which often are in islands. Ice characteristics condition the distribution and activity patterns of pinnipeds; pack ice offers a more constant substrate than fast ice, which varies highly seasonally in extension. Some pinnipeds reproduce in fast ice, such as the hooded seal (*Cystophora cristata*), and the duration of lactation and rearing of their young strongly depend on ice conditions. In general, seasonal changes in oceanographic conditions and ice cover condition the distribution of pinnipeds in the pack ice.

Among the Phocidae, geographical or latitudinal distributions include the Arctic, sub-Arctic, and temperate areas, subtropical and tropical areas, and sub-Antarctic and Antarctic areas. Antarctic seals are the Weddell (*Leptonychotes weddellii*), crabeater (*Lobodon carcinophaga*), leopard (*Hydrurga lep-

tonyx), and Ross (*Ommatophoea rossii*) seals. A sub-Antarctic and Antarctic seal is the southern elephant seal. In the Northern Hemisphere, tropical and subtropical species are the Hawaiian (*Monachus schauinslandi*), Mediterranean (*Monachus monachus*), and the extinct Caribbean (*Monachus tropicalis*) monk seals. Sub-Arctic and temperate-water seals are the gray (*Halichoerus grypus*), harbor (*Phoca vitulina*), and northern elephant seals. Arctic and sub-Arctic seals are the harp (*Pagophilus groenlandicus*), hooded, bearded (*Erignatus barbatus*), ringed ribbon (*Histriophoca fasciata*), spotted (*Phoca largha*), Baikal (*Pusa sibirica*), and Caspian seal. Among phocids, harp, hooded, bearded, ribbon, spotted, Ross, and leopard seals breed in the pack ice; the crabeater and ringed seals breed in pack and fast ice; the southern elephant seal breeds on land and fast ice; the Baikal and Caspian seals on fast ice; the harbor and gray seals on land and ice; and the northern elephant and the monk seals breed on land. Phocids with coastal and continental shelf distribution are the harp, harbor, gray, bearded, ringed, ribbon, spotted, Weddell, crabeater, leopard, and Mediterranean monk seals. Continental slope and oceanic seals are the Hawaiian monk, northern and southern elephant, Ross, and hooded seals. The walrus (*Odobenus rosmarus*, family Odobenidae) breeds on the pack ice and occurs in waters of the continental shelf. All the Otariidae breed and rear their offspring on land. Most of them disperse after breeding and therefore have neritic and oceanic distributions depending on season and reproductive status. Many Otariidae have subtropical or tropical distributions, such as the California (*Zalophus californianus*) and Galapagos (*Z. wollebaeki*) sea lions and the Guadalupe (*Arctocephalus townsendi*) and Galapagos (*A. galapagoensis*) fur seals. The Steller sea lion (*Eumetopias jubatus*) is found from Arctic to temperate waters of the eastern North Pacific. The other sea lions are distributed in tropical and sub-Antarctic waters in the Southern Hemisphere: the Australian (*Neophoca cinerea*), New Zealand (*Phocarctos hookeri*), and South American (*Otaria flavescens*) sea lions. In the Southern Hemisphere, all the fur seals are found in temperate or sub-Antarctic waters: the New Zealand (*Arctocephalus forsteri*), South African (*A. pusillus pusillus*), subantarctic (*A. tropicalis*), Australian (*A. pusillus doriferus*), Juan Fernandez (*A. philippi*), and South American (*A. australis*) fur seals. Only the Antarctic fur seal (*A. gazella*) can be strictly considered sub-Antarctic or Antarctic. In the Northern Hemisphere, the northern fur seal (*Callorhinus ursinus*) is found in the sub-Arctic and temperate North Pacific.

C. Sirenians

All the Sirenia are found in tropical or subtropical waters. The manatees have restricted ranges in different oceans and river systems. The west Indian manatee (*Trichechus manatus*) is found from southern North America and Caribbean to northern South America, in the western Atlantic, and the Amazon manatee (*T. inunguis*) in the Amazon drainage. In the eastern Atlantic, the African manatee (*T. senegalensis*) is found in western Africa, from Senegal to Angola. Manatees are coastal, although they may be found in continental shelf waters, transiting between islands, in the Caribbean. The dugong (*Dugong dugon*) is the most widely distributed sirenian, in the Indian

and the western Pacific oceans, with a preference for shallow coastal bays.

D. Polar Bear and Sea Otter

The polar bear has a circumpolar distribution, mostly above the Arctic circle. It uses coastal, neritic waters, and breeds and rears its offspring on ice. Ice is also important for polar bears as a platform to travel, especially in the ice floes, between foraging areas and areas where they give birth and rear their young and as a substrate to hunt seals. The sea otter is found in the Pacific coasts of North America and Russia, essentially in temperate and sub-Arctic waters. It lives near shore and comes ashore on Aleutian Islands. Its distribution is conditioned by predators (e.g, killer whales) and food availability, such as the prey they usually use in kelp forests (e.g., sea urchins and abalone).

II. Factors Affecting Marine Mammal Distribution

Marine mammal distributions are affected by demographic, evolutionary, ecological, habitat-related, and anthropogenic factors. Demographic factors include the ABUNDANCE, AGE, and sex structure of the marine mammal populations and the reproductive status and life cycle of individuals. Evolutionary factors include morphological, physiological, and behavioral aspects of the species' adaptations. Ecological factors include biological production and use of prey, distribution of prey and predators, and competitors. Habitat includes factors such as water temperature, salinity, density, thermocline depth, and the type of substrate and the bathymetry. Anthropogenic factors are the human effects that alter the natural distribution of marine mammals, including pollutants, human-induced sounds, and incidental and direct kills. Distribution is the product of factors that act in a parallel or interactive way over different scales of space and time on each species, and sometimes on groups of species. As an example, baleen whale distribution depends on their ability to exploit planktonic organisms (evolutionary), the oceanographic characteristics of the water masses where they feed (habitat), and the trophic level they exploit (ecological).

A. Demographic Factors

The dynamics of marine mammal populations can determine distribution changes and patterns. The number of individuals that live in particular areas depends on the capacity of those areas to sustain their biological requirements. In general, the most critical requirements are prey availability and energy. The depletion of food resources by marine mammal populations influences the movement or dispersal to other areas. The age and sex structure of marine mammal aggregations also affect the distribution patterns. Habitat requirements for breeding females or females with offspring are not the same as those of adult males. In the case of odontocetes, females with calves may require coastal areas with locally abundant food resources and protection from predators. Adult males, not having to care for their offspring, are less limited in terms of movements and distribution range. In offshore dolphins, large cohesive aggregations may be required by breeding females for protection in the open ocean and foraging distances will be greater due to patchiness of their prey. In the case of pinnipeds, distributional differences according to age and sex classes and reproductive status are related to the seasonality of their life cycles, their adaptation to aquatic feeding, and their need to periodically return to land to breed. In breeding colonies, individuals will gather seasonally to mate, give birth, and nurse their pups over variable periods of time according to species. After the breeding season, pinnipeds often display age-related differences in habitat use and foraging areas. Dispersal according to age and sex classes is often associated with these characteristics.

B. Evolutionary Factors

All factors related to the secondary aquatic adaptation of marine mammals influence their distribution to some extent. Diving capacities in terms of duration and maximum depths allow particular species to exploit different habitats. In the case of sperm whales or elephant seals, deep diving allows access to prey unavailable to the shallower diving dolphins or porpoises. Hence, their distribution is associated with deep canyons and other deep ocean areas. In sperm whales, this ability also requires complex social systems that ensure the protection of newborns or youngsters, particularly while mothers spend long times underwater in search of prey. Another notable physiological adaptation is thermoregulation, which allows marine mammals to extend their distribution ranges from the warm equatorial waters to the coldest high latitudes. Efficient insulation and body temperature regulation systems allow the polar bear and the sea otter to spend a substantial part of their life at sea and survive in cold waters. The relative inability to regulate body temperature adequately in colder water of neonates is a hypothetical factor that leads baleen whales to migrate from the cold feeding grounds to the warmer calving grounds. Morphological adaptations, such as the feeding apparatus of baleen whales, also influence their distribution. As active filter feeders, they can capture planktonic (e.g., copepods, krill) or schooling (e.g., sand lance, capelin, herring) prey, which are abundant in the particular areas where whales distribute. Finally, the cohesiveness of large dolphin schools and the sensorial integration of individuals allow them to range in offshore areas, find food actively and efficiently, and obtain protection from predators.

C. Ecological Factors

Marine mammal distribution is in great measure related to prey distribution. The ability to exploit different trophic levels and resources classifies different marine mammals from top predators, such as the killer whales, to low-trophic level feeders, such as northern right whales or manatees. Marine mammals can be considered as either specialists or generalists, and these two aspects imply differentiated distribution patterns. Manatees, being specialist feeders, have restricted distributions where sea grass meadows provide continued food. Despite often being categorized as specialists, odontocetes or phocids tend to use a wide range of prey items. Thus, they can be distributed over wider ranges and change their distribution seasonally according to the availability of their prey. The killer whale, as a species, has a broad diet, yet different populations have more specialized diets: transient killer whales feed mainly on pinnipeds and other marine mammals but must range widely to maintain this diet, whereas resident killer whales feed on large fishes such as salmon. In both instances, the distributions of the whales are synchronized to the life cycles of their prey. The transients concentrate seasonally near pinniped rookeries, whereas residents live near the mouth of salmon-spawning rivers. In other cases, marine mammals tend to use the same home range, such as coastal bottlenose dolphins, feeding on different prey species that change their distribution seasonally. In this case, distribution patterns must be studied and interpreted at a finer scale. Interspecific competition is an additional ecological factor determinant of variable distribution. Violent attacks on harbor porpoises by bottlenose dolphins have been reported in their common range in the North Sea. Finally, predation plays an important role in the selection of habitats and distribution areas by marine mammals, especially those of smaller size, such as ringed seals. This species appears to select the fast ice to avoid predation by polar bears.

D. Habitat-Related Factors

Marine mammals are usually found in waters with high densities of principal prey species. These waters are characterized by the physical conditions that facilitate the accumulation of these prey. Relevant oceanographic variates characterizing marine mammal habitats are water temperature, salinity, density, chlorophyl concentration, and thermocline depth. These characteristics are related to upwelling fronts, often related to differences in species distribution. As an example, spinner and spotted dolphins range in the same areas of the eastern tropical Pacific, often traveling in the same schools. They occur in the same overall ocean area as common and striped dolphins, but appear to have preferences for water masses of different oceanographic characteristics (Reilly, 1990). Ocean topography and bathymetry are often related to local oceanographic phenomena that influence marine mammal distribution. Underwater canyons, marine ridges, and irregular topographies concentrate prey for deep divers such as sperm whales or elephant seals. In contrast, mysticetes often have preferences for shallow waters with high topographic variation. In these waters their prey accumulates at frontal interfaces between mixed and stratified waters. The ice is also a critical habitat element for marine mammals; the seasonal and highly dynamic changes of ice cover determine their patterns of change in distribution. It provides shelter during reproduction for pinnipeds, access to seasonally abundant food, and also delimits the distribution ranges of some cetaceans, such as the bowhead whale.

E. Anthropogenic Factors

Human alteration of habitats can change marine mammal distributions significantly. Marine mammals that haul out on land are particularly affected by habitat encroachment by human development. The three species of monk seals have suffered substantial changes in their original distributions, and one of them, the Caribbean monk seal, became extinct because of this. In the case of the Mediterranean monk seal, a major change in habitat preferences occurred as a result of human development, but also of deliberate kills for human uses. The seals changed their haul outs from open beaches to difficult-to-access caves, often with underwater entrance. This has created severe habitat fragmentation. Commercial exploitation has also affected marine mammal distributions greatly. Whale stocks were reduced to the point that many original distribution areas are not used anymore. Overfishing of prey items has led to changes in marine mammal distributions. Pollution of coastal areas has degraded many original marine mammal habitats, thus affecting their original distributions. Human-induced changes in local water temperature have changed the seasonal distribution of the Florida manatee (Reynolds and Odell, 1991). This population previously migrated to warmer waters in winter but now uses the thermal vents in waters close to power plants and has changed migration patterns substantially. Expanding sources of sound in the ecosystem (e.g., large ship traffic, naval experiments) and pollution may also affect marine mammal distribution.

III. Movements and Seasonality

The distribution of marine mammals changes seasonally as their biological and ecological requirements change. Marine mammals respond to changes in the environment, such as in temperature, ice coverage, and prey availability. Daily requirements in terms of energy or protection against predators depend on the reproductive status and the season; these are clearly not the same for females with nursing offspring as for solitary males. Movements are a response to changes in the environment and the biological requirements of a species. In tropical areas, movements are expected to vary according to the patchiness of the environment. Distances covered in short periods of time may vary depending on the conditions, but a very marked seasonality is not commonly found. In high latitudes, changes during the cold winter affect the distribution of marine mammals, their tolerance to physical conditions, and their life history requirements. Thus, seasonality is more marked.

Movements can be classified as migration, dispersal, and daily travel. Migration is the seasonal change between two geographic locations that is related to species reproductive cycle, changes in temperature, and prey availability. Dispersal is the movement from the place of birth to other areas in which individuals reach a feeding area, join a breeding population, or find another group of individuals with which to spend the next stage of its life. The classification of movements may be somewhat arbitrary because marine mammals do not always follow strict periodic patterns. They instead respond to the limitations of the environment in providing constant food or other requirements. Short-scale movements are difficult to detect and must be put in the context of the species life cycle before being classified. A typical example of migration is the one of baleen whales; humpback whales undertake long-distance travels, often thousands of kilometers, between the tropical calving grounds in winter and the high-latitude feeding grounds in summer. In contrast, most Otariidae have dispersal movements from their birth colonies toward different feeding areas or other breeding colonies when they reach sexual maturation. In any case, movement patterns vary among individuals, according to their age, sex, and reproductive condition. A prereproductive young whale may delay its departure from the high-latitude feeding grounds to extend the feeding season, whereas a pregnant female must leave for the low-latitude calving grounds to give birth to its offspring.

A. Cetaceans

Cetaceans spend their entire life in aquatic habitats and are in constant movement. Understanding their seasonal distribution is more difficult than in pinnipeds for technical and logistical reasons: the manipulation and tagging of animals is less efficient and more expensive. Thus, classifying movements as dispersal or migration is even more confusing, except in some well-studied populations of baleen whales. Their life patterns and cycles make the concept of dispersal a little ambiguous, however, because seeming residency, site fidelity, or habitat discreteness may be just apparent, short-term attributes of their distribution. Only migration in large whales is known from long-term studies. Studies on migration range from the examination of catch statistics of whaling operations to the use of modern TELEMETRY technology. Contrary to classic accounts of whale migration, the most recent studies show how movements vary across whale populations and species. Mysticetes appear to have periodical migrations with relatively consistent patterns over the years. Seasonal movements in odontocetes are far less consistent over time, including those of the sperm whale, which has been classified as a migratory species with marked seasonal patterns. In general, as in other marine mammals, factors inducing migration are the biological cycle, greatly determined by reproductive needs, and factors in the environment (e.g., prey availability, changes in water temperature). These factors may trigger the start of seasonal movements, although not all individuals will respond in the same way.

Annual migrations are best known for species with more coastal ranges, such as the gray, right, or humpback whales. However, virtually all mysticete species are known to migrate. No data are available for the pygmy right whale. Most mysticetes have latitudinal migrations, from tropical breeding grounds to high(er) latitude feeding grounds. In breeding grounds, mating and calving take place. Migratory species are right, blue (*Balaenoptera musculus*), fin (*B. physalus*), sei (*B. borealis*), humpback, and gray whales. Bowhead whales also migrate, but their longitudinal movements are equal to or greater than their latitudinal changes, and they never leave Arctic waters. Bryde's (*Balaenoptera edeni*), common minke (*B. acutorostrata*) and Antarctic minke (*B. bonaerensis*) whales, however, have less clear movement patterns. Bryde's whales spend most of the year in warm tropical waters and calving does not have the same marked seasonality seen in other balaenopterids. This indicates a possibly different reproductive cycle, in which both whales feed and mate year round. In this case, whale movements are more similar to those of many odontocete species, in constant search for food, with variable utilization of prey, and different prey types through the year. Among the best known migrations are those of the gray and humpback whales. Gray whales migrate annually from feeding grounds in the Arctic to their calving areas in the lagoons of Baja California in Mexico. Interannual changes in the timing and numbers reaching the different migratory destinations have been observed. The migration of humpback whales is also very well studied, and sperm whales are the best example of long-range migration in odontocetes.

Movements in odontocetes have different scales depending on geographical areas, family, and species. It is generally accepted that most movements are in response to prey availability, and the largest movements, often called migration, occur in oceanic odontocetes. In the eastern tropical Pacific, movements are reported to be wide. Several species of *Stenella* had daily movements of 53 km/day and hundreds of kilometers over months, and these reflected seasonal changes in distribution. Dolphins moved inshore, toward the American continent, in fall and winter and offshore in spring and summer (Perrin, 1975). Other methods, such as line transect surveys in California, have shown how several dolphin species have different patterns of abundance and distribution depending on the season. Pacific white-sided dolphins (*Lagenorhynchus obliquidens*), Risso's dolphins (*Grampus griseus*), common dolphins (*Delphinus* spp.), and northern right whale dolphins (*Lissodelphis borealis*) were less abundant in summer than in winter, and significant north/south shifts in distribution were reported for Dall's porpoises and common and Pacific white-sided dolphins. Significant inshore–offshore differences were found for the northern right whale dolphin. Some dolphin species show variable distribution patterns, such as bottlenose dolphins or killer whales. Difference in patterns has been attributed to different varieties or ecotypes of the same species. In the case of bottlenose dolphins, a well-studied coastal population (Scott *et al.*, 1990) showed a year-round residency with slight seasonal changes within the population home range. It has been argued that dolphins in Florida follow the mullet migrations into the Gulf of Mexico during the fall. Short-term movements have also been observed in a resident bottlenose population in east Scotland. In contrast, Atlantic offshore bottlenose dolphins, described as a possible different form, have wider movements and a broader distribution (Wells *et al.*, 1999).

B. Pinnipeds

Migration is not uncommon in pinnipeds, and the advent of new telemetry has helped describe the migratory movements of several species. Dispersal is very common in pinnipeds and depends on the abundance of prey, its energy content, and the seasonality of prey distribution. In addition, their reproductive cycle mandates that individuals return to land or ice to give birth, nurse, and rear their offspring and molt. Pinnipeds also haul out for resting, thermoregulation, and to escape predators, among other reasons. If the environment provides constant food resources, such as in some tropical areas, there will not be a clear need to disperse. In contrast, pinnipeds living in high latitudes will be more dependent on ice cover, availability of seasonally changing prey, reproduction, and population size. These will create density-dependent effects, such as dispersal and distributional changes. Thus, dispersal can vary with latitude, based on the stability of prey resources.

Phocids appear to migrate more than otariids, as they generally live in higher latitudes, where the environment (e.g., the ice cover) is more variable (Bowen and Siniff, 1999). For otariids that live in tropical areas with a more constant environment, habitat regulates the growth of the standing colonies and conditions of dispersal. Otariids also have longer lactations and rearing periods than seals. Their breeding behavior and requirements in terms of habitat are also different, allowing them to stay longer in the breeding colonies. Periodic events that lead to drastic changes in food availability or other environmental limitations, such as the El Niño southern oscillation (Trillmich and Ono, 1981), also favor dispersal. Among otariids, only the northern fur seal has a well-studied and distinctive migration.

Both elephant seal species and the hooded and the harp seals are good examples of migratory seals. Northern and southern elephant seals spend between 8 and 10 months at sea each year, with long-distance migrations between breeding and molting sites. Both species have two long migration trips between postbreeding and postmolting areas. The northern elephant seal migration, of between 18,000 and 21,000 km, is the longest reported for any mammal.

Harp and hooded seals have interannually variable distribution patterns, dependent on the time of the year, the geographic location, and the density of individuals in the breeding colonies. Harp seals live in colonies in the subarctic pack ice, where breeding takes place. The largest population of this species is in Newfoundland. Individuals from this population start their southward migration in late September along the coast of the Baffin Islands and go eastward through the Hudson Straits, reaching Labrador between October and December. There are variations in migration timing and patterns between age classes.

Gray seals from numerous colonies in the British Isles and gray seal pups disperse widely during the first year. Adult seals show high variability in their movements along the coasts of Scotland, especially in postbreeding periods and after the molt, from March to May. Although long-distance travel by adults occurs, short travel at close range from particular haul outs is more common. Juveniles tend to spend longer periods at sea. (Hammond *et al.*, 1993).

C. Sirenians

The best-studied movements by sirenians are of manatees in waters of Florida. Water temperature is a major determinant of seasonal movements of Florida manatees, and dispersal is higher in warmer months. In winter, manatees tend to aggregate in areas of warmer waters, such as natural freshwater springs or the outfalls of power plants (Reynolds and Odell, 1991).

D. Polar Bear

Seasonal movements in polar bears have been reported in all their distribution range. Long-range movements also occur and are mostly related to the ice cover and extent. PREDATION on seal pups also influences movements, and bears disperse more during pinniped pupping seasons. In summer, when ice melts in many areas, bears move to land, where they remain for a few months, before leaving in November–December. Pregnant females stay longer on land.

IV. Study of Marine Mammal Distribution

The study of distribution depends on each species' habitat and its abundance, so that scale is a significant factor. Distribution changes have to be interpreted in space and time, and different methods are to be used according to species range and density. The distribution of a species occupying an extensive ocean area is best studied by air- or shipboard surveys following systematically placed transects. In surveys, visual and/or acoustic data on species are collected according to predetermined protocols. Oceanographic variates and data on position of individuals can be incorporated in spatial modeling (Hedley *et al.,* 1999). Results of this modeling on repeated surveys can be compared to study seasonal patterns and changes over time (Forney and Barlow, 1998). In species that live in fragmented habitats and that are not abundant, knowledge of the location of animal aggregations is essential. In these cases, the best possible information is obtained from telemetry studies using high-frequency radio tags, satellite-linked radio tags, and geolocation time-depth recorders (Bowen and Siniff, 1999). The use of telemetry devices is also essential in understanding seasonal movements and patterns. The life of the batteries and permanence of the tags in the animals are critical to the duration of the studies. The study of habitat, an integral part of distribution, changes with the species life time because of the above-mentioned factors. It is difficult to monitor a cohort of animals, ideally tagged since birth, because of the long average life span of marine mammals. In practice, the general distribution patterns of a marine mammal population are the sum of the individual specific movements over space and time. Monitoring just a few animals over a restricted time duration (e.g., that of a telemetry device battery) produces partial information on the overall patterns and may show a high variability between individuals. Therefore, inferences at the population level must be made cautiously to avoid biased perceptions of the species distribution.

See Also the Following Articles

Biogeography ▪ Habitat Pressures ▪ Migration and Movement Patterns ▪ Ocean Environment

References

Bowen, W. D., and Siniff, D. B. (1999). Distribution, population biology, and feeding ecology of marine mammals. *In* "Biology of Marine Mammals" (J. E. Reynolds and S. A. Rommel, eds.), pp. 423–484. Smithsonian Institution Press, Washington.

Forney, K. A., and Barlow, J. (1998). Seasonal patterns in the abundance and distribution of California cetaceans, 1991–1992. *Mar Mamm Sci.* **14,** 460–489.

Gaskin, D. E. (1982). "The Ecology of Whales and Dolphins." Heineman Educational Books, London.

Hammond, P. S., McConnell, B. J., and Fedak, M. A. (1993). Grey seals off the east coast of Britain: Distribution and movements at sea. *Symp. Zool. Soc. (Lond.)* **66,** 211–224.

Hedley, S. L., Buckland, S. T., and Borchers, D. L. (1999). Spatial modelling from line transect data. *J. Cetacean Res. Manage.* **1,** 255–264.

Reilly, S. B. (1990). Seasonal changes in habitat and distribution differences among dolphins in the eastern tropical Pacific. *Mar. Ecol. Progr. Ser.* **66,** 1–11.

Reynolds, J. E., III., and Odell, D. K. (1991). "Manatees and Dugongs." Facts on File, New York.

Riedman, M. L. (1990). The Pinnipeds: Seals, Sea Lions, and Walruses. Univ. California Press, Berkeley, CA.

Riedman, M. L., and Estes, J. A. (1988). A review of the history, distribution and foraging ecology of sea otters. *In* "The Community Ecology of Sea Otters" (G. R. Van Blaricom and J. A. Estes, eds.), pp. 4–21. Springer-Verlag, Heidelberg.

Scott, M. D., Wells, R. S., and Irvine, A. B. (1990). A long-term study of bottlenose dolphins in the west coast of Florida. *In* "The Bottlenose Dolphin" (S. Leatherwood and R. R. Reeves, eds.), pp. 235–265. Academic Press, San Diego.

Trillmich, F., and Ono, K. A. (1981). "Pinnipeds and El Niño: Responses to Environmental Stress." Ecological Studies 88. Springer-Verlag, Berlin.

Wells, R. S., Rhinehart, H. L., Cunningham, P., Whaley, J., Baran, M., Koberna, C., and Costa, D. P. (1999). Long distance offshore movements of bottlenose dolphins. *Mar. Mamm. Sci.* **15,** 1098–1114.

Wiig, O., Born, E., and Garner, E. W. (1995). "Polar Bears: Proceedings of the Eleventh Working Meeting of the IUCN/SSC Polar Bear Specialist Group." IUCN, Cambridge, UK.

Diving Behavior

BRENT S. STEWART
*Hubbs–SeaWorld Research Institute,
San Diego, California*

Except for the polar bear (*Ursus maritimus*), all marine mammals feed exclusively in aquatic environments, and mostly in the world's oceans. The depths at which they hunt for and capture prey and the time spent submerged vary among pinnipeds, cetaceans, sea otters, and sirenians as a function of physical and physiological adaptations among these taxa, environmental conditions (e.g., coastal or pelagic, tropical or polar, season), and body size, age, and health of individuals. All are ultimately tied to the sea surface to periodically breathe, yet natural selection has operated to minimize the time needed there and to maximize the amount of time that may be spent submerged hunting and capturing prey. What has become known in recent years is that these animals spend substantial parts of their lives moving within the water column to relatively great depths and some over vast geographic areas in search of food. Among the amphibious pinnipeds, these aquatic foraging bouts may extend, with minor interruptions, for several weeks to several months, punctuated by periods of several days to weeks on land or ice when no feeding occurs when these animals rest, molt, or breed. For the less amphibious sea otters (*Enhydra lutris*), diving and foraging periods may be separated by periods spent sleeping or resting at the sea surface rather than on land. Among the wholly aquatic cetaceans and sirenians, foraging bouts may last several hours or perhaps days, interrupted by periodic resting periods at the sea surface. Individuals of some species, particularly sperm whales and many mysticete cetaceans, may fast during migrations or in particular breeding areas. Although the diving performance and the patterning of individual dives or sequences of dives vary among species, what has become apparent for all marine mammals is that little time is spent at the surface between successive dives to exchange gases (i.e., unload carbon dioxide from tissue and blood and and restore tissue oxygen stores). This allows for sustained, repetitive diving and hunting, and is made possible by physiological adaptions for conserving heat and oxygen and by anatomical adaptations that promote effective movement in the aquatic environment (e.g., reducing drag through streamlining and efficient propulsion mechanisms).

I. Methods of Studying Diving Behavior

The simplest method for studying the diving behaviors of marine mammals is direct observation of the timing and location of appearances of individuals at the sea surface, the number of breaths taken there, and the duration of the animal's disappearance under water before reappearing. With some assumptions and strong inference, much can be deduced about what animals are doing while hidden beneath the ocean surface. Indeed, most early knowledge of diving, feeding, traveling, and migratory behaviors was based on such interpretations.

Other techniques for documenting diving behavior have used radio transmitter and telemetry instruments, operating at different radio frequencies. Sonic transmitters, operating at relatively low frequencies or wavelengths, allow the tracking of animals when they are submerged by placing a microphone (hydrophone) beneath the sea surface to listen for and orient to these signals (Fig. 1). Higher frequencies are used for in-air detection and tracking but generally yield less detailed observations, mostly when an animal reached the surface and how long it spent there. Durations are inferred from periods of radio silence, as transmissions that occur when the animal is submerged will generally be reflected downward from the sea surface and so not be capable of detection in air. When vocalizing underwater, some marine mammals may also be tracked with hydrophones to detect and localize those sounds. All of these techniques require constant tracking and observers must be within a few hundred meters (surface observers) or kilometers (observers in aircraft), as the signals attenuate quickly.

During the past several decades, and in particular during the 1990s, an enormous amount of information has been added to those simple observations due to technological developments and their application to free-ranging marine mammals. For example, in the late 1960s and early 1970s, an encapsulated mechanical photographic device was used in the Antarctic to study the diving patterns of Weddell seals (*Leptonychotes weddellii*). That instrument provided a continuous trace on photographic film of the depth of the seal versus time. The spooled film was pulled at a known rate past a small radioactive particle, which rested on a pressure-sensitive arm. Thus a two-dimensional record was made on the film of depth versus time. From these records came the

Figure 1 *A sonic tag glued to the pelage of an adult male Weddell seal for monitoring its underwater movements in McMurdo Sound, Antarctica. Photo by B. S. Stewart.*

first long-term (about 7 days continuous, based on film capacity) data on the vertical movements of marine mammals in the open ocean. Those instruments, called time-depth recorders (TDRs), were later deployed on a number of species of fur seals and some sea lions to study the effects of variation in body size and environment on the foraging patterns of lactating females. However, because the instruments were rather large and because they were attached with harnesses, they likely had some influence on the recorded durations of dives because of the effects of drag on swimming that they imposed, particularly for fur seals. Other simple instruments used capillary tubes with pressure sensors attached to record the maximum depth of a single dive or the maximum depth achieved during a period of diving.

Mechanical instruments were replaced in the late 1980s with much smaller electronic instruments, armored to keep seawater out under extreme hydrostatic pressures. These instruments could collect and store substantially more data on depth and duration of dives and also had less impact on behavior. Indeed, today most of these instruments weigh less than 50 g and can be glued (Fig. 2) to the hair or fur of pinnipeds for long-term (up to a year) monitoring, attached to the dorsal fin of small cetaceans (Fig. 3), attached to the skin surface with subdermal anchors (Fig. 4) or deeply embedded into the blubber of large whales, or attached with suction cups to the skin of cetaceans for shorter term (up to several days) study. Because these instruments may now also collect data other than just water depth as a function of time (e.g., swim speed, ambient light level, compass bearing, seawater temperature, salinity), they are called time-data recorders. These instruments are generally controlled by small microprocessors that can be pro-

Figure 2 *A satellite-linked radio transmitter (PTT) attached to the head of a southern elephant seal at Marion Island. Photo by B. S. Stewart.*

grammed to record measures of various parameters at particular intervals that are then stored in electronic memory for several months or more. Thus, detailed records (e.g., at 1-sec intervals) of a marine mammal's position in the water column, in addition to other environmental and behavioral data, can be collected continuously for months or more.

Even more recently, technological developments and improvements have involved remote sensing of diving patterns and geographic movements of marine mammals using radio transmitters that communicate with earth-orbiting satellites, most notably the two polar orbiting satellites of the ARGOS

Figure 3 *A satellite-linked radio transmitter (PTT) attached to the dorsal fin of a short-beaked common dolphin. Photo by B. S. Stewart.*

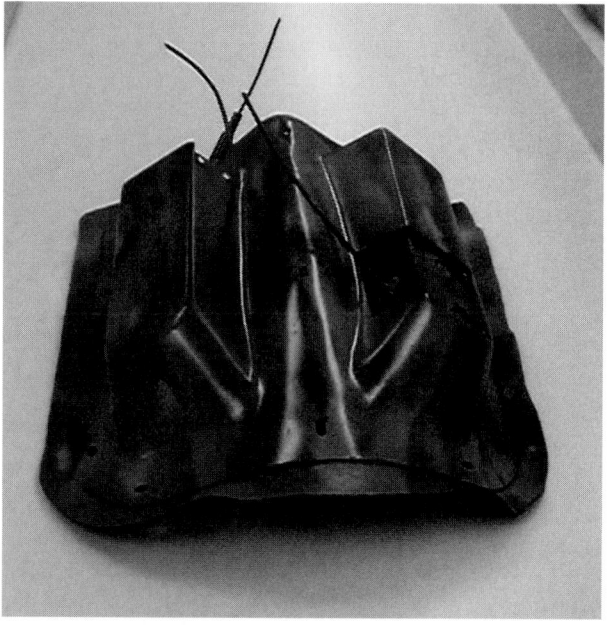

Figure 4 *A saddle package with a satellite-linked data recorder and transmitters custom built for attachment to the skin surface of a California gray whale. Photo by B. S. Stewart.*

Data Collection and Location (DCLS) system. These transmitters are known as platform transmitter terminals (PTTs). They allow animals to be located several times each day. They also allow small amounts of behavioral and environmental data to be transmitted through the DCLS. Further continuing improvement and miniaturization of film and digital video equipment are allowing the underwater diving, social, and hunting behaviors of marine mammals to be visually documented.

Most of what is now known and summarized below on the diving behaviors of marine mammals is based on two-dimensional (depth versus time) data from electronic TDRs, which are occasionally supplemented by geographic locations of the animals at the sea surface. Almost nothing is known of the movements of these animals in three-dimensional ocean space beneath the sea surface. Nevertheless, the seductiveness of representations of a single spatial vector (depth) versus time as a trace in a two-dimensional, linear spatial format led some researchers to infer the geographical form of dives in three-dimensional space. Moreover, some researchers further assigned physiological and behavioral function to those inferred spatial forms. However, the validity of such inferences and conclusions of function is untested, although they are interesting hypotheses for further rigorous inspection.

A substantial amount of information was collected on diving patterns of a number of pinniped species in the 1990s compared to relatively little progress in the study of cetacean diving patterns. The primary reason for the difference in quantity and quality of data between these taxa is principally due to the greater difficulty of keeping instruments attached to cetaceans compared to the long-term attachment of instruments, up to one year, to pinnipeds by gluing them to their hair.

Regardless, the dive patterns of virtually all species were limited to particular times of the year and even to particular classes of individuals (e.g., lactating female pinnipeds). Nearly year-round monitoring of northern elephant seals (*Mirounga angustirostris*) has been the exception. Consequently, any discussion of diving patterns is conditioned on these important constraints. Moreover, whether hunting or feeding occurs whenever animals are submerged and diving has not been confirmed. The incorporation of additional environmental and physiological sensors to TDRs and PTTs will undoubtedly help refine studies of diving patterns to more rigorously evaluate spatial form and function of subsurface movements and to enhance the summaries of dive patterns presented here.

II. Pinnipeds
A. Otariids

California sea lion females (*Zalophus californianus*) dive mostly to depths of around 75 m for about 4 min during summer and then deeper and longer the rest of the year (maximum depth of 536 m and longest dive of 12 min). When at sea for several days at a time and up to 1 to 2 weeks at some seasons, California sea lions dive virtually continually and rest at the surface for only about 3% of the time.

Juvenile Steller (northern) sea lions (*Eumetopias jubatus*) dive to average depths of 21 m (maximum 200 m). Most dives last less than 1 min. They are generally shallower at night and deeper in spring and summer than in winter. Adult females dive deeper than juveniles. Dives are deeper in winter than in autumn.

Southern sea lions (*Otaria flavescens*) dive mostly at night, apparently to the sea bed, where they hunt at depths down to 250 m. While at sea near the Falkland Islands, these sea lions dive virtually continually. Near Patagonia, over half of the time that lactating females are at sea they are diving. Their dives are mostly to depths of 19 to 62 m (maximum of 97 to 175 m) and for 2 to 3 min (maximum of 4.4 to 7.7 min). Diving is continuous during these bouts, and time spent at the surface between successive dives is brief, around 1 min.

Lactating New Zealand (Hooker's) sea lions (*Phocarctos hookeri*) also dive almost continually when at sea, averaging about 7.5 dives per hour, varying little with time of day. Dive depths average about 123 m (maximum of 474 m) and last between 4 and 6 min (maximum of 11.3 min). Most dives are evidently to the sea bed to forage on demersal and epibenthic fish, invertebrates, and cephalopods.

A few lactating Australian sea lion (*Neophoca cinerea*) females were reported to repeatedly forage on the sea bottom (ca. 150 m deep) on the continental shelf of South Australia within 30 km of the coast.

Northern fur seals (*Callorhinus ursinus*) may be at sea continuously for several months or more from autumn through spring, but their diving behavior has not been studied then. Most data come from lactating female fur seals that are foraging near ROOKERIES in the Bering Sea in summer. Then they forage in bouts that mostly occur at night. Seals mostly make shallow dives to depths of 11 to 13 m, lasting around 1 to 1.5 min. These dives tend to be at night when seals are in pelagic habitats.

Depths and durations of dives of Galapagos fur seals (*Arctocephalus galapagoensis*) increase as they get older. Six-month-old seals dive to depths of around 6 m for up to 50 sec and dives occur at all hours. One-year-old seals reach depths of 47 m and durations average 2.5 min. Most of those dives occur at night. When 18 months old, seals are at sea mostly at night, diving continually for periods lasting around 3 min and reaching depths of 61 m.

Lactating Juan Fernandez fur seal (*Arctocephalus philippii*) females dive mostly at night to depths of 50 to 90 m, although most dives are shallower than 10 m. They last, on average, 1.7 to 2.0 min (longest 3.46 min).

Lactating female New Zealand fur seals (*Arctocephalus forsteri*) dive as deep as 274 m, and their longest dives have been measured at around 11 min. Median dive depths are around 5 to 10 m. They occur in bouts with the longest bouts at night. The deepest dives occur around dawn and dusk. Dives are shallowest (30 m) and shortest (1.4 min) in summer and get progressively deeper and longer through autumn (54 m, 2.4 min) and winter (74 m, 2.9 min).

Most dives of female Australian fur seals (*Arctocephalus pusillus doriferus*) are to the sea bed on the continental shelf at depths of 65 to 85 m. The median depth of one foraging male fur seal was 14 m and the median duration of dives was 2.5 min. The deepest and longest dives of that male were 102 m and 6.8 min, respectively, and the seal spent about one-third of its time at sea diving and foraging, with little variation in activity with time of day.

Lactating female subantarctic fur seals (*Arctocephalus tropicalis*) at Amsterdam Island dive predominantly at night. These foraging dives get progressively deeper and longer from summer (10 to 20 m and about 1 min long) through winter (20 to 50 m and about 1.5 min long). The deepest dive recorded is 208 m and the longest is 6.5 min.

Lactating female Antarctic fur seals (*Arctocephalus gazella*) dive mostly at night when they are at sea for periods of 3 to 8 days at a time in summer. These nighttime dives are shallower (about 30 m or less) than dives made during the day (40 to 75 m), closely matching the vertical distribution of krill. Maximum depths and durations of dives have been measured at 82 to 181 m and 2.8 to 10 min, respectively, for individual females. Seals apparently adjust their diving behavior to maximize the proportion of time that they spend at depth. Young pups dive mostly to depths of about 14 m, depending on their body size, for mean durations of 20 sec. Their diving abilities continue to develop during their first couple of months of life, and by the time they are weaned at around 4 months of age, they are able to dive to the same depths and for about the same amount of time as adult females.

B. Odobenids

The diving patterns of the walrus (*Odobenus rosmarus*) are not well studied. It is known, however, that its dives may last 20 min or more, although most may be less than 10 min and exceed 100 m. The longest dive yet recorded lasted about 25 min and the deepest was to 133 m. Most dives are likely shallower than about 80 m, as its benthic prey of mollusks are generally found in relatively shallow coastal or continental shelf habitats. Near northeast Greenland, walruses may be submerged about 81% of the time when they are at sea and are presumably diving and foraging most of that time.

C. Phocids

Phocid seals generally are at sea continually for weeks to months and appear to dive, and perhaps forage, virtually constantly.

Elephant seals are perhaps the best studied of marine mammal divers. The dives of weaned southern elephant seal pups (*Mirounga leonina*) are to about 100 m for about 6 min and they dive virtually continuously when at sea for several months. Heavier pups dive deeper (to ca. 130 m) and longer (ca. 7 min) than smaller pups (88 m and 5 min). Dives of juvenile southern elephant seals last around 15.5 min (maximum of 39 min) to depths averaging 416 m (deepest 1270 m) and they spend about 90% of their several months at sea diving. Intervals between dives are brief, rarely lasting more than 2 min. Adult southern elephant seal dives average 400 to 600 m and 19 to 33 min (deepest 1444 m and 113 min) and also occur continuously while they are at sea for up to 7 to 8 months.

Northern elephant seals also dive continually when at sea for several months or more with only brief periods at the surface (1 to 3 min) between dives. Dives of adults are to modal depths of 350 to 400 m and 700 to 800 m (maximum of 1567 m) and average 20 to 30 min (maximum of 77 min). Depths and durations of dives differ between adult males and females depending on season and geographic location. Generally, these seals feed on pelagic fish and squid, although some seals may also dive to and feed near the sea floor near the coastlines of continents and islands.

Dives of Hawaiian monk seals (*Monachus schauinslandi*) are between 3 and 6 min long and mostly shallow, between 10 and 40 m deep, where the seals forage near the sea bed on epibenthic fish, cephalopods, and other invertebrates. Adults may occasionally dive to greater depths of up to 550 m when foraging outside of the shallow atoll lagoons of the northwestern Hawaiian Islands.

Weddell seals (*Leptonychotes weddellii*) may forage for much, if not most, of the year beneath the unbroken fast ice and the more open pelagic pack ice zones of the Antarctic. Diving and foraging occur in bouts of about 40 to 50 consecutive dives over a several hour period, usually to depths of 50 to 500 m. Dives of young pups are relatively shallow and brief but get progressively deeper and longer as pups age. They plateau when the pups are weaned when about 6 to 8 weeks old. The dives of 1 year olds are somewhat shallower, to around 118 m, compared to adult females (163 m). The deepest recorded is about 750 m and the longest over 73 minutes. Dives are shallower (350 to 450 m) in spring (October to December) than in summer (January; 50 to 200 m), evidently reflecting a shift in preferred hunting depths.

Among the Antarctic pack ice seals, Ross seals (*Ommatophoca rossii*) are also relatively deep divers. One female that was monitored near the Antarctic Peninsula in summer dove exclusively at night, mostly to depths of 110 m (maximum of 212 m) and for about 6.4 min (longest 9.8 min). Diving was continual while the seal was in the water with about 1 min between dives. The deepest dives (175 to 200 m) occurred near twilight and the shallowest (ca. 75 to 100 m) at midnight.

In summer, crabeater seals (*Lobodon carcinophaga*) dive primarily at night and haul out on pack ice during the day, although some diving bouts may last up to 44 hr without interruption. Most dives are 4 to 5 min long to depths of 20 to 30 m, with maximum depths and durations of 430 m and 11 min, respectively. Dives near twilight are deepest and those near midnight shallowest.

Baikal seals (*Pusa sibirica*) apparently dive continually from September through May, when the freshwater Lake Baikal is frozen over, and haul out only infrequently then. Most dives are 10 to 50 m deep in the middle of Lake Baikal where the water depth is around 1000 to 1600 m. Occasionally seals descend to more than 300 m. Dives last between 2 and 6 min but some have been measured at more than 40 min.

Dives of another closely related freshwater seal, the Saimaa seal (*Pusa hispida saimanensis*) of Lake Saimaa in eastern Finland, last about 6 min in spring and increase to about 10 to 11 min by autumn. In summer and autumn, long series of sequential dives lasting more than 10 min each may occur over 3 hr or more. The longest dive recorded is about 23 min, when the seal may actually have been resting on the bottom rather than feeding.

Modal dive depths for breeding age, male ringed seals (*P. hispida hispida*) are 10 to 45 m and for subadult males and postpartum females 100 to 145 m. Durations of dives for adult males are around 4 min and around 7.5 min for adult females.

Harbor seal (*Phoca vitulina*) diving behaviors have been studied in a number of areas throughout their range in the North Pacific and North Atlantic Oceans. Dives in the Wadden Sea (northeast Atlantic) average from 1 to 3 min (maximum of 31 min) with little variation between night and daytime behavior. When in the water, about 85% of their time is spent diving. In the western Atlantic, foraging dives of adult males are mostly deeper than 20 m but are shallower during the mating period, when they are defending aquatic territories or searching for females to mate with instead of foraging. Dives of lactating females are to 12 to 40 m and occur in bouts lasting several hours, mostly during the day. In southern California, dives are as deep as 446 m. Most, however, are to modal depths of 10, 70, or 100 m with an occasional mode at around 280.

Bearded seal (*Erignathus barbatus*) adult females near the coast of Spitzbergen, Norway, dive mostly at night to depths of around 20 m (deepest at 288 m) and for 2 to 4 min (longest 19 min). Nursing pups may dive to around 10 m (maximum of 84 m) for about 1 min (maximum of 5.5 min). Pups spend about 40% of the time that they are in the water diving. Depths and durations of dives increase as the pups age.

Most dives of lactating female gray seals (*Halichoerus grypus*) are to the sea floor and last about 1.5 to 3 min (maximum of 9 min). Most foraging dives of juvenile gray seals in the Baltic Sea are to depths of 20 to 40 m.

Lactating female harp seals (*Pagophilus groenlandicus*) dive about 40 to 50% of the time that they are at sea. Dives average about 3 min (maximum of 13 min) to depths of up to 90 m.

Hooded seals (*Cystophora cristata*) repeatedly dive to depths of 1000 m or more and for 52 min or longer. Most feeding dives appear to be to depths of 100 to 600 m.

III. Cetaceans

A. Odontocetes

Limited data for odontocete cetaceans so far indicate that short-beaked common dolphins (*Delphinus delphis*) may forage at depths of up to 260 m for 8 min or more, although most dives are around 90 m deep, last about 3 min, and are mostly at night. Pantropical spotted dolphins (*Stenella attenuata*) dive to at least 170 m; most of their dives are to 50 to 100 m for 2 to 4 min and most feeding appears to occur at night. Atlantic bottlenose dolphins (*Tursiops truncatus*) near Grand Bahama Island off southeastern Florida often dive to the ocean bottom (7–13 m depth) and burrow into the sediment ("crater-feeding") to catch fish dwelling or hiding there. Long-finned pilot whales (*Globicephalas melas*) dive to over 500 to 600 m for up to 16 min. Northern bottlenose whales (*Hyperoodon ampullatus*) regularly dive to the sea bed at depths of 800 to 1500 m for more than 30 min per dive and occasionally for 2 hr.

Harbor porpoises (*Phocoena phocoena*) near Japan have dived almost continuously when observed for short periods. Maximum dive depths are around 70 to 100 m, although about 70% of dives may be less than 20 m. These porpoises descend to and ascend from depth at greater rates when diving deep than when the dives are shallow. In waters near Denmark, porpoises dive as deeply as 84 m and for up to 7 min from spring through late autumn.

Female white whales (*Delphinapterus leucas*) dive more often between 2300 and 0500 hr than during the day, although males may dive at the same rate at all hours. Dive rates and time spent at the surface decline whereas dives deepen and lengthen from early through late autumn. Most dives are deep (400 to 700 m), with the deepest recorded at 872 m, and last about 13 min on average (maximum of 23 minutes). Dive duration increases with body size.

Narwhals (*Monodon monoceros*) regularly dive to more than 500 m and occasionally deeper than 1000 m, but most dives are to depths of 8 to 52 m and last less than 5 min, although as long as 20 min on occasion. The rate of diving varies between adult males and females. When diving shallow, narwhals descend and ascend relatively slowly (<0.05 m/sec) compared with deeper, longer dives (1 to 2 m/sec) where substantially more time is spent at maximum depth.

Killer whales (*Orcinus orca*) along the northern coast of North Island, New Zealand, dive to the ocean bottom (ca. 12 m depth or less) after stingrays and perhaps burrow into the sediment to catch them.

Sperm whales (*Physeter macrocephalus*) dive to depths of up to 2000 m for 60 min or more. Near Kaikoura, New Zealand, the average duration of dives is about 41 min with about 9 min spent at the surface between dives. Both durations and surface intervals are longer in summer than in winter. Males spend little time at the surface compared to females. Average dive durations have been measured at 36 min near Sri Lanka and about 55 min near the Azores. Sperm whales in the Caribbean were reported to make dives averaging 22 to 32 min during the day (longest 79 min) and 32 to 39 min at night (longest 63 min).

B. Mysticetes

As yet there is no evidence for a taxonomic relationship between body size and maximum dive depths for mysticete cetaceans, although preliminary correlations have been reported between maximum dive durations and body size.

While in shallow coastal lagoons during the spring breeding season, gray whales (*Eschrichtius robustus*) dive for about 1 to 5 min (maximum of 28 min) to average depths of 4 to 10 m (maximum recorded of 20.7 m). It is not clear what the function of these dives may be other than perhaps subsurface resting, as breeding whales are presumed to fast. In the Bering Sea in summer, when whales are feeding, dives average about 3 min.

Fin whales (*Balaenoptera physalus*) in the Ligurian Sea dive repeatedly to depths around 180 m (maximum 474 m) for around 10 min (longest 20 min) while they prey on deep-dwelling krill. Elsewhere, fin whale dives have been reported to last about 5 min near Iceland and about 3 min in the North Atlantic and near Long Island, New York, in summer.

When chased by commercial whalers, dives of blue whales (*Balaenoptera musculus*) lasted up to 50 min. Blue whales off central and southern California otherwise spend about 94% of their time submerged. Dives lasting longer than 1 min are 4.2 to 7.2 min, on average (longest 18 min), and to around 105 m (deepest 150 to 200 m). Dives of pygmy blue whales (*B. m. brevicauda*) have been measured to average 9.9 min (longest 26.9 min).

Humpback whales (*Megaptera novaeangliae*) in Frederick Sound, Alaska, make rather brief (most less than 3 min) and shallow (60 m or less) dives, although some may exceed 120 m on occasion.

When on summer feeding grounds in the Beaufort sea, dives of bowhead whales (*Balaena mysticetus*) last 3.4 to 12.1 min and some are to the relatively shallow sea bed. Dives of calves are very short compared to adults and they also spend more time at the surface between dives. Most dives of juveniles last about 1 min (longest 52 min) to depths of around 20 min. Longer dives, up to 80 min, have been observed for bowhead whales that were harpooned and being chased by whalers. Dives lasting longer than 1 min ("sounding dives") average between 7 and 14 min. Dives made while whales are migrating through heavy pack ice are deeper and longer than those made while in open water. Lactating females dive less often and for shorter periods than other adult whales.

Dives of North Atlantic right whales (*Eubalaena glacialis*) near Cape Cod, Massachussetts, last around 2.1 min.

IV. Other Marine Mammals

Manatees (*Trichechus manatus*, *T. inunguis*, and *T. senegalensis*) feed on floating and submerged vegetation in shallow nearshore habitats so it is unlikely that their dives often exceed 25 to 30 m. Direct observations of free-ranging animals have shown that most dives are less than 5 min, although a few have been timed at more than 20 min. These longer dives may have periods of rest at the bottom rather than feeding activity. Dugongs (*Dugong dugon*) also feed on submerged vegetation, most often in coastal and offshore seagrass beds either on the sea bottom to depths of 20 m or in surface canopies. The longest foraging dives observed are around 6 min, but most have been reported to last only between 2 and 4 min.

Sea otters dive and forage mostly in shallow nearshore waters. Dives may be in bouts lasting several hours during the day and night, interrupted by periods at the surface to groom, process food, or rest. Juvenile males often dive in deeper water, for longer periods, and further from shore than juvenile and adult females. Details on diving sequences, depths, durations, and other parameters are just beginning to emerge from the recent deployment of TDRs in southeast Alaska and in the Gulf of Alaska.

Polar bears (*Ursus maritimus*) are powerful swimmers and probably make some dives while moving among ice floes, the fast-ice edge, or coastlines, but nothing is known of the details of such diving performance. They prey mostly on ringed seals and whale carcasses on the surface of the ice or along shorelines and also on white whales and narwhals that they may attack and kill at the sea surface and then drag out of the water to consume.

See Also the Following Articles

Feeding Strategies and Tactics ■ Swimming ■ Telemetry

References

Baker, J. D., and Donohue, M. J. (1999). Ontogeny of swimming and diving in northern fur seal (*Callorhinus ursinus*) pups. *Can. J. Zool.* **78**, 100–109.

Bengtson, J. L., and Stewart, B. S. (1992). Diving and haulout behavior of crabeater seals in the Weddell Sea, Antarctic during March 1986. *Polar Biol.* **12**, 635–644.

Bengtson, J. L., and Stewart, B. S. (1997). Diving patterns of a Ross seal (*Ommatophoca rossii*) near the eastern coast of the Antarctic Peninsula. *Polar Biol.* **18**, 214–218.

Bowen, W. D., Boness, D. J., and Iverson, S. J. (1999). Diving behaviour of lactating harbour seals and their pups during maternal foraging trips. *Can. J. Zool.* **77**, 978–988.

Brillinger, D. R. and Stewart, B. S. (1997). Elephant seal movements: Dive types and their sequences. *In* "Modelling Longitudinal and Spatially Correlated Data: Methods, Applications and Future Directions," pp. 275–288. Springer, Berlin.

Campagna, C., Fedak, M. A., and McConnell, B. J. (1999). Postbreeding distribution and diving behavior of adult male southern elephant seals from Patagonia. *J. Mamm.* **80**, 1341–1352.

Georges, J. Y., Tremblay, Y., and Guinet, C. (2000). Seasonal diving behavior in lactating subantarctic fur seals on Amsterdam Island. *Polar Biol.* **23**, 59–69.

Hooker, S. K., and Baird, R. W. (1999). Deep-diving behaviour of the northern bottlenose whale, *Hyperoodon ampullatus* (Cetacea: Ziphiidae). *Proc. R. Soc. Lond.* (B) **266**, 671–676.

Horning, M., and Trillmich, F. (1997). Ontogeny of diving behaviour in the Galapagos fur seal. *Behaviour* **134**, 1211–1257.

Jaquet, N., Dawson, S., and Slooten, E. (2000). Seasonal distribution and diving behaviour of male sperm whales off Kaikoura: Foraging implications. *Can. J. Zool.* **78**, 407–419.

Kraft, B. A., Lydersen, C., Kovacs, K. M., Gjertz, I., and Haug, T. (2000). Diving behaviour of lactating bearded seals (*Erignathus barbatus*) in the Svalbard Area. *Can. J. Zool.* **78**, 1408–1418.

Lagerquist, B. A., Stafford, K. M., and Mate, B. R. (2000). Dive characteristics of satellite-monitored blue whales (*Balaenoptera musculus*) off the central California coast. *Mar. Mamm. Sci.* **16**, 375–391.

Martin, A. R., and Smith, T. G. (1999). Strategy and capability of wild belugas, *Delphinapterus leucas*, during deep, benthic diving. *Can. J. Zool.* **77**, 1783–1793.

Mattlin, R. H., Gales, N. J., and Costa, D. P. (1998). Seasonal dive behaviour of lactating New Zealand fur seals (*Arctocephalus forsteri*). *Can. J. Zool.* **76**, 350–360.

Ralls, K., Hatfield, B. B. and Siniff, D. B. (1995). Foraging patterns of California sea otters as indicated by telemetry. *Can. J. Zool.* **73**, 523–531.

Reeves, R. R., Stewart, B. S., and Leatherwood, J. S. (1992). "The Sierra Club Handbook of Seals and Sirenians." Sierra Club Books, San Francisco.

Sjoberg, M., and Ball, J. P. (2000). Grey seal, *Halichoerus grypus*, habitat selection and haulout sites in the Baltic Sea: Bathymetry or central-place foraging. *Can. J. Zool.* **78**, 1661–1667.

Stewart, B. S., and DeLong, R. L. (1995). Double migrations of the northern elephant seal, *Mirounga angustirostris*. *J. Mammal.* **76**, 196–205.

Stewart, B. S., Petrov, E. A., Baranov, E. A., Timonin, A., and Ivanov, M. (1996). Seasonal movements and dive patterns of juvenile Baikal seals, *Phoca sibirica*. *Mar. Mamm. Sci.* **12**, 528–542.

Diving Physiology

GERALD L. KOOYMAN
*Scripps Institution of Oceanography,
La Jolla, California*

Ever since humankind has lived by and gone down to the sea, we have been awestruck by the creatures that make it their home. First we feared them, later we ate them, and now we try to emulate them with humble attempts to set "world" diving records. At present the record for a descent-assisted dive at 133 m during a breath hold is a little over 2 min. Most marine mammals exceed that depth within the first few months of life. Premier divers such as elephant seals (*Mirounga* spp.) and sperm whales (*Physeter macrocephalus*) will occasionally dive to depths beyond a kilometer (Table I). The spectacular abilities of marine birds and mammals to dive deep and for long periods of time are a source of interest and curiosity for marine scientists and amateurs alike.

When marine mammals descend below the sea's surface they leave behind the thin skin of the earth's atmosphere with one of its essential ingredients to all vertebrate life—oxygen. They begin a journey that is incredible in diverse ways. The magnitude of incredulity varies according to the species, but for all, even the most humble of marine mammals, such as the sea otter (*Enhydra lutris*), much if not most of the experience is beyond our imagination. Unlike flying, in which our technology now enables us to fly faster, higher, and further than any bird, bat, or pterosaur ever has or did, marine mammals, particularly those that dive to great depths, explore and exploit a realm that overwhelms much of our technology and which enables us to gain only fleeting glimpses of what their environment is like. Recently we have enlisted the animals themselves to help us discover more about this cold, dark world without oxygen, where awesome hydrostatic pressures always prevail. However, "crittercams" will only give us fleeting glimpses, under very special conditions, with those few species that lend themselves to the attachment of these bulky cameras. Life in the deep blue remains a mystery. So too do the means that enable diving mammals to exploit this habitat.

TABLE I
Distribution and Quantity of Oxygen Stores, Maximum and Routine Diving Depths, and Durations for Some Marine Mammals

Species	Body mass (kg)	Total store (ml/kg)	Lung	% Blood	Muscle	Routine depth (m)	Maximum depth (m)	Routine duration (min)	Maximum duration (min)
Human	70	20	24	57	15	5	133	0.25	6
Weddell seal	400	87	5	66	29	200	700	15	93
Elephant seal	400	97	4	71	25	500	1500	25	120
California sea lion	100	40	21	45	34	40	275	2.5	10
Bottlenose dolphin	200	36	34	27	39		535		
Sperm whale	10000	77	10	58	34	500	2035	40	75

This article discusses some of what is known about adaptations to breath holding and overcoming the crushing effects of pressure. These adaptations are among the most unique among vertebrates. Even after our primordial fish ancestors overcame great obstacles to adapt to the terrestrial environment, and eventually to spread throughout all land habitats of the world, the sea continued to be a rich habitat that would bring great success to those species that exploited it. Some air-breathing vertebrates are doing just that. In fact, this has occurred several times in the history of vertebrates as they reinvaded the sea. The marine reptiles of the Mesozoic were diverse, abundant, and no doubt very capable divers. They had at least one major advantage over marine mammals: a small brain. The brains of mammals require a substantial share of the oxygen being supplied to the body, and it is an obligate need with very little reserve for those times when supply is interrupted. Within 3 min after blood flow and oxygen transport to the human brain are interrupted, there is irreversible damage. This sensitivity of large, complex brains to a grave need for oxygen makes it seem a contradiction that animals who routinely breath hold many times every day are all so smart. Proportionately in terms of brain size relative to body size, several cetacean species have some of the largest brains of mammals. Despite this "handicap," marine mammals have been an extremely successful group that are found in all the world's oceans, in extremely large numbers, and have the biomass of some species matching that of any of the formerly abundant terrestrial mammals of the world.

What is the secret of their success? Some routinely dive to depths of several hundred meters, and a few species may occasionally descend from 1 to 3 km (Table I). Although these depths may seem just a superficial range compared to the ocean limit of 11 km, with an average depth of 3.5 km, the range used by most marine mammals is in the zone of greatest oceanic life. Nevertheless, this region of cold, dark waters requires special adaptations enabling the animal to endure low temperatures and to find prey in the "dark." Marine mammal diving skill provides a dramatic contrast to human capacities. On average, we can dive to a few meters for about 30 sec. The super athletes who make a career of setting records, such as the record breath-hold dive of 133 m, require mechanical aids of weights, pulleys, and drop lines. To extend our depth beyond these few meters, humans have gone to costly extremes in mechanical devices. Most deep submersibles are usually limited to several hundred meters depth, but ALVIN, the workhorse of the scientific submersibles, can go as deep as 4500 m.

Adaptations of marine mammals to the marine environment are diverse in order for them to become successful marine predators. They involve many systems in and out of the body, ranging from external body shape to overcome the high density and viscosity of water to the sensory systems necessary to find their way and to detect prey and predator. Space will allow for discussion of only a few of the numerous adaptations necessary for a successful marine mammal. The following paragraphs discuss adaptations to hypoxia and pressure. These paragraphs address pelagic, offshore deep divers in which the adaptations are the most extreme.

I. Adaptations to Hypoxia
A. Oxygen Stores and Their Distribution

An increased total body O_2 store is considered an essential factor in the breath-hold capacity of diving mammals. The oxygen consumed by body metabolism during a breath hold is stored in three compartments: the respiratory system, the blood, and the body musculature. The theoretical maximum amount of oxygen available in each compartment is a function of several criteria. The respiratory oxygen store is dependent on lung volume and the concentration of oxygen in the lung at the start of a breath hold. The blood and muscle oxygen stores are dependent on blood volume and muscle mass and on the concentration of the oxygen-binding proteins of hemoglobin in blood and myoglobin in muscle. From the measurements of myoglobin concentration in the muscles of many species of divers, it is clear that one of the most consistent hallmarks of oxygen storage in all marine mammals that dive to depth is an elevated myoglobin concentration. This trait is more characteristic of deep divers than any changes in blood volume, hemoglobin concentration, or respiratory volumes. However, increased blood volume and hemoglobin concentration often contribute to elevated oxygen storage.

As the distributions of oxygen stores vary among species, so do the ranges of the total oxygen store (Table I). In humans the total store is 20 ml O_2kg^{-1} body mass, which is about a fifth of the nearly 100 ml O_2kg^{-1} body mass in elephant seals (*Mirounga* spp.). Using the seal as our basic model, it is noted that most of its oxygen is in blood and muscle. The large amount relative to terrestrial animals, using the human average as a standard, is a result of a blood volume 3 times, a hemoglobin concentration 1.5 times, and a myoglobin concentration approximately 10 times the human value. In seals the lung is a minor source of oxygen, as it is in most other marine mammals. It is less than 5% of the total, in part because seals exhale to 50% of their total lung capacity just before diving. Furthermore, at depth the lung is collapsed and does not exchange gas.

B. Cardiovascular Responses

The cardiovascular response to breath holding falls into at least two categories of whether the dive is extended or of routine duration for that species. Measurements of cardiovascular and metabolic responses under these circumstances are very limited for any species and most measurements are from seals. Diving mammals are arrhythmic breathers with pauses between series of breaths. The resting maintenance heart rate is probably most closely reflected in the rate during the respiratory pause or apnea. Using the heart rate during apnea as a basis of comparison for heart rates during a routine dive, heart rates during the dive are lower than the rate of a resting apneusis, and this occurs despite the fact that the mammal is swimming. When an extended dive is performed, the heart rate is even lower than that during routine diving. Because no measurements of blood flow distribution have been directly measured during dives of marine mammals, it is by extrapolation from indirect measures of other organ functions that allude to what may be occurring. During routine dives it is likely that gastric, renal, and hepatic functions are reduced a small amount, but no more than what can be compensated for by higher than normal performance during the short breathing intervals at the surface. Muscle may utilize a small part of the circulating blood oxygen, but it probably relies on its internal store of oxygen bound to myoglobin for much of the aerobic metabolic needs.

Extended dives, those that are three to five times the routine dives, are uncommon. They are most likely to occur because of some urgent need, such as a Weddell seal (*Leptonychotes weddellii*) searching for a new hole under sea ice or an elephant seal hiding at depth to escape notice from a passing pod of killer whales (*Orcinus orca*) near the surface. The cardiovascular response in these extreme cases may be a limitation of blood flow to obligate aerobic tissues, the most conspicuous of which is the brain. Having no internal store of oxygen, and a need to be at full functional capacity, a constant supply of oxygen and other metabolites provided by the blood, as well as transport of waste products of metabolism from the brain, means that constant blood flow is essential. There is a lesser need for transport of oxygen to the heart because of a reduced work load (the slower heart rate) and a small store of internal oxygen. The blood flow to muscle is reduced to a trickle as it draws from the large oxygen store within the muscle and the internal store of glycogen for the production of the high-energy compounds of adenosine triphosphate (ATP). The high concentration of myoglobin in *all* mammals that dive to depths

greater than about 100 m indicates that myoglobin is a key adaptation for diving. Blood flow would be a liability as the affinity of myoglobin for oxygen is much higher than the affinity of hemoglobin for oxygen. Consequently, any flow to muscle that had utilized much of its oxygen store would strip oxygen from the circulating blood and deprive more vital organs, such as the brain, from oxygen. A reduced blood flow to muscle also decreases cardiac output needs and, hence, the work of the heart and its oxygen consumption. Thus, the degree of muscle blood flow reduction during long and short dives is key to understanding the management of oxygen stores. Unfortunately, little is known about this crucial topic. Unlike other organs, muscle is widely distributed in the body, and the vascularity is diffuse. Consequently, it is an intractable problem which has not lent itself to study.

Muscle also has a great capacity for anaerobic metabolism and tolerance for high concentrations of the metabolic end product of lactic acid, which is stored in the form of lactate. Nevertheless, muscle must continue to function for locomotion either continuously, as a Weddell seal swims below the sea ice, or intermittently, as in an elephant seal as it drifts in the depths, but in the end must call upon muscle to provide the locomotion to return to the surface. In contrast, the splanchnic organs may shut down or greatly reduce function until the diving mammal returns to the surface.

C. Metabolic Responses

The cessation of metabolic function in the splanchnic organs reduces metabolic rate substantially, as these organs functioning at normal rates account for nearly 50% of the total resting metabolism of the animal. In addition, the heart beats more slowly and performs less work, which may also be the case for striated muscle. In the cold environment at depth, some tissues may also be cooling, which would result in an additional savings in energy consumption. The final result is to lower the overall metabolic rate to below the resting level during these short and metastable conditions.

D. Anaerobic Metabolism

Dominating the many factors that affect how long an animal may breath hold is the amount of oxygen available and its rate of utilization. Through oxygen-supported metabolic pathways, 18 times more high-energy ATP is produced from glucose than through anaerobic processes. Furthermore, carbon dioxide and water, the end products of oxygen-supported catabolism, are less polluting to the cells and circulation than those of anaerobic catabolism. Finally, nerve cells, especially within the brain, are completely dependent on aerobic metabolism. Therefore, the duration of time an animal may breath hold is most strongly affected by the availability of oxygen, with subsidiary support from anaerobic glycolysis and creatine phosphate catabolism. Although an animal may extend its dive considerably by relying on anaerobic glycolysis, the subsequent recovery is in turn extensive because of the time required to process lactic acid and restore the acid base balance of the cells and circulatory system. For routine dives that occur in sequence over many hours, aerobic metabolism is the only practical option. Oxygen-supported metabolic pathways are also the only means of producing ATP that is derived from catabolism of fat and protein.

E. Aerobic Diving Limit

The only diving mammal in which there has been a detailed correlation of the diving duration and the postdive blood lactate concentration is the Weddell seal (Fig. 1). The source of the lactate has been shown to be from muscle, in which it accumulates rapidly as muscle oxygen is depleted. After the seal surfaces, there is increased blood flow to muscle and much of the lactate is flushed into the circulation and gradually disappears over several minutes. If the seal should dive again before all of the blood lactate is processed, it will continue to decline over the course of the dive unless that dive exceeds what has been termed the aerobic diving limit (ADL). The ADL is defined as the diving duration beyond which there is a net increase in lactate production. This rise in lactate concentration first occurs primarily in muscle and eventually diffuses from the organ into the circulation where it can be measured easily. It has been proposed that this threshold be called the diving lactate threshold (DLT) to avoid confusion about the numerous ways that the ADL has been derived since the first measurements were made on Weddell seals. From those measurements it was also shown that with some reasonable accuracy the ADL could be obtained from the quotient of the O_2 store divided by the metabolic rate, the calculated ADL (cADL). Because this limit predicts basic information about the foraging behavior of diving animals, as well as clarifying physiological responses and models to breath holding, it has been calculated for many diving species. Some of the most intriguing calculations have been made for the elephant seal, a continuous diver that appears to allow no surface time

for recovery from dives exceeding the cADL. There are several possibilities that could resolve this puzzle. A recent computer simulation of oxygen store depletion provides a physiological model of aerobic diving that may clarify this problem and provide direction for further studies.

This comprehensive, numerical model uses as its data source the Weddell seal because there is extensive information on this species. The calculations in the model are based on or derived from available data on cardiac output, O_2 depletion rates of blood and muscle, blood flow patterns in various organs, and the diving metabolic rates that may occur in the Weddell seal. The model demonstrates how the matching or mismatching of oxygen transport and regional oxygen consumption can affect the ADL. This theoretical treatment of the ADL goes a long way in understanding how oxygen must be managed during a dive and, in particular, explains how 31% of the body O_2 store remains unconsumed under the most optimal cardiac output conditions. The model also shows that only 49% of the muscle oxygen provision comes from the internal store during the longest possible aerobic dive. In regard to recovery from an extended dive, the oxygen replenishment rate is much more rapid than the reconversion of lactate to glycogen.

Assuming that the model can be applied to other aquatic species, it may help explain the enigma of the serial dives in the elephant seal that exceed previous cADLs. However, the model does not take into account the influence of creatine phosphate to support the few dives that may appear to exceed the ADL, and for which some have invoked some unusual hypometabolic responses. Because the creatine phosphate concentration is 15–20 mmol kg^{-1} in mammals, this is enough to have a significant effect on the magnitude of the ADL and the production of energy without oxygen, but before measurable amounts of lactate are produced.

II. Adaptations to Pressure

Once a marine mammal descends below the surface, it not only must deal with the lack of oxygen but also with the effects of pressure. This is one of the most imposing physical variables to which vertebrates must adjust. We become especially sensitive to pressure during the most modest dive to depth because our airspaces, such as the middle ear and facial sinuses, make us acutely aware of any difference between the ambient pressure and our internal pressure. More subtle is the effect of pressure on the lung. For humans the lung is an important oxygen store, but in deep-diving mammals the lung is not an important oxygen store. Over a long period of evolution the main function of the vertebrate lung became the exchange of gases between blood and air. During the descent to depth, this function is diminished in deep-diving marine mammals. As the transfer of gases between the lung and blood slackens or ceases, the rise in nitrogen partial pressure within the lung is not matched in the blood. The lack of gas exchange also results in the avoidance of nitrogen narcosis and oxygen toxicity. Even with this adaptation there is still the pure physical effect of hydrostatic pressure on the nervous system. In terrestrial mammals, pressure causes overstimulation or uncoordinated nerve conduction and dysfunction called high-pressure nervous syndrome (HPNS). How

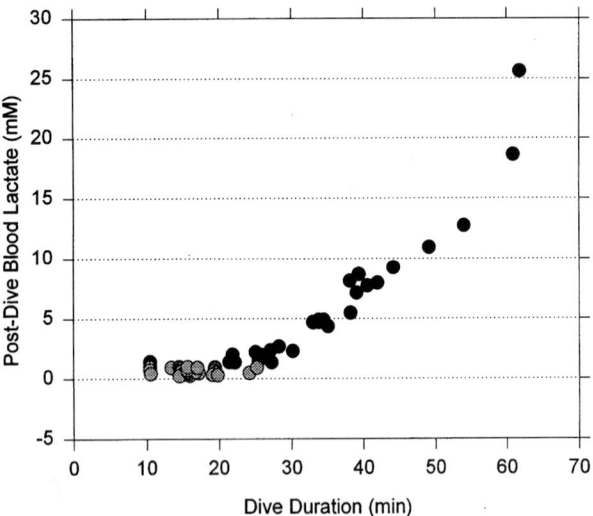

Figure 1 *Peak concentration of lactate in arterial blood after dives of different duration in adult Weddell seals. The inflection represents transitions from completely aerobic dives and is considered the aerobic diving limit or diving lactate threshold. Gray circles reflect blood values of dives in which there is no net production of lactate, and black circles are those in which there was a net production.*

do marine mammals manage to avoid these problems that are manifested in their terrestrial relatives?

The pressure within all airspaces must closely match that of the ambient pressure or the diving animal will suffer damage to the membranes and blood vessels lining the space and a breakdown in normal function. There are at least three major airspaces within most mammals that are liabilities for diving. First are the facial sinuses. Any experienced diver or airline passenger is aware that flying or diving during or soon after having a head cold is a bad idea. The blockage that may ensue during rapid pressure changes can cause extreme pain and serious damage to tissues and blood vessels lining the walls of these cavities. Marine mammals do not have this problem because they have no facial sinuses. Thus, one problem is dealt with by the absence of the airspace.

Similar to all other mammals there is a cavity that forms the middle ear. This is a rigid structure that has little or no compressibility. A pressure differential is prevented, at least for seals and sea lions, because of a complex vascular sinus lining the wall of the middle ear cavity. As the pressure within the middle ear cavity begins to fall below that of the vascular tree, the blood sinuses volume increases. This is a result of the close match between ambient pressure and blood pressure transferred from one fluid (sea water) to another (blood). Another problem is resolved by the reduction of an airspace by hydraulic compression through the vascular system.

Third, the largest airspace of all, and potentially the most problematic, is the lung. Volume pressure curves of the chest wall and lung of the ribbon seal, *Histriophoca fasciata*, show that both the chest and the lung are nearly limitless in the degree of compression collapse that they can tolerate. This must be so for other diving mammals as well, and in less detailed studies it has been shown that there is exceptional compressibility in other seals, sea lions, and dolphins. Dolphins and other toothed whales show the most extreme modifications within the lung among marine mammals, or any other mammal. Most notable is the reinforcement of peripheral airways, the loss of respiratory bronchioles, and the presence of a series of bronchial sphincters. Sea lions also have robust cartilaginous airway reinforcement extending to the alveolar sac, but there are no bronchial sphincters. In seals there is no cartilage in the terminal airway, but the walls are thickened by connective tissue and smooth muscle, which reduces their compliance to less than that of the alveoli. Hence, during compression the more compliant alveoli collapse first and the gases within these alveoli are squeezed into the upper airway spaces.

These airways enable a graded collapse of the lung to occur during a dive to depth. The result is that most of the lung air is forced into the upper airways where gas exchange with the blood ceases. It has been shown that blood PN_2 in seals only rises slightly, no matter how deep the dive. During simulated and actual dives to depth, the P_aN_2 of the northern elephant seal, *Mirounga angustirostris*, peaked at 300 kPa and equilibrated to 200 kPa where it was approximately the same as venous PN_2 (Fig. 2). This was independent of the ambient pressure from 37 to 138 m (1480 kPa). Similar values were obtained for Weddell seals diving voluntarily to depths as great as 230 m (2400 kPa). These small increases in P_aN_2 indicate that lung

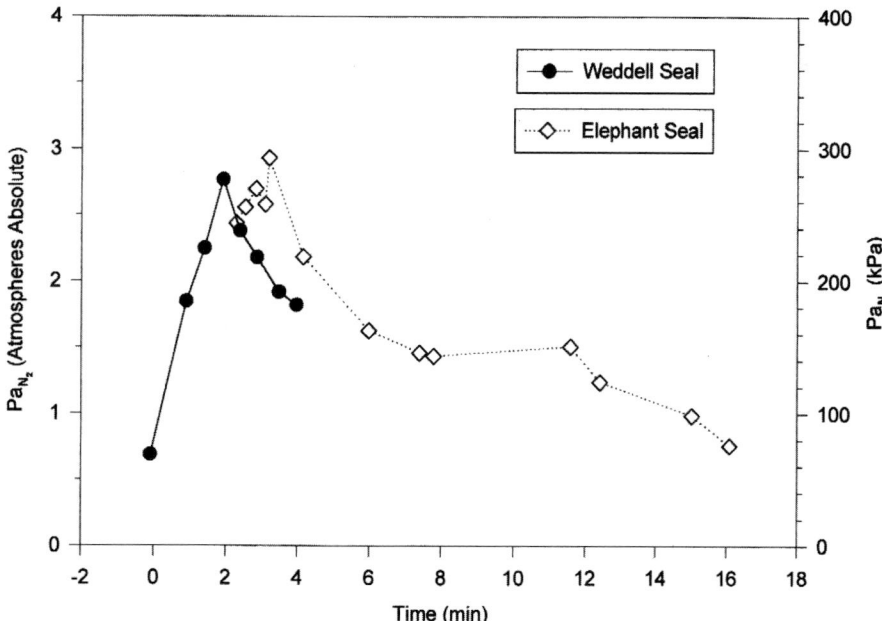

Figure 2 *Arterial N_2 tensions in an elephant seal and a Weddell seal. The elephant seal submersion was simulated in a water-filled hydraulic compression chamber to pressures equivalent to a seawater depth of 272 m. Compression began at zero; at 11 min pressure was released and at 14 min the submersion ended. The Weddell seal made a free dive to 89 m under Antarctic ice. Maximum depth was reached at 5 min, and the dive ended after 8 min.*

collapse in both species occurred between 20 and 50 m. The early occurrence of lung collapse in seals makes the lung almost useless as an O_2 store, but it limits N_2 absorption during the dive. These N_2 values are below the minimum PN_2 of 330 kPa found to be necessary for bubble formation in cats, and it is assumed that a similar threshold for bubble formation prevails in marine mammals. An additional benefit of early lung collapse is that it eliminates the likelihood of nitrogen narcosis. This condition is often experienced by SCUBA divers descending to depths greater than 30 m. At these depths the tissue nitrogen level is at least 399 kPa, greater than the P_aN_2 measured in seals. A final thought is the intriguing condition of elephant seals at sea when they spend 90% of their time underwater and at depths greater than 100 m. At these times the lung does not do what it was originally evolved to do, i.e., to exchange gas with the blood and with the atmosphere. Instead it is collapsed to a solid organ, and the alveoli become unavailable for gas exchange.

Finally, in humans and other nonaquatic animals descending to depths of more than 100 m and at rates of 100 m min^{-1}, the mechanical compression on nervous tissue can cause HPNS. The symptoms are modest to severe tremors throughout the body that can become so severe as to be incapacitating. The range of depths and the rate of descent are modest compared to that of some deep-diving marine mammals. Surely a well-adapted mammal does not experience HPNS, which leads to the compelling question of what the difference is between the neural makeup of a marine mammal that protects it from experiencing HPNS and that of a terrestrial mammal that is susceptible to HPNS. We must conclude that the structure of the nervous system is modified in a unique way for a life at high pressure.

III. Epilogue: Mysteries of the Deep

Our understanding of the physiology of diving in marine mammals is still elementary. These animals have adapted to some of the most extreme conditions on the planet. What we can learn from them about hypoxia and pressure is of much intellectual interest as well as of clinical significance. In addition to the brief summary presented, there is a host of other adaptations not mentioned and in some cases not studied. Some of these are blood and muscle interactions under extreme hypoxia, and acoustical and visual sensing under the extreme conditions of depth where pressure is intense, the cold is penetrating, and light from the surface is at times nil. Many organisms have adapted to a life under these various conditions, but marine mammals commute to the depths, and where they excel is in their adaptability to rapid changes in these extremes as they move from the conditions at the interface of air and water to those of several hundred meters to even a few kilometers beneath the surface.

See Also the Following Articles

Brain Size Evolution ▪ Breathing ▪ Circulatory System ▪ Earless Seals ▪ Endocrine Systems ▪ Health ▪ Musculature

References

Butler, P. J., and Jones, D. R. (1997). Physiology of diving of birds and mammals. *Physiol. Rev.* **77**(3), 837–899.

Davis, R., Fuiman, L. A., Williams, T. M., Collier, S. O., Hagey, W. P., Kanatous, S. B., Kohin, S., and Horning, M (1999). Hunting behavior of a marine mammal beneath the Antarctic fast ice. *Science* **283**, 993–996.

Davis, R. W., and Kanatous, S. B. (1999). Convective oxygen transport and tissue oxygen consumption in Weddell seals during aerobic dives. *J. Exp. Biol.* **202**, 1091–1113.

Kooyman, G. L., and Ponganis, P. J. (1997). The challenges of diving to depth. *Am. Sci.* **85**, 530–539.

Kooyman, G. L., and Ponganis, P. J. (1998). The physiological basis of diving to depth: Birds and mammals. *Ann. Rev. Physiol.* **60**, 19–133.

Kooyman, G. L., Ponganis, P. J., and Howard, R. S. (1999). Diving animals. *In* "The Lung at Depth." Dekker, New York.

Ponganis, P., Kooyman, G. L., and Castellini, M. A. (1993). Determinants of the aerobic dive limit of Weddell seals: Analysis of diving metabolic rates, postdive end tidal PO2 and blood and muscle oxygen stores. *Physiol. Zool.* **66**, 732–749.

Dolphin
SEE *Delphinids, Overview*

Dugong
Dugong dugon

HELENE MARSH
James Cook University,
Townsville, Queensland, Australia

The dugong (*Dugong dugon*) looks rather like a cross between a rotund dolphin and a walrus. Its body, flippers, and fluke resemble those of a dolphin without a dorsal fin. Its head looks somewhat like that of a walrus without the long tusks. Growing to a length of up to about 3 m, the dugong is the only extant plant-eating mammal that spends all its life in the sea. The other sea cows (or sirenians), the three species of manatee, all use fresh water to varying degrees.

I. Diagnostic Characteristics

Dugongs can be difficult to distinguish from dolphins in the wild, especially as they often occur in muddy water. They surface very discreetly, often with only their nostrils showing above the water. Dugongs tend to move more slowly than dolphins.

Adults are gray in color but appear brown from the air or from a boat. Older "scarback" individuals may have a large area of unpigmented skin on the back above the pectoral fins. There is no dorsal fin. The dugong's head is distinctive with the mouth opening ventrally beneath a broad, flat muzzle. The tusks of mature males and some old females erupt on either side of the head. The eyes are small and not prominent. Externally the ears consist of only of small openings, one on either side of the head. The flippers are short and, unlike those of the West Indian and West African manatees, lack nails. There are two mammary glands, each opening via a single teat situated in the "armpit" or axilla. The mammaries are somewhat reminiscent of the breasts of human females, which probably explains the legendary links between mermaids and sirenians. Hindlimbs are absent. Unlike manatees, which have a paddle-shaped tail, the tail of the dugong is triangular like that of a whale (Fig. 1).

II. Distribution and Abundance

The dugong has a large range that spans over 37 countries and includes tropical and subtropical coastal and island waters from east Africa to Vanuatu, between about 26° and 27° north and south of the equator. The dugong's historic distribution was broadly coincident with the tropical Indo-Pacific distribution of its seagrass food plants. It is believed that throughout most of this region outside Australia, the dugong is currently represented by relict populations separated by large areas where it is close to extinction or extinct. The degree to which dugong numbers have dwindled, and their range fragmented, is not known.

Over most of its range, the dugong is known only from incidental sightings, accidental drownings, and the anecdotal reports of fishermen. However, within Australia, extensive aerial surveys have resulted in a more comprehensive knowledge of dugong distribution. A significant proportion of the world's dugongs is found in northern Australian waters from Moreton Bay in the east to Shark Bay in the west. Dedicated aerial surveys of dugong populations in Australian waters indicate that dugongs are the most abundant marine mammal in the inshore waters of northern Australia. Although some areas of suitable habitat have not been surveyed, the population estimates sum to about 85,000 dugongs. This estimate is likely to be an underestimate as the correction for the number of animals that are not available to observers due to water turbidity is probably conservative.

III. Food and Foraging

Dugongs are seagrass specialists, uprooting whole plants when they are accessible, leaving long serpentine furrows bare of seagrass in seagrass meadows. When the whole plant cannot be uprooted, dugongs feed only on the leaves. Dugongs may also selectively forage for the rhizomes of some seagrasses. Dugongs prefer "weedy" or "pioneer" species of seagrass, especially species of the genera *Halophila* and *Halodule*. These seagrasses are low in fiber, high in available nutrients, and easily digested. Experiments simulating dugong grazing indicate that feeding dugongs alter both the species composition and the nutrient qualities of seagrass communities. Dugongs are like farmers cultivating their crops. If dugongs become locally extinct, those seagrass meadows are likely to deteriorate as dugong habitat.

IV. Habitat

Dugongs frequent coastal waters. Major concentrations of dugongs tend to occur in wide shallow protected bays, wide shallow mangrove channels, and in the lee of large inshore islands where there are sizable seagrass beds. Dugongs are also regularly observed in deeper water further offshore in areas where the continental shelf is wide, shallow, and protected. This distribution reflects that of deepwater seagrasses. Dugong feeding scars have been observed at depths of up to 33 m in seagrass beds off northeastern Queensland.

V. Behavior

Knowledge of the social behavior of dugongs is rudimentary. The habits and habitats of dugongs make them difficult to observe, and the lack of distinct size classes or obvious sexual dimorphism limits the data obtained. The only definite long-lasting social unit is the cow and her calf. Most dugongs are sighted in groups of one or two animals. Large aggregations of up to several hundred animals are regularly seen at some locations. Despite this gregariousness, little is known of the structure and function of dugong herds. One of the functions of herds seems to be cultivation grazing, which maintains the seagrass meadow at the stage favored by dugongs as discussed earlier.

As with many other mammals, the mating behavior of dugongs seems to vary with location. Mating herds have been observed at several locations along the Queensland and Northern Territory coasts. In these herds, splashing and fighting precede mating. The presumed female swims, turns, twists, and thrashes as she attempts to escape from her persistent

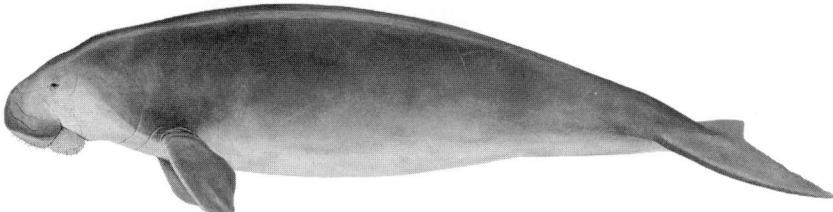

Figure 1 *Gentle vegetarians, dugongs often frequent muddy nearshore and estuarian waters that make it hard to positively identify them. Slow moving and lacking any hint of a dorsal fin, dugongs can be confused with dolphins, especially at a distance. Pieter A. Folkens/Higher Porpoise DG.*

entourage. Members of this entourage engage in bouts of violent fighting before attempting to mount the female in a form of pack rape (Fig. 2).

In contrast, in South Cove in Shark Bay in Western Australia, presumed male dugongs defend mutually exclusive territories in which unique behaviors are displayed in order to attract females. It is not known whether this behavior occurs elsewhere in the dugong's range.

More than 60 dugongs that have been tracked using satellite transmitters. Most movements have been localized to the vicinity of seagrass beds and are dictated by the tide. At the high-latitude limits of their range, dugongs make seasonal movements to warmer waters. In winter in Moreton Bay near Brisbane in eastern Australia, many dugongs regularly make round trips of 15–40 km between their foraging grounds inside the bay and oceanic waters, which average up to 5°C warmer. Some dugongs undertake long-distance movements. In tropical Australia, several dugongs have been recorded making trips of more than 100 and up to 600 km over a few days. Many of these movements have been return trips.

VI. Life History

Dugongs are long-lived with a low reproductive rate, long generation time, and a high investment in each offspring. Like the teeth of other marine mammals, dugong tusks accumulate "growth layer groups" that are used to estimate age, rather like the growth rings of a tree. The oldest dugong whose tusks have been examined for age determination was estimated to be 73 years old when she died. Females do not bear their first calf until they are at least 10 and up to 17 years old. Gestation is approximately 13 months. The usual litter size is one. The calf suckles for 18 months or so, and the period between successive births is very variable; estimates range from 3 to 7 years. Dugongs start eating seagrasses soon after birth and grow rapidly during the suckling period. Population simulations indicate that a dugong population is unlikely to be able to increase more than 5% per year. This makes the dugong highly susceptible to overexploitation by humans.

VII. Interactions with Humans

Dugongs are vulnerable to human impacts because of their life history and their dependence on seagrasses that are restricted to coastal habitats, which are often under pressure from human activities. The sustainable harvest is likely to be in the order 2% of the female population per year. This rate will be lower in areas where their reproductive rate has been shortened by food shortage. Dugongs may be short of food for several reasons, including habitat loss, seagrass dieback, decline in the nutrient quality of available seagrass, or a reduction in the time available for feeding due to boat traffic.

Experience from various parts of northern Australia suggests that episodic losses of hundreds of square kilometres of seagrass are associated with extreme weather events such as some cyclones and floods. For example, ~900 km^2 of seagrass was lost in Hervey Bay in 1992–1993, possibly because of high turbidities resulting from flooding of local rivers and runoff turbulence from a cyclone. Such events can cause extensive damage to seagrass communities through severe wave action, shifting sand, adverse changes in salinity, and light reduction.

Accidental entangling in gill and mesh nets set by commercial fishers is considered a major, but largely unquantified, cause of dugong mortality in many countries. Shark nets set for bather protection have been another source of dugong mortality in Queensland, Australia. Between 1962 and 1999, shark nets set on swimming beaches in Queensland netted some 800 dugongs. In most dugong habitats in Queensland, baited hooks have replaced shark nets. The rate of decline in the catch per beach in the shark nets over this 40-year period averaged about 8% per year. If the shark net catches are a reliable index of the overall decline in the dugong population from all causes, the population of dugongs along the urban coast of Queensland is only a small fraction of that in the 1960s.

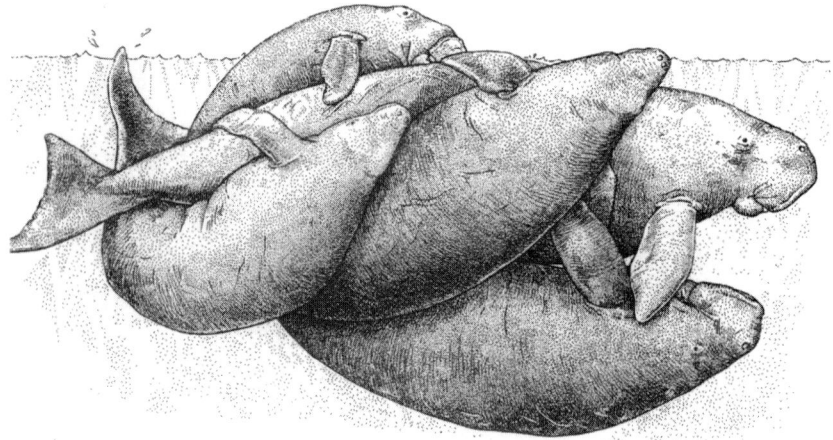

Figure 2 *Part of a mating herd of dugongs. Several males are attempting to embrace the female and mate with her. Drawing by Lucy Smith. Reproduced from Bryden et al. (1998), with permission.*

In response to concerns about declines in dugong abundance on much of the urban coast of Queensland, the Australian and Queensland governments agreed to several measures including a resolution not to issue permits for traditional hunting in this region. A two-tier system of dugong protection areas (DPAs) has been established in the Great Barrier Reef region in which gill and mesh netting is greatly restricted or banned (seven zone A DPAs totaling 2407 km^2) or subject to lesser modifications designed to reduce the probability of a dugong drowning after entangling in a commercial net (eight zone B DPAs totaling 2243 km^2). An additional DPA of 1703 km^2 in which gill and mesh netting practices have been modified was established in Hervey Bay, immediately south of the region.

Dugong meat tastes like beef or pork, and dugong hunting for food and oil was once widespread throughout the dugong's range. Today the dugong is legally protected in most countries. Members of Aboriginal and Torres Strait Islander communities in northern Australia and villagers in Western Province in Papua New Guinea are still permitted to hunt. Australia's indigenous peoples consider dugong hunting to be an important expression of their cultural identity. Dugong oil extracted by boiling the parts of the dugong not used for food, such as the head, is used as a panacea for aches, pains, and many illnesses. A dugong yields about 35% of its body weight in usable meat and fat and on average ~18 liters of oil.

It is increasingly recognized that arrangements to ensure that indigenous hunting of dugongs is sustainable will require the active participation of relevant indigenous communities. The Turtle and Dugong Hunting Management Strategy developed by the Hope Vale community on Cape York is northeastern Australia in association with the Great Barrier Reef Marine Park Authority is a useful model for community-based management of dugong hunting.

VIII. Conclusions

The dugong is a long-lived, slow-breeding animal with a low potential rate of population increase even under ideal conditions. It has a highly specialized diet, feeding selectively on pioneer species of seagrasses. Thus the dugong is dependent on habitats that are under pressure from coastal development.

The World Conservation Union (IUCN) classifies the dugong as vulnerable to extinction at a global scale on the basis of anecdotal evidence of declines in its abundance throughout most of its range. In recent years, there has been an encouraging and widespread increase in interest in dugong biology and management. Nonetheless, the difficulties of managing adverse influences on dugongs in highly populated developing countries emphasize the importance of the remote regions of tropical Australia to dugong conservation.

Acknowledgments

Some of the text in this article has been based on material in Bryden *et al.* (1998) and Marsh *et al.* (1999).

See Also the Following Articles

Indo-West Pacific Marine Mammals ∎ Manatees ∎ Sirenian Life History

References

Bryden, M., Marsh, H., and Shaughnessy, P. (1998). "Dugongs, Whales, Dolphins and Seals: A Guide to Sea Mammals of Australia." Allen and Unwin, Sydney.

Marsh, H., and Lefebvre, L. (1994). Sirenian status and conservation efforts. *Aquat. Mamm.* **20.3,** 155–170.

Marsh, H., Eros, C., Corkeron, P., and Breen, B. (1999). A conservation strategy for dugongs: Implications of Australian Research. *Mar. Freshwat. Res.* **50,** 979–990.

Nishiwaki, M., and Marsh, H. (1985). The dugong [*Dugong dugon* (Muller, 1776)]. "Handbook of Marine Mammals" (S. H. Ridgway and R. J. Harrison, eds.), pp. 1–31. Academic Press, London.

Reynolds, J. E., III and Odell, D. K. (1991). "Manatees and Dugongs." Facts on File, New York.

Reynolds, J. R., and Rommell, S. A., (1999). "Biology of Marine Mammals." Smithsonian Institution Press, VA.

Ripple, J. (1999). "Manatees and Dugongs of the World." Voyageur Press, MN.

Eared Seals
Otariidae

ROGER L. GENTRY
National Marine Fisheries Service,
Silver Spring, Maryland

The Otariidae, or "eared seals," evolved around the rim of the North Pacific Ocean 11–12 million years ago and subsequently spread into the Southern Hemisphere. About three million years ago, sea lions diverged from the fur seals that preceded them. The two are distinguishable by their pelage. Sea lions have a single coat of hair whereas fur seals have stiff outer guard hairs emerging from a dense, thick layer of fine underfur, which is waterproof, thereby providing thermoregulation.

Otariids comprise 16 species (Table I). All breeding sites they use are between 60° S and 55° N latitude. Eight species breed in temperate latitudes, two in subpolar, and two in equatorial regions. The sea lions use only temperate or equatorial regions, but fur seals use all three. No species of otariids mate on ice, but at least two encounter snow during breeding and pup rearing. Four species mate on the shores of deserts where the climate is ameliorated by fog or cool ocean winds. Only two species of fur seals and three species of sea lions still inhabit the North Pacific waters where otariids evolved. Otariids never occupied the North Atlantic Ocean, perhaps because the Central American seaway was closed when the family dispersed into the Southern Hemisphere.

Otariids use both island and mainland sites, but at present, island sites predominate worldwide. Islands offer a combination of freedom from terrestrial predators (including humans), cooling winds for THERMOREGULATION, and closer access to offshore prey concentrations than most mainland sites provide. The preferred substrate for mating is rock or sand, rarely soil or mud. All species mate on open ground, except some small populations of Hooker's sea lions (see Table I for scientific names), which use forested sites.

I. Unique Traits

Otariids are truly amphibious in that they feed at sea but mate and rear their young on land. Therefore, their anatomy strikes a balance between functioning in air and under water.

Unlike other pinnipeds or cetaceans, otariids have an external ear flap (hence their scientific and common names), an air-filled external auditory canal, and a middle ear structure very similar to a terrestrial ear. Otariids hear best in the frequency range of 2–12 kHz. In contrast, dolphins and porpoises that rely on a sophisticated echolocation system to feed hear best from 8 to 90 kHz or higher. This difference suggests that otariids either do not echolocate or do so in a rudimentary way.

Laboratory tests have demonstrated that sea lions have excellent visual acuity at low-light levels. Large eyes, an all-rod retina, and a well-developed reflective layer (*tapeta lucidum*) account for this ability. These features suggest that eared seals feed visually in the dark (at night, in deep water, or both). However, they maintain good visual acuity in air in bright light by being able to close down their pupil to a pinhole. Well-developed, highly-enervated vibrissae may provide tactile cues that combine with visual cues for feeding in the dark.

Locomotion also points out the balance between their terrestrial and aquatic lives. Unlike phocids, otariids can rotate their rear flippers forward and walk or run using all four limbs. They can outrun a human over slippery rocks. Using their chin and their long, strong front flippers, they can climb steeply sloping surfaces, which means they can take advantage of broken, rocky terrain for BREEDING SITES. They swim using the front flippers powered by well-developed pectoral muscles, while the rear flippers trail behind and are used only in turning and stopping. Otariids are quick, graceful swimmers that can leap free of the water and "porpoise" to breathe while swimming fast. They can execute a forward somersault while surfing.

Laboratory tests have also shown that otariids have well-developed cognitive abilities. They can perform complex learning tasks and have excellent memories, abilities that are perhaps related to their role as high-level consumers which depend on patchy prey to survive.

Otariids must reduce heat loss to cold ocean waters, but simultaneously have the ability to lose body heat while on land

TABLE I
Otariids

Common name	Scientific name	Location	Approximate numbers	Trend
Northern fur seal	*Callorhinus ursinus*	Subpolar	1,400,000	Stable
Antarctic fur seal	*Arctocephalus gazella*	Subpolar	3,000,000	Increasing
Sub-antarctic fur seal	*A. tropicalis*	High temperate	>310,000	Increasing
Juan Fernandez fur seal	*A. philippii*	Temperate	18,000	Increasing
Guadalupe fur seal	*A. townsendi*	Temperate	>7,000	Increasing
Cape fur seal	*A. pusillus pusillus*	Temperate	1,700,000	Increasing
Australian fur seal	*A. pusillus doriferus*	Temperate	<60,000	Stable
South American fur seal	*A. australis*	Temperate	<285,000	Increasing
New Zealand fur seal	*A. forsteri*	Temperate	135,000	Increasing
Galapagos fur seal	*A. galapogoensis*	Equatorial	40,000	Fluctuating
Galapagos sea lion	*Zalophus wollebaeki*	Equatorial	40,000	Fluctuating
California sea lion	*Z. californianus*	Temperate	<188,000	Increasing
Japanese sea lion	*Z. japonicus*	Temperate	Extinct	
Steller sea lion	*Eumetopias jubatus*	High temp.	>76,000	Decreasing
Hooker's sea lion	*Phocarctos hookeri*	Temperate	13,000	Stable
Australian sea lion	*Neophoca cinerea*	Temperate	12,000	Stable
Southern sea lion	*Otaria flavescens*	Temperate	275,000	Decreasing

in the sun. A BLUBBER layer (abetted by the underfur in fur seals) reduces heat loss to water. To promote heat loss on land, otariids use shade, immersion in tide pools or spray, flipper waving, or panting. At least one species urinates on its rear flippers for evaporative cooling. Only the flippers are naked, so evaporative cooling from sweat is ineffective. A few animals drink seawater, but usually only at the beginning of fasting at the start of the breeding season. As in VISION and HEARING, the primary adaptations for thermoregulation and water balance seem to be for life at sea, with secondary adaptations for life on land.

II. Diet

Otariids tend to be generalist feeders, taking a wide variety of prey. For example, the northern fur seal takes at least 63 species of prey over its full range. Fur seals and sea lions tend to feed on dissimilar prey. Fur seals often feed in deep water beyond the continental shelf break on small squid or fish, especially myctophids (lantern fish). The Antarctic fur seal takes krill as well as fish, and the New Zealand fur seal supplements fish with rock lobster. Sea lions tend to feed on or near the continental shelf and to take larger or more mature stages of prey than are used by fur seals (e.g., adult halibut). Hooker's and Australian sea lions specialize in squid and octopus, which are hunted on the bottom and under rocks. Several species of sea lions are known to eat the young of other seals, one species exhibits cannibalism, and several are known to eat penguins or other birds. For an as yet unexplained reason, otariids (especially sea lions) intentionally swallow fist-sized rocks.

III. Maternal Strategy

Otariids differ from phocids in that the females of all species continue to feed while they are lactating. Otariid mothers capture prey and within a few days time transfer the energy it contains to their young on shore in the form of milk fat. They make a series of brief, regular nursing visits to shore from birth to weaning, an interval that varies with latitude. In contrast, most phocid mothers gather energy from prey for several months, lay it down as extensive blubber reserves, and then deliver it to their young in a single, prolonged nursing bout that may last a few weeks at most.

Feeding during lactation has widespread implications for otariid natural history. Eared seal mothers are restricted in how long they may be absent for foraging by the limited fasting abilities of their newborn pups. To meet this restriction, mothers must find abundant prey of a given type close enough to the colony that they can commute between the two and still experience a net energetic gain for themselves and their young. The number of sites that meet these conditions and that have proper terrain, cooling winds, absence of predators, and other factors are relatively few. Otariids tend to gather in large numbers (up to hundreds of thousands) on the few sites in the world that meet these specific needs and to form dense aggregations there. Therefore, largely because of maternal commuting, otariids tend to breed in a few large, dense colonies that are most often on islands. Most phocids do not commute for foraging and therefore tend to breed dispersed as pairs or in small numbers on a large number of dispersed sites. Among phocids, only elephant seals (*Mirounga* spp.) form colonies that superficially resemble those of otariids (few sites, large dense groups).

The details of otariid maternal strategies vary considerably according to the local foraging environment. As a generalization, the duration of the maternal feeding trip, duration of nursing visits, and fat content of the milk all increase with increasing distance to feeding locations. However, the amount of milk fat and its delivery schedule are both constrained by the need to wean the young at a particular age. At high latitudes this age is 4 months, which allows mother and young time to migrate to lower latitudes before winter begins. At temperate latitudes the age is 9–12 months, which allows females to bear young annually without having to support two simultaneously. The two species at the equator are not limited by season, but by the rate at which mothers can bring their pups to the size needed for independent foraging. In good years, Galapagos fur seal mothers may achieve this in 12 months. However, when periodic El Niño events disrupt the food supply, they may not achieve it until age 3 years.

IV. Reproductive Adaptations

Otariids and phocids both possess a postpartum estrus. The uterus is Y shaped with two "horns," one of which holds the full-term fetus while the other prepares to receive the new blastocyst soon after birth. In all but one species of otariid, females enter estrus and mate during the 4- to 11-day perinatal nursing period. (The exception is the California sea lion, which mates about 23 days postpartum, well beyond its perinatal nursing period.) Because the perinatal period may be the females' longest single shore visit of the year, estrus at that time gives adult males the longest uninterrupted chance to mate with them. The process is quite efficient; pregnancy rates may exceed 93% for some age groups of females.

Dense gatherings have produced the most striking feature of the otariids, namely SEXUAL DIMORPHISM in body size and appearance (Fig. 1). In each species the adult male is 2 to 4.5 times the size of the female. Looking across species, there is more dissimilarity in the appearance of males as a group than there is among females as a group. These differences in body size and appearance probably result from the dissimilar selection pressures that the two sexes are under. Unlike females, in which fitness is measured by the quality of the single young they bear, male fitness is measured by the number of offspring produced. This means that males attempt to obtain the largest possible number of mates but females do not. Increased size not only gives males an advantage in fights related to obtaining mates, it also allows them to fast longer and thus remain among estrous females longer than males of smaller size. Male otariids weigh from 200 to 1000 kg and can fast for 12 weeks, although a fast of 4 to 6 weeks is more common.

V. Mating System

In all species, males defend space on the land sites that females use for giving birth and nursing their young. In two species (California sea lions and Juan Fernandez fur seals) males also partition some of the aquatic areas where females gather. Defending space gives males exclusive reproductive access to all females on that site. Failure to obtain space among females means having low reproductive success because of the characteristics of estrus (see later). In very few cases do males defend the actual females on a site rather than the boundaries around the site they use.

Defended space in otariids should be referred to as territories, not as leks. The reason is that, except possibly in the California sea lion, otariid females do not actively choose their mates as in a lek system. Instead, females choose their parturition site and then mate with any nearby male when estrus occurs. Females may use the exact same parturition site for 10 years or more, whereas the males that defend these sites change every breeding season or two.

All otariids are polygynous; average adult sex ratios of up to 10 females per male are common. Although the sex ratio at birth is near unity, males have higher mortality rates during maturation than females, and some males that reach adulthood are excluded from mating by the territorial system, which leaves more mating females than males. The average adult sex ratio is difficult to determine because a variable number of females will be at sea on any given day.

Sexual receptivity in female otariids is usually terminated by the physical act of coitus. For that reason, receptivity may last only minutes per year, and most females (85 to 90%) copulate only once. No second estrus is known to occur later in the year, even for females that fail to mate immediately postpartum. The advantage to females of a single copulation is that they can dispense with mating and quickly resume feeding to support their young. The advantage to males is that they can inseminate more females with fewer copulations in a shorter period of time. Most virgin females mate on the same grounds as more mature females, but at the end of the breeding season. For these reasons, males that do not acquire territory on traditional pieces of ground at predictable times of year find many fewer mates than males that do.

Figure 1 *Adult male, female, and newborn Steller sea lions showing the sexual dimorphism for size which characterizes the otariids. Photograph by Roger Gentry.*

Female otariids have an open society that features a loose, size-related dominance system. Larger, older, more experienced females generally have more access to favored rest sites, water, or shade than younger, smaller females. However, no evidence has been found for hierarchies of *individuals*. Also, no evidence has been found for any social bond between adult females, including those between mothers and their previous female offspring. The society is open in the sense that females can enter and leave it frequently without loss of social status. It thus fits with, and may have resulted from, the need of females to forage specific to the needs of their nursing offspring.

Females maintain close spacing while on shore. All sea lions plus the South African fur seal are thigmotactic (seek full body contact with others); spacing in the other species varies seasonally and with the radiant load (less contact on hot days). Despite their close spacing on shore, females are simultaneously aggressive toward each other. Fights that draw blood are uncommon, but females threaten each other frequently in various ways.

The offspring of otariids are precocial. Some can swim on the day of birth but most defer swimming for a month or more. Sexual dimorphism is evident at birth. In one species (northern fur seal) it exists as early in embryonic development as the sexes can be distinguished. During the lactation period, otariid pups tend to form dense aggregations and to avoid contact with adult males, which may bite and kill any pup that approaches them. Pups engage in play bouts that feature many of the components of adult aggressive behavior, sexual behavior, and prey handling in mixed order. These patterns appear to be innate in that pups born in captivity display behavior that they cannot have witnessed. Sneak suckling, especially in starvelings, occurs probably in all otariids. In some species (Steller sea lion), some females that lose their pups may actually foster a foreign pup.

Weaning is difficult to observe and document. It has been studied well in only two species (northern and Antarctic fur seals). In both species, most pups leave shore before their mothers, thereby weaning themselves at 4 months, just prior to migrating. In nonmigratory species, varying degrees of mother–young conflict exist at weaning, depending on environmental conditions, age at weaning, and presence of younger offspring.

Outside the breeding season, when no adult males are present, weaned juveniles may reside among the females and young. Juvenile females may remain there during the breeding season, but juvenile males are usually excluded from breeding areas and may gather on nearby all-male landing areas. In some areas of the world, several species of seals may mix on these landing areas. Juvenile males spend much time play fighting and occasionally make brief running forays into breeding areas or wait offshore to intercept females departing on foraging trips. Copulations sometimes occur in these circumstances.

VI. Foraging Behavior

Otariids usually only dive deeply when foraging. This summary is based on dive records for females, which have been studied more thoroughly than males. Most otariids feed in the water column, sometimes over very deep water. Dive depths measured in the field reflect the vertical distribution of the prey more than the physiological limits of the divers. These depths are usually less than 450 m, and dives to more than 200 m are uncommon. Dive durations are usually less than 12 min, and dives of more than 5–7 min are uncommon. Most dives of otariids (more than 85% for individuals) are aerobic, i.e., the duration of most dives does not exceed the estimated oxygen that animals take down stored in hemoglobin or myoglobin. Diving occurs at all hours, but nighttime diving predominates. The dive depth may change with time of day, suggesting that the otariid is following the daily migration of prey toward the surface at dusk and toward deep water at dawn. There is as yet no evidence of cooperative feeding by otariids (such as group attacks by killer whales), but coordinated group diving has been observed.

VII. Population Trends

Starting in the 16th century, otariids were harvested for furs, hides, blubber, various organs, or, in the case of Steller sea lions, vibrissae (for cleaning opium pipes). By the end of the 19th century, sealers had obliterated many stocks and reduced many species to near extinction. Stocks recovered throughout the 20th century in varying degrees. After a long lag time, the Antarctic fur seal recovered at a rate of 10% per year, the highest known for any otariid population. Recovery for most species was nearer 5–7%. The Japanese sea lion is probably extinct. Some populations (Hooker's sea lion, Australian sea lion, see Table I) are small but apparently stable, whereas others are small but growing (Guadalupe and Juan Fernandez fur seals). Two species (Galapagos fur seal and sea lion) experience intermittent declines and recoveries related to El Niño events. The northern fur seal has been declining at various rates from 1956 to the present. The only otariid species presently of concern to managers is the Steller sea lion, which is rapidly declining for as yet unknown reasons. Worldwide, fur seal populations tend to be increasing faster than sea lion populations and outnumber them by an order of magnitude (nearly 7 million fur seals worldwide compared to just over 600,000 sea lions; Table I). The diet and place of foraging differ between the two groups and may explain these differences. It could be the habit of feeding beyond the continental shelf, like a fur seal, that has accounted for the present rapid growth of the California sea lion population.

See Also the Following Articles

Pinniped Ecology ▪ Pinniped Life History ▪ Rookeries ▪ Sea Lions, Overview

Reference

Riedman, M. (1990). "The Pinnipeds: Seals, Sea Lions, and Walruses." University of California Press, Berkeley, CA.

Earless Seals
Phocidae

MIKE O. HAMMILL
*Department of Fisheries and Oceans,
Mont Joli, Canada*

I. Systematics

The family Phocidae contains the earless or "true" seals. They are distinguished from sea lions and fur seals (family Otariidae) by the absence of external visible ear pinnae, internal testes, a generally larger size, and the inability to draw their hindlimbs forward under their body when on land. This latter character, the absence of tusks, and a notched tongue also distinguish them from the family Odobenidae or walruses.

There had been considerable debate as to whether pinnipeds were diphyletic or monophyletic in origin. The diphyletic view proposed that odobenids and otariids were related to the bears (Ursidae) and that phocids were more closely linked to otters, weasels, and skunks (Mustelidae). However, a reevaluation of morphological evidence and the application of molecular techniques support the monophyletic hypothesis, with pinnipeds descending from arctoid carnivores, a group that includes the bears.

The Phocidae can be divided into two subfamilies: the Monachinae (*Monachus* spp.), with $2n=34$ chromosomes, consisting of the southern phocids, the southern and northern elephant seals (*Mirounga* spp.), and the monk seals; and the Phocinae, or northern seals, with $2n=34$ chromosomes in bearded (*Erignathus barbatus*) and hooded (*Cystophora cristata*) seals and $2n=32$ chromosomes in the remaining seven species (Table I). The separation between these two groups has been confirmed in molecular studies, but the relationships among members within the subfamilies are uncertain. More recent work indicates that the gray seal (*Halichoerus grypus*) may not be sufficiently separated from the ringed seal to warrant its own genus, and a closer relationship between ribbon and harp seals than of either to the *Phoca* group. This suggests that the harp seal should retain its old name: *Pagophilus groenlandicus*. Support has also been found for grouping the ribbon, harp, and hooded together, but the use of karyotypic data as a diagnostic landmark in the phylogenetic analysis may be sufficient to maintain the separation (Table I).

II. Life History

Phocids are found throughout all of the world's major oceans except for the Indian Ocean. Twelve species breed on ice and six species breed on land, including the Caribbean monk seal, which is probably extinct. The gray seal breeds on both land and on ice (Fig. 1). Among the ice-breeding seals, eight species breed primarily on pack ice. Four species breed on land-fast ice. Phocids in the Northern Hemisphere have also colonized freshwater areas; these include the harbor seal (*Phoca vitulina mellonae*) in freshwater lakes of northern Quebec, the ringed seal (*Pusa hispida lagodensis* and *Pusa hispida saimensis*) in Lakes Lagoda and Saimaa in Russia and Finland, respectively, and the Baikal seal (*Pusa sibirica*) in Lake Baikal in Siberia. Ringed seals also frequent Nettelling Lake on Baffin Island in northern Canada.

Pinnipeds are adapted to marine foraging but must haul out on land or ice for parturition and successful rearing of offspring. Marine adaptations include a thick blubber layer for insulation, modifications in limbs and body shape to improve hydrodynamics and agility, and anatomical and physiological changes to improve diving performance. Compared to otariids, phocids generally spend more time at sea, swim more slowly, and dive to deeper depths and for longer periods. Southern elephant seals (*Mirounga leonina*) may dive to 1200 m and remain below the surface for up to 120 min. Other deep-diving phocids include northern elephant seals (*M. angustirostris*) (1500 m), Weddell seals (*Leptonychotes weddellii*) (700 m), and hooded seals (1000 m). Phocids have adopted what is sometimes referred to as a "slow-lane" strategy to reduce energy use during diving. This is achieved through a combination of (1) apnea with exhalation upon initiation of diving to minimize buoyancy and pressure-related problems; (2) an enhanced oxygen-carrying capacity, which is accomplished by a greater blood volume, an increase in red cell mass (haematocrit) within the blood cell volume, a greater hemoglobin concentration in the red blood cells, and possibly a higher content of oxygen-carrying myoglobin in the muscles; and (3) a generally larger body size to maximize oxygen-carrying abilities while minimizing mass-specific energy demands while diving.

The improvements in swimming ability have occurred at the expense of mobility on land or ice, which in turn increases vulnerability to terrestrial predators. Phocids whelp on the ice, isolated islands, or inaccessible beaches, which makes it more difficult for terrestrial predators to approach seals undetected. Some ice-breeding species, e.g., the Baikal seal and the ringed seals, limit further their accessibility to predators such as humans, bears (*Ursus maritimus*), arctic foxes (*Alopex lagopus*), and birds (e.g., *Corvus* sp., *Larus* sp.) by using small caves or lairs under the snow. These lairs also provide shelter from cold ambient temperatures. Current global-warming trends will likely result in the reduction of suitable ice habitat for many phocids, particularly in the more temperate regions of their distribution. This will impact not only seal populations, but also predators that rely on seals as food.

III. Reproduction

Sexual maturity is delayed in phocids. Some females are sexually mature at the age of 3+ years, but the mean age of sexual maturity is normally around 4–6 years, although it may vary with the population size and availability of resources. In northwest Atlantic harp seals the mean age of sexual maturity among females may vary from 5.8 and 4.6 years. These changes have mirrored changes in population size due to exploitation. As the population has increased, the mean age of sexual maturity has also increased. It is currently around 5.4 years. Normally about 80% of the adult females are pregnant, but some interannual variability in adult reproductive rates can occur. Extremely low (~60%) adult reproductive rates have been

TABLE I
Members of the Family Phocidae, General Distribution, and Breeding Habitat

Common name	Latin name	Subspecies	Distribution	Breeding habitat
Northern Hemisphere				
Gray seal	*Halichoerus grypus*		North Atlantic	Land, ice breeder
Harp seal	*Pagophilus groenlandicus*		North Atlantic	Pack-ice breeder
Harbor seal	*Phoca vitulina*	*P.v. vitulina* *P.v. concolor* *P.v. stejnegeri* *P.v. richardsi* *P.v. mellonae*	Atlantic, Pacific Oceans. Arctic regions	Land breeder
Spotted seal	*Phoca largha*		North Pacific, Chukchi Sea	Pack-ice breeder
Caspian seal	*Pusa caspica*		Caspian Sea	Fast-ice breeder
Ringed seal	*Pusa hispida*	*P.h. hispida* *P.h. botnica* *P.h. ochotensis* *P.h. krascheninikovi* *P.h. saimensis* *P.h. ladogensis*	Arctic regions, Baltic Sea	Fast-ice breeder
Hooded seal	*Cystophora cristata*		North Atlantic	Pack-ice breeder
Bearded seal	*Erignathus barbatus*	*E.b. barbatus* *E.b. nauticus*	Arctic	Pack-ice breeder
Baikal seal	*Pusa sibirica*		Lake Baikal, Siberia	Fast-ice breeder
Northern elephant seal	*Mirounga angustirostris*		North Pacific	Land breeder
Ribbon seal	*Histriophoca fasciata*		North Pacific (Chukchi, Bering, and Okhotsk Seas)	Pack-ice breeder
Hawaiian monk seal	*Monachus schauinslandi*		Pacific Ocean (Hawaiian islands)	Land breeder
Mediterranean monk seal	*Monachus monachus*		Mediterranean Sea, Black Sea, Atlantic (NW African coast)	Land breeder
Caribbean monk seal	*Monachus tropicalis*		Caribbean Sea–Gulf of Mexico area	Land breeder
Southern Hemisphere				
Southern elephant seal	*Mirounga leonina*		Sub-Antarctic, Antarctic, southern South America	Land breeder
Weddell seal	*Leptonychotes weddellii*		Antarctic	Fast-ice breeder
Ross seal	*Ommatophoca rossii*		Antarctic	Fast-ice breeder
Leopard seal	*Hydrurga leptonyx*		Antarctic	Pack-ice breeder
Crabeater seal	*Lobodon carcinophaga*		Antarctic	Pack-ice breeder

documented in some years among ringed seals in the Beaufort Sea and Hudson Bay areas of northern Canada. These changes may be related to changes in ice conditions and the availability of food resources.

Males become sexually mature around the same time or slightly later than females, but recently mature males appear to be incapable of defending access to females until 2 or more years after they are sexually mature. Sexually monomorphic species have the longest life expectancies; e.g., ringed seals aged 45 years old have been reported. The life expectancy of sexually dimorphic species is much shorter, particularly among males. In elephant seals, which show perhaps the most extreme level of SEXUAL DIMORPHISM, males are sexually mature at about 5 years of age, but they seldom achieve any rank within the

colony until the age of 8. The greatest reproductive success occurs between the ages of 9–12 years, and males die by the age of 14.

The female phocid reproductive cycle is characterized by parturition, a short lactation period (4–50 days), copulation near the end of lactation, embryonic diapause (~3 months), and active fetal growth (~9 months). In most temperate species, implantation occurs during late summer–early autumn when light levels are decreasing, and pupping occurs during the spring, when light levels are increasing. However, an irregular pattern is seen among gray seals. Gray seals in the United Kingdom give birth during the fall, and implantation occurs during the spring. Implantation of the embryo occurs after the moult, when female energy reserves are at a minimum.

Figure 1 *Gray seals on the ice in Northumberland Strait, Canada. The male is in the foreground, the female in the middle, and the white pup in the background.*

Figure 2 *Newborn hooded seal pup with its mother weighing about 20 kg (A) and a weaned pup 4 days old weighing 44 kg (B). Behind the newborn, note the small balls ("silver dollars") of fetal hair on the ice that were ejected with the placenta at birth.*

Embryonic diapause provides females a mean of terminating reproduction if conditions are poor before her investment becomes too costly. It also leads to synchronization of reproductive activity.

The characteristic phocid lactation strategy consists of building up energy reserves throughout the year, fasting during lactation, and lactating for a short period. The utilization of fat reserves to satisfy their own energy requirements and the costs of providing milk for her pup has favored selection for large body size because energy stores scale to Mass$^{1.0-1.19}$, whereas metabolic requirements scale to Mass$^{0.75}$. Thus an increased body size increases energy-storing capabilities at a greater rate than increasing mass-specific energy requirements. This has led to a larger body size in phocid females than among otariid females (phocid females: mean = 229 kg, median = 141 kg; otariid females: mean = 80 kg, median = 55 kg). The need to build up energy reserves over the year probably adds about 12% to the daily energy requirements of a female phocid, but it also allows the spatial and temporal separation of FEEDING and reproduction.

To minimize the metabolic overhead associated with lactation, phocids have shortened the lactation period to 4–50 days instead of the months seen in otariids and odobenids. This is achieved by remaining beside the pup, providing more opportunities for the pup to suckle, and producing a very fat-rich milk, which increases the energy transfer per volume of milk consumed. In elephant seals, the fat content is very low at the beginning of lactation (~10%), but increases to about 50% fat by midlactation. In harp, gray, and Weddell seals, the milk fat content increases from around 40% to between 50 and 65% fat by midlactation, whereas in hooded seals there is relatively little change in fat content of the milk (55–68%) over the short 4-day lactation period (Fig. 2).

Stability of the whelping habitat and vulnerability to predation appear to be two important factors that have influenced the duration of lactation within the phocid group. Pups on unstable, drifting pack ice nurse for only 4 days in the hooded seal to as much as 30 days among the crabeater and Ross seals (mean = 18 days). In contrast, the longest lactation periods are found among the fast-ice-breeding Weddell, Baikal, and ringed

seals (mean = 57 days). The lengthy lactation period among the very small ringed and Baikal seals may be related to their small size and their relative inability to store and deliver energy quickly to their offspring (Fig. 3). The Weddell seal, which weighs around 450 kg, is almost four times heavier than the diminutive ringed seal. The lengthy nursing period of Weddell seal pups may keep the young away from the fast-ice edges where their exposure to aquatic predators such as leopard seals and killer whales would be greater. The duration of lactation among land-breeding phocids is intermediate to that of the fast-ice and pack-ice animals, with an average of 32 days.

The "fasting strategy" is certainly characteristic of lactation in the largest phocids such as the elephant seals and in the hooded seal. Facultative feeding occurs among female Weddell, bearded, harp, and gray seals, whereas feeding appears to be obligatory among the smaller phocids (≤ 100 kg), such as harbor and ringed seals. Feeding may be necessary because females are unable to store sufficient energy to satisfy their own energy requirements, plus energy required for lactation. The need to continue foraging during lactation means that females

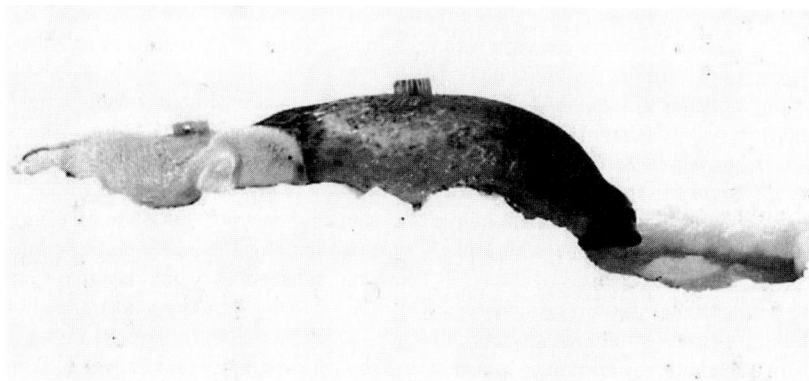

Figure 3 *Female ringed seal with her pup on fast ice in Svalbard, Norway.*

must leave their pups in a safe area while they forage or the pup must accompany the female. In both instances, the spatial separation between whelping sites and foraging areas is limited. In the ringed seal, females scrape out lairs beneath the snow to haul out in. Females leave their pups to feed under the ice but remain close enough to help the pup move to an alternate lair if a predator approaches. Among harbor seals, foraging acivity is restricted during the early stages of lactation. In large groups, as lactation advances, some females leave their pups unattended while they forage. In areas where only small groups are seen, the pups may follow the females over extensive distances of as much as 30 km away from the original haul-out site.

Fewer studies have examined male energy expenditures during breeding, due in part to the difficulties associated with handling large, dangerous, and aggressive animals or their inaccessibility in the case of males that spend much of their time in water. Daily energy expenditures of males during the breeding season are much lower than those of females because they do not incur the costs of milk production. However, once females wean their pups they resume feeding. This contrasts with males, who continue fasting to maximize their access to successively receptive females as the breeding season advances. Elephant, gray, and hooded seals appear to terminate breeding activity when fat levels have declined to levels similar to those observed in females. This suggests that the overall reproductive effort is similar between the two sexes.

At birth, phocid neonates are larger than otariid neonates. When female body size is taken into account, phocid and otariid mass at birth represents about 10% of their mother's mass. The pups are quite lean at birth, with a fat content of 5–8%. Notable exceptions are hooded, harbor, and bearded seals, which are larger relative to other species, with a mass at birth equal to about 12–13% of the maternal mass. Harbor, hooded, and bearded seal pups also differ from otariid and other phocid pups by the presence of a thin blubber layer at birth. Fat content represents 11–14% of the total body mass in harbor and hooded seal pups at birth. Phocid pups gain weight rapidly, achieving a weaning mass two to five times their birth mass in a period of 4–50 days, but mass at weaning is relatively constant across species, being equivalent to 25–30% of the mother's mass. Among hooded, harp, and gray seal pups, 65–75% of the

milk energy is deposited, primarily as fat. Elephant, hooded, harp, ringed, and gray seal pups are 40–50% fat at weaning, whereas species with very active pups, such as Weddell, bearded, and harbor seals, may contain only 34–37% fat at weaning. The rate of mass gain is inversely related to the duration of the lactation period when expressed relative to the female's metabolic mass. The lowest rates of relative mass gain are seen among Hawaiian monk seals, bearded, harbor, and ringed seals, species where the pups are very active and begin entering the water early during lactation. Ice-breeding neonate phocids, such as harp, ringed, ribbon, Caspian, Baikal, and largha seals, are born with a white, relatively long fur called lanugo (Fig. 3). The white color may provide some protective camouflage on the white snow from predators. However, the lanugo may be more effective in its role as an insulator, particularly to the very young pup who has not yet developed a thick layer of BLUBBER for insulation. The structure and color of this fur permits short-wave energy received from the sun to be transmitted through the fur, where it is absorbed by the dark skin. It also traps heat energy radiated from the animal's skin and thus acts like a greenhouse, heating up the air trapped within the fur, but limiting heat loss to the outside air. Among species where the young enter the water very soon after birth (e.g., harbor, bearded, and hooded seals), the young are born with a thin layer of blubber, and the lanugo is shed or moulted within the uterus. This is because blubber acts as a much better insulator in water than fur. Among harbor seals, 5–30% of the pups are born with a white lanugo, depending on the region, but this is quickly replaced by a grayish pelage similar to that of adults. In bearded seals, shedding of the lanugo begins *in utero,* but is completed after birth. In hooded seals, the fetal fur is expelled in small clumps on the ice with the placenta; these are referred to as silver dollars by commercial sealers. At birth the pups are covered by blue (dorsal) and silver (ventral) fur that is much thicker and longer than the adult fur, and at this stage the pups are known as bluebacks (Fig. 2). Hooded seal pups differ from harbor and bearded seals in that they do not enter the water until they are weaned. However, in this species lactation lasts for only 4 days. The remaining species that whelp on the ice in the Northern Hemisphere tend to give birth to pups with a white lanugo, which may afford some protection against surface predators. Gray seals have their pups on both the land and on the ice, but the pups are born with a white lanugo (Fig. 1). Neonates of southern ice-breeding seals, such as Weddell, crabeater (*Lobodon carcinophaga*), leopard (*Hydrurga leptonyx*), and Ross seals (*Ommatophoca rossii*), are born with a gray, brown, or grayish-brown pelage. Southern ice-breeding phocids are not exposed to surface predators other than humans and hence the white lanugo may not be required. Elephant and monk seal pups are born on land with a black pelage.

Pup mortality during the lactation period is normally quite low. Among land-breeding species, trampling and wounds caused by interactions between adults may encourage infection, which may result in death of the pup. This problem is more aggravated in crowded colonies where the number of interactions would increase. In Arctic ice-breeding species, predation by bears, foxes, and birds such as ravens (*Corvus* sp.) and gulls (*Larus* sp.) in the case of ringed seal pups is an important source of pup mortality, although the effects of PREDATION are reduced by the use of lairs. Ringed seal pups are also quite active and are capable swimmers at a very young age. However, swimming incurs a high metabolic cost due to the minimum blubber thickness. Repeated disturbance may affect growth and survival. Surface predators are not present in the southern polar regions, but leopard seals and killer whales are important marine predators. In the pack ice, pups may drown or be crushed as a result of storms causing the breakup and rafting of the ice.

IV. Mating Systems

The study of mating systems among phocids has relied heavily on behavioral observations of animals in the breeding areas. Male reproductive success has been evaluated by a male's ability to monopolize females or by the number of copulations observed. However, DNA techniques have suggested that the evaluation of reproductive success may be more complex, with the existence of alternative mating strategies within a population. In captive harbor seals, behavioral observations during courtship were not reliable indices of paternity, whereas in one population of gray seals, large males sired significantly fewer pups than would otherwise have been indicated from their observed mating opportunities. Females tended to produce several pups fathered by the same male, who in many cases was not the large attendant male.

Phocids breeding on land prefer islands or isolated beaches, where threats from predation are reduced. The combination of habitat limitations and synchronization of reproductive activity encourages the aggregation of females. Males are not involved in caring for the pup. Therefore the best strategy for males is to copulate with as many females as possible, whereas the strategy of females is to successfully rear her pup. Males will attempt to prevent other males from having access to females. In phocids, this may involve defending a geographical area, but if the female moves, the male will follow to defend a new space around the female. The number of females that can be defended is limited by habitat features and the skills of defending males. Large, open beaches are more difficult for a male to control than more topographically irregular sites, where geographic barriers will

aid established males in limiting the approach of intruding males. Reproductive success will also be affected by the fighting and signaling ability of males and how long they can remain beside females without leaving to feed. In male hooded seals, a series of displays involving the inflatable nasal sac and nasal septum are often associated with the approach of other males (Fig. 4). Not all approaches result in combat, suggesting that some signaling occurs. The gradual evolution toward large size observed in females, which permitted the separation of reproduction and feeding, would also have operated on males as well. Larger males could spend more time ashore fasting. Larger males would also be favored over small males in combat, although experience and individual skill development would also be contributing factors. The greatest degree of polygyny seen among phocids is exhibited by the elephant seals, who may defend harems containing upward of 100 females. Southern elephant seal males may weigh up to 3700 kg, whereas northern elephant seal males may reach 2300 kg. In both species the males are typically five to six times larger than the females. Males arrive on the whelping area just before the females and the most successful males remain on the beaches fasting until the last females leave about 3 months later.

In other species the development of polygyny is more variable. Gray seals copulate on land and occasionally in water. Males at 350 kg are about 50% larger than females. In some areas in the British Isles, they control access to up to 8 females, whereas in the open beaches and ice-breeding areas in Canada, males are only able to control access to 1 to 3 females. Less is known about the structure of mating systems of phocids that copulate primarily in water. The development of polygyny is limited by the three-dimensional nature of the marine environment. Among hooded seals, the marked sexual dimorphism observed in many terrestrial mating species is also observed. Male hooded seals weigh up to 440 kg, whereas females reach about 290 kg. The hooded seal male begins to defend one female about midway through the short 4-day lactation period (Fig. 4). Once the pup is weaned, he accompanies the female into the water, mates with her, and then returns to the ice and will attempt to establish himself beside another female. Because pupping occurs over a 2- to 3-week period, males have opportunities for multiple matings during that period. Mating success of a male is affected by his ability to defend access to females against other males and by the amount of time he can spend on the ice before his energy reserves are depleted. Among the remaining phocids that copulate in the water, little difference in body size is seen between males and females; in some cases, such as bearded and Weddell seals, the females appear to be slightly larger. In these cases, smaller size may favor underwater agility. The mating system of harp seals has been referred to as promiscuous, but little information is available. Males do haul out, but no displays or fighting are observed on the ice. Extensive vocal activity within harp seal whelping patches has been recorded, and groups of males are observed often patrolling leads, vocalizing and diving. It is possible that male harp seals are displaying to females, suggesting more of a lek-type system, but there is insufficient data to comment further. Among ringed seals, the presence of predators in the fast ice, such as bears, foxes, and humans, selects against aggregation, whereas the need to con-

Figure 4 *Two male hooded seals fighting on pack ice in the Gulf of St. Lawrence, Canada. Note the inflatable sac on the male on the left.*

tinue feeding during lactation, in an area generally considered to have low productivity, would also be a contributing factor. In this species, some underwater vocal activity has been recorded, but unlike the underwater vocalizations of the widely dispersed bearded seal that can be heard over distances of 25 km, the underwater vocalizations of the ringed seal are relatively weak, limiting their use as a signal to potential mates. However, the males emit a strong odor during the breeding season due to the enlargement and increased secretion activity of sebaceous and apocrine glands in the muzzle region. The strong odor may serve as a signal to inform both females and other males that a breathing hole is used or belongs to a particular male. In harbor seals, the females move with their pups between haul-out sites and foraging areas. Males are unable to defend females or sites against other males. As the time that females will become receptive approaches, males reduce the size of the range that they occupy, but remain in the water, making repeated short dives that are associated with underwater vocal displays. Some males establish themselves near haul-out sites and some near foraging areas used by females, whereas others appear to establish themselves on transit routes between haul-out sites and foraging areas. It has been suggested that a "lekking" type of mating system occurs in this species. In the Antarctic Weddell seal, males appear to defend underwater territories around breathing holes and cracks.

V. Foraging

Many early studies relied on stomach content material from hunted animals, fecal collections, and entrapments in fishing gear to provide information on diving and foraging activity. During the last decade, major technological advances have provided researchers with satellite transmitters, time-depth recorders, stomach temperature probes, and video recorders to study diving and foraging activity.

Phocids feed on a wide variety of prey, including invertebrates, such as amphipods, mysids, squid, and krill, and vertebrate prey, such as fish. Birds have been recorded in the diet of some species such as the leopard seal and harp seal. Leopard seals also prey on other seal species, and cannibalism has been reported in gray seals. DIET composition may change seasonally, geographically, and with age. Newly weaned pups of ringed and harp seals begin foraging on zooplankton in their initial attempts to forage independently and then become more piscivorous as their skills develop. Phocids feed primarily on smaller prey that can be consumed whole, but large prey may be taken. Under certain conditions where prey are very abundant or accessible, e.g., when fish are caught in nets, seals may consume only pieces from fish.

Little is known about factors affecting prey choice. Research has indicated that feeding preferences in harp and harbor seals occur for particular types and sizes of prey that may be independent of local abundance. Harp seals digest capelin more efficiently than most other prey, and throughout their range capelin forms an important component of their diet, whereas other species such as commercially important Atlantic cod (*Gadus morhua*) form only a very minor component in the overall diet. While foraging, phocids must balance their intake of oxygen and the distribution of oxygen for locomotion, body maintenance, and processing of food (specific dynamic action or heat increment of feeding). Their approach to balancing these sometimes conflicting needs will influence their foraging strategy. A seal may process (digest) food while actively swimming and foraging. If it uses only aerobic metabolism, then the consumption of oxygen required to process the food will reduce the amount of time spent diving and collecting food. If the seal attempts to maintain the duration of diving, then the switch to anaerobic conditions will force the animal to rest at the surface until lactic acid levels are reduced. A second strategy is to forage and then spend time resting at the surface, hauled out on the ice or on land until food processing is completed. However, resting at the surface increases vulnerability to surface predators such as sharks or killer whales, whereas returning to shore or ice involves time lost due to transit and may limit the distances that foraging can occur away from haul-out sites. A third strategy involves foraging and then if successful, reducing locomotory costs by drifting during the surfacing phase of the dive, allowing food processing to occur. Many phocids may utilize the first two strategies. Weddell, ringed, harp, and hood seals are often seen hauled out on the ice and occasionally on land. Harbor and gray seals may forage in offshore areas, but rarely spend more than a few days away from haul-out sites. The third strategy appears to be utilized by elephant seals, who spend almost 8–9 months of the year at sea.

Foraging activity has been examined in detail in elephant seals. In northern elephant seals, females are dispersed over a broad geographic area across the northeastern Pacific from the coast to as far west as 150°W, but tend to remain between 44° and 52°N. Foraging occurs both offshore and during transit between inshore and offshore areas. Diurnal changes in diving depths are observed, indicating that they are foraging on vertically migrating prey in the pelagic and mesopelagic environment. Their principal prey are mesopelagic squid and fish. Males utilize a different foraging strategy by foraging little while en route to particular foraging areas along continental margins off the stage of Washington to as far north as the Aleutian Islands. Once on site, repeated, uniform flat-bottomed dives predominate diving, with little diurnal variability, suggesting intensive foraging activity, possibly on benthic prey such as energy-rich elasmobranchs and cyclostomes (sharks, skates, ratfish, hagfish). While at sea, both species dive continuously, spending almost 90% of their total time at sea submerged, leading to the suggestion that they should be called surfacers instead of divers. At the opposite end of the size spectrum, a very different strategy is seen among harbor seals. Although capable of diving to depths of 500 m, harbor seals rarely dive deeper than 65 m and, in some studies, an average of 65% of diving activity occurs at depths of less than 4 m. Visual and telemetry data indicate that harbor seals in some areas spend most of their time very close to the coast in shallow water areas. Foraging distances rarely exceed 50 km away from haul-out sites. Little difference is seen between males and females in dive depths and foraging distances away from haul-out sites outside of the breeding season.

See Also the Following Articles

Mating Systems ■ Pinnipedia ■ Pinniped Life History ■ Sea Lions, Overview

References

Árnason, Ú., Bodin, K., Gullberg, A., Ledje, C., and Mouchaty, S. (1995). A molecular view of pinniped relationships with particular emphasis on the true seals. *J. Mol. Evol.* **40,** 78–85.

Carr, S. M., and Perry, E. A. (1997). Intra- and interfamilial systematic relationships of phocid seals as indicated by mitochondrial DNA sequences. *In* "Molecular Genetics of Marine Mammals" (A. E. Dizon, S. J. Chivers, and W. F. Perrin, eds.), pp. 277–290. Allen Press Inc., Lawrence, KS.

Hochachka, P. W., and Mottishaw, P. D. (1998). Evolution and adaptation of the diving response: Phocids and otariids. *In* "Cold Ocean Physiology" (H. O. Portner and R. C. Playle, eds.), pp. 391–431. Cambridge Univ. Press.

Lydersen, C., and Kovacs, K. M. (1999). Behaviour and energetics of ice-breeding, North Atlantic phocid seals during the lactation period. *Mar. Ecol. Prog. Ser.* **187,** 265–281.

Renouf, D. (1991). "The Behaviour of Pinnipeds." Chapman and Hall, New York.

Boyd, I. L. (1993). "Marine Mammals: Advances in Behavioural and Population Biology." Symposia of the Zoological Society of London No. 66. Clarendon Press, Oxford.

King, J. E. (1983). "Seals of the World." 2nd Ed. Cornell Univ. Press, New York.

Le Boeuf, B. J., and Laws, R. M. "Elephant Seals: Population Ecology, Behavior and Physiology." Univ. of California Press, Los Angeles.

Smith, T. G., Hammill, M. O., and Taugbol, G. (1991). A review of the developmental, behavioural and physiological adaptations of the ringed seal, *Phoca hispida,* to life in the arctic winter. *Arctic* **44,** 124–131.

Echolocation

WHITLOW W. L. AU
University of Hawaii, Kaneohe

Echolocation is the process in which an animal obtains an assessment of its environment by emitting sounds and listening to echoes as the sound waves reflect off different objects in the environment. In a very general sense any animal that can emit sounds may be able to hear echoes from large obstacles (i.e., humans yelling in a canyon); however, this type of process is not considered echolocation. The term echolocation is reserved for a specialized acoustic adaptation by animals who utilize this capability on a regular basis to forage for prey, navigate, and avoid predators. Therefore, echolocating dolphins are often searching for entities that are considerably smaller than themselves and must make fine discrimination of these objects. Over the eon of time, this specialized adaptation has been continually refined under evolutionary pressures.

Echolocation in bats was suspected as early as 1912, but it was not until 1938 when G. Pierce and D. Griffin (1938) provided evidence of bats emitting ultrasonic pulses using an ultrasonic detector that the concept began to gain acceptance. Echolocation in dolphins was suspected around 1947 as was evidenced in the personal notes of A. McBride, the first curator of Marine Studio (later Marineland) in Florida (McBride, 1956). However, it was not until 1960 that Kenneth Norris and colleagues performed the first unequivocal demonstration of echolocation in dolphins by placing rubber suction cups over the eyes of an Atlantic bottlenose dolphin (*Tursiops truncatus*) and observing that the animal was able to swim and avoid various obstacles. Ultrasonic pulses were also detected as the blindfolded dolphin swam and avoided obstacles, including pipes suspended vertically to form a maze.

Since the Norris demonstration, considerable progress has been made in our understanding of the echolocation capabilities of dolphins. Most of the research has been done with the Atlantic bottlenose dolphin, the most common dolphin in captivity. Research in dolphin echolocation can be divided into the following areas: (1) sound production mechanism and propagation in the dolphin's head, (2) sound reception and auditory capabilities, (3) sound transmission and the characteristics of echolocation signals, (4) target detection capabilities, (5) target discrimination capabilities, (6) auditory nervous system function and capabilities, and (7) signal processing modeling. This article addresses each of the first six areas, providing the most recent findings in most cases along with some fundamental capabilities.

I. Sound Production Mechanism and Propagation in the Dolphin's Head

The head of a dolphin shown in Fig. 1 is a very complex structure with unique air sacs and special sound-conducting

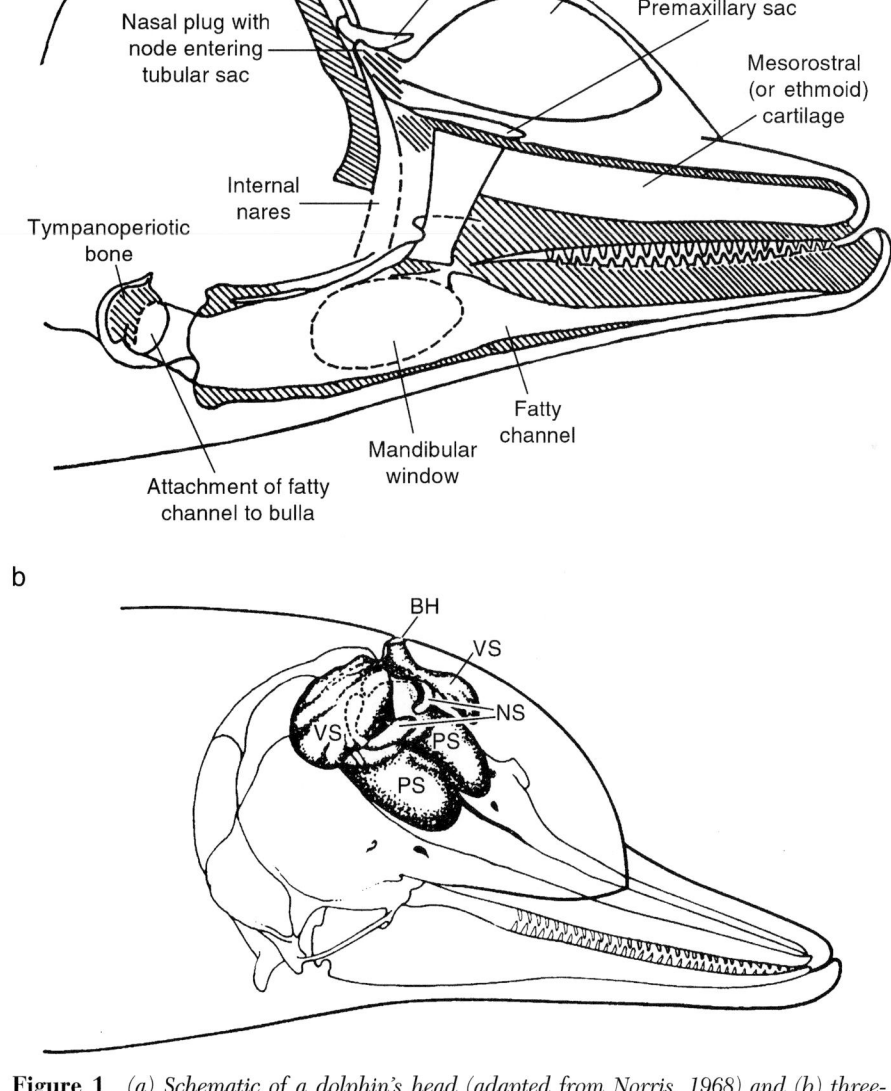

Figure 1   (a) Schematic of a dolphin's head (adapted from Norris, 1968) and (b) three-dimensional diagram of the air sacs in a dolphin's head. PS, premaxillary sac; VS, vestibular sac; NS, nasofrontal (tubular) sac; AS, accessory sac. Adapted from Purves and Pilleri (1983).

fats. Once of the most perplexing issues that has eluded researchers since the discovery of echolocation has been the location and mechanism of sound production in dolphins. In the mid-1980s, Ted Cranford began using modern X-ray computer tomography (CT) and magnetic resonance imaging techniques to study the internal structure within a dolphin's head. These noninvasive techniques allowed Cranford to study the relative position, shape, and density of various structures in the dolphin's head and helped him to conclude that the monkey lip–dorsal bursae (MLDB) region of the dolphin nasal complex was the location of the sound generator. Eventually, Cranford and colleagues were able to obtain high-speed video in 1997, of the phonic lips (previously referred to as the monkey lips) with simultaneous hydrophone observations of echolocation signals. There are two sets of phonic lips, associated with the two nares in the dolphin's nasal complex. Cranford and colleagues have obtained additional high-speed video observation of movements in both sets of phonic lips during the production of echolocation signals and whistles.

The numerical simulation of sound propagation in the head of a dolphin by James Aroyan, then a Ph.D. student at the University of California Santa Cruz, has provided considerable understanding of the role of the air sacs, skull, and melon in the propagation of sounds in a dolphin's head. One of the interesting problems Aroyan considered was that of a plane wave propagating toward the head of a dolphin (as depicted in Fig. 2a) to determine where sounds would focus in the dolphin's head in a similar manner to the process used by geologists to determine the epicenter of an earthquake. He numerically solved the three-dimensional wave equation (also shown in Fig. 2) using a finite-difference technique and a super computer. The density and sound velocity structure of the dolphin's head were estimated from the CT scan results of Ted Cranford. The grid points represent a pictorial illustration of how the head of a dolphin may be mathematically subdivided so that the solution of the wave equation is numerically determined at each grid point. The results for the geometry depicted in Fig. 2a are shown in Fig. 2b, with focal regions at the two auditory bullas and at the MLDB region of the nasal system, supporting Cranford's earlier suspicion of the MLDB being the site of the sound generator for echolocation sounds.

Aroyan then placed a hypothetical sound source at the MLDB region and numerically solved the three-dimensional wave equation as the sound propagated through the dolphin's head into the water. He found that the skull, the various air sacs, and the nonhomogeneous melon all played important roles in forming the beam in which sounds are transmitted into the free field. The specific characteristics of this beam are discussed in Section III. He also showed that if a hypothetical sound source was placed at the larynx, the resulting beam in the free field was not compatible to the actual beam measured

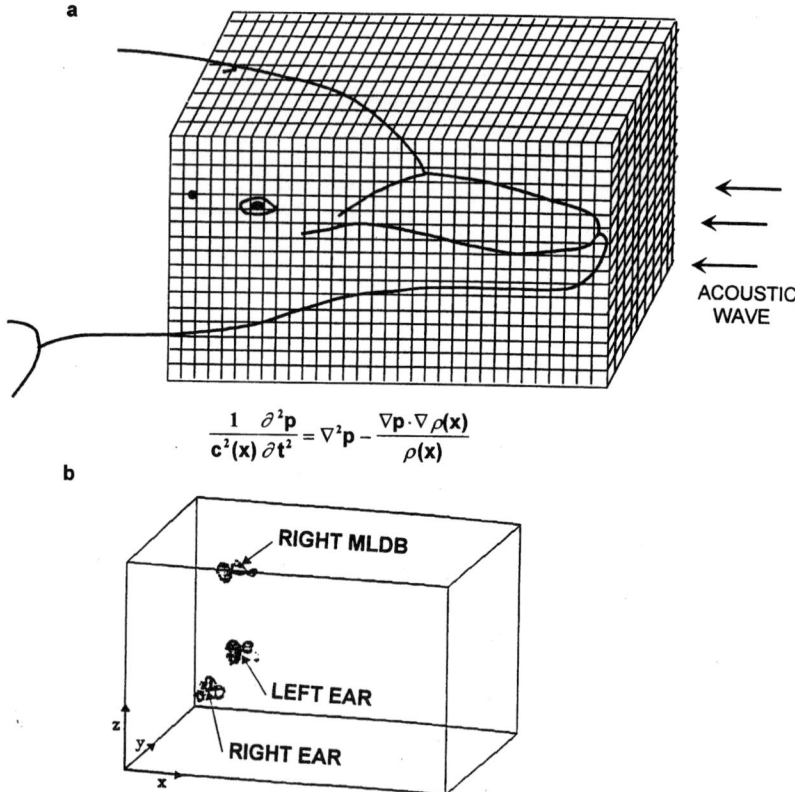

$$\frac{1}{c^2(\mathbf{x})}\frac{\partial^2 p}{\partial t^2} = \nabla^2 p - \frac{\nabla p \cdot \nabla \rho(\mathbf{x})}{\rho(\mathbf{x})}$$

Figure 2 *(a) Configuration of numerical simulation of sound propagation in the head of a dolphin to determine acoustic focal regions and (b) results of numerical simulation for the geometry depicted. After Aroyan (1996).*

for dolphins, therefore essentially eliminating the larynx as a possible site for a sound generator.

II. Sound Reception and Auditory Capabilities

A. Sound Reception Site

Dolphins do not have pinnae and their external auditory meatus is but a pinhole with vestigial fibrous tissues connecting the surface to the tympanic structure. Kenneth Norris was the first to postulate that sound enters the dolphin's auditory system through the thin posterior portion of the mandible (see Fig. 1a) and is transmitted via a fat-filled canal to the tympano-periotic bone, which contains the middle and inner ears. Two electrophysiological measurements, one by Theodore Bullock and colleagues, were conducted that provided evidence to support Norris' theory. However, the acoustic conditions for both experiments were less than ideal: the subjects were confined to small holding tanks and their heads were held near the surface to keep the electrodes from shorting out by the water. The acoustic propagation for such a situation can be extremely variable, with the sound pressure level changing drastically because of multipath propagation, on the order of 10 to 20 dB.

Bertel Møhl and colleagues took a slightly different electrophysiological approach by measuring the brain stem-evoked potential of a bottlenose dolphin that was trained to beach itself on a rubberized mat. A special suction cup hydrophone having a water interface between the piezoelectric element and the skin of the dolphin was positioned at different locations on the dolphin's head. By performing the measurement in air, the point at which sound from the piezoelectric element enters into the dolphin could be firmly established. Acoustic energy will only propagate toward the dolphin's skin, and energy propagating in any other direction will be reflected back at the boundary of the suction cup. Møhl and colleagues positioned the suction cup hydrophone at different locations around the dolphin's head, and at each loca-

tion, the amount of attenuation needed to obtain the evoked potential threshold was determined. Their results are shown in Fig. 3, where the circles indicate the different positions of the suction cup and the number within each circle represents the amount of attenuation needed to achieve threshold. Therefore, a larger number is indicative of a more sensitive region of sound reception. The dashed line indicates the area of the pan bone or mandibular window shown in Fig. 1a. These results indicated that the area just forward of the pan bone area of the dolphin's lower jaw is the most sensitive area of sound reception, which seems to be inconsistent with Norris' pan bone theory. However, the numerical simulations of acoustic propagation by Aroyan suggest that sounds that enter the dolphin's lower jaw just forward of the pan bone actually propagate below the skin surface to the pan bone and enter into the lower jaw through the pan bone.

B. Hearing Sensitivity

The hearing sensitivity of a dolphin was first measured accurately in 1967 by Dr. Scott Johnson in a pioneering experiment. Johnson found that the upper limit of hearing of an Atlantic bottlenose dolphin was 150 kHz. Since Johnson's research, audiograms have been determined for the harbor porpoise (*Phocoena phocoena*), Amazon River dolphin (*Inia geoffrensis*), beluga whale (*Delphinapterus leucas*), false killer whale (*Pseudorca crassidens*), Chinese river dolphin (*Lipotes vexillifer*), Risso's dolphin (*Grampus griseus*), Tucuxi (*Sotalia fluviatilus*), and killer whale (*Orcinus orca*). The audiograms for these odontocetes are shown in Fig. 4.

One of the most remarkable features of these audiograms is the high upper frequency limit of hearing extending beyond 100 kHz. This is rather remarkable when the wide range of sizes of the animals depicted in Fig. 4 is considered. The largest animal represented in Fig. 4 is the killer whale, which weighed about 3600 kg and was about 5 m in length compared to the smallest animal, the harbor porpoise, which typically weighs about 33 kg and is about 1.3 m in length. The typical rule of thumb in mammalian hearing is that larger animals tend to

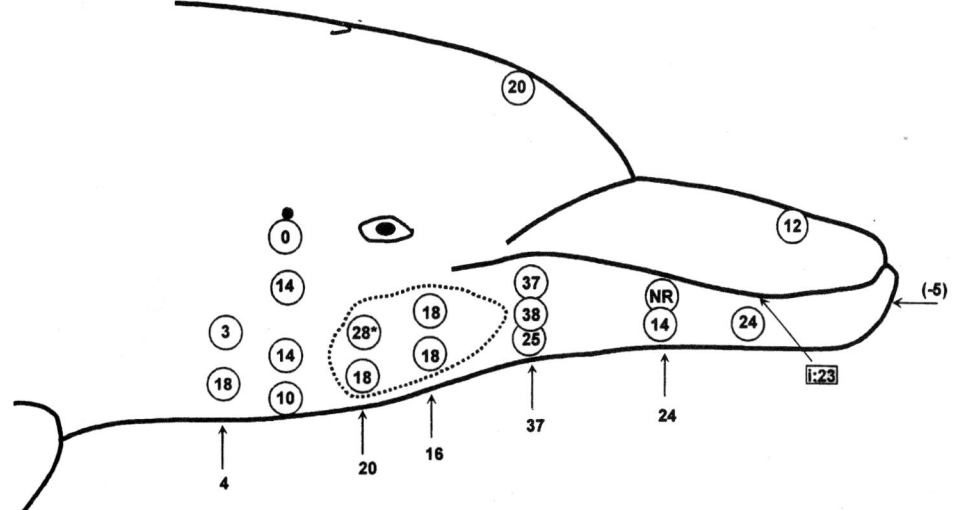

Figure 3 *Relative reception sensitivity of hearing at different locations around a dolphin's head. The higher the number, the more sensitive the region. Adapted from Møhl et al. (1999).*

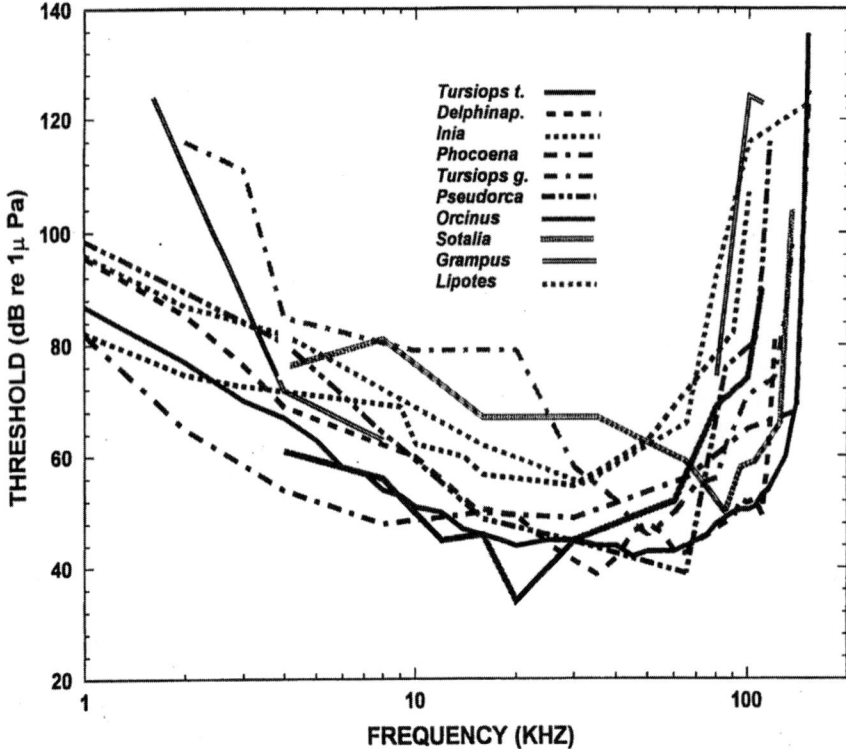

Figure 4 *Audiograms for 10 species of odontocetes.*

have limited high-frequency hearing capabilities. A killer whale is considerably larger than the smallest dolphin, yet its upper limit of hearing is approximately 120 kHz, and may therefore represent an exception to the norm. Another interesting feature of Fig. 4 is the fact that the maximum sensitivity for the different odontocetes has a very similar value, within 10 dB.

Although dolphins do not have pinnaes, the auditory system of the dolphins is directional. Sounds are received best when the source is directly in front of an animal. The receiving beam pattern of a *T. truncatus* in both horizontal and vertical planes is shown in Fig. 5. Several features of the beam patterns are worth pointing out. First, the beam becomes wider as the frequency decreases. Planar transducers also behave in a similar fashion with their beam becoming narrower as the frequency increases. Second, the major axis of the beam in the vertical plane is pointed upward with respect to the teeth line by about 5–10°. Third, the major axis in the horizontal plane is pointing directly in front of the animal parallel to the longitudinal axis of the dolphin. The receiving beam pattern can also be discussed in terms of the spatial variation in the hearing sensitivity of the dolphin. Therefore, the dolphin has the best high-frequency hearing sensitivity when sounds approach from the front and poorer sensitivity as the sound sources move to other locations about the animal's head.

III. Sound Transmission and the Characteristics of Echolocation Signals

There is a distinct difference in the echolocation signals used by odontocetes that produce whistle signals and those that do not whistle. Whistling dolphins project short, almost exponentially decaying signals with durations of 40 to 70 μsec and bandwidths of tens of kHz. Nonwhistling dolphins and porpoises project signals with much durations of 120 to 200 μsec and with narrow bandwidths that are typically less than 10 kHz. An example of a typical echolocation signal produced by an Atlantic bottlenose dolphin is shown in Fig. 6, along with a typical echolocation signal produced by a harbor porpoise (a nonwhistling animal). Whether riverine dolphins produce whistles is still an open question; however, these dolphins emit signals that are of the broadband, short duration variety. Most odontocete species produce whistles, and only a few, such as the harbor porpoise, Commerson's dolphin (*Cephalorhynchus commersonii*), Hector's dolphin (*C. hectori*), Dall's porpoise (*Phocoenoides dalli*), and pygmy sperm whale (*Kogia breviceps*), are known to not.

The amplitudes of the echolocation signals also are very different between whistling and nonwhistling odontocetes. Whistling dolphins, such as *T. truncatus, P. crassidens*, and *D. leucas*, can project echolocation signals with peak-to-peak amplitudes as high as 225 dB re 1 μPa. The center frequency of the signals used by whistling dolphins is affected by the level of the outgoing signal. The center frequency of clicks varies almost linearly with the peak-to-peak amplitude. Nonwhistling dolphins and porpoises, such as *P. phocoena* and *Phocoenoides dalli*, emit signals that normally do not exceed 170 dB re 1 μPa. Peak-to-peak source level measurements for *P. phocoena* by Au and colleagues in 1999, while the animal was performing a target detection task, indicated an average peak-to-peak source level of only 160 dB, which is considerably smaller than the 210–225 dB used by *T. truncatus*,

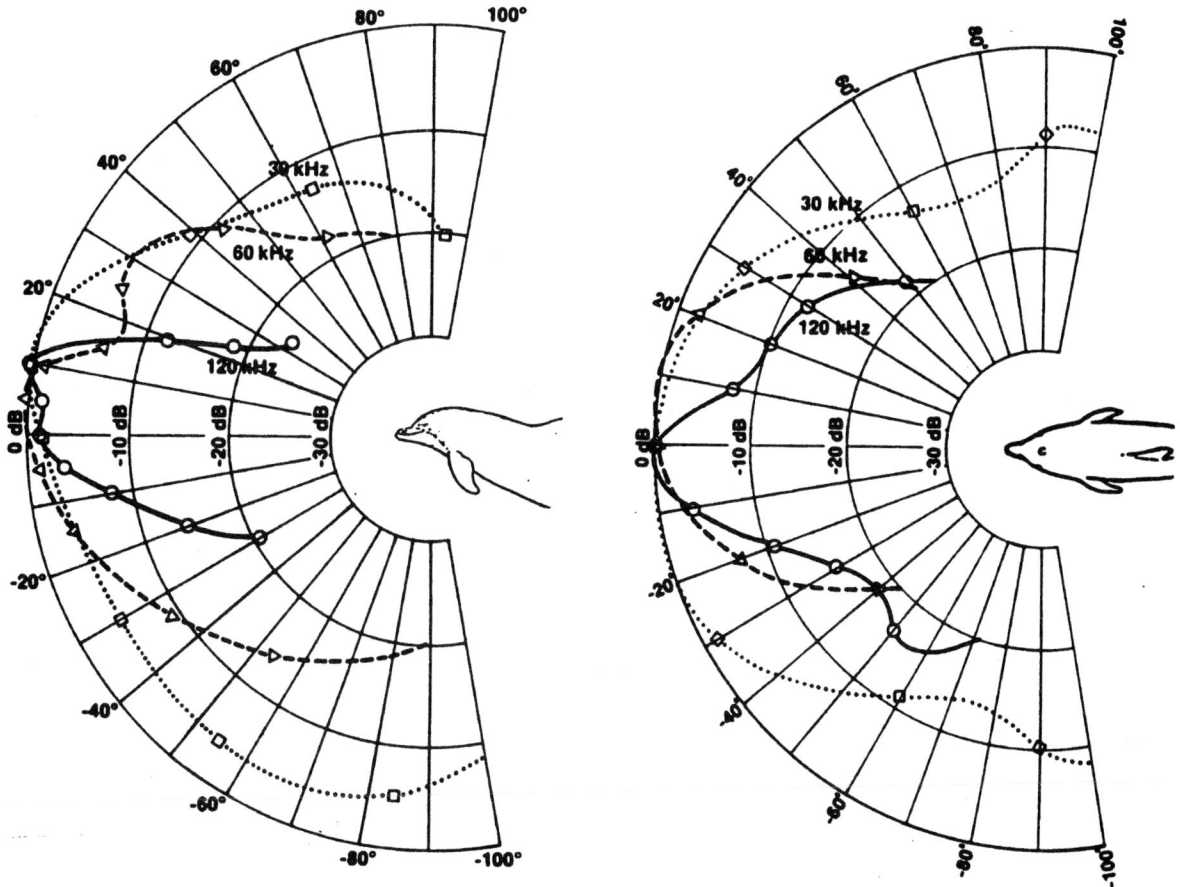

Figure 5 *The receiving beam patterns in the horizontal and vertical planes for different frequencies. Adapted from Au (1993).*

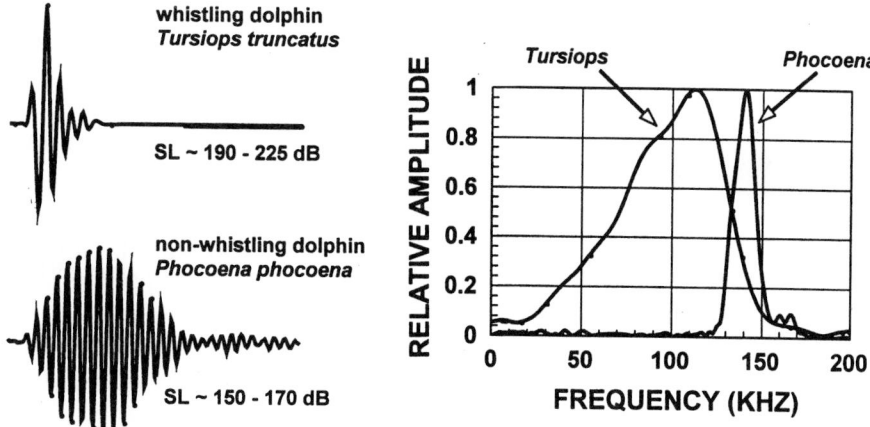

Figure 6 *Typical echolocation signal emitted by* T. truncatus *(a whistling dolphin) and* P. phocoena *(a nonwhistling porpoise). The source level (SL) is the peak-to-peak sound pressure level referenced to 1 μPa at 1 m.*

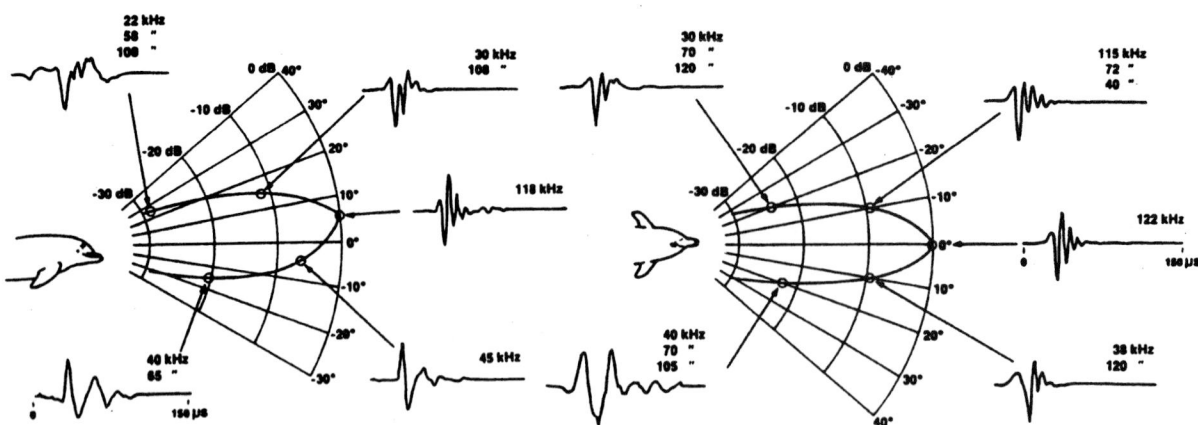

Figure 7 *The transmission beam pattern in horizontal and vertical planes. The signals shown with each beam pattern are all the same signal captured simultaneously by five hydrophones located about the dolphin's head. After Au (1993).*

P. crassidens, and *D. leucas.* The center frequency of the *P. phocoena* signal, which is typically between 120 and 145 kHz, does not depend on the level of the projected sonar signals.

Echolocation signals are projected from a dolphin's head in a beam. An example of the transmitting beam pattern for a *T. truncatus* in both horizontal and vertical planes is shown in Fig. 7. The signal shown at different angles about the animal's head is the same signal captured by an array of hydrophones. Note that only the signal traveling along the major axis of the beam is undistorted. This phenomenon occurs in horizontal and vertical planes. The numbers above each signal are the maxima in the frequency spectra of the signals, in order of descending amplitude. The further away from the major axis of the beam, the more the signal is changed. This property of the beam makes it very difficult to measure echolocation signals in the wild. Occa-sionally, dolphins in the wild may actually swim directly toward a hydrophone so that relatively true measurements can be made.

Beam pattern measurements have also been conducted for *D. leucas, P. crassidens,* and *P. phocoena.* The signals from all of these animals, with the exception of *P. phocoena,* exhibit changes in frequency content when the measuring hydrophone is located away from the major axis. However, in the case of *P. phocoena,* the signals detected by hydrophones located away from the major axis are not distorted, as can be seen in Fig. 8. Distortion does not occur because the signal has a relatively narrow bandwidth.

IV. Target Detection Capabilities

One of the most fundamental properties of a sonar system is its maximum detection range. A simple way to determine the

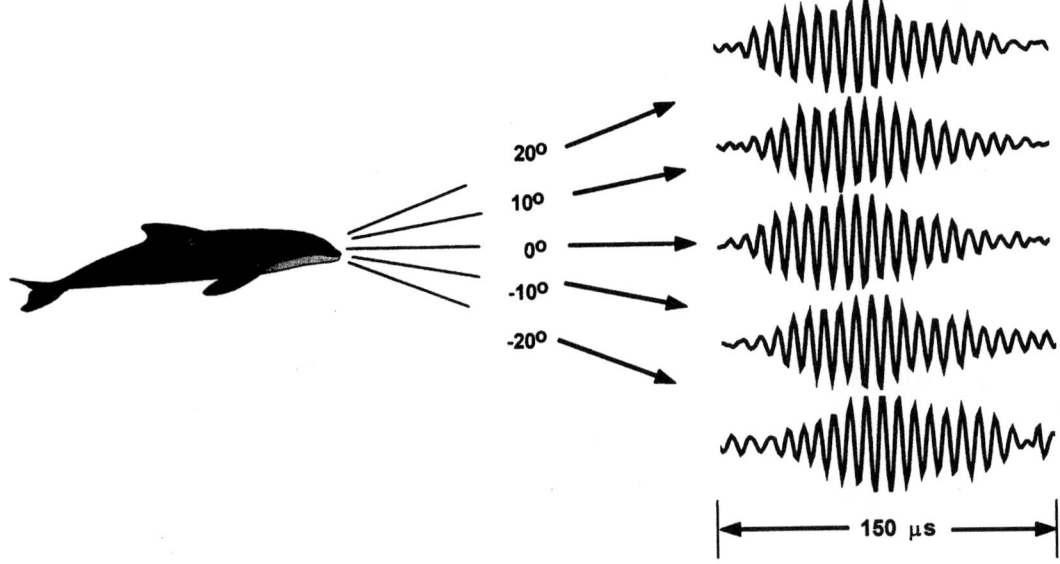

Figure 8 *The waveform of an echolocation signal detected by hydrophones spaced about the head of a* P. phocoena. *After Au et al. (1999).*

maximum target detection range of a sonar is to gradually move a specific target away from the sonar until the target can no longer be detected. Au (1993) used a 7.62-diameter water-filled stainless-steel sphere as the target to determine the maximum detection range of *T. truncatus*. The target was moved progressively away from an echolocating dolphin until the animal could no longer detect its presence. Stringent psychophysical techniques were used and many sessions were conducted in order to stabilize performance and to determine the probability of detection as a function of range. Kastelein and colleagues (1999) used the same type of target (7.62-cm diameter water-filled stainless-steel sphere) as Au and colleagues to determine the sonar detection capability of *P. phocoena*. The results of both experiments are shown in Fig. 9.

The 50% correct detection threshold for the bottlenose dolphin occurred at a range of 113 m. The 50% correct detection threshold for the harbor porpoise was approximately 26 m. An experiment by Au and Snyder took place in Kaneohe Bay, Hawaii, a bay that has a high level of snapping shrimp sounds or noise. Therefore, the bottlenose dolphin was masked by the background noise level. The harbor porpoise target detection experiment was performed at a location in the Netherlands, where the ambient background noise, between 100 and 150 kHz was essentially at sea state 0. Therefore, the harbor porpoise was not masked by background noise and yet its detection threshold was considerably shorter than *Tursiops'*. The difference in the two-way propagation losses for 113 and 26 m is 36 dB. The bottlenose dolphin typically produces clicks that are 50–60 dB greater than that of the harbor porpoise. Therefore, the large difference in the levels of projected signals can account for most, but not all, of the very large difference in the detection threshold ranges of both animals. If the target detection experiment with *T. truncatus* were conducted in a body of water with a low ambient background noise, the dolphin's target detection range would be much longer and, in that case, the difference in the two-way transmission would probably match the difference between the source levels used by the two different species.

V. Target Discrimination Capabilities

There have been many target discrimination experiments involving echolocating dolphins. Unfortunately, in many of these experiments the reflection characteristics of the targets were not measured or were measured with tone-burst signals instead of with a simulated dolphin-like signal. The experiment involving wall thickness discrimination by an echolocating dolphin is one that provided appropriate echo characteristics of the targets (Au, 1993). In this experiment, the dolphin was presented with two hollow aluminum targets separated by 20° at a range of 8 m. The standard target had a wall thickness of 0.63 cm, and the comparison targets had wall thicknesses that were different than the standard by ± 0.8, ± 0.4, ± 0.3, and ± 0.2 mm. All the targets had a length of 12.7 cm. On any given trial, the standard and comparison were introduced into the water separated by ± 20° about the center axis. The dolphin was required to swim into a hoop and echolocate the targets when a screen was lowered out of the way and then touch a paddle that was on the same side of the center line as the standard target. Two sets of targets were available so that the position of the standard could be switched on any given trial.

The results of the wall thickness experiment are shown in Fig. 10. The dolphin performed very well with correct responses in the mid-90 percentile. The animal's correct response performance became progressively worse as the difference in wall thickness decreased. The 75% correct performance threshold was 0.27 mm for the case in which the comparison targets were thinner than the standard target and 0.23 mm when the comparison targets were thicker than the standard.

Echoes from the standard target and the −0.3 mm comparison target are shown in Fig. 11. There are several cues that the animal might have used in order to perform this discrimination. One cue is the difference in the time delay between the first and the second echo component for both the standard and the comparison target. If this cue was used, it suggests that the dolphin could discriminate differences of about 0.5–0.6 μsec.

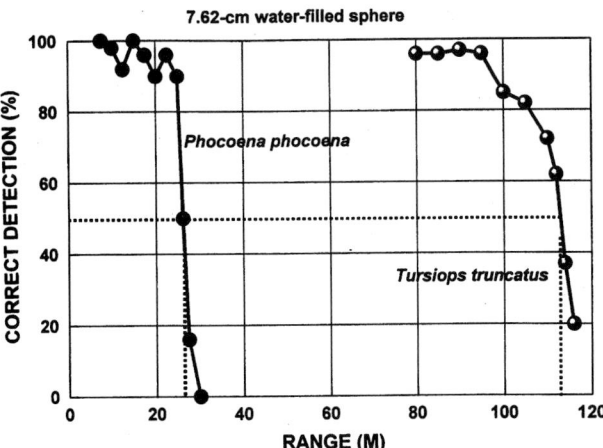

Figure 9 *Target detection performance as a function of range for* T. truncatus *and* P. phocoena. *After Au (1993) and Kastelein (1999).*

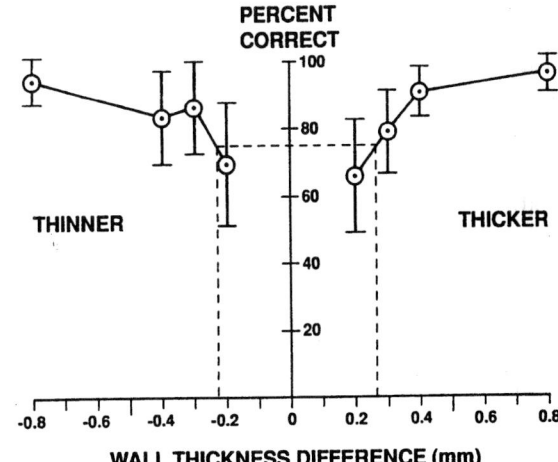

Figure 10 *Performance of an echolocating dolphin in the wall thickness discrimination experiment. Adapted from Au (1993).*

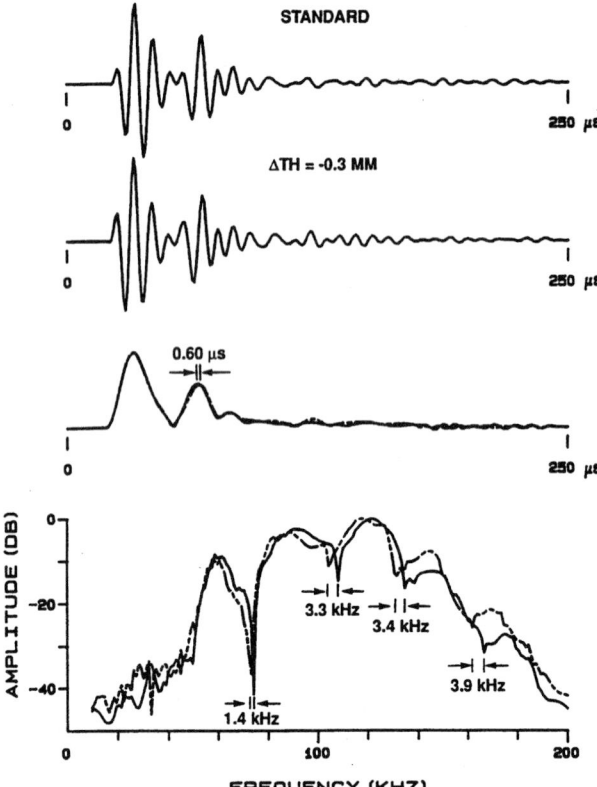

Figure 11 *Echoes from the standard and the −0.3 mm targets. The top two traces are the echo waveforms, the middle trace is the envelope of the echo waveforms overlayed upon each other, and the bottom curve is the spectra of the echoes. Adapted from Au (1993).*

Another cue could be the difference in the time-separation pitch (TSP). When humans are presented with two correlated broadband acoustic signal separated by time *T*, a time-separation pitch equal to $1/T$ can be perceived. TSP stimulii also have a frequency spectrum that is rippled. The third possible cue is the difference in the frequency spectra of the echoes. The frequency spectrum of the echo from the −0.3 mm comparison target is shifted between 2 and 3 kHz from the spectrum of the standard target. If the dolphin was using this cue, it suggests that the animal was able to perceive a frequency difference of 2–3 kHz in the broadband echoes.

VI. Auditory Nervous System Response

The fact that echolocating dolphins can produce signals with peak frequencies between 100 and 135 kHz implies a relatively fast nervous system response because the time between periods of sound transmission and the auditory response can be short, on the order 7 to 10 μsec. The speed of the auditory response can be determined by performing an electrophysiological experiment in which the time difference between the projection of the stimulus and the onset of the brain stem response is measured. This time difference is typically referred to as the

latency of response. A comparison of the latency for a variety of mammals is shown in Fig. 12. In order to fully appreciate Fig. 12, it is important to understand that the brain stem-evoked potential consists of several waves (indicated by Roman numerals) that arrive at the measuring electrodes at slightly different times. From Fig. 12, it is obvious that dolphins have an extremely fast auditory nervous response, even faster than that of a rat. The acoustic stimulus must travel into the inner ear where the cochlea nerves discharge and send the electrical pulses to higher auditory centers and eventually to the brain stem. What is very astonishing is the fact that a rat's head is considerably smaller than that of dolphins, yet dolphins have shorter responses than the rat.

The response time of the auditory system of a dolphin can be estimated by performing an integration-time experiment. Au (1993) performed a target detection experiment using phantom echoes. The phantom echo generator would digitize the outgoing signal, which was detected by a hydrophone 1.5 m in front of the dolphin stationing in a hoop. The digitized signal was stored in the memory of a personal computer, and at the appropriate time delay representing the two-way transit time for a target at 20 m, the "echo," was sent back to the dolphin via a small transducer located 2 m in front of the animal. In the initial phase, a single echo was sent to the dolphin and the dolphin's hearing threshold for the single echo was determined by varying the amplitude of the echo in a staircase fashion. In the second phase, two echoes were separated by a variable spacing that was sent back to the dolphin. The dolphin's threshold was obtained by varying the amplitude of the whole echo in a staircase fashion for various separation times between the two echoes.

The results of the phantom echo experiment are shown in Fig. 13. For an echo consisting of a single click, the threshold is shown on the ordinate of the curve in Fig. 13. Then two echoes were sent back to the animal with a separation time of 50 μsec. The threshold for the two-click echo at 50 μsec was approximately 3 dB lower than for the single click threshold. This was expected because the two-click echo had twice the amount or 3 dB more energy than the single click echo, and

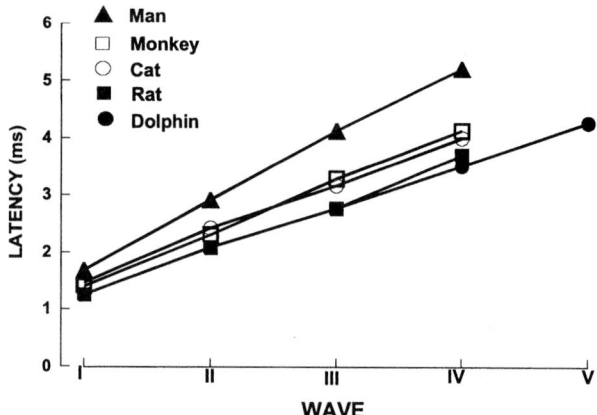

Figure 12 *Brain stem-evoked potential latency for different animals. Adapted from Ridgway (1983).*

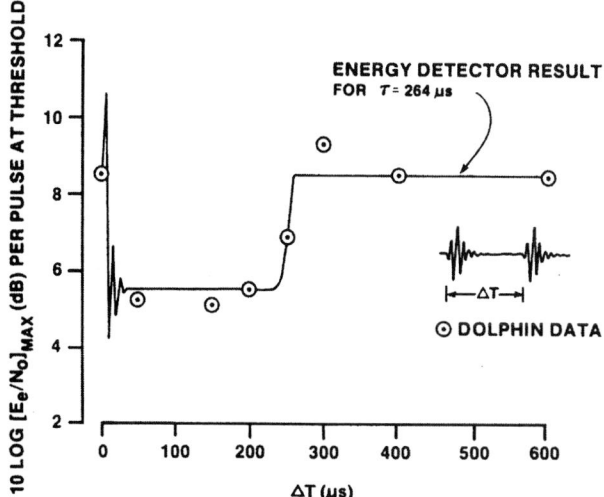

Figure 13 *Results of the phantom echo experiment. After Au (1993).*

the dolphin auditory system, like most mammals, behaves as an energy detector. As the time separation between the two clicks increased to 200 μsec, the dolphin's threshold remained very constant. However, when the time separation increased to 250 μsec and beyond, the dolphin's threshold began to move toward the threshold for a single click echo. The solid line in Fig. 13 represents the output of an energy detector having an integration of 264 μsec. The energy detector response with a 264-μsec integration time best fits the animal results. This integration time is extremely small compared to the integration time of any other mammal.

VII. Conclusions

Dolphins have very keen echolocation capabilities, much keener than any man-made sonar, especially in a shallow water environment. The use of relatively short broadband echolocation signals by whistling dolphins is probably the most important factor in the dolphin's good discrimination capabilities. The broad frequency range of hearing extending over 10 octaves and the good peak sensitivity of 30 to 40 dB re 1 μPa are certainly contributing factors in the dolphin's echolocation capabilities. Another feature of the dolphin's auditory system that contributes to its good echolocation capabilities is the extremely rapid response of its auditory nervous system. The auditory nervous system of the dolphin probably responds faster than that of any other animal if the relative dimensions of the auditory system are taken into account. Finally, dolphins are extremely mobile and can investigate objects at different aspects and angles to maximize the amount of echo information from objects and thus enhance their echolocation capabilities.

See Also the Following Articles

Brain ■ Hearing ■ Song ■ Sound Production

References

Aroyan, J. L. (1996). "Three-Dimensional Numerical Simulation of Biosonar Signal Emission and Reception in the Common Dolphin." Ph.D. Dissertation, Univ. of Calif. Santa Cruz., Santa Cruz, CA.

Au, W. W. L. (1993) "The Sonar of Dolphins." Springer-Verlag, New York.

Au, W. W. L., Kastelein, R. A., Rippe, T., and Schooneman, N. M. (1999). Transmission beam pattern and echolocation signals of a harbor porpoise (*Phocoena phocoena*). *J. Acoust. Soc. Am.* **106**, 3699–3705.

Kastelein, R. A., Au, W. W. L., Rippe, T., and Schooneman, N. M. (1999). Target detection by an echolocating harbor porpoise (*Phocoena phocoena*). *J. Acoust. Soc. Am.* **105**, 2493–2498.

Møhl, B., Au, W. W. L., Pawloski, J. L., and Nachtigall, P. E. (1999). Dolphin hearing: Relative sensitivity as a function of point of application of a contact sound source in the jaw and head region. *J. Acoust. Soc. Am.* **105**, 3421–3424.

Norris, K. S. (1968). The echolocation of marine mammals. *In* "The Biology of Marine Mammals" (H. T. Andersen, ed.), pp. 391–423. Academic Press, New York.

Purves, P. E., and Pilleri, G. (1983). "Echolocation in Whales and Dolphins." Academic Press, London.

Ridgway, S. H. (1983). Dolphin hearing and sound production in health and illness. *In* "Hearing and Other Senses: Presentations in Honor of E. G. Wever (R. R. Fay and G. Gourevitch, eds.), pp. 247–296. Amphora Press, Gronton.

Ecology, Overview

BERND WÜRSIG
Texas A&M University, Galveston

Marine mammals have entered just about all ocean habitats, and several mighty rivers and inshore seas as well. Only the deep abyss is foreign to them, but—remarkably—elephant seals (*Mirounga* spp.), sperm whales (*Physeter macrocephalus*), and several other toothed whales can "easily" dive to depths that exceed 1000 m, where it is cold and dark and where the pressure is 100 times and more what we experience on land. Perhaps just as remarkable is the fact that some of these divers, the pinnipeds, are also able to live on land, where they mate, give birth, and molt.

Morphologic, physiologic, and behavioral adaptations to the environments of marine mammals are largely driven by their food, and the habitats of their prey. Although there are various ways that ecological adaptations can be divided, this article does so by several broad-based general habitat types: open ocean, semipelagic, coastal, and riverine feeding and breeding habitats and—for pinnipeds and the polar bear—their obligatory stint on land to breed.

I. Open Ocean

There are two major types of open ocean marine mammals: "surface dwellers" and "deep divers."

A. Surface Dwellers

Most of the open ocean, or pelagic, smaller toothed whales and dolphins spend their entire lives within about 200 m of the surface. The near-surface environment is low in primary and secondary productivity except in latitudes higher than about 50° north and south of the equator. Therefore, these pelagic cetaceans travel great distances in search of food, often in large herds of hundreds to thousands. The large herds may be for better detection of prey, possible cooperative prey herding, and enhanced detection of predators such as deep water sharks and the larger cousins of dolphins, killer whales (*Orcinus orca*). All of these capabilities may be enhanced by several species traveling together, in so-called multispecies aggregations. An example in the eastern tropical Pacific (ETP), where a dolphin herd may travel over 1000 km in 1 week, is the co-occurrence of spinner (*Stenella longirostris*), pantropical spotted (*S. attenuata*), and common (*Delphinus* spp.) dolphins. These dolphins are slim-bodied (or "sleek"), built for speed and long-distance endurance. They do not have the thick blood (packed with red blood cells) so characteristic of deep divers. Instead, they feed on sporadically encountered near-surface fishes and squid, or at night on animals that rise to within several hundred meters of the surface in association with the deep scattering layer (DSL). Their occurrence in large schools has another function: the school is the social, breeding, and calf care-giving unit, and these nomadic wanderers tend to be within their "complete" society at all times. Exceptions are when young males, for example, may form separate bachelor herds or bands or when adult males move among breeding herds (as in sperm whales).

While several species of baleen whales migrate through deep water, they tend to feed on rich areas of invertebrates and fishes that are found more often close to shore. However, others habitually feed in open ocean waters. As is the case for the surface-dwelling odontocetes, baleen whales most often feed within about 200 m of the surface, as none of them are exceptionally deep divers. Blue (*Balaenoptera musculus*), fin (*B. physalus*), sei (*B. borealis*), and Bryde's and Eden's whales (*B. brydei* and/or *edeni*) are good examples of oft-pelagic, near-surface feeders. Blue and fin whales tend to feed on euphausiid crustaceans, or KRILL; whereas sei and Bryde's whales feed more commonly on shoals of fishes. All of them lunge through their food rapidly. The right (*Eubalaena* spp.) and bowhead whales (*Balaena mysticetus*) often surface-skim feed in the open ocean of productive high latitudes, whereas gray whales (*Eschrichtius robustus*) feed on ampeliscid (tube-dwelling) amphipods. They do so in waters less than 200 m deep, both near shore and far away from land in the Bering, Chukchi, and Beaufort Seas. Although rorquals are built for speed so that they can lunge into food rapidly, right whales and gray whales tend to the more rotund body shape, or chunky.

Many pinnipeds also feed near the surface and, at times, up to several hundred kilometers from shore. The smaller true seals (such as ringed seals, *Pusa hispida*, for example) and all of the eared seals are not deep divers and therefore stay near the surface in those generally higher latitude waters where they find themselves in the open sea. Near-surface feeding pinnipeds are not likely to be out on truly oceanic seas further than several hundred kilometers from land. Northern fur seals

(*Callorhinus ursinus*), however, are often found in deep pelagic waters of the north Pacific.

B. Deep Divers

Many of the larger toothed whales and a few true seals dive "deeply," or below about 500 m. Sperm whales are likely to be the champion divers. They routinely feed at depths around 500 m on fishes and squid, but can also dive to 2,000 m and more in search of the larger truly pelagic squid. Although we know little of the dive capabilities of other deep-diving odontocetes—pilot whales (*Globicephala* spp.), 20 species of beaked whales (family Ziphiidae), dwarf and pygmy sperm whales (family Kogiidae), and the false killer whale (*Pseudorca crassidens*) are good examples—it is likely that all of them are capable of greater than 500-m dives as they feed largely on midsized deep water fishes (often of the family Myctophidae) and squids. Curiously, the largest dolphin-like (or delphinid) cetacean, the killer whale, appears to feed without diving deeply. It is possible, but remains unproved, that some smaller toothed whales can evade killer whales by diving down.

The champion pinniped divers are northern and southern elephant seals (*Mirounga* spp.) as well as the Weddell (*Leptonychotes weddellii*) and probably several other large true seals. They can (but do not often) dive down to 1000 m and beyond. They feed on fishes and squid at these depths, but it has been surmised that at least some deep dives are "resting dives" as the animals conserve energy while their metabolism is largely shut down at depth. Such possible rest (or "sleep") may even help them evade detection by predators such as most active sharks, who are not deep divers, and killer whales. Because elephant seals spend only about 15% of their time at the surface, it is not really correct to call them "divers." Their life is underwater and they are indeed "surfacers" who come up only for life-sustaining air.

All deep divers have adapted physiologically and morphologically for the task. Blood and muscles have changed to hold as much oxygen as possible, and peripheral vasoconstriction and shutting off of nonvital body functions during a dive take place.

II. Semipelagic

Quite a few marine mammals habitually occur in the zone between shallow and deep water, often at the edge of the continental shelf or some other underwater feature. There is high productivity there, caused by upwelling or current systems as sea meets land, and it makes sense that this is a major point of aggregation. Sperm whales off Kaikoura, on the South Island of New Zealand, feed in such a zone near the deep Hikurangi Trench about 10 km out. However, the sperm whales are often within 1–5 km from shore, in productive waters 200 to 600 m in the deep, shore side of the trench. Blue whales of Monterey Bay, California, do so as well, as they take advantage of large stands of krill to enter the area in late summer. Dall's porpoises (*Phocoenoides dalli*) are also found in some abundance in Monterey Bay, not as frequently in very shallow nor very deep waters, but on the edge of the productive Monterey Canyon. Dozens of species and hundreds of geographic examples could be cited as those that occur in such productive "neither nearshore nor open ocean" zones.

Several dolphins are "semipelagic" in another sense. They seek out deep productive waters in areas close to shore so that they can feed in the open sea yet retreat to the shallows, often into bays and inlets or onto expansive shoals during rest. Spinner dolphins of the tropical islands of the Pacific have such a habit. During the day, they rest and socialize within island bays and lagoons, even entering atolls through narrow passes in some areas. It is believed that nearshore rest is to avoid large oceanic swells and trade winds, as well as predation by large oceanic sharks. At night, these dolphins head out to sea, often only 1–5 km from land off these abruptly rising volcanic islands. The dolphins meet the DSL as it comes to within several hundred meters of the surface at night and thus have a food resource available that these only-average divers could not obtain during the day, when the DSL is 600 m or more below. Atlantic spotted dolphins (*Stenella frontalis*) appear to do the same, but have daytime rest over an expansive shallow area: the Grand Banks of the Bahamas, only 6–10 m deep.

Pinniped females that go on foraging dives in between nursing their young on land, such as Galapagos fur seals (*Arctocephalus galapagoensis*), also use the productive shelf and drop-off waters to feed while—in their case—needing to return to land to take care of their young.

III. Coastal

Many marine mammals can be termed "coastal," and because all of the various taxonomic orders and suborders have coastal representatives, one to several examples of each group are given.

The most coastal baleen whale is undoubtedly the gray whale, for it feeds in shallow waters of the Bering Sea, usually but not always near coasts; travels on its immense migration from the Bering Sea to Baja California, Mexico—and back—along the coast; and mates and calves near and in coastal lagoons of the subtropics. It is likely that this rather slow cetacean hugs the coastline for safety (mainly, one surmises, for its young) against shark and killer whale predation. It probably also uses the coast to navigate. It would not be surprising, although present information is not clear on this point, if gray whales use the depth contours, rocky outcroppings of headlands, and other near-coastal features as signs of location as surely as we find our way to and from the supermarket. The coastline also allows them to find clouds of mysids, small aggregating invertebrates, among kelp beds, and to occasionally feed on stands of in-benthic invertebrates while on migration. A second "coastal" animal is the humpback whale (*Megaptera novaeangliae*), for it feeds in bays and inlets, breeds near islands, and only uses deep oceanic waters to get to and from these ends of migration. In the northeast Pacific, humpbacks feed in the fjord-like bays of southern Alaska and breed around the Hawaiian, and Mexican Revillagigedo, islands.

Odontocete cetaceans have many coastal representatives, with the best studied of them being the bottlenose dolphin (*Tursiops truncatus*). While separate populations of this highly adaptable species can exist in deep oceanic waters as well, it is the coastal form that has taken our fancy and makes for one of the better captive animals, presumably because it feels at home in small groups and with confines of cliffs; rocks, bayous, and channels. Bottlenose dolphins variably nose and poke their way among rocks to feed; feed on schooling fishes in the nearshore, at times trapping schools against a beach or cliff; feed on the bottom; and encircle prey as a cooperating group in the open coast sea. Dolphins of the *Cephalorhynchus* genus of southern oceans tend toward coastal living, as do the humpback dolphins of the genus *Sousa*, harbor porpoises (*Phocoena phocoena*), and beluga whales of the arctic (*Delphinapterus leucas*). Interestingly, these animals appear to have some form of fission–fusion society, traveling in subgroups of variable size from day to day. It is likely that they aggregate in small groups for greatest efficiency in hunting and that the social or breeding unit is all of the small groups of an area that get together at some time throughout a year, but never all at once. Most but not all coastal waters are turbid as well, and it may be that echolocation and communicative sounds are particularly well developed in these animals.

Many pinnipeds have coastal representatives, especially for the physically smaller species. California sea lions (*Zalophus californianus*) rest on the shore and feed in the coastal zone, hardly ever venturing further than several kilometers from land. Harbor seals (*Phoca vitulina*) and the two living but highly endangered tropical monk seals (*Monachus* spp.) do so as well.

Sea otters (*Enhydra lutris*), marine otters of Chile (*Lontra felina*), and the sirenians are all coastal shallow-water feeders. Otters feed on invertebrates on the bottom or on kelp-associated fishes. While many populations of sea and marine otters do not frequently haul out on land, they use kelp beds as resting stations and perhaps as a means to hide from sharks and killer whales. The West Indian manatee (*Trichechus manatus*) and the dugong (*Dugong dugon*), the latter largely of the nearshore Indian Ocean, feed on sea grasses and are thereby restricted to the shallows.

IV. Riverine

While the term "marine mammals" is meant for mammals that take all or most of their sustenance from the sea, several species are included that have gone to a largely freshwater environment. Because these have close taxonomic affiliations to several other marine mammals, this inclusion makes sense.

There are several obligate river dolphins: the susu and bhulan (now listed as subspecies within one species, [*Platanista gangetica*]) of the Indian subcontinent; the baiji (*Lipotes vexillifer*) of the Yangtze River of China; and the boto or Amazon river dolphin (*Inia geoffrensis*) that also occurs in the Orinoco basin of South America. These dolphins live their lives in mighty rivers, feeding on invertebrates and fishes, generally in small groups numbering fewer than about six animals. Their eyes have adapted to the less saline environment, and their kidneys do not need to process the salty foods of the ocean. It is likely that they would not survive in salt water. The Amazonian manatee (*Trichechus inunguis*) is also restricted to the extensive freshwater system of the Amazon basin.

In addition to obligate river dolphins and the Amazonian manatee, there are several mammals that are facultative, those who have populations that occur in rivers and those who go in and out of rivers to the adjacent ocean. Of the first type are finless porpoises (*Neophocaena phocaenoides*) that occur throughout nearshore waters of southern Asia and a bit of the Indian Ocean, but have a thoroughly fresh water population in the Yangtze River. Recent work shows that the freshwater form has eyes, skin, and kidneys that are adaptively different from their ocean-going conspecifics. As well, the diminutive tucuxi dolphin (*Sotalia fluviatilis*) occurs nearshore along much of the tropical Atlantic Central and South American coast, but as separate populations in the Amazon River basin. Of the second type of marine mammals, where some go in and out of rivers as members of the same population, are Irrawaddy dolphins (*Orcaella brevirostris*), bottlenose dolphins, belugas, and West Indian and West African manatees (*Trichechus manatus* and *T. senegalensis*). To date, there are no well-defined morphologic or physiologic differences between those who frequent fresh waters more than others, and it is assumed that this wide salinity tolerance is itself an adaptation that allows exploitation of food resources in ecologically diverse realms. Belugas seem to enter rivers more often during a concentrated period of skin sloughing, or molt; these are the only whales known to molt.

Almost all pinnipeds are generally tied to the sea to feed, but a form of the harbor seal and the Asian Lake Baikal (fresh water) and Caspian Sea (somewhat salty) seals (*Pusa* spp.) occur in land-locked areas. They occur in remnants of areas that were once connected to oceans.

V. Life on Land

Polar bears do considerable feeding on land or ice, pinnipeds all need to come to land to give birth, and sea otters do so variably by population. Off California, sea otters give birth in the water, but are usually surrounded and buoyed by *Macrocystis* sp. giant kelp fronds. While some pinnipeds, such as the walrus (*Odobenus rosmarus*) and Weddell seals, mate in water, most do so on solid land or ice, and all females need to come to solid substrates to give birth and to suckle their young. Indeed, a newborn pinniped (and polar bear) is not yet a marine mammal and would become overexposed rapidly and die if it were to be dunked into water. The natal pelt of most true or phocid seals is a downy fur, or lanugo, that holds insulating hair but is not waterproof; they have brown fat, a type of lipid that breaks down rapidly to generate heat; and they instinctively huddle near mother and each other to stay warm.

Pinnipeds have shortened and greatly changed fore and hind flippers, modified beautifully for swift and precise movement in water. However, they have had to compromise their morphology to keep a bit of it—so very necessary for procreation—available for life on land. It is now known that early cetaceans lived a similarly dual existence, with morphologic, physiologic, and likely behavioral compromises to survive in both realms. It is tempting to speculate whether pinnipeds, given another 20 million years of evolution, can make the same total transition to the sea.

See Also the Following Articles

Cetacean Ecology ■ Diving Physiology ■ Ocean Environment ■ Pinniped Ecology

References

Berta, A., and Sumich, J. L. (1999). "Marine Mammals: Evolutionary Biology." Academic Press, San Diego.

Bonner, W. N. (1989). "Whales of the World." Facts on File Press, New York.

Costa, D. P. (1993). The relationship between reproductive and foraging energetics and the evolution of the Pinnipedia. *Symp. Zool. Soc. Lond.* **66,** 293–314.

Estes, J. A., and Duggins, D. O. (1995). Sea otters and kelp forests in Alaska: Generality and variation in a commuity ecological paradigm. *Ecol. Monogr.* **65,** 75–100.

Evans, G. P. H. (1987). "The Natural History of Whales and Dolphins." Facts on File Press, New York.

LeBoeuf, B. J., and Laws, R. M. (eds.) (1994). "Elephant Seals." Univ. of California Press, Berkeley.

Reynolds, J. E., III, and Odell, D. E. (1991). "Manatees and Dugongs." Facts on File Press, New York.

Reynolds, J. E., III, and Rommel, S. A. (eds.) (1999). "Biology of Marine Mammals." Smithsonian Institution Press, Washington, DC.

Stirling, I. (1988). "Polar Bears." Univ. of Michigan Press, Ann Arbor, MI.

Elephant Seals

Mirounga angustirostris and *M. leonina*

MARK A. HINDELL

University of Tasmania, Hobart, Australia

I. Diagnostic Characters

The northern elephant seal (*Mirounga angustirostris*) and the southern elephant seal (*Mirounga leonina*) are the largest pinnipeds. The most striking characteristic of both species is the pronounced SEXUAL DIMORPHISM, with males weighing 8–10 times more than females. Male southern elephant seals have been recorded to weigh 3700 kg, whereas females only weigh between 400 and 800 kg. This makes the elephant seal the most sexually dimorphic of any mammal (Fig. 1). There are other pronounced secondary sexual differences in morphology, all of which are related to the highly polygynous mating strategy of the species. Most notable of these is the large proboscis of the male that plays a key part in dominance displays with other males.

The evolutionary origins of the species are unclear, with estimates of the divergence from a common ancestor ranging from as little as 10,000 years ago to as far back as the Pleistocene. Morphological differences between the species are, however, quite distinct and they are readily differentiated. The southern species is larger, with northern elephant seals rarely reaching more than 2300 kg. In both species, adult females exhibit a considerable range in body weight and there are no clear differences between them in this feature. Adult males of the northern species have a longer proboscis than the southern species. Northern elephant seals also have a more developed chest shield, a region of the neck, chest, and shoulders of thickened and scared skin associated with fighting. In the northern species, this has a distinctive red coloration. Females lack the proboscis and chest

Figure 1 *Adult male (background), adult female, and new-born pup southern elephant seals. Adult males weigh 8–10 times as much as the females, making them the most sexually dimorphic of all mammals.*

shields, but northern females are distinguished by a noticeably narrower and flatter nose than in the southern species.

II. Distribution

Despite their physical similarities, the two species have very different geographic distributions, with at least 8000 km separating them. Southern elephant seals have a more extensive range, with breeding sites on islands scattered right around the sub-Antarctic (Fig. 2). Very occasionally, pups are even born on the Antarctic mainland. Studies have indicated that when not ashore during the breeding season or for their annual molt, southern elephant seals utilize most of the Southern Ocean ranging from waters north of the Antarctic Polar Front (sometimes called the Antarctic Convergence) to the high Antarctic pack ice. There is some separation of feeding areas between the sexes, with males tending to feed in the more southerly waters associated with the Antarctic continental shelf.

Northern elephant seals have a limited breeding distribution, pupping at approximately 15 colonies between Point Raines in northern California to the Baja California Peninsula in Mexico (Fig. 3). Most of the colonies occur on offshore islands, but a small number occur on the mainland coast. As with southern elephant seals, the northern species disperses widely during the nonbreeding phase of its annual cycle. Many individuals travel northward along the North American coast to feed in the Gulf of Alaska and the Aleutian Islands, which is a round trip of more than 10,000 km. This MIGRATION is even more remarkable as many individuals make it twice per year, returning to their southern breeding grounds to molt. The northern species also exhibits sexual differences in foraging areas, with males tending to use the more northerly areas and females heading in a more northwesterly direction and feeding in deep oceanic waters of the North Pacific Ocean.

III. Human Interactions

Today, both species of elephant seal are relatively free of adverse interactions with humans. Southern elephant seals are only rarely captured in the nets of Southern Ocean fishing fleets, and this has never been reported for the northern species. There are some grounds for concern that some large-scale fisheries may be competing with the seals for preferred prey species, but this is difficult to quantify given the current paucity of information on the diet of both species.

However, both species have a long history of direct exploitation by humans as they were hunted extensively during the 1800s for their blubber, which yielded an unusually high-quality oil. In the case of the northern elephant seal, this hunting was so intense that the populations were reduced to a small group breeding on a single island by 1890. The more widespread and southerly distribution of the southern species meant that the exploitation was less intensive. Nonetheless, the seals were reduced dramatically at all of their major breeding sites. The exploitation of southern elephant seals also continued longer than that of northern elephant seals, with commercial operations continuing until 1919 at Macquarie Island and until 1964 at South Georgia.

IV. Geographic Variation

As a direct consequence of the extensive hunting in the 19th century, northern elephant seals passed through an intense genetic bottleneck, which has seen almost all genetic variation removed from the population. The relatively recent expansion onto several islands has not yet resulted in discernible genetic variation between the breeding groups.

In contrast, southern elephant seals have quite a clear genetic structure, with four distinct stocks: the southern Pacific Ocean, the south Atlantic, the southern Indian Ocean, and a small, but increasing population of Peninsula Valdes in Argentina. The integrity of the subpopulations appears to be maintained by the extremely low interchange rates between populations. Although genetically distinct, animals from the subpopulations are indistinguishable from each other in external features.

V. Population Size and Trends

Northern elephant seals are presently undergoing a rapid population increase and range extension, whereas the southern species is currently experiencing a decline in two of its three major populations.

Population declines in the order of 50% have been recorded in both the southern Indian and the southern Pacific Ocean populations since the 1950s and 1960s, whereas the populations in the South Atlantic are stable, or increasing. The current estimated total population size for southern elephant seals is 640,000. The underlying cause of the declines are presently unclear, but are thought to be related to changes in the distribution and abundance of the seal's prey.

Conversely, northern elephant seal populations are currently increasing at a rate of approximately 6% per annum. This is the latest phase in one of the most remarkable population recoveries of any mammal. The total number of northern elephant seals in 1890 is thought to have been less than 100 individuals after 50 years of intensive and indiscriminate hunting by sealers. The last published estimates of the population put the total population at 127,000 in 1991.

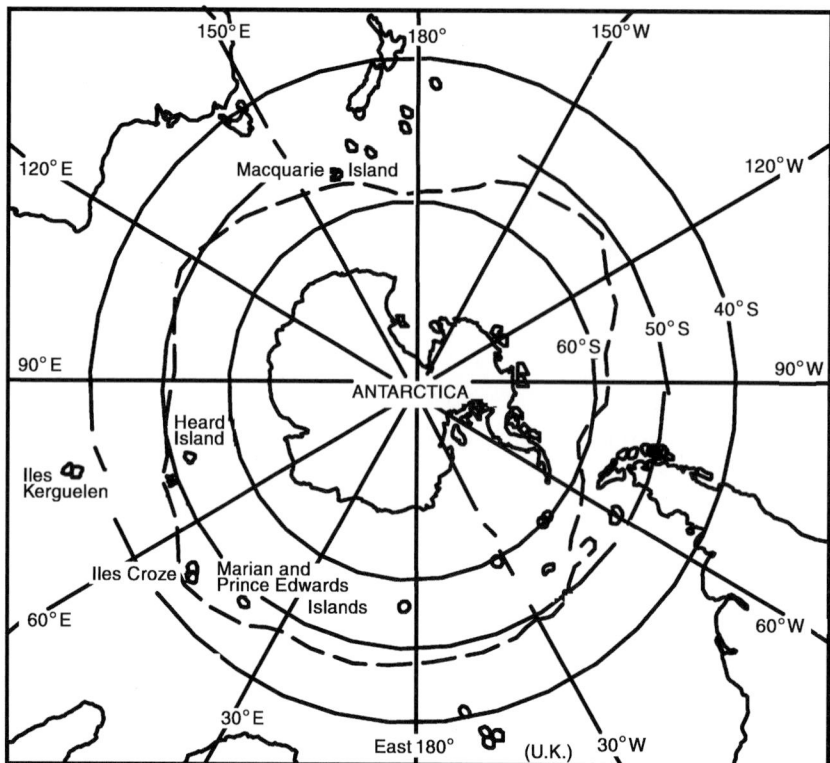

Figure 2 *Map of the Southern Ocean indicating the major southern elephant seal breeding islands.*

VI. Breeding

Both species of elephant seal are highly polygynous, with large, dominant males (alpha males) presiding over large aggregations of females, know as harems. Competition between males for the alpha position is intense and leads to spectacular fights (Fig. 4). Successful males will have almost exclusive ac-

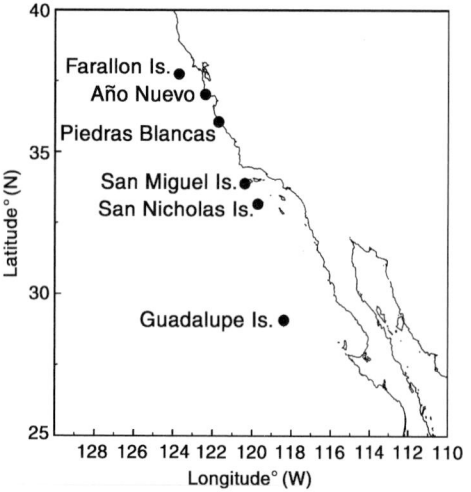

Figure 3 *Map of the West coast of North America indicating the major northern elephant seal breeding sites.*

cess to harems consisting of up to 100 females, and so the reproductive benefits of success are very high. This has led to the evolution of the pronounced secondary sexual characteristics of immense body size and exaggerated proboscis.

The annual breeding cycle begins when the largest males haul out on deserted beaches (in August for *M. leonina* and in December for *M. angustirostris*). Pregnant females then haul out in large numbers, aggregating into harems, and giving birth to their single pup 2–5 days after arriving. The females stay with their pup throughout the ensuing lactation period, never feeding and relying on their thick BLUBBER layer to sustain them and to supply the many liters of milk required by the rapidly growing pup. At birth the pups weigh between 30 and 40 kg in both species, but by the time they wean, southern elephant seal pups weigh approximately 120–130 kg and northern elephant pups weigh approximately 140–150 kg. The difference in weaning weight is due to the slight difference in the duration of lactation in the two species, with southern elephant seals weaning at 23–25 days and northern elephant seals at 26–28 days.

Several days before weaning their pups, the females come into estrus and are mated by the dominant males. Although fertilization takes place at this time, the blastocyst does not implant until several months later. This ability, known as delayed implantation, is common to many pinnipeds. Once the pup is weaned, the females depart to sea, leaving the pups to fend for themselves. The pups spend the next 4 to 6 weeks teaching themselves to swim and hunt, during which time they rely

Figure 4 *Two adult male northern elephant seals fighting during the breeding season. Note the enlarged proboscis and predominant chest shields.*

heavily on the large reserves of blubber that they got from their mothers while suckling. When the pups eventually leave their natal beaches they spend the next 6 months at sea. This is a difficult time for the pups and as many as 30% of them die at this time.

VII. Molt

Monachine seals all have an unusual annual molt, which entails the shedding of epidermal tissue in addition to the hair. The rich supplies of blood required at the body surface for the new skin and hair require the animals to leave the water in order to conserve body heat. The seals therefore spend 3–5 weeks fasting ashore during this time, once again relying on stored blubber to supply their energy requirements.

VIII. Behavior at Sea

Elephant seals spend more than 80% of their annual cycle at sea, feeding intensively to build up the blubber stores required to support them during breeding and molting haul outs. The seals prey on deep-water squid and fish and, as a result, have developed the remarkable ability to dive to depths in excess of 1500 m and for as long as 120 min. While these values are the extremes of those recorded, even the average values are

impressive. Adult females routinely make dives of 20 min and reach depths of 400–800 m. Paradoxically, although the males generally dive for longer, about 30 min, they often do not go as deep. This is a reflection of their tendency to feed over continental shelves, whereas females use deeper open water. Seals spend as much as 90% of the time submerged, the majority of it hunting for food, but other behaviors, such as traveling from place to place, and apparently even resting, take place at depths of more than 200 m.

See Also the Following Articles

Blubber ▪ Mating Systems ▪ Pinniped Evolution

References

Le Boeuf, B. J., and Laws, R. M. (1994). "Elephant Seals: Population Ecology, Behaviour and Physiology." Univ. of California Press, Berkley.
Slade, R. W., Moritz, C., Hoelzel, A. R., and Burton, H. R. (1998). Molecular population genetics of the southern elephant seal *Mirounga leonina*. *Genetics* **149,** 1945–1957.
Stewart, B. S. (1997). Ontogeny of differential migration and sexual segregation in northern elephant seals. *J. Mammal.* **78,** 1101–1116.

Endangered Species and Populations

JOHN E. REYNOLDS III
U.S. Marine Mammal Commission,
Bethesda, Maryland

DOUGLAS P. DEMASTER
National Marine Fisheries Service,
Seattle, Washington

GREGORY K. SILBER
National Marine Fisheries Service,
Silver Spring, Maryland

Eschricht and Reinhardt (1861) and Scammon (1874) warned well over a century ago that certain species or populations of marine mammals had reached dangerously low levels and that survival of these groups was in peril. Since that time, formal and informal efforts have been made to prevent extinction of marine mammals perceived to be in jeopardy. Some of the efforts have been undertaken by a single nation—witness the Marine Mammal Protection Act (MMPA) of 1973, which protects *all* marine mammals in United States waters. Other efforts, such as the administration of the Convention on International Trade in Endangered species of Wild Fauna and Flora (CITES), have been intended to apply on a global scale. As different groups have attempted to develop lists of species and populations with pressing conservation needs, at least four criteria are frequently considered: (a) population size and demography, (b) the extent to which human activities adversely affect the species/population either directly or indirectly, (c) the adequacy and protection of habitat deemed necessary for the survival of the group in question, and (d) the extent of markets and trade for products from the population/species to be conserved. Management agencies, in concert with scientists, have begun developing quantitative criteria that will reduce the subjectivity involved in categorizing species as endangered under the U.S. Endangered Species Act (U.S. ESA) (e.g., IUCN, 1996; Gerber and DeMaster, 1999).

This article focuses on groups of marine mammals that seem to truly require protection in order to survive well into the 21st century and beyond. We provide definitions of criteria under which such groups are classified by some of the more powerful and inclusive legislation, conventions, or other approaches, such as CITES, the MMPA, the U.S. ESA, and the International Union for the Conservation of Nature and Natural Resources' (IUCN; now called the World Conservation Union) "Red List of Threatened Animals" (IUCN, 1996). We describe reasons why some marine mammals are in critical condition today, and we look for lessons and trends that may suggest why some groups recover and others do not. We do not assume, however, that all endangered groups of organisms may be able to recover; in some cases, humans have usurped or modified so much important habitat that a group may always be at risk, regardless of what is done in the future. Finally, we suggest approaches that may be fruitful in the fight to preserve species in the future.

This article deliberately uses the word "group" quite often. Although national legislation such as the U.S. ESA would, by its very name, appear to approach conservation at the species level, the actual intent of the U.S. Congress in passing the legislation was that management be carried out at the stock or population level by local or regional resource managers. Management at the species level can lead to loss of biological diversity if local populations are exterminated, even though the species as a whole appears healthy. The importance of management at the stock or population level becomes clear when one considers the status of the Steller sea lion (*Eumetopias jubatus*). In 1990, the species was designated as threatened under the U.S. ESA. By the late 1990s, however, scientists had demonstrated that (a) there were two discrete stocks of the Steller sea lion; (b) the eastern stock was smaller than the stock occupying the western part of the species range but was stable or even possibly increasing in numbers; and (c) the western stock was in precipitous decline, with counts at some sites dropping by 90% or more since the late 1970s. Accordingly, in 1997 the listing under the U.S. ESA was modified so that the western stock was reclassified as endangered, whereas the eastern stock remained listed as threatened (U.S. Marine Mammal Commission, 1999).

The Steller sea lion example illustrates or reinforces some important points. First, good science is essential for effective management. Second, with endangered marine mammals (and other organisms as well), we need to work to better define what

"distinct population segments" (or "distinct vertebrate populations," e.g., as defined later according to the U.S. ESA) are, so that management and conservation efforts can be focused on the most critical groups and areas. Third, the management process should be adaptive and flexible in order to accommodate new data. Finally, even though many stocks, populations, and even species may be accorded special protection, a few are truly in desperate need of assistance. Unfortunately, due either to lack of scientific information that identifies separate populations or to other reasons, management still occurs at the species level in some cases. Worse, the general public may have serious misconceptions about modern management practices and have often equated the precarious status of a population (e.g., the North Atlantic right whale, *Eubalaena glacialis*) with the status of all species in a taxonomic group (e.g., baleen whales).

Only one marine mammal species has definitely become extinct since Eschricht and Reinhardt (1861) warned of the possible demise of the bowhead whale in the middle of the 19th century: the Caribbean monk seal, *Monachus tropicalis*. Steller's sea cow (*Hydrodamalis gigas*) had already been driven to extinction by the end of the 18th century, and the Japanese sea lion (*Zalophus japonicus*) has probably gone extinct sometime in the last 50 years (Rice, 1998). However, populations of some other species have also become extinct (e.g., the North Atlantic gray whale, *Eshrictius robustus*), and other stocks and even whole species may not survive long into the 21st century. We must change how we value and manage marine mammals if we intend for them to remain a functioning part of the marine ecosystem.

I. Listings of Specially Protected Marine Mammals

Although globally and regionally there are a number of lists of specially protected species, we provide three widely accepted lists of protected marine mammals. Table I lists endangered and threatened (from the U.S. ESA) and depleted (from the U.S. MMPA) species and stocks. Table II lists marine mammals found in Appendix I and Appendix II of CITES. Table III lists those marine mammals categorized by the IUCN as critically endangered (CR), endangered (EN), and vulnerable (V). We did not herein reproduce the IUCN "Red List of Threatened Animals" for all marine mammal species because of spatial constraints.

Because predicting extinction probabilities is fraught with uncertainty, objectively classifying stocks and populations according to their precise level of vulnerability is very difficult. Although all of the criteria used to classify species at risk are credible, we prefer the use of criteria similar to those reported in the IUCN "Red List of Threatened Animals."

It should be noted that of the 128 extant species of marine mammals, only 3 are presently listed by the IUCN as CR [i.e., Mediterranean monk seal (*Monachus monachus*), vaquita (*Phocoena sinus*), and baiji (*Lipotes vexillifer*)]. However, it should further be noted that 8 species are listed as EN and 16 species as VU, while a classification of lower risk has been assigned to 14 species of marine mammals. Finally, status of 38 species of marine mammals has been determined to have insufficient data for classification.

TABLE I
Marine Mammal Species and Populations Listed as Endangered (E) or Threatened (T) under the U.S. Endangered Species Act and Depleted (D) under the Marine Mammal Protection Act[a]

Common name	Scientific name	Status	Range
Manatees and dugongs			
West Indian manatee	*Trichechus manatus*	E/D	Caribbean Sea and North Atlantic from southeastern United States to Brazil; and Great Antilles Islands
Amazonian manatee	*Trichechus inunguis*	E/D	Amazon River basin of South America
West African manatee	*T. senegalensis*	T/D	West Africa coasts and rivers; Senegal to Angola
Dugong	*Dugong dugon*	E/D	Northern Indian Ocean from Madagascar to Indonesia; Philippines; Australia; southern China; Palau
Otters			
Marine otter	*Lutra felina*	E/D	Western South America; Peru to southern Chile
Southern sea otter	*Enhydra lutris nereis*	T/D	Central California coast
Seals and sea lions			
Hawaiian monk seal	*Monachus schauinslandi*	E/D	Hawaiian Archipelago
Caribbean monk seal	*M. tropicalis*	E/D	Caribbean Sea and Bahamas (probably extinct)
Mediterranean monk seal	*M. monachus*	E/D	Mediterranean Sea; northwest African coast
Guadalupe fur seal	*Arctocephalus townsendi*	T/D	Baja California, Mexico, to southern California
Northern fur seal	*Callorhinus ursinus*	D	North Pacific Rim from California to Japan
Western North Pacific Steller sea lion	*Eumetopias jubatus*	E/D	North Pacific Rim from Japan to Prince William Sound, Alaska (east of 144°W longitude)
Eastern North Pacific Steller sea lion	*E. jubatus*	T/D	North Pacific Rim from Prince William Sound, Alaska, to California (east of 144°W longitude)
Saimaa seal	*Phoca hispida saimensis* (= *Pusa hispida saimensis*)	E/D	Lake Saimaa, Finland
Whales, porpoises, and dolphins			
Baiji	*Lipotes vexillifer*	E/D	Changjiang (Yangtze) River, China
Indus river dolphin	*Platanista minor* (= *P. gangetica minor*)	E/D	Indus River and tributaries, Pakistan
Vaquita	*Phocoena sinus*	E/D	Northern Gulf of California, Mexico
Northern offshore spotted dolphin	*Stenella attenuata*	D	Eastern tropical Pacific Ocean
Eastern spinner dolphin	*S. longirostris orientalis*	D	Eastern tropical Pacific Ocean
Mid-Atlantic coastal bottlenose dolphin	*Tursiops truncatus*	D	Atlantic coastal waters, New York to Florida
Northern right whale	*Eubalaena glacialis* (includes *E. japonica*)	E/D	North Atlantic, North Pacific Oceans; Bering Sea
Southern right whale	*E. australis*	E/D	South Atlantic, South Pacific, Indian, and Southern Oceans
Bowhead whale	*Balaena mysticetus*	E/D	Arctic Ocean and adjacent seas
Humpback whale	*Megaptera novaeangliae*	E/D	Oceanic, all oceans
Blue whale	*Balaenoptera musculus*	E/D	Oceanic, all oceans
Finback or fin whale	*B. physalus*	E/D	Oceanic, all oceans
Western North Pacific gray whale	*Eschrichtius robustus*	E/D	Western North Pacific Ocean
Sei whale	*Balaenoptera borealis*	E/D	Oceanic, all oceans
Sperm whale	*Physeter macrocephalus*	E/D	Oceanic, all oceans

[a] As of December 31, 1999 (from Marine Mammal Commission, 1999). Equivalent species names used by Rice (1998) appear in parentheses.

Not all species in need of protection require that protection equally (see also Clapham *et al.*, 1999). The western North Atlantic stock of the humpback whale (*Megaptera novaeangliae*), which numbers more than 10,000 individuals, scarcely merits the same level of protection as do vaquita (perhaps 500–600 individuals left), northern right whale [probably fewer than 300 individuals in the North Atlantic, and probably low 100s of animals in the North Pacific (now recognized as a separate species, *E. japonica*)], Mediterranean monk seal (around 300 left in the wake of an epizootic in 1997, which killed over half of the members of the largest colony in northwest Africa), and baiji (a few tens of animals in isolated stretches of the Yangtze River).

TABLE II
Marine Mammals Listed under Appendices I and II of CITES[a]

Order	Species	Common name	Appendix
Cetacea	*Balaena mysticetus*	Bowhead	I
	Balaenoptera acutorostrata (includes *B. bonaerensis*)	Minke whale	I
	B. borealis	Sei whale	I
	B. edeni	Bryde's whale	I
	B. musculus	Blue whale	I
	B. physalus	Fin whale	I
	Berardius spp.	Beaked whales	I
	Caperea marginata	Pygmy right whale	I
	Eschrichtius robustus	Gray whale	I
	Eubalaena spp.	Right whales	I
	Hyperoodon spp.	Bottle-nosed whales	I
	Lipotes vexillifer	Chinese river dolphin; white flag dolphin	I
	Megaptera novaeangliae	Humpback whale	I
	Monodon monoceros	Narwhal	II
	Neophocaena phocaenoides	Finless porpoise	I
	Phocoena sinus	Vaquita; Gulf of California harbor porpoise	I
	Physeter catodon (= *P. macrocephalus*)	Sperm whale	I
	Pontoporia blainvillei	La Plata River dolphin	II
	Sotalia spp. (= *S. fluviatilis*)	Humpbacked dolphins	I
	Sousa spp.	Humpbacked dolphins	I
Carnivora	*Arctocephalus australis*	Southern fur seal	II
	A. galapagoensis	Galapagos fur seal	II
	A. philippii	Juan Fernandez fur seal	II
	A. townsendi	Guadalupe fur seal	I
	Enhydra lutris nereis	Southern sea otter	I
	Mirounga leonina	Southern elephant seal	II
	Monachus spp.	Monk seals	I
	Ursus maritimus	Polar bear	II
Sirenia	*Dugong dugon* (except in Australia)	Dugong	I
	D. dugon (Australia)	Dugong	II
	Trichechus inunguis	Amazonian manatee	I
	T. manatus	West Indian manatee	I
	T. senegalensis	West African manatee	II

[a]From Federal Register (1999). Equivalent species names used by Rice (1998) appear in parentheses.

We suggest the following as some of the truly critical species and populations, those for which immediate, effective conservation efforts are needed if they are to survive much longer.

Pinnipeds: Mediterranean monk seal; Saimaa (ringed) seal (*Pusa hispida saimensis*)

Cetaceans: Baiji; Indus river dolphin (*Platanista gangetica minor*); vaquita; North Pacific right whale; North Atlantic right whale; several populations of blue whales (*Balaenoptera musculus*); western North Pacific gray whale; Cook Inlet and St. Lawrence River populations of beluga whales (*Delphinapterus leucas*); bowhead whale (*Balaena mysticetus*) populations in the eastern Arctic

Sirenians: Several populations of dugongs (*Dugong dugon*); several populations of West African and West Indian manatees (*Trichechus senegalensis* and *T. manatus*, respectively)

II. Why Do Marine Mammal Populations Become Endangered?

Relative to any species of plants and animals, marine mammals are susceptible to catastrophic reductions in their numbers for a variety of reasons. One of the reasons for this susceptibility relates to the biology of marine mammals: life history attributes that make them extremely vulnerable to overharvest. In addition, many populations of marine mammals are susceptible to extirpation because the value of marine mammal prod-

TABLE III
Marine Mammals Listed as Critically Endangered (CR), Endangered (E), or
Vulnerable (V) by the IUCN (1996)

Order	Species	Common name	Category
Cetacea	*Balaena mysticetus*	Bowhead	E/V
	Balaenoptera borealis	Sei whale	E
	B. musculus	Blue whale	E/V
	B. physalus	Fin whale	E
	Cephalorhynchus hectori	Hector's dolphin	V
	Delphinapterus leucas	Beluga	V
	Eschrichtius robustus	Gray whale	E
	Eubalaena glacialis (includes *E. japonica*)	Northern right whales	E
	Inia geoffrensis	Boto, Amazon river dolphin	V
	Lipotes vexillifer	Baiji, Yangtze river dolphin	CR
	Megaptera novaeangliae	Humpback whale	V
	Neophocaena phocaenoides	Finless porpoise	E
	Phocoena phocoena	Harbor porpoise	V
	P. sinus	Vaquita	CR
	Physeter catodon (= *Physeter macrocephalus*)	Sperm whale	V
	Platanista gangetica (= *P. gangetica gangetica*)	Ganges river dolphin	E
	P. minor (= *P. gangetica minor*)	Indus river dolphin	E
Carnivora	*Arctocephalus galapagoensis*	Galapagos Island fur seal	V
	A. philippii	Juan Fernandez fur seal	V
	A. townsendi	Guadalupe fur seal	V
	Callorhinus ursinus	Northern fur seal	V
	Eumetopias jubatus	Steller seal lion	E
	Halichoerus grypus	Gray seal	E
	Lutra felina	Marine otter	E
	Monachus monachus	Mediterranean monk seal	CR
	M. schauinslandi	Hawaiian monk seal	E
	Phoca caspica (= *Pusa caspica*)	Caspian seal	V
	P. hispida botnica (= *Pusa hispida botnica*)	Baltic seal	V
	P. h. ladogensis (= *Pusa hispida ladogensis*)	Ladoga seal	V
	P. h. saimensis (= *Pusa hispida saimensis*)	Saimaa seal	E
	Phocarctos hookeri	Hooker's sea lion	V
	Zalophus californianus japonicus (= *Z. japonicus*)	Japanese sea lion	Extinct?
	Zalophus californianus wollebaeki (= *Z. wollebaeki*)	Galapagos sea lion	V
Sirenia	*Dugong dugon*	Dugong	V
	Trichechus inunguis	Amazonian manatee	V
	T. manatus	West Indian manatee	V
	T. senegalensis	West African manatee	V

[a]Where more than one classification category is given for a particular species, it means that different populations or stocks of that species are threatened at different levels of severity. Similarly, a particular classification does not necessarily mean that a species is threatened rangewide at that level; the classification may reflect the status of only one population or stock. Equivalent species names listed by Rice (1998) appear in parentheses.

ucts has caused them to be hunted for centuries; they are killed or injured incidentally in a number of fisheries and during other human activities (e.g., boat strikes often kill or seriously injure North Atlantic right whales and West Indian manatees); and human activities such as chemical and acoustic pollution and competition for resources have led to dramatic changes in habitat. The combination of a limited ability to recover from overharvesting and various human impacts has led to the precarious status of several species and many stocks.

A. Life History

Life history attributes are biological characteristics of a species that evolved to maximize the fitness of individual male and female members of the species. Life history attributes include such characteristics as the age at sexual maturity, age-specific survivorship, sex- and age-specific rates of growth, reproductive interval, reproduction rate, and longevity. Stated otherwise, life history attributes dictate the potential for population growth. Biologists coarsely contrast two life history strategies among organisms: the so-called *r* strategists and the *K* strategists (see discussions in Pianka, 1970; Reynolds *et al.*, 2000).

The r strategists tend to be small in size, to reach sexual maturity quickly and produce large numbers of offspring (the letter "*r*" denotes intrinsic rate of increase—a feature maximized by r strategists), to colonize new areas well, to have short life spans, to have populations that fluctuate a great deal over time, and to have extremely high levels of infant mortality, due in part to low levels of parental care. The *K* strategists (which maintain their populations at or around the carrying capacity of the local environment, i.e., at *K*) have the opposite suite of attributes: delayed sexual maturity, long life span, low reproductive rate, large body size, lots of parental care and social behavior, good survival of infants, and poor colonizing ability.

Most species of plants and animals lie somewhere along the continuum between the purely *r*-selected and the purely *K*-selected species, but the life history attributes of marine mammals, as a group, place them *extremely* close to the end of the spectrum—on the *K*-selected side. Consider the highly endangered North Atlantic right whale, for which longevity is several decades (and may approach or even exceed a century as it does in the related bowhead whale), reproduction is very slow (with an average interval between calves of 3–6 years, tending more toward 6 years in the 1990s), and habitat needs are very specific. When right whale populations were reduced to numbers in the tens or low hundreds by overhunting, it became biologically impossible for right whale numbers to rebound quickly. In addition, when the size of the right whale population became very low, the more *r*-selected species' populations could respond by increasing in number relatively rapidly, causing the level of competition for resources to be higher than what would be predicted for an extremely depleted population of right whales. At the same time, humans are degrading the quality of right whale habitat by introducing boats, fishing gear, pollutants, and noise into areas critical to successful foraging and calving. In *K*-selected species, which generally adapt poorly to changing conditions, effects of humans on the environment may severely compromise recovery.

Thus, based on biology alone, right whale populations are not adapted to rapid recovery from overharvest, but biology is not the only reason. A variety of human-related factors (discussed later) also complicate their recovery. We simply note that for marine mammals, their life history strategies provide points of vulnerability.

B. Human Activities and Their Effects

Historically, humans around the world have taken marine mammals for subsistence and for their commercial value. Scammon (1874) was among the first whalers to point out that hunting had caused extreme depletion of a number of marine mammal species, including the gray whale, the bowhead, and the Guadalupe fur seal (*Arctocephalus townsendi*). Many species of marine mammals have been hunted to the extent that their numbers have been reduced—in some cases dramatically. One should not assume that overharvesting only became a serious problem after the advent of relatively modern hunting methods; in fact, elimination of the right whales in the Bay of Biscay by Basque hunters in the 1600s and Steller's sea cow by Russian fur traders a century later indicates that the problem of significant overharvest is nothing new.

The general public tends to perceive commercial whaling as (a) a thing of the past and (b) an activity that focused on the "great whales." Although commercial harvest of marine mammals reached its zenith (or its nadir, depending on one's perspective) some time ago, in fact, commercial harvest of even the great whales continues today [e.g., by Norway in the North Atlantic for minke whales (*Balaenoptera acutorostrata*)] and the harvest of small cetaceans continues to endanger certain species. Another relatively new perspective has been provided by the Russians, who have made available modern whalers' log books that have shown that records of species-specific landings are not always reliable (Danilov-Danil'yan and Yablokov, 1995). It would not be surprising if other whaling nations also misreported some of their harvest data. Thus, at least some species or populations thought to have been protected for decades in the past were actually protected only on paper.

An additional revelation was that some of the whale meat in Tokyo markets in the 1990s was found to be from protected species, such as the humpback whale, *Megaptera novaeangliae* (Baker and Palumbi, 1994). To some people, this suggests recent, illegal takes, although the Japanese government has responded that the meat has been stored, frozen, for decades. Note, also, that meat from whales taken incidentally during gill-net fishing, for example, could be marketed quite legally under INTERNATIONAL WHALING COMMISSION guidelines.

Habitat alteration by humans may exacerbate problems for marine mammals. In two cases involving endangered marine mammals [the western stock of the Steller sea lion (*Eumetopias jubatus*) and the Hawaiian monk seal (*Monachus schauinslandi*) population around French Frigate Shoals], the harvest by humans of these two marine mammals' preferred prey has been suggested as at least one potential cause of population decline.

Humans modify the coastal and marine environments in subtle and diverse ways, making it difficult to tease apart the various possible impacts to marine mammals. In some cases, those impacts can cause sudden, steep reductions in apparently healthy populations to the point where they warrant special protection. One especially telling case involves the die-off of common bottlenose dolphins (*Tursiops truncatus*) along the southeastern coast of the United States in 1987–1988. At least 740 animals died, causing the National Marine Fisheries Service to list that "stock" (quotation marks are used because the stock structure of dolphins in that area is still very uncertain) as depleted under the MMPA. Cause of death was first suggested to be ingestion of brevetoxin; later, scientists suggested that high contaminant loads were involved; and later still, other scientists noted the presence of morbillivirus in preserved tissues of dolphins from the die-off. As noted by Reynolds *et al.* (2000), the precise interplay among the natural toxins, anthropogenic toxicants, viral infections, immune dysfunction, opportunistic infections, and death is still unclear.

The dolphin die-off illustrates that human activities may compromise the health and well-being of *both* individuals and populations by making large numbers of animals susceptible to natural pathogens that might not be harmful to robust individuals. In a relatively large population such as coastal bottlenose dolphins, the problem is serious. For species like the Mediterranean monk seal, which suffered a disastrous die-off in 1997 due to as yet undetermined causes (perhaps either saxitoxin poisoning or morbillivirus, or some combination of these or other causes), the problem becomes extremely critical when an already low population size is further reduced over a matter of weeks.

It is relatively easy to count how many animals are killed by hunters or through incidental take. However, it becomes exceedingly difficult, due to both the variety and magnitude of effects and to potential synergisms, to understand the extent to which chemical and noise pollution, harvest of marine mammal prey by humans, and other effects on local environments may compromise, or at least retard, the recovery of depleted populations of marine mammals.

A final point worth making is that, thus far, we have addressed only local or regional effects of humans on habitats. The adverse effects of global climate change (if they are as profound as some believe and predict), superimposed on all the smaller scale changes induced by our species, paint a bleak picture for *K*-selected marine mammals.

III. Recovery and Nonrecovery of Species and Stocks: Lessons and Trends

Scientists and managers do not yet know why some depleted marine mammal populations increase in abundance, whereas others continue to decline or remain static. Perhaps most notable among those groups to have recovered relatively quickly (or at least to have shown recent increases in population size) are the eastern North Pacific stock of gray whale; the western Arctic bowhead whale; the southern right whale (*Eubalaena australis*); most populations of humpback whales; and some pinniped populations in parts of their range, including southern fur seals (*Arctocephalus* spp.), harbor seals (*Phoca vitulina*), and California sea lions (*Zalophus californianus*). In contrast, northern right whale populations, eastern Arctic populations of the bowhead whale, and Hawaiian and Mediterranean monk seals have either continued to decline or failed to increase in abundance in recent years. Some marine mammal species, such as the sea otter (*Enhydra lutris*) and harbor seal, are declining or are stable in parts of their range while increasing in others. As described earlier, the reasons for different patterns of recovery, where known, probably involve a mix of biological attributes (e.g., life history attributes), naturally caused environmental change, and human-related factors; however, the precise interplay of factors that permits or enhances recovery in certain instances (or prevents it in others) remains unclear.

There are a number of examples of divergent recovery trajectories in (a) closely related species, (b) sympatric and ecologically similar but distantly related species, and (c) populations of the same species occurring in different regions. Examples of these divergent recovery patterns are discussed.

As noted, the eastern North Pacific and western North Atlantic right whale populations are at very low levels, possibly numbering fewer than 500 individuals for the two areas combined. The status of the eastern Arctic bowhead whale (a species closely related to right whales) population is also of considerable concern. In contrast, right whale populations in the Southern Hemisphere have increased at rates estimated at 7–8% per year for a number of years. Southern right whale populations off Australia and New Zealand, South Africa, and South America combined now number at least 7500 individuals. Similarly, the western Arctic bowhead whale population has increased steadily at about 3–4% for at least the past decade, despite removals of 44 to 66 animals per year by Alaskan Native aboriginal hunters, and was estimated to number 8000 individuals in 1993.

Each of the just-mentioned right and bowhead whale populations was reduced to very low numbers by commercial whaling in the 19th and early to mid-20th centuries. However, responses to the cessation (at least on paper) of this human activity have been very different. Inasmuch as these are populations of closely related species, it is likely, although certainly not assured, that they are ecological counterparts (in different habitats) and have generally similar life history patterns. Therefore, major differences in life history patterns or large-scale environmental failure are not likely to be the culprits causing the differences in abundance and population growth. Instead, the differences are most likely related to serious injury and mortality and possibly habitat degradation from human activities. In the North Atlantic right whale population, 46 deaths known to have been caused by collisions with ships or ENTANGLEMENT in fishing gear have occurred since the 1970s (the actual number is almost certainly higher because not all carcasses are reported). Because this population was reduced to low levels by whaling and the species is a consummate *K* strategist, it cannot therefore sustain removals by human activities. Other factors may help explain why certain populations have failed to recover and other populations of marine mammals seem to be recovering. Some of these factors are described.

Before we proceed, however, we note our tendency here to use qualitative terms such as "low" and "very low." Such qualitative terms suggest that a population or species may have approached or reached some level that seriously jeopardizes its long-term survival. However, a population reduced to a few tens of individuals has a substantially lower probability of survival than one in which a few hundred or a thousand individuals survive; for both groups, one could well use descriptive terms such as "very low." Especially in the case of strongly *K*-selected species (such as right whales), survival seems bleak if only tens of animals exist and human-related pressures continue.

Sympatric marine mammal populations of distantly related species can also show divergent population growth rates. The western stock of the Steller sea lion has declined at alarming rates in the past several decades, whereas northern fur seal (*Callorhinus ursinus*) counts in the same region have remained stable for a number of years, albeit at a much lower level than in the past. The causes of the Steller sea lion decline have come under extensive scrutiny—the reasons are not yet completely clear but may in part be related to local competition with a very large commercial fishery in the area. Differences in the lack of growth in the western Steller sea lion stock and those of some sympatric pinniped populations may also be related to ecological shifts (e.g., prey availability) that affect some marine mammals far more than others. In the case of the northern fur seal, the stable but lower than historical population suggests a lower carrying capacity for the species.

In another case, the humpback whale, which occurs sympatrically with right whales in the North Atlantic, is on a positive recovery course (possibly as high as 6–9% per year) relative to the recovery course of the right whales. Although right and humpback whales do not have exactly the same habitat needs and feed at different trophic levels, the recovery of humpback whales suggests an absence of an ecosystem collapse of the North Atlantic marine habitat (as a factor that could impact right whale fecundity) and points further to human impact on the right whale population. Right whales generally occur closer to shore than humpback whales, perhaps exposing the former to greater human impact.

The gray whale provides an interesting example of recovery in one population and a concurrent lack of recovery in a geographically distant population of the same species. The eastern North Pacific gray whale population has grown steadily since the 1970s to levels only slightly less than the lower limit of estimated abundance prior to the initiation of commercial whaling. As a result, this population was removed in 1994 from the list of endangered and threatened wildlife—one of only a handful (but growing number) of endangered species, and the only marine mammal population, to be removed from the list. In contrast, the western North Pacific population of gray whale is very low and remains listed as endangered. Here again, such discrepancies in population growth rates are difficult to interpret. Both populations of gray whales migrate very close to shore and to human population centers along coasts. Presumably, both populations are very similar ecologically and in their life history patterns. The western North Pacific population may be exposed to some heretofore unidentified human activity that adversely affects its growth (such as entanglement in fishing gear, ship strikes, directed and illegal hunting, or pollution or other forms of habitat degradation) or other undetermined selection pressures. Situations analogous to the gray whale scenario exist for other species as well, e.g., sea otters and harbor seals.

A cursory review of the examples just provided suggests that knowledge of the interplay of factors that promote or retard recovery is incomplete. However, where causes of nonrecovery are known, it appears that a common theme involves the adverse effects of human activities.

IV. Improving the Recovery of Species and Stocks

This section makes three key points: (1) it is very unlikely that conservation measures needed to halt the decline of listed species or populations will be effective without understanding the life history of the listed taxon, (2) more information is needed about the effects of human activities on marine mammals, and (3) it is not possible (or at least it becomes dauntingly difficult) to evaluate the effectiveness of recovery efforts or to prioritize the extremely limited funding available for listed species without the establishment of objective criteria for managing and classifying populations under the Endangered Species Act (or other appropriate conventions or legislation).

A. Need for Information on Life Histories of Listed Species

Earlier in this chapter we described listed populations that are still declining (e.g., the western population of Steller sea lion) and listed populations that have remained at very low population levels without showing any signs of recovery (e.g., North Atlantic right whale, vaquita, baiji, the western population of North Pacific gray whale, and both extant species of monk seal). As noted, reasons for the lack of recovery are varied. Therefore, in promoting the recovery of these populations, scientists need, whenever possible, to collect the necessary information on age-specific rates of birth and survival. That is, it is necessary to determine whether recovery is hindered by inadequate reproduction, which could be due to nutritional stress (which, for example, could be caused by COMPETITION WITH COMMERCIAL FISHERIES) or contamination (which, for example, could be related to high levels of pollutants such as organochlorines). Similarly, it is necessary to determine whether there is adequate recruitment into the adult population or high enough levels of adult survival to sustain recovery. For some populations, both reproduction and survival will be found to be inadequate to support a recovery. However, once reliable information on life history parameters is available, it is possible for researchers or managers to step in, implement informed and deliberate strategies, and try to rectify the situation.

For example, levels of recruitment and adult survival in the North Atlantic population of right whale were found to be low because of human-related injury and mortality. In a cooperative effort, the National Oceanic and Atmospheric Administration, U.S. Coast Guard, U.S. Marine Mammal Commission, the International Fund for Animal Welfare, and other agencies and organizations involved in right whale recovery activities worked through the International Maritime Organization to design and implement a system whereby officers of large vessels are required to notify a shore-based station when entering right whale habitat. In return, a message is passed to the ship describing right whale vulnerability to ship strikes and locations of recent whale sightings. This approach seems to be an obvious first step toward a solution.

Similarly, juvenile mortality in Hawaiian monk seals at three of their six breeding colonies has been found to be sufficiently high so as to preclude recovery. Further, it was found that part of this mortality was due to known aggressive male monk seals mobbing and killing adult females and/or pups. In response to this behavior, researchers removed many (but not all) of the adult male monk seals in one area where mobbing was most severe. The result was an immediate and almost total elimination of additional deaths caused by mobbing in that area.

A third example involves the southern sea otter (*E. lutris nereis*). In the late 1970s and early 1980s this population was reported to be in decline. Based on data from observer programs operating in several commercial fisheries along the central California coast, it was determined that the level of incidental mortality of otters in the set gill-net fishery for halibut was not sustainable. Soon after this finding was reported, the state of California passed legislation that moved this fishery offshore, where otter entanglement was thought to be unlikely. Following this management action, the population started to recover.

In summary, it is possible in some cases to take actions that promote the recovery of listed populations. However, without knowledge as to what life history parameters are abnormal, as well as knowledge about the underlying cause of the problem, it is often impossible to know what CONSERVATION measures are appropriate. Unfortunately, the research necessary to determine rates of reproduction and survival in wild populations is both expensive and difficult. In addition, it often take several years of collecting such data to get estimates precise enough to allow scientists to evaluate different hypotheses about why the population is not recovering. Nonetheless, without such information, recovery efforts are severely hampered.

We should also point out that it is not always possible to determine what exactly is causing a listed population to decline (e.g., Steller sea lion) or preventing its recovery (e.g., northern fur seal). Scientists and conservationists can know a tremendous amount about the life history of an affected group, but this knowledge does not always help to direct management efforts. In such cases, the management agencies are required (although they may fall short) to manage in a way that "errs on the side of the animal." This means that managers are required to be cautious in authorizing human activities that are likely to adversely affect a listed species or the habitat upon which it depends. Finally, it should be noted that some populations listed under the U.S. ESA are recovering with little or no assistance from humans. This is typically the situation when a population was overharvested (e.g., many of the large whale populations), subsequently depleted, and then managed in such a way that additional harvest-related mortality was either eliminated or greatly reduced. When the affected population was not too severely reduced, when human-related mortality and serious injury remain low, and when the habitat is still healthy, recovery can occur. This has been observed for populations of gray, right, humpback, and blue whales in many of the world's oceans.

B. Need for Objective Classification Criteria

The U.S. ESA, as amended, requires the various management agencies to develop criteria for determining when a population has recovered to such an extent that it can be removed from the list of endangered and threatened wildlife. Unfortunately, this has not been done for any listed marine mammals. However, efforts are underway by several agencies (U.S. Fish and Wildlife Service, National Marine Fisheries Service, IUCN) to develop objective criteria that can be used to classify the degree to which a particular population is at risk of extinction. Such a task is not easy, in part because the information typically needed in population viability analysis is not available for most populations of marine mammals. For example, many of the available software packages that are used to estimate the likelihood of extinction require estimates of carrying capacity, home range size, and the degree to which environmental catastrophes affect a population. For marine mammals, these data are rarely available. Rather, the only data that can be reliably determined for most marine mammal populations include population size, trends in abundance, human-related removal levels, gross production, and stock structure. Therefore, it is critical that classification criteria be developed that utilize only this suite of information.

Gerber and DeMaster (1999) reported on a generic approach that could be applicable to large whales. In a separate paper (Gerber *et al.*, 1999), this approach was applied to a data set from the already delisted eastern North Pacific gray whale population. The results indicated that with about 10 years of data, collected at a cost of just over $650,000, a minimum data set could be developed that would be adequate for classifying this population under the U.S. ESA. The authors went on to argue that such an expenditure was well worth the cost because (1) U.S. ESA listings should mean something (i.e., when a population has truly recovered it should be delisted), and (2) the cost to society in terms of regulations and management for populations listed under the U.S. ESA can be expensive, especially for marine species. Therefore, where appropriate it is in society's interest to delist populations as soon as possible.

Other classification schemes have been proposed in the past by Mace and Lande (1991) and Mace *et al.* (1992) and more recently by the IUCN (1996). These approaches have considerable merit in that they use a series of "triggers" to classify the degree to which a population is threatened with extinction; any one of these "triggers" can lead to a more conservative classification. These criteria are also based on information that is possible to collect.

The management agencies need to move forward aggressively to adapt methods like those discussed earlier in classifying stocks under the U.S. ESA. It is hoped that by 2010, management agencies will have developed objective criteria that can be used to evaluate whether progress has been made in recovering each listed population and whether each population is still in need of the special protection afforded it under the U.S. ESA. It is worth noting that in its fourth iteration, the revised Florida manatee recovery plan (which is being developed but is incomplete as we write) is likely to include objective, measurable criteria for downlisting.

In the past, reclassification, or at least delisting, has met with considerable resistance from the environmental community. This can be unfortunate. It implies that the members of the environmental community do not trust the management agencies to properly evaluate the extent to which a particular population is threatened with extinction; it may also suggest some distrust in the ability of managers to effectively address problems unless populations or species are listed as endangered or threatened. Resistance to delisting may also mean that the agencies are not doing enough to inform the general public about what it means to be listed under the U.S. ESA (e.g., what degree of threat is tolerable before a species or stock is listed). Finally, if objective criteria for classifying listed groups do not exist, then it is likely that important decisions will be made more often by engaging in emotional debate than by relying on scientific data. This situation needs to be addressed by Congress, which has passed excellent laws concerning protected species but which, unfortunately, has not always appropriated sufficient funds for the responsible agencies to carry out these laws. To some extent, this situation is the fault of the agencies for not being more aggressive and responsive in recognizing the value of developing objective classification criteria. Once objective criteria are developed for at least some of the listed populations and are used either to prioritize extremely limited funding relative to the job that needs to be done or to reclassify populations, there will be a general recognition of the value of this approach by the agencies, Congress, and the involved stakeholders.

Armed with better life history information and clearer, more objective criteria, conservationists may be able to work more effectively to help listed species or stocks recover. As noted earlier, however, lack of recovery in many cases is clearly associated with effects of diverse and intensive human activities. To reinforce an old adage used by Ken Norris (1978), effective management of wildlife ultimately involves effective management of people.

Acknowledgments

We are grateful to D. Laist, R. Mattlin, W. Perrin, J. Powell, R. Reeves, S. Rommel, and R. Wallace for constructive comments on the manuscript; to J. Leiby and J. Quinn for their careful editing; and to B. Brooks and S. Spangle for providing a definition of distinct population segment.

See Also the Following Articles

Baiji ■ Conservation Efforts ■ Extinctions, Specific ■ Hunting of Marine Mammals ■ Mass Die-Offs ■ Otters ■ Population Dynamics ■ Vaquita

References

Baker, C. S., and Palumbi, S. R. (1994). Which whales are hunted? A molecular genetic approach to monitoring whaling. *Science* **265**, 1538–1539.

Brewer, R. (1988). "The Science of Ecology." Saunders, Philadelphia.

Clapham, P. J., Young, S. B., and Brownell, R. L. Jr. (1999). Baleen whales: Conversation issues and the status of the most endangered populations. *Mamm. Rev.* **29**(1), 35–60.

Danilov-Danil'yan, V. F., and Yablokov, A. V. (1995). "Soviet Antarctic Whaling Data (1947–1972)." Center for Russian Environmental Policy, Moscow.

Eschricht, D. F., and Reinhardt, J. (1861). On the Greenland right whale (*Balaena mysticetus*, Linn.) with especial reference to its ge-

ographic distribution and migrations in times past and present, and to its external and internal characteristics. For the Royal Society, R. Hardwicke, London.

Federal Register (1999). Code of Federal Regulations. 50 CFR, section 23.23.

Gerber, L. R., and DeMaster, D. P. (1999). A quantitative approach to Endangered Species Act classification of long-lived vertebrates: Application to the North Pacific humpback whale. *Conserv. Biol.* **13**(5), 1203–1214.

Gerber, L. R., DeMaster, D. P., and Kareiva, P. M. (1999). Gray whales and the value of monitoring data in implementing the U.S. Endangered Species Act. *Conserv. Biol.* **13**(5), 1215–1219.

IUCN (1996). "1996 IUCN Red List of Threatened Animals." IUCN, Gland, Switzerland, and Cambridge, UK.

Mace, G. M., Collar, N., Cooke, J., Gaston, K., Ginsberg, G., Leander-Williams, N., Maunder, M., and Milner-Gulland, E. J. (1992). The development of new criteria for listing species on the IUCN Red List. *Species* **19**, 16–22.

Mace, G. M., and Lande, R. (1991). Assessing extinction threats: Toward a reevaluation of IUCN threatened species categories. *Conserv. Biol.* **5**(2), 148–157.

Norris, K. S. (1978). Marine mammals and man. Pp. 320–338. *In* "Wildlife and America" (H. P. Brokaw, ed.), U.S. Government Printing Office, Washington, DC.

Pianka, E. R. (1970). On r and K selection. *Am. Nat.* **104**, 592–597.

Reynolds, J. E., III, Wells, R. S., and Eide, S. D. (2000). "The Bottlenose Dolphin: Biology and Conservation." Univ. Press of Florida, Gainesville, FL.

Rice, D. W. (1998). "Marine Mammals of the World: Systematics and Distribution." Special Publication No. 4, The Society for Marine Mammalogy, Lawrence, Kansas.

Scammon, C. M. (1874). The marine mammals of the northwestern coast of North America together with an account of the American whale-fishery. 1968 reproduction, Dover Publications, Inc., New York.

U.S. Fish and Wildlife Service (1996). Endangered Species Act of 1973 as amended through the 100th Congress. U.S. Government Printing Office 414-990/50033.

U.S. Marine Mammal Commission (1995). The Marine Mammal Protection Act of 1972 as amended February 1995. Available through Marine Mammal Commission, Bethesda, Maryland.

U.S. Marine Mammal Commission (1999). Annual Report to Congress 1998. Bethesda, Maryland.

Endocrine Systems

DAVID J. ST. AUBIN
Mystic Aquarium, Connecticut

Endocrine systems function to control physiological and metabolic processes, and integrate these functions to meet specific environmental conditions or chronological needs. The effector substances are termed hormones and are typically produced and released into circulation by specialized cells that are usually localized in small glands. Because of their great potency and ability to broadly influence body functions, hormones are regulated by an exquisite set of negative and positive feedback loops that may link several glands (Fig. 1). For

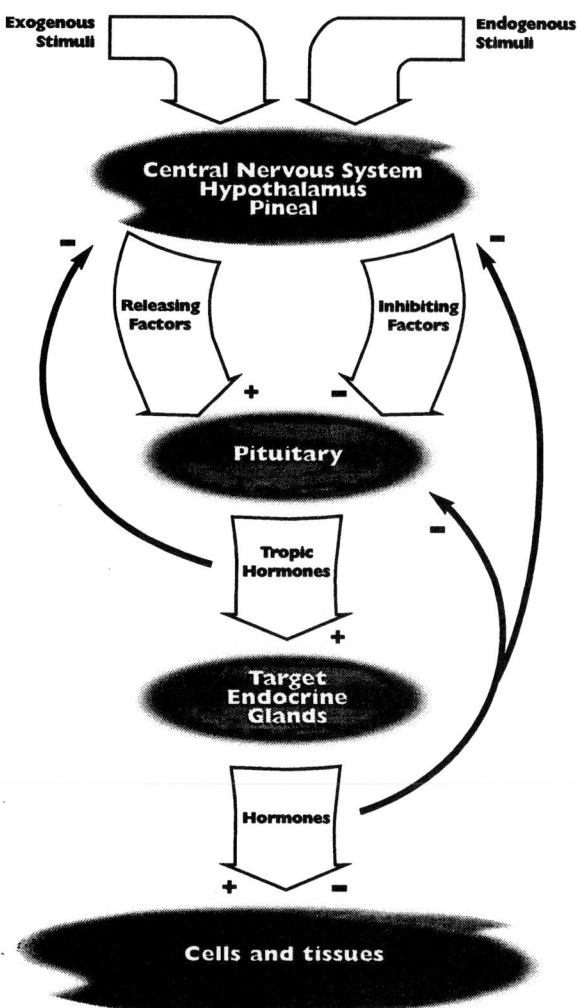

Figure 1 *Schematic representation of stimulatory (+) and inhibitory (−) relationships among principal components of mammalian endocrine systems. Figure by D. Scotti.*

the most part, endocrine systems in marine mammals follow the basic organization and chemical characteristics of other mammals. Nevertheless, it is intriguing to examine how these systems allow marine mammals to meet the peculiar challenges imposed by their environment.

I. Integration of Environmental Cues—Pineal–Hypothalamus–Pituitary

Many species of marine mammals inhabit environments that express drastic seasonal changes in air temperature and light or undergo migrations that produce the same effect. Mysticetes exploit productive cold waters at high latitudes during the long days of summer but retreat to tropical habitats in the fall to bear calves under more forgiving conditions. Day length appears to be an important cue that signals the time to abandon habitats that will soon become ice bound. Pinnipeds, particularly those in polar environments, seasonally partition ac-

tivities such as breeding and molting to take advantage of favorable conditions that will increase the survival of offspring or allow recovery from fasts necessitated by long periods ashore.

Melatonin is the hormone most commonly associated with photoperiodism in mammals. It is produced primarily by the pineal gland (epiphysis) located above the third ventricle of the brain. Daylight suppresses the production of melatonin, which regulates the activity of the hypothalamus, pituitary, and, indirectly, the gonads and thyroid. Unfortunately, the hormone and its activity have received relatively little attention in the marine mammal literature. This deficit is ironic, as the pineal gland of the newborn southern elephant seal, *Mirounga leonina,* can weigh over 9 g, which is the size of the brain of a hamster, the subject of so much of the research on melatonin.

The exceedingly high plasma levels of melatonin in newborn seals draw attention to the important role of this hormone in the survival of animals born under harsh environmental conditions. Concentrations approach 300,000 pmol/l (69,000 pg/ml) in southern elephant seals at birth, but fall steadily during the ensuing 7–10 days. A similar pattern has been observed in harp, *Pagophilus groenlandicus,* gray, *Halichoerus grypus,* Weddell, *Leptonychotes weddellii,* and northern elephant seals, *M. angustirostris.* It has been postulated that the hormone promotes the generation of the active thyroid hormone triiodothyronine (T_3), which in turn accelerates metabolism and provides the heat necessary to withstand the extreme conditions experienced at the time of birth.

Marked seasonality is evident in the size of the pineal in southern elephant seals. The gland is largest in the winter dark period, and the high circulating levels of melatonin at that time presumably suppress gonadal function; hormone levels are uniformly low during the spring breeding season. A circadian rhythm in hormonal secretion is evident during the winter, but equivocal during the summer. In Weddell seals, at least, the cycle is abolished under continuous natural daylight.

The literature on melatonin in cetaceans is even more sparse. The very presence of a discernible pineal gland had been uncertain, but was eventually identified in a number of cetacean species. Extrapineal sources of melatonin, such as from the retina, likely augment the role of the small gland in integrating photoperiod with metabolic functions. One might expect that polar cetaceans such as belugas, *Delphinapterus leucas,* and narwhals, *Monodon monoceros,* need to entrain their endocrine physiology with seasonal changes marked by day length.

Some of the hypothalamic effects of melatonin are transferred to the pituitary gland or hypophysis, a small, compound structure located at the base of the brain. The glandular portion (adenohypophysis) produces a set of hormones that either directly elicit tissue responses (e.g., growth hormone, prolactin, and gonadotropins) or modulate the activity of other endocrine glands [e.g., adrenocorticotropic hormone (ACTH) and thyrotropin or thyroid-stimulating hormone (TSH)]. Early investigators on whaling vessels were impressed by the size of the pituitary in mysticetes (up to 53.5 g in a blue whale, *Balenoptera musculus*), although on a body weight basis, it is unremarkable. Nevertheless, the abundance of the tissue afforded the opportunity for extensive studies to extract and characterize the principal hypophyseal hormones. Fin whale, *B. physalus,* ACTH was found to be identical to that of humans, whereas blue whale TSH more closely resembled that of nonprimates. Studies on beluga whales and bottlenose dolphins, *Tursiops truncatus,* have shown that synthetic ACTH and bovine TSH are capable of eliciting expected responses from the target glands, demonstrating at least some homology in the structure of these hormones among various groups of mammals.

II. Regulation of Metabolism—Thyroid

Early investigators were as impressed by the size of the cetacean thyroid as they were by the pituitary, to the degree that large whales were considered as a possible commercial source for thyroxin (T_4), the principal hormone synthesized by the gland. It was not only the size but the proportions of the thyroid that called attention to the importance of this gland in cetaceans; beluga whales have three times more thyroid tissue per unit body weight than a thoroughbred horse, and bottlenose dolphins have nearly twice as much thyroid as do humans (400 mg/kg vs 250 mg/kg). Because thyroid hormones regulate a wide variety of metabolic processes, particularly those associated with energy metabolism, the finding of relatively high circulating levels of thyroid hormones (TH) in some cetaceans compared with humans and other mammals reinforced long-standing notions of elevated metabolic rates in cetaceans. A more critical analysis of both cetacean metabolic rates and concentrations of free thyroid hormones, which are a better index of available hormone, has revised assumptions about hormonally stimulated metabolism as an adaptation to meet the thermal challenges of the marine environment. In pinnipeds, circulating TH levels and the basal metabolic rate are comparable to those in terrestrial mammals, but, surprisingly, their blood TH are lower than in West Indian manatees, *Trichechus manatus,* a species with a distinctly slower metabolism. Assumptions about the metabolic rate cannot be based solely on circulating TH levels.

Thyroid hormones do exhibit some important dynamics during the course of the development and life history of marine mammals. Circulating levels in neonatal phocid seals are understandably high in view of the need for metabolically derived heat until their lanugo coat gains insulating ability and blubber reserves are established. Thereafter, seasonal fluctuations in the blood levels of T_4 and T_3 have been correlated with changing metabolic needs. Marine mammals are well known for their ability to seasonally manage energy stores, enabling them to fast during migration or breeding. The hormonal modulation of energy consumption by tissues can allow for fat deposition, even when the energy intake is low, or mobilization of fat to sustain accelerated growth while feeding.

Thyroid hormones play a role in an important annual event for pinnipeds—the molt. As in other fur-bearing mammals, TH are seasonally elevated to promote hair growth. Controversy in the literature regarding the degree of association between visible molt and thyroid activity in pinnipeds may derive from the difficulty in recognizing when hair growth is actually stimulated by elevated TH. Hair loss, the overt sign of molting, may be enhanced by increased levels of cortisol at a time when T_4 and T_3 levels are low. Cortisol suppresses the secretion of TSH from the pituitary and inhibits the activation of T_4 to T_3. Fluctuating circulating levels of metabolically potent substances such as cortisol and TH during the molt draw attention to how intensely pinnipeds are physiologically affected at that stage of their annual cycle.

Among cetaceans, the seasonality of TH activity has been described for the beluga whale. The summer period of estuarine occupation is characterized by marked elevation in circulating levels of T_4 and T_3 and by histological evidence of intense cellular activity in the thyroid gland. Colloid reserves are depleted by columnar follicular cells, in contrast to the quiescent appearance of the gland in spring and fall when low cuboidal cells surround abundant stores of thyroid hormone. The implications of this burst of thyroid activity are broad, favoring mobilization of blubber and promoting the effects of other agents such as GH. It also coincides with a unique event in beluga epidermis. Cell production is enhanced, presumably under the influence of TH, and superficial turnover is accelerated by the relatively warm (10–15°C) freshwater environment. Taken together, these events constitute a true molt, lacking only the production of a hair coat to be fully analogous to the process in a pinniped. No comparable transformation has been described in any other cetacean; only the rubbing behavior of killer whales, *Orcinus orca,* on the cobbles of Telegraph Cove, British Columbia, hints of a seasonal pulse in epidermal growth. Studies on circulating levels of TH in Atlantic bottlenose dolphins have not revealed significant annual variation, even though the 15°C range of water temperatures experienced by these animals is equivalent to that encountered by belugas.

Reverse T_3 (rT_3) is a metabolically inactive product of T_4 monodeiodination and typically represents one-third to one-half the circulating concentration of T_3 in humans and other mammals. Cetaceans differ from this balance in T_4 metabolism, showing rT_3 levels that are equivalent or up to three times greater than those of T_3. The ratio is particularly skewed in belugas during the time of enhanced thyroid activity. Circulating levels of rT_3 up to 6.7 nmol/liter (440 ng/dl) are the highest observed in any mammal. Belugas may use this pathway to modulate the effects of markedly elevated levels of T_4, although other feedback mechanisms reducing the secretion of TSH would seem to be more efficient. The sensitivity of deiodinating pathways to increases in circulating levels of cortisol in belugas mimics that in laboratory mammals and is evidenced by coincidental elevation in plasma rT_3 and reduction in T_3.

III. Reproduction—Ovary and Testis

Like the molt, reproduction in most marine mammals is a highly seasonal activity. Most notable in this respect are high-latitude pinnipeds. For animals that are only periodically gregarious, synchrony of the reproductive state is critical. As in other mammals, gonadotropins released from the pituitary drive ovarian and testicular activity, resulting in ovulation and spermatogenesis, respectively, and the associated behaviors of receptivity in females and rut in males. In phocids and otariids, fetal development requires considerably less time than the interval that separates the annual congregation of animals for parturition and breeding. Delayed implantation of the blastocyst therefore serves as a useful strategy to allow parturition to occur at the time of year most appropriate for successful pup rearing. Day length likely serves as a critical cue for implantation, although the sequence of hormonal events that trigger implantation has not yet been documented.

Reproductive cycles in cetaceans are highly variable, with some falling into relatively narrow time frames, such as in belugas, whereas others show very little seasonality. Bottlenose dolphins, which have been studied extensively in captivity, demonstrate some of the variability possible within a species. Some females may be seasonally polyestrous, with up to five cycles per year, whereas others may be anestrous for a year or more. The presence of males has an inconsistent effect, but there is the impression that ovulation may be induced. Testosterone levels in males tend to be higher in spring and fall, roughly coinciding with calving peaks (gestation is approximately 12 months), although individual males show varying patterns from year to year and are capable of impregnating females in almost any month. Seasonal constraints on breeding would appear to be less critical in tropical and subtemperate species than in those exploiting more polar habitats. An exception would be river dolphins exposed to drastic seasonal fluctuations in habitat associated with dry and rainy seasons. Synchrony of reproductive activity, and by inference reproductive hormones, is likely an important consideration.

IV. Fluid and Electrolyte Regulation—Pituitary and Adrenal

For a mammal, the marine environment represents a number of challenges, one of which is to obtain water while eliminating potentially life-threatening salt burdens. Renal adaptations and water balance are considered elsewhere in this volume; this discussion focuses on hormonal mechanisms for maintaining appropriate hydration and electrolyte concentrations. Marine mammals possess all the recognized endocrine systems for managing fluid and electrolytes. Antidiuretic hormone (ADH), also known in some species as arginine vasopressin (AVP) from the posterior pituitary (neurohypophysis), retains water, whereas aldosterone from the zona glomerulosa of the adrenal reclaims sodium. The former would seem to be highly desirable, whereas the latter is of questionable value considering the principal needs of these animals. It is paradoxical that studies on vasopressin have produced a conflicting picture of its role, whereas aldosterone is essential in preventing sometimes fatal hyponatremia in phocid seals.

In fasting northern elephant seal pups, AVP levels in blood decline over time, an unexpected finding in an animal needing to conserve water. Nevertheless, urinary output also falls despite the low circulating levels of AVP. Gray seals show a more typical response, with concurrently rising AVP and urinary osmolality. However, water loading in these animals does not produce the expected decrease in circulating AVP, even though urine volume rises and osmolality decreases. Early investigations in bottlenose dolphins failed to identify a significant role for AVP in this species. Manatees have higher vasopressin levels when in brackish water than in freshwater, presumably as an aid to conserving body water in the former environment.

Despite its apparent redundancy in the marine environment, aldosterone has been detected in all marine mammal species studied. The zona glomerulosa is particularly well developed in young seals, suggesting the need for a fully functional system for salt regulation at birth. Plasma levels in bottlenose dolphins and California sea lions, *Zalophus californianus*, are correlated with changes in renin activity, an important regulator of aldosterone secretion. Sodium restriction causes inconsistent changes in aldosterone in ringed seals, although levels are periodically elevated above those in controls. The freshwater-dwelling Baikal seal, *Pusa sibirica*, is more efficient in sodium conservation than its close marine relative, the ringed seal, *P. hispida*. Aldosterone levels in manatees vary according to their environment, with higher levels favoring sodium conservation when in freshwater. Some species of phocid seals are prone to hyponatremia, apparently when the zona glomerulosa has exhausted its ability to produce aldosterone. Hyponatremia can occur even when aldosterone levels are within the normal range, possibly as a result of a condition known as "aldosterone escape" in which tubular epithelial cells of the kidney lose their sensitivity to the hormone after prolonged stimulation.

Studies have identified atrial natriuretic peptide in several pinniped species. Produced within cardiomyocytes and secreted under the influence of blood volumetric expansion, this hormone results in natriuresis as a mechanism to enhance fluid loss and thereby restore normal blood volume. Its role in managing electrolyte burdens in marine mammals is still poorly understood, although its secretion in the harbor seal, *Phoca vitulina*, at least, appears to be less sensitive to volumetric stimuli than has been found in other mammals. The large intravascular capacity represented by distensible sinuses in these animals likely dampens the atrial stretch required to induce changes in circulating levels of the hormone.

V. Stress Response—Adrenal

Adrenal morphology in marine mammals follows the same pattern of cortical and medullary stratification as in terrestrial mammals, although the cetacean cortex is more highly lobulated. The importance of adrenal function in marine mammals is highlighted by the fact that it is one of the few tissues to which blood supply is maintained during a dive. Cortisol is the principal glucocorticoid in marine mammals; corticosterone is also secreted, but typically represents less than 10% of the circulating level of cortisol. Weddell seals are distinguished by excessively high circulating levels of cortisol, averaging 1600–2200 nmol/l (60–80 μg/dl), at least an order of magnitude greater than in humans and other terrestrial and marine mammals. The underlying cause of this variation is unclear, although it has been determined that concentrations of cortisol-binding proteins are proportionally high in this species, leaving only a small fraction of metabolically available hormone.

Glucocorticoid secretion is under the influence of ACTH from the adenohypophysis (Fig. 2) and has been evaluated using synthetic analogs of ACTH in a variety of marine mammal species. In pinnipeds, the cortical response represented by elevations in circulating levels of cortisol follows that of most terrestrial mammals, resulting in two- to fourfold increases within 15–30 min and persisting for up to 90 min. Cetaceans

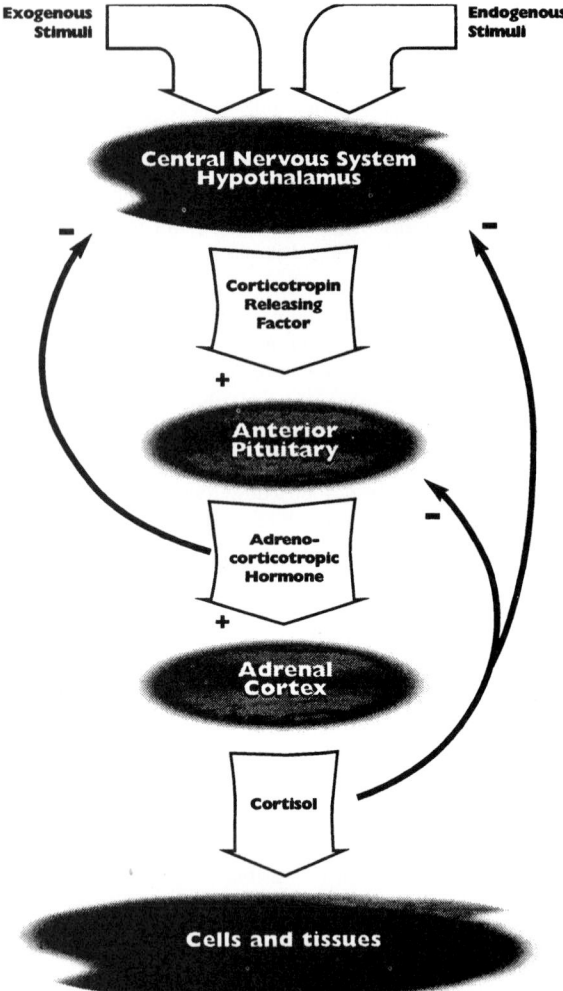

Figure 2 *The secretion of glucocorticoids (e.g., cortisol) from the adrenal cortex is initiated when a stimulated central nervous system produces corticotropin-releasing factor (CRF), which causes the secretion of adrenocorticotropic hormone (ACTH) from the anterior pituitary (adenohypophysis). Increased circulating levels of cortisol inhibit the further secretion of CRF and ACTH, as ACTH also inhibits the production of CRF, thus preventing excessive and prolonged elevation of cortisol in the blood. Figure by D. Scotti.*

show a different pattern. Resting levels of cortisol in belugas and bottlenose dolphins are relatively low, generally in the range of 50–100 nmol/l (2–4 μg/dl), and ACTH stimulation results in only modest elevations, usually less than twofold. Nevertheless, several glucocorticoid-mediated changes, such as alterations in circulating white blood cells (lymphopenia, eosinopenia, and neutrophilia), decreased T_4 and T_3, and elevated rT_3, are observed following ACTH stimulation, suggesting that only mild changes in cortisol are necessary to produce the physiologically appropriate effects. High cortisol availability in plasma or enhanced sensitivity of tissue receptors could account for this fundamental difference.

Adrenocortical stimulation in marine mammals produces an uncharacteristic response in aldosterone secretion. Cetaceans and phocid seals consistently show three- to fivefold increases in circulating levels of aldosterone following ACTH administration; terrestrial mammals, including humans, have less than half that response. One possible explanation lies in the development of hyponatremia in chronically stressed phocid seals. ACTH stimulation in these subjects fails to elicit an aldosterone response, presumably because of exhaustion of the chronically stimulated zona glomerulosa of the adrenal. Aldosterone's participation in the stress response in phocids suggests the need for sodium conservation at such times. The parallel finding of ACTH-stimulated aldosterone secretion in belugas and bottlenose dolphins draws attention to a common need and convergent adaptation to regulate electrolytes during stress. The advantage conferred by this mechanism is that sodium resorption should enhance water conservation, which would be important in an animal that might be deprived, in the short term at least, of dietary water.

The few studies on catecholamine physiology that have been undertaken in marine mammals have focused on the dive response. In hooded and Weddell seals, epinephrine induces splenic contraction, releasing a reservoir of stored erythrocytes and enhancing the aerobic dive capacity of the animal. Both epinephrine and norepinephrine increase during dives longer than a few minutes in Weddell seals and rapidly return to normal levels at the conclusion of the dive. These changes suggest that catecholamines play an integral role in the management of oxygen reserves during diving.

VI. Glucose Regulation—Endocrine Pancreas

Glucose homeostasis is managed as in terrestrial mammals by insulin and glucagon, secreted by pancreatic islet cells. Insulin promotes the uptake of glucose into cells, whereas glucagon has the opposite effect. The literature on these two hormones in marine mammals is sparse. The ratio of insulin to glucagon is typically low, consistent with a metabolism geared to producing glucose from a variety of precursors (gluconeogenesis) rather than deriving it from the small amount of carbohydrates typically found in their diet. During diving, glucose delivery to the brain is critical, and circulating levels of this energy source are maintained by the falling ratio of insulin to glucagon. This also contributes to hyperglycemia, or elevated blood sugar, at the end of the dive.

VII. Summary

As they have with a number of physiological systems, marine mammals have refined basic mammalian endocrine mechanisms to meet the demands imposed by their environment. Unusually high blood levels of melatonin in young seals, particularly elephant seals, act in conjunction with thyroid hormones to stimulate metabolism during the critical early stages of life in cold environments. Distinct seasonal cycles in thyroid, adrenal, and reproductive hormones entrain metabolism, molt, breeding, and parturition to optimize survival. Marine mammals have no unusually developed hormonal mechanisms for sodium excretion or water retention, which would be an asset in their hypertonic saline environment, but do show adaptive increases in aldosterone to conserve sodium when in freshwater. The stress response of pinnipeds and cetaceans includes elevations in aldosterone, perhaps as a means of securing water and electrolyte balance during challenging times.

See Also the Following Articles

Energetics ■ Estrus and Estrous Behavior ■ Female Reproductive Systems ■ Hair and Fur ■ Male Reproductive Systems ■ Osmoregulation

References

Boily, P. (1996). Metabolic and hormonal changes during the molt of captive gray seals (*Halichoerus grypus*). *Am. J. Physiol.* **270**(5 Pt 2), R1051–R1058.

Bryden, M. M. (1994). Endocrine changes in newborn southern elephant seals. *In* "Elephant Seals: Population Ecology, Behavior and Physiology" (B. J. LeBoeuf and R. M. Laws, eds.), pp. 387–397. Univ. of California Press, Berkeley.

Gardiner, K. J., Boyd, I. L., Follett, B. K., Racey, P. A., and Reijnders, P. J. H. (1999). Changes in pituitary, ovarian, and testicular activity in harbour seals (*Phoca vitulina*) in relation to season and sexual maturity. *Can. J. Zool.* **77**, 211–221.

Haulena, M., St. Aubin, D. J., and Duignan, P. J. (1998). Thyroid hormone dynamics during the nursing period in harbour seals. *Phoca vitulina. Can. J. Zool.* **76**, 48–55.

Hochachka, P. W., Liggins, G. C., Guyton, G. P., Schneider, R. C., Staneck, K. S., Hurford, W. E., Creasy, R. K., Zapol, D. G., and Zapol, W. M. (1995). Hormonal regulatory adjustments during voluntary diving in Weddell seals. *Comp. Biochem. Physiol. B Biochem. Mol. Biol.* **112**, 361–375.

Kirby, V. M. (1990). Endocrinology of marine mammals. *In* "Handbook of Marine Mammal Medicine: Health, Disease and Rehabilitation" (L. A. Dierauf, ed.), pp. 303–351. CRC Press, Boca Raton, FL.

Ortiz, R. M., Adams, S. H., Costa, D. P., and Ortiz, C. L. (1996). Plasma vasopressin levels and water conservation in fasting, post-weaned northern elephant seals pups (*Mirounga angustirostris*). *Mar. Mamm. Sci.* **12**, 99–106.

St. Aubin, D. J. (2001). Endocrinology. *In* "CRC Handbook of Marine Mammal Medicine" Second Edition (L. A. Dierauf and F. M. D. Gulland, eds.) pp. 165–192. CRC Press, Boca Raton, FL;

Stokkan, K. A., Vaughan, M. K., Reiter, R. J., Folkow, L. P., Martensson, P. E., Sager, G., Lydersen, C., and Blix, A. S. (1995). Pineal and thyroid functions in newborn seals. *Gen. Comp. Endocrinol.* **98**, 321–331.

Zenteno-Savin, T., and Castellini, M. A. (1998). Plasma angiotensin II, arginine vasopressin and atrial natriuretic peptide in free ranging and captive seals and sea lions. *Comp. Biochem. Physiol. C Pharmacol. Toxicol. Endocrinol.* **119C**, 1–6.

Energetics

Daniel P. Costa
University of California, Santa Cruz

Energetics provides a method to quantitatively assess the effort animals spend acquiring resources, as well as the relative way in which they allocate those resources. Energy flow models are analogous to cost-benefit models used in economics. Costs take the form of energy expended to acquire and process prey, and to maintain body functions. The energetic benefits are manifest as the food energy used for growth and reproduction. By measuring the various avenues of energy transfer, we can determine how animals organize their daily or seasonal activities and how they prioritize their behaviors. Thus, energy flow can be described as what goes into the animal as food and what comes out in the form of growth, waste, or metabolic work. Survival requires a positive balance between the costs of maintenance and the acquisition of food energy. If a marine mammal cannot compensate for decreases in energy acquisition, it must either reduce its overall rate of energy expenditure or utilize stored energy reserves. Conversely, in order to grow and reproduce, animals must obtain more energy than is just needed to survive. Marine mammals undergo profound variations in this dynamic equilibrium as they frequently go into prolonged negative energy balance while fasting during migration, reproduction or molt.

The balance of how energy acquisition and expenditure is achieved differs for individual species and environments. For some species, sea otters (*Enhydra lutris*), sea lions, and fur seals, (Otariidae) very high rates of energy expenditure are met by high rates of energy acquisition. These animals may preferentially live in nearshore environments or upwelling regions where food is abundant. Sirenians represent the opposite extreme. These sluggish marine mammals exhibit comparatively low existence costs and are able to survive on a low-quality diet. They have adapted to a diet of grasses that are in high abundance but are of low quality, energetically. They are able to do this because they live in the climatically benign tropics where maintenance costs may be kept to a minimum.

The seasonal migrations of large cetaceans demonstrate this interrelationship among energetic demand, energy availability, and local productivity. Although maintenance costs may be elevated in the polar regions, the ability to take advantage of the seasonally high productivity associated with the sea ice during the polar summer appears to more than compensate. When confronted with the high energetic costs of reproduction and of winter conditions, the mysticete whales opt for the more benign tropics. Further, their large body size makes the cost of migration extremely low. Prey availability may be low, but so too are the existence costs, especially for a large animal that is able to utilize energy reserves stored in the blubber.

The distribution of energy into the various physiological processes is outlined in Fig. 1. Ingested energy is the total energy consumed by the animal. Metabolic energy (ME) is energy available for maintenance, growth, or reproduction. Energy expended for maintenance includes key processes such as basal metabolism, digestion (heat increment of feeding), thermoregulation, and activity (locomotion, grooming, feeding, etc.).

I. Energy Acquisition

Not all of the ingested material consumed is digestible. Food energy remaining after digestion and elimination of fecal energy is known as apparently digested energy. The apparent digestibility for a DIET of fish ranges from 88 to 97.9%, whereas

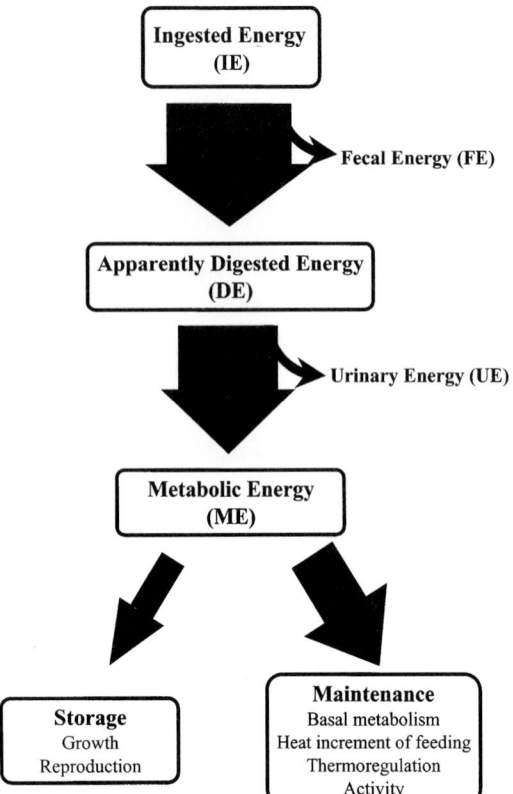

Figure 1 *Pattern of energy flow through a typical mammal.*

invertebrate prey with a high chitin content, such as shrimp, are lower at 72.2%. Plant material is less digestible because it contains cellulose, which is difficult to digest. Although sirenians extract less energy from their food than other marine mammals (84.6%), they are more efficient than other hindgut fermenters, such as horses (45–59%).

Chemical energy is also lost in the urine as urea and other metabolic end products and is defined as urinary energy. Metabolizable energy is the net energy remaining after fecal and urinary energy loss and represents the energy available for growth or reproduction and for supporting metabolic processes such as work (locomotion) and respiration (thermoregulation, maintenance metabolism, heat increment of feeding). The ME for pinnipeds varies from 78.3% for a squid diet to 91.6% for an anchovy diet.

II. Energy Expenditure
A. Cost of Maintenance Functions

Maintenance costs are those associated with homeostasis and include basal metabolism, the heat increment of feeding, thermoregulation, and activity.

1. Basal Metabolism It has generally been assumed that the basal metabolic rates (BMRs) of aquatic mammals are elevated when compared to terrestrial mammals of similar size. The current view of basal metabolic rates of marine mammals is more complex. Many previous studies did not conform to standardized criteria for measurements of basal metabolism. These criteria require that the subjects be adults, resting, thermoneutral, nonreproductive, and postabsorptive. This has been further confused by the expectation that all marine mammals should employ the same metabolic response. Specialization for marine living has occurred independently in three mammalian orders: sirenians, cetaceans, and carnivores. Further, within the carnivores there are three separate transitions to a marine existence: pinnipeds, sea otters, and polar bears, *Ursus maritimus*. Based on this diversity, we might expect different metabolic adaptations among the groups (Fig. 2). Manatees, *Trichechus manatus*, have BMRs lower than values predicted, whereas phocid seals have BMRs equivalent to those of similar-sized terrestrial mammals. Conversely, sea otters, otariids, and odontocetes appear to have BMRs greater than predicted.

The tremendous variation in body composition of marine mammals introduces another complicating factor when attempting to compare basal metabolic rates. Some species such as sirenians and walrus, *Odobenus rosmarus*, have dense bone, whereas seals may be composed of as much as 50% fat. When metabolic rates are expressed relative to body mass, a disproportionate amount of fat or particularly dense bone will lower the apparent metabolic rate. This is due to the low metabolic rates of bone and adipose tissue in comparison to lean tissue. Metabolic rate comparisons are further complicated by the wide variation in individual body composition. Many marine mammals undergo prolonged fasts that are accompanied by profound changes in body composition. Most of the mass change during fasting is due to loss of adipose tissue with a comparatively smaller change in lean tissue. For example, northern elephant seal females, *Mirounga angustirostris*, lose 42% of their initial mass, but of this only 14.9% comes from lean tissue with 57.9% coming from adipose tissue. This results in the overall change in body composition of 39% fat at parturition to 24% fat at weaning. Because lean mass is the primary contributor to whole animal metabolism, the whole body metabolism of the animal is likely to change little even though there has been a major change in its body mass. Such wide variations in body composition within and between marine mammals make comparison of metabolic rates problematic, especially relative to terrestrial animals that typically do not undergo such tremendous variations in body composition.

2. Fur vs Blubber It is important to consider the potential differences in the energy budgets of animals that use fur or blubber for insulation. Nonetheless, the overall time–energy budget of an animal that uses fur (fur seal or sea otter) is fundamentally different from an animal that uses blubber (sea lion, seal, or dolphin). Although fur is not a living tissue, it requires that an air layer be maintained, which is done by frequent grooming. Sea otters spend up to 16% of their day grooming. While sea lions, seals, and dolphins spend no time grooming, they must take in sufficient food to lay down a thick blubber layer, which is a living tissue and must be supplied with blood. Furthermore, blubber serves dual roles as an insulator and as an energy store. During fasting or periods of low

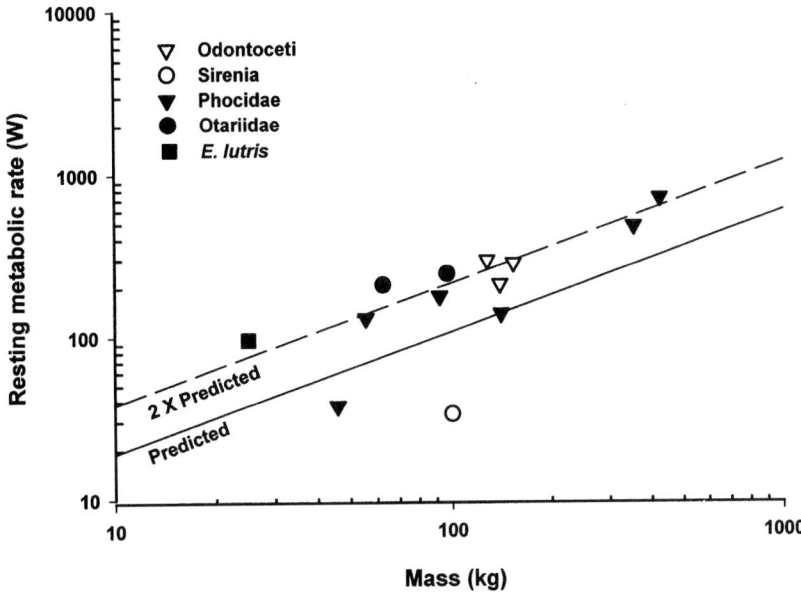

Figure 2 *Resting metabolic rate of marine mammals in relation to body mass. Measurements were made for the animals resting in water. The solid line denotes the predicted metabolic rate for equal-sized terrestrial animals; the dashed line represents two times the predicted levels. Species included are Odontoceti,* Stenella attenuata *and* Tursiops truncatus; *Phocidae,* Phoca vitulina, Pusa hispida, Leptonychotes weddellii, Pagophilus groenlandicus, *and* Haliochoerus grypus; *Otariidae,* Zalophus californianus; *and Sirenia,* Trichechus manatus.

food availability, a marine mammal must balance its utilization of blubber for energy needs with the potential loss of the blubber layer as an insulator.

3. Heat Increment of Feeding   When food is consumed, the animal's metabolic rate increases over fasting levels. The heat increment of feeding (HIF), also known as the specific dynamic action (SDA), may be considered the "tax" that is required to process food energy for conversion to metabolizable energy. The magnitude of energy allocated to HIF varies between 5 and 17% of the ME. In addition, the duration of HIF following a meal will depend on the amount of food consumed and its composition. In many mammals, the HIF is considered excess or waste heat. However, sea otters incorporate the additional heat increment of feeding to reduce their already high thermoregulatory costs. Sea otters are the smallest marine mammal and therefore have a very high rate of heat loss. In order to reduce their already high maintenance costs, sea otters use the heat liberated from activity and HIF to augment their thermoregulatory requirements. While grooming, feeding, and swimming, sea otters use the heat produced from these activities to supplement their thermoregulatory needs. Sea otters typically rest or sleep after feeding and use the HIF to compensate for the decreased heat production from activity. In sea otters, the HIF results in a 54% increase in metabolism that returns to basal levels within 4–5 hr. Sea otters incorporate the heat produced from HIF to augment their thermoregulatory needs while resting (Fig. 3).

B. Cost of Growth and Reproduction

For growth and reproduction to occur, an animal must acquire energy and nutrients in excess of that required for supporting maintenance functions. These additional energetic costs vary with the species of marine mammal, the sex, and reproductive pattern. In pinnipeds, polar bears, and sea otters (and

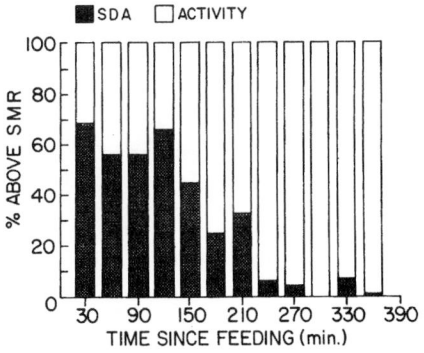

Figure 3 *The thermal budget of sea otters relies on heat production above standard metabolic rate (SMR, equivalent to BMR). The contribution of SDA (or HIF) to the overall metabolism is highest immediately after feeding and reaches zero within 390 min after feeding. As SDA decreases, the animal becomes more active and thus compensates for the decrease in heat production from SDA. From Costa and Kooyman (1984), by permission of University of Chicago Press.*

probably mysticetes and sirenians), the cost of the reproduction in males is limited to the cost of finding and maintaining access to estrous females. Evolution favors a pattern of energy expenditure that maximizes reproductive success in males. Costs associated with reproduction in aquatic and terrestrially breeding males are quite similar when normalized for differences in body mass. A larger body size is preferred in terrestrially breeding male pinnipeds because it confers both an advantage in fighting and allows the male to maintain terrestrial territories longer. In addition, larger animals can fast longer because they have a lower mass-specific metabolic rate than smaller animals. In species that compete for females in the water, we find that the males are comparatively smaller than the species that breed on land. For aquatic breeders, underwater agility is more important than large size when competing for mates. Nevertheless, even in aquatic breeders, large males have an advantage over small males because they do not need to make as many foraging trips and thus have more time to compete for females.

The cost of reproduction for females can be broken down into the energetic requirements of gestation and lactation. The cost of gestation is small relative to the cost of lactation. Even given the considerable variation of maternal investment in marine mammals, there is little variation in fetal mass at birth. Ma-

rine mammals also appear to invest more energy into gestation than terrestrial mammals (Fig. 4). However, as a group, marine mammals exhibit considerable variation in both the duration and the intensity of overall maternal investment (Fig. 5). Phocid seals and mysticete whales have extremely short lactation durations, which are compensated for by a higher rate of energy investment that enables the young to grow rapidly (Fig. 6).

Although phocid pups are weaned early, they still rely on maternally derived energy, stored as blubber, for weeks or months after weaning. The disadvantage of this rapid growth is that most of the mass and energy are stored as fat with proportionately little protein. The advantage of longer lactation is that young get more protein and other nutrients, allowing greater growth of lean tissue. However, longer lactation is energetically more expensive.

1. Variation in Milk Composition   The rapid growth of marine mammal young is made possible by the ingestion of extremely lipid-rich milk. With a few exceptions, most terrestrial animals produce milk that is low in fat; cows and humans produce milk that contains 3.7 and 3.8% milk fat, respectively. Lipid-rich milk allows the mother to transfer high levels of energy in a very short period. Hooded seals, (*Cystophora cristata*), are most impressive, with a 4-day lactation interval and a milk

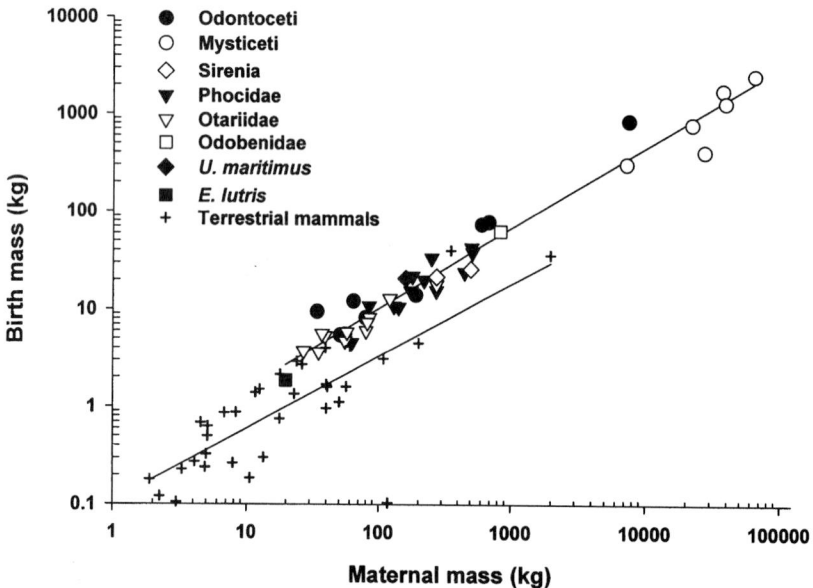

Figure 4   *Birth mass plotted in relation to maternal mass for marine and terrestrial mammals. Species of marine mammals included are Odontoceti,* Inia geoffrensis, Pontoporia blainvillei, Stenella attenuata, Globicephala melas, Physeter macrocephalus, T. truncatus, Phocoena spinnipinnis, P. phocoena, *and* Delphinapterus leucas; *Mysticeti,* Balaenoptera musculus, B. physalus, B. acutorostrata, B. borealis, Megaptera novaeangliae, *and* Eschrichtus robustus; *Phocidae,* Mirounga angustirostris, M. leonina, Cystophora cristata, Phoca vitulina, Pusa hispida, Leptonychotes weddellii, Monachus schauinslandi, Pagophilus groenlandicus, Erignathus barbatus, Lobodon carcinophaga, Histriophoca fasciata, *and* Haliochoerus grypus; *Otariidae,* Arctocephalus gazella, A. forsteri, A. galagagoensis, A. tropicalis, A. pusillis, Callorhinus ursinus, Zalophus californianus, Neophoca cinerea, Eumatopias jubatus, *and* Otaria flavescens; *and Sirenia,* Dugong dugon *and* Trichechus manatus.

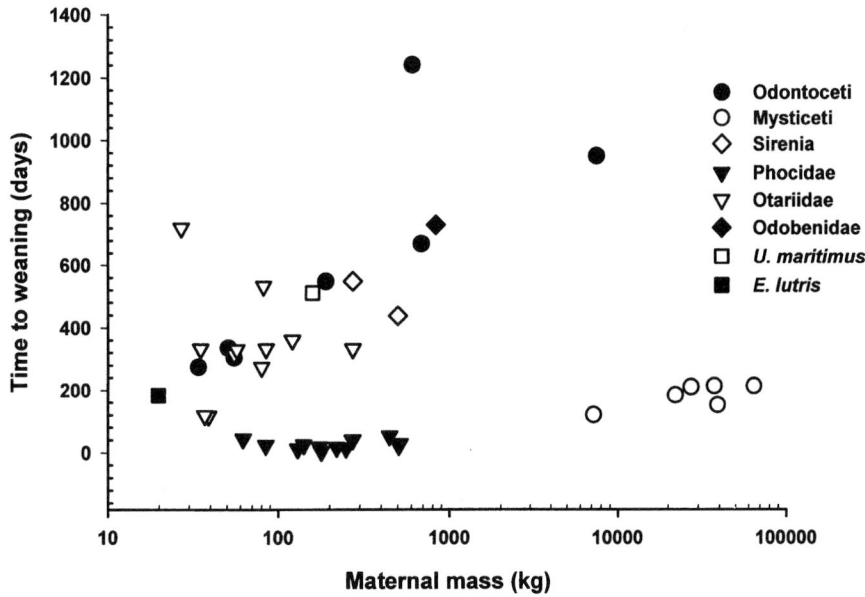

Figure 5 *Time to weaning plotted as a function of maternal mass for marine mammals. Lactation durations of phocid seals and mysticete whales are shorter than all other marine mammals. Species are the same as in Fig. 3.*

fat of 65% lipid. In view of this, it is not surprising that marine mammals with the highest growth rates produce milk with the highest lipid content.

Lactation also enables mothers to optimize the delivery of energy to their young. The energy content of the milk is independent of the type or quality of prey consumed or the distance or time taken to obtain it. Although milk is ultimately derived from the prey consumed, a mother can process, concentrate, or utilize stored reserves to produce milk. For example, some species feed on fish, whereas others feed on fish or squid. However, all of these species provision their offspring with milk of significantly greater energy density than the prey consumed.

2. Body Size and Maternal Resources: The Role of Maternal Overhead Fasting during lactation is a unique component of the life history pattern of many marine mammals. With the

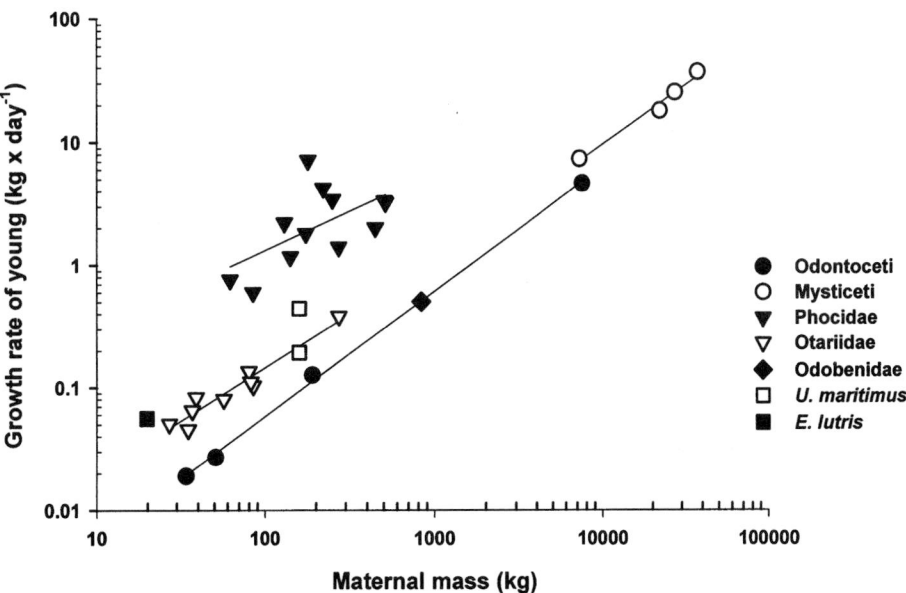

Figure 6 *Growth rate of suckling marine mammals as a function of maternal mass. Lines represent least-square regressions for each taxonomic group. Species are the same as in Fig. 3.*

exception of bears, no other mammal is capable of producing milk without feeding. By undertaking this energetic challenge, mysticetes and pinnipeds are able to separate where and when they feed from where and when they breed. In mysticete whales, this allows them to feed in the highly productive polar regions of the world's oceans but retain the thermal advantage of breeding in the calm sub-tropical regions. Migrating to warmer waters for parturition reduces the thermal demands on the newborn calf and additional thermal savings for the mother.

Among pinnipeds, the separation of feeding from lactation is necessary to allow for terrestrial parturition. Most phocids store sufficient energy reserves for the entire lactation period, whereas all otariids must feed during lactation. Phocid mothers typically remain on or near the rookery continuously from the birth of their pup until it is weaned; milk is produced from body reserves stored prior to parturition. Although some phocids feed during lactation, most of the maternal investment is derived from body stores. Their reproductive pattern is less constrained by the time it takes to travel and exploit distant prey, which may allow utilization of a more dispersed or patchy food resource. By spreading out the acquisition of prey energy required for lactation over many months at sea, northern elephant seal females only need to increase their daily food intake by 12% to cover the entire cost of lactation.

The ability of a marine mammal female to fast while providing milk to her offspring is related to the size of her energy and nutrient reserves and the rate at which she utilizes them. When food resources are far from the breeding grounds, as may occur for some phocids and large mysticete whales, the optimal solution is to maximize the amount of energy and nutrients provided to the young and to minimize the amount of energy expended on the mother. The term "metabolic overhead" refers to the amount of energy a female expends on herself while onshore (seals) or while in the calving grounds (whales). Larger females have a lower metabolic overhead than smaller females. This is because maintenance metabolism scales as mass$^{0.75}$ and fat stores scale as mass$^{1.0}$. As body size increases, energy reserves increase proportionately faster than maintenance metabolism.

3. Energy Investment and Trip Duration   Many phocids fast throughout the lactation interval, whereas otariid females feed intermittently between suckling bouts onshore. Otariid mothers modify the timing and patterning of energy and nutrient investment to optimize energy delivery to their young. Otariid mothers making short feeding trips that provide their pups with less milk energy than mothers that make long trips. In comparison to otariids, phocids may have a reproductive pattern that is better suited for dealing with dispersed or unpredictable prey or prey that is located at great distances from the rookery. However, fasting during lactation places a limit on the duration of investment, which limits the total amount of energy that a phocid mother can invest in her pup.

C. Field Metabolic Rates

A number of approaches have been used to study the metabolic rate of animals at sea. One approach, time budget analysis, sums the daily metabolic costs associated with various activities. Field observations of behavior are coupled with metabolic rate measurements made in captivity. Other methods rely on predictive relationships between physiological variables and metabolic rate. For example, metabolic costs can be assessed indirectly by measurements of changes in body mass and composition, variations in heart rate or ventilation rate, or dilution of isotopically labeled water.

Field metabolic rates (FMR) provide insight into the energetic strategies used by marine mammals. The best data exist for pinnipeds and the bottlenose dolphin, *Tursiops truncatus* (Fig. 7), and indicate that foraging otariids and bottlenose dolphins expend energy at 6 times the predicted basal metabolic level. In contrast, the metabolic rate of diving elephant seals and Weddell, *Leptonychotes weddellii,* seals are only 1.5 to 3 times the predicted basal rate. The lower diving metabolic rate of phocid seals contributes to their superb diving ability.

Field metabolic rates are quite variable both between and within species (Fig. 7). Such variation is thought to be associated with year-to-year changes in both the abundance and the availability of prey. For example, California sea lions, *Zalophus californianus,* significantly increased their foraging effort in response to reductions in prey availability during the 1983 El Niño event. Foraging effort will also vary with the type of prey consumed. Northern fur seals, *Callorhinus ursinus,* foraging on mature pollock during 1981 had lower metabolic rates than fur seals foraging predominately on juvenile pollock during 1982. Surprisingly, bottlenose dolphins, *Tursiops truncatus,* in Sarasota Bay had lower metabolic rates in the winter than in the summer. These dolphins either worked harder to lose heat in the summer or spent more time looking for prey. The importance of the thermal environment on field metabolic rate can also be seen in Galapagos fur seals, *Arctochepalus galapagoensis,* which due to the warm equatorial climate have a substantially reduced FMR compared to other otariids. The field metabolic rate integrates all of the costs incurred over the measurement interval and reflects differences in the foraging intensity as well as the thermal budget of the animal.

1. Energetics of Prey Choice   The amount of work and therefore energy expenditure that an animal puts into locating prey vary as a function of the energy content, availability, and location of the prey. Both size and proximate composition (fat, carbohydrate, protein, and water content) affect the energy content of prey. Prey availability varies as a function of the absolute abundance of prey (amount of prey per unit of habitat) and its distribution in the environment. A predator is more efficient when foraging on prey that is clumped than on prey that is evenly dispersed. Marine mammals forage in areas where prey has been concentrated as a result of oceanographic processes such as eddies, fronts, and upwelling regions associated with bottom topography.

Sea otters provide an excellent example of the factors that determine the energetics of prey choice. In recently occupied areas, sea otters feed on preferred prey items such as clams, abalone, or sea urchins. In such environments they find large, energy-rich, abundant prey that is easy to handle, consume, and digest. In such situations, lower quality prey items (turban snails, sea stars, mussels, chitons) are generally not eaten. These items

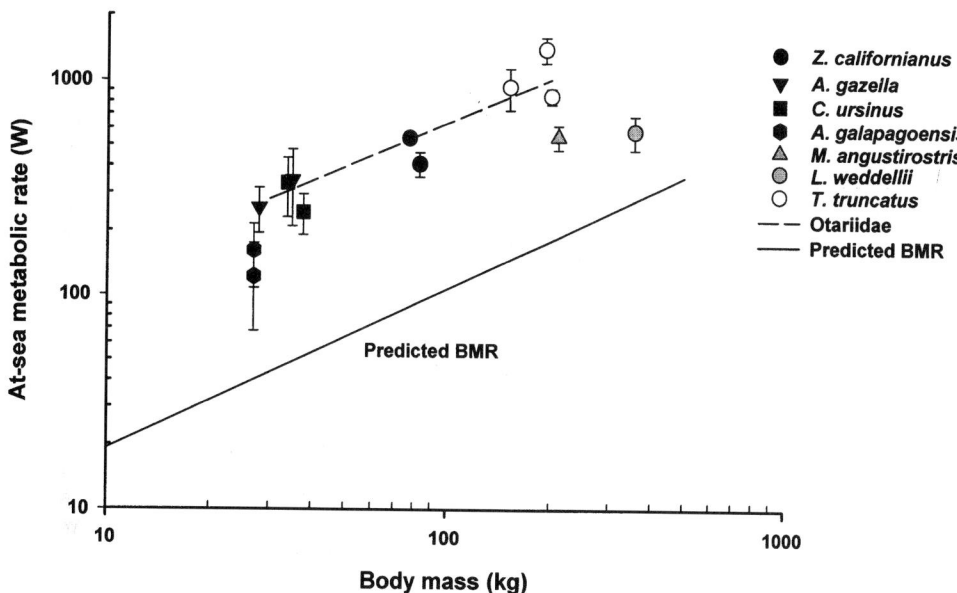

Figure 7 *At-sea metabolic rate measurements determined from the O-18 doubly labeled water method as a function of body mass. Data on* L. weddellii *were measured using open circuit respirometry on seals diving from an ice hole. The solid line represents the predicted basal metabolic rate for a terrestrial animal of equal size; the dashed line is the best-fit linear regression for the otariidae with the exception of data from* A. galapagoensis ($r^2 = 0.53$). *Error bars represent* ± *1 SD. Multiple points for each species reflect measurements taken over different years and show the range of interannual variation within a species.*

may be abundant, but they are energy poor and difficult to eat and digest. As the ABUNDANCE and size of their preferred prey decline, sea otters switch to less preferred, but more accessible prey such as turban snails, kelp crabs, and, in some cases, chitons and sea stars. Some sea otters specialize on different types of prey and are more efficient predators than nonspecialists.

Polar bears represent another example of optimal prey choice and its relation to the prey energy quality. Feeding predominately on ring seals, *Pusa hispida*, polar bears eat the energy-rich blubber layer and leave behind the lean "core" of the carcass. Due to its high lipid content, the blubber has a per unit mass energy content almost 10 times greater than that of the lean tissue of the ring seal. Thus, polar bears consume the most energy-dense part of the ring seal and then move on to find another kill.

2. Variations in Foraging Energetics The relationship between prey availability and reproductive success has been examined for a wide variety of pinniped species. Otariids are more susceptible to variations in nearby food resources because their breeding pattern is linked to continuous prey availability directly offshore, whereas phocids are buffered against fluctuations in prey availability near the rookery because they accumulate the resources they need for lactation over the previous year. Successful reproduction by otariids requires that mothers use a foraging pattern that optimizes the amount of time spent feeding at sea with the amount of milk energy delivered to her pup waiting on the rookery. Studies of female otariids with dependent young show that as food resources decrease, mothers can either increase the time spent foraging at

sea or the intensity of their foraging effort. However, simply increasing the time spent at sea increases the time between visits to the pup. Consequently, a greater proportion of the ingested milk energy is spent on pup maintenance rather than directed to pup growth. Increases in trip duration associated with declining prey resources result in slower pup growth because less milk is delivered over the same time interval.

Phocids are buffered from short-term fluctuations in prey availability due to their unique reproductive pattern. In phocids, reproductive performance (maternal investment) during a given season reflects prey availability over the preceding year and represents the mother's foraging activities over a much larger spatial and temporal scale than the foraging activities of otariids. It follows that the weaning mass of a phocid pup is an indicator of its mothers' foraging success over the previous year, whereas the subsequent postweaning survival of the pup is related to both its weaning mass (energy reserves provided by the mother) and the resources available to the pup after weaning.

See Also the Following Articles

Blubber ▪ Circulatory System ▪ Diving Physiology ▪ Predator-Prey Relationships ▪ Thermoregulation

References

Anderson, S. A., Costa, D. P., and Fedak, M. A. (1993). Bioenergetics. "Antarctic Pinnipeds: Research Methods and Techniques" (R. Laws, ed.), pp. 291–315. Cambridge Univ. Press, Cambridge, England.

Bowen, W. D. (1997). Role of marine mammals in aquatic ecosystems. *Mar. Ecol. Progr. Ser.* **158**, 267–274.

Boyd, I. L. (1998). Time and energy constraints in pinniped lactation. *Am. Nat.* **152**, 717–728.

Boyd, I. L. (1996). Temporal scales of foraging in a marine predator. *Ecology* **77**, 426–434.

Brodie, P. F. (1975). Cetacean energetics, an overview of intraspecific size variation. *Ecology* **56**, 152–161.

Costa, D. P., and Kooyman, G. L. (1984). Contribution of specific dynamic action to heat balance and thermoregulation in the sea otter, *Enhydra lutris. Physiol. Zool.* **57**(2), 199–203.

Costa, D. P., and Williams, T. E. (1999). Marine mammal energetics. *In* "The Biology of Marine Mammals" (J. Reynolds and S. Rommel, eds.), pp. 176–217. Smithsonian Institution Press, Washington, DC.

Costa, D. P. (1991). Reproductive and foraging energetics of high latitude penguins, albatrosses and pinnipeds: Implications for life history patterns. *Am. Zool.* **31**, 111–130.

Costa, D. P. (1993). The relationship between reproductive and foraging energetics and the evolution of the Pinnipedia. Pages 293–314 *in* "Recent Advances in Marine Mammal Science" (I. Boyd, ed.), pp. 293–314. Symposium Zoological Society of London No. 66, Oxford Univ. Press, Oxford, England.

Huntley, A. C., Costa, D. P., Worthy, G. A. J., and Castellini, M. A. (1987). "Approaches to Marine Mammal Energetics." Allen Press, Lawrence, KS.

Lockyer, C. (1978). A theoretical approach to the balance between growth and food consumption in fin and sei whales, with special reference to the female reproductive cycle. *Rep. Int. Whal. Commn.* **28**, 243–249.

Oftedal, O. T., Boness, D. J., and Tedman, R. A. (1987). The behavior, physiology, and anatomy of lactation in the Pinnipedia. *Curr. Mammal.* **1**, 175–245.

Entrapment and Entanglement

JON LIEN
Memorial University,
St. John's, Newfoundland, Canada

Fishermen use a variety of techniques to capture fish. A common method is the use of nets that hang passively, like curtains, in the water and ensnare fish that blunder into them, or make barriers that direct the fish into traps that hold them until they are removed.

Because fishing nets are an unusual barrier, cryptic and hard to detect, they also on occasion catch marine mammals. When nontarget species are accidentally caught in nets they are termed bycatch. Bycatches of some species of marine mammals, such as harbor porpoises (*Phocoena phocoena*), are common in several areas. Because of the strength of modern materials now used in constructing nets, even larger species of cetaceans are sometimes captured incidentally in fishing gear. Such entrapments seriously threaten the North Atlantic right whale (*Eubalaena glacialis*) population. Any species can be captured in nets; however, humpback whales (*Megaptera novaeangliae*), perhaps because of their abundance in coastal water where nets are commonly used or because of the many BARNACLES they carry, seem extremely vulnerable to entanglement in fishing gear.

In the late 1970s, humpback whales were seen in greater numbers in inshore waters of Newfoundland and Labrador as the bait fish capelin, which is their major prey, was seriously depleted offshore on the Grand Banks. The humpbacks moved inshore to feed on spawning capelin, which occur in the same areas where fishermen place their nets. Inevitably this meant trouble. Fishermen began to report whale collisions, which left nets badly damaged. On occasion, whales would actually be caught and held in the nets. About 50% of the animals that were caught died. Because this was a new source of mortality in this recovering humpback population, it was a serious conservation concern.

A program was established by the Whale Research Group of Memorial University of Newfoundland to aid both the whales and the fishermen when incidental captures occurred. If fishermen anywhere along Newfoundland's 17,000 km of coastline accidentally caught a whale they could call the Entrapment Assistance Program by a toll-free phone number. A trained team would be dispatched to release the whale alive and to save as much of the fishing gear as possible. Because there were real benefits for fishermen in minimizing gear losses and lost fishing time, they cooperated very well with the program and whales benefited as well. Humpback mortality as a result of entrapment was reduced to about 10% of the animals that were captured. These deaths occurred before help could reach the animal.

This did not solve the problem. During the 1980s, groundfish populations were being seriously depleting by fishing. To make a living, fishermen responded by adding more nets. Inshore effort by fishing nets increased dramatically. With more barriers, the frequency of collisions by whales, and entrapments, increased. In one fishing season the Entrapment Assistance Program received over 150 reports of entrapped humpback whales. In addition, other species, such as common minke whales (*Balaenoptera acutorostrata*) were also being caught. The program to release whales was extremely busy and effective. However, it was apparent that something was required to prevent collisions.

It was not practical to expect fishermen to stop fishing or to substantially modify where or how they fished. Instead, because cetaceans are acoustic specialists, scientists experimented with electronic devices that could be placed on nets. The hypothesis was that noisier nets would better alert the whales to their presence so they could avoid them. The alarms emitted higher frequency sounds that cetaceans could hear but were not detected by groundfish. Thus fish catches would not be similarly reduced. Such devices were used successfully in areas where the likelihood of collisions was high. Acoustic alarms reduced collisions by about 80%. This was good news for both whales and fishermen.

Other news was not good for fishermen, however. By 1992, groundfish populations were so seriously depleted by fishing that a moratorium on fishing was established. All nets were removed from the water. Collisions and entrapments of whales were reduced to near zero. The moratorium on groundfish fishing continued until 1998 when small quotas were once again established.

The quotas that were established were very small compared to historical levels, so far fewer nets were used. The reduction in the number of nets kept accidents with humpbacks low. However, in addition, quotas were allocated in shares to individual fishermen. Thus, each fisherman did not have to fish competitively but was assured of a fair portion of fish that they could catch when they wanted, usually when they could realize the best prices for fish. In Newfoundland, best prices occur in the fall and most fishing occurred then, a period when whale abundance inshore is relatively low. This combination of lower total fishing effort and a shift in fishing effort to a different season had kept collisions and entrapments to a very low level.

The Entrapment Assistance Program continues to be available to release whales carefully and to aid fisherman in retrieving fishing gear. In areas where whale abundance is high, fishermen continue to place acoustic alarms on nets. However, at present, the incidental capture of humpback whales in nets off Newfoundland and Labrador is a minor problem for both whales and fishermen. Because fisheries are highly dynamic and changing activities that can importantly effect the environment, they must be monitored continuously to ensure the protection of cetacean habitat.

See Also the Following Articles

Fishing Industry, Effects of ■ Harbor Porpoise ■ Humpback Whale ■ Incidental Catches

Reference

Lien, J. (1994). Entrapments of large cetaceans in passive inshore fishing gear in Newfoundland and Labrador (1979–1990). *In* "Gillnets and Cetaceans" (W. F. Perrin, G. P. Donovan, and J. Barlow, eds.), pp. 149–157. Rep. Int. Whal. Comm. Spec. Issue 15.

Estrus and Estrous Behavior

DARYL J. BONESS
National Zoological Park,
Smithsonian Institution, Washington, DC

Estrus is a state of sexual receptivity during which the female will accept the male and is capable of conceiving. This behavioral state is under hormonal regulation involving the ovary and pituitary gland and precedes or coincides with ovulation, i.e., production of an egg. The existing anatomical and physiological evidence suggests that the physiological process underlying estrus in marine mammals is comparable to that in other mammals.

Our knowledge about estrus and estrous behavior in marine mammals is highly variable. Among cetaceans, most of it is derived from studies of the ovaries of animals killed during whaling or collected from beached and stranded specimens. The little behavioral information that is available comes mostly from studies of captive animals and a few observations of free-ranging animals. Among pinnipeds, we know far more from behavioral observations, but this is concentrated on those species that mate on land. There is almost a complete void of information on aquatically mating species, which comprise about half the pinnipeds and the majority of the phocids or true seals. Estrous cycles in marine mammals are relatively long compared to terrestrial mammals, usually a year or more, and are seasonal and synchronous within a species or population.

I. Hormones and Anatomy of Estrus

As noted earlier, the hormonal cycle associated with estrus and reproduction in marine mammals appears to be similar to what happens in other mammals. High plasma concentrations of estrogens at the time of parturition decline rapidly and then begin a sharp rise again (Fig. 1). This estrogen rise, along with a decline in progesterone, which inhibits follicle growth, is likely responsible for rapid follicular growth in the ovary and the onset of estrous behavior.

All marine mammals have a reproductive tract that has two uterine horns (bicornuate) terminating at the ovaries. In pinnipeds, the ovary and horn involved in reproduction alternates between successive reproductive periods. This is not the case in cetaceans, however. Among baleen whales, both ovaries are equally functioning, whereas in many odontocetes one ovary appears to function early in life and the other later in life.

II. Timing of Estrus

In pinnipeds, most of our knowledge about the timing of estrus comes from behavioral observations of mounting and mating behavior. Most species are monestrous, with the exception being the walrus (*Odobenus rosmarus*), which has two estrous periods within a year. The double estrus in walruses is undoubtedly linked to the 15-month gestation. With this long gestation, parturition occurs when there are few reproductively active males available, yet there is a postpartum estrus. The second estrus occurs 9–10 months after birth, during the peak of male sexual activity. Otariid seals show relatively little variation in the timing of estrus, with it occurring early in lactation during what has been called the perinatal period and within 4–14 days following parturition for most species. In a few species (California sea lions, *Zalophus californianus;* Steller sea lions, *Eumetopias jubatus;* and Galapagos fur seals, *Arctocephalus galapagoensis*), estrus occurs after maternal foraging trips begin, which may be as much as a month after parturition. Although a postpartum estrus may be an ancestral condition, the nearly yearlong lactation in otariids, along with the dispersal of males away from BREEDING SITES after mating, would have selected for estrus to be early in lactation and shortly after parturition. Linked to this ecological pattern of spatial separation between breeding on land and feeding at sea is the existence of delayed implantation in seals, which tends to synchronize parturition, and thus estrus. The benefit to otariid females being receptive shortly after parturition would then be having access to the most competitive and highest quality males because they are present on land.

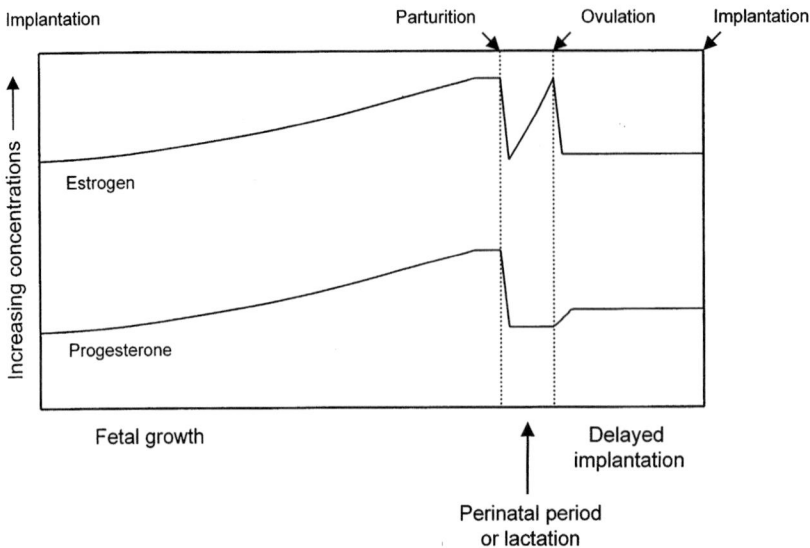

Figure 1   *A schematic of estrogen and progesterone levels in the circulatory system of female seals in relation to the timing of various reproductive events. Modified from Boyd* et al. *(1999).*

In phocid seals the timing of estrus is more variable with respect to parturition, ranging from about 5 days to between 1 and 2 months after parturition. In fact, estrus tends to occur in late lactation or after pups have been weaned in phocids. This accompanies a very different lactation pattern from that of otariids. All phocids have a short lactation ranging from 4 to 60 days. Although those phocids that have extremely short lactation periods will come into estrus within 2 weeks of parturition like most otariids, those species that lactate for a month or more have shifted estrus away from being juxtaposed to parturition if this was the ancestral pattern. Reasons for this difference between otariids and phocids are unclear. It may relate to phocid females being in estrus for longer periods of time, providing for greater opportunities to mate with multiple males and thereby increasing the quality of males fertilizing their offspring through sperm competition.

The duration of estrus in otariids appears to be very short based on observed copulatory behavior. In most species, females are observed mating once, although about 30% of females in some species are seen mating twice (see later for a possible explanation). These multiple matings still appear to occur within a 2-day period. We know nothing about the duration of estrus in the aquatically mating phocids, but for all three of the species that mate on land estrus lasts for several days. At the maximum, it may last up to 13 days in the northern elephant seal (*Mirounga angustirostris*) and 5–7 days in the southern elephant (*M. leonina*) and gray seals (*Halichoerus grypus*). An estrous period several days long is common in the sea otter (*Enhydra lutris*) as well, a species in which females are known to mate with more than one male.

Sea otters have an annual cycle like most pinnipeds. Female sea otters typically become receptive within a few days of weaning their pup and receptivity lasts 3–4 days. It is possible that

ovulation is induced by copulation in this species as other mustelids exhibit induced ovulation. There is evidence of polyestrus in females that are not impregnated successfully during an initial estrus.

The cycles of cetaceans are much more variable than the cycles of other marine mammals. Many of the mysticete whales have a 2-year cycle. For example, the fin whale (*Balaenoptera physalus*) becomes receptive in the winter in low-latitude, warm water, where mating occurs, and then migrates to cooler high latitudes where it feeds through the summer. In the fall it again migrates to low latitudes to give birth, and after a second migration to cooler water for feeding and lactation the female weans the calf and has a period of anestrus for 5–6 months before becoming receptive again, having made another migration to warm, low-latitude waters. The North Atlantic right whale (*Eubalaena glacialis*) may have an even longer interval of 3–4 years between estrous periods, whereas the minke whale (*Balaenoptera acutorostrata*) appears to have an annual cycle.

Odontocetes have an annual breeding season with most mating occurring over a diffuse 2- to 5-month peak for many species studied. However, the estrous cycles of individual animals are not likely to be annual because of multiyear lactation periods for many species, which may inhibit cycling. Lactation periods of 2–3 years are common among odontocetes, Unfortunately, more behavioral studies in which individual animals are followed are needed. However, even this has its drawbacks in that copulatory behavior in some dolphins may have a nonsexual social component that could make it difficult to obtain a clear understanding of the estrous cycle from behavior.

Captive studies have provided some useful insights into the estrous cycle of nonlactating odontocetes. For example, a 3-year study with two common dolphin (*Delphinus delphis*) females and with four bottlenose dolphin (*Tursiops truncatus*) females found somewhat similar results with respect to periods of estrus and anestrus. The common dolphin females exhibit extended anestrus of 1–2 years but also cycled as many as seven times within a year. There was no seasonality to these cycles. The bottlenose dolphin females also had a yearlong anestrus followed by polyestrus, although the maximum number of cycles within a year was only three. Along with this reduced number of cycles compared to the common dolphin the cycles were seasonal, occurring only in April and May.

Sirenians (manatees, *Trichechus* spp.; and dugongs, *Dugong dugon*) and polar bears (*Ursus maritimus*) appear to follow a similar pattern as the cetaceans with a multiyear lactation that results in an estrous cycle of 2 or more years. Among sirenians, after weaning a calf, females may also undergo polyestrus before becoming impregnated. Breeding may not be continuous, as 33 and 56% of a sample of ovaries from manatees and dugongs, respectively, showed that females were neither pregnant nor lactating. Little is known about the length of estrus in sirenians. However, given that males and females are solitary and need to search out each other for reproduction, it is likely that estrus is relatively long and ovulation might be induced. Mating or estrus groups, consisting of one female and multiple males trying to mate with the female, in manatees and dugongs last for variable periods of time, from a few hours to several days. There is anecdotal evidence for polar bears, which also live separate solitary lives except during breeding, to suggest that estrus occurs over multiple days with "courtship behavior" that may be necessary to stimulate ovulation.

III. Estrous Behavior and Signals

Inexperienced observers of the behavior of the two elephant seal species and the gray seal might come to the conclusion that females do not come into estrus, except that they ultimately do mate before departing from the breeding colony. Females generally show no soliciting behavior and appear to protest mounting attempts by males right up to the time females leave. However, upon closer observation, subtle differences in female behavior may become apparent. For example, females may reduce the duration of their protests when they are in estrus, but they may do this selectively to "potentially high-quality" males. In southern elephant seals, males reportedly will move from female to female, placing their head across the female's back and waiting for her reaction before either moving on or subsequently attempting to try to mount her. In northern elephant seals near the end of lactation, a female will not respond vocally to the harem master when he places his foreflipper over her back and she may even show a lordosis response by spreading her flippers and raising her tail end.

Most likely there are cues that signal the sexual status of these phocid females. For example, in the gray seal, attempts by males to mount females are not indiscriminate. They begin about a week before females are first observed copulating and increase in frequency linearly during that period (Fig. 2). As male gray or elephant seal males do not sniff the head or tail end of females prior to attempting to mount them, the existence of olfactory signals seems unlikely, especially in view of

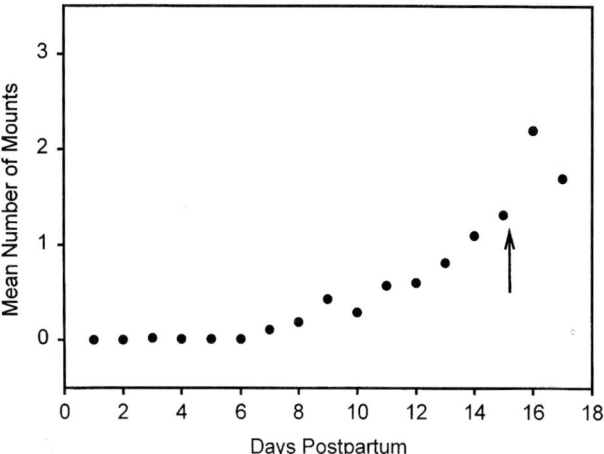

Figure 2 *Mean number of attempted copulations with females by gray seal males as a function of the number of days postpartum for the females. The arrow represents the average time of onset of estrus, defined by the first observed copulation. Modified from Boness and James (1979).*

such sniffing behavior preceding mounting in otariids (see later). One possible explanation for the discriminate mounting by gray seal males could be that they use either the body shape of females that are becoming depleted from fasting or the increased rotundness of their pups as they approach being weaned.

Another interesting example of behavior that likely indicates signaling (probably chemical) of estrus occurs in polar bears. As noted earlier, males and females are solitary and must find each other for reproduction. Males have been observed detecting the tracks of a female and following them up to 100 km to find the female. Manatee and dugong females may increase the range of their movements when they become estrous to increase the likelihood of encountering males.

Our best information about estrous behavior in pinnipeds comes from some elegant experimental work by Roger Gentry in his 19-year study of the northern fur seal (*Callorhinus ursinus*) in Alaska. These studies involved capturing groups of females, controlling their access to males, manipulating the size and age of males to which the females were given access, and using surrogates in place of males. Field observations found that prior to estrus, males engaged in sniffing the nose and mouth of females and that sniffing the tail region sometimes followed this. Investigation of the latter often led to a lordosis response. With this in mind, the studies were designed to determined the role of the male in triggering estrus, terminating it, determining the change in behavior of females, confirming the importance of olfactory cues, and understanding the mechanism of how estrus is terminated.

The following briefly summarizes the findings of Gentry's studies. Females prevented from copulating did not prolong estrus nor recycle, and receptivity lasted an average of 34 hr. A nonestrous female's behavior changed dramatically within as short as a 5-min period, going from being aggressive to a male to emitting low-level estrous vocalization and showing lordosis. The change was reflexive in that either artificial means or a male elicited it. Copulation by a territorial male usually terminated estrous behavior within 15 min, whereas a juvenile or senescent male copulation did not terminate estrus quickly. All copulation led to fertilization, however. Males that actively suppressed the aggressive behavior of females appeared to enhance the speed with which females became receptive. Finally, there appears to be a "vaginal code" based on male size and thrusting pattern for the termination of estrus following copulation.

Very little can be said about the pre-estrous and estrous behavior of cetacean females. In one study of captive spinner dolphins (*Stenella longirostris*), an attempt was made to correlate behavioral and hormonal patterns, examining behaviors that could have a sexual function. The behaviors included genital-to-genital contact, beak-to-genital contact, other-to-genital contact, ventral presentation, chases, and nongenital contact. The only behavior that showed any relationship with hormonal states indicative of estrus was beak-to-genital contact. This involved females inserting her beak into the genital slit of the male, most often during the follicular phase of her estrous cycle.

Our best opportunity to better understand estrus and estrous behavior in cetaceans, sirenians, polar bears, and the aquatically mating pinnipeds will almost certainly come from further studies of captive animals because of our inability to observe individual animals for long periods of time in the wild. For odontocetes, there is the further complication of making inferences about receptivity from behavior alone because of the nonsexual use of sexual behavior in other social contexts.

See Also the Following Articles

Endocrine Systems ■ Female Reproductive Systems ■ Reproductive Behaviors

References

Boness, D. J., and James, H. (1979). Reproductive behaviour of the grey seal (*Halicherous grypus*) on Sable Island, Nova Scotia. *J. Zool. Lond.* **188,** 477–500.
Boyd, I. L. (1991). Environmental and physiological factors controlling the reproductive cycles of pinnipeds. *Can. J. Zool.* **69,** 1135–1148.
Boyd, I. L., Lockyer, C., and Marsh, H. D. (1999). Reproduction in marine mammals. *In* "Biology of Marine Mammals" (J. E. Reynolds III and S. A. Rommel, eds.), pp. 218–286. Smithsonian Institution Press, Washington, DC.
Gentry, R. L. (1998). "Behavior and Ecology of the Northern Fur Seal." Princeton Univ. Press, Princeton, NJ.
Perrin, W. F., Brownell, R. L., and Demaster, D. P. (eds.) (1984). "Reproduction in Whales, Dolphins and Porpoises." International Whaling Commission, Cambridge.
Riedman, M. L., and Estes, J. A. (1990). The sea otter (*Enhydra lutris*): Behavior, ecology and natural history. U. S. Fish and Wildlife Service, Biological Report 90.
Stirling, I. (1988). "Polar Bears." Univ. of Michigan Press, Ann Arbor, MI.
Wells, R. S., Boness, D. J., and Rathbun, G. B. (1999). Behavior. *In* "Biology of Marine Mammals (J. E. Reynolds III and S. A. Rommel, eds.), pp. 324–422. Smithsonian Institution Press, Washington, DC.*

Ethics and Marine Mammals

MARC BEKOFF
University of Colorado, Boulder

The whale in the sea, like the wolf on land, constituted not only a symbol of wildness but also a fulcrum for projecting attitudes of conquest and utilitarianism and, eventually, more contemporary perceptions of preservation and protection.
—Stephen R. Kellert, 1996

I. Humans and Other Animals: Multidimensional Encounters

Humans are a curious lot, and our intrusions, intentional and inadvertent, have significant impacts on other people, nonhuman animal beings ("animals"), plants, water, the atmosphere, and inanimate landscapes. Often our influence is subtle and long term. There are many important and difficult issues that demand serious consideration in discussions of the ethics of how human beings interact with animals. Their complexity is compounded because highly charged subjective opinions and passions run high. This article highlights just how complex and multidimensional these issues are. It is meant to be a starting point for a discussion of different perspectives; none is more important than others. What is important is that we all agree that ethics is an essential element in any discussion of human interactions with other animals.

This article concentrates on human–dolphin and human–whale encounters (e.g., hunting, keeping animals in captivity, swimming programs, using them for entertainment), for ethical questions that arise when considering these types of highly visible interactions can be used as illustrations for human encounters with other marine mammals, including the pinnipeds (seals, sea lions, and walruses), manatees (*Trichechus* spp.), and polar bears (*Ursus maritimus*). There is much public interest in these rendezvous.

Dolphins, whales, and other marine mammals have often been fabricated to be the animals we want them to be. Most descriptions of dolphins and other cetaceans picture them as highly intelligent and capable of experiencing pleasure and pain (sentient) animals with remarkable social and cognitive skills. Indeed, dolphins and other marine mammals seem to fulfill some criteria of "personhood" in that they are alive, aware of their surroundings, sentient, and may have a sense of self. Why, then, do some people feel comfortable intruding into their worlds if it will cause pain and suffering? Toni Frohoff (1998, p. 84) has poignantly noted: "Currently we are walking a fine line in our relationship with cetaceans. The same attraction that motivates us to protect them from harm is also what drives us to be close to them, to have them 'within reach.'" It is because dolphins and other marine mammals are thought to be attractive, harmless, endowed with mystical qualities, or to be of economic value as commodities for show or food that we seek them out. However, we may bring much harm to them in our efforts to include them in our lives, even in ways that do not involve killing them.

Many issues that are pertinent when considering marine mammals are also raised when discussing other mammals such as terrestrial CARNIVORES. Wolves and whales have been among the most persecuted of animals during the last three centuries, wolves because they were feared predators and whales because of their economic value. The near EXTINCTION of both wolves and some whales was important in setting environmental policies. However, while many people swim and wade with dolphins, few if any truly dance or howl with wolves. Thus, the close contact that many humans have with some marine mammals leads to other questions that are unique to these encounters.

A. What Should We Do?

Human impacts on other animals, intentional and inadvertent, are universal. A major question in need of serious debate is should we ever interfere in other animals' lives—when might human interference be permissible? Thus, should we let other animals be and not intentionally interfere in their lives? Or, should we hunt them for food whenever we so wish? Should we hunt only when there are no alternative food sources? Should we interfere in other animals' lives when we have spoiled their habitats or when they are sick, provide food when there is not enough food to go around, provide care to young if a parent does not, stop aggressive encounters, stop predators in their tracks, or translocate individuals from one place to another, including zoos, wildlife and marine theme parks, and aquariums? Should human interests always trump those of other animals? If not, then when should the interests of other animals trump our own?

The question of when humans *should* intrude is a difficult one. However, just because we *can* do something does not mean that we *should* or *have* to do it. Furthermore, just because some intrusions may be *relatively* less malign than others, this line of reasoning places us on a very slippery slope and can, in the end, lead to narrow or selfish anthropocentric claims. Even in situations when humans have good intentions, those intentions are not always enough.

This article discusses some basic principles that underlie the use and exploitation of marine mammals, especially whales and dolphins, presents a few general questions, and discusses some representative examples. Definitive answers to these and other questions are quite elusive, but open discussion can provide guidelines for proactive decision-making. All too often we are left in the position of trying to rectify messy and difficult situations that we have created; proactivity needs to become the *modus operandi* for future actions. For many questions about how animals should be treated by humans there are no "right" or "wrong" answers. However, there are better and worse answers. Perhaps it will turn out that in some cases what we think is the "right" action is not when the big picture is carefully analyzed.

It is essential to remember that even if wild or captive marine mammals develop close social bonds with humans, these animals are socialized or habituated individuals, but they are not domesticated animals. Often people remark that individuals who interact closely with humans have become domesticated.

However, domestication does not happen to an individual during his or her lifetime. Domestication is a long-term evolutionary process during which humans selectively breed animals for desirable traits. "Domesticated" and "socialized" or "habituated" are not synonyms.

B. Human Responsibilities

It generally is accepted that humans have unique responsibilities to other living organisms (and to inanimate environments). Our unique responsibilities stem from our (at least most of us) being moral agents who are responsible for our actions. It usually is assumed that neither animals nor young human infants and mentally impaired adults are moral agents. Rather, they are moral patients, not usually held responsible for their actions. They do not know "right" from "wrong" or "good" from "bad." Nature, too, cannot be good or bad, although the consequences of natural acts can be better (good) or worse (bad). Nature simply is. We do not have to apologize for nature's ways. Nor should her ways—her supposed cruelty and ruthlessness—be used as excuses for what we do to other animals (including humans).

It is important to stress that most, if not all, other animals depend on our goodwill and mercy. Individuals can choose to be intrusive, abusive, or compassionate. We do not have to do something because someone else wants us to it. We do not have to do something just because we can do it. Each of us is responsible for our choices.

II. Animal Rights and Animal Welfare: The Moral Status of Animals

In current discussions about the moral status of animals, there is an obvious "progressive" trend for greater protection for wild and captive animals. (This might be due in part to an increasing number of people moving from farms and rural areas to more urban environments.) This is clearly the case for marine mammals (Kellert, 1999). In a survey of American's perception of marine mammals, most respondents were opposed to commercial WHALING, often for ethical reasons. Concern was also expressed for the commercial exploitation of seals, sea otters, walruses, and polar bears. Most Americans also objected to commercial whaling by native peoples or the resumption of killing gray whales. A majority of Alaskans opposed oil and gas development if it injured or killed marine mammals. There also was an unsuccessful effort prior to the reauthorization of the Marine Mammal Protection Act in 1988 to prohibit any invasive research involving marine mammals unless that research would directly benefit the subject of the research.

In recent years, philosophers and scientists have devoted increasing attention to questions about the moral status of animals. Many people support a position called the *rights* view. To say that an animal has a right to have an interest protected means that the animal has a claim, or entitlement, to have that interest protected even if it would benefit us to do otherwise. Humans have an obligation to honor that claim for other animals (just as they do for humans who cannot protect their own interests). Thus, if a wild dolphin has a right to feed, then humans have an obligation to allow her to do so and not do anything to interfere with her feeding activities. Likewise, if a dolphin has a right to life, she cannot be used in war games, warfare, or other activities in which death is possible.

Animal rights advocates stress that animals' lives are valuable in and of themselves (they have inherent value) and that their lives are not valuable because of what they can do for humans (their utility) or because they look or behave like us. Animals are not property or "things," but rather they are living organisms who are worthy of our compassion, respect, friendship, and support. Animals are not "lesser" or "not as valuable" as humans; they are not property that can be abused or dominated. Human benefits are irrelevant for determining how animals should be treated.

Many people believe that the rights view and the *animal welfare* view are identical. They are not. Animal welfarists focus on individuals' usefulness to humans. They practice *utilitarianism*, in which the general rule of thumb is that the right actions are those that maximize utility summed over all those who are affected by the actions. Often welfarists/utilitarians are called "wise users." They believe that while humans should not abuse or exploit animals, as long as we make the animals' lives comfortable, physically and psychologically, we are taking care of them and respecting their welfare. Welfarists are concerned with the quality of animals' lives. However, welfarists do not believe that animals' lives are valuable in and of themselves. Many conservation biologists and environmentalists are utilitarians who are willing to trade off individuals' lives for the perceived good of higher levels of organization such as populations, species, or ecosystems.

The welfarists' rule of thumb, and it is not a moral rule, is that it is permissible to use animals if the relationship between the costs to the animals and the benefits to the humans is such that the costs are less than the benefits. Welfarists believe that if animals experience comfort, appear happy, experience some of life's pleasures, and are free from prolonged or intense pain, fear, hunger, and other unpleasant states, then we are fulfilling our obligations to them. If individuals show normal growth and reproduction and are free from disease, injury, malnutrition, and other types of pain and suffering, they are doing well. Thus, welfarists argue that using animals in experiments, slaughtering them for human consumption, and using them for treating human disorders [e.g., dolphin-assisted therapy (DAT) programs] are permissible as long as these activities are conducted in a humane way. Welfarists do not want animals to suffer from any unnecessary pain, but they sometimes disagree among themselves about what pain is necessary and what humane care really is. Welfarists agree that the pain and death animals experience are sometimes justified because of the benefits that humans derive. The ends—human benefits—justify the means—the use of animals.

III. Hunting Whales

Whale HUNTING brings to light numerous issues that reflect utilitarian thinking. Whales are frequently viewed as commodities. Whether they are hunted centers on whether it is economical. Rarely are the costs to the individuals entered into the equation. In the past, many people thought that whales were an inexhaustible resource and historically there were few restrictions on killing them. When whale watching became popular, whales were more valuable alive than dead. They went from being a consumptive to a nonconsumptive resource.

Political and sociocultural motives also play a role in whale hunting. Various indigenous people (e.g., the Makah in Washington State) want to be able to hunt whales (in the Makah case, gray whales) because their ancestors did so, because it was part of their cultural heritage. They claim that the tradition of whale hunting defines "who they are."

The revival of aboriginal whaling is controversial in various parts of the world and involves species other than the gray whale. When the target species is on the brink of extinction, few argue that any type of whaling is permissible. Likewise, when killing whales is essential for food, few argue against the practice. However, when the whales are not endangered, people disagree about continuing to kill whales or reestablishing this practice. Some argue that whale hunting is permissible because it is part of the heritage of a given indigenous group (it is cultural revival), whereas others argue there are other cultural practices that are no longer followed and little effort to regain them. Why is whaling hunting so important if it is not essential for sustenance?

Methods of killing whales also are controversial. For example, the Makah used a rifle to kill a whale who had been wounded by a harpoon. A majority of Americans oppose the use of weaponry. Hunting whales produces much pain and suffering. Chasing and stalking individuals compromises their physical and psychological well-being and death usually is not instantaneous, frequently taking upward of 10 min. Furthermore, family groups are broken up. All in all, hunting whales and other animals, including such marine mammals as seals, raises numerous difficult ethical issues.

IV. Keeping Animals in Captivity
A. Swimming with Dolphins

"Swim with dolphin" and "petting pool" programs are very controversial. Such proffered reasons as "it's fun," "aren't the animals cute," or "it's a spiritual experience" are insufficient to justify these practices. Much attention has been given to the question of whether human encounters with dolphins may have negative effects on the dolphins. Human–dolphin interactions may be noisy and stressful. One study reported that captive dolphins showed enlarged adrenals, especially those individuals exposed to humans on numerous occasions. The long-term effects of swim programs on dolphin behavior and well-being still need to be studied systematically, but evidence shows that the stress associated with these programs may have long-term effects on the dolphins.

Swim programs are risky to humans. Dolphins are large and strong animals. While higher risks seem to be experienced more in noncontrolled swims, there are also serious risks in controlled swims that might be fatal. "Controlled" refers to situations when all interactions are directed by trainers, who give the dolphins commands at all times. "Noncontrolled" means interactions are allowed to occur spontaneously.

It is also important to know if DAT programs truly work. While some researchers claim that DAT is an effective therapeutic intervention for several disorders (e.g., depression, autism, cerebral palsy, mental retardation), others disagree. Criticisms center on the use of improper statistical methods and the lack of controlled studies. It often is very difficult to assess experimentally the positive effects of animals on people. In many programs, no other animals, including such domesticated species as dogs, were used as controls to see if they might be as or more effective than dolphins.

Another question that also is important to consider is whether programs that involve interactions with captive dolphins help educate people about these and perhaps other animals. More research is needed to determine if contacts with dolphins actually change people's attitudes about them. Intuitions are not enough. To date, there is no solid evidence that interactive captive programs with dolphins are more effective educationally than noninteractive programs. Indeed, some marine biologists fear that these programs may send the message that it is permissible to take animals from the wild and bring them into captivity and keep them in small tanks where they are bored, deprived, and needlessly die. There also are serious concerns about the fate of dolphins once they are too old or aggressive to partake in swim programs. Yet another concern centers on the possibility that these programs may teach people to expect the same kinds of interactions from free-ranging wild animals.

While there have been attempts to regulate swimming programs, little has actually been accomplished. In the United States, federal regulations controlling these programs, mandated in 1998 (after a delay of 4 years before finalizing them) by the U.S. Department of Agriculture's Animal and Plant Health Inspection Service (APHIS), were suspended in April 1999 soon after they were invoked because, according to their press release: "It has come to our attention that the language in the new regulations may be confusing to some. Therefore, we are suspending enforcement of the regulations in order to take a closer look at the language and make it more understandable." (http://www.aphis.usda.gov/lpa/press/1999/04/dolphin.txt) During this process, there are no regulations for these popular programs. Some people believe the federal regulations were suspended because of pressure from the lucrative industries that exploit dolphins.

B. Petting and Feeding Programs

Petting and feeding programs allow people to pet and feed captive dolphins. Many of these programs may not be adequately supervised or monitored. There are some major concerns with these programs, including that dolphins may be unable to avoid encounters with humans and may be highly stressed, the water in which dolphins and humans interact is often heavily chlorinated and may be unhealthy for dolphins (and humans), dolphins may be fed foreign objects that can harm them, and there seems to be little, if any, education value to these programs. There are few data that speak to these and other concerns and this information is needed to determine if petting and feeding programs can be properly regulated. One of the main questions is whether dolphins can be accessible to people in these programs and still be protected from harm.

While feeding and harassing wild dolphins is illegal in the United States, and there are severe penalties for engaging in these activities, this is not so for other countries. There are documented instances of wild dolphins being fed firecrackers, golf balls, plastic objects, balloons, and fish baits with hooks (so that hooked dolphins can be caught). Provisioning dolphins with fish has been associated with a change in the social behavior of free-ranging bottlenose dolphins (*Tursiops aduncus*) in Monkey Mia, Australia. Dolphins who have been fed also change their foraging behavior and frequent heavily trafficked harbors and marinas. Some get struck by boats. People have also been seriously injured trying to feed wild dolphins. The National Marine Fisheries Service and other organizations are mounting highly visible campaigns to stop the feeding and harassment of wild dolphins. It also has been noted that some problems associated with feeding terrestrial mammals (changes in foraging patterns and hunting skills) are relevant to concerns about the feeding of dolphins.

Clearly, much more information is needed concerning petting and feeding programs for captive and wild animals. Especially needed are data concerning the effects of these programs on dolphin mortality, health, and psychological and emotional well-being. It also is important to counter the possibility that feeding captive dolphins may send the message that it is permissible to feed wild individuals.

C. Zoos: Aquariums and Marine Theme Parks

The existence of zoos, including aquariums and marine theme parks, raises many important and difficult ethical questions. Certainly, numerous people are interested in exotic animals, including marine mammals. Kellert found that a majority of Americans objected to the captive display of marine mammals in zoos and aquariums if there were no demonstrated educational and scientific benefits. They were concerned with the care given to captive individuals. To date, no unequivocal data show that there are any significant educational and scientific benefits that help the animals, despite beliefs that such benefits accrue. An average zoo visitor spends only about 30 sec to 2 min at a typical exhibit and only reads some signs about the animals. A number of surveys have shown that visiting zoos to be entertained was the predominant reason people went to the zoo. In one study at the Edinburgh Zoo in Scotland, only 4% of zoo visitors went there to be educated, and no one specifically stated they went to support conservation. To date, very few empirical data support the notion that much educational information is learned *and* retained that helps the animals in the future. Indeed, some people worry that keeping animals in captivity for humans to view carries the message that it is alright to do so.

Many questions center on how individuals are captured, transported, and kept in various types of captive situations. Animals often are injured and otherwise stressed during capture and transport. Family groups may be broken up and the social structure of populations decimated. The effects of changing the social structure of wild populations are little known. Well-intentioned people often argue that the lives of captive animals are better, of higher quality, than those of wild relatives, but available data for marine mammals suggest that this claim is not well supported. From an ethical perspective, one must consider whether this claim is even relevant, for keeping animals in captivity radically alters numerous behavior patterns that have evolved over millennia. Predation, starvation, and disease are part of what it is to be wild. Is a longer unnatural life in captivity better than a shorter natural life in the wild?

Breeding surplus animals for profit (e.g., polar bears who become the center of media parades and then are moved to other zoos when their resource value or utility is fully exploited) also demands serious discussion. Similarly, the trading, donating, or loaning of unwanted or surplus animals who cannot be released into the wild—treating them as property—also raises numerous ethical questions.

The benefits of keeping marine mammals in captivity, to the animals themselves, remain unknown. Because the social and physical environments of marine mammals are virtually impossible to replicate in captivity, ethical questions arise when these animals are maintained in unnatural environments. There can be little doubt that the quality of life is compromised. In captivity, evolved patterns of foraging, caregiving, and migrating are lost as are natural patterns of social organization (group size and composition). In captivity, for practical reasons, group sizes may be much smaller than those observed in wild relatives. Stereotyped behaviors often result from conditions of captivity, as do self-mutilation and unusually high levels of aggression. Furthermore, individuals often cannot escape from the glaring eye of the public and cannot choose when and where to rest.

There also seems to be higher mortality (spontaneous abortions, still-births) in captive versus wild individuals (especially killer whales, *Orcinus orca*). There also is higher mortality for adult killer whales in captivity. Limited data on annual survival rates suggest that there is high mortality during acclimation to captivity and differences in annual survival rates among different species and age classes within species.

There is little evidence that people leave zoos or aquariums with any long-lasting sentiments or knowledge that benefit either the animals they have seen or their wild relatives. Furthermore, few zoos are engaged in conservation efforts for marine mammals.

Many ethical concerns are also raised because first and foremost, zoos are businesses and their bottom line centers on money. It costs an enormous amount of money to bring marine mammals into captivity and to keep them there. It has been suggested that the money used to capture, transport, and keep animals in captivity would be better used to do research in the wild. Also, much money is spent on public relations and not on the animals themselves. Some feel that the images of nature that are represented to the public are a manufactured corporate point of view that centers more on what the public wants than what is good for the animals. Witness the existence of numerous "Flippers" (the prototypical dolphin) and "Shamus" (the model killer whale), whose lives do not resemble even closely the lives of free-living conspecifics or relatives.

Similar questions are raised when considering research on captive animals. Certainly, information may be gathered about various aspects of their lives (e.g., maternal behavior, self-recognition, social behavior, communication, and cognitive capacities). However, research on captive animals is being increasingly carefully scrutinized by some researchers, philosophers, many universities, and various funding agencies. Some relevant questions include: is it ever permissible to keep individuals in captivity regardless of their utility, is the knowledge that is gained by studying captive individuals justified by keeping them in cages or tanks, and could more reliable data be collected under more natural conditions? Very little still is known about the life histories of most marine mammals. For many people it is the benefits that the captive individuals and other members of their (or other) species might accrue that is central, not the benefits that humans might gain. However, rarely are results used to benefit the animals other than in learning about medical treatments and husbandry to make their lives in captivity better. Rarely do wild individuals benefit from work done on captive relatives.

V. Research Ethics

In addition to questions concerned with how humans treat other animals, the study of ethics also considers questions dealing with such areas as (1) the context of research (where it is done, are local people involved when researchers "go into the field" in countries other than their own, are local customs and beliefs about native fauna respected); (2) scientific integrity (researchers' responsibility for integrity in data collection, analysis, and dissemination); (3) the ownership of data (do data "belong" to any single person or to the team that is involved in their collection, analysis, and dissemination); (4) authorship (whose names should appear on a publication and in what order); and (5) individual responsibility for the integrity of a project as a whole and for the integrity of the results. A good deal of trust is involved in all research, and questions that arise in these general areas require, and are receiving, much attention in the scientific community. Studies of marine mammals often require large teams of people, some of whom have never met, and it is important for all individuals to realize that each is responsible not only for his or her involvement, but for the composite product that is generated.

Another area of concern, some aspects of which are included in this volume, is research methodology (trapping, marking, tracking, and observing animals; experimentally manipulating social groups, food supply, and habitat). Often, human intrusions have major effects on animals' behavior even if they are unintentional. For example, the mere handling of individuals can influence their behavior and their acceptance back into a group, as can fitting individuals with various telemetric devices. Tracking or stalking animals can lead to changes in their activity patterns so that they spend more time avoiding humans than feeding or giving care. Most data come from animals other than marine mammals, and future studies of the effects of various methods are needed. Ethical considerations require that we learn about the effects of research methodologies and attempt to avoid them. In some cases the methods used may preclude collecting data relevant to the questions at hand.

VI. Ecotourism

Ecotourism (whale watching, swimming with wild dolphins, photographing animals, visiting pinniped rookeries), some aspects of which are discussed in this volume, also raises numerous ethical questions concerning human intrusions into the lives of other animals. When, if ever, this activity is justified requires serious debate. People often try to interact with wild marine mammals but do not attempt to pet wild zebras or lions. What principles underlie these differences in attitude?

Whether ecotourism is less intrusive on the lives of marine mammals than various research practices awaits further study. Indeed, there are observations of humans causing seal pups to stampede and being trampled and of humans striking and injuring animals with boats. It is essential to educate the public of possible negative effects of ecotourism.

VII. The Future: Being Proactive

Kellert's study of American perceptions of marine mammals and their management shows clearly that most people support the various goals of the U.S. Marine Mammal Protection Act. Most are willing to "render significant sacrifices to sustain and enhance marine mammal populations and species. . . . These findings clearly indicate that marine mammals possess considerable aesthetic, scientific, and moral support among the great majority of Americans today." (Kellert, 1999, pages iv–v)

It is in the best traditions of science to ask questions about ethics; it is not antiscience to question what we do when we interact with other animals. Ethics can enrich our views of other animals in their own worlds and in our different worlds and help us to see that their lives are worthy of respect, admiration, and appreciation. Indeed, it is out of respect, admiration, and appreciation that many humans seek out the company of whales, dolphins, polar bears, and other marine mammals.

Many ethical issues are extremely difficult to reconcile and generate highly charged and deep emotional and passionate responses. Achieving a win–win situation for animals and humans will be very difficult. However, it is clear that the increasingly detailed attention being given to various sorts of human–marine mammal interactions is showing that there are innumerable negative effects on the lives of the animals. While many negative influences have been anticipated or are not surprising, the severity of human influences has not been fully appreciated. We must be careful not to love these animals to their (or our) deaths. Humans are indeed dangerous to marine mammals and they are dangerous to us.

The study of ethics can also broaden the range of possible ways in which we interact with other animals without compromising their lives. Ethical discussions can help us see alternatives to past actions that have not served us or other animals well. Thus, the study of ethics can be enriching to other animals and to ourselves. If we believe that ethical considerations are stifling and create unnecessary hurdles over which we must jump in order to get done what we want to accomplish, then we will lose rich opportunities to learn more about other animals and also ourselves. Our greatest discoveries come when our ethical relationships with other animals are respectful and not exploitive.

Allowing human interests always to trump the interests of other animals is not the best strategy if we are to solve the numerous and complex problems at hand. We need to learn as much as we can about the lives of wild animals. Our ethical obligations also require us to learn about the innumerable ways in which we influence animals' lives when we study them in the wild and in captivity and what effects captivity has on them. As we learn more about how we influence other animals, we will be able to adopt proactive, rather than reactive, strategies.

The fragility of the natural order requires that people work harmoniously so as not to destroy nature's wholeness, goodness, and generosity. The separation of "us" (humans) from "them" (other animals) engenders a false dichotomy, the result of which is a distancing that erodes, rather than enriches, the possible numerous relationships that can develop among all animal life.

Public education is critical. To disseminate information about what is called the "human dimension," administrators of zoos, wildlife theme parks, aquariums, and areas where animals roam freely should inform visitors of how they may influence the behavior of animals they want to see. Tourism companies, nature clubs and societies, and schools can do the same. By treading lightly, humans can enjoy the company of other animals without making them pay for our interest in their fascinating lives. Our curiosity about other animals need not harm them.

Many marine mammals are closely linked to the wholeness of many ecosystems, and how they fare is tightly associated with how communities and ecosystems fare. By paying close attention to what we do to them, and why we do what we do where and when we do it, we can help maintain the health of individuals, species, populations, and ecosystems.

Acknowledgments

I thank Toni Frohoff, Robert Hofman, Dale Jamieson, Naomi Rose, and Trevor Spradlin for comments on an earlier draft of this paper. Trevor Spradlin kindly sent me voluminous material dealing with human–dolphin interactions, much of which I can only summarize here. Support for summary statements can be found, for the most part, in J. R. Twiss, Jr., and R. R. Reeves (eds.), "Conservation and Management of Marine Mammals," Smithsonian Institution Press, Washington, DC, and in other sections of this encyclopedia. Some sections of this essay are modified from my essay titled "Troubling Tursiops," in Toni Frohof and Brenda Peterson (eds.), "Dolphins: A Literary and Scientific Exploration."

See Also the Following Articles

Captivity ■ Hunting of Marine Mammals ■ Marine Parks and Zoos ■ Whale Watching

References

Beck, A., and Katcher, A. (1996). "Between Pets and People: The Importance of Animal Companionship" (revised edition). Purdue Univ. Press, Lafayette, IN.

Bekoff, M. (ed.) (1998). "Encyclopedia of Animal Rights and Animal Welfare." Greenwood, Westport, CT.

Bekoff, M. (2001). Human–carnivore interactions: Adopting proactive strategies for complex problems. *In* "Carnivore Conservation" (J. L. Gittleman, S. M. Funk, D. W. Macdonald, and R. K. Wayne, eds.). Cambridge Univ. Press, London.

Bekoff, M., and D. Jamieson. (1991). Reflective ethology, applied philosophy, and the moral status of animals. *Perspect. Ethol.* **9**, 1–47.

Bekoff, M., and Jamieson, D. (1996). Ethics and the study of carnivores. *In* "Carnivore Behavior, Ecology, and Evolution" (J. L. Gittleman, ed.), pp. 16–45. Cornell Univ. Press, Ithaca, NY.

Davis, S. G. (1997). "Spectacular Nature: Corporate Culture and the Sea World Experience." Univ. of California Press, Berkeley.

Francione, G. L. (1999). "Introduction to Animal Rights: Your Child or the Dog?" Temple Univ. Press, Philadelphia.

Frohoff, T. G. (1998). In the presence of dolphins. *In* "Intimate Nature: The Bond between Women and Animals" (L. Hogan, D. Metzger, and B. Peterson, eds.), pp. 78–84. Ballantine, New York.

Frohoff, T. G., and Packard, J. M. (1995). Human interactions with free-ranging and captive bottlenose dolphins. *Anthrozoös* **8**, 44–53.

Herzing, D. L., and White, T. I. (1998). Dolphins and the question of personhood. *Etica Anim.* **9/98**, 64–84.

Iannuzzi, D., and Rowan, A. N. (1991). Ethical issues in animal-assisted therapy programs. *Anthrozoös* **4**, 154–163.

Jamieson, D., and Regan, R. (1985). Whales are not cetacean resources. In "Advances in Animal Welfare Science, 1984" (M. W. Fox and L. Mackley, eds.), pp. 101–111. MartinusNijhoff, The Hague.

Kellert, S. R. (1996). "The Value of Life: Biological Diversity and Human Society." Island Press, Washington, DC.

Kellert, S. R. (1999). American perceptions of marine mammals and their management. Humane Society of the United States, Washington, DC (see also D. M. Lavigne, V. B. Scheffer, and S. R. Kellert (1999). The evolution of North American attitudes toward marine mammals. *In* "Conservation and Management of Marine Mammals" (J. R. Twiss, Jr., and R. R. Reeves, eds.), pp. 10–47. Smithsonian Institution Press, Washington, DC.

Kirkwood, J. K., Bennett, P. M., Jepson, P. D., Kuiken, T., Simpson, V. R., and Baker, J. R. (1997). Entanglement in fishing gear and other causes of death in cetaceans stranded on the coasts of England and Wales. *Vet. Rec.* **141**, 94–98.

Marino, L., and Lilienfeld, S. O. (1998). Dolphin-assisted therapy: Flawed data, flawed conclusions. *Anthrozoös* **11**, 194–200.

Nathanson, D. W. (1998). Reply to Marino and Lilienfeld. *Anthrozoös* 11, 201–202.

Nollman, J. (1999). "The Charged Border: Where Whales and Humans Meet." Holt, New York.

Rose, N., and Farinato, R. (1995/1999). "The Case against Marine Mammals in Captivity." Humane Society of the United States, Washington, DC.

Samuels, A., and Spradlin, T. R. (1995). Quantitative behavioral study of bottlenose dolphins in swim-with-dolphin programs in the United States. *Mar. Mamm. Sci.* **11**, 520–544.

Evolutionary Biology, Overview

FRANK CIPRIANO
San Francisco State University, California

Evolutionary biology is a multidisciplinary field devoted to the study of the biological significance of evolution, as well as documentation of its effects. This branch of biology is not *just* the study of evolution, although that is one of its integral components, and may perhaps be better defined as "the work that evolutionary biologists do." This includes a focus on many different aspects of biological processes, informed by an evolutionary perspective, and addressing questions from the microscale of point mutations in DNA to the genealogy relating all living organisms on planet Earth.

I. Components of Evolutionary Biology

From its very beginning, evolutionary biology has been, by necessity, an interdisciplinary field. Four areas of biology were most important in demonstrating the existence and operation of organic evolution; studies of morphology, paleontology, genetics, and ecology have been termed the "four pillars" of evolutionary biology. However, evolution encompasses all aspects of biology, as all living things do and have evolved, and evolutionary biology thus contributes to, and is influenced by, all biological disciplines. In current practice, evolutionary biology depends on an understanding of zoology, botany, geology, physics, analytical chemistry, and biochemistry, as well as some familiarity with aspects of microbiology, physiology, embryology, cell biology, mathematics, statistics, anthropology, archaeology, SOCIOBIOLOGY, psychology, history, and philosophy. Genetics has been the most significant contributor to the development of evolutionary theory after Darwin, and its offspring, molecular biology (co-parented by cell biology), is fundamentally grounded on evolutionary mechanisms. "Biotech," a derivative of molecular biology, genetics, and the pharmaceutical industry, is clearly the most active area of biological investigation today and for the foreseeable future. The synthesis of population genetics and ecology, each also essentially rooted in evolutionary principles, led to the development of the field of population biology, which mainly deals with the interplay among gene frequencies, population demographics, and selection coefficients in a purely mathematical framework. Biological systematics (the study of the relationships among groups of living organisms) is necessarily practiced in an evolutionary framework. A new but very rapidly developing field is nicknamed "evo/devo"—essentially the study of development in an evolutionary framework, which depends on aspects of embryology, biochemistry, molecular biology, and evolution—and which will probably be the next important focus of the "biotech" industry.

Evolution itself can be considered the process (and also the result) of change in heritable aspects of biological populations over time, as observed routinely in the fossil record and in living populations alike. There is incontrovertible evidence for such changes both in the past and currently; that biological evolution happens is a well-documented fact. *Evolutionary theory* is concerned with the details: elucidation of mechanisms that cause and influence evolutionary changes, and evaluation of the relative importance of particular factors. However, what is generally meant by "the" theory of evolution (as developed by Charles Darwin and Alfred Russel Wallace) is a simple, three-part observation that describes evolution resulting from the operation of *natural selection:* (1) members of a species usually have a high reproductive capacity and thus the population will tend to increase in size and its members to compete with each other, (2) individuals of a species differ in a variety of ways, some of which are heritable, and (3) particular individuals possessing some advantageous heritable quality, in competition with other members of their species, will survive and reproduce more often so that their offspring and the advantageous qualities they possess will be more numerous in succeeding generations. This is "a statement of what are held to be the general laws, principles, or causes of something known or observed," and thus a theory in the scientific sense. Criticisms that evolution is "only a theory" are ignoring both the proven fact of evolutionary change and the scientific meaning of *theory*.

Evolutionary biology is currently a significant discipline within the biological sciences because an understanding of evolution is important in understanding the functioning, as well as the history, of morphological structures, physiological mechanisms, cellular processes, and so on (consider, for example, the giraffe's neck or the development of antibiotic-resistant disease bacteria). A search of recent science literature shows that article titles and abstracts containing the word "evolution" account for 1% of the 10 million scientific articles in the Science Citation Index (Institute for Scientific Investigation), about the same number of references as for articles containing the words "chemistry," "physics," or "computer" (Fig. 1). This key word search probably underestimates the influence of evolutionary thought, as an understanding of the evolution of enzymes, proteins, cell-recognition mechanisms, and gene regulation is fundamental to the booming fields of molecular and cell biology and "biotech," although such a perspective may not be reflected in the titles and abstracts of articles (Fig. 1, see entries for "cell," "gene," and "molecular").

Given the fundamental influence of evolution on all living things, one might reasonably claim that *all* biology is evolutionary biology. After all, an interest in the purely mechanistic functioning of the most minor part of a single cell carries with it the implicit question of why and how that part came to be. Without going that far, it seems fair to suggest that it is the *ex-*

plicit asking of such questions that defines the evolutionary biologist: how they go about finding an answer is a matter of individual taste, possibly even of style (and hence the claim in the second sentence of this article)! An evolutionary perspective, applied to a wide variety of biological questions, defines the field (accurately if not precisely) and predicts the breadth of subject matter that is the grist to our mill.

II. Model Systems and the Subject Matter of Evolutionary Biology

Although it was claimed earlier that the study of evolution itself is not the only focus of evolutionary biology, it is certainly one of the most important. A glance through the table of contents of books on evolution and evolutionary biology will reveal that the main areas of interest in evolutionary studies are limited to just a few issues. These include studies of adaptation, tracing evolutionary descent, the history of biological diversity and geographic patterns thereof, debate over the source material of evolution and its causes (selection and/or chance events?), estimates of the rates and directions of evolution, and assessment of the mechanism(s) of SPECIATION. Studies in these areas have produced the body of knowledge of mechanisms that cause, influence, and constrain evolutionary changes referred to earlier as *evolutionary theory*. Such theory is developed in a back-and-forth between observation of living systems, interpretation of patterns thought to indicate the operation of evolutionary laws or principles, erection of formal theories encapsulating those laws or principles, and tests of the theories with further observation and/or experiment.

Crucial to development of our understanding in each of these subject areas is the establishment of appropriate model systems—organisms in an environmental and evolutionary context—some that provide initial data that (according to particular interpreters) illustrate the patterns and others that can be used to test the generality of the theory, for specific issues at appropriate time, distance, and evolutionary-relatedness scales. Biological model systems are analogous to what engineers call a "test bed"—strap a newly designed engine onto a 747, then take her up and see how she flies. . . . The model systems that are the test beds of evolution are often organisms that only a devoted biologist could have much fondness for: fruit flies, yeast, and roundworms. These valiant conscripts to the cause of science serve well because their size, care requirements, and rapid generation time make them ideal experimental subjects. None of these criteria characterize any marine mammal species—even the smallest dolphin is a relatively large mammal, slow-growing, reproducing late in life, and difficult, if not impossible, to breed in captivity. Marine mammals have certainly undergone some interesting evolution—invaders of an alien and difficult habitat that require extensive adaptations of mammalian morphology, physiology, sensory systems, and behavior—but have studies of marine mammals contributed much to the development of evolutionary theory?

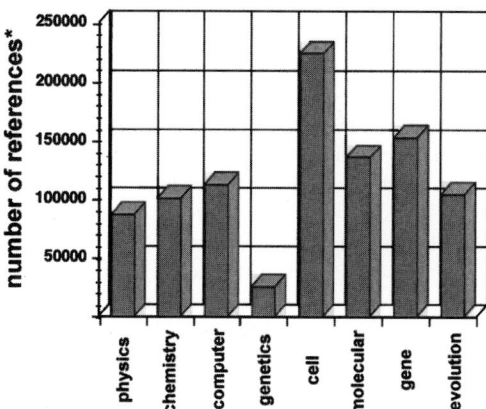

Figure 1 *Results of a literature search for certain keywords (most denoting scientific disciplines) using the online Science Citation Index (© Institute for Scientific Investigation). Bars show the number of articles found containing the key words (in the title, abstract of the article, or listed keywords) indicated along the bottom axis.* °*At the time of this search, the Science Citation Index (expanded) 1988–present contained 10,163,098 articles.*

III. Marine Mammals and Evolutionary Biology

Out of the 10 million entries in the current database of the Science Citation Index, a search for articles containing the term "evolutionary biology" along with "cetacean," "pinniped," "marine mammal," "whale," "dolphin," "seal," or "sea lion" turns up exactly two articles. Not necessarily indicative, given the assertion that evolutionary biology is more a matter of evolutionary perspective applied to biological questions than an explicit focus on evolution alone. This is perhaps supported by searches for "evolutionary biology" along with "Drosophila," "human," "bacteria," or "HIV" (all of which are of major interest to evolutionary biologists), which produced just 59, 53, 8, and 4 articles containing those combinations of terms, respectively. A search for just "evolution" combined with the names of various groups of marine mammals was somewhat more successful, producing a list of several hundred articles.

The literature on marine mammals *and* evolution (not just marine mammal evolution) includes a fairly broad cross section of all the evolutionary topics outlined previously. This literature does include "purely evolutionary" discussion of the time of origin of particular marine mammal groups, from the divergence of the cetaceans from terrestrial mammals about 40 million years ago to the in-progress split between polar bears (*Ursus maritimus*) and brown bears (*U. arctos*). In addition, the divergence time of whales from land mammals is an often-used molecular landmark used to "calibrate" divergence times of other groups.

Although marine mammals are usually unsuitable for direct experimental manipulation (difficult husbandry, political issues, and so on), the analysis of "natural" experiments involving marine mammals has identified particular factors generally important in evolution. Natural experiments are often initiated by the dispersal of species or populations across a wide variety of habitats, or disruption of widely distributed species into separate regions by major environmental changes. Documentation that divergent species or populations use contrasting methods to find prey or avoid predators, find mates, or use resources differently in their particular environments is useful in identifying patterns that can be analyzed for correlation with evolutionary principles. Much attention has been spent on comparative studies made possible by natural experiments of gene evolution, social and breeding systems, morphological and developmental differences, and physiology, both within and between groups of marine mammals. For example, many comparisons have been made within and between seals and sea lions, and very many others between particular marine mammal examples and other terrestrial mammals, or vertebrates in general. Cetaceans, especially the large whales, are often used in such comparisons because both their large size and ancient divergence make them a useful yardstick for comparative studies of gene evolution, physiological adaptations, sensory system functioning, and even disease resistance. Physiologists, life history modelers, and sociobiologists alike have recognized that if they can demonstrate a consistent relationship between factors along a continuum from mice to whales, that says a lot; if particular species are above or below the mouse-to-whale line (for breeding system, brain:body size, and so on), that says a lot too.

In addition, it appears that although marine mammals may not have contributed directly to development of evolutionary theory as model systems, some evolutionary principles are exemplified beautifully by marine mammals. Just about any general biology textbook will illustrate two general principles of evolution with drawings like those shown in Fig. 2. Homology is the similarity between organisms or their parts due to derivation from a common ancestor; Fig. 2a shows that forelimbs from a variety of mammals, including whales, have the same underlying bone structure related to their common ancestry. CONVERGENT EVOLUTION is almost the opposite: similar features found in distantly related organisms; Fig. 2b shows how a common physical regime has resulted in very similar body shapes for three very distant groups of marine vertebrates.

The evolutionary origins of whales, and of pinnipeds, have been subjects of much controversy. Some datasets (and their interpreters) suggested separate origins for toothed vs baleen whales, whereas other interpretations indicated a common origin for all whales. A similar disagreement raged (some will say still rages) over the independent vs common origin of pinnipeds. Although not all proponents are yet satisfied with a resolution of the pinniped origin question, baleen and toothed whales are now generally recognized as resulting from a single origin. These controversies provide clear examples of the difficulty of disentangling homology (identity by descent) from analogy (similarities related to functional convergence), a distinction of broad evolutionary interest. The current debate over the placement of sperm whales (are they more closely related to baleen whales or to other toothed whales?) continues to underscore such difficulties. Detailed investigations of the molecular genetic methodology used to address this question have indicated the importance of certain methodological choices used in the analysis and also the effect of including data from particular, distantly related groups ("outgroups") to the results of such comparisons. Although many different types of genetic comparisons consistently conclude that whales and hippos (*Hippopotamus* spp.) are nearest relatives, morphological comparisons of fossils suggest a different (and now extinct) group of early ungulates ("hooved" mammals and their progenitors) not related to hippos were the progenitors of whales. These difficulties underscore another point important for the consideration of evolutionary biologists—morphology and molecules often do not agree—and how to handle such difficulties is a topic taking up many pages in the current issues of the scientific journals of the field.

Another prime example provided by marine mammals is for the consequences of extreme "population bottlenecks" associated with near extinctions. Northern elephant seals (*Mirounga angustirostris*) were almost completely extirpated by unregulated human hunts and their genetic diversity was greatly reduced. A closely related Southern Hemisphere "sister species" (*M. leonina*) was not depleted so severely, and a higher level of genetic diversity compared to the northern species has been maintained. Loss of genetic diversity has been predicted to result in an increased susceptibility to disease as well as problems associated with inbreeding, but so far northern elephants seals appear to be thriving. Scrutiny of their reproductive status, study of their genetic makeup, and comparisons to their southern relatives and other mammals continue. Similarly, North Atlantic right whales (*Eubalaena glacialis*) were badly depleted by unregulated whaling, and only a small population of about 300 individuals remains off the east coast of North America. Unfortunately, this population continues to decline, and there is some evidence that the low level of genetic diversity remaining is associated with reproductive problems stemming from inbreed-

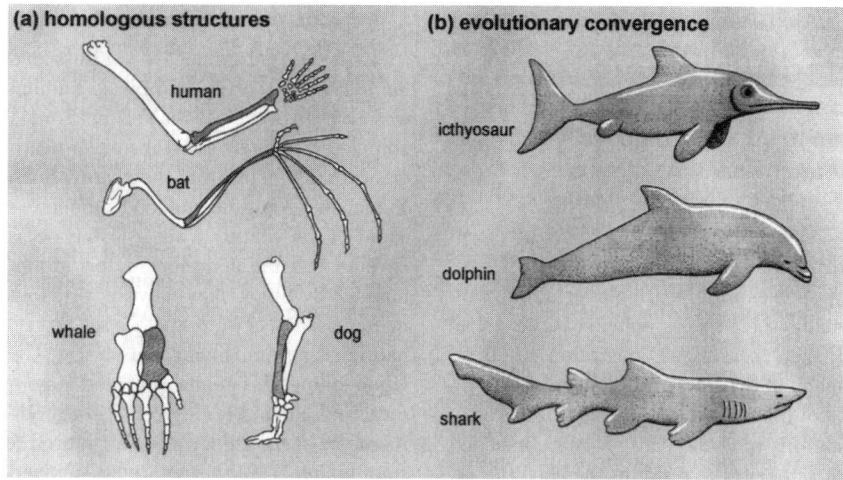

Figure 2 *(a) Mammalian forelimb bone structure showing derivation of differently shaped limbs based on the same underlying (homologous) bone structure. The dark-shaded bone in each illustration is the radius. (b) Examples of evolutionary convergence, where distantly related marine vertebrates have developed similarly streamlined body shapes with biolobed tail, dorsal fins, and pectoral flippers. The icthyosaur is an ancient, now extinct, marine reptile. Both figures redrawn from Minkoff (1983), used by permission.*

ing. Because of the many threats faced by marine mammals and their history of overexploitation by humans, the development of genetic diversity, identification of distinct population units, and conservation status of many species are under intense examination. Thus, details of speciation and molecular divergence mechanisms are being developed with studies of these groups.

Finally, the new insights brought about by evolutionary thinking are currently spurring interesting and integrative work on marine mammals. The details of these studies are starting to feed back into the evolutionary biology literature so that the contributions of marine mammals to the discipline are likely to increase as these areas are explored further.

See Also the Following Articles

Cetacean Evolution ▪ Hippopotamus ▪ Molecular Ecology ▪ Paleontology ▪ Pinniped Evolution ▪ Sociobiology ▪ Sperm Whales, Evolution ▪ Systematics, Overview

References

Árnason, Ú., Gullberg, A., Janke, A., and Xu, X. F. (1996). Pattern and timing of evolutionary divergences among hominoids based on analyses of complete mtDNAs. *J. Mol. Evol.* **43**, 650–661.

Boyd, I. L. (1998). Time and energy constraints in pinniped lactation. *Am. Nat.* **152**, 717–728.

Brandon, R. N. (1996). "Concepts and Methods in Evolutionary Biology." Cambridge Univ. Press, Cambridge.

Caswell, H., Fujiwara, M., and Brault, S. (1999). Declining survival probability threatens the North Atlantic right whale. *Proc. Natl. Acad. Sci. USA* **96**, 3308–3313.

Connor, R. C., Mann, J., Tyack, P., and Whitehead, H. (1998). Social evolution in toothed whales. *Trends Ecol. Evol.* **13**, 228–232.

Fordyce, R. E. (1980). Whale evolution and oligocene southern ocean environments. *Palaeogeogr. Palaeoclimatol. Palaeoecol.* **31**, 319–336.

Futuyma, D. J. (1979). "Evolutionary Biology." Sinauer, Sunderland, MA.

Harvey, P. H., and Page, M. D. (1991). "The Comparative Method in Evolutionary Biology." Oxford Univ. Press, Oxford.

Hasegawa, M., and Adachi, J. (1996). Phylogenetic position of cetaceans relative to artiodactyls: Reanalysis of mitochondrial and nuclear sequences. *Mol. Biol. Evol.* **13**, 710–717.

Hoelzel, A. R. (1994). Genetics and ecology of whales and dolphins. *Annu. Rev. Ecol. Syst.* **25**, 377–399.

Lento, G. M., Hickson, R. E., Chambers, G. K., and Penny, D. (1995). Use of spectral analysis to test hypotheses on the origin of pinnipeds. *Mol. Biol. Evol.* **12**, 28–52.

Lipps, J. H., and Mitchell, E. (1976). Trophic model for the adaptive radiations and extinctions of pelagic marine mammals. *Paleobiology* **2**, 147–155.

Messenger, S. L., and McGuire, J. A. (1998). Morphology, molecules, and the phylogenetics of cetaceans. *Syst. Biol.* **47**, 90–124.

Milinkovitch, M. C., LeDuc, R. G., Adachi, J., Farnir, F., Georges, M., and Hasegawa, M. (1996). Effects of character weighting and species sampling on phylogeny reconstruction. *Genetics* **144**, 1817–1833.

Minkoff, E. C. (1983). "Evolutionary Biology." Addison-Wesley, Reading, MA.

Ridley, M. (1985). "The Problems of Evolution." Oxford Univ. Press, Oxford.

Slade, R. W. (1992). Limited MHC polymorphism in the southern elephant seal: Implications for MHC evolution and marine mammal population biology. *Proc. Royal Soc. Lond. B Biol. Sci.* **249**, 163–171.

Springer, M. S., Amrine, H. M., Burk, A., and Stanhope, M. J. (1999). Additional support for Afrotheria and Paenungulata, the performance of mitochondrial versus nuclear genes, and the impact of data partitions with heterogenous base composition. *Syst. Biol.* **48**, 65–75.

Thewissen, J. G. M., and Fish, F. E. (1997). Locomotor evolution in the earliest cetaceans: Functional model, modern analogues, and paleontological evidence. *Paleobiology* **23**, 482–490.

Van Valen, L. (1968). Monophyly or diphyly in the origin of whales. *Evolution* **22**, 37–41.

exterminated by whalers, this population has rebounded as a result of being completely protected, except for aboriginal subsistence hunting (Reilly, 1984). Another population is found in the western North Pacific, where it migrates between the Okhotsk Sea and southern South Korea. This population has been reduced to very low numbers due to overexploitation and is currently highly endangered. A third population in the western North Atlantic appears to have existed until the 17th century and was probably rendered extinct by early whaling activity (Mead and Mitchell, 1984). Accounts from this period, although somewhat confusing, describe a whale known as the "scrag" whale, which bears a strong resemblance to the gray whale. Earlier Icelandic accounts describe a "sandloegja," which is also felt to have been a gray whale (Fraser, 1970). Subfossil remains of gray whales have been found in eastern North America, and radioactive dating has shown the latest to have been from approximately 1675 A.D. (Mead and Mitchell, 1984). The population went extinct shortly thereafter. Subfossil remains of gray whales have also been found in Europe, but are apparently much older. Therefore, an eastern Atlantic population of gray whales probably also occurred in historical times, apparently going extinct sometime before 500 A.D., quite possibly at the hands of early European whalers (van Deinse and Junge, 1937).

V. Prospects for the Future

Even though few species of marine mammals have gone extinct at the hands of humans, many have come very close. Elephant seals, fur seals, monk seals, walruses, and sea otters all have narrowly escaped extinction. Among cetaceans, the great rorquals, the gray whale, and the right whales were all nearly exterminated. Some have rebounded, some appear to be slowly increasing, others apparently are not increasing and may never recover. The history of sealing and whaling makes for depressing reading. Although these industries have mostly disappeared in today's world (with some notable exceptions), some species of marine mammals are still highly endangered. The Mediterranean monk seal is down to less than 1000 individuals and is being forced into tiny pockets of habitat by an explosion of tourism. The vaquita, a small porpoise from the northern Gulf of California, exists as a small population under pressure from unintentional destruction from fisheries (Rojas-Bracho and Taylor, 1999). The baiji, the river dolphin of the Yangtze River, is the most endangered marine mammal in the world. With probably less than 100 now remaining and its river habitat being massively altered by humans, the future of this species is bleak. Many other species are also endangered, although they are not in as precarious a state as these. As much as may be desired, it will never again be possible to observe the "sea wolves" of Columbus lounging on the tropical beaches of the Caribbean or Steller's sea cow rising out of the northern mists.

See Also the Following Articles

California, Galapagos, and Japanese Sea Lions ▪ Endangered Species and Populations ▪ Monk Seals ▪ Steller's Sea Cow

References

Allen, J. A. (1887). The West Indian seal (*Monachus tropicalis*, Gray). *Bull. Amer. Mus. Nat. Hist.* **2**, 1–34.

Bonner, W. N. (1990). "The Natural History of Seals," p. 197. Facts on File, New York.

Domning, D. P. (1978). Sirenian evolution in the North Pacific Ocean. *In* "University of California Publications in Geological Sciences," Vol. 118, p. 176. University of California Press, Berkeley, CA.

Domning, D. P. (1999). Endangered species: The common denominator. *In* "Conservation and Management of Marine Mammals" (J. R. Twiss, Jr., and R. R. Reeves, eds.), pp. 332–341. Smithsonian Institution Press, Washington.

Ford, C. (1966). "Where the Sea Breaks Its Back," p. 206. Little, Brown, Boston.

Fraser, F. C. (1970). An early 17th century record of the California gray whale in Icelandic waters. *Invest. Cetacea* **2**, 13–20.

Haley, D. (1978). Steller sea cow. *In* "Marine Mammals" (D. Haley, ed.), pp. 236–241. Pacific Search Press, Seattle.

Kenyon, K. W. (1977). Caribbean monk seal extinct. *J. Mammal.* **58**, 97–98.

Kenyon, K. W. (1981). Monk seals. *In* "Handbook of Marine Mammals" (S. H. Ridgway and R. J. Harrison, eds.), Vol. 2, pp. 195–220. Academic Press, London.

King, J. E. (1983). "Seals of the World," p. 240. Cornell Univ. Press, Ithaca, N.Y.

LeBoeuf, B. J., Kenyon, K. W., and Villa-Ramirez, B. (1986). The Caribbean monk seal is extinct. *Mar. Mamm. Sci.* **2**, 70–72.

Maxwell, G. (1967). "Seals of the World," p. 153. Constable, London.

Mead, J. G., and Mitchell, E. D. (1984). Atlantic gray whales, *In* "The Gray Whale" (M. L. Jones, S. L. Swartz, and S. Leatherwood, eds.), pp. 33–53. Academic Press, Orlando.

Nishiwaki, M. (1973). Status of the Japanese sea lion. *In* "Seals," pp. 80–81. New Series Supplementary Paper No. 39. IUCN, Morges, Switzerland.

Reeves, R. R., Stewart, B. S., and Leatherwood, S. (1992). "The Sierra Club Handbook of Seals and Sirenians," p. 359. Sierra Club Books, San Francisco.

Reilly, S. B. (1984). Assessing gray whale abundance: A review. *In* "The Gray Whale" (M. L. Jones, S. L. Swartz, and S. Leatherwood, eds.), p. 203–223. Academic Press, Orlando.

Rice, D. W. (1973). Caribbean monk seal (*Monachus tropicalis*). *In* "Seals," pp. 98–112. New Series Supplementary Paper No. 39. IUCN, Morges, Switzerland.

Rice, D. W. (1998). "Marine Mammals of the World: Systematics and Distribution," p. 231. Special Publication 4. Society for Marine Mammals, Lawrence, KS.

Rojas-Bracho, L., and Taylor, B. L. (1999). Risk factors affecting the vaquita (*Phocoena sinus*). *Mar. Mamm. Sci.* **15**, 974–989.

Smith, F. D. M., May, R. M., Pellew, T. H., Johnson, T., and Walter, K. (1993). How much do we know about the current extinction rate? *Trends Ecol. Evol.* **8**, 375–378.

Stejneger, L. (1887). How the great northern sea cow (*Rytina*) became exterminated. *Am. Nat.* **21**, 1047–1054.

Stejneger, L. (1936). "Georg Wilhelm Steller, The Pioneer of Alaskan Natural History," p. 623. Harvard Univ. Press, Cambridge, MA.

van Deinse, A. B., and Junge, G. C. A. (1937). Recent and older finds of the California gray whale in the Atlantic. *Temminckia* **2**, 161–188.

Ward, H. L. (1887). Notes on the life history of *Monachus tropicalis*, the West Indian seal. *Amer. Nat.* **21**, 257–264.

Whitmore, Jr., F. C., and Gard, Jr., L. M. (1977). "Steller's Sea Cow (*Hydrodamalis gigas*) of the Late Pleistocene Age from Amchitka, Aleutian Islands, Alaska," p. 20. U.S. Government Printing Office, Washington, DC.

False Killer Whale

Pseudorca crassidens

ROBIN W. BAIRD
Dalhousie University,
Halifax, Nova Scotia, Canada

The false killer whale (*Pseudorca crassidens*) is one of the larger members of the family Delphinidae, with adult males reaching lengths of almost 6 m and females reaching up to 5 m. The common name comes not from a similarity in external appearance to the killer whale (*Orcinus orca*), but rather similarity in skull morphology of these two species (Fig. 1). In fact, the two species do not appear to be closely related; based on genetic similarity, false killer whales appear to be most closely related to the Risso's dolphin (*Grampus griseus*), melon-headed whale (*Peponocephala electra*), pygmy killer whale (*Feresa attenuata*), and pilot whales (*Globicephala* spp.). There is some evidence of geographic variation in skull morphology, although no subspecies are currently recognized.

Largely black or dark gray in color (usually with a lighter blaze on the ventral surface between the flippers), they are easily recognizable with their rounded head, gracile shape, small falcate dorsal fin located at the midpoint of the back, and distinctive flippers (with a bulge on the leading edge). False killer whales are slightly SEXUALLY DIMORPHIC, with the melon of males protruding farther forward than in females. Their teeth are large and conical, with 7–11 in each of the upper jaws and 8–12 in each lower jaw.

False killer whales are found in all tropical and warm temperate oceans of the world, and occasional records of their presence in cold temperate waters have also been documented. Although they are typically characterized as pelagic in habits, they do approach close to shore at oceanic islands and are one of the handful of species that regularly mass strands. These oceanic habits have hindered study of this species in the wild, and most of what is known comes from stranded individuals, captive animals, and a few observations of wild groups (usually around oceanic islands). No long-term behavioral or population studies of wild individuals have been undertaken. No estimates of worldwide population size are available, and a few regional estimates based on surveys are imprecise.

False killer whales are considered to be extremely social, usually traveling in groups of 20 to 100 individuals. Details of social organizations are lacking; however, strong bonds between individuals are evident from their propensity to strand en masse (the largest number of individuals mass STRANDING was over 800) and by the affiliative behavior of stranded animals. False killer whales are active during the day, and food sharing in the wild has been recorded on several occasions. The diet of false killer whales appears to be diverse, both in terms of species and size of prey. In general they feed on a variety of oceanic squid and fish, but have also been documented FEEDING on smaller delphinids being released from tuna purse-seines in the eastern

Figure 1 *The quite social false killer whale* (Pseudorca crassidens) *does not resemble the killer whale* (Orcinus orca) *in external appearance. The skulls of the two species are, nevertheless, quite similar. Pieter A. Folkens/Higher Porpoise DG.*

tropical Pacific. One case of predation on a humpback whale (*Megaptera novaeangliae*) calf has also been recorded, and they have been documented attacking sperm whales (*Physeter macrocephalus*). Nonaggressive interspecific associations with bottlenose dolphins (*Tursiops* spp.) have also been reported. Nothing is known about the DIVING BEHAVIOR of this species, and no predators of false killer whales have been reported, although large sharks likely take some individuals.

Life history information comes entirely from stranded individuals. Because the deposition rates of growth layer groups in the teeth have never been calibrated, there is some uncertainty in life history parameters. Both sexes are thought to mature between about 8 and 14 years of age, although there is some suggestion that males may mature later. Maximum longevity has been estimated at 57 years for males and 62 years for females. Calving interval for one population has been reported as almost 7 years, and calving may occur year-round, with a peak in late winter.

A number of types of interactions between humans and false killer whales have been documented. They are the only "blackfish" to regularly bowride on vessels, and in Hawaii they are regularly encountered by commercial whale- or dolphin-watching vessels. They have been maintained in captivity in a number of aquaria around the world, including Japan, the United States, the Netherlands, Hong Kong, and Australia. They have been successfully bred in captivity in several locations, and there they have produced viable interspecies hybrids with bottlenose dolphins. False killer whales are one of several species of odontocetes that occasionally steal fish from both commercial and recreational fishermen, with these types of interactions noted in Japan, Hawaii, the Indian Ocean, and the Gulf of Mexico. Such conflicts have resulted in direct killing in Japan, and small numbers have been occasionally taken in fisheries, both directly and incidentally as bycatch. They are one of a growing list of species that has been recorded ingesting discarded plastic, and high levels of toxins have been documented in tissues collected from stranded animals. It is unknown, however, whether such toxins contribute to immunosuppression in this species.

See Also the Following Articles

Melon-Headed Whale ▪ Pilot Whales ▪ Pygmy Killer Whale ▪ Risso's Dolphin

References

Acevedo-Gutierrez, A., Brennan, B., Rodriguez, P., and Thomas, M. (1997). Resightings and behavior of false killer whales (*Pseudorca crassidens*). *Mar. Mamm. Sci.* **13**, 307–314.

Au, W. W. L., Pawloski, J. L., Nachtigall, P. E., Blonz, M., and Gisner, R. C. (1995). Echolocation signals and transmission beam pattern of a false killer whale (*Pseudorca crassidens*). *J. Acoust. Soc. Am.* **98**, 51–59.

Brown, D. H., Caldwell, D. K., and Caldwell, M. C. (1966). Observations on the behavior of wild and captive false killer whales, with notes on associated behavior of other genera of captive delphinids. *Los Angeles County Mus. Contrib. Sci.* **95**, 1–32.

Koen Alonso, M., Pedraza, S. N., Schiavini, A. C. M., Goodall, R. N. P., and Crespo, E. A. (1999). Stomach contents of false killer whales

(*Pseudorca crassidens*) stranded on the coasts of the Strait of Magellan, Tierra del Fuego. *Mar. Mamm. Sci.* **15**, 712–724.

Notarbartolo-Di-Sciara, G., Barbaccia, G., and Azzellino, A. (1997). Birth at sea of a false killer whale, *Pseudorca crassidens*. *Mar. Mamm. Sci.* **13**, 508–511.

Odell, D. K., and McClune, K. M. (1999). False killer whale *Pseudorca crassidens* (Owen, 1846). *In* "Handbook of Marine Mammals" (S. Ridgway, ed.). Vol 6, pp. 213–243. Academic Press, New York.

Palacios, D. M., and Mate, B. R. (1996). Attack by false killer whales (*Pseudorca crassidens*) on sperm whales (*Physeter macrocephalus*) in the Galapagos Islands. *Mar. Mamm. Sci.* **12**, 582–587.

Purves, P. E., and Pilleri, G. (1978). The functional anatomy and general biology of *Pseudorca crassidens* (Owen) with a review of hydrodynamics and acoustics in Cetacea. *Invest. Cetacea* **9**, 67–227.

Stacey, P. J., and Baird, R. W. (1991). Status of the false killer whale, *Pseudorca crassidens*, in Canada. *Can. Field-Nat.* **105**, 189–197.

Stacey, P. J., Leatherwood, S., and Baird, R. W. (1994). Pseudorca crassidens. *Mammal. Species* **456**, 1–6.

Feeding Strategies and Tactics

MICHAEL R. HEITHAUS AND LAWRENCE M. DILL
Simon Fraser University,
Burnaby, British Columbia, Canada

Marine mammals are found in a wide range of habitats, including the open ocean, coastal waters, rivers, lakes, and even on ice floes and land. They feed on a variety of prey species from aquatic plants to microscopic zooplankton to the largest marine mammals, and a diverse array of strategies and tactics are used to locate and capture these prey. Some marine mammals consume huge numbers of prey items at a time (batch feeding), whereas other attack and consume prey items singly (raptorial feeding). Many marine mammals forage in large groups, whereas others feed alone. This article considers the wide range of marine mammal foraging behaviors and the circumstances and habitats that led to the adoption of particular feeding strategies and tactics.

Before embarking upon a review of marine mammal foraging, it is important to make a distinction between a strategy and a tactic, terms which have specific meanings in the field of behavioral ecology. Put simply, a strategy is a genetically based decision rule (or set of rules) that results in the use of particular tactics. Tactics are used to pursue a strategy and include behaviors (Gross, 1996). Tactics may be fixed or flexible; in the latter case they depend on the condition of the individual or characteristics of the prey or environment. For example, a humpback whale's (*Megaptera novaeangliae*) strategy may be to use a tactic that will maximize energy intake at any particular time. The whale may pursue this strategy by switching between the tactics used to capture fish and those used to catch krill, depending on the relative ABUNDANCE of these two prey types.

Our understanding of marine mammal foraging is hampered by the difficulty of studying these animals. They live in an environment where observations are difficult (often beneath the surface), our presence can disturb their foraging behavior, and feeding events often occur quickly and are easy to miss. Despite this, much is known. This article begins with a review of marine mammal foraging by considering ways that marine mammals find and capture their prey, continues with a discussion of group foraging, and then concludes with a discussion of the causes of variation in feeding strategies and tactics.

I. Finding Prey

The first step in foraging is locating prey. This may be done over many temporal and spatial scales and can involve migrations of thousands of kilometers or switching between habitats separated by only a few meters to forage in prey-rich locations. Then, once a marine mammal is in a prey-rich area, it still must locate prey.

A. Habitat Use

One way that marine mammals can increase their chances of encountering prey is to spend time foraging in those habitats with high prey abundance. Evidence shows that a variety of marine mammals tend to aggregate in areas with high food concentrations. For example, the highest densities of polar bears (*Ursus maritimus*) are found along floe edges and on moving ice, habitats that contain the highest density of seals. Resident killer whales (*Orcinus orca*) are most abundant in Johnston Strait, British Columbia, when salmon migrate through the strait. Also, the distribution of humpback whales in the Gulf of Maine appears to reflect the availability of fish prey, and humpback whale distribution in southeast Alaska may partially be determined by KRILL abundance. When there are a variety of habitats available to marine mammals in a restricted area (such as nearshore environments), a theoretical model predicts that if the main concern of the animals is to maximize energy intake, they should be distributed proportional to the amount of food available in each habitat (e.g., Fretwell and Lucas, 1970). Testing this hypothesis is difficult because marine mammal prey availability is often hard to quantify. However, the distribution of bottlenose dolphins (*Tursiops aduncus*) in Shark Bay, Western Australia, conforms to this hypothesis and matches that of their fish prey in winter months. Humpback and common minke whales (*Balaenoptera acutorostrata*) do not appear to conform to this hypothesis. Instead, they appear to show a threshold response to prey availability, only using a habitat once prey density has reached a particular level. Of course, prey availability is not the only factor that might influence the habitat use of marine mammals, which will be considered in detail later.

B. Migration

When suitable habitats for a marine mammal are widely spread, movements between them are considered migrations. There is thus a continuum between habitat use decisions and migrations. Some migrations appear to be driven primarily by the variation in food availability. Unlike baleen whales, sperm whales (*Physeter macrocephalus*) cannot fast for long periods of time, and female groups use migrations up to 1100 km as part of a strategy for surviving in a variable habitat with low local food abundance and poor foraging success. In fact, this strategy may be the reason that female sperm whales are found in permanent social groups as they may benefit from the experience of old females during migrations.

Migration frequently involves trade-offs between feeding and another factor, such as reproduction. Baleen whales and some pinnipeds feed only for a relatively short period of time in high-productivity, high-latitude waters and then fast during the rest of the year while moving to, and spending time at, low-latitude breeding grounds. For example, northern elephant seals (*Mirounga angustirostris*) forage along the entire North Pacific and then migrate to a few beaches on the California coast to breed. Also, humpback whales reproduce in warm, low-productivity, Hawaiian waters and then move to the more productive waters of the North Pacific to feed during the summer months. There is some evidence that not all individuals migrate each year. Some humpback whales can be seen in the Southeast Alaska feeding grounds year-round, and individuals that do not consume enough prey during the feeding season may forego migration in order to continue feeding.

C. Searching and Diving

The way in which an animal moves through its environment can influence its encounter rate with prey, and many animals exhibit stereotyped search patterns. Marine mammals that forage on concentrated prey may patrol continually through areas where they expect to encounter concentrations. For example, leopard seals (*Hydrurga leptonyx*) will patrol along ice edges where departing and returning penguins congregate, and killer whales patrol nearshore areas in search of seals. When groups of marine mammals forage, they often spread out into widely spread subgroups and/or move forward in a line abreast formation [e.g., dusky dolphins (*Lagenorhynchus obscurus*), pilot whales (*Globicephata* spp.), Risso's dolphins (*Grampus griseus*), bottlenose dolphins (*Tursiops* spp.), killer whales (*Orcinus orca*)]. Spreading out in such fronts may either reduce foraging competition among individuals or increase the probability that prey is detected so the subgroups can converge to feed.

Once a marine mammal has selected a habitat for foraging, it must execute a strategy that will optimize its net energy intake rate, often with respect to trade-offs and constraints. For diving animals this means that it must balance the energetic costs of diving with the energetic gains from foraging. The costs of diving vary greatly among marine mammals. Polar bears, sea otters (*Enhydra lutris*), and most pinnipeds are divers: they spend most of their time above water or have long surface intervals between food-gathering dives. In contrast, most cetaceans and sirenians can best be thought of as surfacers: they spend the majority of their time submerged and make trips to the surface only to breathe.

Theoretical studies of optimal diving suggest that as the depth at which prey are located increases, both dive times and surface times should increase (Kramer, 1988), and the type of a dive a marine mammal executes will depend on the depth and distribution of prey. Some predictions of optimal diving theory are supported by several studies of marine mammals, and both dive times and surface times increase with dive depth in pinnipeds, cetaceans, and sirenians. Because a diving individual should behave in a manner that optimizes its net energy intake (e.g., Kramer, 1988), marine mammals may exceed aerobic limits when the energetic payoff is sufficient.

There is a great deal of variation in the depths to which marine mammals dive. Some, such as sea otters, nearshore odontocetes, and otariids, tend to be shallow divers. Others, including sperm whales, elephant seals, and beaked whales, are extremely deep divers, sometimes foraging over 1000 m from the surface. Some species minimize the depths to which they must dive, and thus the costs, by modifying their diel pattern of foraging. For example, some dolphins and pinnipeds are nocturnal foragers on prey whose daily movements bring them closer to the surface at night [e.g., spinner dolphins (*Stenella longirostris*), northern fur seals (*Callorhinus ursinus*), Antarctic fur seal (*Arctocephalus gazella*)]. The diving tactics of beluga whales (*Delphinapterus leucas*) may be influenced by competition with pinnipeds, who are superior divers. Belugas generally forage over the deepest waters and, because of their body size, are able to gain access to benthic areas that the smaller pinnipeds cannot. Although the time spent at the bottom decreases with increasing depth, belugas compensate by increasing their ascent and descent rates as dive depth increases, a result also found in narwhals.

D. Prey Detection

Marine mammals have many ways to detect their prey, including vision, various types of mechanoreception, echolocation, and hearing. Most marine mammals appear to rely on vision to at least some extent. The large, forward pointing eyes of pinnipeds suggest that vision is an important method for detecting prey. Even species that dive to extreme depths, such as elephant seals, are capable of using vision to find prey in dark waters at their foraging depth. VISION may be less important in other taxa. For example, the river dolphin (*Platanista gangetica*) of the Indian subcontinent has eyes that are greatly reduced and may be mostly blind. Sea otters can use their forepaws to find food and discriminate prey items without the aid of vision, and many pinnipeds are found in turbid waters, making vision a poor method of prey detection. However, they are able to use their vibrissae (whiskers) to detect prey through active touch or through minute water movements caused by their prey.

Odontocete cetaceans have a method of prey detection not available to other marine mammals: echolocation. In controlled situations, odontocetes can detect relatively small objects at a considerable distance. For example, a common bottlenose dolphin can detect a 7.62-cm-diameter sphere from over 100 m. However, it is still unclear how efficient echolocation is under natural conditions. It is likely to be less efficient than suggested by laboratory and controlled experiments (as has been shown for bats) and may vary greatly depending on environmental conditions such as noise. Echolocation appears to be the primary method of prey detection in some essentially blind river dolphins such as the Indian river dolphin, which inhabits murky waters.

Echolocation is not always an effective way to detect prey. While most fish cannot hear echolocation calls, clupeid fish and other marine mammals can. Therefore, odontocetes foraging on prey that can detect their echolocation may have to use tactics other than echolocation for detecting prey. This difference in the ability of potential prey to detect echolocation is reflected in the foraging behavior of fish-eating ("resident") and mammal-eating ("transient") killer whales off British Columbia. Whereas resident whales commonly use echolocation during foraging, transients do not. Also, when transients echolocate, their pulses are of low intensity and are irregular in timing, frequency, and structure, a pattern that may be difficult for prey to detect. Instead of echolocation, mammal-eating killer whales appear to use passive listening to detect their prey. Other marine mammals probably use passive listening opportunistically, especially bottlenose dolphins, which feed on a variety of noisy fish species. Elephant seals and other pinnipeds also have good hearing abilities in water and may use passive listening to find prey.

II. Capturing and Consuming Prey

A diverse array of tactics is used by marine mammals to capture and consume their prey once they have located it (Fig. 1). The most widespread tactic of raptorial predators is to simply chase down individual prey items that they have encountered. However, there are many other more unique tactics employed by marine mammals.

A. Stalking and Ambushing

Marine mammals often hunt prey that are nonsessile, fast-moving, and have good sensory abilities and, thus, could avoid predators if their approach were too obvious. For example, seals can avoid polar bears by diving back through the ice, and penguins can avoid leopard seals by hauling out, as can pinnipeds approached by killer whales near land. When hunting elusive prey, a predator must rely on either stalking or ambushing. A stalking predator attempts to conceal its identity or presence until it approaches its prey close enough for a sudden, successful attack. In contrast, an ambush predator conceals itself and lies in wait, leaving the approach to the prey.

Polar bears use both stalking and ambush methods when hunting seals hauled out on the ice near breathing holes. In terrestrial stalking, bears creep forward and use ice for cover to closely approach their intended prey. Bears also stalk seals by swimming circuitously through interconnected channels or even under the ice, occasionally surfacing through holes to breathe and monitor their prey. However, an ambushing tactic, where a bear lies, sits, or stands next to a breathing hole waiting for a seal to surface, is the most energy-efficient and most commonly used foraging tactic.

Leopard seals also use both stalking and ambush tactics when foraging. Stalking leopard seals may swim under the ice below a penguin and then break through to capture the bird, or they may swim submerged near a fur seal beach and lunge at pups when they get close enough. Alternatively, leopard seals may ambush their prey by hiding between ice flows near a penguin-landing beach. Sea otters will stalk birds by swimming underwater and grabbing them from below, a tactic similar to that used by Steller sea lions (*Eumetopias jubatus*) hunting northern fur seal pups and leopard seals stalking Adelie penguins.

Figure 1 *Whales employ a diverse array of foraging tactics. Pieter A. Folkens/Higher Porpoise DG.*

Another behavior that could be considered stalking is wave riding and intentional beaching used to capture young pinnipeds and penguins near the water's edge. This tactic is commonly used by killer whales and occasionally by Steller sea lions and leopard seals. This may be a particularly dangerous foraging tactic, especially for young killer whales who may not be able to return to the water if they strand too high on the beach.

Some stalking predators make detours that involve moving away from the prey and potentially losing visual contact temporarily before making another approach. Polar bears will make detours from their prey while stalking aquatically, and dolphin subgroups may detour away from a school of fish to attack it from opposing sides. Weddell seals (*Leptonychotes weddellii*) have also been observed making detours when stalking cod under fast ice. These detours allow the seal to remain out of the fish's view and to attack from very close range below the fish.

B. Prey Herding and Manipulation

In order to capture them more efficiently, marine mammals may actively manipulate the behavior of their prey. In other words, marine mammals take advantage of normal prey behaviors to enhance their ability to capture them. These manipulations may help a marine mammal flush prey from hiding, capture an individual prey item, or increase the density of prey aggregations so as to increase the forager's energetic intake rate. Prey herding is a common tactic used by dolphins, porpoises, whales, and pinnipeds and may be considered prey manipulation when they take advantage of natural schooling and flight behavior of their prey. Dolphin and porpoise groups and individuals have been observed herding prey against shorelines or other barriers, reducing the number of escape routes. Dolphins use shorelines for more than herding fish. Common bottlenose

dolphins inhabiting salt marshes are known to form small groups that rush at fish trapped against a mud bank. The wave created by the rapid swim causes fish to strand on the mud bank and the dolphins slide up the bank and pick fish off the mud before sliding back into the water. A similar behavior is performed by both individuals and groups of humpback dolphins (*Sousa plumbea*) foraging around sand banks off Mozambique.

Marine mammals also herd fish in open waters. When schools are at the surface, dolphins may split into groups to attack from different directions, herding the fish into a ball between subgroups. Other times, fish may be herded up from deeper waters and trapped between circling individuals and the surface. During a herding event, individuals swim around the fish school, and below it, preventing its escape. Fish herding in open waters has been reported in many dolphin species, porpoises, and sea lions. Sea lions are also found feeding on schools of fish that are herded to the surface by dusky dolphins (*Lagenorhynchus obscurus*), but it is unclear if the sea lions aid in fish herding. The tactic of herding fish is found in a variety of marine predators, and although there are no reports of prey herding for many species of pelagic dolphins, it is probably a widespread tactic employed by marine mammals feeding on schooling fish.

During a prey-herding event, many different tactics may be used to cause the fish to move into a tight ball and to capture fish in these balls. Splashing at the surface causes fish schools to compact. Dusky dolphins perform leaps at the edge of fish schools that they are herding, and Atlantic spotted dolphins (*Stenella frontalis*) have been observed tail slapping and splashing at the edge of a fish school when it started to break apart or move in a different direction, but the function of these behaviors is still unclear. Killer whales in Norway and humpback whales in the northwest Atlantic also use tail slaps when near schools of prey. Tail flicks by humpbacks may also be used to concentrate schooling euphasid prey in Southeast Alaska, although this may simply be a hydromechanical effect.

Another tactic that marine mammals can use to herd prey is flashing light-pigmented areas of their body toward a fish school. Killer whales herding herring swim under the school and flash their white undersides to keep the school from diving, and a similar behavior has been noted in spotted dolphins. Humpback whales in Southeast Alaska may also use flashes to help concentrate prey by rotating their elongated pectoral flippers while they herd herring, thereby showing the highly visible white undersides.

Fish show strong avoidance responses to bubbles and are reluctant to cross barriers composed of them. Not surprisingly, marine mammals take advantage of this response. The use of bubbles during foraging has been observed in many odontocetes, mysticetes, and pinnipeds. Atlantic spotted dolphins use bubbles to isolate individual fish, pulling them away from the school with the water disturbance created by the passing bubble, so they can be consumed, and Weddell seals blow bubbles into ice crevices where fish are hiding to flush them out. Killer whales also use bubbles to flush prey and blow large bubbles toward rays buried in the sediment, causing them to move. However, bubbles are primarily used to concentrate and contain schools of fish. For example, killer whales blow large bubbles near the surface to keep fish in a tight ball. Humpback whales are the best-known bubble users, and bubble feeding may be conducted by individual whales or in large groups. Whales deploy bubbles in a variety of formations, including columns, curtains, nets, and clouds, with the tactic used dependent on the characteristics of the prey aggregations.

Sound and pressure waves may also be used to manipulate prey behavior. For example, bubble-netting humpback whales in Southeast Alaska produce loud "feeding calls" as they rise to the surface, presumably herding prey up into bubble nets that are meters above the herring schools. Bottlenose dolphins off Australia and Florida use tail slaps known as "kerplunks" while foraging in shallow seagrass habitats. The kerplunk displaces a significant amount of water, creates a plume of bubbles, and causes a low-frequency sound. Kerplunks may cause startle responses in fish and help the dolphin locate and flush their prey, while the bubbles may provide a barrier to contain the fish. Humpback whales in the western Atlantic may flush burrowing fish (sand lance) from the bottom by scraping the substrate with their head and then feed on the fish once they have entered the water column.

C. Prey Debilitation

Marine mammals sometimes debilitate their prey before they consume it. Killer whales attacking mysticetes often swim onto their backs when the prey tries to surface, and in some cases the victim may drown instead of dying from its wounds. While killer whales are herding herring, individuals thrash their tail through the school, stunning fish with the physical impact of their flukes; they then feed on the stunned and injured fish. The whales probably use this tactic because it is energetically more efficient than whole-body attacks. Bottlenose dolphins

strike fish with their tails ("fish whacking") when foraging alone or in groups, sometimes knocking the fish through the air. Also, there is evidence that walruses (*Odobenus rosmarus*) may use their tusks to kill or stun intended seal prey. Sound may also be used to debilitate prey, and several authors have hypothesized that odontocetes may be able to use echolocation to stun or disorient fish, but this hypothesis remains largely untested.

D. Tool Use

"Tool use is the external employment of an unattached environmental object to alter more efficiently the form, position or condition of another object, another organism, or the user itself when the user holds or carries the tool during or just prior to use and is responsible for the proper and effective orientation of the tool" (Beck, 1990). Tool use by marine mammals is reviewed in detail elsewhere in this volume, but several marine mammals make extensive use of tools during foraging and deserve brief mention here.

Sea otters are the best-known marine mammal tool users and will pick up rocks from the bottom and place them on their chest to use as an anvil for crushing mussels, crabs, or urchins, or use them to smash or dislodge abalone off rocks. In some cases, the rocks are retained between foraging dives to be reused. Certain bottlenose dolphins carry sponges on their rostra, apparently as a tool to aid foraging (Fig. 2). Also, there are popular accounts of polar bears throwing blocks of ice at basking seals to injure or trap them, and polar bears in captivity are often observed throwing large objects, raising the possibility of tool use in the wild. Another behavior that might be considered tool use involves killer whales creating waves to wash hauled out seals into the water. For example, in ANTARCTICA, a group of killer whales spotted an adult crabeater seal (*Lobodon carcinophaga*) on a small ice floe. The whales made a rapid swim up to the floe and dove rapidly. The resulting wave tipped the floe and knocked the seal into the water, but predation was not observed. Finally, as described earlier, several marine mammals, including humpback whales, killer whales, bottlenose dolphins, and Atlantic spotted dolphins, use bubbles as a tool to concentrate schooling fish.

Figure 2 *Bottlenose dolphins use sponges as tools to aid in foraging. (Photograph by Michael R. Heithaus).*

E. Benthic Foraging

While most marine mammals pursue their prey in the water column, several species forage on benthic organisms. There are three basic methods that marine mammals use to obtain prey from the bottom: collecting, extracting, and engulfing. Epibenthic prey are simply collected by foraging marine mammals. For example, sea otters collect echinoderms (mostly sea urchins), crabs, and other benthic organisms using their forepaws.

Infaunal prey items must be extracted from the substrate and require the predator to excavate in some manner. Sea otters use their forepaws to dig for clams in soft-sediment areas and may produce large pits over the course of several dives, occasionally surfacing with a clam. Harbor seals (*Phoca vitulina*) dig for prey in sandy habitats with their foreflippers or snouts while narwhals (*Monodon monoceros*) and belugas use water jets to dislodge mollusks buried in the sea floor. Walruses use a combination of tactics to obtain buried bivalves, including digging with their snouts (not tusks) and hydraulic jetting. Walruses make multiple excavations on each dive and have been recorded consuming at least 34 clams on a single dive. Killer whales, in New Zealand, engage in benthic foraging on rays and have been observed pinning them to the bottom and may also be digging for them. Bottlenose dolphins in the Bahamas also dig for infaunal prey ("crater feeding"), and once a burrowing fish has been located, the dolphin will dive into the soft sand and use its flukes to drive deeper, almost up to the flippers, to catch the fish.

If many small infaunal prey items are consumed in a single feeding event, they may be engulfed while still in the sediment. Gray whales (*Eschrichtius robustus*) feeding near the bottom use suction to pull sediment and prey into their mouths and then filter the sediment and water out through their baleen.

F. Batch Feeding

Batch feeding is a tactic employed to consume a large number of prey items in a single feeding event. While mysticetes are obligate batch feeders, some pinnipeds facultatively use this tactic. There are two basic types of batch feeding: skimming and engulfing. Skimmers, most notably the right whales and bowhead whale, swim through concentrations of zooplankton, either at the surface or in the water column, with their mouths open, filtering water through their fine baleen plates, which trap prey.

Engulfers include rorquals and several pinnipeds. These species engulf large amounts of water and prey and then filter the water back through their baleen plates or teeth. Rorquals have a suite of adaptations, including expandable gular pleats and a lower jaw that can disarticulate from the upper jaw, that allow them to engulf huge volumes of water, and fish or crustacean prey, in each feeding attempt. "Lunge feeding" is one of the most common tactics of rorquals feeding near the surface and may take several forms. During a typical lunge, a whale surfaces with its mouth open to capture prey near the surface. Lunge feeding may be done singly or in groups, and in combination with many of the prey concentration tactics.

All Antarctic seals [crabeater, Weddell, Ross (*Ommatophoca rossii*), leopard seals] include zooplankton in their diet, as do some Arctic seals [ringed (*Pusa hispida*), ribbon (*Histriophoca fasciata*), harp (*Pagophilus groenlandicus*), larga (*Phoca largha*), and harbor seals]. Of these, the crabeater seal is the most specialized batch feeder, and zooplankton may comprise up to 94% of its diet. The cheek teeth of crabeater and some other seals are modified for straining krill, which are probably sucked into the mouth when the seal depresses its tongue and are then trapped against the cheek teeth as the water is expelled.

G. Ectoparasitism, Kleptoparasitism, and Scavenging

Predators kill their prey in the course of consuming it (Ricklefs, 1990). While most marine mammal foraging is predatory, there are several ways that animals may forage that do not involve killing their own prey. For example, an animal may gouge mouthfuls of flesh from a "host" without killing it (sometimes referred to as ectoparasitism). Although marine mammals fall victim to such ectoparasites (small sharks), there are no concrete examples of marine mammals using this tactic. However, killer whales may effectively ectoparasitize large whales as some attacks do not kill the victim. Kleptoparasitism (food stealing) has been observed in only otters and polar bears, but may occur within and among other species. Scavenging is a common foraging tactic, but does not appear to be widespread among marine mammals. However, it may be an important tactic for polar bears and some pinnipeds.

H. Herbivory

Sirenians (manatees, *Trichechus* spp., and dugongs, *Dugong dugon*) are the only marine mammals that routinely feed on plants, and both manatees and dugongs may be found foraging individually or in large groups. Manatee feeding appears to be more flexible than that of dugongs, as the former will consume either floating or rooted vegetation and sometimes leaves from overhanging branches or vegetation along banks. Dugongs feed almost exclusively on seagrasses, but may also intentionally consume benthic invertebrates. While manatees tend to crop vegetation, dugongs dig up rhizomes and leave large feeding trails through seagrass beds, which can have a large impact on seagrass biomass, both above and in the sediment (Fig. 3).

I. Prey Preparation and Consumption

While some marine mammal prey can be consumed immediately after capture, others require extensive handling before they are eaten, and some are only partially consumed. Sea otters remove the heads of birds that they capture and strip the muscle from the breast, neck, and legs. Many dolphins and sea lions remove the heads from large fish before consuming them, and bottlenose dolphins will strip flesh from spiny fish (Fig. 4). Head and spine removal may reduce the probability that a predator is injured while consuming prey, but it may also be a mechanism to reduce the intake of bony material that provides no nutritional value. Dolphins do not always consume their prey correctly, and sharp spines have been implicated in mortalities of bottlenose dolphins stranded in Texas and Florida.

Odontocetes cannot chew prey and must spend considerable time handling large prey items. For example, bottlenose dolphins will drag large fish along sandy bottoms until pieces

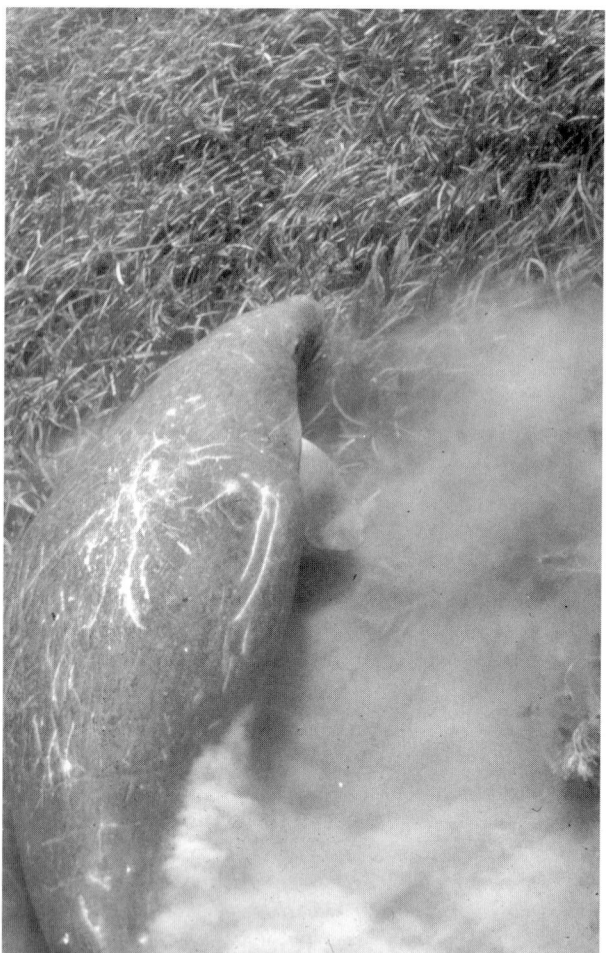

Figure 3 *Dugongs preferentially forage on below-ground portions of many seagrass species, creating a cloud of sediment during foraging activity. (Photograph by Michael R. Heithaus).*

Figure 4 *Not all prey that are captured can be consumed immediately. Large fish, like this one captured by a bottlenose dolphin, may take extensive handling before they can be consumed. (Photograph by Michael R. Heithaus).*

that are small enough to swallow are broken off. Killer whales are well known for their extensive handling of prey, especially pinnipeds, after capture. Killer whales often breach upon sea lion and seal prey as well as slap them with their tails. The function of these behaviors is unclear, but they may tenderize the prey, aid in training calves in hunting techniques, or even debilitate dangerous prey. Selective feeding on energy-rich portions of prey is common in both killer whales and polar bears. Killer whales will selectively eat the blubber and tongue of whales that they kill, and polar bears prefer the blubber and muscle of seals and narwhals over the internal organs.

III. Group Foraging

Many foraging tactics are executed by groups of marine mammals. Sometimes these groups are merely aggregations of animals attracted to the same resource, and there appears to be little interaction among individuals as they pursue prey individually. Other group-foraging behaviors, such as herding of prey, appear to be highly coordinated efforts and may involve animals cooperating with each other to increase their net energy intake rate. It is not always easy to determine whether group-foraging marine mammals are cooperative or not. For example, the echolocation rate of an individual resident killer whale decreases as group size increases, suggesting that there may be information transfer. While this could represent cooperative information sharing, it is also possible that individual whales are parasitizing the information of others, as shown in bats. In many cases it is difficult to assess whether marine mammals are foraging cooperatively because group living may be selected for by factors other than food, and group foraging, whether cooperative or noncooperative, is therefore simply a necessary epiphenomenon. One important consideration in studies of cooperation is whether groups are kin based, as individuals in kin groups are more likely to engage in cooperative behavior to increase their inclusive fitness.

A. Cooperative Foraging, Food Sharing, and Cultivation

Cooperation can be defined as an "outcome that—despite individual costs—is 'good' in some appropriate sense for the members of the group . . . and whose achievement requires collective action" (Mesterson-Gibbons and Dugatkin, 1992). Most cooperation is achieved through a mechanism of by-product mutualism in which an individual acts selfishly to benefit itself, and its actions incidentally benefit other individuals, but not all by-product mutualisms are cooperative. A possible example of by-product mutualism involves bowhead whales (*Balaena mysticetus*) skim feeding in groups with whales staggered in an inverse "V" formation. This formation may aid a whale in prey capture by using adjacent whales as a wall to trap prey or to catch prey that have escaped from the whale in front. These groups sometimes appear to be coordinated, with whales changing direction and leadership. All individuals probably act selfishly, but their presence may benefit other whales.

The just-described definition of cooperation requires three things be shown to support the hypothesis that a group is cooperative. First, individuals acting cooperatively must realize a short-term cost. This cost may include having to share food with other individuals or an opportunity cost by not attacking prey immediately while herding. Next, energy intake rate of individuals benefiting from cooperation must be higher than what they would have gained without cooperation. Finally, collective action must be required for the hunt to be successful. It is worth noting that in cooperative groups not all group members are required to receive equal benefits, and in groups that appear to be cooperative, a number of individuals may be noncooperative (Packer and Ruttan, 1988).

There are many possible examples of cooperative foraging in the marine mammal literature involving mysticetes, odontocetes, pinnipeds, and sirenians. However, most anecdotal accounts of possible cooperative foraging behavior do not provide enough detail to determine whether these groups were truly cooperative. For example, many dolphin species are known to break into subgroups that spread out across a large front when foraging or to travel in line abreast formation. Generally, when one subgroup finds fish, other subgroups join to feed, and this behavior has often been considered cooperative foraging. However, none of the three criteria for cooperation outlined earlier has been shown to apply to these cases. Furthermore, although some authors have assumed that joining subgroups were recruited, they may simply be converging once they determine that another group located food. Fish herding by dolphins, porpoises, whales, and sea lions has been cited as an example of cooperative foraging. In these cases, there does appear to be a cost involved as individuals do not start foraging immediately, but wait until the school has been herded to the surface ("temporary restraint"). Larger groups of dusky dolphins forage on a single fish school for longer periods of time than small groups. Some authors have suggested that this indicates an increase in individual intake and that herding requires collective action. However, it is important to measure individual intake rates because longer foraging durations of large groups may simply be the result of larger schools being herded (and increased time until school depletion) or of increased foraging interference in large groups. More studies are required to support the hypothesis that such groups are cooperative.

Deliberate prey sharing provides strong evidence for cooperative hunting, but must be viewed with caution as some apparent food sharing may represent intense competition for large prey items or kleptoparasitism (e.g., Packer and Ruttan, 1988). Prey sharing has been observed in few marine mammal species, but has been documented in many killer whale populations. Also, an apparent case of prey sharing has been documented in leopard seals when two seals killed penguins, but one individual released its penguin to be consumed by the other.

There are a few other examples of marine mammal foraging that appear to represent cooperative foraging. Leopard seals have been observed hunting in a coordinated fashion, with one seal driving penguins toward a second seal hiding behind an ice flow. The process was repeated several times, and both seals caught penguins each time, sharing prey in one instance. Collective action is required if killer whales are to capture large or swift prey and, in general, larger groups are seen when transient killer whales attack such prey. There may even

be a division of labor during their hunts. Also, there is a cost as prey are divided among group members. Bubble-netting humpback whales feeding on herring in Southeast Alaska represent another potential example of cooperative foraging. In these groups, one whale deploys a bubble net, starting at a depth shallower than the herring schools. The whales then apparently drive the prey up into the bubble net and simultaneously lunge through the herring trapped against the surface. Although there are apparently costs to this behavior and coordination is probably required, no data exist on intake rates in these groups.

One study has suggested that dugongs cultivate seagrass as they forage in large groups moving among seagrass banks. Although dugong grazing changes seagrass communities to stands of more profitable species, for this type of cultivation to occur, cooperation among dugongs would be required. In general, cultivation is favored to evolve only when the individual that gardens realizes the benefits of that action (Branch *et al.*, 1992). This implies both a fixed and a defendable feeding site (Branch *et al.*, 1992). Cooperative cultivation by dugongs is unlikely, as individual dugongs that moved to a previously cultivated area, before the cultivating individuals, would benefit from reduced foraging competition. Also, there do not appear to be any mechanisms to prevent such cheating. A more likely explanation for the observed pattern of dugong foraging is "trap lining" where dugongs rotate among the most profitable seagrass meadows, and the changes in seagrass communities are an incidental byproduct of dugong foraging on rhizomes.

B. Optimal Group Size

The question of why particular group sizes are observed has been raised several times. For some species, group size has been suggested to be that which maximizes the intake rate of individuals in the group (optimal group size). However, this may not generally be the case. When it is difficult for a group to exclude joiners (e.g., when foraging on a large fish school), the observed group size will often be greater than that which maximizes the intake of each group member, as individuals will continue to join a foraging group until the average energy intake in the group approaches that of a solitary forager (stable group size; e.g., Giraldeau, 1988). Also, the benefits of defending resources may be low in large groups, as individuals that do not defend the resource will realize higher intake rates than those individuals that try to defend against joining individuals. Finally, group size is likely to be larger than that which is optimal for foraging considerations if there are other benefits of being in a group (e.g., mating opportunities, protection from predators). Therefore, it is likely that most dolphins feeding on large schools of fish are in groups larger than those that would maximize the energy intake of each group member. However, some marine mammals may be found in groups that are of optimal size for maximizing energy intake. For example, killer whales feeding on marine mammals may be able to regulate group size as individual prey items are easily defended, and groups (which are kin based) may be able to exclude other individuals before foraging commences. This may explain why the modal group size of three individuals observed in foraging transient killer whales is the group size that maximizes its members' energy intake.

IV. Variation in Feeding Strategies and Tactics

Marine mammals show a high degree of variability and flexibility in their foraging tactics. Individuals may be flexible in their foraging tactics depending on their state or circumstances, and this flexibility may lead to variation in foraging tactics among populations, individuals, and age/sex classes. A variation in feeding tactics may also arise from differences in the ways individuals solve cost–benefit trade-offs. Some of these differences among individuals may be genetically based and thus considered strategic variation.

A. Trade-offs

Evolution favors strategies that maximize fitness (usually by maximizing lifetime reproductive success). For example, animals may pursue a strategy that maximizes their expected lifetime energy intake, which may involve a trade-off between maximizing short-term energy intake and minimizing predation risk because habitats that are prey rich are often the most dangerous (e.g., Lima and Dill, 1990; Fig. 5). Therefore, marine mammals may sometimes accept lower energetic returns in order to forage in safe habitats. For example, Indian Ocean bottlenose dolphins in Shark Bay, Western Australia, match the distribution of their prey when their primary predator, the tiger shark, is absent, but shift habitats to forage mostly in low-risk, low-food habitats when sharks are abundant. Also, female polar bears with cubs may preferentially select habitats with lower food abundance to avoid adult males who might kill their cubs. Trade-offs between predation risk to calves and food availability at high latitudes may also be the reason that a strategy that includes seasonal MIGRATION with discrete periods of feeding and fasting evolved in baleen whales.

Trade-offs between feeding and predation may also result in habitat use patterns that vary with behavior. For example, spinner dolphins rest in shallow nearshore coves with sandy bottoms during the day, possibly to reduce the probability of shark attack, and then move offshore to feed on deep scattering layer organisms at night. Similarly, bottlenose dolphins in Shark Bay rest almost exclusively in safer, relatively deep waters, but will often move into higher risk but more productive shallow habitats to feed.

Trade-offs between feeding and reproduction may also influence foraging patterns. For example, most phocid females fast during lactation and must consume sufficient food before the breeding season, whereas female otariids make foraging trips of variable duration throughout lactation (for a review, see Wells *et al.*, 1999).

Prey selection can be viewed as the result of another type of trade-off. Each potential prey item differs in the energy required to capture it and the amount of energy the predator will gain from eating it. This trade-off sometimes results in selective foraging where one prey type is favored over others irrespective of its relative abundance. For example, harp seals always preferentially feed on capelin and select Arctic cod only in nearshore waters. Prey preferences have also been shown in resident killer whales. Off Alaska, resident killer whales prefer coho salmon, whereas those off British Columbia prefer chinook salmon, which are energy-rich but relatively scarce. Prey selection may also take the form of capturing a particular size of prey. For example, harbor seals in Scotland feed primarily on the most abundant fish species, but prefer fish 10–16 cm in length. Changes in the relative costs and benefits of particular prey items may lead to prey switching, which has been observed in some marine mammals.

B. Ontogenetic Variation

There are often distinct differences in the foraging behaviors of marine mammals of different ages. Such differences may be the result of changing physiological or foraging abilities, the relative importance of energy intake and survival at different life history stages, or differences in experience if a learning period is required for the successful use of a particular foraging tactic. Diving by young seals and sea lions is constrained by physiological development and they typically make shorter and shallower dives than adults (e.g., Steller sea lions, Weddell and elephant seals, *Mirounga* spp.). During their first trip to sea, elephant seal pups make a transition from short, shallow dives to a pattern similar to that in adult seals, with longer deeper dives that show diel fluctuations. This transition appears to be related to changes in the physiology of young seals and possibly prey distribution. Young seals of different sizes may adopt different diving tactics. For example, larger yearling Weddell seals engage in relatively shallower dives to forage on benthic prey compared to small yearlings, which make deeper dives to forage on energy-rich prey. However, the cause of this variation is unclear.

Learning plays an important role in the acquisition of foraging tactics in cetaceans. For example, there is a long period of practice required for young killer whales to become adept at using the intentional stranding tactic to capture pinnipeds. This period of learning may involve calves preferentially associating with the female pod members (not necessarily their mother) that engage in this tactic most frequently.

Figure 5 *Foraging decisions made by individuals can be influenced by the presence of predators. Some individuals may forage in areas where they are more likely to be attacked by predators if the energy gain in these habitats is sufficient. (Photograph by Michael R. Heithaus).*

C. Interindividual Variation

Within many marine mammal populations, substantial differences exist among individuals in the foraging tactics that they employ. Northern fur seal females perform two distinct types of foraging dives: shallow dives, which seem to be directed toward vertically migrating prey, and deep dives to feed near the bottom. Shallow dives are made only between dusk and dawn, whereas deep dives occur both at night and during daylight hours. Some individual seals specialize in one dive type or the other, whereas other individuals use a mix of tactics. Southern sea lion individuals differ in their propensity to hunt fur seal pups. In Alaska, only juvenile male Steller sea lions prey upon fur seal pups, whereas in Peru, most hunting is done by just a few adult males, and there are large differences in the success rates of different individuals. Similarly, sea otter predation on birds appears to be largely restricted to a few individuals, and a few individual sea lions have learned to wait at fish ladders and at the mouths of freshwater streams to take advantage of spawning steelhead. Leopard seals also vary in their hunting tactics. For example, a single individual was responsible for all ambushing attacks on Adelie penguins observed in Prydz Bay, Antarctica. Finally, individual variation in the prey species consumed by sea otters may be a result of differences in diving tactics as juvenile males forage further offshore and make longer dives than other age/sex classes.

Cetaceans also show individual variation in feeding tactics. In Shark Bay, many unique tactics, including kerplunking, sponge carrying, and extreme shallow water foraging, are restricted to a small number of individual bottlenose dolphins. It also appears that some dolphins forage only in deep waters, whereas others forage mainly over shallow seagrass habitats or use a mixture of shallow and deep-water tactics. Adult female killer whales perform most of the intentional strandings to catch elephant seals, and within a pod individual females differ in their use of this tactic. Most individual minke whales (*Balaenoptera acutorostrata*) around the San Juan Islands specialize in either lunge feeding or feeding in association with birds. These two tactics are usually observed in different regions with individual whales showing inter- and intraseasonal site fidelity. Individual humpback whales differ in their use of various types of lunge feeding and bubble-netting tactics, which may relate to dietary specializations on either krill or herring and to the distribution of these prey items. Finally, reproductive state may influence the foraging tactics of cetaceans, as lactating female bottlenose, common, and pantropical spotted dolphins (*Stenella attenuata*) consume different prey items than other dolphins.

Cultural transmission appears to be responsible for the acquisition, in humpback whales, of lobtail feeding where a whale slaps the water with its flukes before lunge feeding. Lobtail feeding may have originated with an increase in the importance of sand lance in the whales' diets and then spread as individuals born into the population learned the lobtail feeding tactic rather than adults learning this behavior or new animals recruiting to the area. Cultural transmission of foraging tactics may also maintain the divergence in feeding tactics of sympatric fish-eating and mammal-eating killer whale populations.

Finally, sea otter tool use may be transmitted culturally from mothers to their offspring.

D. Intraindividual Variation

Individual marine mammals can switch among foraging locations and tactics depending on their age, body, condition, group size, and prey distribution and abundance. For example, pinnipeds can change their diving behavior in response to increased foraging costs as seals make shallower dives and dive at a steeper angle to maximize their time at a foraging depth. Individuals that encounter different habitats often switch among tactics depending on their location. For example, humpback whales may switch between foraging in large bubble-netting groups and engaging in individual lunges to capture krill. Sperm whale foraging behavior is linked to foraging success, and foraging is more common when prey availability is high or the energetic cost of capturing prey is relatively low. Also, sea otters change the number of prey items they collect on each foraging dive depending on the average prey size available.

The flexibility of marine mammals is highlighted by their ability to take advantage of human activities. Many odontocetes, pinnipeds, and sea otters have learned to steal fish from nets. Sea lions will even jump into encircling nets to feed or will follow fishing vessels for days to take advantage of the abundant food resources offered by fishing operations. Bottlenose dolphins are well known for foraging behind trawlers and feeding on discarded fish or fish in nets. Some individual bottlenose dolphins also have learned to take advantage of direct handouts of fish offered by people, and many species of odontocetes remove either bait or fish from fishing lines. In the Bering Sea and off southern Brazil, killer whales may damage over 20% of the fish captured by long line fisheries.

Both the diversity of habitats in which marine mammals live and the flexibility of individuals have led to the wide variety of foraging tactics exhibited by the group. However, studies of these tactics are still in their infancy. Future studies should move beyond anecdotal accounts of particular foraging behaviors to systematic investigations of the function and use of particular tactics and the circumstances in which they are employed. Only such detailed studies will improve the ability to predict influences of anthropogenic changes to marine habitats and prey availability on marine mammals and aid in efforts to conserve them.

See Also the Following Articles

Diet ■ Diving Behavior ■ Echolocation ■ Group Behavior ■ Tool Use

References

Beck, B. B. (1980). "Animal Tool Behavior: The Use and Manufacture of Tools by Animals." Garland STPM Press, New York.

Branch, G. M., Harris, J. M., Parkins, C., Bustamante, R. H., and Eekhout, S. (1992). *In* "Plant–Animal Interactions in the Marine Benthos" (D. M. John, S. J. Hawkins, and J. H. Price, eds.), pp. 405–423. Clarendon Press, Oxford.

Evans, P. G. (1987). "The Natural History of Whales and Dolphins." Christopher Helm, London.

Fretwell, S. D., and Lucas, H. L. (1970). On terrestrial behavior and other factors influencing habitat distribution in birds. *Acta Biotheor.* **19**, 16–36.

Giraldeau, L.-A. (1988). The stable group and the determinants of foraging group size. *In* "The Ecology of Social Behavior" (C. N. Slobodchikoff, ed.), pp. 33–53. Academic Press, New York.

Gross, M. R. (1996). Alternative reproductive strategies and tactics: Diversity within sexes. *Trends Ecol. Evol.* **11**, 92–98.

Kramer, D. L. (1988). The behavioral ecology of air breathing by aquatic animals. *Can. J. Zool.* **66**, 89–94.

Lima, S. L., and Dill, L. M. (1990). Behavioral decisions made under the risk of predation: A review and prospectus. *Can. J. Zool.* **68**, 619–640.

Mesterson-Gibbons, M., and Dugatkin, L. A. (1992). Cooperation among unrelated individuals: Evolutionary factors. *Q. Rev. Biol.* **67**, 267–281.

Packer, C., and Ruttan, L. (1988). The evolution of cooperative hunting. *Am. Nat.* **132**, 159–198.

Ricklefs, R. F. (1990). "Ecology." Freeman, New York.

Wells, R. S., Boness, D. L., and Rathburn, G. B. (1999). Behavior. *In* "Biology of Marine Mammals" (J. E. Reynolds and S. A. Rommel, ed.), pp. 324–422. Smithsonian Institute Press, Washington, DC.

Female Reproductive Systems

ROBERT E. A. STEWART

*Department of Fisheries and Oceans,
Winnipeg, Manitoba, Canada*

BARBARA E. STEWART

Sila Consultants, St. Norbert, Manitoba, Canada

The female reproductive system in marine mammals is composed of the basic mammalian reproductive organs: ovary, oviduct, uterus, cervix, vagina, clitoris, and vaginal vestibule. Under control of the endocrine system, these organs are engaged in the reproductive cycle of ovulation, fertilization, implantation, fetal growth, and parturition. Ancillary to reproduction are the mammary glands and lactation. Some variation in anatomy, morphology, and physiology of the reproductive organs, and in reproductive cycles, exists among orders of marine mammals. Species-specific differences within orders also exist, reflecting both phylogeny and the variety of environments inhabited by marine mammals. Greater variation exists in how marine mammals use their basic mammalian anatomy in different marine habitats. This article describes the gross anatomical and morphological characteristics of female reproductive systems and notes the functional adaptations.

I. Anatomy and Morphology

The ovary is the organ where eggs or ova mature and are released during ovulation. Usually, there are two functional ovaries suspended from the abdominal or pelvic cavity by a short mesentery, the mesovarium, which attaches to the dorsal side of the broad ligament. Dugong (*Dugong dugon*) ovaries are also attached to the diaphragm by peritoneal folds that form pouches in the dorsal abdominal wall.

The ovaries are surrounded by the ovarian bursa, a fold of mesosalpinx that forms a peritoneal capsule. There is considerable variation in development and in the extent to which the bursa communicates with the celomic cavity. The ovarian bursa of odontocetes develops *in utero*, whereas in mysticetes it develops after birth. In polar bears (*Ursus maritimus*) and other carnivores, the periovarian space between the ovary and the peritoneal lining of the bursa communicates with the peritoneal cavity by a narrow passage, which may become distended at estrus with fluid of unknown origins. In all marine mammals, and mammals in general, the function of the bursa is to ensure that the ova pass into the oviduct where fertilization occurs.

Marine mammal ovaries vary in size and shape. Quiescent dugong ovaries are small, flattened ovoids or spheres. Ovaries in the Amazonian manatee (*Trichechus inunguis*) are broad and flattened against a short mesovarium. Sea otters (*Enhydra lutris*) have lenticulate, compressed oval ovaries. The odontocete ovary is more or less spherical to ovoid in shape, with a smooth surface in the resting condition, whereas in mysticetes, ovaries are flat and elongated. Phocid ovaries are ovoid and smooth in the resting state. In some species of phocids (e.g., gray seals, *Halichoerus grypus*), fetal hypertrophy of the ovaries exists through hormonal influence of the pregnant female. This condition may be less pronounced in otariids.

Typically, eggs ripen and ovulate alternately between the ovaries in successive reproductive cycles and the ovaries are of similar size. However, in some odontocete cetaceans there is a prevalence of activity in the left ovary (e.g., 70% in pilot whales, *Globicephala*) and the left ovary is larger than the right (Slijper, 1996). The right ovary may become active later in life.

The mammalian ovary is covered by germinal epithelium (Fig. 1) below which lies connective tissue (tunica albuginea) of varying thickness. Germinal epithelium is often invaginated into the tunica albuginea, forming small folds, pits, or subsurface crypts. These invaginations are particularly well developed in pinnipeds and form surface fissures in sea otter ovaries. Below the tunica albuginea is a layer of follicles and corpora that are derived from them. The ovary also contains stromal and connective tissue, interstitial tissue, vascular, nervous, and lymphatic tissues, and embryological remnants. The interstitial cells of cetaceans are less numerous and less prominent than those in some other mammalian orders, such as rodents. Understanding the maturation process of the follicles and the development and subsequent regression of the luteal bodies for each species allows researchers to assess the reproductive status of females (immature, ovulating, etc.).

Follicular maturation (Fig. 1) proceeds through a series of changes characterized by two phases. In the first phase, there is a rapid increase in the size of the oocyte and a slow increase in the size of the follicle. Second, there is slow growth of the oocyte and a rapid increase in the size of the follicle, which can be seen macroscopically. In dugongs, mature follicles may be just visible as translucent bodies or they may protrude from the

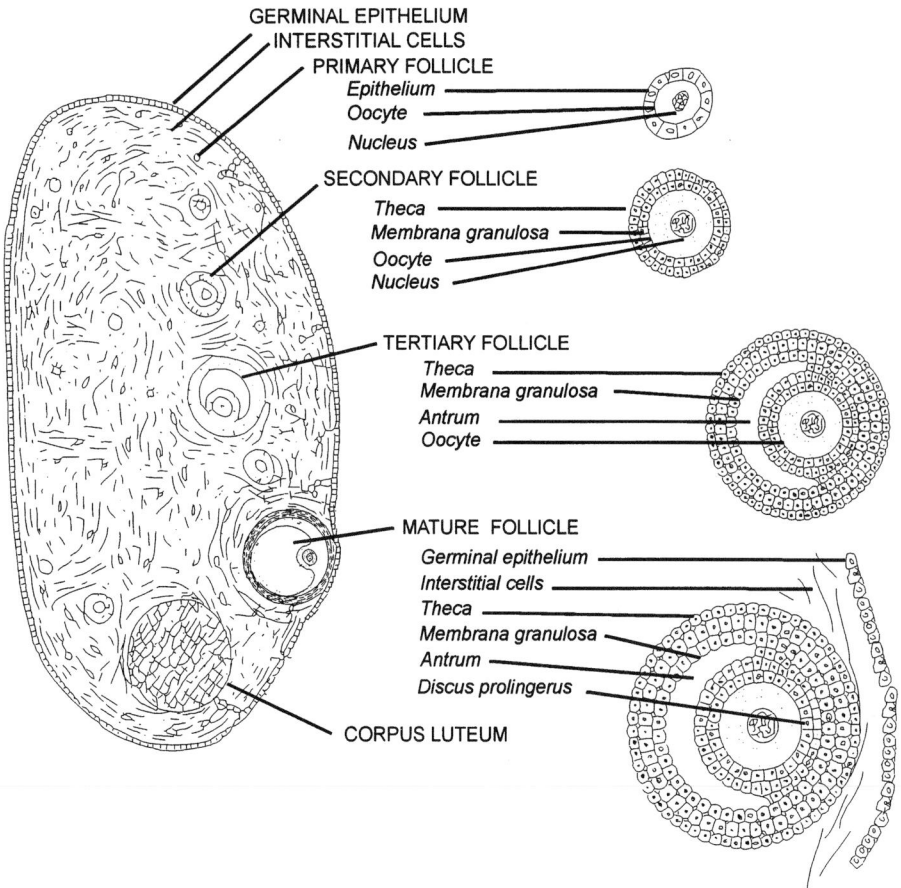

Figure 1 *Main structures of a marine mammal ovary showing stages of follicular development. Modified from Penny and Waern (1965).*

ovarian surface. In West Indian manatees (*Trichechus manatus*), mature follicles appear as large masses of bead-like spherules in the ovary. Similarly, mysticete ovaries may appear grape-like with protruding follicles. Maturing follicles of odontocetes and pinnipeds tend to be more widely dispersed in the ovary.

Oocytes develop within the ovary during fetal development but are dormant until puberty is reached. After puberty, and partly in concert with the annual reproductive cycle, they become primary follicles (Fig. 1) with a single layer of flattened epithelial cells surrounding each oocyte. These follicles lack connective tissue or thecal investment. As primary follicles increase in size they sink deeper into the cortex of the ovary toward the medulla or central area.

Secondary follicles are formed as the single layer of flattened epithelial cells around the oocyte thickens and becomes cuboidal or columnar, forming a distinct membrana granulosa (granular membrane or layer). This granulosa layer quickly becomes several layers thick while being encapsulated in an outer sheath or theca derived from the stroma. The theca divides into two layers. The inner layer, the theca interna, is glandular and well vascularized whereas the outer layer, the theca externa, is composed of connective tissue. The oocyte is now surrounded by a membrane, the zona pellucida, which is jelly-like, contains large amounts of polysaccharides, and lies between the plasma membrane of the oocyte and the granulosa cells.

Tertiary follicles rapidly increase in size due to an increased number of granulosa and thecal cells. Then, one or more cavities form in the granulosa and as the cavities enlarge they coalesce into an antrum, which fills with follicular fluid (liquor folliculi). The fluid-filled follicle is now surrounded by a wall except at the point of attachment of the oocyte.

Mature or Graafian follicles (Fig. 1) contain an oocyte that is surrounded by an irregular cluster of granulosa cells. These granulosa cells eventually form the corona radiata, which in turn is attached to cells forming the discus prolingerus or cumulus oophorus. There is mortality among developing follicles such that only a fraction of those primary follicles that start to develop will ever mature.

Under endocrine control, the mature follicle releases the ovum (ovulation). After ovulation, the corpus luteum (yellow body) develops from cellular components of the follicle. Luteinization is the process of transformation of follicular granulosa cells into luteal cells that contain carotenoid luteins (yellow pigments). There is a significant increase in cell size during luteinization. This volumetric growth is in contrast to the accretional growth of thecal cells in tertiary follicles. The corpus luteum (CL) is considered to be a gland and several types exist, the nomenclature being based on the morphology or function of the CL. For example, a CL of pregnancy develops when ovulation is followed by fertilization; accessory corpora lutea (plural) develop by luteinization of unruptured follicles (common in cetaceans) (Perrin *et al.*, 1984).

If successful fertilization of the ovum occurs, the CL of pregnancy persists through gestation. There are several phases of development in this corpus luteum gravidatitis: a short postovulatory phase when the CL is small, poorly vascularized, and minor changes occur in the luteal cells; a phase during delayed implantation when the CL is smaller than its initial size, vascularization is still poor, and luteal cells show marked cytoplasmic vacuolation; a short phase related to implantation when vacuoles disappear, the CL enlarges with resumed glandular activity, and vascularization increases; a phase postimplantation and during early pregnancy characterized by minor cell adjustments such as fluid accumulation, more obvious intercellular spaces, appearance of small vacuoles, increase of connective tissue, and thickened walls of blood vessels; the phase of the duration of pregnancy when the corpus luteum size is maintained; and the postparturient phase as a corpus albicans (CA). In cetaceans, accessory CLs may also form in the ovary of pregnancy.

In the beluga (*Delphinapterus leucas*), the corpora lutea gravidatitis are about 3–4 cm in diameter and weigh approximately 22 g. CL diameters in blue whales (*Balaenoptera musculus*) average approximately 14 cm; in common minke whales (*B. acutorostrata*) they average about 7 cm and 160 g. In some odontocetes, such as common bottlenosed dolphins (*Tursiops truncatus*), the CL gravidatitis may protrude far out from the general outline of the ovary and is connected by a stalk (pedunculated).

The process of degeneration of the corpus luteum into the corpus albicans is similar regardless of the type of corpus luteum (CL of ovulation, pregnancy, pseudopregnancy, or lactation) that is regressing. There are four patterns of degeneration: fibrohyalin invasion, lipoid degeneration, slow necrobiosis, and fast necrobiosis. Regardless of the regression pattern, glandular elements are lost, lutein granules vanish, and the size of the body diminishes until the white or gray scar-like CA is formed.

In most mammals, corpora albicantia are assimilated either relatively quickly postpartum or after one or two reproductive cycles, as in the sea otter. In marine mammals that cycle every 2 or 3 years, such as the walrus (*Odobenus rosmarus*), the CA may persist for some time. In cetaceans, corpora albicantia are thought to persist throughout a female's lifetime (Perrin *et al.*, 1984), a consequence of the large amount of connective tissue present and its poor vascularization, leading to a slow rate of regression. Some attempts have been made to characterize various types of cetacean corpora albicantia, but a definitive way to distinguish those bodies derived from ovulation from those of pregnancy has not been established.

An ovarian structure that appears in at least some pinnipeds is the hilar rete (Boyd, 1984). These grandular cells are most abundant during delayed implantation, but their function is not clear. They also occur in a number of terrestrial mammals, including carnivores, primates, rodents, and hyraxes.

The oviduct, uterus, and vagina are all derivatives of the Müllerian duct system. The oviduct, fallopian tube, or uterine tube is generally highly convoluted and is enclosed in the ovarian bursa. The anterior end forms a funnel or infundibulum near the ovary and the posterior end of the oviduct enters the uterus. In eutherian mammals with a bicornuate uterus (e.g., cetaceans), the isthmus can be straight or convoluted but it has a thick wall and a narrow lumen. Dugong uterine tubes lack a mesosalpinx and exist as a 4-cm-long cord-like convoluted tube that lies dorsal to the peritoneum. Generally, the oviduct is lined by simple columnar epithelial that are ciliated and have occasional goblet cells. There is an inner circular layer and an outer longitudinal layer.

The uterus classification scheme in mammals is based on progressive fusion of the caudal ends of the oviducts. Four major types are recognized (Table I, Fig. 2), of which three types are represented in marine mammals; no marine mammal has a simplex uterus. All types of uteri are supported by broad ligaments, have two oviducts, and deliver into a single vagina.

The uterine wall has three layers: on the outside, the serous membrane; in the middle, the myometrium, which contains the internal circular muscle and the external longitudinal muscle separated by the vascular layer; and the inner lining of the uterus, the endometrium, composed of an epithelial lining of the lumen, a glandular layer, and some connective tissue. All uterine types exhibit changes in the layers of the uterine wall that precede implantation and development of the placenta. During the luteal phase of the follicle (postovulatory) the endometrium increases in thickness and the glands become extremely branched and convoluted.

Embryonic membranes develop before the embryo implants. The yolk sac forms first as endoderm surrounds the nutrient uterine fluid. The chorion and amnion develop as a double layer, originating from embryonic ectoderm. The innermost layer is the amnion, which forms the fluid-filled amniotic sac encompassing the fetus. The outer layer of cells of this sac wall

TABLE I
Uterine Types

Type of uterus	Uterine horns	Cervix	Example
Duplex	Two completely separate horns	2	Walrus
Bipartite	Two horns separated internally by a septum but sharing a small common area near the cervix	1	Phocids
Bicornuate	Two horns with no internal septum, forming a single body of the uterus	1	Cetaceans, sirenians, mustelids, ursids
Simplex	No horns, one uterine body without compartments	1	Humans

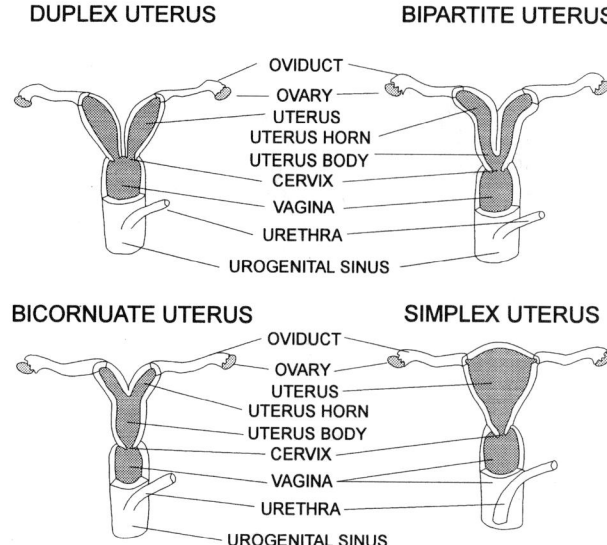

DUPLEX UTERUS **BIPARTITE UTERUS**

OVIDUCT
OVARY
UTERUS
UTERUS HORN
UTERUS BODY
CERVIX
VAGINA
URETHRA
UROGENITAL SINUS

BICORNUATE UTERUS **SIMPLEX UTERUS**

OVIDUCT
OVARY
UTERUS
UTERUS HORN
UTERUS BODY
CERVIX
VAGINA
URETHRA
UROGENITAL SINUS

Figure 2 *Uterus types found in placental mammals. The simplex type is not found in marine mammals. Modified from Romer (1962).*

is the chorion. The allantois develops as an extension of the embryonic hindgut and forms the allantoic cavity. The allantois fuses with the chorion, forming a small round area, the allantochorion. This region becomes the placenta (Fig. 3).

The embryonic membranes persist after the placenta is fully developed. In whales, the amniotic sac and allantois extend into the contralateral uterine horn. The chorion has many folds that mesh with the uterine mucosa. In pinnipeds, the chorion extends beyond the zonary placenta. The large allantois almost completely surrounds the amnion. In sirenians, the allantois is also large and nearly fills the chorionic sac.

The placenta is composed of maternal and embryonic tissue in close union. It allows nutritional, respiratory, and excretory exchange between maternal and fetal circulatory systems by diffusion across the placental membranes. Additionally, the placenta functions as a protective barrier to bacteria and other

large molecules, produces some food materials, and synthesizes hormones required to maintain the pregnancy. The umbilical cord connects the placenta to the ventral surface of the embryo and is formed of mesoderm and blood vessels. This connection is broken at birth.

There are three types of placentas seen in mammals, defined by the fetal tissue that adheres to the uterine wall. They are the chorionic placenta, the yolk-sac placenta, and the chorioallantoic placenta, which is found in all marine mammals. The chorioallantoic placenta is the most advanced placenta type in its ability to provide rapid diffusion between uterine and fetal circulatory systems. As the blastocyst implants and sinks into the uterine endometrium, chorionic villi grow quickly and push further into the endometrium. This process is accompanied by a breakdown of uterine tissue. The degraded debris is called embryotroph. The blastocyst absorbs the nourishing embryotroph until the villi are fully developed and the embryonic vascular system becomes functional.

Although all placentas are derived from both maternal and fetal tissues, the degree of separation of maternal and fetal circulatory systems is variable and is a function of the type of placenta that has developed. Chorioallantoic placentas can be further subdivided based on the fetal and maternal cell layers that are in contact. In epitheliochorial placentas (cetaceans), the epithelium of the chorion is in contact with the uterine epithelium. The villi rest in endometrial pockets. In endotheliochorial placentas (pinnipeds, mustelids, ursids), degradation of the maternal endometrium is more pronounced and the epithelium of the chorion is in contact with the endothelial lining of the uterine capillaries. Hemochorial placentation (sirenians) is characterized by destruction of the endothelium of the uterine blood vessels, allowing blood sinuses to develop in the endometrium. There is direct contact therefore between chorionic villi and maternal blood. Hemoendothelial placentas are not found in marine mammals but occur in some rodents and in lagomorphs.

Placenta shape is characterized by the pattern in which villi are distributed over the chorion. Two shapes occur in marine mammals: diffuse and zonary (Fig. 3). In the diffuse placenta (cetaceans), villi occur over the entire chorion, providing a

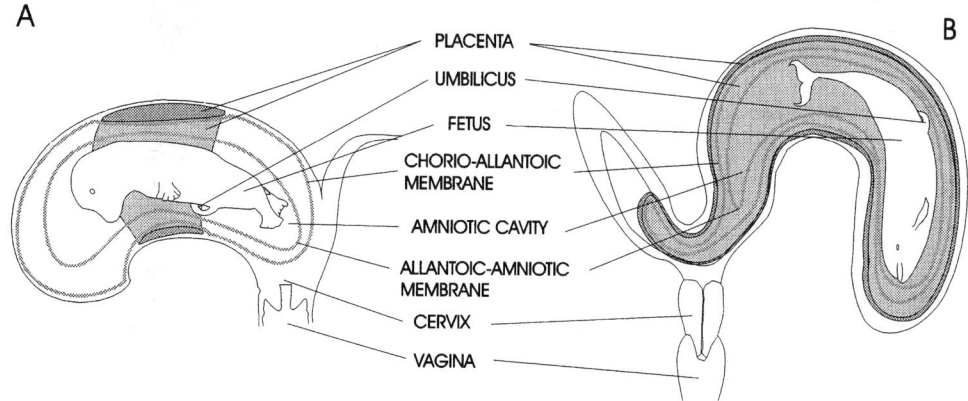

A   B

PLACENTA
UMBILICUS
FETUS
CHORIO-ALLANTOIC MEMBRANE
AMNIOTIC CAVITY
ALLANTOIC-AMNIOTIC MEMBRANE
CERVIX
VAGINA

Figure 3 *Zonary (A) and diffuse (B) placentas found in marine mammals [B modified from Slijper (1966), copyright University of California Press.].*

large surface area for exchange. In a zonary placenta (all other marine mammals), there is a continuous band of villi covering the equator of the chorion. In walrus and sirenians, the zonary placenta leaves detectable scars on the uterus. Both diffuse and zonary placentas are also found in terrestrial mammals.

When the placenta is expelled from the uterus postpartum as afterbirth, a maternal component may be lost. Epitheliochorial placentas have villi that pull out of the uterine pits easily and no endometrium is pulled away. Therefore, no bleeding occurs at birth and this placenta is referred to as nondeciduous. The other types of placentas allow for closer association of maternal and fetal circulatory systems through degradation of the endometrium and extensive intermingling of uterine and chorionic tissue. At birth then, part of the uterine component of the placenta is torn away and bleeding occurs. These are termed deciduous placenta. Bleeding is arrested quickly by collapse of the uterus, contractions of the myometrium constricting blood vessels, and blood clotting. The subsequent development of uterine scars at the bleeding site can result in persistent features that can be used in the interpretation of reproductive history in a female (e.g., walrus; Fay, 1982), although their persistence is variable among species.

The cervix is a well-muscled sphincter that marks the transition between the uterus and the vagina. The West Indian manatee has a rounded cervix. In the dugong, the long, thin-walled vagina has a keratinized shield originating in the vault region. This shield surrounds the cervix and extends along the ventral wall of the vagina. In cetaceans, the cervix is long with a thick wall and a narrow, sinuous lumen. The portion of the cervix projecting into the vagina (portio vaginalis uteri) is species specific in length, e.g., very short in the narwhal (*Monodon monoceros*) but long in the harbor porpoise (*Phocoena phocoena*). Vagina length in pinnipeds equals or slightly exceeds the urogenital canal in length.

Many cetaceans have several folds in the upper part of the vagina that are not found in any other mammal. There is a valve-like arrangement to these 4–12 circular folds, which project distally and look like a chain of funnels with their mouths directed toward the cervix. This configuration may act to retain sperm, but the function of these folds is unclear. Harrison (1969) noted that the folds appeared capable, anatomically, of a pumping action and speculated that they may relate to the formation of vaginal plugs. Vaginal plugs are congealed masses of semen that occur in some mammalian orders. They are thought to assist sperm retention and discourage subsequent matings by competing males. There is some suggestion of vaginal plugs in *Tursiops* and *Delphinus* but little other evidence. Vaginal calculi are masses of organic and inorganic material that have been found in *Delphinus*, *Stenella*, and *Lagenorhynchus*. Calculi contain material identical in composition to mammalian bone and some contain recognizable embryonic bones (Perrin *et al.*, 1984). "Vaginal plugs" therefore may represent small or deteriorating vaginal calculi.

The remainder of the female reproductive system consists of the clitoris, urethra, and vaginal vestibule. For all pinnipeds, the presence of the os clitoridis has been recorded, but it is generally small (<1 cm) and its appearance is irregular. In cetaceans, the clitoris projects from a strand of fibrous connective tissue on the anterior border of the vulva. Dugongs possess a clitoris with a large conical glans that has prominent fissures dividing it into lobes.

The urethra also opens into the vaginal vestibule, draining the bladder, which lies ventral to the vagina. The urethra in pinnipeds opens by a large urinary papilla just caudal to the hymen. Pinnipeds have long urogenital canals compared to other mammals and there is a prominent hymenal fold, which is smaller in otariids. Dugongs have a urethra surrounded by prostate-like tissue, which also surrounds the narrow distal portion of the vagina.

In pinnipeds, the vaginal vestibule opens to the exterior just ventral to the anus in a common furrow. In cetaceans, the vulva is a slit-like aperture with labia majora and labia minora, which may be poorly developed. The slit is positioned just anterior to the anus.

Considerable variation exists in mammary gland configuration and nipple location in marine mammals. Phocids have distinct mammary glands enclosed in connective sheaths lying under the blubber layer. In otariids, the mammary glands coalesce to form a sheet-like layer under the blubber over most of the ventral surface of the body. Cetacean mammary glands are elongate, narrow, flat organs that extend in the subcutaneous connective tissue at both sides of the ventromedial line. They extend from a little posterior of the umbilicus to slightly anterior to the anus.

Cetaceans have two nipples in elongated recesses, one on either side of the midline in the urogenital slit. The nipples become protruded during nursing, and the milk is expelled under pressure, likely due to contraction of either the cutaneous muscles or the myoepithelial cells surrounding the alveoli. Sirenians have two pectoral nipples. Sea otters have only two functional teats on the lower abdomen, compared to six or more in most other mustelids. Polar bears have four teats, two on either side of the midline of the belly slightly posterior of the axillae and two about 15 cm posterior to the anterior pair. Otariids, odobenids, and two genera of phocids (*Erignathus* and *Monachus*) have four teats while the other phocids have two that correspond to the posterior pair of those in otariids. Pinniped nipples are retracted beneath the level of the body surface when a pup is not nursing and become erect during suckling.

II. Reproductive Cycle

In marine mammals, ovulation can be either spontaneous or induced. Spontaneous ovulators release an egg even in the absence of breeding (phocids, cetaceans). Induced ovulators release an egg only in response to coital stimulation (polar bear, sea otter).

The ova normally are fertilized in the oviduct within 24 hr of breeding and the zygote's first cell divisions occur there. At the 8–16 cell stage, movement to the uterus occurs with assistance of muscle contractions in the oviduct. There, further differentiation into the blastocyst and subsequent implantation take place.

Implantation of the blastocyst may be immediate (within 1–2 weeks) or after some protracted delay of several months. This delay, also know as embryonic diapause, occurs in carnivores and pinnipeds and is considered to be obligate. Its duration is species specific. Facultative delayed implantation can occur in other species of mammals if a female is nursing a large litter at the time of insemination (e.g., some marsupials, some insectivores). The length of active gestation is generally related to the body size of the female, and a delay in implantation is thought to allow the young to be born at an advantageous time. For example, harp seals (*Pagophilus groenlandicus*) have a delay of about 3 months followed by an active gestation of about 8.5 months. The delay produces a cycle that is nearly 1 year long, allowing births and breeding to occur during large spring aggregations when pack-ice conditions are suitable. There may be some flexibility in the duration of the delay, and in sea otters, this flexibility may lead to the variation seen in estimates of total gestation, ranging from 6 to 8 months.

Obligate-delayed implantation is characterized by ovulation, fertilization, and differentiation up to the blastocyst stage, which creates a hollow ball of 100–400 cells surrounding a fluid-filled cavity. Further differentiation of the blastocyst then stops and implantation in the endometrium does not occur. The blastocyst is free-floating in the uterus and is covered by a zona pellucida, a noncellular protective layer, for the period of dormancy. Resumption of blastocyst differentiation occurs prior to implantation. During delayed implantation, pregnancy is indicated, macroscopically, by the presence of a corpus luteum and an increased diameter of a uterine horn, which shows marked surface vascularization, smoothening of the endometrial folds, and development of a nidation chamber where implantation will occur.

Usually, implantation of the blastocyst occurs in the uterine horn corresponding to the active ovary (ipsilateral). However, in odontocetes with only one active ovary, there is a tendency for transuterine migration of the blastocyst, which will implant in the other (contralateral) horn. Fetal membranes project into the horn opposite the implantation site. In mysticetes, and most other mammals, there is a slight prevalence (60%) of implanted fetuses in the right horn, reflecting a similar rate of ovarian activity in the right ovary (60%) (Slijper, 1966).

Once implanted, the blastocyst begins to differentiate tissues and organs, and remains in the uterus for the duration of its fetal phase. Nourishment and protection *in utero* allow for relatively high survival rates of the fetus.

At birth, powerful and rhythmic contractions of the uterine myometrium aided by the abdominal muscles expel the fetus. Continued contractions force the placenta from the uterus and vagina. In cetaceans, birth underwater must be rapid to prevent drowning of the neonate. The newborn swims unaided or is pushed to the surface by its mother or attendants to breathe for the first time. Other birthing platforms include land-fast ice (ring seal, *Pusa hispida;* Weddell seal, *Leptonychotes weddellii*), pack ice (harp seal, crabeater seal, *Lobodon carcinophaga*), and terrestrial sites (harbor seal, *Phoca vitulina;* polar bears; otariids). Few births have been observed in sea otters and may take place both on shore and in the water. Siren-ian births are also rarely seen but are thought to occur in shallow water, although there is some evidence they may also calve on low sand bars.

Cetacean and pinniped neonates have relatively large body sizes, approximately 8–10% maternal weight compared to mammals in general (Slijper, 1966; Kovacs and Lavigne, 1992), a benefit to these animals that need to swim at birth or shortly afterward and to maintain homeothermy in cold water. Most marine mammals usually give birth to a single offspring. Twin live births are exceedingly rare in pinnipeds and the sea otter, and have never been documented in cetaceans, although multiple fetuses have been observed. Some twins have been reported among sirenians. It is thought that multiple births are incompatible with the production of newborns that are large relative to maternal body size, as is found in these animals. It may also be difficult for a marine mammal mother to properly tend more than one offspring in the marine environment.

Multiple births are the norm in polar bears; however, ursids have extremely small young, relative to adult female body weight, that are born in dens and emerge with mother after considerable time when substantial postnatal growth has occurred. Most ringed seals and Baikal seals (*Pusa siberica*) are also born in dens, excavated in snow drifts from a hole scratched in the sea ice by the mother.

All marine mammals suckle their young with milk exclusively before a transition to solid food items and complete weaning is made. The period of lactation is again species specific and can be relatively short (4 days in hooded seals, *Cystophora cristata;* 10–12 days in harp seals) or more protracted (up to 2 years in sirenians, some cetaceans, walrus), although the young may start to eat solid food before weaning is completed. Marine mammal milk is typically high in fat (40–50%), high in protein (7–19%), and low in lactose (trace–5%) compared to terrestrial mammals.

Milk may be forcefully ejected from the teats or may be sucked from the teats by the neonate. In cetaceans, forceful ejection of milk is required because neonates cannot suck with their lips and must hold their breath during underwater nursing bouts. Indeed, because young *Tursiops* can only remain underwater for less than a minute, they nurse two or three times an hour over an entire 24-h period. Walruses may suckle young in water, and the teats are surrounded by sphincter-like folds of skin, which suggest that milk may be squirted into the calves' mouth. Young sea otter pups nurse while lying on the female's chest, whereas older ones lie in the water perpendicular to her. Sirenian calves nurse at the surface with their nostrils in the air or just below the surface. All other marine mammals nurse their young on ice or on land and neonates actively suck milk.

Weaning can be abrupt by abandonment (most phocids and otariids) or extend over some time (walrus, sirenians). The extended care of the young during lactation, and sometimes beyond in a period of learning, further increases survivorship of the offspring beyond the high rate of fetal survival. It also increases the efficiency of reproduction in that maternal energy expended toward young results in a high rate of offspring that reach reproductive maturity, consistent with other *K*-selected life history traits.

Age of maturation varies by species and, within a species, can be influenced by environmental factors that affect growth and fattening. Breeding success is often lower in younger breeders, but lifetime reproductive success and the number of descendants produced of those that breed young and survive can be high. Diminished reproductive frequency (reproductive senescence) has been described for some marine mammals (walrus, polar bear, some fur seals and some cetaceans). Although short-finned pilot whales (*Globicephala macrorhynchus*) of advancing age become senescent with no follicular activity, they may still lactate, nursing not only their own previous young but possibly also other young in the pod.

See Also the Following Articles

Endocrine Systems ■ Estrus and Estrous Behavior ■ Male Reproductive Systems ■ Pelvic Anatomy ■ Prenatal Development in Cetaceans ■ Reproductive Behavior

References

Berta, A., and Sumich, J. L. (1999). "Marine Mammals: Evolutionary Biology." Academic Press, San Diego.

Boyd, I. L. (1984). Occurrence of hilar rete glands in the ovaries of grey seals (*Halichoerus grypus*). *J. Zool. (Lond.)* **204,** 585–588.

Boyd, I. L., Lockyer, C., and Marsh, H. D. (1999). Reproduction in marine mammals. *In* "Biology of Marine Mammals" (J. E. Reynolds III and S. A. Rommel, eds.), pp. 218–286. Smithsonian Institution Press, Washington, DC.

Fay, F. H. (1982). Ecology and biology of the Pacific walrus *Odobenus rosmarus divergens* Illiger. North American Fauna, Number 74, Department of the Interior, Washington, DC.

Harrison, R. J., Brownell, R. L., Jr., and Boice, R. C. (1972). Reproduction and gonadal appearances in some Odontocetes. *In* "Functional Anatomy of Marine Mammals" (R. J. Harrison, ed.), Vol. 1, pp. 361–429. Academic Press, London.

Harrison, R. J., and King, J. E. (1965). "Marine Mammals." Hutchinson & Co., London.

Kovacs, K. M., and Lavigne, D. M. (1992). Maternal investment in otariid seals and walruses. *Can. J. Zool.* **70,** 1953–1964.

Leatherwood, S., and Reeves, R. R. (eds.) (1990). "The Bottlenose Dolphin." Academic Press, San Diego.

Penny, D. A., and Waern, R. (1965). "Biology, an Introduction to Aspects of Modern Biological Science." Sir Isaac Pitman (Canada) Limited, Toronto.

Perrin, W. F., Brownell, R. L., Jr., and DeMaster, D. P. (eds.) (1984). "Reproduction in Whales, Dolphins and Porpoises." Reports of the International Whaling Commission Special Issue 6. International Whaling Commission, Cambridge.

Romer, A. S. (1962). "The Vertebrate Body." Saunders, Philadelphia.

Slijper, E. J. (1966). Functional morphology of the reproductive system in Cetacea. *In* "Whales, Dolphins and Porpoises" (K. S. Norris, ed.), pp. 277–319. Univ. of California Press, Berkeley.

van Tienhoven, A. (1968). "Reproductive Physiology of Vertebrates." Saunders, Philadelphia.

Vaughn, T. A. (1972). "Mammalogy." Saunders, Philadelphia.

Zuckerman, L., and Weir, B. J. (eds.) (1977). "The Ovary," Vols. 1 and 2. Academic Press, New York.

Fetal Development

SEE *Prenatal Development*

Filter Feeding

DONALD A. CROLL AND BERNIE R. TERSHY
University of California, Santa Cruz

I. Filter Feeding and the Marine Environment

A fundamental necessity for any organism is acquiring sufficient food for maintenance, growth, and reproduction. This search for food likely drove the return of mammals to the ocean where they were able to exploit highly productive coastal waters. With their return to the sea, marine mammals evolved a number of foraging techniques. Filter feeding, found in mysticete whales and three species of pinnipeds (crabeater seals (*Carcinophaga lobodon*), leopard seals (*Hydrurga leptonyx*), and Antarctic fur seals (*Arctocephalus gazella*), is the most unique of these adaptations for feeding and is not found in any terrestrial mammals.

Filter feeding allows these marine mammals to exploit extremely abundant but small schooling fish and crustaceans by taking many individual prey items in a single feeding event. This adaptation arose in response to the unique patterns of productivity and prey availability in marine ecosystems.

Low-standing biomass and high turnover of small-sized primary producers that respond rapidly to nutrient availability characterize marine food webs. Due to spatial differences in the physical dynamics of marine ecosystems, productivity tends to be more patchy and ephemeral than in terrestrial systems. Consequently, marine grazers (e.g., schooling crustaceans and fish) often occur in extremely high densities near patches of high primary production. Most marine mammals are primary carnivores and feed on these dense, patchily distributed aggregations of schooling prey. The spatial and temporal patchiness of this prey means that marine mammals must often travel long distances to locate prey, and the larger body size of marine mammals likely plays an important role.

Initially, thermoregulatory requirements selected for larger body sizes as mammals returned to the ocean. However, once dependent on marine prey, a large body size also provided a buffer for the patchy and ephemeral distribution of marine prey. Thus, larger individuals could endure longer periods and travel longer distances between periodic feeding events on patchy prey. While adaptive for exploiting patchy prey resources, a consequence of larger body size is a higher average daily prey requirement. For marine mammals that feed on patchy and ephemeral resources, this requires individuals to take in large quantities of prey during the short periods of time it is available.

Filter feeding is a foraging strategy that allows individuals to capture and process large quantities of prey in single feeding events, thus allowing them to acquire energy at high rates when small prey are aggregated. Indeed, for mysticetes, a large body size is probably a prerequisite for attaining a sufficiently large surface area for filter feeding. Thus, the interaction of availability of prey resources, high concentrations of prey in schools, and selection for large body size likely led to the evolution of filter feeding. Ultimately, a large body size and filter feeding allowed some marine mammals to exploit the extremely high densities of schooling prey that develop at high latitudes during the spring and summer, while fasting during the winter when these resources disappear. A large body size provided an energy store for wintering and long-distance migration without feeding.

Due to this dependency on patchy but extremely productive food resources, it is not surprising that filter-feeding whales are believed to have first evolved and radiated in the Southern Hemisphere during the Oligocene at the initiation of the Antarctic circumpolar current (ACC). It is generally agreed that the initiation of the ACC led to cooling of the southern oceans, increased nutrient availability, and thus increased productivity. This increased productivity provided a rich resource of zooplankton that could be exploited effectively through filter feeding.

Present-day filter-feeding marine mammals concentrate their foraging in polar regions and highly productive coastal upwelling regions. The Southern Ocean is still the most important foraging area for filter-feeding marine mammals. Prior to their exploitation by humans, the highest densities of mysticetes occurred in highly productive southern waters. Crabeater seals, Antarctic fur seals, and leopard seals are found primarily in the southern oceans where seasonally dense aggregations of krill develop.

II. Diet, Filter-Feeding Structures, and Prey Capture

All filter-feeding species feed on prey that form dense aggregations (primarily pelagic schooling fish and crustaceans or densely aggregated benthic amphipods). Two feeding adaptations have evolved to allow the exploitation of these dense aggregations: baleen (mysticete whales) and modified dentition (seals).

A. Seals: Diet, Feeding Morphology, and Behavior

Unlike mysticetes, pinnipeds evolved in the Northern Hemisphere where krill was not likely an important component of their DIET. As a result, adaptations for filter feeding are not nearly as extensive in pinnipeds as in mysticetes.

Only three pinniped species regularly filter feed: crabeater seals, leopard seals, and Antarctic fur seals. When filter feeding, all three species feed almost exclusively on Antarctic krill, *Euphausia superba,* in the Southern Ocean where it is abundant and forms extremely dense aggregations. Of the three pinniped species, crabeater seals are the most highly specialized, with krill comprising up to 94% of their diet, whereas krill comprises approximately 33% of the diet of leopard seals and Antarctic fur seals. The most remarkable adaptation for filter feeding in pinnipeds is found in the dentition of crabeater and leopard seals. In both species, elaborate cusps have developed on the postcanines in both the upper and the lower jaws (Fig. 1). Once the mouth closes around a small group of krill, water is filtered out through the cusps, trapping krill

against the insides of the modified teeth. Little detailed information is available on the behavior used by filter-feeding pinnipeds to capture prey. However, data from Antarctic fur seals indicate that they track the diel migration of krill: shallow dives are performed during the night and deeper dives during the day.

B. Mysticetes

1. Diet and Feeding Morphology Most mysticetes feed primarily on planktonic or micronectonic crustaceans (copepods and krill) and pelagic schooling fish found in shallow waters. The gray whale diet consists primarily of benthic apeliscid amphipods, although they can forage on a wide variety of prey, including schooling mysids in some areas. Right (*Eubalaena* spp) and bowhead (*Balaena mysticetus*) whales primarily feed on copepod crustaceans of the genus *Calanus*. All of the rorquals feed on euphausiids (krill) to some extent, and blue whales (*Balaenoptera musculus*) feed almost exclusively on euphausiids. The other rorquals have a more varied diet that includes copepods (sei whales—*B. borealis*) and schooling fish (minke—*B. acutorostrata* and *B. bonaerensis*, Bryde's—*B. edeni,* humpback—*Megaptera novaeangliae,* and fin whales—*B. physalus*).

All mysticetes lack TEETH and instead have rows of baleen plates made of keratin that project ventrally from the outer edges of the palate. Similar to fingernails, the plates grow continuously from the base, but are worn by the movements of the tongue. As the edges of the plates wear, hair-like fibrous strands emerge as fringes. The outer fibers of these fringes are coarser, whereas the inner fibers form a tangled fringe that overlaps with fringes on adjacent baleen plates. Rows of baleen plates form an extended filtering surface along each side of the palate.

The coarseness of the hair-like fibrous fringes, the density of fibers (number of fibers/cm^2), the number of baleen plates, and the length of baleen plates vary among species and are related to the prey species captured in the filtering mechanism. Because gray whales (*Eschrichtius robustus*) feed primarily on sediment-dwelling benthic amphipods, they have the coarsest filtering mechanism, made up of about 100, 5–25 cm-long individual plates with very coarse fibers. This coarse filtering structure allows them to separate amphipods from bottom sediments. In contrast, right whales that feed on small copepods have a fine filtering mechanism composed of more than 350 baleen plates that can exceed 3 m in length. The fibers of right whale baleen are very fine, forming a dense mat capable of capturing copepods that are less than 5 mm long. The strong, flexible, and light characteristics of baleen plates made them commercially important in the 19th century where they served some of the roles of today's plastics.

Mysticetes have evolved three types of filter feeding: sediment straining (gray whales), skimming (right and bowhead whales), and lunging or gulping (rorquals). The morphology of mysticetes reflects these different strategies. Gray whale heads are straight and relatively short and contain short, coarse baleen and their throat regions possess only a few grooves (three to five) in the gular region that allows limited distention for taking in bottom sediment, water, and amphipods. Right and bowhead

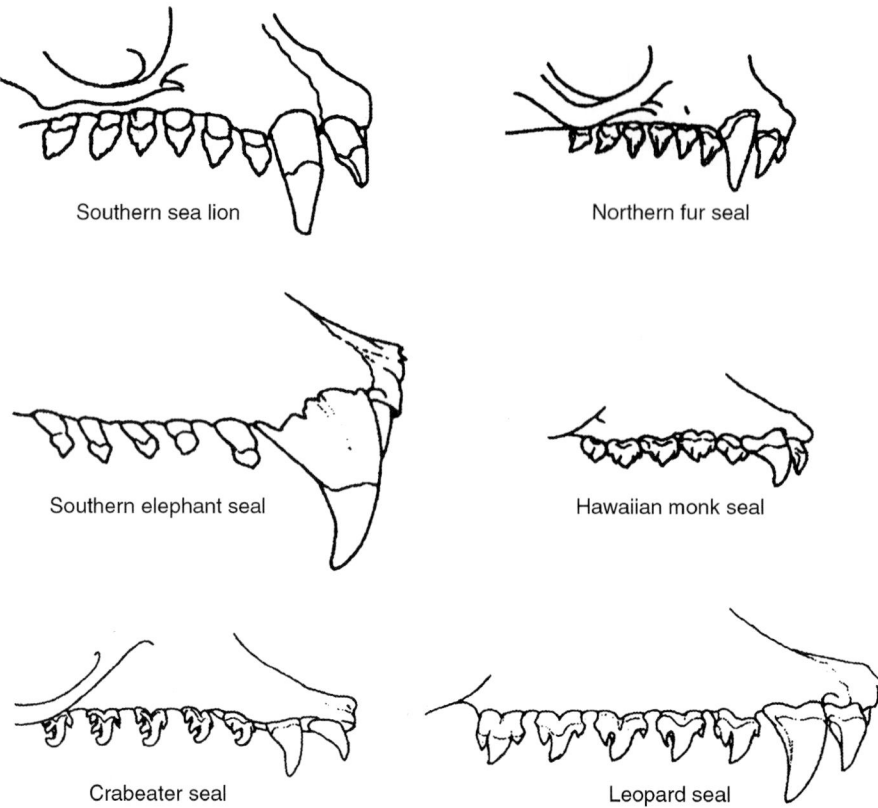

Figure 1 *Dentition patterns in pinnipeds. Note modified cusps in postcanine teeth in filter-feeding crabeater and leopard seals. From Berta and Sumich (1999).*

whale heads have a strongly arched rostrum that allows them to have very long and fine-textured baleen within a relatively blunt mouth. They have no throat grooves for distension and instead feed by swimming slowly and skimming prey from the water. Rorqual heads are large and contain enormous mouths that extend posteriorly nearly half of the total body length. Their mouths contain relatively short baleen that ranges from fine (sei whales) to medium texture (blue, fin, humpback, and minke whales). The heads and bodies of rorquals are much more streamlined than other mysticetes, allowing them to swim rapidly into a prey school to gulp large quantities of water and schooling prey. One of the most remarkable adaptations for feeding is the presence, in rorquals, of 70–80 external throat grooves. During gulping, these grooves open like pleats to allow the mouth cavity to expand up to four times in circumference, taking in a volume of water equivalent to about 70% of the animals' body weight.

2. Feeding Behavior Observations of feeding gray whales in the Arctic and Bering Sea have shown that the whales roll to one side and suck benthic invertebrate prey and bottom sediments, with some distension of the mouth cavity through the expansion of the throat grooves. Water and mud are expelled through the side of the mouth. A similar behavior is used by gray whales that do not migrate as far north where they feed on a variety of benthic invertebrates and schooling myscids.

This benthic foraging behavior creates scrapes of about ½ m deep and 1 by 5 m in shape in the ocean floor, and several studies have shown that the disturbance is an important factor in the ecology of soft-bottom benthic communities of the Arctic and Bering Seas.

Right and bowhead whales forage by skimming with their mouths open through concentrations of crustaceans. As the whale swims, water and prey enter through the front of the mouth and water exits along the sides of the mouth while prey are trapped in the fine baleen (Fig. 2). Whereas right and bowhead whales generally feed singly, at times they may feed alongside one another—a V formation of 14 bowhead whales has been observed.

Rorqual lunge feeding has been described as the largest biomechanical event that has ever existed on earth. Rorquals capture food by swimming rapidly at a prey school and opening the mouth to gulp vast quantities of water and schooling prey (Fig. 3). To maximize the opening, the lower jaw opens to almost 90° of the body axis. This is possible because the lower jaw has a well-developed coronoid process. This process is located where the large temporalis muscle inserts and provides an anchor and mechanical advantage for control of the lower jaw while maximizing the gape for prey capture. It is not developed in other whale species, and a tendinous part of the temporalis muscle, the frontomandibular stay, enhances and strengthens the mechanical linkage between the SKULL and the lower jaw.

Figure 2   *Skim feeding in right and bowhead whales. From Berta and Sumich (1999).*

With the mouth open, the onrush of water and prey is accommodated by the distending ventral pleats. The tongue invaginates to form a hollow sac-like structure (cavum ventrale) that lines the inside of the gular region and the ventral pleats distend fully. After engulfing entire schools of prey, the lower jaw is closed. The muscular tongue and the elastic properties of the ventral walls of the throat act in concert to force water out through the baleen (Fig. 3).

Although the process just described is fundamentally the same in all rorquals, some species exhibit modifications and additional adaptations. Sei whales skim feed in a manner similar to right whales, as well as feeding by lunging. Fin and blue

whales often feed in pairs or trios that have a consistent echelon configuration. Humpback whales have a diverse diet and a wider variety of feeding behaviors. They have been observed bottom feeding and, while feeding on schooling fishes, have been observed to produce a cloud of bubbles and feed cooperatively to assist in prey capture.

Laboratory experiments have shown schooling fish to react to bubbles by aggregating more densely. Humpback whales appear to take advantage of this, as one member of a group of foraging whales produces a net of bubbles. The bubble cloud serves to aggregate and confuse the prey. Members of the group dive below the bubble cloud and surface together—one whale immediately

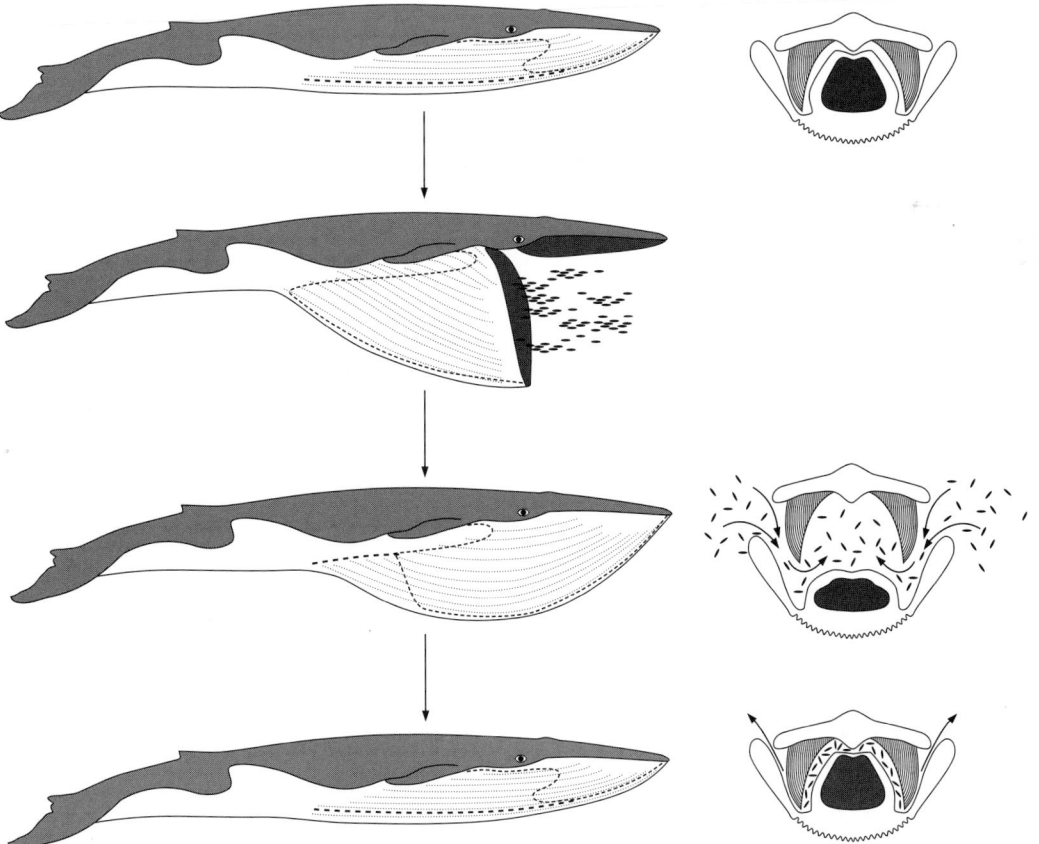

Figure 3   *Lunge feeding in rorqual whales demonstrating expansion of the throat pleats in invagination of the tongue. From Berta and Sumich (1999).*

adjacent to another. Group foraging humpbacks form long-term associations and the location of the whales in the surfacing group appears to be fairly constant through time. Humpbacks thus likely enhance prey capture success by using both bubbles and foraging cooperatively. A variation of bubble cloud feeding has been observed in humpback whales feeding on sand lance off New England. Here the bubble cloud feeding is followed by a tail slap—believed to cause the sand lance to aggregate more densely.

3. Feeding Ecology   All filter-feeding whales exhibit distinct MIGRATION patterns linked to seasonal patterns in prey abundance. Seasonally dense aggregations of prey are probably necessary for successful filter feeding. For example, gray whales undergo the longest migration of any mammal: foraging during the summer and fall in the Bering Sea and Arctic Ocean when dense aggregations of benthic amphipods become available with the seasonal increase in productivity. Humpback whales seasonally migrate from breeding areas to higher latitude foraging areas where schooling fish and krill become seasonally abundant. The timing of coastal migration patterns of the California blue whale appears to be linked to annual patterns in coastal upwelling and krill development patterns.

Studies of the diving behavior and daily movement patterns of right whales have shown that they are found at dense aggregations of copepods, which in turn track oceanographic features such as fronts. Zooplankton densities in regions where right whales foraged in the southwestern Gulf of Maine were approximately three times the mean densities in the region (whale feeding densities averaged 3.1–5.9 g m^{-3}, compared to 1.1–3.6 g m^{-3} where whales were not foraging). In a related study using hydroacoustic surveys, zooplankton densities where right whales were foraging were 18–25 g m^{-3} (compared to 1–5 g m^{-3} where whales were not foraging). Whale diving behavior is related to the depth of prey aggregations. In a year when copepods did not undergo diel migrations, dive depths averaged 12 m, with no dives exceeding 30 m throughout the day and night. In contrast, in a year when copepods showed strong diel shifts in depth (near the surface at night, deeper during the day), whale dive depths were significantly longer during the day.

Rorquals also track seasonal and diel patterns in the abundance and behavior of their prey. In general, the distribution and movement patterns of most rorquals consist of a seasonal migration from high latitudes where foraging takes place to low latitudes where they mate and give birth. However, data from blue whales in the Pacific indicate that feeding also takes place at low-latitude, "upwelling-modified" waters, and data from both the Pacific and the Indian Oceans indicate that some blue whales may remain at low latitudes year-round. Fin and blue whales foraging on krill off the coast of North America concentrate their foraging effort on dense aggregations of krill that are deep (150–300 m) in the water column during the day and may cease feeding when krill become more dispersed near the surface at night.

Rorqual foraging appears to only occur in regions of exceptionally high productivity, often associated with fronts, upwelling centers, and steep topography. It has been estimated that fin whales require prey concentrations of at least 17.5 g m^{-3} to meet daily energy requirements. Krill densities where humpback whales were foraging in Southeast Alaska have been estimated at 910 individuals m^{-3}, and minimum required prey densities for humpbacks were about 50 individuals m^{-3}. Krill densities in schools where blue whales were foraging in Monterey Bay, California, were estimated at 145.3 g m^{-3} compared to an overall mean density of zooplankton of 1.3 g m^{-3} in the area.

III. Summary

Filter feeding in marine mammals is an adaptation that allows individuals to take in large quantities of prey in one mouthful. This is particularly adaptive in marine ecosystems where prey are relatively small and often densely aggregated, but patchy and ephemeral in space and time. Most filter-feeding species feed on schooling fish and crustaceans. The large body size of marine mammals, particularly mysticetes, facilitates filter feeding by providing the ability to have a large filtering area relative to body volume. In addition, a large body size likely provides an energetic buffer for animals that must move long distances between dense prey patches and endure long periods of fasting between foraging events.

See Also the Following Articles

Baleen ■ Crabeater Seal ■ Feeding Strategies and Tactics ■ Krill

References

Berta, A., and Sumich, J. L. (1999). "Marine Mammals: Evolutionary Biology." Academic Press, San Diego.
Bowen, W. D., and Siniff, D. B. (1999). Distribution, population biology, and feeding ecology of marine mammals. *In* "Biology of Marine Mammals" (J. E. Reynolds III and S. A. Rommel, eds.), pp. 423–484. Smithsonian Institution Press, Washington.
Clapham, P. J., and Brownell, R. L. (1996). Potential for interspecific competition in baleen whales. *Rep. Int. Whal. Comm.* **46**, 361–367.
Clapham, P. J., Young, S. B., and Brownell, R. L. (1999). Baleen whales: Conservation issues and the status of the most endangered populations. *Mamm. Rev.* **29**(1), 35–60.
Reidman, M. (1990). "The Pinnipeds: Seals, Sea Lions, and Walruses." Univ. of California Press, Berkeley.
Tershy, B. R. (1992). Body size, diet, habitat use, and social behavior of *Balaenoptera* whales in the Gulf of California. *J. Mammal.* **73**, 477–486.

Finless Porpoise
Neophocaena phocaenoides

MASAO AMANO
*Otsuchi Marine Research Center,
University of Tokyo, Japan*

I. Diagnostic Characters and Taxonomy

The finless porpoise is a small phocoenid cetacean lacking a dorsal fin as the common name suggests (Fig. 1). Instead of a dorsal fin, a ridge runs down the middle of the back. The species has a rounded head without an apparent beak.

Figure 1 *Mother and calf pair of Japanese finless porpoises in captivity (courtesy of the Minamichita Beachland Aquarium).*

Its color is uniformly dark to pale gray and somewhat lighter on the ventral side. Teeth are spatulate as in other phocoenids.

The finless porpoise was originally described by G. Cuvier as *Delphinus phocaenoides* based on a skull supposedly from the Cape of Good Hope. However, there have been no further records of this species from the west coast of Africa and the true type locality is now considered to be the Indian coast. The genus names *Neomeris*, which is not valid since preoccupied by a polychaete, and *Meomeris*, which is a misspelled name and not available, were formerly used (Rice, 1999).

II. Distribution

Finless porpoises inhabit shallow coastal waters and some rivers in the Indo-Pacific region (Fig. 2). Their distributional range is from the Persian Gulf in the west, through the coasts

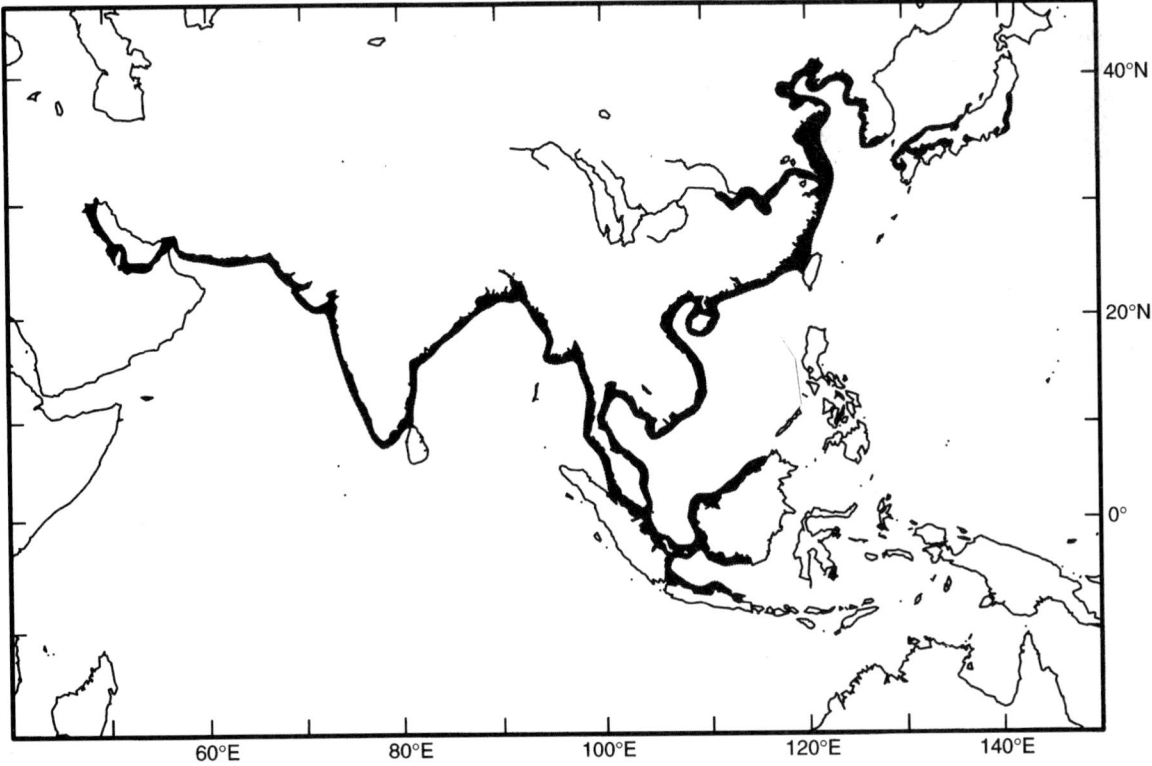

Figure 2 *Distributional range of the finless porpoise.*

Figure 1 *Lateral view of the right side of the fin whale showing asymmetric coloration of the cephalic region. Pieter A. Folkers/Higher Porpoise DG.*

be KRILL composed of the euphausiid *Meganyctiphanes norvegica,* although other species of planktonic crustaceans (*Thysanoessa inermis, Calanus finmarchicus*), schooling fishes such as capelin (*Mallotus villosus*), herring (*Clupea harengus*), mackerel (*Scomber scombrus*), and blue whiting (*Micromesistius poutassou*), and even small squid are also consumed. In the Southern Hemisphere, diet is almost exclusively krill, mostly the euphausiid *Euphasia vallentini* but also other planktonic crustaceans such as *Euphasia superba, Parathemisto gaudichaudii,* or *Calanus tonsus* in smaller proportions. Like other balaenopterids, the fin whale feeds in summer, when an adult

whale is estimated to consume up to 1 ton of euphausiids per day, and fasts in winter (see later).

Because the distribution range and the DIET of fin whales overlap with those of other balaenopterid whales, interspecific competition is likely to occur. This is especially likely in the case of the blue whale (*B. musculus*), which is often found forming mixed schools with fin whales. This association, together with the evolutionary proximity of the two species, makes blue fin hybrids relatively common (see earlier discussion).

Because they are large and swim fast, fin whales do not have significant predators, with the exception of the killer whale

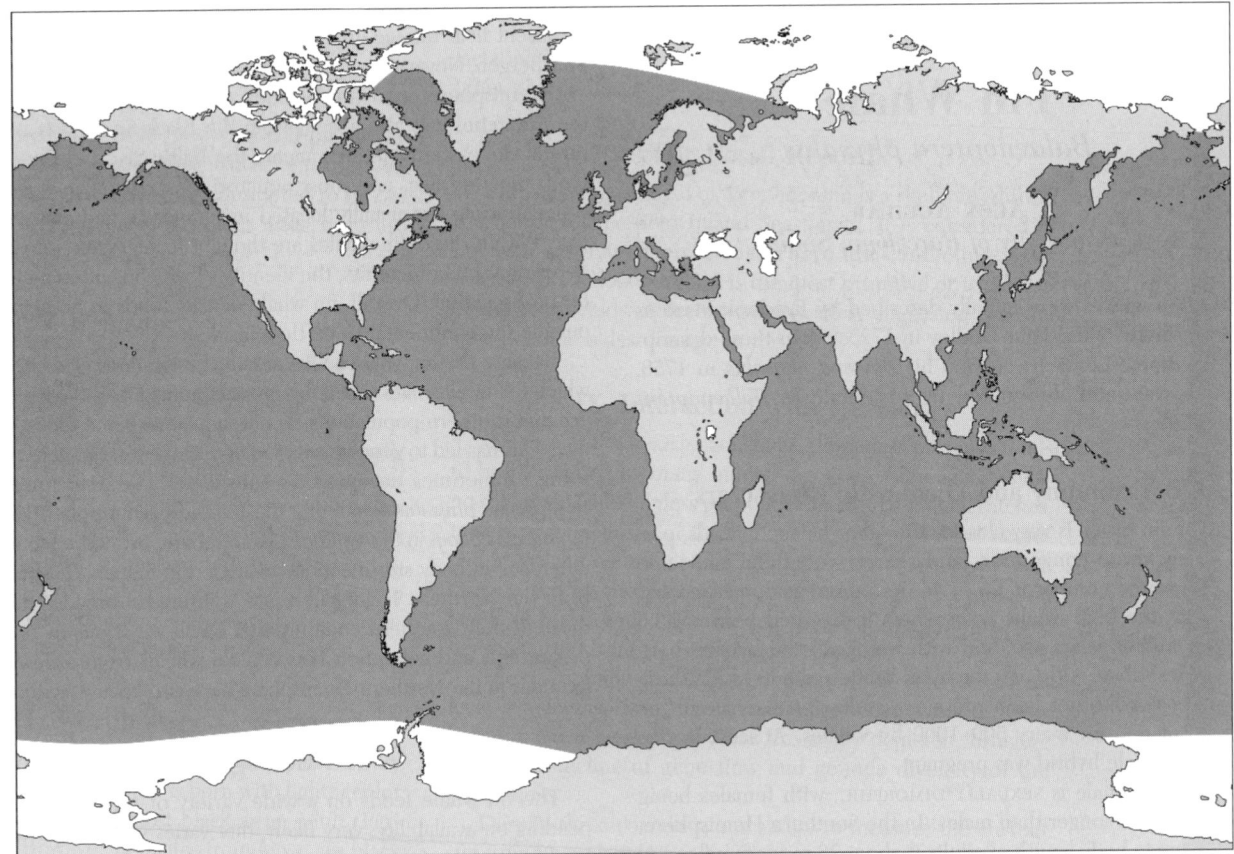

Figure 2 *Distribution of fin whales (dark shaded area).*

prey sp
marine
tional"
fish fro
gear. In
but tole
locales,
bers" a

Mar
diate hc
cial fish
codwor
grypus)
the mus
The res
Ecol
mamma
cial fish
fish spe
mercial
threater

Evid
nipeds,
to remo
actions
seines,
erations
penditu
ing fem:
Fishing
lect foo
wise av:
making
ficult (e
Feec
havior,
seeking
gested t
generati
Mari
during u
during fi
are dam
how the
Mari
fish, ren
tween 1
as in son
of cetac
(*Pseudo*
America

(*Orcinus orca*). In certain regions where this odontocete is abundant, signs of past attacks of killer whales can often be seen on the flippers, flukes, and flanks of fin whales.

IV. Behavior

In the Southern Hemisphere, fin whales engage in north–south seasonal migration, feeding in the summer and breeding and fasting in the winter. A similar latitudinal movement has been generally proposed for the Northern Hemisphere, although in many areas this pattern is not clear. For example, in the North Atlantic, although several summer grounds have been identified in medium to high latitudes (northern Morocco, Gibraltar Straits, northeastern Spain, Scotland, northern and western Norway, Newfoundland, the Faroe Islands, and Iceland), no definite wintering grounds are known. It has been suggested that in this ocean the latitudinal movements of the species are short due to the influence of the Gulf Stream, which would make the higher latitudes suitable as wintering grounds. An alternative explanation is that individuals that concentrate near the coast in the feeding season tend to disperse into open waters during the winter, therefore being more difficult to detect. Indeed, there is evidence that some whales remain in higher latitudes during the coldest months of the year. Also, fin whales may remain at lower latitudes year-round if food is available, as shown by the lack of seasonality of catches in the Gibraltar Straits (36°N).

The migration route apparently follows areas of low geomagnetic intensity and gradient and, similarly to other balaenopterids, not all components of the population move together. Pregnant females are usually the first to initiate the seasonal movement and are soon followed by adult males and resting females. Lactating females and juvenile individuals of both sexes are the last to migrate.

Compared to other large cetaceans, fin whales are relatively fast swimmers. Normal cruising speed is 5–8 knots but may increase up to 15 knots for short bursts. The SWIMMING is usually smooth, and BREACHING is very rare, except when the whale is harassed. Dives are limited to 100–200 m and usually last 3–10 min. The species is not gregarious and its only social bond appears to be that of cows with their nursing calves, a link that vanishes at weaning. School size is small; the fin whale usually swims single or in groups of two to seven, but transitory large aggregations may occur in highly productive areas.

The sounds of fin whales are simple and mostly consist of low-frequency moans and grunts and high-frequency pulses, apparently with a social function. Regional differences have been found between the Gulf of California and several Atlantic and Pacific Ocean regions. Fin whale moans are loud and can be heard for at least tens, probably hundreds of kilometers.

V. Anatomy

The body mass of adult individuals typically ranges from 40 to 50 metric tons in the Northern Hemisphere and from 60 to 80 metric tons in the Southern Hemisphere. A general formula for estimating body weight (W) from body length (L) is $W = 0.0015 L^{3.46}$. If the girth at the level of the navel (G) is available, a more precise formula is $W = 0.0469 G^{1.23} L^{1.45}$.

The relative mass of body tissues varies seasonally according to nutritive condition. Average mass relative to total body weight is $18.4 \pm 3.3\%$ for blubber, $45.3 \pm 4.4\%$ for muscle, $15.5 \pm 2.4\%$ for bone, and $9.8 \pm 2.1\%$ for viscera. The liver is large, usually weighing 230–600 kg. The heart is similar in relative size to that of terrestrial mammals but larger than in odontocetes and weighs 130–290 kg. Kidneys are large and weigh 50–110 kg. The right lung is about 10% heavier than the left, with each one weighing 100–160 kg. The spleen weighs 2–7 kg and sometimes has accessory bodies of smaller size.

The rostrum of the fin whale is sharply pointed, without the lateral curvature typical of blue whales. The zygomatic width is about 50–55% of the condylo-premaxillary length; the width of the rostrum at midlength is approximately 30–35% of its basal width, and the whole skull measures about 20–25% of the total body length. Ribs usually number 16 pairs, with the last pair being smaller and not attached to the vertebral column. The number of vertebrae ranges from 60 to 63, with a typical formula of C, 7; D, 15 (16); L, 14 (13–16); and Ca, 25 (24–27). The sternum is broad and variable in shape, usually in the form of a cross or a trefoil. As in other balaenopterids, scapulae are fan shaped with a convex upper margin. Flippers lack a third digit. The typical digit formula is I, 3–4; II, 6; IV, 5–6, and V, 3–4.

VI. Life History

Growth during the postnatal period is rapid, and 95% of the maximum body size is reached when whales are 9–13 years old. Males grow faster than females but stop growing sooner. Physical maturity, as determined by the degree of ossification of the vertebral column, is reached at about the age of 25 in both sexes. Longevity has not been determined precisely, but individuals of up to 80–90 years old are known.

Reproductive behavior is poorly known, but taking into account the reduced mass of their testes, their mating system probably relies on competition between males for monopolizing females and reduced sperm competition. In the Northern Hemisphere, sexual maturity is attained at an approximate length of 17.5 m in males and 18.5 m in females; corresponding figures for the Southern Hemisphere are 19 m in males and 20 m in females. These lengths typically correspond to an age of about 6–7 years in males and 7–8 years in females.

In the Northern Hemisphere, the mating period is December–February, whereas in the Southern Hemisphere it is May–July. Gestation lasts about 11 months, at the end of which a calf about 6–7 m long and weighing 1–1.5 metric tons is born. Only one calf is usually produced per pregnancy, although twinning has been reported. Weaning occurs when the calf is about 6–7 months old and measures 11–13 m long.

Weaning is followed by a 6-month resting period, at the end of which mating takes place. Therefore, the reproductive cycle is completed in about 2 years. Gross pregnancy rates (number of pregnant females in relation to that of adult females) are typically estimated at 38–49%.

VII. Interactions with Humans

Second in size after the blue whale, the fin whale has been much sought after by whalers; it was one of the target species

when mod
ploitation
ing came
large num
as early as
in abunda
cause of i
other area
tic, the Pa
ing the pe
largest pan
als taken a
tion far ex
one after a
    Fortun
by the In
and 1970s
a low level
the blue v
proved, la
taking plac
tions cease
the specie
10–15 indi
fishery.
    Apart f
threats hav
are uncom
sels somet
tions when
oceanic di
the food cl
icant impa
tremely lo
North Atla
levels of o
erage conc
12 ppm an
whales off
lower, ofte
heavy met;
though usu
North Atla
zinc, 0.2–2
ppm of me
    After th
ther stoppe
timated po
Antarctic s
1000 off W
Norway, 70
7500 in the
ern Medite
and Labrad

See Also

Blue Whale

Figure 1   *California sea lions often move into a purse seine net to start feeding before the net is pursed and continue to do so until all the fish are pumped aboard the fishing vessel. Courtesy of Jon Stern.*

being raised in aquaculture pens. The impact of pinnipeds on fisheries is of particular concern through depredation and gear damage during gill netting on the west coasts of North America, Japan, Britain, Scandinavia, and Chile; through depredation, net damage, and disturbance at fish farms in Britain and Chile; and depredation from trawls, depredation and gear loss from hand lines, and disturbance of purse seining in South Africa (Wickens, 1995). Estimates of the consumption by seals from fishing operations in South Africa show it to be a minimal percentage of fishery catches and a small proportion of the total predation by seals. Preliminary calculations of overall economic losses resulting from these seals' interference show this to be small (0.3%) in comparison with the wholesale value of the catches.

Some pinnipeds converge on areas where anadromous fish stocks aggregate or where the movements of fish are constrained naturally or artificially (bottleneck or "choke points" where salmonids aggregate in response to human-made structures or natural river physiography, such as fish ladders or below falls, respectively). The most thoroughly studied pinniped/salmonid conflict is California sea lion predation on winter steelhead (*Oncorhynchus mykiss*) at Washington State's Ballard Locks. The severe decline in salmon is considered primarily a direct result of human activities (Fraker and Mate, 1999); however, much concern has been voiced that the expanding populations of seals and sea lions may be causing a further decline (or impeding the recovery) of various salmon runs in the Pacific Northwest.

IV. Cetacean Interference with Fisheries

Like pinnipeds, cetaceans interact with a diversity of fisheries. Cetaceans may feed on a fishery's target species, such as killer whales feeding on sablefish (black cod *Anoplopoma fimbria*) in the North Pacific longline fishery, or on fish that are ancillary to the catch, as in the case of bottlenose dolphins feeding on bycatch from trawl fisheries for shrimp and prawn (Fig. 2). Whales and dolphins may interfere with traps or pots, preying on target species, as well as bait. Long-finned pilot whales (*Globicephala melas*) in Newfoundland frequent traps to remove the target species squid. Bottlenose dolphins in Belize have been observed retrieving fish from local, homemade fish traps, whereas this species in the Indian River Lagoon in Florida interacts with the crab pot fishery, apparently to steal bait fish. Bottlenose dolphins and "blackfish" (e.g., killer whales, false killer whales, pilot whales) are notorious fish stealers, and there are widespread reports of catch and gear damage by these species (Fig. 3).

V. Sirenian Interference with Fisheries

Fishermen in Jamaica and Sierra Leone have complained about damage caused to gill nets by "net-robbing" West Indian and West African manatees (*Trichechus manatus* and *T. senegalensis*), respectively. Manatees have been described as stripping the flesh off fish entangled in gill nets and leaving the bones.

VI. Toward Solutions

Fishermen use various means to deter marine mammals in an attempt to safeguard their catches and gear. Lethal methods have been attempted, including shooting at or killing the marine mammal with a variety of objects and methods, sometimes involving poison. Sometimes these practices are illegal. Seals have been persecuted much more intensely than cetaceans.

Culls or control programs for marine mammals have been carried out with the expectation that they will increase fishery yields by reducing marine mammal predation on fish stocks (Earle, 1996). One of the best-known culls occurred at Iki Island in Japan between 1976 and 1982. Thousands of bottlenose dolphins and hundreds of Pacific white-sided dolphins

Figure 2   *Bottlenose dolphins pluck fish from the meshes of shrimp trawl nets. Mesh from nets may be damaged in the process. Courtesy of Dagmar C. Fertl.*

Figure 3 *Killer whales depredate longline fisheries around the world. This killer whale is feeding on a longline-caught bluenose* (Hyperoglyphe antarchia) *in New Zealand. Courtesy of Ingrid Visser.*

(*Lagenorhynchus obliquidens*) false killer whales, and Risso's dolphins (*Grampus griseus*) were killed in response to declines in catches of yellowtail (*Seriola dorsalis*). The annual Canadian commercial hunt for harp seals (*Pagophilus groenlandicus*) and hooded seals (*Cystophora cristata*) in the Gulf of St. Lawrence and Newfoundland is conducted to boost the recovery of over-fished cod stocks. Scientific evaluation of a proposal to cull marine mammals for the purpose of benefiting commercial fisheries needs to consider the complexity of ecological interactions among the marine mammal population(s), the relevant fish stocks, and the fishery/fisheries that catch them.

Lethal methods have not been found to be a consistently effective means of keeping pinnipeds from interacting with fishing operations. The idea is that if problems are caused by a few rogue seals, then removal of these animals should eliminate the problem. However, this method removes individuals that are then often replaced by others.

A diversity of nonlethal methods has been attempted (Jefferson and Curry, 1996). At the most basic level, fishermen throw stones or bait to distract the predator. Other methods used include firing gunshots, using explosives (such as firecrackers and seal bombs), biological sounds (vocalizations from a predator, killer whale), gear modifications (such as mechanical and electronic sound generators), vessel chase, tactile harassment (e.g., rubber bullets), visual signals, and taste aversion (baiting fish using a chemical to induce vomiting). Acoustic "harassment" devices (AHDs) have been widely used to attempt to reduce depredation on fish, especially by pinnipeds (Reeves *et al.*, 1996). Pinnipeds are difficult to deter by acoustic methods, and the acoustic signal of the AHDs over time was a "dinner bell" effect, alerting animals to the presence of a fish pen or trap. New high-intensity AHDs appear to be effective, but have a greater potential for causing hearing damage, as well as affecting nontarget species. In some cases, a problem may

be eased by changing the location of the fishing effort. The most successful mitigation measures appear to be changes to fishing gears or fishing methods where a particular change may reduce or exclude problems, thereby resulting in a permanent solution. Implementation of antipredator cages around fish farms, physical barriers at the entrance of fish ladders, and the change to synthetic twine in gill nets are some examples.

Capture and relocation of "problem" pinnipeds has proven ineffective, with the animals returning to the problem area. California sea lions have been captured at the Ballard Locks and placed in temporary captivity and released after the steelhead run. This proved ineffective in the long term, as did permanent captivity, which eliminates the "problem" sea lions without having to kill them but is limited by the availability of facilities that can hold the sea lions and the costs involved in capturing and holding the animals.

Past efforts have been unsuccessful in finding effective, long-term, nonlethal approaches to eliminating or reducing marine mammal–fishery conflicts. Some nonlethal deterrence measures appear to be effective initially or effective on "new" animals, but become ineffective over time or when used on "new" animals in the presence of "repeat" animals that do not react to deterrence. Further research on the development of new technologies and techniques is needed.

See Also the Following Articles

Competition with Fisheries ▪ Feeding Strategies and Tactics ▪ Incidental Catches ▪ Management ▪ Noise, Effects of ▪ Parasites

References

Beddington, J. R., Beverton, R. J. H., and Lavigne, D. M. (eds.) (1985). "Marine Mammals and Fisheries." George Allen & Unwin, London.

Earle, M. (1996). Ecological interactions between cetaceans and fisheries. *In* "The Conservation of Whales and Dolphins" (M. P. Simmonds and J. D. Hutchinson, eds.), pp. 167–204. Wiley, Chichester.

Fertl, D., and Leatherwood, S. (1997). Cetacean interactions with trawls: A preliminary review. *J. Northwest Atlantic Fish. Sci.* **22,** 219–248.

Fraker, M. A., and Mate, B. R. (1999). Seals, sea lions, and salmon in the Pacific Northwest. *In* "Conservation and Management of Marine Mammals" (J. R. Twiss, Jr., and R. R. Reeves, eds.), pp. 156–178. The Smithsonian Institution Press, Washington, DC.

IUCN. (1981). "Report of the IUCN Workshop on Marine Mammal/Fishery Interactions," La Jolla, California, 30 March–2 April 1981. International Union for the Conservation of Nature and Natural Resources, Gland, Switzerland.

Jefferson, T. A., and Curry, B. E. (1996). Acoustic methods of reducing or eliminating marine mammal-fishery interactions: Do they work? *Ocean Coast. Manag.* **31,** 41–70.

Nedelec, C., and Prado, M. (1990). "Definition and Classification of Fishing Gear Categories." FAO Fisheries Technical Paper, No. 222 (Revision 1).

Northridge, S. P. (1984). "World Review of Interactions between Marine Mammals and Fisheries." FAO Fisheries Technical Paper, No. 251.

Northridge, S. P. (1991). "An Updated World Review of Interactions between Marine Mammals and Fisheries." FAO Fisheries Technical Paper, Supplement to No. 251.

Northridge, S. P., and Hofman, R. J. (1999). Marine mammal interactions with fisheries. *In* "Conservation and Management of Marine

Mammals" (J. R. Twiss, Jr., and R. R. Reeves, eds.), pp. 99–119. Smithsonian Institution Press, Washington, DC.

Powell, J. A. (1978). Evidence of carnivory in manatees (*Trichechus manatus*). *J. Mammal.* **59**, 44.

Reeves, R. R., Hofman, R. J., Silber, G. K., and Wilkinson, D. (1996). "Acoustic Deterrence of Harmful Marine Mammal–Fishery Interactions." Proceedings of a workshop held in Seattle, Washington, 20–22 March 1996. U.S. Dept. of Commerce, NOAA Technical Memorandum NMFS-OPR-10.

Wickens, P. A. (1995). "A Review of Operational Interactions between Pinnipeds and Fisheries." FAO Fisheries Technical Paper, No. 346.

Yano, K., and Dahlheim, M. E. (1995). Killer whale, *Orcinus orca*, depredation on longline catches of bottomfish in the southeastern Bering Sea and adjacent waters. *Fish. Bull.* (*U.S.*) **93**, 355–372.

Fishing Industry, Effects of

SIMON NORTHRIDGE
University of St. Andrews,
Scotland, United Kingdom

The fishing industry probably represents the single area of human activity that has the most profound effects on marine mammals. These effects can be categorized broadly as "operational effects" and "biological effects."

Operational effects include the accidental capture of marine mammals in fishing gear, a problem that has brought more than one marine mammal population to the brink of extinction. Although accidental capture usually results in the death of the animal concerned, there are also instances where marine mammals are injured or affected in some way during fishing operations so that their survival probability or reproductive potential is compromised. Not all operational interactions have a negative effect on marine mammals. In some cases the effect of the fishing operations may be positive for the marine mammal where, for example, they feed on discarded fish or take fish that have been caught before these can be retrieved onto the fishing vessel. In a few cases there are even mutually beneficial collaborative efforts between fishermen and marine mammals, with marine mammals assisting in fish capture and being rewarded with a portion of the catch.

Biological effects encompass all the consequences of the large-scale removal of animal biomass from the marine ecosystem through fishing activities, including, although not limited to, possible competition for resources between fisheries and marine mammals. Competitive interactions can be direct or indirect. Direct competition occurs where the mammal and the fishery are both taking the same kind of fish. Indirect competition includes situations where the fishery and the marine mammal population are taking two different types of fish, but where the removal of one of these fish influences the availability of the other through some competitive or predatory link. Indirect interactions need not be competitive, and sometimes the effect of the fishing industry may be to increase the abundance of marine mammal prey items through indirect ecological interactions. Sometimes fisheries may physically alter a habitat and so change the composition and abundance of the fish community to the detriment or advantage of marine mammals and other predators.

I. Operational Effects

Operational effects cover interactions between fisheries and marine mammals that relate to the mechanical process of fishing. Several fisheries have well-documented problems with unwanted entrapment of marine mammals. In some cases the numbers of animals involved are large enough to seriously endanger the marine mammal populations concerned, and in one or two instances, species have been brought to the verge of extinction through such accidental effects. Examples considered cover gill net fisheries, pelagic trawls, and purse seine fisheries.

A. Gill Net Fisheries

Gill nets are a widely used fishing gear with a long history of use in many parts of the world. Their use has become more widespread since the 1950s or 1960s with the introduction of nylon as a netting material during the 1950s. They represent a fuel-efficient means of fishing and, when set on the seabed, provide a fishing method that can be used to exploit areas of rough ground that cannot be fished easily by towed gear. When used in surface waters, they are usually left to drift with the wind and tide and are effective in targeting dispersed fish schools. They are usually left to fish unattended, and fishing times may range from a few hours to several weeks, but 24 hours would be a typical soak time. It has been suggested that in contrast to the traditional nets that were made of cotton and other natural fibres, the use of stronger nylon twines has contributed to an increased rate of marine mammal ENTANGLEMENT. This, coupled with a dramatic increase in their usage since the 1950s, has led to some serious conservation and animal welfare problems with respect to marine mammals.

Small cetaceans, such as porpoises (Phocoenidae), and some species of seals seem especially prone to becoming entangled in gill nets. In some instances, this does not present a conservation problem. In Britain, for example, gray seals (*Halichoerus grypus*) are frequently caught and drowned accidentally in gill net and tangle net fisheries (see Fig. 1). In a seal tagging program run by the Sea Mammal Research Unit in the North Sea, over 20,000 gray seals have been tagged soon after birth since the 1950s. Returns of tags by fishermen indicate that at least 10% of known subsequent mortalities are due to net entanglement, and at least 1.5% of all pups tagged were recovered dead in fishing nets. Tag loss over the months and years after tagging and failure to return tags from entangled seals are two reasons why this latter figure must be an underestimate of total mortality rates due to entanglement. Despite such mortality in fishing gear, gray seal numbers are increasing in British waters.

In other cases, accidental catches in fishing gear can lead to conservation problems. Concerns have been expressed over the numbers of harbor porpoises (*Phocoena phocoena*) that become entangled in gill nets throughout much of this species' range. Numbers are known to have declined in some areas such as the Baltic Sea, possibly as a result of entanglements. In other areas, including the North Sea, the Celtic Sea, and the Gulf of Maine, the total numbers of porpoises entangled an-

Figure 1 *Young gray seal caught and drowned in a skate tangle net set in the English North Sea. Photo by Dave Sanderson, SMRU.*

nually are thought to be unsustainable and are likely to result in long-term population declines.

A close relative of the harbor porpoise, the vaquita (*Phocoena sinus*), is threatened with extinction through accidental catches in gill nets. In this case, the species has a restricted range, in the upper Gulf of California in Mexico, where there are large numbers of small boats using gill nets and tangle nets to catch a wide variety of fishes. Although the population may number no more than about 600, one estimate suggests that at least 40 animals drown in gill nets every year. This is clearly an unsustainable rate of mortality and this species' future therefore seems bleak.

The North Atlantic right whale (*Eubalaena glacialis*), once one of the more commonly seen whales in the North Atlantic, is now reduced to a population of around 300 animals in that area. The population is declining and it is estimated that at current rates it will be extinct within 200 years. Most of this population migrate along the eastern seaboard of the United States every year, where they too are vulnerable to entanglement in gill nets and also in lobster pot lines. Lobster pots are usually set in "strings" of several pots in a line, with each end of the string marked by a surface-floating buoy attached by a line to the pot string on the seabed. Although only three right whale deaths have been attributed directly to this cause since 1970, evidence of entanglement-related scars on live animals has been identified in around 60% of the population. It has been suggested that some entangled whales die and are dragged to the seabed without being recovered, others may suffer injuries that lead to subsequent death by other causes, while at least two right whales, encumbered with fishing gear, have been fatally wounded in collisions with ships. These high levels of entanglement may therefore be a major factor in the decline in this population of right whales, even though the vast majority of the entanglements are not immediately fatal.

Marine mammal entanglements in gill nets may occur for a variety of possible reasons. Some people maintain that the animals do not detect the netting and swim into it before realizing it is there. It may be that the animals do detect the netting but that they do not recognize it as something dangerous and attempt to swim through it as though it were some natural obstacle, such as seaweed. Another possibility is that the animals are fully aware of the netting, and the danger it poses, but they simply make mistakes and become entangled while feeding close to the net due to inattention.

There has been much attention given to means of reducing the numbers of marine mammals that become caught in gill nets and tangle nets because of the conservation problems that such entanglements represent. So far, the only effective means of reducing bycatch that has been found is the use of pingers. Pingers are small battery-powered devices that emit a brief high-pitched noise every few seconds. They are attached to the float line or lead line of the gill net and are of similar size and shape as a net float so as to avoid tangling the net when it is being set or hauled. They are effective in reducing the entanglement rates of several marine mammal species in gill nets, although exactly why they work is not clear. They have been developed in collaboration with the fishing industry and are currently being used in several major gill net fisheries around the world.

B. Pelagic Trawls

Pelagic trawling is another fishing method that has increased enormously in recent decades. Although trawling dates back for more than a century, this was initially performed with low-opening nets dragged along the seabed. As various aspects of technology have improved, so trawling techniques have been refined, and trawls have been used to catch fish above the seabed and even near the surface of the water. The development in the 1950s of acoustic fish finders and net sounders that enable the skipper to control the position of the net with respect to that of a fish school has been a key technological development.

Initially, during the 1950s, pelagic trawls were used to catch fish like herring that form dense schools. Typical nets might have had an opening of around 2000 m², perhaps 50 m wide and 40 m high. Since then, both nets and trawlers have grown in size, as other pelagic fish species have been targeted, some of which may form much more dispersed schools. Net openings at least 10 times this are now common, with horizontal and vertical openings of up to 200 m. There are numerous records of marine mammals becoming caught in pelagic trawls, sometimes in large enough numbers to present a conservation problem, although the nature and scale of the problem remain obscure in most fishing areas.

Hooker's sea lion (*Phocarctos hookeri*) is endemic to the sub-Antarctic islands of New Zealand, around which a pelagic trawl fishery for squid developed in the 1980s. Observations of fishing activity between 1988 and 1995 suggested annual mortalities of between 20 and 140 Hooker's sea lions, at a rate of about 1 animal every 340 trawl tows. The total population of this species was only around 13,000 animals in the mid-1990s, and the accidental catches were therefore considered significant. There is currently a quota system in operation in this fishery to limit accidental catches.

Dolphins have also been reported caught in pelagic trawl fisheries in several areas, including the United States, Europe, and New Zealand. The capture of dolphins in pelagic trawls appears to be very variable, depending on the area and probably the type of trawl being used. In one U.S. pair-trawl fishery common bottlenose (*Tursiops truncatus*), short-beaked common (*Delphinus delphis*), and Risso's dolphins (*Grampus griseus*) were observed accidentally caught in an average of one in every 5 trawl tows, although this fishery has subsequently been closed. In one New Zealand midwater trawl fishery, rates of common and bottlenose dolphin catches averaged about one animal in every 9 trawl tows observed. In European Atlantic waters, one study recorded common and Atlantic white-sided dolphins (*Lagenorhynchus acutus*) caught in sea bass (*Dicentrarchus labrax*) and albacore tuna (*Thunnus alalunga*) pelagic trawl fisheries at a rate of one in every 10 to one in every 15 trawl tows. In other pelagic trawl fisheries for anchovy (*Engraulus encrasicolus*), pilchard (*Sardina pilchardus*), and mackerel (*Scomber scombrus*) in the same area, no dolphin deaths were recorded.

The reasons why some pelagic trawl fisheries have relatively high levels of accidental dolphin catches and others have low or zero levels are not yet clear. It is probable that such factors as the dimensions of the net, the towing speed and duration, and the foraging activity of dolphins around the nets are important. There are several accounts of dolphins taking advantage of trawling activity by FEEDING on fish escaping through the meshes of the trawl. Such behavior may increase chances of dolphin entanglement. The trend toward increasing net sizes may also have increased the numbers of dolphins caught in some pelagic trawl fisheries in the past few decades, but these interactions remain relatively poorly investigated.

C. Tuna Purse Seine Fisheries

As with gill nets and pelagic trawls, technical innovations during the 1950s enabled the development and expansion of purse seine fisheries around the world. Purse seines are used to catch pelagic fish and work by first encircling a school of fish with a long net, hanging from the surface down to depths sometimes of several hundred meters. The bottom edge of the net can be pursed so as to prevent any escape under the netting once it has been set in a circle. The major technical innovations that allowed this fishery to develop globally were the introduction of nylon as a netting material, enabling much larger and stronger nets to be constructed, and the development of the power block, with which large amounts of netting could be hoisted up out of the water.

American fishermen working in the eastern tropical Pacific Ocean during the late 1950s worked out a way of using dolphin schools in conjunction with purse seine nets to catch yellowfin tuna (*Thunnus albacares*). They discovered that tuna would aggregate under dolphin schools, even if the dolphins were chased. This meant that by using speedboats to corral dolphins they could exploit this behavioral characteristic of the fish to round up tuna schools that would otherwise be invisible below the surface. By corralling a dolphin school and setting a purse seine net around it, a school of tuna would normally also be encircled. The dolphins were not intentionally killed in this activity, but many died as nets were being hauled in and fish were being brought on board. The scale of the fishery meant that in some years hundreds of thousands of dolphins drowned as a result of this fishing activity.

The eastern spinner dolphin stock (*Stenella longirostris orientalis*) was depleted to 44% of its original (pre-1959) level, whereas an offshore pantropical spotted dolphin stock (*Stenella attenuata*) was reduced to about 20% of its pre-exploitation level. From the 1970s onward, in the face of public concerns over the issue, the fishery developed and implemented means of reducing this toll. By encouraging the dolphins out of the net before trying to remove the fish, annual mortalities were reduced to a few thousand per year.

D. Other Types of Fishing Activity

Although tuna purse seines, pelagic trawls, and gill nets have been highlighted as fishing methods where operational interactions have been a cause for concern, many other fishing methods can, under the wrong circumstances, have an impact on marine mammals. Marine mammals have been reported ensnared in the most unlikely types of fishing gear, including the lobster pot lines referred to earlier. In most cases, such interactions occur rarely and the number of animals involved is so small that there is no cause for concern over unsustainable levels of catch. Occasionally, however, circumstances may conspire to cause a problem. At other times the use of certain types of fishing gear may even result in a benefit to the marine mammals and cause a loss to the fishery.

In several parts of the world cetaceans, often killer whales (*Orcinus orca*) and false killer whales (*Pseudorca crassidens*), have been reported to remove fish from hooks during long line fishing operations. This can make fishing in certain areas with long lines unprofitable, and many methods have been tried to eliminate such behavior. There has been remarkably little success reported in trying to prevent such predation, and sometimes the boats involved have had to switch gear or move to other areas.

Normally, interactions between marine mammals and hook and line fisheries have few negative impacts on the marine mammals. In the case of the baiji or Chinese river dolphin (*Lipotes vexillifer*), however, the situation is different. The baiji is endemic to the Yangtze River system in China; the population probably numbers no more than a hundred and is in decline. One of the most important causes of mortality is due to animals becoming ensnared in "rolling hooks." This type of fishing involves using many sharp unbaited hooks on a line set on the bottom of the river to snag bottom-dwelling fish. Dolphins foraging near such hooks sometimes become snagged too, occasionally causing death. Over 50% of recovered river dolphin carcasses had died as a result of such entanglements. The small population size and the restricted distribution of this species mean that even uncommon occurrences such as these can have a highly significant impact on a species.

Other types of fishery where marine mammals actively benefit from fishing activities include trawling, purse seining, and lobster potting. Several species, including bottlenose and white-beaked dolphins (*Lagenorhynchus albirostris*), gray seals, and South African fur seals (*Arctocephalus pusillus*), have been reported to remove fish from fishing gear. Dolphins typically take undersized fish as they come through the cod end of a trawl. Fur seals swim into purse seine nets by climbing over the float line once a school of fish has been encircled and make a meal of the trapped fish. Gray seals have been observed removing bait from baited lobster pots. There are numerous other examples of marine mammals taking advantage of fishing activities in similar ways, often taking marketable fish, which may provoke considerable resentment on the part of the fishing crews.

In a few places in the world, including Burma (Myanmar), Mauritania, and Brazil, fishermen and dolphins have learned to collaborate in the capture of fish, usually by dolphins driving fish toward fishermen waiting with nets. Some of the catch is then given back to the dolphins as a reward.

II. Biological Effects

The more widespread but less well-understood interactions between fisheries and marine mammals are ecosystem-level effects, where fisheries may cause fundamental changes to the species composition of the marine environment.

Every year the fishing industry removes about over 90 million ton of fish and other marine organisms from the world's oceans. Another 20–30 million ton of unwanted animal biomass may be caught but discarded prior to landing every year. It has been suggested that fisheries may account for 8% of the global primary productivity of the oceans, and in some of the more heavily exploited areas as much as 35% of local primary productivity may be required to sustain fishery catch levels. It is clear that such levels of fishing activity are likely to have profound effects on marine ecosystems, especially on the top predators, such as marine mammals, as fish populations are reduced and restructured on a very large scale. In theory, therefore, fisheries may compete with marine mammals by depleting their food.

There is of course another side to this concern. Whereas fisheries may cause a depletion of the food resource for marine mammals, marine mammals are often accused by fishery bodies of consuming large amounts of fish, thereby reducing the amount of food available for people to eat. As a result of this latter concern, in some parts of the world there are frequent calls for marine mammals to be culled as unwanted competitors.

In both cases, however, it has proved extremely difficult to demonstrate any clear competitive interaction between marine mammals and fisheries. This is mainly because of the complexities of the marine ecosystem, which make it very hard to predict how changes in one fish stock will effect either their predators or their prey. Some brief examples will illustrate this point.

A. Increased Food Availability

The North Sea and adjacent areas are among the most heavily fished sea areas of the world, with annual landings of all species of around 2.5 million metric tons. One of the most numerous marine mammals in this region is the gray seal (*Halichoerus grypus*), which feeds on a range of fish species, including Atlantic cod, haddock, whiting, saithe, sandeels, sole, plaice, Atlantic herring, and sprats (*Gadus morhua, Melanogrammus aeglefinus, Merlangius merlangus, Pollachius virens, Ammodytes* spp., *Solea solea, Pleuronectes platessa, Clupea harengus,* and *Sprattus sprattus*). Fisheries also target all of these species, and most of the fish stocks concerned are designated as fully exploited or overexploited. Despite these facts, the gray seal population has expanded at an apparently steady rate over several decades during which fishing pressure in the region has been intense, to reach 108,500 animals in 1994. There are at least two good reasons for this apparent paradox.

First, the most important food of the gray seal is the sand eel (or "sandlance" in the United States). The sand eel fishery takes over 1 million metric tons of sand eels per year and is the single largest fishery (in terms of the amount landed) in the North Sea. Despite fishing pressure, sand eels are still extremely abundant and appear to have increased in abundance since the 1950s. The proposed reason for this is that sand eel numbers have increased in response to massive declines in the abundance of Atlantic herring and mackerel as a result of intense herring and mackerel fishing, especially in the 1960s and 1970s. Although sand eels are an oily fish (and well suited to the seal's diet), unlike herring they are not considered useful for human consumption, and those that are fished are generally reduced to fishmeal. This means that they are a low-value species and have so far escaped the intense levels of fishing pressure applied to others stocks. Being a fairly sedentary species that spends much of its time buried in the sand, they also appear to be well suited to gray seal foraging habits.

Alongside this, although gray seals and fisheries consume other more marketable species in common, gray seals typically consume smaller-sized individuals of around 15 to 20 cm in length, whereas commercial fisheries generally concentrate on fish of 30 cm and larger. As commercial fishing pressure intensifies, larger fish have become scarcer, but smaller fish of the same species may not be affected (or at least not until there are two few large fish left to generate sufficient eggs to replenish the stock). Indeed, in many of the commercially useful gadid (cod family) fishes, small fish are consumed in large numbers by bigger fish of the same species. A reduction of the numbers of larger fish may actually boost the numbers of smaller fish.

It would seem that although gray seals are feeding in an area that is very heavily fished because they exploit a niche that is not directly in competition with fisheries, they are able to thrive. Indeed their population expansion may have been assisted by fishery-induced changes to the species and size structure of the system they inhabit.

B. Decreased Food Availability

Although pinniped numbers may have increased with increasing fishing pressure in one area, this is by no means the norm. In the Gulf of Alaska and the Bering Sea, several pinniped populations have undergone dramatic declines over periods when fishing activity has been increasing. The Pribilof population of Northern fur seals (*Callorhinus ursinus*) declined from 1.25 million animals in 1974 to around 877,000 animals in 1983. At around the same time, harbor seal (*Phoca vitulina richardii*) numbers in the Gulf of Alaska and southeastern

Bering Sea declined, with one major haul-out site at Tugidak Island recording an 85% reduction in numbers between 1976 and 1988. Populations of both of these species now seem to be stable or increasing, but numbers of Steller sea lions (*Eumatopias jubatus*) in western Alaska started to decline in the 1970s and numbers are still declining. Declines of up to 80% have been recorded in some areas.

The reasons for the declines of these three species over much the same time period are not known, but there is general agreement that food availability, especially for younger animals, seems to be a key issue. For all of these three species, Alaska pollock (*Theragra chalcogramma*) is an important prey item. Alaska pollock is also the target of one of the largest single species fisheries in the world. The fishery for Alaska pollock increased greatly during the 1970s with over a million tons being landed annually from the eastern Bering Sea alone throughout most of the 1980s and 1990s.

An obvious explanation is that the fishery has deprived the pinnipeds of their food. A closer inspection, however, reveals that the situation is more complex. Overall, the pollock biomass has stayed remarkably buoyant throughout this time period. Numbers of pollock, especially numbers of the larger or older age classes that are the target of the fishery, have not declined until relatively recently. Smaller pollock are consumed by the pinnipeds and are not targeted by the fishery, but are cannibalized by the larger fish. Numbers of smaller pollock have declined. Furthermore, for Steller sea lions at least, pollock does not appear to be a favored food item. In other parts of the Steller sea lion's range, such as southeastern Alaska, where the population is not in decline, pollock makes a smaller contribution to the diet, after oily fish such as herring.

It may be that the three pinniped species have all suffered from some change in the relative abundance of their preferred diet items. It has not yet been possible to determine whether such ecosystem level changes have been the result of long-term oscillations in oceanographic conditions or whether the pollock fishery has in some way altered the abundance of the pinniped's preferred prey items through the cascading effects of restructuring the pollock population.

C. Other Effects

The effects of fisheries on marine mammals do not necessarily have to be mediated through changes in their food supply. Changes to predation on marine mammals could also arise through the effects of fishing. On the Atlantic coast of Canada, also a heavily fished region, gray seal numbers have also been increasing steadily for more than two decades for unknown reasons. One suggested reason has been that fisheries may have greatly reduced the number of large sharks in coastal waters of Atlantic Canada, and as gray seals are known to be subject to predation by certain sharks, such a reduction might be one of the factors contributing to the increase in gray seal numbers.

Conversely, recent increases in predation on sea otters (*Enhydra lutris*) in the Aleutian Islands have been attributed to behavior changes in killer whales in the region, some of which now seem to be preying more heavily on sea otters than in previous decades. The possibility has been raised that this change in behavior has been caused by a decline in other more favored food items, such as sea lions. Such declines could, as has been suggested earlier, be at least partly due to the effects of fishing on sea lion food items.

All of these hypotheses demonstrate the complex ways in which fishery-induced changes to the marine ecosystem may affect marine mammals, although none has yet proved testable. In almost all cases where some form of competition is perceived, any closer scrutiny of the situation reveals that the complex predatory interrelations of the marine ecosystem make it extremely difficult to predict the results of any proposed management action. The extent to which the fishing industry competes with marine mammals is therefore still very much an open question.

See Also the Following Articles

Competition with Fisheries ■ Entrapment and Entanglement ■ Forensic Genetics ■ Incidental Catches ■ Tuna–Dolphin Issue ■ Vaquita

References

Gosliner, M. (1999). The tuna-dolphin controversy. *In* "Conservation and Management of Marine Mammals" (J. Twiss and R. Reeves, eds.), pp. 120–155. Smithsonian Institution Press, Washington, DC.

Marine Mammal Commission (2000). Annual Report to Congress, 1999. Marine Mammal Commission, Washington, DC.

Mangel, M., and Hofman, R. J. (1999). Ecosystems: Patterns, processes and paradigms. *In* "Conservation and Management of Marine Mammals" (J. Twiss and R. Reeves, eds.), pp. 87–98. Smithsonian Institution Press, Washington, DC.

Merrick, R. L., Cumbley, M. K., and Bryd, G. V. (1997). Diet diversity of Steller sea lions (*Eumatopias jubatus*) and their population decline in Alaska: A potential relationship. *Can. J. Fish. Aquat. Sci.* **54**, 1342–1348.

Northridge, S. P., and Hofman, R. J. (1999). Marine mammal interactions with fisheries. *In* "Conservation and Management of Marine Mammals" (J. Twiss and R. Reeves, eds.), pp. 99–119. Smithsonian Institution Press, Washington, DC.

Northridge, S. P. (1991). An updated world review of interactions between marine mammals and fisheries. FAO Fisheries Technical Paper 251 (suppl. 1). Food and Agriculture Organisation of the United Nations, Rome.

Perrin, W. F., Donovan, G. P., and Barlow, J. (1994). "Cetaceans and Gillnets" (G. P. Donovan, ed.), Report of the International Whaling Commission, Special Issue 15, International Whaling Commission, Cambridge, UK.

Fluking

BERND WÜRSIG
Texas A&M University, Galveston

Fluking is the act of a whale, dolphin, or porpoise (very rarely, a manatee or dugong) raising its tail, or flukes, above the surface of the water during the beginning of a dive. Usually, whales fluke when diving steeply in water deeper than at least two of their own lengths, although fluking can also occur during shallow submergences. There is great variability in fluking behavior by species: humpback (*Megaptera novaeangliae*) and sperm whales (*Physeter macrocephalus*) almost always fluke (Fig. 1) or "fluke out" during the dive; minke whales (*Balaenoptera acutorostrata*) and fin whales (*B. physalus*) do so rarely (Leatherwood and Reeves, 1983). Right (*Eubalaena* spp.), bowhead (*Balaena mysticetus*), and gray whales (*Eschrichtius robustus*) are known to vary the amount and type of fluking depending on whether they are feeding near the surface (no fluking), at moderate depth (occasional fluking), or at depths of 60 m or more ("always" fluking). These species also generally fluke on MIGRATION, during the final dive after a series of "near-surface" dives between respirations. Even smaller toothed whales, such as bottlenose dolphins (*Tursiops truncatus*) and pilot whales (*Globicephala* spp.), at times fluke during deep or at least steeply angled dives (Carwardine *et al.*, 1998).

Large whales fluke as a part of bending their bodies as they angle downward, and the fluke out becomes a natural extension of the animal "rolling" forward and down. In smaller delphinids, however, fluking probably has a distinct advantage during the initiation of the dive: the tail and tailstock (or caudal peduncle) held above the surface provide in-air weight to the body and help propel it downward. The effect of this action is quite similar to human skin divers kicking their feet and legs out of water while bending at the waist as they initiate a dive.

Bowhead whales about to dive steeply often stretch their entire bodies so that head and flukes are at the surface, and the belly or midpart of the body hangs much further below. The spine, in other words, is curved downward. Such a "pre-dive flex" takes about 2 sec and occurs just before the last blow before a dive. The pre-dive flex is almost always predictive of a fluke-out dive, and it and the flukes out indicate steep, generally deep, diving for feeding or migrating (Würsig and Clark, 1993).

While fluking is the term properly reserved for the smooth lifting of the flukes above water during the initial stages of a dive, whales and dolphins lift their flukes out of water in other ways as well. A common form is tail slapping (often termed lobtailing when performed by large whales), consisting of forcefully and often repeatedly slapping the tail onto the surface of the water. This action makes a loud in-air and underwater percussive sound and may well be used to communicate with conspecifics. Tail slapping and breaching often beget more such aerial activities by others, especially when wind and sea states rise; these observations indicate COMMUNICATION and perhaps a social facilitation function of all animals of the area coordinating their states of alertness. Tail slapping may also be used to repel danger as an agonistic display against a conspecific nearby and even to help contain and corral prey near the surface. In humpback whales, tail slapping can be a sign of frustration or anger, as it and breaches are often performed after one male appears to lose during an interactive battle with another one. It is believed that similar evidence of "frustration" has been seen in bowhead, right, and gray whales apparently repulsed by a female. Repeated tail slapping can also be performed "for the fun of it" or as a form of play; in spinner

Figure 1 *A sperm whale (New Zealand) in the act of diving and fluking. Initiation of the dive (top), as the back curves just before lifting the flukes. This fluking sequence shows a particularly high "fluke out" before a fully vertical descent.*

(*Stenella longirostris*), dusky (*Lagenorhynchus obscurus*), and probably other dolphins, this appears to be its main function.

A final form of "flukes out" consists of raising the flukes above water and keeping them there for some time. Dolphins wave their flukes about for at times over 1 min, "headstanding" in apparent play; right, bowhead, and gray whales do so for at times over 10 min. It has been suggested that southern right whales are purposefully sailing by holding their large tails in air during a stiff breeze, but this recreational use of their tails has not been substantiated (Payne, 1995).

Flukes of whales and dolphins have thin trailing edges (similar to the trailing edges of dorsal fins) and are therefore easily tattered or marked by conspecific interactions, getting entangled in lines, or touching other objects. These marks make individual recognition of the flukes possible in those species that habitually fluke out, and fluke-based photographic identification is being practiced for sperm, humpback, and gray whales. In the latter two species, an additional bonus is variable spot or mottling patterns on the flukes (Hammond *et al.*, 1990).

See Also the Following Articles

Diving Behavior ■ Lobtailing

References

Carwardine, M., Hoyt, E., Fordyce, R. E., and Gill, P. (1998). "Whales, Dolphins, and Porpoises." Weldon Owen Press, Sydney, Australia.

Hammond, P. S., Mizroch, S. A., and Donovan, G. P. (eds.) (1990). Individual recognition of cetaceans: Use of photo-identification and other techniques for estimating population parameters. *Rep. Int. Whal. Comn. Spec. Issue* **12**, 43–52.

Leatherwood, S. L., and Reeves, R. R. (1983). "The Sierra Club Handbook of Whales and Dolphins." Sierra Club Books, San Francisco.

Payne, R. S. (1995). "Among Whales." Scribner and Sons, New York.

Würsig, B., and Clark, C. W. (1993). Behavior. *In* "The Bowhead Whale" (J. Burns, J. Montague, and C. Cowles, eds.), pp. 157–199. Allen Press, Lawrence, KS.

Folklore and Legends

ROCHELLE CONSTANTINE
University of Auckland, New Zealand

Folklore and legends are usually traditional stories popularly regarded as the telling of historical events. When in the form of myths, they often involve some form of the supernatural. They have been with us for thousands of years and, because of this, folklore and legends form the basis of many religious beliefs, value systems, and the way we perceive our place in the world and our interaction with other animals. Man has long revered whales and dolphins in legends. For thousands of years they have been aligned with the gods, mythologized, and celebrated in art.

I. Ancient Greece

Some of the earliest legends about dolphins were told in Greek mythology, where it was believed the sun god Apollo assumed the form of a dolphin when he founded his oracle at Delphi on the edge of Mount Parnassus. The ancient Greeks also believed Orion was carried into the sky riding on the back of a dolphin and was gifted three stars by the gods. This constellation is now known as Orion's Belt.

Many cultures, both ancient and recent, revered dolphins and believed them to be messengers from the gods. The pre-Hellenic Cretans appeared to have honored dolphins, and the ancient Greek, Oppian (ca. A.D. 180) wrote of godly intervention in the dolphins' move to the sea. It was believed that by the devising of Dionysus, the Greek god of wine, dolphins exchanged their life on land for life at sea and took on the form of fishes. Even though they changed form, it was believed that they retained the righteous spirit of man and, because of this, they preserved their human thoughts and deeds. Oppian also wrote of dolphins stranding themselves to die so mortals could bury them and thereby remember the dolphin's gentle friendship. This was seen as an example of how magnificent dolphins were.

The close alignment with man meant ancient Greeks held dolphins in extremely high regard and that killing a dolphin was tantamount to killing a person. Both crimes were punishable by death.

II. Romans

Pliny the Elder (A.D. 23–79), a Roman philosopher, told the story of a peasant boy in the Mediterranean Sea who developed a relationship with a solitary dolphin he named Simo. The legend tells of a boy who fed a dolphin and, in return, the dolphin gave him rides across the bay on its back. The boy became ill and died and, according to local knowledge, the dolphin returned to their meeting place for many days until it was believed that it died of a broken heart. Many of the Roman legends involved close human/dolphin contact and may seem fanciful, but in more recent times these bonds between solitary dolphins and humans have been well documented.

III. Art

Drawings carved into rocks in northern Norway of killer whales and other local animals are the earliest known art work portraying dolphins. These drawings have been estimated at 9000 years old. The most detailed and colorful ancient art work was done by the ancient Greek and Minoan (Crete 3000–1500 B.C.) people. Dolphins were portrayed on frescoes, mosaic floors, coins, vases, and in sculpture. One of the earliest known pieces is a dolphin fresco painted ca. 1600 B.C. on the wall of the queen's bathroom in the Minoan palace of Knossos. The Dionysus cup dated 540 B.C. shows the Greek wine god with dolphins and grapes. Coins portraying dolphins have been found in Syracuse, Greece, ca. 480 B.C. and the Romans also had dolphin coins in 2nd-century B.C.

IV. Whales

It appears in many legends that the great whales were not necessarily held in such high regard as the dolphins. Whales were typically described as monsters of the sea, their great size to be feared by all. Oppian (ca. A.D. 180) told of the hunt of a whale; its monstrous size and unapproachable limbs a terrible sight to behold. In biblical times, the story of Jonah and the whale was well known, and it is popular even today. The story tells of Jonah who fled from the lord by boat to Tarshish. When the ship was underway, the lord caused a great storm. In fear of their lives, Jonah asked the mariners to cast him into the sea so the lord would again make the sea calm and spare the mariners lives. Once Jonah was in the sea, however, the lord prepared a "great fish" to swallow him. He was in the belly of the whale for 3 days and 3 nights where he prayed and vowed salvation to the lord. Upon his vow the lord spoke to the whale and it vomited Jonah onto dry land and spared his life. Although today we know that it is unlikely that this event truly occurred, the story displayed the power of the lord and what he was capable of doing to those who defied him.

In his 1851 novel "Moby Dick," Herman Melville described a white sperm whale (*Physeter macrocephalus*) of uncommon magnitude, capable of great ferocity, cunning, and malice. Melville's novel summarized the fears of Yankee whalers that the tables would be turned and the whale would become the attacker.

Not all folklore portrays whales as fearsome beasts. Maori folklore of the Ngati Porou people tells of their ancestors being carried safely across the Pacific to New Zealand on the back of a whale. The Ngai Tahu people consider the sperm whales off the coast of the South Island as taonga (treasures). If a whale strands, prayers are said in order to return its spirit to Tangaroa, the Maori god of the sea. After this, the lower jawbone is removed for ceremonial carving and placement on the marae (the tribes' traditional meeting grounds).

The north Alaska Inuit people have for over 1000 years relied on whale products for their survival. As with many traditional HUNTING societies, ceremonies accompany the hunt that assure good luck, and many hunters take charms or amulets to ensure their luck and safety. Some believe the skull of the dead whale must be returned to the sea in order to assure the immortality and reincarnation of the whale, thereby protecting the future hunting success.

V. Legends and Folklore around the World

A. Haida

The Haida people of northwestern North America tell of an evil ocean people who used killer whales (*Orcinus orca*) as canoes. The Haida turned a chief into a killer whale and they believe that this whale now protects them from attacks by the ocean people.

B. Tlingit

The Tlingit (pronounced "Kling-kit") people of southeastern Alaska immortalize killer whales in their beliefs and folklore. Images of killer whales appear in many of their masks, carvings, totems, and blankets. At gatherings, the Tlingit tell stories, including one about the origin of killer whales. They believe a man from the Seal people carved many killer whales

from wood but only the one carved from yellow cedar would swim. The legend says he carved many more from cedar and they swam up the inlet where he taught them how to hunt and what to hunt for. He also taught them not to hurt people. The Tlingit in return do not hunt killer whales and they believe that because of this the killer whales look after them (Fig. 1).

C. Australian Aborigines

On Mornington Island in the Gulf of Carpentaria, northern Australia, a tribe of Aborigines have been in direct communication with Indian Ocean bottlenose dolphins (*Tursiops aduncus*) for thousands of years. They have a medicine man who calls the dolphins and "speaks" to them telepathically. By these communications he assures that the tribes' fortunes and happiness are maintained.

D. Amazonians

Many people who live on the banks of the Amazon River believe that river dolphins or botos (*Inia geoffrensis*) have the ability to transform themselves into handsome young men in order to woo women during fiestas and times of ceremony. So strong is this belief that some children are believed to have been fathered by these dolphins.

E. Japan

A Japanese legend tells of a gigantic whale who challenges a sea slug to a race after boasting that he is the greatest animal in the sea. The sea slug accepts and arranges for his friends to wait at different beaches along the chosen course. On the day of the race, the whale surges ahead, but when he arrives at the first beach he is astonished to find the sea slug already there. So he challenges it to another race, only to have the sea slug win again. This happens many times until the whale admits defeat. This legend is analogous to the European legend of the tortoise and the hare but shows the Japanese peoples close relation to the sea and its inhabitants and their use in teaching moral lessons.

Perhaps such legends and folklore serve the purpose of helping people understand their past or to help society learn

Figure 1 *Killer whale images on the front of a Tlingit house. Photograph courtesy of Alaska State Library.*

valuable lessons. In many societies today we revere whales and dolphins, and this will continue to develop our folklore into the future.

See Also the Following Articles

Inuit and Marine Mammals ■ Popular Culture and Literature ■ Whaling, Early and Aboriginal

References

Melville, H. (1851). "Moby Dick."
Unsworth, B. (1996). "Classic Sea Stories." Random House, London.

Food Sources
SEE *Diet*

Forensic Genetics

C. SCOTT BAKER AND GINA M. LENTO
University of Auckland, New Zealand

Molecular genetics provides a powerful new tool for the CONSERVATION and management of marine mammals: the forensic identification of products derived from hunting, strandings, and fisheries bycatch. Such products include soft tissue such as meat, organs, BLUBBER, skin, and blood, as well as teeth, bone, baleen, and hair. Although the species origins of these products may be impossible to determine on the basis of appearance, they contain DNA that can be sequenced and compared to sequences from known specimens. With a complete library of reference sequences, a product of unknown origin can be attributed in most cases to 1 of the (approximately) 80 cetacean species or 40 pinniped species or subspecies. If a comprehensive archive of tissue is maintained as part of a regulated hunt, it is possible to trace the origins of a product to a specific individual.

The molecular methods used in marine mammal forensic genetics have their origins in efforts to isolate and amplify DNA from ancient biological material (Pääbo, 1988). With the advent of the polymerase chain reaction (PCR) and "universal" primers (Kocher *et al.*, 1989), DNA can now be recovered from almost any biological source, even products that have been preserved, cooked, or canned. The interpretation of DNA sequences for the identification of species and geographic origins, however, is also dependent on basic research in species-level phylogenetic relationships and on the genetic structure of populations (Baker and Palumbi, 1995). The application of forensic genetics to marine mammals has, in turn, highlighted inadequacies in the basic taxonomy of cetaceans and pinnipeds and the recognition or definition of subspecies and management units (i.e., "stocks") below the species level (see later).

The forensic use of molecular genetic methods is of particular interest to the INTERNATIONAL WHALING COMMISSION (IWC), as it attempts to develop a revised management scheme for the regulation of any future commercial whaling, and to the Convention on International Trade in Endangered Species of Wild Fauna and Flora (CITES), as it attempts to implement a verifiable system for controlling trade in cetacean products. Progress in the molecular monitoring of commercial markets in whale products is described later as an example of this application. Other applications of forensic genetics to marine mammals include identifying strandings and fisheries bycatch, particularly for poorly described species such as beaked whales (Dalebout *et al.*, 1998; Henshaw *et al.*, 1997), and monitoring of trade in pinniped penises marketed as aphrodisiacs (Malik *et al.*, 1997). These methods are finding similar applications in the conservation and management of other taxa, most notably the HUNTING and bycatch of turtles (Bowen, 1995) and international trade in caviar (DeSalle and Birstein, 1996).

I. Phylogenetic Identification

Species-level identification of marine mammals has relied primarily on the phylogenetic reconstruction of DNA sequences from the control region or cytochrome *b* gene of the mitochondrial (mt) genome. In general, mtDNA offers two important advantages over traditional genetic markers such as allozymes. First, because of its maternal inheritance and absence of recombination, the phylogenetic relationship of mtDNA sequences reflects the history of maternal lineages within a population or species. (If hybridization is encountered, nuclear markers are required to identify the paternal species; see later.) Second, all else being equal, the effective population size of mtDNA genomes is one-fourth that of autosomal nuclear genes and its rate of random genetic drift is proportionately greater. This results in more rapid differentiation of mtDNA lineages among populations, compared to nuclear genes, and consequently greater sensitivity in the detection of recent historical demographic or speciation events. The ability to detect population differentiation is also enhanced by the rapid pace of mtDNA evolution, which is generally estimated to be 5 to 10 times faster than nuclear coding DNA in most species of mammals. The control region of the mtDNA does not code for a protein or RNA and, in the absence of these constraints, accumulates mutational substitutions more rapidly than other regions. For this reason, it has become the marker of choice for most studies of the population structure of cetaceans and pinnipeds. The cytochrome *b* gene, a protein region of the mtDNA, has also been used widely in species-level systematics and, in some cases, for the population structure of marine mammals (Lento *et al.*, 1997). Because of the large number of reference sequences available on public databases such as GenBank and EMBL, both loci have been used for species-level identification of marine mammals.

The basic steps involved in the phylogenetic identification of an unknown specimen or product are illustrated in Fig. 1. First, mtDNA is extracted from the product in question, such as a flensed piece of skin and blubber from a commercial market. Second, a fragment of the mtDNA control region is amplified from the product via PCR: a cyclic, *in vitro* enzymatic reaction

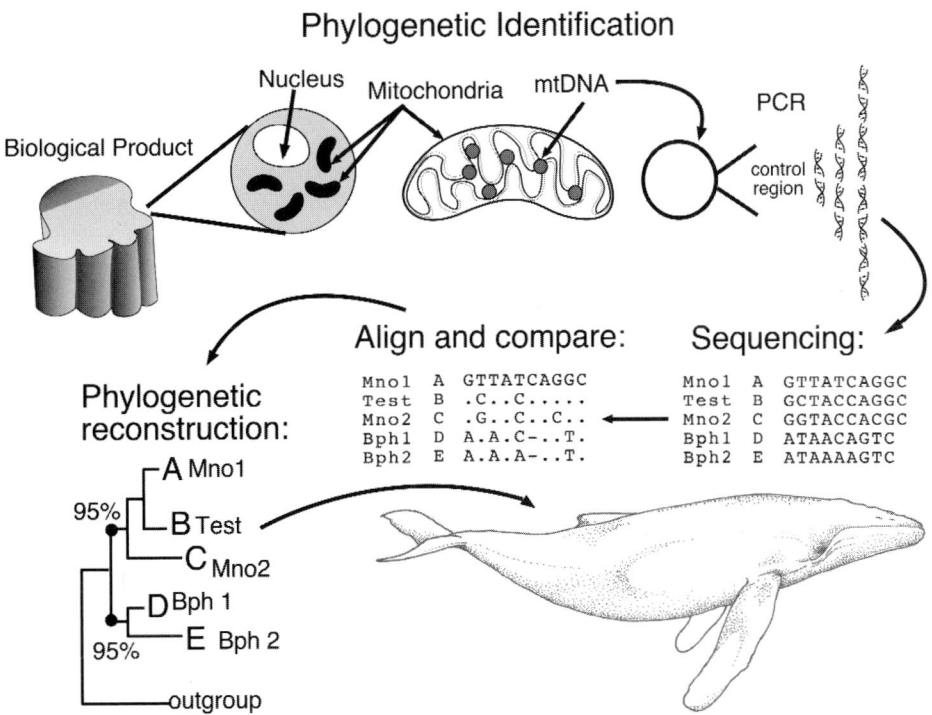

Figure 1 *The basic steps involved in the phylogenetic identification of a whale product using nucleotide sequences amplified by PCR from the mtDNA control region.*

that results in the exponential replication of a small targeted fragment of DNA (usually less than 1000 bp). Third, the exact nucleotide sequence of the amplified fragment is determined using a dideoxy-terminator sequencing reaction, followed by electrophoresis through an acrylamide gel. In many laboratories, this step is now automated with a computer-assisted, laser scanner. Fourth, the sequence of the product, now referred to as the "test sequence," is aligned to and compared with the sequences from reference samples. Finally, the sequence from the product is grouped, by phylogenetic reconstruction, with the most closely related reference sequences. The reconstruction is usually represented as a "tree," with closely related sequences forming neighboring branches. This allows a hierarchical comparison to establish, first, the suborder and family derivation using a small number of reference sequences from a large number of species. A close relationship or match with a "reference" sequence provides evidence for identification of the species origin of the product. One or more "outgroups" (i.e., distantly related species) are used to protect against a misclassification error. The strength of support for an identification or phylogenetic grouping is evaluated by "bootstrap" resampling of sequence data. The relative support for a grouping or branch in the tree is shown as the percentage agreement from a large number (>500) of bootstrap simulations.

As a conservative approach to forensic identification, a species identification should be considered "confirmed" only if the test sequence is "nested" within the range of type sequences for a given species. This is necessary because the molecular systematics of some marine mammals, particularly cetaceans, are

not fully described (see later). If a test sequence is intermediate between two groups of reference sequences, rather than nested within one or the other, it could be a related species or subspecies not included in the reference database.

When a large set of reference sequences is available from the range of a single species, it is possible to use "intraspecific" variation to evaluate the geographic origin of a sample. In many cases, the MANAGEMENT of marine mammals is based on geographic populations or "stocks" (Dizon *et al.*, 1992). Catch quotas and limits of incidental mortality from fisheries bycatch are usually set according to such stock definitions, as well as according to species. Hunting may be allowed in an abundant stock but prohibited in another stock of the same species that is depleted from past exploitation. However, the ability to identify or estimate the stock origins of a specimen or product is determined by the genetic distinctiveness of the recognized stocks, as well as by the comprehensiveness of the reference samples. For some (but not all) species of baleen whales and a few pinnipeds, it is possible to identify the origin of a product to oceanic population and to geographic stock within oceans (see later).

An alternative to the phylogenetic identification of an unknown specimen or product is individual identification by DNA "fingerprinting" or "profiling" using nuclear markers. This is only possible for the control of whale products originating from a regulated hunt or bycatch. As in human forensic genetics, a combination of highly variable nuclear markers can be used to establish individual identity with high probability (or to exclude identity with certainty). The DNA profiles of each individual whale can be stored on a searchable electronic database, forming a "register" of all products intended for the market. If the register is comprehensive (fully diagnostic), a match with a market product would confirm the legality of the product. A product that did not have a match in the register would be illegal. Further genetic investigation would be required to determine the species and geographic origin of illegal products.

II. "Portable Polymerase Chain Reaction"

In some cases, the application of forensic genetics to international conservation problems has involved an additional technical step required by international law. Under the requirements of the CITES, it is illegal to transport native products derived from most marine mammals without both importation and exportation permits from the respective countries of origin and destination. The processing of such permits is a lengthy affair in many countries, and for politically sensitive requests either nation can deny a permit, effectively terminating the research. However, CITES regulations do not apply to "synthetic" DNA created during amplification by PCR, assuming that all "native" product has been removed (Jones, 1994). To comply with these regulations, it has been necessary to assemble a portable PCR laboratory and conduct the initial PCR amplifications from whale products on site in the Japan and South Korea (Baker and Palumbi, 1994). Once the amplified DNA was isolated from the native product, it could be transported internationally for sequencing and final analysis. In this way, the "portable PCR" has become an important tool for international conservation genetics.

III. Monitoring Commercial Markets in Whale, Dolphin, and Porpoise Products

In recognition of historic patterns of overexploitation, the IWC voted in 1982 to impose a global moratorium on commercial whaling. Although the moratorium took effect in 1986, whaling never actually stopped. IWC member nations continue to hunt some species of whales for scientific research or for aboriginal and subsistence use. Whales killed for scientific research can legally be sold to domestic consumers and traded to other member nations of the IWC, thereby sustaining a commercial market for meat, skin, blubber, and other whale products. Small cetaceans are also hunted or taken as fisheries bycatch and sold for consumption in many parts of the world. Although the IWC regulates only hunting of large whales, international trade in all cetaceans is subject to CITES. When some species are protected by an international prohibition against hunting or trade but similar species are not, it is crucial to identify the origin of products that are actually sold in retail markets.

In an effort to monitor the sale and trade of cetaceans products, molecular methods have been used to identify the species and geographic derivation of products sold in two countries with active commercial markets: Japan and the Republic of (South) Korea. Whale meat is widely available in retail markets of both countries despite the international moratorium on commercial whaling. In 1995, high-quality whale meat sold for up

to $30/kg in Korea and $460/kg in Japan (Chan *et al.*, 1995). Japan sustains a legal market for whale products by killing several hundred minke whales in the Southern Hemisphere and 100 minke whales in the North Pacific Ocean each year under an exemption to the moratorium for scientific research. South Korea has no program for scientific hunting but reports a substantial fisheries bycatch of cetaceans each year, including up to 128 minke whales from coastal waters (Mills *et al.*, 1997). It is assumed that products from this unregulated incidental mortality are sold in local markets, but their international trade is prohibited by CITES.

IV. Species, Stock, and Individual Identification

At least 11 independent reports have been published or formally presented to the IWC documenting the molecular genetic identification of nearly 1000 products from commercial markets (Dizon *et al.*, 2000). About 50% of all products were identified as Antarctic minke whale (*Balaenoptera bonaerensis*), the target of scientific hunting by Japan in the Antarctic. About 20% were identified as North Pacific minke whale (*Balaenoptera acutorostrata scammoni*), the target of Japanese scientific hunting in the North Pacific. The remainder were found to be small cetaceans (including porpoises, family Phocoenidae; and a variety of dolphins, family Delphinidae), beaked whales (family Ziphiidae), sperm whales (family Physeteridae/Kogiidae), and other protected baleen whales. The latter group included fin (*B. physalus*), blue (*B. musculus*) or blue/fin hybrids, Bryde's (*B. edeni*), sei (*B. borealis*), and humpback whales (*Megaptera novaeangliae*). A few products sold as "whale" were found to be sheep or horse.

Detailed comparisons of mtDNA sequences have provided information on the stock derivation of some commercial products. For example, the North Pacific minke whale forms at least two stocks with marked genetic differences (Goto and Pastene, 1997): the "J" stock found in the Sea of Japan/East Sea and the "O" stock found in the North Pacific to the east of Japan. Although the "O" stock is subject to legal scientific hunting by Japan and is reported to be relatively abundant, the "J" stock was depleted by commercial hunting before 1986 and is considered a "protection stock" by the IWC. Using molecular methods, market surveys from 1993 to 1999 showed that a large proportion of products from Korea and a substantial proportion from Japan were derived from the protected "J" stock (Baker *et al.*, 2000). Although one source of these products is fisheries bycatch, the possibility of illegal hunting of this stock cannot be excluded.

The detailed analysis of a Japanese market product identified initially as a blue whale shows the importance of maintaining archived tissue from a regulated hunt for verification of trade records. The mtDNA sequence for this product matched closely with the published sequence of a blue/fin hybrid killed during a scientific whaling program by Iceland. Because mtDNA is inherited maternally, it cannot, by itself, identify a product as a hybrid. Subsequent comparison of variable nuclear DNA markers from tissue archived during the Icelandic whaling program confirmed that this product was derived from this hybrid individual (Cipriano and Palumbi, 1999).

V. Taxonomic Uncertainties

A final, and critical, consideration in using phylogenetic methods for species identification is the assumption that the taxonomy of the group in question is complete. If this basic biological information is lacking, questions about the adequacy of a reference database and the genetic distinctiveness of recognized species cannot be answered. Surprisingly, taxonomic uncertainties remain a problem among the baleen whales, even though this group includes only a handful of widely distributed, formerly abundant species. For example, two products purchased on Korean markets in 1994 grouped with reference sequences from both the sei and Bryde's whales but did not group closely with either of these two recognized species (Baker *et al.*, 1996). As a result, it seems likely that these products originated from an unrecognized species or subspecies of Bryde's whales. To be fully confident in all identifications, worldwide surveys are needed for nuclear and mtDNA variation among each of the 11 recognized species of baleen whales, particularly the Bryde's/sei complex and minke whales. Such surveys would provide the basis for an objective evaluation of current distinctions among species, subspecies, morphological forms, and populations or stocks.

See Also the Following Articles

Classification ▪ Molecular Ecology ▪ Stock Identity

References

Baker, C. S., Cipriano, F., and Palumbi, S. R. (1996). Molecular genetic identification of whale and dolphin products from commercial markets in Korea and Japan. *Mol. Ecol.* **5**, 671–685.

Baker, C. S., Lento, G. L., Cipriano, F., and Palumbi, S. R. (2000). Predicted decline of protected whales based on molecular genetic monitoring of Japanese and Korean markets. *Proc. R. Soc. Lond. B* **267**, 1191–1199.

Baker, C. S., and Palumbi, S. R. (1994). Which whales are hunted? A molecular genetic approach to monitoring whaling. *Science* **265**, 1538–1539.

Baker, C. S., and Palumbi, S. R. (1995). Population structure, molecular systematics and forensic identification of whales and dolphins. *In* "Conservation Genetics: Case Histories from Nature" (J. Avise and J. L. Hamrick, eds.), pp. 10–49. Chapman and Hall, New York.

Bowen, B. (1995). Tracking marine turtles with genetic markers. *BioScience* **45**, 528–534.

Chan, S., Ishihara, A., Lu, D. J., Phipps, M., and Mills, J. A. (1995). Observations on the whale meat trade in East Asia. *TRAFFIC Bull.* **15**, 107–115.

Cipriano, F., and Palumbi, S. R. (1999). Genetic tracking of a protected whale. *Nature* **397**, 307–308.

Dalebout, M. L., Van Helden, A., Van Waerebeek, K., and Baker, C. S. (1998). Molecular genetic identification of southern hemisphere beaked whales (Cetacea: Ziphiidae). *Mol. Ecol.* **7**, 687–694.

DeSalle, R., and Birstein, V. J. (1996). PRC identification of black caviar. *Nature* **381**, 197–198.

Dizon, A., Baker, C. S., Cipriano, F., Lento, G., Palsboll, P., and Reeves, R. (2000). "Molecular Genetic Identification of Whales, Dolphins,

and Porpoises: Proceedings of a Workshop on the Forensic Use of Molecular Techniques to Identify Wildlife Products in the Marketplace." La Jolla, CA, USA, 14–16 June 1999.

Dizon, A. E., Lockyer, C., Perrin, W. F., Demasters, D. P., and Sisson, J. (1992). Rethinking the stock concept: A phylogeographic approach. *Conserv. Biol.* **6**, 24–36.

Goto, M., and Pastene, L. A. (1997). Population structure in the western North Pacific minke whale based on an RFLP analysis of the mtDNA control region. *Rep. Int. Whal. Commn.* **47**, 531–538.

Henshaw, M. D., LeDuc, R. G., Chivers, S. J., and Dizon, A. E. (1997). Identifying beaked whales (family Ziphiidae) using mtDNA sequences. *Mar. Mamm. Sci.* **13**, 487–495.

Jones, M. (1994). PCR products and CITES. *Science* **266**, 1930.

Kocher, T. D., Thomas, W. K., Meyer, A., Edwards, S. V., Pääbo, S., Villablanca, F. X., and Wilson, A. C. (1989). Dynamics of mitochondrial DNA evolution in animals: Amplification and sequencing with conserved primers. *Proc. Natl. Acad. Sci. USA* **86**, 6196–6200.

Lento, G. M., Haddon, M., Chambers, G. K., and Baker, C. S. (1997). Genetic variation, population structure and species identity of Southern Hemisphere fur seals, *Arctocephalus* spp. *J. Hered.* **88**:202–208.

Malik, S., Wilson, P. J., Smith, R. J., Lavigne, D. M., and White, B. N. (1997). Pinniped penises in trade: A molecular genetic investigation. *Conserv. Biol.* **11**, 1365–1374.

Mills, J., Ishirhara, A., Sakaguchi, I., Kang, S., Parry-Jones, R., and Phipps, M. (1997). "Whale Meat Trade in East Asia: A review of the Markets in 1997." TRAFFIC International, Cambridge.

Pääbo, S. (1988). Ancient DNA: Extraction, characterization, molecular cloning, and enzymatic amplification. *Proc. Natl. Acad. Sci. USA* **86**, 1939–1943.

Phipps, M., Ishihara, A., Kanda, N., and Suzuki, H. (1998). Preliminary report on DNA sequence analysis of whale meat and whale meat products collected in Japan. *TRAFFIC Bull.* **17**, 91–94.

Fossil Record

R. Ewan Fordyce
University of Otago, Dunedin, New Zealand

The fossil record of marine mammals extends back more than 50 million (M) years (Fig. 1). Hundreds of species are known, with Cetacea numerically and taxonomically dominant and globally widespread, followed by Sirenia in mainly warm settings. Seals, sea lions, and relatives (Pinnipedia) are locally abundant, but other marine carnivores (otters, amphicyonids) are rare and a marine sloth unique. Fossils occur in marine strata from nearshore to deep-ocean settings and, occasionally, freshwater habitats. Remains vary from near-complete skeletons through skulls and teeth to single and usually undiagnostic bones. The taxonomic framework, although not always firm, is adequate to review the diversity and spatiotemporal distribution of fossil marine mammals. Standard zoological techniques are used in taxonomy, classification, and analysis of function, whereas routine geological techniques are used to date fossils in terms of relative and absolute time scales and to interpret sedimentary environments. The fossil record

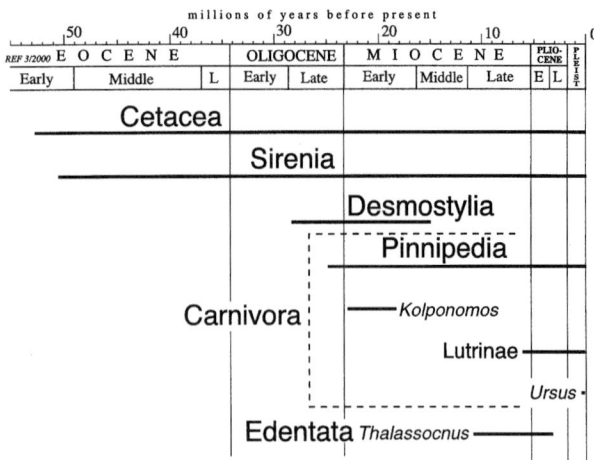

Figure 1 *Summary of geological age ranges for the main groups of fossil marine mammals.*

shows patterns of evolution and extinction that link strongly with environmental change in the oceans.

I. Occurrence, Environment, and Age

Fossil marine mammals occur in sedimentary rocks. Originally, remains accumulated in mud, silt, sand, or gravel, which, as flesh decayed, was buried and turned to rock through compaction and/or deposition of cementing minerals. Sedimentary rocks are recognized as discrete formations (genetically unified bodies of strata) and are named formally, e.g., the Calvert Formation, Maryland. Marine mammals come from strata including sandstone, mudstone, limestone, greensand, and phosphorite, most of which are marine rocks now exposed on land. Rare fossils have been recovered from the sea floor. Because broadly similar rock types may form at different times and places, sedimentary rocks must be dated to establish their time relationships.

Two correlated time scales, relative and absolute, are used in discussing the fossil record. The relative time scale has named intervals (epochs; Fig. 1) in an agreed international sequence: Eocene, Oligocene, Miocene, Pliocene, and Pleistocene. These epochs are usually subdivided into early, middle, and late. Stages (e.g., Aquitanian of Fig. 2) may provide finer subdivision. Typically, time intervals are recognized by a distinct suite of age-diagnostic fossils. The most reliable dates are based on oceanic microfossils with short time ranges, such as foraminifera, to allow correlation between ocean basins. Because of compounded errors of long-distance correlation, ages are rarely accurate to within 1 M years and many fossils can be placed only roughly within a stage. Beyond the relative time scale, absolute dates in millions of years are needed to understand rates of processes. Absolute dates are obtained usually from radiometric analysis of grains of volcanic rock interbedded with strata that also contain age-diagnostic fossils.

II. Taxonomic Framework

Fossil marine mammals are classified on the evidence of skeletons. No other useful body parts preserve, and fossils have not yet produced biomolecules useful in molecular taxonomy. SKULLS are by far the most versatile and thus important elements in classification, but teeth are taxonomically useful in most groups, and other bones (vertebrae, limb elements) have been used at times.

III. Cetacea

Fossils show that cetacean history extends back over 50 M years (Fig. 2). The earliest cetaceans—Archaeoceti—were small amphibious species that lived in fresh and brackish waters in the warm subtropical Tethys seaway between Eurasia, India, and Africa. By about 40 million years before the present (Ma), archaeocetes included large and fully marine species that had spread to temperate latitudes. The earliest modern or crown group whales, species in the filter-feeding group Mysticeti, appeared at about the Eocene/Oligocene boundary, ~34–35 Ma. Odontocetes—echolocating toothed whales and dolphins—probably radiated about the same time. Odontocetes diversified dramatically to become the most speciose group of marine mammals. Most extant cetacean families were established before the end of the Miocene, 5–10 Ma, but no living species has a history clearly longer than 2 M years.

A. Archaeoceti

Knowledge of basal whales has expanded dramatically since the early 1980s, giving new insights into cetacean phylogeny, ecology, and distribution. As noted in Kellogg's classic monograph, for many years, basal archaeocetes were known only from *Protocetus atavus* (Mokattam Formation, Middle Eocene, ~46.0 Ma; Egypt-Tethys), which was represented by a skull and uncertainly related postcranial skeleton. Since the 1980s, new finds by Gingerich, Thewissen, and others, especially in the eastern Tethys, have greatly increased diversity at the level of species, genus, and family. Basal archaeocetes are placed in the Pakicetidae, typified by the small *Pakicetus inachus*, from nonmarine red beds of the Kuldana Formation (~49–49.5 Ma), Pakistan. Skull structure (Fig. 3B) indicates limited underwater hearing capabilities, and the teeth are more simple than those of many later forms. *Pakicetus* has been cited as evidence that the earliest cetaceans radiated slowly in productive shallow waters of the Tethys seaway between Asia and India. The pakicetid (and cetacean) record has been extended back to ~53.5 Ma, based on a fragmentary jaw of *Himalayacetus subathuensis* from India. Oxygen isotopes from *Himalayacetus* indicate marine rather than nonmarine habits. Other pakicetids include species of *Ichthyolestes* and *Nalacetus,* also from the eastern Tethys.

The family Protocetidae, now expanded beyond *Protocetus,* is an Early to Middle Eocene grade for species in which the skull has an enlarged supraorbital shield, the mandible has a large mandibular foramen, and hindlimbs are reduced; they lack the complex teeth and pterygoid sinuses of younger cetaceans. Some protocetids occur in the western Tethys to western central Atlantic, including the large protocetid *Eocetus* (Egypt, North Carolina), *Pappocetus* (Nigeria; possibly North Carolina), *Natchitochia* (Louisiana), and apparently *Protocetus* (Texas). Protocetids have a high diversity in the eastern Tethys, judging from the range of teeth from Pakistan and India, although few skulls are known. *Babiacetus* (~43.5 Ma) is known

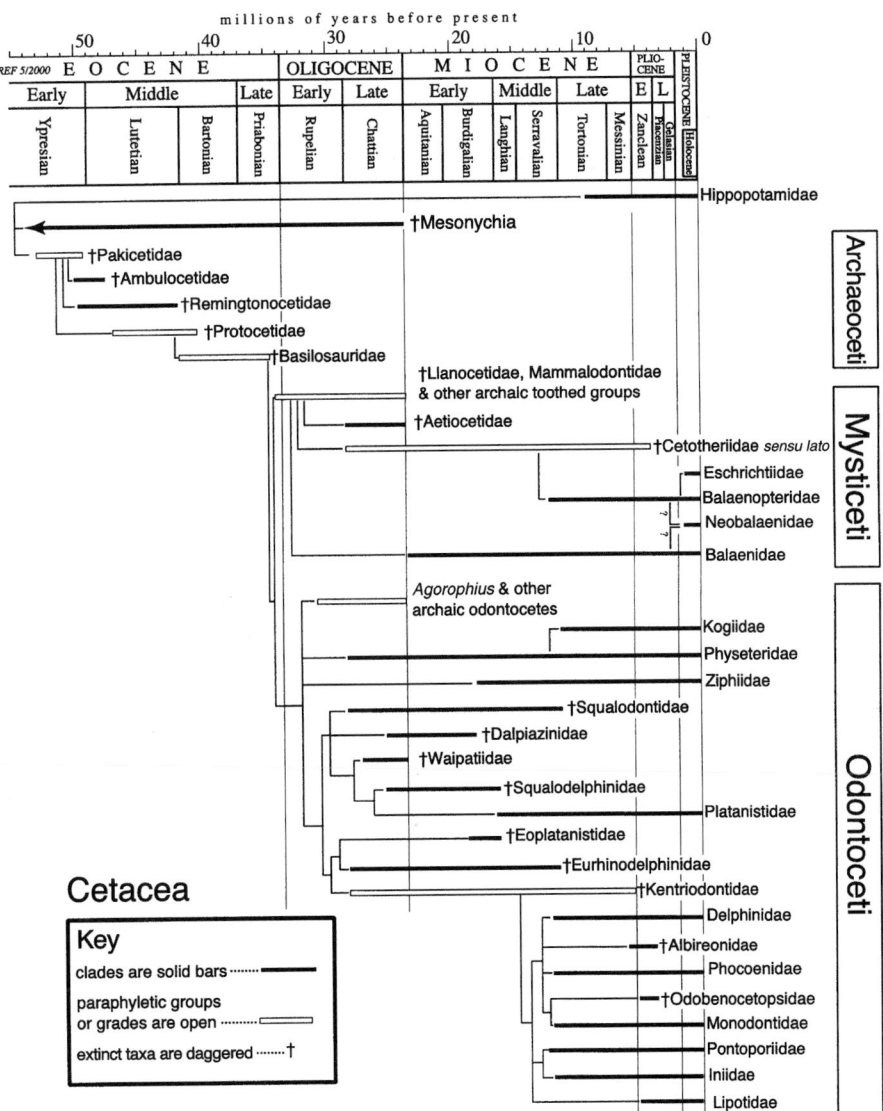

millions of years before present

Figure 2 *Geological age ranges of major groups of Cetacea. Time scale shows absolute time, Epochs (e.g., Eocene, Oligocene) and their subdivisions (e.g., early, middle), and stages (e.g., Priabonian, Rupelian). Bars show age ranges for family level cetacean taxa, either clades or, where cladistic classification is lacking, grades. Accuracy of ranges varies between different groups and different time intervals. Inferred relationships follow Uhen and others for archaeocetes, and Muizon, Fordyce and Muizon, Barnes, Heyning, and Waddell et al. for odontocetes and mysticetes. From Fordyce, R. E. (2000). Cetacea. In "Encyclopedia of Life Sciences." Nature Publishing Group, London. www.els.net*

from teeth and jaws, whereas the slightly older (45.5–46 Ma) *Gaviacetus, Takracetus,* and *Indocetus* are represented by partial skulls. More complete is *Rodhocetus kasrani* (Domanda Formation, 46–46.5 Ma; Pakistan), in which the skeleton is strikingly intermediate between that of mesonychians (the terrestrial putative ancestors of cetaceans) and later whales. Cetacean features include the short neck vertebrae and more posterior vertebrae adapted for dorsoventral oscillation, but *Rhodocetus* retains a femur and sacrum. *Rodhocetus kasrani* is

from deep rather than shallow water deposits, implying early colonization of offshore habitats. Another reputed protocetid, *Georgiacetus vogtlensis* (McBean Formation, 40–41 Ma; Georgia), has somewhat elaborate cheek teeth, a pterygoid sinus in the skull base, and a reduced link between sacrum and pelvis; because there are otherwise basilosaurid features, *Georgiacetus* is perhaps better placed in the Basilosauridae.

Two other archaeocete families are reported only from the Early and Middle Eocene of the eastern Tethys. First,

Archaeoceti

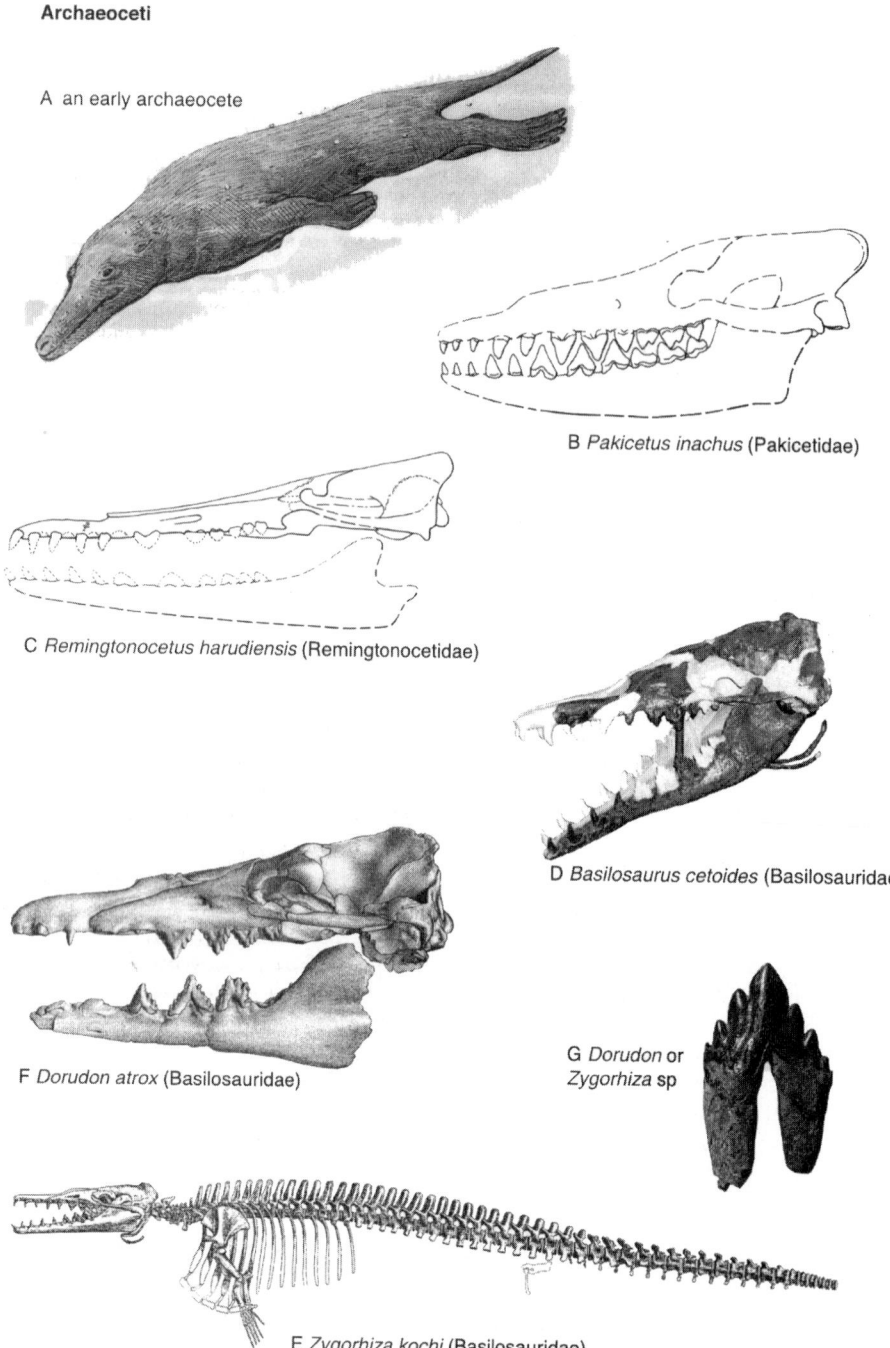

A an early archaeocete

B *Pakicetus inachus* (Pakicetidae)

C *Remingtonocetus harudiensis* (Remingtonocetidae)

D *Basilosaurus cetoides* (Basilosauridae)

F *Dorudon atrox* (Basilosauridae)

G *Dorudon* or *Zygorhiza* sp

E *Zygorhiza kochi* (Basilosauridae)

Figure 3 *Archaeocete cetaceans. (A) Reconstruction of a pakicetid-grade archaeocete, by C. Gaskin, from the Geology Museum, University of Otago. (B) Skull and mandibles of* Pakicetus inachus *(Eocene, Pakistan), lateral view, after Gingerich and Russell. (C) Skull and mandibles of* Remingtonocetus harudiensis *(Eocene, India), lateral view, after Kumar and Sahni; mandibular form is speculative. (D) Skull and mandibles of* Basilosaurus cetoides *(Eocene, Alabama), oblique lateral view of specimen in the U.S. National Museum of Natural History. (E) Skeleton of* Zygorhiza kochi *(Eocene, Alabama and Mississippi), reconstruction based on composite specimens, from Kellogg (1936); (F) Skull and mandible of* Dorudon atrox *(Eocene, Egypt), lateral view, slightly modified from Andrews. (G) Tooth of* Dorudon *or* Zygorhiza *species indeterminate (Eocene, New Zealand), medial view.*

Ambulocetus natans (Kuldana Formation, 48.0–49.0 Ma; Ambulocetidae) includes a substantially complete skeleton with a long-snouted skull and well-developed fore- and hindlimbs. *Ambulocetus* perhaps swam using pelvic paddling and dorsoventral undulations of the tail, comparable in style to some modern otters. A crocodile-like mode of predation in water is possible, but locomotion on land was probably clumsy. Second, the family Remingtonocetidae encompasses specialized long-snouted Middle Eocene species of *Remingtonocetus* (Fig. 3C), *Andrewsiphius*, and allies, known from at least partial skulls (~43.5–45 Ma). Despite previous suggestions, remingtonocetids seem unrelated to the later radiation of odontocetes.

Basilosaurids, from the later Middle and Late Eocene, are the oldest cetaceans known beyond the Tethys. They are typified by the 15-m-long *Basilosaurus cetoides* (Fig. 3D), first described and named as a fossil reptile from Louisiana (Jackson Formation, Late Eocene, ~36–39 Ma; western North Atlantic). The large size of *Basilosaurus* and its elongate vertebrae are specialized features used to recognize a subfamily Basilosaurinae. The latter include *Basilosaurus isis* (~39 Ma; Egypt, central Tethys), which has small but functional hindlimbs of ungulate-like character. Large later Eocene archaeocetes, presumably basilosaurines, have been reported from scattered localities worldwide (e.g., Northeastern Atlantic, proto-Southern Ocean, Southwest Pacific), indicating an expanding range for Cetacea.

The second subfamily of basilosaurids, the Dorudontinae, is a grade that includes smaller, more generalized, and somewhat dolphin-like species of *Dorudon, Pontogeneus, Zygorhiza* (Fig. 3E), *Saghacetus, Ancalecetus*, and perhaps *Georgiacetus*. These genera are rather similar to one another and are diagnosed on size, tooth form, and limb form. *Dorudon* is known from two species, the others one each. The typical species *D. serratus* is fragmentary, but others include some magnificent fossils (e.g., *Dorudon atrox*, Birket Qarun Formation, 39–40 Ma: Egypt; Fig. 3F). These formally named species are from the Tethys and western Central Atlantic, but tantalizing referred specimens (Fig. 3G) occur in other widely separated localities pointing, as for basilosaurines, to an early geographic spread.

Basilosaurids differ from more basal archaeocetes in having cheek TEETH with complex denticles and expanded basicranial air sinuses. These features, which indicate more sophisticated feeding and hearing capacities, link basilosaurids closely with early odontocetes and mysticetes. Several dorudontines are equally plausible sister taxa to the Odontoceti + Mysticeti, but basilosaurines seem too specialized, in terms of large size and elongate vertebral bodies, to be directly ancestral to living cetaceans. No positively identified archaeocetes are known from Oligocene or younger rocks.

B. Mysticeti

Since the 1960s, the fossil record of mysticetes has expanded to reveal diverse toothed and toothless Oligocene species, which effectively "bridge the gap" between archaeocetes and baleen-bearing mysticetes such as cetotheres, balaenopterids, and right whales. Pivotal here is *Aetiocetus cotylalveus* (Aetiocetidae; Yaquina Formation, Late Oligocene; Fig. 4A) from Oregon.

Initially, this small cetacean was identified as an archaeocete because it has teeth, but other features, including the flattened triangular rostrum, indicate that it is an archaic mysticete. Aetiocetids are moderately diverse in their known range (North Pacific) and include species of *Chonecetus* and *Morawanocetus* from Japan. Because aetiocetids are Late Oligocene only, they are probably relict basal mysticetes that persisted after more crown-ward baleen-bearing mysticetes had appeared.

Older archaic toothed fossil mysticetes are more problematic. The enigmatic *Llanocetus denticrenatus* (Llanocetidae) was based on a fragmentary large toothed jaw and a brain cast from the La Meseta Formation, Eocene–Oligocene boundary (~35 Ma) of Seymour Island, Antarctica (Fig. 4B). Fragments of unnamed small toothed mysticetes have been described from the basal Oligocene of New Zealand (Southwest Pacific–marginal proto-Southern Ocean) (Fig. 4C). Also from the margins of the proto-Southern Ocean are two other notable toothed species: the specialized short-snouted *Mammalodon colliveri* (latest Oligocene or earliest Miocene, ~23–24 Ma, Australia; Mammalodontidae) and the enigmatic large *Kekenodon onamata* (Late Oligocene, ~27–28 Ma, New Zealand; Kekenodontidae). Other toothed Cetacea formerly identified as archaeocetes and odontocetes from New Zealand, Australia, and France probably also belong in the Aetiocetidae, Llanocetidae, Mammalodontidae, and Kekenodontidae. Most are archaeocete-like, with broad-based rostra and otherwise subtle mysticete characters, and they are more widespread and diverse than previously suspected. Their teeth were probably used in filter feeding, perhaps supplemented by baleen of which no trace has been reported. These fossils represent early branches in mysticete evolution, some of which persisted as relict taxa.

Remains of toothless and baleen-bearing mysticetes are relatively abundant in Miocene and younger strata worldwide, and some fossils (e.g., species of *Mauicetus* and *Cetotheriopsis*) occur in Late Oligocene sequences back to 29–30 Ma. Many of the fossils, particularly those older than Late Miocene (~12 Ma), lack the distinguishing skull features of right whales, balaenopterids, and gray whales and have been placed in the family Cetotheriidae. As commonly used, the Cetotheriidae is a grade or "waste basket" family (paraphyletic and probably polyphyletic), including several different lineages of archaic mysticetes. More than 20 genera have variously been placed in the group. Some Miocene cetotheres are clearly close to balaenopterids, differing mainly in the more primitive structure of the frontal bone over the eye, and these fossils may indeed be on the lineage leading to rorquals. A few of the latter, e.g., *Idiocetus* and *Plesiocetus*, have been classified alternatively in the Cetotheriidae or Balaenopteridae.

Strictly, cetotheres are typified by the Middle Miocene (12–13 Ma) *Cetotherium rathkii* from Ukraine (Parathethys), which is known only from an incomplete skull rather different in structure from those of living mysticetes (Fig. 4D). For example, the upper jaw (rostrum) thrusts back into bones of the braincase with a sharp narrow triangular apex, almost obscuring the nasal bones. The Pliocene *Herpetocetus* from the Yushima Formation of Japan shows a similar structure. Eventually, the

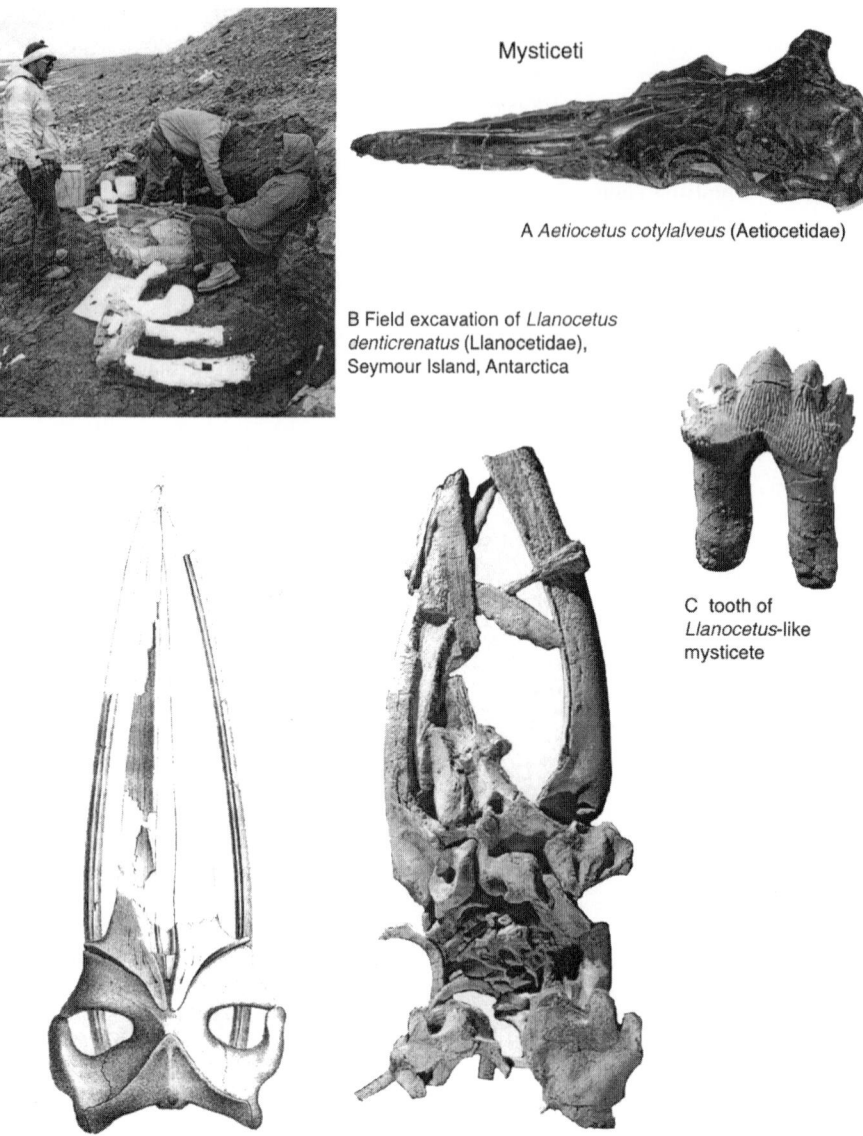

Mysticeti

A *Aetiocetus cotylalveus* (Aetiocetidae)

B Field excavation of *Llanocetus denticrenatus* (Llanocetidae), Seymour Island, Antarctica

C tooth of *Llanocetus*-like mysticete

D *Cetotherium rathkii* (Cetotheriidae)   E "Cetotheriidae" species undescribed

Figure 4 *Mysticete cetaceans. (A) Skull of* Aetiocetus cotylalveus *(Oligocene, Oregon), oblique dorsolateral view of holotype. (B) Field site showing excavation of ribs of the archaic mysticete* Llanocetus denticrenatus *(Eocene/Oligocene boundary, Antarctica). (C) Tooth of* Llanocetus-*like archaic mysticete (?Llanocetidae) (Oligocene, New Zealand), from Fordyce. (D) Skull and mandibles of* Cetotherium rathkii *(Miocene, Ukraine), dorsal view, holotype skull and uncertainly associated mandibles, from Van Beneden and Gervais. (E) Broken skull, mandibles, and associated elements of an undescribed* Mauicetus-*like "cetothere" (Oligocene, New Zealand), dorsal view; specimen in Geology Museum, University of Otago.*

Cetotheriidae will be defined as a clade for *Cetotherium rathkii* and its close relatives, although this will leave many other archaic mysticetes (e.g., Fig. 4E) uncertainly placed. This matter, the exact family level identity of many fossil mysticetes, is a key problem area in cetacean phylogeny.

Fossil balaenopterids, like their living relatives, have a distinctive skull structure in which the frontal bone above the eye is de-pressed to house the large muscles that close the lower jaw. Fossils such as *Megaptera miocaena* (Sisquoc Formation, California, northeast Pacific) indicate that *Megaptera* (subfamily Megapterinae, humpback whales) had diverged from the *Balaenoptera* lineage (subfamily Balaenopterinae, rorquals) by the Late Miocene (~12 Ma). Other Late Miocene and younger records of *Megaptera*, such as *M. hubachi* (Figs. 5A and 5B), are known. Although

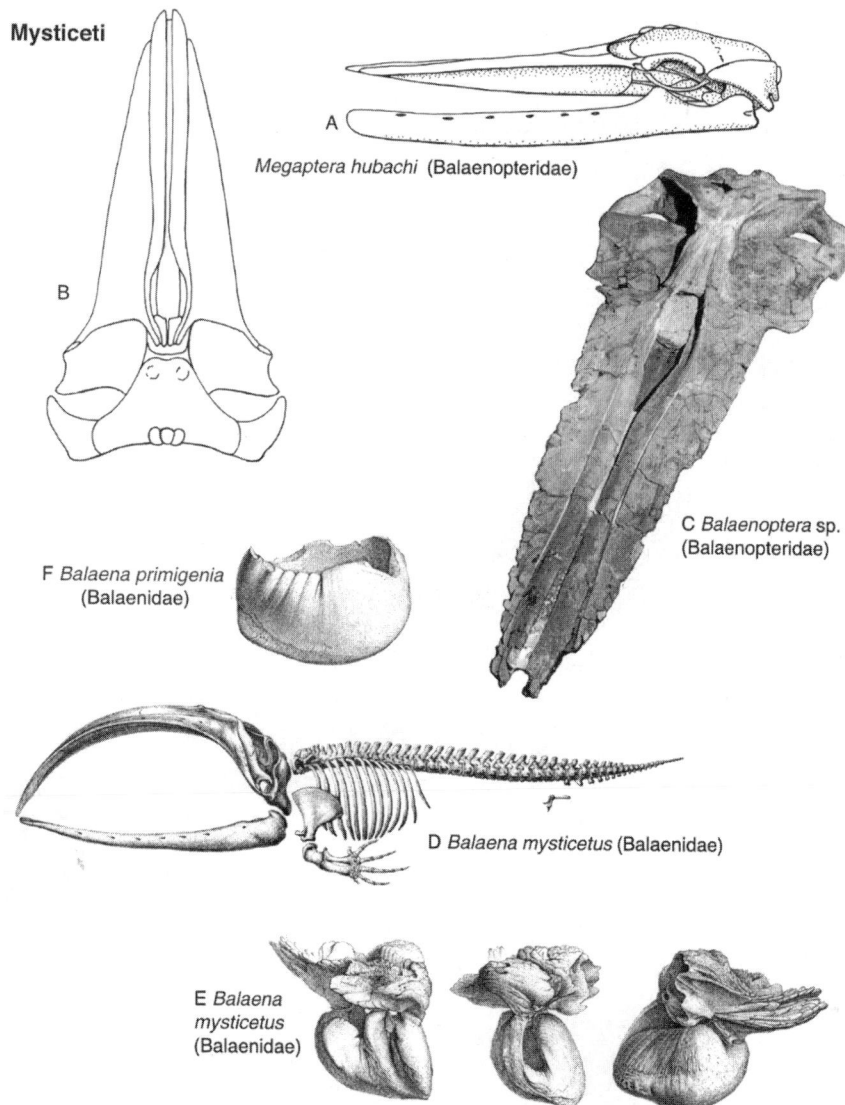

Mysticeti

A

Megaptera hubachi (Balaenopteridae)

B

C *Balaenoptera* sp.
(Balaenopteridae)

F *Balaena primigenia*
(Balaenidae)

D *Balaena mysticetus* (Balaenidae)

E *Balaena*
mysticetus
(Balaenidae)

Figure 5 *Mysticete cetaceans. Skull and mandibles of* Megaptera hubachi *(Pliocene, Chile), after Dathe (1983, Zeitschrift für geologische Wissenschaften 11): (A) skull, dorsal view; and (B) skull and mandibles, lateral view. (C) skull of an undescribed species of* Balaenoptera *(Pliocene, New Zealand), oblique dorsal view; specimen in Museum of New Zealand. Skeleton and ear bones of* Balaena mysticetus *(extant, Arctic), from Van Beneden and Gervais (1868–1880). (D) lateral view of skeleton and (E) lateral (left), anterior (middle), and internal (right) views of ear bones, with periotic above and tympanic bulla below. (F) isolated tympanic bulla of* ?Balaena primigenia *(Pleistocene?, Britain), internal view, from Van Beneden and Gervais.*

Megaptera is rather divergent from *Balaenoptera*, this does not necessitate a much older split between these two groups. For *Balaenoptera* and close relatives, the oldest described fossils are also Late Miocene (~12 Ma), with less certain Middle Miocene records. There are many records of later Miocene, Pliocene, and Pleistocene *Balaenoptera* fossils (Fig. 5C).

The oldest fossil gray whale, from the Pleistocene (~0.5 Ma), gives no obvious clue to the origins of *Eschrichtius robustus*. Fossils do not support the notion of links between the gray whale

and cetotheres; equally, they do not discount relationships with *Balaenoptera*. Geologically young gray whale fossils indicate that these animals occurred in the North Atlantic.

Mysticetes diversified greatly in the Oligocene, before about 25 Ma, yet the earliest described right whale (Balaenidae) is the Early Miocene *Morenocetus parvus* (~20 Ma) from Patagonia. *Morenocetus* anchors the Balaenidae (represented by living *Balaena* and *Eubalaena*; Figs. 5D and 5E) as the oldest family of living mysticetes. Jaws are not known for *Morenoce-*

tus, but other aspects of the skull fit the balaenid pattern well. The later Miocene record of right whales is patchy, with only fragmentary fossils reported, but better Pliocene (2–5 Ma) and Pleistocene (<2 Ma) records include many nominal species (e.g., *B. primigenia*, Fig. 5F) and other published records of partial skeletons. A nearly complete skeleton of a large Early Pliocene (4–5 Ma) balaenid close to *Balaena mysticetus* is known from the Yorktown Formation, Atlantic Coastal Plain. There is no published fossil record of Neobalaenidae to indicate the origins of the pygmy right whale, *Caperea marginata*, from its presumed balaenid ancestors.

C. Odontoceti

Odontocetes are much more diverse in terms of taxa and structure than mysticetes. The oldest named species that are well dated are Late Oligocene (<30 Ma). Reportedly older species (Early Oligocene, ~30–34 Ma) are undescribed or of less certain age. The most archaic described odontocete is *Archaeodelphis patrius*, based on a fragmentary skull of uncertain origin and possible Oligocene age. Uniquely, this enigmatic species barely shows evidence of nasofacial muscles, which, in other odontocetes, are implicated in echolocation. Despite its reputedly ancestral position, because its skull base is somewhat specialized, *Archaeodelphis* is perhaps not directly on a lineage leading to living species. Almost as archaic is *Xenorophus*, a bizarre Late Oligocene genus containing one or two species with specialized facial structures superimposed on a rather primitive skull (Fig. 6A).

There is not yet consensus about the widely discussed Agorophiidae, a group previously linked with the squalodontids. Strictly, the family includes only the type species of *Agorophius* (*Agorophius pygmaeus*; Cooper Marl, Late Oligocene, >24 Ma, South Carolina), for which most of the holotype skull (Fig. 6B) is lost. Some paleocetologists prefer a grade family Agorophiidae, including *Archaeodelphis*, *Xenoro-phus*, and varied fragmentary archaic odontocetes. There is no evidence at present to regard the Agorophiidae as a basal clade of odontocetes.

Among living odontocetes, sperm whales (*Physeter macrocephalus*, Figure 7A-B; and *Kogia* spp.) represent a basal radiation, yet sperm whales have a poor early fossil record. The fragmentary *Ferecetotherium kelloggi* is reportedly Late Oligocene (23+ Ma; Maikop? Formation), providing the earliest record for a living family of odontocetes. Early Miocene sperm whales, such as *Diaphorocetus* and *Idiorophus* (Figs. 7D and 7E) from the South Atlantic, show the characteristic basined facial bones of later sperm whales, but retain a more primitive braincase, which links them firmly with other odontocetes. Many named species of fossil sperm whales are based on isolated teeth (e.g., species of *Scaldicetus*; Fig. 7C), which, however, reveal little about the actual animal. The oldest Kogiidae are later Miocene species from the eastern Pacific, including *Praekogia* from Isla Cedros (Almejas Formation, ~6 Ma), and the large narrow-skulled *Scaphokogia* from the Pisco Formation of Peru.

Beaked whales, Ziphiidae, have a disappointing early record based on isolated and worn rostra ("beaks" or upper jaws) but few diagnostic skull remains. The oldest reported ziphiid is *Squaloziphius emlongi* (Clallam Formation, latest Oligocene or earliest Miocene, ~23 Ma, Washington), which, equally, may be a eurhinodelphinid. Other supposed ziphiids actually represent the platanistoid dolphin group Squalodelphinidae. Most of the described ziphiid fossils are resistant isolated rostra of Middle Miocene age or younger (<16 Ma). Many of these are named, mostly as species of *Belemnoziphius*, *Choneziphius* (Fig. 7E), and *Mesoplodon*, despite the problem of determining exactly what sort of ziphiid they are. Ziphiid beaks and ear bones (e.g., Figs. 7H–7J) are common cetacean fossils at unconformities and among deep sea dredgings. Perhaps the most

Odontoceti

A *Xenorophus sloani* (Odontoceti incertae sedis)

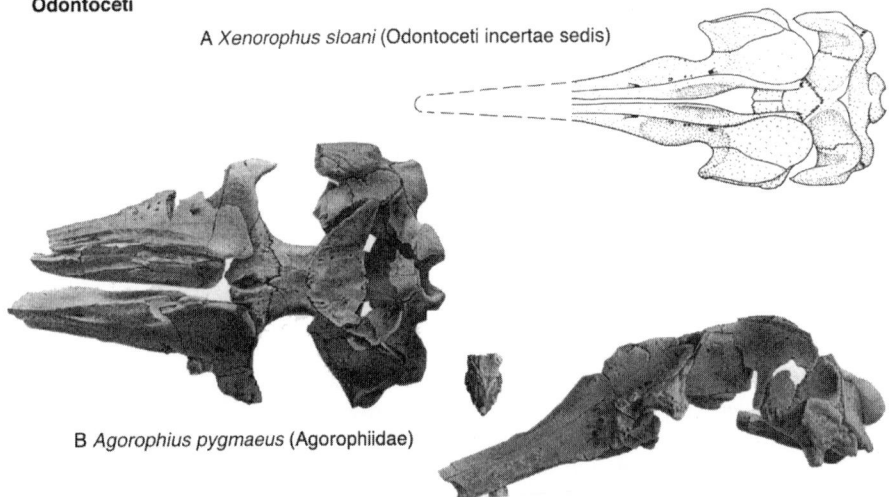

B *Agorophius pygmaeus* (Agorophiidae)

Figure 6 *Odontocete cetaceans. (A) Skull of* Xenorophus sloani *(Oligocene, South Carolina), dorsal view, after Kellogg (1923) and after Whitmore and Sanders (1977, Systematic zoology 25). (B) Skull and teeth of* Agorophius pygmaeus *(Oligocene, South Carolina), dorsal and lateral views of skull and detail of one tooth, from True.*

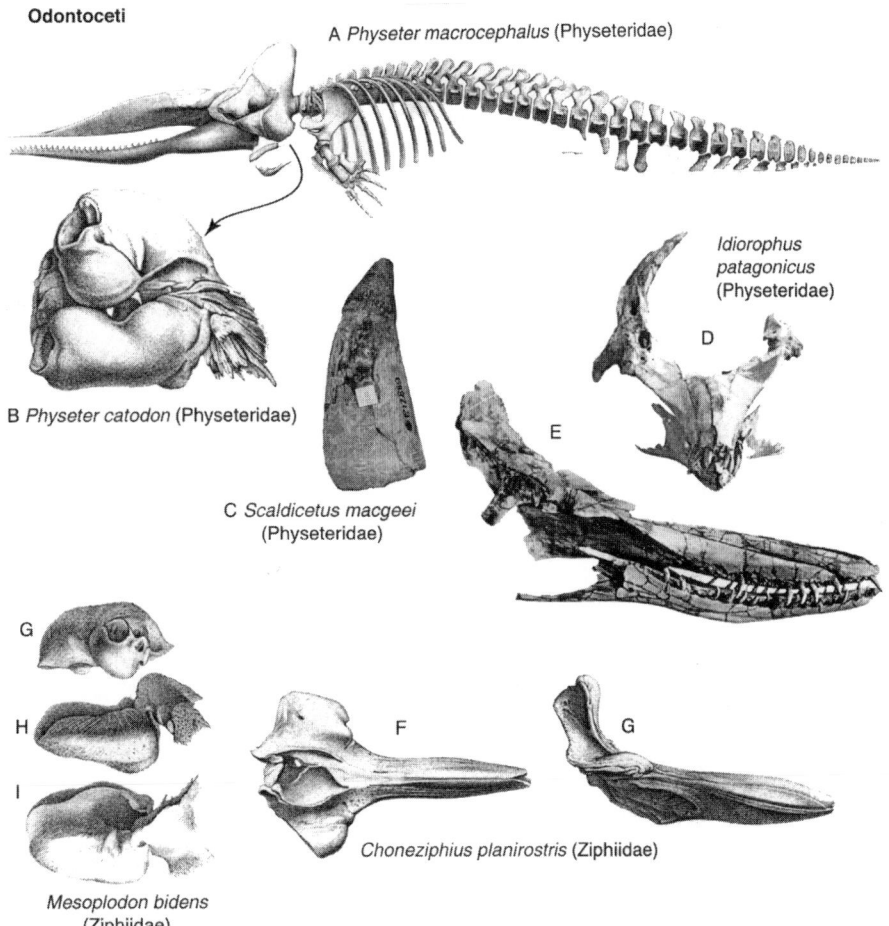

Odontoceti

A *Physeter macrocephalus* (Physeteridae)

Idiorophus patagonicus (Physeteridae)

D

E

B *Physeter catodon* (Physeteridae)

C *Scaldicetus macgeei* (Physeteridae)

G

H

I

F

G

Choneziphius planirostris (Ziphiidae)

Mesoplodon bidens (Ziphiidae)

Figure 7 *Odontocete cetaceans. Skeleton and ear bones of* Physeter macrocephalus *(extant, cosmopolitan), from Van Beneden and Gervais: (A) lateral view of skeleton and (B) internal view of ear bones, periotic above and tympanic bulla below. (C) Tooth of* Scaldicetus macgeei *(Pliocene, Australia). Skull of* Idiorophus patagonicus *(Miocene, Patagonia), from Lydekker (1894), (D) anterior view and (E) lateral view. Incomplete skull of* Choneziphius planirostris *(Miocene–Pliocene, Belgium), from Van Beneden and Gervais: (F) dorsal view and (G) lateral view. Earbones of* Mesoplodon bidens *(extant, Atlantic), from Van Beneden and Gervais: (H) periotic, internal; (I) tympanic bulla, internal; and (J) tympanic bulla, dorsal.*

unusual occurrence of a fossil cetacean is that of a large ziphiid in freshwater Miocene strata of Kenya.

Living platanistids are known only from one species of *Platanista* from rivers in India and Pakistan, but fossil relatives are diverse. Some extinct long-beaked marine dolphins from the western North Atlantic, including *Zarhachis* (Early–Middle Miocene, ~14–16+ Ma) and *Pomatodelphis* (Late Miocene, 6–9 Ma), have facial crests, which place them close to *Platanista* in the Platanistidae. The marine *Allodelphis* (Jewett Sand, Early Miocene, ~20 Ma; California) is a more archaic possible platanistid that lacks facial crests. Clearly, platanistids invaded freshwater late in their history.

Related to the platanistids is the extinct and more archaic family Squalodelphinidae, which is based on the long-beaked *Squalodelphis* (Early Miocene, Mediterranean). Southern squalodelphinids include several species of *Notocetus* (Fig. 8A)

from the latest Oligocene and Early Miocene (~18–24 Ma; New Zealand and Patagonia) bordering the Southern Ocean; formerly, some were identified wrongly as beaked whales.

A small heterodont dolphin with a slightly asymmetrical skull, *Waipatia maerewhenua* (Late Oligocene, ~25–26 Ma; New Zealand), typifies a family Waipatiidae. *Waipatia* (Figs. 8B and 8C) shows skull features that indicate an ability to echolocate. Other previously enigmatic odontocetes, such as *Sulakocetus* and *Microcetus* from the Tethys-North Atlantic, may also be waipatiids. These animals probably lie close to the base of the platanistoids.

Probably the best known of the platanistoids are the shark-toothed dolphins, Squalodontidae, which are geographically widespread medium to large odontocetes with long rostra and conspicuous, robust, triangular heterodont teeth (Figs. 8D and 8E). Squalodontids were long implicated in the phylogeny of

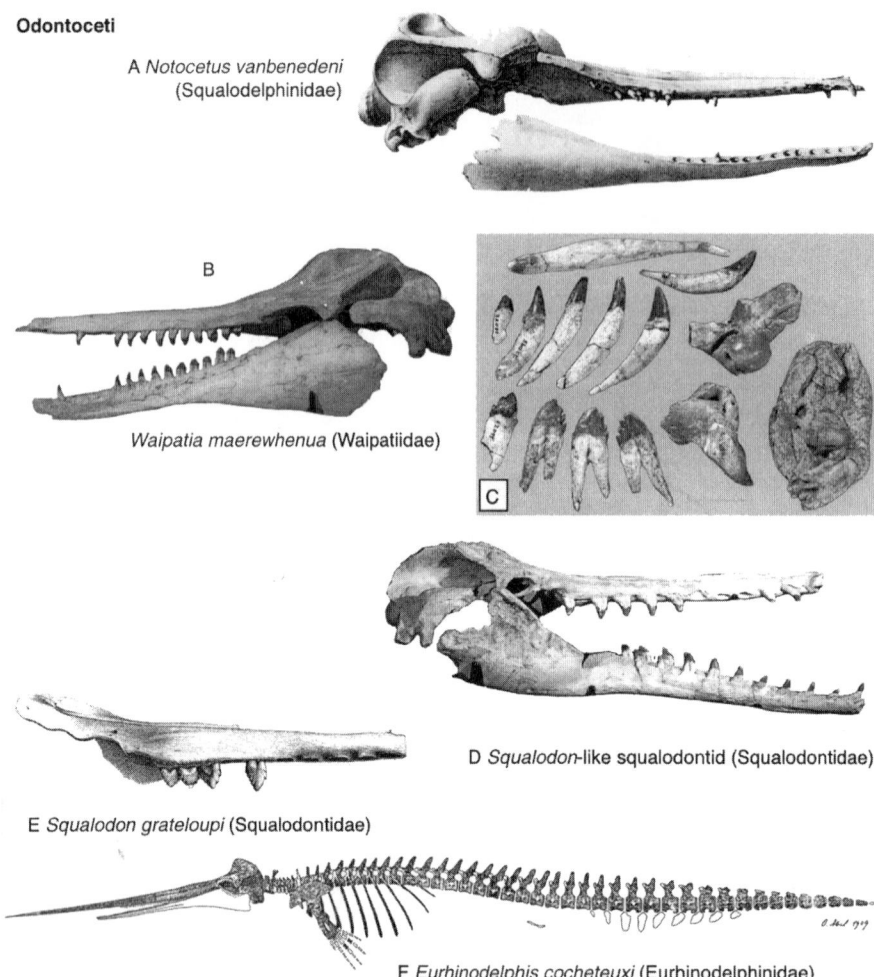

Odontoceti

A *Notocetus vanbenedeni*
(Squalodelphinidae)

B

Waipatia maerewhenua (Waipatiidae)

C

D *Squalodon*-like squalodontid (Squalodontidae)

E *Squalodon grateloupi* (Squalodontidae)

F *Eurhinodelphis cocheteuxi* (Eurhinodelphinidae)

Figure 8 *Odontocete cetaceans. (A) Skull and mandibles of* Notocetus vanbenedeni *(Miocene, Patagonia), from Lydekker.* Waipatia maerewhenua *(Oligocene, New Zealand): (B) skull, lateral view; and (C) teeth and ear bones. (D) Skull and mandibles of* Squalodon-*like squalodontid (Oligocene, New Zealand), lateral view. (E) Incomplete skull of* Squalodon gratelupi *(Miocene, France), lateral view, from Van Beneden and Gervais. (F) Skeleton of* Eurhinodelphis cocheteuxi *(Miocene, Belgium), lateral view, from Abel (1909).*

living odontocetes, but recently have been included among the platanistoids. Squalodontids are based on *Squalodon gratelupi* (Early Miocene, ~20 Ma, eastern North Atlantic; Fig. 8E) and other clearly related species. Other supposed species of *Squalodon*, which were named from fragments, apparently represent other families of odontocetes or even mysticetes. Probably, the squalodontids include only *Squalodon, Kelloggia, Eosqualodon,* and *Phoberodon.* The broad-beaked *Prosqualodon* (latest Oligocene–Early Miocene, ~18–~24 Ma, marginal Southern Ocean) may represent a related separate family, Prosqualodontidae.

Species in the extinct family Eurhinodelphinidae (sometimes called Rhabdosteidae) have dramatically long rostra and quite specialized skulls. The early record is patchy, with Oligocene forms restricted and poorly known. This, and problems of interpreting skull bones, means that relationships are obscure; they could lie with delphinoids or, equally, platanistoids. Early and Middle Miocene eurhinodelphinids such as *Eurhinodelphis* (Fig. 8F) and *Agyrocetus* are widespread. One Late Oligocene species occurs in freshwater strata of central Australia. Eurhinodelphids are related to another extinct group, the Early Miocene Eoplatanistidae.

Early true dolphins represent the grade family Kentriodontidae (Delphinoidea). Kentriodontids are geographically widespread small to medium-sized animals with largely symmetrical skulls, including *Kentriodon* (Fig. 9A) and *Hadrodelphis* (both Calvert Formation, Early to Middle Miocene, 14–16+ Ma). Rare Late Oligocene (>23 Ma) kentriodontids indicate ancient origins for the lineage leading to delphinids. The geologically youngest kentriodontids were contemporaneous with early delphinids from the Late Miocene (~10–12 Ma), animals characterized particularly by markedly asymmetrical skulls.

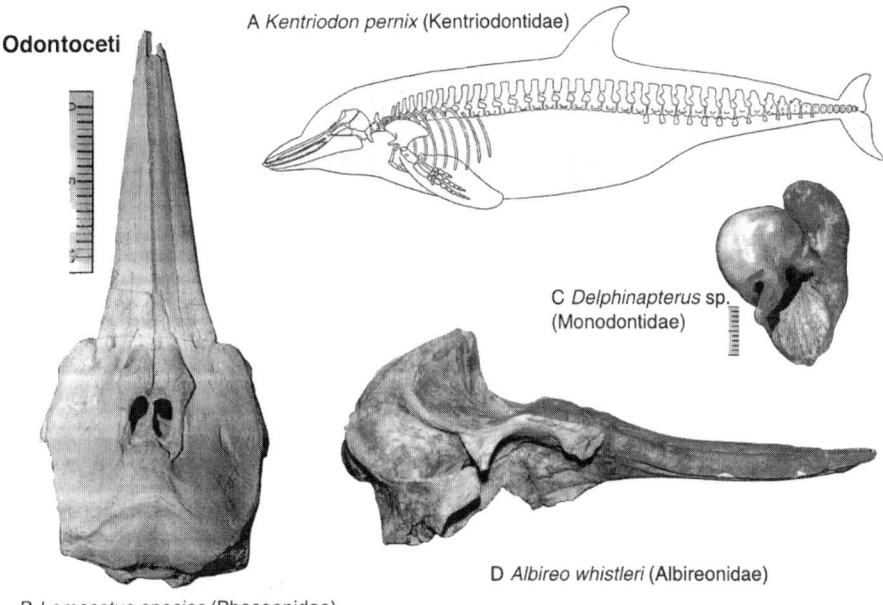

Figure 9 *Odontocete cetaceans. (A) Skeleton of* Kentriodon pernix *(Miocene, Maryland), lateral view of speculative reconstruction, from Kellogg (1927). (B) Skull of* Lomacetus gins-burgi *(Miocene, Peru), dorsal view. (C) Periotic of* Delphinapterus *sp. (Pliocene, North Carolina), ventral view. (D) Skull of* Albireo whistleri *(Miocene, Isla Cedros), lateral view.*

(Supposedly older records of delphinids, >12 Ma, appear incorrect). Delphinids are important components of, especially, Pliocene assemblages (2–5 Ma). For example, fossils from northern Italy include skulls, teeth, and ear bones referred to species of *Stenella, Tursiops, Orcinus,* the *Tursiops*-like *Hemisyntrachelus,* and the extinct *Astadelphis.*

Like delphinids, porpoises (Phocoenidae) and white whales (Monodontidae) range back to the Late Miocene, although no clear ancestors have been identified. The record of porpoises is better than that of monodontids, originating with *Salumiphocaena stocktoni* from the eastern North Pacific (Monterey Formation, 10–11 Ma; California). Fossil ear bones and skulls indicate a high diversity for phocoenids in the latest Miocene and earlier Pliocene of the eastern Pacific (e.g., *Lomacetus* and *Piscolithax* spp.; Fig. 9B). A more sparse record of monodontids reveals that this group was also important in warm equatorial waters of the eastern Pacific until well into the Pliocene, with fossils such as *Denebola* present in Peru and Cedros Island. In the Atlantic, Early Pliocene (~4.5 Ma) *Delphinapterus* occurs in North Carolina (Yorktown Formation; Fig. 9C), and there are reports of geologically young (Late Pleistocene, <0.5 Ma) *Delphinapterus* and *Monodon* from midlatitude North Atlantic shores.

While most of the Late Miocene and younger fossil delphinoids are of modern aspect, two extinct families are known from the latest Miocene and earlier Pliocene of the eastern Pacific. The Albireonidae is based on the latest Miocene porpoise-like *Albireo whistleri* (Fig. 9D) from Cedros Island, whereas two species of tusked "walrus whales" (*Odobenocetops*; Odobenocetopsidae) are from the Pisco Formation, Peru,

The latter are uncannily like the living walrus, *Odobenus rosmarus,* in skull form.

Several groups of river dolphin lie close to the Delphinoidea. Among members of these groups, pontoporiid fossils occur in freshwater strata in South America and Mio-Pliocene marine rocks bordering the eastern North Pacific and western North Atlantic. The small short-beaked *Brachydelphis mazeasi* (Pisco Formation, ~12–15 Ma) from Peru represents a distinct subfamily (Pontoporiidae: Brachydelphininae) related to the long-beaked Peruvian fossil *Pliopontos littoralis* (~4–5 Ma) and the living franciscana (*Pontoporia blainvillei*). Marine long-beaked species of *Parapontoporia* from Isla Cedros (Almejas Formation, ~5 Ma) and southern California (San Diego Formation, ~2–3 Ma) may be related to *Pontoporia,* although ear bone anatomy points to relationships with living baiji (*Lipotes vexillifer*). The problematic *Prolipotes,* based on a fragment of possibly Miocene lower jaw from freshwater deposits in China, is too incomplete to confirm a relationship with *Lipotes.* Many fossils have been placed in the family Iniidae, with the living *Inia geoffrensis,* but most of these belong elsewhere. Fossil iniids include the South American Late Miocene ("Mesopotamiense" horizon, >5 Ma) *Ischyrorhynchus* and probably *Saurodelphis* and may include the marine *Goniodelphis hudsoni* (Early Pliocene, 4–5 Ma) from Florida.

IV. Marine Carnivora

The record of marine Carnivora starts about the Oligocene/Miocene boundary (~25 Ma). Pinnipeds are most speciose and geographically widespread aquatic carnivores, well represented in northern temperate regions. Related closely is the fossil marine amphicyonid, *Kolponomos.* The only marine ursid, the polar bear, *Ursus maritimus,* has no significant fossils and probably originated only a few tens of thousands of years ago. Otters have a meager but notable fossil record.

A. Pinnipeds: Seals, Walruses, Fur Seals, and Sea Lions

As with Cetacea, the fossil record of seals, sea lions, walruses, and relatives has expanded greatly in recent decades, particularly through finds around the North Pacific. Fossil pinnipeds are less diverse than the Cetacea in terms of species, genera, and families, and the geological record is shorter, extending back to the late Oligocene (~25 Ma) (Fig. 10). Furthermore, the lack of consensus about family level taxonomy and indeed about pinniped monophyly versus diphyly complicates a review of these animals; a genus, or different species on one genus, may be placed in one of several different taxa, some of which are acknowledged grades, and others of which are clades of varying family level rank (e.g., family or subfamily).

Pinniped diphyly was favored strongly from the 1960s to 1980s (Fig. 10A). Of the two living groups and their fossil relatives, the eared seals—fur seals and sea lions—supposedly originated from a bear-like stock in the North Pacific. The walrus was placed tentatively with the eared seals to form the group Otarioidea (or Otariidae). In contrast, the earless or true seals (Phocidae, or Phocoidea) were thought to be a distinct and different group related to mustelids and originating in the North Atlantic. By the 1990s, some anatomically based cladistic analyses of pinnipeds concluded that seals and relatives are monophyletic, with the walrus and Phocidae being the most crown-ward taxa, and Otariidae more basal in the group Pinnipedimorpha (Fig. 10B). Classification is still volatile, and communication is hampered because taxonomic names have quite different meanings to different workers. The cladistic methods seem clear and repeatable, supporting the notion of monophyly. Conversely, a good case has been made on morphological grounds that some key characters used in cladistics are ambiguous, arguably including significant reversals and convergences, undermining claims of monophyly. Further, the monophyletic pattern plotted against time (Fig. 10C, lower) produces long "ghost lineages," whereas the diphyletic pattern reveals long-ranging paraphyletic groups (Fig. 10C, upper). Some recent molecular taxonomies have corroborated anatomical cladistics and recognized the living pinnipeds as monophyletic, but have supported "traditional" views in placing the walrus closer to otariids than to phocoids. There is not yet a consensus.

The small marine carnivore *Enaliarctos mealsi* is an archaic species based on skulls and a skeleton from the Jewett Sand of California (about Oligocene/Miocene boundary; ~23 Ma) (Fig. 11A). The species typifies the Enaliarctinae (or, alternatively, Enaliarctidae), is widely regarded as a basal otariid, and may lie at the stem of all pinnipeds. Other enaliarctines of comparable age have been named from fossils from California and Oregon (e.g., *Pteronarctos goedertae,* Astoria Formation, Middle Miocene, ~16 Ma) (Fig. 11B), and an undetermined *Enaliarctos*-like species occurs in Japan (late Early Miocene, ~17

Ma). Also from California, and roughly contemporaneous with *Enaliarctos,* is *Pinnarctidion bishopi,* an archaic pinniped also initially placed in the Enaliarctinae. A second species, *P. rayi,* from coastal Oregon (Early Miocene, >19 Ma) includes skulls and postcranial bones, which, alternatively, have been allied with Phocidae.

Fur seals and sea lions, [Otariidae (or Otariinae)] appeared by the Late Miocene (>9 Ma) in the North Pacific, as shown by *Thalassoleon* and *Pithanotaria* (Fig. 11C). The two living otariid groups, the fur seals and sea lions, have a short fossil record around the Pacific and Southern Ocean. Fossil northern fur seals (arctocephalines) include the small Late Pliocene (2–3 Ma) *Callorhinus gilmorei* (Japan, California, Baja, California), which is close to the living North Pacific *C. ursinus.* A larger, southern fur seal is the Late Pliocene (~2 Ma) *Hydrarctos lomasiensis* from Peru, which is possibly allied to living fur seals in the genus *Arctocephalus.* Among sea lions (otariines), North Pacific fossil *Eumetopias* is known from teeth and some postcranial bones of latest Pliocene age (~2.5 Ma, Japan), and there are reports of Pleistocene *Zalophus.* In the south, the extinct *Neophoca palatina* is based on a single skull of middle Pleistocene age (~0.5+ Ma, New Zealand).

The diverse cluster of medium to large extinct pinnipeds known as allodesmines is typified by the rare *Allodesmus kernensis* and *A. kelloggi* (Fig. 11D) and more common *A. gracilis* (Fig. 11E) from Sharktooth Hill, California (Middle Miocene, 13–14 Ma). Allodesmines range temporally from the early Middle to Late Miocene, and widely across the North Pacific (Japan, and Washington to Baja California, Mexico). At least eight species have been described in four genera (*Allodesmus, Brachyallodesmus, Atopotarus,* and *Megagomphos*). They were large-eyed, deep-diving animals that swam by forelimb propulsion. *Allodesmus* and its relatives have been placed variously (as a family or a subfamily) in the Otariidae or alternatively close to the true seals, Phocidae.

The single species of living walrus is the sole survivor of a once diverse group. Walrus origins may lie with the Early Miocene (16–22 Ma) genus *Desmatophoca,* which contains two large rare species from the Northeast Pacific (Fig. 11F). Alternatively, *Desmatophoca* has been allied with the phocid radiation. Another group implicated in walrus origins is the North Pacific Miocene Imagotariinae. Of these, the basal *Neotherium* was small, but most imagotariines were medium to large archaic sea lion-like forms. Middle Miocene (~13–16 Ma) fossils include species of *Prototaria* (Japan) and *Proneotherium* (Oregon); *Imagotaria* (California) is a representative Late Miocene form. Though on the walrus lineage, these were not particularly mollusk eaters but probably had a generalized piscivorous diet.

The North Pacific group Dusignathinae includes extinct forms regarded variously as walrus-like sea lions ("pseudowalruses") or, alternatively, true walrus relatives (with sister group status in Fig. 11C, upper and lower). These animals, including *Dusignathus* and the very large *Gomphotaria* (known from a nearly complete skeleton), range from Late Miocene (5–8 Ma) to Late Pliocene (2–3 Ma). Dusignathines differ from odobenines in that both the upper and the lower canines are enlarged as tusks.

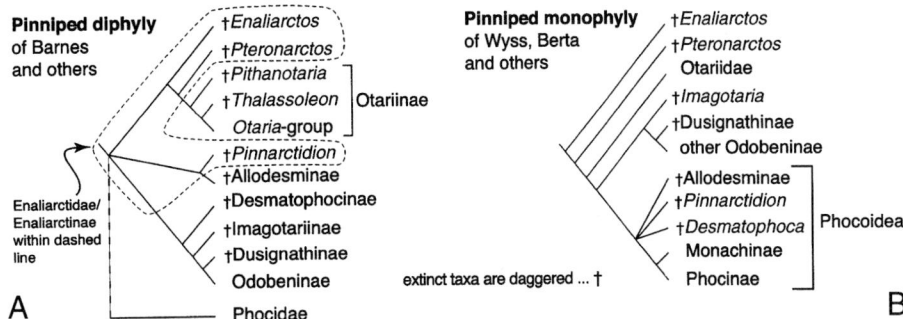

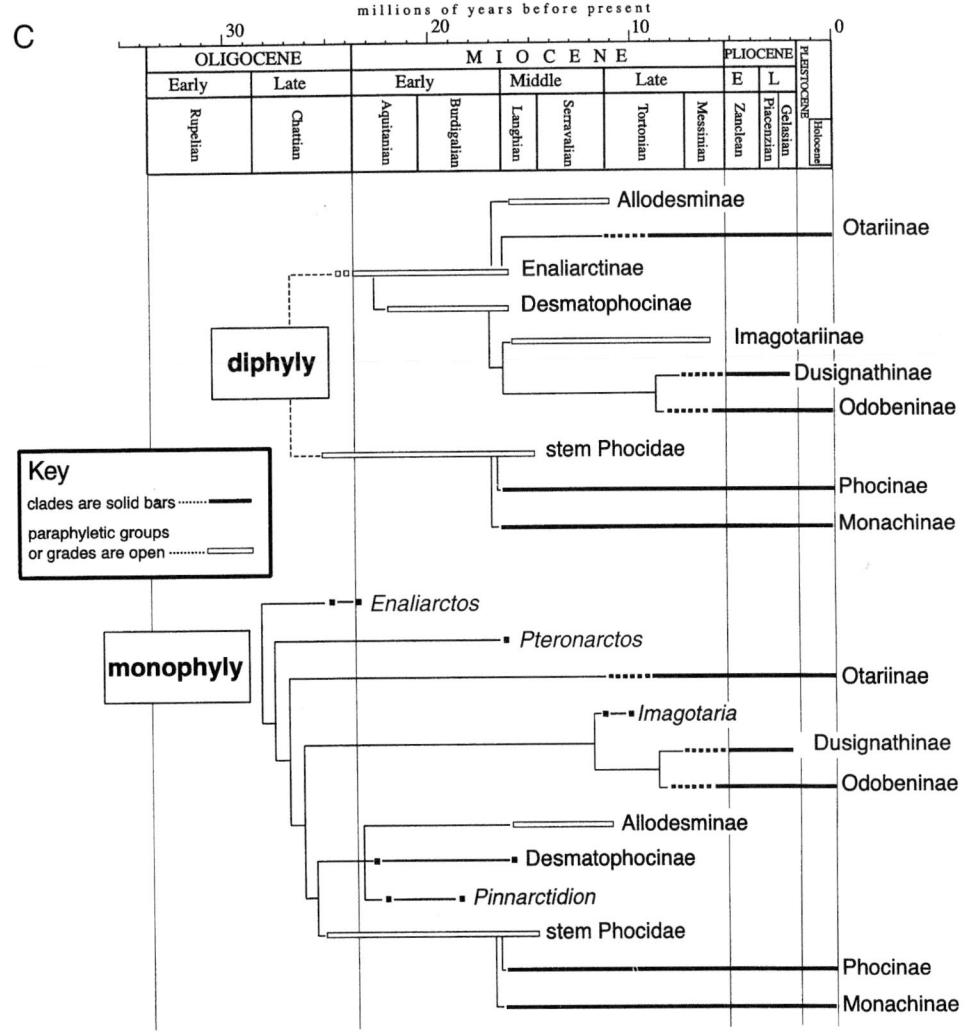

Figure 10 *Different hypotheses of relationships among pinnipeds, and geological age ranges of major groups of pinnipeds. (A) One concept of pinniped diphyly, based on Barnes and others. (B) One concept of pinniped monophyly, based on Berta and others. (C) Geological age ranges of major groups of pinnipeds, plotted against a standard time scale (for time scale details, see Fig. 2). Hypotheses of diphyly (upper) or monophyly (lower) are used to show inferred relationships. Bars show age ranges for taxa, which are mostly families and subfamilies but sometimes genera. Clades are in-filled bars, grades or paraphyletic or stem groups are open bars, and genera or key species within genera are dots.*

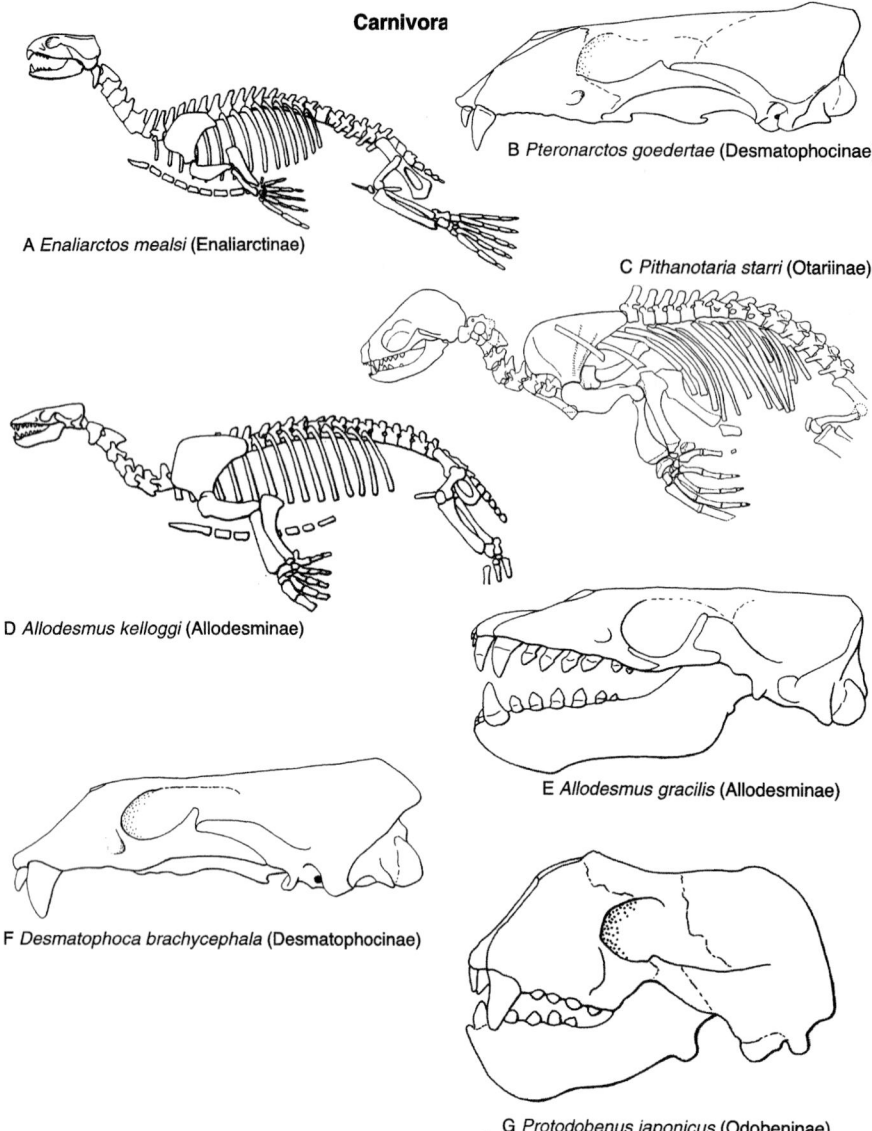

Carnivora

A *Enaliarctos mealsi* (Enaliarctinae)

B *Pteronarctos goedertae* (Desmatophocinae)

C *Pithanotaria starri* (Otariinae)

D *Allodesmus kelloggi* (Allodesminae)

E *Allodesmus gracilis* (Allodesminae)

F *Desmatophoca brachycephala* (Desmatophocinae)

G *Protodobenus japonicus* (Odobeninae)

Figure 11 *Pinniped carnivores. (A) Skeleton of* Enaliarctos mealsi *(Oligocene/Miocene boundary, California), lateral view, after Berta et al. (B) Skull of* Pteronarctos goedertae *(Early Miocene, California), lateral view, after Barnes (1989, Contributions in science, Nat. Hist. Mus. of LA County 403). (C) Skeleton of* Pithanotaria starri *(Late Miocene, California), lateral view, after Kellogg (1925). (D) Skeleton of* Allodesmus kelloggi *(Middle Miocene, California), lateral view, after Mitchell. (E) Skull and mandibles of* Allodesmus gracilis *(Middle Miocene, California), lateral view, after Barnes and Hirota (1995, The Island Arc 3[4]). (F) Skull of* Desmatophoca brachycephala *(Early Miocene, California), lateral view, after Barnes. (G) Skull and mandibles of* Protodobenus japonicus *(Early Pliocene, Japan), lateral view, after Horikawa (1995, The Island Arc 3[4]).*

Many extinct genera have been placed with the large, tusked living walrus, *Odobenus rosmarus*, in the Odobenidae. Late Quaternary fossils (<1 Ma) of walrus are known around the margins of the North Atlantic and North Pacific, with *Odobenus* sp. also reported from the Late Pliocene (2–2.7 Ma) of the North Pacific. Other extinct species of odobenine (large-tusked) walruses lived in northern waters outside polar regions, in sit-

uations known to be warmer than today. For example, the Late Pliocene *Valenictus chulavistensis*, which is related closely to *Odobenus*, is known from skulls and associated skeletons from southern California (San Diego Formation, 2–3 Ma), whereas teeth referred to the extinct *Trichecodon huxleyi* have been reported from the earlier Pliocene (Bone Valley Formation, 4–5 Ma) of Florida and Pleistocene (Red Crag, <2 Ma) of Britain, south of the inferred range of the living walrus. *Protodobenus japonicus* (Fig. 11G), from Japan, may be as old as 5 Ma and thus is the oldest reported odobenine walrus.

In contrast to the various classifications of otariids and odobenids, true seals (Phocidae) are well unified by specialized swimming features. This unity might mislead one into thinking that phocid paleontology and phylogeny are straightforward. The fossil record of phocids is patchy, and few basal or archaic taxa are known. The oldest reported specimens, from South Carolina, are, surprisingly, from the Late Oligocene (>23 Ma). If the concept of pinniped monophyly is correct, with the Pacific Oligo/Miocene *Enaliarctos* at the stem of pinnipeds and phocids much further crown-ward, this early Atlantic record for phocids has major implications for the geography and timing of pinniped evolution. Phocids are a negligible component of North Pacific fossil faunas (with sporadic *Phoca*-like fossils from the Pleistocene and, rarely, Pliocene), but are significant in the Atlantic and, lesserly, the Southern Ocean.

The typical phocid genus *Phoca* has been used widely, and often wrongly, for diverse fossil seals and even some fossil cetaceans, but *Phoca* in the strict sense is not certainly older than Pleistocene. *Phoca* and its living relatives comprise a group Phocini (or Phocinae), which also includes fossils as old as Middle Miocene (14–16+ Ma). For example, the Middle Miocene presumed phocine *Leptophoca lenis*, which was based on an isolated humerus, is known from many skeletal elements from the Calvert Formation (~15 Ma; USA). *Leptophoca, Prophoca,*

Cryptophoca, and related genera are all from the Atlantic shores of North America and Europe or from the Paratethys, which was an ancient shallow and even brackish northeastern extension of the Mediterranean Sea.

Fossils from a second group of phocid seals, the Monachinae, include a scatter of *Monachus*-like genera based on fragmentary North Atlantic fossils, such as *"Monotherium"* from the Calvert Formation (~15 Ma) related to living Monk seals (Monachini of some authors). Other excellent Late Miocene to Early Pliocene specimens are allied to living *Lobodon carcinophaga* (Fig. 12A) in the Lobodontini. Lobodontines include *Piscophoca pacifica* and the elongate, slender-bodied *Acrophoca longirostris* (both Pisco Formation, Peru, eastern Pacific) and, in the Atlantic, the Late Miocene *Properiptychus argentinus* (Argentina). From the margins of the Southern Ocean come the well-known *Homiphoca capensis* (earliest Pliocene, ~5 Ma) (Figs. 12A and 12B) from South Africa and more fragmentary *Homiphoca*-like fossils from southern Australia. The only notable fossil of a living Antarctic lobodontine seal is a jaw identified as *Ommatophoca rossi* from New Zealand (latest Pliocene, >2 Ma; southwest Pacific) (Fig. 12D).

B. Amphicyonidae

The extinct Early Miocene carnivore *Kolponomos* (Amphicyonidae) is known from two North Pacific species from the Clallam and Nye Formations of Oregon and Washington. The skull is massive, with forward-directed eyes, a narrow snout,

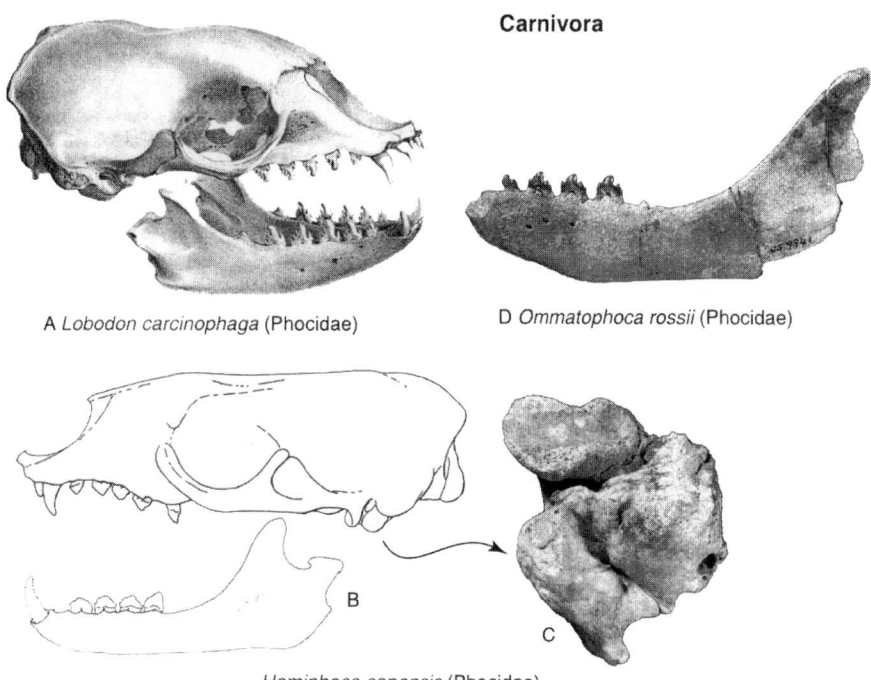

Carnivora

A *Lobodon carcinophaga* (Phocidae)

D *Ommatophoca rossii* (Phocidae)

Homiphoca capensis (Phocidae)

Figure 12 *Pinniped carnivores. (A) Skull and mandibles of* Lobodon carcinophaga *(extant, Antarctica), lateral view, from Gray (1846).* Homiphoca capensis *(earliest Pliocene, South Africa): (B) skull and mandibles of lateral view, after de Muizon and Hendey (1980, Annals of the South African Museum 82[3]) and (C) right temporal bone (ear region), ventral view. (D) Mandible of* Ommatophoca rossii *(latest Pliocene, New Zealand), lateral view.*

and broad low-crowned teeth, which indicate a crushing feeding habit, probably on shelled invertebrates (Fig. 13A). Other than the skull, the skeleton is known poorly, but was presumably bear like. *Kolponomos* is interpreted as amphibious, not a strong swimmer, and living nearshore. It has been placed close to the base of the pinnipeds.

C. Sea Otters (Mustelidae: Lutrinae)

Sea otters are known as fossils from around the North Pacific and North Atlantic. Despite their presence in marine strata, none of the species was clearly an obligate marine mammal. A key diagnostic feature is blunt molar teeth, which indicate a durophagous (shell-crushing) diet. The giant Late Miocene (~7 Ma) otter, *Enhydritherium terraenovae*, is known from fragmentary specimens from marine strata in California and Florida and from an informative articulated skeleton from freshwater strata in Florida (Fig. 13B). The latter represents an animal with powerful neck muscles, forelimbs that probably helped in swimming, and hindlegs suitable for terrestrial locomotion. In younger fossils, such as *Enhydra macrodonta* (Late Pleistocene, California), limb structure indicates hindlimb swimming, with forelimbs used to manipulate food.

V. Tethytheres: Sirenians and Desmostylians

Tethytheres are herbivorous, mostly large mammals that include elephants, sirenians, and the extinct Desmostylia. Sirenians and desmostylians have a significant marine fossil record.

A. Sirenia

Living sirenians are obligate aquatic mammals that dwell in shallow subtropical and tropical waters, both marine and freshwater. The fossil record, which is mostly from the Northern Hemisphere (Fig. 14), shows that the two living families (Trichechidae and Dugongidae) have a long and diverse history. Two extinct families of sirenians are based on fossils. The fossil record of sirenians, as documented extensively by Domn-

ing, is judged reasonably complete and probably gives a reliable guide to the evolution of the group.

The oldest and most archaic sirenian is *Prorastomus sirenoides* (sole member of the Prorastomidae), named by the famous anatomist Richard Owen for a skull, mandible, and vertebra from the Early and Middle Eocene (~50 Ma) of Jamaica. A few other specimens are known. This occurrence in the western tropical Atlantic is surprising, as tethytheres are thought to have evolved in the eastern Tethys. *Prorastomus* is generally intermediate in structure between other tethytheres and later Sirenia, although perhaps it is not directly ancestral to any known later sirenian. Its notable sirenian features include the inflated rostrum, pachyostotic skull (with dense bone), retracted enlarged nares, and five premolars.

Before the end of the Eocene, sirenians occupied warm waters from the western tropical Atlantic through the Tethys to the western Pacific. Many of the fossils represent species of *Protosiren* (extinct family Protosirenidae; Middle–Late Eocene), distinguished from *Prorastomus* by details of the ear and a more down-turned rostrum. *Protosiren* occurs in Egypt and India and presumably ranged through the middle Tethys. Some specimens retained short hindlimbs and were probably amphibious rather than completely aquatic. Protosirenids are plausible structural ancestors for the two extant families: Dugongidae and Trichechidae. Despite the abundance of protosirenid fossils and the presence of effectively a circumglobal equatorial seaway (Tethys–Caribbean–Pacific), there are no reports of Eocene or Early Oligocene sirenians from outside Indonesian west Pacific.

The earliest dugongids—the halitheriines (paraphyletic subfamily Halitheriinae; Eocene–Pliocene)—were contemporaneous with protosirenids, and overlap in range, being known from Middle Eocene to Oligocene rocks in Egypt (e.g., *Eosiren*, Fig. 15A). Dugongids are characterized by a loss of hindlimbs and by changes in the ear region of the skull, but relationships among archaic forms are not fully resolved. Fossils indicate that, apart from the hydrodamalines (see later), the family is subtropical–tropical.

In the Oligocene, halitheriine dugongids occurred beyond the shrinking Tethys sea, with significant fossils of *Halitherium* in Eu-

Carnivora

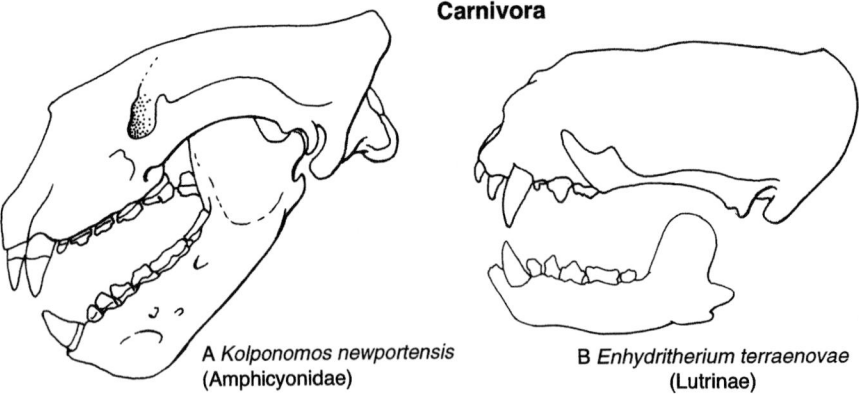

A *Kolponomos newportensis*
(Amphicyonidae)

B *Enhydritherium terraenovae*
(Lutrinae)

Figure 13 *Other marine carnivores. (A) Skull and mandibles of* Kolponomos newportensis *(Early Miocene, Oregon), lateral view, after Tedford* et al. *(1994, Proc. San Diego Mus. Nat. Hist. 29). (B) Skull and mandibles of* Enhydritherium terraenovae *(Late Miocene, Florida), lateral view, after Lambert (1997).*

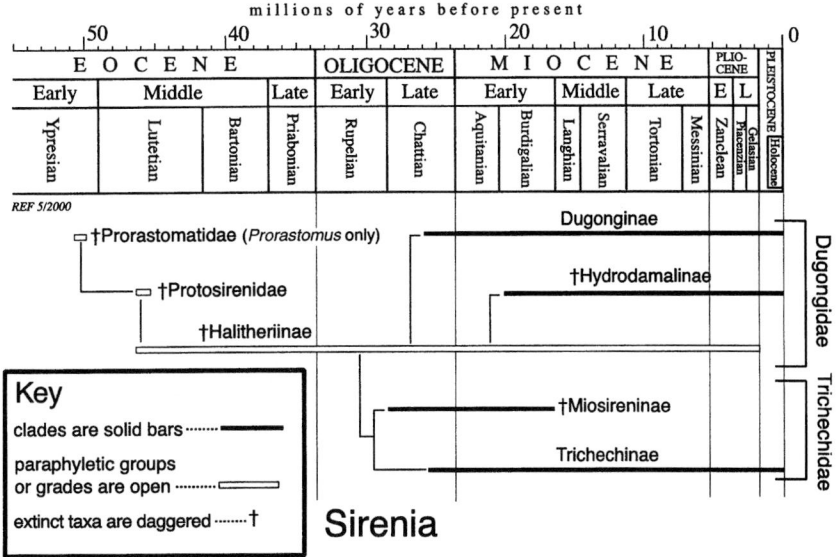

millions of years before present

Figure 14 *Geological age ranges of families and subfamilies within the Sirenia plotted against a standard time scale (for time scale details, see Fig. 2). Clades are in-filled bars, and grades or paraphyletic or stem groups are open bars. Inferred relationships follow the work of Domning (1996).*

rope and the western North Atlantic. *Halitherium* (Germany; Fig. 15B) includes some of the very few Early Oligocene marine mammals known. Geologically, younger halitheriines include species in the widespread Early–Middle Miocene *Metaxytherium* (Fig. 15C), characterized by strongly down-turned snouts and small upper incisor tusks, and interpreted as generalized bottom feeders. *Metaxytherium* is reported from Europe (eastern Atlantic/Mediterranean), the western Atlantic, and the North to equatorial Pacific. Presumably, Pacific halitheriines, with their significant Miocene record, originated from animals that spread westward through the Central American seaway. One halitheriine, *Metaxytherium crataegense*, is reported both from the western Atlantic and the eastern tropical Pacific. Older Pacific records of dugongids are fragmentary and enigmatic; they include an indeterminate Late Oligocene dugongid from Japan and an unnamed Early Miocene (18–25 Ma) halitheriine from Oregon.

In the Early to Middle Miocene, a significant new lineage of dugongids, the hydrodamalines, radiated in the North Pacific. This group, characterized primarily by a large body size, probably arose within *Metaxytherium*. Early fossils include two species of *Dusisiren*, while the cold-tolerant *Hydrodamalis* is known from the circum-North Pacific Late Miocene to Pliocene *H. cuestae* and its descendant, *Hydrodamalis gigas*. The latter, Steller's sea cow, was exterminated by humans; it was a huge animal that, unlike other sirenians, lived in exposed cold high-latitude waters where it fed on kelp. Steller's sea cow has significant Quaternary records.

The dugong lineage arose by the Late Oligocene, when the Dugonginae branched away from the halitheriines. The oldest and most archaic dugongines (*Crenatosiren*) and greatest known diversity are in the western North Atlantic–Caribbean region, and perhaps the group originated there. Dugongines are, however, known as fossils from the eastern North

Atlantic–Mediterranean (*Rytiodus*; Early Miocene) and the eastern North Pacific (*Dioplotherium*; Early–Middle Miocene). Beyond a few fragments of Pleistocene age, there are no notable fossils for the living western tropical Pacific *Dugong dugon*; however, phylogenetic analyses indicate a significant long but unrecorded history for the lineage leading to this species.

Manatees (Trichechidae), characterized by a secondarily reduced rostrum, probably evolved from *Prototherium*-like dugongs in the later Eocene or Early Oligocene. The oldest trichechids are the European Miosireninae–the Late Oligocene *Anomotherium* and Early Miocene *Miosiren*. By the Middle Miocene (~14 Ma), archaic manatees, species of *Potamosiren* (Trichechinae), occupied fresh waters in South America and since then have been restricted mainly to inshore waters. The Late Miocene to Pliocene (~4–6 Ma) *Ribodon*, from the lower La Plata basin, shows the distinctive manatee-like character of supernumerary molar teeth, interpreted as an adaptation for feeding on abrasive freshwater grasses. In turn, aquatic plants perhaps evolved in response to changing South American drainages caused by the uplift of the Andes. An Early Pliocene (~5 Ma) *Ribodon* is known from North Carolina at about the time that dugongs disappeared from the Caribbean–western North Atlantic, and fossils close to *Trichechus* appear in the Late Pliocene (2–3 Ma). Such fossils indicate that manatees ecologically replaced dugongs in the Caribbean–western North Atlantic, although whether by competition is uncertain. There is no fossil record to reveal the dispersal of *Trichechus* eastward to Africa.

B. Desmostylia

Desmostylians are large extinct North Pacific marine mammals that were first described, in the 1880s, on the basis of isolated molar teeth with distinctive bundles of high columnar

Tethytheres - Sirenia and Desmostylia

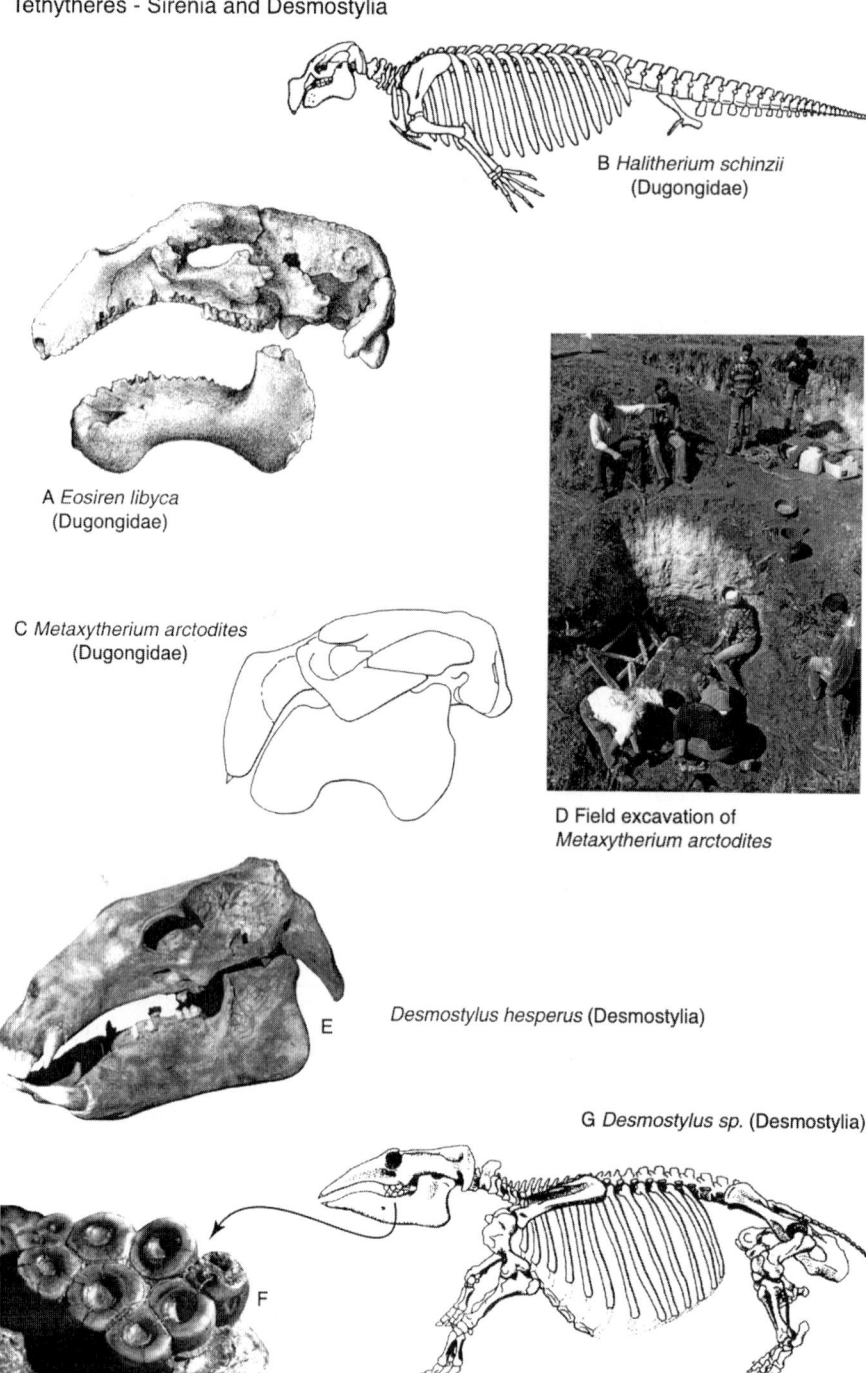

A *Eosiren libyca*
(Dugongidae)

B *Halitherium schinzii*
(Dugongidae)

C *Metaxytherium arctodites*
(Dugongidae)

D Field excavation of
Metaxytherium arctodites

Desmostylus hesperus (Desmostylia)

G *Desmostylus sp.* (Desmostylia)

Figure 15 *Sirenians and desmostylians. (A) Skull of* Eosiren libyca *(Eocene, Egypt), lateral view, from Andrews (1906). (B) Skeleton of* Halitherium schinzii *(Oligocene, Germany), lateral view, from Romer (1945). (C) Skull of* Metaxytherium arctodites *(Miocene, Baja California Sur), lateral view, after Aranda-Manteca et al. (1994, Proc. San Diego Mus. Nat. Hist. 29). (D) Field site showing excavation of holotype of* Metaxytherium arctodites *(Miocene, Baja California Sur).* Desmostylus hesperus *(Middle Miocene, Japan): (E) skull, oblique lateral view; (F) cheek tooth, oblique occlusal view; and (G) skeleton, lateral view, after Inuzuka et al. (1995, The Island Arc 3[4]).*

cusps (*Desmostylus;* Fig. 15F). In 1953, they were placed in their own group, which is the only extinct order of marine mammals. Currently they are regarded as tethytheres allied with, but distinct from, Sirenia. Desmostylians have been interpreted as amphibious, HIPPOPOTAMUS-like, shallow-water mammals that fed on algae and seagrasses. They are known from subtropical to temperate coasts of North America and Japan, but did not range into the South Pacific.

The most archaic desmostylian is *Behemotops,* with a single species (*B. proteus*) based on mandibles and teeth of Late Oligocene age (~25–29 Ma) from Washington. *Behemotops* also occurs in Hokkaido, Japan. Tooth form in *Behemotops* is reminiscent of a group of terrestrial Asian tethytheres, the archaic proboscidians called Anthracobunidae.

Among the four other recognized genera, *Cornwallius* (Late Oligocene) and *Vanderhoofius* (Middle Miocene) are known only from western North America. Both species of *Desmostylus* (Early to Middle Miocene) (Figs. 15E–15G), which are distinguished on the basis of teeth and skulls, occur on both the eastern and the western coasts of the Pacific. Similarly, *Paleoparadoxia* (Early to Middle Miocene), with three to four species, also occurs on both Pacific coasts. Desmostylians have not, however, been reported from the Southern Hemisphere.

Unifying features for desmostylians include a skull with a long broad muzzle and prominent tusk, dorsally protruding eyes, a shortened neck, a broad sternum, and many details of the postcranial skeleton. The matter of body stance is contentious; it has been reconstructed, using skeletal form and fossilized posture, as reptilelike, with the limbs extended laterally. Alternatively, stance has been interpreted as the familiar quadrupedal form of mammals, with the body well off the ground and limbs more or less under the body. Irrespective of stance, desmostylians were probably large and slow moving on land and swam much like polar bears.

C. Edentata

One of the most unusual fossil occurrences is that of abundant ground sloths from marine Pliocene strata (~4 Ma) of Peru. The extinct marine sloth, *Thalassocnus natans,* is from a rich assemblage of marine vertebrates. Although the sloth shows no clearly obligate aquatic features, its structure would allow swimming comparable to that of otters. Because the adjacent Peruvian coast was a desert, the sloth probably ate seaweeds or seagrasses.

See Also the Following Articles

Archaeocetes, Archaic ▪ Baleen Whales, Archaic ▪ Basilosaurids ▪ Beaked Whales, Overview ▪ Cetacean Evolution ▪ Desmostylia ▪ Pinniped Evolution ▪ Sirenian Evolution

References

Barnes, L. G., Domning, D. P., and Ray, C. E. (1985). Status of studies on fossil marine mammals. *Mar. Mamm. Sci.* **1**(1), 15–53.
Barnes, L. G., Inuzuka, N., and Hasegawa, Y. (eds.) (1995). Evolution and biogeography of fossil marine vertebrates in the Pacific realm. *Island Arc* **3**(4), 243–537.
Berta, A., and Demére, T. (eds.) (1994). Contributions in marine mammal paleontology honoring Frank C. Whitmore, Jr. *Proc. San Diego Mus. Nat. Hist.* **29**, 1–268.
Domning, D. P. (1996). Bibliography and index of the Sirenia and Desmostylia. *Smith. Contrib. Paleobiol.* **80**, 1–611.
Fordyce, R. E., and Barnes, L. G. (1994). The evolutionary history of whales and dolphins. *Annu. Rev. Earth Planet. Sci.* **22**, 419–455.
Fordyce, R. E. (2000). Cetacea. In "Encyclopedia of Life Science." Nature Publishing Group, London. www.els.net
Kellogg, A. R. (1936). A review of the Archaeoceti. *Carnegie Inst. Wash. Publ.* **482**, 1–366.
Lambert, W. D. (1997). The osteology and paleoecology of the giant otter *Enhydritherium terraenovae.* *J. Vertebr. Paleontol.* **17**(4), 738–749.
Rice, D. W. (1998). "Marine Mammals of the World: Systematics and Distribution." Society for Marine Mammalogy, Lawrence, KS.
Thewissen, J. G. M. (ed.) (1998). "The Emergence of Whales: Evolutionary Patterns in the Origin of Cetacea." Plenum, New York.

Fossil Sites

R. EWAN FORDYCE
University of Otago, Dunedin, New Zealand

Fossil marine mammals—Cetacea, Sirenia, Desmostylia, Pinnipedia, and other aquatic carnivores—are known from hundreds of sites worldwide (Fig. 1). In the 1800s most localities were in temperate northern latitudes, but since 1900 finds have spanned from modern tropics to poles, in both north and south and on all major continents. Usually, sites preserve marine sedimentary rocks, which have been exposed on land through sea-level fall and/or uplift, followed by erosion. There are negligible records (dredgings) from the deep ocean, but there are some important freshwater sites for secondarily nonmarine species. Clearly, fossils give only a rough and indirect guide to former distributions in ancient oceans. Sites vary from rich localized concentrations at sites a few tens of meters across, to scattered occurrences across many kilometers, which become significant at the regional level, and they range in age from Eocene to Pleistocene (Fig. 2). The case studies in this article, given in sequence from oldest to youngest, span all the major time intervals and oceans.

I. Role of Geological Processes

Marine mammal history has been affected by geological changes in oceans and climates during the last 50+ million (M) years. These changes ultimately reflect global tectonic processes. Continents are now relatively more emergent than for much of the past 50 M years, with less continental shelf and less extensive shallow continental sea than in the past. Most continents preserve coast-parallel strips of ancient marine rock now exposed on land. These may be extensive and a notable source of cetacean fossils (e.g., the Atlantic Coastal Plain of the eastern United States) or limited (e.g., most of Africa). Sometimes extensive shallow epicontinental seas onlapped the continents, as

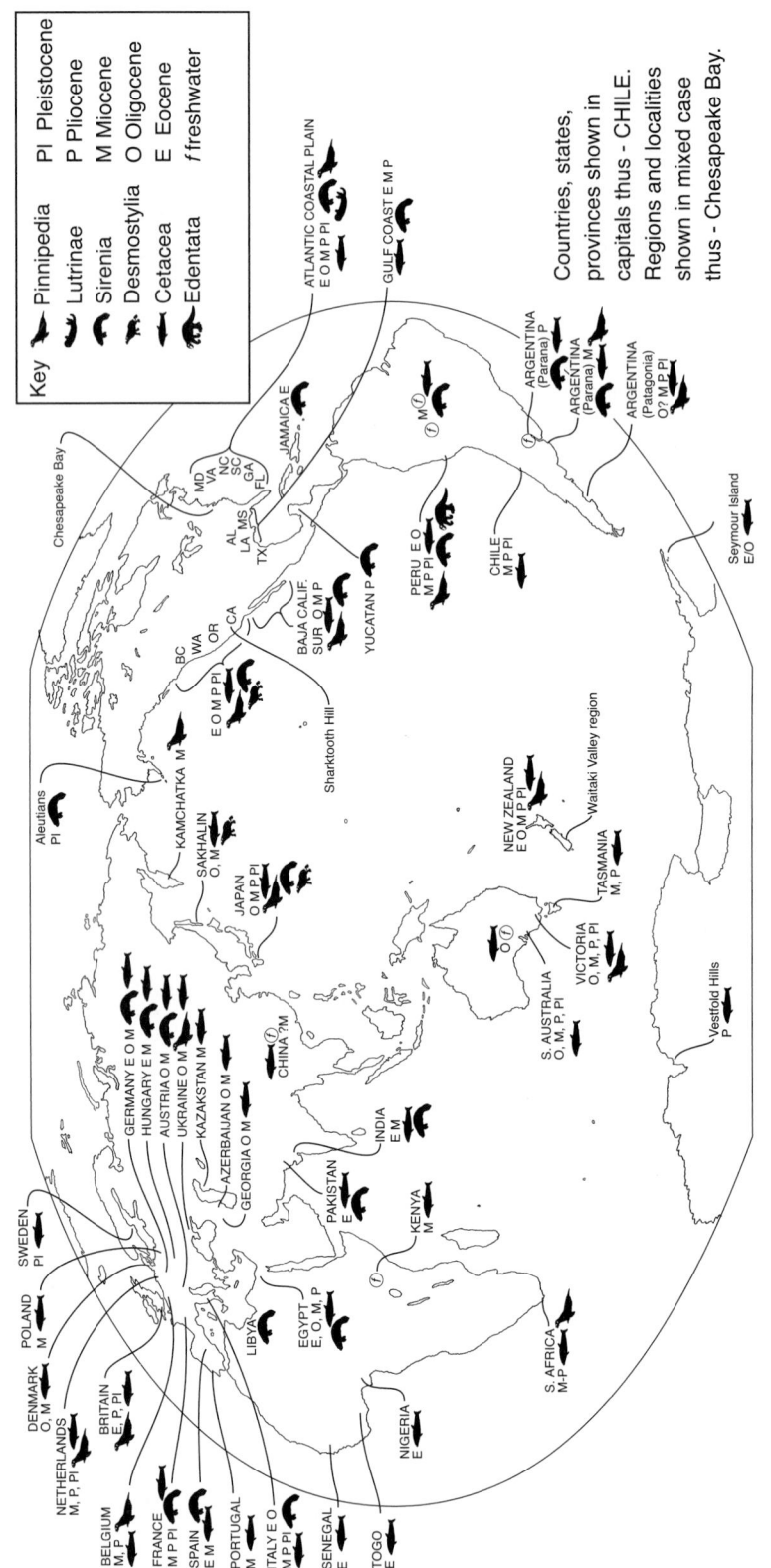

Figure 1 *Selected localities for fossil marine mammals.*

472

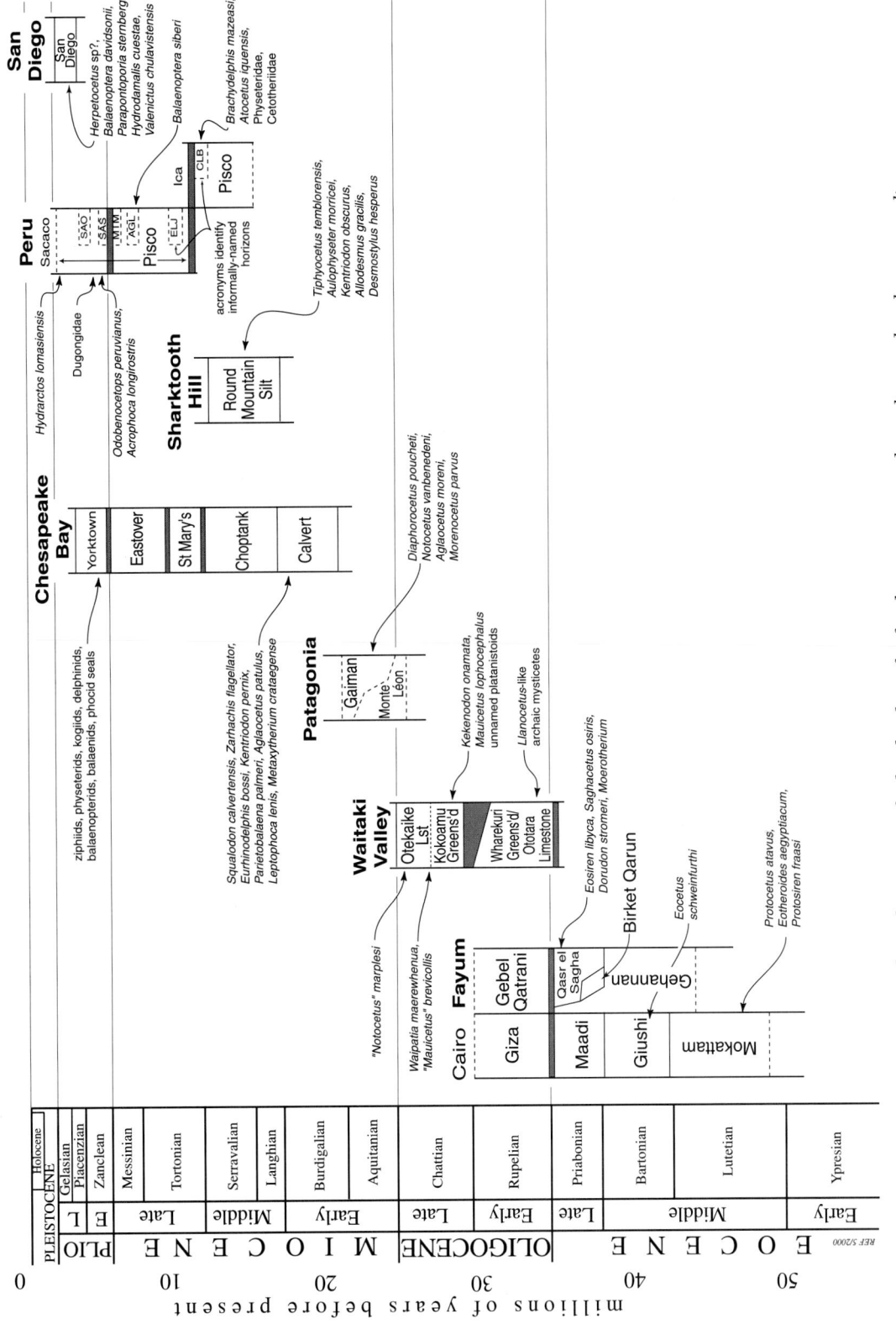

Figure 2 *Geological age ranges for key localities for fossil marine mammals, as discussed under case studies.*

in northern Europe and the Paratethys (see later). Major drops in sea level occurred about 30 million years before the present (Ma) and, associated with widespread glaciation and global cooling, since 2 Ma (major fluctuations).

When the first marine mammals appeared, the extensive shallow Tethys Sea stretched from the Pacific to about the modern Mediterranean. By the end of the Eocene, India had moved northward to collide with Asia, closing much of the Tethys. More western remnants of the Tethys, through what is now southern Eurasia, were eliminated in the Miocene, when Africa collided with Eurasia. Later, the Mediterranean dried out completely about 6 Ma, with dramatic consequences for the biota.

In the south, Australia moved north away from Antarctica opening part of the Southern Ocean by the end of Eocene time (34 Ma). Later, Antarctica and South America separated in the Oligocene (>23 Ma) to open the Drake Passage, which allowed the west to east flow of a newly developed Circum-Antarctic Current. This current isolated Antarctica thermally and probably allowed the Antarctic ice cap to expand, global climates to cool, and global oceans to become more heterogeneous. Australia continued to drift north, so that in about Middle Miocene (~15 Ma) it closed the Indopacific seaway between Australia and Asia and restricted equatorial circulation between the Indian and the Pacific Oceans. In the middle Pliocene (~3–4 Ma), the Panama Seaway closed, cutting Caribbean–Pacific links. The closure of the Panama Seaway correlates closely with the start of Northern Hemisphere continental glaciation.

II. Global Summary of Localities

Important localities occur in marine sequences around the modern Mediterranean, which is a remnant of the formerly extensive Tethys sea and the now-vanished Paratethys. Cetaceans, pinnipeds, and sirenians are notable. Italy has many sites of Pliocene to Oligocene age, whereas the most significant localities along the southern Mediterranean are in the Egyptian Eocene (see later). Paratethyan localities to the northeast include some in Austria (e.g., Miocene cetaceans), Hungary, (e.g., Eocene sirenians), Slovakia (e.g., Miocene pinnipeds), Croatia, Romania, and Ukraine and several in the Caucasus mountains and borders of the Caspian Sea, including Georgia (e.g., Oligocene cetaceans), Azerbaijan, and Kazakhstan.

Eastern North Atlantic faunas have come from Miocene–Pleistocene and, rarely, Eocene–Oligocene sequences bordering the North Sea, in Denmark, northern Germany, Poland, Sweden (e.g., Pleistocene cetaceans), the Netherlands, Belgium, North Sea dredgings, and Britain (e.g., Pleistocene cetaceans and pinnipeds). Eocene to Pliocene fossils from the western North Atlantic include many for Cetacea, Sirenia, pinnipeds, and sea otters in strata from the Atlantic Coastal Plain (from Delaware to Florida; see later) and Gulf Coast. The Caribbean has a few important sites (sirenians; Yucatan, Jamaica). Southwest Atlantic fossils from Argentina include cetaceans, sirenians, and pinnipeds from Oligocene–Miocene marine and Miocene–Pliocene nonmarine strata (see later). A few sites in the eastern tropical Atlantic have produced Eocene cetaceans (Nigeria, Senegal, Togo); the tip of South Africa has

rich Pliocene bone beds, including cetaceans and pinnipeds; and cetacean bones have been dredged from near Cape Agulhas.

Bordering the Indian Ocean, Eocene Tethyan species from Pakistan and India represent ancient cetaceans, and less common sirenians. There are scattered reports of younger Cetacea. From Kenya comes a Miocene apparent freshwater beaked whale (Ziphiidae), and ziphiid rostra have been dredged from off Western Australia but, otherwise, there are no other significant described faunas from around the Indian Ocean.

A few regions around the Pacific, which was the largest ocean during cetacean history, have received concentrated attention. Japanese Neogene cetaceans, sirenians, desmostylians, and pinnipeds are well documented, and studies of Oligocene species are underway. There are only sporadic records of fossil marine mammals from further north in the Pacific (Sakhalin and the Aleutian chain). From the eastern North Pacific (Mexico to British Columbia; see later), Miocene–Pleistocene assemblages, including cetaceans, desmostylians, and pinnipeds, are well documented, but of hundreds of known Oligocene marine mammals, only a few species are described. Rich assemblages of Neogene marine mammals come from Peru (see later), with a few records from Chile. New Zealand assemblages, from the Southern Ocean margin, span from the Eocene to Quaternary, including important Oligocene material (see later). A scattered Oligocene to Neogene record from Australia also hints at the composition of Southern Ocean faunas. One Paleogene and one Pliocene site are known from Antarctica.

III. Eocene: Mediterranean/Tethys (Northern Egypt)

Eocene strata in northern Egypt, near Cairo, and southward at Fayum have produced fundamentally important archaeocete cetaceans and sirenians. Although assemblages are modest in terms of diversity, fossils are generally well preserved and may be quite abundant. The Cairo sequence, at Gebel Mokattam, is slightly older than that of Fayum. Recent work by Gingerich interpreted the fossiliferous strata in terms of changing sea levels interacting with shallow marine deposition at a passive continental margin. The sequences include marked unconformities caused by a lowered sea level. Paleoenvironments range from nonmarine (riverine; which has, however, produced sirenians) through estuarine, active shoreface, barrier bar, and shallow shelf.

At Gebel Mokattam, Cairo, the oldest unit comprises about 120 m of limestones of the Mokattam Formation, deposited in a shallowing marine shelf setting from 48 to ~41 Ma; this unit produced the widely cited archaic cetacean *Protocetus atavus* and early sirenians, including *Eotheroides aegyptiacum* (named by famous anatomist Richard Owen) and *Protosiren fraasi*, all about 46 Ma. From the overlying shallow marine Giushi Formation (30 m of limestone with shale, ~41 to 37 Ma) comes the archaeocete *Eocetus*. Significant marine mammals have not been described from younger units of Gebel Mokattam, but there are important assemblages in the Fayum.

In the south, Fayum marine strata span the Middle and Late Eocene (~43–43 Ma). Gehannan Formation limestone has yielded fossils of the sirenians *Eotherioides* and *Protosiren*, and more or less articulated skeletons of the archaeocetes *Dorudon* (formerly *Prozeuglodon*) *atrox* and *Basilosaurus isis*, the latter known to have hindlimbs. Archaeocete skeletons are abundant at horizons interpreted as representing low stands of sea level and *Moeritherium*, an estuarine proboscidean, is present. The top of the Gehannan Formation varies in age according to locality. It is succeeded in places by barrier beach sands of the Birket Qarun Formation, a unit that has many archaeocete skeletons in the base. The youngest marine rocks at Fayum are lagoonal strata of the Qasr el Sagha Formation (~35–37 Ma). It has produced the sirenian *Eosiren libyca*, two dorudontine archaeocetes—the small *Saghacetus osiris* and larger *Dorudon stromeri*—and *Moeritherium*. Above this marine sequence, a sirenian has been found in Oligocene riverine strata of the Gebel Qatrani Formation.

The two Egyptian localities are important because they produced some of the earliest recognized archaic cetaceans and sirenians. The specimens—many of which are types (basic standards of reference)—include some of the youngest archaeocetes known, being close in age to the oldest crown-group Cetacea from about the Eocene/Oligocene boundary.

IV. Oligocene: Southwest Pacific (Waitaki Valley, New Zealand)

Thin marine Oligocene strata (23–33 Ma) (Fig. 3) in North Otago and South Canterbury, around the Waitaki Valley area, are a source of early "modern" Cetacea. Important localities include those near Oamaru, Duntroon, Wharekuri, and Hakataramea Valley, spanning some tens of kilometers. Notable earlier work is that of B. J. Marples on archaic mysticetes referred to *Mauicetus*. Most of the abundant modern finds are undescribed.

The Wharekuri Greensand (~12 m) and equivalents such as the Ototara Limestone have produced some of the few Early Oligocene (~33 to ~30 Ma) cetaceans known worldwide. These units are truncated by a major unconformity caused by sea level fall at ~30 Ma and are succeeded by the Kokoamu Greensand, a thin (usually <5 m), burrowed, shelly, sediment-starved unit. In turn, Greensand grades up over several meters into the massive bioclastic Otekaike Limestone (~50 m thick), which spans the later Late Oligocene (~26 Ma) to earliest Miocene (~23 Ma). Sediments accumulated mainly in a quiet setting below storm wave base in mid to outer shelf depths. Associated vertebrates include penguins, sharks, and bony fish, but no other marine mammals. Macroinvertebrates are locally abundant and probably caused the bioerosion seen on many bones. Cetaceans occur both as isolated bones and as skeletons (Fig. 4). The source units are often cemented, producing resistant outcrops so that excavations need pneumatic tools, are limited in scope, and mostly recover partial rather than whole skeletons.

Cetaceans from the Wharekuri Greensand include two presumed mysticetes, a large species showing severe vertebral pathology ("spina bifida" and periostitis), and a juvenile or neonate. Elsewhere, fragments from the Ototara Limestone include a small *Llanocetus*-like toothed archaic mysticete. Mysticetes from the overlying Kokoamu Greensand include enigmatic and rare archaic toothed forms such as *Kekenodon*

Figure 3 *Outcrop of Earthquakes Marl, Kokoamu Greensand, and Otekaike Limestone, Waitaki Valley region, New Zealand. Otekaike Limestone here is about 10 m thick. Photo by R. E. Fordyce.*

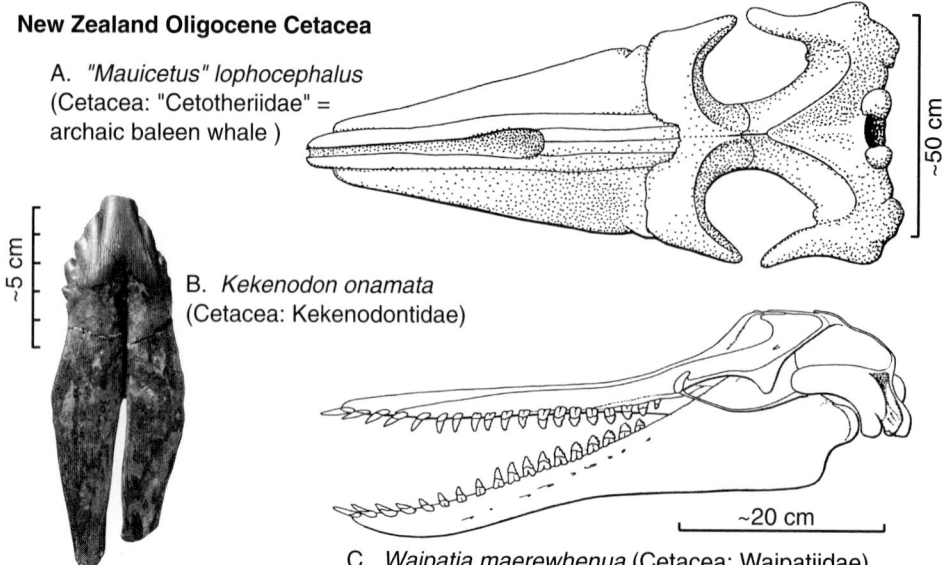

New Zealand Oligocene Cetacea

A. *"Mauicetus" lophocephalus*
(Cetacea: "Cetotheriidae" =
archaic baleen whale)

~50 cm

~5 cm

B. *Kekenodon onamata*
(Cetacea: Kekenodontidae)

~20 cm

C. *Waipatia maerewhenua* (Cetacea: Waipatiidae)

Figure 4   *Oligocene Cetacea from New Zealand. (A) Skull of cetacean* "Mauicetus" *lopho-cephalus, dorsal view, after Marples. (B) Tooth of cetacean* Kekenodon onamata; *photo by R. E. Fordyce. (C) Skull and mandible of cetacean* Waipatia maerewhenua, *lateral view, after Fordyce and Barnes (1994).*

onamata (formerly misidentified as an archaeocete), *Mammalodon*-like species, and an unnamed species with teeth reminiscent of the odontocete *Squalodon*. More common baleen-bearing whales include the enigmatic *Mauicetus lophocephalus*, which is one of the older global records of a toothless mysticete, and other fossils more closely matching cetotheres in the traditional sense. Odontocetes from the lower to middle Kokoamu Greensand mostly belong in the Platanistoidea. They include species of Squalodontidae and provisionally identified Waipatiidae, Squalodelphinidae, and Dalpiazinidae; all are unnamed new species. There are no true dolphins (e.g., Kentriodontidae) from lower in the Greensand.

Cetaceans appear less common but are better preserved in the uppermost Kokoamu Greensand and overlying Otekaike Limestone. No species identified from the underlying Greensand is known to occur in the Limestone. Mysticetes include unnamed species of *"Mauicetus"* and other cetotheres in the broad sense. Odontocetes are mainly rare but well-preserved platanistoids, including new species of Squalodontidae, the small *Waipatia* (Waipatiidae), *"Microcetus"* and *Notocetus* (both Squalodelphinidae), and undescribed Dalpiazinidae. Tantalizing fragments of an archaic delphinoid (not a kentriodontid) and a sperm whale are known. There are notable absences, including the mysticete group Aetiocetidae, archaic odontocetes comparable to *Agorophius* and *Xenorophus*, Eurhinodelphinidae, and beaked whales (Ziphiidae).

Overall, small to medium-sized mysticetes dominate assemblages, with small odontocetes also conspicuous. The shallow broad seaway could have been a breeding ground for mysticetes from the recently developed Southern Ocean ecosystem south of New Zealand. Why platanistoids are common but delphinoids and sperm whales are rare is not clear. Perhaps the

shallow seaway was only rarely inhabited by more oceanic or deep-diving species of delphinoids and sperm whales.

The assemblages reinforce the idea that the Oligocene was a time of structural/ecological experiment. Species were scattered more evenly among diverse family level taxa. In contrast, for extant cetaceans, a few families account for most of the species diversity. Some of the better-preserved fossils will help resolve cladistic relationships of extant Cetacea, thus providing an independent standard against which to compare molecular classifications. Revised classifications should also help quantitative zoogeography. Meanwhile, qualitative comparisons suggest a faunal composition more similar to that of the Caucasus than of the Atlantic Coastal Plain or northeast Pacific. Only archaic-toothed mysticetes hint at trans-Pacific links.

V. Miocene: Southwest Atlantic (Patagonia)

Strata of the "Patagonian" marine stage in Santa Cruz, Chubut, and Rio Negro Provinces of southern Argentina have produced Early Miocene Cetacea, which include some of the oldest described representatives of modern families (Fig. 5). These fossils are from the Gaiman and Monte Leon Formations and represent the Leonian local stage, Early Miocene, and possibly latest Oligocene. The strata represent shallow-water settings, in which age-diagnostic microfossils are uncommon, and exact international correlations are uncertain. The Chubut River region has provided the main described material. Other marine mammals, e.g., Sirenia, pinnipeds, are unknown. More than a century of basic taxonomy includes work by Moreno, Lydekker, True, Cabrera, and Kellogg, with significant modern contributions from Cozzuol and de Muizon. Conse-

Patagonian Early Miocene Cetacea

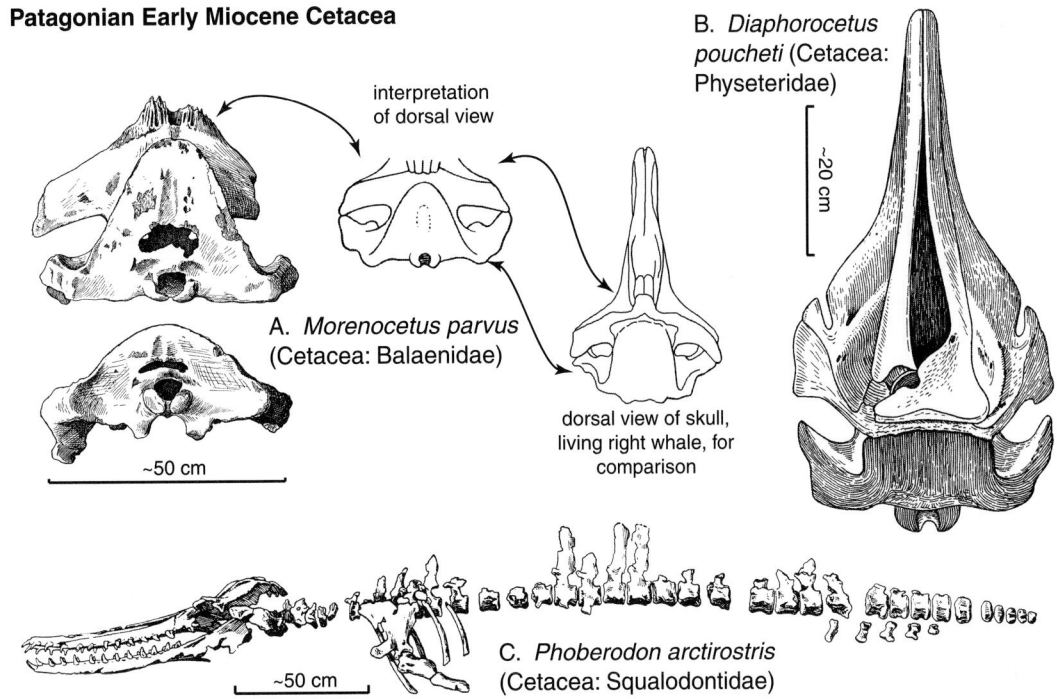

Figure 5 *Miocene Cetacea from Patagonia. (A) Skull of cetacean* Morenocetus parvus, *dorsal and posterior views and interpretive dorsal view (all from Cabrera, 1926), and interpretive view of skull of a living right whale (after Van Beneden and Gervais, 1868–1880). (B) Skull of cetacean* Diaphorocetus poucheti, *dorsal view, from Raven and Gregory (1937). (C) Skeleton of cetacean* Phoberodon arctirostris, *lateral view, from Cabrera.*

quently, the fossils have been cited widely and are entrenched in the literature.

Among the Early Miocene Cetacea, the small *Morenocetus* is the world's oldest described right whale. The rostrum is not known, but other skull features indicate that *Morenocetus* is a balaenid. Of less certain relationships are the "cetotheres" *Aglaocetus* and *"Plesiocetus,"* also known from skulls. Skull form is quite different between these mysticetes, indicating quite different habits.

Odontocetes are notably more diverse and include some of the best-known early sperm whales: the small *Diaphorocetus* and the much larger *Idiorophus*. Both are archaic in appearance, retaining upper teeth with obvious enamel and a narrow anterior on the rostrum. Archaic platanistoid odontocetes are significant, including rather complete material for the large and long-beaked shark-toothed dolphin *Phoberodon* (Squalodontidae) and the short-beaked *Prosqualodon*. *Prosqualodon* is also present in New Zealand and Australia; it has been variably placed in the Squalodontidae or its own group Prosqualodontidae, in turn dubiously allied with the true dolphins (Delphinida). A third group of platanistoids, the Squalodelphinidae, is represented by *Notocetus*, in turn important in refining the higher classification of the Platanistoidea. *Notocetus*-like odontocetes also occur in New Zealand pointing, as for *Prosqualodon*, to a Southern Ocean distribution. The distribution pattern for *Notocetus* and squalodelphinids also includes the North Atlantic and Mediterranean. The eurhinodelphinid *Argyrocetus* shows extreme lengthening of the rostrum; this

genus occurs in contemporaneous strata in California. True dolphins (Delphinida) are rare, represented by the kentriodontid *Kentriodon*.

Patagonian cetaceans indicate the start of "modern" ecological structuring (e.g., deep-diving sperm whales, skim-feeding balaenids). They also reveal distribution patterns (e.g., circum-Southern Ocean) seen among living cetaceans.

VI. Miocene: Northwest Atlantic (Chesapeake Bay)

Flat-lying strata of the Chesapeake Group exposed around the western shores of Chesapeake Bay, in Maryland (Fig. 6) and Virginia, have been an important source of Miocene and Pliocene marine mammals for over 150 years. Fossil cetaceans, sirenians, and phocid seals are present (Fig. 7). Early studies on these fossils were made in the 1800s by the pioneering vertebrate paleontologists Harlan, Leidy, and Cope. Later, F. W. True (early 1900s) and A. R. Kellogg (1920s to 1960s) produced such detailed descriptions of taxa such as *Delphinodon*, *Squalodon* and *Parietobalaena* that the Chesapeake fossils have become international standards of comparison. Significant collections are held in the Smithsonian Institution.

Marine mammals occur in five formations of the Chesapeake Group: Calvert (Early to Middle Miocene) (Fig. 6), Choptank (Middle Miocene), St. Mary's (later Middle Miocene), Eastover (Late Miocene), and Yorktown (Pliocene). These are mainly fine-grained and shallow-water deposits

Figure 6 *Outcrop of Calvert Formation, north of Scientists Cliffs, Maryland. Photo by R. E. Fordyce.*

deposited in climate regimes that ranged from subtropical early in the Miocene (~17 Ma) to near-modern in the Pliocene (~4.5 Ma). Strata are soft so that fossils can be collected easily. The fossils are often preserved well, with fine sutural detail present and only limited crushing. Material ranges from single worn bones to nearly complete skeletons.

Fossils from the Calvert Formation are important in revealing faunal composition around the Early to Middle Miocene boundary. Among the diverse cetaceans, odontocetes include species in three families of Platanistoidea: the shark-toothed dolphins, Squalodontidae (*Squalodon*), Squalodelphinidae (*Phocageneus, Notocetus*), and Platanistidae (*Zarhachis, ?Pomatodelphis*). At least 2 and possibly more than 10 species of the long-snouted enigmatic Eurhinodelphinidae occur (*Eurhinodelphis, "Rhabdosteus"*). True dolphins—Delphinoidea—are placed in the archaic grade family Kentriodontidae (*Kentriodon, Macrokentriodon, Delphinodon, Liolithax,* and *Hadrodelphis*). There is a small sperm whale, Physeteridae (*Orycterocetus*), and various extinct dolphins of uncertain affinities (*Tretosphys, Araeodelphis,* and *Pelodelphis*). Five genera of mysticetes represent the paraphyletic and possibly polyphyletic family Cetotheriidae (*Parietobalaena, Mesocetus, Diorocetus, Aglaocetus,* and *Pelocetus*). Other marine mammals are markedly rare in the Calvert Formation. Phocid seals, known mostly from isolated elements but also from rare partial skeletons, form the basis for the phocine *Leptophoca* and a species of the monachine *Monotherium*. Sea cows include the extinct dugongid sirenian *Metaxytherium crataegense,* which also occurs in the Pacific (Montera Formation, Peru), indicating movement presumably through the Central American Seaway.

Marine mammal assemblages from the Chesapeake Group reveal marked faunal change over time. Archaic cetaceans from the Calvert Formation, such as eurhinodelphinids, squalodontids, and squalodelphinids, are rare or absent in overlying (younger) units, whereas extant families such as the Delphinidae and Balaenopteridae become significant by the Pliocene. Concurrently, species level diversity drops, and ecological shifts are indicated particularly by the absence of squalodontids and long-snouted eurhinodelphinid dolphins.

South of Chesapeake Bay, strata of the lower Yorktown Formation (Early Pliocene, ~4.5 Ma) at Lee Creek Mine, North Carolina, have produced thousands of isolated marine mammal bones, expanding the fauna beyond the Yorktown at Chesapeake Bay. Fossils include ziphiids, physeterids, kogiids, delphinids, balaenopterids, balaenids, and phocid seals. Unusual elements include monodontids (now found only at high latitudes) and pontoporiids (now restricted to the South Atlantic). Similar assemblages occur in the San Diego and Pisco Formations (see later).

Strata of Chesapeake Group, especially the Calvert Formation, are one of the world's richest sources of fossil marine mammals. The abundance of young animals could reflect favorable conditions for calving or, equally, a high mortality for young individuals.

VII. Miocene: Northeast Pacific (Sharktooth Hill, California)

The Sharktooth Hill bone bed (Fig. 8) of Kern County, in the foothills of the Sierra Nevada, California, is one of the most important horizons for marine mammal fossils in the Pacific Basin. The assemblage includes more than 100 marine vertebrates documented thoroughly by Kellogg, Mitchell, Barnes, and others. Cetaceans, pinnipeds, rare desmostylians, and terrestrial vertebrates occur abundantly in, and less commonly above, a thin (10–30 cm) dense, and geographically widespread layer near the top of the Round Mountain Silt (Temblor Formation, Middle Miocene). This bone bed formed over thousands of years, about 13–14 Ma; it represents a mix of taxa from different environments. Most fossils are isolated bones, but sometimes there are natural associations of bones from one individual. Preservation varies; bones may be finely preserved or eroded, and sometimes with marks caused by scavengers or predators.

Cetaceans dominate the assemblage. Mysticetes comprise four cetotheres (*Aglaocetus, Parietobalaena, Peripolocetus,* and *Tiphyocetus*) known from partial skulls and a possible right whale (undetermined Balaenidae). Among odontocetes, sperm whales (Physeteridae) include the moderate-sized *Aulophyseter*. The taxonomy of the smaller odontocetes is less certain, for Kellogg based several new species and genera on isolated periotics, which, in some cases, have been linked to later-discovered SKULLS. The latter represent a long-beaked platanistid-like dolphin (*"Squalodon" errabundus*) and species in the archaic delphinoid group Kentriodontidae (*Kentriodon, Kampholophos*). Several named species of odontocete are still enigmatic (e.g., species of *Oedolithax* and *Lamprolithax*). In terms of ecological composition, cetaceans are comparable to faunas in Californian waters today.

Otariid (or otarioid) pinnipeds are significant and include the types for some species or genera. They are the small ar-

Calvert Formation early Middle Miocene marine mammals

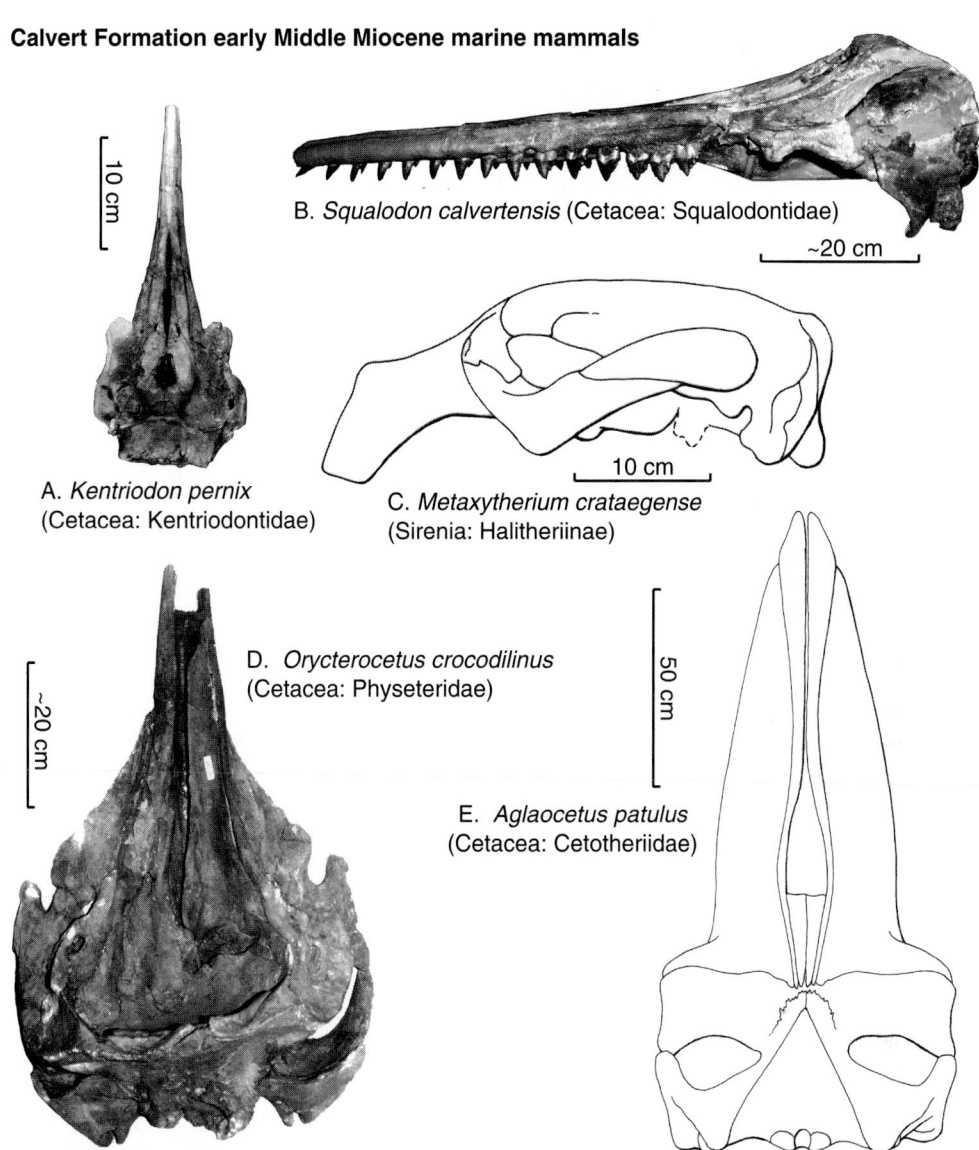

B. *Squalodon calvertensis* (Cetacea: Squalodontidae)

~20 cm

10 cm

A. *Kentriodon pernix*
(Cetacea: Kentriodontidae)

C. *Metaxytherium crataegense*
(Sirenia: Halitheriinae)

10 cm

D. *Orycterocetus crocodilinus*
(Cetacea: Physeteridae)

~20 cm

50 cm

E. *Aglaocetus patulus*
(Cetacea: Cetotheriidae)

Figure 7 *Miocene marine mammals from Chesapeake Bay. (A) Skull of cetacean* Kentriodon pernix, *dorsal view (paratype), photo by R. E. Fordyce. (B) Skull of cetacean* Squalodon calvertensis, *lateral view, photo by R. E. Fordyce. (C) Skull of sirenian* Metaxytherium crataegense, *lateral view, after Kellogg. (D) Skull of cetacean* Orycterocetus crocodilinus, *dorsal view, photo by R. E. Fordyce. (E) Skull of cetacean* Aglaocetus patulus, *dorsal view, after Kellogg (1968, U.S. Nat. Mus. Bull. 247).*

chaic imagotariine *Neotherium,* three species of the large *Allodesmus* (including the rare type-species *A. kernensis* and more common *A. gracilis*), the large rare *Pelagiarctos,* and two unnamed "desmatophocines." No phocids have been identified. Of other marine mammals, sirenians have not been reported, but fragmentary desmostylians occur.

There are conspicuous faunal similarities with the slightly older assemblage from the Calvert Formation of Maryland and Virginia, western North Atlantic; for example, *Parietobalaena* and *Kentriodon* are reported for both places. Differences are also marked; phocid seals and squalodontid and eurhinodelphinid odontocetes are absent from the Sharktooth Hill bone bed.

VIII. Mio-Pliocene: Southeast Pacific (Peru)

Since the early 1980s, de Muizon has documented diverse later Miocene and Pliocene marine mammals (Fig. 9), including some quite surprising ecotypes, from the Pisco Formation of the southern coast of Peru. Rich localities are around Sacaco,

Figure 8 *Outcrop of Round Mountain Silt, Sharktooth Hill, California. L. G. Barnes is looking at cetacean remains, including part of a cetothere. Photo by R. E. Fordyce.*

near Lomas, where well-preserved skeletons of cetaceans and pinnipeds occur, and Ica, to the north. Pisco strata include sandstone, siltstone, and occasional conglomerate, sometimes volcanic rich, with marine invertebrates and bone-rich horizons. Settings are intertidal, barrier bar, and lagoonal; these environments were mostly protected from the open Pacific, perhaps by rocky promontories. The sequence of vertebrate faunas is not in doubt (Cerro la Bruja at the base, Sacacao at the top), but age determination is difficult because the Pisco Formation mostly lacks fossils useful in international correlation.

Odontocete cetaceans are the most diverse marine mammals. The oldest productive Pisco horizon, Cerro la Bruja, may be of late Middle Miocene age (>11 Ma); if so, its odontocetes include an early record for the Pontoporiidae or franciscanas (the peculiarly short-beaked *Brachydelphis*). Other younger Miocene horizons provide early records of porpoises (Phocoenidae: *Australithax*, *Lomacetus*, and *Piscolithax*) and pygmy sperm whales (Kogiidae, narrow-skulled *Scaphokogia*), accompanied by archaic dolphins (Kentriodontidae: *Atocetus*, *Belonodelphis*, and *Incacetus*), beaked whale (Ziphiidae), and sperm whales (Physeteridae). Younger assemblages, from Pliocene horizons within the Pisco Formation, are similar in taxonomic structure, containing pontoporiids (*Pliopontos*), porpoises (*Piscolithax*), ziphiids (*Ninoziphius*), sperm whales (Physeteridae), and kentriodontids, but also more modern dolphins (Del-

phinidae) and an unnamed beluga-like species (Delphinapterinae). The most peculiar Pliocene odontocetes are two species of the extinct tusked *Odobenocetops*. These animals, reported only from the Pisco Formation, have secondarily lost many of the distinctive facial features of odontocetes.

Mysticeti from the Pisco Formation include two cetotheres and up to six balaenopterids for which few details are published. Species briefly described and named by Pilleri include the small cetothere, *Piscobalaena*, the larger *Piscocetus* (Cetotheriidae?), and an extinct *Balaenoptera* (Balaenopteridae). Several specimens of balaenopterids preserve baleen plates *in situ* on the skull.

Up to nine species of phocid seals (Phocidae) are present. Two lobodontines, the *Monachus*-like *Piscophoca pacifica* and long-skulled *Acrophoca longirostris*, are known from well-preserved fossils, including articulated skeletons. From the relatively barren upper Pisco Formation comes the unique specimen of an extinct fur seal, *Hydrarctos* (Otariidae: Arctocephalinae), which is probably Late Pliocene. There is a single record of a sirenian, probably close to *Dugong*, from Early Pliocene lagoonal deposits. Despite the unexpected occurrence, it seems certain that the extinct Pisco sloth *Thalassocnus* really was a marine mammal. Fossil sloths occur in the five main levels at Sacaco, they are abundant, there are no other putative land mammals in the vertebrate assemblage, and the adjacent coast was a desert.

Pliocene assemblages from the Pisco Formation include cetaceans and pinnipeds similar to those from the Yorktown Formation of the Chesapeake Group, North Atlantic, indicating contact through the Central American Seaway before the uplift of Panama. Several genera also occur at Isla Cedros, Baja California Sur. When considered with roughly contemporaneous assemblages from elsewhere in the eastern Pacific (Isla Cedros; also San Diego) and contrasted with modern communities, it seems that there must have been considerable faunal turnover late in the Pliocene or in the Pleistocene.

IX. Plio-Pleistocene: Northeast Pacific (San Diego, California)

Well-preserved geologically young marine mammals—from the later Pliocene and earlier Pleistocene—are rare, but important in revealing marine mammal ecology before the dramatic climate shifts and sea level change of the Pleistocene glaciations. One notable sequence is the ~84 m of San Diego Formation at and near San Diego, southern California. This sandy to gravelly unit was deposited late in the Pliocene (2–3 Ma) and possibly Pleistocene (>1.5 Ma) in settings mostly from shoreface to mid- and outer shelf. The lower finer strata with abundant fossils are marine, whereas the coarser and sparsely fossiliferous upper strata are partly nonmarine. The formation contains many marine mammals that are now extinct, including some whose descendants live in quite different settings today.

Pinnipeds, Cetacea, and Sirenia are present, including many complete enough to identify to species level. The pinnipeds include Otariidae (an extinct species of *Callorhinus* and an unnamed genus) and Odobenidae—the extinct long-tusked walrus *Valenictus*. *Dusignathus*, also present, has been allied variously with walruses or identified as a "pseudo-walrus"; there are no phocids.

Peru - Pisco Formation Miocene and Pliocene marine mammals

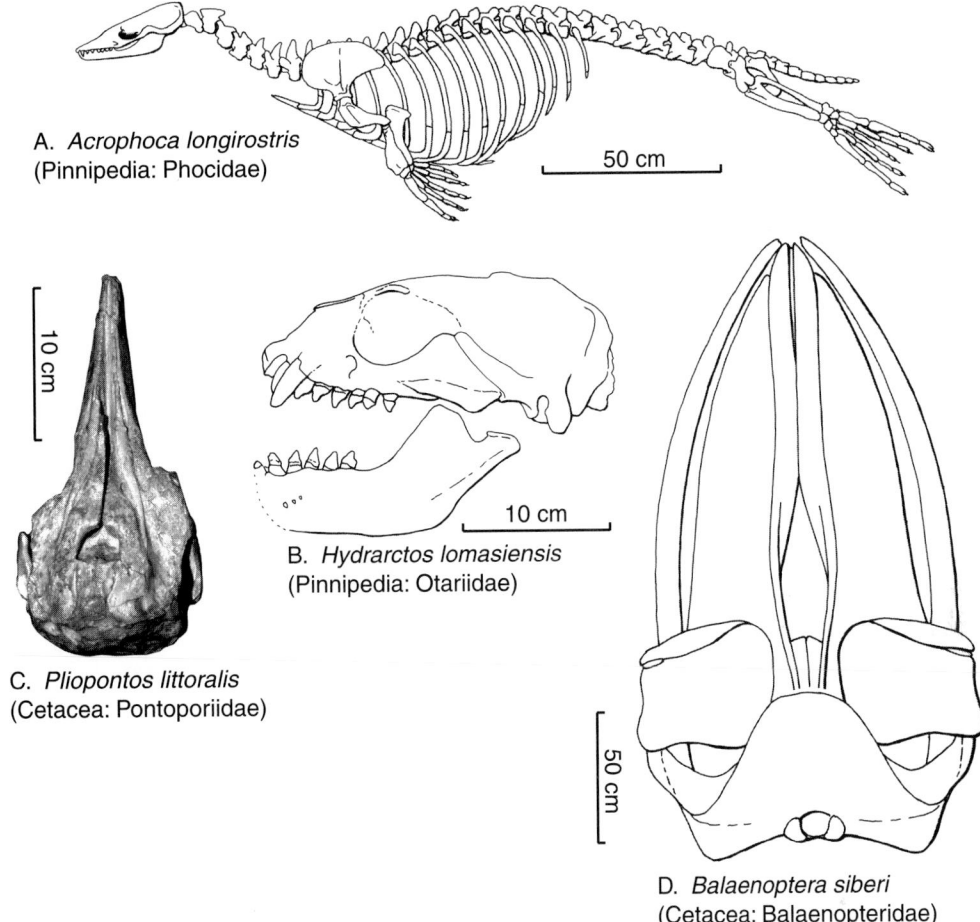

A. *Acrophoca longirostris*
(Pinnipedia: Phocidae)

50 cm

10 cm

B. *Hydrarctos lomasiensis*
(Pinnipedia: Otariidae)

C. *Pliopontos littoralis*
(Cetacea: Pontoporiidae)

50 cm

D. *Balaenoptera siberi*
(Cetacea: Balaenopteridae)

Figure 9   *Miocene and Pliocene marine mammals from Peru. (A) Skeleton of pinniped Acrophoca longirostris, lateral view, after de Muizon (1985). (B) Skull of pinniped Hydrarctos lomasiensis, lateral view, after de Muizon (1978, Bull. d'Inst. Fr. d'Etud. Andines 7). (C) Skull of cetacean Pliopontos littoralis, dorsal view, photo by R. E. Fordyce. (D) Skull and mandibles of cetacean Balaenoptera siberi, dorsal view, after Pilleri (1989).*

Three families of mysticetes include two species of the cetothere *Herpetocetus*, providing one of the last records of Cetotheriidae *sensu stricto*. The extinct minke whale *Balaenoptera davidsonii* is one of five rorquals (Balaenopteridae), and there are two right whales (Balaenidae). Among odontocetes, the long-beaked *Parapontoporia* has been regarded as one of the last marine Pontoporiidae or, alternatively, a marine relative of the endangered Yangtsze dolphin *Lipotes*. Other odontocetes are two porpoises (Phocoenidae), a beluga-like animal (Delphinapterinae), and two dolphins (Delphinidae, including *Stenella* or *Delphinus*). The huge sirenian *Hydrodamalis cuestae* appears to be a direct ancestor to the recently exterminated *Hydrodamalis gigas,* Steller's sea cow, of the cold North Pacific.

These fossils imply a major shift in geographic range and/or ecology in geologically recent times: walruses (now only one species, of *Odobenus*) and belugas currently live in cold north-

ern waters, as *Hydrodamalis*, also did until a few hundred years ago. Descendants of *Parapontoporia* have left the eastern Pacific, and cetotheres are extinct.

See Also the Following Articles

Baleen Whales, Archaic ■ Basilosaurids ■ Delphinoids, Evolution ■ Extinctions, Specific ■ Kentriodontidae ■ Sirenian Evolution ■ Sperm Whales, Evolution

References

Berta, A., and Deméré, T. (eds.) (1994). Contributions in marine mammal paleontology honoring Frank C. Whitmore, Jr. *Proc. San Diego Mus. Nat. Hist.* **29**, 1–268.

Barnes, L. G., Inuzuka, N., and Hasegawa, Y. (eds.) (1995). Evolution and biogeography of fossil marine vertebrates in the Pacific realm. *Island Arc* 3[4 (for 1994)], 243–537.

Cozzuol, M. A. (1996). The record of the aquatic mammals in southern South America. Münchner Geowissenschaftliche Abhandlungen. *Geol. Paläontol.* **30**, 321–342.

Fordyce, R. E., and Barnes, L. G. (1994). The evolutionary history of whales and dolphins. *Annu. Rev. Earth Planet. Sci.* **22**, 419–455.

Gingerich, P. D. (1992). Marine mammals (Cetacea and Sirenia) from the Eocene of Gebel Mokattam and Fayum, Egypt: Stratigraphy, age and paleoenvironments. University of Michigan papers on paleontology **30**, 1–84.

Kellogg, A. R. (1931). Pelagic mammals from the Temblor formation of the Kern River region, California. *Proc. Cal. Acad. Sci. Ser. 4* **19**, 217–297.

Mitchell, E. (1965). "History of Research at Sharktooth Hill, Kern County, California." Kern County Historical Society and County of Kern.

Muizon, C., de and De Vries, T. J. (1985). Geology and paleontology of late Cenozoic marine deposits in the Sacaco area (Peru). *Geol. Rundschau* **74**, 547–563.

Repenning, C. A. (ed.) (1977). Symposium: Advances in systematics of marine mammals. *Syst. Zool.* **25**(4), 301–436.

Franciscana
Pontoporia blainvillei

ENRIQUE A. CRESPO
Centro Nacional Patagónico,
Puerto Madryn, Argentina

I. Diagnostic Characters and Comments on Taxonomy

The franciscana (*Pontoporia blainvillei*) is also known as the La Plata River dolphin. In Uruguay and Argentina it is called *franciscana* whereas in Brazil it is called *toninha* or *cachimbo*. While both this species and the Yangtze river dolphin, *Lipotes vexillifer*, were until recently regarded as of the family Pontoporiidae, the franciscana is now the sole member of this family. The franciscana is the only one of the four river dolphin species living in the marine environment. It is one of the smallest dolphins and has an extremely long and narrow beak and a bulky head. The franciscana is brownish to dark gray above turning lighter on the flanks and belly (Fig. 1). The number of teeth in the upper and lower jaws ranges from 53 to 58 and from 51 to 56, respectively.

II. Distribution/Range Map

The species is endemic in southwestern Atlantic waters. Based on the distribution of sightings and catches, the franciscana lives in a narrow strip of coastal waters beyond the surf to the 30 m isobath (Fig. 2). The complete range known for the franciscana extends from Itaúnas (18°25′S, 39°42′W) in Espirito Santo, Brazil, to the northern coast of Golfo San Matías (41°10′S) in northern Patagonia, Argentina.

III. Geographic Variation

Skull morphology, genetic markers, and parasites have been used to identify stocks. The existence of two potential populations was tested by means of the differences in skull morphology. A northern (smaller) form was proposed between Rio de Janeiro and Santa Catarina, and a southern (larger) form for Rio Grande do Sul, Uruguay, and Argentina. The existence of differences between populations was confirmed some years later using mitochondrial DNA from samples collected at Rio de Janeiro and Rio Grande do Sul. It was found that six exclusive haplotypes were present in the northern population and 5 in the southern one, indicating some degree of segregation between the stocks. Gastrointestinal parasites were also used as bioindicators in order to study the existence of stocks in franciscana. The PARASITES seem to indicate segregation into two functional or ecological stocks between southern Brazil–Uruguay and Argentina. Three species of parasites were recommended as biological tags [*Hadwenius pontoporiae*, *Polymorphus* (*P*) *cetaceum* and *Anisakis typica*]. On the basis of this information, at least three stocks or populations could exist.

IV. Ecology

Little is known about the northern stock or population, between Espirito Santo and Santa Catarina, a region that is under the influence of the Brazil tropical current. Between southern Brazil and Golfo San Matías, the franciscana lives in a transition zone in which the surface circulation of the southwestern Atlantic is dominated by the opposing flows of subtropical and sub-Antarctic water masses. The coastal marine ecosystem is characterized by continental runoffs with a high discharge of high-nutrient river flows (e.g., Lagoa dos Patos, Río de la Plata). Juvenile sciaenids, the most important prey of the franciscana, are typically associated with those continental runoffs and the influence of subtropical shelf waters. The franciscana feeds mostly near the bottom on fishes of several families, such as sciaenids, engraulids, batrachoidid, gadids, carangids, and atherinids. However, sciaenids account for most of the fish species. The DIET also includes squids, octopus, and shrimps. The franciscana feeds on the most abundant species in the region and seems to change its diet according to seasonal prey fluctuations. A comparison of results between two studies carried out 15 years apart in Rio Grande do Sul showed shifts in prey composition in which important prey of the former period were depleted in artisanal fisheries. Among predators, remains of franciscanas were found in stomach contents of killer whales and several species of sharks.

V. Behavior

Very little is known about the behavior of free-ranging franciscanas, in part because they are difficult to observe in the wild and in part as a consequence of low sighting effort. The franciscana is thought to be solitary or not gregarious. Herd size may range from 2 to 15 individuals. In aerial surveys carried out in southern Brazil with the objective of estimating abundance, 37 sightings gave a mean herd size of 1.19 (SD: 0.47, range: 1–3). A study of wild behavior at Bahía Anegada in southern Buenos Aires Province showed a seasonal pattern with cooperative FEED-

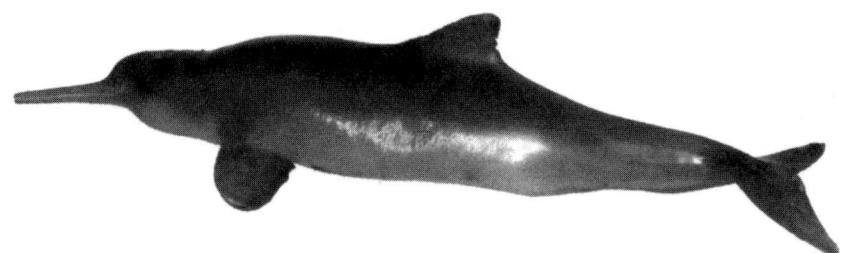

Figure 1 *Lateral view of a 137.5-cm adult female franciscana.*

ING, with traveling activities increasing during winter and high tide. The mean swimming speed was estimated in 1.3 m/sec (±0.09) with a maximum of 1.8 m/sec, and mean dive duration was estimated in 21.7 sec (±19.2) (range from 3 to 82 sec).

VI. Notable Anatomy, Physiology, Life History

Females are larger than males. Adult females range between 137 and 177 cm in total length whereas males range between 121 and 158 cm. The weight of mature females ranges between 34 and 53 kg and that of males ranges between 29 and 43 kg. Neonates in Uruguay range in size between 75 and 80 cm, whereas in southern Brazil they range between 59 and 77 cm (some of the smaller neonates could be near-term fetuses). Neonates weigh around 7.3 to 8.5 kg. Age at sexual maturity was estimated to be 2.7 years, and the gestation period is between 10.5 and 11.1 months. Females gave birth around November and lactation lasts for 9 months. However, calves take solid food around the third month, weighing between 77 and 83 cm. Mating seems to occur in January and February. The calving interval is around 2 years; nevertheless, few females

Figure 2 *Distribution range of the franciscana dolphin in the southwestern Atlantic Ocean. The shaded area represents the area inside the approximate 30-m isobath.*

were found lactating and pregnant at the same time. Reproductive capacities and life span are low for the species, which is a problem for the population to sustain the mortality rates caused by fisheries. Longevity has been estimated close to 15 years for males and 21 for females, fairly low when compared to most of the small cetaceans. Few individuals attain ages over 10 years. Three types of acoustic signals have been recorded, including low, high, and ultra high frequency clicks.

VII. Fossil Record

Three records have been related to the franciscana and assigned to the family Pontoporiidae: *Brachidelphis mazeasi,* a middle miocene fossil from the Pisco Formation (Perú); *Pontistes rectifrons,* a late miocene fossil found in the Paraná Formation (Argentina); and *Pliopontos littoralis,* a pliocene fossil closely related to the living species described for the Pisco Formation (Perú).

VIII. Interactions with Humans

Incidental catches in gill nets, mostly of juvenile individuals, became a serious problem for the species throughout its distribution range probably since the end of World War II. At that time, many artisanal fisheries for sharks developed in the region for vitamin A production, which was exported to Europe. During the 1970s, gill net mortality in Uruguay was estimated at above 400 individuals/year and fell to around 100 individuals/year in the last few years for economic reasons. Nevertheless, minimum mortality rates were always estimated at several thousands of individuals throughout the distribution range. At present, higher mortality rates are shown by the fisheries at Rio Grande do Sul and Buenos Aires Province, where no less than 700 and 500 are, respectively, incidentally taken. The estimated mortality for the whole distribution range could be no less than 1500 individual per year. It is not known if these mortality rates are sustainable. There has been only one attempt to estimate abundance, carried out at Rio Grande do Sul, where around 42,000 individuals were estimated to be in 64,000 km^2 between the coast and the 30-m isobath. The upper limit of the ABUNDANCE ESTIMATION cannot sustain the lowest estimates of mortality. Therefore, more precise estimates are needed along with conservation measures in order to preserve the species. Other threats to the franciscana include habitat degradation. A large proportion of the distribution range is subject to pollution from several sources, especially the agricultural use of land and heavy industries between São Paulo in Brazil and Bahía Blanca in Argentina. The coastal zone is also intensely used for boat traffic, tourism, and artisanal and industrial fishing operations.

See Also the Following Articles

Incidental Catches ■ River Dolphins

References

Andrade, A., Pinedo, M. C., and Pereira, J., Jr. (1997). The gastrointestinal helminths of the franciscana, *Pontoporia blainvillei,* in southern Brazil. *Rep. Intl. Whal. Comm.* **47,** 669–674.

Aznar, F. J., Raga, J. A., Corcuera, J. and Monzón, F. (1995). Helminths as biological tags for franciscana (*Pontoporia blainvillei*) (Cetacea, Pontoporiidae) in Argentinian and Uruguayan waters. *Mammalia* **59**(3), 427–435.

Barnes, L. G. (1985). Fossil pontoporiid dolphins (Mammalia: Cetacea) from the Pacific Coast of North America. *Contr. Sci. Nat. Hist. Mus. L. A. County* **363,** 1–34.

Bassoi, M. (1997). Avaliação da dieta alimentar de toninha, *Pontoporia blainvillei* (Gervais & D'Orbigny, 1844), capturadas acidentalmente na pesca costeira de emalhe no sul do Rio Grande do Sul. Dissertação de Bacharelado, Fundação Universidade do Rio Grande. Rio Grande - RS.

Bordino, P., Thompson, G., and Iñiguez, M. (1999). Ecology and behaviour of the franciscana dolphin *Pontoporia blainvillei* in Bahía Anegada, Argentina. *J. Cetacean Res. Manage.* **1**(2), 213–222.

Brownell, R. L., Jr. (1975). Progress report on the biology of the franciscana dolphin *Pontoporia blainvillei* in Uruguayan waters. *J. Fish. Res. Board. Can.* **32**(7), 1073–1078.

Brownell, R. L., Jr. (1984). Review of reproduction in platanistid dolphins. *Rep. Intl. Whal. Comm.* (special issue 6), 149–158.

Busnel, R. G., Dziedzic, A., and Alcuri, G. (1974). Études préliminaires de signaux acoustiques du *Pontoporia blainvillei* Gervais et D'Orbigny (Cetacea, Platanistidae). *Mammalia* **38,** 449–459.

Corcuera, J., Monzón, F., Crespo, E. A., Aguilar, A., and Raga, J. A. (1994). Interactions between marine mammals and coastal fisheries of Necochea and Claromecó (Buenos Aires Province, Argentina). *Rep. Intl. Whal. Comm.* (special issue 15), 283–290.

Crespo, E. A., Harris, G., and Gonzalez, R. (1998). Group size and distributional range of the franciscana *Pontoporia blainvillei.* *Mar. Mamm. Sci.* **14**(4), 845–849.

Cozzuol, M. A. (1996). Contributions of southern South America to vertebrate paleontology. *Münch. Geowiussenschaftliche Abhandlungen* **30,** 321–342.

Kasuya, T., and Brownell, R. L., Jr. (1979). Age determination, reproduction and growth of franciscana dolphin *Pontoporia blainvillei.* *Sci. Rep. Whale Res. Inst.* **31,** 45–67.

Muizon, C. de (1988). Les Vertebrés fossiles de la Formation Pisco (Pérou) Triosieme partie: Les Odontocétes (Ceacea, Mammalia) du Miocene. Recherche sur les Grandes Civilisations, Institut Française d'Etudes Andines. *Mémoire* **78,** 1–244.

Pérez Macri, G., and Crespo, A. (1989). Survey of the franciscana, *Pontoporia blainvillei,* along the Argentine coast, with a preliminary evaluation of mortality in coastal fisheries. *In* "Biology and Conservation of the River Dolphins" (W. F. Perrin, R. L. Brownell, Jr., K. Zhou, and J. Liu, eds.), pp. 57–63. Occasional Papers of the IUCN Species Survival Commission (SSC) 3.

Pinedo, M. C. (1982). Analises dos contudos estomacais de *Pontoporia blainvillei* (Gervais and D'Orbigny, 1844) e *Tursiops gephyreus* (Lahille, 1908) (Cetacea, Platanistidae e Delphinidae) na zona estuarial e costeira de Rio Grande, RS, Brasil. M.Sc. Thesis, Universidade do Rio Grande do Sul, Brasil.

Pinedo, M. C. (1991). Development and variation of the franciscana, *Pontoporia blainvillei.* Ph.D. Thesis, University of California, Santa Cruz.

Pinedo, M. C., Praderi, R., and Brownell, R. L., Jr. (1989). Review of the biology and status of the franciscana *Pontoporia blainvillei.* *In* "Biology and Conservation of the River Dolphins" (W. F. Perrin, R. L. Brownell, Jr., K. Zhou, and J. Liu, eds.), pp. 46–51. Occasional Papers of the IUCN Species Survival Commission (SSC) 3.

Praderi, R., Pinedo, M. C., and Crespo, E. A. (1989). Conservation and management of *Pontoporia blainvillei* in Uruguay, Brazil and Argentina. *In* "Biology and Conservation of the River Dolphins" (W. F. Perrin, R. L. Brownell, Jr., K. Zhou, and J. Liu, eds.), pp. 52–56. Occasional Papers of the IUCN Species Survival Commission (SSC) 3.

Secchi, E. R., Ott, P. H., Crespo, E. A., Kinas, P. G., Pedraza, S. N. and Bordino, P. (2001). A first estimate of franciscana (*Pontoporia* blainvillei) abundance off southern Brazil. *Journal of Cetacean Research and Management* 3(1), 95–100.

Secchi, E. R., Zerbini, A. N., Bassoi, M., Dalla Rosa, L., Moller, L. M., and Roccha-Campos, C. C. (1997). Mortality of franciscanas, *Pontoporia blainvillei*, in coastal gillnets in southern Brazil: 1994–1995. *Rep. Intl. Whal. Comm.* 47, 653–658.

Secchi, E. R., Wang, J. Y., Murray, B. Roccha-Campos, C. C., and White, B. N. (1998). Populational differences between franciscanas, *Pontoporia blainvillei*, from two geographical locations as indicated by sequences of mtDNA control region. *Can. J. Zool.* 76, 1622–1627.

Fraser's Dolphin
Lagenodelphis hosei

M. LOUELLA L. DOLAR
Tropical Marine Research, San Diego, California

Fraser's dolphin was described in 1956 based on a skeleton collected by E. Hose from a beach in Sarawak, Borneo in 1895. F. C. Fraser gave it the genus name *Lagenodelphis*, due to what appeared to him as similarity of the skull to those of *Lagenorhynchus* spp. and *Delphinus delphis*. The external appearance of this species was not known until 1971 when specimens were found in widely separated areas: near Cocos island in the eastern tropical Pacific, South Africa, and southeastern Australia.

I. Characters and Taxonomic Relationships

Fraser's dolphin is easily identified by its stocky body, short but distinct beak, and small, triangular or slightly falcate dorsal fin—the flippers and flukes are also small (Fig. 1). The color pattern is striking and varies with age and sex. For example, a distinct black stripe that extends from eye to anus is absent or faint in juveniles, wider and thicker in adult males, and variable in adult females. The same is true with the facial stripe or "bridle": it is absent in calves, variable in females, and extensive in adult males, where it merges with the eye-to-anus stripe to form a "bandit mask." Color pattern in the genital region may also be sexually dimorphic. The back is brownish gray, the lower side of the body is cream colored, and the belly is white or pink. Other features that appear to vary with age and sex are dorsal fin shape and the postanal hump. With some variability, the dorsal fin is slightly falcate in calves and females and more erect or canted in adult males. Similarly, the postnatal hump is either absent or slight in females and young of both sexes and well developed in adult males. From a distance, the eye-to-anus stripe makes Fraser's dolphin look similar to the striped dolphin, *Stenella coeruleoalba*. However, the distinctive body shape of Fraser's dolphin rules out confusion with other species. The largest male recorded was 2.7 m long and the largest female 2.6 m with males over 10 years old significantly larger than females. Large males could weigh up to 210 kg. Based on a limited number examined,

it is tentatively proposed that Fraser's dolphins in the Atlantic are larger than those in the Pacific. Fraser's dolphin belongs to the subfamily Delphininae. Based on cytochrome b mtDNA sequences, it is more closely related to *Stenella, Tursiops, Delphinus*, and *Sousa* than it is to *Lagenorhychus*. Morphologically, the skull structure shows close similarity with that of the common dolphin, *D. delphis*, in terms of the presence of deep palatal grooves and with those of *S. longirostris, S. coeruleoalba*, and the clymene dolphin *Stenella clymene* in several other characteristics.

II. Distribution and Ecology

Fraser's dolphin is a tropical species, distributed between 30°N and 30°S. Strandings outside this limit, such as in southeastern Australia, Brittany, and Uruguay, are considered unusual and are probably influenced by temporary oceanographic events. It is typically an oceanic species, except in places where deep water approaches the coast such as in the Philippines, Indonesia, and Lesser Antilles, where Fraser's dolphins can be observed within 100 m from shore. In the eastern tropical Pacific, they were observed to occur at least 15 km offshore and, mostly, on high seas approximately 45–110 km from the coast where water depth was between 1500 and 2500 m. In the Sulu Sea, Philippines, highest sighting rates were in waters >500 m and up to 5000 m, although some animals were observed in shallower waters adjacent to the continental shelf. In the Gulf of Mexico, sightings have been around 1000 m depth, and the animals appear to be more common in the Gulf than anywhere else in the North Atlantic. Affinity to deep waters can be explained by the type of prey eaten by Fraser's dolphins, which is composed of mesopelagic fish, crustaceans, and cephalopods. It is also suggested that Fraser's dolphins feed selectively on larger prey that inhabit deeper waters. In the eastern tropical Pacific and the Sulu Sea, the most common fishes in the diet are the Myctophidae and Chauliodontidae, and the most common crustaceans are the Oplophoridae. Cephalopods were not reported eaten by Fraser's dolphins in the eastern tropical Pacific, but this group comprised a significant amount of the DIET in the Sulu Sea animals, i.e., about 30% by volume. Based on prey composition, it was hypothesized that Fraser's dolphins in the eastern tropical Pacific feed at two depth horizons: the shallowest level of no less than 250 m and the deepest no less than 500 m. In the Sulu Sea, Fraser's dolphins appear to feed from near surface to at least 600 m. Examination of myoglobin (Mb) concentrations in the skeletal muscles of Fraser's dolphin support the distribution and feeding habits of this species. The value averages at 7.1 g Mb 100^{-g} muscle and is comparable to those of the very best divers such as the Weddell seal (*Leptonychotes weddellii*), bottlenose whale (*Hyperoodon ampullatus*), and sperm whale (*Physeter macrocephalus*). However, in South Africa and in the Caribbean, Fraser's dolphins were observed FEEDING near the surface.

Although no PREDATION has been reported, Fraser's dolphins may be preyed upon by killer whales (*Orcinus orca*), false killer whales (*Pseudorca crassidens*), and large sharks. Cookie cutter sharks (*Isistius brasiliensis*) are thought to inflict circular wounds.

An external PARASITE, *Xenobalanus* sp., and internal parasites (*Phyllobothrium delphini, Monorhygma grimaldi, Anisakis*

Figure 1 *Fraser's dolphins in the eastern tropical Pacific. Courtesy of R. L. Pitman.*

simplex, Tetrabothrius sp., *Bolbosoma* sp., *Strobicephalus triangularis, Campula* sp., and *Stenurus ovatus*) have been observed in Fraser's dolphins.

III. Behavior and Life History

Fraser's dolphins often swim in tight fast-moving schools of 100 to 1000 individuals with the members of the school "porpoising in low-angle, splashy leaps" and have been reported to swim away from vessels in the eastern tropical Pacific. In the Philippines they were observed to ride the bow if the boat ran at less than 3 knots but were often displaced by melon-headed whales (*Peponocephala electra*).

In the eastern tropical Pacific and the Gulf of Mexico, Fraser's dolphins are often found together with melon-headed whales. Although also sometimes seen with melon-headed whales (when in relatively shallow waters), Fraser's dolphins in the Sulu Sea are often seen with short-finned pilot whales, *Globicephala macrorhynchus*. Other species Fraser's dolphins are seen with are the false killer whale, Risso's dolphin (*Grampus griseus*), spinner dolphin, pantropical spotted dolphin (*S. attenuata*), bottlenose dolphin (*Tursiops truncatus*), and sperm whale. In the western tropical Indian Ocean, Fraser's dolphins were not seen with any other species.

A sample of 108 dolphins from a school captured in Japan showed a mixed-age group and a ratio of 1:1 between males and females; the oldest was estimated to be 17.5 years old. Males reach sexual maturity at about 7–10 years at 220–230 cm and the females at 5–8 years at 210–220 cm. Mating may be promiscuous. The annual ovulation rate is about 0.49, and the gestation period is about 12.5 months. The calving interval is approximately 2 years; in Japanese waters, calving appears to peak in spring and fall. Limited samples from South Africa suggest that calving occurs in summer. Length at birth is estimated to be about 100–110 cm.

IV. Interaction with Humans

Fraser's dolphins are caught in drive nets in Japan and by harpoon in Lower Antilles, Indonesia, and (before they became protected) in the Philippines. They are also caught incidentally in purse seines in the eastern tropical Pacific and the Philippines, in traps nets in Japan, in gill nets in South Africa and Sri Lanka, in antishark nets in South Africa, and in drift nets in the Philippines.

See Also the Following Articles

Coloration ■ Delphinids, Overview

References

Amano, M., Miyazaki, N., and Yanagisawa, F. (1996). Life history of Fraser's dolphin, *Lagenodelphis hosei*, based on a school captured off Pacific coast of Japan. *Mar. Mamm. Sci.* **12,** 199–214.

Balance, L. T., and Pitman, R. B. (1998). Cetaceans of the tropical western Indian Ocean: Distribution, relative abundance, and comparisons with cetacean communities of two other tropical ecosystems. *Mar. Mamm. Sci.* **14,** 428–459.

Dolar, M. L. L. (1994). Incidental takes of small cetaceans in fisheries in Palawan, central Visayas and northern Mindanao in the Philippines. *Rep. Int. Whal. Commn. Spec. Issue* **15,** 355–363.

Dolar, M. L. L. (1999). "Abundance, Distribution and Feeding Ecology of Small Cetaceans in the Eastern Sulu Sea and Tañon Strait, Philippines." Ph.D. Dissertation, University of California, San Diego.

Dolar, M. L. L., Leatherwood, S., Wood, C., Alava, M. N. R., Hill, C., and Aragones, L. V. (1994). Directed fisheries for cetaceans in the Philippines. *Rep. Int. Whal. Commn.* **44,** 439–449.

Dolar, M. L. L., Suarez, P., Ponganis, P., and Kooyman, G. L. (1999). Myoglobin in pelagic small cetaceans. *J. Exp. Biol.* **202**, 227–236.

Jefferson, T. A., and Leatherwood, S. (1994). *Lagenodelphis hosei. Mamm. Spec.* **470**, 1–5.

Jefferson, T. A., Leatherwood, S., and Weber, M. A. (1993). "FAO Species Identification Guide: Marine Mammals of the World." FAO, Rome.

Jefferson, T. A., Pitman, R. L., Leatherwood, S., and Dolar, M. L. L. (1997). Developmental and sexual variation on the external appearance of Fraser's dolphins (*Lagenodelphis hosei*). *Aquat. Mamm.* **23**, 145–153.

Le Duc, R. G., Perrin, W. F., and Dizon, A. E. (1999). Phylogenetic relationships among the delphinid cetaceans based on full cytochrome b sequences. *Mar. Mamm. Sci.* **15**, 619–648.

Perrin, W. F., Leatherwood, S., and Collet, A. (1994). Fraser's dolphin, *Lagenodelphis hosei* Fraser, 1956. *In* "Handbook of Marine Mammals" (S. H. Ridgway and R. Harrison, eds.), Vol. 5, pp. 225–240. Academic Press, San Diego.

Perrin, W. F., Best, P. B., Dawbin, W. H., Balcomb, K. G., Gambell, R., and Ross, G. J. B. (1973). Rediscovery of Fraser's dolphin *Lagenodelphis hosei. Nature* **241**, 345–350.

Perryman, W. L., Au, D. W. K., Leatherwood, S., and Jefferson, T. A. (1994). Melon-headed whale, *Peponocephala electra* Gray, 1846. *In* "Handbook of Marine Mammals" (S. H. Ridgway and R. Harrison, eds.), Vol. 5, pp. 363–386. Academic Press, San Diego.

Robison, B. H., and Craddock, J. E. (1983). Mesopelagic fishes eaten by Fraser's dolphin, *Lagenodelphis hosei. Fish. Bull. U.S.* **81**, 283–289.

Wade, P. R., and Gerrodette, T. (1993). Estimates of cetacean abundance and distribution in the eastern tropical Pacific. *Rep. Int. Whal. Comm.* **43**, 477–493.

Watkins, W. A., Daher, M. A., Fristrup, K., and Notobartolo di Sciara, G. (1994). Fishing and acoustic behavior of Fraser's dolphin (*Lagenodelphis hosei*) near Dominica, southeast Caribbean. *Carib. J. Sci.* **30**, 76–82.

Würsig, B., Jefferson, T. A., and Schmidly, D. J. (2000). "The Marine Mammals of the Gulf of Mexico." Texas A & M Univ. Press, College Station, TX.

Fur

SEE *Hair and Fur*

G

Gastrointestinal Tract

JAMES G. MEAD
National Museum of Natural History,
Smithsonian Institution, Washington, DC

The gastrointestinal tract consists of all structures derived from the primitive gut tube and distal to the esophagus, including the stomach, small intestine, large intestine, and those accessory structures that have formed from that part of the gut (liver, gallbladder, pancreas, hepatopancreatic duct, anal tonsils). The posterior boundary is the lower part of the anal canal where the mucous membrane of the gut ends and the epidermis begins. This article follows the terminology of Chivers and Langer (1994).

The anatomy of the gastrointestinal tract has long fascinated workers. Grew (1681) is the earliest worker who dealt with that topic exclusively. Tyson (1680), in his marvelous treatment of the anatomy of the harbor porpoise (*Phocoena phocoena*), went extensively into the gastrointestinal tract. Owen dissected the dugong (*Dugong dugon*) (1838) and then summarized all the information on the digestive system of mammals in his magnum opus on comparative anatomy (1868). William Turner did extensive studies of the stomach of cetaceans, which are summarized in his catalog of the specimens of marine mammals in the Anatomical Museum of the University of Edinburgh (1912). Langer (1988) and Reynolds and Rommel (1996) did a very good treatment of the gastrointestinal tract of the sirenians.

Measurements of the gastrointestinal tract, both in terms of length and volume, are extremely difficult due to the elasticity of the organs. At death the muscles lose their tonus and the length and volume can double or triple (Slijper, 1962).

The parts of the gastrointestinal tract are described starting with the stomach and progressing distally. The anatomy of each part is treated in sequence according to the following classification: Pinnipedia (Phocidae, Otariidae, Odobenidae); Sirenia (Dugongidae, Trichechidae); Cetacea-Odontoceti [Delphinoidea (including Phocoenidae and Monodontidae), Platanistoidea (including Platanistidae, Iniidae, Pontoporiidae, and Lipotidae),

Physeteroidea, Ziphiidae]; and Mysticeti (Balaenopteridae, Balaenidae, Eschrichtiidae, Neobalaenidae). The major features of the gastrointestinal tract are summarized in Table I.

I. Major Organs

A. Stomach

The stomach is a series of compartments starting with the cardiac, fundic, and ending with the pyloric. The boundary of the stomach with the esophagus is determined by the epithelial type: stratified squamous for the esophagus and columnar for the stomach. The distal boundary is marked by the pyloric sphincter.

The stomach is suspended by the mesogastrium, which, in development, becomes complexly folded and differentiated into the greater and lesser omenta.

1. Pinnipedia The stomach in pinnipeds is relatively uncomplicated when compared to the rest of marine mammals. The stomach in the California sea lion (*Zalophus californianus*) consists of a simple cardiac chamber into which the esophagus enters, followed by a narrowing into the pyloric chamber. There is a prominent pyloric sphincter. The pyloric end of the stomach is strongly recurved onto the cardiac portion. The stomach in the southern sea lion (*Otaria flavescens*) and Weddell seal (*Leptonychotes weddellii*) does not differ from the California sea lion. The stomach of the walrus (*Odobenus rosmarus*), although it is not described in any detail, does not appear to differ markedly from that of the other pinnipeds.

Pinnipeds seem to follow the carnivore plan of a relatively simple single-chambered (monolocular), nonspecialized stomach.

2. Sirenia The stomach in the dugong is moderately complex. Externally it is a simple oval organ with the esophageal opening in the center. Internally, there is a ridge (gastric ridge) that divides the stomach into two compartments: the cardiac and pyloric portions. There is development of a powerful sphincter up to 4 cm thick at the esophageal/gastric junction (Owen, 1868). The stomach walls are highly muscular. The cardiac gland is roughly spherical and about 15 cm in diameter in adults. The cardiac gland opens into the first compartment, where the esophagus also opens. The mucosa in the cardiac

TABLE I
Comparative Morphology of the Gastrointestinal System of Marine Mammals

Taxon	Stomach						Small intestine					Large intestine		Accessory organs				
	Type	Forestomach	Main stomach[a]	Connecting chambers	Pyloric stomach	Cardiac gland	Duodenum	Duodenal ampulla	Duodenal diverticula	Jejunum[b]	Ileum	Cecum	Colon	Liver	Gall-bladder	Pancreas	Hepato-pancreatic duct	Anal tonsils
Pinnipedia																		
Phocid	Unilocular	Absent	Present	Absent	Absent	Absent	Present	Absent	Absent	Undiff.	Undiff.	Present	Present	Multilobed	Present	Present	Absent	Absent?
Otariid	Unilocular	Absent	Present	Absent	Absent	Absent	Present	Absent	Absent	Undiff.	Undiff.	Present	Present	Multilobed	Present	Present	Present	Absent?
Odobenid	Unilocular?	Absent?	Present?	Absent?	Absent	Absent?	Present	Absent	Absent?	Undiff.	Undiff.	Present	Present	Multilobed	Present	Present	Present	Absent?
Sirenia																		
Dugonid	Unilocular	Absent	Present	Absent	Present	Present	Present	Present	Present	Undiff.	Undiff.	Hyper.	Present	Multilobed	Present	Present	Absent	Absent?
Trichechid	Unilocular	Absent	Present	Absent	Present	Present	Present	Present	Present	Undiff.	Undiff.	Hyper.	Present	Multilobed	Present?	Present	Absent	Absent?
Cetacea																		
Mysticete																		
Balaenopterid	Plurilocular	Present	Present	Present	Present	Absent	Present	Present	Absent	Undiff.	Undiff.	Present	Present	Bilobed	Absent	Present	Present	Absent?
Eschrichtiid	Plurilocular	Present	Present	Present	Present	Absent	?.	?.	Absent	?.	?.	Present?	Present?	Bilobed	Absent	Present	Present	Present
Balaenid	Plurilocular	Present	Present	Present	Present	Absent	?.	?.	Absent	?.		Absent	Undiff.	Bilobed	Absent	Present	Present	Present?
Neobalaenid	Plurilocular	Present	Present	?.	Present	Absent	?.	?.	Absent	?.	?.	Present?	Present?	Bilobed	Absent	Present	Present?	Absent?
Odontocete																		
Delphinoid																		
Delphinid	Plurilocular	Present	Present	Present	Present	Absent	Present	Present	Absent	Undiff.	Undiff.	Absent	Undiff.	Bilobed	Absent	Present	Present	Absent
Phocoenid	Plurilocular	Present	Present	Present	Present	Absent	Present	Present	Absent	Undiff.	Undiff.	Absent	Undiff.	Bilobed	Absent	Present	Present	Absent?
Monodontid	Plurilocular	Present	Present	Present	Present	Absent	Present	Present	Absent	Undiff.	Undiff.	Absent	Undiff.	Bilobed	Absent	Present	Present	Absent?
Platanistoid	Plurilocular	Variable	Hyper.	Variable	Variable	Absent	Present	Present	Absent	Undiff.	Undiff.	Variable	Undiff.	Bilobed	Absent	Present	Present	Variable
Physeteroid	Plurilocular	Present	Present	Present?	Present	Absent	Present	Present	Absent	Undiff.	Undiff.	Absent	Undiff.	Bilobed	Absent	Present	Present	Present?
Ziphiid	Plurilocular	Absent	Variable	Hyper.	Present	Absent	Present	Present	Absent	Undiff.	Undiff.	Absent	Undiff.	Bilobed	Absent	Present	Present	Absent?

[a] hypertrophied.
[b] undifferentiated.

gland is packed with gastric glands that are distinguishable from the glands in the main stomach compartment. The glands consist of chief and parietal cells at a ratio of 10:1. The mucosa in the cardiac glands is arranged in a complex plicate structure. The pyloric aperture is in the second compartment. The cardiac region of the stomach extends for several centimeters from the esophageal junction. The stomach is lined by gastric glandular epithelium with a particular abundance of goblet cells and mucus-secreting gastric glands.

The stomach of the dugong appears to be modified to secrete mucus to aid in lubricating the ingested material and prevent mechanical abrasion to the mucosa. It is interesting that the salt content of the dugong diet is high; sodium is about 30 times and the chloride about 15 times that of terrestrial pasture plants.

The stomach in the recently extinct *Hydrodamalis gigas* (Steller sea cow) was apparently very large. According to Steller, it was 6 feet long and 5 feet wide when distended with masticated seaweed.

The stomach in the manatees (*Trichechus* spp.) is very similar to that in the dugong. The stomach is divided by a muscular ridge into cardiac and pyloric regions. A single cardiac gland opens into the cardiac region of the stomach.

3. Cetacea   The cetacean stomach is a diverticulated composite stomach (pleurolocular), consisting of regions of stratified squamous epithelium, fundic mucosa, and pyloric mucosa.

The stomach, as typified by a delphinid, consists of four chambers (Fig. 1). These have been referred to by a variety of anatomical terms: forestomach (first, esophageal compartment, paunch), main stomach (second, cardiac, fundus glandular, proximal), connecting chamber (third, fourth, "narrow tunneled passage," "conduit ètroit," intermediate, connecting channel, connecting division), and pyloric stomach (third, fourth, fifth, pyloric glandular, distal).

A. FORESTOMACH

There has been discussion about the homologies of the forestomach in Cetacea. It is lined with stratified squamous epithelium, like the esophagus, and there was reason to believe that it was just an esophageal sacculation. Embryological work in the common minke whale (*Balaenoptera acutorostrata*) demonstrated that the forestomach was formed from the stomach not the esophagus and that it was homologous to the forestomach of ruminants.

Odontocetes

Delphinoid: The forestomach is lined with stratified squamous nonkeratinized epithelium. The epithelial lining is white in freshly dead animals and is thrown into a series of longitudinal folds when empty. The forestomach was often referred to as the "paunch" in older literature. Like the other chambers in the stomach, it is variable in size. It is pyriform and on the order of 30 cm long in an adult *Tursiops truncatus* (280 cm total length). The forestomach is highly muscular but has no glandular functions. The forestomach/main stomach is a wide opening (3 to 5 cm in adult *Tursiops*) in the wall of the forestomach near the esophageal end. The forestomach functions as a holding cavity analogous to the crop of birds or the forestomach of ungulates. Because the communication with the main stomach

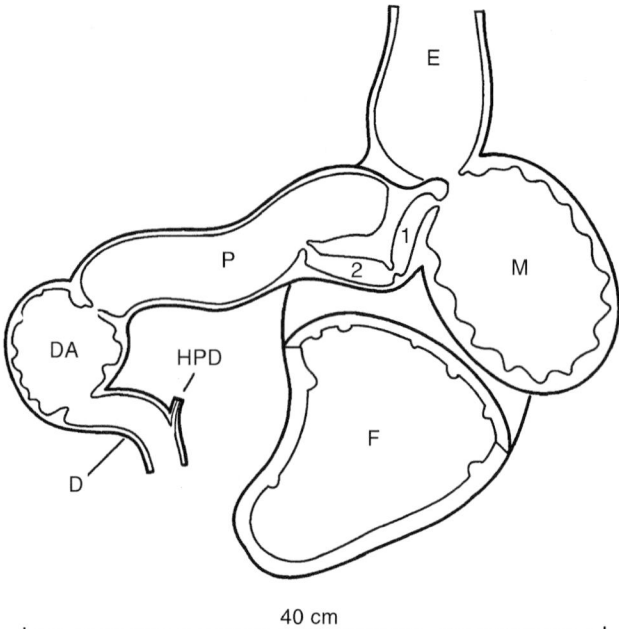

Figure 1   *Stomach of a spinner dolphin,* Stenella longirostris, *ventral view. D, duodenum; DA, duodenal ampula; E, esophagus; F, forestomach; HPD, hepatopancreatic duct; M, main stomach; P, pyloric stomach; 1, 2, compartments of connecting chambers. After Harrison et al. (1970).*

is so wide, there is a reflux of digestive fluids from the main stomach and some digestion takes place in the forestomach.

The same general relationships hold in *Phocoena, Delphinapterus leucas,* and *Monodon monoceros.*

Platanistoid: The forestomach is unusual in *Inia geoffrensis* and *Platanista gangetica* in that the esophagus runs directly into the main stomach and the forestomach branches off the esophagus. In the two other families of platanistoids, the forestomach is lacking entirely.

Physeteroid: The forestomach is present in *Physeter macrocephalus*. It was a compartment about 140 and 140 cm, lined with yellowish-white epithelium in a 15.6-m male.

Ziphiidae: The forestomach is absent in all ziphiids.

Mysticetes

The forestomach is present in all species of mysticetes.

B. MAIN STOMACH   The main stomach has a highly vascular, glandular epithelium that is grossly trabeculate. The epithelium of the main stomach is dark pink to purple. The main stomach secretes most of the digestive enzymes and acids and where digestion commences. It has also been known as the fundic stomach. It is present in all cetaceans.

Odontocetes

Delphinoid: The main stomach is approximately spherical and on the order of 10–15 cm in adult *Tursiops*. The same general relationships hold in *Phocoena, Delphinapterus,* and *Monodon.*

Platanistoid: In *Platanista* there is a constricting septum of the main stomach that forms a small distal chamber, through which the digesta must pass.

Lipotes vexillifer presents an unusual situation in having three serially arranged main stomach compartments. The second and third compartments are very much smaller than the first and are topographically homologous with the connecting chambers. However, they are lined by epithelium that has fundic glands, typical of the main stomach.

Physeteroid: There is nothing remarkable about the main stomach of physeteroids.

Ziphiid: Some ziphiids develop a subdivision in their main stomach. There is an incipient constriction in the main stomach of *Berardius bairdii* and *Mesoplodon bidens* that divides the stomach into two compartments. The connecting chambers exit off the second compartment. Another type of stomach modification has occurred in *Mesoplodon europaeus* and *M. mirus*, where a large septum has developed, forming a blind diverticulum in the main stomach. An additional septum has developed in the diverticulum in *Mesoplodon europaeus* subdividing it.

Mysticetes

There is nothing remarkable about the main stomach in mysticetes.

C. CONNECTING CHAMBERS The connecting chambers, also called the connecting channel, the intermediate stomach, and the third stomach are present in all Cetacea. They are lined with pyloric epithelium and are easily overlooked in dissections. They are small in most cetaceans but have been developed greatly in ziphiids. Because of their proliferation in ziphiids, where they seem to function as something more than channels between the main and pyloric stomachs, their name was changed from connecting channels to connecting chambers.

Odontocetes

Delphinoid: The connecting chambers in a typical delphinoid consist of two narrow compartments lying between the main stomach and the pyloric stomach. The diameter of the connecting chambers is 0.8 cm in adult *Tursiops* and the combined length is 7–9 cm. The epithelial lining is very similar to pyloric stomachs. In some species the compartments are simple serially arranged; in others they may have diverticulae.

The same general relationships hold in *Phocoena*, *Delphinapterus*, and *Monodon*.

Platanistoid: They occur in all the species of platanistoids, with the exception of *Lipotes*. In that species the compartments lying between the main stomach and the pyloric stomach (second and third compartments of the main stomach) are lined with epithelium containing fundic and mucous glands in the first compartment and fundic glands in the second compartment. This would make them subdivisions of the main stomach. The connecting chambers appear to be absent in *Lipotes*.

Physeteroid: Although none of the works that describe the sperm whale stomach mention the connecting chambers, there is no reason to assume that they are absent.

Ziphiid: The connecting chambers in ziphiids are globular compartments, ranging in number from 2 to 11 (Fig. 2). They are separated by septa and communicate by openings in the septa. The openings are either central or peripheral in the septa. The connecting chambers are lined with pyloric epithelium. The connecting chambers in specimens of adult *Mesoplodon* (ca. 5 m long) are about 10 cm in diameter.

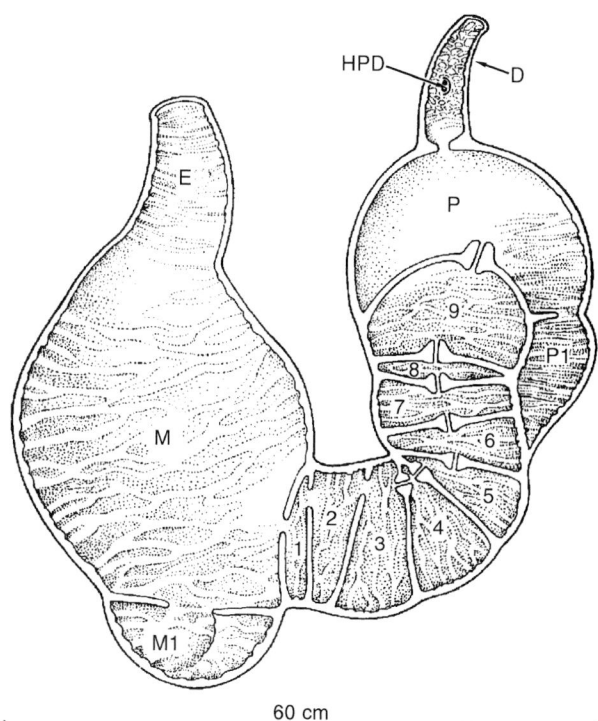

60 cm

Figure 2 *Stomach of a Gulf Stream beaked whale,* Mesoplodon europaeus, *dorsal view. D, duodenum; DA, duodenal ampulla; E, esophagus; HPD, hepatopancreatic duct; M, main stomach; M1, accessory main stomach; P, pyloric stomach; P1, accessory pyloric stomach; 1, 2, etc., compartments of connecting chambers. Drawing by Trudy Nicholson.*

Mysticetes

Balaenopterid: Many workers have described the connecting chambers in a number of species of *Balaenoptera* (blue, *B. musculus*; fin, *B. physalus*; sei, *B. borealis*; minke, *B. acutorostrata* and *B. bonaerensis*). The connecting chambers in common minke whales are 10 to 30 cm in length.

Balaenid: The inflated connecting chambers in an 8.5-m female *Balaena mysticetus* were 5 cm in diameter and 17 cm combined length. The presence of connecting chambers was not mentioned in dissections of right whales.

Eschrichtiid: The connecting chambers are relatively large in a newborn *Eschrichtius robustus*.

Neobalaenid: There are no data on the connecting chambers in *Caperea marginata*.

D. PYLORIC STOMACH

Odontocetes

Delphinoid: The pyloric stomach in delphinoids is a simple tubular cavity lined by typical mucous-producing pyloric glands. The epithelium is in many ways similar to the epithelium of the small intestine. The pyloric stomach is about 20 cm long and 4 cm in flat diameter in an adult *Tursiops*.

The same general relationships hold in *Phocoena*, *Delphinapterus*, and *Monodon*.

Platanistoid: The pyloric stomach in *Platanista gangetica* is a single chamber about 12 cm long and contains abundant large tubular pyloric glands. The pyloric stomach is comparable in *Inia* and *Pontoporia*, but differs markedly in *Lipotes*. In that species it is differentiated into a proximal bulbous compartment and a smaller distal compartment. The epithelial lining in *Lipotes* is similar to all other Cetacea.

Physeteroid: Available data on the pyloric stomach of physeterids are scanty. The pyloric compartment is present and there is no reason to assume that it is different from the rest of the cetaceans.

Ziphiid: The pyloric stomach in a newborn *Ziphius* is a simple spherical compartment that measures about 10 cm in diameter. It is lined with smooth pyloric epithelium and communicates with the duodenum through a strong pyloric sphincter. This is also the case in *Hyperoodon* spp., *Tasmacetus*, and some species of *Mesoplodon* (*M. densirostris*, *M. hectori*, and *M. stejnegeri*).

In *Berardius bairdii* the main pyloric compartment has expended in volume to where it is nearly the size of the main stomach and has developed a small distal accessory chamber. The pyloric compartments are in series, with the accessory chamber communicating with the duodenum.

In all other species of *Mesoplodon* examined to date (*M. bidens*, *M. europaeus*, and *M. mirus*), a blind diverticulum has developed. The diverticulum comes off the proximal side of the pyloric stomach and lies along the distal connecting chambers. The accessory pyloric stomach communicates with the pyloric stomach through a wide opening.

Mysticetes

Balaenopterid: In all of the balaenopterid species examined (*B. acutorostrata*, *B. borealis*, *B. musculus*, and *B. physalus*), the pyloric stomach is smaller than the stomach. The pyloric stomach contains 8.5 to 12.1% of the total inflated stomach volume (18–391).

Balaenid: The balaenids appear to be similar to the balaenopterids.

Eschrichtiid: In a dissection of a newborn *Eschrichtius*, the pyloric compartment seemed to be comparable to that in balaenopterids.

Neobalaenid: There are no data available on the pyloric stomach in *Caperea*.

B. Small Intestine

The small intestine starts at the pyloric sphincter. Digestion continues in the small intestine where absorption of the nutrients takes place. The duodenum derives its names from its length of about 12 inches (30 cm) in humans. It has no mesentery and its wall has longitudinal folds. The hepatic and pancreatic ducts open into the duodenum.

The longitudinal folds of the duodenum become circular (plicae circulares) in the jejunum. The jejunum starts when the small intestine becomes suspended by a mesentery.

Differentiation between the jejunum and the ileum is not sharp. The distal portions of the ileum contain longitudinal folds, which differentiates it from the proximal portion of the jejunum, which contains circular folds. The distal boundary of the ileum is sharp. The diameter of the intestine increases at the ileocolic orifice. This orifice is usually provided with a sphincter permitting partial closure.

The small intestine is characterized by the presence of absorptive villi in the mucosa.

1. Pinnipedia

Phocid: The demarcation between pylorus and duodenum is not sharply marked by position of the duodenal (Brunner's) glands in *Leptonychotes weddelli*. The duodenum is 1 or 2 feet in length. Small plicae circulares and short irregular villi were present in the duodenum. Jejunum and ileum are hard to differentiate.

Phoca vitulina has a small intestine of "great length": 40 feet (~12 m) in a seal 3 feet (91 cm) long (snout → end of flippers). An adult male *Mirounga leonina* (4.80 m sl) had a small intestine length of 202 m.

Otariid: *Otaria flavescens* lacks plicae circulares, with villi being arranged on delicate transverse linear folds. *Eumetopias jubatus* has a small intestine length of 264 feet (~80 m).

Odobenid: Owen (1853) described the intestine in passing in his description of a young walrus. The small intestine was 75 feet (~23 m) long, the cecum was 1.5 inches (3.8 cm), and the large intestine was 1 foot (~30 cm) in length.

2. Sirenia

The duodenum of the dugong and manatees has two duodenal diverticula that are crescentic in shape and about 10–15 cm long, measured in the curve. They communicate via a common connecting channel with the duodenum. The lining of the diverticulae is similar to the pyloric region of the stomach and contains mucous glands. The duodenum is about 30 cm in length, similar to other medium-sized mammals. Both the duodenum and the diverticulae contain prominent plicae circulares. There is a weak sphincter at the distal end of the duodenal ampulla. The length of the small intestine is from 5.4 to 15.5 m, four to seven times the body length of the animal. The small intestines were about 1 inch (2.5 cm) in diameter in the juvenile that Owen dissected. Brunner's glands are present in the duodenum and diverticulae. Paneth cells are absent in contrast to most domestic terrestrial herbivores. The diverticulae appear to enlarge the surface of the proximal duodenum, which would allow a larger volume of digesta to pass from the stomach at one time.

3. Cetacea

Odontocetes

Delphinoid: In delphinoids there is no cecum and no marked differentiation between large and small intestines. Intestine length ranged between 8.85 and 16.80 m in specimens of *Tursiops*, *Delphinus delphis*, and *Stenella* spp. (total lengths from 160 to 230 cm). There is a duodenum about 30 cm long, but differentiation between a jejunum and ileum is lacking. Examination of the small intestines by light microscopy revealed a lack of well-developed villi in delphinids.

Platanistoid: The length of the small intestine in a 204-cm *Inia geoffrensis* was 4.15 m. The duodenum was approximately 20 cm long. The jejunum was not differentiated from the ileum. A prominent longitudinal fold began at the opening of the hepatopancreatic duct in the duodenum and continued throughout the small intestine. The small intestine varied in diameter from 0.7 to 0.8 cm. The small intestine graded into a "smooth-walled portion" that was 1 cm in diameter. The authors were unable to tell where the small intestine ended and the large intestine began. The "smooth-walled portion" was 80 cm in length and graded into the colon distally.

There are no plicae circulares or typical villi in the intestine of *Pontoporia*. A distinct uninterrupted longitudinal fold occurs in the small intestine.

There were abundant plicae circulares in the proximal part of the intestine in *Platanista gangetica,* changing to longitudinal folds in the last meter or two. There was a prominent cecum and an ileocolic sphincter.

The small intestine in most platanistids is extremely long. The ratio of small to large intestine length is between 50 and 60% in *Pontoporia,* 50% in *Inia,* but only around 9% in *Platanista.*

Physeteroid: The total intestinal length in adult sperm whales can range up to 250 m. The plicae circulares were unusual in that they appeared to be spiral, giving the impression of a spiral valve in sharks. There is no cecum and the transition between the small and the large intestine is gradual.

Ziphiid: It is said that the combined intestinal length of *Hyperoodon* is six times the body length. There is a unique vascular rete (mirabile) intestinale associated with the large and small intestine in at least *Ziphius cavirostris* and *Berardius* spp. There is no cecum.

Mysticetes

Balaenopterid: The mean ratio of the length of the small intestine to body length in minke whales (*Balaenoptera acutorostrata*) was rather small (3.92) and averaged 36 m in length. The minke whale possessed a duodenal ampulla, but there was no indication of differentiation of the jejunum and ileum.

Balaenid, Eschrichtiid, and Neobalaenid: Nothing is available on the anatomy of the small intestine for balaenids, eschrichtiids, and neobalaenids.

C. Large Intestine

The large intestine consists of the colon (ascending, transverse and descending), cecum, vermiform appendix, rectum, anal canal, and anus. The cecum is a diverticulum off the proximal end of the colon, near the ileocolic juncture. The vermiform appendix is the narrowed apex of the cecum. The sigmoid colon of humans is that portion of the distal end of the colon that has its mesocolon extended so that it is free in the brim of the pelvis.

The colon is suspended by a mesentery (mesocolon). The colon functions to absorb water and consolidate fecal material. Most mammalian colons have their longitudinal muscle fibers arranged into bands called taenia coli. The colon is formed into sacculations, which are called haustra. The rectum is the straight portion of the large intestine that transverses the pelvis. The anal canal is the specialized terminal portion of the large intestine. The anal canal has many lymph nodes and glands that reflect the difficulties of fecal excretion. The anus represents the end of the gastrointestinal tract. The anal sphincter controls excretion of fecal wastes.

1. Pinnipedia In pinnipeds the cecum is short and blunt or round and an appendix is not present. The large intestine is relatively short and not much larger in diameter than the small intestine. No taenia coli, plicae semilunares, haustra, and appendices epiploicae are present.

Phocid: The colon is about 6 feet (183 cm) long in an adult *Leptonychotes.* The colon grades into the rectum, which begins at the pelvic inlet and ends at the anal canal. Throughout the length of the rectum the lining is thrown into large irregular transverse rectal plicae. Toward the distal portion of the rectum, the plicae become organized into five longitudinal anal columns that continue into the anal canal. The anal canal is much smaller in diameter than the rectum. Small coiled tubular rectal glands were present. The anal canal ends where the mucosa changes into a pigmented cornified stratified squamous epithelium (epidermis). Circumanal glands, coiled tubular structures, representing modified sweat glands, were confined exclusively to this region. In *Leptonychotes* there was no evidence of other anal glands, sacs, or scent glands.

2. Sirenia The cecum in the dugong is conical and was about 6 inches (~15 cm) long and 4 inches (~10 cm) wide at the base in the half-grown specimen. A sphincter is present in the ileocecal juncture. There is no constriction between the cecum and the colon. The epithelial lining of the cecum is smooth and its walls are muscular.

The colon in the dugong is thinner walled than the small intestine and is between 4 and 11 times the total body length. There are no taeniae coli. The lining of the colon is smooth, with the exception of irregular folds that are present at the wider terminal portion. The lining of the rectum is provided with longitudinal folds, which become finer and more numerous in the anal canal. The lining of the anus is grayer and harder than the lining in the rectum. The anal canal is about 5 cm long. At the distal end of the canal the longitudinal folds become higher and terminate in globular swelling, which occlude the lumen and which have been termed "anal valves."

The cecum in manatees is very pronounced and unusual in shape. It is an oval body about 20 cm in diameter and has two horn-shaped appendages that can reach up to 15–20 cm in length.

3. Cetacea
Odontocetes

Delphinoid: As was stated earlier, there is no cecum and no marked differentiation between large and small intestines in delphinoids.

Platanistoid: The colon in a 204-cm *Inia geoffrensis* was 40 cm long, followed by a 5-cm rectum and a 3-cm anal canal. The proximal and distal portions of the colon were 1 and 1.5 cm in diameter, respectively.

There is a pronounced cecum that is 5 to 9 cm long in *Platanista gangetica.* The large intestine is short, 60 cm in adults. Lengths of the large intestine (cecum, colon, rectum, and anus) in four specimens that ranged between 76 and 127 cm total length ranged from 25.5 to over 58 cm (the 127-cm specimen was lacking the cecum). There was no trace of taenia coli.

There is no cecum in *Pontoporia.* The longitudinal fold in the small intestine of *Pontoporia* becomes two distinct longitudinal folds. Taenia and haustra coli were not found.

Physeteroid: In large adult *Physeter* the large intestine can be up to 26 m long. The mucosa of the large intestine in *Physeter* is not folded. There is no cecum. The diameter of the descending colon is increased markedly in *Kogia* spp.

Ziphiid: There is no cecum and the transition between large and small intestines is gradual.

Mysticetes

Mysticetes have a very short cecum except in the case of right whales where the cecum is absent. There is a marked difference between the diameter of the large and small intestines in right whales (*Eubalaena* spp.). There is one mention of taeniae and haustrae coli in the blue whale where the taeniae consisted of three longitudinal muscular bands.

Balaenopterid: The mean ratio of the large intestine to body length in common minke whales was 40%. The mean ratio of cecum length to body length was 4%; the cecum varied between 30 and 50 cm.

Balaenid, Eschrichtiid, and Neobalaenid: There appears to be no specific data for the large intestine of balaenids, eschrichtiids, and neobalaenids.

II. Accessory Organs

A. Liver

The liver is derived from a diverticulum of the embryonic duodenum. That diverticulum also gives rise to the gallbladder, which serves as a reservoir for hepatic secretions. The liver expands to become the largest internal organ. The liver functions in the storage and filtration of blood, in the secretion of bile, which aids in the digestion of fats, and is concerned with the majority of the metabolic systems of the body.

1. Pinnipedia   The liver is multilobed in pinnipeds; up to seven or eight lobes have been reported in *Otaria*.

2. Sirenia   The diaphragm has become oriented in the dorsal plane instead of the transverse plane, as it is in other mammals. The liver in the dugong and manatees is flattened against the dorsally oriented diaphragm. The liver is composed of four lobes: the normal central, left and right and the fourth, Spigelian lobe that lies on the dorsal border of the liver and is closely associated with the vena cava.

3. Cetacea   The liver in cetaceans is divided into two lobes by a shallow indentation. Occasionally a third intermediate lobe develops. Because relative liver weights are more than would be expected, the tentative conclusion is that cetaceans, particularly odontocetes, may have increased metabolic rates.

B. Gallbladder

The gallbladder is located on the posterior side of the liver where the hepatic duct issues. The developmental origins of the pancreas, liver, and gallbladder are related. The gallbladder stores and concentrates the bile that is secreted by the liver.

1. Pinnipedia   The gallbladder is present universally in pinnipeds and tends to be pyriform and located in a fossa of one of the subdivisions of the right lobe of the liver.

2. Sirenia   The gallbladder is small in the dugong and is strongly sigmoid in shape. It lies on the ventral surface of the central lobe where the falciform and round ligaments attach. It does not appear to be described in manatees.

3. Cetacea   The gallbladder is absent in all members of the order Cetacea. Part of this lack is compensated for by an increased capacity for storage in the hepatic and bile duct system.

C. Pancreas

The pancreas develops out of outgrowths of the embryonic duodenum. It consists of two developmental bodies, the dorsal and ventral pancreas, which may empty into either the hepatic duct or directly into the duodenum. The pancreas secretes enzymes that are discharged into the duodenum and insulin that is discharged directly into the blood.

The pancreas in marine mammals appears to have no remarkable differences from that in other mammals.

D. Hepatopancreatic Duct

The hepatopancreatic duct represents the developmental fusion of the hepatic, ventral, and dorsal pancreatic ducts. The hepatopancreatic duct opens into the duodenum through the major duodenal papilla.

1. Pinnipedia

Phocids: In *Phoca vitulina* and *Leptonychotes* the pancreatic duct joined the common bile duct before either of them come into contact with the duodenum. In this case they coalesce to form a hepatopancreatic duct, which empties into the duodenum.

Otariids and Odobenids: In *Otaria* and *Odobenus* the pancreatic duct and common bile duct empty separately into the duodenum. In the case of *Odobenus* the two ducts coalesce within the walls of the duodenum and enter via a common opening.

2. Sirenia   A hepatopancreatic duct does not form in the Sirenia; the common bile and pancreatic ducts open separately into the duodenum.

3. Cetacea   The bile duct and pancreatic ducts coalesce into a hepatopancreatic duct in all known cetaceans.

E. Anal Tonsils

Tonsils are bodies of organized lymphatic tissues around crypts used to communicate to the lumen of whatever system they are in. They are most common in the digestive system where everyone is familiar with the tonsils at the back of the throat. In some marine mammals, clusters of lymphatic tissue that fall into the definition of tonsils occur at the other end of the digestive system, in the anal canal.

1. Pinnipedia and Sirenia   There is no mention of anal tonsils in pinnipeds or sirenians. There is enough description of the anal canal to be relatively certain that they do not occur.

2. Cetaceans

Odontocetes

Delphinoid: Anal tonsils have not been reported in delphinoids.

Platanistoid: Anal tonsils were found in the anal canal of *Platanista gangetica* and *Stenella coeruleoalba*. Lymphoid tissue was found in the anal canal of *Inia geoffrensis* that was not organized into structures that could be called tonsils. There was no trace of lymphatic tissue in the anal canal of *Pontoporia blainvillei*.

Physeteroid: Similar structures were also found in some sperm whales in South Africa. These were noticed upon external examination of the whales, so they were at the distal end of the anal canal. There were no comparable data on other whales so their findings may have represented abnormal swelling in normal lymphatic tissue.

Ziphiid: No anal tonsils have been reported in ziphiids.

Mysticetes

Eschrichtiid: Anal tonsils were found in the gray whale. These tonsils consisted of masses of lymphatic tissue that communicated with the anal canal via crypts. They lie near the boundary of the anal canal with the rectum, 30 to 40 cm from the anal orifice.

Balaenopterid, Balaenid, and Neobalaenid: No anal tonsils have been reported.

See Also the Following Articles

Anatomical Dissection: Thorax and Abdomen ■ Diet ■ Energetics

References

Chivers, D. J., and Langer, P. (eds.) (1994). "The Digestive System in Mammals: Food, Form and Function." Cambridge Univ. Press, Cambridge.

Green, R. F. (1972). Observations on the anatomy of some cetaceans and pinnipeds. *In* "Mammals of the Sea: Biology and Medicine" (S. H. Ridgway, ed.), Chapter 4, pp. 247–297. Charles C. Thomas, Springfield.

Grew, N. (1681). Musaeum regalis societatis or a catalogue & description of the natural and artificial rarities belonging to the Royal Society and preserved at Gresham College . . . whereunto is subjoyned the comparative anatomy of stomachs and guts. London, 4to, il. 7, 386 pp., l. 1 + il. 2, pp. 1–43, 31 pls.

Harrison, R. J., Johnson, F. R., and Young, B. A. (1970). The oesophagus and stomach of dolphins (*Tursiops, Delphinus, Stenella*). *J. Zool.* **160,** 377–390.

Langer, P. (1988). "The Mammalian Herbivore Stomach: Comparative Anatomy, Function and Evolution." Gustav Fischer, Stuttgart.

Langer, P. (1996). Comparative anatomy of the stomach of the Cetacea. Ontogenetic changes involving gastric proportions—mesenteries—arteries. *Zeitschr. Säugetierkunde* **61,** 140–154.

Olsen, M. A., Nordboy, E. S., Blix, A. S., and Mathieson, S. D. (1994). Functional anatomy of the gastrointestinal system of northeastern Atlantic minke whales (*Balaenoptera acutorostrata*). *J. Zool.* **234,** 55–74.

Owen, R. (1838). "On the Anatomy of the Dugong," pp. 28–46. Proceedings of the Zoological Society of London.

Owen, R. (1866–1868). "On the Anatomy of Vertebrates," Vols. I–III. Longmans, Green and Co., London.

Reynolds, J. E. III, and Rommel, S. A. (1996). Structure and function of the gastrointestinal tract of the Florida manatee, *Trichechus manatus latirostris. Anat. Rec.* **245,** 539–558.

Slijper, E. J. (1962). "Whales." Hutchinson and Co., London.

Turner, W. (1912). The marine mammals in the Anatomical Museum of the University of Edinburgh. Macmillan, London.

Tyson, E. (1680). Phocoena, or the anatomy of a porpess, dissected at Gresham College. B. Tooke, London.

Genetics, Overview

PER J. PALSBØLL
University of California, Berkeley

Genetics constitute the study of heredity and variation of inherited characteristics. In the case of genetic analyses of natural animal populations at the level of organisms or above (e.g., populations or phyla), most studies draw their inferences from relative differences in consanguinity (i.e., kinship or relatedness). However, in the case of natural populations, we usually possess little or no prior knowledge as to the exact degree of relatedness among the individuals that we are comparing (whether from the same or different species). Hence, the primary task becomes to obtain an accurate estimate of the degree of relatedness among individuals sufficient for the purposes of the specific study.

In principle, the relative degree of relatedness among organisms is estimated from, and positively correlated with, the proportion of shared inherited characters. It is possible to use any trait of an organism toward this end; however, the further removed from the locus encoding the trait (i.e., the DNA itself), the higher the chance that external factors may have altered the phenotypic expression of a trait. Hence, while relatedness can be estimated from morphological characters and a single morphological character might represent many loci, the phenotypic expression might be influenced by extrinsic factors, such as environmental or physiological variables, to an unknown extent. In contrast, the composition of most cellular components is not susceptible to such extrinsic variables and thus the interpretation of the observed variation can be directly linked to the state of the encoding locus, the genotype. This explains why biochemical/molecular methods were so readily adopted and applied to estimate genetic and phylogenetic relationships when these first emerged in the mid-1960s. Until the 1980s the biochemical/molecular methods applied to natural populations were mainly indirect, e.g., the most widely employed method, isozyme electrophoresis, detects differences in the overall electric charge of enzymes due to amino acid substitutions. An important limitation of isozyme analysis is that the proportion of the genome, which encodes for detectable proteins, is very small, and only a subset of amino acid substitutions will yield a change in the overall electric charge of the enzyme. In addition, homoiotherm organisms (birds and mammals) have a reduced level of isozyme variation compared to poikilotherm animals and plants.

Despite these limitations, a large number of studies have been conducted based on isozyme electrophoresis, providing novel and valuable insights. Interested readers should consult the works of Wada and Danielsdóttir, both of whom have undertaken extensive isozyme-based studies of various species of cetaceans.

The most basic source of genetic data, the nucleotide sequence of the genome itself, became accessible in a practical manner due to a series of technical advances during the 1980s culminating with the development of the polymerase chain reaction (PCR) by Mullis and co-workers in 1987. The PCR technique permits simple *in vitro* amplification of any specific nucleotide sequence if the nucleotide sequence of the flanking regions is known. Once amplified, the exact nucleotide sequence of the locus is readily determined. Today, PCR-based analyses of DNA sequences are the predominant methods used in genetic studies of marine mammals, which is why this article relies on examples based on the analysis of DNA sequences rather than isozymes or morphological characters.

I. Obtaining Tissue Samples

A prerequisite for DNA-based methods is, naturally, DNA. The most common source is samples of soft tissue from which the DNA subsequently is extracted. Soft tissue samples are readily available from dead animals, e.g., stranded or killed specimens. However, it is often scientifically or ethically desirable to obtain samples from free-ranging, live animals. The advantage of PCR-based techniques is that only a minute amount of target DNA is required. Adequate amounts of DNA are contained in skin biopsies, sloughed skin, hair, and even feces, which can be collected from free-ranging marine mammals with relative ease.

The sensitivity of PCR-based methods also enables the use of historical samples, such as hair from old furs, baleen, or even dried blood obtained from old log books. However, the quality of DNA extracted from such historical samples is usually inferior and obtained in much lower concentrations than DNA extracted from current samples. The same is usually true for DNA extracted from fecal or similar degraded samples. The low concentration and often highly degraded DNA obtained from such samples necessitate additional precautionary measures to prevent contamination as well as repeated analyses to ensure that a correct genotype is obtained.

Tissue samples from free-ranging animals can be collected by invasive and noninvasive techniques, each with their respective advantages and disadvantages. Invasive techniques, such as the collection of skin biopsies, enable a directed sampling scheme. This implies that, conditions permitting, skin biopsies can be collected from individuals relevant to the specific objective and a biopsy can be linked to a specific individ-

ual. Multiple biopsy systems have been developed to collect skin biopsies from marine mammals, all principally consisting of a delivery unit, such as a crossbow or gun, and a projectile unit, usually an arrow. The projectile unit carries the biopsy tip and a stop to limit the depth of penetration, which may act as a float as well. The biopsy tip is typically a simple hollow tube of stainless steel with one or more barbs retaining the sample. Systems of various kinds and ranges have been developed, the most recent is the long-range system developed by Dr. Finn Larsen at the Danish Institute for Fisheries Research with which a skin biopsy was collected from a blue whale, *Balaenoptera musculus*, at a distance of approximately 70 m (~210 feet, see Fig. 1). Skin biopsies from pinnipeds or smaller odontocetes are usually collected when the animals haul out on land or bow ride using a pole onto which a biopsy tip is mounted. Invasive sampling techniques are at times viewed as intrusive and thus undesirable. In order to investigate such concerns, data have been collected during biopsy sampling in order to detect possible adverse effects. To date the only discernible effects appear to be short term and may be equally attributable to the close approach of the boat necessary to collect a sample.

The alternative, noninvasive sampling methods are usually of a more opportunistic and random nature, which may prohibit the pursuit of some research objectives. For cetaceans, the most common kind of noninvasive sample is sloughed skin. The outer epidermis in cetaceans differs from most other mammals by the lack of dead keratinized cells and consists mostly of live cells complete with nuclei and mitochondria: the organelles that host the two cellular genomes. There is considerable variation among cetacean species in terms of the amount and how often they slough skin. Sperm whales have been observed to slough massive amounts of skin, whereas other species, such as fin and minke whales, rarely slough any skin. The main disadvantage when collecting samples such as sloughed skin is the opportunistic nature of the samples and the difficulty in linking a specific sample to a particular individual during multi-individual sightings, which may influence the pursuit of some objectives. In addition, the quality and quantity of DNA extracted from such samples are more variable than those obtained from skin biopsies.

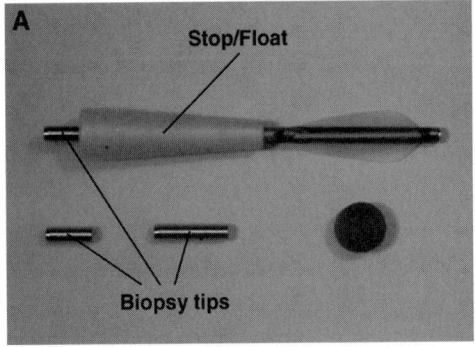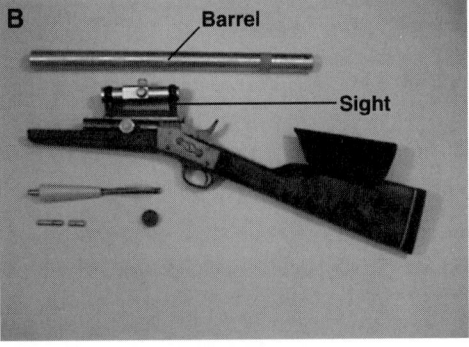

Figure 1 *The "Larsen" long-range skin biopsy system. (A) The projectile unit with biopsy tips and concave stop, which acts as a float as well. (B) The delivery system (a Remington rolling block system rifle), complete with barrel and sighting aid. Courtesy of Dr. Finn Larsen.*

Genomic DNA has also been extracted successfully from fecal plumes collected in the water column from dugong and dolphins, which contain epithelial cells from the intestinal tract. Among pinnipeds, the most common type of noninvasive samples is fecal samples, typically collected from haul-out sites on land. In bears, sloughed hair has proven an excellent source of noninvasive samples where the DNA is extracted from the root cells. In the case of bears, a simple, yet highly effective sampling scheme has been utilized based on "hair traps" with scent lures to attract bears and barbed wire that passively collects hair samples.

Samples are usually preserved by freezing with or without conservation buffer. Commonly used conservation solutions are 70–96% ethanol or distilled water saturated with sodium chloride and 20% dimethyl sulfoxide (DMSO).

II. Commonly Analyzed Loci

As mentioned earlier, genetic analyses of different taxa, e.g., individuals, populations, or species, are basically about estimating relative degrees of consanguinity. Put simply, the higher the proportion of shared traits/characters between two taxa, the higher degree of relatedness or, in the case of nucleotide sequences, the more mutations (i.e., differences in the nucleotide sequence) at the same locus separating two different taxa, the less related.

In principle, two kinds of mutations are observed in nucleotide sequences: single nucleotide substitutions or insertions/deletions of one or more nucleotides.

The latter kind of mutation is common at microsatellite loci. The most commonly analyzed microsatellite loci are dinucleotide repeats (e.g., GT), which are more common than tri- and tetranucleotide microsatellite loci in the mammalian genome. Most sequence changes at microsatellite loci consist of additions or deletions of one or more repeats. This kind of mutation is likely due to single-strand slippage, which subsequently results in misalignment during DNA replication. This mode of mutation is termed a stepwise mutation model (Fig. 2). Mutation rates at microsatellite loci are often high and have been estimated at 10^{-4}–10^{-5}, which is severalfold higher than that observed for single nucleotide substitutions. The high mutation rate typically yields multiple alleles at each locus and consequently high levels of heterozygosity. Microsatellite loci are thus well suited as genetic markers in the estimation of close relationships, such as parent–offspring relations. In contrast, microsatellite loci are less well suited to estimate more distant relationships due to high levels of homoplasy generated by the high rate and mode of mutation (insertion/deletion of repeats). Alleles at a microsatellite locus will differ solely by the number of repeats, and two copies of the same allele (i.e., the same number of repeats) may be allozygous or autozygous (Fig. 2). This aspect has to be taken into account during data analysis, and several estimators of genetic divergence have been developed specifically for microsatellite loci. However, the stepwise mutation mode introduces additional variance in the estimation of genetic divergence, which in turn reduces the precision. While the probability of homoplasy is low among closely related individuals, such as members of the same population, it increases with genetic divergence and thus poses more of a problem at distant relationships. Other mutational constraints appear to occur at microsatellite loci as well, such as a limit on the number of repeats and rare multirepeat mutations, both which affect the feasibility of microsatellite loci to estimate distant evolutionary relationships, such as divergent populations or different species.

The rate of single nucleotide substitutions is typically severalfold lower than that of single-strand slippage at microsatellite loci. The lower mutation rate implies that single nucleotide substitutions are less prone to homoplasy and thus in many ways are better suited than microsatellite loci to estimate more distant evolutionary relationships. However, the rate of single nucleotide substitutions differs among and within loci due to varying (often unknown) selective pressures. An example is codons in exons. In most cases a single amino acid is encoded by at least four different codons. The different codon sequences encoding for the same amino acid typically differ at the third position, at times on the first, and only rarely at the second codon position. Hence, nucleotide substitutions at the third position are usually synonymous and not subject to selective constraints. In contrast, the majority of nucleotide substitutions at the first and second codon positions are nonsynonymous. The selective constraints are thus higher at the first and second codon positions, and the substitution rate is usually lower than that of the third codon position. Because of the different selective pressures relative to codon position, phylogenetic analyses usually stratify nucleotide sequence data according to codon position. There are, however, multiple exceptions to this rule of thumb.

In mammals the vast majority of the genome does not encode for proteins and is thus presumably under little or no selection pressure. However, the large variations in mutation rates among such noncoding DNA sequences indicate the existence of selective constraints acting on these DNA sequences

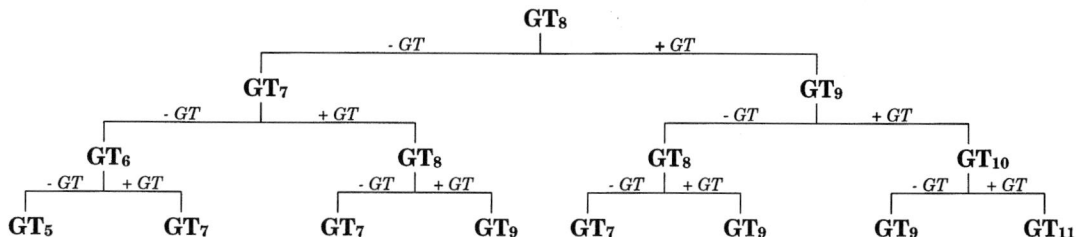

Figure 2 *The stepwise mutation mode at microsatellite loci. +/− (GT) denotes a mutation by single-strand slippage, i.e., addition or deletion of a single GT repeat unit.*

as well. Possible explanations are aspects such as chromosome pairing during meiosis, replication and transcription rates, and chromosomal stability.

A prerequisite for the estimation of the relative degree of genetic divergence among taxa is a model of the underlying evolutionary mechanisms. One important assumption in most evolutionary models is the absence of homoplasy. The commonly employed infinite-site mutation model assumes that mutations always occur at a new site in the nucleotide sequence. The infinite-allele model differs slightly in that multiple mutations at the same position can occur, but no allozygous alleles have identical nucleotide sequences. The consequence of either model is that identical nucleotide sequences all are assumed to be autozygous. While these idealized models probably are applicable to closely related taxa, multiple mutations do occur at the same position, especially at fast evolving nucleotide sequences, such as the commonly analyzed mitochondrial control region.

The earlier mentioned variance in mutation rates among loci is in fact an advantage as it enables the researcher to pick loci with mutation rates appropriate for the level of divergence under study. Usually the goal is to uncover sufficient amounts of variation to facilitate accurate estimations, while keeping the amount of homoplasy as low as possible.

Mammalian cells, like all eucaryotic cells, contain two different genomes; the cell nucleus harbors a paternal and maternal set of chromosomes, and the mitochondria, in the cell cytoplasm, possess a small genome, a circular DNA molecule of approximately 16,500 nucleotides in length in cetaceans and pinnipeds. During formation of the zygote, the sperm cells do not seem to contribute any mitochondria in mammals, although rare cases of paternal leakage of mitochondrial DNA have been reported. Thus in principle and for all practical purposes, the offspring inherits only the maternal mitochondrial genome.

III. Analyses of Individuals

In the case of marine mammals, genetic methods have been applied to identify individuals and parent–offspring relations as well as full siblings for a number of different purposes.

Identifying marine mammals by traditional tagging methods is often not feasible in most species. In many instances, marine mammals are simply too large, have too wide ranges, and live in a too dense medium to make traditional tagging practical. Tag attachments are usually relatively short-lived, in part because of the significant drag caused by the water unless attached to solid structures, such as the tusk of a male narwhal. While individual identification from natural markings has been applied successfully to a number of marine mammal species, this approach is limited to species with sufficient levels of natural variation among individuals.

In comparison, individuals from most species can be identified by "genetic fingerprinting," even species with much reduced levels of genetic variation, such as northern elephant seals, *Mirounga angustirostris*. Palsbøll and co-workers (1997) set out to verify if "genetic tagging" was feasible for a wide-ranging cetacean species. Their study included 3068 skin biopsy samples collected over a period of 8 years (from 1988 to 1995) from humpback whales, *Megaptera novaeangliae*, across the North Atlantic. Each humpback whale was identified by its composite genotype collected from six hypervariable microsatellite loci. The main issue in individual identification from a genetic profile is the probability of identity. The probability of identity is estimated readily for all degrees of consanguinity, ranging from unrelated individuals to parent–offspring pairs, from the population allele frequencies and decreases with the number of loci genotyped. The difficulty lies in determining the proportion of each kind of relationship among the collected samples, which in turn determines the expected number of individuals that have identical genetic profiles by chance. While the probability of identity is positively correlated with the degree of consanguinity, the proportion of pairs of a specific degree of relation decreases with consanguinity. In the case of the humpback whale study mentioned earlier, the probability of identity and expected numbers of different individuals with identical composite genotypes were estimated for unrelated individuals only, first for each maternally related FEEDING aggregation and subsequently for the entire population. The expected number of pairs of different individuals with identical genetic profiles by chance in the total sample of 3068 samples was estimated at less than one. Consequently, skin biopsy samples with identical genetic profiles were inferred as originating from the same individual. In total, 698 such samples with duplicate genetic profiles were detected. In a few cases, samples had been collected from the same individual humpback whale as far apart as 7500 km. The overall pattern of resightings within and among sampling areas was in agreement with data based on two decades of sighting records of individual humpback whales from their natural markings. The genetic "tags" were also used to estimate the abundance of humpback whales on the breeding grounds in the West Indies using mark-recapture techniques. Because the sex of each individual whale had been determined by genetic analysis as well, separate estimates of male and female abundance were calculated. Unexpectedly, the study yielded a significantly higher estimate of males at 4894 (95% confidence interval, 3374–7123) relative to that of females at 2804 (95% confidence interval, 1776–4463). The reason for this apparent underrepresentation of females on the breeding range (the sex ratio among calves and all whales on the feeding grounds has previously been estimated at 1:1) could not be resolved on the basis of data collected during the study. However, the authors suggested either spatial or temporal segregation among females as the source of the difference between the two abundance estimates.

An aspect of marine mammal biology where genetic methods are especially useful is the determination of parentage, e.g., to study breeding strategies and to assign individual reproductive fitness. Paternal reproductive success can be assessed in several ways, either by determination of specific parentage or by the level of paternal variation among the offspring. The former approach is relatively straightforward, as individuals that are related as parent and offspring will have at least one allele in common at each locus. However, as is the case for individual identification (see earlier discussion), two individuals that are not related as parent and offspring may also share at least one allele at each locus by chance. The probability that two individuals not related in parent–offspring manner share one or two alleles at each locus by chance decreases with the number and variability of loci genotyped. Hence, confident assignment of parentage requires that a relatively large number of variable loci are genotyped. In addition to a sufficient number of genetic markers, an adequate set of samples is required in order to ensure that parent and offspring pairs are contained among the samples collected. To date, only a few studies have attempted assignment of paternity in marine mammals, e.g., in gray seals, *Halichoerus grypus,* or harbor seals, *Phoca vitulina,* by analysis of either microsatellite loci or "multilocus" DNA fingerprinting as in the case of the northern elephant seal where genetic diversity is exceptionally low.

Hoelzel and co-workers compared the reproductive success of northern and southern, *M. leonina,* male elephant seals estimated as the proportion of pups fathered by the alpha male in his own harem. Previous behavioral observations indicated a higher level of competition for matings among male northern elephant seals compared to male southern elephant seals, leading to the hypothesis that northern elephant seal alpha males on average are less successful than their southern conspecifics. The genetic analysis corroborated this hypothesis, finding that southern elephant seal alpha males sired a significantly higher proportion of pups in their own harem than did northern elephant seal alpha males.

Multilocus DNA fingerprinting differs from microsatellite analysis mainly by the fact that the alleles from multiple loci are detected simultaneously. The simultaneous detection of multiple loci prevents the assignment of individual alleles to loci, which is why the degree of relatedness usually is estimated from the proportion of bands shared between individuals. However, the relationship between the degree of band sharing and relatedness is not straightforward, which is why the degree of band sharing is usually calibrated with a sample of individuals of known relationship, i.e., parent–offspring pairs.

Amos and co-workers (1993) employed multilocus fingerprinting as well as microsatellite loci to study the pod structure of long-finned pilot whales, *Globicephala melas.* Long-finned pilot whales are found in groups known as pods. Pilot whale pods appear to consist of mature animals as well as immature animals, presumably calves of the mature females. However, genetic analyses revealed that adult males within a pod were also closely related to mature females in the same pod, indicating that males stay within their natal pod, even after they become mature. Genetic analyses further revealed that mature males had not sired the calves in their own pod. Curiously, calves of the same cohort in a pod shared paternal alleles, indicating that a single or few closely related males sired calves of the same age. The authors proposed that mature males leave their natal pod briefly and mate with receptive females when pods meet during the breeding season. This hypothesis would explain why no males were found to have sired calves in their own pod, as well as the observation of few paternal alleles among calves belonging to the same cohort within a pod. Mature males of different ages within a pod would then also be maternally related.

Individual-based analyses like examples just given have the potential to address new issues with genetic methods that pre-

viously were not feasible. Traditional population genetic analyses (see later) yield evolutionary estimates of genetic divergence and may thus be of limited relevance to contemporary management and CONSERVATION issues. However, identifying individuals and parent–offspring relations provides a "real time" insight into population structure and dispersal at a time scale relevant to management and conservation purposes.

IV. Analyses of Populations

A large number of genetic studies of marine mammals have been undertaken for the purpose of identifying population structure and mechanisms of intraspecific evolution. In practical terms the aim is to determine if individuals belonging to the same partition are more closely related to each other than with individuals from other partitions, which is expected if partitions represent different taxonomic units (e.g., pods, population, or species). In numerical terms, this objective translates into estimation of the degree of genetic heterogeneity among subpopulations. The degree of genetic heterogeneity among subpopulations is traditionally estimated as the relative increase in homozygosity due to population subdivision, e.g., Wright's *F* statistics. The increase in homozygosity due to population structure is a product of random genetic drift. Random genetic drift denotes the oscillation in allele frequencies resulting from sampling each new generation from the parental generation. If we assume that mating is random within each subpopulation with respect to the locus under study (which is likely to be the case in most instances) and the absence of any selection, the offspring generation can then be viewed as a random sample of the parental alleles. As with any random sampling process, such sampling is subject to stochastic variation, i.e., alleles are not resampled in exactly the same proportions as those found in the parental generation and the allele frequencies will thus oscillate between generations. The long-term consequence of random genetic drift in a finite-sized subpopulation is that all but one allele will be lost from the subpopulation, in the absence of introduction of new alleles by gene flow and mutation. In other words, due to random genetic drift, alleles are lost from a subpopulation (increasing the homozygosity) at a rate depending on the rate of introduction of new alleles by either mutation and/or gene flow from other subpopulations. Because the process is random, it follows that different alleles will in/decrease in frequency due to random genetic drift in different subpopulations. Overall the effect of random genetic drift is that we find more homozygotes among the sampled individuals (collected from more than one sub-population) than expected from the overall allele frequencies *if* our sample contains individuals from a single random mating subpopulation. Gene flow will homogenize the allele frequencies among subpopulations by transferring alleles from one subpopulation to others. If there are no major fluctuations in effective population size, gene flow, or mutation rates, an equilibrium state is reached where the divergence in allele frequencies caused by random genetic drift and mutation is equivalent to the rate of homogenization due to gene flow. Even very low levels of gene flow (e.g., 10 individuals per generation) among subpopulations will homogenize allele frequencies among subpopulations to an extent that no effect of random genetic drift and mutation can be detected. Neither the mutation rate nor the effective population size is usually known in natural populations. For instance, two populations may have a similar level of genetic variation (e.g., estimated as the heterozygosity) but differ in terms of population sizes and mutation rates. For instance, the degree of heterozygosity estimated among samples collected from a small population at loci with high mutation rates may be similar to that estimated from a large population at loci with low mutation rates. As the level of genetic variation depends on the combination of effective population size and mutation rate (and these are typically unknown), it is common to simply combine both in the composite parameter $\theta = 4N_e\mu$, where N_e denotes the effective population size and μ the mutation rate. The advantage of this approach is that the composite parameter θ can be estimated from population genetic data, i.e., from the number of alleles, heterozygosity, polymorphic nucleotide positions, and the variance in allele size (for microsatellite loci). Comparisons of estimates of θ can be used to draw inferences regarding differences in mutation rates among loci within single populations or differences in effective population size among populations as well as estimates of genetic divergence.

As mentioned earlier, many population genetic studies of marine mammals have employed analysis of microsatellite loci. In addition, the nucleotide sequence of the maternally inherited mitochondrial control region is usually determined as well. The mitochondrial control region constitutes the only major noncoding region of the mitochondrial genome, with mutation rates well above the remainder of the mitochondrial genome. Usually the sequence of the first 200–500 nucleotides in the mitochondrial region is determined, which constitutes the most variable part of the mitochondrial control region. Because the mitochondrial genome is maternally inherited, any results from this locus estimate only the degree of maternal relation among samples. Most microsatellite loci, however, are of autosomal origin and thus inherited in a Mendelian manner.

The different mode of transmission of the mitochondrial and nuclear genome implies that each may reflect a different evolutionary relationship for the same set of samples. Palumbi and Baker investigated this aspect in 1994 in a study of humpback whales. In addition to mitochondrial control region sequences, the study also included data collected from the first intron in the nuclear protein-encoding locus actin. A phylogenetic analysis of actin intron I allele nucleotide sequences revealed the existence of two main evolutionary lineages with no apparent geographic affinities. The two lineages could be distinguished by digestion with the restriction endonuclease *Mnl*I basically defining a system of two alleles. This two-allele system was subsequently employed in the analysis of samples collected off Hawaii and western Mexico, both winter breeding grounds for eastern North Pacific humpback whales. While the distribution of mitochondrial control region alleles was highly heterogeneous between the same two population samples (the Hawaiian sample being almost monomorphic), no significant level of heterogeneity was detected in the distribution of the two actin intron I alleles. These "contrasting" results, i.e., little/no gene flow at the mitochondrial locus but indications of high levels of gene flow at the nuclear actin intron I locus, were

interpreted as the result of male-mediated gene flow, different rates of random genetic drift at each of the two genomes, or a combination of both. However, a subsequent study by Baker and co-workers also revealed significant levels of heterogeneity at nuclear loci (mainly microsatellite loci) among humpback whale samples collected off California and Alaska, which winter off Mexico and Hawaii, respectively. The simplest explanation for the seemingly discrepant outcome of the two just-mentioned studies is likely an increase in statistical power due to larger sample sizes and the inclusion of additional nuclear loci in the analysis (actin intron I as well as four microsatellite loci). However, the results do not rule out the possibility of some contribution from male-biased gene flow to the level of heterogeneity. More work is necessary before any affirmative conclusions can be reached.

The issue mentioned earlier, i.e., different degrees of male and female gene flow, is highly relevant when studying marine mammals. This has been clearly demonstrated in several population genetic analyses of species such as the North Atlantic humpback whales as well as northern right whales, *Eubalaena glacialis*, and belugas, *Delphinapterus leucas*. Specifically, North Atlantic humpback whales summer at several high-latitude feeding grounds off the eastern sea border of North America, west Greenland, Iceland, Jan Mayen and Bear Island in the Barents Sea. Humpback whales from these distinct feeding grounds all appear to congregate on common winter grounds in the West Indies. The winter constitutes the breeding and mating season. Calves are born during the winter and follow the mother during the spring MIGRATION to a high-latitude feeding ground and later on the autumn migration back to the West Indies. At the end of their first year the calves separate from their mother. The calf will, however, continue to migrate back to the same high-latitude feeding ground in subsequent summers to which it went with its mother during the first summer. The population genetic consequence of this maternally directed migration pattern is that North Atlantic humpback whale summer feeding grounds can be viewed as a single panmictic population with respect to nuclear loci, but structured in terms of mitochondrial loci. The latter is due to the maternal transmission of the mitochondrial genome in combination with the maternally directed site fidelity to the high-latitude summer feeding grounds. Nuclear alleles are exchanged when humpback whales from different summer feeding grounds mate in the West Indies. However, the calves only inherit their maternal mitochondrial genome and thus there is in principle no exchange of mitochondrial DNA among summer feeding grounds if calves keep returning to their maternal high-latitude summer feeding ground. Several population genetic studies have analyzed North Atlantic humpback whales and, in conclusion, found what was expected from the explanation just described. However, low levels of heterogeneity have also been detected at nuclear loci when comparing western and eastern North Atlantic high-latitude summer feeding grounds, indicating that some eastern North Atlantic humpback whales may winter and breed elsewhere than in the West Indies.

On a more detailed spatial scale, Hoelzel and co-workers determined the genotype at multiple microsatellite loci and the nucleotide sequence in the variable part of the mitochondrial control region in samples collected from pods of killer whale, *Orca orcinus*, observed in Puget Sound in the northeastern Pacific. Two kinds of killer whale pods are found in Puget Sound: resident and transient. The latter pods spend only part of the year in Puget Sound. While the resident pods seem to feed almost exclusively on fish, the diet of transient pods is mainly composed of marine mammals. The two kinds of pods also differ in the average number of individuals and vocalizations. Genetic analysis revealed significant levels of heterogeneity between resident and transient killer whales not only at the mitochondrial locus but at nuclear loci as well. This result was inferred as evidence of restricted gene flow between two different kinds of foraging specialists, and in fact it might be that this feeding specialization drives the genetic divergence between the two sympatric groups of killer whales.

All the just-mentioned examples assume the absence of selection, but one could well envision natural selection affecting the degree and distribution of genetic variation among and within subpopulations.

One such possibility is the sperm whale, *Physeter macrocephalus*, among which very low levels of variation have been detected in the mitochondrial control region on a worldwide scale. This observation prompted Whitehead to propose selectively advantageous cultural transmission in matrilineal whale species as the cause of the low levels of variation at maternally inherited mitochondrial loci. The basic principle proposed by Whitehead is that long-term association between females and their offspring facilitate an efficient cultural transmission of behavioral traits, e.g., feeding behaviors. If a maternal lineage adopts more efficient behaviors that are selectively advantageous, which in turn increases that lineage's reproductive success, such maternal lineage will eventually increase in proportion within the population. The model is similar to the genetic inheritance of selectively advantageous traits, i.e., natural selection, the only difference being that transmission across generations is cultural as opposed to inheritance. Because the mitochondrial genome is transmitted maternally, it will thus "hitchhike" along with the maternal cultural transmission of advantageous behavioral traits. The study reported low levels of genetic variation at mitochondrial loci in species which were classified as matrilineal by the author, i.e., species with pods, presumably consisting of females and their offspring, such as pilot whales (*Globicephala* spp.) and sperm whales. In contrast, the nucleotide diversity was on average 10-fold higher in species classified in the study as nonmatrilineal. Using computer simulations, the author demonstrated that the maternal cultural transmission of advantageous behavioral traits could indeed reduce the nucleotide sequence variation at mitochondrial loci significantly if the cultural transmission was efficient and the selective advantage was relatively high ($\sim$0.1). While there was no objection to the hypothesis that cultural transmission of advantageous behavioral traits might occur in cetaceans, others have pointed toward other evolutionary models, such as continued selection and fluctuating population sizes, as equally compatible with the observed data collected from sperm whales.

The environment inhabited by marine mammals is relatively devoid of physical barriers in comparison to the terrestrial environment. In addition, many marine mammal species have wide ranges and thus there is a high potential for dispersal. Despite this, most genetic studies of marine mammals have detected population structure in the distribution of genetic variation within as well as between ocean basins. The lack of physical barriers to dispersal indicates that intrinsic factors may play a role in generating population structure, such as foraging specialization and maternally directed site fidelity. Even for species where no obvious behaviors limiting dispersal have been observed, population genetic structure was detected, such as in the case of polar bears, *Ursus maritimus*, and fin whales, *Balaenoptera physalus*. In these two instances, it appears that the availability of prey is, at least in part, responsible for generating population genetic structure. In the case of polar bears, Paetkau and co-workers (1999) analyzed 16 different microsatellite loci in a total of 473 polar bears collected from all areas of the Arctic. The study detected a pattern of genetic divergence among subpopulations that was consistent with the distribution of active annual sea ice, which in turn relates to the abundance of ringed seals, which is their main prey. The study of North Atlantic and Mediterranean Sea fin whales by Bérubé (1998) was based on analyses of mitochondrial control region sequences as well as six microsatellite loci in each of 309 specimens. The population structure revealed by the genetic analyses was consistent with an isolation-by-distance model, which could be explained by a distribution, described as a "patchy-continuum" previously suggested by Sergeant and based on the overall distribution of prey. Interestingly, the fin whale study also revealed the possible effect of major geological events, in this case glaciation, on the present-day levels and distribution of genetic variation. The frequency distribution of mitochondrial nucleotide sequences suggested that the fin whale population in the western North Atlantic had undergone rapid expansion in abundance most probably from a small postglacial founder population (Fig. 3).

V. Analyses of Interspecific Relationships

A well-founded phylogenetic description of marine mammals is fundamental to our understanding of the unique evolution and adaptations observed in this group of mammals. Phylogenetic studies have been conducted at several levels, e.g., among cetaceans as well as at higher levels, such as the relation of cetaceans to ungulates.

The latter question has attracted much attention as molecular data are emerging complementing earlier morphological estimates of the phylogenetic affinities of marine mammals. Results emerging from the molecular data are, at the moment, inconsistent with the morphological findings as well as among the different molecular data sets themselves with regard to the branching order in several parts of the evolutionary tree. There are multiple possible explanations for such incongruence, such as incomplete taxonomic sampling, inadequate model of change (molecular and morphological), insufficient choice and number of outgroups, as well as incomplete fossil records. As mentioned earlier, the level of homoplasy increases with genetic divergence, which complicates the interpretation of nucleotide sequence data. Instead of DNA nucleotide sequences, the more

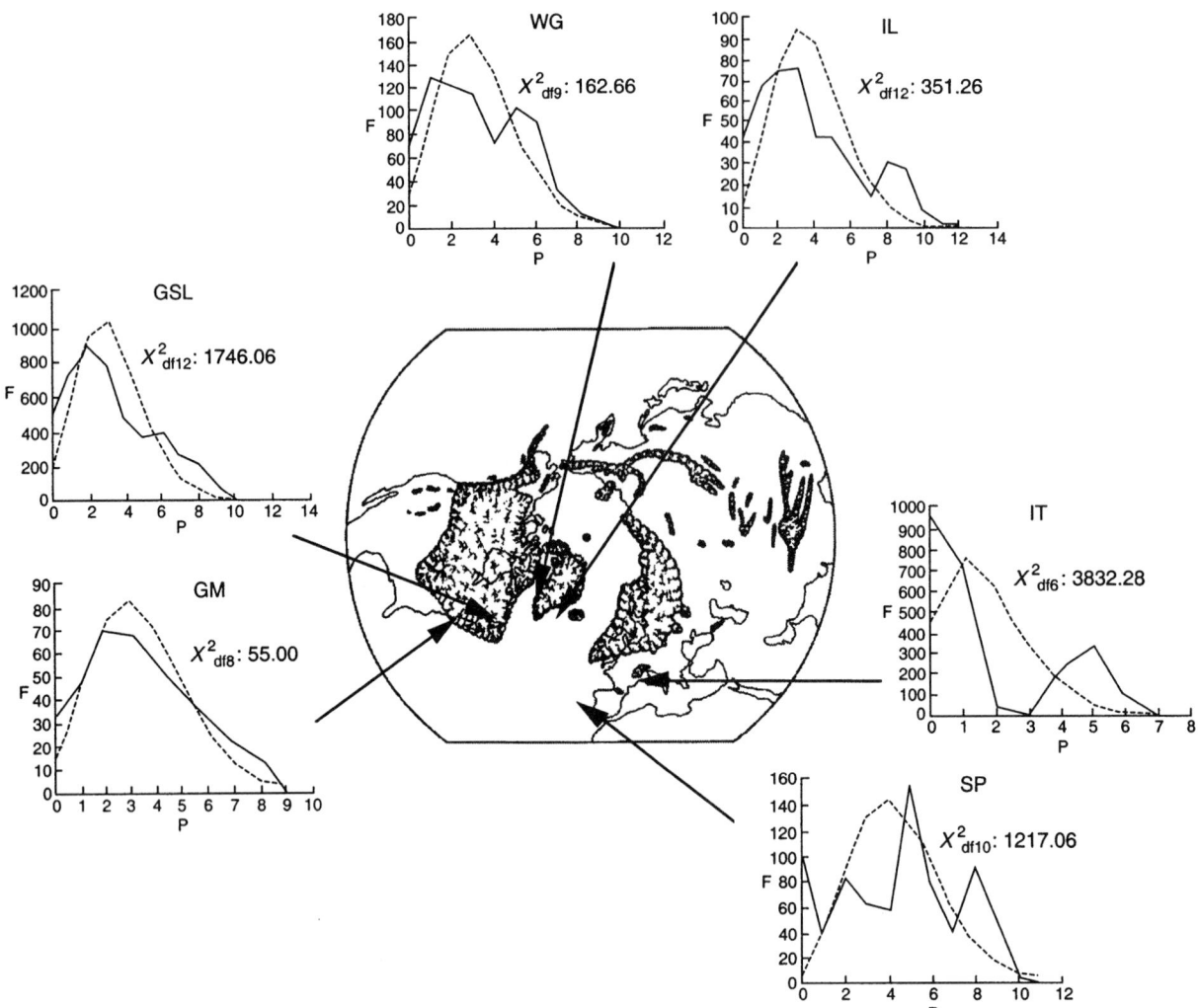

Figure 3 *Indication of postglacial expansions on western North Atlantic fin whale,* Balaenoptera physalus, *populations from genetic data. Observed (solid line) and expected (dashed line) frequency distributions of pairwise differences among mito-chondrial control region nucleotide sequences in North Atlantic fin whale populations under a model of exponential expansion (see text for details). A close match between the observed and the expected distribution suggests that the samples were obtained from an exponential expanding population. The marked areas on the map of the Northern Hemisphere indicate the presence of solid ice sheets during the last Pleistocene glaciation. Map from Pielou, "Biogeography." Reprinted by permission of John Wiley & Sons, Inc. Copyright © 1979 John Wiley & Sons, Inc.*

common sort of data employed in phylogenetic analyses, Shi-mamura and co-workers mapped the presence or absence of retroposons, termed short interspersed elements (SINEs), at different locations in a number of ungulate and cetacean species. SINEs are in many ways thought to be ideal phyloge-netic characters as they presumably are inserted into the host genome in a random and irreversible manner, i.e., a very sim-ple mutation model devoid of many of the problems, such as homoplasy, codon position, transition/transversions ratio, and the like, which introduce variability in analyses of single nu-cleotide substitutions. The SINE-based study found support

for the notion that Artiodactyla is a paraphyletic group in that cetaceans did not constitute a sister group but originate within Artiodactyla (Fig. 4). Earlier studies based on a sequence analy-sis of nuclear loci encoding milk proteins by Gatesy (1998) also arrived at the same conclusion, i.e., that artiodactyls are a pa-raphyletic group, also from the position of the cetacean branch. The paraphyly of Artiodactyla was subsequently supported in a comprehensive phylogenetic estimation conducted by Gatesy involving data from several nuclear and mitochondrial loci. Given the highly specialized cetacean morphology, compar-isons of morphological characters with terrestrial mammals are

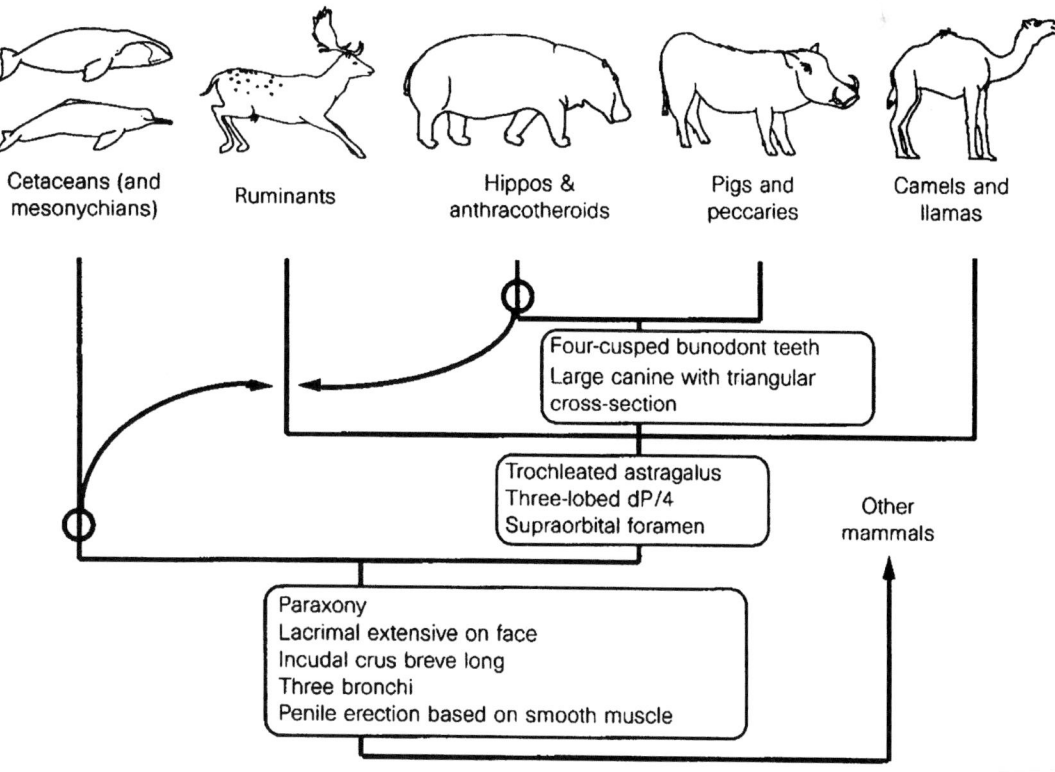

Figure 4 *Changes to the traditional artiodactyl phylogeny suggested by the findings of Shimamura* et al. *(1997). See text for details. Reprinted by permission from* Nature 388, *622–623. Copyright © 1997 Macmillan Magazines Ltd.*

not a trivial matter either and may in part account for the observed discrepancies between the morphological and molecular approaches.

A perhaps more controversial study is that of Milinkovitch and co-workers (1993) who estimated the phylogenetic position of the sperm whales within Cetacea from mitochondrial nucleotide sequences. Conventional taxonomy based on morphological characters places this distinct and old lineage of cetaceans among the odontocetes, as sperm whales share many morphological characters with other odontocetes, the presence of TEETH and ECHOLOCATION being the most obvious traits. In contrast, the study by Milinkovitch and co-workers found that sperm whales were significantly more closely related to the baleen whales than to the remainder of the odontocetes. The result of this study has since been the subject of numerous additional analyses and reanalyses and, in many ways, has become a case study of phylogenetic estimation. These additional analyses have showed that estimation of taxonomic relationships from nucleotide sequences is sensitive to aspects such as choice of outgroups, taxonomic sampling, sequence alignment, and long branches. Subsequent analyses based on nuclear and mitochondrial loci by Gatesy showed a strong support among nuclear genes for the traditional odontocete affinity of the sperm whales and less strong support for the alternate view suggested by the mitochondrial nucleotide sequences.

In both of the instances just described, the lack of congruence among the different approaches and loci demonstrates that our understanding is still far from satisfactory and that additional work is necessary before we have a more thorough and definitive understanding of the evolution of this highly specialized group of mammals and the underlying molecular mechanisms employed in our inferences.

See Also the Following Articles

Forensic Genetics ■ Hybridism ■ Molecular Ecology ■ Stock Identity

References

Amos, B., Schlötterer, C., and Tautz, D. (1993). Social structure of pilot whales revealed by analytical DNA profiling. *Science* **260**, 670–672.

Baker, C. S., Medrano-Gonzalez, L., Calambokidis, J., Perry, A., Pichler, F., Rosenbaum, H., Straley, J. M., Urban-Ramirez, J., Yamaguchi, M., and Von Ziegesar, O. (1998). Population structure of nuclear and mitochondrial DNA variation among humpback whales in the North Pacific. *Mol. Ecol.* **7**, 695–707.

Bérubé, M. (1998). Evolution, genetic structure and molecular ecology of the North Atlantic fin whale, *Balaenoptera physalus* (Linnaeus,

1758). *In* "Department of Natural Resource Sciences, Macdonald Campus," p. 184. McGill University, Ste-Anne de Bellevue.

Clapham, P. J., Palsbøll, P. J., and Mattila, D. K. (1993). High-energy behaviors in humpback whales as a source of sloughed skin for molecular analysis. *Mar. Mamm. Sci.* **9**, 213–220.

Gatesy, J. (1998). Molecular evidence for the phylogenetic affinities of Cetacea. *In* "The Emergence of Whales" (J. G. M. Thewissen, ed.), pp. 63–111. Plenum Press, New York.

Hoelzel, A. R., Dahlheim, M., and Stern, S. J. (1998). Low genetic variation among killer whales (*Orcinus orca*) in the eastern North Pacific and genetic differentiation between foraging specialists. *J. Hered.* **89**, 121–128.

Hoelzel, A. R., Le Boeuf, B. J., Reiter, J., and Campagna, C. (1999). Alpha-male paternity in elephant seals. *Behav. Ecol. Sociobiol.* **46**, 298–306.

Lambertsen, R. H. (1987). A biopsy system for large whales and its use for cytogenetics. *J. Mammal.* **68**, 443–445.

Milinkovitch, M. C., Orti, G., and Meyer, A. (1993). Revised phylogeny of whales suggested by mitochondrial ribosomal DNA sequences. *Nature* **361**, 346–348.

Mowat, G., and Strobeck, C. (2000). Estimating population size of grizzly bears using hair capture, DNA profiling, and mark-recapture analysis. *J. Wildl. Manag.* **64**, 183–193.

Mullis, K. B., and Faloona, F. (1987). Specific synthesis of DNA in vitro via a polymerase-catalyzed chain reaction. *Methods Enzymol.* **155**, 335–350.

Paetkau, D., Amstrup, S. C., Born, E. W., Calvert, W., Derocher, A. E., Garner, G. W., Messier, F., Stirling, I., Taylor, M. K., Wiig, O., and Strobeck, C. (1999). Genetic structure of the world's polar bear populations. *Mol. Ecol.* **8**, 1571–1584.

Palsbøll, P. J., Allen, J., Bérubé, M., Clapham, P. J., Feddersen, T. P., Hammond, P., Hudson, R. R., Jørgensen, H., Katona, S., Larsen, A. H., Larsen, F., Lien, J., Mattila, D. K., Sigurjónsson, J., Sears, R., Smith, T., Sponer, R., Stevick, P. T., and Øien, N. (1997). Genetic tagging of humpback whales. *Nature* **388**, 676–679.

Palsbøll, P. J., Clapham, P. J., Mattila, D. K., Larsen, F., Sears, R., Siegismund, H. R., Sigurjónsson, J., Vasquez, O., and Arctander, P. (1995). Distribution of mtDNA haplotypes in North Atlantic humpback whales: The influence of behaviour on population structure. *Mar. Ecol. Progr. Ser.* **116**, 1–10.

Palumbi, S. R., and Baker, C. S. (1994). Contrasting population structure from nuclear intron sequences and mtDNA of humpback whales. *Mol. Biol. Evol.* **11**, 426–435.

Reed, J. Z., Tollit, D. J., Thompson, P. M., and Amos, W. (1997). Molecular scatology: The use of molecular genetic analysis to assign species, sex and individual identity to seal feces. *Mol. Ecol.* **6**, 225–234.

Shimamura, M., Yasue, H., Oshima, K., Abe, H., Kato, H., Kishiro, T., Goto, M., Munechika, I., and Okada, N. (1997). Molecular evidence from retroposons that whales form a clade within even-toed ungulates. *Nature* **388**, 666–670 (1997).

Valsecchi, E., Palsbøll, P., Hale, P., Glockner-Ferrari, D., Ferrari, M., Clapham, P., Larsen, F., Mattila, D., Sears, R., Sigurjónsson, J., Brown, M., Corkeron, P., and Amos, B. (1997). Microsatellite genetic distances between oceanic populations of the humpback whale (*Megaptera novaeangliae*). *Mol. Biol. Evol.* **14**, 355–362.

Wada, S., and Numachi, K. I. (1991). Allozyme analyses of genetic differentiation among the populations and species of the *Balaenoptera*. *Rep. Intl. Whal. Comm. Spec. Issue* **13**, 125–154.

Whitehead, H. (1998). Cultural selection and genetic diversity in matrilineal whales. *Science* **282**, 1708–1711.

Genetics for Management

ANDREW E. DIZON
Southwest Fisheries Science Center,
La Jolla, California

Certain kinds of genetic information are particularly well suited to assist in designing strategies to protect human-impacted marine mammals. What sort of genetic information is required depends on the particular CONSERVATION goals wildlife managers seek to achieve when protecting specific species or populations within species. For example, is the goal to prevent EXTINCTION of the species as a whole or to prevent extirpation of local, but not necessarily genetically unique, populations? For most developed nations, these goals are codified in laws presumably reflecting, at least in democratic societies, the will of the public. To achieve these goals, managers often choose between controversial and conflicting strategies, such as limits on the species and numbers of marine mammals that can be incidentally killed during certain fishing operations. Relaxed limits favor the fishermen but may put a population of marine mammals at risk; stringent limits are less risky but may put an unsupportable burden on fishermen by restricting their fishing options. Obviously, the kind and quality of biological data, genetic or otherwise, informing this choice are critical. Decisions have to be based on the current scientific information available or will be challenged in the courts. While most scientific information on impacted populations is of value, certain kinds of information are much more important for the management process. If only limited data are available, inappropriate decisions can be made, eventually imperiling the population needing protection in the first place. Biological data on marine mammals, especially cetaceans, are difficult and consequently expensive to obtain. By consuming limited conservation funds, even good but irrelevant studies can impede the conservation effort. To ensure that genetic studies proposed are relevant for MANAGEMENT needs requires an understanding of the policy (the conservation goals) before doing the science (the information gathering).

Currently, management-oriented genetic studies use primarily (1) microsatellite loci within the 3×10^8 or so base pairs (bp) of the mammalian nuclear genome or (2) DNA sequence data from a portion of the 1.6×10^4 bp of the mitochondrial genome; the sub-sequence is also known as a haplotype (Fig. 1). Mitochondrial (mt)DNA is a multicopy, circular, cytoplasmic DNA that, in marine mammals, is inherited intact from the mother. In contrast, microsatellites are part of the nuclear genome and are inherited biparentally. They are short stretches of repeated DNA that are distributed abundantly in the nuclear genome and show exceptional variability in most species. In addition, gender-specific markers have been developed, and sex can be determined genetically. Finally, there is potential of measuring a variety of condition-related parameters (e.g., stress, maturity, pregnancy, spectral sensitivity, age) by examining the DNA or an expressed product of the DNA (i.e., the protein) in a small piece of skin.

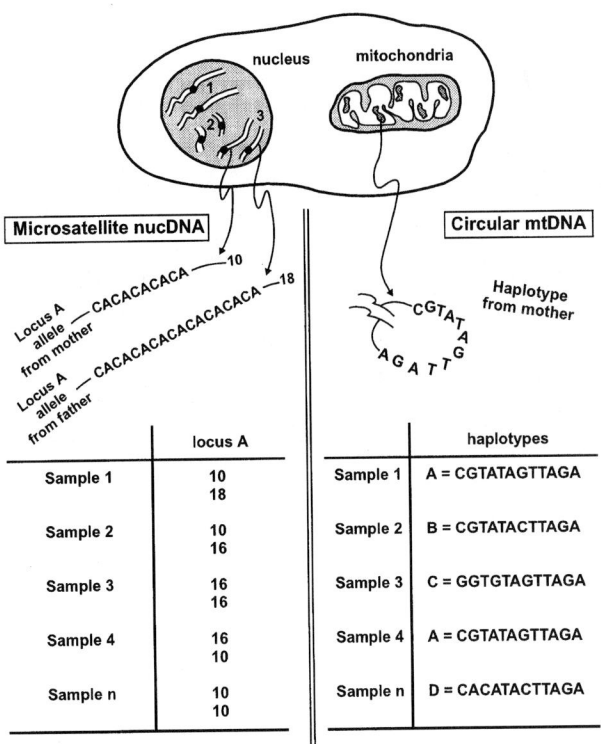

Figure 1 *Data for management genetic studies primarily consist of microsatellite DNA, mitochondrial DNA, or both. Microsatellites are short tandem repeats (two, three, or four base repeats) of base pairs, e.g., CACACACACA . . . , ATGATGATB . . . , or GATAGATAGATA. . . . Microsatellite data consist of n pairs of alleles at m number of microsatellite loci within the 3×10^9 or so base pairs (bp) of the mammalian nuclear genome. There is estimated to be a microsatellite region every 3000 or so base pairs. Microsatellites are part of the nuclear genome and are inherited biparentally. Mitochondrial data consist of n subsequences of base pairs (haplotypes) at some locus within the 1.6×10^4 bp of the mitochondrial DNA genome. mtDNA is a multicopy cytoplasmic DNA that, in vertebrates, is inherited intact from the mother. Each mitochondria may have 5–10 DNA molecules, and there may be from 100 to 1000 mitochondria per cell. The three pairs of paired lines within the nucleus represent chromosomes; the 10 and 18 bp represent two alleles at a microsatellite locus located on the long arms of chromosome 3. For mitochondrial DNA, the arrow leading out of mitochondria shows a sequenced portion of 12 bp of the 16,000-bp molecule. For example, sample 1 is heterozygous at microsatellite locus A having a pair of alleles that have five and nine CA repeats. Sample 1 also possesses an "A"-type mitochondrial haplotype that, for example, differs by 2 bp from the "C"-type. For actual studies, the number of microsatellite loci examined might range from 4 to 10, and the size of the mitochondrial sequence examined might range from 350 to 1200 bp.*

One advantage that genetic analyses have over "whole animal" studies is that data are easier to collect and few constraints are put on the quality of a sample or its origin. DNA is a relatively tough molecule, and adequate samples can be obtained from tiny amounts of a variety of tissues such as skin, blood or blood stains, hair follicles, placenta, excrement, baleen, modern or ancient bone, or, in some circumstances, formalin-preserved tissues. For instance, adequate amounts of mtDNA from ca. 1000-year-old bowhead whale (*Balaena mysticetus*) bones from the Chukchi Peninsula have been obtained. For live animals, projectile biopsying (crossbow, firearm, or lance) has been used successfully for all but the smallest and shyest cetaceans. Harbor porpoises (*Phocoena phocoena*) have proven particularly elusive. However, Fig. 2 is a photograph of a crossbow biopsy being taken of a highly endangered North Pacific right whale (*Eubalaena japonica*) at a range of 70 m.

I. The "Conservation Unit"

Today, defining the population segment on which to focus conservation efforts is the primary use of genetic information. The U.S. Marine Mammal Protection Act of 1972 (MMPA), the Endangered Species Act of 1973 (ESA), and the Revised Management Procedure of the INTERNATIONAL WHALING COMMISSION (IWC) all direct that management efforts must be focused on populations below the species level. Although other countries have not necessarily established laws codifying the conservation unit, biologists are generally in agreement that species comprise a collection of semi-isolated populations (i.e., species-wide panmixia is the exception) and that those semi-isolated populations should be the focus of management. However, the devil is in the details, and there is much controversy on the precise definition of these units. Besides having obvious biological consequences for getting the groupings correct, there can be economic ones as well. For instance, quotas on harvest or incidental take are calculated as some allowable fraction of the overall ABUNDANCE within the chosen conservation unit. A small conservation unit is the most biologically risk averse because quotas are then necessarily small, and there is the greatest likelihood that removals will be equally distributed over the whole unit. However, a large conservation unit is the most economically risk averse because the quotas are larger, and there is the potential that excessive removals in one part of the range (the sink) will be compensated for by immigration from outside the exploited region (the source).

Policy tries to provide managers with guidance to balance conservation and economic issues by defining the management unit. For instance, the ESA seeks to prevent the extinction of distinct population segments that are evolutionarily unique. The policy addresses last-ditch efforts to rescue populations whose abundances are so low, or whose abundances will become so low in the near future, that if something is not done immediately, they will likely go extinct. These so-called evolutionarily significant units (ESU) are defined in the statues as (1) being "substantially" reproductively isolated from other population segments of the same species and (2) representing an important component in the evolutionary legacy of the species. The first criterion speaks to the rate of exchange between the population segment and other segments. The second speaks to the time the population segment has been isolated. In contrast, the MMPA seeks to maintain viable populations across their historical ranges at 50% of their historical population size. This act addresses maintenance of abundance. The

Figure 2 *Crossbow biopsy being taken of a highly endangered North Pacific right whale at a range of 70 m. Visible at about 40% of the distance from the bow of the boat and the whale is the cross-bow projectile. A genetic sample was obtained and analyzed successfully. Photo by Wayne Perryman, Southwest Fisheries Science Center, La Jolla, CA.*

MMPA conservation units could be characterized as demographically significant units (DSUs) to contrast them with ESUs. Some use the term "management unit" to refer to a DSU, but because both DSUs and ESUs are management units in the strict sense, it is important to distinguish them.

Genetic data are useful for defining both. However, the policy goals are different and, consequently, the details of genetic studies directed toward either must take slightly different approaches.

A. The Evolutionarily Significant Unit

Because the ESA is concerned with conservation units that are characterized as being "evolutionarily" different, the genetic methodology employed must be sensitive to evolutionary distances between taxa. Indeed, the traditional academic use of genetic data is employed to reconstruct common ancestry and to group taxa based on common ancestry. No restriction is based on the taxon level examined (subspecies, species, genus, family, etc.) save that the taxa are assumed to be reproductively isolated and that sufficient time has passed so that measurable genetic differences have accrued between every individual in one taxa and every individual in another. For higher level taxonomic relationships, the grouping derives *a priori* from a particular classification based on morphological distinctiveness. For groupings below the species level, the grouping derives *a priori* from geographical clustering; some have termed this phylogeography to contrast it to traditional phylogenetics.

Regardless, the key to ESU status is still reproductive isolation and time. Using DNA sequence data to test these *a priori* groupings to see if they are genetically accurate, an investigator demonstrates that all the individuals of each *a priori* stratum fall into exclusive genetic clusters. If so, ESU status can be presumed for the groupings. The evidence addresses the policy that protection should be offered to a population segment that is first of all "substantially" reproductively isolated. If they

were not isolated, it would be impossible to demonstrate the presence of exclusive genetic clustering.

If animals are commonly moving between groups and interbreeding, the groups would not be reproductively isolated from one another and would share genetic material. As a result, the genetic analysis would not find unique groupings of individuals corresponding to each population, and no ESUs could be defined. However, if the individuals of *a priori* defined groups do cluster into exclusive genetic groups, that indicates they have been reproductively isolated from one another for a significant period of time and do represent at ESU. As such, they are likely following unique evolutionary pathways and each must be conserved independently. The genetic evidence is usually presented in the form of a branching diagram representing the evolutionary pathways leading to mutually exclusive genetic clusters (Fig. 3A).

B. The Demographically Significant Unit

Consider, however, if the individuals in the sample fail to fall into exclusive genetic clusters that are congruent with the *a priori* classification. For example, what is happening if some of the individuals sampled in the northern hemisphere cluster genetically with those in the south (Fig. 3B)? This situation can be the result of (1) insufficient time having elapsed from when the populations were split to purge ancestral genes from the populations, (2) a degree of gene flow exists or has existed recently (e.g., a few adventuresome individuals immigrated to the south or vice versa to breed), or (3) a combination of the two. It also means that the populations under consideration do not meet ESU criteria. Nevertheless, the populations may be genetically distinguishable if there are significant frequency differences in alleles or haplotypes between the groups. These populations would be characterized as DSUs and the definition would pertain to an intermediate situation between complete, long-term isolation of the ESU and free gene flow between geographically distinct populations (panmixia).

A Phylogeographic Concordant Clustering

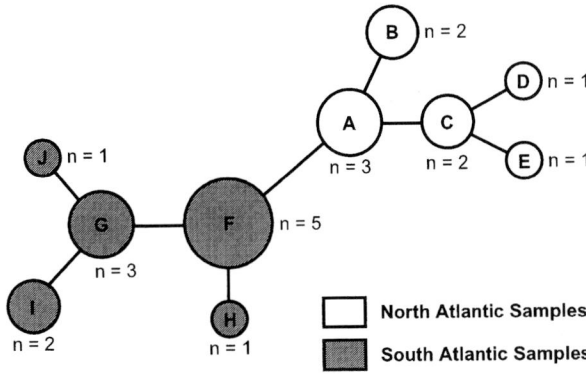

B Phylogeographic Modal Clustering

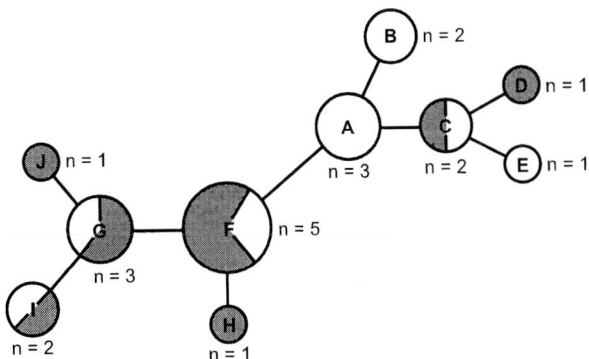

Figure 3 *Hypothetical genetic evidence representing two different evolutionary histories presented in the form of branching diagrams representing the evolutionary pathways leading to haplotypes observed in a sample of marine mammals. The size of circles is proportional to the number of individuals in the sample exhibiting the particular haplotype, and each haplotype differs from a connected neighbor by a 1-bp difference. (A) North Atlantic and South Atlantic stock has been isolated for a sufficient amount of time so that there are no haplotypes common to both. Geographic strata are concordant with genetic ones. (B) The isolation of the two stocks is (1) recent so that common haplotypes (C, F, G, and I) have not yet been purged via genetic drift from the North Atlantic, the South Atlantic, or both or (2) the isolation is incomplete, and there is a degree of continual interchange between the stocks. Even though the geography and the genetics are not strictly concordant, the distribution of haplotypes within each of the two stocks in this example is modally different.*

It is in the range of dispersal rates between the virtual isolation of the ESU and complete panmixia where the interpretation of genetic information requires an understanding of policy. The logical thread goes as follows: The MMPA establishes, albeit somewhat obliquely, that populations be maintained at 50% of their historical capacity as functioning elements of their ecosystems. This is interpreted to mean that adequate population levels shall be maintained *across their historical ranges*. It would forbid management action that resulted in extirpation in one portion of the range, although such extirpation would not reduce the overall species abundance to below 50% of historical levels.

What happens if anthropogenic mortality occurs at different levels in different parts of the range, i.e., there is heavy incidental take in the southern part of the range because it overlaps with a gill net fishery, but none at all in the central and northern part of the range? For example, consider a temperate, coastal species that inhabits waters from northern California through Canada, the Aleutian Peninsula, to Japan. Due to the large distances involved, distinct habitat differences, and the coastal behavior of this species, complete panmixia is not very likely and some population structure is, i.e., dispersal between certain population segments is reduced. Say samples are available from each of five putative population groupings (defined *a priori*) in the U.S. Pacific northwest waters. An extensive genetic analysis using both mtDNA and microsatellites is performed, and initial analyses using phylogenetic methods demonstrate no striking genetic clustering concordant with the geographic groupings. However, proximal populations were observed to share haplotypes and microsatellite, and a χ^2 analysis showed that significant frequency differences for the mtDNA haplotypes and for many of the microsatellite loci distinguish the populations. The inference here is that dispersal is sufficiently limited among the five populations so that some genetic differentiation has occurred among them. The populations are isolated but cannot be considered ESUs because the "evolutionary legacy" criterion is not met. They should be considered DSUs because dispersal between them is sufficiently reduced to warrant managing them separately (e.g., establishing individual quotas for incidental take for each population).

This recommendation can actually be made with confidence because of the shape of the curve that relates genetic differentiation and dispersal (Fig. 4). The strength of the result is reflected in the left-hand portion of the graph: genetic differentiation is detectable only when exchange rates between the putative populations are virtually nonexistent from a demographic or management point of view. This is in the range of a few dispersers per generation. However, the weakness of genetic analyses comes from how rapidly genetic differentiation falls as dispersal increases only slightly. Genetic differentiation disappears at dispersal rates that still would be considered nonsignificant from a demographic point of view, say a few percent per *year*. In other words, it is very difficult to demonstrate statistically significant genetic differentiation if dispersal between strata is more than a few dispersers per generation.

So by demonstrating genetic differentiation, the geneticist has confidently demonstrated demographically insignificant exchange rates. The management consequences are that any anthropogenic mortality within the strata must be compensated for by production from within rather than dispersal from adjacent, perhaps less impacted, units. Under this circumstance, which is actually common in coastal populations, mistakenly assuming that adjacent populations will serve as a source for the losses within the impacted population can result in destruction of the impacted population and failure to maintain it as a functioning element of its ecosystem. Disregarding the geneticist's recommendation may mean that the manager will have failed to meet a policy goal stipulated in the MMPA.

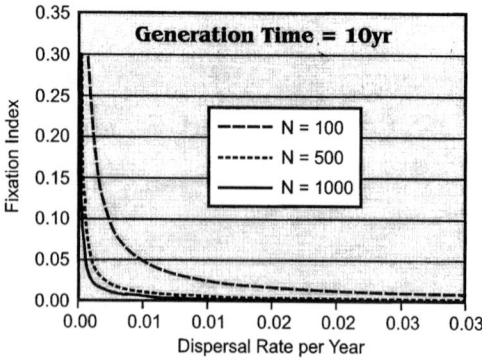

Figure 4 *The idealized relationship between the degree of genetic differentiation expressed as a fixation index, dispersal rate expressed as the average dispersal rate year, and population size expressed as the number of breeding animals or breeding females in the case of mtDNA analyses (effective population size). The fixation index ranges between 1 (no common alleles or haplotypes) to 0 (no differences in allelic or haplotypic distribution). Demographically insignificant rates of exchange (e.g., 1% per year) in anything but the smallest effective population sizes probably result in an inability to subdivide populations with any degree of statistical confidence. Perhaps more importantly, because the curve is so flat at this point and higher, genetic data have little resolution to accurately estimate dispersal rate in this range.*

However, it is not a "symmetrical" situation. What happens when genetic evidence fails to establish significant demographic isolation between units? A manager may be tempted to use this negative evidence to infer, because there was no evidence of population subdivision and hence restricted dispersal, that the putative populations could be coalesced into one larger management unit. Coalescence of two or more small populations into one larger MU would allow the manager to establish a larger incidental take quota and avoid the inevitable economic and political consequences of restricting fishing effort to reduce the incidental fishing mortality. The manager argues that high levels of take in one localized portion of the range (the sink) will be compensated for by production in and dispersal from less exploited portions of the range (the source).

This would turn out to be an appropriate decision if the failure to find evidence of population subdivision was due to demographically high levels of exchange between the exploited and the unexploited regions. However, the decision may have serious biological consequences if the failure to find genetic differences was simply because the experimental design of the genetic study lacked statistical power to discriminate subdivision (e.g., too few samples tested, too little portion of the genome tested, or an insufficiently variable portion of the genome tested). In reality, although undetected, the populations were demographically isolated, and it would be unlikely that adjacent populations could replenish losses due to incidental take in the exploited region. Because exchange between populations may be high enough to prevent detection genetically but not high enough for demographic replenishment, fail-

ure to discriminate the subdivision genetically should not at present be used as a scientific rationale for coalescing smaller populations into larger management units.

II. Focusing on the Individual

In the previous section, the focus was on a *population* of animals united by some characteristic, e.g., geographic locale. In this section, the focus is on the individual and what information genetic studies can provide to management.

A. Illegal Traffic and Trade

Two sorts of questions are usually asked: Did sample X come from the same individual as sample Y? Microsatellite analysis is used to establish an individual's genetic fingerprint; this is also known as genotyping. The second question is what is the provenance of sample X, i.e., what species or geographic population characterizes the sample? For this, sequence analyses are generally employed.

Question 1 is much like placing crime suspects at the crime scene via something the suspect has left behind (e.g., clothing fibers, fingerprints, hair, DNA), and genotyping is a highly reliable means of answering it. The genetic profile of a piece of meat in a market of unknown provenance could be compared with the genetic profiles in a database of "legally" harvested whales or, alternatively, the sample could be compared with the genetic profiles in a database of biopsied, protected ones.

Question 2 is more general and deals with establishing that the sample came from an animal that belonged to a certain group (taxa). Genetic analyses can help determine whether a given market sample came from a proscribed or a permitted taxon. For example, a particular market sample is humpback whale (*Megaptera novaeangliae*). The unknown sample is compared genetically with samples whose taxon identity is known. Because the genetic differences between taxa above the species level are so large, assignment analyses are almost infallible (e.g., Did the sample come from a whale or a cow?). In most situations, assignment is accurate at the species level (e.g., Did the sample come from a minke whale *Balaenoptera acutorostrata/ bonaerensis*, or a blue whale *B. musculus*?). However, there are exceptions, such as discriminating species among the genera *Delphinus*, *Stenella*, and *Tursiops*. Accurate assignment of an individual sample to its geographic origin is very difficult [e.g., Did the sample come from a gray whale (*Eschrichtius robustus*) harvested off the eastern Pacific Ocean or from the Okhotsk Sea?]. While there are exceptions to this rule, in general there is an increasing level of difficulty in distinguishing provenance of an individual sample, the lower the taxonomic division.

B. Other Uses of Individual-Oriented Genetic Information

Genetic mark-recapture methods based on genotyping can be substituted for traditional tagging methods, i.e., Discovery tags, for estimating population size, dispersal rate, and MIGRATION pathways. The management value of such data is obvious. However, small population sizes are necessary to ensure a high frequency of "recaptures." Large-scale mark-recapture studies based on molecular techniques are impractical because a recapture can only be recognized after a biopsy is taken and analyzed. As a result, a large number of expensive analyses would have to be done to ensure an adequate number of recaptures. Besides reidentification of individuals, genotyping can be used to reliably identify parent–offspring relationships, although large numbers of microsatellite loci must be examined to do this accurately. It is probably worth the effort because by doing so, dispersal can be measured over two generations rather than over the lifetime of single individuals. For conservation decisions, inter- rather than intragenerational movement is probably a more important parameter than movements of a single individual. Another important demographic parameter that emerges from a study of parent–offspring relationships is the fraction of mature animals enjoying reproductive success. In other words, what is the particular breeding structure of the population?

Finally, determining gender provides a means to examine geographical segregation by sex and whether males or females are the dispersers. It is a common situation with many marine mammal species that females tend to be strongly philopatric, returning year after year to specific feeding or BREEDING SITES. Female philopatry can be demonstrated by examining genetic population subdivision separately in males and females. If only females are strongly philopatric, mtDNA subdivision should be apparent among the females but not the males. If there are some data on age, it is sometimes possible to demonstrate that the likelihood of dispersal increases with age of the males. When males are the dispersers but not females, microsatellite subdivision should be nonexistent because the males of breeding age serve as a "conduit" to homogenize the alleles between populations. There are policy implications in demonstrating female philopatry. While this sort of population structuring would not qualify the population as an ESU, it does qualify them as a DSU worthy of management. If the animals from a particular feeding or breeding area are extirpated (males and females both), recolonization will not likely take place. The strongly philopatric females from other breeding or feeding grounds would not recolonize the depopulated region, and the dispersing males would not likely return to an area with no females. Thus, if policy deliberately excluded populations based on female philopatry, there could likely arise a situation where harvest could reduce or fragment ranges.

C. The Hidden Power of Molecular Genetics

In addition to providing answers to population subdivision, dispersal, individual identities, and breeding behavior, molecular genetic analyses present a previously unexploited opportunity for gaining understanding of marine mammals via remote, nonlethal sampling. Some of these data can have direct relevance for management. Consider that a skin sample contains the entirety of the individual's genetic blueprint. The ability to read this blueprint is progressing at an astounding rate, and although most of the progress is within the human genome, around 70% of the cetacean genes are homologous, and tools developed for medical research can be utilized for marine mammals. Two new approaches will be described briefly in which analyses of skin DNA and its expressed products can be used to gain understanding regarding the biological characteristics of populations and individuals.

DNA sequence information extracted from the genes of skin cells can provide data about expressed characteristics of other tissues or organs. Sequencing visual pigment genes from skin is a good example. With collateral data about visual performance of particular photoreceptors via behavioral or physiological testing, it is possible to extrapolate from the DNA sequence to the spectral sensitivity. Understanding the visual abilities of cetacean could aid in the design of fishing nets with increased color contrast, making them more visible to marine mammals, thereby reducing ENTANGLEMENT rates. A second approach is to directly examine expressed proteins within the skin itself, asking what proteins are up- or downregulated in skin under certain conditions. A good example is stress. Changing environmental conditions can perturb homeostasis, causing cellular stress and triggering a molecular stress response. At our laboratory, using immunohistochemical staining of thin skin sections, to date, 15 stress-responsive proteins (SRPs) have been identified whose expression is increased by 10-fold in stressed dolphins and whales in comparison to unstressed control animals (Fig. 5). The SRPs examined are widespread in the proliferating epithelial cells of the epidermis, and their induction is conserved in different species, genders, ages, and stressors. The procedure employed is simple and can be used for screening the large numbers of skin specimens necessary for correlating the presence of cellular stress with various environmental and anthropogenic factors. The management value is clear: Monitoring cellular stress in representative components of the marine ecosystem could provide an early warning system, allowing timely intervention in the case of habitat alteration. Cetaceans are top predators sensitive to many forms of environmental and anthropogenic stress, making them highly suitable as stress-reporting marine species.

III. Conclusion

While examination of genetic material offers unparalleled insights into all biological aspects of an animal's life, certain sorts of genetic information provide data that are directly relevant to the management process. The most important is the definition of the conservation unit. By common sense and by law in many countries, this unit is created out of the understanding that the vast majority of species (marine mammal or otherwise) are not panmictic. Species are subdivided geographically into isolated and semi-isolated groupings. Genetic analyses can measure this directly and provide the main avenue whereby the geneticist can provide information to facilitate decision making. Other genetic information on impacted populations is certainly of high value. This article has provided some examples. Regardless of the sort of genetic information collected, to ensure that genetic studies and information will be useful for management requires a clear understanding of the conservation policy that the studies are designed to help implement.

See Also the Following Articles

Conservation Efforts ■ Forensic Genetics ■ Molecular Ecology ■ Stock Identity

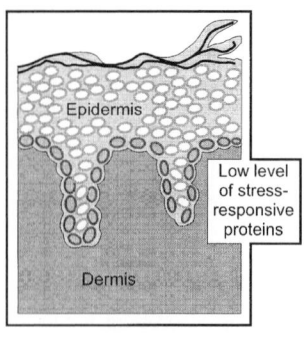

Molecular Analysis of Stress

Using Projectile Skin Biopsies

Methods: During stress, many stress-responsive molecules are mobilized in skin and elsewhere. These proteins can be visualized via immuno-histochemical staining of these tissue sections. Technique works especially well with fresh, frozen, or formulin-preserved skin.

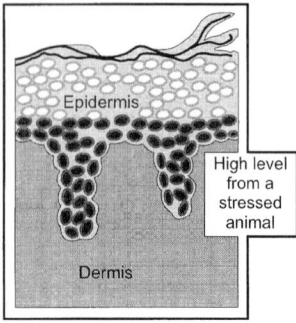

Results: High levels of 22 stress-responsive proteins were identified in skin of bottlenose dolphins and belugas that were independently assessed as highly stressed. Matched control animals had one to two orders of magnitude lower levels of the stress-responsive proteins.

Figure 5 *Immunohistochemical staining of skin sections to examine stress in marine mammals. Rapidly proliferating epithelial cells in the basal portion of the epidermis, in dolphins and whales exposed to stressful conditions, have higher concentrations of stress-responsive proteins (shaded cells) than cells proliferating under nonstressful conditions. These higher concentrations of stress-responsive proteins are maintained as the epithelial cells are pushed to the surface of the skin and shed by new cell growth underneath. The process takes approximately 2 months. Thus the skin provides a "record" of the approximate intensity of stress that has occurred over that period. Only stress that has occurred continuously for about a half a day can be detected presently, making the test insensitive to acute stress such as short bouts of high-intensity exercise.*

References

Amos, W. (1997). Marine mammal tissue sample collection and preservation for genetic analyses. *In* "Molecular Genetics Marine Mammals" (A. Dizon, S. Chivers, and W. Perrin, eds.), pp. 107–116. The Society for Marine Mammalogy, Lawrence, KS.

Avise, J. C. (1998). The history and purview of phylogeography: A personal reflection. *Mol. Ecol.* **7**, 371–380.

Dizon, A., Baker, C. S., Cipriano, F., Lento, G., Palsbøll, P., and Reeves, R. (eds.) (2000). Molecular genetic identification of whales, dolphins, and porpoises. *In* "Proceedings of a Workshop on the Forensic Use of Molecular Techniques to Identify Wildlife Products in the Marketplace." U.S. Department of Commerce, NOAA Technical Memorandum, NOAA-TM-NMFS-SWFSC-286.

Hillis, D. M., Moritz, C., and Mable, B. K. (eds.) (1996). "Molecular Systematics," 2nd Ed. Sinauer Associates, Sunderland, MA.

Moritz, C. (1994). Defining "evolutionarily significant units" for conservation. *TREE* **9**, 373–375.

O'Corry-Crowe, G. M., Suydam, R. S., Rosenberg, A., Frost, K. J., and Dizon, A. E. (1997). Phylogeography, population structure and dispersal patterns of the beluga whale, *Delphinapterus leucas,* in the western Nearctic revealed by mitochondrial DNA. *Mol. Ecol.* **6**, 955–970.

Palsbøll, P. (1999). Genetic tagging: Contemporary molecular ecology. *Biol. J. Linn. Soc.* **68**, 3–22.

Taylor, B. L., and Dizon, A. E. (1999). First policy then science: Why a management unit based solely on genetic criteria cannot work. *Mol. Ecol.* **8**, S11–S16.

Waples, R. S. (1991). Pacific salmon, *Oncorhynchus* spp., and the definition of "species" under the endangered species act. *Mar. Fish. Rev.* **53**, 11–22.

Genital Anatomy

SEE *Female Reproductive Systems; Male Reproductive Systems*

Geographic Variation

WILLIAM F. PERRIN
*Southwest Fisheries Science Center,
La Jolla, California*

I. The Nature of Geographic Variation

Mammals vary from place to place, in size, shape, coloration, osteology, and genetic features, including chromosomes, enzymes, and DNA sequences. They also vary in sounds produced, other behavior, life history, parasites, contaminant loads, biochemical features such as fatty acids, and other characters. This article focuses on geographic variation in morphology. When morphological variation and range are discontinuous, i.e., the populations or metapopulations are allopatric and can be diagnosed from one or, more commonly, a few characters, they are usually recognized as species, with the inference that they have diverged irrevocably in their evolutionary paths. When this is not true and groups differ from each other on average (modally) rather than absolutely, the variation is considered to be *geographic variation* within a species, and the form is recognized as a *subspecies, race,* or *geographic form* or *variant.*

Mammal species tend to vary geographically most in those features that vary most within a population. If, as for most mammals, body size varies broadly within a population, then geographic variants will usually differ in average body size. In another example, odontocete cetaceans are unusual among mammals in that they vary greatly in the number of TEETH and vertebrae within a species, and, as expected, these features differ sharply between geographic forms (Fig. 1).

Geographic variation may be *discordant,* i.e., the geographic pattern may differ among characters. For example, spinner dolphins (*Stenella longirostris*) in the eastern tropical Pacific vary differentially in color pattern, dorsal fin shape, and SKULL features (Perrin *et al.,* 1991), creating a complex mosaic of subspecies (see later) and varying zones of intergradation, depending on what feature is being looked at.

It is likely that most geographic variation in morphology (and the underlying genetic basis) in marine mammals is due to differential selection (ecological divergence) rather than genetic

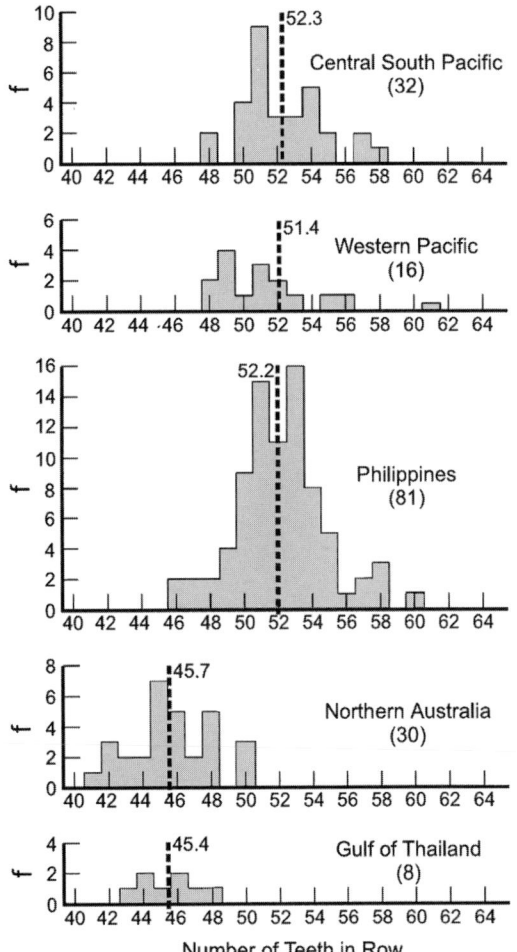

Figure 1 *Geographic variation in the number of teeth in spinner dolphins from five regions. Average and frequency distribution are shown; sample sizes are shown in parentheses. By permission from Perrin* et al. *(1999).*

drift. By saying that two populations belong to the same species, we are implying that there is, or recently has been, gene flow between them. Populations can diverge sharply morphologically in the presence of even substantial gene flow if the ecologically engendered differential selection is strong enough (Orr and Smith, 1998). However, modeling studies have indicated that social behavioral characteristics, such as female phylopatry and polygynous breeding systems, both common in marine mammals, can lead to the sequestering of variation due to drift within populations; this may accelerate evolutionary divergence (Storz, 1999).

Neutral genetic differences can accumulate across populations due to drift, and markers for this geographic variation are used extensively in defining marine mammal populations for purposes of assessment and management.

II. Subspecies

Subspecies are formally named or otherwise recognized geographic variants within a species. Subspecies are currently

recognized for 28 of the 127 or so marine mammal species (Table I).

The situation for subspecies is not as tidy as might be implied from Table I. Many of these subspecies were poorly described and may prove to be invalid; others may turn out to be full species. Some probably include multiple distinctive populations that deserve subspecific status but are as yet poorly understood. As indicated, some geographic forms have been recognized but not yet named. Some workers would disagree with certain of these subspecific designations and perhaps recognize others. Many additional subspecies were described in the past but have since been discounted (Rice, 1998). As for all of taxonomy, progress in classification at this level (*beta taxonomy*) is uneven and iterative.

III. Cetaceans

For the odontocetes, in every case where adequate samples of specimens from different regions have been available for examination, geographic variation has been found (Perrin and Brownell, 1994; Stacey and Leatherwood, 1997; Archer and Perrin, 1999; Amano and Miyazaki, 1992), so it can be expected to be universal.

Body size tends to be larger in open waters than in closed seas. For example, in the contiguous eastern North Atlantic, Mediterranean, and Black Sea, the short-beaked common dolphin, *Delphinus delphis,* is largest in the North Atlantic, smallest in the Black Sea, and intermediate in average size in the Mediterranean (Perrin, 1984). The bottlenose dolphin, *Tursiops truncatus,* shows the same pattern, being largest in the open Atlantic and smallest in the Black Sea, as does the beluga, *Delphinapterus leucas,* in the Canadian Arctic (Stewart and Stewart, 1989; Doidge, 1990). Body size also varies inshore/offshore and between riverine and marine populations. In the eastern tropical Pacific, the coastal subspecies of the pantropical spotted dolphin, *Stenella attenuata graffmani,* is on average larger than the offshore form, *S. a. attenuata.* It also has larger teeth; it may prey on larger tougher, benthic fish species, whereas the offshore form feeds primarily on small epipelagic fishes and squids. The pattern is repeated in the Atlantic spotted dolphin; the coastal form is larger than the offshore form in the Gulfstream (Fig. 2). However, in the bottlenose dolphin in the western North Atlantic, the pattern is reversed; the offshore form is larger than the coastal form (Hoelzel *et al.,* 1998) (in correlation with different stomach contents and parasite loads). The riverine form of the tucuxi, *Sotalia fluviatilis fluviatilis,* is smaller than the marine form, *S. f. guianensis.* Variation in size can also be latitudinal; short-beaked common dolphins in the eastern Pacific are longest in the Central Stock off Central America and shorter to the north and south (Perryman and Lynn, 1993). While it has been suggested that some of this variation in body size could be ecophenotypic (due, for example, to differential nutrition across areas of varying productivity), it is thought to most likely be determined genetically.

The dorsal fin is another feature that varies markedly with region in some odontocetes. A dramatic example of this can be seen in the tropical Pacific; whereas the fin in spinner dolphins

TABLE I
Currently Recognized Subspecies[a,b]

Species	Subspecies
Cetaceans	
Balaenoptera acutorostrata	*B. a. acutorostrata* (North Atlantic)
	A. a. scammoni (North Pacific)
	B. acutorostrata subsp. (Southern Hemisphere)
B. borealis	*B. b. borealis* (North Atlantic and North Pacific)
	B. b. schlegellii (South Hemisphere)
B. physalus	*B. p. physalus* (North Atlantic and North Pacific)
	B. p. quoyi (Southern Hemisphere)
B. musculus	*B. m. musculus* (North Atlantic and North Pacific)
	B. m. indica (northern Indian Ocean)
	B. m. brevicauda (Southern Hemisphere)[c]
	B. m. intermedia (Southern Hemisphere)[c]
Platanista gangetica	*P. g. gangetica* (Ganges and Brahmaputra)
	P. g. minor (Indus River system)
Inia geoffrensis	*I. g. geoffrensis* (Amazon below Bolivia)
	I. g. boliviensis (Rio Madeira, Bolivia)
	I. g. humboldtiana (Orinoco River system)
Cephalorhynchus commersonii	*C. c. commersonii* (Southern South America)
	C. commersonii subsp. (Kerguelen Islands)
Sotalia fluviatilis	*S. f. fluviatilis* (Amazon River system)
	S. f. guianensis (western Atlantic)
Stenella attenuata	*S. a. attenuata* (pelagic tropical waters)[d]
	S. a. graffmani (eastern Pacific coastal)
S. longirostris	*S. l. longirostris* (pelagic tropical waters)
	S. l. orientalis (eastern Pacific offshore)
	S. l. centroamericana (eastern Pacific coastal)
	S. l. roseiventris (inner Southeast Asia)
Lagenorhynchus obscurus	*L. o. obscurus* (southern Africa)
	L. o. fitzroyi (southern South America)
	L. obscurus subsp. (New Zealand)
Globicephala melas	*G. m. melas* (North Atlantic)
	G. m. edwardii (Southern Hemisphere)
	G. melas subsp. (Japan)
Neophocaena phocaenoides	*N. p. phocaenoides* (Indian Ocean to Southern China Sea)
	N. p. sunameri (western North Pacific)
	N. p. asiaeorientalis (Yangtze River)
Phocoena phocoena	*P. p. phocoena* (North Atlantic)
	P. p. vomerina (eastern North Pacific)
	P. phocoena subsp. (western North Pacific)
Phocoenoides dalli	*P. d. dalli* (North Pacific)
	P. d. truei (Kurile Peninsula, northern Japan)
Carnivores	
Arctocephalus pusillus	*A. p. pusillus* (southern Africa)
	B. p. doriferus (Australia)
A. australis	*A. a. gracilis* (southern South America)
	B. a. australis [Falkland Islands (Malvinas)]
Odobenus rosmarus	*O. r. rosmarus* (Atlantic Arctic)
	O. r. laptevi (Kara Sea to east Siberia)
	O. r. divergens (Pacific Arctic)
Erignathus barbatus	*E. b. barbatus* (Atlantic Arctic)
	E. b. nauticus (Laptev Sea to Pacific Arctic)
Phoca vitulina	*P. v. concolor* (western North Atlantic)
	P. v. mellonae (freshwater, eastern North America)
	P. v. vitulina (eastern North Atlantic)
	P. v. stejnegeri (western North Pacific)
	P. v. richardii (eastern North Pacific)

(continues)

TABLE I (*Continued*)

Species	Subspecies
Pusa hispida	*P. h. hispida* (Arctic Ocean and Bering Sea)
	P. h. botnica (Baltic Sea)
	P. h. lagodensis (Lake Ladoga, Russia)
	P. h. saimenis (freshwater lakes in Finland)
	P. h. ochotenis (Sea of Okhotsk)
Halichoerus grypus	*H. g. grypus* (western and eastern Atlantic)
	H. g. macrorhynchus (Baltic Sea)
Pagophilus groenlandicus	*P. g. groenlandicus* (North Atlantic)
	P. g. oceanicus (White and Barents Seas)
Ursus maritimus	*U. m. maritimus* (Atlantic Arctic)
	U. m. marinus (Pacific Arctic)
Enhydra lutris	*E. l. lutris* (western North Pacific)
	E. l. kenyon (Aleutians to Washington)
	E. l. nereis (California to Mexico)
Sirenians	
Trichechus manatus	*T. m. manatus* (South American mainland)
	T. m. latirostris (southeastern United States)
Dugong dugon	*D. d. dugon* (Indian and western Pacific Oceans)
	D. d. hemprichii (Red Sea)

[a]From Rice (1998) and Perrin *et al.* (1999).
[b]Approximate ranges in parentheses.
[c]Relative winter (breeding) ranges of *B. m. brevicauda* and *B. m. intermedia* unknown.
[d]Combines "subsp. A" and "subsp. B."

in Hawaii and the South Pacific is slightly falcate and subtriangular, typical of the species around the world, in large adult males in the far eastern Pacific (*Stenella longirostris orientalis* and *S. l. centroamericana*) the fin is canted forward, with a convex posterior margin (Perrin, 1990, 1998; Fig. 3). Animals in a broad zone of HYBRIDIZATION or intergradation are intermediate. A similar variation is present in short-beaked common dolphins; large adult males from the equatorial offshore eastern Pacific have more erect, triangular dorsal fins than in other regions. In both species the more erect (or forward-canted) dorsal fin is correlated with the development of a post-anal ventral hump (of unknown function).

The color pattern also varies within a species. In the *truei* form of Dall's porpoise (*Phocoenoides dalli truei*) in the western Pacific the ventrolateral white field is greatly enlarged from that in *P. d. dalli*. The just-described geographic variation in dorsal fin shape in the spinner dolphin is correlated with a variation in color pattern; the dorsal overlay in the eastern spinner is extensive and dark, obscuring the cape and giving the animal a monochromatic rather than a tricolor appearance. In killer whales in the Antarctic, the cape is visible; in other areas of the world it is not (Evans *et al.*, 1982). The degree of spotting in the Atlantic spotted dolphin (*Stenella frontalis*) varies from intense along the U.S. east coast to slight or none in animals in the offshore Gulf Stream to the northeast (Perrin *et al.*, 1987).

The most extensive studies of geographic variation in odontocetes have dealt with cranial features, characters that can be measured in collections of museum specimens. Within a species, variation has been found to be greatest in elements involved in feeding: size and number of teeth, length and breadth of the rostrum, and size of the temporal fossa. This implies that much

geographic variation must be associated with trophic ecology: available forage, foraging techniques, and competition. Cranial variation in the offshore spotted dolphin has been shown to be correlated with environmental parameters such as water depth, solar insulation, sea-surface temperature, surface salinity, and thermocline depth (Perrin *et al.*, 1994), and the distribution of two forms of the spinner dolphin in the eastern Pacific is associated with different water masses (Fiedler and Reilly, 1994). Different geographic forms or subspecies may also exhibit different patterns in life history parameters, such as age and size at attainment of sexual maturity, fecundity, and survival, but these differences can be due to differential population status as well as genetic factors (Chivers and DeMaster, 1994).

Mysticetes have not been as well studied because of their large size and a paucity of MUSEUM series of specimens (Rice, 1998). Subspecies and populations have been recognized mainly on the basis of DISTRIBUTION and, more recently, genetic differences. No adequate comparisons of the recognized populations of bowhead whales, *Balaena mysticetus*, have been carried out; the same is true for the two extant populations of the gray whale, *Eschrichtius robustus*, and the several populations of humpback whales, *Megaptera novaeangliae*. It is only in the rorquals that progress has been made in documenting geographic variation in morphology; this has been due to the availability of large series of specimens in factory-ship whaling operations. A dwarf form of the minke whale, *Balaenoptera acutorostrata*, exists in the Southern Hemisphere. Minke whales from the Sea of Japan ("J Stock") and Pacific coast of Japan ("O Stock") may differ modally in body proportions and baleen and flipper coloration (Kato *et al.*, 1992). Small coastal and large offshore forms of Bryde's whale, *B. edeni*, have been

Figure 2 *Geographic variation in shape and color pattern in spinner dolphins from the eastern and central Pacific: (top) Hawaii, (bottom) far eastern Pacific, and (middle) intermediate form from region between far offshore in eastern Pacific. By permission from Perrin (1990).*

described from South Africa and Japan. A pygmy "Bryde's whale" from Southeast Asia (Perrin *et al.*, 1996) may belong to a different species (which may carry the name *B. edeni;* if it does, the name *B. brydei* would probably then apply to the "ordinary" Bryde's whale). The pygmy blue whale, *B. musculus brevicauda,* is shorter (by 2 m in the North Pacific; Gilpatrick *et al.*, 1998) and heavier than other blue whales. (Thus the heaviest animal on earth is called "pygmy.")

IV. Carnivores

While for some small cetaceans, efforts to find genetic markers concordant with geographic morphological variation have failed (e.g., Dizon *et al.*, 1991, for *Stenella longirostris*), the reverse is true for pinnipeds; genetic differences or reproductive isolation have been found between populations that cannot be distinguished morphologically in *Arctocephalus forsteri, Zalophus californianus, Eumetopias jubatus,* and *Pusa hispida* (Rice,

1998; Hoelzel, 1997; Loughlin, 1997). The reasons for this may be polygyny and strong phylopatry (promoting accumulation of neutral variation due to drift) combined with relatively uniform ecological selection (promoting morphological homogeneity) over the range of the species.

Like cetaceans, pinnipeds can vary geographically in body size (e.g., *Arctocephalus tropicalis* among Amsterdam, Gough, and Marion Islands; *Odobenus rosmarus* between the Atlantic and Pacific Arctic; *Mirounga leonina* between Macquarie Island and South Georgia; *Pusa hispida* between the Baltic and the Sea of Okhotsk; Rice, 1998), coloration (*Phoca vitulina* between different islands off California; Yochem *et al.*, 1990), and cranial features (*Phoca vitulina* among several subspecies; *Pusa hispida* between pack ice and shore-fast ice and between the freshwater populations; *Halichoerus grypus* between the two sides of the Atlantic and the Baltic; *Histriophoca fasciata* between the western and eastern parts of the Bering Sea; and *Monachus monachus* between the Mediterranean and Atlantic; Rice, 1998).

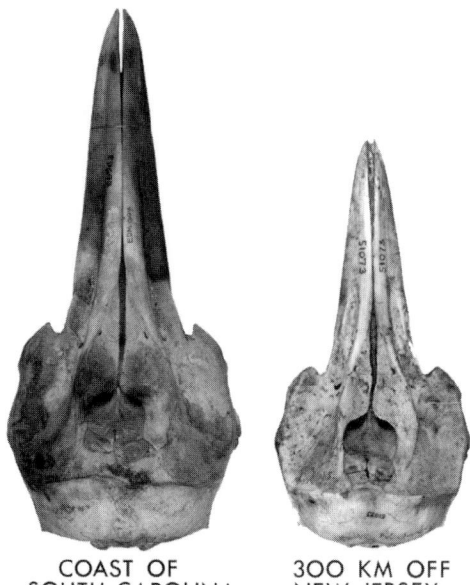

COAST OF
SOUTH CAROLINA

300 KM OFF
NEW JERSEY

Figure 3 *Geographic variation in adult skull size in the At-lantic spotted dolphin: larger size in coastal waters than in off-shore waters. (Large skull, 461 mm long; small skull, 360 mm).*

The three subspecies of sea otter are distinguished on the basis of body size and cranial characters (Wilson *et al.,* 1991) and the two subspecies of polar bears on the basis of skull size (Rice, 1998).

V. Sirenians

A molecular study of *Trichechus manatus* (García-Rodríguez *et al.,* 1998) found variation not in accordance with the presently recognized subspecies based on cranial characters; *T. m. latirostris* is closely linked to the Caribbean population of *T. m. manatus,* whereas the phylogenetic distances among the Caribbean, Gulf of Mexico, and South American populations of *T. m. manatus* are comparable to that between *T. manatus* and the Amazonian manatee, *T. inunguis.* As for many other marine mammal taxa, the taxonomy is ripe for revision based on both morphological and molecular characters.

Geographic variation in cranial morphology has been found within *Dugong d. dugon,* between Australia and Tanzania and between the Gulf of Carpentaria and Queensland in Australia (Rice, 1998).

See Also the Following Articles

Biogeography ■ Coloration ■ Genetics for Management ■ Morphology, Functional ■ Speciation ■ Species

References

Amano, M., and Miyazaki, N. (1992). Geographic variation and sexual dimorphism in the skull of Dall's porpoise, *Phocoenoides dalli. Mar. Mamm. Sci.* **8,** 240–261.

Archer, F. I., II, and Perrin, W. F. (1999). *Stenella coeruleoalba. Mamm. Species* **603,** 9.

Chivers, S. J., and DeMaster, D. P. (1994). Evaluation of biological indices for three eastern Pacific dolphin species. *J. Wildl. Manage.* **58,** 470–478.

Dizon, A. E., Southern, Š. O., and Perrin, W. F. (1991). Molecular analysis of mtDNA types in exploited populations of spinner dolphins (*Stenella longirostris*). *Rep. Int. Whal. Commn. Spec. Iss.* **13,** 183–202.

Doidge, D. W. (1990). Age-length and length-weight comparisons in the beluga, *Delphinapterus leucas. In* "Advances in Research on the Beluga Whale" (T. G. Smith, D. J. St. Aubin, and J. R. Geraci, eds.). *Can. Bull. Fish. Aquat. Sci.* **224,** 59–68.

Evans, W. E., Yablokov, A. V., and Bowles, A. E. (1982). Geographic variation in the color pattern of killer whales. *Rep. Int. Whal. Commn.* **32,** 687–694.

Fiedler, P.C., and Reilly, S. B. (1993). Interannual variability of dolphin habitats in the eastern tropical Pacific. II. Effects on abundances estimated from tuna vessel sightings, 1975–1990. *Fish. Bull. U.S.* **92,** 451–463.

García-Rodríguez, A. I., Bowen, B. W., Domining, D., Mignucci-Giannoni, A. A., Marmontel, M., Montoya-Ospina, R. A., Morales-Vela, B., Rudin, M., Bonde, R. K., and McGuire, P. M. (1998). Phylogeography of the West Indian manatee (*Trichechus manatus*): How many populations and how many taxa? *Mol. Ecol.* **7,** 1137–1149.

Gilpatrick, J. W., Perryman, W. L., Brownell, R. L., Jr., Lynn, M. S., and DeAngelis, M. L. (1997). Geographic variation in North Pacific and Southern Hemisphere blue whales (*Balaenoptera musculus*). *IWC Sci. Comm.* SC/49/O9, 32. Available from International Whaling Commission, The Red House, 135 Station Road, Impington, Cambridge CB4 9NP, UK.

Hoelzel, A. R. (1997). Molecular ecology of pinnipeds. *In* "Molecular Genetics of Marine Mammals." *Soc. Mar. Mamm. Spec. Pub.* **3,** 147–157.

Hoelzel, A. R., Potter, C. W., and Best, P. B. (1998). Genetic differentiation between parapatric "nearshore" and "offshore" populations of the bottlenose dolphin. *Proc. R. Soc. Lond. B* **265,** 1177–1183.

Houck, W. J., and Jefferson, T. A. (1999). Dall's porpoise *Phocoenoides dalli* (True, 1885). *In* "Handbook of Marine Mammals" (S. H. Ridgway and R. Harrison, eds.), Vol. 3, pp. 443–472. Academic Press, London.

Kato, H., Kishiro, T., Fujise, Y., and Wada, S. (1992). Morphology of minke whales in the Okhotsk Sea, Sea of Japan and off the east coast of Japan, with respect to stock identification. *Rep. Int. Whal. Commn.* **42,** 437–453.

Loughlin, T. R. (1997). Using the phylogeographic method to identify Steller sea lion stocks. *In* "Molecular Genetics of Marine Mammals." *Soc. Mar. Mamm. Spec. Pub.* **3,** 159–171.

Orr, M. R., and Smith, T. B. (1998). Ecology and speciation. *TREE* **13,** 502–505.

Perrin, W. F. (1984). Patterns of geographical variation in small cetaceans. *Acta Zool. Fennica* **172,** 137–140.

Perrin, W. F. (1990). Subspecies of *Stenella longirostris* (Mammalia: Cetacea: Delphinidae), *Proc. Biol. Soc. Wash.* **103,** 453–463.

Perrin, W. F. (1998). *Stenella longirostris. Mamm. Species* **599,** 7.

Perrin, W. F., Akin, P. A., and Kashiwada, J. V. (1991). Geographic variation in external morphology of the spinner dolphin *Stenella longirostris* in the eastern Pacific and implications for conservation. *Fish. Bull. U.S.* **89,** 411–428.

Perrin, W. F., and Brownell (1994). A brief review of stock identity in small marine cetaceans in relation to assessment of driftnet mortality in the North Pacific. *Rep. Int. Whal. Commn. Spec. Iss.* **15,** 393–401.

Perrin, W. F., Dolar, M. L. L., and Robineau, D. (1999). Spinner dolphins (*Stenella longirostris*) of the western Pacific and Southeast Asia: Pelagic and shallow-water forms. *Mar. Mamm. Sci.* **15,** 1029–1053.

Perrin, W. F., Dolar, M. L. L., and Ortega, E. (1996). Osteological variation of Bryde's whales from the Philippines with specimens from other regions. *Rep. Int. Whal. Commn.* **46,** 409–413.

Perrin, W. F., Mitchell, E. D., Mead, J. G., Caldwell, D. K., Caldwell, M. C., van Bree, P. J. H., and Dawbin, W. H. (1987). Revision of the spotted dolphins, *Stenella* spp. *Mar. Mamm. Sci.* **3,** 99–170.

Perrin, W. F., Schnell, G. D., Hough, D. J., Gilpatrick, J. W., Jr., and Kashiwada, J. V. (1994). Reexamination of geographic variation in cranial morphology of the pantropical spotted dolphin, *Stenella attenuata,* in the eastern Pacific. *Fish. Bull. U.S.* **92,** 324–346.

Perryman, W. L., and Lynn, M. S. (1993). Identification of geographic forms of common dolphin (*Delphinus delphis*) from aerial photogrammetry. *Mar. Mamm. Sci.* **9,** 119–137.

Rice, D. W. (1998). Marine mammals of the world. *Soc. Mar. Mamm. Spec. Pub.* **4,** 231.

Stacey, P. J., and Leatherwood, S. (1997). The Irrawaddy dolphin, *Orcaella brevirostris:* A summary of current knowledge and recommendations for conservation action. *Asian Mar. Biol.* **14,** 195–214.

Stewart, B. E., and Stewart, R. E. A. (1989). *Delphinapterus leucas. Mamm. Species* **336,** 8.

Storz, J. F. (1999). Genetic consequences of mammalian social structure. *J. Mamm.* **80,** 553–569.

Wilson, D. E., Bogan, M. A., Brownell, R. L., Jr., Burdin, A. M., and Maminov, M. K. (1991). Geographic variation in sea otters, *Enhydra lutris. J. Mamm.* **72,** 22–36.

Yochem, P. K., Stewart, B. S., Mina, M., Zorin, A., Sadovov, V., and Yablokov, A. V. (1990). Non-metrical analyses of pelage patterns in demographic studies of harbor seals. *Rep. Int. Whal. Commn. Spec. Iss.* **12,** 87–90.

Geological Time Scale

ELLEN M. WILLIAMS
Northeastern Ohio Universities College of Medicine, Rootstown

Much of our concept of time is measured by our daily and yearly activities, but time in terms of Earth history is infinitely longer. Rather than increments of minutes, hours, or even years, geological time ticks by at millions and billions of years. The calendar of Earth's existence is the geological time scale; a measure developed over the last four centuries by many geoscientists and paleontologists.

I. Methods for Deriving the Time Scale
A. Lithostratigraphy

The geological time scale is based on principles of lithostratigraphy that allow geologists to decipher a geologic history from a sequence of strata. These include the principle of superposition, developed in the late 17th century by Nicolaus Steno, which states that in any sequence of undisturbed strata, the oldest layers lie at the bottom and progressively become younger toward the top. In practical application, if a geologist can identify the bottommost and topmost beds in a sequence, he or she might also be able to infer the relative ages of those

beds. Steno also founded the principle of original horizontality (under the influence of gravity, sediment will settle horizontal or parallel to the earth's surface) and lateral continuity (as originally deposited, strata will extend laterally thinning at the basin margin), which further aid in the identification of beds and their relationship to one another.

In traveling through the world's mountain belts, it is apparent that not all beds lie in a horizontal position, nor do they lie in undisturbed sequences. Therefore, the concepts and principles set forth by James Hutton and Charles Lyell in the 18th and 19th centuries broadened Steno's original principles. Hutton's concept of unconformities (a break or hiatus in the geologic record as represented by an erosional surface) not only allows stratigraphers to determine age relations within one sequence of strata, but also to correlate unconformities across great distances. This was followed by Lyell's principle of crosscutting relationships, which showed that any geologic feature that crosscuts another is younger than that feature. For the concept of inclusions, Lyell established that fragmentary bits (inclusions) of rock contained within another rock body are older than the enclosing mass. Taken together, the work of Lyell, Hutton, and Steno provide a lithostratigraphical basis for determining the ages of beds within and across outcrops. Figure 1 shows a geological history obtained from applying those stratigraphical principles and concepts.

B. Biostratigraphy

The physical position of rock units is not the only tool for deciphering the Earth's geochronology, however. At the turn of the 18th century, William Smith noted that certain rock units often contained a particular assemblage of fossils. He used these assemblages to trace beds over short distances and also to correlate beds of different lithologies that contained the

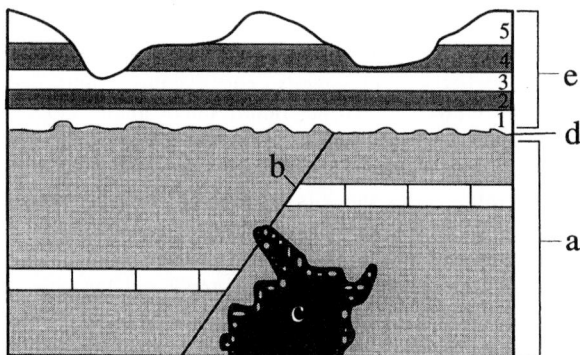

Figure 1 *Geologic crosssection and associated relative geochronology obtained by applying stratigraphic principles. Chronologically from oldest to youngest are (a) deposition of light gray and block units; (b) faulting (principle of crosscutting relationships); (c) intrusion of igneous body with inclusion of light gray unit (principle of crosscutting relationships and Lyell's concept of inclusions); (d) erosional surface (unconformity); and (e) deposition of horizontal strata (layer 1 being oldest, 5 youngest; principle of superposition). Quantitative dating methods such as radiometric or paleomagnetic dating would provide actual ages for some of the section.*

same fossils. These observations are the foundation of the principle of biologic succession, which stipulates that unique faunas characterize different periods in Earth's history, and that the identification of similar faunas in disparate deposits will allow geologists to infer chronological correlations between the deposits. Sixty years later, Charles Darwin would place Smith's biologic succession in an evolutionary context.

The use of fossils as time markers has advanced since Smith's time, and we now recognize that both assemblages of fossils and individual genera are useful in geochronologic determinations. The most effective time markers are fossils known as index or guide fossils. These taxa have a short geologic existence but are geographically widespread or cosmopolitan. Consequently, these fossils can narrow the possible age limits for the strata in which they are found and simultaneously be correlated across great distances. By using multiple overlapping ranges of index fossils, even smaller increments of time can be determined. The discipline that has developed from fossil-based stratigraphic correlations is biostratigraphy.

C. Magnetostratigraphy

A more recent method for deriving a time scale is the use of paleomagnetic analysis. The Earth's geomagnetic poles are inclined 11° from Earth's rotational axis, and studies have shown that periodically these poles reverse, therefore our "North" today was our "South" 30,000 years ago. These reversals occur roughly every half million years but are interspersed with shorter reversals at erratic intervals (such as the one 30,000 years ago). The history of reversals is recorded in both sedimentary and igneous deposits. For igneous rocks, pole orientation is recorded when iron oxide minerals such as magnetite (Fe_3O_4) align themselves parallel to the poles by crystallization in an igneous body. In the case of a cooling lava, alignment begins as the temperature nears 500° C. The magnetic declination preserved in the igneous rock on complete crystallization is known as thermoremanent magnetization. For sedimentary deposits, such as those on the ocean floor, grains of iron oxide will continually align themselves with the poles until lithification occurs; this is termed depositional remanent magnetization. The history of reversals as recorded in lava flows from around the world in combination with known radiometric dates have allowed scientists to construct a time scale based on ancient magnetism, the paleomagnetic time scale.

D. Radiometric Dating

The time scale framework established by stratigraphic and fossil studies has improved significantly over the last century with the advent of methods that allow for obtaining a quantitative geochronology. Within 10 years of the discovery of radioactivity, the physicist Ernest Rutherford suggested in 1905 that radioactive decay might be used to measure the actual age of a rock, and thus radiometric dating was born.

Decay occurs by particle emission from an atom, which results in the formation of another atom with a different composition, known as the daughter product. For example, the decay of uranium-238 will initially produce thorium-234 as its daughter product, and this process will continue until a stable configuration is obtained; in the case of uranium-238, the stable product is lead-206. Because each individual isotope has a particular mode and the length of decay that is unique, it is possible to determine the age of a particular specimen by measuring the amount of parent product versus daughter product by a mass spectrometer. For computational ease, the most useful increment to measure decay is half-life, the time required for half of the original quantity of radioisotope to decay. Half-lives are also unique for individual isotopes; for instance, carbon-14, which decays to nitrogen-14, has a half-life of 5730 years, and therefore is useful in dating recent organic materials. Whereas, uranium-238 has a half-life of 4468 million years, the isotope with the longest half-life is rubidium-87 (daughter product of strontium-87) with 48,800 million years.

II. Geochronologic Units

By applying stratigraphic and biostratigraphic principles, early geologists and paleontologists began construction of a relative geological time scale. This was based on the identification and correlation of chronostratigraphic units that represent the rocks deposited during a specific time interval. Chronostratigraphic units were then translated to geochronologic units representing increments in time. Figure 2 shows the time scale in its current form. It has been revised significantly over the last century with the use of quantitative methodologies and better stratigraphic constraint.

The largest division of the time scale is the Eon; in succession from oldest to youngest, there are three eons: the Archean, Proterozoic, and Phanerozoic. Collectively, the Archean and Proterozoic are referred to as the Precambrian and represent the largest span of Earth's history (approximately 87%). The rocks of the Archean and Proterozoic are predominantly igneous and metamorphic, although they do also contain the earliest records of bacteria, algae, and multicellular organisms.

The Phanerozoic, spanning 570 millions years ago (Ma) until today, is divided into progressively smaller units; the largest being the era, followed by the period, epoch, and finally age. The boundaries of these units indicate pronounced changes in fauna and flora or other geologic events. The oldest era, the Paleozoic, lasted 325 million years and is marked by a proliferation and diversification of shelly marine invertebrates in the early part (Cambrian through Silurian) and by bony and cartilaginous fishes, amphibians, reptiles, and mammal-like reptiles (therapsids) in the later part (Devonian through Permian).

The Paleozoic ended in a catastrophic mass extinction. Numerous marine invertebrates (rugose corals, spiny productid brachiopods, two orders of bryozoans, and many echinoderms) disappeared as well as over 70% of Permian amphibians and reptiles. It is hypothesized that the union of the supercontinent Pangaea with its associated climatic changes may be responsible for the Permian mass extinction.

The Mesozoic era spans 180 million years of geologic history and contains three periods: the Triassic, Jurassic, and Cretaceous. During this time, two new vertebrate classes evolved, the birds and mammals, and other groups experienced impressive radiations. The Mesozoic also marked the breakup of Pangaea and greater development of the Tethys Seaway. However, the Mesozoic is probably best known for the amazing reptile

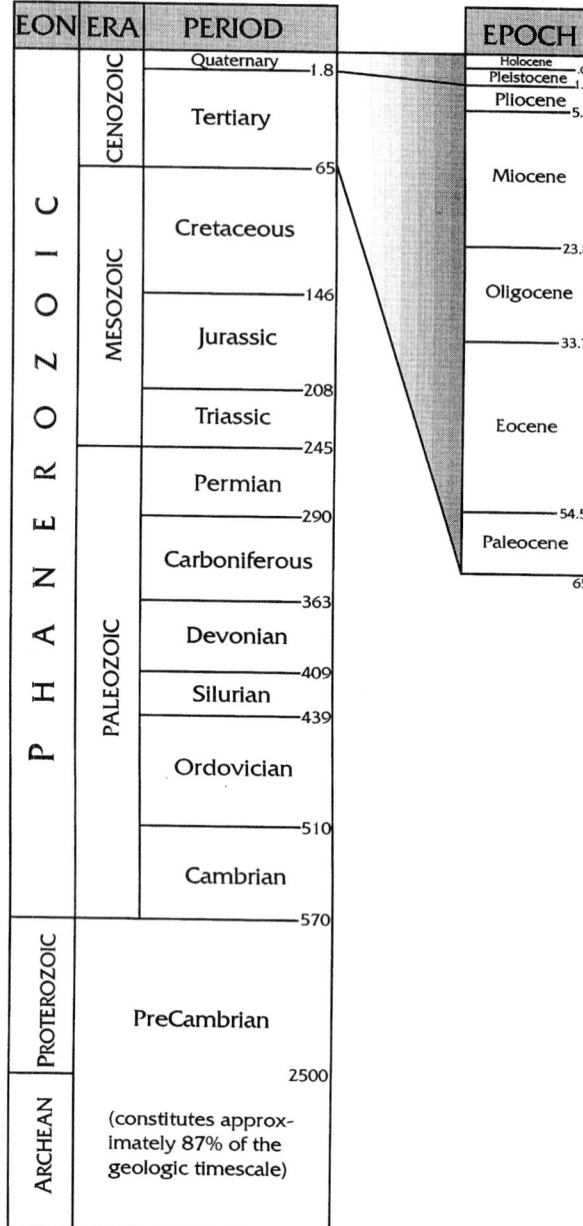

Figure 2 *Geological time scale. Dates along the right margin are millions of years before present. Geochronologic ages are not shown. Modified from Harland* et al. *(1990) and Berggren* et al. *(1996).*

the K–T boundary (Cretaceous–Tertiary boundary) and set the stage for the rapid diversification of all mammalian families during the Cenozoic era, the age of mammals.

The Cenozoic is composed of two periods, the Tertiary and Quaternary; the latter being the period in which we live today. These periods are additionally divided into epochs with the Paleocene, Eocene, Oligocene, Miocene, and Pliocene falling within the Tertiary, and the Pleistocene and Holocene within the Quaternary. Immense fossil collections have been amassed for the Tertiary and Quaternary, and comparison with studies of known environmental and geographic change has allowed scientists to paint a clearer picture of this portion of Earth history than any other period.

Like many other mammal groups, marine mammals evolved and radiated during the early to middle part of the Tertiary. The first sirenians appear in the early Eocene with the primitive genus *Prorastomus*. They underwent a radiation in the middle Eocene followed by a gradual decline through the Miocene into the earliest Pliocene. Three genera are known in recent times (*Trichechus*, *Hydrodamalis*, and *Dugong*). The fossil group Desmostylia appeared in the early Oligocene with the youngest representatives present in the middle to late Miocene.

Among cetaceans, the oldest representatives, archaeocetes, originate in the early Eocene, whereas the suborder Mysticeti appears in the late Eocene and Odontoceti in the early Oligocene. Within odontocetes, most of the major modern families are established by the middle to late Miocene and among mysticetes by the earliest to middle Miocene with the exception of eschrichtiids and neobalaenids, which appear in the Pleistocene. The younger geologic appearance of eschrichtiids and neobalaenids may not indicate their origin, but rather it may point to gaps in the fossil record.

For marine carnivorans, the oldest pinnipeds are late Oligocene in age. Although phocids appear in the middle Miocene, they are most abundant and diverse in recent times. Desmatophocids have a much shorter geological existence from the early to middle Miocene. Like phocids, the oldest odobenids are Miocene in age, although they enjoy their greatest diversity in the late Miocene–early Pliocene with only one species, *Odobenus rosmarus*, present today. The earliest otariids are late Miocene in age and, like phocids, the otariids are most species rich in recent times.

See Also the Following Articles

Fossil Record ■ Fossil Sites ■ Origins of Marine Mammals ■ Speciation

References

Berggren, W. A., Kent, D. V., Swisher, C. C., and Aubrey, M.-P. (1996). A revised Cenozoic geochronology and chronostratigraphy. *In* "Geochronology, Time Scales and Global Stratigraphic Correlation" (W. A. Berggren, D. V. Kent, M.-P Aubry, and J. Hardenbol, eds.), pp. 129–212. SEPM Spec. Publ. No. 54.

Carroll R. L. (1988). "Vertebrate Paleontology and Evolution," p. 689. Freeman, New York.

Currie, P. J., and Padian, K. (1997). "Encyclopedia of Dinosaurs," p. 869. Academic Press, San Diego.

faunas that dominated terrestrial and aquatic realms worldwide. The most spectacular representatives of this time are the dinosaurs.

Like the Paleozoic, the Mesozoic ended in biotic crisis with approximately 25% of all known families of animals eliminated, including dinosaurs, pterosaurs, and marine reptiles such as ichthyosaurs, plesiosaurs, and mosasaurs, as well as numerous invertebrates. This extinction occurred across what is known as

Gradstein, F. M., Agterberg, F. P., Ogg, J. G., Hardenbol, J., van Veen, P., Thierry, J., and Huang, Z. (1996). A Triassic, Jurassic, and Cretaceous time scale. *In* "Geochronology, Time Scales and Global Stratigraphic Correlation" (W. A. Berggren, D. V. Kent, M.-P. Aubry, and J. Hardenbol, eds.), pp. 95–126. SEPM Spec. Publ. No. 54.

Harland, W. B., Armstrong, R. L., Cox, A. V., Craig, L. E., Smith, A. G., and Smith, D. G. (1990). "A Geologic Time Scale 1989," p. 263. Cambridge Univ. Press, Cambridge, UK.

Press, F., and Siever, R. (1985). "Earth," 4th Ed., p. 656. Freeman, New York.

Gestation

SEE *Female Reproductive Systems*

Giant Beaked Whales

Berardius bairdii and *B. arnuxii*

TOSHIO KASUYA

Teikyo University of Science and Technology, Uenohara, Japan

These two species are the largest animals of the family Ziphiidae, and their taxonomic status is still unsettled. They live in cohesive schools, feed on deep-water bottom fish, and are known to have curious life history characteristics.

I. Taxonomy

Arnoux's beaked whale, *Berardius arnuxii* Duvernoy, 1851, was described using a skull from New Zealand. A specimen of similar characters found in the Bering Sea was the basis for another species, Baird's beaked whale, *B. bairdii* Stejneger, 1883.

Characters once thought to distinguish the two species are now considered invalid, beside a few slight differences, i.e., smaller adult size (8.5–9.75 m vs 9.1–11.1 m) and possible differences in flipper size and in the shape of nasal bones and vomer. Such minor differences throw doubt on the validity of the separate species (Fig. 1).

II. Skeleton and Internal Morphology

Condylobasal lengths of skulls of adult Arnoux's beaked whales range from 1174–1420 mm, and those of Baird's beaked whale are 1343–1524 mm. Other measurements in percentage of condylobasal length are (both species combined): length of rostrum 60.7–69.5%, width of rostrum at base 64.4–82.3%, and breadth across zygomatic processes of squamosals 47.1–56.5%. Nasal bones are large but do not overhang the superior nares. Among Ziphiidae, their SKULL is the least bilaterally asymmetrical and the crest formed by nasals, premaxillae, maxillae, and frontals at the skull vertex is the least developed.

A pair of large TEETH erupt on the anterior end of the lower jaw at around sexual maturity and abrade rapidly. The tooth is flat, triangular in shape (about 8×8×3 cm) and has elements of rudimental enamel, thin dentine, massive secondary dentine filling the pulp cavity, and thick cementum that covers the root.

The vertebral formula of 3 Arnoux's beaked whales was C7, T10-11, L12-13, Cd17-19, total 47-49, and that of 49 Baird's beaked whales off Wadaura, Japan, C7, T9-11, L12-14, Cd17-22, total 47-52 (n=48.9), most of which (41) had either 48 or 49 vertebrae. There are five phalanges in the manus.

The stomach lacks an esophageal compartment, and the glandular stomach has up to nine segments. The caecum is absent. The nasal tract has three pairs of sacs.

III. External Morphology

The entire body is dark brown, and the ventral side is paler and has irregular white patches. Tooth marks of conspecifics are numerous on the back, particularly on adult males.

Figure 1 *Giant beaked whales in the genus* Berardius *are distributed disjunctly. (A)* B. arnuxii *occurs in waters around the Antarctic, reaching northward to the shores of the Southern Hemisphere continents. (B)* B. bairdii *ranges across the northern Pacific from Japan, throughout the Aleutians, and southward along the coast to the southern tip of Baja California (see Fig. 2). Despite this widely separated occurrence, evidence for the distinctiveness of these two taxa remains equivocal.*

They are the least SEXUALLY DIMORPHIC among the family (Table I). The body is slender and has a small head, a low falcate dorsal fin, and small flippers that fit into the depressions on the body. A pair of throat grooves and some accessory ones contribute to expand the oral cavity at suction feeding. The equation $W = (6.339 \times 10^{-6}) L^{3.081}$ expresses the relationship between body weight (W, in kg) and body length (L, in cm) off Japan.

The blowhole is crescent shaped with the concavity directing anteriorly. The melon is small and its front surface is almost vertical, from which a slender rostrum projects.

IV. Distribution

Arnoux's beaked whales inhabit vast areas of southern oceans, except the tropics, from the Ross Sea at 78°S to Sao Paulo (24°S), northern New Zealand (37°S), South Africa (31°S), and southeastern Australia (29°S) (Fig. 2).

Baird's beaked whales inhabit the temperate North Pacific and adjacent seas, mainly deep waters over the continental slope. The northern limits are at Cape Navarin (62°N) in the Bering Sea and in the central Okhotsk Sea (57°N), where they occur even in shallow waters.

On the American side they occur north of northern Baja California (30°N), but there are records from the southern Gulf of California.

The southern limits on the Asian side are at 36°N on the Japanese coast in the Sea of Japan and at 34°N on the Pacific coast. They occur year-round in the Okhotsk Sea and Sea of Japan, including the drift ice area of the former.

Off the Pacific coast of Japan, the whales appear in May in waters over the continental slope at depths of 1000–3000 m and their numbers increase toward summer when hunting commences and then decrease toward October. During this period they are almost absent in waters further offshore. This reflects their food preference and DIVING ability. Their wintering ground is unknown.

V. Biology of Baird's Beaked Whales off Pacific Japan

When traveling, they form tight schools of 1–30 individuals ($n = 5.9$). Schools of 2–9 individuals constitute 64% of the encounters and singletons 14% (Fig. 3).

Diving lasts up to 67 min ($n = 18.2$), 39% last less than 11 min, 27% 11–20 min, and 18% 21–30 min. Time at surface is 1–14 min ($n = 3.9$) and tends to be greater after a longer dive. During surface schooling, individuals blow continuously while swimming slowly and are easily identifiable from vessels.

Age is determined using growth layers in the teeth. The gestation time is unknown, although 17 months is suggested from interspecies relationship among toothed whales. Neonates are about 4.6 m. Females first ovulate at age 10–15 years when they are 9.8–10.7 m and live to about 54 years. Ovulation occurs throughout life at an average rate of once every 2 years. There is no evidence of significant postreproductive life.

The testis is histologically mature when it weighs 1.5 kg, which corresponds to age 6–11 years or a body length of 9.1–9.8 m, but continues growing until age 30, when it reaches 3–9 kg.

TABLE I
External Measurements of Baird's Beaked Whales off Wadaura, Pacific Coast of Japan

Measurement[a]	Males			Females		
	n	Mean	SD	n	Means	SD
1. Body length	20	1038	29.4	14	1049	28.1
2. Tip of lower jaw	18	0.8	0.16	13	0.7	0.15
3. Base of melon	20	5.6	0.56	14	5.4	0.51
4. Center of eye	20	9.2	0.48	14	9.2	0.51
5. Anterior insertion, flipper	20	16.1	0.83	14	15.8	0.70
6. Umbilicus, center	20	46.4	3.80	13	45.8	4.54
7. Genital orifice, center	20	64.7	2.27	14	69.8	2.20
8. Anus, center	20	72.3	2.09	14	73.3	2.41
9. Tip of dorsal fin	20	71.3	1.62	12	73.1	2.50
10. Anterior insertion, tail flukes	10	91.9	0.53	4	91.8	4.65
11. Flipper, length	20	12.4	0.49	14	12.0	0.05
12. Flipper, width	20	4.3	0.23	14	4.1	0.19
13. Dorsal fin, height	14	2.9	0.39	10	3.0	0.27
14. Girth at middle	4	51.9	4.93	2	48.3	6.06

[a]Body length is in centimeters, and other figures are length from tip of upper jaw expressed as percentage of body length. Body lengths of samples are 1002–1120 cm for males and 1015–1110 cm for females.

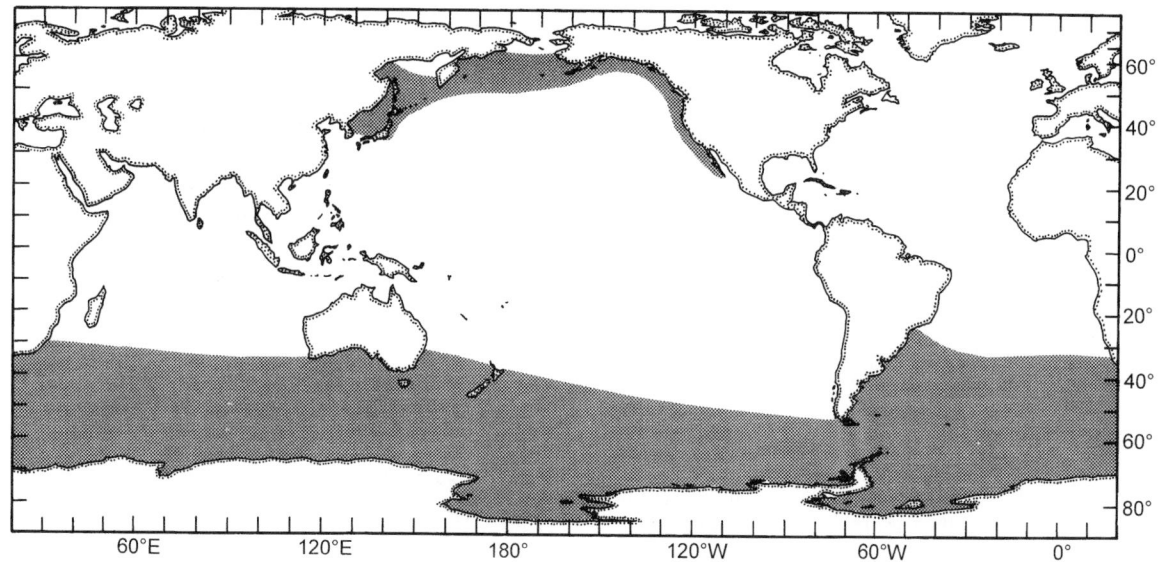

Figure 2 *Distribution of giant beaked whales in the North Pacific* (Berardius bairdii) *and Southern Hemisphere* (B. arnuxii) *(shaded areas). Precision of our current knowledge is greater in the former waters.*

Males live to about 84 years. Lack of behavioral data inhibits judgment of the age at which males begin to participate in reproduction.

Physical maturity, determined from the fusion of vertebral epiphyses to the centrum, is attained before 15 years and within 5 years from sexual maturity. Mean body lengths of samples 15 years or older are 10.45 m (SD=0.31, n=22) in females and 10.10 m in males (SD=0.35, n=66).

The male proportion is 44% at age 3–9 years and then increases with age to reach 100% at 55 years and over. Seventeen percent of samples of both sexes were old males (55–84 years). Such a sex ratio imbalance is common among whaling samples from Japan (Sea of Japan, Okhotsk Sea, and Pacific), Russia (Kuril and Aleutian Islands), and Canada (off Vancouver Island) and is believed to reflect a lower natural mortality rate of males.

The proportion of females among sexually mature individuals is 23%. This is only partially improved to 37% with the assumption that males attain reproductive capacity at 30 years when testicular growth ceases. The selective benefit for such male longevity is unknown.

Figure 3 *A school of Baird's beaked whales off Boso coasts, Japan.*

VI. Feeding, Parasites, and Predators

Squid beaks have been found in the stomach of an Arnoux's beaked whale.

Off Japan and California, Baird's beaked whales feed mainly on benthic fishes (Moridae and Macrouridae) and cephalopods, but occasionally on pelagic fish such as mackerel, sardine, and saury. Cyamids attach on teeth and skin, stalked BARNACLES on teeth, and diatoms on skin. Wounds attributable to the cookiecutter shark, *Isistius brasiliensis,* are common off Japan. Scars of killer whale teeth are common on flippers and tail flukes, suggesting PREDATION. Internally, they are heavily parasitized in the stomach, liver, blubber, and kidney, with extensive kidney pathology due to the nematode *Crassicauda giliakiana.*

VII. Abundance and Exploitation

ABUNDANCE is estimated only for Japanese waters: 5029 for the Pacific coast, 1260 for the eastern Sea of Japan, and 660 for the southern Okhotsk Sea, with 95% confidence intervals of about 50% on both sides of the mean.

Exploitation has not been reported for the Arnoux's beaked whale. HUNTING of Baird's beaked whales by USSR, Canada, and the United States was of low level. Hunting in Japan started in the early 17th century at the entrance of Tokyo Bay. The annual catch was less than 25 before 1840 and then declined. In 1891, whaling cannons were introduced and the operation moved to outer seas. After World War II, the fishery, expanded to the entire northern Pacific, reported a catch of over 300, and subsequently declined. Now the industry operates with a quota of 8 for the Sea of Japan, 2 for the southern Okhotsk Sea, and 52 for the Pacific coasts.

See Also the Following Articles

Beaked Whales, Overview ■ Japanese Whaling

References

Balcomb, K. C., III (1989). Baird's beaked whale *Berardius bairdii* Stejneger, 1883: Arnoux's beaked whale *Berardius arnuxii* Duvernoy, 1851. *In* "Handbook of Marine Mammals" (S.H. Ridgway and R. Harrison, eds.), Vol. 4, pp. 261–288. Academic Press, London.

Brownell, R. L. (1974). Small odontocetes of the Antarctic. *In* "Antarctic Mammals" (V. C. Bushnell, ed.), Antarctic Map Folio Series 18, pp. 13–19. American Geographical Society, Washington, DC.

International Whaling Commission (1992). Report of the Subcommittee on Small Cetaceans. *Rep. Int. Whal. Commn.* **42,** 108–119.

International Whaling Commission (1994). Report of the Subcommittee on Small Cetaceans. *Rep. Int. Whal. Commn,* **44,** 178–228.

Kasuya, T. (1977). Age determination and growth of the Baird's beaked whale with a comment on the fetal growth rate. *Sci. Rep. Whales Res. Inst.* **29,** 1–20.

Kasuya, T. (1986). Distribution and behavior of Baird's beaked whales off the Pacific coast of Japan. *Sci. Rep. Whales Res. Inst.* **37,** 61–83.

Kasuya, T., Brownell, R. L., Jr., and Balcomb, K. C., III (1997). Life history of Baird's beaked whales off the Pacific coast of Japan. *Rep. Int. Whal. Commn.* **47,** 969–979.

Kasuya, T., and Miyashita, T. (1997). Distribution of Baird's beaked whales off Japan. *Rep. Int. Whal. Commn.* **47,** 963–968.

McCann, C. (1975). A study of the genus *Berardius* Duvernoy. *Sci. Rep. Whales Res. Inst.* **27,** 111–137.

Reeves, R. R., and Mitchell, E. (1993). Status of Baird's beaked whale, *Berardius bairdii. Can. Field-Nat.* **107,** 509–523.

Ross, G. J. B. (1984). The smaller cetaceans of the south east coast of South Africa. *Ann. Cape Prov. Mus. (Nat. Hist.)* **15**(2), 173–410.

True, F. W. (1910). An account of the beaked whales of the family Ziphiidae in the collection of the United States National Museum, with remarks on some specimens in other American museums. *Bull. U.S. Nat. Mus.* **73.**

Gray Seal
Halichoerus grypus

AILSA HALL
*University of St. Andrews,
Scotland, United Kingdom*

The gray seal is a highly successful marine predator of the Northern Hemisphere. This precocious and inquisitive marine mammal has had a changing relationship with humans, being exploited for its fur and blubber in the early last century but now being protected under national and international legislation. It continues to be a controversial species in relation to its interactions with fisheries where it often comes into conflict with humans for resources. However, it has sustained the vagaries of this relationship, and its population is increasing in almost every area it is found.

I. Diagnostic Characteristics

The gray seal is the only member of the genus *Halichoerus.* Its species name, *grypus,* means hook nosed, referring to the Roman nose profile of the adult male. *Halichoerus* means sea pig in Greek. This species exhibits SEXUAL DIMORPHISM with the mature males weighing between 170 and 310 kg and adult females between 100 and 190 kg. Individuals from the population in the western Atlantic are significantly larger than those from the eastern Atlantic; males can weigh over 400 kg and females over 250 kg. Genetic studies suggest that the western and eastern Atlantic populations are distinct and diverged approximately 1 million years ago (Boskovic *et al.,* 1996).

II. Selected Morphological and Physiological Data

Morphological differences between the sexes can be seen in Fig. 1. The neck and chest of the male are wrinkled and often scarred, whereas females are much sleeker. Both have the convex nose and wide muzzle, which are very pronounced in the male. Many of the females are gray in color with a distinctive cream/off-white background and markings, particularly around the neck, with generally a dark back and light underside. Males are more uniformly dark when mature, but subadults can have similar cream-colored patches on the neck and the side of the

Figure 1 *Adult male (top) and female (bottom) gray seals. The female has distinct markings on the fur, whereas the male is more uniform in color and larger than the female.*

face. Females mature at between 3 and 5 years old and males around 6 years, although they are probably not socially mature until 8 years old.

III. Distribution and Geographic Variation

Figure 2 shows the geographic range of the gray seal. Breeding rookeries are on remote uninhabited islands or on fast ice. The single biggest island-breeding colony is on Sable Island (85,000, increasing at almost 12% per annum). Other major sites in the western Atlantic are in the Gulf of St. Lawrence (69,000). The northeast Atlantic population in Iceland was estimated to be 11,600 in 1987 with approximately 3000 in Norway, 2000 in Ireland, and between 1000 and 2000 in the White Sea. The Baltic population is estimated at approximately 5000 animals. The British population is surveyed annually and is cur-rently approximately 110,000 animals, increasing at about 6% per annum.

IV. Life History and Ecology

The females give birth, on land or on ice, to a single white-coated pup between September and March. The earliest breeding colonies are those in the south of the United Kingdom and Ireland. Further north around the British Isles the breeding season is later, between October and November. In Canada, peak pupping is not until January and in the Baltic it occurs in late February–early March.

At birth the pup weighs between 11 and 20 kg and, over the lactation period, lasting an average of 18 days, can quadruple in weight to over 40 kg. The mothers' milk is very fat-rich (around 50–60% lipid) and is mobilized from her blubber stores. The pup's white coat, known as the lanugo, is shed at weaning. The pup then undergoes a postweaning fast on land for a period lasting between 10 and 28 days, during which it loses approximately 0.5 kg per day. The reason for this fasting period is not fully understood, but physiological changes during this time suggest that it is related to the development of diving ability.

Toward the end of lactation the female comes into estrus and mates. On some colonies there may be as many as 10 females to 1 male, whereas on rookeries, where access is not restricted by narrow gullies, the sex ratio may be 2 females to 1 male. Males compete for access to females but do not defend discrete territories, and matings may occur in the water as females return to the sea, as well as on land. Females fast during the breeding season and may lose up to 40% of their initial body weight during the breeding season, as they do not feed during this time. The gestation period is 8 months, and to achieve a 12-month breeding cycle the fertilized egg is not implanted until 4 months after conception. This occurs around the time of the annual molt when animals spend longer time hauled out on land. Gray seals generally return to their natal site to breed and show a high degree of site fidelity, often returning to within meters of their previous pupping sites (Pomeroy *et al.*, 1994).

Gray seals feed on a variety of fish species and cephalopods (Hammond *et al.*, 1994a,b). However, a large proportion of their diet is sand eels or sand lance (*Ammodytidae*), which can make up over 70% of the diet at some locations and in some seasons. Other prey include whiting, cod, haddock, saithe, and flatfish (plaice and flounder). They are largely demersal or benthic feeders, and foraging trips lasting between 1 and 5 days away from a haul-out site are frequently focused on discrete areas that are within 40 km of a haul-out site (McConnell *et al.*, 1999). On average, gray seal dives are generally short, lasting between 4 and 10 min with a maximum recorded duration of about 30 min. Typically, in the United Kingdom, animals dive down to the sea bed, which is largely 60 m in depth, falling to 200 m in some areas, although dives at depths >300 m have been recorded.

See Also the Following Articles

Hunting of Marine Mammals ■ Pinniped Life History ■ Rookeries

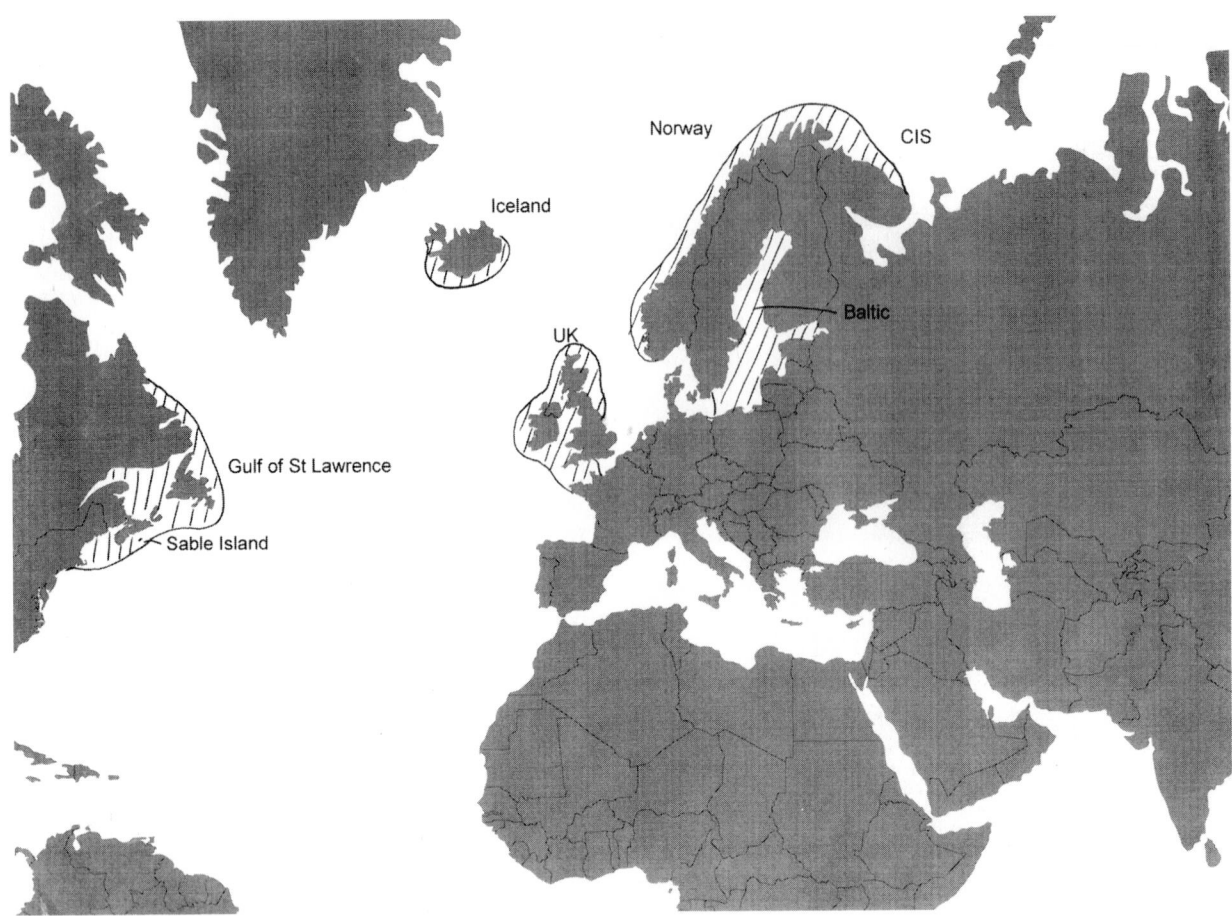

Figure 2 *Map showing the geographic distribution of the gray seal.*

References

Bonner, N. (1994). "Seals and Sea Lions of the World." Blandford, London.

Boskovic, R., Kovacs, K. M., Hammill, M. O., and White, B. N. (1996). Geographic distribution of mitochondrial DNA haplotypes in grey seals (*Halichoerus grypus*). *Can. J. Zool.* **74,** 1787–1796.

Hammond, P. S., Hall, A. J., and Prime, J. H. (1994a). The diet of grey seals around Orkney and other island and mainland sites in northeastern Scotland. *J. Appl. Ecol.* **31,** 340–350.

Hammond, P. S., Hall, A. J., and Prime, J. H. (1994b). The diet of grey seals in the Inner and Outer Hebrides. *J. Appl. Ecol.* **31,** 737–746.

McConnell, B. J., Fedak, M. A., Lovell, P., and Hammond, P. S. (1999). Movements and foraging areas of grey seals in the North Sea. *J. Appl. Ecol.* **36,** 573–590.

Pomeroy, P. P., Anderson, S. S., Twiss, S. D., and McConnell, B. J. (1994). Dispersion and site fidelity of breeding female grey seals (*Halichoerus grypus*) on North Rona, Scotland. *J. Zool. Lond.* **233,** 429–447.

Reijnders, P., Brasseur, S., Van der Toor, J., Van der Wolf, P., Boyd, I., Harwood, J., Lavigne, D., and Lowry, L. (1993). "Seals, Fur Seals, Sea Lions and Walrus." Status Survey and Conservation Action Plan, IUCN/SSC Seal Specialist Group, IUCN, Gland, Switzerland.

Riedman, M. (1990). "The Pinnipeds: Seals, Sea Lions and Walruses." University of California Press, Berkeley.

Gray Whale
Eschrichtius robustus

MARY LOU JONES AND STEVEN L. SWARTZ
*Southeast Fisheries Science Center,
Miami, Florida*

The family Eschrichtiidae includes a single known genus and species, the gray whale, which now is found only in the North Pacific and Pacific Arctic Oceans (though it once lived in the North Atlantic until the 17th or early 18th century). Grays are by far the most coastal of all the great whales, and inhabit primarily inshore or shallow, offshore continental-shelf waters. They tend to be nomadic, highly migratory, and are tolerant of climatic extremes. Each year, they make the longest migration of any whale (up to 15,000–20,000 km round trip) largely without FEEDING, traveling along nearshore routes between a summer feeding zone of high productivity in Arctic or subarctic waters and a winter breeding zone in temperate or subtropical southern waters. Unlike other mysticetes, the gray is primarily a bottom feeder and influences the topography of the seabed in the Arctic (from sucking its prey out of the sediments). It is the only whale to bear its young in warm, shallow, coastal areas and lagoons.

There are two populations. The western North Pacific population (or Korean-Okhotsk) migrates along the coast of Asia. It was hunted to the verge of extirpation and is extremely rare. Another much larger, eastern North Pacific population (or California-Chukchi) migrates along the coast of North America and eastern Siberia (Fig. 1). It too was severely overexploited in the latter half of the 19th and early 20th centuries, but, following protection from commercial WHALING, has increased to about 26,600 (in 1999). The resilient eastern North Pacific gray whale is the only cetacean population that, following severe depletion, has sufficiently recovered under protection from commercial whaling to be removed from the list of endangered species. The western Pacific population, however, remains listed as critically endangered.

An active but gentle species, as long as they are not molested, the gray whale had a reputation for ferocity among the old whalers, who dubbed it "devilfish" for its habit of crashing into and staving in boats when harpooned or in defense of its young. Despite it being the trickiest and most dangerous prey, early whalemen developed a special affection for grays and found them to be the most interesting and intelligent of all the great whales. Grays seemed to learn quickly the dangers from whaling and performed a remarkable array of evasive maneuvers. They were admired for their fierce protection of their young and habit of giving assistance or "standing by" an injured companion, often reaching self-sacrificing measures. When attacked, they showed a power of resistance and tenacity of life that distinguished them from all other cetaceans. Today, many people have come to value gray whales more highly as a living resource than as one to slaughter, and they have become a WHALE-WATCHING phenomenon. Their coastal habits make them the most accessible of all the mysticetes and they can be seen most easily, often from shore. Gray whales are unusually sportive; breaching, spyhopping, LOBTAILING, and mating extravaganzas are essential elements of their migratory and breeding-grounds repertoires. Their willingness to allow whale watchers to stroke them is an added attraction, and grays are now known as "friendly" whales.

I. Systematics

A. Evolutionary History and Classification

No fossils of a direct gray whale ancestor have been found. The family Eschrichtiidae is known only from the Recent and from a single Pleistocene specimen about 50,000–120,000 years old found in California, and a less certain one from Alaska. A long-held theory proposed that gray whales could have evolved from the Cetotheridae, an extinct family of whales dating back some 38 million years, and could be their closest living relatives. Due to the lack of any fossil remains linking the modern gray whale to the far more ancient cetothere, some challenge that view and are unwilling to link them to any of the known early whales. A highly distinctive species, the gray whale has been placed in its own family: Eschrichtiidae (Ellerman and Moirrison-Scott, 1951) (=Rhachianectidae, Weber, 1904).

Most experts have considered the gray whale to be more closely related to the rorquals (Balaenopteridae) than to the right

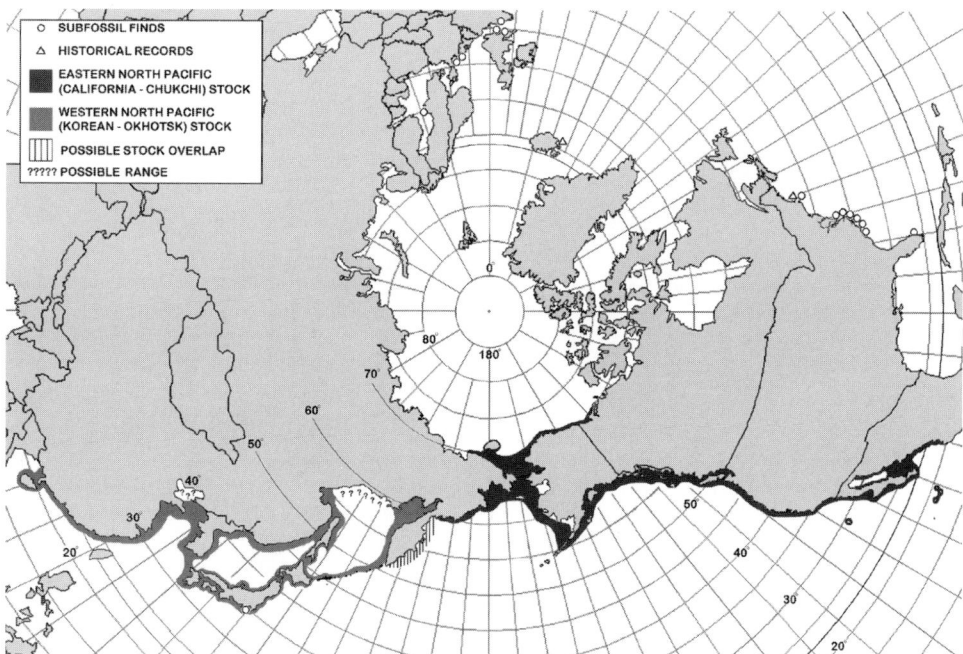

Figure 1 *Known distribution, historic and current, of the gray whale. The eastern North Pacific population (black) has recovered from depletion. The western North Pacific population (gray) remains critically endangered. The Atlantic gray whale is extinct and is known from subfossil finds (circles).*

whales (Balaenidae). Others have given it an intermediate position between the two. However, for the four modern families of baleen whales commonly recognized (Balaenopteridae: rorquals, or fin whales; Balaenidae: bowhead and right whales; Eschrichtiidae: the gray whale; and Neobalaenidae: the pygmy right whale), the pattern of phylogenetic relationships at the base of baleen whale divergence is unresolved. With respect to gray whales, analyses of their position within the Mysticeti conflict. Molecular studies position gray whales within the balaenopterids, while analyses based on morphology and including fossil and extant taxa differ in suggesting grays are linked with the balaenids and the pygmy right whale. Moreover, some biologists place the gray whale as a subfamily of Balaenopteridae.

B. Names

The gray whale has many English names, first applied by 19th century whalers. *Scrag* was used by old whalers on the Pacific coast of North America because they identified it with a whale called a scrag that was taken in the Atlantic Ocean in the 17th and 18th centuries. *Devil fish* and *hard head* were derived from the often violent reaction of the grays that commonly smashed boats with their heads and flukes when harpooned. *Mud digger* and *mussel digger* referred to the bottom feeding of the whales. *Gray* and *gray back* characterized its color. *Okhotsk* and *Korean* denoted the western population's feeding and presumed breeding grounds, and *Chukchi* and *California*, the feeding and breeding grounds for the eastern population (also the whaling grounds).

As for its scientific name, the generic name *Eschrichtius* (Gray, 1865) was given to honor a 19th century Danish zoologist, Daniel Eschricht; and the specific name *robustus* (Lilljeborg, 1861) is from the Latin for "oaken" or "strong." The gray whale first became known to science not through observations of living animals but through the discovery of subfossil skeletal remains from Europe where it had long been extinct. Conspecificity cannot be proven by purely anatomical data, but the SKELETON of the gray is distinctive and no anatomical difference has been found between extinct Atlantic and extant Pacific populations (or between the eastern and western populations of the Pacific) that would justify separating them on the basis of species, or even subspecies. Thus, the odd situation exists where the remains from the extinct Atlantic population serve as the type specimen for in the Pacific Ocean (=*Eschrichtius gibbosus* Erxleben, 1777).

II. Description

The gray whale is a robust, slow-moving whale with a flexible body, more slender than the right whales and more stocky than most rorquals (Fig. 2). This species is readily identified by the mottled gray color of the skin with numerous lighter patches scattered all over the body (although color may vary from gray-brown to slate-black). Grays have more external PARASITES and epizoites than any other cetacean. The barnacle, *Cryptolepas rhachianecti,* thought to be host specific, has been found on beluga whales (*Delphinapterus leucas*). As larvae, BARNACLES are free swimming but soon settle onto calves and adults alike,

Figure 2 *The narrow head of the gray whale is usually covered with patches of barnacles and whale lice (top left). The blow is heart-shaped and 3–4 m high (top right). Instead of a dorsal fin, grays have a low hump followed by a series of bumps (bottom left). The flukes are over 3 m wide, frequently bear scars from the teeth of killer whales, and are often lifted before a deep dive (bottom right).*

Figure 3 *Dense clusters of barnacles surrounded by whale lice develop shortly after birth. Barnacles leave white scars on the whale's skin, which slowly repigments over time.*

eventually forming large colonies that are deeply embedded in the skin. Grays also host three species of WHALE LICE (they are cyamids, not insects) that feed on skin and damaged tissue: *Cyamis scammoni* and *Cyamus kessleri* occur only on grays, whereas *Cyamus ceti* also lives on other whales (Fig. 3). The lice cling by the thousands in areas of reduced water flow, such as around barnacle clusters, blowholes, and folds of skin, and swarm into wounds. In the breeding lagoons, schools of topsmelt (*Atherinops affinis*) symbiotically clean lice and sloughing skin from the whales. Much of the whale's mottled appearance comes from the parasites or scars from previous infestations and abrasions. By photographing the skin pigmentation patterns on the backs and sides, it is possible to identify individual animals, which is important to the study of gray whales.

The gray whale's relatively short, narrow head is triangular (in top view) and moderately curved downward (in side view) (Fig. 4). It is encrusted with patches of barnacles and associated whale lice, particularly on top. Widely spaced bristles sprout from small dimples on the upper and lower jaw (no other whale has so many hairs); these short bristles, linked with sensory cells, are extra noticeable on calves. The skull comprises only about 20% of the total skeletal length. A unique feature is the presence of paired occipital tuberosities on the posterior part of the skull. Small eyes, with eyelids, are located just behind the corners of the mouth. Directly above them, on top of the head, is a pair of blowholes (nostrils). Barely visible, the ear opening is a tiny hole just behind the eye. The narrow upper jaw has 130 to 180 baleen plates hanging down on each side, separated in the front of the snout. A gray's BALEEN is the shortest (5–40 cm), thickest, and coarsest of all mysticetes and is cream-white to pale yellow. The lower jaw is broad, with a keel-like protuberance in front, and slightly arched. On the throat there are two to seven (commonly three) short, deep creases that stretch open and allow the mouth to expand a little during feeding, but they do not extend beyond the throat region and are insignificant compared with the many long ventral grooves found in balaenopterids.

Gray whales lack a dorsal fin but have a low hump followed by a series of 8–14 small bumps (knuckles) along the top of the tail stock. The ventral part of the body is smooth, without any longitudinal grooves. The paddle-shaped flippers are up to 200 cm long. Tail flukes are over 3 m wide on adults, with smooth trailing edges and a deep median notch. The flippers and flukes are often marred with tooth scars from killer whales (*Orcinus orca*). Unique to this species is a cyst-like structure (10–25 cm in diameter) beneath a swelling on the ventral surface of the tail stock, which may be similar to sebaceous glands of land mammals, or function as a track-laying scent gland, although its exact function

Figure 4 *Gray whales commonly spyhop, lifting the head vertically above the water. The head is narrow and triangular when viewed from the top (left), and they have from two to seven short creases on the throat (right), rather than the long, ventral throat grooves found in balaenopterids.*

is unknown. Grays, which survive in extremely cold water for about half of the year during the feeding season, are insulated with a layer of BLUBBER averaging 15 cm thick beneath the skin and can tolerate a great drop in their skin surface temperature to only a degree or so above that of the surrounding water.

Newborn grays (calves) average 4.6 to 4.9 m long and weigh about 920 kg. The sex ratio is parity at all ages. They reach puberty at anywhere from 6 to 12 years of age (average is 8), at a mean length of 11.7 m in females (called cows) and 11.1 m in males (called bulls). Adults weigh 16,000 to 45,000 kg and stop growing at about 40 years, when the average female is 14.1 m long and the average male is 13.0 m. The largest female recorded was 15 m, and the largest male 14.6 m long. Although adult females are slightly bigger than males, there is no significant difference in their appearance (the distance from the genital slit to the anus is wider in males). The maximum, as well as average, life span is unknown (age is calculated from growth layers in the waxy ear plugs that fill the auditory canal). One large female was estimated to have been 75–80 years old when she was killed and she was pregnant.

III. Ecology and Behavior
A. Social Organization

The gray whale is not a highly social species. Individuals may associate with many conspecifics, but they do not appear to form stable pairs or groups and come together for only part of the year during migration and on the winter breeding range. The only persistent social bond known is between a mother and a calf, which disappears at weaning. Now and then, short-term associations lasting several days or weeks are reported, but their significance is still a mystery to us. Very little research into the social organization of the gray whale has been done. It is possible that they communicate even over large distances, sending and receiving acoustic signals. No territoriality, dominance, or overt aggression toward conspecifics has been reported.

On the summer feeding grounds, grays are usually widely spaced, solitary (commonly pregnant females), or in pairs, and less often in small groups of 3–5, although many may be in proximity in the patchily distributed food-rich areas. Larger aggregations in tens or even hundreds can occur in a particularly rich feeding area but are likely related to a mutually available mass of food rather than to social cohesion or interaction (these aggregations fluctuate constantly). Occasionally, some grays stop feeding to form groups of 30–40 or 100–400 animals that engage in bouts of social activity (lasting 1–4 days) reminiscent of courting and mating; however, their function is unknown. During migration, singles, pairs, and trios are most common but grays sometimes form transient groups of up to 16 individuals.

On the winter breeding grounds, large aggregations of mothers with young and courting/mating whales are common, but are in constant flux (1000 or more will crowd into the largest breeding lagoon). Initially, mothers with neonates have little interaction with other mothers and calves, although many are concentrated in the nursery areas of the breeding grounds. When calves are 2–3 months old, however, they often form highly interactive social groups. In these encounters, mothers and young cavort en masse, rolling about on top of each other,

rubbing and touching from head to flukes, and often emitting huge bursts of underwater air bubbles. Groups last from a few minutes to over 3 hr and involve up to 40 individuals at a time, with many others coming and going, and may play a role in the social development of the calf.

Overall, there is a low degree of cooperation among gray whales, except limited examples of joint defense against killer whale attacks and assistance or support behavior, mainly for the aid of the young and especially in the calving areas. This is evidenced by adults coming to the aid of a mother whose calf is in trouble. Standing by (whales in a pair or group assisting, supporting, or staying with an injured companion) also occurs occasionally among adults in times of distress.

B. Feeding

Gray whales do almost all of their feeding during summer and fall when they are in higher latitudes, where they forage on the ocean floor in shallow waters over continental shelves (4–120 m deep). They are adapted to exploit the tremendous seasonal abundance of food that results as the Arctic pack ice (sea ice that is unattached to land) retreats in spring, exposing the sea to the polar summer's continuous daylight, which triggers an enormous bloom of microorganisms in the water down to the sea floor. Unlike other baleen whales, the gray is mainly a bottom feeder and sucks small invertebrates and crustaceans out of the sand and mud. Their distribution in the feeding grounds coincides with the concentrations of these bottom-dwelling prey. As the summer advances and the edge of the pack ice recedes and uncovers more of the feeding grounds, the whales move. They feed heavily from about May through October, gaining enough stores of fat to sustain them during fasting or greatly reduced intake of food during the rest of the year, when the polar feeding grounds are ice covered and they migrate south to warm winter breeding grounds.

During about 5 months of intensive feeding in Arctic waters, an adult will consume roughly 170,000 kg of food. By the time the grays return to the feeding grounds (5 to 6 months later) they will have lost up to 30% of their body weight and must single-mindedly forage to replenish their fat reserves. The highest energy costs during migration are incurred by pregnant or lactating cows. For cows, the cost of reproduction includes the ENERGETIC requirements for gestation (producing a calf) and lactation (nursing young until weaning), which is far greater. During summer and fall, pregnant cows put on 25–30% more weight than other gray whales (exclusive of fetus).

An extraordinary aspect of the gray whale's feeding ecology is its apparent dietary flexibility. Over 80 species of prey have been identified, reflecting its opportunistic approach to foraging. On the summer feeding grounds, grays primarily consume benthic gammaridean amphipods (shrimp-like crustaceans that live on or buried in the sediment). Amphipods from four families account for about 90% of the food, but depending on the feeding area, 1 of 7 species is usually dominant. Four are from the family Ampeliscidae (*Ampelisca macrocephala*, *A. eschrichti*, *Byblis gaimardi*, *Haploops* sp.). They are tube builders that live in dense colonies or "tube mats" in the upper few centimeters of sea floor sediments. Overall, the amphipod *A. macrocephala* (up to 33 mm long) is probably the most commonly taken species (and occurs in concentrations as high as 23,780/m^2 in the Chirikov Basin in the Bering Sea). The other three species are from separate families: Haustoriidae (*Pontoporia femorata*), Lysianassidae (*Anonyx nugax*), and Atylidae (*Atylus bruggeni*), which are mobile scavenging amphipods that rove freely over the seafloor in search of prey. In some areas, polychaete tube worms (*Travisia forbesi*) are their main food. Planktonic prey items eaten in the peripheral feeding areas south of the main feeding grounds occur in swarms or schools and include mysids, crab larvae, red crab, mobile amphipods, herring eggs and larvae, squid, megalops, and bait fish. Some plant material also occurs in their stomachs.

To bottom feed, grays roll to one side, bringing the head parallel with the seabed, sweep the side of the mouth close over the bottom a few centimeters above it, and open the jaws slightly to suck sediment containing prey into the mouth (which has flexible lips) (Fig. 5). Water, sand, and mud are strained through the comb-like baleen, leaving the food trapped on its inner margin. The suction might be created by retracting the large, strongly muscled tongue (weighing 1400 kg). The grays move slowly along the bottom, sucking up infauna in pulses, and surface with clouds of sediment (called mud plumes) streaming from the mouth. Mud plumes mark the meandering path of the feeding whales. Seabirds feed on prey brought to the surface in the plumes.

Grays impact their feeding grounds more than any other cetacean. Bottom feeding leaves mouth-sized depressions or "feeding pits" in the sea floor, from which the top layers of sediment are removed. Foraging is a major source of physical disturbance to the benthic community and plays an important role in the rate of turnover of the epibenthos. In some areas of the Arctic, over 40% of the seafloor is pock-marked with feeding pits. It is thought that by clearing space in the bottom, whales open areas for recolonization, succession, and maturing of the prey community, thus promoting the growth and diversity of life on the seafloor. Periods of nonuse are presumed to correspond to rapid recovery of the habitat. However, if the resource is overutilized and the area is stripped, it could be a one-way street leading to the permanent loss of amphipod communities and changing feeding patterns. In this way, gray whales are an integral part of the coastal community and participate in a dynamic feedback loop, termed "niche construction," whereby their feeding activities function to shape their ecological niche through alteration of the benthos.

In addition to bottom feeding, grays also occasionally feed by surface skimming and engulfing planktonic prey out of the

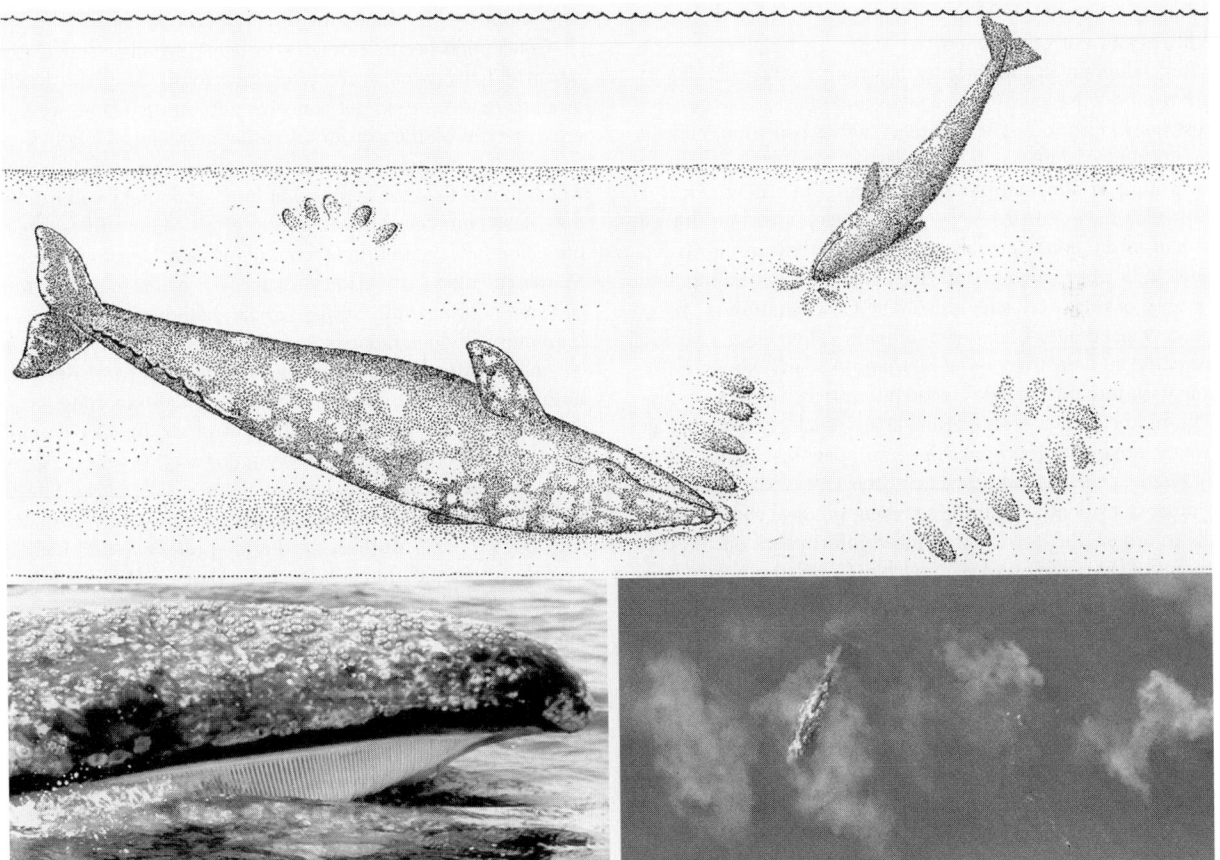

Figure 5 *A bottom feeding gray whale swims on one side to suck prey from the seafloor, creating mouth-sized depressions or feeding pits (top). The cream-white baleen plates are the coarsest, shortest, and fewest of any mysticete (bottom left). Sieving prey-laden sediments through the baleen creates billowing mud plumes (bottom right).*

water column. Zooplankton are only known to be utilized outside of the principle feeding grounds, in peripheral feeding areas throughout the migratory range. Instead of traveling the entire distance to the feeding grounds, some whales spend the summer feeding along the coast in other parts of their range. Also, whales destined for the summer grounds sometimes stop to feed periodically on the way if the opportunity arises. The importance of peripheral feeding areas is unclear. With three modes of feeding (benthic suction, engulfing, and skimming) the gray has perhaps a greater range of foraging techniques than any other of the great whales.

C. Reproduction

Gray whales are thought to have a promiscuous mating system: males and females do not form long-term pair associations and both sexes may copulate with several partners during the same breeding season. Because multiple inseminations can occur, it is proposed that *sperm competition* may be taking place in gray whale fertilization (sperm from two or more males compete to fertilize the ovum within a female). Adult males have relatively large testes weight (averaging 38 kg in mating season) to body weight ratios and presumably produce large quantities of sperm. In this mating strategy, copulating males attempt to dilute or displace the sperm of other males to increase the likelihood of being the male to fertilize the female. Their fibroelastic penis reaches 170 cm in length and is erected by muscle fibers and not vasodilation.

Reproduction in gray whales is strongly seasonal. The female reproductive cycle lasts 2 years and consists of the onset of estros (the period of sexual receptivity), ovulation, conception, gestation, lactation, and an anestros period. Most females bear young in alternate years, although some may rest 2 or more years between calves. In general, each year one-third to one-half of the adult females are birthing (they are not receptive to bulls after calving) and the remainder are mating, with a reversal of roles each successive year. Cows continue to breed at an advanced age. Bulls mate annually. They have a peak of spermatogenetic activity in late autumn or early winter, correlating with the time females come into estrus. Some sexual behavior on the feeding grounds and among males occurs that apparently serves nonreproductive social purposes.

Lengthy courting (precopulatory) behavior is part of the mating process, evidently requiring sufficient physical contact by the bulls to arouse the cows, but detailed information on the constituents of courtship is not yet available. Copulation occurs belly to belly. Pairs or trios of whales sometimes court and mate quite gently together. More often (or perhaps just more readily visible), there is a high level of activity, marked by whales rolling, touching, splashing, and cavorting energetically, at which times bulls with extended penises can be seen (Fig. 6). While some nudging and pushing may take place to get close to a cow, bulls do not appear to fight to keep others away. Bulls outnumber available cows by as much as two to one, leading to the belief of a ménage à trois mating group by early naturalists. In fact, although trios are common, so are pairs and groups of various sizes that can blossom into a giant free for all involving as many as 20 consorting adults at a time. The large groups constantly

Figure 6 *Gray whales mate with multiple partners, often in large, energetic courting groups (top). Newborn calves have more uniformly dark skin and are supported on their mothers' backs for their first few breaths (bottom).*

fluctuate, with some participants departing while others join in as if stimulated by the sexual activity of the initial core group.

Conception occurs primarily in late November and December while the whales are migrating south from the feeding grounds, but some do not conceive until in the winter assembly area, or even on the northward spring migration. Length of gestation is disputed, but is generally thought to last 11 to 13 months, which means that newly pregnant females do not give birth until they have completed the following year's southward fall migration. The birth season lasts from about late December to early March (median birth date: January 27), when most near-term cows are in or near the calving grounds, although some calves are born during the migration from California south. Cows bear a single calf, unattended, and provide sole parental care. Reports of births cite head-first deliveries, with the cow supporting her calf at the surface for the first few breaths of air. Initially, its movements are uncoordinated, but swimming soon steadies (Fig. 6).

A mother's bond with her calf is especially close. She displays an unusual degree of affection, often gently stroking it with her flippers. Mothers are highly protective and will fight fiercely to defend their young from danger. While in the lagoons and on migration, calves stay close to and almost touching their mothers. They drink about 189 liters of rich milk (about 53% fat and 6% protein) each day and grow rapidly, reaching 8.7 m when weaned. The calf remains dependent on its mother until it is weaned in the summer feeding area, at about 7–8 months of age, and perhaps 1 or 2 months longer, when they have solid food in their stomachs but remain with their mother. It is thought that calves begin to forage during the latter stages of nursing and thus may gain some experience while still with the mother. After the calves are weaned, around August, cows are anestrous for 3 to 4 months. Then in November to December a new mating period begins.

D. Sensory Perception

1. Acoustics Once reported to be almost silent, it is now known that gray whales are soniferous both day and night. They create a variety of phonations that sound like rasps, croaks, snorts, moans, groans, grunts, pops, roars, quick series of clicks, belches, and metallic knocks and bongs. These low-frequency broadband signals range from about 100 Hz to 4 kHz, but may go up to 12 kHz. The most prevalent sounds for whales feeding in the Arctic and those in the breeding lagoons are pulsive signals, usually emitted in bursts, that sound like a series of metallic knocks (broadband pulses, from about 100 Hz to 2 kHz). Tonal moans are the most common phonation from migrating whales. Some behaviors may also serve an acoustic function. Grays expel huge bursts of air bubbles underwater (explosive exhalations). These emissions are often released in profusion from the blowholes in social settings. Occasionally, large quantities of air are released from the sides of the mouth as the whale swims by, producing a spectacular display of bubbles. The functions are still obscure, but the joint effect of the acoustic and visual components could create a potent short-range communication signal. Other behaviors that may have an acoustic function include percussive jaw claps, head slaps, backslaps, BREACHING, flipper slapping, and LOBTAILING.

Gray whales are not known to echolocate by means of high-frequency click trains as odontocetes do. However, some low-frequency click-like sounds resembling ECHOLOCATION (which enables a whale to detect objects by listening to the reflected echo of its own sound pulses) have been recorded. These sounds are very tentatively proposed as evidence for primitive echolocation aptitudes that may serve a long-distance function limited to large targets (such as whales) or to detecting broad topographical or oceanic features useful for orientation and navigation. The theory of echolocation in gray whales, however, is as yet unsubstantiated.

Whalers have long stood in awe of the gray's sensitivity to sound. Even the water disturbance by an oar may put a whale to flight. The relatively low upper limit of the frequency range of their vocalizations suggests that they may hear well into the low sonic or infrasonic regions (below the range of human hearing, frequencies lower than 18 Hz). The use of mostly low-frequency sounds is thought to be an adaptive strategy whereby gray whales circumvent the high levels of natural background noise prevalent in their coastal environment by producing sounds that are generally at frequencies below it. Unfortunately, much of the man-made noise in the ocean also occurs in the lower frequency range and has a high level of output, which could interfere with or mask the gray whale's sounds or possibly damage their hearing. Gray whales appear to try to get around some man-made noise by increasing their call types, calling rates, and the loudness of calls to enhance signal transmission and reception.

2. Other Sensory Perception Gray whales can see moderately well both in air and water, but color VISION is probably weak. The position of the eyes suggests that they have stereoscopic vision forward and downward permitting efficient estimation of distance. The eyes are adapted for heightened sen-

sitivity to dim light and for improving contrast and resolution underwater. Grays have retained some sense of smell but are micronosmic at best. In water, the nares are almost always closed (but whales may smell the air as they breathe). The sense of touch is well developed. Some taste buds occur at the back of the tongue, and the possibility of chemoreception through taste has been conjectured.

E. Sleep

It is not known if gray whales sleep. Whales on migration have not been observed to stop to rest for long periods of time. One exception is mothers and calves, which stop to nurse and rest during the north migration. In polar regions during summer when daylight is continuous, most gray whales remain active continually, usually foraging or moving between feeding areas, although occasionally a few resting animals are seen. It is speculated that grays, like some delphinids, may rest one hemisphere of the brain at a time (presumably essential to a voluntary breather). In the breeding grounds, there are more obvious indications that grays sleep, particularly near-term pregnant females and those with neonates. They rest, barely awash, floating at or just beneath the surface, with head and flukes hanging down, for up to an hour, and raise the head to breathe periodically in a slow rhythmic pattern.

F. Swimming, Breathing, and Diving

Overall, gray whales are relatively slow but steady swimmers on migration, although speeds vary from the beginning to the end of the route and there are periods of wandering, resting, milling, feeding, and breeding activity in addition to directed travel along the way. They make the southward trip from the feeding to the breeding grounds in an average of 55 days, swimming at about 7–9 km/hr, and cover a distance of about 144–185 km/day. On the north migration, grays move at a slower speed, averaging 4.5 km/hr (88–127 km/day), and may socialize and feed more, which effectively slows their diurnal rate of migration. Mothers and calves travel up to 96 km/day. Speed of directed travel is about the same as that of other whales, but mothers and calves pause to rest and nurse along the migration. When pursued, grays may reach about 13 km/hr but can only maintain this pace for a few hours. Speed under duress can surge to 16 km/hr, at least for short bursts (avoiding predators). Interestingly, gray whales are very efficient swimmers. They travel mostly at speeds that minimize their energy expenditure and maximize their range, and swim at depths that minimize total drag, important factors in successfully covering the long migratory distances they travel.

Gray whales usually are not exceptionally long or deep divers. The pattern of BREATHING between dives can vary greatly for different activities, with grays averaging only 3% of the time at the surface. When migrating, whales typically remain submerged during traveling–dives for 3 to 5 min during which they may travel 300 m. They surface to blow three to five times at intervals of 15 to 30 sec during a series of short, shallow, surface dives showing only a small portion of the back. The bushy spout is 3 to 4 m high (Fig. 2). Following the terminal blow in a series, traveling whales typically lift their flukes into the air (fluke up) to begin the next traveling dive. During prolonged dives, they may remain submerged 7 to 10 min (or longer) and travel 500 m or more before resurfacing to breathe. Usually, the longer the dive, the greater the number of blows, as the need to reoxygenate the system is greater. Their maximum known dive depth is 170 m.

Breathing and DIVING BEHAVIOR on the feeding and breeding grounds is more variable than on migration. When foraging on summer feeding grounds in shallow coastal waters of 50–60 m, grays dive to the bottom by submerging almost vertically and lifting their flukes above water, and stay under for 5 to 8 min while swimming very slowly. In the breeding lagoons, about 50% of the dives are less than 1 min and 99% are less than 6 min, whereas dives longer than 12 min are associated with resting animals. Mothers, for example, typically float at or slightly below the surface for periods of up to an hour and then submerge for 5 to 10 min, or up to 26 min. When evading detection, grays often surface cautiously, exposing only the blowholes, exhale quietly without a visible blow, and sink silently beneath the surface (called snorkeling).

The species is active at the surface; spyhopping (raising the head vertically out of the water), breaching (leaping vigorously into the air), and other aerial behaviors (head stands with tails in the air, flipper slaps on the surface, etc.) are commonly performed by adults and older calves, especially on the breeding grounds (Fig. 7, also Fig. 4). Throughout their range, grays often appear to "play" and surf in or near the breakers and shallow water along shore. Some grays regularly rub themselves on beaches and sandbars on the breeding grounds and on the rubbing beaches off Vancouver Island. They also rub on pebble beaches and rocks in the Arctic, leaving behind shed barnacles. Some enter brackish water in fjords, coastal lagoons, and the mouths of rivers and emerge cleaned of barnacles and lice. Gray whales are noted to frequent places so shallow that they appear to be lying on the bottom. Occasionally during the ebb tide, some are stranded (apparently unharmed) until the incoming tide refloats them.

G. Friendly Whale Behavior

Gray whales exhibit a sense of curiosity that appears early in life as calves investigate and "play" with floating objects such as balls of kelp and small logs. The whales, including mothers and calves, frequently approach whale-watching skiffs, particularly on the breeding grounds. Behaviors include stationing alongside the skiffs, rubbing against them, bumping, lifting, and blowing bubbles beneath the boats, and allowing the passengers to pet and stroke them (Fig. 8). This activity is popularly termed "friendly" behavior. In the lagoons, these curious grays seem to be initially attracted to the sounds made by the motors of the skiffs, which fall within the same frequency range as gray whale vocalizations. Since the first encounter with a friendly whale at the calving lagoons in the 1970s, friendly whales have become commonplace there and are also encountered to a lesser degree along the migratory route and even in the Bering Sea.

H. Predators and Mortality

Killer whales are the only predator of gray whales (besides humans), although several species of sharks, including the great white shark (*Carcharodon carcharias*) and tiger shark (*Galaeocerdo cuvier*), scavenge on carcasses and might kill a small number of calves. Pods of killer whales cooperatively pursue grays, especially calves and juveniles, and seem to attack by repeatedly ramming along their sides, grasping the flukes and flippers to immobilize and drown them, and trying to open their mouths to bite into the tongue. Killer whales have frequently been reported feeding on the tongues of gray whales and then leaving the carcasses as carrion. Sometimes grays turn on their back and slash out with a powerful tail to ward off the swift wolf-like packs of killer whales. Oddly, if cornered, they may go into "shock," floating motionless at the surface, stomach up, while killer whales bite at the tongue and flippers. Rakemark scars from teeth are often seen on living whales, indicating that many successfully ward off an attack. Some attacks may also represent practice or play by killer whales. A reduced risk of PREDATION from killer whales (more abundant at high lati-

Figure 7 *Gray whales breach frequently while migrating and on their winter breeding grounds. One animal was observed to breach 40 consecutive times.*

Figure 8 *A "friendly" gray whale cow and calf allow whale watchers to pet them (note the tip of the mother's lower jaw in the foreground).*

tudes in colder coastal seas) might be a primary benefit to females leaving polar waters to give birth in the subtropics. However, predation pressure does not appear to be a significant determinant in the gray whale's social organization.

Other known causes of gray whale mortality include ship collisions, ENTANGLEMENT in fishing gear (particularly gill nets) and man-made debris, and whaling (legal aboriginal takes and poaching). Also, calves are sometimes severely struck by whales involved in courting/mating groups, which could result in accidental fatalities. No infectious diseases have been reported. Internal parasites occur but are not known to cause death. In 1999, 2000, and 2001 an unexplained, severe deterioration was seen in the physical condition and health status of some individuals in both eastern and western populations (gray whales were unusually thin, or emaciated). In the eastern population, mortality was unusually high, and some whales appeared to have died from starvation.

IV. Distribution, Migration, and Status
A. North Atlantic Population(s) (Extinct)

The gray whale once existed on both sides of the North Atlantic. Complete and partial skeletons of grays that are subfossils (not yet mineralized) have been found on the east coast of the United States (from New Jersey to Florida) and in the eastern Atlantic from the Baltic coast of Sweden, the Netherlands, Belgium, and the Channel coast of England, the most recent dated from about 1650 A.D. (see Fig. 1). In the western Atlantic, the gray whale is thought to have migrated all along the Atlantic seaboard from Florida to Canada. The youngest North American specimen is from colonial times about 1675 A.D., whereas the oldest are around 10,000 years old. The European gray whale may have disappeared around 500 A.D., but there is a credible record for Iceland in the early 17th century. Evidently, based on written accounts, the last few gray whales in the Atlantic were exterminated by the late 17th or early 18th century, apparently by early Basque, Icelandic, and Yankee whalers. The disappearance of grays from both sides of the Atlantic coincides with the development of WHALING, supporting the idea that overhunting in Europe, Iceland, and North America was responsible for, or at least contributed to, its demise.

B. Western North Pacific Population (or Korean-Okhotsk)

1. Distribution and Migration   Historic records suggest that the western North Pacific population of gray whales formerly occupied summer feeding grounds in the Okhotsk Sea as far north as Penzhinskaya Bay and south to Akademii and Sakhalinskiy Gulfs on the west and the Kikhchik River on the east (see Fig. 1). In autumn, the whales migrated south along the coast of eastern Asia from the Tatarskiy Strait to South Korea (passing Ulsan from late November to late January) to winter breeding grounds suspected to be along the coast of Guangdong Province and around nearby Hainan Island in southern China. The southern-most record was from the east coast of Hainan Island. The long-held belief that the western grays spent the winter on the south coast of Korea was unfounded. It was proposed that an additional migration corridor led down the east coast of Japan to winter breeding grounds in the Seto Inland Sea (where

calving occurred) in southern Japan, but this is largely unsubstantiated. In spring, it is assumed that the whales undertook a reverse migration, passing back through the Sea of Japan to reach their summer feeding habitat in the Okhotsk Sea.

Today, the number of gray whales inhabiting the above region is severely reduced. Currently their only known summer–fall feeding ground is off the northeastern coast of Sakhalin Island, Russia. The winter calving and mating grounds are unknown, but may be in coastal waters of the South China Sea.

2. Exploitation and Population Status   The western North Pacific gray whale was considered to be extirpated, or nearly so, during the 20th century but is known to survive today as a tiny remnant population. It is one of the most endangered and little-known whale populations in the world. This group was hunted intensely during the past three centuries, but its decline can be largely attributed to modern commercial whaling off Russia, Korea, and Japan between 1890 and 1960. Preexploitation abundance is unknown. Whaling pressure from the Japanese hand-harpoon fishery was underway by the 16th century. Japanese whalers continued to take grays in the 17th, 18th, and 19th centuries. A branch of the population speculated to have bred in the Seto Inland Sea of Japan was gone by 1900. Beginning in the 1840s, American and European whalers took grays in the Okhotsk Sea and western North Pacific until the early 20th century. The last major whaling period occurred between 1910 and 1933, when about 1400 whales were harvested by Japanese and Korean whalers. The fishery dwindled as the whales ran out, and many authorities thought the population was exterminated. However, catch records for 67 whales taken from the Korean coast from 1948 to 1966 indicated that some western grays remained. From 1967 to 1975, a few were continuously caught. Sightings along the coast of Korea, Japan, China, and Russia after that were rare.

During the 1990s, a small number of gray whales were found feeding during summer and fall in the Okhotsk Sea, mostly along northeastern Sakhalin Island (in Russian waters north of Japan), emphasizing its importance as a feeding ground. The population size of western gray whales was estimated to be about 100 individuals in 1999 and less than 100 in 2001. The World Conservation Union listed this population as critically endangered in 2000. Some believe it is likely that the population is below a critical size sufficient for recovery and may soon become extinct; others suggest that it may be increasing slowly. There are no data from the population's southern range off China, North Korea, South Korea, or Japan, and research is needed. It is generally agreed that the western and eastern gray whales are discrete geographical populations. Recent genetic work has documented pronounced differences between them (implying negligible levels of gene flow) and indicates that the eastern and western gray whales can be genetically differentiated at the population level.

C. Eastern North Pacific Population (or California-Chukchi)

1. Distribution and Migration   From the end of May through September, most of the eastern North Pacific population is on its summer feeding grounds in the shallow, continental shelf waters of the Bering Sea and Chukchi Sea (between Alaska and Siberia), the Beaufort Sea (east to 130°W), and the east Siberian Sea (west to 178°30′E) (Fig. 1). The range reaches its northern limit at 69°N at the edge of the zone of close pack ice (to Wrangel Island in some years). Access to the vast feeding ground is controlled by the seasonal formation, disintegration, and drift of ice (for 5–6 months it is ice covered). Gray whales are widely dispersed throughout much of the region, but the major feeding areas where they occur in greatest ABUNDANCE are the northcentral and northwestern Bering Sea, as well as the western and southwestern Chukchi Sea. Although many of the feeding areas have not been studied, those that have are underlain by dense, infaunal amphipod communities. A highly preferred habitat is the Chirikov Basin (between St. Lawrence Island and Bering Strait). It contains one of the largest and most productive amphipod beds in the world and extends over 40,000 km². Apparently, whales do not forage in the coastal waters on the eastern side of the Bering and Chukchi Seas, which is consistent with the lack of benthic amphipod infauna in that portion of the continental shelf. As a rule, grays are distributed in shallow waters near shore and rarely go beyond 50 km offshore, although they also aggregate on shallow flats a great distance from shore (up to 180 km). The habitat utilized averages 38 to 40 m in depth, and from 1% to 7% ice cover, but can be as great as 30%. The grays are constantly moving; their DISTRIBUTION varies yearly, and even monthly, as a result of constant ranging between feeding areas. Their foraging areas also support the largest number of bottom-feeding marine mammals in the world, including walruses (*Odobenus rosmarus*), bearded seals (*Erignathus barbatus*), and sea otters (*Enhydra lutris*).

The departure of grays from the northern feeding grounds in late summer and fall is cued primarily by shortening photoperiods and ultimately necessitated by advancing ice formation over feeding areas as the Arctic summer draws to a close. Some turn southward as early as mid-August and begin the long migration extending 7500–10,000 km to the breeding grounds, depending on where they are on the feeding range. Starting in September, grays leave the Beaufort and east Siberian Seas and converge into the Chukchi Sea. In October and November, whales move south out of the Chukchi Sea into the Bering Sea. Then, whales travel southeast and exit the Bering Sea via Unimak Pass, Alaska (in the Aleutian Islands), the easternmost prominent corridor between the Bering Sea and the North Pacific Ocean. Some pass through as early as October, others as late as January, but 90% leave from mid-November to late December. Females in late pregnancy go first, followed by other adults and immature females, and then immature males.

Once through Unimak Pass, the whales travel along the coast of North America down to central California. The migration is spread out all along the coast of Canada and the United States. The main body of the population arrives in central California by mid-January and takes about 6 weeks to pass. Beyond Point Conception, California, the majority take a more offshore route across the southern California Bight, through the Channel Islands, and reencounter the coast in northern Baja California. When the last of the southward migrants reach central California in February, they begin to overlap with the first of the northward migrants returning to the feeding grounds.

From January to early March (through May for some cows and calves), most of the population is in the winter assembly area, which extends from about central California (Point Conception) southward along the west coast of the Baja California Peninsula and continues around Cape San Lucas to the southeastern shore of the Gulf of California off Sonora and Sinaloa, Mexico, Historically, a few continued on to Guadalupe Island, whereas others reached the Revillagigedo Islands. Although a few calves are born off California, most are born along the open coast and in the calving lagoons and bays of Baja California and mainland Mexico. The principal calving areas (with 85% of the calves) are Scammon's Lagoon (Laguna Ojo de Liebre), Black Warrior Lagoon (Laguna Guerrero Negro), San Ignacio Lagoon (Laguna San Ignacio), and the Magdalena Bay complex (from Boca de las Animas to Bahía Almejas), all on the outer coast of the Baja California Peninsula. A few calves are also born on the mainland coast of Mexico at Yavaros in Sonora, and Bahía Reforma in Sinaloa.

The breeding lagoons penetrate far into desert regions through narrow entrances marked by lines of whitewater over barrier sand bars. Except for mothers and calves, however, the vast majority of gray whales in Baja California are outside the lagoons and estuaries in Bahía Sebastián Vizcaíno and Bahía de Ballenas and along the coastline, milling, courting, and wandering along the coast. Courting whales in general do not remain in the lagoons for extended periods. Rather, they are constantly passing and repassing into and out of them, and roving to other areas of the winter assembly grounds, leading to a high turnover of courting whales and subadults in the lagoons. The activity of the grays continues unabated day and night. Cows with newborns seek the quiet, inner reaches of the lagoons early in the season, away from harassment by courting whales concentrated in the areas around the lagoon entrances and outside along the outer coast, where much rolling, splashing, and sexual play can be seen. However, cows with calves also move into the ocean (often at night) and then return during darkness in morning hours, and some visit other lagoons within a season. As the consorting adults start their north migration, the mothers and calves essentially abandon the inner lagoon nurseries and occupy the area near the lagoon entrances. Some cows return to the same lagoon in successive years to bear their young, whereas others rear calves in various lagoons in different years.

The spring migration north to the Arctic feeding grounds begins in mid-February. It retraces the route of the fall migration, but is not as concentrated or as fast. Newly pregnant females migrate first, returning soonest to the Arctic to feed in preparation for the high energetic cost of gestation and lactation. They are followed by anestrous females, adult males, and then immatures. Last to migrate are the mothers and calves; they remain in the breeding area 1–1.5 months longer than most grays while the calves strengthen and grow. The first journey to the Arctic is a time of particular danger for the calves, which are occasional targets of killer whales. Cows and calves tend to travel extremely close to shore (90% are within 200 m) and are mostly alone or in pairs. Northbound whales funnel into the Bering Sea through Unimak Pass from March through June.

The north migration culminates in the dispersal of gray whales throughout their Arctic feeding grounds, which is extended in time and closely related to the ice condition (spring melt). The earliest arrivals generally reach St. Lawrence Island by May as ice recedes north or when leads or polynyas (a large area of water in pack ice that remains open throughout the year) are extensive. The main core of the population usually arrives in the Bering Strait by the end of May, where they are distributed along the cracks of ice throughout areas free of pack ice. One part of the population moves southward along the Asiatic coast and another passes through the Bering Strait into the Chukchi Sea where the whales split off in two directions: east toward the Alaska Peninsula and west toward the Chukotka Peninsula. Another smaller route possibly runs toward the Asian coast, along the Aleutian and Commander Islands. By June, grays are common in the northern Bering Sea in ice-free years, and through the Bering Strait into the southern Chukchi Sea during summer and autumn, as well as into the northeastern Chukchi and Beaufort Seas. By August and September, the ice has retreated north an average of 480 km into the Chukchi Sea. Their eastern distribution in the Beaufort Sea is limited by pack ice, as is their western distribution in the Chukchi and east Siberian Seas.

The vast majority of gray whales go to the northern feeding grounds; however, a small but perhaps increasing number do not migrate the entire distance and spend the summer feeding along the coast from Baja California to British Columbia. These whales (called seasonal residents) join the southbound migrants again in early winter. Areas where they have been observed out of season in Mexico include Bahía San Quintin and Cabo San Lorenzo, on the Pacific coast of Baja California, and Bahía de Los Angeles in the Gulf of California.

2. Exploitation and Population Status Native peoples of North America and Siberia have taken gray whales from the eastern North Pacific population for thousands of years, and a few groups continue to hunt them today. The impact of aboriginal whaling was relatively slight, however, compared to the wholesale slaughter of this population by the first American and European commercial whalers to hunt them in the Pacific. In 1846, they discovered the winter breeding grounds of the gray whale, and commercial harvests began soon thereafter in the lagoons of Baja California, then along the migration route, and spread to the feeding grounds in the Bering Sea. From 1846 to 1874, it is estimated that a minimum of 11,390 grays (not including calves) were taken. From its inception, the relentless 19th century whaling, mainly by American whalers, devastated the population. The hunt in the breeding range was largely concentrated on the cows and calves that were easily killed in the crowded lagoons and bays. Because most of the cows carried fetuses, or would have been impregnated, or had calves that were killed or died of starvation, the reproductive capacity of the population was reduced greatly. By 1900, the once abundant population was thought to be nearly extinct, and whaling all but stopped due to lack of quarry. The attention of the whalers turned to other species, allowing the gray (perhaps a few thousand remained) a brief respite before the advent of modern whaling.

With the introduction of floating factory ships on the west coast of North America in 1905, the hunting of gray whales resumed. A few were taken off Baja California and California in 1919, but mostly between 1925 and 1929. About 48 were taken annually in the Bering Sea from 1933 to 1946. All together, at least 1153 were taken from the remnant population, mainly by Norwegian, Russian, Japanese, and United States vessels. Only fear of EXTINCTION led to their official protection in 1946, except for an aboriginal harvest of about 160 whales each year that have been taken legally by Siberian Eskimos, and also a few by Alaskan natives. Since receiving protection, and the end of research harvests of about 316 grays in the 1960s, the population has increased steadily (by 2.5% per year). Based on the most recent survey (in 1997–1998), the eastern North Pacific population was estimated to be 26,600, possibly exceeding the 1846 preexploitation abundance, which most experts place at between 15,000 and 24,000. There have been indications, however, that the population is approaching, or possibly exceeding, its carrying capacity and may have become food limited (large decreases in amphipod biomass have been linked to increased predator pressure from gray whales and to detrimental effects of global warming in the Arctic). If this is correct, we can expect the gray whale population to level off or even decline.

V. Conservation and Management

A. Legal Protection

Gray whales received partial protection from commercial whaling in 1931 under the Convention for the Regulation of Whaling (which was largely ineffectual). The major whaling nations, Japan and the former Soviet Union, were not signatories to this agreement. They continued to take grays until 1946, when they joined 15 other countries and ratified the International Convention for the Regulation of Whaling, which established the INTERNATIONAL WHALING COMMISSION (IWC). The IWC was intended to provide for the proper conservation of whale stocks and thus make possible the orderly development of the whaling industry. Although it failed in its primary mission, one of its first actions was to officially halt commercial whaling for gray whales in 1946, while allowing native subsistence harvests and scientific collections. Nevertheless, there were violations of the agreement by member nations of IWC, as well as pirate whaling (whaling that is practiced by fleets that acted beyond any national jurisdiction). In 2000, Russian scientists revealed that "literally at every sighting" this prohibited species was illegally killed by the former Soviet Union from 1961 to 1979, and whaling statistics were falsified.

Gray whales were listed as endangered under the U.S. Endangered Species Conservation Act in 1969. Further protection was given by the Marine Mammal Protection Act of 1972 and the U.S. Endangered Species Act of 1973. Under the protection afforded by these and other measures, the eastern population of gray whales recovered. In 1994, it was removed from the List of Endangered and Threatened Wildlife and Plants (under the U.S. Endangered Species Act) when the population numbered 21,000. The population was also downlisted in the World Conservation Union's "1996 IUCN Red List of

Threatened Animals," from "endangered" to "lower risk: conservation dependent." However, changes to the listing of the eastern North Pacific gray whale had no bearing on the status of the western North Pacific gray whale population, which is still critically endangered.

There is no allowable commercial take of any gray whales. The IWC quota for the years 1998–2002 of 140 eastern grays annually (with an overall total of 620 in five seasons) is in response to the catch requested by the Russian Federation for its native people. It also includes an annual quota of five whales requested by the United States to satisfy the Makah Indian tribe's tradition of whaling in Washington state. No grays have been allocated to Alaskan native hunters since 1991. Further protection for eastern gray whales was given by Mexico in 1972 when two of the principal breeding lagoons, Black Warrior Lagoon and Scammon's Lagoon, were declared the world's first whale sanctuaries. The same status was extended to San Ignacio Lagoon in 1979. All lie within the Vizcaíno Desert Biosphere Reserve, Mexico's largest refuge, and entrance into the lagoons is regulated. Currently, not only is it illegal to hunt gray whales, it is also illegal to harm, harass, or even cause behavioral changes without special permits.

B. Concerns

Recently there has been a major shift in the physical environment of the Arctic region with wide-ranging effects on the biota, which may have a deleterious impact on gray whales. Over the past 20 to 30 years, there has been a trend of decreasing sea ice concurrent with increased sea surface temperatures due to global warming. Primary productivity has decreased an estimated 30–40% since 1976. Major declines of marine mammal, fish, and bird populations have occurred in the Arctic's Bering Sea. Although the effects of climate warming on grays are unknown, there are indications that the depression in primary production may lead to reductions in the benthic prey communities on which they feed. Increased predation from the growing population of whales themselves also appears to be stressing the amphipod populations. The eastern North Pacific gray whales may be expanding their summer range in search of additional feeding grounds. Moreover, it is hypothesized that the increase in gray whale mortality in 1999 and 2000 included some whales that were starving. A substantial reduction in food resources, through anthropogenic or natural causes, could have long-term effects on the future health, growth, and stability of the gray whale population.

The region of the Okhotsk Sea around Shakalin Island holds large reserves of oil and gas and is currently being developed jointly by Russian, Japanese, and U.S. companies; oil drilling and production activities plus increased shipping and aircraft traffic may cause physical habitat damage or disturb or displace the highly endangered western Pacific population of gray whales on their only known feeding ground.

Gray whales are intimately related to the coastal habitats in which they have evolved, and it is the dynamic nature of coastal regions that has shaped their unique life history and behavior. It is also precisely their coastal habits that place them in direct conflict with humans. It is not enough to stop overharvesting the whales, we must also protect their critical habitat and allow them living space. They cannot avoid exposure to our intensive coastal development, POLLUTION, vessel traffic, military activities, noise, and industrial activities associated with increased exploration and development of continental shelf, oil, and gas resources over virtually their entire range. Additional concerns include disturbance from ecotourism along migration routes and within the calving grounds, entanglement in fishing gear (particularly gill nets), ship strikes, pollution from salt extraction facilities in Mexico's gray whale refuges, and commercial developments in the breeding area of Magdalena Bay, Mexico. In a world where the human population is expected to double in the next century, the pervasive effects of the population explosion will lead to additional regional and global environmental problems and further approbation of living space and resources that the gray whale requires to sustain itself.

See Also the Following Articles

Diving Physiology ■ Endangered Species and Populations ■ Migration and Movement Patterns ■ Reproductive Behaviors

References

Andrews, R. C. (1914). Monographs of the Pacific Cetacea. I. The California gray whale (*Rhachianectes glaucus* Cope). *Mem. Am. Mus. Nat. Hist.* **1**(5), 227–287.

Braham, H., and Donovan, G. P. (eds.) (in press). "Special Issue on the Gray Whale." *The Journal of Cetacean Research and Management.* Special Issue No. 3, Cambridge.

Darling, J. D., Keogh, K. E., and Steeves, T. E. (1998). Gray whale (*Eschrichtius robustus*) habitat utilization and prey species off Vancouver Island, B.C. *Mar. Mamm. Sci.* **14**(4), 692–720.

Dedina, S. (2000). "Saving the Gray Whale: People, Politics, and Conservation in Baja California." University of Arizona Press, Tucson.

Henderson, D. A. (1972). "Men and Whales at Scammon's Lagoon." Dawson's Book Shop, Los Angeles.

Jones, M. L., Swartz, S. L., and Leatherwood, S. (eds.) (1984). "The Gray Whale (*Eschrichtius robustus*)." Academic Press, Orlando.

Le Boeuf, B. J., Pérez-Cortés M., H., Urbán Ramírez, J., Mate, B. R., and Ollervides U. F. (2000). High gray whale mortality and low recruitment in 1999: Potential causes and implications. *J. Cetacean Res. Manage.* **2**(2), 85–99.

Rice, D. W. (1998). "Marine Mammals of the World, Systematics and Distribution." Society for Marine Mammalogy, Special publication No. 4, Lawrence, KS.

Rice, D., and Wolman, A. A. (1971). "Life History and Ecology of the Gray Whale (*Eschrichtius robustus*)." American Society of Mammalogy, Special Publication No. 3.

Rugh, D. J., Muto, M. M., Moore, S. E., and DeMaster, D. P. (1999). Status review of the eastern north Pacific stock of gray whales. U.S. Dep. Commer., NOAA Tech. Memo. NMFS-AFSC-103.

Scammon, C. M. (1874). "The Marine Mammals of the Northwestern Coast of North America." John H. Carmany and Co., San Francisco.

Tomilin, A. G. (1957). Mammals of the U.S.S.R. and adjacent countries. Vol. IX. Cetacea. Akad. Nauk, SSSR, Moscow (transl. by Israel Program for Sci. Transl., Jerusalem, 1967).

Weller, D. W., Würsig, B., Bradford, A. L., Burdin, A. M., Blokhin, S. A., Minakuchi, H., and Brownell, R. L., Jr. (1999). Gray whales off Sakhalin Island, Russia: Seasonal and annual occurrence patterns. *Mar. Mamm. Sci.* **15**(4), 1208–1227.

Group Behavior

ALEJANDRO ACEVEDO-GUTIÉRREZ
University of California, Santa Cruz

Many animals spend part or all of their lives in groups. A group may be viewed as any set of individuals, belonging to the same species, which remain together for a period of time interacting with one another to a distinctly greater degree than with other conspecifics. Thus, the study of group living is the study of social behavior, and marine mammals societies can be remarkably diverse (Fig. 1).

Groups can be classified based both on the amount of time individuals interact with each other and on the benefits that individuals receive. Schools last for periods of minutes to hours whereas groups last months to decades. Aggregations (or non-mutualistic groups) do not provide a larger benefit to individuals than if they were alone, whereas groups (or mutualistic groups) do provide such a benefit to their members. Aggregations are formed because a nonsocial factor, e.g., food, attracts individuals to the same place, and groups are formed because they provide a benefit to their members.

I. Group Living

There appear to be three conditions under which group living will evolve: the benefits to the individual outweigh the costs, the costs outweigh the benefits but strong ecological constraints prevent dispersal from the natal territory, and the area where the group lives can accommodate additional individuals at no cost.

A. Benefits and Costs of Group Living

Group living is explained in terms of benefits to the individual group members—direct fitness, kin selection, reciprocity, and mutualism—and to the groups themselves—group selection. Kin selection and reciprocity are perhaps the arguments employed most frequently to explain benefits of group living. For instance, kin selection explains the generalities of cooperative breeding in mammals and birds and the evolution of cooperation among male chimpanzees. However, these traditional explanations are, in some cases, inadequate and some behaviors are best explained in terms of direct fitness. Delayed direct benefits to the subordinate male explain the occurrence of dual-male courtship displays in long-tailed manakins, and direct benefits from the early detection of danger explain the sentinel behavior of meerkats.

Reduction of predation and increase of foraging efficiency and the number of individuals that can be supported by the available local resources have been typically viewed as important factors shaping group living. However, the reduction of predation and increased foraging efficiency can be accomplished through a myriad of different mechanisms (Table I). In addition, these two benefits are sometimes inadequate to explain group living: in African lions, female-grouping patterns are best explained as facilitating the cooperative defense of cubs against infanticidal males and the defense of territory against other females, not as increasing foraging efficiency. Group living can also impose several costs to individuals (Table II). The magnitude of these costs may be important in shaping group living; it has been suggested that differences in group size between primate and carnivore species may be related to the differences in their costs of locomotion. Marine mammals reduce the costs of locomotion by developing energy-conserving swimming behaviors such as routine transit speeds, wave riding, porpoising, and gliding. It has been hypothesized that this reduced cost of locomotion, coupled with a lack of restriction to a particular territory, has allowed some populations of killer whales (*Orcinus orca*), and possibly long-finned pilot whales (*Globicephala melas*), to develop societies in which females and males remain with their natal group for life (Fig. 2). In this manner, males traveling with their mothers can have large home ranges and thus find potential mates.

Studies have documented novel strategies followed by individuals living in groups, including the complexity of intra- and intergroup interactions. Female African lions cooperate to defend their territory from intruders; however, some individuals consistently lead the approach, whereas other individuals lag behind without being punished by the leaders. One potential explanation for this tolerance is that females need to defend their territories against other groups and their success depends in part on the number of defending females, even if some individuals

Figure 1 *(Top) Blue whales are usually found alone or in small numbers. Photo by B. Tershy. (Bottom) South American sea lions aggregate in large numbers during the breeding season. Photo © A. Acevedo-Gutiérrez.*

TABLE I
Benefits of Group Living

Reduction of predation
    Enhanced ability to detect predators: sensory integration[a]
    Enhanced ability to deter predators, even larger than group members[a]
    Enhanced ability to escape, including predator confusion and coordinated evasion behavior[a]
    Reduced individual probability of being selected as prey
        By associating with conspecifics: dilution effect[a]
        By hiding behind conspecifics: selfish herd[a]

Allocation of time to other activities
    Reduced individual vigilance time
        Because of group vigilance (many eyes)[a]
        Because of decreased individual predation risk[a]
    Increased foraging time for mothers by having babysitters[a]

Enhanced detection and capture of prey
    Foraging in risky, but profitable, areas.
    Finding prey or reducing variation in food intake through cooperative searching: sensory integration[a]
    Following more knowledgeable animals in the group to a food source: information transfer[a]
    Following other species with more specialized senses to a food source[a]
    Joining resources uncovered by others, also known as conspecific attraction, kleptoparasitism, area copying, scrounging, or tolerated theft[a]
    Acquisition of innovative feeding behaviors from another group member
        Social learning through social facilitation (contagion of motivational states)
        Directing attention to particular locations or objects: local enhancement
        Imitation of knowledgeable tutors[a]
        Information sharing and cultural transmission[a]
    Increased diversity and size of prey that is captured
        Due to more individuals foraging[a]
        Due to prey flushed by movements of group members[a]
        Due to individuals with different skills or abilities foraging together: skill pool effect
    Increased food intake as a result of communal foraging[a]
    Lower risks of injury while hunting[a]

Acquisition or defense of resources
    Large groups defend, occupy, or displace small groups from better territories
    Large groups acquire or defend localized food sources, including carcasses, from conspecifics or other species[a]

Improved reproduction
    Caring and protection of offspring[a]
    Learning to be a parent[a]
    Finding mates in isolated or vast areas[a]
    Enhanced reproductive synchrony[a]
    Enhanced survival when there is prevention of dispersal to neighboring territories
    Males benefit from cooperative displays, subdominant males receive the payoff later in time
    Males in large groups gain access to females[a]

Reduction of parasitism
    When number of hosts in a group increases more rapidly than the number of mobile parasites, reduced individual probability of being
        parasitized by associating with conspecifics: dilution effect

Other
    Huddling to survive cold temperatures[a]

[a]Suggested or documented benefits in marine mammals.

never lead the charge. Complex social behaviors have also been reported in marine mammals. In a breeding colony of gray seals (*Halichoerus grypus*) at the island of North Rona, Scotland, a few large males monopolize matings on the breeding beaches; however, females over the years give birth to full siblings not sired by the dominant male. Fathers of the pups are nondominant males that mate with the same females in different seasons. Thus behavioral polygyny and genetic fidelity seem to operate si-

multaneously in this colony. It has been suggested that this strategy of partner fidelity is maintained in the population because it may diminish aggressive interactions between dominant males and thus reduce the pup mortality originated by these clashes. Male Indian Ocean bottlenose dolphins (*Tursiops aduncus*) in Shark Bay, Australia, exhibit very complex levels of alliances that are only matched by humans. Males form strong and stable bonds for over 10 years with one or two other males, and males

TABLE II
Costs of Group Living

Increased predation
 Large groups more attractive to predators
 Larger groups more likely to be detected by a predator: encounter effect

Reduced foraging efficiency
 Increased amount of food needed for group[a]
 Increased energy spent, distance traveled, or area covered to find food for group[a]
 Increased conspicuousness: prey able to detect predators sooner than if predators are alone
 Reduction in food intake due to sharing of prey, scramble competition, scrounging, and individual discrepancies in foraging success
 Reduction in food intake due to interference by the behavior of other individuals[a]
 Reduce ability to learn innovative foraging skills due to scroungers in the group

Increased conflicts for resources due to the presence of more conspecifis or other species
 Individuals from other groups or species following: social parasitism[a]
 Individuals from other groups or species attracted to feeding parties: local enhancement[a]

Reduced reproduction, increased competition for mates or other limited resources
 Individual discrepancies in number or quality of mates obtained[a]
 Extrapair copulations and loss of fertilizations to other members of group[a]
 Increased intraspecific competition for limited resources[a]
 Increased infant mortality[a]
 Increased risk of exploitation of parental care by conspecifics
 Theft of nest material

Increased risk of infection
 Increased contagious parasitism
 Increased disease transmission

[a]Suggested or documented costs in marine mammals.

in these first-order alliances cooperate to form aggressively maintained consortships with individual females. Each first-order alliance forms moderately strong bonds with one or two other alliances (these second-order alliances do not endure for more than a few years) and males cooperate to take or defend females from other alliances. A different strategy is for males to form a

Figure 2 *In certain populations, male and female killer whales remain with their natal groups throughout their lifetime. Photo by B. Tershy.*

large but loose superalliance that competes with the smaller and more stable first-order or second-order alliances. Members of the superalliance split into smaller alliances of pairs and trios that are constantly changing but that are always composed of males from the superalliance. These pairs and trios join conflicts involving members of the superalliance and are always victorious. It is hypothesized that the large size of the superalliance allows individuals to compete with the smaller alliances and that the fluidity of individual associations within the superalliance allows males to maintain affiliative bonds.

B. Female Social Behavior

One previously neglected area of research is the study of female social behavior. Because females and males frequently have different interests, female relationships are important in understanding social evolution independently of the behavior of males. For instance, dominant female chimpanzees have a higher reproductive success than subordinate ones, apparently because they are able to establish and maintain access to good foraging areas, competing in extreme cases as intensely as males. Females may also influence behaviors that affect the interests of males. Female bird song appears to have evolved in part to compete for males; however, this behavior has the potential consequence of preventing polygyny by deterring rival females.

The study of females is also essential to understand group living in marine mammals. Captive female common bottlenose dolphins (*Tursiops truncatus*) maintain dominance hierarchies and also compete aggressively against each other. However, unlike chimpanzees, it is unknown if female dominance hierarchies in free-ranging dolphins translate into differences in reproductive success. The preference and fidelity of female gray seals at North Rona toward nondominant males undermine the polygynous strategy of dominant males and result in a different mating system from that inferred by behavioral observations. The large number of females in colonies of certain pinnipeds, such as elephant seals (*Mirounga* spp.), has permitted the existence of alloparenting and the appearance of a distinct suckling strategy by calves: milk stealing. Male and female sperm whales (*Physeter macrocephalus*) have different grouping strategies: females aggregate in complex groups whereas adult males are less social. The function of the female groups is to provide care for calves that are too young to follow their mothers during their deep foraging dives (Fig. 3). It has been suggested that alloparenting in sperm whales reduces the period in which the calf is unaccompanied and thus provides protection from predators and also perhaps provides communal nursing. Thus key features of the sperm whale society are explained solely by the behavior of females.

II. Social Behavior of Marine Mammals

There are several differences between terrestrial and marine environments that have allowed the evolution of distinctive strategies in marine mammals. Drag, heat loss, and density of the water generate differences in scaling and costs of locomotion, allowing many marine mammals to have large body sizes and large home ranges. Sound is the form of energy that best propagates in water and, not surprisingly, marine mammals employ it for social COMMUNICATION and many species employ ECHOLOCATION to navigate. Marine mammals must find food that is for the most part dispersed and patchy, thus they appear to have no territories outside of the breeding season. Due to the global effects of the atmosphere and the ocean in the marine environment, marine mammals are affected by both global and local processes as exemplified by the impact of El Niño–Southern Oscillation events on different populations.

A. General Strategies

Sirenians, sea otters (*Enhydra lutris*), and polar bears (*Ursus maritimus*) are solitary animals that have few social interactions beyond mating and mother/offspring pairs. The time that these pairs remain together is 1 to 1.5 years in sirenians, 5 to 7 months in sea otters, 2.5 years in polar bears. When a female becomes receptive, sirenians form aggregations that have as many as 17 males physically competing for access to the female or defending display territories. During the breeding season, male sea otters establish territories that include the areas occupied by several females, whereas male polar bears mate with only one partner because females have a dispersed distribution.

Most pinnipeds aggregate in colonies during the breeding season. A major factor influencing the size of these colonies is the distribution of habitat available for parturition. All pinnipeds give birth out of the water and thus the areas favored for parturition are oceanic islands, ice, or isolated mainland regions not easily accessible to terrestrial predators. When available space is limited, females become densely aggregated in large colonies that favor mating systems in which males defend either aggregations of females or areas occupied by females, or aggregate and display before aggregations of females (Fig. 4). When parturition space is dispersed, females are isolated, males usually have access to only one female, and no colonies are formed. The strongest association found in pinnipeds is formed by a mother and her offspring and lasts from less than 1 week to almost 3 years, depending on the species. Pinnipeds haul out together outside of the breeding season. Although this nonreproductive social behavior is poorly known, evidence shows that it increases vigilance for predators in harbor seals (*Phoca vitulina*). It is believed that hauling out together also allows pinnipeds to rest, avoid predators, molt, or warm themselves. For instance, walruses (*Odobenus rosmarus*) in large numbers may decrease the rate of body heat loss, particularly in calves, when on land or ice.

Figure 3 *Alloparenting behavior apparently allows sperm whale mothers to make deep foraging dives. Photo by B. Tershy.*

Figure 4 *Pinniped males are able to monopolize access to clustered females during the breeding season. Photo © A. Acevedo-Gutiérrez.*

The complexity of cetacean societies appears to be related to the amount of time invested in lactating and in rearing the calf after weaning. Baleen whales are found in schools of varying size, from single individuals to more than 20 whales. Pairs of mothers and their offspring form stable associations that last less than 1 year. It is currently unclear if long-term associations exist among adult whales. Most females give birth every 2 to 3 years and have the potential to produce more than 20 calves throughout their lifetime. Schools of baleen whales have been observed in both feeding and breeding grounds. For instance, feeding humpback whales (*Megaptera novaeangliae*) forage alone, in aggregations, or as a group, depending on prey type, whereas aggregations of breeding males display acoustically or compete directly for access to females. Odontocetes are the most social marine mammals and have different types of societies as suggested by the large variation in school size among species (Table III). Short-term associations between adults characterize several porpoises (Phocoenidae). Associations between mothers and their offspring last 8 to 12 months. Females breed every 1 or 2 years and may give birth to 15 calves or more during their life span. In medium-sized delphinids, such as bottlenose dolphins, associations between adults are varied: they last a short amount of time in some individuals and several years in others; long-term associations appear to be common between males in certain populations. Calves remain with their mothers 2 to 11 years (Fig. 5). Females give birth at least every 3 years and may produce close to 10 calves throughout their lifetime. Baird's beaked whales (*Berardius bairdii*) apparently employ a novel social strategy. Males live longer than females and thus there is an excess of mature males over females. It has been hypothesized that these traits indicate a society in which males provide significant parental care by rearing weaned calves: protecting them from predators and teaching them foraging skills. In the case of the sperm whale and large-sized delphinids (pilot whales, *Globicephala* spp., and some populations of killer whales), females appear to spend their entire lives within their natal group, forming strong matrilineal societies. Females usually breed every 3 to 6 years and may give birth to about 5 calves throughout their lifetime, more in the case of long-finned pilot whales. Females may live over 20 years past their reproductive years. It has been suggested that this strategy allows old females to transmit and store cultural information and provide alloparental behavior. In the case of short-finned pilot whales (*G. macrorhynchus*), it is possible that nonreproductive females even provide alloparental nursing. Male killer whales in some populations, and perhaps male long-finned pilot whales, remain in their natal group for life but mate with females from other groups when they meet. Male sperm whales, and perhaps male short-finned pilot whales, leave their nursing group when they reach puberty; however, the former, after they have reached their late 20s, roam among nursery groups looking for mates, whereas the latter appear to join other nursery groups. It has been suggested that adult male short-finned pilot whales remain in the nursery group that they join and engage in few clashes with other males because they are able to engage in nonreproductive mating with old

TABLE III
School Sizes of Odontocetes

Species[a]	Average school size	Maximum school size
Phocoenoides dalli (7)	2.3–7.4	5–500
Neophocaena phocaenoides (1)	2.0	13
Phocoena phocoena (6)	1.2–5.7	15–100
P. sinus (1)	1.9	7
P. spinipinnis (1)	4.5	10
Cephalorhynchus commersonii (1)	6.9	110
Lissodelphis borealis (2)	9.9–110.2	60–2000
Delphinus sp. (4)	46.8–385.9	650–4000
Grampus griseus (9)	6.3–63	20–500
Lagenodelphis hosei (1)	394.9	1500
Lagenorhynchus acutus (1)	53.2	?
L. obliquidens (3)	10.8–88	50–6000
L. obscurus (3)	9.5–86	24–1000
Sotalia fluviatilis (1)	2.5	10
Sousa plumbea (1)	6.6	25
Stenella attenuata (5)	26.0–360.0	148–2400
S. clymene (1)	41.6	100
S. coeruleoalba (3)	60.9–302	500–2136
S. frontalis (2)	6.0–10.0	50–65
S. longirostris (4)	37.6–134.1	95–1700
Steno bredanensis (2)	14.7–40.0	53
Tursiops aduncus (2)	10.2–140.3	80–1000
T. truncatus (29)	3.1–92.0	18–5000
Feresa attenuata (1)	27.9	70
Globicephala macrorhynchus (2)	12.2–41.1	33–230
G. melas (3)	9.3–84.5	220
Orcinus orca (10)	2.6–12.0	5–100
Pseudorca crassidens (1)	18	89
Peponocephala electra (2)	135.3–199.1	400
Delphinapterus leucas (3)	3.8–32.9	100–500
Monodon monoceros (1)	3	50
Inia geoffrensis (2)	1.6–2.0	8–10
Lipotes vexillifer (1)	3.4	10
Platanista gangetica (1)	1.4	3
Kogia sima (1)	1.7	?
Physeter macrocephalus (6)	3.7–22.1	17–120
Berardius bairdii (1)	7.2	25
Ziphius cavirostris (1)	2.3	7

[a]Values in parentheses indicate number of studies.

Figure 5 *Bottlenose dolphin calves remain with their mothers for up to 11 years. Photo © A. Acevedo-Gutiérrez.*

females, as apparently occurs in bonobos. A recent hypothesis that remains to be tested suggests that the cultural transmission of behaviors, i.e., learned behaviors passed onto family members, is being conserved within matrilineal groups of odontocetes, affecting the course of genetic evolution. This modification of the course of genetic evolution through culture has thus far been documented only in humans.

B. Foraging

Increased foraging efficiency is considered to be one of the principal roles of group living in cetaceans. However, thus far transient killer whales provide the only clear example supporting the argument that marine mammals live in groups because of foraging benefits. Transient killer whales live in the Pacific

Northwest and prey on harbor seals and other small marine mammals. Individuals maximize their caloric intake if they feed in groups of three, which is the size of the group in which they live. The small size of these groups is apparently maintained by the departure of all female offspring and all but one male offspring from their natal group.

Two benefits of group living through foraging efficiency are the ability to search for prey as a group and to forage communally (Fig. 6). Searching for prey as a group allows individuals to combine their sensory efforts, which should be an advantage when prey has a dispersed and patchy DISTRIBUTION. Communal foraging allows individuals to combine efforts to pursue and capture prey. This behavior has been reported in dolphins, baleen whales, including fin whales (*Balaenoptera physalus*) and bowhead whales (*Balaena mysticetus*), and pinnipeds, such as fur seals and sea lions. However, in some instances it is unclear whether individuals combine efforts to pursue and capture prey or merely aggregate in an area where food is concentrated. A particular type of communal foraging behavior, termed prey herding, has been observed when feeding on shoaling fish. Individuals encircle shoals of fish and thus create a tight, motionless ball of prey from which they can grab individual fish; in some cases, individuals release bubbles to further tighten the ball of prey. This herding of prey has been well described in humpback whales, dusky dolphins (*Lagenorhynchus obscurus*), and killer whales. However, it has also been reported in other species, such as bottlenose dolphins, common dolphins (*Delphinus* spp.), Clymene dolphins (*Stenella clymene*), and Atlantic spotted dolphins (*S. frontalis*). It is difficult to document this behavior, and only one study has quantified the success of cetaceans in herding prey. This study noted that common bottlenose dolphins off Costa Rica seldom herd prey into a ball when it is scattered, rather they pursue the individual fish. Perhaps evasive maneuvers of prey or interference by hunters from other species feeding on the same prey prevents dolphins from herding.

Figure 6 *Communal foraging allows dolphins to combine pursuing efforts. Photo © A. Acevedo-Gutiérrez.*

C. Predation

The reduction of PREDATION is considered to be another principal function of group living in cetaceans. Certain shark species and some large delphinids attack cetaceans, and calves suffer higher mortality than adults. However, pinnipeds apparently also form groups in response to predation. Walruses sometimes form groups lasting throughout the year in the water and on haul-out sites. It has been suggested that this may be a female strategy for pup defense against predation by polar bears.

Thus far no conclusive evidence shows that group living in cetaceans is driven because of benefits in reduction of predation, although it has been suggested that this could be the case in sperm whales. Nonetheless, group living may provide several benefits to reduce predation. Groups are able to mob and chase away predators, as has been observed in humpbacked dolphins (*Sousa* spp.) when attacked by a shark. It is believed that other dolphins also employ this antipredatory strategy. Sperm whales, and perhaps humpback whales, employ the marguerite formation, in which adults surround young individuals by having their heads toward the center (horizontal formation) or toward the surface (vertical formation). Adults have their flukes toward the periphery and employ them to slap at predators, which in the majority of observations have been killer whales.

Group living appears to be related to food and predation in terrestrial and marine mammals. Thus it has been argued that the variation of group sizes among dolphin species is related to food availability, to prey habitat, or to the need to defend against predators. Data have been compiled on the average school size from 24 species of the family Delphinidae (Table III). Because definitions of school vary among researchers, the author has attempted to make values comparable by selecting only studies with at least 30 observations throughout a season and defined schools as the number of individuals engaged in similar activities regardless of distance between them. The author averaged the values from species belonging to the same genus and related them to crude measures of predation pressure and prey habitat, measures obtained from the literature. Results indicate that regardless of the body weight of the species, average school sizes are larger when predation pressure is high rather than low and when prey is found in open rather than in enclosed waters (Fig. 7). School sizes are largest when both pressure from predators is high and prey lives at depth in oceanic waters. Thus it appears that the average school size of dolphins is related both to the predation they experience and to the habitat where their prey lives. In turn, reproductive success might also be related to these two environmental factors. The reproductive success of female bottlenose dolphins is highest in shallow waters, either because calves and their mothers are able to detect and avoid predators or because prey density is highest.

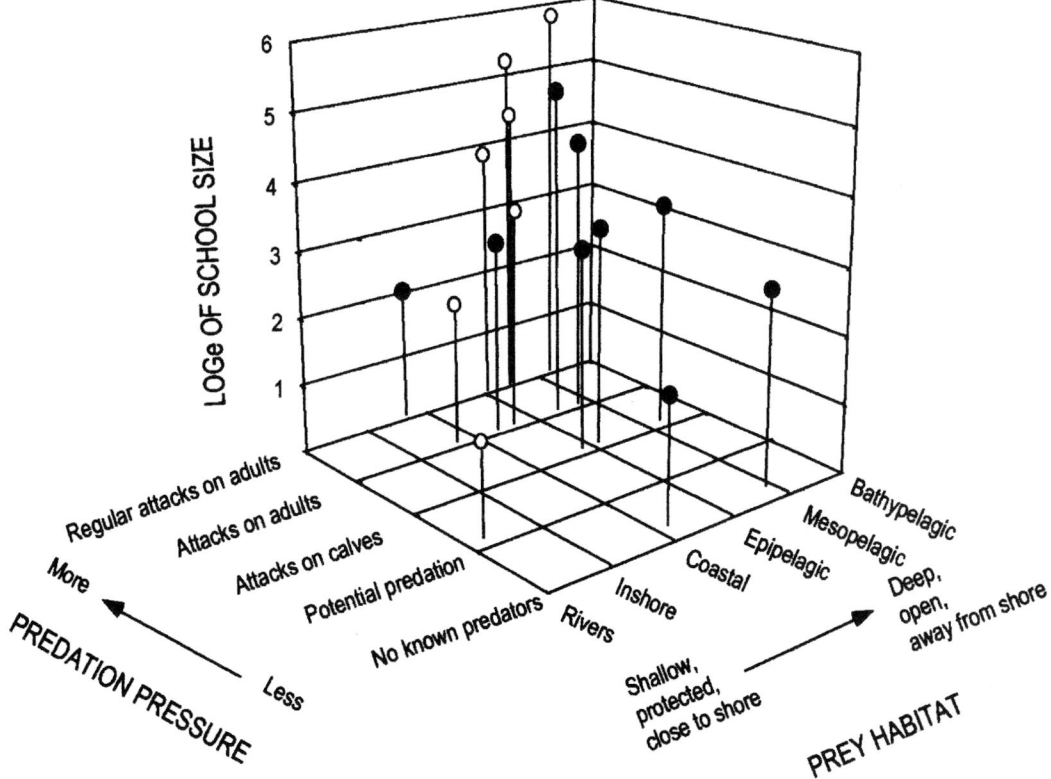

Figure 7 *Relationship among predation pressure, prey habitat, and average school size of 16 genera of the family Delphinidae. Open circles indicate small species (females weigh less than 150 kg) and solid circles indicate large species (females weigh more than 150 kg).*

III. Conclusion

Group living involves benefits and costs, and the resultant society represents a balance between the different interests of all group members. The aquatic environment has allowed marine mammals to pursue complex and sometimes unique social strategies. At the same time, the basic needs of finding food, ensuring reproduction, and evading predators are also found in terrestrial environments. This convergence provides interesting parallels between the social strategies of marine mammals and those of terrestrial mammals: chimpanzees and bottlenose dolphins, elephants and sperm whales. Although many questions about the group behavior of marine mammals remain unanswered, long-term studies and research on poorly known species will provide a more profound understanding of their societies.

See Also the Following Articles

Feeding Strategies and Tactics ▪ Mating Systems ▪ Predation on Marine Mammals ▪ Sociobiology

References

Alexander, R. D. (1974). The evolution of social behavior. *Annu. Rev. Ecol. System.* **5**, 325–383.

Berta, A., and Sumich, J. L. (1999). "Marine Mammals: Evolutionary Biology." Academic Press, San Diego.

Connor, R. C., Heithaus, M. R., and Barre, L. M. (1999). Superalliance of bottlenose dolphins. *Nature* **397**, 571–572.

Connor, R. C., Mann, J., Tyack, P. L., and Whitehead, H. (1998). Social evolution in toothed whales. *Trends Ecol. Evol.* **13**, 228–232.

Evans, P. G. H. (1987). "The Natural History of Whales and Dolphins." Facts on File, New York.

Kasuya, T. (1995). Overview of cetacean life histories: An essay in their evolution. *In* "Developments in Marine Biology" (A. S. Blix, L. Walløe, and Ø. Ultang, eds.), Vol. 4, pp. 481–497. Elsevier Science, Amsterdam.

Krebs, J. R., and Davies, N. B. (eds.) (1978–1997). "Behavioural Ecology: An Evolutionary Approach." 1st-4th Eds. Blackwell, Oxford.

Mann, J., Connor, R. C., Tyack, P. L., and Whitehead, H. (eds.) (2000). "Cetacean Societies: Field Studies of Dolphins and Whales." Univ. of Chicago Press, Chicago.

Renouf, D. (ed.) (1991). "The Behaviour of Pinnipeds." Chapman & Hall, London.

Reynolds, J. E., and Rommel, S. A. (1999). "Biology of Marine Mammals." Smithsonian Institution Press, Washington, DC.

Ridgway, S. H., and Harrison, R. H. (1981–1999). "Handbook of Marine Mammals," Vols. 1–6. Academic Press, San Diego.

Riedman, M. (1990). "The Pinnipeds: Seals, Sea Lions, and Walruses." Univ. of California Press, Berkeley.

Whitehead, H. (1998). Cultural selection and genetic diversity in matrilineal whales. *Science* **282**, 1708–1711.

Habitat Pressures

PETER G. H. EVANS
University of Oxford, United Kingdom

Like other animals, marine mammals may have preferred locations in which they spend the majority of time or where they engage in particular important life history activities such as giving birth, calf rearing, or feeding. The array of physical and oceanographic features that typify those locations forms the habitat of a species or local population. Often these are difficult to define. An ice-breeding seal clearly depends on pack ice upon which to give birth and that constitutes its breeding habitat, and a gray whale may seek out a sheltered tropical lagoon to calve, but for a large open-ocean baleen whale such as the blue whale, identifying its habitat requirements for breeding can be a difficult task. The same applies to feeding habitats: manatees and dugongs, for example, require specific habitats such as shallow seagrass beds for feeding, but oceanic dolphins may range the high seas in pursuit of shoaling fish, making it hard to identify whether they have specific habitat requirements. Human activities impinge upon the lives of marine mammals if they damage or destroy those habitats that may be important to them. Our knowledge of habitat pressures facing marine mammals is therefore limited to particular species and especially to locations nearshore where animals have been studied more intensively and their ecological requirements better defined.

Habitats formed by eddies, thermoclines, and fronts may shift from one locality to another during the life span of a marine mammal, leading to shifts in their geographic distributions. Habitats determined by geomorphological features, such as depth, topography, and available haul-out or den sites [in the case of pinnipeds and polar bears (*Ursus maritimus*), respectively], are relatively stable over time in relation to location. Strong site fidelity may lead a population to have difficulty adjusting to changes in local food availability.

Habitat pressures on marine mammals from anthropogenic influences may be grouped into five categories: (1) physical damage to their environment: a river or seabed and its constituent communities; (2) contamination from chemical pollutants; (3) direct removal of important prey through fisheries; (4) disturbance from human activities either by the introduction of sound into the environment or through ship strikes; and (5) physical and oceanographic effects from global climate change.

I. Physical Damage

Human population pressures frequently lead to direct changes to coastal and riverine environments. Estuaries are turned into industrial harbors, wetlands are drained for agricultural purposes or for tourism, and coastal waters are modified often irreversibly by dredging of the seabed and input of a wide variety of pollutants. Some of the most obvious detrimental changes to a habitat come from the alteration of rivers inhabited by particular dolphin species. Water is often taken out of rivers for other uses, such as for drinking, flood control, or irrigation agriculture. In Pakistan, for example, most of the annual flow of the Indus River is diverted into canals, and this, along with dam construction, has resulted in the Indus river dolphin (*Platanista gangetica minor*) losing probably at least half of its historical range. Dams modify water flow and affect the sedimentation of rivers; they also block traditional movement patterns of marine mammals, which can lead to population fragmentation. The construction of large dams (such as the Ghezouba Dam and the Three Gorges Dam) along the Yangtze Kiang river system thus poses serious threats to the already endangered Chinese river dolphin or baiji (*Lipotes vexillifer*), which now numbers only a few hundred (or fewer) individuals. It may also restrict movements of more widespread species such as the Amazonian manatee in Brazil.

On land, one of the greatest habitat pressures leading to mass extinctions of fauna and flora is that of deforestation, particularly in the tropics. In the 1980s, Latin American countries are estimated to have eliminated 7.4 million hectares of tropical forests annually, with Brazil sustaining the greatest annual loss with 3.2 million hectares per year. This deforestation directly affects the freshwater habitats of the boto or Amazon river dolphin (*Inia geoffrensis*), as well as the Amazonian manatee (*Trichechus inunguis*).

After centuries of direct exploitation, pinnipeds have largely sought sites remote from human activities to give birth to their pups. They therefore are less likely to experience direct physical damage to those breeding habitats.

II. Chemical Pollution

Nearshore environments in particular are exposed to a potential wide range of pollutants as a result of industrial and agricultural activities. Those pollutants may concentrate in the food web and either degrade the habitat by removing important prey populations or cause health deficiencies in the local populations of marine mammal species. Although high levels of potentially damaging pollutants have been detected frequently in marine mammals, particularly seals and coastal small cetaceans inhabiting nearshore environments, direct causal links with health status have rarely been demonstrated. Baltic ringed (*Pusa hispida*) and gray (*Halichorus grypus*) seals during the 1970s had lesions of the reproductive system attributed to high PCB and DDT levels in their tissues. By the late 1980s–1990s, as levels in those pollutants declined, the proportion with lesions had declined substantially, along with an increase in their pregnancy rate. In an experimental study with harbor seals (*Phoca vitulina*), females fed with fish from the heavily polluted Dutch Wadden Sea had poorer reproductive success than those fed less contaminated fish from the North Atlantic. The effects were attributed to PCBs or their metabolites, and seals with the highest PCB intake were found to have reduced blood levels of thyroid hormones and vitamin A, both of which are known to be important in reproduction, including spermatogenesis.

Belugas (*Delphinapterus leucas*) in the highly polluted St. Lawrence estuary in North America had a high prevalence of tumors that had been attributed to carcinogenic compounds, such as polycyclic aromatic hydrocarbons, and other toxic compounds, such as PCBs. These were thought to account for low reproductive success in this population. However, although both sets of compounds occurred at high levels in this population, it remains difficult to demonstrate a direct link, and the population in fact appears to have increased in the 20 years since hunting ceased in 1979.

Harbor porpoises (*Phocoena phocoena*) inhabiting the continental shelf around the British Isles on postmortem are significantly more likely to be diseased when they have high hydrocarbon concentrations. Mass mortalities of striped dolphins (*Stenella coeruleoalba*) in the Mediterranean, common bottlenose dolphins (*Tursiops truncatus*) in the eastern United States, harbor seals in the North and Baltic Seas, and Baikal seals (*Pusa sibirica*) in Lake Baikal all showed significantly high concentrations of PCBs, which was thought to have reduced resistance to disease, thus making these populations more susceptible to virus infection.

Despite examples like these of apparent links between contamination and health status, the biological significance and the nature of effects remain uncertain, and it has been impossible to demonstrate conclusively that demographic changes to a population can be attributed to pollution. The only exceptions are where POLLUTION can be shown to lead directly to mortality. After the *Exxon Valdez* went aground in Price William Sound, Alaska, in 1989, releasing large volumes of crude oil, several thousand sea otters (*Enhydra lutris*) and about 300 harbor seals died as a result of their oiled pelts losing vital insulation properties.

III. Competition with Fisheries

Habitats comprise animal and plant communities in an often complex web of interaction. When one or more members of the community are removed in large numbers this can have repercussions throughout the food web, altering PREDATOR–PREY RELATIONSHIPS and competition for resources. Following the intense exploitation of large baleen whales in the Southern Ocean during the first half of the 20th century, it was estimated that their overall biomass was reduced from 43 million tons to about 6.6 million tons, and that this made available a "surplus" of about 153 million tons of KRILL. These massive changes to the food web of the Southern Ocean had important effects on the remaining members, with individual whales growing faster, reaching sexual maturity at an earlier age, and exhibiting increased pregnancy rates. Similar changes in life history parameters were seen in other marine species, such as the Antarctic crabeater seal (*Lobodon carcinophaga*) and several seabird species.

During the 20th century, fisheries around the world intensified to such an extent that major changes in fish stocks have been observed for many species. Rarely, however, is it possible to show that prey depletion has reduced the numbers of a particular marine mammal species. Many marine mammals have catholic diets and appear to respond by switching prey. The relative ease of capture and nutritive contents of different prey species may vary, but it has scarcely ever been possible to demonstrate that these have affected reproductive or survival rates, and hence led to a decline in that population. More often than not, the species appears to respond by shifting its DISTRIBUTION.

On both sides of the North Atlantic, fishing activities have reduced the stocks of Atlantic mackerel and herring markedly, resulting in other fish (upon which they prey), such as sand lance, sprat, and gadoid species, becoming very abundant locally. Not only did some cetacean species such as harbor porpoise and humpback whale (*Megaptera novaeangliae*) switch their diet to include those prey in greater amounts, but some also showed geographic shifts in distribution. Gray seals, feeding largely on sand lance, increased in number in the North Sea at around 7% per year, whereas right whales (*Eubalaena glacialis*), feeding largely on plankton (the prey of sand lance) in the northwest Atlantic, showed local declines. When some local sand lance and sprat populations crashed a few years later, further changes were witnessed. In the Gulf of Maine, for example, fish-eating humpback and fin whales (*Balaenoptera physalus*) were replaced by plankton-eating right and sei (*B. borealis*) whales, harbor porpoises moved nearer to shore, and Atlantic white-sided dolphins (*Lagenorhynchus acutus*) became abundant and white-beaked dolphins (*L. albirostris*) rare.

In the Bering Sea and Gulf of Alaska, substantial declines in the numbers of Steller sea lions (*Eumetopias jubatus*), harbor seals, and northern fur seals (*Callorhinus ursinus*), as well as several species of fish-eating birds, have occurred since the 1970s. Although other factors may also be involved, most of these declines have been attributed to a decline in food availability resulting from the development of the walleye pollock fishery, a key prey species for many of these marine mammals following the demise of local herring stocks. Similarly, the collapse of productivity of the Barents Sea ecosystem, brought on partly from excessive fishing mortality, has had far-reaching effects on a range of species from seabirds to marine mammals.

IV. Disturbance

Sounds are introduced into marine and freshwater environments from a wide variety of sources: motor-powered vessel traffic of various sizes; active sonar for object detection, including fish finding; seismic exploration and subsequent drilling and production for oil and gas; explosions from military exercises and ocean science studies; and marine dredging and construction. Most of the sounds produced are concentrated between 10 and 500 Hz frequency. However, speedcraft of various types generate noise mainly between 2 and 20 kHz by cavitation of the propellor, and sidescan and military sonar generate sounds between 2 and 500 kHz.

Among cetaceans, baleen whales appear to have rather different hearing sensitivities to those of toothed whales and dolphins. The former are thought to be most sensitive at low frequencies below 5 kHz, and the latter above 10 kHz. Thus baleen whales are likely to be most vulnerable to large vessels, oil and gas activities, marine dredging, and construction, whereas toothed whales and dolphins may be more susceptible to recreational speedboats and most forms of active sonar.

Changes in behavior (e.g., movement away from the sound, increased dive times, clustering behavior) are often recorded in the vicinity of loud sounds. Few experimental studies have been conducted to test the nature and duration of negative responses. One such study in relation to low-frequency regular ATOC (acoustic thermometry of ocean climate project) sound pulses was conducted west of California. Aerial surveys showed no significant differences in numbers of marine mammals of any species between control and experimental surveys, but humpback and sperm whales (*Physeter macrocephalus*) were on average further from the sound source during the experimental periods. Although many other studies have reported negative reactions, there is very little information concerning the long-term impact of sound disturbance. In Hawaii, humpback whale mothers with their calves are thought to have shifted their distribution offshore in response to the high volume of recreational traffic. Whale and seal watching itself can impose pressures on marine mammals, disturbing seals from haul-out or breeding sites, and whales (and dolphins) from favored feeding areas.

In addition to those indirect effects where sound disturbance may interfere with or frighten marine mammals, there is some evidence that loud sounds can cause physical damage. Temporary or permanent shifts in hearing thresholds may occur that could affect auditory acuity, and postmortem examination of humpback whales found dead in the vicinity of drilling operations has revealed ear damage.

Powered vessels pose an obvious threat to marine and freshwater mammals through direct damage. Collisions have been reported in a wide variety of species, and in some, such as the Florida manatee (*Trichechus manatus latirostris*) and the North Atlantic right whale, they are regarded as the major threat to their survival. With the advent of high-speed ferries in many parts of the world, ship strikes are being reported with increasing frequency, affecting especially the slower swimming species such as sperm and pilot whales.

V. Climate Change

As a result of emissions by humans of substances that deplete the ozone layer, our increasing use of hydrocarbons for energy and fuel, and large-scale deforestation and desertification, the world is experiencing climate change such that it is predicted that, in the next hundred years, temperatures will rise by 1.0–3.5°C and the overall sea level will rise by anywhere from 15 to 95 cm. Obvious consequences will be the melting of polar ice, drowning of coastal plains, and changes to shallow seas. Other less direct implications include an increase in the frequency and velocity of storms, and more extreme seasonal fluctuations in local climate (including, for example, El Niño Southern Oscillation events). Shifts in areas of primary productivity may lead to distributional changes for many marine mammal species, but some, such as the polar bear (*Ursus maritimus*), land-breeding pinnipeds, and coastal cetaceans and sirenians, may find it difficult to adjust to the loss of important feeding or breeding habitats. Already there is concern that less stable ice in some parts of the Arctic has reduced the availability of ringed seals (*Pusa hispida*) to polar bears, thus reducing the breeding success of the bears, which in those areas depend on this species for food.

During recent El Niño events, there has been reproductive failure in many seabird populations and some colonies of fur seals. During the 1982 El Niño, for example, all Galapagos fur seal (*Arctocephalus galapagoensis*) females lost their pups due to starvation. However, many pelagic toothed whales and dolphins, being less tied to a particular locality, simply shifted their distributions: short-finned pilot whales (*Globicephala macrorhynchus*), for example, left southern Californian waters following the departure of a species of squid, their main prey. Such changes can affect other members of the ecosystem. When the squid returned some years later, the temporarily vacant niche became occupied by another cetacean species, the Risso's dolphin (*Grampus griseus*).

Despite the many pressures on their habitats, marine mammals appear to be remarkably resilient, often living in highly modified coastal and riverine environments. Of course, because demographic changes may be slow and difficult to detect, we rarely know whether these are having negative effects. In the case of small local populations of endangered species such as the North Atlantic right whale, vaquita (*Phocoena sinus*), various river dolphins, monk seals (*Monachus* spp.), and manatee populations, the dangers of habitat pressures are all to obvious. However, even for other species, a precautionary approach would be prudent, and there is hope for the establishment of protective areas where human activities can be zoned.

References

Bjørge, A. (2000). How persistent are marine mammal habitats in an ocean of variability? Habitat use, home range and site fidelity in marine mammals. *In* "Marine Mammals: Biology and Conservation" (P. G. H. Evans and J. A. Raga, eds.). Plenum Press/Kluwer Academic, London.

Evans, P. G. H. (1996). Human disturbance of cetaceans. *In* "The Exploitation of Mammals: Principles and Problems Underlying Their Sustainable Use" (N. Dunstone and V. Taylor, eds.), pp. 376–394. Cambridge Univ. Press, Cambridge.

International Whaling Commission (IWC) (1997). Report of the IWC workshop on climate change and cetaceans. *Rep. Int. Whal. Comn.* **47**, 293–313.

Kenney, R. D., Payne, P. M., Heinemann, D. W., and Winn, H. E. (1996). Shifts in northeast shelf cetacean distributions relative to trends in Gulf of Maine/Georges Bank finfish abundance. *In* "The Northeast Shelf Ecosystem: Assessment, Sustainability, and Management" (K. Sherman, N. A. Jaworski, and T. J. Smayda, eds.), pp. 169–196. Blackwell, Oxford.

Loughlin, T. R. (ed.) (1994). "Marine Mammals and the Exxon Valdez." Academic Press, San Diego.

O'Shea, T. J. (1999). Environmental contaminated and marine mammals. *In* "Biology of Marine Mammals" (J. E. Reynolds III and S. A. Rommel, eds.), pp. 485–563. Smithsonian Institution Press, Washington, DC.

Reeves, R. R., Chaudhry, A. A., and Khalid, U. (1991). Competing for water on the Indus plain: Is there a future for Pakistan's river dolphins? *Environ. Con.* **18**, 341–350.

Reeves, R. R., and Leatherwood, S. (1994). Dams and river dolphins: Can they co-exist? *Ambio* **23**, 172–175.

Reeves, R. R., and Smith, B. D. (1999). Interrupted migrations and dispersal of river dolphins: Some ecological effects of riverine development. *CMS Tech. Ser. Pub.* **2**, 9–18.

Reijnders, P. J. H., Aguilar, A., and Donovan, G. P. (eds.) (1999). Chemical pollutants and Cetaceans. *J. Cet. Res. Manage.* Special issue 1.

Reynolds, J. E., III (1999). Efforts to conserve manatees. *In* "Conservation and Management of Marine Mammals" (J. R. Twiss, Jr., and R. R. Reeves, eds.), pp. 267–295. Smithsonian Institution Press, Washington, DC.

Richardson, W. J., Greene, C. R., Jr., Malme, C. I., and Thomson, D. H. (eds.) (1995). "Marine Mammals and Noise." Academic Press, San Diego.

Rosas, F. C. W. (1994). Biology, conservation and status of the Amazonian manatee *Trichechus inunguis*. *Mamm. Rev.* **24**(2), 49–59.

Trillmich, F., and Dellinger, T. (1991). The effects of El Niño on Galapagos pinnipeds. *In* "Pinnipeds and El Niño: Responses to Environmental Stress" (F. Trillmich and K. A. Ono, eds.), pp. 66–74. Springer Verlag, Berlin.

Tynan, C. T., and DeMaster D. P. (1997). Observations and predictions of Arctic climate change: Potential effects on marine mammals. *Arctic* **50**(4), 308–322.

Würsig, B., and Evans, P. G. H. (2000). Cetaceans and humans: Influences of noise. *In* "Marine Mammals: Biology and Conservation" (P. G. H. Evans and J. A. Raga, eds.). Plenum Press/Kluwer Academic, London.

Hair and Fur

PAMELA K. YOCHEM AND BRENT S. STEWART
*Hubbs-SeaWorld Research Institute,
San Diego, California*

I. Structure and Function

The presence of hair is one of the characteristics that distinguishes mammals from other vertebrates. Hair consists of keratinized epidermal cells, formed in hair follicles located in the dermal layer of the skin. Pinniped and sea otter (*Enhydra lutris*) hairs are flattened in cross section rather than round as in other carnivores. This is evidently an adaptation for enhancing streamlining of the body and reducing drag during swimming. Pinnipeds and sea otters have diffuse smooth muscle in their dermis, but they lack true arrector pili muscles. Pinnipeds, sea otters, and polar bears (*Ursus maritimus*) possess sebaceous glands and sweat glands, but these are absent in cetaceans and sirenians. Cetacean skin is hairless except for a few vibrissae or bristles occurring mostly on the rostrum or around the mouth. These are usually lost before or soon after birth. Sirenians have widely scattered hairs. The integument of pinnipeds, sea otters, and polar bears generally has two layers of hair. The outer protective layer consists of long, coarse, guard hairs and the inner layer is composed of softer intermediate hairs or underfur. Polar bear, sea otter, and otariid guard hairs are medullated (having a sheath), whereas phocid and walrus hairs (*Odobenus rosmarus*) are not. The hairs typically grow in groups or clumps, with a single guard hair emerging cranial to one or more underfur hairs. Each hair grows from a separate follicle, but the underfur follicles feed into the guard hair canal so that all hairs in a particular clump emerge from a single opening in the skin. Some pinnipeds have a relatively sparse hair coat [walrus, elephant seals (*Mirounga* spp.), and monk seals (*Monachus* spp.) with a single guard hair per hair canal], whereas others have a lush, thick coat (fur seals, with dozens of underhair or fur follicles feeding into each guard hair canal). Sea otters have the densest fur of any mammal, with approximately 130,000 hairs/cm^2, about twice as dense as that of northern fur seals (*Callorhinus ursinus*).

The appendages of some pinnipeds and the pads of sea otters are hairless, allowing these species to readily lose excess body heat by conduction to the environment. Although most marine mammals rely on blubber for insulation, a layer of air trapped within the hair or fur serves as the primary insulator in fur seals and sea otters and keeps the skin dry when the animals are submerged. Sea otter pelage is coated with squalene, a hydrophobic lipid that aids in waterproofing the fur.

II. Molt

Many phocid seals possess a white lanugo coat *in utero*; this may be lost before birth or may persist for several weeks (as in some arctic and antarctic species). This pelage provides insulation for neonates of ice-breeding seals until they develop a

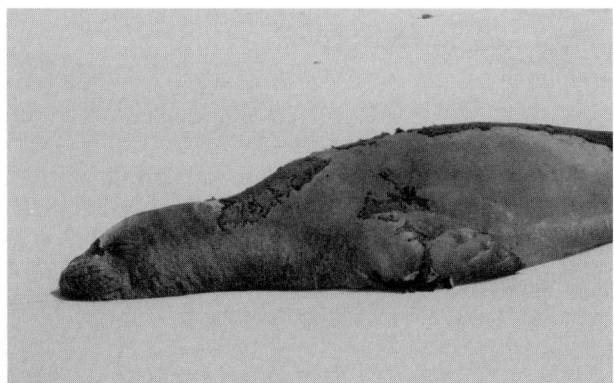

Figure 1 *Catastrophic-type molt in the Hawaiian monk seal* (Monachus schauinslandi) *where the upper epidermis and hairs are shed in large patches within a few weeks. Photograph by B. S. Stewart.*

blubber layer and also may serve as camouflage or protective coloration. Other examples of distinct neonatal pelage include the wooly black coat of elephant seals, which is replaced by a silvery hair coat after the pup is weaned, and the fluffy buff-colored pelage of sea otter pups, which persists for several months. The signals for initiation and control of the annual pelage cycle are not known for most species but are thought to include endocrine (thyroid, adrenal, and gonadal hormones), thermal, and nutritional influences. Molt is generally seasonal, beginning shortly after breeding. Sea otters may molt year round, although more hairs are generally replaced in summer than in winter. The duration of molt in pinnipeds ranges from a very rapid and "catastrophic" shedding of large patches of epidermis and associated hairs (elephant seals, monk seals) (Fig. 1) to the more gradual pattern seen in otariids, with hairs replaced over several months.

See Also the Following Articles

Blubber ■ Energetics ■ Pinniped Physiology ■ Streamlining ■ Thermoregulation

References

Ashwell-Erickson, S., Fay, F. H., and Elsner, R. (1986). Metabolic and hormonal correlates of molting and regeneration of pelage in Alaskan harbor and spotted seals (*Phoca vitulina* and *Phoca largha*). *Can. J. Zool.* **64**, 1086–1094.

Fay, F. (1985). *Odobenus rosmarus. Mammal. Species* **238**, 1–7.

Kenyon, K. W. (1969). The sea otter in the eastern Pacific Ocean. North Am. Fauna Ser., U. S. Dept. Int., Fish Wildl. Serv. No. 68.

Ling, J. K. (1970). Pelage and molting in wild mammals with special reference to aquatic forms. *Quart. Rev. Biol.* **45**, 16–54.

Ling, J. K. (1974). The integument of marine mammals. *In* "Functional Anatomy of Marine Mammals" (R. J. Harrison, ed.), Vol. 2, pp. 1–44. Academic Press, London.

Pabst, D. Ann, Rommel, S. A., and McLellan, W. A. (1999). The functional morphology of marine mammals. *In* "Biology of Marine Mammals" (J. E. Reynolds II and S. A. Rommel, eds.), pp. 15–72. Smithsonian Institution Press, Washington, DC.

Reeves, R. R., Stewart, B. S., and Leatherwood, J. (1992). "The Sierra Club Handbook of Seals and Sirenians." Sierra Club Books, San Francisco.

Scheffer, V. (1964). Hair patterns in seals (Pinnipedia). *J. Morphol.* **115**, 291–304.

Sokolov, V. E. (1982). "Mammal Skin." University of California Press, Berkeley.

Williams, T. D., Allen, D. D., Groff, J. M., and Glass, R. L. (1992). An analysis of California sea otter (*Enhydra lutris*) pelage and integument. *Mar. Mamm. Sci.* **8**, 1–8.

Harbor Porpoise
Phocoena phocoena

ARNE BJØRGE AND KRYSTAL A. TOLLEY
Institute of Marine Research, Bergen, Norway

The harbor porpoise (*Phocoena phocoena*) is a small odontocete inhabiting the temperate and boreal waters of the Northern Hemisphere. The harbor porpoise derives its common English name from the Latin for pig (*porcus*) and is sometimes referred to as the "puffing pig" in parts of Atlantic Canada. Harbor porpoises have a short, stocky body resulting in a rotund shape, an adaptation that helps them limit heat loss in the cold northern climes. On average, adult females reach 160 cm in length and 60 kg. Males are smaller than females, growing only to about 145 cm and 50 kg (Fig. 1). The largest recorded size for this species was from a female who was over 200 cm and 70 kg.

I. Distribution and Variation

The dorsal side of the harbor porpoise is dark gray, while the chin and underbelly are a contrasting white, which sweeps up to the midflanks in a mottled pattern. Gray stripes originate

Figure 1 *Two immature harbor porpoises swim side by side in the murky waters of the Irish sea. The characteristic stripe running from the mouth to the flipper is clearly visible, as is the mottled gray pattern on the flanks. Photograph by Florian Graner.*

on each side of the head near the back of the mouth and run back toward the flippers. The small triangular-shaped dorsal fin makes this species easily recognizable at sea, as does its characteristic swimming pattern of several short, rapid surfacings followed by an extended dive of several minutes. Occasionally, harbor porpoises are known to "log," or lie at the surface, while they move their head about in a sweeping motion.

Harbor porpoises are distributed throughout the coastal waters of the North Pacific, the North Atlantic, and the Black Sea (Fig. 2). Although sightings of porpoises occur in deep waters between land masses, they generally prefer the shallower inshore waters of the continental shelves. Porpoises in each ocean basin are reproductively isolated, resulting in the classification of harbor porpoises into two subspecies: *Phocoena phocoena vomerina* in the Pacific Ocean and *Phocoena phocoena phocoena* in the Atlantic Ocean, plus a third possible subspecies, *Phocoena phocoena relicta,* in the Black Sea. These subspecies differ from each other morphologically as well as genetically, as their cranial size and proportions are different. Pacific harbor porpoises have smaller skulls but longer jaws than Atlantic porpoises. Although the fossil record containing porpoises is poor, recent genetic investigations have made it possible to reconstruct the most probable relationships among the porpoises.

Early morphological studies suggested that harbor porpoises are related to Burmeister's porpoise (*Phocoena spinipinnis*) and the vaquita (*Phocoena sinus*), and therefore these three species have been placed in the same genus. However, genetic information suggests that the harbor porpoise's closest relative is the Dall's porpoise (*Phocoenoides dalli*), a species endemic to the Pacific Ocean.

II. Ecology and Behavior

Harbor porpoises are normally found in small groups (1–3 animals) often consisting of at least one mother–calf pair. Groups of 6–8 animals are not uncommon and, on rare occasions, harbor porpoises may form much larger aggregations. Their SWIMMING and surfacing movements are quick, but harbor porpoises rarely leap out of water. When surfacing, their dorsal side is exposed for a few seconds; due to their short body length and blunt shape, their movement at the surface resembles a forward roll. Often, harbor porpoises can be seen resting at the surface with their body tilted slightly backward.

Porpoises are fish and squid feeders, often foraging near the sea bottom in waters less than 200 m in depth. Bottom-dwelling fishes and small pelagic schooling fishes with high lipid content

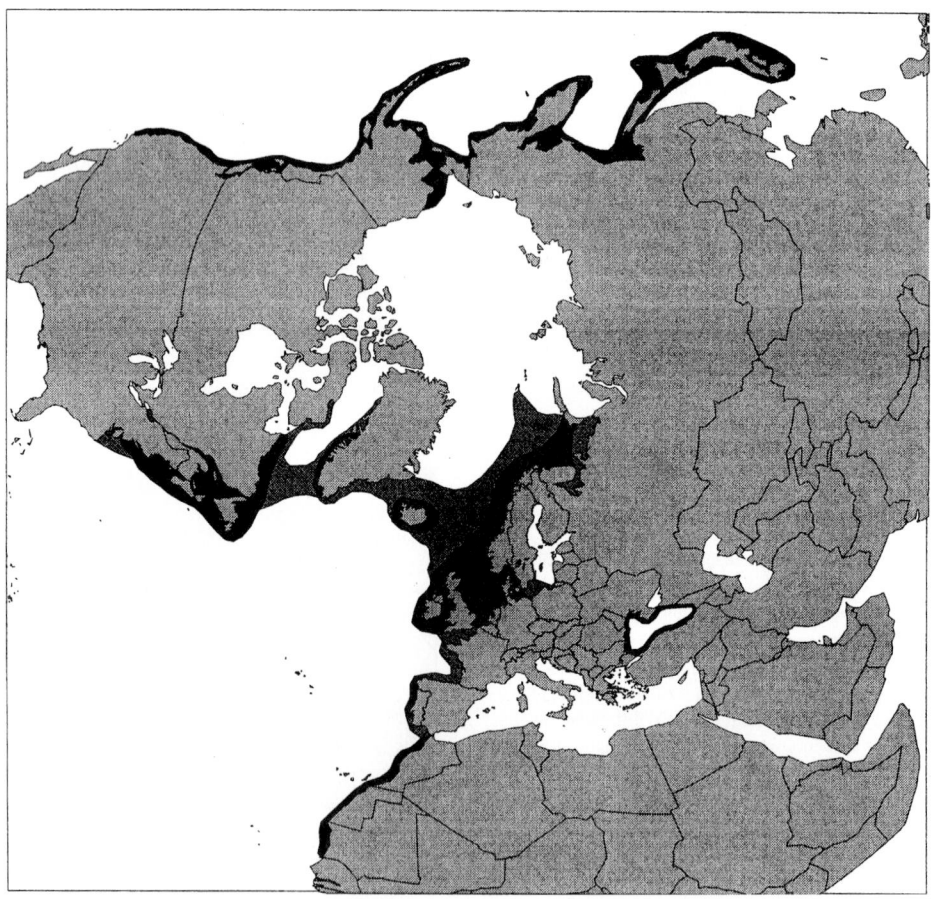

Figure 2 *The primary worldwide distribution of the harbor porpoise* (Phocoena phocoena) *is shown in dark gray. The secondary distribution of harbor porpoises is shown in light gray.*

such as herring, sprat, and anchovy are common prey items. Occasionally, porpoises occur in deeper waters where their diet includes midwater fish such as pearlsides.

Some studies have demonstrated that porpoises may reside within an area for an extended period of time. However, onshore/offshore migrations and movements parallel to the coast are thought to occur. These movements may mirror seasonal changes in DISTRIBUTION and availability of important prey species, and some studies indicate that these movements may correspond to underwater ridges and banks. In some coastal areas, porpoises migrate offshore to avoid ice during winter, but in the long and narrow fjords of Norway, where porpoises live year-round, fresh water input from rivers may freeze within a few hours. Under such circumstances porpoises can be fatally trapped.

III. Anatomy, Physiology, and Life History

Harbor porpoises have spade-shaped teeth, a characteristic that distinguishes them from the dolphin family, which have conical teeth. A characteristic feature of harbor porpoises are tubercles, or small hard bumps on the leading edge of the dorsal fin. The function of the tubercles is not yet known. Harbor porpoises have extremely thick blubber, an adaptation that aids in THERMOREGULATION in cold waters. Calves have thicker blubber and are more rotund than adults, providing them with an excellent capacity to conserve heat. Dive telemetry data have shown that porpoises can dive to depths of at least 220 m. The majority of dives usually last just over a minute, although dives of over 5 min have been recorded.

Harbor porpoises have an average life span of about 8–10 years, although some have been documented to have lived longer than 20 years. They become sexually mature between 3 and 4 years but are not physically mature until they are 5 (males) to 7 (females) years old. Harbor porpoises exhibit reproductive seasonality, whereby all porpoises give birth in a contracted calving season lasting only a few weeks. The calving season varies from region to region, but in most areas, calving takes place within a time period from May to August. The gestation period is approximately 10.5 months. Calves are weaned before they reach 1 year but begin to take small solid food items (e.g., euphausiids) when they are just a few months old. Calves are usually about 70–75 cm and 5 kg at birth but grow rapidly in their first year with males reaching about 120 cm and females 125 cm in length.

Mating takes place approximately a month and a half after the calving season. In the Atlantic, most females produce a new calf every year, but in the Pacific it appears that the calving interval may be every other year. Harbor porpoises likely have a promiscuous mating system, whereby each individual mates with several other individuals. Further, they are thought to be "sperm competitors" because males produce large quantities of sperm, presumably in order to mate with several females in an effort to reproduce. The testes undergo an exceptionally large seasonal change in size where they enlarge up to 800 g just prior to the mating season but regress in winter to a total weight of about 200 g.

IV. Human Interactions

Because harbor porpoises inhabit coastal waters, they are affected by the activities of modern man. POLLUTION, NOISE, ship traffic, and overfishing of prey species are just a few of the human-induced disturbances on this species. In the past, harbor porpoises were harvested for their meat and blubber, but at present, porpoises are given legal protection in nearly every country. However, legal protection does not protect them against accidental deaths in fishing nets, and entanglement in fishing nets is the most significant human-induced threat to porpoises. In many areas this incidental mortality may exceed sustainable levels, and modifications in fishing practices are urgently needed to ensure the long-term survival of some porpoise populations. In some areas, knowledge of porpoise movements and habits has aided in setting fishing regulations designed to help protect this species.

Although porpoises may be able to detect fishing nets through the use of echolocation, they do not always echolocate and may blunder into nets without ever detecting them. The use of specialized highly visible nets may help porpoises detect fishing nets. "Pingers," or devices that emit warning sounds, also may be attached to nets to make known the location of fishing nets. However, testing of these special nets and devices over a long period of time is required to verify if these are adequate solutions to this problem.

See Also the Following Articles

Blubber ■ Echolocation ■ Porpoises, Overview

References

Barlow, J., Hill, P. S., Forney, K. A., and DeMaster, D. P. (1998). U.S. Pacific Marine Mammal Stock Assessments: 1998. NOAA Technical Memorandum NOAA-TM-NMFS-SWFSC-258. U.S. Department of Commerce. Available at: http://www.ntis.gov/

Berta, A., and Sumich, J. L. (1999). "Marine Mammals: Evolutionary Biology." Academic Press, San Diego.

Bjørge, A., and Donovan, G. P. (eds.) (1995). "Biology of the Phocoenids." Report of the International Whaling Commission, Special Issue 16. Cambridge.

Gaskin, D. E. (1982). "The Ecology of Whales and Dolphins." Heinemann Educational Books, London.

Leatherwood, S., and Reeves, R. R. (1983). "The Sierra Club Handbook of Whales and Dolphins." Sierra Club Books, San Francisco.

Read, A. J. (1999). Harbour porpoise Phocoena phocoena (Linneaus, 1758). In "Handbook of Marine Mammals," Vol. 6, pp. 323–355. Academic Press, London.

Read, A. J. (1999). "Porpoises." Colin Baxter Photography Ltd., Granton-on-Spey, Scotland.

Read, A. J., and Hohn, A. A. (1995). Life in the fast lane: The life history of harbor porpoises from the Gulf of Maine. Mar. Mamm. Sci. 11, 423–440.

Reynolds, J. E., and Rommel, S. A. (eds.) (1999). "Biology of Marine Mammals." Smithsonian Institution Press, Washington.

Rosel, P. E., Haygood, M. G., and Perrin, W. F. (1995). Phylogenetic relationships among the true porpoises (Cetacea: Phocoenidae). Mol. Phylogenet. Evol. 4, 463–474.

Westgate, A. J., Read, A. J., Berggren, P., Koopman, H. N., and Gaskin, D. E. (1995). Diving behaviour of harbour porpoises, Phocoena phocoena. Can. J. Fish. Aqu. Sci. 52, 1064–1073.

Harbor Seal and Spotted Seal

Phoca vitulina and *P. largha*

JOHN J. BURNS
Fairbanks, Alaska

The harbor seal is also widely known as the common seal. It occurs over a great latitudinal range and in many different coastal and insular habitats around the rims of both the North Atlantic and the North Pacific regions. Spotted seals, in contrast, occur only in seasonally ice-covered seas of the Western Hemisphere. The name larga seal is sometimes used for the spotted seal and is derived from *largha*, which is part of the scientific name. These two sibling species are the most closely related members of the subfamily *Phocinae* and are fascinating examples of adaptations to vastly different environments. Most harbor seals occur in habitats that are sea ice free throughout the year, or at least where their coastal haulout sites are clear of sea ice during the breeding season. Spotted seals utilize sea ice during the breeding season. In this context, it is important to distinguish between sea ice and freshwater icebergs calved from tidewater glaciers. Both species are of medium size. In some areas of the North Pacific their distributions overlap.

I. General Appearance

Based on external appearances, harbor and spotted seals older than weaned pups are not readily distinguishable from each other. Body size of spotted seals falls within the range of that for all but the largest harbor seals.

The pelage pattern and coloration of harbor seals is variable (Fig. 1). Background color ranges from yellowish or yellowish-gray (light phase) to blackish (dark phase). Light-phase seals are usually more pale on the flanks and belly than on the back, are covered with small black spots, and often show small pale rings, usually on the slightly darker dorsum. Dark phase harbor seals also have dark spots that are largely masked by the background COLORATION. Usually the dark seals show obvious light rings, especially on the dorsum. Seals of intermediate coloration are common. Throughout their broad range there are regions within which a particular pelage type predominates. Ungava seals are of the dark phase, as are most western Pacific harbor seals. Spotted seals are more uniform in color and pattern (Fig. 2). They tend to resemble light-phase harbor seals, which has contributed to the confusion about these two species.

II. Diagnostic Characters

There are morphological, ecological, and behavioral differences between harbor and spotted seals. The breeding habitat of harbor seals is coastal and insular. They give birth mainly on shore ROOKERIES, although in some parts of Alaska and Greenland they utilize icebergs calved from tidewater glaciers in protected fjords. Spotted seals use seasonal sea ice, mostly far

Figure 1 *Adult harbor seals from the eastern Pacific region. The uppermost seal is of the dark color phase.*

Figure 2 *A female spotted seal with pup in the Bering Sea. The pup is nearly weaned and is beginning to shed lanugo around its hind flippers.*

from shore. During the breeding season, harbor seals occur in herds with no obvious social organization. Spotted seals occur as widely scattered adult pairs, usually with a pup (triads). In the areas where they occur together, harbor seals breed about 2 months later than spotted seals (reproductive separation). The pelage of newborn harbor seals is like that of adults because the lanugo is shed before birth (*in utero*). Occasionally, especially in the northern parts of their range or in the case of premature pups, the lanugo is retained for up to a few days after birth. The pups usually enter the water shortly after birth, often within an hour. Spotted seal pups retain their whittish woolly lanugo, which is important for thermoregulation, for about 4 weeks. After the lanugo is shed the pelage resembles that of adult animals (Fig. 3). They remain on the ice during the nursing period and are abruptly weaned (abandoned).

There are no individual cranial measurements that consistently separate harbor from spotted seals. As with body size,

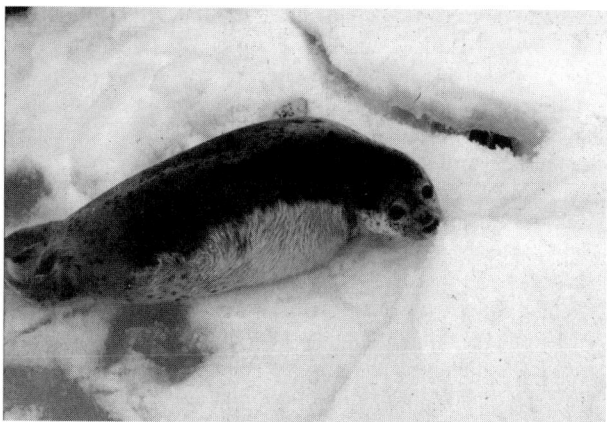

Figure 3 *A weaned spotted seal pup that has almost completely shed its lanugo.*

skull measurements are within the range of those in harbor seals. The ratios of some measurements are useful for differentiating between the two species. Those ratios include jugal length/condylobasal length; nasal length from maxillo-frontal suture/condylobasl length; and interorbital width/mastoid width. Several nonmetrical characters, used in combination, do permit differentiation between the species. In harbor seals the skull is more massive, the bullae are more flattened and angular, the premolar teeth of adults are mostly obliquely set (straight in spotted seals), the posterior margin of the jugal bone is mostly angular (as opposed to rounded), the glenoid fossa is more flattened and angular, the bony process of the external auditory meatus is mostly straight and blade like (as opposed to mostly blunt and rounded), the shape of the posterior edge of the bony palate is mostly acute (as opposed to mostly rounded), and the hyoid arch is mostly incomplete, having abbreviated stylohyals that are not attached to the bullae (as opposed to complete and attached in spotted seals). Unfortunately, none of these diagnostic characteristics are useful for differentiating live seals under field conditions.

Very experienced observers can distinguish between these two seals, even those with similar pelage, based on behavior when hauled out together on land, on general facial features of adults, and on behavior when frightened into the water.

III. Distribution and Movements

The distribution of harbor and spotted seals is shown in Fig. 4. Harbor seals occur over a latitudinal range from about 30°N to 80°N in the eastern Atlantic region and about 28°N to 62°N in the eastern Pacific region. They have the broadest distribution and occur in most different habitats of any other pinniped. Although the centers of abundance (greatest numbers of breeding animals) are in the northern temperate zone, breeding colonies of these seals occur north or south of that zone, depending on the presence of required environmental conditions created by regional oceanographic and climatic conditions. The high latitude distribution in the Atlantic region is due to relatively warm oceanographic features, including the so-called North Water in Baffin Bay (eastern Canada–West Greenland), and the strong influence of warm water carried across the Atlantic to northern Europe by the Gulf Stream and associated gyres.

There are five presently recognized subspecies of harbor seals: *P.v. vitulina* Linnaeus, 1758; *P.v. concolor* (DeKay, 1842); *P.v. mellonae* (Doutt, 1942); *P.v. richardii*[1] (Grey, 1864); and *P.v. stejnegeri* Allen, 1902. The spotted seal, *P. largha* Pallas, 1811, is considered to be a monotypic species. The different subspecies of *P. vitulina* were originally recognized on the basis of geographical separation and skeletal morphology. Recent studies of their genetics sustain those conclusions. Boundaries

---

[1]Editorial protocol for this book requires the nomenclature of Rice (1998), which is *P.v. richardii*. The correct nomenclature, in the authors opinion, is *P.v. richardsi* in accordance with the explanation in Shaughnessy and Fay (1977). The person in whose honor this subspecies of seal was named was Capt. Richards, not Capt. Richard.

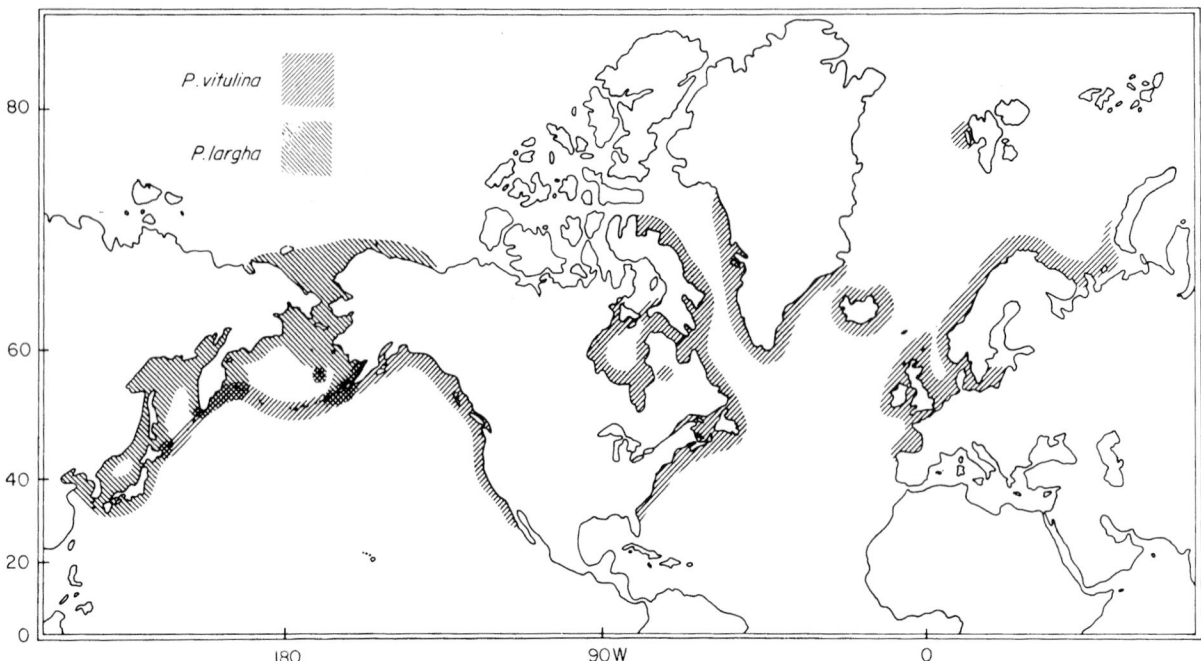

Figure 4 *The distribution of harbor seals (five subspecies) and spotted seals.*

between the eastern and the western subspecies within both the North Atlantic and North Pacific oceans are not known with certainty.

In the eastern Atlantic, *P.v. vitulina* normally occurs from the French coast bordering the English Channel, throughout the North Sea and northward to Finmark on the Barents Sea, including into the southern Baltic Sea and waters of Ireland and Great Britain. Stragglers occur to Portugal in the south and to the eastern Barents Sea in the northeast. The northernmost breeding population (here assumed to be *P.v. vitulina*) is in western Svalbard (Spitsbergen) at 78°30′N.

The boundary between *P.v. vitulina* of the eastern Atlantic and *P.v. concolor* of the western Atlantic is not known. However, harbor seals extend across the North Atlantic as a series of widely separated populations that occur at Svalbard, the Faeroe Islands (uncertain), Iceland, southern East Greenland, and West Greenland northward to about Upernavik (72°N). In Greenland the seals are considered to be *P.v. concolor,* as are those in most of eastern North America.

In the western Atlantic region the normal range of *P.v. concolor* extends from about 40°N (New Jersey) to about 73°N (northern Baffin Island, Canada), including into Hudson Bay and southern Foxe Basin. Stragglers have occurred as far south as Florida. The resident freshwater seal of the Ungava Peninsula in eastern Canada, *P.v. mellonae,* was first described and recognized as a separate subspecies by Doutt (1942), mainly on the basis of skull features and apparent isolation. It occurs in several drainage systems that empty into eastern Hudson Bay, where *P.v. concolor* is found. The subspecific designation of the Ungava seals was in doubt for several reasons, including their close proximity to saltwater harbor seals; the fact that in general harbor seals occur frequently in rivers and lakes; and be-

cause the freshwater drainages in which it occurs flow into Hudson Bay. However, passage to and from salt water is blocked by numerous obstacles resulting from isostatic uplifting (rebound) of the peninsula since the last Pleistocene glaciation. The distinct status of this rare freshwater seal has been upheld on the basis of genetic differences.

In the North Pacific region the distribution of harbor seals extends from Cedros Island near the west-central coast of Baja California, Mexico (about 28°N), northward to the Gulf of Alaska and southeastern Bering Sea, across the entire Aleutian Ridge (the Aleutian and Commander islands) to the Kamchatka Peninsula of eastern Russia, southward in the Kuril Islands and beyond to Hokkaido island in northern Japan. *P.v. richardii* is the subspecies of the eastern North Pacific region, and *P.v. stejnegeri* occurs in the western Pacific. The boundary between these two subspecies is currently thought to be in the western Alaska Peninsula–eastern Aleutian Islands, although uncertainty about that question still exists. The northernmost pupping colonies in the Pacific region are in Prince William Sound, Alaska, at about 61°13′N. That is some 1920 km farther south than the northernmost breeding group in the Atlantic region.

Great distances separate the Atlantic and Pacific forms. There are no breeding colonies between Baffin Island in northeastern Canada and the Pribilof Islands of southeastern Bering Sea, nor between northern Norway and the Pribilof Islands.

Seasonal and annual movements of harbor seals are quite varied depending on the environments in which they occur. They are usually considered to be relatively sedentary, with a high degree of fidelity to one or a few haul-out sites. This view, although perhaps applicable to some populations, is a gross oversimplification. It is now recognized that they move, in some cases quite extensively. Generalizations are inappropriate in view of this seals' wide distribution and differences in stock sizes, population dynamics, and the varied environments they occupy. In most instances, some individuals are likely more sedentary and show stronger site fidelity than others. Kinds of movement include migrations, juvenile dispersal, seasonal shifts, shifts related to breeding activity, responses to seasonal habitat exclusion, responses to acute or chronic disturbance, and immigration/emigration, occasionally on a relatively large scale.

The spotted seal was, until recently, considered to be a subspecies of the harbor seal. It is now recognized as a distinct species that includes several widely separated breeding populations. The centers of abundance during the breeding season are mainly in the temperate/sub-arctic boundary regions. The seal is well adapted to exploit the "front" and broken ice zones of seasonal sea ice that overlies continental shelves during winter and spring. Spotted seals resort to haul outs on land during ice-free seasons of the year. There are great seasonal expansions and contractions of range, commensurate with the annual cycle of sea ice advance and retreat. Their distribution in all areas is most restricted during the period of maximum ice cover. They occur in the Bering, Chukchi (in summer), Beaufort (in summer) and Okhotsk seas, Tartar Strait, the Sea of Japan and the northern Yellow Sea/Bo Hai (Bohai Sea), and adjacent embayments that border the Korean Peninsula and China. The most southern breeding populations (about 38°N) are in the Sea of Japan and the Yellow Sea. Their occurrence at these southern latitudes is because of a cold winter climate, dominated by the so-called Siberian high pressure system, that results in a limited sea ice cover during midwinter.

In all areas, as the seasonal ice cover recedes and disintegrates, spotted seals expand their range and haul out on land. Some animals of the population that winters and pups on ice in the Bering Sea migrate northward into the Chukchi and Beaufort seas during the ice-free months. Their summer–early autumn distribution extends as far north as 71°30′N near Point Barrow, Alaska, and to about 70°N on the northern shores of Chukotka, Russia. Thus, the total range of the Bering Sea population extends over 15° of latitude, or about 1665 km.

Spotted and harbor seals are sympatric (have overlapping ranges) in the southeastern and southwestern Bering Sea, on the Kamchatka Peninsula, in the Kuril Islands, and northern Japan. Similarities in general appearance and occurrence on land (sometimes in close proximity) have long contributed to the confusion about these two different species.

IV. Population Size

Populations of harbor and spotted seals fluctuate in size due to both natural and anthropogenic causes, including hunting, incidental taking, competition for food with commercial fishers, habitat alteration, disturbance, protective measures, diseases, climate regime shifts, and other factors. Some populations are small and isolated, persisting in what may be marginal habitat. These may be the ones most vulnerable to changes in environmental factors and to direct exploitation. In general, direct exploitation has now been reduced greatly. Most popula-

tions are currently protected from HUNTING except under terms of special licenses or in areas where they are taken by indigenous peoples for subsistence purposes. Population estimates are fragmentary and, in many cases, outdated.

Estimates of the size of harbor seal populations provide useful indications of regional abundance and further illustrate this seals' coastal and insular distribution, primarily in the northern temperate zone. In the middle 1980s there were perhaps 98,000 eastern Atlantic harbor seals. By then populations had recovered after prolonged and sometimes intensive hunting and control programs. The largest numbers were and still are around the rim of the North Sea and Iceland. Areas of greatest abundance were in Great Britain (up to 47,000), Iceland (28,000), the Wadden Sea (10,000), and Kattegat/Skagerrak (6000). The smallest known populations are in the Baltic Sea (perhaps 200) and around Svalbard (500 to 600 in 1990). In 1988 a large proportion of some populations died from a viral epidemic: up to 48% in parts of southeastern Great Britain and an estimated 60% in the Wadden Sea and Kattegat/Skagerrak. These affected populations recovered rapidly and by 1992 there were an estimated 7250 seals in the Wadden Sea and perhaps 5200 in Kattegat/Skagerrack.

The number of western Atlantic harbor seals is not known, although in 1993 there were reported to be between 40,000 and 100,000, of which 30,000 to 40,000 were in Canadian waters. In some parts of northern Canada and Greenland these seals are harvested for meat and for their beautiful hides, which are made into clothing and other articles of Native handicrafts. The population in the United States was about 4700. There is no estimate of the number in Greenland. Fewer than 60 were harvested there in 1989. Apparently they were never abundant and have declined or disappeared in several locations.

Estimates (unreliable) of the number of Ungava seals range from 120 to perhaps 600 animals. The actual number is probably closer to the lower value. This includes the entire subspecies. Ungava seals are considered to be possibly endangered, vulnerable, and rare because of a lack of information, a very limited range, low numbers, and potential threats from proposed development in the region. This subspecies may well be a relict in habitat that has been altered drastically and unfavorably by natural geological processes during Holocene and Recent times.

In the eastern Pacific region, harbor seals are abundant, although as elsewhere numbers have fluctuated greatly. There is no estimate for the number in Mexican waters. From California to southeastern Alaska, these seals have increased greatly during the past three decades. Estimates during the mid-1990s were: California, 34,500; Oregon and coastal Washington, 29,900; inland waters of Washington, 13,800; British Columbia, 100,000 (with increases of about 12.5%/year through 1993); and southeastern Alaska, 35,000. In south-central and southwestern Alaska the population trend has been the opposite of that farther south, with an 85% decline in the Gulf of Alaska between 1976 and 1988. These opposite trends in the southern and northern parts of their range are apparently mainly responses to a major oceanic regime shift that began about 1976, although factors such as protection, continued hunting and incidental taking are also involved. At present the estimated number in the Gulf of Alaska region, including Prince William Sound, is about 20,000 animals. In southwestern Alaska, primarily Bristol Bay, there are an estimated 15,000 seals not including animals in the Aleutian Islands, which are herein considered to be a different subspecies.

The range of the western Pacific harbor seal extends across the Aleutian Ridge to Asia. These seals are predominately of the dark color phase, they tend to occur in very small groups (as opposed to large aggregations), and they mainly occupy rocky islands and shorelines. Regional estimates of numbers are: Aleutian Islands, about 3400 (in 1994); Commander Islands, 1500; Kamchatka Peninsula, 200; Kuril Islands, 1900; and northern Japan, 300. Estimates for the latter four areas are from the early 1990s. They are classified as rare in Japanese and Russian waters and are now protected.

There are no reliable recent population estimates for the spotted seal, except perhaps in the Bohai Sea. Indirect and anecdotal information suggests that since the early 1980s the population in the Bering Sea has declined, perhaps due to changing climate and therefore more unfavorable sea ice conditions and changed food-web dynamics. This is also in accord with trends of harbor seals in the northern part of their Pacific range. For the Bering Sea population, estimates (educated guesses) were 200,000–250,000 in the early 1970s and 100,000–135,000 in the 1980s. The actual current size of the population is unknown, although these seals remain of common occurrence and are important to Native subsistence hunters in coastal areas of the Bering and Chukchi seas. Fewer than 5000 per year are thought to be harvested annually. Vessel-based commercial hunting by the Russians ended in 1995. The population in the Okhotsk Sea was placed at 130,000 in 1982. There is essentially no subsistence hunting there and, as in the Bering Sea, commercial harvests have stopped. The current size of populations in the Sea of Japan and Tartar Strait is not known. In the Bohai Sea, including Liaodong Bay, there were an estimated 4500 in 1990. This compares with estimates of >7000 in the 1930s; >8100 in 1940; and 2269 in 1979 (after a period of intensive harvesting). These seals were accorded protection from hunting in the 1980s.

V. Behavior and Life History

As already noted, harbor and spotted seals are superficially quite similar in appearance. Harbor seals haul out mainly on land, although in some areas of mainland Alaska and Greenland they use icebergs calved from tidewater glaciers. Also, in the northern parts of their range, where labile sea ice occurs to or very near shore, they haul out on it until the land sites are accessible, usually long before the pupping season. They use haul outs throughout the year, although most frequently and in greatest numbers during the pupping and molting seasons. Regardless of season, haul-out activity is strongly affected by the stage of the tide, air temperature, wind speed, precipitation, and time of day. They lay close to the water when hauled out and usually flee when disturbed, although habituation is not uncommon near large human population centers, if they are not harassed unduly. The substrate at natural haul outs on land is diverse and includes mud flats, sand and gravel bars and beaches, rocks, glacial icebergs, and occasionally sea ice. Depending on the region, haul outs can be on lakes, rivers, estuaries, bays, ocean shorelines, islands, islets, ledges, and any other setting where the seals can rest, undisturbed, with immediate access to deep water. They may, on occasion, haul out on man-made structures such as docks, floats, and log rafts.

Spotted seals use sea ice starting with its formation in autumn. They often concentrate in large numbers on the early ice that forms near river mouths and estuaries (freshwater freezes at a higher temperature then seawater) and feed on autumn spawning fishes. As the ice thickens, becomes attached to land, and extends farther from shore, spotted seals move seaward into the drifting ice. Their association is mainly with the highly labile marginal areas and they migrate (southward in the Chukchi/Bering sea region) to maintain an association with that habitat. During the cold weather of winter they rarely haul out. Peak haul out on the ice is during the pupping and molting season.

As the sea ice cover retreats and disintegrates in late spring–early summer, spotted seals again move shoreward and, in the Bering Sea, northward. Again, large aggregations can often be seen close to shore on the last remnants of former shorefast ice and on ice flushed from rivers and estuaries. At this time of year they feed extensively on the dense schools of spawning herring and smelt. They haul out on shore when the ice is gone. Between haul-out bouts on land, some seals travel long distances in the open sea, even between Alaska and Siberia, and use multiple haul outs (as they also do when migrating). Shore haul outs are mostly on isolated mud, sand or gravel beaches, or on rocks close to shore. They are often on river bars, tidal flats, and barrier islands. Spotted seals are especially vigilant on land, where they may be subjected to attack from a variety of predators. Their association with sea ice starts again as soon as it begins to form in the autumn.

A. Food Habits

Feeding forays of harbor seals can be close to haul-out sites or many miles distant, either along the coast (including rivers) or seaward. They are capable of FEEDING at considerable depths (to 500+ m) and are generalists that prey mainly on abundant and easily available foods, with diets varying by season and region. There are long-term changes in foods that are associated with environmental changes, and therefore dynamic changes in the abundance of different prey species. Primary food items are small to medium size fishes (or age classes), such as various members of the codfish family, hake, mackerel, herring, sardines, smelts, shad, capelin, sandlance, sculpins, a variety of flatfishes, salmonids, and many others. Their propensity for cod, salmons, and other commercially important species has resulted in long-standing conflicts with fishermen in many areas. Cephalopods (squid and octopus) are usually reported as being next important after fishes, followed by crustaceans, including mainly shrimps and crabs. Several studies have reported that shrimp may be particularly important to recently weaned pups. Although there is great diversity in foods, a few items usually comprise the majority of seasonal diets in an area. As examples, in Atlantic Canada, 23 different food items were identified but 4 accounted for 84% of the estimated biomass of prey consumed. In the Gulf of Alaska, fishes comprised 73.8% of the diet and 27 different species were eaten. The four most important foods were pollock, cephalopods, capelin, and flatfishes. In the western Aleutian Islands the main food items, at least in 1958 and 1962, were Atka mackerel (*Pleurogrammus monopterygius*) and octopus. The main foods of the Ungava seals are thought to be resident brook and lake trout. Seals in Lake Iliamna, Alaska, feed on the variety of salmonids (charr, trout, and salmon) that occur there in large numbers.

Spotted seals are also generalist feeders that primarily utilize similar types of abundant fishes, crustaceans, and cephalopods. Because they have a pelagic distribution in winter–spring and a different coastal and pelagic distribution during ice-free months, there are major seasonal and regional differences in food habits. Additionally, there are age-related differences. Most reports about food habits are based on seals examined during spring (mainly April and May) when they are associated with sea ice. A few samples are from animals collected in the coastal zone during autumn, and there are anecdotal observations of summer feeding, especially in areas where subsistence and commercial fishing activities occur. There are few data from the late autumn and winter months, although in the Bering and Okhotsk seas these seals occur where pollock, herring, eelpout, flounders, shrimp, and crabs are abundant.

Independent feeding by spotted seal pups begins about 10 to 15 days after they are abruptly weaned. During the time of fasting and early independent feeding they live on their accumulated fat reserves and lose between 18 and 25% (sometimes up to 30%) of their weight. The first food consumed is frequently small amphipods or euphausiids. Abundant schooling fishes are the main foods of older seals and, in the Okhotsk Sea, occurred in 89% of seals 1–4 years old and 70% of seals >5 years old. Cephalopods were next in importance, followed by decapods. Amphipods were still consumed by the 1–4 year olds but were not found in older animals. The frequency of occurrence of cephalopods was higher in older age animals. Spotted seals were reported to feed more in the morning and evening than at other times of the day. During spring the main food items in the Bering Sea, depending on the region, were pollock, arctic cod, sand lance and capelin. In the Okhotsk Sea, pollock were most important. In Peter the Great Bay (Sea of Japan), the dominant fishes were saffron cod, flounders, and rockfish and in Tarter Strait they were saffron cod, flounders, and salmon. In all areas crustaceans and cephalopods were also important. There has been little sampling in coastal habitats during summer when anadromous and coastal spawning fishes such as charr, salmon, capelin, smelt, herring, flounders, saffron cod, and other species are abundant. According to traditional local knowledge, those foods are utilized intensively by the seals.

B. Size

The average length of harbor seals varies among populations. The smallest and largest seals occur in the North Pacific region and therefore they bracket the size of animals from other regions. Those from the northern Gulf of Alaska, members of the *P.v. richardii* complex, are the smallest. The average standard length and weight of adult males from that area is about 160 cm and 87 kg, while that of adult females is about 148 cm and 65 kg. Newborn pups average 82 cm and 10 kg. The largest seals are of the *P.v. stejnegeri* complex from northern Japan. Length and weight of adult males range from 174 to 186 cm and 87 to 170 kg and that of adult females from 160 to 169 cm and 60 to 142 kg. newborn pups were up to 98 cm and 19 kg.

Spotted seals are about the same size as most harbor seals and there are slight differences among populations. Adult males from the Bering Sea range from 161 to 176 cm and 85 to 110 kg. Adult females are 151 to 169 cm and 65 to 115 kg. Near-term fetuses and newborn pups from the Okhotsk Sea are 78 to 92 cm long and 7 to 12 kg. Healthy pups usually double and sometimes triple their birth weight during the 3- to 4-week nursing period.

C. Reproduction

In general, female harbor seals reach sexual maturity at ages 3 to 4 years and physical maturity by age 6 or 7. Males obtain sexual maturity at 4 to 5 and physical maturity by 7 to 9. The maximum life span is around 35 years, although few animals live that long in the wild. All harbor seal populations have a similar reproductive cycle. However, over their very broad range the specific timing of events varies. Depending on the region in question, births occur in late spring or summer. Within a specific population the peak of pupping can change slightly over time, apparently in response to significant environmental shifts. Additionally, there is some interannual variability. In general the pupping season extends over a period of up to about 10 weeks, within which there is about a 2-week peak. Females bear a single pup, although twinning has been recorded. In most regions, pups are born on land, usually between the high and the low tide water lines. In some parts of Alaska (and presumably also in Greenland), pups are born on floating icebergs calved from tidewater glaciers in protected fjords.

Newborn pups can and do enter the water, often being forced to do so by tidal inundation of birth sites or because of disturbance by birds scavenging afterbirth. Mother–pup bonding is a critical phase of behavior within the first hour of birth, as mutual recognition is required to locate and/or remain with each other on rookeries and in the water. A young pup often clings to its mother's back in the water. Mothers feed during the approximately 4-week nursing period (some reports indicate as long as 6 weeks). Pups start to catch their own food during the late stages of the nursing period.

Mating occurs in the water at about the time that pups are weaned, although females mating for the first time or that have not given birth in a specific year may breed outside of the peak period of the postparturient animals. There is intermale competition for receptive females and no obvious social organization during the breeding season. As with all other pinnipeds, fertilization is followed by a prolonged period of delayed implantation (embryonic diapause) that lasts about 2.5 months, after which the embryo implants and resumes development. The total gestation period, from fertilization to birth, is about 10.5 months. In most populations, pregnancy rates exceed 85%; in other words, most sexually mature females bear a pup every year.

To put the timing of this generalized reproductive cycle into a regional context, the peak period of pupping can be used as the benchmark event. In Europe, most pupping occurs during late June and early July. In most of eastern Canada and Greenland, births are mainly during mid-May to mid-June, slightly later at higher latitudes. However, the Ungava seals reportedly pup during late April or early May. There are considerable differences among populations of the Pacific region. Births occur during early February in Mexican waters; in March–April in southern California; in May along the outer Washington coast; between late June and September in Puget Sound and southern British Columbia; during May to late June in northern British Columbia, most of Alaska and Japan; and early June to late July (peak around July 1) in the Aleutian and Pribilof Islands.

Spotted seals have the same basic reproductive cycle as harbor seals, although the timing of events is directly related to the most favorable sea ice conditions at the time of birth and weaning. Pups are born exposed on the ice and, during the first 2 or 3 weeks, are more like land mammals. They spend most of the time on ice floes until weaned. Unseasonably early disintegration of ice is thought to result in a high mortality of nursing pups. During early life the dense coat of lanugo provides the required insulation for maintaining body heat, although that important function is assumed by the rapidly increasing blubber layer acquired during the 4-week nursing period. At weaning, most pups are heavier than at any other time during their first year of life. They are so fat and buoyant that they are poor divers. This large energy store provides sustenance during the early stages of adjustment to independent life.

Timing of the birth period has evolved to coincide with the average period of greatest extent and stability of the seasonal sea ice cover and varies by region. Weaning, which is abrupt, coincides with the normal seasonal onset of ameliorating spring weather and disintegration of the seasonal ice cover. The use of sea ice as a platform on which to bear and nurture pups is central to the ecology of spotted seals. These events (birth, dependence during the nursing period, weaning, and early independence) are more restricted in time than is the case with harbor seals. Pups are born earlier in the more southerly parts of this species' range. In the Yellow Sea the peak period is during late January; in the Sea of Japan it occurs during February and March; and in both the Okhotsk and Bering seas the peak is during the first half of April. Mothers feed during the nursing period, although the pups remain on the ice, sometimes wriggling over brash ice to move between closely adjacent ice floes.

Mating occurs at about the time that pups are weaned, and most females breed annually. The MATING SYSTEM is quite different than that of harbor seals. Spotted seals begin to form pairs early during the pupping season. They are considered to be annually monogamous and territorial. Triads consisting of a female, her pup, and an attending male can be seen lying on the ice. These triads are widely spaced, although there are regions of high abundance. Females frequently attend their pups on the ice, especially during the early nursing period, and the males stay with the females. Pairs that include an adult female that did not pup are also formed. In the Bering and Okhotsk seas, such pairs are seldom seen on the ice in early April (prior to the molt in adults), probably because there is no pup to attend. As with all aspects of the breeding cycle, there is a latitudinal element to the mating period that relates to the

regional seasonality of sea ice, and the timing of birth and lactation.

Nursing and recently weaned pups remain on the ice flows without benefit of snow lairs. Their only protection from wind is that provided by their mothers or the shelter of ice ridges. The exposed and relatively immobile pups are not subjected to significant predation by polar bears or arctic foxes because the labile marginal ice in which they occur during spring is well south of the normal range of those predators. Polar bears and arctic foxes do not occur in the Okhotsk Sea of farther south.

D. Molt

In harbor seals the molt generally occurs during midsummer to early autumn, within 2 or 3 months of the pupping season. During the molt, seals haul out more frequently than at any other time of the year except for the pupping season. There are differences in timing among age and sex cohorts. Usually yearlings begin and end the molt earliest, followed by subadults, then adult females, and last, adult males. There is overlap among these general age groups. Throughout their extensive range the molt occurs after cessation of the breeding season. Accordingly, it occurs later in the year in the late breeding populations of harbor seals such as those in Europe, British Columbia, and Puget Sound.

Spotted seals of the Okhotsk and Bering seas molt mainly in late spring. Pups, as mentioned, have the color and pelage pattern of adults after their lanugo is shed. Older seals begin the molt after the breeding season and show an overlapping age-related sequence similar to that of harbor seals. The period of intensive molt is during May and June, during which time the sea ice is retreating rapidly and deteriorating. In areas where the ice disappears early, or in minimal ice years, the molt is completed on shore haul outs.

VI. Mortality Factors

Excessive overexploitation, large-scale die-offs due to epizootic diseases, and natural long-term population changes are known to occur in harbor seals. As examples, in late 1979 and 1980 an estimated 500 seals died along the New England coast from an influenza virus of avian origin. Another less severe disease-caused die-off occurred in the same area in 1982. The largest known incident of mass deaths occurred in northern Europe, during 1988–1989, when an estimated 18,000 harbor seals died due to a viral infection that spread rapidly among some colonies in the North Sea region. In all areas the populations had previously reached high levels after cessation of control programs. In the eastern North Pacific south of the northern Gulf of Alaska there has been a sustained long-term increase in numbers. Farther north they declined about 85% between 1976 and 1988. These changes in the eastern Pacific are probably mainly related to natural large-scale climatic regime shifts now known to have occurred. Nothing is known about natural fluctuations in spotted seals, although it is probable that they have also been affected by climate change and therefore changes in the carrying capacity of their more remote environment.

The reported predators of harbor seals include killer whales, sharks, Steller sea lions, eagles, ravens, and gulls. Spotted seals are preyed on by those same animals and also by walruses, polar and brown bears, wolves, and arctic foxes. Humans hunt both species of seals, especially in the north, and also occasionally take them incidentally in the course of other activities, particularly fishing.

VII. Conservation Concerns

There are similar conservation concerns relevant to both harbor and spotted seals. The general problem of pollution from military, agricultural, and/or industrial activities (including coastal and off-shore oil and gas development) is of particular concern because of its direct and indirect effects on seals and the foods they eat. Oil spills are a chronic problem. Recent disease outbreaks may have been intensified because of suppressed immune responses caused by a variety of pollutants. Hunting may still be an important factor in limiting or reducing some of the small breeding populations of harbor seals in Greenland and northern Canada. Fishing activities can affect both species adversely by causing incidental mortalities and by competing for fish the seals depend on for food. FISHERY INTERACTIONS are probably limiting any increase of the small populations of harbor seals in northern Japan and parts of Greenland. Development projects can alter or eliminate important habitat or displace seals by increased disturbance near haul outs. The Ungava seal may be particularly vulnerable to proposed hydroelectric projects.

Climate change, specifically warming, will have major impacts on harbor and spotted seals. The contentious aspect of that important issue is the extent to which natural change is being exacerbated by anthropogenic effects. Climate has changed many times in the past and has been an important force affecting zoogeography, population fluctuations, extirpation, and EXTINCTION. Global warming is definitely causing later formation and earlier breakup and reducing the extent and thickness of seasonal sea ice. It might well result in an increase of suitable habitat for harbor seals in the north and an overall decrease of spotted seal habitat, especially in the southern parts of its range.

See Also the Following Articles

Earless Seals ▪ Mass Die-Offs ▪ Migration and Movement Patterns ▪ Pinniped Ecology ▪ Skull Anatomy

References

Barlow, J., Brownell, R. L., Jr., DeMaster, D. P., Forney, K. A., Lowry, M. S., Osmek, S., Ragen, T. J., Reeves, R. R., and Small, R. J. (1995). "U.S. Pacific Marine Mammal Stock Assessments." NOAA Technical Memorandum NMFS-SWFSC-219.

Bigg, M. A. (1981). Harbour seal *Phoca vitulina* Linnaeus, 1758 and *Phoca largha* Pallas, 1811 (1981). *In* "Handbook of Marine Mammals" (S. H. Ridgway and R. J. Harrison, eds.), Vol. 2, pp. 1–27. Academic Press, London.

Boulva, J., and Mclaren, J. A. (1979). Biology of the harbor seal, *Phoca vitulina*, in eastern Canada. "Bulletin of the Fisheries Research Board of Canada," Bulletin 200. Ottawa, Canada.

Burns, J. J., Fay, F. H., and Fedoseev, G. A. (1984). Craniological analysis of harbor and spotted seals of the North Pacific region. *In* "Soviet–American Cooperative Research on Marine Mammals" (F. H.

Fay and G. A. Fedoseev, eds.), Vol. 1, pp. 5–16. NOAA Technical Report NMFS 12.

Dong, J., and Shen, F. (1991). Estimates of historical population size of harbour seal in Liaodong Gulf. *Mar. Sci.* **3**, 40–45.

Doutt, J. K. (1942). A review of the genus *Phoca. Ann. Carnegie Mus.* **29**, 61–125.

Ebbesmeyer, C. C., Cayan, D. R., McLain, D. R., Nichols, F. H., Peterson, D. H., and Redmond, K. T. (1991). 1976 step in the Pacific climate: Forty environmental changes between 1968–1975 and 1977–1985. *In* "Proceedings Seventh Annual Pacific Climate Workshop," Asilomar, CA., pp. 115–126. California Department of Water Research.

Geraci, J. R., St. Aubin, D. J., Barker, I. K., Webster, R. G., Hinshaw, V. S., Bean, W. J., Ruhnke, H. L., Prescott, J. H., Early, G., Baker, S. S., Madoff, S., and Schooley, R. T. (1982). Mass mortality of harbor seals: Pneumonia associated with influenza A virus. *Science* **215**, 1129–1131.

Gjertz, I., and Børset, A. (1992). Pupping in the most northerly harbor seal (*Phoca vitulina*). *Mar. Mamm. Sci.* **8**, 103–109.

Heide-Jørgensen, M.-P., and Härkönen, T. J. (1988). Rebuilding seal stocks in the Kattegat-Skagerrak. *Mar. Mamm. Sci.* **4**, 231–246.

Hoover, A. A. (1988). Harbor seal, *Phoca vitulina. In* "Selected Marine Mammals of Alaska: Species Accounts with Research and Management Recommendations" (J. W. Lentfer, ed.), pp. 125–187. Marine Mammal Commission, Washington, DC.

Kenyon, K. W. (1965). Food of harbor seals at Amchitka Island, Alaska. *J. Mammal.* **46**, 103–104.

King, J. E. (1983). "Seals of the World," 2nd Ed. Comstock Publishing Associates, Ithaca, NY.

Lowry, L. F., and Frost, K. J. (1981). Feeding and trophic relationships of phocid seals and walruses in the eastern Bering Sea. *In* "The Eastern Bering Sea Shelf: Oceanography and Resources" (D. W. Hood and J. A. Calder, eds.), Vol. 2. University of Washington Press, Seattle, WA.

Lowry, L. F., Frost, K. J., Davis, R., DeMaster, D. P., and Suydam, R. S. (1998). Movements and behavior of satellite-tagged spotted seals (*Phoca largha*) in the Bering and Chukchi seas. *Pol. Biol.* **19**, 221–230.

Naito, Y. (1974). The hyoid bones of two kinds of harbour seals in the adjacent waters of Hokkaido. *Sci. Rep. Whales Res. Inst.* **26**, 313–320.

O'Corry-Crowe, G. M., and Westlake, R. L. (1997). Molecular investigations of spotted seals (*Phoca largha*) and harbor seals (*P. vitulina*), and their relationship in areas of sympatry. *In* "Molecular Genetics of Marine Mammals" (A. E. Dizon, S. J. Chivers, and W. F. Perrin, eds.) pp. 291–304. Special Publications Number 3, The Society for Marine Mammalogy.

Olesink, P. F., Bigg, M. A., and Ellis, G. M. (1990). Recent trends in abundance of harbour seals, *Phoca vitulina*, in British Columbia. *Can. J. Fish. Aquat. Sci.* **47**, 992–1003.

Pitcher, K. W. (1980). Food of the harbor seal, Phoca vitulina richardii in the Gulf of Alaska. *Fish. Bull.* **78**, 544–549.

Quakenbush, L. T. (1988). Spotted seal, *Phoca largha. In* "Selected Marine Mammals of Alaska: Species Accounts with Research and Management Recommendations" (J. W. Lentfer, ed.). Marine Mammal Commission, Washington, DC.

Reijnders, P. J. H., Brasseur, S., van der Toorn, J., van der Wolf, P., Boyd, I., Harwood, J., Lavigne, D., Lowery, L., and Stuart, S. (eds.) (1993). "Seals, Fur Seals, Sea Lions and Walruses: Status of Pinnipeds and Conservation Action Plan." IUCN, Gland, Switzerland.

Rice, D. W. (1998). "Marine mammals of the World. Systematics and Distribution." Special Publication No. 4. The Society for Marine Mammalogy, Lawrence, KS.

Shaughnessy, P. D., and Fay, F. H. (1977). A review of the taxonomy and nomenclature of North Pacific harbour seals. *J. Zool. Lond.* **182**, 385–419.

Small, R. J., and DeMaster, D. P. (1995). "Alaska Marine Mammal Stock Assessments 1995." NOAA Technical Memorandum NMFS-AFSC-57.

Smith, R. J. (1999). "The Lacs des Loups Marins Harbour Seal, *Phoca vitulina mellonae* Doutt 1942: Ecology of an Isolated Population." Ph.D. Thesis, The University of Guelph, Guelph, Ontario, Canada.

Teilmann, J., and Dietz, R. (1994). Status of the harbour seal, *Phoca vitulina*, in Greenland. *Can. Field Nat.* **108**, 139–155.

Temte, J. L., Bigg, M. A., and Wiig, O. (1991). Clines revisited: The timing of pupping in the harbour seal (*Phoca vitulina*). *J. Zool. Lond.* **224**, 617–632.

Tikhomirov, E. A. (1966). Certain data on the distribution and biology of the harbor seal in the Sea of Okhotsk dring the summer-autumn period and hunting it. *In* "Soviet Research on Marine Mammals in the Far East." Izvestia TINRO **58**, 105–115.

Withrow, D. E., and Loughlin, T. R. (1995). Abundance and distribution of harbor seals (*Phoca vitulina richardii*) along the Aleutian Islands during 1994. *In* "Marine Mammal Assessment Program, Status of Stocks and Impacts of Incidental Take 1994," pp. 173–205. National Marine Mammal Laboratory, Seattle, WA.

Figure 1 *Adult female harp seal with "whitecoat" pup. Photo by N. Lightfoot.*

Harp Seal
Pagophilus groenlandicus

DAVID M. LAVIGNE
International Fund for Animal Welfare, Guelph, Ontario, Canada

The harp seal is one of the most abundant and best known of all pinniped species. Commonly referred to by scientists in recent decades by the Latin name *Phoca groenlandica*, the Greenland seal (following Burns and Fay 1970), it is more correctly named *Pagophilus groenlandicus* (Erxleben, 1777)—the ice lover from Greenland (Rice, 1998). Its most popular common name, the harp seal, comes from the black, wishbone marking found on the backs of adult animals, which is reminiscent of the musical instrument of the same name (Fig. 1). Another common name—the saddleback seal—refers to the same marking. The faces of adults are also black whereas the remainder of the body appears silvery-gray when dry. Young pups, which have a characteristic white pelt at birth, are known as "whitecoats" (Fig. 1).

Adult harp seals are about 1.7 m in length, with females being marginally smaller, on average, than males. Adults weigh about 130 kg early in the pupping season, but their mass (weight) varies considerably throughout the year and from year to year.

I. Distribution and Populations

Harp seals inhabit the North Atlantic and Arctic Oceans from northern Russia in the east to Newfoundland and the Gulf of St. Lawrence (Canada) in the west (Fig. 2). Although some authors recognize two subspecies (see Rice, 1998), it is more common to refer to three distinct populations (Reijnders *et al.*, 1993; Reeves *et al.*, 1992), based on small morphological, genetic, and behavioral differences. One population, which is found largely in the Barents Sea, reproduces on the "East Ice" in the White Sea off the coast of northwest Russia. (This population is designated by some Russian scientists as *P.g. oceanis*.) A second population lives off the east coast of Greenland and breeds on the "West Ice" near the island of Jan Mayen. The third lives in the northwest Atlantic off the east coast of Canada and breeds in two locations: on the "Front" off the coast of Newfoundland and Labrador, and in the Gulf of St. Lawrence (Fig. 2). (The latter two populations are assigned to *P.g. groenlandicus* by some scientists.)

The northwest Atlantic harp seal population is the largest of the three. The Canadian government now estimates that the population stabilized in the late 1990s (Department of Fisheries and Oceans, 2000). Pup production in 1999 was estimated at 998,000 ± 200,000 (mean ± 95% confidence limits). The year 2000 population was estimated to number 5.2 million animals (95% confidence interval = 4.0–6.5 million).

The most recent assessment for the West Ice population estimated 1994 pup production and population size to be 59,000 and 286,000, respectively (Anonymous, 1994).

A 1998 aerial survey in the White Sea found pup production on the East Ice to be on the order of 300,000–400,000, higher than previously thought (see Anonymous, 1999). These results are consistent with a total population size of about 1.5 to 2.0 million animals.

The harp seal is a highly gregarious and migratory species that lives in close association with pack ice. Its annual range (Fig. 2) is essentially defined by the southern and northern limits of pack ice and is largely coincident with the subarctic region of the North Atlantic (Dunbar, 1968).

II. Growth and Reproduction

From late February to mid-March, female harp seals congregate near the southern limits of their range (Fig. 2) to give birth to their pups. Each female gives birth to a single pup (although twin fetuses have been reported). Newborn pups weigh

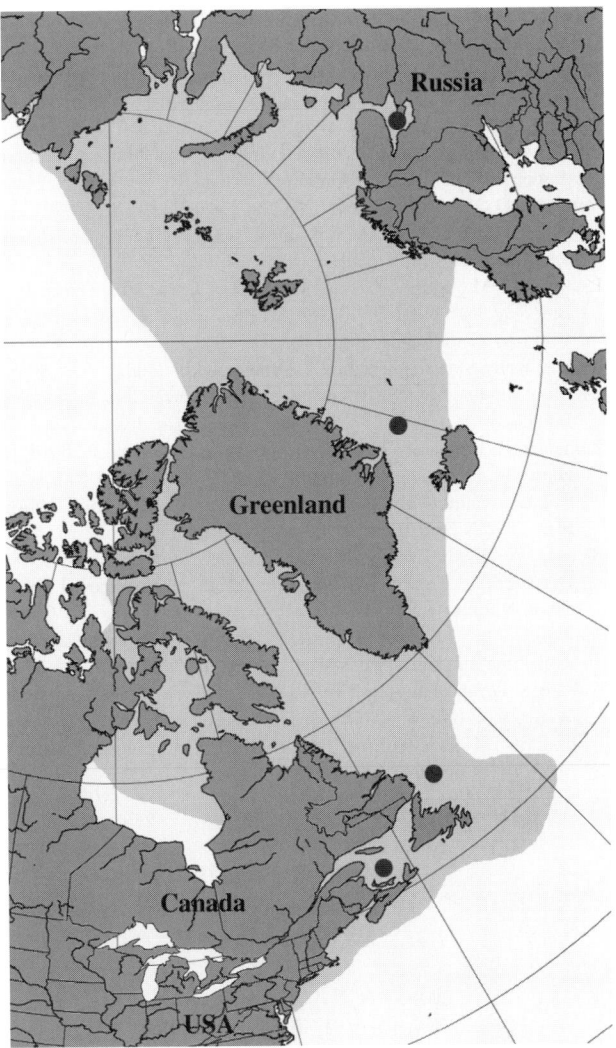

Figure 2 *Harp seal distribution in the North Atlantic Ocean. The four black circles indicate pupping areas (see text for details).*

Figure 3 *Molted harp seal pup or "beater," aged about 4 weeks. Photo courtesy of International Fund for Animal Welfare.*

about 11 kg at birth and lack the thick insulating layer of BLUBBER found in older seals. They are nursed on a fat-rich milk for about 12 days, during which time they deposit a thick (ca. 5-cm) blubber layer and grow about 2.2 kg per day. During this time they pass quickly through a number of recognizable developmental stages (Lavigne and Kovacs, 1988; Hannah, 2000). At weaning the pups weigh on average 36 kg. At this stage they are known as "graycoats" because their spotted, gray juvenile pelage has grown in and can be seen beneath the white neonatal coat. Shortly thereafter, the white coat becomes loose and, within a few days, it begins to fall out. Once the white coat is completely molted, exposing the black-spotted, silvery-gray pelt of the young harp seal, the animal is called a "beater" (Fig. 3).

At the age of about 13–14 months, young harp seals undergo their second molt. It is at this time that the "beater" pelt is replaced by a similar spotted pelt and the animals are renamed "bedlamers." They will retain this spotted pelt through successive annual molts until the spots start to disappear and

the dark, harp-shaped pattern of the adult coat begins to emerge. Older harp seals with a combination of the spotted bedlamer pelt and the distinct adult "harp" are called "spotted harps." The transition from the bedlamer pelt to the adult pelt begins with the onset of sexual maturity, at the age of 4 or 5 years (possibly later in males). Most male harp seals develop the black "harp" marking abruptly, whereas in females the transition is more gradual and may take many years. Indeed, some female harp seals never lose all their spots or develop a complete "harp." Any seal with the complete harp and black face may be aged anywhere from about 5 to 30 years—the life expectancy of the species.

Weaning in harp seals is abrupt. The adult females simply leave their pups on the ice and turn their attention to mating and the production of next year's pup. Mating usually occurs in the water (Lavigne and Kovacs, 1988), but has also been photographed on the ice (F. Bruemmer, personal communication). The fertilized egg, which results, divides several times, forming a spherical embryo, which floats freely in the womb for more than 3 months before being implanted in the wall of the uterus. This type of suspended development—known as delayed implantation—ensures that all females give birth to their pups at the same time each year when short-lived pack ice is available as a whelping platform.

Meanwhile, the weaned pups remain on the ice and undergo a postweaning fast, during which they may lose up to half of their body mass as they draw on their thick blubber layer for sustenance. The entire fast may last upward of 6 weeks, but they eventually enter the water and begin feeding on their own.

Following the breeding season, older harp seals congregate once again on pack ice farther north to undergo their annual molt. Following the molt, in which the pelt and surface layers of skin are replaced, the animals MIGRATE to summer FEEDING areas in subarctic and arctic waters to the north. All three populations exhibit similar patterns of annual migrations, although the timing of specific events, such as pupping, mating, and molting, varies slightly among populations.

III. Ecology

By pinniped standards, harp seals are modest divers. The average maximum dive depth is 370 m and the average dive duration is about 16 min (Schreer and Kovacs, 1997).

Harp seals exhibit catholic feeding habits, which vary with age, season, location, and year. While at least 67 species of fin fish and 70 species of invertebrates have been recorded in their stomachs over the past 50 years (Wallace and Lawson, 1997), harp seals tend to concentrate on smaller fishes, such as capelin (*Mallotus villosus*), arctic cod (*Boreogadus saida*), and polar cod (*Arctogadus glacialis*), and a variety of invertebrates, including euphausiids (*Thysanoessa* sp.). Contrary to many popular reports, harp seals rarely eat commercially important Atlantic cod, *Gadus morhua*. Harp seals are themselves prey for polar bears (*Ursus maritimus*), killer whales (*Orcinus orca*), and sharks (e.g., *Somniosus microcephalus*), but their major predator would appear to be *Homo sapiens*.

IV. Conservation Status

All three populations of harp seals have been hunted by humans for centuries; all three have undergone documented declines in numbers as a result of overexploitation; and all three populations continue to be hunted today. The commercial hunt for northwest Atlantic harp seals remains the largest hunt for any marine mammal population in the world, with over 460,000 animals acknowledged killed in each year between 1996 and 1999 (Department of Fisheries and Oceans, 2000, also see Johnston *et al.*, 2000). The seals are hunted in Canada during and after the spring breeding season in what was once called "the greatest hunt in the world." The same population is also hunted off the west coast of Greenland during the summer months.

Overexploitation, particularly in the northwest Atlantic (Johnston *et al.*, 2000), and an expanding and largely unregulated trade in seal products (especially seal skins, seal oil, and penises) remain potential threats to the species. Other potential threats include continued proposals to cull harp seal populations, ostensibly to benefit commercial fisheries; reduced food availability due to human overfishing or climate change; incidental catches in fishing gear; and environmental contaminants (Reijnders *et al.*, 1993).

See Also the Following Articles

Earless Seals ■ Hair and Fur ■ Hunting of Marine Mammals ■ Pinniped Physiology

References

Anonymous (1994). Report of the Joint ICES/NAFO Working Group on Harp and Hooded Seals. International Council for the Exploration of the Sea. C. M. 1994/Assess:5.

Anonymous (1999). Report of the Joint ICES/NAFO Working Group on Harp and Hooded Seals. Advisory Committee on Fishery Management. International Council for the Exploration of the Sea C. M. 1999/ACFM:7.

Burns, J. J., and Fay, F. (1970). Comparative morphology of the ribbon seal, *Histriophoca fasciata*, with remarks on the systematics of Phocidae. *J. Zool.* (*Lond.*) **161**, 363–394.

Department of Fisheries and Oceans (2000). Northwest Atlantic harp seals. DFO Science Stock Status Report E1-01.

Dunbar, M. J. (1968). "Ecological Development in Polar Regions: A Study in Evolution." Prentice-Hall Inc., Englewood Cliffs, NJ.

Hannah, J. (2000). "Seals of Atlantic Canada and the Northeastern United States," 2nd Ed., revised. International Marine Mammal Association Inc., Guelph, Ontario, Canada.

Johnston, D. W., Meisenheimer, P., and Lavigne, D. M. (2000). An evaluation of management objectives for Canada's commercial harp seal hunt, 1996–1998. *Conserv. Biol.* **14**, 729–737.

Lavigne, D. M., and Kovacs, K. M. (1988). "Harps and Hoods: Ice-Breeding Seals of the Northwest Atlantic." Univ. of Waterloo Press, Waterloo, Ontario, Canada.

Reeves, R. R., Stewart, B. S., and Leatherwood, S. (1992). "The Sierra Club Handbook of Seals and Sirenians." Sierra Club Books, San Francisco.

Reijnders, P., Brasseur, S., van der Toorn, J., van der Wolf, P., Boyd, I., Harwood, J., Lavigne, D., and Lowry, L. (1993). "Seals, Fur Seals, Sea Lions and Walrus: Status Survey and Conservation Action Plan." IUCN Seal Specialist Group, Gland, Switzerland.

Rice, D. W. (1998). "Marine Mammals of the World: Systematics and Distribution." The Society for Marine Mammalogy, Special Publication Number 4.

Schreer, J. F., and Kovacs, K. M. (1997). Allometry of diving capacity in air-breathing vertebrates. *Can. J. Zool.* **75**, 339–358.

Wallace, S. D., and Lawson, J. W. (1997). A review of stomach contents of harp seals (*Phoca groenlandica*) from the Northwest Atlantic: An update. IMMA Technical Report 97–01. International Marine Mammal Association, Inc., Guelph, Ontario, Canada.

Health

JOSEPH R. GERACI AND
VALERIE J. LOUNSBURY
National Aquarium in Baltimore, Maryland

The health of any animal is affected by age, behavior, and environment. Like terrestrial species, marine mammals are subject to infection, injury, and metabolic disturbances. Our understanding of marine mammal health has been impeded not only by the difficulties inherent in studying these species in the wild, but also by their unique biology. In recent decades, the challenge has been compounded by human impacts on the health of marine mammals and their environment.

I. Adaptations to Life at Sea

Cetaceans, sirenians, pinnipeds, and sea otters, although taxonomically distant, have evolved similar biological mechanisms to cope with a marine existence. These functional adaptations include strategies for controlling body temperature, diving, maintaining salt and water balance, and promoting reproductive success—adaptations vital to health and survival.

A. Thermal Balance

Except for some areas in the tropics, the sea is almost always cooler than mammalian body temperature. Even a few degrees difference is enough to drain a mammal's thermal reserves, as water steals body heat about 20 times faster than air. To cope with this drain, marine mammals have generally evolved mechanisms to conserve body heat. Of these, blubber has arguably been the key to evolutionary success. This coat of fat provides cetaceans and certain pinnipeds with mechanical protection, warmth, buoyancy, nutrients when food is scarce, and fresh water in reserve. Otariid pinnipeds have thinner blubber and less body fat than phocids or the walrus (*Odobenus rosmarus*) and are thus less tolerant of cold and depend to a certain extent on their pelage for insulation. This is especially true for otariid pups, which may not acquire an adult coat or adequate fat until they are about 3 months old and, in the meantime, are prone to hypothermia when they become wet. Species with less blubber rely on other strategies. The sea otter (*Enhydra lutris*) depends entirely on a high metabolic rate (and high caloric intake) to generate heat and on its dense, well-groomed fur to prevent heat loss. The living sirenians, with low metabolic rates and little ability to control surface heat loss, are functionally restricted to tropical and subtropical waters.

Environmental temperature has more than a subtle bearing on health. To survive in a cold climate, a marine mammal must be robust, appropriately insulated, and have all surface heat control mechanisms operating. If not, the only recourse is to increase the metabolic rate and either eat more or borrow fat from vital fat reserves. Ironically, as a last measure to conserve heat, pinnipeds and sea otters may haul out on land where the prospect of feeding is hopeless.

Overheating is rarely a problem for a marine mammal in water. In a warming environment, a whale may eat less and metabolize blubber, which effectively reduces insulation, and shed excess heat by increasing blood flow to the skin, particularly of the extremities. Losing heat on land is not as easy. A wet seal or sea otter may get some comfort from evaporative cooling, but once it is dry, it depends mostly on circulatory and behavioral adaptations (e.g., seeking shade, sleeping, moving to the surf zone) to avoid hyperthermia. These strategies work to a point. A sea otter out of water can become distressed at air temperatures as low as 10°C and die within hours at 21°C. A cetacean stranded on a sunny beach can literally cook inside its own blubber.

B. Breathing and Diving

Marine mammals forage at all depths. While sea otters and sirenians may have little need to dive deeply or for more than just a few minutes, some species of phocids and odontocete cetaceans make prolonged dives to 2000 to 3000 feet or more. Such deep diving is made possible by a suite of adaptations for coping with pressure and potentially deadly nitrogen, and for storing and utilizing oxygen efficiently. During a prolonged dive, circulation to the skin and viscera may almost cease, allowing oxygen to be channeled to organs that need it most, such as the heart and brain.

Deep dives may require a shift to anaerobic metabolism. Such dives may be useful for escape and exploration but are costly in terms of time and energy. For most animals, survival depends on obtaining sufficient prey within the depth and time limits imposed by aerobic diving capacity, which in turn depends on the species and the size, age, and health of the individual. Because of their relatively greater capacity to store oxygen, large animals tend to be better divers. It is not surprising that juveniles may find it difficult to reach prey that is easily accessible to adults.

C. Salt and Water Balance

The osmotic concentration of the sea is nearly four times greater than that of mammalian body fluids. Chemical equilibrium thus favors loss of body fluids into the sea and encroachment of salts into the animal. Marine mammals have evolved a number of strategies to prevent this from happening: (1) external surfaces are impermeable to seawater; (2) body water is highly conserved—sweat glands are either reduced or absent and the kidneys efficiently concentrate urine; (3) they drink little seawater and acquire most of their fresh water from food (water makes up about 70% of a fish, 80% of a squid, and over 90% of aquatic plants, and each gram of dietary fat or metabolized blubber yields close to its weight in fresh water). In pinnipeds and cetaceans, the physiological response to stress is also designed to conserve water. Secretion of aldosterone, a hormone produced by the adrenal cortex, promotes the resorption of sodium from the kidney, thereby drawing water back into the body. Maintaining electrolyte balance thus depends on adequate blubber, well-functioning kidneys, proper hormonal balance, and a healthy, intact epidermis.

D. Strategies for Rearing Young

The social, physical, and biological conditions that together create a healthy environment are especially critical during the period of an animal's life when it depends totally on its mother. Vulnerabilities can often be predicted on the basis of species, location, patterns of maternal care, and environmental conditions. For example, many pinnipeds have evolved strategies to ensure breeding opportunities for animals that may be dispersed for much of the year, with the result that pupping and mating occur at predictable times and locations. While an effective strategy for the population, consequent crowding on rookeries can increase the risk of injury, infection, and disease transmission for individuals.

II. What Can Go Wrong?

Body systems work together, each complementing the others. Impairment of one system can disturb the entire equilibrium, leading to secondary problems, which then become a threat. For example, blubber is a source of energy, insulation, water reserves, and buoyancy. If it becomes depleted because food is unavailable, the animal may eventually be unable to rest at the surface, maintain body heat, forage, escape predators, or keep up with a group. The ensuing stress may open the door to infection, further reducing the chance of survival.

Injuries and illnesses are not always apparent and are often detected only after analyzing blood or tissue samples from a living animal or dissecting a dead one. Even careful study might not reveal serious biochemical and physiological conditions. Stress is poorly understood and difficult to quantify. What little is known about the process in marine mammals shows that it can disrupt thyroid and adrenal cortical function, water and electrolyte balance, and metabolism and reproduction, and can decrease circulating levels of certain blood cells, perhaps reducing resistance to parasitic infection and compromising immune responses.

A. Reproductive Failure and Death of the Newborn

An orderly, coordinated progression of biological and behavioral factors is required for an animal to reproduce successfully. Weakness or disruption at any point can result in reproductive failure, evident as abortion, stillbirth, premature birth, or death of the newborn. The causes of reproductive failure are often obscure, particularly in species that cannot be studied from shore.

In some species, the risk of abortion or stillbirth appears to be greater for first-time mothers. Young mothers are also often smaller and give birth to smaller offspring, which are more vulnerable to hypothermia and injury. The health and nutritional condition of any mother, regardless of age or size, affects the fetus. Decreased prey abundance, such as associated with El Niño events, may be associated with decreased fertility, increased abortions, and reduced pup production. Infectious disease [e.g., morbillivirus in harbor seals (*Phoca vitulina*) in Europe] may also lead to increased premature births and abortions. In fact, a rise beyond the expected level of reproductive failure may be one of the first signs of an environmental disruption, such as a viral epidemic, reduced prey stocks, or high levels of certain anthropogenic contaminants.

A successful birth is no guarantee of prolonged survival. Neonates that are weak at birth or suffer from serious congenital defects soon die. Healthy neonates may, for one reason or another, be abandoned by their mothers or face an early death if the mothers fail to provide proper care, whether due to illness, injury, disturbance, or simply to inexperience.

B. Starvation

Marine mammals spend much of their time searching for food of the appropriate type, size, and quality to satisfy needs that may vary seasonally and with age. Some animals, e.g., dependent young, the sick, and the very old, can starve even when food is plentiful. Many factors determine how long an animal can survive without food: its age, fat reserves, metabolic rate, energy demands, and general health. Large animals with low metabolic rates survive longer than those with high energy demands, such as small species, newborn, and growing pups. Baleen whales may feed very little for 6 to 8 months of the year, but a sea otter without food for even 2 days can die from gastroenteritis and shock. Starvation is a major cause of death in pinniped and sea otter pups.

Throughout the period of dependency, a young animal's survival hinges on the health of its mother. Before giving birth, a phocid seal or baleen whale must develop ample fat reserves to carry it through a period of fasting or reduced feeding during lactation. The pup or calf born of a malnourished mother is at risk from the moment of birth and its longevity is compromised early in its development. The young of species in which females feed continuously during lactation face a different threat. A bottlenose dolphin (*Tursiops truncatus*) calf depends on the state of its mother's nourishment throughout what may be a year or more of nursing. An otariid pup risks starvation if a shift in prey abundance forces its mother to spend longer periods away from the rookery.

Weaning frees a young animal from dependence to face the challenge of providing for itself. Manatees (*Trichechus* spp.) and some cetaceans and otariids remain with their mothers long enough to learn foraging skills. Not so for all sea otters. Newly independent juveniles, handicapped by high metabolic demands and inexperience, and at a time when erupting TEETH create problems with chewing, often starve. Females may be especially vulnerable because they tend to remain in established areas of the range even when prey become depleted.

Depletion of food stocks, whether from overexploitation, overfishing, or climatic or oceanographic fluctuation, can affect entire populations. Food scarcity in one area may cause some animals to move elsewhere. When food abundance changed during the El Niño of 1982–1983, California sea lions (*Zalophus californianus*) moved northward, and many northern fur seals (*Callorhinus ursinus*) may have emigrated from San Miguel Island to rookeries in the Bering Sea. Some animals are unable or unwilling to make such excursions, e.g., females with pups, territorial males, or populations in remote ranges. When fish disappeared from surface waters around the Galapagos Islands during the 1982–1983 El Niño, widespread starvation among the islands' fur seal population (*Arctocephalus galapagoensis*) soon followed.

Starving animals eventually die—some quickly, as would a pup deprived of milk or a sea otter overcome with hypothermia and exhaustion. Others die after a period of illness triggered by malnutrition and mediated by factors such as hypothermia, dehydration and electrolyte imbalance, hormonal disturbances, and infection by parasites and opportunistic pathogens. Some starving seal pups may ingest whatever is nearby—gravel, stones, or grass—and consequently die of an impacted stomach.

While a sudden shortage of prey may cause outright starvation of large numbers of animals, the more subtle effects of nutritional stress, including low productivity and decreased juvenile survival, may prove equally damaging to a population.

C. Direct Environmental Effects

Extreme weather conditions can take a toll on all age classes. Among Florida manatees (*Trichechus manatus latirostris*), intense or prolonged cold weather can cause mortality equivalent to 1.5–2% of the estimated population, with the greatest impact on juveniles. Storms hitting a crowded pinniped rookery at the peak of breeding season can be disastrous: pups become hypothermic, are battered on rocks or drowned, are separated from their mothers and starve, or become victims of adult aggression. Unusual ice conditions can be hazardous for cold-water species. Sea otters trapped out of water by heavy ice soon die of starvation, stress, and shock. An untimely freeze in polar waters can trap cetaceans in ice, where they may ultimately suffocate or starve. Ice-breeding seals can drown or be crushed in large numbers if their ice floes are broken up by storms or unseasonably warm temperatures.

D. Trauma

For most marine mammals, the risk of injury is continual, whether from natural sources, such as storms, predators, and aggressive encounters, or human activities, such as fishery operations and recreational boating. For example, injuries are common on pinniped ROOKERIES, where pups are often trampled accidentally or attacked by adults, fall into gullies or crevices, or wash off unprotected beaches into pounding surf. Adults can be victims as well, as bulls compete for territories and females, and females compete for space.

Historically, commercial hunting had serious impacts on certain species or stocks of marine mammals. Today, more animals die in accidents; interaction with fishers is a leading cause of death and injury. Pelagic odontocetes die in purse seines and drift nets, coastal cetaceans and pinnipeds in gill net and trawl fisheries, and some river dolphins by fishing methods that employ electricity and explosives. Marine mammals thought to compete with commercial operations may be killed deliberately.

Entanglement in discarded net fragments, ropes, packing bands, monofilament line, and other debris is a risk for many species. The effects on populations vary; some suffer no appreciable impact, whereas others may be seriously threatened. For the individual, the problem is always serious. An animal that does not drown immediately may escape with fractures and internal injuries or may carry net fragments, ropes, or bands that increase drag, impede swimming ability, or become snagged. A seal pup growing into its packing-band "collar" will eventually die, either from suffocation or from deep cuts and infection.

Coastal dwellers are vulnerable to injury from a variety of human activities. For example, many dugongs (*Dugong dugon*) in Queensland (Australia) waters have died in shark nets set to protect public beaches. Right whales (*Eubalaena glacialis*) in the northwest Atlantic and manatees in Florida are injured or killed by collisions with vessels at rates that jeopardize these populations.

E. Predation

There are times in a marine mammal's life when it draws the attention of predators. Probably the easiest meal is a small, inexperienced animal that can be found in a particular place on schedule—criteria often met by young pinnipeds, whether on land or ice or at sea. As some examples, arctic foxes and polar bears break into ringed seal (*Pusa hispida*) birth lairs to take pups and, sometimes, their mothers. Steller sea lions (*Eumetopias jubata*) on the Pribilof Islands eat young northern fur seals that venture into the water, whereas southern sea lions (*Otaria flavescens*) raid South American fur seal (*Arctocephalus australis*) rookeries, driving away the adults and killing pups. Leopard seals (*Hydrurga leptonyx*) consume large numbers of crabeater seals (*Lobodon carcinophaga*) from the time the weaned pups leave the safety of the ice until they are several months old and large enough to escape attack. Killer whales (*Orcinus orca*) patrol some pinniped rookeries and may work vigorously to wash a seal into the water or even chase one onto the beach; other pods may attack baleen or sperm (*Physeter macrocephalus*) whales. Sharks pose a danger to many species or populations, including sea otters along California and Hawaiian monk seals (*Monachus schauinslandi*).

The impact of a predator can extend beyond its effect on the individual prey. Killing a pregnant mother with a dependent young removes not one, but three animals from the population. A female northern elephant seal (*Mirounga angustirostris*) may recover rapidly from a shark attack, as many seem to do, but may be less able to nurse her pup and is unlikely to mate in the compressed breeding season. In this case, a single attack, while only injuring the mother, may have cost the population two pups.

F. Parasites

Almost all marine mammals are infected by parasites by the time they are weaned or shortly afterward. Most of these parasites have evolved with their hosts and, under normal circumstances, cause little damage to otherwise healthy animals. Among these are the amphipods and copepods that eat bits of whale skin, seal lice that normally occur in small numbers and consume insignificant amounts of blood, and gastrointestinal helminths ("stomach worms"). Others are harmful enough to affect the well-being of individuals and even segments of a population. For pinnipeds, these include heartworms, some lungworms, and the hookworm *Uncinaria lucasi;* and in cetaceans, the nematodes *Crassicauda* spp. (in the mammary glands, cranial sinuses, and kidneys) and the trematodes *Nasitrema* spp. (in the cranial sinuses) and *Campula* spp. (in the liver and pancreas). However, any parasite can become destructive when the mechanisms that maintain the host–parasite balance break down, as they do when an animal is ill or starving. Prolonged stress, by retarding wound healing and destroying protective blood cells, can set the stage for a parasite to do the most harm. Debilitated animals that come ashore often suffer from serious parasitic conditions.

An animal's parasite burden can offer clues to its overall health and to changes in its environment, such as alterations in prey abundance. Seal lice are transmitted and proliferate on the animal only on land. A heavy infestation requires that the animal be on shore a long time, one sign that it may be ill. A fast-swimming odontocete offers barnacles little opportunity to attach; the presence of species such as *Lepas* sp. or *Xenobalanus* sp. on a dolphin's flukes or dorsal fin suggests that the animal has been moving unusually slowly, a common sign of illness. Differences in parasite fauna can indicate differences in feeding habits. For example, walruses feeding on benthic invertebrates have few if any nematodes in their stomachs, whereas those that eat fish have more. The relationship between diet and parasitism is predictable enough that variations in parasite burden may be used to distinguish populations and help identify segregated social groups.

G. Microorganisms

Microorganisms of all kinds—bacteria, fungi, and viruses among them — abound in the sea. Some are of the types found on land and in land dwellers; others, including certain *Vibrio* bacteria, thrive only in aquatic habitats. Like terrestrial mammals, marine mammals harbor many organisms that may be regarded as normal. Few of these are necessarily pathogenic, meaning they do not always cause infectious disease, but some are more threatening than others. The fine line between infection and infectious disease depends on both the virulence of the organism and the susceptibility of the host, which is determined by the history of previous contact with the organism and the health of the animal's immune system. Age is also a factor. A very young animal may be protected by maternal antibodies, which protect it against organisms with which the mother has earlier come into contact. The pup or calf then develops its own active immune capability, which affords increasing protection until its declining years, when immune function once again weakens. For these reasons the very young and the very old are more likely to acquire infections. Of course, natural and human-related stresses may compromise immune function in animals of all ages.

1. Bacteria The nature and severity of bacterial infections can be influenced by the animal's behavior and age, and environmental conditions. Habitat also plays a determining role. A phocid pup born on clean sand is less likely to acquire a serious navel infection as it drags its unhealed umbilicus across the rookery than a pup born in areas fouled by feces, stagnant water, or decaying vegetation. For pups in fouled environments, bite wounds provide another route for infection by bacteria such as *Streptococcus* sp. and *Corynebacterium* sp.

Infections are sometimes predictable. During molt, seals slough skin and hair. In northern elephant seals the process is exaggerated and large sheets of epidermis are lost; many animals, particularly yearlings, come ashore with skin infections during this time. Weddell seals (*Leptonychotes weddellii*), which use their teeth to maintain breathing holes in ice, and sea otters that feed on hard-shelled prey grind down their teeth to such an extent that they develop abscesses and bone infection.

A few bacteria are inherently pathogenic. Leptospirosis, caused by the spirochete *Leptospira* sp., occurs in domestic and wild animals worldwide. Infection in California sea lions has caused kidney disease in juvenile and subadult males and abortion in females. *Mycobacteria* of the complex associated with tuberculosis (*M. bovis, M. tuberculosis*) are of growing concern. An outbreak in a captive colony of New Zealand fur seals (*Arctocephalus forsteri*) and Australian sea lions (*Neophoca cinerea*) was the first indication that this disease, subsequently found in free-ranging pinnipeds from Australia, New Zealand, and South America, may be endemic in certain wild populations. Bacteria representing an apparently new strain or species of *Brucella* have been found in a number of marine mammal species, primarily from the North Atlantic and Artic oceans. Although implications for marine mammals are uncertain, infection in terrestrial mammals commonly leads to abortion.

The impact of an infection on animal health depends on the organ involved. An isolated abscess in a muscle may have little apparent effect, while a similar infection in the lung can be seriously debilitating. Bacterial pneumonia, often associated with lungworms, can be serious enough to cause death or stranding. Infections that increase metabolic stress or disturb water and electrolyte balance, such as gastroenteritis with vomiting and diarrhea, can be rapidly fatal.

2. Mycotic Infections Fungal organisms rank low on the list of primary pathogens of marine mammals. They tend to infect animals that are weakened, perhaps by other chronic debilitating disease. Infections are usually acquired from soil-, dust-, or water-borne fungi and enter the body through the skin or by inhalation. A wide variety of organisms have been isolated from marine mammals, including *Candida, Aspergillus, Coccidioides, Blastomyces, Histoplasma, Fusarium, Nocardia,* and *Loboa.*

Lobomycosis, a skin infection caused by the yeast *Loboa loboi,* has an unusual range. The disease occurs in free-ranging and captive bottlenose dolphins from Florida waters and in tucuxi (*Sotalia fluviatilis*) in South America. Curiously, other than in cetaceans, Lobo's disease occurs only in people inhabiting low-lying wetlands of Central and South America.

Coccidioidomycosis, generally a rather innocuous fungal disease of domestic animals, was until recently considered rare in marine mammals. What might be described as outbreaks of infection in California sea lions and sea otters between 1986 and 1994 coincided with a rise in human infections, attributed to environmental conditions that favored the growth of *Coccidioides imitis.*

3. Viruses First recognized in the late 1960s, viral infections in marine mammals have emerged as the greatest cause of large-scale mortality. To spread rapidly, a virus requires a naive host population of a minimum density, which can arise either through population growth or changes in social behavior. Once infected, a migrating or wandering animal may carry the virus to new habitats.

More than 450 harbor seals (*Phoca vitulina*) died in a disease outbreak in New England during the winter of 1979–1980. The cause was found to be an influenza virus of avian origin that had infected the seals, probably as they hauled out on the rookeries of Cape Cod. Seals of all ages developed pneumonia, which forced many out of the water and onto crowded beaches where the virus could spread easily from seal to seal by aerosol transmission. This was the first marine mammal die-off of demonstrated viral origin.

More devastating were the outbreaks of morbillivirus infection that swept through populations of pinnipeds and cetaceans in Europe in the late 1980s and early 1990s. The series of epidemics began with the outbreak of canine distemper virus infection that killed thousands of Baikal seals (*Pusa sibirica*) in 1987–1988. A related morbillivirus (phocine distemper virus) killed about 17,000 harbor seals and a few hundred gray seals (*Halichoerus grypus*) in Europe in 1988–1989. Between 1990 and 1992, another morbillivirus (dolphin morbillivirus) killed thousands of striped dolphins (*Stenella coeruleoalba*) in the Mediterranean Sea. Infected animals developed pneumonia, fever, and neurological disorders associated with encephalitis. The immunosuppressive effect of these viruses led to the development of secondary, often overwhelming, infections by bacteria, fungi, and other viruses.

Studies since 1988 indicate that morbillivirus infection, often without recognized illness, is common in many marine mammal species and may have occurred in many North Atlantic marine mammal populations prior to the European epidemics. The outbreaks in European seals may have been the result of viruses, perhaps introduced by infected migrating harp seals (*Pagophilus groenlandicus*) entering naive populations that were dense enough to support transmission. The epidemic in striped dolphins showed that the brief periods during which dolphins surface in unison, and perhaps some behaviors underwater, may be enough for a viral infection to spread rapidly from one cetacean to another.

The morbillivirus outbreaks offered clues to past events, such as the unexplained die-off of crabeater seals along the Antarctic Peninsula in 1955. Serological studies have tentatively linked that event to morbillivirus, perhaps transmitted from sled dogs. The 1990–1992 epizootic in striped dolphins suggested that morbillivirus infection, which was observed in some bottlenose dolphin carcasses examined during a die-off along the U.S. mid-Atlantic coast in 1987–1988, may have played an important role in that event. Indeed, retrospective studies present strong evidence that morbillivirus outbreaks have occurred sporadically in coastal bottlenose dolphin populations along the southeast United States since the early 1980s.

A number of viruses are associated with less serious health conditions. Poxviruses, for example, commonly cause skin lesions in pinnipeds and cetaceans; pox disease often appears in conjunction with other illnesses or stress. Herpesviruses are also common in cetaceans and pinnipeds and, although not usually serious, have been associated with fatal pneumonia and hepatitis in harbor seal pups and encephalitis in one stranded harbor porpoise (*Phocoena phocoena*). Calicivirus infection is common among many marine mammals in the North Pacific; clinical disease, which in California sea lions appears as vesicular lesions on the skin of the flippers and mouth, may accompany stress, debilitation, or other infectious conditions, particularly leptospirosis.

Numerous other viruses have been found in marine mammals, many without recognized effect. The list will undoubtedly grow, as will our understanding of their significance, as investigators become more alert to the presence and effects of viruses, and as techniques to isolate and identify them continue to improve.

H. Metabolic Disorders

Metabolic processes sometimes break down. Environmental and biological factors that control hormonal regulation may fail to become synchronized, demands on the system may be overtaxing, and organ function, under the influence of a genetic clock, deteriorates with age and illness. The animal becomes incapacitated, but the underlying reason may be evident only at the molecular level and therefore is difficult to detect. Not surprisingly, little is known about metabolic diseases in aquatic species.

In marine mammals, salt and water balance is regulated in part by the adrenal gland. Aldosterone, secreted from the adrenal cortex, normally acts on the kidney tubules to conserve sodium and thereby maintain salt and water balance. In pinnipeds, conditions that lead to prolonged stress, including molt, malnutrition, and disease, can exhaust the gland of aldosterone, resulting in loss of sodium from the body, a condition known as hyponatremia. Affected animals lose their appetite, become weak and disoriented, and eventually die. Aldosterone features in the stress response of cetaceans as well, only it does not become depleted and the animals do not develop hyponatremia.

Quite the contrary, in severe stress following a stranding, a cetacean may eventually begin to drink seawater and develop a condition of salt overload, or fatal hypernatremia, that dehydrates tissues, including the brain.

I. Tumors

Marine mammals develop all kinds of tumors, from benign lipomas that are little more than fatty lumps in the great whales to highly malignant lymphomas in young seals. As studies on marine mammals have increased, so have the numbers and variety of tumors reported.

In other mammals, tumors have been associated with a variety of factors, including hormones, viruses, congenital and hereditary defects, and physical and chemical agents. Establishing these links has generally required years of investigation on large populations and a systematic consideration and elimination of other possible contributors. These requirements are difficult to meet in marine mammal studies. Hence it may never be possible to prove the assumption that environmental contaminants are responsible for the unusually high incidence of tumors in beluga whales (*Delphinapterus leucas*) in the St. Lawrence River, however tempting the link. One study may be more fruitful. Recent investigations suggest that a virus may be responsible for the high incidence of urogenital tract carcinomas observed in stranded California sea lions.

J. Biotoxins

Certain species of dinoflagellates and algae produce toxins that accumulate in some fishes and invertebrates, eventually poisoning animals further up the food chain. Prior to the late 1980s, such biotoxins had been suspected, but not proven, to play a role in several events involving marine mammals. For example, ciguatoxin, a dinoflagellate neurotoxin, was implicated in the illness of about 50 Hawaiian monk seals on Laysan Island in 1978; the weak, lethargic seals eventually became emaciated, suffered from severe parasitic infections, and died. Fourteen humpback whales (*Megaptera novaeangliae*) died in Cape Cod Bay (Massachusetts) in the winter of 1987 after eating mackerel containing saxitoxin, a neurotoxin that even in minute quantities causes respiratory paralysis.

In 1988, brevetoxin, a neurotoxin produced by the dinoflagellate *Gymnodinium brevis*, the organism responsible for "red tides," was implicated in a die-off of several hundred Atlantic bottlenose dolphins along the U.S. mid-Atlantic coast. Although the extent of the role of brevetoxin in that event remains unclear, this toxin has since been linked to the mortality of bottlenose dolphins and Florida manatees in the Gulf of Mexico, where red tides are a recognized threat to the manatee population. Red tide outbreaks in southwest Florida in 1983 and 1996 killed about 37 and 155 manatees, respectively; these animals died from the acute and chronic effects of ingestion of toxins and inhalation of toxic brevetoxin aerosols.

In 1998, California sea lions along central California were poisoned by domoic acid, a neurotoxin produced by the diatom *Pseudo-nitzschia* sp. It caused convulsions, loss of coordination, and vomiting. While more than half of the stranded sea lions died, others were brought to rehabilitation centers and recovered. The similarity of this event to previous episodes in

California sea lions and northern fur seals in the same area suggests that blooms of this diatom could have an effect on discrete populations.

Marine mammals may be particularly susceptible to the neurological action of biotoxins for several reasons: (1) during a dive, blood is channeled to the heart and brain, effectively concentrating toxin there, and away from the liver and kidney where it is normally metabolized and excreted; (2) a short period of disorientation may be enough to impede an animal's ability to reach the surface for a vital breath of air; and (3) animals that remain in the area of a bloom may be subject to the cumulative effects of toxins ingested or toxic aerosols inhaled over a period of days or weeks.

K. Strandings

Stranding is defined as having run aground. The term here describes any marine mammal that falters ashore ill, weak, or simply lost. Most animals die at sea and only a fraction reach the shore. Those that do generally reflect the age, sex, and density of the animals in the area. Any change in the expected profile may be a signal that something unusual is happening, such as a toxic event, a disease outbreak, intensive local fisheries operations, or a change in prey abundance.

Pinnipeds and, to a lesser extent, sea otters normally spend time ashore, but only those unwilling or unable to return to sea are considered stranded. These would include pups that become separated from their mothers prematurely or fail to make a successful transition to independence. Most strand in the vicinity of the rookery, although some may stray far from their normal range. Other than in spring, when pups come ashore in large numbers, and in the absence of unusual events such as disease outbreaks, pinnipeds normally strand alone.

Many cetaceans that strand singly are debilitated in some way. Some offshore species strand with characteristic illnesses. Short-beaked common dolphins (*Delphinus delphis*) along California, for example, develop parasite-related brain damage, and dwarf (*Kogia sima*) and pygmy (*K. breviceps*) sperm whales along the U.S. Atlantic and Gulf coasts come ashore with impacted stomachs after ingesting plastic bags and other debris.

A mass STRANDING can be defined as two or more cetaceans, excluding mother–calf pairs, that come ashore alive at the same time and place. Highly social species of odontocetes [e.g., sperm whales, pilot whales (*Globicephala* sp.), false killer whales (*Pseudorca crassidens*), and Atlantic white-sided dolphins (*Lagenorhynchus acutus*)] are the most probable victims. Many explanations have been proposed, but the only common link seems to be the strong social nature of these species. Once one or more animals strand, for whatever reason, the compulsion to stay together brings others ashore.

A stranded animal's chances of surviving diminish by the hour. Sea otters and pinnipeds risk hyperthermia, injury from terrestrial predators, and starvation. A cetacean has difficulty shedding heat even in cold weather, and a larger one may develop respiratory fatigue and distress as the chest cavity is compressed under its own weight. Within a few hours of stranding, some cetaceans begin to show evidence of shock or vascular collapse, which leads to poor circulation and impaired organ function. The onset of shock further impairs the whale's health and may prevent its recovery, even if it is returned to sea in what appears to be good condition.

L. Habitat Alteration and Disturbance

Marine mammals have adapted over millions of years to the often harsh conditions of the marine environment. In the past few decades, environmental change has proceeded at a rate far exceeding the slow pace of evolution. How well can marine mammals cope with urban and industrial wastes, coastal dredging, undersea construction, vessel traffic, and NOISE? As with other influences on health, the effects—if they can be determined with any degree of certainty—will vary depending on species, sex, age, individual tolerance and behavior, and a host of other factors.

1. Contaminants As long-lived predators at the top of the food chain, marine mammals accumulate contaminants in their tissues. The concentrations and distribution within tissues depend on the type of contaminant and the animal's age and sex. Because most compounds accumulate over time, older animals generally have more. Fat-soluble substances, such as the persistent DDT, PCBs, and related organochlorines, reside in fatty tissues like BLUBBER, liver, and brain; heavy metals are found in liver, but also distribute in muscle, kidney, and other organs. Pregnant and lactating females produce milk using stored fat and the chlorinated hydrocarbons that came with it. While the suckling offspring loads up with contaminated milk, the female depletes her stores and, over time, has proportionally less and less than a male of equivalent size and age. What concentrations are eventually harmful to the male, to the female as she loads and unloads the compounds with each reproductive cycle, or to the pup that may be even more sensitive? What happens to an animal of any age that becomes ill, stops eating, and uses stored fat, which releases these potentially toxic compounds into the bloodstream where, in increasingly higher concentrations, they are carried to other tissues?

As yet, no clear picture emerges, and broad differences in effects among species continue to invite speculation. In Baltic seals, organochlorine levels seem to be associated with low pregnancy rates and uterine pathology, as well as a disease complex characterized by metabolic disorders, hormonal imbalance, cranial bone lesions, and reduced immune function. The nature of marine mammals and the environment they live in pose serious challenges to conducting investigations that require tight controls and sophisticated technology. Meanwhile we rely on empirical observations and preliminary studies that offer clues. Experimental studies are, nevertheless, yielding data supporting the link between exposure to certain chlorinated hydrocarbons and impaired immune function in at least some species. A better understanding of the influence of contaminants on susceptibility to infectious disease will likely emerge from continued laboratory investigations.

2. Oil Spills Oil spills are visible and unsightly, and sea otters show us how quickly fatal one can be. The 1989 *Exxon Valdez* incident in Prince William Sound, Alaska, was dramatic and beyond the proportion of other spills that have affected marine mammals. Until that event, relatively few marine mammals were known to have been killed by oil.

The impact of spilled oil depends on its composition, environmental conditions, and the species involved. During the first few hours or days after a spill, low molecular weight fractions are the most acutely toxic. They irritate and harm tissues, especially the sensitive membranes of the eyes and mouth; they can be ingested during feeding or when a fouled animal is grooming; or their vapors can be inhaled and damage the lungs. Light fractions are absorbed into the blood where they can attack the liver, nervous system, and blood-forming tissues. Sea otters caught in the *Exxon Valdez* spill showed signs of lethargy, respiratory distress, and diarrhea, and evidence of liver damage, kidney failure, and endocrine imbalance. Between 3500 and 5500 otters were estimated to have died. Three hundred harbor seals also died; many had brain lesions, probably resulting from inhalation of vapors from fresh oil.

Evaporation of the low molecular weight fractions leaves heavy residues and thick, foamy emulsions called mousse. By sticking tenaciously to vital insulating hairs of sea otters, polar bears, and some species of pinnipeds (e.g., fur seals), these substances can destroy the animals' ability to maintain thermal balance. The sea otter is especially vulnerable because its entire existence depends on a well-groomed hair coat.

Except for the sea otter, there is no real evidence that marine mammals ingest much oil. They may be able to deal with small quantities of fresh oil or that premetabolized by their prey because they, like other mammals, have the liver enzymes required to metabolize and excrete such compounds.

3. Ingesting Debris Some marine mammals become entangled in fishing nets and debris. Others are as likely to ingest various types of discarded items and trash that enter the oceans—mostly from land sources—at a rate of over 6 million metric tons each year. Florida manatees, for example, face increasing risks of ingesting fishing line and hooks, wire, plastic bags, and other rubbish trapped in floating mats of vegetation. Some cetaceans, including pygmy sperm whales and some beaked whales, share a tendency to ingest plastics. Some items are small and inconsequential; others may block or perforate the gastrointestinal tract, leading to slow starvation or sudden death.

4. Other Disturbing Influences Habitat degradation can take many other forms: prey depletion, nutrient enrichment that leads to toxic algal blooms, underwater drilling noise, heavy vessel traffic, and disturbance of pupping or calving areas, to name a few. The potential range of effects is immense. A boat traveling through one of Florida's canals might collide with a manatee and kill it or raise the turbidity and inhibit the growth of water plants that are vital to its diet. Individuals might respond to food shortage or disturbance by moving to marginally suitable environments, e.g., northward to colder waters, where risks of cold stress are increased. A harp seal wandering far from its normal range following a collapse of prey stocks might introduce a pathogenic virus into a susceptible population. A sudden, unusual noise near a crowded pinniped rookery might cause animals to panic and stampede, trampling or abandoning their young.

Other reactions to disturbances may be more subtle. In terrestrial mammals, intense noise alone can cause disorders ranging from long-term hearing loss to physiological stress, hypertension, hormonal imbalance, and lowered resistance to disease. Such effects are nearly impossible to document in marine mammals. It can be assumed that animals are generally unlikely to become habituated physiologically to any disturbances that are associated with threatening situations.

III. The Future

We have a growing understanding of the range of pathogens in the sea and the mechanisms marine mammals have evolved to counter their effects. Except for the inevitable rise of new disease agents and the discovery of old ones, the elements of this endless tug-of-war are unlikely to change. Here, humans are only observers. However, the expression of illness, whether in an individual or a population, is governed by dynamic environmental conditions, and some of these are within our ability to control. Responsible stewardship of the oceans and coastal waters may, by that reckoning, be the key to marine mammal health in the future.

See Also the Following Articles

Cetacean Physiology ▪ Circulatory System ▪ Diving Physiology ▪ Mass Die-Offs ▪ Parasites ▪ Pathology ▪ Pollution and Marine Mammals ▪ Stranding

References

Anderson, D. M. (1994). Red tides. *Sci. Am.* **271,** 62–68.

Baker, J. R. (1984). Mortality and morbidity in grey seal pups (*Halichoerus grypus*): Studies on its causes, effects of environment, the nature and sources of infectious agents and the immunological status of pups. *J. Zool. (Lond.)* **203,** 23–48.

Duignan, P. J., House, C., Odell, D. K., Wells, R. S., Hansen, L. J., Walsh, M. T., St. Aubin, D. J., Rima, B. K., and Geraci, J. R. (1996). Morbillivirus infection in bottlenose dolphins: Evidence for recurrent epizootics in the western Atlantic and Gulf of Mexico. *Mar. Mamm. Sci.* **12,** 499–515.

Elsner, R. (1999). Living in water: Solutions to physiological problems. *In* "Biology of Marine Mammals" (J. E. Reynolds and S. A. Rommel, eds.), pp. 73–116. Smithsonian Institution Press, Washington, DC.

Geraci, J. R., Harwood, J., and Lounsbury, V. J. (1999). Marine mammal die-offs: Causes, investigations, and issues. *In* "Conservation and Management of Marine Mammals" (J. R. Twiss, Jr., and R. R. Reeves, eds.), pp. 367–395. Smithsonian Institution Press, Washington, DC.

Geraci, J. R., and Lounsbury, V. J. (1993). "Marine Mammals Ashore: A Field Guide for Strandings." Texas A&M University Sea Grant Publications, Galveston, TX.

Geraci, J. R., and St. Aubin, D. J. (1987). Effects of parasites on marine mammals. *Int. J. Parasitol.* **17,** 407–414.

Geraci, J. R., and St. Aubin, D. J. (eds.) (1990). Summary and conclusions. *In* "Marine Mammals and Oil: Confronting the Risks," pp. 253–256. Academic Press, San Diego.

Hall, A. J. (1995). Morbilliviruses in marine mammals. *Trends Microbiol.* **3,** 4–9.

Harwood, J., and Hall, A. J. (1990). Mass mortality in marine mammals: Its implications for population dynamics and genetics. *Trends Ecol. Evol.* **5,** 254–257.

Higgins, R. (2000). Bacteria and fungi of marine mammals: A review. *Can. Vet. J.* **41,** 105–116.

Kanwisher, J., and Ridgway, S. H. (1983). The physiological ecology of whales and porpoises. *Sci. Am.* **248,** 110–120.

Laist, D. W., Coe, J. M., and O'Hara, K. J. (1999). Marine debris pollution. *In* "Conservation and Management of Marine Mammals" (J. R. Twiss, Jr., and R. R. Reeves, eds.) pp. 342–366. Smithsonian Institution Press, Washington, DC.

Loughlin, T. R. (ed.) (1994). "Marine Mammals and the *Exxon Valdez.*" Academic Press, San Diego.

O'Shea, T. J. (1999). Environmental contaminants and marine mammals. *In* "Biology of Marine Mammals" (J. E. Reynolds and S. A. Rommel, eds.), pp. 485–563. Smithsonian Institution Press, Washington, DC.

Trillmich, F., Ono, K. A., Costa, D. P., DeLong, R. L., Feldkamp, S. D., Francis, J. M., Gentry, R. L., Health, C. B., Le Boeuf, B. J., Majluf, P., and York, A. E. (1991). The effects of El Niño on pinniped populations in the Eastern Pacific. *In* "Pinnipeds and El Niño: Responses to Environmental Stress" (F. Trillmich and K. A. Ono, eds.), pp. 247–270. Springer-Verlag, Berlin.

Van Bressem, M.-F., Van Waerebeek, K., and Raga, J. A. (1999). A review of virus infections of cetaceans and the potential impact of morbilliviruses, poxviruses and papillomaviruses on host population dynamics. *Dis. Aquat. Org.* **38,** 53–65.

Hearing

J. G. M. THEWISSEN
Northeastern Ohio Universities College of Medicine, Rootstown

Vision and smell are senses that lose much of their usefulness in the watery medium where marine mammals live. As a result, hearing is generally more important in marine mammals than it is in land mammals. Hearing is a matter of life and death in odontocete cetaceans because they echolocate: they emit high-frequency sounds and determine the shape of their surroundings on the basis of the reflections of those sounds. Echolocation requires a sophisticated sound production organ (located in the nose) and sophisticated hearing. However, the physics of waterborne sound are very different than airborne sound (for a summary, see Denny, 1993). These two factors combined, the importance of well-developed hearing and the differences between sound in air and in water, have caused the organ of hearing to undergo a number of important changes. Some of these changes are well understood and occur across vertebrates that straddle the water–land boundary (Fay and Popper, 1985; Lombard and Hetherington, 1993), whereas others are the focus of intense investigations and controversies.

Sadly, sound is also important to marine mammals in another way. Dolphins may die if their middle ear is infected by nematodes because they are unable to detect prey. Loud sounds similarly may deprive whales and dolphins of their auditory abilities and may be a source of mortality.

I. Anatomy

The organ of hearing of all mammals consists of three parts: the external ear, the middle ear, and the inner ear (Fig. 1A). In land mammals, the external ear is basically an air-filled tube (the external auditory meatus) attached to a structure that funnels sound (the pinna). The tube ends medially at the tympanic membrane (eardrum), suspended by a bone called the ectotympanic. This tube fills with water when an animal enters the water. Medial to the tympanic membrane is the middle ear. It is an air-filled cavity, containing three small bones (ear ossicles) with two muscles attached to them. The bones are called malleus (hammer), incus (anvil), and stapes (stirrup) and form a chain contacting the tympanic membrane on one side (malleus) and the inner ear on the other (stapes). The middle ear cavity is air filled and is connected to the pharynx by means of the auditory tube (Eustachian tube), it does not fill with water when an animal is submerged. The inner ear is encased in a thick bone cover (the petrosal bone), it consists of a series of cavities in this bone that are filled with two kinds of fluids (perilymph and endolymph). For hearing, the most important of these cavities has the shape of a snail shell and is called the cochlea. It houses a long, hollow, rolled-up organ that mainly consists of three ducts that run the full extent of the cochlea. The scala vestibuli starts at the vestibule of the inner ear (where the oval window opens) and extends to the apex of the cochlea. There the scala vestibuli is connected to a second duct, the scala tympani, which extends from the apex to the base of the cochlea and ends at the round window. Both of these ducts are filled with perilymph. A third duct, the cochlear duct, is filled with endolymph and extends the length of the cochlea sandwiched between the two scalae. The floor of the cochlear duct is the basilar membrane in which a long row of neurons is implanted. The neurons of the organ of Corti are part of cranial nerve VIII (vestibulo-cochlear nerve) and pass auditory information to the brain.

Sea otters and polar bears have ears that are very similar to those of their land relatives. Pinnipeds and sirenians have ears that are, in gross features, similar to those of land mammals, although there are some important modifications (such as the absence of the pinna in phocids and sirenians). In cetaceans, the middle and inner ears still retain clearly recognizable anatomical features of land mammals, but the sound pathway through the external ear has been completely modified.

The cetacean external ear (Fig. 1B) retains an external auditory meatus supported by cartilages, but the duct is narrow to the point where it is not patent and there is no pinna. Instead, the cetacean mandible is involved in sound reception. The posterior part of the odontocete mandible, behind the teeth, has a thin lateral wall (sometimes called the pan bone). This part of the mandible is medially concave and houses a large fat pad. This fat pad extends anteriorly into the mandibular foramen, which is enormous, and continues into the area ventral to the teeth. Posteriorly, this fat pad touches the lateral bony wall of the ectotympanic. This wall is extremely thin, and the attachment area for the tympanic membrane in cetaceans is reduced greatly. The ectotympanic has a joint (syndesmosis) with the petrosal (also called periotic), but the two bones combined have very limited connections to the remainder of the skull, unlike most mammals. In delphinids, this connection is limited to one small piece of cartilage; in all other odontocetes, the tympano-periotic has a larger connection; and in mysticetes,

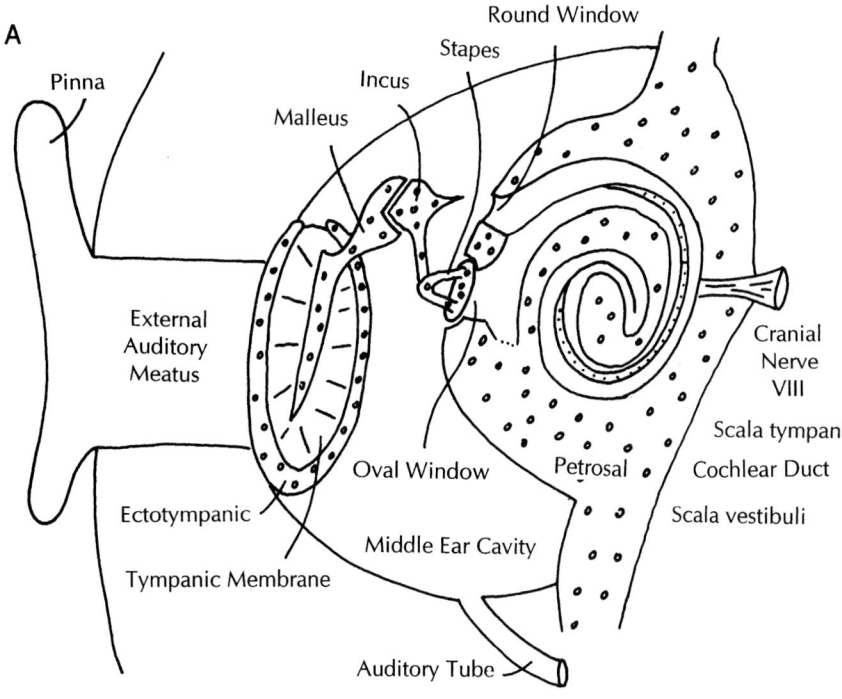

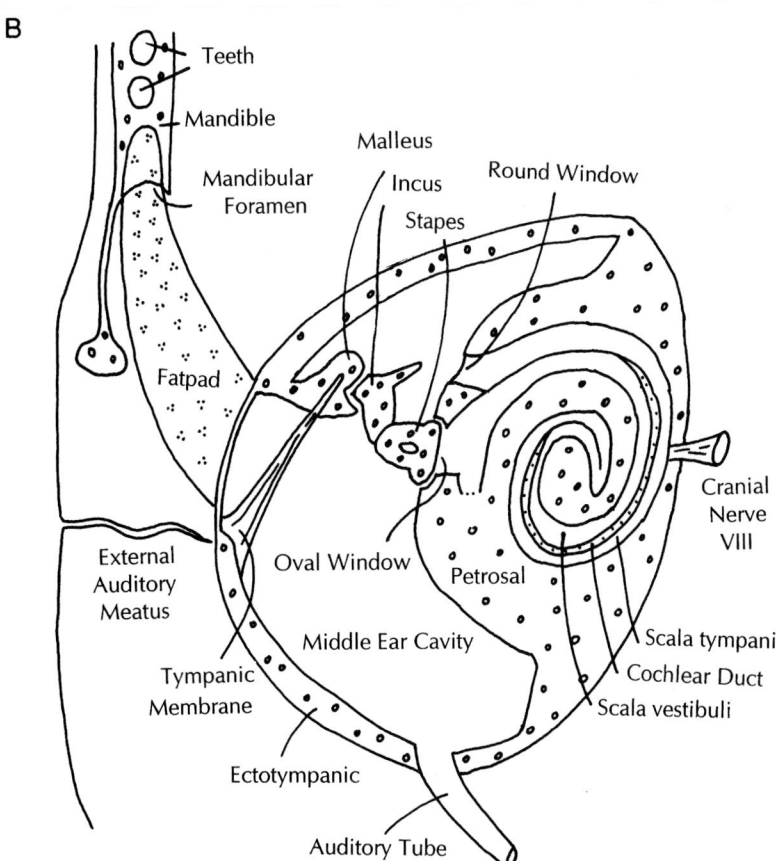

Figure 1 *Diagram of the ear in a generalized mammal (A) and a cetacean (B). Pinnipeds and sirenians have auditory systems similar to A.*

the complex is integrated into the skull similar to land mammals. The entire tympano-periotic of odontocetes is surrounded by spongy tissue that is filled with air spaces.

The tympanic membrane of odontocetes (often called tympanic ligament) is not the flat, more or less circular structure of land mammals. Instead, it is a strand of elastic tissue attaching to a small circular ectotympanic ring and then narrowing into an elastic tissue strap that attaches to the malleus (not unlike a folded-in umbrella). A similar structure occurs in mysticetes, but here an additional piece of the tympanic membrane forms a large, blunt protrusion into the external auditory meatus, often referred to as the glove finger. The ear ossicles of all cetaceans are highly pachyosteosclerotic and very different in shape from those of land mammals (Fig. 2). All cetaceans, except *Kogia* spp., have ossicles that are similar in overall form, and many of these morphological traits go back to the earliest Eocene cetaceans. In mysticetes, the stapes is much longer than in odontocetes. In *Kogia* spp., the malleus is fused to the ectotympanic in a unique shape. Sirenians and phocid pinnipeds (but not otariids and odobenids) also have pachyosteosclerotic ossicles, but they retain the shapes of the ossicles of their land relatives.

Middle ear muscles of cetaceans are reduced greatly. Suspended in the middle ear cavity is a plexus of veins and arteries collateral to the internal carotid artery. The internal ear of cetaceans is similar to that of land mammals, except that its proportions are different and that there are great differences among cetaceans (see Ketten, 1992).

The organ of balance in mammals consists mainly of three canals that run in the petrosal in a circular fashion. They are located immediately posterior to the cochlea and are also filled with perilymph and are innervated by cranial nerve VIII. Whereas the size of the cochlea scales closely to body size in mammals, the semicircular canals scale with body size in all mammals except cetaceans. In all modern cetaceans, the semicircular canals are much smaller than would be expected for their body size.

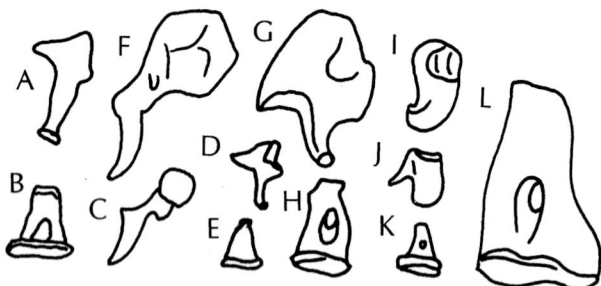

Figure 2 *Auditory ossicles of marine mammals; all are left ossicles in similar views and to scale. (A and B) Ursus maritimus, incus, and stapes. (C–E) Eumetopias jubatus, malleus, incus, and stapes. (F–H) Lobodon carcinophaga, malleus, incus, and stapes. (I–K) Delphinus sp., malleus, incus, and stapes. (L) Dugong dugon, stapes. A–E represent more or less primitive morphologies for mammals, whereas all others are modified to various degrees. Modified after Doran (1876).*

II. Functional Morphology

Sound consists of waves of vibrations of the molecules that constitute air or water (in the case of marine mammals). The ear amplifies the sounds and translates them into neural impulses. Left and right ears together also gather directional information.

The pathway of sound in pinnipeds and sirenians is not significantly different from that of land mammals (Fig. 1A). Sound is funneled to the tympanic membrane, which starts oscillations of this membrane. These vibrations are transmitted to the manubrium of the malleus, which leans against the tympanic membrane. The vibrations are then passed along the chain of ossicles, eventually causing the foot plate of the stapes to pump in and out of the oval window. Ossicles function as an amplifier in two ways. First, sound energy that arrives at the tympanic membrane is concentrated on the much smaller area of the foot plate of the stapes. Second, the amplitude of the vibrations is enlarged by a lever-arm system: small vibrations of the long in-lever (the manubrium of the malleus) are transmitted to a much shorter out-lever (the crus breve of the incus). These two mechanisms result in higher pressures at the stapedial foot plate then at the tympanic membrane. Higher pressures are necessary to start vibrations in the dense fluid of the inner ear (perilymph). As such, the middle ear matches the acoustic properties of the air in the external auditory meatus to those of the perilyph of the inner ear and is technically often described as an impedance matcher.

The vibrations that are caused by the stapes set up standing waves in the fluids of the scala vestibuli. The standing waves are transmitted to the endolymph in the cochlear duct and these stimulate the hair cells on the basilar membrane. The hair cells fire electric impulses that are carried to the brain. Different frequencies are perceived by the stimulation of different sections of the basilar membrane: specific hair cells are receptors for specific frequencies. Low frequencies are perceived near the apex of the cochlea, whereas high frequencies are perceived near the base of the cochlea.

The inner ear of cetaceans functions in the same way as that of land mammals, but the external and middle ears are very different. The area that is most sound sensitive on a dolphin's head is not the ear, but the skin over the lower jaw (Fig. 3). From here, sounds are transmitted through the bone of the mandible and through the fat pad to the bony wall of the middle ear. What happens here is controversial, but the best functional middle ear model for odontocetes has been proposed by Hemilä *et al.* (1999). A model in which the entire ectotympano-ossicle complex rotates around an axis through the malleus correlates well with experimental data for hearing at low frequencies. A model in which four bony units (malleo-incus, stapes, ectotympanic, and periotic) are connected by springs that mainly allow translations predicts higher frequency data well.

The cetacean middle ear (and that of some pinnipeds as well) also contains a plexus of arteries and veins. It is possible that this plexus represents an adaptation for deep diving and that it can be inflated to reduce the airspace in the middle ear cavity, thus increasing its pressure to match ambient pressure.

The inner ear of cetaceans is structurally and functionally similar to that of land mammals, but differs in the details. Ket-

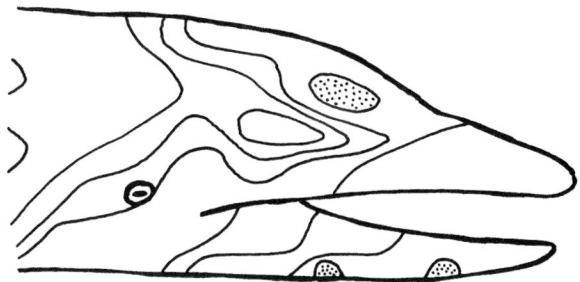

Figure 3 *One of the earliest determinations of sound sensitivity of the dolphin head. A hydrophone producing a sound of 65 kHz was pressed on different areas of this* Stenella attenuata, *resulting in this map of areas with similar sensitivities. The greatest sensitivity to sound (stippled areas) was on the lower jaw, with a lesser maximum on the forehead. The external auditory meatus (not shown, but located posterior to the eye) does not represent a maximum of sound sensitivity. Redrawn after Bullock et al. (1968). (See also Fig. 3, Echolocation).*

ten (1992) distinguished two types of odontocete cochleas mainly on the basis of the shape and size of the basilar membrane. The differences are related to specific frequency ranges. Mysticetes have a different cochlea that is adapted to low-frequency sound reception.

The middle ear of pinnipeds and sirenians contains the same elements as in land mammals, but the ossicles of phocids and sirenians are greatly enlarged. This pachyosteosclerosis does not occur in odobenids and otariids.

III. Measures of Hearing

Hearing can be described by a number of quantifiable variables. Components of hearing include localizing ability (directional hearing), spatial resolution, and frequency discrimination (for a review, see Wartzok and Ketten, 1999). One of the most useful measures of hearing is the minimum audible intensity, which varies as a function of the frequency of the sound. The resulting plot is called an audiogram (Fig. 4). Sound intensities in these plots are usually indicated in decibels (dB), a relative and nonlinear measure that requires a reference intensity. To make matters even more complicated, it is customary to use different reference pressures for in-air and underwater measurements. Frequency is measured in hertz (Hz). The organ of hearing operates over five orders of magnitude, and frequency is therefore usually plotted on a logarithmic scale. Although straightforward to understand and display, determining audiograms and other

Figure 4 *Audiograms of cetaceans (A) and pinnipeds (B) for waterborne sound; (C) for airborne sound. The X axis is frequency on a logarithmic scale in Hz. The Y axis represents pressure level at minimum audible sound (in dB, with 1 A as a reference for waterborne sound, and SPL for airborne sound). Hatched areas represent envelopes containing data for multiple specimens. The source for A and B is Wartzok and Ketten (1999) and for C is Fay (1988).*

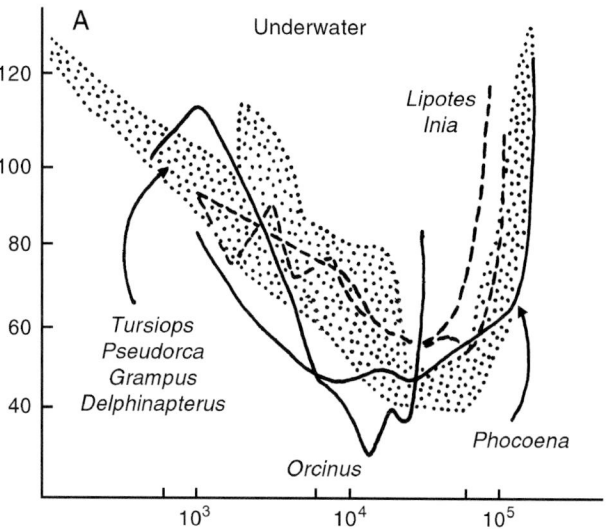

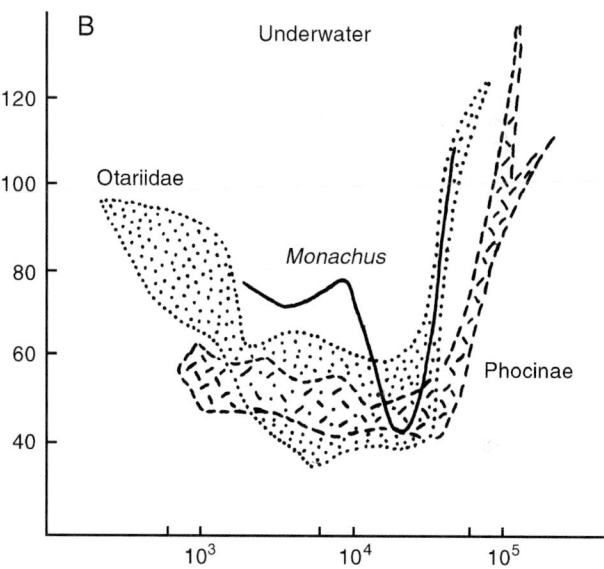

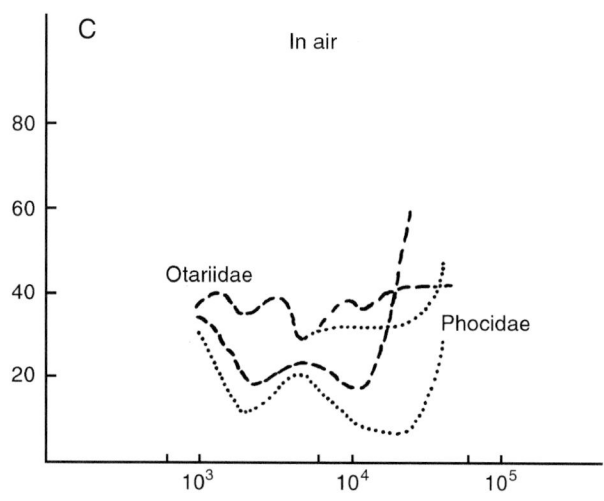

measures of hearing is usually technically very hard, and results are partly dependent on factors outside of the control of the investigator (such as the motivation of the investigated animal if the audiogram is based on behavioral responses). For cetaceans, audiograms are only available for five families of odontocetes, four of which are represented by a single individual. Sample sizes are thus small, although some generalizations can be made.

The delphinids *Tursiops* spp., *Pseudorca crassidens*, and *Grampus griseus* and the monodontid, *Delphinapterus leucas* show audiograms that have a fairly typical shape. For low frequencies (left side of Fig. 4A), minimum audible intensity decreases with increasing frequency. Auditory sensitivity peaks between 11,000 and 18,000 Hz (lowest part of the data envelope), and at higher frequencies, sensitivity is reduced. The peak in odontocete hearing is well above the frequency of best hearing in humans (2000 Hz, but variable with age) as a result of the use of high frequencies in echolocation. Optimal hearing in *Phocoena* spp. and *Orcinus orca* occurs at frequencies that are lower than for the delphinoids mentioned earlier. For *Phocoena*, the lower optimal frequency has been related to its particular cochlear shape (Ketten, 1992). *Orcinus* is a delphinid of large body size, and, in general, ears of animals with larger body sizes are tuned to lower frequencies, which is related to their longer cochlear duct (the size of the cochlea scales with body weight). The lower optimal frequency in *Orcinus* can also be understood in another way. The size of objects to be discerned poses limits on the echolocation frequencies that can be used in such a way that, to discern smaller objects, higher echolocation frequencies need to be used. *Orcinus* hunts larger prey than other delphinids and hence can use lower frequency signals. The freshwater dolphins *Inia geoffrensis* and *Lipotes vexillifer* have overall curves that are similar to those of the generalized delphinids but are less sensitive than in the generalized group. Several explanations are possible; one of them holds that these species hunt in small-scale river environments, where only prey close to the predator can be pursued. In open sea settings, prey that is further away can be pursued successfully. Echolocation signals and echos attenuate with distance, and hence a closer object can be discerned with a receiver that is less sensitive.

Underwater audiograms differ among different pinnipeds (Fig. 4B). Significantly, phocines (*Phoca* spp., *Pusa* spp., and *Pagophilus groenlandicus*) consistently show better high-frequency sensitivity than otariids. The only monachine (*Monachus* spp.) for which an audiogram is available lacks the high-frequency sensitivity of the other phocids and is similar to the otariids in this respect. The differences between phocines and otariines do not occur in in-air audiograms (Fig. 4C), suggesting that pinnipeds, means of sound transmission differ in air and in water. The curious bimodal sensitivity peak in two individuals also remains to be explained and may be the result of the compromise that the pinniped ear represents, a compromise between hearing in air and in water.

See Also the Following Articles

Brain ■ Echolocation ■ Morphology, Functional ■ Noise, Effects of ■ Skull Anatomy ■ Sound Production ■ Vision

References

Bullock, T. H., Grinnell, A. D., Ikezono, E., Kameda, K., Katsuki, Y., Nomoto, M., Sato, O., Suga, N., and Yanagisawa, K. (1968). Electrophysiological studies of central auditory mechanisms in cetaceans. *Zeitschr. Physiol.* **59,** 117–156.

Denny, M. W. (1993). "Air and Water, the Biology and Physics of Life's Media." Princeton University Press, Princeton, NJ.

Doran, A. H. G. (1876). Morphology of the mammalian Ossicula auditus. "Transactions of the Linnean Society," Series 2, Zoology, 1:371–497, pl. 58–64.

Fay, R. R. (1988). "Hearing in Vertebrates, a Psychophysics Databook." Hill-Fay Associates, Winnetka, IL.

Fay, R. R., and Popper, A. N. (1985). The octavo-lateralis system. *In* "Functional Vertebrate Morphology" (M. Hildebrand, D. M. Bramble, K. F. Liem, and D. B. Blake, eds.), pp. 291–316. Harvard Univ. Press, Cambridge, MA.

Hemilä, S., Nummela, S., and Reuter, T. (1999). A model of the odontocete middle ear. *Hear. Res.* **133,** 82–98.

Ketten, D. R. (1992). The marine mammal ear: Specializations for aquatic audition and echolocation. *In* "The Biology of Hearing" (D. Webster, R. Fay, and A. Popper, eds.), pp. 717–754. Springer-Verlag, New York.

Lombard, R. E., and Hetherington, T. E. (1993). The structural basis for hearing and sound transmission. *In* "The Skull" (J. Hanken and B. K. Hall, eds.), Vol. 3, pp. 241–302. University of Chicago Press.

Wartzok, D., and Ketten, D. R. (1999). Marine mammal sensory systems. *In* "Biology of Marine Mammals" (J. E. Reynolds and S. A. Rommel, eds.), pp. 117–176. Smithsonian Press, Washington, DC.

Heat Balance

SEE *Thermoregulation*

Hippopotamus

JOHN GATESY
University of Wyoming, Laramie

Hippopotamuses are stocky, large, semiaquatic, hoofed mammals that are classified as members of the family Hippopotamidae (order Artiodactyla). There are two extant species. The endangered pygmy hippopotamus, *Choeropsis liberiensis*, inhabits the dense forests of western Africa. The much larger common hippopotamus, *Hippopotamus amphibius* (see Fig. 1), has a broader geographic range that includes many of the water systems of sub-Saharan Africa.

I. Anatomy and Behavior

Due to its secretive nature, field observations of the pygmy hippopotamus are rare, so little is known of the ecology and natural history of this species. The animal is knockwurst-shaped, weighs approximately 150–250 kg, is basically hairless, and has a broad muzzle with prominent canine TEETH. *Choeropsis* is thought to be more solitary and less aquatic than *Hippopotamus*.

Figure 1 *Possible phylogenetic relationships among Cetacea (top) and artiodactyl ungulates—hippopotamuses (middle) and ruminants (bottom). Molecular data support the relationship on the left, and gross anatomical comparisons imply the relationship on the right. Molecular data suggest that cetaceans and hippopotamuses share aquatic specializations that may be further evidence of their close phylogenetic kinship. Whale photo is by Howard Rosenbaum (WCS), hippopotamus photo is by G. Amato (WCS), and bison photo is by C. Hayashi.*

The huge common hippopotamus ranges in size from approximately 1000 to 3500 kg. Like the pygmy hippopotamus, *Hippopotamus amphibius* is nearly hairless. The skin lacks sebaceous glands and is underlain by a thick layer of insulating fat. The limbs are short and stubby, and the skull is huge with the orbits and nostrils positioned dorsally. *Hippopotamus* individuals spend a large proportion of their lifetimes in rivers, lakes, or swamps. They are capable swimmers and are able to stay submerged for over 5 min. At dusk, the common hippopotamus advances to the land to graze on grasses and at dawn returns to the safety of the water to loll about, digest, and rest. *H. amphibius* is highly gregarious and, under some environmental conditions, this species concentrates into large herds of over 100 individuals. Groups of 10 to 20 animals are more typical. The males are territorial and battle viciously when competing for mates, but a variety of ritualized social behaviors limit intermale conflicts to some degree. Because of their enormous size and dangerous gape, adult common hippopotamuses have few natural enemies beside humans.

II. Fossil Record

The paleontological record of Hippopotamidae is quite rich in the Pleistocene and the Pliocene, the last five million years of geologic history. In fact, during some time intervals, five or more hippopotamid species were contemporaneous. The geographic distribution of hippopotamids was also more extensive in the not so distant past; hippopotamids ranged into Europe, Asia, North Africa, and Madagascar. In the Pleistocene, there were hippopotamuses in Great Britain! So the current complement of two extant species is a mere relic of past hippopotamid diversity.

Despite their relative abundance in the Plio-Pleistocene, definitive hippopotamids only extend back into the fossil record approximately 15 million years. The earliest described genus, *Kenyapotamus*, is found at several localities in East Africa (Pickford, 1983), but the evolutionary history of hippopotamids prior to 15 million years ago is obscure and may only be illuminated by further fossil discoveries.

III. Phylogeny of Hippopotamids

Semiaquatic hippopotamuses are not classified as marine mammals, the main focus of this encyclopedia. However, hippopotamuses recently have been implicated as close evolutionary relatives of cetaceans (whales, dolphins, and porpoises) and are thus relevant to this volume. The phylogenetic origins of Hippopotamidae are not clearly characterized, but among living species, mammalogists have traditionally aligned hippopotamuses closest to pigs and peccaries with ruminating ungulates (cattle, antelopes, deer, giraffes, chevrotains, and camels) considered their next closest evolutionary kin (Fig. 2a; Gentry and Hooker, 1988 and references therein). Based on morphological similarities, hippopotamuses and these other hoofed mammals

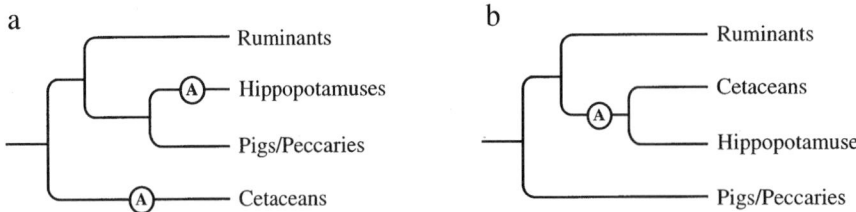

Figure 2 *Two phylogenetic hypotheses of hippopotamus origins that have different implications for the evolution of aquatic features. Acquisitions of aquatic traits shared by hippopotamuses and cetaceans are represented by open circles with "A" inside. (a) If hippopotamuses group within Artiodactyla, even-toed ungulates, and are only distantly related to Cetacea, aquatic specializations were derived independently in hippopotamuses and whales. (b) If cetaceans are the closest living relatives of hippopotamuses, shared aquatic specializations of these taxa may have been acquired only once in their common ancestors.*

are classified as members of the mammalian order Artiodactyla, even-toed ungulates. Hooves in which the axis of symmetry runs between the third and fourth digits, the so-called paraxonic condition, characterize all species in this group.

According to the majority of the gross anatomical evidence, cetaceans are only distantly related to hippopotamids (Fig. 2a). Therefore, the common aquatic specializations of cetaceans and hippopotamuses traditionally have been interpreted as convergences. In this case, independent adaptation to an aquatic environment was hypothesized to have driven the parallel evolution of grossly similar characteristics, such as near hairlessness, a thick layer of insulating fat, the lack of sebaceous glands, the lack of scrotal testes, and the ability to nurse offspring underwater. These features are common to all extant cetaceans and one or both extant species of hippopotamus.

Recently, a radically different interpretation of hippopotamid phylogeny has been presented. Examination of diverse genetic data suggests that cetaceans, not other artiodactyls, are the closest relatives of hippopotamids (Fig. 2b; Gatesy, 1998; Ursing and Árnason, 1998; Nikaido *et al.*, 1999). In this scheme, whales are simply highly derived members of "Artiodactyla." Therefore, at least some of the common aquatic specializations of hippopotamuses and whales could be interpreted as further evidence of a close evolutionary relationship between these mammals (Fig. 2b). The recent discovery of fossil whales with fully functional hindlimbs and paraxonic feet (Thewissen *et al.*, 1994) lends further credibility to close phylogenetic ties between whales and artiodactyls. If genetic data portray an accurate picture of hippopotamid origins, the seemingly ungainly hippopotamus may offer additional clues to the evolutionary derivation of the most graceful of marine mammals, the cetaceans (Fig. 2).

See Also the Following Articles

Archaeocetes, Archaic ■ Artiodactyla ■ Cetacean Evolution

References

Gatesy, J. (1998). Molecular evidence for the phylogenetic affinities of Cetacea. *In* "The Emergence of Whales: Evolutionary Patterns in the Origin of Cetacea, Advances in Vertebrate Paleobiology" (J. Thewissen, ed.), pp. 63–111. Plenum Press, New York.

Gentry, A., and Hooker, J. (1988). The phylogeny of the Artiodactyla. *In* "The Phylogeny and Classification of the Tetrapods" (M. Benton, ed.), Vol. 2, pp. 235–272. Clarendon Press, Oxford, England.

Nikaido, M., Rooney, A., and Okada, N. (1999). Phylogenetic relationships among cetartiodactyls based on insertions of short and long interspersed elements: Hippopotamuses are the closest extant relatives of whales. *Proc. Natl. Acad. Sci. USA* **96**, 10261–10266.

Pickford, M. (1983). On the origins of Hippopotamidae together with descriptions of two new species, a new genus, and a new subfamily from the Miocene of Kenya. *Geobios* **15**, 193–217.

Thewissen, J., Hussain, S., and Arif, M. (1994). Fossil evidence for the origin of aquatic locomotion in archaeocete whales. *Science* **263**, 210–212.

Ursing, B., and Árnason, Ú. (1998). Analyses of mitochondrial genomes strongly support a hippopotamus-whale clade. *Proc. R. Soc. Lond. B* **265**, 2251–2255.

History of Marine Mammal Research

Bernd Würsig

Texas A & M University, Galveston

If research is the gathering of knowledge, then we can think of marine mammal research to have gone on as long as humans have gazed at whales spouting offshore and seals pupping on beaches. But early observations of nature were largely tied up with myths about animals and legends of their capabilities. A common theme appears to have been the changing of humans to dolphins and whales, and the reverse. This theme is recognized in remaining legends of Australian aborigine "dream time," baiji (*Lipotes vexillifer*) and boto (*Inia geoffrensis*) river dolphin folklore (Zhou and Zhang, 1991; Sangama de Beaver and Beaver, 1989, respectively), tales of the god-like killer whales (*Orcinus orca*) of Pacific Northwest indigenous tribes (McIntyre, 1974), and many more.

Nevertheless, some early writings show remarkable insights in marine mammal biology. Well over 2000 years ago, scholars of China's Han Dynasty described baiji as related to marine dolphins, implying that those were known to intellectuals of the time. These descriptions, in the annotated dictionary "Er-Ya," survive to this day. Even earlier, the Greek philosopher/scientist Aristotle (384–322 B.C.) differentiated between baleen and toothed whales, and described both types in some detail. It is unfortunate but totally understandable in hindsight that he classified cetaceans as fishes, a practice still present in Britain's "Royal Fishes" as all whales and dolphins belonging by law to the Crown. The Roman writer/lawyer/admiral Pliny the Elder (23–79 A.D.) published a book on dolphins and whales 400 years after Aristotle's time as part of Pliny's 37-volume "Natural History."

Not much scientific inquiry or thought was conducted between Roman times and the western Renaissance, and knowledge, at least written knowledge, of marine mammals languished as well. The modern progression of marine mammal research can perhaps best be described as occurring in four general (and not mutually exclusive) phases: (1) morphological description from beach-cast specimens and fossils, (2) descriptions of behavior and anatomy as gathered during hunting and whaling activities, (3) studies of physiology and behavior in captivity, and (4) studies of ecology, behavior, and physiology in nature. These phases follow a rough chronology, with morphology and systematics the main topics pre-1900s, hunting-related morphological and behavioral research mainly from the 1850s to the 1970s, scientific captive animal descriptions beginning around 1950, and more ecologically oriented descriptions in nature beginning around the 1970s. All phases are ongoing, with electronic devices promising to elevate in-field research on marine mammal lives to a new level of sophistication. A very readable recent account of the history of marine mammal studies is found in Berta and Sumich (1999). This volume lists some of the major deceased marine mammal researchers of the past, with their annotated classic works in the field.

Pierre Bélon was probably the first "modern" marine mammal author since Pliny's time. He published accurate descriptions and woodcuts of whales, dolphins, and seals (Belloni, 1553), and these (and also, unfortunately, the less accurate ones) were much-copied by others in the next two centuries.

The real burst of marine mammal knowledge did not come until later, however. And then it came suddenly, in tune with 18th century awakening of scientific thought in the western world. While many authors could be mentioned, three early contemporaries did much to advance cetacean descriptions, taxonomy, and systematics. These were the French zoologist La Cépède (1804) and the Cuvier brothers. Georges Cuvier, who arguably founded modern evolutionary theory, wrote on many topics, including cetaceans; whereas his less-famed brother Frederic published two important works on cetaceans (F. Cuvier, 1829, 1836). These three were followed by the Belgian zoologist Van Beneden in the latter half of the 19th century, with work mainly consisting of compilations of information on fossil whales, and by a host of fine morphologists, taxonomists, systematists, and evolutionary historians in the

20th century (summaries are provided by Rice, 1998; Pabst *et al.*, 1999; Reynolds *et al.*, 1999; and Thewissen, 1998). While much of the earlier work centered on cetaceans, the British zoologist John Gray described both seals and whales in the British Museum (Gray, 1866), and the American zoologist Joel Allen wrote excellent monographs on whales, pinnipeds, and sirenians (Allen, 1880).

Yamase (1760) began the science of marine mammalogy in Japan at about the same time as serious studies began in the west. He presented accurate figures and descriptions of the external morphology of six toothed and seven baleen whale species and distinguished them from fishes. His work was brought to the west in a marine mammal section of "Fauna Japonica" by Siebold (1844). Otsuki began to describe the internal anatomy of cetaceans of Japan in 1808, but his manuscript remains unpublished.

A second major phase of information gathering, often linked intricately with that just described, involved descriptions of animals as related to HUNTING and WHALING. Morphological information was at the core of these descriptions, but behavior and the basic society structure of whales and pinnipeds—of course much of the time affected by the hunting activities themselves—were recorded as well. One of the earliest accurate accounts consisted of German-born and Russian-naturalized Georg Steller's descriptions of pinnipeds and the soon-after extinct Steller's sea cow (*Hydrodamalis gigas*), the largest and only cold-water sirenian known (originally published in Latin in 1751, but republished in English as Steller, 1899). Quite a few books related especially to whaling were produced, but perhaps the most enduring one from the 19th century was by the North American whaling captain Charles Scammon, who wrote with feeling and accuracy on behavior and life history habits of marine mammals of the North Pacific (Scammon, 1874). In the 20th century, one of the most famous works largely relying on whaling-accumulated data consists of Everhard Slijper's book "Whales and Dolphins" (published in English in 1976). A very readable account of whaling and the literature derived from whaling can be found in "Men and Whales" by Richard Ellis (1991).

While whaling, sealing, and other forms of direct hunting are much abated today as compared to in the 1960s, there are still powerful low-level, oft-indigenous hunts, especially in protein-poor areas of the world (Perrin, 1999). As a result, data are being accumulated and analyzed on morphology, GENETICS, taxonomy and SYSTEMATICS, life history, prey patterns, and so on. Excellent recent information has become available from results of hunting on (for example) pilot whales (*Globicephala* spp.), oceanic dolphins (especially of the genus *Stenella*), bowhead whales (*Balaena mysticetus*), sperm whales (*Physeter macrocephalus*), and several seal, fur seal, and sea lion species (summaries in Berta and Sumich, 1999; Reynolds and Rommel, 1999; and Twiss and Reeves, 1999).

A third major research avenue has come about as a result of keeping marine mammals in captivity. Attempts to do so in the early part of the last century usually resulted in the animals' untimely deaths—due to poor water, incorrect or tainted food, disease, and intraanimal aggression in confined spaces. Facilities that housed marine mammals simply replaced dead ones by more captures from nature. However, especially since the 1970s, amazing strides in husbandry have been made for all marine mammals (except large whales), and the better aquaria now keep—and breed—animals very well. Unfortunately, there are still many "primitive" facilities, especially in less-developed parts of the world. At present, there are representatives of all major taxonomic groups in captivity, as show animals and for research: toothed whales and dolphins (only two baleen whales, each time young gray whales, *Eschrichtius robustus,* have been kept); pinnipeds of all types, but especially California sea lions (*Zalophus californianus*); sirenians (mainly the Caribbean manatee, *Trichechus manatus,* and the dugong, *Dugong dugon,*); and polar bears (*Ursus maritimus*) and sea otters (*Enhydra lutris*).

Only through holding animals in controlled situations have researchers learned that dolphins echolocate (Au, 1993); that all marine mammals exhibit reduced heart and general metabolic rates during dives (Ridgway, 1972; Pabst *et al.,* 1999); and that both dolphins and sea lions have remarkably advanced cognitive capabilities (Tyack, 1999). Furthermore, it is now fully appreciated that while pinnipeds and cetaceans are finely tuned underwater swimmers and divers with superbly evolved methods of breath holding, avoiding or reducing lactic acid depth during long submergences, and navigating in dark and cold waters, there is no secret "magic" to their energetic capabilities (Costa and Williams, 1999).

One major misstep from studies in captivity took place: the American John Lilly avowed in the 1960s that his research on bottlenose dolphins (*Tursiops truncatus*) proved that these popular show animals have an intelligence superior even to that of the brightest dogs and chimpanzees, and likely equal to that of humans (Lilly, 1967). Careful studies by others have shown that dolphins are undeniably "smart" (intelligence is very difficult to define and compare, but has something to do with well-developed flexibilities of behavior and of innovative learning), but that there is no reason to believe that dolphins fare better in this "intelligence/cognition" sphere than many other highly social mammals (Herman, 1980, 1986; Tyack, 1999; Wells *et al.,* 1999).

While the study of marine mammals as derived dead from nature and live from captivity continues and grows, a relatively new approach has become the major research avenue since the 1970s. This consists of researchers going out into nature to observe the animals in their own milieu; as the animals associate with conspecifics; eat and are being eaten; and mate, give birth, and raise their young. We are learning more about the lives of these generally social creatures as they face storms, heavy years of sea ice, seasons of poor food resources (e.g., caused by "El Niño" southern oscillation climatic events), giant parasite infestations, adoring but noisy boatloads of whale-watching tourists, crowded shipping lanes, and habitat degradation near shore and in mighty rivers. This information on ecology of marine mammals is vital if we are to help protect them and their natural ecosystems from the depredations of overfishing, habitat POLLUTION by chemicals, heavy metals, and noise; and the very real possibility of global climate change and whole-scale habitat destruction due to the effects of ozone depletion and global warming (Tynan and DeMaster, 1997).

Studies in nature often rely on visual or photographic recognition of individual whales, dolphins, and pinnipeds, often with the help of tags or color marks but also by natural markings (Hammond *et al.,* 1990). Researchers have described movement patterns by tracking animals with surveyor's transits from shore, and from shore and vessels by small radio tags placed on their bodies (Würsig *et al.,* 1991). Since the early 1990s, satellite tags that relay position information to earth-orbiting satellites have become smaller, less expensive, and ever more popular. As a result, we know that northern elephant seals (*Mirounga angustirostris*) swim and dive into deep oceanic waters for months at a time, humpback whales (*Megaptera novaeangliae*) take rapid zigzag courses between their mating and feeding grounds, North Atlantic right whales (*Eubataena glacialis*) undergo previously unsuspected jaunts between Greenland and New England during the feeding summer, and much more (Wells *et al.,* 1999). Tags are being fitted not only with depth-of-dive measuring and telemetering devices, but also with ways to ascertain swimming velocity, angles of dives, water and skin temperature, individual sound production, heart rate, and, in the future, other physiological measures. Recent advances in small and low-light capable video camera/record systems are even giving data on SWIMMING, socializing, and feeding behavior directly from the animals underwater (Davis *et al.,* 1999).

Physiological research, previously entirely within the realm of captivity, is more and more possible with innovative or sophisticated techniques in nature. Samples of stool, urine, blood, and even mother's milk are being collected from pinnipeds resting on land or ice. Trained dolphins have been released at sea, commanded to dive, and then told to exhale into a funnel to ascertain oxygen consumption values and to station themselves so that blood can be drawn. This interaction between animals in captivity and nature is especially fruitful for physiological research. Small darts have been developed that are fired from a cross bow or pneumatic pistol and that obtain skin and blubber samples for analyses of genetics (Dizon *et al.,* 1997), toxin loads, and blubber energy content for relative measurements of health within and between populations. Skin samples of breaching whales have been successfully (and in a totally benign fashion) collected from the water and genetically sampled for gender, social grouping, and population data. A technique has been developed to harmlessly "skin-swab" bow-riding dolphins, also for genetic analysis (Harlin *et al.,* 1999).

In response to an apparent increase in marine mammal strandings and the emergence of new marine mammal diseases in recent years, studies of wild marine mammal disease and ocean chemical contaminants are on the increase. While studies in nature have yielded data on the presence of deadly viruses and contaminant levels in tissues of beached and dying marine mammals (Aguilar and Borrell, 1997), they have provided little insight into immune defense against disease or the biochemical consequences of contaminants. For example, species-specific biomarkers have been developed to assess the dolphin immune system (Romano *et al.,* 1999). Because they are readily available for long-term studies requiring serial sampling of tissues and health and reproductive histories, captive marine mammals afford unique opportunities to provide basic insight into the relationships among contaminants, the immune system, and animal health. Once they are developed and tested on animals in captivity, biomarkers can be used with wild marine mammal populations to assess contaminant exposures and their possible effects on immune systems and neurologic responses (Ridgway and Au, 1999), as well as on reproductive success (Ridgway and Reddy, 1995), growth, and development.

The sensitive hearing of marine mammals has led to concerns that intense sound or noise pollution generated by humans could impede COMMUNICATION, cause stress, or damage hearing. Marine mammal hearing studies currently underway should help to define mitigation criteria for the effects of human-generated sound in the ocean (Schlundt *et al.*, 2000), and ultimately allow us to find a balance between the ecological needs of marine mammals and the role the ocean plays in commerce, exploration, travel, and defense.

Ever-more sophisticated electronic and biochemical techniques are being developed to study the lives of marine mammals. However, the "tried and true" methods of looking at fossil bones, dissecting and describing pathologies of a net-entangled animal or one cast on shore after a storm, safely and carefully experimenting with animals in captivity, and the dogged gathering of behavioral information by binoculars and notebook are by no means passé. We are, in this new 21st century, in a vibrant phase of marine mammal research, and we see a very bright future for ever-more knowledge being gathered within our field.

References

Aguilar, A., and Borrell, A. (eds.) (1997). Marine mammals and pollutants: An annotated bibliography. Foundation for Sustainable Development, Barcelona.

Allen, J. A. (1880). History of North American pinnipeds: A monograph of the walruses, sea-lions, sea-bears and seals of North America. *U.S. Geol. Surv. Terr. Misc. Publ.* **12**; 1–785.

Au, W. W. L. (1993). "The Sonar of Dolphins." Springer Verlag, New York.

Belloni, P. (1553). "De Aquatibilis (Book Two)." Stephan Press, Paris.

Berta, A., and Sumich, J. L. (1999). "Marine Mammals: Evolutionary Biology." Academic Press, San Diego.

Costa, D. P., and Williams, T. M. (1999). Marine mammal energetics. *In* "Biology of Marine Mammals" (J. E. Reynolds III and S. A. Rommel, eds.), pp. 176–217 Smithsonian Institution Press, Washington, DC.

Cuvier, F. (1829). Cétacés. *In* "Histoire Naturelle des Mamifères." Roret Press, Paris.

Cuvier, F. (1836). "De l'Histoire Naturelle des Cétacés." Roret Press, Paris.

Davis, R. W., Fuiman, L. A., Williams, T. M., Collier, S. O., Hagey, W. P., Kanatous, S. B., Kohin, S., and Horning, M. (1999). Hunting behavior of a marine mammal beneath Antarctic fast ice. *Science* **283**; 993–996.

Dizon, A. E., Chivers, S. J., and Perrin, W. E. (eds.) (1997). "Molecular Genetics of Marine Mammals." Special Publ. No. 3, The Society for Marine Mammalogy, Allen Press, Lawrence, KS.

Ellis, R. (1991). "Men and Whales." Knopf Press, New York.

Gray, J. E. (1866). "Catalog of Seals and Whales in the British Museum" 2nd Ed. British Museum Press, London.

Hammond, P. S., Mizroch, S. A., and Donovan, G. P. (eds.) (1990). "Individual Recognition of Cetaceans: Use of Photo Identification and Other Techniques to Estimate Population Parameters." Intl. Whal. Comm. Special Issue No. 12, Cambridge Univ. Press, Cambridge.

Harlin, A. D., Würsig, B., Baker, C. S., and Markowitz, T. M. (1999). Skin swabbing for genetic analysis: Application on dusky dolphins (*Lagenorhynchus obscurus*). *Mar. Mamm. Sci.* **15**; 409–425.

Herman, L. M. (1980). Cognitive characteristics of dolphins. *In* "Cetacean Behavior: Mechanisms and Functions" (L. M. Herman, ed.), pp. 363–429. Wiley-Interscience Press, New York.

Herman, L. M. (1986). Cognition and language competencies of bottlenosed dolphins. *In* "Dolphin Cognition and Behavior: A Comparative Approach" (R. J. Schusterman, J. A. Thomas, and F. G. Woods, eds.), pp. 221–252. Lawrence Erlbaum Press, Hillsdale, NJ.

La Cépède, Compte de. (1804). "Histoire Naturelle des Cétacés." Paris.

Lilly, J. C. (1967). "The Mind of the Dolphin." Doubleday Press, New York.

McIntyre, J. (1974). "Mind in the Waters." Charles Scribner's Sons, New York.

Pabst, D. A., Rommel, S. A., and McLellan, W. A. (1999). The functional morphology of marine mammals. *In* "Biology of Marine Mammals" (J. E. Reynolds III and S. A. Rommel, eds.), pp. 15–72. Smithsonian Institution Press, Washington, DC.

Perrin, W. F. (1999). Selected examples of small cetaceans at risk. *In* "Conservation and Management of Marine Mammals" (J. R. Twiss, Jr. and R. R. Reeves, eds.), pp. 296–310. Smithsonian Institution Press, Washington, DC.

Reynolds, J. E., Odell, D. K., and Rommel, S. A. (1999). Marine mammals of the world. *In* "Biology of Marine Mammals" (J. E. Reynolds III and S. A. Rommel, eds.), pp. 1–14. Smithsonian Institution Press, Washington, DC.

Reynolds, J. E., and Rommel, S. A. (1999). "Biology of Marine Mammals." Smithsonian Institution Press, Washington, DC.

Rice, D. W. (1998). "Marine Mammals of the World: Systematics and Distribution." Special Publ. No. 4, The Society for Marine Mammalogy. Allen Press, Lawrence, KS.

Ridgway, S. E. (1972). "Mammals of the Sea: Biology and Medicine." Charles H. Thomas Press, Springfield, IL.

Ridgway, S. E., and Au, W. W. L. (1999). Hearing and echolocation: Dolphin. *In* "Encyclopedia of Neuroscience" (G. Adelman and B. Smith, eds.), 2nd Ed. pp. 858–862. Springer-Verlag, New York.

Ridgway, S., and Reddy, M. (1995). Residue levels of several organochlorines in *Tursiops truncatus* milk collected at varied stages of lactation. *Mar. Pollut. Bull.* **30**; 609–614.

Romano, T. A., Ridgway, S. H., Felton, D. L., and Quaranta, V. (1999). Molecular cloning and characterization of CD4 in an aquatic mammal, the white whale, *Delphinapterus leucas*. *Immunogenetics* **49**; 376–383.

Sangama de Beaver, M., and Beaver, P. (1989). "Tales of the Peruvian Amazon." AE Publications, Largo, FL.

Scammon, C. M. (1874). "The Marine Mammals of the North-Western Coast of North America Described and Illustrated Together with an Account of the Whale-Fishery." John. H. Carmany, San Francisco, CA.

Schlundt, C. E., Finneran, J. J., Carder, D. A., and Ridgway, S. H. (2000). Temporary shift in masked hearing thresholds (MTTS) of bottlenose dolphins, *Tursiops truncates*, and white whales, *Dephinapterus leucas*, after exposure to intense tones. *J. Acoust. Soc. Am.* **107**, 3496–3508.

Siebold, P. F. von. (1842). Fauna Japonica: Les Mammiferes Marins, Batavia Press, Jakarta.

Slijper, E. J. (1976). "Whales and Dolphins." Univ. of Michigan Press, Ann Arbor, MI.

Steller, G. W. (1899). The beasts of the sea. *In* "The Fur Seals and Fur Seal Islands of the North Pacific Ocean" (D. S. Jordan, ed.), pp. 179–218. U.S. Government Printing Office, Washington, DC.

Thewissen, J. G. M. (ed.) (1998). "The Emergence of Whales: Evolutionary Patterns in the Origin of Cetacea." Plenum Press, New York.

Twiss, J. R., Jr., and Reeves, R. R. (1999). "Conservation and Management of Marine Mammals." Smithsonian Institution Press, Washington, DC.

Tyack, P. L. (1999). Communication and cognition. In "Biology of Marine Mammals" (J. E. Reynolds III and S. A. Rommel, eds.), pp. 287–323. Smithsonian Institution Press, Washington, DC.

Tynan, C. T., and DeMaster, D. P. (1997). Observations and predictions of Arctic climatic change: Potential effects on marine mammals. *Arctic* **50**; 308–322.

Wells, R. S., Boness, D. J., and Rathbun, G. B. (1999). Behavior. *In* "Biology of Marine Mammals" (J. E. Reynolds III and S. A. Rommel, eds.), pp. 324–422. Smithsonian Institution Press, Washington, DC.

Würsig, B., Cipriano, F., and Würsig, M. (1991). Dolphin movement patterns: Information from radio and theodolite tracking studies. *In* "Dolphin Societies: Discoveries and Puzzles" (K. Pryor and K. S. Norris, eds.), pp. 79–111. Univ. of California Press, Berkeley, CA.

Yamase, H. (1760). Geishi [Natural History of Whales]. Osaka Shorin, Osaka, Japan.

Zhou, K., and Zhang, X. (1991). "Baiji, the Yangtze River Dolphin, and Other Endangered Animals of China." Yilin Press, Nanjing, China.

Hooded Seal

Cystophora cristata

KIT M. KOVACS
Norwegian Polar Institute, Tromsø

I. Description and Distribution

The hooded seal is a large, northern phocid. It is silver-gray in color with irregular black spots covering most of the body (Fig. 1); the face is usually solid black. Adult males are about 2.5 m long and weigh an average of 300 kg, although some reach over 400 kg. Adult females are considerably smaller than males, measuring approximately 2.2 m long and weighing an average of 200 kg. Hooded seal pups are approximately 1 m long when they are born and weigh about 25 kg. They are blue-black on their backs and silver-gray on their bellies. The common name for hooded seal neonates "blue-back" comes from this distinctive coloration pattern. The most unique feature of the species is the prominent two-part nasal ornament of sexually mature males that gives the species its common name. When relaxed, this nasal appendage hangs as a loose, wrinkled sac over the front of males' noses. However, when they clamp their nostrils shut and inflate the sac it becomes a large, tight, bilobed "hood" that covers the front of the face and top of the head. Adult males also have a very elastic nasal septum that they can extrude through one of their nostrils as a big membranous pink balloon. These two rather bizarre structures are secondary sexual characters that males use to display to females and to other males during the breeding season.

Figure 1 *Hooded seal mother–pup pair with an attending male in the background (top). Blue-back hooded seal newly weaned, 4 days old (middle). Hooded seal male with nasal septum extruded and hood somewhat inflated (bottom).*

Hooded seals are a migratory species with a range that encompasses a large sector of the North Atlantic (Fig. 2). They follow an annual movement cycle that keeps them in close association with drifting pack ice. During the spring the adults concentrate for breeding purposes in three locations: one group forms off the east coast of Canada, which is split into two whelping patches, one in the Gulf of St. Lawrence and the other north of Newfoundland—an area known as the Front; a second group congregates in the Davis Strait; and a third comes together on the West Ice, east of Greenland. Some weeks after breeding the animals move into traditional molting areas on the southeast coast of Greenland, near the Denmark Strait, or in a smaller patch that is found along the northeast coast of Greenland, north of Jan Mayen. After the annual molt, hooded seals disperse broadly for the summer and autumn months, preferring areas along the outer edges of pack ice.

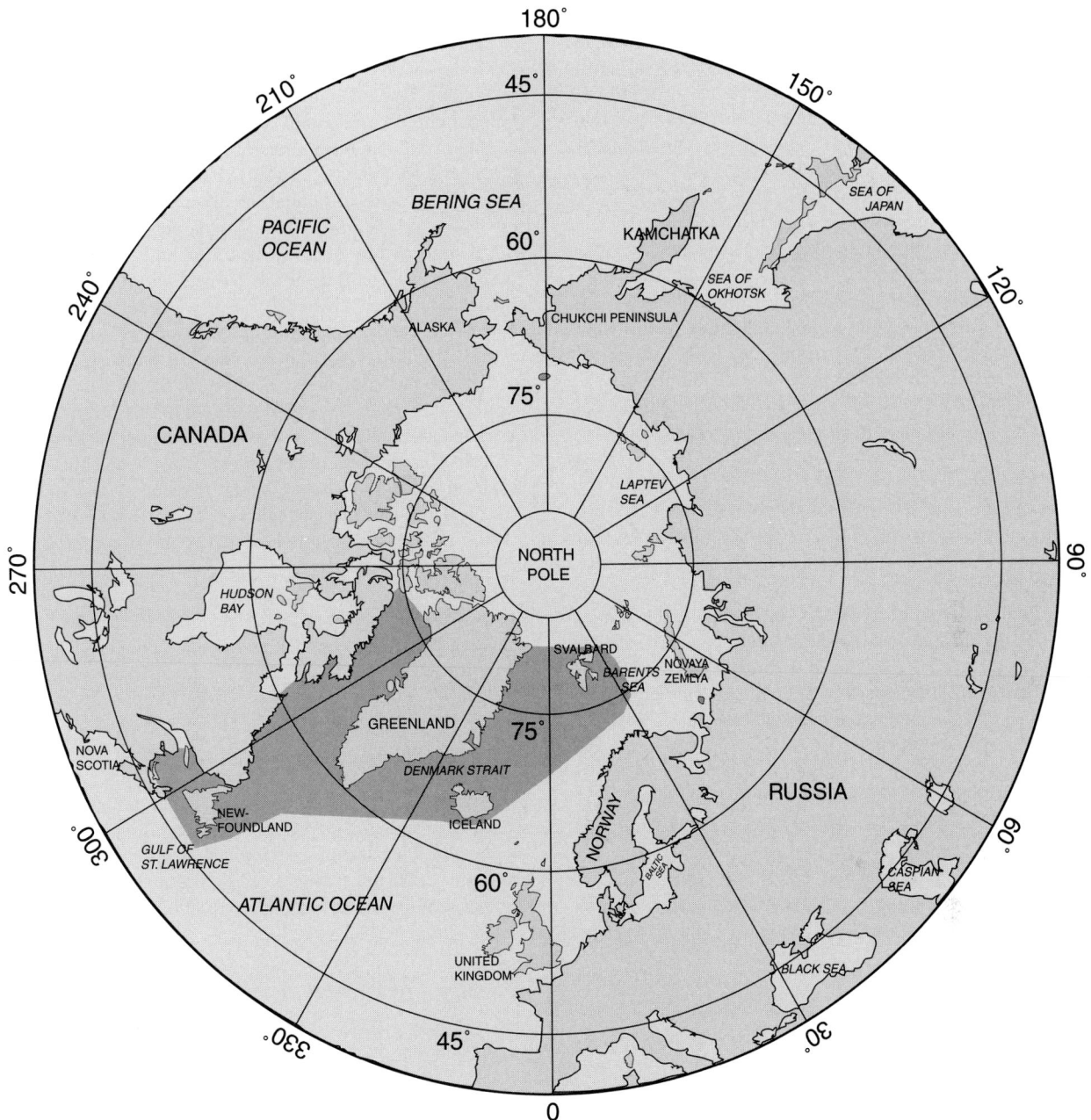

Figure 2 *Map showing the distribution of hooded seals (dark-shaded area).*

Records of hooded seals being found outside their normal range are not uncommon; young animals in particular are great wanderers. Juveniles have been found as far south as Portugal and Florida in the Atlantic Ocean and in California on the Pacific side.

II. Behavior and Ecology

Hooded seals are solitary animals outside the breeding and molting seasons. Even during these two annual phases when they do aggregate into loose herds, they are very aggressive with one another and do not tolerate close contact beyond the mother–offspring bond or a short male–female pairing period. Their vocal repertoire is quite simple, as would be expected for a species that is not highly social. The breeding season occurs in late March. It is short, lasting only 2–3 weeks in a given area. Females give birth in loose pack ice areas, preferring quite thick first-year ice floes for whelping. They space themselves out within the herd at intervals of 50 m or more when ice conditions permit, but the form of the herd and interfemale distances are highly variable, depending on the ice conditions. Mothers attend their pups continuously during the 4-day long period of lactation. Pups are born in a very advanced developmental state, having already shed their grayish-white embryonic first coat of hair and having already accumulated a thin layer of subdermal BLUBBER. During the incredibly short nursing period, pups drink up to 10 l of milk per day that contains an average fat content of 60%. This energy-rich diet allows pups to more than double their birth mass during the few days that they are cared for by their mothers; they gain over 7 kg per day. Pups are weaned weighing 50–60 kg. During the time when mothers are with their pup, a male often attends the pair. Males compete with one another to maintain positions close to a female. The battles are often bloody. When a mother is ready to leave her offspring, the attending male accompanies her to the water where mating takes place. Hooded seals have delayed implantation of the embryo, for up to 4 months, similar to many other seal species. Males will return to the whelping area after mating with a female to resume mate searching. Individual males have been recorded with up to eight females in one breeding season.

The pups remain alone on the ice for some days or weeks before going to the water and learning to swim, dive, and forage. During their time on the ice they fast, using body reserves stored in their substantial blubber layer to fuel their energy needs. When they do start to eat, pups feed on krill and other invertebrates initially until they have sufficient aquatic skills to capture fish. Little is known about juvenile hooded seals. They are seen only infrequently among adult breeding or molting aggregations. It is assumed that they spend much of their time at sea and in isolated Arctic pack ice areas. When they reach sexual maturity, at an age of 3 for most females and 4 or more for males, they join the species-typical annual migratory cycle. Hooded seal live for 25–30 years.

Adult hooded seals can dive to depths of over 1000 m and can remain underwater for periods of up to almost an hour. They fed on a variety of deep-water fishes, including Greenland halibut (*Reinhardtius hippoglossoides*) and a variety of redfish species (*Sebastes* spp.), as well as squid. Herring (*Clupea harengus*), capelin (*Mallotus villosus*), and various gadoid fishes, including Atlantic cod (*Gadus morhua*) and Arctic cod (*Boreogadus saida*), have also been found in hooded seal stomachs.

Polar bears (*Ursus maritimus*) are natural predators of hooded seals, but human exploitation is likely the greatest source of mortality. Killer whales (*Orcinus orca*) are also a likely predator, although this has never been documented conclusively.

III. Exploitation and Conservation Status

Hooded seals have been commercially exploited for centuries, usually in conjunction with hunts whose primary target was the more abundant harp seal. Norway, Russia, Denmark–Greenland, Great Britain, and Canada have taken part in the commercial harvesting of hooded seals. Pre-World War II hunting was done for oil and leather, but improved techniques for handling furs meant that the blue-back pelt was the most financially lucrative product of the hooded seal har-

vest following the war. Adults continued to be taken for oil and leather production, but the numbers were reduced because the market demand for these products dropped. Because adult females remain on the ice to defend their pups against hunters, many adult females were killed. Regulations limiting the killing of mothers have become more restrictive in recent decades and relatively few females are now taken in the whelping patches. Annual catches of hooded seals have always varied dramatically, depending largely on ice conditions at the time of breeding. In years of high harvests, up to 150,000 animals have been taken in the North Atlantic. Seal management in international waters was put under the auspices of the International Commission for the Northwest Atlantic Fisheries (ICNAF), with Canada, Norway, and Denmark being voting members in the early 1960s. Documented population declines of hooded seals led to the introduction of quota management during the 1970s. A bilateral agreement for East-Atlantic harvesting between Norway and Russia was also formulated. Following Canada's declaration of a 200-mile economic zone in the late 1970s, Norway and Canada also created a bilateral agreement and ICNAF was transformed into the Northwest Atlantic Fisheries Organization (NAFO). Under this agreement, Canada and Norway cooperate extensively with information exchange regarding hooded seal abundance estimates and commercial hunting quota revisions. The small population of hooded seals breeding in the Gulf of St. Lawrence is currently protected from harvesting, as is the Denmark Strait molting concentration. The European economic community banned the importation of all seal products in 1985. This had a marked effect on the value and market place for hooded seal pelts. Subsistence harvesting of hooded seals takes place in Arctic Canada and in Greenland.

The global population size of hooded seals is very difficult to estimate because this species is difficult to survey. The total population size is almost certainly in excess of half a million animals.

See Also the Following Articles

Earless Seals ■ Energetics ■ Mating Systems ■ North Atlantic Marine Mammals

References

Bowen, W. D., Oftedal, O. T., and Boness, D. J. (1985). Birth to weaning in 4 days: Remarkable growth in the hooded seal, *Cystophora cristata. Can. J. Zool.* **63**, 2841–2846.

Folkow, L. P., and Blix, A. S. (1995). Distribution and diving behaviour of hooded seals. *In* "Whales, Seals, Fish and Man" (A. S. Blix, L. Walløe, and Ø. Ulltang, eds.), pp. 193–202. Elsevier Science, Amsterdam.

Kovacs, K. M., and Lavigne, D. M. (1996). *Cystophora cristata.* Mamm. *Spec.* **258**, 9.

Lavigne, D. M., and Kovacs, K. M. (1988). "Harps and Hoods." Univ. of Waterloo Press.

Lydersen, C., and Kovacs, K. M. (1999). Behaviour and energetics of ice-breeding, North Atlantic phocid seals during the lactation period. *Mar. Ecol. Prog. Ser.* **187**, 265–281.

Hourglass Dolphin
Lagenorhynchus cruciger

R. NATALIE P. GOODALL
Centro Austral de Investigaciones Científicas,
Tierra del Fuego, Argentina

The hourglass dolphin, *Lagenorhynchus cruciger*, is the world's southernmost small dolphin. An oceanic species found in sub-Antarctic and Antarctic waters, it often bow rides ships and accompanies larger cetaceans but is one of the least studied cetaceans. Only three specimens were collected in the 136 years from the discovery of the species to 1960. Our knowledge of the biology of this species rests on 20 specimens (only 6 of them complete), 4 stranding observations and sightings at sea.

I. Taxonomy

The name *Delphinus cruciger* was based on a drawing from a sighting in the South Pacific in 1820. Synonyms include *D. albigena, D. bivittatus, Electra clancula, E. crucigera, D. superciliosus, Phocoena crucigera, P. d'Orbignyi, Lagenorhynchus wilsoni, L. latifrons,* and *L. Fitzroyi* (with *L. australis* and *L. obscurus*). The accepted combination, *L. cruciger,* was made by Van Beneden and Gervais in 1880. Common names have included the *crucigere,* the *albigena, grindhval,* sea skunks, and springers; the name in Spanish is *delfín cruzado.*

II. Diagnostic Characters
A. Pigmentation

The hourglass dolphin is mainly black or dark with two elongated lateral white areas, in some animals joined with a fine white line, which give it its common name (Fig. 1). The forward patch extends onto the face above the eye, which is within the black surface but outlined with a large dark eye spot with a point forward and a thin white line. The dark pigment of the lips is of varying shape; a gape to flipper stripe may be gray to tan, beige, or even rose. One animal had a white half-moon mark outlining the blowhole. On the side below the white flank patch, there is a lobe of white projecting forward, which may form a sharp point, a blunt, curved shape, or a hook. The flank patches on some animals almost meet on the upper tail stock. Part of the underside of the flippers is white. The ventral region is generally white, with some dark areas forward from the tail stock to the genital region. The pigmentation of juveniles has not been described.

B. Size and Shape

The hourglass is a rather stocky dolphin with a large, recurved dorsal fin that is variable in shape from erect to hooked. The tail stock is often keeled. Total lengths (*n*=9) range from 142 to 187 cm. Females (*n*=5) measure 142–183 cm, males (*n*=3) 163–187. This is probably not the total range of length for the species. Weights are known for only three specimens.

Figure 1 *Hourglass dolphins possess two lateral white areas along the flank that are united by a thin white line that resembles an hourglass. Schools of individuals can swell from the small group of 3 pictured, commonly around 6 or 7, and rarely up to 60 individuals. Photo by Robert Pitman.*

Females of 163.5 and 183 cm weighted 73.5 and 88 kg. A 174-cm male weighed 94 kg.

III. Distribution

The hourglass dolphin is circumpolar in the Southern Ocean, in both Antarctic and sub-Antarctic waters, from about 45°S south to fairly near the ice pack (Fig. 2). Exceptional sightings were at 36°14′S in the South Atlantic and 33°40′S in the South Pacific off Valparaiso, Chile. The southernmost sighting was at 67°38′S in the South Pacific. Most specimens were found between 45 and 60°S, the northernmost from New Zealand and the southernmost from the South Shetland Islands. Sightings and specimens are plotted in Fig. 2.

IV. Ecology
A. Habitat

This dolphin is pelagic and circumpolar in the Southern Hemisphere on both sides of the Antarctic Convergence and northward in cool currents associated with the West Wind Drift. Recorded water temperatures range from −0.3 to 13.4°C. Although oceanic, sightings of this dolphin are often made near islands and banks. Sightings reflect observer effort, with most in the Drake Passage, reflecting ship traffic between South America and the Antarctic Peninsula.

B. Prey and Predators

The stomachs of five specimens from different oceans have been examined; one was empty. Prey items included unidentified small fish, the fish *Krefftichtys andersonii* (Mycophidae) of about 2.4 g and a length of 55 mm; small squid, including some from the families Onychoteuthidae and Enoploteuthidae; and crustaceans. They often feed in large aggregations of sea birds and in plankton slicks.

No predators are known, although killer whales and leopard seals are possibilities.

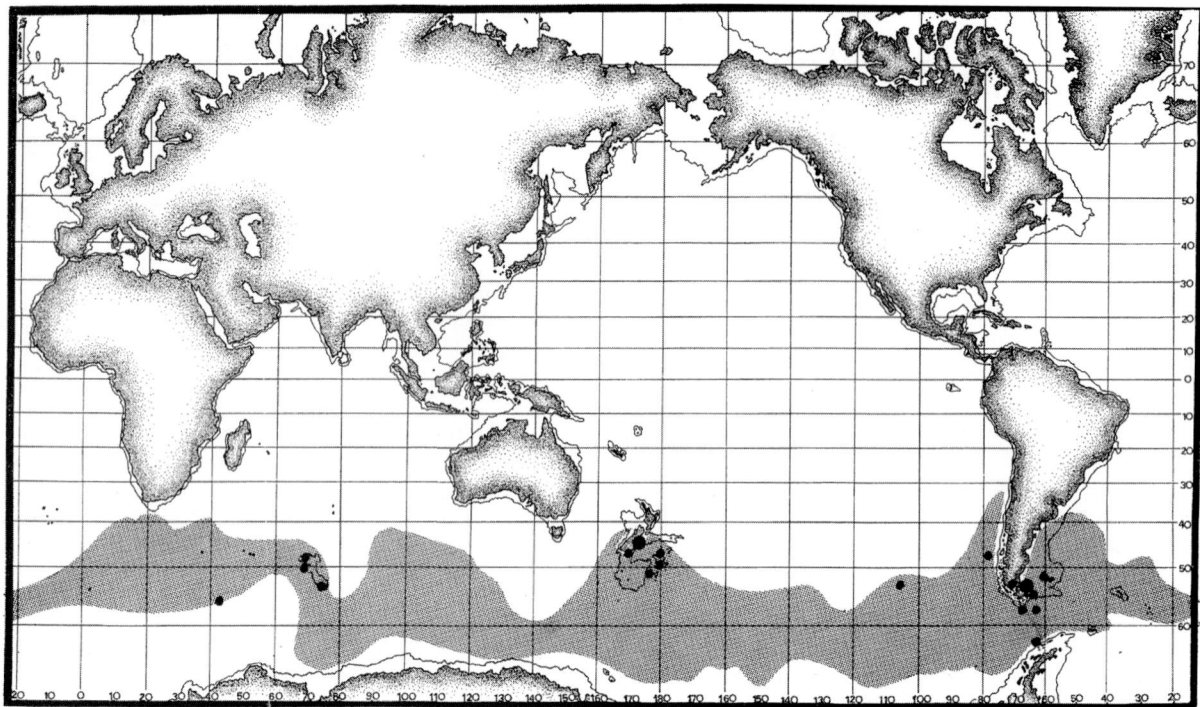

Figure 2 *Distribution of the hourglass dolphin compiled from incidental and dedicated sighting surveys and the location of specimens from 1824 to 1997. Small circles indicate single specimens, and large circles represent three specimens off New Zealand and four at Tierra del Fuego.*

V. Behavior

Nothing is known of the migratory movements of this species. Hourglass dolphins are rapid swimmers with a forward, plunging movement. They commonly bow-ride ships, especially in rough weather. These dolphins were found so often with fin whales (*Balaenoptera physalus*), that whalers used them as "spotters" in finding whales. They have also been seen with sei and minke whales (*B. borealis* and *B. bonaerensis*), large bottlenose whales (*Hyperoodon* and *Berardius*), pilot whales (*Globicephala melas*), and southern right whale dolphins (*Lissodelphis peronii*).

During the Southern Hemisphere minke whale assessment cruises conducted by the INTERNATIONAL WHALING COMMISSION, school sizes ranged from 1 to 60 animals (mean 7). Other studies reported mean sizes of 4 and 5.7 animals.

The sounds of this species have not been recorded.

VI. Internal Anatomy

The condylobasal lengths of 11 skulls ranged from 316 to 370 mm. Visible teeth numbered 26–34 upper and 27–35 lower in each jaw. Vertebral count is CV7, Th12–13, L18–22, and Ca 29–33 for a total of 69–72 (*n*=7). The first two cervicals are fused (*n*=4). The vertebrae of *L. cruciger* are smaller than those of *L. australis*, but slightly larger than those of *L. obscurus* and are similar to the latter in shape. There are 12–13 ribs (*n*=7). One specimen had seven pairs of sternal ribs, another eight. The phalangeal formula (*n*=5) is I=2–3, II=8–11, III=6–8, IV=2–4, and V=0–2.

The internal organs were examined for one specimen and organ weights are known for another. The intestine length of one specimen was 18 m.

VII. Life History
A. Growth and Reproduction

Very little is known of growth and reproduction in this species. A 163.5-cm female was sexually immature and one of 183 cm was pubescent. Males of 174 and 187 cm were sexually mature. Based on fusion of the vertebral epiphyses, the 163.5-cm female was physically immature. The 183-cm female was subadult. In males, an animal of 163 cm was subadult, one of 187 cm was nearly mature, and one of 174 cm was physically mature. Nothing is known of the young, times of birth, and reproduction rates; only three calves, seen in January and February, have been reported. No studies on aging have been published.

B. Parasites and Disease

Nematodes (*Anisakis* sp.) were reported in the stomachs and intestines of two animals. The largest animal known, a male, had long-standing gastric trauma with extensive peritonitis. No other diseases have been reported.

VIII. Interactions with Humans

Several hourglass dolphins have been taken for scientific study. No other directed catches are known. The only INCIDENTAL

CATCHES reported were three females from New Zealand and a drift net catch in the southern Pacific Ocean. No animals have been kept in CAPTIVITY.

See Also the Following Articles

Antarctic Marine Mammals ■ Delphinids

References

Brownell, R. L., Jr. (1999). Hourglass dolphin, *Lagenorhynchus cruciger* (Quoy and Gaimard, 1824). *In* "Handbook of Marine Mammals; the Second Book of Dolphins and Porpoises" (S. H. Ridgway and R. Harrison, eds.), vol. 6, pp. 121–135.

Gazitúa, F., Gibbons, J., and Cárcamo, J. (1999). Descripción de un ejemplar de delfín cruzado, *Lagenorhynchus cruciger* (Delphinidae), encontrado en el Estrecho de Magallanes. *Anales Instit. Patagónica Se. Cs. Nat. (Chile)* **27**, 73–82.

Goodall, R. N. P. (1997). Review of sightings of the hourglass dolphin, *Lagenorhynchus cruciger,* in the South American sector of the Antarctic and sub-Antarctic. *Rep. Int. Whal. Commn.* **47**, 1001–1013.

Goodall, R. N. P., Baker, A. N., Best, P. B., Meyer, M., and Miyazaki, N. (1997). On the biology of the hourglass dolphin, *Lagenorhynchus cruciger* (Quoy and Gaimard, 1824). *Rep. Int. Whal. Commn.* **47**, 985–999.

International Whaling Commission (1997). Report of the sub-committee on small cetaceans. Annex H. *Rep. Int. Whal. Commn.* **47**, 169–191.

Kasamatsu, F., Hembree, D., Joyce, G., Tsunoda, L., Rowlett, R., and Nakano, T. (1988). Distribution of cetacean sightings in the Antarctic: Results obtained from the IWC/IDCR minke whale assessment cruises, 1978/9 to 1983/84. *Rep. Int. Whal. Commn.* **38**, 449–487.

Kasamatsu, F., and Joyce, J. (1995). Current status of odontocetes in the Antarctic. *Antarc. Sci.* **7**(4), 365–379.

LeDuc, R. G., Perrin, W. F., and Dizon, A. E. (1999). Phylogenetic relationships among the delphinid cetaceans based on full cytochrome *B* sequences. *Mar. Mamm. Sci.* **15**(3), 619–648.

Van Waerebeek, K., Goodall, R. N. P., and Best, P. G. (1997). A note on evidence for pelagic warm-water dolphins resembling *Lagenorhynchus. Rep. Int. Whal. Commn.* **47**, 1015–1017.

Humpback Dolphins
Sousa chinensis, S. plumbea, and *S. teuszi*

GRAHAM J. B. ROSS
Australian Biological Resources Study, Canberra

The names given to the humpback dolphins, genus *Sousa,* reflect both their distribution (Fig. 1) and their distinctive humped appearance when they surface (Figs. 2 and 3). These coastal animals occur in small groups from northwestern Africa to northern China and Australia and are subject locally to incidental mortality in nets. Their biology is poorly known and their classification is open to discussion, not least because study material across the range is limited.

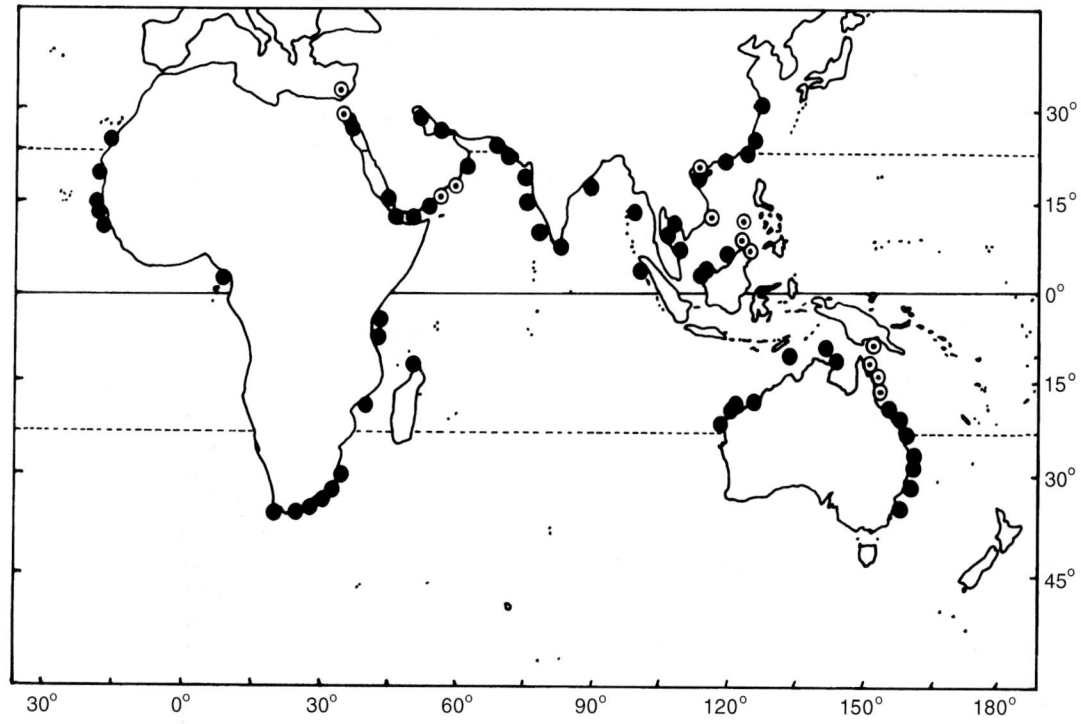

Figure 1 *Distribution of* Sousa *worldwide. Closed and open circles represent specimens and sightings, respectively. After Kerem and Goffman (2000), Ross* et al. *(1994, 1995), and Smith* et al. *(1997).*

Figure 2 *Subadult* Sousa *in Richards Bay, southern Africa, showing the humped back and ridged dorsal fin base typical of animals westward of Sri Lanka.*

I. Description

Humpback dolphins are medium sized and robust in form. The melon, moderate in size, is slightly depressed and, in profile, slopes gradually to an indistinct junction with the long, narrow snout. Neonates have vibrissae. The gape is straight. The broad flippers are rounded at the tip, and the flukes are broad and full, with a deep median caudal notch. Dorsal and ventral ridges on the caudal peduncle are well developed in African and Indian Ocean populations. The distinctive dorsal fin comprises a thickened base supporting a thinner, fin-shaped upper component; its overall form varies geographically (Figs. 2 and 3). Its length varies in relation to body length. Total length reaches a maximum of 2.5–2.8 m in different parts of the distribution. South African males and females are sexually dimorphic; mean lengths and weights for fully grown males are 2.70 m and 260 kg compared to 2.40 m and 170 kg in females. The few available data suggest that Arabian Gulf animals may also be dimorphic in length. The biology of any population is poorly known. These dolphins live close to shore in depths of 25 m or less. Characteristic features of the skull include a long,

Figure 3 *Adult female* Sousa chinensis *and her calf off North Lantau Island, Hong Kong, showing the total loss of gray body pigment during growth. Photo © L. Porter/SWIMS.*

narrow rostrum, strengthened by raised premaxillary bones and increasingly compressed toward the tip (Fig. 4), large temporal fossae on which the jaw muscles insert, and pterygoid bones that are separated in the midline by up to 11 mm. A broad gap exists between the posterior margin of the maxillary bones and the supraoccipital crest of the skull. The mandibular symphysis is long, and each jaw bears 27 to 38 teeth, wedge-shaped at their base.

II. Distribution

Atlantic humpback dolphins occur widely along the West African coast, between southern Morocco (23°54′N) and Cameroon. Indo-pacific humpback dolphins are distributed continuously along the coast from False Bay, South Africa to the South China Sea, including the Red Sea, Arabian Gulf, the Indian subcontinent, Gulf of Thailand, Malacca Straits and northern Borneo, and the coast of China to the Yangtze River (31°50′N). At least one animal reached the Mediterranean via the Suez Canal (Fig. 1). Australian animals reach 25°S on the west coast, extending to 34°S facilitated by the warm eastern boundary current; similarly, those resident in southern South Africa live at 34°S in water temperatures of 15–22°C.

III. Taxonomy

The taxonomy of the genus is not well established, partly because sample sizes for morphological and genetic comparisons are small for most populations. Three species were recognized by Rice (1998): *Sousa chinensis* (Osbeck, 1765), *S. plumbea* (G. Cuvier, 1829), and *S. teuszii* (Kukenthal, 1892). Other nominal species include *S. lentiginosa* (Owen, 1866) and *S. borneensis* (Lydekker, 1901). However, recent morphological studies, supported somewhat equivocally by genetic analyses, indicate that there is a single, variable species for which the name *S. chinensis* has priority.

IV. Regional Differences

Regional differences occur in external proportions, especially in body length, snout length, and the length of the dorsal fin base, although data on the mean or maximum body length exist for few populations. Animals in northeastern Australia and Hong Kong waters attain 2.3–2.5 m in length; Indian dolphins are generally shorter than 2.6 m; however, four animals exceeded 3.0 m in length. Beak length varies from about 6% of total length in West African animals to about 10% in northeast Africa. Dorsal fin height varies only in relation to total length, rather than regionally. In dolphins from Sri Lanka westward, the base of the dorsal fin forms a distinct elongate mid-dorsal ridge with a small, falcate upper component (Fig. 2). The basal part includes vascular structures similar to those in the dorsal fin of other delphinids, perhaps assisting in THERMOREGULATION, and may attain 39% of body length in southern African animals. Eastward of Sri Lanka, the fin base comprises a broad, thickened pad that rises slightly above its surroundings surface and supports a larger, almost triangular upper component with a rounded tip (Fig. 3). The few specimens known from the east

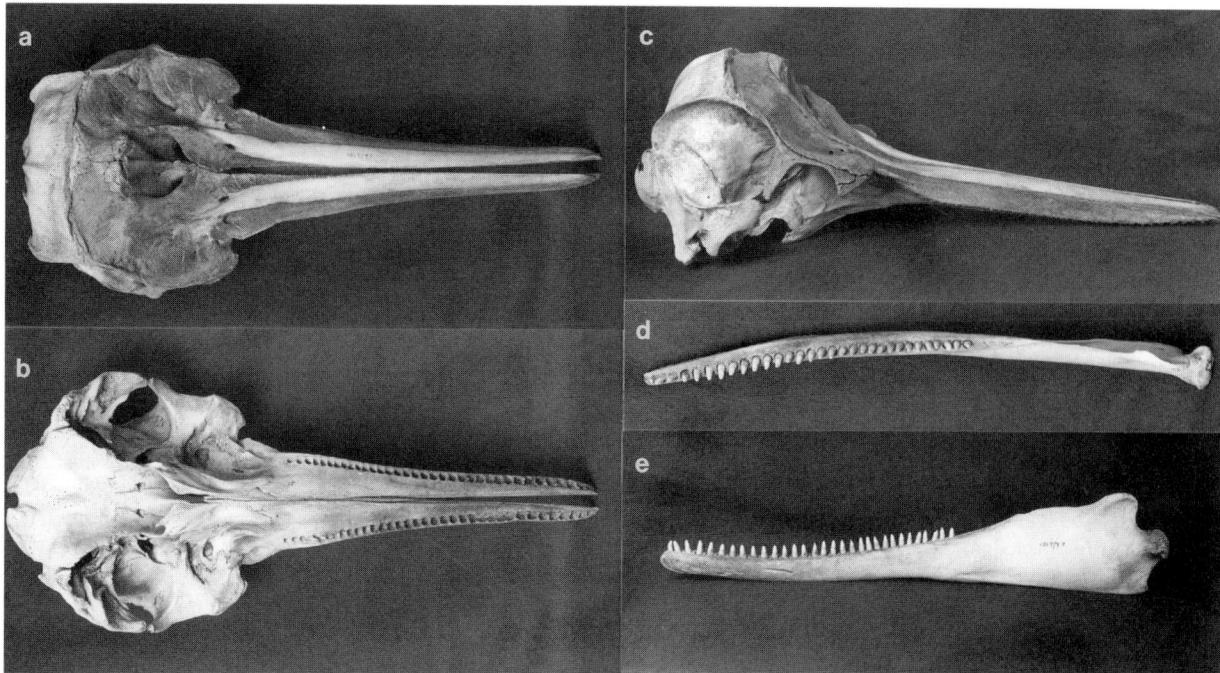

Figure 4 *Dorsal (a), ventral (b), and lateral (c) views of the skull, and dorsal (d) and lateral (e) views of the mandible of an adult male* Sousa *from South Africa. Photo by G. Ross, after Ross* et al. *(1994).*

coast of India suggest that the transition from one form of fin to the other occurs in this region.

Color varies greatly, with age and location, in both the timing and extent of the loss of the gray background color to become white (pink when flushed) and the development of spotting on the flanks and back. Born with a typical neonatal dark-grayish pigmentation above and paler gray below, animals lose gray pigmentation at different rates. Thus populations on the coast of China pale earlier than any other, turning white or pink within a few years of birth, a process that takes considerably longer in Australian and other populations to the west. Dolphins from eastern India to the Arabian Gulf coastline become spotted with white as mature adults, and there is no such pigment loss in any African population other than a pale fin mark in some southern African adults. Black spotting over the head and body develops in adults of all populations from the Arabian Gulf eastward.

Skull morphology is similar in all populations, apart from lower tooth counts, a shorter mandibular symphysis, and a broader cranium in West African animals. However, such low tooth counts may reflect small sample size, as the well-sampled southern African population varies greatly (30–38 teeth). The median number of TEETH per jaw increases eastward from 28 or 29 in West African animals to 36 or 37 teeth in north Indian Ocean populations and declines to 32 or 33 teeth in Southeast Asian and Australian animals. Regional differences occur in the ear bones of different populations. The range of vertebral formulae in South African animals was C7 T11–12 L9–12 Cd 20–24 = 49–52. Vertebral counts in humpback dolphins further east are similar to those of the South African sample (49–53), and West African humpback dolphins have 52–53 vertebrae.

V. Habitat

Humpback dolphins rarely occur in waters more than 25 m deep throughout their range. Saline, often turbid, waters in mangrove channels, embayments, and tropical river deltas or over shallow banks form important habitats. Occasionally they occur offshore, such as in the Great Barrier Reef, usually associated with islands or reefs, or in reef lagoons, such as Ningaloo Reef, Western Australia. Along the high-energy southern African coast, where sandbars limit access to rivers, humpback dolphins generally occur less than 1 km offshore, often close to the surf zone; conversely, in southern China, dolphins may swim up rivers for tens of kilometers.

VI. Behavior

Humpback dolphins swim slowly at about 5 km/hr, surfacing briefly at intervals of up to a minute. Longer dives may last up to 5 min. Typically they avoid boats and rarely bow ride. When approached, they generally dive, split up into small groups or single animals, and often change course underwater, reappearing unexpectedly some distance away. When a dolphin surfaces, the beak or occasionally the whole head is typically raised clear of the water and the body is arched, humping the back and perhaps exposing the flukes, before sounding. In one study, movement, feeding, social activities, and resting accounted for 49, 27, 15, and 8% of observation time, respectively. Single animals or pairs are generally adult; immatures tend to associate with groups containing more than one adult. Calves form about 10% of the population. Group size is generally 4 to 7 (range 2 to about 25). Body contact, displays during SWIMMING, leaping and chasing, or activities focused on inanimate or animate objects form part of the behavioral repertoire.

VII. Biology

These dolphins feed primarily on fish and cephalopods in shallow reef, littoral, or estuarine-associated habitats. Senegalese dolphins follow the rising tide into mangrove channels to feed, returning with the ebb tide, and South African dolphins spend more time feeding on the rising tide. Temporary beaching by dolphins to retrieve bonefish deliberately washed onto exposed sandbanks has been reported. Fishes in the families Haemulidae, Sciaenidae, Sparidae, Mugilidae, and Clupeidae and cuttlefishes (Sepiidae) are the most important prey across the range of *Sousa*. Notably, the first three of these families produce underwater sounds, which may assist dolphins to detect their prey.

Clear evidence of migration in humpback dolphins has not been found, although Senegalese animals may shift northward in summer. Humpback dolphins on the southern Cape coast of South Africa appear to be resident, and those in southern China are present throughout the year.

The effects of predation on humpback dolphins are uncertain. Off Natal, some 35% of these dolphins captured in nets bear healed shark bites, suggesting that shark attack is a significant cause of mortality, as this sample merely represents the survivors of such attacks. The few records of PARASITES include the whalelouse *Syncyamus aequus* and the nematode *Halocercus* (?) *pingi*.

Sound production and reception are vital to humpback dolphins in the often murky habitat they occupy. They produce clicks comprising highly directional single pulses repeated in series at variable rates between 10 and 500 Hz, apparently used for echolocation. Whistles and screams are frequency-modulated sounds: the former are produced singly or in series and are of variable length (milliseconds to seconds), whereas screams have a harmonic structure and occur in series. These may be important in communication, as their high-frequency components exceed the frequency range of sounds produced by fish and crustaceans and ambient environmental NOISE. The cochlear nerve in *Sousa* is specialized with large nerve fibers, permitting rapid transfer of information to the brain; the largest of some 77,000 fibers in this nerve are 50 μm in diameter, the greatest observed in any vertebrate.

Reproductive and other life history data are minimal for all populations. Calves are about 1 m long at birth. Off South Africa, births occur throughout the year, two-thirds of which are in summer; females mature at about 10 years old and males about 3 years later, assuming one growth layer group in teeth per year. The length at maturity is unknown. Data from other populations are minimal. The fat content of humpback dolphin milk is about 10%, similar to that of bottlenose dolphins but lower than that of common dolphins. Copulation occurs with one dolphin inverted below its partner, lasting 20–30 sec. Observations of dolphins rising vertically belly to belly in a vertical position in the Arabian Gulf and the Indus delta have been ascribed to mating behavior.

VIII. Relationships

The relationships of *Sousa* with other delphinoid genera are equivocal. Air sinus structure suggests that *Sousa* and *Steno* are closely related, whereas earbone structure places *Cephalorhynchus* with these two genera as a closely related group comprising the subfamily Sotaliinae. More recent genetic studies indicate that *Sousa* is a member of the Delphininae. No fossils of *Sousa* are known.

IX. Human Influence

Humpback dolphins from South African, Australian, and Thai populations have survived in captivity for periods from 3 months to over 30 years. In southern Queensland, up to seven free-ranging humpback dolphins visit Tin Can Bay regularly, where they are fed fish by visitors. Humpback dolphins are particularly susceptible to the effects of human activities in the coastal zone, especially fishing. Incidental catches in nets are reported for West Africa, northeastern Africa, the Arabian Gulf, Indus delta, and southwestern India. Off Natal, shark nets set to protect bathing beaches caught at least 67 humpback dolphins between 1980 and 1989, from a population of unknown size, although clearly small. Catch data for humpback dolphins in Australian protective nets are unknown. Changes in the Indus delta and other tropical regions through construction, drainage, and destruction of mangroves strongly affect the prime habitat of humpback dolphins.

Organochlorine residues are evident in several populations. In three southern Indian animals, total DDTs and PCBs in blubber reached 11,000–14,000 ppm and 920–1800 ppm, respectively. Off Natal, total DDT levels in blubber ranged from 59 to 243 ppm and PCBs 1000 to 130,000 ppm, the highest residue level for any marine mammal of that region. Mercury pollution levels of up to 0.9 ppm in Hong Kong *Sousa* present a potential threat to this population.

The conservation status of almost all populations of humpback dolphins throughout their range is uncertain, primarily because the monitoring of mortality rates is minimal in most regions and determining population size is difficult, even locally, although all appear to be small. Simple population estimates for the Indus delta and the Saloum delta, Senegal, are 500 and 100 animals, respectively. Density in Moreton Bay, southeast Queensland, is estimated at 0.1 dolphin/km^2.

See Also the Following Articles

Delphinids ■ Geographic Variation ■ Incidental Catches ■ Sound Production

References

Barros, N. B., and Cockcroft, V. G. (1991). Prey of humpback dolphins (*Sousa plumbea*) stranded in eastern Cape Province, South Africa. *Aqu. Mamm.* **17**(3), 134–136.

Cockcroft, V. G., Leatherwood, S., Goodwin, J., and Porter, L. J. (1997). The phylogeny of humpback dolphins genus *Sousa*: Insights through mtDNA analyses. Paper SC/49/SM25 presented to the IWC Scientific Committee, September 1997, Bournemouth.

Connell, A. D. (1994). Pollution and effluent disposal off Natal. *Lect. Notes Coast. Estuarine Stud.* **26**, 226–251.

Corkeron, P. J., Morisette, N. M., Porter, L., and Marsh, H. (1997). Distribution and status of hump-backed dolphins, *Sousa chinensis*, in Australian waters. *Asian Mar. Biol.* **14**, 49–59.

Gao, G., and Zhou, K. (1995). Fiber analysis of the vestibular nerve of small cetaceans. *In* "Sensory Systems of Aquatic Mammals." (R. A.

Kastelein, J. A. Thomas, and P. E. Natchigall, eds.), pp. 447–453. De Spil Publishers, Woerden.

Huang-Zongguo, Liu-Wenhua, Zheng-Chengxing, Lin-Ruicai, and Cai-Jialiang (1997). Chinese white dolphin (*Sousa chinensis*) in Xiamen Harbor. 1. Appearance and internal organs. *J. Oceanogr. Taiwan Strait* **16**(4), 473–478.

James. P. S. B. R., Rajagpolabn, M., Dan, S. S., Fernando, A. B., and Selveraj, V. (1989). On the mortality and stranding of marine mammals and turtles at Gahirmatha, Orissa from 1983 to 1987. *J. Mar. Biol. Assoc. India* **31**(1-2), 28–35.

Jefferson, T. A., Curry, B. E., Leatherwood, S., and Powell, J. A. (1997). Dolphins and porpoises of West Africa: A review of records (Cetacea: Dephinidae, Phocoenidae). *Mammalia* **61**, 87–108.

Kerem, D., and Goffman, O. and Spanier, E. (2001). Sighting of a single hump-backed dolphin (*Sousa* sp.) along the Mediterranean coast of Israel. *Mar. Mamm. Sci.* **17**, 170.

Parsons, E. C. M. (1998). Trace metal pollution in Hong Kong: Implications for the health of Hong Kong's Indo-Pacific hump-backed dolphins (*Sousa chinensis*). *Sci. Total Environ.* **214**, 175–184.

Peddemors, V. M., de Muelenaere, H. J. H., and Devchand, K. (1989). Comparative milk composition of the bottlenosed dolphin (*Tursiops truncatus*), humpback dolphin (*Sousa plumbea*) and common dolphin (*Delphinus delphis*) from South African waters. *Comp. Biochem. Physiol.* **94A**(4), 639–641.

Rice, D. W. (1998). "Marine Mammals of the World: Systematics and Distribution." The Society for Marine Mammalogy, Lawrence, KS.

Ross, G. J. B., Heinsohn, G. E., and Cockcroft, V. G. (1994). Humpback dolphins *Sousa chinensis* (Osbeck 1765), *Sousa plumbea* (G. Cuvier, 1829) and *Sousa teuszii* (Kukenthal, 1892). *In* "Handbook of Marine Mammals." (S. H. Ridgway and R. Harrison, eds.), Vol. 5, pp. 23–42. Academic Press, London.

Ross, G. J. B., Heinsohn, G. E., Cockcroft, V. G., Parsons, E. C. M., and Porter, L. J. (1995). Revision of the taxonomy of humpback dolphins, genus *Sousa*. Abstract in Proceedings of the Symposium on the Biology and Conservation of Small Cetaceans in Southeast Asia, 26–30 June 1995, Dumaguete, Philippines. (Working Document UNEP/SEA 95/WP19).

Schultz, K. W., and Corkeron, P. J. (1994). Interspecific differences in whistles produced by inshore dolphins in Moreton Bay, Queensland, Australia. *Can. J. Zool.* **72**, 1061–1068.

Smith, B. D., Jefferson, T. A., Leatherwood, S., Dao Tan Ho, Chu Van Thuoc, and Le Hai Quang (1997). Investigations of marine mammals in Vietnam. *Asian Mar. Biol.* **14**, 145–172.

Tanabe, S., Subramanian, A., Ramesh, A., Kumaran, P. L., Miyazaki, N., and Tatsukawa, R. (1993). Persistent organochlorine residues in dolphins from the Bay of Bengal, South India. *Mar. Pollu. Bull.* **26**, 311–316.

Humpback Whale
Megaptera novaeangliae

PHILLIP J. CLAPHAM
*Northeast Fisheries Science Center,
Woods Hole, Massachusetts*

The humpback whale (Fig. 1) is one of the best known and easily recognizable of the large whales. It is known for its frequent acrobatic behavior and its occasional tendency to approach vessels. In recent years, thousands of humpback whales have been identified individually from natural markings (notably the pattern on the ventral surface of the tail flukes), and as a result, much has been learned about the biology and behavior of this species.

I. Characters and Taxonomic Relationships

At close range, humpback whales are easily distinguished from any other large whale by their remarkably long flippers, which are approximately one-third the length of the body. The flippers are ventrally white and can be either white or black dorsally depending on the population and the individual; flippers of North Atlantic humpbacks tend to be white, whereas those in the North Pacific are usually black (Fig. 1). The body color is black dorsally, with variable pigmentation on the underside (black, white, or mottled). The head and jaws have numerous knobs called tubercles, which are also diagnostic of the species. The dorsal fin is small but highly variable in shape, ranging from low (almost absent) to high and falcate. Like all rorquals, humpbacks have a series of ventral pleats running back from the tip of the lower jaw, in this species to the umbilicus. The tail is usually raised during a dive; the underside exhibits a pattern that is unique to each individual, which ranges from all white to all black. The presence of white on the ventral surface, and the prominent serration of the trailing edge, distinguishes humpbacks from other whales that "fluke" while DIVING, such as right, bowhead, blue, gray, and sperm whales.

Adult female humpback whales are typically 1 to 1.5 m longer than males. Maximum reliably recorded adult lengths are in the 16- to 17-m range, although 14–15 m is more typical (Clapham and Mead, 1999). Calves are 3.96 to 4.57 m at birth and approximately 8–10 m at independence (Clapham *et al.* 1999), which occurs at the end of the calf's natal year. There are no easily observable differences between male and female humpbacks. Females possess a grapefruit-sized lobe at the rear of the genital slit; this lobe is absent in males (Glockner-Ferrari and Ferrari, 1990). In addition, the spacing between the genital slit and the anus is considerably greater in males.

The SKULL of the humpback whale is easily distinguished from that of other baleen whales by the narrowness of the rostrum relative to the zygomatic width. The humpback has between 270 and 400 baleen plates on each side of the mouth. The plates are usually black, although those close to the tip of the jaw are sometimes white or partly white.

The genus *Megaptera* is monotypic and is one of two genera in the family Balaenopteridae (the "rorquals"). No subspecies are recognized. The binomial *Megaptera novaeangliae* derives from the Greek for "big wing" (*mega* + *pteron*) and the Latin for "New England," which was the origin of the specimen used by Borowski in his description of the species in 1781.

II. Distribution and Ecology

Humpback whales are found in all oceans of the world (Fig. 2). They are a highly migratory species, spending spring through fall on feeding grounds in mid- or high-latitude waters, and wintering on calving grounds in the tropics, where they do not

Figure 1 *The long flippers of humpback whales are white, whereas the dorsum of the body is usually black. The display of the flipper is a small part of a spectacular behavioral repertoire that includes tail slashing, breaching, and other behaviors. Pieter A. Folkens/Higher Porpoise DG.*

eat (Dawbin, 1966). Humpback whales are typically found in coastal or shelf waters in summer and close to islands or reef systems in winter. Some documented MIGRATORY MOVEMENTS of this species represent the longest known migration of any mammal, being almost 5000 miles one way (Palsbøll *et al.,* 1997). Not all humpbacks migrate every year, although the sex/age class of nonmigratory animals remains unclear. Remarkably, the purpose of migration remains unknown; it may reflect a need to maximize energetic gain by exploiting pulses of productivity in high latitudes in summer and then gaining thermodynamic advantages by overwintering in warm water in winter. The only nonmigratory population is that residing in the Arabian Sea, where monsoon-driven productivity in summer permits the whales to remain in tropical waters year-round (Mikhalev, 1997).

In the North Atlantic, humpbacks return each spring to specific feeding grounds in the Gulf of Maine, Gulf of St. Lawrence, Newfoundland, Labrador, Greenland, Iceland, and Norway. Fidelity to these areas is strong and is determined by where a calf was taken by its mother in the former's natal year. Recent genetic analysis has indicated that this fidelity is maintained on an evolutionary time scale in at least Iceland and Norway (Palsbøll *et al.,* 1995; Larsen *et al.,* 1996). Despite this fidelity, whales from all feeding grounds migrate to the a common breeding area in the West Indies, where they mate and calve (Katona and Beard, 1990). Historically important breeding areas in the Cape Verde Islands and the southeastern Caribbean appear to be utilized by relatively few whales today.

In the North Pacific, there are at least four separate breeding grounds in Hawaii, coastal Mexico, offshore Mexico (Revil-

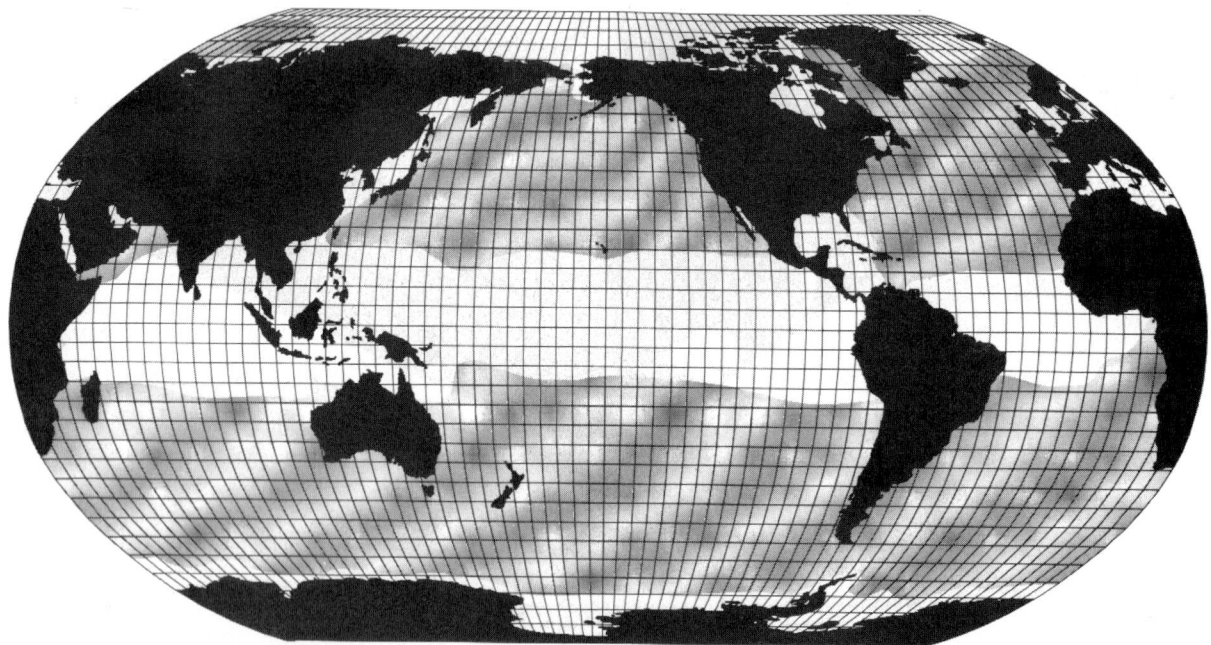

Figure 2 *The distribution of the humpback whale* (Megaptera novaeangliae) *illustrates a common pattern that, for the most part, excludes tropical warm waters in summer (Arabian humpbacks are an exception). The humpback is especially well known as a migratory species, tending to feed and mate at higher latitudes but calving at lower latitudes. From Perry* et al. *(1999); adapted from Johnson and Wolman (1984).*

lagigedos Islands), and Japan (Calambokidis *et al.*, 1997). Whales from these wintering areas migrate primarily to Alaska, California, possibly the Bering Sea, and the western North Pacific, respectively. However, crossover is not unknown and some transoceanic movements have been recorded (e.g., British Columbia to Japan and back).

In the Southern Hemisphere, humpbacks feed in circumpolar waters around the Antarctic and migrate to relatively discrete breeding grounds in tropical waters to the north. Six populations or "management areas" are recognized by the International Whaling Commission in the Southern Hemisphere, but some movement between these areas is very likely.

The humpback whale has a generalist diet, feeding on euphausiids and various species of small schooling fish. The latter include herring (*Clupea* spp.), capelin (*Mallotus villosus*), sand lance (*Ammodytes* spp.), and mackerel (*Scomber scombrus*). Humpbacks appear to be unique among large whales in their use of bubbles to corral or trap schooling fish. Whales blow nets, clouds, or curtains of bubbles around or below schools of fish and then lunge with mouths open into the center of the bubble structure (Jurasz and Jurasz, 1979; Hain *et al.*, 1982). As with other balaenopterids, the ventral pleats expand when a humpback is feeding, allowing the animal to increase the capacity of its mouth greatly.

Rake-mark scars from TEETH attest to the fact that humpbacks are commonly attacked by killer whales (*Orcinus orca*). However, it seems likely that fatal attacks are largely confined to very young calves, and predation does not appear to be a significant effector in the social organization of the humpback (Clapham, 1996).

III. Life History and Behavior

Breeding in humpback whales is strongly seasonal. Females come into estrus in winter, at which time testosterone production and spermatogenesis also peak in males (Chittleborough, 1965). The gestation period is about 11 months, with the great majority of calves born in midwinter. Calves probably begin to feed independently at about 6 months of age, but nursing likely continues in many animals until shortly before independence at about a year of age. Sexual maturity is reached in both sexes on average at 5 years. Interbirth intervals in females are most commonly 2 years, although annual calving is not unknown (Clapham and Mayo, 1990; Glockner-Ferrari and Ferrari, 1990). Although multiple fetuses have been recorded in dead pregnant females, living twins or multiplets are unknown.

The social organization of the humpback is characterized by small unstable groups, and individuals typically associate with many companions on both feeding and breeding grounds (Clapham, 1996). Longer-term associations (those lasting days or weeks) are occasionally recorded, but their basis is unclear. There appears to be no territoriality in this species.

In winter, male humpback whales sing long, complex songs, the primary function of which is presumably to attract females. All whales in a given population sing essentially the same song, and although the form and content of all songs change over time, the whales somehow coordinate these changes. Males also compete very aggressively for access to females (Tyack and Whitehead, 1983), and the resulting "competitive groups" can last for hours and involve tail slashing, ramming, or head butting. Males

may also form coalitions, but further research is required to assess the significance and composition of such alliances.

In part because of the prominent male display aspect (i.e., singing behavior), the mating system has been compared to a lek (Mobley and Herman, 1985), although it does not possess the rigid territoriality common to such systems. Males almost certainly remain in breeding areas longer than females and attempt to obtain repeated matings, whereas newly pregnant females return quickly to higher latitudes (Dawbin, 1966) where they will feed for many months in order to prepare for the considerable energetic cost of lactation.

Humpbacks are well known for their often spectacular aerial behaviors. These include breaching, LOBTAILING, and flippering. Such behaviors occur at all times of year and in a variety of contexts, and it is clear that they perform a range of functions. These may include play, COMMUNICATION, parasite removal, and expression of excitement or annoyance.

IV. Conservation Status

The humpback whale was heavily exploited by the whaling industry for several centuries. Because of its coastal distribution, it was often the first species to be hunted in a newly discovered area. This century, some 200,000 humpbacks were slaughtered in the Southern Hemisphere alone; of these, more than 48,000 were taken illegally by the Soviet Union (Yablokov *et al.*, 1998). It is quite likely that more than 90% of the animals in some populations were killed during the most intensive periods of exploitation. As a result, the humpback is considered an endangered species. Despite this, most studied populations appear to be making a strong recovery. The North Atlantic population has been estimated at 10,400 animals (Smith *et al.*, 1999) and the North Pacific at 6000–8000 (Calambokidis *et al.*, 1997). Strong population growth rates have been reported for many areas, ranging from 6.5% in the Gulf of Maine to more than 10% in some Southern Hemisphere populations (IWC, 1999). Commercial whaling for humpbacks officially ended worldwide in 1966, although the Soviets continued to hunt this species for some years afterward. Small aboriginal hunts for humpbacks still occur in a couple of locations, and many more whales die from entanglement in fishing gear or collisions with ships. However, none of these impacts appears to be significant at the population level, and the outlook for this once overexploited species appears good in most areas.

See Also the Following Articles

Breaching ▪ Entrapment and Entanglement ▪ Song

References

Calambokidis, J., *et al.* (1997). Population abundance and structure of humpback whales in the North Pacific basin. Final report to Southwest Fisheries Science Center, La Jolla, CA.

Chittleborough, R. G. (1965). Dynamics of two populations of the humpback whale, *Megaptera novaeangliae* (Borowski). *Aust. J. Mar. Freshw. Res.* **16**, 33–128.

Clapham, P. J. (1996). The social and reproductive biology of humpback whales: An ecological perspective. *Mamm. Rev.* **26**, 27–49.

Clapham, P. J., and Mayo, C. A. (1990). Reproduction of humpback whales, *Megaptera novaeangliae,* observed in the Gulf of Maine. *Rep. Int. Whal. Comm. Spec. Issue* **12,** 171–175.

Clapham, P. J., and Mead, J. G. (1999). *Megaptera novaeangliae. Mamm. Species* **604,** 1–9.

Clapham, P. J., Wetmore, S. E., Smith, T. D., and Mead, J. G. (1999). Length at birth and at independence in humpback whales. *J. Cetacean Res. Manage.* **1,** 141–146.

Dawbin, W. H. (1966). The seasonal migratory cycle of humpback whales. *In:* Whales, Dolphins and Porpoises (K. S. Norris, ed.), pp. 145–170. Univ. of California Press, Berkeley, CA.

Glockner-Ferrari, D. A., and Ferrari, M. J. (1990). Reproduction in the humpback whale (*Megaptera novaeangliae*) in Hawaiian waters, 1975–1988: The life history, reproductive rates and behaviour of known individuals identified through surface and underwater photography. *Rep. Int. Whal. Comm. Spec. Issue* **12,** 161–169.

Hain, J. H. W., Carter, G. R., Kraus, S. D., Mayo, C. A., and Winn, H. E. (1982). Feeding behaviour of the humpback whale, *Megaptera novaeangliae,* in the western North Atlantic. *Fish. Bull.* **80,** 259–268.

IWC (1999). Report of the Scientific Committee. *J. Cetacean Res. Manage.* **1**(Suppl.)

Johnson, J. H., and Wolman, A. A. (1984). The humpback whale, *Megaptera novaeangliae. Mar. Fish. Rev.* **46,** 30–37.

Jurasz, C. M., and Jurasz, V. P. (1979). Feeding modes of the humpback whale, *Megaptera novaeangliae,* in Southeast Alaska. *Sci. Rep. Whales Res. Inst. Tokyo* **31,** 69–83.

Katona, S. K., and Beard, J. A. (1990). Population size, migrations and feeding aggregations of the humpback whale (*Megaptera novaeangliae*) in the western North Atlantic Ocean. *Rep. Int. Whal. Comm. Spec. Issue* **12,** 295–305.

Larsen, A. H., Sigurjónsson, J., Øien, N., Vikingsson, G., and Palsbøll, P. J. (1996). Population genetic analysis of mitochondrial and nuclear genetic loci in skin biopsies collected from central and northeastern North Atlantic humpback whales (*Megaptera novaeangliae*): Population identity and migratory destinations. *Proc. R. Soc. Lond. B* **263,** 1611–1618.

Mikhalev, Y. A. (1997). Humpback whales *Megaptera novaeangliae* in the Arabian Sea. *Mar. Ecol. Progr. Ser.* **149,** 13–21.

Mobley, J. R., and Herman, L. M. (1985). Transience of social affiliations among humpback whales (*Megaptera novaeangliae*) on the Hawaiian wintering grounds. *Can. J. Zool.* **63,** 763–772.

Palsbøll, P. J., Allen, J., Bérubé, M., Clapham, P. J., Feddersen, T. P., Hammond, P., Jørgensen, H., Katona, S., Larsen, A. H., Larsen, F., Lien, J., Mattila, D. K., Sigurjónsson, J., Sears, R., Smith, T., Sponer, R., Stevick, P., and Øien, N. (1997). Genetic tagging of humpback whales. *Nature* **388,** 767–769.

Palsbøll, P. J., Clapham, P. J., Mattila, D. K., Larsen, F., Sears, R., Siegismund, H. R., Sigurjónsson, J., Vásquez, O., and Arctander, P. (1995). Distribution of mtDNA haplotypes in North Atlantic humpback whales: The influence of behavior on population structure. *Mar. Ecol. Progr. Ser.* **116,** 1–10.

Perry, S. L., DeMaster, D. P., and Silber, G. K. (1999). The great whales: History and status of six species listed as endangered under the U.S. Endangered Species Act of 1973. *Mar. Fish. Rev.* **61,** 1–23.

Smith, T. D., Allen, J., Clapham, P. J., Hammond, P. S., Katona, S., Larsen, F., Lien, J., Mattila, D., Palsbøll, P. J., Sigurjónsson, J., Stevick, P. T., and Øien, N. (1999). An ocean-basin-wide mark-recapture study of the North Atlantic humpback whale (*Megaptera novaeangliae*). *Mar. Mamm. Sci.* **15,** 1–32.

Tyack, P., and Whitehead, H. (1983). Male competition in large groups of wintering humpback whales. *Behaviour* **83,** 1–23.

Yablokov, A. A., Zemsky, V. A., Mikhalev, Y. A., Tormosov, V. V., and Berzin, A. A. (1998). Data on Soviet whaling in the Antarctic in 1947–1972 (population aspects). *Russ. J. Ecol.* **29,** 38–42.

Hunting of Marine Mammals

Randall R. Reeves
Okapi Wildlife Associates, Hudson, Quebec, Canada

Ancient middens testify to the importance of marine mammals in the lives of early maritime people around the world. Many of the bones and bone fragments found in such sites probably came from animals that were scavenged from beaches. However, ingenious methods of capturing pinnipeds, sirenians, and cetaceans eventually were developed, and the archaeological refuse came to signify past hunting. The rewards were tempting—large amounts of nutritious meat and fat, hides, ivory, sinews for sewing, and bones for making household implements or weapons. These products eventually came to have high commercial value, fueling global whaling and sealing industries in modern times.

Among the marine mammals, no taxonomic group has been entirely spared from hunting pressure. However, some species have been hunted more intensively than others. The great whales (the sperm whale *Physeter macrocephalus* and the baleen whales) have been sought for their oil, meat, and baleen; pinnipeds for their oil or pelts; sea otters (*Enhydra lutris*) for their furs; and sirenians mainly for their flesh and skins. In contrast, some dolphin populations have hardly been hunted at all, and they remained secure until the advent and proliferation of unselective fishing methods, which result in the incidental killing of nontarget organisms. Marine mammals have also been hunted with the intention of reducing their predation on valued resources such as fish, crustaceans, or mollusks (Northridge and Hofman, 1999). This culling, often implemented through government-sponsored bounty programs, is similar to that directed at wolves, mountain lions, and other predators in parts of North America, with the outspoken support of ranchers and sport hunters.

I. Hunting of Whales, Dolphins, and Porpoises

People in the Arctic were hunting bowhead whales (*Balaena mysticetus*) as long ago as the middle of the first millennium, and western Europeans were taking right whales (*Eubalaena glacialis*) by the beginning of the second (Ellis, 1991; McCartney, 1995). While the technology and culture of subsistence whaling spread within the Arctic and sub-Arctic from the Bering Strait region, the Basques were responsible for the development and spread of commercial WHALING (see Section V). From its beginnings in the Bay of Biscay, this whaling eventually reached all of the world's oceans and involved people of many nationalities. Modern whaling, characterized by engine-driven catcher boats and deck-mounted harpoon cannons firing explosive grenades, began in Norway in the 1860s (Tønnessen and Johnsen, 1982). A key feature of modern whaling was that it made possible the routine capture of any species, including the blue whale (*Balaenoptera musculus*), fin whale

(*B. physalus*), and other fast-SWIMMING balaenopterines. In the first three-quarters of the 20th century, factory ships from several nations (e.g., Norway, the United Kingdom, Germany, Japan, the United States, and the Soviet Union) operated in the Antarctic, the richest whaling ground on the planet. At its pre-War peak in 1937–1938, the industry's 356 catcher boats, associated with 35 shore stations and as many floating factories, killed nearly 55,000 whales, 84% of them in the Antarctic.

Commercial whaling declined in the 1970s as a result of conservationist pressure and depletion of whale stocks. The last whaling stations in the United States and Canada were closed in 1972, and the last station in Australia ceased operations following the 1978 season. By the end of the 1970s, only Japan, the Soviet Union, Norway, and Iceland were still engaged in commercial whaling. With the decision by the INTERNATIONAL WHALING COMMISSION (IWC) in 1982 to implement a global moratorium on commercial whaling, Japan and the Soviet Union made their final large-scale factory-ship expeditions to the Antarctic in 1986/1987, and Japan stopped its coastal hunt for sperm whales (*Physeter macrocephalus*) and Bryde's whales (*Balaenoptera edeni*) in 1988. Iceland closed its whaling station in 1990 and shortly thereafter withdrew its membership in the IWC. Contrary to the widespread belief that commercial whaling had ended, however, Norway and Japan continued their hunting of minke whales through the 1990s. By formally objecting to the moratorium decision, Norway reserved its right to carry on whaling. Thus, Norwegian whalers continued to kill more than 500 northern minke whales per year in the North Atlantic. Using a provision in the whaling treaty that allows member states to issue permits to hunt protected species for scientific research, Japan continued taking more than 400 southern minke whales in the Antarctic and 100 northern minke whales in the western North Pacific each year. In 1999, the Icelandic parliament approved the resumption of a shore-based commercial hunt. Because the main incentive for commercial whaling in the 1990s was Japanese demand for whale meat, Norway and Iceland were eager to reopen the international trade in whale products. Norway continued to stockpile meat and blubber in anticipation that the trade ban under the Convention on International Trade in Endangered Species of Wild Fauna and Flora would be lifted.

The hunting of smaller cetaceans has generally been confined to coastal waters and conducted on a smaller, or at least localized, scale. There are, however, some examples of large, well-organized hunts (Mitchell, 1975). Fishermen in the Faroe Islands have continued to kill hundreds, and in some years well over a thousand, long-finned pilot whales (*Globicephala melas*) and Atlantic white-sided dolphins (*Lagenorhynchus acutus*) in a drive fishery that is centuries old. The driving method involves a number of small boats that herd the animals into shallow water where they can be killed with lances, long knives, or firearms. There has also been a long-standing drive fishery in Japan, taking a variety of delphinid species, most notably striped dolphins (*Stenella coeruleoalba*). The Japanese hunt for small cetaceans involves other methods as well, including the hand-harpooning of dolphins and Dall's porpoises (*Phocoenoides dalli*) and the use of harpoon guns to take short-finned pilot whales (*G. macrorhynchus*) and other medium-sized cetaceans.

A large commercial hunt for dolphins (*Delphinus delphis* and *Tursiops truncatus*) and harbor porpoises (*Phocoena phocoena*) was conducted in the Black Sea from the 19th century through the mid-1960s (the Soviet Union banned dolphin hunting in 1966), and this hunting with rifles and purse seines continued in the Turkish sector until at least 1983.

Aboriginal hunters in Russia, the United States (Alaska), Canada, and Greenland kill several tens of bowhead whales, 100–200 gray whales (*Eschrichtius robustus*), and many hundreds of white whales (*Delphinapterus leucas*), narwhals (*Monodon monoceros*), and harbor porpoises (Greenland only) each year (Fig. 1). This hunting is primarily for food, and the products are generally consumed locally or sold within proscribed markets (see Section V). In recent years, aboriginal whalers in Washington State (USA), British Columbia (Canada), and Tonga (a South Pacific island nation) have expressed interest in reestablishing their own hunts for large cetaceans. In fact, in the spring of 1999, the Makah tribe in Washington took their first gray whale in more than 50 years.

Figure 1 *Adult white whales (belugas) killed by Eskimos in Kasegaluk Lagoon near Point Lay on the Chukchi Sea coast of Alaska in July 1993. Canoes powered by outboard motors are used to drive the whales toward shore before killing them with rifles (top). The flukes, flippers, and skin with adhering blubber (locally called maktak) are saved as a delicacy (bottom). Courtesy of Greg O'Corry-Crowe.*

II. Pinnipeds

Sealing began in the Stone Age when people attacked hauled-out animals with clubs (Bonner, 1982). Later methods included the use of harpoons thrown from skin boats and gaff-like instruments for killing pups on ice or beaches. Traps and nets were used as well. The introduction of firearms transformed the hunting of pinnipeds and caused an alarming increase in the proportion of animals that were killed but not retrieved, especially in those hunts where the animals were shot in deep water before first being harpooned. This problem of "sinking loss" also applies to many cetacean hunts.

In addition to their meat and fat, the pelts of some seals, especially the fur seals and phocids, are of value in the garment industry. Markets for oil and sealskins fueled commercial hunting on a massive scale from the late 18th century through the early 20th century (Busch, 1985). The ivory tusks and tough, flexible hides of walruses (*Odobenus rosmarus*) made these animals exceptionally valuable to both subsistence and commercial hunters. At least 10,000 walruses are killed every year, most of them by the native people of northeastern Russia, Alaska, and northeastern Canada. The killing is accomplished mainly by shooting. The meat, BLUBBER, and skin are eaten by people or fed to dogs, while the tusks are either used for carving or sold as curios. Native hunters in the circumpolar north also kill more than a hundred thousand seals each year, mainly ringed seals (*Pusa hispida*) but also bearded (*Erignathus barbatus*), ribbon (*Histriophoca fasciata*), harp (*Pagophilus groenlandicus*), hooded (*Cystophora cristata*), and spotted seals (*Phoca largha*) (Fig. 2). Seal meat and fat remain important in the diet of many northern communities, and the skins are still used locally to make clothing, dog traces, and hunting lines. There is also a limited commercial export market for high-quality sealskins and a strong demand in Oriental communities for pinniped penises and bacula. The sale of these items, along with walrus and narwhal ivory, white whale and narwhal skin (maktak), and polar

bear (*Ursus maritimus*) hides and gall bladders, has helped offset the economic losses in some local hunting communities caused by the decline in international sealskin markets.

The scale of commercial sealing, like that of commercial whaling, has declined considerably since the 1960s. It continues, however, in several parts of the North and South Atlantic. After a period of drastically reduced killing in the 1980s, the Canadian commercial hunt for harp and hooded seals has been reinvigorated, at least in part as a result of governmental subsidies. An estimated 350,000 harp seals were taken by hunters in eastern Canada and West Greenland in 1998 (Lavigne, 1999). A few tens of thousands of molting pups are clubbed to death on the sea ice, but the vast majority of the killing is accomplished by shooting.

Norwegian and Russian ships continue to visit the harp and hooded seal grounds in the Greenland Sea ("West Ice") and Barents Sea ("East Ice"), taking several tens of thousands of seals annually. Also, thousands of South African fur seals (*Arctocephalus pusillus pusillus*) and South American fur seals (*A. australis*) have been taken annually in southwestern Africa and Uruguay, respectively. These hunts are centuries old, having been driven initially by commercial markets for skins and oil and, more recently, by the Oriental demand for seal penises and bacula. Also, especially in Africa, the hunt has been justified as a response to concerns about competition between seals and fisheries.

III. Sirenians

Sirenians have been hunted mainly for their delectable meat and blubber and their strong hides. The Steller's sea cow (*Hydrodamalis gigas*) was hunted to extinction within about 25 years of its discovery by European sea otter and fur seal hunters. Much like tortoises on tropical islands, the sea cows were easy to catch and provided local sustenance to ship crews, enabling the men to carry on their pursuit of fur, oil, and other valuable resources. Local people in West Africa and Central and South America used manatee hides to make shields, whips, glue, and plasters for dressing wounds. Large-scale commercial killing of Amazonian manatees (*Trichechus inunguis*) to supply mixira (fried manatee meat preserved in its own fat) took place in Brazil from the 1780s to the late 1950s, and manatee hides were in great demand for making heavy-duty leather products and glue between 1935 and 1954 (Domning, 1982). Dugongs (*Dugong dugon*), like manatees, have long been a prized food source for seafaring people throughout their extensive Indo-Pacific range (Nietschmann, 1984). It is impossible to make a reasonable guess of how many manatees are killed by villagers each year in West Africa and South America, but the total (three species, combined) is probably in the thousands. Hunting of dugongs continues in much of their range, including areas where the species is almost extinct.

Sirenians have been captured using many different methods, apart from simply stalking and lancing or harpooning them from boats, or setting nets to enmesh them. People in West Africa and South America developed ingenious fence traps and drop traps to catch manatees. These could be baited to attract the animals or just placed strategically to take advantage of

Figure 2 *Greenlanders butchering a bearded seal in Wolstenholme Fjord in June 1988. Bearded seals are especially prized by native people of the circumpolar north because of their tasty meat and tough, flexible hide. The hide is used to make leather lines, boat and tent covers, boot soles, and various other items. Courtesy of Steve Leatherwood.*

their natural movements through constricted channels. Dugong hunters in some areas used underwater explosives to kill their prey. In Torres Strait between Australia and New Guinea, portable platforms were set up on seagrass beds, and the hunter waited there overnight for opportunities to spear unsuspecting dugongs as they grazed.

IV. Sea Otters and Polar Bears

Sea otters have been cursed by the luxuriance of their pelts, which are among the most desirable of all mammalian furs. They were hunted remorselessly to supply the Oriental market from the 1780s onward—until very few were left and protection came in 1911. As otters were depleted in a region, hunting efforts there would be redirected at fur seals. Although anchored nets were sometimes used to catch sea otters (Kenyon, 1969), most of the hunting was conducted by men in boats, using lances initially and rifles later on. In California, otters were sometimes shot by men standing on shore, and in Washington, shooting towers were erected at the surf line and Indians were employed to swim out and retrieve the carcasses (Busch, 1985). Aboriginal people in Alaska are still allowed to hunt sea otters as long as the furs are used locally to make clothing or authentic handicraft items. The reported annual kill during the mid- to late-1990s was in the range of 600–1200.

Polar bears have always been prime targets of Eskimo hunters, and non-Eskimo sport hunters have taken large numbers of bears as trophies (Stirling, 1988). At least several hundred polar bears are still killed each year, most of them by Eskimos for meat and the cash value of their hides. In Canada, the hunting permits allocated to native communities are often sold to sport hunters, on condition that a local guide accompany the hunter and that only the head and hide be exported. These expeditions generally involve dogsled travel, thus reinvigorating a traditional mode of winter transportation while at the same time creating a need for more hunting—to obtain fish and marine mammal meat to feed to the dogs. Today, polar bears are killed almost exclusively by shooting them with high-powered rifles, but in the past they were also hunted with baited set-gun traps in Svalbard. A small number of polar bears are killed each year in self-defense.

V. Market (Commercial) vs Subsistence (Household-Use) Hunting

An important, although often problematical, distinction has been made between hunting for profit and hunting as a means of survival. This distinction is more than academic. The nature and degree of regulation have often depended on how a given hunter's enterprise was classified. The dichotomy between "commercial" and "subsistence" exploitation has had particular meaning in the context of the worldwide regulation of whaling. The IWC recognizes "aboriginal subsistence" whaling as a special category and has traditionally exempted certain groups of whalers from regulation. Similarly, many national and multilateral restrictions on sealing have applied only or primarily to industrial operations and not to "aboriginal" hunters hunting for "subsistence" (e.g., the U.S. Marine Mammal Protection Act and the North Pacific Fur Seal Convention).

Initially, the reasoning behind such special treatment was that these hunters used less destructive or wasteful gear and methods, and served only local, relatively small markets. However, those criteria are now called into question as aboriginal hunters have adopted modern weaponry and mechanized transport, and have increasingly chosen to sell their produce for cash. Some products, notably sealskins and the ivory obtained from walruses and narwhals, enter a global marketplace.

Anthropologists argue that the term "subsistence" should be broadly defined and not exclude cash-based exchanges when these occur within a context that emphasizes local production and consumption. They point to the fact that modern Eskimos, for example, are simply adapting to a changing world by hunting marine mammals with rifles, outboard motors, and snowmobiles. Only by selling skins, tusks, and, in the case of polar bears, their own services as hunting guides are these traditional hunters able to obtain the cash needed to live comfortably while continuing to be engaged in a domestic mode of production, providing highly esteemed and nutritious food for their home communities. Indeed, the IWC still considers Greenland whaling for minke and fin whales to be "aboriginal subsistence" whaling even though most of the whales are killed with deck-mounted harpoon guns firing explosive grenades and the meat and other products enter a country-wide, cash-based exchange network (Caulfield, 1997). At the same time, the IWC has resisted Japan's efforts to have "small-type coastal" whaling, which also serves a domestic but cash-based market, reclassified as something other than commercial whaling.

The difficulty of distinguishing commercial from subsistence hunting is not unique to situations involving marine mammals. Similar issues have arisen in relation to the trade in "bush meat" in Africa, Asia, and the Neotropics. Unregulated hunting is incompatible with the concept of sustainability. Considering the enormous increases in killing power afforded by firearms and mechanized transport, together with rapid human population growth and the attendant rise in resource consumption, we are long past a time when racial or cultural entitlement can be allowed to preclude a vigorously enforced management regime based on conservation principles.

VI. Future Hunting

For two reasons, the hunting of marine mammals in the foreseeable future is unlikely to approach the scale at which it was pursued throughout the 19th and much of the 20th century. First, the populations of many species remain far below earlier levels. Even if some recovery is achieved, the environmental carrying capacity has almost certainly declined in many instances. Considering the low productivity of these relatively large, long-lived animals, it is unrealistic to expect their aggregate biomass to return to "pristine" levels in a world so thoroughly transformed by human endeavor. Second, attitudes toward marine mammals have changed considerably, and any initiative to expand the scope or scale of hunting is subject to public scrutiny as never before. Many people, particularly in the United States, Europe, and Australasia, are morally opposed to the killing of cetaceans, if not all marine mammals. While this certainly does not mean that hunting will stop altogether, it does make it more likely that hunters will need to prove that their enterprise is both sustainable (within the productive capacity of the affected animal population) and humane.

See Also the Following Articles

Conservation Efforts ▪ Incidental Catches ▪ Inuit and Marine Mammals ▪ Polar Bear ▪ Steller's Sea Cow ▪ Whaling

References

Bonner, W. N. (1982). "Seals and Man: A Study of Interactions." Univ. of Washington Press, Seattle.

Busch, B. C. (1985). "The War against the Seals: A History of the North American Seal Fishery." McGill-Queen's University Press, Kingston, Ontario.

Caulfield, R. A. (1997). "Greenlanders, Whales, and Whaling: Sustainability and Self-determination in the Arctic." Univ. of New England Press, Hanover, NH.

Domning, D. P. (1982). Commercial exploitation of manatees *Trichechus* in Brazil c. 1785–1973. *Biol. Conserv.* **22**, 101–126.

Ellis, R. (1991). "Men and Whales." Alfred A. Knopf, New York.

Kenyon, K. W. (1969). "The Sea Otter in the Eastern Pacific Ocean." United States Department of the Interior, Bureau of Sport Fisheries and Wildlife, North American Fauna No. 68.

Lavigne, D. M. (1999). Estimating total kill of northwest Atlantic harp seals, 1994–1998. *Mar. Mam. Sci.* **15**, 871–878.

McCartney, A. P. (ed.) (1995). "Hunting the Largest Animals: Native Whaling in the Western Arctic and Subarctic." Canadian Circumpolar Institute, University of Alberta, Edmonton.

Mitchell, E. (1975). "Porpoise, Dolphin and Small Whale Fisheries of the World: Status and Problems." International Union for Conservation of Nature and Natural Resources, Gland, Switzerland. IUCN Monograph No. 3.

Nietschmann, B. (1984). Hunting and ecology of dugongs and green turtles, Torres Strait, Australia. *Natl. Geogr. Res. Rep.* 1976 Projects, **17**, 625–651.

Northridge, S. P., and Hofman, R. J. (1999). Marine mammal interactions with fisheries. *In* "Conservation and Management of Marine Mammals" (J. R. Twiss, Jr., and R. R. Reeves, eds.), pp. 99–119. Smithsonian Institution Press, Washington, DC.

Stirling, I. (1988). "Polar Bears." Univ. of Michigan Press, Ann Arbor, MI.

Tønnessen, J. N., and Johnsen, A. O. (1982). "The History of Modern Whaling." Univ. of California Press, Berkeley.

Hybridism

MARTINE BÉRUBÉ
University of California, Berkeley

Speciation is assumed to be a function of genetic divergence caused by reproductive barriers between gene pools (i.e., different populations). Traditionally, species are either *allopatric* or *sympatric*. In the case of sympatric species mechanisms such as temporal segregation, behavioral differences, and gametic incompatibilities ensure reproductive isolation. Hybridization denotes the successful mating between two individuals from different and reproductively isolated gene pools accepted as species. Hybridization is observed frequently among higher plants, but only rarely among vertebrates. Within mammals, hybrids have been recorded in a number of marine as well as terrestrial species. The evolutionary consequences of such hybrids vary depending on the frequency, the degree of genetic differences between the parental species, mating system, and the ecological circumstances.

The examination of hybrids has always attracted much attention, as such incidences and their frequency might provide clues to reproductive behavior, dispersal capabilities, and phylogenetic relationship of species. As might be expected, hybrids are more common within genera where the different species have similar life histories and habitat requirements. When the frequency of hybridization is low, the fitness of the hybrids is generally low as well and hybrids usually are nonviable or sterile and thus do not represent a threat to the genetic constitution of the parental species. However, as the frequency of hybridization increases, so may the number of viable and reproductive hybrids, which in turn might cause the breakdown of previous reproductive barriers between the two species. One evolutionary consequence of such a scenario is introgression. A recent well-documented example of this is the high incidence of coyote genes in what morphologically seem to be gray wolves observed in North America (Lehman *et al.*, 1991). The ultimate evolutionary consequence of introgression is the extinction of the species whose genome is being replaced by the other.

With regard to marine mammals, a total of some 61 cases of alleged hybridization have been described; 40 within Cetacea and 21 within the pinniped Carnivora. Putative hybrids have been observed in captivity as well as in the wild. Most of the marine mammal hybrids reported so far have only been described morphologically. However, molecular techniques have been applied in some cases in order to confirm the identity of hybrids and identify the parental species.

I. Evidence of Mating between Species

Although the theoretical expectation is that male and female genital morphology evolves continuously and thus makes interspecific mating difficult or impossible, attempts of interspecific mating have been observed between pinniped species where no hybrids have yet been reported. Such mating appears to be aggressive, and usually the heterospecific male is much larger than the female. Often the female does not survive such a mating (Miller *et al.*, 1996).

This kind of aggressive interspecific mating was first observed between a male gray seal (*Halichoerus grypus*) and a female harbor seal (*Phoca vitulina*) (Wilson, 1975). Later reports of such aggressive behaviors include mating between (i) a male New Zealand sea lion (*Phocarctos hookeri*) and a dead female New Zealand fur seal (*Arctocephalus forsteri*) (King, 1983), (ii) a south American sea lion (*Otaria flavescens*) and a South American fur seal (*A. australis*) (Miller *et al.*, 1996), (iii) a female California sea lion (*Zalophus californianus*) and a male Steller sea lion (*Eumetopias jubatus*) (Miller *et al.*, 1996), (iv) and finally between southern elephant seals (*Mirounga leonina*) and Australian fur seals (*A. pusillus*) (Miller *et al.*, 1996).

The aggressive mating by sea lions with heterospecific females has been interpreted as "excess of violent sexual selection" (Miller *et al.*, 1996). This aggressive behavior seems to be widespread in the family Otariidae (eared seals), and possibly the number of hybrids is much higher than reported to date.

II. Reported Occurrences of Hybridization in Captivity

Among captive cetaceans, 25 hybrids have been identified, all within the suborder Odontoceti (toothed whales). All hybridizations occurred between seven species of the Delphinoidea superfamily, where the bottlenose dolphin, *Tursiops truncatus*, was one of the parental species in all cases (see Table I, Fig. 1). The majority of these hybrids have not survived. However, a first-generation hybrid between a bottlenose dolphin (*T. truncatus*) and a false killer whale (*Pseudorca crassidens*) has given birth twice after mating with a bottlenose dolphin (Duffield, 1998). When these occurrences were reported in 1998, one of the two calves from this second generation was still alive. Finally, within the pinnipeds, hybridization in captivity has been observed within the two families Phocidae (earless seals) and Otariidae (eared seals) (see Table I).

III. Reported Occurrences of Hybridization in the Wild

Probably the most impressive occurrences of hybridization among marine mammals are those identified within the suborder Mysticeti (baleen whales). A total of 11 hybrids among baleen whale species have been reported so far; all of these were captured during commercial whaling operations. In all incidences the parental species involved were a blue (*Balaenoptera musculus*) and a fin whale (*B. physalus*). The first report of such anomalous baleen whales was in 1887 by A. H. Cocks (1887), who recorded 6 hybrids, or "Bastards," along the Lapland coast. However, this number is likely to be an underestimate as the author mentioned that sometimes hybrids were entered in their records as a fin whale instead of "Bastard." Later, Doroshenko (1970) reported a hybrid between a blue and fin whale, taken in 1965 off Kodiak Island (in the Gulf of Alaska), identified from its exceptional but intermediate morphological traits.

More recently, three anomalous baleen whales, one female and two males, caught during the Icelandic whaling operations between 1983 and 1989 were described morphologically as fin/blue whale hybrids. The parental species of these specimens were later confirmed by molecular analyses of the maternally inherited mitochondrial genome as well as Mendelian-transmitted nuclear genes (Árnason *et al.*, 1991; Spilliaert *et al.*, 1991). Interestingly, the female Icelandic fin/blue whale hybrid was in her second pregnancy. Molecular analyses of the fetus found that it was the result of a mating between the hybrid mother and a male blue whale. Finally, a fin/blue whale hybrid caught off northwest Spain in 1984 was identified morphologically, and subsequent molecular analyses found the maternal species to be a blue whale and the paternal species a fin whale (Bérubé and Aguilar, 1998).

Within the Odontoceti, the first three hybrids described were from a stranding on the West Coast of Ireland in

TABLE I
Reported Occurrences of Captive Hybridization

Families involved	Species	Parental role	Method of detection	Reported number of hybrids	Reference
Delphinidae	*T. truncatus* × *G. griseus*	Dam Sire	Morphological and molecular	13	Shimura *et al.* (1986), Sylvestre and Tasaka (1985)
	T. truncatus × *D. delphis*	Dam Sire	Morphological	2	Duffield (1998)
	T. truncatus × *D. capensis*	Dam Sire	Morphological	1	W. Perrin, personal communication
	T. truncatus × *P. crassidens*	Dam Sire	Morphological	6	Duffield (1998), Nishiwaki and Tobayama (1982)
	S. bredanensis × *T. truncatus*	Dam Sire	Morphological	1	Dohl *et al.* (1974)
	G. macrorhynchus × *T. truncatus*	Dam Sire	Morphological	2	Duffield (1998)
Phocidae	*P. hispida* × *H. grypus*	Dam Sire	Morphological	1	King (1983)
Otariidae	*C. ursinus* × *Z. californianus*	Dam Sire	Morphological	1	Duffield (1998)
	Z. californianus × *A. pusillus*	Dam Sire	Morphological	1+	King (1983)
	Z. californianus × *O. flavescens*	Dam Sire	Morphological	1	King (1983)

Blacksod Bay (Fraser, 1940). Morphological analysis concluded that the three stranded specimens were hybrids from matings between bottlenose and Risso's dolphins (*Grampus griseus*). The occurrence of as many as three hybrid individuals in the same stranding, each a cross of the same parental species, is highly unusual given the overall low rate of hybridization among cetaceans per se. For the same reason, Fraser himself first thought the hybrids to be a novel species rather than hybrids. Since the stranding in Blacksod Bay, only a single incidence of hybridization in the wild has been reported within the family Delphinidae. This specimen was caught by fishermen off the Peruvian coast and determined to be a hybrid between common (*Delphinus capensis* or *delphis*) and dusky dolphin (*Lagenorhynchus obscurus*) based on its morphology (Reyes, 1996).

In 1990, an anomalous whale skull was collected in Disko Bay at west Greenland. The morphological characteristics of this skull were intermediate between those of adult narwhal (*Monodon monoceros*) and beluga (*Delphinapterus leucas*), and the authors hypothesized that the specimen was likely a narwhal–beluga hybrid (Heide-Jøgensen and Reeves, 1993).

The most recent case of hybridization reported within Odontoceti is a female fetus recovered from a dead Dall's porpoise (*Phocoenoides dalli*). Morphological and molecular analyses determined the fetus as a cross between a Dall's and a harbor porpoise (*Phocoena phocoena*) (Baird *et al.*, 1998).

The most common occurrence of hybridization in pinnipeds is between the sub-antarctic (*Arctocephalus tropicalis*) and the Antarctic fur seal (*A. gazella*) (Table II). Based upon the population estimates for each of the two species and the frequency of hybrids, the magnitude of hybridization has been estimated to represent 9.3 and 0.1% of the *A. gazella* and *A. tropicalis* populations, respectively (Kerley, 1983).

IV. Evolutionary Implications of Hybridization

The evolutionary significance of hybridization is not known, but hybridization does provide an opportunity for gene flow between otherwise isolated gene pools, e.g., exchange of adaptive traits. Among marine mammals, hybridization has been shown to occur between a variety of species (Tables I and II). However, the overall rate of hybridization appears to be quite limited, and no cases of introgression have yet been identified. The apparent scarcity of hybrids may not be a true reflection of the actual rate, i.e., it is possible that hybrids simply are overlooked or not reported (during commercial or subsistence whaling) in order to avoid sanctions for killing protected species (e.g., blue whales). Furthermore, the identification of hybrids so far has relied primarily on morphological characters, which usually require that the specimen be killed. However, the introduction of nonlethal methods (Lambertsen, 1987) to obtain the necessary tissue for molecular methods as skin biopsies from free-ranging cetaceans makes it a simple task to identify hybrids today.

Marine mammals are genetically relatively similar. In comparison, the level of genetic divergence between the fin and the blue whale is similar to that observed among human (*Homo sapiens*), chimpanzee (*Pan troglodytes, P. paniscus*), and gorilla (*Gorilla gorilla*) (Árnason and Gullberg, 1993). Even within all cetaceans, where mysticetes and odontocetes probably diverged some 40 million years ago, nearly all species have the same num-

TABLE II
Reported Occurrences of Hybridization in the Wild

Families involved	Species involved	Parental role	Method of detection	Reported number of hybrids	Reference
Balaenopteridae	*B. physalus* × *B. musculus*	Sire and dam / Sire and dam	Morphological and molecular	11+	Árnason *et al.* (1991), Bérubé and Aguilar (1998), Cocks (1887), Doroshenko (1970), Spilliaert *et al.* (1991)
Delphinidae	*T. truncatus* × *G. griseus*	?	Morphological	3	Fraser (1940)
Delphinidae	*D. capensis* × *L. obscurus*	?	Morphological	1	Reyes (1996)
Monodontidae	*D. leucas* × *M. monoceros*	?	Morphological	1	Heide-Jørgensen and Reeves (1993)
Phocoenidae	*Phocoenoides dalli* × *Phocoena phocoena*	Dam / Sire	Morphological and molecular	1	Baird *et al.* (1998)
Phocidae	*C. cristata* × *Pagophilus groenlandicus*	Dam / Sire	Morphological and molecular	1	Kovacs *et al.* (1997)
Otariidae	*A. gazella* × *A. tropicalis*	?	Morphological	15	Kerley (1983)
Otariidae	*O. flavescens* × *A. australis*	?	Morphological	?	Miller *et al.*, (1996)

Figure 1 *Hybridization can occur in both natural and captive settings. Instances are more common between closely related taxa, but the ability to hybridize is not always indicative of close phylogenetic affinity. Pictured here is a hybrid between* Tursiops *and* Steno. *Photo by S. Leatherwood.*

ber of chromosomes ($2n=44$; a few have $2n=42$) and similar karyotypes. Among the pinnipeds, more variation in chromosome number has been detected; the number of diploid chromosomes can be from 32 to 36 (Árnason, 1990). The relatively similar genetic background and often sympatric existence (in feeding or breeding range) among closely related marine mammals would seem to favor hybridization. However, as mentioned earlier, hybridization is rare, and between species where several hybrids have been observed (such as the fin and the blue whale), the genetic integrity of the parental species appears intact.

It has been argued that hybrids of the heterogametic sex (males in mammals with a single X and Y chromosome) were most likely to be sterile or nonviable (Haldane, 1922). Since then, evolutionary geneticists have been looking to test this "Haldane's" rule. Among cetaceans, specifically the family Mysticeti, the only two male blue/fin whale hybrids examined to date were both sexually immature despite their relatively high age (Árnason *et al.*, 1991). Although consistent with Haldane's rule, which has been supported in a number of terrestrial mammals, the small sample size makes it impossible to assess with certainty if the rule applies to marine mammals as well.

The incidence of anomalous marine mammals reported so far has shown that hybridization does occur in captivity as well as in natural settings. Some of these hybrids (mainly in captive animals) have been carried to term and survived. However, only a single case has produced viable calves (unfortunately, the time of birth was not mentioned), of which only one was reported still alive in 1998 (Duffield, 1998). In the case of captive animals, it is difficult to assess if the seemingly low viability of the offspring is related to their hybrid origin or to the general low rate of survival in observed captive-born cetaceans (Van Gelder, 1977). Nonetheless, the observed occurrences of viable and fertile hybrids in captivity suggest that such could

happen in the wild. However, to date no offspring of a hybrid has been observed alive in the wild. Whether this is due to our limited ability to detect such hybrids or if indeed (as the lack of introgression indicates) that such viable offspring from hybrids are rare is still an open question.

See Also the Following Articles

Baleen Whales ■ Captive Breeding ■ Genetics, Overview ■ Mating Systems ■ Reproductive Behavior ■ Speciation

References

Árnason, A., and Gullberg, A. (1993). Comparison between the complete mtDNA sequences of the blue and the fin whale, two species that can hybridize in nature. *J. Mol. Evol.* **37**, 312–322.

Árnason, Ú. (1990). Phylogeny of marine mammals: Evidence from chromosomes and DNA. *In* "Chromosomes Today" (K. Fredga, B. A. Kihlman, and M. D. Bennet, eds.), Vol. 10, pp. 267–278.

Árnason, Ú., Spillaert, R., Palsdottir, A., and Arnason, A. (1991). Molecular identification of hybrids between the two largest whale species, the blue whale (*Balaenoptera musculus*) and the fin whale (*B. physalus*). *Hereditas* **115**, 183–189.

Baird, R. W., Willis, P. M., Guenther, T. J., Wilson, P. J., and White, B. N. (1998). An intergeneric hybrid in the family Phocoenidae. *Can. J. Zoo,* **76**, 198–204.

Bérubé, M., and Aguilar, A. (1998). A new hybrid between a blue whale, *Balaenoptera musculus,* and a fin whale, *B. physalus:* Frequency and implications of hybridization. *Mar. Mamm. Sci.* **14**, 82–98.

Cocks, A. H. (1887). The fin whale fishery of 1886 on the Lapland coast. *Zoologist* **11**, 207–222.

Dohl, T. P., Norris, K. S., and Kang, I. (1974). A porpoise hybrid: Tursiops X Steno. *J. Mamma.* **55**, 217–221.

Doroshenko, N. V. (1970). A whale with features of fin whale and blue whale. *Tinro* **70**, 255–257.

using the exceptional resolving capabilities of the human eye. Additional rigor is often incorporated into the process through the use of multiple judges for difficult final identifications. Computer-assisted matching is becoming increasingly important as catalogs are now incorporating many thousands of individuals, and as contributions to centralized catalogs are being made by numerous researchers in widely dispersed locations.

Other kinds of "natural markings" that are being used increasingly are genetic markers from skin biopsy samples. Molecular analyses of small samples allow determination of gender and individual identification from genotypes provided by microsatellite loci. This technique was developed for large-scale use during an ocean basin-wide study of humpback whales in which photographs were used to identify 2998 individual whales and microsatellite loci were used to identify 2015 whales (Smith *et al.*, 1999). Based on the results of these initial studies, molecular techniques hold a great deal of promise for studies of a variety of cetaceans.

B. Temporary Markings

Natural temporary markings include skin lesions on parts of the body visible to researchers (Wilson *et al.*, 1999) and soft-bodied barnacles that attach to dorsal fins, for example. Such markings can be useful for distinguishing between otherwise unmarked animals within a group, but their changeable nature make them less reliable for accurate identifications over long periods. Skin lesions may take weeks to months to fully heal and disappear, but their characteristics change during the healing process. Soft-bodied barnacles favor dorsal fin tips for attachment, leading to low variability in positioning, thus minimizing their value for identification.

Anthropogenic temporary markings have been found to be of limited utility with cetaceans (as reviewed by Scott *et al.*, 1990). Remotely applied paint and tattoos have been tested with small cetaceans, and in all cases the animals were either not reidentified or the markings disappeared within 24 hr due to skin sloughing. In some cases, zinc oxide-based, brightly colored sun protection ointments have been applied to dolphins' dorsal fins prior to release. These have allowed for the short-term identification of animals otherwise lacking in distinctive marks, and transfer of colors between animals can indicate social interactions.

C. Scarring and Branding

Dorsal fin notching has been attempted in a few cases with killer whales, bottlenose dolphins, pantropical spotted dolphins (*Stenella attenuata*), and spinner dolphins (Scott *et al.*, 1990). Notching provides the same kinds of features used in the photographic identification of natural marks. Such notching requires capturing the animals, which also provide opportunities to learn the sex and age of the marked dolphin, as well as other biological information. One report indicated minor but persistent bleeding as a result of notching, but this has not been reported by others.

Freeze branding, using metal numerals 5–8 cm high applied to the animals' body or dorsal fin for 10–20 sec, has been used safely and successfully with a variety of small cetaceans, including bottlenose dolphins, spinner dolphins, short-beaked

common dolphins (*Delphinus delphis*), Pacific white-sided dolphins, short-finned pilot whales, false killer whales (*Pseudorca crassidens*), Amazon River dolphins, and rough-toothed dolphins (*Steno bredanensis*) (Irvine *et al.*, 1982; Scott *et al.*, 1990). Freeze-brand application typically results in little or no reaction by dolphins, but minor skin lesions may occur if brands are applied for too long. Readable white marks usually appear within a few days (Fig. 4). Freeze brands fade over time, but the marks can often still be identified for many years in good-quality photographs even if they are not readily visible in the field. Fading appears to be age related, with brands disappearing more rapidly and more completely on younger animals but remaining readable on adults for as long as 11 years or more (Irvine *et al.*, 1982; Scott *et al.*, 1990).

D. Attachment Tags

The use of attachment tags for identification purposes (rather than telemetry, covered elsewhere in this volume), including

Figure 4 *(Top) Dorsal fin of a 2-year-old female bottlenose dolphin showing a fresh freeze brand ("7") above a year-old freeze brand and a 1-year-old roto tag. (Bottom) Dorsal fin of the same bottlenose dolphin at 5 years of age showing 3- and 4-year-old freeze brands, a naturally acquired notch at the top of the leading edge of the fin, and a notch formed from loss of a roto tag on the trailing edge of the fin. Photographs by Sarasota Dolphin Research Program.*

Discovery tags, spaghetti tags, button tags, and roto tags, has been reviewed by Scott *et al.* (1990). Discovery tags are numbered metal cylinders shot into the BLUBBER from whaling ships or research vessels. The tags have been used primarily with baleen and sperm whales and are recovered when the whales are captured and rendered, providing information on two points within the animals' range. Tagging was initiated in 1932 and continued until the whaling moratorium in 1985. More than 20,000 Discovery tags have been used, but return rates have been low, typically below 15%. Smaller versions of these tags have been used with small whales without notable success, and use with cetaceans less than 4.6 m long has been discouraged because of risk of serious injury.

Streamer or spaghetti tags, originally developed for fish tagging, are colored vinyl-covered strands of wire cable of variable length with steel or metal dart tips that are applied with either a crossbow or a jab stick, with the intent of anchoring the tip between blubber and muscle. Thousands of these tags have been applied to dolphins, porpoises, and belugas, especially in association with the tuna seine net fishery in the eastern tropical Pacific Ocean. Because of poor retention and high risk of injury to the animal, use of spaghetti tags with small cetaceans has been discouraged for many years (Irvine *et al.,* 1982).

Dorsal fins or ridges are commonly used for tag attachment because of their structure, prominence, and regularity of appearance above the water's surface. Button tags, typically numbered and colored fiberglass or plastic disks or rectangular plates designed after the Peterson disk fish tags, have been applied to several species of small cetaceans, including bottlenose dolphins, pantropical spotted dolphins, spinner dolphins, common dolphins, Pacific white-sided dolphins, belugas, and harbor porpoises (Evans *et al.,* 1972; Scott *et al.,* 1990). Usually, button tags are attached through the dorsal fin by means of one or more plastic (especially delrin) or stainless-steel bolts or pins that connect the tag halves on each side of the fin. Although some button tags have lasted for several years on pelagic dolphins, inshore animals often lose the tags within weeks or months, often by breaking them through rubbing on the shallow sea floor. Use of button tags has been largely discontinued due to poor tag retention and the potential for injury to the animals (Irvine *et al.,* 1982).

Small plastic cattle ear tags, or rototags, clipped through the trailing edges of dorsal fins have proved successful for identifying small cetaceans in the field, including bottlenose dolphins, pantropical spotted dolphins, spinner dolphins, common dolphins, rough-toothed dolphins, Pacific white-sided dolphins, short-finned pilot whales, and harbor porpoises (Fig. 4a; Norris and Pryor, 1970; Scott *et al.,* 1990). Typically, a small hole is made in the thin tissue of the trailing edge using a sterile technique, and the tag is clipped through the fin with special pliers. Although the written markings are too small to be read at a distance, the number of tags, color, and position on the fin provide a useful degree of variation. Rototags have remained in position of periods of years, although often they are lost within months. Rototag halves may separate, leaving a healed hole in the fin, or they migrate through the trailing edge of the fin, leaving a small healed notch; both pose minimal risks to the animals but offer continuing identification features. Barnacle and

algae fouling and pressure necrosis are infrequent problems. As a modification of this technique, small VHF radio transmitters have been attached to rototags for short-term tracking (up to 30 days), with a modification involving the use of a corrosible nut system to release that tag at that time.

Other attachment techniques, such as the use of tethers or plastic-coated wires or polypropylene or soft rubber tubing, have proved to be ineffective and injurious to the animals when attached to the caudal peduncle. Tag loss rates have been high, and abrasions were frequently noted.

II. Pinnipeds
A. Natural Markings

Natural body markings have been used in only a few studies of pinnipeds such as gray seals (*Halichoerus grypus*), northern elephant seals (*Miromga angustirostris*), Steller sea lions (*Eumetopias jubatus*), Hawaiian monk seals (*Monachus schauinslandi*), harbor seals (*Phoca vitulina*), and California sea lions (*Zalophus californianus*). Yochem *et al.* (1990) examined pelage patterns of harbor and larga (*Phoca largha*) seals to distinguish between populations and individuals. Using black and white photographs they scored the presence or absence of spots, clarity of spots, relative density of spots, complexity of spots, presence of rings, and spacing of rings in selected body areas (especially sides of the head, neck, and chest). Hiby and Lovell (1990) described a computer-aided matching system for screening a library of digitized natural mark photographs of gray seals. Their system created a three-dimensional model to locate features on the seal's body, especially using the side of the neck. For most pinniped species, studies using natural markings are hampered by a lack of distinctive markings and the large numbers of individuals or pack ice distributions of many species (Erickson *et al.,* 1993). Most pinniped researchers have resorted to the use of artificial markings and tags for individual identification.

B. Temporary Markings

Techniques for temporary markings of pinnipeds include paints, dyes, bleaches, and pelage clippings (Erickson *et al.,* 1993). These techniques offer the advantages of often being able to be applied without having to restrain the animals and permitting remote identification without disturbance. However, these marks are typically lost upon moulting, precluding the continuity of identification beyond a single season. A variety of paints (marine, highway, rubber-based, quick-drying cellulose, aerosol sprays, and house paint) have been used to mark seals and sea lions. Paints have been applied from brushes or rollers on poles and from plastic bags thrown at the animals. Quick-drying paint has proved relatively effective, with a useful lifespan of about 1 month on average. Northern fur seals (*Callorhinus ursinus*) have been marked successfully for 2–12 months with a fluorescent plastic resin, naptha-based paint. This technique apparently results in the matting of guard hairs, which then break off, leaving an outline of the mark. High-gloss marine enamel applied from aerosol cans to mark Hooker's sea lions (*Phocarctas hookeri*) has resulted in markings lasting 3 months, even after the animals have been at sea. Carbon dioxide-powered paint guns firing small capsules have proved less effective for marking elephant seals due to reliability problems and the small size of the marks.

Dyes have been used with several species of pinnipeds, especially light-colored species (Erickson *et al.*, 1993). Successful dying usually occurs when permanent dyes are used and when the animals are dry and remain out of the water for a period of time following application. Colored dyes and black Nyanzol D have lasted 3–4 months on gray seals, harbor seals, and California sea lions. The addition of alcohol to Nyanzol D leaves a more distinct marker because it dissolves fur oils and also prevents the dye solution from freezing. Yellow picric acid in a saturated alcohol solution has been used with gray seals, with results that last through pup molting, appearing on the adults as well. This solution can be applied from a backpack tree sprayer to wet or dry seals. Fluorescent dye mixed with small quantities of epoxy resin has also been used with success. In some cases, such as southern elephant seals (*Mirounga leonina*), dyes have been less successful.

Bleach offers a very effective and sometimes longer-lasting alternative to paints and dyes (Erickson *et al.*, 1993). Many of the bleach solutions can be applied to sleeping animals via a squeeze bottle, thus minimizing risk, effort, and disturbance. Commercially available products such as Lady Clairol Ultra Blue dye in combination with various chemicals have been used most often, resulting in a white or cream-colored mark that is most visible on dark pelages. Combinations resulting in thicker consistency allow for distinct lines. Bleach marks on elephant seals last until moult, sometimes for 6 months, and have lasted for two seasons on fur seals. Combinations of bleaches and dyes have also been used in some cases, such as northern elephant seals (Fig. 5).

Hair clipping is somewhat more difficult than the previous techniques, but effective when the underfur is a different color from the guard hairs (Erickson *et al.*, 1993). This technique involves clipping or singeing the pelage to create a distinctive mark. It has been used with success with northern fur seals, Steller sea lions, and Antarctic fur seals (*Arctocephalus gazella*).

Figure 5 *Bleach markings on a northern elephant seal. "Bilbo" is marked in black dye for identification through the summer molt and in bleach for the winter breeding season. Photo by C. J. Deutsch.*

C. Scarring and Branding

Punch marks and amputations have been used extensively with fur seals, with poor success and concerns about injury to the animals (Erickson *et al.*, 1993). Initial efforts to mark northern fur seals and Antarctic fur seals by punching holes in flippers in unique combinations of numbers and positions found this technique to be unreliable due to healing and occlusion. Hair on the flippers of phocids seals precludes utility with these species. Flipper notching was also found to be unreliable due to tissue regrowth. Although ear notching was used successfully for cohort marking in northern fur seals, it is no longer used because of concerns regarding interference with diving abilities.

Both hot branding and freeze branding have been used with great effect with pinnipeds (Erickson *et al.*, 1993). Hot brands have been used since 1912 with thousands of northern fur seals, Cape fur seals (*A. pusillus*), southern elephant seals, Weddell seals (*Leptonychotes weddellii*), grey seals, and leopard seals (*Hydrurga leptonyx*). Some marks have remained readable for up to 20 years. The technique seems best suited to colonial seals due to the bulky nature of the branding tools and heat source. Typically, brands are heated to red hot and are applied with firm, even pressure for 2–7 sec, depending on whether the hair has been clipped. Brands are applied to the upper saddle, middle back, or upper shoulder to optimize sightability.

Freeze branding differs from hot branding in that it involves the selective killing of pigment-producing cells through contact with a super-cooled metal numeral or symbol (typically 5 cm high) (Erickson *et al.*, 1993). Brands are cooled with liquid nitrogen or a dry ice and alcohol mixture and applied for about 20 sec to an area where hair has been removed. Correct freeze brand application results in a nonpigmented pelage mark, ranging from dark (elephant seals, California sea lions) to pink (California sea lions; walrus, *Odobenus rosmarus*). Freeze branding has had mixed success. Many freeze brands on pinnipeds have been found to repigment within 1–2 years, perhaps as a result of excessive branding. Readable brands have been obtained for elephant seals (up to a year, discernible for 3 years), California sea lions (readable for 1.5 years, discernible for up to 4 years), walrus (readable for many years), and Australian sea lions, *Neophoca cinerea* (legible on flippers for 7 years, on flanks for 4 years).

D. Attachment Tags

Plastic or metal attachment tags are used more widely than any other kind of individual identification system with pinnipeds (Erickson *et al.*, 1993). Monel or stainless-steel tags such as those used to mark livestock are the most common metal tags. These metal strap tags are self-piercing and are attached by means of special pliers to the trailing edge of the fore flippers of otariids and to the interdigital web of the hind flippers of phocids. Typically, the tags are stamped with an organization address and serial number. Thousands of metal tags have been attached to phocids. Retention rates on phocid seals are low, with postattachment tears and cuts sometimes becoming infected. Hundreds of thousands of metal tags have been attached to otariids, with similar poor results.

The use of plastic tags is now much more common than metal tags for identifying pinnipeds (Fig. 6). Two kinds of plastic bags are used commonly: rototags and Allflex tags. Both consist of

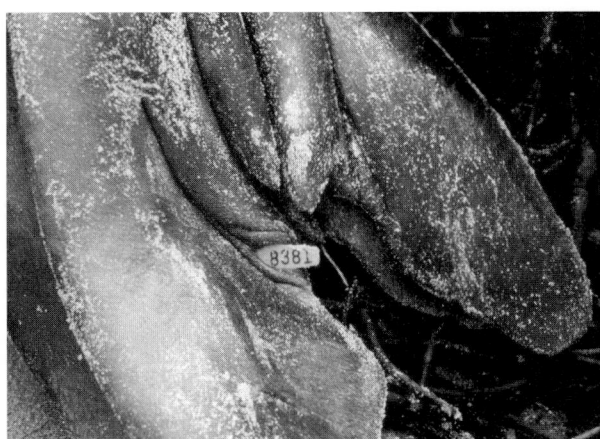

Figure 6 *Flipper tag on a northern elephant seal. Photo by B. J. LeBoeuf.*

self-piercing male and female elements that are applied with special pliers, as with metal tags. Plastic tags are available in a variety of colors, leading to more than 300 unique color combination possibilities. The visibility of both metal and plastic attachment tags can be enhanced through the use of streamer markers, such as nylon cloth strips reinforced with vinyl, which may last for a year or more.

Tagging success with both metal and plastic tags is less than desired. Loss rates of the two kinds of plastic tags are variable, but tend to be lower than for metal tags, about 10% annually. However, the long-term durability of metal tags is better than plastic. Wounds from metal tags are more common than for plastic.

III. Sirenians

A. Natural Markings

The process of developing new techniques and applying existing technology to studies of sirenians has been reviewed by G. Rathbun in Wells *et al.* (1999). Natural marks, including deformities and scars, have been used to identify individual manatees since the 1950s. Among the marks that have proven most useful for individual identification are the scars from collisions with boats, especially propeller scars. Most manatees (*Trichechus manatus*) in Florida waters bear scars from boat collisions, often from more than one event. Boat scars occur over all parts of the manatee's body, but especially the dorsal surface and paddle, where notches may be cut by propellers (Fig. 7). Individual identification progressed from sketches of marks to surface and underwater 35-mm photography. Photography allowed for the tracking of changes in identifying characteristics through time and for distinguishing between manatees with similar markings. Technological advances have resulted in photographic images of scar patterns being saved, cataloged, and searched with the assistance of computers.

B. Temporary Markings

No widely accepted techniques currently exist for temporarily marking sirenians. Paint, flipper bands, and harnesses

have been tested, but have been found to be ineffective (Irvine and Scott, 1984). "Paintstiks," oil-based crayon-like markers, have remained visible for 3–7 days during field tests, although rubbing eventually smears or removes them. Aerosol paint was short-lived and application startled the animals and polluted the water.

C. Scarring and Branding

Although not intentional, the most widely used features for identifying individual manatees are propeller scars. In recent years, scientists have begun cutting small notches in the paddles of manatees. The positions of the notches around the paddle are coded to provide information on cohorts. Freeze branding is also used with manatees that have been captured or rehabilitated on occasion, with some success (Irvine and Scott, 1984). Although most brands fade with time, some have remained readable at distances of 15 m for as long as 4 years. Success may vary with whether the manatees are shedding, as well as season, water temperature, and salinity.

D. Attachment Tags

Lacking dorsal fins, sirenians provide few opportunities for tag attachment. As described for cetaceans, spaghetti tags have been tested with manatees (Irvine and Scott, 1984). These 20-cm-long plastic streamer tags attached to a metal dart have been applied with either a lance or a crossbow, attempting to anchor the tag about 2 cm below the skin. Spaghetti tags demonstrated poor retention and caused abscesses on some manatees.

The most effective technique for tag attachment involves a break-away "belt" looped around the animal's peduncle. This belt is designed to minimize chafing, break away if it should become snagged on an obstacle in the environment, and carry a floating VHF or satellite-linked radio transmitter at the end of a tether. Each transmitter float is color coded to allow for individual identification visually. The tethers can be replaced by swimmers as necessary.

Passive integrated transponders, or PIT tags, have been implanted in nearly every Florida manatee that has been handled in recent years (Wright *et al.*, 1998). These glass-encapsulated microchips are about the size of a rice grain. They are implanted

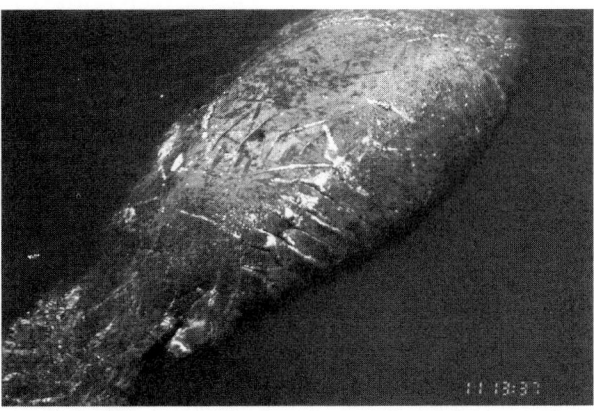

Figure 7 *Identifying scars from boat collisions on a Florida manatee. Photo by J. K. Koelsch.*

subcutaneously at a depth of about 3.5 cm, dorsal and caudal to the ear, and medial to the scapula. A small incision is made, and the tag is inserted via a 12-gauge needle. Each is programmed with a unique identification code that is activated by a hand-held scanner when it passes nearby. PIT tags are relatively easy to implant, last a long time, are reusable, rarely infect the animals, have an unlimited number of potential codes, and allow for easy data recording and transfer, but they suffer from the fact that they must be scanned from no more than 15 cm away and that the receivers are not waterproof.

See Also the Following Articles

Coloration ■ Hair and Fur ■ Population Dynamics ■ Telemetry

References

Altmann, J. (1974). Observational study of behavior: Sampling methods. *Behaviour* **49**, 227–267.

Bigg, M. (1982). An assessment of killer whale (*Orcinus orca*) stocks off Vancouver Island, British Columbia. *Rep. Int. Whal. Comm.* **32**, 655–666.

Erickson, A. W., Bester, M. N., and Laws, R. M. (1993). Marking techniques. *In* "Antarctic Seals" (R. M. Laws, ed.), pp. 89–118. Cambridge Univ. Press, Cambridge.

Evans, W. E., Hall, J. D., Irvine, A. B., and Leatherwood, J. S. (1972). Methods for tagging small cetaceans. *Fish. Bull.* **70**(1), 61–65.

Hammond, P. S., Mizroch, S. A., and Donovan, G. P. (1990). Report of the workshop on individual recognition and the estimation of cetacean population parameters. *In* "Individual Recognition of Cetaceans: Use of Photo-Identification and Other Techniques to Estimate Population Parameters" (P. S. Hammond, S. A. Mizroch, and G. P. Donovan, eds.), pp. 3–17. Report of the International Whaling Commission, Special Issue 12, Cambridge.

Hiby, L., and Lovel, P. (1990). Computer aided matching of natural markings: A prototype system for grey seals. *In* "Individual Recognition of Cetaceans: Use of Photo-Identification and Other Techniques to Estimate Population Parameters" (P. S. Hammond, S. A. Mizroch, and G. P. Donovan, eds.), pp. 57–61. Report of the International Whaling Commission, Special Issue 12, Cambridge.

Irvine, A. B., and Scott, M. D. (1984). Development and use of marking techniques to study manatees in Florida. *Florida Sci.* **47**(1), 12–26.

Irvine, A. B., Wells, R. S., and Scott, M. D. (1982). An evaluation of techniques for tagging small odontocete cetaceans. *Fish. Bull.* **80**, 135–143.

Katona, S., Baxter, B., Brazer, O., Kraus, S., Perkins, J., and Whitehead, H. (1979). Identification of humpback whales from fluke photographs. *In* "Behavior of Marine Mammals: Current Perspectives in Research" (H. E. Winn and B. Olla, eds.), pp. 33–44. Plenum Press, New York.

Norris, K. S., and Pryor, K. W. (1970). A tagging method for small cetaceans. *J. Mammal.* **51**(3), 609–610.

Payne, R., Brazier, O., Dorsey, E. M., Perkins, J. S., Rowntree, V. J., and Titus, A. (1983). External features in southern right whales (*Eubalaena australis*) and their use in identifying individuals. *In* "Communication and Behavior of Whales" (R. Payne, ed.), pp. 371–445. Westview Press, Boulder, CO.

Scott, M. D., Wells, R. S., Irvine, A. B., and Mate, B. R. (1990). Tagging and marking studies on small cetaceans. *In* "The Bottlenose Dolphin" (S. Leatherwood and R. R. Reeves, eds.), pp. 489–514. Academic Press, San Diego.

Smith, T. D., Allen, J., Clapham, P. J., Hammond, P. S., Katona, S., Larsen, F., Lien, J., Matilla, D., Palsbøll, P. J., Sigurjónsson, J., Stevick, P. T., and Øien, N. (1999). An ocean-basin-wide mark-recapture study of the North Atlantic humpback whale (*Megaptera novaeangliae*). *Mar. Mamm. Sci.* **15**(1), 10–32.

Wells, R. S., Boness, D. J., and Rathbun, G. B. (1999). Behavior. *In* "Biology of Marine Mammals" (J. E. Reynolds, III and S. A. Rommel, eds.), pp. 324–422. Smithsonian Institution Press, Washington, DC.

Wells, R. S., and Scott, M. D. (1990). Estimating bottlenose dolphin population parameters from individual identification and capture-release techniques. *In* "Individual Recognition of Cetaceans: Use of Photo-Identification and Other Techniques to Estimate Population Parameters" (P. S. Hammond, S. A. Mizroch, and G. P. Donovan, eds.), pp. 407–415. Report of the International Whaling Commission, Special Issue 12, Cambridge.

Wilson, B., Arnold, H., Bearzi, G., Fortuna, C. M., Gaspar, R., Ingram, S., Liret, C., Pribanic, S., Read, A. J., Ridoux, V., Schneider, K., Urian, K. W., Wells, R. S., Woods, C., Thompson, P. M., and Hammond, P. S. (1999). Epidermal diseases in bottlenose dolphins: Impacts of natural and anthropogenic factors. *Proc. R. Soc. Lond. B* **266**(1423), 1077–1083.

Wright, I. E., Wright, S. D., and Sweat, J. M. (1998). Use of passive integrated transponder (PIT) tags to identify manatees (*Trichechus manatus latirostris*). *Mar. Mamm. Sci.* **14**(3), 641–645.

Würsig, B., and Jefferson, T. A. (1990). Methods of photo-identification for small cetaceans. *In* "Individual Recognition of Cetaceans: Use of Photo-Identification and Other Techniques to Estimate Population Parameters" (P. S. Hammond, S. A. Mizroch, and G. P. Donovan, eds.), pp. 43–55. Report of the International Whaling Commission, Special Issue 12, Cambridge.

Würsig, B., and Würsig, M. (1977). The photographic determination of group size, composition, and stability of coastal porpoises (*Tursiops truncatus*). *Science* **198**, 755–756.

Yochem, P. K., Stewart, B. S., Mina, M., Zorin, A., Sadavov, V., and Tablakov, A. (1990). Nonmetrical analyses of pelage patterns in demographic studies of harbor seals. *In* "Individual Recognition of Cetaceans: Use of Photo-Identification and Other Techniques to Estimate Population Parameters" (P. S. Hammond, S. A. Mizroch, and G. P. Donovan, eds.), pp. 87–90. Report of the International Whaling Commission, Special Issue 12, Cambridge.

Illegal and Pirate Whaling

ROBERT L. BROWNELL, JR.
Southwest Fisheries Science Center,
La Jolla, California

A. V. YABLOKOV
Center for Russian Environmental Policy, Moscow

Illegal whaling occurs in contravention of national laws or internationally agreed quotas, seasons, area restrictions, and other limitations, whereas "pirate whaling" refers to unregulated whaling conducted outside the aegis of the International Whaling Commission, usually under a flag of convenience. Such activities can lead directly to depletion of whale stocks through overexploitation. Furthermore, the lack of catch data, or the reporting of falsified data, can lead to serious error in assessment of the size and status of stocks and erroneous management advice ultimately contributing to their collapse.

I. Illegal Whaling

Known instances of illegal whaling were conducted by several nations. Because the offenses of the USSR were the most egregious, most of this discussion will focus on what is now known of Soviet activities. The Soviet Union conducted massive illegal whaling and falsification of data over a period of decades, with catastrophic consequences for whale CONSERVATION and development of the science of whale MANAGEMENT. The USSR commenced pelagic whaling in the North Pacific with the *Aleut* in June 1933 (Tønnessen and Johnsen, 1982). This floating factory operation continued through the 1967 season. After this date, four additional Soviet factory ships conducted pelagic whaling operations in the North Pacific. For more than a decade following the end of World War II, the USSR operated a single whaling factory ship (*Slava*) in the Antarctic. Beginning in 1959, the Soviets began expanding their whaling operations, adding one new factory ship in each of the next three Antarctic seasons (*Sovietskaya Ukraina* in 1959/1960, *Yuri Dolgorukiy* in 1960/1961, and *Sovietskaya Rossiya* in 1961/1962). This expansion occurred at a time when there was extensive discussion at the IWC about declining stocks and other countries were decreasing their whaling operations. The USSR voted against the drastic reductions in catch quotas required to meet scientific recommendations and against implementation of an International Observer Scheme (IOS), both of which were eventually put in place. After the breakup of the Soviet Union, a number of Russian and Ukrainian biologists who had served as scientists aboard Soviet factory ships and knew of the existence of accurate but unreported catch statistics decided to collect them and make them available to the world scientific community. This section summarizes information on Soviet illegal activities during the two major phases of its whaling in the Southern Hemisphere and North Pacific and briefly notes recent information on illegal whaling activities by other nations.

A. Southern Hemisphere

A summary of the disparity between USSR catch data reported to the IWC and actual takes is given in Table I (modified from Yablokov *et al.*, 1998). These data concern catches made by Soviet fleets working in, or en route to or from Antarctic waters. The period concerned is from the beginning of postwar Soviet whaling in 1947 until the introduction of international observers in the 1972/1973 whaling season, when most illegal activities ceased. During this period, unreported catches totaled 102,335 whales, almost half (44%) of which were humpbacks (*Megaptera novaeangliae*). However, 11,397 animals (primarily fin whales, *Balaenoptera physalus,* which were then an unprotected species) were actually overreported; this was done to conceal the massive illegal catches of pygmy blue (*B. musculus brevicauda*), sei (*B. borealis*), humpback, and southern right whales (*Eubalaena australis*).

Massive falsification of geographic catch data began in 1959 and was practiced by all four Soviet Antarctic whaling fleets. The primary areas for illegal catches were numerous sections of the South Atlantic, the Indian Ocean (including as far north

TABLE I
Comparison of Southern Hemisphere USSR Catch Data (1947–1972)[a]

Species	Reported	Actual	Disparity
Blue whale	3,887	3,681	206
Pygmy blue whale	10	9,215	−9,205
Sei whale	29,751	53,366	−23,615
Fin whale	52,860	42,889	9,971
Bryde's whale	10	1,457	−1,447
Minke whale	1,246	384	862
Humpback whale	2,820	45,831	−43,011
Southern right whale	4	3,368	−3,364
Sperm whale	50,715	72,372	−21,657
Killer whale	482	124	358
Southern bottlenose whale	10	17	−7
"Others"	0	29	−29
Total	141,795	232,733	−90,938

[a]Reported to the IWC with numbers actually taken. Modified from Yablokov *et al.* (1998) and Tormosov *et al.* (1998).

as the Arabian Sea), and the southwestern Pacific (Yablokov, 1994; Zemsky *et al.*, 1995a,b). IWC regulations prohibited the taking of baleen whales (mysticetes) north of latitude 40°S, although the killing of sperm whales (*Physeter macrocephalus*), killer whales (*Orcinus orca*), and bottlenose whales (*Hyperoodon planifrons*) in this region was permissible. Therefore, the Soviets used the pretext of hunting toothed whales to exploit mysticete populations in the prohibited area. Soviet whaling fleets bound for the Antarctic began search and catcher operations immediately after leaving the Suez Canal or after passing either Gilbraltar or Portuguese waters.

Virtually all biological data reported to the IWC were "corrected" to disguise the extensive illegal catches. Because of a prohibition on killing mothers and calves, all such cases were either unreported or were concealed with false data. Thus, a fin whale mother and calf might be reported as "two sei whales," whereas a catch of four female sperm whales would be reported as "two males." It has been estimated that at least 80% of all officially reported Soviet data on length, weight, sex ratio, reproduction, and maturational state are false.

The scale of the Soviet catches partly explains the apparent failure to recover that has been evident in some mysticete populations despite their supposedly protected status. The Soviets killed 12,896 blue whales in the Southern Hemisphere, of which more than 9200 were unreported pygmy blues killed after the IWC accorded protected status to both subspecies in the 1965/1966 season. Catches were made over a wide area, including the Indian Ocean north of the equator. Thus, these populations were reduced to much lower levels than the 800–1600 (blue) and 10,000 (pygmy blue) that were estimated at the time. Although recent estimates of 500 and 5000 for current Antarctic populations of blue and pygmy blue whales, respectively, are not statistically robust, it is apparent that the population sizes are smaller than would be expected following three decades of protection.

Humpback whales were even more seriously impacted by illegal catches. In addition to the huge number (45,831) of unreported catches made by the Soviets, it is known that additional illegal takes were made in Antarctic waters by the *Olympic Challenger*, a pirate factory ship owned by Aristotle Onassis (Tønnessen and Johnsen, 1982), discussed later. The Australian biologist Graeme Chittleborough (1965) asserted that large discrepancies in calculated mortality coefficients for humpbacks could be explained only by the occurrence of extensive illegal catches, a view that is now validated by the Soviet catch data reported here. Many of the Soviet catches were made from the management division known as Antarctic Area V (Dawbin, 1966), which explains the collapse of shore whaling station operations in eastern Australia and New Zealand in the 1960s. There is now evidence of strong population growth in some Southern Hemisphere humpback populations (Bannister, 1994; Paterson *et al.*, 1994; Findlay and Best, 1996), but the status of other stocks remains unclear.

Additional illegal Soviet pelagic whaling occurred in the Arabian Sea on humpback (Mikhalev, 1997a), blue, Bryde's (*Balaenoptera edeni*), and sperm whales during the 1960s (Mikhalev, 1996, 1997b; personal communication). This whaling occurred while the whaling fleets were en route to the Antarctic whaling grounds. Biological data from humpback catches made in November 1966 off Oman and India have resolved a longstanding issue regarding the identity of this population, which is unique among humpbacks in that it resides in tropical waters year-round (Mikhalev, 1997a). However, the status of this tropical population of humpbacks remains uncertain.

Southern right whales have always been protected under IWC regulations; this status dates from a League of Nations agreement in 1935. Recovery (increasing populations) is today apparent in only 4 of the 13 putative populations of this species (Best, 1993), although five of the remaining nine stocks are considered impossible to monitor on a regular basis. True data show that the Soviets made large (3368) unreported catches of this species between 1950 and 1971 (Tormosov *et al.*, 1998). Many of these takes were made around remote islands or in midoceanic areas such as Campbell Island, Crozet, Kerguelen, Tristan da Cunha, and the central Indian Ocean.

B. North Pacific

A single Soviet whaling factory operated in the North Pacific between 1933 and 1967, a small vessel named the *Aleut*. Four additional Soviet factory ships later operated in the North Pacific. A new ship, the *Sovetskaya Rossiya* (build as the sister ship to the *Sovetskaya Ukraina*), operated for four seasons (1962–1965) and then again for three more seasons in 1973, 1978, and 1979. The *Slava*, after working many years in the Southern Hemisphere, worked in the North Pacific for four seasons (1966–1969). Two sister whaling factory ships were built specifically for the North Pacific (*Vladivostok* and the *Dalniy Vostok*); both started operations in 1963. The *Vladivostok* operated through 1978 and the *Dalniy Vostok* through 1979. The main areas of operations for these two fleets were the Bering Sea, Gulf of Alaska, and other more southern parts of the North Pacific.

The whales in the North Pacific did not fare any better than those in the Southern Hemisphere. However, the available records are not as good. Doroshenko (2000a) and others reported numerous illegal catches of North Pacific right whales (*Eubalaena japonica*) in both the western and eastern North Pacific. Brownell and colleagues (2000) provided some data on massive illegal catches of sperm whales in the North Pacific. Soviet pelagic whaling operations in the North Pacific also illegally took blue, humpback, and gray whales (*Eschrichtius robustus*) (Doroshenko, 2000a). A summary of the inconsistencies between data reported by the USSR to the IWC and actual catches is in Table II.

It is known that illegal whaling on a similarly large scale was also conducted by the USSR throughout the Northern Hemisphere, although relevant catch data have yet to be analyzed. In light of these revelations in both the Southern Hemisphere and the Northern Hemisphere, it is clear that current views regarding the status and recovery potential of virtually all affected whale populations worldwide need revision. This is a long-term project that will continue for the next decade or longer.

At the 50th IWC meeting in Oman in 1998, the IWC took action on the Scientific Committee's (SC) concern about the falsified Soviet whaling data, mainly for sperm whales, by adopting the SC recommendation to remove official USSR Southern Hemisphere whale catches from the IWC database.

The USSR was not alone in the illegal harvest of whales. Recent evidence on the falsification of catch statistics has been reported in various North Pacific coastal-based operations conducted by the Japanese (Kasuya, 1999; Kasuya and Brownell, 1999, 2001; Kondo, 2001). Suspicions about illegal reporting in Japanese operations are not new and have been presented in the past (Kasuya and Miyazaki, 1997). The scale of these activities, however, was much smaller and the consequences less severe than in the case of the Soviet illegal whaling. Sperm whales catches between 1954 and 1964 were 1.4 to 3 times greater than the numbers Japan reported to the IWC. The total true catches of Bryde's whales taken during the final years of commercial land-based whaling (1981–1987) off the Bonin Islands were 1.6 times

TABLE II
Comparison of North Pacific Commercial USSR Catch Data (1961–1972)[a]

Species	Reported	Actual	Disparity
Blue whale	517	1,205	−688
Fin whale	10,613	8,621	+1,992
Sei whale	9,048	4,177	+4,871
Bryde's whale	775	714	+61
Humpback whale	3,043	6,793	−3,750
Gray whale	0	138	−138
Right whale	0	508	−508
Bowhead whale	0	133	−133
Total	23,996	22,289	+1,707

[a]Reported to the IWC, with numbers actually taken.

the numbers Japan reported to the IWC. Fin whales were reported taken illegally by the Republic of Korea in the 1980s.

During the 1990s and the following decade, numerous reports appeared regarding the sale of "illegal whale products" from protected whales collected in the Japanese market (Baker *et al.*, 2000). It is argued that Japan's scientific whaling program (since 1989) has acted as a cover for undocumented or illegal products from various protected species (fin, sei, humpback, and gray). While this is possible, there are no available data to support the occurrence of any large-scale illegal whaling during the 1990s. The most parsimonious explanation for the whale products from protected species is that they are from (1) whales killed before the 1986 IWC moratorium on commercial whaling, (2) past scientific hunts by Iceland or Norway, (3) by-catches from Japanese fisheries, and (4) STRANDINGS in Japan.

II. Pirate Whaling

As noted earlier, unregulated whaling conducted under the flags of non-IWC member nations has contributed to the depletion of some whale stocks. The most famous of these operations was that conducted by interests in Norway and Japan from 1968 to 1979 in the North and South Atlantic under the flags of Somalia, Cyprus, Curaçao, and Panama. Meat from the whales was shipped to Japan for human consumption. The *Run* operated mainly in the South Atlantic from January 1968 to February 1972. It was renamed the *Sierra* in 1972 and expanded major operations to the North Atlantic in 1975, where it continued taking whales until it was rammed and sunk by the *Sea Shepherd* (a privately operated vessel dedicated to interference with commercial whaling) in 1979 off Portugal (Watson, 1979). The *Tonna* joined the *Sierra* in December 1977 and operated until July 1978, when it foundered during processing of a large whale on deck. The *Cape Fisher*, later renamed the *Astra*, operated briefly as a processing vessel for the *Sierra* in 1979.

The Sierra Fishing Agency submitted its catch statistics to the Bureau of International Whaling Statistics in Norway until 1976, when the practice was discontinued because of a perceived lack of credibility of data. Data for the remaining years of operation were destroyed, but some information was salvaged through interviews with former crew members (Best, 1992). The catches included blue, fin, sei, Bryde's, humpback, and minke (*Balaenoptera acutorostrata* and/or *B. bonaerensis*). Large catches of fin whales totaling hundreds were made off the coasts of Spain and Portugal (the IWC's "Spain–Portugal–British Isle Management Area") after 1976.

Another notorious episode of pirate whaling occurred in the Southern Hemisphere from 1951 to 1956 by the factory ship *Olympic Challenger* and its fleet of 12 catcher boats (Tønnessen and Johnson, 1982). The Olympic Whaling Company, an affiliate of the Pacific Tankers Co. of New York, was financed by the Greek-born Argentine citizen Aristotle Onassis. The ownership of the vessel, a converted tanker, was later transferred to the Olympic Whaling Company S.A. in Montevideo, Uruguay. The captain was German and the expedition manager Norwegian. The factory ship and some of the catcher vessels flew the Panamanian flag and the remainder of the catchers the Honduran flag. Neither Panama nor Honduras were members of the IWC at the time, so the whaling

operations were completely unregulated. The expedition took thousands of whales in the Antarctic South Pacific sector and off Chile, Peru, and Ecuador, including blue, humpback, sei, right, and sperm. Catch data were reported to the International Bureau of Whaling Statistics, but these have been shown to incorporate extensive falsification of numbers, species, and sizes of whales caught (Barthelmess *et al.*, 1997). As noted earlier, these unregulated catches in combination with later illegal catches by Soviet fleets contributed in a major way to the catastrophic decline of whales in the Southern Ocean, particularly the humpback.

See Also the Following Articles

Humpback Whale ■ Hunting of Marine Mammals ■ International Whaling Commission ■ Japanese Whaling ■ Stock Assessment

References

Baker, C. S., Lento, G. M., Cipriano, F., Dalebout, M. L., and Palumbi, S. R. (2000). Scientific whaling: Source of illegal products for market? *Science* **290**, 1695.

Bannister, J. (1994). Continued increase in humpback whales off western Australia. *Rep. Int. Whal. Commn.* **44**, 309–310.

Barthelmess, K., Kock, K.-H., and Reupke, E. (1997). Validation of catch data of the *Olympic Challenger's* whaling operations from 1950/51 to 1955/56. *Rep. Int. Whal. Commn.* **47**, 937–940.

Best, P. B. (1992). Catches of fin whales in the North Atlantic by the M. V. *Sierra* (and associated vessels). *Rep. Int. Whal. Commn.* **42**, 697–700.

Best, P. B. (1993). Increase rates in severely depleted stocks of baleen whales. *ICES J. Mar. Sci.* **6**, 93–108.

Brownell, R. L., Jr., Yablokov, A. V., and Zemsky, V. A. (2000). USSR pelagic catches of North Pacific sperm whales. *In* "Soviet Whaling Data" (1949–1979) (A. V. Yablokov and V. A. Zemsky, eds.), pp. 123–130. Center for Environmental Policy and the Marine Mammal Council, Moscow.

Chittleborough, R. G. (1965). Dynamics of two populations of the humpback whales, *Megaptera novaeangliae* (Borowski). *Aust. J. Mar. Freshw. Res.* **16**, 33–128.

Dawbin, W. H. (1966). The seasonal migratory cycle of humpback whales. *In* "Whales, dolphins and porpoises" (K. S. Norris, ed.), pp. 145–170. Univ. California Press, Berkeley and Los Angeles.

Doroshenko, N. V. (2000a). Soviet catches of humpback whales (*Megaptera novaeangliae*) in the North Pacific. *In* "Soviet Whaling Data" (1949–1979) (A. V. Yablokov and V. A. Zemsky, eds.), pp. 48–95. Center for Environmental Policy and the Marine Mammal Council, Moscow.

Doroshenko, N. V. (2000b). Soviet whaling for blue, gray, bowhead and right whales in the North Pacific, 1961–1979. *In* "Soviet Whaling Data" (1949–1979) (A. V. Yablokov and V. A. Zemsky, eds.), pp. 96–103. Center for Environmental Policy and the Marine Mammal Council, Moscow.

Findlay, K. P., and Best, P. B. (1996). Estimates of the numbers of humpback whales observed migrating past Cape Vidal, South Africa, 1988–1991. *Mar. Mam. Sci.* **12**, 354–370.

Kasuya, T. (1999). Examination of the reliability of catch statistics in the Japanese coastal sperm whale fishery. *J. Cetacean Res. Manage.* **1**, 109–122.

Kasuya, T., and Brownell, R. L., Jr. (1999). "Additional Information on the Reliability of Japanese Coastal Whaling Statistics." IWC Sci. Comm. unpub. meeting doc. SC/51/O7. Available from IWC, The Red House, 135 Station Rd, Impington, Cambridge, CB4 9NP, UK.

Kasuya T. and Brownell, R. L., Jr. (2001). "Illegal Japanese coastal whaling and other manipulation of catch records." IWC Sci. Comm. unpub. meeting doc. SC/53/RMP/24. Available from IWC, The Red House, 135 Station Rd, Impington, Cambridge, CB4 9NP, UK.

Kasuya, T., and Miyazaki, N. (1997). Cetacea and Sirenia. *In* "Red List of Japanese Mammals" (T. Kawamichi, ed.), pp. 139–187. Bunichi Sogo Shyuppan, Tokyo. [In Japanese with English summary.]

Kondo, I. (2001). "Rise and Fall of Japanese Coastal Whaling." Sanyosha, Tokyo. [In Japanese.]

Mikhalev, Y. A. (1996). "Pygmy Blue Whales of the Northern-Western Indian Ocean." IWC Sci. Comm. unpub. meeting doc. SC/48/SH30, 30 pp. Available from IWC, The Red House, 135 Station Rd, Impington, Cambridge CB4 9NP, UK.

Mikhalev, Y. A. (1997a). Humpback whales *Megaptera novaeangliae* in the Arabian Sea. *Mar. Ecol. Prog. Ser.* **149,** 13–21.

Mikhalev, Y. A. (1997b). "Bryde's Whales of the Arabian Sea and Adjacent Waters." IWC Sci. Comm. unpub. meeting doc. SC/49/SH30, 10 pp. Available from IWC, The Red House, 135 Station Rd, Impington, Cambridge CB4 9NP, UK.

Paterson, R. A. and Cato, D. H. (1994). The status of humpback whales *Megaptera novaeangliae* in east Australia thirty years after whaling. *Biol. Cons.* **70,** 135–142.

Tønnessen, J. N., and Johnsen, A. O. (1982). "The History of Modern Whaling." [Translated from the Norwegian by R. I. Christophersen] University of California, Berkeley.

Tormosov, D. D., Mikhalev, Y. A., Best, P. B., Zemsky, V. A., Sekiguchi, K., and Brownell, R. L., Jr. (1998). Soviet catches of southern right whales, *Eubalaena australis,* 1951–1971: Biological data and conservation implications. *Biol. Cons.* **86,** 185–197.

Watson, P. (1979). Pirate whaler smashed. *Defenders* **54**(6), 363338.

Yablokov, A. V. (1994). Validity of whaling data. *Nature* (*Lond.*) **367**(6459), 108.

Yablokov, A. V., Zemsky, V. A., Yu, A., Mikhalev, Tormosov, V. V. [sic], Berzin, A. A. (1998). Data on Soviet whaling in the Antarctic in 1947–1972 (population aspects). *Russ. J. Ecol.* **29,** 38–42. [Translated from Ekologiya 1, 43–48 (1998)]

Zemsky, V. A., Berzin, A. A., Mikhalev, Y. A., Tormosov, D. D. (1995a). Soviet Antarctic pelagic whaling after WWII: Review of actual catch data. *Rep. Int. Whal. Commn.* **45,** 131–135.

Zemsky, V. A., Berzin, A. A., Mikhalev, Y. A., Tormosov, D. D. (1995b). "Soviet Antarctic Whaling Data." Centre for Environmental Policy, Moscow.

Incidental Catches

SIMON NORTHRIDGE
University of St. Andrews,
Scotland, United Kingdom

Marine mammals sometimes get caught up and killed in fishing operations. In many cases these deaths are entirely unintended by the fishermen concerned and are incidental to the main fishing operation. They are therefore referred to as incidental catches. Sometimes they are also referred to as "bycatches," although this term is also used to described the capture of some species that, while not the main target of a fishery, still have some value and may therefore be landed. Incidental catches are generally unwanted and discarded.

Incidental catches of marine mammals have probably occurred for as long as people have been putting nets and lines into the water. Most species of marine mammal that occur in places that are heavily fished have been recorded caught in at least one type of fishing gear. Most types of fishing gear have been reported to ensnare marine mammals at one time or another. Some captures seem to defy reason. Large whales, for example, may become caught in a single lobster pot line, whereas porpoises get caught in simple fish traps that they are able to find their way into, but not out of. Others catches are easier to comprehend, as when trawls with openings of several hundred meters in circumferences scoop whole schools of dolphins from the sea.

In the past, and indeed in many parts of the world today, unintentional captures of marine mammals might be treated as a bycatch and landed for consumption. During the latter half of the 20th century, however, fishing technology has changed faster and more completely than ever before, which has led to a reappraisal of the issues surrounding bycatch and incidental catch. Nets have become larger and stronger, numerous new fishing techniques have been devised, and fishing intensity throughout the world has increased dramatically, nearly trebling marine fishery landings over a period of just 40 years.

Such developments have had unintended negative impacts on nontarget species, including marine mammals, so that incidental catches have now become a critical issue for some marine mammal populations. Marine mammals generally reproduce slowly, and their populations are not able to withstand much additional nonnatural mortality. The removal of just 1% of the population per year may be more than a marine mammal population can sustain in the longer term. For this reason, many nations now legislate to protect marine mammal populations from deliberate or accidental exploitation, and there are several international agreements with the same aim.

Legislation to protect marine mammals from excessive mortality has resulted from a variety of case studies that have uncovered unsustainable levels of incidental capture. Several of these case have become widely publicized and have generated considerable public attention and debate.

I. Examples

A. Eastern Tropical Pacific Tuna Purse Seine Fishery

The first interaction to be recognized as a serious concern for the conservation of marine mammals was the large-scale capture of pelagic dephinids (mainly *Stenella* and *Delphinus* species) in the U.S. tuna purse seine fishery of the eastern tropical Pacific Ocean (ETP). Tuna boat skippers learned that they could catch large tuna by herding dolphin schools with speedboats and then surrounding them with long, deep, purse seine nets. Fishermen were exploiting the curious fact that in the ETP (and some other places), large yellowfin tuna (*Thunnus albacares*) will school under and follow dolphin schools. Once the dolphins and associated tuna are surrounded, the nets can be "pursed," whereby the bottom end of the net is closed off, thereby catching the tuna. At this point the dolphins can still surface to breathe within the encircled net and could escape by jumping over the floats. Pelagic delphinids, however, seem to find it difficult to escape from such an enclosed situation, and many became trapped and died under folds of the surrounding purse seine or simply fainted and died.

This fishing technique was begun in the 1950s, but was not recognized as a potential conservation problem until the early 1970s, when a monitoring program was established. During much of the 1960s and up to 1972, annual mortalities are thought to have ranged between 200,000 and 500,000. Thereafter a variety of efforts were made to reduce the kill, but tens of thousands of dolphins were still being killed annually throughout most of the 1980s. Pantropical spotted dolphins (*Stenella attenuata*) were the most frequently killed species, and numbers of this species in the ETP were more than halved over the 1960s and 1970s. Populations of other species were also severely impacted.

Largely as a result of public pressure, and the introduction of "dolphin safe" tuna retailing, this practice has now been greatly reduced. New techniques have been devised by the skippers to ensure that a very high proportion of the dolphins used in this way to catch tuna are encouraged to escape from the nets before the fish are removed. Under a training and monitoring scheme run by the Inter-American Tropical Tuna Commission, dolphin mortalities had been reduced to between 2000 and 4000 animals per year by the mid-1990s. Efforts continue to reduce these figures further still.

Throughout the world, since the discovery of the effect of the ETP tuna fishery on dolphin populations, it has become clear that there are numerous other fisheries in which marine mammals are being killed in large numbers. In some cases, populations or species have been threatened with extinction. Two of the most severe cases are those concerning the baiji (*Lipotes vexillifer*) and the vaquita (*Phocoena sinus*).

B. The Baiji

The baiji, otherwise known as the Chinese river dolphin, inhabits the middle and lower parts of the Yangtse River system in China. The total population size is not known, but is thought to be a few hundred at most and is declining. The major source of mortality for this species appears to be snagging in "rolling hook" fishing lines. These are lines equipped with many closely set, sharp, unbaited hooks designed to snag fish foraging on the river bed in the same areas as the Baiji. In one study, 45% of all known Baiji deaths were attributed to snagging in rolling hooks.

C. The Vaquita

The vaquita is a species of porpoise restricted to the upper part of the Gulf of California in Mexico. Population studies suggest that only around 600 animals remain, that numbers are declining, and that the species is in critical danger of extinction. Again, the major source of mortality is incidental catches in fishing operations, in this case gill net for sharks and other large fishes. Gill nets are simple long panels of netting that are set to stand vertically in the water with floats along their top and a weighted rope on their bottom. Depending on the amount of weight added, they either sit on the seabed floating upward or they float at the surface hanging down. They are left to ensnare fish that happen to swim into them, but also catch marine mammals by entangling them. Annual vaquita mortalities in gill net fisheries are estimated at around 40–80 per year, which is clearly an unsustainable level of mortality given the size of the population.

D. New Zealand Sea Lion

Another species that has been threatened in a similar way, but by an entirely different sort of fishery, is the New Zealand sea lion (*Phocarctos hookeri*). These sea lions are restricted to New Zealand's sub-Antarctic islands, mainly the Auckland Islands. The population is thought to have been much reduced by commercial sealing activities in the 19th century. The total population size was estimated at around 13,000 animals in the mid-1990s, although it has been further reduced by a mass die-off in 1998. A trawl fishery for squid was started around the Auckland Islands in 1979, which subsequently resulted in significant numbers of sea lions being drowned in the large trawl nets, raising concern that the population might be reduced further or even threatened with extinction. The New Zealand government implemented an observer scheme in the late 1980s to monitor the numbers being killed and it now sets incidental catch limits every year in order to prevent unsustainable levels of mortality. Once the annual limit is reached the fishery must stop fishing.

II. Causes for Concern

Although the just-described examples are perhaps the most extreme cases, there are numerous others around the world where significant numbers of marine mammals are killed incidentally in fishing operations. It is usually the smaller species and those that occur in continental shelf waters where most fishing occurs that are impacted most heavily.

Incidental catches do not always impact on entire species. In many instances, marine mammal species may be widespread and in little danger of overall extinction. Nevertheless, incidental catches may be frequent enough to reduce or eliminate a local population (Fig. 1). This is the case for the harbor porpoise (*Phocoena phocoena*). While they are in no imminent danger of extinction as a species, in several areas, including the Gulf of Maine off the U.S. northeast coast, incidental catch rates are or have been high enough to push local populations into decline.

In other parts of its range, including the English Channel and the Baltic Sea, harbor porpoises have all but disappeared. While the causes of these disappearances are not known, fishing is intense in both areas and incidental catch rates in adjacent areas are known to be high enough to be unsustainable.

Throughout the world, small inshore species such as the harbor porpoise are known to be victims of incidental catches in fishing operations, but the level of such catches and the likely impact remain unknown. Monitoring incidental catch rates and estimating population sizes are both very expensive. A significant issue in this regard is that there does not need to be a very large number of incidental kills for the total effect to be significant. When a marine mammal population numbers in the hundreds or even the tens of thousands, a few individuals to a few hundred individuals taken per year may be enough to generate a population decline. Furthermore, even when the marine mammal population is much larger, if the fishery is also large, significant catches can occur while still remaining unknown. Generally speaking, incidental catches are rare events. Typically, a capture event only occurs in one or two out of every hundred fishing operations. Such low levels may remain unnoticed, although the aggregate effect over a large number of vessels and operations may be significant. Such low levels of capture also make monitoring more difficult.

Although most attention worldwide has focused on the potential conservation issues that incidental catches of marine

Figure 1 *A harbor porpoise entangled in a cod gillnet in the North Sea, one of several thousand dying this way every year in European gillnet fisheries. Photo by Nigel Godden/Sea Mammal Research Unit.*

mammals raise, animal welfare considerations are also a concern. Whereas some incidental capture of marine mammals in fishing operations is an inevitable consequence of fishing, in some nations at least, any large-scale fatalities of marine mammals are publicly unacceptable regardless of whether they are sustainable at a population level.

III. Attempts to Resolve the Problem

Most of the numerically significant incidental catches of marine mammals tend to be in static fishing gear, mainly gill nets. Despite the attention focused on this subject in recent years, it is still not known how or why marine mammals actually become caught in such nets. It is not known, for example, whether mammals are attracted to nets by curiosity or by the presence of trapped fish, whether they do not notice the netting, or whether they simply do not understand the potential consequence of swimming into it. Despite our ignorance, some progress has been made toward resolving the problem.

One potential solution to the problem of marine mammal capture in gill nets has been developed in North America. Pingers, or acoustic beacons, exploit the sensitive hearing of marine mammals by emitting an intermittent, short, high-pitched noise that most fish cannot hear but that appears to repel or warn off marine mammals. Attached at regular intervals along the length of a gill net, these 16-cm-long cylindrical devices have been shown to reduce the numbers of marine mammals (mainly harbor porpoises, but also dolphins and sea lions) caught by up to 90% (Kraus *et al.*, 1997). Pingers were first developed in Canada, and their use is now mandatory in several U.S. fisheries.

Pingers certainly appear to be useful but there are still some concerns about their use. As they are a recent technological innovation, marine mammals may become habituated to their noises and start to ignore them. If, as seems to be the case, pingers displace animals from an area, and if they are used to the very large numbers that would seem to be necessary, then it is also possible that marine mammals may become excluded from parts of their foraging habitats. Pingers rely on batteries, and they rely on people replacing those batteries to ensure that they continue to work. This can be an expensive and time-consuming operation that many people might eventually prefer to forget about. Finally, the pingers themselves are expensive, so that the cost of equipping a net with pingers may exceed the cost of the net. In many less-developed countries, it is unlikely that they will ever become widely used for this reason alone.

Issues with mobile fishing gear are somewhat different. There are or have been several initiatives worldwide that aim to keep marine mammals out of towed fishing gear. In the ETP tuna fishery referred to earlier, special techniques and nets have been developed to help dolphins to escape from the purse seine net once the net has fully encircled the school of tuna. During the "backdown procedure" the skipper reverses the vessel and is able to sink a part of the net floatline under the water, enabling the dolphins to escape. This part of the net is also made up with a smaller meshed panel (the Medina panel), reducing the chances of dolphins becoming entangled as they escape. Similarly, in New Zealand, special marine mammal escape devices have been designed and used in squid trawls. A large grid is placed near the rear of the net, set at a 45° angle to the vertical plane. Fish can pass through the grid, but larger animals such as sea lions are forced upward and out the net through an escape hatch.

In general, the incidental capture of marine mammals is caused by a combination of fishing technique or gear design and the behavior of the marine mammal. Resolving problematic interactions therefore involves some combination of change to fishing gear use or design and the manipulation of marine mammal BEHAVIOR. Very little is known about the behavior of marine mammals in relation to fishing gear, especially in the context of incidental capture. In part this is because of the difficulties of studying marine mammals underwater, but it is also because of the rarity of such events, which makes observing their occurrence very difficult. Finding solutions to the problem is therefore a slow and arduous process.

Most fishing practices and gear designs have been adopted by fishing communities because they are effective in catching fish, and making changes may therefore reduce the profitability of a fishery. Effective mitigation measures therefore need to be devised in collaboration with the fishing community in order to minimize the adverse impacts on fish catches, but they may also require a legislative approach to ensure compliance or equability within a fishery. In this respect, managing the incidental capture of marine mammals may be seen as part of a much more wide-ranging and ongoing problem of managing a global industry that, in the last 50 years, has outgrown its resource base.

See Also the Following Articles

Entrapment and Entanglement ■ Fisheries, Interference with ■ Tuna–Dolphin Issue

References

Gosliner, M. L. (1999). The tuna-dolphin controversy. *In* "Conservation and Management of Marine Mammals" (J. Twiss and R. Reeves, eds.), pp. 120–155. Smithsonian Institution Press, Washington, DC.

Kraus, S. D., Read, A. J., Solow, S., Baldwin, K., Spradlin, T., Anderson, E., and Williamson, J. (1997). Acoustic alarms reduce porpoise mortality. *Nature* **388**, 525.

Northridge, S. P., and Hofman, R. J. (1999). Marine mammal interactions with fisheries. *In* "Conservation and Management of Marine Mammals" (J. Twiss and R. Reeves, eds.), pp. 99–119. Smithsonian Institution Press, Washington, DC.

Perrin, W. F. (1999). Selected examples of small cetaceans at risk. *In* "Conservation and Management of Marine Mammals" (J. Twiss and R. Reeves, eds.), pp. 296–310. Smithsonian Institution Press, Washington, DC.

Perrin, W. F., Donovan, G. P., and Barlow, J. (eds.) (1995). Gillnets and Cetaceans. Report of the International Whaling Commission, Special Issue 15, Cambridge.

Wickens, P. A. (1995). A review of operational interactions between pinnipeds and fisheries. FAO Fisheries Technical Paper 346. Food and Agriculture Organisation of the United Nations, Rome.

Indo-Pacific Beaked Whale

Indopacetus pacificus

ROBERT L. PITMAN
Southwest Fisheries Science Center, La Jolla, California

The Indo-Pacific (or Longman's) beaked whale (Ziphiidae) is one of the least known extant cetaceans. Its existence is known only from two skulls (Fig. 1), and it

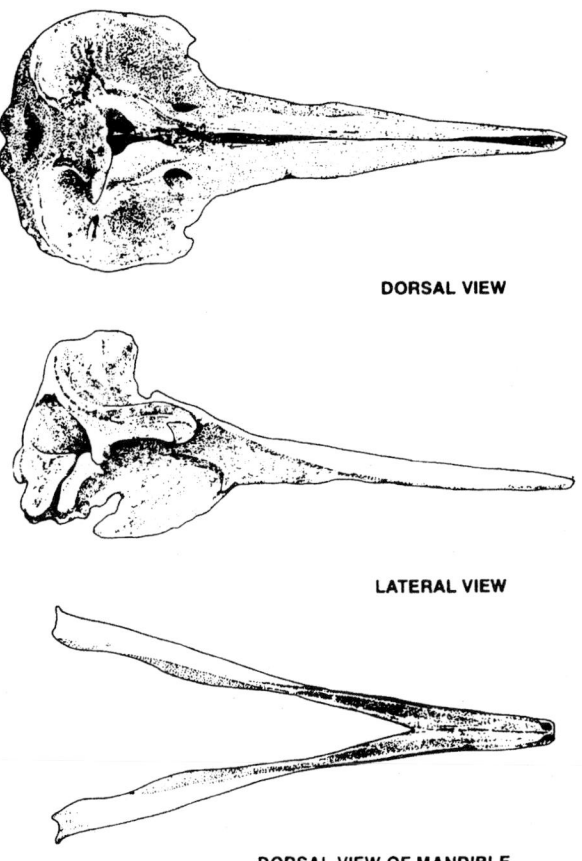

DORSAL VIEW

LATERAL VIEW

DORSAL VIEW OF MANDIBLE

Figure 1 *Skull of* Indopacetus pacificus. *Courtesy of the Food and Agriculture Organization of the United Nations; from Jefferson et al. (1993), Marine Mammals of the World, FAO, Rome.*

has never been identified in the flesh, alive or dead. It was originally described in 1926 as *Mesoplodon pacificus* from a beach-worn skull collected in Queensland, Australia, in 1882. The validity of the species was initially challenged by researchers who variously suggested that it was a subspecies of True's beaked whale (*Mesoplodon mirus*) or an adult female southern bottlenose whale (*Hyperoodon planifrons*). J. C. Moore later refuted these allegations and recognized it as a valid species. This was confirmed by the subsequent discovery of a second specimen, another skull, found on the coast of Somalia in 1955 (but not reported until 1968). After further study of both specimens, Moore found them sufficiently distinct to warrant a new genus, *Indopacetus*. The five diagnostic characters he cited for distinguishing *Indopacetus* from *Mesoplodon*, and the other living genera of beaked whales, were (1) a single pair of teeth alveoli (sockets), apical in the mandible, that become progressively shallower in the adult male to at least 30 mm; (2) frontal bones occupying an area of the synvertex equal to or greater than that of the nasal bones; (3) almost no posterior process of the premaxillary crest; (4) a deep horizontal groove in the maxillary bone just above, and about half as long as, the orbit; (5) about midlength in the beak, a swelling of the lateral margins so that the beak does not grow

Figure 2   *(a) A presumed cow and calf pair of an unidentified bottlenose whale from the eastern tropical Pacific that is probably* Indopacetus pacificus; *notice the pigmentation pattern of the calf in the foreground. (b) Same cow/calf pair showing the crease between melon and beak of the adult. Courtesy of Bernard Brennan.*

narrower throughout its entire length. The validity of the genus *Indopacetus* has not been universally accepted, however, and some workers include the species in *Mesoplodon*.

All that can be surmised about the natural history of *Indopacetus* is that, based on the two STRANDING records, it appears to inhabit tropical waters of the Indo-Pacific, and based on skull size the total length may be 7.0–7.6 m, making it a fairly large beaked whale. Although the two teeth that erupt only in adult males have never been examined, based on alveoli shape they appear to be oval in cross section and point forward.

Indopacetus pacificus is perhaps the largest animal left on the planet that has not been positively identified alive in the wild, but there is reason to believe that it may not be as rare as the meager stranding record indicates. Recently, the existence of a large and distinctive, but as yet unidentified, species of beaked whale has been confirmed from 45 sightings scattered from the eastern tropical Pacific to the western tropical

Indian Ocean. The calf of the so-called "tropical bottlenose whale" is dark with a pale melon that has a posterior margin confluent with the blowhole; it also has a whitish flank patch that extends high up on the sides of the animal and a small white "ear spot" (Fig. 2a). The adult is a large beaked whale (length estimates range from 7 to 8 m), which has a crease between the melon and the beak (Fig. 2b) identifying it as a bottlenose whale. These attributes are not referable to any known species of ziphiid, but are consistent with what little is known about *I. pacificus*, and the suggestion was made that the tropical bottlenose whale may well be that species. The question will remain unresolved until a suitable specimen becomes available.

See Also the Following Articles

Beaked Whales ■ Mesoplodont Whales

References

Moore, J. C. (1968). Relationships among the living genera of beaked whales with classifications, diagnoses, and keys. *Field. Zool.* **53**, 206–298.

Moore, J. C. (1972). More skull characters of the beaked whale *Indopacetus pacificus* and comparative measurements of austral relatives. *Field. Zool.* **62**, 1–19.

Pitman, R. L., Palacios, D. M., Brennan, P. L. R., Brennan, B. J., Balcomb, K. C., III, Miyashita, T. (1999). Sightings and possible identity of a bottlenose whale in the tropical Indo-Pacific: *Indopacetus pacificus? Mar. Mamm. Sci.* **15**, 531–549.

Indo-West Pacific Marine Mammals

PETER RUDOLPH AND CHRIS SMEENK
National Museum of Natural History,
Leiden, The Netherlands

The Indo-West Pacific is defined here as the tropical and subtropical waters of the Indian Ocean, from about 35°S off South Africa to the Red Sea and Persian Gulf and from Australia and Southeast Asia to about 30°N (Fig. 1).

The Indo-West Pacific probably offers the greatest diversity of marine mammal species in the world. Within it live representatives of 10 of the 14 families of the order Cetacea, with more than 40 of the 83 species treated as such by Rice (1998), as well as one member of the order Sirenia. We have not included species of the order Carnivora (sea lions, walruses, and seals), nearly all of which normally live in higher latitudes. Only the ranges of the Cape and Tasmanian fur seals *Arctocephalus pusillus pusillus* and *A. p. doriferus,* the south Australian fur seal *A. forsteri,* and the Australian sea lion *Neophoca cinerea* (family Otariidae) include the southernmost part of the area considered, in South Africa and southern Australia, respectively.

I. Endemic Taxa

Many species occurring in the Indo-West Pacific have a cosmopolitan or pantropical distribution, and several Northern and Southern Hemisphere species extend their range to within the confines of the area. However, a relatively large number (11) of species or currently recognized subspecies are endemic to the Indo-West Pacific, although the taxonomic position and distribution of some forms are still insufficiently known. Most of these occur mainly in shelf and/or fresh-water ecosystems; *Tursiops aduncus* is also found in deep oceanic waters. Three seem to be largely confined to shelf areas: *Balaenoptera edeni* (if correctly defined), *Delphinus tropicalis* (whether a valid species or

a subspecies of *D. capensis*), and *Stenella longirostris roseiventris;* five have a decidedly coastal, estuarine, or even partly riverine distribution: *Sousa plumbea, S. chinensis, Orcaella brevirostris, Neophocaena phocaenoides,* and the sirenian *Dugong dugon;* and two are true river dolphins: *Platanista gangetica* (including *P. g. minor*) and *Lipotes vexillifer.* In addition, at least one oceanic species—*Mesoplodon ginkgodens*— is endemic to deep waters of the (sub)tropical Indo-Pacific at large and the same may hold true for *Indopacetus pacificus,* although the IDENTIFICATION and DISTRIBUTION area of the latter are still somewhat uncertain.

II. Zoogeography

The shelf areas of the Indo-West Pacific show a high primary productivity, the result of monsoon-related currents and strong upwelling. Although the shallow waters of Southeast Asia and Australia would seem to constitute a barrier between the Indian and Pacific Ocean basins, the deep passages through the eastern Indo-Malayan Archipelago offer suitable dispersal routes for oceanic species, most of which, including sperm whales, *Physeter macrocephalus,* have indeed been recorded from the deeper straits and seas between the islands. Not considering the river dolphins, none of the marine mammals in the area is restricted to either the Indian Ocean or the Pacific side of the archipelago, with the exception of the western form *Sousa plumbea,* which, if valid, probably is the result of a recent vicariant speciation within the genus.

The endemic marine mammal taxa of the Indo-West Pacific shelf waters differ in the extent of their present range. *B. edeni, S. longirostris roseiventris* (as far as known at present), and *O. brevirostris* seem restricted to the Sunda and Sahul Shelves and adjacent waters; all other species also occur further west and north in the Indian and Pacific Ocean, respectively. In the east, the deep waters of the Pacific seem to form a barrier to further dispersal, although the dugong has penetrated beyond the continental shelves, occurring as far as Micronesia and, as a vagrant, to Fiji; the eastern confines of *T. aduncus* are not known. In the north, *Neophocaena* has extended its range to Korea and Japan. In the southwest, *S. plumbea* and *T. aduncus* occur as far as South Africa. Here, the cold waters of the Cape apparently have prevented their dispersion into the Atlantic Ocean, although the geographic history of the West African *Sousa teuszii* is not clear. In the northwest, *S. plumbea* occurs in the Suez Canal; this species, as well as the dugong, has even strayed into the Mediterranean Sea.

III. Annotated Species Account
A. Rorquals, Family Balaenopteridae

Most baleen whales undertake extensive seasonal migrations between cold, productive summer feeding grounds in temperate or high latitudes and winter mating and calving areas in tropical or warm temperate waters, suggesting that there is only little mixing of Southern and Northern Hemisphere populations. Baleen whales in the Indo-West Pacific, particularly the Indian Ocean, are poorly known.

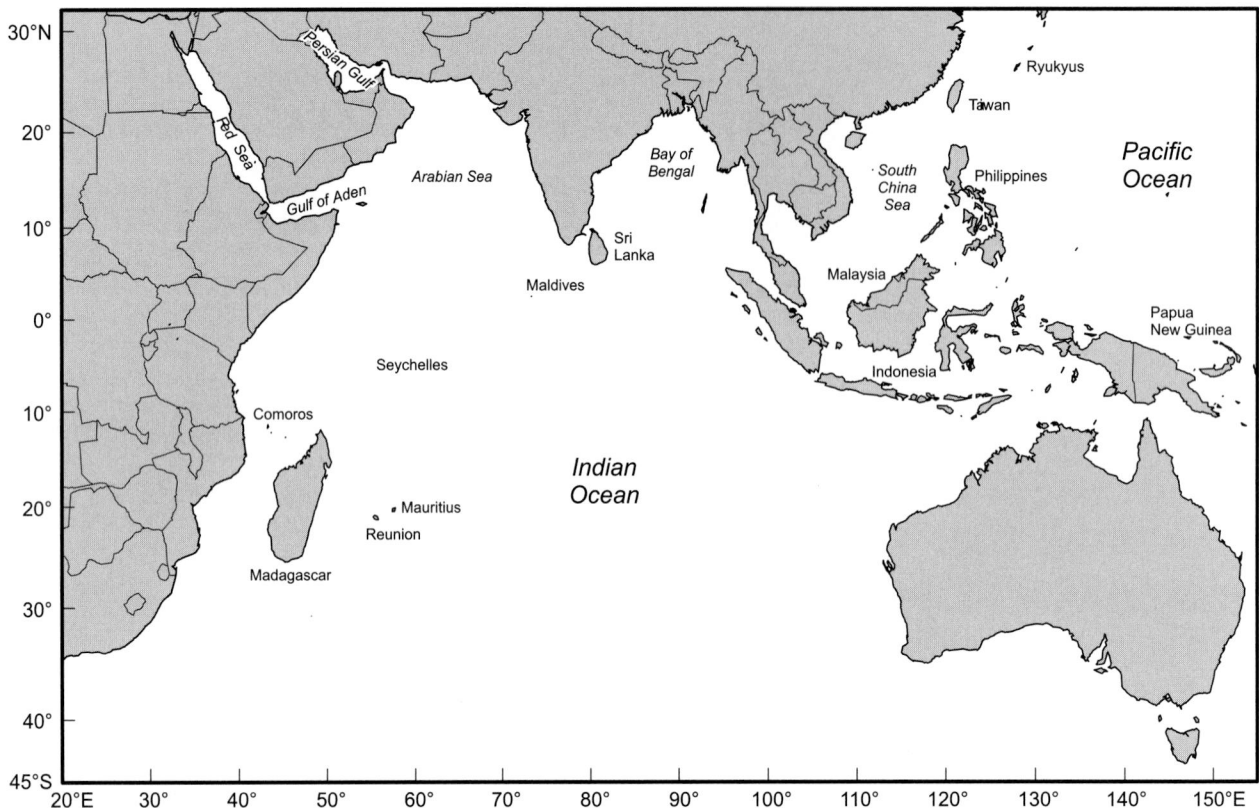

Figure 1 *The Indo-West Pacific is home to a wide variety of marine mammals, including many endemic taxa.*

The modern WHALING industry, following the invention of the explosive grenade harpoon, was initially based mainly on rorquals, although sperm and right whales were also taken. In the Indian Ocean, whaling stations existed in South Africa, Mozambique, and western Australia. Although most pelagic whaling occurred below 40°S, it was practice for whalers to take baleen whales as well as sperm whales in tropical waters on their passage to and from the Antarctic. Newly revealed data on illegal Soviet-Russian whaling operations between 1947 and 1972 have shown that baleen and sperm whales were taken in the central and northern Indian Ocean, near Madagascar and Australia, as well as in Indonesian waters. These included blue whales (*Balaenoptera musculus*) (caught mainly after 1965, when the species had already received protection from the IWC); humpback whales (*Megaptera novaeangliae*) off Oman, Pakistan, and India; as well as southern right whales (*Eubalaena australis*) around the Crozet and Kerguelen Islands and in the central Indian Ocean. Japanese whalers caught 232 Bryde's whales (*Balaenoptera brydei* and *B. edeni*) in the western Indian Ocean, south of Java and the Lesser Sunda Islands, and near the Solomon Islands during the seasons 1976/1977–1978/1979. Relatively small-scale, directed catches of baleen whales are today only known from the Philippines and from Lamakera on Solor Island, Indonesia. In the Philippines, baleen whales (mostly Bryde's whales) were or are still hunted off Pamilacan, Bohol, and Camiguin Islands. The fishery, now legally prohibited, at Pamilacan Island was seasonal and opportunistic. It started in January and ended in June, with most whales taken in April and May. The animals were caught using a hook of stainless steel, which was driven into the whale by one of the hunters jumping on the whale's back and using his weight to drive the hook in.

Humpback whale, *Megaptera novaeangliae*

Growing evidence shows that discrete populations of humpback whales remain in the Arabian Sea, including the Gulf of Oman and the Gulf of Aden, and in the Bay of Bengal year-round, with at least some of these animals both feeding and breeding there. Biological examination of 238 humpback whales caught illegally in these waters in November 1966 by Soviet-Russian whalers showed that they differed significantly from Antarctic humpbacks in size, COLORATION, body scars, and pathology (Mikhalev, 1997). The song structure of humpback whales recorded off Oman is also different from that in the North Pacific and North Atlantic. Southern Hemisphere humpback whales, which feed in Antarctic waters, winter near Mozambique, around Madagascar, off northwest and northeast Australia, and elsewhere in the tropical Pacific. In Southeast Asian waters, humpback whales have been reported from Vietnam, the Philippines, the South China Sea, and the waters around Taiwan; probably, these are animals that spend the summer in the Northwest Pacific. The occurrence in the Indo-

Malayan Archipelago has yet to be confirmed. In the 19th and 20th centuries, humpback whales were pursued off Mozambique, Madagascar, and northwestern Australia.

Minke whale, *Balaenoptera acutorostrata*

The North Pacific population, sometimes regarded as the subspecies *B. acutorostrata scammoni*, occurs in summer south to at least the East China Sea and the central Pacific to about 30°N; its winter range extends into (sub)tropical waters. A southern population, the "dwarf minke whale," is known to winter as far north as South Africa and northern Australia; the relationships of these animals with other stocks are not known. In the tropical Indian Ocean, minke whales have been reported year-round from the Red Sea, Gulf of Aden, Arabian Sea, Persian Gulf, and the seas around Sri Lanka. Reports of the species from Vietnam, the Indo-Malayan Archipelago, and the Philippines have yet to be confirmed.

Antarctic minke whale, *Balaenoptera bonaerensis*

The winter range of this Southern Hemisphere species is known to extend into tropical waters; it has been found north to about 7°S. Animals recorded from the Indian Ocean possibly belong with this species.

Bryde's whale, *Balaenoptera brydei*
Sittang whale, *Balaenoptera edeni*

These whales, which until recently were considered conspecific, have a primarily tropical and warm-temperate distribution. The larger form, the "true" Bryde's whale, ranges from the Cape of Good Hope north to the Red Sea and Persian Gulf, the Indo-Malayan Archipelago and South China Sea, south to Shark Bay and off Queensland, Australia. Inshore and offshore populations have been reported from South Africa and Japan, differing in body size, external appearance, feeding ecology, and reproductive behavior.

A distinct small form, the Sittang whale or pygmy Bryde's whale, has been identified in the eastern Indian Ocean and tropical West Pacific; in southeast Asia it has been found from Burma (Myanmar) to the southern Philippines, the South China Sea and the Solomon Islands, and apparently is restricted to coastal and shelf waters. Molecular evidence from allozyme and mtDNA analyses, as well as osteological comparisons, indicates that the small form does not belong genetically with the larger Bryde's whales and almost certainly constitutes a separate species, to which the name *B. edeni* may apply, if the holotype of the nominal species, which is from the mouth of the Sittang River, Burma, proves to be of that form (see species account).

Sei whale, *Balaenoptera borealis*

The sei whale appears to be uncommon in tropical waters. None of the published new records in the northern Indian Ocean is convincing. Sei whales have routinely been confused with Bryde's or fin whales by observers who were not familiar with all three species. Sei whales from the Southern Hemisphere generally winter as far north as South Africa and Western Australia (to about 25°S). Japanese scouting and research

vessels reported sighting 10 sei whales south of Sumatra (5–10°S) during November in the period 1974/1975–1984/1985. In Southeast Asia its occurrence has been reported for the South China Sea (although this needs confirmation) and by one specimen from Java. The Javan animal is the holotype of the putative subspecies *B. borealis schlegelii*.

Fin whale, *Balaenoptera physalus*

The fin whale also seems to be uncommon in tropical waters. A North Pacific population is thought to winter from the Sea of Japan south to the Philippine Sea, with concentrations in the East China and Yellow Seas. Southern Hemisphere fin whales, which may belong to the southern subspecies *B. physalus quoyi*, winter in the Indian Ocean off South Africa, Madagascar, and Western Australia. Fin whale sightings reported by Japanese vessels in the period 1974/1975–1984/1985 were concentrated in two longitudinal areas: one west of 50°E and the other at 70–100°E. The northern limits of the species' range were about 20–25°S in November/December and 40–45°S in March. There are several published reports of fin whales from the northern Indian Ocean, Persian Gulf and off the Seychelles. There are a few records from Indonesian and Philippine waters.

Blue whale, *Balaenoptera musculus*

The blue whale occurs throughout the Indo-West Pacific. It has been recorded from around the Seychelles, from the Arabian Sea and Persian Gulf, and from Sri Lanka, and the Bay of Bengal. Strandings and sightings of blue whales in Southeast Asia have been reported from the Indo-Malayan Archipelago and southern China. In the Indo-Malayan Archipelago they have been recorded for the months of June, August, September, and December and probably occur there year-round. The "pygmy blue whale" *B. musculus brevicauda*, known from subantarctic waters of the Indian, southeast Atlantic, and southwest Pacific Oceans, may winter in tropical waters.

It is not known whether the blue whales found in the northern Indian Ocean and the Arabian Sea, which reside in tropical waters year-round, constitute a separate population. The name *B. musculus indica* has been given to this stock, but its distinguishing features are poorly defined. Animals observed in the Arabian Sea, around Sri Lanka and in the Indo-Malayan Archipelago have been tentatively identified as pygmy blue whales. However, the type specimen of *B. musculus indica*, collected in Burma (Myanmar), measured 25.6 m and an individual stranded on Java 27 m, whereas the largest pygmy blue whale ever taken measured only 24.1 m.

The year-round occurrence of blue and humpback whales in the northern Indian Ocean seems to be related to the high primary productivity of this region, the result of a combination of monsoon-related currents and strong upwelling.

Two other species of baleen whale occur in the Indian Ocean. The southern right whale *Eubalaena australis* (family Balaenidae) extends its range north to southern Mozambique and Western Australia. In the 19th and early 20th centuries, right whales were taken in the Indian Ocean between 30° and

40°S, around Crozet, Kerguelen, and Amsterdam Islands and on their calving grounds off South Africa and southeastern Australia. The pygmy right whale *Caperea marginata* (family Neobalaenidae) has been recorded north to nearly 33°S off southwestern Australia. The critically endangered western population of the gray whale *Eschrichtius robustus* (family Eschrichtiidae), known as the Korea/Okhotsk Sea stock, winters off southern China, with records from the Yellow, East China, and South China Seas, as far south as Hainan. It has been suggested that breeding and calving grounds may exist off Guangdong Province in southern China, but this should be investigated.

B. Sperm Whales, Family Physeteridae

Sperm whale, *Physeter macrocephalus*

The sperm whale occurs in all oceans of the world. Concentrations such as found in the traditional whaling grounds seem to be associated with oceanographic fronts, steep bottom topography, and high productivity. Whereas adult males reach temperate or even polar waters and return to lower latitudes to breed, females and immatures are restricted to tropical and warm-temperate seas. The sperm whale occurs throughout the deep waters of the Indian Ocean, including the Gulf of Aden and Gulf of Oman, and in the West Pacific; its main distribution is well known from records of 19th-century whalers. The species is sometimes found in shallow waters. The deep passages and seas between the islands of the eastern Indo-Malayan Archipelago have been supposed to be a migration route of sperm whales between the Indian and Pacific Ocean.

In the western Indian Ocean, exploitation of sperm whales began at about 1800 on the whaling grounds near the Cape of Good Hope and in later years extended east and north to Madagascar, Mozambique, and the "Zanzibar grounds" (Comoros, Seychelles, and the east African coast), Arabia, and west of Sri Lanka. In the eastern Indian Ocean and western Pacific, sperm whales were taken south of Sumatra and Java, in the Timor Sea and off Western Australia, in the Moluccan and Sulu Seas and north of New Guinea, and off the east coast of Japan. The fishery declined for economic reasons toward the close of the 19th century. Today, sperm whales are only hunted in a subsistence whale fishery in Indonesia. The whalers of Lamalera on Lembata Island specialize in catching sperm whales and other toothed whales. The animals are caught by using hand harpoons from open rowing boats called "peledang." During the whaling season, the boats search an area of up to a few kilometers south of the coast. When a whale is approached, the harpooner leaps from a small platform on the bow and adds his weight to drive the harpoon into the whale's back, similar to the technique used by Philippine whalers, who hunt baleen whales.

C. Pygmy and Dwarf Sperm Whale, Family Kogiidae

Our knowledge of the distribution of the pygmy sperm whale *Kogia breviceps* and dwarf sperm whale *K. sima* is sketchy and mainly based on specimens stranded or caught in fishing nets. Both species occur in tropical and warm-temperate seas, with the pygmy sperm whale extending its range into slightly colder waters. They are difficult to observe at sea and probably are much more common than sighting records would suggest. Although their diets overlap, prey composition indicates that the pygmy sperm whale has a more oceanic distribution, whereas the dwarf sperm whale prefers continental shelves and shelf edges.

In the Indo-West Pacific, both species have been recorded from South Africa north to Oman, east to Australia and the Indo-Malayan Archipelago, and north to southern Japan. Both are killed accidentally in Sri Lanka; the dwarf sperm whale is caught in directed fisheries and incidentally in gill net and seine fisheries in the Philippines. In Indonesia, the dwarf sperm whale is taken by subsistence whalers of Lamalera on Lembata Island.

D. Beaked Whales, Family Ziphiidae

Twelve species of beaked whales have been recorded from the Indo-West Pacific. Only four of these are distributed in tropical waters. The other eight normally live in temperate or cold waters of the Southern Hemisphere and occasionally migrate or stray into lower latitudes. All are oceanic species, mainly known from stranded animals.

Cuvier's beaked whale, *Ziphius cavirostris*

This species is the most widespread of the beaked whales, occurring worldwide in tropical and warm-temperate waters. It has been reported from South Africa, the Comoros and Seychelles, the Arabian Sea, Sri Lanka, Australia, the Indo-Malayan Archipelago, Taiwan, the Philippines, and southern Japan.

Ginkgo-toothed whale, *Mesoplodon ginkgodens*

This species is distributed in tropical and warm-temperate waters of the Indian and Pacific Oceans. In the Indo-West Pacific it has been found in Sri Lanka, the Strait of Malacca, the Indo-Malayan Archipelago, off Taiwan, and in southern Japan, with one record from the northern Yellow Sea.

Blainville's beaked or dense-beaked whale, *Mesoplodon densirostris*

This species occurs in tropical and warm-temperate waters of all oceans. It has been reported from South Africa, Mauritius, the Seychelles, Australia, the Philippines, China, Taiwan, and Japan.

"Tropical bottlenose whale," *Indopacetus pacificus*

Large, unidentified beaked whales that look much like *Hyperoodon* have been observed and photographed on many occasions in the Indian and Pacific Oceans, between about 20°S and 35°N. Records from the Indo-West Pacific are from the Gulf of Aden and the Arabian Sea, Sri Lanka, the Indo-Malayan Archipelago, the Philippines, and north to southern Japan. Pitman *et al.* (1999) presented good evidence that they represent Longman's beaked whale *I. pacificus*, so far only known from two skulls of stranded animals, in Somalia in 1955 and in northeastern Australia in 1882.

Two species—the strap-toothed whale *Mesoplodon layardii* and the scamperdown whale *M. grayi*—occur at least seasonally within the southern confines of the area considered. The remaining six are irregular visitors: Arnoux's beaked whale *Berardius arnuxii*, Tasman beaked whale *Tasmacetus shepherdi*, southern bottlenose whale *Hyperoodon planifrons*, Hector's beaked whale *Mesoplodon hectori*, True's beaked whale *M. mirus*, and Andrew's beaked whale *M. bowdoini*. A skull of *H. planifrons* (the holotype of the species) was found in the Dampier Archipelago, Western Australia, at about 20°S.

E. River Dolphins, Families Platanistidae and Lipotidae

Two species of river dolphins are represented in the Indo-West Pacific region: the Indian river dolphin or susu *Platanista gangetica* (family Platanistidae) and the Yangtze dolphin or baiji *Lipotes vexillifer* (family Lipotidae).

Two subspecies of the Indian river dolphin are recognized: the Ganges dolphin *P. g. gangetica* in the Ganges/Brahmaputra River system of India, Bangladesh, Nepal, and possibly Bhutan, and the Indus dolphin *P. g. minor*, formerly common throughout the Indus River and its tributaries, but now only found in a restricted area in Pakistan. Both subspecies are regarded as endangered, particularly the Indus dolphin with an estimated population in the low hundreds, suffering from habitat degradation (pollution, construction of dams) and mortality in fishing gear.

Of all living cetaceans, the Yangtze dolphin in China probably is the closest to EXTINCTION, with an estimated number of less than 100 animals remaining. Baiji are found in the middle and lower course of the Yangtze River and may enter its tributary lakes during periods of flooding.

F. Dolphins, Family Delphinidae

Rough-toothed dolphin, *Steno bredanensis*

This species mainly inhabits deep, warm oceanic waters. In the Indo-West Pacific it ranges from South Africa to the Gulf of Aden and Arabian Sea, east to Australia, the Indo-Malayan Archipelago and the Philippines, and north to southern Japan. Small numbers are killed in gill nets off Sri Lanka.

Humpback dolphins, genus *Sousa*

Humpback dolphins inhabit tropical and subtropical coastal waters of the Indo-West Pacific and often occur in mangroves, river deltas, and estuaries. They ascend far up the main rivers of Asia, e.g., the Ganges and Yangtze.

Rice (1998) recognizes two species in the Indo-West Pacific. The Indian humpback dolphin *S. plumbea* occurs from South Africa and Madagascar to the Red Sea, including the Suez Canal, and from the Persian Gulf to the western Bay of Bengal.

The Pacific humpback dolphin *S. chinensis* ranges from the Strait of Malacca, Gulf of Thailand, and the northwest coast of Borneo to southern China. It also occurs in northern Australia, down the east coast to about 34°S. Humpback dolphins are reported to decline in the two areas where they have been studied in some detail. Off South Africa they seem to be killed in antishark nets at an unsustainable rate and it has been suggested that they are declining in Hong Kong waters. They are caught for oceanaria along the coast of Thailand. Incidental catches are known from shallow-water fisheries throughout its range.

Bottlenose dolphins, genus *Tursiops*

Bottlenose dolphins are distributed throughout the Indo-West Pacific, where they occupy a wide range of habitats. Some populations inhabit bays, lagoons, and estuaries, whereas others live in more open coastal or even oceanic waters. The taxonomy of the genus is largely unresolved. Until recently, most authors have recognized one cosmopolitan species, *T. truncatus*. In the Indo-West Pacific, two types can be distinguished: a larger form that is very similar to *T. truncatus* and one that is more slender, has a longer beak, and often a spotted abdomen. To the latter, the name *T. aduncus* is commonly applied, although that species is not very well defined. Morphological and osteological studies as well as molecular genetic analyses of mitochondrial and nuclear DNA have confirmed that the two forms are indeed specifically distinct (Wang *et al.*, 1999, 2000 a, b; Hale *et al.*, 2000). Based on an analysis of complete mtDNA cytochrome B sequences, LeDuc *et al.* (1999) concluded that the smaller form may even belong in the genus *Stenella*. Most authors assume that *T. aduncus* is the more common species throughout the Indo-West Pacific, ranging from South Africa to the Red Sea and Persian Gulf, east to eastern Australia and New Caledonia, the Indo-Malayan Archipelago, and north to the Ryukyu and perhaps Ogasawara

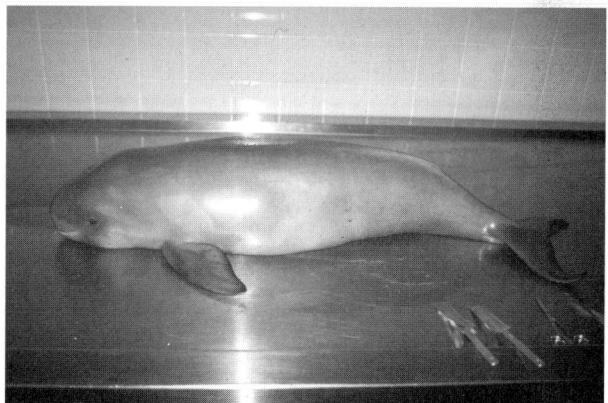

Figure 2 *The finless porpoise (top) and the Indopacific humpback dolphin (bottom) are endemic to the Indo-West Pacific. Photographs by Thomas Jefferson.*

(Bonin) Islands. It occurs mainly, but not exclusively, in coastal waters.

It is not clear to what extent the larger species occurs in the area. Thirteen bottlenose dolphins from east of Mauritius and near the Maldives were identified by genetic analysis as *T. truncatus*. Animals of this type have also been reported from South Africa, the Seychelles, the Red Sea, Oman, southern and eastern Australia, and from the Northwest Pacific, south to Hong Kong, Taiwan, and the Philippines. It occurs in both oceanic and coastal waters.

Incidental and directed takes of *Tursiops* sp. occur in several fisheries in the Indo-Pacific. Directed catches have been reported for Sri Lanka, Thailand, the Philippines, and Taiwan. A large incidental take has occurred in a Taiwanese gill net fishery off Australia, with perhaps over 2000 animals killed per year.

Pantropical spotted dolphin, *Stenella attenuata*

The pantropical spotted dolphin occurs throughout the tropical and subtropical Indo-West Pacific, from South Africa to the Red Sea and Persian Gulf, east to Australia, the Indo-Malayan Archipelago, and the Philippines, and north to southern Japan. Mixed groups of *S. attenuata* and *S. longirostris* have been reported from the western Indian Ocean, Red Sea, and the Indo-Malayan Archipelago, often associated with concentrations of seabirds and tuna. The spotted dolphin is taken in harpoon fisheries in the Philippines, Laccadive Islands, Sri Lanka, and Indonesia. A former drive fishery at Malaita in the Solomon Islands took several hundred animals per year in the 1960s. The species is incidentally caught in gill net fisheries in Pakistan, Sri Lanka, the Philippines, and northern Australia.

Spinner dolphin, *Stenella longirostris*

The spinner dolphin ranges throughout the tropical and subtropical Indo-West Pacific, from South Africa to the Red Sea and Persian Gulf, east to Australia, the Indo-Malayan Archipelago and the Philippines, and north to southern Japan.

Perrin *et al.* (1999) identified two forms of spinner dolphin in the Indo-West Pacific based on external, osteological, and behavioral differences, as well as habitat preferences. They distinguish the "dwarf spinner dolphin" as a subspecies, *S. l. roseiventris*. This form has been found in shallow inner waters of Southeast Asia (Gulf of Thailand, Borneo, Moluccan Sea, and northern Australia). Dwarf spinner dolphins are smaller, with fewer vertebrae and teeth than the pelagic form *S. l. longirostris*. The latter inhabits the Indian Ocean and West, Central, and South Pacific. Dwarf spinner dolphins feed mainly on benthic and coral reef fishes and invertebrates, whereas the larger form primarily feeds on mesopelagic fish and squid. Off Oman, two types of spinner dolphin are recognized, one of them consisting of very small, dark-colored and often pink-bellied animals (Van Waerebeek *et al.*, 1998). The taxonomic position of the latter is still unresolved. Spinner dolphins are taken in harpoon fisheries in Sri Lanka, the Philippines, and Indonesia. Incidental catches in fishing gear have been reported for Pakistan, India, Sri Lanka, the Philippines, and northern Australia.

Striped dolphin, *Stenella coeruleoalba*

Although the striped dolphin is primarily a warm water species, its range extends into higher latitudes than that of spotted and spinner dolphin. The species generally seems to be restricted to oceanic waters and approaches the shore only where deep water occurs close to the coast. The striped dolphin has been reported from South Africa to the southern Red Sea, east to Australia, and from the Philippines north to Japan. Records from the Indo-Malayan Archipelago have not yet been confirmed. INCIDENTAL CATCHES occur in gill nets in the northern Indian Ocean (e.g., Sri Lanka).

Common dolphins, genus *Delphinus*

Common dolphins are distributed in warm-temperate and tropical waters throughout the Indo-West Pacific. Morphological and genetic analysis has revealed the existence of at least two species: the short-beaked common dolphin *D. delphis* and the long-beaked common dolphin *D. capensis*.

In the Indian Ocean, only the occurrence of long-beaked animals has as yet been confirmed; *D. capensis* has been reported from South Africa and Madagascar. In the northwest Pacific, this form may occur as far south as Taiwan.

In coastal waters of the northern Indian Ocean and southeast Asia, a very long-beaked form exists, which by some authors is regarded as a separate species: *D. tropicalis*. Its taxonomic status is still unclear, but it exceeds *D. capensis* in rostral length and the average number of teeth. This form has been documented for the Arabian Sea, Gulf of Aden and southern Red Sea, Gulf of Oman and Persian Gulf, Pakistan and western India, the Gulf of Thailand, Gulf of Tonkin, and the South China Sea, north as far as Taiwan. Reports that it occurs sympatrically with *D. delphis* or *D. capensis* off Oman seem erroneous. The situation around Taiwan is not clear.

The short-beaked common dolphin *D. delphis* has been reported from various areas in the West Pacific: in the northwest from Taiwan and southern Japan, in the southwest from New Caledonia and Australia, particularly the south coasts.

Fraser's dolphin, *Lagenodelphis hosei*

Fraser's dolphin was described in 1956 from a skeleton found on a beach in Sarawak, Borneo, prior to 1895. The species occurs primarily in deep tropical and subtropical waters. In the Indian Ocean it has been reported from South Africa and Madagascar, the Seychelles, southern India and Sri Lanka, Borneo (type specimen) and the deeper waters of the eastern Indo-Malayan Archipelago, the Philippines, China, Taiwan, and southern Japan. In Australia, strandings have occurred south to about 38°S. In Indonesia and the Philippines, Fraser's dolphin has been observed in mixed groups with various other dolphin species. Directed catches have been reported for the Philippines and Indonesia.

Genus *Lagenorhynchus* [= *Sagmatias*]

Two species of this genus occur marginally within the confines of the Indo-West Pacific. The (North) Pacific white-sided dolphin *L.* [*S.*] *obliquidens* has been found as far south as Taiwan. The dusky dolphin *L.* [*S.*] *obscurus* in some years enters the coastal waters of southern Australia.

Risso's dolphin, *Grampus griseus*

Risso's dolphin is distributed throughout tropical and temperate seas, particularly seaward of steep continental shelf edges. In the Indian Ocean it is found from South Africa to the Red Sea and Persian Gulf, east to the Bay of Bengal, the deeper waters of the eastern Indo-Malayan Archipelago and Australia, and throughout the West Pacific. Risso's dolphins are known to be directly taken off Palawan in the Philippines and in the Indonesian whale fishery of Lembata and Solor Islands. Between 1983 and 1986, 241 individuals were reported landed in the Sri Lanka gill net fishery, but the actual numbers taken here may be about 1300 animals per year.

Melon-headed whale, *Peponocephala electra*

The melon-headed whale is mainly found in deep tropical and subtropical, oceanic waters. In the Indo-West Pacific it has been recorded from the Seychelles to the Arabian Sea and Bay of Bengal, Australia, the deep waters of the Indo-Malayan Archipelago, the Philippines, Gulf of Thailand, Taiwan, and southern Japan. There are a few records from South Africa at 34°S and from southern Australia. Mass STRANDINGS are known from the Seychelles, Indonesia, Australia, and Japan. There is a direct catch of melon-headed whales in the harpoon fishery off Pamilacan Island in the Philippines and off Lembata and Solor Islands in Indonesia.

Pygmy killer whale, *Feresa attenuata*

The pygmy killer whale is mainly found in deep tropical and subtropical, oceanic waters. In the Indo-West Pacific it has been recorded from South Africa, to the Gulf of Aden and Gulf of Oman, east to Australia, the Indo-Malayan Archipelago, the Philippines, Taiwan, and southern Japan. Incidental and directed catches have been reported for Sri Lanka, with an estimated 300–900 animals taken annually. Although not confirmed by a specimen yet, the species may also be taken in the subsistence whale fishery in Indonesia.

False killer whale, *Pseudorca crassidens*

The false killer whale is found primarily in deep tropical to warm-temperate waters. It occurs throughout the Indian Ocean from South Africa to the Red Sea and Persian Gulf, east to Australia, the Indo-Malayan Archipelago, and the Philippines, north to the Yellow Sea and southern Japan. In Australia, mass strandings occur relatively often, about once every 2.5 years since 1970, on the average involving about 100 individuals. Mass strandings have also been reported from Tanzania and Sri Lanka. In ancient times, false killer whales were hunted for their ivory in the Arabian Sea. Although not reported yet, the species might be taken in the subsistence hunt in Indonesia. Incidental catches have been reported for South Africa, India, Sri Lanka, and northern Australia.

Killer whale, *Orcinus orca*

The cosmopolitan killer whale occurs throughout the Indian Ocean, including the Gulf of Aden and Red Sea, and in the West Pacific. In the Indian Ocean there are records for all months and all latitudes. The existence of three migratory stocks in the southern Indo-Pacific has been postulated: near the coast of Africa, off the west coast of Australia, and in oceanic waters of the central Pacific. Their distribution may be linked to the movements of rorquals, particularly minke whales, on which killer whales occasionally feed.

Usually, killer whales were only secondary targets of whalers, but some have been taken by whalers operating from Durban in South Africa and probably by Soviet-Russian pelagic whaling operations in the 1970s, although the localities of the catches were not reported to the IWC. Killer whales are caught in the whale fishery at Lembata Island, Indonesia, where the landing of 24 animals was recorded between 1960 and 1994. Catches in net fisheries, though rare, have been reported for Sri Lanka.

Pilot whales, genus *Globicephala*

Two species are recognized, with largely parapatric distributions. The short-finned pilot whale *G. macrorhynchus* mainly occurs in tropical and subtropical waters. In the Indo-West Pacific it occurs from South Africa to the Red Sea and Gulf of Oman, east to Australia, the Indo-Malayan Archipelago, the Philippines, and north to Japan. Mass strandings have been reported for India, the Indo-Malayan Archipelago, and Australia. The long-finned pilot whale *G. melas* has a bipolar distribution in temperate waters. Southern animals are distinguished as *G. m. edwardii*. In the Indo-West Pacific, its northern limits overlap the range of the short-finned pilot whale off the Cape Province in South Africa and off southern Australia; occasionally, the species strays further north.

Short-finned pilot whales are taken in coastal fisheries off Sri Lanka and Pakistan. They are caught in subsistence whale fisheries in Indonesia and the Philippines.

Irrawaddy dolphin, *Orcaella brevirostris*

This species has distinct riverine, estuarine, and coastal populations. It is distributed discontinuously in muddy, shallow coastal waters of the Bay of Bengal, Strait of Malacca, the Indo-Malayan Archipelago, Gulf of Thailand, and northern Australia. It enters the systems of the Ganges/Brahmaputra, Irrawaddy, Mekong, and several other rivers, with populations living as far as about 1300 km upstream in the Irrawaddy and 1000 km up the Mekong/Sekong. Along the Asian mainland, the Irrawaddy dolphin occurs in northeastern India, Bangladesh, Burma (Myanmar), Thailand, Malaysia and Singapore, Cambodia, Laos, and Vietnam. In the Indo-Malayan Archipelago it has been found on northeastern Sumatra, the Seribu Islands, Java, and in many places on Borneo: Kendawangan, Kumai Bay, and the Barito River in the south, the Kajan and Mahakam Rivers and their major tributaries and lakes in the eastern, central, and northern part of the island, and from Sarawak, Brunei, and Sabah in the northwest. In 1978 the population in the Mahakam River was estimated to number at least 100–150 animals, whereas an unpublished account from 1993 reported only 68 individuals. A 1997 survey in the Mahakam River did not give a population estimate (Kreb, 1999), but in 1999 the numbers in this river were estimated at not more than 50. The species has further been found in southwestern Sulawesi and in northwestern and southern New Guinea and the Philippines. In northern Australia it ranges from Point Cloates in Western Australia to Gladstone in Queensland. Incidental catches are known for many areas.

G. Porpoises, Family Phocoenidae

Finless porpoise, *Neophocaena phocaenoides*

The finless porpoise is distributed along a narrow band of shallow water along the coasts of southern and eastern Asia. The species occurs in inshore waters, mangrove zones, and delta areas, including the lower reaches of the major river systems such as the Indus, Ganges/Brahmaputra, and Mekong. Three subspecies are reasonably well differentiated: *N. p. phocaenoides* occurs from the Persian Gulf to the South China Sea and southern part of the East China Sea, and in the Indo-Malayan Archipelago east to Java; *N. p. sunameri* is found along the coast of northeastern China, Korea, and southern Japan. The only population exclusively inhabiting fresh water, *N. p. asiaeorientalis*, occurs in China in the lower and middle course of the Yangtze River and adjacent lakes, ranging over an area of almost 1670 km. Some populations seem to be seriously depleted, mainly due to incidental mortality in fishing gear and to habitat degradation.

H. Dugongs, Family Dugongidae

Dugong, *Dugong dugon*

Dugongs live in tropical and subtropical coastal waters of the Indo-West Pacific. Two subspecies are recognized: *D. d. hemprichii* occurs throughout the Red Sea and *D. d. dugon* is distributed discontinuously from Mozambique to the Gulf of Aden and from the Persian Gulf east to Australia. Its range includes many islands in the western and northern Indian Ocean, the Indo-Malayan Archipelago, the Philippines, north to southern China, Taiwan (where probably extinct), and the Ryukyu Islands, east to Guam, Palau, Yap, Pohnpei, the Bismarck Archipelago, Solomon Islands, Vanuatu, New Caledonia, and, as a vagrant, to Fiji. The waters of Papua New Guinea and northern Australia are the most important stronghold for the species. Dugongs live in areas where there are large quantities of seagrasses (families Potamogetonaceae, Zosteridae, and Hydrocharitaceae). In many areas, the species has been reduced to widely separated relict populations, mainly by overhunting. Although the dugong has become protected over most of its range, direct hunting for food and indirect catches in fishing gear are still substantial in many areas (east Africa, India, Sri Lanka, the Indo-Malayan Archipelago, the Philippines and Australia), although few data on numbers taken are available.

For a review of work in the Indian Ocean, see Leatherwood and Donovan (1991). A compilation of recent publications and information on cetaceans in southeast Asia is found in Perrin *et al.* (1996) and Smith and Perrin (1998).

See Also the Following Articles

Endangered Species and Populations ■ North Pacific Marine Mammals ■ Ocean Environment

References

Andersen, M., and Kinze, C. C. (1999). Annotated checklist and identification key to the whales, dolphins, and porpoises (order Cetacea) of Thailand and adjacent waters. *Nat. Hist. Bull. Siam Soc.* **47,** 27–62.

Balance, L. T., and Pitman, R. L. (1998). Cetaceans of the western tropical Indian Ocean: Distribution, relative abundance, and comparisons with cetacean communities of two other tropical ecosystems. *Mar. Mam. Sci.* **14,** 429–459.

Baldwin, R. M., Gallagher, M., and Van Waerebeek, K. (1999). A review of cetaceans from waters off the Arabian Peninsula. *In* Fisher, M., Ghazanfar, S. A., and Spalton, A. (eds.). "The Natural history of Oman," pp. 161–189. Backhuys Publishers, Leiden.

Hale, P. T., Barreto, A. S, and Ross, G. J. B. (2000). Comparative morphology and distribution of the *aduncus* and *truncatus* forms of bottlenose dolphin *Tursiops* in the Indian and Western Pacific Oceans. *Aquatic Mammals* **26,** 101–110.

Kreb, D. (1999). Observations on the occurrence of Irrawaddy dolphin, *Orcaella brevirostris*, in the Mahakam River, East Kalimantan, Indonesia. *Z. Säugetierkunde* **64,** 54–58.

Leatherwood, S., and Donovan, G. P. (eds.) (1991). "Cetaceans and Cetacean Research in the Indian Ocean Sanctuary." Marine Mammal Technical Report Number 3, UNEP, Nairobi.

Leatherwood, S., and Reeves, R. R. (eds.) (1989). Marine mammal research and conservation in Sri Lanka 1985–1986. Marine Mammal Technical Report Number 1, UNEP, Nairobi.

LeDuc, R. G., Perrin, W. F., and Dizon, A. E. (1999). Phylogenetic relationships among the delphinid cetaceans based on full cytochrome B sequences. *Mar. Mam. Sci.* **15,** 619–648.

Mikhalev, Y. A. (1997). Humpback whales *Megaptera novaeangliae* in the Arabian Sea. *Mar. Ecol. Prog. Ser.* **149,** 13–21.

Perrin, W. F., Dolar, M. L. L., and Alava, M. N. R. (eds.) (1996). "Report of the Workshop on the Biology and Conservation of Small Cetaceans and Dugongs of Southeast Asia." East Asian Seas Action Plan, UNEP, Bangkok.

Perrin, W. F., Dolar, M. L. L., and Robineau, D. (1999). Spinner dolphins (*Stenella longirostris*) of the western Pacific and Southeast Asia: Pelagic and shallow-water forms. *Mar. Mam. Sci.* **15,** 1029–1053.

Pitman, R. L., Palacios, D. M., Brennan, P. L. R., Brennan, B. J., Balcomb III, K. C., and Miyashita, T. (1999). Sightings and possible identity of a bottlenose whale in the tropical Indo-Pacific: *Indopacetus pacificus?* *Mar. Mam. Sci.* **15,** 531–549.

Rice, D. W. (1998). "Marine Mammals of the World: Systematics and Distribution." Special Publication Number 4, The Society for Marine Mammalogy, Lawrence KS.

Rudolph, P., Smeenk, C., and Leatherwood, S. (1997). Preliminary checklist of Cetacea in the Indonesian Archipelago and adjacent waters. *Zool. Verh. Leiden* **312,** 1–48.

Smith, B. D., and Perrin, W. F. (eds.) (1998). "Asian Marine Biology 14." Hong Kong Univ. Press, Hong Kong.

Van Waerebeek, K., Gallagher, M., Baldwin, R., Papastavrou, V., and Al-Lawati, S. M. (1998). Morphology and distribution of the spinner dolphin, *Stenella longirostris*, rough-toothed dolphin, *Steno bredanensis* and melon-headed whale, *Peponocephala electra*, from waters off the Sultanate of Oman. *J. Cetacean Res. Manage.* **1,** 167–177.

Wang, J. Y., Chou, L.-S., and White, B. N. (1999). Mitochondrial DNA analysis of sympatric morphotypes of bottlenose dolphins (genus: *Tursiops*) in Chinese waters. *Molecular Ecology* **8,** 1603–1612.

Wang, J. Y., Chou, L.-S., and White, B. N. (2000a). Osteological differences between two sympatric forms of bottlenose dolphins (genus *Tursiops*) in Chinese waters. *J. Zool., Lond.* **252,** 147–162.

Wang, J. Y., Chou, L.-S., and White, B. N. (2000b). Differences in the external morphology of two sympatric species of bottlenose dolphins (genus *Tursiops*) in the waters of China. *J. Mammalogy* **81,** 1159–1165.

Zhou, K., Leatherwood, S., and Jefferson, T. A. (1995). Records of small cetaceans in Chinese waters: A review. *Asian Mar. Biol.* **12,** 119–139.

Infanticide and Abuse of Young

CLAUDIO CAMPAGNA
Centro Nacional Patagónico,
Puerto Madryn, Argentina

The killing and abuse of young by conspecifics is a widespread phenomenon. Parental and nonparental infanticide have been reported in almost 100 species of mammals, most of which are terrestrial (Hausfater and Hrdy, 1984; Parmigiani and vom Saal, 1994). Infant killing can be the direct outcome of a violent interaction or can result from the indirect neglect of a young or an accident. This article focuses on violent, nonparental forms of infanticide in aquatic mammals. Parental killing in this group is apparently restricted to the indirect effects of maternal neglect (see Le Boeuf and Campagna, 1994) and will not be treated here. Infant abuse is a much more common behavioral occurrence than infanticide. It may imply active violence or passive neglect, and it does not necessarily involve the intended death of the victim. Death in the context of abuse is usually perceived as accidental, a byproduct that often follows a process of infection and starvation (Le Boeuf and Campagna, 1994). Infant or young refers to a lactating or recently weaned pup, calf, or cub.

Except for otariids and phocids, data on killing and abusing young are sparse for aquatic mammals. Infanticide is an event that may pass unobserved or unreported. Spotty research coverage, with some species being well known and others virtually unstudied, suggests that the relevance and diversification of abuse and killing of young may be more widespread than reported here. Explanations of the well-documented cases of abuse and infanticide in aquatic mammals rarely support the adaptive hypotheses that would account for similar episodes in terrestrial species.

I. Abuse and Killing of Young by Males

Violent behavior toward young was described in four out of seven sea lion species (with the closely related *Zalophus californianus, Z. japonicus* and *Z. wollebacki* being the exceptions). Subadult and juvenile males of the South American sea lion, *Otaria flavescens,* abduct (seize), abuse, and kill pups during the breeding season (Campagna *et al.,* 1988). The BEHAVIOR was observed in coastal Patagonia (Campagna *et al.,* 1988), Uruguay (Vaz Ferreira, 1965), Chile (H. Paves Hernandez and C. Espinoza, personal communication), Peru (Harcourt, 1993; P. Majluf and K. Soto, personal communication), and the Falkland Islands (C. Duck and D. Thompson, personal communication). At Punta Norte, Península Valdés, Argentina, 163 successful abductions were recorded in four breeding seasons. More than 20% of the 400 pups born each season were abducted by males. In a typical abduction, a juvenile or subadult male approached the breeding area alone or as part of a group raid (Campagna *et al.,* 1988), dashed toward a pup, and grabbed it. The pups were then abducted away from the breeding group and some were carried out to sea (11% of the abductions), whereas others were released and held close to the abductor. Pups that attempted to escape were shaken violently from side to side, tossed in the air, held crushed against rocks, or submerged. Males defended their abducted pup from other males. Some abductors mounted pups, fully covering them with their massive bodies. About 6% of the pups abducted and 1.3% of the pups born during a season died as a consequence of physical abuse. Dead pups showed tooth puncture wounds and extensive hematomas.

Australian (*Neophoca cinerea*) and Hooker's (*Phocarctos hookeri*) sea lions abduct and abuse pups in a similar fashion described for *O. flavescens,* with the important difference that adult Hooker's cannibalize the killed pups. Adult male Australian sea lions grab pups that may be alone or with the mother and bite, shake, and toss them several times (Higgins and Tedman, 1990). Eight attacks recorded in two breeding seasons resulted in four dead pups (5% of the pups observed) and accounted for 19% of pup mortality in the rookery (Higgins and Tedman, 1990). Adult and subadult male Hooker's sea lions grab pups by the neck, violently thrash them from side to side, and sometimes carry them out to sea and drown them (Wilkinson *et al.,* 2000). Adult abductors were also observed eating pups. Opportunistic observations on Hooker's sea lions report males abducting pups on two occasions and eating them on nine occasions (Wilkinson *et al.,* 2000). After thrashing the victim repeatedly from side to side, males bit the flesh off the carcass and consumed it. This is the only otariid species for which cannibalism has been described. Immature males do not apparently kill pups, although they may try to keep them under control and occasionally mount them.

Steller sea lions, *Eumetopias jubatus,* may trample or crush pups or push them over a cliff as an indirect consequence of territorial disputes. In some instances, however, pups are killed as a direct violent action by males (B. Porter, personal communication).

Episodes of violent behavior toward pups are rare or absent among fur seals. Juvenile male northern fur seals, *Callorhinus ursinus,* occasionally abduct conspecific pups in a context that suggests a form of mate substitution (R. Gentry, personal communication). Male Antarctic fur seals, *Arctocephalus gazella,* rarely respond to pups, even to the extent that they will fail to respond if they happen, apparently accidentally, to lie on top of a pup (I. Boyd, personal communication). Pups may be killed accidentally by males if they come between fighting individuals, a relatively common occurrence in otariids.

Among the other pinnipeds, infant abuse and killing were described in at least four phocids and the walrus. Male northern and southern elephant seals, *Mirounga* spp., of different age classes, kill suckling pups and weanlings (Le Boeuf and Campagna, 1994). Pups are trampled accidentally by bulls in the context of male–male competition and may then die of internal injuries. Weaned pups are abused by pubertal males that attempt to mate with them and, in the process, injure and kill individuals of both sexes (Rose *et al.*, 1991). At the time of departure, 30–50% of northern elephant seal (*M. angustirostris*) weaned pups show signs of having been mounted by a male that range from neck bites, scraps, cuts, and puncture wounds to deep gashes exposing BLUBBER and profuse bleeding. An adult southern elephant seal, *M. leonina*, male from the Patagonian colony of Península Valdés killed and apparently ate pups (J. C. López, personal communication). He grabbed weaned pups from the beach, dragged them out to sea, kept them underwater until struggling ceased, and then tore off chunks and consumed them. The cannibal returned to the same place at least during two consecutive breeding seasons and killed dozens of weanlings. Male gray seals, *Halichoerus grypus*, occasionally shake, toss, bite, mount, and kill pups (D. Boness and P. Pomeroy, personal communication). There is also evidence of cannibalism in this species (Bédard *et al.*, 1993; Kovacs, 1996). An adult male was involved in the killing and eating of pups during three breeding seasons. In a similar modality to the southern elephant seal cannibal, the gray seal male grabbed his victims by the hind flippers, dragged them into the water, and drowned them. He later tore off chunks of the pup's body with a biting–shaking action and consumed the blubber, skin, and muscle. Hawaiian monk seals (*Monachus schauinslandi*) males mount pups, suffocate, or drown them (Hiruki *et al.* 1993; M. Craig, personal communication). Some individuals persist in this behavior and may kill many pups. Finally, adult male, female, and immature walruses, *Odobenus rosmarus*, can jab a pup with their tusks and cause lethal injuries.

In summary, adult, subadult, and juvenile males of several pinniped species injure, abuse, or kill suckling and recently weaned pups in the following contexts: (a) accidentally, often as an indirect outcome of trampling and crushing during dominance, female defense, and territorial disputes; (b) as a direct or indirect consequence of misdirected sexual assault, such as during abductions and abuse by pubertal males; (c) as a direct or indirect consequence of misdirected aggressive behavior with no clear sexual component, such as attack of pups by territorial males not associated to mounting, herding, or harassing; and (d) as an apparent source of food (cannibalism). The age class involved in the abuse and killing varies with the species. In Australian and Hooker's sea lions, adults are the most aggressive toward pups, but subadult and juvenile males also sequester pups and engage in biting, mounting, and holding them underwater. In the South American sea lion, subadult and juvenile males do most of the abductions; adults are rarely involved in pup abuse. Among phocids, young males seem to be involved more often

in abuse than adults; adults may cause pup death or injury as an epiphenomenon of male–male competition.

Reports of violence toward young in the rest of the aquatic mammals are rare. Male polar bears, *Ursus maritimus*, occasionally kill and eat cubs, a behavior that is apparently generalized throughout the Arctic (Taylor *et al.*, 1985). Indirect evidence suggests infanticide in the bottlenose dolphin, *Tursiops truncatus* (Patterson *et al.*, 1998). Stranded dolphin calves were found with internal injuries that included contusions around the head and thorax, bone fractures, and lacerated organs compatible with violent behavior. The interactions that may have caused the death of the calves were not observed. However, an adult dolphin was observed to interact violently with a dead conspecific calf, and dolphins were also seen to chase and hit harbor porpoises, *Phocoena phocoena*, hard enough to toss them into the air. Stranded harbor porpoises had evidence of trauma similar to that reported for the stranded dolphin calves. Additional indirect evidence of conspecific killing in *T. truncatus* is available for a population of the southeast Virginia coast (D. Dunn, personal communication). Nine bottlenose dolphins within their first year of life, thus still dependent on their mothers, stranded with multiple skeletal fractures, hematomas, organ lacerations, contusions, and hemorrhages, indicating multidirectional trauma. External signs of trauma were absent, an observation incompatible with predation, boat strike, and fisheries interactions, but similar to antemortem injuries reported for harbor porpoises and dolphins (Patterson *et al.*, 1998).

II. Abuse and Killing of Young by Females

Adult pinniped females repel alien young in the context of aggressive protection of resources intended for their own pup. In the northern elephant seal, females aggressively reject alien pups that approach them (references in Le Boeuf and Campagna, 1994). They shake, throw, and viciously bite unrelated pups. Attacks may be violent enough to cause extensive wounds or fractures, with subsequent infection and death. Orphans attempt to nurse from any female, thus being particularly vulnerable to attack and injury.

An unusual behavior involving females and resulting in the death of unrelated pups was described for the South American sea lion rookery at Islas Ballestas (Peru; K. Soto, personal communication). During the 1997–1998 El Niño breeding season, virtually all pups born starved to death. The following year, only about one-quarter of the females gave birth. These mothers had to defend their newborn pups from the sustained attempts of neighboring females to abduct the latter. It often occurs that otariid females close to parturition attempt to bring alien pups near them. However, the particular breeding context of the post-El Niño year resulted in an unusually high incidence of a behavior that may be related to confusing alien pups with their own pup. Abductions occurred at a rate of one pup every 2.7 hr of observation. Females did not nurse the abducted pups, which were later abandoned. Those that failed to reunite with their mother died from starvation or were killed by young males. Almost 300 female abductions resulted in 11 pup deaths, and the incidence of pup mortality due to male abductions increased from the regular 1.5 to 8.0% of the pups born.

III. Male Violent Behavior toward Mature Females and Interspecific Pups

From a behavioral standpoint, abuse and killing of conspecific young by male pinnipeds resemble male violent behaviors directed toward mature females of the same species and toward females and pups of other species. Attributes that allow males to physically overpower competitors would also promote aggressive sexual behaviors related to achieving access and maintaining control of breeding females. For example, adult and subadult *O. flavescens* males abduct females from established harems (Campagna *et al.*, 1988). Abductions involve grabbing, tossing, herding, mounting, and biting. Some females are badly injured and killed in the process. Male harassment of conspecific females may be relatively common in phocids; it has been reported for both species of elephant seals: the Hawaiian monk seal and the gray seal (Mesnick and Le Boeuf, 1991; Hiruki *et al.*, 1993; Boness *et al.*, 1995).

Strong and large pinniped males with an indiscriminate sexual urge often injure and kill females of other species. Males killing interspecific females during mating attempts were reported in all sea lions (Miller *et al.*, 1996). *O. flavescens* males kill *A. australis* females, and *E. jubatus* kill *Z. californianus* females and even males. Mating attempts with dead females of the same and of a different species occur in some otariids and phocids, such as the South American sea lion and the elephant seal. Abnormal escalation of aggressive sexual behaviors may lead to instances such as a Steller sea lion male killing at least 84 California sea lion females and 12 males over three seasons (see references in Miller *et al.*, 1996) or a southern elephant seal male killing more than 100 *A. pusillus* breeding female over successive breeding seasons (Best *et al.*, 1981).

Sea lion predation of pups of other otariid species typically involves grabbing a pup by the neck, shaking it from side to side, tossing and recovering, dragging it to sea, submerging and drowning, biting off flesh, and consuming it. It has been described for at least three species. Steller sea lion prey on northern fur seal neonates (pups under 5 months of age; Gentry and Johnson, 1981). Adult South American sea lions prey on South American for seal *Arctocephalus australis,* pups (Harcourt, 1993). Hooker's sea lions, a species for which cannibalistic behaviors have been described (Wilkinson *et al.*, 2000), prey on New Zealand fur seals, *A. forsteri,* and on Antarctic and subantarctic fur seals, *A. tropicalis.*

In the South American sea lion, interspecific predation and conspecific abuse may be particulary associated. Juvenile and subadult sea lion males abduct and kill *A. australis* pups but do not consume them. Interspecific pup abduction was observed in Peru (Harcourt, 1993) and in Uruguay (Vaz Ferreira and Bianco, 1987), where sea lion and fur seals live sympatrically. Males grab a fur seal pup, take it to a neighboring beach and toss and shake the pup. However, instead of killing and eating the pup, as adult male behaving as predators would do, these younger males defend them from other sea lions, mount them, and behave as they do with pups of their own species. Fur seal pups may be killed as an indirect consequence of violent treatment, but are not consumed by their abductors.

In summary, pup killing in some species (e.g., Steller sea lions) is more common in the context of interspecific predation, whereas in others (e.g., South American and Australian sea lions) it occurs more often in a sexual or aggressive social context. In general terms, pup abuse follows a similar pattern as female abuse, with the most aggressive species toward pups being also violent toward conspecific and interspecific females.

IV. Adaptive Meaning of Infant Abuse and Killing

Several hypotheses have been proposed to account for infanticide in terrestrial mammals (Hausfater and Hrdy 1984): (1) exploitation or predation, young are killed for nutritional benefits; (2) resource competition, adults kill unrelated young to increase access to food or breeding space for themselves or their offspring; (3) sexual selection, males kill unrelated offspring to achieve reproductive access to females; (4) parental manipulation, a parent reduces litter size by eliminating all or part of a litter; and (5) social pathology, a maladaptive behavior. Adaptive explanations for the killing of young in aquatic mammals have been suggested for bottlenose dolphins (sexual selection; Connor *et al.*, 1996) and Hooker's sea lion (cannibalism; Wilkinson *et al.*, 2000).

Cannibalism is exceptional among aquatic mammals, and social pathology would be involved in cases such as the cannibal adult male gray seal and the subadult male southern elephant seal described earlier. However, cannibalism in *P. hookeri* was suggested to fit the food resource hypothesis. Several males kill and eat pups in a fashion similar to that described for the same species preying on fur seals. Pups are easy targets for males and may supply calories in excess of the daily energy requirement of a male, as has been suggested to explain the predation of *O. flavescens* on *A. australis* (Harcourt, 1993). Cannibalism in polar bears appears to occur as carrion feeding and as attacks by males on cubs. There is also evidence of a polar bear male feeding on an adult female (I. Stirling, personal communication), but this is a rare observation of difficult interpretation.

Most instances of infant abuse and killing in pinnipeds are better understood as epiphenomena of indiscriminate sexual and aggressive behaviors (Le Boeuf and Campagna, 1994). Social context, SEXUAL DIMORPHISM, and sexually selected behaviors would set the context for the occurrence of injury and death of young. Pinniped colonies are often dense, parental investment is limited to females, males are large relative to pups and females, and male movements are frequent in the vicinity of pups. At times during the breeding season, pups may be the most abundant age class in a rookery, increasing the opportunity of social interaction. Reproductive females are aggressive toward conspecifics in general and alien pups in particular. Female aggressive behavior in this context would be explained by the cost of producing milk for individuals that are fasting while nursing. A large proportion of the breeding males do not have sexual access to females and males have an indiscriminate sexual behavior. Pups, particularly those close to being weaned, may be almost as large as young mature females and are often confused as females. South American sea lion and northern and southern elephant seal males would kill pups in the context of misplaced sexual behavior. Abducted *O. flavescens* pups, for example, are treated as female substitutes, perhaps a practice of herding or harem keeping (Campagna *et al.*, 1988). Pups born in a harem are not likely to be the offspring of the dominant male, as they were sired the previous season. Behavioral mechanisms that can protect pups from direct and indirect violence (e.g., being crushed during male fights) would not then be under selective pressure. Infanticide in the Australian sea lion would be the consequence of misdirected aggression. It was suggested that territorial males may perceive pups as a threat. After killing a pup, males return to their usual position in the territories (Higgins and Tedman, 1990).

It is not yet clear to what extent the abuse and killing of conspecific pups may have on a common evolutionary substrate with violent behaviors directed to mature females of the same or other species and toward young of other species. Examples among otariids suggest that a circular gradation may exist from simple predation to infanticide to cannibalism. Steller sea lions kill pups of other species as predators but rarely or never abuse conspecific pups as abductors; South American sea lions prey on pups of other species when adults and abduct (but do not eat) pups of the same and other species when young; and Hooker's sea lions abduct, abuse, kill, and eat conspecific and interspecific pups. It remains to be determined if this progression is actual or deceptive. It is possible, however, that in the behavioral similarities among these phenomena may underlay a key to understanding the evolutionary origin of abuse.

See Also the Following Articles

Aggressive Behavior ■ Predation on Marine Mammals

References

Bédard, C., Kovacs, K., and Hammill, M. (1993). Cannibalism by grey seals, *Halichoerus grypus,* on Amet Island, Nova Scotia. *Mar. Mamm. Sci.* **9,** 421–424.

Best, P. B., Meyer, M. A., and Weeks, R. W. (1981). Interactions between a male elephant seal *Mirounga leonina* and Cape fur seals *Arctocephalus pussilus. S. Afr. J. Zool.* **16,** 59–66.

Boness, D. J., Bowen, W. D., and Iverson, S. J. (1995). Does male harassment of females contribute to reproductive synchrony in the grey seal by affecting maternal performance? *Behav. Ecol. Socibiol.* **36,** 1–10.

Campagna, C., Le Boeuf, B. J., and Cappozzo, H. L. (1988). Pup abductions, and infanticide in southern sea lions. *Behaviour* **107,** 44–60.

Connor, R. C., Richards, A. F., Smolker, R. A., and Mann, J. (1996). Patterns of female attractiveness in Indian Ocean bottlenose dolphins. *Behaviour* **133,** 37–69.

Gentry, R. L., and Johnson, J. H. (1981). Predation by sea lions on Northern fur seal neonates. *Mammalia* **45,** 423–430.

Harcourt, R. (1993). Individual variation in predation on fur seals by southern sea lions (*Otaria byronia*) in Peru. *Can. J. Zool.* **71,** 1908–1911.

Hausfater, G., and Hrdy, S. B. (1984). "Infanticide: Comparative and Evolutionary Perspectives." Aldine Press, New York.

Higgins, L. V., and Tedman, R. A. (1990). Effect of attacks by male Australian sea lions, *Neophoca cinerea,* on mortality of pups. *J. Mammal.* **71,** 617–619.

Hiruki, L. M., Gilmartin, W. G., Becker, B. L., and Stirling, I. (1993). Wounding in Hawaiian monk seals (*Monachus schauinslandi*). *Can. J. Zool.* **71,** 458–468.

Kovacs, K. (1996). Grey seal cannibalism. *Mar. Mamm. Sci.* **12,** 161.

Le Boeuf, B. J., and Campagna, C. (1994). Protection and abuse of young in pinnipeds. *In* "The Protection and Abuse of Young in Animals and Man" (S. Parmigiani and F. vom Saal, eds.), pp. 257–276. Harwood Academic, Chur, Switzerland.

Mesnick, S. L., and Le Boeuf, B. J. (1991). Sexual behavior of northern elephant seals. II. Female response to potentially injurious encounters. *Behaviour* **117,** 262–280.

Miller, E., Ponce de León, A., and DeLong, R. L. (1996). Violent interspecific sexual behavior by male sea lions (Otariidae): Evolutionary and phylogenetic implications. *Mar. Mamm. Sci.* **12,** 468–476.

Parmigiani, S., and vom Saal, F. (eds.) (1994). "The Protection and Abuse of Young in Animals and Man." Harwood Academic, Chur, Switzerland.

Patterson, I. A. P., Reid, R. J., Wilson, B., Grellier, K., Ross, H. M., and Thompson, P. M. (1998). Evidence for infanticide in bottlenose dolphins: An explanation for violent interactions with harbour porpoises? *Proc. R. Soc. Lond. Biol. Ser.* **265,** 1167–1170.

Rose, N. A., Deutsch, C. J., and Le Boeuf, B. J. (1991). Sexual behavior of male northern elephant seals. III. The mounting of weaned pups. *Behaviour* **119**(3-4), 171–192.

Taylor, M. K., Larsen, T., and Schweinsburg, R. E. (1985). Observations of intraspecific aggression and cannibalism in polar bears (*Ursus maritimus*). *Arctic* **38,** 303–309.

Vaz-Ferreira, R. (1965). Comportamiento antisocial en machos subadultos de *Otaria hyronia* ("lobo marino de un pelo"). *Rev. Facult. Human. Ciencias. Montevideo* **22,** 203–207.

Vaz Ferreira, R., and Bianco, J. (1987). Acciones interespecificas entre *Arctocephalus australis y Otaria flavescens. Rev. Museo Argent. Ciencias Nat. Zool.* **14,** 103–110.

Wilkinson, I. S., Childerhouse, S. J., Duignan, P. J., and Gulland, F. M. D. (2000). Infanticide and cannibalism in the New Zealand sea lion, *Phocarctos hookeri. Mar. Mamm. Sci.* **16**(2), 494–500.

Intelligence and Cognition

BERND WÜRSIG

Texas A & M University, Galveston

Dolphins and sea lions are wonderful crowd pleasers in oceanaria: they leap, toss balls, swim through hoops or other obstacles, and vocalize on demand. In nature, they race toward boats, surf in the bow wave, and perform amazing acrobatics for—it seems—the pure joy of it. They are highly social, communicate, enjoy contact with humans, and appear to spend much of their time playing. It is therefore easy to understand why one of the most common questions asked by nonmarine mammal researchers is: "They are very intelligent, are they not?" This question is an excellent one, for it forces us to attempt to analyze what we mean by intelligence and how marine mammals might fit our definition of the concept.

Intelligence and cognition go hand in hand. The former refers to the mental capabilities of a human or nonhuman animal and usually is described by assessing problem-solving skills. The latter refers to the information processing within the animal and may be inferred by an analysis of how it appears to plan an action or alter it based on past experience. A "more intelligent" animal responds to an environmental stimulus faster or more accurately than the "less intelligent" one; the "more cognitive" action or animal may indicate more insight and more awareness of the problem than the "less cognitive" one. Unfortunately, past determinations of the concepts tended to be biased by our own human problem-solving skills and sensory systems and, to large degree, still are. However, we now know that indicators of intelligence can even be very different for different human societies or cultural backgrounds, i.e., within species. Can we say that the nature-living Australian aborigine who scores very low on an "intelligence" test designed with problem-solving questions of our modern industrial/electronic society is less intelligent than the student who takes the test in the industrialized world? If we answer "yes," we should be forced to "take the test" on the Aborigine's terms, perhaps by coming up with solutions of survival in the alternately extremely hot and cold, rugged, and food-poor outback. Similarly, it is not reasonable to study intelligence in dolphins and sea lions by asking them to solve problems relative to our linguistic communication or hand manipulation skills (in cognitive psychology, this is called the comparative approach). It is also unreasonable to compare "intelligences" of river dolphins with those of oceanic species by asking them to solve the same problems of space or objects.

An alternative to the comparative approach of describing intelligence and cognition is often called the "absolute method." It involves an attempt to find out how an animal thinks about things. Thinking is defined as mental manipulation of the internal representation of the external world, the stimulus. The cognitive animal is influenced to change its internal manipulations in part by past experience, and the more adept animal does this better than the "less intelligent" one. While it is difficult to judge mental processes, approximate tests and observations to do so have been devised and will be described later on.

One important window into intelligence and cognition for social species (and all marine mammals show a reasonable to very high level of sociality) is certainly communication. The individuals and species that communicate among each other in sophisticated and at times novel, interactive ways are likely the "more intelligent" (by, in this case, the prime criterion of communication) than those whose communication may be structured more rigidly or less complicated. The great U.S. ethologist Donald Griffin has argued persuasively that communication is a major "window into the mind," not only of humans, dolphins, and other mammals, but of ants and honeybees as well (Griffin, 1981). He went on to postulate that it may be more parsimonious to explain the dance language of bees by considering them to be aware of their actions than it is to consider them reacting to complicated chains or sets of stimuli in unthinking ("noncognitive") fashion. This intriguing idea is not yet widely accepted by behavioral researchers and cognitive ethologists. However, most researchers now accept the possibility of "intelligences" and cognition in nonhuman animals, potentially very different in operating modes from our own, and not testable by traditional comparative approaches.

I. Brain Size and Characteristics

A BRAIN is needed to think and to have the chance of being aware (as a modern book, we need to mention the "brain" of artificially intelligent computers as well). Within a particular taxonomic group, larger and more complex brains tend to show a crude relationship to greater flexibility of behavior, adaptiveness to novel situations, and communication skills, i.e., intelligence. The relationship is imperfect, however, and is notoriously difficult to measure. For example, the entire brain has usually been used for descriptions of size and relative complexity, but there are motor, body function, and sensory parts of the brain that have very little to do with storing, processing, and integrating aspects of memory and thought (the latter occur only in the cerebrum).

Large mammals tend to have larger brains than small ones so brain size to body size ratios have been devised. One of these is the encephalization quotient (EQ), championed by Jerison (1973) and accepted by many researchers, albeit with often slightly different forms of calculation. The EQ is the ratio of brain mass observed to the brain mass predicted from an allometric equation of brain mass/body mass ratio of mammals as a whole. Therefore, an EQ of 1 means that the animal has an "average" brain size. It has been found for terrestrial mammals that EQs tend to be higher for those species that have few offspring, delayed physical and sexual maturity, long parental care, and generally high behavioral complexity (as estimated by degree of sociality and amount of behavioral flexibility). Examples are primates and social carnivores such as cats and canids. Within the primates, EQs tend to be higher for those in the categories just mentioned than for others, demonstrating that meaningful life history–brain size comparisons can be made at least in that group. Some aspects of general intelligence appear to be correlated with those higher EQs, from tree lemurs at the low end of the scale to the great apes at the pinnacle. Nevertheless, the very concept of EQ represents a general statement for potential comparison within or between taxa, but does not represent a fundamental phenomenon per se.

Polar bears (*Ursus maritimus*), sea otters (*Enhydra lutris*), and pinnipeds have EQs around 1, as predicted by the overall regression line of brain weight to body weight among mammals. Their brains tend to weigh between 0.1 and 0.3% of their bodies. In other words, there is nothing unusual in brain size of these mammals relative to their terrestrial carnivore cousins. Because brains are energetically expensive, it has been postulated that those of pinnipeds that dive to great depths and hold their breaths for long periods of time might be smaller. At first glance, this appears to be the case for such divers as Weddell (*Leptonychotes weddellii*) and elephant seals (*Mirounga* spp.), but an analysis by Worthy and Hickie (1986) showed that brain size and dive capability have no clear relationship.

Dolphins and whales have large brains but not all have large brain to body weights or EQs. The sperm whale has the largest brain on earth, weighing about 8 kg. This brain is in a body that weighs about 37,000 kg, however. The brain is only 0.02% of the weight of the body, or one-fifth of the size ratio of the smallest-brained pinnipeds. However, at large sizes, a straight-line allometric comparison is probably not fair by any measure, and perhaps the body of the sperm whale (*Physeter macrocephalus*) simply does not need relatively much brain mass for muscle movement, skin sensation, visceral action, and so on. The common bottlenose dolphin (*Tursiops trumcatus*), however, has a brain weighing 1.6 kg in a body that weighs about 160 kg, making it—at 1% of body weight—one of the largest relative brains on earth. This competes only with several other dolphins, great apes, and humans (whose brains are about 1.5 kg in a 65-kg body, or about 2.3%) (Table I).

Baleen whales, like sperm whales, have large absolute brains (about 7 kg in an 80,000-kg fin whale, *Balaenoptera physalus*), but none have brain to body weight ratios as large as even the relatively small ones of the sperm whale. Sirenians have neither absolute nor relatively large brains, with the Caribbean manatee (*Trichechus manatus*) having a 300-g brain in a 750-kg body (0.04% of body weight). It has been postulated that the sirenian, a herbivore, increased body size to house a large gut for processing low-energy food, and a concomitant increase in brain size was not needed to support this size. Similarly, the huge size of baleen whales allows them to have huge mouths and to fast for extended periods. Again, this is a very different

allometric growth than that of a cow, for example, that is "simply" scaled up in size from a sheep.

Brain weight/body weight relationships are of general interest and have some relationship to relative information-processing capabilities. However, a larger absolute or relatively sized brain than that of another animal does not necessarily serve a "smarter" animal. The concept of intelligence is not a linear one; because there are so many "intelligences" depending on measure or the describer's concept of what is important, intelligence is not definable in absolute terms. All of the marine mammals have well-developed cerebrums. The brains of toothed whales have especially high amounts of neocortical folding and therefore high surface areas (Fig. 1). This quality is believed to be related to thought processes and behavioral flexibility. Whereas polar bears, sea otters, and pinnipeds show a general "terrestrial carnivore" level of folding, baleen whales and sirenians have very smooth cerebrums, with minimal surface areas. Nevertheless, the internal structure of whale and sirenian cerebrums is as well developed as those of other social mammals, and there is no reason to believe that these animals are "dumber" than others based on brain size and gross morphology. Perhaps their ways of finding and securing food, without the need of sophisticated hunting strategies as by toothed whales and carnivores, coupled with some aspects of their communication and society interactions, simply do not require the elaborate neocortical folding seen in many other mammals.

TABLE I
Brain and Body Weights of Some Marine Mammals as Compared to Humans[a]

Species	Brain weight (g)	Body weight (ton)	(Brain weight/body weight) × 100
Pinnipeds			
Otariids			
Northern fur seal (*Callorhinus ursinus*)	355	250 (male)	0.142
California sea lion (*Zalophus californianus*)	363	101	0.359
Southern sea lion (*Otaria flavescens*)	550	260	0.211
Phocids			
Bearded seal (*Erignathus barbatus*)	460	281	0.163
Gray seal (*Halichoerus grypus*)	320	163	0.196
Weddell seal (*Leptonychotes weddellii*)	550	400	0.138
Leopard seal (*Hydrurga leptonyx*)	542	222	0.244
Walrus (*Odobenus rosmarus*)	1020	600	0.170
Odontocetes			
Common bottlenose dolphin (*Tursiops truncatus*)	1600	154	1.038
Short-beaked common dolphin (*Delphinus delphis*)	840	100	0.840
Pilot whale (*Globicephala* sp.)	2670	3,178	0.074
Killer whale (*Orcinus orca*)	5620	5,448	0.103
Sperm whale (*Physeter macrocephalus*)	7820	33,596	0.023
Mysticetes			
Fin whale (*Balaenoptera physalus*)	6930	81,720	0.008
Sirenian			
Florida manatee (*Trichechus manatus latirostris*)	360	756	0.047
Human	1500	64	2.344

[a]Modified from Berta and Sumich (1999). Pinniped data from original sources listed in Bryden (1972), Spector (1956), Sacher and Staffeldt (1974), Bryden and Erickson (1976), and Vaz-Ferreira (1981); cetaceans from Bryden and Corkeron (1988); and sirenians from O'Shea and Reep (1990).

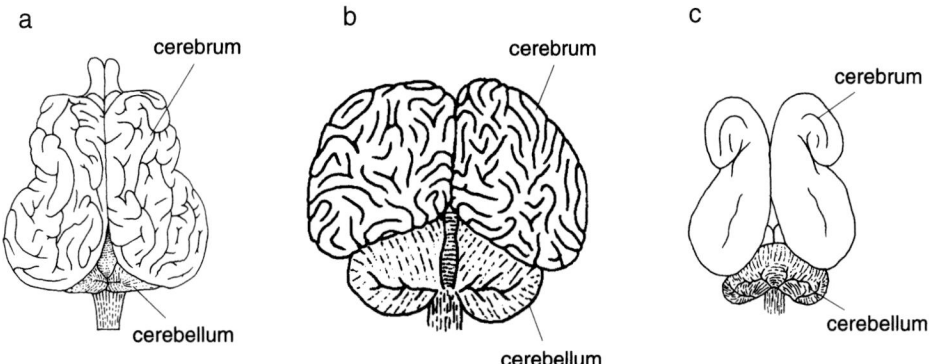

Figure 1. *Comparison of pinniped,* Otaria flavescens *(a); cetacean,* Tursiops truncatus *(b); and sirenian,* Dugong dugon *(c) brain, dorsal views. Illustrated by P. Adam. From Berta and Sumich (1999).*

While much more work on brain size and sensory capabilities needs to be done, it is known that toothed whales and dolphins, who echolocate and use sounds intensively for communication, have well-developed auditory processing lobes. Pinnipeds and especially polar bears, however, have well-developed areas for processing smell.

While brain size and complexity issues used to dominate our thinking about relative intelligence, it is becoming apparent that these can give only vague indicators of complexity of thought. It is likely that brains are structured more along lines of how an animal interacts with others and with its ecology. Higher brain function is a complex mixture of sensory inputs; processing, storing, and reactions to stimuli; innovation; and retrieval and use of previously stored events. Our inability to find clear links of these with measures of brain size and aspects of gross complexity may simply be because of the relatively primitive state of cognitive science, or it could be that clear "all-encompassing" rules of relationships simply do not exist. Promising avenues for future brain studies are noninvasive electrobiological and chemobiological studies from remote sensing of brain tissue while it is undergoing particular tasks. The findings to come from such work will make our present discussions of brain function seem very primitive indeed.

II. Learning

We know that dolphins and sea lions do marvelously complex things in captivity, but we also know that most of these behaviors have been reinforced from existing simpler ones and shaped into that dramatic leap to catch a fish. It is positive reinforcement behavior, or operant conditioning, that is at work; the animal gets a food or other reward for having done a good job. Typically, a sea lion or dolphin reward is one to three small fish per performed action. This is not unlike "training" a cat to run into the kitchen when it hears the sound of a can opener or the guppies in a home aquarium all aggregating near the top when a drawer with dried shrimp is opened. Operant conditioning can be performed on just about all animals on earth, and only speed of learning and some aspects of the amount of behavioral shaping can be indicators of a measure of "smart-

ness" or relative intelligence. The animals learn, but there is not necessarily insight to their learning.

A. Language Studies

It has long been known that dolphins have squeaks and whistles that appear to be used for communication. In captivity, bottlenose dolphins at times appear to imitate or mimic human and other sounds. These observations led an early dolphin communication researcher, John Lilly (e.g., Lilly, 1961), to attempt to communicate with dolphins by teaching them human speech. The results were a total failure, with not one clearly definable mimicked human sound; although dolphins are quite good at matching the staccato rhythm, in the form of bursts of sounds emitted in air (or underwater), of human speech. Dolphins obviously do not have the vocal apparatus to produce human speech and may not have the neural wiring for it either. Nevertheless, Lilly's association with dolphins did not stop him from postulating that dolphins have great "extraterrestrial" intelligence. He used their large brains and their purported friendliness as arguments, but could not muster communicative interactions with humans as a part of the argument. Unfortunately, his popular writings have swayed countless laypersons, and a substantial "cult" of believers in extremely high dolphin intelligence and sophisticated human–dolphin communication, even at the nonverbal extrasensory level, has evolved. No other scientists have made similar claims, but the unscientific nature of Lilly's assertions deterred many others from studying dolphin and whale communication, and early on addressing intelligence and cognition in an obviously behaviorally flexible taxonomic order of mammals. By the way, some seals and beluga whales (*Delphinapterus leucas*) do have the ability to mimic human sounds, and one now-deceased harbor seal, *Phoca vitulina*, ("Hoover") at the New England Aquarium used to delight visitors with his rendition of simple sentences mimicked from a human pool cleaner, replete with the pool cleaner's Maine harbor-side accent. This ability does not indicate greater intelligence than in other seals and toothed whales who do not mimic. Instead, the ability (generally found in male pinnipeds) may relate to the way the animals use natural sounds in order to work out dominance relations for mating access to females and for other social interactions.

Two researchers who were not scared off by the unfounded claims of John Lilly and who nevertheless began language communication research, were Lou Herman of the University of Hawaii and Ron Schusterman, now of the University of California at Santa Cruz. Their studies began in the 1970s and are still ongoing, with a cadre of graduate students and postdoctoral researchers.

Lou Herman's work represents the only truly pioneering language study conducted on dolphins. As in some of the successful studies with chimpanzees, who like dolphins also cannot utter sophisticated human sounds, Herman uses a modified form of sign language, with volunteers' arms at poolside "talking to" common bottlenose dolphins. This is thus a gestural, not vocal, language. While Herman and his team have delved into many fascinating aspects of dolphin abilities, the basic study goes somewhat like this. Teach a dolphin a simple sentence, such as "fetch ball hoop," to indicate taking the ball from the hoop and bringing it to poolside. Once this command, reinforced by operant conditioning, is perfected, then the dolphin is presented with new, untrained challenges. Perhaps it is asked to "fetch hoop ball," or either hoop or ball or both objects are replaced with novel items never before put into this context. It is clear that dolphins quickly grasp the basic concept of "object 1," "object 2," and "command" and act correctly a large percentage of the time. These sentence structures have been made more complicated, with similarly positive results. The dolphins are reasonably good at syntactic structure, and they also seem to be able to conceptualize general categories of items. In others words, the ball used in training can be substituted successfully by another ball, and a gestural symbol ("word") can be made to refer to an item very specifically or to be more general, just as in human word use (Herman, 1986).

Ron Schusterman has repeated many of Herman's studies and invented other experiments of his own, but with California sea lions (Zalophus californianus). His results are essentially the same: sea lions are also adept at learning and extrapolating from human-like syntactic structure (Schusterman and Krieger, 1986). Interestingly, the conclusions drawn by these two fine researchers are quite different, indicating the state of knowledge and vibrant nature of the field of animal language and cognition. Herman interprets his findings as the animals using language. "Fetch hoop ball" represent a verb, a direct object, and an indirect object. Schusterman, however, states that there is no reason to believe that the animal perceives this interaction as anything more than an action command and that the linguistic concept "verb" need not enter into the equation. It is true that human children, for example, do not learn language in the structured operant conditioning style as performed here. Instead, we learned (mainly) from people talking around us and from acquiring words and syntactic rules as we went along. It was not until language was already well formed that we were required in school to understand syntactic structure by diagramming or labeling the parts of sentences.

Language acquisition learning in dolphins and sea lions has taught researchers much about imitation, learning, and mental processing abilities. It is undeniable that dolphins learn the basic concepts very rapidly (sea lions a bit less rapidly) and faster than most mammals except for chimpanzees and humans. This by itself indicates a high level of that nebulous and poorly defined "intelligence." However, whether these studies can be called language, or whether that is even an important question, is open to debate. We humans have taken human syntax and foisted it on nonhuman species. Nevertheless, the animals have done remarkably well with what they were given. Perhaps they can do even better as they communicate among each other with signs and symbols and emotive content for which they have evolved.

B. Inventive Dolphins

Pinnipeds, sea otters, polar bears, and sirenians show elements of learning and play in captivity, but do not show the same kind of quick thinking or innovation as do some dolphins. However, most work has been done with dolphins, so there is some element of bias. Nevertheless, bottlenose dolphins and rough-toothed dolphins (Steno bredanensis), both with very large brains, are known as "the best" of performers in oceanaria. It is not clear whether these animals adjust better to captivity than others or whether they are innately more behaviorally flexible than others.

One interesting story of behavioral flexibility comes from a study done on two rough-toothed dolphins at Sea Life Park, Hawaii, in the mid-1960s. Karen Pryor, then head trainer at Sea Life Park, introduced a new demonstration into her on-stage performance with one of her dolphins named Malia. The intent was for Pryor to demonstrate to the audience how a previously unconditioned behavior could be reinforced by operant conditioning. In order to do so, she could not use a previously trained repertoire, but each day had to choose a simple behavior (such as a particularly high surfacing or loud blow) that the animal did and then reinforce it. After several days of this, Malia "spontaneously" recognized that "only those actions will be reinforced which had not been reinforced previously" (Pryor et al., 1969). In order to receive rewards rapidly (or for the pure fun of it), Malia "began emitting an unprecedented range of behaviors, including aerial flips, gliding with the tail out of the water, and 'skidding' on the tank floor" (Pryor et al., 1969). None of these behaviors had been shaped, none had even been seen before in the basic repertoire of dolphin behaviors at Sea Life Park! Pryor and her colleagues then repeated the work with an untrained female rough-toothed dolphin named Hou in order to assess experimentally whether creativity could be induced by operant conditioning in another dolphin and how long it would take. The experiment succeeded splendidly, and in a few trials, Hou was also presenting a new "act" after each one that received an operant reward.

Pryor et al. (1969) discussed their results very cautiously and reminded the reader that such training for novelty can probably be successful in horses and perhaps even pigeons as well. Many students of animal behavior and intelligence agree and are content to explain the development of novel behavior as simply a trained response. However, others have taken the experimental results further and suggested that much more insight than normal is required for the animal to "learn to learn" (the great philosopher Gregory Bateson called this "deuterolearning") and that the relatively quick manner in which dolphins "caught on" confirms their high intelligence. By the way, similar nonverbal training of reinforcement for novel behaviors has also been conducted for humans; the humans took about as long to realize what was being trained as did the dolphins (Maltzman, 1960). For Hou and for the humans, there was a period of strong frustration (even anger, in the humans) where they had not "caught on." They would be reinforced for a behavior, do it, and then not be rewarded for it ever again. It took some time for the "realization" to come that they then needed to exhibit a new behavior to get a reward. Once realized, the humans expressed great relief at having figured out "the problem," whereas the dolphins raced around the tank excitedly and displayed more and more novel and body-twisting behaviors— to the obvious delight of the researchers.

An interesting observation about dolphins is that they—at least bottlenose dolphins—readily recognize images of humans and of themselves in mirrors and on television screens. Herman *et al.* (1990) were able to elicit correct answers from the televised image of a human giving sign-based directions, even to the point where only white-gloved hands were shown going through the signaling motions. This demonstrates that the animals were able to use representations of the gestural instructions. Several investigators have shown dolphins mirrors and real-time video images of themselves; the dolphins react to the images with curiosity and playfulness, moving their rostrums rapidly and following their own eye movements. Furthermore, the reactions to video images of other dolphins appear to indicate that the viewing animals recognize different individuals on the screen, including themselves. This indicates a "sense of self" and has been described as an important insight into cognition. Interestingly, chimpanzees and other apes do not have this innate capability to see images on a flat screen as representations of themselves, others, or humans. They can be taught to process the images meaningfully, but only after prolonged exposure.

III. Behavioral Complexity in Nature

A. Carnivores and Sirenians

Most marine mammals are highly social, and we would expect that they have sophisticated ways of communicating with each other by showing innovative and variable behaviors in the face of social strategies and interactions. However, the less social species are obviously also behaviorally complex. Examples are polar bears and sea otters. Polar bears have a large repertoire of "sneaking up" on their generally ice-bound prey. They move against the wind, come from the side of the sun glare, and use ice obstructions and stealth in order to surprise their prey. It has been reported that in captivity, they figure out rapidly how to unlatch (and unhinge) doors in order to escape or to move from pen to open enclosure. Sea otters are tool users, prying mussels and abalone from the substrate with rocks or stones they keep cached in an armpit while not in use. At the surface, they retrieve the tool in order to break open their shellfish food; at times using the rock as a hammer and at times laying it on their stomach and using it as an anvil. Individual sea otters have preferred methods of tool use, implying learning and innovation. Polar bears and sea otters are obviously "bright," but few

behavioral studies or systematic investigations of learning have been conducted.

Pinnipeds are also behaviorally adept, and—as we have seen—sea lions can learn tricks and some aspects of language in captive training settings. They are all social mammals, especially while hauled out on land in order for males (of most species) to work out dominance relations with each other and for females to mate, give birth, and take care of their altricial (not well developed) young. Vocalizations, body postures, and smell are important aspects of communication. In the sea, most pinnipeds are less social (with the walrus being a strong exception), but they likely use more individualized but sophisticated strategies for finding and securing enough prey to survive. We expect that the animals need to periodically adapt to different types of prey, learn which could be physically harmful or poisonous, and learn how to detect and avoid large sharks, killer whales, and leopard seals. Many pinnipeds do not take their young out to sea with them, and therefore all learning to hunt and to survive needs to be without substantial help from more experienced adults. The author suspects, but has no proof for, that the brains of pinnipeds are adapted for relatively quick self-learning to survive and are less adapted or structured for social communication except as that needs to develop for procreation. Polar bears, sea otters, and sirenians would be an exception, although while generally less social then other marine mammals, mothers take prolonged care of their young while the young develop feeding and other skills. We assume, but again have no direct proof for this assertion, that the young learn more easily and completely in the presence of their mother.

B. Baleen Whales

Baleen whales are social creatures, especially during mating times. Vocal communication is extremely important to them, with drum-like sounds of gray whales (*Eschrichtius robustus*), long low-frequency moans of blue whales (*Balaenoptera musculus*), short low-frequency grunts of fin whales, and the rich repertory of groans, moans, and scream-like sounds of the right (*Eubalaena* spp.) and bowhead (*Balaena mysticetus*) whales. Whereas all whales appear to produce sounds, the most elaborate (and best-studied) sounds are the songs of male humpback whales, which likely serve as a male–male (intrasexual) dominance signal, male–female (intersexual) mating advertisement, or both. The songs are copied from listening to each other, are long and complicated, and must require reasonably formidable powers of learning and memory. Baleen whales on the mating grounds also sort out dominance relationships in either aggressive (humpback) or more gentle but highly maneuvering surface-active groups of gray whales, right whales, and bowhead whales. In the latter grays and right-bowhead groups, it is likely that multiple males allow each other to inseminate a particular female and practice a form of sperm competition instead of physical competition to increase the chances of fathering a young. It is also likely, although behavioral researchers have gathered only incomplete glimpses of the possibility, that female whales make it more difficult for some males than others to mate with them, thereby performing mate choice of preferred partners. If true, it must be important for females to gauge the relative "goodness" of males from the complicated matrix of social sounds and close-up interactions that present themselves. In right whales and bowhead whales, an adult female has only one young every 2 to 5 years. The calf gestates in her body for 1 year and then is nursed for another. This low reproductive rate means that she must take very good care of the young to attempt to assure its survival, and researchers would not be surprised at all to find that she also wants to choose the father of her young with care.

Baleen whales tend to be less social on the feeding grounds, although recent behavioral research indicates that at least some long-term bonds of affiliation persist between breeding and feeding grounds. This does not appear to be the norm, however. Generally, blue whales, humpback (*Megaptera novaeangliae*), gray whales, right whales, and bowhead whales (these five are the best-studied baleen whale species) aggregate at particular areas because of food concentrations. An aggregation due to an outside stimulus is not necessarily a social unit, although it can result in one. Some social interactions do occur, and it is even likely that the whales are paying close attention to each other in order to detect perhaps new or better feeding opportunities somewhere else. As well, blue whales often lunge into their food in tandem, apparently so as to provide a wall next to each other toward which fast-moving krill will not escape. Bowhead and right whales will swim in staggered formations of "echelons," side by side, apparently for the same purpose (Würsig, 1988).

The winner in the baleen whale feeding complexity department must surely be the amazing humpback whale. Humpback whales lunge into their fish food, alone and in coordinated groups up to an observed 22 animals. They are not merely aggregated in such a case, but all lunge (from below and toward the surface) at essentially the same time, coming to the surface within about 6 sec of each other. Apparently, although this is not yet proved, there is a vocal signal at the beginning of these highly coordinated lunges. One whale signals and others follow. Hitting the prey, a huge fish or bait ball, at one time presumably allows for each mouth to be better filled in the resultant prey's confusion than if one or a few mouths attacked. Humpback whales also flick their tails at prey and then circle to engulf it; they flick their long foreflippers forward as their mouths open, presumably to flash the white undersides of these flippers at the prey and to herd it more efficiently into the mouth. Finally, they release a stream of bubbles from their blowholes while circling around the prey and upward. The rising bubble screen forms an effective net around the prey, and the humpback (alone or with several others) then lunges toward the surface in the center of the "net," filling its capacious mouth with concentrated prey. It is unclear how flexible the several feeding behaviors are, but it is certain that several need social coordination. It is also likely that young humpbacks need to learn and perfect the techniques, and we assume that social learning is the major vehicle to do this.

C. Toothed Whales

Toothed whales are highly social creatures, except for older adult male sperm whales who tend to be loners, some lone killer whales (*Orcinus orca*), and an extremely ("aberrant") low level of singles in many species of dolphins. Some of the deep-ocean beaked whales may be loners as well, but we have no good data on this point. Whereas most species are social, there are very different forms. Hector's dolphins (*Cephalorhynchus hectori*), harbor porpoises (*Phocoena phocoena*), and river dolphins tend to occur in small groups of up to a dozen animals, rarely more. We surmise that in at least some of these dolphins, individuals know each other well. Pantropical spotted (*Stenella attenuata*) and striped (*S. coeruleoalba*) dolphins of the open ocean, however, travel in "herds" of thousands of animals. While there appear to be subgroups with at least some interindividual fidelity, it is very unlikely that all members of the herd know each other; some may never even meet each other. However, the herd acts as a coordinated unit, traveling at the same speed (which must be near the speed of the slowest animals), turning in essential unison, often diving in synchronized fashion. If a disturbance occurs along a flank or somewhere below, e.g., a shark zooms out of depth, there is rapid information transfer from animal to animal so that the group cascades away from the perceived danger in coordinated fashion. The information transfer is so rapid that we assume that animals are aware not only of their nearest neighbors, but are "looking beyond others," by sight when possible and probably also by echolocation. This is sort of a chorus line effect, where dancers coordinate their movements better by not merely paying attention to their nearest neighbors, but by anticipating the wave of raised legs, for example, as it (the wave) approaches. As well, it is likely that each dolphin pays attention to the vocalizations and movements of others nearby and thus integrates response information in what the great cetacean researcher Ken Norris called a sensory integration system for dolphins (Norris *et al.*, 1994). Jerison (1986) used the idea of shared echolocation among dolphins to postulate that the animals share sensory inputs in a way that might synergistically enhance an expanded sense of "self." A human analogy would be if several people of a group could know their world and their place in it better by sharing neural data of aggregate visual systems. Jerison postulated that this potential sharing of echolocation data might itself account, at least in part, for large dolphin brains, but we have no direct information on this provocative point.

Coordination of group movements and activities need not be a matter of high intelligence and cognition, of course; and sensory awareness and a collective sensory integration system are well developed in schooling fishes, flocking birds, and so on. Instead, we might do better to look at the complexities of social interactions to gain "a window into the dolphins' minds" (after Griffin, 1981). Alas, we do not yet know very much about details of communication in delphinid cetaceans, but we do know enough to call it "complex."

Dolphins in a group are constantly aware of each other. A flipper touch here, a glance there, a slow echolocation-type click, a whistle. They interact by all sensory modalities available to them. We guess (and it is only a guess) that they are constantly gauging each other, deciding dominance/subservience relationships, seeking the comforting presence of relatives or those that they have found to be helpful in previous encounters, and avoiding those that might be aggressive. We know that there are at least occasions of political intrigue. Indian Ocean bottlenose dolphin (*Tursiops aduncus*) males of Shark Bay, southwestern Australia, have a strong tendency to form alliances to kidnap females. They apparently do so to gain access to reproductive females—access that might otherwise not be available because these males may not be of sufficiently high dominance status or would not be chosen by the females. Interestingly, super alliances of two or more alliances form in order to steal females from another male alliance (Connor *et al.*, 2000). Richard Connor is presently attempting to find out whether males that cooperate together in this fashion are more often closely related or whether alliances are formed along lines of friendship more than kinship.

Toothed whales appear often to be structured along matriarchal (female-based) lines. Sperm whales, killer whales, pilot whales (*Globicephala* spp.), and bottlenose dolphins (of at least several populations) have close ties between mother and female young even after weaning, and in sperm and pilot whales, these ties appear to last for life. This means that potential cultural transmission of knowledge, from generation to generation, is expected to flow especially efficiently along female lines. Mom teaches young, young teaches its children, and so on. In a society of relatively resident killer whales of the U.S. and Canadian Pacific Northwest, female and male offspring stay within the pod for life. This society is thus socially "closed." However, females mate with males outside, and the males mate with females of neighboring pods. Each pod is therefore reproductively matriarchal. These societies of relatively stable long-lived individuals are likely to develop behavioral cultures of their own. We have some evidence: killer whale pods have individually distinctive sound repertoires, or dialects. Individuals of pods can recognize each other easily as of that pod. It is likely, but not proved, that individuals also recognize each other as individuals by sound. In the more open but still matriarchal societies of at least one population of common bottlenose dolphins, studied by Randy Wells and colleagues, of Sarasota, Florida, male offspring develop signature whistles (individually distinct sounds) more like those of their mothers than do female offspring. The moms and female offspring stay together as daughters mature. The sons, however, leave the natal group, roam elsewhere, and only now and then interact with their natal groups as adults. It is hypothesized that the similar signature sounds of moms and sons may provide an efficient means of recognition and thereby inbreeding avoidance (Sayigh *et al.*, 1995). Signature whistles are also copied by dolphins who are answering the original whistler. This rapid imitation may serve as a societal binding mechanism. It has been postulated that basic greetings and verbal recognition were prerequisite to the development of human language. Dolphins have the signature recognition portion of this capability (Janik, 2000).

Sound has been studied recently relative to kin and others of a society, and much more sound-based learning is likely to come to light as studies progress. This is to be contrasted with the relatively stereotyped sounds of the great apes, for example, that do not change much with age or social association (but then, in all fairness, apes are generally less vocally communicative and more visually based than cetaceans). However, it is also likely that not only vocal evidence for learning and social transmission will come to light. We have some hints, and only hints, here as well. Killer whales of Patagonia, Argentina,

beach themselves in order to take sea lion (*Otaria flavescens*) and elephant seal pups. The beaching maneuver requires great skill, as the predator needs to gauge exactly where the prey is on the beach, after having seen the prey only from a distance and through murky nearshore waters, beyond the surf zone through which it needs to make its rush to the beach. As well, it needs to beach with such velocity and angle as to be certain that the spilling waves will allow it to reach deep water again. Killer whale adults have been described as making sham rushes at a beach and then waiting along the sides while young killer whales attempt the maneuver, usually clumsily and ineffectively, again and again. Now and then, the adult makes an intervening rush and then retreats to the side again. This behavior was pointed out as probable teaching by Argentine killer whale researchers Juan Carlos and Diana Lopez (1985) and has been studied in greater detail and verified since. It is unclear how well youngsters would learn beaching "on their own," but it is likely that it is transmitted culturally, as killer whale beaching behavior is found in only several populations worldwide. In Galveston Bay, Texas, certain female bottlenose dolphins and their young follow shrimp boats much more so than others, even maneuvering into the shrimp nets to take live fish and then wriggling out again while the shrimper is underway. This activity, video taped underwater, requires skill and dexterity to avoid being entangled in the fishing gear. The dolphins who exhibit this behavior do so "with ease," whereas others do not fish at all in this manner. Again, we wonder whether cultural learning and societal transmission of knowledge is important here. While culture has been explored in birds and nonhuman primates, very little has been written on this subject for marine mammals (but see Whitehead, 1998).

Even on an hour-to-hour basis, dolphins of a group are likely to be coordinating their activities superbly well. While there are many potential examples (and each behavioral observer has his or her favorite ones), the author prefers one that he and his wife have studied for some time. Dusky dolphins (*Lagenorhynchus obscurus*) of the shallow waters of Patagonia, Argentina, coordinate activities to corral fish schools. It appears, and much more work is needed to properly describe the individual behaviors, that dolphins (circling while vocalizing, tail swiping, and blowing bubbles) surround the prey ball and thereby cause it to tighten. They also herd the prey ball to the surface and then use the surface as a wall through which the prey cannot escape. Interestingly, dolphins do not appear to feed until the prey has been tightened and is at the surface. There may be a form of "temporary restraint," with all animals working toward the common good of getting the prey secured. This coordinated activity stands in stark contrast to taking individual advantage of the prey by grabbing a mouthful here or there and causing the prey to scatter and escape. As an example, sea lions that enter the area work on their own and are highly disruptive to the herding efforts of the dolphins. While we still need to look at the details of this behavior, to see whether kin, for example, help each other more often, we assume that much communication, learning, and individual trust need to go into such coordination. It is likely, but unknown at present, that animals know each other well enough as to have preferred "working" partners and have mechanisms for detecting and effectively ostracizing those cheaters who do not help or are disruptive at critical phases of prey gathering (Würsig *et al.*, 1989).

Such activities require individual recognition, concepts of strategies for dealing with different behaviors of fish schools, coordination, memory of past events, and potential teaching or at least learning from others; in short, considerable behavioral sophistication and flexibility. In New Zealand, far from Argentina and in a different deep-water environment, most dusky dolphins feed not on schooling fishes but on mesopelagic ("midwater, deep ocean") fishes at night. It has been found that a small subsegment of the dolphin population—the same individuals on a regular basis—travels to bays where dolphins take seasonal advantage of fish stocks to herd prey into tight balls as described earlier for Argentina. Because it is apparently the same animals doing so year after year, we believe that there might be cultural transmission of information here as well; only some have learned (or care) to take advantage of this particular foraging style.

A final example of at-times sophisticated-seeming behavior is certainly play. Almost all young mammals play, and this has been interpreted as gaining skills necessary to survive. It certainly seems like much fun as well. In dolphins and a handful of other mammals, adults habitually engage in behavior that is difficult to rationalize as anything but play. Dusky dolphins pull on the legs of floating birds, and individuals of several species perfect the balancing of pieces of kelp or other objects on their rostrums, flippers, dorsal fin, and tail. Play is not the purview of only dolphins, however. Adult baleen whales, sea lions, sea otters, and polar bears play with objects, at times for up to an hour or more. Play seems less common in phocid seals, but the imitative sounds of "Hoover" the harbor seal may have represented a form of vocal play. Play seems more rare (or absent?) in wild adult sirenians, but then long-term studies underwater have not been conducted.

IV. Conclusions

Marine mammals are not of one taxonomic group and live in many varied ways; we therefore are not surprised to find that they have different brain sizes and ways of adapting to their ecologies, social structures, and behaviors. Because all use marvelously adaptive behaviors to help them survive, they are all "smart." However, such a general definition is not very satisfying. The polar bear, sea otter (and marine otter, *Lontra felina*, of Chile), pinnipeds, sirenians, and baleen whales all have behavioral characteristics and ways of living that might refer to "intelligences" not all that different from terrestrial mammals. Several of the dolphins (not all) stand out as being exceptionally large brained and behaviorally sophisticated; they are quick learners in captivity and have social structures and behaviors that appear to be highly complex and variable. While much of the large brains of these odontocetes may well be taken up by the neural processing required for ECHOLOCATION and other senses, as has often been speculated, it is highly likely that a large part of it also deals with relationships, learning, and long-term memory of events (Schusterman *et al.*, 1986).

Much more needs to be learned about dolphin whistle and click communication. However, it does not seem likely that

their combinations of whistles and clicks can be termed "language" in the sense of putting sets of (for example) whistles together as referential communication for different objects or constructs (ideas). Instead, vocalizations seem to carry emotive content, signature information, and may well serve as an important tool for binding social relationships (Janik, 2000). Nevertheless, there are certain to be surprises to be gained from studies on delphinid communication as more information is gleaned. One important avenue for exploration is the extent to which communication and behavior have been transmitted from generation to generation, resulting in distinct cultures in such animals as sperm whales, killer whales, and several species of dolphins (Whitehead, 1998).

While we think of dolphin and other marine mammal "intelligences" and cognitive processes and realize what marvelous animals they are, it is also fair to contemplate their limits. Dolphins are beautifully tuned to the environments in which they have evolved for millions of years, but they do not necessarily have the capability to make behavioral extrapolations that seem to us very simple. A prime example is the fear (or mental incapability) of wild dolphins to leap over obstructions. This has been a major problem for the tuna purse seining industry—dolphins caught in a net could easily all leap to freedom as the net is pursed. They do not do so because it is not in their repertoire to do so and are caught (and at times entangled and killed) as a result. Only dolphins trained to leap over nets will do so or some animals that seem to have "accidentally" (perhaps the most innovative ones?) discovered the capability in nature. This article ends on this theme of focused mental capabilities because it illustrates two related points: (1) dolphins are not those "super-intelligent" beings as claimed by some aspects of the news media and many books and films and (2) dolphins are indeed "intelligent" for those things that they need to solve and interact with in their natural world, but their natural world is very different from ours.

See Also the Following Articles

Behavior, Overview ▪ Brain Size Evolution ▪ Communication ▪ Group Behavior ▪ Language Learning

References

Connor, R. C., Read, A. J., and Wrangham, R. (2000). Male reproductive strategies and social bonds. *In* Cetacean Societies: Field Studies of Dolphins and Whales" (J. Mann, R. C. Connor, P. L. Tyack, and H. Whitehead, eds.). University of Chicago Press, Chicago, IL.

Griffin, D. R. (1981). "The Question of Animal Awareness: Evolutionary Continuity of Mental Experience." Rockefeller Univ. Press, New York.

Herman, L. M. (1986). Cognition and language competencies of bottlenosed dolphins. *In* "Dolphin Cognition and Behavior: A Comparative Approach" (R. J. Schusterman, J. Thomas, and F. G. Wood, eds.). Lawrence Erlbaum Associates, Hillsdale, NJ.

Herman, L. M., Morrel-Samuels, P., and Pack, A. A. (1990). Bottlenosed dolphin and human recognition of veridical and degraded video displays of an artificial gestural language. *J. Exp. Psychol. Gen.* **119,** 215–230.

Janik, V. M. (2000). Whistle matching in wild bottlenose dolphins (*Tursiops truncatus*). *Science* **289,** 1355–1357.

Jerison, H. J. (1973). "Evolution of the Brain and Intelligence." Academic Press, New York.

Jerison, H. J. (1986). The perceptual world of dolphins. *In* "Dolphin Cognition and Behavior: A Comparative Approach" (R. J. Schusterman, J. Thomas, and F. G. Wood, eds.) Lawrence Erlbaum Associates, Hillsdale, NJ.

Lilly, J. C. (1961). "Man and Dolphin." Doubleday Press, New York.

Lopez, J. C., and Lopez, D. (1985). Killer whales (*Orcinus orca*) of Patagonia, and their behavior of intentional stranding while hunting nearshore. *J. Mammal.* **66,** 181–183.

Maltzman, I. (1960). On the training of originality. *Psychol. Rev.* **67,** 229–242.

Norris, K. S., Würsig, B., Wells, R. S., and Würsig, M. (1994). "The Hawaiian Spinner Dolphin." University of California Press, Berkeley.

Pryor, K. W., Haag, R., and O'Reilly, J. (1969). The creative porpoise: Training for novel behavior. *J. Exp. Anal. Beh.* **12,** 653–661.

Sayigh, L. S., Tyack, P. L., Wells, R. S., Scott, M. D., and Irvine, A. B. (1995). Sex differences in signature whistle production of freeranging bottlenose dolphins, *Tursiops truncatus. Behav. Ecol. Sociobiol.* **36,** 171–177.

Schusterman, R. J., and Krieger, K. (1986). Artificial language comprehension and size transposition by a California sea lion (*Zalophus californianus*). *J. Comp. Physiol.* **100,** 348–355.

Schusterman, R. J., Thomas, J. A., and Wood, F. G. (eds.) (1986). "Dolphin Cognition and Behavior: A Comparative Approach." Lawrence Erlbaum Associates, Hillsdale, NJ.

Whitehead, H. P. (1998). Cultural selection and genetic diversity in matrilineal whales. *Science* **282,** 1708–1711.

Worthy, G. A. J., and Hickie, J. P. (1986). Relative brain size of marine mammals. *Am. Nat.* **128,** 445–459.

Würsig, B. (1988). The behavior of baleen whales. *Sci. Am.* **256**(4), 102–107.

Würsig, B., Würsig, M., and Cipriano, F. (1989). Dolphins in different worlds. *Oceanus* **32,** 71–75.

International Whaling Commission

GREGORY P. DONOVAN
*International Whaling Commission,
Cambridge, United Kingdom*

The International Whaling Commission (IWC) is the intergovernmental body established in 1946 to conserve whale stocks and regulate whaling. Membership is open to any sovereign state. There were 40 member nations (Table I) in the year 2000.

I. Historical Background

Whaling cannot be put forward as an example of the successful sustainable management of a renewable resource. From the start of the "commercial" exploitation of whales, the story was usually one of eventual overexploitation. Modern commercial whaling began with the invention of the explosive harpoon

TABLE I
List of Members of the IWC in August 2000[a]

Contracting government	Date of entry into force
Antigua and Barbuda	1982
Argentina	1960
Australia	1948
Austria	1994
Brazil	1974
Chile	1979
People's Republic of China	1980
Costa Rica	1981
Denmark[b]	1950
Dominica	1992
Finland	1983
France	1948
Germany	1982
Grenada	1993
Guinea	2000
India	1981
Ireland	1985
Italy	1998
Japan[c]	1951
Kenya	1981
Republic of Korea	1978
Mexico	1949
Monaco	1982
Netherlands	1977
New Zealand	1976
Norway[d]	1960
Oman	1980
Peru	1979
Russian Federation[b]	1948
St. Kitts and Nevis	1992
St. Lucia	1981
St. Vincent and The Grenadines[b]	1981
Senegal	1982
Solomon Islands	1993
South Africa	1948
Spain	1979
Sweden	1979
Switzerland	1980
United Kingdom	1948
United States[b]	1948

[a]Note some nations have left and subsequently rejoined. The year of entry applies to their most recent adherence.
[b]Engaged in aboriginal subsistence whaling.
[c]Whaling under scientific permit.
[d]Commercial whaling.

combined with the development of steam-powered catcher boats in the 1860s. This allowed whalers to take the faster swimming rorquals (e.g., blue and fin whales, *Balaenoptera musculus* and *B. physalus*). The promise of large numbers of whales caused whalers to investigate the Antarctic, and the first whaling station was established on South Georgia in 1904 and took 195 whales. By 1913, there were 6 true land stations and 21 floating factories that had to be moored in suitable harbors; the total catch was 10,760 whales. The invention of the stern slipway in 1925 allowed vessels to operate in offshore waters and by 1930/1931, 41 factory ships took over 37,000 whales. This overproduction led to a catastrophic decline in the price of whale oil.

It was the fear of low prices rather than the fear of overexploiting whale stocks that was the driving force behind early moves to limit catching. Despite attempts under the auspices of the League of Nations to establish some international control, the production agreements negotiated among themselves by the whaling companies produced the first effective limitation of catches in the early 1930s.

World War II caused a world shortage in the supply of fats, and several nations had their eyes on profits from pelagic whaling. It was in this light, and with the experience gained in developing international agreements just before the war, that discussions were held in London in 1945 and in Washington in 1946 on the international regulation of whaling.

II. Establishment of the International Whaling Commission

The International Convention for the Regulation of Whaling was signed at the 1946 conference. It was a major step forward in the international regulation of natural resources, as it was one of the first to place "conservation" at the forefront. The convention was established "to provide for the proper conservation of whale stocks and thus make possible the orderly development of the whaling industry." This was a laudable aim, but finding the difficult balance between "conservation" and the development of the whaling industry has dominated the history of the IWC.

An important feature of the convention was that it established a mechanism whereby regulatory measures included in the Schedule to the convention (catch limits, seasons, size limits, inspections, etc.) could be amended when necessary by a three-quarters majority of members voting (excluding abstentions).

The convention also formally assigned importance to the need for scientific advice, requiring that amendments to the regulations "shall be based on scientific findings." To this end, the commission established a scientific committee comprising scientists nominated by member governments (and later invited experts when appropriate).

Despite this, there are aspects of the convention that have attracted criticism. For example, any government can "object" to any decision with which it does not agree within a certain time frame. This (along with the right of nations to unilaterally issue permits to catch whales for scientific purposes) has led to accusations that the IWC is "toothless." However, it should be recognized that without these provisions, the convention would probably have never been signed.

From a management perspective, a more serious flaw was that the IWC could neither restrict operations by numbers or nationality nor allocate quotas per operation. Although it may be questioned whether the IWC could have agreed to national quotas or numbers of vessels, certainly if such limitations had been reached, this would have reduced the management problems associated with increasing numbers of vessels chasing limited quotas.

The convention formally established the International Whaling Commission. The IWC comprises one commissioner from each government who has one vote and may be accompanied by one or more experts and advisers.

III. The IWC before 1972

Perhaps the most serious problem of early management was the use of the blue whale unit (BWU). In terms of oil yield, one blue whale was considered equal to 2 fin, 2.5 humpback (*Megaptera novaeangliae*), or 6 sei (*B. borealis*) whales. In 1945, a catch limit of 16,000 BWU was set (suggested by three scientists as being a "reassuring" value in the middle of their estimate of 15–20,000). The flaw in the BWU system is apparent: it allows catching of depleted species below levels at which catching that species alone would be economically unviable. This is apparent from catch data up to the 1970s, which reveal that as blue whale catches declined, so fin whale catches (the next largest species) increased until they too were overexploited and sei whale catching began.

The lack of national quotas resulted in an "Olympic" system, where it became a race to catch as many whales as possible before the total quota was reached, leading to waste during processing and the use of increasing numbers of catcher boats (129 in 1946/1947 and 263 in 1951/1952). This neither made economic sense nor encouraged conservation.

As early as 1952, it was recognized that the catch quota was too high. The difficulty was in getting all the whaling nations to agree to a reduction; if one nation objected, then all objected. This was the start of a difficult period for the IWC, trying to match the evidence of science against the needs of the industry. At one stage, both the Netherlands and Norway withdrew from the commission and its survival seemed in doubt. After considerable argument and controversy, by 1971/1972, the catch limit had been reduced to 2300 BWU, and certain species, including the blue and humpback whales, had been protected from commercial whaling.

IV. A Period of Change: 1972 to the "Moratorium"

In 1972, the UN Conference on the Human Environment called for an increase in whale research, a 10 year "moratorium" on commercial whaling and a strengthening of the IWC. Although proposals for 10 year moratoria were subsequently tabled at the IWC, they failed to reach the required three-quarters majority, largely because the IWC scientific committee believed that management on a stock-by-stock basis (Antarctic catches were first set by species in 1972) was the most sensible approach; if required, each stock could be independently protected. The UN resolution was, however, taken seriously by the IWC. By 1976, a permanent secretariat had been established in Cambridge, an international decade of cetacean research had been declared, and a management procedure [the new management procedure (NMP)] had been adopted. The NMP was aimed at bringing all stocks of whales to an optimum level at which the largest number of whales can be taken consistently [the maximum sustainable yield (MSY)] without depleting the stock. It also gave complete protection to stocks at 54% of their estimated preexploitation size, i.e., well before they became endangered. The NMP was regarded as a major step forward in the management of whaling. It appeared to take the issue of catch limits largely out of the hands of the politicians and place them in those of the scientific committee. In addition, from 1973, the long-awaited international observer scheme was in operation (from 1973) that aimed at ensuring that new catch limits were enforced.

A major feature over this period was the increase in IWC membership. In 1963, there were 18 member nations, of which only 4 were nonwhaling countries; in 1978, there were 17, of which 8 were nonwhaling, and by 1982, membership was 39. Of the 13 whaling nations, 3 had only aboriginal/subsistence operations (Denmark, the United States, and St. Vincent and the Grenadines).

The 1979 meeting was a turning point in the commission's history. Doubts had been expressed by some over (1) the theoretical and practical application of the NMP and (2) the morality of whaling, irrespective of the status of the stocks. At that meeting, a proposal to end pelagic whaling for all species except minke whales was adopted and a sanctuary was declared for the Indian Ocean outside the Antarctic. Whereas the onus in the past had been for positive evidence of a decline in stocks before a reduction in catch limits was agreed, positive evidence was now required if a catch limit was to be set. By 1982, a *Schedule* amendment was adopted that implemented a pause in commercial whaling (or to use popular terminology, a "moratorium") from 1986. Originally, four whaling nations, Japan, Norway, Peru, and the USSR, lodged objections to this decision, although Peru and Japan subsequently withdrew theirs. In the year 2000, only Norway carries out commercial whaling under an objection.

One obvious question to ask as the IWC's moratorium came into effect was whether the commission had been a success? At one level the answer must be no; indeed, it could be argued that it had been a disaster. For example, in the Antarctic, the most important area to the IWC initially, (i) blue and fin whales had been reduced to at best 5 and 20% of their original numbers, and possibly much less, respectively; hardly a good example of "conservation of whale resources"; (ii) the 1983/1984 catch was 6655 minke whales (*Balaenoptera bonaerensis*), a species not considered worth catching in 1947/1948 when the catch in BWU was 25 times greater; hardly "the orderly development of the whaling industry."

So, had the IWC achieved anything? First of all, while it is easy with current levels of knowledge to criticize the IWC's performance, it has to be said that modern whaling had not resulted in the extinction of any species; IWC actions, while insufficient, were better than nothing. Since the 1970s, the trend has been very much toward conservative catch limits based on

scientific advice, to a degree probably unparalleled in any fisheries commission. It has been argued by some that this trend reached unreasonable limits with the introduction of the "moratorium." It is indicative of the inherent problems within the commission that the same decision is hailed by some as its greatest success and by others as its most abject failure.

V. The Commission Today

Since 1976, the IWC has had a full-time secretariat (of 15–20 people) with headquarters in Cambridge, UK. Each year, the annual meeting of the commission is held, either by invitation in any member country, or in the United Kingdom. The scientific committee (comprising up to 120 scientists) meets in the 2 weeks immediately before the main commission meeting and may hold special meetings during the year. The information and advice it provides form the basis upon which the commission develops the regulations for the control of whaling.

A. Management Issues

The primary function of the IWC is the conservation of whale stocks and the management of whaling. In addition to commercial whaling, the IWC has recognized the discrete nature of aboriginal subsistence whaling, and allowed aboriginal catches from stocks that have been reduced to levels at which commercial whaling would be prohibited.

1. Commercial Whaling At the outset of its discussions on the work to be carried out after the moratorium came into place, the scientific committee recognized the need to develop management procedures that did not repeat past mistakes and recognized the limitations of both data it had and data it was likely to obtain. Clearly, it was not acceptable to try out experimental management procedures in the wild. Apart from the serious consequences of getting it wrong, on long-lived species such as whales, it would take a considerable time to assess whether it really worked. The approach adopted was therefore to use computer simulations of whale populations over a long (100-year) period.

The most important part of any development process is the determination of *management objectives*. These were set by the commission and can be summarized as: (1) catch limits should be as stable as possible; (2) catches should not be allowed on stocks below 54% of the estimated carrying capacity (as in the NMP); and (3) the highest possible continuing yield should be obtained from the stock. The highest priority was given to the second objective.

After 8 years of intense work, the committee developed a procedure for determining safe catch limits that required knowledge of only two essential parameters: estimates of current abundance taken at regular intervals and knowledge of past and present catches. Intensive testing of the procedure against numerous assumptions and problems had been undertaken; some of these are summarized in Table II.

The way in which catch limits are calculated from the required information is specified by the *Catch Limit Algorithm* (CLA). This is a "feedback" procedure; as more information accumulates from sighting surveys (and catches if taken), then the estimates of necessary parameters are refined. In this way, the

TABLE II
Examples of Trials

Several different population models and associated assumptions

Different starting population levels, ranging from 5 to 99% of the "initial" population size

Different MSY levels, ranging from 40 to 80%

Different MSY rates, ranging from 1 to 7% (including changes over time)

Various levels of uncertainty and biases in population size

Changes in carrying capacity (including reduction by half)

Errors in historic catch records (including underestimation and underreporting by half or more)

Catastrophes (irregular episodic events when the population is halved)

Various frequencies of surveys

procedure monitors itself constantly. Catch limits are set for periods of 5 years. The CLA was initially tested on the assumption that it is applied to known biological stocks. To date, testing for specific species and areas has only been carried out for minke whales in the North Atlantic and Southern Hemisphere. Without such trials, catch limits will be zero under the *Revised Management Procedure* (RMP). It is clear that for very many populations, such as blue whales in the Southern Hemisphere, it will be a very long time before catches would be allowed under the RMP.

The CLA plus the rules about, *inter alia*, stock boundaries, allocation of catches to small areas, and what to do if many more of one or other sex are caught form the RMP. The RMP sets a standard for the management of all marine and other living resources. It is very conservative, and this is a reflection of the relative priorities assigned to the objectives, the level of uncertainty in the information on abundance, productivity and stock identity of whale stocks, and the fact that many years are required before the CLA refines its estimates of the required parameters.

Although these scientific aspects were adopted by the IWC in 1994, its actual implementation is a political decision. The IWC will not set catch limits for commercial whaling until it has agreed and adopted a complete *Revised Management Scheme* (RMS). Any RMS will also include a number of nonscientific issues, including inspection and enforcement, and perhaps humaneness of killing techniques. The importance of an international inspection scheme was highlighted by the discovery of widespread falsification of catch data by Soviet whaling operations prior to 1972.

2. *Aboriginal Subsistence Whaling* Aboriginal subsistence whaling is permitted by Denmark [Greenland: fin and minke (*Balaenoptera acutorostrata*) whales], the Russian Federation [Siberia: gray (*Eschrichitius robustus*)] and bowhead [(*Balaena mysticetus*) whales)], St. Vincent and the Grenadines (Bequia: humpback whales) and the United States (Alaska: bowhead and gray whales). It is the responsibility of the committee to provide scientific advice on safe catch limits for such stocks. With the completion of the RMP, the scientific committee began the process of developing a new procedure for the management of

aboriginal subsistence whaling (AWMP) that takes into account the different objectives for the management of such whaling as compared to commercial whaling. The commission will be establishing an aboriginal whaling scheme that comprises scientific and logistical (e.g., inspection/observation) aspects of the management of all aboriginal fisheries. The scientific component will comprise some general aspects common to all fisheries and an overall AWMP within which there will be common components and case-specific components. The first components are expected to be completed in 2002.

3. *Scientific Permit Whaling* A major area of discussion since the moratorium has been the issuance of permits by national authorities for the killing of whales for scientific purposes. The right to issue them is enshrined in Article VII of the convention (that furthermore requires that the animals be utilized once scientific data have been collected), and prior to 1982, over 100 permits had been issued by a number of governments, including Canada, the United States, USSR, South Africa, and Japan. Since the "moratorium," Japan, Norway, and Iceland have issued scientific permits as part of their research programs. The discussion has centered on accusations that such permits have been issued merely as a way around the moratorium decision contrasted by claims that the catches are essential to obtain information necessary for rational management and other important research needs. All proposed permits have to be submitted for review by the scientific committee, but the ultimate responsibility for their issuance lies with the member nation. Only Japan has issued scientific permits for the year 2000 [400 ± 10% Antarctic minke whales in the Antarctic, the 12th full-scale survey of a 16-year program; 100 northern minke whales, 50 Bryde's whales (*Balaenoptera edeni*), and 10 sperm whales (*Physeter macrocephalus*) in the western North Pacific]. As in previous years, a majority of the commission members urged Japan to refrain from issuing the permits.

4. *Small Cetaceans* It can be argued that no species of large whale is endangered by whaling today and will not be by any resumption of whaling under the RMS. Threats to those species, such as the North Atlantic right whale (*Eubalaena glacialis*), which remain severely reduced, do not include direct hunting. The most seriously threatened cetaceans (by direct hunting and incidental captures in fisheries) are a number of species and populations of the smaller cetaceans. At present, there is no single international body responsible for their conservation and management. There is considerable disagreement within the IWC as to whether the present convention is sufficient to allow the IWC to assume such a role. Fortunately, there is general agreement that the IWC scientific committee can consider the status of small cetaceans and provide advice to governments even though the IWC cannot set management regulations; it is to be hoped that governments respond individually and collectively. It remains a matter of some urgency that an international agreement or series of regional agreements be reached to ensure the conservation of small cetaceans.

5. *Whale Watching* The IWC has become involved (in a monitoring and advisory capacity) with aspects of the manage-

ment of whale watching as one type of sustainable use of cetacean resources. It has adopted a series of objectives and principles for managing whale watching proposed by the scientific committee.

B. Other Scientific Issues

The commission funds and acts as a catalyst for a good deal of cetacean research. In the year 1999/2000 some $400,000 was allocated to scientific research in addition to the IWC-related work undertaken by individual member governments. One major program is a series of Antarctic cruises to estimate abundance that has been carried out since 1978. These are now called Southern Ocean Whale and Ecosystem Research (SOWER) circumpolar cruises and include a component dedicated to blue whales.

With increasing awareness that detrimental environmental changes may threaten whale stocks, the IWC has accorded priority to research on the effects of such changes on cetaceans. While the RMP addresses such concerns adequately, the scientific committee has agreed that the species most vulnerable to such threats would be those reduced to levels at which the RMP, even if applied, would result in zero catches. It has developed two major research programs: one (POLLUTION 2000+) on the effects of chemical pollutants on cetaceans and another on the effects of climate change and ozone depletion (SOWER 2000). It is also increasing collaboration and cooperation with governmental, regional, and other international organizations working on related issues.

The work carried out by the IWC scientific committee is recognized worldwide. The commission has increasingly published scientific reports and papers; this culminated in the launch of the *Journal of Cetacean Research and Management* in 1999.

C. Politico-ethical Issues

Of prime consideration from both a scientific and an ethical viewpoint is the possibility of extinction of any population due to whaling. No population of whales is currently under threat of extinction from whaling, and it is clear that any acceptable management procedure will ensure that this cannot happen. However, this presumes an acceptance that whales are a natural resource to be harvested. While this is certainly the stated position of many members of the IWC, it is not universally accepted. A wide range of opinions have been expressed, ranging from the belief that whales are such a "special" group of animals that they should not be killed under any circumstances, through the view that they should not be killed commercially as whale products are not essential, to the view that the whales are a natural resource like any other.

In this regard, the question of humane killing has once more arisen within the IWC, with some nations stating that even if a safe management procedure is adopted, catch limits should not be set unless a "satisfactorily humane" killing method is available. This subject has been addressed several times during the history of the IWC, and the commission has been active in promoting work on more humane killing techniques for both commercial and aboriginal subsistence whaling. However, obtaining agreement on what comprises a "satisfactorily humane" technique will not be simple. In particular, in the case of aboriginal subsistence whaling, arguments of tradition and culture can clash with the adoption of modern technology.

VI. Conclusion

Many of the "politicoethical" issues listed are linked to questions of culture and freedom; they are complex and almost inevitably will not be resolved unanimously. There is clearly a divergence of opinion within the IWC on such matters to an extent unparalleled in any similar organization. It is, for example, difficult to think of any fisheries organization where some of the members believe it is immoral to catch fish under any circumstances. This is not the place to enter into a philosophical debate over the rights of nations or groups of nations to impose their moral values on others, but merely to point out the necessity of such a debate and the need for a degree of compromise if the IWC is not going to fragment, with potentially serious consequences for the world's cetaceans.

See Also the Following Articles

Ethics and Marine Mammals ∎ Illegal and Pirate Whaling ∎ Management ∎ Whale Watching ∎ Whaling, Modern

References

Allen, K. R. (1980). "Conservation and Management of Whales." Univ. of Washington Press, Seattle and Butterworth & Co., London.

Donovan, G. P. (1992). The International Whaling Commission: Given its past, does it have a future? In "Whales: Biology–Threats–Conservation" (J. J. Symoens, ed.), pp. 23–44. Royal Academy of Overseas Sciences, Brussels, Belgium.

Gambell, R. (1977). Whale conservation: Role of the International Whaling Commission. *Mar. Policy,* 301–310.

International Whaling Commission (1950–1998). *Rep. Int. Whal. Comm.* **1–48.**

International Whaling Commission (1999–). *J. Cetacean Res. Manage.* **1–.**

International Whaling Commission (1999–). *Annu. Rep. Int. Whal. Comm.* www.iwcoffice.org.

Tønnessen, J. N., and Johnsen, A. O. (1982). "The History of Modern Whaling." C. Hurst & Co., London.

Inuit and Marine Mammals

STEPHEN A. MACLEAN
Texas A&M University, Galveston

GLENN W. SHEEHAN
Barrow Arctic Science Consortium, Alaska

ANNE M. JENSEN
Ukpeagvik Iñupiat Corporation Science Division, Barrow, Alaska

Inuit is a northern Alaskan term meaning "people" that has come to include the native "Eskimo" peoples of Chukotka, northern Alaska, Canada, and Greenland (Fig. 1). Inuit represent one extreme of the hunter–gatherer paradigm,

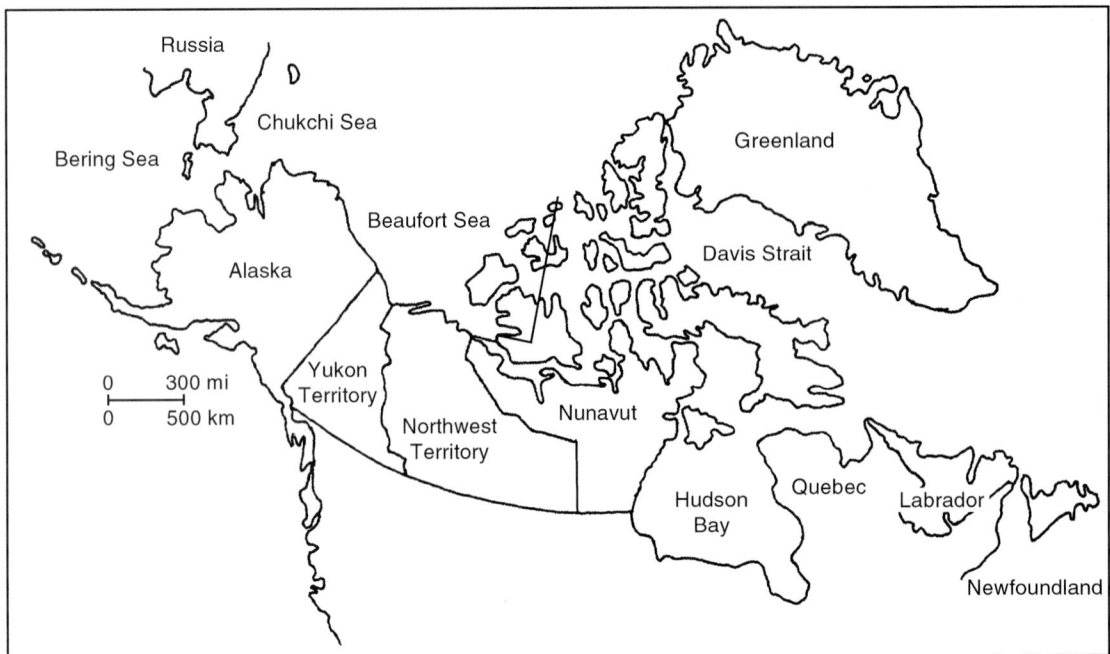

Figure 1 *Coastal Arctic inhabited by the Inuit. Redrawn from Freeman* et al. *(1998).*

relying almost exclusively on hunting to thrive in one of Earth's harshest environments, the Arctic. Most Inuit hunting has focused on marine mammals, with the bowhead whale (*Balaena mysticetus*) making up a central part of the harvest. Whaling was important to Inuit from Alaska to Greenland and underwrote the formation and survival of permanent sedentary villages on Alaska's arctic coast.

Inuit have depended on hunting marine mammals and caribou for thousands of years. The Birnirk culture (A.D. 400–800) was the first to successfully incorporate whale hunting into their subsistence regime. Whaling was completely integrated into the succeeding Thule culture starting around A.D. 800. Around A.D. 1000, Thule folk and their whaling culture spread out of Alaska and into Canada and Greenland.

The ancestral Inuit tool kit employed raw materials from hunted species plus some worked stone and driftwood. Their technology depended heavily on compound tools made from several types of raw materials and incorporating several parts. A harpoon might employ a driftwood shaft, a foreshaft made from caribou antler, a socket piece from walrus (*Odobenus rosmarus*) bone, a finger rest made from walrus ivory, lashings made from caribou sinew, a head made from whale bone, a blade made from slate, a line made from walrus hide, and a sealskin float.

The harpoon head toggled, or turned, 90° once it was thrust into the animal, preventing withdrawal. As the head toggled, the shaft fell away, leaving a hide cord running from the head back to the hunter or to a float. The float was a sealskin with all but one of its orifices sewn shut. The remaining orifice was used to inflate the float through an ivory inflation nozzle, which was then plugged with a piece of driftwood. The float marked the prey's location and slowed it down, tiring it as it attempted to swim or dive.

The first commercial whalers to enter the northern sea near Greenland in the 14th century found Inuit hunting bowhead whales from umiat (skin-covered driftwood framed boats), using compound harpoons with toggling heads. By the early 17th century, Greenlandic Inuit were severely impacted by commercial whaling, which decimated the whale stocks, perhaps even eliminating the Svalbard stock upon which the east Greenlanders seem to have depended. In Canada, much commercial WHALING for the European trade came to be shore based and carried out by local Inuit crews, entailing major alterations to Inuit lifestyles compounded by the destruction of the whale stocks.

Westerners first reached northern Alaska in 1826. However, Inuit lifestyles there were relatively unaltered by contact with the West until the second half of the century, when depredation of the bowhead whale stocks by commercial whaling and the spread of European diseases had disastrous consequences for the Inuit.

Inuit clothing was superior to Western cold weather gear and was often sought by Yankee whalers in Alaskan waters. Entire Inuit families were often hired to travel aboard commercial whaling ships in the Arctic; women skin sewers made and mended clothing for the crew while the men hunted with the Yankees. By the late 19th century, Yankee whalers also adopted the Inuit toggling harpoon head (Bockstoce, 1986).

The Inuit diet relied upon meat and blubber from whales, seals, and polar bears (*Ursus maritimus*). Caribou meat was eaten with seal oil or whale oil. Inland Inuit relied upon traded seal oil for a critical part of their dietary intake (Sheehan, 1997). Skins for boats came from seals and walruses. These, along with caribou and birds, also provided skins for clothing. Whale and seal oil provided fuel for lamps, the only source of heat other than body heat in houses.

In Alaska, driftwood semisubterranean houses incorporated long entrance tunnels made up of whale bones, while in areas of Canada and Greenland, where driftwood was scarce, even the houses were constructed with whale bones, or with stone and bone. The only prehistoric qargi, or whalers' ceremonial house, that has been excavated in north Alaska was made almost entirely of whale bones.

Pokes (seal skins) filled with seal oil were used to preserve meat. Prehistorically in Alaska, i.e., prior to 1826 and even past the middle of the 19th century, seal oil and whale oil pokes were major trade items from coastal areas (Maguire, 1988). Return trade from inland Inuit was primarily caribou skins for clothing and blankets. The economy left nothing to waste, with dog teams consuming old clothing as well as any of the harvest not used directly by the Inuit.

Whaling provided a dependable food surplus to the prehistoric coastal Alaskan communities, allowing them to organize their lives around the whale hunt (Sheehan, 1997). This whaling culture was successful for a thousand years. Whaling remains the organizing focus of Inuit life today in northern Alaska and is still an important part of Inuit ideology in other parts of the Arctic. Marine mammal hunting continues to underpin Inuit subsistence activities and social interactions.

I. Precontact Whaling

It is commonly believed that indigenous whaling developed in the Bering Sea and Bering Strait region about 2000 years ago with the Okvik and Old Bering Sea cultures. An increase in the diversity and complexity of tools used for hunting marine mammals took place from approximately 100 B.C. to A.D. 600. This suggests an increased dependence on large whales and other marine mammals (Stoker and Krupnik, 1993). There appear to be two significant differences between the early groups that hunted whales but did not rely upon them and later groups that were dependent for their survival on the whale hunt. One of these differences was technological, the other social.

The introduction of drag float technology may have transformed whale hunting from a "status" activity resulting, when lucky, in a "windfall," into a "normal" activity resulting in a regular and substantial payoff. Transformation of the umialik (whaling captain) from a temporary hunt leader into a permanent political leader responsible for distributing the whaling surplus throughout the community allowed the population to thrive and grow. The combination of technological and social change culminated in the period of the Punuk and Thule cultures starting at A.D. 800.

Although it is generally agreed that widespread large whale hunting did not occur until the Thule culture spread across North America to Greenland, whaling may have developed independently in several areas at different times. The earliest of these may be the Maritime Archaic tradition of Labrador and Newfoundland, dating from approximately 3000 B.C. The Maritime Archaic is believed to be one of the earliest cultures to use the toggling harpoon head. Møbjerg (1999) reported that the Saqqaq culture of Greenland's west coast, part of a broader Arctic small tool tradition, which stretches across the North American Arctic, may have been hunting baleen whales as early

as 1600–1400 B.C. One of the most interesting cases is the old whaling culture of Cape Kruzenstern, near Kotzebue Sound in Alaska, which appeared suddenly around 1800 B.C. but disappeared shortly thereafter. These people used large lance and harpoon points, possibly to hunt for baleen whales. The abundance of whale bones in the area suggests that whaling was practiced, but there is no evidence that the technology was passed to later cultures (Giddings, 1967).

The Thule whaling culture developed in northwestern Alaska around A.D. 800 and spread very quickly across arctic Alaska and Canada as far as Labrador and Greenland within a few hundred years. The rapid spread of the Thule whaling culture was perhaps influenced by a period of climatic warming. The warmer weather may have resulted in seasonally open water across the entire coast from northwest Alaska to eastern Canada and Greenland, making Pacific and Atlantic populations of whales contiguous and more numerous. These conditions would encourage the expansion of a shore-based whaling culture.

The climate of the far north did not remain warm and stable for long. Colder weather and a resulting increase in expanse and duration of ice cover reduced the distribution and numbers of whales in the Arctic, with a concomitant reduction in the geographic range that could sustain a whaling-focused economy, and made reliance on whales risky in areas that were more marginal. Thule people who could no longer succeed in whaling focused more heavily on smaller marine mammals and other small game. Some parts of the central Canadian Arctic were depopulated.

The climatic variations resulted in dramatic changes to the Thule whaling culture throughout its range. The remnant Thule cultures gave rise to the contemporary Inuit cultures of present-day Canada, Greenland, and Alaska. In Alaska, whalers were able to continue their primary reliance on whale hunting by clustering in large permanent villages at points of land, where every spring they could rely on currents and geography to place them within walking distance of nearshore leads in the ice. Whales followed the leads as they went north for the summer. The leads became the foci of the whale harvest, supplemented by fall whaling in open water, as the whales passed the points on their way south.

II. Mysticetes
A. Bowhead Whale, agviq

The bowhead whale is the largest animal hunted by any prehistoric or historic hunter–gatherer society. Adults reach at least 20 m and weigh 50,000 kg or more. The slow moving, blubber-rich whale is a particularly suitable target, as it often travels close to shore in predictable migration patterns.

The advent of commercial whaling and the consequential contact with Europeans forever changed the patterns of indigenous bowhead whaling. Commercial whalers reduced bowhead populations to levels too low to support a subsistence hunt in most of the whales' range. The Chukotkan natives continued bowhead whaling until the late 1960s when Soviet authorities replaced the shore-based hunt with a catcher-based hunt, primarily for gray whales (*Eschrichtius robustus*). In

1997, the INTERNATIONAL WHALING COMMISSION (IWC) allotted a quota of five bowheads to Chukotkan natives. With assistance and training by Alaskan whalers, the Chukotkan Inuit have begun to hunt bowhead whales again. One whale was landed in 1997 and another in 1998. The Canadian Inuit ceased traditional bowhead hunting around World War I due to low whale numbers and active discouragement by the Canadian government. In 1991, the Canadian Inuit at Aklavik, in the Mackenzie River delta, landed a bowhead for the first time since the early 20th century. An unsuccessful hunt was carried out in 1994 and a successful hunt in 1996. Greenlandic Inuit hunted bowheads for many centuries before commercial whaling depleted the Atlantic stocks nearly to extinction. Greenlandic Inuit were employed by Danish commercial whalers from the late 18th century until 1851, when depleted bowhead numbers brought a halt to commercial hunts. Currently the bowhead whale is hunted under the quota system in northern Alaska, in the villages of Savoonga, Gambell, Little Diomede, Wales, Kivalina, Point Hope, Wainwright, Barrow, Nuiqsut, and Kaktovik, along the Bering, Chukchi, and Beaufort Seas.

After commercial whaling ceased in the early 20th century, Alaskan Inuit returned to a strictly subsistence bowhead hunt. Bockstoce (1986) estimated that an average of 15–20 whales were landed each year from 1914 to 1980. After 1970 there was a significant increase in the number of bowheads landed in Alaska. This was a result of a combination of factors. There was an increase in cultural awareness by Native Americans in general and Alaska Natives in particular, brought about by the passage of the Alaska Native Lands Claim Settlement Act in 1971. The discovery of oil in Prudhoe Bay in 1968 and the construction of the Trans-Alaska pipeline provided significant cash input into the economy of northern Alaska, which prompted a large increase in the number of whaling captains. The position of whaling captain in northern Alaskan Inuit whaling communities has always been one of great respect and authority. Traditionally, only those hunters who demonstrated great hunting success and respect for customs rose to the position of whaling captain. The expense of obtaining whaling gear limited the number of crews and ensured that only experienced whalers rose to the position of captain. The influx of money and employment in the 1970s resulted in a doubling of the whaling crews in northern Alaska from 44 in 1970 to 100 in 1977. The number of whales landed also increased from an average of 15/year to about 30/year from 1970 to 1977. There was also a large increase in the number of whales struck but lost and presumably killed.

The increase in the number of struck but lost whales, combined with an estimate from the IWC that only 600–2000 bowheads remained in the Arctic, prompted the IWC to call for a total ban on bowhead whaling. The Inuit reacted strongly to this ban. They formed the Alaska Eskimo Whaling Commission (AEWC), composed of whaling captains from each whaling village. In 1978 the AEWC, through the U.S. delegation to the IWC, negotiated a quota of 12 bowheads landed or 18 struck for the 9 Alaskan whaling villages. Since then the IWC has established quotas for Alaskan whalers, and the AEWC has distributed strikes to the 10 Alaskan whaling villages (Little Diomede joined AEWC in 1992). Research paid for and conducted through the AEWC and the North Slope Borough (NSB, the regional government in northern Alaska) Department of Wildlife Management indicates that the Eskimo whaling captains were correct when they asserted that there were many more whales than the IWC estimated. Careful censuses of the Bering–Chukchi–Beaufort Seas bowhead population have shown that the bowhead population in the western Arctic actually numbers around 8000 and is increasing about 3% per year. In consequence, the number of strikes allotted to Alaskan whalers has also increased. In 1997, a block quota was set for the years 1998–2002. The quota of 280 whales to be landed during that period includes five whales to be taken per year in Chukotka.

Alaskan Inuit hunt bowhead during the spring and fall migration. In spring, bowheads migrate from wintering grounds in the Bering Sea north through the Bering Strait to feeding areas in the eastern Beaufort Sea. The whales move along open leads in the ice created when drifting pack ice shears away from the grounded, shore-fast ice. These leads occur in predictable places along the Alaskan coast. Bowheads begin the migration north from the Bering Sea in late March through early April and pass the whaling villages of Gambel and Savoonga soon thereafter. The whales pass by Pt. Barrow from mid-April to early June and arrive in the eastern Beaufort Sea in May. Bowheads begin the fall migration across the central Beaufort Sea in early September and pass Alaska's north coast from mid-September to early October. Some whales may continue across the northern Chukchi Sea arriving in Chukotka in November, and others may move southward, likely crossing the central Chukchi Sea.

Equipment used in the modern whale hunt is a combination of precontact technology and tools adopted from Yankee whalers. The boat used for the hunt is a skin-covered frame called an umiaq. The frame was traditionally made of driftwood lashed with baleen with some whale bone fittings, but now is made from prepared lumber. The cover is made from the skins of bearded seals or walrus hunted the previous summer. The skins are left to ferment, which softens the skin and allows the hair to be stripped off easily. The skins are sewn together using a special waterproof stitch and stretched over the frame using rawhide thongs or, more recently, jute or nylon line. The average umiaq in Barrow (Fig. 2) requires six bearded seal skins for the cover, is 6.5–8.5 m long, 1.5–1.8 m across the beam, and weighs approximately 160 kg when dry (Stroker and Krupnik, 1993). The skins are usually replaced every 1 or 2 years, depending on their condition. In some places, aluminum or wooden boats powered with outboard motors have replaced the umiat (plural of umiaq). However, in areas where heavy ice is often encountered, umiat are still used because they are easier to move across and through heavy ice. During fall whaling in Barrow, Nuiqsut, and Kaktovik and during spring whaling in areas where leads are wide and whales travel farther from the lead edge, aluminum or fiberglass boats powered with outboard motors are used.

Weapons used for hunting are essentially the same equipment used by commercial whalers at the end of the 19th century. The darting gun and shoulder guns were introduced by Yankee whalers soon after the Civil War and were adopted by

Figure 2 *Umiaq (skin boat) and harpoon used by Eskimo whalers in Barrow, Alaska. Photo by Steve A. MacLean.*

Inupiat hunters in the last decades of the 19th century (Bockstoce, 1986). The harpoon consists of a wooden shaft 1.5–2 m long tipped with a detachable steel harpoon with a brass toggling head attached to a float with 55 m of strong nylon line. The harpoon is tipped with a plunger trigger-driven gun that fires an 8-gauge, brass bomb simultaneously with the harpoon strike. A second darting gun that resembles the harpoon but without the toggling head harpoon is used to deliver a second bomb. Heavy brass shoulder guns are also used to fire bombs from distances greater than can be attempted with the darting gun. The brass-encased bombs are charged with penthrite, which replaced black powder in 1998. Penthrite bombs deliver a sudden concussion and kill by shock rather than laceration and tissue damage. This reduces the number of whales that are struck but lost. Other equipment includes flensing tools hand made of steel blades (often from hand saws) attached to long wooden handles, heavy-duty block and tackle to haul the whale onto the sea ice, an aluminum or fiberglass boat used to chase and retrieve a whale after a strike is made from the umiaq, and snowmobiles used to tow equipment to and from camp and to carry meat and maktak back to the village.

Preparations for whaling begin well before the whales arrive. Male members of the crew clean weapons and the ice cellar for storing meat and build sleds and other equipment needed for the camp on the ice. The wives of the captain and crew members sew a new skin cover for the umiaq frame. When the skins are dry, the umiaq is lashed to a sled for the wait until a lead opens.

Sometime before the arrival of the first bowheads the captain will decide where to place his camp. One or several "roads" are built across the ice to the selected sites. The roads are built to smooth the route across the maze of pressure ridges on the ocean ice. Smoothing the route eases the task of hauling sled loads of meat and maktak in the event of a successful hunt and provides a quick escape route if ice conditions become unsafe. Stakes with colors or symbols are often placed along the roads. Camps are located on the ice edge, often in "bays" in anticipation of whales swimming under projecting points and surfacing in those bays, or on points that provide good views of approaching whales.

Inuit believed, and many continue to believe, that whales give themselves willingly to hunters worthy of their sacrifice. Traditionally, many taboos governed activities in whaling camps, and these taboos were strictly followed to ensure a successful hunt. Tents, sleeping gear, and cooking were prohibited in camps. Most taboos have been dispensed with, but traditions still govern activity in camps. One tent is set up in camp to allow crew members to sleep in short shifts and for cooking meals. The tent is placed away from the lead and to the right of the boat to prevent approaching whales from seeing the camp. The umiaq is kept ready at the water's edge with a smooth ramp cut into the edge so that it can be launched silently. The harpoon and darting gun are positioned in the bow of the umiaq with the line from the harpoon neatly coiled on the bow. The weapons, lines, and floats are always kept on the right side of the boat, and the strike is always made over the right side of the boat to prevent ENTANGLEMENT in the line. At least one crew member remains on watch at all times, scanning the lead for any sign of an approaching whale.

When a whale comes within range and is determined suitable, the umiaq is launched silently with the harpooner ready in the bow. Two to five paddlers are situated along each side of the umiaq, with a steersman in the stern to steer the umiaq toward the whale. The umiaq is paddled silently, with all crew members stroking in unison. The steersman directs the umiaq to where he or the captain hopes the whale will surface next. The harpooner strikes the whale from as close as possible, often from point-blank range. The preferred target is the postcranial depression just forward of the back. A hit here will often kill the whale instantly. If this target is not available, the spine, heart, or kidney regions are targeted. As soon as the whale is struck, the float is thrown overboard on the starboard (right) side. If possible, a senior crew member other than the harpooner will fire the shoulder gun to plant another bomb into the whale. Other crews, alerted by VHF radio, quickly converge on the site of the strike in aluminum boats powered by outboard motors and may fire another bomb into the whale in an attempt to kill it quickly. Aluminum boats are much faster than umiat and help ensure that a struck whale will not be lost.

Immediately after the whale is killed the captain of the crew that first struck the whale says a prayer (to the Christian God). The prayer is often broadcast over VHF radio and is the first signal of a successful hunt to villagers waiting on shore. The whale's pectoral flippers are then lashed together and the flukes may be removed to reduce drag. A long line is attached to the caudal section forward of the flukes and all available boats attach to the line, with the successful crew at its head, to tow the whale tail first to the butchering site on the ice. Word of the successful hunt is sent to the village by snowmobile, and the whaling flag of the successful crew is raised over the captain's home. Many members of the community then travel to the butchering site to help with hauling the whale onto the ice and butchering it.

At the butchering site a large block and tackle is attached to the ice and used to haul the whale onto the ice. Every available crew member and community member hauls on the free end of the line running through the block and tackle, pulling on commands from the whaling captains. If the whale is too large to haul onto the ice, some butchering may commence in the water. The tongue or SKULL may be removed to ease the task of hauling the carcass onto the ice. Butchering begins as quickly as possible after the whale is hauled onto the ice because the thick blubber layer retards heat loss and the meat in an unbutchered whale quickly spoils. The whale is butchered according to strict customs governing the distribution of shares (Fig. 3). Parts of the whale are reserved for the captain of the crew that struck the whale. Most of that portion will be shared with the community at feasts and festivals that occur throughout the year. Additional shares are divided among the successful crew and the crews that assisted in killing, towing the whale to the butchering site, hauling the whale onto the ice, and butchering. Individuals not representing a crew are also offered shares of meat and maktak. A group of 20–25 people can butcher an average size bowhead in 6 or 7 hr. No shares are distributed until the butchering is complete. Traditionally, following butchering some skulls were rolled into the ocean to allow the spirits of the whales to enter other bodies and again be hunted. The spirit of the whale would remember that the captain treated it well and so sacrifice itself to that captain again. Other skulls were brought ashore and placed at the beginning of the tunnels that led to the entrances of villagers' semisubterranean homes. These symbol-

ically placed skulls suggested that as you entered the home you also entered the world of the whale. The prehistoric qargi or whalers' ceremonial house was built entirely of whale parts to represent a complete whale (Sheehan, 1990). Today, some skulls are not returned to the ocean but are taken ashore where they are cleaned and displayed in the village. The remainder of the skeleton is left on the ice for gulls, foxes, and polar bears.

Bowhead maktak, served boiled fresh, or raw and frozen, is the most prized food in the Arctic. Shares of meat and maktak are widely distributed among family and neighbors, often to family members living in cities who would not receive traditional foods otherwise. Meat is eaten raw and frozen, boiled, or fermented in blood. Many internal organs are also eaten. The kidney, intestines, and heart are boiled. The huge tongue of the bowhead is considered a delicacy when boiled. BALEEN was traditionally used to make toboggans, for lashing of umiaq frames, for bird snares, and to make fish nets and seal nets that could easily be freed of the ice that forms on nets immediately as they are removed from the water. A simple snap of the net broke off the ice from this resilient material. Now baleen is crafted into artwork and sold.

On the day following butchering, the captain of the successful crew opens his home to the community in celebration. All comers are offered food and drink. In early June the umiat of the successful whaling crews are hauled off the ice in ceremonies (apugauti). Once again, the captain supplies food and drink to all who attend. Nalukataq, the formal whaling festival, takes place in June. Each successful crew will have their own nalukataq, or several crews will hold one together. At nalukataq, the members of successful crews distribute the majority of the

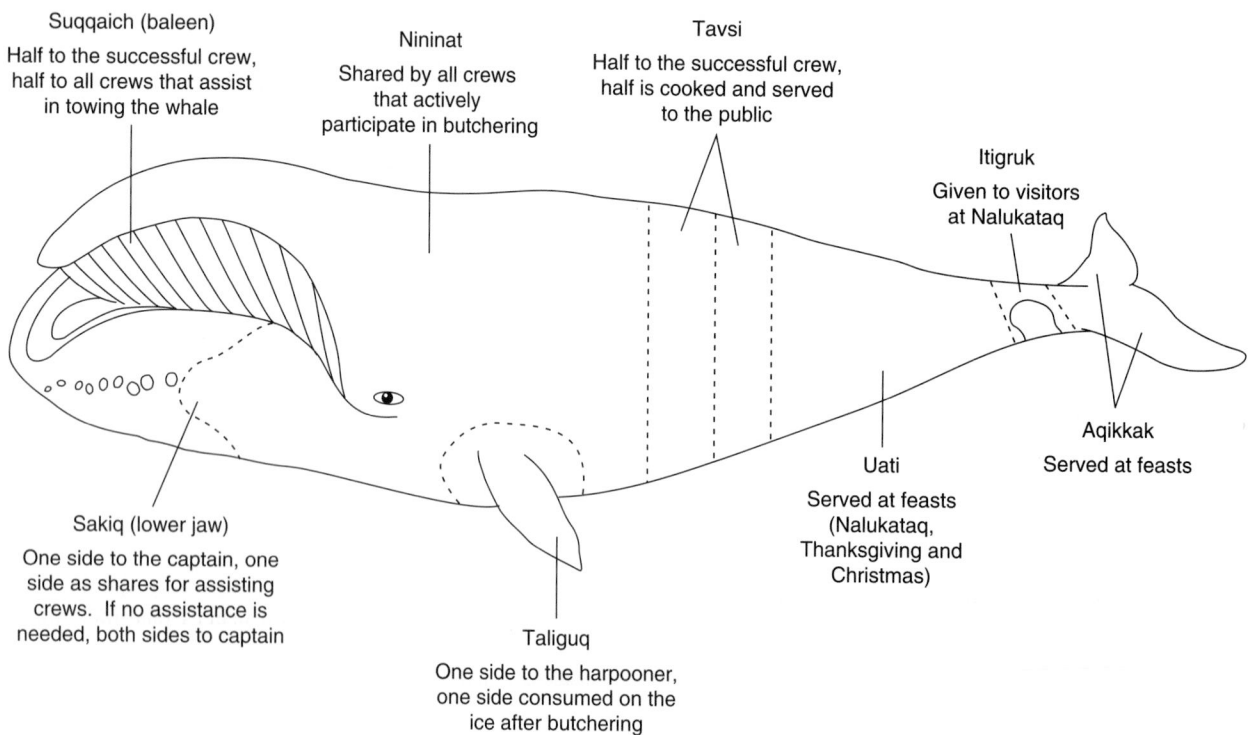

Figure 3 *Division of bowhead whale shares in Barrow, Alaska. From Harry Brower, Jr.*

meat and maktak reserved for the community. The captain and crew also distribute other foods collected during the year, such as caribou meat and soup, duck soup, goose soup, and many other traditional foods. Fruit and candy are also distributed, and coffee, tea, and soft drinks are served throughout the day.

After the food is distributed, the blanket toss begins. Skins from the successful umiaq are removed from the boat and resewn to form a blanket with rope handles along the edge. Community members climb onto the blanket, one at a time, and are thrown into the air by people pulling on the handles in unison. The objective is to jump as high and as many times as possible without falling. Members of successful crews will often climb onto the blanket with bags of candy to fling to the crowd while jumping. After the blanket toss a traditional dance is usually held in the community center. Each successful crew and their families will dance by themselves, but most dances are open to anyone. Nalukataq is one of the most joyful times in the village, and the traditional dance that is the culmination of nalukataq can last late into the night.

B. Gray Whale (*Eschrichtius robustus*)

Only the Chukotkan Inuit of the Russian Far East regularly hunt gray whales. Historically, Chukotkan Inuit hunted both bowhead and gray whales from shore-based stations. The traditional shore-based hunt was banned by the Soviets and replaced by a catcher boat-based hunt in 1954 (Freeman *et al.*, 1998). As a result, the cultural traditions were lost and few people now remember traditional hunting methods. The Soviet catcher boat *Zvyozdnyi* last hunted in 1992 (Freeman *et al.*, 1998). After the catcher boat stopped whaling, the villagers began to hunt marine mammals again to supplement dwindling food supplies.

The return to traditional, shore-based whaling was a difficult and costly endeavor. The lack of equipment and knowledge had serious consequences in several villages. Hunters from the village and Nunlingran died in several hunting accidents, and one whaling boat from Sireniki was sunk, killing all aboard. However, the hunters from seven Chukotkan villages landed 51 gray whales in 1994 (Freeman *et al.*, 1998). In Lorino, several experienced marine mammal hunters were able to teach younger hunters the proper use of harpoons, spears, and rifles. Hunters from Lorino landed 38 gray whales in 1994. Several other villages solicited aid from Lorino, and with training from experienced hunters began to successfully hunt gray whales. The hunt is now sanctioned and controlled by the IWC, with a quota of 120 gray whales landed in Chukotkan villages from 1998 to 2002. Gray whale hunting has again become an important part of Chukotkan Inuit cultural and dietary lives.

Gray whale hunting is carried out in the summer when gray whales move into the Bering Sea from their wintering grounds. Whaling is conducted from shore stations using skin boats (baidara) or wooden whaling boats. The harpoon–spear is a special whaling implement traditionally used by the Inuit of Chukotka (Freeman *et al.*, 1998), consisting of a wooden shaft with a detachable metal spear that is attached to a line with a small float. Each boat carries 7 to 10 of the metal spears and one wooden shaft. The spear is thrown by hand and the metal spear detaches from the wooden shaft. The wooden shaft is re-

trieved from the water, fitted with another harpoon–spear, and the whale is approached again. The harpooner aims for the back of the whale, trying to hit the main blood vessels or vital organs. Once harpoons have been set, the whales are shot with large-caliber rifles. This form of hunting is often dangerous. Gray whales are known to fight aggressively. Two boats are used to ensure the hunters' safety. The hunters also try to take small or medium sized whales.

Gray whales are taken for their meat and blubber. The meat and maktak are eaten frozen, thawed and raw, or boiled. Oil is rendered from the blubber and used as food by itself or added to edible roots, willow leaves, and other vegetables.

In northern Alaska during the early historic period, commercial trade for baleen from bowhead whales created wealth that allowed people to increase the number of dogs in their teams. As a consequence, some gray whales were hunted primarily to feed sled dogs, although some hunters also found the meat to be very tasty. Gray whales are no longer hunted in Alaska.

C. Humpback Whales (*Megaptera novaeangliae*) and Fin Whales (*Balaenoptera physalus*)

Greenlandic Inuit hunted humpback whales from skin boats in much the same way they hunted bowhead whales. Humpback whales are slow-swimming whales, and the techniques used for bowhead whales were successful for humpbacks as well. Although Greenlandic bowhead hunting ceased in the mid 19th century, humpback whaling continued until the 1980s.

In the 1920s, changing sea ice conditions caused food shortages among the Greenlandic Inuit who could no longer catch seals or humpback whales using traditional means. The Danish government operated a steel catcher boat, the *Sonja*, with a Danish crew from 1924 to 1949. The *Sonja* was able to catch larger and faster-swimming whales. In 1927 the *Sonja* caught 22 fin whales, 9 humpbacks, 7 blue whales, and 2 sperm whales. The meat was provided to Inuit of western Greenland and the blubber was shipped to Denmark, where it was rendered to oil and sold. In 1950, the *Sonja* was replaced with the larger *Sonja Kaligtoq*. From 1954 onward, the whales were taken to a single flensing station where meat and maktak were frozen for distribution and sale throughout Greenland. In addition to the government catcher boats, in 1948 some local fishermen began installing harpoon cannons on their boats and hunting whales. Fin and humpback whales were taken to the community where meat and maktak were sold. In the late 1980s the IWC eliminated the humpback whale quota, so fin and minke whales are currently the only baleen whales that are hunted in Greenland.

D. Minke Whales (*Balaenoptera acutorostrata*)

Minke whales have been hunted in Greenland since 1948. The minke whale hunt is now controlled by quotas set by the IWC and administered by the Greenland Home Rule Authority. The variable quotas consider the socioeconomic, cultural, and nutritional needs of the people and the regional abundance of whales. In the 1990s the quota varied from 110 to 175 per year. Minkes are hunted in summer and fall when ice conditions permit.

Hunts from fishing boats and small skiffs are opportunistic. Hunts take place from fishing boats whenever whales are sighted or from skiffs when enough small boat hunters are available. Whalers on fishing boats use deck-mounted harpoon cannons, whereas those aboard skiffs use hand-thrown harpoons and rifles. In each case the whales are towed back to the community for flensing and distribution. Shares are distributed to the vessel owner and crew members, and a large share is reserved for the boat. Little personal share of meat or maktak is sold, but the boat share is sold to contribute to the cost of operating a commercial fishing boat. In the small skiff hunt, shares are divided equally among the participants of the hunt and those helping with the flensing.

III. Odontocetes

A. Beluga Whales (*Delphinapterus leucas*)

Beluga whales are hunted across their range in Chukotka, Alaska, Canada, and Greenland. Ancestors to the modern Inuit were involved in beluga hunting as early as 5500 years ago in Alaska (Freeman *et al.*, 1998). The techniques used by the ancestral Inuit are the same as those used in Alaska, Canada, and Greenland before contact with commercial whalers. Entire communities were involved in a collective whale hunt or drive. A shaman typically guided the hunt, which was led by a distinguished hunter from one of the communities involved. Freeman *et al.* (1998) quoted an elder from Escholtz Bay, Alaska, describing a traditional drive from around 1870: "They made a line and moved together. They hollered, splashed their paddles, waved their harpoons to scare them into real shallow water. . . . When a hunter got a beluga, he ties it to his qayaq (kayak) and brought it to shore; if he get two, he'd tie one on each side. . . . If wind came up while men were out hunting, women would take umiaqs (skin boats) off the racks and go to help those hunters who were towing two belugas. People always helped together when they landed and pulled those beluga on the shore." Friesen and Arnold (1995) determined that beluga whales were a focal resource for precontact Inuit of the Mackenzie delta, constituting up to 66% of their meat. Two or more hunters would cooperate in a beluga hunt. The whales were approached by hunters in kayaks who threw harpoons attached to sealskin floats. After the whale tired, it was lanced in the heart with a blade attached to one end of the kayak paddle. In some locations, hunters in kayaks working cooperatively would drive belugas into shallow water where they were killed. In northern Greenland, and possibly elsewhere, belugas were hunted at large cracks in the ice where the whales congregated to breathe.

In the 18th, 19th, and early 20th centuries, Canadian Inuit were hired by commercial whaling enterprises to hunt belugas. Skins and blubber from the belugas were shipped to European markets. The Inuit hunters kept the meat and some of the maktak and received trade goods, which often included wooden boats.

Methods changed with the introduction of rifles, fiberglass and aluminum boats, and outboard motors. Today, hunters in Alaska use one of four methods to hunt belugas: harpooning or shooting from the ice edge in spring, shooting from motorized boats in open water, netting, or driving the whales into shallow water. Ice edge hunting occurs during the northward migration, sometimes concurrently with bowhead whaling. Belugas can also be shot directly from shore if the migrating whales are close enough, as happened in Barrow in 1997. Open water hunting is common in summer and fall when the ocean is free of ice. Netting occurs at headlands where predictable movement patterns make netting practical. Shallow water drives are most common in shallow bays and estuaries, such as Pt. Lay and Wainwright, Alaska.

Sealskin kayaks were last used to hunt belugas in the 1960s in communities in northern Quebec and the Belcher Islands. Now hunters use skiffs or freighter canoes powered with outboard motors. Harpoons with detachable heads attached to floats are still used, although now floats are made from man-made materials rather than seal skins. Rifles (.222 to 30.06 caliber) are used to kill the whales after harpoons have been attached.

Belugas are the most commonly and widely taken whale species in Canada (Freeman *et al.*, 1998). Beluga maktak is highly prized by Canadian Inuit. After a successful hunt the meat and maktak are distributed to family members and neighbors according to traditional customs. In some communities a successful hunt is announced over community radio and all community members are invited to collect a share. Because beluga maktak is so highly prized, very little of it is sold for redistribution through retail outlets in the Canadian Arctic. Beluga maktak is usually eaten raw and fresh, although some now deep-fry it. The meat is usually air dried before being eaten. In some communities, sausages are made by placing meat in sections of intestine that are lightly boiled before being dried or smoked. Beluga oil was used for lamp oil, softening skins, and cleaning and lubricating guns and other equipment.

Beluga hunting in Greenland has followed a history similar to hunting of other larger whale species. For many centuries, local hunters supplied meat and maktak to meet community needs. In colonial times, beluga blubber and oil became an important trade commodity. As a result, the Greenland Trade Department established commercial beluga drives and hired local hunters to carry out the hunt. Commercial drives continued until the 1950s when the European market for whale oil disappeared. Commercial whale drives reappeared in the 1960s when improved coastal communication and refrigeration made it possible to transport beluga meat and maktak from northern hunting communities to southern Greenland. Today, belugas are hunted with rifles (30.06 caliber to 7.62 mm) from small boats. Typically, kayaks and motorized skiffs are used to hunt belugas, often singly or in pairs, but sometimes a larger number of small boats cooperate to hunt belugas swimming together. Meat and maktak are distributed throughout the community, including sale at the local market, and in retail stores throughout Greenland.

Beluga hunting in Russia only occurs in a few villages in Chukotka, and the numbers taken are small. Belugas in Russia are associated with the distribution of fish, especially arctic cod and arctic char. Hunting occurs opportunistically when belugas are encountered during other activities. Hunting occurs either from shore or from the ice edge. Hunters hide behind hummocks of ice and shoot the whales with rifles (7.62 or 9 mm). Meat is dried, frozen, boiled, or fried. Maktak is eaten raw, fresh, boiled, or fried. The skin is used for boot soles, belts, and lines. The oil is used with fish and salad plants. Historically, beluga oil was traded for reindeer meat and skins, although when Soviet state-run fur farms were operating the oil was sold to the farms (Freeman *et al.*, 1998).

B. Narwhal (*Monodon monoceros*)

Narwhals have been hunted in Greenland and eastern Canada for centuries, and may have brought the Greenlandic Inuit in close contact with the Norse in Greenland beginning in the 10th century. Narwhal ivory was bartered among Inuit long before European contact. Narwhal tusks were highly valued by European traders in the Middle Ages, who sold the tusks in Europe mislabeled as unicorn horn, sometimes for their weight in gold. The royal throne of Denmark, made in the 15th century, is made almost entirely of narwhal ivory. Narwhal tusks were the basis of trade between Greenlandic Inuit and Europeans from the 10th through the 19th centuries and were important to Canadian Inuit after the collapse of commercial bowhead whaling in the late 19th century. Inuit in Greenland and Canada used the tusks to create durable and functional tools, especially harpoon foreshafts.

Narwhals were hunted from kayaks either along the flow edge, in ice cracks, or in open water. Near ice, the narwhals were harpooned and hauled ashore. In open water, hunters worked together to drive the narwhals into shallow water where they were killed. Another method was to station hunters with rifles on cliffs who would shoot the whales as they swam by. Several hunters in kayaks would wait offshore and harpoon the whales once they were shot. Now, hunting in Canada takes place with small skiffs, rifles, and harpoons attached to floats. Narwhal hunting in northern Greenland is still accomplished with kayaks. Five-meter skiffs or 10- to 12-m cutters are used in southern Greenland, although occasionally narwhals are shot from shore or netted.

Maktak from narwhals is prized and is eaten fresh raw or aged. Narwhal oil was considered of higher quality than seal oil and was used in lamps for heat and light. The tusk remains the most highly prized product from narwhal. Today tusks are used for artwork or sold. Narwhal ivory sold for an average of $100 per foot (30 cm) in 1997 (Freeman *et al.*, 1998). Narwhal meat was used to feed hunters' dog teams.

C. Other Small Cetaceans

Small numbers of other cetaceans are taken in eastern Canada and Greenland. The principal species taken in Canada are common bottlenose dolphin (*Tursiops truncatus*) and harbor porpoise (*Phocoena phocoena*). In Greenland, killer whales (*Orcinus orca*), long-finned pilot whales (*Globicephala melas*), northern bottlenose whales (*Hyperoodon ampullatus*), harbor porpoise, white-beaked dolphins (*Lagenorhynchus albirostris*), and Atlantic white-sided dolphins (*Lagenorhynchus acutus*) are taken.

IV. Pinnipeds

A. Ringed Seal (*Pusa hispida*), natchiq, Bearded Seal (*Erignathus barbatus*), ugruk, and Harp Seal (*Pagophilus groenlandicus*)

Seals are probably the most widely distributed, abundant, and reliable food resource available to coastal Inuit popula-tions. Ringed seals are available near shore for much of the year. Bearded seals are also important, although less abundant and less widely available than ringed seals. They are important not only for their meat, but also as a source of raw materials, particularly their hides (Jensen, 1987). Harp seals are seasonally very abundant in certain areas of Greenland and eastern Canada, and were taken when present. Ribbon seals (*Histriophoca fasciata*), Larga seals (*Phoca largha*), and harbor seals (*Phoca vitulina*) are only occasionally encountered. All of these pinnipeds are hunted in similar ways and have been combined for the following discussion.

Natchiq are ice adapted. They are hunted at breathing holes, in subnivean lairs, on drift ice, and in open water. Other seals are not as ice adapted as the natchiq. They can also be hunted on drift ice and in open water. Harbor seals and Larga seals tend to stay away from ice if it is present in significant amounts. Ugruk are common on ice pans and commonly hunted on pans or in open water. Harbor seals tend to be more common than natchiq in more southerly areas (southern Greenland, Labrador), although they have been regarded as shy and also potentially aggressive. Harp seals were generally taken from kayaks in open water or when hauled out on offshore drift ice, although they could be harpooned from shore or from the ice edge under certain circumstances.

Traditionally, natchiq were hunted at breathing holes on the ice, at pupping dens, while basking in the sun, by netting at the breathing hole, from the ice edge, or from boats in open water. Breathing-hole hunting was most common, as the ocean is ice covered for much of the year. Ringed seals carve out and maintain breathing holes in the ice throughout the winter. In flat ice the breathing holes may be visible from the surface, but often they are covered with snow, and practically invisible. Ringed seals maintain numerous breathing holes, so there was never any guarantee that a seal would visit the hole where the hunter was waiting.

Breathing-hole hunting was a difficult, cold endeavor and is no longer practiced to any great extent anywhere in the Arctic. Boas (1964) presented an excellent description of pre-rifle seal hunting methods and equipment. A hunter would first locate a breathing hole with the use of one of his sled dogs. Once the hole was found, the hunter set up his equipment around the hole. The hunter sat on an ice block with his feet resting on a piece of fur or stood on the fur with his harpoon in his hand or at his side and waited for the seal to arrive at the breathing hole. There was never any way to determine how long the hunter would have to wait. If the village needed food, it was not uncommon for hunters to wait 24 hr or longer for a seal to arrive. Now, more efficient and less strenuous methods are preferred.

When a seal arrives at a breathing hole, the first breath is a short, shallow sniff for any sign of danger. If the seal does not detect danger, the next breath will be deeper. On this second breath, the hunter thrust his harpoon straight down the hole striking the seal on the head or neck. The toggling head detached, preventing the seal from escaping. The seal was killed and the breathing hole enlarged to pull the seal through. Once rifles became available, seals were shot when they came to the hole, then immediately harpooned to prevent the seal from drifting away or sinking.

After the breeding season, seals enlarge their breathing holes located on large areas of flat ice so they can climb out and bask in the sunshine. Traditionally, Inuit had several methods for hunting seals at this time, described in detail in Nelson (1969) and Boas (1964). A hunter might simply wait near one of the holes for a seal to surface. The water within the hole pulsates when a seal arrives at its hole. When the seal broke the surface of the water, it was speared or shot. Occasionally, hunters placed lines with several hooks along the wall of a breathing hole to catch seals backing into the water after surfacing.

Another traditional seal hunting technique required great stealth and skill. The hunter emulated the behavior of a seal, sliding along the ice on his side, often with a piece of sealskin beneath him to reduce friction and keep his clothing dry. Often hunters would scrape the ice with seal claws attached to a piece of wood to mimic the scratching sound made by resting seals. A skilled hunter could approach very close to a seal basking in the sun. In this way hunters were often able to kill 10–15 seals in 1 day. In a variant of this method, the hunter pushed a small sled with a white shield that hid him from the seals.

Seals could also be netted at their breathing holes. Netting was done at night to prevent the seal from seeing and avoiding the net. This also reduced the hunters' vision and exposed the hunter to many dangers. Four holes were cut around a breathing hole and the net lowered into the water to approximately 10 feet. Seals generally approach breathing holes along the surface, so they did not encounter the net. When the seals dove from the hole, they dove straight down and became entangled in the net. Seal netting was discontinued in the 1960s.

In spring, pregnant ringed seals hollow a natal den in the snow covering one of their breathing holes. Hunters again use one of their dogs to find the dens. The hunter cut a small hole in the wall of the den through which he could watch for the return of the mother seal. When the seal returned, the hunter jumped through the snow between the seal and its hole, trapping it. Prior to the introduction of rifles the seals were killed with a spear or club; later they were shot through the wall of the den.

Traditionally, ice-edge hunting was accomplished with a small harpoon that was thrown at seals swimming near the edge. A line was attached to the harpoon to retrieve struck seals. Hunters were limited by how far they could accurately throw the harpoons, usually 10–20 feet. The introduction of the rifle changed the nature of seal hunting. Hunting seals from the ice edge using rifles is easier and more efficient than breathing-hole hunting, and the range of the hunters has been increased greatly by the rifles. The increased range brought about two new inventions specifically for use in ice-edge rifle hunting: the retrieving hook (manaq or manaqtuun) and a small skin boat (umaiggaluuraq). The manaq consists of a rope up to 200 feet long, attached to a piece of wood with four hooks extruding from the sides. A float is attached to keep the hooks afloat for winter hunting (when seals float after being shot), and a sinker is attached for summer hunting to retrieve seals that sink to the bottom. Once a seal is shot, the hunter grabs his manaq to retrieve the seal from the water. The line is coiled and held in the left hand, while the right hand holds the line 3–5 feet

from the hooks. The hook is thrown beyond the seal, the line is slowly drawn in until the hooks are near the seal, a sharp tug sinks the hooks into the hide, and the seal is carefully pulled to the ice edge.

The umaiggaluuraq (literally "small umiaq") is 7–10 feet long and 36–40 inches wide (Nelson, 1969). Two bearded sealskins are used to cover a wooden frame. Once a hunter shoots a seal, he pulls the boat to the ice edge, often with the help of another hunter to prevent damage to the skins by dragging the boat. The boat is rowed to the seal with two short oars lashed to the gunwales. When the hunter reaches the seal, he tows it back to the ice edge with a small hook and line.

Open water hunting and hunting of seals basking on drift ice became most popular after the introduction of rifles. Before rifles were introduced, hunters occasionally harpooned seals from kayaks, but only in calm water. After rifles and outboard motors became readily available, several men would hunt together from a single umiaq. The hunters were often members of the same whaling crew using the captain's boat. Seals were shot with rifles ranging from 22 to 30.06 caliber and harpooned. Now, aluminum boats have replaced skin boats, but the same methods are used. Open water hunting from aluminum boats is currently the most popular way to hunt both the ringed and the bearded seal in northern Alaska. Harpoons are still used in the Yukon-Kuskokwim Delta because people feel that shot seals sink too quickly. In Greenland, certain areas still forbid motorized boats in the hunt, although they may be used to travel to the hunting area.

B. Walrus, aiviq

Walruses are almost always associated with pack ice and are only hunted when the pack ice is close to shore. They do haul out on shore in certain locations, although this has become less common. Nelson (1969) reported that hunters in Wainwright, Alaska, only traveled offshore as far as land was still visible on the horizon. However, Spencer (1959) reported that hunters in Barrow often traveled 50 to 100 miles into the ocean to find walruses. The distances traveled are probably dependent on the proximity of the pack ice to shore and undoubtedly changed with the introduction of outboard motors.

Hunting walruses was, and remains, a collective hunt. The size of the walrus and the logistics of butchering and transporting the meat back to the village make it necessary for several hunters to work cooperatively. Traditionally, walruses were hunted using large harpoons similar to the harpoons used in bowhead whaling. Long lines, often made of walrus skin, were attached to the harpoons and fastened to a large piece of ice or were held by the hunter who used a smaller spear to drive the end of the line into the ice. Walruses were harpooned while they were lying on the ice. When the harpooned walrus dove, the line prevented it from escaping. When the walrus tired, it was killed with a lance through the heart. Occasionally, walruses were hunted from umiaqs when they were encountered away from the pack ice. In those circumstances, floats were attached to the line or the line was fastened to the umiaq. The walrus was killed with a lance once it tired. Nelson (1969) summarized an elder recounting one traditional method of hunting walruses in which two hunters harpooned two walruses facing opposite directions. The lines from the two harpoons were quickly tied together, and the walruses pulled against each other until they tired enough to be killed with lances through the neck. Now, large rifles are used instead of harpoons, but the methods used to approach the walruses are the same. When a walrus herd is sighted, the ice surrounding the herd is evaluated. There must be enough ice-free water to allow approach and to allow sufficient time for the killed walrus to be butchered before ice closes in.

Walruses are approached slowly with the outboard running. Generally, walruses are approached to within 10 feet before they are shot. All hunters shoot at the same time and continue the volley until enough have been taken or the herd escapes into the water. Walruses must be shot in the brain or the anterior portion of the spinal cord to ensure a kill. Walruses will not float once killed, so any dead or seriously wounded walruses that fall into the water are considered struck and lost. Fay *et al.* (1994) reported that up to 42% of walruses struck in Alaskan hunts from 1952 to 1972 were lost. Wounded walruses are often dangerous, and Nelson (1969) recounted several instances in which wounded walruses damaged boats. In fact, walruses can be so aggressive that they have disrupted mail delivery by kayak and even forced the abandonment of a settlement in Greenland.

Walrus flippers "ripened" in seal oil are considered a delicacy in much of the Arctic. Select portions of meat are eaten, but the bulk of the walrus was used to feed the hunters' dog teams. The skin, bones, and especially the tusks were the most valuable parts of the walrus. Walrus skins often replaced bearded seal skins on umiaqs in places where bearded seals were not abundant. Walrus skins were also used to create strong lines that were attached to harpoons used in seal, walrus, and whale hunting. The bones of walruses were used to make tools, and the ivory tusks were often used to make harpoon points and foreshafts. Now, ivory is used in artwork and much is sold to generate a cash income.

V. Polar Bears, nanuq

Polar bears are found throughout the Arctic and are hunted through much of their range. Polar bears remain on the pack ice for most of the year, and most hunting takes place during the winter on the pack ice. Polar bears are also taken opportunistically when they are encountered on land or in open water.

Polar bear hunting is considered one of the most dangerous hunting activities and successful hunters often enjoy high status in village communities. Traditionally, single hunters using spears, lances, or knives hunted polar bears. Boas (1964) and Nelson (1969) both described polar bear hunts before the introduction of rifles. In the Canadian and Greenlandic Arctic, it was common to release dogs to chase the bears and tire them. Once the bears stopped, they were approached on foot and killed with lances or spears. Dogs were not used commonly in Alaska, but were released if the bears were on young, unsafe ice. Spears and lances were quickly given up once rifles became available.

Hunting for polar bears is now nearly always done on the sea ice, and hunters often travel far offshore to find bears. Walking used to be the preferred method of transportation because it offered the advantages of a silent approach and the ability to hide quickly among the ice hummocks and ridges. Now, snowmobiles are preferred. With snowmobiles, hunters can pull sleds to transport the meat and hide back to the village, eliminating the need to drag the hide and then return with dogs to transport the meat.

Hunters usually find tracks rather than finding the animal itself. From the tracks hunters can tell the size of the animal, its direction and speed, and how long ago the bear passed. Tracks are followed until the bear is sighted. The hunter can then either move quickly to overtake the bear or move ahead to wait in ambush. In either case, it is important to get as close to the bear as possible to ensure a lethal shot. Wounded polar bears are dangerous and sometimes attack the hunter. If the bear is in a position that the hunter cannot reach, the hunter will sometimes try to lure the bear closer by mimicking a sleeping seal. Once the bear stalks close enough, the hunter picks up his rifle and shoots. Sometimes hunters leave seal blood or blubber on the ice and return to the area later to see if any bears have been lured by the smell. When bears venture close to villages or whaling camps they are almost always shot.

Polar bears are hunted for both their meat and their hides, which are divided among the village according to local tradition. In Greenland, the person who sights the bear becomes its "owner" regardless of whether they participate in the hunt. Any other people who shoot the bear or touch it before it is killed also receive shares of the bear. In Alaska and Canada, shares were traditionally distributed widely within the village. A young hunter's first bear was shared among all the people in the hunting party or was distributed to the elders in the village if he was hunting alone. Now, the shares are distributed less formally, but meat is usually shared with family members and others outside the family. The successful hunter usually keeps the hide.

Polar bear meat is prized by many people in the Arctic. Meat is always well cooked to prevent trichinosis, and the liver is never eaten due to high concentrations of vitamin A. In Alaska the sale of polar bear hides is prohibited by the Marine Mammals Protection Act of 1972. Hides are used for clothing such as boots, mittens, or trim for parkas and also for sleeping mats when camping on the ice. In Greenland, polar bear skins were used for warm hunting pants, but now all skins are sold to Greenland's trading department. Since 1994, polar bear hunters in Greenland have been able to sell bear meat to restaurants and hotels.

VI. Conclusion

Inuit and their ancestors have hunted marine mammals for thousands of years. The technology and techniques of hunting marine mammals evolved in a culture intimately associated with the sea and the creatures that inhabit it. In modern times, the technology and techniques of hunting marine mammals have changed, but many traditions and beliefs remain. Marine mammal hunting provides access to status within the community and a sense of self-worth for a generation of Inuit struggling to cope with the burdens of cultural assimilation. It is the traditions and beliefs that are necessary for marine mammal hunting to remain an important part of the Inuit people's subsistence and cultural lives.

See Also the Following Articles

Beluga Whale ■ Bowhead Whale ■ Folklore and Legends ■
Hunting of Marine Mammals ■ Polar Bear

References

Boas, F. (1964). "The Central Eskimo." University of Nebraska Press, Lincoln.

Bockstoce, J. R. (1986). "Whales, Ice, and Men: The History of Whaling in the Western Arctic." University of Washington Press, Seattle.

Fay, F. H., Burns, J. J., Stoker, S. W., and Grundy, J. S. (1994). The struck-and-lost factor in Alaskan walrus harvests, 1952–1972. *Arctic.* **47**(4), 368–373.

Freeman, M. R., L., Bogoslovskaya, R. A., Caufield, I., Egede, I. I., Krupnik, I. I., Stevenson, M. G. (1998). "Inuit, Whaling, and Sustainability." AltaMira Press, Walnut Creek, CA.

Friesen, T. M., and Arnold, C. D. (1995). Zooarchaeology of a focal resource: Dietary importance of beluga whales to the precontact Mackenzie Inuit. *Arctic* **48**(1), 22–30.

Giddings, J. L. (1967). "Ancient Men of the Arctic." Alfred A. Knopf, New York.

Jensen, A. M. (1987). Patterns of bearded seal exploitation in Greenland. *Études/Inuit/Stud.* **11**(2), 91–116.

Maguire, R. (1988). "The Journal of Rochfort Maguire, 1852–1854: Two Years at Point Barrow, Alaska, Aboard *H.M.S. Plover* in the Search for Sir John Franklin" (J. R. Bockstoce, ed.), 2 vols. Works issued by the Hakluyt Society, Second Series No. 169. The Hakluyt Society, London.

Møbjerg, T. (1999). New adaptive strategies in the Saqqaq culture of Greenland, c. 1600–1400 BC. *World Archaeol.* 30(3), 452–465.

Nelson, R. K. (1969). "Hunters of the Northern Ice." University of Chicago Press, Chicago.

Sheehan, G. W. (1990). Excavations at Mound 34. *In* "The Utqiagvik Excavations (E. S. Hall, Jr., ed.), Vol. 2, pp. 181–325, 337–353. The North Slope Borough Commission on Iñupiat History, Language and Culture, Barrow, Alaska.

Sheehan, G. W. (1997). "In the Belly of the Whale: Trade and War in Eskimo Society." Aurora, Alaska Anthropological Association Monograph Series—VI, Anchorage, Alaska.

Spencer, R. F. (1959). "The North Alaskan Eskimo: A Study in Ecology and Society." Smithsonian Institution Press, Washington, DC.

Stoker, S. W., and Krupnik, I. I. (1993). Subsistence whaling. *In* "The Bowhead Whale" (J. J. Burns, J. J. Montague, and C. J. Cowles, eds.). Academic Press, Lawrence, KS.

Irrawaddy Dolphin
Orcaella brevirostris

PETER W. ARNOLD
*Museum of Tropical Queensland,
Townsville, Australia*

The Irrawaddy dolphin *Orcaella brevirostris* (Owen in Gray, 1866) is a coastal Indo-west Pacific species that also occurs in major river systems of Southeast Asia. Until recently, it was known mainly through the classic monograph by Dr. John Anderson, Superintendent of the Indian Museum, Calcutta from 1865 to 1886.

I. Systematics

The Irrawaddy dolphin resembles the beluga *Delphinapterus leucas* (Pallas, 1776) in general appearance and certain anatomical features, such as the tympanoperiotic earbones. This led some specialists to consider it a tropical relative of the beluga, placing it in the variably defined family Monodontidae or Delphinapteridae. It has also been placed in a family of its own. Recent morphological and genetic studies consistently place it in the family Delphinidae. Genetic studies suggest that its closest relative may be the killer whale *Orcinus orca* (Linnaeus, 1758).

The common name refers to its occurrence in the Irrawaddy (Ayeyarwady) River, Burma (Myanmar). The freshwater population was originally described as a separate species *Orcaella fluminalis* (Anderson in Gray, 1871); however, studies provide little support for this distinction. Some morphological and genetic studies do suggest a potential taxonomic separation between animals found throughout Southeast Asia and those in Australia and Papua New Guinea.

II. Description

The head is broadly rounded, with no sign of a beak. There is an indentation behind the head, forming a distinct neck crease. The head is very mobile and can be deflexed almost 90° to the body axis. The dorsal fin, over halfway along the back, is small—hence the alternative name "snubfin dolphin" (Fig. 1). A midventral crease runs along the belly from level with the flippers to the genital slit. The flippers are broad and paddle-like; they are highly mobile and may be rotated forward through a wide angle.

The color pattern may vary regionally. Animals have been described as "dark gray" but Australian animals are lighter colored, with a subtle three-tone pattern (white belly, gray cape along the back, and lighter gray to brown along the sides).

Maximum recorded length of the Irrawaddy dolphin is 2.75 m, but it averages 2.1 m in length and 115–130 kg mass. Males appear to be about 15% longer than females.

There are 16–20 TEETH in each half of the upper jaw and 15–19 teeth in each lower row. In dolphins from the Mahakam River system, East Kalimantan, the teeth may not erupt from the gums.

III. Distribution

The species is widely distributed in coastal waters from the Madras coast of India as far north as Calcutta (22°N) to the northeast coast of Australia almost to the Queensland–New South Wales border (27°S). In between it is recorded from Bangladesh, Myanmar, Thailand, Kampuchea, Malaysia, Singapore, Brunei, Indonesia, the Philippines (unpublished records of Dr. A. A. Yaptinchay), Vietnam, and Papua New Guinea (Fig. 2).

It occurs to at least 1300 km upstream in the Irrawaddy River. There are also records of sightings throughout the Mekong River, from its mouth in Vietnam upstream to the border between Cambodia (Kampuchea) and Laos (Lao Peoples' Democratic Republic). Landlocked populations occur in the Lao PDR and in East Kalimantan, Indonesia.

Figure 1 *These animals clearly show the rounded head and small dorsal fin characteristic of the Irrawaddy dolphin. The squirting of water, which has been reinforced by training in these captive dolphins, occurs naturally in freshwater populations. The function is unknown. Courtesy of Isabel L. Beasley.*

IV. Ecology and Behavior

The Irrawaddy dolphin has a generalized diet, taking a wide variety of fishes as well as cephalopods (squid, cuttlefish, octopus) and shrimps. Food is taken both near the bottom and in open water. Captive animals were fed about 4–8% of their body weight daily, but this may exceed their metabolic needs. Phonations at about 60 kHz were recorded from Mahakam River dolphins held captive in Indonesia. These signals are apparently used for echolocation but vision also appears to be good. There is no basis for referring to this species as a "blind dolphin," as occurs in some of the general literature.

In the Northern Hemisphere, mating is reported from December to June. The gestation period has been estimated at 14 months. Newborn animals or near-term fetuses have been reported from April to June; births occurred in captive Indonesian animals from December to July. The newborn dolphin is about 1 m long and has a mass of 10–12 kg. Captive animals started taking fish at about 6 months and were fully weaned by 2 years of age. A small number of animals from northeastern Australia, aged using dentinal growth layer groups in teeth, reached near adult size at 4 to 6 years. Maximum life span was considered to be about 30 years.

Group size is usually 2 to 3, but may reach 10–15 animals. Surfacing is generally unobtrusive, with a low roll showing little of the back. Given this, and the small size of the dorsal fin, the Irrawaddy dolphin is often easy to miss in the field. However, it may raise the flukes above water and wave or slap the flippers, as well as spyhop or breach. The maximum dive time was recorded at 12 min but animals more usually submerge for about 2 to 3 min.

Although Anderson reported dolphins riding the bow of steamers in the Irrawaddy River, recent observations suggest lack of interest in or even avoidance of large power vessels. However, there are reports from Burma and Thailand, dating to the last century, of Irrawaddy dolphins driving fishes into fishermen's nets. Such cooperative fishing has been confirmed in the Irrawaddy River.

V. Conservation Status

Irrawaddy dolphins may be killed for their oil or captured for display in public aquaria. Both practices appear to be limited but may affect local populations. Legislation protecting dolphins is in place in most countries where Irrawaddy dolphins occur, but enforcement is a problem given the extensive (and poorly patrolled) range of the species. Dolphins may be killed in those areas where explosive fishing is still used. Of wider concern is accidental capture in fishing nets or nets used to control the number of sharks off popular swimming beaches. Concern has been raised especially about habitat reduction, with possible fragmentation of dolphin populations, and degradation. Suggested effects include reduced food supplies (from habitat degradation and overfishing), increased industrial and urban pollution of coastal sites, and disturbance from increased vessel traffic. Such potential threats are largely unquantified. There are few reliable population estimates from anywhere within the range of this species, and in most conservation plans it is listed as "insufficiently known." However, with the exception of northern Australia, the Irrawaddy dolphin has been reported in low numbers throughout its range, with suggestions of population declines in several Southeast Asian countries.

See Also the Following Articles

Delphinids, Overview ■ Indo-West Pacific Marine Mammals ■ River Dolphins

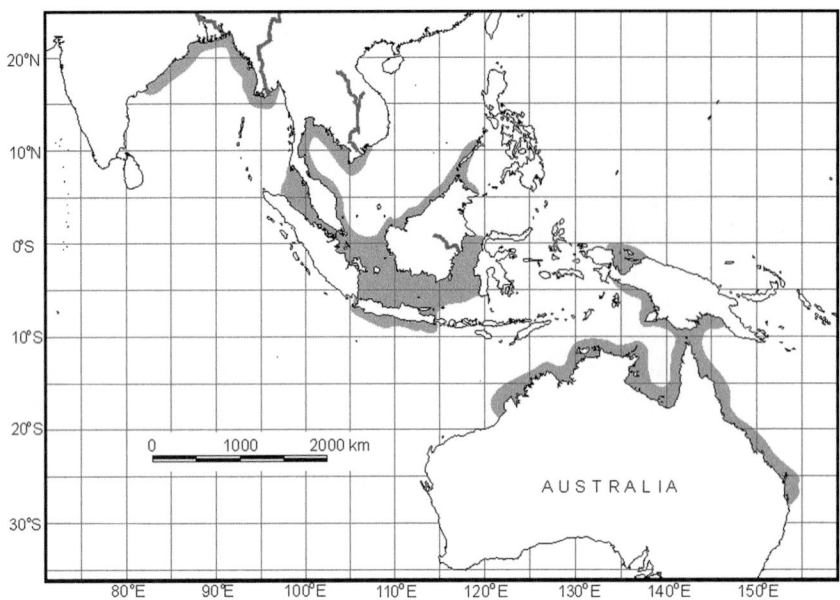

Figure 2 *Distribution of Irrawaddy dolphin, based largely on information in Stacey and Leatherwood (1997) but including subsequent published records. More information is needed to substantiate the apparent gap between confirmed localities within Southeast Asia and those in Australia/Papua New Guinea.*

References

Anderson, J. (1879). Anatomical and zoological researches comprising an account of the zoological results of two expeditions to western Yunnan in 1868 and 1875; and a monograph of the two cetacean genera *Platanista* and *Orcella*. B. Quaritch, London.

Arnold, P. W., and Heinsohn, G. E. (1996). Phylogenetic status of the Irrawaddy dolphin *Orcaella brevirostris* (Owen in Gray): A cladistic analysis. *Mem. Queensl. Mus.* **39**, 141–204.

Marsh, H., Lloze, R., Heinsohn, G. E., and Kasuya, T. (1989). Irrawaddy dolphin *Orcaella brevirostris* (Gray, 1866). *In* "Handbook of Marine Mammals" (R. J. Harrison and S. Ridgway, eds.), Vol. 4, pp. 101–118. Academic Press, New York.

Stacey, P. J., and Arnold, P. W. (1999). *Orcaella brevirostris. Mamm. Species* **616**, 1–8.

Stacey, P. J., and Leatherwood, S. (1997). The Irrawaddy dolphin, *Orcaella brevirostris:* A summary of current knowledge and recommendations for conservation action. *Asian Mar. Biol.* **14**, 195–214.

Japanese Whaling

Toshio Kasuya

*Teikyo University of Science and Techology,
Uenohara, Japan*

Whaling is a fishing activity that targets whales, but the term does not often fit because of the ambiguity of *whale,* sometimes construed to exclude small cetaceans. This is also true in Japanese whaling. This article adopts the broadest meaning for the term Japanese whaling to include activities of hunting any cetaceans in Japanese territory, by Japanese companies, or by any companies known to be sponsored by them.

I. Subsistence Whaling

Numerous bones of gregarious dolphins in a site of the Jomon Era (10,000 B.P.–200 B.C.) on the Noto coasts, Sea of Japan, suggest the presence of a drive fishery. Other sites of similar antiquity on the Pacific coasts of central and northern Japan and on the coasts of northern Kyushu facing the Sea of Japan/East China Sea revealed remains of small cetaceans and detachable harpoon heads. The Okhotsk Sea culture of Hokkaido in the 5–14th centuries left skeletons, harpoons, and drawings depicting whale harpooning. Ainu people on Uchiura Bay, Pacific coast of southern Hokkaido, opportunistically hunted minke whales in the late 19th century using aconite-poisoned detachable harpoon heads and floats.

Skeletal remains from these sites represent at least 13 species of cetaceans, i.e., North Pacific right whale (*Eubalaena japonica*), common minke whale (*Balaenoptera acutorostrata*), sei whale (*B. borealis*), humpback whale (*Megaptera novaeangliae*), sperm whale (*Physeter macrocephalus*), false killer whale (*Pseudorca crassidens*), long-finned pilot whale (*Globicephala melas*), Pacific white-sided dolphin (*Lagenorhynchus obliquidens*), common dolphin (*Delphinus* sp.), common bottlenose dolphin (*Tursiops truncatus*), Dall's porpoise (*Phocoenoides dalli*), harbor porpoise (*Phocoena phocoena*), and unidentified beaked whales (Ziphiidae) (e.g., Kasuya, 1975). Differentiation of whales hunted from those stranded, however, is often difficult.

II. Traditional Commercial Whaling

On the coasts of Noto and northern Kyushu, whales were hunted until the end of the 19th century by placing small trap nets at whale passages. Recorded takes of harpooned whales in a harbor at Ine, Sea of Japan, included 167 humpback, 148 fin, and 40 right whales in 1656–1913. A village next to Ine took small cetaceans in the same way. A similar fishery was also recorded in the 14th century at villages on Tsushima Island, off northern Kyushu. A cooking recipe in 1489 recommended whale meat for noble guests.

Records of "harpoon whaling" started in the 1570s at Morosaki at the entrance of Mikawa Bay, a bay attached to Ise Bay that opens to the Pacific (Fig. 1). The whalers first used light harpoons with a detachable head and line. Harpoons with fixed head and lancing followed. The winter operation continued to the early 1800s and took gray (*Eschrichtius robustus*) and humpback whales for oil and meat.

This was soon transmitted eastward to Katsuyama (35°05′N) at the entrance of Tokyo Bay for Baird's beaked whales (*Berardius bairdii*) and survived until the late 19th century. It was also transmitted westward to the nearby Ise and Kii areas (before 1606), Shikoku (1624), northern Kyushu (1630s), and Nagato (around 1672).

Harpoon whaling on the Ise and Kii coasts mostly ceased before 1770. A whaling group at Taiji, Kii, was an exception. They modified old harpoon whaling learned from Morosaki in 1606 into new "net whaling" in 1677. During the whaling season (winter and spring), harpoon boats waited offshore for a signal from the spotters on cliffs. With signals of flags and smoke indicating species, number, and position of whales, they drove the whales toward shore where net boats waited, which placed nets in front of the whales to entangle them. Then the procedures of harpoon whaling followed, i.e., harpooners threw harpoons (of fixed head) and then lances followed. When a whale became weak, a harpooner swam to the whale to tie ropes through holes made near the blowholes and the back of the body to prevent the carcass from sinking (Fig. 2). Boats on each side of the whale towed it, using these ropes and additional ropes that surrounded the body, to the beach for flensing. This method was again transmitted to Shikoku (1681) and northern Kyushu (1684) (Fig. 1). The preference was to harpoon calves first and then their mothers to secure both with ease.

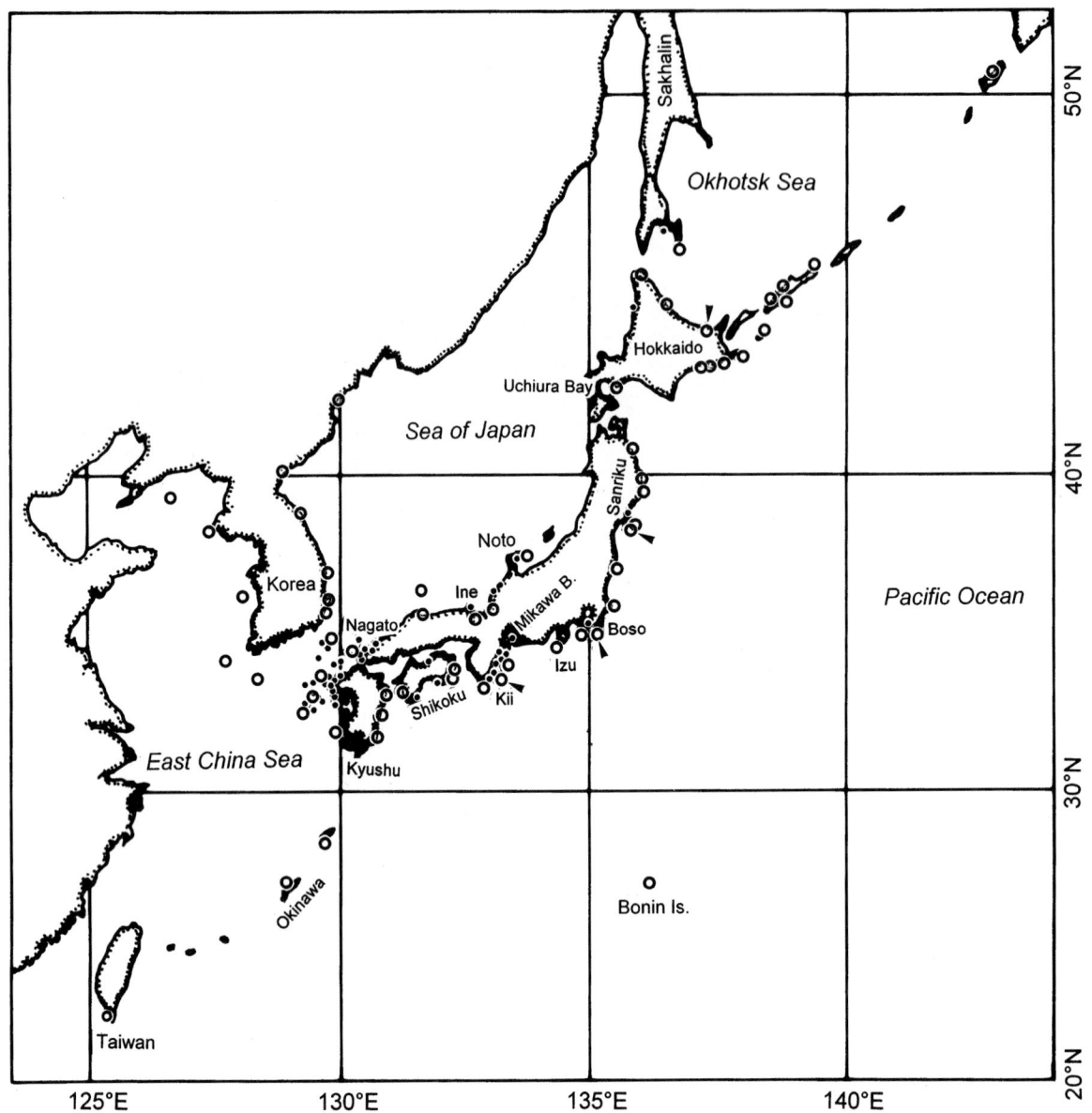

Figure 1 *Location of major land stations used by Japanese whaling. Closed circles represent harpoon or net whaling, open circles represent large-type whaling during pre- and postwar periods, and arrows indicate head land station currently in use by small-type whaling.*

Catches declined in the late 19th century, and some groups started modern Norwegian-type whaling, whereas others attempted to improve their traditional method. A few net whalers moved to new grounds in Hokkaido and southern Sakhalin and took gray whales.

Meat and most of the blubber were sold for human consumption fresh or salted. Oil was extracted from chopped bones and some BLUBBER and was used for light, for human consumption, and as a pesticide in rice paddies nationwide in Japan.

A Tsuro group in Shikoku in the late 1890s used 15 harpoon boats, 2 whale towers, and 14 net boats. The total full-time workers was 356, including 10 whale spotters, 12 flensers, 2 car-

penters, 1 cooper, and 2 blacksmiths. Another group off northern Kyushu recorded 587 workers in the early 1800s. Each group took low tens of whales yearly (Table I). Annual expenditures of 12,423–15,864 yen made a profit of 987–25,640 yen (mean 9778) for a group at Kawajiri, Nagato, during 1884–1893.

III. Modern Coastal Whaling

A Russian, A. Dydymov, started modern Norwegian-type whaling in the western North Pacific in 1889 using a land station east of Vladivostok, Russia. In 1891, the Pacific Whaling Company was established at Vladivostok and operated from the

Figure 2 *A scene of net whaling (from Oyamada, 1832). A harpooner is climbing on a humpback whale to attach a line to the animal. Boats are ready to kill the animals with lances.*

Korean to Sakhalin coasts. Large amounts of whale meat sold by Russians at Nagasaki stimulated the Japanese to begin similar operations. After several attempts that caught the first whales in 1898 and survived only for a short period, Nihon Enyo Gyogyo (Japan Far Seas Fishery), founded in 1899 at Senzaki, Nagato, established modern whaling in Japan using Norwegian gunners. The company expanded the business and renamed themselves Toyo Gyogyo (Oriental Fishery) in 1904, absorbing

other whaling companies. In 1908 a total of 12 modern whaling companies operated using 28 catcher boats. Toyo Gyogyo and five others merged in 1909 to form new Toyo Hogei (Oriental Whaling), which owned 20 land stations (3 in Korea) and 21 whale catcher boats. Six others remained independent. Data on whales taken by them are not available by species, but comparison with later records throws some light on the species and cetacean fauna during earlier whaling (Table II).

TABLE I
Number of Whales Taken by Japanese Net Whaling at Kawajiri (Tada, 1978) and Tsuro (Yamada, 1902)[a]

Seasons	Humpback	Right	Gray	Blue	Fin	Bryde's	Total
Kawajiri, Nagato							
1698–1737	391 (9.8)	105 (2.6)	60 (1.5)		22 (0.6)		518 (13.0)
1738–1840[b]	304 (5.4)	113 (2.0)	72 (1.3)		3 (0.1)		492 (8.8)
1845–1889	198 (5.0)	39 (1.0)	37 (0.9)		131 (3.3)		405 (10.1)
Tsuro, Shikoku							
1849–1865	209 (12.3)	19 (1.1)	101 (5.9)	5 (0.3)	0 (0)	35 (2.1)	369 (21.7)
1874–1890	108 (6.4)	21 (1.2)	82 (4.8)	24 (1.4)	9 (0.5)	41 (2.4)	285 (16.8)
1891–1896	26 (4.3)	2 (0.3)	18 (3.0)	18 (3.0)	5 (0.8)	31 (0.5)	100 (16.7)

[a]Average annual catches are in parentheses.
[b]Records for 47 years in this period are missing.

TABLE II
Change in Whaling Grounds and Species Hunted off Japan[a]

Season	East Korea	NW/Kyushu to Nagato	SE/Kyushu and Shikoku	Kii	Boso	Sanriku	Hokkaido
1899/0	15	—	—	—	—	—	—
1900/1	42	—	—	—	—	—	—
1901/2	58	—	—	—	—	—	—
1902/3	89	—	—	—	—	—	—
1903/4	182	—	—	—	—	—	—
1904/5	336	—	—	—	—	—	—
1905/6	294	4	—	1	74	22	—
1906/7	378	—	198	199	32	88	—
1907/8	236	47	289	248	160	217	—
1908/9	244	59	126	381	56	297	—
1909/0	—	—	—	58	—	96	—
Total	1874	110	613	887	322	720	—
1911							
Blue	1	4	64	—	177	—	54	0
Fin	183	281	7	—	31	—	394	66
Humpback	5	14	4	—	25	—	8	3
Sei/Bryde's	0	12[b]	13[b]	—	87[c]	—	260[d]	1
Gray	119	2	0	—	0	—	0	0
Right	0	0	1	—	1	—	0	0
Sperm	0	0	4	—	9	—	149	0
Total	308	313	93	—	330	—	865	70

[a]Statistics for July 1899–April 1910 are from Akashi (1910) and those for 1911 from Kasahara (1950).
[b]Bryde's whales.
[c]Mostly Bryde's whales.
[d]Mostly sei whales.

In November 1909 the Japanese government placed hunting of sperm and baleen whales other than minke whales under its control and limited catcher boats to 30. This was further decreased to 25 (1934–1963) and to 7 (1977–) in several steps. On June 8, 1938, Japan enacted the protection of certain whales, i.e., cows accompanied by calves and whales below minimum size limits. However, it allowed the taking of right and gray whales, and the size limits, particularly for blue whales (*B. musculus*), were smaller than those existing in international agreements. In November 1945, Japan adopted international regulations of the time and joined the International Convention for the Regulation of Whaling of 1946 in April 1951 (before the peace treaty).

Postwar coastal whaling started in September 1945 and continued until March 1988 by five major companies using a maximum of 20 land stations (Fig. 2). This whaling was called large-type whaling to distinguish it from small-type whaling established in December 1947. The fishing season and land stations used changed over time. The last season of the fishery (1987/1988) used a land station at Bonin Islands for 317 Bryde's whales (*B. edeni*), and four stations at Yamada and Ayukawa (both in Sanriku), Wadaura (Boso), and Taiji (Kii) took 188 sperm whales.

Japan started a national sperm whale quota in 1959. The quota by North Pacific whaling counties (Canada, Japan, USA, USSR) replaced this starting in 1971. The four countries set quotas for fin (*B. physalus*), sei, and Bryde's whales in 1969; these were replaced by quotas of the International Whaling Commission (IWC) in 1972. The IWC prohibitions by species and dates of enforcement were blue (1965), humpback (1966), and fin and sei whales (1976).

Maximum annual catches by species since 1911 and their dates are 300 blue (1911); 1043 fin, 160 humpback, and 155 gray whales (1914); 14 right whales (1932); 1035 sei whales (1959); 504 Bryde's whales (1962); and 3747 sperm whales (1968). Whaling companies manipulated coastal statistics, particularly for sperm whales (Kasuya, 1999) and Bryde's whales.

IV. Pelagic Whaling in the Antarctic
A. Before World War II

In 1934, Nihon Hogei ("Japan Whaling," renamed from "Toyo Hogei" in 1934, which merged with Nihon Suisan in 1937) purchased the Norwegian factory ship *Antarctic* (9600 tons) and five catcher boats for £55,000 (¥900,000), which was three times the profit from one Japanese Antarctic fleet in

1937/1938. On the way to Japan the *Antarctic* and three catcher boats operated from December 1934 and took 213 whales. This was the first Japanese Antarctic operation. This company built a second fleet in 1936/1937 and a third one in 1937/1938.

Hayashikane Shoten, the antecedent of Taiyo Gyogyo (Ocean Fisheries), sent the *Nisshinmaru* fleet in the 1936/1937 season and had a second fleet in 1937/1938. Kyokuyo Hogei (Polar Sea Whaling) sent the *Kyokuyomaru* fleet in the 1938/1939 season. Thus the total Japanese Antarctic operation increased to six factory ships in 5 years.

The Japanese government enacted regulations of pelagic whaling on June 8, 1938, including (1) a fishing season from November 1 to March 15; (2) protection of gray whales and right whales (*Eubalaena* spp.), except for the North Pacific north of 20°N; (3) cows accompanied by calves; (4) minimum size limits; (5) processing within 36 hr; and (6) full utilization of the catch. This differed from international agreements of the time in a season about 6 weeks longer and the blue whale size limit about 1.4 m shorter, allowing taking blue whales that migrated earlier in the season. Implementation of international agreements in the industry had to wait new regulations in September 1946.

The main product of these operations was whale oil for export. The government strictly limited the importation of Antarctic whale meat until the 1939/1940 season to protect coastal whaling. Some of the whale oil of the last two seasons (1939/1940 and 1940/1941) was landed in northern Korea and was exported to Germany via Siberia.

B. Postwar Operations

In order to feed the starving Japanese population, the General Headquarters of the Allied Forces (GHQ) issued a permit for Antarctic whaling in August 1946. Taiyo Gyogyo converted an oil tanker into *Nisshinmaru No. 1* and Nihon Suisan another vessel to *Hashidatemaru*. These fleets caught 932 BWU, or 6% of the world catch of the 1946/1947 season, and produced 12,260 tons of whale oil and 22,167 tons of other edible products. The total was 36.9 tons/BWU, almost double the maximum prewar production of 19.0 ton/BWU. Whale meat became an important product of Japanese whaling. Kyokuyo Hogei sent the *Baikarumaru* fleet for sperm whales only in 1951/1952, before it returned to the Antarctic in 1956/1957.

While world Antarctic fleets recorded an increase from 9 (1946/1947) to 21 (1960/1961–1961/1962) and a subsequent decline, the decline of Japanese fleets was slightly slower, i.e., from a peak of 7 fleets in 1960/1961–1964/1965 to 1 in 1977/1978–1986/1987. In 1956, Japan purchased a foreign fleet to expand its operation. The objective of purchases changed in 1962, when Japan got a quota allocation of 41% and a further increase was permitted with fleet purchase. Out of 9 fleets purchased by Japan in the postwar period, 4 were for their quotas.

The total Japanese fleet and number of workers involved varied by quota and species hunted. The *Nisshinmaru No. 1* fleet in the 1950/1951 season, when it processed 631 blue and 1014 fin whales, had 348 persons on the factory ship, 604 on three freezing and salting vessels, and 197 on nine catcher boats. The total was 1149.

Takes of significant numbers of sei whales started in 1949/1950 and reached a maximum of 11,310 in 1965/1966, and that of minke whales (*B. acutorostrata* subsp. and *B. bonaerensis*) started in 1971/1972 and reached 3950 in 1976/1977.

Three Antarctic whaling companies split off their whaling sections to merge them into a new company, Nihon Kyodo Hogei (Japan Union Whaling), in 1976. The new company sent two fleets to the Antarctic in 1976/1977 and one in 1977/1978–1986/1987. The last two seasons were operated under objection to the IWC moratorium on commercial whaling.

Southern humpback whales were completely protected as of 1963/1964, "true" blue whales as of 1963/1964, all southern blue whales as of 1964/1965, fin whales as of 1976/1977, sei whales as of 1978/1979, and sperm whales as of 1981/1982.

V. Pelagic Whaling in the North Pacific
A. Before World War II

The *Tonanmaru* fleet was sent out in the 1940 and 1941 seasons by Hokuyo Hogei (Northern Sea Whaling), established jointly by three whaling companies, and caught 74 blue, 659 fin, 114 humpback, 9 sei, 333 sperm, 58 gray, and 4 North Pacific right whales in the two seasons off southern Kamchatka and in the Bering and Chukchi Seas.

B. Postwar, off the Bonin Islands (Ogasawara Islands)

Whaling had been operated in 1927–1944 using land stations on the Bonin Islands for humpback, Bryde's, and sperm whales, but the 1945 permit of the GHQ to whale off Bonin Islands prohibited the use of land stations. Therefore, Taiyo Gyogyo converted a navy vessel to a factory ship and whaled in March–April 1946. The number of fleets and companies involved subsequently varied by season; the last fleet was sent out in 1951. In 1952, pelagic whaling started in the northern North Pacific, and operations off the Bonin Islands ceased.

In the six seasons they took 923 Bryde's, 606 sperm, and 29 other whales. Only 20 humpback whales were taken because of offshore operations.

Subsequent whaling by Nitto Hogei, Nihon Hogei, and Sanyo Hogei was operated for Bryde's whales from a land station on Hahajima Island, Bonin Islands, in 1981–1987.

C. Northern North Pacific

The Peace Treaty came into effect in April 1952, and Japan sent out the *Baikarumaru* fleet to the North Pacific. The fleets increased to two in 1954 and three in 1962, and then decreased to the fleet of Kyodo Hogei (1976–1979). The operation was a joint venture of most of the Japanese whaling companies; Kyokuyo Hogei, Nihon Suisan, and Taiyo Gyogyo were the major ones. Factory ships and quotas changed frequently. The IWC ban on pelagic whaling for species other than minke whales came into effect in the 1979/1980 Antarctic and 1980 northern summer season.

The *Miwamaru*, a whale catcher–factory ship, operated in the 1973–1975 seasons for 279 minke and 6 Baird's beaked whales.

The first national quota of 350 BWU was for a 1-year test operation by one of the two fleets in 1954. This was followed by a blue whale quota of 70 (1955–1961) or 60 (1962–1965) and quotas of about 800 BWU (1957–1964) and 1001 BWU (1965–1968) for species other than blue whales. Sperm whale quotas were from 1500 to 1800 (1957–1961), 2460 to 2700 (1962–1965), and 3000 (1965–1968).

The North Pacific whaling countries set quotas by species in 1969, which were followed by quotas of the IWC (as of 1971 for baleen whales and 1972 for all large whales). The IWC has protected blue and humpback whales since 1966 and fin and sei whales since 1976.

VI. Whaling under Foreign Jurisdiction

Since 1957, Japanese whaling has expanded into foreign territories, presumably for new whale stocks, efficient vessel allocation, unregulated operation, or for new business opportunities.

A. Taiwan, 1957–1959

Taiwan was outside the ICRW. Kyokuyo Hogei whaled for two seasons jointly with a local company using a land station in southern Taiwan but took only 29 humpbacks and a sperm whale. The Taiwanese partner operated for a few additional years.

B. Okinawa, 1958–1965

Under supervision of the U.S. military, the Ryukyu government governed the Ryukyu Islands from the end of World War II to 1972 when the islands were returned to Japan. Hand harpoon fishermen at Okinawa Island, the Ryukyu Islands, took humpback whales using harpoon guns beginning around 1950. In 1958 the Ryukyu government introduced IWC regulations. Only Nago whalers and two local companies got the new licenses. Two Japanese whaling companies, Taiyo Gyogyo and Nitto Hogei, offered crew and catcher boats to them. The land stations were at Nago, Sashiki, and Itoman.

In addition to catches of 52 humpback whales by the Nago group (1950–1957), the three groups took 788 humpback, 31 sperm, and 1 Bryde's whale in 1958–1965.

C. Brazil, 1959–1984

Two groups whaled off Brazil, each inviting a Brazilian partner. The Taiyo group whaled in 1960–1963 from a land station at Cabo Frio, and the Nichirei group in 1959–1984 from Costina. Catches were mostly sei whales in 1959–1964 (3214 whales in the six seasons) and then shifted to minke whales with a maximum recorded catch of 1036 in 1975. Sperm, fin, and blue whales were also taken (in decreasing order). Some of the supposed sei whales were Bryde's whales.

D. Canada, 1962–1972

Japanese whalers operated jointly with Canadian partners off Newfoundland and Vancouver Island. The Taiyo group operated in 1962–1967 using a land station at Coal Harbor, Vancouver Island, and caught mostly sei (2153), sperm (1108), and fin (837) whales, but some blue and humpback whales were also taken.

Off Newfoundland, the Kyokuyo group operated in 1966–1972 using a land station at Dilldo, and the Taiyo group in 1967–1972 using a Williams Port station. Their catch was mostly fin whales (1168) and a few humpback, sei, and sperm whales. The Canadian government closed commercial whaling in 1973.

E. South Georgia, 1963/1964–1965/1966

Two Japanese expeditions operated using South Georgian land stations leased from the United Kingdom. Their total catches were 1273 fin, 919 sei, and 218 sperm whales. Under international pressure, the United Kingdom agreed at the 1966 IWC conference to voluntarily retain the South Georgian catches at or below the level of the 1964/1965 seasons. This terminated Japanese expeditions.

F. Chile, 1964–1968

Chile, Peru, and Ecuador jointly regulated whaling in their territorial waters until 1979, when Chile and Peru joined the ICRW.

Nitto Whaling and its local partner whaled using one or two land stations and took 516 blue, 582 fin, 1061 sei, and 1221 sperm whales. The catch of blue whales occurred only in 1965 and 1966. Some of the supposed sei whales were probably Bryde's whales.

G. Peru, 1967–1985

A local company sponsored by Nihon Kinkai Hogei (Japan Coastal Whaling) whaled using a station at Paita. The operation ended in March 1985. The fishing season lasted almost 12 months of the year, with occasional interruption of 1 or 2 months in winter. The total catch was 291 fin (1968–1977), 3408 Bryde's (1973–1983), 232 sei (1973–1978), 2304 Bryde's or sei (1968–1972), and 14,331 sperm whales (1968–1981) (Valdivia *et al.*, 1984).

H. Philippines, 1983–1984

A local company whaled for two seasons using the *Faith No. 1*, the renamed *Miwamaru* catcher–factory ship of Japan. One of the Japanese sponsors had taken part in an earlier *Miwamaru* operation in Japan. A take of 9 Bryde's whales in 1983 and 47 Bryde's whales was reported with production of 277 tons of meat in 1984. The operation ended due to Japanese rejection of meat import and new regulation for her nationals concerning participation in foreign whaling. The local company operated in 1985 and took 40 Bryde's whales.

VII. Small-Type Whaling

This is defined as a whaling activity that takes minke whales and toothed whales other than sperm whales using a vessel and whaling cannon below a certain size limit. This fishery started around the start of the 20th century, e.g., the Baird's beaked whale fishery off the Boso coast introduced Greener's harpoon guns in 1892 and Taiji fishermen 20-mm five-barrel harpoon guns for pilot whales in 1904. The fishery was placed under control of the Minister of Agriculture and Forestry in December 1947. Before this the operation was unregulated except for the Boso coast where the Baird's beaked whale fishery required a license from Chiba Prefecture.

About 20 vessels operated the fishery off northern Kyushu, Kii, Boso, and Sanriku before World War II. The number increased to 53 in 1942 and 80 in 1950 and then it declined rapidly to 9 in 1970, 4 in 1988, and 5 since 1992. Conversion from several small vessels to one larger vessel contributed to the earlier decline. During the war the vessel size was 5–20 tons. The size limit was 30 tons in 1947, 40 tons in 1963, and is now 50 tons. The maximum caliber of harpoon gun changed from 40 mm (1947) to 50 mm (1952–present) (Fig. 3). Other regulations included the prohibition of killing cows accompanied by calves and a fishing season of 6 months. The vessels usually leave port in the morning and return in the evening.

This fishery had no quota until 1977 and took minke, Baird's beaked, pilot, and killer whales (*Orcinus orca*). Dolphins and porpoises were also taken. IWC set a quota for minke whales for the seasons 1978–1987. The government of Japan set a national quota for Baird's beaked whales at 40 (1983–1987), 60 (1988), 54 (1989–1998), and 62 (1999–present).

The Japanese government maintains that target species of this fishery other than minke whale are outside the IWC competence, thus the decision of IWC to cease commercial whaling does not prohibit take of these species. Currently five catcher boats operate using four land stations: Abashiri on the Okhotsk Sea coast of Hokkaido, Ayukawa on the Sanriku coast, Wadaura on the Boso coast, and Taiji on the Kii coast. Their quota is 62 Baird's beaked whales, 100 short-finned pilot whales (*Globicephala macrorhynchus*) (50 for each of the two populations), and 30 Risso's dolphins (*Grampus griseus*).

VIII. Dolphin and Porpoise Fisheries

In Japan, dolphins and porpoises are taken by drives, hand harpoon, and small-type whaling.

At least 52 villages have operated dolphin drive fisheries since the 14th century on the Sea of Japan and Pacific coasts, but the number declined throughout the 19th and 20th centuries. When it was placed under the license system of the prefecture governments in 1982, only four groups acquired license. Currently, two groups, at Futo on the Izu coasts and Taiji on the Kii coasts, operate drive fisheries with quotas of about 3000 dolphins of six species.

The fishermen drive schools of gregarious dolphins into harbor using several fast boats. Other equipment used is a cone-shaped steel disk welded to one end of a 2-m-long steel pipe. The cone is placed underwater and the other end of the pipe in the air is hammered to scare dolphins acoustically.

Harpoon fisheries started in prehistoric time (see earlier discussion), but large-scale commercial hunts began possibly in 1920s when motor-driven vessels became common in coastal fisheries. Dolphins and porpoises are harpooned when they come to bow ride. An electric shocker is usually connected to the hand harpoon with a detachable head.

This fishery came under the control of regional fishery coordination committees in 1989. Currently, about 400 vessels of Hokkaido, Sanriku, Boso, Kii, and Okinawa operate with a quota of about 18,000 dolphins and porpoises. Okinawa hunters use crossbows; their efficiency is superior to hand harpoons for pilot whale hunts. The fishing season is variable among locations.

Quotas for drive and harpoon fisheries are 400 short-finned pilot whales of southern form, 50 false killer whales, 1280 Risso's dolphins, 1100 bottlenose dolphins, 950 pantropical spotted dolphins (*Stenella attenuata*), 725 striped dolphins (*Stenella coeruleoalba*), and 17,700 Dall's porpoises (two populations).

IX. Research Whaling

During 1956–1979, Japan issued several permits to take whales for research purpose based on Article 8 of the ICWR. The hunt accompanied the operation of ordinary commercial whaling. It killed a relatively small number of whales or lasted for only a few seasons. These features are different from the research whaling of the later period.

The ban on commercial whaling by the IWC came in effect in the 1985/1986 Antarctic season and 1986 coastal season. Japan withdrew its objection to this IWC decision on July 1, 1986, taking effect from May 1, 1987 (pelagic), October 1, 1987 (coastal minke and Bryde's whales), and April 1, 1988 (coastal sperm whales). In November 1987, Nihon Kyodo Hogei

Figure 3 *A small-type whaling vessel in search of pilot whales off the Sanriku coast, October 1985.*

dissolved. Half of the staff formed Nihon Kyodo Senpaku (Japan Union Shipping) to operate vessels acquired from Nihon Kyodo Hogei, and some others and Geirui Kenkyusho (Whales Research Institute) merged to establish Nihon Geirui Kenkyusho [(Japan) Institute of Cetacean Research (ICR)].

ICR started to take whales for scientific purposes in the 1987/1988 Antarctic season and 1994 North Pacific season. In the 1998 and 1999 seasons it took about 400 Southern Hemisphere and 100 North Pacific minke whales using a factory ship and catcher boats chartered from Nihon Kyodo Senpaku. Proceeds of products from these operations, about 4 billion yen/year (US$=100–110 yen), and subsistence and contract of about 1 billion yen from the Japanese government are used to finance activities of the institute. ICR expanded the project to Bryde's and sperm whales in the North Pacific in 2000.

See Also the Following Articles

Illegal and Pirate Whaling ■ International Whaling Commission ■ Whaling, Traditional

References

Akashi, K. (1910). "Honpono Noruweshiki Hogeishi (History of Norwegian-Type Whaling in Japan)." Toyohogei, Osaka. [In Japanese.]

Hashiura, Y. (1969). "Kumano Taijiura Hogeishi (History of Whaling in Taiji, Kumano)." Heibonsha, Tokyo, [In Japanese.]

Hattori, T. (1887–1888). "Nihon Hogei Iko (Japanese Whaling Miscellanea)." Dainihon Suisankai, Tokyo. [In Japanese.]

Hawley, F. (1960). Miscellanea Japonica. II. Whales and Whaling in Japan. Vol. 1, Part 1. Privately published by the author.

Itabashi, M. (1987). "Nanpyoyo Hogeishi (History of Antarctic Whaling)." Chuo Koron, Tokyo. [In Japanese.]

Kasahara, A. (1950). Nihon Kinkaino Hogei Gyoto Sono Shigen (Whaling and Whale Resources around Japan). *Rep. Inst. Nihon Suisan* **4**, 1–103. [In Japanese.]

Kasuya, T. (1975). Past occurrence of *Globicephala melaena* in the western North Pacific. *Sci. Rep. Whal. Res. Inst.* **27**, 95–110.

Kasuya, T. (1999). Examination of reliability of catch statistics in the Japanese coastal sperm whale fishery. *J. Cetacean Res. Manage.* **1**(1), 109–122.

Kasuya, T., and Kishiro, T. (1993). Review of Japanese dolphin drive fisheries and their status. *Rep. Int. Whal. Commn.* **43**, 439–452.

Maeda, K., and Teraoka, Y. (1952). "Hogei (Whaling)." Isana Shobo, Tokyo.

Matsubara, S. (1896). "Nihon Hogeishi (History of Japanese Whaling)." Fishery Association of Japan, Tokyo. [In Japanese.]

Ohsumi, S. (1975). Review of Japanese small-type whaling. *Fish. Res. Bd. Can.* **32**(7), 1111–1121.

Omura, H. (1950). Whales in the adjacent waters of Japan. *Sci. Rep. Whales Res. Inst.* **4**, 27–113.

Omura, H. (1984). Nihonkaino Kujira (Whales in the Sea of Japan). *Geiken-Tsushin* **354**, 65–73. [In Japanese.]

Omura, H., Matsuura, Y., and Miyazaki, I. (1942). "Kujira (Whales)." Suisansha, Tokyo. [In Japanese.]

Oyamada, Y. (1832). Insanatori Ekotoba (Whaling in Words and Pictures). Tatamiya, Edo. (in Japanese). [Translated into English in *Invest. Cetacea* **14**(Suppl.), 1–119 under wrong title Yugiotoru Eshi and wrong spelling of the author Y. Yamada.]

Tada, H. (1978). "Meijiki Yamaguchiken Hogeishino Kenkyu (Study of Whaling History of Yamaguchi in Meiji Era)." Matsuno Shoten, Tokuyama. [In Japanese.]

Tato, K. (1985). "Hogeino Rekishito Shiryo (Whaling History and Data)." Suisansha, Tokyo. [In Japanese.]

Tokuyama, N. (1992). "Taiyo Gyogyo Hogei-jigyono Rekishi (History of Whaling Enterprise of Taiyo Gyogyo)." Privately published by the author.

Tonnessen, J. N., and Johnsen, A. O. (1982). "The History of Modern Whaling." Hurst, London.

Valdivia, J., Landa, A., and Ramirez, P. (1984). Peru, progress report on cetacean research 1982–83. *Rep. Int. Whal. Commn.* **34**, 223–228.

Yamada, S. (1902). "Tsuro Hogeishi (History of Whaling at Tsuro)." Tsuro Whaling, Kochi, [In Japanese.]

Jumping

SEE *Leaping Behavior*

Kentriodontidae

Susan D. Dawson
*University of Prince Edward Island,
Charlottetown, Canada*

Kentriodontidae is the most diverse family of archaic delphinoids in terms of morphology, geographic range, and temporal range. Unfortunately, while they are the best-studied archaic delphinoids, there are still many unanswered questions about this group, including the very basic question of whether kentriodontids are monophyletic (Muizon, 1988). The grade-level family Kentriodontidae has essentially been defined based on its members holding an intermediate position between primitive odontocetes and modern delphinoids. Additionally, no computer-aided cladistic analysis has described the relationships of taxa attributed to the Kentriodontidae.

I. History

Kentriodontids have a long history in the science of paleontology. The specimen now known as *Lophocetus calvertensis* was first described by Harlan in 1842 as *Delphinus calvertensis* (this is an early example of a fossil odontocete being taxonomically lumped with a modern genus). The family takes its name from *Kentriodon pernix*, described by Kellogg in 1927. Slijper named the subfamily Kentriodontinae in 1936 and considered it to be within the family Delphinidae. Barnes (1978) articulated the modern concept and classification of the family Kentriodontidae as distinct from but related to modern delphinoids. Subsequent studies (Muizon, 1988; Dawson, 1996) have been based on that classification.

II. Morphology

Archaic delphinoids resemble modern delphinoid species in general body morphology. Hindlimbs are absent, forelimbs are modified to form paddle-like flippers, and the rostrum is elongated to form a pronounced snout or beak in most species. Differences between archaic and modern delphinoids are apparent in features such as SKULL proportions and morphology,

development of basicranial sinuses, tooth number and morphology, and vertebral number and proportions. Each of these features has primitive and derived character states, which are distributed among archaic and modern delphinoids.

Among modern delphinoids, there is a range of variation of flipper shapes, from the narrow flippers of the common dolphins (*Delphinus* spp.), to the wider flippers of the beluga whale (*Delphinapterus leucas*), to the wide ovoid flippers of the killer whale (*Orcinus orca*). Forelimb bones are known from only a few species of archaic delphinoids, making prediction of fossil flipper morphology tenuous. The humerus, radius, and ulna are known from *Hadrodelphis*, and the articular surfaces of these bones suggest that the flipper was wide, although without the bones of the manus it is impossible to predict the exact width.

Morphology of the cervical vertebrae is known from a few species (such as *Hadrodelphis calvertense*) and indicates that the cervical vertebrae of archaic delphinoids are less compressed along the cranio-caudal axis than the cervical vertebrae of modern delphinids and phocoenids. In these modern species, the cervical vertebral bodies are extremely thin and may be fused together, making the neck relatively rigid and very short. The increased length of the cervical vertebral bodies in archaic forms suggests that the neck was more elongate and mobile in these species, perhaps comparable to modern beluga whales. Different kentriodontid taxa also exhibit a range of vertebral centrum proportions in other regions of the vertebral column. While no species has lumbar vertebral centra elongated to the extreme extent found in some archaeocetes, some species such as *Kentriodon pernix* do have more elongated vertebral bodies than species of *Atocetus*. Differences in vertebral centrum proportions may have functional significance, a concept which has been explored by Buchholtz (1998). Individuals with relatively long vertebral centra may have had more flexible vertebral columns and may have been slower, more maneuverable swimmers, such as the modern beluga whale. Individuals with shorter, more compressed vertebral centra may have more rigid vertebral columns and may be faster swimmers, such as the modern Dall's porpoise.

Kentriodontids have been traditionally divided into three groups (Table I). Kentriodontines include the genera *Kentriodon*, *Delphinodon*, and *Macrokentriodon*. These genera have a cranial vertex morphology which is relatively flat and tabular. There is broad exposure of the nasal and frontal bones at the

TABLE I
Classification of Archaic and Modern Delphinoids, with Temporal and Geographic Distributions[a]

Order Cetacea Brisson, 1762
  Suborder Odontoceti (Flower, 1864)
    Superfamily Delphinoidea (Gray, 1821)
      Family Kentriodontidae (Slijper, 1936)
        Subfamily Kentriodontinae (Slijper, 1936) Late Oligocene–
          Middle Miocene; eastern and western North Pacific,
          western South Pacific, western South Atlantic, western
          North Atlantic oceans)
            Kentriodon Kellogg, 1927
            Macrokentriodon Dawson, 1996
            Delphinodon (True, 1912)
        Subfamily Lophocetinae Barnes, 1978 (Middle Miocene–
          Late Miocene; eastern North Pacific, western North
          Atlantic oceans)
            Lophocetus (Harlan, 1842)
            Liolithax Kellogg, 1931
            Hadrodelphis Kellogg, 1966
        Subfamily Pithanodelphinae Barnes, 1985 (Middle
          Miocene–Late Miocene; eastern North Pacific, eastern
          South Pacific, eastern North Atlantic oceans)
            Pithanodelphis Abel, 1905
            Atocetus Muizon, 1988
      Family Albireonidae Barnes, 1984 (Late Miocene–Early
        Pliocene; eastern North Pacific ocean)
          Albireo Barnes, 1984
      Family Monodontidae Gray, 1821 [Late Miocene–Recent;
        eastern North Pacific (Late Miocene–Pliocene), circumpolar
        Arctic]
      Family Delphinidae Gray, 1821 (Late Miocene–Recent; all
        oceans)
      Family Phocoenidae (Gray, 1825) (Late Miocene–Recent; all
        oceans)

[a]Kentriodontid genera of uncertain affinity are not included in this classification.

vertex, but the nasal bones are not enlarged or inflated, nor is the vertex significantly elevated. Lophocetines include the genera *Lophocetus, Liolithax,* and *Hadrodelphis* and have nasal bones that are more elongate and the cranial vertex slightly to significantly elevated. Pithanodelphines (including the genera *Pithanodelphis* and *Atocetus*) have large, inflated nasal bones. With the exception of some pithanodelphines, kentriodontids demonstrate the primitive condition of cranial symmetry; this feature is in contrast to modern delphinids, in which the cranial vertex is markedly asymmetrical and the left premaxilla does not contact the left nasal bone.

While primitive odontocetes retained heterodont teeth, archaic delphinoids (as well as other Oligocene–Miocene odontocetes such as eurhinodelphids) mark the transition to the polydont, homodont dentition found in modern delphinids (Fig. 1). There is variation in the number of TEETH in different species, but all kentriodontids have a tooth count that is increased beyond the generalized mammalian condition. Kentriodontids do not have the large, triangular, double-rooted cheek teeth of the Squalodontidae. Instead, all teeth are conical in shape and have

a single root. Modern phocoenids also have a homodont dentition with single-rooted teeth, but the crowns of these teeth are spatulate rather than conical (Fig. 1A). There is no evidence of this tooth morphology in any kentriodontid, nor is it present in the earliest phocoenids. *Kentriodon* demonstrates an unusual morphology, with the most anterior premaxillary teeth elongate and procumbent, oriented in a horizontal rather than vertical plane (Fig. 1C). Some kentriodontid taxa, such as *Hadrodelphis* and *Kampholophos,* exhibit variation within the tooth row, such that more posterior teeth have a shelf-like projection on the lingual surface and a series of small cuspules (Fig. 1D). Anterior teeth are smooth and conical. This slightly heterodont condition is also found in modern Iniidae, the boto of the Amazon River (*Inia geoffrensis*), and is part of the evidence that led paleontologists to hypothesize relationships between archaic delphinoids and river dolphins. Further investigations of odontocete relationships, however, have focused on more diagnostic cranial characters than dentition and confirmed the relationship of kentriodontids with modern delphinoids.

Modern and archaic delphinoids exhibit wide variation in rostrum length (Fig. 1). Species such as the long-beaked *Hadrodelphis* have a rostrum that is approximately two-thirds the total length of the skull (Fig. 1D). Short-beaked species such as *Delphinodon dividum* have a rostrum that is only one-half the total skull length. Although there is also a range of variation between genera, monodontids and phocoenids tend to have shorter rostra than delphinids (Fig. 1B), and all three modern families tend to have shorter rostra than long-beaked kentriodontids. The short rostra of phocoenids (Fig. 1A) may be associated with paedomorphic modifications postulated for this group (Fordyce and Barnes, 1994). Longer rostra are found in the river dolphins, including platanistids, pontoporiids, lipotids, and iniids. No living odontocete possesses a rostrum as long as that of the bizarrely long-beaked eurhinodelphids, which had a rostrum four-fifths (or more) the total length of the skull.

III. Taxonomic, Temporal, and Geographic Diversity

The fossil record of modern delphinids, phocoenids, and monodontids extends to at least the Late Miocene (Table I). As a family, the modern Delphinidae are cosmopolitan in distribution, with individual genera having a more restricted range. The earliest fossil delphinids are reported from eastern North Pacific deposits of the Late Miocene. Barnes (1990) summarized the evolutionary history of the bottlenose dolphin, *Tursiops truncatus*. The genus is known from at least the early Pliocene, and possibly from the latest Miocene. Fossil species of *Tursiops* are widely distributed, having been reported from Atlantic, Pacific, and Mediterranean deposits.

Modern phocoenids are widely distributed in the Atlantic, Pacific, and Indian Oceans, but are not found in all regions of those oceans. The earliest Phocoenidae are known from the Late Miocene. Early phocoenid fossils are reported from the western coasts of North and South America, and this group may have originated in the Pacific basin.

Extant monodontids are restricted in DISTRIBUTION and diversity when compared to fossil members of this family. There

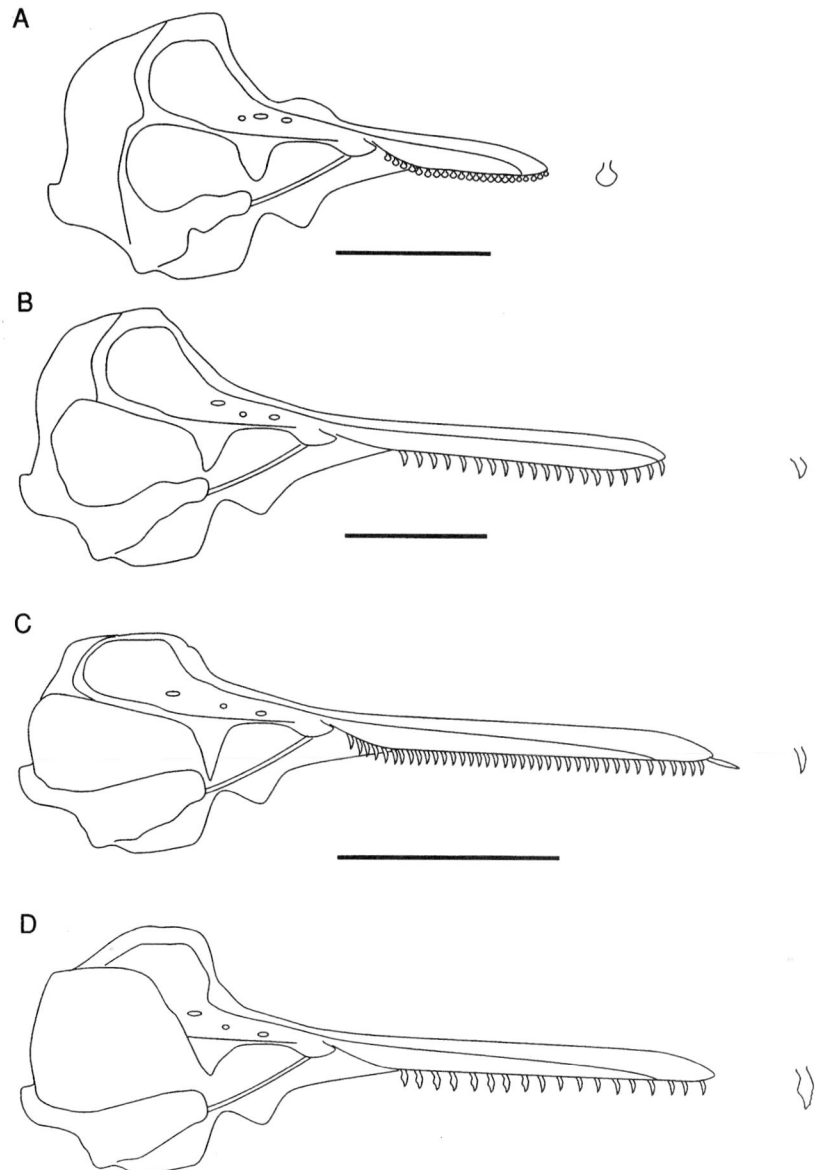

Figure 1 *Skull and tooth morphology of modern and fossil delphinoids. Scale bar: 10 cm. (A) Family Phocoenidae,* Phocoena phocoena; *(B) Family Delphinidae,* Tursiops truncatus; *(C) Family Kentriodontidae,* Kentriodon pernix *(fossil, reconstructed); and (D) Family Kentriodontidae,* Hadrodelphis calvertense *(fossil, reconstructed).*

are two living monodontids, the beluga (*Delphinapterus leucas*) and the narwhal (*Monodon monoceros*), both of which are found only in circumpolar waters of the Arctic. The earliest fossil monodontids are reported from the Late Miocene. Fossil monodontids showed greater taxonomic diversity and had a much different geographic distribution; they are reported from the Late Miocene and Pliocene of California and Baja Mexico.

The family Kentriodontidae is often described as cosmopolitan in geographic distribution, with a temporal range from the Late Oligocene to Late Miocene. However, because the group may not be monophyletic, the distribution of "Ken-

triodontidae" may be a spurious and irrelevant question. It is perhaps more meaningful to first examine the distribution of individual taxa and the taxonomic distribution of species within genera.

A. Interpreting the Fossil Record of Kentriodontids

All studies of fossil organisms are necessarily limited by the quality of available specimens. This problem is particularly acute with kentriodontids, as there is a lack of described specimens as well as a lack of specimens represented by diagnostic, comparable material. While it is preferable for holotype specimens to consist of a skull and associated periotic, several kentriodontid taxa have been named based on isolated periotics or teeth. Analyzing the distribution of individual species of kentriodontids is particularly problematic due to the paucity of specimens. Most taxa are known only from the holotype material, making it difficult to account for individual variation; e.g., eight genera of kentriodontids are reported from North America, but only two of these genera are known from more than one skull (Dawson, 1996). There is little evidence of an individual species being known from more than one ocean basin or region. The sole published example of one species described from more than one basin or region are two specimens referred to as "aff. *Delphinodon dividum*" from the Middle Miocene of California and Japan; both of these specimens consist of isolated periotics, making this designation problematic.

B. Diversity of Kentriodontid Species

The distribution of genera both geographically and temporally is slightly less problematic. There are 16 genera included in the family; of these only 5 genera have more than one published species. The proper generic assignment of at least two of these species is questionable. There are two species formally assigned to the genus *Hadrodelphis: H. calvertense* and *H. poseidoni*. *Hadrodelphis poseidoni*, however, is questionable because it is based only on isolated teeth. The genus *Lophocetus* has also undergone considerable revision. There are currently two species formally assigned to the genus *Lophocetus*. Both species, *L. calvertensis* and *L. repenningi*, are represented by a skull with periotics. However, Muizon has questioned the generic assignment of *L. repenningi*, suggesting that it does not belong in that genus. Kellogg originally assigned *Liolithax pappus* to the genus *Lophocetus*, but Barnes' 1978 revision of the kentriodontids placed it in its currently accepted genus. The type species of the genus *Liolithax* is *L. kernensis* Kellogg 1931, which is described on the basis of isolated periotics. No cranial or postcranial material has been formally referred to *L. kernensis*.

The two species of the genus *Atocetus* offer a well-founded example of temporal and spatial diversity of a kentriodontid genus. *A. nasalis* is reported from the eastern North Pacific basin, whereas *A. iquensis* is reported from the eastern South Pacific. Both species are described from several diagnostic specimens.

The genus *Kentriodon* is the most diverse of the kentriodontids, with three named species and at least five undescribed species mentioned in the literature. It is the oldest described kentriodontid genus, reported from the Late Oligocene to the Middle Miocene; it is not reported from the Late Miocene (Ichishima *et al.*, 1994). *Kentriodon* also has the widest geographic range, reported from the eastern and western North Pacific, eastern and western South Pacific, western North Atlantic, and western South Atlantic. It has not been reported from the eastern North Atlantic, although most kentriodontids reported from this region are Late Miocene in age.

See Also the Following Articles

References

Barnes, L. G. (1978). A review of *Lophocetus* and *Liolithax* and their relationships to the delphinoid family Kentriodontidae. *Nat. Hist. Mus. L.A. County Bull.* **28**, 1–35.

Barnes, L. G. (1990). The fossil record and evolutionary relationships of the genus *Tursiops*. *In* "The Bottlenose Dolphin" (S. Leatherwood and R. R. Reeves, eds.), pp. 3–26. Academic Press, San Diego.

Berta, A., and Sumich, J. L. (1999). "Marine Mammals: Evolutionary Biology." Academic Press, San Diego.

Buchholtz, E. A. (1998). Implications of vertebral morphology for locomotor evolution in early Cetacea. *In* "The Emergence of Whales: Evolutionary Patterns in the Origin of Cetacea" (J. G. M. Thewissen, ed.), pp. 325–352. Plenum Press, New York.

Dawson, S. D. (1996). A description of the skull and postcrania of *Hadrodelphis calvertense* and its position within the Kentriodontidae. *J. Verteb. Paleontol.* **16**(1), 125–134.

Fordyce, R. E., and Barnes, L. G. (1994). The evolutionary history of whales and dolphins. *Ann. Rev. Earth Planet. Sci.* **22**, 419–455.

Ichishima, H., Barnes, L. G., Fordyce, R. E., Kimura, M., and Bohaska, D. J. (1994). A review of kentriodontine dolphins: Systematics and biogeography. *Island Arc* **3**, 486–492.

Muizon, C. de (1988). Les relations phylogenetiques des Delphinida. *Ann. Paleontol.* **74**(4), 159–257.

Kidney, Structure and Function

CAROL A. BEUCHAT
University of Arizona, Tucson

The marine environment presents mammals with unique physiological challenges. Most obviously, the ocean's high salt content and the inavailability of sources of fresh drinking water make it highly desiccating. The kidney is the sole osmoregulatory organ in mammals, so it must excrete excess minerals in urine while at the same time conserving water to prevent dehydration. Moreover, it must maintain osmotic homeostasis of the body fluids even while diving, when depressed cardiovascular function may compromise many physiological processes. To meet the special needs of mammals living in the sea, the kidneys differ in some significant and interesting ways from those of most terrestrial species.

I. General Structure of the Mammalian Kidney

In all mammals, the fundamental functional units are the millions of nephrons that transform plasma filtered from the blood into urine, which contains metabolic toxins and excess minerals that are then excreted from the body. Macroscopically, the kidney is organized into an outer tissue layer, the cortex, that surrounds an inner region, the medulla. In the simplest kidneys, such as those of rodents, the medulla forms a single cone-shaped papilla (unipapillary kidney). In other species, the kidney can have multiple papillae, forming a "multipapillary" kidney. This type of kidney is also referred to as "multireniculate," with the reniculi corresponding to the medullary papillae and their associated cortical tissue (Sperber, 1944).

Multireniculate kidneys can be of two types (Oliver, 1968). In a "compound" multireniculate kidney, the cortex is continuous and encloses all of the papillae. The urine produced by each of the papillae is drained through the collecting ducts (the terminal segment of the nephron) into a funnel-like structure, the pelvis, that surrounds all of the papillae. This type of kidney is found in only a few species of mammals, including humans and beavers.

The "discrete" multireniculate kidney is actually a single organ composed of many miniature kidneys, the reniculi, enclosed in a sheet of connective tissue and peritoneum (the renal capsule) to form a contiguous renal mass (Fig. 1). Each of the reniculi typically resembles a simple unipapillary (or bipapillary) kidney, with its own blood supply and renal pelvis (Figs. 2 and 3). The microcirculatory organization and arrangement of the nephrons in each reniculus are essentially the same as those in a simple unipapillary kidney. Urine produced by each of the reniculi drains from the pelvis into ureteral tubules that connect to the main ureter, which carries the urine to the bladder. The reniculi can number in the hundreds or even thousands in each of the paired kidneys. Discrete multireniculate kidneys are found in almost all marine mammals, including cetaceans, pinnipeds, otters, and polar bears (*Ursus maritimus*), but in only a few nonmarine species.

II. Structure of the Kidneys of Marine Mammals

A. Cetaceans and Pinnipeds

The division of the multireniculate kidney into discrete, independent units is developed to the greatest degree in cetaceans and pinnipeds, which are considered here together because of their general similarities in renal morphology. There is no published information about the renal structure of the walrus (*Odobenus rosmarus*), although the regular harvesting of this animal historically and even in the present would seem to provide abundant opportunity for study.

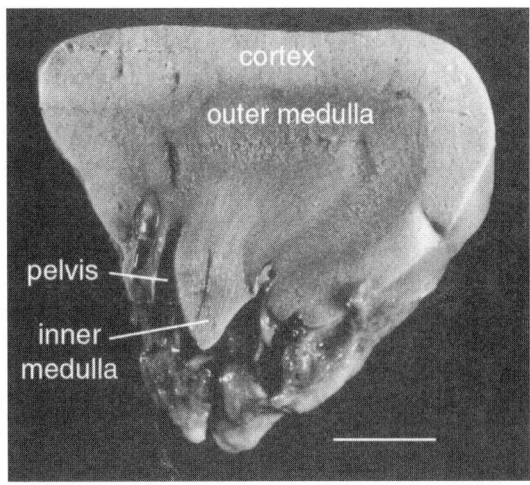

Figure 2 *A unipapillary reniculus from the kidney of the Cape clawless otter. Like the typical mammalian kidney, there is an outer covering of cortical tissue, and the medulla is divided into outer and inner regions. The inner medulla forms a conical papilla that contains the longest loops of Henle. Urine drains from the nephrons into the pelvis, and from there to the bladder via the ureter. Scale bar: 2.5 mm. Modified from Beuchat (1999).*

In cetaceans and pinnipeds, the renicular lobulation of the kidney is clearly evident as grooves on its surface. In some species, the reniculi are rather loosely packed within the renal capsule and are separated by considerable connective tissue (perirenal fascia), whereas in others the packing is tight and the reniculi are more hexagonal than round, minimizing the space between them. Especially in cetaceans, they can be arranged in grape-like clusters on branches of the ureter (Abdelbaki *et al.*, 1984).

Although most reniculi in cetaceans and pinnipeds are usually unipapillary, reniculi with two or three medullary pyramids are not uncommon, and reniculi with as many as seven pyramids have been noted in the North Atlantic right whale,

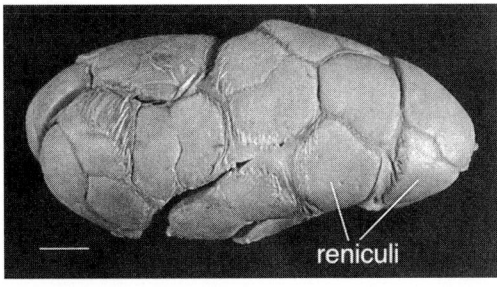

Figure 1 *The discrete multireniculate kidney of the Cape clawless otter* (Aonyx capensis) *with its covering of perirenal fascia. The outlines of the reniculi are evident on the surface of the kidney. Scale bar: 10 mm. Modified from Beuchat (1999).*

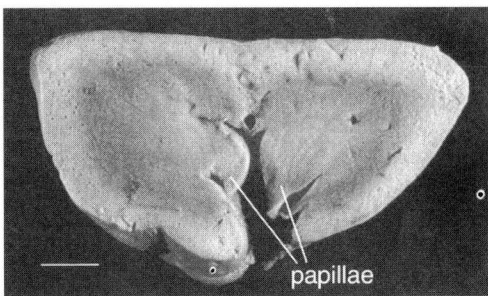

Figure 3 *A bipapillary reniculus from the kidney of the Cape clawless otter. The cortex is continuous, surrounding both medullae. The two medullary papillae are functionally separate, but the urine drains into a common pelvis. Scale bar: 3 mm. Modified from Beuchat (1999).*

Eubalaena glacialis (Maluf, 1989). In these multipapillary reniculi, which themselves resemble a compound multireniculate kidney, the medullary tissue forming each papilla appears to be functionally discrete, whereas the cortex can either have divisions corresponding to each papilla or can be continuous, forming a capsule that encloses all the papillae within the reniculus.

In the typical bean-shaped, unipapillary kidney of terrestrial mammals, the renal artery enters the kidney, and the renal vein and ureter exit, from the hilus, which is located at an indentation on the concave medial surface of the kidney. In cetaceans and pinnipeds, the comparable vessels (i.e., renicular artery and vein) are similarly arranged in the individual reniculus. In some pinnipeds, the entire multireniculate kidney is likewise bean shaped, with a hilus-like medial indentation where the renal artery enters and the renal vein and ureter exit. In other pinnipeds and most of the cetaceans, the renal mass is elongate, and the renal artery and vein enter toward the anterior end and the ureter exits more caudally.

A unique feature of the kidneys of cetaceans and pinnipeds is the presence of a conspicuous sporta perimedullaris, a fibromuscular sheet that separates the cortex and the medulla. First described in cetaceans, it was for a time thought to be unique to this group. Subsequently, however, it has been found to be well developed in both cetaceans and pinnipeds. Because it is present but poorly developed and inconspicuous in nonmarine mammals, it appears to be a feature of the kidney that is related to a marine existence.

The sporta perimedullaris lies at the corticomedullary junction and has perforations that allow penetration of tubular and vascular elements from the cortex into the medulla. In longitudinal cross section of the reniculus, the sporta perimedullaris usually appears as short ribbons of connective tissue and smooth muscle fibers. The function of the sporta perimedullaris remains unknown, but the presence of smooth muscle fibers suggests that, like the renal pelvis, it may be spontaneously contractile, perhaps assisting in expulsion of the urine from the kidney.

B. Otters

There are published accounts of renal structure for three species of otters (Sperber, 1944; Beuchat, 1996, 1999): the sea otter (*Enhydra lutris*), which is exclusively marine; the river otter (*Lutra lutra*), which occurs only in fresh water; and Africa's Cape clawless otter (*Aonyx capensis*), which like the manatee inhabits both freshwater and marine environments. Despite their very different osmoregulatory requirements, all three species have discrete multireniculate kidneys that most resemble those of pinnipeds. Lobulation of the kidney is readily apparent externally, and both unipapillary and bipapillary reniculi have been noted in the Cape clawless otter (Figs. 1–3).

C. Polar Bears

Nothing is known about the renal anatomy of the polar bear, the only ursid that can be considered truly at home in the sea. However, discrete multireniculate kidneys have been described in all other bear species examined (i.e., grizzley, *U. arctos;* sun bear, *U. malayanus;* and sloth bear, *Melursus ursinus;* Sperber, 1944). Studies of kidney function in polar bears are lacking.

D. Sirenians

The most exclusively marine of the recent sirenians is the extinct Steller's sea cow (*Hydrodamalis gigas*). The only descriptions of its kidney are from publications by Steller from the 1750s in which he notes that the kidney resembled "those of seal and sea otter," with numerous reniculi (Sperber, 1944; Maluf, 1989). From this the kidney would seem to be of the discrete multireniculate type. There were no studies of renal function in Steller's sea cow before its extinction.

The dugong (*Dugong dugon*) is also highly marine. Its kidney, however, is unlike that of any other marine mammals, bearing a greater resemblance to the crest-type kidney of the dog. It is elongated but lacks superficial grooves and internal division into reniculi. There is a continuous cortex and medial hilus. The medulla is divided into cranial and caudal regions, each forming a long crest with girdle-like dorsal and ventral extensions that are suggestive of transverse lobulation.

Manatees are found in a range of salinities from seawater to freshwater. The Amazonian manatee (*Trichechus inunguis*) is restricted to freshwater. The West Indian manatee (*Trichechus manatus*) and the West African manatee (*T. senegalensis*) are the only mammals besides the Cape clawless otter that naturally occur in both freshwater and marine environments. Sperber (1944) described the kidneys of *T. manatus* and *T. senegalensis* as appearing "to represent a transitional stage between kidneys with undivided cortex and divided medulla, and the renculi kidneys." In these species, the kidney seems to most closely resemble a compound multireniculate design, with 6–11 medullary papillae. Although a continuous cortex surrounds all of the medullae, there are folds of cortex (plicae corticales) and septa that separate adjacent medullary regions, producing the external appearance of lobulation in the adult (Maluf, 1989). The papillae are not conical, forming instead at their tips a concave surface that protrudes into a large, muscular renal pelvis. There is no inner medulla. The kidney is bean shaped, with the renal artery and vein and the ureter located together at the indentation of the hilus.

III. Size of the Kidney

The mass of the kidney increases with body size in mammals (Beuchat, 1996). In marine mammals that have discrete multireniculate kidneys, the number of reniculi in each kidney increases as well. In the harbor porpoise (*Phocoena phocoena*), for example, the kidneys of an adult (body length = 1.6 m) weigh 150–200 g and contain roughly 300 reniculi that range in weight from 0.15 to 0.6 g each. In an 11-m right whale, one of the kidneys can weigh 32 kg and contain 5400 reniculi, each with an average weight of 2.6 g.

In general, the kidneys of cetaceans and pinnipeds are larger than those of similarly sized terrestrial mammals (Beuchat, 1996). This may be a consequence of the division of the kidney into reniculi, with the connective tissue surrounding each reniculus accounting for at least part of the difference in mass.

IV. Urinary-Concentrating Ability

All marine mammals examined to date can produce urine that is at least as concentrated as seawater (1000 mosM), and most can do substantially better than this. Measurements from cetaceans (six species) range from 1081 to 1737 mosM and those from pinnipeds (five species) and the sea otter are even higher, from 2000 to 2750 mosM (Beuchat, 1996). The highest value recorded from a manatee is 1100 mosM.

Although these osmolalities are far less than those achieved by some desert rodents, in which the urine can exceed 8000 mosM, they are surprisingly high considering the structure of the kidney in most marine mammals. The countercurrent multiplier theory predicts that the concentrating ability should increase with increasing length of the loop of Henle. In species with discrete multireniculate kidneys, the nephrons have very short loops of Henle because the individual reniculi are much smaller than an appropriately sized nonmultireniculate kidney would be for that species (Beuchat, 1996). How marine mammals produce such concentrated urine with such short loops of Henle remains a paradox.

See Also the Following Articles

Morphology, Functional ■ Ocean Environment ■ Osmoregulation

References

Abdelbaki, Y. Z., Henk, W. G., Haldiman, J. T., Albert, T. F., Henry, R. W., and Duffield, D. W. (1984). Macroanatomy of the renicule of the bowhead whale (*Balaena mysticetus*). *Anat. Rec.* **208,** 481–490.

Beuchat, C. A. (1996). Structure and concentrating ability of the mammalian kidney: Correlations with habitat. *Am. J. Physiol.* **271,** R157–R179.

Beuchat, C. A. (1999). Kidney structure of a euryhaline mammal, the Cape clawless otter (*Aonyx capensis*). *S. Afr. J. Zool.* **34,** 163–165.

Maluf, N. S. R. (1989). Renal anatomy of the manatee, *Trichechus manatus,* Linneaus. *Am. J. Anat.* **184,** 269–286.

Oliver, J. (1968). "Nephrons and Kidneys: A Quantitative Study of Developmental and Evolutionary Mammalian Renal Architectonics." Harper & Row, New York.

Sperber, I. (1944). Studies on the mammalian kidney. *Zool. Bidrag Fran Uppsala* **22,** 249–432.

Killer Whale

Orcinus orca

JOHN K. B. FORD
*Fisheries and Oceans Canada,
Nanaimo, British Columbia, Canada*

With its striking black and white markings and cosmopolitan range, the killer whale (*Orcinus orca*), or orca, is one of the most easily recognized and widely distributed of all cetaceans. Although the species has long been held in high regard by many aboriginal maritime cultures, other societies feared the killer whale as a reputedly ruthless and dangerous predator, and the animals were commonly vilified and persecuted. Attitudes toward killer whales have fortunately improved over the past few decades. The species has been an admired display species in aquaria for over 30 years and has been featured in numerous movies, documentaries, and other forms of popular media. It has become the focus of commercial whale-watching operations in several regions. Long-term field research using photo-identification of individuals from natural markings has resulted in certain populations of killer whales being among the best studied of any cetacean species (Fig. 1).

I. Characters and Taxonomic Relationships

The killer whale is one of the most distinctive of the odontocete cetaceans and is unlikely to be confused with any other species. It is a large dolphin, attaining maximum body lengths of 9.0 m in males and 7.7 m in females. Maximum measured weights are 3810 kg for a 6.7-m female and 5568 kg for a 6.75-m male (Dahlheim and Heyning, 1999). In addition to sexual dimorphism in size, mature males develop disproportionately larger appendages than females (Fig. 2). This includes the pectoral flippers, tail flukes (the tips of which curl downward in males), and dorsal fin, which is erect in shape and may attain a height of 1.8 m in males. At birth, neonate killer whales are approximately 2–2.5 m long and weigh approximately 200 kg.

The most distinguishing feature of the killer whale is its striking coloration. Killer whales are generally black dorsally and white ventrally. Above and behind the eye on each lateral side of the whale's head is a conspicuous, elliptically shaped white patch, referred to as the postocular patch. On the posterior lateral sides of the whale, the ventral white region continues dorsoposteriorly to form flank patches that extend almost half-way to the dorsal ridge. At the posterior base of the dorsal fin is a gray-pigmented area of variable shape termed the "saddle patch." In neonates, the normally white pigmented areas on

Figure 1 *Left side view of an adult male killer whale showing representative dorsal fin nick and saddle-patch scars that are used in photographic identification of individuals. Photo by G. Ellis.*

Figure 2 *Adult male killer whale breaching near San Juan Island, Washington. Note large pectoral flippers and tall dorsal fin typical of mature males. Photo by C. Emmons.*

the body have an orange hue, and the saddle patch is indistinct or absent for the first year of life. Considerable variation exists among killer whale populations and individuals in the size and shape of white and gray patches. In some populations, particularly in the Southern Hemisphere, killer whales have a faint gray pigmentation over much of their body and a black dorsal "cape" anterior to the dorsal fin.

The skull of the killer whale can be distinguished from other odontocetes by its shape, size, dental formula, and large teeth. Typically, 10–12 (up to 14) teeth are found per row, with teeth usually up to 10 cm in length. Upper and lower teeth interlock when the jaws are closed, which may result in considerable wear along their anterior and posterior facets. In some populations, extreme wear of the tooth crowns has been observed, even in young individuals, which may relate to diet.

Taxonomically, the killer whale is the largest species of the family Delphinidae. The genus *Orcinus* is considered monotypic, although two species, *O. nanus* and *O. glacialis*, were independently proposed for a population of purportedly small individuals in the Antarctic. These new species designations have not received general acceptance. The morphological and ecological features described in Antarctic killer whales likely represent the kinds of population-specific variations that may be typical of the species over much of its range (Hoelzel *et al.*, 1998; Barrett-Lennard, 2000).

II. Distribution and Abundance

The killer whale is second only to humans as the most widely distributed mammal in the world. They are found in all oceans and most seas, but are generally most commonly observed in coastal, temperate waters, especially in areas of high productivity. Notable concentrations occur in waters along the northwestern coast of North America, around Iceland, and along the coast of northern Norway. In the Antarctic, killer whales are commonly found up to the pack ice edge in many areas and may extend well into ice-covered waters. In the Canadian Arctic, killer whales are rarely seen in the vicinity of pack ice, but do visit the region during the open water season in later summer. Information on the species' distribution in most tropical and offshore waters is limited, but numerous scattered records attest to its widespread, if rare, occurrence.

Because of its general scarcity and sporadic occurrence in most regions, the killer whale is a difficult species to census. Photo-identification studies in nearshore waters of the northeastern Pacific Ocean, from the eastern Aleutian Islands to California, have yielded a total population count of approximately 1500 whales (Ford *et al.*, 2000). Similar studies off northern Norway have identified 445 whales (Similä, 1997). No reliable global population estimate is available for the species, although some rough estimates have been given for some ocean regions. Intensive vessel sighting surveys for cetaceans in the eastern tropical Pacific have resulted in an estimate of about 8500 killer whales in an area of 19 million km^2. Ship-board cetacean surveys in the Antarctic have yielded a rough estimate of 70,000 killer whales (Dahlheim and Heyning, 1999).

III. Ecology

The DIET of killer whales comprises an extremely diverse array of prey species, which can vary widely both within and among regions. It is the only cetacean that routinely preys upon marine mammals, with attacks or kills documented for over 35 different species (Jefferson *et al.*, 1991). Mammalian taxa that have been recorded as prey of killer whales include other cetaceans—both mysticetes and odontocetes—pinnipeds, sirenians, mustelids, and, on rare occasions, ungulates. A variety of fish species are also important prey of killer whales, notably salmon (*Oncorhynchus* spp.), herring (*Clupea harengus*), cod (*Gadus* spp.), tuna (*Thunnus* spp.), and various sharks and other elasmobranchs (Dahlheim and Heyning, 1999; Saulitis *et al.*, 2000; Ford *et al.*, 1998; Visser, 1999). Other animals noted as killer whale prey include squid, octopus, sea turtles, and sea birds.

In the northeastern Pacific, there is remarkable dietary specialization in different sympatric populations of killer whales, and a growing body of evidence suggests that similar degrees of specialization may also exist in other regions. Long-term photo-identification studies in British Columbia, Washington, and Alaska have shown that two different populations inhabit the same coastal waters, yet maintain social isolation from each other (Bigg *et al.*, 1990). These populations differ in genetic structure, morphology, behavior, distribution patterns, and ecology. One population, referred to as residents, are fish specialists, whereas the other, termed transients, are primarily mammal hunters.

Residents show strong seasonal movements associated with the coastal migrations of salmon. Observational studies of residents and analyses of stomach contents from beach-cast carcasses have determined that salmon is their predominant prey for at least half the year and that preference is shown for the largest or fattiest available species [chinook (*Oncorhynchus tshawytscha*) in British Columbia and coho (*O. kisutch*) in

Prince William Sound, Alaska; Ford *et al.*, 1998; Saulitis *et al.* 2000] (Fig. 3). Squid (*Gonatopsis borealis*) and a variety of nonsalmonid fish species are also eaten by residents (Ford *et al.*, 1998). There is no evidence that marine mammals are consumed, although porpoises and seals are harassed and killed by residents on rare occasions. Foraging groups of residents typically ignore marine mammals in their vicinity and seldom elicit avoidance responses from those species (Jefferson *et al.*, 1991; Saulitis *et al.*, 2000). The distribution and diet of residents in winter and early spring are poorly known.

Transients show relatively little seasonal change in ABUNDANCE and DISTRIBUTION, most likely because their preferred prey species are present year-round in coastal waters. In British Columbia, harbor seals (*Phoca vitulina*) are the primary prey of transients, although Steller sea lions (*Eumetopias jubatus*), California sea lions (*Zalophus californianus*), harbor porpoises (*Phocoena phocoena*), Dall's porpoises (*Phocoenoides dalli*), and Pacific white-sided dolphins (*Lagenorhynchus obliquidens*) are also important (Ford *et al.*, 1998) (Fig. 4). Larger cetaceans are seldom attacked and killed by transients in coastal waters of British Columbia, although this is common off California. Transients have been observed to harass and occasionally eat a variety of species of seabirds in British Columbia and southeastern Alaska. In Prince William Sound, Alaska, transients feed mostly on harbor seals and Dall's porpoises, but seabird predation has not been recorded (Saulitis *et al.*, 2000). Transients have not been observed to eat any species of fish, and no fish remains have been found in the stomachs of stranded transients.

Such extreme dietary specialization in sympatric populations is without precedent in mammals. These specializations likely evolved gradually over a long period by means of increasingly refined and successful foraging strategies that were learned by

Figure 4 *Transient killer whale ramming a Steller sea lion during an attack in Blackfish Sound, British Columbia. Photo by J. Borrowman.*

individuals and passed across generations. Effective foraging for the very different types of prey of residents and transients may require such divergent skills and tactics that lifestyles dependent on one or the other prey type are mutually exclusive. Foraging specializations may have played a role in the historical separation of ancestral resident and transient groups, and over time the two populations became socially and eventually reproductively isolated. Residents and transients are highly distinct in both mitochondrial and nuclear DNA composition (Hoelzel *et al.*, 1998; Barrett-Lennard, 2000). A third sympatric population has been documented in coastal waters off British Columbia south to California (Ford *et al.*, 2000). Provisionally termed "off-shores," these whales are seldom encountered in protected inshore waters and have not been observed mixing with either the resident or the transient population. They form a genetically distinct group, although they are more closely related to residents than to transients (Barrett-Lennard, 2000). The dietary habits of this population are so far unknown.

Populations of killer whales in other regions may also be highly specialized in feeding habits. In the Antarctic, two sympatric populations have been reported that differ morphologically and ecologically, with one population preying primarily on marine mammals and the other on fish. Off the northern coast of Norway, a population of killer whales moves seasonally in relation to the migration pattern of its principal prey, herring (Similä, 1997). It is likely that populations with dietary specializations exist wherever sufficiently abundant and reliable prey resources are available to sustain them year-round. In other regions, more opportunistic foraging strategies may be expected. For example, in the sub-Antarctic Crozet Islands, killer whales feed seasonally on southern elephant seal (*Mirounga leonina*) pups, but also forage for fish. Groups of killer whales have been observed to attack and kill large baleen whales and sperm whales in various locations (Jefferson *et al.*, 1991), but it is not known whether these groups specialize on such prey items.

The killer whale has no natural predators other than humans.

Figure 3 *Resident killer whale with freshly killed salmon in Haro Strait, Washington. Photo by K. Soloman.*

IV. Life History

Most detailed information on reproduction, mortality, and other life history parameters of killer whales has been derived from long-term photo-identification studies of resident killer whales in British Columbia and Washington (Olesiuk *et al.*, 1990). The reliability and completeness of this information are due to the extremely stable social structure of residents, where emigration from the natal group does not take place and mortalities can be documented reliably (see next section for more detail). It is not known whether these life history parameters are typical of other populations or regions.

Studies of captive whales indicate that sexually mature females have periods of polyestrous cycling interspersed with noncycling intervals of 3–16 months. The gestation period is 15–18 months. In resident killer whales, births may take place in any month, although most are in October–March. Neonate mortality is high, with an estimated 43% dying in the first 6 months (Olesiuk *et al.*, 1990). Calves are nursed for at least a year, but may start taking solid food from the mother while still nursing. Typical age at weaning is not known, but is likely between 1 and 2 years of age.

Females give birth to their first viable calf at between 11 and 16 years of age (mean of 15 years) (Olesiuk *et al.*, 1990). Intervals between viable calves average about 5 years (range 2–14 years). Females have an average of 5.35 viable calves over a 25-year reproductive life span, which ends at approximately 40 years of age. Females then become reproductively senescent for an average period of 10 years, but this postreproductive period may extend to more than 30 years. Mean life expectancy for females (calculated at age 0.5 years, following the period of high neonate mortality) is estimated to be approximately 50 years, and maximum longevity is 80–90 years. Males attain sexual maturity at about 15 years of age, as indicated by a rapid growth of the dorsal fin. Males continue to grow until they reach physical maturity at about 21 years of age. Mean life expectancy for males (calculated at age 0.5 years) is estimated to be about 29 years, with maximum longevity about 50–60 years. Mortality curves for both males and females are U-shaped, although the male curve is narrower.

V. Social Organization

Killer whales are usually observed traveling alone or in groups of up to about 50 individuals. Reports of larger groups likely involve temporary aggregations of smaller, more stable social units. Long-term photo-identification studies have provided information on the social organization of the species in several regions of the world. The most detailed of these are studies in coastal British Columbia, Washington, and Alaska, particularly for the resident population (Bigg *et al.*, 1990; Matkin *et al.*, 1999; Ford *et al.*, 2000). Resident societies can be arranged into a number of groupings based on maternal genealogy, social association, and acoustical relationship. The basic social unit of residents is the matriline, which is a highly stable group of individuals linked by maternal descent. A typical matriline is composed of a female, her sons and daughters, and the offspring of her daughters. Because females may live to 80–90 years of age, and females have their first viable calf at about 15 years of age, a matriline may contain as many as four generations of maternally related individuals. Some matrilines contain only one generation, which can result if a female dies and leaves only sons or daughters that have no young of their own. The bonds among members of a matriline are extremely strong, and individuals are seldom seen apart from the group for more than a few hours. No permanent dispersal of individuals has been observed from a resident matriline.

The next level of social organization in resident killer whales is the pod, which is a group of related matrilines that shared a common maternal ancestor in the recent past. Matrilines within pods are thus more closely related to one another than to matrilines in other pods. Pods are less stable than matrilines, and member matrilines may travel apart for periods of weeks or months. However, these matrilines still tend to travel more often with others from their pod than with matrilines from other pods (Ford *et al.*, 2000). The majority of pods are composed of one to three matrilines. Resident pods in British Columbia, Washington, and Alaska contain a mean of 18 whales (range = 2–49).

Above the pod is the clan, which is a level of social structure defined by the acoustic behavior of pods. It is composed of pods that have similar vocal dialects (see Section VII). All pods within a clan have most likely descended from a common ancestral pod through a process of growth and fragmentation along matrilines. Thus, the related dialects of clan members seem to reflect the common matrilineal heritage of the pods. Those pods with very similar dialects are probably more closely related, and have split more recently, than those with more different features in their dialects. Clans are sympatric, and pods from different clans frequently travel together. Clan membership is occasionally—but usually not—reflected in patterns of association. It is not clear how clans are related to each other, as they have no acoustical features in common, nor is the origin of clans known.

The top level of structure in a resident society is the community, which is made up of pods that regularly associate with one another. The community is thus defined solely by association patterns rather than maternal genealogy or acoustic similarity. Pods from one community have not been seen to travel with those from another, although their ranges may partly overlap. Three communities of residents have been identified in coastal waters of British Columbia, Washington, and Alaska: southern (83 whales in 3 pods, 1 clan), northern (214 whales in 16 pods, 3 clans), and southern Alaskan (237 whales in 11 pods, 2 clans) (Matkin *et al.*, 1999; Ford *et al.*, 2000).

Social organization in mammal-eating transient killer whales is not as well known as in residents (Ford and Ellis, 1999). Like residents, the basic social unit is the matriline, but unlike residents, offspring may disperse from matrilines for extended periods or permanently, either as juveniles or as adults. As a result, transient matrilines tend to be smaller than those of residents, and lone individuals, particularly males, are often observed. Small group sizes of transients appear to reflect the marine-mammal foraging specialization of this population (Baird and Dill, 1996; Ford and Ellis, 1999). Association patterns of transient matrilines are very dynamic, and they do not form consistent groupings of matrilines equivalent to resident pods.

All transient groups in a community can be linked together via this network of associations. All members of a transient community share a related call repertoire, as in a resident clan. However, regional differences exist in the vocal repertoire of transients.

Social organization based on matrilineal descent may be typical of killer whales globally. In other regions where long-term photo-identification studies have been undertaken, close and prolonged associations of mothers and offspring are commonly seen (e.g., Norway, Crozet Islands, Argentina). Temporal persistence of these bonds may be a primary variable determining group sizes and structure.

VI. Activity States and Behavior

The activity states of killer whale groups are of four basic types: foraging, traveling, resting, and socializing (Ford, 1989; Saulitis *et al.*, 2000). Minor differences in definitions and classification criteria of activities by different researchers make detailed comparisons difficult, but general patterns are evident. Foraging and traveling are the predominant activity states noted in all populations, although the proportions of the activity budget dedicated to these activities vary. Mammal-eating transients in coastal waters of the northeastern Pacific spend the great majority of their time (approximately 90–95%) foraging and traveling, whereas fish-eating residents spend only about 60–70% of their time doing so, at least during summer when salmon is abundant. Residents spend considerably more time resting and socializing than transients. Fish-eating killer whales in northern Norway have activity budgets very similar to those of northeastern Pacific residents (Similä, 1997).

A. Foraging

Behavior patterns of foraging killer whales vary considerably among populations and prey types. Groups of salmon-hunting residents often disperse over large surface areas while foraging, with members moving at roughly the same speed (mean = 6.0 km/hr) and direction. Foraging episodes are typically 2–3 hr in duration, but may last up to 7 hr. Individual salmon are pursued, captured, and eaten by single animals or small subgroups, usually a mother and juvenile offspring. Norwegian killer whales feed on herring in a coordinated manner referred to as "carousel feeding" (Similä, 1997). Using percussive actions such as tail lobbing, releasing blasts of bubbles, and flashing the white ventral side of their bodies, the whales herd herring into a tight ball close to the surface. The whales then stun fish by striking the edges of the ball with their tail flukes and eat the debilitated prey.

Transient killer whales in the northeastern Pacific typically forage in smaller groups than fish-eating killer whales. Transient groups hunt harbor seals in groups averaging three to four individuals (Baird and Dill, 1996; Ford *et al.*, 1998; Saulitis *et al.*, 2000), usually close to shore and near seal haul-out sites. While foraging, transients remain acoustically very quiet apparently to avoid detection by potential prey and possibly to locate prey by passive listening (Barrett-Lennard *et al.*, 1996). Harbor seals are killed and shared among group members relatively quickly compared to Steller sea lions or California sea lions, which may take over 2 hr to kill and consume. Sea lions

are usually rammed or butted with the whales' heads, and slapped repeatedly with tail flukes, until the animal is debilitated sufficiently to be taken underwater and drowned. When hunting porpoises or dolphins, transients forage in slightly larger groups averaging five members, which spread out in open water in a rough line abreast. The whales single out an individual porpoise, chase it until it tires, and then ram it or jump on it to complete the kill (Ford *et al.*, 1998). Larger schools of Pacific white-sided dolphins are often driven by transients into confined bays where individual dolphins are trapped against the shore and killed (Ford and Ellis, 1999).

A variety of specialized tactics have been described for killer whales hunting marine mammals in other regions. In Patagonia, Argentina, killer whales hunt southern sea lion (*Otaria flavescens*) and elephant seal (*Mirounga leonina*) pups in the shallows along sloping pebble beaches, and often intentionally strand themselves temporarily in the process (Fig. 5). The whales hunt cooperatively and share their prey after capture. Killer whales in the Crozet Islands hunt elephant seal pups in a manner similar to those in Patagonia, and adults appear to teach this technique to their offspring (Guinet and Bouvier, 1995). In the Antarctic, killer whales have been observed to locate seals hauled out on ice pans by spyhopping and then dislodging them from the pan with a wave created by the whale's fast approach. Attacks on baleen whales or sperm whales often involve groups of 10–20 killer whales, which work together in a coordinated manner to subdue the prey. Individuals will attempt to grasp the tail flukes or pectoral flippers to immobilize the larger whale, while others attack the head and blowhole area evidently to prevent the whale from breathing. Once killed, often just the tongue, lips, and BLUBBER are consumed (Jefferson *et al.*, 1991). Because most baleen whales sink upon death, killer whales may only be able to feed extensively on carcasses of whales killed in shallow waters (Guinet *et al.*, 2000).

Figure 5 *Female killer whale catching a southern sea lion pup in Patagonia, Argentina. Photo by J. Ford.*

B. Traveling

Traveling killer whales move in a single direction at a consistent, fast pace, with no evidence of foraging or FEEDING. Groups often travel in a line abreast, with synchronized dives and surfacings. Resident killer whales have been documented to travel at speeds of over 20 km/hr (mean = 10.4 km/hr; Ford, 1989).

C. Resting

When resting, resident killer whales usually form a line abreast, often with individuals grouped tightly together (Fig. 6). Group diving and surfacing become closely synchronized and regular, with longer dives of 2–5 min duration separated by three or four short, shallow dives. The rate of forward progression is slow compared to foraging and traveling, and resting groups may stop altogether and rest motionless at the surface for several minutes (Ford, 1989; Similä, 1997).

D. Socializing

Socializing activity includes a wide range of physical displays and social interactions. Aerial behaviors are frequent and may include spyhops, breaches, flipper slaps, tail lobs, and head stands. Juveniles often chase each other, roll and thrash at the surface, and engage in various other forms of play behavior, including playing with objects such as kelp or jellyfish. Sexual interactions involving penile erections are commonly observed, predominantly in all-male play groups. Some individuals may rest quietly at the surface while other pod members actively socialize. Beach rubbing is a common behavior observed during socializing in some populations. Residents belonging to the "northern" community in British Columbia visit certain beaches repeatedly to rub their bodies on smooth pebbles in shallow water (Ford, 1989). Interestingly, "southern" community residents do not share this behavioral tradition and have not been seen to rub at any location.

VII. Sound Production

Like most delphinids, killer whales are highly vocal. They produce a wide variety of clicks, whistles, and pulsed calls for ECHOLOCATION and social signaling. Studies of resident killer

Figure 6 *Resting pods of resident killer whales in Prince William Sound, Alaska. Photo by C. Matkin.*

whales in British Columbia have documented vocal variations associated with activity state and group identity (Ford, 1989, 1991). Vocal exchanges among foraging resident whales are dominated by highly stereotyped, repetitive discrete calls from a repertoire averaging 12 call types (range 7–17 call types) per pod. Resting activity is usually associated with greatly reduced vocal activity, and occasional use of certain calls heard predominantly, but not exclusively, in such contexts. Socializing whales use mainly whistles and nonrepetitive, variable pulsed calls, and aberrant versions of discrete calls. Excitement or motivational levels of vocalizing individuals are reflected in minor variations in pitch and duration of discrete calls.

Call repertoires of resident killer whale pods have features that are distinct, forming systems of group-specific dialects. The entire call repertoire appears to be shared by all pod members. Some portions of a pod's call repertoire may be shared with certain other pods, whereas other portions may be unique. Levels of similarity in these group-specific dialects appear to reflect the degree of relatedness of different pods better than do patterns of travel association. Dialects most likely are learned by young whales by mimicking their mother and siblings, and are retained in the matriline due to the lack of individual dispersal. Divergent variations in dialects among related matrilines likely accompany the gradual fission that leads to pod formation. Dialects likely provide an acoustic means of maintaining group identity and cohesion, and may serve as indicators of relatedness that help in the avoidance of inbreeding between closely related whales (Ford, 1991; Barrett-Lennard, 2000). Dialects have also been documented within a community of pods of killer whales in northern Norway, and likely exist elsewhere.

Mammal-eating transient killer whales in British Columbia and Alaska have greatly reduced vocalization rates compared to residents. Transients are generally silent when foraging. This includes the use of echolocation, which in one study was heard 27 times less often from foraging transients than foraging residents (using an index adjusted for group size; Barrett-Lennard et al., 1996). Transients are more likely than residents to use individual (or "cryptic") clicks rather than click trains, presumably to avoid alerting potential prey to their approach. When vocal, transients off the coasts of southeastern Alaska to California produce a number of calls that are shared among all groups in the community. Certain other calls seem exclusive to transient groups in different portions of this range. Group-specific dialects as seen in resident pods are not evident, presumably due to the reduced stability of social structure in transients (Ford and Ellis, 1999).

VIII. Conservation Status

Globally, the killer whale is listed by the IUCN as "lower risk: conservation dependent," meaning that the species, although presently not considered at risk, could become so should existing conservation programs be discontinued and exploitation expanded. Historically, killer whales in several regions have been the target of directed fisheries, culling, and persecution. An average of 43 whales per year were taken by Japanese whalers in their coastal waters during 1946–1981, mostly for human consumption. This fishery is no longer active. Norwegian whalers took an average of 56 whales per year during

1938–1981 in a government-subsidized hunt aimed at reducing killer whale numbers to reduce competition for other fisheries. The killer whale meat from this fishery was used only for animal consumption. A small number of killer whales were taken annually by Soviet whalers in the Antarctic, with the exception of a large take of 916 animals in the 1979/1980 season. No significant directed hunt for killer whales continues today.

Killer whales have long been feared as dangerous predators or vilified as perceived or real threats to fisheries in many regions, and were often harassed or shot opportunistically. Although much reduced, some such persecution continues today. Killer whales have been shot by fishermen in Alaska and possibly other regions to prevent the whales from taking fish from long-line fishing operations (Matkin *et al.*, 1999). There is evidence that populations in coastal waters of British Columbia and Washington state were already depressed from shootings when a live-capture fishery developed there in the mid-1960s. Killer whales became highly sought for public display in aquaria following the first successful capture and display of the species at Vancouver in 1964. During 1964–1977, 63 whales were taken in this fishery to supply aquaria in many parts of the world (Olesiuk *et al.*, 1990). During the late 1970s to mid-1980s, live captures shifted to the waters of Iceland, where over 50 whales were taken. The improved success of captive breeding during the past decade has reduced the need for capture from wild populations, although periodic live captures continue.

Other CONSERVATION concerns include direct effects of oil spills and other forms of toxic pollution of killer whale survival. The *Exxon Valdez* oil spill in Alaska was strongly correlated with the subsequent loss of 14 whales from a pod that was seen swimming through light oil slicks early in the spill, although it was not possible to directly attribute the deaths to this cause. Oil spills may also have indirect effects on killer whales by reducing prey abundance. Being at high trophic levels in the food web, killer whales are susceptible to bioaccumulation of organochlorine pollutants. Levels of PCBs in resident and, in particular, transient killer whales in British Columbia and Washington state have been shown to be among the highest observed in any cetacean and are higher than levels found to affect the health of European harbor seals. It is not known whether there is a direct impact of PCBs on health in these killer whales, although such effects as immunosuppression and reduced reproductive success are possible (Ross *et al.*, 2000).

Other potential impacts of human activities on killer whale status are reduced prey availability and disturbance caused by vessel traffic. As an example, many stocks of salmon, the principal prey of residents killer whales, have declined significantly in British Columbia and Washington State as a result of overfishing, degradation of spawning grounds, and reduced ocean survival. Vessel disturbance is of particular concern in areas of intensive whale watching, although many forms of boat traffic have the potential to affect whales. The physical presence of moving boats near killer whales can disrupt their activities, particularly resting. Underwater noise from vessels has the potential to interfere with social or echolocation signals, or to mask passive acoustic cues that may be important in finding prey.

It is possible that many of these potential impacts on killer whales are, when taken alone, insufficient to negatively affect killer whale survival. However, there is a potential for more serious cumulative effects that could displace killer whales from critical habitats or result in reduced survival.

See Also the Following Articles

Captivity ■ Delphinids, Overview ■ Marine Parks and Zoos ■ Predation on Marine Mammals ■ Whale Watching

References

Baird, R. W., and Dill, L. M. (1996). Ecological and social determinants of group size in transient killer whales. *Behav. Ecol.* **7**, 408–416.

Barrett-Lennard, L. G. (2000). "Population Structure and Mating Patterns of Killer Whales (*Orcinus orca*) as Revealed by DNA Analysis." Ph.D. Dissertation, University of British Columbia, Vancouver, BC.

Barrett-Lennard, L. G., Ford, J. K. B., and Heise, K. A. (1996). The mixed blessing of echolocation: Differences in sonar use by fish-eating and mammal-eating killer whales. *Anim. Behav.* **51**, 553–565.

Bigg, M. A., Olesiuk, P. F., Ellis, G. M., Ford, J. K. B., and Balcomb III, K. C. (1990). Social organization and genealogy of resident killer whales (*Orcinus orca*) in the coastal waters of British Columbia and Washington State. *Rep. Int. Whal. Commn. Spec. Issue* **12**, 383–405.

Dahlheim, M. E., and Heyning, J. E. (1999). Killer whale *Orcinus orca* (Linnaeus, 1758). *In* "Handbook of Marine Mammals" (S. H. Ridgway and R. Harrison, eds.), pp. 281–322. Academic Press, San Diego.

Ford, J. K. B. (1989). Acoustic behaviour of resident killer whales (*Orcinus orca*) off Vancouver Island, British Columbia. *Can. J. Zool.* **67**, 727–745.

Ford, J. K. B. (1991). Vocal traditions among resident killer whales (*Orcinus orca*) in coastal waters of British Columbia. *Can. J. Zool.* **69**, 1454–1483.

Ford, J. K. B., and Ellis, G. M. (1999). "Transients: Mammal-Hunting Killer Whales of British Columbia, Washington, and Southeastern Alaska." UBC Press and Univ. of Washington Press, Vancouver, BC and Seattle, WA.

Ford, J. K. B., Ellis, G. M., and Balcomb, K. C. (2000). "Killer Whales: The Natural History and Genealogy of *Orcinus orca* in the Waters of British Columbia and Washington." UBC Press and Univ. of Washington Press, Vancouver, BC and Seattle, WA.

Ford, J. K. B., Ellis, G. M., Barrett-Lennard, L. G., Morton, A. B., Palm, R. S., and Balcomb III, K. C. (1998). Dietary specialization in two sympatric populations of killer whales (*Orcinus orca*) in coastal British Columbia and adjacent waters. *Can. J. Zool.* **76**, 1456–1471.

Guinet, C., Barrett-Lennard, L. G., and Loyer, B. (2000). Co-ordinated attack behaviour and prey sharing by killer whales at Crozet Archipelago: Strategies for feeding on negatively-buoyant prey. *Mar. Mam. Sci.* **16**, 829–834.

Guinet, C., and Bouvier, J. (1995). Development of intentional stranding hunting techniques in killer whale (*Orcinus orca*) calves at Crozet Archipelago. *Can. J. Zool.* **73**, 27–33.

Hoelzel, A. R., Dahlheim, M. E., and Stern, S. J. (1998). Low genetic variation among killer whales (*Orcinus orca*) in the eastern North Pacific, and differentiation between foraging specialists. *J. Heredity* **89**, 121–128.

Jefferson, T. A., Stacey, P. F., and Baird, R. W. (1991). A review of killer whale interactions with other marine mammals: Predation to co-existence. *Mamm. Rev.* **21**, 151–180.

Matkin, C. O., Ellis, G., Olesiuk, P., and Saulitis, E. (1999). Association patterns and inferred genealogies of resident killer whales, *Orcinus orca*, in Prince William Sound, Alaska. *Fish. Bull.* **97**, 900–919.

Matkin, C., Ellis, G., Saulitis, E., Barrett-Lennard, L., and Matkin, D. (1999). Killer whales of Southern Alaska. North Gulf Oceanic Society, Homer, Alaska.

Olesiuk, P. F., Bigg, M. A., and Ellis, G. M. (1990). Life history and population dynamics of resident killer whales (*Orcinus orca*) in the coastal waters of British Columbia and Washington State. *Rep. Int. Whal. Comm. Spec. Issue* **12**, 209–242.

Ross, P. S., Ellis, G. M., Ikonomou, M. G., Barrett-Lennard, L. G., and Addison, R. F. (2000). High PCB concentrations in free-ranging Pacific killer whales, *Orcinus orca:* Effects of age, sex and dietary preference. *Mar. Poll. Bull.* **40**, 504–515.

Saulitis, E., Matkin, C., Barrett-Lennard, Heise, K., and Ellis, G. (2000). Foraging strategies of sympatric killer whale (*Orcinus orca*) populations in Prince William Sound, Alaska. *Mar. Mamm. Sci.* **16**, 94–109.

Similä, T. (1997). "Behavioral Ecology of Killer Whales in Northern Norway." Dr. Scient Thesis, Norwegian College of Fisheries Science, University of Tromsø, Tromsø, Norway.

Visser, I. (1999). Benthic foraging on stingrays by killer whales (*Orcinus orca*) in New Zealand waters. *Mar. Mamm. Sci.* **15**, 220–227.

Figure 1 Euphausia superba. *Courtesy of I. Everson, British Antarctic Survey.*

Krill

ROGER P. HEWITT AND JESSICA D. LIPSKY
*Southwest Fisheries Science Center,
La Jolla, California*

Euphausiids, or krill, have long been recognized as a critical element of the natural economy of the world's oceans (Sars, 1885; Brinton, 1962; Marr, 1962; Mauchline and Fisher, 1969; Mauchline, 1980). Early fishery biologists repeatedly stressed the importance of various species of euphausiids as food for exploited fish and whale stocks (Lebour, 1924; Hickling, 1927; Hjort and Rund, 1929). Norwegian whalers referred to the euphausiids found in large numbers in the stomachs of whales caught in the North Atlantic as *stor krill* (or large krill, referring to *Meganyctiphanes norvegica*) and *smaa krill* (or small krill, referring to *Thysanoëssa inermis*); the word "krill" is now used in reference to euphausiids in general (Mauchline and Fisher, 1969). Laws (1985) estimated that 190 million tons of Antarctic krill (*Euphausia superba*) were consumed annually by baleen whales in the Southern Ocean prior to their exploitation. It is estimated that the current populations of whales, birds, pinnipeds, fish, and squid consume 250 million tons of Antarctic krill annually (Miller and Hampton, 1989). Of the 85 species of krill, Mauchline and Fisher (1969) listed only nine of primary importance in terms of their distribution range, biomass, and dominance in the diets of vertebrate predators. They noted that these species constitute a large fraction of the plankton where they are found and that their biomasses are largest at high latitudes. In addition to their numbers, the habit of euphausiids to form large swarms makes them particularly important as prey to marine vertebrates (Fig. 1).

The krill species considered most important to the trophodynamics of marine ecosystems are *M. norvegica*, *T. raschii*, and *T. inermis* in the North Atlantic Ocean; *E. pacifica*, *T. in-*ermis, *T. raschii*, *T. longipes*, and *T. inspinata* in the North Pacific Ocean; and *E. superba*, *E. crystallorophias*, and *T. macrura* in the Southern Ocean. Mauchline and Fisher (1969) list another seven species of importance in more restricted geographical areas and/or seasons: *Nyctiphanes couchii* in the North Atlantic Ocean; *T. spinifera* and *E. similis* in the North Pacific Ocean; *N. capensis* near the southern part of Africa; *N. australis* and *Pseudoeuphausia latifrons* from Western Australia to New Zealand; and *E. vallentini* in the Southern Ocean. In addition, the following seven species are often cited in predator diet samples from restricted locales and time periods: *Nematocelis megalops* in the North Atlantic; *E. recurva*, *E. lucens*, *E. hemigibba*, *T. gregaria*, *E. spinifera*, and *N. megalops* from Western Australia to New Zealand; *E. recurva*, *E. lucens*, and *T. gregaria* near the southern part of Africa; and *N. simplex* along the western coast of North America (Fig. 2).

While many of these species are broadly dispersed, they exhibit their highest densities in areas of enhanced seasonal primary and secondary production. These areas include eastern boundary currents, coastal and oceanic upwelling regions, and sea ice edge zones, as well as estuaries, fjords, and small-scale eddies where physical mechanisms may enhance the aggregation of krill. It is not surprising therefore to find krill predators, including baleen whales and crabeater seals, concentrated in these areas as well (Fig. 3).

Krill species differ in their geographic distribution, body size (ranging from less than 1 cm to 14 cm), and longevity (ranging from less than 1 year to as many as 10 years) but share many other characteristics that contribute to their importance as prey for baleen whales. Furthermore, baleen whales have not shown strong species or size selectivity among krill when foraging in an area where more than one species and/or developmental stage is present. Krill are therefore described here in general terms with species-specific reference only where appropriate.

I. General Morphology and Life History

The body plan of krill (Fig. 4) is divided into two main regions: the cephalothorax and the abdomen. The cephalothorax,

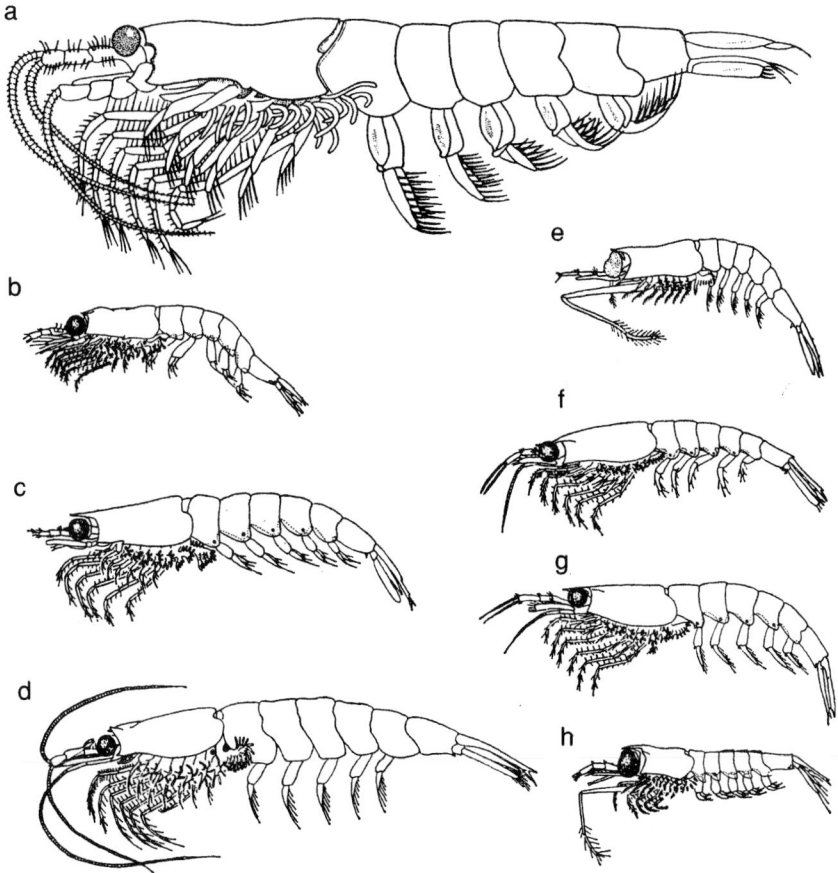

Figure 2 *Scale drawing of the eight most important krill species:* Euphausia superba *(a),* E. Pacifica *(b),* E. crystallorphias *(c),* Meganyctiphanes norvegica *(d),* Thysanoëssa macrura *(e),* T. inermis *(f),* T. raschii *(g), and* T. longipes *(h). From Mauchline and Fischer (1969).*

a fused head and thorax, contains the internal organs, including the digestive system, the heart, and gonads. It is about one-third of the body length and is covered by a thin shell or carapace. The muscled abdomen is made up of six segments ending with a telson and two pair of uropods, which together form a fan shape at the tail. At the head there are a pair of eyes and two pair of antennae with tactile and olfactory sensors; excretory organs open near the second set of antennas. The mouth is made up of several parts whose function is to filter, macerate, and manipulate food prior to ingestion. Six to eight pairs of limbs are connected to the thorax and are used to filter particles out of the water and pass them to the mouth. Unlike decapod crustaceans (crabs, lobsters, prawns, shrimps) the gills of krill are exposed, hanging below the carapace. The first five abdominal segments each have a single pair of limbs (pleopods) attached, which are used for swimming; the sixth abdominal segment has no appendages. On a mature adult male the first pair of pleopods is modified to form a petasma, which is used during copulation to clasp and transfer spermatophores to the female. The thelycum, or female copulatory organ, is located on the anterior underside of the thorax near the opening of the oviducts.

The exoskeletons of krill are translucent, allowing a view of the internal organs, including the heart, stomach, and hepatopancreas, which is often colored dark green or red. Krill are also luminescent with light-emitting photophores located at the bases of their pleopods, near the thelycum, close to the mouth and in the eye stalks. The photophores are a deep red color but emit electric blue light in the water. Many species are also pigmented with red chromatophores that expand when the animal is stimulated. As a result, swarms of krill often appear to be bright red, particularly when under attack by a predator. The guano of krill-eating birds is often pink in color, and the feces of krill-eating marine mammals are characteristically dark red.

As krill mature sexually, males elaborate packets of sperm called spermatophores and females develop clusters of eggs or broods. During spawning the male grasps the female with his petasmae and transfers spermatophores to her body where they adhere in the vicinity of her thelycum. Among the various species of krill, brood size ranges from tens of eggs to several thousand, and some species have been observed to spawn several broods during a single breeding season. When a female releases a brood of eggs, they are fertilized by spermatozoa now liberated from the spermatophores. For some species the

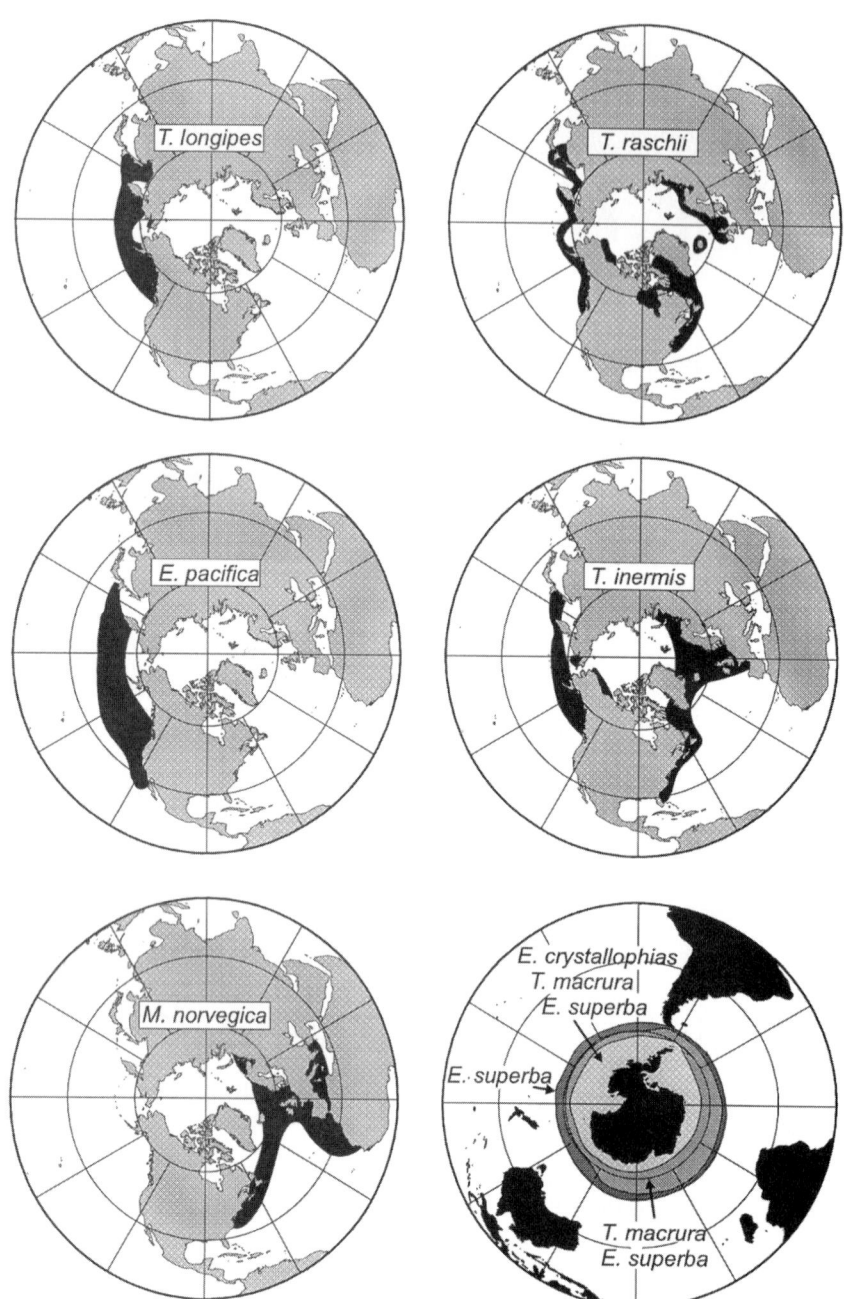

Figure 3 *Northern and Southern Hemisphere maps showing dispersion of important krill species. Redrawn from Mauchline and Fischer (1969).*

female carries the fertilized eggs in brood pouches until they hatch, thereby protecting them from predation. For most species, however, eggs are released into the open sea. In some cases the eggs are neutrally buoyant, but often they are heavier than water and sink before hatching into nauplius larvae, which in turn develop and molt through a series of larval stages, each resembling the adult morphology more than the previous stage. In the case of *E. superba*, a brood of 10,000 fertilized eggs may be released by a single female in a near-surface swarm of spawning adults; the eggs sink to depths of greater than 1000 m, incubate, and hatch. The nauplius has no swimming appendages and continues to sink as it grows, molts, and gives rise to more advanced larval forms. Once it can swim the larva begins its ascent into the surface waters, progressing through several more molts and ultimately emerging as a calyptosis larva. Calyptoses continue to eat, grow, and molt through additional stages in preparation for the winter when food is less available. Sometime in the late winter or early spring the larvae finally

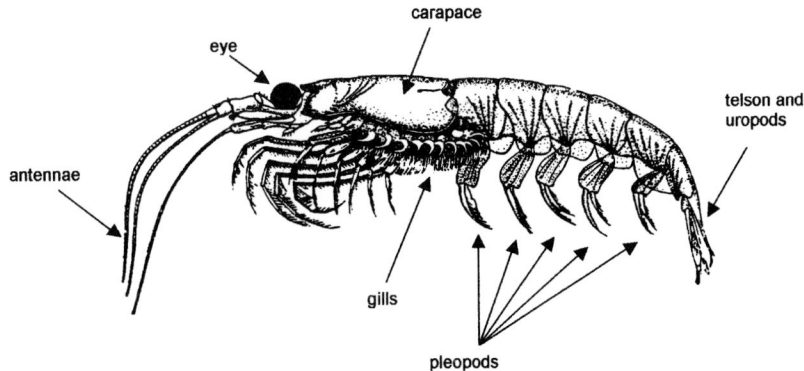

Figure 4 *General krill body plan.*

metamorphose into juvenile krill, but it may be as long as another year before they are ready to spawn themselves. In the case of *E. pacifica*, this process is compressed to a few months with spawning occurring during the spring and recruitment into the adult population occurring during the fall.

Except for rich fat stores invested in developing eggs, larval and postlarval krill do not elaborate high levels of fat reserves. Consequently, they must eat constantly in order to offset the energy costs of swimming, growth, and reproduction. In addition, krill periodically shed their exoskeletons throughout their life, adding substantially to their energy requirements. Krill are generally thought to be FILTER-FEEDING herbivores, grazing on phytoplankton in the surface layers of the ocean. Many species, however, are reported to be omnivorous, filtering and/or capturing copepods and other small zooplankton. *E. superba* has been observed in the cavities and cracks on the underside of winter sea ice, presumably feeding on interstitial ice algae. Krill growth and reproductive activity have been directly linked to available food supplies. Negative growth and regression of sexual characteristics have been observed in several species and related to lowered availability of food.

II. Swarming

Krill are heavier than water and must swim continually in order to maintain their position. They aggregate into dense swarms, which can take on a variety of shapes from discreet balls to extensive layers. The swarms may range from 1 m thick to several tens of meters and may extend horizontally tens of meters to several thousand meters. Individual animals appear to be in constant movement, and a sharp gradient in density is often observed at the periphery of a swarm. Within the swarm, volumetric densities may range up to several thousand animals per cubic meter. Near the shelf break surrounding islands in the southwest Atlantic sector of the Southern Ocean it is not uncommon to observe large swarms of *E. superba*, each estimated to contain several thousand tons of krill. Most krill species migrate vertically each day, moving into the upper waters at night and dispersing and moving downward just before dawn and aggregating into denser concentrations. It is generally thought that this behavior is the result of a trade-off between avoiding predation (dense swarms deep in the water during the day) and

maximizing feeding efficiency (dispersed individuals in the more particle-rich surface water at night). Although this is a regular pattern, vertical migration behavior varies between species and within a species depending on location and season. Daytime surface swarms have been observed for several species and often contain reproductively mature individuals.

The highest densities of krill have been reported near areas of strong vertical mixing and enhanced primary production. These include coastal upwelling zones, ocean frontal boundaries, and topographic features that interrupt or modify currents such as continental shelf breaks, underwater canyons and escarpments, and seamounts. Krill swarms also tend to aggregate in areas of water flow discontinuity such as eddies and sheer zones between opposing currents.

III. Recruitment Variability

Recruitment of young animals into adult euphausiid population is highly variable in space and time. The production of spawn and the survival of larvae may vary widely within the distribution range of a species as well as between reproductive events. In his review of euphausiid life histories, Siegel (2000) noted that most species reduce their growth phase and extend their reproductive phase toward the center of their distribution ranges. Closer to their distribution limits, krill put more time into growth and less into reproduction.

There is no apparent relationship between stock size and the production of new recruits for most species studied. Relatively large adult stocks can produce few new recruits, and small adult stocks are capable of producing enough new recruits to increase the stock abundance severalfold. The intensity of spawning, survival of eggs and larvae, and the rate of growth have been shown to vary widely between years for several species, resulting in large year-to-year variability in abundances. Interannual variability in the abundances of *E. pacifica* and *N. simplex* off the west coast of North America has been estimated as 10-fold, 25-fold for *T. inermis* in the Barents Sea, and 5- to 60-fold for *M. norvegica* at different parts of its range in the North Atlantic.

Recruitment success is affected by exogenous factors, which act to enhance adult reproduction, survival of eggs, and growth of larvae. The best documented of these is the influence of coastal upwelling, which enhances primary production and the subsequent growth and maturation of young krill. Temperature affects the incubation rate of eggs and the growth rate of larvae, exposing them to longer or shorter periods of predation. Fluctuations in currents may also transport animals into unfavorable areas. Near the Antarctic Peninsula, *E. superba* spawn earlier in the spring and for a longer period following winters of extensive sea ice development; their larvae enjoy a higher survival rate if sea ice is extensive during the following winter (Loeb *et al.*, 1997). Four to 5-year cycles are apparent in the seasonal extent of sea ice and the recruitment of krill in this region of the Southern Ocean. The postulated affect of seasonal sea ice is to provide a refuge, access to a wintertime food source (ice algae), and to inhibit rapid springtime population growth of a potential competitor to krill, *Salpa thompsoni* (Loeb *et al.*, 1997). *Salpa thompsoni* is a pelagic tunicate and obligate filter feeder, which requires open water access to springtime phytoplankton blooms in order to reproduce.

Figure 5 *Blue whale lunging through a subsurface krill layer, from an animation. Courtesy of 4.2.2 LTD., Bristol, U.K., for National Geographic Television.*

IV. Foraging Tactics of Baleen Whales and Crabeater Seals

The two characteristics of euphausiids described earlier—an immediate response, in terms of individual growth and reproductive output, to favorable conditions and highest densities in predictable locales—allow efficient exploitation of krill by baleen whales.

In general, baleen whales migrate between high-latitude summer feeding grounds and low-latitude winter breeding and calving grounds. Exceptions are bowhead whales (*Balaena mysticetus*), which are restricted to Arctic regions, and Bryde's whales (*B. edeni*), which usually range from subtropical to temperate waters. Blue, fin, and sei whales (*Balaenoptera musculus, B. physalus*, and *B. borealis*) tend to migrate in offshore waters while gray (*Eschrichtius robustus*), right (*Eubalaena* spp.), and humpback (*Megaptera novaeangliae*) whales tend to use a more coastal MIGRATION route. Adult whales are thought to feed less during migration than immature or undernourished animals. Off the western coast of North America, blue, fin, Bryde's, and humpback whales have been observed feeding on euphausiids aggregated along underwater escarpments and canyons during both winter and summer. The location and timing of whale foraging follow the appearance of high densities of euphausiids and tend to progress from south in the winter to north in the summer. In recent years, aggregations of euphausiids and foraging whales have been a predictable event in the Gulf of California during late winter, near underwater seamounts and canyons off northern California in the summer, and along the shelf break surrounding the Channel Islands in the fall.

Actively feeding whales have been observed to lunge through surface swarms of krill, engulfing large quantities of water and distending their bellies, before expelling the water and extruding as much as several hundred kilograms of krill (Fig. 5). Similar feeding behavior on subsurface swarms has been inferred from acoustic records of krill layers superimposed with dive tracks simultaneously recorded by instruments attached to foraging whales. There are many reports of humpback and fin whales herding and concentrating their prey before lunging through an aggregation of krill. Bryde's and minke whales (*B. acutorostrata* and *B. bonaerensis*) have also been observed gulping large quantities of aggregated euphausiids. Foraging by right and bowhead whales has been described as skimming a continuous stream of water rather than gulping; this behavior may be more efficient with dispersed prey (Nemoto, 1970). Sei and gray whales appear to use both methods.

Despite their name, crabeater seals (*Lobodon carcinophaga*) eat very little other than krill. They are found in the sea ice zone in the Southern Ocean and constitute 50% by number (75% by weight) of the world pinniped population. Crabeaters have lobed cusp TEETH with spaces between them. It is presumed from the shape of the mouth, tongue, an spacing between the teeth that crabeater seals engulf a portion of an aggregation of krill and then strain the water similar to a baleen whale. Crabeaters tend to feed at night when krill are in the upper layers and more dispersed than during the day.

V. Marine Mammal Diets and Euphausiid Consumption by Ocean Basin

A. North Pacific

Blue whales in the eastern North Pacific, foraging from British Columbia to the Californias, feed principally on three species of krill in the California Current; *E. pacifica* and *Thysanoëssa spinifera*, the more inshore species, which is replaced by *Nyctiphanes simplex* moving south (Fig. 6). Fin whales have been observed feeding from the Gulf of California to the northern parts of the Bering Sea from April to September, respectively. During the late winter and spring, fin whales feed on *N. simplex* in the southern portion of their foraging range; moving north in the summer they feed on *E. pacifica, T. raschii, T. longipes*, and *T. inermis*. Fin whales have also been observed to feed on copepods, with the change in prey type related to changes in local relative densities of prey. During the summer months, sei whales consume a variety of euphausiid species, including *T. gregaria, E. pacifica, E. recurva, E. diomedeae, E. tenera, T. inermis, T. spinifera, N. difficilis*, and *N. gracilis*.

Figure 6 *A blue whale swimming through a surface swarm of krill. Courtesy of John Calambokidis, Cascadia Research.*

South of Japan less than 2% of the sei whale diet has been reported to consist of fish, whereas prey species consumed near the Aleutian Islands include copepods, amphipods, decapod crustaceans, fishes, and squid. In comparison to blue whales and fin whales, sei whales appear to be more opportunistic feeders, willing to switch prey type more readily in response to local availability. Bryde's whales have been observed consuming *E. similis, N. difficilis,* and *T. gregaria* as well as amphipods, copepods, and fish in the western Pacific and both euphausiids and fish in the Gulf of California. Humpback whales have been observed foraging on euphausiids, including *E. pacifica, T. raschii, T. longipes,* and *T. spinifera,* from Southeast Alaska to Baja California, although a substantial part of their diet includes clupioid fish as well. Bowhead whales forage primarily on *T. raschii* and *T. inermis* in the Bering and Beaufort Seas during summer and fall, although copepods, mysids, and amphipods also form a part of their diet. Minke whales have been observed foraging on euphausiids, but appear to prefer fish throughout the North and northeastern Pacific. Gray whales are thought to consume primarily benthic amphipods in the Bering Sea during the summer months, although there are reports of gray whales consuming *T. raschii* in the Bering Sea and *E. pacifica* off northern California. Prey selectivity among ringed seals (*Pusa hispida*) appears to be dependent on seasonality and location. Ringed seals have been reported to eat *T. raschii, T. longipes,* and *T. inermis* in offshore waters in the Northern Hemisphere in spring and summer when krill abundance is greatest; in the winter they consume Arctic cod and other fish species in inshore waters.

B. South Pacific

Bryde's whales have been observed feeding on *E. diomedeae, E. recurva,* and *T. gregaria* and occasionally fish in the Coral Sea (western South Pacific) during the austral spring. In the eastern South Pacific, Bryde's whales consume euphausiids during the austral summer between 35° and 40° south latitude. Humpback whales have been observed off the east and west coasts of Australia feeding on euphausiids, including *E. hemmigibba, P. latifrons,* and *E. spinifera.*

C. North Atlantic

Fin whales feed primarily on *M. norvegica, T. inermis,* and *T. raschii* during the summer months, switching between prey species in response to local availability. Minke whales consume *T. inermis* and *M. norvegica* in the North Atlantic where euphausiids form a much larger portion of the diet than in the Pacific. North Atlantic right whales (*Eubalaena glacialis*) feed primarily on copepods, although consumption of euphausiids has been observed particularly when associated with copepods. Harp seals (*Pagophilus groenlandicus*) feed on a variety of prey, including decapods, amphipods, euphausiids, and pelagic fishes; however, newly weaned pups and young seals have been reported to feed mainly on *Thysanoëssa* species.

D. Indian Ocean

Fin and minke whales have been observed feeding on euphausiids in the southwest Indian Ocean during their spring and fall migrations to and from the Southern Ocean; prey species include *E. recurva, E. lucens, T. gregaria, E. spinifera, N. capensis,* and *E. diomedeae.* Bryde's whales forage on these euphausiids species as well in the southwest Indian Ocean. Near Durban, South Africa humpback whales have been observed feeding on *E. recurva* and *T. gregaria,* and a single pygmy blue whale (*B. musculus brevicauda*) was reported feeding on *E. recurva* and *E. diomedeae.* Sei whales were observed to consume euphausiids, as well as copepods, amphipods, pteropods, and fish.

E. Southern Ocean

Fin and minke whales consume several species of krill in the Southern Ocean throughout the austral summer. Species preference appears to be related to local availability with *T. macrura* and *E. vallentini* more prevalent in the diets of animals foraging in open waters and *E. frigida* and *E. crystallorophias* more prevalent near the continental shelf and ice edge regions. The numerically dominant euphausiid in the Southern Ocean, *E. superba,* is consumed in all areas. Southern right whales (*E. australis*) have been observed foraging on *E. superba* in the Atlantic sector of the Southern Ocean. Humpback whales have been frequently observed foraging on *E. superba* in bays and fjords along the Antarctic Peninsula. Crabeater seals consume *E. superba* and *E. crystallorophias* in the sea ice zone and in coastal fjords and bays; Antarctic silverfish have been reported as seasonal constituents of their diet but krill has been estimated to provide over 90% of their prey requirements. Much smaller portions of the diets of leopard (*Hydrurga leptonyx*), Ross (*Ommatophoca rossii*), Antarctic (*Arctocephalus gazella*), and subantarctic (*A. tropicalis*) fur seals have been reported to be composed of krill.

From these observations, some generalizations may be drawn: (1) blue and fin whales appear to have a higher preference for euphausiids than minke, humpback, or bowhead whales; (2) sei and Bryde's whales appear to be more opportunistic feeders; (3) gray whales and northern right whales prefer prey other than euphausiids but will consume them; (4) crabeater seals have a higher preference for euphausiids than other seals in the Southern Ocean; and (5) ringed and harp seals in the Northern Hemisphere include euphausiids in their diets during certain times of the year and life cycle.

Gross estimates of the consumption of euphausiids by marine mammals are summarized in Tables I and II. Estimates of stock

TABLE I
Consumption of Euphausiids by Whales

Ocean basin	Whale species	Abundance	Average body weight (tons)	Summertime ingestion rate (10^3 kcal/day)	Feeding period (days)	% krill in diet	Krill consumed (10^3 tons)
North Pacific	Blue whale	3,000–4,000	69.2	2136	180	100	1,240–1,654
	Fin whale	14,600–18,600	42.3	1452	180	80	3,282–4,182
	Sei whale	13,000–13,000	19.9	805	180	80	1,621–1,621
	Bryde's whale	34,500–45,500	13.2	584	180	40	1,559–2,056
	Minke whale	30,000–32,000	5.3	284	180	70	1,153–1,229
	Humpback whale	5,000–6,000	31.8	1161	180	60	674–809
	Bowhead whale	8,000–10,000	80.0	2392	180	80	2,963–3,704
	Gray whale	25,000–27,000	25.0	962	180	5	233–251
	Northern right whale	400–600	55.0	1784	180	25	35–52
						Total	12,760–15,559
North Atlantic	Blue whale	750–1,300	69.2	2136	180	100	310–538
	Fin whale	45,000–50,000	42.3	1452	180	80	10,117–11,241
	Sei whale	9,000–13,000	19.9	805	180	80	1,122–1,621
	Minke whale	120,000–182,000	5.3	284	180	70	4,610–6,992
	Humpback whale	10,000–11,000	31.8	1161	180	60	1,349–1,483
	Northern right whale	300–350	50.0	1656	180	25	24–28
						Total	17,532–21,903
South Hemisphere	Blue whale	600–800	83.0	3708	120	100	287–383
	Pygmy blue whale	2,000–6,000	68.9	3205	120	100	827–2,481
	Fin whale	10,000–20,000	48.0	2415	120	100	3,116–6,232
	Sei whale	35,000–45,000	17.5	1096	120	80	3,960–5,091
	Bryde's whale	78,000–108,000	13.2	879	120	40	3,538–4,899
	Minke whale (two species)	650,000–950,000	7.0	535	120	100	44,858–65,561
	Humpback whale	15,000–16,000	26.5	1517	120	100	2,936–3,131
	Southern right whale	6,500–7,500	55.0	2687	120	100	2,253–2,600
						Total	61,775–90,379

abundances were obtained from working papers and reports of the INTERNATIONAL WHALING COMMISSION, reports from the U.S. National Marine Fisheries Service, and the primary literature. In some cases, no reliable estimates are available and broad ranges were used. Daily ingestion rates for baleen whales during the feeding season were estimated from energetic requirements as a function of body weight following Sigurjónsson and Vikingsson (1997). A daily ingestion rate for seals was estimated as 7% of body weight. Average body weights, the percentages of euphausiids in the diets, and the caloric value of euphausiids (0.93 kcal/g) were taken from the primary literature. The duration of the feeding season was assumed to be 180 days for Northern Hemisphere baleen whales and seals, 120 days for Southern Hemisphere baleen whales and Antarctic and sub-Antarctic fur seals, and 335 days for crabeater, leopard, and Ross seals.

Although Tables I and II are based on several simplifying assumptions, some general conclusions may be drawn. Total consumption of euphausiids by marine mammals is on the order of 10–20 million tons per year in the North Pacific, 15–25 million tons per year in the North Atlantic, and 125–250 million tons per year in the Southern Hemisphere, with the bulk of the latter portion consumed in the Southern Ocean. In the North Atlantic, the largest portion is consumed by fin whales

followed by minke whales. In the North Pacific, consumption is distributed more evenly, with fin and bowhead whales consuming the most, followed by blue, sei, Bryde's, and minke whales, all of which consume similar portions. In the Southern Hemisphere, comparable proportions of euphausiids are consumed by crabeater seals and baleen whales. Of the estimated total krill consumption by baleen whales in the Southern Ocean, minke whales (two species) consume approximately two-thirds. Crabeater seals consume more krill than any other marine mammal population in the world.

These crude calculations suggest that baleen whales consume a substantial amount of euphausiids; moreover, their food requirements must have been several times more prior to commercial whaling. Unfortunately, there is little information on which to judge whether krill production was higher prior to the onset of commercial whaling or whether other krill predators (e.g., crabeater seals) benefited as a result of the decline in baleen whale stocks.

What is more apparent is that krill abundance can vary dramatically over relatively short periods of time and that baleen whales have adapted to this variability. Their size and ability to accumulate substantial energy stores allow them to integrate over large distances and periods of time in their search for food.

TABLE II
Consumption of Euphausiids by Seals

Ocean basin	Seal species	Abundance	Average body weight (tons)	Summertime ingestion rate (10^3 kcal/day)	Feeding period (days)	% krill in diet	Krill consumed (10^3 tons)
Northern Hemisphere	Ringed seal	6,000,000–7,000,000	75	5.3	180	25	1418–1654
	Harp seal	100,000–400,000	130	9.1	180	25	41–176
						Total	1458–1830
Southern Hemisphere	Crabeater seal	15,000,000–30,000,000	220	15.4	335	94	72,742–145,484
	Leopard seal	300,000–500,000	275	19.3	335	37	716–1,193
	Antarctic fur seal	1,000,000–1,500,000	50	3.5	120	50	210–315
	Sub-antarctic fur seal	300,000–500,000	85	6.0	120	50	107–179
	Ross seal	125,000–225,000	175	12.3	335	10	51–92
						Total	73,826–147,263

Their longevity allows them to spread reproductive effort over several years. It is reasonable to expect, however, that the supply of euphausiids will not be sufficient in all years to meet total energy requirements and that reproductive success and population growth among krill-dependent baleen whales may vary from year to year in response to the availability of their prey.

VI. Anthropogenic Effects

The production of euphausiids can be very sensitive to environmental conditions. This raises two concerns with regard to the influence of human activities. The first is that highly productive euphausiid populations may be able to sustain large fisheries. The second is that climatic change (whether man-induced or not) may affect the frequency of environmental conditions that are favorable for reproductive success. Because these are relatively recent developments, we cite three studies as entries into a larger body of literature.

Fisheries on euphausiids have the potential of being the largest in the world. In their review of krill fisheries, Nicol and Endo (1999) described the harvest of *E. pacifica* off the coasts of Japan and western Canada, *T. inermis* off the coasts of Japan and eastern Canada, *E. nana* off the coast of Japan, *T. raschii* and *M. norvegica* off the coast of eastern Canada, and *E. superba* in the Southern Ocean. In recent years, the harvest of *E. pacifica* off Japan (ca. 60,000 tons per year) and *E. superba* in the Scotia Sea region of the Southern Ocean (ca. 80,000 tons per year) comprised over 90% of the world harvest of euphausiids. Nicol and Endo (1999) noted that these yields are well within their theoretical potentials, although expansion of the coastal fisheries is unlikely because of ecological, economic, and political considerations. They further note, however, that as conventional fisheries decline and demand for krill as aquaculture feed increases, fishing pressure is likely to shift to *E. superba* in the Southern Ocean where current harvests are far below current estimates of sustainable yields.

Evidence suggests that the production of euphausiids may be affected by long-term climatic change. Warming of the surface waters of the California Current since the mid-1970s has been ac-companied by a reduction in the depth of the thermocline, reduced nutrient input via coastal upwelling, reduced primary production, and an overall decrease in macrozooplankton biomass by as much as 80% (Roemmich and McGowan, 1995). Euphausiids are the dominant taxa in the macrozooplankton fauna of the California Current and have shown decreased abundances during warm (El Niño) years and increased abundances during cold (La Niña) years. A 50-year warming trend in the Antarctic Peninsula region has been associated with a decrease in the annual production of sea ice. Loeb *et al.* (1997) correlated the reproductive success of *E. superba* with the wintertime extent of sea ice and suggested that the warming trend may cause a decrease in the frequency of strong year classes of Antarctic krill, a decrease in the mean population abundance of krill, and a change in the carrying capacity of vertebrate krill predators in the region.

See Also the Following Articles

Baleen Whales ▪ Diet ▪ Filter Feeding ▪ Plankton ▪ Predator–Prey Relationships

References

Bargmann, H. E. (1945). The development and life-history of adolescent and adult krill, *Euphausia superba*. *Disc. Rep.* **23**, 103–176.

Boden, B. P., Johnson, M. W., and Brinton, E. (1955). The Euphausiacea (Crustacea) of the North Pacific. *Bull. Scripps Inst. Ocean.* **6**(8), 287–400.

Brinton, E. (1962). The distribution of Pacific euphausiids. *Bull. Scripps. Inst. Ocean.* **8**, 51–270.

Einarsson, H. (1945). Euphausiacea. I. North Atlantic species. *Dana Rep.* **27**, 1–185.

Hickling, C. F. (1927). The natural history of the hake. *Fish. Invest. Lond. Ser. II* **10**, 1–100.

Hjort, J., and Rund, J. T. (1929). Whaling and fishing in the North Atlantic. *Rapp. P.-v. Reun. Cons. Int. Explor. Mer.* **41**, 107–119.

Laws, R. M. (1985). The ecology of the Southern Ocean. *Am. Sci.* **73**, 26–40.

Lebour, M. V. (1924). The Euphausiidae in the neighborhood of Plymouth and their importance as herring food. *J. Mar. Biol. Assoc. U.K.* **13**, 810–846.

Loeb, V., Siegel, V., Holm-Hansen, O., Hewitt, R., Fraser, W., Trivelpiece, W., and Trivelpiece, S. (1997). Effects of sea-ice extent and krill or salp dominance on the Antarctic food web. *Nature* **367,** 897–900.

Marr, J. W. S. (1962). The natural history and geography of the Antarctic krill (*Euphausia superba* Dana). *Disc. Rep.* **32,** 33–464.

Mauchline, J. (1980). The biology of mysids and euphausiids. *Adv. Mar. Biol.* **18,** 1–681.

Mauchline, J., and Fischer, L. R. (1969). The biology of euphausiids. *Adv. Mar. Biol.* **7,** 1–454.

Miller, D. G. M., and Hampton, I. (1989). Biology and ecology of the Antarctic krill. *Biomass Sci. Ser.* **9,** 1–166.

Nemoto, T. (1970). Feeding pattern of baleen whales in the ocean. *In* "Marine Food Chains" (J. H. Steele, ed.), pp. 241–252. Univ. Calif. Press, Berkeley.

Nicol, S., and Endo, Y. (1999). Krill fisheries development, management and ecosystem implications. *Aquat. Living Resour.* **12**(2), 105–120.

Roemmich, D., and McGowan, J. (1995). Climatic warming and the decline of zooplankton in the California Current. *Science* **267**(5202), 1324–1326.

Sars, G. O. (1885). Report on the Schizopoda collected by H.M.S. "Challenger" during the years 1873–1876. *Voyage H.M.S. "Challenger"* **13**(37), 1–128.

Siegel, V. (2000). Krill demography, life history, and aspects of population dynamics. *Can. J. Fish. Aqu. Sci.* **57**(Suppl.1), 130–150.

Sigurjonsson, J., and Vikingsson, G. A. (1997). Seasonal abundance of and estimated food consumption by cetaceans in Icelandic and adjacent waters. *J. Northw. Atl. Fish. Sci.* **22,** 271–287.

Tattersall, W. M. (1908). Crustacea. VII. Schizopoda. National Antarct. Exped., 1901–1904. *Nat. Hist. (Zool.)* **4,** 1–42.

Language Learning

Louis M. Herman
University of Hawaii, Honolulu

No single trait has been linked more closely with the human species than language. However, the definition of language and its uniqueness as a human trait continue to be areas of study and debate. Some, such as the linguist Noam Chomsky, take an evolutionary discontinuity position professing that language is a highly unique adaptation supported by special modifications of the brain that appear only in humans. Others, such as the anthropologist Barbara King, favor a continuity position, which suggests that language must have its roots in earlier hominoid adaptations for communication and that some of these adaptations may still be extant in modern ape species.

I. Human Language and Ape Language

The work on teaching various language-like systems to apes by Beatrice and Alan Gardner, David Premack, Duane Rumbaugh, and others, beginning in the mid-1960s and continuing throughout the decade of the 1970s, seemed to provide a genuine link between human and ape in fundamental language competency. This early work reported that chimpanzees were able to learn to use and understand not only words but also sentences. Sentences give human language its great communicative power through the infinite variety of meanings that can be constructed by the recombination of words. To understand a sentence the human listener must take account of both the meaning of the words and their grammatical relationships to one another, as governed by word order or other syntactic devices. This early work on teaching language to apes was thrown into disarray, however, by additional studies and criticisms from other researchers, such as Herbert Terrace and Carolyn Ristau. These researchers argued that the putative "sentences" produced by the apes were largely an artifact of context, imitation, or cue-ing. In particular, although sequences of symbols were indeed produced by the apes, the sequences had no syntactic structure that enhanced, explained, or modified meaning.

Until recently, the work with apes was focused on language production and paid scant attention to language comprehension. Investigators attempted to teach the apes to produce language—where words were represented by gestures, keyboard symbols, or other types of artificial symbols. These investigators assumed that if the ape produced a word, or series of words, that it therefore understood what the word or sequence represented. They also assumed that the ape could understand those same words or sequences when produced by the human partner. These assumptions, when finally tested, proved false. It was shown, instead, that comprehension did not automatically flow from language production. The preeminence of comprehension in language development, only recently appreciated in the ape language field, has long been emphasized among those studying child language. Language comprehension by young children develops earlier than language production, and even into adulthood comprehension vocabularies exceed speaking vocabularies.

Recent work with bonobo chimpanzees, pioneered by Sue Savage-Rumbaugh, has emphasized language comprehension and has progressed well beyond the earlier ape language studies. The bonobos have shown an ability to learn to understand instructions given in spoken English sentences. Together with some of her earlier work, Savage-Rumbaugh has shown that chimpanzees can learn to appreciate the symbols (words) of the language "referentially." The understanding that words *refer* to things or events in the real world is one of the key characteristics of human language. Among other things, referential understanding enables us to discuss things that are not immediately present or that happened at a different place or time.

II. Dolphins and Language
A. Natural Language?

Dolphins (including the common bottlenose dolphin *Tursiops truncatus*) produce various types of sounds, including clicks, burst-pulse emissions, and whistles. Clicks are used for ECHOLO-CATION, the dolphin's form of sonar. Through echolocation, the dolphin can examine its world through sound by listening to the echoes returning from objects struck by the clicks. Burst-pulse sounds may indicate the dolphin's emotional state, ranging from pleasure to anger. However, these type of vocalizations have been little studied and much remains to be learned about them. Whistles may be used for COMMUNICATION, but it is still an open question as to whether, or how much, whistle communication is intentional versus unintentional (e.g., rapidly repeated whistling may be elicited by stress, without any specific intention to convey that emotional state to others). During the 1960s, researchers attempted to determine whether the whistle vocalizations might be a form of language. Investigators recorded whistles from many dolphins in many different situations, but failed to demonstrate sufficient complexity in the vocalizations to support anything approaching a human language system. Some of the early work instead pointed to the stereotypy of the whistles from individual dolphins, leading David and Melba Caldwell to suggest that the whistle functioned principally as a "signature," with each individual dolphin producing a unique signature. Presumably, this enabled that individual to be identified by others. Other researchers have noted, however, that there can be a great deal of flexibility in the whistle. Douglas Richards, James Wolz, and Louis Herman, at the Kewalo Basin Marine Mammal Laboratory at the University of Hawaii, reported a study showing that a bottlenose dolphin could use its whistle mode to imitate a variety of sounds generated by a computer and broadcast underwater into the dolphin's habitat. Peter Tyack later reported that one dolphin could imitate another's whistle, thereby possibly referring to or calling that individual. As was noted earlier, referring symbolically to another individual, or to some other object or event in the environment, is one of the basic characteristics of a language. However, we still do not know to what extent the dolphin's whistles may be used to refer to things other than themselves or another dolphin. This is a fruitful area for additional study, however.

Although the evidence strongly suggests that dolphins do not possess a natural language, like the case for apes, it is still important and informative to study whether dolphins might nevertheless be able to learn some of the fundamental defining characteristics of human language. Any demonstration of language-learning competency by dolphins would bear on questions of the origins of human language, shifting the emphasis from the study of precursors in other hominoid species to common convergent characteristics in ape and dolphin that might lead to advanced communicative and cognitive capacities.

B. Early Attempts at Teaching Language to Dolphins

From the mid-1950s to the mid-1960s, John Lilly promoted the idea that bottlenose dolphins might possess a natural language. He based this supposition on this species' exceptionally large brain with its richly developed neocortex. He reasoned that the large brain must be a powerful information processor having capabilities for advanced levels of intellectual accomplishment, including the development of a natural language. He set about to uncover the supposed language. Failing in that quest, he then attempted, also without success, to teach human vocal language (English) to dolphins he maintained in his laboratories. Dolphins have a rich vocal repertoire, but not one suited to the production of English phonemes. The procedures used by Lilly and data he obtained were presented only sketchily, making any detailed analysis of his efforts at teaching language moot.

In the mid-1960s, Duane Batteau developed an automated system that translated spoken Hawaiian-like phonemes into dolphin-like whistle sounds that he projected underwater into a lagoon housing two bottlenose dolphins. He then attempted to use these sounds as a language for conveying instructions to the dolphins. A major flaw in his approach, however, was that individual sounds were not associated with individual semantic elements, such as objects or actions, but instead functioned as holophrases (complexes of elements). For example, a particular whistle sound instructed the dolphin to "hit the ball with your pectoral fin." Another sound instructed the dolphins to "swim through a hoop." Unlike a natural language, there was no unique sound to refer to *hit* or *ball*, or *hoop*, or *pectoral fin*, or any other unique semantic element. Hence, there was no way to recombine sounds (semantic elements) to create different instructions, such as "hit the hoop (rather than the ball) with your pectoral fin." After several years of effort, the dolphins were able to learn to follow reliably the holophrastic instructions conveyed by each of 12 or 13 different sounds. However, because of the noted flaw in the approach to construction of a language, the experiment failed as a valid test of dolphin linguistic capabilities.

C. Kewalo Basin Dolphin Language Studies

The work on dolphin language competencies by Louis Herman and colleagues at the Kewalo Basin Marine Mammal Laboratory in Honolulu was begun in the mid-1970s and emphasized language comprehension from the start. These researchers, working principally with a bottlenose dolphin named Akeakamai housed at the laboratory, constructed a sign language in which words were represented by the gestures of a person's arms and hands. The words referred to objects in the dolphin's habitat, to actions that could be taken to those objects, and to relationships that could be constructed between objects. There were also location words, *left* and *right*, expressed relative to the dolphin's locations, that were used to refer to a particular one of two objects having the same name, e.g., *left hoop* vs *right hoop*. Syntactic rules, based on word order, governed how sequences of words could be arranged into sentences to extend meaning. The vocabulary of some 30 to 40 words, together with the word-order rules, allowed for many thousands of unique sentences to be constructed. The simplest sentences were instructions to the dolphin to take named actions to named objects. For example, a sequence of two gestures glossed as *surfboard over* directs the dolphin to leap over the surfboard, and a sequence of three gestures glossed as *left Frisbee tail-touch* directs the dolphin to touch the Frisbee on her left with her tail. More complex sentences required the dolphin to construct a relationship between two objects, such as taking one named object to another named object or placing one named object in or on another named object. To interpret relational sentences correctly, the dolphin had to take account of both word meaning and word order. For example, a sequence of three gestures glossed as *person surfboard fetch* tells the dolphin to bring the surfboard to the person (who is in the water), but *surfboard person fetch*, the same gestures rearranged, requires that the person be carried to the surfboard. By incorporating *left* and *right* into these relational sentences, highly complex instructions could be generated. For example, the sequence of five gestures glossed as *left basket right ball in* asks the dolphin to place the ball on her right into the basket on her left. In contrast, the rearranged sequence *right basket left ball in* means the opposite, "put the ball on the left into the basket on the right." The results, published by Louis Herman, Douglas Richards, and James Wolz, showed that the dolphin was proficient at interpreting these various types of sentences correctly, as evidenced by her ability to carry out the required instructions, including instructions new to her experience. These were the first published results showing convincingly an animal's ability to process both semantic and syntactic information in interpreting language-like instructions. Semantics and syntax are considered core attributes of any human language.

Ronald Schusterman and Kathy Krieger tested whether a California sea lion (*Zalophus californianus*) named Rocky might be able to learn to understand sentence forms similar to those understood by the dolphin Akeakamai. Rocky was able to carry out gestural instructions effectively for simpler types of sentences requiring an action to an object. The object was specified by its class membership (e.g., "ball") and, in some cases, also by its color (black or white) or size (large or small). In a later study, Schusterman and Robert Gisiner reported that Rocky was able to understand relational sentences requiring that one object be taken to another object. These reports suggested that the sea lion was capable of semantic processing of symbols and, to some degree, of syntactic processing. A shortcoming of the sea lion work, however, was the absence of contrasting terms for relational sentences, such as the distinction between "fetch" (take to) and "in" (place inside of or on top) demonstrated for the dolphin Akeakamai. Additionally, unlike the dolphin, the sea lion's string of gestures were given discretely, each gesture followed by a pause during which the sea lion looked about to locate specified objects before being given the next gesture in the string. In contrast, gestural strings given to the dolphin Akeakamai were without pause, analogous to the spoken sentence in human language. Further, Rocky did not show significant generalization across objects of the same class (e.g., different balls), but unlike the dolphin seemed to regard a gesture as referring to a particular exemplar of the class rather than to the entire class. Thus, although many of the responses of the sea lion resembled those of the dolphin, the processing strategies of the two seemed different, and the concepts developed by the sea lion appeared to be more limited than those developed by the dolphin.

D. Akeakamai's Knowledge of the Grammar of the Language

As a test of Akeakamai's grammatical knowledge of the language she had been taught, Louis Herman, Stan Kuczaj, and Mark Holder constructed *anomalous* gestural sentences. These were sentences that violated the syntactic rules of the language or the semantic relations among words. The researchers then studied the dolphin's spontaneous responses to these sentences. For example, the researchers compared the dolphin's responses to three similar gestural sequences: *person hoop fetch, person speaker fetch,* and *person speaker hoop fetch*. The first sequence is a proper instruction; it violates no semantic or syntactic rule of the learned language. It directs the dolphin to bring the hoop to the person, which the dolphin does easily. The second sequence is a syntactically correct sequence but is a semantic anomaly inasmuch as it directs the dolphin to take the underwater speaker, firmly attached to the tank wall, to the person. The dolphin typically rejects sequences like this by not initiating any action. The final sequence is a syntactic anomaly in that there is no sequential structure in the grammar of the language that provides for three object names within a sequence. However, embedded in the four-item anomaly are two semantically and syntactically correct three-item sequences: *person hoop fetch* and *speaker hoop fetch*. The dolphin in fact typically extracts one of these subsets and carries out the instruction implicit in that subset by taking the hoop to the person or to the underwater speaker.

These different types of responses revealed a rather remarkable and intelligent analysis of the sequences. Thus, the dolphin did not terminate her response when an anomalous initial sequence such as *person speaker* was first detected. Instead, she continued to process the entire sequence, apparently searching backward and forward for proper grammatical structures as well as proper semantic relationships, until she found something she could act on, or not. This analytic type of sequence processing is part and parcel of sentence processing by human listeners.

E. Understanding of Symbolic References to Absent Objects

Louis Herman and Paul Forestell tested the dolphin Akeakamai's understanding of symbolic references to objects that were not present in the dolphin's habitat at the time the reference was made. For this purpose, they constructed a new syntactic frame consisting of an object name followed by a gestural sign glossed as *"question."* For example, the two-item gestural sequence glossed as *basket question* asks whether a basket is present in the dolphin's habitat. The dolphin could respond *yes* by pressing a paddle to her right or *no* by pressing a paddle to her left. Over a series of such questions, with the particular objects present being changed over blocks of trials, the dolphin was as accurate at reporting that a named object was absent as she was at reporting that it was present. These results gave a clear indication that the gestures assigned to objects were understood referentially by the dolphin, i.e., that the gestures acted as symbolic references to those objects.

F. Interpreting Language Instructions Given through Television Displays

The television medium can display scenes that are representations of the real world, or sometimes of imagined worlds. As viewers, we understand this and often respond to the displayed content similarly to how we might respond to the real world. We of course understand that it is a representation and not the real world. It appears, however, that an appreciation of television as a representation of the real world does not come easily to animals, even to apes. Sue Savage-Rumbaugh wrote in her book, "Ape Language," that chimpanzees show at most a fleeting interest in television, and that from their behavior it was not possible to infer that they were seeing anything more than changing patterns or forms. Her own language-trained chimpanzee subjects, Sherman and Austin, only learned to attend and to interpret television scenes after months of exposure in the presence of human companions who reacted to the scenes by exclaiming or vocalizing at appropriate times. Louis Herman, Palmer Morrel-Samuels, and Adam Pack tested whether the dolphin Akeakamai might respond appropriately to language instructions delivered by a trainer whose image was presented on a television screen. Akeakamai had never been exposed to television of any sort previously. Then, for the first time, the researchers simply placed a television monitor behind one of the underwater windows in the dolphin's habitat and directed Akeakamai to swim down to the window. On arriving there she saw an image of the trainer on the screen. The trainer than proceeded to give Akeakamai instructions through the familiar gestural language. The dolphin watched and then turned and carried out the first instruction correctly and also responded correctly to 11 of 13 additional gestural instructions given to her at that same testing session. In further tests, Akeakamai was able to respond accurately even to degraded images of the trainer, consisting, for example, of a pair of white hands moving about in black space. The overall results suggested that Akeakamai spontaneously processed the television displays as representations of the gestural language she had been exposed to live for many years previously.

III. Implications

The results of the language comprehension work with the bonobo chimpanzee and the dolphin Akeakamai show many similarities, especially in the receptivity of the animals to the language formats used and in their proficiency at responding to sequences of symbols. The dolphin has been tested in more formal procedures than the bonobo, leading to a fuller understanding of the dolphin's grammatical competencies than has been attained for the chimp. Findings with the bottlenose dolphin are in keeping with many other demonstrations of the cognitive abilities of this species. The advanced cognitive abilities of apes are also well documented. An early summary by Herman (1980, p. 421) still seems appropriate to accommodate the convergent cognitive and language-learning abilities of ape and dolphin: "The major link that cognitively connects the otherwise evolutionarily divergent (dolphins) . . . and primates may be social pressure—the requirement for integration into a social order having an extensive communication matrix for promoting the well-being and survival of individuals. . . . Effective functioning in such a society demands extensive socialization and learning. The extended maturational stages of the young primate or dolphin and the close attention given it by adults and peers . . . provide the time and tutoring necessary for meeting these demands. In general, high levels of parental care and high degrees of cortical encephalization go together. . . . It

is not difficult to imagine that the extensive development of the brain in (dolphins) . . . and the resulting cognitive skills of some members of this group, have derived from the demands of social living, including both cooperation and competition among peers, expressed within the context of the protracted development of the young. These cognitive skills may in turn provide the behavioral flexibility that has allowed the diverse family of (dolphins) . . . to successfully invade so many different aquatic habitats and niches."

See Also the Following Articles

Intelligence and Cognition ■ Sound Production ■ Training

References

Caldwell, M. C., and Caldwell, D. K. (1965). Individualized whistle contours in bottlenose dolphins (*Tursiops truncatus*). *Nature (London)* **207,** 434–435.

Chomsky, N. (1972). "Language and Mind." Harcourt Brace Jovanovich, New York.

Gardner, B. T., and Gardner, R. A. (1971). Two-way communication with an infant chimpanzee. *In* "Behavior of Nonhuman Primates" (A. M. Schrier and F. Stollnitz, eds.), Vol. 4, pp. 117–184. Academic Press, New York.

Herman, L. M. (1980). Cognitive characteristics of dolphins. *In* "Cetacean Behavior: Mechanisms and Functions" (L. M. Herman, ed.), pp. 363–429. Wiley Interscience, New York.

Herman, L. M. (1986). Cognition and language competencies of bottlenose dolphins. *In* "Dolphin Cognition and Behavior: A Comparative Approach" (R. J. Schusterman, J. Thomas, and F. G. Wood, eds.), pp. 221–251. Lawrence Erlbaum Associates, Hillsdale, NJ.

Herman, L. M. (1989). In which Procrustean bed does the sea lion sleep tonight? *Psychol. Rec.* **39,** 19–50.

Herman, L. M., and Forestell, P. H. (1985). Reporting presence or absence of named objects by a language-trained dolphin. *Neurosci. Biobehav. Rev.* **9,** 667–691.

Herman, L. M., Kuczaj, S., III, and Holder, M. D. (1993). Responses to anomalous gestural sequences by a language-trained dolphin: Evidence for processing of semantic relations and syntactic information. *J. Exp. Psychol. Gen.* **122,** 184–194.

Herman, L. M., Morrel-Samuels, P., and Pack, A. A. (1990). Bottlenose dolphin and human recognition of veridical and degraded video displays of an artificial gestural language. *J. Exp. Psychol. Gen.* **119,** 215–230.

Herman, L. M., Richards, D. G., and Wolz, J. P. (1984). Comprehension of sentences by bottlenose dolphins. *Cognition* **16,** 129–219.

Herman, L. M., and Tavolga, W. N. (1980). The communication systems of cetaceans. *In* "Cetacean Behavior: Mechanisms and Functions" (L. M. Herman, ed.), pp. 149–209. Wiley Interscience, New York.

Herman, L. M., and Uyeyama, R. K. (1999). The dolphin's grammatical competency: Comments on Kako (1998). *Anim. Learn. Behav.* **27,** 18–23.

King, B. J., and Shanker, S. G. (1997). The expulsion of primates from the garden of language. *In* "Evolution of Communication" (S. Wilcox, B. King, and L. Steels, eds.), Vol. 1, pp. 59–99. John Benjamins Publishing Company, Philadelphia.

Lilly, J. C. (1967). "The Mind of the Dolphin: A Nonhuman Intelligence." Doubleday, New York.

Premack, D., and Premack, A. (1983). "The Mind of an Ape." W. W. Norton & Company, New York.

Richards, D. G., Wolz, J. P., and Herman, L. M. (1984). Vocal mimicry of computer generated sounds and vocal labeling of objects by a bottlenose dolphin, *Tursiops truncatus*. *J. Comp. Psychol.* **98,** 10–28.

Ristau, C. A., and Robbins, D. (1979). Language in the great apes: A critical review. *In* "Advances in the Study of Behavior" (J. F. Rosenblatt, R. B. Hinde, C. Beer, and M-C Busnel, eds.), Vol. 12, pp. 141–255. Academic Press, New York.

Rumbaugh, D. M. (1977). "Language Learning by a Chimpanzee: The Lana Project." Academic Press, New York.

Savage-Rumbaugh, E. S. (1986). "Ape Language: From Conditioned Response to Symbol." Columbia Univ. Press, New York.

Savage-Rumbaugh, E. S., Murphy, J., Sevcik, R. A., Brakke, K. E., Williams, S. L., and Rumbaugh, D. M. (1993). Language comprehension in ape and child. *Monogr. Soc. Res. Child Dev.* **58,** No. 3–4.

Schusterman, R. J., and Gisiner, R. (1988). Artificial language comprehension in dolphins and sea lions: The essential cognitive skills. *Psychol. Rec.* **34,** 3–23.

Schusterman, R. J., and Krieger, K. (1984). California sea lions are capable of semantic comprehension. *Psychol. Rec.* **38,** 311–348.

Terrace, H. S., Petitto, L. A., Sanders, R. J., and Bever, T. G. (1979). Can an ape create a sentence? *Science* **206,** 891–902.

Tyack, P. L. (1986). Whistle repertoires of two bottlenose dolphins, *Tursiops truncatus:* Mimicry of signature whistles? *Behav. Ecol. Sociobiol.* **18,** 251–257.

Leaping Behavior

BERND WÜRSIG
Texas A&M University, Galveston

Dolphins and whales leap above water in seemingly exuberant displays of sheer joy. While play may at times be a cause of leaping, there appear to be multiple leap types and reasons, not totally understood.

When large whales leap, the activity is generally termed breaching. Whales appear to breach to communicate to others, due to a high activity level (or state of alertness), or—at times—from apparent frustration or anger after a social interaction with one or more other whales (Whitehead, 1985). Breaching may also occur for the pure "joy" or "fun" of it, but these potential motivations are difficult for researchers to assess. Leaping by the smaller toothed whales may have similar general functions, as well as several of the more specific ones outlines here.

I. Basic Description

Leaping clear of the water is always an energetic and usually a highly acrobatic feat. First, to clear the water the cetacean needs to attain a rapid forward speed and momentum, near the limit of its SWIMMING capability. It generally bends its body abruptly to exit the water, then twists the body midair to reenter the water in some structured fashion. Even a noisy "belly flop" after a leap has been designed as such, as multiple similar leaps of the same animals demonstrate. Reentering the water can be head first, creating minimal splash and NOISE. It can consist of a side, back, or belly splash, resulting in a welter of white water and foam and a considerable percussive (splash) noise in-air and underwater. Finally, there is the "showy" acrobatic leap that consists of spins, somersaults, and various in-air twists. Frame-by-frame analysis of high-speed photographs shows that dolphins control these acrobatics to within split-second timing, affecting muscle movements that allow them to perform the same leap and reentry onto the water again and again. In human terms, a well-trained gymnast or pool diver comes to mind. Leaps tend to last for 1 to 2 sec, depending on the acrobatics being performed and the size of the leaping individual.

II. The Head-First Reentry Leap

There are three main variations of this leap that tend to create little water disturbance or noise upon reentry. One consists of a "stationary" leap, where the animal comes steeply from depth, usually greater than three body lengths. It leaps in-air, breathes, and tucks its body into a bend to reorient the head downward, then rapidly descends into depth at or very near the original exit point. This leap appears to be executed for the animal to leave whatever it is doing at depth for a minimal time, breathe, and use the in-air weight of its body to regain its position. The need for such an efficient mechanism to breathe becomes clear when we consider that dolphins feeding or mating at depth, for example, essentially need to interrupt these activities to obtain life-sustaining air. If they can do so rapidly, all the better. The stationary leap is performed singly by dolphins herding food fish below the surface (Würsig, 1979), but often in twos or threes during socializing (Norris *et al.,* 1994) (Fig. 1).

A second head-first reentry leap consists of rapid swimming just below the surface, a very abrupt bend of the body to exit the water, and then a long arcuate in-air leap that may propel the dolphin forward by up to three times its own body length. While the reentry is head first, there is nevertheless some splashing of water due to the rapidity of the action, kicked up by the body as it exits and again as it enters. This is the "running leap" of dolphins moving at speed, a form of high porpoising.

Figure 1 *A head-first reentry of a dusky dolphin.*

Dolphins propel themselves underwater with several powerful but rapid tail beats and then "sail" through the air, a medium 800 times less dense than water. There is considerable drag generated by crossing the air/water interface, but for an animal that needs to come to the surface to breathe anyway (such as penguins and dolphins), travel efficiency increases above a particular speed by leaping rather than swimming. For a 2.5-m-long dolphin, the crossover speed from swimming to leaping is about 4.6 m/sec, or 16.6 km/hr (Au and Weihs, 1980). Above about 4 m body length (and concomitant body weight), high porpoising is no longer as easy, although killer whales (*Orcinus orca*) moving very rapidly may leap in this manner for short periods of time (Fig. 2).

The third head-first reentry leap is designed to gain height. Dolphins, often in twos or threes, leap as high as three times their own body lengths above the surface of the water, usually but not always reentering the water head first. A 2.5-m male pantropical spotted dolphin (*Stenella attenuata*)—spotted dolphins are the champion high leapers—thus leaps about 7 m into the air, or the equivalent of over two apartment stories high. While these leaps may be performed largely for "fun," they may also serve the function of seeing to greater distance by gaining height. Dusky dolphins (*Lagenorhynchus obscurus*) leap in this fashion just before high porpoising toward feeding aggregations with flocking birds some kilometers away (Würsig and Würsig, 1980) (Fig. 3).

III. The Noisy Leap

When dolphins or whales fall back onto the water with a broad side of the body, they invariably create a large splash and a percussive slap sound. Frame-by-frame analysis of high-speed photography shows that there are actually two splashes: one is created as the animal falls onto the water surface and initiates a crater of water underneath it and the other is the secondary splash (and slap sound) produced as the crater collapses upon itself. This cavitation is particularly dramatic for breaching whales, but is associated with almost all noisy leaps.

Dolphins that noisy-leap exit the water in similar fashion as in head-first reentry leaps, but twist the body to reenter with back, side, or belly first. Many noisy leaps end with the dolphin

Figure 3 *A high leap of a dusky dolphin.*

merely falling back onto the water surface. Others are particularly designed to have the animal reenter in a predetermined fashion, and high-speed photography shows subtle tail, flipper, head, or other body readjustments even split seconds before reentering and appearing to be structured to force the body onto the water with a maximal intensity of splash. These observations have led to speculation that noisy leaps are structured for omni-directional communication among dolphins and whales. Indeed, noisy leaps tend to occur more often in higher wind states (when near-surface ambient noise greatly increases), and this observation fits with the hypothesis of communication. Noisy leaps also occur around the periphery of near-surface schools, and in that case, the percussive slaps, as well as the underwater bubble clouds formed by dolphins reentering the water, may serve to frighten fish and cause them to school or aggregate more tightly. Dolphins may at times also stun or debilitate fish prey with the slaps of noisy leaps (as well as with tail slaps), but there is no detailed information on this possibility (Fig. 4).

IV. The Acrobatic Leap

Some dolphins are especially showy for at least some of their leaps, with spins, somersaults, combinations of flips, head twists, extra tail kicks in-air, and so on. These leaps are almost always associated with an obviously high level of social activity in a school or pod, as evidenced by social rubbing, sexual activity, and a cacophony of whistle and other sounds. Acrobatic leaps usually occur in bouts, with one dolphin (or whale) leaping at least several times. The more social the group, the more leaping dolphins and the longer the individual bouts. These leaps appear graceful to our human eyes and appear particularly structured to be enjoyed in the making and the viewing, like art. However, this may not be the case; we simply do not know.

Spinner dolphins (*Stenella longirostris*) spin by rotating their body rapidly around the long axis up to six times (usually

Figure 2 *Dusky dolphins porpoising.*

Figure 4 *A side slap of a dusky dolphin.*

Figure 5 *A spinner dolphin spinning.*

two to four times) before falling back into the water. They do so in both vertical spins and horizontal fashion (Norris and Dohl, 1980). Members of the genus *Lagenorhynchus*, such as Pacific white-sided (*L. obliquidens*) and dusky dolphins, are probably the most aerially acrobatic of all dolphins and whales, with somersaults, twists, and various inventive bends and contortions (Brownell and Cipriano, 1999). Individuals also have the longest bouts of any of the dolphins (some whale breach bouts are as long), with up to 36 somersaults having been counted in one dolphin in one about 5-min duration. Interestingly, *Lagenorhynchus* spp. individuals will "never" change leap type during a bout. If a dusky dolphin begins a backwards somersault with a half twist to the left and a tail kick just before reentering the water, it will continue this same leap, with absolutely no noticeable variation, during that leap bout. Toward the end of the bout, it will tire, muscle action will slow, and the leap will be slightly imperfect. It then quits and breathes while resting at the surface for several minutes. Later, in a different bout, the same individual will leap differently, demonstrating that it knows more than one leap type.

Acrobatic leaps tend to be noisy, but are not structured specifically to make noise. They are structured to be acrobatic, and it is difficult to imagine that they occur for anything but the "fun" (or the art) of it. A more scientifically acceptable explanation may be that acrobatic leaps are not merely an outgrowth of a high level of social activity, but are themselves a call for social activity. Acrobatic leap types may thus serve a social facilitation function that helps to coordinate members of a school or pod. Such facilitation may be especially useful to animals that coordinate finding and aggregating of food and that may need to establish and maintain delicate balances of social and sexual hierarchies. One argument against this stands out: dolphins leaping acrobatically are not being watched by others. They perform their show above the surface while, at any one time of a leap, most or all others are below. Acrobatic leaps may create somewhat different splash sounds from other more simple leaps, but this is not known (Figs. 5 and 6).

V. Conclusions

The descriptions in this article are not to be thought of as complete explanations of the various—and variable—leaps of cetaceans. There is much that we do not yet know, but it is certainly fair to say that surface-active cetaceans, being large-brained social mammals, may have many different reasons for their actions. One long-existing guess for leaping may have some truth to it as well. This is the suggestion that leaping, and the attendant splashing onto the water's surface, may be an attempt to dislodge parasites or other biological hitchhikers. Indeed, there is evidence for Hawaiian spinner dolphins, that individuals with remoras attached to their sides or bellies leap more than those that do not, and the author has seen remoras fly off spinning spinner dolphins, even before the dolphins hit the water. To get rid of potentially bothersome (and drag-inducing) freeloaders may be a partial explanation, but it cannot be the entire explanation. Leaping occurs for a variety of reasons, having to do with locomotion,

Figure 6 *A dusky dolphin somersaulting.*

rapid breathing, seeing in air, communicating, possibly scaring prey, social–sexual displaying and facilitation, removal of ectoparasites, and—most certainly—fun.

A discussion of leaping would be incomplete without mention of other than cetacean marine mammals. Indeed, otariid pinnipeds also high-porpoise (probably for the same hydrodynamic efficiency consideration as for dolphins), and especially fur seals and sea otters leap at the surface by rapidly rolling around their own axes. This activity serves to aerate the extremely fine, long, and dense pelage of these marine mammals that use air for insulation. Leaping in pinnipeds and sea otters at times may also consist of play activity, but there is no further information on this point

See Also the Following Articles

Breaching ■ Communication ■ Playful Behavior ■ Speed ■ Swimming

References

Au, D. and Weihs, D. (1980). At high speeds dolphins save energy by leaping. *Nature* **284,** 548–550.

Brownell, R. L., Jr., and Cipriano, F. (1999). Dusky dolphins, *Lagenorhynchus obscurus. In* "Handbook of Marine Mammals, Vol. 6." (Ridgway and Harrism, eds.), pp. 85–104. Academic Press, San Diego.

Norris, K. S., and Dohl, T. P. (1980). Behavior of the Hawaiian spinner dolphin, *Stenella longirostris. U.S. Fishery Bulletin* **77,** 821–849.

Norris, K. S. Würsig, B., Wells, R. S., and Würsig, M. (1994). "The Hawaiian Spinner Dolphin." Univ. of California Press, Berkeley.

Wells, R. S., Boness, D. J., and Rathbun, G. B. Behavior. (1999). *In* "Biology of Marine Mammals" (J. E. Reynolds III and S. A. Rommel, eds.), pp. 324–422. Smithsonian Press, Washington, DC.

Whitehead, H. (1985). Why whales leap. *Sci. Am.* (March).

Würsig, B. (1979). Dolphins. *Sci. Am.* **240,** 136–148.

Würsig, B. and Würsig, M. (1980). Behavior and ecology of the dusky dolphin, *Lagenorhynchus obscurus,* in the South Atlantic. *U.S. Fish. Bull.* **77,** 871–890.

Leopard Seal
Hydrurga leptonyx

TRACEY L. ROGERS
University of Sydney, Australia

Leopard seals are large seals, displaying sexual dimorphism where the females are larger than the males. Females grow up to 3.8 m and weigh up to 500 kg, whereas males grow up to 3.3 m and weigh up to 300 kg.

I. Distribution

The main leopard seal population remains within the circumpolar Antarctic pack ice but the seals are regular, although not abundant, visitors to the sub-Antarctic islands of the southern oceans and to the southern continents. The most northerly leopard seal sightings are from the Cook Islands. Juveniles appear to be more mobile, moving further north during the winter. Because it does not need to return to pack ice to breed, the leopard seal can escape food shortages during winter by dispersing northwards. Every 4 to 5 years the number of leopard seals on the sub-Antarctic islands oscillates from a few to several hundred seals. The periodic dispersal could be related to oscillating current patterns or resource shortages in certain years.

The leopard seal population is estimated to be 222,000 to 440,000. During summer, leopard seals breed on the outer fringes of the pack ice where they are solitary and sparsely distributed. Their density is inversely related to the amount of pack ice available to the seals as haul-out platforms. Pack ice cover varies with the season, from a maximum between August and October to a minimum between February and March. Population densities are greatest in areas of abundant cake ice (ice floes of 2 to 20 m in diameter) and brash ice (ice floes greater than 2 m in diameter), whereas they are least in areas with larger floes. Densities range from 0.003 to 0.151 seals/km², and there is an age-related difference in their spatial behavior. Due to intraspecific aggression there is a greater degree of spatial separation among older seals.

II. Diet

Leopard seals take a diverse range of prey, including fish, cephalopods, sea birds, and seals. Different food sources are used when available or when opportunities to take other more sought after prey are few. Krill makes up the largest proportion of their diet, particularly during the winter months when other food types are not abundant. At this time the leopard seals must compete directly with krill-feeding specialists, such as the crabeater seal (*Lobodon carcinophaga*) and Adélie penguin. This is believed to be a time of potential food shortage and causes some juvenile leopard seals to move north from the pack ice during the austral winter. The leopard seal is responsible for more predation on warm-blooded prey than any other pinniped. Leopard seals capture and eat juvenile crabeater seals in particular, but also prey on Weddell (*Leptonychotes weddellii*), Ross (*Ommatophoca rossii*), southern elephant (*Mirounga leonina*), sub-Antarctic and Antarctic fur seals (*Arctocephalus tropicalis* and *A. gazella*) and southern sea lions (*Neophoca cinerea* and *Phocarctes hookeri*). Newly weaned crabeater seals are the most vulnerable and are taken from November to February. Crabeater seal survivors bear characteristic parallel paired scars from leopard seal attacks, and approximately 78% of adult crabeater seals display such marks. The teeth of the leopard seal have a dual role; the large recurved canines and incisors are designed for gripping and tearing prey, whereas the upper and lower tricuspid (three cusped) molars interlock to provide an efficient krill sieve.

III. Reproductive Biology

Male leopard seals are sexually mature by 4.5 years and females by 4 years of age. Females give birth to their pups and wean them on the ice floes of the Antarctic pack ice. Males do not remain with the females; only mother–pup groups are observed on ice floes. Length at birth is about 120 cm, with rapid growth through the first 6 months postpartum. Births are believed to occur from October to mid-November and mating

Figure 1 *Leopard seal in threatening posture.*

from December to early January, after the pups have weaned. Lactation is believed to last for up to 4 weeks. Mating in the wild has been observed rarely, but captive seals mount only when in the water. There is a period of delayed implantation from early January to mid-February. Implanted fetuses are found after mid-February when the corpus luteum (glandular structure in the ovary) has begun to increase in size and the corpus albicans (scar from ovarian glandular structure) from the previous pregnancy has continued to regress.

Acoustic behavior is important in the MATING SYSTEM of the leopard seal. Leopard seals become highly vocal prior to and during their breeding season (Fig. 1). Lone male leopard seals vocalize for long periods each day, from early November through January. Female leopard seals also use long-distance acoustic displays during the breeding season. However, female seals vocalize for a brief period only from the beginning of estrus until mating, presumably to advertise sexual receptivity. The calls of the leopard seal are at low-to-medium frequencies and so powerful that they can be heard through the air–water interface and felt through the ice.

See Also the Following Articles

Antarctic Marine Mammals ■ Krill ■ Sexual Dimorphism

References

Bonner, N. (1994). "Seals and Sea Lions of the World." Blandford, London.

Erickson, A. W., Siniff, D. B., Cline, D. R., and Hofman, R. J. (1971). Distributional ecology of Antarctic seals. *In* "Symposium on Antarctic Ice and Water Masses" (G. Deacon, ed.), pp. 55–76. Heller and Sons, Cambridge.

Gilbert, J. R., and Erickson, A. W. (1977). Distribution and abundance of seals in the pack ice of the Pacific Sector of the Southern Ocean. *In* "Adaptations within Antarctic Ecosystems" (G. A. Llano, ed.), pp. 703–740. Smithsonian Institution, Washington, DC.

Kooyman, G. L. (1981). Leopard seal (*Hydrurga leptonyx* Blainville, 1820). *In* "Handbook of Marine Mammals" (S. Ridgway and R. Harrison, eds.), Vol. 2, pp. 261–274.

Laws, R. M. (1984). Seals. *In* "Antarctic Ecology" (R. M. Laws, ed.), Vol. 2, pp. 621–715. Academic Press, London.

Lowry, L. F., Testa, J. W., and Calvert, W. (1988). Notes on winter feeding of crabeater and leopard seals near the Antarctic Peninsula. *Polar Biol.* **8,** 475–478.

Rogers, T., Cato, D. H., and Bryden, M. M. (1996). Behavioral significance of underwater vocalizations of captive leopard seals, *Hydrurga leptonyx. Mar. Mamm. Sci.* **12,** 414–427.

Rogers, T. L., and Bryden, M. M. (1997). Density and haul-out behavior of leopard seals (*Hydrurga leptonyx*) in Prydz Bay, Antarctica. *Mar. Mamm. Sci.* **13(2),** 293–302.

Rounsevell, D., and Pemberton, D. (1994). The status and seasonal occurrence of leopard seals, *Hydrurga leptonyx,* in Tasmanian waters. *Aust. Mammal.* **17,** 97–102.

Siniff, D. B., and Bengtson, J. L. (1977). Observations and hypothesis concerning the interactions among crabeater seals, leopard seals, and killer whales. *J. Mammal.* **58,** 414–416.

Siniff, D. B., and Stone, S. (1985). The role of the leopard seal in the tropho-dynamics of the Antarctic marine ecosystem. *In* "Antarctic Nutrient Cycles and Food Webs" (W. R. Siegfried, P. R. Condy, and R. M. Laws, eds.), pp. 555–559. Springer-Verlag, Berlin.

Stirling, I., and Siniff, D. B. (1979). Underwater vocalizations of leopard seals (*Hydrurga leptonyx*) and crabeater seals (*Lobodon carcinophagus*) near the South Shetland Islands, Antarctica. *Can. J. Zool.* **57,** 1244–1248.

Testa, J. W., Oehlert, G., Ainsley, D. G., Bengtson, J. L., Siniff, D. B., Laws, R. M., and Rounsevell, D. (1991). Temporal variability in Antarctic marine ecosystems: Periodic fluctuations in the phocid seals. *Can. J. Fish. Aquat. Sci.* **48,** 631–639.

Literature

SEE *Popular Culture and Literature*

Lobtailing

NATHALIE JAQUET
Texas A&M University, Galveston

Lobtailing is performed by the majority of cetacean species. It consists of slapping either the ventral or the dorsal side of the tail flukes against the water (Fig. 1) any number of times, from a single tail slap to over a hundred depending on the species and the context. It is a purposeful behavior, happening mainly in social and/or foraging contexts. As is the case with many social displays, lobtailing seems to be contagious and is often accompanied by breaches. The significance of lobtailing is still a mystery and may vary among species and contexts. Although its is probable that lobtailing is a form of communication, scientists are still far from having uncovered the exact functions of lobtailing.

I. Who Lobtails?

Although a large number of cetacean species have been observed lobtailing at least once, quantitative information on lobtailing rate is only available for a very few species. What is clear, however, is that there is a very high variability in the occurrence of lobtailing among species. Some species of large whales, such as sperm whales (*Physeter macrocephalus*), humpback whales (*Megaptera novaeangliae*), right whales (*Eubalaena australis, E. japonica* and *E. glacialis*), and gray whales (*Eschrichtius robustus*), are commonly seen lobtailing. However, lobtailing is much less common in minke (*Balaenoptera acutorostrata* and *B. bonaerensis*), sei (*B. borealis*), Bryde's (*B. edeni*), blue (*B. musculus*), and fin (*B. physalus*) whales. Large differences in the lobtailing rate also exist among species of small cetaceans: porpoises and river dolphins are seldom observed lobtailing whereas lobtailing is very common in most species of Delphinidae. For the least known species of cetaceans (e.g., many species of beaked whales and some poorly known species of dolphins), lobtailing frequencies are unknown.

However, despite the paucity of quantitative information available on lobtailing rates, it seems that, in general, species with complex social systems have a higher lobtailing rate than species that are more solitary. This is consistent with what has been found for some closely related behavior such as breaching.

II. How?

Lobtailing is performed differently by large and small cetaceans. Baleen whales and sperm whales lobtail when vertically in the water column with their tail flukes high above the water, the stock is then bent and the flukes are slapped forcefully on the water surface (Fig. 2). While lobtailing, these whales tend to stay almost stationary. However, small delphinidae usually lobtail while lying horizontally in the water, either on their belly or

Figure 2 *Lobtailing sperm whale: (top) the stock is bent and (bottom) the flukes are slapped forcefully against the water surface.*

Figure 1 *Lobtailing sperm whale. This is the first lobtail of a sequence of eight consecutive lobtails and thus the flukes tend to be still very high above the water.*

on their back, thus slapping either the ventral or the dorsal side of their flukes. Forward movement is sometimes associated with lobtailing. In both cases, lobtailing results in the production of a loud noise that propagates underwater for up to several hundred meters. Lobtails are often executed in sequences and a particular individual may lobtail over 100 times in a row.

III. When Does Lobtailing Occur?

Lobtailing occurs in a wide range of circumstances, and the circumstances may vary depending on the species. However, for all species, most lobtailing activities occur in either social and/or foraging contexts. Furthermore, for all species, it seems that lobtailing is very seldom performed by a lone animal, which would not be either in visual and/or in acoustic contact with at least another conspecific. It is also very seldom performed by resting animals. As it is difficult to generalize the occurrence of lobtailing for all species of cetaceans, a few specific examples are given.

In sperm whales, the occurrence of lobtailing seems to be strongly correlated with the complexity of the social situation. Sperm whales show a considerable sexual segregation in social organization: females and immatures form cohesive long-term family units of about 12 individuals, whereas bachelor and mature males form either loose aggregations or are found singly or in pairs. Females and young are often observed lobtailing, whereas this behavior occurs only very seldom in bachelor and mature males (Table I). Lobtailing occurs more often when a group of females and immatures is socializing than when it is foraging, and there is a good correlation between the occurrence of lobtailing and the occurrence of breaching. As it is generally a contagious phenomenon, it is rare to see only a single individual lobtailing.

Similarly, in humpback whales, lobtailing occurs more often when they are engaged in mating and calving (in winter) than when they are foraging (in summer), and there is also a strong correlation between lobtailing and breaching. However, in some populations of humpback whales, lobtailing is also associated with foraging activities. While surface feeding in the southern Gulf of Maine, humpbacks often lobtail one to three times before releasing a bubble cloud and ending the feeding event by lunging in the middle of the cloud.

In bowhead whales (*Balaena mysticetus*), lobtailing is associated with other socializing activities and mainly occurs in social–sexual groups, especially in the fall. Tail slapping is also used as an aggressive act toward conspecifics.

In dusky dolphins (*Lagenorhynchus obscurus*), lobtailing seems to be associated with surface feeding in large groups. Significantly more lobtails occur in the 15 min preceding a feeding bout and during a feeding bout than at any other time. In winter, when the dolphins tend to feed more individually, very little aerial BEHAVIOR is observed.

IV. Why Lobtail?

The function of lobtailing has been subject to much speculation and is not yet clearly understood. It is likely, however, that lobtailing has multiple functions, and therefore that this question does not have a single answer.

It has often been suggested that lobtails produce a loud percussive noise that can be heard underwater for long distances and thus that lobtailing may serve as a nonvocal acoustic signal. However, as the tail slap is produced at the surface, the underwater intensity of the noise is likely to be limited. Furthermore, measurements of received levels a few hundred meters from a lobtailing bowhead whale showed that the noises created by the tail slaps were much less intense than their calls.

As shown in the previous sections, lobtailing seems to occur mainly in species with a complex social organization and/or social contexts. Therefore, despite having limited underwater noise propagation, lobtailing is still likely to have some COMMUNICATION function. It could be an attention-getting signal, as suggested for

---

TABLE I
Rate of Observing Lobtails per High-Quality Fluke Identification Photographs of Sperm Whale[a]

	Identification photographs (=~indication of vessel proximity to each class)	No. of lobtails observed	Rate of observing lobtails (No. lobtails/No. ID)
Gulf of California Females and immature	92	90	0.98
Galápagos Females and immature	1551	2848	1.84
Kaikoura Bachelor males	521	31	0.06
Scotian Shelf Males	62	2	0.03
Galápagos Mature males	66	0	0
Gulf of California Mature males	10	0	0

[a]Data from Waters and Whitehead (1990) and N. Jaquet, unpublished data.

Hector's dolphins, or communicate presence to a conspecific. It could also signal danger or precede synchronous dives as observed for spinner dolphins. In bowhead whales, it has been shown that tail slapping is at times used as an aggressive act toward conspecifics. In many species of dolphins and in humpback, bowhead, and southern right whales, it has also been suggested that lobtailing is a reaction to annoyance. In some species of dolphins, lobtailing may also inform nearby schoolmates that a school of fish has been found and thus act as a recruitment method.

Lobtailing also seems to play a role in foraging. In the case of humpback whales, it has been suggested that the lobtails associated with surface feeding were creating a frightening disturbance, causing near-surface fish to school more tightly. In dolphin schools (e.g., dusky dolphins), lobtailing and motorboating may be used in cooperative foraging to keep surface-herded fish from escaping laterally.

Any behavior that cannot be classified easily and for which the functions is unclear is often described as "play behavior." It is not possible to rule out that lobtailing also has a play function, but it seems unlikely that it is one of its major functions as, in most species, calves do no seem to lobtail more often than adults.

See Also the Following Articles

Breaching ▪ Leaping Behavior

References

Norris, K. S., Würsig, B., and Wells, R. S. (1994). Aerial behavior. In "The Hawaiian Spinner Dolphin" (K. S. Norris, B. Würsig, R. S. Wells, and M. Würsig, eds.), pp. 103–121. Univ. of California Press, Berkeley.

Pryor, K. (1986). Non-acoustic communicative behavior of the great whales: Origins, comparisons, and implications for management. Rep. Int. Whal. Comm. Spec. Issue 8, 89–96.

Waters, S., and Whitehead, H. (1990). Aerial behaviour in sperm whales. Can. J. Zool. 68, 2076–2082.

Weinrich, M. T., Schilling, M. R., and Belt, C. R. (1992). Evidence for acquisition of a novel feeding behaviour: Lobtailing feeding in humpback whales. Megaptera novaeangliae. Anim. Behav. 44, 1059–1072.

Würsig, B., and Würsig, M. (1980). Behavior and ecology of the dusky dolphin, Lagenorhynchus obscurus, in the South Atlantic. Fish. Bull. 77, 871–890.

Würsig, B., Dorsey, E. M., Richardson, W. J., and Wells, R. S. (1989). Feeding, aerial and play behavior of the bowhead whale, Balaena mysticetus, summering in the Beaufort Sea. Aqu. Mamm. 15, 27–37.

Locomotion, Terrestrial

ANDRÉ R. WYSS
University of California, Santa Barbara

Each of the major clades of marine mammals stems from different terrestrial origins—cetaceans sharing a common ancestry with ungulates, pinnipeds with carnivorans, and sirenians and desmostylians with paenungulates (elephants and kin). The degree to which these groups have become specialized for an aquatic existence varies dramatically, with some modern forms spending a considerable portion of their lives on land (pinnipeds), and others none (cetaceans and sirenians); their ability to locomote terrestrially varies accordingly. For those clades whose extant members are exclusively aquatic, the fossil record provides insights into the terrestrial locomotory capabilities of early transitional forms.

I. Pinnipedia and Kin

Being the geologically youngest group of marine mammals, pinnipeds evince the most obvious signs of their terrestrial ancestry. The limbs of pinnipeds are substantially shortened, reducing the terrestrial agility of these animals relative to their carnivoran allies. Even so, pinnipeds retain considerable mobility on land, but the way in which that mobility is achieved varies remarkably with the group. Pinnipeds can be categorized simple mindedly as "wrigglers" or "walkers." In phocids, during progression on land, the hind flippers remain outstretched posteriorly above the ground, with forward motion being achieved through lurching. Weight is borne not by the flippers but alternately by the abdomen and pelvic region. After the latter is drawn forward and planted, earthworm-like extension of the torso pitches the remainder of the animal forward (Tarasoff et al., 1972). By contrast, in otariids and the walrus (Odobenus rosmarus), the hind legs are capable of being turned forward, with the soles and the palms contacting the ground in a more typically mammalian fashion. These drastically different modes of terrestrial locomotion influenced the now-refuted notion of multiple pinniped origins and were long argued to substantiate an exclusive grouping of otariids and walruses. Because use of the limbs in terrestrial locomotion is obviously a primitive feature, however, it bears neither on the question of a close otariid walrus relationship nor on the relationship of phocids to other pinnipeds.

Although typically playing little role in phocid terrestrial locomotion, the fore flippers are sometimes employed to a minor extent. In phocines the fore flippers can be used to help drag the body forward, whereas in elephant seals they provide balance. The crabeater seal (Lobodon carcinophaga) uses alternate strokes of its fore flippers and a sinuous flapping of its hindquarters to move rapidly on ice (O'Gorman, 1963).

In otariids, walking is accomplished with the abdomen held clear of the ground, with weight being borne equally by all four flippers (Beentjes, 1990). The elongated digits of the anterior limb point posterolaterally, with a sharp bend between the manus and the forearm occurring between the rows of carpals. The hindlimbs are bent forward at the ankle, with the toes pointing anterolaterally. As in all pinnipeds, the left and right knees are incorporated within the body wall, severely limiting the forward–backward excursion of the limbs (much as long, tight skirts do in humans). During a slow walk the fore flippers are moved alternately, whereas the hind flippers move in unison in some species and alternately in others. A gallop moves otariids more quickly, with the fore flippers being moved forward in unison and then the hind flippers. Walruses locomote in a similar manner, but most of the weight is carried by the abdomen (which rests on the substrate), with forceful lunges propelling the animals forward (Gordon, 1981).

The early pinnipedimorph *Enaliarctos* (approximately 23 million years in age) is known from a beautifully preserved skeleton, including flipper-footed limbs (Berta *et al.*, 1989). Nothing of its postcranial anatomy suggests that *Enaliarctos* moved any differently on land than do modern otariids. Similarly, the extinct desmatophocids and allodesmids probably walked in the manner of otariids and the walrus.

II. Cetacea and Kin

Apart from sliding onto decks at marine theme parks, and snatching the occasional penguin or pinniped from water's edge, modern cetaceans are completely inept on land. Such was not always the case, however, as spectacular fossil finds have revealed whale progenitors retaining four fully formed appendages. While these early fossils—often termed "archaeocetes"—are nearly universally termed cetaceans, the latter name is not yet defined formally. Cetacea is probably best used, however, to designate the least inclusive clade encompassing mysticetes and odontocetes, in which case "archaeocetes" are not cetaceans by definition (despite being obviously closely related to cetaceans). The recently described *Ambulocetus* (meaning walking whale) from roughly 52-million-year-old river deposits in Pakistan is the most remarkable of these transitional animals (Thewissen *et al.*, 1994, 1996).

In contrast to cetaceans, the elbow, wrist, and digital joints remained fully functional in *Ambulocetus*. The hands sprawled laterally during terrestrial locomotion; movements of the forelimb were probably reminiscent of those seen in otariids. Contrasting again with modern forms, hindlimbs remained fully developed in *Ambulocetus*. On land, the short, stout femur was apparently rotated laterally, causing the enormous feet to be directed laterally. Hindlimb motion likely mimicked that of otariids, with propulsion stemming from extension of the lower back.

III. Desmostylia

Desmostylians inhabited coastal waters of mainly the northern Pacific during Oligocene and Miocene time. Inasmuch as desmostylians retain four stout but otherwise normally proportioned limbs, which are not modified into paddles, and their remains are found in near-shore marine sediments, the group is generally considered to have been amphibious, and therefore capable of locomoting terrestrially. The stance of these animals is highly peculiar, however, with strongly inturned fore and hind feet. Fusion of the radius and ulna severely curtailed rotation of the forelimb; details of foot posture in group remain somewhat controversial.

IV. Sirenia

Although sirenians do not venture onto land, there is no doubt that their forebears did. Eocene outgroups to Sirenia show limited but persuasive evidence that these animals retained functional hindlimbs, particularly the hip structure in *Protosiren* (Domning *et al.*, 1994). This stage in the ancestry of Sirenia thus records the quadrupedal (and likely amphibious) antecedents to fully aquatic members of the clade.

See Also the Following Articles

Eared Seals ■ Earless Seals ■ Morphology, Functional ■ Sirenian Evolution ■ Swimming

References

Berta, A., Ray, C. E., and Wyss, A. R. (1989). Skeleton of the oldest known pinniped, *Enaliarctos mealsi. Science* **244,** 60–62.

Beentjes, M. P. (1990). Comparative terrestrial locomotion of the Hooker sea lion (*Phocarctos hookeri*) and the New Zealand fur seal (*Acrtocephaus forsteri*): Evolutionary and ecological implications. *Zool. J. Linn. Soc.* **98,** 307–325.

Gordon, K. R. (1981). Locomotor behaviour of the walrus (*Odobenus*). *J. Zool.* (*Lond.*) **195,** 349–367.

O'Gorman, F. (1963). Observations on terrestrial locomotion in Antarctic seals. *Proc. Zool. Soc. Lond.* **141,** 837–850.

Savage, R. J. G., Domning, D. P., and Thewissen, J. G. M. (1994). Fossil Sirenia of the East Atlantic and Caribbean region. V. The most primitive known sirenian, *Prorastomus sirenoides* Owen, 1855. *J. Vertebr. Paleont.* **14,** 427–449.

Tarasoff, F. J., Bisaillon, A., Piérard, J., and Whitt, A. P. (1972). Locomotory patterns and external morphology of the river otter, sea otter, and harp seal (Mammalia). *Can. J. Zool.* **50,** 915–929.

Thewissen, J. G. M., Hussain, S. T., and Arif, M. (1994). Fossil evidence for the origin of aquatic locomotion in archaeocete whales. *Science* **263,** 210–212.

Thewissen, J. G. M., Madar, S. I., and Hussain, S. T. (1996). *Ambulocetus natans,* an Eocene cetacean (Mammalia) from Pakistan. *Courier Forschungsinstitut Senckenberg* **191,** 1–86.

Lutrinae

WARREN FITCH
University of Calgary, Alberta, Canada

Otters (Lutrinae) are amphibious members of a group of mammals called the order Carnivora, family Mustelidae. This family also includes the weasels, badgers, and skunks. Technically, otters are defined according to the following suite of characters: (1) upper fourth premolar with a hypocone or hypoconal crest; (2) first lower molar with a strongly basined hypoconal crest; (3) entoconid reduced to a low crest or altogether absent; and (4) numerous aquatic adaptations, which are examined in more detail later.

I. Modern Diversity

Thirteen is currently agreed upon as the number of living species. There is less agreement on what to name these species of otters. Table I lists the 13 extant species recognized by Wozencraft (1989, 1993) and approximately where they are found.

Which genus name is used does not affect the number of species, just how they are hypothesized to be related. The BIOGEOGRAPHY is important in determining possible

TABLE I
Lutrinae

Genus species	Common name	Location
Aonyx capensis	Cape clawless otter	Central and southern Africa
A. cinerea	Oriental small-clawed otter	India to Indonesia, Southeast Asia, south China
A. congica	Congo otter (Zaire clawless otter)	Central Africa
Enhydra lutris	Sea otter	Coastal North Pacific
Lutra canadensis[a]	River otter	North America
L. felina[a]	Marine otter	West coast of most of South America
L. longicaudis[a]		Mexico, Central and northern two-thirds of South America
L. lutra	European otter	Widespread in Eurasia–Ireland to Japan, South to Sri Lanka
L. maculicollis	Spotted-neck otter	Most of sub-Saharan Africa
L. perspicillata[b]	Smooth-coated otter	South Asia–Afghanistan to Southeast Asia
L. provocax[a]	Southern river otter	Patagonian region of South America
L. sumatrana	Sumatran otter (hairy-nosed otter)	Southeast Asia and Indonesia
Pteronura brasiliensis	Giant river otter	Eastern and northern South America

[a]*Lontra* rather than *Lutra* appeared for these species in Wozencraft (1993). See Species List.
[b]*Lutrogale* rather than *Lutra* appeared in Wozencraft (1993).

relationships and evolutionary history. Fossils, though, have to be considered too.

II. Evolutionary History

The first mustelid fossil is from the late Eocene in Europe. Many different kinds of mustelids are known from the early Miocene (23 mya) of both Europe and North America. Lutrines (otters) are present by the late Miocene (16–18 mya) in the holarctic region. Forms considered to be probable lutrines are known from much earlier in the Miocene. The term "probable lutrines" means that we are not sure if they are otters because of a lack of information. Many early fossil specimens are known only from fragmentary data.

Because otters are placed in the Mustelidae, we can easily use the mink (*Mustela vison*) as a model of the possible morphology and ecology of the earliest proto-otter. An evolutionary scenario that leads us to the otters would read something like this: The ancestral lutrine probably lived in an area with a considerable amount of shallow water where many fish and invertebrates lived close to shore. It was terrestrial in habits but was able and willing to occasionally make use of the abundant food in the water. It had little specialized anatomy for aquatic locomotion or for food capture. Gradually, over the course of generations, this proto-otter tended to spend more time in the water, and slight alterations of the morphology began to appear. What are these alterations? They are the adaptations we now identify in modern otters. SWIMMING by the earliest otter-like form would have employed a terrestrial gait much like the

mink still uses. Most likely, it started out hunting fish or other vertebrates in the shallows. Fish eating is considered primitive in lutrines (Berta and Morgan, 1986), and FEEDING on invertebrates evolved later. Piscivory (feeding on fish) involves using the mouth for capture of the prey. The portion of the brain of otters associated with facial sensitivity is larger in most otters compared with other mustelids. This would improve the capture success when using the head to capture prey. A few species of otter that feed on invertebrates such as crayfish show an enlarged area for tactile sensation in the forepaws, which is in line with their increased use in prey capture.

Body shape also changed. The mink shows no special shape adaptations. Gradually the proto-lutrine took on a more fusiform (streamlined) shape. The neck became shorter, the front legs also got shorter, and the hindlimb lengthened mostly through the elongation of the metatarsals and phalanges. These elongate digits on the hindlimb became encased in webbing. These changes went along with changes in the gait. Two types of locomotion developed to produce greater speed in the water. The limbs were used in a joint thrust and recovery stroke (simultaneously, both hind or both front legs would pump and then recover) and the tail was employed to add thrust. In some species, such as the giant river otter, the tail is used for high speed and the limbs just for maneuvering. The adaptations to life in the water reach their extreme in the sea otter. In essence, the sea otter embodies the story of otter evolution. In this species we see the completion of the evolutionary process that turned a land-dwelling mustelid into an animal quite at ease in its marine environment.

See Also the Following Articles

Dental Morphology, Evolution of ▪ Mustelidae ▪ Osmoregulation ▪ Otters

References

Berta A., and Morgan, G. S. (1985). A new sea otter (Carnivora: Mustelidae) from the Late Miocene and Early Pliocene (Hemphillian) of North America. *J. Paleontol.* **59,** 809–819.

Berta, A., and Sumich, J. L. (1999). "Marine Mammals: Evolutionary Biology." Academic Press, New York.

Estes, J. A. (1989). Adaptations for aquatic living by carnivores. *In* "Carnivore Behavior, Ecology and Evolution" (J. L. Gittleman, ed.), pp. 242–282. Cornell Univ. Press, Ithaca, NY.

Ewer, R. F. (1973). "The Carnivores." Cornell Univ. Press, Ithaca, NY.

Hunt, R. M., Jr. (1996). Biogeography of the order Carnivora. *In* "Carnivore Behavior, Ecology and Evolution" (J. L. Gittleman, ed.), Vol. 2, pp. 485–541. Cornell Univ. Press, Ithaca, NY.

Koepfli, K.-P., and Wayne, R. K. (1998). Phylogenetic relationships of otters (Carnivora: Mustelidae) based on mitochondrial cytochrome *b* sequences. *J. Zool. Lond.* **246,** 401–416.

Taylor, M. E. (1989). Locomotor adaptations by Carnivores. *In* "Carnivore Behavior, Ecology and Evolution" (J. L. Gittleman, ed.), pp. 382–409. Cornell Univ. Press, Ithaca, NY.

Wozencraft, W. C. (1989). Appendix: Classification of the Recent Carnivora. *In* "Carnivore Behavior, Ecology and Evolution" (J. L. Gittleman, ed.), pp. 569–594. Comstock Publishing Associates, Ithaca, NY.

Wozencraft, W. C. (1993). Order Carnivora. *In* "Mammal Species of the World" (D. E. Wilson and D. M. Reeder, eds.), 2nd Ed., pp. 309–313. Smithsonian Institution Press, Washington, DC.

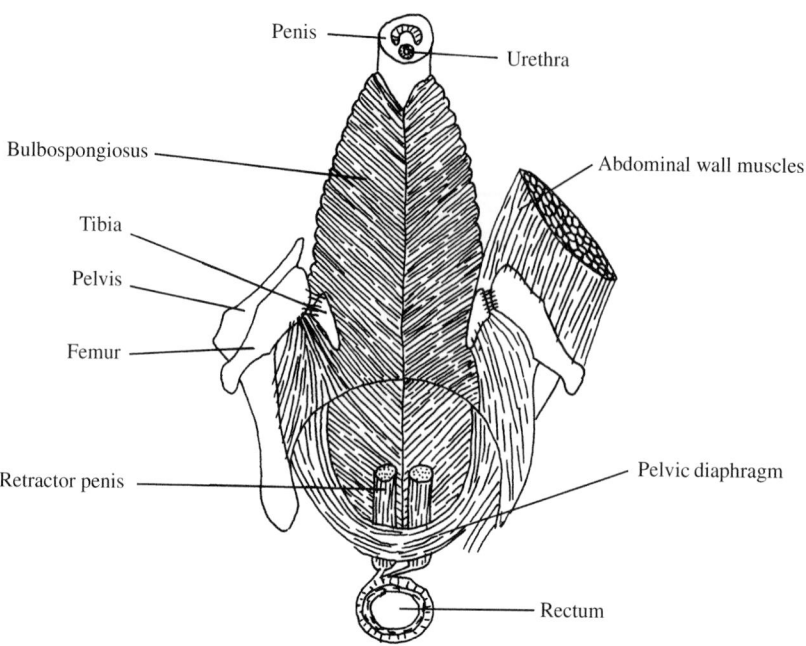

Figure 3 *Pelvis and male genitals of bowhead whale* (Balaena mysticetus) *in ventral view (top of drawing is anterior). This figure shows the close relation between the internal hindlimb bones (pelvis, femur, and tibia), the rectum, and the penis. Retractor penis and bulbospongiosus are penile muscles. (Modified after Struthers, 1893.)*

increase around the time of sexual maturity, making it a useful diagnostic tool. A seasonal pattern of circulating testosterone concentrations exists with elevated concentrations during the breeding season (typically in spring, but a few species are autumnal or multiseasonal breeders). In species with a short, tightly synchronized breeding season, testosterone concentrations are increased for 1 to 3 months at the start of the season but decline to baseline levels after breeding ends. Seasonality is also apparent in most male marine mammals in increased size of the testes

and accessory reproductive glands (even muscles in some cetaceans) and increased spermatogenesis. Increased size of the testes is due to an increased diameter of the seminiferous tubules and epididymes, resulting in increases in the volume of sperm. Spermatogenesis usually lags behind testosterone production, as production of testosterone by testicular Leydig cells is necessary for germ cell differentiation in the seminiferous tubules.

A few marine mammals, such as dugongs and sea otters, lack a distinct breeding season. A few older male dugongs that

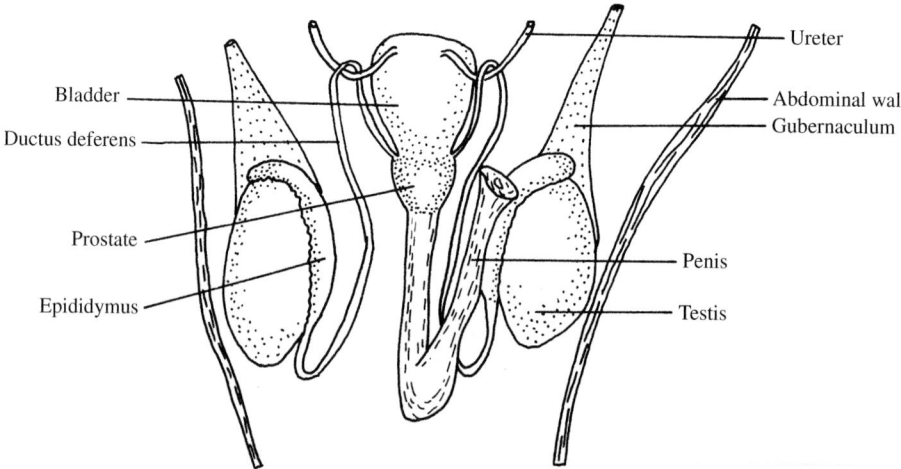

Figure 4 *Male reproductice tract of an otariid in ventral view (top is anterior). (Modified after Boyd et al., 1999.)*

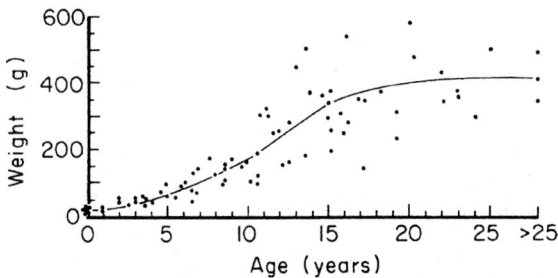

Figure 5 *Weight of nonspermiogenic testes of walrus in relation to age. (Modified from Fay, 1982.)*

were examined were found to be aspermic, suggesting long periods of sterility.

During seasonal quiescence, spermatogenesis ceases, although the testes retain relatively large seminiferous tubules with spermatocytes present. Shrinkage of anterior pituitary cells that produce gonadotrophins is thought to be ultimately responsible for the seasonal testicular regression.

V. Effects of Environmental Factors

Some environmental factors simply cue physiological events, whereas others have the potential to hasten or disrupt the system. The three most important are photoperiod, nutrition, and social factors.

Seasonal reproduction requires that males have adequate numbers of viable sperm when the females enter estrus. Hence the spermatogenic cycle must be initiated months before breeding. Photoperiod is the most commonly cited environmental cue for synchronizing reproductive processes in both males and females; it appears to function months before the breeding season begins. The pineal gland is responsible for the neuroendocrine communication of photoperiod to the rest of the body. Melatonin secretion, which is activated during short photoperiods, acts to relay photoperiodic cues to the target organs. In many species, melatonin is inhibitory to the gonadotrophic-releasing hormones (GnRH) that stimulate testosterone production and spermatogenesis. Thus reproductive processes in most species are stimulated during increasing daylength (i.e., spring). Conversely, increased melatonin concentrations due to a decreased photoperiod lead to inactivation of the testes.

Both sea otters and sirenians lack a defined breeding season, and the Australian sea lion (*Neophoca cinerea*) has a nonannual, nonseasonal reproductive pattern. Spermatogenesis in these species may be continuous. The lack of correlation between testicular activity and season in the dugong may relate to the absence of a pineal gland. No published studies have accounted for the lack of a defined breeding season in sea otters.

There is little published information of nutritional effects on the reproductive biology of male marine mammals. A high plane of nutrition is known to advance the onset of puberty in females and could be expected to have the same effect in males. It is also safe to assume that the plane of nutrition of an individual male will affect its position in a dominance hierarchy. For species in which there can be severe natural impacts on food resources, adult males may have lower blubber thickness during years of poor feeding, resulting in reduced stamina during the breeding season. Although the functional or mechanistic nature of the nutrition–reproduction relationship remains unclear, it can safely be concluded that the measurement of body condition and its effects on various reproductive events, especially during natural environmental perturbations, will continue to be important areas of marine mammal research.

Physiology and the environment influence the development of mating systems by affecting the relative distribution and availability of males and females, thereby altering the reproductive success of an individual male. After sexual maturation, serum testosterone concentrations may vary independent of testis weight, indicating that social factors play a role in reproductive processes. It is not uncommon to find captive situations with cetaceans that are of the same age but at very different reproductive states (i.e., one or more males remain sexually immature much longer than the others). Changing the social structure in an enclosure will often stimulate puberty in those lagging behind in sexual development.

VI. Effects of Pharmacological Agents

The most common reason for prescribing pharmacological agents is to reduce fertility. The three species for which this has been needed in captivity are the common bottlenose dolphin (*Tursiops truncatus*), the California sea lion (*Zalophus californianus*), and the harbor seal (*Phoca vitulina*). All of these can be prolific breeders in captivity, and the need regularly arises to control numbers in some facilities. Until recently, physical separation and contraception of females were the only practical methods. Now GnRH agonists and antiandrogens are being used, with varying success.

A second reason to prescribe pharmacological agents is to suppress aggression among males. The need to control behavior in the captive setting is obvious, especially with adult male bottlenose dolphins during the breeding season. It is less obvious but equally if not more important in the management of the Hawaiian monk seal (*Monachus schauinslandi*), a declining species in which males attempt mass matings, usually with a breeding-aged female, sometimes to the point of killing her.

GnRH agonists work by stimulating the anterior pituitary to release GnRH, which stimulates the testes to produce testosterone and initiate spermatogenesis. Paradoxically, the pituitary quickly becomes refractory and ceases its production of GnRH, which inhibits the testes. Injections of GnRH agonists have been used with some success with harbor seals and effectively decrease circulating testosterone concentrations to prepubertal levels in Hawaiian monk seals. Antiandrogens have been tried unsuccessfully with bottlenose dolphins.

Marine top predators are likely targets for xenobiotic compounds that act either as estrogens or antiandrogens. The most common of these are the polychlorinated biphenyls (PCBs) and dichlorodiphenyltrichloroethanes (DDTs). These compounds bioaccumulate up the food chain, making marine mammals highly susceptible to their biological effects. Male marine mammals continue to accumulate organochlorines throughout their lives, whereas females tend to reduce their body burden via transplacental transfer and lactation. The range of PCB concentrations reported for arctic marine mammals is highest in the walrus (*Odobenus rosmarus*), although the absolute concentrations are highest in polar bears. The effects of organochlorines on male reproductive physiology have not been well studied, as most research has focused on females. The known effects are pathologies related to structural changes and thickening of tubules in organs such as the kidneys, adrenals, and reproductive tract. The most striking possible example has been the occurrence of pseudohermaphroditic polar bears with a normal vaginal opening, a small penis with baculum, and no Y chromosome. The syndrome is hypothesized to be due to either excessive androgen secretion by the mother or endocrine disruption from environmental pollutants. The impacts of all the detected pathologies are unknown. However, there are widespread reports that xenobiotic compounds are also strongly immunosuppressive, rendering contaminated animals more vulnerable to bacterial and viral infections. Experimental studies using minks (mustelids such as the sea otter) indicate that the enzymatic pathways that metabolize steroids are disrupted, but the detailed biosynthetic pathways of the organismal response have not been elucidated.

See Also the Following Articles

Baculum ∎ Captive Breeding ∎ Endocrine Systems ∎ Female Reproductive Systems ∎ Sociobiology ∎ Territorial Behavior

References

Atkinson, S. (1997). Reproductive biology of seals. *Rev. Reprod.* **2**, 175–194.

Atkinson, S., Becker, B. L., Johanos, T. C., Pietraszek, J. R., and Kuhn, B. C. S. (1994). Reproductive morphology and status of female Hawaiian monk seals fatally injured by adult male seals. *J. Reprod. Fertil.* **100**, 225–230.

Berta, A., and Sumich, J. L. (1999). "Marine Mammals: Evolutionary Biology." Academic Press, San Diego.

Desportes, G., Saboureau, M., and Lacroix, A. (1994). Growth-related changes in testicular mass and plasma testosterone concentrations in long-finned pilot whale *Globicephala melaena*. *J. Reprod. Fertil.* **102**, 245–252.

Green, R. F. (1972). Observations on the anatomy of some cetaceans and pinnipeds. *In* "Mammals of the Sea, Biology and Medicine" (S. H. Ridgway, ed.), pp. 247–297. Charles C. Thomas, Springfield, IL.

Harrison Matthew, L. (1950). The male urigenital tract in *Stenella frontalis* (G. Cuvier) [=*S. attenuata*]. *Atlantide Rep.* **1**, 223–247.

Kenagy, G. J., and Trombulak, S. C. (1986). Size and function of mammalian testes. *J. Mammal.* **75**, 1–22.

Kita, S., Yoshioka, M., and Kashiwagi, M. (1999). Relationship between sexual maturity and serum and testis testosterone concentrations in short-finned pilot whales (*Globicephala macrorhynchus*). *Fish. Sci.* **65**, 878–883.

Laws, R. M., and Sinha, A. A. (1993). Reproduction. *In* "Antarctic Seals" (R. M. Laws, ed.), pp. 228–267. Cambridge Univ. Press, Cambridge.

Marsh, H., Heinsohn, G. E., and Glover, T. D. (1984). Changes in the male reproductive organs of the dugong (Sirenia, Dugongidae) with age and reproductive activity. *Aust. J. Zool.* **32**, 721–742.

Morejohn, G. V. (1975). A phylogeny of otariid seals based on morphology of the baculum. *Rapp. Pro.-verb. Reun. Cons. Int. Expl. Mar.* **169**, 49–56.

Ommanney, F. D. (1932). The urino-genital system of the fin whale (*Balaenoptera physalus*). *Disc. Rep.* **5**, 363–466, pls. 2–3.

Pabst, D. A., Rommel, S. A., and McLellan, W. A. (1998). Evolution of thermoregulatory function in cetacean reproductive systems. *In* "The Emergence of Whales" (J. G. M. Thewissen, ed.), pp. 379–397. Plenum Press, New York.

Pabst, D. A., Rommel, S. A., and McLellan, W. A. (1999). The functional morphology of marine mammals. *In* "Biology of Marine Mammals" (J. E. Reynolds, III and S. A. Rommel, eds.), pp. 15–72. Smithsonian Institution Press, Washington.

Ramsay, M. A., and Sterling, I. (1986). On the mating system of polar bears. *Can. J. Zool.* **64**, 2142–2151.

Ramsay, M. A., and Sterling, I. (1988). Reproductive biology and ecology of female polar bears (*Ursus maritimus*). *J. Zool. Lond.* **214**, 601–634.

Slijper, E. J. (1966). Functional morphology of the reproductive system in Cetacea. *In* "Whales, Dolphins, and Porpoises" (K. S. Norris, ed.), pp. 277–319. Univ. of California Press, Berkely.

Mammalia

J. G. M. THEWISSEN
Northeastern Ohio Universities College of Medicine, Rootstown

The animal kingdom is divided into approximately 15 phyla. These phyla represent large groupings of species descended from a single, ancient ancestor. Humans, as well as marine mammals, belong to the phylum Chordata and its largest subdivision, the Vertebrata. Vertebrates are animals with a segmented backbone consisting of a large series of similar bony elements, the vertebrae. Vertebrates are further divided into a number of classes, and mammals (Mammalia in Latin) is one of these (the others are amphibians, reptiles, birds, and several classes of fishes).

I. Mammalian Characteristics

All mammals are derived from a single ancestor that lived in the Late Triassic, approximately 220 million years ago. Scientists distinguish Mammalia from Mammaliformes (Rowe, 1988), but that distinction is not important for marine mammals and we can treat those terms as synonyms here. The first mammal was different from contemporary reptiles by having a lower jaw (half of the mandible) that consisted of a single bone holding both the teeth and the joint with the SKULL. In all nonmammals (e.g., birds and reptiles), the dentary does not take part in the mandibular joint, and instead a bone called the articular forms the joint with a skullbone called the quadrate. The articular and quadrate did not disappear when mammals evolved, instead they became part of the sound transmission mechanism of the ear and are called malleus (hammer) and incus (anvil) in mammals.

II. Marine Mammals

Many other features characterize modern mammals (Fig. 1), but for most of these there are exceptions, and marine mammals are commonly among the exceptions. Most modern mammals, for instance, can be characterized as air-breathing, endotherm (warm-blooded) vertebrates with hair that chew

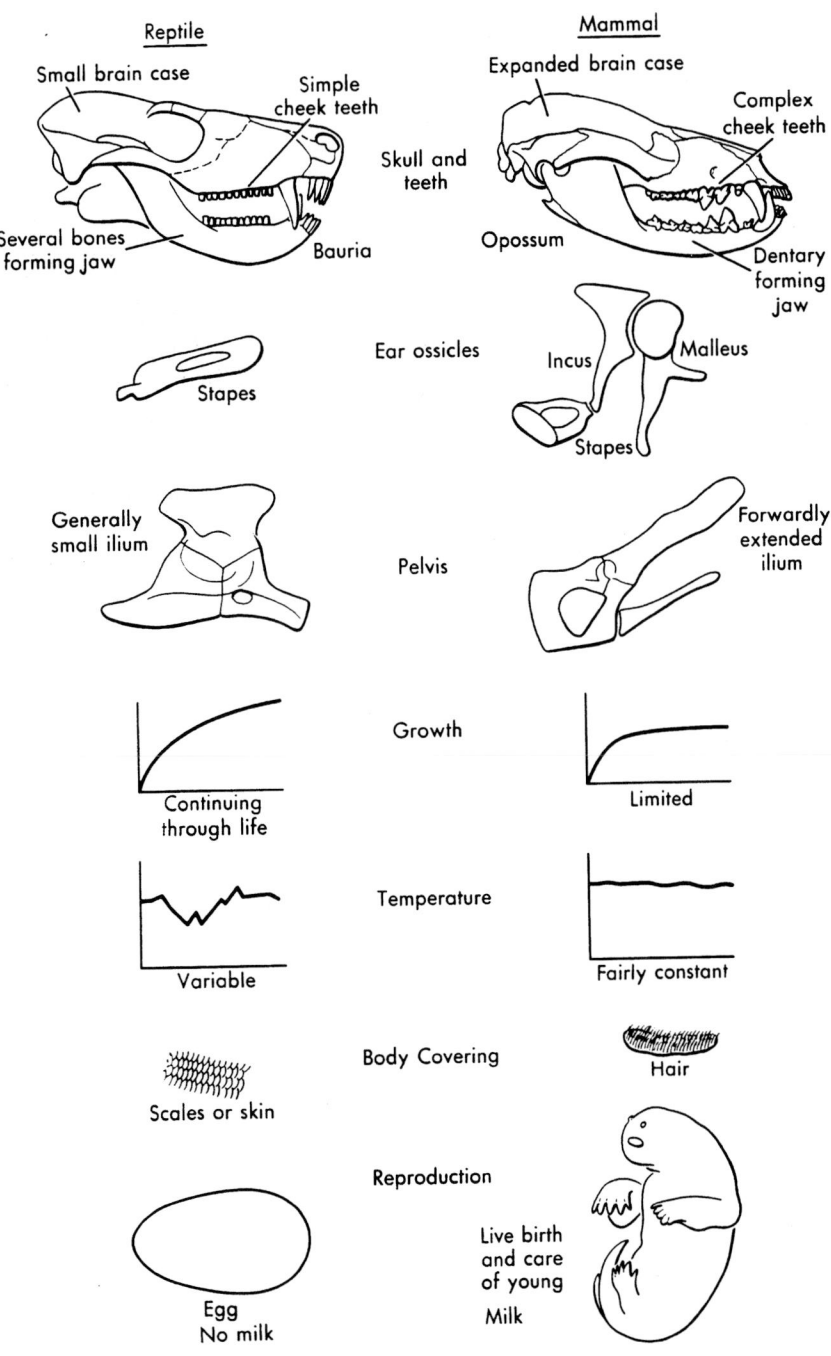

Figure 1 *Some of the characters that distinguish mammals from other vertebrates. From Colbert, E. H., and Morales, M. (1991). "Evolution of the Vertebrates," 4th Ed. Copyright © John Wiley & Sons, Inc. Reprinted by permission of Wiley-Liss, Inc., a subsidiary of John Wiley & Sons, Inc.*

their food with complex TEETH and that feed their young milk produced by mammary glands of the female.

Most modern mammals have a battery of strongly varied teeth that occlude well and are used for chewing. Mysticete whales have no teeth and many modern odontocete whales and pinnipeds lack the complex morphology of the molars. Odonto-

cetes do not chew their food and there is no fine occlusion. Most mammals have hair that covers the entire body, but not whales and sirenians. There are commonly two generations of teeth (deciduous and adult) in an individual mammal, but modern toothed whales only have a single generation. The skull articulates with the vertebral column by means of two convex joints called occipital condyles, which are present in all marine mammals. Nearly all mammals have seven neck vertebrae, but the manatees only have six. Mammals commonly have two pairs of limbs, each with five digits; modern whales and sirenians lack external hindlimbs and digits are commonly not visible externally.

See Also the Following Articles

Classification ▪ Systematics, Overview

References

Colbert, E. H., and Morales, M. (1991). "Evolution of the Vertebrates," 4th Ed. Wiley, New York.
Rowe, T. (1988). Definition, diagnosis, and origin of Mammalia. *J. Vertebr. Paleontol.* **8,** 241–264.
Vaughan, T. A. (1986). "Mammalogy," 4th Ed. Saunders, Philadelphia.

Management

JAY BARLOW
*Southwest Fisheries Science Center,
La Jolla, California*

Management refers to those regulations, laws, treaties, and policies that govern human interactions with marine mammals. Marine mammal management may promote a wide variety of human objectives: conservation of marine mammal populations for their intrinsic value, maintenance of marine mammal populations for human exploitation, protection of human health interests, humane treatment of captive animals, reduction of direct or competitive interference with commercial fisheries, and so on. This article concentrates on the general approaches used for marine mammal management.

I. Management Units

"Management unit" refers to the group of animals that is the target of some management action. It may refer to a colony, subpopulation, population, or species. The term "stock" has traditionally been used instead of "management unit," but this term has evolved to be synonymous with both "population" and "management unit," so, to avoid confusion, "management unit" is preferred. The appropriate definition of a management unit depends on the management objective. Laws to prevent the extinction of a species might have a species or subspecies as a management unit; however, the likelihood of achieving this management objective may be increased by managing on the basis of populations. Laws may not always explicitly define management units, but the stated goals of that law may give some clues as to how the term should be interpreted. For example, if the goal is to maintain marine mammal populations as functional elements of their ecosystems (one of the goals of the U.S. Marine Mammal Protection Act, MMPA), management units might necessarily be smaller than the entire population to ensure than range

contractions would not prevent the attainment of this goal (Taylor, 1997). Knowledge of population structure is critical to defining management units. Population structure has been studied using tagging, radio and satellite tracking, allozymes, DNA fingerprinting, DNA sequencing, photo-identification, morphometrics, and chemical markers. Most of these methods are limited: they can only show that two samples differ and thus that population structure is present, but they cannot be used to demonstrate that population structure is absent. There is almost always some uncertainty in deciding how finely to divide management units, and one of the current challenges in marine mammal management is dealing with this uncertainty.

II. Methods of Marine Mammal Management

A. Traditions, Taboos, and Practices

Prior to modern times, management took the form of culturally enforced practices. Ancient Greeks, natives in the Amazon Basin, and many sea-going cultures held dolphins in especially high regard and had proscriptions against killing or eating dolphins. Monk seals (*Monachus monachus*) were considered by the early Greeks to have prophetic powers and to be protected by Poseiden; however, the popular views toward this species included antipathy and hostility. The societies that did harvest whales and seals (including Inuit and Aleut cultures) often had elaborate rules that determined who could hunt these animals and when they could be hunted. It is not known whether traditions and taboos were important in conserving marine mammals, but there is no evidence of marine mammal EXTINCTIONS caused by humans prior to that of the Steller's sea cow (*Hydrodamalis gigas*) in the 1700s. Traditions based on superstitions have been increasingly ignored as human populations have increased (Johnson and Levigne, 1999).

B. Harvest Bans

The most common method of protecting marine mammals from overexploitation has been a complete ban on harvesting. Most often, this has been practiced after a catastrophic decline has already occurred. Gray whales (*Eschrichtius robustus*), northern elephant seals (*Mirounga angustirostris*), and Guadalupe fur seals (*Arctocephalus townsendi*) were protected by Mexico after their near extinction. A complete ban on whaling for gray whales and right whales (*Eubalaena* spp.) was instituted early in the history of international whale management. Australia, Mexico, New Zealand, South Africa, and the United States have banned the commercial harvest of all marine mammal species in their waters. The European Union members of ASCOBANS (Agreement on the Conservation of Small Cetaceans of the Baltic and North Seas) have agreed to ban the intentional harvest of all small cetaceans. Exceptions to harvest bans are commonly made for aboriginal or subsistence harvests and for incidental mortality pursuant to other commercial enterprises.

C. Age/Sex Limitations on Harvests

Age and/or sex limitations on harvest are commonly employed in the management of terrestrial species. The U.S. North Pacific Fur Seal Act of 1910 outlawed the harvest of northern fur seal (*Callorhinus ursinus*) females and pups. This, together with the provisions of the North Pacific Fur Seal Treaty of 1911, effectively reversed the marked declines of the populations that breed on islands in the Bering Sea. Because many marine mammals do not exhibit marked sexual dimorphism (as do fur seals), similar regulations are not practical for all species. The 1931 Convention for Regulation of Whaling and later regulations of the INTERNATIONAL WHALING COMMISSION (IWC) prohibited the commercial harvest of dependent calves and their mothers. Because most whales have a 2- or 3-year reproductive cycle and are nursing for only 6–12 months, females were not protected for the majority of time. Minimum size limits were also established for various whale species. These and other regulations were not effective in preventing the depletion of most of the world's whale populations.

D. Seasonal Area Closures

The seasonal closure of certain areas or all areas is another common practice in wildlife management. The 1931 Convention for the Regulation of Whaling established a closed season for factory ships in Antarctic waters from April 7 to December 8. Seasonal area closures have also been used to reduce the number of gill net entanglements of Hector's dolphins (*Cephalorhynchus hectori*) in New Zealand and harbor porpoises (*Phocoena phocoena*) in the U.S. Gulf of Maine.

E. Restrictions on Methods and Fishing Gear

Regulations may limit the methods by which marine mammals are killed. The Fur Seal Treaty of 1911 eliminated the at-sea pelagic harvest of northern fur seals (*Callorhinus ursinus*), which were commonly considered to be wasteful (many carcasses could not be recovered) and which were more difficult to monitor. Methodological restrictions are not limited to direct, intentional harvests. Many gear restrictions have been applied to reduce marine mammal bycatch in commercial fishing operations. Finer mesh panels (Medina panels) were added to tuna purse seine nets in the eastern tropical Pacific to reduce dolphin ENTANGLEMENT. The use of acoustic warning devices (pingers) is mandated to reduce cetacean bycatch in several U.S. fisheries, and additional experiments with pingers are currently being conducted worldwide. Regulations may also address how a particular gear is used; the adoption of a "backdown" procedure greatly reduced the mortality of dolphins in tuna purse seines. In addressing marine mammal bycatch problems, restrictions sometimes take the form of a complete ban on a particular gear type. In 1989 the states and territories of the South Pacific banned the use of large-scale (> 2.5 km), drift gill nets in their exclusive economic zones, and in 1992 the United Nations General Assembly extended this ban to all international waters. There have been several attempts to ban drift gill nets in the Mediterranean. In the United States, Florida, Louisiana, Texas, and California have banned gill nets in all or part of their waters in response to marine mammal and other bycatch issues.

F. Quota-Based Restrictions

The most direct method to manage removals from a wild population is to set a limit on the number of animals that can be taken in a given time period (usually 1 year). Quota-based management was first applied to the directed harvest of marine mammals and was later adapted to regulation of bycatch. This method requires some method for estimating annual mortality,

such as from a mandatory program placing observers on whaling or fishing vessels. Whaling on the high seas has been regulated with quotas since 1931, but early quotas were designed only to limit oil production and were based on a "blue whale unit" [the oil equivalent of one blue whale (*Balaenoptera musculus*) being two fin whales (*B. physalus*), six sei whales (*B. borealis*), etc.]. The lack of species- or population-specific whale quotas lasted until 1972 and is widely blamed for the near extinction of most large whale populations. The failure of IWC to effectively manage whaling resulted in an international moratorium on commercial whaling that started in 1986 and continues today. Since 1986, the IWC has devised and adopted a revised management procedure that incorporates a new, well-tested catch limit algorithm (CLA) for setting population-specific quotas. Aboriginal subsistence whaling continues under population-specific quotas that are based on biological considerations and on "cultural and nutritional needs." Quotas were first used in 1976 to limit bycatch in the U.S. tuna purse seine fishery to 78,000 dolphins per year. The U.S. quotas gradually decreased to 20,500 by 1981, but, like the blue whale unit, still had not adequately addressed species- and population-specific conservation concerns. The gradual conversion of the tuna purse seine fishery from a U.S. industry in 1970 to a largely international fleet by 1990 further complicated conservation efforts. The Inter-American Tropical Tuna Commission (and several nongovernmental organizations) have negotiated with its signatory nations to impose vessel-specific quotas on total dolphin bycatch (1993) and stock-specific quotas (2000). Although management of dolphin mortality in the tuna fishery has remained a special case, the United States has adopted a more general approach to setting stock-specific quotas on the maximum allowable levels (potential biological removal, PBR) of human-caused mortality for marine mammal populations in its exclusive economic zone. The PBR approach is like the IWC's CLA in that it sets allowable removal rates that are conservative in the face of uncertainty but which can increase as uncertainties are resolved. New Zealand uses a similar approach to setting annual bycatch limits for Hooker's sea lions (*Phocarctos hookeri*) and has closed its squid trawl fishery when this limit was exceeded. Several countries, including the United Kingdom, are investigating similar approaches to setting limits on fishery bycatch.

G. Market Monitoring and Trade Restrictions

Enforcement of laws on the high seas is often difficult or impossible; therefore, market monitoring and international trade restrictions may be necessary to prevent the illegal harvest and marketing of protected marine mammal. The Convention on International Trade in Endangered Species (CITES) is the primary implement for international trade restrictions and currently bans all trade in whale products, including some species that are not considered "endangered" but whose meat might be confused with that from endangered species. Genetic methods now can distinguish between all species, and CITES is under pressure from pro-whaling countries to lift the "look-alike" ban on nonendangered whales. Surreptitious market surveys by nongovernmental organizations and subsequent genetic analyses have shown that Japanese and Korean markets contain a wide variety of cetacean products, many of which are mislabeled and some of which may have been illegally imported (some cetaceans taken within EEZ waters of those countries and minke whales taken under "scientific whaling" can be legally marketed). Because marine mammal products can be extremely valuable, there will be a strong incentive to cheat. Some IWC member countries are insisting that a system of market monitoring precedes the resumption of commercial whaling, possibly by genetically "fingerprinting" every legally taken whale.

H. Treatment of Wild and Captive Animals

Marine mammals, especially cetaceans, are regarded by many cultures as deserving special treatment by humans. These attitudes may stem from their similarities to humans (large brain, play behavior, etc.), from their representation in popular media, or from the endangered status of some species. Whatever the reason, the special treatment is often evident in national laws that afford more protection for marine mammals than for similar terrestrial mammals. For example, the U.S. MMPA prohibits "harassment" of marine mammals (defined as any pursuit, torment, or annoyance that has the potential to disrupt the natural behavioral patterns of the animal) unless a specific permit is obtained. Virtually any research on marine mammals (except passive observation) has a potential for harassing the subject and therefore requires an MMPA permit. National laws are also frequently implemented to regulate the public display of marine mammals to ensure that adequate space and care are provided to those animals. Some laws and regulations are expressions of public concern for individual animals (rather than concern about species or populations) and are derived more from the animal rights movement than from a conservation ethic, but this distinction is not clear in many cases. Stranding programs that rehabilitate beached animals may aid individuals and, for endangered species, the survival of the species.

I. Marine Sanctuaries

There is a long-standing and growing interest in the use of protected areas or sanctuaries as a management tool for marine species. The first marine mammal refuge [for pinnipeds and sea otters (*Enhydra lutris*) on Afognak Island, Alaska] was established in 1892, but most have been established since 1975. Protected areas are a useful management tool because the concept is so simple (easy to understand and to enforce). For marine mammals, established sanctuaries and protected areas are taxonomically limited: the Indian Ocean and Southern Ocean whale sanctuaries (established by the IWC) protect only large whales, the Irish whale and dolphin sanctuary (established by Ireland) protects only cetaceans, and the Banks Peninsula sanctuary (established by New Zealand) was designed to protect only Hector's dolphins. The level of protection varies between sanctuaries; the Irish whale and dolphin sanctuary does not prevent porpoise and dolphin bycatch in commercial fisheries (although the existence of the sanctuary has focused efforts on reducing bycatch). The utility of protected areas as a management tool depends critically on characteristics of the animals they are designed to protect (residency patterns, home ranges, mating strategies) and on the size of the protected area. The

enormous Southern Ocean whale sanctuary (generally, all waters south of 40° S but excluding the Indian Ocean sanctuary) is currently recognized as being too small to effectively protect its whales (which migrate out of this area during the southern winter). In contrast, small protected areas are quite effective in sheltering breeding colonies of pinnipeds or essential warm spring habitats of manatees. To conservation biologists, a "marine protected area" refers to an area of complete protection at all ecosystem levels. Existing marine protected areas are too small to afford much protection for marine mammal species, although they may protect some critical habitat.

J. Pinniped Control Programs

The recovery of many pinniped populations from a legacy of hunting and near extermination is one of the success stories in marine mammal management, but this recovery is hardly viewed as a success by fishermen and aquaculturists who share their waters. Even conservationists are faced with a dilemma in some situations, such as when California sea lions (*Zalophus californianus*) (protected, but now numbering approximately 200,000) are threatening the survival of a depleted stealhead run in Washington State. Laws protecting marine mammals can and have been modified to deal with such small-scale problems by authorizing the lethal or captive removal of specific problem animals. In some areas, the use of acoustic harassment devices (AHDs) has been authorized to deal with the economic loss to seals by aquaculture facilities or commercial fishermen. Although some of these "fixes" appear to be successful in the short term, their long-term utility is questionable and there is concern about the impact of AHDs on other elements of the ecosystem. Even more controversial are programs designed to reduce entire pinniped populations by culling. Government-sanctioned culling programs to improve fisheries have been practiced in many countries, including Norway and the United States. In Canada, the high annual quota on harp seals (*Pagophilus groenlandicus*) and hooded seals (*Cystophora cristata*) is justified, in part, as a means to reduce seal predation on depleted cod stocks. This approach has been criticized on theoretical grounds because it oversimplifies ecosystem interactions; pinnipeds may feed on a commercially important fish species, but may also feed on predators of that species. Management of culling programs would typically fall under national regulations, but the IUCN Marine Mammal Action Plan has established a protocol to evaluate culling proposals.

K. Ecosystem Management

Ecosystem management refers to approaches ranging from simply considering the impact of a management decision on other elements of the ecosystem to the simultaneous optimization of management strategies to meet management goals of all elements of an ecosystem. There are no examples of the latter approach, although Norway and the signatory nations of the Convention for the Conservation of Antarctic Marine Living Resources are pursuing this goal by promoting multispecies considerations in the management of marine mammal, fish, and seabird resources. Although it is unarguably true that improvements can be made in resource management by considering ecosystem interactions, it is also true that predicting the implications of even a simple ecosystem perturbation is far beyond our current capabilities. Significant progress in implementing ecosystem management may be left to future generations.

III. Trends in Marine Mammal Management

In recent years, there has been a movement toward management procedures that determine quotas for allowable harvests or incidental mortality based on rigid formulae. Both the IWC's CLA and the U.S. PBR approaches are based on formulae that estimate the allowable removals from a management unit based on measurable attributes (such as estimated population size, population growth rates, catch histories, and the precision of the various estimates that are used). The advantage is that all parties can reach a *priori* agreement on the management objectives and on the rules that will be used to reach those objectives without divisive arguments about the effect on anyone's quota. Biological data are inherently imprecise and full of other uncertainties. For both CLA and PBR approaches, computer simulation studies were used to "tune" the quota formulae to achieve their goals even in the presence of imprecision and bias in available data (Wade, 1998). With the increasing emphasis on rigid quota-based management, debates about management practices are changing. Instead of concentrating on which values of biological parameters and which analytical models should be used, managers are now more concerned with how management units should be defined.

Coincident with the movement toward rigid formula-based quota schemes is an increasing reliance on direct approaches to measuring population parameters and a decreasing reliance on industry statistics, such as catch per unit effort. Advances in survey methodology (line-transect and mark-recapture) have greatly improved our ability to estimate the size of cetacean populations. Photo-identification studies, combined with mark-recapture analysis, have refined our understanding of marine mammal life history. Observer programs have increased the reliability of bycatch and harvest estimates. Satellite tagging and the recent revolution in molecular biology have contributed to an explosion of new information on the structure of marine mammal populations. Although all these recent trends promote the potential for effective marine mammal management, the real impediment to effective management is now the lack of collective willpower to implement regulations and to enforce existing regulations.

There has been increasing interests in applying the "precautionary principle" in marine mammal management. In the face of uncertainty, management decisions should be made to minimize the damage caused by being wrong. In most resource protection issues, there are two types of damage: the damage caused to a regulated industry by providing more marine mammal protection than is needed and the damage done to the populations and the industry by providing too little protection. A look at the catastrophic history of marine mammal management illustrates the disastrous economic and ecological results of management approaches that are not precautionary. One way to add precaution is to reverse the legal burden of proof to ensure that any action will *not* adversely affect the population before that action is permitted. Clearly, the future challenge is how to make marine mammal management appropriately precautionary.

See Also the Following Articles

Conservation Biology ■ Genetics for Management ■ Population Status and Trends ■ Stock Identity ■ Whaling, Modern

References

Blix, A. S., Walløe, L., and Ulltang, Ø (eds.) (1994). "Whales, Seals, Fish, and Man." Elsevier, Amsterdam.
Johnson, W. M., and Lavigne, D. M. (1999). "Monk Seals in Antiquity." Netherlands Commission for International Nature Protection.
Taylor, B. L. (1997). Defining "population" to meet management objectives for marine mammals. *In* "Molecular Genetics of Marine Mammals" (A. E. Dizon, S. J. Chivers, and W. F. Perrin, eds.), pp. 49–65. Society for Marine Mammology.
Twiss, J. R., Jr., and Reeves, R. R. (eds.) (1999). "Conservation and Management of Marine Mammals." Smithsonian Institution Press, Washington, DC.
Wade, P. R. (1998). Calculating limits to the allowable human-caused mortality of cetaceans and pinnipeds. *Mar. Mamm. Sci.* **14**(1), 1–37.

Manatees
Trichechus manatus, T. senegalensis, and *T. inunguis*

JOHN E. REYNOLDS III
Eckerd College, St. Petersburg, Florida

JAMES A. POWELL
*Florida Marine Research Institute,
St. Petersburg*

The manatees (order Sirenia; family Trichechidae; subfamily Trichechinae) represent one of the most derived groups of extant mammals. Although ancestral forms were terrestrial, descendant forms have occupied aquatic habitats since the Eocene Epoch, providing the group with a long period of time over which to evolve. Apart from their suite of unusual morphological attributes (adaptations) associated with their herbivory and aquatic habitat, manatees have many behavioral and life history traits that are similar to those of other mammals. For most aspects of species biology, the Florida manatee is the best-studied taxon, and without data to the contrary, scientists assume that other manatees may be similar to the Florida subspecies.

I. Scientific and Common Names

Trichechus inunguis, Natterer, 1883: Amazonian manatee
T. manatus, Linnaeus, 1758: West Indian manatee
T. m. manatus: Antillean manatee
T. m. latirostris: Florida manatee
T. senegalensis, Link, 1795: West African manatee

II. Conservation Status

The various manatee species are protected by laws specific to the countries they occupy, but enforcement of these laws is generally minimal. Several countries (e.g., United States, Mexico, Belize, Guatemala, Ivory Coast, Cameroon, and Nigeria) have created manatee reserves and sanctuaries, and others (Brazil and Peru) have "protected" areas that include important manatee habitat.

The following list provides examples of broader-scale protection efforts.

Trichechus inunguis: CITES: Appendix I
World Conservation Union (IUCN): listed as vulnerable
United States, Endangered Species Act (ESA): listed as endangered
T. manatus latirostris: CITES: Appendix I
World Conservation Union (IUCN): listed as vulnerable
United States, Endangered Species Act (ESA): listed as endangered
T. manatus manatus: CITES: Appendix I
World Conservation Union (IUCN): listed as vulnerable
United States, Endangered Species Act (ESA): listed as endangered
T. senegalensis: CITES: Appendix II
World Conservation Union (IUCN): listed as vulnerable
United States, Endangered Species Act (ESA): listed as threatened
Protected under Class A, African Convention for the Conservation of Nature and Natural Resources

III. Name Derivations

The generic name of the manatees, *Trichechus,* comes from the Greek words *trichos* (hair) and *ekhō* (to have), referring to the sparse body hairs and abundant facial hairs and bristles. *Inunguis* refers to the lack of nails on the pectoral flippers of the Amazonian manatee. At least two possible origins for *manatus* have been suggested: It could refer to the hand (*manus*), as manatees sometimes use their front, or pectoral, limbs to push food into their mouths. More likely, the term comes from the Carib Indian word *manati*, which means woman's breast, perhaps referring to the fact that the manatee's mammary glands are located in the axillary region in approximately the same anatomical location as the breasts of a human female; this particular anatomical feature contributed to the association of the manatee with the mythical mermaid. *Senegalensis* denotes that the West African species is found along the coast of Senegal, although it also occurs in waters of other west–central African countries.

Vernacular names for the manatees vary by region. In English-speaking areas, they are typically referred to as sea cows; similarly, in German, a manatee is referred to as a *seekuh* or *manati*, in Dutch as a *zeekoe*, in French as a *lamantin*, in Spanish as a *manati* or *vaca marina*, and in Portuguese as a *peixe-boi*, or ox fish. In some west African countries, a manatee is called a "*mammy wata*," which refers to a water deity. Diverse indigenous names are also in use in Africa and South America.

IV. Diagnostic Characters
A. Diagnostic Characters of the Modern Trichechidae

Unlike some marine mammals (e.g., pinnipeds, polar bears, and sea otters), manatees and the other living member of the order Sirenia, the dugong (*Dugong dugon*), are totally aquatic. They inhabit shallow waterways and feed primarily on plants, a diet that makes the sirenians unique among modern marine mammals.

Although manatees do not dive to great depths or for prolonged periods as many cetaceans and pinnipeds do, they are anatomically well adapted to aquatic habitats (Fig. 1). They lack hind (pelvic) limbs; have reduced, paddle-like front (pectoral) limbs; have fusiform (streamlined, spindle-shaped) bodies with few external protuberances and thick, tough skin; and are very large (an adaptation that facilitates heat conservation). Their heads are somewhat streamlined, and the nostrils are located on the dorsal side of the muzzle. A dorsal fin is lacking. Internally, manatees have extremely thick, heavy (pachyosteosclerotic) bones and an unusual arrangement of the diaphragm and lungs that facilitates buoyancy control. Manatees, like other marine mammals, have sensory and other adaptations that enhance diving, osmoregulatory, and thermoregulatory abilities.

Unusual adaptations accommodate the manatee's herbivorous diet. These include (1) enlarged lips (especially the upper lip) equipped with prehensile as well as tactile vibrissae and moved by a muscular hydrostat; (2) the presence of supernumerary (polydont) molariform cheek teeth that are replaced via horizontal migration along the jaws throughout the lifetime of each manatee; and (3) a greatly expanded gastrointestinal tract (specialized for hindgut fermentation, as in horses and elephants) with several unusual gross and microscopic features.

Manatees differ sharply from their close relative, the dugong. Manatees have a rounded fluke, whereas dugongs have split flukes similar to those of cetaceans. Dugongs have tusks, which manatees lack, and the mode of tooth replacement in the two differs. The rostrum of the dugong is much more sharply deflected downward than the rostrum in any manatee species. In addition, dugong skin is smoother than is the case for West African and west Indian manatees.

B. Diagnostic Characters of the Individual Species

The West Indian manatee is the largest living sirenian, with individuals approaching 1500 kg in weight and 4 m in length. Females tend to be somewhat larger than males, but body size cannot be used to determine either the sex or the age of an individual. West Indian manatees are euryhaline (can tolerate both salt and freshwater) but may require periodic access to fresh water to drink. West African manatees are generally very similar to West Indian manatees in terms of their size, general body form, and habitat, but the West African manatee has a blunter snout, somewhat protruding eyes, and a slightly more slender body. The Amazonian manatee is the smallest trichechid, measuring about 3 m long or less and weighing less than 500 kg. Its "rubbery" skin is smoother than that of its con-

Figure 1 *Although manatees may not swim as fast or dive as deep as some cetaceans do, manatees have the fusiform bodies, the reduced or absent limbs, and the powerful locomotory fluke that cetaceans also have. This particular animal has been fitted with a belt attached to a floating canister containing telemetry equipment. Photograph by Patrick Rose.*

geners, and it lacks nails at the tips of the pectoral flippers, which are proportionately longer than in the other species. In addition, white or pink belly patches are common. The Amazonian manatee is confined to freshwater habitats.

Manatee species also vary in the degree of rostral deflection, corresponding to the predominate location in the water column of food plants in their natural habitats. West African manatees have the least deflected snouts, and Florida manatees the most deflected.

V. Distribution and Abundance

All extant manatees occupy subtropical and tropical waters (Figs. 2–5).

A. West Indian Manatee, *T. manatus*

This species occupies coastal and riverine habitats from the mid-Atlantic region of the United States, throughout the wider Caribbean Sea and Gulf of Mexico, and into coastal parts of northeastern and central–eastern South America. The Florida manatee, *T. m. latirostris*, occurs from eastern Texas to Virginia in the summer, but occupies waters of Florida and southeastern Georgia year-round (Fig. 2). Although its distribution is not continuous, the Antillean manatee, *T. m. manatus*, occupies the remainder of the species' range, from southwestern Texas to South America. It occupies the waters of 19 countries (Fig. 3). The range of the Antillean manatee may overlap with that of the Amazonian manatee around the mouth of the Amazon River (Fig. 4).

Scientists estimate that there may be 3300 or more Florida manatees. Some recent analyses of population trends of manatees occupying different regions of the species' range suggest that the population grew through the 1980s and early 1990s but leveled off in at least some locations during the mid-1990s. As the 20th century ended, other analyses suggested that the population may be relatively stable or may even be increasing slightly in some regions, but the statistical uncertainty associated with data and models leaves open the possibility that the overall population may be declining.

The number of Antillean manatees is unknown. During recent aerial surveys of Belizean waters and waters of southern Quintana Roo, Mexico, more than 400 manatees were counted. The corridor between Belize and southern Mexico is considered to be a last stronghold for the subspecies.

B. Amazonian Manatee, *T. inunguis*

This species occupies freshwater habitats throughout the drainage of the Amazon River and its tributaries, including rivers and lakes in Brazil, Peru, Ecuador, and Colombia (Fig. 4). There are no reliable population estimates.

C. West African Manatee, *T. senegalensis*

This species is found in the coastal and riverine waters of nearly two dozen countries in central and West Africa, from Senegal to Angola (Fig. 5). Manatees inhabit the upper reaches of the Niger River to Guinea and occur throughout the inland delta of Mali. Manatees in the upper Niger River are cut off from the sea by cataracts and a hydroelectric dam. Manatees inhabit two tributaries of Lake Chad, the Logone and Chari rivers, but are not found in the lake itself. The Logone and Chari rivers do not communicate with the sea; during times when water levels were higher, manatees in these rivers probably were able to mix with other manatees by moving through interconnecting

Figure 2 *Range of the Florida manatee,* T. m. latirostris *(map produced by Leslie Ward).*

lakes to the Benue River, a tributary of the Niger River. There are no reliable population estimates.

VI. Geographic Variation/Subspecies

The only species for which subspecies have been identified is the West Indian manatee. The two subspecies differ most obviously in their skeletal (especially skull) morphology.

However, Garcia-Rodriguez *et al.* (1998) examined mitochondrial DNA (mtDNA) control regions from 86 individual west Indian manatees from 8 different locations. They found 15 different haplotypes that could be clustered into three, rather than two, distinctive lineages for the species. These authors also noted for three presumed West Indian manatees from Guyana that their mtDNA haplotypes were more consistent with that of the Amazonian manatee.

Clearly, the systematics of manatees is a topic that requires additional study.

VII. Ecology

All manatees are herbivores, and as hindgut digesters (like horses and elephants), they can subsist on low-quality forage. Because they are such large mammals, manatees would be expected to have a low weight-specific metabolic rate, but their metabolism is 20 to 30% lower than one would expect. The best-studied species, the West Indian manatee, consumes more than 60 species of plants (almost exclusively angiosperms) and may ingest a mass of food that approximates up to 7% of its body weight each day. In some locations, 50 to 90% of the plant biomass may be eaten or uprooted by grazing animals, but the overall effects on local plant productivity of manatee feeding are not well understood. The dugong has been described as a cultivation grazer, and the manatee may serve the same role.

Although manatees subsist primarily on plants, they also consume flesh. They have been reported to consume fish entangled in nets and tunicates have been found in large numbers in some manatee stomachs. Of course, the plants manatees consume have epiphytic organisms growing on their leaves.

The DISTRIBUTION of the Florida manatee is influenced by temperature, and perhaps, by access to fresh water to drink. In cold weather, manatees tend to migrate south and/or seek refuge at natural and artificial warm-water refugia (Fig. 6). The distribution of other manatee species or subspecies appears to be governed to at least some extent by the availability of water and suitable habitat during the wet and dry seasons. Antillean manatees, for example, may move upstream in coastal rivers during the wet season, when water levels are high, and return to lower reaches of rivers during the dry season. Amazonian manatees occupy lakes during the dry season, when rivers and streams dry up. Because the lakes are murky and lack bottom vegetation, manatees may fast during the dry season for up to 200 days when water levels drop and shoreline vegetation is no longer available for them to eat.

The habitat requirements of West African manatees are similar to those of West Indian manatees. Although manatees along the coast of Africa tend to move up rivers and out of estuaries during the dry season, they can occasionally be found in any aquatic habitat. In the upper reaches of the Niger River and some other large rivers, West African manatees, like Amazonian manatees, may remain in lakes during the dry season, when water levels drop, and stay there until waters rise and they can move back into the rivers.

PREDATION on manatees has not been well documented, but it appears that they have few natural enemies and that preda-

Figure 3 *Range of the Antillean manatee,* T. m. manatus *(map produced by Leslie Ward).*

tion levels are very low. Anecdotal reports suggest that crocodilians and sharks may account for some manatee mortality in different parts of the world; in Florida, such reports are reinforced by the rare presence on living manatees of wounds caused by alligators or sharks. Especially, but not exclusively, when aggregated in lakes during the dry season, Amazonian manatees may also be preyed upon by jaguars.

Natural factors that have been documented to kill large numbers of Florida manatees include cold weather and red tides.

VIII. Behavior

Manatees feed on bottom vegetation, plants in the water column, and floating or shoreline vegetation. Their flexible pectoral flippers and prehensile lips, which are equipped with bristles, are used to push vegetation into the mouth. The most striking and well-documented migrations occur in Florida in response to cold weather. Aggregations of more than 300 manatees occur at each of several natural and artificial sources of warm water in winter (Fig. 6). On very cold mornings, the majority of Florida manatees may be found at a few warm-water refugia scattered along the coast. Florida manatees disperse widely in warm weather. Amazonian manatees also migrate seasonally, from rivers and streams in the wet season to deeper waters, such as lakes, in the dry sea-

son. In some areas, West African and Antillean manatees show movement patterns similar to those of Amazonian manatees in response to low water and lack of freshwater flow.

Mating herds, composed of a female in estrus and a consort of several (up to 22) males, may remain together for periods of up to a month, typically outside the winter season. The cow is receptive to mating for only a day or two during that time. Although "fighting" does not occur, the males vigorously push and shove one another to gain access to the female. Females mate with several males during the estrous period. The MATING SYSTEM is an example of scramble competition polygamy (specifically polyandry).

Cues males use to locate estrous females are not known, but it is possible that the males detect olfactory, gustatory, and acoustic signals produced by the females. Males tend to have larger home ranges than females do; thus, wandering males may routinely encounter a number of females. Females in estrus have a larger ranger of movement than nonestrous females.

Calves tend to stay close to their mothers for some time after birth. Weaning generally occurs when the calf is 1–2 years old, although calves up to 4 years old may still nurse. When traveling, calves swim parallel and close to their mothers, presumably in a position where they are easy to protect, where COMMUNICATION is facilitated, and where hydrodynamic drag is minimized. During the long period of maternal care, calves

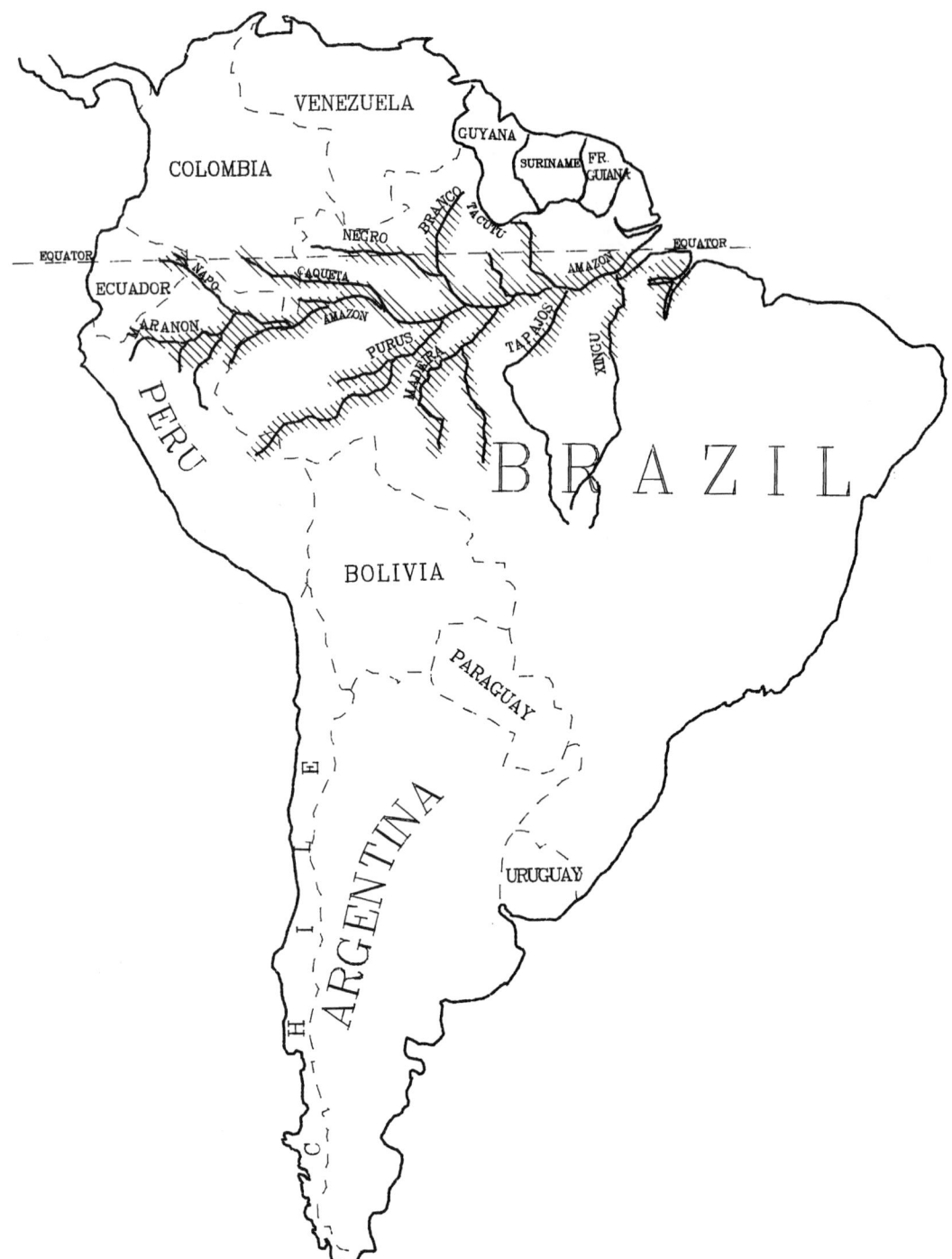

Figure 4 *Range of the Amazonian manatee,* T. inunguis *(map produced by Leslie Ward).*

apparently learn the locations of important resources such as warm-water discharges or fresh water. The learning process causes groups of manatees, including young animals that have recently become independent from their mothers, to use the same areas year after year.

Details of the social structure of manatees are lacking. Florida manatees appear to have a simple fission–fusion soci-

ety in which individuals come together in a series of temporary groups. G. B. Rathbun has stated that although such societies may be relatively unstructured, the lack of social structure is not the same as a lack of social complexity.

Communication among manatees appears to involve acoustic signals (squeaks and squeals, mostly in the 3- to 5-kHz range), tactile cues (rubbing and "kissing"), visual cues, and

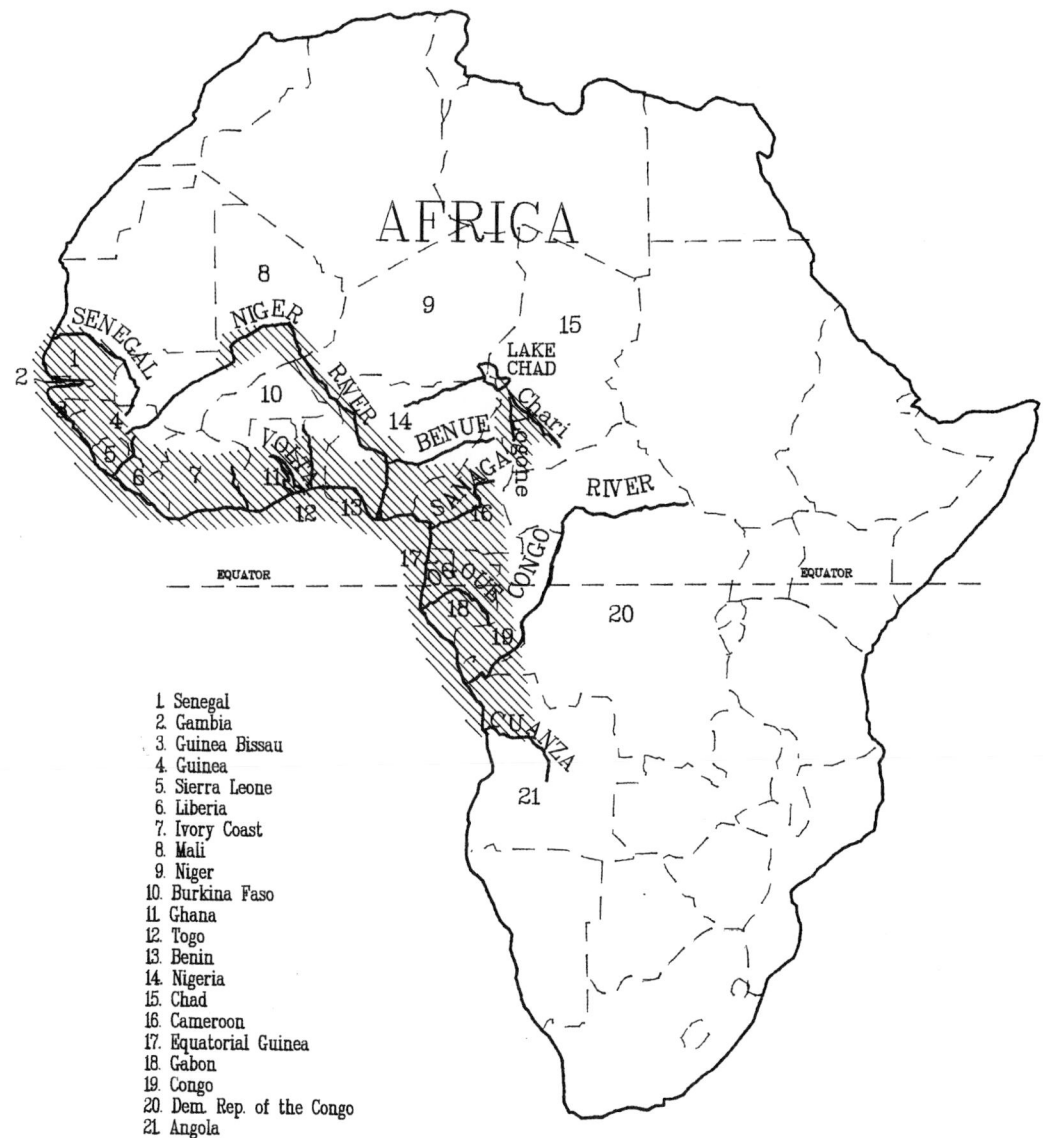

1. Senegal
2. Gambia
3. Guinea Bissau
4. Guinea
5. Sierra Leone
6. Liberia
7. Ivory Coast
8. Mali
9. Niger
10. Burkina Faso
11. Ghana
12. Togo
13. Benin
14. Nigeria
15. Chad
16. Cameroon
17. Equatorial Guinea
18. Gabon
19. Congo
20. Dem. Rep. of the Congo
21. Angola

Figure 5 *Range of the West African manatee,* T. senegalensis *(map produced by Leslie Ward).*

possibly chemical cues (suggested by repeated use of "rubbing posts" and by individuals mouthing one another).

Manatees appear to play. Body surfing and follow the leader have been observed.

IX. Anatomy/Physiology/Life History

Manatees have a suite of unusual adaptations. We note here some features of a variety of organ systems.

The bones, especially the ribs, are dense and heavy (osteosclerotic), and the ribs and some other bones are swollen (pachyostotic). As in other marine mammals, the long bones are shortened and the phalanges in the pectoral appendage are more elongated than are those of "typical" terrestrial mammals. Hyperphalangy, however, does not occur. The first digit is reduced and the fifth digit is enlarged. Pelvic limbs are absent,

although vestigial pelvic bones that are sexually dimorphic remain embedded in hypaxial musculature. Erythropoiesis (formation of red blood corpuscles) and granulopoiesis (formation of certain white blood cells) occur primarily in the vertebral bodies. The SKULL is elongated but not telescoped. Nares and nasal bones have migrated dorsally. The zygomatic arch, which abuts the periotic bone on each side of the skull, is relatively light and porous and is permeated with oil. However, compared to the bones of the zygomatic arches of most marine mammals, the bones of the manatee's arch are huge, reflecting their importance as an attachment for powerful chewing muscles.

The muscle color is of several shades, from almost white to red, apparently due to differences in myoglobin concentration in particular muscles or muscle groups. Axial muscles to the fluke are extremely powerful. The panniculus muscles are very

Figure 6 *Florida manatees aggregate in large numbers around natural and artificial sources of warm water in winter. In this photograph, approximately 230 manatees huddle in the discharge of the Riveria power plant. Photograph by John Reynolds and Florida Power & Light Company.*

well developed, as are muscles to the pectoral flipper (presumably to facilitate both dexterity and strength).

The skin is extremely heavy and thick and may provide some of the ballast needed for buoyancy control. Body hairs are sparse (approximately 1 every cm^2); sweat glands are lacking; and nerve plexuses associated with some hairs suggest that the hairs are important in detecting pressure or in tactile communication. Instead of having one layer of BLUBBER as many other marine mammals do, manatees have alternating layers of panniculus muscle and blubber (somewhat resembling bacon).

The lungs are long and unlobed and occupy virtually the entire dorsal region of the trunk. Manatees exchange about 90% of the air in their lungs in a single breath. The branching pattern of the bronchi is monopodial, and the terminal airways are reinforced with cartilage. The diaphragm is large and powerful, is located in a horizontal (coronal) plane, is constructed as two independent hemidiaphragms, and is instrumental in maintaining buoyancy control.

The large intestine is enormous (more than 20 m long in large animals), a feature that is not surprising in a hindgut fermenter. The stomach has a large accessory organ of digestion (the cardiac gland), and the capacious duodenum has two prominent diverticulae. The cecum is small and bicornuate. The life-long, horizontally oriented tooth replacement is a very unusual feature and may be an adaptation to facilitate the consumption of the gritty plant material that manatees consume. Histology of the various portions of the gastrointestinal tract shows some unusual cellular arrangements. The accessory organs of digestion (liver, pancreas, salivary glands) are unremarkable. Manatees have taste buds but no vomeronasal organ.

The heart is not unusual except for a persistent interventricular cleft, the presence of notable amounts of cardiac fat, and the large amount of pericardial fluid. Circulatory adaptations (retia, arteriovenous anastomoses, countercurrent heat exchangers) facilitate overall heat conservation, while also allowing for the cooling of the reproductive organs and the nervous tissues.

The brain is small (the encephalization quotient for *T. manatus* is 0.27), and the cerebral hemispheres lack extensive convolutions. Notably large trigeminal (cranial nerve V) and facial (cranial nerve VII) nerves are associated with the facial vibrissae.

The uterus is bicornuate. The ovaries are rather flattened and diffuse, and in mature individuals, the ovaries have numerous corpora. The penis and testes are located inside the body wall. The testes are relatively small, but the seminal vesicles are very large. The testes abut the kidneys along the caudal part of the diaphragm.

The kidneys are lobular, are located on the ventral surface of the caudal part of the diaphragm, and are often encapsulated in fat. Their microscopic structure suggests an ability by manatees to produce concentrated urine and therefore to go for prolonged periods without access to fresh water.

Manatees can remain submerged for more than 20 min but generally dive for much shorter periods of time (2–3 min or less). Because the plants manatees consume grow close to the surface where sunlight is available, dives are usually shallow.

Scientists have historically suggested that temperatures below about 19° C induce sufficient stress to cause at least some manatees to seek warm water as a refuge. Some recent evidence suggests that this temperature may be a little high and that 17° C is perhaps a more realistic point at which stress occurs. Even though scientists may be uncertain of the precise point at which thermal stress occurs, it is clear that both chronic and acute exposure to low temperatures may cause death.

The extent to which manatees physiologically *need* fresh water is unclear. It is clear, however, that Florida manatees *like* fresh water to drink. Functional morphology suggests that the kidney should be able to produce hyperosmotic urine and be able to rid the body of excess salt following seawater ingestion.

Manatees, like other marine mammals, are *K* strategists when compared to most other animals. In some ways, however, manatees appear to be less *K* selected if the comparison group is just the marine mammals. Table I provides life history information on Florida manatees.

X. Evolution and Fossil Record

Sirenians probably arose in the Old World (Eurasia and/or Africa) not later than the early Eocene Epoch, 50–55 million years ago. The oldest fossils are from Jamaica. Within a few million years (i.e., in the middle Eocene, 45–50 million years ago), several genera of sirenians existed. Peak diversity of sirenians occurred during the Oligocene and Miocene Epochs (5–35 million years ago).

The first truly manatee-like (i.e., trichechine) sirenian was *Potomosiren*, fossils which are about 15 million years old (Miocene of Colombia). During the Pliocene Epoch (about 2–5 millions years ago), trichechids also inhabited the Amazon Basin and the Caribbean. The Amazonian trichechids gave rise to the Amazonian manatee, and the Caribbean trichechids are thought to have given rise to the West Indian and West African manatees, which are sister taxa.

Due at least in part to their dense bones, sirenians in general are well preserved in the fossil record, but true manatees are rare until the Pleistocene.

Various lines of evidence (e.g., genetic analyses, electrophoresis of serum proteins, and morphological studies) suggest that the order Sirenia (manatees and dugong) is most closely related to a group of mammalian orders called the Paenungulata. The extant paenungulates include the elephants (order Proboscidea) and hyraxes (order Hyracoidea). The sirenians appear to be most closely related to elephants and the extinct, HIPPOPOTAMUS-like desmostylians.

XI. Interactions with Humans

Humans have interacted with the various manatee species in a number of ways, most of them harmful to the manatees. The following information includes both well-documented and presumed interactions.

Manatees have historically been hunted throughout their ranges. In Florida, HUNTING pressure has virtually ceased within the past few decades, although animals are occasionally still taken illegally for meat. The best-documented and most extreme example of manatee hunting occurred in Brazil from 1935 to 1954, when between 80,000 and 140,000 Amazonian manatees were killed for their meat and hides. Primary products included *mixira* (fried manatee meat preserved in its own

---

TABLE I
Estimated Population Traits of the Florida Manatee Based on Long-Term Life History Research[a]

Trait	Description
Maximum life expectancy	60 years
Gestation period	11–13 months
Litter size	1
Percentage of twins	1.79% at Blue Spring; 1.4% at Crystal River
Sex ratio at birth	1:1
Calf survival to year 1	0.60 at Blue Spring; 0.67 at Crystal River
Annual adult survival	90% on Atlantic coast; 96% at Crystal River and Blue Spring
Earliest age of first reproduction: ♀	3–4 years
Mean age of first reproduction: ♀	5 years
Earliest onset of spermatogenesis	2 years
Proportion of adult ♀♀ pregnant	0.33 of salvaged carcasses; 0.41 of living animals at Blue Spring
Proportion of nursing first-year calves during winter season	0.36 (mean)
Mean period of calf dependency	1.2 years
Mean interbirth interval	2.5 years
Period of highest number of births	May–September
Period of highest frequency of mating herds	February–July

[a]From Lefebvre and O'Shea (1995), supplemented by some more recent data.

fat), uncooked meat, lard, and the tough hides, which could be used for a range of products, including whips, shields, and machine belts. Although the market for hides diminished after 1954, several thousand manatees were killed each year through the late 1950s, and probably beyond.

In certain countries such as Peru and Ecuador (Amazonian manatee) and possibly in some West African countries, military patrols hunt manatees, or hire local hunters to catch manatees, for food.

Manatees are also hunted for reasons other than the products they provide. In Sierra Leone, the Mende people hunt manatees, in part, to reduce the number of manatees and thereby to keep them from tearing fishing nets, destroying fish that have been netted, and plundering rice fields.

However, local traditions may work in favor of manatees and prevent their harvest in particular areas. In the Korup region of Cameroon, for example, villagers fear manatees and have no taste for the meat, so they generally do not hunt the animals.

An interesting presumed effect of manatee hunting in tropical America and West Africa is that some manatees have become nocturnal and/or crepuscular.

Manatees are captured accidentally in fishing gear (crab pot lines, trot lines, fishing nets) in the United States and other countries. The extent of serious injury or mortality is unknown.

Collisions with boats and barges account for about 25% of all manatee mortality in the United States (Fig. 7). The number rose at a rate of about 7.5% per year between 1976 and 1996, and currently more than 70 animals die annually in this way. The number of registered boats in Florida alone exceeds 750,000. Based on observations of scarred animals, collisions with boats appear to be occurring with increasing frequency in other parts of the world, but the extent to which those collisions kill manatees outside the United States is unknown. Also unknown is how seriously boat-inflicted injuries debilitate manatees and affect reproduction, without causing immediate death (Fig. 8).

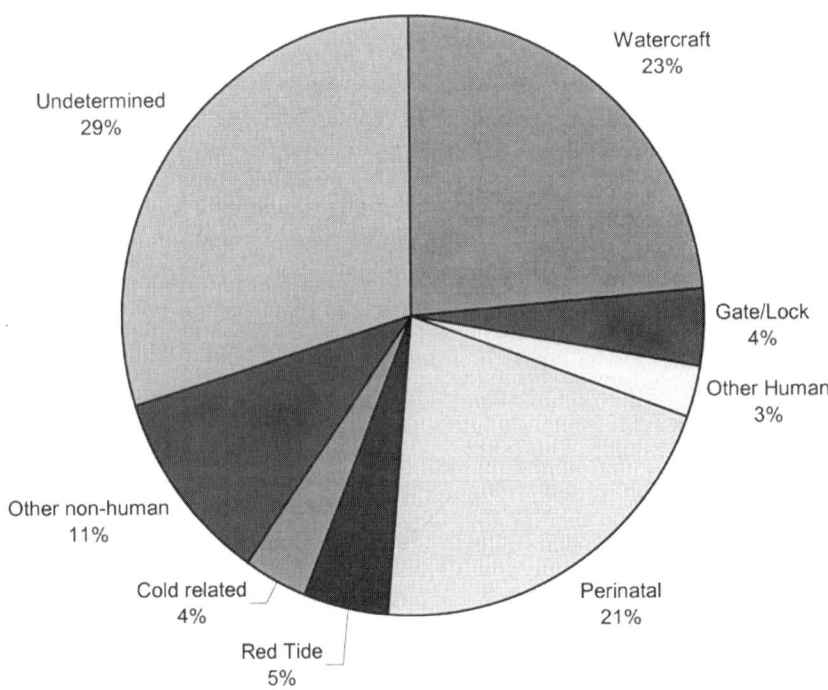

Figure 7 *Categories of manatee mortality in Florida. This pie chart shows manatee mortality categories based on 3501 carcasses recovered or reported to federal or state agencies from 1974 through 1998. The highest percentage of deaths remains undetermined (n = 1043) and includes unrecovered (n = 99), badly decomposed (n = 588), and other carcasses (n = 423) that were not badly decomposed but for which cause of death could not be assigned. Total human-related mortality is high (n = 1065, 31% of total) and includes watercraft-related deaths (n = 828), trauma or drowning caused by canal locks or flood gates (n = 145), and other human-related factors such as entanglement (n = 92). Non human-related causes of death (n = 658, 19% of total) are related to cold exposure (n = 124), red tide outbreaks (n = 164), or other factors (n = 92). Perinatal mortality (n = 735) refers to the death of a small animal (≤ 150 cm long) for which cause of death cannot be determined; perinatal mortality can be either human or nonhuman related. Watercraft-related mortality continues to rise annually and is the single highest known category of death. Produced by James Powell.*

Figure 8 *Although several dozen Florida manatees die each year because of collisions with watercraft, many animals survive such encounters, albeit with considerable pain and disfigurement. The extent to which reproduction and longevity of survivors are compromised is unknown. Photograph by Sirenia Project, U.S. Geological Survey.*

The propeller scars and increased turbidity caused by boats negatively affect the health and distribution of sea grasses and other vegetation eaten by manatees. Boats also make noise, which may affect manatee distribution, habitat use, and energetics. Boats can, therefore, affect manatees both indirectly, by contributing to diminished food resources, and directly, by disturbing, injuring, or killing them. In Florida, manatees sometimes become trapped in flood control structures and canal locks and die. About 4% of the manatees known to die between 1974 and 1996 were crushed and drowned in flood gates or canal locks. Increasingly, scientists and environmentalists at the national, state, and regional levels are concerned about the effects of POLLUTION both on the health of individual animals and on the status of populations. Levels of certain chemical pollutants have been assessed in some marine mammal tissues, but the effects of these chemicals are unclear. In only one case did scientists experimentally demonstrate a clear cause-and-effect relationship between a toxicant and reproductive impairment (in harbor seals). Based on toxicological studies of laboratory animals, scientists suspect that chemical pollution is harming the endocrine and immune systems of at least some stocks of marine mammals, but this has yet to be demonstrated.

Scientists have found elevated levels of copper in the tissues of Florida manatees from certain locations, but levels of other metals and of organochlorines have been unremarkable. No other toxicants have been examined, and the effects of elevated copper levels on manatees are unknown.

In Central America, runoff of pesticides and herbicides and the ingestion of plastic debris have been suspected of causing the death of Antillean manatees. In the Amazon basin, the water pollution associated with mining activities may be harming manatees.

NOISE pollution is a problem of which people around the world have become increasingly aware, but about which few data exist. Many scientists suspect that noise pollution negatively affects manatees, but studies examining this relationship are needed. Underwater noise of anthropogenic origin has been demonstrated or suggested to cause some marine mammals to vary their normal patterns of habitat use and to expend more energy than usual in order to avoid disturbance, and these behavioral changes could logically be expected in manatees too. In the coastal waters of Florida, where more than 750,000 boats are registered and many additional boats are also found, any cumulative effects of anthropogenic noise are a real cause for concern. Even in Costa Rica, where boat traffic is sparse compared to that in Florida, hunters and scientists have noted that manatees react to the noise of approaching boats and that manatee distribution is inversely related to the amount of boat traffic.

Behavioral and anatomical evidence suggests that manatees hear boat motors, but a recent audiogram suggests otherwise. Nevertheless, the suggestion (unwise in our opinion) has been made to put noise makers or acoustic alarms on boats to alert manatees. Not only may such devices be unnecessary if manatees do, indeed, already hear boats, but because their use would greatly increase existing levels of underwater noise, they would most probably disturb or harm not only manatees but many other organisms as well.

Disturbance can occur in ways besides those associated with boat traffic or boat noise as described earlier. At Crystal River and nearby springs in Florida, tourists gather in large numbers to swim with manatees when the animals aggregate in winter. Although most people behave responsibly, some stand on, ride, or tie ropes to the manatees. Such behavior by humans could cause manatees to avoid seeking refuge at warm-water springs in winter, which could lead to even higher than usual manatee mortality in very cold weather. Disturbance of females accompanied by calves could lead to abandonment of a calf, contributing to escalating levels of "perinatal" mortality.

In Southern Lagoon, Belize, and other locations, ecotourism focusing on manatees has developed. Although the financial benefits to local residents may be significant, such activities should be carefully planned to minimize the negative effects on the manatees residing in and using the resources of such locations.

As noted earlier, disturbances from hunting pressure have apparently induced nocturnal or crepuscular behavior in manatees in certain parts of the world.

We have discussed many of the harmful ways that humans have affected manatees—noise, chemical contamination, boat traffic, and ecotourism, for example—but in at least one way, we humans may be helping manatees. Most biologists feel that thermal discharges from power plants and other sources have provided winter habitat that has helped the populations of manatees in Florida to survive and even to grow in at least some areas. However, these plants have finite lifetimes, and manatee dependence on them creates a long-term dilemma for managers. What will the manatees do if warm water is no longer available at a spot where they have learned to depend on it? Industry (primarily Florida Power & Light Company) and the U.S. Fish and Wildlife Service have initiated discussions to attempt to solve this problem.

Another way in which people have helped manatees is by the introduction of exotic aquatic plants into Florida's waterways. Such plants as *Eichhornia crassipes* (water hyacinths) and *Hydrilla verticillata* have proliferated to such an extent that they provide important and abundant food resources for manatees in certain regions of the state. In fact, the exotics displace native vegetation and may grow so luxuriantly that they create navigation problems in some waterways; manatees in Florida have been suggested as possible economical weed-clearing agents, a role that they fill well in some canals in Guyana.

Other human-related habitat modifications have not been helpful to manatees. Dams or other structures prevent manatees from pursuing normal MIGRATION routes along rivers in South America and West Africa. Finally, the eradication of millions of hectares of rain forest each year in Amazonia cannot help but negatively affect *all* species occupying that area because of factors such as reduced productivity, siltation, and changes in hydrological cycles.

Manatees are maintained in nearly 20 different facilities worldwide. They breed in several of the facilities. In certain facilities in Florida, injured or diseased manatees are routinely rehabilitated and released back into the wild, thereby assisting wild populations.

Without entering the debate about the appropriateness of captivity, we simply note here that facilities that display manatees to the public provide important venues for educating people about manatees and their conservation, as well as for conducting basic and applied research on the different species.

Acknowledgments

We thank Holly Edwards, Judy Leiby, Bill Perrin, Sentiel Rommel, and an anonymous reviewer for their valuable comments on the manuscript. We are also grateful to Leslie Ward for producing the species range maps and to Patrick Rose, Florida Power & Light Company, and the Sirenia Project (U.S. Geological Survey) for permission to use slides.

See Also the Following Articles

References

Domning, D. P. (1982). Evolution of manatees: A speculative history. *J. Paleontol.* **56**, 599–619.

Domning, D. P. (1996). Bibliography and Index of the Sirenia and Desmostylia. *Smith. Contrib. Paelobiol.* **80**, 611.

Domning, D. P., and Hayeck, L. C. (1986). Interspecific and intraspecific morphological variation in manatees (Sirenia: *Trichechus*). *Mar. Mamm. Sci.* **2**, 87–144.

Garcia-Rodriguez, A. I., Bowen, B. W., Domning, D., Mignucci-Giannoni, A. A., Marmontel, M., Montoya-Ospina, R. A., Morales-Vela, B., Rudin, M., Bonde, R. K., and McGuire, P. M. (1998). Phylogeography of the West Indian manatee (*Trichechus manatus*): how many populations and how many taxa. *Molecular Evolution* **7**, 1137–1149.

Hartman, D. S. (1979). "Ecology and Behavior of the Florida Manatee." Special Publication No. 5, American Society of Mammalogists.

Lefebvre, L. W., and O'Shea, T. J. (1995). Florida manatees. *In* "Our Living Resources: A Report to the Nation on the Distribution, Abundance, and Health of U.S. Plants, Animals, and Ecosystems" (E. T. LaRoe, *et al.*, eds.), pp. 267–269. U.S. Department of the Interior, National Biological Service, Washington, DC.

Lefebvre, L. W., O'Shea, T. J., Rathbun, G. B., and Best, R. C. (1989). Distribution, status, and biogeography of the West Indian manatee. *In* "Biogeography of the West Indies" (C. A. Woods, ed.), pp. 567–610. Sandhill Crane Press, Gainesville, FL.

O'Shea, T. J., Ackerman, B. B., and Percival, H. F. (eds.) (1995). "Population Biology of the Florida Manatee" U.S. Department of the Interior, National Biological Service, Information and Technology Report 1.

Powell, J. A. (1996). The distribution and biology of the West African manatee (*Trichechus senegalensis* Link, 1795). Report to the United Nations Environment Programme, Nairobi, Kenya.

Reeves, R. R., Stewart, B. S., and Leatherwood, S. (1992). "The Sierra Club Handbook of Seals and Sirenians." Sierra Club Books, San Francisco.

Reynolds, J. E., III (1999). Efforts to conserve the manatees. *In* "Conservation and Management of Marine Mammals" (J. R. Twiss, Jr., and R. R. Reeves, eds.), pp. 267–295. Smithsonian Institution Press, Washington, DC.

Reynolds, J. E., III, and Odell, D. K. (1991). "Manatees and Dugongs." Facts on File, New York.

Reynolds, J. E., III, and Rommel, S. A. (eds.) (1999). "Biology of Marine Mammals." Smithsonian Institution Press, Washington, DC.

Rosas, F. C. (1994). Biology, conservation and status of the Amazonian manatee, *Trichechus inunguis. Mamm. Rev.* **24**, 49–59.

U.S. Fish and Wildlife Service (1996). "Florida Manatee Recovery Plan." Second Revision. Prepared by the Florida Manatee Recovery Team for the U.S. Fish and Wildlife Service, Atlanta, Georgia.

Marine Parks and Zoos

DANIEL K. ODELL

SeaWorld Inc., Orlando, Florida

LORAN WLODARSKI

SeaWorld Florida, Orlando

I. The History of Zoological Parks

Humans have held wild animals in captivity for hundreds if not thousands of years. The earliest zoos were not meant for the average citizen but for the elite, as wealthy rulers collected unusual animals for their enjoyment. Slowly these private collections turned public such as when the animals gathered at Schloss Schönbrunn, Vienna, Austria, first opened to the public in 1765 (Courcy, 1999). The park is considered the first modern zoo. Polar bears (*Ursus maritimus*) and various pinnipeds were probably among the first marine mammals to be held by humans. Polar bears may have been held since about 1060 and harbor porpoises (*Phocoena phocoena*) since perhaps as early as the 1400s, but the majority of marine mammals seen more commonly in marine facilities were not displayed until the late 1800s and early 1900s (Reeves and Mead, 1999). Many types of cetaceans have never been displayed and some have only been seen recently in marine parks. Killer whales (*Orcinus orca*), for example, were first displayed in 1961.

II. Zoos and Marine Parks Worldwide

How many zoos, aquaria, and marine zoological parks exist worldwide? More specifically, how many of these facilities display marine mammals? It is likely that no one has an exact count. New facilities are being built, some facilities close, and some facilities change the animals that they have on display. The Conservation Breeding Specialist Group (CBSG) of the IUCN maintains a global zoo directory on its worldwide web page <www.cbsg.org/gzd.htm>. The directory lists nearly 1800 institutions but does not give information on marine mammals in the collections. A similar list is published in each issue of the International Zoo Yearbook. The yearbook also includes a list of zoo and aquarium associations with postal, phone/fax, and email addresses. Kisling (2001) is the most recent review of zoo and aquarium history and includes an admittedly incomplete listing of over 900 zoos and aquaria worldwide. For example, 230 facilities are listed for the United States; 140 for Japan; 56 for India; and 155 for Asia (Table I). A survey by Couquiaud-Douaze (1999) of facilities holding cetaceans lists 166 institutions in 42 countries located on all continents except Antarctica (Table I). If pinnipeds and polar bears were added, the number of institutions would grow considerably. For example, a 1995 survey (Andrews *et al.*, 1997) of the United States and Canada listed 109 facilities that held 1460 marine mammals, including 11 species of cetaceans, 11 species of pinnipeds, the sea otter (*Enhydra lutris*), and the Florida manatee (*Trichechus*

TABLE I

Worldwide Counts of Zoos and Aquariums with Counts of Those Holding Cetaceans[a]

Region/Country	Cetacean facilities[b]	Total facilities[c]
Africa		21
Egypt	1	
South Africa	2	
Asia/Pacific		155
Australia	3	10
China	6	
French Polynesia	1	
India		56
Indonesia	1	17
Japan	40	140
Korea	1	
New Zealand	1	
Taiwan		1
Thailand	2	
North America		
Bermuda	1	
Canada	3	
Bahamas	2	
USA	38	230
Central and South America		
Argentina	2	31
Bolivia		1
Brazil		73
Central America		8
Chile		2
Colombia	2	1
Cuba	3	
Guyana		1
Honduras	1	1
Mexico	9	1
Peru	1	1
Venezuela	3	4
Europe and Middle East		
Austria	0	3
Bahrain	1	
Belgium	2	3
Bulgaria	1	
Cyprus	1	
Czech Republic and Slovakia		15
Denmark	1	4
Finland	1	2
France	2	7
Germany	4	42
Hungary		9
Israel	1	11
Italy	5	4
Lithuania	1	
Malta	1	
Poland		16
Portugal	2	1
Russia and former Soviet Union	3	31
Spain	8	3

(continues)

[a] The counts given are incomplete but do give an idea of the relative distribution and abundance of zoos and aquariums.
[b] From Couquiaud-Douaze (1999).
[c] From Kisling (2000).

manatus latirostris). Polar bears were not included.

Zoological parks can be found through web sites maintained by regional organizations. The following list is representative and not necessarily all inclusive: American Zoo and Aquarium Association <www.aza.org>; Australian Regional Association of Zoological Parks and Aquariums <www.arazpa.org.au>; Canadian Association of Zoos and Aquariums <www.caza.ca>; European Association of Zoos and Aquaria <www.eaza.net>; Japanese Association of Zoological Gardens and Aquariums <www.jazga.or.jp>; PanAfrican Association of Zoological Gardens, Aquariums and Botanical Gardens <www.paazb.org>; South East Asian Zoos Association <www.seaza.org>; World Zoo Organization <www.worldzoo.org>; Zoological Society of London <http://www.zsl.org/ioz/research/>; and Zoos Worldwide <www.zoos-worldwide.com>.

III. Challenges

As zoos and aquaria developed, there is no question that the quality of the facilities has improved dramatically, especially over the past several decades and especially in developed countries. Enclosure and pool sizes have increased and have gone from caged to cageless exhibits. Governments have enacted (or are considering) legal standards for the care, maintenance, and display of captive animals (e.g., the United States Animal Welfare Act) and these standards are continually evolving. Marginal facilities, including many traveling or temporary exhibits, have been eliminated. However, on a worldwide basis, there remains much room for improvement. These improvements often have a high financial cost and there is often an unavoidable trade-off with funding for other human activities (e.g., health care). If facilities cannot provide proper care for their animals, perhaps they should be closed and the animals relocated to other facilities. Unquestionably, it is in the best interests of all zoo and aquariums staffs to provide the best possible care for the animals in their charge. Whether institutions are formally "for profit" or "not for profit," it still takes large amounts of money to build, operate, maintain, and expand exhibits. Captive breeding is becoming more and more important for marine mammal facilities. These programs often require additional, separate facilities (e.g., maternity pools) and additional animals. Another challenge is the acquisition of high-quality marine mammal food (primarily fish and squid) on a reliable basis. As fish stocks are depleted around the world, marine mammal facility managers must plan accordingly. Some are considering the development of a mass-produced fish replacement for marine mammal food. Such a product could be produced as needed and would not require the storage of a year's supply.

A (perhaps) more difficult challenge deals with the ethics of keeping wild animals (or, for that matter, any animal) in captivity for any purpose (Mench and Kreger, 1996). Worldwide, one will encounter any number of groups dedicated to the elimination of facilities holding cetaceans for any purpose. Interestingly, one seems to see little, if any, opposition to holding pinnipeds. Marine parks and zoos are often targeted by protests when new exhibits are proposed or opened or when new animals are acquired. The effectiveness of these protests and similar activities remains unclear. Ultimately, each person will have to reach her/his own conclusions on the ethics of keeping animals in captivity. One can only ask that people seek factual information (zoos and aquariums have an obligation to provide the best available information on the animals in their charge), even if they choose to ignore it, before making their personal decisions.

IV. Research

The mission statements of most zoological parks and aquaria probably include "recreation, education, conservation, and research" in one form or another. Whether the institution is private, for-profit; private, not-for-profit; or public and supported to some extent by taxes (i.e., public funds), the recreation component is the most visible followed by the education component. The extent to which these institutions are involved in research and conservation programs varies and is, to some extent, dependent on financial resources. However, even the smallest of institutions (in size or in financial resources) can participate in local or multipartnered national or international research and conservation projects.

Research on wild marine mammals, especially cetaceans, is often expensive and subject to the vagaries of environmental conditions (i.e., weather) among other things. Modern technology (radio and satellite tags, time-depth recorders, GPS tags, "critter cams," hydrophone arrays, etc.) have made huge contributions to our knowledge of free-ranging marine mammals. However, short of "Star Trek Tricorder" technology, there is still much that cannot be learned from wild animals that can be learned from captive marine mammals. Behavior, including acoustic emissions, can be observed and recorded 24 hr/day if desired. Animals can be trained to hold position for body measurements, collection of body fluids (blood, urine, milk), various medical procedures (i.e., ultrasound examinations), and collection of expired gases. Animals can be trained for a variety of visual, acoustic (hearing, ECHOLOCATION), locomotion, and learning studies. The birth, growth, and development (behavioral and physiological) of offspring can be detailed. Therefore, studies on captive animals are not a replacement for, but a supplement to, studies on free-ranging animals, and the results must be applied with the limits of these "laboratory" studies in mind.

One measure of the involvement of zoos and aquaria in research can be obtained from annual reports of individual institutions and regional organizations. For example, the American Zoo and Aquarium Association's (AZA) annual report on conservation and science for 1996–1997 (Hodskins, 1998) lists over 1100 publications of all types (abstracts, magazine articles, journal articles, etc.) for all animal groups produced by the AZA member institutions during that time period.

V. Education

Marine parks, zoos, and aquaria offer a wide variety of education program [in-park graphics, exhibit narrations, behind-the-scenes tours, curriculum-specific programs for various age and grade levels from preschool through college, camp programs, classroom programs, off-site outreach programs, and, in the electronic age, satellite television and internet (worldwide web) offerings]. In fact, United States facilities holding marine mammals for public display are required by federal law to have an approved education program.

The world we live in might be remarkably different if everyone could travel to the plains of Africa to view cheetahs (*Acinonyx jubatus*) stalking their prey, to view diminutive Humboldt penguins (*Spheniscus humboldti*) basking in the South American shorelines, or perhaps see meandering Florida manatees slowly grazing on vegetation along the coasts of the state of Florida. However, for most people around the planet, such encounters will never occur. In a 1995 Roper poll, 87% of those who participated stated that their only opportunity to see wild animals came from visiting zoological facilities (Roper Starch Worldwide, 1995). Given the fact that most people cannot afford the time and money of a jet-set lifestyle, zoological parks are vital links to connect mankind with the plethora of animals on the planet.

Zoological parks alert people to the increasing threats these animals face. For example, because most people will not travel to view wild cheetahs in Africa, they may not see how these creatures are slaughtered for their hides, how Humboldt penguins are disappearing due to the mining of their guano (feces) deposits where they nest, or how Florida manatees are highly endangered thanks to an ever-increasing presence of humans in their habitat. Zoological facilities may be entertaining, but education, research, and conservation are now cornerstones of major parks. The same Roper poll revealed that 92% of those surveyed agreed that zoological parks are vital educational resources.

Although approximately 71% of the planet is covered by oceans, this realm and its inhabitants remain a mystery to a majority of people. Marine life parks helps (1) educate the public about the seas and (2) clear up long-rooted misconceptions about ocean animals. A prime example of this is how killer whales were perceived in the past and how they are viewed today.

Some cultures, like aboriginal tribes of the Pacific Northwest, respected killer whales, although several major whaling countries feared these animals. Indeed, the name *Orcinus* is probably derived from Orcus, an ancient mythological Roman god of the netherworld—a reference to the ferocious reputation of this animal. In 1835, Hamilton wrote that the killer whale ". . . has the character of being exceeding voracious and warlike. It devours an immense number of fishes of all sizes . . .,

when pressed by hunger, it is said to throw itself on every thing (sic) it meets with . . ." In modern civilization, many still envisioned killer whales as terrifying threats to humans, with a 1973 United States Navy diving manual warning that killer whales "will attack human beings at every opportunity." In the not too distant past, governments such as Japan, Greenland, Canada, and the United States sanctioned the use of lethal force to be used against killer whales. Killer whales were hunted for commercial use and despised by whalers who would ". . . often carry a rifle expressly for the Killer's benefit," according to Bennett in this book "Whaling in Antarctica." In 1961, a killer whale was displayed publicly for the first time and afterward the perception of these animals began to change. Coupled with a growing environmental awareness in the 1960s, public sentiment rallied to protect cetaceans like killer whales from hunting. Cetaceans are now protected by various national laws and international agreements, and killer whales are generally perceived in a positive way thanks in part to the educational programs of zoological parks.

VI. Conservation

Conservation programs are linked inextricably with both research and education programs. Zoos, aquaria, and marine parks that hold marine mammals can incorporate conservation messages (i.e., do not feed or swim with wild dolphins or manatees; proper field etiquette and trash disposal) into static graphics and show and exhibit scripts and narrations, as well as in classroom programs and the electronic media. The American Zoo and Aquarium Association's annual report on conservation and science for 1996–1997 (Hodskins, 1998) listed over 1200 conservation projects in which the AZA's 185 members were involved during that time period.

Facility staff experienced in handling marine mammals can provide advice and assistance to field workers. Facilities located near an endangered species' habitat can assist with rescue and rehabilitation of sick or injured animals [e.g., monk seals (*Monachus* spp.) in Hawaii and the Mediterranean/eastern Atlantic; manatees in Florida; sea otters in Alaska and California; and Steller sea lions (*Eumetopias jubatus*) in Canada and Alaska]. Unfortunately, no facility is large enough to handle a blue (*Balaenoptera musculus*) or right whale (*Eubalaena* spp.), although SeaWorld California had remarkable success with an orphan gray whale (*Eschrichtius robustus*) calf (Antrim *et al.*, 1998). It may even be possible for facilities to start breeding colonies of severely endangered marine mammals [e.g., the baiji (*Lipotes vexillifer*)] if field conservation efforts prove inadequate. In addition, institutions can make direct monetary contributions to conservation programs and encourage their visitors to make their own contributions to bona fide programs.

See Also the Following Articles

Captivity ■ Conservation Efforts ■ Ethics and Marine Mammals

References

Andrews, B., Duffield, D. A., and McBain, J. F. (1997). Marine mammal management: Aiming at the year 2000. *IBI Rep.* **7**, 125–130.

Antrim, J., McBain, J., and Parham, D. (1998). Rehabilitation and release of a gray whale calf: J.J.'s story. *Endang. Spec. Update* **15**(5), 84–89.

Bennett, A. G. (1932). "Whaling in Antarctica." Henry Holt Publisher.

Couquiaud-Douaze, L. (1999). "Dolphins and Whales: Captive Environment Guidebook." National University of Singapore.

de Courcy, C. (1999). The origin and growth of zoos. *Endanger. Spec.* **1**(1), 16–19.

Hamilton, R. (1835). Mammalia: Whales. *In* "The Naturalist's Library" (W. Jardine, ed.), Vol. 26, pp. 228–232. W. H. Lizars, Edinburgh.

Hodskins, L. G. (ed.) (1998). AZA Annual Report on Conservation and Science 1996-97. Volume II. Member Institution Conservation and Research Projects. American Zoo and Aquarium Association, Silver Spring, Maryland.

Mench, J. A., and Kreger, M. D. (1996). Ethical and welfare issues associated with keeping wild animals in captivity. "In Wild Mammals in Captivity: Principles and Techniques" (D. G. Kleiman, M. E. Allen, K. V. Thompson, and S. Lumpkin, eds.), pp. 5–15. University of Chicago Press, Chicago.

Reeves, R. R., and Mead, J. G. (1999). Marine mammals in captivity. "In Conservation and Management of Marine Mammals" (J. R. Twiss, Jr. and R. R. Reeves, eds.), pp. 412–436. Smithsonian Institution Press, Washington, DC.

Roper Starch Worldwide (1995). Public attitudes toward zoos, aquariums and animal theme parks. Roper Starch Worldwide, New York.

Mass Die-Offs

John Harwood
University of St. Andrews,
Scotland, United Kingdom

The term "mass die-off" has been used rather imprecisely in the scientific literature. In general, it is an event that involves the death of many hundreds of individuals in a relatively short interval (usually 1–2 months). Mass die-offs of rare species may involve smaller numbers of individuals, but the use of this term can be justified if a large proportion of the population is involved in the die-off. STRANDINGS of groups of social cetaceans, such as pilot whales (*Globicephala macrorhynchus* and *G. melas*) and false killer whales (*Feresa attenuata*), should probably not be regarded as mass die-offs. However, large numbers of dead bodies may wash up along a short section of coast during a mass die-off, and this may, initially, resemble a stranding event. Marine mammals spend most of their lives at sea and only a fraction of the number of individuals that die during a mass die-off are likely to be observed. As a result, the true magnitude and effect of a mass die-off cannot be assessed from a simple body count. The scale of the die-off is usually best estimated by comparing abundance estimates made before and after the event.

I. Diagnosis

The fact that marine mammals spend most of their lives at sea makes it not only difficult to determine how many individuals have died during a mass die-off, but also difficult to diagnose the cause. Many days may elapse between the death of an individual and the recovery of its carcass. To make matters worse, many die-offs occur along remote stretches of coastline where access is difficult, further increasing the time between death and examination. The problem is further complicated because many of the agents that can cause mass die-offs also reduce resistance to diseases. Exposure to these agents may then allow other pathogens, which would otherwise by relatively harmless, to contribute toward an individual's death.

Despite these problems, it has proved possible to identify with some confidence the causes of several of the mass die-offs that were observed in the last 20 years of the 20th century. Three factors appear to be particularly important: infectious diseases, naturally occurring toxins, and environmental events. Several of these factors may act together, and the effects of any one of them can be amplified by anthropogenic factors (such as previous exposure to potentially toxic chemicals).

II. Disease

Disease of one kind or another is a frequent cause of mortality in marine mammal populations, but highly infectious disease organisms (particularly viruses) can cause the death of large numbers of animals in a very short period. An influenza virus probably caused the death of at least 450 harbor seals (*Phoca vitulina*) along the New England coast of the United States in 1979–1980. The family of viruses that is most often associated with disease-induced die-offs, however, is the morbilliviruses. Measles is the most familiar virus in this family. Canine distemper virus (probably contracted from domestic dogs or farmed mink) caused the death of several thousand Baikal seals (*Pusa sibirica*) in the Russian Federation in 1987–1988. The closely related phocine distemper virus caused the death of 18,000 harbor seals in the North Sea during 1988, and a dolphin morbillivirus caused the death of several thousand striped dolphins (*Stenella coeruleoalba*) in the Mediterranean Sea between 1990 and 1992. The death of more than 740 bottlenose dolphins (*Tursiops truncatus*) along the Atlantic coast of the United States in 1987 was initially attributed to poisoning by algal toxins (see later), but more recent evidence suggests that this was also caused by dolphin morbillivirus. Morbilliviruses tend to suppress their host's immune system, thus increasing the risk of secondary infection by a wide range of disease agents. These secondary infections are often the final cause of death for an individual infected with a morbillivirus, which can make it difficult to diagnose the real cause of a die-off.

III. Toxins

Some species of single-celled algae (notably diatoms and dinoflagellates) produce poisonous compounds (known as phycotoxins), which can accumulate in any fish or invertebrates animals that eat them. When environmental conditions are particularly suitable, these organisms multiple rapidly, creating "blooms." The resulting high concentrations of phycotoxins can cause mass mortalities of fish and fish predators. The best documented event of this kind involving marine mammals was the death of over 400 California sea lions (*Zalophus californianus*) along the central California coast during May and June 1998. This coincided with a bloom of an algal diatom, which is known to produce domoic acid, a dangerous neurotoxin, in the same area. Domoic acid was detected in northern anchovies, which are plankton feeders and a well-known prey of sea lions, and in the body fluid of sick sea lions. These sea lions also showed many of the neurological symptoms commonly associated with domoic acid poisoning.

Toxins produced by dinoflagellate protozoa have been implicated in the deaths of Hawaiian monk seals (*Monachus schauinslandi*) in 1978, in the death of 14 humpback whales (*Megaptera novaeangliae*) in Cape Cod Bay in 1987, and in the death of large numbers of manatees (*Trichechus manatus*) in Florida in 1982 and 1996.

In May and June 1997 the bodies of over 100 Mediterranean monk seals (*Monachus monachus*) were found along a short stretch of the West African coast near the border between Mauritania and the former Western Sahara. Initial investigations revealed that at least some of these individuals had been infected with a morbillivirus, which most closely resembled dolphin morbillivirus, and the mass die-off was originally attributed to this agent. However, most of the seals died quickly with few, if any, overt signs of disease. This was very different from what had been observed in other morbillivirus-induced events. Subsequent analysis provided evidence of the presence of several phycotoxins in dead seals and high concentrations of a dinoflagellate known to produce at least one of these phycotoxins in local coastal waters. It is possible that both agents were involved in the die-off. On the basis of the evidence currently available, it is not possible to say with any confidence that either was responsible.

IV. Environmental Effects

Unusual environmental conditions can cause high levels of mortality, particularly among young animals. For example, severe storms coupled with unusually high tides during the winter of 1982–1983 resulted in the death of up to 80% of all northern elephant seal (*Mirounga angustirostris*) pups born at some California colonies.

Even more dramatic effects can be caused by changes in oceanographic conditions. For example, El Niño southern oscillation events can dramatically alter the availability of prey species around marine mammal colonies. The severe 1982–1983 El Niño event had wide-ranging effects on fur seal and sea lion colonies throughout the eastern Pacific. Its effects were even evident in populations of seals in the Antarctic.

An intrusion of low oxygen content water into the coastal waters of Namibia in early 1994 resulted in a massive reduction in the availability of fish. Colonies of Cape fur seals (*Arctocephalus pusillus pusillus*) in Namibia suffered the highest levels of pup mortality ever observed: approximately 120,000 pups had died by the end of May 1994. There was also very high mortality among those subadult males that remained on the breeding grounds.

V. Anthropogenic Effects

Deliberate killing by humans can cause mass die-offs of marine mammals, but the effects of such activities are dealt with elsewhere. However, human activities can also result in mass die-offs through indirect effects. The most obvious of these is the exposure of marine mammals to harmful chemicals that are spilled or discharged into the marine environment. Although marine mammals are probably less vulnerable to the effects of oil spills than seabirds, species such as the sea otter (*Enhydra lutris*), which rely on their dense fur for insulation, can be seriously affected. Indeed, it is estimated that 3500 to 5500 sea otters died in 1989 after the tanker *Exxon Valdez* spilled 42 million liters of oil in Prince William Sound, Alaska. Other chemicals may have a more insidious effect. A wide range of man-made halogenated organic compounds are preferentially soluble in fat and can accumulate at high concentrations in the blubber of predatory marine mammals. These compounds can lower resistance to disease, and individuals with high tissue levels may be particularly vulnerable during mass die-offs caused by disease agents. High organochlorine levels may well have been a contributory factor to mortality during the morbillivirus epidemics in the North Sea and Mediterranean.

A combination of environmental factors and fishing activity may also result in die-offs. For example, large numbers of harp seals (*Pagophilus groenlandicus*) appeared off the north coast of Norway between 1985 and 1988, probably as a result of the collapse of the stocks of capelin, an important prey species, in the Barents and Norwegians Seas. Many of these seals became entangled in fishing nets and subsequently drowned. The Norwegian government compensated fishermen for the damage this caused, and these compensation statistics provide an estimate of the minimum number of seals that drowned. Compensation for 79,000 seals was paid in 1987 and 1988.

VI. Effects on Populations

Determining the effect of mass die-offs on the marine mammal population is often difficult because baseline data on the size and status of the affected population are lacking. However, their impact can be substantial. For example, the morbillivirus-induced die-off in the North Sea killed approximately 40% of the harbor seal population, and local mortality rates were as high as 60%. The mass die-off of Mediterranean monk seals in 1997 killed 70% of the local population and about one-third of the world population of this species.

Some marine mammal populations have shown a remarkable ability to recover from the effects of die-offs. The North Sea harbor population returned to its pre-epidemic level within 10 years of the seal die-off, although some local populations (e.g., those along the eastern seaboard of England) are still depleted. However, species whose populations are already small (such as the Mediterranean monk seal, whose world population was less than 1000 individuals at the time of the die-off) may be reduced to such low levels that they are more susceptible to other problems associated with low population size (such as increased levels of inbreeding and the loss of genetic diversity through genetic drift). As a result, their risk of extinction may be increased substantially.

VII. Future Trends

There are a number of grounds for predicting that the frequency of mass die-offs will increase during this century. Certainly marine mammals populations, like many other species, will be exposed ever more frequently to novel pathogens as a result of the increased movement of humans and their domestic animals who can act as vectors for these agents. Global warming is also likely to lead to new movement patterns, which will increase the exchange rate of these pathogens.

Exposure to toxins, particularly those produced by dinoflagellate algae, is also likely to increase. Levels of nutrients and minerals, which are normally in short supply, are periodically increased in coastal waters by large-scale run-off of rainwater from agriculture land. This occurs more frequently now because of modern drainage techniques and can create conditions that are favorable to algal blooms. In addition, dinoflagellates are particularly well adapted to transportation in the ballast water of large ships because, when conditions are unfavorable, they become encased in a protective cyst. As a result, many species that historically had a very restricted distribution now have a global one. For example, *Gymnodinium catenatum* (one of the species implicated in the mass die-off of Mediterranean monk seals) was, until recently, confined to the east coast of the United States, but since 1970 it has been recorded in Japan and Australia, as well as off the west coast of Africa.

See Also the Following Articles

Health ▪ Pathology ▪ Pollution and Marine Mammals

References

Geraci, J., Harwood, J., and Lounsbury, A. (1999). Marine mammal die-offs: Causes, investigations and issues. *In* "Conservation and Management of Marine Mammals" (J. Twiss and R. Reeves, eds.), pp. 367–395. Smithsonian Institute Press, Washington, DC.

Harwood, J. (1998). What killed the Mediterranean monk seals? *Nature* **393**, 17–18.

Scholin, C., Gulland, F., Doucette, G., Benson, S., Busman, M., Chavez, F., Cordaro, J., Delong, R., De Vogelaere, A., Harvey, J., Haulena, M., Lefebvre, K., Lipscomb, T., Loscutoff, S., Lowenstine, L., Marin, R., Miller, P., McLellan, W., Moeller, P., Powell, C., Rowles, T., Silvagni, P., Silver, M., Spraker, T., Trainer, V., and Van Dolah, F. (2000). Mortality of sea lions along the central California coast linked to a toxic diatom bloom. *Nature* **403**, 80–83.

Mating Systems

SARAH L. MESNICK
Southwest Fisheries Science Center,
La Jolla, California

KATHERINE RALLS
National Zoological Park, Smithsonian Institution,
Washington, DC

Animal mating systems are diverse, complex, and variable. The key to understanding this variation in mating systems is the realization that individuals behave so as to maximize their lifetime reproductive success and that males and females maximize reproductive success in different ways. In general, males tend to maximize reproductive success by mating with as many females as possible to increase the number of offspring they sire. In contrast, females can produce only a limited number of young and tend to maximize their reproductive success by producing and successfully rearing high-quality offspring. Males and females are thus subject to different selective pressures and do not necessarily cooperate. They may, in fact, be in direct conflict over mating arrangements or try to deceive one another. Together, the mating strategies of males and females define the mating system of a species. This article presents an overview of male and female mating strategies and describes how the different groups of marine mammals solve the problem of finding mates.

Mating systems have traditionally been categorized on the basis of three criteria: the number of partners each sex copulates with during the breeding season, whether the male and female form pair bonds, and how long these bonds are maintained. We focus on a single criterion, the number of mating partners for each sex during a breeding season, to facilitate comparisons among taxa. Thus, we define four mating systems: *monogamy,* in which each individual has a single partner; *polygyny,* in which some males have two or more partners; *polyandry,* in which some females have two or more partners; and *polygynandry,* in which some males and some females have multiple partners.

In marine mammals, as in most other mammals, maternal gestation and lactation provide the majority of nutrient requirements for developing young and males often contribute nothing but sperm to their offspring. Consequently, males are free to devote the bulk of their time and energy to competing for access to receptive females. Thus, theory predicts that most marine mammals should be polygynous.

The potential for polygyny, and the extent to which that potential is realized, is determined to a great extent by the degree to which receptive females are aggregated in space and time. The distribution of females, in turn, is determined by phylogenetic constraints (such as the retention of terrestrial birthing in pinnipeds) and by ecological and social conditions. Among the most important of these conditions are the distribution of resources necessary for breeding, predation pressure, and the costs and benefits of group living. Several types of polygynous mating systems have been identified in marine mammals. Males may defend resources that are vital to females, such as parturition sites (resource defense polygyny), or they may defend females directly either simultaneously (female defense polygyny) or sequentially (sequential female defense polygyny). They may aggregate on traditional display sites and advertise for females (lekking) or they may search widely and spend little time with females except to mate (roving).

It is becoming increasingly clear that females of many species mate with more than one male during a breeding season. Polygynandry is the most accurate term for these mating systems where at least some members of both sexes mate with more than one individual. However, the reproductive success of a female marine mammal does not vary with the number of times she mates per season, as she usually gives birth to only a single offspring and therefore can be inseminated by only one male. Thus, some marine mammals have polygynandrous mating systems but polygynous fertilization patterns.

I. Male Mating Strategies

Male competition for access to mates takes at least three general forms: aggressive interactions to limit the access of other males to females (contest competition), competition to disperse and find sexually receptive females (scramble competition), and competition in courtship to be chosen by the female (mate choice competition). Each of these behavioral strategies has, in turn, generated a number of structures and physiological adaptations. For example, marine mammal males are often distinguished by large body size, big canines, or tusks that can be used as weapons in combat with other males (Fig. 1). In most mammalian species, males disperse more widely than females, and there is increasing evidence that the pattern holds in some marine mammals [e.g., Dall's porpoise (*Phocoenoides dalli*), Escorza-Treviño and Dizon (2000); beluga (*Delphinapterus leucas*), O'Corry-Crowe *et al.* (1997)]. Males are typically more active in courtship and are the more conspicuous and ornamented sex. They may attempt to entice and attract females through visual, acoustic, and pheromonal displays (Fig. 2).

More recently, several additional forms of male competition have been described. Males may attempt to outcompete other males by producing higher quality or greater quantities of sperm or by removing other male's sperm (sperm competition). When a male cannot monopolize access to females on his own, males may cooperate and form alliances. These alliances effectively act as a single male in competing for access to females and have been described for Indian Ocean bottlenose dolphins from Shark Bay, Australia (*Tursiops aduncus;* Connor *et al.* 1996), and common bottlenose dolphin from Sarasota Bay, Florida (*T. truncatus;* Wells *et al.,* 1987). The existence of male alliances is suspected in a growing number of other marine mammal species. Males may also form consortships with females in which the male attempts to associate and copulate with a female during the presumptive fertile period. In common bottlenose dolphins, such consortships often correlate with the later birth of offspring (Wells *et al.,* 1999). In species in which males are larger than females, possess dangerous weapons, and aggressively pursue copulation, some males may forcibly coerce the female to

Figure 2 *A singing adult male humpback whale (*Megaptera novaeangliae*). Singing by humpback males presumably acts to attract females, although whether songs contain cues to mate quality remains in dispute. Singing may also function to space males in a breeding area or to aid in the establishment of dominance hierarchies. © Flip Nicklin (Minden Pictures).*

mate [forced copulation; northern elephant seals (*Mirounga angustirostris*), Le Boeuf and Mesnick (1990)]. Moreover, some males "cheat." They may sneak copulations when alpha bulls are distracted [northern elephant seals; Le Boeuf (1974)] or abduct females from the territories of dominant males [kleptogyny; northern fur seals (*Callorhinus ursinus*), Gentry (1998)].

These forms of competition need not be mutually exclusive. It is likely, based on better-studied taxa, that individual males within a marine mammal population will utilize different strategies depending on their age, size, dominance rank, and the number and quality of available females. Although competition between males is obvious in many species of marine mammals, the possibility that males of some species may also prefer to mate with particular females is an area that deserves further attention.

II. Female Mating Strategies

Females of most marine mammals produce only a single offspring at a time. The interbirth interval ranges from 1 year in most pinnipeds and small cetaceans to 5, 6, or even 7 years in larger toothed whales such as sperm whales (*Physeter macrocephalus*),

Figure 1 *Adult male northern elephant seals (*Mirounga angustirostris*) fight for positions in a dominance hierarchy that confers access to receptive females. Photo by Sarah L. Mesnick.*

killer whales (*Orcinus orca*), and short-finned pilot whales (*Globicephala macrorhynchus*). Females can thus produce only a limited number of young and must maximize their reproductive potential by successfully rearing high-quality young. To give offspring a competitive edge, females can enhance their fitness by choosing males that offer resources or genetic benefits. This choice may occur either pre- or postcopulation (the latter may lead to sperm competition). At present, precopulation mate choice has been studied in only a few pinniped species and virtually nothing is known about postcopulatory choice by female marine mammals.

In other well-studied taxa, such as birds, females are highly discriminating in their choice of sexual partners. Moreover, most females choose in a similar way so that a few males achieve many copulations and many other males none. Females may choose among potential mates directly (based on resources, size, strength, dominance, or display) or indirectly (by mating with the winner of contests for access to females). Some marine mammal females actively seek out particular males and mate. For example, in California sea lions (*Zalophus californianus*), some females change pupping locations from one year to the next to remain with a territorial male who changed territory location (Heath, 1989). Females may also incite male–male competition. By protesting male sexual advances loudly, female northern elephant seals instigate fights among males and subsequently mate with the winner of these battles.

It is difficult to establish the presence of female choice and even more difficult to determine why females choose particular mates or to quantify the benefit to females of exercising choice. Direct benefits to the female, in the form of nutritional resources to the female or parental care, are not known to exist in marine mammals. Females can, however, benefit by choosing males with higher-quality territories, which provide better parturition or thermoregulatory sites, or by choosing males that give protection from harassment by subordinate males, which provides uninterrupted time for lactation and reduces vulnerability to aggression from other conspecific males. Females can also benefit by discriminating among potential mates on the basis of indirect (genetic) benefits. These include choosing males of the correct species, males with immunologically compatible genes, males with "good genes" who can produce offspring of higher quality. Females can also choose males with better fertilization ability or virility.

Females may make very different decisions regarding which males they associate with, which males they mate with, and which male sires their offspring. In land-breeding pinnipeds, for example, a female may reside with one dominant or territorial male during lactation but later leave this male to copulate with another male elsewhere (extraterritorial copulation). In some marine mammal species, such as the bottlenose dolphin, sexual behavior is a frequent and important component of nonreproductive social life and has little to do with fertilization. As with males, female strategies need not be mutually exclusive and it is likely that different females will utilize different strategies depending on their age, dominance rank, and the number and quality of available mates.

III. Mating Systems in Different Groups
A. Pinnipeds

For all pinnipeds studied to date, data support, or are highly suggestive of, a polygynous mating system. Pinnipeds are predisposed to polygyny because they give birth on land, which results in the spatial clustering of females, and have an annual birthing cycle, which results in reproductive synchrony among females. The degree of polygyny varies both within and among species with the extent of reproductive synchrony and spatial clustering. Most species have a peak availability of receptive females lasting about 1 month, but the availability of receptive females ranges from 10–15 days in harp (*Pagophilus groenlandicus*) and hooded (*Cystophora cristata*) seals to a period of several months for species that breed in tropical habitats such as monk seals (*Monachus* spp.) and Galapagos sea lions (*Zalophus wollebaeki*).

Variation in the degree of spatial clustering within and among species is due to a variety of factors, including the spatial distribution of suitable breeding sites, whether mating takes place on land or at sea, the intensity of male harassment, predation pressure and/or thermoregulatory needs. Polygyny and sexual dimorphism are generally much more extreme in species that mate on land than in those that mate in the water.

1. Otariids Otariid females feed during lactation. Lactation is energetically costly so females must raise their young on sites near highly productive marine areas. Because these sites are limited, females typically occur in dense aggregations (numbering from a few individuals to more than a thousand) on beaches or rocky shelves on islands. Mating occurs on land, although evidence of at least some mating at sea exists for a few species [e.g., Juan Fernández fur seal (*Arctocephalus philippii*) and the California sea lion]. The combination of dense female aggregations and terrestrial mating gives some males the opportunity to monopolize mating with many females. Sexual dimorphism among otariids is correspondingly extreme; males are on average three times, and sometimes up to six times, heavier than females and have other traits favored in physical combat over females: large canines, thick chests, and dense manes.

The northern fur seal is among the most polygynous of the otariids: a single male at the St. George Island rookery mated with 161 females and hundreds of males had no copulations at all (Gentry, 1998). The lowest levels of polygyny probably occur in the Juan Fernández fur seal, the South American sea lion (*Otaria flavescens*), the Galapagos fur seal (*Arctocephalus galapagoensis*), and Hooker's sea lion (*Phocarctos hookeri*), in which the ratio of sexually active adults ranges from two to six females per male (Boness *et al.*, 1993).

Male otariids typically defend territories containing resources needed by females—parturition and thermoregulatory sites—rather than individual females (Fig. 3). However, female defense has been demonstrated in at least one otariid, the South American sea lion. The two types of polygyny are difficult to distinguish and are not necessarily mutually exclusive. There is some evidence suggestive of lekking in three species (the California sea lion, the South American fur seal (*Arctocephalus australis*), and Hooker's sea lion), although this interpretation remains controversial.

A male's ability to acquire and defend a territory depends on his size and age, his ability to compete with other males, and his ability to fast during his tenure (contest competition). Un-

Figure 3 *A Steller sea lion* (Eumetopias jubatus) *territory. Adult males defend resource-based territories that encompass female parturition and thermoregulatory sites. Females choose among males in a surprisingly consistent way. As a result, some males holding territories never or rarely mate, while a few males mate with many females. Photo by Robert L. Pitman.*

der most circumstances, the boundaries of territories are fixed and are delineated by breaks in the topography. Males use a species-specific threat display when defending the boundaries of their territory. A male that secures a territory will probably, but not necessarily, mate with many of the females that give birth on his territory. Climate and rookery topography also play important roles in determining a male's mating success. Those males defending territories containing access to the water, tide pools, or shade acquire a disproportionately large number of females.

Most otariid bulls fast while maintaining their territories, sometimes for the entire 2- to 3-month breeding season [e.g., Steller sea lions (*Eumetopias jubatus*) and northern fur seals]. Some males return to the same territory in subsequent years whereas others move to new territories or are not seen again. Territorial males may try to herd females to prevent them from leaving their territories, but in most species, females determined to leave generally can. The males of some species, however [e.g., northern fur seal and the South American and Australian (*Neophoca cinerea*) sea lion], are able to prevent females from leaving their territories by threats, herding, and sometimes physical aggression leading to injury.

The importance of male courtship displays in otariids is not well understood. For example, it is not known whether male displays, such as the incessant barking of male California sea lions, are used as threat displays for males, or as displays for females, or both. Alternative male mating strategies are widespread and generally thought to be practiced by subadult or subordinate males. These include gang raids by groups of nonterritorial males (up to 40) to abduct or mate with females in the main breeding territories (South American and Australian sea lions), males stealing females from the territories of their neighbors (kleptogyny; northern fur seals), and males trying to sneak copulations (several species). How successful these strategies are in inseminating females is not known.

Female mating strategies are less well understood than male mating strategies, but several lines of evidence suggest that females exercise more choice among males than previously suspected. Female otariids choose which territory to haul out in and usually, but not always, move freely in and out of a male's territory. Estrus occurs within 1 to 2 weeks postpartum in all but one species (California sea lion, about 21–27 days). When it is time to mate, females may leave the male's territory in which they have given birth and mate with another male. This has been documented through behavioral observations in California sea lions and South American fur seals and by paternity studies in fur seals on Macquarie Island.

Climate, ROOKERY topography, and the intensity of male harassment influence the ability of females to exercise mate choice. Females of species breeding in hot climates have more opportunity for mate choice due to their frequent thermoregulatory movements between their birthing site and the water. Intense male herding restricts female choice and may injure females. Female northern fur seals are thought to successfully reduce the risk of injury from males by forming dense aggregations and competing for central locations within these groups, which minimizes contact with males, and by acting submissive around males. In this species, females do not appear to choose males directly. Rather, by gathering on traditional mating grounds, the result is that males fight and females subsequently mate with the winners of these contests. Female otariids may also directly solicit and initiate copulation from males. In Steller sea lions, for example, females gain the sexual attention of males by lateral neck swings, dragging of the hindquarters, and sinuous movements of the female's body against the male's body. While females tend to direct most solicitation behavior toward the older "proven" territorial males, the extent of female choice remains unclear. Multiple mating is known in 30% of otariids studied (Boness *et al.*, 1993) and suggests an important and variable role across species for sperm competition and mate choice.

2. Phocids Most phocid females fast during a short and concentrated lactation period, utilizing energy stored as fat before parturition. Because phocid females are not dependent on concentrated marine production during lactation like otariid females, they can mate in more dispersed locations. Moreover, in 16 of the 19 phocid species, the majority of mating takes place in the water near or after the end of lactation. Females of many of these species give birth on ice and do not aggregate as densely as those of terrestrially breeding species. Thus, during the breeding season, females are dispersed spatially (solitary or dispersed in small- to moderate-sized well-spaced colonies) and mobile during mating. Because males have less opportunity to defend and mate with multiple females, aquatically mating phocids are less polygynous and sexually dimorphic than otariids.

Moreover, the breeding season is short and, in species that breed on ice, mating takes place when temperatures are well below freezing. Reverse sexual dimorphism, with females larger than males, occurs in several species. Large female size may help a mother provide greater quantities of fat-rich milk to her pup and protect her from low polar temperatures. Small size in males is thought to facilitate agility underwater, where males may defend territories and mate with females. Nevertheless, aquatically mating species are considered to be slightly or moderately polygynous. Mating takes place within a few days of the weaning of the pup.

We have limited knowledge of male and female mating strategies in most aquatically mating phocids. In some cases, males defend the lactating female and her vicinity directly, a strategy akin to roving and sequential defense of a single female or a small group of females. A typical group consists of a female and her pup and an adult male who may have to wait before the female comes into estrus and is receptive to mating. Presumably, the male will mate with the female when she enters the water after weaning her pup. This system can be described as sequential polygyny, as males may leave after mating to search for another receptive female. It occurs in crabeater seals (*Lobodon carcinophaga*), spotted seals (*Phoca largha*), and hooded seals. Other males may surround these "triads" and they may compete for access to the female, typically with threats and sometimes bloody fights.

In some cases, males appear to defend aquatic territories (called "maritories") off the beach or ice where females reside. Males spend considerable time in these territories giving vocal and visual display. This characterizes such phocids as bearded (*Erignathus barbatus*), harp, and Weddell (*Leptonychotes weddellii*) seals and, in some cases, harbor seals (*Phoca vitulina*). A male may mate with any receptive female that enters his territory. Genetic studies of harbor seals on Sable Island show that male success is moderate to low, with most males fertilizing one or no females and the maximum number of females fertilized for any male being five (Coltman *et al.*, 1998).

Males of many aquatically mating species are thought to use visual and acoustic displays to threaten other males and to attract females. "Eerie but melodious" songs have been described for male bearded seals and "trills," "knocks," "buzzes," and "chirps" for male Weddell seals. Male hooded seals make numerous sounds underwater and also produce sounds in air as they inflate and deflate their hood and red nasal sac. In ringed seals (*Pusa hispida*), there is much social interaction and sniffing between males and females (males have a strong odor during the breeding season), which raises the question of whether there may be preferred mates. Virtually nothing is known of female mating strategies among aquatically mating phocids.

The northern and southern (*Mirounga leonina*) elephant seal and some populations of the gray seal (*Halichoerus grypus*) are unusual among phocids in that mating takes place on land. These species exhibit a form of female defense polygyny. Males maintain a position near a receptive female or females and attempt to exclude other males from their vicinity. In elephant seals, males use visual and acoustic threats as well as physical fighting to compete for dominance in a social hierarchy that confers access to females. Polygyny in elephant seals is extreme; at the Año Nuevo rookery in California, as few as five males may be responsible for 48 to 92% of the copulations observed during a breeding season (Le Boeuf, 1974). The lifetime reproductive success of most males is nil or low. Many die before reaching breeding age and higher-ranking males prevent some of those that survive from breeding.

Genetic analyses confirm that the proportion of pups sired by alpha males is consistent with that expected from observed mating success in southern elephant seals but show that behavioral observations overestimate the success of some northern elephant seal alpha males (Hoelzel *et al.*, 1999). The relatively lower success of northern elephant seal males was probably due in part to the behavior of the Año Nuevo females, which copulate more frequently, the greater success of non-alpha males, and/or reduced fertility of specific alpha males. Female elephant seals may exercise mate choice by competing for central positions in harems where dominant males reside and by inciting male–male competition and subsequently mating with the winner of these battles.

Mating behavior among the geographically widespread gray seal is difficult to categorize. Gray seal females do not cluster as tightly and are more mobile in the colony than elephant seal females. Dominant males maintain their proximity to females by using visual threat displays and occasional fights to deter other males. In the Scottish Islands, behavioral observations suggest a classical polygynous system. Genetic studies, however, reveal that many fathers spend little time at shore, that some pairs of seals show partner fidelity, and that dominant males to not father as many offspring as behavioral observations would suggest (Worthington Wilmer *et al.*, 2000).

3. Walrus Walruses (*Odobenus rosmarus*) have the most elaborate courtship displays of all pinnipeds. Walruses show marked sexual size dimorphism and are thought to be strongly polygynous. Atlantic walruses in the Canadian High Arctic exhibit a mating system that resembles female defense polygyny. Pacific walruses in the Bering Sea may have a lek-like mating system. Groups of males cluster around females, which form dense aggregations on pack ice. Males are aggressive toward one another and produce intricate visual and vocal displays, consisting of barks, whistles, growls, and underwater bell-like sounds. The massive tusks of the male walrus also appear to play an important role as a symbol of rank (to threaten other males) and as a visual signal to females, who may choose among males partly by the size of their tusks.

B. Cetaceans

1. Odontocetes In contrast to pinnipeds, which are relatively sedentary and clustered during the breeding season, female odontocetes are mobile and dispersed. This has two important consequences for male mating strategies: males have less opportunity to control access to aggregated females and less assurance of paternity. It is not surprising, therefore, that the basic mating strategy of male odontocetes appears to be one of searching for receptive females and spending little time with them other than to mate. It is likely that mate guarding, or monopolization of females long enough to ensure conception, also occurs, although the phenomenon has been well documented only in bottlenose dolphins.

Female mating strategies in odontocetes are little understood. Given their mobility and three-dimensional habitat, it is generally thought that females are able to exercise choice by outmaneuvering males or by rolling belly-up. Observational and hormonal evidence suggests that females of several species copulate frequently both during and outside the breeding season and may be polyestrous. Frequent copulation may function to induce sperm competition, aid in assessing future mates, or help to establish social bonds with potential future partners. In many odontocete species, sexual behavior is an important component of nonreproductive social interactions and often has little to do with fertilization, making it difficult to infer mating strategy from incomplete observations. Perhaps the best example is the intriguing "wuzzling" behavior of Hawaiian spinner dolphins (*Stenella longirostris longirostris*). Wuzzling refers to interweaving masses of caressing and copulating dolphins of both sexes and all ages, which are especially common in the summer months, when many females come in estrus. Is the behavior social? Sexual? Both?

We know little about mating systems in the vast majority of odontocetes. However, there are substantial data on bottlenose dolphins, sperm whales, and killer whales. A mating system of female defense or sequential defense polygyny has been suggested for Indian Ocean bottlenose dolphins. Males form stable coalitions of a few males that may work alone or with other closely associated coalitions to form temporary consortships with individual females, often through aggressive herding (Shark Bay, Australia; Connor *et al.*, 1996). Male common bottlenose dolphin individuals and members of long-term pair bonds form temporary consortships with females without obvious aggressive herding (Sarasota Bay, Florida; Wells *et al.*, 1999). The extent to which this sequential female defense strategy is successful is uncertain, however, as individual females can cycle multiply and associate with several males during the season in which they conceive. These behaviors may facilitate female mate choice and promote sperm competition. In another location, male common bottlenose dolphin apparently do not form alliances or aggressively herd females, although single males may accompany groups of females throughout the breeding season (Moray Firth, Scotland; Wells *et al.*, 1999). Among sites, the level of male bonding may be inversely related to male body size and the degree of sexual dimorphism (Tolley *et al.*, 1995). At sites where animals are small, males may form alliances to gain and maintain access to females; where animals are large, males can do this on their own.

Most sperm whales in the Galápagos Islands appear to rove between groups of females searching for potential mates. One or more large mates may attend a group of females (sometimes simultaneously) for short periods of time ranging from a few minutes to several hours (Fig. 4). Rather than herding female groups, females have been observed to alter course and speed so that they could join a large male hundreds of meters away. Males did not interact aggressively with each other within female groups, despite several accounts in the literature of males fighting outside of groups. Given the apparent roving strategy of males, the role of the tremendously large nose of the male sperm whale and its possible use as a sound-generating organ remains unclear. The loud clicks may function in male–male competition or advertisement to attract females (Cranford, 1999).

Pods of resident killer whales in the Pacific Northwest are frequently observed associating with one another in the summer months when prey (and observer) abundance is high. In these multipod groups, there is much sexual activity among all pod members, young and old alike. Because no dispersal of either males or females occurs from resident pods, it is thought that mat-

Figure 4 *Adult male sperm whales* (Physeter macrocephalus) *rove among female groups searching for receptive individuals and staying with each group for only a few hours at a time.* © *Flip Nicklin (Minden Pictures).*

ing takes place during these encounters. Considering that the entire pod engages in these encounters, it is likely that their function is both sexual and social. Similarly, genetic analyses of long-finned pilot whales captured in a Faroese fishery indicate that males remain in their natal groups but do not mate within them (Amos *et al.*, 1993). Young were sired by males not captured with the group, implying that pilot whales must mate when two or more groups meet or when adult males pay brief visits to other groups.

Very little is known about mating systems in the remaining species of toothed whales. However, we can infer something about the mating strategies of these species from the type and degree of sexual dimorphism and its association with other characteristics, such as bodily scarring and relative testis size. For example, testis size ranges dramatically among odontocete species, from less than 0.05% [several *Mesoplodon* species, the franciscana (*Pontoporia blainvillei*), the baiji (*Lipotes vexillifer*), and sperm whale] to 5% or greater [harbor porpoise (*Phocoena phocoena*), finless porpoise (*Neophocaena phocaenoides*), and dusky dolphin (*Lagenorhynchus obscurus*)]. These data suggest the importance of sperm competition in several odontocete species, especially among some of the delphinids and porpoises. The importance of mate choice competition, attempts to entice and attract females through elaborate displays, is suggested by differences between the sexes in song and exaggerated visual signals such as the postanal hump or en-larged dorsal fins. At present, sexually dimorphic acoustic signals are known only in sperm whales. However, because odontocetes produce a wide range of sounds, acoustic displays are likely to occur in several other species as well. The importance of contest competition for access to mates is suggested by sexual dimorphism in size, weaponry (teeth and tusks), and the presence of scarring of conspecific origin (tooth rakes). Sperm whales, the beaked whales, narwhal, and bottlenose whale exhibit these traits.

2. Mysticetes　Among the mysticetes, substantial data on breeding behavior exist only for the humpback, (*Megaptera novaeangliae*), right (*Eubalaena* spp.), and gray (*Eschrichtius robustus*) whales. Even in these species, virtually nothing is known about female behavior. The humpback whale has been studied most intensively. Male humpbacks adopt one or more of three primary strategies: display by singing long, complex songs; direct competition with other males for females in "competitive groups"; and escort of females, including those with newborn calves. Males escorting females are most likely waiting for mating opportunities or guarding females after copulation. Two secondary strategies, roving and sneaking, have also been suggested. The relative importance and success of each of these strategies are unknown.

Female humpbacks sometimes aggressively reject subadult males and they may incite competition among males. Although molecular analysis of paternity has shown that females are mated by different males between years (Clapham and Palsbøll, 1997), it is unknown whether females mate multiply within a given breeding season. Singing by male humpbacks is an intriguing phenomenon, as songs change over time, yet all members of a population sing essentially the same song at any one time. Singing by humpback males presumably acts to attract females, although whether songs contain cues to mate quality remains in dispute. Singing may also function to space males in a breeding area or to aid in the establishment of dominance hierarchies. Whether the aggregation and displaying of humpback whales at specific sites constitutes lekking also remains controversial.

Little else is known about the mating systems or other balaenopterid whales. Blue (*Balaenoptera musculus*) and fin (*B. physalus*) whales seem to be widely dispersed during the winter breeding season. Male fin whales have a patterned call, which has been termed a breeding display. The question of whether male blue whales have specific calls that may function as mating displays is currently under investigation.

Right whales show sexual activity throughout the year, although calving is strongly seasonal. Because the gestation period is 1 year and there is no evidence of diapause, mating leading to conception presumably occurs primarily in the winter. The function of sexual activity during other seasons is unknown. Observations of multiple male right whales mating with single females, together with the huge (1 ton!) testes, strongly suggest that sperm competition is a principal mating strategy in these species, and also probably in bowhead (*Balaena mysticetus*) and gray whales (Brownell and Ralls, 1986). The level of aggression in male–male interactions in these species is low compared to that observed in humpback whales, data consistent with the predominance of sperm competition as a mating strategy.

C. Sirenians, Sea Otters, and Polar Bears

Male manatees (*Trichechus* spp.) and dugongs (*Dugong dugon*) tend to be solitary and search for potential mates by roaming over large areas that include the home ranges of several females. Groups of males sometimes follow and try to mate with a single female, forming a "mating herd." In both manatees and dugongs, males in these herds threaten and fight with each other but it is still unknown whether this behavior is a form of scramble competition or is more akin to a type of lekking. In Shark Bay, Australia, dugongs associate in a more classical kind of lekking, with several males patrolling exclusive areas and engaging in activities usually indicative of both male competition and mate attraction, including acoustic signaling. In both manatees and dugongs, the mating season extends over several months and sexual dimorphism is slight. Interbirth intervals are at least 2 years and may be as much as 5 in some cases.

Female sea otters (*Enhydra lutris*) typically give birth annually. Births generally peak in the spring, although females in warmer areas may give birth in any month. Adult males are larger than females. Male sea otters establish territories, usually overlapping one or more female home ranges, that contain food resources and sheltered resting places. Males may defend territories seasonally or all year. Other males congregate in groups outside of the areas occupied by territorial males. Courtship and mating, as are typical for many mustelid species, are rough and females may be injured or killed by males. Copulation occurs with both the male and the female on their backs near the water's surface. The male grasps the female's head or jaws, including the nose, in his own jaws. Recently mated female typically have red, swollen noses. After mating, the pair may stay together for a few days in which they feed, groom, play, and rest in close company.

Polar bears (*Ursus maritimus*) are highly sexually dimorphic and polygynous; adult males may be over twice as heavy as adult females. Female polar bears have extensive home ranges, and males travel over large areas when searching for mates. Males apparently fight among themselves for access to females. Specific courtship behaviors are lacking or are yet undescribed. The largest and strongest males apparently do most of the mating, while other males sometimes wait in the distance. Polar bears are notable among marine mammals in that they are the only species in which females give birth to multiple young (one or two is the most common litter size and rarely three or four). The interbirth interval is about 3 years.

IV. Mating System Studies and the Future

This is an exciting time for the study of marine mammal mating systems as technical advances such as the use of molecular markers and underwater acoustic and visual recording devices are providing new insights and making it possible to investigate previously inaccessible species. The results of molecular studies have confirmed some hypotheses regarding marine mammal mating systems and refuted others. Paternity analyses in a number of species are revealing that dominant males are not as successful in siring offspring as expected from behavioral observations alone. These results suggest that female choice plays a more important role than was previously suspected. Female mating strategies, such as the incitation of male–male competition, extraterritory copulation, and the promotion of sperm competition by frequent mating, are gaining increasing attention. As our understanding of the physiology of female receptivity grows, we will be better able to interpret both female and male mating behavior. At the same time, our increasing ability to hear and see underwater will enable us to tap into the little known realm of underwater acoustic and visual displays.

See Also the Following Articles

Aggressive Behavior, Intraspecific ■ Breeding Sites ■ Courtship Behavior ■ Sexual Dimorphism ■ Territorial Behavior

References

Amos, B., Schlötterer, C., and Tautz, D. (1993). Social structure of pilot whales revealed by analytical DNA profiling. *Science* **260**, 670–672.

Boness, D. J., Bowen, W. D., and Francis, J. M. (1993). Implications of DNA fingerprinting for mating systems and reproductive strategies of pinnipeds. *Symp. Zool. Soc. Lond.* 66–93.

Brownell, R. L., and Ralls, K. (1986). Potential for sperm competition in baleen whales. *In* "Behavior of Whales in Relation to Management" (G. P. Donovan, ed.), Special Issue 8, pp. 97–112. Reports of the International Whaling Commission, Cambridge.

Coltman, D. W., Bowen, W. D., and Wright, J. M. (1998). Male mating success in an aquatically mating pinniped the harbor seal (*Phoca vitulina*), assessed by microsatellite DNA markers. *Mol. Ecol.* **7**, 627–638.

Clapham, P. J., and Palsbøll, P. J. (1997). Molecular analysis of paternity shows promiscuous mating in female humpback whales (*Megaptera novaeangliae*, Borowski). *Proc. R. Soc. Lond. B* **264**, 95–98.

Connor, R. C., Richards, A. F., Smolker, R. A., and Mann, J. (1996). Patterns of female attractiveness in Indian Ocean bottlenose dolphins. *Behavior* **133**, 37–69.

Cranford, T. W. (1999). The sperm whale's nose: Sexual selection on a grand scale? *Mar. Mamm. Sci.* **15**(4), 1133–1157.

Escorza Treviño, S., and Dizon, A. E. (2000). Phylogeography, intraspecific structure and sex-biased dispersal of Dall's porpoise, *Phocaenoides dalli*, revealed by mitochondrial and microsatellite DNA analyses. *Mol. Ecol.* **9**.

Gentry, R. L. (1998). "Behavior and Ecology of the Northern Fur Seal." Princeton Univ. Press, Princeton, NJ.

Goldsworthy, S. D., Boness, D. J., and Fleischer, R. C. (1999). Mate choice among sympatric fur seals: Female preference for conphenotypic males. *Behav. Ecol. Sociobiol.* **45**, 253–267.

Heath, C. B. (1989). "The Behavioral Ecology of the California Sea Lion." Ph.D. Thesis, University of California, Santa Cruz.

Hoelzel, A. R., Le Boeuf, B. J., Reiter, J., and Campagna, C. (1999). Alpha-male paternity in elephant seals. *Behav. Ecol. Sociobiol.* **46**, 298–306.

Le Boeuf, B. J. (1974). Male-male competition and reproductive success in elephant seals. *Am. Zool.* **14**, 163–176.

Le Boeuf, B. J., and Mesnick, S. L. (1990). Sexual behavior of male northern elephant seals. I. Lethal injuries to adult females. *Behaviour* **116**(1-2), 143–162.

O'Corry-Crowe, G. M., Suydam, R. S., Rosenberg, A., Frost, K. J., and Dizon, A. E. (1997). Phylogeography, population structure and dispersal patterns of the beluga whale *Delphinapterus leucas* in the western Nearctic revealed by mitochondrial DNA. *Mol. Ecol.* **6**, 955–970.

Tolley, K. A., Read, A. J., Wells, R. S., Urian, K. W., Scott, M. D., Irvine, A. B., and Hohn, A. A. (1995). Sexual dimorphism in wild bottlenose dolphins (*Tursiops truncatus*) from Sarasota, Florida. *J. Mammal.* **74**(4), 1190–1198.

Wells, R. S., Boness, D. J., and Rathbun, G. B. (1999). Behavior. *In* "The Biology of Marine Mammals" (J. E. Reynolds III and S. Rommell, eds.), pp. 324–422. Smithsonian Institution Press, Washington.

Wells, R. S., Scott, M. D., and Irvine, A. B. (1987). The social structure of free-ranging bottlenose dolphins. *Curr. Mammal.* **1**, 247–305.

Worthington, Wilmer J., Overall, A. J., Pomeroy, P. P., Twiss, S. D., and Amos, W. (2000). Patterns of paternal relatedness in British grey seal colonies. *Mol. Ecol.* **9**, 283–292.

Melon-Headed Whale
Peponocephala electra

WAYNE L. PERRYMAN
Southwest Fisheries Science Center,
La Jolla, California

The melon-headed whale is one of a group of small, dark-colored whales that are often referred to as "blackfish." It is only recently that much has been known about these little whales because they generally occur far offshore and in many areas they avoid approaching vessels.

I. Characters and Taxonomic Relationships

The melon-headed whale (Fig. 1) is mostly dark gray in color with a faint darker gray dorsal cape that is narrow at the head and dips downward below the tall, falcate dorsal fin. A faint light band extends from the blowhole to the apex of the melon. A distinct dark eye patch, which broadens as it extends from the eye to the melon, is often present and gives this small whale the appearance of wearing a mask. The lips are often white, and white or light gray areas are common in the throat region and stretching along the ventral surface from the leading edge of the umbilicus to the anus. At sea, this species is difficult to distinguish from the pygmy killer whale (*Feresa attenuata*). It differs externally from the pygmy killer whale by having a more pointed or triangular head and sharply pointed pectoral fins. Both of these characters are difficult to recognize at sea unless these small whales are seen from above. Experienced observers often rely more on behavioral than physical characters to separate these two blackfish in the field. In stranded specimens, the melon-headed whale can be distinguished from all other blackfish by its high tooth count, 20 to 26 per row, compared to generally less than 15 teeth per row for pygmy killer whales.

Melon-headed whales are about 1 m in length at birth (Bryden *et al.*, 1977) and continue to increase in length until they are 13 to 14 years old. Asymptotic length for males (2.52 m) is greater than for females (2.43 m), and males also have comparatively longer flippers, taller dorsal fins, and broader tail flukes (Best and Shaughnessy, 1981; Miyazaki *et al.*, 1998). In addition, some males exhibit a pronounced ventral keel that is found posterior to the anus. The longest specimen reported was a 2.78-m female that stranded in Brazil (Lodi *et al.*, 1990). A 2.64-m male that stranded in Japan weighing 228 kg is the heaviest specimen reported (Miyazaki *et al.*, 1998).

The SKULL of the melon-headed whale is typically delphinid in shape, with the exception of a very broad rostrum and deep antorbital notches. It is similar to the skull of the common bottlenose dolphin (*Tursiops truncatus*), but the TEETH of the melon-headed whale are much smaller and appear more delicate. The high tooth count of this species separates its skull from those of other small beakless whales.

The melon-headed whale is a member of the subfamily Globicephalinae where it is closely allied with the very similar pygmy killer whale and the larger pilot whales (*Globicephala melas* and *G. macrorhynchus*). Investigations regarding the interrelations of these species have yet to produce definitive results.

II. Distribution and Ecology

Melon-headed whales are found worldwide in tropical to subtropical waters. They have occasionally been reported from higher latitudes, but these sightings are often associated with incursions of warm water currents (Perryman *et al.*, 1994). They are most often found in offshore, deep waters, and nearshore sightings are generally from areas where deep oceanic waters are found near the coast. Squids appear to be the preferred prey of this species, but small fish and shrimps have also been found in their stomachs (Jefferson and Barros, 1997).

III. Behavior and Life History

Melon-headed whales are most often found in large aggregations, a behavior that separates them from the very similar pygmy killer whale. They are often seen in large mixed aggregations with Fraser's dolphin (*Lagenodelphis hosei*). They have also been sighted in mixed herds with spinner dolphins (*Stenella longirostris*) and common bottlenose dolphins (Dolar, 1999). Although they are reported to flee from approaching vessels in the eastern Pacific, it is not uncommon for melon-headed whales to briefly ride the bow wave of passing ships in other areas. They may bow ride for longer periods if the vessel slows to a speed of a knot or less.

Mass STRANDINGS of melon-headed whales have been reported on several occasions; the cause of the strandings is unknown. In two strandings from Japan and one in Brazil, the specimens had high loads of internal PARASITES, which might have caused some animals to strand. It has also been suggested that mass strandings of these highly social animals may be caused by a panic response in the school when a few members accidentally strand (Miyazaki *et al.*, 1998).

IV. Interactions with Humans

When captured live and transferred to aquariums, melon-headed whales have not thrived and have been difficult to train. They have been aggressive toward keepers and have caused in-

Figure 1 *Melon-headed whales,* Peponocephala electra, *occur around the world in subtropical and tropical waters. Photographed by R. L. Pitman in the Gulf of Mexico.*

juries by ramming individuals with their heads or raking them with their teeth. In Hawaiian waters, melon-headed whales have approached divers in an aggressive manner, swimming rapidly and opening and closing their jaws causing an audible clapping sound. Swimmers should be cautious if entering the water around these small whales.

Melon-headed whales are taken in small numbers in harpoon and drift net fisheries in the Philippines (Dolar, 1994), Indonesia, Malaysia, and in the Caribbean near the island of St. Vincent. Schools of melon-headed whales have been taken in the drive fishery operated from the port of Taiji, Japan. On rare occasions, a member of this species is taken in the purse seine fishery for yellow-fin tuna in the eastern tropical Pacific. Because most of these fisheries are not extensively monitored, the effect of these direct and incidental takes on local populations is unknown.

See Also the Following Articles

Pilot Whales ■ Pygmy Killer Whales

References

Best, P. B., and Shaugnessy, P. D. (1981). First record of the melon-headed whale *Peponocephala electra* from South Africa. *Ann. South Afr. Mus.* **83**, 33–47.

Bryden, M. M., Harrison, R. J., and Lear, R. J. (1977). Some aspects of the biology of *Peponocephala electra* (Cetacea: Delphinidae). I. General and reproductive biology. *Aust. J. Mar. Freshwat. Res.* **18**, 703–715.

Dolar, M. L. (1999). "Abundance, Distribution, and Feeding Ecology of Small Cetaceans in the Eastern Sulu Sea and Tañon Strait, Philippines." Ph.D. Thesis, University of California San Diego.

Jefferson, T. A., and Barros, N. B. (1997). *Peponocephala electra.* *Mamm. Spec.* **553**, 1–6.

Lodi, L. F., Siciliano, S., and Capistrano, L. (1990). Mass stranding of *Peponocephala electra* (Cetacea Glopicephalinae) on Pirancanga Beach, Bahia, Brazil. *Sci. Rep. Cetacean Res. Inst.* **1**, 79–84.

Miyazaki, N., Yoshihiro, F., and Iwata, K. (1998). Biological analysis of a mass stranding of melon-headed whales (*Peponocephala electra*) at Aoshima, Japan. *Bull. Natl. Sci. Mus. Tokyo Ser. A* **24**, 31–60.

Perryman, W. L., Au, D. W. K., Leatherwood, S., and Jefferson, T. A. (1994). Melon-headed whale—*Peponocephala electra* (Gray, 1846). In "Handbook of Marine Mammals" (S. H. Ridgway and R. Harrison, eds.), Vol. 5, pp. 363–386. Harcourt Brace, London.

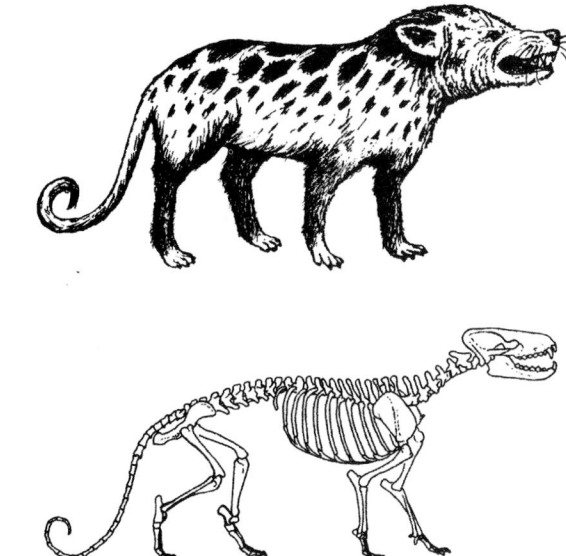

Figure 1 *Skeleton of one of the better known mesonychian fossils,* Mesonyx *(bottom), and an artist's (Luci Betti) reconstruction of how it might have looked (top). This animal was approximately the size of a large dog.*

Mesonychia

MAUREEN A. O'LEARY
State University of New York, Stony Brook

Mesonychians are an extinct group of four-footed land mammals that lived in the Early Tertiary and are recognized by paleontologists to be the closest relatives of whales. This hypothesis has, however, been difficult to reconcile recently with the molecule-based hypotheses that cetaceans may be most closely related to hippos. Mesonychians were unique because they possessed hooves like many plant-eating mammals, but sharp teeth like many carnivorous mammals. Mesonychians obtained a relatively wide distribution throughout the globe; their fossils are found in North America, Europe, and Asia in rocks dating from the early Paleocene through the end of the Eocene, an interval of about 30 million years. Despite a relatively wide geographic DISTRIBUTION, mesonychian fossils remain some of the rarest elements of early Tertiary faunas. Most species have been described only from jaws and teeth but several are also well known from skulls and postcrania. The first appearance of mesonychians in the fossil record precedes the first appearance of whales by approximately 10 million years. Unlike whales, mesonychians became completely extinct around the end of the Eocene. Mesonychians are significant because they provide scientists with a hypothesis of how whale ancestors may have looked before they left land for a life in water (Fig. 1).

I. Origins and Relationships

Mesonychians were first named for specimens discovered in North America in 1874. Since then about 20 genera of varying sizes have been recognized, none of which contains many species (Table I). Mesonychians were initially thought to belong to the order Creodonta, an extinct group of mammals closely related to the CARNIVORA. This was proposed because mesonychians are similar to both creodonts and carnivorans in having tall, pointed lower molar teeth (Fig. 2). Paleontologist Leigh Van Valen argued in 1966, however, that because mesonychians also have hooves and certain other features of the skull, their anatomy more closely resembles that of hoofed mammals. Hoofed mammals include, among other species, artiodactyls, perissodactyls, and all of their extinct relatives. Van Valen argued that as mesonychians evolved they diverged from other hoofed mammals, which are primarily herbivorous with short, square teeth and independently developed dental similarities resembling those of certain meat-eating animals like carnivorans. The teeth of mesonychians also resemble the teeth of early whales, also thought to have been carnivorous, not only because their living descendants are carnivorous but because of the sharp pointy shape of their teeth. Mesonychians thereby became an important fossil intermediate to link a carnivorous group like whales to living and extinct hoofed mammals, which primarily eat plants.

Van Valen's hypothesis has since been corroborated by phylogenetic analyses of mammals that show that mesonychians are the extinct species mostly closely related to whales and that artiodactyls are the living mammals most closely related to this whale + mesonychian group. The features that mesonychians and whales share are primarily those of the dentition (Fig. 2) and the skull. Mesonychians, whales, and artiodactyls, however, all share cranial and postcranial features in common, including possession of a paraxonic foot (Fig. 3). Paraxonia is a specially evolved condition in these mammals in which the weight of the body is transmitted along an imaginary line between digits

TABLE I
Mesonychian Taxa and Their Stratigraphic and Geographic Ranges[a]

Genus	Stratigraphic range	Geographic range
Yantanglestes	Paleocene	Asia
Hukoutherium	Paleocene	Asia
Dissacusium	Paleocene	Asia
Ankalagon	Paleocene	North America
Sinonyx	Paleocene	Asia
Dissacus	Paleocene–Eocene	North America, Asia, Europe
Pachyaena	Paleocene–Eocene	North America, Asia, Europe
Jiangxia	Paleocene	Asia
Mongolonyx	Eocene	Asia
Harpagolestes	Eocene	Asia
Hessolestes	Eocene	Asia
Synoplotherium	Eocene	North America
Mesonyx	Eocene	North America, Asia
Guilestes	Eocene	Asia
Mongolestes	Eocene	Asia
Hapalodectes	Paleocene–Eocene	North America, Asia
Hapalorestes	Eocene	North America
Metahapalodectes	Eocene	Asia
Lohoodon	Eocene	Asia
Honanodon	Eocene	Asia

[a] From McKenna and Bell (1997).

three and four (Fig. 3). Digit one (which is equivalent to the thumb or big toe in a human) is reduced such that it is not weight bearing and in some animals it is completely lost.

When generating phylogenetic analyses of Cetacea on the basis of DNA it is difficult to evaluate the full impact of mesonychians because they are completely extinct and information about their genes remains unknown. Mesonychians may nevertheless have been a very pivotal group.

II. Anatomy and Function

Study of the best-preserved skeletons of mesonychians, primarily the genera *Dissacus*, *Pachyaena*, *Mesonyx*, and *Sinonyx*, indicates that these animals evolved the ability to run fast relative to their Early Tertiary contemporaries. Paleontologists surmise this from the structure of mesonychian vertebral columns, limbs, ankles, and feet. Many of the joints of these mammals have evolved to restrict motion of the limbs to flexion and extension as is typical of many cursorial animals. In so doing many mesonychians sacrificed having a wide range of mobility of their joints for having increased joint stability within a limited range of mobility. This condition may be advantageous for an animal that is moving at high speeds across a terrestrial substrate. Study of the vertebral column of the mesonychian genus *Pachyaena* indicates that it was a stiff-backed runner, meaning that its vertebral column did not exhibit much motion side to side or up and down during run-

ning. This feature characterizes many large-bodied hoofed mammals, and observation of this functional similarity is another shared feature of mesonychians and hoofed mammals. These functional characters are perhaps best understood in the genus *Pachyaena* (Fig. 4), which includes some of the larger mesonychian species, and which in one species is estimated to have had a body weight of approximately 400 kg. The earliest whales, therefore, may have evolved into fully aquatic animals by modifying a body that had originally evolved for running on land.

The dentition of mesonychians is simplified such that they have lower premolars and molars that resemble each other in exhibiting three main cusps (Fig. 2). Upper teeth are simplified and triangular in all taxa. The morphology of the lower dentition resembles that seen in living piscivorous mammals such as seals and toothed whales, and some paleontologists have argued that mesonychians may have been piscivorous also. Mesonychians had a chewing mechanism that was restricted largely to orthal motions (up and down as opposed to side to side). Carnivorans chew in a similar fashion; however, the mesonychian chewing mechanism appears to have been less tightly interlocking than that of carnivorans as evidenced by the variable position of tooth-wear facets on mesonychian molars (Fig. 2).

Studies of endocasts (molds made from the inside of the skull to estimate brain size and shape) of mesonychians indicate that these animals had more specialized BRAINS than con-

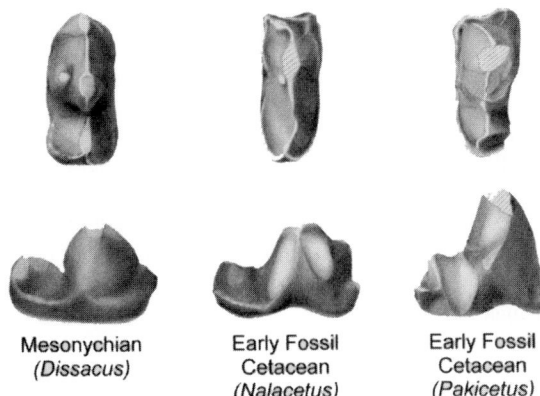

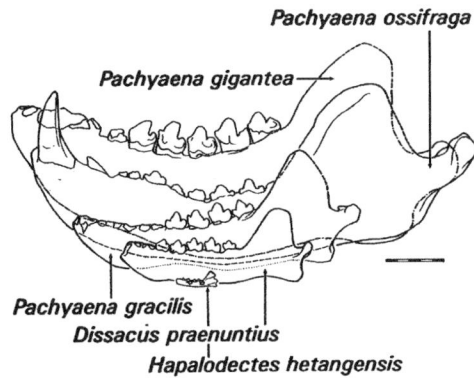

Figure 2 *Right lower molars of a mesonychian and primitive whales for comparison. The top row of teeth is shown from a view of the chewing surface, and the bottom row of teeth is shown from a view from the cheek side. Reprinted with permission from O'Leary and Uhen (1999).*

Figure 4 *Lower jaws of four different mesonychian species indicating the range of size variation in this group. The species Dissacus praenuntius was approximately the size of an average dog. Scale bar: 5 cm. Reprinted with permission from O'Leary and Rose (1995).*

temporaneous hoofed mammals and carnivorans. In particular, mesonychians had reduced olfactory lobes, suggesting a decreased reliance on the sense of smell.

III. Paleoecology

Because of their strange combination of hoofed mammal and carnivoran characteristics, mesonychians have no close modern analogue, something that makes reconstruction of their paleoecology particularly challenging. In the Early Tertiary (Paleocene and Eocene) faunas in which mesonychians are found, they are among the largest predators and some of the largest mammals. Carnivorans did not reach the size of mesonychians until millions of years later. It was mesonychians and not carnivorans that filled

the role of pursuit predators in the Eocene. Paleontologists continue to grapple with the question of why mesonychians have a cursorially modified skeleton. Was it important for defensive (escape) or offensive (attack) behavior? The diet of mesonychians has also been difficult to determine, with suggestions ranging from omnivory (a varied diet) to piscivory, molluscivory (mollusks), and carnivory. By further researching both phylogenetic and functional questions, paleontologists may better understand why whales left land and returned to water.

See Also the Following Articles

Artiodactyla ■ Hippopotamus ■ Perissodactyla

References

Carroll, R. L. (1988). "Vertebrate Paleontology and Evolution." Freeman, New York.

Janis, C. M., and Wilhelm, P. B. (1993). Were there mammalian pursuit predators in the Tertiary? Dances with wolf avatars. *J. Mamm. Evol.* **1**, 103–125.

Luo, Z., and Gingerich, P. D. (1999). Terrestrial Mesonychia to aquatic Cetacea: Transformation of the basicraniuim and evolution of hearing in whales. *Univ. Mich. Papers Paleontol.* **31**, 1–98.

McKenna, M. C., and Bell, S. K. (1997). "Classification of Mammals above the Species Level." Columbia Univ. Press, New York.

O'Leary, M. A., and Rose, K. D. (1995). Postcranial skeleton of the early Eocene mesonychid *Pachyaena* (Mammalia, Mesonychia). *J. Vertebr. Paleontol.* **15**, 401–430.

O'Leary, M. A., and Uhen, M. D. (1999). The time of origin of whales and the role of behavioral changes in the terrestrial–aquatic transition. *Paleobiology* **25**, 534–556.

Szalay, F. S., and Gould, S. J. (1966). Asiatic Mesonychidae (Mammalia, Condylarthra). *Bull. Am. Mus. Nat. Hist.* **132**, 1–173.

Van Valen, L. (1966). The Deltatheridea, a new order of mammals. *Bull. Am. Mus. Nat. Hist.* **132**, 1–126.

Zhou, X., Sanders, W. J., and Gingerich, P. D. (1992). Functional and behavioral implications of vertebral structure in *Pachyaena ossifraga* (Mammalia, Mesonychia). *Contrib. Mus. Paleontol. Univ. Mich.* **28**, 289–319.

Zimmer, C. (1998). "At the Water's Edge." Free Press, New York.

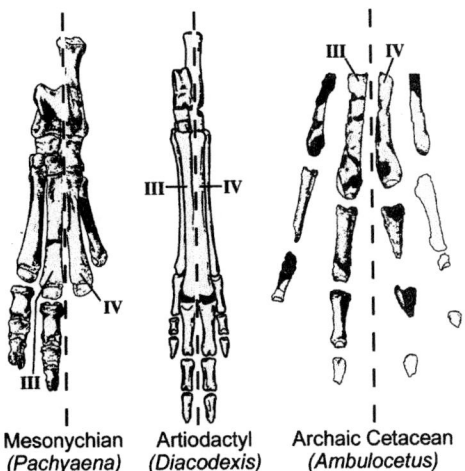

Figure 3 *Feet of a mesonychian, a fossil artiodactyl, and a fossil cetacean all shown as if looking down on the top surface of the foot. All exhibit a paraxonic foot, i.e., one in which the weight of the body passes along an imaginary line (dotted) between digits three and four and in which the foot is largely symmetrical.*

Mesoplodont Whales
Mesoplodon spp.

ROBERT L. PITMAN
Southwest Fisheries Science Center,
La Jolla, California

The genus name *Mesoplodon* (Greek: *mesos,* middle: *hopla,* arms; *odon,* tooth; i.e., armed with a tooth in the middle of the jaw) was coined by Gervais in 1850. Ziphiids (beaked whales), including mesoplodonts, appeared suddenly in the FOSSIL RECORD in the lower Miocene (26 Mya), are well represented by the upper Miocene (5 Mya), and their diversity has declined steadily since then. Currently there are 13 recognized species in *Mesoplodon* (Table I), making it by far the largest cetacean genus, and it is likely that new species are yet to be discovered. In the eastern Pacific alone, two new species have been described (*M. peruvianus* in 1991 and *M. bahamondi* in 1995), and the description of a third is in preparation. Morphological and molecular genetic analyses confirm species limits within the group. Based on morphology, *M. bowdoini* and *M. carlhubbsi* have been suggested as possible subspecies, but genetic analyses indicate species level differences. Although there

are no currently recognized subspecies, *M. mirus* has (apparently) disjunct populations in the North Atlantic and the Southern Ocean that have markedly different color patterns, suggesting at least a subspecific level of divergence. Another ziphiid, *Indopacetus pacificus,* up until very recently known only from two skulls, has often been included within *Mesoplodon,* but is now known to be generically distinct.

Historically, mesoplodont species have been diagnosed by features of the skull, relying mainly on the length of the rostrum, and the shape, size, and placement of teeth, especially of adult males (Fig. 1). However due to anatomical similarities (especially of females and young), specimens are often misidentified, even by experts, and molecular genetic analyses have recently become important for identifying individual specimens. Genetic techniques are also uncovering "cryptic" species. For example, geneticists recently reported a new species of mesoplodont that appears to be morphologically nearly indistinguishable from *M. hectori.*

I. Description

Mesoplodonts are small whales, ranging in size from 3.9 (*M. peruvianus*) to 6.2 (*M. layardii*) m; there are too few data to determine if there are consistent size differences between males and females, although in at least some cases, females appear to be slightly larger, as is the case for most ziphiids. The

TABLE I
Living Species of *Mesoplodon*[a]

Latin name	English name(s)	Length (m)	Distribution
M. hectori	Hector's beaked whale; New Zealand beaked whale	4.4	Circumglobal in temperate waters of Southern Hemisphere
M. mirus	True's beaked whale	5.3	Warm temperate North Atlantic and southern Indian Ocean
M. europaeus	Gervais' beaked whale; Antillean beaked whale; Gulf Stream beaked whale	5.2	Warm temperate and tropical waters of North Atlantic
M. bidens	Sowerby's beaked whale; North Atlantic beaked whale; North Sea beaked whale	5.5	Temperate North Atlantic from Europe to North America
M. grayi	Gray's beaked whale; Haast's beaked whale; scamper-down whale; small-toothed beaked whale	5.6	Circumglobal in temperate waters of Southern Hemisphere
M. peruvianus	Pygmy beaked whale; Peruvian beaked whale; lesser beaked whale	3.9	Eastern Pacific from northern Mexico to northern Chile
M. bowdoini	Andrew's beaked whale; deep-crest beaked whale	4.7	Known only from strandings in Australia and New Zealand
M. bahamondi	Bahamonde's beaked whale	?	One specimen from Juan Fernandez Island, Chile
M. carlhubbsi	Hubbs' beaked whale; arch-beaked whale	5.3	Temperate north Pacific from California to Japan
M. ginkgodens	Ginkgo-toothed whale; Japanese beaked whale	4.9	Tropical and warm temperate waters of the Indian and Pacific oceans
M. stejnegeri	Stejneger's beaked whale; Bering Sea beaked whale; saber-toothed whale	5.2	Sub-Arctic and temperate north Pacific from California to Japan
M. layardii	Layard's beaked whale; strap-toothed whale; long-toothed beaked whale	6.2	Circumglobal in temperate and sub-Antarctic southern waters
M. densirostris	Blainville's beaked whale; dense beaked whale	4.7	Circumglobal in tropical and warm temperate waters

[a]Lengths are maxima.

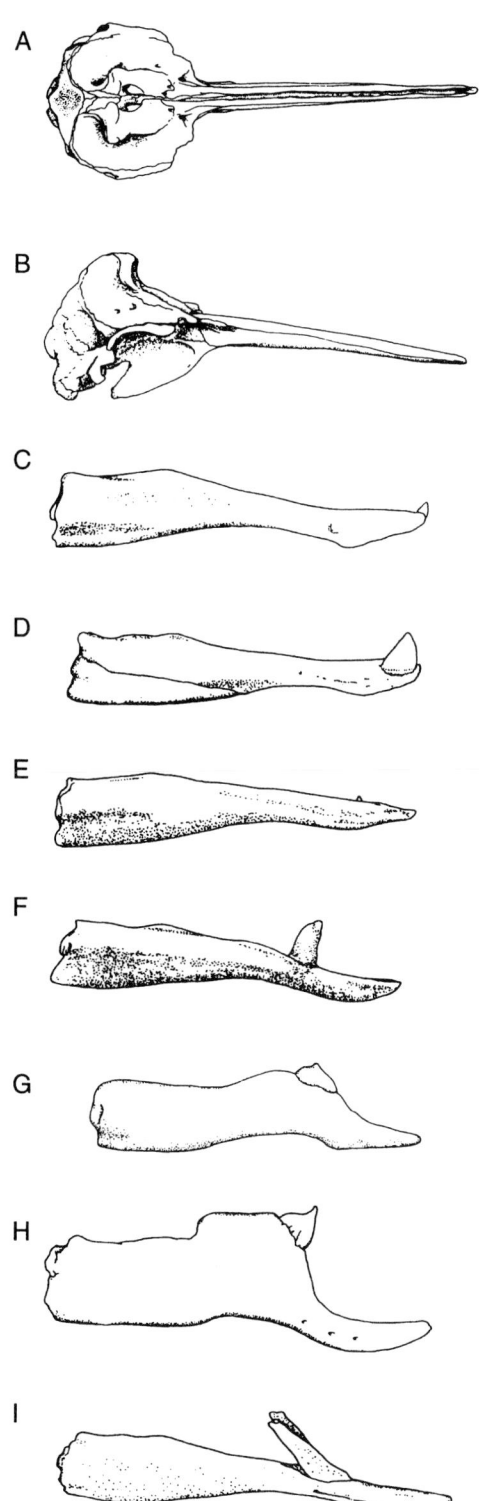

body is spindle shaped, with a small, usually triangular dorsal fin, located approximately two-thirds of the way back on the body. The flippers are small and narrow and fit into pigmented depressions in the body. The unnotched flukes are usually straight across the trailing edge or even slightly convex. A single pair of external throat grooves is present between the mandibles that apparently assists in suction feeding (see later). The head is small and tapered. The melon is small and blends without a crease into the beak. The blowhole is a half circle with the ends pointed forward and not always symmetric. Beak length is variable depending on the species, ranging from short (e.g., *M. densirostris*) to very long (e.g., *M. grayi*) (Fig. 2).

Most species show three sexually dimorphic traits, all of which relate to male aggressive behavior at the onset of sexual maturity: (1) only adult males have functional TEETH, (2) only adult males have extensive secondary ossification of the mesorostral canal, and (3) only adult males show extensive and conspicuous body scarring. These features are discussed later.

Males apparently use their teeth for intraspecific fighting with other males to establish breeding hierarchies. In fully mature males, a single pair of teeth erupts from the mandibles and projects up, outside of the mouth, along both sides of the rostrum. Depending on the species, the teeth are located anywhere from the tip of the lower jaw (apical) to about halfway back along the jaw, and they vary markedly in size and shape (Fig. 1). In most species, the teeth are laterally compressed, although in *M. mirus*, which has the most apical teeth, they are oval in cross section. The tooth usually has a sharp denticle on top for inflicting wounds, but this can be worn down smooth in older males. In some species, the tooth is raised up on a high bony arch in the lower jaw (e.g., *M. densirostris*); in other species, the mandible is relatively flat and teeth of varying length are surrounded and supported mainly by gum tissue.

A hypothesis of evolutionary trend for mesoplodonts proposes that the occurrence of apical teeth is a primitive condition

Figure 1 *Selected* Mesoplodon *skulls and mandibles. (A) Dorsal view of* M. grayi *skull. (B) Lateral view of* M. grayi *skull. Lower jaws of (C)* M. mirus, *(D)* M. hectori, *(E)* M. europaeus, *(F)* M. stejnegeri, *(G)* M. ginkgodens, *(H)* M. densirostris, *and (I)* M. layardii. *Jaws are all of adult males except E. Modified after Jefferson* et al. *(1993).*

Figure 2 *Gray's beaked whale is unusual among the* Mesoplodon *species in having a long beak like a dolphin. This photo was taken in Antarctic waters by Richard A. Rowlett.*

and more posterior locations are derived. Teeth positioned further back on the mandible apparently allow animals to attack more forcefully with their rostrum with less risk of damage to the mandibles or teeth. However, teeth further back along the lower jaw need to be elevated or elongated so that they are not obstructed by the rostrum, which gets wider and deeper toward the base. These factors have probably contributed to the marked variation in tooth size and placement within this group (Fig. 1). As an extreme example, *M. layardii* has some of the most bizarre teeth of any known animal. They are long (to at least 34 cm) and curl back over the rostrum so that they sometimes overlap each other, clamping the jaws nearly shut (Fig. 3). These teeth have dorsally projecting denticles and, judging by the amount of scarring on adult males, are still effective for intraspecific fighting. However, in many individuals, tooth wear results from the rostrum rubbing against the inner sides of the teeth. This clearly indicates that jaw movement is impaired, and the adaptive significance of tooth shape in this species has never been adequately explained.

The mesorostral canal is a narrow groove in the midline of the upper rostrum that is filled with cartilage in most cetaceans. This cartilage is continuous with the mesethmoid and homolo-

gous to the nasal septum of terrestrial mammals. In adult male mesoplodonts, the cartilage is displaced by expansion of the vomer from below, which is composed of extremely dense bone, in some cases more dense than elephant ivory. When males attack, they make contact with the top of their rostrum and use their teeth with the mouth in the closed position. It has generally been assumed that the heavily reinforced rostrum was selected to allow fighting males to be more forceful with their attacks while reducing the possibility of damage to the rostrum. A recent study, however, suggested that the bone of the rostrum, although very dense, was too brittle to provide mechanical reinforcement and that it might, for example, have a hydrostatic role to assist in deep diving.

Color patterning among mesoplodonts is poorly known because animals at sea are sighted so infrequently and the few fresh animals that do strand on beaches lose their colors quickly. Most species are so similar, especially females and juveniles, that even on the rare occasion when animals are clearly seen at sea, few can be identified. This applies to stranded animals as well. Live animals are usually a nondescript gray or brown dorsally and somewhat paler ventrally. Most have no distinguishable overall pattern, although some are quite distinctively marked. For example, *M. mirus* (Southern Hemisphere form) has an all white tail stock, dorsal fin, and flukes; *M. layardii* has a very distinctive black and white pattern to the head, face, and beak, and adult males of an unidentified mesoplodont from the eastern tropical Pacific (probably *M. peruvianus*) have a broad white swathe across the body that forms a conspicuous chevron when viewed from above.

Most mesoplodonts exhibit sexually dimorphic COLORATION, with adult males patterned more conspicuously than females. This can be due either to pigment deposition (e.g., *M. carlhubbsi*) or, more often, to adventitious coloration from scarring. Most scars on mesoplodonts are attributable to tooth rake injuries from other males or wounds from cookie-cutter shark bites (see later). All mesoplodonts, except apparently *M. ginkgodens*, form white scar tissue over external wounds so that the body retains a permanent visual record of any injuries. As a result, adult males in groups of mesoplodonts are usually easily recognizable by their prominent scarring (Fig. 4), and it has been suggested that this may have evolved as a social signal for indicating male "quality." In addition to heavy scarring, adult males (and, to a lesser extent, females) of some species have white pigment patches that may serve to highlight important anatomical landmarks, including the beak tip, head, lips, and genital area.

II. Distribution and Zoogeography

Mesoplodonts are so difficult to approach and identify under normal conditions that there are several species that have never been identified alive in the wild (e.g., *M. bowdoini*, *M. hectori*, *M. bahamondi*, *M. europaeus*, *M. ginkgodens*). Consequently, nearly everything known about their distribution and abundance has been inferred from relatively infrequent stranding events. As a group they are widespread, occurring in all of the world's oceans except for the coldest waters of the Arctic and Antarctica. They normally inhabit deep ocean waters

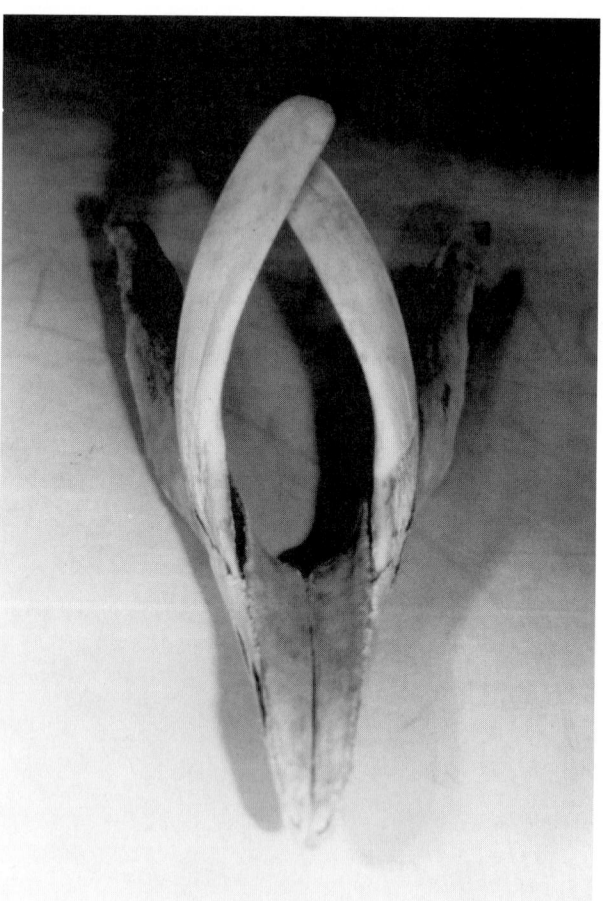

Figure 3 *Lower jaw of an adult male* Mesoplodon layardii *showing how the teeth wrapped around the upper jaw in the living animal.*

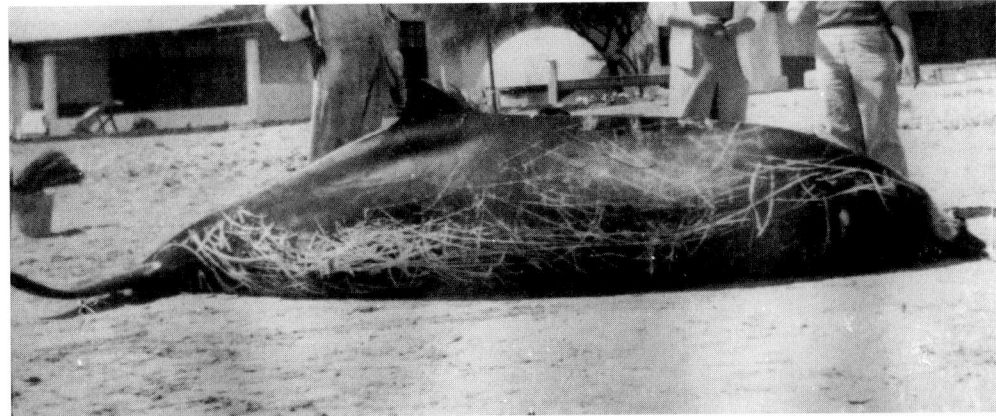

Figure 4 *A stranded specimen of an adult male* Mesoplodon *(probably* M. peruvianus*) from Paracas, Peru. Note white scarring over most of the body from fighting with other males. Photo by E. Link, courtesy of J. Mead, Smithsonian Institution.*

(>2000 m) or continental slopes (200–2000 m), and only rarely stray over the continental shelf. The distribution of most species tends to be somewhat localized (limited to single ocean basins; Table I), although *M. densirostris* is found in all tropical and warm temperate oceans and is perhaps the most widespread mesoplodont. Seasonality of stranding records suggests that at least some high-latitude species (e.g., *M. layardii*) may undergo some limited migration to lower latitudes during the local winter.

III. Food and Feeding

Based on stomach contents of stranded animals, mesoplodonts feed primarily on mesopelagic squid (e.g., *Histioteuthis* spp., *Taonis* spp., *Gonatus* spp.), although some mesopelagic fish may also be taken (at least some of these fish, however, are probably secondary, i.e., squid, prey). Most prey are probably caught at depths of 200 m or more. A reduced dentition among various species of odontocetes (toothed cetaceans) is generally interpreted as an adaption for FEEDING on squid. Most squid-feeding cetaceans, including mesoplodonts, are thought to be suction feeders: prey is sucked in and swallowed whole. Large muscles at the back of the tongue (hyoglossus and styloglossus) are anchored to an enlarged hyoid bone; this allows the tongue to be retracted in a piston-like manner while the throat pleats allow distention of the mouth floor, creating the necessary vacuum for sucking in prey. Because ziphiids in general no longer need their teeth for feeding this has freed them up to evolve for other purposes, i.e., as fighting weapons. It has been hypothesized that white pigmentation on the anterior floor of the mouth of many beaked whale species may serve as an attractant for BIOLUMINESCENT squid.

IV. Behavior

Almost nothing is known about mesoplodont behavior, partly because they are so rarely sighted, but also because their behavioral repertoire at the surface appears to be very limited and stereotyped. The most commonly reported behavior has been slow swimming, usually away from a vessel, and often a mile or more away. When undisturbed, they roll quietly several times at the surface and then dive from 20 to over 45 min at a time. Most groups surface simultaneously and within a few body lengths of each other, indicating that some communication is probably going on as they forage in total darkness of the deep ocean. While at the surface, either traveling or stationary, individuals in groups usually remain within a couple body lengths of each other. BREACHING (leaping out of the water) has been recorded on only a very few occasions. Long-beaked forms (e.g., *M. grayi, M. bidens,* and *M. layardii*) often bring their beak up out of the water at a 45° angle when they surface. Normally there is no visible blow, and none are known to lift their flukes when diving. Tail slapping has been reported once. Male mesoplodonts are assumed to use their erupted teeth as tusks to fight with each other, and although none of these battles have ever even reported by human observers, the extensive scarring found among males of this group suggests that it is of frequent occurrence (Fig. 4).

V. Other Life History Notes

Because of their shy nature, far offshore habitat, and apparent rarity, very little is known about the biology of mesoplodonts, and nearly everything that is known has come from the examination of stranded animals. As in all cetaceans, females give birth to a single calf. Mean body length for calves at birth has ranged from 2.1 m for *M. europaeus* to 2.5 m for *M. carlhubbsi*, representing from 40 to 48% of the adult female body lengths (these are neither the largest nor the smallest mesoplodonts). The smallest mesoplodont calf reported to date is 1.9 m (*M. hectori*), although *M. peruvianus* will probably prove to be smaller. There is no information on gestation or lactation periods.

Longevity data for mesoplodonts are virtually nonexistent, although they may be quite long lived: a count of tooth layers in a specimen of *M. europaeus* suggests that it was at least 27 years old. A female *M. densirostris* estimated to be 9 years old (based on tooth layer counts) had just recently become sexually mature. Very little has been recorded in the way of diseases, parasites, or commensals. Osteomyelitis has been reported twice. Endoparasites recorded have included cestodes and occasional heavy infestations of nematodes (*Crassicauda* sp.) in the kidneys. Ectoparasites recorded include *Penella* sp. (a parasitic copepod) and cyamids. The erupted teeth of males often have stalked BARNACLES (*Conchoderma* sp.) attached to them; bunches of these often appear as "tassels" on the teeth of live animals at sea. A pseudo-stalked barnacle, *Xenobalanus* sp., is sometimes attached to the flukes and dorsal fins. Almost nothing is known about mesoplodont vocalizations. Stranded animals on beaches have been reported to make cow-like sounds but there are no known recordings of vocalizations at sea. Because they almost certainly use sounds to coordinate deep diving behavior among herd members and locate prey in total darkness, it is likely that they do vocalize.

Mesoplodonts occur in small groups typically ranging in size from 1 to 6 animals, although groups of up to 10 have been reported, and a mass stranding of 28 *M. grayi* occurred in New Zealand. Mean school size for 125 *Mesoplodon* sightings (including at least three different species) from the eastern tropical Pacific was 3.0, with 2 being the most common group size. Although mixed groups of adult males with females and calves have been observed at sea, there is some evidence from both sightings and strandings data that there may be some segregation by sex or age class at times. Predators of mesoplodonts probably include killer whales and large sharks, although direct observations are lacking. Mesoplodonts often have white oval scars (diameter to about 8 cm) caused by the bites of cookie-cutter sharks (*Isistius* sp.). These are small (to about 50 cm), mesopelagic sharks that feed by snatching mouthfuls of flesh off larger fish and cetaceans. Although individual mesoplodonts are often riddled with scores of healed bite wounds, these do not appear to contribute to mortality.

VI. Status and Conservation

So few mesoplodonts have been reliably identified at sea that it is impossible to accurately determine the population status of any species, although based on stranding data, at least some species may not be as rare as the sighting records indicate. *M. grayi* and *M. layardii* appear to be widespread and fairly common in the Southern Ocean, as is *M. densirostris* in tropical oceans. These may be the most abundant mesoplodonts. Most species, however, appear to be neither numerous nor widespread, and some may be quite rare (e.g., *M. bowdoini*, *M. hectori*). The large number of species in this group suggests a high rate of endemism with naturally small populations and restricted ranges. Although there has never been any directed fishery for mesoplodonts, a few are occasionally harpooned opportunistically by whalers, and unknown, but potentially significant, numbers are killed by high seas drift nets and long line fishing gear. Only stranded specimens have ever been kept in captivity and these have usually died within a few days (usually from preexisting conditions).

See Also the Following Articles

Aggressive Behavior, Intraspecific ■ Beaked Whales, Overview ■ Cetacean Life History

References

Heyning, J. E. (1984). Functional morphology involved in intraspecific fighting of the beaked whale, *Mesoplodon carlhubbsi*. *Can. J. Zool.* **62**, 1645–1654.

Jefferson, T. A., Leatherwood, S., and Webber, M. A. (1993). FAO Species Identification Guide. Marine mammals of the world. FAO, Rome.

Mead, J. C. (1989). Beaked whales of the genus *Mesoplodon*. Pages 349–430 in S. H. Ridgway and R. J. Harrison, eds. "Handbook of Marine Mammals." Vol. 4. Academic Press, London.

Moore, J. C. (1968). Relationships among the living genera of beaked whales with classifications, diagnoses, and keys. *Field. Zool.* **53**, 209–298.

Reyes, J. C., Mead, J. G., and Van Waerebeek, K. (1991). A new species of beaked whale *Mesoplodon peruvianus* sp. n. (Cetacea: Ziphiidae) from Peru. *Mar. Mamm. Sci.* **7**, 1–24.

Migration and Movement Patterns

S. JONATHAN STERN
Florida State University, Tallahassee

Migration is a critical part of the life history strategies of a diverse group of organisms. An optimal strategy may include the need to move from one location to another and back in some systematic fashion. Migration is the large-scale movement between different parts of the home range, with some energy allocation to support movement or time to meet reproductive needs.

Migration underscores an individual's need for some resource such as food or mates. An individual has a home range that is a function of relative body size and mobility as well as a variety of other factors. In general, large animals need large home ranges. As home range increases in size, an individual experiences variability in environmental conditions. In the marine environment, two parcels of water 10 cm apart are more similar on average than two parcels of water 1000 km apart. Accordingly, waters of higher latitude have generally higher productivity than those of lower latitudes. In addition, a single parcel of water may have similar oceanographic conditions 2 days apart but may differ significantly between winter and summer. Length of the productive season is shorter at higher latitudes due to decreased sunlight in winter, in addition to other oceanographic changes. Thus resources are variable in space as well as time. In general, one part of a home range may be very different with respect to the availability of resources than another. Disparity in resource availability results in the necessity to move between places in the home range.

Resources for marine mammals include food, mates, and space. Mates may only be seasonally receptive or available, or sexes may have a different spatial distribution in a particular season. Space is a variable resource because not all habitats are suitable. In territorial species, once a territory is occupied, other individuals of the same species are excluded. In nonterritorial species, crowding often occurs, and while habitat is nearby, and apparently suitable, there is a tendency to form a crowd. Space for hauling out may not be available at all times due to covering by tides once or twice a day.

Given that prey resources are variable in space and time, one strategy is to stay in one area and tough out the hard times by somehow reducing the effects of variability. This leads to formation of denser fur, thicker BLUBBER layer, or adopting a new strategy such as hibernation. Female polar bears (*Ursus maritimus*) hibernate over winter, giving birth and feeding cubs until emerging from the den in spring. Other segments of polar bear populations do not hibernate, but make large-scale movements in search of food. One benefit of staying in one location is that an individual does not face any energetic cost of moving or potentially adverse conditions along the way. For cetaceans, the cost of moving long distances while on migration is probably not very different than moving around in one location. Some pinnipeds may conserve energy by hauling out or lying at the surface for extended periods. The second strategy is to move to another part of their home range where conditions are more favorable, such as a higher resource density, or where environmental conditions are better. For example, Caribbean manatees (*Trichechus manatus*) move in relation to changing water temperature on a seasonal basis. They prefer waters warmer than 68°F and move in relation to these waters. Migration occurs in populations where some parts of a home range may have more and/or better resources of one type, while another part of the home range has more and/or better resources of another type than the first. The general pattern is movement between feeding and reproductive grounds or haul-out site.

For marine mammals, migration was assumed to occur based on the seasonal occurrence of large numbers of a particular species at different locations. However, migration can only be proven by Lagrangian studies, involving marking individuals in some fashion on one migratory destination and recapturing them on another. Whaling provided the first real evidence for migration in large whales. Numbered darts were fired into dorsal blubber and muscle. If that individual was killed during subsequent whaling operations, tagging and killing dates and locations could be compared and some assessment of movements could be made. Some pinnipeds were marked with numbered flipper tags. Censuses were conducted on a number of haul outs. Movement was documented as tagged individuals moved between haul-out sites. Movement and migration patterns have been described in varying levels of detail using photo-identification and satellite telemetry. For example, locations of individual northern elephant seals (*Mirounga angustirostris*) sent by satellites every 2 days provide valuable information on movement and migration to and from areas of high food productivity, inferred by persistent signals from a relatively confined location, such as over a seamount.

I. Terminology

Defining migration is the first step to discussing it because not all large-scale movements are migration. This is not an easy task, as literature on migration does not have a consistent definition. How do we distinguish between daily movements of a seal that returns to the same haul out every low tide from a whale that swims from the Antarctic ice edge to the waters off Brazil? Movement encompasses a hierarchy of displacements ranging from thousands of kilometers, encompassing thousands of surfacings. The following terms describe a variety of types of movement of increasing scale (Fig. 1). A step is a relevant distance moved, such as distance between sites of long dives. Kinesis refers to changes in turning or movement rates. Kinesis can often result in station keeping, where an individual maintains a relative position through relatively frequent turns. Foraging is a search for resources within a patch. A patch is an area within which resources are randomly distributed. Commuting occurs between adjacent patches. After searching patches in a region, animals can then move to another region, which is subject to a different set of local oceanographic conditions. Movement between regions is referred to as ranging. For example, fin whales (*Balaenoptera physalus*) make an overall east to west displacement through the Gulf of Alaska during a feeding season.

Migration is persistent and more or less rectilinear movement, presumably between two different parts of the home range, each with its own resources or use. For marine mammals, these destinations are areas for feeding/breeding/birth/lactation and, in addition, for pinnipeds, molting. General mysticete migration patterns are shown in Fig. 2. This movement occurs on a seasonal basis, with the majority of a population, sex, or age class undertaking the same overall movement pattern. Upon arrival at one destination, behavior changes to maintain relative location. For molting, breeding, and lactating pinnipeds, this is achieved by hauling out on land or, if on FEEDING grounds, floating in one spot or clusters of deep dives in a limited area.

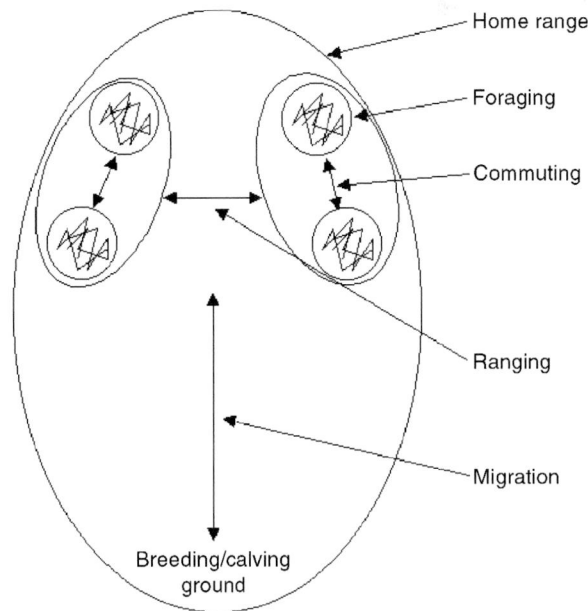

Figure 1 *Different type of movements are described in the text.*

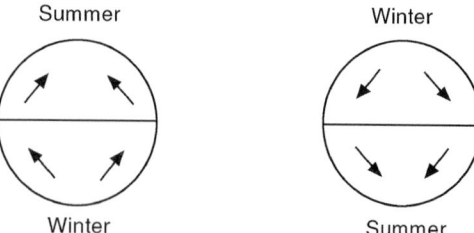

Figure 2 *Generalized migration patterns of baleen whales in the Northern and Southern Hemispheres. Whales spend summers on productive feeding grounds and then migrate to the winter breeding/calving grounds in warmer waters. Most species of whales are believed to feed little, if at all, while away from their feeding grounds.*

For cetaceans, station keeping involves changing the direction of travel relatively frequently.

One characteristic of marine mammal migration is seasonal changes in energy allocation and storage. This results in an increased fat store to support reproduction. Based on stomach contents analysis of whales killed on winter grounds, mysticetes are thought to feed little on migration routes and breeding grounds, living off stored blubber. However, recent evidence suggests that feeding occurs on the winter grounds, at least opportunistically. Some phocids such as elephant seals (*Mirounga* spp.) migrate long distances. They feed continuously to store energy to support time spent on breeding and molting haul outs. Stored energy is used for lactation in females and for mating and agonistic displays in males. Elephant seals lose much of their fat when hauled out and need to begin storing energy as rapidly as possible upon reentry into the water. Otariids feed daily around haul outs, although males may fast while defending a territory.

In this discussion, migration and movement focus in the horizontal dimensions (x and y). Displacement in the vertical dimension (z) is trivial by comparison, although is not a trivial part of foraging ecology. For marine mammals, migration is movement considerably greater than 2 hr, the maximum dive time recorded for a marine mammal and the maximum depth of a dive.

The following scenarios are not true migrations, but are often labeled as such in literature. Seasonal movement may be a response to changing prey distribution. The occurrence of some groups of killer whales (*Orcinus orca*) in the inland waters of the Pacific Northwest correlates with the seasonal migration of salmon (*Onchyrhynchus* spp.). Because these fish ultimately go upstream to die, the whales must find other prey during the winter. It is not clear if they feed on other species during the winter or go offshore to find other salmon schools. Gray seals (*Halichoerus grypus*) move to distinctly different areas on a seasonal basis to feed in productive areas in the Northwest Atlantic. Polar bears seasonally roam over large home ranges, searching for available food. Other large-scale movements are also not truly migration. For example, movement may be in relation to shifting environmental conditions, such as the seasonal advance or retreat of an ice edge. While geographic lo-

cation of an individual changes, it is essentially maintaining itself in the same general environment.

Dispersal is not migration, as there is no return to the original area. Dispersal is colonization of new or recolonization of historic breeding habitats. For example, breeding sites of northern elephant seals were historically on islands off the mainland of California, likely due to the presence of terrestrial predators on the mainland. In association with postexploitation recovery and the decline of terrestrial predators, elephant seals returned to all historic breeding islands and invaded new sites on the mainland. Another example of a dispersal event was observed off California in response to the 1982/1983 El Niño Southern Oscillation event. Common bottlenose dolphins (*Tursiops truncatus*) moved from southern California to San Francisco with the northward advance of warmer waters. This group of dolphins remained after the warm waters retreated back south.

True migrating species have distinct adaptations. Because energy is a limiting resource, migration can be examined from the perspective of energy acquisition (Table I). Fasting species have different physiological and energy intake characteristics than nonfasting species. Fasting during some part of the year requires intense feeding during some other part of the year in order to store energy as well as to meet metabolic needs at the time.

Species that fast either spend their entire lives or part of the year in productive polar, subpolar, and temperate waters where they can feed intensely to form a blubber layer. Other than ice-breeding seals, migrating marine mammals have feeding grounds located pole-ward of the breeding areas.

II. Why Migrate?

Any strategy has associated costs and benefits. Natural selection weighs the costs and benefits of migrating; if the benefits are greater than the costs, a species will migrate. Different areas have different proximate currencies, with energy acquisi-

---

TABLE I
Energy Acquisition

Fasting
    Mysticetes[a] (except bowhead whales)
    Elephant seals[a]
    Some male otariids[a] (on breeding grounds)
    Harp seals (*Pagophilus groenlandicus*)[a]
    Hooded seals (*Cystophora cristata*)[a]
    Polar bears (females can hibernate)
    Most phocids
    Recently weaned pinnipeds (some exploratory swimming around
      haul-out site)
Nonfasting
    Odontocetes
    Female otariids
    Some phocids
    Polar bears
    Sirenians
    Otters

[a]A true migratory species.

tion important in the feeding areas and reproductive success important in the breeding/calving areas. Feeding success, however, has a direct influence on reproductive success, the ultimate currency, for survival of the genes of an individual. For cetaceans, migration occurs for food and reproduction. Pinnipeds migrate for these reasons as well as for molting.

A. Pinniped Patterns

Pinnipeds are central-place foragers, moving from haul-out sites to feeding grounds at varying distances. Large-scale movement away from a haul out reduces pressure on local resources and has other benefits as well. Because predators may congregate around haul outs, an individual can reduce its chance of being killed by a predator by reducing the number of entrances to and exits from the water. Dive profiles of elephant seals suggest that while they feed along an entire migration route, there are areas with a higher frequency of deep dives, suggesting intense, localized feeding. Some of these highly productive areas are associated with seamounts. These seals make deep foraging dives for deep-water prey. The areas around their haul outs are seasonally productive and support a number of mysticete species. Therefore, productivity is not the key; rather it is productivity of a certain type of prey that makes migration beneficial. Some otariids make daily excursions from haul-out sites in search of food. For a lactating pinniped, large movements increase the time spent away from a pup, resulting in less rapid pup growth. While increased entrances and exists from the water increase the probability of being killed by a predator, the sheer number of other seals entering and exiting the water reduces the per capita probability of attack.

B. Mysticete Migration

The feeding grounds of baleen whales are in productive cold waters so it is clear why they migrate to these areas. A question remains as to why they must migrate to warm waters for reproduction. Four reasons have been suggested for migration to warm water breeding and calving areas. The first is to minimize thermal stress on calves. This is likely not a problem for a newborn calf. Smaller mammals with less insulation are able to survive in those conditions. Because of its large body size, a calf is not thermally stressed.

The second reason to move to warm waters is resource tracking, i.e., following prey. By definition, this is not migration. While an individual, school, or population is moving, movement results in reduced resource variability, as prey are always in the vicinity.

Killer whale predation on calves has been a suggested reason for mysticete migration. By migrating to warm, relatively killer whale-free waters to give birth, calf mortality would be reduced.

The final reason to migrate is essentially an evolutionary holdover: individuals migrate because their ancestors did. The evolutionary holdover hypothesis includes feeding and reproduction into a life history strategy. Intense feeding leads to energy storage as the short-term goal that maximizes reproductive success, the ultimate, evolutionary goal. Natural selection favors individuals that are successful at migrating, feeding, and reproducing.

C. Pleistocene Conditions

The productive waters of the higher latitudes have changed significantly over the past 20,000 years. These changes likely had a profound influence on regional productivity, affecting migration routes, destinations, and foraging areas of many cetaceans, pinnipeds, carnivores (bears, otters) and sirenians (Steller's sea cow, *Hydrodamalis gigas*). Increased ice extent and land emergence of the Pleistocene made current feeding grounds unavailable to gray whales (*Eschrischtius robustus*), bowhead whales (*Balaena mysticetus*), beluga whales (*Delphinaptrus leucas*), narwhals (*Monodon monocerous*), walruses (*Odobenus rosmarus*), Northern fur seals (*Callorhinus ursinus*), and other seals. The North Atlantic north of 45° was an ice-bearing polar sea with conditions similar to the Antarctic Convergence, resulting in a larger, more productive sea than at present. The distance between productive cold water and warmer water was much less, resulting in relatively little or no distance between feeding and breeding/calving areas (Fig. 3). In the Southern Ocean, an equator-ward shift in isotherms was in response to a northward extension of ice. The Antarctic convergence was 5° north of its current position.

D. Present Oceanographic and Migratory Conditions

At the glacial maxima, cold, productive waters were closer to the equator than at present. At these latitudes, the number of hours of sunlight per day were not as seasonally extreme as at the poles. Assuming that mid and lower latitude waters are relatively unchanged from Pleistocene conditions, the only difference would be the retreat of cold waters toward the poles. As cold, productive waters retreated toward the poles, sunlight for photosynthesis became more variable over the course of a year, leading to intense seasonal peaks in production followed by reduced production in winter. Over time, whale distribution followed the pole-ward retreat of fronts of productive oceans.

This explains the summer distribution of baleen whales, but does not address the need to migrate to warm waters for reproductive activities. Many species of marine mammals, and mammals in general, have highly seasonal reproductive strategies in order to time births relative to optimal environmental conditions. Day length is a cue for seasonal breeding in a number of birds and mammals. Photoperiod is an important seasonal cue because it is invariant from year to year. This means that the timing of reproduction and migratory movements can be the same from one year to the next. The advantage is that an individual can maximize its use of seasonal prey resources as well as seasonally available mates.

Both circadian (daily) and circannual (yearly) cycles use light as a cue: however, the specific cues from light, or *zeitgebers*, vary. Circadian signals are dawn and dusk, whereas circannual signals are perceived as the ratio of number of light to dark hours in a 24-h period. Thus both cycles are used for seasonal cues. Other cues may act as secondary synchronizers, although these, such as food availability, are more variable.

The pineal gland is responsible for time keeping in birds and mammals via the production of melatonin, as well as other compounds. The number of hours of darkness per 24 hr is "counted" by the biosynthesis of melatonin, which is produced

Figure 3 *Post-Pleistocene distribution of the polar front in the North Atlantic Ocean. Lines represent the southern extent of sea surface temperature relating to the polar front. Numbers associated with each line are thousands of years before present. Whale movement from warm-water calving/breeding waters to cold productive feeding waters would have been similarly truncated toward the equator in the past. The equator-ward extension of cold water meant that productive seasons were probably longer in the past than at present due to increased hours of daylight.*

more in hours of darkness. At seasonal scales, as winter approaches, the hours of daylight decrease and hours of darkness increase. In a given 24-hr period, the amount of melatonin produced increases, suppressing gonadal activity. In many mammals the breeding season of females corresponds to periods of decreasing daylight per 24 hr. Increasing daylight after the winter solstice is responsible for triggering estrus. Pineal glands are exceptional in size in polar species such as the Weddell seal (*Leptonychotes weddellii*), northern fur seal, and walrus, species that live where day length is most variable.

As productive waters retreated pole-ward, cetacean distribution shifted accordingly, changing the overall lighting regime from more or less equal hours of dark and light to one with more hours of darkness in winter, with resultant gonadal suppression. In addition, production decreases in winter so there is no benefit to stay in colder waters.

Individuals need to move to a lighting regime that switches on the gonads. In migrating animals, a shift in the lighting regime on the feeding grounds may trigger migration toward the equator. As an individual approaches the equator, the hours of daylight per 24 hr increase. In addition, after December 21, the fewest hours of daylight in the Northern Hemisphere, and June 21, the fewest hours of daylight in the Southern Hemisphere, the number of hours of daylight per 24 hr begins to increase. The rate change in the lighting regime reaches its maximum at the equinoxes. Together, these result in reduced melatonin levels and restored gonadal activity, which triggers mating behavior. This may be a triggering mechanism in the evolution of migratory behavior in mysticetes.

Testosterone and its metabolites trigger migratory behavior in some animals. Male California sea lions (*Zalophus califor-*

nianus) spend the bulk of the year hauled out in large bachelor groups. By June, they have left their haul-outs in central and northern California, Oregon, Washington, and Alaska and migrated to the Channel Islands off southern California. Here, breeding occurs as males set up territories and defend females against other males with whom they spend most of the year in relatively peaceful coexistence. Once migration back to the feeding grounds is triggered in July, an individual is exposed to increasing hours of daylight per 24 hr as it swims pole-ward. One benefit from such a signal is that it is invariant from year to year.

Curiously, the peak migration of gray whales is variable from year to year. The initiation of the southward migration to the breeding grounds may be linked to a change in foraging success or some other environment factor, such as the formation of sea ice. Individuals may be able to override these signals. Extended feeding seasons for baleen whales occur on occasion as prey are uncharacteristically available in late full. For example, blue whales (*Balaenoptera musculus*) were observed feeding on euphausiids in late November in Monterey Bay, California. Whales generally leave the area in late September.

In some cases, not all members of a migratory species actually migrate in a given year. For example, in every location where minke whales (*B. acutorostrata* and *B. bonaerensis*) feed, individuals are observed in winter.

III. Orientation and Navigation

The mechanisms of orientation, plotting their location at any time, and navigation, directing movement from one location to another, are not known. Individuals are often seen in the same locations from one year to the next. In the interim, they have traveled thousands of kilometers, indicating that marine mammals use some type of cues for orientation and navigation between migratory destinations. Organisms tend to meander if they lack orientation and navigation cues. Therefore, marine mammals must know where they are at a given time (orientation) and where they are going next (navigation).

At the initiation of migration, a direction must be selected. Advancing ice may simply eliminate certain directions as a choice, displacing individuals toward the equator. In higher latitudes, changes in sea conditions influence prey availability, which may also trigger the migratory response.

Cues may vary over time and the course of migration. For example, once migration is initiated, the only cue necessary is which overall direction to travel: north, south, east, or west. Celestial navigation has been suggested as one mechanism of navigation. In the north/south directions, the relative location of the sun in the sky can be monitored. This may be as simple as "keep sunrise on the left side when migrating to the breeding ground and on the right side when migrating to the feeding ground" or as complicated as estimating latitude as a function of position of the sun. Navigation by star location has also been suggested as a mechanism.

Another possible large-scale cue is the direction of a major current if an animal is moving against it. Near the equator in the Northern Hemisphere, western boundary currents move from south to north, while eastern boundary currents move from north to south. Coastal processes and minor currents cause the deformation of major currents at higher latitudes, resulting in the formation of gyres and eddies. Migrating whales may use these currents for a free ride.

Magnetite in the brains of some species has been implicated as a mechanism by which individuals could track changes in the earth's magnetic field. Mass STRANDINGS often occur at the same location. These locations may have anomalies in the local earth's magnetic field, which cause whales to become disoriented and strand. A tantalizing example of the possibility of using magnetic cues is seen in humpback whales (*Megaptera novaeangliae*) migrating from Hawaii to Southeast Alaska. Tracks were within 1° of magnetic north.

At smaller scales, other cues could be used. For example, while mysticetes do not have a true sense of smell, they do have a well-developed Jacobsen's organ. This may allow them to "taste" differences in water mass composition. For example, freshwater from ice melt or riverine input might provide a "taste trail" to a rich feeding ground, as a lens of fresh water floats on denser salt water.

Routes to and from feeding and breeding grounds may be variable or essentially a retracing of the migratory path. Male humpback whales migrate south from the Gulf of Maine relatively far offshore, while the return trip is much closer to shore. Gray whales along the west coast of North America probably migrate between calving lagoons in Baja California and feeding grounds in the Bering, Chukchi, and Beaufort Seas by following contours of the coastline. Gray whales migrate along the same nearshore corridor in both migratory directions. There is spatial or temporal segregation based on age class or gender. Northern elephant seals disperse from haul-out sites into the Gulf of Alaska; however, males go further north than females.

In cases where a baleen whale species has a sufficiently long lactation period, offspring can learn migration routes and feeding locations from their mothers. In instances where offspring are weaned prior to reaching the feeding grounds, offspring are left to make exploratory migrations in hopes of finding suitable feeding grounds. This may be one reason that populations of some minke whales are segregated according to age class.

IV. Physiology of Migration

Marine mammals have a suite of physiological adaptations for energy allocation and fasting, as well as storage and mobilization of fats to and from blubber stores. Fats are the most important energy source, as lipids hold more energy per unit weight than other forms. Lipids are also less bulky than protein or carbohydrates because no water is required for storage. During feeding on productive feeding grounds, hyperphagia promotes increased lipid synthesis, fat uptake, and rate of fatty acid synthesis. Storage of fats occurs when the supply in blood exceeds metabolic demand. Mobilization of fats occurs when the demand for energy in blood exceeds supply.

However, little is known about hormonal activity in relation to migratory movements. Prolactin is a hormone responsible for promoting milk production and lactation in mammals. It has the effect of increased fattening in birds. If there were a similar effect on mammals, it would be of importance. For pinnipeds, where lactating females leave pups for one to a few days, increased fat storage for milk is vital. A lactating female mysticete on the feeding ground would store fat relatively faster. This would not only provide for milk for the offspring, but also help in restoring the female's blubber layer for subsequent migration back to the breeding/calving grounds.

V. Effects of Migration

Marine mammals may have localized seasonal effects on their feeding grounds. For example, nutrient recycling may be enhanced locally in bays and inlets used by groups of feeding humpback whales in Alaska. Benthic feeders such as gray whales and walruses represent a disturbance mechanism to these communities. Disturbance opens areas for colonization and settlement of pelagic larvae of benthic species. Other coadaptations occur in barnacles living on gray whales that have timed spawning activities to coincide to when whales are in the calving lagoons. Whales, barnacles, and thus released gametes of barnacles are concentrated in limited areas.

VI. Metapopulation–Removal Migrations

Stock boundaries are delineated by mark and recapture analysis, genetic, biochemical, demographic differences, or other techniques. These differences give the impression of little or no flux of individuals between stocks. Evidence from photographic identification studies of humpback whales in the North Pacific Ocean suggests that some differences may be artificial as individuals move between stocks. Within an ocean basin, stocks can be viewed as a metapopulation, which is a number of populations connected by the dispersal of individuals between them. If a species is distributed into populations spread over a sufficiently large area, environmental conditions are more or less independent between areas; therefore a catastrophe in one area will not affect other populations. Dispersal reduces the risk of population extinction by minimizing the effects of chance environmental changes or changes in population demography. Further, genetic heterozygosity is maintained and the population is less likely to exhibit genetic problems.

A Steller sea lion (*Eumetopias jubatus*) metapopulation has been explored in Alaska. In the Aleutian Islands and Gulf of Alaska, these sea lions have declined by 50% since the 1960s. However, Steller sea lions from Southeast Alaska south to Oregon have remained stable or slightly increased during this time. Evidence from Alaska and the Aleutian Islands suggests that fragmentation will occur, with rookeries being reduced in size and eventually becoming extinct. One reason for hope would be if the population in Southeast Alaska became a source of dispersers into the Aleutian population. The Mediterranean monk seal (*Monachus monachus*) experienced a recent population decline with habitat fragmentation throughout its range. Large expanses of unsuitable habitat separate major pupping sites, with little chance of dispersal between the two remaining large populations, although each separate population may be viable over time.

VII. Migration, Movement, and the Future

One of the main reasons to construct models of movement and migration patterns is to develop descriptive and predictive models to study the effects of changing environmental conditions. Migration in marine mammals evolved within the context of constantly changing environmental conditions. Species had to adapt to deal with novel situations and conditions. A polar front in the Pleistocene as described earlier retreated at a rate that allowed individuals to adapt to its changing distribution. Climatologists predict that global temperatures will increase by as much as 4.5°C in the next century. While it is clear that marine mammals are capable of adapting to changing environments, they might not be able to adapt at a rate commensurate with that of the change in environmental conditions in the near future. The potential implications are profound, and the environmental effects are not entirely clear.

Global warming will likely have variable effects depending on latitude, with polar and temperate regions affected to a greater extent than more tropical areas. Because these areas represent feeding grounds for migratory as well as resident species, understanding these effects is of considerable importance. The effect on marine mammals will likely be through changes in the distribution of resources in space and time. Key to survival will be how individuals respond to changes in resource distribution over space and time, and how this affects reproductive success.

See Also the Following Articles

Behavior, Overview ■ Breeding Sites ■ Cetacean Ecology ■ Distribution ■ Pinniped Ecology

References

Baker, R. R. (1978). "The Evolutionary Ecology of Animal Migration." Holmes and Meier Publishers.

Berta, A., and Sumich, J. L. (1999). "Marine Mammals: Evolutionary Biology." Academic Press, San Diego.

Dingle, H. (1996). "Migration: The Biology of Life on the Move." Oxford Univ. Press.

Lockyer, C. L., and Brown, S. G. (1981). The migration of whales. *In* "Animal Migration" (D. J. Aidley, ed.), pp. 105–137. Cambridge Univ. Press, Cambridge.

McMullough, D. R. (1996). "Metapopulations and Wildlife Conservation." Island Press.

Reidman, M. (1990). "Pinnipeds: Seals, Sea Lions, and Walruses." University of California Press.

Stern, S. J. (1998). Field studies of large mobile organisms: Scale, movement and habitat utilization. *In* "Ecological Scale: Theory and Applications" (D. Peterson and V. T. Parker, eds.), pp. 289–308. Columbia Univ. Press.

Mimicry

PETER L. TYACK
*Woods Hole Oceanographic Institution,
Massachusetts*

The words "mimicry" and "imitation" often have a negative connotation in English of being an unoriginal fake. A mimic is often an annoying copy cat, while an imitation can be a second-rate copy of a more valuable original. However, mimicry and imitation are based on special cognitive abilities that are rare among animals and that form the basis of culture. Humans learn most cultural traits—from the words in our language to the way we prepare food or hold a tool—through observational learning and imitation. We must learn thousands of these cultural traits through imitation before we can make an original contribution to our culture.

Imitation is a form of social learning—it requires an animal to observe a "demonstrator" performing a behavior and then to be able to perform that behavior itself (Galef, 1988; Whiten and Ham, 1992). Many psychologists distinguish between vocal learning, in which an animal modifies the sounds it produces based on the sounds it hears, and motor imitation, which involves an animal watching a posture or movement of another animal and then copying that movement—"monkey see, monkey do." Some also distinguish between motor imitation of something like clapping, where the animal can watch its hands in the same way that it watches the demonstrator, versus facial gestures or whole body movement, where the actor cannot receive sensory input about its own performance that directly parallels the observation of the demonstrator.

There is strong evidence for vocal learning and imitation among a variety of animals, including marine mammals and birds (Janik and Slater, 1997). Perhaps the simplest evidence involves vocal mimicry where an animal demonstrates the ability to produce sounds after exposure to model sounds that were not part of its pre-exposure repertoire. Animals such as parrots (Todt, 1975; Pepperberg, 2000), starlings (West *et al.*, 1983), and a harbor seal (*Phoca vitulina*) named Hoover (Ralls *et al.*, 1985) have demonstrated abilities to imitate the sounds of human speech. Other animals such as dolphins have been trained to imitate acoustic features of artificial sounds (Richards *et al.*, 1984).

Humpback whales, *Megaptera novaeangliae*, have not been kept in captivity where one could most easily study imitation of manmade sounds, but their songs have a structure that cannot be explained by any mechanism other than vocal imitation. At any one time, most singing humpback whales within a population sing songs that are similar (Tyack, 1999). These songs change week by week, month by month, and year by year, and individual whales have been shown to track these changes. Humpback songs have been recorded for decades, with no suggestion that the songs repeat. This suggests that humpback whales learn to produce the current song and to track the progressive changes that make up such a distinctive feature of this signal.

Vocal imitation has also been reported for the natural sounds of bottlenose dolphins (*Tursiops truncatus*). Most dolphins studied in either captive or wild settings develop an individually distinctive signature whistle. Dolphins can imitate the whistles of social partners (Tyack, 1986). Three male Indian Ocean bottlenose dolphins (*T. aduncus*) that formed a strong social bond were reported to modify their whistles over 3 years as the bond formed, such that all three converged on a shared whistle (Smolker and Pepper, 1999). The functions of this imitation are currently not known, but similar imitation has been suggested to function as a name for reference (Tyack, 1999), as a threat (Janik and Slater, 1997), or as an affiliative signal (Smolker and Pepper, 1999).

It has been controversial whether animals can perform motor imitation (Galef, 1988). This stems in large part from the difficulty of proving that a display of which the demonstrator was capable was not part of the pre-exposure repertoire of the animal. Ethologists expect that many animal displays represent fixed action patterns that are inherited. If an action pattern is simply triggered by sensing a conspecific performing the same action, that does not demonstrate observational learning. There are many anecdotes about motor imitation in marine mammals that are difficult to explain via any mechanism other than observational learning. Tayler and Saayman (1973) provided some of the most interesting examples. They reported captive bottlenose dolphins swimming with postures and motor patterns similar to those of seals, turtles, fishes, and penguins that were housed in the same pool. These postures and swimming patterns are so awkward and different from normal dolphin locomotion that it is scarcely credible that they represent anything other than learned behaviors. The most striking example of imitation involved a calf Indian Ocean bottlenose dolphin that observed through an underwater window a human blowing out a cloud of cigarette smoke. The calf swam over to its mother, suckled, swam back to the window, and expelled a mouthful of milk into a cloud that looked similar to exhaled tobacco smoke! This kind of anecdote clearly suggests that it would be worth conducting careful experimental tests of motor imitation in cetaceans.

A good teacher can shape our behavior in ways that may look like imitation, but only demand associative learning on the part of the student. Animal trainers can shape the behavior of animals in the same way. If wild animals were to train one another in this way, this could create a faulty appearance of imitation, but this is thought not to be a problem for there is little evidence that one animal will train another (Caro and Hauser, 1992). However, there are some indications of what might be called teaching in cetaceans. Tyack and Sayigh (1997) provided suggestive evidence for possible teaching of signature whistles in wild bottlenose dolphins, and Guinet and Bouvier (1995) suggested that killer whales (*Orcinus orca*) teach the young how to strand in order to catch pinnipeds on the beach. Rather than being an alternative to passive observational learning, teaching appears at least in our own species to function in tandem with observational learning. This potential synergy between teaching and imitation would be most likely to benefit highly social animal groups, such as carnivores, primates, and cetaceans, in which cultural traditions for foraging provide a strong selective advantage. These kinds of observations in wild cetaceans, coupled with careful experimental tests with captive cetaceans, suggest that cetaceans are promising subjects for the study of mimicry in the development of cultural traits in animals.

See Also the Following Articles

Culture in Whales and Dolphins ■ Intelligence and Cognition ■ Song

References

Caro, T. M., and Hauser, M. D. (1992). Is there teaching in nonhuman animals? *Q. Rev. Biol.* **67**, 151–174.

Galef, B. G., Jr. (1988). Imitation in animals: History, definitions, and interpretation of data from the psychological laboratory. *In* "Social Learning: Psychological and Biological Perspectives" (T. Zentall and B. G. Galef, Jr., eds.), pp. 3–28. Lawrence Erlbaum Associates, Hillsdale, NJ.

Guinet, C., and Bouvier, J. (1995). Development of intentional stranding hunting techniques in killer whale (*Orcinus orca*) calves at Crozet Archipelago. *Can. J. Zool.* **73**, 27–33.

Janik, V. M., and Slater, P. J. B. (1997). Vocal learning in mammals. *In* "Advances in the Study of Behavior" (P. J. B. Slater, J. S. Rosenblatt, C. T. Snowdon, and M. Milinski, eds.), Vol. 26, pp. 59–99. Academic Press, New York.

Pepperberg, I. M. (2000). "The Alex Studies: Cognitive and Communicative Abilities of Grey Parrots." Harvard Univ. Press, Cambridge, MA.

Ralls, K., Fiorelli, P., and Gish, S. (1985). Vocalizations and vocal mimicry in captive harbor seals, *Phoca vitulina. Can. J. Zool.* **63**, 1050–1056.

Richards, D. G., Wolz, J. P., and Herman, L. M. (1984). Vocal mimicry of computer-generated sounds and vocal labeling of objects by a bottlenosed dolphin, *Tursiops truncatus. J. Comp. Psychol.* **98**, 10–28.

Smolker, R., and Pepper, J. W. (1999). Whistle convergence among allied male bottlenose dolphins (Delphinidae, *Tursiops* sp.). *Ethology* **105**, 595–617.

Tayler, C. K., and Saayman, G. S. (1973). Imitative behavior by Indian Ocean bottlenose dolphins (*Tursiops aduncus*) in captivity. *Behaviour* **44**, 277–298.

Todt, D. (1975). Social learning of vocal patterns and modes of their application in gray parrots (*Psittacus erithacus*). *Zeit. Tierpsychol.* **39**, 179–188.

Tyack, P. (1986). Whistle repertoires of two bottlenosed dolphins, *Tursiops truncatus*: Mimicry of signature whistles? *Behav. Ecol. Sociobiol.* **18**, 251–257.

Tyack, P. L. (1999). Communication and cognition. *In* "Biology of Marine Mammals" (J. E. Reynolds III and J. R. Twiss, Jr., eds.), Vol. 1, pp. 287–323. Smithsonian Press, Washington, DC.

Tyack, P. L., and L. S. Sayigh (1997). Vocal learning in cetaceans. *In* "Social Influences on Vocal Development" (C. Snowdon and M. Hausberger, eds.), pp. 208–233. Cambridge Univ. Press, Cambridge.

West, M. J., Stroud, A. N., and King, A. P. (1983). Mimicry of the human voice by European starlings: The role of social interaction. *Wilson Bull.* **95**, 635–640.

Whiten, A., and Ham, R. (1992). On the nature and evolution of imitation in the animal kingdom: Reappraisal of a century of research. *In* "Advances in the Study of Behavior" (P. J. B. Slater, J. S. Rosenblatt, C. Beer, and M. Milinski, eds.), Vol. 20, pp. 239–283. Academic Press, New York.

Minke Whales

Balaenoptera acutorostrata and
B. bonaerensis

WILLIAM F. PERRIN AND
ROBERT L. BROWNELL, JR.
*Southwest Fisheries Science Center,
La Jolla, California*

Minke whales are the smallest of the rorquals (family Balaenopteridae) and historically the last to be targeted by commercial whaling in the Antarctic as the greater whales were successively depleted.

I. Characters and Taxonomic Relationships

Until relatively recently, only one species of minke whale was thought to exist (common minke whale); all minke whales were referred to *B. acutorostrata* (Lacépède, 1804) (e.g., Stewart and Leatherwood, 1985). Morphological and genetic evidence of a second species accumulated through the last half of the 20th century, and the Antarctic minke whale, *B. bonaerensis* (Burmeister, 1867) came to be fully recognized in the late 1990s (Rice, 1998; IWC, 2001), although a few workers still treat the minke whales as consisting of one species (e.g., Reynolds *et al.*, 1999; Stewart, 1999) whereas others are withholding judgment about the best taxonomic arrangement pending further studies (e.g., Kato and Fujise, 2000). The two species are partially sympatric in the Southern Hemisphere, where a small form (unnamed subspecies) of the common minke whale, the dwarf minke whale, is much smaller than the Antarctic minke whale and possesses the distinct white flipper mark that is characteristic of the species in the Northern Hemisphere and absent in the Antarctic species (Fig. 1). The two species also differ in relative size and shape of several cranial features (Arnold *et al.,*

1987; Zerbini and Simões-Lopes, 2000; Kato and Fujise, 2000) and in mitochondrial DNA sequences (Pastene *et al.,* 1994).

Rice (1998) recognized three subspecies of the common minke whale: the North Atlantic minke whale (*B. a. acutorostrata*), the North Pacific minke whale (*B. a. scammoni,* formerly *B. a. davidsoni*), and the unnamed Southern Hemisphere dwarf minke whale. The dwarf minke whale is genetically closer to the North Atlantic than to the North Pacific form.

The common name comes from Norway. One story has it that a hapless whale spotter named Meincke identified a minke whale as a blue whale and thereafter small rorquals were called "Minkie's whale." Other common names that have been applied to the minke whales include "lesser rorqual," "little piked whale," "sharp-headed finner," and "lesser finback." Somewhat confusingly, the Antarctic minke is also called "ordinary minke whale" in IWC literature.

In both species, the rostrum is very narrow and pointed and there is a single ridge on the head. The dorsal fin is relatively tall and falcate and is located relatively far forward on the posterior one-third of the body (in comparison to the larger rorquals). The average length of the common minke whale in the North Atlantic at physical maturity has been estimated variously at about 8.5–8.8 m in females and 7.8–8.2 m in males (Horwood, 1990). One estimate for the North Pacific is 8.5 in females and 7.9 in males. Female Antarctic minke whales are estimated to average 9.0 m at maturity and males 8.5 m. The dwarf minke whale of the Southern Hemisphere is on average about 2 m shorter than the Antarctic minke whale (Kato and Fujise, 2000). The white flipper mark of the common minke whale extends up onto the shoulder in the dwarf subspecies of the Southern Hemisphere (Fig. 1). The BALEEN is white in the northern subspecies of the common minke whale but appears dark-gray or brown posteriorly in the dwarf subspecies due to a narrow dark fringe. In the Antarctic minke whale, the baleen plates are black on the left beyond the first few plates and on the right they are white in the first third and black in the rear two-thirds of the row. The baleen filaments in both species are coarser than in the sei whale but finer than in fin, blue, and bryde's whales, about 3.0 mm in diameter. The SKULL (Fig. 2) is larger in *B. bonaerensis* than in both northern and southern forms of *B. acutorostrata* (Fig. 3). The modal number of vertebrae in both species is about 49, ranging from 46 to 51.

II. Distribution and Ecology

In the North Atlantic, the common minke whale is found in summer as far north as Baffin Bay in the Canadian Arctic, Denmark Strait, and Svalbard in the Barents Sea (Brownell *et al.,* 2000). The wintering grounds are poorly known but extend at least to the Caribbean in the west and the Straits of Gibralter in the east. Affinities of minke whales reported from farther south to Senegal are unknown. In the North Pacific, the summer range extends to the Chukchi Sea. In the winter, common minke whales are found south to within 2° of the equator, although those south of central Baja California, Mexico, in the eastern North Pacific are of unknown relationship to the whales farther to the north. In the Southern Hemisphere, the distribution of the dwarf subspecies is poorly known. It extends as far north as 11°S in the western Pacific off Australia and 7°S off South America in the Atlantic nearly

Figure 1 *(Top) Dwarf minke whale* (Balaenoptera acutorostrata *unnamed subspecies*), *(middle) Antarctic minke whale* (B. bonaerensis), *and (bottom) North Pacific minke whale* (B. acutorostrata scammoni). *Photos courtesy of Hidehiro Kato.*

year round. In the far south, it is seasonally sympatric with the Antarctic minke whale on the FEEDING grounds during austral summer and occurs off South Africa during the fall and winter. Where sympatric with the Antarctic minke whale, it tends to occur in shallower, more coastal waters over the continental shelf.

Antarctic minke whales are found from 55°S to the ice edge during the austral summer, some occurring in the loose ice pack. Some have been recorded to overwinter in the Antarctic. During the austral winter, most retreat to breeding grounds at midlatitudes: 10–30°S in the Pacific between 170°E and 100°W, off northeastern and eastern Australia, off western South Africa, and off the northeastern coast of Brazil. In these areas their distribution is primarily oceanic, beyond the continental shelf break. One specimen is known from Suriname in the Northern Hemisphere.

Both species of minke whales are catholic feeders but specialize with season and area. In the North Atlantic, reported diet items include sand lance, sand eel, salmon, capelin, mackerel, cod, coal fish, whiting, sprat, wolffish, dogfish, pollack, haddock, herring, euphausiids, and copepods (Stewart and Leatherwood, 1985); krill (euphausiids) are important off West Greenland, whereas capelin and cod are dominant prey in eastern Newfoundland. In the North Pacific, major food items include euphausiids, Japanese anchovy, Pacific saury, and walleye pollack.

In the Antarctic, dwarf minke whales feed mainly on myctophid fishes but also on some euphausiids (Kato and Fujise, 2000), whereas Antarctic minke whales feed mainly on euphausiids.

Killer whales (*Orcinus orca*) prey on minke whales of both species. By one Russian estimate, Antarctic minke whales make up 85% of the diet of killer whales in the Southern Ocean (Stewart and Leatherwood, 1985).

III. Behavior and Life History

Common minke whales are noted for their curiosity about ships, often coming from afar to cross the bow or run with the vessel for minutes or hours. Their sudden appearance on the bow or alongside has startled many an observer. They are difficult to spot at a distance because of their small inconspicuous blow and brief surfacing behavior. Antarctic minke whales are reported to be easily approachable while feeding. Dwarf minke whales in the Great Barrier Reef region of eastern Australia readily approach and stay with divers and are the subject of a WHALE-WATCHING tourist operation.

Single animals are often seen and groups are usually small, consisting of two or three individuals, although larger aggregations of up to 400 may form on occasion in high latitudes.

Figure 2   *Dorsal (left), ventral (right), and lateral views of a skull of a dwarf minke whale* (Balaenoptera acutorostrata *unnamed subspecies*). *Photos courtesy of Hidehiro Kato.*

Minke whale sounds recorded in the North Atlantic included grunts, thumps, and frequency downsweeps ranging to 200 Hz. Similar downsweeps have been recorded in the Ross Sea (Stewart and Leatherwood, 1985).

Differential migration by sex and age leads to segregation by sex and breeding condition. Mating behavior has not been directly observed. Breeding is diffusely seasonal in the common minke whale in the Northern Hemisphere, with calves of 2.4–2.7 m appearing approximately 10 months after conception. Lactation lasts 45 months. Age at attainment of sexual ma-

turity has been estimated at about 7 years in males and 6 years in females. The pregnancy rate among adult females in some populations approaches 100%, suggesting an annual reproductive cycle. Little is known of the life history of the dwarf minke whale, but limited data available suggest similarity with that of the northern forms. The Antarctic minke whale also exhibits similar life history parameters (Horwood, 1990). Age at attainment of sexual maturity is 8 years in males and 7 to 8 years in females (although it may have been higher when overall whale densities were much higher earlier in the 20th century). Preg-

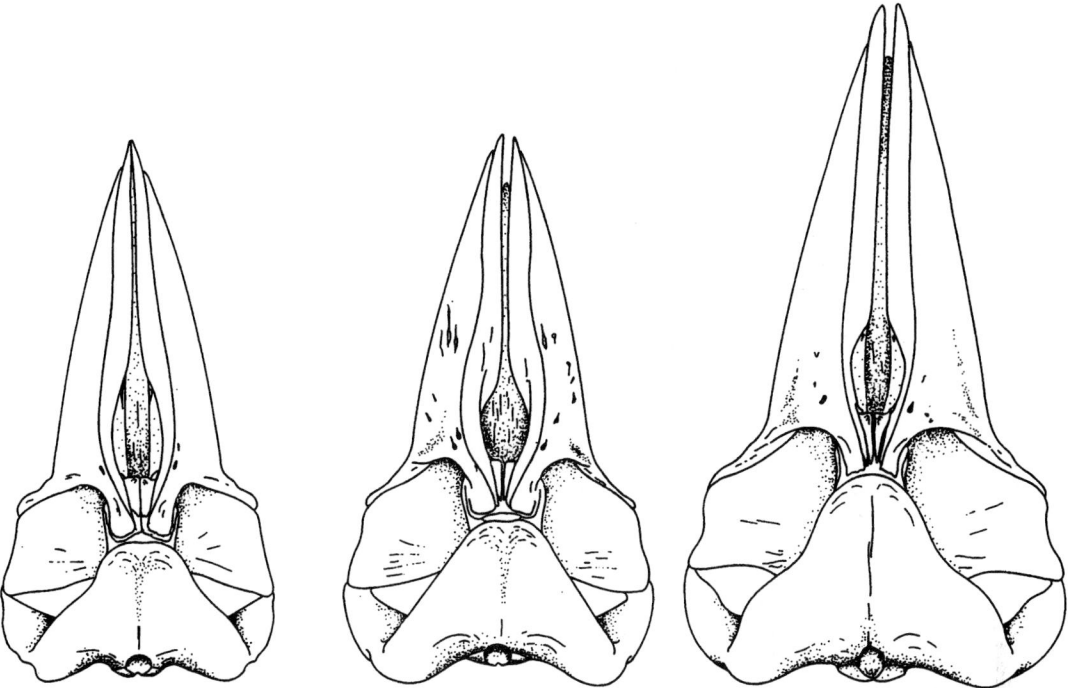

Figure 3 *Dorsal views of skulls (left to right) North Pacific minke whale* (Balaenoptera acutorostrata scammoni), *dwarf minke whale* (B. acutorostrata *unnamed subspecies*), *and Antarctic minke whale* (B. bonaerensis).

nancy rates remain at or near 90% for most of the year, again suggesting an annual cycle on average. Peak births are in July and August. During the feeding season, mature females are found closer to the ice than immature females, and immature males are more solitary than mature males.

IV. Conservation Status

The minke whales in the Southern Ocean were largely ignored in the early days of modern industrialized whaling because of their small size, but as the larger rorquals (blue, fin, and sei) were successively depleted, attention turned to the still abundant minkes in the early 1970s. After 1979, only minke whales were allowed by the IWC to be taken in factory-ship operations. Annual catches in the Antarctic ranged to about 8000 (details on all world catches are given in Horwood, 1990) by Japan and the USSR. Hundreds were also taken from land stations in Brazil and small numbers in South Africa. As of the 1985/1986 Antarctic season, all commercial whaling was banned under an IWC moratorium. Minke whale stocks in the Southern Hemisphere have been thought to be still in good condition and stable, with estimates (both species combined) in the neighborhood of 750,000. However, the results of most recent surveys have thrown these estimates into doubt; the estimates have been abandoned and a new assessment is planned for the year 2001 (IWC, 2001). Meanwhile, Japan has continued to take over 400 Antarctic minke whales annually under a research permit issued under the terms of the whaling convention.

Common minke whales have also been exploited commercially in the North Pacific and North Atlantic, in both land-based and pelagic whaling operations. Some stocks were depleted and became fully protected under IWC regulations, including the west Greenland, northeastern North Atlantic, and Sea of Japan–Yellow Sea–east China Sea stocks. The main whaling nations involved were Norway and Japan, with catches also by Korea, China, and the USSR. Catches from land stations in Japan continued until 1987. In the late 1990s, Norway recommenced commercial whaling on minke whales in the North Atlantic under an objection to the 1986 moratorium; the take in 2000 was 487 out of a self-imposed quota of 655. In the North Pacific, Japan began taking minke whales under a scientific research permit in 1994; the nationally established limit for 2000 was 100 animals [together with 50 Bryde's whales and 10 sperm whales (*Physeter macrocephalus*)].

Aboriginal subsistence whaling is exempt from the IWC moratorium on commercial whaling, and localized whaling for North Atlantic minke whales has continued under this provision in west Greenland. In 1999, 165 minke whales were landed. The impact of this whale fishery on the west Greenland stock is not known.

In summary, most minke whale stocks are in better condition than most stocks of the other large whales, but questions remain about the status of some populations and the effects of continued whaling.

See Also the Following Articles

Antarctic Marine Mammals ■ Bow-Riding ■ International Whaling Commission ■ Killer Whale ■ Whaling, Early and Aboriginal

References

Arnold, P., Marsh, H., and Heinsohn, G. (1987). The occurrence of two forms of minke whales in the east Australian waters with a description of external characters and skeleton of the diminutive or dwarf form. *Sci. Rep. Whales Res. Inst. Tokyo* **38**, 1–46.

Brownell, R. L., Jr., Perrin, W. F. Pastene, L. A., Palsbøll, P. J., Mead, J. G., Zerbini, A. N., Kasuya, T., and Tormosov, D. D. (2000). Worldwide taxonomic status and geographic distribution of minke whales (*Balaenoptera acutorostrata* and *B. bonaerensis*). Int. Whal. Commn meet. doc. SC/52/O27, 1-13. Available from IWC, 135 Station Road, Impington, Cambridge, CB4 9NP, UK.

Horwood, J. (1990). "Biology and Exploitation of the Minke Whale." CRC Press, Boca Raton, FL.

IWC (2001). Report of the Scientific Committee. *J. Cetacean Res. Manage.* 3(Suppl.).

Kato, H., and Fujise, Y. (2000). Dwarf minke whales; morphology, growth and life history with some analyses on morphometric variation among the different forms and regions. Int. Whal. Commn meet. doc. SC/52/OS3, 1-30. Available from IWC, 135 Station Road, Impington, Cambridge CB4 9NP, UK.

Pastene, L. A., Fujise, Y., and Numachi, K. (1994). Differentiation of mitochondrial DNA between ordinary and dwarf forms of southern minke whale. *Rep. Int. Whal. Commn.* **44**, 277–281.

Reynolds, J. E., III, Odell, D. K., and Rommel, S. A. (1999). Marine mammals of the world. *In* "Biology of Marine Mammals" (J. E. Reynolds III and S. A. Rommel, eds.), pp. 1–14.

Rice, D. W. (1998). Marine mammals of the world. *Soc. Mar. Mamm. Spec. Pub.* **4**, 1–231.

Stewart, B. S. (1999). Minke whale *Balaenoptera acutorostrata*. *In* "The Smithsonian Book of North American Mammals" (D. E. Wilson and S. Ruff, eds.), pp. 246–247. Smithsonian Press, Washington, DC.

Stewart, B. S., and Leatherwood, S. (1985). Minke whale *Balaenoptera acutorostrata* Lacépède, 1804. *In* "Handbook of Marine Mammals" (S. H. Ridgway and R. Harrison, eds.), Vol. 3, pp. 91–136. Academic Press, San Diego.

Zerbini, A. N., and Simões-Lopes, P. C. (2000). Morphology of the skull and taxonomy of southern hemisphere minke whales. Int. Whal. Commn meet. doc. SC/52/OS10, 1-28. Available from IWC, 135 Station Road, Impington, Cambridge, CB4 9NP, UK.

Molecular Ecology

A. Rus Hoelzel
University of Durham, United Kingdom

From a practical point of view, molecular ecology broadly refers to ecological studies that employ the tools of molecular biology. However, studies in this young field are often more specifically about the genetic diversity and structure of populations in an ecological context. Just as behavioral ecology explores the evolutionary consequences of the interaction between organisms and their environment from a behavioral perspective, molecular ecology explores the same interaction in the context of gene flow, genetic drift, and natural selection. Mammals in the marine environment have adapted to the challenges of aquatic life with some dramatic changes in anatomy, physiology, and BEHAVIOR. This is especially relevant to molecular ecology when characteristics of the marine habitat, or behavior enabled by these adaptations, affect patterns of dispersal, reproductive behavior, and demographics.

For example, while many cetacean species can move great distances and have broad tolerance for different habitats with respect to breeding and foraging, pinnipeds are tied to suitable terrestrial habitat for breeding (e.g., isolated from predators), but must forage at sea. Characteristics of the resource exploited by pinnipeds mean that some populations can forage near breeding grounds, whereas others must travel great distances in search of prey. Resources can also limit the distribution of suitable breeding grounds. These factors affect population genetic diversity and structure through an impact on dispersal range and the "effective size" of populations. The effective population size (N_e) is the size of an ideal population (with random mating and unaffected by processes such as mutation and selection) that would show the same rate of decay in genetic diversity as the observed population. Population size can be very large at a breeding colony where the resource is local and abundant (and therefore genetic diversity can be high), as seen especially for some otariid species. Phocid seals, however, often travel great distances on foraging excursions, and breeding colonies are often smaller than for otariids.

Genetic diversity decays more quickly over time in populations with smaller effective population size. Within populations, reproductive strategy can impact the level of diversity, as reproductive skew (such as polygynous mating) reduces N_e. The potential for polygyny has been suggested to depend on various aspects of resource exploitation and, in pinnipeds, on the consequences this has for the clumping of females (see Boness, 1991). If females are clumped and not too synchronous in estrus, males can monopolize the mating of multiple females, to the exclusion of other males (polygyny). Most of the otariid species are polygynous, but only a few of the phocid seals.

Another important factor influencing genetic diversity is demographic history, especially for species that have been the subject of intensive hunting (e.g., the right whales *Eubalaena* spp., gray whale *Eschrichtius robustus,* bowhead whale *Balaena mysticetus,* sea otter *Enhydra lutris,* and numerous species of pinnipeds). Elephant seals provide a good illustration of both the potential role of polygyny and the impact of excessive hunting on the loss of genetic diversity. There are two closely related species, the northern (*Mirounga angustirostris*) and southern (*M. leonina*) elephant seal. Behavioral and genetic studies have shown that these species are among the most polygynous of mammals (e.g., see LeBoeuf, 1972; Hoelzel *et al.,* 1999).

In the 19th century, elephant seals were exploited heavily for their blubber in both hemispheres. The southern species retained relatively large population numbers, but the more accessible northern species was forced through a severe population bottleneck (Bartholemew and Hubbs, 1960; Hoelzel *et al.,* 1993). Molecular genetic variation in the northern elephant seal is now low at mtDNA, allozyme, immune system, and repetitive DNA loci (see review in Hoelzel, 1999), consistent with predictions based on simulation models, given the severity of the bottleneck (Hoelzel *et al.,* 1993; Hoelzel, 1999). By comparing post-population-bottleneck genetic diversity with demographic simulation models and historic data, Hoelzel *et al.* (1993) estimated the severity of the population bottleneck to be less than 30 seals over a 20-year period, or a single-year bottleneck of less than 20 seals.

Simulation studies illustrate the role of polygyny in further reducing genetic variation during the period of recovery (Hoelzel, 1999). A survey of 54 allozyme loci at an average of 99 individuals per locus revealed no variation (Bonnell and Selander, 1972; Hoelzel *et al.*, 1993). The estimated bottleneck of 20 seals would not have been sufficient to eliminate variation at these loci in a monogomous species, but polygynous mating results in high variance in male reproductive success, which increases the impact on diversity. The reason is simply that relatively few males are contributing to the gene pool. Figure 1 illustrates how a bottleneck of less than 20 seals can account for the loss of allozyme diversity when the observed level of polygyny is taken into account. Only those simulations based on polygynous mating (Fig. 1, open bars) predict a level of post-bottleneck diversity that is low enough to be consistent with the measured levels of diversity.

Dispersal (and hence gene flow) among populations can be restricted by geographic barriers (such as continents) or by habitat specialization and energetic considerations. However, some marine mammal species exhibit dispersal over vast distances. For example, the sperm whale (*Physeter macrocephalus*) shows little genetic variation even among different ocean basins (Lyrholm *et al.*, 1999). Other species have been shown to be composed of geographically isolated populations that are differentiated genetically, such as Pacific vs Atlantic populations of humpback whales (*Megaptera novaeangliae;* Baker *et al.*, 1994)

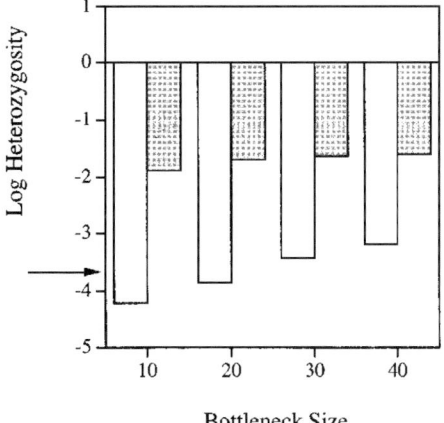

Figure 1 *This chart illustrates for the northern elephant seal the predicted impact of population bottlenecks of various sizes (along the* x *axis) on a measure of genetic diversity (heterozygosity) following the bottleneck. Shaded bars are based on simulations that assume monogamous mating, whereas open bars are based on simulations assuming observed levels of polygyny. No allozyme heterozygosity was found in the modern population, but the arrow indicates the average heterozygosity that would have been seen if just one individual had been heterozygous at just one locus (after Hoelzel, 1999).*

and walruses (*Odobenus rosmarus;* Cronin *et al.*, 1994). This "allopatric" pattern of differentiation is common among both marine and terrestrial species. What is more unusual, but perhaps quite important for the evolution of population structure in marine mammals, is the partitioning of the environment in a way that shows differentiation among sympatric (living in the same geographic region) and parapatric (neighboring regions) populations. This can happen when populations that differentiated in allopatry come back together or by processes that lead to choosing particular mates in preference to others ("assortative mating") in sympatry.

One mechanism for differentiation in sympatry is called "resource specialization" and has been described in detail for various terrestrial and aquatic species (see Smith and Skulason, 1996). It means that individuals of a species specialize on habitat or prey choice. Note, however, that these specializations can only lead to genetic differentiation if they also lead to assortative mating. Various studies show genetic differentiation between resource specialists in dolphin species. For example, the killer whale (*Orcinus orca*, the largest dolphin) travels in highly stable social groups (called pods). Pods, and populations of pods, tend to specialize on prey resources, with some focusing on marine mammal prey and others on fish prey. Sympatric populations of marine mammal and fish specialists in the eastern North Pacific were found to be differentiated genetically (Hoelzel and Dover, 1991; Hoelzel *et al.*, 1998a). The level of differentiation was great enough to suggest the possibility of two "cryptic" (morphologically indistinguishable) killer whale species, differing especially in foraging behavior. However, a global survey of genetic diversity did not support that interpretation (Hoelzel and Dover, 1991; A. R. Hoelzel *et al.*, unpublished results).

A number of dolphin species inhabit both coastal and offshore environments, which can differ with respect to the type and distribution of potential prey. Especially in regions where there is upwelling, the habitat in the marine littoral zone can be very different from the offshore habitat. Several studies of nearshore and offshore dolphin populations indicate intraspecific morphological and, in some cases, genetic distinctions. For example, nearshore and offshore forms of the pantropical spotted dolphin (*Stenella attenuata*) can be distinguished by tooth and jaw structure (Douglas *et al.*, 1984), and the Atlantic spotted dolphin (*S. frontalis*) is also found in nearshore and offshore populations. Two species of common dolphin (*Delphinus delphus* and *D. capensis*) have been classified by the length of the beak. In this case, ranges overlap, and both forms are sometimes found in the nearshore habitat. Rosel *et al.* (1994) found genetic differentiation between these forms.

The best-known example is that of the common bottlenose dolphin (*Tursiops truncatus*), which occurs in coastal and offshore populations throughout its range. In the eastern North Pacific, nearshore and offshore forms were originally classified as two different species—*T. gilli* (the nearshore form) and *T. nuuanu*—through a reappraisal recognizing extensive overlap in morphotypes later reclassified both as *T. truncatus*. In the western North Atlantic the nearshore and offshore forms have been described in some detail and show both morphometric (Mead and Potter, 1995) and genetic differentiation (Hoelzel *et al.*, 1998b). There were consistent differences between the two types in feeding behavior, with the nearshore form feeding primarily on coastal fish, whereas the offshore form concentrated on deep-water squid (Mead and Potter, 1995). The genetic distinction indicated low levels of gene flow (or no gene flow in the recent past) between the two populations (Hoelzel *et al.*, 1998b).

Marine mammals are highly mobile and, in many cases, show seasonal differences in DISTRIBUTION, such as the annual migration between breeding and feeding sites seen in baleen whales. This is an important consideration for the identification of populations for protection and MANAGEMENT, especially when breeding "stocks" mix on feeding grounds where they may be hunted (see review in Hoelzel, 1998). For example, common minke whale (*Balaenoptera acutorostrata*) populations on either side of Japan (off Korean and in the western North Pacific) are differentiated genetically at both allozyme (Wada, 1991) and mtDNA loci (Goto and Pastene, 1996). Both studies found evidence of seasonal mixing on feeding grounds to the north in the Okhotsk Sea. In another example involving minke whales, in this case the Antarctic species, *B. bonaerensis*, a temporal mixing of two genetically differentiated populations from Antarctic management areas IV and V was described based on mtDNA variation (Pastene *et al.*, 1996).

In general, the pattern and degree of genetic differentiation among populations are not predicted easily by geographic patterns for marine mammal species and instead depend on a complex interaction between life history and habitat. Even for the highly mobile and pelagic species, such as the minke whales, there can be considerable genetic differentiation among regional populations and, for some species (such as the killer whale), among populations in sympatry. What we know of the molecular ecology of these species emphasizes the need for more data and a careful consideration of the mechanisms affecting patterns of diversity within and among populations.

See Also the Following Articles

Forensic Genetics ■ Genetics, Overview

References

Baker, C. S., Slade, R. W., Bannister, J. L., Abernethy, R. B., Weinrich, M. T., Lien, J., Urbán, J., Cockeron, P., Calambokidis, J., Vasquez, O., and Palumbi, S. R. (1994). Hierarchocal structure of mitochondrial DNA gene flow among humpback whales world-wide. *Mol. Ecol.* **3**, 313–327.

Bartholomew, G. A., and Hubbs, C. L. (1960). Population growth and seasonal movements of the northern elephant seal, *Mirounga angustirostris*. *Mammalia* **24**, 313–324.

Boness, D. J. (1991). Determinants of mating systems in the Otariidae (Pinnipedia). *In* "Behaviour of Pinnipeds" (D. Renouf, ed.), pp. 1–44. Chapman & Hall, London.

Bonnell, M., and Selander, R. K. (1974). Elephant seals: Genetic variation and near extinction. *Science* **184**, 908–909.

Cronin, M. A., Hills, S., Born, E. W., and Patton, J. C. (1994). Mitochondrial DNA variation in Atlantic and Pacific Walruses. *Can. J. Zool.* **72**, 1035–1043.

Goto, M., and Pastene, L. A. (1996). Population genetic structure in the western North Pacific minke whale examined by two independent RFLP analyses of mitochondrial DNA. Report to the International Whaling Commission (SC/48/NP5).

Hoelzel, A. R. (1998). Genetic structure of cetacean populations in sympatry, parapatry, and mixed assemblages: Implications for conservation policy. *J. Hered.* **89**, 451–458.

Hoelzel, A. R. (1999). Impact of population bottlenecks on genetic variation and the importance of life history; a case study of the northern elephant seal. *Biol. J. Linn. Soc.* **68**, 23–39.

Hoelzel, A. R., Dahlheim, M. E., and Stern, S. J. (1998a). Low genetic variation among killer whales in the eastern North Pacific, and genetic differentiation between foraging specialists. *J. Hered.* **89**, 121–128.

Hoelzel, A. R., and Dover, G. A. (1991). Genetic differentiation between sympatric killer whale populations. *Heredity* **66**, 191–195.

Hoelzel, A. R., Halley, J., O'Brien, S. J., Campagna, C., Arnbom, T., Le Boeuf, B. J., Ralls, K., and Dover, G. A. (1993). Elephant seal genetic variation and the use of simulation models to investigate historical bottlenecks. *J. Hered.* **84**, 443–449.

Hoelzel, A. R., Le Boeuf, B. J., Reiter, J., and Campagna, C. (1999). Alpha male paternity in elephant seals. *Behav. Ecol. Sociobiol.* **46**, 298–306.

Hoelzel, A. R., Potter, C. W., and Best, P. (1998b). Genetic differentiation between parapatric 'nearshore' and 'offshore' populations of the bottlenose dolphin. *Proc. Roy. Soc. B.* **265**, 1–7.

Le Boeuf, B. J. (1972). Sexual behaviour in the northern elephant seal, *Mirounga anustirostris*. *Behaviour* **41**, 1–25.

Lyrholm, T., Leimar, O., Johanneson, B., and Gyllensten, U. (1999). Sex-biased dispersal in sperm whales: Contrasting mitochondrial and nuclear genetic structure of global populations. *Proc. Roy. Soc. B* **266**, 347–354.

Mead, J. G., and Potter, C. W. (1995). Recognizing two populations of the bottlenose dolphin (*Tursiops truncatus*) off the Atlantic coast of North America; morphologic and ecologic considerations. *IBI Rep.* **5**, 31–44.

Pastene, L. A., Goto, M., Itoh, S., and Numachi, K. I. (1996). Spatial and temporal patterns of mitochondrial DNA variation in minke whales from Antarctic areas IV and V. *Rep. Int. Whal. Comm.* **46**, 305–314.

Rosel, P. E., Dizon, A. E., and Heyning, J. E. (1994). Genetic analysis of sympatric populations of common dolphins (genus *Delphinus*). *Mar. Biol.* **119**, 159–167.

Smith, T. B., and Skulason, S. (1996). Evolutionary significance of resource polymorphisms in fishes, amphibians, and birds. *Annu. Rev. Ecol. Syst.* **27**, 111–133.

Wada, S. (1991). Genetic distinction between two minke whale stocks in the Okhotsk Sea coast of Japan. *Rep. Int. Whal. Comm.* (SC/43/Mi32).

Monk Seals
Monachus monachus, M. tropicalis, and *M. schauinslandi*

WILLIAM G. GILMARTIN
Hawaii Wildlife Fund, Volcano, Hawaii

JAUME FORCADA
Southwest Fisheries Science Center, La Jolla, California

The genus *Monachus* includes two endangered phocid species that live in the world's tropical and subtropical seas of the Northern Hemisphere: the Mediterranean monk seal (*M. monachus*) and the Hawaiian monk seal (*M. schauinslandi*). The numbers of these reclusive seals and a third species, the extinct Caribbean monk seal (*M. tropicalis*), once widely distributed around the Caribbean (Timm *et al.*, 1997) and last seen in 1952, have all declined due to killing by humans, habitat loss from coastal development and other activities, and prey resource competition with fisheries.

I. Characters and Taxonomy

Monk seals originated in the North Atlantic, with the Hawaiian species of probable Caribbean origin at least 15 million years ago across the Central American seaway. Unique among pinnipeds in some primitive, unspecialized skeletal and vascular anatomy, the Hawaiian monk seal is known as the "most primitive of living seals" (Repenning and Ray, 1977).

Mediterranean monk seals are uniformly black at birth with a conspicuous white ventral patch unique to each individual and distinct in shape by sex (Fig. 1). Newly molted seals of all ages are silvery gray dorsally and lighter ventrally, with the ventral patch pattern persisting through life. Juveniles have a medium to dark gray pelage, similar to that of adult females but with less COLORATION disruptions caused by scarring. Near the age of 4, adult males become almost uniformly black, with their white ventral patch; coloration in females is more variable, but not as dark as males dorsally and the ventral fur is also lighter. Adult males may also be slightly longer than females (Samaranch and González, 2000). Adult lengths are 2.3–2.8 m and weights are 240–300 kg.

Figure 1 *Three species of monk seals once existed; today only two species survive. The two living species are widely separated geographically, the Mediterranean monk seal being in the Mediterranean (Monachus monachus: top) and the Hawaiian monk seal occurring in Hawaiian waters (M. schauinslandi: bottom). Interestingly, Mediterranean males are larger than conspecific females and Hawaiian males are smaller than conspecific females. The Caribbean monk seal (M. tropicalis: center) is now extinct and, as a result, little is known of its biology. (Pieter A. Folkens/Higher Porpoise DG).*

Hawaiian monk seals are also black at birth, with some showing small white patches at various sites. These seals are also silvery gray following molt, with the fur color changing in juveniles to a yellow brown prior to the next molt and darkening through the year in adults. Hawaiian seals show no differences in fur coloration by sex. Adult females may attain a slightly larger size than males. Adult lengths are 2.1–2.4 m and weights are 170–240 kg.

II. Distribution and Ecology

Historically, the Mediterranean monk seal inhabited the entire Mediterranean Basin and the southeastern North Atlantic, from the Azores Islands to near the equator. The current distribution is severely contracted and fragmented (Aguilar, 1999), with the largest population (ca. 250–300 seals) in the Eastern Mediterranean, on the islands in the Aegean and Ionian Seas, and along the coasts of Greece and Turkey. Only two breeding populations are known in the Atlantic, one at the Cap Blanc peninsula in the Western Sahara and Mauritania (ca. 100 seals) and a smaller group in the Desertas Islands at the Madeira Islands (ca. 20 seals). Sightings are rare now in other areas within the historical range.

Hawaiian monk seals occur only in the Central Pacific in the mostly uninhabited northwestern Hawaiian Islands. A small breeding population also inhabits the main Hawaiian Islands, and monk seals have been reported on rare occasions south of the Hawaiian Archipelago at Johnston Atoll, Wake Island, and Palmyra Atoll. Hawaiian monk seals have a high fidelity to their island of birth; only about 10% of seals born at any of the major breeding islands will move to another island during their life. Very low genetic diversity is also evident in this species, which now includes 1300–1400 individuals.

Both monk seal species consume a highly diverse diet of diurnally and nocturnally active fish, octopus, squid, and lobster (Marchessaux, 1989; Goodman-Lowe, 1999). Proportions of these prey species in the diet vary by location, season, and age of the seals. Although most of the prey species are benthic in the coral reef ecosystem, a few are pelagic. Hawaiian monk seals have a broad prey base of at least 40 species. They forage within their resident atolls and along the fringing reefs; at one site, where the population is food stressed, some individuals forage at reefs 60–200 km from their breeding islands. Foraging commonly occurs near the breeding atolls and at sea mounts to depths of 100 m.

III. Life History and Behavior

Both monk seal species have protracted reproductive seasons, and copulation occurs in the water (Kenyon and Rice, 1959; Marchessaux, 1989; Ragen and Lavigne, 1999). Some female Mediterranean monk seals may be reproductively active at 3 years and Hawaiian monk seals at 4 years. However, in the Hawaiian seal, the mean is perhaps 6–7 years and maturity is as late as 10–11 years in females at a site where prey abundance is low, showing nutritional status is a critical factor in maturation. Mediterranean monk seals give birth year round, but mostly in the summer through early winter months. Monk seal births in Hawaii usually occur from February to August, peaking in April–June, but births are known in all months. At birth, Mediterranean monk seal pups weigh between 15 and 26 kg and gain weight fast during a suckling time of up to 4 months, with the females mixing feeding trips with pup attendance. Hawaiian monk seal pups are born at about 16 kg and are weaned at 6 weeks, after attaining a weight of 70–100 kg. Hawaiian females do not forage during the lactation period. After weaning, pups of both species survive on their fat reserves while they acquire prey-catching skills.

Females of both species have four functional mammary glands, and while they can give birth to single pups in consecutive years, they will also skip some years. Hawaiian females exhibit an average 381-day interval between annual births; thus these females give birth later each year until a year in which they do not give birth, and then they cycle earlier the following year. Some females give birth on a more interrupted schedule.

When Hawaiian monk seal female–pup pairs are near each other, accidental exchanges of the pups between the females can occur. Serious consequences result when the exchanged pups have suckled for very different times and are of very different size. The mothers will still wean their foster nursing pups after about 6 weeks of lactation, leaving one pup larger than normal size and one smaller, with the latter's survival chances highly compromised.

A mating tactic of some nondominant male Hawaiian monk seals is an attack by a group of these males (a few to over 20) on an adult female or an immature seal on some occasions. The attention of the dominant male in consort with the female becomes distracted by one or more of the challengers in the group, allowing the others to then breed with the female during bouts that may last over 3 hr in the water. The repeated and prolonged biting on the back of the female by males attempting copulation results in extensive trauma and tissue and fluid loss, often leading to her death. This detrimental behavior occurs primarily at seal colonies where the adult sex ratio is skewed toward males.

Mediterranean monk seals typically haul out to rest, molt, and give birth in protected sea caves throughout their range. They haul out on open beaches rarely, although this was believed to be a normal behavior for the species before the extensive exploitation of the last centuries. Hawaiian monk seals commonly haul out on open sandy beaches and will use rocky shorelines as well. Sea caves are rare in Hawaii.

Both the hair and the epidermis are sloughed and replaced during the annual molt in monk seals. This type of molt is similar to elephant seals, but unlike all other seals. In the Hawaiian monk seal the actual observed molting period is about 10 days, but based on an observed high proportion of time ashore before and after the hair–skin sloughing period, the entire physiological process is probably much longer.

The maximum known age for Mediterranean monk seals is 44 years and 30 years for a male Hawaiian seal.

IV. Conservation Status

Both living species of monk seals have been impacted greatly by human activities, from direct killing to competition for prey with fisheries and incidental disturbance of seals due to human presence on or near hauling and breeding beaches. Extinction of *M. tropicalis* in the 1950s and the international endangered status of the two remaining *Monachus* species result from this high sensitivity of the genus to direct and indirect human interactions.

The Mediterranean monk seal has been exploited since ancient times, and a significant decline in all of its range occurred in the second half of the 20th century. The total population was thought to be between 600 and 1000 individuals in the 1970s but at present is estimated at 350–450. Pup mortality increases sharply for Hawaiian monk seals when human beach use of a preferred habitat forces females to give birth at unsuitable pup-rearing sites (Kenyon, 1972; Gerrodette and Gilmartin, 1990). Both monk seals also interact with fisheries—bottomfish and longline hooks have been observed in Hawaiian seals, and Mediterranean seals have been entangled in active fishing gear and shot. ENTANGLEMENT in marine debris is a threat to both monk seals. Usually affecting pup and juveniles, the frequency of observations of entangled Hawaiian monk seals is increasing and, in many cases, where the debris is not removed, the seals are likely to be seriously injured or die.

Although the monk seal colony on the Western Sahara coast had been characterized by high adult survival rates, it also exhibited high pup mortality and very low recruitment, yet this colony had been considered the most viable population of the species. However, during May–July 1997, the size of this colony was tragically reduced by two-thirds due to a large-scale mass mortality event (Forcada *et al.*, 1999). The probable cause was a phytoplankton paralytic toxin; a morbillivirus was also implicated. The age structure of the surviving population was severely altered because juvenile mortality was insignificant compared to that of adults. This high mortality event severely compromised the recovery potential of the species in the Atlantic.

Hawaiian monk seal numbers have also been reduced due to human activities (Kenyon, 1972; Gerrodette and Gilmartin, 1990). Currently, however, all of the major breeding islands of the Hawaiian monk seal occur within federal and state government refuges where access is controlled and the foraging habitat is partially protected and managed to reduce fishery interactions (Wexler, 1993). The Hawaiian seal has a well-organized research and recovery effort, with guidance provided by a "recovery team" of scientists.

Recovery actions (Ragen and Lavigne, 1999) have included disentanglement of seals and two successful programs to enhance female survival. Underweight female pups were collected, rehabilitated, and then released back to the wild during the 1980s and early 1990s. In another effort, adult male seals that were killing females were captured and relocated to areas remote from the main breeding populations (Winning, 1998). Both projects contributed to population recovery in some colonies during the 1990s.

The remaining populations of both monk seal species are highly vulnerable to random catastrophic events such as die-offs due to introduced disease, the effects of inbreeding depression and low genetic variability, human disturbance, and competition with fisheries. The stability of the extant populations relies on high adult female survival rates. Fortunately, the Hawaiian monk seal population is moderately buffeted from anthropogenic pressures by the isolation and protected status of its major breeding habitat. The Mediterranean monk seal is not as fortunate. While a few protected areas have been established for the Mediterranean seal by Greece in the Aegean and by Portugal at the Desertas Islands in the Atlantic, only an immediate and significant reduction in anthropogenic pressures on the Mediterranean species and a range-wide coordinated recovery effort will avoid its extinction in the 21st century.

See Also the Following Articles

Competition with Fisheries ■ Endangered Species and Populations ■ Mass Die-Offs

References

Aguilar, A. (1999). "Status of Mediterranean Monk Seal Populations." RAC-SPA, United Nations Environment Program, Aloès Editions, Tunis.

Forcada, J., Hammond, P. S., and Aguilar, A. (1999). Status of the Mediterranean monk seal *Monachus monachus* in the western Sahara and the implications of a mass mortality event. *Mar. Ecol. Prog. Ser.* **188,** 249–261.

Gerrodette, T., and Gilmartin, W. G. (1990). Demographic consequences of changed pupping and hauling sites of the Hawaiian monk seal. *Conserv. Biol.* **4,** 423–430.

Goodman-Lowe, G. D. (1999). The diet of the Hawaiian monk seal, *Monachus schauinslandi*, from the Northwestern Hawaiian Islands during 1991–1994. *Mar. Biol.* **132,** 535–546.

Kenyon, K. W. (1972). Man versus the monk seal. *J. Mammol.* **53,** 687–696.

Kenyon, K. W., and Rice, D. W. (1959). Life History of the Hawaiian monk seal. *Pacific Sci.* **13,** 215–252.

Marchessaux, D. (1989). "Recherche sur la Biologie, l'Ecologie et le statut de Phoque moine, *Monachus monachus.*" GIS Posidonie Publ., ISBN No2-905540-13-3, Marseille, France.

Ragen, T. J., and Lavigne, D. M. (1999). The Hawaiian monk seal: Biology of an endangered species. *In* "Conservation and Management of Marine Mammals" (J. Twiss and R. Reeves, eds.), Vol. II, pp. 224–245. Smithsonian Institution Press, Washington, DC.

Repenning, C. A., and Ray, C. E. (1977). The origin of the Hawaiian monk seal. *Proc. Biol. Soc. Wash.* **89,** 667–688.

Samaranch, R., and González, L. M. (2000). Changes in morphology with age in Mediterranean monk seals (*Monachus monachus*). *Mar. Mamm. Sci.* **16,** 141–157.

Timm, R. M., Salazar, R. M., and Townsend Peterson, A. (1997). Historical distribution of the extinct tropical seal, *Monachus tropicalis* (Carnivora: Phocidae). *Conserv. Biol.* **11,** 549–551.

Wexler, M. (1993). A monk on their backs. *Natl. Wild.* **31,** 44–49.

Winning, B. (1998). The roller coaster ride of the Hawaiian monk seal. *California Wild* **51,** 30–35.

Morphology, Functional

CHRISTOPHER D. MARSHALL
University of Washington, Seattle

Function [is] the dynamic aspect of structure—structure changing in time.
—Picken

Functional morphology is a diverse field of biology that integrates anatomy, biomechanics, physiology, and behavior. It is the study of structure and its relationship with function; it is a way of viewing the world. The functional morphologist asks, "How does it work? The aquatic realm presents serious physical and physiological challenges for mammals. How mammals have overcome these challenges and adapted to the marine environment is the focus of marine mammal functional morphology. While our knowledge of the anatomy of marine mammals is extensive and continues to grow quickly, our understanding of the function of anatomical structures, organ systems, and functional complexes is incomplete. This article considers body size as an aquatic adaptation and then reviews major organ systems of marine mammals for which function is well established. Innovative solutions to aquatic challenges (i.e., buoyancy, thermoregulation, locomotion, diving, and feeding) are the focus. Because different groups have evolved different solutions for overcoming these challenges, this survey of marine mammal functional morphology is comparative in nature. Where appropriate, marine mammal function is contrasted with that of terrestrial mammals. Although the dog is one of the best known mammals anatomically, valuable comparisons can also be drawn using other subjects of veterinary anatomy, such as the horse and the ox.

"Marine mammal" is a general name for a diverse collection of nonrelated mammals that have returned to the sea for all or a portion of their lives. Generally, five groups of marine mammals are recognized: Cetacea (whales, dolphins, and porpoises), Pinnipedia (seals, sea lions, and walruses), Sirenia (manatees and dugongs), sea otters (Mustelidae), and polar bears (Ursidae). Only one ursid and two mustelids have exploited this habitat. Several mustelid species spend a majority of their time in freshwater, but only the sea otter (*Enydra lutra*) and the marine otter (*Lontra felina*) make the sea their primary habitat.

I. Body Size, Thermoregulation, and the Aquatic Environment

Thermoregulation is of utmost importance for marine mammals. Because the thermal conductivity of water is 25 times greater than air, marine mammals lose heat 25 times faster than terrestrial mammals. Like all mammals, marine mammals are warm-blooded and must maintain a constant body temperature. This can be a difficult task considering that many marine mammals inhabit the near-freezing waters of the polar oceans. One relatively simple solution to reduce heat loss is to increase body size. Large body size is advantageous because of the simple physical relationships among body size, surface area, and volume. As an animal gets larger and the linear body dimension (L_b) increases, surface area (SA) increases proportionally ($\propto$) to the second power, and volume (V) increases proportionally to the third power.

$$\text{a. } SA \propto L_b^2 \qquad \text{b. } V \propto L_b^3 \qquad (1)$$

These relationships can be clearly demonstrated by the following example. Imagine that the body of an animal is represented by a cube with 2-cm edges. Such a cube has a surface area of 4 cm and a volume of 8 cm. If we double the dimensions of the cube (4-cm edges), the surface area of the new cube would be 4^2 cm^2 or 16 cm^2, and the volume would increase to 4^3 cm or 64 cm^3. Small animals have large SA to V ratios, whereas large animals have a low SA:V. Heat is lost at the interface with the environment; a lower SA:V means lower heat loss. Compared to terrestrial mammals, marine mammals are large. Polar bears (*Ursus maritimus*) are the largest species in their family and sea otters are twice as large as their largest terrestrial relative. Aquatic life frees mammals from the constraints of gravity, allowing larger body sizes to be attained. Rorquals (whales of the family Balaenopteridae) have exploited this freedom; the blue whale (*Balaenoptera musculus*) is the largest living animal. In addition to increasing body size, marine mammals can decrease heat conduction through various adaptations of the integument and cardiovascular systems.

II. Integument System

The mammalian integument includes the skin, hair, associated glands (sebaceous and apocrine), claws, and digital pads. Mammalian skin is multifunctional. It provides protection from trauma and microbial intrusion, prevents desiccation, provides sensory information to the central nervous system (touch, pressure, vibration, temperature, and pain), and aids in thermoregulation.

A. Thermoregulatory Function of the Integument

The dense coat of hair (pelage) of mammals provides a layer of insulation. This adaptation is unique to mammals and has allowed them to move into cooler climates and to be active at night. Most mammals have two insulative types of hair: overlying guard hairs and an underlying layer of fine hair often called woolly hair (underfur). Hair keeps mammals warm by trapping and maintaining a layer of air close to the skin. Muscles spanning from the epidermis to guard hair follicles (arrector pilli muscles) can raise or lower the hair shaft, altering the thickness of the trapped air layer. The insulative power of the pelage is correlated with fur depth and hair density. As we have seen, heat loss or heat conduction (Q) is related to surface area. It is also related to the difference between body and environmental temperatures, and the physical material that heat must pass through (e.g., skin or air). The thicker the insulating layer (e.g., fur depth), the lower the heat loss. These relationships are summarized by the following equation:

$$Q = K\,SA[(T_b - T_e)]/L \qquad (2)$$

where SA is surface area, T_b is body temperature, T_e is environmental temperature, L is the distance between T_b and T_e, and k is the thermal conductivity constant of the material heat is being transferred through (e.g., skin and air). Thermal conductivity is simply a measurement of how easily heat flows through a material. Increasing fur depth effectively increases the distance (L) between T_b and T_e, which slows heat loss.

To be an effective insulator, hair must be kept dry. This is accomplished by a thin coating of oil on the hair shafts. The oil is secreted by sebaceous glands at the base of each hair follicle. The dense guard hairs and large amount of oil on the pelage of marine mammals maintain a layer of air even when an animal is completely submerged. The sea otter's pelage is mostly a dense undercoat with relatively few guard hairs; groups of 60–80 underfur hairs surround each guard hair (Kenyon, 1969). It is also the densest pelage of any mammal, approximately 130,000 hairs/cm^2, but ranging as high as 164,000 hairs/cm^2 on certain regions of the body. Its sebaceous glands secrete squalene (a lipid) instead of oil, which repels water more effectively. Frequent grooming is necessary to maintain the insulative properties of the pelage. Fur seals also have a dense, nonwettable pelage; hair density of the northern fur seal (*Callorhinus ursinus*) ranges up to 60,000 hairs/cm^2. The hair density of cats, dogs, and humans is 16,000 to 32,000 hairs/cm^2, 200–9000 hairs/cm^2, and 100,000 total, respectively. Sea otters and fur seals are particularly susceptible to oil spills. Oil penetrates to the skin, which results in loss of the thermal air layer.

To increase streamlining and reduce drag, some marine mammals have secondarily lost their hair and rely on blubber for thermoregulation. BLUBBER is simply the enlargement of the hypodermis and associated adipose tissue. The insulation provided by blubber is lower than fur; fat transmits heat three to five times faster than a dry, high-quality pelage (Schmidt-Nielsen, 1990). According to Eq. (2), Q can also be reduced by increasing the distance (L) between T_b and T_e, by changing k, or both. The insulative properties of blubber depend strongly on both thickness and lipid content. Blubber can be quite thick. Right whale (*Eubalaena* spp.) blubber can be as thick as 50 cm in some regions (Slijper, 1962). Thick blubber increases the distance between the body cavity and the water. The percentage of lipid content of the adipose tissue is seasonally variable and species specific, but can vary between 9 and 82% of wet blubber weight. Increasing the lipid content of the blubber (which is independent of blubber thickness) effectively changes the thermal conductivity constant (k).

B. Modification of the Integument to Reduce Drag

Seawater is 60 times more viscous than air and its density is three orders of magnitude greater. The increased viscosity and density of seawater resists movement through it; this resistance is called drag. The integument of marine mammals is designed to reduce drag. Energy saved by this reduction can be used for other activities, such as feeding and reproducing. Alternatively, reduced drag may allow individuals to swim faster or further for the same amount of invested energy. There are two types of drag to consider: frictional and inertial. Frictional drag occurs as fluid moves past the body. Water in contact with the skin or hair does not move; this region is called the boundary layer (Vogel, 1994). The fluid above this layer is forced to shear or slide past the boundary layer. A gradient is created in which thin layers of fluid move at different speeds around the body. Layers further away from the boundary layer move faster than those close to it. Frictional drag is proportional to the surface area in contact with the water. Marine mammals with hair usually lack arrector pilli muscles. This may allow their hair to lie flatter (although with some loss of thermal protection), which would create a smoother, low-drag profile compared to terrestrial mammals that swim (e.g., muskrats, *Ondatra* spp.). To decrease frictional drag further, some marine mammals have secondarily lost their hair. Smooth surfaces create less frictional drag than rough surfaces. The exceptionally smooth skin in some marine mammals is accomplished by high turnover rates of outer cell layers. The bottlenose dolphin (*Tursiops truncatus*) replaces the outermost epidermal cells every 2 hr. This high turnover rate also keeps the surface free of algae, barnacles, and other settling organisms. As marine mammals move through the water, pressure is generated. This pressure generates another form of drag, inertial drag. Water pressure is highest near the head and lowest at the widest point of the body. Fluid flows from areas of high pressure to low pressure. Fluid departs the body at the widest point and results in a wake trailing the animal. The size of the wake is proportional to the amount of inertial drag generated. Inertial drag can be reduced by delaying the departure of the fluid from the body for as long as possible. This is accomplished by a streamlined, fusiform body. This shape results in a smaller wake and therefore a smaller inertial drag. Blubber is important in STREAMLINING and sculpting the contour of body shape. The goal of body sculpting is to reduce both types of drag.

C. Buoyancy Function of the Integument

A thick, dense, insulating pelage has the disadvantage of being buoyant. Increased buoyancy is also a potential problem associated with thick blubber. Materials that are less dense than water (blubber, lungs) float, whereas materials that more dense than water (bone, muscle) sink. Archimedes' principle states: the buoyant force on a body immersed in fluid is equal to the weight of the fluid displaced by that object. If body weight is less than the weight of displaced water, then the animal experiences an upward force, buoyancy (*B*). If body weight is greater than the weight of displaced water, then the animal experiences a downward force, gravity, and sinks. Proper buoyancy is critical to survival. The relationship between tissue density and weight of displaced water is shown by the expression:

$$B = (\rho_b - \rho_w)Vg, \tag{3}$$

where ρ_b is density of the body, ρ_w is density of water, V is volume of the body, and g is the force of gravity. Marine mammals are generally more dense than water and tend to sink. They can alter their buoyancy by altering ρ_b. Buoyancy can be increased by a reduction of the density of heavy materials (e.g., bone), an increase in the volume of organs that are less dense than seawater (e.g., lungs), or some combination thereof. The integument of marine mammals that use blubber for insulation may impose an additional buoyant force on the animal because blubber is less dense than seawater. It has been shown that air trapped by pelage imposes an additional buoyant force in an aquatic opossum (*Chironectes minimus*). Sea otters likely experience a buoyant force from their pelage as well. Energy not expended toward maintaining neutral buoyancy can be used for other activities. Preliminary evidence for sirenian integument suggests that it may be heavier than water and may function as ballast.

III. Musculoskeletal System

The musculoskeletal system is responsible for movement of the body. It consists of muscle, bone, tendons, ligaments, and joints. It is of primary importance in locomotion, feeding, and respiration. It also provides support and protection for internal organs and aids in sculpting the shape of the body. Marine mammals locomote in diverse ways, which is reflected in the design and adaptations of their musculoskeletal system.

A. Leverage

Muscles are contractile elements that are attached to bones and cross bony joints by tendons; shortening of muscles produces movement at the joint. This design is based on leverage. Levers are simple devices that transmit forces from one place to another using a pivot. For example, a person might want to move a large rock that is too heavy to pick up. With the aid of a rigid beam and a smaller rock, one could build a simple lever by placing the tip of the beam under the large rock and pivoting the beam on top of the small rock. The large rock can be moved by imposing force on the beam at the end opposite to the rock. This force is known as the in force (F_{in}), the small rock is the fulcrum, and the force generated to move the load is the out force (F_{out}). The distance between the fulcrum and the in force is the in-lever arm (L_{in}), and the distance between the fulcrum and the load is the out-lever arm (L_{out}). The directions of F_{in} and F_{out} are called the lines of action. In the simplest case, the lines of action of F_{in} and F_{out} are parallel to each other and at right angles to the lever. Torque (τ) is the rotational movement of the lever. It is the product of a force and its lever arm; therefore, each lever has an in torque and an out torque:

$$\text{a. } \tau_{in} = F_{in}L_{in} \qquad \text{b. } \tau_{out} = F_{out}L_{out} \tag{4}$$

When $F_{out}L_{out} = F_{in}L_{in}$, no motion occurs. If $F_{out}L_{out} > F_{in}L_{in}$, the direction of the force is toward F_{out} and the load is moved. If $F_{out}L_{out} < F_{in}L_{in}$, the direction of the lever moves toward F_{in}. Changes in the distance between the fulcrum and the load change the effort needed to move the rock. Mechanical advantage (*MA*) is the ratio of the in-lever arm to out-lever arm and also the ratio of the out force to the in force:

$$MA = L_{in}/L_{out} = F_{out}/F_{in} \tag{5}$$

The SI (Système Internationale) unit for force is the newton (N). A lever system in which a F_{in} of 5 N results in a F_{out} of 10 N would have a mechanical advantage of 10/5 or 2. Greater mechanical advantage is attained with levers that have a long L_{in} and short L_{out}. Consider the forelimb (Fig. 1), which extends by the contraction of the triceps. The triceps spans posteriorly from the scapula and humerus (arm), across the joint of the proximal limb to a bony process of the forearm (ulna) called the olecranon. The contraction of the triceps contributes the F_{in}, the joint is the fulcrum, the olecranon process is the in lever, and the forearm is the out lever. Notice that the length of the olecranon (L_{in}) is longer and the forelimb is shorter (L_{out}) in the sea lion compared to the dog (Fig. 1). Another component to levers is velocity (*v*). Each lever arm possesses a velocity (v_{in} and v_{out}). The length of the lever arms influences the velocity such that $v_{in}L_{out} = v_{out}L_{in}$. This is the opposite of the force-lever arm relationship ($F_{out}L_{out} = F_{in}L_{in}$). Therefore, animals that can flex their distal limbs quickly (a high v_{out}) must

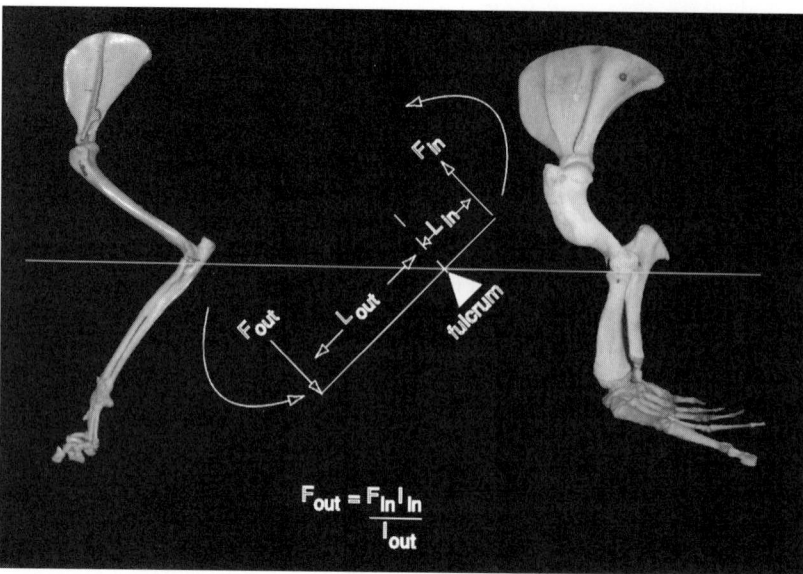

Figure 1 *Forelimb lever. (Left) Forelimb of a dog. (Right) Forelimb of a Steller sea lion* (Eumetopias jubatus). *The horizontal line indicates the pivot point or fulcrum in both forelimbs. The humerus (arm) lies just above the line. The ulna and radius of the forearm lie just below the line. Note the difference in length of the in-lever arm between the dog and the sea lion.*

possess a short L_{in}. However, this reduces mechanical advantage, which increases with increasing L_{in}. There is a trade-off between mechanical advantage and velocity. Associated structures of the musculoskeletal system such as sesamoid bones, bursae, and synovial tendon sheaths function to increase the surface area of attachments, protect tendons that pass over bony prominences, redirect contractile forces for greater effective applied force during movement, and reduce friction at joints.

B. Locomotion

The postcranial skeleton consists of the vertebral column (axial skeleton) and appendages (appendicular skeleton). The skeletons and musculature of cetaceans and sirenians have been modified greatly compared to the dog (Howell, 1930; Slijper, 1962, Evans, 1993; Rommel, 1990; Domning, 1977, 1978; Pabst, 1993, 1999). The sacrum, sacral vertebrae, and hindlimbs have been lost and only rudiments of the pelvis remain. The forelimbs are modified hydrofoils (wings) that assist in aquatic locomotion. Pinnipeds still possess hindlimbs, but differences occur between the more aquatically adapted phocids (seals, family Phocidae) and otariids (sea lions, family Otariidae). Compared to other marine mammals, polar bears and sea otters have not modified their limbs greatly.

Thrust is an important function of the postcranial musculoskeletal system; it can be increased by generating large in forces (F_{in}) and high mechanical advantage. A larger F_{in} can be obtained by increasing the number of muscle fibers in parallel; the force that a muscle can produce is directly proportional to its cross-sectional area. In obligate marine mammals such as cetaceans and sirenians, thrust is primarily generated by oscil-

lation of their axial skeleton. The remaining marine mammals generate thrust in water with their appendages. Otariids locomote using their forelimbs whereas phocids and sea otters generally use their hindlimbs. Walruses (*Odobenus rosmarus*) and polar bears use all four limbs to produce thrust.

The vertebral column of cetaceans is similar to that of the dog with several modifications. The cervical vertebrae are compressed anterioposteriorly; normally the first and second vertebrae, if not all cervical vertebrae, are fused. This compression shortens the neck and aids in streamlining the body. The cervical and thoracic vertebrae of cetaceans are greater in height (dorsoventral) than in length (anterioposterior), which stiffens the anterior body. The lumbar and caudal vertebrae are longer than they are high, which allows a greater range of motion in the dorsal–ventral direction. The lower vertebral column functions as a variably flexible beam that is bent dorsally by epaxial muscles and ventrally by hypaxial and abdominal muscles. Some marine mammals have modified the spines of their vertebrae (Figs. 2A–2C). Compared to the dog, the spinous and transverse processes of cetacean thoracic and lumbar vertebrae are long, which increases the surface area for the attachment of large epaxial and hypaxial muscles (Fig. 2B) and increases mechanical advantage. Only the transverse processes of sirenian lumbar (Fig. 2C) and caudal vertebrae are enlarged; their vertebral column likely functions in a manner similar to cetaceans, but with decreased power and a smaller range of motion.

Flukes at the end of the vertebral column deliver forces produced by muscles to the water, whereas the dorsal and pectoral fins are used for stability and maneuverability. Fins, flippers, and flukes can be thought of as hydrofoils (Vogel, 1994). They are streamlined in cross section. The rounded leading edge

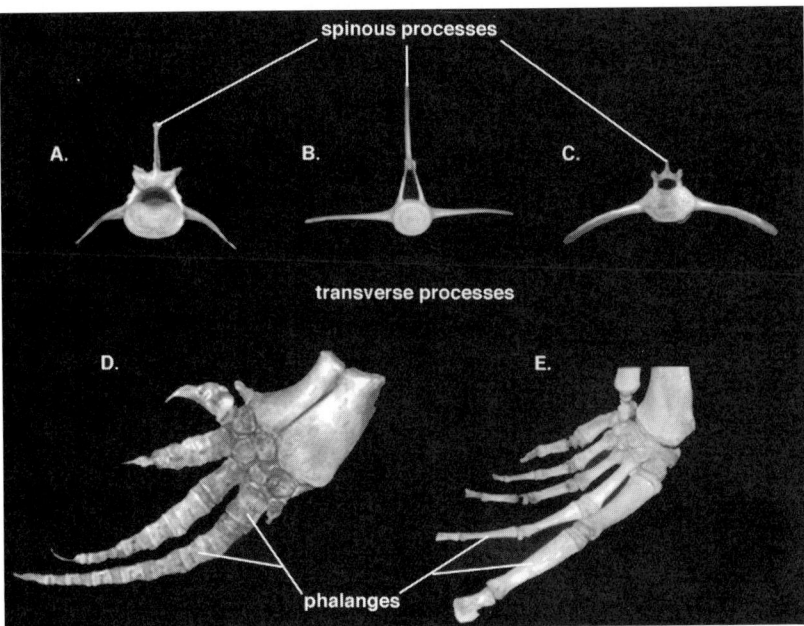

Figure 2 *Vertebrae and flippers. Lumbar vertebra of a dog (A), bottlenose dolphin (B), and a West Indian manatee (C). Note the length of the spinous and transverse processes. (D and E) Manus of a bottlenose dolphin* (Tursiops truncatus) *and Steller sea lion, respectively. Note that the increase in flipper length is achieved by hyperphalangy in the dolphin and elongation of the phalangeal bones in the sea lion.*

tapers to a trailing edge much like an airplane wing. The dorsal fins and flukes are not supported by skeletal elements. However, the pectoral fins of cetaceans and the flippers of other marine mammals are supported by bones of the forelimb and manus (hand). The bones of the forelimb are flat and form the basis of the pectoral fin shape; the radius is located on the leading edge of the flipper and the ulna on the trailing edge (Figs. 2D and 2E). The aspect ratio (AR) is the ratio of the length of the fin (span) to its width (cord) parallel to the fluid flow. The AR of the fin affects the magnitude of lift and drag forces produced. Hydrofoils with a high AR produce large lift forces with minimal drag forces, whereas low AR hydrofoils produce small lift forces and greater drag. Dolphins have a moderate-to-high AR relative to other aquatic vertebrates, a design that may lower the energetic cost of swimming. The high AR in dolphins is accomplished by elongation of the manus by hyperphalangy (increased number of phalanges or finger bones per digit; Fig. 2D), which is unusual among mammals. The bones of the manus and the forelimb are encased in connective tissue and skin and are relatively immobile. The mobility of the entire pectoral fin varies among cetaceans. For example, bottlenose dolphins have relatively immobile flippers that maintain stability during locomotion, whereas the long pectoral fins of humpback whales (*Megaptera novaeangliae*) are very mobile and specialized for high maneuverability. The pinniped forelimb is also flattened and the humerus is short. As in dolphins, the forelimb is rigidly immobilized by connective tissue and skin. The entire limb is mobile and important in maneuvering during locomotion. Forelimbs of sirenians can be used alone for slow locomotion and to

move the body across the bottom during feeding. As mentioned previously, otariid and phocid locomotion are divergent. Otariids swim using their forelimbs as hydrofoils to generate lift (thrust) and their hindlimbs as a rudder (Hildebrand, 1988; Fish, 1996). Their scapulae (shoulder blades), which are modified compared to the dog and other marine mammals, have a prominent spine and an accessory spine that subdivides the supraspinous fossa. The humerus possesses large tuberosities, and the ulna has a long olecranon process (Fig. 1). This increased surface area is indicative of larger brachial (arm) muscles. Larger muscles have a greater cross-sectional area and are capable of producing more force. These specializations increase the mechanical advantage not only by increasing F_{in}, but also by increasing the in-lever arm and decreasing the length of the out-lever arm. Although pinnipeds do not exhibit hyperphalangy, the individual metacarpals (hand bones) and phalanges of otariids are elongated (Fig. 2E). The increased length of the foreflipper increases the AR of the entire forelimb. On land, otariids are able to rotate their hindlimbs beneath the body and ambulate as tetrapods; they are remarkably agile. The forelimbs bear much of the weight. The entire palmar surface of the elongated manus is placed on the ground and each forelimb is alternately moved when walking. The head and neck are swung side to side to facilitate the movement of their bulky flippers. The larger transverse processes of the cervical vertebrae reflect the greater behavioral repertoire of head movement during land-based locomotion. Transverse and spinal processes of the lumbar vertebrae among otariids are small and differ little from those of the dog.

Thrust in phocids is generated by the hindlimbs using short, alternate, medial strokes toward the midline (King, 1983; Fish, 1996). During each power stroke the hindflippers expand, increasing their surface area. The hindflippers narrow during lateral recovery strokes, reducing drag. The medial–lateral motions of the hindlimbs are amplified by the simultaneous side-to-side movement of the posterior vertebral column. The transverse processes of lumbar vertebrae are enlarged for greater surface area for muscle attachment, although not to the extent seen in cetaceans and sirenians. Attachment of large hindlimb muscles, which presumably have a larger cross-sectional area compared to otariids, increase F_{in}. Additionally, the ilium (part of the pelvis) of phocids is angled such that the medial face is oriented more anteriorly, increasing the surface area for axial swimming muscle attachment. The femur of phocids is short and flattened. As with the humerus of otariids, the short femur decreases the out-lever arm, which increases the mechanical advantage. The ball and socket of the femur–pelvic joint is shallow and several ligaments are absent. These modifications increase the range of motion. The phocid forelimb lacks the specialization of the otariid forelimb; an accessory scapular spine is absent and the manus is not elongated. The forelimbs are held close to the body and used only for changes in direction. The shorter neck of phocids produces a more streamlined body than otariids. The shorter neck is a result of the slight retraction and greater curvature of the cervical vertebral column, relatively smaller neck musculature, and differences in sculpting effects of blubber. The length of the cervical region relative to the entire vertebral column is the same proportionally in phocids, otariids, and dogs.

Walruses incorporate aspects of both otariid and phocid locomotion. Thrust is produced by the hindlimbs but the forelimbs also contribute. The cross-sectional area of the hindlimb muscles of walruses is relatively larger than that of sea lions and therefore walruses are able to produce a greater F_{in}. The femur is short compared to phocids. Their forelimb musculature is not as enlarged as that of sea lions. The walrus scapula is intermediate between that seen in otariids and phocids; it does not possess an additional scapular spine, but the supraspinous fossa is enlarged compared to phocids.

The sea otter swims on the surface by undulation of its hindlimbs and tail in the vertical plane (Howell, 1930). Although no mechanical studies of sea otter swimming are available, the muscles of the hindlimb are large, the femur is short, and the digits and phalanges are elongated. These characteristics suggest an emphasis on hindlimb propulsion.

Polar bears are excellent shallow divers. They use all four limbs in a paddle stroke much like a dog. Their appendages are not modified for increased thrust and little information exists regarding their mechanics.

C. Musculoskeletal System and Buoyancy

Buoyancy can be modified by altering the density of body tissues. Bone is the densest material of the body and a likely place to investigate for buoyancy modifications. One might imagine that the skeleton of cetaceans would be denser than terrestrial mammals to offset the buoyant nature of the blubber. The long bones of cetaceans are solid and lack a medullar canal; this condition could increase the density of the forelimb. However, relative skeletal density data of aquatic and terrestrial mammals are not well established. Sirenians have clearly modified their skeletal density. These marine mammals possess a heavily mineralized, dense, and thickened skeleton compared to terrestrial and aquatic mammals. Sirenian skeletons are commonly described as exhibiting pachyostosis (thickening of the bone). In addition to increased cross-sectional area, cancellous (spongy) bone has been replaced by compact (hard) bone, a condition known as osteosclerosis. The sirenian condition is best described as pachyosteosclerosis. This condition is not pathological and may result from altered functions of the endocrine system. Typically, mammalian long bones contain a marrow-filled cavity, the primary center for red and white blood cell, and platelet production (collectively called hemopoiesis). Both sirenians and cetaceans lack this medullar canal; the site of hemopoiesis is in the vertebrae. The increased size and density of the sirenian skeleton is hypothesized to function as ballast and aid in buoyancy compensation.

D. Skull Structure

Marine mammals display an amazing amount of divergence and diversity in their SKULL structure compared to their terrestrial counterparts. Their skull morphologies transcend many aspects of their biology and are linked intimately with their behavioral ecology. The hallmark of mammals is the differentiation of teeth and complex patterns of mastication. However, many marine mammals have secondarily modified or lost these typical mammalian characteristics.

The skull of modern cetaceans is perhaps the most derived among mammals. In contrast to the dog, cetaceans have drastically remodeled the rostrum, nares, cranium, ear bones (petrosal bones), and mandible (Fig. 3). The maxilla and premaxilla elongate the facial region, but the cranium is shortened due to the overlapping (telescoping) of cranial bones and is dome like in appearance. This disparity of facial length vs cranial length is variable among cetaceans [e.g., the short blunt face of pilot whales (*Globicephala* spp.) and the long narrow rostra of river dolphins]. The concomitant shortening of the cranial vault with the elongation of the rostrum results in the posterior migration of the nares and associated structures to the dorsum of the skull, posterior to the orbits. The position of the nares is important during locomotion and allows quick and efficient gas exchange without breaking "stride."

Odontocete mandibles apparently perform two functions: sound reception and food acquisition. The mandibular bone is relatively thin and the medullar canal is greatly expanded and filled with fat. Its posterior–medial surface is absent, exposing the medullar canal and mandibular fat. Due to shortening of the cranium, the end of the mandible lies in close proximity to the ear bones (Fig. 3B). The mandibular fat extends posteriorly and attaches to the ear bone. It is hypothesized that returning sound (generated by echolocation) is picked up by the mandible and delivered to the ear along the length of the mandibular fat. In all modern delphinoids the ear bones are completely separated and detached from the rest of the skull, suspended by a series of five ligaments, and surrounded by a mucous foam-like substance. These ligaments acoustically isolate the ear bones

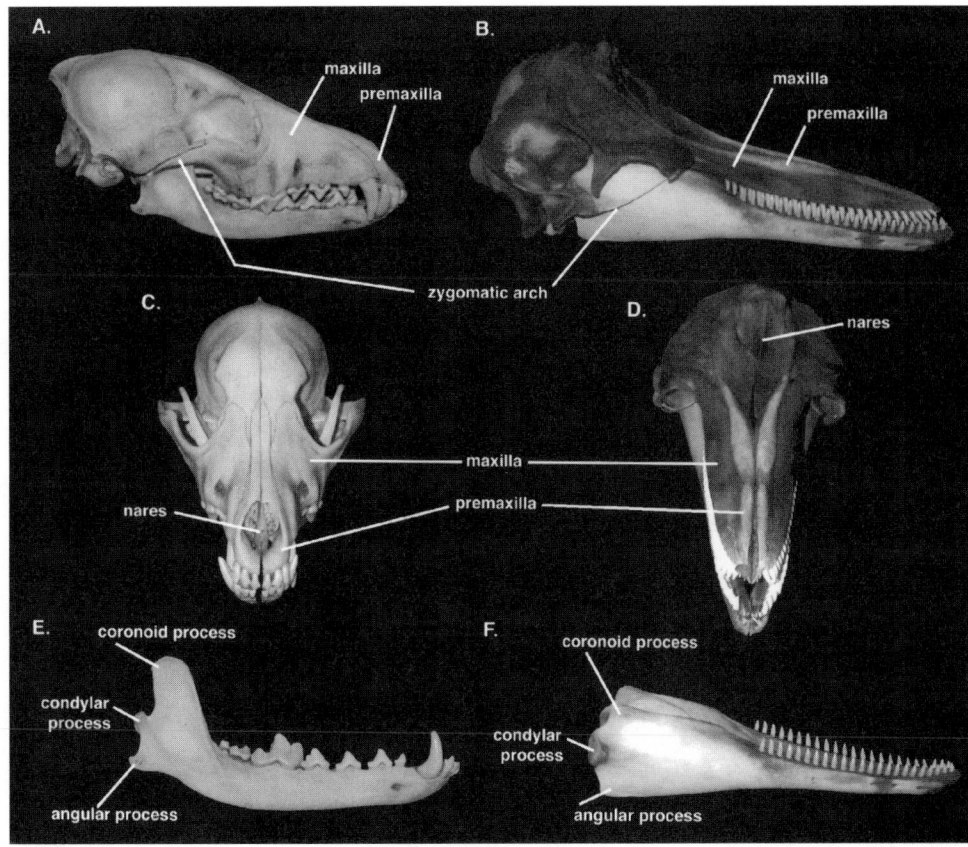

Figure 3 *Comparison of the skull morphology of a dog (A) with a typical odontocete, the bottlenose dolphin (B). Note the elongation of the premaxillary and maxillary bones, thin zygomatic arch, and simple mandible.*

from the skull. This would permit a single acoustic pathway to the ear by way of the mandible. Odontocete jaws can be virtually edentulous (ziphiids) or filled with several hundred simple, single-rooted homodont TEETH (plantanistids). The condylar process of the mandible (articulation of the mandible with the skull) is simple (Fig. 3F) and allows for only a simple dorsal–ventral motion; most mammals can also move their jaws side-to-side (herbivores excel at this). In general, delphinid muscles of mastication (temporalis, masseter, and pterygoids) are relatively small. The pterygoid muscles seem to be the dominant muscle group—a condition found elsewhere only in anteaters. The killer whale (*Orcinus orca*) is an exception; this delphinid has a dominant temporalis, perhaps related to the requirements of taking large prey. Compared to the dog, skull attachments for the temporalis and masseter are also reduced (i.e., zygomatic arch, and coronoid and angular processes; Figs. 3E and 3F).

The facial region of mysticetes dominates the skull even more so than in odontocetes (Fig. 4D). The maxilla arches higher above the mandible compared to odontocetes. Among mysticetes the height of the maxillary arch in balaenids (whales of the family Balaenidae) is distinctively greater than in balaenopterids. The jaws are edentulous; teeth have been replaced by baleen—these sheets of keratinized epidermis (much like

hair, nails, and horns) are used to filter water. The mysticete mandible is loosely articulated with the skull, and the mandibular symphysis (anterior region of the mandible where right and left sides join) is not fused. These features allow the mandible to open at nearly 90° to the long axis of the body. As in delphinids, the muscles of mastication of mysticetes and their bony attachments are also relatively small (Slijper, 1962). Clearly the oral apparatus of cetaceans is not designed for mastication.

The sirenian skull is also derived relative to other mammals. The facial region also dominates the skull; however, whereas cetaceans have evolved a relatively enlarged maxilla, sirenians exhibit pronounced and expanded premaxillary bones (Figs. 4G and 4H). This is particularly true of dugongs (*Dugong dugon*; Fig. 4H). The large anterior bones of both manatees and dugongs increase the surface area for the attachment of large facial muscles that form a muscular hydrostat and are involved in feeding; the large narial basin possibly allows for greater movement of these facial muscles. Sirenians are unique in that they are the only herbivorous marine mammals in existence. Unlike cetaceans, sirenians do masticate. The temporalis muscle is well developed but the masseter muscle is relatively simple (unlike terrestrial herbivores). The coronoid process of the mandible is expanded and modified. The condylar process

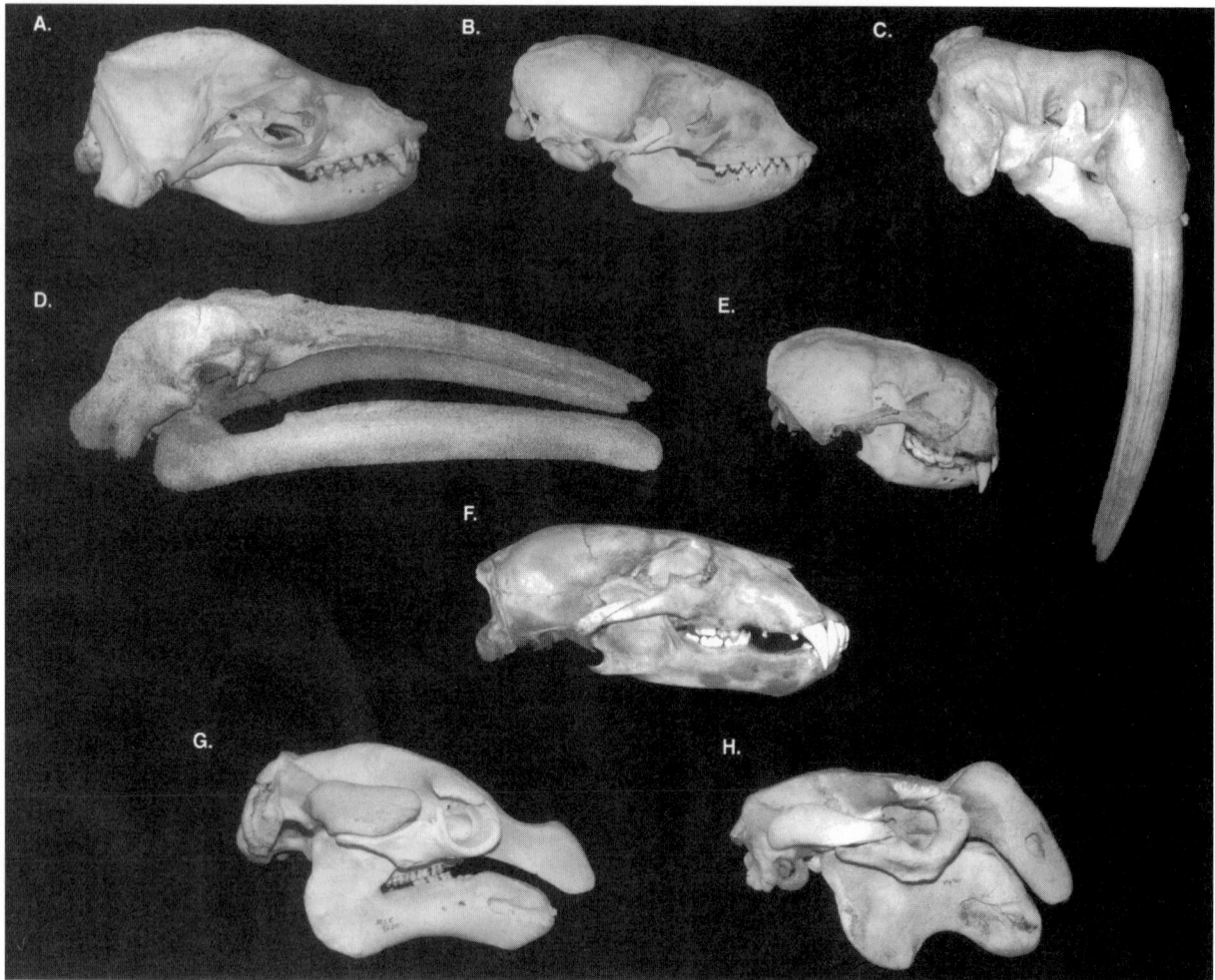

Figure 4 *Diversity of skull morphology among marine mammals: (A) California sea lion, (B) harbor seal, (C) walrus, (D) gray whale, (E) sea otter, (F) polar bear, (G) West Indian manatee, and (H) dugong.*

is small and flat. The mandibular fossa is shallow, allowing for great mobility of the tympano-mandibular joint. The mandible is large and heavy; only cheek teeth are present. An unusual skull character is the massive pterygoid process, which may provide a second jaw articulation (ptyergoid-mandibular) that could allow the lower jaw to rotate and produce transverse (side-to-side) movements during mastication (Domning, 1978). Such an articulation would displace the fulcrum of the lower jaw from the condyles to the pterygoid processes; an unusual situation among mammals.

Pinnipeds represent a diverse group of marine mammals with varying life histories. However, some generalizations are possible. Among marine mammals, their skulls are most similar to that of the dog (Figs. 3A, 4A, and 4B), with notable exceptions such as walruses (Fig. 4C). Pinnipeds are characterized by a large rounded cranium, short snout, large orbits, and narrow interorbital distance. In general, the skulls of otariids are less variable than those of phocids, although some species show sexually dimorphic characteristics. In small phocids the orbits (and therefore the eyes) encompass a greater proportion

of the skull than in larger phocids. The narial basin of elephant seals (*Mirounga* spp.) and hooded seal (*Cystophora cristata*) skulls is enlarged; as in sirenians, it is presumed that this allows greater movement of their mobile proboscises. Pinnipeds that employ suction feeding [walruses, bearded (*Erignathus barbatus*), crabeater (*Lobodon carcinophaga*), ringed (*Pusa hispida*), and harp (*Pagophilus groenlandicus*) seals] tend to have short, wide rostra with jaws that have scoop-like anterior ends and a long mandibular symphysis or a mandible in which the ventral borders are angled toward the oral cavity. The skull of the walrus differs significantly from other pinnipeds. Most of these differences are linked with its specialized feeding and ecology and the presence of tusks (the dominant feature of the skull, Fig. 4C). The maxillary bones are enlarged and adapted to accommodate and anchor the tusks to the skull. The short, wide rostrum is advantageous for benthic (bottom) feeding and increases the surface area for highly mobile whiskers. The posterior head is flat and broad, providing a large surface area for the attachment of neck flexor muscles. The enlargement of the maxillary bones to anchor the tusks to the skull and large regions for attachment of neck muscles are important for hauling-out behavior. Walruses commonly use their tusks to pull and lift their bodies from the water. Hence the derivation of their Latin name *Odobenus* (tooth walker).

IV. Digestive System

The digestive system is responsible for acquiring food, its mechanical and chemical processing, absorption of nutrients, and excretion of waste products. Methods of acquiring food are quite variable among mammals and include behaviors such as manipulating, shearing, or grasping food with the anterior teeth, lips, tongue, or oral appendages. Acquiring and processing food are largely dependent on adaptations of the skull and associated structures. Because ingestion occurs at the interface between the environment and the body, it is the only portion of food processing that can be seen. Therefore, acquisition of food comprises a large proportion of our current understanding of feeding mechanics in marine mammals.

A. Food Acquisition

1. Cetacean Feeding Cetaceans have developed some of the most specialized and varied FEEDING mechanisms among mammals. This should not be a surprise, as cetaceans exhibit an amazing amount of ecological diversity and inhabit a diverse number of habitats ranging from the tropics to the polar regions. These mechanisms can be generalized to four categories that are specific to the major cetacean groups. Mysticetes are generally categorized as skimmers and engulfers, while the feeding modes of odontocetes involve grasping and suction of prey.

A. MYSTICETES All mysticetes possess BALEEN and most feed on plankton. However, the diet of some rorquals is predominantly fish and squid, whereas gray whales (*Eschrichtius robustus*) feed on marine amphipods and fish in addition to plankton. The method in which prey are captured differs significantly between balaenids and balaenopterids. In general, balaenids are skimmers and balaenopterids are engulfers.

As mentioned previously, the maxilla of balaenids has a greater arching curvature. This accommodates the taller, narrower, and greater number of baleen plates relative to balaenopterids. Baleen plates have a straight outer edge; the inner edge is rounded and wider at the top. A fringe of hair lines the inner edge of each plate, and multiple plates result in the intertwining of the hairs into a woven mat, which can strain plankton when water is passed through. The greater number and height of balaenid baleen plates increase the surface area for straining. Balaenids feed by SWIMMING slowly at the surface with their mouths slightly open. Water flows into the mouth, through the baleen plates where plankton are filtered, and out the corner of the mouth. Periodically the mouth is closed and plankton are removed from the baleen by the tongue and ingested. A model of hydrodynamic flow through the oral cavity of a bowhead whale suggests that the unique morphological structures and actions of the oral apparatus, such as the subrostral gap, orolabial sulcus, curvature of baleen, mandibular rotation, and lingual mobility, permit the steady flow of water through the baleen and may improve the efficiency of filtration.

Instead of skimming the water, balaenopterids engulf large quantities of water, which is then strained through the baleen.

This requires several unique adaptations, including numerous throat grooves that allow for an expansion of the gular region (ventral region between the two bodies of the mandible) and a frontomandibular stay apparatus (Fig. 5). This feeding strategy relies on the momentum of the body to operate. Despite their enormous size, rorquals are slender, streamlined, and faster than the relatively bulky and slow-swimming balaenids; body shape is linked with feeding strategy. As whales approach prey, the mouth is opened during active swimming. The forward momentum and sudden increase of inertial drag force the mouth open and cause the throat grooves to expand. Throat grooves in balaenopterids are pleats of blubber that undergo large-scale deformation to provide an expansive cavity (cavum ventrale) that enables these whales to engulf enormous quantities of water. The throat blubber of fin whales (*Balaenoptera physalus*) can expand to as much as 4 times its resting length circumferentially and up to 1.5 times its resting length longitudinally. This extensibility is a direct result of the material properties of the grooved blubber. The small and weak mandibular articulation with the skull is not capable of preventing the mandible from being detached from the head during the forceful engulfment event. The frontomandibular stay apparatus involves a strong ligament extending from the supraorbital process of the frontal bone to the coronoid process of the mandible (Lambertsen *et al.*, 1995; Fig. 5). This innovation allows the mouth to open to 90° while protecting the jaw joint from overextension

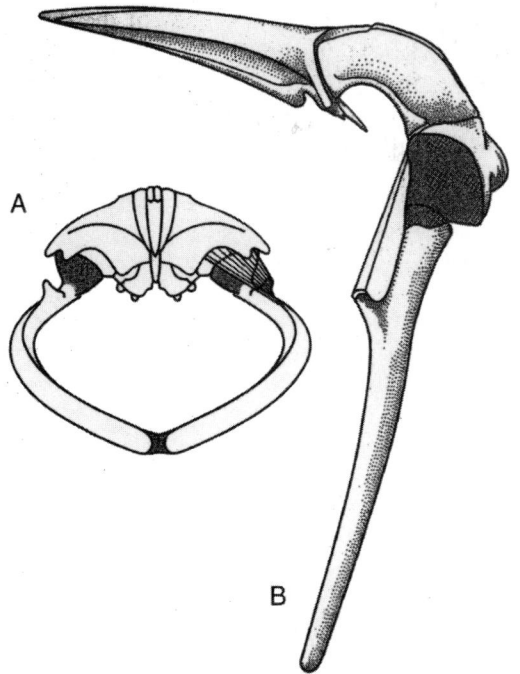

Figure 5 *Frontomandibular stay apparatus of Balaenopteridae. (A) Frontal view. Dark areas represent nonbony, flexible connective tissue at mandibular symphysis and the jaw joint. (B) Lateral view. Note the location of the fibrous connection between the skull and the mandible that prevents overextension of the jaw. From Lambertsen et al. (1995), by permission.*

and damage. The stay apparatus also allows each side of the mandible to rotate ventrodorsally along its long axis (the dorsal surface rotates medially, the ventral surface rotates laterally). This is possible due to the loose articulation of the mandible with the skull and the unfused and flexible mandibular symphysis. Mandibular rotation further increases the amount of water the whale is able to engulf and may assist in the expansion of the throat grooves. After the mouth is closed, water is forced through the baleen and plankton, fish, or other prey are ingested. This water movement is powered by the retraction of the elastic throat grooves, contraction of muscles deep to the grooved blubber, and the return of the tongue to its original position.

The oral apparatus of gray whales possesses many characteristics of both balaenids and rorquals. Instead of filtering water, gray whales filter sediment for marine amphipods. Feeding is not completely understood but is accomplished with the animal swimming on its side and using its jaws to excavate long troughs on the sea floor. The baleen of gray whales is unusually thick and sturdy, and it is presumed that amphipods are strained through it in a manner similar to balaenids straining plankton from water. Gray whales also skim surface water for plankton in the same manner as balaenids but are capable of engulfing prey in a manner similar to balaenopterids. This diversity in feeding modes allows for dietary flexibility so that alternate food sources can be used when prey distribution changes.

B. ODONTOCETES   Although the odontocete oral apparatus is not as derived as that of mysticetes, it is still greatly modified compared to terrestrial mammals. One reason is that the mandible plays a dual role in feeding and sound reception. The number of teeth is variable; some species may possess 200–300 homodont teeth (e.g., river dolphins) within long pincher-type jaws, whereas other odontocetes are characterized by a drastic reduction in tooth number, tooth function, and possess blunt rostra. These odontocetes are suspected of using suction as a primary method of feeding [e.g., pilot whales (*Globicephala* spp.), beaked whales (Ziphiidae), and sperm whales (*Physeter macrocephalus*)].

Although anatomical data and anecdotal information regarding feeding in cetacean abound, little functional data exist for feeding mechanics among odontocetes. The delphinid jaw is often called a clap trap, which refers to the grasping method of prey capture. The elongation of the skull and mandible results in a lower mechanical advantage but increases the velocity of jaw closure. The modifications of the teeth, a relatively diminished masticatory muscle mass, and a low mechanical advantage are modifications for quickly grasping and capturing prey, which are then swallowed whole without mastication.

Another odontocete feeding strategy is suction feeding, which requires a mechanism that lowers the intraoral pressure relative to that of the surrounding water. An increase in the volume of the oral and pharyngeal cavities can create this pressure difference. Suction feeding in terrestrial vertebrates generates intraoral pressure differences by the retraction and depression of a large, piston-shaped tongue. These animals typically have a suite of anatomical features such as a smooth, flat palate and a robust hyoid apparatus with large lingual and hyoid musculature (e.g., genioglossus, styloglossus and hyoglossus). Cetaceans

thought to use suction to feed display many of these morphologies and probably generate negative intraoral pressures in the same way as other vertebrates. Throat grooves may allow for greater expansion of the gular region and would assist in generating greater negative pressures. Much of the evidence for suction feeding among odontocetes is anecdotal. One published account describes belugas sucking a 50-cent coin (about 11.2 grams) off the bottom of their tank from 4 inches away. Similar observations have been made for pilot and killer whales. Sucking has been observed in live-stranded beaked whale (*Mesoplodon carlhubbsi*) calves; investigative palpation suggested that the motion originated in the region of the throat grooves. Manipulation of ziphiid cadavers demonstrated that the tongue could be retracted easily toward the hyoid apparatus by the extrinsic tongue muscles. Retraction of the tongue by manipulation of these muscles resulted in distention of the throat grooves.

2. Sirenian Feeding   The most striking characteristic of sirenian feeding is the mobility and use of the lips and associated vibrissae. Vibrissae are specialized hairs that transmit tactile information from the environment to the central nervous system. The follicles of these tactile hairs have a blood-filled sinus that surrounds the hair shaft; within the walls of this sinus are numerous mechanoreceptors. It is thought that deflections of the hair shaft are amplified and transmitted to these mechanoreceptors by the blood sinus. All sirenians possess muscular snouts that are covered by short tactile hairs and modified vibrissae. The lips of West Indian manatees (*Trichechus manatus*) move groups of the modified vibrissae or bristles to grasp vegetation and introduce it into the mouth (Hartman, 1979; Marshall *et al.*, 1998). These bristles are homologous with mystacial and mental vibrissae of the dog but are short, thick, and robust. The use of vibrissae by sirenians to manipulate objects in their environment is a departure from the classical sensory function of mammalian vibrissae. For example, many other mammals use vibrissae to detect tactile cues. Pinnipeds employ whisking movements for more directed tactile exploration. The sniffing behavior and related vibrissal movement during exploration by rodents involve sweeping of the mystacial vibrissae forward and backward in conjunction with protraction and retraction of the rhinarium and head. The modification of manatee bristles to actively manipulate food and other objects appears to be unique to sirenians.

Deflection of the rostrum is related to sirenian feeding ecology (Figs. 4G and 4H). Dugongs are seagrass specialists. Their rostrum is strongly downturned (~70°), which is advantageous for benthic feeding. The Amazonian and West African manatee (*T. inunguis* and *T. senegalensis*) possess the least deflected rostra (~30° and 26°, respectively). The Amazonian manatee is a feeding specialist on floating plants and meadows of the freshwater Amazonian tributaries. The neck musculature of Amazonian manatees differs from other sirenians and allows for greater extension of the neck. The diet of West African manatees is not well known, but is likely similar due to the similar habitat. West Indian manatees have rostral deflections that are intermediate between those of dugongs and Amazonian and West African manatees. West Indian manatees are generalist feeders and can feed anywhere in the water column.

Manatees possess an unusual dentition and form of tooth replacement. Only cheek teeth, which number six to eight at any one time, are present. Teeth erupt in the back of the mouth and move anteriorly as they wear. The bony septa between the teeth are reabsorbed in front of a tooth and redeposited behind it. This allows the teeth to move through the bone of the mandible. As teeth wear they progress forward. When the teeth reach the anteriormost portion of the tooth row and are completely worn, the roots are absorbed and the tooth is shed. A new tooth erupts in the posterior tooth row that continues the conveyer-like process. Manatees are apparently able to produce an unlimited number of cheek teeth. Dugongs possess simplified peg-like teeth that are open rooted. Enlarged horny pads on the upper and lower palate play an important role in mechanical reduction and processing of sea grasses. Incisors are present but erupt only in males. These are often referred to as tusks but are not homologous with the canine tusks of walruses.

3. Pinniped Feeding Prey capture by pinnipeds has not been widely investigated. However, vibrissae appear to be important for prey capture and discrimination. Vibrissal tactile discrimination by California sea lions (*Zalophus californianus*) and harbor seals (*Phoca vitulina*) has been shown to be as sensitive as the hands of monkeys (Dehnhardt and Ducker, 1996). Vibrissae in harbor seals form a hydrodynamic receptor system that is tuned to the frequency of water movement made by fish (Dehnhardt, 1998). The number of teeth is reduced in pinnipeds compared to the dog and the teeth are more uniform, changes presumed to relate to eating fish. At least two species, the leopard seal (*Hydrurga leptonyx*) and crabeater seal (*Lobodon carcinophaga*), have specialized teeth. The leopard sea feeds primarily on large vertebrates, such as penguins and other seals. The distinctive cheek teeth possess three long shearing cusps. Crabeater seals actually feed on KRILL, not crabs. The cusps of their cheek teeth are complicated and modified to form a sieve. The seal swims into a krill patch with mouth open, sucking in water. Water is drained through the sieve and the krill are consumed. Walruses excel at suction feeding. Powerful intraoral pressures are generated by the piston-like tongue and design of the oral cavity. The rostrum of walruses is broad and covered with hundreds of sensitive vibrissae used to detect shellfish. The use of suction for feeding is likely more prevalent among pinnipeds. The masticatory apparatus of most pinnipeds appears to be similar to other carnivores.

4. Sea Otter and Polar Bear Feeding Sea otters forage on the bottom in waters as deep as 40 m. Their diet is varied, but shellfish and urchins comprise a large portion. Otters use their forepaws to excavate clams and to pry shellfish and urchins from the rocky substrate, sometimes using tools. Food is usually consumed at the surface, and behavioral observations suggest that otters do not use their teeth underwater, even when feeding on fish. Upon surfacing, fish are killed by a bite to the head. A rock or some other tool is usually carried in a flap of skin in the axilla region (under the arm) and is used to pound open shellfish. The spines of urchins are simply bitten off and the test (shell) of the urchin is crushed with the cheek teeth. Their cheek teeth are broad, flat, and covered with thick enamel. The shearing cusps of the carnassial teeth have been lost; sea otters are adapted for crushing their food (Kenyon, 1969).

Very little is known about the feeding mechanics of polar bears. Polar bears grasp their prey with their mouths and break the neck or skull with their large masticatory muscles and robust dentition. Their masticatory apparatus appears to resemble that of other bears.

B. Mechanical and Chemical Digestion

In general, the stomach of cetaceans is four chambered (although the number of chambers is variable) and bears a striking resemblance to that of the ox and other ruminants, despite the fact that cetaceans are carnivores. The morphology of the cetacean stomach is more likely related to phylogeny than to function.

The sirenian gastrointestinal tract is of interest because of this group's herbivorous diet. All sirenians are hindgut fermenters. The stomach of West Indian manatees is simple (one chamber) with a prominent muscular ridge that projects into the stomach lumen, partially dividing it into anterior and posterior regions. Both manatees and dugongs possess a discrete digestive accessory gland, the gastric or cardiac gland, which contains parietal and chief cells. This gland protrudes from the greater curvature of the stomach wall. The segregation of digestive acid-secreting cells from the rest of the stomach is an unusual feature among mammalian digestive systems. Although its function is unclear, this segregation may protect acid-secreting cells, and therefore the rest of the stomach, from the abrasive nature of the sirenian diet. The small intestine of West Indian manatees is long and may exceed 20 m. In both West Indian manatees and dugongs, a pair of blind pouches, or diverticula, extend from the expanded portion of the duodenum (duodenal ampulla). The mucosal glands of the duodenum are unusual in that they secrete an acidic mucous instead of an alkaline mucous. This represents an unusual feature among mammalian digestive tracts; the function of acidic mucous in the duodenal region of the small intestine is not understood. It is interesting to note that two other mammals, the koala (*Phascolarctos cinereus*) and the wombat (*Vombatus hirsutus*), also possess a discrete cardiac or gastric gland and secrete acidic mucous in the duodenum. Except for its length, the rest of the small intestine does not differ significantly from that of other mammals. The large intestine of West Indian manatees is also long (>20 m) and wide (~15 cm). At the junction of the small and large intestines is a large cecum with two diverticula. Histology of the region, volatile fatty acid (VFA) analysis, and the existence of large populations of symbiotic microbes all point to the proximal large intestinal and cecum as sites of cellulose breakdown in both manatees and dugongs.

V. Cardiovascular System

Several important morphological modifications of the cardiovascular system exist in marine mammals. These include (1) the blood supply to the brain, (2) arterial–venous anastomoses, (3) countercurrent heat exchangers, and (4) modifications to the venous system and circulation. The cardiovascular system of marine mammals is modified to solve three critical physiological problems of life at sea: conserving oxygen, protecting oxygen-sensitive tissues during deep dives, and protecting body tissues from the cold environment (thermoregulation).

A. Diving Adaptations of the Cardiovascular System

Among marine mammals foraging for food requires diving. For marine mammals that engage in prolonged deep dives (cetaceans and pinnipeds), efficient oxygen use translates to a greater percentage of time foraging at depth. Some of the greatest modifications of the mammalian cardiovascular system occur among cetaceans and pinnipeds. One possible way to increase dive time is to take more oxygen on a dive. This is exactly what some marine mammals do. One might think that this is done by increasing lung capacity. With the exception of the sea otter, marine mammals have not greatly increased lung size relative to body size. In fact, many deep divers have smaller-than-predicted lungs for mammals of their body weight. Instead cetaceans and pinnipeds store oxygen in blood and muscle tissue (oxygen binds to the molecules hemoglobin and myoglobin, respectively). To facilitate oxygen storage, these mammals possess a greater volume of blood, a higher density of red blood cells (higher hematocrit), and a greatly increased concentration of muscle myoglobin relative to terrestrial mammals. Dives can be prolonged further by efficient use of these extended oxygen stores. Stereotypical changes in the cardiovascular circulation called the diving reflex function to conserve oxygen. Most diving mammals (including humans) exhibit bradycardia (slowing of the heart rate) and peripheral vasoconstriction during a dive. Vasoconstriction increases peripheral resistance and therefore maintains blood pressure during bradycardia. During a dive, blood flow is decreased or shut off to tissues that are tolerant of hypoxic conditions. For example, blood supply to the diaphragm, pancreas, liver, and skeletal muscle is reduced to approximately one-twentieth of normal circulation while blood flow to the heart and brain is maintained. The brain and the heart are profoundly sensitive to hypoxic conditions; even a few minutes of ischemia can cause severe damage or death. Blood flow to the intestines, which hold up to 50% of the normal blood volume, is completely shut off. Although muscles can utilize the oxygen bound to myoglobin, this supply is quickly used up and cannot explain the long dive durations observed. Muscle tissue can work anaerobically. Pinnipeds have a high tolerance for lactic acid buildup, a byproduct of anaerobic metabolism and the cause of muscle fatigue during exercise.

B. Blood Supply to the Brain

In the dog and most terrestrial mammals, blood supply to the BRAIN is provided by two bilateral arteries: the internal carotid and the vertebral arteries. The vertebral artery enters the skull through the foramen magnum and the internal carotid artery enters at the base of the skull. Both arteries join a circular anastomosing vasculature at the base of the brain called the circle of Willis. From this circle, cerebral arteries emerge to supply specific regions of the brain. The function of the circle of Willis is not known but is thought to provide collateral blood supply in the event of vessel blockage and to dampen pulses of pressure emanating from the heart that could damage the delicate brain tissue.

The blood supply to the brain in cetaceans is unusual. The internal carotid artery does not reach the brain and they do not possess vertebral arteries (Walmsley, 1938; Slijper, 1962). Instead, the cetacean brain receives blood from a series of vessels called dorsal intercostal arteries. In the dog, arteries branch off the thoracic aorta to supply the regions between each rib (dorsal intercostal arteries) and the muscles on the dorsal side of the vertebral column (epaxial muscles). In cetaceans, the dorsal intercostal arteries form a dense network of small anastomosing arteries called rete mirabile (literally "wonderful net"). These retia are extensive and form the main blood supply to the brain. The functional significance of the retial system in cetaceans is not known but many functional hypotheses exist. The most popular hypothesis is that the retia dampen pulse pressures from the heart and protect the delicate tissues of the brain, as does the circle of Willis in terrestrial mammals. A histological investigation of the rete mirabile in the foramen magnum of bowhead whales (*Balaena mysticetus*) reported an exceptionally thick smooth muscle layer (tunica media) of retial arteries. Previous investigators have reported an absence of nerve fibers within the rete mirabile in other cetaceans. In the bowhead, diffuse nerve fibers were not observed; however, ganglion-like complexes were found within the thick smooth muscle layer. This finding suggests that retia mirabile are under neural control and blood flow may be modulated by vasoconstriction. Physiological data have demonstrated that in a retial system composed of hundreds of muscular anastomosing vessels, the mean blood pressure remains unchanged but periodic increases in pressure are reduced.

C. Arteriovenous Anastomoses

Arteriovenous anastomoses (AVAs) are direct connections between arterioles and venules that allow blood to bypass capillary beds. Blood flow continues directly from the arterial system to the venous system without gas and solute exchange with the tissues. Although these structures usually occur at the level of the capillary beds, they also connect larger vessels further away from the capillaries. Cetaceans and pinnipeds make extensive use of AVAs in regions of blood supply to the integument. These structures are particularly important for thermoregulation and oxygen conservation. Decreased blood flow to the skin, accomplished by shunting blood through an AVA, reduces heat loss through the skin. Conversely, increased blood flow to the skin can bypass the insulation of the blubber and dump excessive heat to the environment. Arteriovenous anastomoses are also used to conserve oxygen by shunting blood away from oxygen-insensitive organs, such as the intestines, during dives.

D. Countercurrent Heat Exchangers

A countercurrent heat exchanger (CCHE) is an organization of arteries and veins that transfers heat from the arterial system to the venous system; the function of this vascular organization is to control the amount of heat lost to the environment. These structures are found in many terrestrial and aquatic vertebrates in regions of the body that are poorly insulated. In marine mammals, they are particularly well developed in locations that have little to no blubber, such as the dorsal and pectoral fins of cetaceans, limbs of pinnipeds and sea otters, and pectoral flippers of sirenians. In cetaceans, arteries that supply the fins and flukes (central arteries) are surrounded by numerous veins (circumarterial veins); heat from warm arterial blood is transferred to cool venous blood returning to the heart (Fig. 6). The amount of heat transferred to veins is proportional to the length of the CCHE. Arterial blood reaching the periphery is significantly cooler and therefore less heat is lost to the environment [the difference between T_b and T_e has been reduced; see Eq. (2)]. Cool venous blood traveling from the periphery is warmed and therefore does not significantly reduce core body temperature. Countercurrent heat exchangers can be bypassed by dilation of the central artery, which compress and collapse the circumarterial veins. Venous blood flow is then rerouted to veins further away from the central artery and nearer to the surface of the skin (Fig. 6). This allows excessive heat to be dumped to the environment; these regions are often called thermal windows.

Countercurrent heat exchangers are especially important in the reproductive system of marine mammals. The testes of terrestrial mammals typically lie outside the abdominal cavity in the scrotum. This is because sperm production requires tem-

peratures 1.1°C cooler than core body temperature. Sperm production at core body temperature results in malformed, nonfunctional spermatozoa. The testes of many marine mammals are intraabdominal, which increases streamlining. Not only do these marine mammals testes experience core body temperatures, but testes of cetaceans, sirenians, phocids, and walruses are located near large swimming muscles, which can generate heat one to two orders of magnitude greater than resting muscles. Bottlenose dolphins possess robust CCHEs that are adjacent to the testes. Veins carrying cooled blood from the dorsal fin and tail flukes directly supply a venous plexus that surrounds the arteries supplying the testes. The juxtaposition of a venous and arterial plexus regulates the temperature of the region near the testes. Even during exercise, the CCHE is able to maintain the testes at a lower temperature. In male phocids, anastomoses that span from the foot to the pelvis allow cool venous return through the gluteal, pelvic, and pudendoepigastic veins. This cool venous return is separate from the warmer venous return generated by the hindlimb CCHE and is directed at the venous plexus, which lies medial to the testes. Male West Indian manatees appear to possess a CCHE similar to that found in cetaceans and phocids.

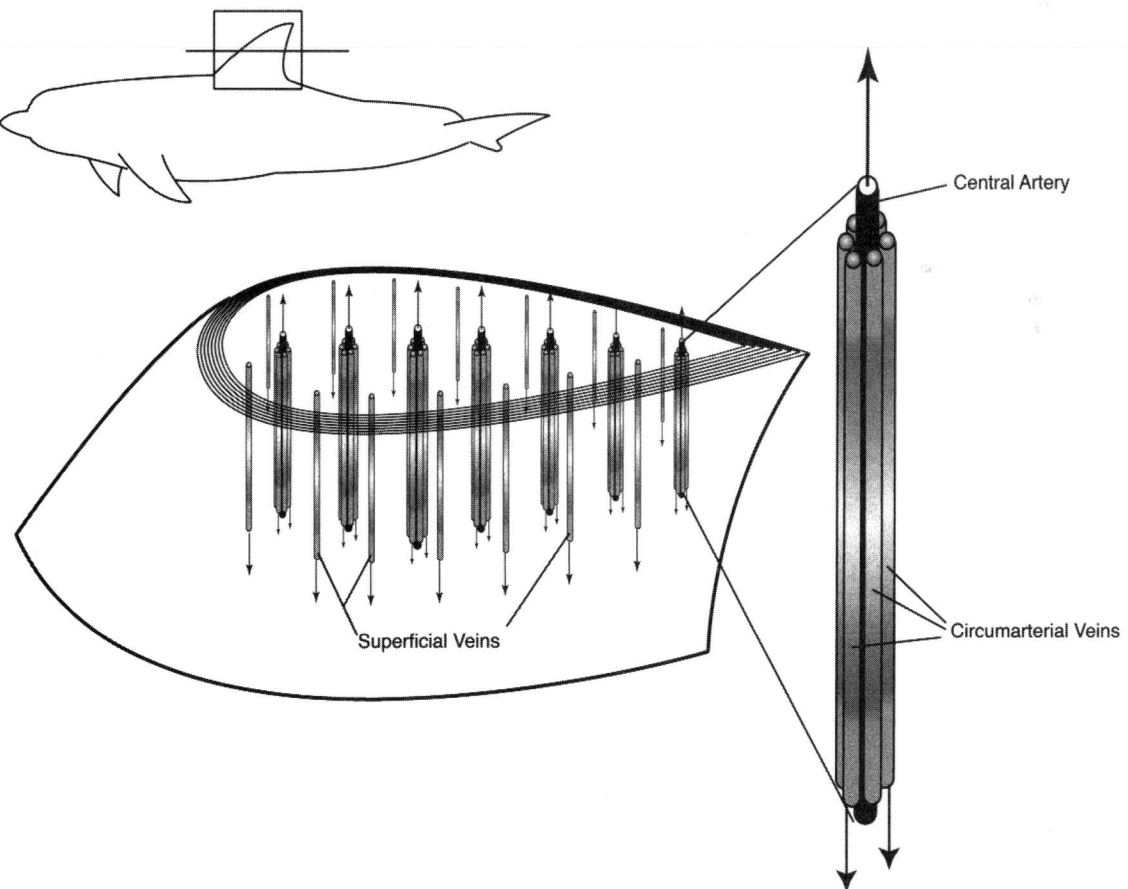

Figure 6 *Schematic of a countercurrent heat exchanger (CCHE) in the dorsal fin of a bottlenose dolphin. Note the circumarterial veins surrounding each central artery, the positions of CCHEs within the dorsal fin, and the peripheral location of the superficial veins not associated with the CCHE.*

Female marine mammals face a similar problem during pregnancy. Elevated abdominal temperatures could harm fetal development. In addition, the mammalian metabolic rate of the fetus is about twice that of maternal tissues; the fetus produces heat. Data from experimental terrestrial mammals indicate that heat is transferred to the abdomen and then to the external environment through the thermal window of the ventral abdomen. Because of their thick blubber, female marine mammals do not have this option. Cetacean and phocid females possess CCHEs that may regulate the temperature of the reproductive organs and abdomen. The system in cetaceans involves the juxtaposition of the lumboposterior venous plexus with the uterovarian arterial plexus in a manner similar to the cetacean testicular CCHE. The system in female pinnipeds involves cool venous return from the three anastomoses in the hindlimb directed toward the abdominal wall plexus that lies lateral to the uterus. Thus, cool venous return could lower the regional temperature of the uterus.

E. Modifications of the Venous System

The venous systems of pinnipeds and cetaceans are modified and more complex relative to the dog. These modifications are related to diving as well as thermoregulation. Much of our current knowledge of cardiovascular function in marine mammals has been derived from *in vivo* angiographies of a few phocid species during experimentally elicited diving responses (Ronald *et al.*, 1977).

Due to the lack of data on a wide range of pinniped species, one must be careful not to overgeneralize cardiovascular function to all pinnipeds. Important differences exist among phocids, walruses, and otariids. For example, in those otariids investigated, the major cranial venous drainage is accomplished largely by the external jugular vein, and secondarily by the vertebral veins. As in dogs, otariids have small extradural veins. However, the major intracranial drainage in phocids is through the modified extradural intravertebral veins (Fig. 7); the external jugular vein is reduced in size.

1. Abdominal Venous Plexus and the Hepatic Sinus   The extradural vein in phocids spans from the cranium to the sacrum. Many peripheral veins drain into it along its entire length. In the neck, extradural veins communicate extensively with vertebral veins (cervical vertebral system) and provide a major connection to the heart. In the thorax, the extradural vein communicates with intercostal veins and therefore indirectly with the azygos vein (another major route of return to the heart, Fig. 7). Near the sacrum, branches of the extradural vein communicate directly with the posterior vena cava (PVC) and indirectly with the PVC through a large bilateral abdominal wall plexus. The PVC is the major venous drainage of the abdomen and bifurcates just below the level of the kidneys. Anterior to the kidneys, the right and left branches of the PVC join. At the level of the liver, large hepatic veins join the PVC to form an enlarged sinus called the hepatic sinus. As the PVC passes through the diaphragm and into the thorax, it becomes the anterior vena cava (the major venous drainage of the thorax). A band of striated muscle from the diaphragm encircles the PVC to form a sphincter as it passes into the thorax (Fig. 7). This caval sphincter is innervated by the right and left phrenic nerves. *In vivo* stimulation of the right phrenic nerve produced a strong contraction of this band of striated muscle that occluded the PVC. Constriction of the PVC would severely limit (if not stop) most venous return from the abdomen to the heart. *In vivo* angiography of harbor and harp seals at rest and during forced dives gives us some idea of the function of their venous system. Shortly after the beginning of the dive, smooth muscles surrounding the cervical vertebral system contract, forcing venous blood further posteriorly. Blood from the intestinal arteries is shunted away into the venous system (via the portal system) by AVAs. The large quantity of still oxygenated blood in the portal system is pumped into the PVC by the contraction of smooth muscle surrounding the veins. Almost no blood is present in the intestinal vessels during a dive. The hepatic sinus expands to accommodate the large influx of oxygenated blood. The bifurcation of the PVC also accommodates the increased blood volume. The caval sphincter prevents

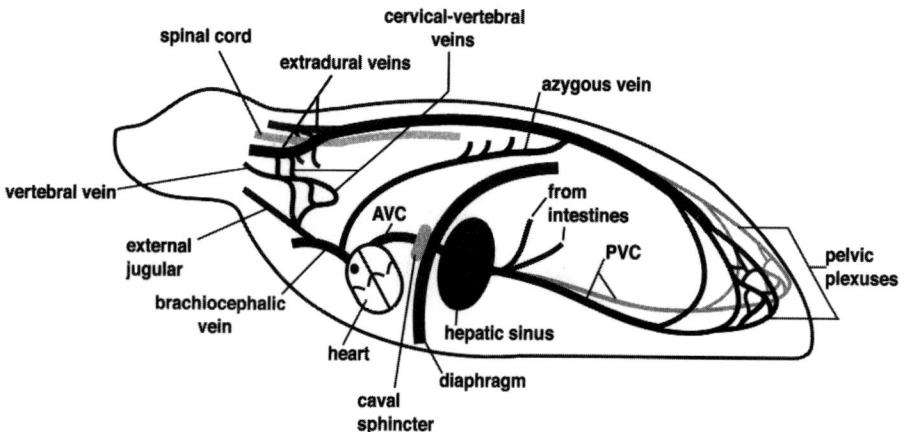

Figure 7   *Schematic of a simplified phocid venous system (after King, 1983). AVC, anterior vena cava; PVC, posterior vena cava. Note the bifurcation of the PVC and the locations of the extradural veins near the spinal cord, hepatic sinus, and caval sphincter.*

movement of this blood to the heart. As the heart expands and a bolus of blood from the anterior vena cava enters the heart, the caval sphincter synchronously opens momentarily, and a bolus of oxygenated blood from the PVC enters the anterior vena cava. Deoxygenated blood draining from the brain flows posteriorly down the extradural vein and is held in the lateral abdominal wall plexus. In this way, deoxygenated venous blood is segregated from oxygenated venous blood. At the end of the dive the circulation returns to normal; the cervical vertebral system opens and flow is reestablished to the heart. Flow from the head to the abdominal venous plexus (through the thoracic epidural vein) reverses and drains venous blood by way of the azygous vein and cervical vertebral route. The caval sphincter opens and blood held in the PVC and abdominal lateral wall plexus is quickly flushed into the anterior vena cava and to the heart. Blood flow returns to the intestines.

The otariid vascular system is not as well studied as the phocid system. However, there are several important differences. A hepatic sinus and caval sphincter are present in both otariids and walruses, but the caval sphincter is not as well developed as in phocids. Phocids have a higher hematocrit, greater blood volume, and twice the myglobin concentration in muscle than otariids and walruses. Based on current knowledge, it appears that the phocid cardiovascular system is more specialized for diving.

The venous vasculature of cetaceans is also highly derived compared to that of the dog. A large extradural spinal vein runs posteriorly down the spinal cord in close association with the spinal, cervical, thoracic, and lumbar rete mirabile. This spinal vein has many communicating branches with the anterior and posterior vena cava, as well as with an elaborate lateral abdominal wall plexus. The PVC is bifurcated and communicates directly with the pelvic venous plexus. Odontocetes also possess a hepatic sinus and caval sphincter; the presence of such structures in mysticetes in uncertain. The venous vasculature of odontocetes bears striking similarities to the phocid vasculature.

VI. Respiratory System

The function of the respiratory system is to deliver air to the alveoli in the lungs and facilitate gas exchange. Additionally, the respiratory system protects the delicate tissues of gas exchange from physical and pathological damage, permits vocalization, and houses the olfactory receptors.

A. Aquatic Respiratory Adaptations

One of the most obvious traits of the respiratory system in cetaceans is the position of the nares. The dorsal position allows for quick and convenient exhalation and inhalation with little to no interruption during locomotion. Cetaceans lack conchae (turbinate bones of the nasal cavity), which allows for a rapid and forceful exhalation and inhalation. As the nasal passage emerges from the top of the skull, an adipose cushion sits within the entrance like a thick plug (nasal plug). In its relaxed state, the nasal plug occludes the nasal passage. Strong muscles spanning the skull pull the plug forward and open the blowhole. The nasal plug prevents water from entering the respiratory system, even at high pressures experienced when diving at

depths. Sirenians possess similar paired nasal plugs. Pinnipeds nares are not as specialized. Annular muscles encircle the opening of each nostril to keep them closed. Opening the nares requires voluntary relaxation of these muscles.

The larynx of marine mammals is generally similar to terrestrial mammals. In most mammals, the epiglottis, that most anterior cartilage of the larynx (responsible for keeping food out of the airway), rests above the soft palate of the mouth. Many mammals are able to remove and reinsert the tip of the epiglottis above the soft palate. The larynx of odontocetes is highly modified compared to that of the dog. The odontocete epiglottis and other laryngeal cartilages are elongated; this part of the larynx is aptly named the goosebeak. The goosebeak projects far into the nasal pharynx; at the laryngeal end of the nasal pharynx a strong muscular sphincter creates a tight seal around the goosebeak. This condition completely separates the foodway from the airway so that odontocetes are obligate nasal breathers, much like the horse. This modification prevents water from the oral cavity from entering the airway but may also be implicated in supplying air to specialized nasal diverticula (air sacs) for sound generation used for echolocation.

B. Respiratory Adaptations to Diving

Air is composed of 79% nitrogen and 21% oxygen. Under pressure all gases will go into solution. Remove the pressure and gas comes out of solution, sometimes very quickly. This is what happens when a carbonated drink is opened. Nitrogen in the lungs of a scuba diver at depth enters the bloodstream in the same way because the air in the lungs is compressed by water pressure. Water pressure forces nitrogen into the bloodstream and peripheral tissues. The longer a scuba diver spends at depth, the more nitrogen enters the bloodstream. If the scuba diver ascends slowly, the nitrogen will come out of solution slowly and the diver can exhale the gas. However, if the scuba diver ascends rapidly, nitrogen comes out of solution faster than can be possibly exhaled. Gas bubbles collect in respiratory tissues and in the tissues surrounding the peripheral circulatory system, such as the joints. This phenomenon is called decompression sickness (the bends) and can result in serious injury or death. Marine mammals do not get the bends. Cartilaginous rings of the conduction tract extend to the level of the alveoli in marine mammals. At depth, the rib cage is compressed and the lungs of cetaceans and pinnipeds collapse under pressure, but the conducting tract does not occlude. Air from the collapsed alveoli is forced into the conduction system. Without these structures, air would be trapped at the respiratory surfaces and nitrogen could enter the bloodstream. Water pressure decreases upon ascent, allowing air to expand and reenter the alveoli. This source of oxygen could be important to the animal at the end of a prolonged dive. Most pinnipeds do not fill their lungs prior to a dive and many actually have smaller than predicted lung size relative to mammals of the same body size (e.g., Weddell seal, *Leptonychotes weddellii*). Small lung size circumvents the additional expenditure of energy to overcome the increased buoyancy of filled lungs. For deep diving marine mammals, the lungs are not a source of oxygen storage; instead oxygen is stored in the blood (by binding to hemoglobin) and in muscle (by binding to myoglobin).

VII. Reproductive System

The reproductive system of marine mammals differs little from that of terrestrial mammals. The morphology of the organs and glands of this system reveals much about the phylogenetic relationships among marine and terrestrial mammals. Many of the aquatic specializations of the reproductive system involve thermoregulation of the testes and the developing fetus, which have been discussed. Cetaceans and sirenians, being obligate marine mammals, give birth to large-bodied precocial young. Nursing underwater requires certain adaptations of both mother and calf to be successful. Cetacean mammary glands are paired and lie within the subcutaneous connective tissues on each side of the ventral midline just anterior and adjacent to the genital slit. The nipples (teats) open into the genital slit and can be extruded. Milk flow appears to be a voluntary act on the part of the mother and may be initiated by the contraction of abdominal muscles. Dolphins do not possess muscular lips as do other mammals; calves must grasp the teat between the palate and tongue. The scalloped end of the tongue creates a seal around the teat to facilitate suckling. The paired mammary glands of sirenians are located in the axilla (armpit), as in elephants. Calves clamp on the posterior side of the flipper with their muscular mouths and create a tight seal to nurse; rehabilitating manatee calves are capable of suckling from a bottle and it is presumed that calves use suction when nursing from their mothers. Observations of wild mother–calf pairs report the movement of the lips in a rhythmic motion during nursing, which indicated suckling.

VIII. Concluding Remarks

The diversity of morphological adaptations for life in the aquatic environment by marine mammals attests to the power of natural selection. Marine mammals provide a tangible means to investigate the range of possibilities of mammalian form and function. As marine mammals have garnered more scientific attention recently, previous gaps in our knowledge regarding the diversity of their biology are beginning to fill in. Increased emphasis on integrative, functional research and expanding technological tools will surely provide additional fascinating insights into their functional morphology.

Acknowledgments

I thank Sarah Cox, Roger Reep, and Susan Herring for editorial assistance. I thank the Burke Museum of Natural History for photographic use of their specimens. Financial support was provided by NIH Grants T32 DC00033 and F32DE05731-01A1.

See Also the Following Articles

Anatomical Dissection: Thorax and Abdomen ■ Circulatory System ■ Dental Morphology, Evolution of ■ Diving Physiology ■ Gastrointestinal Tract ■ Musculature ■ Thermoregulation

References

Dehnhardt, G. (1998). Seal whiskers detect water movements. *Nature* **394,** 235–236.

Dehnhardt, G., and Ducker, G. (1996). Tactual discrimination of size and shape by a California sea lion (*Zalophus californianus*). *Anim. Learn. Behav.* **24,** 366–374.

Domning, D. P. (1977). Observations on the myology of *Dugong dugon* (Muller). Smithsonian Contributions to Zoology Number 226, Washington, DC.

Domning, D. P. (1978). The myology of the Amazonian manatee (*Trichechus inunguis*) (Natterer) (Mammalia: Sirenia). *Acta Amazon.* **8**(Suppl. **1**), 1–81.

Evans, H. E. (1993). "Miller's Anatomy of the Dog," 3rd Ed. Saunders, Philadelphia.

Fish, F. E. (1996). Transitions from drag-based to lift-based propulsion in mammalian swimming. *Am. Zool.* **36,** 628–641.

Hartman, D. S. (1979). Ecology and behavior of the manatee in Florida. Special Publication Number 5, American Society of Mammalogists.

Hildebrand, M. (1988). "Analysis of Vertebrate Structure." Wiley, New York.

Howell, A. B. (1930). "Aquatic Mammals: Their Adaptations to Life in the Water." Dover Publications, New York.

Kenyon, K. W. (1969). The sea otter in the eastern Pacific Ocean. *North Am. Fauna* **68,** 1–352.

King, J. E. (1983). "Seals of the World," 2nd Ed. Cornell Univ. Press, Ithaca, NY.

Lambertsen, R., Ulrich, N., and Straley, J. (1995). Frontomandibular stay of Balaenopteridae: A mechanism for momentum recapture during feeding. *J. Mammal.* **76,** 877–899.

Marshall, C. D., Huth, G. D., Edmonds, V. M., Halin, D. L., and Reep, R. L. (1998). Prehensile use of perioral bristles during feeding and associated behaviors of the Florida manatee (*Trichechus manatus latirostris*). *Mar. Mamm. Sci.* **14,** 274–289.

Pabst, D. A. (1993). Intramuscular morphology and tendon geometry of the epaxial swimming muscles of dolphins. *J. Zool. (Lond.)* **230,** 159–176.

Pabst, D. A., Rommel, S. A., and McLellan, W. A. (1999). The functional morphology of marine mammals. *In* "The Biology of Marine Mammals." Smithsonian Press, Washington, DC.

Rommel, S. A. (1990). Osteology of the bottlenose dolphin. *In* "The Bottlenose Dolphin." (S. Leatherwood and R. R. Reeves, eds.), pp. 29–49. Academic Press, San Diego.

Ronald, K., McCarter, R., and Selley, L. J. (1977). Venous circulation in the harp seal. *In* "Functional Anatomy of Marine Mammals." (R. J. Harrison, ed.), Vol. 3, pp. 235–270. Academic Press, London.

Schmidt-Nielsen, K. (1990). "Animal Physiology: Adaptation and Environment," 4th Ed. Cambridge Univ. Press, Cambridge.

Slijper, E. J. (1962). "Whales." Basic Books, New York.

Walmsley, R. (1938). Some observations on the vascular system of a female fetal finback. *Contrib. Embryol.* **164,** 109–178.

Vogel, S. (1994). "Life in Moving Fluids: The Physical Biology of Flow," 2nd Ed. Princeton Univ. Press, Princeton.

Musculature

J. G. M. Thewissen
Northeastern Ohio Universities College of Medicine, Rootstown

The muscular system of mammals was designed on a single blueprint; there is a remarkable constancy of muscles and associated nerves from the most agile bat to the fastest antelope, and the largest whale. Details differ, and most of these differences reflect adaptations to specific demands of the environment. This article presents an outline of muscular anatomy of marine mammals, emphasizing how pinnipeds, cetaceans, and sirenians differ from terrestrial mammals. General summaries of cetacean muscles can be found in Slijper (1936, 1962), and of manatees in Domning (1978). Howell (1930) discussed locomotor morphology of all marine mammals.

I. Cranial Muscles

In the cranial muscles, marine mammals differ from terrestrial mammals in the arrangement of their eye muscles, facial muscles, their masticatory muscles, and the muscles of the palate, pharynx, larynx, and tongue.

The facial muscles in land mammals are attached to the skin of the face and moderate facial expressions. In all marine mammals, the facial muscles are involved in closing of the nose opening (or blowhole) to prevent the entry of water during diving. In sirenians, the most important of these muscles insert on the mobile snout and are involved in the manipulation of food. In cetaceans, the facial muscles are greatly rearranged and are positioned around the airsac system on the forehead (Purves and Pilleri, 1983). As such, they are greatly involved in the production of sound. In many mysticetes, facial muscles also extend between the two halves of the mandible where they assist in squeezing gulps of ingested water through the baleen.

One particular facial muscle, the buccinator, is unusual in that it is not near the surface of the skin in land mammals. It forms the wall of the cheek and gives the cheek a rigid wall when suction is produced. As such it is critical for nursing young. In cetaceans, partly as a result of the long snout, this muscle cannot give the cheek a rigid wall. Nursing females assist young in suckling by actively squirting milk into their mouths by the contraction of special skin muscles overlying the mammary gland.

The masticatory muscles of pinnipeds are similar to those of terrestrial carnivores, and sirenian masticatory muscles are not unlike those of herbivores. In cetaceans, the temporal muscle is greatly reduced and the muscles used to close the jaws are the pterygoids and masseter. Unlike land mammals, in which these muscles direct lateral movements of the lower jaws, simple closing of the jaws is their main function in cetaceans.

The muscles of the throat of most marine mammals do not differ greatly from those of land mammals. The throat of odontocete cetaceans (Fig. 1) is more specialized than that of other marine mammals. The larynx of odontocetes is elongate and its epiglottis projects far anteriorly, reaching the back of the palate and extending into the nasopharyngeal duct. The walls of the nasopharyngeal duct, including the soft palate, consist of a strong annular muscle that encloses the epiglottis and seals the lumen of the larynx functionally from the pharynx.

The tongue in land mammals is mostly made of muscle, and its large size in baleen whales is remarkable. In blue whales (*Balaenoptera musculus*), the tongue is the size of an adult elephant and makes up 2.5% of the whale's entire weight. However, rorqual tongues are not very muscular, consisting mainly of fat and connective tissue.

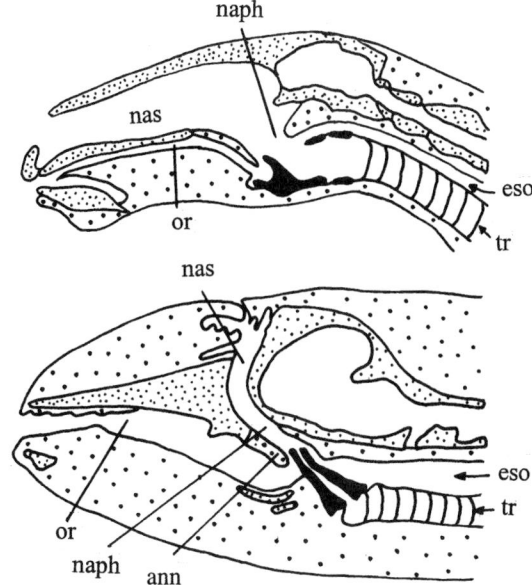

Figure 1 *Midline sections through the head of a horse (top) and common porpoise (*Phocoena phocoena, *bottom*) showing the unique shape of the throat and larynx in odontocete cetaceans. Air in all mammals passes from the nasal cavity (nas) to the nasopharyngeal duct (naph), to the larynx, and to the trachea (tr). Food in all mammals passes from the oral cavity (or), to the pharynx (throat), and to the esophagus (eso). The laryngeal cartilages (colored black) form a spout in odontocetes that fits into the nasopharyngeal duct and can be closed tightly by means of annular muscles (ann). This closure causes a tight separation of the air and food passages. Modified after Slijper (1962).*

II. Axial Muscles

The neck muscles in cetaceans are unremarkable because the neck is short. The neck is long in pinnipeds and may be very muscular. It functions in balancing the body during locomotion and powers the blows that males deal their conspecifics.

Muscles extending along the back and tail are the main muscles of propulsion in cetaceans. Epaxial muscles extend along the dorsal side of the transverse processes of the vertebrae. These muscles contract and cause dorsal concavity of the back and tail, pulling the fluke up in the upstroke. These muscles, especially the multifidus and longissimus, are enormous (Fig. 2). The upstroke in cetaceans is powered mainly by the longissimus and extensor caudae lateralis (Pabst, 1993). The latter muscle inserts directly on the dorsal surface of the vertebrae of the fluke, but the longissimus exerts its power by attaching to a subdermal sheath of tendons (Fig. 2) that attaches on spinous and transverse processes along most of the back of the cetacean. It is through the connections of this sheath to the terminal tail vertebrae that the fluke is moved, allowing muscular force to be distributed evenly along the caudal peduncle. The multifidus does not insert on this sheath, instead attaching to the posterior thoracic vertebrae and the lumbar vertebrae. Its main function appears to be to stiffen the back, providing a

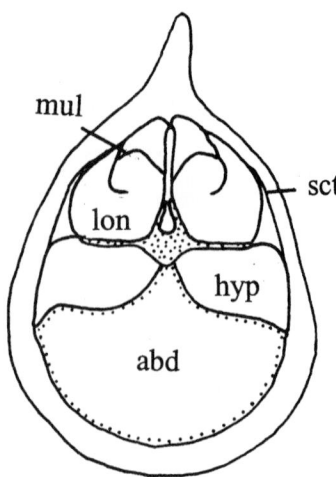

Figure 2 *Cross section through the lumbar region of a bottlenose dolphin* (Tursiops truncatus). *Note the large development muscles, which are larger in this cross section than the abdominal cavity (abd). Dorsal to the vertebra (stippled) are epaxial muscles (mul, multifidus; lon, longissimus) and ventral to the vertebra are hypaxial (hyp) muscles. Muscles are closely associated to the subdermal connective tissue sheath (sct). Modified after Pabst* et al. *(1993).*

stable platform of origin for longissimus. The longissimus is also large in sirenians and is probably important in their upstroke. Lumbar epaxial muscles are also important in powering the undulatory movements of *Enhydra lutris* (sea otter).

Epaxial and hypaxial muscles are large in phocids, where they are used to produce the side-to-side movements that propel the body in SWIMMING. Among the larger of these muscles is the iliocostalis.

III. Muscles of Thorax, Abdomen, and Pelvis

A large superficial skin muscle, the cutaneous trunci (sometimes called panniculus carnosus) covers much of the thorax and abdomen in many mammals. In sirenians, this muscle is especially large and assists in the downstroke of the tail. In cetaceans, part of this muscle is specialized and overlies the mammary gland. It can compress the gland (Fig. 3) and squirt milk into the mouth of nursing young.

The downstroke of the fluke in cetaceans is mainly powered by muscles attaching to the ventral side of the thoracic and lumbar vertebrae and inserting, via a tendon sheet, to the ventral side of the caudal vertebrae and chevron bones (Pabst, 1983). These muscles are large and are commonly called hypaxialis lumborum. In sirenians, specialized tail muscles called sacrocaudalis ventralis lateralis and medialis produce depression of the tail.

Unusual among mammals is the muscle system associated with the penis of cetaceans. Just like in most mammals, erection in cetaceans is not under muscular control. However, unlike most mammals, retraction of the penis into a pouch on the body of the cetacean is caused by contraction of the retractor penis muscles (see Fig. 1, Male Reproductive Systems). These muscles also occur in artiodactyls but are absent in other mammals.

IV. Forelimb Muscles

The forelimb of cetaceans is mainly involved in steering and does not provide propulsive force during rectilinear swimming. Shoulder movement are mainly adduction and abduction; flexion and extension are limited. The shoulder of cetaceans allows less mobility than that of most terrestrial mammals. The clavicle is absent, and tight muscles anchor the scapula to the thorax. These muscles include pectoralis, rhombdoids, serratus ventralis, and latissimus dorsi. A large additional muscle, the trapezius, occurs in most mammals, but is absent in cetaceans. At the scapulo-humeral joint, the deltoid is a strong abductor, and the latissimus dorsi probably the main adductor, assisted by the subscapularis. The joint between scapula and humerus is a ball and socket joint in cetaceans, as in all mammals, but there are no flexible synovial joints below the cetacean shoulder. Ligamentous connections at the elbow, wrist, and fingers allow for some elastic mobility. A few muscles in cetaceans (e.g., triceps) insert distal to the elbow, but there are no muscle bellies in the forearm and hand.

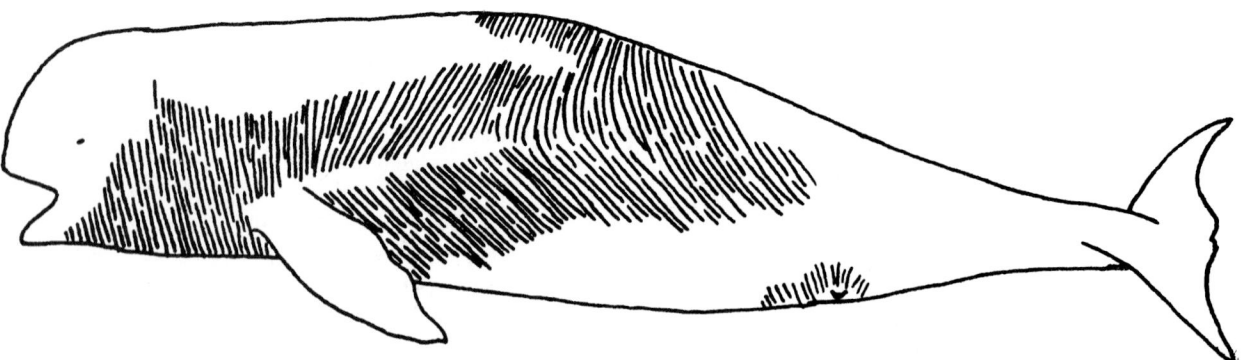

Figure 3 *The Cutaneous Trunci muscle of finless porpoise* (Neophocaena phocaenoides). *Note the muscular tissue overlying the mammary gland (ventrally, near the tail). After Howell (1930).*

Just like cetaceans, the forelimb of phocid and odobenid pinnipeds does not provide much of the propulsive force when swimming. It is, however, important in land locomotion. In contrast, otariid pinnipeds use their forelimb as the main propulsor during swimming, and the forelimbs also have an important role in land locomotion.

All pinnipeds lack a functional clavicle, and the shoulder is loosely attached to the chest. Of the shoulder muscles, the pectoralis and latissimus dorsi are the largest and probably provide most propulsive force during swimming in otariids (Fig. 4). The forearm and wrist of otariids contain synovial joints, although mobility at the wrist is reduced. The flippers that form the hands of otariids lack extensive musculature.

The forearm and hand of phocids are relatively mobile, unlike those of otariids. In the northern phocids (phocines), the hands are used in terrestrial locomotion and have well-developed muscles.

Modern Sirenia do not use their forelimbs for propulsion, but retain synovial joints at the shoulder, elbow, and wrist. The hands are used in manipulating food and retain many of the muscles that are present in land mammals. In the extinct *Hydrodamalis gigas* (Steller's sea cow), there were no wrist and hand bones and, consequently, no hand muscles.

V. Hindlimb Muscles

There are no hindlimb muscles in modern cetaceans, although they were developed in Eocene forms. A rudiment of the pelvis and sometimes the femur occurs in some modern cetaceans, but its main purpose appears to be the attachment of retractor penis.

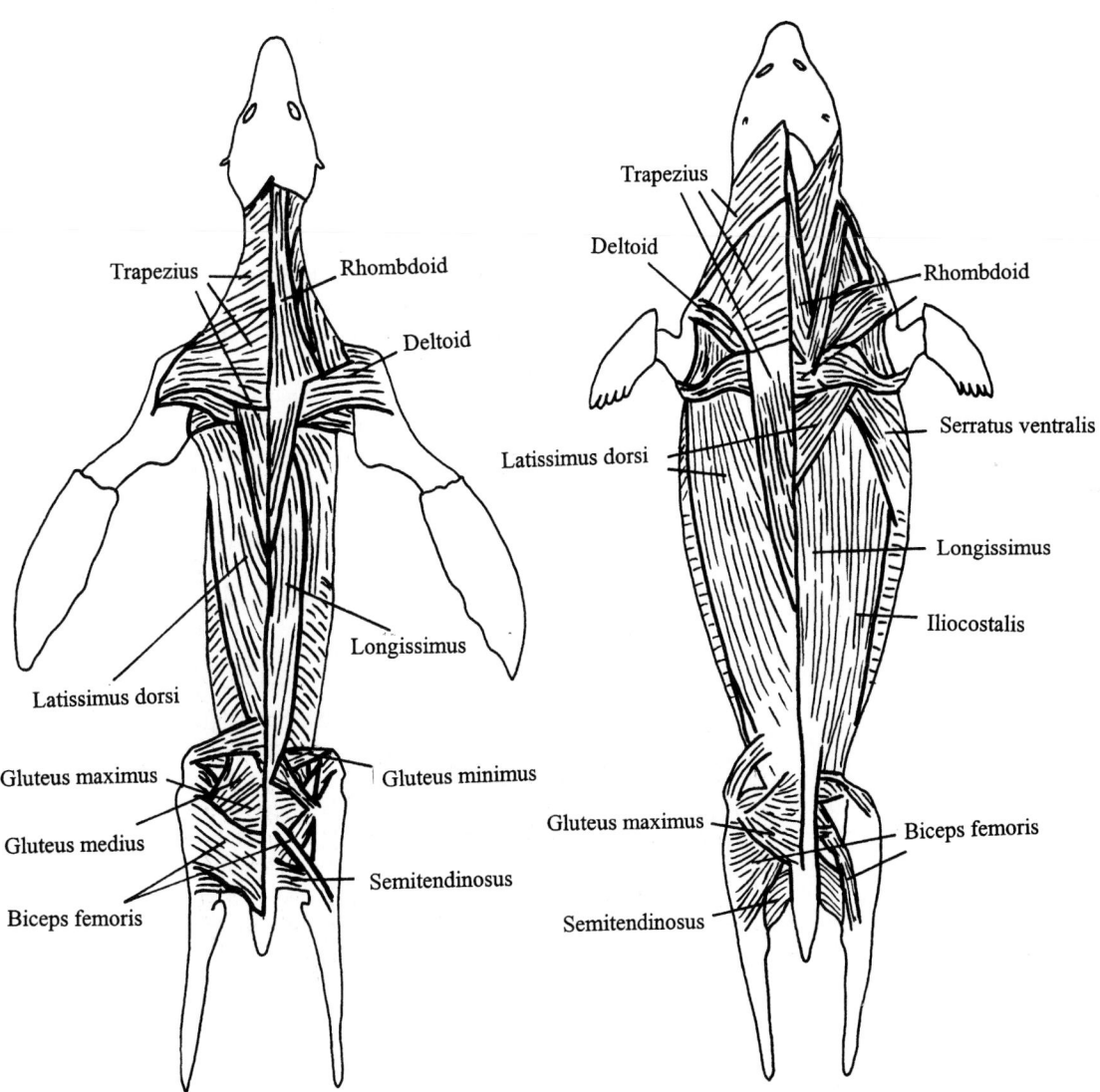

Figure 4 *Dorsal views of a partially dissected sea lion* (Zalophus, *left*) *and a harbor seal* (Phoca, *right*): *superficial muscles on left side of the animal, deeper muscles on the right. Modified after Howell (1930).*

Pelvic bones were also known in early sirenians, but these bones and all of the hindlimb musculature were lost in the extant forms.

Hindlimbs are well developed in the pinnipeds. Phocidae and Odobenidae use their hindlimbs in aquatic locomotion, making adduction/abduction movements with an inverted foot. Phocidae do not use their hindlimbs to support the body while on land. Otariidae trail their hindlimbs during swimming, but use them during locomotion on land. Odobenidae also support their body with their hindlimbs while on land.

In the phocids, the hindlimb flexors (e.g., the hamstrings) are reduced, whereas the extensors are supported by changed insertion of the adductors and obturator externus, which also serve in extension (Fig. 4). Muscles below the knee are present in all pinnipeds, but those crossing the heel are stronger in phocids than in otariids.

See Also the Following Articles

Anatomical Dissection: Thorax and Abdomen ▪ Locomotion, Terrestrial ▪ Male Reproductive Systems ▪ Sound Production ▪ Vision

References

Domning, D. P. (1978). The myology of the Amazonian manatee, *Trichechus inunguis* (Natterer) (Mammalia, Sirenia). *Acta Amazon.* **8**, Suppl. 2, 1–80.

Howell, A. B. (1930). "Aquatic Mammals, Their Adaptations to Life in the Water." C. C. Thomas, Baltimore.

Pabst, D. A. (1993). Intramuscular morphology and tendon geometry of the epaxial swimming muscles of dolphins. *J. Zool. Lond.* **230**, 159–176.

Purves, P. E., and G. E. Pilleri. (1983). "Echolocation in Whales and Dolphins." Academic Press, London.

Slijper, E. J. (1936). Die Cetaceen, vergleichend-anatomisch und systematisch. *Cap. Zool.* **6 and 7**, 1–590.

Slijper, E. J. (1962). "Whales." Basic Books, New York.

Museums and Collections

JOHN E. HEYNING
*Natural History Museum of Los Angeles County,
California*

The integrative approach to studying biology is similar to constructing a jigsaw puzzle—each discipline and data set contribute in a meaningful way to understand the whole. Individual pieces may contribute more or less to the picture, but nonetheless all pieces are important. In biology, each discipline contributes its own unique set of pieces to the puzzle of a species' unique biology. Research in museums has historically focused on specimen-oriented disciplines and thus has contributed to these suites of puzzle pieces. Specimens are potential sources of data for the disciplines of systematics, paleontology, morphology, histology, genetics, pathology, life history, parasitology, toxicology, and biochemistry. In addition, museums serve as important forums of informal learning for the visitors that peruse the exhibits or engage in an educational program.

I. Biodiversity and Systematics

Perhaps the most fundamental among the specimen-oriented disciplines is the study of biodiversity, the defining of species and populations within species. Most marine mammalogists working within museums in the 19th and early 20th centuries spent their hours primarily describing new species from the vast array of specimens unloaded from some recent voyage of exploration so characteristic of that time. For instance, from the numerous marine mammals specimens collected by the Southern Hemisphere expeditions of the HMS *Erebus* and *Terror* during the years 1839–1843, John Gray of the British Museum (Natural History) described numerous new species, including the Ross seal (*Ommatophoca rossii*), the crabeater seal (*Lobodon carcinophaga*), the pygmy right whale (*Caperea marginata*), and the Chilean dolphin (*Cephalorhynchus eutropia*). While the heyday of prolific new species description peaked a century ago, the need for the ongoing study remains very relevant today. Several new species (or resurrected old species) have been defined within recent years, and most populations are just now being understood.

The classical approach of using morphology to define species continues to be relevant. However, analyses of molecular genetic data provide us with additional new tools to help define populations, species, and the relationship among species. Exemplary of this is the recent discovery of a new species of beaked whale. In the mid-1970s, several STRANDINGS occurred of a small species of beaked whale along a restricted section of southern California coastline. Because these specimens morphologically resembled the Southern Hemisphere species *Mesoplodon hectori*, scientists tentatively assigned these California animals to that taxon. A graduate student from New Zealand investigating beaked whale phylogeny sampled the DNA from these specimens along with many others held in museums, including the type specimen of *M. hectori* catalogued into the British Museum over 100 years ago. To her astonishment, these California specimens clustered nowhere near specimens of *M. hectori* from the Southern Hemisphere (Dalebout *et al.*, 1998), hence providing evidence that they represented a new species hithertofore undescribed!

Determining the evolutionary relationships, or phylogeny, among this diversity of species, both living and extinct, is the study of systematics. Systematics provides an evolutionary framework that becomes the foundation for the comparable biological approach. Phylogenies can be constructed using a variety of data sets, morphological, molecular, and fossils—all of which reside primarily within museums. Hence, researchers today can infer past events from phylogenetic reconstructions of evolutionary relationships. Most modern systematists use a philosophical approach called cladistics. The basic tenets of cladistics are quite simple: organisms are deemed to be related based on shared derived characters called synapomorphies. Derived characters are defined as having arisen in the common ancestor of the taxa and subsequently passed onto their descendant taxa.

Museums have a long-term commitment to house specimens for research. Thus material collected in the 1700s and 1800s is still available for scientific inquiry today. For many species, it is only through the accumulation of specimens and data over several decades, even over a century, that we can obtain the samples sizes needed to begin to understand even the basic biology of these species. For systematic studies it is crucial to examine a large series of specimens (Fig. 1). In order to define species or populations, one must first know the limits of variation—individual, ontogenetic, SEXUAL DIMORPHISM—in order to ascribe that the observed variation is due to limited genetic exchange.

II. Morphology

How can a blue whale engulf up to 70 tons of water? Why doesn't a narwhal break its tusk? How can a dolphin cool its testes so that spermatogenesis can occur? All these questions require the detailed examination of anatomical structures. This in turn requires that some specimens are readily available. Some studies are limited to hard parts and can be answered by examining osteological material. However, studies of soft anatomy require that these structures be preserved. For most organisms, storage of the whole beast can be accomplished easily by plunking the specimen into a jar of formalin and/or alcohol. Preservation for future study of a good-sized dolphin, let alone a whale, presents far more of a logistical challenge. As the immense specimens typically need be dissected without preservation, the task can be demanding, as these large, oil-laden mammals produce a rich organic bouquet as they decompose. Fortunately, there is now a renaissance of morphological work requiring innovative ways of preserving and studying cetacean anatomy.

III. History of Museum Research

The first large collections of marine mammals had their genesis in the grand museums of Europe. Baron von Cuvier amassed and published on a very important collection in the early 1800s, which now resides in the Museum National d'Histoire Naturelle in Paris. By the mid-1800s, the British Museum of Natural History (now the Natural History Museum, London) had built major collections as the British Empire explored the world. Two of the preeminent marine mammalogists of this era, William Henry Flower and John Edward Gray, increased our knowledge considerably by studying the specimens within this venerable museum. Aside from the collections amassed from expeditions, museums in Britain had a distinct advantage for growing their collections. In 1324, stranded whales and dolphin were declared "Royal Fishe" and therefore property of the Crown. The original intent of this decree was to ensure that an economically valuable stranded fresh whale would enhance the coffers of the government. An unforeseen benefit was that the majority of strandings were of the economically nonvaluable uneatable variety and therefore available for government supported museums. Hence, the first stranding program began (Fraser, 1977). This original decree and subsequent museum-oriented mindset was passed along to the then British colonies. These former colonies now have museums with major collections including those in Australia, New Zealand, South Africa, and the United States.

Marine mammals as museum specimens are difficult to acquire, store, and maintain. As a result, there are very few large collections for researchers to use. Of the largest collection of land mammals, well over one dozen have more than 100,000 specimens. The majority of specimens in these collections are

Figure 1 *A series of pilot whale skulls* (Globicephala *spp.). Series such as these allow biologists to define species and to understand populations within species. Defining these biological units is crucial to conservation biology among other disciplines.*

the taxonomically diverse and numerically abundant rodents and bats. In striking contrast, less than a dozen or so museums have collections of marine mammals numbering over a mere 1000 or so. These include the National Museum of Natural History (Smithsonian); Natural History Museum of Los Angeles County; National Science Museum, Tokyo; The Natural History Museum London; National Museum of New Zealand; American Museum of Natural History; California Academy of Sciences; South Australian Museum; Museum National d'Histoire Naturelle, Paris; and South African Museum.

For a specimen to be of greatest utility for answering questions, it needs to have as much associated data with it as possible. Such archives provide context for the additional data collected by scientists. Originally, museum curators collected only skulls or skeletons along with occasional sketches of the living beast. Early in this century, following the lead set by the systematic collection of data from whaling stations, museum workers began documenting more data from each specimen. As the number of questions regarding marine mammal biology have increased concurrent with new analytical tools to address these questions, far more is being collected. Now it is not uncommon to collect the complete skeleton, frozen tissues, measurements, fluid-preserved tissues, photographs, and notes.

IV. Public Display

Over the past century and a half, museums have served an increasing role as important centers for the public to learn about the natural world. Accurately mounted exhibits can convey great biological detail and grand-scale presence that would be difficult for the public to ever experience in the wild.

Many museums have also capitalized on the immensity of whales to create exhibit icons, most notably the articulated skeleton of a large whale (Fig. 2) or a model of a living blue whale (*Balaenoptera musculus*). In 1907, a model created from a 74.4-foot blue whale went on display at the American Museum of Natural History. Subsequently, the British Museum of Natural History erected its own blue whale model measuring some 88 feet in length. In the early 1960s, the Smithsonian unveiled their 92-foot model. Not to be outdone, the American Museum christened their new and anatomically more accurate 94-foot (28.7 m) model in 1969!

Museums hold collections in the public trust so that they are available to scholars in perpetuity. Thus they serve as guardians of the tangible evidence of the past and the archivists of our current natural heritage. In addition, museums serve as important centers at which the public can learn.

See Also the Following Articles

History of Marine Mammal Research ■ Paleontology ■ Systematics, Overview

References

Conover, A. (1996). The object at hand. *Smithsonian* **27**(7), 28, 30, 31.
Dalebout, M. L., van Heldsen, A., van Waerebeek, K., and Baker, C. S. (1998). Molecular genetic identification of southern hemisphere beaked whales (Cetacea: Ziphiidae). *Mol. Ecol.* **7**, 687–694.

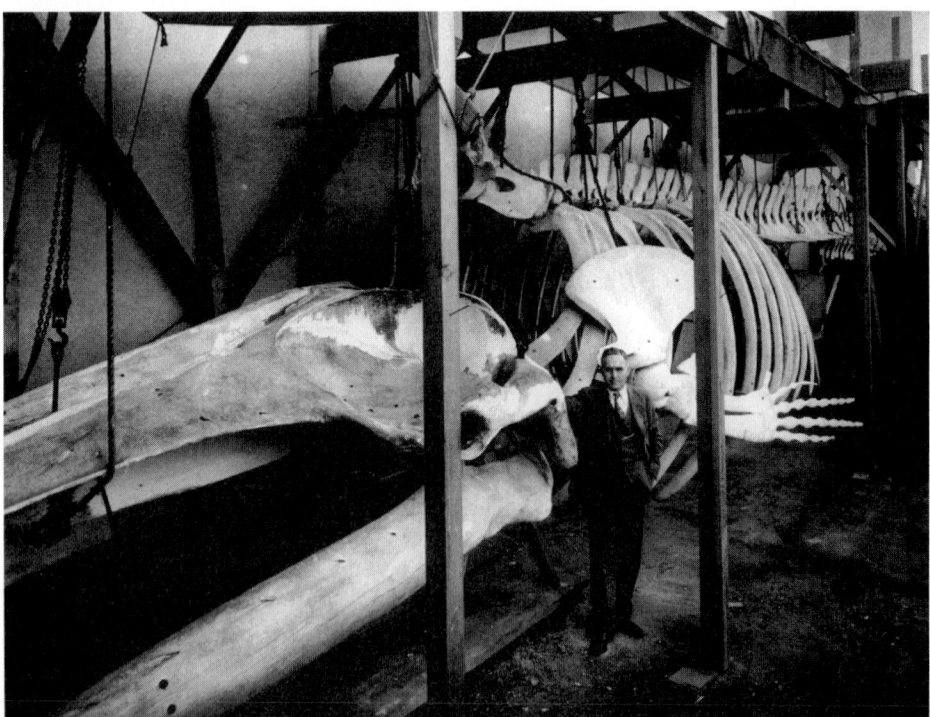

Figure 2   *The mounting of a fin whale* (Balaenoptera physalus) *skeleton in the 1930s at the Natural History Museum of Los Angeles County.*

Fraser, F. C. (1977). Royal fishes: The importance of the dolphin. *In* "Functional Anatomy of Marine Mammals" (R. J. Harrison, ed.), Vol. 3, pp. 1–44. Academic Press, London.

Heyning, J. E. (1991). Collecting and archiving of cetacean data and specimens. *In* "Marine Mammal Stranding in the United States: Proceedings of the Second Marine Mammal Stranding Workshop" (J. E. Reynolds and D. K. Odell, eds.), pp. 69–74. NOAA Technical Report NMFS 98.

Van Gelder, R. G. (1970). Whale on my back. *Curator* **12**(2), 95–119.

Mustelidae

RONALD E. HEINRICH
Ohio University, Athens

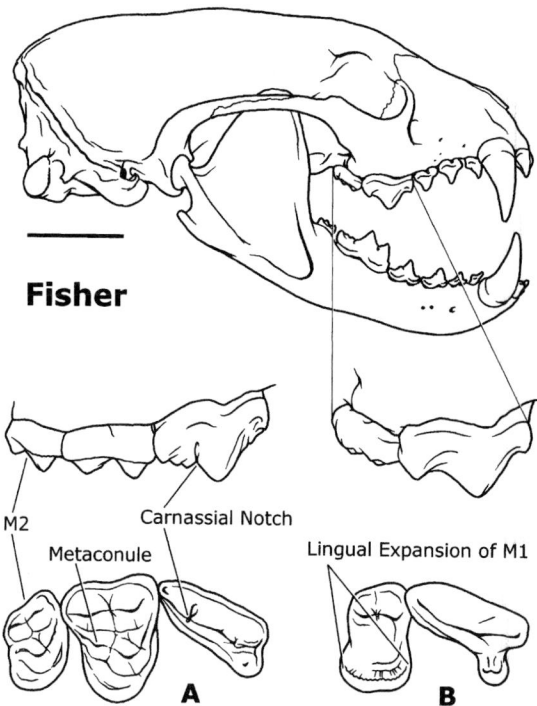

Figure 1 *Characteristics of upper carnassial (P4) and molar (M1) teeth in mustelids. The dentitions of a canid (gray fox, A) and mustelid (fisher, B) are compared in lateral (top) and occlusal views (bottom). Eumustelids have lost the second molar, the metaconule on M1 and the carnassial notch on P4, and expanded the lingual surface of M1. Scale bar = 2 cm.*

Mustelidae is one of the most successful families of the order Carnivora; its 23 genera and 65 extant species of weasels, badgers, otters, and skunks have a nearly worldwide distribution. These small to medium-sized mammals tend to have long and slender bodies, exhibit sexual dimorphism (males are generally 25% larger than females), and possess well-developed anal scent glands. Several cranial and dental characters, including flooring of the suprameatal fossa (an expansion into the roof of the auditory canal), reduction/loss of the upper second molar (M2), reduction/loss of the metaconule and lingual expansion of the upper first molar (M1), and loss of the carnassial notch on the last upper premolar (P4), have been argued to be synapomorphies of a monophyletic Mustelidae (Fig. 1). Recent molecular analyses, however, suggest that skunks are distantly related to other mustelids and occupy a phylogenetic position as sister taxon to Musteloidea, a clade consisting of Mustelidae (minus skunks) and Procyonidae (raccoons and their relatives).

I. Palaeomusteloids, Early Eumustelids, and Origins of Lutrinae

The earliest mustelid-like fossils are assigned to *Mustelavus* from the late Eocene of western North America and *Mustelictis* and *Plesictis* from the early Oligocene of France. Because these animals retain upper and lower second molars and an upper carnassial notch, they are conservatively placed in an undifferentiated musteloid stem group. Others have argued, however, that the small size of the second molars and reduction of the metaconule are indicative of mustelid affinities and place all three genera in the family Mustelidae (Baskin, 1998). An increased diversity of palaeomusteloids occurred in the late Oligocene–early Miocene of North America and Europe, and although most of these taxa have little to do with later mustelid evolution, two European genera, *Paragale* and *Plesiogale*, lose the carnassial notch, exhibit some flooring of the suprameatal fossa, and show rudimentary lingual expansion of the upper first molar suggesting an ancestral relationship to eumustelids (Hunt, 1996). Although eumustelids evolved in Eurasia, the earliest representatives of this group are found in the early

Miocene of North America, suggesting a late Oligocene immigration event, with immigration events into Africa and South America occurring in the early Miocene and late Pliocene, respectively. By the late Miocene there is considerable mustelid diversity in the fossil record and all of the extant mustelid subfamilies are represented.

One enigmatic taxon that has been included with the palaeomusteloids by some workers and as an incertae sedis member of Arctoidea [Musteloidea + Ursidae (bears) + Pinnipedia] by others is the late Oligocene *Potamotherium*. Known from a number of complete and well-preserved skeletons, this aquatically adapted otter-like genus exhibits a combination of musteloid and pinniped characters that have suggested to some that it may be transitional between mustelids and phocids (Tedford, 1976), although others have argued that these morphological similarities are convergent. The earliest widely accepted fossil otter is the early to mid-Miocene genus *Mionictis* known from dental remains in North America, Europe, and China. Diversification of the major lutrine clades likely occurred by the mid-Miocene (Koepfli and Wayne, 1998), although fossil records for most lineages tend to be Plio-Pleistocene in age. One exception to this is the fossil record of the sea otter (*Enhydra lutris*), the most marine and one of the most morphologically derived of all otters. Remains of *Enhydritherium* and *Enhydriodon*, the consecutive outgroups to

Enhydra, respectively, have been described from throughout the late Miocene, sharing with the living sea otter loss of upper and lower first premolars, a short robust jaw, and lower first molars with low, inflated cusps (Berta and Morgan, 1985).

II. Phylogenetic Relationships of Musteloids

A number of recent morphologic and molecular analyses have addressed phylogenetic relationships both among genera that have been referred to Mustelidae and between mustelids and other extant members of the monophyletic suborder Caniformia, which includes canids (dogs), procyonids, pinnipeds, and ursids. While most of these analyses have supported a traditionally held view that Procyonidae and Mustelidae are sister taxa, and that Ursidae and Canidae represent successive outgroups (Flynn *et al.,* 1988), the position of Pinnipedia has proven more difficult to resolve, with morphological evidence

supporting a sister taxon relationship to Ursidae and molecular data tending to support a sister group relationship to Musteloidea (Fig. 2A).

Among the 23 genera and five subfamilies attributed to Mustelidae, there is strong support, both molecular and morphologic, for the monophyly of otters (Lutrinae) and skunks (Mephitinae), little support for Mustelinae monophyly, and no support for badger (Melinae) monophyly (weasels, martens, wolverines, among others). Lutrinae appears to be nested well within the mustelid clade (Fig. 2B), and several studies have suggested that skunks are the sister taxon to otters (Wyss and Flynn, 1993). Ribosomal protein sequence data (Dragoo and Honeycutt, 1997), however, argue for a phylogenetic position of skunks as sister taxon to Musteloidea (Fig. 2B), implying that characters such as loss of the second molars and upper carnassial notch evolved more than once among musteloids. Consensus on musteloid phylogeny will require more analyses that combine molecular and morphologic data, and that include important fossil taxa such as *Mustelavus* and *Potamotherium,* as well as the early procyonid *Amphictis.*

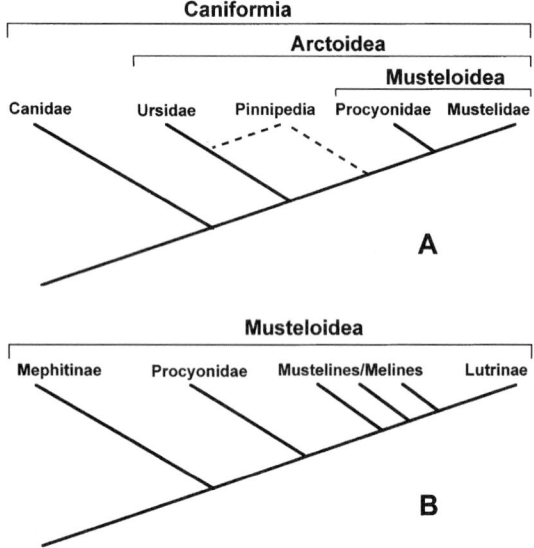

Figure 2 *Cladograms depicting the relationship of mustelids to other extant members of the carnivoran suborder Caniformia (A) and one hypothesis of phylogenetic relationships among musteloids based on molecular data where skunks (Mephitinae) are not part of a monophyletic Mustelidae. Dotted lines (A) show alternative hypotheses for pinniped relationships, and the musteline/meline designation (B) is used to indicate the paraphyletic nature of these subfamilies.*

See Also the Following Articles

Carnivora ■ Otters

References

Baskin, J. A. (1998). Mustelidae. *In* "Evolution of Tertiary Mammals of North America" (C. M. Janis, K. M. Scott, and L. L. Jacobs, eds.), Vol. 1, pp. 152–173. Cambridge Univ. Press, Cambridge.

Berta, A., and Morgan, G. S. (1985). A new sea otter (Carnivora: Mustelidae) from the Late Miocene and Early Pliocene (Hemphillian) of North America. *J. Paleontol.* **59,** 808–819.

Dragoo, J. W., and Honeycutt, R. L. (1997). Systematics of mustelid-like carnivores. *J. Mammal.* **78,** 426–443.

Flynn, J. J., Neff, N. A., and Tedford, R. H. (1988). Phylogeny of the Carnivora. *In* "The Phylogeny and Classification of the Tetrapods" (M. J. Benton, ed.), Vol. 2, pp. 73–115. Clarendon Press, Oxford.

Hunt, R. M., Jr. (1996). Biogeography of the order Carnivora. *In* "Carnivore Behavior, Ecology, and Evolution" (J. L. Gittleman, ed.), Vol. 2, pp. 485–541. Cornell Univ. Press, Cornell.

Koepfli, K.-P., and Wayne, R. K. (1998). Phylogenetic relationships of otters (Carnivora: Mustelidae) based on mitochondrial cytochrome *b* sequences. *J. Zool. Lond.* **246,** 401–416.

Tedford, R. H. (1976). Relationship of pinnipeds to other carnivores (Mammalia). *Syst. Zool.* **25,** 363–374.

Wyss, A. R., and Flynn, J. J. (1993). A phylogenetic analysis and definition of the Carnivora. *In* "Mammal Phylogeny: Placentals" (F. S. Szalay, M. J. Novacek, and M. C. McKenna, eds.), pp. 53–73. Springer-Verlag, New York.

Narwhal

Monodon monoceros

M. P. Heide-Jørgensen
*Greenland Institute of Natural Resources,
Nuuk*

In 1758, Linnaeus used the scientific name *Monodon monoceros* for the whale with one tooth and one horn. Together with the close relative the white whale or beluga, *Delphinapterus leucas*, the narwhal now forms the two-species family of Monodontidae (Fig. 1).

I. External Appearance and Dentition

Newborn narwhals are evenly gray or dark-brownish gray. While nursing for 1–2 years, the coloration changes gradually to a dark background color with white patches that give a mottled appearance. When adult, the animals are completely mottled on the dorsum but with increasing white fields on the ventral side. Old adult males only maintain a narrow dark-spotted pattern on the top of the back, whereas the rest of the body is white. Unlike in other cetaceans, the tail flukes are concave in fully grown narwhals and a low ridge replaces the dorsal fin.

The most conspicuous feature of the narwhal is the up to 3-m-long spiraled tusk. Six pairs of maxillary and two pairs of mandibulary teeth are present in early narwhal embryos, but only two maxillary pairs persist and develop. Of these the two anterior teeth develop into an elongated tooth that is the start of the tusk. The other two TEETH remain vestigial. In males, the left of the two elongated teeth grows and protrudes through the maxillary bones and skin of the rostrum of the whale. During growth the tusk spirals and grooves to the left. In males, the right of the elongated maxillary teeth and in females both maxillary teeth remain inside the SKULL, sometimes just protruding through an opening in the maxillary bone. Irregularities in the development of tusks are frequently seen: females sometimes attain a tusk, males occasionally have no tusk, and narwhals with two tusks (so-called "double tuskers") are not rare.

There is great variability in the shape and dimensions of the protruding tusk. Some are fairly straight and others corkscrew like; some are thin and fragile, whereas others are short and thick. The largest tusk measured was 267 cm, but a full-grown male usually carries a tusk of about 200 cm. Tusks are sometimes broken, and there are records of tusk from another narwhal sitting inside the broken tip. The purpose of the tusk has been much disputed, but because both females and males without tusks thrive, tusks do not seem critical for survival. The tusk is more likely a secondary sexual character that is related to the hierarchy of male narwhals. Displays and crossing of tusks are

Figure 1 *The narwhal,* Monodon monoceros, *occurs in the remote North Atlantic and Arctic Oceans and is conspicuous with a long tusk in males, usually formed from one tooth in the left upper jaw. Pieter A. Folkens/Higher Porpoise Design Group.*

frequently seen on narwhal summering grounds and it is likely that this activity determines social rank.

II. Fossil Records

There are several records of narwhal fragments from Pleistocene deposits in England and Germany. Bones found along the Russian Arctic coasts—both on the mainland and on the Russian Arctic Islands—also suggest a different occurrence of narwhals before or during the most recent glaciation. In Canada, bone remains from early postglacial times have also been found both north (Ellesmere Island) and south (Gulf of Saint Lawrence) of present narwhal distribution.

III. Distribution and Abundance

The main reason the narwhal remained a legendary animal for so long may be because of its preference for remote and inaccessible habitats, usually in areas over deep water that is covered with heavy pack ice during dark winter months. Europeans did not visit most of these areas until the 19th century, and even though Inuit hunters traded the tusks with whalers, precise descriptions were lacking.

The narwhal essentially inhabits the Atlantic sector of the Arctic Ocean with few records of stragglers from the Pacific sector (Fig. 2). During the last glaciation, narwhals were restricted to the North Atlantic but with the retreating ice they inhabited the archipelago of the Canadian High Arctic, northern Hudson Bay, Davis Strait, Baffin Bay, the Greenland Sea, and the Arctic Ocean between Svalbard and Franz Josef Land. Today, low numbers of narwhals are found offshore in deepwater areas of the Eurasian sector of the Arctic Ocean, where they are seen most frequently around Franz Josef Land and Svalbard. The northernmost recordings of narwhals are from the area between 84°N and 85°N northeast of Franz Josef Land at 70–80°E. In the Greenland Sea, narwhals are widely distributed in the pack ice but probably in low numbers. Along the coast of East Greenland, narwhals are found during the open water season in fjords from 65°N to 81°N, with particularly large concentrations in Scoresby Sound and Kangerlussuaq. No complete ABUNDANCE ESTIMATES are available from any of the Northeast Atlantic areas, but in 1983 an estimate of 300 narwhals was derived from a survey in Scoresby Sound.

In West Greenland, narwhals visit coastal areas in northwest Greenland (Inglefield Bredning and Melville Bay) during summer and central West Greenland during autumn (Uummannaq) and winter (Disko Bay). Up to 4000 narwhals have been counted in Inglefield Bredning in August and 3000 in Disko Bay in March. Offshore, narwhals are abundant in the heavy consolidated pack ice in northern Davis Strait and Baffin Bay from late November through May, and the number of narwhals wintering in this area has been estimated at 35,000 whales.

During ice break-up, narwhals move into the Canadian High Arctic through Lancaster Sound and Pond Inlet. They visit the fjord systems of Eclipse Sound, Admiralty Inlet, Prince Regent Inlet, and Peel Sound during the open water season from June to September. The abundance in these areas was estimated at 18,000 narwhals in 1984. With the formation of fast ice in October, narwhals move east toward Baffin Bay and Davis Strait.

In northern Hudson Bay and Foxe Basin, an apparently isolated group of 1300 narwhals persists. It is believed that they move south to Hudson Strait in winter.

IV. Migration

Narwhals follow the DISTRIBUTION of the ice and move toward coastal areas in summer when these are ice free. During freeze-up the coastal areas are abandoned and the narwhals move offshore. In winter they stay in very heavy consolidated pack ice, usually in leads or holes in 10/10 of ice. When ice breaks up in the spring, narwhals penetrate north through narrow leads and open water channels.

Movements from summer through winter have been monitored by tracking of narwhals instrumented with satellite-linked radio transmitters attached to the tusk of males. At summering grounds in West Greenland and Canada, narwhals moved back and forth between glacier fronts, offshore areas, and neighboring fjords. When fast ice formed the whales moved out to deeper water, usually up to a 1000-m water depth. In October the whales moved southward toward the edge of the continental shelf where the water depth increases over a short distance from 1000 to 2000 m. This slope was also used as a wintering ground, and even though the whales seemed stationary in this area, they still conducted shorter movements along this steep slope. Narwhals tracked from Canada and West Greenland were within a few kilometers from each other at these wintering grounds at the deep slope at the edge of the continental shelf in central Baffin Bay. The importance of this area as a wintering ground has also been confirmed by aerial surveys. No satellite trackings of whales have been conducted so far in spring and early summer. The mean swimming speed of traveling narwhals is 5 km/hr.

V. Growth in Length and Weight

Length at birth is approximately 160 cm. The tusk erupts at a body length of 260 cm and attains a length of 150 cm at sexual maturity. Body length at sexual maturity is around 360 and 420 cm for females and males, respectively. Mean length and weight at physical maturity are around 400 cm and 1000 kg for females and 475 cm and 1600 kg for males.

No reliable methods are available for estimating the age of narwhals; both the protruding tusk and the embedded teeth contain distinctive growth layers in both dentine and cementum, but with increasing age the growth layers apparently collapse and become unreadable. Also, there is no empirical way to determine how many growth layers are deposited annually. So far, no narwhals have been kept successfully in captivity.

VI. Reproduction

The gestation period of the narwhal is subject to some uncertainty, as mating probably occurs in inaccessible areas in March–May. Calving seems to occur in July–August in both Greenland and Canada, and with a mating season early in spring, this implies a gestation period of 13–16 months. Lactation lasts 1–2 years, and females are generally believed to calve every 3 years, but data supporting this seem inadequate.

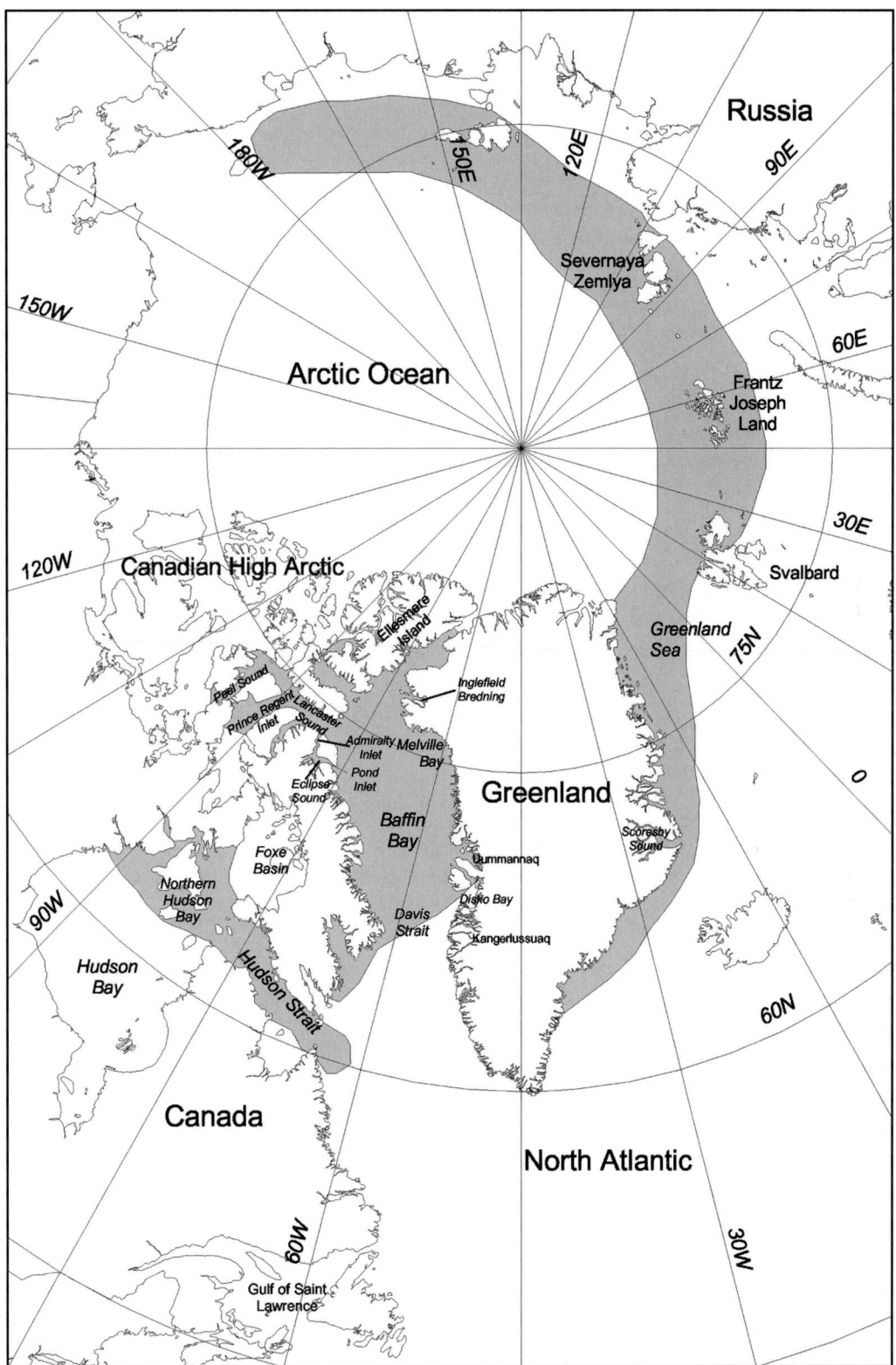

Figure 2 *Distribution of the narwhal.*

VII. Feeding

The DIET of narwhals has been studied in Greenland and Canada and they typically prey on high Arctic fish species such as polar cod, *Boreogadus saida,* and Arctic cod, *Arctogadus glacialis;* both are pelagic species that are often associated with the underside of ice. Narwhals also take demersal species that are found at great depths, such as Greenland halibut, *Reinhardtius hippoglossoides;* redfish, *Sebastes marinus;* and bottom-dwelling cephalopods. In some areas the narwhals seem to be feeding exclusively on schools of squids, *Gonatus fabricii,* which can be found at variable depths. Apparently little FEEDING takes place during the open water season in August.

VIII. Population Structure

Narwhals are usually found in small groups of 5–10 whales migrating together. Sometimes larger herds are formed that consist of several smaller groups often all on a directional movement along a coastline or toward the head of a fjord. The narwhal groups are usually segregated with adult males in separate groups and females with calves sometimes together with immature males. Mixed groups occur especially in large herds, but single animals, particularly males, can also be found.

Studies of mitochondrial DNA have revealed a low level of nucleotide and haplotype diversity in narwhals. This is probably the result of a rapid expansion of the population after the last glaciation from a small founding population. Despite the low variation in narwhal mtDNA, there are still genetic differences between narwhals from different areas. Not so surprisingly, narwhals from East Greenland are different from those inhabiting Baffin Bay, but more surprising was the distinctness of narwhals at two summering grounds (Inglefield Bredning and Melville Bay) and one autumn ground (Uummannaq) in West Greenland. Apparently, narwhals have annual fidelity to certain summer and autumn feeding localities, but the extent of mixing on the wintering grounds is unknown.

IX. Diving Behavior

Data on narwhal diving have been collected from whales instrumented with satellite transmitters in both Canada and Greenland. Narwhals are able to dive to depths exceeding 1000 m, and the deepest dive recorded was 1164 m. However, the dives are usually completed within 20 min and never exceed 25 min, so the whales only have a short time at the bottom as ascent–descent rates for deep dives are 2 and 1 m/sec for shallow dives. Narwhals apparently reduce their diving activity during autumn and early winter and make more deep dives.

X. Vocalizations

Narwhals are known to make a variety of noises. Clicks that are believed to be used for echolocation have been measured to have their maximum amplitudes at 48 kHz with rates of 3–10 clicks/sec. Faster click rates of 110-150 clicks/sec had maximum amplitudes at 19 kHz. Whistles or pure tones in frequencies from 300 Hz to 18 kHz have also been recorded and they are suspected to serve as social signals among the whales.

XI. Unicorn Myth and Systematics

The narwhal is the animal behind the legend about the mysterious unicorn: a horse-like creature with a spiraled horn protruding from the forehead. The horn was supposed to have healing abilities, and the wild and shy animal could only be captured with a virgin as bait. Based on narwhal tusks that were brought south from Arctic coasts, this was essentially how narwhals were perceived in western civilization until the 17th century when the first descriptions of a fish-like sea monster appeared.

XII. Human Effects and Interactions

Narwhals have never been a target for commercial WHALING probably because of their skittishness and the difficulties involved in catching them. Inuit hunters in Greenland and Canada hunt narwhals for their valuable tusks and the highly prized skin that is considered a delicacy throughout the Inuit communities. The harvest level was on average 550 and 280 during 1993–1995 in Greenland and Canada, respectively, and it is considered small relative to the population size in most areas; however, depending on the population structure, some subpopulations may be overexploited.

Narwhals have high levels of some organochlorines and heavy metals where at least the first are of anthropogenic origin. Possible effects of these contaminants have not been studied in narwhals.

Because of their prevalence for high-density pack ice, narwhals are susceptible to climatic changes that influence the water currents and thereby ice formation in the Arctic. Whether it is naturally occurring or human-induced climate changes, narwhals may become entrapped or lose access to important feeding areas if ice conditions change.

XIII. Ice Entrapments

A peculiar feature of the natural history of narwhals is their susceptibility to being entrapped in ice. Because of their preference for heavy pack ice, large schools of narwhals are occasionally caught in ice that freezes rapidly during intense cold, thereby preventing the whales from getting enough air to breathe. This happens particularly often in areas where unpredictable ice conditions persist due to the mixing of warm and cold water masses of variable strength, e.g., Disko Bay in West Greenland. Large numbers of narwhals may succumb during such an ice entrapment, and in January 1915, more than 1000 narwhals died in a well-known ice entrapment in Disko Bay. If the whales are discovered, Inuit hunters may also prey upon them, using the word "sassat" for the event.

See Also the Following Articles

Beluga Whale ■ Folklore and Legends ■ Inuit and Marine Mammals

References

Dietz, R., and Heide-Jørgensen, M. P. (1995). Movements and swimming speed of narwhals, *Monodon monoceros,* equipped with satellite transmitters in Melville Bay, Northwest Greenland. *Can. J. Zool.* **73**, 2120–2132.

Dietz, R., Heide-Jørgensen, M. P., Glahder, C., and Born, E. W. (1994). Occurrence of narwhals (*Monodon monoceros*) and white whales (*Delphinapterus leucas*) in East Greenland. *Meddr Grønland Biosci.* **39**, 69–86.

Ford, J. K. B., and Fisher, H. D. (1978). Underwater acoustic signals of the narwhal (*Monodon monoceros*). *Can. J. Zool.* **56**, 552–560.

Hay, K. A., and Mansfield, A. W. (1989). Narwhal *Monodon monoceros* Linnaeus, 1758. *In* "Handbook of Marine Mammals" (S. D. Ridgway and Sir R. Harrison, eds), Vol. 4, pp. 145–176, Academic Press, San Diego.

Heide-Jørgensen, M. P., and Dietz, R. (1995). Some characteristics of narwhal, *Monodon monoceros*, diving behaviour in Baffin Bay. *Can. J. Zool.* **73**, 2106–2119.

Miller, L., Pristed, J., Møhl, B., and Surlykke, A. (1995). The click-sounds of narwhals (*Monodon monoceros*) in Inglefield Bay, Northwest Greenland. *Mar. Mamm. Sci.* **11**(4), 491–502.

Palsbøll, P., Heide-Jørgensen, M. P., and Dietz, R. (1997). Distribution of mt DNA haplotypes in narwhals, *Monodon monoceros*. *Heredity* **78**, 284–292.

Siegstad, H., and Heide-Jørgensen, M. P. (1994). Ice entrapments of narwhals (*Monodon monoceros*) and white whales (*Delphinapterus leucas*) in Greenland. *Meddr Grønland Biosci.* **39**, 151–160.

Neoceti

R. Ewan Fordyce
University of Otago, Dunedin, New Zealand

Neoceti is the taxonomic group containing the two living clades of Cetacea (Odontoceti and Mysticeti), but excluding Archaeoceti. The two living groups are quite disparate, each distinguished by a unique combination of anatomical and ecological attributes. The Odontoceti (toothed whales, dolphins, porpoises) are echolocating macropredators, whereas Mysticeti (baleen whales) are filter feeders. Ancient cetaceans from Oligocene times (25 to 30+ Ma) also show skull structures indicative of ECHOLOCATION in odontocetes and of filter feeding in mysticetes, emphasizing the early divergence of feeding habits. Apart from the feeding apparatus, however, basal odontocetes and mysticetes are much more similar to one another than are their modern descendants. Similarities include some evolutionary novelties (synapomorphies) of the skull, which are not seen in archaeocetes. Thus, odontocetes and mysticetes are regarded as sister taxa, forming a clade variably termed crown group Cetacea, or Neoceti, or Autoceta. Basal odontocetes and mysticetes also show marked similarities with archaic cetaceans (Archaeoceti), pointing clearly to an origin within the archaeocete family Basilosauridae.

I. Changing Concepts of Names

The name Cetacea was first used in a modern sense by Brisson in 1762 for genera and species of living whales, dolphins, and porpoises. Until the mid-1800s, high-level classification was based on superficial features, with no implication that patterns among living cetaceans had arisen by evolution. Formal subdivisions for the living baleen whales and toothed cetaceans were proposed by W. H. Flower for Mysticeti (1864) and Odontoceti (1867), with the implication that these were real groups (in modern terms, clades). The discovery of fossils broadened the concept of Cetacea in the early to mid-1800s. Initially, most fossils were recognized as related to living species, and modern generic names (e.g., *Delphinus, Balaena*) were applied to some of these. The discovery of the archaic Eocene whale *Basilosaurus* in the 1830s eventually led in 1883 to naming of a formal group of archaic cetaceans, the Archaeoceti. Thus, the concept of Cetacea was expanded to include three suborders: one extinct (Archaeoceti) and two living (Odontoceti and Mysticeti).

II. The Monophyly of Odontoceti

Odontocetes include 71–72 living species in the families Physeteridae, Kogiidae, Ziphiidae, Platanistidae, Delphinidae, Phocoenidae, Monodontidae, Iniidae, Pontoporiidae, and Lipotidae. Strictly defined, the Odontoceti comprises the most recent common ancestor of all living species, plus all the descendants of that ancestor. Such an ancestor probably lived in latest Eocene or Early Oligocene times. In practice, fossil and recent odontocetes are distinguished by osteological features, particularly in the skull. For example, above the eye, a large supraorbital process in each maxilla rises posteriorly over the frontal, usually forming a voluminous facial fossa in which open dorsal infraorbital foramina for nerves and blood vessels; in living species, this fossa forms the origin for the nasofacial muscles, which manipulate diverticula or sacs in the soft nasal passages. In turn, the diverticula probably help produce echolocation sounds. Where the rostrum passes into the facial fossa, a vertical antorbital notch forms a path for the facial nerve, which supplies the nasofrontal muscles. Anteriorly, each side of the rostrum in front of the bony nares has a premaxillary sac fossa, premaxillary foramen, and usually premaxillary sulci; in living species, these are also implicated in sound generation in the nasal passages. Below the face, the infraorbital process is vestigial or absent, and the most posterior tooth lies far forward of the antorbital notch. In the ear region on the skull base, the parietal and squamosal roof the periotic so that this ear bone no longer contributes to the floor of the braincase. Finally, all odontocetes that are well-preserved show evidence of a small middle sinus extending laterally from the ear toward the jaw joint in the glenoid cavity.

The only serious challenge to odontocete monophyly has come from recent molecular analyses, which placed the sperm whale, *Physeter macrocephalus*, closer to living mysticetes than to other odontocetes. Such a relationship would make the Odontoceti paraphyletic. Reanalyses, based both on anatomy and on molecules, confirmed that the sperm whale lies within the Odontoceti and that the group is a clade.

III. The Monophyly of Mysticeti

Mysticetes include 13–14 living species in the families Balaenidae, Neobalaenidae, Balaenopteridae, and Eschrichtiidae. Mysticete monophyly has never been in serious doubt (cf. *Physeter* and odontocete monophyly). Strictly, the Mysticeti comprises the most recent common ancestor of all living species, plus

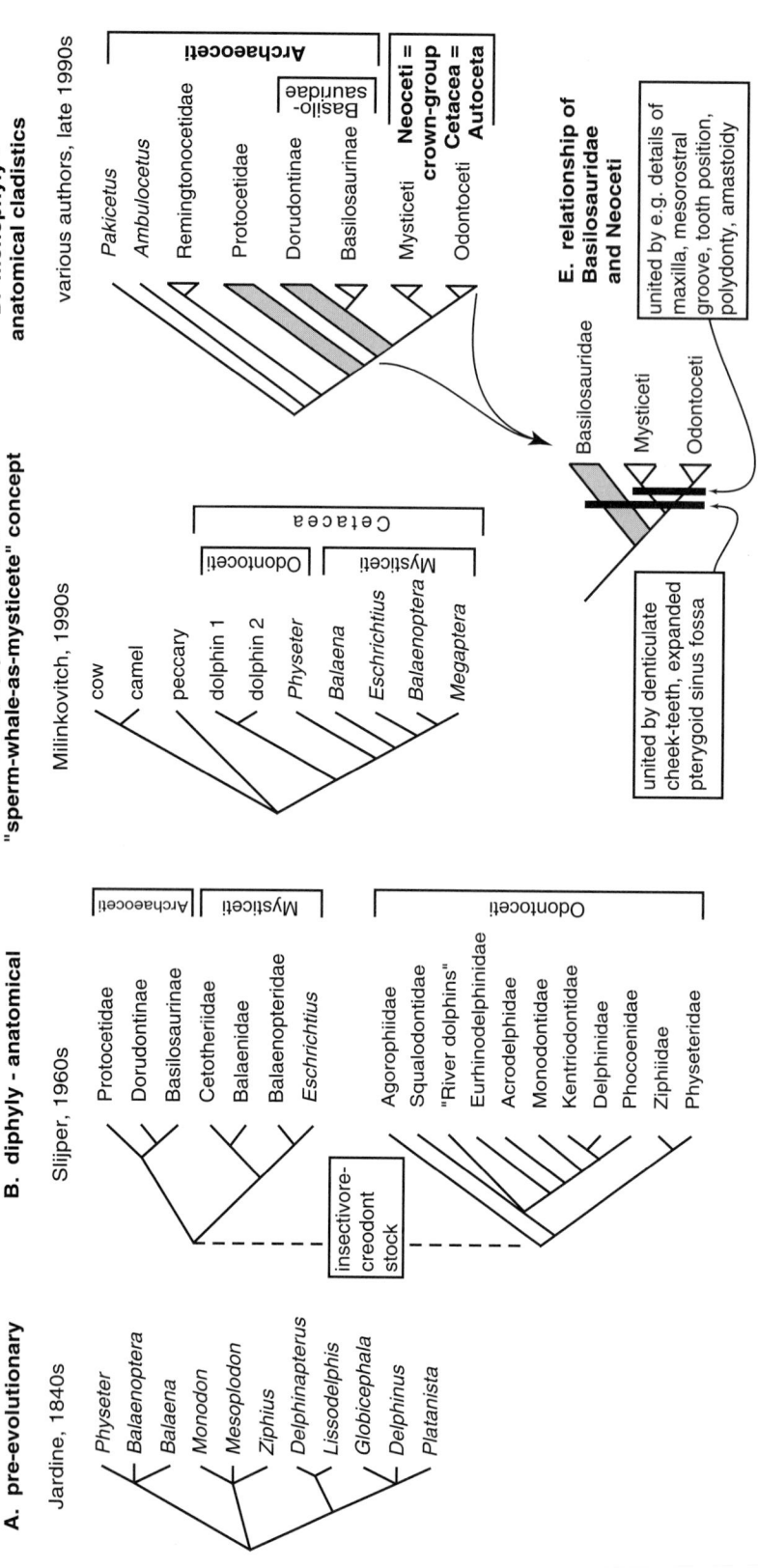

Figure 1 *Changing concepts of the Cetacea. (A) Preevolutionary classification as used by Jardine and others, early to mid-1800s. Species are clustered on the basis of sometimes superficial features. Genealogical relationships are not particularly implied. (B) Widely held concept of cetacean diphyly, as used by Slijper in the 1960s, and supported by many other cetologists (based on Slijper). (C) Widely cited but now abandoned concept of relationships between Odontoceti and Mysticeti showing the sperm whale, Physeter macrocephalus, as more closely related to living mysticetes than to other odontocetes (based on Milinkovitch). (D) Current concept of Cetacea showing crown group Cetacea (Neoceti) with two sister taxa, Odontoceti and Mysticeti, and a cluster of progressively more stemward archaeocete groups (based on Uhen and others). Grade taxa are shown with gray infill. (E) Summary of relationship between later archaeocetes, the Basilosauridae, and Odontoceti + Mysticeti.*

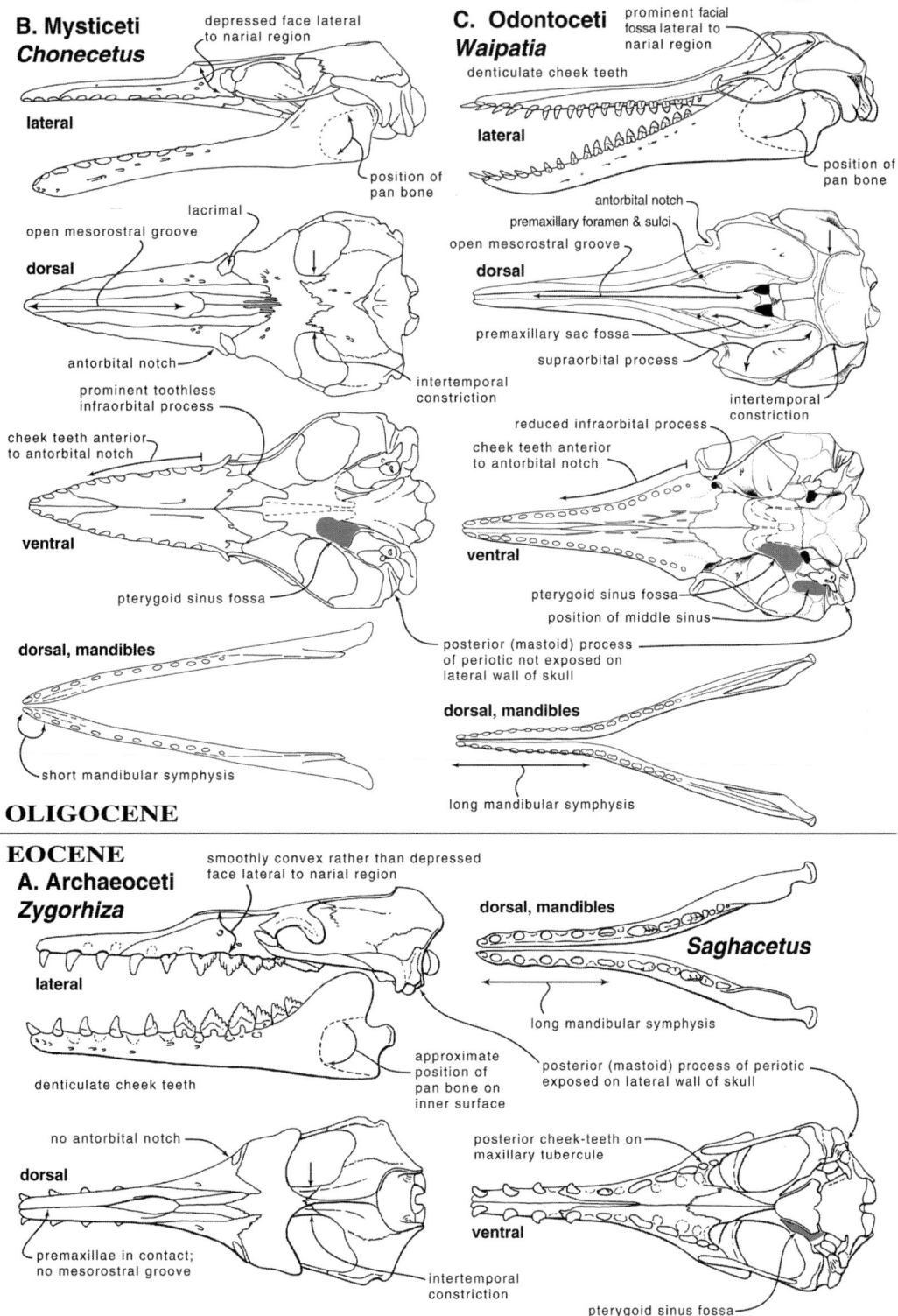

Figure 2 *Morphological similarities and differences among later archaeocetes and basal odontocetes and mysticetes. (A) Archaeocete. Lateral, dorsal, and ventral views of the archaeocete skull show the dorudontine Zygorhiza kochii (Basilosauridae: Dorudontinae; Priabonian, latest Eocene), based on Kellogg (1936). Dorsal view of the archaeocete mandible shows the dorudontine Saghacetus osiris (Basilosauridae: Dorudontinae; Priabonian, latest Eocene), based on Stromer (1908). (B) Archaic mysticete. Lateral, dorsal, and ventral views of skull and dorsal view of mandible show Chonecetus geodertorum (Mysticeti: Aetiocetidae; Chattian, Late Oligocene), based on Barnes et al. (1995), with the addition of some features. Figures of teeth are not available. (C) Basal platanistoid odontocete. Lateral, dorsal, and ventral views of skull and dorsal view of mandible show Waipatia maerewhenua (Odontoceti, Waipatiidae; Chattian, Late Oligocene), based on Fordyce (1994), with modifications, including the addition of some features. Some teeth are in situ on the original fossil, but for simplicity are not shown in other than the lateral view.*

all the descendants of that ancestor. Plausible ancestors of latest Eocene and Early Oligocene age are known. Unlike living species, fossil mysticetes are not recognized by the presence of baleen; although this probably occurred in most extinct species, it preserves rarely. Rather, osteological features, particularly in the skull, distinguish the Mysticeti. The rostrum is relatively large, with thin edges and a flat broad lower surface. The main bones in the rostrum (vomer, premaxilla, and maxilla) are sutured loosely with each other and, posteriorly, with the cranium. The lacrimal is also loosely sutured, between the frontal and a prominent transverse preorbital ridge on the maxilla. Loose sutures between the feeding apparatus and cranium account for the common loss of the rostrum in fossil mysticetes; perhaps such sutures function in skull kinesis during filter feeding. Ventrally, the maxilla is usually toothless, forming an origin for baleen, but in archaic forms the maxilla may have teeth that lie well forward of the orbit (cf. archaeocetes). Posteriorly, the maxilla extends toward the orbit, forming a prominent infraorbital plate below the frontal. Finally, the halves of the mandible are joined by ligaments at a short symphysis. Other putative diagnostic features of the skull are seen in most, but not all, mysticetes, as noted later for archaic forms.

IV. The Monophyly of Odontoceti + Mysticeti

In the later 1800s and indeed until the 1960s, the known archaeocetes and fossil mysticetes and odontocetes seemed rather divergent from one another. Slijper, Kellogg, and other influential cetologists doubted a close relationship between odontocetes and the other two groups and were uncertain about mysticete origins among the archaeocetes. Thus, the two living groups of cetaceans were regarded as diphyletic, of different ancestry. They were sometimes classified as distinct orders.

From the 1970s to 1990s, several major advances overturned the notions of diphyly, and ultimately changed cetacean nomenclature. The fossil record of Eocene archaeocetes and of early (especially Oligocene) odontocetes and mysticetes expanded markedly, helping to bridge the structural and stratigraphic "gap" among the three groups. It became clear that evolution is not always slow and gradual and that major structural change can occur in short geological intervals. Developments in the deep sea drilling project led to much improved geological correlation, helping to date and clarify evolutionary sequences. Molecular and biochemical approaches to phylogeny indicated close relationship between odontocetes and mysticetes. The rise of cladistics (phylogenetic systematics) also clarified many concepts of relationship and nomenclature.

The Odontoceti and Mysticeti are now widely regarded as forming a clade or monophyletic group, equivalent to the Cetacea in the sense of Brisson and, indeed, of many modern systematics. Strictly, this group is crown group Cetacea; it includes all descendants, living and extinct, of the most recent common ancestor of Odontoceti + Mysticeti.

Odontocetes and mysticetes do share bony features not seen in archaeocetes, supporting their sister group relationship. In both, the posterior portion of the maxilla is at least slightly concave, rather than smoothly convex, and carries one or more dorsal infraorbital foramina that open dorsally rather than anteriorly. On the rostrum, an open mesorostral groove extends far anteriorly so that the premaxillae have little or no contact in the midline. The posterior-most teeth in odontocetes and toothed mysticetes lie anterior to the antorbital notch. Most (not all) basal species are polydont, with more than the usual mammalian number of cheek teeth, and a tooth succession is unknown. Compared with basilosaurid archaeocetes, the zygomatic process of the squamosal is more robust and anteriorly produced, with a more delicate jugal. Finally, odontocetes and mysticetes are amastoid, with the posterior (mastoid) process of the periotic not exposed laterally on the skull wall.

Odontocetes and mysticetes are often identified as having a "telescoped" skull in which bone positions have moved dramatically relative to familiar mammalian landmarks such as the nose and eye. However, "telescoping" is a wide-ranging term applied to at least four different functional shifts involving both the facial region and the braincase. It should not be cited to support the monophyly of odontocetes and mysticetes.

V. Primitive Features in Basal Odontocetes and Mysticetes

Some early fossil odontocetes and mysticetes have features similar or even identical to those seen in some basilosaurid archaeocetes. Such fossils include, among odontocetes, species of *Xenorophus*, *Mirocetus*, and *Archaeodelphis*, whereas mysticetes include species of Aetiocetidae, *Mammalodon*, *Llanocetus*, and some unnamed taxa. (Perhaps some of these are stem rather than crown group members; e.g., *Archaeodelphis* possibly represents a stem odontocete.) The most obvious basilosaurid-like features of archaic odontocetes and mysticetes are the prominent intertemporal constriction, formed by elongate parietals dorsally on the braincase, and heterodont teeth. In all, the anterior teeth have single roots and simple crowns clearly distinct from two- or three-rooted cheek teeth with complex denticulate crowns. Multiple denticles on the crown are an evolutionary novelty linking basilosaurids, odontocetes, and mysticetes. Further, the posterior mandibular cheek teeth in archaic odontocetes have a distinctive anterior vertical groove, as in basilosaurids. (The loss of the last upper molar, M3, in basilosaurids, has been used to dismiss a basilosaurid origin for odontocetes and mysticetes. However, a widely variable tooth complement in the latter and the likelihood that polydonty involved an increase in the number of mid- to posterior cheek teeth would allow a basilosaurid origin.)

Other parts of the feeding apparatus are revealing. In basilosaurids and basal odontocetes and mysticetes, the mandible has a large mandibular fossa (reduced in more crownward mysticetes), and the temporal muscle has a distinct vertical origin on the frontal. (This origin changes dramatically in most odontocetes, becoming overridden by facial bones and, in mysticetes, migrating over the orbit.)

In all groups, the complex of foramina in the orbit are not tightly clustered (as in living species) but are scattered anteroposteriorly. Ventrally, in the skull base, an enlarged subspherical pterygoid sinus fossa is formed by alisphenoid and pterygoid. (Such a fossa is absent in Protocetidae and more basal archaeocetes and becomes more elaborate in more crowned Odontoceti and Mysticeti.) Common features in the ear region include a lack of fusion between periotic and tympanic bulla (fusion occurs in later mysticetes), laterally compressed processes on the periotic (becoming more inflated in many later odontocetes), and a rather low squat tympanic bulla (becoming smoothly rounded in later mysticetes, but more elevated and delicate in later odontocetes). Other basilosaurid features include persistent postparietal foramina in the braincase and prominent exoccipital condyles.

VI. The Paraphyly of Archaeocetes

Because odontocetes and mysticetes (= crown group Cetacea) arose from among basilosaurid archaeocetes, the suborder Archaeoceti is paraphyletic. Archaeocetes form an artificial cluster of cetaceans that lack the features of Odontoceti or Mysticeti. (Many cladists would not recognize archaeocetes as a formal group and would use the term "stem group Cetacea" as an alternative to Archaeoceti.) Note that this expands the concept of Cetacea beyond the crown group, and indeed expands it beyond the concept used by Brisson.

See Also the Following Articles

Archaeocetes, Archaic ■ Skull Anatomy ■ Systematics ■ Teeth

References

Barnes, L. G., Kimura, M., Furusawa, H., and Sawamura, H. (1995). Classification and distribution of Oligocene Aetiocetidae (Cetacea; Mysticeti) from western North America and Japan. *Island Arc* **3**(4), 392–431.

Fordyce, R. E. (1994). *Waipatia maerewhenua*, new genus and new species (Waipatiidae, new family), an archaic Late Oligocene dolphin (Cetacea: Odontoceti: Platanistoidea) from New Zealand. *Proc. San Diego Mus. Nat. Hist.* **29**, 147–176.

Fordyce, R. E., and Barnes, L. G. (1994). The evolutionary history of whales and dolphins. *Ann. Rev. Earth Planet. Sci.* **22**, 419–455.

Heyning, J. E., and Mead, J. G. (1990). Evolution of the nasal anatomy of cetaceans. *In* "Sensory Abilities of Cetaceans" (J. Thomas and R. Kastelein, eds.), pp. 67–79. Plenum, New York.

Kellogg, A. R. (1936). A review of the Archaeoceti. *Carnegie Inst. Wash. Publ.* **482**, 1–366.

Milinkovitch, M. C. (1997). The phylogeny of whales: A molecular approach. *Soc. Mar. Mammal.* **3**, 317–338.

Miller, G. S. (1923). The telescoping of the cetacean skull. *Smith. Miscell. Collect.* **76**(5), 1–70.

Nikaido, M., Rooney, A. P., and Okada, N. (1999). Phylogenetic relationships among cetartiodactyls based on insertions of short and long interpersed elements: Hippopotamuses are the closest extant relatives of whales. *Proc. Natl. Acad. Sci. USA* **96**(18), 10261–10266.

Rice, D. W. (1998). Marine Mammals of the World: Systematics and Distribution." Society for Marine Mammalogy, Lawrence, KS.

Uhen, M. D. (1998). Middle to Late Eocene basilosaurines and dorudontines. "The Emergence of Whales" (J. G. M. Thewissen, ed.), pp. 29–61. Plenum, New York.

Whitmore, F. C., and Sanders, A. E. (1977). Review of the Oligocene Cetacea. *Syst. Zool.* **25** (for 1976), 304–320.

New Zealand Sea Lion
Phocarctos hookeri

NICHOLAS J. GALES
Australian Antarctic Division,
Tasmania

*P*hocarctos hookeri (Gray, 1844) is named in honor of Sir James Hooker, who was the botanist with the British Antarctic expedition of 1839–1843. It is monotypic and is one of the world's seven extant sea lions.

New Zealand sea lions, like all otariids (eared seals), have marked sexual dimorphism; adult males are 240–350 cm long and weigh 320–450 kg and adult females are 180–200 cm long and weigh 90–165 kg. At birth, pups are 70–100 cm long and weigh 7–8 kg; the natal pelage is a thick coat of dark brown hair that becomes dark gray with cream markings on the top of the head, the nose, tail, and at the base of the flippers. Adult females' coats vary from buff to creamy gray with darker pigmentation around the muzzle and the flippers. Adult males are blackish-brown with a well-developed black mane of coarse hair reaching the shoulders (Fig. 1).

I. Distribution and Abundance

The New Zealand sea lion is endemic to New Zealand and is one of the most regionally localized and rare of the world's pinnipeds. It is classified as a threatened species by the International Union for the Conservation of Nature (IUCN). Prior to human arrival in New Zealand, *P. hookeri* was more widespread and probably more abundant than today. While the pristine breeding range of *P. hookeri* included almost all New Zealand coastal, island, and subantarctic territory, a combination of subsistence hunting by Maori and commercial sealing by Europeans have virtually reduced the current breeding range to two groups of sub-Antarctic islands. Currently more than 95% of New Zealand sea lion pups are born at three sites in the remote Auckland Islands; the remaining pups being born

Figure 1 *Adult female New Zealand sea lion (right) and subadult male New Zealand sea lion.*

at nearby Campbell Island (Fig. 2). Occasionally, single pups are born at regular sea lion haul-out sites on the South Island of New Zealand (Otago Peninsula) and around Stewart Island. Total pup production during the 1995/1996 breeding season was estimated to be about 2800; estimates of absolute abundance based on this pup production are 12,500 (95% CI: 11,100–14,000) (Gales and Fletcher, 1999). While variations in pup production have been recorded, the population appears to have been stable for the past few decades.

II. Habitat

New Zealand sea lions breed and haul out on a diverse range of terrestrial habitats, including sandy beaches, reef flats, grass and herb fields, dense bush and forests, and solid bedrock. Each site generally has easy access to relatively protected waters (Fig. 3). The sympatric New Zealand fur seal (*Arctocephalus forsteri*) selects rockier, more exposed sites, and the two species rarely interact on land. The marine habitats of New Zealand sea lions have only been described for lactating females from the Auckland Islands. Here they forage on benthic habitats in the waters of the adjacent Auckland Island shelf. Other age and sex classes

disperse more widely; in particular, adult and subadult males, tagged at the Auckland Islands, have been seen at Macquarie Island and around the southern parts of New Zealand.

III. Breeding Behavior

The breeding behavior of *P. hookeri* is typical of that of a polygynous otariid. Territorial males begin to assemble and defend physical territories at breeding rookeries in late November. Pregnant females begin to arrive in early December and aggregate into harems of up to 25 animals attended by a single dominant bull. Challenges from peripheral males are regular and the tenure of territorial males is short. There are no published studies to elucidate the behavioral mechanisms by which female movement within the rookeries is determined. Females give birth soon after arrival; the pupping season lasting about 35 days. By mid-January most territorial bulls have departed, the harems break up, and females and pups disperse to occupy surrounding areas.

Postparturient females exhibit estrus 7–10 days after the birth of their pup and are mated by the territorial bull. Soon after they depart on their first foraging trip. Trip durations average 1.7 days, interspersed with an average of 1.2 days ashore feeding the

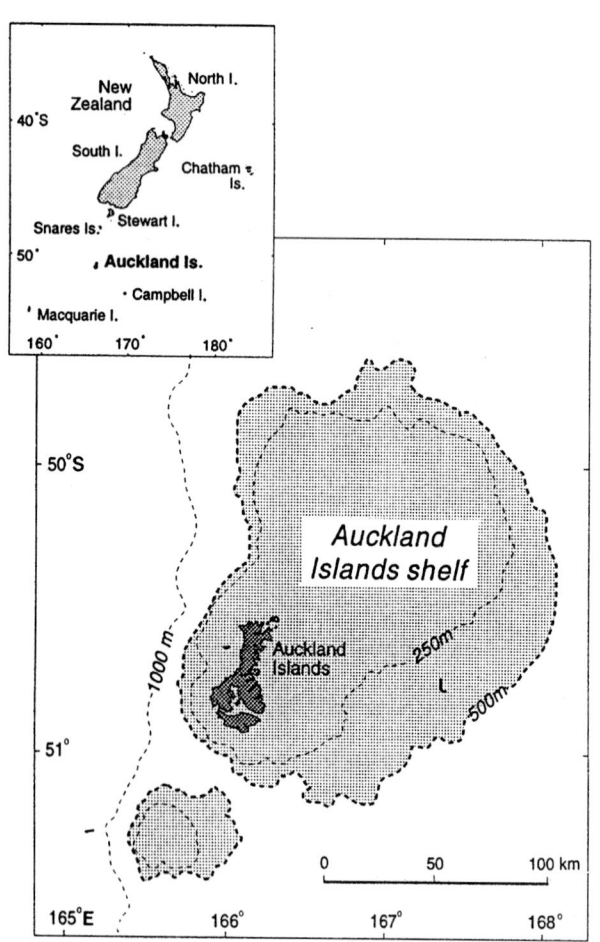

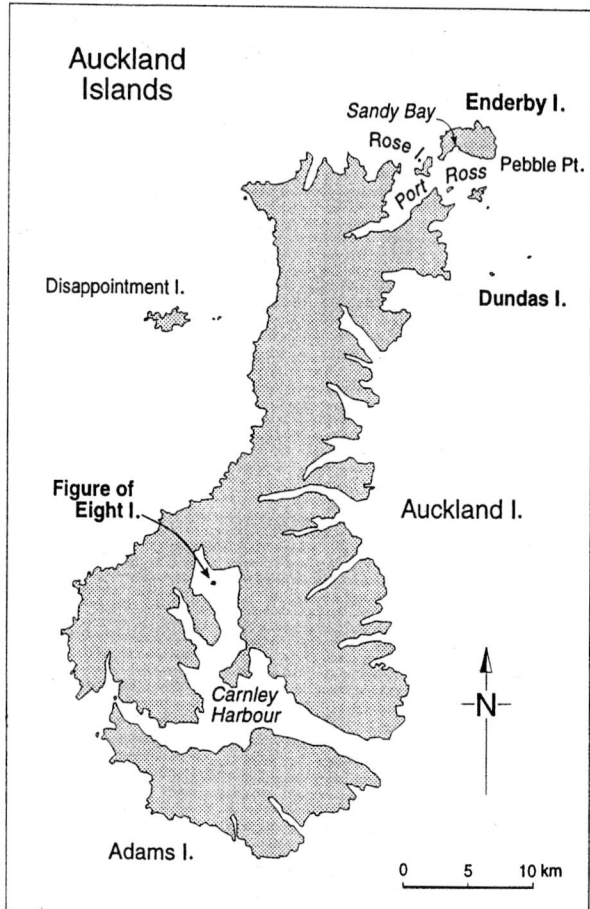

Figure 2 *Distribution map showing breeding colonies of the New Zealand sea lion,* Phocarctos hookeri.

Figure 3 *Typical breeding aggregation of New Zealand sea lions,* Phocarctos hookeri, *showing pups and adult females in the foreground and adult males fighting in the background.*

pup. Unattended pups gather into groups during maternal absence. At least 10% of pups die during the pupping season. While the causes of this early mortality has not been studied, it is thought to be caused principally by starvation, parasitism, trauma, and disease. INFANTICIDE and cannibalism by bulls have been observed. A high frequency of cows simultaneously nursing more than one pup (6% of observations) has been reported for *P. hookeri* during the first few weeks of pup rearing. This fostering behavior may be associated with kin selection in small populations. Lactation lasts about 10 months (Gales, 1995).

Females are thought to become sexually mature at 3 years or older and produce their first pup a year later. The duration of an assumed embryonic diapause has not been measured. Pregnancy and pupping rates also remain unknown. Males are reported to be sexually mature at 5 years but do not hold territories for a further 3–5 years. Survivorship and longevity have not been measured.

IV. Disease and Predation

Little is known of the normal disease status of New Zealand sea lions. The highly localized distribution of this species makes it particularly vulnerable to the effects of epizootic disease, and indeed in January 1998 an unusual mortality event occurred. At least 53% ($n = 1600$) of the pups of the year died, as well as many juveniles and adults. A principal cause has not been determined, but it is thought to be the result of bacterial infection. A probable new species of bacteria has been identified that may prove to be the main pathogen. The sea lion population was also found to have been exposed previously to phocine distemper virus, a virus that has resulted in many marine mammal deaths worldwide. An undefined suite of environmental factors (including a strong ENSO event, or El Niño) that stressed the sea lion population and decreased its immunity may have caused the event.

Sharks are likely to be the most significant predator of sea lions; with recent and healed bite wounds being a feature of many animals at the Auckland Islands. Occasional visitors such as leop-

ard seals (*Hydrurga leptonyx*) have been observed to eat small sea lions, and killer whales (*Orcinus orca*) may also be a predator.

V. Foraging Ecology

Sea lions at the Auckland Islands forage on a wide variety of prey, with benthic and pelagic organisms being represented. Thirty-three taxa have been identified from analyses of identifiable remains in scats and regurgitations, with fish comprising 59%, cephalopods 22%, and crustaceans 15% of the remains found. The six most abundant prey items [in decreasing order of abundance: opalfish (*Hemerocoetes* species), octopus (*Enteroctopus zelandicus*), munida (*Munida gregaria*), hoki (*Macruonus novaezelandiae*), oblique-banded rattail (*Coelorhynchus aspercephalus*), and salps (*Pyrosoma* species)] accounted for 90% of the total prey items. The diet of male sea lions on the Otago Peninsula has been found to represent a similar range of prey. Given the problems of bias associated with quantifying diet from scat and regurgitate analysis, further work utilizing newer, more precise tools such as fatty acid analysis would be most instructive.

New Zealand sea lions are the deepest and longest diving of the otariids. At sea they dive at a mean rate of 7.5 dives/hr and spend 45% of the time submerged. They undertake dives to a mean of 123 m (median 124 m, maximum >500 m) and spend an average of 3.9 min (median 4.33 min, maximum 11.3 min) on each dive >6 m. Almost half of the dives exceed the calculated *aerobic dive limit,* leading to a hypothesis that the dive behavior of *P. hookeri* reflects either successful physiological adaptation to exploiting benthic prey or a marginal foraging environment in which diving behavior is close to physiological limits (Gales and Mattlin, 1997). It has been shown subsequently that New Zealand sea lions have indeed equipped themselves physiologically for deep diving by having the largest blood volume of any otariid (Costa *et al.,* 1998). Behavioral adaptations, such as "burst and glide" diving, also appear to be used by this species to maximize the time available for foraging on the benthos, while still maintaining an effective energy budget. They have also been shown to be operating at what is likely to be close to their physiological maximum, as the gains in diving performance have been made with O_2 storage increases but not through a significant decrease in their at-sea metabolic rate (Costa and Gales, 2000).

VI. Interactions with Humans

New Zealand sea lions are subject to incidental drowning in squid and other trawl fisheries that operate around the Auckland Islands. The number of sea lions killed in the squid fishery (estimated to range from 17 to 141 per year for the period 1988–1997) has been the cause of serious concern and has led to a number of management measures. These include the imposition of a 12 nautical-mile marine mammal sanctuary around the Auckland Islands in which commercial fishing is prohibited; the deployment of government observers on trawlers to record the incidence of marine mammal bycatch (7–32% of tows observed each year for 1988–1997; bycatch rate varied from 0.6 to 3.8% of tows during this period); a delay in the opening of the seasonal fishery until February 1 each year; a voluntary code of practice for the industry, which aims to reduce the chance of bycatch; and the imposition by the New Zealand government of a maximum allowable level of fishing related mortality (MALFIRM), the reaching of which leads to the early closure of the fishery for that season. This MALFIRM (set at the beginning of each season on the basis of an approved model at about 60–80 sea lions) was exceeded each year between 1995 and 1998, and the fishery was closed early in the latter 3 years. In efforts to mitigate the sea lion/fishery interaction, specially designed escape devises are now being tested.

Another identified danger to New Zealand sea lions is tourism (principally sub-Antarctic, but also on the Otago Peninsula). These interactions are regulated via a limited-entry permit system on the sub-Antarctic islands and by behavioral protocols for tourists here and elsewhere. Other impacts, such as POLLUTION, ENTANGLEMENT, and direct killing, are not thought to be significant. There are no *P. hookeri* held in captivity. Under the auspices of a draft population management plan, the New Zealand government aims to monitor and research the sea lion population, mitigate threatening processes and remove the sea lion from its threatened status within 20 years (from 1999).

See Also the Following Articles

Diving Physiology ▪ Eared Seals ▪ Management ▪ Rookeries

References

Costa, D. P., and Gales, N. J. (2000). Foraging energetics and diving behavior of lactating New Zealand Sea lions *Phocarctos hookeri*. *J. Exp. Biol.* **203,** 3655–3665.

Costa, D. P., Gales, N. J., and Crocker, D. E. (1998). Blood volume and diving ability of the New Zealand sea lion, *Phocarctos hookeri*. *Phys. Zool.* **71,** 208–213.

Gales, N. J. (1995). Hooker's sea lion recovery plan (*Phocarctos hookeri*). Threatened Species Recovery Plan Ser. No. 17, New Zealand Department of Conservation, Wellington.

Gales, N. J., and Mattlin, R. H. (1997). Summer diving behaviour of lactating New Zealand sea lions, *Phocarctos hookeri*. *Can. J. Zool.* **75,** 1695–1706.

Gales, N. J., and Fletcher, D. J. (1999). Abundance, distribution and status of the New Zealand sea lion, *Phocarctos hookeri*. *Wildl. Res.* **26,** 35–52.

Noise, Effects of

BERND WÜRSIG
Texas A&M University, Galveston

W. JOHN RICHARDSON
LGL Ltd., King City, Ontario, Canada

W‍hen we humans dunk our heads underwater, the ocean seems relatively silent. This misconception occurs because our ears are optimized to hear in air and have poor sensitivity in the much denser medium of water. In reality, the oceans are full of sounds. Natural sources of underwater sound include breaking waves and surf, rain striking the sea surface, ice cracking and groaning in the higher latitudes, and the distant rumble of storms and earthquakes. Besides these physical sources, there is also a rich biological repertoire. There are sounds of snapping shrimp, grunting fishes, squeaking and popping sirenians, and the amazingly varied vocalizations of pinnipeds and cetaceans. Walruses (*Odobenus rosmarus*) display by knocks and mews; bearded seals (*Erignathus barbatus*) emit elaborate trills during their breeding season; toothed whales whistle, send bursts of staccato-like click trains, and echolocate; and large whales moan, groan, and sing for group cohesion, sexual displays, and COMMUNICATION (Tyack, 2000). Some researchers suspect that strong low-frequency sounds of certain baleen whales may also function as active sonar, helping them to navigate across wide open ocean spaces or around ice, or to locate silent conspecifics.

Unfortunately, the industrialized world has created other sources of noise underwater (Fig. 1). There is motorized shipping, underwater blasting, and offshore drilling, dredging, and construction. These activities produce underwater sounds incidentally, not purposefully. Several other types of underwater sounds are created purposefully: fathometers and sonars of many types operating at frequencies ranging from very high to low; air gun pulses for oil and gas exploration; pingers used to locate underwater equipment and to alert marine mammals to the presence of fishing nets; acoustic harassment devices (such as seal bombs) to keep marine mammals away from mariculture facilities; and sounds used for ocean science measurements (such as ATOC). Fish and marine mammals have evolved with the rich physical and biological cacophony of nature and are presumably well adapted to those sounds. However, most anthropogenic (human-generated) sounds first appeared in the past 100 years or so and are increasing in intensity and geographical extent decade by decade (Gisiner *et al.*, 1999; Jasny, 1999).

I. Importance of Sound to Marine Mammals

Marine mammals rely on underwater sound for communicating, finding prey, avoiding predators, and probably navigating. Other senses are available to them, but sound is the most important one at distances or in environments where the senses of touch, taste, and sight are not available. It is unclear how much sea otters (*Enhydra lutris*) and polar bears (*Ursus maritimus*) rely on sounds underwater. However, it is well known that pinnipeds (sea lions, fur seals, seals, and walruses), sirenians (manatees and the dugong, *Dugong dugon*), and cetaceans (dolphins, porpoises, and whales) use sound both passively, when listening to the environment, and actively, when communicating. The toothed whales also echolocate to find prey, detect predators, and maneuver in the environment.

The acoustic frequencies that are most important vary with the type of marine mammal. Baleen whales tend to use lower frequencies of sound: usually below 1 kHz and reaching down into the infrasonic range (<20 Hz) in fin and blue whales (*Balaenoptera physalus* and *B. musculus*). Toothed whale communication is mainly above 1 kHz, and the echolocation sounds of most species are at very high frequencies, above 20 to 30 kHz.

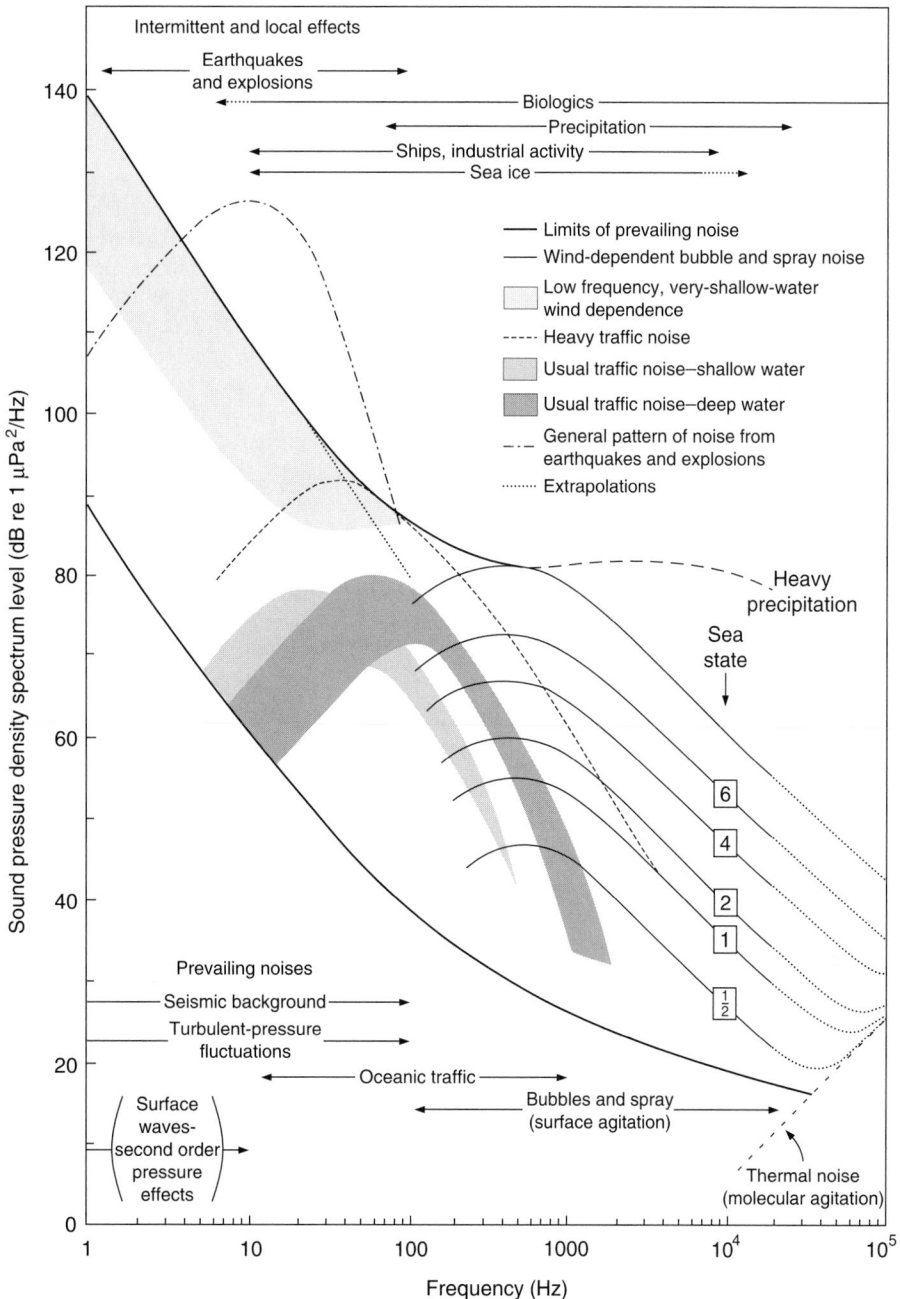

Intermittent and local effects

Earthquakes and explosions

Biologics

Precipitation

Ships, industrial activity

Sea ice

Limits of prevailing noise

Wind-dependent bubble and spray noise

Low frequency, very-shallow-water wind dependence

Heavy traffic noise

Usual traffic noise—shallow water

Usual traffic noise—deep water

General pattern of noise from earthquakes and explosions

Extrapolations

Heavy precipitation

Sea state

6

4

2

1

½

Prevailing noises

Seismic background

Turbulent-pressure fluctuations

Oceanic traffic

Surface waves-second order pressure effects

Bubbles and spray (surface agitation)

Thermal noise (molecular agitation)

Sound pressure density spectrum level (dB re 1 μPa^2/Hz)

Frequency (Hz)

Figure 1 *General types of natural and human-made sounds in the world's oceans (adapted from Wenz, 1962).*

The latter sounds are above the frequency range audible to humans ("ultrasonic"). Pinnipeds and sirenians tend to be intermediate between baleen and toothed whales with respect to the frequencies of their calls (mainly at 0.5–10 kHz) and optimum hearing range. The optimum frequencies for pinnipeds and sirenians are similar to those for humans.

While sounds of sirenians, some pinnipeds, and some small odontocetes are weak and audible only within a few tens of meters, most cetacean sounds are much stronger (Richardson

et al., 1995). Received sound levels and maximum detection distances depend not only on the strength of the sound at the source ("source level"), but also on the frequency and on physical properties of the environment through which the sound propagates. Nevertheless, smaller dolphins and porpoises can be heard to several hundred meters. Killer whale (*Orcinus orca*) "screams," social sounds of pilot whales (*Globicephala* spp.) and the staccato click sounds of sperm whales (*Physeter macrocephalus*) reach to several kilometers. Communication sounds of many odontocetes and mysticetes have source levels of 160 to 180 dB re 1 μPa at 1 m distance. The clicks of bottlenose dolphins and sperm whales can be much more intense, >220 dB in the same units (Au *et al.*, 2000; Møhl *et al.*, 2000). Echolocation clicks of some smaller odontocetes can have similarly high peak levels, but are very brief and do not contain much energy. In a few species of seals, e.g., bearded and Weddell seals (*Leptonychotes weddellii*), the males appear to use complicated tonal sounds as advertising displays to warn off other males (and possibly to attract females); these sounds can also be quite intense. These seals, which defend underwater territories (or "maritories"), are exceptional among the pinnipeds, however.

A. Potential Types of Noise Effects

Anthropogenic sounds can result in a variety of effects whose consequences to marine mammals can range from nil to severe, depending on the type and received level of the sound:

Tolerance (No Overt Response): Mammals exposed to audible levels of man-made sounds often exhibit no obvious responses, continuing their normal activities without moving away. A marine mammal hearing a sound does not always react overtly to it. However, responses can be subtle so a detailed behavioral and physiological study is needed before it is legitimate to conclude that there is no response. Animals might tolerate noise in order to remain in a preferred location, such as a feeding area, even if the noise caused stress or other inconspicuous effects.

Changes in Behavior or Activity: Alterations in behavior are common when marine mammals are exposed to man-made sounds. Sometimes the effects are subtle, discernible only by detailed observation and statistical analysis, e.g., changes in surfacing/respiration/dive cycles. More conspicuous effects include changes in activity, e.g., from resting or feeding to alert, facing toward the noise source, and so on.

Avoidance Reactions: Upon exposure to strong man-made sounds, marine mammals engaged in feeding, social interactions, or other "normal" activities often interrupt those activities and swim away. When migrating whales approach a noise source, they typically deflect a few degrees off their "normal" course and swim to one side or the other of the noise source.

Masking: Masking is the process whereby sounds of interest to a listener are obscured by interfering sounds. If those sounds are important to the listener (e.g., predator calls), masking could have a serious effect (Richardson *et al.*, 1995).

Hearing Impairment: Animals (including humans) exposed to strong sounds can incur a reduction in their hearing sensitivity. This impairment is often temporary, provided the sound levels are not too high or too prolonged. However, repeated exposure to strong sounds, and even a single brief exposure to an extremely strong sound (e.g., nearby explosion), can cause permanent hearing impairment.

Nonauditory Physiological Effects and Stress: In humans, exposure to very strong underwater sounds can cause resonances in lung cavities and other types of nonauditory physiological effects. Chronic exposure to strong noise may also at times cause stress reactions. These phenomena are almost entirely unstudied in marine mammals.

B. Noise Definition, Categories, and Levels

The term "noise" indicates an unintentional or unwanted sound, possibly disagreeable or noxious. We tend to think of noise as sound we do not like, but this is a subjective impression; a sound that is noise to one person (or animal) may be an important signal (or music) to another. For example, the songs of humpback whales (*Megaptera novaeangliae*) are undoubtedly important to them. We do not know whether these strong and—in some areas—seasonally persistent sounds are regarded as "music" or as noise by pilot whales and bottlenose (*Tursiops* spp.), spinner (*Stenella longirostris*), and spotted dolphins (*S. attenuata* and *S. frontalis*) that are attempting to communicate in the presence of this background of whale song. However, industrial sounds are likely to be perceived by marine mammals as noise. Nevertheless, some anthropogenic noises may themselves be important cues to mammals: a boat is approaching, take evasive action; a seismic vessel is active, do not get too close, etc. Thus, noise can have an important signaling function. In human terms, it is probably good that we hear an approaching truck or train while we are on foot or on a bicycle nearby.

Sound strength can be measured near the source or at some greater distance. By convention, source levels usually are reported for a standard distance of 1 m. The received sound level tends to diminish with increasing distance. In interpreting quoted sound levels, it is essential to know whether the value is a source or received level and, for the latter, the distance from the source. The received levels 10, 100, and 1000 m from a source are often about 20, 40, and 50–70 dB less than the corresponding source level. A further complication is that sound levels are measured in many different and sometimes difficult-to-convert units. Decibel (dB) values are meaningless unless the reference level is also specified. This is typically 1 μPa for underwater sound and 20 μPa for airborne sound, but other references are sometimes used, especially in the older literature. Even when the distance from source and the reference level are stated, other complications can make it difficult or impossible to compare acoustical measurements from different studies. For example, simple comparisons of levels in water vs air are usually of doubtful validity.

The levels quoted in this article concern underwater sound measured in dB re 1 μPa and, except for values shown in Fig. 1, are overall broadband levels (i.e., including a wide range of frequencies). In the case of source levels, the levels are standardized to a distance of 1 m.

One way of categorizing anthropogenic noises is whether they are transient or continuous. Transient noises are generated by helicopters or planes passing by, sonar pings, explosions, seismic surveys, and ocean acoustic studies. The source (and received) levels of some of these transient sources can be high (Richardson *et al.*, 1995). However, because these sounds are brief and intermittent, the average sound level over an extended period is lower than the peak level. Transient sounds may have less effect on animal behavior and hearing as compared with continuous sounds having a similar received level.

The strongest transient sounds in the sea come from explosions, which produce shock waves as well as very strong (but brief) underwater sounds. Nominal source levels can be on the order of 270–280 dB—higher than for any other anthropogenic sound. The shock waves from explosions can injure, stun, and even kill nearby fishes, sea turtles, humans, and marine mammals. The distances within which death or injury will occur depend on the size and depth of the explosion, the type of animal, and other factors. Aside from explosions, some of the strongest sources of underwater sound are seismic surveys (typically using air gun arrays to produce low-frequency noise pulses) and military search sonars operating at medium and low frequencies. Air gun arrays and sonars can produce transient sounds with effective source levels of 230 dB or more. Sounds received from these sources diminish rapidly with increasing distance, but often are detectable as much as 100 km away. There are many other types of transient anthropogenic sounds with a wide variety of levels and characteristics.

Continuous sounds are caused by drill ships, dredging, and vessels of all sizes underway. Of these sound producers, large tankers, container ships, and icebreakers working in ice are among the strongest sources. Their source levels are on the order of 200 dB or more. Although not as high as for some transient sources, this level is sustained without gaps. Much of this sound energy is at infrasonic frequencies, which are apparently important to at least some of the baleen whales, but may be irrelevant to dolphins that have little or no hearing sensitivity at those frequencies. Much of the shipping is concentrated along defined shipping lanes, where sound levels will wax and wane as individual ships pass. However, at least in the Northern Hemisphere, the overall din of vessels pervades not only the lanes themselves, but major parts of the ocean basins involved (Fig. 2).

The strongest components of sound from many of the major anthropogenic sources are below 1 kHz, in the same frequency band as is used by most of the baleen whales for their communication calls. For seismic surveys and large ships, most sound energy is below 200 Hz. A recently acknowledged source of strong low-frequency sound is the low-frequency active (LFA) sonar used to detect quiet submarines. The U.S. Navy's LFA projectors produce a beam of intense sound (source level up to about 240 dB) at frequencies in the 100- to 500-Hz band. However, many of the strongest military sonars operate in the midfrequency (a few kilohertz) range. Likewise, the sounds from outboard engines operating at high speed, snowmobiles traveling on ice, and acoustic harassment devices at mariculture facilities are in the low kilohertz range. Sounds at medium and especially at low frequencies tend to propagate for longer distances than sounds at high frequencies (e.g., above 10 kHz). At high frequency, sound is absorbed into seawater, and the absorption rate increases rapidly with increasing frequency. Hence, high-frequency pinger and sonar sounds (including dolphin clicks) generally diminish to low or undetectable levels within 1 km or less and thus do not present potential problems except at close range.

C. Animals of Concern

Because most, if not all, marine mammals rely on underwater sound for various purposes, any strong anthropogenic sounds at relevant frequencies might have an effect. Animals attuned to different sound frequencies will be most affected by different types of sounds. Mid- and high-frequency sounds as produced by small vessels and mid- and high-frequency sonars are likely to affect dolphins, porpoises, pinnipeds, and sirenians. In contrast, large whales are more susceptible to lower frequency sounds such as those from large vessels, drilling operations, and sounds specifically designed to propagate long distances through the water and/or bottom, such as LFA sonar, marine seismic exploration, and ocean tomography (e.g., ATOC).

Although it was once thought that nearshore animals are likely to be most vulnerable because of the concentration of industrial activities in nearshore waters, this is no longer as clear. Marine mammals on the high seas may be affected as well, given the occurrence in deep water of shipping, military operations, acoustic oceanography projects, and (increasingly) oil industry operations. In general, it is probable that mammals of the southern oceans are still affected less often and less intensively by anthropogenic sounds than in the northern oceans, simply because of the preponderance of human occupation, industry, and shipping in the north. However, underwater noise is now everywhere and has the potential to affect all animals that can hear it and that need to communicate through it.

II. Effects on Hearing

Most, if not all, marine mammals have evolved special adaptations to hear well underwater (Au *et al.*, 2000). However, the adaptations that make their ears sensitive to pressure fluctuations in the water may also increase the risk of damage from exposure to strong waterborne sound and shock waves.

The most drastic effects come from shock waves, which can damage not only the ear but also other organs, in extreme cases causing death. Explosions produce mechanical and pressure trauma effects that can cause profound physical injury or direct death of marine mammals. It has also been suggested that some deaths of beaked whales (Ziphiidae) in the Mediterranean Sea, the Bahamas, and elsewhere were attributable to strong noise from sonar trials. The evidence is circumstantial and inconclusive for the Mediterranean. Preliminary accounts of the Bahamas incident indicate that sonar sounds may have caused injury to the auditory system and associated structures, although not enough to kill the animals directly.

Sounds that are not strong enough to cause profound physical injury or outright death may nonetheless impair hearing, resulting in either temporary or permanent threshold shifts (TTS and PTS, respectively). TTS occurs in humans and animals subjected to intense sounds for short periods of time or to less intense sounds for longer times. The temporary loss of hearing acuity by humans listening to a rock concert is the classic example. Mild TTS has been demonstrated in bottlenose dolphins (*Tursiops truncatus*) and beluga whales (*Delphinapteras leucas*) exposed to a single 1-sec pulse of strong sound (192–201 dB) and in dolphins and various pinnipeds exposed to lower sound levels for 20–50 min (Kastak *et al.*, 1999; Schlundt *et al.*, 2000). Hearing thresholds returned to preexposure levels within about 12 hr, and these brief exposures caused no long-term hearing impairment.

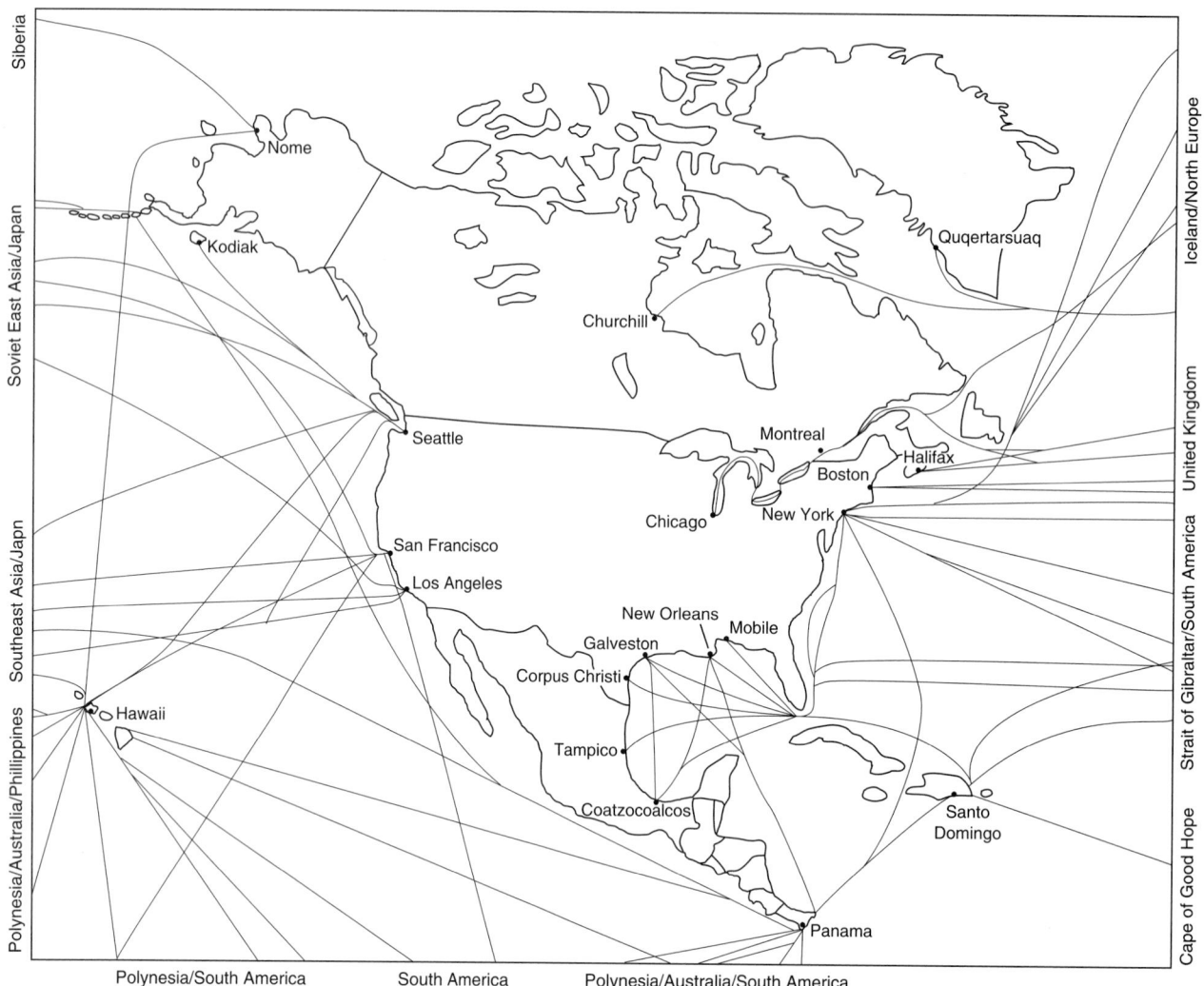

Figure 2 *Major international shipping lanes in North American waters. From Jasny (1999) and U.S. Department of State.*

Temporary hearing impairment could be deleterious to animals that rely strongly on their abilities to detect sound. In addition, data on the sound levels and durations at which TTS begins are useful in identifying situations when there is concern about permanent hearing damage (PTS). However, it is not known how much additional exposure (higher levels and/or longer durations) would be necessary before PTS would occur. In general, permanent damage typically results from a loss of sensory hair cells within the inner ear. The loss is most pronounced in the part of the inner ear responsible for detecting sound at the frequency of the injurious stimulus. In terrestrial mammals, lost hair cells do not grow back, and this is likely the case for marine mammals as well. (Interestingly, at least partial regeneration occurs in birds and fishes.) Older bottlenose dolphins in captivity are known to have reduced hearing sensitivity (especially at the higher frequencies), presumably at least in part because of the cumulative effects of sound (Ridgway and Carder, 1997). This loss is similar to the progressive hearing loss in humans, or presbycusis.

The initial TTS results from small odontocetes confirm that sound levels necessary to cause TTS are correlated with the duration of exposure. More data are needed to quantify this relationship and to determine the TTS thresholds for repeated sounds such as seismic and sonar pulses. Sound levels necessary to cause TTS and PTS in baleen whales and sirenians are entirely unknown.

III. Nonauditory Physiological Effects

Even less is known about potential effects of strong noises on nonhearing physiology than on hearing. Human and animal studies show that strong sounds can affect the vestibular system (and thus sense of balance), air sinuses and adjacent tissues, neural transmission (skin tingling in divers), and reproductive functions. Studies of some terrestrial animals exposed to strong noise have shown reduced sperm production, menstrual irregularities, abortions, and stillbirths. Most of these drastic effects come from intensive low-frequency sounds and attendant physical vibrations of body tissues. It is not known whether the low-frequency underwater sounds of ATOC, LFA, or seismic exploration could elicit similar effects in marine mammals. If so, these effects would probably be limited to animals close to the sound source.

It is possible that strong sound could cause bubbles to form in blood or tissues, and that animals might succumb to a lodging of these bubbles in the BRAIN and elsewhere (Crum and Mao, 1996). This situation is analogous to bubble formation in human divers when they return too rapidly from depth—the condition known as the bends.

Stress is one possible outcome of exposure to sounds that are disturbing to animals. Stress could, in theory, inhibit normal social interactions, FEEDING, reproduction, longevity, and a host of other important functions (Curry, 1999). However, noise-induced stress is not well understood even in humans and terrestrial mammals, and there are essentially no data on noise-induced stress in marine mammals. Research is needed.

IV. Effects on Behavior

A. General

Short-term behavioral and avoidance reactions to noise have been studied more than damage to hearing and other physiological effects, to the point that there are now at least some data on noise-induced disturbance in every group of marine mammals. For most marine mammals, it is relatively easy to observe some aspects of their DISTRIBUTION and behavior, as they need to come to the surface to breathe. Nevertheless, much of their time is spent below the waves, and our brief glimpses of them at the surface can present a biased view. Also, short-term reactions (or lack thereof) are not necessarily a good indicator of long-term effects, and the latter have rarely been studied directly. This section does not attempt to review the numerous studies on behavioral and distributional responses to underwater noise; for more details, see Richardson *et al.* (1995).

Mammals often appear to become alert to novel sounds when they are of low intensity, but this alertness quickly wanes if there is no danger connected with the sound. Thus, the droning of ships or drilling platforms in the distance often appears to be ignored due to habituation or sensory adaptation. Some mammals may avoid the immediate area around the noise source, but remain within the area where the sounds are at least faintly audible. When sounds are tolerated, it is possible (though unproven) that they may elicit stress or other unseen physiological reactions.

Marine mammals often react more dramatically when received sound levels are high (indicative of a strong or nearby source) or when they are increasing (indicative of an approaching source). Bowhead whales (*Balaena mysticetus*) migrating toward a drill ship or a marine seismic operation typically deflect their course so that their closest point of approach as they pass the noise source is at least 20 km away. At that distance, the received level of the strongest air gun pulses was near 130 dB, averaged over pulse duration. Migrating gray whales (*Eschrichtius robustus*) show similar deflections of their migration route as they approach a simulated seismic operations or LFA sonar, but they seem to tolerate higher sound lev-

els than bowheads. These deflections by bowhead and gray whales are not a sudden fright reaction, but an edging away—analogous to a person walking along a sidewalk, seeing a disturbance ahead, and crossing to the sidewalk on the other side of the road. Also, at least in bowheads, there is a decrease in durations of surfacings and dives, and number of breaths per surfacing. In other words, the whales cycle through their surfacing/dive repertoire more rapidly. There must be at least subtle commensurate changes in activities such as feeding, socializing, or rest.

Cetacean responses to noise are quite variable, depending on the circumstances of exposure and the activities of the animals at the time. There are anecdotal reports from Heard Island in the Southern Ocean, and from the Gulf of Mexico, that sperm whales are sometimes sensitive to air gun pulses at even greater distances than bowheads. In contrast, observers on seismic vessels near the United Kingdom found that sperm whales did not show strong avoidance when the air guns were operating. The beluga whale is another example of a species whose responses to anthropogenic noise are highly variable. Belugas sometimes tolerate exposure to large fleets of fishing vessels, but in other circumstances flee when exposed to faint sounds from approaching ships 35–50 km away (Finley *et al.*, 1990).

A noise that is associated with danger, such as a catcher boat of a whaling fleet or a purse-seine vessel that encircles dolphins to catch the tuna underneath, can trigger strong avoidance reactions at distances of 10 km or more. Reactions may become stronger upon repeated exposure to aversive stimuli; in this case, "sensitization" is said to occur. At least some dolphins seem to distinguish vessels based on their sounds and react differently to boats that habitually harass the animals (such as aggressive tour operators or researchers who tag or collect biopsy samples) vs boats that approach slowly and carefully. In the latter case, the marine mammals may have learned that the vessel is not harmful—a case of behavioral adaptation.

In contrast to the extreme examples just cited, some marine mammals tolerate exposure to high levels of sound. Seals and sea lions attracted to locations where prey fish are concentrated often tolerate exposure to high levels of noise from acoustic harassment devices designed to disperse pinnipeds. Also, during the Heard Island study, hourglass dolphins (*Lagenorhynchus cruciger*) approached to within several hundred meters of a powerful source of low-frequency sound (57 Hz; source level 209–220 dB), whereas pilot (*G. melas*), southern bottlenose (*Hyperoodon planifrons*), and minke whales (*Balaenoptera* sp.) were seen less often during sound transmissions than during silent periods (Bowles *et al.*, 1994). The hearing systems of the small hourglass dolphins probably were not very sensitive to low-frequency sounds; they may have been curious about but not discomfited by the noise. However, it is also noteworthy that, when visible near the surface, they would not have been exposed to levels much above 160 dB given the nature of the sound field around this particular sound source. It is possible that they exhibited a vertical avoidance response, staying near the surface simply because the noise was less strong there.

Marine mammals sometimes approach and tolerate a sound source for their own benefit. Bottlenose dolphins that feed on prey stirred up by shrimp fishers know the sounds of all aspects of the trawling and net-lifting operations; they move from vessel to vessel according to stage of trawling, even from several kilometers away. Killer whales and several pinniped species of the northeast Pacific similarly react to net lifting like the gong of a dinner bell. Many species of dolphins and some species of sea lions and fur seals approach vessels to bow ride or wake ride near the vessel, apparently "for fun."

While this discussion centers mainly on underwater sounds, in-air noises can affect pinnipeds on land, occasionally with serious consequences. A low-flying aircraft, the sudden honk of an automobile horn, or the noisy approach of humans on foot can cause animals on land to stampede into the water. If this occurs on a birthing/nursing beach, adults sometimes trample pups in their rush to escape the perceived danger; however, it is not always clear whether sound or sight (and, at times, substrate vibration) is more responsible for the stampede and loss of pups.

B. Communication

Because sound is such an important sensory modality for almost all marine mammals, it stands to reason that noise has the potential to disrupt the efficiency of communication (and echolocation). Indeed, there is some evidence for this. Some animals fall silent when they perceive danger (or a novel sound). During the aforementioned Heard Island study, sperm and pilot whales were heard 24% and 8% (respectively) of 1137 min when there was no sound, but none were heard to vocalize during 2202 min with sound transmissions. Although these data suggest a strong effect of noise on communication, they are not yet sufficient to indicate at what received levels the whales became quiet (Bowles *et al.*, 1994).

Acoustic reactions to intense sounds vary among species and with the activity of the animals when they perceived the sound. Belugas are vocally active when they detect ship sounds, but narwhals fall silent. Dusky dolphins (*Lagenorhynchus obscurus*) resting in small groups fall silent when they perceive danger. However, those same dolphins will continue their social whistling and burst-pulse "chatter" when they are engaged in high-energy social and sexual activities. They seem less easily disturbed at these higher energy times.

Masking of sounds useful to animals by stronger anthropogenic noise is another important phenomenon. We humans have the ability to converse with each other even at noisy parties and can separate useful speech from an amazing background din (the well-known "cocktail party" phenomenon). Marine mammals can probably do likewise, for they have evolved in a world where physical and biological sounds other than their own abound. Nevertheless, studies of captive dolphins and pinnipeds have confirmed that, in marine as well as terrestrial mammals, strong man-made noise can physically "drown out" other sounds at similar frequencies. The noise reduces the maximum distance at which one animal can hear calls from another animal or some other environmental sound that may be important (Richardson *et al.*, 1995). There are indications (not fully proven) that gray whales, belugas, and bottlenose dolphins sometimes shift the primary frequencies of their sounds to reduce overlap with the frequencies of background noise. If so, this suggests that some cetaceans may be able to reduce the negative effects of noise masking, but this suggestion needs further study.

C. Social Structure

Marine mammals are generally very social. If noise disrupts social structure, and if the effect is sufficiently strong and long-lasting, then it could be detrimental to the well-being and survival of individuals and perhaps ultimately populations. Noise disturbance can cause some degree of social disruption, but the consequences are largely unknown. When a supply ship comes within 2–4 km of a group of bowhead whales, the whales scatter in all directions. They are "socially disrupted" for as much as a few hours. However, we have no idea how important this disruption may be to the well-being and productivity of the whales. We can guess that they will be harmed if this kind of disruption happens day after day to animals that feed more efficiently when together than when apart. Social disruption also includes the accidental separation of a nursing mother from her pup or calf. When this happens, the probability of pup or calf survival may be reduced substantially.

We do not know to what extent the masking of communication sounds by noise can cause social disruption. Blue whales and fin whales can sometimes hear each other over distances of several hundred kilometers. If there are important social interactions over those long distances (not proven), these interactions may be impeded by the masking effects of background noise, especially if that noise is continuous. For example, these whales may not be able to find each other as efficiently for feeding and mating purposes given the masking effect of the man-made noise. If whale societies are mediated in large part by acoustic contact, as we suspect (Wells *et al.*, 1999; Tyack, 2000), then background noise that diminishes the distances over which whales can communicate also diminishes the spatial scale of their societies.

D. Habitat Use

While short-term disruption of behavior by noise is known to occur, it is less clear how this may translate to overall use of habitat and to long-term disturbance. It was mentioned earlier that migrating bowhead whales "edge away" from seismic exploration and drill ships, in most cases maintaining a minimum distance of about 20 km. It is unlikely that this results in any major harm to the animals. However, if there were a similar and prolonged displacement from a localized area of important feeding habitat, then the potential for harm would be great. At this point, we know that some whales move away from sources of strong noise, but we do not know whether this significantly reduces their yearly food intake or causes other important disruption.

Similarly, human disturbance of island spinner dolphins has affected their use of nearshore bays that appear to be important to them for daytime rest (after feeding in deep water at night). It is possible to argue that this reduced use of safe havens might impact their survivability as individuals and therefore as populations. However, to date, we have no information on such a population-wide effect.

The concept of habitat available vs habitat disturbed is an important one in CONSERVATION BIOLOGY. The purist would argue that any habitat disrupted is too much, and compromises the natural world. The pragmatist argues that we humans are so overpoweringly disruptive that we must be content with keeping at least a minimum of area available for the animals in question to survive. In the case of marine mammals, that means keeping some critical habitats relatively free of strong noises that create unacceptable disruption in short-term behavior, long-term behavior, aspects of physiology, and hearing. The task is a difficult one and needs further research, monitoring, mitigation, and political enforcement to have a chance of success.

V. Research and Potential Mitigation

The first step in mitigation is to develop a better knowledge base. Detailed discussions of research needs can be found in NRC (1994, 2000) and Richardson *et al.* (1995). We need better information on auditory sensitivity, especially of large whales, and the levels and frequencies of noises that cause TTS, PTS, and nonhearing physiological disruptions in all marine mammals. Researchers have acquired the technical capability to study noise-induced stress and other aspects of physiology, with miniature data loggers and telemetry transmitters attached to animals under field conditions. This is a relatively new avenue of research for marine animals and is likely to grow in the near future. Studies of short-term behavioral disruptions are worthwhile, but it is long-term behavioral, social, communication, and habitat disruption that are of greatest need of study. It is our challenge to monitor the health of marine mammals as related not only to anthropogenic sound pollution, but also to other aspects of habitat degradation.

Potential mitigation avenues are many. Naval vessels are already engineered to reduce sound emissions, e.g., by physical decoupling of rotating machinery from the hull, propeller design, and emission of screens of air bubbles. This knowledge needs to be taken into the private sector, which is responsible for the vast majority of shipping noise. Bubble screening has been shown to strongly decrease the noise from a stationary pile-driving activity in Hong Kong waters, and this technique could be effective in dampening industrial sounds in many other situations (Würsig *et al.*, 2000). A controversial technique that is often employed is "ramping up" of sounds in order to alert animals of impending strong noise, essentially to chase them away from the zone of most influence. This approach probably is effective for some species and situations (e.g., baleen whales vs airguns). However, in other situations it may attract curious marine mammals into a zone of danger. Further study is needed to determine the situations when ramping up is a useful mitigation technique.

In projects where sound is emitted purposefully, e.g., marine seismic exploration, ocean science studies, and navy LFA sonar, refinements in equipment and operational procedures may be possible that will reduce the exposure of marine mammals to noise. Some efforts have already been made to reduce source levels to the minimum that will be effective and to improve beam forming so as to reduce sound radiation in unnecessary directions. Also, these (and some other) noisy activities are amenable to regulation as to where and when they are used, especially as pertains to the seasonal MIGRATION of sensitive baleen whales. It is common for noisy projects to be restricted to certain times of year and/or to certain areas in order to reduce impacts on marine mammals.

It is up to scientists to help provide badly needed information on disturbance reactions and zones of potential influence, and up to politicians and regulators to write and enforce legislation to curb the uncontrolled proliferation of ocean noise. Combined efforts by all will be needed in order to provide adequate protection for marine mammals while avoiding unnecessary restrictions on worthwhile human activities.

See Also the Following Articles

Behavior, Overview ■ Echolocation ■ Habitat Pressures ■ Hearing ■ Pollution and Marine Mammals ■ Sociobiology

References

Au, W. W. L., Popper, A. N., and Ray, R. R. (eds.) (2000). "Hearing by Whales and Dolphins." Springer-Verlag, New York.

Bowles, A. E., Smultea, M., Würsig, B., DeMaster, D. P., and Palka, D. (1994). Relative abundance and behavior of marine mammals exposed to transmissions from the Heard Island Feasibility Test. *J. Acoust. Soc. Am.* **96**(4), 2469–2484.

Crum, L. A., and Mao, Y. (1996). Acoustically enhanced bubble growth at low frequencies and its implications for human diver and marine mammal safety. *J. Acoust. Soc. Am.* **99**(5), 2898–2907.

Curry, B. E. (1999). Stress in mammals: The potential influence of fishery-induced stress in dolphins in the Eastern Tropical Pacific Ocean. NOAA Technical Memo, National Marine Fisheries Service, SWFSC, La Jolla, CA.

Finley, K. J., Miller, G. W., Davis, R. A., and Greene, C. R. (1990). Reactions of belugas, *Delphinapterus leucas,* and narwhals, *Monodon monoceros,* to ice-breaking ships in the Canadian high arctic. *Can. Bull. Fish. Aquat. Sci.* **224**, 97–117.

Gisiner, R., Cudahy, E., Frisk, G., Gentry, R., Hofman, R., Popper, A., and Richardson, W. J. (1999). Proceedings/Workshop on the effects of anthropogenic noise in the marine environment, 10–12 February 1998. Office of Naval Research, Arlington, VA. Available at www.onr.navy.mil/sci_tech/personnel/cnb_sci/Proceed.pdf

Jasny, M. (1999). Sounding the depths: Supertankers, sonar, and the rise of undersea noise. Natural Resources Defense Council Publications, New York. Available at www.nrdc.org/wildlife/marine/sound/sdinx.asp

Kastak, D., Schusterman, R. J., Southall, B. L., and Reichmuth, C. J. (1999). Underwater temporary threshold shift induced by octave-band noise in three species of pinniped. *J. Acoust. Soc. Am.* **106**(2), 1142–1148.

Møhl, B., Wahlberg, M., Madsen, P. T., Miller, L. A., and Surlykke, A. (2000). Sperm whale clicks: Directionality and source level revisited. *J. Acoust. Soc. Am.* **107**(1), 638–648.

NRC (1994). Low-frequency sound and marine mammals: Current knowledge and research needs. U.S. Nat. Res. Counc., Ocean Studies Board, Committee on low-frequency sound and marine mammals (D. M. Green, H. A. DeFerrari, D. McFadden, J. S. Pearse, A. N. Popper, W. J. Richardson, S. H. Ridgway, and P. L. Tyack). Natl. Acad. Press, Washington, DC.

NRC (2000). Marine mammals and low-frequency sound: Progress since 1994. U.S. Nat. Res. Counc., Ocean Studies Board, Committee to Review Results of ATOC's Marine Mammal Research Program (A. N. Popper, H. A. DeFerrari, W. F. Dolphin, P. L. Edds-Walton, G. M. Greve, D. McFadden, P. B. Rhines, S. H. Ridgway, R. M. Seyfarth, S. L. Smith, and P. L. Tyack). Natl. Acad. Press, Washington, DC.

Richardson, W. J., Greene, C. R., Jr., Malme, C. I., and Thomson, D. H. (1995). "Marine Mammals and Noise." Academic Press, San Diego, CA.

Ridgway, S. H., and Carder, D. A. (1997). Hearing deficits measured in some *Tursiops truncatus,* and discovery of a deaf/mute dolphin. *J. Acoust. Soc. Am.* **101**(1), 590–594.

Schlundt, C. E., Finneran, J. J., Carder, D. A., and Ridgway, S. H. (2000). Temporary shift in masked hearing thresholds of bottlenose dolphins, *Tursiops truncatus,* and white whales, *Delphinapterus leucas,* after exposure to intense tones. *J. Acoust. Soc. Am.* **107**(6), 3496–3508.

Tyack, P. L. (2000). Functional aspects of cetacean communication. *In* "Cetacean Societies: Field Studies of Dolphins and Whales" (J. Mann, R. C. Connor, P. L. Tyack, and H. Whitehead, eds.), pp. 270–307. University of Chicago Press, Chicago, IL.

Wells, R. S., Boness, D. J., and Rathbun, G. B. (1999). Behavior. *In* "Biology of Marine Mammals" (J. E. Reynolds III and S. A. Rommell, eds.), pp. 324–422. Smithsonian Institution Press, Washington, DC.

Wenz, G. M. (1962). Acoustic ambient noise in the ocean: spectra and sources. *J. Acoust. Soc. Am.* **34**, 1936–1956.

Würsig, B., Greene, C. R., Jr., and Jefferson, T. A. (2000). Development of an air bubble curtain to reduce underwater noise of percussive piling. *Mar. Environ. Res.* **49**(1), 79–93.

North Atlantic Marine Mammals

Gordon T. Waring and Debra L. Palka
*Northeast Fisheries Science Center,
Woods Hole, Massachusetts*

Marine mammals are a diverse, widespread, and significant component of North Atlantic marine ecosystems. Four of the five commonly recognized marine mammal taxa reside in the North Atlantic: cetaceans (mysticetes, baleen whales; and odontocetes, toothed whales, dolphins, and porpoises), sirenians (manatees), pinnipeds, (seals and walruses), and polar bears (Table I). A fifth taxon (marine and sea otters) and sea lions and fur seals (family Otariidae) have not inhabited the North Atlantic since at least the late Pleistocene. The systematics of marine mammals are still being disputed. However, with the advent of molecular, biochemical, morphological and TELEMETRY studies, the exact number of extant species may be settled. Marine mammals occupy all North Atlantic marine regimes, tropical to polar, although species-specific ranges exist and distribution patterns are not uniform. The large-scale, non-random distribution of marine mammals is influenced by oceanographic features, whereas smaller-scale distributions are influenced by factors such as the animal's physiology, behavior, and ecology. Over geologic time scales, the diversity and ecology of North Atlantic marine mammals reflect adaptation to a dynamic aquatic environment. As elsewhere, North Atlantic marine mammal populations have been significantly impacted by human interactions. Some species have been, and continue to be, harvested for subsistence and commercial use and for their cultural value. Overexploitation has resulted in EXTINCTION [e.g.,

Caribbean monk seal (*Monachus tropicalis*), Atlantic gray whale (*Eschrichtius robustus*)] and significant population declines (e.g., North Atlantic right whale, *Eubalaena glacialis*) and has also likely caused significant ecological "changes" (e.g., reduction of top predators and competitive interactions). Indirect mortality (e.g., fishery bycatch, pollution) has adversely affected numerous species [e.g., harbor porpoise (*Phocoena phocoena*), bottlenose dolphin (*Tursiops truncatus*), beluga whale (*Delphinapterus leucas*), gray seal (*Halichoerus grypus*)].

I. Physical Environment

The physical characteristics of the North Atlantic ecosystem (Fig. 1) critically influence marine mammal distribution. Although the ocean basin provides marine mammals with an open pathway that extends from the equator northward to the Arctic and includes adjacent bodies of water (e.g., Gulf of Mexico, Caribbean Sea, North Sea, Norwegian Sea, and Bay of Biscay), the North Atlantic has many different ecosystems. Some adjacent seas, such as the Baltic and Mediterranean, are more isolated from the open ocean and form separate ecosystems. In open ocean, water masses define tropical to polar ecosystems that are influenced by circulation patterns of the major ocean currents, such as the Gulf Stream, Greenland current, and North Equatorial current. There are broad continental shelf ecosystems defined by basins, banks, channels, ice, submarine canyons, and volcanic islands. Sea mounts and the mid-Atlantic Ridge also define important ecosystems. These types of oceanographic features influence productivity to concentrate prey and create high-use marine mammal habitats.

II. Distribution and Habits

Baleen whales are widely distributed in the North Atlantic, with individual species exhibiting preferences for certain ecosystems. Some preferences are temperature driven. For example, bowhead whales (*Balaena mysticetus*) only occupy polar waters, whereas Bryde's whales (*Balaenoptera edeni*) are only found in tropical waters. Other preferences are more topography driven. For example, right, humpback (*Megaptera novaeangliae*), fin (*Balaenoptera physalus*), and minke (*B. acutorostrata*) whales prefer continental shelf ecosystems, whereas blue (*B. musculus*), sei (*B. borealis*), and Bryde's whales are associated with shelf-edge and deeper oceanic water. Large whales, however, are highly mobile and seasonally may occupy different habitats. Baleen whales, except bowhead and Bryde's whales, undergo the most extensive seasonal migrations of all North Atlantic marine mammals, migrating between warm low-latitude breeding grounds in winter and cold high-latitude feeding grounds in summer. North Atlantic humpback whales exemplify this migratory behavior. In summer, humpback whale stocks feed in Iceland, Greenland, Newfoundland, Gulf of St. Lawrence, and Gulf of Maine/Scotian Shelf and then spend winter on breeding grounds in the Caribbean Sea. Molecular genetic studies indicate that the feeding stocks are matrilineal groups of related individuals. There is little evidence of recent genetic exchange between North Atlantic and South Atlantic populations of baleen whales, due largely to seasonal differences in the migration paths of the two populations.

TABLE I
North Atlantic Marine Mammal Taxa[a]

Scientific name	Common name	Scientific name	Common name
Order Cetacca		*S. frontalis*	Atlantic spotted dolphin
Suborder Mysticeti (baleen whales)		*S. longirostris*	Spinner dolphin
Family Balaenidae (right whales)		*Steno bredanensis*	Rough-toothed dolphin
Balaena mysticetus	Bowhead whale	*Tursiops truncatus*	Common bottlenose dolphin
Eubalaena glacialis	Northern right whale		
		Family Phocoenidae	
		Phocoena phocoena	Harbor porpoise
Family Balaenopteridae (rorquals)		Order Sirenia (manatees and	
Balaenoptera acutorostrata	Common minke whale	dugongs)	
B. borealis	Sei whale	Family Trichechidae (manatees)	
B. edeni	Bryde's whale	*Trichechus manatus*	West Indian manatee
B. musculus	Blue whale	Subspecies	
B. physalus	Fin whale	*T. m. latirostris*	Florida manatee
Megaptera novaeangliae	Humpback whale	*T. m. manatus*	Antillean manatee
Suborder Odontoceti (toothed whales)		Order Carnivora	
Family Physeteridae		Suborder Pinnipedia	
Physeter macrocephalus	Sperm whale	Family Phocidae (true seals)	
		Subfamily Phocinae (northern phocids)	
Family Kogiidae		*Cystophora cristata*	Hooded seal
Kogia breviceps	Pygmy sperm whale	*Erignathus barbatus*	Atlantic bearded seal
K. sima	Dwarf sperm whale	*Halichoerus grypus*	Gray seal
		Pagophilus groenlandicus	Harp seal
Family Monodontidae		*Pusa hispida botnica*	Baltic seal
Delphinapetrus leucas	White whale/beluga whale	*P. hispida hispida*	Arctic ringed seal
Monodon monoceros	Narwhal	*Phoca vitulina concolor*	Western Atlantic harbor seal
		P. vitulina vitulina	Eastern Atlantic harbor seal
Family Ziphiidae (beaked whales)			
Hyperoodon ampullatus	Northern bottlenose whale	Subfamily Monachinae (southern phocids)	
Mesoplodon bidens	Sowerby's beaked whale	*Monachus monachus*	Mediterranean monk seal
M. densirostris	Blainville's beaked whale	*M. tropicalis[b]*	Caribbean monk seal
M. europaeus	Gervais' beaked whale		
M. mirus	True's beaked whale	Family Odobenidae	
Ziphius cavirostris	Cuvier's beaked whale	*Odobenus rosmarus*	Walrus
Family Delphinidae (ocean dolphins)			
Delphinus delphis	Short-beaked common dolphin	Suborder Fissipedia	
		Family Ursidae	
Delphinus capensis	Long-beaked common dolphin	*Ursus maritimus*	Polar bear
Feresa attenuata	Pygmy killer whale		
Globicephala macrorhynchus	Short-finned pilot whale		
G. melas	Long-finned pilot whale		
Lagenodelphis hosei	Fraser's dolphin		
Grampus griseus	Risso's dolphin		
Lagenorhynchus acutus	Atlantic white-sided dolphin		
L. albirostris	White-beaked dolphin		
Orcinus orca	Killer whale		
Peponocephala electra	Melon-headed whale		
Pseudorca crassidens	False killer whale		
Sousa teuszii	Atlantic hump-backed dolphin		
Stenella attenuata	Pantropical spotted dolphin		
S. clymene	Clymene dolphin		
S. coeruleoalba	Striped dolphin		

[a]From Jefferson *et al.* (1993)
[b]Extinct.

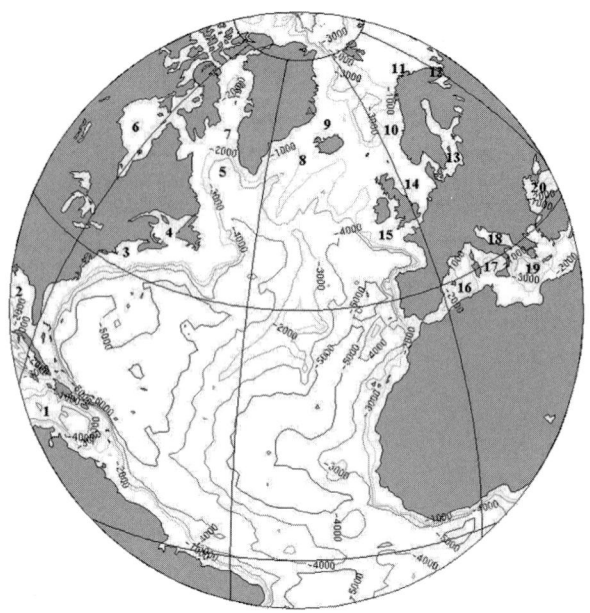

North Atlantic

Figure 1 *Bodies of water in the North Atlantic. Depth contours in meters. 1, Caribbean Sea; 2, Gulf of Mexico; 3, Gulf of Maine; 4, Gulf of St. Lawrence; 5, Labrador Sea; 6, Hudson Bay; 7, Davis Strait; 8, Denmark Strait; 9, Greenland Sea; 10, Norwegian Sea; 11, Barents Sea; 12, White Sea; 13, Baltic Sea; 14, North Sea; 15, Celtic Sea; 16, Mediterranean Sea; 17, Tyrrhenian Sea; 18, Adriatic Sea; 19, Aegean Sea; 20, Black Sea. Produced using software written by Dr. Pawlowicz, University of British Columbia, Canada.*

Odontocetes also occupy nearly all marine ecosystems in the North Atlantic, with individual species exhibiting preferences for particular ecosystems. Continental shelf species found in cool temperate to subpolar waters are harbor porpoises, Atlantic white-sided and white-beaked dolphins (*Lagenorhynchus acutus* and *L. albirostris*) and long-finned pilot whales (*Globicephala melas*), and two Arctic species, narwhal (*Monodon monoceros*) and beluga whales. Continental shelf break/pelagic species found in temperate to cooler waters include bottlenose (offshore and coastal forms), short-beaked common (*Delphinus delphis*), Risso's (*Grampus griseus*), striped (*Stenella coeruleoalba*), and Atlantic spotted (*S. frontalis*; coastal form) dolphins, sperm (*Physeter macrocephalus*) and northern bottlenose (*Hyperoodon ampullatus*) whales, and Cuvier's (*Ziphius cavirostris*), Blainville's (*Mesoplodon densirostris*), Sowerby's (*M. bidens*), and True's (*M. mirus*) beaked whales. The range of northern bottlenose and Sowerby's beaked whales extends into sub-Arctic waters. Continental shelf break/pelagic species found in warm temperate to tropical waters are pantropical spotted (*Stenella attenuata*), Atlantic spotted (offshore form), spinner (*S. longirostris*), Clymene (*S. clymene*), rough-toothed (*Steno bredanensis*), Atlantic hump-backed (*Sousa teuszii*), and Fraser's (*Lagenodelphis hosei*) dolphins and melon-headed (*Peponocephala electra*), false killer (*Pseudorca crassidens*), pygmy killer (*Feresa attenuata*), short-finned pilot (*G.*

macrorhynchus), pygmy sperm (*Kogia breviceps*), dwarf sperm (*K. sima*), and Grevais' beaked (*M. europaeus*) whales. Within warm-temperate to tropical water mass habitats, bottom topography and frontal boundaries are important characteristics that define cetacean distribution. Unlike baleen whales, only a few odontocetes (e.g., sperm and long-finned pilot whales) are known to undergo long-range seasonal migrations. Stock structure is largely unknown, except for a few nearshore continental shelf species (e.g., harbor porpoise, beluga). Some oceanic odontocetes likely move between North and South Atlantic waters (e.g., pantropical spotted dolphin, Cuvier's beaked whale).

North Atlantic seals (phocids) include both Northern and Southern hemisphere species (Table I). Northern phocids [harbor (*Phoca vitulina*), and gray seals (*Halichoerus grypus*)] are widely distributed in boreal to polar waters. The ice seals [hooded (*Cystophora cristata*), bearded (*Erignathus barbatus*), harp (*Pagophilus groenlandicus*), and ringed (*Pusa hispida*) seals] pup on ice and have seasonal migrations that are strongly associated with seasonal ice fluctuations. Bearded, hooded, and harp seals also utilize pelagic habitats. Ranges change; for example, since the 1990s, the winter/spring distributions of hooded and harp seals extended southward into northeast U.S. coastal waters. Harbor seals are the most widely distributed species, occupying cool temperate to Arctic North Atlantic waters. Gray seals have a discontinuous distribution in cold temperate to sub-Arctic coastal waters. Southern phocids include the Mediterranean (*Monachus monachus*) and Caribbean (extinct) monk seals. The Mediterranean monk seal is primarily found in the Mediterranean, adjacent seas, and along northwestern Africa. The Caribbean monk seal previously inhabited the Caribbean Sea and southern portion of the Gulf of Mexico. Stock structure for North Atlantic seals is well defined.

Cetaceans and phocid seals constitute the largest component of North Atlantic marine mammal fauna. Additional species include walruses (*Odobenus rosmarus*), polar bears (*Ursus maritimus*), and manatees (*Trichechus manatus*). Walruses and polar bears have a circumpolar distribution. Both species are usually associated with ice habitats but also spend time on coastal land areas. The Florida and Antillean manatees have a tropical to subtropical distribution. The Florida manatee is found in coastal waters of the Gulf of Mexico and southeastern United States. The Antillean manatee is distributed from northern Mexico to central Brazil and throughout the islands of the Caribbean.

III. Feeding

The taxonomic division of cetaceans into Odontoceti and Mysticeti reflects their different feeding strategies. Baleen whales are strainers who largely feed on planktonic or micronektonic crustaceans and/or relatively small pelagic fish by using visual or passive acoustic techniques. Toothed whales are graspers who capture fish, squid, and other species by hunting using sight, sound, or active ECHOLOCATION. Pinnipeds and polar bears are carnivores. Pinnipeds consume primarily fish and invertebrates, and some species occasionally eat sea birds, seals, or small whales. Polar bears prey primarily on seals and sometimes feed on fish and other small mammals. In contrast, manatees are herbivores, grazing in shallow waters on vegetation using primarily their sense of touch.

IV. Human Impact

Centuries of human activities have affected all North Atlantic marine mammal populations. Prehistoric people hunted coastal marine mammals for subsistence use, and in some areas (e.g., Bequia, Canada, Greenland) aboriginal hunting still exists. Early subsistence hunting, however, was likely insignificant compared to commercial whaling that began in Europe during the 10th century. By the beginning of the 18th century, European whalers depleted bowhead and right whale stocks in the eastern North Atlantic. American whalers also depleted right and humpback whale stocks in coastal waters off the American colonies. Whalers then hunted bowhead and right whales in the western North Atlantic, from Greenland to the Gulf of St. Lawrence. Depletion of these stocks initiated pelagic whaling for sperm and humpback whales in waters as far south as the Caribbean. Modern whaling, as we know it today, began in the late 19th century when Norwegians invented the explosive harpoon and converted from sail to steam vessels. This allowed whaling to expand to the faster swimming blue, fin, and sei whales. By the 1920s the North Atlantic large whales were depleted, and so whaling activities were redirected into Antarctic waters. Commercial whaling depleted most of these stocks as well. In 1946 the International Convention for the Regulation of Whaling was signed to provide for the CONSERVATION of whale stocks. However, North Atlantic whaling continued until the 1987 INTERNATIONAL WHALING COMMISSION moratorium was enacted. Following the moratorium, fin and minke whales were still taken for subsistence in West Greenland, and Norway continued a small minke whale fishery. Recently, Norwegian minke whaling has increased. Despite the many years since whaling of most species has stopped, some of the North Atlantic large whales (in particular the North Atlantic right whale) have not yet recovered. This is probably due to slow growth, low reproductive rates, and other human interactions.

Commercial exploitation of smaller cetaceans began in the 14th century when the Danes initiated organized hunts of Baltic Sea harbor porpoises. Although these hunts ended in the mid-20th century, there are still very few harbor porpoises in the Baltic Sea. In the 1500s, the Faeroese initiated a pilot whale drive fishery that continues to this day. Examples of hunts during the early to mid-1900s include Norwegian hunts of minke, killer, northern bottlenose, and pilot whales, American bounty hunts on harbor porpoises, and Newfoundland pilot whale drives. The Newfoundland fishery had to stop recently due to local depletion. Small-cetacean hunts occurring today are small-scale subsistence fisheries, such as for harbor porpoises in Greenland and belugas in Canada, Greenland, and Russia. It is unknown whether these stocks can sustain these removals.

North Atlantic walrus populations were similarly exploited. In the early 1600s, England initiated walrus hunting around Spitzbergen, Jan Mayen, and Norway. Russian, European, and Canadians joined in to expand the hunts further northward. As a result, these walrus populations were severely depleted by the 19th century and they have not yet recovered.

Seals were first commercially hunted for oil and blubber in Europe and Newfoundland. In these areas, large-scale commercial hunts for seal skins started in the early 18th century, focusing on harp and hooded seals, although bearded, ring, gray, and harbor seals were also taken. By the late 1800s, hunting expanded to Greenland for harp, hooded, and ringed seals. During the World Wars, hunting slowed down, allowing some populations to recover. However, sealing resumed immediately afterward. During the 1960s, killing methods raised public opinion against sealing, which then prompted MANAGEMENT actions and quotas to reduce hunting. The largest reduction began in 1983, particularly in Canadian waters, when the European community enacted a ban on the importation of seal skins. However, since 1996, the level of Canadian harp sealing has resumed to pre-1983 levels because new markets for skins and meat have opened up.

Long-standing conflicts between humans and seals have occurred because seals impact economically valuable fishery resources. Impacts include seals preying on fish species, and seals, particularly gray and harbor seals, infecting many North Atlantic fish species with seal (or cod) worm, *Pseudoterranova decipiens*. These issues have initiated seal bounty programs in Europe and North America, which resulted in regional extirpation (e.g., northeast United States, Baltic Sea), of some gray and harbor seal stocks. Although bounty programs have either ended or been greatly reduced, ecological interactions between seals and fisheries remain a management challenge in the North Atlantic.

Following World War II, technological improvements in fishing gear and vessels led not only to the expansion of coastal and high seas fisheries, but to the incidental mortality of thousands of marine mammals and rapid depletion of fish resources. By the 1970s, the elevated levels of marine mammals takes, particularly dolphins in the eastern tropical Pacific tuna purse seine fishery, instigated management and conservation measures that were aimed at reducing incidental takes of marine mammals in fisheries (e.g., U.S. Marine Mammal Protection Act of 1972). Over the past two decades, national and international measures have aimed to improve fish stocks and to monitor and reduce fishery-related impacts on marine mammals (e.g., 1991 U.N. Resolution 46/125, ending high seas driftnet fisheries). Unfortunately, marine mammal mortality still occurring in many fisheries threatens some marine mammal populations in the North Atlantic, such as right whales, bottlenose and common dolphins, harbor porpoises, and Mediterranean monk seals.

Environmental contaminants potentially pose a threat to the health of marine mammals. Contaminant levels can become toxic in marine mammals because most feed at high trophic levels and so accumulate low levels of toxins from their contaminated prey. Numerous studies have documented the presence of organochlorine and heavy metals in tissues of marine mammals. The debate is: are these levels dangerous? Potential deleterious biological effects of these contaminants include immunosuppression, endocrine disruption, and reproductive and pathological disorders. Documented cases of deleterious effects include the reproductive failure that has been linked to organochlorine levels in seals from the Baltic and Wadden Seas and to beluga whales from the St. Lawrence Estuary. It has been suggested that some of the large-scale die-off events that have killed thousands of seals and dolphins in northern Europe, the Mediterranean, the U.S. east coast, and Gulf of Mexico are due, at least in part, to high levels of organochlorines or toxic metals (e.g., cadmium, mercury). Epizootic events and toxic algal blooms have also caused large-scale die-offs. For example, in 1988 an epizootic in the Kattegat-Skagerrak (body of water bordering Denmark, Sweden, and Norway) killed approximately 56% of the harbor seal population in those waters. In winter 1987/1988, 14 humpback whales died in the vicinity of Cape Cod after consuming Atlantic mackerel containing a dinoflagellate saxitoxin. However, in nearly all cases, it has not been possible to demonstrate a direct link between death and contaminants. Other types of potentially dangerous environmental contaminants include oil spills and acoustic disturbances because these may cause behavioral modifications, prey displacement, or direct mortality.

V. Status

The current status of North Atlantic marine mammal populations is tightly linked to the population's biological characteristics and their long history of interacting with human activities. Most populations are no longer commercially hunted, but some are still severely depleted (e.g., North Atlantic right whales). Human activities, such as hunting, incidental takes, and environmental contaminants, continue to directly and indirectly adversely impact marine mammals. Conservation and research programs, particularly for small cetaceans, are highly variable among countries. Because most marine mammal populations are mobile, the only way to assess the status of and conserve these populations is to ensure that scientific research and conservation programs are effective ocean wide.

See Also the Following Articles

Cetacea, Overview ■ Distribution ■ Fishing Industry, Effects of ■ Hunting of Marine Mammals ■ Pinnipedia

References

Aguilar, A., and Borrell, A. (1995). Pollution and harbour porpoises in the Eastern North Atlantic: A review. *In* "Biology of the Phocoenids" (A. Bjørge and G. P. Donovan, eds.), pp. 231–244. Report of the International Whaling Commission (Special Issue 16), Cambridge.

Baker, S. C., and Palumbi, S. R. (1997). The genetic structure of whale populations: Implications for management. *In* "Molecular Genetics of Marine Mammals" (A. E. Dizon, S. J. Chivers, and W. F. Perrin, eds.), pp. 117–146. Special Publication 3. The Society for Marine Mammalogy, Lawrence, KS.

Bowen, W. D., and Siniff, D. B. (1999). Distribution, population biology, and feeding ecology of marine mammals. *In* "Biology of Marine Mammals" (J. E. Reynolds III and S. A. Rommel, eds.), pp. 423–484. Smithsonian Institution Press, Washington, DC.

Boyd, I. L., Lockyer, C., and Marsh, H. D. (1999). Reproduction in marine mammals. *In* "Biology of Marine Mammals" (J. E. Reynolds III and S. A. Rommel, eds.), pp. 218–286. Smithsonian Institution Press, Washington, DC.

Gambell, R. (1999). The International Whaling Commission and the contemporary whaling debate. *In* "Conservation and Management of Marine Mammals" (J. R. Twiss, Jr., and R. R. Reeves, eds.), pp. 179–198. Smithsonian Institute Press, Washington, DC.

Geraci, J. R. (1989). Clinical investigation of the 1987–88 mass mortality of bottlenose dolphins along the U.S. central and south Atlantic coast. Final report to National Marine Fisheries Service, U.S. Navy Office of Naval Research, and U.S. Marine Mammal Commission, Washington, DC.

Jefferson, T. A., Leatherwood, S., and Webber, M. A. (1993). "Marine Mammals of the World." FAO Species Identification Guide, Food and Agriculture Organization of the United Nations, Rome.

Katona, S. K., and Kraus, S. D. (1999). Efforts to conserve the North Atlantic right whale. *In* "Conservation and Management of Marine Mammals" (J. R. Twiss, Jr., and R. R. Reeves, eds.), pp. 311–331. Smithsonian Institute Press, Washington, DC.

Kinze, C. C. (1995). Exploitation of harbour porpoises (*Phocoena phocoena*) in Danish waters: A historical review. *In* "Biology of the Phocoenids" (A. Bjørge and G. P. Donovan, eds.), pp. 141–244. Report of the International Whaling Commission (Special Issue 16), Cambridge.

Northridge, S. P., and Hofman, R. J. (1999). Marine mammal interactions with fisheries. *In* "Conservation and Management of Marine Mammals" (J. R. Twiss, Jr., and R. R. Reeves, eds.), pp. 99–119. Smithsonian Institution Press, Washington, DC.

O'Shea, T. J. (1999). Environmental contaminants and marine mammals. *In* "Biology of Marine Mammals" (J. E. Reynolds III and S. A. Rommel, eds.), pp. 485–536. Smithsonian Institution Press, Washington, DC.

Reijnders, P. J. H. (1986). Reproductive failure in common seals feeding on fish from polluted coastal waters. *Nature* **324**, 456–457.

Rice, D. W. (1998). "Marine Mammals of the World: Systematics and Distributions." Special Publication 4. The Society for Marine Mammalogy, Lawrence, KS.

Rosel, P. (1997). A review and assessment of the status of the harbor porpoise (*Phocoena phocoena*) in the North Atlantic. *In* "Molecular Genetics of Marine Mammals" (A. E. Dizon, S. J. Chivers, and W. F. Perrin, eds.), pp. 209–226. Special Publication 3. The Society for Marine Mammalogy, Lawrence, KS.

Sahrhage, D., and Lundbek, J. (1992). "A History of Fishing." Springer-Verlag, Berlin.

Waring, G. T., Palka, D. L., Clapham, P. J., Swartz, S., Rossman, M. C., Cole, T. V. N., Hansen, L. J., Bisack, K. D., Mullin, K. D., Wells, R. S., Odell, D. K., and Barros, N. B. (1999). U.S. Atlantic and Gulf of Mexico Marine Mammal Stock Assessments–1999. NOAA Technical Memorandum NMFS-NE-153.

North Atlantic, North Pacific, and Southern Right Whales

Eubalaena glacialis, E. japonica, and *E. australis*

ROBERT D. KENNEY
University of Rhode Island, Narragansett

Right whales occur in temperate to subpolar latitudes of all of the world's oceans. North Atlantic [(*Eubalaena glacialis* (Müller, 1776)] right whales were the first target of commercial whaling and the first whale species to be seriously depleted, and today the two Northern Hemisphere species are the rarest large whales in the world. North Atlantic and North Pacific [(*E. japonica* (Lacépède, 1818)] right whales remain critically depleted even after more than six decades of international legal protection and may still be facing the threat of EXTINCTION. Right whale populations off the eastern United States, Argentina, South Africa, and Australia have been the focus of intensive long-term research and may be some of the most intensively studied large whales. Nevertheless, there are many areas where our knowledge is sadly lacking and where additional research is critically needed.

I. Systematics and Nomenclature

A. Common Names

Conventional wisdom holds that the common name "right whale" comes from English whalers, who designated this as the "right" (i.e., correct) whale to hunt because it occurred near shore, swam slowly enough to be caught from a small boat propelled by sails or oars, floated when dead, and yielded large amounts of valuable oil and BALEEN. Alternatively, it may have derived from the sense of right whale meaning "true whale," as later formally recognized in the Latin generic name *Eubalaena*. Other common names in English include black right whale and black whale.

B. Systematics

While Northern and Southern Hemisphere right whales have long been treated as two species, it was recommended in 1998 that all right whales should be considered as a single species, *Balaena glacialis*. In June 2000, the INTERNATIONAL WHALING COMMISSION's scientific committee reviewed right whale taxonomy. Although morphological differences are nearly absent, the most recent genetic evidence demonstrates that right whales of the North Atlantic, North Pacific, and Southern Ocean possess fixed, unique genetic patterns indicating complete and long-established isolation. Genetic data also show that North Pacific and southern [*E. australis* (Desmoulins, 1822)] right whales are more closely related to one another than either is to North Atlantic right whales, making the traditional combination of North Atlantic and North Pacific animals as "northern right whales" a systematics error. Based on the evidence, the committee recommended that the three populations be considered as separate species. They also recommended retention of the generic name *Eubalaena*, published by J. E. Gray (see the accompanying box) to separate right whales from bowhead whales (*Balaena mysticetus*), the only other living species included in the family Balaenidae. However, while there are no currently accepted genetic criteria for differentiation at the generic level, systematic classifications are scientific hypotheses, subject to revision after further study. The few authors who have looked critically at morphological or molecular differences between right whales and bowheads have concluded that there is little evidence supporting their separation at the generic level and that those differences which do exist are less than those between species of *Balaenoptera*. In addition, the known fossil species of *Balaena* cannot be shown as more closely allied with either right whales or bowheads.

Taxonomic Rules, J. E. Gray, and Right Whale Names

There is a set of very specific rules, the International Code of Zoological Nomenclature, for determining the correct scientific name to apply to any particular taxon. One important aspect of the code is the rule of priority—when determining the valid name for any species (or genus or higher level category), the first published name applied should be used. For example, the original descriptions of all three right whale species included them in the genus *Balaena*, a genus name published by Linnaeus in 1758. If biological evidence continues to support classifying right whales as a distinct genus from bowheads, the name *Eubalaena*, published by J. E. Gray in 1864, is the oldest name available applicable only to the right whales and therefore the valid name by the rule of priority.

John Edward Gray (1800–1875) was an English zoologist and an important figure in the history of cetacean taxonomy. He began his career at about age 15 as a volunteer insect collector, received a temporary appointment at the British Museum in 1824 to catalog reptiles, and was Keeper of Zoology there from 1840 until his retirement only 2 months before his death. In his years at the museum, he published well over 1000 papers, including half of the 200 catalogs issued by the museum during his time. *Eubalaena* is not the only cetacean name he coined; 15 of the 85 currently accepted names of extant cetacean species, 15 of the 40 generic names, and 11 of the 14 family names were authored by Gray. One might conclude that he was particularly knowledgeable about cetacean taxonomy. However, he worked at what was likely the world's most influential museum of his day and apparently had a penchant for creating new species, genera, and higher taxa. He has been called one of the most notorious taxonomic "splitters" of all time. He also authored at least four other generic names for right whales, and in a single publication in 1871 he counted six genera and 14 species of right whales.

II. Description

Right whales have an extremely robust body form, bordering on rotund, with the girth at times exceeding 60% of total body length. The body is mostly black, sometimes with irregular white ventral patches. Some individuals may have a mottled appearance, and calves may sometimes be lighter colored. There is no dorsal fin. The pectoral flippers are large, broad, and blunt, and the flukes are very broad (up of 40% of body length), black on both dorsal and ventral surfaces, deeply notched, and smoothly tapered to the tips. Calves are 4.5–6 m long at birth; typical adults are 13–16 m. North Pacific right whales attain larger maximum sizes than the other species, up to 18 m and over 100 metric tons.

The head is relatively large, comprising about one-quarter to one-third of the total body length. The upper jaw is somewhat arched, and the margin of the lower lip forms a very pronounced curve. There are 200–270 baleen plates on each side

Are Whale-Lice Parasites?

Two species are said to be symbiotic when they live intimately associated with each other, with the exact nature of the relationship depending on the fitness costs and benefits incurred by each species. In the symbiosis involving right whales and cyamids, it can be assumed that the cyamid benefits from the relationship by having a place to live and a ready food supply. If the right whale also benefits, the association is defined as mutualism; if the whale receives neither benefits nor costs, the association is commensalism; and if the whale is harmed, the association is parasitism. Most sources refer to cyamids as parasites or ectoparasites. It is known that they do feed on the whale's skin, however, there is no evidence that their presence or feeding causes any harm to the whale. Similarly, while it is possible to construct various hypotheses for potential benefits provided to the whale by the cyamids, there is no evidence to test or confirm these theories. It therefore seems that the conservative course would be to consider cyamids as ectocommensals of right whales until convincing evidence demonstrating otherwise has been shown.

of the upper jaw. The baleen plates are relatively narrow and 2–2.8 m long, with very fine fringing hairs.

The most conspicuous external characteristics of right whales are the callosities on the head. These are irregular patches of thickened, keratinized tissue, which are inhabited by dense populations of specialized amphipod crustaceans, known as cyamids or "whale lice" (see accompanying box). At least three species of whale lice occur on right whales: *Cyamis gracilis, C. ovalis,* and *C. erraticus.* In southern right whales the callosities are also inhabited by barnacles, *Tubicinella* sp. The callosities occur at the tip of the snout (called the "bonnet" by whalers), on the lower lips and chin, above the eyes, and in front of and behind the blowholes. The callosities are congenital and not caused externally, as their beginnings are present in fetuses and neonates, but the pattern is not fully developed and colonized by cyamids until the whale is at least several months old. The callosity patterns are unique to individuals and are therefore extremely useful as a natural "tag," which allows repeated identification of individuals by photographs (Fig. 1).

The closest relative of the right whale is the bowhead whale. Bowheads are somewhat longer and substantially stouter, have relatively larger heads (about 40% of body length) with a more arched appearance, have much longer baleen plates (up to 5.2 m long), and completely lack callosities.

III. Distribution and Abundance

Right whales are found in the middle latitudes of both the Northern and Southern Hemispheres, between approximately 20° and 60°S and N latitudes (Fig. 2). There are three geographically isolated populations currently recognized as separate species: in the North Atlantic, North Pacific, and Southern Ocean. The populations are kept separated by Arctic ice and warm equatorial waters so that there is no interchange between populations, and apparently has not been for millions of years.

A. North Atlantic Right Whales

The historical range of North Atlantic right whales apparently extended as far south as Florida and northwestern Africa and as far north as Labrador, southern Greenland, Iceland, and Norway. The traditional hypothesis has been that there were separate stocks with little interchange on the western and eastern sides of the basin; however, recent evidence is casting some doubt on that model. Analysis of some 19th century whaling specimens in European museums shows that they do not differ genetically from living western individuals, and there have been one or two right whales seen in the eastern North Atlantic in recent years that were known individuals from the western stock. It is possible that the structure of a right whale population is that a particular ocean basin is inhabited by a single breeding population without long-term genetic isolation of stocks, but where return to traditional habitats learned from the mother (matrilineal habitat fidelity) maintains shorter term separation between two or more subsets of the population.

The present range of western North Atlantic right whales, from Florida to Nova Scotia with very occasional occurrence beyond those limits, is much reduced from its historical extent. The best estimate of present abundance is about 300 animals. In the

Figure 1 *Right whales are in the genus* Eubalaena *and include three species. Individuals can be identified with varying success using callosity patterns on the head. Southern right whales* (E. australis), *pictured here, are represented by multiple stocks, and some of these stocks appear to be both healthy and increasing, in contrast to the much reduced abundance and distribution of the North Atlantic right whale. Pieter A. Folkens/Higher Porpoise DG.*

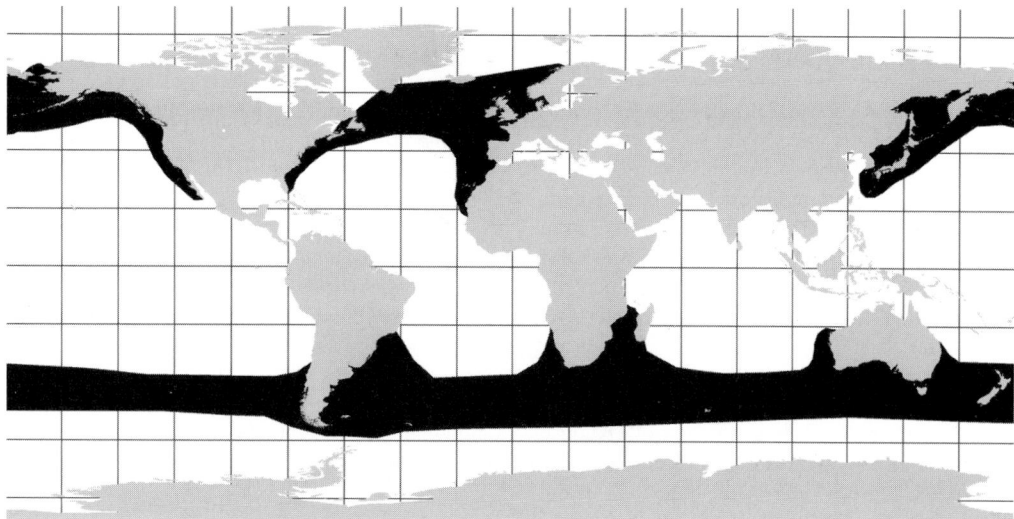

Figure 2 *Worldwide range of right whales* (Eubalaena glacialis, E. japonica, *and* E. australis)*. Much of what is shown here is relatively speculative based on sparse historical records of whaling catches and the available recent sightings. Most recent data come from areas relatively nearshore, with few or no data for the pelagic areas between the known coastal habitats.*

eastern North Atlantic, there have been only a handful of right whale sightings in the last few decades. It is not known whether these represent a small remnant eastern stock or whether some or all of them are individuals from the known western population.

B. North Pacific Right Whales

The historical range in the North Pacific was similarly much more extensive than today. Right whales occurred from Japan and northern Mexico north to the Sea of Okhotsk, Bering Sea, and Gulf of Alaska. Recent sightings are extremely rare, primarily in the Sea of Okhotsk and eastern Bering Sea. There are no reliable estimates of abundance, and there may be even fewer whales than in the North Atlantic. There are also insufficient genetic or resighting data to address whether there is support for the traditional separation into eastern and western stocks.

C. Southern Right Whales

Right whales are known from several areas of the Southern Ocean. Multiple stocks have been hypothesized, including Argentina/Brazil, South Africa, east Africa/Mozambique, western Australia, southeastern Australia, New Zealand, and Chile. Additional stocks have been hypothesized for the central Indian Ocean, the Campbell and Auckland Islands in the southwestern Pacific, and Tristan da Cunha in the central South Atlantic. There have also been suggestions of even finer stock structuring, e.g., between Argentina and Brazil in the western South Atlantic or between Namibia and South Africa in the eastern South Atlantic. Preliminary genetic analysis shows incomplete separation between eastern and western South Atlantic right whales. Further genetic studies should shed light on the degree of gene flow and stock separation in southern right whales and on the possible role of matrilineal habitat fidelity in structuring the population.

Right whale populations in Argentina, South Africa, and Australia are presently the largest and the best studied. The total abundance of those three stocks is currently estimated at about 7000 animals, and all three stocks appear to be healthy and increasing at 7–8% annually.

IV. Ecology
A. Prey

Right whales feed entirely on zooplankton, especially on large calanoid copepods (crustaceans approximately the size of a grain of rice). At times they also feed on smaller copepods, krill (larger shrimp-like crustaceans), pteropods (tiny planktonic snails), or the planktonic larval stages of BARNACLES and other crustaceans. The details of their diet likely differ between regions, e.g., it is likely that krill comprise a higher proportion of the DIET in southern right whales. It is also likely that right whales can be somewhat opportunistic, FEEDING on any prey of a size that can be filtered efficiently by the baleen, which does not swim strongly enough to escape, and which is concentrated into sufficiently dense patches to trigger feeding behavior. For example, there have been recent observations of North Atlantic right whales in the Bay of Fundy feeding on aggregations of salps.

B. Habitat

Right whales migrate annually between high-latitude feeding grounds and low-latitude calving and breeding grounds. These are substantial differences in the locations where most research has been conducted between the Northern and the Southern Hemispheres; therefore, there is often a lack of directly comparable information for different populations.

1. Feeding Grounds Right whale feeding takes place in spring, summer, and fall in higher latitude feeding grounds, where ocean temperatures are cooler and overall biological productivity is much higher. The best known right whale feeding grounds are in the western North Atlantic. These habitats are in nearshore and shelf waters, where some combination of bottom topography, water column structure and stratification, and currents acts to physically aggregate zooplankton into extremely dense concentrations. The densest zooplankton concentrations measured in the North Atlantic were found by sampling near right whales. There are probably also offshore feeding grounds, in locations not yet known, based on historical whaling records and on the fact that some known whales are often missing from the known habitats for months or years at a time. There must also have been other feeding grounds in the past, when the range of North Atlantic right whales was much more extensive.

Feeding grounds for the other species of right whales are much more poorly known. In the North Pacific, based on historical whaling records and the few recent sightings, the principal feeding grounds were most likely in the Sea of Okhotsk, central and eastern Bering Sea, and Gulf of Alaska. All of these feeding areas are much more pelagic or offshore than in the well-studied North Atlantic habitats. In the Southern Ocean, right whale feeding grounds are also apparently mostly in offshore, pelagic regions. Southern right whale feeding grounds are likely to be found associated with areas of extremely high productivity; limited sighting data available show most whales in the regions between the subtropical and Antarctic convergences.

2. Calving Grounds   Calving in right whales occurs during winter. Where the calving grounds are known, they are in shallow coastal regions or bays. The only known current calving ground in the western North Atlantic is in coastal waters near Georgia and northeastern Florida. It has been speculated that other coastal areas, including Delaware Bay and Cape Cod Bay, may have been calving grounds before the population was depleted by whaling. It has been noted, for example, that Cape Cod is similar topographically to Peninsula Valdés in Argentina and is located at about the same latitude. In the eastern North Atlantic, Cintra Bay in northwestern Africa is believed to have been a historical right whale calving ground. It is possible that areas near the Azores and Madeira, as well as the Bay of Biscay, were also calving grounds. In the North Pacific, no right whale calving grounds have ever been discovered. In the Southern Hemisphere, shallow coastal waters and bays in many areas are currently known to be southern right whale calving areas or hypothesized to have been calving grounds historically, including Argentina, Brazil, Falkland Islands, Tristan de Cunha, Namibia, South Africa, Mozambique, Kerguelen Island, Australia, New Zealand, Auckland Islands, and Chile.

3. Breeding Grounds   Breeding or mating also occurs during the winter. Because of the 3-year female reproductive cycle, breeding can take place geographically distant from calving. In the western North Atlantic, the location of the majority of the population during the winter is not known, and adult males are nearly absent from the calving ground. Breeding must occur wherever the adult population spends the winter, but it is not known whether there are specific, distinct winter habitats or whether the whales are broadly dispersed across wide regions of the North Atlantic.

In southern right whales, at least some mating occurs in or near the calving grounds, although there may be small-scale segregation of breeding adults from females with calves. In Argentina, because females are observed infrequently in these breeding groups in the year prior to calving, it is possible either that the mating which actually leads to conception occurs in some other, unknown habitat or that receptive females only visit coastal waters for very brief periods.

4. Learning and Habitat Use   Circumstantial evidence suggests that learning is an important component of habitat selection in right whales. Calves apparently learn the locations of feeding grounds by accompanying their mothers during the first year of life and then return to those same habitats for the rest of their lives. This pattern of matrilineal habitat fidelity seems to be common in migratory whale species; resighting and genetic data demonstrate that it is responsible for population structuring in North Atlantic humpback whales.

C. Predators

Potential predators of right whales include killer whales and large sharks. There have been a few direct observations of killer whale attacks on right whales; more common are records of right whales in the North Atlantic, South Atlantic, and North Pacific with scarring patterns on the flukes that appear to be tooth rakes or missing pieces of the flukes. Predators may be more likely to attack calves or juveniles than adults. Predation may have been one of the selective pressures leading to the evolution of right whales' use of shallow coastal habitats for calving, particularly since at least white sharks are known to often attack their prey from below.

V. Behavior

A. Aerial Behaviors

Right whales are observed to frequently perform highly energetic behaviors at or above the surface of the water. These aerial behaviors include BREACHING (jumping partly or almost completely above the surface), LOBTAILING (violently slapping the surface with the flukes), and flippering (slapping the surface with a pectoral flipper). The functions of these behaviors are not known. They all produce very loud sounds, which may sometimes have a communicative and/or aggressive function. Right whales in Argentina and South Africa have been observed to lift their flukes above the surface, where the flukes act like a sail and allow the wind to push the whale horizontally. This "tail-sailing" behavior has not been reported in other habitats.

B. Feeding Behavior

Right whales are "skimmers" (Fig. 3). They feed by swimming forward with the mouth agape. Water flows into the opening at the front, and out through the baleen, straining their prey from the water. Feeding can occur at or just below the surface, where it can be observed easily, or at depth. At times, right whales apparently feed very close to the bottom, because they are observed to surface at the end of an extended dive with mud on the head. Typical feeding dives last for 10–20 min.

C. Reproductive and Social Behavior

Courtship in right whales often involves aggregations of whales termed "surface-active groups." These are usually centered around

Figure 3 A "skimming" North Atlantic right whale feeding on plankton near the surface. Photograph by William Watkins.

a single female and may involve large numbers of males; groups of more than 20 animals have been observed. Often the female is belly-up at the surface, while the males stroke her with their flippers or attempt to push her under. There is some evidence that the female initiates the interaction by vocalizing. In the North Atlantic, surface-active groups are observed in all seasons, even though calving is highly synchronous and restricted to winter. Therefore much of the observed activity does not lead to fertilization and may serve a social function. The female may simply use the interactions to assess male quality for later mating. The interactions between males in the group generally involve very little of the violence and aggressiveness seen in humpback whales. One theory is that right whales engage in sperm competition, where the volume of semen is important in displacing the sperm of other males mating with the same female. Right whale testes may be the largest of any animal, at 2 m long and 500 kg each.

D. Sound Production

Right whale vocalizations are primarily low-frequency moans, groans, belches, and pulses. Most acoustic energy produced is below 500 Hz, with some sounds up to 1500–2000 Hz. The functions of these sounds are not well understood. Hypothesized functions include maintenance of contact between separated individuals, threats or other aggressive signals, and social signals, including their possible involvement in surface-active group behavior.

VI. Life History

A. Age at Maturity

Information on the age at maturity in right whales is not available from whaling data as it is for other whale species taken by 20th century industrial whaling. The information must be derived from photoidentification studies which track known individuals from birth. The youngest mature female in the western North Atlantic was 4 at maturity and 5 at first calving. From both North Atlantic and Southern Hemisphere data, the average age at first calving is closer to 9 or 10 years. Age at maturity is not yet known for males, as there is no external method for identifying paternity. Newer genetic studies may be better able to identify fathers of calves and begin to provide data for age of maturity in males.

B. Growth

Growth in right whales is relatively rapid from birth to weaning at about age 1, by which time the calf approximately doubles in body length to 9–11 m. Available data on growth after age 1 are not entirely consistent. For example, some studies indicate that growth also can be relatively rapid in year 2, by which time total length may reach 12–13 m, and thereafter is much slower. However, South African right whales apparently grow little between ages 1 and 4. Growth after age 1 is likely to be dependent on feeding success. The western North Atlantic female that matured at age 4 remained with her mother well into her second year, possibly growing much faster than the typical rate by nursing for a longer period. Right whales are believed to reach sexual maturity at body lengths of 13–16 m.

C. Life Span

There are very few data on the longevity of right whales. Aging baleen whales is an extremely difficult problem. Japanese attempts to use the wax plug found in the auditory canal from the North Pacific right whales taken for research in 1956–1968 were not successful. The oldest known right whale to date was in the North Atlantic. A mother–calf pair near Fort Lauderdale, Florida, was pursued and shot at by fishermen on March 24, 1935. The calf was killed, but the mother escaped. A photograph of her published in the New York Herald Tribune at the time was later matched to photographs taken in Cape Cod Bay (by pioneering right whale researcher William E. Schevill) in April 1959. She was also photographed in 1980, 1985, and 1992. On August 13, 1995 she was photographed offshore, with a large, gaping wound on the head apparently caused by a ship strike. It is unlikely she could have survived that injury. Assuming she was at least 10 years old in 1935, she would have been at least 70 years old in 1995. Recent research on bowhead whales suggests that they may live even longer.

D. Reproduction

The typical reproductive cycle in mature female right whales is 3 years between births. The gestation period is approximately 1 year, and weaning occurs at about 1 year of age. The female then takes a third year to replenish her energy stores, although it is possible for a female who has been especially successful at feeding to skip the resting year and calve after only a 2-year interval (one case observed in the North Atlantic, and at least one in Argentina). An alternative explanation for an observed 2-year interval is calf mortality soon after birth and subsequent avoidance by the mother of the high energetic demands of lactation; documented twice in Australia. This presumes that the mother–calf pair is sighted during the brief interval between birth and death of the calf. Otherwise what would be observed is an apparent 5-year interval, of which 25 were recorded in the North Atlantic between 1980 and 1998. Calving has been observed very rarely; in other instances, known females have been sighted in the calving ground both before and after the calf was born.

VII. Fossil Record

Five fossil species of *Balaena* have been described from deposits of late Miocene, Pliocene, or Pleistocene age (2–10 million years old) from Europe and North America: *B. affinis*,

B. etrusca, B. montalionis, B. primigenius, and *B. prisca.* The last of these is similar enough to modern bowheads (*B. mysticetus*) that it may, in fact, not be a separate species. There is then a long gap in the balaenid FOSSIL RECORD to *Morenocetus parvus,* the oldest known member of the family, found in early Miocene (23 million years old) deposits in South America.

VIII. Interactions with Humans

A. Whaling

North Atlantic right whales were the first whales to be harvested commercially by the Basques along the Atlantic coast of western Europe as early as the 11th century. The whales were killed primarily for oil, which was sold across Europe, as the technology of the time did not permit preservation and transportation of meat. As populations nearest shore were depleted, Basque whaling expanded to more distant waters, reaching eastern Canada by 1530. Basque whaling in Canada was centered in the Strait of Belle Isle between Labrador and Newfoundland and took 300–500 whales per year at its peak. Catches were declining by 1610–1620 and ended in 1713, by which time they had taken as many as 40,000 whales, including both right whales and bowheads.

Local shore-based right WHALING in North America began soon after the establishment of permanent colonies during the early 17th century. Peak catches were in the early 18th century (e.g., 86 in Nantucket, Massachusetts in 1726; 111 in Long Island, New York in 1707), and right whales in western North Atlantic waters may have been effectively extinct as a basis for a commercial fishery by the middle of the 18th century. The familiar Yankee whaling industry soon developed as a high-seas fishery targeting sperm whales; however, the whalers continued to opportunistically take any right whales encountered. Yankee whaling (including ships from several European nations) spread to the South Atlantic by 1775, into the South Pacific in 1789, and into the North Pacific by 1820. The Japanese had also begun their own shore-based fishery, which took some coastal migrant right whales, in the late 16th century.

The traditional high-seas Yankee whale fishery finally ended in the early 20th century, when it was replaced by modern industrial whaling. Total right whale catches (although records are not complete) were at least 38,000 in the South Atlantic, 39,000 in the South Pacific, 1300 in the Indian Ocean, 15,000 in the North Pacific, and at least a few hundred in the North Atlantic. Some shore-based whaling in the eastern United States persisted into the 1920s, but it was minor, with only 8 taken in Long Island after 1900. In the North Atlantic, the last episode of intensive right whaling was in the late 19th and early 20th century off Norway, Iceland, and Scotland, and the last right whales taken were at Madeira: 1 in 1952 and 2 in 1967. All right whale populations worldwide were protected from commercial whaling in the 1930s by the first International Convention for the Regulation of Whaling. However, the Japanese took 23 North Pacific right whales in the 1940s and 13 more under special scientific research permits between 1956 and 1968, some illegal takes of right whales along the coast of Brazil were reported in the 1950s, there was a significant amount of illegal Soviet taking of right whales in the North Pacific and Southern Ocean into the 1960s, and it is possible that there has been illegal right whaling elsewhere in the world.

B. Ship Strikes

The most significant human-related source of mortality at present in western North Atlantic right whales is collision with large ships. Between 1970 and 1999, 16 right whales were known to have been killed by ships, and 2 others were last seen with serious and probably fatal injuries. There are probably additional mortalities that are never discovered because the carcasses are lost at sea. Ship collisions may be less of a mortality factor in other oceans, where right whales spend less time in nearshore habitats or where the level of industrial development is lower, although at least three probable ship-strike mortalities have been recorded in recent years off the Brazilian coast.

C. Entanglements

The second most important human-related mortality factor in western North Atlantic right whales is incidental capture in commercial fishing gear. The gear involved is fixed gear (set in one location rather than towed behind a vessel), including sink gill nets, drift nets, and a variety of pot and trap fisheries. Since 1970, three right whales are known to have been killed by ENTANGLEMENTS and eight others were seriously injured but disappeared and probably died. It is not always known whether entanglements occur in actively fishing gear or in gear that has been lost, damaged, or moved by storms or vessels (often termed "ghost" gear). There are few data on entanglement mortalities in other populations.

Entanglement seems to be very common in right whales. Many entanglements involve the tail, where the leading edges of the flukes begin, and leave characteristic scars afterward. Over 60% of whales in the western North Atlantic carry such scars, and some individuals have been entangled two or three times. Entanglements are therefore often not lethal. They may be more dangerous in younger animals, who might grow into a relatively benign entanglement until it becomes life-threatening.

D. Climate Change

Right whales are feeding specialists, with a relatively narrow range of acceptable prey species and requiring prey to be concentrated in exceptionally high densities. The development of right whale feeding grounds is closely linked to physical phenomena such as water structure, currents, and temperature. This may make right whales more sensitive than other species to impacts from global climate change. Any possible impacts may be increased because of matrilineal fidelity to their feeding grounds, and possibly a relatively low ability to locate new feeding grounds when conditions change. For example, right whales have never reoccupied the habitat in the Strait of Belle Isle where Basque whaling killed many thousands.

E. Other Human Impacts

There are a number of other potential human impacts on right whales:

1. Habitat loss due to high levels of human activity is mentioned frequently as a possible impact. Right whales no longer occur in Delaware Bay, eastern United States; Table Bay, South Africa; Wellington Harbor, New Zealand; or Derwent River, Tasmania. However, a plausible alternative explanation is that they were extirpated by whaling and have never reoccupied the habitat due to matrilineal habitat fidelity.

2. Pollution is another potential impact that is mentioned frequently but where evidence is sparse. Oil spills may be a bigger threat to right whales than to other baleen whales because their very fine baleen might be fouled more easily. Blubber samples show a presence of toxic contaminants, but at lower levels than in cetaceans that feed at higher trophic levels. A recent concern is that some contaminants may act as hormone mimics, affecting reproduction, or as immune system suppressants.

3. Man-made noise may have the potential for interfering with acoustic communication, particularly since the major noise source, shipping, is also concentrated in the lower frequencies.

4. Effects of intensive commercial fisheries may alter ecosystem structure, increasing the abundance of other species that feed on zooplankton, particularly small fishes with lower economic value than the larger species harvested by FISHERIES.

5. The long-term effects of extreme population depletion by whaling might include reduced genetic diversity and associated health and reproductive problems.

See Also the Following Articles

Bowhead Whale ■ Callosities ■ Filter Feeding ■ Species ■ Whale Lice

References

Bannister, J. L., Brownell, R. L., Best, P. B., and Donovan, G. P. (eds.), (2001). "Journal of Cetacean Research and Management, Special Issue 2." International Whaling Commission, Cambridge, UK.

Brownell, R. L., Best, P. B., and Prescott, J. H. (eds.) (1986). "Right Whales: Past and Present Status: Reports of the International Whaling Commission, Special Issue 12." International Whaling Commission, Cambridge, UK.

Cummings, W. C. (1985). Right whales *Eubalaena glacialis* (Müller, 1776) and *Eubalaena australis* (Desmoulins, 1822). *In* "Handbook of Marine Mammals" (S. H. Ridgway and R. Harrison, eds.), Vol. 3, pp. 275–304. Academic Press, San Diego.

Katona, S. K., and Kraus, S. D. (1999). Efforts to conserve the North Atlantic right whale. *In* "Conservation and Management of Marine Mammals" (J. R. Twiss and R. R. Reeves, eds.), pp. 311–331. Smithsonian Institution Press, Washington, DC.

Omura, H., Ohsumi, S., Nemoto, T., Nasu, K., and Kasuya, T. (1969). Black right whales in the North Pacific. *Sci. Rep. Whales Res. Inst.* **21**, 1–96.

Payne, R. (1995). "Among Whales." Dell Publishing, New York.

Rice, D. W. (1998). "Marine Mammals of the World: Systematics and Distribution." Special Publication Number 4. Society for Marine Mammalogy, Lawrence, KS.

Northern Fur Seal
Callorhinus ursinus

ROGER L. GENTRY
*National Marine Fisheries Service,
Silver Spring, Maryland*

The northern fur seal, *Callorhinus ursinus,* inhabits the North Pacific Ocean and Bering Sea (Fig. 1). *Callorhinus* is the oldest living genus of the family Otariidae, with origins 2–5 million years ago. *Callorhinus* means "beautiful nose," and *ursinus* means "bear-like"; sea bear was a former common name. This species has a short rostrum and therefore lacks the dog-like profile of the other fur seal genus, *Arctocephalus.* Females are small (45 kg) and gray-brown with a light underbelly; males are larger (200–250 kg) and vary from black to reddish (Fig. 2). The young (pups) are generally black with a light belly. About 91% of the northern fur seal population breeds at the Pribilof Islands of Alaska (74%) or at the Commander Islands of Russia (17%). The remainder breeds at the Kurile Islands and Robben Island (Russia), Bogoslof Island (Alaska), and the Farallone and San Miguel Islands (California). The species predates some of the islands on which it now breeds so some past population redistribution has occurred. It is primarily a subpolar species whose adaptations do not preclude breeding at a lower latitude. Bones in kitchen middens suggest that breeding colonies once occurred at several mainland sites in the United States.

Northern fur seals have been studied longer than most other marine mammals. George Wilhelm Steller first described the species in 1749 based on his observations in 1742 when he was a member of Vitus Bering's expedition that discovered the Commander Islands, Russia. Heavy exploitation of this colony (for pelts to be made into garments) began almost immediately after discovery and expanded considerably in 1786 when the Pribilof population was discovered. To protect the populations from overexploitation, the Russians prohibited the killing of females, which was one of the first known wildlife management actions. Despite this protection, major declines in population size occurred three times since 1742, each one associated with killing for pelts or other management actions. After one such decline (1911), the North Pacific Fur Seal Treaty was enacted, which involved four nations (Japan, Russia, United States, and Britain for Canada). It ended permanently in 1985. The population declined from 1956 to about 1983, was apparently stable at about 1.5 million animals until 1990, and declined at 2% per year until 2000.

If the commercial kill for pelts at times affected the population size, it was also responsible for a vigorous research program that made this among the best known marine mammal species. Population records extend back to at least 1865, and data are especially good between 1911 and 1985 when comparative research at the major land breeding sites was carried out under the treaty mentioned earlier. Presently research is conducted under a bilateral environmental agreement with Russia.

I. Natural History

This species is among the most pelagic of pinnipeds; females spend all but about 35 and males all but about 45 days per year at sea. From November to about March each year, the species remains north of the transition zone between the Oyashio and the Kuroshio currents (about 35°N latitude) without coming ashore. In March and April, animals gather along the continental shelf breaks and begin to migrate north to their respective breeding islands, apparently so that females can rear their young while feeding on greater (but more seasonal) food concentrations than are available at lower latitudes. The young

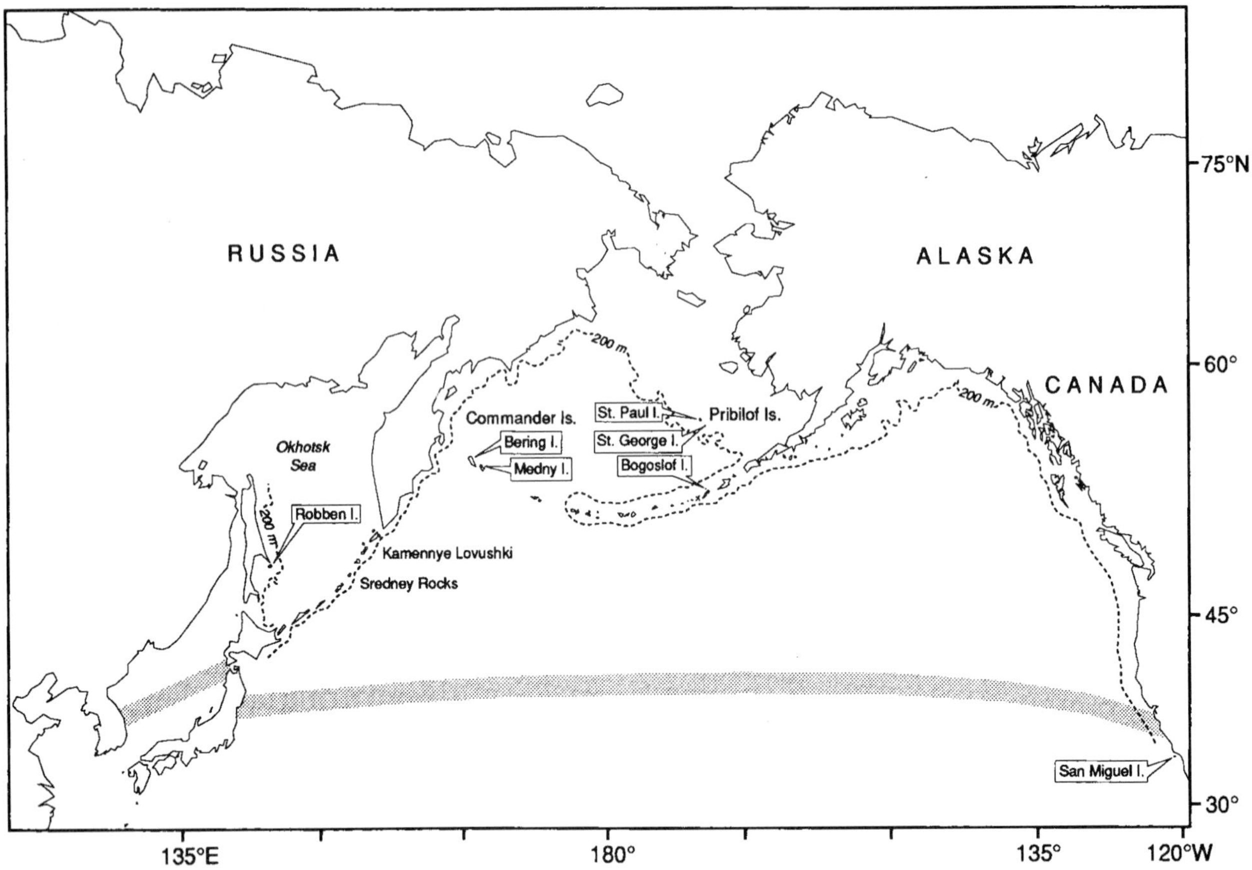

Figure 1 *Distribution of the northern fur seal.*

Figure 2 *Adult male and female northern fur seals. Photo by V. B. Scheffer.*

are born mainly in July, are nursed for about 4 months, wean in October or November just as winter arrives, and begin their southward migration with the rest of the population. The only animals that do not migrate are those that breed at San Miguel Island and the adult males of all other breeding colonies. It may be the larger mass of adult males that allows them to overwinter in cold northern waters unlike smaller animals that migrate to the southern end of the range.

This species is not under intense competition for breeding space by other species of otariids. California sea lions (*Zalophus californianus*) compete with them for space only at San Miguel Island. Steller sea lions (*Eumetopias jubatus*) may rest among northern fur seals (and occasionally eat their young at sea) but do not usually mate on the same sites. The Guadalupe fur seal (*Arctocephalus townsendi*) does not presently share any breeding islands with the northern fur seal. Furthermore, northern fur seals do not often compete with other otariids for food. They take small fish and squid over deep water, unlike California and Steller sea lions. They feed along the shelf break, but not upon the shelf like California sea lions. Guadalupe fur seals probably take many of the same prey items as the northern fur seal, but the Oyashio/Kuroshio transition zone effectively separates the two fur seal species at sea.

Individuals show strong preferences for using specific land sites at particular times of the year. Despite having many square kilometers of suitable land available for giving birth, individual females bear their young within 8–10 m of the same site in successive years. Males will defend only one territorial location for their reproductive lifetime. This tie to specific sites may be an extreme form of philopatry (mating on the site of their own birth). Furthermore, females and males tend to arrive on a date that is specific to them as individuals, despite the fact that pupping and mating in the population occur over a 6-week period. On the beach, the tens of thousands of breeding animals may superficially resemble a large, amorphous mass, but in reality it is composed of individuals each with fairly specific preferences for breeding at a given location and time of year.

In timing and location, the breeding aggregations of this species are highly predictable because of the site fidelity of individuals. Thirty-one BREEDING SITES (formerly called "rookeries") now exist. One of these ("North" breeding site on Bering Island, Commander Islands, Russia) has been occupied yearly since at least 1742. Breeding sites do not change much in size or shape from year to year. Northern fur seals colonize new islands and new beaches at a low rate compared to other fur seals. The onset and duration of the pupping/mating season are quite stable and vary little because of weather or climatic conditions.

The young leave land with modest fat reserves and little foraging experience at 4 months of age. At sea they must immediately learn to find and capture solid food alone while migrating in the appropriate direction. The difficulty of these tasks is reflected in the fact that some 60% or more of the pups born per year fail to survive to age 2 years when their age mates first return to land. Despite this seemingly low survival rate, the species is evolutionary quite successful; it numbered approximately 2.5 million animals at its peak in the mid-1950s.

The ecological role of the species is a high-level consumer, taking a variety of fish, cephalopods, crustaceans, and an occasional bird. Seventy-five different prey species have been identified in northern fur seal stomachs. Juvenile pollock, Atka mackerel, capelin, eulachon, herring, and several species of squid predominate. They take these prey items at relatively shallow depths (100–200 m for females, less than 400 m for males) compared to those species of marine mammals that can dive to a thousand meters or more. Northern fur seals feed mostly at night on prey that migrate vertically. A major component of the northern fur seal's diet is lantern fish (myctophids), which have no commercial value. Northern fur seals are preyed upon by large sharks, killer whales, and (for young only) Steller sea lions.

II. Reproduction

The species is SEXUALLY DIMORPHIC; males are at maximum 4.5 times larger than females. A size difference exists as early in embryonic growth as the sexes can be identified, and it accelerates at the onset of sexual maturity, about 3–5 years of age. Females mate almost immediately after becoming sexually mature and begin to produce one young per year. Males do not become socially mature (large enough to gain a territory among

females) until they are 8–9 years old. Like other otariids, the sex ratio at birth of northern fur seals is nearly 1:1, but due to greater age-specific mortality rates in males than in females, and delayed mating in males, mating is polygynous. The average adult sex ratio is about 9 females per male, but it may increase to 60:1 when males are killed for pelts.

Females have the highest pregnancy rates between ages 8 and 13 years, with rates sometimes exceeding 93% of an age class. Females can produce offspring yearly to about age 22 years, although few survive to that age. Contrary to sexual selection theory, older, larger females do not preferentially give birth to male pups, but larger mothers give birth to larger pups. Twinning is rare. For unknown reasons, females in the Commander Islands population have their first pup on average a year younger than females in the Pribilof population.

Females usually enter estrus and mate about 5.3 days after parturition. About 85% of the females have only a single copulation per estrus; the other 15% have no more than two. The reason for this low number of copulations is that the physical act of coitus terminates receptivity. Males can trigger the onset of estrus by interacting with females (the Whitten effect). However, if no male is present, females enter estrus spontaneously and become nonreceptive again after about 36 hr if they fail to copulate. No second estrus occurs that year.

Sexually receptive females mate indiscriminately with any male that approaches them. When no adult males are present (as in captivity), females will mate with a male of any size or age, and they adopt a stereotyped posture indicating sexual receptivity when touched in the pelvic area by humans, other females, or even their pups. Most matings are between adult males and females not because females reject juvenile males, but because adult males exclude them from the breeding areas. In contrast to females, males may mate 115 times or more per year, depending on their size and the location of their territory. The reproductive life span of males averages 1.5 seasons, significantly shorter than for females. Ten seasons is the maximum territorial occupancy recorded for any male northern fur seal.

III. Mating System

Males partition the pupping areas used by females into spaces called territories. They defend the boundaries of these territories (not the females therein) using mainly vocal and postural threats and rarely fighting. Males arrive on territory about 1 month before the females and spend their time maintaining boundaries and fasting. Fasts last on average 38 to 42 days for the population, but uninterrupted fasts of 80 days have been recorded for individuals. Throughout the main breeding season (July), males spend most of their time interacting with females and mating. Each male has exclusive reproductive access to all females in his territory. They abandon their territory in late July and are replaced by large subadult males that perform perhaps 20% of the year's copulations, including with all the virgin females that arrive after pupping has ended. Males are highly specific in the area they will defend, maintaining at least some of the same boundaries from year to year. Very few defend a territory that is more than 10 m distant from the previous year's territory. Males will persist in defending a territory from year to year even if it contains few females.

Females arrive usually 1 day before parturition occurs. They select a parturition site without apparent regard to the male that defends that site. Most births occur at night. Females remain with the pup in one location and defend it from other females until estrus and mating occur. Usually 1–24 hr after mating they depart on the first of a series of foraging trips. When they return they locate their pups by vocal exchanges and probably smell. Females will not suckle any pup but their own. Orphans and starvelings sneak-suckle, but are often bitten and tend to die of wounds and infection.

Females use the same parturition site from year to year to the greatest extent possible given the number of females on land, date, and other factors. Females prefer the company of other females. They are reluctant to be the sole female in a male's territory probably because such males bite females to keep them from moving. By living in a group, females reduce the number of male/female contacts and thus reduce the risk of injury. Females will abandon a favored birthing site if reaching it means being along in a territory with an adult male.

Despite seeking the close company of others, females are quite aggressive toward near neighbors. They bite and frequently threaten females around themselves, generally do not allow others to lie in full body contact (like sea lions do), and appear to lack identifiable social bonds with any other female. Females are therefore gregarious but not social. This combination allows each animal to enter and leave the aggregation without loss of social status in the usual sense of the term. The largest females win contests over resting sites, but there is no evidence of a stable social hierarchy based on individual identity.

Active female mate choice is not apparent. Females mate at nearly the same site for many years whereas the male defending that site as part of his territory (and having sexual access to all females on it) changes on average every 1.45 seasons. That is, females do not move around in an apparent attempt to find the highest ranking, most fit male. Furthermore, as stated earlier, females appear unable to discriminate high and low ranking males (or even their own species) when they are receptive. Females mate with the most fit males in the population, but by a more indirect means than individual choice. By arriving predictably at a given location and time of year and not deviating from this pattern, females become a predictable resource over which males compete. The winners of this competition remain as potential mates. Females thus indirectly acquire as potential mates the class of males that is highest ranking, but they accomplish this without having made a choice among individual males. This appears to explain why females are uniformly receptive to all males when they are in estrus.

IV. Foraging Behavior

Many aspects of foraging, such as duration of trips to sea, vary according to the local foraging environment. For example, trips to sea are shorter at the Commander Islands, which have a narrow continental shelf, than at the Pribilof Islands, which have a wide shelf. Some aspects of foraging behavior, such as

the duration of trips to sea, depth of dives, number of dives per day, and others, change seasonally in response to changes in the behavior of prey and the energy needs of the pup. Other aspects of foraging behavior, such as the duration of the deepest dives and the recovery times from these dives, do not change seasonally and may be determined physiologically. Females tend to specialize in one of three types of dive patterns [shallow (<50 m maximum), deep (100–200 m maximum), and mixed (a combination of the other two)] and to use the same pattern for several years. Furthermore, they tend to feed at about the same location on different trips to sea. Deep dives occur on the shelf and apparently represent feeding near or at the bottom. Shallow divers change depth at dawn and dusk, suggesting that the prey are migrating vertically. Females using the deep and shallow dive patterns have different fatty acids in their blubber, indicating that different prey bases are being exploited.

V. Ontogeny

As in other pinnipeds, the young of northern fur seals are precocial. They can swim on the day of birth but do not voluntarily do so until they are a month old. While their mothers are away foraging, pups withdraw to unoccupied parts of the breeding grounds to avoid being bitten by adult females. There they join other pups in dense "pods" where they sleep and play until their mothers return. Because adult males actively avoid trampling pups, the pups are actually safer from the much larger males than they are from adult females. Intentional infanticide by adult males is unknown in this species. Newborn pups can perform many of the components of adult behavior. For example, a 5-day old male pup can mount a female correctly and make coordinated, pelvic thrusting motions without ever having observed mating BEHAVIOR in adult males.

About 10% of pups die on shore before weaning. Starvation, trauma, hookworm, and several infectious diseases account for most of the deaths.

Males are on average 0.6 kg heavier than females at birth and remain larger until weaning. The growth rate during suckling is slow (1.1–1.3% of maternal mass per day) compared to phocids. The reason is that unlike phocids, northern fur seal pups fast each time their mothers are way foraging, and maintenance metabolism is costly given the cold temperatures, wind, and rain that the pups face. Pups are weaned at about 40% of adult female mass. Most pups wean themselves and depart from land before their mothers.

The migratory pathways, distribution, diet, and behavior of pups at sea are largely unknown because they are small and difficult to instrument and recapture. It is known that they migrate through Aleutian passes into the North Pacific Ocean in November of their birth year and that some of them are seen on the coasts of Washington, British Columbia, and Japan by the following January. Pups return to land during the summer of their second or third year, usually to the island of their birth. Juvenile females usually land on the breeding grounds, but juvenile males are excluded from these areas by adult males and instead come ashore on nearby all-male landing areas. They remain around the breeding island for 80 days or so, during which time they make periodic feeding trips to sea, visit all the landing areas available, and spend time in play bouts where they perfect the movement patterns used in fighting. Many juvenile males makes brief incursions onto the breeding grounds where they visit sites that some of them will later defend as territories, which suggests philopatry. The landing areas used by juvenile males were the sites where sealers used to stage the kill for pelts. Despite the fact that killing occurred on some of these sites weekly during the breeding seasons for over two centuries, fur seals did not learn to avoid them.

VI. Physiology

The northern fur seal reduces heat loss to cold ocean waters by means of dense underfur that traps small air bubbles. This fur stays dry even when in water. While at sea, and especially after deep dives, fur seals spend long periods grooming the fur and restoring its air content. The variable part of their thermoregulatory system is their naked rear flippers, which are the longest and thinnest of any pinniped. To reduce heat loss to cold water they hold one front flipper and both rear flippers in the air, often in an arch or "jug handle." To increase heat loss on land in warm weather they fan their rear flippers, keep them damp, or pant. The fur is such an effective insulator that snow falling on sleeping animals does not melt or make the animals shiver.

Fur seals are apneustic breathers, meaning they hold their breath on inspiration even in their sleep. This type of breathing is the rule in diving mammals. The longest dive northern fur seals can make without resorting to anaerobic metabolic pathways is about 3.7 min for a 45-kg female. They require about 20 min to recover from such a dive. During the dive their heart rate decreases dramatically as in other diving mammals. While SWIMMING at high speed northern fur seals often leap free of the water to breath, thus making a "porpoising" motion.

Both vision and hearing are highly effective in air and underwater. In bright light their pupil closes to a small pinhole showing a brown iris. In dim light the pupil opens wide to gather more light. An open pupil and a reflective layer in the retina contribute to good visual acuity in dim light where most feeding occurs. The reflective layer is visible as an "eye shine" if a light is trained on animals at night.

See Also the Following Articles

Breeding Sites ■ Mating Systems ■ Thermoregulation ■ Vision

References

Allen, J. A. (1880). History of the North American pinnipeds, a monograph of the walruses, sea-lions, sea-bears and seals of North America. U.S. Geol. Geogr. Surv. Terr., Misc. Publ. 12:16+785.

Gentry, R. L. (1998). "The Northern Fur Seal." Princeton Univ. Press, Princeton, NJ.

Reidman, M. (1990). "The Pinnipeds." University of California Press, Berkeley.

Scheffer, V. B. (1958). "Seals, Sea Lions, and Walruses: A Review of the Pinnipedia." Stanford Univ. Press, Stanford, CA.

North Pacific Marine Mammals

SERGIO ESCORZA-TREVIÑO
*Southwest Fisheries Science Center,
La Jolla, California*

I. North Pacific Marine and Freshwater Biomes

The vastness and diversity of the North Pacific Ocean are reflected in the richness of its marine mammal community. Sixteen of the world's 36 species of pinnipeds, 50 of the 85 species of cetaceans, and 2 of the 5 species of sirenians have been reported to occur in the North Pacific, in addition to the polar bear (*Ursus maritimus*) and the sea otter (*Enhydra lutris*). Most of these species are also found in other parts of the world, as is the case of most balaenids and delphinids, many ziphiids, and some otariids, phocids, and phocoenids. However, a large proportion of the species found in the North Pacific are endemic to its marine or riverine ecosystems: 9 pinnipeds, 10 cetaceans, 1 sirenian, and the sea otter.

The North Pacific Ocean ranges from about 80°W to 130°E, covering about 60% of the earth's circumference, and from the Arctic Ocean to the Equator (Fig. 1). The North Pacific encompasses a great number of peripheral basins, as different as the highly evaporative and relatively small Gulf of California (also known as Sea of Cortés) in the east, the large and epicontinental Bering Sea in the north, or the complex region of small, semienclosed seas and shallow shelves around the Indo-Pacific Archipelago in the west, where the Pacific and the Indian oceans meet. In addition, a number of large, complex river systems exist that extend thousands of kilometers upstream, as is the case of the Yangtze River in China.

The geographic DISTRIBUTION of mammal species in the ocean depends on a number of factors, among which temperature, depth, and productivity tend to be the most important. Some species, such as the killer whale (*Orcinus orca*) or the sperm whale (*Physeter macrocephalus*), are considered cosmopolitan. Others, like the vaquita (*Phocoena sinus*) or the now extinct Steller's sea cow (*Hydrodamalis gigas*), have or had very limited ranges. Many species are circumglobal but limited to particular climatic zones. For example, some species are pantropical, inhabiting low-latitude waters in all the world oceans, whereas others have antitropical (or bipolar) distributions. Species such as the ringed seal (*Pusa hispida*) and polar bear have been sighted as far north as the North Pole. Others can range hundreds of kilometers up the great rivers of both sides of the Pacific Ocean, either permanently or on a seasonal basis.

The North Pacific is dominated by a large subtropical gyre (Fig. 1). This North Pacific central gyre flows clockwise, bounded to the west by the Kuroshio Current, to the north by the North Pacific Current, to the east by the California Current, and along the south by the North Equatorial Current. To the north of the

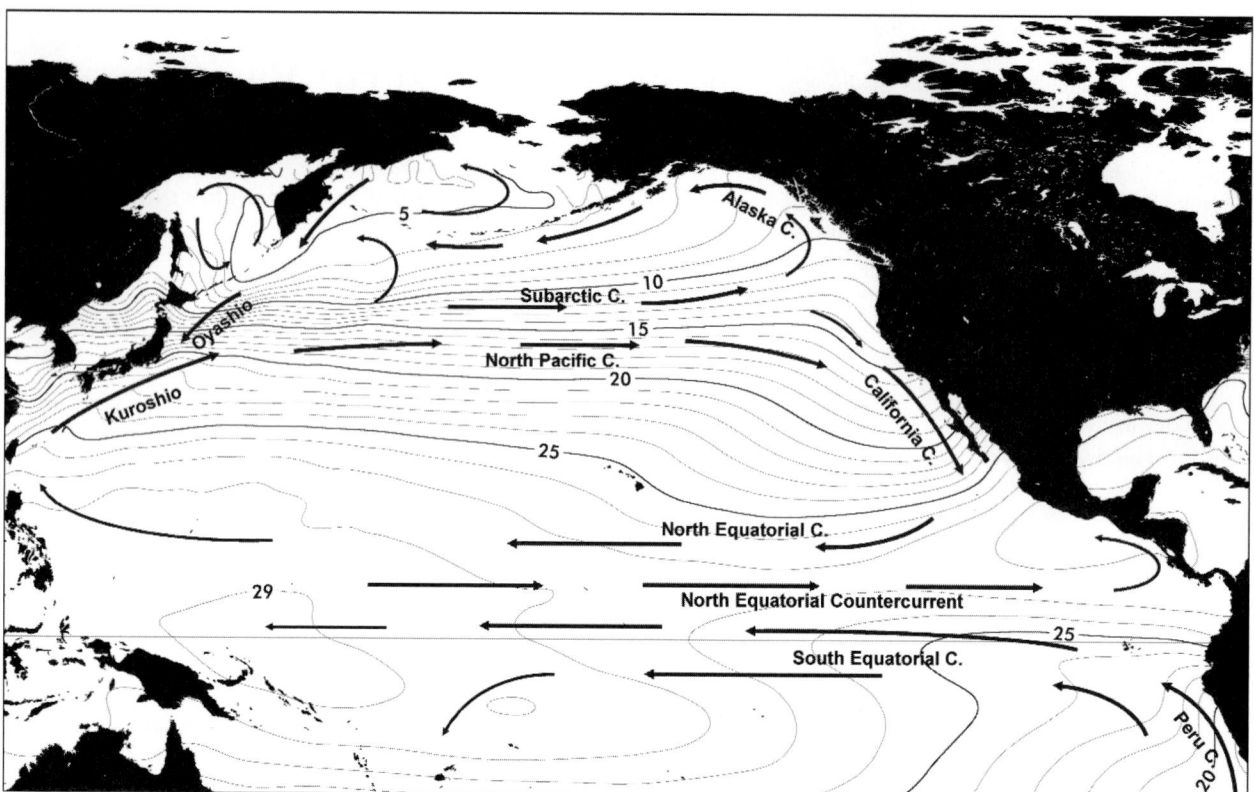

Figure 1 *Annual mean surface temperatures and principal currents of the North Pacific Ocean (graphic by Paul Fiedler).*

North Pacific central gyre, the cold Oyashio Current flows along the Kamchatka Peninsula and forms the western boundary of a counterclockwise subarctic gyre. The Alaska Current flows counterclockwise along the southeastern coast of Alaska and the Aleutian Peninsula. The convergence zone of these subarctic gyres and the central gyre, known as the Subarctic Boundary, crosses the western and central North Pacific at about 42°N and marks the steepest change in the abundance of cold-water vs warm-water species. To the south of the central gyre, the equatorial current system consists of the North Equatorial Countercurrent between the North and the South Equatorial Currents.

An important differentiating characteristic between the Pacific and other ocean basins is the strong effects of El Niño—Southern Oscillation (ENSO) events. Two distinct modes of Pacific circulation exist, depending on the variability of trade winds stress. When the trade winds weaken and are replaced by westerlies at low latitudes in the western Pacific (an ENSO event), the entire heat balance at low latitudes is perturbed, a situation that can last from a few months to 1 or 2 years. When this occurs, the major oceanographic features usually found in the eastern tropical Pacific can weaken or even disappear completely, resulting in atypical distributions for many marine mammal species.

II. Marine Mammals of Cold Water Marine Ecosystems

Most of the endemic species of marine mammals in the North Pacific Ocean inhabit its cold temperate, subarctic, or arctic wa-

ters (Table I). Among the pinnipeds, in the family Otariidae, we can find several of these endemic species. In fact, of the seven species of otariids present in the North Pacific, five are endemic to the North Pacific, whereas the other two, because they are present in the Galapagos archipelago, are also represented in the South Pacific. The Guadalupe fur seal (*Arctocephalus townsendi*) now breeds only on Guadalupe Island, off Baja California. The northern fur seal (*Callorhinus ursinus*) is a pelagic species that ranges from the Sea of Okhotsk and southern Bering Sea to the Sea of Japan and northern Baja California. The California sea lion (*Zalophus californianus*) includes two geographical divisions: one on the Pacific coast and one in the Gulf of California. Some animals have also been sighted on both sides of the Atlantic, but represent escapees from captive facilities. Along the western side of the Pacific, the Japanese sea lion (*Zalophus japonicus*) is, very probably, extinct. The last confirmed sighting took place 50 years ago. The Steller sea lion (*Eumetopias jubatus*) inhabits the coastal and immediate offshore waters of the cool temperate North Pacific.

Of the eight species of phocids (seals) present in the North Pacific Ocean, four are endemic: the spotted seal (*Phoca largha*), the ribbon seal (*Histriophoca fasciata*), the Hawaiian monk seal (*Monachus schauinslandi*), and the northern elephant seal (*Mirounga angustirostris*). The former two inhabit the pack ice zone of the arctic North Pacific, although the ribbon seal becomes pelagic during the summer. The Hawaiian monk seal, which probably represents the most primitive of the living phocids, inhabits the northwestern chain of the Hawaiian Islands,

TABLE I
Marine Mammal Species Present in the North Pacific Ocean by Taxa and Waters They Inhabit[a]

Taxa	Cold	Warm	River
Order Sirenia			
Family Dugongidae			
Dugong dugon (dugong)		X	
***Hydrodamalis gigas* (Steller's sea cow)**	X		
Order Carnivora			
Family Ursidae			
Ursus maritimus (polar bear)	X		
Family Mustelidae			
***Enhydra lutris* (sea otter)**	X		
Family Otariidae			
***Arctocephalus townsendi* (Guadalupe fur seal)**	X		
A. galapagoensis (Galapagos fur seal)		X	
***Callorhinus ursinus* (northern fur seal)**	X		
***Zalophus californianus* (California sea lion)**	X		
***Z. japonicus* (Japanese sea lion)**	X		
Z. wollebaeki (Galapagos sea lion)		X	
***Eumetopias jubatus* (Steller sea lion)**	X		
Family Odobenidae			
Odobenus rosmarus (walrus)	X		
Family Phocidae			
Erignatus barbatus (bearded seal)	X		
Phoca vitulina (harbor seal)	X		
***Phoca largha* (spotted seal)**	X		
Pusa hispida (ringed seal)	X		
***Histriophoca fasciata* (ribbon seal)**	X		
Cystophora cristata (hooded seal)	X		
***Monachus schauinslandi* (Hawaiian monk seal)**	X		
***Mirounga angustirostris* (northern elephant seal)**	X		
Order Cetacea			
Family Balaenidae			
***Eubalaena japonica* (North Pacific right whale)**	X		
Balaena mysticetus (bowhead whale)	X		
Family Eschrichtiidae			
***Eschrichtius robustus* (gray whale)**	X		
Family Balaenopteridae			
Megaptera novaeangliae (humpback whale)	X	X	
Balaenoptera acutorostrata (minke whale)	X	X	
B. borealis (sei whale)	X	X	
B. brydei (Bryde's whale)	X	X	
B. edeni (small type Bryde's whale)	X	X	
B. musculus (blue whale)	X	X	
B. physalus (fin whale)	X	X	
Family Physeteridae			
Physeter macrocephalus (sperm whale)	X	X	
Family Kogiidae			
Kogia breviceps (pygmy sperm whale)	X	X	
K. sima (dwarf sperm whale)	X	X	
Family Ziphiidae			
Ziphius cavirostris (Cuvier's beaked whale)	X	X	
***Berardius bairdii* (North Pacific bottlenose whale)**	X		
Indopacetus pacificus (Longman's beaked whale)		X	
Mesoplodon sp. A		X	
M. hectori (Hector's beaked whale) (?)	X		
M. peruvianus (pygmy beaked whale)	X		
***M. carlhubbsi* (Hubbs' beaked whale)**	X		
M. ginkgodens (ginkgo-toothed whale)		X	

(continues)

TABLE I (*Continued*)

Taxa	Cold	Warm	River
M. stejnegeri (saber-toothed whale)	X		
M. densirostris (Blainville's beaked whale)		X	
Family Lipotidae			
Lipotes vexillifer (baiji)			X
Family Monodontidae			
Delphinapterus leucas (beluga)	X		
Monodon monoceros (narwhal)	X		
Family Delphinidae			
Steno bredanensis (rough-toothed dolphin)		X	
Sousa chinensis (Pacific hump-backed dolphin)		X	X
Tursiops truncatus (Common bottlenose dolphin)	X	X	
T. aduncus (Indian Ocean bottlenose dolphin)		X	
Stenella attenuata (pantropical spotted dolphin)		X	
S. longirostris (spinner dolphin)		X	
S. coeruleoalba (striped dolphin)	X	X	
Delphinus delphis (short-beaked common dolphin)	X	X	
D. capensis (long-beaked common dolphin)	X	X	
Lagenodelphis hosei (Fraser's dolphin)		X	
Lagenorhynchus obliquidens (Pacific white-sided dolphin)	X		
Lissodelphis borealis (northern right whale dolphin)	X		
Grampus griseus (Risso's dolphin)	X	X	
Peponocephala electra (melon-headed whale)		X	
Feresa attenuata (pygmy killer whale)		X	
Pseudorca crassidens (false killer whale)	X	X	
Orcinus orca (killer whale)	X	X	
Globicephala melas (long-finned pilot whale)	X		
G. macrorhynchus (short-finned pilot whale)	X	X	
Orcaella brevirostris (Irrawaddy dolphin)		X	X
Family Phocoenidae			
Neophocaena phocaenoides (finless porpoise)	X	X	X
Phocoena phocoena (harbor porpoise)	X		
P. sinus (vaquita)	X		
Phocoenoides dalli (Dall's porpoise)	X		

[a]Endemic species are shown in bold.

whereas the northern elephant seal, whose population numbers once dwindled down to a few tens, ranges throughout the northeastern Pacific, from the Aleutian Islands to the Gulf of California. Other seals present in the North Pacific are the bearded seal (*Erignatus barbatus*) and the ringed seal (*Pusa hispida*), which are both arctic species, the harbor seal (*Phoca vitulina*), whose range extends from the Bering Sea to Baja California, and the hooded seal (*Cystophora cristata*), present only as a rare vagrant into the Beaufort Sea and Bering Sea and even to southern California. The only living species of the family Odobenidae, the walrus (*Odobenus rosmarus*), also inhabits the arctic North Pacific.

There are many cetacean species present in the higher latitudes of the North Pacific. Among the Mysticeti, three families are found. The balaenids are represented by both the bowhead whale (*Balaena mysticetus*), which inhabits the pack ice zone of the Arctic, and the migrant and endemic North Pacific right whale (*Eubalaena japonica*), whose range extends from the Bering Sea to California in the east and Taiwan in the west. The only living representative of the family Eschrichtiidae, the gray whale (*Eschrichtius robustus*), occurs only in the North Pacific,

as the North Atlantic populations are extinct. At present, the western, or "Korean," gray whale population is in very low numbers, whereas the eastern, or "California," population is considered to have fully recovered from past exploitation and ranges from subarctic waters to Baja California. Five species of rorquals are found in the cold waters of the North Pacific. The humpback whale (*Megaptera novaeangliae*) inhabits all the oceans of the world, from high-latitude summer FEEDING grounds to tropical winter grounds. The sei whale (*Balaenoptera borealis*) is also present in all oceans, but in more temperate waters than other rorquals. The common minke whale (*Balaenoptera acutorostrata*) can be found from the Chukchi Sea almost to the equator. The blue whale (*Balaenoptera musculus*) lives almost everywhere in the world, from the tropics to the arctic pack ice. The fin whale (*Balaenoptera physalus*), also almost worldwide in distribution, can be sighted in the North Pacific from the Okhotsk Sea, Bering Sea, and Chukchi Sea to Taiwan, Hawaii, and Baja California (including the Gulf of California).

The cosmopolitan sperm whale is present from the equator to the edges of the pack ice. The ranges of the two species of the family Kogiidae are somewhat smaller, but they are present in temperate and tropical waters around the world. The pygmy sperm whale (*Kogia breviceps*) inhabits oceanic waters, whereas the dwarf sperm whale (*Kogia sima*) lives over the continental shelf and slopes.

The family Ziphiidae is well represented in the North Pacific. Cuvier's beaked whale (*Ziphius cavirostris*) lives in all temperate and tropical waters around the world, whereas the North Pacific bottlenose whale (*Berardius bairdii*) is an endemic species, present only in the temperate and subarctic North Pacific. Also endemic are Hubbs' beaked whale (*Mesoplodon carlhubbsi*), in temperate waters, and the saber-toothed whale (*Mesoplodon stejnegeri*), which ranges from subarctic waters to the cold temperate North Pacific. The pygmy beaked whale (*Mesoplodon peruvianus*) is a recently discovered species, known only from the Gulf of California and Peru. Hector's beaked whale (*Mesoplodon hectori*), although considered to be circumglobal in the temperate waters of the Southern Hemisphere, can also be regarded as a North Pacific marine mammal because individuals have stranded and been sighted off southern California (although recent genetic analyses suggest that these may have been of a different, undescribed species).

Both members of the family Monodontidae are also present in the North Pacific Ocean. The beluga (*Delphinapterus leucas*) inhabits the Arctic Sea and spreads into its adjacent seas, mainly in shallow shelf waters. The narwhal (*Monodon monoceros*) is present as a vagrant in the Chukchi Sea and Bering Sea.

Several delphinids found in the cold North Pacific are widespread species that mainly inhabit warm waters and are discussed in the next section. In addition, the cosmopolitan killer whale is present from equatorial regions to the polar pack ice. The long-finned pilot whale (*Globicephala melas*) represents a special case. This species is bipolar in temperate waters and, although it is common in the North Atlantic, there are no historical records of living animals in the North Pacific. However, skulls were recovered at two archeological sites in Japan, in areas now occupied by the short-finned pilot whale (*G. macrorhynchus*). Two other species present in this area are endemic to the North Pacific: the northern right-whale dolphin (*Lissodelphis borealis*), in subarctic and temperate waters, and the Pacific white-sided dolphin (*Lagenorhynchus obliquidens*) (Fig. 2), also in cool temperate areas. Other species of *Lagenorhynchus* occupy this same niche in the North Atlantic and Southern Hemisphere. However, studies have indicated that the genus might be polyphyletic, and the Pacific white-sided dolphin might indeed represent a different genus. If this were indeed the case, the correct scientific name for the Pacific white-sided dolphin would be *Sagmatias obliquidens*.

Four species of porpoises are present in the North Pacific. Dall's porpoise (*Phocoenoides dalli*) is endemic to the cold temperate waters of North Pacific. The recently discovered and critically endangered vaquita (*Phocoena sinus*), with only a few hundred individuals left, is endemic to the northern Gulf of California. Its closest relative (Burmeister's porpoise, *Phocoena spinipinnis*), curiously, does not live in the North Pacific, but in southern latitudes, indicating that the vaquita evolved independently in the Gulf of California when it became trapped there after crossing the equator during the last cooling period. The other two species present are the harbor porpoise (*Phocoena phocoena*), restricted to shallow coastal temperate waters, and the finless porpoise (*Neophocaena phocaenoides*), which inhabits coastal waters along the mainland of southern and southeastern Asia and Japan, as well as living up the Yangtze River system, up to 1600 km inland.

A member of the order Sirenia, the only herbivorous marine mammals, used to be present in the cold waters of the

Figure 2 *The Pacific white-sided dolphin* (Lagenorhynchus obliquidens), *thought closely related to a North Atlantic congener, may be a separate lineage and deserve a different genus name. Evidence on either side of this question can be interpreted in different ways. This dolphin is pelagic, is found in large herds, and occurs in colder waters of the Northern Pacific. Photo courtesy of NMFS.*

North Pacific. The Steller's sea cow, endemic to shallow sub-arctic waters of this ocean, has been extinct since 1768. Hopes for the survival of the species went up briefly in 1962, but the alleged sighting has been discredited.

Two other marine mammals in the North Pacific do not belong to the "traditional" marine mammal groups. One is the sea otter, which is endemic to the North Pacific and inhabits the cold temperate and subarctic zones. This species dipped to near EXTINCTION at the end of the 19th century and beginning of the 20th century. The other is the polar bear, which inhabits the pack ice regions of the Arctic Ocean and contiguous seas and adjacent coastal areas.

III. Marine Mammals of Warm Water Marine Ecosystems

The marine mammals that inhabit these ecosystems tend to be also present south of the equator, or even in other oceans. Among the otariids, two Pacific species occur only in tropical waters. The Galapagos sea lion (*Zalophus wollebaeki*) was originally confined to this archipelago, but a small rookery now exists on La Plata Island (01°S), which is why the species cannot be considered endemic to the North Pacific. The Galapagos fur seal (*Arctocephalus galapagoensis*) is also endemic to this equatorial archipelago.

Many of the cetaceans present in warm waters are widespread species, with wide latitudinal ranges, and have already been referred to in the previous section (Table I). This is the case of most rorquals, except for Bryde's whale (*Balaenoptera brydei*), which prefers tropical and warm temperate waters, and the small-type or pygmy Bryde's whale (*Balaenoptera edeni*), which lives in coastal and shelf waters of the eastern Indian Ocean and the western Pacific. Among the ziphiids, some are found exclusively in warm water. Such is the case of the gingko-toothed whale (*Mesoplodon ginkgodens*), which lives in tropical and warm temperate areas of the Indian and Pacific Oceans, and Blainville's beaked whale (*Mesoplodon densirostris*), also with a tropical and warm temperate distribution, but around the world. Most beaked whales, especially the genus *Mesoplodon,* are rarely spotted at sea and, even when they are, their identification is extremely difficult. It is very probable that some species have not been described yet. One such undescribed species of *Mesoplodon* has been sighted in the eastern tropical Pacific (*Mesoplodon* sp. A), and it is possible that large unidentified "tropical bottlenose whales" seen in the tropical Indian and Pacific Oceans are indeed Longman's beaked whales (*Indopacetus pacificus*), although this will remain a mystery until specimens are collected.

In addition to the widespread delphinids mentioned in the Section II, several species typically found in warm waters can also range into temperate waters. Probably the most familiar is the common bottlenose dolphin (*Tursiops truncatus*), which is present in temperate and tropical zones of all oceans. The species usually inhabits coastal areas and shallow offshore banks, although there are pelagic populations that live far offshore. The type specimen of the Indian Ocean bottlenose dolphin (*Tursiops aduncus*) comes from the Red Sea, but some *aduncus*-like *Tursiops* have been recorded from the western

Northern Pacific (Taiwan). The striped dolphin (*Stenella coeruleoalba*) is also found worldwide in temperate and tropical waters, as are Risso's dolphin (*Grampus griseus*) and the false killer whale (*Pseudorca crassidens*). Other species are characteristic of the tropical and warm temperate waters around the world and thus are also found in the North Pacific. Such is the case of the rough-toothed dolphin (*Steno bredanensis*), the pantropical spotted dolphin (*Stenella attenuata*), the spinner dolphin (*Stenella longirostris*), Fraser's dolphin (*Lagenodelphis hosei*), the melon-headed whale (*Peponocephala electra*), the pygmy killer whale (*Feresa attenuata*), and the short-finned pilot whale (*Globicephala macrorhynchus*). Both the short-beaked common dolphin (*Delphinus delphis*) and the long-beaked common dolphin (*Delphinus capensis*) (Fig. 3) inhabit tropical and warm temperate waters worldwide. However, while the former lives offshore, the latter is a coastal species. The Pacific hump-backed dolphin (*Sousa chinensis*) is distributed discontinuously in coastal waters of the western Pacific, as well as river basins. It is also present in the southern Pacific.

The dugong (*Dugong dugon*), the only sirenian found in warm North Pacific waters, has a widespread but discontinuous range along the continental coasts among the islands of the Indian and western Pacific oceans.

IV. Aquatic Mammals of River Ecosystems

The river basins, systems, and their mouths in southeastern Asia provide a unique habitat for several cetacean species (Table I). The most important hydrographic system for marine mammals is the Yangtze River, and several species can be found here. The baiji (*Lipotes vexillifer*) lives exclusively in fresh water and is endemic to this basin, from its estuary upstream for 1600 km. The Pacific hump-backed dolphin (*Sousa chinensis*) can ascend as far as 1200 km up this river system as well. The finless porpoise (*Neophocaena phocaenoides*) inhabits the coastal waters along the mainland of southern and southeast-

Figure 3 *The long-beaked common dolphin* (Delphinus capensis) *inhabits the North Pacific but is also found in disjunct populations in other warm waters around the world. The short-beaked common dolphin* (D. delphis) *is smaller. Common dolphins were not distinguished as separate species until the mid-1990s. Photo by R. S. Wells, Chicago Zoological Society.*

ern Asia and Japan and also ranges into the middle and lower reaches of the Yangtze River and its tributaries. Finally, the Irrawaddy dolphin (*Orcaella brevirostris*) is distributed discontinuously in shallow and murky waters at the mouth of rivers around the Asian mainland and Australia, as well as in river systems, such as the Mekong.

V. Conservation Issues

The problems faced by North Pacific marine mammals are common to marine mammals in other areas of the world: pollution, habitat degradation, fishery interactions, directed takes, and lack of sufficient resources for successful protection and MANAGEMENT. Although the hunting of marine mammals for oil, meat, or fur has been largely reduced by international agreements, some species are still targeted. The active harpoon fishery for the North Pacific endemic Dall's porpoise is still active and causes the largest fishery-related mortality for any cetacean species in the world. A large number of animals from several species are taken incidentally each year by net ENTANGLEMENTS, fishery bycatch, or boat strikes. Two North Pacific marine mammal species have been driven to extinction by humans in the recent past: the Japanese sea lion and the Steller's sea cow. Several other North Pacific species are dangerously close to following in their wake, as is the case for the vaquita, the baiji, the western population of the gray whale, and the North Pacific right whale. However, the recovery of other North Pacific species gives us reason for hope. The northern elephant seal, driven down to a few tens of individuals by hunting in the early 1900s, now numbers in the tens of thousands. The eastern gray whale, whose current population size is similar to its preexploitation stock size, was taken off the U.S. endangered species list in 1995. With continued protection, two species that once faced extinction are now showing positive signs of recovery: the sea otter and the Guadalupe fur seal. Because of their mobility and vast geographic ranges, the distributions of most marine mammal species extend across borders and international waters. For these reasons, effective management and conservation strategies depend greatly on international agreements that coordinate objectives and efforts.

See Also the Following Articles

Arctic Marine Mammals ▪ Biogeography ▪ Extinctions, Specific

References

Favorite, F., Dodimead, A. J., and Nasu, K. (1976). "Oceanography of the Subarctic Pacific Region." International North Pacific Fisheries Commission, Vancouver, Canada.

Green, G. A., Bonnell, M. L., Balcomb, K. C., Bowlby, D. E., Grotefendt, R. A., and Chapman, D. G. (1990). Synthesis of information on marine mammals of the eastern North Pacific, with emphasis on the Oregon and Washington OCS area. *In* "Oregon and Washington Marine Mammal and Seabird Surveys: Information Sythesis and Hypothesis Formulations" (OCS Study MMS 89-0030) (J. J. Bruggeman, ed.). U.S. Department of the Interior, Minerals Management Service, Pacific OCS Region, Washington, DC.

Leatherwood, S., Reeves, R. R., Perrin, W. F., and Evans, W. E. (1988). "Whales, Dolphins, and Porpoises of the Eastern North Pacific and Adjacent Arctic Waters: A Guide to Their Identification. Dover Publications, New York.

Longhurst, A. (1998). "Ecological Geography of the Sea." Academic Press, San Diego.

Nishiwaki, M. (1967). "Distribution and Migration of Marine Mammals in the North Pacific Area." Bulletin of the Ocean Research Institute University of Tokyo, No. 1. Tokyo, Japan.

Osborne, R. W., and van Beever, H. (1979). "Sightings of Marine Mammals in Eastern North Pacific Coastal Water." Ocean Research and Education Society, Boston, MA.

Rice, D. (1998). "Marine Mammals of the World: Systematics and Distribution" Society for Marine Mammalogy. Special Publication Number 4, Society for Marine Mammology, Lawrence, KS.

Seed, A. (1971). "Toothed Whales in Eastern North Pacific and Arctic Waters. Pacific Search Books, Seattle, WA.

Seed, A. (1972). "Baleen Whales in Eastern North Pacific and Arctic Waters." Pacific Search Books, Seattle, WA.

O

Ocean Environment

Paul C. Fiedler
*Southwest Fisheries Science Center,
La Jolla, California*

Marine mammals live, feed, and reproduce in a vast, three-dimensional fluid environment—the ocean. Air breathing necessitates frequent attendance to the sea surface for all marine mammals, while pinnipeds cannot reproduce in the water and have thus retained a close tie to land or ice. Despite these ties to the boundaries of the ocean environment, oceanography is an important part of the study of marine mammals. Habitat and distribution of marine mammals are affected by the physical and chemical properties of the water through which they swim and communicate, the topography of the ocean bottom where they feed, the physical state of the ocean surface where they breathe and haul out, and numerous factors influencing the distribution of food organisms.

I. Surface Temperature, Salinity

Temperature at the earth and ocean surface generally decreases from the equator to the poles (Fig. 1), but local processes complicate this simple pattern. Upwelling brings cold water to the surface along the equator and along eastern boundaries between oceans and continents. Ocean currents move cold water equatorward along these eastern boundaries and warm water poleward along western boundaries.

Salinity of the surface waters of the open ocean varies between 32 practical salinity units (psu) in the subarctic Pacific and 37 psu in subtropical gyres. This variation affects the density of seawater and thus the distribution of mass and the resulting thermohaline circulation. Salinity variations over this range have little or no physiological effect on marine mammals, but can influence the availability of food organisms through effects on stratification and circulation. At the coastal and polar limits of the ocean and in marginal seas, processes such as local precipitation and evaporation, river runoff, and ice forma-

tion can result in salinities less than 10 and greater than 40 psu. Marine mammals have adapted to tolerate even these extremes.

II. Surface Currents, Winds

Surface waters of the ocean are constantly in motion due to waves, tides, and currents. Currents move water across ocean basins and thus alter the distribution of temperature, salinity, and, indirectly, food organisms. Ocean currents are driven by energy from the sun, both by changes in the distribution of mass due to heating and cooling and by wind forcing at the surface. The distribution of solar radiation and rotation of the earth set up a basic pattern of winds consisting of (1) easterly trade winds over low latitudes, (2) westerlies over midlatitudes, and (3) polar easterlies. Heating and cooling of land masses and the seasonal cycle of solar energy input alter this basic pattern.

The surface circulation of most of the area of the world's oceans consists of subtropical gyres, which move anticyclonically (Fig. 2). Equatorial currents move from east to west, western boundary currents (Gulf Stream, Kuroshio) move warm water from near the equator toward the poles, eastward currents (Antarctic Circumpolar, North Pacific) move water back across the oceans on the poleward side of the gyres, and eastern boundary currents (California, Peru, Canary, Benguela) move cold water toward the equator.

The strength of these currents varies seasonally. Eastern and western boundary currents spin off eddies, on the order of 100 km in diameter, which move both water and prey organisms and alter the distribution of favorable habitat for marine mammals. Eastern boundary currents are sites of seasonal coastal upwelling, when equatorward winds move surface water offshore and deep nutrient-rich water is brought to the surface. Westward equatorial currents are also sites of upwelling, due to divergence of surface water caused by the rotation of the earth.

III. Vertical Structure

Physical (and biological) variability in the ocean environment is generally much greater in the vertical than in the horizontal dimension. Water temperature may be nearly constant for a few meters or tens of meters below the surface in the mixed layer but then decreases rapidly with depth in the thermocline. Vertical changes in salinity may contribute to this

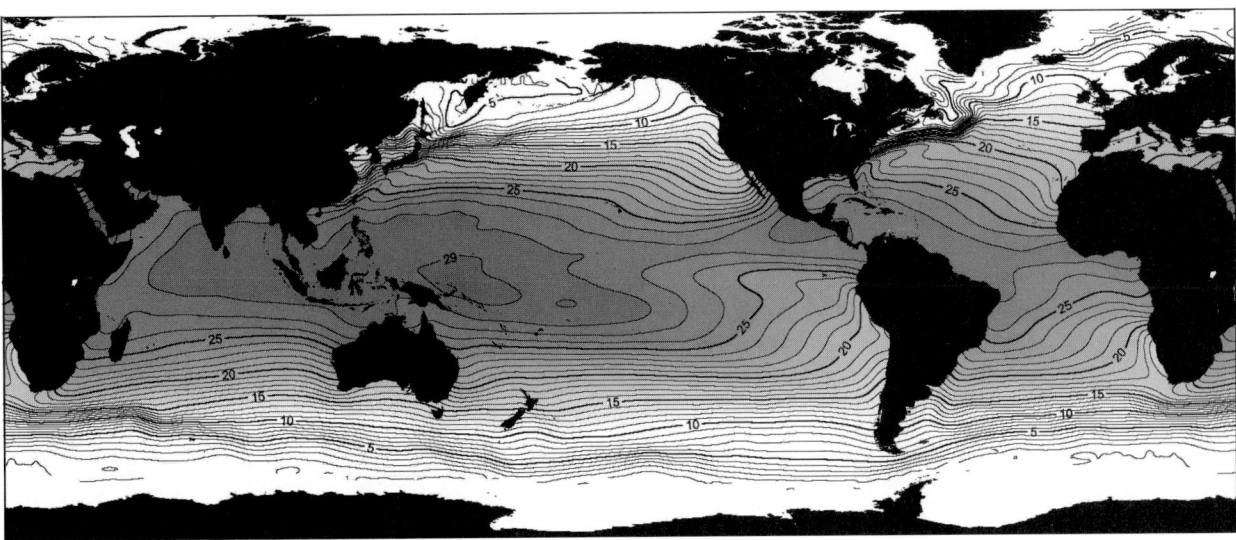

Figure 1 *Mean sea surface temperature (°C).*

stratification, especially in polar seas. Stratification influences productivity because deeper, colder water has higher nutrient concentrations than the warmer water near the surface. Stratification also influences the distribution of food organisms.

IV. Productivity

Biological productivity is the rate of production of living matter; oceanographers often speak of "production," which is the variable standing stock of living organisms and is closely linked to productivity. Primary productivity in the ocean is the rate of production of plants (phytoplankton) and is often limited by light and nutrient availability, both of which are controlled by physical oceanographic processes. Light intensity at the sea surface varies with latitude and time of year, as well as time of day. Thus, photosynthesis occurs in polar seas only during summer. Light intensity decreases rapidly with depth, so

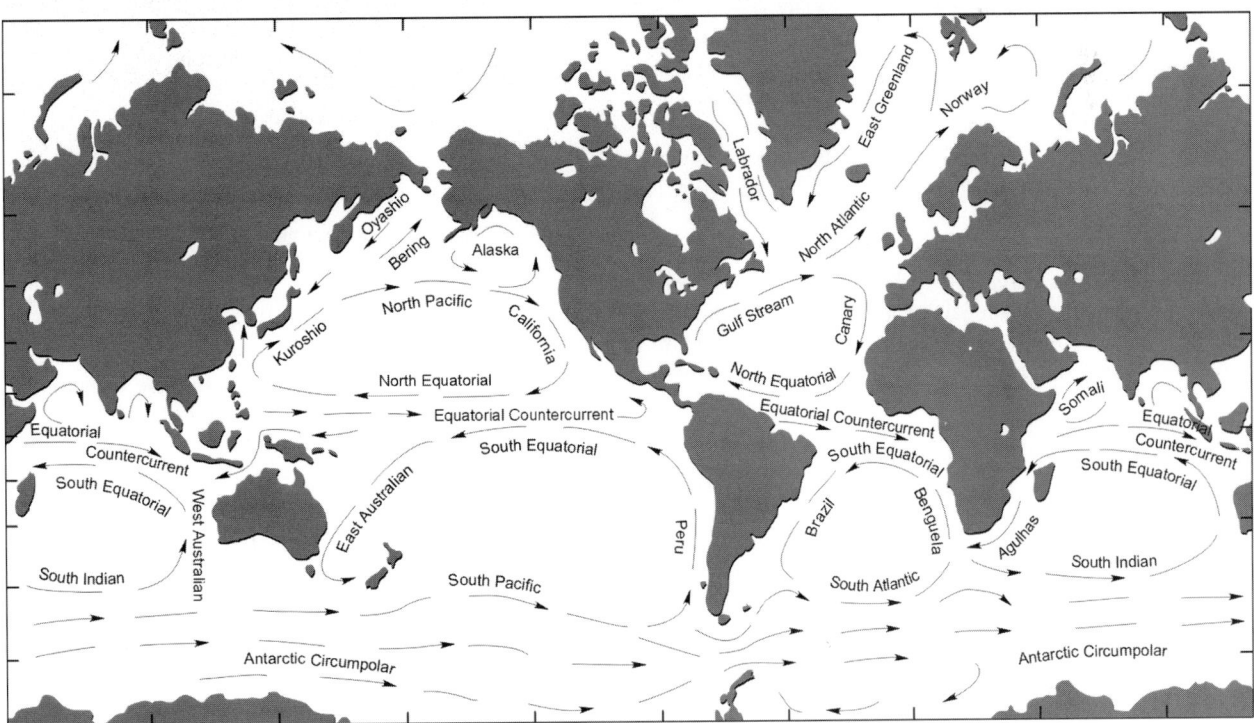

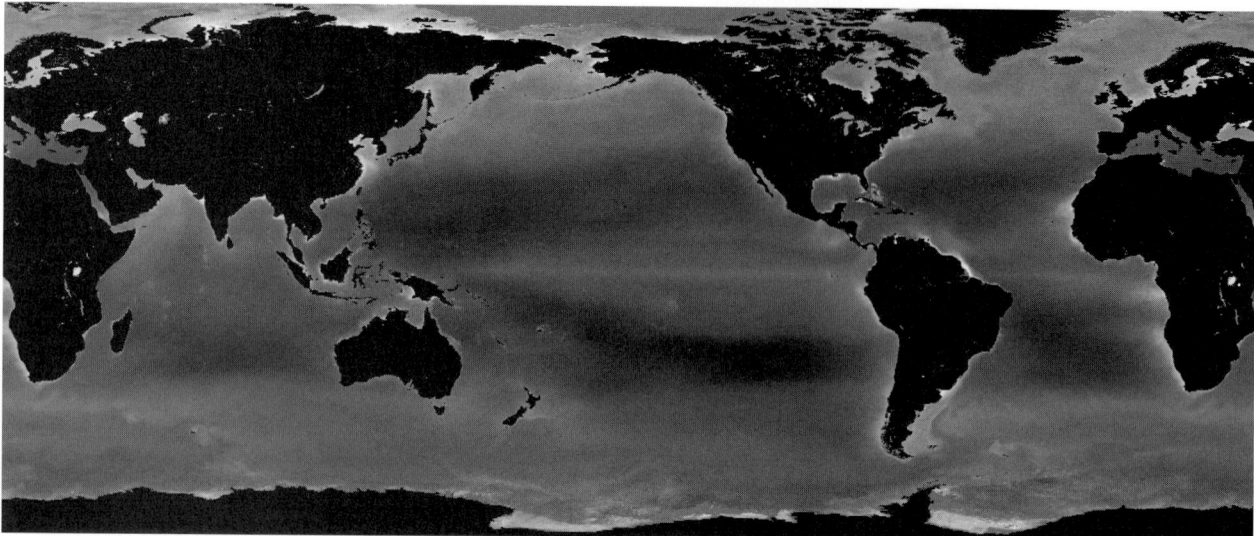

Figure 3   *Mean phytoplankton pigment concentration from SeaWiFS satellite data, 1998–1999, increasing from darker to lighter grays; uniform light gray near the poles indicates missing data.*

that photosynthesis is possible within only a few meters or tens of meters of the sea surface (Fig. 3).

Plant growth also requires nutrients (nitrate, phosphate, and, for some phytoplankton, silicate). The importance of micronutrients—notably iron—in some regions of the open ocean, such as the Southern Ocean around Antarctica where many baleen whales feed, has been recognized in recent years. Nutrient availability is affected by oceanographic processes including stratification, wind and tidal mixing, circulation, and upwelling.

Marine mammals consume zooplankton, fish, squid, and even other marine mammals. Therefore, distribution and foraging are not directly determined by primary production. Food chains involving marine mammals are as short as diatoms–krill–baleen whales. Sperm whales (*Physeter macrocephalus*), however, consume large deep-living squid and are at least four steps removed from primary production at the sea surface.

V. Ice

Ice is an important component of the habitat of migratory and endemic marine mammals in both the Arctic and the Southern Ocean. Ice cover varies seasonally with temperature and is also subject to the influence of winds and currents. There must be some open water in any ice-covered habitat utilized by marine mammals to allow access to both air and food. Ice on the sea surface is found in a variety of types (pack ice, icebergs, shore ice, fast ice, drift ice, new ice) and forms (small floes, brash ice, pancake ice, etc.).

Sea ice in the Arctic is tightly packed except at the ice edge. Much of the Arctic Ocean is permanently ice covered. The Antarctic has a broad zone of looser pack ice with many internal leads. Most of this ice melts during summer. Polynyas are areas within the pack ice that are almost always clear of ice. Providing both access to the water and locally enhanced productivity, they are relatively more important in the Arctic than in the Antarctic.

VI. Temporal Variability

The ocean environment varies seasonally, especially at higher latitudes where ice closes accessibility to most species during winter. Seasonal variability in wind-driven upwelling causes significant changes during the year even in the tropics. The ocean environment varies between years as well. The cycle in the ocean–atmosphere system known as the El Niño Southern Oscillation (ENSO) or simply "El Niño" causes major changes in the tropics every 3 to 5 years. Both seasonal and interannual variability occur in all marine mammal habitats. However, the relative magnitude of variability at each scale varies (Fig. 4).

Marine mammals have experienced climate change—variations in the ocean environment over periods of hundreds to thousands of years—throughout their evolutionary history. Concern is now focused on the possible effects of rapid global warming caused by mankind's input of excess carbon dioxide into the atmosphere. Scientists generally, although not universally, agree that by 2100 the global average temperature will increase by 1 to 3°C, the sea level will rise by several tens of centimeters, and sea ice coverage will be reduced. Changes in environmental variability on shorter time scales, such as the frequency or magnitude of El Niño events, the amplitude of seasonal cycles, or the frequency or intensity of storms, are possible. Regional changes in precipitation, circulation and upwelling, and UV-B (ultraviolet) radiation are expected.

VII. Effects on Life History and Function

Temperature can be a critical factor in the energy budget of warm-blooded mammals. However, marine mammals live and breed in the polar ice and even small, young harp seals (*Pagophilus groenlandicus*) are able to tolerate freezing temperatures. Some baleen whale species tend to be slightly larger in the cold Southern Ocean than in the warmer North Pacific. Adaptations, including an insulating BLUBBER layer and an ability to reduce

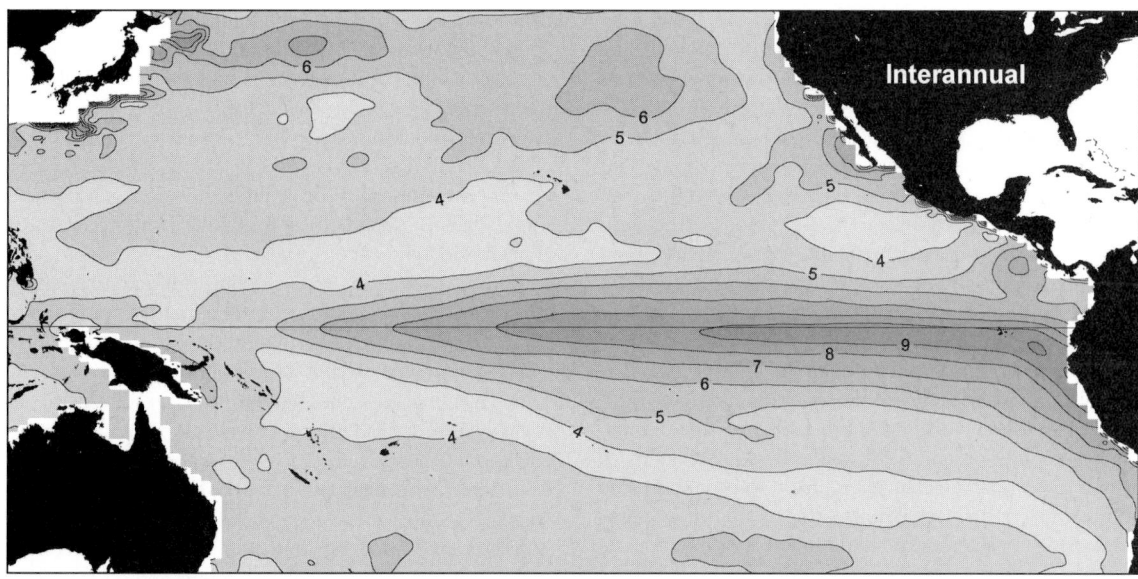

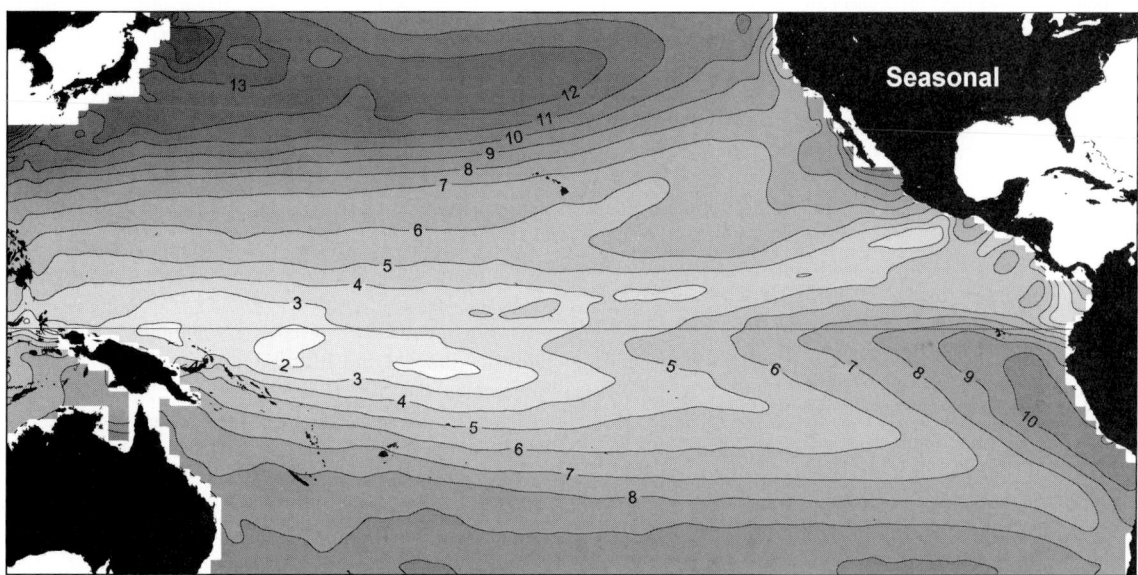

Figure 4 *Approximate interannual and seasonal ranges in sea surface temperature in the temperate and tropical Pacific in °C.*

blood circulation to peripheral parts of the body, have reduced the direct influence of temperature on distribution.

It has long been believed that large whales migrate from summer polar FEEDING grounds to warmer tropical breeding grounds because of adverse effects of low temperature on growth and survival of neonates. However, this hypothesis is no longer supported by considerations of energy intake, insulation by blubber, and heat loss. Even though the energy cost of migration may be very low for large whales, adult whales do not optimize their energy budget by migrating to warm tropical or temperate waters.

Cetaceans use underwater sound in several ways. Odontocetes use high-frequency sound to locate prey individuals and patches over short distances. Mysticetes use low-frequency sig-

nals for navigation and COMMUNICATION. While the effects of temperature, salinity, and pressure on the propagation of sound in water are well known, it is not known whether oceanographic variability can affect the use of sound by marine mammals. If blue and fin whales (*Balaenoptera musculus* and *B. physalus*) use the low-frequency sound channel of the deep ocean to acoustically "visualize" their environment, variations in water properties very likely distort or obscure the image they can obtain.

VIII. Effects on Feeding

Marine mammals are generally apex predators, or at the top of the food chain, although some of them are preyed on by killer whales (*Orcinus orca*) and leopard seals (*Hydrurga leptonyx*). Because they are motile, social, and intelligent, they are able to take advantage of prey in locally dense patches. Like all aquatic and terrestrial predators, marine mammals cannot forage efficiently on average concentrations of prey. It has been shown that right whales (*Eubalaena* spp.) and humpback whales (*Megaptera novaeangliae*) require concentrations of prey that are more dense than the maximum densities observed by net sampling. This is an indication of the limitations of oceanographic sampling, as well as the paucity of prey in the ocean environment. Sampling by acoustic backscattering from echo sounders and by photography with tethered cameras has not revealed small patches of prey of the required densities.

Prey consumed by marine mammals include crustaceans (copepods, euphausiids or krill, amphipods, shrimp), cephalopods (squid), and schooling fish (herring, capelin, cod, mackerel, myctophids or lanternfish, and others). Availability of these prey in dense patches has been linked to oceanographic features, including bathymetry, fronts, eddies, and primary productivity. Except for bathymetry, these features or characteristics of the ocean environment vary over time.

Many baleen whales feed on dense patches of zooplankton or smaller fishes. The distribution of feeding North Atlantic right whales (*E. glacialis*) can be predicted by surface temperature and bathymetric variables, including depth and slope. These are the simplest oceanographic variables to measure, but they explain the distribution of whales only indirectly through effects on prey. Zooplankton patches are available in productive coastal waters, where upwelling and wind mixing results in high primary production, and in polar seas, where high primary productivity results from summer insolation and enhanced phytoplankton growth at the melting ice edge and even under the ice. The patches are often located at bathymetric features such as the shelf edge, islands, and seamounts, because bathymetry affects circulation and production and serves as a cue for aggregation of krill and fish. Blue whales feed on KRILL at the ice edge in the Antarctic, but also at the shelf edge off California (Fig. 5).

Along the northeast United States continental shelf, cetaceans tend to frequent distinct regions based on food preferences. Piscivores—humpback, fin, and minke (*B. acutorostrata*) whales and bottlenose (*Tursiops truncatus*), Atlantic white-sided (*Lagenorhynchus acutus*), and common dolphins (*Delphinus delphis*)—are most abundant over shallow banks in the western Gulf of Maine and midshelf east of Chesapeake Bay. Planktivores—right, blue, and sei (*B. borealis*) whales—are most abundant in the western Gulf of Maine and over the western and southern portions of Georges Bank. Teuthivores (squid eaters)—sperm and long-finned pilot (*Globicephala melas*) whales and Risso's dolphins (*Grampus griseus*)—are most abundant at the shelf edge. Most of these patterns can be related to the availability of specific prey. Right whales feed on dense patches of copepods in the vicinity of the Great South Channel. Humpbacks and fin whales feed on a small schooling fish, the American sand lance, which is very abundant on Stellwagen Bank and Jeffreys Ledge in the Gulf of Maine and is also abundant off Chesapeake Bay.

Cetaceans are distributed over distinct depth ranges along the continental slope in the northwestern Gulf of Mexico. Deep-diving teuthivores, such as Risso's dolphins, sperm whales, and

beaked whales (*Ziphiidae*), are found near temperature fronts in deep water, where their squid prey may aggregate. Sperm whales are more abundant in regions of high zooplankton biomass and steep topography in the Pacific Ocean, where deep-diving squid are aggregated. Most of the 19th-century sperm whaling grounds in the Pacific were in regions of coastal or oceanic upwelling where primary production is high. However, the occurrence of some whaling grounds in unproductive waters, such as the region around Tahiti, and the absence of catches in productive waters, such as the California upwelling, indicate that other factors affect the availability of prey to whales and/or whales to whalers.

In the Antarctic, both Antarctic fur seals (*Arctocephalus gazella*) and minke whales (*Balaenoptera bonaerensis*) are found in the marginal ice zone, where primary productivity and, presumably, prey availability are enhanced. Winter pack ice extends out beyond the narrow Antarctic continental shelf and over the deep ocean. Minke whales, beaked whales, and several species of seals live in this ice and feed on krill (the euphausiid *Euphausia superba*), myctophid fish, and squid. As the pack ice retreats in summer, many of these winter residents and other predators, including large baleen whales, male sperm whales, and killer whales follow the ice edge to feed on abundant plankton and nekton. Right whales, feeding on smaller copepods not concentrated at the ice edge, remain in open waters north of the retreating ice.

Ice is an important habitat component for breeding and shelter of pinnipeds. Antarctic ice floes in the marginal ice zone are a refuge for fur seals (*Arctocephalus gazella*) from predation by killer whales. In contrast, crabeater seals (*Lobodon carcinophaga*), are found in the interior of the pack ice where larger, more stable floes provide BREEDING SITES.

In the Arctic, gray (*Eschrichtius robustus*), bowhead (*Balaena mysticetus*), and beluga whales (*Delphinapterus leucas*) migrate into the Chukchi and Beaufort Sea north of Alaska to feed in summer and fall. Bottom-feeding gray whales feed in shallow, ice-free waters; piscivorous belugas feed in deep water near the ice edge; and planktivorous bowhead whales feed closer to the coast where ice cover is less. These distribution differences are clearly related to prey availability. River plumes formed by summer runoff into the Arctic have been shown to limit the distribution of prey and thus determine the distribution of foraging whales.

IX. Effects on Distribution and Migration

The distribution of baleen whales on summer feeding grounds and movement on these feeding grounds are associated with oceanographic features, including fronts between water masses, eddies along these fronts, eddies caused by circulation around islands and capes, and oceanic and coastal upwelling. Prey availability is enhanced by nutrient input and enhanced primary production, or by convergence and aggregation of plankton. Japanese whalers have exploited concentrations of (1) sei whales off Japan where the warm Kuroshio current meets the cold Oyashio, (2) fin whales in a cold cyclonic eddy and in a mixing zone between cold coastal waters and warmer Bering Sea waters in the western subarctic Pacific, (3) fin whales at shelf edge fronts and upwelling zones off islands in the eastern Bering Sea, and (4) blue and fin whales at the polar front north of the Antarctic pack ice.

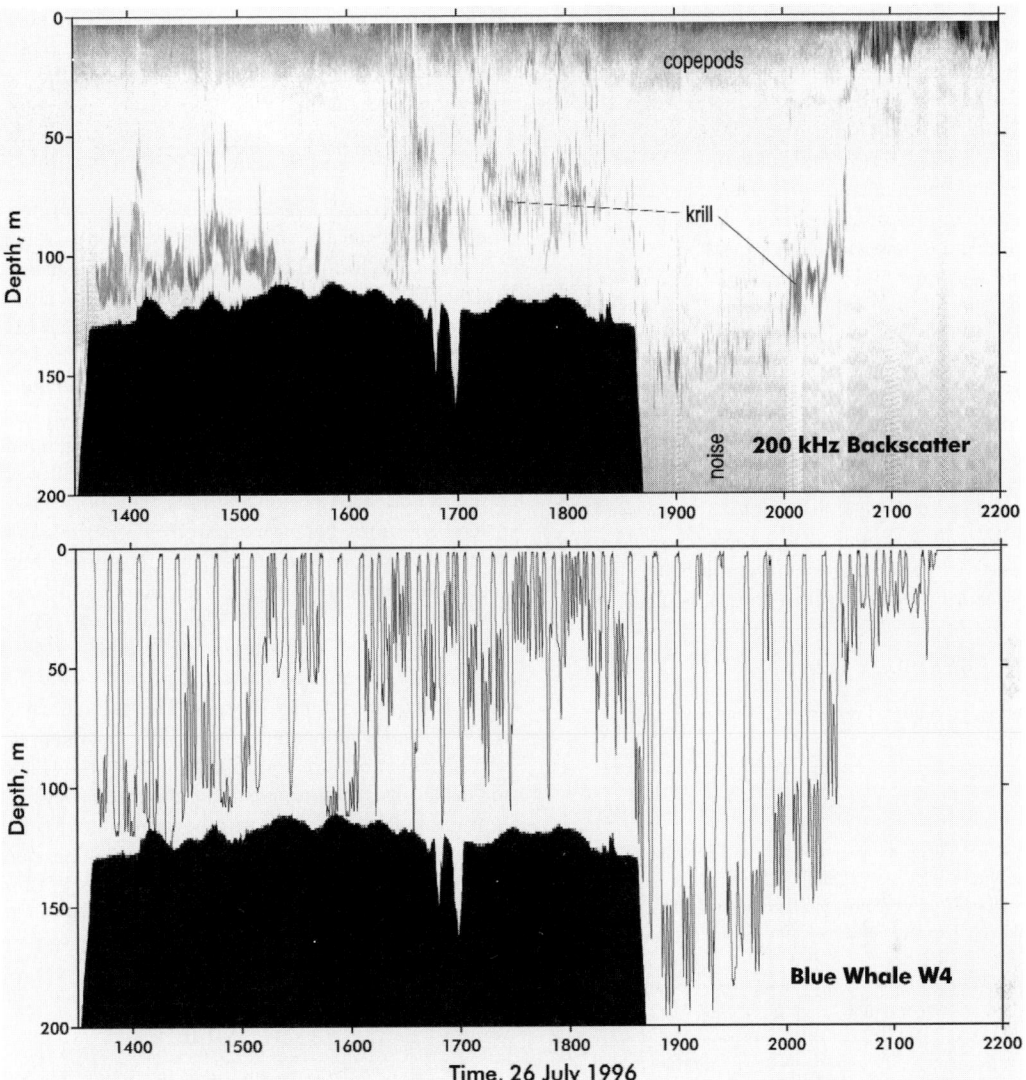

Figure 5 *Track of a blue whale (bottom) diving into and feeding on a layer of krill (top) at the shelf edge off Santa Rosa Island, California. Reprinted from* Deep-Sea Research II, *45, Fiedler, P. C., et al.,* Blue whale habitat and prey in the California Channel Islands, *pp. 1781–1801, Copyright 1998, with permission of Elsevier Science.*

Relatively little is known about the migration routes of most baleen whales between summer feeding grounds and winter breeding grounds. Gray whales migrating south along the California coast remain near the coast and apparently use bottom topography for navigation. Humpback whales have been tagged off Hawaii and observed to linger at seamounts in the middle of the North Pacific on the way to summer feeding grounds in the Gulf of Alaska. Seamounts are known to be sites of aggregation of fish. These whales may be feeding, regrouping, or simply navigating.

X. Effects of Climate Change

Direct effects of temperature increase are unlikely for most marine mammals because of their mobility and thermoregula-

tory ability. However, direct effects could occur for populations at the limits of their range or with physical limits to migration. For example, the vaquita (*Phocoena sinus*) is a rare species trapped at the warm northern end of the Gulf of California. Other species, such as the beluga whale, with low skin pigmentation and living in a polar region subject to atmospheric ozone depletion, may suffer direct effects of increased UV-B radiation.

Indirect effects on marine mammal populations will occur through alterations of habitat characteristics and distribution, and prey availability. Global warming is predicted to be greatest at high latitudes in the Northern Hemisphere. The resulting reduction of ice cover in the Arctic will affect ice-associated species through changes in the distribution of critical habitat such as polynyas and the ice edge. Seasonal and geographical shifts in prey availability in both the Arctic and the Antarctic may have a variety of indirect effects: nutritional status and reproductive success, geographic range, and timing or patterns of migration. Bowhead whales and narwhals (*Monodon monoceros*), and especially beluga whales, which feed close to the ice edge, will be affected by reduced ice extent in the Arctic. Warming could also reduce the extent of pack ice in the Antarctic and thus affect the distribution and abundance of krill. Declining krill abundance in the region of the Antarctic Peninsula during the early 1990s has been linked to low winter sea ice extent.

Global warming is expected to be less intense at lower latitudes. However, reduction in wind-driven coastal and equatorial upwelling is possible and may reduce prey availability for marine mammals. A decline in zooplankton biomass off California since 1951 has been linked to warming of surface waters, increased stratification, and reduced mixing and upwelling of nutrients. Odontocetes that depend on squid, such as sperm whales, may be affected more adversely because cephalopod populations are known to be highly variable. A rising sea level will change the lagoons and shelf areas where gray whales and humpbacks breed, but past changes in these breeding sites are known to have occurred over longer periods of geologic time.

See Also the Following Articles

Diet ▪ Distribution ▪ Habitat Pressures ▪ Migration and Movement Patterns

References

Gaskin, D. E. (1982). "The Ecology of Whales and Dolphins." Heinemann Educational Books Ltd., London.

International Whaling Commission (1997). Report of the IWC workshop on climate change and cetaceans. *Rep. Int. Whal. Commn.* **47**, 293–319.

Kenney, R. D., and Winn, H. E. (1986). Cetacean high-use habitats of the northeast United States continental shelf. *Fish. Bull.* **84**(2), 345–357.

Tynan, C. T., and DeMaster, D. P. (1997). Observations and predictions of arctic climatic change: Potential effects of marine mammals. *Arctic* **50**(4), 308–322.

Odobenocetops

CHRISTIAN DE MUIZON

National Museum of Natural History, Paris, France

One of the most unexpected recent discoveries in marine mammal paleontology is that of a new family and genus of tusked cetacean—Odobenocetopsidae, *Odobenocetops*—from the early Pliocene beds of the Pisco Formation of Peru (Muizon, 1993; Muizon *et al.*, 1999). As its name indicates, the new marine mammal presents several similarities with the living walrus, *Odobenus rosmarus*. However, *Odobenocetops* is clearly a cetacean and presents a startling and unprecedented example of convergence with the walrus in its cranial morphology and inferred feeding habits (Muizon,

1993; Muizon *et al.*, 1999). *Odobenocetops* is known from two species: *O. peruvianus*, Muizon, 1993, and *O. leptodon*, Muizon, Domning, and Parrish, 1999, with the former being slightly older than the latter.

I. Descriptive Anatomy

Odobenocetops (Figs. 1–3) has lost the elongated rostrum of the other cetaceans. Instead, the rostrum is short, rounded, and blunt and formed by the premaxillae, which are greatly enlarged. They have large, asymmetrical ventral alveolar sheaths holding sexually dimorphic tusks. The right tusk of the male is large and can reach 1 m or more in length. The left tusk is approximately 25 cm long, of which a few centimeters only were erupted. Both tusks are straight. In the female, both tusks approach the size of the left tusk of the male, although the right tusk is slightly larger than the left. Premaxillary sheath and tusks are oriented posteroventrally. The bony nares have displaced anteriorly (when compared to other odontocetes). The palate is very deep and arched, and its anterior border is U to V shaped. *Odobenocetops* lacks maxillary teeth. The orbits face anterolaterally and dorsally. The portions of the frontal and maxillae, which cover the temporal fossae in other odontocetes, have been reduced and narrowed in such a way that the temporal fossae are opened dorsally. The glenoid cavity is an anteroposteriorly elongated groove. The occipital condyles are very salient and more convex than in any other Pliocene or living cetaceans. The periotic and tympanic have the characteristic morphology observed in other delphinoid cetaceans. The mandible is unknown. The postcranial skeleton is poorly known. The atlas is shortened anteroposteriorly as in other odontocetes, and the forelimb has the basic pattern of that of other odontocetes. The length of the body could have ranged from 3 to 4 m.

II. Affinities

Despite its extremely modified morphology, several characters of the skull clearly relate *Odobenocetops* to the delphinoid cetaceans. Cetacean characters of *Odobenocetops* include the presence of large air sinuses in the auditory region connected to well-developed pterygoid sinuses; the thickened and pachyostotic tympanic bulla; and the dorsal opening of the narial fossae. Odontocete affinities of *Odobenocetops* are attested by maxillae covering the supraorbital processes of the frontals, although they are partially withdrawn secondarily; the dorsoventral expansion of the pterygoid sinuses; the presence of large premaxillary foramina (in *O. peruvianus* only); and the asymmetry of premaxillae and maxillae in the facial region. The sigmoid morphology of the dorsomedial view of the involucrum and the presence of a medial maxilla–premaxilla suture at the anterolateral edge of each narial fossa are delphinoid characters (Muizon, 1988). Several characters of the pterygoid, alisphenoid, and temporal fossa indicate close affinities of *Odobenocetops* and the Monodontidae (Muizon, 1993; Muizon *et al.*, 2000). In view of the exceptional specializations of *Odobenocetops*, it was referred to as a new odontocete family, the Odobenocetopsidae, regarded as the sister group of the Monodontidae.

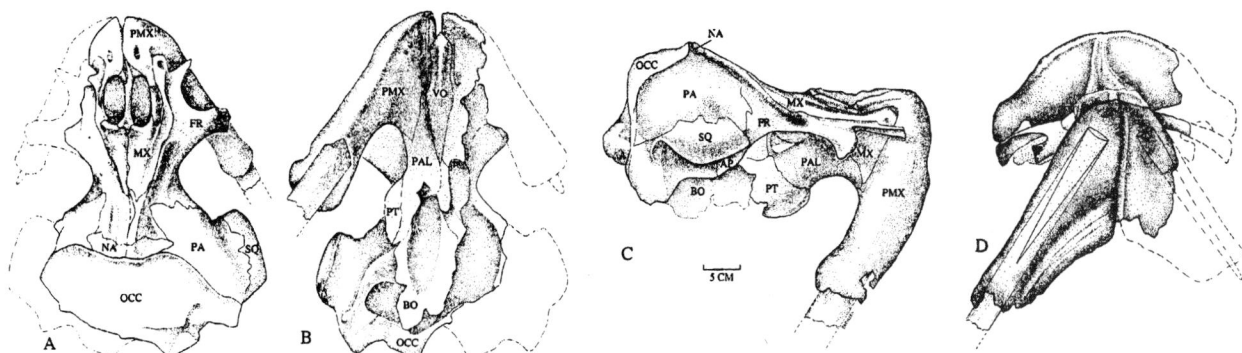

Figure 1 Odobenocetops peruvianus: *Skull of a male in dorsal (A), ventral (B), lateral (C), and anterior (D) views. Scale bar: 5 cm. AP, alisphenoid; BO, basioccipital; FR, frontal; MX, maxilla; NA, nasal; OCC, occipital; PA, parietal; PAL, palatine; PMX, premaxilla; PT, pterygoid; SQ, squamosal; VO, vomer. Modified from Muizon (1993), reprinted with permission from* Nature *(365: 745–748) copyright (1993) Macmillan Magazines Ltd.*

The occurrence of tusks in *Odobenocetops* is a convergence with *Monodon* as in the latter genus the large tusk of the male is implanted in the left maxilla, whereas in *Odobenocetops* it is implanted in the right premaxilla. Consequently, the tusks of *Monodon* and *Odobenocetops* are not homologous.

III. Functional Anatomy and Habits
A. Feeding Adaptations

Odobenocetops and the living walrus show several morphogical convergences: a large and deep palate; strong development of a wide, blunt snout, which is highly vascularized and innervated with strong muscular insertions, suggesting the presence of a powerful tactile upper lip; the presence of tusks; and a reduction of maxillary dentition (postcanine maxillary teeth are absent in the fossil walrus *Valenictus chulavistensis*). The living walrus feeds mainly on benthic invertebrates (e.g., *Mya, Clinocardium, Tellina, Spisula, Macoma*) (Fay, 1982), sucking out the siphon and foot of bivalves or gastropods shells, using

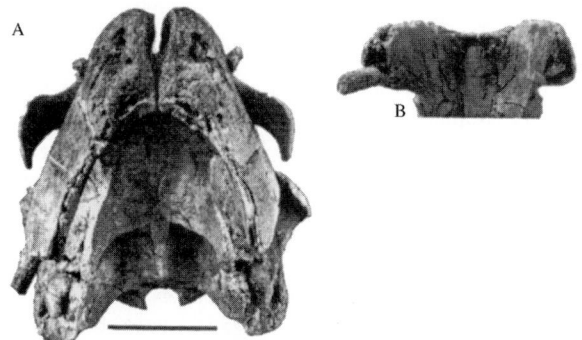

Figure 2 Odobenocetops peruvianus: *skull of a female in anteroventral view (A); ventral view of the anterior region of the palate of the same skull (B). Scale bar: 5 cm. From Muizon et al. (1999), reprinted with permission from the* Comptes Rendus de l'Académie des Sciences, Série IIA, *329: 449–455) copyright (1999) Elsevier.*

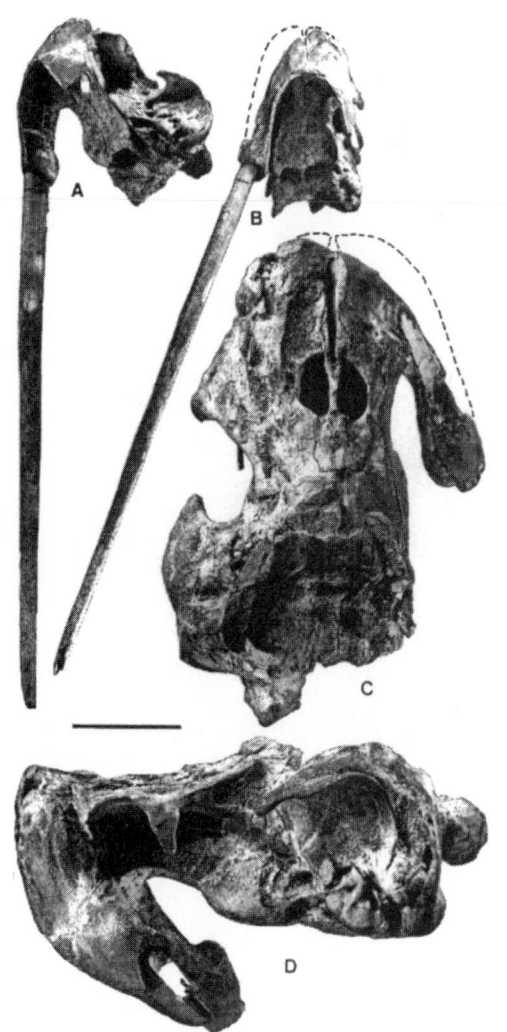

Figure 3 Odobenocetops leptodon: *skull of a male in left ventrolateral (A), anteroventral (B), dorsal (C), and left lateral (D) views. Scale bar: 20 cm (A, B); 10 cm (C, D). From Muizon et al. (1999), reprinted with permission.*

the tongue as a piston powered by the very large lingual retractors and depressors. Their mouth works like a powerful vacuum pump. Given the anatomy of its palate, it is likely that *Odobenocetops* presented a similar feeding adaptation. This adaptation was probably even more efficient in *O. leptodon* because its palate is longer than in *O. peruvianus*. *Odobenocetops* therefore probably fed on benthic invertebrates (bivalves and gastropod mollusks and/or crustaceans), which are abundant in the Pisco Formation at Sacaco. Because of the great vascularization of the inferred upper lip, it is possible that, as in the walrus, *Odobenocetops* had vibrissae, which would have had an important tactile role in the search for food. When feeding on the sea floor, the body of *Odobenocetops* was probably held in an oblique position as is observed in the living walrus (Fig. 4).

B. Vision vs Echolocation

VISION and HEARING were probably slightly different in the two species of *Odobenocetops*. Because of the shortening of the rostrum and the very short space between the premaxillary foramina and the nares, it is probable that the premaxillary air sacs in *O. peruvianus* were absent, unlike in other odontocetes. Furthermore, a melon was also probably absent or very reduced. In contrast, in *O. leptodon*, fossae for premaxillary sacs are clearly

Figure 4 *Reconstruction of* Odobenocetops leptodon *(males) in swimming and feeding position. The third individual in the background is in ventrolateral view from the left side in order to show the erupted small left tusk. From Muizon* et al. *(1999), reprinted with permission. Painting by Mary Parrish (Smithsonian Institution, Washington, DC).*

present on the premaxillae, just anterior to the nares as are observed in other odontocetes. Furthermore, the dorsal surface of the rostrum is longer and wider and there is enough space for a small melon anterior to the nares. The cetacean melon is related to echolocation and, consequently, *O. peruvianus* was probably devoid of echolocating ability, whereas some echolocating ability may have been present in *O. leptodon*.

The anterodorsal edge of the orbit (i.e., the anterior border of the supraorbital process of the frontal) in *O. peruvianus* is deeply notched, allowing good anterodorsal binocular vision. In contrast, this region of the frontal in *O. leptodon* is only slightly concave, which reduced binocular vision.

The probable lack of ECHOLOCATION ability in *O. peruvianus*, was probably compensated for by good anterodorsal binocular vision. As in the walrus, vision was especially useful when the animal was searching for food on the sea flood with the body in an oblique position. In *O. leptodon*, limited binocular vision was compensated for by echolocation ability as in other odontocetes.

C. Tusk, Position and Function

The posteroventrally oriented tusks are certainly one of the most unusual characteristics of *Odobenocetops*. Because of its great length and slenderness, the right tusk of the male was probably very fragile and it was unlikely that the animal swam with the tusk held at a 45° angle with the axis of the body. Comparison of the occipital condyles to the anterior articular facets of the atlas indicates that, when swimming, the dorsal plane of the skull was oriented anterodorsally and that the tusk was approximately parallel to the axis of the body. When feeding, the large tusk was dragged on the sea floor, and because of the oblique position of the body, the tusk was held at an angle of 20° to 40° with the axis of the body (Fig. 4). In other respects, the convexity of the occipital condyles of *Odobenocetops* indicates that its neck was extremely mobile, allowing a degree of head movement approaching 90° or more.

The narwhal tusk is mainly used in social activities. The orientation and length of the large tusk of *Odobenocetops* suggest that this tooth was very fragile and was unlikely to have been used to exert forces (in digging or fighting). The probable SEXUAL DIMORPHISM in the size of the tusks of *O. peruvianus* points to a nonviolent social role (Muizon *et al.*, 2001). In addition, the tusks of walruses and the premaxillary tusks and sheaths of *Odobenocetops* may play a role in bottom feeding as orientation guides to keep the mouth and any vibrissae (if present) in constant position relative to the substrate.

Although suction feeding has been increasingly recognized among odontocetes (e.g., *Tursiops, Globicephala, Phocoenoides, Phocoena, Monodon, Delphinapterus, Physeter, Mesoplodon, Berardius*), the extreme modifications of *Odobenocetops* are unique among the Cetacea and suggest concomitant expansion of the adaptive breadth of the order. *Odobenocetops* occupied, in the Southern Hemisphere, an ecological niche otherwise unknown among odontocetes that was similar to that of the odobenine walruses, which were widespread in the Northern Hemisphere during the Pliocene.

See Also the Following Articles

Cetacean Evolution ▪ Convergent Evolution ▪ Narwhal ▪ Walrus

References

Fay, F. H. (1982). "Ecology and Biology of the Pacific Walrus," *Odobenus rosmarus divergens* Illiger." U.S. Department of Interior; Fish and Wildlife Service. North American fauna, Number 74, 1–279.

Muizon, C. de (1988). Les relations phylogénétiques des Delphinida. *Ann. Paléontol.* **74**, 115–183.

Muizon, C. de (1993). Walrus-like feeding adaptation in a new cetacean from the Pliocene of Peru. *Nature* **365**, 745–748.

Muizon, C. de, Domning, D. P., and Ketten, D. R. (2001). *Odobenocetops peruvianus,* the walrus-convergent delphinoid (Cetacea, Mammalia) from the lower Pliocene of Peru. *Smith. Contrib. Paleobiol.,* in press.

Muizon, C. de, Domning, D. P., and Parrish, M. (1999). Dimorphic tusks and adaptive strategies in the Odobenocetopsidae, walrus-like dolphins from the Pliocene of Peru. *C. R. Acad. Sci. Paris Sci. Terre Planètes* **329**, 449–455.

Origins of Marine Mammals

MICHAEL A. TAYLOR
*National Museums of Scotland,
Edinburgh, United Kingdom*

W hales, pinnipeds, and sirenians are examples of a major evolutionary phenomenon: the dramatic transformation of terrestrial amniotes (reptiles, birds, and mammals) to a more or less fully aquatic lifestyle. Indeed, marine mammals are far from unique, for this has happened several tens of times in evolutionary history. The whales are ecological replacements for the extinct ichthyosaurs, which were reptiles as fish-like as whales. The sea lions have a body form and swimming method broadly similar to those of the penguins, the extinct great auk (*Pinguinus impennis*) and plesiosaurs. Other examples include the plotopterid birds, mosasaurs (aquatic lizards), at least two lineages of marine crocodilians, and the mollusc-eating placodont reptiles that were analogous to the modern sea otter (*Enhydra lutris*). Yet, however many times it has happened, it is still a huge transition from a terrestrial animal, adapted to live, feed, and breed on land, to an animal such as a whale, which is so notoriously fish-like that its relationships puzzle nonbiologists. Indeed, how whales could have arisen from land mammals was a question routinely posed by opponents of Darwin's "Origin of Species," and he himself regarded it as a crucial issue.

The problem can be approached today by asking what phylogenetic route the transition took: what are the actual relationships? Then what can we say about the animals that were the actual intermediates? What did they look like and in what habitats did they live? What was, in fact, the ecological route of the transition? And, in parallel, does the evidence make functional sense that corroborates the hypothesis? This article focuses on general issues common to all lineages.

I. How the Transition Can Be Studied

It is a central tenet of Darwinian evolutionary theory that relatively gradual changes can lead to major transitions. Under natural selection, a species can gradually change if this is advantageous (more correctly, advantageous through increasing the frequency of the individual gene alleles under selection). The degree of adaptation need not be perfect. An animal only has to be sufficiently well adapted to gain some advantage; even quite a poor degree of adaptation can be enough. For a land animal, the occasional ability to swim, however poorly, can be critical in escaping predators or locating new food resources. Furthermore, adaptation is "fuzzy": animals have some flexibility in how they use their adaptations, allowing natural selection to start driving change in either direction. There is almost no such thing as a purely land mammal, for most mammals can at least swim in an emergency. This provides the potential basis for the evolution of a more aquatic lifestyle.

The transition from a land animal to, say, a whale will have taken place by a relatively gradual process with many intermediate forms. For this transition from land to sea to be possible, natural selection should be able to act along the whole range of the spectrum, and each intermediate form across the spectrum should be reasonably well adapted. In other words, there should be no insurmountable adaptive barrier in the way. (The available evidence on the origin of marine mammals is not sufficiently fine-grained for worthwhile discussion of the fine detail of this transition—for instance, whether morphological change was truly gradual or was concentrated in bursts during the formation of new species, as the hypothesis of punctuated equilibria predicts.)

The first line of evidence for demonstrating the viability of a transition is direct, from actual fossils of the transitional forms (or more probably close relatives). They may not (yet) be available, but are enormously valuable when they are. Early whales are an excellent example.

The second line of evidence for the viability of a transition is the existence of a spectrum of biological analogues, unrelated species that demonstrate a complete range of adaptations from fully terrestrial to fully marine. Their survival is evidence that their various lifestyles, and therefore the stages in the transition, are viable.

The third type of evidence is to some extent an independent test. It comprises theoretical biomechanical analyses that confirm that particular hypotheses of evolutionary transitions are functionally possible. Conversely, a hypothesis that required a functionally implausible intermediate form would thereby be considerably weakened.

II. Environment of the Transition

What resources drove natural selection during the land–sea transition? The most likely one is food. Less obvious but also very important are shelter from predators and shelter from extremes of temperature. Different resources occur in different environments; e.g., seaweed occurs on the shore and in seawater, but basking in the sun may only be practical on land. It is likely that the transition was driven by the move between environments in order to exploit resources.

The fact that the transition is driven by nondirectional natural selection has the major implication that selection and therefore evolution can in principle simply halt, or reverse direction, for greater or lesser periods, within any one lineage. This has undoubtedly happened. Indeed, certain forms are truly amphibious, relying on a highly specific combination of land and water resources to enable a lifestyle that is otherwise impossible. For instance, the HIPPOPOTAMUS feeds on land at night, but rests in the water during the day, whereas the river otter (*Lutra lutra*) forages in water but comes out to warm up, rest, and breed. To speak of these as being stuck at an intermediate stage of the evolution of aquatic forms risks missing the truth, which is that such amphibious forms are highly adapted in their own right. Crocodilians likewise evolved a highly efficient amphibious freshwater and estuarine lifestyle, which remains the main crocodilian way of life. However, some crocodilian lineages became fully marine and others reverted to a fully terrestrial lifestyle. Suggestions that some modern terrestrial groups are secondarily terrestrial, derived from amphibious or aquatic forms, are therefore entirely reasonable in principle and can only be assessed on actual evidence. Such suggestions have been made for snakes, proboscideans, and human beings, for instance.

What was the environment of the transition from land to sea? There are various possible environments where land, fresh water, and sea variously meet:

1. Freshwater river
2. Lake or other body of standing fresh water
3. Estuary with at least some saltwater influence
4. Sandy open seashore
5. Rocky open seashore
6. Unusual habitats, especially highly productive ecosystems rich in food; modern examples include mangrove swamps, the Florida Everglades, and the regularly flooded lowland riverine forests of the Amazon Basin. It must not be forgotten that, quite apart from the effects of human activity today, ecosystems in the past may have been very different, even without allowing for the "wild card" of unpredictably unusual habitats without any modern analogue.

These different environments pose varying advantages and problems to animals that exploit them. Freshwater rivers and lake shores, however, offer good opportunities for amphibious forms such as hippopotamuses, which rest in the water but feed on land at night. However, because of the risk of stranding during occasional droughts, only the largest rivers are suitable for fully aquatic animals such as odontocetes. Conversely, rocky shores offer plenty of fish and cephalopods to animals such as river otters, and molluscs and sea urchins to animals such as sea otters. However, life in salt water poses problems of salt excretion and THERMOREGULATION. River otters lack blubber and use air-filled fur for insulation. They risk death if their coats are not washed in fresh water, as the salt water mats their coats when it dries. They are therefore dependent on the availability of standing pools of fresh water on shore.

An example of a hypothesis of simple transition from land to sea through a single habitat, rocky seashores, can be formulated using modern analogues:

1. Primarily terrestrial carnivore/carrion feeder, not normally a swimmer, but foraging on the shore above water, e.g., stoat (*Mustela erminea*).
2. Feeding facultatively on rocky shores above water and in rock pools, and possibly diving for food. Some swimming ability, but not much more than most terrestrial mammals, e.g., American mink (*Mustela vison*).
3. Shallow-water swimmer, foraging routinely in inshore water, for example, benthic invertebrates, remaining on land otherwise (e.g., for resting, thermoregulation, and breeding). Swimming by forceful kicks of webbed hind feet linked to flexion and extension of back, supplemented by dorsoventral undulation of a flattened tail, e.g., river otter (pursuit predator) or the archaic whale *Ambulocetus* (probably an ambush predator).
4. Shallow-water swimmer and diver, remaining in the water for much of the time and specializing in aquatic food. Swimming somewhat more specialized for aquatic movement, notably the development of bone ballast in buoyancy control, which slows the animal, e.g., the sea otter.
5. Fully specialized swimmer, feeding in open water but perhaps resting and breeding on beaches inaccessible to land predators, with increased development of a caudal fin to provide most or all of the thrust, e.g., archaic protocetid whales such as *Rodhocetus* and the living gray seal (*Halichoerus grypus*) (in which the hind feet are modified to the functional equivalents of caudal fins).
6. Fully aquatic, not coming to land at all, e.g., short-beaked common dolphin (*Delphinus delphis*).

This testable hypothesis is given as an illustrative example based solely on carnivorous mammals on rocky temperate coastlines, and alternatives can be developed for other groups and other environments. Nevertheless, note that this hypothesis contains two critical elements:

- Ecological plausibility—these are stages in a gradual spectrum, exploiting resources as they become newly accessible
- Functional plausibility—again, these are stages in a gradual spectrum, with a clear explanation for how the development of caudal fin propulsion might occur, which can be reinforced by biomechanical analyses

Second, there may be no one single environment for the transition. Because the transition may take place in stages, natural selection can operate in successively different ways depending on the environment of the time. This has major implications, as discussed later.

It is also possible, and even likely, that the transition from land to water moved through a series of habitats, not just one. This may be important in easing the direct transition from land to sea. For instance, the rocky shore model given earlier does not offer any obvious answers to the problem of the origin of sirenians, which are marine herbivores. Individual red deer (*Cervus elaphus*) facultatively eat intertidal marine algae or "seaweed," and interestingly the North Ronaldsay breed of domestic sheep, which has long been confined to the shore of North Ronaldsay in the Orkney Islands, is now so adapted to a diet containing at least some algae that it cannot survive on grass alone. However, there are no living specialized mammalian grazers on algae comparable to (say) the Galápagos marine iguana (*Amblyrhynchus cristatus*). This might be explained on two grounds: (1) that the transition did go through a rocky shore phase, and it just happens that no analogues are alive today, in which case the extinct Pliocene aquatic sloth *Thalassocnus* may be one such analogue; or (2) that the transition took place somewhere else, presumably in fresh water or estuaries because sandy beaches are devoid of edible plants. This leads to another possible spectrum, again with some modern analogues, but this time moving through geographical as well as niche space:

1. Riverine animal, walking (but not normally swimming) in water and feeding on land, e.g., hippopotamus, or land animal wading in fresh water to feed on aquatic plants.
2. More aquatic, swimming, feeding on freshwater plants in lakes and rivers.
3. Aquatic, swimming form, feeding in estuaries.
4. Fully aquatic, feeding in estuaries and nearshore, e.g., manatees (*Trichechus* spp.).
5. Fully aquatic, feeding largely or wholly on marine plants, e.g., dugong (*Dugong dugon*) and Steller's sea cow (*Hydrodamalis gigas*).

Note that this offers one explanation of how land animals can survive in the salt-rich "desert" of the sea, by way of intermediate forms able to cope with increasingly brackish water (although alternatively a marine mammal might be derived from desert-living land mammals, which are already adapted to retaining body water, e.g., by excreting urine much saltier than the blood).

The environment, or environments, of transition also bias the fossil record. Even if an animal dies, escapes scavenging, and is fossilized, it is not necessarily buried where it lived in life. For instance, water currents might have moved the carcass. To summarize a complex situation, corpses in estuaries and rivers are apt to be swept out to sea. Carcasses of animals that lived on rocky coasts, which are being eroded, tend not to be preserved *in situ* as there is no sediment to bury them in. Carcasses on sandy coasts also tend not to be preserved because these coasts are relatively high-energy environments that disrupt the corpse. If the carcasses are buried, this may be because they have drifted offshore and sunk to the sea floor, or possibly because open sea animals have been stranded on the shore and preserved there. Only animals in rivers and lagoons have much chance of being preserved where they lived, and even then riverine animals are quite likely to be washed away downriver. These biases are borne in mind by competent paleontologists and can be partly mitigated by functional analysis (which would reveal anomalies such as an amphibious animal in an open-sea deposit).

III. Functional Issues

The functional problems involved in the transition from land to water, such as swimming, FEEDING, and sensory physiology, are mostly discussed elsewhere in other entries. Here is a case study, focusing on buoyancy control, and showing how this one function interacts with several others, such as swimming and feeding.

Land mammals breathe air and may also rely on air-filled fur for heat insulation, especially in temperate and polar regions. This air makes swimming and diving difficult if the buoyancy provided by air in the lungs (and any fur) is not precisely compensated by the denser parts of the body such as bone. This wastes at least part of the swimming effort, which goes to staying up or down in the water rather than thrusting the animal forward. Animals well adapted to aquatic life have proper adaptations to buoyancy control separate from swimming, which can thus be optimized for forward movement. However, locomotion interacts with respiration here. Air in large lungs is useful as an oxygen store, but only for animals that do not swim fast (because the added bulk slows them down) or dive deeply (because larger lungs compress more with depth, making buoyancy changes even greater, and because of the risk of the "bends" if high-pressure air in the lungs is allowed to remain in respiratory exchange with the bloodstream). To compensate for the air in their lungs, marine mammals may have stones (gastroliths) in the stomach or massively enlarged bony skeletons for ballast. Alternatively, they may reduce the size of their lungs and replace air-filled fur with naked skin underlain by an insulating layer of subcutaneous fat or BLUBBER. In practice, of course, all functions and organs are interdependent. For instance, sirenians can only tolerate their large lungs and correspondingly massive skeletons because they feed on plants that do not swim away, whereas odontocetes have to be much faster swimmers to catch their prey, and their smaller lungs are coordinated with the evolution of much lighter skeletons. These changes appear to occur in predictable patterns in the evolution of marine mammals and their reptilian and avian analogues:

1. Land animal, relatively buoyant in water, e.g., stoat.
2. Amphibious animal with little or no bone ballast, because it slows it too much on land, but remaining too buoyant in water for efficient locomotion, e.g., American mink, river otter, polar bear (*Ursus maritimus*), desmostylians, *Thalassocnus* (sloth), *Ambulocetus* (primitive whale), and Galápagos marine iguana. Tied to living on land (or ice) for most of the time.
3. Primarily aquatic form with bone ballast in the form of larger, or denser, bony skeletons. Swimming can be decoupled from buoyancy control and made more efficient, e.g., sea otter and some primitive whales. Typical of animals living close inshore. Nile crocodile (*Crocodylus niloticus*) uses gastroliths but is not quite so fully aquatic.
4. Further evolution into more specialized fully aquatic forms, three different groupings being identifiable depending on locomotor and feeding adaptations, which drive selection in different ways, even reversing it:
   a. Slow-swimming predators (or herbivores) on sessile food such as plants or benthic invertebrates: full development of bone ballast and large lungs that act as air stores and give rapid loss of buoyancy during a dive as the increasing water pressure compresses the lungs, enabling the animals to become negatively buoyant and walk on the bottom, e.g., placodonts (extinct reptiles feeding on hard-shelled benthic invertebrates) and sirenians. This is an

essentially shallow water adaptation, linked to inshore and coastal waters.

  b. Predators on mobile food—an apparently intermediate body plan of underwater fliers with hydrofoil-shaped swimming limbs, using gastroliths and/or bone ballast for buoyancy control: plesiosaurs, penguins, and sea lions. Also permitting open sea lifestyles, but (in penguins and sea lions) allowing them to retain hindlimbs relatively unmodified for use when breeding on land (it is not clear what plesiosaurs did, as their two pairs of hydrofoil "wings" are essentially identical). Also permits open sea lifestyles, although these animals are for some reason more coastal in lifestyle and distribution than the cetaceans.

  c. Predators on mobile food—fast swimming with caudal fins: selection for improved speed and acceleration leads to reversal of selection to give extensively lightened skeletons and smaller lungs (which mean less change of net buoyancy with depth, making deep diving more efficient); e.g., ichthyosaurs and cetaceans. Permits open sea lifestyles.

IV. The Problem of the Intermediate

The problems of functional change are not a simple matter of moving from terrestrial to aquatic adaptations. "Amphibious" forms have to tolerate an intermediate level of adaptation, develop a doubly functional anatomy (which may itself involve a level of compromise), or remain poorly adapted in one of the two environments. This is not in itself a problem if the benefits compensate for the disadvantages, as it is better to be an inefficient swimmer (in terms of energy cost of locomotion) with access to abundant aquatic food than to starve on land.

However, a net increase in energy intake is not sufficient. Animal species do not normally exist in isolation, but have to compete with one another for resources. It is not clear how intermediate forms can survive at all under these circumstances. At any one time and place an intermediate form will always risk being outcompeted by a land specialist or by a sea specialist, if not both. For instance, an intermediate form will be a slower and less energy-efficient swimmer than a specialist marine form, and a slower and less efficient runner than a specialist land form. River otters are not quite as nimble as comparable land mustelids, and sea otters are more aquatic but still clumsier on land. The dilemma can only be resolved if, as seems likely, intermediate forms rely on "refuges" free of competition. These refuges might be real physical ones, in space and time, or virtual ones, in ecological "niche space." Possible refuges and examples of animals exploiting them include:

1. Ecological isolation in an area inaccessible to land or water predators and competitors: possibly sea otters in rocky shallow water.
2. Specialization in a dietary resource not exploited by similar animals: sea otters feeding on hard-shelled shallow water invertebrates.
3. Time averaging of predator avoidance, e.g., getting into the water when a land carnivore is around, but getting out when a shark arrives. The risk that both will be around at the same time so that escape becomes impossible is a multiple of the two separate probabilities that each will be around, and is therefore smaller than either.
4. Time averaging of food or other resource, e.g., hippopotamuses feeding on land at night, but retreating to water during the day.
5. Combining complementary resources not available to pure land or pure water specialists, e.g., using land to bask in the sun to gather heat, but wading and diving in cold water to feed on algae, as adult Galápagos marine iguanas do (perhaps only relevant to those mammals that are small enough and live in cold enough sea water to have problems with thermoregulation, notably river otters on northern temperate seacoasts).
6. Geographical isolation to escape land predators, perhaps on oceanic islands, e.g., Galápagos marine iguana.
7. Isolation in evolutionary time. Modern marine mammals evolved as replacements for the Mesozoic marine reptiles, the last of which went extinct in the mass extinction at the end of the Cretaceous. The mammals were thus sheltered from competition by these direct analogues in a way that would not be the case today (at least before human interference). However, this is only a partial explanation, as early Tertiary marine mammals had to contend with predatory fishes and crocodiles in fresh water and in the sea, together with land predators.

It is entirely possible that more than one type of refuge was exploited during the evolution of any one marine mammal lineage, perhaps even at the same time.

V. The Limits of Adaptation

So far the discussion has been based on what is evolutionarily, ecologically, and functionally plausible—roughly, what is permitted by the ecological context, the laws of physics, and the need for gradual transformation through a series of equally viable ancestors. However, the adaptation of marine mammals is also limited by phylogenetic constraint: the concept that an evolutionary lineage's change is limited in direction or in scope by inherited features. Indeed the need to evolve along a continuous spectrum of viable forms is in itself such a constraint, as suggested by the repeated patterns in the evolution of buoyancy control outlined earlier. But does the inherited body plan itself pose constraints? A good example seems to be marsupial "whales," or rather their nonexistence, which at first sight seems surprising. Convergent evolution with placental mammals has generated, for instance, marsupial "mice," "cats," and "wolves," so why not "seals" and "whales"? This is probably because marsupial young are born at a very immature stage compared to placental mammal young and develop within the mother's pouch or hanging from the teat so that they would drown the moment their mother went swimming. Marsupials cannot therefore easily evolve aquatic forms. The only well-attested semiaquatic marsupial is the yapok or water opossum (*Chironectes minimus*), which forages for prey such as fish and crustaceans in fresh water. In the female, the waterproof pouch can be closed off with a sphincter to keep water out, and the young can tolerate the ensuing low oxygen levels for many minutes. However, this is an impractical solution for a lifestyle involving swimming for more than brief periods. The marsupial mode of reproduction therefore appears to be a phylogenetic constraint impeding the development of fully aquatic lifestyles beyond the semiaquatic stage. Because placental mammals carry their young for much longer periods, they have been able to bypass this constraint.

Indeed, it can be argued that many marine mammals (and marine amniotes more generally) are, or were, not fully aquatic. The pinnipeds, the penguins, marine turtles, and most seabirds (for instance), remain critically dependent on breeding on land or on floating ice, which has major impacts on their biology. They have to retain at least some ability to move on land, compromising full adaptation to life in water. Even so, they are extremely vulnerable to land predators when they go ashore to breed. They therefore have to swim or fly, sometimes over global distances, to suitable rookery areas in sometimes inhospitable environments. The spectacular concentrations sometimes resulting at these ROOKERIES provide dense populations of potential mates and competitors, which may drive intense competition between males, selecting for massive sexual dimorphism, as in elephant seals (*Mirounga* spp.). Plainly something about their various modes of reproduction prevents young being born live, or if it is born live, from surviving in water. These animals therefore still occupy an evolutionary refuge of the kind outlined earlier in the form of predator-free, land breeding areas.

Whales and sirenians do give birth to live young in water and have therefore—or concurrently—lost the ability to move on land, freeing them to become entirely adapted to life in water, just as the ichthyosaurs did more than a hundred million years before. Whales and sirenians, then, seem to be the only truly marine mammals with no remaining link to the land that their ancestors left millions of years before. Nevertheless, the contrast between ichthyosaurs and dolphins reveals evidence of phylogenetic constraint in what are superficially functionally optimal animals. They are superficially very similar, under the same selective pressures of a fully aquatic lifestyle, with fish-like bodies with dorsal and caudal fins and long narrow snouts for snapping at fish and cephalopods underwater. They even give, or gave, birth to live young underwater (fossils of gravid female ichthyosaurs have been found). Still, they differ in ways that reflect their differing ancestries. Ichthyosaurs were (presumably) derived from lizard-like land reptiles that ran by using lateral undulations of their bodies to increase the forces and movements of their legs, and hunted by sight rather than smell. Ichthyosaurs accordingly had tails that beat from side to side with vertically oriented caudal fins. They also had huge eyes for seeing in deep or silt-laden water. In contrast, dolphins were ultimately derived from small nocturnal insectivorous mammals rather shrew-like in habits, with high-frequency hearing and rudimentary sonar (as in rats and blind people) and whose running was enhanced by dorsoventral flexion and extension of the body. Accordingly, dolphins swim with up and down beats of their tails, which bear horizontally oriented caudal fins. Instead of the huge eyeballs of ichthyosaurs, the domed foreheads of dolphins contain the sonic lenses and huge brains needed to send sonar pulses and process the resulting data.

Thus even the most adapted of the marine mammals show the signs of their origins.

See Also the Following Articles

Breathing ▪ Evolutionary Biology ▪ Locomotion, Terrestrial ▪ Mammalia ▪ Morphology, Functional ▪ Swimming

References

Benton, M. J. (2000). "Vertebrate Palaeontology," 2nd Ed. Blackwell Science, Oxford.

Callaway, J. M., and Nicholls, E. L. (eds.) (1997). "Ancient Marine Reptiles." Academic Press, San Diego, CA.

Carroll, R. L. (1987). "Vertebrate Paleontology and Evolution." Freeman, San Francisco.

Kemp, T. S. (1999). "Fossils and Evolution." Oxford Univ. Press, Oxford.

Maddock, L., Bone, Q., and Rayner, J. M. V. (eds.) (1994). "Mechanics and Physiology of Animal Swimming." Cambridge Univ. Press, Cambridge.

Taylor, M. A. (2000). Functional significance of bone ballast in the evolution of buoyancy control strategies by aquatic tetrapods. *Hist. Biol.* **14**, 15–31.

Thewissen, J. G. M. (ed.) (1998). "The Emergence of Whales: Evolutionary Patterns in the Origin of Cetacea." Plenum Press, New York.

Thewissen, J. G. M., and Fish, F. E. (1997). Locomotor evolution in the earliest cetaceans: Functional model, modern analogues, and paleontological evidence. *Paleobiology* **23**, 482–490.

Vogel, S. (1994). "Life in Moving Fluids," 2nd Ed. Princeton Univ. Press, Princeton, NJ.

Osmoregulation

DANIEL P. COSTA
University of California, Santa Cruz

An organism can be thought of as a large volume of fluid surrounded by a body wall. Mammals must maintain both the concentration and the volume of this internal fluid within a very narrow range and can only tolerate minor deviations. Even though most marine mammals live in an aquatic medium, the animals' internal fluid composition differs from the ambient environment and therefore requires active processes to maintain it. Osmoregulation describes the way in which the internal water and electrolyte concentration of this internal environment is maintained. When animals feed they take in both water and electrolytes that must be excreted. While they gain water from metabolizing food, they lose water through evaporation when they breathe to obtain the oxygen necessary for metabolism. Maintenance of a constant internal environment requires that whatever comes into the animal must equal what goes out. The easiest way to understand osmoregulation is to account for the ways water and electrolytes enter and leave the organism (Fig. 1). For example, if a dolphin consumes a

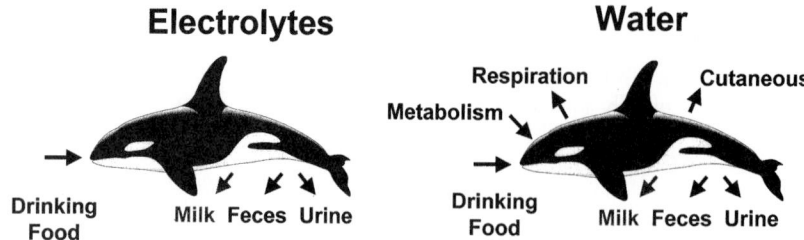

Figure 1 *Schematic of ways by which water and electrolytes enter and leave a marine mammal. Excretion of electrolyte and water as milk occurs only when females are lactating.*

large volume of water and electrolytes, it must have the capability to excrete an equivalently large volume in the feces and urine, through BREATHING, and in milk during lactation. Conversely, if a seal on the beach does not have access to food or water, it must be able to survive on the water produced from metabolism and have mechanisms in place to reduce water loss. Following the relative rates of water and electrolyte input and output helps us understand the mechanisms that marine mammals use for osmoregulation.

I. Water and Electrolyte Ingestion

Water and electrolytes enter the animal through the ingestion of food and water. Water that is consumed as water, i.e., water contained in the food or actively drunk, is called preformed water. Compared to terrestrial mammals, marine mammals consume a water-rich diet of fish and marine invertebrates, which are composed of between 70 and 80% water. Prey are also composed of electrolytes and nitrogen that require water for excretion by the kidney. Marine mammals face different osmoregulatory problems depending on the type of prey that they consume. For example, the ratio of water to electrolytes is quite different between vertebrate and invertebrate prey. The internal fluid of vertebrates is one-third the electrolyte concentration of seawater, whereas the internal fluid of invertebrates is essentially the same as seawater. Thus, a dolphin eating squid will get almost three times as much electrolytes than if it consumed fish. Furthermore, an animal such as a manatee with access to fresh water can drink fresh water to flush electrolytes, whereas an oceanic dolphin can only drink seawater. Water is also produced as a by-product of metabolism; this is called metabolic water production (MWP). Because the amount of metabolic water produced varies with the fuel oxidized, different diets produce varying amounts of metabolic water. For example, 1.07 g of water is generated for every gram of fat oxidized, 0.56 g of water per gram of carbohydrate, and only 0.39 g of water per gram of protein.

II. Water and Electrolyte Output

Both water and electrolytes are excreted in the urine and feces, whereas only water is lost through evaporation. Water is lost via evaporation both across the skin (cutaneous water loss) and through the lungs (respiratory evaporative water loss). Because marine mammals lack sweat glands there is no loss of electrolytes (salts) across the skin. Unlike sea birds and marine reptiles, marine mammals lack specialized glands to excrete salts. All salt excretion is through the kidney, and marine mammals have developed a specialized kidney to handle the large volume of electrolytes and water they process.

III. Do Marine Mammals Drink Seawater?

In most cases, marine mammals can derive sufficient water from their diet so that they do not need to ingest seawater. Measurements of the water, electrolyte, and nitrogen intake, coupled with measurements of evaporative, urinary, and fecal water loss, suggest that a feeding seal can get all of the water it needs from its prey (through both preformed and metabolic water). This is due to the high water content of the prey coupled with the low evaporative water loss of an animal living in a marine environment.

Do animals drink seawater when they become osmotically stressed in environments where the evaporate water loss is high? To determine whether a marine mammal can gain fresh water by drinking seawater we need to know whether the animal can excrete urine that is more concentrated than seawater. The more concentrated the urine, the greater the amount of "fresh water" that can be derived from the ingestion of seawater. A simple calculation can show how much water is gained or lost relative to the concentrating ability of the kidney (Table I). For example, if a humpback whale, *Megaptera novaeangliae*, consumed 1000 ml of seawater and its kidney had the ability to excrete urine with a chloride concentration of 820 mmol $\times$ liter^{-1}, it could gain 350 ml of fresh water. Whereas humans, who cannot produce urine as concentrated as seawater, would lose 350 ml of fresh water for every liter of seawater they consumed. The maximum urine concentrating ability of marine mammals and a few terrestrial mammals is presented in Table II.

So, do marine mammals drink seawater? Many species of marine mammals have the capacity to drink seawater, but they do not always do so. Fur seals and sea lions have been observed drinking seawater while fasting on the rookery. However, visual observations are not quantitative and do not allow an assessment of the relative importance of seawater ingestion to water balance. Isotopically labeled water and/or electrolytes have been used to quantify seawater drinking in a variety of marine

TABLE I
Differences in Urine Concentrating Ability of Humpback Whale, *M. novaeangliae,* and Human
Given to Show a Gain or Loss of Body Water after Ingestion of 1 Liter of Seawater

	Seawater consumed (volume ml)	Cl− concentration (mmol liter^{-1})	Max urine concentration (mmol liter^{-1})	Urine volume produced (ml)	Water balance gain or loss (ml)
Whale	1000	535	820	650	+350
Human	1000	535	400	1350	−350

mammals (Table III). In these studies, the amount of water and/or electrolytes consumed in the food was added to that produced by metabolism and compared to the total amount of water and/or electrolytes that passed through the animal is measured by isotopic tracers. Using these methods, investigators found that sea otters, *Enhydra lutris,* and bottlenose dolphins, *Tursiops truncatus,* that were feeding and Galapagos fur seals, *Arctocephalus galapagoensis,* short-beaked common dolphins, *Delphinus delphis,* and short-finned pilot whales, *Globicephala melas,* that were fasting all consumed substantial amounts of seawater. In contrast, feeding and fasting harbor seals, *Phoca vitulina,* feeding northern fur seals, *Callorhinus ursinus,* and fasting Antarctic fur seals, *A. gazella* all had negligible amounts of seawater ingestion. Fasting northern elephant seal pups, *Mirounga angustirostris,* fast for up to 3 months without any measurable ingestion of seawater. The need to drink seawater varies with climate and habitat. For example, Galapagos fur seal females drank during the perinatal fast, whereas female Antarctic fur seals did not. The need to drink seawater in the Galapagos is most likely due to the in-

TABLE II
Maximum Urine Chloride Concentration and Maximum
Osmolarity Measured for Marine Mammals Compared to
Values of Representative Terrestrial Mammal

	Cl concentration (mEq liter^{-1})	Osmolarity (mOsm liter^{-1})
Balaenoptera borealis	370	1340
B. physalus	390	
B. musculus	340	1340
Megaptera novaeangliae	820	
Tursiops truncatus	632	2458
Zalophus californianus	760	2223
Enhydra lutris	555	2130
Human, *Homo sapiens*	400	1230
White rat, *Rattus rattus*	760	2900
Camel, *Camelus dromedarius*	1070	2800
Sand rat, *Psammonys obesus*	1920	6340
Seawater	535	1000

creased evaporative water loss associated with a warm tropical environment.

IV. Relative Reductions in Water Loss

As described earlier, marine mammals do not need to drink seawater because they have reduced their evaporative water loss. Amazingly, northern elephant seals can fast for months without access to food or water. The only water available to fasting seals is MWP from the oxidation of fat and protein in their tissue. Remember that a positive water balance requires that water input equals water output. This requires that water lost in the urine, feces, and from evaporation be equal to or less than water produced from metabolism (MWP). How then do elephant seals, and probably other seals and sea lions, reduce their water loss?

A. Cutaneous Water Loss

Given their aquatic lifestyle, marine mammals have very low evaporative cutaneous water loss. In water there would be no evaporative water loss, and on land, because pinnipeds lack sweat glands, their evaporative loss is quite low. However, common dolphins and harbor porpoises, *Phocoena phocoena,* appear to lose a substantial amount of water across their skin surface. Common dolphins lose as much as 4 liter H_2O day^{-1} or 70% of their total water intake. It may be that seawater ingestion is necessary to make up for the water lost across the skin.

B. Respiratory Evaporative Water Loss

Endotherms lose water through respiration by the simple physics of warming and saturating the air they breathe. Ambient air is inhaled, warmed, and humidified to core body temperature. For example, air fully saturated (100% relative humidity) with water at 10°C contains 10 mg H_2O per liter of air, whereas fully saturated air in the lungs at 37°C contains 40 mg H_2O per liter of air. Unless there is a mechanism to recover water, a seal would lose 30 mg of H_2O for every liter of 10°C air it inhaled.

Marine mammals employ a few tricks to reduce the water lost through respiration. The first is to breathe periodically, i.e., to inhale, hold their breath, and then exhale. This is called apneustic breathing. Apneustic breathing increases the amount of oxygen extracted per liter of air inhaled. Whereas terrestrial animals typically extract 4% oxygen per breath, marine mammals can extract as much as 8% per breath. This allows marine mammals to breathe less frequently and thereby lose less water because they make

TABLE III
Rate of Seawater Ingestion Measured Using Isotopic Tracer Techniques in Marine Mammals

	Body mass (kg)	Rate of seawater consumption		
		ml kg^{-1} day^{-1}	ml day^{-1}	Proportion of total water influx (%)
Globicephala melas	605	4.5	2720	n.a.
Tursiops truncatus				
Feeding	198	37.5	7420	68.8
Delphinus delphis				
Fasting	57	12.5	700	17
Arctocephalus gazella				
Fasting	39.4	1.0	39	15
A. galapagoensis				
Fasting	37.4	18.3	684	84
Callorhinus ursinus				
Fasting	23	1.8	41	2.0
Phoca vitulina				
Feeding	29.4	3.0	137	9.2
Fasting	28.6	1.3	37	7.3
Enhydra lutris				
Feeding	24.3	62	1507	23

fewer respirations to obtain an equivalent amount of oxygen. Pinnipeds, sea otters, and polar bears, *Ursus maritimus*, reduce their respiratory evaporative water loss further by employing a nasal countercurrent heat exchanger. It has been suggested that dolphins, which lack nasal turbinates, recover respiratory water through the adiabatic cooling associated with explosive respirations.

Nasal Countercurrent Heat Exchanger Marine mammals, rodents, and desert ungulates have small passageways in their nasal passages that allow them to recover water vapor and heat that was added to the air at inhalation. Nasal turbinates are composed of very small passageways that allow intimate contact between the inhalant air and the nasal membranes (Fig. 2).

As the cold air passes across the small nasal passage, it is warmed and water evaporates. Heat and moisture are transferred from the nasal passage to the air so that by the time it leaves the nasal turbinate it is warmed and humidified to body temperature. In the process of warming the inhaled air, the membranes lining the nasal passages have cooled. On the following exhalation the warm moisture-laden air is cooled as it passes over the cool membranes. As the air temperature declines water vapor condenses and is recovered in the nasal passage (Fig. 3).

C. Fecal Water Loss

Although there are no direct measurements, fecal water loss of feeding cetaceans is probably quite high. Fecal water loss in

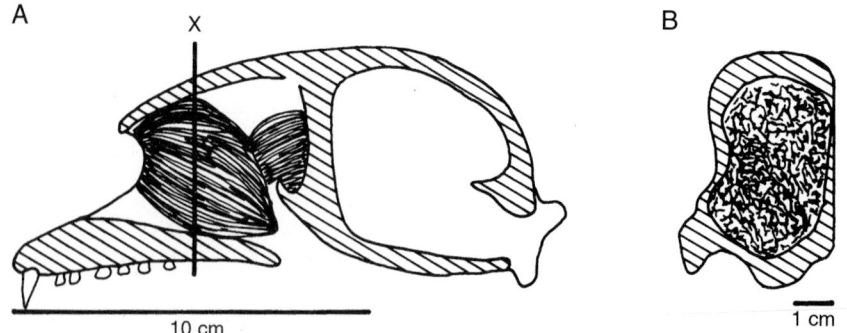

Figure 2 (A) Sagittal section of a weanling elephant seal skull showing the nasal turbinates. (B) Cross section through one-half of the skull at line "X" in A. From Huntley et al. (1984). Reproduced with permission of the Company of Biologists, Ltd.

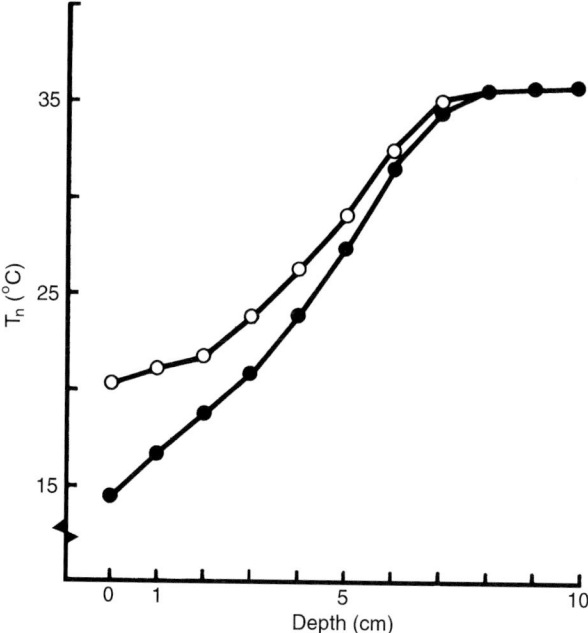

Figure 3 *Temperature at 1-cm intervals within the nasal passage of a weanling elephant seals where the ambient air temperature was 15°C (○) and 5°C (●). From Huntley et al. (1984). Reproduced with permission of the Company of Biologists, Ltd.*

pinnipeds feeding on fish is comparable to that of terrestrial carnivores. However, it is not clear how marine mammals that ingest seawater avoid the laxative effect of $MgSO_4$. Fasting animals have negligible fecal water loss, as their fecal production is quite small.

TABLE IV
Water, Lipid, and Protein Content of Marine Mammal Milk Compared to Human and Cow

	% water	% lipid	% protein
Balaenoptera musculus	45.4	41.5	11.9
B. acutorostrata	60.4	24.4	13.6
Physeter macrocephalus	64.5	24.4	9.1
Tursiops truncatus	69.6	15.3	11.5
Arctocephalus galapagoensis	58.5	29.4	12.1
Callorhinus ursinus	44.3	41.5	14.2
Neophoca cinerea	64.7	25.8	9.5
Mirounga angustirostris	36.6	54.4	9.0
Cystophora cristata	33.7	61.4	4.9
Halichoerus grypus	36.6	52.2	11.2
Human	87.6	3.8	1.2
Cow	87.3	3.7	3.3

D. Urinary Water Loss

The rate and amount of water lost in the urine are directly related to both the urine concentrating ability of the kidney and the hydration state of the animal. The kidney ultimately regulates the water and electrolyte state of the animal. When there is a surplus of water, the kidney produces diluted urine, whereas during periods of water stress, the kidney excretes concentrated urine. The kidney must be able to excrete metabolic end products in the form of urea and all excess electrolytes with the water that remains after cutaneous, respiratory, fecal, and in some cases, lactational water loss. While at sea, marine mammals either get all of their water from their prey or they drink seawater. This requires the processing of large urine volumes at moderate to high urine concentrations, and most marine mammals (cetaceans, pinnipeds, sea otters) have a specialized lobulate or reniculate kidney that enables them to do this.

Pinnipeds, such as the northern elephant seal, undergo prolonged fasts on land without access to water. These animals are able to stay in water balance by a combination of low rates of evaporative water loss coupled with low rates of urine production. Elephant seals utilize fat almost entirely (96–98%) for their metabolism while fasting. Fat oxidation produces only CO_2 and H_2O, whereas the oxidation of protein results in CO_2, H_2O, and urea. Urea is the end product deamination of amino acids and requires water to be excreted by the kidney. Therefore, fat is not only an efficient way to store energy, it is also economical with respect to water balance.

V. Water Balance during Reproduction

Many female pinnipeds do not have access to water while they suckle their young and thus could become dehydrated during lactation. However, marine mammal milk is high in lipids and low in water compared to terrestrial mammals (Table IV). This has the advantage of providing the young with the maximum amount of energy with minimal loss of water from the mother. This is likely an advantageous by-product of the energetics of marine mammal lactation and not a derived adaptation for water balance. It is important to note that pups also do not have access to water and therefore must be capable of maintaining their water balance entirely from water contained in the milk and that derived from its oxidation.

See Also the Following Articles

Energetics ▪ Habitat Pressures ▪ Kidney, Structure and Function

References

Adams, S. H., and Costa, D. P. (1993). Water conservation and protein metabolism in northern elephant seal pups during the postweaning fast. *J. Comp. Physiol. B.* **163**, 367–373.

Costa, D. P. (1982). Energy, nitrogen, and electrolyte flux and seawater drinking in the sea otter *Enhydra lutris. Physiol. Zool.* **55**, 35–44.

Costa, D. P., and Trillmich, F. (1988). Mass changes and metabolism during the perinatal fast: A comparison between Antarctic and Galapagos fur seals. *Physiol. Zool.* **61**, 160–169.

Coulombe, H. N., Ridgway, S. H., and Evans, W. E. (1965). Respiratory water exchange in two species of porpoise. *Science* **149**, 86–88.

Depocas, F., Hart, J. S., and Fisher, H. D. (1971). Sea water drinking and water flux in starved and in fed harbor seals, *Phoca vitulina*. *Can. J. Physiol. Pharma.* **49**, 53–62.

Gentry, R. L. (1981). Seawater drinking in eared seals. *Comp. Biochem. Physiol.* **68**, 81–86.

Hui, C. A. (1981). Seawater consumption and water flux in the common dolphin *Delphinus delphis*. *Physiol. Zool.* **54**, 430–440.

Huntley, A. C., Costa, D. P., and Rubin, R. D. (1984). The contribution of nasal countercurrent heat exchange to water balance in the northern elephant seal, *Mirounga angustirostris*. *J. Exp. Biol.* **113**, 447–454.

Pilson, M. E. Q. (1970). Water balance in California sea lions. *Physiol. Zool.* **43**, 257–269.

Otters

JAMES A. ESTES AND JAMES L. BODKIN
U.S. Geological Survey, Santa Cruz, California

The otters (Mustelidae; Lutrinae) provide a unique look into the evolution of marine living by mammals. This is because most extant marine mammals have been so highly modified by long periods of selection for life in the sea that they bare little resemblance to their terrestrial ancestors. Marine otters, in contrast, are recent expatriates from terrestrial and freshwater habitats, and some species still live in both environments. Contrasts within this group, and among the otters, terrestrial mammals, and the more highly adapted pinnipeds and cetaceans potentially offer deep insight into mammalian adaptations to life in the sea. Among the marine mammals, sea otters also provide the clearest understanding of predation and ocean ecosystem function. This is due in part to serendipitous opportunities provided by history and in part by the relative ease with which shallow coastal systems where sea otters live can be observed and studied. These two qualities of the otters are what make them interesting to marine mammalogy. Thus, our contribution to this volume on the marine mammals is built around these themes.

I. Evolution and Phylogeny

Mustelids radiated from primitive arctoid carnivores at the Eocene/Oligocene border. Early lutrine phylogenies were based on the morphology of fossil and extant species, from which the otters were viewed as a monophyletic group that diversified into three clades: the fish eaters (*Lutra, Lontra,* and *Pteronura*), crab eaters (*Aonyx*), and the sea otter (*Enhydra*). However, the otters probably have been under strong selection for parallel or CONVERGENT EVOLUTION, thereby confounding efforts to understand phylogenetic relationships based on morphology. Distinctive features of the three purported clades might thus have resulted from differences among their common ancestors or selective divergence resulting from different prey (fishes vs invertebrates) or habitats (the ocean vs fresh water). Patterns of brain form and function exemplify this problem in the otters.

Architectural and functional differences among otter brains correlate with their principal foraging modes: invertebrate vs fish feeding. Sensory and motor function in the brains of all mammals map mediolaterally along the prefrontal gyrus. The fish-eating otters, which require precise sensory/motor function of the mouth and facial area, have well-developed proximal regions of the prefrontal gyrus. In contrast, the invertebrate feeding otters (*Enhydra* and *Aonyx*), which require precise sensory/motor function of their forelimbs for prey capture, have more highly developed lateral regions of the prefrontal gyrus. But are these features primitive or derived? If primitive, then they might accurately reflect phylogeny; otherwise, they surely do not.

Nucleotide sequence analysis of the mitochondrial cytochrome *b* gene was used to disassociate the confounding effects of adaptation in constructing a lutrine phylogeny for 9 of the 13 extant species (Fig. 1). These data demonstrate that earlier phylogenies based solely on morphology were grossly inaccurate. The molecular analysis indicates three primary clades, including: (1) the North American river otter (*Lontra canadensis*), neotropical otter (*L. longicaudus*), and marine otter (*L. felina*); (2) the sea otter (*Enhydra lutris*), Eurasian otter (*Lutra lutra*), spotted-necked otter (*Lutra maculicollis*), cape clawless otter (*Aonyx capensis*), and small-clawed otter (*A. cinerea*); and (3) the giant otter (*Pteronura brasiliensis*). Fundamental life history differences (e.g., seasonal vs aseasonal reproduction; direct vs delayed implantation) between Eurasian and North American river otters, species once thought to have formed only after the late-Pleistocene isolation of Asia and North America, make much more sense under the molecular phylogeny. The pattern of long terminal branches and short internal branches in the phylogenetic tree further suggests rapid evolutionary radiation of the otters. Estimates of divergence time indicate that the clades containing *Pteronura* and *Lontra* had separated by the late Miocene. This phylogeny also suggests that sea otters, the only fully marine otter and the most distinctive of all otter species in terms of their morphology, physiology, and behavior, diverged recently but have taken on these different characters because of strong selection imposed by life in the sea. Lutrines appear most closely related to the extant *Mustela* (mink, weasels, and polecats), although whether the otters originated from one or multiple radiations of this group remains uncertain.

II. Marine Otters

At least 6 of the 13 extant otter species are fully or partially marine living (Fig. 2). All of the marine living species or populations occur at high latitudes. Sea otters (North Pacific Ocean) and the chungungo (*Lontra felina*; southeast Pacific Ocean) are the only lutrines that feed exclusively in the sea. Sea otters range from the northern Japanese archipelago, across the rim of the Pacific to about central Baja California, Mexico. Chungungos range along the west coast of South America from Peru to southern Chile. Sea otters typically rest and give birth at sea where chungungos only enter the sea briefly to hunt. Chungungos often go ashore to eat their prey, rest, give birth, and rear their young in dens formed in rocky areas just above high

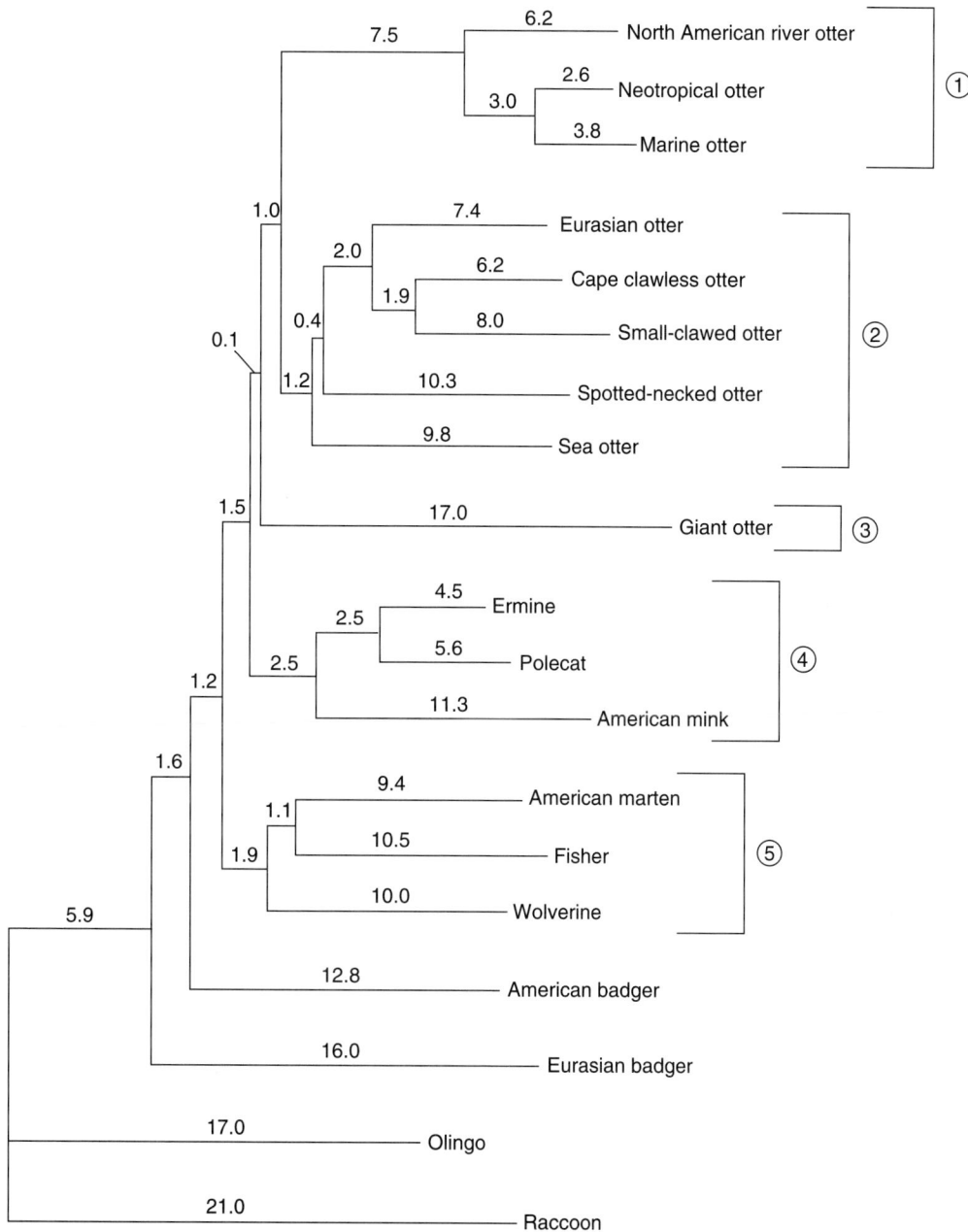

Figure 1 *Phylogeny of the otters. Marine otter is* Lontra felina; *Olingo is* Bassaricyon. *For numbers, see text. From Koepfli and Wayne (1998).*

water. Sea otters occupy a broad range of habitats, from protected bays to exposed outer shores, whereas chungungos occur only along exposed shorelines. Marine living populations of *Lontra provicax* occur in the protected inner waters of coastal Chile and southern Argentina. Marine living populations of North American otters occur northward from about San Francisco in the Pacific and Martha's Vineyard in the Atlantic. Marine living populations of Eurasian otters have a roughly similar latitudinal distribution (from about Portugal northward to northern Scotland and Scandinavia in the North Atlantic; from Japan into Russia in the North Pacific). Marine living populations of cape clawless otters occur in the southernmost regions of South Africa. Tropical lutrines, in contrast, rarely enter the sea.

While marine living species or populations of otters occur only at high latitudes, freshwater living otters occur over a much broader latitudinal range: from the most poleward ice-free environments to the tropics in the northern and southern hemispheres of both the Old and the New worlds. These distributional differences between marine- and freshwater-living otters probably relate to latitudinal differences in production between

Figure 2 *Geographical distributions of marine-living otters.*

freshwater and coastal marine habitats, and the fact that otters have secondarily entered the sea from their primitive freshwater environments. At low latitudes, freshwater production exceeds that of the ocean, whereas at high latitudes this pattern is reversed. This production gradient may have drawn the primitively freshwater living otters into the sea at high latitudes. The most compelling evidence for the proposal is provided by Hans Kruuk's comparative study of freshwater and marine living populations of Eurasian otters in northern Scotland. Kruuk found that marine living populations are able to meet their energy requirements by 2–3 hr of fishing per day, whereas those living in freshwater must double to triple the time investment in foraging to meet their energy requirements. Remarkably different population densities of marine and freshwater living otters further indicate that production and food availability are superior in coastal marine habitats. Reported densities for Eurasian and North American river otters in rivers and streams at high latitudes range between about 1 and 5 individuals per 10 km. Marine living populations of these same species are about fivefold greater, and sea otter densities range from 60 individuals per 10 km shoreline in California to >300 individuals per 10 km shoreline in the Aleutian Islands. These data indicate that otters at high latitudes maintain higher population densities in the sea than they do in freshwater and that freshwater habitats are only marginally capable of producing enough food to meet their energy requirements.

Different population densities of otters in freshwater and marine environments appear to have influenced the evolution of social behavior. Low-density freshwater environments afford little or no opportunity for males to compete for females; consequently, all of the freshwater living otters appear to have monogamous or promiscuous mating systems. In contrast, the elevated production and resulting high female densities may have driven sea otters to their strongly polygynous mating system.

III. Status and Trends

Except for the sea otter, there is little current information on the status and trends of marine living otter populations. The chungungo, listed in IUCN's "Red Data Book," is believed by some authorities to be threatened with EXTINCTION. Illegal harvesting for fur, habitat destruction in the form of deforestation, mining, and pollution, and competition with fisheries are thought to be the species' main threats. Chungungos are rare from Peru southward through central Chile, but there are conflicting reports on their abundance from Chiloe Island southward to Cape Horn.

North American river otters are common to abundant along much of the west coast of North America. Their abundance along the Atlantic coast of North America is uncertain, and information on population trends is lacking for both areas. Eurasian otters are common in marine habitats of Europe and Russia, especially from Norway to Scotland in the eastern North Atlantic. However, as for marine living river otters in North America, information on the status and trends of these populations is lacking. Coastal populations in Asia are practically unknown. The status and trends of marine living populations of southern river otters and cape clawless otters are unknown or unreported.

IV. The Sea Otter

A. History

Species resembling modern sea otters (based on body size and dentition) had arisen by the Miocene. There are two recognized lineages, one of which led to the extinct *Enhydriodon* and the other of which led to *Enhydritherium* and presumably to modern sea otters, *Enhydra* spp. Members of both lineages possessed large flattened molars for crushing the exoskeletons of their invertebrate prey. Fossil remains of these early sea otters are known from North America, Eurasia, and Africa. The distribution of modern sea otters, which apparently arose during the late Pliocene or early Pleistocene, was restricted to the north Pacific Ocean (Fig. 3). One extinct species, *E. macrodonta*, is described from the late Pleistocene of California.

Aboriginal maritime hunters apparently depleted sea otter populations, as evidenced by the character of faunal remains in Aleut kitchen middens. However, judging from the large numbers of otters encountered by European fur hunters at the time

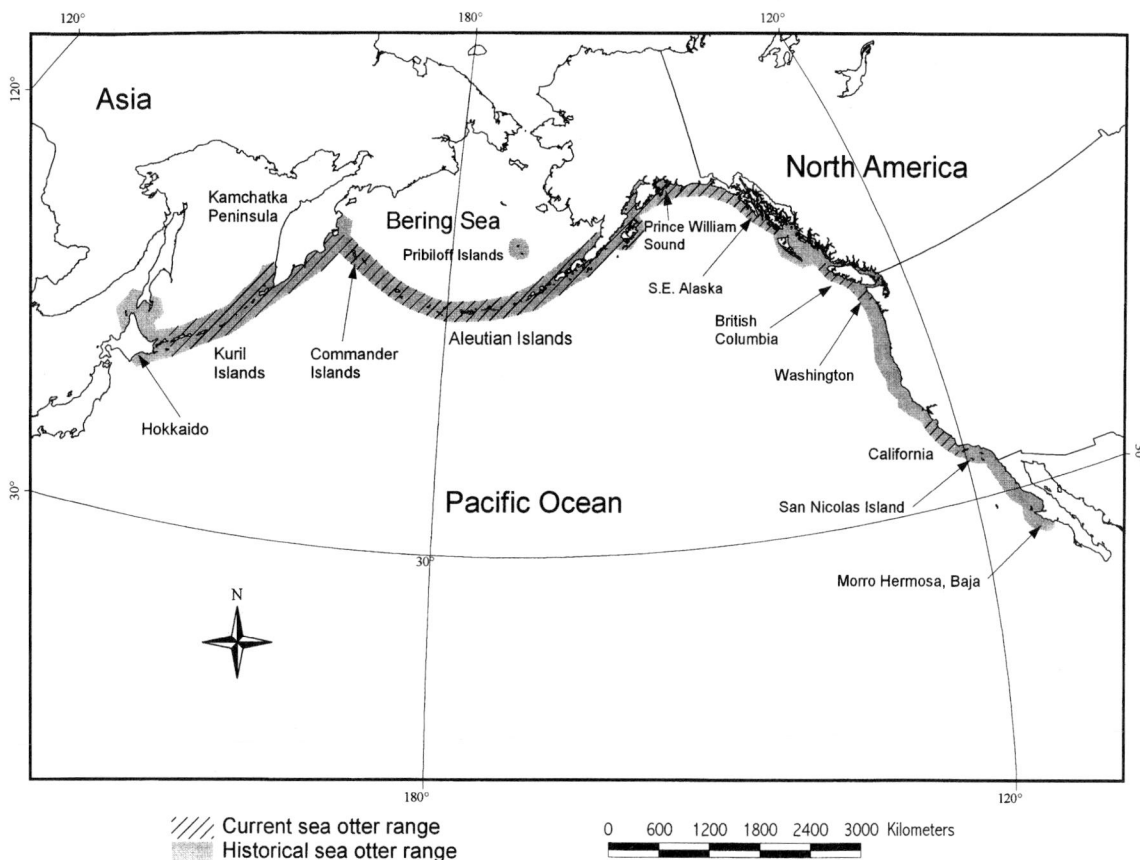

Figure 3 *Historical and current distribution of sea otters in the North Pacific Ocean. Modified from Kenyon (1969).*

of their first contact with aboriginal peoples, these reductions probably were limited to village sites. Sea otter populations subsequently were hunted to near extinction between 1750 and 1900 in the Pacific maritime fur trade. About a dozen remnant colonies in total containing no more than a thousand or so individuals survived into the early 20th century. Following protection in 1911, these colonies increased and several had again reached historical levels by the late 1930s or early 1940s. Reintroductions were undertaken in the 1960s and 1970s to reestablish sea otters in Southeast Alaska, British Columbia, Washington, and Oregon, and sea otters were relocated to San Nicolas Island (southern California) in 1987. Most of the translocated populations increased at rates of 17–20% per year, about the theoretical maximum (R_{max}) for sea otters. The California sea otter population has increased at a much lower rate: about 5% per year. Although precise reasons for the slow population growth in California sea otters are uncertain, the general cause is elevated mortality rather than reduced fertility. All or most of the sea otter's range in Asia is now reoccupied. Several populations in the Kuril and Commander islands had recovered to historical levels by the 1970s or 1980s.

While sea otter populations generally have increased in number and range during the 20th century, there are exceptions. In apparent response to increased killer whale predation, sea otter populations across the western and central Aleutian archipelago declined by roughly 90% during the 1990s. The geographical extent of this decline is uncertain. Several thousand sea otters were killed by the *Exxon Valdez* oil spill in 1989, and while the overall impact of this event on the population is unclear, chronic detrimental effects appear to have persisted through the 1990s. The California sea otter population, which increased slowly throughout most of the 20th century, has since declined by about 12%. The reintroduced population at San Nicolas Island, after declining sharply because of emigration, has remained roughly stable at 15–25 individuals since the early 1990s. Reintroduced sea otter populations in Southeast Alaska, British Columbia, and Washington all continue to increase in numbers and range.

B. Morphology and Physiology

Little is known about vision and hearing in sea otters. However, they are difficult to approach from upwind, thus suggesting sensitive olfaction. Sea otters appear to sense estrous females and territorial males by moving to a downwind position from other animals. Adult male sea otters may locate estrous females by following water-borne scents. Upon entering a group of conspecifics, sea otters often perform a ritualized greeting that appears to involve scent recognition. They have a complex vocal repertoire and at least one vocal modality, the scream, is recognizably distinct among individuals.

The high thermal conductivity of water presents severe challenges to aquatic mammals. This is especially so for small species (because of high surface/volume ratios) at high latitudes or in areas of intense upwelling (because of the cold water). The high capacity for heat loss probably explains why so few marine mammal species are small. As the smallest fully aquatic mammal, sea otters face an especially severe thermal challenge because they spend all or most of their lives immersed in cold water. This challenge has somehow been met by increasing heat production or reducing heat loss. Like other mustelids, the basal metabolic rate of sea otters is well above that predicted from the Kleiber curve, and sea otters gain additional heat by the specific dynamic activity (SDA) of digestion. Sea otters lack the BLUBBER layer that insulates cetaceans and pinnipeds, instead depending on their dense fur. Although fur in air is a superior insulator to blubber, it has three disadvantages in water: high maintenance costs, the inability to regulate heat flow, and compressibility at depth.

In order to maintain the fur's insulation, sea otters must groom almost continuously, an activity that consumes up to 10% of their time. Because fur (in contrast to blubber) is an inflexible insulator, sea otters require some means of facilitating heat loss during exercise. This apparently is accomplished mainly through the enlarged rear flippers, which are sparsely furred and high vascularized. The sea otter's flippers also may absorb heat from solar radiation. The compressibility of fur causes it to lose volume, and hence insulative value, with increased depth. The rate of heat loss is doubled at a depth of about 10 m (the equivalent of two atmospheres) and is essentially unrestricted during deep dives. Water conservation also presents a challenge to sea otters because they feed primarily on marine invertebrates, which are isotonic to seawater. They meet this challenge with a large and efficient kidney.

Like all marine mammals, sea otters must dive to feed. This poses three problems: oxygen debts are incurred, with associated changes in CO_2 concentration and pH; lactate buildups occur during prolonged, anaerobic dives; and rapid and extreme pressure forces nitrogen from the lungs into blood and other tissues, resulting in nitrogen toxicity and "the bends." Although sea otters do not dive to such extreme depths as many other marine mammals, they are considerably better divers than other otter species, being able to attain depths in excess of 100 m. The respiratory system is appropriately modified for deep diving. For instance, the tracheal length–width ratio is less in sea otters than in river otters, thus permitting more rapid and complete air exchange before and after a dive. Like the pinnipeds, sea otters have cartilaginous airways that empty directly into their alveoli, thus ensuring patency until alveolar compression collapse during deep dives. Sea otters also have comparatively large lungs that provide both oxygen storage and increased buoyancy. Although the sea otter's blood hemoglobin concentration is similar to that of most terrestrial mammals and less than that of most phocid seals, oxygen–hemoglobin affinity in the sea otter is relatively high, thus increasing their blood–oxygen storage capacity.

C. Locomotion

Locomotor efficiency of the semiaquatic otters is reduced on both land and in the water compared with fully terrestrial or fully aquatic mammals (Fig. 4). Among the lutrines, sea otters possess the most extreme modifications for aquatic propulsion. These include enlarged, flipper-like hindlimbs, a loosely articulated skeleton that permits increased flexibility, reduced forelimb length, and an increased tendency toward body movement and away from paddling in swimming.

The sea otter has retained a limited degree of terrestrial mobility while exhibiting morphological adaptations for improved aquatic locomotion. They rarely leave the water in some areas, whereas in others they haul out regularly. In rare instances, when sea ice prevents foraging north of the Alaska Peninsula, sea otters travel many kilometers inland in an apparent attempt to reach the Pacific Ocean.

Two types of terrestrial LOCOMOTION—walking and bounding—follow the pattern typical of terrestrial carnivores. In walking, the general motion is one of forward movement of alternate limbs in a rolling gait parallel to the long axis of the body. The head and neck are held close to the ground and the back may be horizontal, with the abdominal region contacting the ground, or the back may be arched. In some instances, sea otters pull themselves along the substrate with their fore paws. Running, described in *Lontra* spp. as a rapid forward movement in the same pattern as walking, has apparently been lost in the sea otter. Rapid movement of the sea otter on land is accomplished through bounding, where both forelimbs are moved forward simultaneously followed by simultaneous forward movement of the hindlimbs. During the bounding motion the back is usually highly arched. A maximum velocity on land of about 5 m/sec can be achieved in short bursts.

There are three forms of aquatic locomotion. Two of these are used predominantly while on the sea surface. One is accomplished by a sweeping motion of the tail and is used for

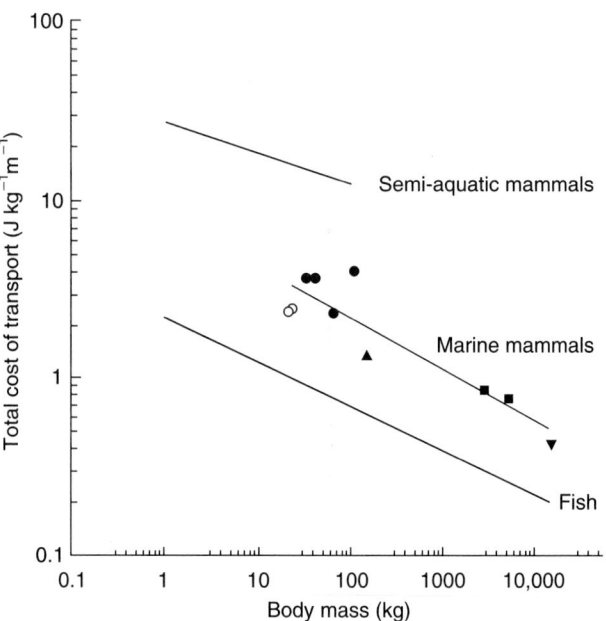

Figure 4 *The cost of transport for terrestrial, semiaquatic, and full aquatic mammals. From Williams (1999).*

slow movement while feeding, grooming, and maintaining its position while resting. The other, a paddling motion, consists of vertical thrusts and recovery of the hindlimbs while in either a supine or a pronate position. This type of movement is typically used during long-distance travel and may be interspersed with submerged travel, grooming, or foraging behavior. Paddling velocities range from about 0.5 to 1.0 m/sec.

A third means of aquatic locomotion in sea otters is accomplished by craniocaudal thrusts of the pelvic limbs, often including bending of the lumbar, sacral, and caudal regions for increased speed. This movement is used in diving. While swimming, the posterior margins of the hind flippers approximate the lunate pattern and undulating movement of the caudal fin of cetaceans. The sea otter loses swimming efficiency in the resistance and turbulence during the recovery stroke and through spaces between the flippers and tail.

Travel velocities over short distances (<3 km) range from about 0.5 to 0.7 m/sec, with a maximum of about 2.5 m/sec. Estimated average rates of travel over long distances (10–75 km) range from 0.16 to 1.5 m/sec. Submerged swimming speeds average about 1.0 m/sec.

D. Diving

Diving occurs during grooming, traveling, and foraging activities. Grooming dives usually occur before feeding or resting periods and are of short duration and shallow depth. Because locomotion is more efficient underwater than on the surface, otters frequently make relatively long (30–60 sec), shallow dives while traveling between resting and FEEDING areas.

The characteristics of foraging dives have been described at numerous locations within the sea otter's range. Two general types of data have been obtained: direct visual observations and remotely acquired information through radiotelemetry. Dive attributes from visual observations include dive duration, surface intervals between dives, and approximate water depths at the estimated dive locations. These data are inherently biased against animals foraging further from shore. Thus, estimates of average and maximum dive duration from animals instrumented with radio transmitters are substantially longer than those obtained visually.

Mean SWIMMING speeds during descent and ascent in foraging dives average about 1 m/sec. In California, average dive times vary with age and sex, with maximum values estimated from juvenile males and minimum values from adult females with dependent pups. The differences likely reflect differential habitat utilization, with juvenile males foraging in deeper water offshore and females with pups foraging in shallower water nearshore. Maximum reported dive durations are 246 sec in California and 260 sec in Alaska. Dive times and their interceding surface intervals correlate with water depth, although the deepest dives are not necessarily associated with maximum dive times. Surface intervals are highly correlated with prey size and type, with the longest intervals allied with the largest prey, thus reflecting associated increases in handling and consumption times. Sea otters commonly dive to depths exceeding 40 m in the Aleutian Islands, and there is one record of a sea otter drowned in a crab pot set in 91 m of water.

Information obtained using time–depth recorders (TDRs) from Southeast Alaska further indicates individual- and gender-related differences in mean and maximum foraging dive depth distributions. Generally, adult females dive to shallower depths than adult males, although some females regularly dive to depths exceeding 60 m. Most adult males foraged at depths between 40 and 60 m, although several repeatedly dove to depths exceeding 60 m. Maximum respective dive depths were 76 and 100 m for females and males, respectively.

E. Feeding Ecology

With a few notable exceptions, sea otters forage on sessile or slow-moving invertebrates from soft sediments, rocky reefs, and kelp canopy habitats (Table I). Prey are captured mainly with the forepaws rather than in the mouth. VISION probably is used to capture highly motile prey, such as fish. However, the fact that sea otters forage at night, in highly turbid water, and at great depths suggests that vision is not a requisite sensory modality for successful feeding.

Sea otters use their enlarged molariform teeth to crush the exoskeletons of their invertebrate prey. Tool use has developed in both the sea otter and Cape clawless otter as a further aid to foraging. This is accomplished by using rocks or other hard objects to break open the exoskeletons of their invertebrate prey. Tool use is unknown in all other mammals, except primates. Tool use renders any marine invertebrate, regardless of size or shell thickness, vulnerable to sea otter predation.

Foraging for infauna requires excavation of sediments, accomplished by digging with the forepaws while offsetting the effects of neutral buoyancy with the hind flippers. Excavations for burrowing clams may be up to 1 m in depth and require displacement of large amounts of sediments. Sea otters also obtain prey by pirating it from conspecifics. Territorial males most commonly employ this method by stealing food from females that forage within their territories.

While more than 150 prey species are known to be consumed by sea otters, only a few of these predominate in the diet, depending on location, habitat type, season, and length of occupation. In California, otters foraging over rocky substrates and in kelp forests mainly consume decapod crustaceans, gastropod and bivalve molluscs, and echinoderms. In protected bays with soft sediments, otters mainly consume infaunal bivalves (*Saxidomus nuttalii* and *Tresus nuttalltii*), whereas along exposed coasts of soft sediments, *Tivela stultorum* is a common prey. Important prey in Washington include crabs (species of *Cancer* and *Pugettia*), octopus (*Octopus* sp.), the intertidal clam (*Prototheca* sp.), sea cucumbers (*Cucumaria miniata*), and the red sea urchin (*S. franciscanus*). Less frequently recovered prey include the gumboot chiton (*Cryptochiton stelleri*) and the purple sea urchin (*S. pupuratus*). The predominantly soft sediment habitats of southeast Alaska, Prince William Sound, and Kodiak Island support populations of clams that in turn are the primary prey of sea otters. Throughout most of southeast Alaska, burrowing bivalve clams (species of *Saxidomus, Prototheca, Macoma,* and *Mya*) predominate in the sea otters diet, accounting for 50% of the identified prey. In Prince William Sound and Kodiak, clams account for 34–100% of the otter's prey. Mussels (*Mytilus trossulus*) apparently become more important as the length of occupation by sea otters increases, ranging from 0% at newly occupied sites at Kodiak

TABLE I
Prey Items of the Sea Otter[a]

Prey	California	Oregon	Southeast Alaska	Aleutian Islands	Commander Islands	Prince William Sound	Kodiak archipelago	Washington	Shumagian Islands
Echiura									
Echiurus echiurus	.	.	X	.	.	X	X	.	.
Urechis caupo	X	.	.	.	.	.	.	.	.
Sipuncula	.	.	.	.	.	.	.	X	.
Nemertea									
Emplectonema sp.	.	.	.	X	.	.	.	.	.
Annelida									
Polychaeta									
Arenicola sp.	.	.	.	X	.	.	.	.	.
Eudistylia polymorpha	X	.	.	.	.	.	.	.	.
Eudistylia sp.	.	.	.	.	.	X	X	.	.
Nereis sp.	.	.	.	X	.	X	.	.	.
Nereis vexillosa	X	.	.	.	.	.	.	.	.
Arthropoda									
Crustacea									
Cirripedia									
Thoracica									
Balanus cariosus	.	.	.	.	X	.	.	.	.
B. nubilus	X	.	X	.	.	.	.	X	.
Balanus sp.	.	.	X	.	.	.	.	.	.
Lepas anatifera	.	.	.	.	X	.	.	.	.
Malacostraca									
Isopoda									
Idotea sp.	.	.	.	X	.	.	.	.	.
Isopod (unidentified)	.	.	.	X	.	.	.	.	.
Amphipoda									
Amphipod (unidentified)	.	.	.	X	.	.	.	.	.
Gammarus sp.	.	.	.	.	X	.	.	.	.
Decapoda									
Blepharipoda occidentalis	X	.	.	.	.	.	.	.	.
Cancer antennarius	X	X	.	.	.	.	.	X	.
C. gracilis	X	.	.	.	.	.	.	.	.
C. magister	X	X	X	.	.	X	X	X	X
C. oregonensis	.	.	.	X	.	X	.	.	.
C. productus	X	X	.	.	.	.	.	X	.
Cancer sp.	.	.	X	X	.	.	.	X	.
Chionecetes bairdi	.	.	X	X	.	.	.	.	.
C. opillio	.	.	.	.	X	.	.	.	.
Cryptolithodes sitchensis	X	.	.	.	.	.	.	.	.
Dermaturus mandtii	.	.	.	X	.	.	.	.	.
Emerita analoga	X	.	.	.	.	.	.	.	.
Hapalogaster cavicauda	X	.	.	.	.	.	.	.	.
H. grebnitzkii	.	.	.	.	X	.	.	.	.
Hemigrapsus sp.	X	.	.	.	.	.	.	.	.
Hyas coarctatus	.	.	.	.	X	.	.	.	.
Lopholithodes foraminatus	X	.	.	.	.	.	.	.	.
Loxorhynchus crispatus	X	.	.	.	.	.	.	.	.
Oregonia gracilis	.	.	X	.	.	.	.	.	.
Pachygrapsus crassipes	X	.	.	.	.	.	.	.	.
Paguristes sp.	X	.	.	.	.	.	.	.	.
Pagurus gilli	.	.	.	X	.	.	.	.	.
P. hirusutiusculus	.	.	.	X	X	.	.	.	.
Pagurus sp.	.	.	.	X	.	.	.	.	.
Panulirus interruptus	X	.	.	.	.	.	.	.	.
Paralithodes camtchaticus	.	.	X	.	.	.	.	.	.
Placetron wosnessenski	.	.	.	X	.	.	.	.	.
Pleuroncodes planipes	X	.	.	.	.	.	.	.	.
Pugettia producta	X	X	.	.	.	.	.	X	.
P. richii	X	.	.	.	.	.	.	.	.

(continues)

TABLE I *(Continued)*

Prey	California	Oregon	Southeast Alaska	Aleutian Islands	Commander Islands	Prince William Sound	Kodiak archipelago	Washington	Shumagian Islands
Pugettia sp.	·	·	·	·	·	·	·	X	·
Sclerocrangon boreas	·	·	·	X	·	·	·	·	·
Telmessus cheiragonus	·	·	X	X	·	X	X	·	·
Mollusca									
Gastropoda									
Astraea gibberosa	X	·	·	·	·	·	·	·	·
A. undosa	X	·	·	·	·	·	·	·	·
Argibuccinium oregonensis	·	·	·	X	·	·	·	·	·
Buccinium sp.	·	·	·	X	·	·	·	·	·
Calliostoma sp.	X	·	·	·	·	·	·	·	·
Crepidula adunca	X	·	·	·	·	·	·	·	·
Fusitriton oregonensis	·	·	X	·	·	X	·	·	·
Neptunia sp.	·	·	X	·	·	X	·	·	·
Haliotis cracherodii	X	·	·	·	·	·	·	·	·
H. kamtschatkana	·	·	X	·	·	·	·	·	·
H. rufescens	X	·	·	·	·	·	·	·	·
H. walallensisa	X	·	·	·	·	·	·	·	·
Haliotis sp.	·	X	·	·	·	·	·	·	·
Lottia gigantea	X	·	·	·	·	·	·	·	·
L. ochracea	·	·	·	·	X	·	·	·	·
Megathura crenulata	X	·	·	·	·	·	·	·	·
Natica clausa	·	·	·	X	X	·	·	·	·
Notoacmaea persona	·	·	·	·	·	X	·	·	·
Polinices lewisii	X	·	·	·	·	·	·	·	·
Tectura spp.	·	·	·	X	·	·	·	·	·
Tegula brunnea	X	X	·	·	·	·	·	·	·
T. funebralis	X	·	·	·	·	·	·	·	·
T. montereyi	X	·	·	·	·	·	·	·	·
T. pulligo	X	·	·	·	·	·	·	·	·
Tegula sp.	·	X	·	·	·	·	·	·	·
Thais sp.	·	·	·	X	·	·	·	·	·
Bivalvia									
Chlamys sp.	·	·	·	·	·	X	·	·	·
Clinocardium ciliatum	·	·	·	X	·	·	·	·	·
Clinocardium facanum	X	·	·	·	·	·	·	·	·
C. nuttallii	X	·	X	·	·	X	X	·	·
Entodesma navicula	·	·	X	·	·	·	·	·	·
Gari californica	X	·	X	·	·	X	·	·	·
Hiatella arctica	·	·	·	·	·	X	·	·	·
Hinnites giganteus	X	·	·	·	·	·	·	·	·
H. multirugosus	·	X	·	·	·	·	·	·	·
Humilaria kenerlia	·	·	X	·	·	X	X	·	X
Liocyma viridis	·	·	·	X	·	·	·	·	·
Macoma incongrua	·	·	·	·	·	X	·	·	·
Macoma inquinata	·	·	·	·	·	X	X	·	·
M. nasuta	·	·	X	·	·	·	·	X	·
Macoma sp.	·	·	·	X	·	·	X	·	·
Mactromeris polynyma	·	·	X	·	X	·	·	·	·
Modiolus modiolus	X	·	X	X	X	X	X	·	·
Musculus niger	·	·	·	·	X	·	·	·	·
M. vernicosa	·	·	·	·	X	·	·	·	·
Musculus sp.	·	·	·	·	·	·	·	·	X
Mytilus californianus	X	X	X	·	·	·	·	X	·
M. trossulus	X	·	X	X	X	X	X	·	·
Mya arenaria	·	·	X	·	·	X	X	·	·
M. truncata	·	·	X	·	X	X	X	·	·
Panopea generosa	·	·	X	·	·	·	·	·	X
Pecten beringianus	·	·	·	X	·	·	·	·	·
P. islandica	·	·	·	X	·	·	·	·	·
Pododesmus cepio	X	·	·	·	·	·	·	·	·
P. macroschisma	·	·	X	X	X	·	·	·	·
Protothaca staminea	·	X	X	·	·	X	X	·	X

(continues)

TABLE I (*Continued*)

Prey	California	Oregon	Southeast Alaska	Aleutian Islands	Commander Islands	Prince William Sound	Kodiak archipelago	Washington	Shumagian Islands
Protothaca sp.	X	X	.	.	.	.	.	X	.
Saxidomus giganteus	.	.	X	.	.	X	X	X	X
S. nuttalli	X	X	.	.	.	.	.	.	.
Saxidomus sp.	.	X	.	.	.	.	.	.	.
Serripes groenlandicus	.	.	X	X	.	X	.	.	.
Siliqua patula	X	.	.	.	X	.	.	.	.
Solen sicarius	X	.	.	.	.	.	.	.	.
Spisula hempelli	X	.	.	.	.	.	.	.	.
Tagelus californianus	X	.	.	.	.	.	.	.	.
Tivela stultorum	X	.	.	.	.	.	.	.	.
Tresus capax	.	.	X	.	.	X	X	X	.
T. nuttallii	X	.	.	X	.	.	.	.	.
Venericardia paucicostatus	.	.	.	X	.	.	.	.	.
Voluplopsius beringi	.	.	.	.	X	.	.	.	.
Polyplacophora									
Callistochiton crassiocostatus	X	.	.	.	.	.	.	.	.
Cryptochiton stelleri	X	X	X	X	X	.	.	X	.
Ischnochiton sp.	X	.	.	.	.	.	.	.	.
Mopalia sp.	.	.	.	X	.	.	.	.	.
Schizoplax brandtii	.	.	.	X	.	.	.	.	.
Tonicella marmorea	.	.	.	X	.	.	.	.	.
T. ruber	.	.	.	X	.	.	.	.	.
Cephalopoda									
Loligo opalescens	X	.	.	.	.	.	.	.	.
Octopus sp.	X	X	X	X	.	X	X	X	.
Polypus sp.	.	.	.	.	X	.	.	.	.
Echinodermata									
Echinoidea									
Dendraster excentricus	X	.	.	X	.	.	.	.	.
Stronglyocentrotus drobachiensis	.	.	X	X	X	X	.	.	.
S. franciscanus	X	X	X	.	.	.	.	X	.
S. polyacanthus	.	.	.	X	X	.	.	.	.
S. purpuratus	X	X	X	.	.	.	.	X	.
Asteroidea									
Asterina miniata	X	.	.	X	.	.	.	.	.
Ceramaster sp.	.	.	.	X	.	.	.	.	.
Evasterias troschelii	.	.	.	.	.	X	.	.	.
Henricia sp.	.	.	.	X	.	.	.	.	.
Leptasterias sp.	.	.	.	X	.	.	.	.	.
Pisaster brevispinus	X	.	.	.	.	.	.	.	.
P. giganteus	X	.	.	.	.	.	.	.	.
P. ochraceus	X	X	.	.	.	.	.	X	.
Pycnopodia helianthoides	X	.	X	.	.	.	X	X	.
Solaster	.	.	X	.	.	.	.	.	.
Ophiuroidea									
Brittle stars (unidentified)	.	.	X	X	.	.	.	.	.
Gorgonocephalus eucnemis	X	.	X	.	.	.	.	.	.
Holothurioidea									
Cucumaria miniata	.	X	.	.	.	.	.	X	.
C. piperata	X	.	.	.	.	.	.	.	.
C. fallax	.	.	X	.	.	.	.	.	.
Cucumaria sp.	.	.	X	X	.	X	.	.	.
Parastichopus sp.	X	.	.	.	.	.	.	X	.
Chordata									
Ascidiacea									
Styela montereyensis	X	.	.	.	.	.	.	.	.
Tunicata	.	.	.	X	.	.	.	.	.
Pisces									
Ammodytes hexapterus	.	.	.	X	X	.	.	.	.
Anoplopoma fimbria	.	.	.	X	X	.	.	.	.
Aptocyclus ventricosus	.	.	.	X	X	.	.	.	.
Cottidae species	X	.	.	X	.	.	.	.	.

(*continues*)

TABLE I (Continued)

Prey	California	Oregon	Southeast Alaska	Aleutian Islands	Commander Islands	Prince William Sound	Kodiak archipelago	Washington	Shumagian Islands
Cyclopterichthys glaber	.	.	.	X	.	.	.	.	.
Embiotocidae species	X	.	.	.	.	.	.	.	.
Gadus morhua	.	.	.	X	X	.	.	.	.
Gymnocanthus pistilleger	.	.	.	X	.	.	.	.	.
Hexagrammos superciliosus	.	.	.	X	.	.	.	.	.
Hexagrammos sp.	X	.	.	.	X	.	.	.	.
Hemilepidotus hemilepidotus	.	.	.	X	.	.	.	.	.
H. jordani	.	.	.	.	X	.	.	.	.
Lepidopsetta bilineata	.	.	.	.	X	.	.	.	.
Mallotus villosus	.	.	.	.	X	.	.	.	.
Oncorhynchus nerka	.	.	.	.	X	.	.	.	.
Pleurogrammus monoterygius	.	.	.	X	.	.	.	.	.
Theragra chalcograma	.	.	.	.	X	.	.	.	.
Ophiodon elongatus (egg mass)	.	X						X	
Aves									
Anatidae									
Anas crecca	.	.	.	X	.	.	.	.	.
Melanitta perspicillata	X	.	.	.	.	.	.	.	.
Gaviidae									
Gavia immer	X	.	.	.	.	.	.	.	.
Laridae									
Larus sp.	X	.	.	.	.	.	.	.	.
Phalacrocoracidae									
Phalacrocorax sp.	X	.	.	.	.	.	.	.	.
Podicipedidae									
Aechmophorus occidentalis	X	.	.	.	.	.	.	.	.

[a] Updated from Riedman and Estes (1990).

to 22% in long occupied areas. Crabs (*C. magister*) were once important sea otter prey in eastern Prince William Sound, but apparently these have been depleted by otter foraging and thus are no longer eaten in large numbers. Sea urchins are minor components of the sea otter's diet in Prince William Sound and the Kodiak archipelago. In contrast, the sea otter's diet in the Aleutian, Commander, and Kuril Islands is dominated by sea urchins and a variety of fin fish (including hexagrammids, gaddids, cottids, perciformes, cyclopterids, and scorpaenids). Sea urchins tend to dominate the diet of low-density sea otter populations whereas fishes are consumed in populations near equilibrium density. For unknown reasons, fish are rarely consumed by sea otters in regions east of the Aleutian Islands.

Sea otters also take advantage of episodically abundant prey. Examples include squid (*Loligo* sp.) and pelagic red crabs (*Pleuroncodes planipes*) in California and smooth lumpsuckers (*Aptocyclus ventricosus*) in the Aleutian Islands. Pelagic red crabs appear in coastal waters of southern and central California during strong El Niño events, and vast numbers of lumpsuckers appear episodically in coastal waters of the western and central Aleutian Islands to spawn. Sea otters, on occasion, attack and consume sea birds, including teal, scoters, loons, gulls, grebes, and cormorants.

Dietary diversity increased through time as otter populations recolonized new habitats and grew toward resource limitation. Similar patterns of increased dietary diversity have been chronicled in the Aleutian Islands, Prince William Sound, and California. These changes are probably the consequence of otters reducing the abundance of their preferred prey.

Studies of marked sea otters in California have shown extreme individual variation in diet and foraging behavior. Most otters specialize on one to three prey types. This individual variation does not appear to be directly influenced by prey availability as different individuals often consume different prey at the same time and place. Dietary patterns, which appear to be inherited matrilineally, persist for years and may be lifelong characters of individuals. The causes and consequences of individual foraging patterns remain uncertain.

F. Community Ecology

1. Food Web Effects In 1960, Hairston, Smith, and Slobodkin proposed that "the world is green" because predators regulate herbivore populations, in turn releasing plants from regulation by herbivory. Sea otters and kelp forests provided early empirical support for this hypothesis. The sea otter's role in structuring kelp forest communities was discovered by contrasting islands in the Aleutian archipelago at which the species was present in large numbers or absent, a serendipitous consequence of the Pacific maritime fur trade. Shallow reef habitats at islands with abundant sea otters had few sea urchins and well-developed kelp forests, whereas abundant sea urchins had destroyed the kelp forests at islands lacking sea otters (Fig. 5). The explanations for this pattern is a straightforward consequence of what has since come to be known as a "trophic cascade." That is, sea urchin populations are regulated by sea otter PREDATION, in turn allowing the kelp forest to flourish in the absence of significant herbivory. When otters were removed,

Figure 5 *Alternate community states in areas with (kelp forests) and without (sea urchin barrens) sea otters. Photographs were taken of reef habitats at about 10-m depths in the western Aleutian Islands.*

sea urchins increased to such levels that deforestation occurred. Similar patterns between areas with and without sea otters have since been documented in Southeast Alaska, British Columbia, Washington, and California. Time series studies of areas in which sea otters have become reestablished or declined provide further evidence that the presence or absence of sea otters maintains these alternative states of community organi-

zation. Similar trophic cascades resulting from the effects of apex predators are now known for many systems.

Numerous indirect effects of the sea otter–urchin–kelp trophic cascade are known or suspected (Fig. 6). Most of these relate to the role of kelp as habitat and a source of production for other species of consumers in coastal food webs. The kelps (order Laminariales) and other species of macroalgae are extremely productive, in large measure because none of the essential ingredients for photosynthesis (light, water, CO_2, and nutrients) are limiting in shallow coastal waters at high latitudes. Thus, systems with and without sea otters (and thus with and without well-developed kelp forests) vary substantially in total productivity. This fact was confirmed using the naturally occurring stable carbon isotopes to measure the relative photosynthetic contributions of macro- vs microalgae across islands of the Aleutian archipelago with and without sea otters. Total production was estimated to be three to four times greater where sea otters were present. Further tests of the otter-productivity hypothesis were done by outplanting recently settled barnacles and mussels from a common source to intertidal and subtidal habitats at islands with and without sea otters. Growth rates were two to five times greater in otter/kelp-dominated systems.

The limiting effects of otters on sea urchins and other herbivorous invertebrates reduce the disturbance from herbivory, thus enhancing the strength of competitive interactions among kelp species. This effect has been demonstrated in the Aleutian Islands, Southeast Alaska, British Columbia, and central California. Further indirect effects on higher trophic level consumers result from increased production, altered habitat, or competition with otters for common food resources. The production effects are so strong that they almost certainly influence numerous co-occurring species. Bald eagle (*Haliaetus leucocephalus*) and harbor seal (*Phoca vitulina*) densities are substantially greater on islands of the Aleutian archipelago that also support abundant sea otter populations. Radically different DIETS and patterns of foraging behavior occur in glaucous winged gulls (*Larus glaucescens*) between sites with and without sea otters. Sea otters also appear to influence the population densities of benthic feeding sea ducks. Common eiders (*Somateria mollissima*), which feed mainly on sea urchins, mollusks, and other benthic invertebrates, occur at higher densities where otters are rare or absent than where they are abundant. Otters also influence predator–prey interactions between sea stars and mussels in the Aleutian archipelago. The biomass of sea stars at Attu Island declined by roughly two orders of magnitude following the spread of sea otters into Massacre Bay along the island's southeastern shore, in turn substantially increasing the survival rates of mussels and barnacles. This example is of particular interest because the influence of sea star predation on mussel beds (enhanced species diversity by the predation-induced reduction in competitive exclusion by the dominant mussels) is one of the most well-known and influential paradigms in ecology. The influence of sea otters on kelp beds even appears to elevate the environmental carrying capacity for otters themselves. Although sea otters limit the abundance of their benthic invertebrate prey, coastal fish populations are enhanced by increased production and the habitat

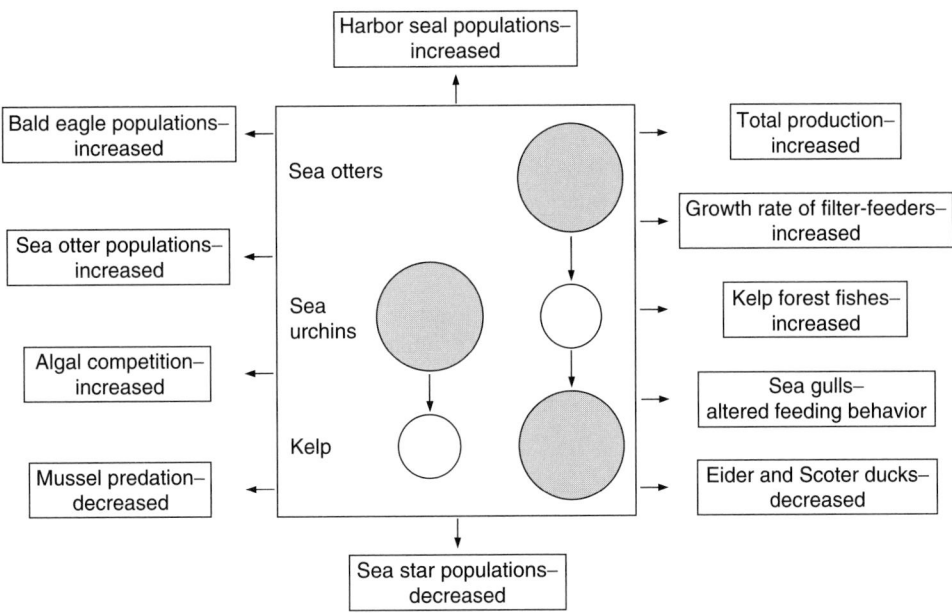

Figure 6 *A schematic of trophic cascades in kelp forest systems with and without sea otters (central box). Known or suspected indirect effects of these interactions are represented at the perhipheries.*

provided by kelp forests. Besides elevating the equilibrium density, this interaction may stabilize sea otter populations because the fishes are less vulnerable than benthic invertebrates to population regulation by sea otter predation.

The top-down influences of apex predators through trophic cascades have been incorporated into a conceptual model relating the strength of plant–herbivore interactions to trophic complexity (Fig. 7). The model predicts that plant–herbivore interactions should be strong in food chains with an even number of trophic levels (e.g., two, four, and so on) and relatively weak for odd-numbered food chains (e.g., one, three, five, and so on). The accumulated evidence from three decades of research on sea otters and kelp forests is highly consistent with that view.

2. Evolutionary Forces Strong species interactions maintained over sufficiently large scales of space and time should inevitably lead to selective responses in the interacting species. This expectation, coupled with predicted differences in the strength of plant–herbivore interactions between odd- vs even-numbered food webs, has led to a view that kelp forests in the North Pacific Ocean evolved in the absence of intense herbivory. While the kelps have no FOSSIL RECORD, a variety of evidence indicates that they radiated in the North Pacific Ocean during the late Cenozoic in concert with sea otters, their recent ancestors, and other groups of benthic feeding marine mammals. The hypothesis that predators decoupled the coevolution of kelps and their herbivores was evaluated by contrasting various features of marine plants, invertebrate herbivores, and plant–herbivore interactions between North Pacific and Australasian kelp forests. Australasian kelp forests were chosen for this contrast because they lack predators of comparable effect to sea otters. Because kelp forest systems in the North Pacific

and Australasia, respectively, evolved as odd-numbered (three) vs even-numbered (two) systems, the coevolution of plant defenses and herbivore resistance to those defenses were hypothesized to be strong in Australasia and weak in the North Pacific.

Marine algae were known to use secondary metabolites as their chief defense against herbivory. Therefore, the initial test of this evolutionary hypothesis was to compare the diversity and concentration of secondary metabolites between North Pacific and Australasian seaweeds. This comparison demonstrated that phlorotannins (the principal secondary metabolites in brown algae) were roughly an order of magnitude more concentrated in southern hemisphere kelps, in some cases approaching 20% dry

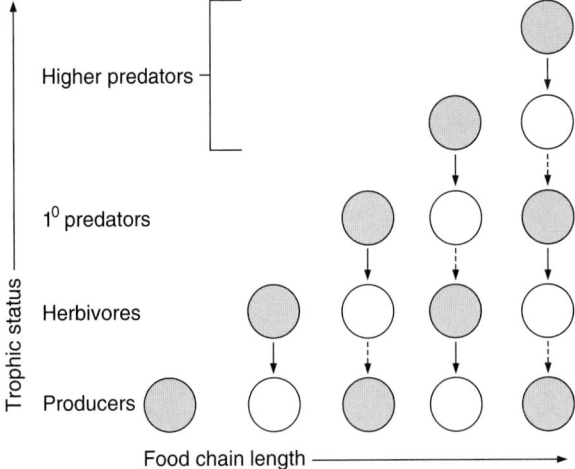

Figure 7 *Fretwell's (1987) model of alternating plant–herbivore interaction strength with increasing food chain length.*

weight. A second prediction of the hypothesis was that herbivore resistance to algal secondary metabolites is stronger in Australasian than in North Pacific species. This prediction was tested by extracting the metabolites from both North Pacific and Australasian seaweeds and measuring their strength of feeding deterrence to Pacific and Australasian herbivores. Regardless of their source, the strength of deterrence by phlorotannins was dramatically greater on North Pacific herbivores, thus suggesting that resistance to these compounds had evolved in southern hemisphere herbivores. This perspective and supporting data suggest that sea otters, their recent ancestors, and perhaps other predators have decoupled the coevolution of plant–herbivore interactions in North Pacific kelp forests, thus explaining why North Pacific kelp forests were so vulnerable to destructive grazing following the demise of sea otters and other benthic feeding predators.

G. Population Biology

1. Genetics Sea otters provide unique opportunities for the study of marine mammal population and conservation genetics. The sea otter's dependency on benthic/demersal habitats and limited diving capacity restrict its distribution to shallow coastal environments. Its small home ranges (generally <20 km of coast) probably restrict gene flow. Extensive human harvests reduced and fragmented sea otters into a small number of isolated colonies, from which all current populations are derived. Additionally, translocations from either one or two remnant populations have resulted in viable reestablished populations in Washington, British Columbia, and Southeast Alaska. Pleistocene glacial advances and retreats likely influenced genetic exchange. Isolation and extinctions caused by ice sheets extending over large coastal areas probably resulted during glacial maxima, whereas high levels of gene flow must have occurred as these areas were recolonized during periods of glacial retreat. Both morphological and genetic examinations of sea otters indicate some level of population structuring prior to 18th and 19th century fur harvests.

Sea otter skull size declines from Russia, across Alaska, and into California, and variation in skull morphology forms the basis for the three currently designated subspecies: Russia, *E. l. lutris;* Alaska, *E. l. kenyoni;* and California, *E. l. nereis.* Mitochondrial DNA (mtDNA) haplotype frequencies identify at least four distinct grouping of sea otters: California, Prince William Sound, the Kodiak/Aleutian/Commander Islands, and the Kuril Islands. The magnitude of difference in current mtDNA haplotype frequencies suggests at least some level of genetic differentiation among these groupings existed prior to the population declines and isolation that occurred between 1750 and 1900. The extant California population probably represents a monophyletic mtDNA lineage that contains two unique mtDNA haplotypes.

The extent to which long-term evolutionary processes and recent human exploitation have contributed to modern genetic differences among and within the subspecies of sea otters is unknown. The degree of difference in mtDNA haplotypes suggests that there has been little gene flow across the range of the species from California to the Kuril Islands of Russia. However, low levels of mtDNA sequence divergence in sea otters

across their range also suggest that major phylogenetic breaks or long-term barriers to gene flow do not exist.

There is concern over the potential for loss of genetic diversity stemming from severe population bottlenecks experienced by sea otter populations over the past two centuries. However, theoretical analyses suggest that as little as 23% of the original genetic variation was lost from the California population because of this bottleneck effect. Subsequent empirical studies show that a high proportion of diversity was maintained during the period of recent isolation among populations, thus supporting this theoretical result.

Reintroduced and remnant sea otter populations provide the opportunity to study the effects of the duration and magnitude of bottlenecks on genetic diversity and population growth rates. In sea otters, mtDNA haplotype diversity is significantly correlated with both minimum population size and the number of years a population remained at that minimum size. No relationship was detected between genetic diversity and population growth rates, although translocated populations demonstrated significantly higher annual growth rates (λ = 1.18–1.24) than remnant populations (λ = 1.06–1.09). Genetic diversity is greater in translocated populations that were derived from two source populations compared with those derived from a single source population. Haplotype frequencies in populations with estimated founding sizes of 4 and 28 animals (Washington and British Columbia) differed from the source populations, probably signaling the effect of genetic drift. This interpretation is supported by the fact that haplotype frequencies from sea otters in Southeast Alaska, a translocated population with an estimated founding size of 150 individuals, did not differ from the source populations.

As sea otters continue to reoccupy former habitat, and currently isolated populations become contiguous, we may be afforded the opportunity to view the process of genetic exchange across the species' range. Genetic differences currently observed among geographically isolated populations may diminish, reducing current levels of genetic population structure.

2. Demography As in all sexually reproducing species, sea otter populations are regulated by age- and sex-specific rates of reproduction and survival. These life history patterns in sea otters are more similar to those of the pinnipeds (with whom they share the ocean as a common environment) than they are to the other lutrines (with whom they share a more recent common ancestor). Perhaps the most remarkable life history feature of sea otters is that they almost invariably conceive and give birth to a single young, a character shared with other marine mammals but not other lutrines (Fig. 8).

Female sea otters typically become sexually mature at 3 years, occasionally earlier. The reproductive cycle is normally 1 year, with roughly 6 months from conception to birth and another 6 months from birth to weaning. Substantial learning occurs during this latter period. Primiparous females usually fail to successfully wean their pups, probably because they have not yet learned to be good mothers. Adult females apparently enter estrus within several hours to a few days after mother/pup separation (either from weaning or death). The majority of preweaning deaths occur shortly after birth. Thus, in many areas there

Figure 8 *A female sea otter with her pup.*

is a biannual peak in births, with the primary peak occurring in spring or early summer and the secondary peak (by females who failed in their previous cycle) in fall or early winter. Females continue to reproduce throughout life, with little or no evidence for either reproductive senescence or adjustments of fertility rate to environmental variation.

Population regulation in sea otters occurs largely or exclusively through variation in age- and sex-specific mortality rates. The high rates of population growth that have been observed in parts of Alaska, Canada, and Washington (17–20% per year, near the theoretical R_{max}) could only be realized if the probability of mortality from birth to senescence were very low. As growing populations become limited by resource availability, the mortality rate in young otters increases greatly. Data from the Aleutian Islands and central California indicate that about half of the pups born fail to reach weaning age. The probability of mortality during the next 6 months, while as yet unmeasured, also seems to be relatively high. Mortality rates from about 1 year of age to physiological senescence (10–15 years) are low, even in food-limited populations. Thus, the principal mechanism of population regulation in sea otters appears to be starvation-induced mortality early in life.

Although the sex ratio at birth is about 50:50, the postweaning mortality rate is greater in males than in females, thus resulting in a female-biased adult sex ratio. The elevated male mortality probably is caused by aggressive interactions with adult males, thus forcing juvenile males to disperse into lower-quality habitats. These mortality patterns are reflected in the age and sex composition of beach-cast carcasses at locations where populations are at or near equilibrium density. In California, the age composition of beach-cast sea otter carcasses more closely resembles the age structure of living populations, thus indicating an elevated mortality rate of prime-age animals. Because of the generally similar age-specific birth rates across all sea otter populations, this elevated mortality in prime-age animals is responsible for a depressed growth rate in the California sea otter population.

H. Behavior

The behavioral ecology of sea otters strongly reflects life in the sea. Vigorous grooming is the species' behavioral hallmark.

Sea otters probably spend more time and energy grooming their fur than any other mammal. This is accomplished by rubbing, rolling, blowing, and splashing. Grooming is necessary for cleaning and replenishing air to the under fur.

Sea otters are unusual among both carnivores and marine mammals in the generally small size of their home ranges, which typically includes no more than several miles of shoreline. Both males and females occasionally move longer distances for uncertain reasons. Extralimital sightings in central Baja California and near Wrangel Island in the Chuchi Sea demonstrate that, on occasion, individual sea otters move hundreds of miles.

Adult male sea otters maintain territories that in California average about 40 hectares. Adult females apparently move freely among these territories, but the territory holder aggressively excludes juvenile males. The precise function of male territoriality in sea otters is unclear, although ultimately territories serve to increase reproductive success. Adult males frequently harass females with large pups in an apparent effort to force separation, thus inducing the female to enter estrus and bear his offspring. Copulation occurs repeatedly during brief consorts, after which the males and females separate. A male grasps the female's nose in his mouth and rolls vigorously on the surface of the ocean to achieve intromission. Distinctive nose scars in adult females often result from this behavior. In severe cases, trauma to the nose and facial region may result in death to the female. Some males seem especially prone to such brutality. Upon killing their mate, these males usually continue to copulate with the corpse, sometimes for days following her death. Adult male sea otters have been observed attempting to copulate with young harbor seals, in two such instances killing the seals during the process.

This sea otter's polygynous mating system (and the resulting high male libido) likely evolved in response to their unusually high population density, thus promoting male competition for females as the limiting resource in sexual reproduction. Polygynous mating systems are typical of all otariids and some phocids in temperate latitude systems but apparently are rare or absent in other species of otters. As is true for other polygynous species, male sea otters provide no parental care.

I. Conservation and Management

1. Asia The status of sea otter populations throughout most of Asia is uncertain. Sea otters currently occupy all of their historic range in Russia and have been reported from northern Japan. However, present-day abundance is believed to be less than during the preharvest period, as areas of low density occur along the Kamchatka Peninsula and northern Japan. Sea otters recolonized Bering Island, in the Commander Islands, in about 1983. This population continued to increase until 1990, at which time it declined by about 40%. The population has remained stable since that time at about 3500 individuals. Systematic surveys of sea otter populations have not been done throughout most of Asia as recent governmental change has resulted in dramatic reductions in support of CONSERVATION and research programs.

2. Alaska Remnant and reintroduced sea otter populations experienced rapid and widespread recovery during most of the

20th century, probably due in large measure to the superabundant prey resources that developed in their absence. In some areas these growing populations eventually attained levels where food or other resources became limiting. These patterns of recovery and growth have been heralded as one of conservation's great success stories. However, there are new concerns over the sea otter's long-term welfare.

A. OIL SPILLS AND OTHER FORMS OF MARINE POLLUTION.   The *Exxon Valdez* oil spill killed large numbers of sea otters—the minimum mortality estimate was 750 individuals. Moreover, in the 10 years since the *Exxon Valdez* spill, sea otter populations in oil-impacted areas have recovered at an annual rate of 3.3% per year, far below the growth rate observed in this same population prior to the spill. This had led some to conclude that chronic effects from the *Exxon Valdez* spill continue in Prince William Sound.

In the early 1990s, high levels of PCBs were discovered in sea otters from the Aleutian archipelago. These now appear to be associated with past military activities in the area. Although concentrations measured in sea otter tissues exceed those known to be harmful in mink, presently there is no other evidence that they are having detrimental effects on sea otters.

B. HUMAN HARVESTS.   Alaska natives are permitted to harvest sea otters for traditional purposes. Presently, most harvested animals are taken from Southeast Alaska and the Kodiak archipelago. There are no restrictions on take and large numbers are currently being removed from some populations. Sustainable harvest levels and take quotas need to be developed from life table data and the spatial ecology of individual otters. Quotas also should take into account the status and genetic structure of populations, and the availability of food and space.

C. DECLINE IN CENTRAL ALEUTIANS.   During the 1990s the abundance of sea otters in the west-central Aleutian Islands has declined by about 80–90%. Available evidence indicates that these declines were caused by increased predation by killer whales. Killer whales may have switched from consuming pinnipeds to sea otters following the collapse of Steller sea lion and harbor seal populations in western Alaska during the late 1970s and 1980s. While the geographic extent of the sea otter declines appears to be large, additional surveys are needed to determine the range and magnitude of decline. As the result of these declines, sea urchin populations have increased, followed by a dramatic increase in the rate of herbivory and subsequent deforestation of the nearshore community (Fig. 9).

3. British Columbia   The sea otter population in British Columbia continues to grow and expand its range. Conflicts with various shellfisheries are the primary concern of resource managers, whereas others believe that the population is still precariously small. Sea otter harvests have been proposed but at present are not permitted under either national or provincial law.

4. Washington/Oregon   Like British Columbia, the Washington sea otter population continues to expand in numbers and range. From the relocation site near La Push on the outer coast, otters have spread around Cape Flattery eastward through the Straits of Juan de Fuca. These animals now occupy tribal waters of the Makah Indians and are rapidly depleting red urchin populations, which until recently supported a tribal fishery. Sea otters are presently absent from the coast of Oregon.

5. California   California sea otters have long been the objects of debate and controversy. Fishery interests have maligned sea otters because they compete with humans for various commercial and recreational resources. Conservationists, however, have been concerned with the small size and slow growth rate of the California sea otter population. Increased trafficking of oil tankers and outer continental shelf oil development along the California coast have heightened concern that a spill might reduce or even extinguish the California sea otter population. For these reasons, the California sea otter population was listed as "Threatened" in 1977 under the U.S. Endangered Species Act. The U.S. Fish and Wildlife Service adopted a recovery plan for the California Sea Otter in 1981. Congress subsequently established Public Law 625 in an effort to both enhance the California sea otter population and protect shellfisheries. In 1987, under provisions of this law, sea otters were reintroduced to San Nicolas Island in southern California. The otter population at San Nicolas Island dwindled to about 15 individuals over the next several years, after which it remained relatively stable through the 1990s, despite the birth of at least 50 pups. Reasons for the growth failure of the population at San Nicolas Island are uncertain.

For most of the 20th century, the California sea otter maintained a slow but steady increase in numbers and range. There was a period of decline from the mid-1970s to the early 1980s, apparently resulting from increased losses in a coastal set net fishery. The population resumed growth almost immediately following the implementation of protective measures. The California sea otter population again began a gradual decline in the mid-1990s. Reasons for the current decline are uncertain. Reasonable possibilities include increased pollution and disease, incidental losses in fishing gear, and food resource limitation.

During this most recent period of numerical decline, the California sea otter's range expanded southward, thus resulting in large numbers of otters moving south of Pt. Conception into the "no-otter zone" established by PL 625. Shellfishery interests are presently demanding that these otters be removed, whereas conservation groups are demanding that something be done to curtail the ongoing population decline. The U.S. Fish and Wildlife has yet to decide how they will respond to this unfortunate dilemma.

6. Mexico   Except for occasional wanders from central and southern California, sea otters are currently absent from the waters of Baja California.

V. Concluding Remarks

After having been hunted to the brink of extinction, sea otters have recovered dramatically. While the status of other marine living otters is less certain, these animals appear better off than their freshwater-living counterparts, many of which are in a very bad way because of human exploitation and various kinds of habitat destruction. The relative well-being of marine otters is thought to result from habitat destruction being less problematic

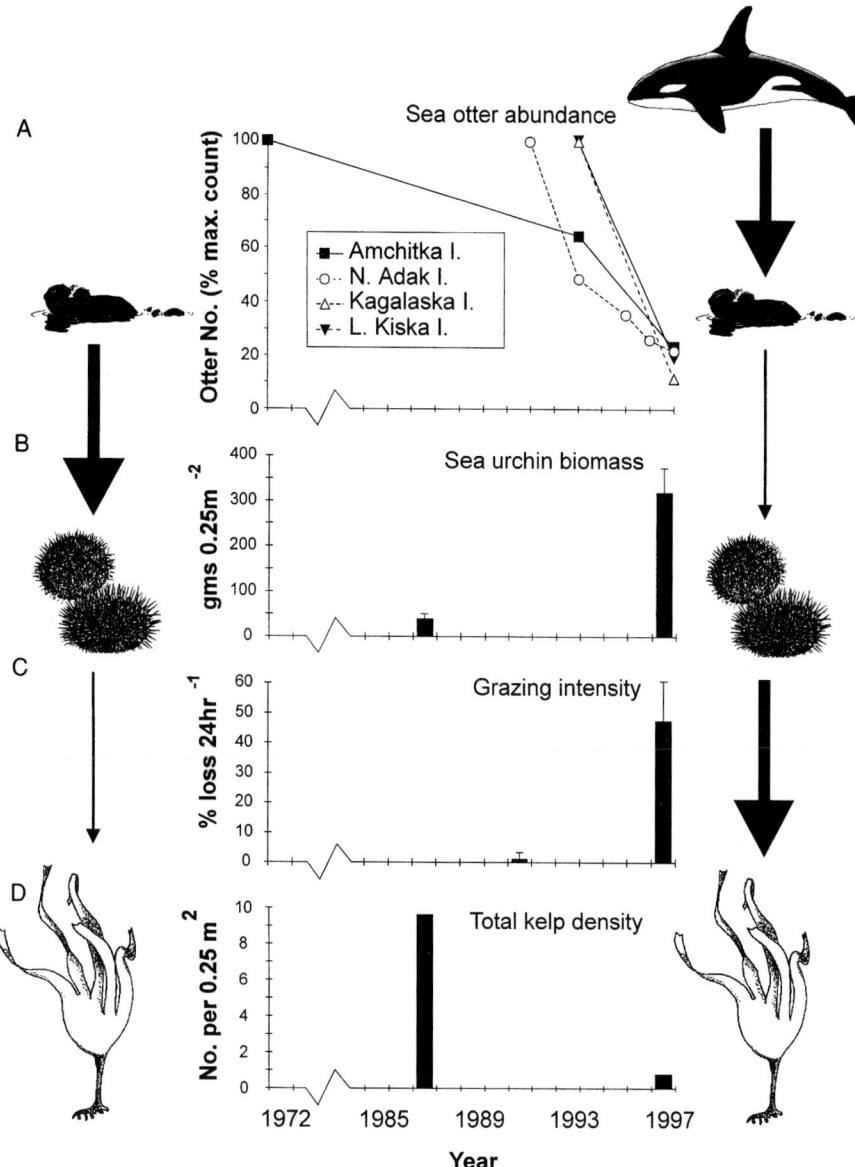

Figure 9 *Changes in sea otter abundance over time at several islands in the Aleutian archipelago, and concurrent changes in sea urchin biomass, grazing intensity, and kelp density measured from kelp forests at Adak Island. Error bars indicate 1 SE. The proposed mechanisms of change are portrayed in the margins: the one on the left shows how the kelp forest ecosystem was organized before the sea otter's decline and the one on the right shows how this ecosystem changed with the addition of killer whales, an apex predator. Heavy arrows represent strong trophic interactions; light arrows represent weak interactions. Reprinted with permission from Science 282, Estes et al., Killer whale predation on sea otters linking coastal with oceanic ecosystems, pp. 473–476. Copyright 1998 American Association for the Advancement of Science.*

in the sea than it is on land. While that may be true, coastal marine environments increasingly are the recipients of terrestrial pollution and are being overexploited by numerous fisheries. Furthermore, scientists are beginning to understand that the coastal zone is linked in important and complex ways with both the open sea on its one side and land on the other. The fact that people live in disproportionately high densities along the land–sea margin is a direct threat to coastal marine ecosystems and many of the species that depend on these systems. All of this does not bode well for the future of marine living otters. Overall, sea otters are far more abundant today than they were 50 years ago. However, the recent and ongoing declines of otter populations in Alaska and California may be among the first signs from these animals that coastal oceans are in peril.

In the process of recovery, expanding otter populations have come into conflict with fishery interests that developed during their absence. Similar conflicts surround other carnivores, but in few cases is the effect so apparent as it is with sea otters and shellfisheries. These conflicts strongly testify to the role apex predators often play in regulating their prey populations. Numerous other species and ecosystem-level processes are also influenced by the indirect effects of sea otter predation. Again, these examples have helped lead to the growing realization that apex predators often play complex but important roles in the maintenance of biodiversity, and thus that any affective strategy for biodiversity conservation must be sufficient to include viable populations of apex predators.

Acknowledgments

We thank Ronald Jameson for providing unpublished dietary information from Washington, Tim Tinker for preparing the maps, and especially Stacey Reese for editorial and technical assistance. Support for this work was provided by the U.S. Geological Survey.

See Also the Following Articles

Hunting of Marine Mammals ▪ Mustelidae ▪ Population Dynamics ▪ Territorial Behavior ▪ Thermoregulation

References

Berta, A., and Morgan, G. S. (1986). A new sea otter (Carnivora: Mustelidae) from the Late Miocene and Early Pliocene (Hemphilian) of North America. *J. Paleontol.* **59,** 809–819.

Bodkin, J. L., Ballachey, B. E., Cronin, M. A., and Scribner, K. T. (1999). Population demographics and genetic diversity in remnant and translocated populations of sea otters. *Conserv. Biol.* **13**(6), 1378–1385.

Duggins, D. O., Simenstad, C. A., and Estes, J. A. (1989). Magnification of secondary production by kelp detritus in coastal marine ecosystems. *Science* **245,** 170–173.

Estes, J. A. (1989). Adaptations for aquatic living in carnivores. *In* "Carnivore Behavior Ecology and Evolution" (J. L. Gittleman, ed.), pp. 242–282. Cornell Univ. Press.

Estes, J. A., and Duggins, D. O. (1995). Sea otters and kelp forests in Alaska: Generality and variation in a community ecology paradigm. *Ecol. Monogr.* **65,** 75–100.

Estes, J. A., Jameson, R. J., and Johnson, A. M. (1981). Food selection and some foraging tactics of sea otters. *In* "Worldwide Furbearer Conference Proceedings" (J. A. Chapman and J. D. Pursley, eds.), pp. 606–641. Frostburg, Maryland.

Estes, J. A., Tinker, M. T., Williams, T. M., and Doak, D. F. (1998). Killer whale predation on sea otters linking coastal with oceanic ecosystems. *Science* **282,** 473–476.

Fretwell, S. D. (1987). Food chain dynamics: The central theory of ecology? *Oikos* **50,** 291–301.

Hairston, N. G., Smith, F. E., and Slobodkin, L. B. (1960). Community structure, population control and competition. *Am. Nat.* **94,** 421–425.

Kenyon, K. W. (1969). The sea otter in the eastern Pacific Ocean. North American Fauna No. 68. Bureau of Sport Fisheries and Wildlife, Dept. of Interior, Washington, DC.

Koepfli, K. P., and Wayne, R. K. (1998). Phylogenetic relationships of otters (Carnivora: Mustelidae) based on mitochondrial cytochrome b sequences. *J. Zool.* **246**(4), 401–416.

Riedman, M. L., and Estes, J. A. (1990). The sea otter (*Enhydra lutris*): Behavior, ecology and natural history. U.S. Fish Wildlife Serv. Biol. Rep. **90**(14).

Williams, T. M. (1999). The evolution of cost efficient swimming in marine mammals: Limits to energetic optimization. *Phil. Trans. R. Soc. Lond. B* **354,** 193–201.

Pacific White-Sided Dolphin and Dusky Dolphin

Lagenorhynchus obliquidens and *L. obscurus*

KOEN VAN WAEREBEEK
Peruvian Centre for Cetacean Research, Lima

BERND WÜRSIG
Texas A&M University, Galveston

Because of their morphological similarities, the stocky dusky dolphin *Lagenorhynchus obscurus* from the Southern Hemisphere and the Pacific white-sided dolphin *Lagenorhynchus obliquidens* from the northern Pacific Ocean have long been recognized as phylogenetically closely related species despite the absence of a FOSSIL RECORD. Several researchers have gone so far as to suggest that the latter could almost equally well be regarded as a subspecies of the dusky dolphin. A close scrutiny of morphological and life history parameters, however, does not support this premise. Recent cytochrome *b* sequence analysis is consistent with the "sister-species" hypothesis, and the divergence date is estimated at 1.9–3.0 million years ago (Cipriano, 1997).

I. Pacific White-Sided Dolphin

Described in 1865 by Theodore Gill from three skulls collected in California, the boldly colored Pacific white-sided dolphin is black or dark gray on the back and posterior sides, as well as on the short snout, the leading edge of the tall dorsal fin, and the pointed flippers. The light gray thoracic patch is sharply delineated from the white belly by a thin dark line, in contrast with the dusky dolphin, which lacks this line and a sharp demarcation. Gray, linear dorsal flank blazes, often called "suspender stripes," project forward from the grayish flank patches along the back and disappear above the eyes (Fig. 1). The average adult size is 2.1–2.2 m with a body weight of 75–90 kg. Exceptionally, a length of 250 cm (males) and 236 cm (females) can be reached (Walker *et al.*, 1986) and a maximum weight of 181 kg, which is at least 50% heavier than the dusky dolphin.

A. Distribution

The mostly pelagic Pacific white-sided dolphin has a primarily temperate distribution across the North Pacific, from Taiwan to the Kurile and Commander Islands on the west and from 20–21°N to 61°N on the east, as well as more or less continuously across the North Pacific. Some seasonal shifts occur; while more common in coastal waters during fall and winter, these dolphins move offshore during spring and summer, in rough synchrony with movements of anchovy and other prey (Leatherwood *et al.*, 1984; Walker *et al.*, 1986).

No subspecies are recognized but either two or three populations are distinguished from the northeast Pacific and another two forms seem to exist in Japanese waters. *L. ognevi* named by Sleptsov in 1955 is a synonym of *L. obliquidens*.

Figure 1 *The Pacific white-sided dolphin* (Lagenorhynchus obliquidens) *is usually found in pelagic waters of the North Pacific. Photo by Robert L. Pitman.*

B. Behavior

These acrobatic dolphins often form large herds, averaging some 90 individuals, but groups of more than 3000 dolphins have been reported. Some segregation occurs based on sex and age. They readily associate with many other cetacean species and avidly BOW RIDE. Generalist feeders, off California and Japan, *L. obliquidens* mostly prey on lantern fishes (Myctophidae), anchovies, hake, and squid (Fitch and Brownell, 1968); off British Columbia they feed on herring, salmon, cod, shrimp, and capelin (Heise, 1997a). Males and females become sexually mature at 170–180 and 175–186 cm, respectively. The calving season peaks during the summer, the gestation period is 1 year, and females reach sexual maturity at an estimated 7.5 years old (Heise, 1997b).

C. Abundance and Exploitation

Pacific white-sided dolphins are still fairly numerous; however, like for most delphinids, no total abundance estimates are available. On the continental shelf off northern and central California in the mid-1980s, a population of 85,000 dolphins was proposed, but a 1991–1993 estimate for the same general area amounted to only 11,200 (Barlow and Gerrodette, 1996).

While it is not taken in drive fisheries, Japanese fishermen for decades have harpooned at times large numbers. A single company, for example, landed 697 animals in May and June 1949. Nonetheless, precise catch statistics were not compiled until fairly recently, and the long-term effect of the exploitation on the western population has not been evaluated. At least tens of Pacific white-sided dolphins are taken each year incidentally to a variety of coastal fisheries off western North America. The now outlawed multinational pelagic driftnet fisheries in the North Pacific may have caused significant mortality, but reliable figures have not been published. Pacific white-sided dolphins have been maintained in captivity in the United States, Japan, and Canada, but survival rates have been low.

II. Dusky Dolphin

The dusky dolphin was described by John W. Gray in 1828 from a stuffed skin and a single skull shipped to the British Museum from the Cape of Good Hope. This smallish dolphin can be recognized by its short beak and the bluish black to dark gray of the dorsal field contrasting with the white belly, as well as the light gray of the thoracic patch and two-pronged flank patch (Fig. 2). The dark lips and eyepatch also stand out. The falcate dorsal fin is two-toned with a dark leading edge. Unlike in *L. obliquidens,* the linear dorsal flank blaze does not extend farther anteriorly than about midbody. Heavily pigmented specimens are found off Peru and Argentina. The holotype of *Delphinus fitzroyi* (Waterhouse, 1838) caught off Argentina from Darwin's ship *Beagle* was such a melanized form. Both males and females off Peru reach sexual maturity at about 175 cm, while the largest two known individuals out of many hundreds measured 211 and 205 cm, respectively. They rarely exceed 100 kg in weight. No significant dimorphism is present except that adult males, much as in Pacific white-sided dolphins, have a significantly bigger and more strongly curved dorsal fin, presumably a secondary sexual characteristic.

Figure 2 *The dusky dolphin* (L. obscurus), *which may be closely related to its northern congener, the Pacific white-sided dolphin, occurs in Southern Hemisphere waters in both the Atlantic and the Pacific. Photo by K. Van Waerebeek.*

Gestation lasts for some 13 months. In Peru and New Zealand, most births occur in winter but a few neonates appear during other seasons. In Argentina, summer is the prime birth season. In Peru, sexual maturity for females is estimated at 4.3–5 years and for males at 3.8–4.7 years. The lower values, which are the most recent, possibly indicate a density-dependent response to heavy exploitation (Chávez-Lisambarth, 1998). Female Patagonian dusky dolphins mature at 6.3 years of age (Dans *et al.,* 1997).

A. Distribution

Dusky dolphins are distributed around South America, from northern Peru south to Cape Horn and from southern Patagonia north to about 36°S, including the Falkland Islands; off southwestern Africa from False Bay to Lobito Bay, Angola; and in New Zealand waters, including off Chatham and Campbell Islands. Populations of unknown size inhabit waters surrounding oceanic islands of the Tristan da Cunha archipelago in the mid-Atlantic, the Prince Edward Islands, and Amsterdam Island in the southern Indian Ocean (Van Waerebeek *et al.,* 1995). Dusky dolphins have been confirmed from southern Australia but seem to be rare there. Their coastal habits explain the discontinuous DISTRIBUTION across the temperate southern ocean, reproductive isolation, and cranial differences at a subspecific level (Brownell and Cipriano, 1999). Dusky dolphins from southwest Africa and New Zealand are some 8–10 cm shorter than the ca. 185 cm average adult body size of Peruvian specimens of both sexes (Van Waerebeek, 1993).

B. Behavior

Although dusky dolphins can move over great distances—a range of 780 km is confirmed—no well-defined seasonal MIGRATION patterns are apparent. However, the Argentina and New Zealand populations exhibit inshore–offshore movements both on a diurnal and on a seasonal scale. Dusky dolphins over the extensive continental shelf of Patagonia forage cooperatively on small schooling fishes during the day. On the east coast of New Zealand, where deep oceanic waters reach close to shore, animals typically feed at night on prey associated with the deep scattering layer (Würsig *et al.*, 1997). Some of these same dolphins, identified by natural marks on their bodies, have been found to feed cooperatively during daytime in the Marlborough Sounds on the northern edge of the South Island of New Zealand, reinforcing a generally held belief that these animals adapt behaviorally relative to habitat and prey availability patterns. Prey items include a variety of smaller fish; however, the diet everywhere tends to be dominated by anchovies, hakes, and several squid species. Surface feeding activity typically happens in large groups accompanied with extensive aerial display. It is believed that such aerial activity helps synchronize cooperative foraging and after-feeding social (including sexual) activities (Würsig *et al.*, 1989). Anomalous dentine deposition during El Niño years off Peru may reflect low foraging success. Killer whales (*Orcinus orca*) and some sharks are the only confirmed predators; dusky dolphins will enter very shallow water to avoid killer whales.

C. Abundance and Exploitation

In the absence of reliable ABUNDANCE ESTIMATES, the species' status is indeterminate. Unknown numbers are caught in gill nets in New Zealand waters, although current catches have dropped from those of the 1970s and 1980s. Off Argentina, from 1982–1994, a variable number of Patagonian dusky dolphins, typically a few hundred each year, died in midwater trawls. Arguably of biggest concern are fishery-related mortality levels in Peruvian coastal waters, where both directed and accidental entanglements in artisanal drift nets, as well as harpooning, have killed thousands each year since 1985. In the 1991–1993 period, an estimated 7000 animals per annum were captured, an exploitation thought to be unsustainable. It is believed but not confirmed that this level of human-caused mortality has dropped off in recent years, partly due to conservation legislation and partly because of the depletion of the population. Only a few individuals have been exhibited in aquaria in New Zealand, South Africa, and Australia, largely because dusky dolphins adapt rather poorly and have failed to reproduce successfully in captivity.

See Also the Following Articles

Delphinids, Overview ■ Speciation

References:

Barlow, J., and Gerrodette, T. (1996). Abundance of cetaceans in Californian waters based on 1991 and 1993 ship surveys. NOAA Tech. Memorandum NOAA-TM-NMFS-SWFSC-233.

Brownell, R. L., Jr., and Cipriano, F. (1999). Dusky dolphin, *Lagenorhynchus obscurus. In* "Handbook of Marine Mammals, Vol. 6." (Ridgway and Harrison, eds.), pp. 85–104. Academic Press, San Diego.

Chávez-Lisambarth, L. E. (1998). Age determination, growth and gonad maturation as reproductive parameters of dusky dolphin *Lagenorhynchus obscurus* (Gray, 1828) from Peruvian waters. Thesis Doctor rer. nat., University of Hamburg.

Cipriano, F. (1997). Antitropical distributions and speciation in dolphins of the genus *Lagenorhynchus*: A preliminary analysis. *In* "Molecular Genetics of Marine Mammals," *Soc. Mar. Mammal Spec. Pub.* 3, pp. 317–338.

Dans, S. L., Crespo, E. A., Pedraza, S. N., and Alonso, M. K. (1997). Notes on the reproductive biology of the female dusky dolphins (*Lagenorhynchus obscurus*) off the Patagonian coast. *Mar. Mamm. Sci.* **13**(2), 303–307.

Fitch, J. E., and Brownell, R. L., Jr. (1968). Fish otoliths in cetacean stomachs and their importance in interpreting feeding habits. *J. Fish. Res. Board Can.* **25**, 2561–2574.

Heise, K. (1997a). Diet and feeding behaviour of Pacific white-sided dolphins (*Lagenorhynchus obliquidens*) as revealed through the collection of prey fragments and stomach content analyses. *Rep. Int. Whal. Comm.* **47**, 807–815.

Heise, K. (1997b). Life history and population parameters of Pacific white-sided dolphins (*Lagenorhynchus obliquidens*). *Rep. Int. Whal. Comm.* **47**, 817–825.

Leatherwood, S., Reevers, R. R., Bowles, A. E., Stewart, B. S., and Goodrich, K. R. (1984). Distribution, seasonal movements, and abundance of Pacific white-sided dolphins in the eastern North Pacific. *Sci. Rep. Whales Res. Inst. Tokyo* **35**, 128–157.

Van Waerebeek, K. (1993). External features of the dusky dolphin *Lagenorhynchus obscurus* (Gray 1828) from Peruvian waters. *Estudios Oceanol.* **12**, 37–53.

Van Waerebeek, K., van Bree, P. J. H., and Best, P. B. (1995). On the identity of *Prodelphinus Petersii* Lütken, 1889 and records of dusky dolphins *Lagenorhynchus obscurus* (Gray, 1828) from the Indian and southern mid-Atlantic Oceans. *S. Afr. J. Mar. Sci.* **16**, 25–35.

Walker, A. W., Leatherwood, S., Goodrich, K. R., Perrin, W. F., and Stroud, R. K. (1986). Geographical variation and biology of the Pacific white-sided dolphin, *Lagenorhynchus obliquidens*, in the northeastern Pacific. *In* "Research on Dolphins" (M. M. Bryden and R. J. Harrison, eds.), pp. 441–465.

Würsig, B., Cipriano, F., Slooten, E., Constantine, R., Barr, K., and Yin, S. (1997). Dusky dolphins (*Lagenorhynchus obscurus*) off New Zealand: Status of present knowledge. *Intl. Whal. Comm. Sci. Rep.* 1996.

Würsig, B., Würsig, M., and Cipriano, F. (1989). Dolphins in different worlds. *Oceanus* **32**(1), 71–75.

Paenungulates

PASCAL TASSY
National Museum of Natural History, Paris, France

The name Paenungulata, i.e., "near-ungulates," was coined in 1945 by the American paleontologist George Gaylord Simpson to group seven orders of placental mammals in a major taxon ranked as a superorder. This grouping included three orders—Sirenia (sirenians or sea cows), Proboscidea (elephants), and Hyracoidea (hyraxes, also called "dassies")—and four extinct orders—Pantodonta (pantodonts), Dinocerata (uintatheres), Pyrotheria (pyrotheres), and Embrithopoda (arsinoitheres). It must be emphasized that Simpson included an extinct group in the Sirenia, the desmostylians, as suborder Desmostyliformes (in modern times usually considered an order, DESMOSTYLIA).

I. The Extant of Paenungulates

The contents and validity of the Paenungulata are controversial. Simpson himself acknowledged the fact that the superorder was "possibly polyphyletic." Probably amphibious, they were not as specialized to aquatic life as sirenians. Proboscideans are terrestrial but may have had an aquatic origin and hyracoids are terrestrial and arboreal (Fig. 1). The extinct orders were terrestrial and nearly all were unfamiliar herbivorous animals already diverse in the early Tertiary. Pantodonts were largely distributed in America and Eurasia, the largest forms being no larger than a cow. Uintatheres were frequent in North America and Asia and reached the size of living rhinoceroses and superficially resemble them with strange paired horns on both the front and the back of the skull. Pyrotheres were restricted to the early Tertiary of South America and developed features paralleling the elephants: the incisors were in the shape of short tusks and the nasal openings were retracted, indicating the presence of a trunk. Arsinoitheres (Fig. 2), first described in Africa (in the early Oligocene of Egypt), were also present in Eurasia, although rare. They reached the size of living rhinos. They possessed a gigantic pair of horns on the nasal bones and a much smaller one on the frontal bones.

According to Simpson, the "nucleus" of the Paenungulata was the association of proboscideans, sirenians, hyracoids, and the extinct arsinoitheres. Today it can be said that the nucleus of the nucleus is the grouping of the proboscideans and sirenians, now called the Tethytheria. The relationship of the tethytheres to other mammalian orders is much debated and, as a consequence, the content of Paenungulata remains an enigma.

However, most modern authors agree that pantodonts, uintatheres, and pyrotheres are not related to elephants and sirenians and are therefore not paenungulates. This leaves the question: Are Proboscidea, Sirenia, Hyracoidea, and Embrithopoda monophyletic and should they be called collectively Paenungulata? When only extant species are considered, this question can be phrased as: Are hyracoids closely related to elephants and sirenians?

II. Hyracoids

The order Hyracoidea includes eight species (the exact number is still debated), classified into three genera: *Procavia*

or rock hyrax (Fig. 3), *Dendrohyrax* or tree hyrax, and *Heterohyrax* or bush hyrax. They live in tropical and subtropical habitats of Africa. *Heterohyrax* is known in east Africa, and *Dendrohyrax* is distributed in central and western Africa. *Procavia* is largely distributed in Africa, but ranges north into Lebanon. Today hyracoids are small animals the size of a rabbit. Some fossil hyracoids were very large, such as tapir-sized *Megalohyrax eocaenus* found in the late Eocene and Oligocene of the Fayum area in Egypt or *Pliohyrax graecus* from the Miocene of Eurasia. The earliest fossil hyracoids are known from middle or lower Eocene sites from Algeria and associated large (*Titanohyrax*) and small (*Microhyrax*) forms.

Extant hyracoids have feet with three toes. The third is the largest, and the first and fifth toe are absent. The hands have four fingers; the third is the largest, the fifth is reduced, and the first is lacking. The pattern of hands and feet is "serial," meaning that the bones of the carpus and tarsus are stacked into separate columns. This is a derived condition compared to the generalized mammalian condition, which is alternate, where the bones of carpus and tarsus are stacked in overlapping rows, like bricks in a wall. Hyracoids also display a structure considered primitive for ungulates: the presence of a central bone located between the trapezoid and magnum. Aspects of the hyracoid ankle are unique. A strong medial process (malleolus medialis) of the tibia articulates into a deep fossa situated on the astragalus medial to the main tibial articular facet. The astragalus has a flat distal articular surface for the navicular bone.

The gait is plantigrade. The sole of the foot has a soft elastic pad that allows cohesion on irregular substratum, such as trunks and branches. Sudorific glands keep the pad moist. The molars are lophodont. The first incisors are enlarged. The canines show an unusual shape: they are premolariform and nearly identical to the premolars so that canines were long understood as first premolars.

The cranium shows unique specializations, such as the postorbital bar (processus zygomaticus) made by both parietal and frontal bones. Also the hyoid apparatus is unique in the fact that the lingual process is formed by the hypohyal and not, as found elsewhere (i.e., perissodactyls and bovids), the basihyal.

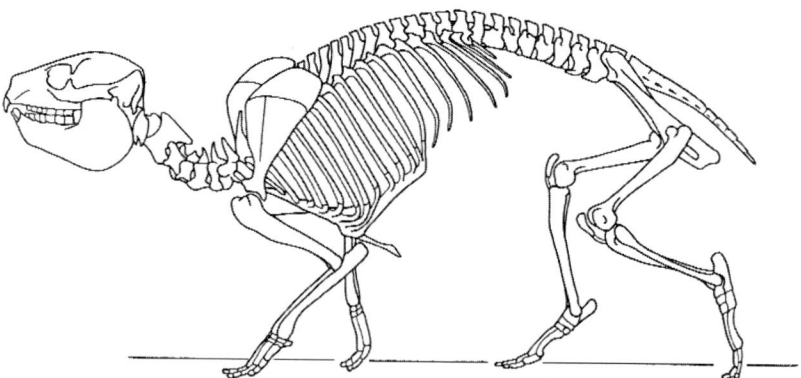

Figure 1 *Skeleton of the hyracoid* Procavia capensis. *Courtesy of M. S. Fischer. Length: 50 cm.*

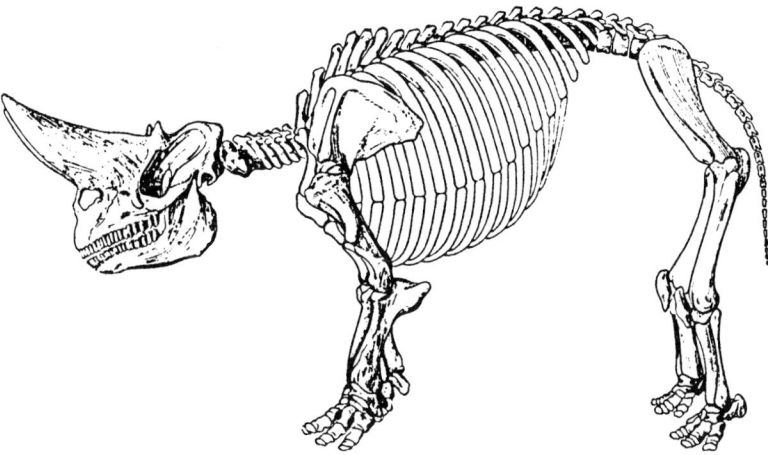

Figure 2 *Skeleton of the embrithopod* Arsinoitherium, *Fayûm, Egypt. Length: 3.50 m. From Andrews.*

Studies have shown that alleged primitive features, such as the plantigrady, were in fact derived secondarily. Ontogenetic development of the limbs shows that hyraxes are basic unguligrads.

III. Pantomesaxonian Interrelationships

Several authors classify extant perissodactyls, hyracoids, sirenians, and proboscideans in a clade called Pantomesaxonia (also named Altungulata). On the other hand, paenungulate characters that support a sister group relationship between hyracoids and tethytheres come from placentation and fetal membrane development and from the locomotor apparatus. The peculiar structure of hyracoid fetal membranes is also known in tubulidentates (the aardvark) and could be associated to a more inclusive group. The serial arrangement of carpals and tarsals is a classical character uniting paenungulates. However, this condition is also known in nonpaenungulate ungulates, such as *Phenacodus*, a genus known from the late Paleocene up to the late Eocene, and usually considered related to Perissodactyla (living tapirs, rhinoceroses, and horses). Among tethytheres, desmostylians do not show a taxeopod condition; hence a reversal is to be assumed for desmostylians. Other characters are clearly homoplastic, such as the amastoidy (lack of mastoid exposure on the cranium), which is also present in various taxa, such as rhinocerotids and pigs. The primitive condition is a reduced exposure (a possible pantomesaxonian character), which occurs in tapirs and in an extinct genus from the late Paleocene of China, *Radinskya*. The short face shared by living hyracoids and tethytheres is a homoplastic feature, as more primitive extinct hyracoids possessed a long face. On the other hand, different characters support sister group relationships between Hyracoidea and Perissodactyla. These include a flat distal articular surface on the astragalus for the navicular; the hoof with mesial side with horned fused lamellae; the course of the sternoscapularis muscle. The presence of an eustachian sac—a pharyngeal dilatation of the eustachian tube—has also been cited as a synapomorphy for hyracoids and perissodactyls. It is known in hyracoids, tapirs, and horses, but the black rhino *Diceros bicornis* lacks it. The eustachian sac changes the course of the internal carotid artery by pushing it away from the tympanic area of the osseous bulla. Because the course of the internal carotid is extrabullar in *Diceros*, this position is perhaps a reversal, an indication of a former possession of the eustachian sac in rhinocerotids. These characters support monophyly of hyracoids and perissodactyls and refute the monophyly of paenungulates. However, no molecular data support a sister group relationship between hyracoids and perissodactyls. On the contrary, various molecular analyses bring hyracoids closer to tethytheres with different exact topologies: sister group of tethytheres, or closer to sirenians, or closer to proboscideans.

In conclusion, although many morphological classifications and phylogenies suggest monophyly of pantomesaxonians, the question of monophyly or paraphyly of paenungulates is still

Figure 3 *The rock hyrax* Procavia capensis *in its type area (Table Mountain, Cape Town, South Africa). Photo by P. Tassy.*

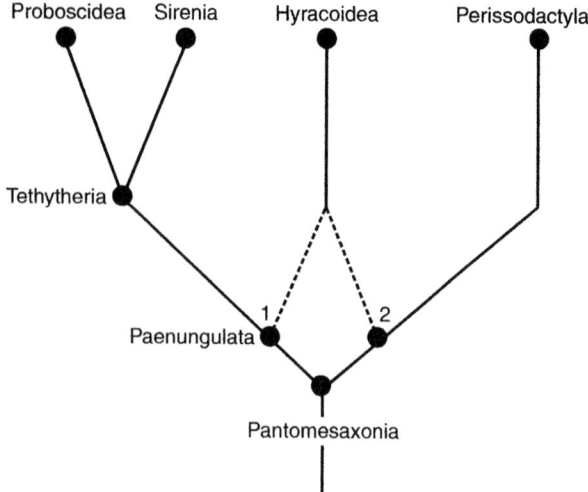

Figure 4 *Interrelationships of pantomesaxonians. If Hyracoidea is related to Perissodactyla (hypothesis 2) and not to Tethytheria (hypothesis 1), the Paenungulata (Proboscidea, Sirenia, and Hyracoidea) becomes paraphyletic.*

not satisfactorily resolved (Fig. 4). As a matter of fact, the recent classification of Mammalia by McKenna and Bell (1997) proposes a new name for extant elephants, sirenians, and hyracoids, i.e., for the nucleus of paenungulates sensu Simpson; the Uranotheria, introduced as an order made of three suborders: Hyracoidea, Tethytheria, and extinct Embrithopoda.

See Also the Following Articles

Perissodactyla ▪ Sirenian Evolution ▪ Tethytheria

References

Fischer, M. S. (1986). Die Stellung der Schliefer (Hyracoidea) im phylogenetischen System der Eutheria. *Cour. Forsch.-Inst. Senckenberg* **84**, 1–132.
Fischer, M. S., and Tassy, P. (1993). The interrelation between Proboscidea, Sirenia, Hyracoidea, and Mesaxonia: The morphological evidence. *In* "Mammal Phylogeny: Placentals" (F. S. Szalay, M. Novacek, and M. C. McKenna, eds.), pp. 217–234. Gustav Fischer, New York.
Gaudin, T. J., Wible, J. R., Hopson, J. A., and Turnbull, W. D. (1996). Reexamination of the morphological evidence for the cohort Epitheria (Mammalia, Eutheria). *J. Mamm. Evol.* **3**, 31–79.
McKenna, M. C., and Bell, S. K. (1997). "Classification of Mammals above the Species Level," p. 631. Columbia Univ. Press, New York.
Novacek, M. J. (1992). Fossils, topologies, missing data, and the higher level of phylogeny of eutherian mammals. *Syst. Biol.* **41**, 58–73.
Simpson, G. G. (1945). The principles of classification and a classification of mammals. *Bull. Am. Mus. Nat. Hist.* **85**, 1–350.
Springer, M. S., Amrine, H. M., Burk, A., and Stanhope, M. J. (1999). Additional support for Afrotheria and Paenungulata, the performance of mitochondrial versus nuclear genes, and the impact of data partitions with heterogeneous base composition. *Syst. Biol.* **48**, 65–75.
Thewissen, J. M. G., and Domning, D. P. (1992). The role of phenacodontids in the origin of the modern orders of ungulate mammals. *J. Vertebr. Paleontol.* **12**, 494–504.

Paleontology

J. G. M. THEWISSEN
Northeastern Ohio Universities College of Medicine, Rootstown

Paleontology is the study of ancient life and, as such, it is rooted in observations of fossils. Fossils are traces of living beings that are older than 10,000 years. In the case of marine mammals, the remains most commonly preserved as most fossils are teeth and bones. However, there are many other kinds of fossils: tracks and burrows of an animal in petrified soil, mummified pieces of skin, or fossilized excrements called coprolites.

I. Fossilization

Fossilization is the process that makes a fossil out of a living organism. In marine mammals, the first step in fossilization usually starts with the death and burial of its body. This step is studied by a branch of paleontology called taphonomy. In order for fossilization to continue, this initial stage needs to be short, avoiding destruction of the remains by scavengers, waves, or other biological and physical elements. After burial, the remains are subject to diagenesis: a series of processes that act on the cadaver after burial and before exposure on the earth's surface. There are many diagenetic processes, common ones include chemical alteration of the bones by minerals dissolving in circulating water, deformation of the fossil by high pressures deep in the Earth's crust, and precipitation of minerals in cavities in the fossil bones. Diagenesis may last millions of years; it ends when the fossil is brought to the surface by forces that move blocks of sediments around and expose them to erosion.

In general, fossilization is a rare event, and even though an ancient ocean may have abounded with marine mammal life, very few of these organisms are usually preserved as fossils. However, there are taphonomic processes that concentrate marine mammal remains. Waves in a shallow sea, for instance, may wash dead marine mammals onto the beach, making them abundant in the fossilized beach deposit.

II. Collecting and Curation

Collecting marine mammal fossils can be easy when established sites, known to be rich in fossils, are visited. More commonly, however, it takes a thorough understanding of the local geology to discover localities that produce marine mammals. Geological maps are studied to determine the location of areas with sedimentary rocks that record a marine environment. Aerial photos or an actual survey of these areas is necessary to find out whether these sedimentary rocks are actually exposed and accessible. These areas are then subjected to a thorough inspection by eye while walking or crawling over the locality, a process called prospecting. Individual, dispersed fossils can then be gathered; this is called surface collecting. When rich concentrations of fossil bones occur, buried fossils can be exposed by digging or chiseling. This method of fossil collecting is called quarrying and may involved specialized manual skills and tools, power tools, and glues to harden fossils that disintegrate easily. Delicate fossils sometimes need to be fortified by encasing them in plaster or synthetic casts called jackets to ensure that they are not damaged during transportation. A third collecting method is called screen washing and is usually used only to collect small fossil fragments that are in relatively unconsolidated sediment. Large amounts of sediment are sieved, either on dry screens or by washing them through sieves with water. Sediment falls through the sieves, whereas fossils stay on them and can be gathered. Taking copious notes is an essential part of collecting fossils for scientific purposes. The topographic location where the fossil is found needs to be written down and marked on maps, and geological data about the sediments in which the fossil was found have to be noted. Without such notes, fossils lose much of their scientific value. After collection, fossils are returned to a fossil preparation laboratory, where sediment is removed and fossils are hardened with special glues. These procedures require considerable technical skills and insight into the physical and chemical properties of fossils and sediments. The last step in collecting fossils is curation, where fossils are assigned a unique museum number that can be used as an identifier in publications and as a guide to find the field records concerning the specimen.

Although preliminary identifications and study start in the field when a fossil is collected, a thorough investigation into the biology of a fossil marine mammal cannot be undertaken until the specimen is curated. Several natural history museums have large collections of fossil marine mammals, and these collections are studied by scientists who can use them for comparison with newly collected specimens. Common ways to study a fossil include making observations on its shape and taking measurements and attempting to identify the species that it pertained to. There is no limit to the number of means of study, and paleontologists have made many advances by applying methods and techniques originally developed in other sciences. Some of these new methods to study fossils include CT scan, stable isotope analysis, and the analysis of DNA sequences of fossils.

The last step of any scientific investigation, including paleontology, is the publication of the specimen in a scientific journal. This makes the existence of the new specimen known to the scientific community.

See Also the Following Articles

Fossil Sites ▪ Geological Time Scale ▪ Museums and Collections

Reference

Raup, D. M., and Stanley, S. M. (1978). "Principles of Paleontology," 2nd Ed. Freeman, San Francisco.

Pantropical Spotted Dolphin
Stenella attenuata

WILLIAM F. PERRIN
*Southwest Fisheries Science Center,
La Jolla, California*

The pantropical spotted dolphin (*Stenella attenuata*) is a common resident of all tropical seas and has become familiar to many as the main species involved in the "tuna/dolphin problem" that has taken the lives of millions of dolphins over the last four decades.

I. Characters and Taxonomic Relationships

This dolphin can be identified externally by its long beak sharply demarcated from the melon, slender body, strongly falcate dorsal fin, and (in adults) spots (Fig. 1). The newborn calf is unspotted. Dark spots begin to appear ventrally in large juveniles. Near-adult animals and some young adults have large discrete or overlapping spots both above (light) and below (dark). In adults, the ventral spots fuse and fade to a medium gray, and the dorsal light spots intensify, sometimes to the point of making the animal appear nearly white above. The light spots sweep up behind the dorsal fin. The underlying pattern (observable in calves and juveniles) consists of a dark cape sweeping over the eye to maximum depth below the dorsal fin, a very light to dark gray lateral/ventral field, a narrow well-defined eye stripe to the apex of the melon, a dark band of varying definition extending from the lower corner of the mouth to the flipper, and dorsoventral division of the peduncle into darker upper and lower lighter halves. The last can also be seen in adults. The tip of the beak is white in adults. Details of coloration vary regionally. The large coastal spotted dolphin of the eastern tropical Pacific, *S. g. graffmani*, is extremely heavily spotted, whereas animals around Hawaii tend to be lightly spotted (Perrin and Hohn, 1994).

Sexually mature adults examined range from 166 to 257 cm ($n > 1650$) and weigh up to 119 kg (a 257-cm male from Panama Bay), with a wide GEOGRAPHIC VARIATION. Males are on the average slightly larger than females in body size and most skull characters.

This species may be confused in the tropical Atlantic with the endemic Atlantic spotted dolphin, *S. frontalis*, which is of similar size and may be seen in the same area. A distinguishing characteristic of *S. frontalis* is a light spinal blaze that sweeps up through the cape toward the dorsal fin, but this may be almost absent in some individuals. The dorsoventral division of the peduncle present in *S. attenuata* is absent in *S. frontalis*. The ground pattern is three-part, with a distinct cape, lateral field, and ventral field, as opposed to the two-part pattern in *S. attenuata*. Adults may have dark ventral spots on a very light

Figure 1 *Adult pantropical spotted dolphin in the eastern tropical Pacific. Photo by Robert L. Pitman, from Perrin and Hohn (1994).*

ground; this is not seen in adults of *S. attenuata* (ventral spots are medium gray and obscured by fusion).

The skull (35.6–46.0 cm long in 183 adults) overlaps with that of *S. frontalis* in tooth counts and all measurements; some specimens can be identified to species only through multivariate analysis (Perrin and Hohn, 1994). The skull in both species has a long narrow rostrum with no palatal grooves, about 35–50 teeth in each row, medium-sized rounded temporal fossae, convergent premaxillae, and arcuate mandibles. However, vertebral counts determined to date have been nonoverlapping: 74–84 (*n* = 75) vs 67–72 (*n* = 52) in *S. frontalis*.

The pantropical dolphin is a member of the delphinid subfamily Delphininae *senus stricto* (LeDuc *et al.*, 1999). In a cladistic phylogenetic analysis based on cytochrome *b* mtDNA, it shares a strongly supported polytomic clade with *S. clymene* and *S. coeruleoalba* (sister species), *S. frontalis*, *Delphinus* spp., and *Tursiops aduncus* (to the exclusion of *T. truncatus*). In skull characters it is similar to *S. frontalis*, *T. truncatus*, and *T. aduncus* vs another coherent series of species composed of *S. longirostris*, *S. clymene*, and *S. coeruleoalba*. However, these latter groupings are not supported by results of the molecular studies to date. Some of the similarities may be plesiomorphies, similarities due to the retention of primitive character states.

II. Distribution and Ecology

Distribution is worldwide in tropical and some subtropical waters, from roughly 30–40°N to 20–40°S (Jefferson *et al.*, 1993). It ranks first in abundance among cetaceans (second only to bottlenose dolphins, *Tursiops truncatus*) in the deeper waters of the Gulf of Mexico (Würsig *et al.*, 2000), second in the eastern tropical Pacific and the Sulu Sea, but only sixth in the tropical Indian Ocean (Balance and Pitman, 1998; Dolar, 1999). In the eastern Pacific, the offshore form, *S. a. attenuata*, inhabits the tropical, equatorial, and southern subtropical water masses, being most abundant in waters underlain by a sharp thermocline at depths of 50 m or less, a surface temperature

over 25°C, and salinities less than 34 parts per thousand. In these areas, pantropical spotted dolphins commonly occur in large multispecies aggregations, including spinner dolphins (*S. longirostris*) and yellowfin tuna (*Thunnus albacares*). A large coastal subspecies, *S. a. graffmani*, occurs in the eastern tropical Pacific from Mexico to northern Peru; this form is replaced ecologically in the Atlantic by a large coastal form (unnamed) of the endemic Atlantic spotted dolphin, *S. frontalis* (Perrin and Hohn, 1994).

Prey of the offshore form include mainly small epipelagic fishes, squids, and crustaceans (Robertson and Chivers, 1997). Flying fish are a major diet item in some regions. The diet in the eastern Pacific overlaps greatly with that of the yellowfin tuna, *T. albacares*. Diet of the large coastal form is unknown but may include larger and tougher benthic fishes. Predators include the killer whale (*Orcinus orca*) and sharks, probably the pygmy killer whale (*Feresa attenuata*), and possibly the false killer whale (*Pseudorca crassidens*) and the short-finned pilot whale (*Globicephala macrorhynchus*). Parasites may cause direct or indirect mortality.

III. Behavior and Life History

Why spotted dolphins associate closely with tuna in the eastern Pacific is unknown, although it is suspected to be related to foraging efficiency at some level (Perrin and Hohn, 1994). Other suggested reasons for the association involve physiological efficiency or protection from predators. Immature and subadult males and females tend to form smaller schools or join larger spinner dolphin schools not associated with tuna. The breeding system is unknown but may be promiscuous as in the spinner dolphin. Spotted dolphins exhibit a wide variety of aerial behavior (but not spinning); juveniles (identifiable as such by their lack of spots) make especially high vertical leaps. In areas where they are not harpooned or pursued by purse seiners, they readily come to vessels to ride the bow wave and sometimes can be observed closely for long periods. Burst swimming speed exceeds 12 knt (about 22 km/hr), and dives of up to 3.4 min have been observed.

Gestation is 11.2–11.5 months (Perrin and Hohn, 1994). Length at birth is 80–85 cm. Females reach sexual maturity at 9–11 years and males at 12–15 years. The calving interval is about 2–3 years but varies with population status. The length of nursing also varies greatly with population status, ranging from 1 to 2 years. Breeding is diffusely seasonal, with multiple peaks in some regions.

IV. Interactions with Humans

The species has not been kept successfully in captivity. Direct and incidental catches have been substantial. Japanese drive and harpoon fisheries take spotted dolphins for human consumption; the catch in 1 year (1982) was 3799 (Perrin and Hohn, 1994), and annual catches in 1994–1997 ranged from 23 to 449 (International Whaling Commission, 1999). Tuna fishermen seek out aggregations of spotted dolphins and tuna in the eastern tropical Pacific and set their nets on them to capture the tuna. Abundance of the northeastern stock of the offshore form has been reduced to a fraction of its original size by incidental kill in tuna purse seines since the early 1960s. Directed fisheries also exist or have existed in the Philippines, Laccadive Islands in the Indian Ocean, Solomon Islands, Indonesia, and St. Helena in the South Atlantic (Perrin and Hohn, 1994), and bycatches also occur in fishing nets of various types in the Philippines, India, Australasia, western North Pacific, coastal Peru and Ecuador, and Taiwan (Perrin *et al.*, 1994, 1996). The impacts of most of these takes on the populations have not been assessed.

See Also the Following Articles

References

Balance, L. T., and Pitman, R. B. (1998). Cetaceans of the tropical western Indian Ocean: Distribution, relative abundance, and comparisons with cetacean communities of two other tropial ecosystems. *Mar. Mamm. Sci.* **14**, 429–459.

Dolar, M. L. L. (1999). "Abundance, Distribution and Feeding Ecology of Small Cetaceans in the Eastern Sulu Sea and Tañon Strait, Philippines." Ph.D. dissertation, University of California, San Diego, CA.

International Whaling Commission (1999). Report of the Scientific Committee. *J. Cetacean Res. Manage.* **1**(Suppl.), p. 284.

Jefferson, T. A., Leatherwood, S., and Webber, M. A. (1993). "FAO Species Identification Guide: Marine Mammals of the World," p. 320. FAO, Rome.

LeDuc, R. G., Perrin, W. F., and Dizon, A. E. (1999). Phylogenetic relationships among the delphinid cetaceans based on full cytochrome *b* sequences. *Mar. Mamm. Sci.* **15**, 619–648.

Perrin, W. F., Dolar, M. L. L., and Alava, M. N. R. (eds.) (1996). "Report of the Workshop on the Biology and Conservation of Small Cetaceans and Dugongs of Southeast Asia, Dumaguete, Philippines, 27–30 June 1995." UNEP(W)/EAS WG.1/2, United Nations Environment Programme, Bangkok, p. 101.

Perrin, W. F., Donovan, G. P., and Barlow, J. (eds.) (1994). Cetaceans and gillnets. *Rep. Int. Whal. Commn.* (*Spec. Issue 15*), p. 629.

Perrin, W. F., and Hohn, A. A. (1994). Spotted dolphin *Stenella attenuata. In* "Handbook of Marine Mammals" (S. H. Ridgway and R. Harrison, eds.), Vol. 5, pp. 71–98. Academic Press, San Diego, CA."

Robertson, K. M., and Chivers, S. (1997). Prey occurrence in pantropical spotted dolphins, *Stenella attenuata,* from the eastern tropical Pacific. *Fish. Bull. U.S.* **95**, 334–348.

Würsig, B., Jefferson, T. A., and Schmidly, D. J. (2000). "Marine Mammals of the Gulf of Mexico." Texas A&M Univ. Press, College Station, TX.

Parasites

J. Antonio Raga, F. Javier Aznar, Juan A. Balbuena, and Mercedes Fernández
University of Valencia, Spain

B eyond their sanitary or economic importance, parasites are an integral part of the biosphere. Parasites are extremely diverse and of great importance to the health of their host. However, they are usually small and go unnoticed in biological studies. Parasites are also so pervasive that they virtually infect every free-living organism, potentially influencing, among other things, host behavior and population size, food web dynamics, and community structure. These effects are usually undesirable when human health or economy are at stake but confer parasites a paramount importance in many ecological and evolutionary studies.

This article presents an overview of the diversity of marine mammal parasites. Its aim is to explain concisely what they are and how they have become associated with their hosts. Other aspects, such as the impact of parasites on marine mammal populations, are covered only briefly as these topics are dealt with extensively elsewhere (Aznar *et al.*, 2001). Under the term "parasite," we will consider protozoans and metazoans (helminths and arthropods) which have adopted this trophic strategy.

I. Parasite Diversity

A. Cetaceans

Protozoan parasites have been reported rarely in marine mammals, perhaps due to insufficient sampling or because these organisms are actually rare. In fact, the number of protozoons detected and sampling effort seem clearly related. Three groups of coccidians occur in cetaceans: *Cystoisospora delphini* in bottlenose dolphins, *Tursiops truncatus*, causing enteritis; *Toxoplasma* sp., in bottlenose dolphins and striped dolphins, *Stenella coreuleoalba*, associated with toxoplasmosis; and *Sarcocystis* sp. in toothed whales, with little or no pathologic effect. The life cycles of these coccidians have not been elucidated, but they should be similar to those of their terrestrial counterparts. Regarding ciliates, *Haematophagus megapterae* is found attached to the baleen plates of humpback whales, *Megaptera novaeangliae,* feeding on blood cells, and *Kyaroikeus cetarius* (Fig. 1b) (belonging to a genus exclusive to toothed whales) is sessile and dwells in the lungs of wild bottlenose dolphins. The latter has also been reported in captive toothed whales. Its transmission is believed to be direct, requiring perhaps repeated exposure to be successful. Finally, one unidentified flagellate and the sarcodinan *Entamoeba* sp. have been recorded in the contents of the colon of the bowhead whale, *Balaena mysticetus.*

Four families of flukes occur typically in cetaceans: Campulidae occur in whales and dolphins; Brauninidae and Pholeteridae occur in dolphins, and Notocotylidae occur in whales. In terms of diversity and geographic extension, the family Campulidae is the most important. It comprises some 41 species, 35 of which are distributed among the majority of cetacean families. Species of *Campula, Oschmarinella,* and *Zalophotrema* inhabit the hepatic and pancreatic ducts of TOOTHED WHALES; *Hadwenius,* the intestine; *Nasitrema* (Fig. 1c), the air sinuses; and *Hunterotrema,* the lungs. Baleen whales harbor species of *Lecithodesmus* in the bile ducts. The life cycles of the Campulidae are not known, but all digenean flukes possess heteroxenous cycles. The ciliated miracidium larvae of *Nasitrema* have eyespots and are active swimmers, whereas those of *Oschmarinella rochebruni* lack eyespots. These differences may indicate that the life cycle of the former occurs near the surface, whereas that of the latter occurs in deep water. The widespread occurrence of campulids in fish- or squid-eating cetaceans suggests that these prey may act as second intermediate or transport hosts. The family Pholeteridae includes two species, one of which, *Pholeter gastrophilus,* parasitizes many toothed whales (mainly delphinids) in the world. These flukes live in the wall of the main and pyloric stomachs, more rarely the duodenum, sometimes generating extensive fibrosis. Another fluke, with a strange ovoid morphology, *Braunina cordiformis* (Brauninidae), attaches to the stomach and the duodenal ampulla of small toothed whales. Baleen whales harbor, in the large intestine, several species of the genus *Ogmogaster* (Notocotylidae) (Fig. 1d), which have been found in many areas (Antarctic, Pacific, Atlantic, and Mediterranean). The life cycles are unknown, but information derived from other notocotylids suggests that the larval stage known as cercaria might not need a second intermediate host, encysting directly on the crustaceans that serve as prey for whales.

Cetaceans are infected with adult tapeworms of the families Tetrabothriidae (16 species of 23 in marine mammals) and Diphyllobothriidae (11 of 48 in marine mammals). These worms have small to very large sizes and occupy the intestine, from the duodenum (*Tetrabothrius forsteri*) to the terminal colon and rectum (*Strobilocephalus triangularis*). Tetrabothriidae have diversified morphologically in cetaceans [the genera *Trigonocotyle* (Fig. 1g), *Strobilocephalus, Priapocephalus,* and *Tetrabothrius;* only the latter two have representatives in baleen whales]. The diphyllobothriids belong to the genera *Diphyllobothrium* (Fig. 1f) (in toothed whales, more rarely in baleen whales), *Diplogonoporus* (in baleen whales and the sperm whale, *Physeter marcocephalus*), *Hexagonoporus* (in the sperm whale), and *Plicobothrium* (typical from pilot whales, *Globi-*

cephala spp., rarely other dolphins). In general, tapeworms of marine homeotherms use zooplanktonic crustaceans as first intermediate hosts. Euphasiids act as intermediate hosts, and fishes as transport hosts, in the life cycle of some tetrabothriids. In contrast, the known cycles of members of the Diphyllobothriidae involve both copepods and fishes as intermediate hosts. Plerocercoids (larval stages) of the family Phyllobothriidae are encysted in the blubber and mesenteries of a number of cetacean species, being called *Phyllobothrium delphini* (Fig. 1h) and *Monorygma grimaldii,* respectively. Because the adult stages are not known, it is difficult to ascertain whether these designations actually correspond to species or to complexes of species. The definitive hosts are probably sharks that become infected by either preying or scavenging on cetaceans. This raises interesting questions about the evolution of these life cycles because no other tetraphyllideans use mammals as intermediate or transport hosts.

Among the nematodes, the family Anisakidae is probably the most successful in terms of potential for colonizing hosts in many environments. Studies of this family have revealed the existence of several complexes of sibling species. The anisakids of cetaceans belong to the genera *Pseudoterranova, Contracaecum,* and particularly, *Anisakis.* The latter is found mainly in cetaceans, hence its vernacular name of "whale-worm." Species of *Anisakis* (five defined morphologically, more based on isozyme analyses) have been reported in at least 35 species of marine cetaceans and occur also in river dolphins. Whaleworms favor the forestomach, where larvae can provoke ulcers by attaching in aggregates. The life cycle of *Anisakis simplex* is the best documented. Eggs shed with the feces hatch and release a free-living larva that is subsequently ingested by planktonic crustaceans. Fish and squid become infected by feeding on these crustaceans. Cetaceans acquire larva by consumption of infected prey. The worms molt to the adult stage and mate in the stomach, where females release the fertilized eggs. The large Crassicaudidae nematodes (a family containing about 11 species exclusive of cetaceans) are found in the kidneys and urigenital organs, placenta, mammary glands, muscles, and pterygoid sinuses, sometimes causing extensive damage. Species of *Crassicauda* (Fig. 1l) infect both whales and dolphins, whereas the up to 3.75-m-long *Placentonema gigantissima* dwells in the placenta of sperm whales. Life cycles are largely unknown. It has been speculated that *C. boopis* might infect fin whale calves, *Balaenoptera physalus,* either by ingestion of larvae shed in their mothers' urine or by prenatal (transplacental) transmission. Pseudaliids (about 17 species exclusive to toothed whales) are distributed among the genera *Pseudostenurus, Pharurus, Torynurus, Stenurus, Halocercus, Pseudalius,* and *Skrjabinalius* (Fig. 1k). They occur mainly in the lungs, air sinuses, and heart of phocoenids and monodontids, secondarily delphinids. Prenatal transmission of *Halocercus* in common bottlenose dolphins has been suggested. However, several authors consider that the life cycles of pseudaliids should also involve intermediate and transport hosts.

Only two genera of acanthocephalans (family Polymorphidae), closely allied to forms infecting aquatic birds, occur in the intestine and occasionally the stomach of marine mammals.

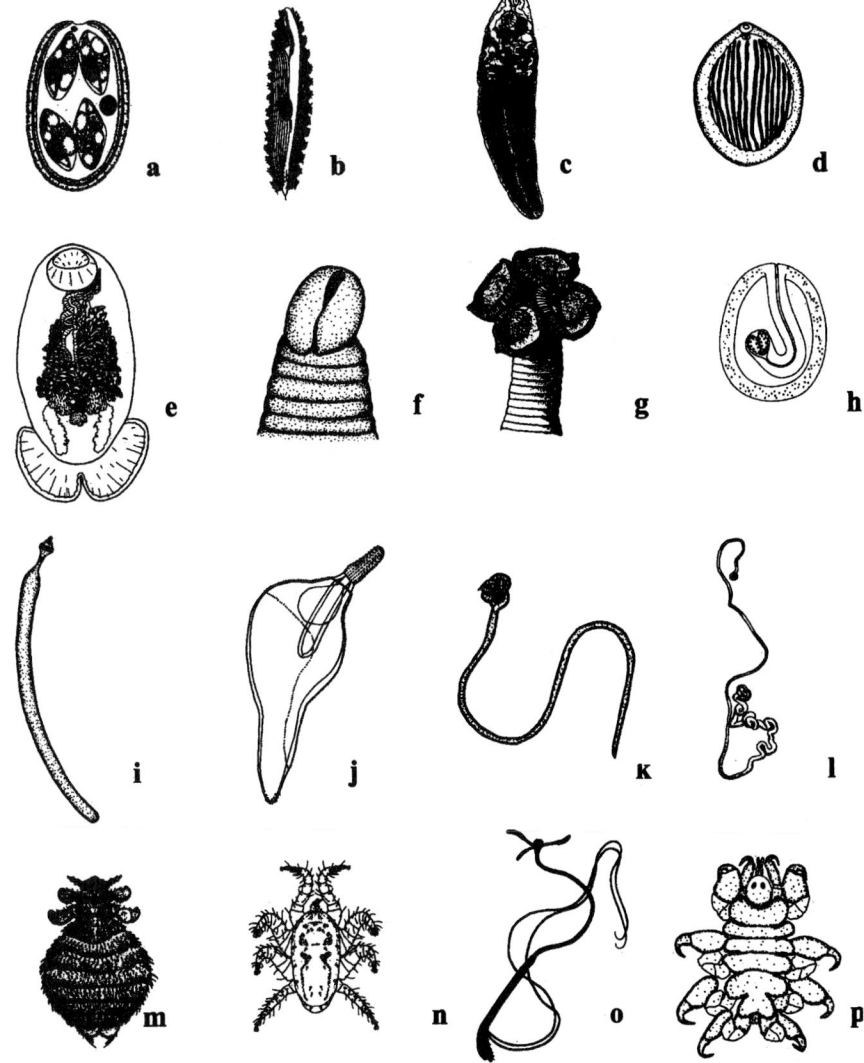

Figure 1 *Some representatives of marine mammal parasites. Protozoons (a) Eimeria phocae from harbor seal, (b)* Kyaroikeus cetarius *from common bottlenose dolphin; digeneans: (c)* Nasitrema *sp. from long-finned pilot whale, (d)* Ogmogaster antarcticus *from fin whale, (e)* Taprobanella bicaudata *from dugong; cestodes: (f)* Diphyllobothrium *sp. from long-finned pilot whale, (g)* Trigonocotyle globicephalae *from long-finned pilot whale, (h)* Phyllobothrium delphini *(plerocercoid) from short-beaked common dolphin; acanthocephalans: (i)* Bolbosoma capitatum *from false killer whale, (j)* Corynosoma australe *from leopard seal; nematodes: (k)* Skrjabinalius guevarai *from striped dolphin, (l)* Crassicauda grampicola *from Risso's dolphin; insects: (m)* Echinophthirius horridus *from harbor seal; acari: (n)* Orthohalarachne attenuatta *from walrus; crustaceans: (o)* Pennella balaenopterae *from fin whale, (p)* Syncyamus aequus *from striped dolphin. Not to scale. Reproduced with permission from (a) Hsu* et al. *(1974)* J. Parasitol. *60, 399; (b) Sniezek* et al. *(1995)* J. Euk. Microbiol. *42, 262; (c) Raga* et al. *(1994)* Rep. Int. Whal. Comm. *44, 372, courtesy of the International Whaling Commission; (e) Blair (1981).* Aus. J. Zool. Suppl. Ser. *81, 26; (f) Balbuena and Raga (1993) Diphyllobothriids (Cestoda: Pseudophyllidea) from the long-finned pilot whale* Globicephala melas *(Traill, 1809) off the Faroe Islands, with comments on the taxonomy of* Diphyllobothrium *Cobbold, 1858 species from cetaceans.* Syst. Parasitol. *26, 121, Fig. 2, © 1993 Kluwer Academic publishers, with kind permission from Kluwer Academic Publishers; (h) Testa and Dailey (1977)* Bull. South. Calif. Acad. Sci. *76, 103; (j) Zdzitowiecki (1991). Antarctic acanthocephala. Synopses of the Antarctic benthos, Vol. 3, p. 63, Koeltz Scientific Books; (l) Raga (1987)* Vie et Millieu, *Vol. 37, p. 216; (m) Scherf (1963) Ein Beitrag zur Kenntnis zweier Pinnipedierläuse (*Antarctophthirus trichechi *Boheman und* Echinophthirius horridus *Olfers).* Z. Parasitenk. *23, Fig. 13a, © Springer-Verlag; (n) Webb* et al. *(1985)* J. Parasitol. *71, 389; (p) Raga (1988)* Crustaceana *54, 150, by permission of Brill NV.*

Whales and dolphins are the typical hosts for *Bolbosoma* (Fig. 1i) (some nine species). Pelagic euphasiids and copepods are postulated to act as intermediate hosts, and fishes as transport hosts. A few species of *Corynosoma* have also been described in cetaceans, but the bulk of diversity in this group is found in pinnipeds (see later).

Cetaceans harbor a specific, heterogeneous, and rather diverse fauna of parasitic and nonparasitic crustaceans; interestingly, such intimate associations are rare in sirenians and do not occur on pinnipeds. Whale lice (Cyamidae, about 26 species) (Fig. 1p) have diversified extensively and exclusively among dolphins and whales, becoming one of the few groups of parasitic amphipods. Whale lice attach to the skin (especially on natural openings, wounds, and scars), feeding on epidermic tissue and algal filaments. The life cycle takes place completely on the hosts and, therefore, females have developed a four-plated pouch that keeps the eggs and weak-legged young from being washed away. Transmission is by bodily contact. Likewise, around 6 *Pennella* species (Pennellidae) (Fig. 1o) have become established in cetaceans. The life cycles of these copepods are complex: after two free-swimming stages, they develop, mate, and attain sexual maturity on a cephalopod intermediate host and then females seek a suitable definitive host, e.g., cetaceans, where they burrow on the host's tissue (mainly on the back and belly) to get their head anchored, feeding on blood and body fluids.

B. Sirenians

Four species of Coccidia have been described in sirenians: *Eimeria manatus, E. nodulosa,* and *Toxoplasma gondii* in the West Indian manatee, *Trichechus manatus,* and *E. trichechi* in the Amazonian manatee, *T. inunguis.* The species of *Eimeria* have been detected as oocysts in feces. Nothing is known about their life cycles, but the oocyst of *E. nodulosa* bears peculiar nodules that are thought to serve for the attachment in the aquatic vegetation. Thus the cycle might be direct.

The metazoan parasitic fauna of manatees and dugongs is restricted to digeneans and nematodes, but it is fairly diversified, given the extant number of host species, and very specific. Except *Nudacotyle undicola,* which belongs to a genus typical from nonmarine mammals, all species and genera recorded so far (2 genera, 3 species of nematodes; 16 genera, 22 species of digeneans) are exclusive to sirenians (3 out of 6 families also are). Interestingly, the dugong, *Dugong dugon,* exhibits a distinct helminth fauna with greater diversity than the manatees (17 vs 7 species). The reason for this difference is a matter of speculation. It might simply result from differences in research effort. However, dugongs inhabit the Tropical Pacific, whose richer marine fauna may offer a larger selection of intermediate hosts and, perhaps more importantly, dugongs have remained longer in the marine habitat of ancestral sirenians. In contrast, the manatees entered freshwater habitats, which probably impoverished their helminth fauna.

The parasites of sirenians occur in a variety of sites within the host. The roundworms *Paradujardinia halicoris* and *Heterocheilus* spp. inhabit the stomach, more rarely the small intestine. Flukes exhibit a notable morphological variety, e.g., *Taprobanella bicaudata* (Fig. 1e), and are found in diverse lo-

cations. Species of Paramphistomidae (4), Rhabdiopoeidae (4), and Nudacotylidae (1) and most of the Opisthotrematidae (6) occur in different sites in the digestive system, including the stomach, pyloric ceca, duodenum, ileum, colon, pancreas, and liver. In some cases they form cysts, e.g., *Faredifex clavata* (Rhabdiopoeidae) or *Lankatrema mannarense* (Opisthotrematidae). In contrast, *Opisthotrema* (2), *Cochleotrema* (2), and *Pulmonicola* (1) species dwell in the ear system or the lungs, whereas *Labicola elongata* (Labicolidae) occurs in the upper lip of dugongs. The life cycles have not been determined for any of these parasites. Some authors have suggested that the nematodes of sirenians use crustaceans as intermediate hosts that would be consumed incidentally while feeding on vegetation. Alternatively, the nematode eggs might be ingested directly when the hosts feed on contaminated vegetation. Similar conjectures have been advanced for digeneans. *Chiorchis fabaceus* is thought to infect manatees through the incidental ingestion of snails containing metacercariae (larvae), whereas the cercariae of *Lankatrematoides gardneri* use a mollusk as an intermediate host, and perhaps then the larvae escape and encyst on aquatic vegetation, awaiting to infect dugongs.

C. Pinnipeds

The most common protozoons found in pinnipeds throughout the world are coccidians. Species of *Eimeria* have been detected in the intestine of seals, sometimes causing severe disorders. *Eimeria phocae* (Fig. 1a) is typical from harbor seals, *Phoca vitulina,* in the northwestern Atlantic. Although its life cycle is not clear, experimental evidence shows that oocysts sporulate in feces incubated in air, but not when suspended in seawater; this may suggest that transmission occurs on land. Another six species of *Eimeria* occur in Weddell seals, *Leptonychotes weddellii,* and crabeater seals, *Lobodon carcinophaga.* Species of *Sarcocystis* have been reported in harbor seals in California, northern fur seals, *Callorhinus ursinus,* in Alaska, and leopard seals, *Hydrurga leptonyx,* in Antarctica. *Isospora miroungae* and *Cystoisospora israeli* have been described in young Antarctic southern elephant seals, *Mirounga leonina,* and South African fur seals, *Arctocephalus pusillus,* respectively. Finally, toxoplasmosis associated with *Toxoplasma gondii* has been observed in a harbor seal pup. In addition to coccidians, the flagellate *Giardia* sp. has been found in ringed seals, *Pusa hispida,* from the western Arctic region of Canada. In fact, they have been considered as potential reservoir hosts for giardiasis in the Arctic.

Flukes are represented by 6 families. Within the Campulidae, 4 out of 5 species of the genus *Orthosplanchnus* infect Arctic and Subarctic phocids, and walruses, *Odobenus rosmarus,* whereas *O. antarcticus* is found in Antarctic seals. *Zalophotrema hepaticum* is associated with California sea lions, *Zalophus californianus,* in the Northeastern Pacific. Campulids of pinnipeds live in the bile ducts, gall bladder, and, rarely, the intestine of their hosts. The family Heterophyidae is widespread in pinnipeds from the Northern Hemisphere. About 10 intestinal species have been recorded so far, from the genera *Cryptocotyle, Phagicola, Rossicotrema, Galactosomum, Pricetrema,* and *Phocitrema.* Species of the two latter genera are particularly noticeable, as they are associated more specifically with pinnipeds (in the Pacific region). Nothing is known of the life cycles of these heterophyids. Inferences made from other species strongly suggest that a gastropod would function as first and various species of fish as second intermediate hosts. Five species of opistorchiids from the genera *Opistorchis*, *Metorchis*, and *Pseudamphistomum* inhabit the bile ducts of seals from the Northern Hemisphere. Their life cycles are unknown, but, from other opistorchiids, it might be inferred that fish act as second intermediate hosts. Microphallids (3 species) from the genera *Microphallus* and *Maritrema* are intestinal flukes. For instance, *Microphallus orientalis* has been reported in immense numbers in walruses and bearded seals, *Erignathus barbatus*, from the Barents Sea. The majority of microphallid larvae encysts in benthic crustaceans, which may explain their occurrence in seals that feed on bottom invertebrates. Species of the genera *Echinostoma* and *Mesorchis* (Echinostomatidae) have been reported incidentally in seals. Finally, *Ogmogaster antarcticus* (Notocotylidae) is known from Weddell and crabeater seals. These hosts likely become infected when feeding on benthic invertebrates (see earlier discussion).

The tapeworm fauna of pinnipeds is rich. The majority of species belong to the family Diphyllobothriidae (37 species out of 48 in marine mammals). Species of *Diphyllobothrium* form the major component, being distributed in pinnipeds worldwide. In contrast, species of the genera *Baylisia*, *Baylisiella*, and *Glandicephalus* are exclusive to Antarctic seals, and those of *Diplogonoporus*, to boreal pinnipeds. Insights of the life cycles of diphyllobothriids have already been discussed. One genus of Tetrabothriidae, *Anophryocephalus* (7 species) is associated with Arctic and sub-Arctic pinnipeds. Euphasiids appear to be intermediate hosts, and fishes transport hosts in the life cycle of these parasites.

Pinnipeds harbor a diverse nematode fauna that comprises 6 families: Ancylostomatidae (2 species), Dipetalonematidae (2 species), Trichinellidae (1–2 species), Filaroididae (4 species), Crenosomatidae (1 species), and Anisakidae (14 species). To a great extent, this parasitic fauna is related to taxa found in land carnivores. Thus the cycle of some groups is land dependent and resembles that of their terrestrial counterparts. For instance, the eggs and the first three larval stages of *Uncinaria lucasi* (Ancylostomatidae) develop in the soil on the otariid rookeries in the North Pacific. Third-stage larvae infect adult northern fur seals through the skin (especially the flippers), migrating to host tissues, particularly the ventral blubber and mammary glands. Larvae use the milk of adult female seals to infect the pups, where worms become adult in the intestine. Transmission occurs only during the breeding period; infective larvae survive the severe winter on the rookery, awaiting the arrival of the fur seals in early June. The heartworm, *Dipetalonema spirocauda* (Dipetalonematidae), likely requires an athropod vector (a louse) to trigger development and to transmit the microfilaria to North Atlantic seals, although this cycle has yet to be proven. Transplacental or transmammary transmission to pups has been also suggested to occur. The species of *Trichinella* (Trichinellidae) are typical tissue parasites of terrestrial mammals. However, as discussed in the next section, a natural cycle of trichinosis involving pinnipeds seems to exist in the Arctic. With regard to nematodes with aquatic cycles, experimental research has demonstrated that the lungworm *Parafilaroides decorus* (Filaroididae) infects fishes that

consume feces from Californian sea lions, contaminated with the larvae. These coprophagous fishes are, in turn, preyed upon by the sea lions. Recent experimental studies of the lungworm *Otostrongylus circumlitus* (Crenosomatidae), a species parasitizing boreal seals, have shown that fishes seem to be used as intermediate hosts. By far the most widespread nematode group in seals is the Anisakidae, dwelling in the stomach and duodenum. Pinnipeds are primary hosts for species of *Pseudoterranova* and *Phocascaris*, whereas species of the genus *Contracaecum* appear primarily in aquatic birds, having some representatives in phocids and otarids. In addition, there are records of *Anisakis simplex* maturing in some phocid species, particularly the gray seal, *Halichoerus grypus*. The complex life cycle of the sealworm, *Pseudoterranova decipiens*, is the best known. Experimental studies have shown that the sealworm is able to infect a wide variety of invertebrates. Under natural conditions, however, it utilizes only benthic and epibenthic organisms as intermediate and transport hosts. The exploitation of benthonic food webs can be traced to the early development of the parasite. The free-living larva emerging from the egg has negative buoyancy and adheres to the substrate by its caudal extremity. This behavior favors ingestion by benthic copepods, which, in turn, are consumed by benthic macroinvertebrates. At this point, the larvae have molted twice and are ready to infect the seal, but benthophagous fish, or their demersal predators, can be used as transport hosts, enhancing transmission.

Acanthocephalans of the genus *Corynosoma* (Fig. 1j) (Polymorphidae, some 20 species in pinnipeds) exhibit a protracted history of association with pinnipeds. They have a cosmopolitan distribution, appearing in the intestine of most species of seals, sea lions, and walruses. The complete life cycle for *C. strumosum* and *C. pseudohamanni* has been inferred from field collections. Nearshore amphipods acquire the cystacanth larvae, which, following ingestion, encyst in the body cavity of several fish species, awaiting seals to prey on these fish. Species of *Bolbosoma* have also been reported incidentally in Arctic seals.

Two main arthropod groups are associated with pinnipeds: sucking lice (Echinophthiriidae) (Fig. 1m) and mites (Halarachnidae) (Fig. 1n). The entire life cycle of sucking lice is spent on a single host and, therefore, these insects rely on direct contact for transmission. Each individual host must provide acceptable ecological conditions, such as hairs to glue the eggs. Echinophthiriids (four genera, nine species) are associated with all pinniped groups (otariids, odobenids, phocids) worldwide. They are especially adapted to the hosts' ecological conditions and are transmitted most likely on land. Mites (six species) inhabit the respiratory tract and belong to two genera: *Halarachne* (in phocids) and *Orthohalarachne* (in otariids and odobenids). The life cycle takes place in the same individual host and comprises four stages: a larva, two nymphal stages, and the adult. Transmission occurs on the terrestrial environment when active larvae are transferred by nose contact or are sneezed from the nostrils of infested animals. Acari typical of terrestrial mammals also have some representatives in pinnipeds. *Dermacentor rosmari* is the only species within the Ixodidae. These ticks appear between the fingers of walruses, especially in the hind legs on the skin, in the Arctic waters of Russia; nymphal stages are not known. *Demodex zalophi*

(Demodicidae) occurs in the hair follicles of the flippers and genital region of California sea lions. Each follicle usually contains one female and four males. All stages of development take place in the hair follicle, usually in the distal portion of the duct of the sebaceous gland.

D. Sea Otter

A total of 17–18 parasite species have been reported from sea otters, *Enhydra lutris*, throughout their range. The majority are acquired directly from sympatric pinnipeds: *Orthosplanchnus fraterculus*, *Pricetrema zalophi*, *Phocitrema fusiforme*, *Diplogonoporus* sp., *Diphyllobothrium phocarum*, *Pseudoterranova decipiens*, *P. azarasi*, *Corynosoma strumosum*, *C. villosum*, and *Halarachne miroungae*. Other parasites appear to derive from marine birds, namely *Microphallus pyrum* (Microphallidae), and three species of *Polymorphus* (Polymorphidae) occurring as immature. *Anisakis* larvae have also been recorded. Apparently, the only parasite specific to the sea otter is *Corynosoma enhydri*, the largest species in this genus.

E. Polar Bear

Polar bears, *Ursus maritimus*, have been relatively little analyzed for parasites. To our knowledge, the only helminth recorded thus far in wild animals is *Trichinella* spp. (Trichinellidae).

II. Patterns and Processes in Host–Parasite Associations

To understand how marine mammals and their parasites have become associated, we have to discuss first the principles that regulate the outcomes of host–parasite interactions. The evolutionary fate of every parasite species ineluctably depends on that of its hosts. To survive, parasites must track their hosts, switch to its hosts, or go extinct. Host tracking should result in congruent phylogenies of hosts and parasites (i.e., a perfect cospeciation pattern). However, incongruence often occurs because parasites speciate or go extinct whereas their hosts do not. Another reason for incongruence is host switching (also called "host capture"), i.e., parasites colonize phylogenetically unrelated hosts through ecological mechanisms. Colonization deserves more detailed comments because of its importance for the development of parasite faunas in marine mammals (Fig. 2).

The likelihood that a parasite successfully colonizes a new host depends on two factors: the probability of encounter and the compatibility between the new host and the parasite. Encounters depend on behavioral, ecological, or biogeographical factors. For instance, the transmission of sirenian parasites appears to be linked to sea vegetation or the organisms associated

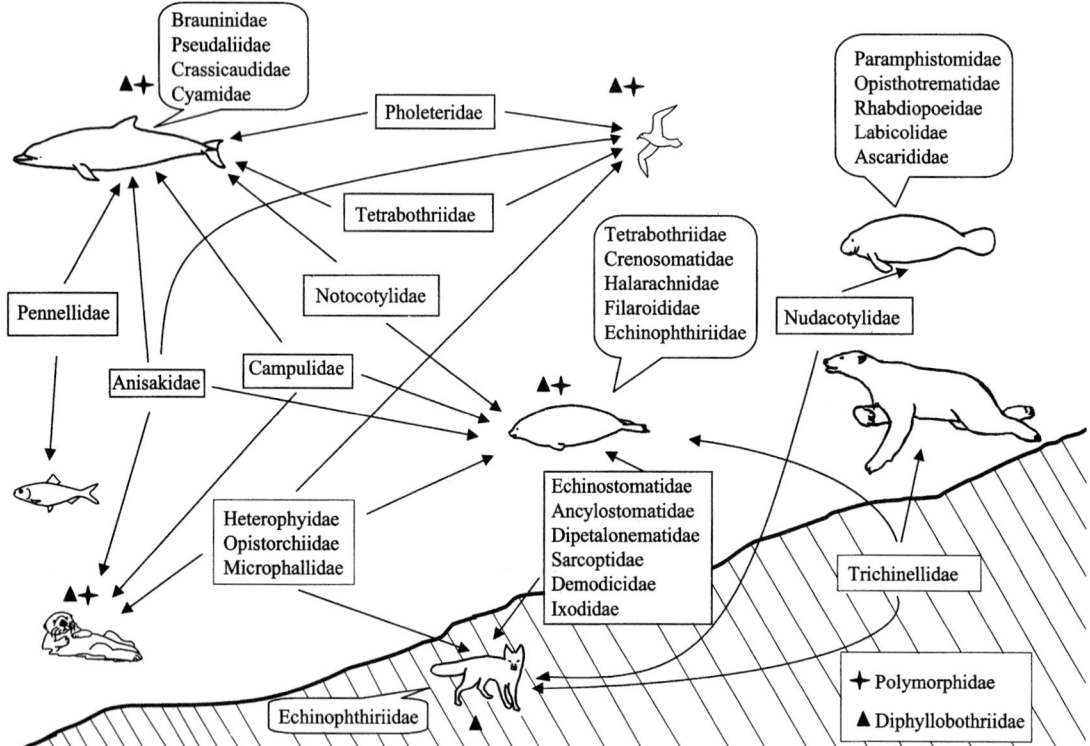

Figure 2 *Sharing patterns of parasitic helminth and arthropod genera. For simplicity, only the parasite families to which these genera belong are displayed. Balloons enclose families with genera exclusive to each host group, and rectangles those with genera common to several host groups (as indicated by arrows). For example, within the Tetrabothriidae, the genus Tetrabothrius is shared between marine birds and cetaceans, whereas the genus Anophryocephalus is exclusive to pinnipeds. Therefore, the Tetrabothriidae appear as both exclusive of pinnipeds and shared between birds and whales.*

to it. Therefore, the chances of infection to fish-eating marine mammals are very low. In this context, most marine mammal parasites exploit food webs for transmission and cannot exert strong host selection, being consumed by both "right" and "wrong" hosts. Whether a host is "right" or "wrong" depends on filter compatibility, i.e., those morphological, physiological, or immunological barriers that preclude that the parasite becomes established, matures, or reproduces in/on the contacted host. For instance, lice on pinnipeds require hair and a suitable surrounding climate to survive. Therefore, these lice have little opportunity to establish on cetaceans or sirenians. In another example, species of *Corynosoma* are mainly associated with pinnipeds worldwide, but their larvae frequently end up in sympatric seabirds or cetaceans. However, for unknown physiological and/or immunological reasons, these larvae rarely mature in nonpinniped hosts.

When encounters with "wrong" hosts are fortuitous, their ecological and evolutionary significance is minimal. For example, sea otters from California seldom acquire larvae of *Polymorphus* from aquatic birds. However, if encounters repeat predictably throughout time, true accommodation or SPECIATION of parasites in formerly "wrong" hosts can be promoted, providing that parasites are eventually able to overcome the compatibility filter. This is obviously easier as the target host is phylogenetically closer to the original host. However, we should bear in mind that the probability of encounter is always of paramount importance, e.g., carnivorous marine mammals have more parasitic taxa in common with piscivorous marine birds than with sirenians (Fig. 2).

A typical case of accommodation is that of the ecological facultative hosts, i.e., suitable hosts for a parasite species that originated elsewhere. For example, the tapeworm *Anophryocephalus ochotensis* originated in the Steller sea lion, *Eumetopias jubatus*, by colonization from *Anophryocephalus* spp. of seals. However, *A. ochotensis* also infects and reproduces now in sympatric northern fur seals. Facultative hosts are particularly important when they sustain a significant portion of a parasite population.

A. General Hypothesis on the Origin of Associations

Based on the above concepts, we can elaborate a general hypothesis on the origin of host–parasite associations. Let us consider the terrestrial ancestors of each marine mammal group (cetaceans, sirenians, pinnipeds, the sea otter, and the polar bear). These ancestors certainly harbored parasites in a variable number, but the subsequent host transitions from the land to the sea likely posed a barrier for these terrestrial parasites to track their hosts. In other words, parasites were phylogenetically constrained to either precisely adjust their life cycle and/or physiology to the new environment or go extinct. There are beautiful examples of such biological adjustment (e.g., the life cycle of *U. lucasi*, see earlier discussion), but extinction is thought to have occurred often, particularly in marine mammals, the cetaceans and sirenians, that eventually severed their ties with land completely.

When the ancestors of marine mammals made it to the sea, they became literally immersed in an ocean of infective stages of marine parasites. Then the mechanisms favoring colonization began to work. Some host captures involved marine mammals and parasites from nonmammalian hosts. However, due to compatibility limitations, these episodes should have been rare as compared to the parasite exchange between marine mammals themselves. An example can illustrate the intricacies of colonization events. The Campulidae, a family of digeneans occurring in marine mammals worldwide, is the putative sister group of flukes occurring in fish as adult. Apparently, the campulid ancestors amplified a cycle that formerly ended up in these fish by including their predators (some unknown ancestor of toothed whales) as new definitive hosts. This initial host switch opened the possibility of subsequent coevolution (host tracking) of campulids and toothed whales and, of course, of new host-switching events. Not surprisingly, the latter involved other marine mammals: baleen whales and pinnipeds acquired campulids from toothed whales, and sea otters from pinnipeds.

We therefore postulate that most parasite taxa infecting marine mammals have a marine origin, with coevolved products from land having remained in hosts with a permanent or close contact with this realm. The evidence available confirms that the arthropods and intestinal helminths of cetaceans probably rose from marine ancestors, whereas a sizeable part of parasites of pinnipeds has a clear terrestrial affiliation. To test this prediction thoroughly, we need robust phylogenetic hypotheses dealing with all parasite taxa.

B. Parasite Exchange in Ecological Scenarios

We have learned from the arguments described earlier that the distribution of parasites in marine mammals can be largely understood from patterns of exchange, either past or present-day, within communities of marine vertebrates. This section analyzes with more detail some major features of these patterns within each specific ecosystem.

1. Terrestrial Ecosystems For obvious reasons, we expect that most monoxenous parasites disappeared when the terrestrial ancestors of marine mammals entered the aquatic environment. In fact, whale lice, coming from free-living amphipods, appear to be the only aquatic forms that rely exclusively on direct transmission. However, some parasites of pinnipeds and polar bears, derived putatively from terrestrial counterparts, have cycles constrained to develop on land (interestingly, this does not occur with sea otters). Most of these land parasites have monoxenous life cycles and, to a great extent, are host specific. For instance, Echinophthiriid lice and halarachnid mites likely coevolved with the ancestor(s) of pinnipeds and are largely restricted to these hosts today. In this context, the occurrence of *Halarachne miroungae* in sea otters should be regarded as fortuitous. Perhaps the most interesting case is that of *Trichinella* species. These extremely generalist parasites occur worldwide in terrestrial CARNIVORES and other mammals. Bearded and ringed seals, walruses, and polar bears are known to serve as hosts of these nematodes in the Arctic. The infections can evidently be traced to a terrestrial origin, but patterns of transmission are intricate and have not been proven definitively. Crustaceans and birds have been suggested to acquire *Trichinella* larvae when feeding on mammal carcasses. Then, seals would be infected by direct ingestion of these crustaceans or while feeding on grounds contaminated with bird feces. The nematode might then be transmitted to polar bears and some walruses that feed on seals. Polar bears could also acquire *Trichinella* independently by consuming carrion of terrestrial mammals.

2. Freshwater Ecosystems   River dolphins, Baikal seals, *Phoca sibirica*, and manatees have secondarily colonized freshwater habitats, raising the interesting question of what parasites they are expected to harbor. Either parasites tracked these hosts from the marine realm or colonized them in freshwater or terrestrial habitats. The species of *Hunterotrema* and *Anisakis* infecting the Amazonian river dolphin, *Inia geoffrensis*, exemplify the first possibility. In contrast, many other parasites have been acquired more recently in nonmarine habitats. For instance, the fluke *Nudacotyle undicola*, from West Indian manatees, belongs to a genus whose species infect freshwater and terrestrial mammals, and Baikal seals, *Pusa sibirica*, are infected with the tapeworm *Diphyllobothrium dendriticum*, a typical freshwater species that usually infects aquatic birds and land mammals. Similarly, an echinostomatid fluke, *Ruffetrema indirae*, has been described in the Indian river dolphin, *Platanista gangetica*. Relatives of this fluke infect terrestrial birds and mammals. However, the Indian river dolphin infections seem accidental because no mature eggs were found in any worm.

Some parasites of these mammals have an uncertain origin. The digenean *Chiorchis fabaceus* and nematodes of the genus *Heterocheilus* are common to species of manatees from both freshwater and coastal habitats, but we do not know where these associations originated. The nematode *Contracaecum lobulatum* is exclusive to Indian river dolphins, and species of *Contracaecum* are typical of piscivorous freshwater and marine birds, pinnipeds, and cetaceans. Where might *C. lobulatum* have come from? In any case, river dolphins and manatees have a very poor, specific helminth fauna. Although we should bear in mind that these hosts have been poorly analyzed for parasites, the evidence suggests that parasite exchange with sympatric hosts seems to occur often at ecological time.

3. Coastal Ecosystems   Coasts constitute probably the most important realm in terms of historical and current parasite exchange between carnivorous marine mammals, and between these hosts and other vertebrates. Perhaps the best illustration is the observation that the parasite fauna of sea otters is made up of parasites from either sympatric pinnipeds or seabirds. The influx of parasites through coastal food webs puts many parasites in contact with incidental hosts, in which, depending on the compatibility filter, parasites are able or not to reproduce at ecological time. Examples of apparent failure are, for example, *Pseudoterranova decipiens* in sea otters, *A. simplex* in many seals, or most *Corynosoma* spp. in nonpinniped hosts. In contrast, some parasites are notoriously unspecific with respect to the choice of their final hosts: heterophyid, opistorchiid, and microphallid digeneans found typically in terrestrial mammals or aquatic birds infect and reproduce in pinnipeds and the sea otter (rarely coastal cetaceans) in neritic and littoral waters. Most of these records are occasional,

but in the Caspian seal, *Pusa caspica*, they represent an important portion of the helminth fauna: 16 of the 21 helminth species recorded so far are generalist parasites found in sympatric aquatic birds or terrestrial carnivores.

There are also many examples of recent accommodation or speciation of parasites in new hosts (although, in most cases, we are not certain of which were the donor and the target hosts). For instance, the campulid *Orthosplanchnus fraterculus* infects and reproduces readily in walruses, bearded seals, and sea otters, which share a similar diet on benthic invertebrates (sea otters probably acquired this fluke from pinnipeds). Likewise, diet similarity seems to be responsible for the occurrence of *Ogmogaster antarcticus* in fin whales, blue whales, Weddell seals, and crabeater seals in Antarctic waters. There are also many examples of apparent parasite speciation associated with recent colonization in the coastal environment. Species of *Corynosoma* in cetaceans (e.g., *C. cetaceum* in dolphins from the Southern Hemisphere), sea otters (*C. enhydri*), and seabirds (*C. shakletoni* in the Gentoo penguin *Pygoscelis papua*) appear to constitute independent host-switching events from *Corynosoma* of pinnipeds.

Finally, some cases represent more archaic colonization events followed by secondary diversification within the new association. It is postulated that the genera *Anophryocephalus* (Tetrabothriidae) and *Orthosplanchnus* (Campulidae), typical from pinnipeds of the Arctic and sub-Arctic waters, arose during the Quaternary from ancestors infecting toothed whales. After the initial host capture, a complex history of parasite diversification apparently occurred associated to pinnipeds themselves. One of the most striking examples is that of *Contracaecum* spp. in pinnipeds: it is believed that boreal seals and austral otariids independently acquired species of *Contracaecum* from seabirds.

4. Pelagic Ecosystems   Despite the appearance of continuity between coastal and oceanic domains, the parasitic fauna of pelagic marine mammals is distinctive and comparatively poorer. Several factors appear to contribute to this singularity. First, some parasitic groups are underrepresented. The diversity of digeneans is probably constrained because they almost exclusively require gastropods or bivalves as first intermediate hosts, which are abundant in coastal habitats but not in pelagic waters. Indeed, the Campulidae are among the few flukes with representatives in pelagic marine mammals. Second, host communities are generally species poor as compared to coastal ones; therefore, the possibility of parasite exchange is decreased (note, for example, that there is little chance for terrestrial influence). Third, infective stages are much more "diluted" in the pelagic environment. Transmission rates are thus lowered, also reducing the likelihood of host captures.

Some parasitic groups infecting marine mammals are predominantly pelagic and probably originated in this ecosystem. Except for species of the genus *Anophryocephalus*, most tetrabothriid cestodes occur in pelagic birds and cetaceans. Seabirds have been considered as initial hosts for ancestral tetrabothriids, from which these tapeworms would have switched to cetaceans (seabirds, dolphins, and whales share the genus *Tetrabothrius*). There are two other interesting cases of host switching. Pelagic cetaceans seem to have acquired species of *Pennella* from oceanic fishes. There are even records of the same species, *P. crassicornis*, parasitizing both cetaceans and fishes. However, females of *Pennella* can actively select their definitive hosts, in contrast to the majority of examples discussed thus far. How did this host switch to cetaceans happen? However, some species of *Phyllobothrium* and *Monorygma* (Cestoda) appear to use cetaceans (mostly pelagic) as intermediate hosts. Evidently, a shift in the ancestral life cycle of these tapeworms had to occur as other relatives within the class Tetraphyllidea use only invertebrates and fish.

III. Effects and Applications

A. Parasitosis

The severity of the parasite-induced damage is related to the type of parasite, its abundance, the host's health status, and the concurrence of other pathogenic agents. The effects of parasites usually have little relevance on host health, such as only local reactions produced by the proboscis of *Bolbosoma* or *Corynosoma*. In other cases, the lesions can be more important, such as ulcers and hemorrhages caused by *Anisakis* and *Pseudoterranova*. On occasion, parasites can compromise the health of its host and even lead to its death: *Nasitrema* flukes can provide neuropathies that have been related to both single and mass strandings. Lungworms of cetaceans and pinnipeds cause bronchitis and pneumonia that may lead to mortality, particularly among the youngest individuals. Sometimes, the direct action of the parasite might not be severe, but propitiates viral, bacterial, and other parasite infections, e.g., sucking lice may transmit heartworms while feeding on their hosts.

Several authors have suggested that some parasites contribute significantly to marine mammal POPULATION DYNAMICS by affecting either host reproduction or survival. For instance, the mammary glands of many females belonging to a herd of Atlantic white-sided dolphins, *Lagenorhynchus acutus*, suffered from mastitis caused by *Crassicauda* nematodes. The parasites altered the secretory tissue, affecting the quality and quantity of the milk, which would ultimately compromise the survival of the calves and the reproductive output of the herd. In other instances, parasites can lead to direct mortality. *Crassicauda* sp. causes cranial damage in spotted dolphins, *Stenella attenuata*, from the eastern Tropical Pacific. A study of dolphins caught in a tuna purse seine fishery showed an increase of lesions among individuals up to 5 years old. In contrast, in animals older than 8 years, lesions diminished by 12.3% annually. Assuming that the lesions were irreversible, that there were no reinfections, and that dolphins with lesions had the same probability of being caught as those without, the mortality rate attributable to the parasite was estimated as 1.1%. Since the annual natural mortality rate for the Pacific spotted dolphin population is 10–13%, the study suggests that *Crassicauda* sp. accounts for 11–14% of that rate.

In some cases, parasites might regulate marine mammal populations. A density-dependent relationship between parasite-induced mortality and population size has been shown for the hookworm, *U. lucasi*, in northern fur seals from the North

Pacific. These nematodes constitute the most important mortality factor among newborns by causing diarrhea, anemia, and intense intestinal hemorrhages. The population size of pups born in St. Paul Island (Alaska) peaked around 1940 and declined more or less steadily until the present. Data of hookworm mortality from previous surveys showed a decrease in pup numbers, which means that recent mortality is less than in the past. These data suggest a density-dependent relationship and, therefore, a possible regulatory effect.

B. Economic and Public Health Importance

The parasites of marine mammals can infect valuable animals in aquaria, causing lesions and diseases. For this reason, expensive prophylactic measures must be used routinely. The most important economic impact, however, is due to anisakid nematodes whose larvae use commercial fish and squid for transmission to marine mammals. These larvae also have public health repercussions (see Box), but from an economic perspective, the problem is cosmetic, rendering the fish unappealing to consumers. In 1982, the losses caused by *Pseudoterranova decipiens* in eastern Canada were valued $20 million, only in processing of cod fillets.

The parasites with repercussions on public health are those that can infect humans with food (Box 1), either by eating meat from marine mammals or, more frequently, raw fish or squid containing living infective stages.

C. Natural Tags

Many parasites are useful natural markers of biological and environmental phenomena because their transmission is linked intimately to features of the host and the ecosystem. Despite some limitations, marine mammal parasites have proven themselves as biological indicators of phylogeny, local migration, distribution, disease, stock identity, and social behavior of their hosts.

Regarding host behavior, differences in the occurrence of two whale louse (Cyamid) species on sperm whales were interpreted as evidence of spatial segregation between the bulls and the rest of the population off South Africa. Assuming that males leave their natal herd at attainment of sexual maturity, parasite information suggested that this should occur at 12 m of length. This estimation was validated later on by analyses of gonadal tissue. Another example is the comparison of intestinal helminth abundance between pods of North Atlantic pilot whales, *Globicephala melas*. Monte Carlo simulations revealed that adult males were more difficult to allocate into their pods than other individuals. This suggests male exchanges between pods, a conclusion that agrees with genetic studies that adult males do not breed within their natal pods.

Parasites transmitted through the food web have also provided information on past and present host feeding grounds. This has served to reveal geographical differences between areas for Antarctic whales, larga seal (*Phoca largha*), populations on both sides of the North Pacific, inshore and offshore forms of bottlenose dolphins in both the east coast of the United States and Peru, and geographical segregation of franciscanas, *Pontoporia blainvillei*, south and north of the La Plata River Estuary, between Argentina and Uruguay.

Main Parasitic Zoonoses

Trichinosis occurs mainly among people from the Arctic due to the consumption of raw or undercooked meat, particularly of polar bear, walrus, and some seals. *Trichinella* larvae are found in the striate muscle of many mammals, encapsulated inside individual nurse cells. When the muscles are ingested by another mammal, the larvae are released into the small intestine, where they penetrate the mucosa and develop to adult stage. The females give birth to numerous larvae, which migrate through the circulatory system to the skeletal muscles. The damage to the host is due to both penetration of adult females in the mucosa, migration of juveniles, and penetration in the muscle and nurse cell formation. Infections in humans can be fatal.

Diphyllobotriosis is caused by tapeworms of the genera *Diphyllobothrium* and *Diplogonoporus*. Human infections occur by ingestion of plerocercoids (larval stages) encysted in fish muscles. These larvae can develop into the adult stage in the human intestine. Usually the infections are not severe, but, in some cases, can lead to pernicious anemia due to a B_{12} vitamin deficit. Cases of marine-transmitted diphyllobothriosis are particularly common in Japan and Peru due to the ingestion of raw fish dishes such as ceviche.

Anisakidosis is produced by anisakid nematodes, particularly those of the genus *Anisakis*. Infections in humans occur when the larvae are eaten with either raw or lightly cooked fish or squid. The larvae cannot develop to the adult stage in the digestive tract of humans, but can cause considerable damage to the gastric or intestinal wall. They may produce ulcers and eventually peritonitis and other severe pathologies. Although anisakidosis has been traditionally common in Asian countries, especially in Japan, the popularity of raw fish dishes, as sushi, has resulted in human infections worldwide. Allergic reactions due to antigens released by the worms in the fish have been reported both among consumers and workers in fish-processing plants. This puts the problem of anisakidosis under a new light because the antigens are thermoresistant. Thus, common prophylactic methods to kill the larvae, such as cooking or freezing, cannot prevent allergies.

See Also the Following Articles

Health ▪ Stock Identity ▪ Whale Lice

References

Aznar, F. J., Balbuena, J. A., Fernández, M., and Raga, J. A. (2001). Living together: The parasites of marine mammals. *In* "Marine Mammals: Biology and Conservation" (P. G. H. Evans and J. A. Raga, eds.). Kluwer Academic/Plenum Publishers, New York.

Balbuena, J. A., Aznar, F. J., Fernández, M., and Raga, J. A. (1995). The use of parasites as indicators of social structure and stock identity of marine mammals. *In* "Whales, Seals, Fish and Man" (A. S. Blix, L. Walløe. and Ø. Ulltang, eds.), pp. 133–139. Elsevier Science, Amsterdam, The Netherlands.

Beck, C., and Forrester, D. J. (1988). Helminths of the Florida manatee, *Trichechus manatus latirostris*, with a discussion and summary of the parasites of sirenians. *J. Parasitol.* **74**, 628–637.

Dailey, M. D. (1985). Diseases of mammalia: Cetacea. *In* "Diseases of Marine Animals" (O. Kinne, ed.), Vol. 4, Part 2, pp. 805–847. Biologische Anstalt Helgoland, Hamburg, Germany.

Delyamure, S. L. (1955). "Helminth Fauna of Marine Mammals (Ecology and Phylogeny)." Akademiya Nauki SSSR, Moscow. [Translated by Israel Program for Scientific Translation, Jerusalem, 1968.]

Geraci, J. R., and St. Aubin, D. J. (1987). Effects of parasites on marine mammals. *Int. J. Parasitol.* **17**, 407–414.

Hoberg, E. P. (1995). Historical biogeography and modes of speciation across high-latitude seas of the Holarctic: Concepts for host-parasite coevolution among the Phocini (Phocidae) and Tetrabothriidae (Eucestoda). *Can. J. Zool.* **73**, 45–57.

Lauckner, G. (1985). Diseases of mammalia: Carnivora. *In* "Diseases of Marine Animals" (O. Kinne, ed.), Vol. 4, Part 2, pp. 645–682. Biologische Anstalt Helgoland, Hamburg, Germany.

Lauckner, G. (1985). Diseases of mammalia: Pinnipedia. *In* "Diseases of Marine Animals" (O. Kinne, ed.), Vol. 4, Part 2, pp. 683–793. Biologische Anstalt Helgoland, Hamburg, Germany.

Lauckner, G. (1985). Diseases of Mammalia: Sirenia. *In* "Diseases of Marine Animals" (O. Kinne, ed.), Vol. 4, Part 2, pp. 795–803. Biologische Anstalt Helgoland, Hamburg, Germany.

Margolis, L., Groff, J. M., Johnson, S. C., McDonald, T. E., Kent, M. L., and Blaylock, R. B. (1997). Helminth parasites of sea otters (*Enhydra lutris*) from Prince William Sound, Alaska: Comparisons with other populations of sea otters and comments on the origin of their parasites. *J. Helminthol. Soc. Wash.* **64**, 161–168.

Oshima, T., and Kliks, M. (1987). Effects of marine mammal parasites on human health. *Int. J. Parasitol.* **17**, 415–421.

Raga, J. A. (1997). Parasitology of marine mammals. *In* "Marine Mammals, Seabirds and Pollution of Marine Systems" (T. Jauniaux, J. M. Brouqueqneau, and F. Coignoul, eds.), pp. 67–90. Presses de l'Université de Liège, Liège, Belgium.

Raga, J. A., Balbuena, J. A., Aznar, F. J., and Fernández, M. (1997). The impact of parasites on marine mammals: A review. *Parasitologia* **39**, 293–296.

Smith, J. W. (1999). Ascaridoid nematodes and pathology of the alimentary tract and its associated organs in vertebrates, including man: A literature review. *Helminthol. Abstr.* **68**, 49–96.

Parental Behavior

JANET MANN
Georgetown University, Washington, DC

Parental behavior in pinnipeds, sirenia, sea otters, polar bears, and cetaceans shares several features: paternal care is virtually absent, gestation and lactation periods are typically long, females give birth to and nurse one offspring at a time (polar bears excepted), and many marine mammals fast during the early stages of lactation. In sum, marine mammal mothers invest extensively and exclusively in single offspring; this article reviews the diversity and nature of that investment.

Marine mammal maternal strategies vary in important respects. Polar bears (*Ursus maritimus*), sea otters (*Enhydra lutris*), and all three families of pinnipeds—Odobenidae (walrus, *Odobenus rosmarus*), Phocidae (earless or "true seals"), and Otariidae (eared seals, sea lions, fur seals)—give birth on land or ice. Twenty-three species of pinnipeds breed on land and 13 breed on ice. Cetacean and sirenian females give birth in the water; this pattern favors precocial swimming and diving. Among many pinnipeds, maternal care is largely restricted to milk transfer, whereas the prolonged association characteristic of many cetaceans, sirenians, otters, polar bears, and some pinnipeds also involves protection and potentially extensive information transfer.

I. Feeding, Lactation, and Patterns of Association

Lactation strategies in marine mammals generally depend on trade-offs among foraging, predation risk, and reproduction. This trade-off is exemplified by many marine mammal species that fast during the early stages of lactation. That females forgo feeding by breeding on land (i.e., pinnipeds) or in warm coastal waters (i.e., baleen whales) suggests that benefits, such as reduced predation risk and rapid energy transfer from mother to offspring, outweigh the costs of fasting. Larger bodied mammals can withstand fasting for longer periods than smaller, thus able to afford longer fasting periods devoted to offspring care.

Fasting and lactation coincide in many marine mammal species and only rarely in terrestrial mammals. Early-weaning marine mammals tend to have milk that is high in fat, investing heavily in offspring for a shorter period. Late-weaning marine mammals tend to have lower fat milk (although still much higher than for terrestrial mammals). This pattern is generally true for comparisons between phocids and otariids, or toothed and baleen whales (see Oftedal, 1997; Table I). Other factors, such as the development of pup or calf foraging skills, also contribute to late weaning ages and prolonged association (see Section IV).

Phocids tend to fast and remain near the ROOKERY until their pup is weaned; they rely on fat stores to nurse offspring. Phocid maternal strategies are generally characterized as *fasting*. A few phocids feed during lactation, notably the harp (*Pagophilus groenlandicus*), harbor (*Phoca vitulina*), ringed (*Pusa hispida*), and Weddell (*Leptonychotes weddellii*) seals. Most remarkable is the hooded seal (*Cystophora cristata*), which breeds on pack ice and nurses her pup for only 4 days, transferring approximately 748,000 J or 178,657 kcal to the pup in that time.

Predator and prey distributions are likely to influence breeding habitat (i.e., pack ice, fast ice, beach, cave, and water) and lactation length. Whether breeding habitat and lactation length are related is difficult to determine using the comparative method because species breeding in similar habitats tend to be close phylogenetically (and are thus nonindependent). However, some general comparisons are possible. Pack or floe ice is unstable compared to fast ice and could constrain the duration of lactation. Some species with protective lairs or caves [arctic Baikal ringed seal (*Pusa sibirica*), and Mediterranean monk seal (*Monachus monachus*)] wean late for phocids, at 2–3 months. Body size and fasting ability also influence the duration of nursing. Phocid

TABLE I
Parameters Related to Marine Mammal Maternal Care

Taxon	Birth location (land, ice, water)	Sexual dimorphism (high/moderate vs low)	Mother fasts during lactation	Milk fat (%)	Duration of lactation (months)	Association postweaning?
Pinnipedia						
Otariidae	L	H	Yes	25–53	4–30	No?
Phocidae	L, I	H, L	Yes	47–61	<3	No
Odobenidea	L, I	H	No	—	24	Yes?
Sirenia	W	L	No	—	>12	No?
Ursidae (polar bears)	L	L	Yes	17–25	>12	No?
Mustelidae (sea otters)	L	L	No	—	6–12	Yes
Cetacea						
Platanistoidea	W	L	No	13	<12	Yes
Delphinoidea	W	H, L	No	22–30	12–48	Yes
Ziphoidea	W		No	18–24	—	—
Physeteroidea	W	H	No	24	>12	Yes
Balaenidae	W	L	Yes	22	<10	No
Neobalaenidea	W	L	Yes	—	<6	No
Eschrichtiidae	W	L	Yes	53	<6	No
Balaenopteridae	W	L	Yes	24–40	<10	No

females tend to be larger than otariid females and can thus fast longer.

Otariid mothers leave their young after the first week to hunt for 1–14 days before returning to nurse their pups. When the mother is hunting, the pup is typically fasting. She returns for 1–3 days and fasts while she feeds her pup. The duration of her absence is correlated positively with the milk she provides per pup visit. If food is abundant, then mothers make shorter foraging trips than when food is scarce and feed their young more often. Thus, prey abundance is correlated positively with maternal body condition and pup growth rate. Some otariids, such as sea lions, nurse their pups for 6–11 months. The Galapagos fur seal (*Arctocephalus galapagoensis*) nurses her pup for as long as 3 years.

As pups grow, otariid mothers may spend three-fourths of their time at sea, but each foraging trip tends to vary in duration. Phocids are adapted to store energy and fast for the entire lactation period, whereas otariids must supplement stores with some food. Thus otariid mothers must find breeding breaches that are also near productive feeding grounds. Beach breeding can provide safety to pups, but also constrains how far and how often females can venture to hunt. Otariid pups grow more slowly than phocid pups, but tend to be weaned at larger weights, 40% of maternal mass compared with 30% in phocids. When food abundance is high, otariid mothers can transfer more milk to their offspring than when food is scarce. Phocid females concentrate on feeding offspring rather than expend energy traveling to and from feeding areas. In sum, most otariids wean their young within the year, but later than most phocids.

Of all pinnipeds, walruses appear have the longest average period of maternal care, nursing their young for up to 3 years. It is not known whether walrus mothers fast during the initial stage of lactation. They give birth on ice, but the mother nurses her pup in the water. Known for their sociability, females and pups spend most of their time in female herds. Daughters stay in their mothers' herd. Sons leave the herd when 2–3 years old and join all-male herds.

Similar to the fast vs feed pattern that distinguishes lactating phocids and otariids, the larger baleen whales fast for long periods and toothed whales little or not at all. Baleen whales have higher fat milk and nurse for shorter periods than toothed whales. Mysticete mothers, who typically migrate to warmer waters to birth and rear young, can devote their time and energy almost exclusively to milk transfer, much like phocid mothers. Like pinnipeds, baleen whales are characterized by annual breeding and feeding cycles, including cases of postpartum estrus. Among baleen whales and pinnipeds, females are often attractive to males soon after giving birth. Thus, intense competition between males over access to females could interfere with maternal care, but detailed behavioral studies are needed.

Little is known about the nature of the baleen whale mother–calf relationship. Humpback (*Megaptera novaeangliae*) and right whales (*Eubalaena* spp.) are the best-studied baleen whales, likely due to their tendency to visit protected coastal areas to breed. Interactions between more than one mother–calf pair are infrequent, especially between humpback mother–calf pairs, which appear to actively avoid each other.

The explanation for avoidance is unclear. Association between mother–calf pairs may disrupt the development of maternal or calf recognition. Similar to some of the large ungulate species, mothers may avoid unrelated calves to avoid milk theft. Anecdotal reports of Southern right whale (*E. australis*) behavior indicate that calves occasionally attempt to nurse from another female. With such a tremendous energetic output, milk transfer may be difficult to inhibit if the "wrong" calf attempts to nurse. Additionally, fasting in warm water likely reduces the metabolic demands on the mother. (Warm water would also reduce metabolic demands on the calf.) Fasting terrestrial mammals convert body fat to fatty milk more easily than mammals that eat during lactation, and this pattern is likely to favor fasting in baleen whales that invest in rapid growth of a large calf. Consistent with this, fasting mammals (terrestrial and marine), including baleen whales, have milk low in carbohydrate, protein, and water, but very high in fat. The baleen whale mother can thus fatten in the feeding grounds, transfer that energy efficiently to her calf in the warmer breeding grounds, and assure a safer migration to higher latitudes. Larger calf body size would likely reduce predation risk and other somatic costs (e.g., metabolic). Four balaenopterids [blue, (*Balaenoptera musculus*), fin, minke, (*B. acutorostrata* and *B. bonaerensis*), and sei (*B. borealis*) whales] tend to wean early, before or soon after reaching the feeding grounds (at 6–7 months). Humpback, right, bowhead (*Balaena mysticetus*), and gray (*Eschrichtius robustus*) whales continue to nurse offspring to older ages and are typically feeding during the latter stage of lactation. Although bowhead whales migrate, they do not migrate to warmer waters to calve and apparently do not fast. Given this pattern, bowhead whales are predicted to have milk lower in fat, nurse offspring for longer periods, and have slower growth rates compared to other baleen whales. Existing data are too limited to make explicit comparisons.

Little is known about mother–calf behavior and associations during migration. Observations of Pacific humpback whales off the coasts of Australia and of gray whales off the coasts of North America indicate that adult males often accompany mothers and calves during the migration. This may reduce killer whale (*Orcinus orca*) predation risk, although the males are likely there to mate with the female, not protect the calf. Several studies have indicated that up to one-third of humpback whale calves bear killer whale tooth marks by the time they reach the feeding grounds with their mothers. Killer whale attacks on gray and humpback whale calves have been observed, but there is no evidence to suggest that males aid calves during such attacks.

Odontocete life history patterns are slow relative to baleen whales. The baleen whale usually reaches reproductive maturity in 4 or 5 years, but the typical odontocete takes 10 years or more, despite a much smaller body size. Only the smallest odontocetes, such as porpoises, tend to reach reproductive maturity by age 5. With a relatively long period of immaturity and calf vulnerability, odontocetes may benefit more by stable patterns of group living compared to mysticetes.

Odontocetes do tend to have stable groups and patterns of association compared to mysticetes. In contrast to the baleen pattern, odontocete mother–calf pairs tend to associate with each other and stable associations with female kin are common. Adult male associations with mothers and young calves are more variable. For example, adult male killer whales consistently associate with female kin and their offspring. Female sperm whales (*Physeter macrocephalus*) and young spend most of their time with female kin and rarely associate with adult males. Shark Bay bottlenose dolphin (*Tursiops aduncus*) adult males associate with mothers and newborns (0–3 months) for only 5% of the time, although females are generally in larger groups during this period than when the calf is older. Larger group sizes may afford protection to the calf by reducing predation risk or harassment from conspecifics.

Close mother–calf association is characteristic of many toothed whales and dolphins, particularly in the delphinid family. Delphinid calves typically swim alongside the mother for the first few weeks in what is known as "echelon position." Within a few months, they begin to swim under the mother regularly in what is known as "infant position." In contrast, humpback and right whale calves tend to swim alongside the mother throughout development and separations are not common during lactation. Bowhead calves occasionally ride on the backs of their mothers during migrations. Odontocete mothers are presented with particular challenges baleen mothers do not face. Because they feed throughout lactation and hunt single prey, the mother must accelerate or dive deeply to hunt, often leaving her calf alone at the surface. Young calves are not very proficient divers, and they breathe more often, thus they sometimes "wait" at the surface while the mother forages below. This is especially pronounced in deep-diving species such as sperm whales. Mothers may dive for as long as between 30 and 50 min. The calf, often alone at the surface, has a tendency to swim toward any whale that comes up first. Some form of communal care in this deep-diving species may be an important selective force favoring matrilineal sperm whale units. A lone sperm whale mother would be at a distinct disadvantage if she had to abandon her vulnerable calf at the surface for long periods as she hunted for squid at depth. Among pantropical spotted dolphins (*Stenella attenuata*), lactating females feed on flying fish rather than squid (preferred by nonlactating females), which necessitates diving and separation from their calf. In bottlenose dolphins (*Tursiops* spp.), most mother–calf separations occur during the mother's foraging. When foraging, bowhead mothers and calves may separate 1–2 km for 30 min or more.

II. Protection and Predation

Otariids tend to breed on predator-free islands. Phocid pups can be vulnerable on fast ice, where polar bears and killer whales may feed on them. In the North Pacific, killer whales frequently catch harbor seal pups on their first forays into the ocean. South American sea lion (*Otaria flavescens*) pups are picked off beaches by killer whales. Great white sharks (*Carcharon charcharias*) are common predators of elephant seal pups (*Mirounga* spp.) and a variety of other pinniped prey. Mammal-eating killer whales will also prey on most pinniped species they encounter, including the largest species, Steller sea lions (*Eumetopias jubatus*). Maternal protection of pups

from predation is not well documented. Adult male and female pinnipeds sometimes pursue predators, such as killer whales and sharks, but nothing about this behavior suggests that they are protecting offspring per se.

Mothers and other female whales and dolphins do protect offspring from predatory attacks. Sperm whale females may form a ring "marguerite formation" around a calf and sometimes they place their heads to the center and use their powerful tails to deter a predator. At other times they will face out in a circle, with their tails facing the calf. Indian Ocean bottlenose dolphin mothers and close associates have also defended calves from predation attempts by tiger sharks (*Galeocerdo cuvieri*) (Mann and Barnett, 1999). Gray whale mothers have been observed placing themselves between killer whales and their calves (Black *et al.*, 1999).

Bottlenose dolphin infanticide has been documented from retrieved carcasses at two Atlantic Ocean sites: the Moray Firth in Scotland and Virginia, United States. The perpetrators of calf killing are not known. If the pattern is similar to that found widely in felids, ursids, primates, and rodents, then it is likely that unrelated adult males are the perpetrators and might gain a reproductive advantage by monopolizing and mating with the mother of the victim. Polar bear mothers, like other ursid females, are known for their protectiveness and fierce defense of cubs from infanticidal males.

Among land-breeding SEXUALLY DIMORPHIC pinnipeds, offspring are occasionally killed incidentally by large patrolling males (e.g., elephant seals and sea lions). These killings appear to be by accident, e.g., when the male is preoccupied with defending the breeding beach. However, males also abduct and force matings on pups, sometimes resulting in the pup's death. Among ice-breeding pinnipeds, or when mating occurs in the water, pups are less vulnerable to patrolling males. In these species, sexual dimorphism is also less pronounced and the risk of being squashed is lower than for land-breeding species.

III. Mother–Offspring Recognition

Mutual vocal and olfactory recognition is common in a number of pinniped species, especially among otariids. This might be expected given that otariid mothers and pups often separate during the mother's foraging trips and the reunion depends on mutual recognition through repeated calling. Otariid pups are also fairly mobile, so the previous location of her pup would not be a reliable cue. In addition, otariids breed colonially, complicating the task of finding the pup or mother among hundreds. Phocid mothers would not necessarily need a recognition system if they remain close to their offspring, but some species that separate during lactation (e.g., harbor seals, *Phoca vitulina*) exhibit vocal recognition. After feeding trips, it is common for mothers to inspect several pups before finding her own.

Sea otter vocalizations have been compared to pinnipeds in structure and complexity. There is some evidence for short-range COMMUNICATION and vocal recognition between mothers and pups. Olfactory cues may also play a role.

Cetaceans have little or no sense of smell, but likely use vocal communication for individual identification. Little is known about mysticete mother–calf recognition and only slightly more is known in odontocetes. Delphinids produce a diverse array of sounds, including echolocation clicks, burst-pulse sounds, and whistles. Dolphin calves are born able to whistle and will whistle often during the first days of life. Bottlenose dolphins and potentially other delphinids exhibit signature whistles, individually distinctive whistles in the first years of life. In bottlenose dolphins, simple whistles are present at birth, but the calf develops a more distinctive contour with age (although the calf may continue to produce simple whistles as well). Mechanisms influencing the development of SIGNATURE WHISTLE contours are not known. Although the calf's signature whistle is not present at birth, the young calf may learn to identify his or her mother's signature whistle soon after birth and mothers may be able to recognize the calf's simple whistle. Field studies suggest that whistles mediate natural separations and reunions between mothers and calves. Calves tend to whistle more than mothers, perhaps indicating the calf's greater motivation to reunite with the mother than vice versa. Experimental work with captive and temporarily captured animals suggests that dolphins can recognize the whistles of others. In the marine environment, depth (water pressure) and other habitat features may alter sound enough to make voice recognition difficult; thus, selection may favor individually distinctive whistle contours in an environment where individuals join and leave each other and frequently lose visual contact. Few studies have sufficiently examined cetacean mother–calf communication in captive or wild settings to elaborate further.

IV. Maternal Influence on Offspring Behavioral Development

Mothers influence offspring behavioral development by the experiences they provide, including migration and navigation, communication, social interactions, and foraging. To secure the transition from dependence to independence, marine mammal offspring must acquire necessary survival skills. Although data are limited, maternal influence on the development of foraging skills has received the most attention.

Some types of marine mammal foraging may not require learning, but other foraging skills clearly require experience. Marine mammal young may develop foraging skills by independent (nonsocial) learning or by social learning, including stimulus or local enhancement (e.g., exposure to foraging areas with the mother), observation of maternal foraging or prey caught, and rarely pedagogy. Teaching requires that the "teacher" modifies his or her behavior in the presence of a naïve observer (pupil) at some cost or without benefit to the "teacher," which results in or facilitates acquisition of some skill or knowledge by the "pupil" (Caro and Hauser, 1992). As such, a pup or calf can "learn" from another animal, without teaching occurring. There are good examples of extensive social and nonsocial learning in Cetacea, but only a few possible examples of teaching. There are no examples of teaching in pinnipeds and little evidence for learning from the mother. In sea otters, learning by observing from the mother appears to be important, but teaching probably does not occur.

Pinniped pups theoretically could learn some foraging skills by following and observing the mother. However, there is little

evidence to suggest this. Phocid mothers fast for most of lactation and leave the rookery before their pups do. Otariid pups could accompany their mothers on foraging trips, but most studies find little evidence of this. Pup absence from the rookery often coincides with maternal absence (a pup does not leave the rookery while the mother is present), but mother and pup travel and forage independently at sea. Otariid pups have more opportunities to develop swimming, diving, and hunting skills while relying principally on their mothers nutritionally. In the late-weaning Galápagos fur seal, pups catch their own prey for a year or longer before weaning. Some phocid pups gain considerable diving and swimming experience independently before weaning [e.g., bearded (*Erignathus barbatus*) and ringed seals (*Pusa hispida*)]. Some harbor seals change haul-out sites during lactation. The pup, then, may follow the mother 30 km or more. Although research with captive pinnipeds indicates that they are adept learners, including vocal learners, there is little to suggest that learning in the context of the mother–pup relationship is particularly important except for mother–pup recognition.

Several studies clearly document that otter pups (*Lutra lutra* and *Enhydra lutris*) learn from their mothers. Mothers take their pups on foraging trips and the pup watches the mother as she dives and retrieves prey. Mothers share their catch with pups, and in sea otters, pups even develop the specific tool-using strategy of the mother (use of rocks, soft-drink bottles, or other objects) to open shellfish or other prey (Lyons and Estes, 1989). It takes time for pups to develop adequate diving and foraging skills, which may help explain the long dependence period in this small marine mammal. Other MUSTELIDS typically wean offspring at much younger ages (i.e., 2–3 months).

To become independent, cetacean calves must be able to forage successfully on their own. They must also catch their first prey on their own. Unlike other hunting species (e.g., felids, canids), cetacean mothers generally do not share prey with young. One exception is the killer whale, where prey sharing between mother and offspring has been well documented. Despite many thousands of hours of observation on bottlenose dolphin calves in Shark Bay by the author, prey sharing has not been observed. Delphinids exhibit a diverse range of foraging strategies and there is some evidence that these are learned. Whether calves learn foraging skills by simple trial and error, social learning, i.e., observing the mother and other conspecifics, or other mechanisms is generally not known. Shark Bay bottlenose dolphin calves begin capturing prey within 4–6 months, even though they continue nursing 3 years or more. Female bottlenose dolphin calves in Shark Bay, Australia, adopt the sponge-foraging tactic of their mother, and those without sponging mothers do not develop the trait. This unusual form of tool use, which is adopted by only a small subset of the population, may be transmitted socially (and culturally) from mother to offspring. Humpback whale juveniles in the North Atlantic have learned new foraging strategies, likely due to shifts in prey density and distribution. These tactics spread rapidly through the population, although data suggest that older animals generally did not adopt the behavior. The humpback example may provide evidence for learning (horizontal transmission), but not from the mother (vertical transmission).

The diversity of foraging strategies that humpback whales exhibit (e.g., bubblenet, cooperative lunge feeding) may help explain their relatively longer nursing period compared to other mysticetes; i.e., the duration of lactation may, in part, be determined by the ability of young animals to forage independently. This might also help explain the longer nursing periods of odontocetes relative to mysticetes. The odontocete's task of capturing single, elusive, and specialized prey is likely to require greater foraging skill than the gulp-feeding techniques characteristic of mysticetes.

Killer whales may provide the only reasonable cetacean example of teaching, but more observations are needed. Experienced beach-feeding whales (who beach to capture pinnipeds) actively help younger animals develop this risky technique by nudging them to shore, partially wounding prey for young to manipulate, and assisting their departure from the beach.

V. Parity

For many mammals, parity (number of pregnancies/live births experienced) is correlated negatively with offspring mortality. For example, sea otter pups born to older, more experienced mothers tend to have lower mortality than pups born to younger mothers. First-time (primiparous) elephant seal mothers have lower weaning success than experienced multiparous mothers. However, a number of factors may contribute to this difference, including maternal mass (young mothers are smaller than old mothers), experience, dominance, and placement of pups in the rookery. Young, subordinate females relegated to the peripheral area of a breeding beach may have more difficulty relocating their pup and their pup may be more vulnerable to harassment and death.

First-born dolphin calves in captivity have higher mortality than later-born calves, but the cause(s) for the difference is not known. Both maternal experience and body size may be factors. Patterns of first-born mortality in wild populations are not sufficiently well documented to draw conclusions.

VI. Sex-Biased Investment

Several pinniped studies have investigated whether, consistent with the predictions of Trivers and Willard (1973), mothers in good condition preferentially invest in sons over daughters. According to the Trivers-Willard model, in polygynous species, maternal investment is likely to have marginally greater genetic payoff in terms of grandoffspring if the mother can influence offspring quality and thus produce a "super-son" compared to a "super-daughter" because males have greater variance in reproductive success than females. This depends on whether mothers can confer benefits, such as increased body size, to their offspring. Biased investment generally takes two forms: sex ratio and differential investment or maternal care.

Data on biased maternal care or sex ratios are equivocal for both phocids and otariids. The sex ratio does not covary with maternal mass, a good indicator of maternal condition. For example, Northern elephant seal (M. angustirostris) males are born heavier than females and young mothers are less likely to successfully wean sons compared to daughters; however, there

is little evidence that mothers differentially invest in sons compared to daughters. Males gain most of their size during the 3- to 5-year growth spurt, long after weaning. Maternal expenditure may not contribute significantly to male reproductive success. Gray seal data are similarly equivocal. Otariid males are born slightly heavier and appear to grow faster than female pups, but none of the pinniped data provides clear evidence that sons exact greater reproductive costs on mothers than daughters.

In cetaceans, much less is known about biases in either sex ratio or maternal care. However, one study found that humpback whales were more likely to give birth to sons than daughters following the close of a long (3-year) birth interval compared to a short (1- to 2-year) one, although they hypothesized that mothers were in better condition after a 3-year interval they were thus more likely to give birth to sons. However, the actual condition of mothers is unknown. Notably, adult female baleen whales are slightly larger than adult males. It thus remains unclear what benefit mothers in good condition might gain by biasing the sex ratio toward sons if larger body size relative to females or other males is unlikely to give them an advantage. Available growth data suggest that weanling males are slightly smaller than weanling females. Other potential cases supporting Trivers and Willard are findings suggesting that male teenage sperm and short-finned pilot whales (*Globicephala macrorhynchus*) are occasionally still nursing, but female offspring nurse no later than age 7. Because both species are markedly sexually dimorphic, these examples may be more likely to fit the Trivers–Willard model.

VII. Mother–Offspring Conflict and Weaning

In a number of pinnipeds, weaning can be initiated by either mother or pup. Otariid pups may leave the rookery while their mother continues to return from foraging trips. Phocid pups are typically weaned by the mother. She leaves the rookery permanently. Likely, the proximate cue for both the mother's and the pup's departure from land or ice is hunger. The mother leaves when her fat stores are depleted. The pup leaves when it is no longer receiving enough milk to sustain itself. Among some otariids, such as the Galápagos fur seal, the older sibling (1–2 years) may compete with the newborn for milk, providing one of the few examples of clear sibling competition among marine mammals. The size of the older sibling appears to influence the degree of competition with the younger sibling.

Little is known about the weaning process in whales or dolphins. Weaning may be fairly sudden in some baleen whales that separate feeding and breeding. Some baleen species wean during the migration to higher latitudes or soon after, but other species wean toward the end of the first year. Weaning in toothed whales is likely more gradual than in mysticetes. Evidence for mother–calf conflict comes from studies of Indian Ocean bottlenose dolphins in Shark Bay and southern right whales in Patagonia. At both sites, mothers are less responsible for maintaining proximity to their calves as they age and mothers appear to use the sea floor to prevent calves from nursing. A right whale mother may also roll belly up to prevent her calf from nursing.

VIII. Association Postweaning

Like humans and a few other mammals, mother and offspring may continue to associate postweaning and sometimes for life. Evidence for strong mother–offspring association in pinnipeds is weak. Phocids wean abruptly and separate. Among otariids, some postweaning association is possible but difficult to study. Many otariid and walrus females stay in or return to the same breeding areas, and thus may well recognize each other and interact. Few data on this are available.

However, among cetaceans, high group stability and patterns of preferential mother–daughter association in fission-fusion societies (fluid patterns of group association, often with some underlying stability) indicate that strong bonds may persist. For example, resident killer whale male and female offspring remain with their mothers for life, a pattern unique to this species. Pilot whale daughters remain with their mothers postweaning, but sons may disperse temporarily or permanently. Sperm whale daughters remain in their matrilineal unit for life, much like elephants. Bottlenose dolphin daughters associate with mothers more postweaning than sons. In contrast, few daughter or son associations postweaning persist in mysticete whales. Among delphinids, the significance of such long-term kin associations is not known, but given the diversity of prey, the complexity of some delphinid social relationships, and other selective pressures on odontocetes, continued association is likely to offer the benefits of group living. Such benefits include protection from predators or conspecifics and sharing of information and/or tasks, such as calf care.

See Also the Following Articles

Breeding Sites ■ Infanticide and Abuse of Young ■ Mimicry ■ Reproductive Behavior

References

Bachman, K. C., and Irvine, A. B. (1979). Composition of milk from the Florida manatee, *Trichechus manatus latirostris*. *Comp. Biochem. Physiol.* **62A**, 873–878.

Bernard, H. J., and Hohn, A. A. (1989). Differences in feeding habits between pregnant and lactating spotted dolphins (*Stenella attenuata*). *J. Mammal.* **70**, 211–215.

Black, N., Ternullo, R., and Schulman-Janiger, A. (1999). Behavior and ecology of transient kills in Monterey Bay, California. "13th Bienniel Conference on the Biology of Marine Mammals" (Abstract p. 17), Wailea Maui, Hawaii.

Caro, T. M., and Hauser, M. D. (1992). Is there teaching in nonhuman animals? *Q. Rev. Biol.* **67**, 151–174.

Clutton-Brock, T. H. (1991). "The Evolution of Parental Care." Princeton Univ. Press, Princeton, NJ.

Costa, D. P., and Crocker, D. E. (1999). Seals. *In* "Encyclopedia of Reproduction." Academic Press, San Diego.

Gazo, M., Aparicio, F., Cedenilla, M. A., Layna, J. F., and Gonzalez, L. M. (2000). Pup survival in the Mediterranean monk seal (*Monachus monachus*) colony at Cabo Blanco Peninsula (Western Sahara-Mauritania). *Mar. Mamm. Sci.* **16**, 158–168.

Janik, V. M., and Slater, P. J. B. (1997). Vocal learning in mammals. *Adv. Study Behav.* **36**, 59–99.

Le Boeuf, B. J., Condit, R., and Reiter, J. (1989). Parental investment and the secondary sex ratio in Northern elephant seals. *Behav. Ecol. Sociobiol.* **25**, 109–117.

Le Boeuf, B. J., and Laws, R. M. (eds.) (1994). "Elephant Seals: Population, Ecology, Behavior and Physiology." University of California Press, Berkeley.

Lyons, K. J., and Estes, J. A. (1989). Individual variation in diet and the question of optimal feeding behavior in the female California sea otter. 8th Biennial Conference on the Biology of Marine Mammals (Abstract p. 40), Pacific Grove, California.

Mann, J., and Barnett, H. (1999). Lethal tiger shark (*Galeocerdo cuvier*) attack on bottlenose dolphin (*Tursiops* sp.) Calf: defense and reactions by the mother. *Mar. Mamm. Sci.* **15**, 568–575.

Mann, J., Connor, R. C., Barre, L. M., and Heithaus, M. R. (2000). Female reproductive success in wild bottlenose dolphins (*Tursiops* sp.): Life history, habitat, provisioning, and group-size effects. *Behav. Ecol.* **11**(2), 210–219.

Mann, J., and Smuts, B. B. (1998). Natal attraction: Allomaternal care and mother–infant separations in wild bottlenose dolphins. *Anim. Behav.* **55**, 1097–1113.

Mann, J., and Smuts, B. B. (1999). Behavioral development in wild bottlenose dolphin newborns (*Tursiops* sp.). *Behaviour* **136**, 529–566.

McShane, L. J., Estes, J. A., Riedman, M. L., and Staedler, M. M. (1995). Repertoire, structure, and individual variation of vocalizations in the sea otter. *J. Mammal.* **76**, 414–427.

Oftedal, O. T. (1993). The adaptation of milk secretion to the constraints of fasting bears, seals, and baleen whales. *J. Dairy Sci.* **76**, 3234–3246.

Oftedal, O. T. (1997). Lactation in whales and dolphins: Evidence of divergence between baleen- and toothed-species. *J. Mamm. Gland Biol. Neoplasia* **2**, 205–230.

Oftedal, O. T., Bowen, W. D., and Boness, D. J. (1993). Energy transfer by lactating hooded seals, nutrient deposition in their pups, during the four days from birth to waning. *Physiol. Zool.* **66**, 412–435.

Reidman, M. L., Estes, J. A., Staedler, M. M., Giles, A. A., and Carlson, D. R. (1994). Breeding patterns and reproductive success of California sea otters. *J. Wildl. Manage.* **58**, 391–399.

Sayigh, L. S., Tyack, P. L., Wells, R. S., and Scott, M. D. (1990). Signature whistles of free-ranging bottlenose dolphins *Tursiops truncatus*: Stability of mother-offspring comparisons. *Behav. Ecol. Sociobiol.* **26**, 247–260.

Smolker, R. A., Mann, J., and Smuts, B. B. (1993). The use of signature whistles during separations and reunions among wild bottlenose dolphin mothers and calves. *Behav. Ecol. Sociobiol.* **33**, 393–402.

Smolker, R. A., Richards, A. F., Connor, R. C., Mann, J., and Berggren, P. (1997). Sponge-carrying by Indian Ocean bottlenose dolphins: Possible tool-use by a delphinid. *Ethology* **103**, 454–465.

Trillmich, F. (1996). Parental investment in pinnipeds. *In* "Parental Care: Evolution, Mechanisms and Adaptive Significance" (J. S. Rosenblatt and C. T. Snowdon, eds.), pp. 533–577. Academic Press, San Diego.

Trivers, R. L., and Willard, D. E. (1973). Natural selection of parental ability to vary sex ratio offspring. *Science* **179**, 90–92.

Watt, J. (1993). Ontogeny of hunting behaviour of otters (*Lutra lutra* L.) in a marine environment. *Symp. Zool. Soc. Lond.* **65**, 87–104.

Whitehead, H., and Mann, J. (2000). Female reproductive strategies of cetaceans: Life histories and calf care. *In* "Cetacean Societies: Field Studies of Dolphins and Whales" (J. Mann, R. C. Connor, P. L. Tyack, and H. Whitehead, eds.), pp. 219–246. The University of Chicago Press.

Würsig, B., and Clark, C. (1993). Behavior. *In* "The Bowhead Whale" (J. Burns, J. Montague, and C. Cowles, eds.), Chap. 5, pp. 157–199. Special Pub. No. 2, Society for Marine Mammalogy.

Würsig, B., Dorsey, E., Fraker, M., Payne, R., and Richardson, J. (1985). Behavior of bowhead whales, *Balaena mysticetus*, summering in the Beaufort sea: A description. *Fish. Bull.* **83**(3), 357–377.

Würsig, B., Koski, W. R., and Richardson, W. J. (1999). Whale riding behavior: Assisted transport for bowhead whale calves during spring migration in the Alaska Beaufort Sea. *Mar. Mamm. Sci.* **15**, 204–210.

Pathology

Daniel F. Cowan
University of Texas Medical Branch, Galveston

This article focuses on naturally occurring diseases of marine mammals and how their behavioral, physiologic, and anatomical adaptations to life spent mainly or entirely in the water influence those disease processes. Pathology is the study of diseases, including their mechanisms, manifestations, and diagnosis. One of the fundamental principles of pathology is that *every disease is a reaction to injury.* This means that an organism responds to an injury within its anatomical and physiological capacity, and that the way the organism responds, i.e., the way the disease process shows itself, will be determined or at least influenced by the organism's adaptation to its environment. Thus, a fish can develop an infectious disease of the gills, but cannot get pneumonia, a disease of the lungs. This seems obvious enough, but it may easily be forgotten when making assumptions about marine mammals. For example, the presence or absence of a thick coat of hair makes a difference in the way mercury is handled by an animal. In hairy species, a large portion of the mercury burden accumulated in the diet is eliminated by binding to HAIR, which is later shed. Cetaceans, however, which may be exposed to at least as much environmental mercury as their hairy fellow marine mammals, the pinnipeds, otters, and polar bears, have no hair to bind mercury and must find some other way of dealing with it. Their protective mechanism appears to be to combine it with selenium, rendering it insoluble and inert, and tucking it away in storage sites. This seems to serve the cetacean well, but it can be totally misleading to the researcher who finds large tissue burdens, even "lethal levels" of mercury in a healthy dolphin, and immediately casts about for a polluter to blame.

"Injury" is understood broadly to mean any noxious influence, which may include physical trauma; action of physical agents such as heat and cold; infection; intoxication; metabolic disease; nutritional deficiency; genetic disorders; and developmental malformations. To this list some would add "stress," the often (but not always) injurious effects of the attempts of an organism to adapt to environmental influences not directly injurious themselves. This idea allows us to think of concepts of, for example, deleterious effects of excess population density (crowding), even in the presence of an adequate food supply. We can also think of an animal living at the margin of its nutritional support, with enough food available to support day-to-day functions, but unable to respond to an event, such as pregnancy, exposure to unaccustomed cold, and exposure to toxins, among many others, that places an increased energy demand on it. In such circumstances, the factors that comprise immune resistance may be too feeble to prevent heavy parasitic infestation or infection. With this perspective, we can readily understand that the environment in which an organism lives has a profound, if not determining, influence on the things that make the organism ill or cause it to die. By studying the specific diseases and patterns of occurrences of diseases in a group of an-imals, we can gain insight not only into the hazards of the environment, but also the basic physiology of the organism. We also have to consider that animals differ even within species and that they may differ in their reactions because of experience so that two animals that seem nearly identical might exhibit wide differences in their response to apparently identical environmental situations or stressors.

Because environments vary, it may be expected that patterns of disease might vary among populations of a single species of animal living in different places, and because physiology and anatomy vary, patterns of disease may also vary among different species living in the same environment. Indeed, this phenomenon is well known in the world of experience. Human populations living in temperate climates and industrialized societies have very different patterns of disease than people living in tropical agricultural societies, even though both populations would be susceptible to the same noxious agents if only they were exposed to them. Conversely, dogs, cats, and humans all living in the same house will have their own separate infectious diseases, and only rarely do they infect each other.

From the foregoing, it can be seen that study of the diseases of wild marine mammals can offer insights into anatomy, physiology, and environment that might be gained in no other way. Disease represents an interaction between an individual animal and its environment, which demands an understanding of complex, often intricate processes involving several different organisms. For example, it is not enough to simply measure the level of a toxin in water to gauge its effects on animals living in the water, as the toxin itself can be changed not only by the presence of other chemicals in the water or by bacterial action, but also by sequential processing through metabolic systems of different organisms and ultimately by the physiology and chemistry of the marine mammal itself. Observations indicate, for example, that adaptation to environmental mercury over millennia makes it possible for dolphins to tolerate tissue burdens of mercury that would be fatal to cattle (Turnbull *et al.,* 1998). Animals with high tissue mercury burdens may show no sign of the tissue lesions associated with mercury toxicity in land animals (Siebert *et al.,* 1999). Diseases may appear as a secondary effect of some primary phenomenon; e.g., a population with inadequate food supplies may become debilitated, with lowered resistance, and so be overwhelmed by parasites that ordinarily are held in check. In this instance, the load of parasites is obvious, the lowered resistance to parasitism may be inferred, and the lack of appropriate nutrition recognized by other factors, such as body condition, serum protein levels, and measurements of specific nutrients in body fluids.

It is remarkable that until relatively recent times, almost nothing was known about the diseases of free-ranging marine mammals, and even today not much is known about the diseases of sea otters (*Enhydra lutris*), polar bears (*Ursus maritimus*), manatees (*Trichechus* spp), and walruses (*Odobenus rosmarus*) in the wild. Even though many tens of thousands of large whales were taken in the whaling industry, only a few reports of pathology resulted. Simpson and Gardner (1972) compiled a detailed discussion of the histology and histopathology of marine mammals based mainly on their own experience and an appraisal of the scattered reports then available in the literature. This work remained the standard resource and reference for two decades and is still very valuable. For practical purposes, systematic study of the pathology of marine mammals in the wild is limited to pinnipeds, dolphins, and, to a lesser degree, manatees and did not begin until the mid-1960s. The current literature contains many reports of findings in single animals, or small groups of animals, but relatively few investigations resulting from long-term, detailed studies. The state of the art and science of pathology of marine mammals is such that new and important observations may still be made using dissection and light microscopy, techniques developed and applied to many species in the 19th century, supported by the latest developments in molecular biology and molecular diagnostics.

I. Sources of Information

All marine mammals are protected by law in most developed countries with notable exceptions being Canada and Japan, which allow commercial hunting of some species. In countries with protective laws, access to marine mammals is carefully limited and experimentation, with rare exceptions, is prohibited. This means that apart from animals maintained in marine aquaria and under the care of specialized veterinarians, all information is derived from animals that strand on beaches, are killed accidentally as by-catch in the fishing industry, or taken by approved subsistence hunters, who may make arrangements with researchers for access. Not much scientific use is made of animals in commercial hunting operations. Apart from the occasional stranding near a population center or accidental death associated with fishing, it is rare for an animal to be found in an undecomposed condition by a trained observer with appropriate equipment and supplies. Therefore, what we know is based mainly on case reports, often incomplete, and a few studies involving at most a few dozen animals, often of several different species. Stranding and rehabilitation centers are the main source of information. These obviously can describe only what they see, which is mainly coastal species and the occasional pelagic animal that is cast up on the beach.

Despite these limitations, observations made in stranding centers may be the source of powerful insights into the state of the free-ranging, otherwise inaccessible wild populations. Phocine distemper virus, a morbillivirus responsible for the deaths of at least 18,000 harbor seals (*Phoca vitulina*) in Europe in 1988, was first isolated from stranded seals in that year. Phocine herpes virus (PhHV1) was identified in 1985. The highly toxic domoic acid, a product of the diatom *Pseudonitzschia australis*, was recognized as a cause of death in over 400 California sea lions (*Zalophus californianus*) examined in a rehabilitation center in 1998.

It is not clear just how representative beach-stranded animals are of the population at large, but for the time being this issue is of lesser importance than simply building a reliable database of descriptions of pathological findings and relating these to age, sex, reproductive status, incidence, and type of trauma sustained, identification of infectious diseases, toxin and parasite burdens, and so on. Continuing, detailed pathological studies of stranded dolphins are currently taking place in a number of areas, most notably on the North Sea coast of

Germany, the St. Lawrence River and estuary in Quebec, coastal New Zealand, the United Kingdom, Italy, Peru, the Canary Islands, and the Texas and Florida coasts of the Gulf of Mexico, among other locations. Pinnipeds are studied in central California, and manatees in Florida. Important but non-continuing studies of strandings have been done in southern California, in which the focus was solitary strandings, and in the eastern United States, involving mass strandings. Mortality events in which unusually large numbers of deaths of marine mammals, sometimes accompanied by substantial mortality of fish, seabirds, and turtles, occur over a period of days to months, are under a special program conducted by the U.S. National Marine Fisheries Service.

II. Differences between Pathology of Strandings and By-Caught Animals

Spontaneous strandings of single animals are believed to be caused by sickness or some other impairment, such as injury, and are therefore a selected and perhaps nonrepresentative element of the population from which they come, whereas by-catch animals are snatched from their daily lives and may be presumed to be healthy, or at least representative of the general condition of the population. Diseases seen in these may be a better indicator of the disease status of the population or of disease in an early stage, whereas strandings might represent a late stage. This is mostly presumption, based on logic rather than observation, although accumulating experience suggests it is true.

"By-catch" is the term used to refer to animals that are caught by accident during the course of trying to catch something else. In the case of marine mammals, it usually means ENTANGLEMENT of the animal in fishing nets. Commonly this involves one or a few animals, but in the early days of the purse-seine tuna fishery in the eastern tropical Pacific Ocean, it might mean hundreds or even thousands of animals. At least two episodic studies of pathology have been conducted on this group of animals, and a continuing study of by-catch is taking place in northern Germany.

III. Mass Strandings vs Mortality Events and Solitary Strandings

Mass strandings, the more or less sudden appearance on a beach of large numbers of whales or dolphins, remains a problem to explain. Various theories have been put forth, such as bad luck of animals swimming inshore in being caught by an ebbing tide. This presupposes several things: complex bottom topography, a gently sloping bottom, a rapidly moving tide, failure of echolocation in shallow water, interference in echolocation by extraneous noise, and perhaps distraction of attention by feeding. Other theories implicate an impaired leader who misjudges the tide and leads the herd into a strand or, more likely, fails to lead them away from the beach at the right time. One or two observers even see similarities to lemming migrations and postulate a mass suicide. All of these theories lack supporting experimental evidence. What is known from the few pathological studies of mass-stranded animals is that substantial disease is found, mainly parasitism, which would have been present for some time. Some of this may have been severe, but most is of a tolerable level. If the disease causing the strand was present yesterday or the day before, why did they strand today?

Geraci and St. Aubin (1977) examined the naturally occurring diseases in 41 of a herd of approximately 150 stranded Atlantic white-sided dolphins, *Lagenorhynchus acutus*. The most prevalent lesions were associated with parasites: mastitis associated with the nematode *Crassicauda grampicola* and biliary and pancreatic fibrosis associated with biliary flukes *Oschmarinella laevicaecum*. In these conditions, the parasite enters the duct system and provokes inflammation and scarring. Other parasites commonly found were *Phyllobothrium delphini* and *Monorygma grimaldi* cysts in the blubber and abdomen, *Crassicauda* sp. in the fascia, all of which were encysted, and long-term infestations. A roundworm, *Stenurus globicephalae*, was present in the cranial air sinuses and lungs, where it caused minor inflammation, and the tapeworm *Tetrabothrius forsteri* was found in the duodenum. These findings are all typical of wild dolphins in the North Atlantic and cannot be invoked to explain the stranding. This is generally the story of mass strandings; many clues, but little definitive evidence.

"Mortality event" is the term applied to excess deaths (over that known to occur in an average period of time) in a relatively limited geographic area. The difference between a mass stranding and a mass mortality is the time scale and circumstances. Mass stranding is an event of a day or so, whereas mass mortality evolves slowly over days, weeks, or months. In contrast to mass strandings, animals involved in a mortality event are very likely to suffer from infectious disease (e.g., Schulman *et al.*, 1997) affecting some segment of the local population or a toxic event such as the brevitoxicosis associated with red tide (e.g., Bossart *et al.*, 1998) or domoic acid toxicity associated with bloom of a particular diatom (Scholin *et al.*, 2000). One obvious difference between an infection and an environmental toxicosis is that the infection, typically viral, tends to be limited to one kind of animal, whereas the toxicosis may affect mammals, birds, fish, and turtles alike. Epizootic infection can occur among terrestrial animals as well, but an environmental biotoxicosis is a uniquely aquatic event.

Solitary strandings of dolphins, which may be taken as representative of coastal marine mammal strandings, reflect a variety of causes, from trauma (boat strikes), intra- and interspecific aggression, accidents, and a wide variety of disease causes (Haubold *et al.*, 1999). This tends to apply more to juvenile and adult animals, whereas young calves seem to strand more for social reasons, such as separation from the mother, and although debilitated, are less likely to be sick. This observation does not apply to pinniped pups, however, which tend to be suffering from infection, septicemia, dehydration, starvation, or trauma when found stranded.

IV. Parasitism

Infestation by parasites is nearly universal in wild animals, and marine mammals are no exception. Indeed, the parasites found in pinnipeds and cetaceans are so widespread and bear such a consistent relationship to their hosts and their environments that they have been used as "tags" to study specific mammal populations (Dailey and Otto, 1982). Parasitism is probably the best recognized disease factor in free-ranging marine mammals, and the variety of lesions caused may be illustrated by a few examples.

Parasitism is by far the leading cause of pathology associated with stranding in wild pinnipeds and cetaceans, affecting the BRAIN and lungs as well as the GASTROINTESTINAL TRACT, liver, and pancreas. Large (baleen) whales may suffer KIDNEY damage from *Crassicauda*, and seals may carry heavy, often fatal, burdens of heart worm. While parasites occur in other sites, such as in the blubber and under the peritoneal membranes, they are not usually of any particular significance and apparently are not an important mortality factor.

The great majority of small cetaceans, approaching 100%, have nematode lung worms. These seem to be relatively innocuous as long as they remain in the airways, but provoke an inflammatory reaction on entering the alveolar spaces, the distal sites of air exchange. These delicate tissues lack the defenses of the airways, which are covered by a cell layer adapted to a passageway and have a thin coat of protective mucin. The parasite in the alveolus, perhaps associated with bacteria, provokes the formation of small abscesses, which rarely may rupture into the pleural space, but typically subside into fibrous nodules (granulomas), which may mineralize. At this stage, they are inert and probably of no further consequence, except as a marker of past events. In some delphinids, notably the harbor porpoise, *Phocoena phocoena*, in the North Sea, the airways may be filled to occlusion by large nematodes of several species (Siebert *et al.*, 1995).

The effect of parasites in the lungs illustrates very well the influence of adaptation of the lung to diving on the disease process. The cetacean lung differs from the lung of a typical land mammal in that the diving mammal has structural adaptations designed to keep air in the gas exchange spaces, the alveoli, while also tending to eliminate or reduce "dead space." Dead space is the functional compartment of the lung in which gas exchange does not take place; for practical purposes, the airways. In diving, compression of the animal by surrounding water, which increases with depth, forces incompressible blood into any compressible space. This is familiar to human divers as the "squeeze effect." In the lung, the alveoli can be collapsed by compression of the chest wall, but the airways are held open rigidly by cartilage bars. It appears that the reduction of airway dead space and the prevention of the forcing of alveolar air back into airways are accomplished by the action of a series of muscular valves or sphincters within the airway. All of this extra tissue (compared with the lung of a terrestrial mammal) makes the cetacean lung dense and heavy. When an agent such as a lungworm, bacteria, or aspirated contaminated water reaches the distal airways and provokes an inflammatory reaction, it appears that the muscular sphincters go into spasm, preventing movement of air, and evacuation of secretions.

The effect of the peculiar anatomy of the cetacean lung is to cause nearly all infections to form abscesses with focal destruction of tissue. The density of the lung tissues in general prevents spread of the inflammation beyond the local focus. In contrast, bacterial infections of the terrestrial mammalian lung may resolve without destruction of tissue but may spread in an unconfined fashion to involve large regions of the lung.

Delphinids of many species harbor flukes of the genes *Nasitrema* in the pterygoid air sinuses of the skull. These sinuses connect with the external environment by way of the nasal passages and by way of the eustachian tubes, with the specialized structures of the ear, the ossicles or acoustic bullae. Bullae are connected to the brain by the staticoacoustic nerve, concerned with both balance and hearing, which passes into the skull within a dense fibrous sheath. Characteristics of the anatomy of the air sinus, the acoustic bullae, and the staticoacoustic nerve permit invasion by the worm into the subarachnoid space of the brain, in which the spinal fluid circulates. Taking advantage of this space, they migrate over the surface of the brain until they reach a point of penetration, at which they burrow through the cortex or gray matter deep into the white matter, laying large numbers of eggs (ova). This migration produces destructive tracts or galleries in the white matter, with hemorrhage and necrosis. It is not clear what induces the migration of the parasite into the brain, as it is a dead end for the worm as well as damaging to the host. It is a complex problem, seen mainly in beach stranded dolphins in both the Atlantic and the Pacific Oceans. Dolphins of the same species caught as bycatch in the eastern tropical Pacific tuna fishery also have infestations of the same fluke in the air sinuses, but apparently without invasion of the nervous system (Walker and Cowan, 1981). Why this difference should occur is not known. It is possible that a third factor is involved, such as some agent in the inshore water that influences the worm, i.e., makes it "sick" and disoriented. There is clearly a differential species susceptibility among dolphins, as the air sinus infestation occurs with great frequency in stranded Atlantic bottlenose dolphins, *Tursiops truncatus*, but without nervous system invasion.

Flukes of several species (*Campula rochebruni, C. oblongata*, among others) infest the bile duct and pancreatic duct of dolphins and porpoises. The effect may be a relatively low-grade irritation or inflammatory reaction ranging up to chronic active inflammation with fibrosis and occlusion of the duct. This results in inflammation and fibrosis of the affected organs, a form of hepatitis (cholangitis), and pancreatitis. Serious disease is relatively uncommon but may be life-threatening when it occurs.

The stomach and intestine of marine mammals are the frequent site of infestation with nematodes, trematodes, and cestodes. A light infestation may be relatively innocuous, but occasionally the stomach or intestine may be perforated by parasites, resulting in peritonitis. In some geographic areas, gastric ulcers associated with the nematode *Anisakis simplex* are common in stranded cetaceans (Abollo *et al.*, 1998). Sea lions may carry heavy burdens of hookworm.

Massive infestation of the heart and great vessels of elephant seals by nematodes (*Otostrongylus circumlitus*) is described from strandings on the central California coast (Gulland *et al.*, 1997). Heavy burdens in the right atrium, right ventricle, and pulmonary artery were associated with pulmonary thromboembolism (blood clots in the lungs) and pulmonary arteritis, an important cause of mortality. Death of infested juvenile elephant seals (*Mirounga angustirostris*) before the parasite reaches maturity suggests that the host–parasite association is relatively recent. Similar heartworm infestation occurs in seals in the Atlantic Ocean. This disease is very much like the heart worm infestation of dogs seen in parts of the United States.

V. Neoplasia

"Neoplasia" is the process of the formation of autonomous new growths in a tissue. The mass of tissue formed is called a tumor. Some of these are benign, meaning limited in capacity to harm, and some are malignant (cancer), having the ability to invade adjacent tissues and to set up colonies in remote organs and tissues. Some tumors have been shown to be caused by viruses, some by chemicals, and a rare few are associated with parasitic infestation. For most, a definitive "cause" is not known; we can only speak of "associations."

Once thought to be rare in marine mammals, neoplasms were found to occur at an incidence of 2.5% in a large surveyed population of marine mammals (Howard *et al.*, 1983a). As of 1987, there were probably only 41 confirmable reports of tumors in cetaceans. Then, in 1994, 21 additional tumors, some benign and some malignant, were reported in 12 of 24 animals from the small, isolated, and highly contaminated population of beluga whales (*Delphinapterus leucas*) in the St. Lawrence River in Quebec (De Guise *et al.*, 1994). Seven of these animals had more than one tumor; one had as many as three tumors, two malignant. One was as young as 1.5 years, another 3.5 years. The ages ranged up to over 29 years. This particular environment, the estuary of the St. Lawrence River, suffers heavy industrial pollution, with the waters containing polycyclic aromatic hydrocarbons. High concentrations of organochlorines, heavy metals, and benzo[a]pyrene were found in the tissues of the belugas. Thirty-seven percent of all tumors reported from cetaceans to that time were found in this small population of belugas, suggesting the direct carcinogenic effect of the pollutants or an impairment of resistance to the development of tumors.

Gulland *et al.* (1996) reported an incidence of 66 transitional cell carcinomas in a population of 370 (18%) California sea lions (*Zalophus californianus*) examined over a 15-year period. Transitional cell carcinomas are a particular type of cancer that arise from the lining epithelium or membrane of part of the urogenital tract. The original site and cause of this extraordinary incidence of a particular tumor type was a mystery for a number of years due to the advanced state of the disease when recognized. Typically, it had spread to involve most of the lower abdominal lymph nodes and viscera, and the primary site could not be determined. Environmental pollution with a variety of industrial chemicals was speculated to be the cause. Using a variety of modern analytical techniques, strong evidence has been found that the cause of the malignancy is infection with a gamma herpesvirus and not chemical pollution (Lipscomb *et al.*, 1999). This virus is implicated in the etiology of several animal cancers.

These two studies are very instructive in that they emphasize that a presumed low incidence of tumors in wild populations may be merely an artifact of not looking. When populations of wild animals are studied carefully over a long time, clusters of disease may be revealed. They also illustrate the value of modern technology in evaluating cause. In the case of transitional cell carcinomas, the cause, a herpesvirus, is demonstrated as well as it can be, whereas tumors in belugas are associated with pollution on epidemiological grounds but have not been proven conclusively to be caused by pollution.

VI. Infectious Disease

In their life in the oceans, marine mammals are exposed to a very great array of infectious agents, including dozens, if not hundreds, of species of bacteria, viruses, fungi, and protozoa. Because these are not distributed randomly, actual infection will vary with features such as location, water temperature, contamination from terrestrial sources, river effluents, food species, and exposure to other marine mammals. Most organisms are resisted successfully, and a state of more or less peaceful coexistence is maintained. Occasionally, however, an individual's defenses are so weakened or breached that one of these microorganisms can gain effective entry and set up a disease state. Sometimes an otherwise perfectly healthy animal comes into contact with a particularly virulent strain of organism or simply a novel one to which there is no natural or acquired immunity conferred by previous exposure. Under these circumstances, the organism can sweep through a population, causing epizootic disease with high death rates. Some isolated, small populations may be seriously threatened with extinction by new infections.

Bacterial, viral, and mycotic (fungal) infections are important morbidity and mortality factors in marine mammals, often appearing in outbreaks, such as one that occurred in the endangered New Zealand sea lion (*Phoctarctos hookeri*), involving both adults and pups (Gales *et al.*, 1999). At least 53% of the pups born during the 1998 breeding season (total number was not known) had died within the first 2 months of life, with acute necrotizing inflammation of the blood vessels of the skin and lungs in adults and pneumonia and abscesses in the pups. The causative agent was determined to be a pleomorphic gram-negative bacterium, most likely *Campylobacter* sp., an organism not previously associated with this kind of mortality.

Several viral diseases are well known among pinnipeds, including seal pox virus and the San Miguel calcivirus, which produce contagious crusted lesions on the skin and oral mucosa. These are serious but not fatal diseases, much like human chicken pox. Seals, sea lions, and dolphins are known to be infected with herpesviruses, and dolphins are susceptible to a pox virus, which produces transient "pinhole" lesions of the skin. Rabies virus is uncommon, but has been observed in ringed seals. Herpesvirus is familiar to most people in the context of the infections known as shingles and cold sores. This virus occurs in many types and strains, some relatively innocuous, with others capable of producing severe disease of many organs, including the brain, lungs, liver, and heart as well as the skin. Several outbreaks of herpes infection have been recognized among marine mammals, usually producing skin lesions and sometimes pneumonia, as well as the spectacular incidence of cancer in California pinnipeds.

Morbilliviruses, which comprise a large group of viruses that cause measles in humans, and distempers in dogs and many other species also produce respiratory disease, immune deficiency, and neurological injury in seals, dolphins, porpoises, and a wide range of other cetaceans.

Marine mammals are quite susceptible to infection with morbillivirus. Morbillivirus infections, which were not documented in aquatic mammals until 1988, have caused at least five epizootics in these species since the early 1990s (Kennedy, 1998). Disease has been recognized in seals from Europe (*P. vitulina*) and Siberia (*Pusa sibirica*), striped dolphins (*Stenella coeruleoalba*) in the Mediterranean Sea, and bottlenose dolphins in the northern Atlantic Ocean and the Gulf of Mexico, and seropositivity, indicating previous exposure to the virus, has been found in many species of seal, dolphin, porpoise, and whale from the Antarctic to the Arctic circle, in Florida manatees (*T. manatus*), and in polar bears in Russia and Canada. The disease in marine mammals is much like distemper in dogs, with destructive and inflammatory lesions in the brain, gastrointestinal tract, lungs, and lymph nodes, with immune suppression, and frequently superinfection of immune-impaired animals with fungi. Lung lesions include bronchointerstitial pneumonia, with filling of alveolar spaces by exudates, hyaline membranes formed by protein exudate covering gas exchange surfaces, and hemorrhage. Brain lesions are typically in the form of nonsuppurative encephalitis (nonpus-forming inflammation of the brain), with neuronal and glial necrosis, microgliosis, the brain equivalent of scar, and focal demyelination, or loss of the insulating covering of nerve processes. Necrosis of the cerebral cortex is also sometimes found (Kennedy, 1998).

It is not entirely clear whether the observed increase in cases of the morbillivirus distempers is due to actual spread of the infection or to improved case finding permitted by clear descriptions of the lesions in the literature, accompanied by the development of advanced methods of laboratory diagnosis.

Many bacterial species are found in marine waters and may be recovered from marine mammals, either as primary pathogens or as part of a complex normal flora. Bacterial infection is thought to be the main cause of disease and death in marine mammals, especially in captivity (Howard *et al.*, 1983b). Certain marine organisms are known to produce severe or fatal infections. These include the halophilic or salt water *Vibrios*, of which there are many species, and *Edwardsiella tarda*. These organisms are common in the marine environment and are frequently encountered in cetaceans and less often in pinnipeds. The exact means by which these organisms are acquired by marine mammals is not known, but experience with humans indicates that they can be inoculated directly, infect wounds, and be ingested with food items, such as shellfish. It is reasonable to assume these routes of entry in marine mammals as well.

Clostridia are obligate anaerobic bacilli, ubiquitous in the environment in soil, sewage, marine sediments, decaying animals and plant products, and the intestinal tracts of many animals. More than 80 species are known. Some species are potent toxin producers, causing botulism and gas gangrene. Many species of *Clostridia* have been cultured from blood, lesions, and intestinal tract of stranded dolphins in the Gulf of Mexico, but are less common in California pinnipeds. Some *Clostridia* may be merely part of the normal intestinal flora, whereas some species have been recovered from lungs in cases of pneumonia.

Many species of bacteria have been recovered from cultures of respiratory tract, kidney, and intestinal tract, as well as lesions of captive and stranded marine mammals. The majority of these species are known to be pathogenic or potentially pathogenic. There may be a differential distribution of bacteria in different kinds of marine mammals, as in cetaceans *Vibro, Clostridia, Pseudomonas,* and *Edwardsiella* tend to predominate, whereas in pinnipeds from the California coast, for example, the major pathogens encountered in sea lions, elephant seals, and harbor seals are organisms usually associated with the intestine, mainly *Escherichia coli, Klebsiella pneumoniae, K. oxytoca, Proteus* spp., *Pseudomonas* spp., *Enterococcus* spp., and *Salmonella.* Leptospirosis occurs in harbor seals, California sea lions, and northern fur seals (*Callorhinus ursinus*).

Brucella infection of the placenta with abortion has been reported in dolphins. The organism, *Brucella delphini,* appears to be readily transmissible among dolphins and has also been cultured from the lung of a bottlenose dolphin at necropsy. Brucella infection occurs in other cetaceans and seals. A substantial percentage of marine mammal serum samples (about 30%) react positively on tests used to detect antibody to *Brucella* spp., and a number of *Brucella* isolates have been obtained from marine mammals. However, only *B. delphinus* has been associated with reproductive failure.

Mycobacteria include organisms that cause tuberculosis and leprosy. They are hardy organisms and may produce infection across a wide array of warm- and cold-blooded animals. *Mycobacterium marinum,* originally described from fish, was first recognized as a human pathogen in 1951. *M. marinum* infection has been transmitted to a handler by a dolphin bite. The lesions, when localized, resemble abscesses, but are very slow to heal. Infection of deeper tissues and organs, such as heart valves, brain, eye, and joints, is very serious. *Mycobacterium bovis,* the agent of bovine tuberculosis, was transmitted from seals to their trainer. It caused similar lesions in both. Six cases of tuberculosis were observed over a 4-year period in stranded sea lions and fur seals from the coast of Argentina (Bernardelli *et al.,* 1996). Disease was widespread, involving lungs, lymph nodes, liver, spleen, pleura, and peritoneum. The lesions were typical granulomatous inflammation, from which organisms belonging to the *Mycobacterium tuberculosis* complex were isolated. This organism complex was similar to both *M. bovis,* the agent of bovine tuberculosis, and *M. tuberculosis,* the agent of human tuberculosis, but with enough molecular differences to be judged different from both. This suggests that the seals and sea lions did not acquire their infection from contact with humans or cattle. Organisms associated with the pinniped cases all had similar genetic features, suggesting that seal tuberculosis in that geographic area is caused by organisms belonging to a distinct grouping within the *M. tuberculosis* complex. An instance of disseminated *Mycobacterium chelonei* infection in a manatee (*Trichechus inunguis*) has been reported, and cases of cutaneous mycobacteriosis in a manatee and its handler have been attributed to *M. chelonei.* This organism was first identified in a reptile.

Marine mammals are subject to infection with a wide variety of fungi and filamentous bacteria (*Actinomyces, Nocardia, Dermatophilus,* which produce lesions similar to those caused by fungi) (Migaki and Jones, 1983; Reidarson *et al.,* 1999). Some of these are opportunistic, meaning that they occur in the context of debility or other disease, and some are primary pathogens. Both opportunists and pathogens are species that are familiar in human and veterinary medicine and are not peculiar to the marine environment.

Animals suffering immune suppression from morbillivirus disease may suffer severe, disseminated fungal infection, frequently from molds such as *Aspergillus* spp., as a terminal event. These organisms may produce disease in one of two patterns: superficial infections of skin that cause mild disease or indolent, very chronic processes; and the deep or systemic mycoses, which produce severe pyogranulomatous lesions. A pyogranulomatous lesion is a mixture of granulomatous inflammation, which is produced by persistent infection, with a pus-forming acute inflammation, indicative of a more active injury. Deep infections may involve any organ, especially the lungs and respiratory tract, and the brain. They may produce bulky lesions, which displace normal tissues ("mass lesion") as well as destroy tissues. Some organisms tend to permeate tissues and invade and block blood vessels.

Dermatophytosis, ringworm-like lesions, occur in manatees and pinnipeds, caused by the same organisms that cause ringworms in land animals and humans. Because infection with fungi requires spores from the environment rather than the vegetative stages found in marine mammals, direct transmission from animal to animal seems unlikely. Lobo's disease, caused by a fungus, *Loboa loboi* (syn. *Lacazia loboi*), is a very unusual disease recognized only in dolphins and humans and no other species. One instance of transmission from dolphin to human has been reported. Lobo's disease (once called keloidal blastomycosis) is a skin infection producing chronic, treatment-resistant, thick nodular swellings of the superficial dermis and epidermis, occasionally with ulceration. While the lesions in humans and dolphins are quite similar, there are subtle morphological differences in the organisms in the lesions, which may represent separate species (Haubold *et al.,* 2000). In humans, Lobo's disease is a disease of the Central and South American tropics, while in dolphins it ranges from the Gulf of Mexico, mainly Florida, to South America.

A number of protozoal infections, similar to those seen in land animals, are known to occur in marine mammals. These include toxoplasmosis, infection with *Toxoplasma gondii,* found in a West Indian manatee; pinnipeds; stranded Atlantic bottlenose dolphins, *Tursiops truncatus;* a Pacific spinner dolphin, *Stenella longirostris;* beluga whales, *Delphinapterus leucas;* and a killer whale, *Orcinus orca.* The animals were found in both Atlantic and Pacific Oceans, were captive and free ranging, and the lesions ranged from incidental to disseminated and fatal. The mode of transmission in these animals is not known. *Cryptosporidium* morphologically, immunologically, and genetically indistinguishable from those obtained from infected calves have been recovered from feces of California sea lions (*Zalophus californianus*), suggesting that the sea lion could serve as a reservoir for the environmental transmission of this organism.

Giardia have been isolated from fecal material from harp seals (*Pagophilus groenlandicus*), gray seals (*Halichoerus grypus*), and harbor seals in eastern Canada waters and ringed seals (*Phoca hispida*) from western arctic Canada and from California sea lions from northern coastal California. The widespread occurrence of these organisms in both terrestrial and marine mammals suggests transmission via rivers and streams, and perhaps carrier species taken as food.

VII. Chronic Diseases of Undetermined Cause

Several chronic diseases of unknown cause have been recognized in dolphins and small whales stranding along the Texas coast of the Gulf of Mexico. These are arthritis, pulmonary angiomatosis, and amyloidosis.

A high incidence of arthritis of the synovial joints (neck and flipper joints) was found in several cetacean species, including 10 of 49 (20%) bottlenose dolphins (*T. truncatus*), 1 of 2 (50%) striped dolphins, 1 Fraser's dolphin (*Lagenodelphis hosei*), and 1 pygmy sperm whale (*Kogia breviceps*) (Turnbull and Cowan, 1999a). Some of these cases were associated with infection of the joint, whereas others showed features typical of degenerative or osteoarthritis, commonly thought of as a disease of aging. Two such cases, however, were in immature animals. The severity of disease ranged from a relatively minor loss of articular cartilage to complete destruction and fusion of the joint.

Angiomatosis is a newly recognized and bizarre disease, to date reported only in bottlenose dolphins from the Texas coast (Turnbull and Cowan, 1999b). This disease was first recognized in 1991 in a minor form but has since increased in incidence and severity to involve all adult bottlenose dolphins. It is characterized by the proliferation of small blood vessels in the lungs to form small clusters. It involves all parts of both lungs equally and to the same degree and is not associated with inflammation. These vascular clusters progressively enlarge and fuse. In advanced stages of the disease, the walls of the new vessels thicken and lumens are reduced. Proliferation of vessels in the bronchial-lining mucosa erodes the bronchial cartilages and severely constricts airway diameter. The normally very thin and pliable pleura is similarly involved and becomes thick, stiff, and opaque, reaching a thickness of 3 mm. The effect on respiratory physiology has not been measured, but the gross and microscopic appearance suggests marked impairment of lung ventilation and restriction of circulation. One animal with advanced disease had massive thickening of the right ventricle of the heart, suggesting a marked increase in pulmonary artery blood pressure (pulmonary hypertension). Lymph nodes associated with the lungs in advanced cases may also show a pronounced vascular proliferation, and hemangiomas (tumors of blood vessels) are a common finding in the lymph nodes and an occasional one in the lungs. The cause of this disease is not known.

Amyloidosis, a disease characterized by deposition of a dense amorphous waxy proteinaceous material in the interstitial tissues of various organs, occurs in bottlenose dolphins stranding along the Texas coast of the Gulf of Mexico at an incidence of about 20%, a remarkable figure considering that the disease has not been reported from any other cetacean species (Cowan, 1995). This deposition occurs consistently in the kidneys, but also in small arteries in the spleen, heart, and lungs and around the acini of the thyroid gland and the palatal salivary gland. In the amounts present, the amyloidosis likely did not result in death, but experience in humans and other animals is that it can progress to organ failure. Several causes and associations are known for amyloidosis in humans and other animals, including chronic infection and disorders of the lymphoreticular system, but the cause of dolphin amyloidosis remains obscure.

VIII. Conclusions

There is much to be learned from the study of the pathology of marine mammals. Work by many observers since the mid-1970s has done much to reveal the causes of sickness and death among these heretofore mysterious animals. We are coming to understand how similar the diseases of marine mammals can be to those of terrestrial mammals and how different they can be. This is well illustrated by the evolution of inflammatory processes in the lung, which are strongly influenced by the special anatomy of the diving lung, and by the phenomenon of cerebral parasitism, which appears to depend not only on the special anatomy of the nasal passages, air sinuses, the specialized inner ear, and their relations to the brain, but also exposure to a very specific parasite.

Recognition of similarities in tissue structure and diseases can be treacherous if we approach the study of the diseases of marine mammals as if they are just like land animals, only wet. What we see, or think we see, may mean something very different than it would mean if we saw it in a land animal.

A great deal has been learned about the anatomy and physiology of pinnipeds and cetaceans in particular, and especially about cetaceans, who have taken typical mammalian systems, inherited from what many believe to be terrestrial forbears, back into the water and made them work there. The adaptive changes in the anatomy of the lungs in particular, which influence the development of disease processes in that organ, are plain and obvious. These are the adaptations we can see and are relatively easy to puzzle out. We are only now gaining insight into the unseen adaptations, such as the metabolic management of mercury, a protoplasmic poison. The vast majority of mercury (and most other toxic metals) in the environment is from natural, not industrial, sources and comes from rocks and the soil. It is leached out by rain and reach streams, rivers, and ultimately the OCEANS. These metals have always been present in the environment, perhaps not in the levels now recognized in some geographic regions. However, the adaptive experience of animals who spend their entire lives in the water and whose food consists of other organisms that spend their entire lives in the water has endowed them with startling abilities. These natural toxins cannot be avoided; they must be dealt with within the metabolic capacities of the animal who is inevitably exposed to them. Who could have predicted that a dolphin can carry a burden of mercury in its tissues sufficient to kill a cow four times over and still be healthy? The next task is to better understand these unseen adaptations so that we do not misread or misunderstand the message that Nature is sending us about the state of our oceans, written in the tissues of marine mammals.

See Also the Following Articles

Health ■ Mass Die-Offs ■ Parasites ■ Stranding

References

Abollo, E., López, A., Gestal, C., Benavente, P., and Pascual, S. (1998). Long-term recording of gastric ulcers in cetaceans stranded an the Galician (NW Spain) coast. *Dis. Aquat. Organ.* **32,** 71–73.

Bernardelli, A., Bastida, R., Louteiro, J. D., Michelis, H., Romano, M. I., Cataldi, I., and Costa, E. F. (1996). Tuberculosis in sea lions and fur seals from the southwestern Atlantic coast. *Rev. Sci. Office Int. Epizoo.* **15,** 985–1005.

Bossart, G. D., Baden, D. G., Ewing, R. Y., Roberts, B., and Wright, S. D. (1998). Brevitoxicosis in manatees (*Trichechus manatus latirostris*) from the 1996 epizootic: Gross, histologic and immuno-histochemical features. *Toxicol. Pathol.* **26,** 276–282.

Cowan, D. F. (1995). Amyloidosis in the bottlenose dolphin *Tursiops truncatus. Vet. Pathol.* **34,** 311–314.

Dailey, M. D., and Otto, K. A. (1982). Parasites as biological indicators of the distributions and diets of marine mammals common to the eastern Pacific. N.O.A.A. National Marine Fisheries Service, Southwest Fisheries Center, La Jolla, CA.

De Guise, S., Lagac, B., and Land, P. (1994). Tumors in St. Lawrence Beluga whales (*Delphinapterus leucas*). *Vet. Pathol.* **31,** 444–449.

Gales, N., Duignan, P., Childerhouse, S., and Gibbs, N. (1999). New Zealand sea lion mass mortality event, January/February 1998: (1) Descriptive epidemiology. Society for Marine Mammalogy, 13th Biennial Conference, Wailea, Maui, Hawaii, Nov. 29–Dec. 3, 1999 Abstracts, p. 63.

Geraci, J. R., and St. Aubin, D. J. (1977). Pathologic findings in a stranded herd of atlantic whitesided dolphins, *Lagenorhynchus obliquidens acutus* Proceedings, 8th Annual Conference, International Association for Aquatic Animal Medicine, Boston, MA.

Gulland, F. M. D., Beckmen, K., Burek, K., Lowenstine, L., Werner, L., Spraker, T., Dailey, M., and Harris, E. (1997). Nematode (*Otostrongylus circumlitus*) infestation of northern elephant seals (*Mirounga angustirostris*) stranded along the central California coast. *Mar. Mamm. Sci.* **13,** 446–459.

Gulland, F. M. D., Trupkiewicz, J. G., Spraker, T. R., and Lowenstine, L. J. (1996). Metastatic carcinoma of probable transitional cell origin in 66 free-living California sea lions (*Zalophus californianus*), 1979 to 1994. *J. Wildl. Dis.* **32,** 250–258.

Haubold, E. M., Cooper, C. W., Jr., Wen, J., McGinnis, M. R., and Cowan, D. F. (2000). Comparative morphology of *Lacazia loboi* (*syn Loboa loboi*) in dolphins and humans. *Med. Mycol.* (in press).

Haubold, E. M., Turnbull, B. S., Clark, L., and Cowan, D. F. (1999). Cycles and trends of pathology of stranded cetaceans in the western gulf of Mexico from 1991–1999. Society for Marine Mammalogy, 13th Biennial Conference, Wailea, Maui, Hawaii, Nov. 29–Dec. 3, 1999 Abstracts, p. 78.

Howard, E. B., Britt, J. O., and Simpson, J. G. (1983a). Neoplasms in marine mammals. *In* "Pathobiology of Marine Mammals" (E. B. Howard, ed.), Vol. II, pp. 95–107. CRC Press, Boca Raton, FL.

Howard, E. B., Britt, J. O., Matsumoto, G. K., Itahara, R., and Nagano, C. N. (1983b). Bacterial diseases. *In* "Pathobiology of Marine Mammals" (E. B. Howard, ed.), Vol. I, pp. 69–118. CRC Press, Boca Raton, FL.

Kennedy, S. (1998). Morbillivirus infections in aquatic mammals. *J. Comp. Pathol.* **119,** 201–225.

Lipscomb, T. P., Scott, D. P., Gulland, F. M. D., Lowenstine, L. J., King, D. P., Hure, M. C., Stott, J. L., and Garber, R. (1999). Metastatic carcinoma of california sea lions: Evidence of genital origin and association with gamma herpesvirus infection. Proceedings, 30th Annual Conference, International Association for Aquatic Animal Medicine, Boston, MA.

Migaki, G., and Jones, S. R. (1983). Mycotic diseases in marine mammals. *In* "Pathobiology of Marine Mammals" (E. B. Howard, ed.), Vol. II, pp. 1–27. CRC Press, Boca Raton, FL.

Reidarson, T. H., McBain, J. F., Dalton, L. M., and Rinaldi, M. G. (1999). Diagnosis and treatment of fungal infections in marine mammals. *In* "Zoo and Wild Animal Medicine: Current Therapy" (M. Fowler and E. Miller, eds.), 4th Ed., pp. 478–485. Saunders, Philadelphia.

Scholin, C. A., Gulland, F., Doucette, G. J., Benson, S., Busman, M. Chavez, F. P., Cordaro, J., DeLong, R., De Vogelaere, A., Harvey, J., Haulena, M., Lefebvre, K., Lipscomb, T., Loscutoff, S., Lowenstine, L. L., Marin, R., III, Miller, P. E., McLellan, W. A., Moeller, P. D. R., Powell, C. L., Rowles, T., Silvagni, P., Silver, M., Spraker, T., Trainer, V., and Van Dolah, F. M. (2000). Mortality of sea lions along the central california coast linked to a toxic diatom bloom. *Nature* **403,** 80–84.

Schulman, F. Y., Lipscomb, T. P., Moffett, D., Krafft, A. E., Lichy, J. H., Tsai, M. M., Taubenberger, J. K., and Kennedy, S. (1997). Histologic, immunohistochemical, and polymerase chain reaction studies of bottlenose dolphins from the 1987–1988 United States Atlantic coast epizootic. *Vet. Pathol.* **34,** 288–295.

Siebert, U., Joiris, C., Holsbeek, L., Benke, H., Failing, K., Frese, K., and Petzinger, E. (1999). Potential relation between mercury concentrations and necropsy findings in cetaceans from German waters of the North and Baltic Seas. *Mar. Pollut. Bull.* **38,** 285–295.

Siebert, U., Weiss, R., Lick, R., Benke, H., and Frese, K. (1995). Pathology of the respiratory tract of harbour porpoises (*Phocoena phocoena*) from German waters of the North and Baltic Seas. Society for Marine Mammalogy, 11th Biennial Conference, Orlando, FL.

Simpson, J. G., and Gardner, M. B. (1972). Comparative microscopic anatomy of selected marine mammals. *In* "Mammals of the Sea; Biology and Medicine" (S. Ridgway, ed.), pp. 298–418. C. C. Thomas, Springfield, IL.

Turnbull, B. S., and Cowan, D. F. (1999a). Synovial joint disease in wild cetaceans *J. Wildl. Dis.* **35,** 511–518.

Turnbull, B. S., and Cowan, D. F. (1999b). Angiomatosis, a newly recognized disease in atlantic bottlenose dolphins (*Tursiops truncatus*) from the Gulf of Mexico *Vet. Pathol.* **36,** 28–34.

Turnbull, B. S., Cowan, D. F., Ramanujam, V. M. S., and Alcock, N. W. (1998). Do dolphins have protective mechanisms against mercury toxicity? Proceedings 29th Annual Conference, International Association for Aquatic Animal Medicine, San Diego, CA.

Walker, W. A., and Cowan, D. F. (1981). Air sinus parasitism and pathology in free ranging common dolphins (*Delphinus delphis*) in the eastern tropical Pacific. N.O.A.A., National Marine Fisheries Service, Southwest Fisheries Center, La Jolla, CA.

Peale's Dolphin
Lagenorhynchus australis

R. NATALIE P. GOODALL
Centro Austral de Investigaciones Científicas,
Tierra del Fuego, Argentina

Peale's dolphin (*Lagenorhynchus australis*) is common in inshore waters of southernmost South America, but because it seldom strands, its life history is not well known. Nearly a century passed before the first descriptions of its pigmentation and skull were discovered to be of the same species.

Accounts of only five specimens had been published by 1974; many of the 80 specimens now available were beach pick-ups.

I. Taxonomy

This species was first described as *Phocaena australis* Peale, 1848 and *Sagmatias amblodon* Cope, 1866. Although lumped with *Delphinus obscurus* and *L. cruciger* at different times, the present combination *Lagenorhynchus australis* was proposed by Kellogg in 1941. *Tursio chiloënsis* is a synonym. New work with DNA shows that the genus *Lagenorhynchus* needs revision and that the genus name of one Northern Hemisphere species and the three Southern Hemisphere SPECIES should revert to *Sagmatias*. Although there have been several common names in English, Peale's dolphin is now standard. *Delfin austral* (southern dolphin) is the common name in Argentina and Chile, although *llampa* is sometimes used in the latter.

II. Diagnostic Characters

A. Pigmentation

Peale's dolphins are dark gray or black on the dorsal surface, with two areas of lighter pigmentation on the sides (Fig. 1). A curved white-to-gray flank patch angles forward from the vent, narrowing to a single line ending below or in front of the dorsal fin. The posterior curves of the flank patch almost meet above the tail stock. The larger thoracic patch is light to medium gray, outlined with a narrow dark line on its lower surface. Both patches may be flecked with darker gray. A double black eye ring extends forward onto the snout, which is inconspicuous. The black chin or throat patch varies individually in the shape of its posterior border, usually extending backward on the sides to leave a white point in the middle. The white flipper patch is also delineated with gray and may extend onto the lower part of the flipper. The flippers and dorsal fin are dark with lighter posterior edges. Flippers in older animals may have a series of small knobs on the leading edge. The ventral surface behind the throat patch is white, with a few dark streaks in the genital area. Young animals are lighter gray than adults and have less definition between thoracic and flank patches than in adults.

Peale's dolphins can be confused with dusky dolphins, *L. obscurus,* through much of their range. The latter are usually a lighter gray and have white on the sides of the face and two light lines running forward from the flank patch.

B. Size and Shape

This is a stocky dolphin with the barest indication of a beak. Total length for 35 specimens ranged from 98 to 218 cm. Females ($n = 20$) measured 130–210 cm, males ($n = 9$) from 138 to 218 cm, and animals of unknown sex ($n = 5$) from 172 to 213 cm. The collections to date probably do not represent the total size range for the species. The heaviest animal ($n = 5$), a sexually mature female, weighed 115 kg.

III. Distribution

Peale's dolphin is a southern South American species commonly found from 59°S (the Drake Passage south of Cape Horn) northward to Valdivia, Chile (about 38°S), on the west coast and Golfo San Jorge, Argentina (44°S), on the east, including the Falkland (Malvinas) Islands, with exceptional records to 33°S in the southeastern South Pacific and 38°S in the southwestern South Atlantic (Fig. 2). Sightings of animals similar to *L. australis* in 1988 near the Cook Islands, in tropical waters thousands of miles from its normal distribution, have been considered an anomalous occurrence by some authors or perhaps a new species by others.

IV. Ecology

A. Habitat

Peale's dolphins occupy two major habitats; open coasts over shallow continental shelves to the north and deep, protected bays and channels to the south and west. In the channels, this is an "entrance animal" found in tide rips at the entrance to fjords. They often swim shoreward of or through channels in kelp beds (*Macrocystis pyrifera*). Although their distributions overlap, Peale's dolphins are usually coast hugging, while the similarly pigmented dusky dolphin occurs a few miles seaward. There is no information on abundance, stocks, or population size.

B. Prey, Predators, and Parasites

Peale's dolphins are noted for feeding in the kelp, where divers have observed them taking small octopus. They also feed

Figure 1 *Peale's dolphin,* Lagenorhynchus australis. *Pieter A. Folkens/Higher Porpoise DG.*

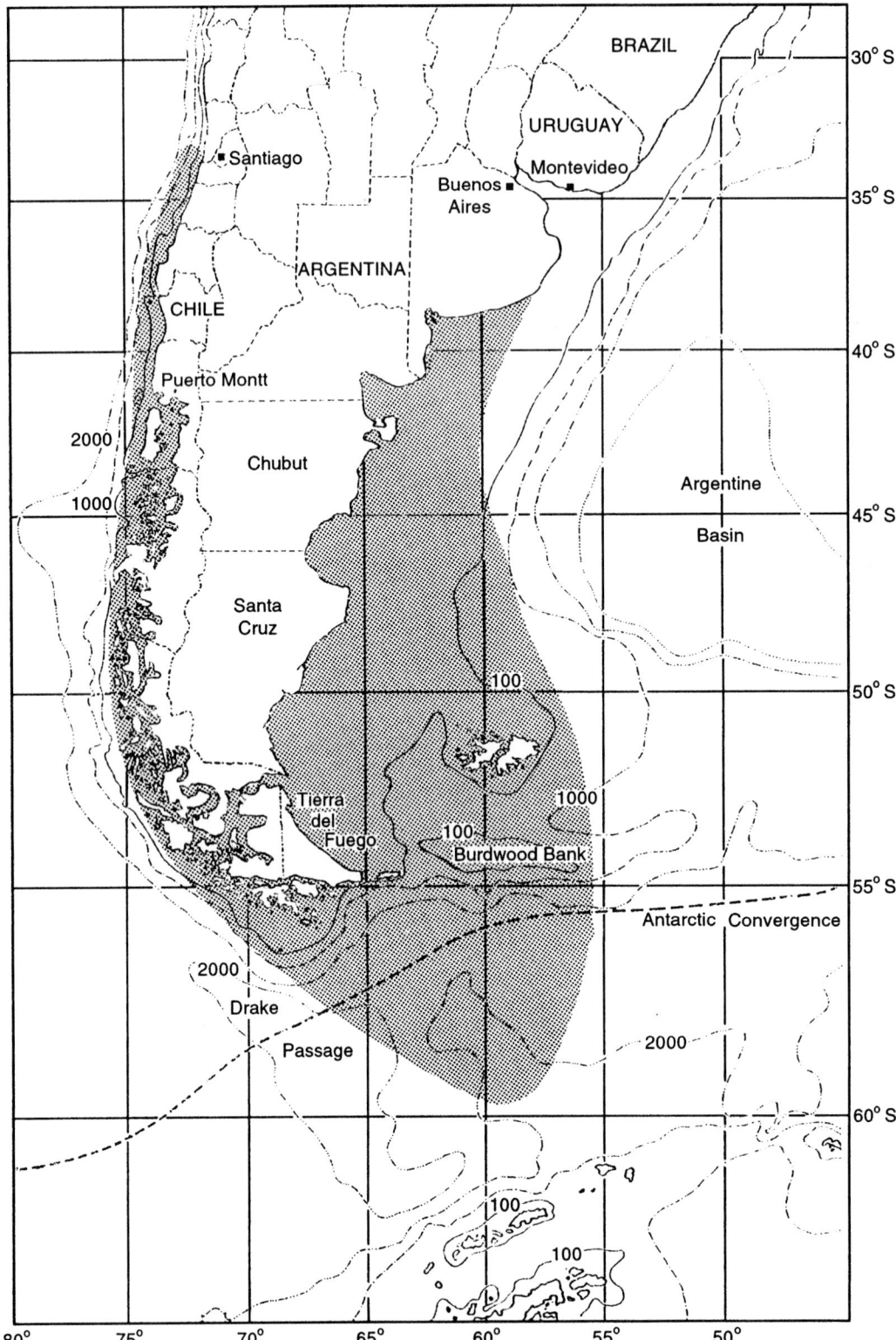

Figure 2 *The distribution of Peale's dolphin, based on sightings and strandings from 1839 to 1997.*

in open waters beyond the kelp on fish, often using the "sunburst" formation for herding. Few stomachs have been examined ($n = 16$), and those only from the southwestern South Atlantic. About 20 prey taxa have been identified, mainly demersal and bottom fish, octopus, and squid species common over the continental shelf or in kelp beds.

No predators but humans are known, although killer whales and leopard seals are a possibility. Parasite fauna are under study; *Anisakis simplex* was present in three females.

V. Behavior
A. Migration

Near Isla Chiloé and in the Strait of Magellan, resident groups have been noted throughout the year, although more animals were present during summer. In southern Tierra del Fuego, animals also seem to move inshore during summer, possibly following fish migrations. It is not known if the dolphins in different parts of the range belong to different populations or stocks.

B. Swimming and Schooling

Peale's dolphins are most often seen swimming slowly in or near the kelp. Dive times range from 3 to 157 sec, with an average of 28 sec ($n = 723$), with three short dives followed by a longer one. They commonly BOW RIDE, with much head movement, rolling, BREACHING, spy hopping and spinning. They produce a wide splash when surface swimming, earning the name "plough-share" dolphins. They associate closely with other dolphins, especially *Cephalorhynchus commersonii*.

Group size is usually small, from 2 to 5 animals, but aggregations of up to 100 have been seen. Little else is known of their behavior.

C. Sounds

Pulsed sounds were recorded in the Chilean channel region, revealing broadband clicks at 5–12 kHz and narrowband clicks at 1- to 2-kHz bandwidths. Whistle-like squeals were not recorded, but may have been above the limits of the recording equipment used.

VI. Internal Anatomy

The condylobasal lengths of 27 skulls ranged from 352 to 359 mm. The number of teeth is variable, with up to 37 upper and 34 lower teeth in each jaw. The mouth is unusual in having a wide lip outside the tooth line. The vertebral count is CV7 with the first two fused, T14, L13–16, Ca 31–34 for a total of 66–70, normally 67–68. The Peale's dolphin is larger and more robust than the other two southern *Lagenorhynchus*, with larger vertebrae. The phalangeal number is I = 2–3, II = 6–9, III = 5–6, IV = 2–4, V = 0–2.

Few tissues have been examined. Skin thickness was 1–2 mm, and blubber thickness was 15–25 mm. The gonads and uterus of three specimens have been described, and organ weights are known for three others. Eye pigments were reported for one animal. Residues of organochlorine contaminants were found in the blubber of one animal.

VII. Life History
A. Growth and Reproduction

There is little information on reproduction. A female of 185 cm was sexually immature, one of 193 cm was pubescent, and one of 210 cm was mature. There is no information on sexual maturity in males. Calves have been reported from spring through fall, October to April.

Physical maturity, on the basis of epiphyseal fusion, has been recorded for 24 specimens. Neonates measured 98–130 cm, juveniles 138–176 cm, subadults 142–210 cm, and adults 190–199 cm. The oldest animal was a physically mature female with 13 GLGs (growth layer groups) in the teeth.

VIII. Interactions with Humans
A. Captivity and History

No Peale's dolphins have been kept in CAPTIVITY. Prehistorically, Peale's dolphins were exploited for food by the canoe peoples of the channels of southernmost South America; remains have been found in kitchen middens dated at 2500 and 6000 years before present. More recently, a few dolphins were taken for scientific research.

B. Fisheries

1. Chile Peale's dolphin have been heavily exploited for crab (*centolla*, *Lithodes antarctica*) bait in southern Chile since at least the early 1970s, with highest exploitation in the 1980s. Overfishing, resulting in greatly reduced populations of crabs, has led to other types of fishing that do not need bait. The full extent of the exploitation of dolphins for crab bait at present is unknown, but is thought to be less than formerly. Although common, Peale's dolphins are rarely taken incidentally in the northern part of their range between Valdivia and San Antonio. A few dolphins have been caught in anti-pinniped nets near the salmon pens in the Isla Chiloé area.

2. Argentina A few Peale's dolphins were taken for crab bait in eastern Beagle Channel in the late 1970s and early 1980s, but this has not continued. A small incidental take occurs in shore-set gill nets off Tierra del Fuego, but fishermen claim that Peale's dolphins are usually strong enough to fight and release themselves from the net. Likewise, a few Peale's dolphins die in off-shore fishing activities south of Golfo San Jorge, but not as many as other species. The extent of this exploitation is unknown; it represents a potential danger that should be monitored.

See Also the Following Articles

Hourglass Dolphin ■ Pacific White-Sided Dolphin and Dusky Dolphin

References

Brownell, R. L., Jr. (1999). Peale's dolphin, *Lagenorhynchus australis* (Peale 1848). *In* "Handbook of Marine Mammals; the Second Book of Dolphins and Porpoises" (S. H. Ridgway and R. Harrison, eds.), Vol. 6, pp. 105–120.

de Haro, J. C., and Iñiguez, M. A. (1997). Ecology and behaviour of the Peale's dolphin, *Lagenorhynchus australis* (Peale, 1848), at Cabo Vírgenes (52°30′S, 68°28′W), in Patagonia, Argentina. *Rep. Int. Whal. Commn.* **47**, 723–727.

Goodall, R. N. P., de Haro, C., Fraga, F., Iñiguez, M. A., and Norris, K. S. (1997). Sightings and behavior of Peale's dolphins, *Lagenorhynchus australis*, with notes on dusky dolphins, *L. obscurus*, off southernmost South America. *Rep. Int. Whal. Comm.* **47**, 757–775.

Goodall, R. N. P., Norris, K. S., Schevill, W. E., Fraga, F., Praderi, R., Iñiguez, M. A., and de Haro, C. (1997). Review and upate on the biology of the Peale's dolphin, *Lagenorhynchus australis. Rep. Int. Whal. Commn.* **47**, 777–796.

Goodall, R. N. P., Schiavini, A. C. M., and Fermani, C. (1994). Net fisheries and net mortality of small cetaceans off Tierra del Fuego, Argentina. *Rep. Int. Whal. Comm.* (Special Issue 15), 295–304.

International Whaling Commission (1997). Report of the sub-committee on small cetaceans. Annex H. *Rep. Int. Whal. Commn.* **47**, 169–191.

Leatherwood, S., Grove, J.S., and Zuckerman, A. E. (1991). Dolphins of the genus *Lagenorhynchus* in the tropical South Pacific. *Mar. Mamm. Sci.* **7**(2), 194–197.

LeDuc, R. G., Perrin, W. F., and Dizon, A. E. (1999). Phylogenetic relationships among the delphinid cetaceans based on full cytochrome B sequences. *Mar. Mamm. Sci.* **15**(3), 619–648.

Lescrauwaet, A.-K. (1997). Notes on the behaviour and ecology of the Peale's dolphin, *Lagenorhynchus australis,* in the Strait of Magellan, Chile. *Rep. Int. Whal. Commn.* **47**, 747–755.

Lescrauwaet, A.-K., and Gibbons, J. (1994). Mortality of small cetaceans and the crab bait fishery in the Magallanes area of Chile since 1980. *Rep. Int. Whal. Comm.* (Special Issue 9), 103–118.

Schiavini, A. C. M., Goodall, R. N. P., Lescrauwaet, A.-K., and Koen Alonso, M. (1997). Food habits of the Peale's dolphin, *Lagenorhynchus australis.;* review and new information. *Rep. Int. Whal. Commn.* **47**, 827–833.

Van Waerebeek, K., Goodall, R. N. P., and Best, P. G. (1997). A note on evidence for pelagic warm-water dolphins resembling *Lagenorhynchus. Rep. Int. Whal. Commn.* **47**, 1015–1017.

Pelvic Anatomy

Peter J. Adam
University of California, Los Angeles

With the development of tail flukes for producing propulsion in whales, manatees (*Trichechus* spp.), and dugongs (*Dugong dugon*), the pelves and hind limbs became vestigial structures that now associate only loosely with the spine. The major role of the pelvic apparatus in these forms, when present, is to serve as attachment points for muscles acting on the genitalia and the abdominal body wall. Marine carnivores, which still maintain close ties with the terrestrial environment, have not had such a dramatic reduction in the pelvis and hind limb structures. Both pinnipeds and the sea otter (*Enhydra lutris*) have united the toes to form flippers. Phocids, which use the hind limbs to generate swimming thrust and which cannot rotate the hind feet under the body while on

land, have highly modified hind limbs. Phocid adaptations include increasing the surface areas available for muscles that flex the leg, modification of limb muscles to aid in undulatory movements of the spine, and a general increase in the muscle mass operating on the hind limb (in particular the muscles acting to flex the limb).

I. Cetaceans

The known fossil record documenting cetacean evolution shows a progressive reduction and loss of hind limb skeletal elements and disassociation of the pelvic girdle from the vertebral column as whales became less dependent on nearshore environments and developed tail flukes to generate swimming thrust. This trend is most marked with the origin of the basilosaurine whales, in which the tibia and fibula became fused with each other and tarsal elements co-ossified into a single immobile mass. Basilosaurines also mark the point during which the pelves became disassociated from the vertebral column. Among modern forms, only vestiges of the hind limb skeleton can be found, and these are contained within the body wall. Mysticetes typically posses only vestigial pelves, whereas the occurrence of hind limb and pelvic elements is more variable among both individuals and species of odontocetes. When present, the pelves bear little resemblance to those of terrestrial mammals and, when undeveloped, may exist only as a band of connective tissue connecting spinal muscles to those of the genitalia and abdominal wall. If present as a bony element, each pelvis is typically cigar or sickle shaped, with only the pelvic bone proper contributing to its structure (Fig. 1). Atavistic femora and occasional tibiae have been described from numerous (mysticete and odontocete) taxa, although occurrence of these elements is infrequent. Hind limb buds are present during early embryogenesis of all whale species documented so far, although the mesodermal cells that usually form the internal limb structures die or are reallocated to other functions as limb buds are resorbed later in ontogeny. Retention of a rudimentary pelvis is associated with the attachment of numerous muscles acting on the reproductive organs of both sexes. In males, the pelvis is usually larger relative to that of females. It serves as the site of origin for muscles acting on the genitals and anal region (e.g., the penis retractor and levator ani muscles) and may also serve as a site of attachment for posterior fibers of the rectus abdominis muscle. When present, the pelvis is isolated from the spine (sacral vertebrae are absent) but maintains a soft tissue attachment to the hypaxial spinal musculature. Rearrangements of spinal and pelvic muscles in association with tail-based locomotion have led to considerable controversy over specific identities of muscles in these regions.

II. Sirenians

The evolutionary loss of the hind limb in sirenians closely parallels that of cetaceans (Fig. 1), with modern forms possessing only a vestigial pelvis composed of iliac and ilium bones. In dugongs, each pelvis is long and stick like in appearance, and the ilium and ischium are of subequal length, fusing by 5 years of age in both sexes. In manatees, the pelves are more

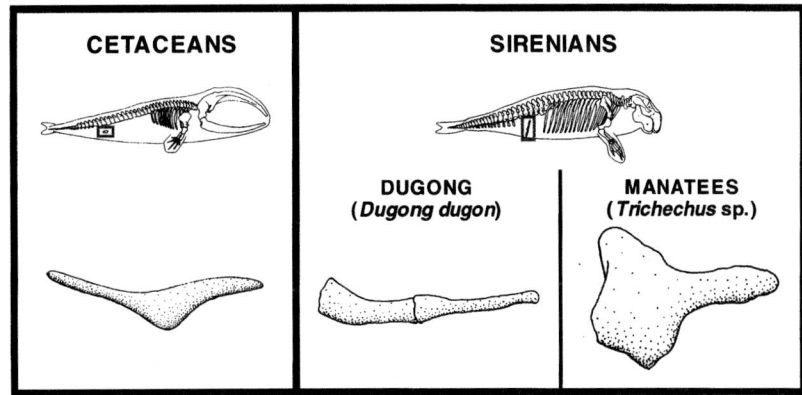

Figure 1 *Line drawings of the right pelvis of a cetacean (left) and sirenians (right) in lateral view (anterior toward the right). Position and orientation of the pelvis are indicated by boxes on the skeletal outlines.*

plate like and cross shaped in lateral view. The ischium is the largest portion of the manatee pelvis, with the ilium forming a small cap on the anterior surface of the bone complex. As in whales, sirenian pelves lack osseous attachment to the vertebral column. In dugongs, the pelves join with the anterior caudal vertebrae by an aponeurosis thought to be homologous with the m. coccygeus, as well as by the retractor ischii and ischiococcygeus muscles to caudal chevron BONES. The pelves serve as the origin for muscles acting on the genital organs (e.g., in females, the constrictor vulvae, constrictor vestibuli, and urethralis muscles) as well as some muscles inserting into the skin of the abdominal region [e.g., the transversus abdominis (in part)]. The atavistic appearance of femora has been reported for manatees.

III. Marine Carnivores

A. Pinnipeds

The earliest known pinniped, *Enaliarctos* from the late Oligocene of California, possessed a well-developed hind flipper, and intermediate stages in the anatomical progression from a limb used for terrestrial locomotion to one specialized for swimming are undocumented. Anatomical adaptations of the hind limbs of extant pinnipeds largely reflect strategies adopted by each family for swimming and terrestrial locomotion. Phocid seals and walruses (*Odobenus rosmarus*) primarily use the hind limbs to generate thrust while swimming and have relatively more muscle mass in the pelvic region relative to the pectoral region and forelimb. Otariids propel themselves with the forelimbs and have relatively lower pelvic muscle mass. On land, otariids and walruses are able to rotate their hind feet under their body and progress with modified walking motions; phocids lack the ability to rotate their feet under their body and move along the ground with undulatory movements of the body.

Externally, the hind limbs of pinnipeds extend beyond the body contour from the approximate middle or end of the crus. In walruses and phocids the middle digit is the shortest, and digits increase in length both laterally and medially, giving the flipper a crescent shape (more marked in phocids). A thin, extensible interdigital webbing stretches between adjacent digits in these forms. In otariids, the interdigital areas are occupied by thick layers of connective and other tissues, making the hind flipper a much more rigid structure. Emargination of the distal interdigital regions of the flipper confers a scalloped shape to its trailing edge. Claws are reduced in all pinnipeds, although those of the middle three digits tend to be better developed than those of the first and fifth digits. Claws are positioned terminally in phocids and subterminally in walruses, but are located considerably farther proximally in otariids due to the development of distal cartilaginous rods on the ungual phalanges. The presence of these cartilaginous extensions gives the ungual phalanges an hourglass shape and roughened distal ends. The plantar surface of otariid and walrus flippers is hairless, with moderately developed foot pads related to their ambulatory terrestrial locomotion. Foot pads are lacking on phocids. With the exception of interdigital regions, which are hairless or have only sparse hair, the dorsal pedal surface typically has a hair density subequal or slightly lower than that of the body.

Departures of the skeletomuscular anatomy from the condition observed in typical terrestrial carnivores are most prevalent in phocid seals due to their highly modified (undulatory) terrestrial locomotion and specialized hind limb swimming. The iliac region of the phocid pelvis is expanded laterally, particularly among phocines. This confers a mechanical advantage to the gluteus muscle complex, which inserts onto the greater trochanter of the femur and functions to flex the leg against the water. The ischiopubic region of phocid pelves, posterior to the acetabulum, is elongate relative to otariids and the walrus. This increases the surface area available for attachment of the strong muscles acting to medially flex the leg during the power stroke of the swimming cycle (e.g., the adductor, gracilis, gemelli, obturatorius, and semimembranosus muscles). The ischial tuberosity, often misidentified as the "ischial spine," is greatly enlarged in phocids, but undeveloped in otariids and the walrus. It serves as the site of origin for the biceps femoris muscle, which inserts broadly onto the tibia. The orientation and widening of the biceps femoris in phocids indicate that it is primarily responsible for lifting the hind limb off the ground during terrestrial locomotion, as well as medially flexing the limb during swimming. The pinniped femur is short and stout, and the distal condyles are inclined relative to the long axis of the shaft. The fovea capitis of the femoral head is lacking. This indicates the loss of the teres ligament, which normally maintains the femoral head within the acetabulum of the pelves in terrestrial mammals that have weight-bearing hip joints. In phocids, the lesser femoral trochanter is either reduced or absent, and the two muscles typically inserting onto it have undergone major changes from their usual orientation and function: (1) the iliacus muscle inserts onto the more distal femoral epicondylar crest or proximal tibia and (2) the psoas major muscle, arising from the posterior thoracic, lumbar, and sacral vertebrae, inserts onto the medial surface of the ilium and thus aids in lateral undulation of the spine during swimming rather than acting on the limb. Proximally, the tibia and fibula of most

pinniped species become fused prior to maturity. The posterior tibial fossa is deep in phocids, reflecting enlargement of the tibialis caudalis muscle, which originates from this region and inserts onto the tarsus and first metatarsal, acting to plantar flex the pes during the swimming power stroke. The tendon of the flexor hallucis longus muscle passes over a posterior projection of the astragalus that is unique to phocids, limiting dorsal flexion of the pes. This, in combination with the elongated isciopubis (which limits anteroventral bending of the spine when phocids are on land), limits the ability of phocids to assume a four-legged stance. Additionally, the tibioastragalar joint of phocids is nearly spherical and would unlikely bear the weight of the animal. This is in contrast to the more rigid, hinge-like joint found in other pinnipeds and terrestrial carnivores. Inserting tendons of the large plantar flexing muscles (i.e., the flexor digitorum longus, flexor hallucis longus, and flexor digitorum superficialis muscles) often combine together in a com-

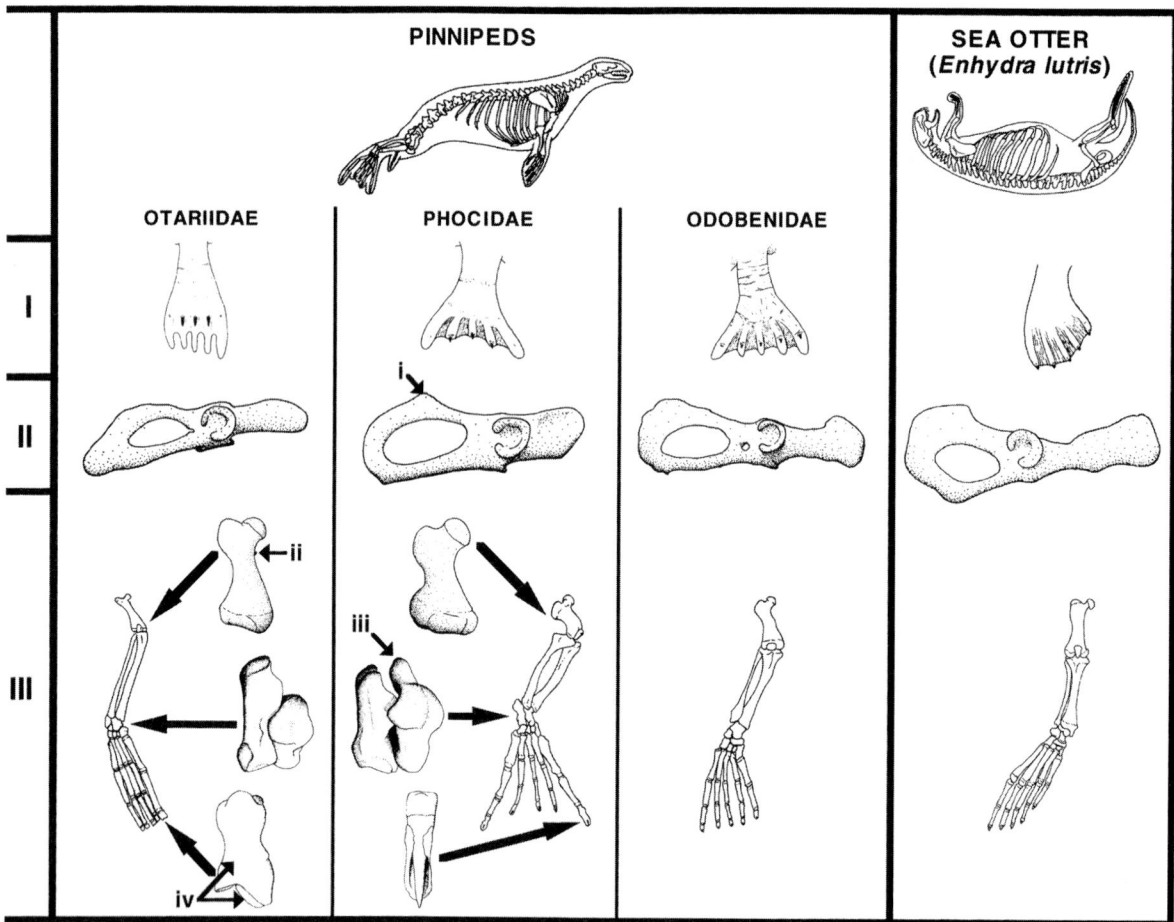

Figure 2 *Line drawings of external flipper morphology (row I), right pelvis (row II, lateral view, anterior is toward the right), and right hind limb skeleton [row III, anterior (dorsal) view] of representative species of pinnipeds (left; Otariidae, Callorhinus; Phocidae, Monachus; and Odobenidae, Odobenus) and the sea otter (right). Indicated features of the otariid and phocid hind limb are as follows: i, enlarged ischial tuberosity of phocids; ii, presence of a lesser femoral trochanter in otariids and walrus (absent or reduced in phocids); iii, posterior projection of the phocid astragalus, over which the tendon of the flexor hallucis longus muscle passes; and iv, first ungual phalanx of the otariid pes showing the lack of a claw and distal roughened surface to which cartilage is attached in life.*

plex manner at the level of the tarsals. A united tendon of these muscles sends branches out to the digits, although slips extending to the first and fifth digits tend to be larger in phocids. The pedal formula of all pinnipeds is 2-3-3-3-3, the primitive condition for all mammals. In phocids and the walrus, metatarsals and phalanges of the first and fifth digits are more robust than those of the middle three digits. This is associated with hind limb swimming, where both digits may act as leading edges of the flipper during the complex power stroke. Metatarsal–phalangeal and interphalangeal joints tend to be of the tongue-and-groove type in phocids and hinge-like in otariids and the walrus. However, all pinnipeds have reduced trochleation of the metatarsals and phalanges (Fig. 2).

Available evidence indicates that blood supplying the crus and pes of phocids passes primarily through the external iliac, femoral, and sapheneous arteries. Maintenance of this primitive condition is related to the efficiency of the system for supplying oxygen and nutrients to the heavily used hind limb musculature. In contrast, otariids have adapted the blood vessels such that most of the blood supplying the distal limb regions passes through the internal iliac and gluteal arteries. The passage of blood via this route is believed to enhance heat dissipation. The walrus is intermediate to these conditions, although detailed description of its anatomy is lacking. The presence of a circulatory countercurrent heat exchange system (*rete mirabile*) in the hind limbs has been reported for several phocid and otariid species. Anatomy of the spinal nerves serving the hind limb of pinnipeds is poorly known. Available evidence suggests that the lumbosacral plexus has shifted posteriorly by one vertebra, being composed of ventral rami arising from the third through fifth lumbar, first through third sacral, and first caudal vertebrae. Division of this plexus into lumbar and sacral plexi is not possible.

B. Polar Bears and Sea Otters

Adequate descriptions of polar bear (*Ursus maritimus*) pelvic and hind limb anatomy have not yet been made, but there is little indication that the morphology of this species has diverged appreciably from that of other species of *Ursus*. Departures of sea otter hind limb anatomy from that of other mustelids, however, are seen more readily. Externally, the leg is enclosed within the loose body skin to the approximate level of the ankle. The digits are bound together by interdigital webbing, although the fourth and fifth digits are bound more closely together than other adjacent digital pairs. The sea otter is unusual in that in overall length the digits decrease in size from the fifth to the first: V > IV > III > II > I. While swimming, sea otters use the hind feet to generate thrust and sweep the leg through the water such that the fifth digit forms the leading edge of the pes. The hair densities for the ankle and interdigital webbing have been estimated at 107,000 and 3300 hairs/cm^2, respectively, compared to a density of 125,000 hairs/cm^2 for the back. Pads are present on the phalangeal portion of each toe and are variably found ventral to the metatarsals. As with pinnipeds, the fovea capitis is absent from the femur, marking the absence of the teres ligament. The biceps femoris muscle inserts onto the middle of the tibia and maintains the leg in a posterior position. The flexor digit V muscle is very large in the sea otter (relative to other mustelids). This enlargement corresponds to the use of the lateral surface of the pes to lead during the power stroke of the limb. The remaining hind limb anatomy of the sea otter corresponds well with that of terrestrial mustelids (Fig. 2).

See Also the Following Articles

Locomotion, Terrestrial ■ Musculature ■ Skeletal Anatomy ■ Swimming

References

Berta, A., and Ray, C. E. (1990). Skeletal morphology and locomotor capabilities of the archaic pinniped *Enaliarctos mealsi. J. Vertebr. Paleontol.* **10,** 141–157.

Bisaillon, A., and Pièrard, A. (1981). Ostéologie de morse de l'Atlantique (*Odobenus rosmarus*, L., 1758) ceintures et membres. *Zentralblatt Veterinärmedizin. Reihe C Anat. Histol. Embryol.* **10,** 310–327.

Domning, D. P. (1977). Observations on the myology of *Dugong dugon* (Miller). *Smith. Contrib. Zool.* **226,** 1–57.

Domning, D. P. (1991). Sexual and ontogenetic variation in the pelvic bones of *Dugong dugon. Mar. Mamm. Sci.* **7,** 311–316.

Fay, F. H. (1974). Comparative and functional anatomy of the vascular system in the hind limbs of the Pinnipedia. Transactions of the First International Theriological Congress (Nauka Publishers, Moskow) **1,** 166–167.

Gambarjan, P. P., and Karapetjan, W. S. (1961). Besonderheiten im Bau des Seelöwen (*Eumetopias californianus*), der Baikalrobbe (*Phoca sibirica*) und des Seeotters (*Enhydra lutris*) in Anpassung an die Fortbewegung im Wasser. *Zool. Jahrbucher (Abteilung Anat. Ontog. Tiere)* **79,** 123–148.

Kaiser, H. E. (1974). "Morphology of the Sirenia: A Macroscopic and X-Ray Atlas of the Osteology of Recent Species. Karger, Basel, Switzerland.

Miller, W. C. S. (1888). The myology of the Pinnipedia. In "Report on the Scientific Results of the Voyage of H. M. S. Challenger during the Years 1873–76 (C. W. Thomson and J. Murray), Vol. 26. Order of Her Majesty's Government.

Muizon, C. de (1981). Une interprétation functionelle et phylogénétique de l'insertion du psoas major chez les Phocidae. *Comp. Ren. Acad. Sci. (Paris)* **292,** 859–862.

Nakanishi, T., Yamamoto, M., and Suenaga, Y. (1978). Comparative anatomical studies on the nerves and muscles of the posterior limb of the northern fur seal and cat. *Okajimas Fol. Anat. Japon.* **54,** 317–340.

Schulte, H. W., and Smith, M. de F. (1918). The external characters, skeletal muscles, and peripheral nerves of *Kogia breviceps* (Blainville). *Bull. Am. Mus. Nat. Hist.* **38,** 7–72.

Sedmera, D., Misek, I., and Klima, M. (1997). On the development of cetacean extremities. I. Hind limb rudimentation in the spotted dolphin. *Eur. J. Morphol.* **35,** 25–30.

Tarasoff, F. J. (1972). Comparative aspects of the hind limbs of the river otter, sea otter and seals. In "Functional Anatomy of Marine Mammals" (R. J. Harrison, ed.), Vol. 1, pp. 333–359. Academic Press, New York.

Tarasoff, F. J., Bisaillon, A., Pièrard, J., and Whitt, A. P. (1972). Locomotory patterns and external morphology of the river otter, sea otter, and harp seal (Mammalia). *Can. J. Zool.* **50,** 915–929.

Uhen, M. D. (1998). Middle to Late Eocene basilosaurines and dorudontines. In "The Emergence of Whales: Evolutionary Patterns in the Origin of Cetacea" (J. G. M. Thewissen, ed.), pp. 29–61. Plenum Press, New York.

Perissodactyla

J. G. M. Thewissen
*Northeastern Ohio Universities College of Medicine,
Rootstown*

I. Overview

The perissodactyls (odd-toed ungulates) are an order of land mammals that comprise, in modern times, horses (including zebras and asses), rhinoceroses, and tapirs. Perissodactyls are mainly characterized by their feet. There is an odd number of toes, and the axis of symmetry of each fore- and hindfoot runs through the middle (third) digit (this is called mesaxony). The third digit is the largest in all perissodactyls and is the only digit in horses. The toes immediately adjacent to the third digit (the second and fourth) are similar to each other and smaller than the central toe. Rhinos have three toes per foot and tapirs have three on their hindfeet and four on the forefeet. The fourth toe in a tapir is medial to and smaller than any other toe. All toes bear a hoof in perissodactyls, just like in the even-toed ungulates (artiodactyls), where there are two or four toes per foot and the axis of symmetry runs between the third and the fourth digit (paraxony).

II. Relation to Marine Mammals

The earliest perissodactyl fossils known date from the early Eocene, approximately 52 million years ago, and are close temporally to the earliest cetaceans and sirenians. Perissodactyls are related to marine mammals. Studies of morphological data (Thewissen and Domning, 1992; Fischer and Tassy, 1993) commonly find that perissodactyls are close relatives to sirenians, proboscideans, and hyracoids. Molecular data disagree with that, instead finding that certain insectivores are related more closely to sirenians (Stanhope *et al.*, 1998). A minority opinion is that perissodactyls are closely related to whales, but most evidence instead points to a close relationship between whales and artiodactyls.

See Also the Following Articles

Artiodactyla ■ Sirenian Evolution

References

Fischer, M. S., and Tassy, P. (1993). The interrelation between Proboscidea, Sirenia, Hyracoidea, and Mesaxonia. *In* "Mammal Phylogeny, Placentals" (F. S. Szalay, M. J. Novacek, and M. C. McKenna, eds.), pp. 217–235. Springer-Verlag, New York.

Stanhope, M. J., Waddell, V. G., Madsen, O., de Jong, W., Hedges, S. B., Cleven, G. C., Kao, M. and Springer, M. (1998). Molecular evidence for multiple origins of Insectivora and for a new order of endemic African insectivore mammals. *Proc. Natl. Acad. Sci. USA* **95**, 9967–9972.

Thewissen, J. G. M., and Domning, D. P. (1992). The role of phenacodontids in the origin of the modern orders of ungulate mammals. *J. Vertebr. Paleontol.* **12**, 494–504.

Pilot Whales

Globicephala melas and *G. macrorhynchus*

Paula A. Olson and Stephen B. Reilly
*Southwest Fisheries Science Center,
La Jolla, California*

Pilot whales are large dolphins, among the largest members of the family Delphinidae. They are widely distributed in the world's oceans. Two species are recognized: *Globicephala melas* (long-finned pilot whale) and *G. macrorhynchus* (short-finned pilot whale). Notable aspects of pilot whale natural history include sexual dimorphism and mass strandings. Pilot whales are among the most gregarious of the cetaceans.

The name "pilot whale" originated with an early theory that a school is piloted by a leader. Other common names for these whales include pothead (after the whales' bulbous melon) and blackfish (a term also used for melon-headed whales, pygmy killer whales, and false killer whales). The genus name, *Globicephala*, is derived from the Latin word *globus*, meaning round ball or globe, and the Greek word *kephale*, meaning head. *Melas* is a Greek word for black. *Macrorhynchus* is likewise derived from Greek words: *macro*, meaning enlarged, and *rhynchus*, meaning snout or beak. For many decades, *malaena* was used as the trivial name for the long-finned pilot whale. However, in 1986 the name was revised to *melas*.

I. Physiology
A. Morphology

Long-finned and short-finned pilot whales are difficult to distinguish at sea (Fig. 1). The morphological differences between the two species are subtle: length of flippers, differences in skull shape, and number of teeth. On average the pectoral flippers of long-finned pilot whales are one-fifth the body length, whereas on short-finned whales they are one-sixth the body length. However, overlap exists between the two species. Long-finned flippers exhibit a noticeable "elbow" whereas short-finned pectorals have a more curved appearance.

Adult pilot whales reach an average length of approximately 6 m. Males are larger than females. Most pilot whales appear black or dark gray in color. The body is robust with a thick tailstock. The melon is exaggerated and bulbous, and there is either no beak or a barely discernible one. A wide, broad-based falcate dorsal fin is set well forward on the body. The flippers are long, slender, and sickle shaped. A faint gray "saddle" patch may be visible behind the dorsal fin as well as a faint postorbital blaze (Fig. 2). A gray midventral line extends anteriorly into an anchor-shaped chest patch and widens posteriorly into a genital patch. Calves are paler than adults.

Because most size, shape, and color pattern distinctions between the two species are so variable, the shape of the skull is the only definitive characteristic for identification to species. The long-finned pilot whale has a narrower skull, with the premaxil-

Figure 1 *Long-finned and short-finned pilot whales exhibit similar external morphology, including a large, bulbous melon and a broad-based dorsal fin. Photo by J. Carretta.*

lae leaving uncovered 1 cm of the lateral borders of the maxillae (Fig. 3A). There are 9 to 12 teeth in each row. The skull of the short-finned pilot whale is shorter and broader, and the premaxillae cover the maxillae (Fig. 3B). There are 7 to 9 teeth in each row. Pilot whales have notably fewer teeth than most other delphiniids. This is similar to the evolutionary reduction of teeth found in Risso's dolphins (*Grampus griseus*) and sperm whales (*Physeter macrocephalus*), two other heavy squid consumers.

B. Sexual Dimorphism

Pilot whales exhibit striking sexual dimorphism in size, similar to that observed in sperm whales and killer whales (*Orcinus orca*). Adult males are longer than females, develop a more pronounced melon, and have a much larger dorsal fin (Fig. 4). The function of sexual dimorphism in pilot whales is unknown, although several have been hypothesized. The males' enlarged features may be used for display to other males or females or for increased agility when maneuvering for mate access or for herding females. The males' large size may aid in defense of their school from attacks by killer whales or sharks.

C. Disease

Pilot whales in the western Atlantic have been affected by the morbilliviruses that have plagued other marine mammals in recent decades. Although to date no large-scale outbreak of disease has been reported in pilot whales, high percentages of both species sampled during the 1980s and 1990s carried virus-neutralizing antibodies. It appears that most individuals are immune. Due to their wide-ranging movements and their propensity to mix with other species, pilot whales may act as a vector for morbilliviruses in other cetaceans, such as bottlenose dolphins (*Tursiops truncatus*).

II. Distribution

Pilot whales are wide ranging (Fig. 5). Generally, short-finned pilot whales have a tropical and subtropical distribution and long-finned pilot whales are distributed antitropically. There is little overlap in the range of the two species. Areas of marginal overlap include the temperate waters of the North and South Atlantic, in the Pacific off the coast of Peru, and off South Africa. Pilot whales are found in both nearshore and pelagic environments.

Figure 2 *Full-body view of the short-finned pilot whale (top) and the long-finned pilot whale (bottom). Drawings courtesy of P. Folkens.*

A

B

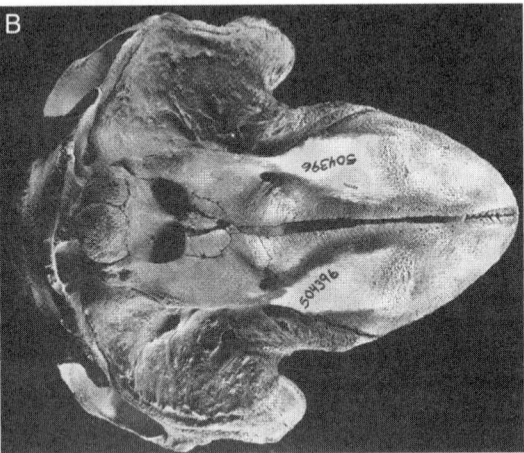

Figure 3 *(A) Dorsal view of the skull of a long-finned pilot whale (G. melas). (B) Dorsal view of the skull of a short-finned pilot whale (G. macrorhynchus). Note the differences in the shape and length of the rostrums and the degree to which the maxillae are covered by the premaxillae.* G. melas *photo courtesy of J. Heyning, Los Angeles County Museum of Natural History.* G. macroryhnchus *photo courtesy of C. Potter, USNM; photo by C. Clark.*

A. Geographical Variation and Subspecies

Long-finned pilot whales inhabit the cold temperate waters of both the North Atlantic and the Southern Ocean. The two populations are isolated. There are slight morphological differences between the populations and they are accorded subspecies status: *G. melas melas* in the North Atlantic and *G. melas edwardii* in the Southern Hemisphere.

In the North Atlantic, the range of *G. melas melas* includes the waters of Greenland (Denmark), Iceland, and the Barents Sea south to the Tropic of Cancer. This species is present in the western Mediterranean Sea. *G. melas edwardii* is circumglobal in the Southern Hemisphere, with records as far north as 14°S in the Pacific and further south than the Antarctic Convergence. Long-finned pilot whales do not occur in the North Pacific.

Short-finned pilot whales are found worldwide in tropical, subtropical, and warm temperate waters. Their northern range in the Atlantic extends to the midcoast area of the United States and to France. Short-finned pilot whales are not found in the Mediter-

Figure 4 *A group of male and female pilot whales. Males are larger than females and develop exaggeratedly wide dorsal fins. Photo by J. Carretta.*

ranean. Latitude 25°S marks the southernmost record for the Atlantic and the Pacific coasts of South America. Elsewhere in the Pacific, the range of the short-finned pilot whale continues north to Japan and to the west coast of the United States.

Pacific short-finned pilot whales in higher latitudes are generally larger than those in lower latitudes. Two distinct populations of short-finned pilot whales are found off northern and southern Japan. These populations exhibit morphological differences in external and cranial features. The populations are segregated geographically and genetically. However, their exact taxonomic status is still undetermined and for the time being they are both classified as *G. macrorhynchus*.

B. Abundance

Estimates of abundance for pilot whales have generally been undertaken in response to management issues. Survey areas are typically determined by management goals rather than natural population boundaries. Most of the quantitatively derived estimates of abundance are for nearshore populations. Estimates using line-transect methods have been made for Newfoundland/Labrador (Canada), the northeast Atlantic, the Antarctic, the U.S. West Coast, northern and southern Japan, and the eastern tropical Pacific (Table I).

III. Fossil Record

The extant delphinid groups (to which *Globicephala* belongs) appeared in the Middle to Late Miocene Epoch. Fossils of the genus *Globicephala* dating from the Pleistocene Epoch have been uncovered in Florida (*G. baerreckeii*). Odontocete remains from the Pliocene Epoch uncovered in Tuscany, Italy, have been designated as *Gobicephala? eturiae*.

Although the long-finned pilot whale does not currently inhabit the North Pacific, skulls dated to the 8th to 12th centuries have been recovered in Japan. Short-finned pilot whales now inhabit those waters.

IV. Ecology

Pilot whales are found on the continental shelf break, in slope waters, and in areas of high topographic relief. In gen-

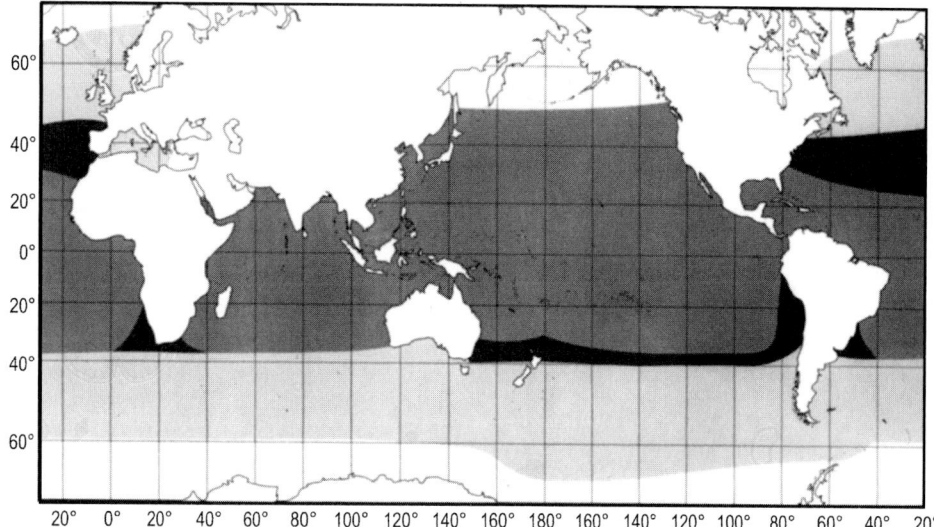

Figure 5 *Worldwide distribution of pilot whales. The distribution of long-finned pilot whales is indicated by light shading and the distribution of short-finned pilot whales by medium shading. Dark shading indicates regions where the species overlap.*

eral, pilot whales are nomadic, but resident populations have been documented in a few locations such as coastal California and Hawaii. Seasonal inshore/offshore movements of pilot whales are related to the distribution of squid, their favorite prey. Studies in Newfoundland and California correlated the seasonal abundance of pilot whales with spawning squid.

The pilot whale diet consists primarily of squid, with lesser amounts of fish. Fish prey in the northwest Atlantic include Atlantic cod (*Gadus morhua*), Greenland turbot (*Rheinhardtius hippoglossoides*), Atlantic mackerel (*Scomber scombris*), Atlantic herring (*Clupea harengus*), hake (*Urophycis* spp.), silver hake (*Merluccius bilinearis*), and spiny dogfish (*Squalus acanthias*). Pilot whales in the northeast Atlantic have been known to take Atlantic Argentine (*Argentina silus*) and blue whiting (*Micromesistius poutassou*).

The ecosystem changes brought about by a strong El Niño event in 1982–1983 affected the distribution of pilot whales off southern California dramatically. With the influx of warm water during the El Niño, squid did not spawn as usual in the area. Pilot whales were virtually absent from the region that year and remained so for another 9 years. It is not known where the whales went during this time or whether the whales sighted there now are returning or are new individuals.

V. Life History

Pilot whales share several features of life history with other large odontocetes: long life span, delayed maturity, different rates of maturation for males and females, seasonal MATING, and the production of a single calf in multiyear intervals.

TABLE I
Estimates of Pilot Whale Abundance Using Line-Transect Methods

Species/Subspecies	Area	Population	Reference
Globicephala melas melas	Newfoundland/Labrador	6,731–19,603	R. Hay (1982). *Rep. Int. Whal. Commn.* **32**, 475–480.
G. melas melas	Northeast Atlantic	778,000	S. T. Buckland *et al.* (1993). *Rep. Int. Whal. Commn.* Spec. Issue 14, pp. 33–49.
G. melas edwardii	Antarctic	200,000	F. Kasamatsu and G. G. Joyce (1995). *Antarct. Sci.* **7**, 365–379.
G. macrorhynchus	West coast U.S.	970	J. Barlow, *et al.* (1997). NOAA Tech. Memo. NMFS-SWFSC-248.
G. macrorhynchus	Northern Japan	5,300	T. Miyashita (1986). Int. Whal. Comm. Scientific Committee Meeting Document SC/38/SM17.
G. macrorhynchus	Southern Japan	53,000	T. Miyashita (1986). Int. Whal. Comm. Scientific Committee Meeting Document SC/38/SM17.
G. macrorhynchus	Eastern tropical Pacific	160,200	P. W. Wade and T. Gerrodette (1993). *Rep. Int. Whal. Commn.* **43**, 477–493.

Short-finned pilot whales have a slower growth rate than long-finned pilot whales and reach a shorter body length. The pattern of growth for both species is similar. Rapid neonatal growth is followed by a less rapid but continual growth phase during the juvenile years. Growth slows even further after the attainment of sexual maturity and ceases some years later. Short-finned females become sexually mature at 9 years, males at age 13 to 16. Long-finned females reach sexual maturity at 8 years, males at about 12. Males become socially mature, e.g., mate successfully, several years after reaching sexual maturity.

In the Northern Hemisphere, mating generally occurs in spring or early summer and parturition in summer or early autumn. An exception to this is the population of pilot whales off northern Japan. Their peak breeding season is autumn with parturition in winter. Gestation is estimated to be 12 months for long-finned pilot whales and 15–16 months for short-finned whales.

The birth interval in pilot whales is one of the longest of all the cetaceans. Lactation lasts for at least 3 years, often longer. Such an extended lactation period probably serves a social rather than nutritional purpose in later years. The capability of a short-finned female to lactate years past final ovulation has been reported. This supports the theory that females may invest more in present offspring as their likelihood to bear more offspring diminishes. Sizable numbers of postreproductive females have been found in populations of both species.

Estimates of natural mortality for pilot whales are based on data from the directed fisheries in the Faeroe Islands (Denmark) and Japan. These data indicate that male pilot whales have a higher mortality rate at all ages than those of females. Females are known to be longer-lived. Females live past 60 years; males reach 35–45 years.

VI. Behavior

A. Social Organization

Pilot whales are highly social and are usually found in schools, or pods, averaging 20–90 individuals. A variety of GROUP BEHAVIORS have been documented. Commonly reported are traveling or foraging in a loose chorus-line formation or collective logging at the surface.

The social structure of pilot whale pods is similar to that of killer whales. Pilot whales form stable pods composed of individuals with close matrilineal associations. All age and sex classes are included, although there is a female bias in adults. The groups are stable; pilot whales grow to maturity in their natal group and most remain there for life. Genetic evidence supports the theory that males breed outside their family group.

Pilot whales are polygynous. Huge aggregations of pilot whales are occasionally reported and it is believed that males move between family groups to mate during these temporary aggregations. This type of social structure where adult males stay with their female kin and mate elsewhere is unusual among mammals.

B. Sound Production

Pilot whales echolocate with a precision similar to that of bottlenose dolphins. Pilot whales also vocalize, the primary purpose probably being to maintain contact among school members. Vocalizations are more complex with active behavior and simpler with calmer behavior.

Significant differences have been found between the calls of long-finned and short-finned pilot whales. The calls of long-finned pilot whales are of a lower frequency and a narrower frequency range than those of short-finned pilot whales. The mean frequency for long-finned pilot whales is 4480 Hz, for short-finned pilot whales the mean frequency is 7870. Short-finned pilot whales were also found to have distinct group-specific call repertoires, as would be expected for a species with stable matrilineal kinship groups.

C. Interspecific Associations

Pilot whales are often observed in mixed species aggregations. They are most commonly sighted in association with common bottlenose dolphins (*Tursiops truncatus*), but have also been seen with short-beaked common dolphins (*Delphinus delphis*), Atlantic white-sided dolphins (*Lagenorhynchus acutus*), Pacific white-sided dolphins (*Lagenorhynchus obliquidens*), striped dolphins (*Stenella coeruleoalba*), Fraser's dolphins (*Lagenodelphis hosei*) killer whales, fin whales (*Balaenoptera physalus*), sperm whales, and gray whales (*Eschrichtius robustus*). There are accounts of pilot whales behaving aggressively toward humpbacks (*Megaptera novaeangliae*), sperm whales, common dolphins, and dolphins of the genus *Stenella*. Pilot whales have also been reported carrying the carcasses of dead California sea lions (*Zalophus californianus*) and towing a human diver.

D. Mass Stranding

Pilot whales are one of the most frequently reported cetaceans in mass strandings. Strandings of pilot whales are dramatic events because they usually involve groups of animals and because live whales that are assisted back into the ocean often return to the beach. Pilot whales strand singly as well as in groups; often these animals are diseased.

Curiously, most of the whales in a group event do not show any pathology. It is not known why apparently healthy animals strand, although there are a variety of hypotheses. The whales may become confused or trapped in shallow areas; geomagnetic anomalies may disorient whales if they are using the earth's magnetic field for navigation; or if an ill animal strands it may be followed by members of its pod. The strong social bonds within a pilot whale pod are likely to play a role in stranding events, whatever the other underlying reasons may be.

VII. Interactions with Humans

A. Directed Catches

Because of their cohesive social structure, pilot whales are susceptible to herding by humans. Meat, blubber, and oil are the desired products. Historically, there have been a number of directed fisheries for pilot whales. In the North Atlantic, drive fisheries for long-finned pilot whales were conducted in Newfoundland, Cape Cod, Norway, Iceland, Orkney Islands (Scotland), Hebrides Islands (Scotland), Greenland, and the Faeroe Islands. The intensive drive fishery in Newfoundland (1947–1971) is estimated to have taken 54,000 animals and to have reduced the local population substantially. The population may be recovering, but more information is needed. Fisheries for short-finned pilot whales have operated in the Caribbean and in Japan.

The drive fishery in the Faeroe Islands and the drive and harpoon fisheries in Japan continue today. These fisheries have been in existence for several hundred years. In 1996, the catch of long-finned pilot whales in the Faeroes was 1524 animals; in 1997, Japan reported a catch of 347 short-finned pilot whales.

B. Bycatch

The INCIDENTAL BYCATCH of cetaceans in fisheries is a worldwide phenomenon. Most bycatch goes unreported because this information is not recorded in many countries. Pilot whales are particularly susceptible to entanglement in driftnets. The effect of such mortality on pilot whale populations is unknown.

In northeast U.S. waters, pilot whales have been taken incidentally in a variety of fisheries: mackerel and squid trawls, pelagic drift gill nets, pelagic long lines, and pelagic pair trawls. Some of these fisheries are now closed, and none of the current fisheries exceed the allowable annual take for pilot whales under U.S. law.

Pilot whales off California and Mexico are taken incidental to driftnet fisheries targeting swordfish and sharks. It is likely that these takes are from the same population of short-finned pilot whales, and the per-set mortality rate reported from both countries is similar. In the United States the take exceeds the allowable annual limit and a take reduction plan has been implemented. Prior to the El Niño of 1982–1983, pilot whales were taken incidentally in the Californian squid purse seine fishery. Pilot whale redistribution in response to El Niño is the likely reason no mortality was reported for this fishery during the following years. Currently the squid fishery is not monitored, but there have been anecdotal reports of pilot whales seen near squid fishing operations in recent years.

C. Contaminants

Long-finned pilot whales from both sides of the north Atlantic carry high levels of organochlorine contaminants (pesticides such as DDT and PCB) in their tissues. Concentrated organochlorines may impair reproduction or increase susceptibility to disease. Studies are continuing on the effects these compounds have on marine mammals.

Accumulations of cadmium and mercury are also present in the tissues of long-finned pilot whales from the Faeroe Islands. As top predators, pilot whales are a repository of these heavy metals accumulating through the marine food chain. Pilot whales seem unusually tolerant to elevated levels of these metals, as studies have not yet revealed a major toxicity problem in these species.

D. Captivity

Numerous short-finned pilot whales have been captured for training and/or display in the United States and other parts of the world. Whales have been taken from northeast United States, California, Hawaii, and Japan by means of direct capture, incidental to drive fisheries, or stranding salvage. The most recently published North American marine mammal display census dates from 1990. At that time, four pilot whales were on exhibit in United States and Canadian aquaria. Stranded pilot whales have occasionally been kept for rehabilitation and subsequently released.

See Also the Following Articles

Delphinids, Overview ▪ Echolocation ▪ Sexual Dimorphism ▪ Stranding

References

Amos, B., Schlotterer, C., and Tauz, D. (1993). Social structure of pilot whales revealed by analytical DNA profiling. *Science* **260**, 670–672.

Bernard, H. J., and Reilly, S. B. (1999). Pilot whales. *In* "Handbook of Marine Mammals" (S. H. Ridgeway and R. Harrison, eds.), Vol. 6, pp. 245–279. Academic Press, London.

Donovan, G. P., Lockyer, C. H., and Martin, A. R. (eds.) (1993). "Biology of Northern Hemisphere Pilot Whales." *Rep. Int. Whal. Commn*, Spec. Issue No. 14, Cambridge.

Duignan, P., House, C., Geraci, J. R., Early, G., Copland, H. G., Walsh, M. T., Bossart, G. D., Cray, C., Sadove, S., St. Aubin, D. J., and Moore, M. (1995). Morbillivirus infection in two species of pilot whales (*Globicephala* sp.) from the western Atlantic. *Mar. Mamm. Sci.* **11**, 150–162.

Rendell, L. E., Matthews, J. N., Gill, A., Gordon, J. C. D., and Macdonald, D. W. (1999). Quantatative analysis of tonal calls from five odontocete species, examing interspecific and specific variation. *J. Zool. Lond.* **249**, 403–410.

Reynolds, J. E., III, and Rommel, S. A. (eds.) (1999). "Biology of Marine Mammals." Smithsonian Institution Press, Washington, DC.

Pinnipedia, Overview

ANNALISA BERTA
San Diego State University, California

Pinnipeds have always been understood to represent a group distinct from other aquatic mammals. They are recognized as members of the mammalian order Carnivora and include three monophyletic lineages; the Otariidae (fur seals and sea lions), the Odobenidae (the walruses), and the Phocidae (true or earless seals). Pinnipeds comprise slightly more than one-fourth (28%) of the species level diversity of marine mammals. Thirty-three to 37 living species of pinnipeds are distributed throughout the world: 18 phocids, 14–18 otariids, and the walrus. One additional species of modern phocid and a modern otariid are reported extinct in historical time (Rice, 1998).

I. Systematics and Distribution

A. Otariidae: Fur Seals and Sea Lions

Of the two groups of seals, otariids are characterized by the presence of external ear flaps or pinnae, and for this reason they are often called "eared" seals (Fig. 1). Otariids can turn

a   b

Figure 1 *Representative otariids, (a) southern sea lion,* Otaria flavescens, *and (b) South African fur seal,* Arctocephalus pusillus, *illustrating pinna. Note also the thick, dense fur characteristic of fur seals. Males are shown behind smaller females. Illustrations by P. Folkens. From Berta and Sumich (1999).*

their hind flippers forward and use them to walk. Otariidae typically are divided into two subgroupings: Otariinae (sea lions) and Arctocephalinae (fur seals). Five living genera and seven species of sea lions are recognized, occurring in both the Northern and Southern Hemispheres: *Eumetopias jubata* (northern sea lion), *Neophoca cinerea* (Australian sea lion), *Otaria flavescens* (southern sea lion), *Zalophus californianus* (California sea lion), *Z. japonicus* (Japanese sea lion), *Z. wollebacki* (Galapagos sea lion), and *Phocarctos hookeri* (New Zealand sea lion; Fig. 2). Fur seals, named for their thick, dense fur, are divided into two genera. The genus *Arctocephalus,* or southern fur seals, consists of eight species: *A. australis* (South American fur seal), *A. forsteri* (New Zealand fur seal), *A. gazella* (Antarctic fur seal), *A. galapagoensis* (Galapagos fur seal), *A. philippii* (Juan Fernandez fur seal), *A. pusillus (A. p. pusillus*

South African fur seal and *A. p. doriferus* Australian fur seal), *A. townsendi* (Guadalupe fur seal), and *A. tropicalis* (sub-Antarctic fur seal). The genus *Callorhinus* is monotypic; a single extant species is recognized, *C. ursinus*. All of the fur seals except the northern and Guadalupe fur seals are found in the Southern Hemisphere. The northern fur seal is found in sub-Arctic waters of the North Pacific, with the exception of a small population on San Miguel Island off California (Fig. 3).

The monophyly of otariids is based on a combination of cranial, postcranial, and soft anatomical characters (Berta and Sumich, 1999). Relationships among otariids based on morphology conflict with those based on molecular data, although as yet the latter is based on an incomplete taxonomic sample (Fig. 4). Morphologic data indicate that only the sea lions (Otariinae) are monophyletic with a sister group relationship

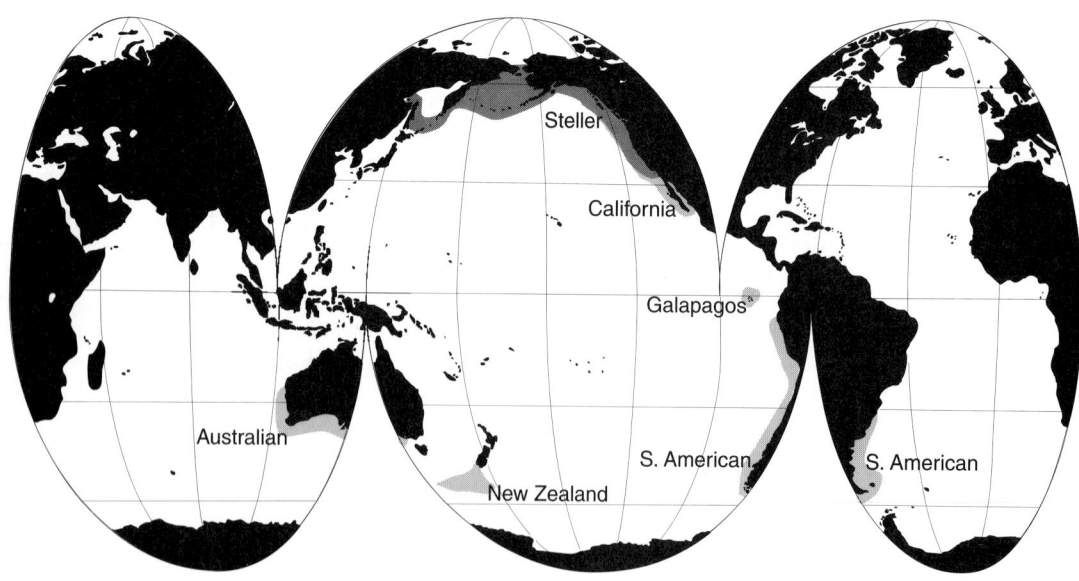

Figure 2 *Distribution map of sea lions. From Berta and Sumich (1999) based on Riedman (1990).*

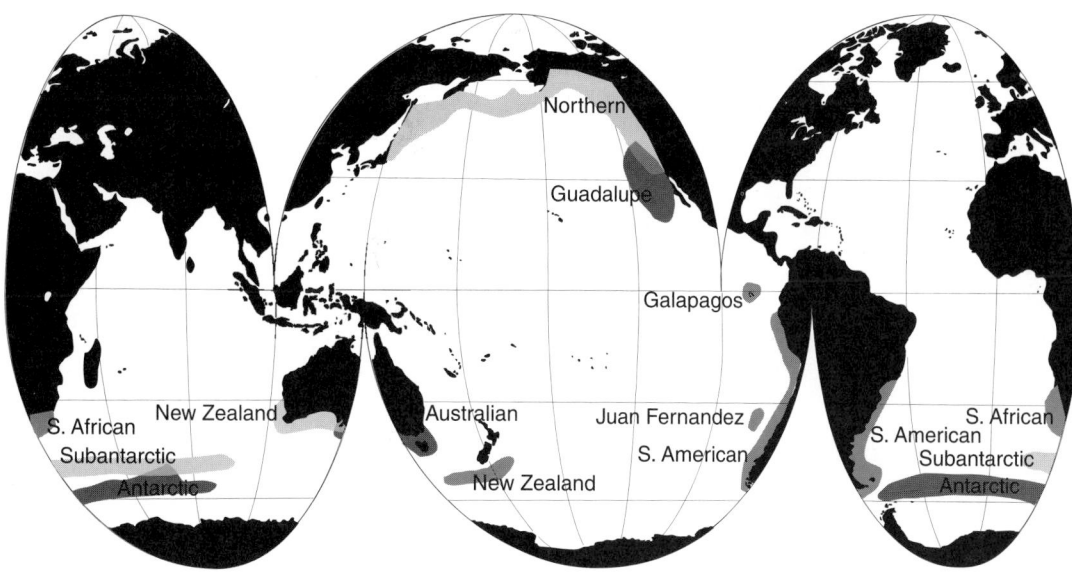

Figure 3 *Distribution map of fur seals. From Berta and Sumich (1999) based on Riedman (1990).*

suggested with only some fur seals (i.e., *Arctocephalus* spp.). Thus far, morphologic data have been unable to resolve relationships among *Arctocephalus* species. Relationships among sea lions based on morphology suggest a pairing of *Otaria* and *Phocarctos* with *Neophoca*, *Eumetopias*, and *Zalophus* as successive outgroups. *Callorhinus* and the extinct taxon *Thalassoleon* are positioned as sequential sister taxa to this clade. Molecular studies have revealed that neither fur seals nor sea lions are monophyletic. In support of these results is evidence for hybridization of various species (e.g., *Arctocephalus gazella/A. tropicalis*, *Zalophus californianus/Otaria flavescens*). Additionally, the violent sexual behavior by male sea lions toward females of different species may have resulted in more hybridization and introgression that has been typically recognized for the evolutionary history of otariids.

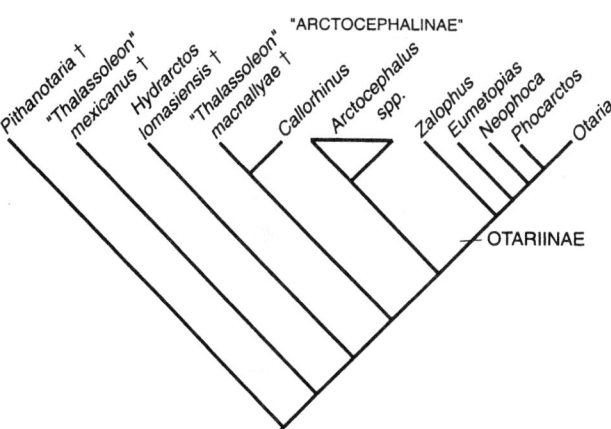

Figure 4 *Phylogeny of Otariidae based on morphologic data showing monophyletic Otariinae and paraphyletic "Arctocephalinae." From Berta and Sumich (1999).*

One modern species of sea lion is probably extinct, *Zalophus japonicus* (Japanese sea lion); its last well-documented record of occurrence was in 1951.

B. Odobenidae: Walruses

Although tusks are arguably the most characteristic feature of the modern WALRUS, *Odobenus rosmarus*, a rapidly improving fossil record indicates that these unique structures evolved in a single lineage of walruses and "tusks do not a walrus make." The living walrus is the sole survivor of a what was once a diverse radiation of at least 10 genera and 13 species of walruses that lived from the early Miocene to the end of the Pliocene (Deméré, 1994). Two subspecies of the modern walrus are recognized: *O. r. rosmarus* (Atlantic walrus) and *O. rosmarus divergens* (Pacific walrus). Pacific walruses are more abundant, are larger, and have longer tusks than Atlantic walruses (Fay, 1981). Walruses inhabit the Northern Hemisphere in areas with pack ice over shallow water of the continental shelf (Fig. 5). Like phocids, walruses lack external ear flaps. A unique feature of members of the modern walrus lineage are enlarged upper canine tusks that function primarily in breeding and social contexts. Walrus locomotion combines elements of phocid and otariid locomotion.

Monophyly of the walrus family is strongly supported, although there is controversy regarding whether walruses are related more closely to otariids or to phocids (Berta and Sumich, 1999). The modern walrus belongs to one of two monophyletic clades of walruses, the Odobeninae, and is most closely related to *Valenictus,* from the late Miocene to late Pliocene of southern and central California (Fig. 6). Several uniquely derived features of *Valenictus* indicate that it diverged from a common ancestry with *Odobenus* prior to pursuing its own evolutionary path toward toothlessness.

A survey of genetic variation among Atlantic and Pacific populations of the walrus suggests separation of the subspecies

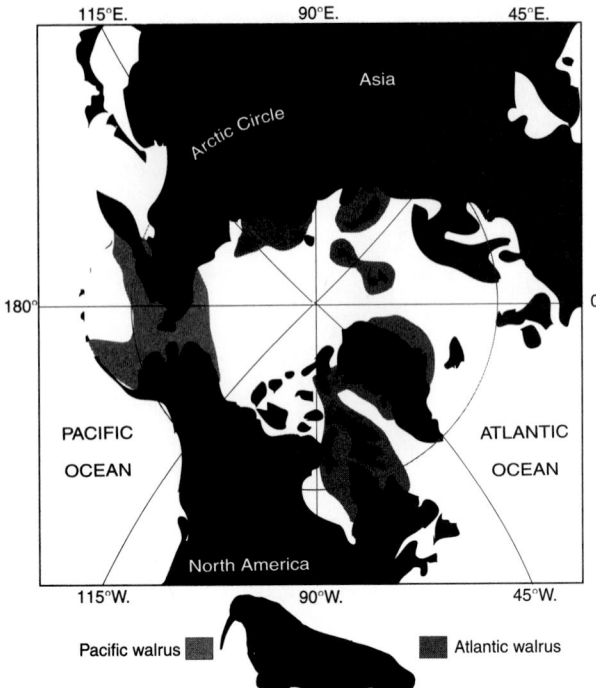

Figure 5 *Distribution of modern walrus subspecies,* Odobenus rosmarus divergens *(Pacific walrus) and* Odobenus rosmarus rosmarus *(Atlantic walrus). From Berta and Sumich (1999).*

about 500,000–750,000 years ago, supporting the suggestion that *Odobenus* evolved in the Pacific and reached the North Atlantic early in Pleistocene time (Cronin *et al.*, 1994).

C. Phocidae: Seals

The second major grouping of seals, the phocids, often are referred to as the "true" or "earless" seals for their lack of visible ear pinnae, a characteristic that readily distinguishes them

from otariids as well as the walrus (Fig. 7). Among the most distinguishing phocid fur seals is their inability to turn the hindlimbs forward to support the body, resulting in a peculiar crawling locomotion on land. Phocids inhabit both Northern and Southern Hemispheres, although they are largely restricted to polar and subpolar regions (Fig. 8). Among pinnipeds, phocids are unique in their ability to survive in estuarine and freshwater habitats (e.g., Caspian and Baikal seals inhabiting land-locked lakes).

Traditionally, phocids have been divided into two or four major subgroups (including the Monachinae, Lobodontinae, Cystophorinae, and Phocinae). Only the Phocinae is universally accepted as being monophyletic (Berta and Sumich, 1999; Fig. 9).

The "Monachinae" subgroup of "southern seals" typically includes *Monachus* (monk seals), *Mirounga* (elephant seals), and the Lobodontini (Antarctic seals). Monophyly of the Monachinae is disputed (Wyss, 1988; Bininda-Emonds and Russell, 1996). Basal phocids have been identified as being *Monachus* spp. (not recognized as a monophyletic taxon by all workers) or alternatively as *Mirounga*. Three species of *Monachus* have been described: *M. schauinslandi* (Hawaiian monk seal), *M. monachus* (Mediterranean monk seal), and the recently extinct *M. tropicalis* (Caribbean monk seal). According to Wyss (1988), *Monachus* is a paraphyletic taxon, and based on morphology of the ear region, *M. schauinslandi* is basal to the other two species (Fig. 9a). Bininda-Emonds and Russell (1996) argue that *Monachus* is monophyletic and, on the basis of other morphologic characters, position *M. monachus* as the basal taxon (Fig. 9b). Molecular sequence data for species of *Monachus* are currently available only for *M. schauinslandi* and thus offer no resolution of relationships among this species group. Elephant seals, named for their enlarged proboscis, are represented by two species: *Mirounga angustirostris* (northern elephant seal) and *Mirounga leonina* (southern elephant seal).

There is also disagreement regarding monophyly of the Lobodontini, which include *Leptonychotes weddellii* (Weddell seal), *Lobodon carcinophaga* (crabeater seal), *Hydrurga lep-*

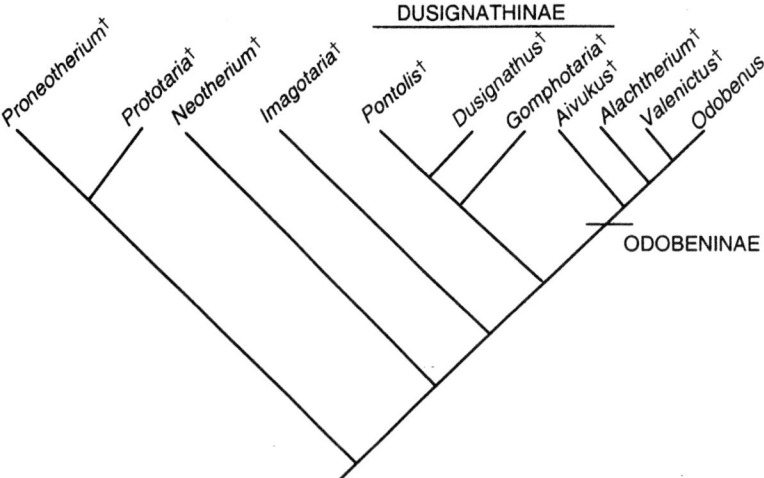

Figure 6 *Phylogeny of Odobenidae. From Deméré (1994) and Deméré and Berta (2001).*

a b

c d

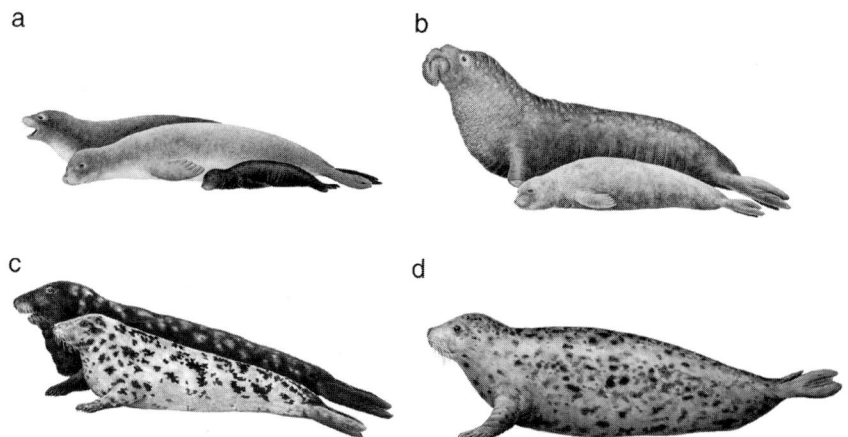

Figure 7 *Representative "monachines" (a) Hawaiian monk seal,* Monachus schauins-landi, *and (b) northern elephant seal,* Mirounga angustirostris, *and phocines (c) harbor seal,* Phoca vitulina, *and (d) gray seal,* Halichoerus grypus. *Males are shown behind smaller females. Illustrations by P. Folkens. From Berta and Sumich (1999).*

tonyx (leopard seal), and *Ommatophoca rossii* (Ross seal). Although Wyss' data support monophyly of this group, relationships among taxa were unresolved. According to Bininda-Emonds and Russell (1996) the Lobodontini are paraphyletic. Their most parsimonious tree (bootstrap analysis) supports a *Lobodon–Monachus* pairing with *Ommataphoca, Leptonychotes,* and *Hydrurga* forming successive outgroups to this clade (Fig. 9b).

Traditionally the Phocinae subgroup of "northern seals" includes *Erignathus barbatus* (bearded seal), *Cystophora cristata* (hooded seal), *Halichoerus grypus* (gray seal), *Phoca* (including among others harbor and spotted seal), *Pusa hispida* (ringed seal), *Histriophoca fasciata* (ribbon seal), and *Pagophilus groenlandicus* (harp seal). Wyss' cladistic analysis supports *Erignathus* and *Cystophora* as successive sister taxa to the Phocini (*Pusa, Histriophoca, Pagophilus, Halichoerus,* and *Phoca*). Although Bininda-Emonds found support for the basal position of *Cystophora,* this was not the case for *Erignathus,* which moved to a sister group relationship with *Pagophilus + Histricophoca* (Fig. 9b). There is disagreement about recognition of the Phocini; paraphyly of several taxa has been proposed based on both morphologic and molecular data. According to Arnason *et al.* (1995) molecular data, Phocini are divided into an earlier diverging lineage consisting of *Histriophoca* and *Pagophilus* and the other composed of various species of *Phoca* (rendering the genus paraphyletic) and *Halichoerus* (Fig. 9c). Bininda-Emonds and Russell (1996) acknowledge weak support and poor resolution among Phocini.

The harbor seal (*Phoca vitulina*) has the most extensive geographic distribution of any seal, with a range spanning over 16,000 km from the east Baltic, west across the Atlantic and Pacific Ocean to southern Japan. The population structure of the harbor seal studied by Stanley *et al.* (1996) revealed that populations in the Pacific and Atlantic Oceans are highly differentiated. Mitochondrial data are consistent with the ancient isolation of populations in both oceans coincident with the development of continental glaciers and extensive sea ice. In the Atlantic and Pacific Oceans, populations appear to have been established from west to east, with the European populations showing the most recent common ancestry.

II. Anatomy and Physiology

Pinniped aquatic specializations include their STREAMLINED shape, reduced external ear pinnae, paddle-like limbs and feet, small tail, and genital organs and mammary glands that are retracted beneath the skin. In comparison to most terrestrial carnivorans, pinnipeds are large, which helps conserve warmth. Pinnipeds, particularly phocids show tremendous diversity in size, ranging from the smallest pinniped, the Baikal seal, *Pusa sibirica,* reaching a length of just over a meter and a weight of 45 kg to the largest pinniped, the elephant seals, nearly 5 m in length (adult males) and up to 3200 kg (Bonner, 1990). Pinnipeds are ecologically diverse with habitats ranging from shelf to surface waters in tropical and polar seas, with some species living in freshwater lakes, whereas others move freely between rivers and the ocean.

Phocids and the walrus have lost much of their HAIR (fur) and are characterized by thick layers of blubber under the skin. Otariids, especially fur seals, have retained a thick fur coat. Color patterns in the pelage of pinnipeds occur almost exclusively among phocids. Ice-breeding seals (e.g., ribbon seal, harp seal, hooded seal, ringed seal, crabeater seal, Weddell seal, and leopard seal) show contrasting dark and light or disruptive color patterns (Fig. 10). The uniform coloration of some pinnipeds (e.g., white harp seal pups) allows them to blend readily into their arctic environment. Pinnipeds come ashore for birthing and molting. All phocid seals undergo an annual molt. Fur seals and sea lions instead renew their pelt gradually all year.

Vibrissae, or whiskers, are stiff hairs that occur on the face. Most prominent are the mystacial whiskers, which range in size from the short stiff bristles of the walrus to the very long, fine bristles of fur seals (Fig. 11). Vibrissae function as sensitive

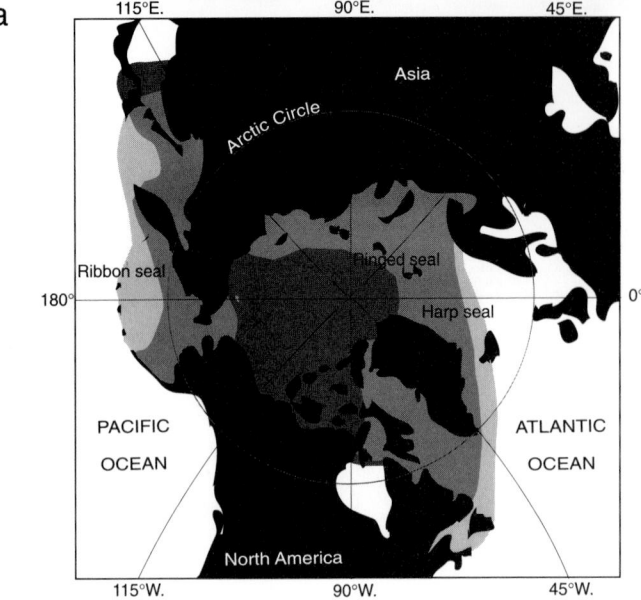

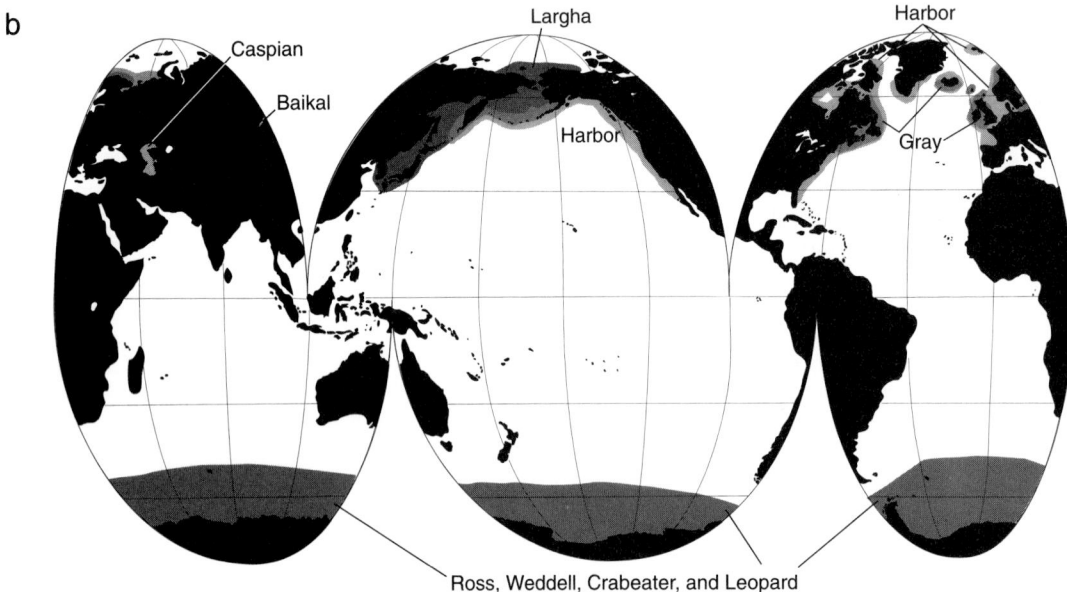

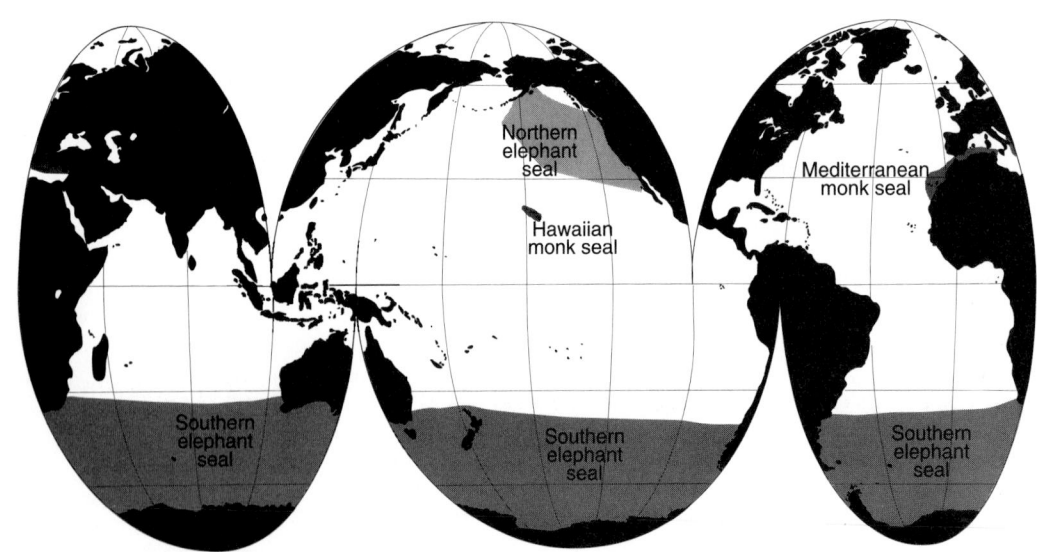

Figure 8 *Distribution of some Arctic phocines. (a), Antarctic lobodontine and phocines (b), and "monachines" (c). From Berta and Sumich (1999) based on Riedman (1990).*

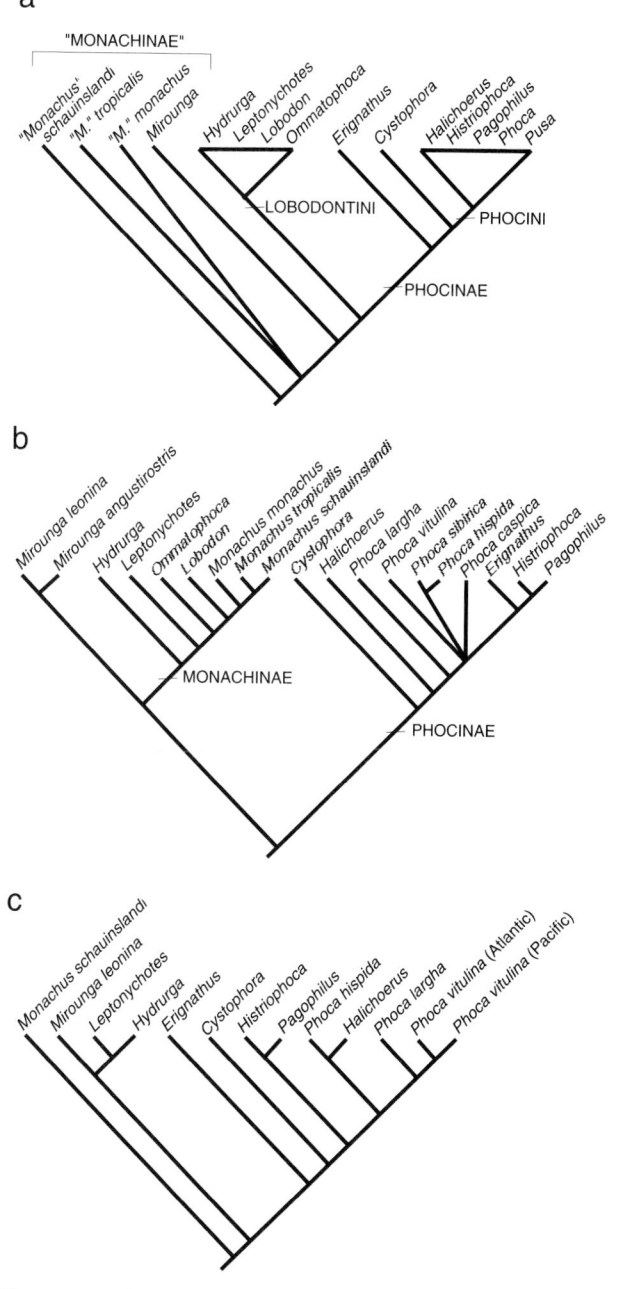

ter. Among these solutions are a spherical body and a resultant decreased surface-to-volume ratio and increased insulation (thick BLUBBER or fur). In addition, heat exchange systems occurring in the flippers, in the fins, and (as discovered recently) in the reproductive tracts of pinnipeds conserve body heat.

Among modern pinnipeds, aquatic and terrestrial locomotion are achieved differently. Three distinct patterns of SWIMMING are recognized: (1) pectoral oscillation (forelimb swimming) seen in otariids where the forelimbs are used in a "flapping" manner to produce thrust, (2) pelvic oscillation (hindlimb swimming) seen in phocids where the hindlimbs are the major propulsors, and (3) a variant of pelvic oscillation exhibited by the walrus where the hindlimb are the dominant propulsive force and the forelimbs are used as rudders or

Figure 9 *Alternative phylogenies for Phocidae based on morphologic data (a) from Wyss (1988) and Berta and Wyss (1994); (b) from Bininda-Emonds and Russell (1996); and (c) molecular data from Arnason et al. (1995).*

touch receptors. Research on the Baltic ringed seal (*P. hispida*) has shown that they have exceptionally well-developed vibrissae, which help them find their way in the dark and often cloudy water beneath the ice (Hyvarinen, 1995). A single vibrissae of the Baltic ringed seal contains more than 10 times the number of nerve fibers typically found in that of a land mammal.

Pinnipeds, like other marine mammals, have evolved ways to accommodate the immense heat loss that occurs in the wa-

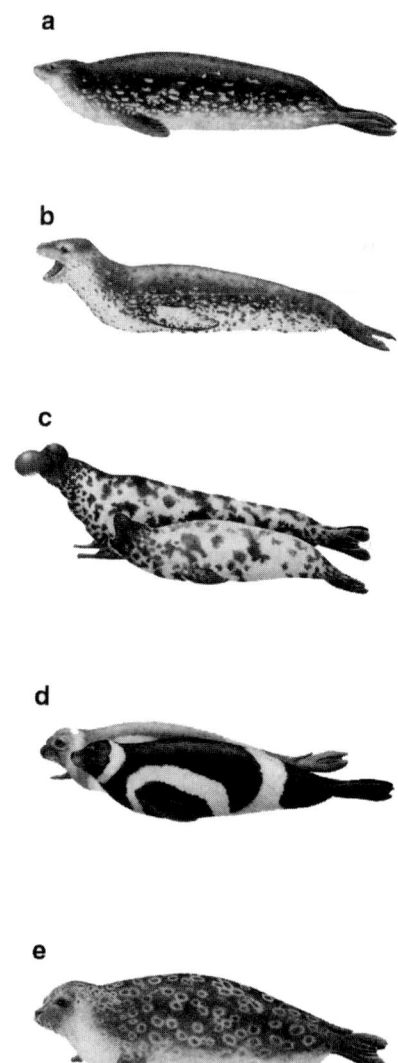

Figure 10 *Examples of phocid pelage patterns: (a) Weddell seal, Leptonychotes weddellii; (b) leopard seal, Hydrurga leptonyx; (c) hooded seal, Cystophora cristata, male in back; (d) ribbon seal, Histriophoca fasciata; and (e) ringed seal, Pusa hispida. Illustrations by P. Folkens, in Berta and Sumich (1999).*

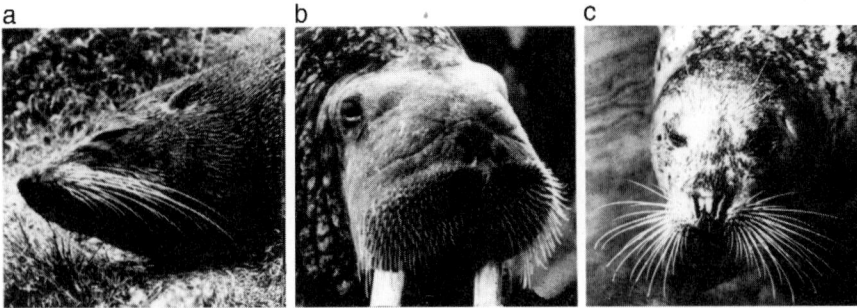

Figure 11 *Heads of various pinnipeds showing facial vibrissae: (a) New Zealand fur seal,* Arctocephalus forsteri, *(b) walrus,* Odobenus rosmarus, *and (c) Pacific harbor seal,* Phoca vitulina richardii. *From Ling (1977) in Berta and Sumich (1999).*

paddles. There is a major difference in locomotion on land between phocids on the one hand and otariids and walruses on the other. The inability of phocids to turn the hindlimbs forward results in forward progression by vertical undulations of the trunk, which do not involve the hindlimbs. In walruses, as in otariids, the hindlimbs can be rotated forward in terrestrial locomotion.

Pinnipeds are carnivores; most are generalists feeding predominantly on fish and squid. Several pinnipeds, notably crabeater and leopard seals, have highly modified cheek teeth with complex cusps to trap and strain krill. Leopard seals also possess well-developed canines for preying on birds and other pinnipeds. Walruses are specialists that feed almost exclusively on clams using a suction-feeding strategy in which the muscular tongue acts as a piston, creating low pressure in the mouth cavity. Some pinnipeds, e.g., elephant seals, rival gray whales (*Eschrichtius robustus*) in the distances traveled in migration (18,000–21,000 km) to forage offshore between breeding seasons. The advent of microprocessor-based time and depth recorders and satellite telemetry has enabled the documentation of details of the foraging behavior of these deep-diving seals.

Among pinnipeds are found the most extraordinary of marine mammal divers. Average dives of small species such as the Ross seal are just under 10 min in duration, increasing to over 1 hr for the northern elephant seal and the Weddell seal. Maximum depths vary from less than 100 m in the Guadalupe fur seal to more than 1500 m in northern elephant seal males. Some seals (in addition to sperm whales *Physeter macrocephalus*, sea turtles, and some penguins) are "incredible diving machines" with unique ways of budgeting their oxygen supply and responding to pressure.

Sounds produced by pinnipeds include air-borne and underwater vocalizations. Air-borne sounds vary from grunts, snorts, or barks identified as either mother–pup calls or threat calls among seals to the distinctive gong-like sounds produced by male walruses striking throat pouches with their flippers as part of a courting display during the breeding season. Pinnipeds produce a variety of underwater sounds that appear related to breeding activities and social interactions. Among these are the whistles, trills, chirps, and buzzes of Weddell seals that are used in territorial defense. These contrast with the soft lyrical calls of leopard seals that may be related to their solitary social system.

In contrast to toothed whales, pinnipeds have not been found to use ECHOLOCATION in their natural surroundings.

The pinniped eye is adapted for VISION both above and under water. The spherical lens, thick retina, and the well-developed tapetum lucidum increase light sensitivity. With the exception of the walrus, which has small eyes, seals and sea lions have large eyes in relation to body size. The question of whether pinnipeds have color vision is still debated, although behavioral experiments and the presence of both rods and cones in the retina have been documented in some species (e.g., California sea lion, spotted seal, walrus).

III. Behavior

Unlike other marine mammals, pinnipeds differ in their need to return to land (or to ice) to give birth. Many pinnipeds (e.g., elephant seal) are extremely polygynous, with successful males mating with dozens of females in a single breeding season. Species that are polygynous tend to breed in large colonies on land where males compete for breeding territories (in otariids) or establish dominance hierarchies (in elephant seals). Because these males must compete for access to females, associated with extreme polygyny is the strong SEXUAL DIMORPHISM seen in elephant seals, including large body size (adult males as much as five times as large as females), elongated proboscis, enlarged canine teeth, and thick skin on the neck. Other phocids (e.g., Weddell seal, harp, ringed, ribbon, bearded, hooded) mate in the water or on ice and show a reduced level of polygyny, which is explained in part by the difficulty in gaining access to females in unstable environments such as pack ice.

Pinnipeds are characterized by sexual bimaturity with females reaching sexual maturity before males. In polygynous species, males require several years of physical maturation following sexual maturity before they successfully compete for access to females. Gestation in most species of pinnipeds averages between 10 and 12 months; walruses have the longest gestation period of 16 months. Most species regulate their reproductive cycle by delayed implantation (from 1.5 to 5 months). Delayed implantation prolongs birth until conditions are more favorable for offspring survival. Pinniped females of all species give birth to a single pup. In most species, pupping occurs in spring or summer.

The maternal behaviors and lactation strategies of pinnipeds are influenced by their breeding habitat whether on ice or land. Most phocids exhibit a fasting strategy where females fast completely and remain out of water for the duration of a relatively short lactation, ranging from less than 1 week in hooded seals to almost 8 weeks in the Weddell seal. It has been suggested that the unstable nature of pack ice has selected for the extremely short lactation periods of some ice-breeding phocid seals. To compensate for the brevity of lactation, the milk produced by these species is energy dense, with a fat content up to 60% in some species (i.e., hooded and harp seals). Rapid pup growth is ensured by the richness of the milk. In the foraging cycle strategy of most otariids, mothers fast for only a few days following the birth of pups. Then the mothers begin foraging trips at sea, leaving their pups for a few days at a time. The lactation period of otariids are longer, ranging from several months to more than 1 year, and milk is generally less energy dense than in phocids (e.g., milk fat content averaging between 24 and 40%). Walruses exhibit a variant of the otariid strategy, termed the aquatic nursing strategy, in which walrus pups accompany their mothers on foraging trips into the water. The length of lactation in walruses is the longest among pinnipeds, lasting 2 or 3 years.

Among generalizations that can be made about pinniped longevity is that females, especially those of polygynous species, tend to live longer than males. In many cases, males do not survive even to the delayed age of sexual maturity. Seals have been known to pup successfully at 24–25 years and live as long as 40 years or more. Significant factors implicated in the natural mortality of pinnipeds include disease (especially morbilliviruses), PREDATION (e.g., white sharks, killer whales), starvation, and PARASITES.

See Also the Following Articles

Eared Seals ■ Earless Seals ■ Locomotion, Terrestrial ■ Reproductive Behavior

References

Árnason, Ú., Bodin, K., Gullberg, A., Ledje, C., and Mouchaty, S. (1995). A molecular view of pinniped relationships with particular emphasis on the true seals. *J. Mol. Evol.* **40**, 78–85.

Berta, A., and Sumich, J. L. (1999). "Marine Mammals: Evolutionary Biology," Academic Press, San Diego, CA.

Berta, A., and Wyss, A. R. (1994). Pinniped phylogeny. *Proc. San Diego Soc. Nat. Hist.* **29**, 33–56.

Bininda-Emonds, O. R. P., and Russell, A. P. (1996). A morphological perspective on the phylogenetic relationships of the extant phocid seals (Mammalia: Carnivora: Phocidae). *Bonn. Zool. Monogr.* **41**, 1–256.

Bonner, W. N. (1990). "The Natural History of Seals." Facts on File, New York.

Cronin, M. A., Hillis, S., Born, E. W., and Patton, J. C. (1994). Mitochondrial DNA variation in Atlantic and Pacific walruses. *Can. J. Zool.* **72**, 1035–1043.

Deméré, T. A. (1994). The family Odobenidae: A phylogenetic analysis of fossil and living taxa. *Proc. San Diego Soc. Nat. Hist.* **29**, 99–123.

Deméré, T. A., and Berta, A. (2001). A re-evaluation of *Proneotherium repenningi* from the middle Miocene Astoria Formation of Oregon and its position as a basal odobenid (Pinnipedia: Mammalia). *J. Vertebr. Paleontol.* **21**(2), 279–310.

Fay, F. H. (1981). Walrus: *Odobenus rosmarus. In* "Handbook on Marine Mammals" (S. H. Ridgway and R. J. Harrison, eds.), Vol. 1, pp. 1–23. Academic Press, New York.

Hyvarinen, H. (1995). Structure and function of the vibrissae of the ringed seal (*Phoca hispida* L.). *In* "Sensory Systems of Aquatic Mammals" (R. A. Kastelein, J. A. Thomas, and P. E. Nachtigall, eds.), pp. 429–445. De Spil Publishers, Woerden, The Netherlands.

Miller, E. H. (1996). Violent interspecific sexual behavior by male sea lions (Otariidae): Evolutionary and phylogenetic implications. *Mar. Mamm. Sci.* **12**(3), 468–476.

Rice, D. W. (1998). Marine Mammals of the World. Special Publication 4, Society of Marine Mammalogy, Allen Press, Kansas.

Riedman, M. (1990). "The Pinnipeds: Seals, Sea Lions and Walruses." University of California Press, Berkeley.

Stanley, H., Casey, S., Carnahan, J. M., Goodman, S., Harwood, J., and Wayne, R. K. (1996). Worldwide patterns of mitochondrial DNA differentiation in the harbor seal (*Phoca vitulina*). *Mol. Biol. Evol.* **13**(2), 368–382.

Wyss, A. R. (1988). On "retrogression" in the evolution of the Phocinae and phylogenetic affinities of the monk seals. *Am. Mus. Novit.* **2954**, 1–38.

Pinniped Ecology

W. Don Bowen
Department of Fisheries and Oceans, Dartmouth, Nova Scotia, Canada

Carrie A. Beck and Deborah A. Austin
Dalhousie University, Halifax, Nova Scotia, Canada

Ecology deals with the interactions between individuals and their environment. In this context, environment is taken broadly to include other organisms and the physical environment. These interactions take place at various spatial and temporal scales and influence both the abundance and the distribution of individuals. However, ecology is also a historical science in that the patterns we see today reflect past events and phylogenetic relationships. Thus, processes acting on both evolutionary and ecological time scales have undoubtedly influenced many of the characteristics of pinniped ecology we see today. Pinnipeds are large, long-lived, aquatic mammals exhibiting delayed sexual maturity and reduced litter size; a single precocial offspring is the norm. As such, they share many of the demographic features of other large mammals. Population numbers do not change dramatically from year to year, and numbers are most sensitive to changes in adult survival, followed by juvenile survival and fecundity (Eberhardt and Siniff, 1977). We assume that these characteristics are under selection and that variability in foraging success affects survival probability and reproductive performance of individuals. Nevertheless, uncertainty about historical condition and influences limits the extent to which present characteristics can be interpreted as evolutionary adaptations. Inevitably, discussions of pinniped ecology and other aspects of pinniped biology will overlap. This article focuses on five aspects of pinniped ecology: abundance, distribution, reproduction, foraging, and the ecological roles of pinnipeds in aquatic ecosystems.

I. Abundance

Despite interest in the ecology of pinnipeds, the abundance of many species is poorly known. The abundance of species of commercial importance (past, northern fur seals, *Callorhinus ursinus;* or present, harp seals, *Pagophilus groenlandicus*) is generally better known than for those species that have not been exploited. The accuracy and precision attached to the estimates of abundance vary greatly due to the difficulty in designing good census procedures or to the lack of effort to obtain good estimates. Good estimates of abundance are important because both abundance and trends in abundance are perhaps the most useful indicators of the status of populations.

Commercial exploitation decimated many pinniped species, in some cases to levels nearing extinction (e.g., northern elephant seals, *Mirounga angustirostris*). Over the past several decades or more, some species have recovered or are continuing to recover. Thus, the present abundance of heavily exploited species may not be a good measure of their preexploitation numbers. Pinniped species range over 4 orders of magnitude in abundance, from the crabeater seals (*Lobodon carcinophaga*) at about 12 million (probably the most abundant marine mammal in the world) to the Mediterranean monk seal (*Monachus monachus*) at probably fewer than several hundred individuals (Reijnders *et al.*, 1993). Phocid species are generally more abundant than otariids, with 15 of 18 phocid species numbering greater than 100,000 individuals compared with only 8 of 14 otariid species (Bowen and Siniff, 1999). The reasons for this difference are not entirely clear. Over the past 100 years, both families have been exploited commercially and subjected to other human factors that might have influenced abundance. More likely, the greater abundance of phocids is the result of their greater use of high-productivity areas in temperate and polar waters than is the case in most otariid species. The three most abundant otariids, the northern fur seal, the Antarctic fur seal (*Arctocephalus gazella*), and the Cape fur seal (*A. pusillus pusillus*), all forage in seasonally productive, high-latitude ecosystems, a characteristic shared with the most abundant phocid species (i.e., the ringed seal *Pusa hispida*, the harp seal, and the crabeater seal).

Abundance is determined by the movement of individuals (in and out of the population), births, and deaths. These processes are influenced by both direct and indirect human activities, as well as by ecological factors such as PREDATION, food supply, breeding habitat, disease, competition with other species, and environmental variability. In the absence of human effects, combinations of these ecological factors are thought to regulate the abundance of a population about a level known as carrying capacity. Although the concept of carrying capacity has a long history in ecology, with the possible exception of the Weddell seal (*Leptonychotes weddellii*), there is little evidence that populations of pinnipeds fluctuate about some long-term average

level. This lack of evidence may be an artifact of the effects of previous exploitation on population trends and the fact that few populations have been surveyed over many decades. For example, a number of pinniped populations recovering from human exploitation (e.g., Antarctic fur seal, gray seal *Halichoerus grypus*, harbor seal *Phoca vitulina*) have increased at rates in excess of 12% per year over several decades or more. At Sable Island, Canada, the number of gray seal pups born each year has increased exponentially, with a doubling time of about 6 years, for more than 40 years. It is also clear that pinniped populations may decline rapidly as a result of epizootics, such as the phocine distemper virus that killed large numbers of harbor seals in the North Sea, and during short-term extreme changes in ocean climate, such as El Niño (see later).

Ecosystem structure and function are influenced by both top-down (i.e., consumer driven) and bottom-up (i.e., producer driven) processes. Pinnipeds likely exert top-down control on some ecosystems through predation (see Section IV) and are affected by changes in food available brought about by changes in primary and secondary productivity.

An example of a top-down ecosystem perturbation affecting pinniped abundance occurred in the Southern Ocean in the early 1900s. The overexploitation of some species of seals and whales led to an enormous uncontrolled "experiment" in this cold-water ecosystem. A high biomass of Antarctic KRILL is the cornerstone of the Southern Ocean food web, accounting for half of the total zooplankton biomass. Of the six species of pinniped that inhabit the Southern Ocean, crabeater seals, Antarctic fur seals, and leopard seals (*Hydrurga leptanyx*) feed mainly on krill, whereas southern elephant seals (*Mirounga leonina*) and Ross seals (*Ommatophoca rossii*) consume mainly cephalopods and Weddell seals eat primarily fish. Krill is also the main food resource for the resident large baleen whales [blue (*Balaenoptera musculus*), sei (*B. borealis*), minke (*B. bonaerensis*), humpback (*Megaptera novaeanglioe*), fin (*B. physalus*), and southern right (*Eubalaena australis*)]. As the cetacean biomass declined from exploitation by more than 50% between 1904 and 1973, an estimated 150 million t of krill were released annually to the remaining predators (Laws, 1985). The abundance of krill-eating species of pinniped, such as the crabeater seal and the Antarctic fur seal, increased substantially following the massive cetacean exploitation.

Bottom-up effects on top predators such as pinnipeds also can occur rapidly over the course of months. Perhaps the most dramatic example of this occurs during El Niño. El Niño events occur approximately every 4 years in the eastern tropical Pacific, resulting in reduced upwelling and a decrease in primary and secondary productivity. During a severe El Niño, the effects of reduced food availability on seabirds and marine mammals can be quite pronounced. Galapagos fur seals (*Arctocephalus galapagoensis*) have an unusually long lactation period of approximately 2 years that is thought to have evolved to buffer young against minor El Niño events. However, during the severe El Niño event between August 1982 and July 1983, pup production of Galapagos fur seals was only 11% of previous years, and no pups survived past the first 5 months. Adult females responded by increasing the foraging trip length, while most adult males did not appear on the BREEDING SITE and were unable to hold territories during the breeding season (Trillmich and Ono, 1991).

Top-down and bottom-up effects at times may work in concert to influence the abundance of pinnipeds. Between the 1950s and early 1970s, intensive harvesting in the Bering Sea and the Gulf of Alaska critically reduced the populations of large whales, flatfishes, herring, and other primary consumers of krill and zooplankton. It is believed that the resulting increase in the availability of krill and other zooplankton, coupled with a regime shift in the ocean climate, favored some species such as walleye pollock and not others. This resulted in a change in the relative abundance of both fish and invertebrate assemblages. There was also a regime shift in ocean climate during the 1970s that likely contributed to changes in the biological productivity of these areas. Although the consequence of these changes on marine mammals remains uncertain, both harbor seal and Steller sea lion (*Eumetopias jubatus*) numbers declined dramatically following these changes, resulting in the Steller sea lion being declared an endangered species. A probable explanation for this cycle of events involves a combination of environmental change affecting producers (bottom-up effect) and human exploitation of predators resulting in changes to the ecosystems that have been detrimental to pinnipeds (Bowen and Siniff, 1999).

II. Distribution

Fundamentally, pinniped distributions reflect the need to give birth on solid substrate (land or ice) and to feed at sea. Within these broad constraints, the distribution of pinnipeds is affected by physical (e.g., ice cover, location of remote islands) and biological (e.g., productivity, abundance of predators) characteristics of habitat, demographic factors (e.g., population size, age, sex, and reproductive status), morphological and physiological constraints, and human effects (e.g., disturbance). Although each of these factors may influence distribution, combinations of factors are generally responsible for the distribution patterns observed. Pinniped distribution is also three dimensional, where the third dimension is water depth and the underlying bathymetry. Although a complete understanding of pinniped distribution must consider this three-dimensional world, this aspect of pinniped behavior is discussed elsewhere.

Pinniped species have a restricted and generally patchy distribution in most aquatic environments: estuaries and continental shelves (e.g., gray seals); tropical seas (e.g., monk seals, Galapagos fur seals); the deep ocean (e.g., elephant seals); both Arctic (e.g., ringed seals) and Antarctic polar seas (e.g., crabeater seals, Antarctic fur seals); and freshwater lakes (e.g., Baikal seals, *Pusa sibirica*) (King 1983). It is important to note, however, that our understanding of the distribution of most species is based primarily on the location of breeding colonies. We know less about where most species forage at sea such that our view of the overall distribution of most species is incomplete. For example, based on the location of breeding colonies, northern elephant seals range from Baja California to central California. However, satellite TELEMETRY studies show that this species forages over broad areas of the North Pacific Ocean for much of the year.

If we examine the distribution of pinniped breeding colonies, it is clear that we see patterns that reflect the evolutionary history of pinnipeds and the distribution of resources. At large scales, both sea lion and fur seal distributions reflect their origins in the Pacific Ocean. Northern fur seals and Steller sea lions are widely distributed along both sides of the North Pacific Ocean. The California sea lion (*Zalophus californianus*) and Japanese sea lion (*Z. japonicus*) are or were endemic to opposite sides of the North Pacific, and the Galapagos sea lion occurs at the equator. The four other species of sea lions occupy colonies along the west coast of South America, southern Australia, and New Zealand. With the exception of the northern fur seal and Guadalupe fur seal, the other six species of fur seals occur in these tropical or subtropical southern water, but also extend into the cool, nutrient-rich waters of the South Atlantic and Indian Oceans. Sea lion and fur seal breeding colonies are usually located on remote islands near areas of high biological productivity (e.g., northern fur seals, Antarctic fur seals), which provide both protection from mainland predators and nearby food sources. These conditions are particularly important for lactating females.

Species of the family Phocidae are widely distributed in biologically productive temperate and polar seas. Although most abundant in the North Atlantic and Antarctic Oceans, a reflection of their evolutionary origins in the Atlantic basin, phocid species have circumpolar distributions in both the Arctic Ocean (e.g., ringed seal, bearded seal, *Erignathus barbatus*) and the Antarctic Ocean (e.g., Weddell seal, crabeater seal) as well as a broad distribution in the North Pacific Ocean [e.g., harbor seal, larga seal (*Phoca largha*), ribbon seal (*Histriophoca fasciata*)]. Several endangered species also occur in tropical waters (Hawaiian and Mediterranean monk seals, *Monachus* spp.).

Pinnipeds must return to a solid substrate (land or ice) to give birth, rear their offspring, and in many species to molt. For most species, these requirements result in seasonal changes in distribution. In the case of species that breed on pack ice, such as harp and hooded (*Cystophora cristata*) seals and the walrus (*Odobenus rosmarus*), seasonal changes in ice cover virtually guarantee some change in the distribution of individuals. This may partly explain why 7 of 13 (54%) species of pinnipeds that give birth on ice (i.e., most phocid seals and the walrus) are migratory compared to only 4 of 20 (20%) species that give birth on land (2 of 6 phocids, 2 of 14 otariids; Bowen and Siniff, 1999). However, this difference also may be partly explained by the variable quality of data on the at-sea distribution of pinnipeds.

Migration appears to be a common feature of the ice-breeding phocid species, but this behavior is perhaps best documented in the northern elephant seal. This land-breeding species shows extreme sexual size dimorphism, with males being about five times heavier than females. Northern elephant seals undertake the longest known migration and some of the deepest dives reported for a mammal (Stewart and DeLong, 1993). Individual elephant seals make two long-distance migrations of 18,000 to 21,000 km between breeding and molting sites in California and pelagic foraging areas in the North Pacific. Using the California current as a corridor to areas further north, northern elephant seals leave the breeding beaches in southern California for northern offshore foraging areas. The first migration occurs following the breeding season, in which adult male and female elephant seals travel an average of 11,967 and 6289 km, respectively, and remain at sea for an average of 124 and 73 days. After the molt, the seals depart on a second migration. Females are at sea for approximately twice as long as males and cover an average distance of 12,264 km compared to an average of 9608 km by males. males migrate farther north than females, with most males traveling as far as the northern Gulf of Alaska and the eastern Aleutian Islands. These sex differences in foraging distribution, and perhaps diet, may have evolved to reduce competition between females and the much larger males of this species.

III. Reproductive Ecology

The reproductive ecology of pinnipeds varies considerably, reflecting differences in body size, geographic distribution, and habitats used by individual species. Despite this diversity, there are common features that reflect their common ancestry as terrestrial carnivores, and their subsequent adaptation to a predominately aquatic lifestyle.

As noted previously, a conserved trait of their terrestrial ancestry is the requirement for all pinniped species to give birth to their offspring on a solid substrate (land or ice). However, pinnipeds must feed at sea, often some distance from the breeding grounds. This spatial and temporal separation of parturition from aquatic foraging is thought to have played a large role in shaping the mating and lactation strategies of pinnipeds. Three general strategies have evolved to deal with the conflict between at-sea foraging and terrestrial parturition (see later); however, the requirement for terrestrial parturition has likely contributed to some common features of pinniped reproduction, such as birth synchrony.

In most pinniped species, reproduction is seasonal and highly synchronous (e.g., harp seals). The evolution of reproductive synchrony is often associated with seasonal resource availability. In ice-breeding species (e.g., harp and hooded seals), the timing of reproduction is linked to the seasonal availability of sea ice. Seasonal changes in prey abundance and environmental conditions can also influence the timing of parturition and mating. The Hawaiian monk seal (*Monachus schauinslandi*) displays only weak synchrony in reproduction. In this species, births extend over a 6-month period. Given the less variable tropical habitat of this species, reproductive synchrony may not have been under strong selection relative to the species in more variable temperate and polar environments. Subtropical populations of California and Galapagos sea lions and Galapagos fur seals also show slightly less temporal synchrony of reproduction relative to more temperate populations (Boness, 1991).

Other common features of pinniped reproduction include postpartum mating and delayed implantation. These two characteristics of pinniped reproduction also appear to reflect the terrestrial ancestry of the taxa with both features occurring in many modern terrestrial carnivores. However, selection for postpartum mating may have continued as pinnipeds adapted to their aquatic environment. Given the wide-ranging and dispersed distribution of pinniped species during the at-sea foraging season, the aggregation of individuals at pupping colonies may have offered one of the few predictable opportunities for male and females to interact.

Another common feature of pinniped reproduction is the production of a single, precocious offspring; litters of two are rare. Offspring are born with their eyes open and begin to vocalize within minutes of birth. Neonates are also able to move short distances to their mother and to begin suckling shortly after birth. Harbor seal females produce extremely precocial offspring that are capable of swimming and diving with their mothers within an hour of birth (Bowen, 1991).

A. Mating Systems

Within the Pinnipedia, mating systems range from extreme polygyny (e.g., northern fur seals) to sequential defense by males of individual females. The mating system of individual species is closely associated with the dimensionality and stability of the habitat used and the distribution of females at parturition. Broadly speaking, species can be grouped as land-breeding and aquatic-breeding species.

1. Land-Breeding Species　Land-breeding pinniped species include all fur seals and sea lions, northern and southern elephant seals, and the gray seal. These species colonize oceanic islands and coastal areas to give birth and mate. The aggregation of individuals during the breeding season has been attributed to the fact that oceanic islands are relatively rare and unevenly dispersed such that the availability of suitable pupping sites may limit the distribution of females (Boness, 1991). Predation may also select for female clustering, with females being less vulnerable to terrestrial predators and/or harassment by conspecific males when in large groups (dilution effect). Aggregation of females within a stable, two-dimensional habitat has led to the evolution of a polygyny in these species, with males defending either resources needed by females (e.g., birth and thermoregulatory sites in otariid species) or the females themselves (e.g., elephant seals and gray seals). By competing with and limiting the access of other males to females, successful males mate with multiple females, thus increasing their reproductive success. The degree of polygyny in land-breeding pinniped species ranges from extreme in the northern fur seal and elephant seals where one male may mate with 16–100 females to moderate (6–15 females) in gray seals, Hooker sea lions (*Phocarctos hookeri*), and the Galapagos fur seal (Le Boeuf, 1991).

As in other polygynous species, land-breeding pinniped species are sexually size dimorphic. Males in these species can be much larger than females and often show other secondary sex characteristics. These dimorphic characteristics are the result of sexual selection for traits that increase an individual's ability to monopolize and defend resources needed by females or females themselves. Large body size, and concomitant large body energy stores in the form of subcutaneous blubber, permits dominant males to fast and thus remain ashore during the period when females become receptive. The most extreme example of sexual size dimorphism in pinnipeds occurs in elephant seals where males are 5–6 times heavier than females in the northern species and up to 10 times heavier than females in the southern species.

2. Aquatic-Breeding Species Walruses and all other phocid seals [Weddell, Ross, crabeater, leopard, bearded (*Erignathus barbatus*), hooded, ringed, Baikal, Caspian (*Pusa caspica*), larga, harp and ribbon] give birth on pack ice or fast ice and mate in the water. Although Hawaiian monk seals and harbor seals give birth to their offspring on land, they too mate in the water. In species where pups are born on ice, females tend to be more widely distributed, although access to breathing holes in the ice may promote clumping in some species (e.g., walrus and Weddell seals). This broader distribution of females, on an unstable habitat, limits the number of females a male can monopolize at any given time and as a result these species typically show reduced levels of polygyny (e.g., harbor seals; Coltman *et al.*, 1999). The fact that mating occurs in the water, a fluid three-dimensional environment, may also limit the ability of males to monopolize females, resulting in reduced levels of polygyny.

Wells *et al.* (1999) suggested that the mating strategies used by ice-breeding species could be classified as scrambling (males search for receptive females and move on to the next); sequential defense (males sequentially defend single females through mating); and lekking (males aggregate and attract females using displays). At present, there is insufficient information on the breeding behavior of most aquatic breeding species to draw firm conclusions about the type of mating system used. Data on the mating behavior of these species are limited to that which can be observed on ice prior to copulation. For example, observational data suggest that hooded seals utilize a sequential defense form of mating system whereby males compete with one another to defend a single mother and her pup on the ice. The male then remains with the pair until the pup is weaned and then enters the water with the female, presumably to mate. The application of new methods, including genetic paternity assessment and animal-borne video, will be needed to clarify the mating strategies used by these species.

In species that mate aquatically, there may be less selective advantage for males to be larger than females because of the limited ability of males to monopolize females in this environment. As a consequence, in most of these species, male and females are of similar size and in some cases females are larger than males. For example, male Weddell seals are slightly smaller than females and it has been suggested that smaller size makes the male more agile during underwater mating activities (Le Boeuf, 1991). Underwater vocalizations also appear to be an important component of the mating behavior in aquatically mating pinniped species. For example, in Pacific walruses, which exhibit a lekking mating system, males perform complex underwater visual and vocal displays in small groups next to female haul-out sites to attract females. Male Weddell, harbor, harp, hooded, and bearded seals also produce a range of underwater vocalizations during the breeding season that may be used to attract females or to establish underwater territories or display areas.

B. Lactation Strategies

Male pinnipeds do not participate in the care of offspring. Thus, parental care is the exclusive responsibility of the female. Female care involves the transfer of energy-rich milk to the pup and protection from conspecifics and terrestrial predators (Bowen, 1991). In some species (e.g., the walrus), females may also teach their young to forage, as young accompany mothers on foraging trips during the lactation period. Female pinnipeds have dealt with the temporal and spatial separation of energy acquisition (aquatic foraging) from high levels of energy expenditure (terrestrial lactation) in different ways, resulting in the three basic lactation strategies: a foraging cycle, fasting, and aquatic nursing. Until recently, each of the pinniped families was thought to use only one of these strategies (otariids, the foraging cycle; phocids, fasting; and odobenids, aquatic nursing) and the evolution of these differing strategies was presumed to reflect phylogeny. However, studies have indicated that this traditional view is an oversimplification (Boness and Bowen, 1996; Boyd, 1998). For example, harbor seal females exhibit a foraging cycle strategy previously thought to occur only in otariid species.

1. Foraging Cycle All otariids and some of the smaller phocid species (e.g., harbor seals) exhibit this lactation strategy. Females come ashore for parturition with a moderate level of stored body energy. After giving birth, females remain onshore and fast while attending and nursing their young for a perinatal period ranging from a few days to a week. After this initial provisioning period, females leave their pups and return to sea to feed. These trips range from less than 1 day in some species to as long as 14 days in others, depending on the distance to the foraging location and prey abundance. Females then return to land to nurse their pup, after which they repeat the cycle until the pup is weaned. The lactation period in otariid species is quite long, ranging from 4 months to 3 years (Bowen, 1991). Females of these species are considered income breeders, relying on current food intake to support both their own metabolic needs and the energetic cost of milk production. The milk produced by female otariids is relatively energy dense (24–40% fat) compared to terrestrial mammalian species. Pup growth rates are rather low, ranging from 0.06 kg/day in Galapagos fur seals to 0.38 kg/day in Steller sea lions (Boness and Bowen, 1996).

Harbor seals, a phocid species, also exhibit this lactation strategy—alternating short foraging trips to sea (7–10 hr) with terrestrial nursing. The harbor seal is a relatively small phocid species, with females weighing approximately 84 kg at parturition. Given the small quantity of body energy that these females are able to store, female harbor seals are forced to make regular foraging trips to acquire sufficient energy to wean their pups successfully. Compared to otariid species, the length of the lactation period in harbor seals is much shorter (24 days) and the milk produced by females has a relatively higher fat content (50%). Consequently, pup growth rate is higher in harbor seals relative to otariid species (0.6 kg/day). Foraging cycles during lactation may also occur in ringed seals and other relatively small phocid species. Evidence shows that the females of two medium size phocids, the Weddell seal and the harp seal, may also forage during the lactation period. However, the intensity of foraging and the degree to which successful weaning of offspring relies on these foraging trips are not clearly understood.

2. Fasting Strategy In the larger-bodied phocid species, females fast during lactation. Females arrive at the breeding site with large energy stores in the form of adipose tissue (i.e., blubber). In the western Atlantic, for example, gray seal females arrive at Sable Island weighing an average of 210 kg. Of this body mass, 32% or 67 kg is fat. After parturition, females fast for the entire lactation period (e.g., 16 days in the case of gray seals) using their stored energy to support the ENERGETIC cost of milk production and their own maintenance metabolism. For this reason, female phocids are considered capital breeders—having stored energy often months before it is needed. The lactation period in phocids is much shorter than in otariid species, ranging from 4 days in hooded seals to 60 days in Weddell seals. Maternal body size, metabolic rate, and the fasting ability of females may play an influential role in determining the duration of lactation both within and between species (Boness and Bowen, 1996; Boyd, 1998).

Another characteristic feature of the phocid fasting strategy is the production of extremely high-fat milk, ranging from 47% fat in southern elephant seals to 61% fat in hooded seals. This energy-dense milk results in a high rate of offspring growth, ranging from 1.4 kg/day in the Hawaiian monk seal to 7.1 kg/day in the hooded seal (Bowen, 1991). Weaning occurs abruptly when mothers return to the sea to feed. Pups often fast for weeks following weaning, living off their accumulated fat stores before entering the water and beginning to forage independently.

3. Aquatic Nursing The walrus is the only pinniped species that exhibits aquatic nursing. Just prior to parturition, pregnant females separate from the herd and give birth to their offspring alone on pack ice. New mothers remain on the ice fasting for the first few days postpartum, relying on stored body energy accumulated prior to parturition. Subsequently, females and their young return to the herd to forage. Walrus pups suckle in the water for between 2 and 3 years on relatively low-fat milk (24.1%). As with otariids, weaning is gradual. Young walruses begin to feed on benthic organisms as early as 5 months of age and likely gain valuable foraging experience from their mothers over the remainder of lactation. At weaning, female offspring are assimilated into the mother's herd, whereas male offspring join other male groups.

Lactation strategies are often viewed from the female's perspective. This seems reasonable, but in long-lived species such as pinnipeds, females may trade-off investment in current offspring against investment in future offspring. This may lead to conflicts between females and their offspring over the level of investment received. The transition from nursing pup to nutritionally independent juvenile usually occurs without parental supervision in pinnipeds. This transition is arguably the most important period of a pinnipeds' young life. As offspring size affects subsequent survival, we should expect that offspring would attempt to obtain as much milk as they can during lactation. Thus, the nutritional requirements and physiological abilities of individual offspring must also play a role in shaping lactation strategies. For example, the fasting ability of offspring constrains the duration of foraging trips by female fur seals and sea lions.

IV. Foraging

Successful foraging is essential for survival and reproduction and is therefore a critical determinant of fitness. Pinnipeds are among the largest vertebrate carnivores in marine ecosystems and yet the foraging behavior of these apex predators is poorly understood. As noted earlier, pinnipeds inhabit diverse environments, consequently they forage at highly varied spatial and temporal scales and in doing so they exploit a wide range of prey.

A. Methods

As pinnipeds generally feed underwater at remote locations, ecologists rely upon indirect methods to gain insight into their foraging behavior and diets. Very high frequency (VHF) radio tags have been used to study the at-sea locations of coastal species such as harbor seals. Acoustic tags have been used to track the underwater movements of gray seals. More recently, microprocessor-based, time-depth recorders (TDRs) have been used to collect information on dive duration, depth, frequency, and temporal distribution and to calculate the at-sea locations of pinnipeds using solar navigation equations. However, the use of TDRs is limited by the need to recover the instrument to retrieve the stored information and therefore only those species that can be reliably recaptured are used in TDR studies. In contrast, satellite-linked, time-depth recorders transmit collected data on diving parameters and surface positions to polar-orbiting satellites operated by Service ARGOS. This technology has broadened the range of species that have been studied, but the expense of using satellite-linked tags often places limits on the number of individuals studied.

Although we have learned a great deal from the use of location telemetry and dive recorders, these studies have provided little insight into the feeding success rate of pinnipeds. Recent work has demonstrated that estimates of feeding success can be determined using stomach temperature telemetry and animal-borne video. The body temperature of a marine prey is often considerably lower than that of its pinniped predator, thus the stomach temperature of the predator should drop following prey ingestion. This approach has been used successfully on free-living harbor seals and several other species. When combined with information on the diving behavior and movement patterns in the same individual, stomach telemetry can provide new insights into the spatial and temporal patterns of foraging success relative to foraging effort. Animal-borne video technology (Marshall, 1998) has taken our understanding of foraging behavior and diet one step further by providing direct observations of the way in which pinnipeds search and capture prey and how foraging behavior changes as a function of prey type. These video images, coupled with data on swim speed, diving characteristics, environmental conditions (such as sea temperature), and energy expenditure, promise to revolutionize our understanding of pinniped foraging ecology.

Determining the diet of marine mammals also requires the use of indirect methods. The most common methods rely on the recovery and identification of hard prey structures that are resistant to digestion from the stomach, intestine, or feces of individual animals. Sagittal otoliths, cephalopod beaks, bones, scales, invertebrate exoskeletons, and shells can be used to determine the species consumed and, in some cases, to estimate the size and age of the prey. Fecal samples are increasingly being used for this purpose because they are less expensive to collect, a high proportion of samples contain identifiable prey, and estimates of diet are less affected by differential rates of digestion than estimates from stomach samples (Bowen and Siniff, 1999). Although the use of hard parts to estimate the diet of pinnipeds is common, this method is subject to a number of biases, which may limit the value of results. First, stomach and fecal contents only provide an estimate of the diet near the point of collection, and as a result, offshore diets cannot be sampled easily. This may seriously bias the diet of wide-ranging species such as elephant seals, harp seals, northern fur seals, and Juan Fernandez fur seals (*Arctocephalus philippii*). Second, hard parts are often eroded during digestion or digested completely such that prey size may be seriously underestimated and prey identification may not be possible. Finally, perhaps the most serious disadvantage is that dietary analysis based on hard parts is strongly biased against soft-bodied or small prey with fragile structures.

Inevitably our understanding of the diets of pinnipeds is tied to the development of new methods. Fatty acid signature analysis is a relatively new method, which has been developed to study marine mammal foraging and diet (Iverson, 1993). Lipids in marine ecosystems are diverse and characterized by long-chain polyunsaturated fatty acids that originate in unicellular phytoplankton. In monogastric carnivores, such as pinnipeds, ingested fatty acids with a carbon chain length greater than 14 are deposited in body tissues in a predictable way. As a result, the fatty acid composition or signature of the predator reflects the fatty acid composition of prey species consumed (Iverson, 1993). By comparing the reference signature of various prey species to the fatty acid signature of the predator (obtained from blubber tissue or milk), diet composition can be estimated both qualitatively and quantitatively. The use of fatty acid signature analysis eliminates the dependence on recovery of hard parts and integrates the diet over a period of weeks to months such that the location of sampling becomes less important.

Stable isotope ratios of carbon and nitrogen found in the muscle, skin, vibrissae, or blood of pinnipeds and other predators are also being used to investigate diet. These ratios reflect a composite of prey species eaten over a broad time scale. By examining the levels of $^{15}N/^{14}N$ found in body tissues, scientists can determine the trophic level at which the pinnipeds fed. The carbon isotope ratio ($^{13}C/^{12}C$) has been found to vary geographically and thus the level of carbon isotope in the predator's tissues provides insight into foraging location. Although this technique is useful in determining trophic level and foraging location, it does not permit the specific diet composition of individuals to be assessed.

B. Diet

A large number of prey species have been identified in the diet of various pinniped species, leading to the view that pinnipeds are generalist predators. This is consistent with the expectation that large, wide-ranging predators consume more types of prey, as their environment becomes patchier. However, in most cases a relatively small number of taxa account for the majority of food eaten (Bowen and Siniff, 1999). For example,

gray seals on the Scotian Shelf, Canada, consumed 24 different taxa; however, only two to four species accounted for over 80% of the energy consumed depending on the time of year.

Fish and cephalopod species are the main prey types eaten by pinnipeds (Table I). However, crustaceans also appear to account for a substantial portion of prey consumed by some species. Crustaceans are a major prey of harp seals in the North Atlantic and of ringed seals and bearded seals in the Bering Sea. In three Antarctic species, Antarctic fur seals, crabeater seals, and leopard seals, krill accounts for up to 50% of the diet. Unlike most pinnipeds, which generally feed on mobile prey (e.g., fishes, cephalopod molluscs, and crustaceans) in pelagic and benthic habitats, the walrus feeds almost exclusively on sessile benthic invertebrates in soft-bottom sediments.

Several pinniped species are also known to feed on other pinnipeds (Bowen and Siniff, 1999). Male southern fur seals appear to commonly feed on young South American fur seals (*A. australis*). Steller sea lions are known to prey on a variety of pinniped species, including harbor seals, ringed seals, bearded seals, young northern fur seals, and larga seals. Walruses prey on larga seals, ringed seals, and young bearded seals.

The diet composition and foraging behavior of pinnipeds are influenced by a number of factors. The ecology and behavior of prey species clearly play a role in shaping the foraging strategies of pinnipeds. Research on male harbor seal foraging behavior at Sable Island, Canada, using animal-born video found that prey behavior affected both capture technique and profitability of different prey types. Other studies have shown that between-year differences in the diet composition of harbor seals were correlated with differences in the distribution and abundance of herring and sprat, two important prey species.

Intrinsic factors, such as age and sex, may also play a role in the diet composition of individuals within pinniped species. Given that pinnipeds are long-lived predators, their individual foraging tactics and behavior may change over time to reflect increased physiological capabilities and learning. For example, harbor seal pups feed on pelagic prey such as herring and squid, whereas the diet of adults is dominated by benthic species. Similarly, the contribution of benthic prey (e.g., crabs, clams, and sculpins) to the diet of bearded seals increases with age. Age-specific differences in diet composition have also been found in southern elephant seals and harp seals.

Diet composition may also differ between sexes in pinniped species that exhibit sexual size dimorphism (e.g., northern and southern elephant seals, otariid species). Due to the relationship between basal metabolic rate and mass, larger individuals require more total energy per unit time than smaller individuals. Oxygen storage capacity also increases with body mass due to the larger blood pool in which to store oxygen and the larger muscle (myoglobin) mass. In addition, larger animals have a slower mass-specific metabolic rate such that they utilize their larger oxygen stores at a slower rate relative to smaller individuals. Thus, larger individuals are capable of longer, deeper foraging dives. These physiological attributes may allow, or require, males (the larger sex) to pursue different prey types (potentially higher quality prey) than females. Although theory suggests that the diets of males and females may differ, cur-

TABLE I
Major Prey of Selected Pinnipeds

Species	Location	Main prey
Gray seal	Eastern Canada	Sand lance, flatfishes, Atlantic cod
	United Kingdom	Sand lance
Harbor seal	Eastern Canada	Herring, Atlantic cod, pollock, squid
	Western Canada	Pacific hake, Pacific herring
	Sweden	Atlantic cod, sole, herring, sand lance
Harp seal	Northwest Atlantic	Arctic cod, herring, capelin
	White Sea/East ice	Capelin, sand lance, herring
Hooded seal	Greenland	Greenland halibut, redfish, Gadidae
Ringed seal	Bering Sea	Saffron cod, Arctic cod, shrimp
Ribbon seal	Bering Sea	Pollock, eelpout, Saffron cod
Bearded seal	Bering Sea	Shrimp, crab, clam
Northern elephant seal	California	Cephalopods, Pacific whiting
Southern elephant seal	Heard/Macquarie Island	Squid, pelagic and benthic fishes
	Heard Island	Squid, pelagic fishes
Leopard seal	Southern Ocean	Krill, cephalopods, penguins, seals
Northern fur seal	North Pacific	Anchovy, herring, capelin, sand lance
	Bering Sea	Pollock, capelin, herring, squid
South African fur seal	Benquela Current	Anchovy, hakes, squid
Antarctic fur seal	South Georgia	Krill, cephalopods, fish
Sub-Antarctic fur seal	Gough Island	Squid
Australian fur seal	Tasmania	Squid
South American fur seal	Peru	Sardine, southern anchovy, jack mackerel
Jan Fernandez fur seal	Alejando Selkirk Island	Myctophid fishes, squid
New Zealand fur seal	New Zealand	Octopus, squid, barracuda
Steller sea lion	Gulf of Alaska	Pollock, herring, squid
California sea lion	California	Northern anchovy, Pacific whiting, squid

rently there are few studies to test this hypothesis. However, studies on southern elephant seals do indicate sex differences in diet.

C. Foraging and Diving Behavior

The foraging ecology of pinnipeds and other air-breathing vertebrates is constrained by the need to surface for oxygen. Dive duration is constrained by the interplay between the amount of oxygen that can be stored and the rate at which the diver expends oxygen. Thus, it is inevitable that patch use, and the resulting distribution of foraging in time and space, will be influenced by the physiological constraints. Other factors, such as prey density and depth, may play an important role in how pinnipeds forage within these physiological constraints.

Foraging pinnipeds dive repeatedly with relatively short surface intervals between dives; this cluster of dives is called a dive bout. In general, dive bouts are thought to indicate foraging within a prey patch, particularly in otariid species. Theoretically, divers should organize their behavior for optimal patch use. To organize their behavior in this way, divers should opti-

mize both the time budget of the dive cycle (dive duration and surface interval) and the number of dive cycles to repeat. Both of these factors will influence the amount of prey caught and the energy and time consumed during the dive bout. However, there may be a trade-off between prey depth and profitability such that prey items that might be exploited when closer to the surface are less likely to be exploited as the depth of that prey increases.

Empirical tests of optimal foraging theory and optimal patch use in diving pinnipeds are uncommon, largely due to the difficulty and expense of studying these wide-ranging predators and their prey. However, it appears that some otariids feed near the surface on vertically migrating prey, such as krill, to maximize energetic efficiency.

Phocids are generally better suited for deep diving and for longer periods of time than are their otariid and odobenid counterparts. This is largely because phocids have a larger blood volume and larger myoglobin content in the muscles and thus store more oxygen per unit of body mass. Phocids also dive in continuous bouts and are known to spend up to 90% of their time in the water submerged. Thus, unlike otariids and odobenids, phocid seals live at depth, returning periodically to the surface to breathe. Although diving behavior is often considered to be synonymous with foraging in otariids, dive shape analysis in phocids demonstrates that diving may also be used for travel, predator avoidance, and sleep (Wells *et al.*, 1999).

The function of different types of dives has been investigated through the analysis of the two-dimensional profile (time vs depth) of individual dives (i.e., dive shape) and swimming speed during diving. Foraging dives are those in which time spent at the bottom of the dives is a significant fraction of the total dive duration. In northern elephant seals and Weddell seals, these dives are often to similar depths over time, suggesting that the seals are exploiting prey patches that remain at a constant depth and are dense enough to maintain high encounter rates. In contrast, dives characterized by a middle segment of slow downward drift are thought to be associated with the digestion of food in female northern elephant seals. Although dive shape undoubtedly contains information about the behavior of individuals, animal-borne video has revealed that different behaviors can be represented by the same dive shape, thus limiting the inferences that can be drawn from shape alone.

D. Factors Affecting Foraging Ecology

Pinnipeds are important consumers of marine resources; however, for most species little is known about how they interact with the biotic and abiotic features of marine ecosystems. Knowledge of the spatial behavior of pinnipeds is important because spatial patterns can fundamentally affect the nature and dynamics of species interactions. These interactions largely determine the distribution of foraging. Within the ocean, food is distributed in patches and this distribution can be strongly influenced by the physical properties, such as water temperature and the availability of nutrients. For example, the distribution and migratory patterns of northern elephant seals correspond with the location of three dominant water masses of the North Pacific. The localized biological productivity in these water masses and associated fronts result in a high abundance of cephalopods, an important food of this species.

Seasonal changes in prey distribution and abundance can also influence pinniped foraging patterns. Reduced prey availability leads to changes in foraging behavior that include increased trip duration, trip distance, and increased foraging effort. For example, Antarctic fur seals increase their times at sea, northern fur seals increase diving effort, and California sea lions use both tactics during periods of limited prey resources.

E. Spatial and Temporal Scales of Foraging

The relative mobility, range, and body size of an animal affects the resolution at which it recognizes environmental heterogeneity. For example, a relatively small-bodied, central place forager, such as a lactating harbor seal, would identify resource patches at a smaller meso-scale than a highly mobile animal, such as a gray seal. To understand the relationship of an organism to its environment, one must understand the interactions between the intrinsic scales of heterogeneity within the environment and the scales at which the organism can respond to this heterogeneity. Scale issues are critical for effective conservation and management of pinnipeds because of shifts in habitat use and dispersal over ontogeny and a relatively long life span.

A large body size and the capacity for storing large amounts of fat in the form of BLUBBER enable some species of pinnipeds to feed irregularly and thus to exploit distant foraging locations and patchy resources. In contrast, smaller pinnipeds, such as Antarctic fur seals, perceive environmental heterogeneity at a more local scale. For example, fur seals forage at two spatial distributions: (1) fine scale, represented by short (<5 min) travel durations between patches and (2) meso-scale, represented by longer periods of travel (>5 min) (Boyd, 1996). Similarly, based on fatty acid signature analysis, harbor seals appear to demonstrate meso-scale partitioning of their foraging habitat in Prince William Sound, Alaska. Fatty acid signatures obtained from harbor seal blubber biopsies differed within the Sound at a spatial scale of about 40–50 km, and at a smaller scale of 9–25 km, reflecting fine-scale differences in diet between haul-out sites (Iverson *et al.*, 1997).

Although the patch structure of an environment is expressed in both space and time, temporal variation in predator behavior is likely to provide an insight into the spatial distribution of a highly dynamic prey source that may be difficult to track in other ways. For example, in the Antarctic Ocean, krill is distributed patchily and is the major prey resource of lactating Antarctic fur seals. By using the diving behavior of females obtained from TDR records, it is possible to track the way in which fur seals respond to within season and interannual variation in prey patchiness and abundance. Over a 5-year period, changes in the distribution of travel durations between diving bouts suggested that the spatial distribution of krill swarms varied between years. Although their foraging behavior did not indicate that there was a reduction in the number of krill patches, reduced pup growth rates suggested that patches were of poorer quality, and thus the females had difficulty meeting lactation needs. To compensate, females spent a greater amount of time at each patch, thereby maximizing their average rate of energy intake (Boyd, 1996).

To maximize fitness during years of reduced prey abundance, pinnipeds must be sufficiently plastic in their foraging strategies to compensate for added foraging costs. To determine the temporal scales at which predators make these behavioral decisions, Boyd and colleagues simulated increased foraging costs in Antarctic fur seals by adding an extra drag to lactating females, thereby increasing energy expenditure. At the scales of individual dives, the treatment group made shorter, shallower dives than the control (no extra drag added) seals. It appeared that diving behavior was adjusted to maximize the proportion of time spent at the bottom of dives. At the scale of diving bouts, there was no variation between the two groups in bout frequency and duration, or the time spent diving. However, at the scale of complete foraging cycles, the time spent at sea was significantly longer in the treatment group, yet there was no difference in pup growth rate between control and treatment groups.

In contrast to otariids, most phocid seals are able to fast throughout much of the breeding season due to their large body size and corresponding energy stores. As a result, behavioral responses of phocids to changes in food availability between years may be more flexible, resulting in less severe effects on their population dynamics. Still, a change to a less profitable prey or increased foraging effort may still have energetic consequences that result in impacts at the population level. In the Moray Firth, Scotland, clupeid fishes are the dominant prey of harbor seals. In years when clupeids are absent from inshore waters, seals travel further to feed and use alternative prey. As a consequence, the seals showed evidence of reduced body condition, suggesting that there were energetic consequences to this change in diet. Between-year differences in survival rates suggest that temporal variation in prey abundance and resulting diets also have consequences for the dynamics of phocid populations (Thompson *et al.*, 1996).

V. Role of Pinnipeds in Aquatic Ecosystems

Although pinnipeds are one of the more visible components of the marine ecosystems, our understanding of their ecological roles is surprisingly limited. Still, there is some evidence that pinnipeds may have important effects on the structure and function of some ecosystems (Bowen, 1997). Given that pinnipeds are large, long-lived animals that are often present in considerable numbers, we might expect some species to exert top-down control on ecosystems. However, conclusive studies are lacking, largely due to the difficulty of conducting manipulative experiments in the ocean, the fact that interactions occur at quite different spatial and temporal scales, and the inherent indeterminancy in the behavior of complex marine systems.

Despite the long-standing debate over the ecological interactions between pinnipeds and commercial fisheries, there is little understanding of the impact of pinniped predation in these situations. For example, the recent collapse of the Atlantic cod stock on the eastern Scotian Shelf has fueled debate over the impact of gray seal predation, both in causing the decline and in preventing early recovery. Model results indicated that gray seal predation accounted for only 10 to 20% of the estimated mortality caused by the fishery and therefore was unlikely to have played an important role in the decline.

One example of top-down control exerted by a pinniped species comes from a study of lakes in northern Quebec. Lower Seal Lake has a population of land-locked harbor seals and, compared to nine neighboring lakes without seals, supports a different fish community. The relative abundance of lake trout (*Salvelinus fontinalis*) was greater in the nine lakes without seals, whereas brook trout (*S. namaycush*) was the dominant species in Lower Seal lake. Compared to lake trout in neighboring lakes, those in Lower Seal Lake were on average smaller, younger, grew more rapidly, and matured earlier, all of which represent life history characteristics that are associated with heavy exploitation. Although based on strong inference rather than direct empirical evidence, it appears that seal predation was responsible for both the changes in community structure and the life history traits of fish species in Lower Seal Lake (reviewed in Bowen, 1997).

Pinnipeds may also play a role in structuring benthic communities. Walruses disturb bottom sediments during feeding. By selectively feeding on older individuals of a few species of bivalve mollusks, walruses may be responsible for structuring the benthic fauna. Ingestion and defecation by walruses may result in substantial redistribution of bottom sediments, which may favor colonization of some species. In addition, during the process of feeding, walruses produce many pits and furrows in the soft sediments. Thus, walrus feeding appears to affect community structure in three ways: by providing food for scavengers such as sea stars and brittle stars, by providing habitat under discarded bivalve shells, and by reducing the abundance of macroinvertebrates in feeding pits compared to surrounding sediments. Nonetheless, the effects of walrus feeding behavior on macrobenthic assemblages over periods greater than a few months and at larger spatial scales remain greatly unknown.

VI. Conclusions

Our understanding of the ecology of pinnipeds has increased dramatically over the past several decades, but advances have been rather uneven. For example, the lactation strategies of pinnipeds are reasonably well understood, but many aspects of foraging ecology and the ecological role of pinnipeds in aquatic ecosystems remain elusive. As in all areas of science, our ability to measure the system under study influences the rate of progress profoundly. New types of data loggers, telemetry, and methods to estimates the diet of free-ranging pinnipeds will undoubtedly play a prominent role in advancing our understanding. However, we should not underestimate the importance of collaborative research involving ecologists, oceanographers, and population and ecosystem modelers.

See Also the Following Articles

Diet ■ Distribution ■ Ecology, Overview ■ Female Reproductive Systems ■ Parental Behavior

References

Boness, D. J. (1991). Determinants of mating systems in the Otariidae (Pinnipedia). *In* "Behaviour of Pinnipeds" (D. Renouf, ed.), pp. 1–44. Chapman and Hall, London.

Boness, D. J., and Bowen, W. D. (1996). The evolution of maternal care in pinnipeds. *BioScience* **46,** 645–654.

Bowen, W. D. (1991). Behavioural ecology of pinniped neonates. *In* "Behaviour of Pinnipeds" (D. Renouf, ed.), pp. 66–127. Chapman and Hall and Cambridge Univ. Press, Cambridge.

Bowen, W. D. (1997). Role of marine mammals in aquatic ecosystems. *Mar. Ecol. Prog. Ser.* **158,** 267–274.

Bowen, W. D., and Siniff, D. B. (1999). Distribution, population biology, and feeding ecology of marine mammals. *In* "Biology of Marine Mammals" (J. E. I. Reynolds and S. A. Rommel, eds.), pp. 423–484. Smithsonian Press, Washington, DC.

Boyd, I. L. (1996). Temporal scales of foraging in a marine predator. *Ecology* **77,** 426–434.

Boyd, I. L. (1998). Time and energy constraints in pinniped lactation. *Am. Nat.* **152,** 717–728.

Coltman, D. W., Bowen, W. D., and Wright, J. M. (1999). A multivariate analysis of phenotype and paternity in male harbor seals, *Phoca vitulina*, at Sable Island, Nova Scotia. *Behav. Ecol.* **10,** 169–177.

Eberhardt, L. L., and Siniff, D. B. (1977). Population dynamics and marine mammal management policies. *J. Fish. Res. Board Can.* **34,** 183–190.

Iverson, S. J. (1993). Milk secretion in marine mammals in relation to foraging: Can milk fatty acids predict diet? *Symp. Zool. Soc. Lond.* **66**, 263–291.

Iverson, S. J., Frost, K. J., and Lowry, L.F. (1997). Fatty acid signatures reveal fine scale structure of foraging distribution of harbor seals and their prey in Prince William Sound. *Mar. Ecol. Prog. Ser.* **151**, 255–271.

King, J. E. (1983). "Seals of the World." Comstock Publishing Associates, Ithaca, NY.

Laws, R. M. (1985). The ecology of the Southern Ocean. *Am. Sci.* **73**, 26–40.

Le Boeuf, B. J. (1991). Pinniped mating systems on land, ice and in the water: Emphasis on the Phocidae. *In* "Behaviour of Pinnipeds" (D. Renouf, ed.), pp. 45–65. Chapman and Hall, Cambridge.

Marshall, G. J. (1998). Crittercam: An animal-borne imaging and data logging system. *Mar. Technol. Soc. J.* **32**, 11–17.

Reijnders, P., Brasseur, S., van der Torn, J., van der Wolf, Boyd, I., Harwood, J., Lavigne, D., and Lowry, L. (1993). "Status Survey and Conservation Action Plan: Seals, Fur Seals, Sea Lions, and Walrus." IUCN, Gland, Switzerland.

Stewart, B. S., and Delong, R. L. (1993). Seasonal dispersal and habitat use of foraging northern elephant seals. *Symp. Zool. Soc. Lond.* **66**, 179–194.

Thompson, P. M., Tollit, D. J., Greenstreet, S. P. R., MacKay, A., and Corpe, H. M. (1996). Between-year variations in the diet and behaviour of harbour seals, *Phoca vitulina*, in the Moray Firth; causes and consequences. *In* "Aquatic Predators and Their Prey" (S. P. R. Greenstreet and M. L. Tasker, eds.), pp. 44–52. Blackwell, Oxford.

Trillmich, F., and Ono, K. A. (1991). "Pinnipeds and El Niño: Responses to Environmental Stress." Springer-Verlag, Berlin.

Wells, R. S., Boness, D. J., and Rathbun, G. B. (1999). Behavior. *In* "Biology of Marine Mammals" (J. E. I. Reynolds and S. A. Rommel, eds.), pp. 324–422. Smithsonian Press, Washington, DC.

Pinniped Evolution

ANNALISA BERTA
San Diego State University, California

The name Pinnipedia was first proposed for fin-footed carnivores more than a century ago. Pinnipeds—fur seals and sea lions, walruses, and seals—are one of three major clades of modern marine mammals, having a fossil record going back at least to the late Oligocene [27 to 25 millions of years before present (Ma)]. The earliest pinnipeds were aquatic carnivores with well-developed paddle-shaped limbs and feet. A North Pacific origin for pinnipeds has been hypothesized; the group subsequently diversified throughout the world's oceans.

I. Pinniped Ancestry: Origin and Affinities

There has long been debate about the relationship of pinnipeds to one another and to other mammals. The traditional view, also referred to as diphyly, proposes that pinnipeds originated from two CARNIVORE lineages, an odobenid (walrus) plus otariids (fur seals and sea lions) grouping affiliated with URSIDS (bears) and phocids (seals) being related to mustelids (weasels,

skunks, otters, and kin) (Fig. 1A). The current view supported by cladistic analysis of both morphologic and molecular data supports pinnipeds as monophyletic (having a single origin). Although the hypothesis presented here positions ursids as the closest relatives of pinnipeds, it is acknowledged that there is difficulty separating the various lineages of arctoid carnivores (mustelids, procyonids, and ursids) at their point of divergence (Fig. 1B). Other hypotheses support pinnipeds as either allied with MUSTELIDS or as having an unresolved arctoid ancestry.

There is, however, still disagreement about relationships among pinnipeds. Most of the controversy lies in the debate as to whether odobenids are most closely related to phocids or to otariids. Most morphologic data support a link between phocids and the odobenids, whereas molecular data favor a link between odobenids and otariids.

II. Divergence of Major Lineages

The broad pattern of evolution within pinnipeds shows divergence of five major lineages. These include the three extant lineages—Otariidae, Phocidae, and Odobenidae—and two extinct groups—the Desmatophocidae and a basal lineage *Enaliarctos* (Berta and Sumich, 1999; Fig. 2). At times the Odobenidae have been included in the Otariidae, although cladistic analysis currently supports these as distinct monophyletic groups.

Within Pinnipedimorpha (living pinnipeds plus their fossil allies) are included archaic pinnipeds *Enaliarctos* and *Pteronarctos* + pinnipeds (Fig. 2). Berta and Wyss (1994) established a cladistic diagnosis of the group based on 18 synapomorphies of the skull, teeth, and postcrania. Unequivocal derived features include large infraorbital foramen, anterior palatine foramina anterior of maxillary-palatine suture, upper molars reduced in size, lower first molar metaconid reduced or absent, humerus short and robust, deltopectoral crest on the humerus strongly developed, and digit I on the manus and digit I and V on the pes emphasized. The basal taxon *Enaliarctos* from the late Oligocene and early Miocene (27–18 Ma; Fig. 3) of California is known by five species, one represented by a nearly complete skeleton (Fig. 4). *Enaliarctos* was a small, fully aquatic pinnipedimorph with shearing teeth (as is typical of most terrestrial carnivorans), flexible spine, and fore- and hindlimbs modified as flippers. Several features of the hindlimb suggest that *Enaliarctos* was capable of maneuvering on land, although it probably spent more time near the shore than extant pinnipeds. Ecologically, the earliest pinnipedimorphs were coastal dwellers that likely fed on fish and other aquatic prey. A later diverging lineage allied more closely with pinnipeds than with *Enaliarctos* is *Pteronarctos* from the late Miocene (19–15 Ma) of coastal Oregon. *Pteronarctos* is recognized as the earliest pinniped to have evolved the unique maxilla diagnostic of modern pinnipeds. The maxilla of pinnipeds makes a significant contribution to the orbital wall. This differs from the condition in terrestrial carnivores in which the maxilla is limited in its posterior extent by contact of several facial bones (jugal, palatine, and/or lacrimal).

Fur seals and sea lions (EARED SEALS), the Otariidae, are diagnosed by frontals that extend anteriorly between the nasals,

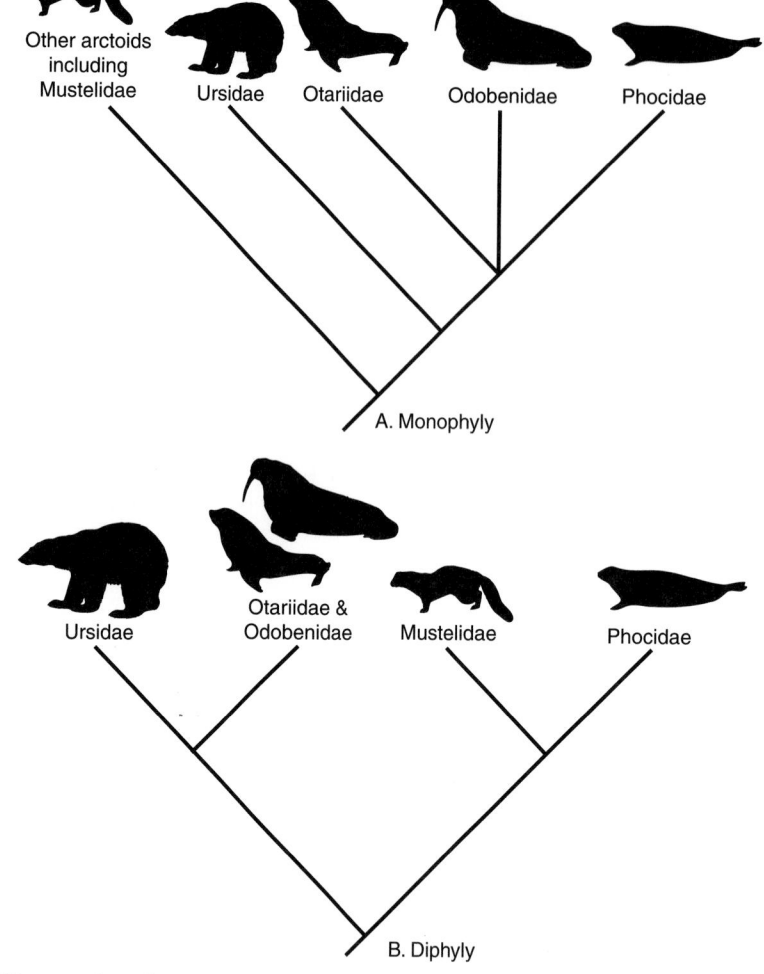

Figure 1 *Alternative hypotheses for relationships among pinnipeds. (A) Monophyly with ursids as the closest pinniped relatives. (B) Diphyly in which phocids and mustelids are united as sister taxa as are otariids, odobenids, and ursids. From Berta and Sumich (1999).*

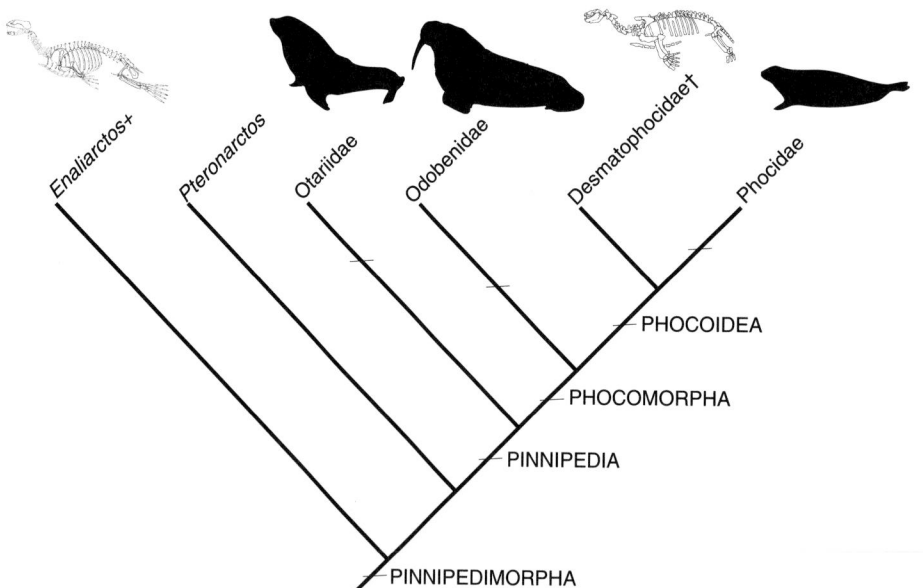

Figure 2 *A cladogram depicting the relationship of the major clades of pinnipeds. From Berta and Sumich (1999).*

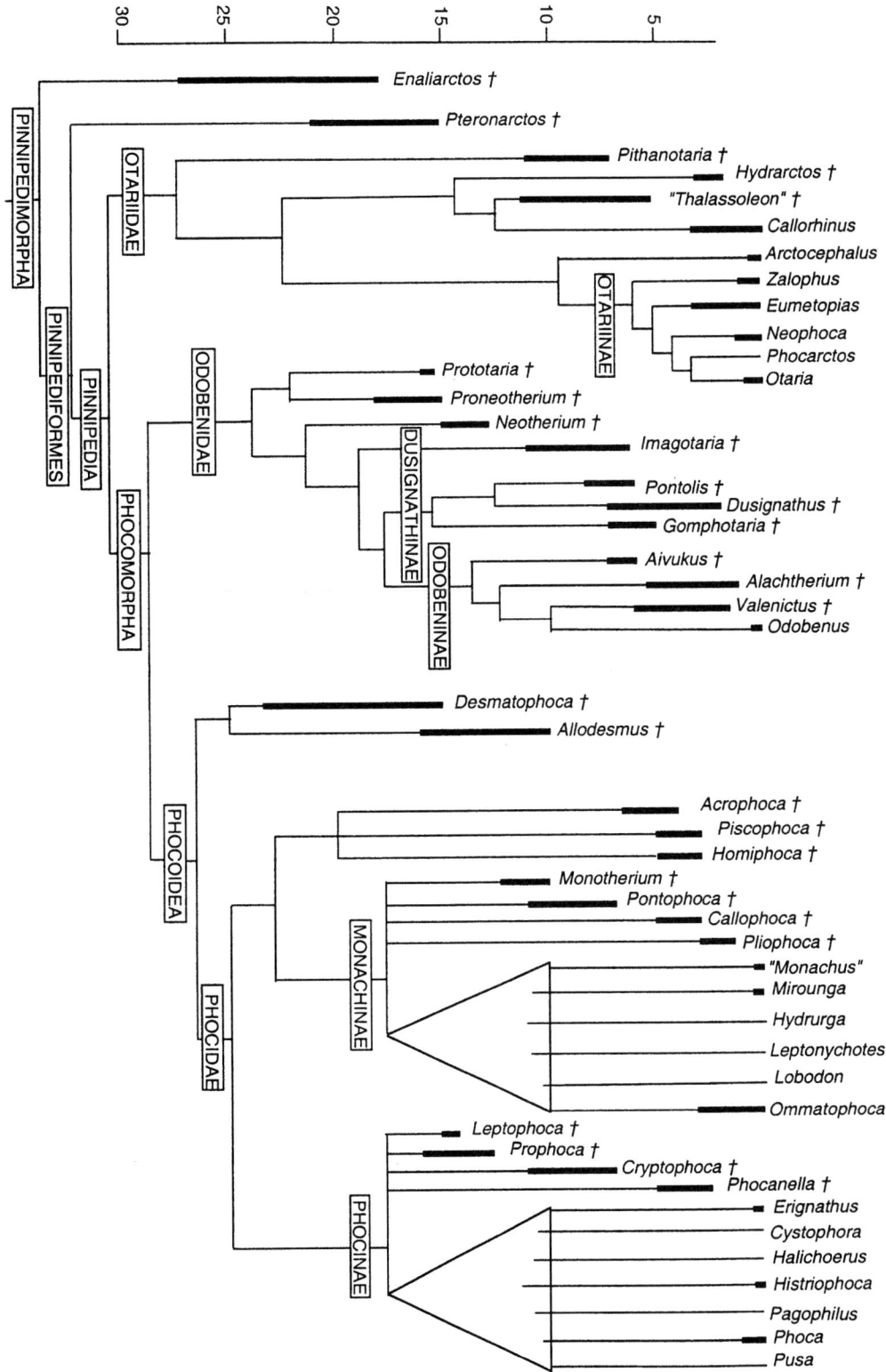

Figure 3 *Chronologic ranges of major lineages and genera of extinct and living pinnipeds (time scale in millions of years). Thick lines show stratigraphic ranges of taxa and thin lines indicate their phylo-genetic relationships. Branching points are not necessarily correlated with the time scale. Note phocid interrelationships are shown as unresolved due to conflicting molecular and morphologic data sets.*

Figure 4 *The pinnipedimorph* Enaliarctos mealsi *showing pinnipedimorph synapomorphies described in the text. From Berta and Ray (1990).*

large and shelf-like supraorbital process of the frontal, secondary spine dividing the supraspinous fossa of the scapula, uniformly spaced pelage units, and the presence of a trachea with an anterior bifurcation of the bronchi (Berta and Sumich, 1999; Fig. 5). Although otariids are often divided into two subfamilies, Otariinae (sea lions) and Arctocephalinae (fur seals), only the sea lions are believed to have descended from an exclusive common ancestor, at least according to morphologic data. The otariids are the earliest diverged pinniped lineage, originating approximately 11–12 Ma (Fig. 3) and including the poorly known *Pithanotaria* from the late Miocene of California and three species of *Thalassoleon* described from the late Miocene of California, Mexico, and Japan. *Thalassoleon* is distinguished from *Pithanotaria* in its larger size and lack of a thickened ridge of tooth enamel at the base of the third upper incisor (Fig. 6). An extinct species of the Northern fur seal genus, *Callorhinus*, has been described from the late Pliocene of southern California, Mexico, and Japan. One fossil fur seal from the Pleistocene of Peru of the southern fur seal genus *Arctocephalus* has been referred to a new subgenus, *Arcto-*

cephalus (*Hydrarctos*). Cladistic analysis of otariids by Berta and Deméré (1986) recommended that *Hydrarctos* be elevated to generic status. Several extant species of *Arctocephalus* have a FOSSIL RECORD extending to the Pleistocene in South Africa and North America (California). The fossil record of modern sea lions is poorly known. The following taxa are reported from the Pleistocene: *Neophoca* (New Zealand), *Eumetopias* and *Zalophus* (Japan), and *Otaria* (Brazil).

Otariids appear to have originated in the North Pacific (Fig. 7). Sea lions diverged as a distinct lineage between 3 and 5 Ma. Both fur seals and sea lions dispersed to the Southern Hemisphere within the past 3 Ma. The diversification of otariids in the Southern Hemisphere has been attributed to several factors, including increased levels of primary productivity in colder waters, which at this time reached to the equator, and island hopping assisted by favorable currents. Sea lion diversification proceeded subsequent to the isolation of populations in the Northern and Southern Hemispheres of the Pacific basin, leading to divergence of the California, Japanese, and Galapagos sea lions (*Zalophus*) and Steller's sea lion (*Eumetopias*) in the

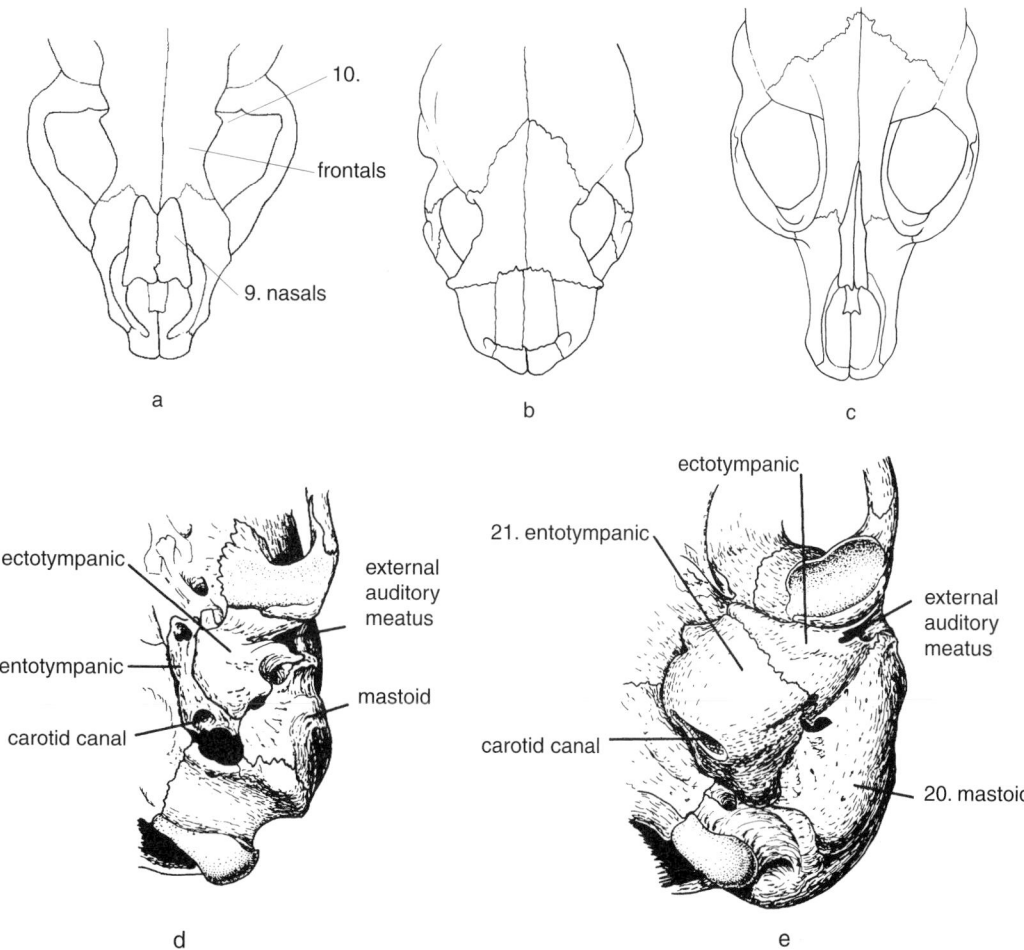

Figure 5 *Skulls/ventral view of ear regions of (a, d) otariid, (b) walrus, and (c, e) phocid illustrating otariid synapomorphies: frontals extend anteriorly between nasals (contact between these bones is transverse in walruses or V-shaped in phocids) and phocid synapomorphies: pachyostotic mastoid bone (not seen in other pinnipeds) and greatly inflated ectotympanic bone. From Berta and Sumich (1999).*

north and the Australian sea lion (*Neophoca*) and southern sea lion (*Otaria*) and the New Zealand sea lion (*Phocarctos*) in the south.

Walruses or Odobenidae are diagnosed as a monophyletic group by the presence of a broad, thick pterygoid strut, fourth upper premolar with a strong posterolingually placed protocone shelf, lower first molar with the talonid heel absent, and a calcaneum with a prominent medial tuberosity (Fig. 6). Another purported odobenid synapomorphy (i.e., construction of the antorbital process by the maxilla and frontal bones) is now known to be primitive (Deméré and Berta, 2001). Morphologic study of the evolutionary relationships among walruses has identified the following taxa: *Prototaria, Proneotherium, Neotherium, Imagotaria,* dusignathines, and odobenines; the latter includes the modern walrus.

Fossil walruses first appear in the early Miocene (16–18 Ma; Fig. 3) fossil record with *Prototaria* in Japan and *Proneotherium* in North America (Oregon). These archaic walruses are characterized by unenlarged canines and narrow, multiple rooted

premolars with a trend toward molarization (Fig. 8), adaptations suggesting retention of the fish diet hypothesized for archaic pinnipeds rather than evolution of the specialized mollusc diet for the modern walrus. Portions of the axial and hindlimb skeleton of *Proneotherium* preserve evidence of a number of aquatic adaptations, including a laterally flexible spine, a broad, shortened femur, and a paddle-shaped foot (Deméré and Berta, 2001). In addition to *Proneotherium* and *Prototaria,* two monophyletic clades of walruses are recognized that diversified in the late Miocene (Fig. 3). Dusignathine walruses, which include *Dusignathus, Pontolis,* and *Gomphotaria,* developed enlarged upper and lower canines. *Gomphotaria pugnax,* the most completely known dusignathine, is distinct cranially and dentally in its possession of large, procumbent upper lateral incisors and canines; the latter with deeply fluted roots and a small orbit. Odobenines, which include *Aivukus, Alachtherium, Valenictus,* and *Odobenus,* evolved the enlarged upper tusks seen in the modern walrus. Unique among pinnipeds is the toothlessness of *Valenictus chulavistensis* from the

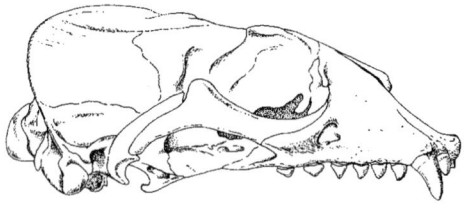

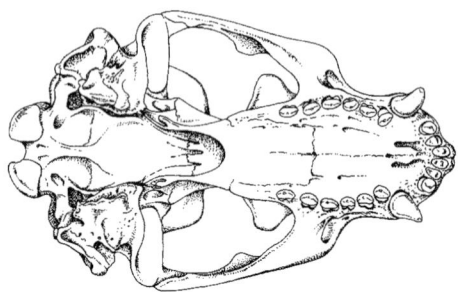

Figure 6 *Skull of an early otariid,* Thalassoleon mexicanus, *from the late Miocene of western North America in lateral (top) and ventral (bottom) views. Original 25 cm long. From Repenning and Tedford (1977).*

late Pliocene of California, presumably an adaptation for suction feeding (Demére, 1994).

It has been proposed that the modern walrus lineage (odobenine walruses) entered the Caribbean from the Pacific via the Central American Seaway (between 5 and 8 Ma) and dispersed northward into the North Atlantic (Fig. 7). Less than 1 Ma the living genus *Odobenus* returned to the North Pacific through the Arctic Ocean. Alternatively, on the basis of a new, earlier record of the modern walrus lineage form Japan, it has been suggested that this lineage may have evolved in the North Pacific and dispersed instead to the North Atlantic through the Arctic during the Pleistocene.

The earless seals, the Phocidae, are diagnosed as a monophyletic group by pachyostotic mastoid region, greatly inflated entotympanic bone, complete absence of the supraorbital process of the frontal, strongly everted ilia, and lack of an ability to draw the hindlimbs forward under the body due to a massively developed astragalar process and greatly reduced calcaneal tuber (Wyss, 1988; Fig. 5). Phocids have traditionally been divided into two or four major subgroupings, there is considerable disagreement regarding which are monophyletic. Most molecular and morphologic evidence supports "monachines" (or Monachinae) as early diverging phocids. The Phocinae are universally accepted as monophyletic, although relationships among taxa within this group differ among workers.

Although an earlier, less well-documented record of phocids from the late Oligocene of South Carolina exists (Koretsky and Sanders, in press), there is undisputed evidence for both the "monachine" and phocine seal lineages from the middle Miocene (approximately 15 Ma) on both sides of the North Atlantic. The phocine lineage is represented by *Leptophoca,*

and the "monachine" lineage is represented by *Monotherium* (Fig. 3). Despite the fact that a number of fossil "monachines" have been described, not all are known by comparable elements. In addition to *Monotherium* among the better known taxa from eastern Europe are *Pontophoca* from the middle Miocene and *Callophoca* from the early Pliocene; both taxa are under study by Koretsky. Several archaic lobodontine (?phocine) seals (i.e., *Arcophoca* and *Piscophoca*) represented by complete skeletons are known from the late Miocene and/or early Pliocene of South America (Peru) and *Homiphoca* from South Africa (Fig. 3). *Acrophoca* is unique among phocids with its long, slender skull, flexible neck, and elongated body (Fig. 9). Although these fossil taxa have been referred to the lobodontines, a subgrouping of "monachines," new discoveries, as well as restudy of material previously referred to these taxa, suggest that they may in fact be more closely related to phocines. The fossil record of extant "monachines" is poorly known and includes only *Monachus* and *Mirounga* from the late Pleistocene and the lobodontine *Ommatophoca* from the late Pliocene (Fig. 3). With regard to fossil phocines among the best known taxa are *Prophoca* (middle Miocene) in the eastern North Atlantic, *Cryptophoca* (late Miocene) in the Black Sea region, and *Phocanella* (early Pliocene) in both the eastern and the western North Atlantic (Fig. 3). Extant phocine genera with a fossil record include *Phoca* from the late Pliocene and *Erignathus* and *Histriophoca* from the late Pleistocene (Fig. 3).

The purported first appearance of phocids in the North Atlantic suggests that the common ancestor of phocids had migrated to the North Atlantic, either northward through the Arctic Basin or southward through the Central American Seaway (Fig. 7). Support for a southern route is based on the hypothesized close relationship of phocids and the extinct desmatophocids, the latter occurring as far south as Mexico, and the fact that the Bering land bridge blocked access to the Arctic through much of the late Oligocene and early Miocene. The BIOGEOGRAPHIC history of monk seals (*Monachus* spp.) is conflicting, as there are differing hypotheses for their phylogenetic position relative to other phocids. Alternative routes show dispersal in both directions through the Central American seaway followed by diversification in the colder waters of the Southern Hemisphere to produce the lobodontine seal fauna of the Antarctic today.

The biogeographic pattern for phocine seals is no less complicated given the different phylogenetic hypotheses proposed. Although it was suggested earlier that phocines were a Northern Hemisphere radiation, a considerable diversity of phocine seals is known from the Southern Hemisphere during the late Miocene and/or early Pliocene (assuming that *Acrophoca, Homiphoca,* and *Piscophoca* are phocines rather than lobodontine "monachines"). In addition, other phocine lineages appear to have been isolated in the Paratethys Sea (northern arm of the Tethys Sea stretching across the area now occupied by the Black, Caspian, and Aral Seas of Eurasia) and the North Atlantic during the late Miocene and Pliocene. Several dispersal routes for phocines seem likely. One hypothesized dispersal involved an initial migration from the Paratethys Sea into the

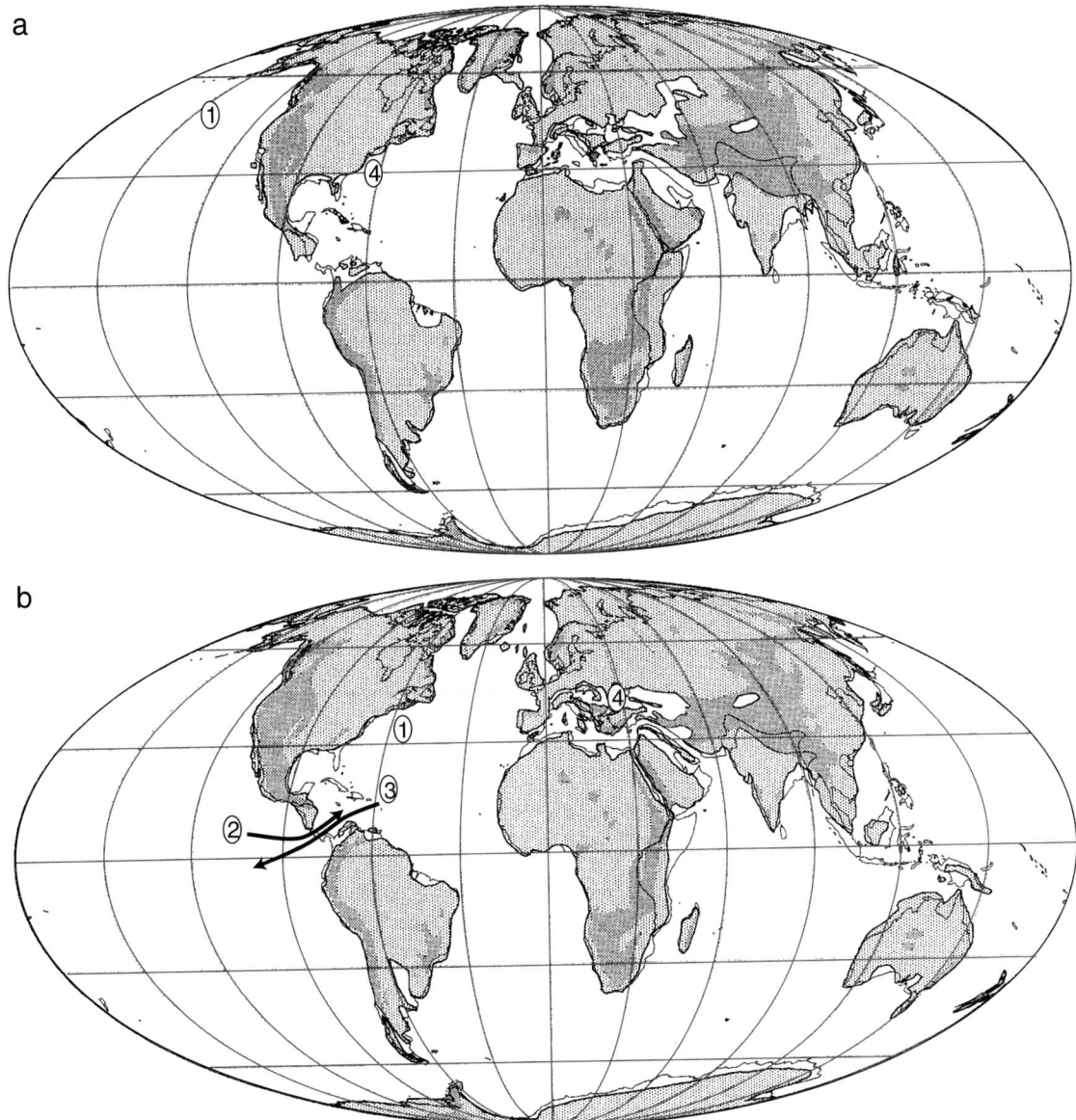

Figure 7 *Reconstruction of continents, ocean basins, and paleocoastlines in the (a) early Miocene (20 Ma) (1, early records of archaic pinnipeds, odobenids, and desmatophocids) and (b) middle Miocene (12 Ma) (1, early well-documented phocids; 2, dispersal of "monachines" and odobenids to Atlantic; 3, dispersal of phocines to South Pacific; and 4, isolation of phocines in remnants of Paratethys Sea and in North Atlantic). From Berta and Sumich (1999); base map from Smith et al. (1994).*

Arctic Basin, followed by an eastward migration to give rise to modern *Phoca/Pusa*. In this scenario the land-locked Baikal seal (*Pusa sibirica*) gained access to Lake Baikal from the Arctic via large lakes at the southern margin of the Siberian ice sheet. A second land-locked species, the Caspian seal (*Pusa caspica*) remained in the Caspian Sea as an isolated remnant of Paratethys as judged by the presence of fossils similar to living Caspian seals in this location. A second dispersal involved a westward migration of the remaining phocines from the North

Atlantic to their current ranges. The hooded seal (*Cystophora cristata*) occurs at high latitudes of the Atlantic basin and apparently never dispersed successfully to the Pacific. The bearded seal (*Erignathus barbatus*) is presently confined to the Arctic and sub-Arctic around the North Atlantic, but Pleistocene records extend as far south as Portugal. Another hypothesis argues for a North Atlantic origin for all phocines with glacial events causing speciation. For example, cyclical fluctuations in glacial maxima (with concomitant variations in

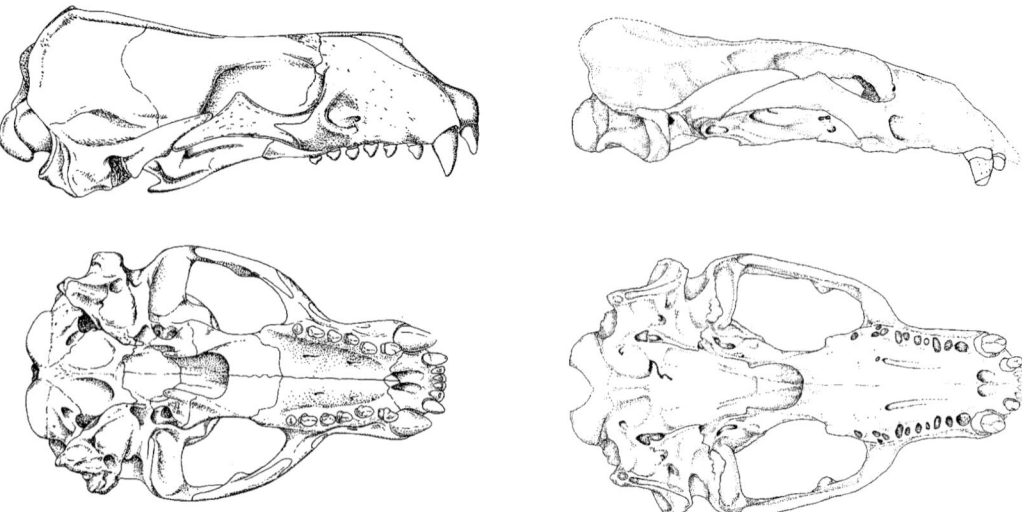

Figure 8 *Lateral and ventral views of skulls of fossil odobenids. (Left)* Imagotaria downsii *from the Miocene of western North America. Original 30 cm long. From Repenning and Tedford (1977). (Right)* Prototaria planicephala *from the Miocene of Japan. Original 27 cm. From Kohno et al. (1995).*

sea level) through the Pleistocene mediated range expansions of *Phoca* spp., ultimately leading to the isolation of populations in refugial centers (i.e., Arctic, Okhotsk, Aleutian) and the divergence of populations (e.g., ribbon seal, *Histriophoca fasciata*, ringed seal, *Pusa hispida*, larga seal, *Phoca larga*, harbor seal, *Phoca vitulina*, harp seal, *Pagophilus groenlandicus*).

An extinct family of archaic pinnipeds, the desmatophocids, are characterized by elongate skulls, relatively large eyes, mortised contact between two cheekbones, and bulbous cheek teeth (Fig. 10). Cladistic analysis has identified the Desmatophocidae, which includes two genera, *Desmatophoca* and *Allodesmus*, as the common ancestors of phocid pinnipeds. This clade, phocids + *Allodesmus* and *Desmotophoca*, termed the Phocoidea, is supported by four synapomorphies, including premaxilla–nasal contact reduced, squamosal–jugal contact mortised and marginal process below ascending ramus well developed (Deméré and Berta, in press; Fig. 2). This interpretation differs from previous work that recognized desmatophocids as otarioid pinnipeds (a paraphyletic group that includes walruses but excludes phocids). Desmatophocids are known from the early and middle Miocene (23–10 Ma) of the western United States and Japan

(Fig. 3). Newly reported occurrences of *Desmatophoca* confirm the presence of sexual dimorphism and large body size in these pinnipeds (Deméré and Berta, in press). *Allodesmus* is a diverse taxon (as many as six species have been named) with taxa informally divided into "broad headed" and "long headed" subgroups. The previous suggestion that *Allodesmus* retained a mobile proboscis, much like the modern elephant seal, seems unlikely on anatomical grounds. *Allodemus*, known by several complete skeletons, retains several features consistent with forelimb propulsion but also displays adaptations for hindlimb SWIMMING (Adam and Berta, 1998).

Acknowledgments

I thank Thomas Deméré for helpful discussions of pinniped biostratigraphy and Irina Koretsky for providing unpublished information on phocid seals.

See Also the Following Articles

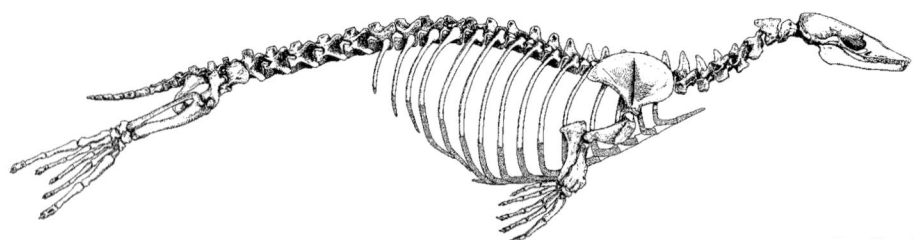

Figure 9 *Skeleton of an archaic phocid,* Acrophoca longisrostris, *from the Miocene of Peru. From Muizon (1981).*

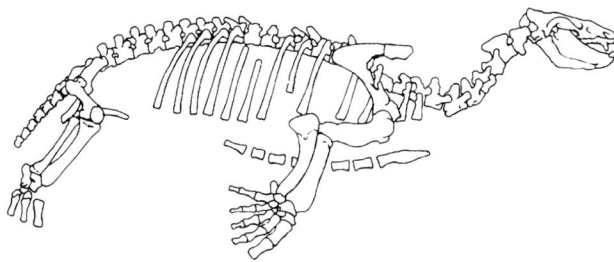

Figure 10 *Skeleton of the desmatophocid,* Allodesmus kernenesis, *from the Miocene of western North America. Original 2.2 m long. From Mitchell (1975).*

References

Adam, P. J., and A. Berta (1998). Using a phylogenetic paradigm to test hypotheses of the evolution of locomotor styles in the Pinnipedimorpha (Mammalia). PaleoBios, 1998 Conference Abstracts Supplement, 1A.

Berta, A., and Deméré, T. A. (1986). *Callorhinus gilmorei* n. sp. (Carnivora: Otariidae) from the San Diego Formation (Blancan) and its implications for otariid phylogeny. *Trans. San Diego Soc. Nat Hist.* **21**(7), 111–126.

Berta, A., and Ray, C. E. (1990). Skeletal morphology and locomotor capabilities of the archaic pinniped *Enaliarctos mealsi. J. Vertebr. Paleo.* **10**(2), 141–157.

Berta, A., and Sumich, J. L. (1999). "Marine Mammals: Evolutionary Biology." Academic Press, San Diego, CA.

Berta, A., and Wyss, A. R. (1994). Pinniped phylogeny. *Proc. San Diego Soc. Nat. Hist.* **29**, 33–56.

Deméré, T. A. (1994). The family Odobenidae: A phylogenetic analysis of fossil and living taxa. *Proc. San Diego. Soc. Nat. Hist.* **29**, 99–123.

Deméré, T. A., and Berta, A. (in press). The Miocene pinniped *Desmotophoca oregonensis* Condon, 1906 (Mammalia: Carnivora) from the Astoria Formation, Oregon. *Smith. Contrib. Paleobiol.*

Deméré, T. A., and Berta, A. (2000). A re-evaluation of *Proneotherium repenningi* from the middle Miocene Astoria Formation of Oregon and its position as a basal odobenid (Pinnipedia: Mammalia). *J. Vertebr. Paleontol.* **21**(2), 279–310.

Kohno, N., Barnes, L. G., and Hirota, K. (1995). Miocene fossil pinnipeds of the genera *Prototaria* and *Neotherium* (Carnivore; Otariidae; Imagotariinae) in the North Pacific Ocean: Evolution, relationships and distribution. *Island Arc* **3**, 285–308.

Koretsky, I. A., and Sanders, A. E. (in press). Pinniped bones from the late Oligocene of South Carolina: The oldest known seal (Carnivora: Phocidae). *Smith. Contrib. Paleobiol.*

Mitchell, E. D. (1975). Parallelism and convergence in the evolution of the Otariidae and Phocidae, Conseil international de l'Exploration de la Mer. *Rapp. Proc-Verbaux Reunions* **169**, 12–26.

Muizon, C. de (1981). Les Vertebres Fossiles de la Formation Pisco (Perou). Part 1. Recherche sur les Grandes Civilisations, Mem. No. 6, Instituts Francais d'Etudes Andines, Paris.

Repenning, C. A., and Tedford, R. H. (1977). Otarioid seals of the Neogene. *Geol Surv. Prof. Pap. (U.S.)* **992**, 1–93.

Smith, A. G., Smith, D. G., and Funnell, B. M. (1994). "Atlas of Mesozoic and Cenozoic Coastlines." Cambridge Univ. Press, Cambridge.

Wyss, A. R. (1988). On "retrogression" in the evolution of the Phocinae and the phylogenetic affinities of the monk seals. *Am. Mus. Novit.* **2924**, 1–138.

Pinniped Life History

IAN L. BOYD
*University of St. Andrews,
Scotland, United Kingdom*

The life history of an individual is the pattern of allocation of resources to maintenance, growth, and reproduction throughout its lifetime. Life history analysis attempts to explain the scheduling of the allocation process throughout an organism's life. It assumes implicitly that it is appropriate to classify individuals by age because this is a major component of the independent variable representing time that is used to examine variation in resource allocation. However, we know that other properties of an individual, such as its body condition or foraging skill, are also important variables that affect reproduction and, ultimately, fitness.

Most life history studies involving pinnipeds have assumed that age is the main force in pinniped life histories when, in fact, age per se may have relatively little to do with influencing fitness. It is a paradox of life history studies that they are, by definition, time-based approaches to examining variation in the fitness between individuals when time itself probably has less biological importance than other factors. One such factor in pinnipeds is body size, long recognized as a determinant of sexual maturity in pinnipeds. Age at sexual maturity in pinnipeds can be expressed as a decreasing function of growth rate. Expressed at the level of populations, this is interpreted to mean that individuals within pinniped populations that are at a level well below the environmental carrying capacity would experience higher growth rates and would, therefore, become sexually mature at an earlier age (Bengtson and Laws, 1985). This was an implicit acknowledgment that age was not the operant factor in pinniped life histories and was at best secondary to the size of the energy reserves of an individual. Nevertheless, despite the considerably greater difficulties that exist with measuring age in pinnipeds than there are with measuring body size (e.g., mass or some other suite of morphometrics), age has continued to be used as the primary independent variable in life history studies.

I. Characteristics of Pinniped Life Histories

Pinniped life histories are characterized by three main features: (1) by mammalian standards, pinnipeds have high annual survival rates, giving potential longevities in the order of 2–4 decades; (2) the average age at sexual maturity is delayed by 2–6 years depending on the species (Table I); and (3) each adult female normally produces a maximum of one offspring per reproductive cycle. Variations on this theme at the level of individuals and species can provide insight into the evolution of life histories in pinnipeds.

Pinniped life histories are assumed to have evolved to maximize the genetic fitness of individuals. This occurs in pin-

TABLE I
Demographic Parameters Used to Describe Life Histories of Pinnipeds[a]

Species	Mean female body mass (kg)	Mean male body mass[b] (kg)	Pup survival rate	Adult female survival rate	Adult male survival rate	Mean age at first parturition (years)	Mean pregnancy rate[c]	Reference
Mirounga angustirostris	393–425	—	0.88	0.69–0.77[d]	—	3–4	0.80	Huber et al. (1991); Reiter and Le Boeuf (1991); Le Boeuf et al. (1994)
M. leonina	400–500	2100	0.98	0.67–0.88	0.50–0.83	4–5	0.88	McCann (1985); Hindell (1991); Galimberti and Boitani (1999)
Leptonychotes weddellii	350–425	—	0.80–0.92	0.76–0.85[e]	—	6–8	0.46–0.79	Testa and Siniff (1987); Testa (1987); Testa et al. (1990); Hastings and Testa (1998)
Lobodon carcinophaga	220	—	0.21[f]	0.90–0.97	—	2.5	0.95–0.98	Boveng (1993)
Pagophilus groenlandicus	100–140	—	—	—	—	4.8	0.82–0.97	Bowen et al. (1981); Kjellqwist et al. (1995)
Pusa hispida	40–50	—	0.84[f]	0.86	—	6–8	0.88	Smith (1987)
Halichoerus grypus	160–190	—	0.66[f]	0.93	—	5–7	0.80–0.98	Harwood and Prime (1978); Boyd (1985); Pomeroy et al. (1999)
Eumetopias jubatus	250	—	0.78[f]	0.84–0.93	—	4–5	0.63	York (1994)
Callorhinus ursinus	29–39	97–165	0.80–0.96	0.86–0.89	0.70[g]	3–4	0.69–0.72	Wickens and York (1997)
Arctocephalus townsendi	49	—	—	—	—	—	—	Wickens and York (1997)
A. galapagoensis	—	64	0.85–0.91	0.85	0.68	5	—	Wickens and York (1997)
A. philippii	—	—	0.92–0.95	—	—	—	—	Wickens and York (1997)
A. pusillus pusillus	57	247	0.65–0.80	0.88[h]	0.70	4	0.71	Wickens and York (1997)
A. pusillus doriferus	76	—	0.85	—	—	4	0.73	Wickens and York (1997)
A. forsteri	—	—	0.40–0.92	—	—	5	0.67	Wickens and York (1997)
A. australis	35–58	—	0.53–0.90	—	—	3	0.80–0.82	Wickens and York (1997)
A. tropicalis	36	—	0.85–0.96	—	—	5	0.79–0.84	Wickens and York (1997)
A. gazella	45	188	0.69–0.96	0.83–0.92	0.50	3	0.68–0.77	Wickens and York (1997)

[a]Rates are expressed per year. Data for fur seals are summaries from tables in Wickens and York (1997); otherwise the original sources are given. Data for male mass were not included if no demographic data were available.
[b]Sexually and socially mature individuals.
[c]Pregnancy and birth rate are assumed to be equivalent.
[d]Juvenile survival rates fall within the same range.
[e]Juvenile survival >1 year old ~0.70.
[f]Survival in first year.
[g]Values for juvenile males aged 4 months–2 years are 0.20–0.50; those for males aged 2–5 years are 0.75–0.90.
[h]Probably negatively biased because of the inclusion of juveniles.

nipeds within the constraints of a semiaquatic existence and has most probably led to the relatively narrow range of life histories we observe within the taxon. All pinnipeds rely to some degree on ice or land for reproduction, particularly the processes of birth and lactation. Many interacting variables have led to the evolution of pinniped life histories, including the joint and sometimes conflicting needs to avoid predation, to forage with maximum efficiency, and to choose a mate of high quality.

By mammalian standards, pinnipeds are animals with a large body size. However, in terms of their demography and their investment in reproduction, pinnipeds do not appear to differ greatly from other mammals after body size has been taken into consideration. There are also no obvious relationships between body size and life history variables at the species level within the pinnipeds (Table I), although, as we shall see, this is not the case for variation between individuals within species.

Large body size has a cost in that relatively large amounts of resources are invested in tissue growth and maintenance and it takes a relatively long time to reach a body size capable of supporting reproduction. There is also a need to produce precocial young that can defend themselves against predation from an early age or that can forage independently of their mothers within days to weeks of birth. This necessitates greater investment in individual offspring and limits the number of young that can be produced at a single reproductive attempt. It also means that the rate of reproduction (number of young born per unit time) is relatively low. The combination of high investment in growth, causing a delay in sexual maturity, and low reproductive rates, even when sexually mature, means that pinnipeds must have relatively high longevities (low rates of mortality). Without this combination of demographic variables individuals could not, on average, replace themselves during their lifetimes.

II. Methods for Examining Life Histories

Life histories are represented most concisely by demographic models based on empirical measurements of survival and fecundity rates. Demographic variables for pinnipeds are summarized in Table I. Amongst the 36 species of pinnipeds, some form of demographic information is available for most species, but as seen from Table I, there are very few for which there could be said to be complete information, and, in almost all of these, information is mainly available for females. Very little is known about the life histories of male pinnipeds. It is also perhaps a little misleading to represent these demographic variables in terms of species, as many vary as much between different populations of the same species as they do between the species themselves. Averaging across populations also has the disadvantage that it obscures the variation in life histories between individuals. Therefore, while life histories may, in practice, often be examined at the level of populations using demographic parameters, it is an important tenet of life history analysis that it is based on the demography of individuals. This distinguishes life history analysis from the study of population dynamics, which normally deals with individuals as if they are all identical.

The most complete information about life histories for any population of pinnipeds comes from Weddell seals (*Leptonychotes weddellii*) at McMurdo Sound, Antarctic (Hastings *et al.*, 1998), and northern elephant seals (*Mirounga angustirostris*) from Año Neuvo or the Farallon Islands, California (Reiter and Le Boeuf, 1981; Sydeman *et al.*, 1991). These studies were based on the long-term mark-recapture of individuals. Similar studies have been carried out on Antarctic fur seals (*Arctocephalus gazella*) (Boyd *et al.*, 1995) and gray seals (*Halichoerus grypus*) (Pomeroy *et al.*, 1999). Mark-recapture is probably the only way to examine life histories in pinnipeds to provide the quality of data necessary to understand the complex interactions between factors that influence fitness. However, such studies can only be undertaken in special circumstances where there is particularly easy access to the study population. In most cases, information about population life histories has been derived from cross-sectional samples of populations based on one-off or sequential culls that were often

part of a commercial harvest (Fowler, 1990; Bowen *et al.*, 1981). Although some of the disadvantages of this method may be offset by the advantages of a large sample size, it has the potential to lead to misinterpretation of the pattern of life history. Some of these problems are discussed.

III. Constraints on Life Histories

Pinniped life histories have evolved under a combination of factors that are broadly based around the need for animals to balance their energy budgets. These include the constraints involved with (1) being homeothermic in water that is 25 times more conductive than air and (2) the high temporal and spatial variability in the distribution of resources within the marine environment. Phylogeny may also be seen as a constraint in that the ancestors of pinnipeds may not have possessed an ideal range of characteristics (physiological, anatomical, social, or distributional) for exploiting the marine environment. Therefore, current pinniped life histories may be constrained by difficulties with inherent mechanisms.

An example of such a constraint is the apparent necessity for a terrestrial (or pagophilic) phase during the reproductive cycle. This may be a consequence of the occupancy by ancestral pinnipeds mainly of temperate and polar marine habitats in which small neotates may have difficulty with thermoregulating in cold water, thereby necessitating terrestrial living for young neonates. Pinnipeds may have been locked into this form of reproductive cycle from an early stage in their evolution.

The constraint of the terrestrial phase in reproduction has brought with it other social and life history consequences. The necessity for mothers to find suitable terrestrial habitat (including ice) for parturition has more or less isolated, both spatially and temporally, the reproductive process from the feeding grounds. Species that exploit distant, unpredictable food sources require larger body mass than those that exploit food that is present at relatively close range to the pupping location. This is because there will be a critical duration over which a pup can be left without feeding and with low risk of starvation. If mothers cannot forage profitably during lactation within this critical duration, it is necessary for mothers to carry with them at parturition most of the food reserves required to raise their pup to independence (Boyd, 1998).

The extreme seasonality of food availability in higher latitudes has also led to extreme seasonality of reproduction, resulting in spatially and temporally synchronized reproduction. It is possible that both sexes have used this to affect greater mate choice, which has produced polygynous, highly competitive mating systems. These combined factors have led, in most species, to an annual cycle of reproduction.

IV. Costs vs Benefits of Reproduction

Even though individuals may have the option to reproduce annually, longitudinal studies show that they do not always exercise this option. Even when individuals do reproduce, they may adjust the amount of resources they supply to their offspring. The reasons for this are centered on the decisions that individuals must make during their life times in order to maximize their fitness, often measured in terms of number of offspring produced across their whole lifetime and not just one reproductive cycle.

There are obvious fitness gains from reproduction, but there are also costs involved. For example, in Antarctic fur seals (*Arctocephalus gazella*), reproduction in any year carries with it a 40% greater chance of dying in the following year. It also carries a similar cost in terms of reduced probability of breeding in the following year (Boyd *et al.*, 1995). In northern elephant seals (*Mirounga angustirostris*), mothers that reproduce for the first time at age 3 incur greater costs, in terms of reduced survival, than those that breed first at age 4 (Reiter and Le Boeuf, 1991). Female gray seals (*Halichoerus grypus*) that expend more on their offspring in 1 year also have reduced reproductive success in the following year (Pomeroy *et al.*, 1999). Thus, female pinnipeds must find a solution of how best to allocate energy between growth/maintenance and reproduction that optimizes the balance between fitness costs and benefits of reproduction. Those individuals that achieve the optimum balance will have greatest lifetime fitness. How pinnipeds make investment decisions in order to optimize this balance is unknown. In reality, few individuals may actually achieve the optimum, especially in variable environments, but natural selection favors those individuals that make investment decisions that approach the optimum most closely.

V. Age at First Reproduction

All pinnipeds experience a delay of several years in the time taken to reach sexual maturity (Table I). Several studies have shown that the age at first reproduction is not constant. In harp seals (*Pagophilus groenlandicus*) it is negatively related to population size (Bowen *et al.*, 1981), implying that the age at which individuals mature is density dependent [although see Trites and York (1993)]. Further evidence for a shift in age at sexual maturity with population size exists for crabeater seals (*Lobodon carcinophaga*) (Bengtson and Laws, 1985). The speed with which the change occurs shows that this is not an effect mediated by natural selection for individuals with different life history patterns, rather it is almost certainly driven by changes in the growth rates of individuals as population density and, by implication, per capita food availability changes. Consequently, the mean age at sexual maturity in a population may simply be a reflection of the mean growth rate.

Among northern elephant seals, females tend to begin breeding at age 3 or 4. The fitness of individuals that begin to breed at age 4 is greater than those that begin at age 3 because there is a cost, in terms of reduced survivorship, for those that began breeding at age 3 (Reiter and Le Boeuf, 1991). In Antarctic fur seals there is a similar disadvantage to breeding at an earlier age (Lunn *et al.*, 1994), although, for those individuals that survive, there is no subsequent effect on reproduction through the remainder of life.

These results suggest how age at sexual maturity can be determined by natural selection. In northern elephant seals and Antarctic fur seals there appears to be a trade-off between the fitness costs of breeding early in life and the fitness gains from early reproduction. Although, on average, individuals that begin breeding at age 3 have lower survival, it is possible that those

that breed at age 3 and survive have increased fitness mainly because they have, on average, one more reproductive attempt than those that begin breeding at age 4. Animals may opt to take a greater risk by breeding first at age 3 but with the prospect of greater ultimate lifetime fitness. For the trade-off between breeding first at age 3 or age 4 to operate and be evolutionarily stable, both strategies must have equal median lifetime fitness.

VI. Variations in Measures of Fitness

Strictly speaking, fitness should be measured in terms of the number of grandchildren that are produced by an individual. However, no study of pinnipeds has been able to do this, so a variety of fitness indices are used. The simplest and least informative of these is fertility rate, followed by weaning rate, proportion of offspring surviving their first year, and proportion of offspring surviving to reproductive age. There are specific examples of each of these measures from studies of pinnipeds.

Fertility rates in pinnipeds are normally in excess of 0.8 (Table I) and, given other vital rates in pinniped demography, they normally have to be of this order for populations to have the potential to grow. Longitudinal studies of individual pinnipeds show that most females experience fallow reproductive cycles in their lifetimes (Lunn *et al.*, 1994). It remains unclear if the observation of declining fertility with increasing age in cross-sectional samples of pinniped populations reflects senescence of individuals. The observation could equally be caused by greater survival rate, and therefore greater representation in older age classes, of individuals with intrinsically low reproductive rates.

Like age at sexual maturity, fertility is probably linked to the attainment of a critical minimum body condition at a specific stage of the reproductive cycle. In fact, physiologically, there may be virtually no difference between the process of puberty and the seasonal recrudescence of the reproductive system, so the two processes could be considered to be controlled by a common mechanism.

Fertility rates are influenced by previous experience of reproduction. In northern elephant seals, it appears that most females that miss a breeding attempt compensate for this by having a higher probability of weaning a pup in the following year, although, early in the reproductive life span, the opposite effect has been observed, i.e., individuals that miss a reproductive cycle have low success in the following year. Therefore, offspring quality may be affected by previous reproductive experience. Antarctic fur seals are significantly less likely to reproduce in a year following a reproductive attempt.

Weaning rates are affected by both age and previous experience of reproduction in northern elephant seals. It appears that although weaning rates increase initially with experience, these begin to decline later in life. This may represent a cumulative cost of reproduction that is manifest as senescence. However, it is still uncertain if this effect is an artifact of sampling caused by greater longevity in individuals that tend to skip reproduction more frequently or invest a smaller proportion of their energy reserves in their offspring.

In Weddell seals (*Leptonychotes weddellii*) offspring survival to age 1 and reproductive age both increase with maternal age and experience and, for male offspring, in relation to maternal body length (Hastings and Testa, 1998). Again, this suggests that those individuals that were able to invest more resources in their offspring, by virtue of their larger size and greater experience (perhaps reflecting the occupancy of better habitat), had enhanced fitness.

VII. Comparing Males and Females

Because females are the limiting sex and because it is much more difficult to study reproductive success in males, more attention has been focused on female than on male pinnipeds. Nevertheless, males may invest large amounts of their energy reserves in reproduction. In general, males have shorter life expectancies than females (shown by lower annual survival rates in Table I), but it is not clear how this is influenced by the investment in reproductive effort. Investment theory would suggest that the shorter life expectancy of males is because of their preparedness to take greater risks with their survival. The potential gains from reproduction, in terms of offspring, in males that are successful competitors because they make a large investment are greater than for females that are restricted to producing a single offspring per season.

There is also confusion in the literature about when males become sexually mature. The age at physiological maturity in males is probably similar to that of females, but many authors make a distinction between physiological and social maturity, which is defined by the age at which individuals are capable of competing for matings. Recent genetic evidence (e.g., Amos *et al.*, 1993) is casting doubt on some of the former interpretations of what social maturity actually means because the pattern of mating success in males often does not follow the pattern suggested by the observed social structure. In the near future, we may have to revise our views of the life history patterns of male pinnipeds.

VIII. Optimal Life Histories: Modeling the Way Forward

Life history analysis in pinnipeds is fraught with difficulties. Longitudinal studies in which individuals are studied throughout their lifetimes can only be carried out on a narrow range of accessible populations and they are expensive and logistically complex to maintain over the time periods (usually decades) required to achieve useful results. Cross-sectional studies are extremely limited in what they can tell us about the dynamics of life histories, and commercial harvests, the usual source of these data, are a thing of the past. We have to find a new way forward.

To date, almost all studies of pinniped life histories have been empirically based and, as pointed out in this description, they have highlighted the interactive nature of parameters such as longevity and reproductive rate. A modeling framework is required in order to allow these interactions to be investigated, to make better use of the data sets that already exist, and to identify critical gaps in the empirical data.

If a pinniped is to maximize its lifetime fitness F, then it must choose the optimal allocation of resources to reproduc-

tion through its lifetime. Thus, $F = f_1 + f_2 + f_3 \ldots f_n$ where f_a is the fitness contribution from year a in the life of the pinniped, which lasts n years. We know that there are certain functional relationships between maternal size or condition and the probability that she will reproduce or survive. If we assume that the relationship between offspring condition and its ultimate fitness is asymptotic, then, up to a certain level, the more energy that a female delivers to her offspring the greater will be her fitness. If the energy delivered to an offspring (e_a) is a proportion p of the energy available to the mother, then from what we know of the growth patterns and the energetic efficiencies of pinnipeds, it is possible to estimate the energy available for reproduction throughout the life span of an average individual. By setting rules that an individual will only reproduce if it has a sufficient excess of energy above that required for maintenance, we may be able to investigate the life history patterns in different environments as well as the effects of stochastic variability in food availability on life histories.

Many of the dynamic relationships described here should become explicit in the results of such an energy-based life history model. Similarly, such a model could help the interpretation of some of the cross-sectional population data in the context of dynamic life history processes. This type of approach seems to be essential if progress is to be made in pinniped life history analysis and for the full implications of life history analysis to be realized. Because the mechanism underlying population trajectories is the sum of individual life histories, understanding the environmental factors that affect life histories is fundamental to understanding population and species viabilities.

See Also the Following Articles

Cetacean Life History ■ Population Dynamics ■ Sirenian Life History ■ Sociobiology

References

Amos, W., Twiss, S., Pomeroy, P. P., and Anderson, S. S. (1993). Male mating success and paternity in the grey seal, *Halichoerus grypus*: A study using DNA fingerprinting. *Proc. R. Soc. Lond. B* **252**, 199–207.

Bengston, J. L., and Laws, R. M. (1985). Trends in crabeater seal age at sexual maturity: An insight into Antarctic marine interactions. *In* "Antarctic Nutrient Cycles and Food Webs" (W. R. Siegfried, P. R. Condy, and R. M. Laws, eds.), pp. 667–675. Springer, Berlin.

Boveng, P. L. (1993). "Variability in a Crabeater Seal Population and the Marine Ecosystem Near the Antarctic Peninsula. Unpublished Ph.D. thesis. Montana State University, Bozeman, MT.

Bowen, W. D., Capstick, C. K., and Sergeant, D. E. (1981). Temporal changes in the reproductive potential of female harp seals (*Pagophilus groenlandicus*). *Can. J. Fish. Aqu. Sci.* **38**, 495–503.

Boyd, I. L. (1985). Pregnancy and ovulation rates in grey seals (*Halichoerus grypus*) on the British coast. *J. Zool. Lond.* **205**, 265–272.

Boyd, I. L. (1998). Time and energy constraints in pinniped lactation. *Am. Nat.* **152**, 717–728.

Boyd, I. L., Croxall, J. P., Lunn, N. J., and Reid, K. (1995). Population demography of Antarctic fur seals: The costs of reproduction and implications for life-histories. *J. Anim. Ecol.* **64**, 505–518.

Fowler, C. W. (1990). Desity dependence in northern fur seals (Callorhinus ursinsus). *Mar. Mammal Sci.* **6**, 171–195.

Galimberti, F., and Boitani, L. (1999). Demography and breeding biology of a small, localized population of southern elephant seal (*Mirounga leonina*). *Mar. Mamm. Sci.* **15**, 159–178.

Harwood, J., and Prime, J. H. (1978). Some factors affecting the size of the British grey seal populations. *J. App. Ecol.* **15**, 401–411.

Hastings, K. K., and Testa, J. W. (1998). Maternal and birth colony effects on survival of Weddell seal offspring from McMurdo Sound, Antarctic. *J. Anim. Ecol.* **67**, 722–740.

Hindell, M. A. (1991). Some life-history parameters of a declining population of southern elephant seals, *Mirounga leonina. J. Anim. Ecol.* **60**, 119–134.

Huber, H. R., Rovetta, A. C., Fry, L. A., and Johnston, S. (1991). Age-specific natality of northern elephant seals at the South Farallon Islands, California. *J. Mammal.* **72**, 525–534.

Kjellqwist, S. A., Haug, T., and Øritsland, T. (1995). Trends in age-composition, growth and reproductive parameters of Barents Sea harp seals, *Phoc groenlandica. ICES J. Mar. Sci.* **52**, 197–208.

Le Boeuf, B. J., Morris, P., and Reiter, J. (1994). Juvenile survivorship of northern elephant seals. *In* "Elephant Seals: Population Ecology, Behavior, and Physiology" (B. J. Le Boeuf and R. M. Laws, eds.), pp. 121–136. University of California Press, Berkeley.

Lunn, N. J., Boyd, I. L., and Croxall, J. P. (1994). Reproductive performance of female Antarctic fur seals: The influence of age, breeding experience, environmental variation and individual quality. *J. Anim. Ecol.* **63**, 827–840.

McCann, T. S. (1985). Size, status and demography of southern elephant seal (*Mirounga leonina*) populations. *In* "Sea Mammals in South Latitudes: Proceedings of a Symposium of the 52nd ANZAAS Congress in Sydney, May 1982" (J. K. Ling and M. M. Bryden, eds.), pp. 1–17. South Australian Museum, Northfield.

Pomeroy, P. P., Fedak, M. A., Rothery, P., and Anderson, S. S. (1999). Consequences of maternal size for reproductive expenditure and pupping success of grey seals at North Rona, Scotland. *J. Anim. Ecol.* **68**, 235–253.

Reiter, J., and Le Boeuf, B. J. (1991). Life history consequences of variation in age at primiparity in northern elephant seals. *Behav. Ecol. Sociobiol.* **28**, 153–160.

Reiter, J., Panken, K. J., and Le Boeuf, B. J. (1981). Female competition and reproductive success in northern elephant seals. *Anim. Behav.* **29**, 670–687.

Smith, T. G. (1987). The ringed seal, *Phoca hispida*, of the Canadian Western Arctic. *Can. Bull. Fish. Aqu. Sci.* **216**, 81.

Sydeman, W. J., Huber, H. R., Emslie, S. D., Ribic, C. A., and Nur, N. (1991). Age-specific weaning success of northern elephant seals in relation to previous breeding experience. *Ecology* **72**, 2204–2217.

Testa, J. W. (1987). Long-term reproductive patterns and sighting bias in Weddell seals (*Leptonychotes weddelli*). *Can. J. Zool.* **65**, 1091–1099.

Testa, J. W., and Siniff, D. B. (1987). Population dynamics of Weddell seals (*Leptonychotes weddelli*) in McMurdo Sound, Antarctic. *Ecol. Monogr.* **57**, 149–165.

Testa, J. W., Siniff, D. B., Croxall, J. P., and Burton, H. R. (1990). A comparison of reproductive parameters among three populations of Weddell seals (*Leptonychotes weddellii*). *J. Anim. Ecol.* **59**, 1165–1175.

Trites, A. W., and York, A. E. (1993). Unexpected changes in reproductive rates and mean age at 1st birth during the decline of the Pribilof northern fur seal (*Callorhinus ursinus*). *Can. J. Fish. Aqu. Sci.* **50**, 858–864.

Wickens, P., and York, A. E. (1997). Comparative population dynamics of fur seals. *Mar. Mamm. Sci.* **13**, 241–292.

York, A. E. (1994). The population dynamics of northern sea lions, 1975–1985. *Mar. Mamm. Sci.* **10**, 38–51.

Pinniped Physiology

DANIEL E. CROCKER
Sonoma State University, Rohnert Park, California

DANIEL P. COSTA
University of California, Santa Cruz

Pinnipeds are unique among mammals because they feed in the marine environment and reproduce on land or ice, requiring a spatial and temporal separation of feeding from lactation. Seals stay at sea for weeks and often months at a time, yet they must spend considerable amounts of time on land. The amphibious nature of pinniped life has necessitated a wide range of physiological adaptations to life in water and on land. Pinnipeds must meet the physiological challenges of marine existence using specialized adaptations that still facilitates existence on land. This life history requires a remarkable plasticity of physiology. Broad categories of physiological adaptation include (1) aquatic locomotion, (2) apnea and diving physiology, (3) sensory physiology, (4) osmoregulation, (5) thermoregulation, (6) fasting physiology, and (7) lactation physiology.

Pinnipeds have had to overcome numerous problems associated with moving efficiently in the dense aquatic medium, and this adaptation has reduced their ability to move about on land. Otariids have hindflippers that can be turned under the body for terrestrial LOCOMOTION, whereas phocid seals cannot turn their hindflippers under the body and instead use lunging movements to get around on land.

Perhaps the most complex suite of adaptations required for making a living in the ocean is the physiology associated with breath-hold diving to foraging depths. In addition to adaptations for dealing with great pressures, pinnipeds exhibit physiological adaptation for apnea, increased oxygen storage, bradychardia, hypoperfusion, hypometabolism, and neuronal and hormonal control of cardiac and spleen function.

The sensory systems of pinnipeds enable them to successfully navigate, forage, and communicate in a variety of environments. Seals hear and see relatively well both in the air and underwater. Because the behavior of sound and light in water is markedly different than that in air, this again requires plasticity in their sensory physiology. Ultimately, sensory physiology must provide the appropriate visual and auditory information to facilitate social interactions on land, while allowing detection and capture of prey and detection and avoidance of predators at sea. Adaptations include well-developed underwater directional hearing and visual sensitivity at low light levels.

Living in salt water poses osmoregulatory problems for pinnipeds. In addition, pinnipeds must stay in water balance during periods on shore during which they may fast completely from food or water. Because animals also lose water for evaporate cooling, osmoregulatory strategies are linked to thermoregulation.

Pinnipeds are exposed to a remarkably variable range of environmental temperatures. They are able to tolerate frigid ocean temperatures at depth as well as high amounts of thermal radiation encountered when hauled out on land. Adaptations that help pinnipeds retain heat in the ocean environment, such as thick blubber or dense fur, may also promote overheating on land. Adaptations that may play a role in thermoregulation include large body size, blubber or dense fur, countercurrent heat exchange systems, and possibly high metabolic rates.

I. Fasting Physiology
A. Lipid Utilization and Protein Sparing

Many pinnipeds fast for extended periods during their breeding season or during molting (Table I). Mating, giving birth, nursing pups, and, for some species, molting all require long periods of time on land. This is particularly true of phocid seals, which undergo voluntary periods of prolonged fasting twice a year. Adult male pinnipeds may abstain from food or water for as long as 3 months while maintaining a territory or competing for dominance rank on the breeding ROOKERY. In many phocid species, females fast completely from food and water for over a month while delivering tremendous amounts of energy to their pups as milk. Offspring are weaned abruptly and, in many species, the pup then undergoes an extended postweaning fast before departing to sea and initiating foraging. This postweaning fast may be an important developmental time relative to the diving physiology of the offspring. In most cases, these extended fasts are associated with behaviors or processes resulting in considerable energy expenditure (e.g., combat, mating, lactation, molting). Adult animals may lose as much as 35–57% of stored body reserves during these periods (Fig. 1).

The lengths of these voluntary fasts may vary considerably. Fasts can last as long as 3 months in breeding males of both otariids and phocids. Otariid females alternate short onshore periods with foraging trips. These fasts can last from several days to 1–2 weeks postpartum. In phocid seals that fast throughout lactation, the fasting duration can be as short as 3–5 days in hooded seals (*Cystophora cristata*), whereas northern ele-

phant seals (*Mirounga angustirostris*) nurse a pup for 23–30 days after an additional 1–2 weeks fasting before parturition. Unlike most other groups of animals that undergo natural fasts, activity levels remain high. Males expend energy in territorial interaction, dominance interactions, and MATING BEHAVIORS. Females expend energy in agonistic encounters for breeding space, interaction with males, and for milk synthesis. Pups are also active during their fasts, making daily movements into the water and often exhibiting high movement rates.

Despite this high level of energy expenditure, seals are able to minimize the depletion of lean body mass, with the bulk of energy reserves coming from adipose tissue. Within the first weeks of the fast, rates of mass loss in nonlactating animals decrease markedly and then remain relatively stable and low for the remainder of the fast. This is accomplished primarily through a reduction in metabolic rate during the fast. This decline is evident in some species on a whole body basis as well as when corrected for changing body size and composition. The key adaptation for extended fasting appears to be the ability to spare protein while fasting, thereby reducing vital organ damage. This stage of fasting, called stage II fasting, is characterized by substantial decreases in blood urea nitrogen levels and urinary excretion of nitrogenous wastes. These characteristics are evident throughout the fasts of phocid seals (Fig. 2). This decreased protein degradation is reflected in reduced absolute and proportional declines in the use of protein reserves. Protein contributes as little as 1–6% of total energy utilization by the end of the fast. For example, at the beginning of the postweaning fast, northern elephant seals meet around 4% of their energy needs through protein catabolism. By the end of the postweaning fast this value has declined to around 1%.

Nonesterified fatty acids (NEFA) provide the majority of the animal's energy needs during long-term fasts. Increases in both turnover and plasma concentrations of NEFA have been demonstrated in several species. Reported NEFA values are greater than those reported for any other animal (as high as 3.1

TABLE I
Duration of Natural Fasts for Pinnipeds Exhibiting Extended Fasts during Breeding

Females		Males	
Crabeater seal	~4 weeks	Crabeater seal	~4 weeks
Gray seal	2.5–3 weeks	Gray seal	3–8 weeks
Hawaiian monk seal	5–6 weeks	Hooded seal	~4 weeks
Hooded seal	1.5–2 weeks	Leopard seal	Unknown
Leopard seal	Unknown	Northern elephant seal	2–3 months
Northern elephant seal	5 weeks	Ross seal	Unknown
Ross seal	Unknown	Southern elephant seal	2–3 months
Southern elephant seal	4 weeks	All fur seals	~2.5 months
Weddell seal[a]	6–7 weeks	All sea lions[b]	~2.5 months

[a]Females enter the water frequently and some may feed.
[b]California sea lion males enter the water periodically to feed. Fasting duration in the species is ~2 weeks.

Figure 1 *An adult female northern elephant seal early (right) and late (left) in the lactation period. Northern elephant seal females lose between 35 and 45% of body mass during breeding.*

mM) and increase over the fast in some species. Plasma glycerol levels show similar increases and probably are an important substrate for gluconeogenesis. There is some evidence that seals can selectively utilize reserves from different parts of the body (i.e., blubber reserves vs core tissues) during different phases of the fast. Ketone bodies (HBA) accumulate somewhat during the fast in weaned elephant seals and gray seals, *Halichoerus grypus,* and Weddell seals, *Leptonychotes weddellii,* and subsequently decline rapidly as the end of the fast is approached. This suggests that ketone bodies may contribute to energy metabolism during long-term fasting, although levels are significantly lower than nonfasting adapted species and never reach levels affecting acid–base balance or causing ketosis. Particularly striking in this regard are data from lactating northern elephant seals, who despite the aforementioned high and increasing NEFA levels over the fast, exhibit consistently low HBA values across lactation (0.03 to 0.13 mM). It is also important to note that increasing HBA levels were only demonstrated in juvenile animals of relatively small body size.

Stage III fasting or terminal starvation occurs when 30–50% of total body protein has been used. In nonfasting adapted species, this is associated with a decrease in lipid utilization and a decline in circulating ketone bodies. Evidence for entrance into stage III fasting in seals has been equivocal. The increase and subsequent decline in ketone bodies at the end of the fast in some fasting pups would suggest entry into stage III. However, only two studies have demonstrated increases in protein catabolism following the period of effective protein sparing. When considering the protein utilization by lactating females, it is important to include the loss of body protein for milk synthesis. One study on northern elephant seals demonstrated reductions in protein sparing with the depletion of lipid reserves that, together with the nutrient demands of milk synthesis, moved females close to the 30% value of body protein loss considered extreme in humans (Fig. 3). It may be that in normal, voluntary fasts, stage III fasting is never reached, with seals departing to sea before this point. Blubber reserves also play an important role in thermoregulation, and blubber depletion for energy needs is limited by the need to thermoregulate.

B. Hormonal and Fuel Regulation

Studies on hormonal and fuel regulation during fasting have suggested that seals may exhibit the protein conservation and high lipid utilization of stage II fasting throughout their lives. Fat is an important energy source throughout development and life,

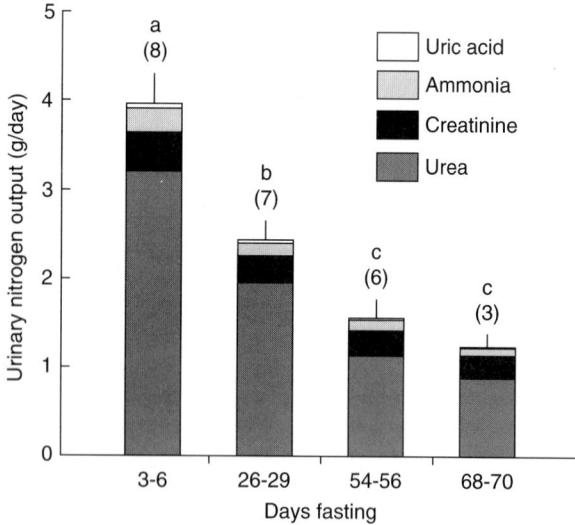

Figure 2 *Changes in daily urinary nitrogen excretion in fasting elephant seal pups. Letters denote significant differences between periods (P < 0.05). Samples sizes are in parentheses. From Adams and Costa,* J. Comp. Physiol. B *165, Water conservation and protein catabolism in northern elephant seal pups during the post weaning fast. Copyright 1993 Springer-Verlag.*

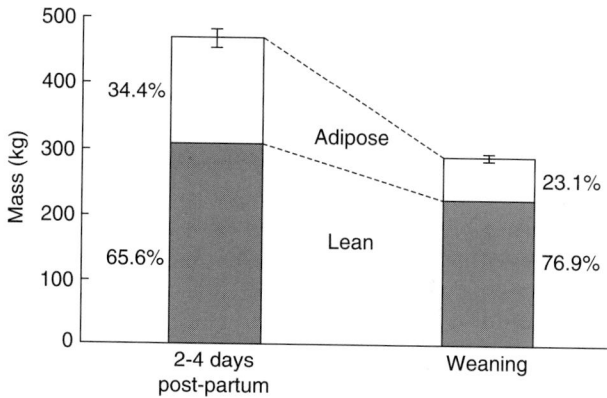

Figure 3 *Changes in maternal mass and body composition over the lactation period in female northern elephant seals. On average, females lose 27% of total body protein stores during lactation.*

including high-fat milk, high-fat fish, and body fat stores. Studies of the role of insulin in glucose regulation in suckling pups have suggested that seals may be preadapted for fasting. Low insulin concentrations, impaired glucose clearance, and low insulin to glucagon ratios exhibited in both feeding and fasting pups contribute to the mobilization of lipids from body stores. In general, fasting pups are hyperglycemic and hypoinsulemic. Slow recovery of baseline blood glucose levels after insulin injection suggests that blood glucose concentrations are not closely regulated in a typical fashion by the release of the hormones insulin and glucagon. In elephant seals, insulin to glucagon ratios were less than one, which is correlated with high rates of hepatic gluconeogenesis in other mammals. The direct contribution of glucose to the total metabolic rate is less than 1% in seals during extended fasts. Fatty acid oxidation studies confirmed that lipid is the main energy source and suggest that glycerol liberated by lipolysis may provide substrate to meet the animal's glucose needs. Studies on elephant seals, harbor seals (*Phoca vitulina*), and gray seals suggest that glucose turnover rates are typical of mammals, although glucose carbon is predominantly recycled and oxidation rates are low. Recycling protein and glucose carbon may serve as an important shuttle mechanism for carbon (e.g., synthesis of nonessential amino acids).

Currently, our understanding of the fasting physiology of pinnipeds is biased toward studies on juvenile animals, especially the postweaning fast. Investigations have demonstrated dramatic differences in fasting physiology between lactating and molting females as well as fasting pups. The pressures of nutrient delivery for milk synthesis may have significant impacts on the metabolic strategies used by females during extended fasts. Very little is known about fasting in adult males, who potentially have some of the highest metabolic rates while fasting. Future studies can benefit from interspecific comparisons of fasting physiology relative to natural fasting durations. Even more instructive may be intraspecific comparisons among sexes and age classes during development, lactation, molting, and breeding. These comparisons will help demonstrate how

the fasting biochemistry of animals responds to the varying energy and nutrient demands of breeding and lactation.

C. Renal Physiology during Fasting

Water balance during fasting is maintained by the input of metabolic-derived water from lipid catabolism. Significant reductions in urinary water loss contribute to the maintenance of water balance (Fig. 4). The low rate of protein oxidation and an efficient urinary-concentrating mechanism in pinnipeds reduce urinary water loss during fasting. Early work on harbor seals demonstrated reductions in glomerular filtration rates (GFR) associated with fasting and hyperfiltration after feeding. Subsequent investigations have been equivocal, leaving it unclear whether the mechanism underlying reduced urine flow is decreased glomerular filtration or increased tubular resorption. Investigations on weaned northern elephant seal pups have revealed no correlation between plasma levels of vasopressin and urinary concentrating ability. Similar investigations have demonstrated a fivefold increase in plasma aldosterone concentration during the first 5 weeks of fasting that then began to decrease. This pattern mirrors reported changes in urine osmolarity, which increases through the eighth week of the fast and then declines. This suggests an important role for aldosterone in regulating urine concentration by its action on sodium resorption in the collecting duct.

Investigations on lactating adult northern elephant seals have demonstrated dramatic increases in GFR across the fast and suggest that these elevated rates could be an adaptation to increasing the efficiency of urea excretion during reduced urine flow. This mechanism reduces residency time and passive resorption of urea in the collecting tubules. The efficiency of urea excretion in lactating females declined from 49 to 38% over lactation, suggesting that with a declining urine flow and stable plasma urea concentration, increased GFR is necessary to increase urea excretion as protein catabolism increases.

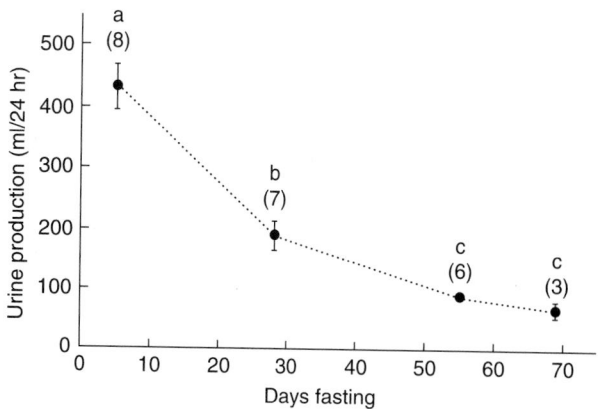

Figure 4 *The decline in daily urine production of elephant seal pups progressing through the postweaning fast. Letters denote significant differences between periods (P < 0.05). Samples sizes are in parentheses. From Adams and Costa, J. Comp. Physiol. B 165, Water conservation and protein catabolism in northern elephant seal pups during the post weaning fast. Copyright 1993 Springer-Verlag.*

II. Immune System

One aspect of pinniped physiology that warrants further investigation is the immune system. For the polygynous species that breed on land, BREEDING SITES are potentially pathogenic environments. Many animals with open wounds (as high as 90% of adult males in some species) are packed closely together on a substrate soiled with excrement and rotting corpses. Neonates with poorly developed immune systems are born directly onto this substrate. Investigations have suggested depressed humoral immunity in neonates and that rapid postpartum increases in humoral immunity are the result of *de novo* synthesis, with a reduced role for passive maternal transfer. Interestingly, some studies have suggested increases in humoral immune response in adult females during breeding.

III. Lactation Physiology

The physiology of lactation in pinnipeds is significantly impacted by constraints resulting from the temporal separation of foraging and parental investment. Pinnipeds have evolved two general lactation strategies to manage pup provisioning within these constraints. After a short perinatal fast, otariids alternate foraging trips with suckling bouts. Initial milk production is synthesized from maternal reserves, whereas subsequent milk nutrients are derived from resources acquired while foraging. Phocids, particularly the larger species, fast during a brief but energy-intensive lactation during which nutrients for milk synthesis are derived exclusively from maternal tissues. Some smaller phocid seals, such as harbor seals or ring seals, *Pusa hispida*, forage during lactation. Pinnipeds consistently produce lipid-rich milk, independent of lipid content of the diet. Even more amazing is that lactation, an energetically expensive period, occurs while the female is fasting. Long-term fasting is characterized by

protein sparing, reductions in metabolic rate, and reductions in water loss for urea nitrogen excretion. Studies on nonlactating fasting phocids have shown that protein stores are spared with the bulk of energy demands being supplied by the oxidation of fatty acids. In contrast, lactation is characterized by dramatic increases in metabolism and significant transfer of nutrients and water to the mammary gland for the synthesis and secretion of milk. The general metabolism of the lactating female is reorganized in a way that ensures the appropriate nutrients are partitioned to the mammary gland. In nonfasting animals, lactation is accompanied by increased levels of food consumption and digestion, with accompanying increases in the absorptive capacity of the gastrointestinal tract. In fasting phocids, regulatory mechanisms override protein and energy-sparing mechanisms to make the nutrients necessary for milk synthesis available at the expense of body nutrient reserves. The high demands of lactation coupled with complete abstinence from food and water present a complex regulatory problem. An investigation in northern elephant seals suggested that changes in the energy demands of milk synthesis across lactation may impact fasting physiology and ultimately limit the period of parental investment. From this perspective, lactating phocid females may be one of the best examples of homeorhesis "orchestrated changes for the priorities of a physiological state," found in nature.

Studies have suggested changes in milk composition and the nutrient requirements of milk synthesis across lactation in pinnipeds. These patterns are controlled by hormonal and biochemical changes. Of these changes, those that impact mobilization of adipose tissue stores, metabolism of lipids, and utilization of lipids by the mammary gland are the most significant. These changes are also important as pinnipeds transition from periods of nutrient deposition and mobilization for milk synthesis. Decreased insulin levels remove the strong antilipolytic effects of this hor-

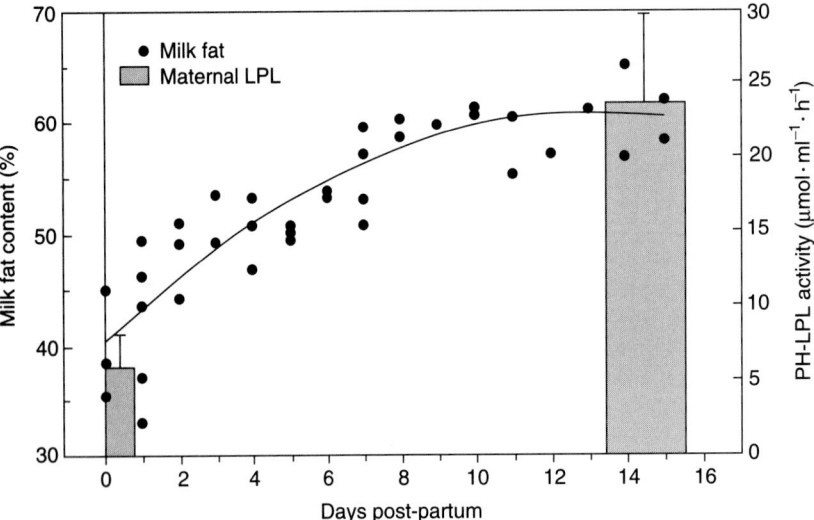

Figure 5 *Changes in maternal postheparin LPL activity in relation to changes in milk fat over lactation in gray seals. From Iverson et al., J. Comp. Physiol. 165, Lipoprotein lipase activity and its relationship to high milkfat transfer during lactation in grey seals. Copyright 1995 Springer-Verlag.*

mone. Cortisol and other glucocorticoids influence lipid metabolism directly and indirectly. Cortisol stimulates hormone-sensitive lipase in adipose tissue and antagonizes the actions of insulin. Hormone-sensitive lipase activity increases lipid mobilization from adipose tissue. Lipoprotein lipase (LPL) is the primary enzyme involved in directing triglycerides mobilized from tissue stores to tissues for utilization. LPL is bound to tissues and facilitates the hydrolyzation of triglycerides, allowing uptake of fatty acids by the tissue. Under normal conditions of insulin release, LPL activity in tissues increases and triglyceride is cleared from the blood. During fasting, LPL activity in adipose tissue decreases, while hormone-sensitive lipase activity in adipocytes increases. The general pattern found in lactation is a decrease in adipocyte LPL activity before parturition and an increase in mammary gland LPL activity. The hormone prolactin is believed to be primarily responsible for LPL regulation in lactation. Investigations on harbor seals and gray seals have suggested a similar pattern. General activity levels of LPL increased 10-fold over lactation in these species and were significantly higher than levels found in humans (Fig. 5). The dramatic increase in milk lipid content early in lactation in some phocids seals may in part be explained by a significant increase in mammary gland LPL activity.

Very little work has been done on the physiology underlying milk secretion in pinnipeds. The release of milk fat globules occurs by apocrine secretion, in which the apical portion of the cell membrane is sloughed off. The high fat contents of pinniped milk suggest significant increases in the amount of membrane and its turnover. Pinnipeds may partially reduce this requirement by utilizing larger fat globules that require smaller amounts of membrane loss per unit lipid secreted. Data on mammary gland size in phocids have been equivocal, suggesting increased size relative to body mass in some species but not others. In any case, phocid seals appear to be particularly efficient at mobilizing and transporting nutrients to the mammary gland, perhaps by reducing the levels of *de novo* synthesis of milk lipids occurring at the mammary gland. This efficiency must in turn be matched by rapid and efficient digestion and assimilation of milk lipids by the offspring.

See Also the Following Articles

Cetacean Physiology ■ Diving Physiology ■ Endocrine Systems ■ Energetics ■ Health ■ Kidney, Structure and Function ■ Osmoregulation

References

Adams, S. H., and Costa, D. P. (1993). Water conservation and protein metabolism in northern elephant seal pups during the post-weaning fast. *J. Comp. Physiol. B* **163**, 367–373.

Castellini, M. A., and Rea, L. D. (1992). The biochemistry of natural fasting at its limits. *Experientia* **48**, 575–582.

Iverson, S. J. (1993). Milk secretion in marine mammals in relation to foraging: Can milk acids predict diet? *In* "Recent Advances in Marine Mammal Science" (I. Boyd, ed.), pp. 263–292. Symposium Zoological Society of London No. 66, Oxford Univ. Press, Oxford.

Iverson, S. J., Hamosh, M., and Bowen, W. D. (1995). Lipoprotein lipase activity and its relationship to high milk fat transfer during lactation in grey seals. *J. Comp. Physiol. B* **165**, 384–395.

Kirby, V. L., and Ortiz, C. L. (1994). Hormones and fuel regulation in fasting elephant seals. *In* "Elephant Seals: Population Ecology, Behavior, and Physiology" (B. J. Le Boeuf and R. M. Laws, eds.), pp. 374–386. University of California Press, Los Angeles.

Wartzok, D. Physiology of behavior in pinnipeds. (1981). *In* "Behavior of Pinnipeds" (D. Renouf, ed.), pp. 236–299. Chapman and Hall, London.

Pirate Whaling
SEE *Illegal and Pirate Whaling*

Plankton

AKITO KAWAMURA
Kyoto, Japan

Plankton is the collective name given to the assemblage of free-swimming or suspended microscopic organisms considered too small to move independently of ocean currents. Large animals that are able to disperse under their own power are called nekton. The distinction between plankton and nekton, however, is sometimes blurred. For example, larger animals that are capable of limited self-propulsion, such as jellies and salps, are often included in the plankton. Large euphausiids, such as Antarctic krill, whose ability to actively determine and maintain their position is poorly understood, have been referred to as either macroplankton or micronekton. Phytoplankters are plants, and zooplankters are animals.

I. Phytoplankton

Phytoplankton consists of microscopic unicellular plants and forms the basis of marine ecosystems; nearly all life in the sea derives from the solar energy fixed in photosynthesis by these plants. Two factors control phytoplankton growth: light irradiance and nutrients. Light is only available in the top layers of the oceans, variably down to about 200 m or less, whereas nutrients are more abundant in deeper layers. Evolution of small size has enabled phytoplankton to absorb scarce nutrients through maximizing the ratio of surface area to volume. Small size, down to 2 μm, also confers high buoyancy and a low sinking rate, keeping the cells near the surface. Grazers on the phytoplankton must also be extremely small in order to be able to feed on them. As well as being the "grass of the sea," phytoplankton comes into cetology in one of the common names of the blue whale (*Balaenoptera musculus*), "sulphurbottom," which refers to a yellowish-brown layer of diatoms (single-celled epiparasitic algae) accumulating on the whale while it feeds in cold polar waters.

II. Zooplankton

Zooplankton consists of animals from several taxonomic groups from Protozoa to Vertebrata and is a main source of food for many marine mammals. Carnivorous, omnivorous, and herbivorous zooplankters have been found in the stomach of baleen whales; three groups of crustaceans are the most important: copepods, euphausiids, and amphipods. These and other planktonic animals have developed a wide variety of specialized mechanisms and techniques for feeding on phytoplankton and suspended particulate matter, including appendicular nets and guiding whorls in copepods, finely structured appendages in euphausiids, and ciliary movements in pteropods such as *Limacina* and *Cavolinia*. "Basket feeding" in the Antarctic krill, *Euphausia superba*, is a kind of mass raptorial feeding. Nektonic animals have also developed filters by modifying gill rakers into functional sieves, e.g., as in the basking shark (*Cetorhinus maximus*) and whale shark (*Rhincodon typus*). Buccal teeth with well-developed accessory cusps also function as sieves in crabeater seals (*Lobodon carcinophaga*) and leopard seals (*Hydrurga leptonyx*), which feed on krill. The most highly derived filtering system among the vertebrates is that of baleen in whales, which enables them to utilize varying sizes of minute particles suspended in the viscose water medium.

III. Plankton and Whales

Whales can feed efficiently on zooplankters because they occur in dense swarms. Their patchy distribution and aggregation may be partially passive and due to oceanographic features such as Langmuir circulation but may also involve active processes on their part. When Antarctic krill swarm densely at the surface, the water is discolored with brick-red patches due to their carotenoid pigmentation. A swarm is usually a species-specific phenomenon, as in birds or fishes. In a particular feeding ground in the subantarctic area, sei whales (*Balaenoptera borealis*) and southern right whales (*Eubalaena australis*) feed on copepods (*Calanus tonsus*) that swarm at a density of 21,000–23,000 indi-viduals/m^3, or 29–35 g/m^3. The background density in the same region is fewer than 1/m^3 (Kawamura, 1974). Based on acoustic data, Antarctic krill has been found to swarm at a density of some 60,000 individuals/m^3 or more or $\geq$33 kg/m^3 (Miller and Hampton, 1989). The baleen whales exploit these swarms, harvesting immense amounts of biomass in a process that could be compared to seining pelagic fish schools. The concentration of biomass by swarming zooplankton is incredible; it has been estimated that the energy represented by the phytoplankton in 5 million gallons of seawater is equal to that in only 1 pound of black beans (Fraser, 1969), yet baleen whales may efficiently collect some 4% of body weight daily by exploiting swarms of zooplankton that feed on the phytoplankton (Fig. 1).

The stomach contents of a whale usually consist of zooplankton of a single species, one of only a limited number of major species (Table I), depending on location and species of whale. This is due to the swarming habit, which does not occur in all planktonic crustaceans (Kawamura, 1980a). Sometime, however, two or more species may be found in a single stomach due to accidental engulfment or to feeding on swarms of different species. The dominant copepods in the North Pacific are *Eucalanus bungii* and *Metridia pacifica* and in the Southern Ocean *Calanus propinquus*, *Calanoides acutus*, and *M. gerlachei*, but these are not fed on by whales because they do not swarm. In temperate and tropical waters, the Bryde's whale (*Balaenoptera edeni*) feeds on the swarming euphausiids *Pseudeuphausia latifrons*, *Thysanoessa gregaria*, *Euphausia recurva*, and *E. diomedeae*. No swarming copepods have been found in the Indian Ocean, despite high species diversity in that region. Balaenopterid and balaenid whales each show similar food habits in different oceans and seas but exhibit different food preferences (Bowen and Siniff, 1999); the former feed by "swallowing" (engulfing) large

Figure 1 *Krill* (Euphausia superba *Dana*) *gushing out from a fin whale stomach in the Antarctic Ocean. Note uncontaminated composition with other prey items.*

TABLE I
Zooplankton Species Representing Major Components of the Stomach Contents of Whalebone Whales

Feeding ground	Zooplankton species
Antarctic	*Euphausia superba, E. crystallorophias*
Sub-Antarctic	*Euphausia vallentini, Thysanoessa macrura, Nyctiphanes australis,[b] Calanus tonsus, C. simillimus, Clausocalanus laticeps, Drepanops pectinatus, Parathemisto gaudichaudii, Munida gregaria[c]*
Northern North Pacific and Bering Sea	*Thysanoessa raschii, T. inermis, T. longipes, Nematoscelis magalops*
	Neocalanus cristatus, N. plumchrus, N. flemingeri, Limacina helicina
Sub-Arctic and transitional North Pacific	*Euphausia pacifica, E. recurva, Thysanoessa spinifera, Calanus pacificus, Sergestes similis*
Pacific Mexican waters	*Pleuroncodes planipes*
North Atlantic	*Meganyctiphanes norvegica, Thysanoessa inermis, Calanus finmarchicus sl., Temora longicornis*
Tropical Eastern Indian Ocean	*Euphausia diomedeae, E. sibogae, Pseudeuphausia latifrons, Thysanopoda tricuspidata*
Coral Sea, and Kermadec	*Euphausia diomedeae, Euphausia recurva, Thysanoessa gregaria*

[a]Adapted from Kawamura (1974, 1980a,b, 1982) and Gaskin (1982).
[b]New Zealand waters.
[c]Patagonian waters.

TABLE II
A Rough Summary of the Number of Species Found in the Stomachs of Baleen Whales[a]

Feeding ground	Taxa						
	Euphausiacea	Copepoda	Amphipoda	Mysidacea	Decapoda	Pisces	Squid
Southern Ocean	12	7	1	—	2	14	3
South Africa[b]	8	20	7	—	1	10	—
Australia and New Zealand	4	2	1	—	1	2	—
South Pacific and Coral Sea	4	—	—	—	—	—	—
Brazilian waters	6	—	—	—	—	2	—
Eastern tropical Indian Ocean	3	—	—	—	—	1	—
Temperate Indian Ocean	2	—	—	—	—	1	—
Northern North Pacific and Bering Sea	6	4	1	1	2	10	>1
Far eastern seas	3	8	—	—	—	16	2
Temperate and tropical Pacific	8	3	1	—	1	12	3
California and Mexico	3	1	—	—	1	1	1
Japan and Ryukyus	6	2	—	—	—	7	1
Northeastern Atlantic	4	2	2	—	—	15	—
Northwestern Atlantic	2	2	—	—	—	5	(1)[c]

[a]The anomalously high number of copepod species in South African waters may reflect supplemental feeding en route to and from the northern breeding ground. Adapted from Kawamura (1980b).
[b]Five pteropod species could be added.
[c]Doubtful because no scientific name is known, but "squid."

amounts of water, which allows them to also feed on squid and fish of suitable size as well as euphausiids, whereas balaenids feed largely by "skimming" for copepods (Mitchell, 1974). However, the bowhead whale (*Balaena mysticetus*) is also known to feed near the bottom on planktonic amphipods in the Arctic. Gray whales (Eschrichtiidae: *Eschrichtius robustus*) feed mainly on benthic amphipods in the Bering and Chukchi Seas but also consume planktonic crustaceans and fish eggs and larvae in the southern parts of their range (Darling *et al.*, 1998).

IV. Plankton and Seals

Seals also feed on plankton, but to a lesser extent than baleen whales, with the degree depending on species, age, breeding condition, and season. Some Antarctic seals eat euphausiids. Two species (*Euphausia superba* and *E. crystallorophias*) were found in 1 of 48 Weddell seals (*Leptonychotes weddellii*), 94 of 100 crabeater seals, 37 of 159 leopard seals, and 9 of 21 Ross seals (*Ommatophoca rossii*) (Laws, 1984). The high proportions for crabeater and leopard seals reflect the specialization of their dentition for straining. In addition, Ross and Weddell seals feed on benthic and near-bottom planktonic amphipods, isopods, and mysids. In leopard seals (*Hydrurga leptonyx*), the proportion of krill in the diet decreases with age; krill were found in the stomachs of 87% of juveniles (vs 23% overall). Larga seals (*Phoca largha*) and ice-breeding harbor seals (*P. vitulina*) in the Okhotsk Sea feed largely on fish, but pups of harbor seals, ribbon seals (*Histriophoca fasciata*), and ringed seals (*Pusa hispida*) feed on swarming euphausiids, such as *Thysanoessa inermis* and *T. raschii*, which appear in association with the pack ice of their habitat (Kato, 1982). This planktonic food is important after weaning until the seals become able to catch fish or squid.

V. An Efficient (but Vulnerable?) Ecology

The marine ecosystem is a world of size-dependent structures, based at the bottom on suspended organic particles of phytoplankton and zooplankton. Annual global marine primary production of about 2×10^{10} tons of carbon has been estimated to produce some 2.4×10^8 tons of fish after passing through up to five trophic levels (Ryther, 1969). The baleen whales shorten this chain, as do other top krill-eating predators in the Antarctic such as seals and penguins. Solar energy is transferred to large animal production with greater efficiency than in any other ecosystem. A 30-m blue whale can consume vast numbers of 4–5-cm krill. Similarly, a 0.3-cm copepod is available to a 13- to 16-m sei whale. The ratio of predator length to prey length is as great as 8300:1. These are the most extreme cases of size disparity between predator and prey in marine ecosystems (Berta and Sumich, 1999). Relatively few species of prey are involved (Table II), e.g., only 17–20 of the 85 or so known species of euphausiids and only 7–10 of the approximately 1800 known species of copepods (Mauchline, 1998). This may indicate ecological vulnerability of the baleen whales, as they have few options for substitution of food species. The distribution and abundance of the largest baleen whale stocks are/were determined by the distribution of their favored densely swarming planktonic prey (Nemoto and Kawamura, 1977). The most marked case is that of the highly stenophagous Southern Ocean blue whale, which feeds exclusively on *E. superba* (Clapham *et al.*, 1999).

See Also the Following Articles

Baleen ■ Bioluminescence ■ Cetacean Ecology ■ Diet ■ Krill ■ Predator–Prey Relationships

References

Berta, A., and Sumich, J. L. (1999). "Marine Mammals: Evolutionary Biology." Academic Press, San Diego, CA.

Bowen, W. D., and Siniff, D. B. (1999). Distribution, population biology, and feeding ecology of marine mammals. *In* "Biology of Marine Mammals" (J. E. Reynolds III and S. A. Rommel, eds.), pp. 423–484. Smithsonian Institution Press, Washington, DC.

Clapham, P. J., Young, S. B., and Brownell, R. L., Jr. (1999). Baleen whales: Conservation issues and the status of the most endangered populations. *Mam. Rev.* **29**, 35–60.

Darling, J. D., Keoghi, K. E., and Steeves, T. E. (1998). Gray whale (*Eschrichtius robustus*) habitat utilization and prey species off Vancouver Island, B. C. *Mar. Mamm. Sci.* **14**, 692–720.

Fraser, J. H. (1969). Plankton resources. *In* "The Encyclopedia of Marine Resources" (E. F. Frank, ed.). Van Nostrand Reinhold.

Gaskin, D. E. (1982). "The Ecology of Whales and Dolphins." Heinemen, London.

Kato, H. (1982). Food habits of largha seal pups in the pack ice area. *Sci. Rep. Whales Res. Inst. Tokyo* **34**, 123–136.

Kawamura, A. (1974). Food and feeding ecology in the southern sei whales. *Sci. Rep. Whales Res. Inst. Tokyo* **26**, 25–144.

Kawamura, A. (1980a). Food habits of the Bryde's whales taken in the South Pacific and Indian Oceans. *Sci. Rep. Whales Res. Inst. Tokyo* **32**, 1–23.

Kawamura, A. (1980b). A review of food of balaenopterid whales. *Sci. Rep. Whales Res. Inst. Tokyo* **32**, 155–197.

Laws, R. M. (ed.) (1984). Seals. *In* "Antarctic Ecology," Vol. 2, pp. 621–715. Academic Press, London.

Mauchline, J. (1998). The biology of calanoid copepods. *Adv. Mar. Biol.* **33**, 1–701.

Miller, D. G., and Hampton, I. (1989). Biology and ecology of the antarctic krill (*Euphausia superba* Dana): A review. *BIOMASS Handbook Ser.* **9**, 1–166.

Mitchell, E. (1974). Trophic relationships and competition for food in Northwest Atlantic whales. *Proc. Can. Soc. Zool. Ann. Mtg.* **1974**, 123–133.

Nemoto, T., and Kawamura, A. (1977). Characteristics of food habits and distribution of baleen whales with special reference to the abundance of North Pacific sei and Bryde's whales. *Rep. Int. Whal. Commn. (Special Issue)* **1**, 80–87.

Ryther, J. H. (1969). Photosynthesis and fish production in the sea. *Science* **166**, 72–76.

Playful Behavior

BERND WÜRSIG

Texas A&M University, Galveston

Play consists of actions performed for no other apparent purpose than their own enjoyment. However, it is recognized that play occurs in young animals to learn motor and social skills needed to survive. Play as "enjoyment" may have evolved simply because something enjoyable will be sought after, and if needed actions of learning are enjoyable, they will be done. Tussling sibling brown bears, rolling and cuffing each other, are obviously playing. But it has long been a truism that such play in the proximal sense is vital in learning self-defense and in establishing rules of association. Play tends to decrease in frequency as mammals become older and does not often occur in adults. Obvious exceptions are some primates and cetaceans, although as behavioral studies gather details, researchers are learning that play in adults of other species is actually more common than described previously. Play may also be an attempt to relieve boredom, and we would expect play to be especially well developed in the larger brained, behaviorally flexible mammals (Goodall, 1990).

I. Motor Imitation

Many marine mammals seem to be especially good at imitating the actions of their conspecifics or of individuals of other species. Thus, untrained dolphins in oceanaria have been described as performing a colleagues' trained repertoire essentially flawlessly when called upon to do so. Apparently, the dolphins learned the motor actions simply by observation (Pryor, 1995). While this by itself is not play, the capability of imitation is often expressed as play: dolphins have imitated a diver's movements of cleaning the pool; as well as the grooming and swimming movements of seals and other pool inhabitants. The dolphins would generally approach the object of imitation, slow their own travel to approximate that of the slower coinhabitant, and then move their body in exaggerated imitation of movements of the other individual. The human diver, alternately bending and straightening at the waist as he cleaned the aquarium tank with a rubber scraper, was imitated by the dolphin bobbing its head and neck up and down in rapid and jerky fashion. At the same time, the dolphin released clouds of bubbles from its blowhole in synchrony with the bubbles of the diver's air regulator and made a squeaking sound in an apparent attempt to reproduce the squeaks of the rubber on glass. The author has seen similar behavior in nature, with a bottlenose dolphin (*Tursiops truncatus*) adult in the Bahamas imitating a particularly clumsy tourist who had difficulty in descending below the water as she propelled herself with her skin dive flippers and by rotating her arms. The dolphin matched her speed, alternately bobbed at the surface and descended in unison with the woman, jerked its peduncle and tail back and forth in unnatural fashion, while at the same time rotating its short front flippers as if they were flailing arms. The effect looked highly hilarious, and it would be difficult to rationalize the behavior as anything but a bit of malicious fun, or play.

II. Vocal Imitation

While motor or physical imitation seems to be mainly in the purview of TOOTHED WHALES, other marine mammals also practice vocal imitation. This imitation may be an outgrowth of learning one's own species-specific (and perhaps group or pod-specific, as in killer whales, *Orcinus orca*) vocalizations, but the capability can then become a method of play. In the 1970s, a captive male harbor seal (*Phoca vitulina*), named Hoover by his caretakers, was capable of imitating the voice of a human worker who frequented his area, complete with a New England accent and a bit of a drunk-sounding slur (Ralls *et al.*, 1985).

Beluga whales (*Delphinapterus leucas*) can also imitate human sound and will at times use these imitations in apparent mischievous play. While dolphins cannot imitate human sounds very precisely (they seem to lack the vocal capability, not the intelligence, to do so), they will easily imitate clicks, whistles, barks, scrapes, and squeaks (as the fore-mentioned window-washing sounds) that can occur in an aquarium with other animals. A particularly readable account of imitation, innovation, intelligence, and cognition can be found in Tyack (1999).

III. Examples of Play

When we meet marine mammals underwater, we are apt to be subjects of intense curiosity. Sea lions, fur seals, and harbor seals will dash around us, pirouette in front of us, and gaze at us. Interestingly, these same animals ignore us or become wary if we approach too close to them when they are hauled out on shore. When in their watery milieu, however, fear is gone. Manatees and dugongs are similarly attracted to humans underwater, except in those places where they are hunted. This curiosity can turn to play. Just about every researcher who dives with marine mammals has tales of sea lions tugging on his or her flippers; manatees and dugongs mouthing an arm or leg in apparent playfulness; and dolphins playing "the leaf game," where they present the human with a bit of flotsam and wait or return until it is given back to them (Johnson and Norris, 1994).

However, although many people prefer to think that dolphins and humans have an especially play-prone bond, dolphins (and pinnipeds) are known to interact in playful fashion with many other animate and inanimate objects in their environment. Thus, they may "tease" a sea turtle by mouthing and pulling on its tail or legs or they may swim beside a like-sized shark, imitating every movement that the shark makes. Some species of dolphins are also known for adroitly balancing pieces of flotsam on their jaws, flippers, dorsal fin, or tail or carrying a piece of algae, plastic, or other pliable object in a manner to keep it balanced on an appendage simply by force of the forward movement of the body. This activity tends to take place when the group of animals is being highly social, usually with much sexual activity as well. The dolphin playing with the inanimate object, however, tends to be alone, apparently transferring its sense of socializing to playing with the object (Fig. 1).

While apparent play is often clouded by the possibilities of curiosity or learning, the author has witnessed several clear-cut examples of play in nature, detailed in the next three sections.

A. The Mischievous Dusky Dolphin

When dusky dolphins (*Lagenorhynchus obscurus*) have fed and socialized, they are in a very playful mood. They race toward a boat from kilometers away, attempting to ride the bow wave of the vessel. They leap in acrobatic somersaults and flips that appear to show sheer exuberance and orient toward, bite at gently, and otherwise manipulate objects in their environment. There is much social/sexual interaction and apparent play, often in twos and threes, but lone animals carry pieces of kelp on their rostrum or melon and anterior edges of flippers, dorsal fin, or fluke. Most of these behaviors could be interpreted as being a part of learning or honing skills, but the author has repeatedly seen nonequivocal play, as follows.

Figure 1 *A dusky dolphin* (Lagenorhynchus obscurus) *playing with kelp in Patagonia, Argentina.*

After dolphins have fed on schooling anchovy near the surface, marine birds who took advantage of the aggregated fish ball sit on the water, digesting. These are of a variety of species, but almost always include kelp gulls (*Larus dominicanus*) and brown-hooded gulls (*L. maculipennis*), yellow or red legs, respectively, dangling below a round white rump. The sight seems almost irresistible for some dolphins, who slowly circle one of these birds, approach it at an angle and from the rear, very gently open their jaws around one or both dangling appendages, and then rapidly but not totally close the mouth and surge forward and down. This action drags the bird below water by about 30 to 60 cm before the mouth is reopened and the bird literally pops back to the surface. The bird frantically flutters and preens before it flies off. The dolphin meanwhile slowly swims off, at times to attempt the same "trick" with another hapless gull. At all times (this has been seen about one dozen times to date), the bird is unharmed, suggesting particular finesse as the dolphin closes its pointy-toothed jaws and surges forward. The intent appears to be to surprise, certainly not to harm, perhaps like a child sneaking up behind a person to pop a balloon. The same attempts have been seen with dolphins orienting toward the dark feet of Magellanic penguins (*Spheniscus magellanicus*) bobbing at the surface, but the penguins, ever aware of their environment below, simply paddle rapidly ahead or dive before the dolphin can complete its action.

B. Creating a Bow Wave

Dolphins ride or surf on oceanic waves and those created by ships. They also ride on the fleeting bow waves created by rapidly surging large whales, and the dolphins appear to "work" particularly hard to get the whales to surge forward. A (generally small) group of dolphins will swim rapidly to the sides and front of the head of a baleen or sperm whale, close to the eyes. When the whale surges forward abruptly in an apparent attempt to evade the dolphins, the dolphins surf the one steep white-water wave so created during that surge. It is believed

that the activity is not pleasing to the whale, as the whale exhales forcefully during the surge, with a sonorous "snort" that indicates aggression or anger in other contexts. In right whales (*Eubalaena* spp.), the loud in-air "dolphin blow" of a snorting, surging, whale alerts researchers that *Lagenorhynchus* sp. or bottlenose dolphins are playing with the leviathans. The snort is so loud that on a calm sea it can be heard for a distance of several kilometers. One whale is "good for" anywhere from 5 to 10 surges. It then tires or decides to give up evading the dolphins, and the sport is over for the time or needs to be reinitiated with another whale nearby.

C. Balancing Bowhead Whales

Baleen whales engage in the surface activities of BREACHING, tail lobbing, flipper slapping, and holding the tail above the surface of the water. While much of this is certainly play, at least at times, the percussive nature of these activities possibly aids in communication and may also represent outgrowths of anger or frustration. Holding the tail out of the water for many minutes at a time, a habit of some right and bowhead whales (*Balaena mysticetus*), may feel good as a stiff breeze touches the skin and may even be a form of recreational "sailing" with the tail (as suggested by Payne, 1995).

Bowhead whale adults (approximately 18 m long) have been seen interacting with large logs (up to 10 m long) on summer–fall feeding grounds in the Beaufort Sea. They nudge and propel the log, handle it with their flippers and tail, and attempt to push it under water. Several of these activities seem similar to surface social interactions during sexual activity, and it could be argued that log handling is play useful to developing physical social skills. However, the most dramatic part of log handling is balancing the object, quite adroitly, on the back or belly. The author has seen log balancing (by an adult female) wherein the whale rolled gently sideways to compensate for the action of large oceanic swells rolling past the whale's body. The whale was so adept at this balancing that she could briefly keep both ends of the huge log suspended in air, a feat perhaps not unlike a trained sea lion balancing a ball on its snout. It is difficult to imagine how this activity could be anything but play or an "artistic" attempt at perfecting a difficult task (Würsig *et al.*, 1989; Wells *et al.*, 1999) (Fig. 2).

D. The Darker Side of Play

While apparent mischievous behavior of pulling on gull or turtle legs or of inciting whales to become aggravated could be potentially AGGRESSIVE or harmful play, it probably is not. However, some interspecies behaviors are definitely harmful and it is unclear whether they represent play or merely aggression. Short-finned pilot whales (*Globicephala macrorhynchus*) have been described aggressively and for almost 2 hr swimming around and into a pod of obviously frightened sperm whales (*Physeter macrocephalus*) (Weller *et al.*, 1996). Because pilot whales are not known to feed on the much larger sperm whales, it is possible that this action occurred only "for the fun of it," although other explanations, such as perhaps attempting to get whales to regurgitate squid gathered at depth, cannot be ruled out. A large male pilot whale in Hawaii carried a human diver into depth and held onto her with his mouth for tens of seconds. While he could easily have bit into her and killed her, his hold-

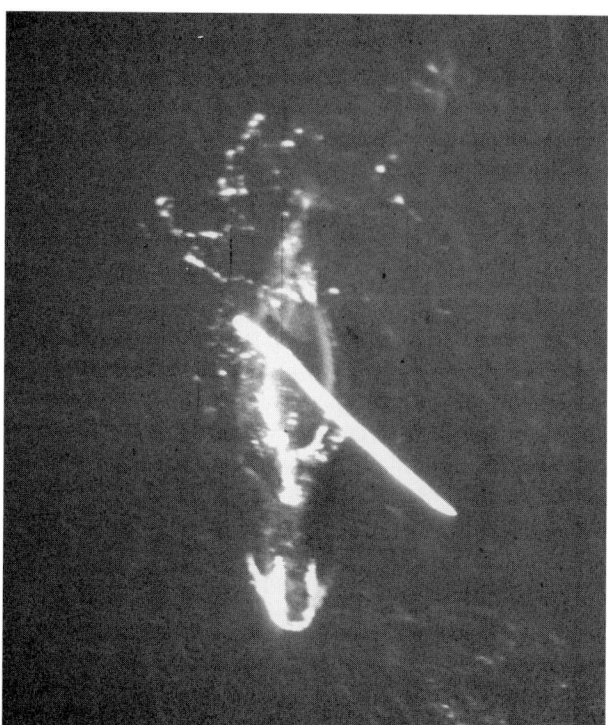

Figure 2 *A bowhead whale playing with a log.*

Shane, S. H., Tepley, L., and Costello, L. (1993). Life-threatening contact between a woman and a pilot whale captured on film. *Mar. Mamm. Sci.* **9**, 331–335.

Tyack, P. L. (1999). Communication and cognition. *In* "Biology of Marine Mammals" (J. E. Reynolds III and S. A. Rommel, eds.), pp. 287–323. Smithsonian Institution Press, Washington, DC.

Weller, D. W., Würsig, B., Whitehead, H., Norris, J. C., Lynn, S. K., Davis, R. W., Clauss, N., and Brown, P. (1996). Observations of an interaction between sperm whales and short-finned pilot whales in the Gulf of Mexico. *Mar. Mamm. Sci.* **12**, 588–594.

Wells, R. S., Boness, D. J., and Rathbun, G. B. (1999). Behavior. *In* "Biology of Marine Mammals" (J. E. Reynolds III and S. A. Rommel, eds.), pp. 324–422. Smithsonian Institution Press, Washington, DC.

Würsig, B., Dorsey, E. M., Richardson, W. J., and Wells, R. S. (1989). Feeding, aerial and play behaviour of the bowhead whale, *Balaena mysticetus*, summering in the Beaufort Sea. *Aqu. Mamm.* **15**, 27–37.

Polar Bear
Ursus maritimus

IAN STIRLING
Canadian Wildlife Service,
Edmonton, Alberta, Canada

The polar bear, *Ursus maritimus* (Phipps, 1774), is, on average, the largest of the eight species of bears found worldwide and is completely white. Typically, the body is stocky, lacks a shoulder hump, and has a longer neck in relation to the rest of the body than other ursids. Other English common names for the species are white bear, ice bear, and Nanuk. The taxonomy is order Carnivora, family Ursidae, genus *Ursus*, and species *U. maritimus*. There are no subspecies.

Adult males typically measure 200 to 250 cm in length from the tip of the nose to the tip of the tail and weigh 400–600 kg, although some individuals may reach about 800 kg. The total length of adult females is 180–200 cm and they normally weigh 200–350 kg while pregnant females may occasionally exceed 500 kg. There is some GEOGRAPHIC VARIATION in size and growth rates of bears in different populations. Polar bears are plantigrade and have five toes on each foot, with nonretractable claws. The forepaws are large and oar-like, as an adaptation for SWIMMING. The hind legs are not used while swimming. Females normally have four functional mammae, although supernumary nipples have also been reported. The dental formula is i 3/3, c 1/1, p 2-4/2-4, and m 2/3. The vitamin A content of the liver ranges is about 15,000 unit per gram and is toxic to humans.

I. Distribution

Polar bears are distributed in ice-covered waters throughout the circumpolar Arctic. About 20 individual populations are currently recognized, varying in size from a few hundred to a few thousand bears in each, with a world population estimate (in 1997) of 22,000–27,000. Although small numbers of animals may be found in the permanent multiyear pack ice of the central arctic basin, their preferred habitat is the annual ice over continental

ing her was relatively gentle (but firm), and play appears to have been the reason (Shane *et al.*, 1993). The woman survived only because she was an expert swimmer and diver and did not panic.

Bottlenose dolphins, the staple of show dolphins in over 100 aquaria worldwide, are known for at times aggressive play. In Scotland, a group of bottlenose dolphins is known to attack and fling about harbor porpoises (*Phocoena phocoena*), not for food and possibly simply for the sport of it. Bottlenose dolphins and harbor porpoises of the area also do not appear to compete with each other for space or food, but it is unknown whether more than a few ("rogue") dolphins are involved in this particularly gruesome sport that has killed at least one dozen harbor porpoises at this writing.

See Also the Following Articles

Bowhead Whale ■ Intelligence and Cognition ■ Mimicry ■ Pacific White-Sided Dolphin and Dusky Dolphin

References

Goodall, J. (1990). "Through a Window: My Thirty Years with the Chimpanzees of Gombe." Houghton Mufflin, Boston, MA.

Johnson, C. M., and Norris, K. S. (1994). Social behavior. *In* "The Hawaiian Spinner Dolphin" (K. S. Norris, B. Würsig, R. S. Wells, and M. Würsig, eds.), pp. 243–286. Univ. of California Press, Berkeley, CA.

Payne, R. (1995). "Among Whales." Scribner and Sons, New York.

Pryor, K. (1995). "On Behavior: Essays and Research." Sunshine Books, North Bend, WA.

Ralls, K., Forelli, P., and Gish, S. (1985). Vocalizations and vocal mimicry in captive harbor seals, *Phoca vitulina. Can. J. Zool.* **63**, 1050–1056.

shelves of the continents and islands around the coastline of the Arctic Ocean where overall biological productivity and densities of ringed seals [Pusa hispida (= Phoca hispida)], their primary prey, are greatest (Fig. 1). The southern limit of their distribution in winter varies with the extent of seasonal pack ice in the Bering, Labrador, and Barents seas. In areas where the ice melts completely in summer, such as Hudson Bay or southeastern Baffin Island in Canada, polar bears spend several months on shore fasting on their stored fat reserves until freeze-up in the fall.

II. Ecology

Regional concentrations and seasonal movement patterns of polar bears are influenced primarily by the type and distribution of sea ice and by the density and distribution of seals. From freeze-up in the fall until breakup in the spring, polar bears prefer coastal areas and inter-island channels lying over the continental shelf, especially active ice areas associated with shore leads or the floe edge. The size of home ranges of bears living within the Canadian Arctic Archipelago is generally smaller than those of polar bears dispersed in open ocean areas such as the Chukchi Sea. In general, polar bears are less abundant in areas of extensive multiyear ice and in the immediate vicinity of polynyas with overwintering walrus (Odobenus rosmarus) populations, probably because the density of ringed seals is lower there.

Polar bears that have continuous access to sea ice hunt throughout the year. Their hunting techniques and rates of success change with the seasons and vary in different areas. The most common method of hunting is to lie beside a breathing hole waiting for a seal to surface and breathe. Contrary to popular myth, they do not cover their noses with a paw when stalking a seal, although when stalking, they do keep their heads low while walking slowly and steadily toward potential prey. The largest proportion of a polar bear's annual caloric intake for the year occurs in spring and early summer, at which time newly

Figure 1 Ursus maritimus, *the polar bear, is aptly named, as the species is often observed miles from the nearest land on polar pack ice and swimming between ice floes where they hunt ringed seals and sometimes bearded seals. Photo by François Gohier.*

weaned ringed seal pups are 50% fat by wet weight. The specific sea ice habitats most hunted by polar bears in spring are stable shore-fast ice with deep snowdrifts along pressure ridges, which are suitable for ringed seal birth lairs and breathing holes; the floe edge where leads are wide (>1 km); and areas of moving ice with seven-eighths or more of ice cover. After breakup of the annual ice in late spring to early summer, hunting success is reduced and, in areas where open water prevails from late summer through autumn, polar bears seek onshore retreats where they simply fast on their stored fat reserves until freeze-up.

The principal prey species of polar bears are ringed seals and, to a lesser degree, bearded seals (Erignathus barbatus). Ringed seals maintain their breathing holes from freeze-up in the fall to breakup in the spring by abrading the ice with the heavy claws on their foreflippers. Many winter breathing holes are located on the last cracks to close over in the fall and bears are able to locate them by smell, even under a meter or more of compacted wind-blown snow. In areas where wind, water currents, or tidal action causes the ice to continually crack and subsequently refreeze, seals are apparently more accessible to polar bears and the bears are able to hunt them there more successfully. In places where the snow cover in the fiords is deep in spring, large numbers of ringed seals give birth to their pups in subnivean liars where they are hunted by polar bears of all ages and sex classes, but especially females with newborn cubs. Bearded seals concentrate where natural cracks and polynyas form through the winter because it is easier to breathe there. Polar bears are also known to occasionally prey on walruses, belugas (Dephinapterus leucas), narwhals (Monodon monceros), waterfowl, and seabirds.

Most maternity denning takes place in snowdrifts on land in coastal areas. In western Hudson and James bays, however, pregnant females must enter maternity dens prior to when suitable snowdrifts form in most years so they dig dens in frozen peat in small banks along the edges of lakes or streams. Individual females show fidelity to denning areas, although not to individual den sites. In the western Beaufort Sea north of Alaska, a large proportion of the adult females den 200 km or more offshore in the multiyear ice of the Beaufort Sea.

III. Behavior and Physiology

Polar bears breed in April and May. Males are twice the size of females, which probably reflects the intense competition between them for mating with available females. Because females keep their cubs for 2.5 years in most areas, they are normally available for mating no more often than once every 3 years. Thus the functional sex ratio of breeding age animals is a minimum of 3 males:1 female. Polar bears have induced ovulation, which means a female must mate many times before she will ovulate and can be fertilized. Polar bears also have delayed implantation so the fertilized egg ceases developing after the first few divisions and remains dormant in the uterus until implantation in mid-September to mid-October, depending on latitude. The young are born in a snow den about 2 months after implantation and weigh about 0.6 kg at birth. The eyes of newborn cubs are closed and their body hairs are so fine they have mistakenly been reported to be hairless at birth. Cubs are nursed by the female until they are large enough to leave the maternity den in late March or early April, by which time they weigh 10–12 kg. In most areas, cubs are weaned at 2.5 years of age, after which the female mates again. Females breed for the first time at 4–5 years of age and continue to have cubs to the end of their lives. About two-thirds of the litters are twins, while single cub litters are the next most common, and triplets are infrequent, although not rare. The largest litters and heaviest cubs are born to adult females in their prime, about 8–18 years of age, whereas younger and older females tend to have single cub litters more often. In populations that are not overharvested, females live into their mid-20s, although some reach their early 30s. Males generally reach their early 20s with occasional animals reaching the late 20s. In general, bears living in areas that are more biologically productive are larger and reproduce earlier than bears in areas that are less productive.

Polar bears have a remarkable ability to store large amounts of fat during periods when prey (mainly seals) are available and then fast for protracted periods when food is not available. In Hudson Bay, where the annual ice melts completely by mid-July and does not reform until mid-November, all bears in the population must fast for at least 4 months on their stored fat, whereas pregnant females do not feed for 8 months, during which period they must support themselves as well as nurse their young so the cubs can grow large enough to withstand the rigors of the arctic environment. Only pregnant females go into dens for the winter; bears of all other age and sex classes remain on the ice and hunt throughout the winter except briefly during the coldest or most inclement periods when they may occupy temporary dens in the snow. Whenever food is unavailable, the metabolism of a polar bear automatically slows to a hibernation-like physiological state, in which energy is consumed at a lower rate. This change occurs after about 7–10 days of not eating and can occur at any season of the year, rather than only in the autumn prior to winter denning, like other species of bears. Thus, because the digestibility of fat by a polar bear is about 98%, it has the ability to store large amounts of fat quickly when food is available and then switch to a more efficient metabolic state for as long as necessary when food is unavailable.

IV. Fossil Record

Polar bears are thought to have originated from brown bears (*Ursus arctos*), possibly in the vicinity of Siberia during the glacialogical advances of the late Pleistocene a million or so years ago. Polar bears and brown bears are capable of interbreeding in zoos and the young are fertile. However, in the wild, there is little overlap in the habitats during the breeding season. The molars and premolars of polar bears are more jagged and sharper than those of other bears, reflecting their rapid evolutionary shift toward carnivory from the flatter grinding teeth of their more omnivorous relatives. The oldest known fossil is from London, England, and is less than 100,000 years old.

V. Interactions with Humans

Polar bears are important to the culture and economy of aboriginal people in Alaska, Canada, Greenland, and Siberia, who harvest 8–900 bears annually. Population research is ongoing in most areas to try to ensure that harvests are undertaken at sustainable levels. Thus, the survival of polar bears is not currently threatened by hunting, but this must still be regulated carefully because they have a low reproductive capability and would require 20 or more years to recover from an overharvest. Many contaminants in the Arctic marine ecosystem are lipophilic and bioaccumulate as they move up in the food chain. Because polar bears are at the top of the ecological pyramid, and live predominantly on fat, they are capable of achieving high levels of contaminant loading in some areas. It is not yet known whether these contaminants are high enough to interfere with reproductive functions or the immune system. In some areas, such as western Hudson Bay, long-term studies have demonstrated that climatic warming is causing the ice to break up earlier, which is correlated with decreasing condition and reproductive rates of polar bears there. In more recent years, ecotourism to view polar bears, especially at Churchill, Manitoba, has increased markedly with positive local economic benefits and a significant increase in the worldwide appreciation of the bears themselves.

See Also the Following Articles

Arctic Marine Mammals ■ Bearded Seals ■ Ringed, Caspian, and Baikal Seals ■ Ursidae

References

Amstrup, S. C., and Gardner, C. L. (1994). Polar bear maternity denning in the Beaufort Sea. *J. Wildl. Manage.* **58**, 1–10.

Derocher, A. E., Garner, G. W., Lunn, N. J., and Wiig, Ø. (eds.) (1998). "Polar Bears." Proceedings of the Twelfth Working Meeting of the IUCN/SSC Polar Bear Specialist Group, 3-7 February 1997, Oslo, Norway. IUCN Species Survival Commission Paper 19.

Derocher, A. E., Nelson, R. A., Stirling, I., and Ramsay, M. A. (1990). Effects of fasting and feeding on serum urea and serum creatinine levels in polar bears. *Mar. Mamm. Sci.* **6**, 196–203.

Derocher, A. E., and Stirling, I. (1998). Maternal investment and factors affecting offspring size in polar bears (*Ursus maritimus*). *J. Zool. (Lond.)* **245**, 253–260.

Ferguson, S. H., Taylor, M. K., Born, E. W., Rosing-Asvid, A., and Messier, M. (1999). Determinants of home range size for polar bears (*Ursus maritimus*). *Ecology Letters* **2**, 311–318.

Norstrom, R. J., Belikov, S. E., Born, E. W., Garner, G. W., Malone, B., Olpinski, S., Ramsay, M. A., Schliebe, S., Stirling, I., Stishov, M. S., Taylor, M. K., and Wiig, Ø. (1998). Chlorinated hydrocarbon contaminants in polar bears from eastern Russia, North America, Greenland, and Svalbard: Biomonitoring of Arctic pollution. *Arch. Environ. Contam. Toxicol.* **35**, 354–367.

Paetkau, D., Amstrup, S. C., Born, E. W., Calvert, W., Derocher, A. E., Garner, G. W., Messier, F., Stirling, I., Taylor, M., Wiig, Ø., and Strobeck, C. (1999). Genetic structure of the world's polar bear populations. *Mol. Ecol.* **8**, 1571–1585.

Prestrud, P., and Stirling, I. (1994). The International Polar Bear Agreement and the current status of polar bear conservation. *Aquat. Mamm.* **20**, 1–12.

Ramsay, M. A., and Dunbrack, R. L. (1986). Physiological constraints on life history phenomena: The example of small bear cubs at birth. *Am. Nat.* **127**, 735–743.

Ramsay, M. A., and Stirling, I. (1988). Reproductive biology and ecology of female polar bears (*Ursus maritimus*). *J. Zool. Lond.* **214**, 601–634.

Stirling, I. (1988). "Polar Bears." University of Michigan Press, Ann Arbor, MI.

Stirling, I. (1990). Polar bears and oil: Ecologic effects. *In* "Synthesis of Effects of Oil on Marine Mammals" (J. R. Geraci and D. J. St. Aubin, eds.), pp. 223–234. Academic Press, San Diego, CA.

Stirling, I., and Lunn, N. J. (1997). Environmental fluctuations in arctic marine ecosystems as reflected by variability in reproduction of polar bears and ringed seals. *In* "Ecology of Arctic Environments" (S. J. Woodin and M. Marquiss, eds.), pp. 167–181. Blackwell, Oxford.

Stirling, I., Lunn, N. J., and Iacozza, J. (1999). Long-term trends in the population ecology of polar bears in western Hudson Bay in relation to climatic change. *Arctic* **52**, 294–306.

Stirling, I., and Øritsland, N. A. (1995). Relationships between estimates of ringed seal and polar bear populations in the Canadian Arctic. *Can. J. Fish. Aquat. Sci.* **52**, 2594–2612.

Pollution and Marine Mammals

PETER J. H. REIJNDERS
Alterra, Marine and Coastal Zone Research, Den Burg, The Netherlands

ALEX AGUILAR
University of Barcelona, Spain

Awareness of the threat of environmental contaminants to marine mammals is widespread. High concentrations of certain compounds in the tissues of these animals have been associated with organ anomalies and impaired reproduction and immune function, as shown by large die-offs among seal and cetacean species. This has prompted alertness about the impact of pollution and stimulated research into the relationship between observed effects and pollutants. However, a clean cause and effect relationship between the residue levels of contaminants and the observed effects has been demonstrated in only a few studies. In the absence of evidence, this might elicit a serious backlash because concerns expressed are easily interpreted as fear mongering. This could lead to inertia in taking appropriate management measures, which is undesirable from a conservation as well as an environmental management perspective.

The main reasons for the lack of proof of the impact of pollution on marine mammals are the difficulty or impossibility of experimenting in laboratory conditions with these animals and the frequent occurrence of confounding factors that hamper the establishment of cause–effect relationships. Examples of these factors are the fact that pollution always occurs as a mixture of a large number of chemical compounds, the lack of data on biological variables influencing tissue levels, quality of samples usually analyzed, the limited information on pathology and occurrence of disease in the specimens studied, the absence of reliable population data, and the lack of information on the influence of other detrimental factors, such as the impact of fisheries and of other human-related sources of disturbance.

I. Substances of Concern

In general, the concept of pollution incorporates many different substances to which marine mammals are exposed and might affect their health adversely. These include chemical compounds, oil pollution-derived substances, marine debris, sewage-related pathogens, excessive amounts of nutrients causing environmental changes, and radionuclides. The influence of oil and petroleum-derived compounds, such as polycyclic aromatic compounds, of marine debris, of sewage-related pathogens, and of nutrients-related changes, such as the occurrence of biotoxins, has not been the subject of focused research in marine mammalogy. As a consequence, data on these pollutants, either as concentrations in tissue of the affected marine mammals or as effects on them, are extremely limited. This article therefore only addresses pollution caused by chemical substances.

Traditionally, most laboratories tended to routinely analyze organohalogenated compounds such as DDT, DDE, DDD, polychlorinated biphenyls (PCBs), lindane, dieldrin, endrin, hexachlorobenzene (HCB), heptachloro-epoxide (HEPOX), and mirex, and trace elements such as mercury, lead, selenium, and cadmium. Some laboratories, able to use more sophisticated equipment, have also analyzed polychlorinated dibenzodioxins (PCDDs) and polychlorinated dibenzofuranes (PCDFs). Such a narrow approach brings the risk of overlooking the impact of other, poorly known compounds. However, the monitoring of all known synthetic organic chemicals and their metabolites currently in use would require analysis of about 300,000 compounds. Therefore, criteria have to be developed to identify priority compounds on which to focus monitoring. Criteria for the identification of these compounds should include the level of production and release into the environment, bioaccumulation potential, and toxicity. Examples of "novel" compounds that fall into the category of priority compounds are organotins, polybrominated biphenyls (PBBs), and polybrominated diphenyl ethers (PBDEs).

Because research funds are limited, another issue to be addressed is the choice between monitoring pollutant concentrations or investigating their effects. The latter option is in our view clearly preferable. If an effect is observed, more focused research for the responsible compounds can follow.

Taking into account the two elements discussed earlier and without ignoring the potential impact of other compounds, it is currently accepted that a list of compounds of highest priority should include all organohalogens usually referred to as persistent organic pollutants (POPs), particularly including PCBs, DDTs, PCDDs, HCB, dieldrin, endrin, mirex, PCDFs, PBBs, PBDEs, polycyclic aromatic hydrocarbons (PAHs) and phenols, and metals, particularly including their organic forms such as methyl-mercury and organotins.

II. Pollution from an Environmental Perspective

Pollution is only one of the many environmental factors that influence the health status of marine mammals. The assessment of the impact of pollution on marine mammals has therefore to be undertaken in a holistic perspective, considering also the potential of pollutants to interfere with their ability to recuperate from stress caused by other environmental forces. As an example, PCBs could cause immune suppression in a given seal population without directly leading to an increased mortality. However, if such a population is exposed to an introduced virus, the extent of a resulting epidemic is likely to be much aggravated.

Marine ecosystems are complex, and environmental forces operating on populations are often multifactorial and produce synergistic or cumulative effects. Therefore, it will be complicated to attribute a given effect to a single factor. To illustrate the complexity of unraveling the impact of pollution, we discuss here some of the environmental factors, natural or anthropogenic, that influence the resilience of marine mammals to pollution.

A. Prey Depletion

Natural environmental variations such as redistribution of planktonic organisms may bring changes in distribution, abundance, or recruitment of the species that constitute the food of marine mammals. However, depletion of prey may often also be caused by overfishing by commercial fisheries. Depending on the extent of the depletion, marine mammals may respond to the reduced supply of prey either by switching to other species, or by temporarily moving to another area. Frequently, however, they undergo an impoverishment of their body condition and their recruitment rates become lower. The resilience of animals in such populations/stocks is affected negatively, potentially increasing the detrimental impact of pollutants.

B. Habitat Disturbance

Habitat may be disturbed by a wide range of human activities, including recreation, construction works, and many others. For pinnipeds, sea otters (*Enhydra lutris*), and some coastal cetaceans, the physical alteration of the litoral, including the mere presence of humans and their associated infrastructure, may be a significant detrimental factor. Noise pollution is a particular source of concern because many marine mammals rely on sound emission and detection for finding their prey, communicating, and navigating. Activities producing noise-related disturbance include shipping, boating, military maneuvers, seismic testing, and oil and gas drilling.

C. Disease

Natural factors influence the incidence of disease. For example, a shift in distribution of prey species may lead to an increased parasite infestation rate likely to affect the resilience of populations to pollutants. Although the incidence of infectious disease in marine mammals is poorly known in general terms, morbillivirus epizootics that have recently affected pinnipeds as well as cetaceans have elicited extensive research on the effects of viral diseases on marine mammal populations.

Deadly bacterial diseases are generally considered to be secondary to other conditions such as viral disease, parasitic infection, or trauma. However, like some pollutants, bacteria can also interfere with reproduction, as was demonstrated by the finding of *Brucella* organisms in porpoises and dolphins.

In marine mammals, similarly to other better studied groups of vertebrates, disease and the toxic effect of pollutants are often interrelated. This relationship is discussed in more detail in the next section, although it should be mentioned here that diseases can affect metabolic systems and, consequently, alter physiological functions. Chronically diseased females, for example, usually have a poor reproductive performance, as do females affected by some pollutants.

D. Overall Environmental Changes: Global Warming, Ozone Depletion

Albeit the potential effect of global changes on marine mammals has been little investigated and its consequences are considered less imminent than those caused by other factors, this matter certainly deserves concern. It is predicted that the global rise in temperature will alter marine communities and their productivity, cause a sea level rise, reduce ice cover, and modify rainfall and water current systems. The consequences for marine mammals are unclear, but undoubtedly those alterations will affect their behavior and distribution. An increased incidence of epizootics among pinnipeds is also postulated, as higher densities as a result of increased haul-out behavior will result in a higher transmission rate of infectious agents.

Despite the longer term character of these threats, changes in the distribution and behavior of marine mammals caused by climate variation should be monitored to detect potential relationships at an early stage.

The examples just mentioned show clearly that studying the impact of pollution on marine mammals requires a multidisciplinary approach. Therefore, we advocate assessing pollution impact not in an isolated way, but always in relation to other environmental factors.

III. Factors in Assessing Pollution Impact

Two sources of information may warn that pollution might affect a given population: high tissue pollutant concentrations in the members of that population and changes in the biological parameters of the population such as physiological condition and changes in reproductive or mortality rates. The latter are often derived from population monitoring and/or pathological investigations. However, a number of biological factors and inconsistencies in the sampling and analytical procedures seriously hamper the establishment of such relationships, sometimes even leading to spurious interpretations of environmental data.

A. Biological Factors Affecting Variability in Pollutant Levels

Some persistent chemicals are bioaccumulative and their concentrations in living organisms undergo a progressive amplification through food chains, a process called biomagnification. However, the increase at each trophic level is usually considerably higher than the 10-fold increase predicted by ecological models. Biomagnification, defined as the ratio of concentrations of a compound in the predator to its prey, can be altered significantly, and often much increased, by a number of variables, such as the route of exposure, the physical and chemical properties of the compound, the metabolic capacity of the predator, and its physiological constitution.

B. Diet

Diet composition is a key factor determining resultant tissue concentrations. Because baleen whales feed on planktonic crustaceans and are thus situated lower in the food web, their tissue organochlorine (OC) concentrations are almost invariably lower than those in the top-predator toothed whales living in the same ecosystem.

Within a population of the same species, OC levels can also differ because of variation in diet. For example, juvenile pinnipeds often exploit different food resources than adults, and in many species of cetaceans and pinnipeds, adult males prey on different species than adult females. In some marine mammals, differences may even be associated with reproductive status: the diet of lactating females of some dolphin species is different from pregnant or resting females. Also, the geographical region where food is consumed is critical: during most of the year, male sperm whales (*Physeter macrocephalus*) occupy different geographical regions than females and, as a consequence, their pollutant profiles are quite dissimilar. Differences in diet are also assumed to have an influence on the tissue concentrations of PAHs in marine mammals. Levels of these compounds in marine mammals are generally low, although they tend to be higher in cephalopod-eating marine mammals than in those relying on fish. The explanation appears to be that the ability of fish to metabolize PAHs is better than that of cephalopods.

Tissue levels of metals also appear to be related to the feeding habits and region of exposure. Cadmium, copper, and zinc levels are higher in cetaceans that feed primarily on squid than in those feeding on fish. This is attributed to the ability of squid to retain these elements selectively. Intraspecific differences in tissue metal concentrations have also been linked to segregation in feeding areas; the levels of lead in kidney and muscle tissue of long-finned pilot whales (*Globicephala melas*) and white beaked dolphins (*Lagenorhynchus albirostris*) occurring during summer in the same areas are much different because they segregate geographically—and feed—during the winter.

C. Age and Sex

The tissue concentration of a pollutant in a marine mammal is a function of the difference between the intake rate and the metabolization and excretion rates. OCs have been found to correlate positively with age; levels are relative low in young animals, increase until a certain age, and then either continue to increase or reach a plateau level or decrease. The leveling off or decreasing phase is different for males and females, as is addressed later on. Factors that influence the age-related pattern of accumulation of organochlorines are detoxification ability and the feeding rate. The capacity for detoxification is low in young animals and improves with age; thus the initial increase during the juvenile stage is slowed down by improved metabolization and excretion rates. The resulting leveling off of tissue concentrations is enhanced further by reduced feeding rates in adults.

Superimposed on these is the effect of reproduction in females. OCs, as most lipophilic compounds, cross the placenta and reach the fetus, although not all chemicals do it at equal rates. For example, the lower chlorine-substituted (lower weight) congeners of PCBs are transported more easily than higher chlorinated ones. In addition to placental transport, OCs are also transferred from mother to offspring through milk. Higher chlorinated OCs are transferred less efficiently from the lipid tissue of the mother to her milk and hence to the suckling calf or pup. This process obviously does not start until the females reach sexual maturity and become pregnant for the first time. Therefore, the first pregnancy marks the start of the leveling off or decrease phase in females. There are differences among species and compounds. Moreover, this reproductive discharge in females is not uniform and depends on the characteristics of the reproductive cycle of the species and the physicochemical properties of the compound. The transfer during lactation is much higher than that occurring through deposition in the tissues of the calf or pup during pregnancy. In cetaceans, the discharge of PCBs, expressed as percentage transferred in relation to maternal tissue load, ranges from 5 to 96% during lactation and from 4 to 6% during pregnancy. In pinnipeds, the ranges are 23–81 and 1–10%, respectively. Not surprisingly, the length of the lactation period significantly influences the proportion of the OCs' load transferred to the offspring. It has been estimated that this proportion ranged from 3 to 27% in fin whales (*Balaenoptera physalus*), with a lactation period of around 7–8 months, whereas it was around 80% in bottlenose dolphins (*Tursiops truncatus*) and 72–91% in striped dolphins (*Stenella coeruleoalba*), two species in which lactation lasts about 14 months. Irrespective of the amount transferred, the reproductive discharge results in lower levels of lipophilic pollutants in reproductively active females as compared to males of the same age. However, there are some exceptions to the general rule. In Antarctic minke whales (*Balaenoptera bonaerensis*), levels of PCBs and DDT were found to be higher in immature males than in mature males as a result of a shift in diet caused by adult migration to less polluted areas. In the North Atlantic, adult female sperm whales are more polluted than males of comparable age because they feed on more polluted species and are distributed year-round in regions where pollutant loads are higher.

Age-related variation in tissue concentrations of trace elements is less homogeneous. Mercury, cadmium, selenium, and lead increase with age, somewhat more steeply in females compared to males. There is no clear leveling off for any element except for lead, in which a slower increase has been observed at an older age. Because these elements are not lipophilic, reproductive transfer does not affect their loads in females. It has been suggested that the higher levels of those elements found in females compared to males may be related to differences in metabolic pathways linked to hormone cycles.

Information on other trace elements is scarce. Copper and zinc show no increase with age. In fact, concentrations in newborns are higher than in adults, which is attributed to an age-related decrease in absorption and retention of these essential elements.

D. Nutritive Condition

Nutritive condition affects the volume of fat in the body and its lipid composition. In some cetaceans and pinnipeds, blubber lipid richness may decrease from 90% in a female near term to 30–35% in females just having weaned their offspring.

Although less impressive, males also show changes in blubber layer thickness during the reproductive season. Apart from this reproduction-related change, seasonal variation may also be significant. Variation in blubber layer thickness is lower in toothed whales compared to baleen whales. In some pinnipeds, independently of the reproduction-related changes, blubber layer thickness can vary by as much as 50% (taking the maximum thickness as a reference). This variation has implications for the dynamics of lipophilic contaminants. Because lipids are mobilized more readily from the blubber than lipophilic pollutants, lipid metabolization typically results in an increase in the residue levels. However, it has been found that the increase is less than a kinetic concentrative model would predict. It has been suggested that the more polar fraction of the pollutants is mobilized more readily through the enhanced metabolization and excretory capacity stimulated by a rise in tissue pollutant concentrations subsequent to lipid metabolization.

It is unclear to what extent changes in nutritive condition affect tissue concentrations of nonlipophilic compounds. Changes in mass and composition of tissues where chemicals (e.g., heavy metals in liver and kidney) are likely to accumulate influence the dynamics of these pollutants, but data on these processes in marine mammals are lacking.

Body growth in young animals also influences tissue level of pollutants. In both pinnipeds and cetaceans it has been found that dilution of contaminants occurs in the early stages of growth due to the rapid deposition of blubber and the amassing of liver and kidney tissue. Calculations of tissue concentrations on a lipid basis instead of a fresh weight basis can partially account for such variation, but it does not account for variation in the qualitative composition of the lipid fraction, which is also likely to affect the retention ability of the tissue.

E. Body Size

The influence of body size on variation in the accumulation pattern of pollutants is somewhat complex. Generally, elimination rates of xenobiotic compounds per unit of bodyweight are related inversely to body weight, a trend that also holds for the activity of detoxifying enzymes. Both would tend to favor accumulation of higher pollutant levels in larger animals. Contrary to that effect, the metabolic rate is inversely correlated to body size. Because metabolic rate is correlated with pollutant intake, a higher pollutant accumulation can be expected in smaller species. The influence of metabolic rate has been found to outweigh the countereffect of elimination and detoxifying activities. The concentration factor in a marine mammal is largely dependent of its daily rate of food consumption—inversely related to body size—and the mean concentration of pollutant in its prey. Small animals therefore carry generally higher loads of pollutants relative to their body weight than larger animals.

Variation in body size is more dramatic in cetaceans than in pinnipeds. Some dolphin and porpoise species weigh, when adult, about 30–40 kg, whereas the larger whales can weigh more than 150,000 kg. The range in adult pinnipeds varies from 50 to 4000 kg. An example of variation in pollutant levels between two species of different size is that of two krill-eating Atlantic baleen whale species, in which differences in tissue pollutant levels were explained by differences in body mass. It has been proposed that in species sharing the same waters, the effect of body mass on tissue concentration outweighs that of the small differences in diet or other biological traits.

F. Body Composition

The distribution pattern of pollutants in the body of an animal depends largely on the physical and chemical properties of the substances involved. For example, much work has been carried out to investigate the influence of the position of H atoms on the biphenyl ring in all PCBs, which largely determines the possibilities for their metabolization by marine mammals.

Because lipophilic pollutants accumulate in fatty tissue, about 70–95% of lipophilic pollutants end up in the blubber, which in marine mammals is the largest fat compartment. The chemical composition of the blubber also influences pollutant concentrations. In species with thick blubber, pollutants are stratified in the different layers and significant differences may be found between inner and outer strata. Therefore, the whole blubber layer must be sampled to obtain a representative picture of the individual's load. Mercury, cadmium, zinc, and other heavy metals accumulate mostly in the liver and kidney, and lead accumulates predominantly in bone tissue.

G. Analysis and Sampling

One of the major handicaps in assessing temporal and spatial trends of contaminants in marine mammals is the poor comparability of data. This holds partly for heavy metals, but it is definitely critical for analyses of OCs. The analytical techniques used, and their accuracy, have changed considerably over time and also vary between different laboratories. This greatly hinders comparison between studies undertaken by different laboratories or time periods. Significant improvement in standardizing procedures has been achieved in the last decade through intercalibration exercises. Quality assurance and quality control are of utmost importance, but this also holds for the sampling procedures. To avoid contamination by the packaging material, clean glass or aluminium foil should be used to preserve samples for OCs analyses, and plastic bags should be used to preserve samples for heavy metal analyses. Each sample should be accompanied by the appropriate biological data and, if possible, also with a detailed pathological examination to reveal the incidence of alterations in reproductive biology, early development, and occurrence of diseases. Detailed field and laboratory protocols taking these considerations into account have to be developed before embarking on any ecotoxicological study.

IV. Impacts of Pollution on Marine Mammals

Numerous studies have suggested that exposure to pollutants has an impact on marine mammal populations, mainly on reproduction and mortality. However, in most of these studies, the existence of confounding factors prevents reaching conclusive results and only a few have actually succeeded in demonstrating such a relationship.

The effects of pollution, either observed or suggested, can be grouped conveniently under three categories: impaired reproduction, indirect mortality, and direct mortality.

A. Impaired Reproduction

OCs, particularly PCBs, have been demonstrated to be responsible for impaired reproduction in the harbor seal (*Phoca vitulina*). This conclusion was reached by means of a feeding experiment in which 12 female harbor seals were fed DIETS low in OCs and 12 females received a diet high in OCs, particularly PCBs and DDE. The conclusion was that reproductive success was significantly lower in the more polluted diet group: 4 pups were born instead of 10 born in the control group. The latter figure is similar to what is normally found in free-ranging harbor seals. In addition, the analysis for estradiol-17β and progesterone in blood samples from these seals revealed that reproductive failure occurred at the implantation stage, as such failure was accompanied by low levels of oestradiol-17β. A plausible explanation of this effect is that PCBs impaired the enzymatic metabolism, lowering the circulating levels of estradiol, which in turn led to imperfect endometrial receptivity and prevented successful implementation of the blastocyst.

Elevated OC concentrations have been associated with reproductive impairment in gray seals (*Halichoerus grypus*) and ringed seals (*Pusa hispida*) in the Baltic Sea and in California sea lions (*Zalophus californianus*). Female Baltic gray and ringed seals exhibited uterine occlusions and stenosis, leading to partial or complete sterility; concentrations of OCs were higher in affected animals than in normally reproducing females. It has been proposed that pregnancy was interrupted by PCBs (or PCB-metabolites), followed by the development of pathological disorders. Epidemiological studies on the involved populations strongly support the hypothesis that PCBs or their metabolites, i.e., methyl sulfones, are responsible for the observed reproductive impairment. This has been apparently confirmed by the fact that the incidence of pathological conditions in younger but mature age classes decreased. OC levels in seals as well as other Baltic biota declined sharply between 1970 and 1980. However, unequivocal evidence for a cause–effect relationship has not been provided, although this stage of proof is probably as far as one can get with the constraints of this type of field research.

The case of the California sea lion is even more complex. Initially, still births and premature pupping were attributed to high OC (PCBs and DDE) concentrations. Later studies demonstrated that pathogenic disease agents could also have been responsible. These confounding factors prevented reaching a clear-cut conclusion on the causative role of pollution.

The proof for reproductive disorders in cetaceans caused by specific pollutants is even weaker than for pinnipeds. Impaired reproductive performance caused by PCBs has been suggested in beluga whales (*Delphinapterus leucas*) in the St. Lawrence River. In 2 out of 120 examined belugas, hermaphroditism was observed. However, the pathological studies were not conclusive, and the lack of sound population data which with to compare the observed findings made it impossible to reach a conclusion on the actual role of pollutants on such abnormalities.

Low levels of testosterone were associated with high levels of PCBs and DDE in Dall's porpoises (*Phocoenoides dalli*). However, the biological significance and underlying mechanism are unclear because both variables are age related; further studies are needed to clarify the potential involvement of pollutants.

Abnormal testes, i.e., transformed epididymal and testicular tissue, were observed in North Pacific minke whales (*Balaenoptera acutorostrata*). A possible relation with high levels of OCs has been suggested, but not proved.

B. Disease

Numerous pathological disorders, including skull lesions (paradentitis, osteoporosis, exostosis), cortical adenomas, hyperkeratosis, nail malformations, uterine stenosis and occlusions, uterine tumors (leiomyomas), and colonic ulcers, have been observed in Baltic gray and ringed seals and, to a lesser extent, in harbor seals. Pathological and epidemiological investigations revealed that the observed symptoms were part of a disease complex called hyperadrenocorticism, a disease syndrome associated with high levels of PCBs and DDT and their metabolites. Contrary to reproductive impairment, it is not possible to evaluate conclusively which of these substances elicit a response in seals because of crossed or synergistic effects. As in the case of reproductive disorders, the prevalence of uterine lesions, adrenocortical hyperplasia, and skull bone lesions was found to decrease following a decline of DDT and PCBs in Baltic biota. Conversely, however, the incidence of uterine leiomyoma in Baltic seals has not changed to date. Of even more concern is the increasing incidence during recent years of colonic ulcers in young Baltic gray seals, indicating an increasingly compromised immune system in these animals. DDT tissue levels in these animals decreased strongly between 1969 and 1997, annually by 11–12%, but PCB levels decreased during the same period at a much lower pace, only 2–4% annually. This may suggest a role of PCBs and/or their metabolites in the observed pathologies, although the potential effect of novel, unknown compounds cannot be excluded.

Some studies have shown direct evidence of the immunotoxicity of OCs. Reduced immune responses were correlated with high levels of PCBs and DDT in *in vitro* immune function assays with peripheral blood lymphocytes from free-ranging bottlenose dolphins. In an experiment with captive harbor seals, *in vitro* and *in vivo* immune function tests showed lower immune function related to higher dietary concentrations of OCs. While these two studies show that OCs adversely interfere with immune function, the toxicological and biological significance unfortunately remains unclear.

It has been suggested that lowered immunocompetence induced by contaminants aggravated the die-offs of bottlenose dolphins in the Gulf of Mexico (1990, 1991, 1993) and on the east coast of the United States (1987–1988), striped dolphins in the Mediterranean Sea (1990–1992), harbor seals in the North Sea (1988), Baikal seals (*Pusa sibirica*) in Lake Baikal (1987–1988), and Caspian seals (*P. caspica*) in the Caspian Sea (2000). In most cases the mortalities were ultimately caused by a morbillivirus infection, but exposure to high levels of OCs was proposed to have played a key role by facilitating viral transmission and increasing the susceptibility of individuals to the disease. However, it has been difficult to conclude on the etiology of these mortalities. Different studies have tried to establish links between die-offs and pollution. In the case of the striped dolphin morbillivirus epizootic, animals killed by the disease carried significantly higher PCB concentrations than survivors. This finding could be explained by (1) immune suppression caused by PCBs, leading to higher mortality of the more polluted individuals; (2) mobilization of pollutants stored in depot tissues thinned by the disease; or (3) changes in physiological functions of the affected individuals, leading to increased PCB concentrations.

In two other studies, levels of organochlorines were related to mortality. In one study, OC levels in seals that died during the morbillivirus outbreak were compared with those in surviving seals. In the other study, OC concentrations in harbor porpoises (*Phocoena phocoena*) that died from physical trauma were compared with animals known to have succumbed to an infectious disease. Both studies were inconclusive in establishing a direct cause–effect relationship between pollutants and susceptibility to disease because of the existence of confounding factors such as heterogeneous body condition between the groups compared. A follow-up study on harbor porpoises from England and Wales has been more conclusive. In this study, PCB concentrations in blubber from animals that died due to physical trauma (e.g., bycatch) were compared with those from animals that died because of an infectious disease. A significant association was demonstrated between blubber PCB concentrations and mortality due to infectious disease, suggesting a causal relationship with chronic PCB exposure. Here, again, the possibility of additive or synergistic effects of other contaminants must be considered.

Other ecotoxicological studies point toward other effects of pollutants on marine mammals. It has been proposed that OCs produced thyroid hormone and vitamin A deficiency in at least harbor seals. Thyroid hormones are important in the structural and functional development of sex organs and the brain, both intra-uterine and postnatal. A vitamin A deficiency may lead to increased susceptibility to microbial infections and retarded growth, as appeared to be indicated by the significant lower birth weights of pups born in the more contaminated dietary group of the captive harbor seal study discussed earlier.

Another noteworthy example of impaired health status possibly caused by pollution is the case of the St. Lawrence beluga population. A range of pathological conditions have been documented in this population, particularly a high prevalence of tumors and digestive tract and mammary gland lesions. High tissue levels of OCs, lead, and mercury have been found in these animals. The establishment of a cause–effect relationship between contaminants and the observed effects in this population is hampered by the possible adverse role of other environmental factors, such as previous overhunting, high levels of noise pollution, and overall habitat destruction. Any of these factors has the potential for causing most of the observed conditions and the population's small size and slow recovery.

C. Direct Mortality

There is no record of any acute chemical poisoning event affecting marine mammals, apart from one case that affected harbor seals: a small colony had been acutely poisoned by an accidental discharge of mercury-contaminated agricultural disinfectant and several deaths occurred.

D. Endocrine-Disrupting Chemicals

Concern has been expressed about xenobiotic-induced endocrine disruption in wildlife and humans. Adverse effects of contaminants on mammalian wildlife through the modulation of endocrine systems are documented predominantly in fish-eating (aquatic) mammals. Indeed, a large number of xenobiotics with endocrine-disrupting properties, such as OCs, have been detected in marine mammal tissue. In previous sections, reproductive and nonreproductive effects, including possible links with the functioning of the immune system, have been discussed in relation to these pollutants. Except for the reproductive toxicity in harbor seals and Baltic seals, evidence of a causal link between endocrine disruption and observed effects is weak or nonexisting. Most often neither a positive proof nor a dismissal, simply a negative endocrine-like effect, could be provided. The reasons for lack of proof are the unavailability of reliable population data, the potential interaction between the many pollutants present, the role of disease agents and other environmental factors, the lack of biomarkers to assess endocrine effects, and the little research on early development in marine mammals.

V. Species Vulnerability

The impact of pollution on marine mammals can occur throughout the entire chain from exposure, uptake, metabolism, and excretion. Concentration in prey is a determining factor. Generally, coastal species are exposed to higher environmental levels than more pelagic species, and species occurring in industrialized (including intensive agricultural) areas usually have higher pollutant levels compared to animals in less developed regions. Among marine mammals, coastal seals and dolphins usually carry the highest tissue residue levels. Superimposed on that is the preferred trophic level of feeding. In the same water mass, species feeding at lower trophic levels are exposed to lower levels of pollutants compared to species feeding higher in the food chain. This is why pollutant levels are almost always lower in baleen whales than in toothed whales. Exceptions to this pattern have been discussed earlier, e.g., for metals in species feeding on squid rather than on fish.

As mentioned earlier, females get rid of pollutants through reproduction. Species that reach sexual maturity at a younger age are at an advantage compared to those that start reproducing at an older age. Early reproduction is also positive for the offspring. The amount of pollutants descendants receive is lower if mothers initiate reproduction activities early because they have not yet built up high tissue concentrations. Similarly, an earlier onset of sexual senescence is a disadvantage in this respect because it halts the discharging process. A protracted lactation period is clearly beneficial for reproductive females because the amount of lipophilic pollutants that they transfer is high. This obviously depends on the time they start to feed again because then the pollutant uptake will counterbalance the discharge. However, the protracted lactation period may have adverse effects on the offspring because the milk is often more polluted than the food that decendants will consume once weaned. It is unclear how this resolves at the population level, i.e., whether the benefit for the reproductive female is higher or lower than the costs for the offspring.

A factor likely to lead to differential vulnerability between species is body size. Small species generally have higher levels of pollutants relative to their body weight than those of larger body mass.

Metabolization is another operative factor in this context. The P450 enzyme system is the main physiological tool for metabolizing OCs. For example, this system can be induced by PCBs, mediated by the arylhydrocarbon (Ah) receptor, which is found in mammals and birds. The metabolic ability, however, is not uniform among marine mammals. Overall, cetaceans have a lower metabolization capacity, as measured by phenobarbital (PB) and methylcholantrene (MC) types of activity. Initially, all cetaceans were thought to lack the (PB) type of enzyme. However, research has shown that several dolphin species possess at least some microsomal PB type of enzyme. Still, their PB and MC type of metabolic activity is usually lower than that of pinnipeds and terrestrial species. At a more specific level, ringed seals and harbor porpoises seem to have metabolic capacities intermediate between those of other seals and cetaceans. In conclusion, apart from the more apparent cetacean–pinniped difference in metabolic capacity, sharp differences also exist among species within any given taxa.

The critical question in this respect is, however, whether a low activity of PB-type and/or MC-type enzymes renders cetaceans more vulnerable to pollution, as has been suggested repeatedly. This may not automatically be the case. For example, PCBs can potentially elicit toxicity in at least two ways: as parent compounds (persistent congeners) and as metabolized congeners. The persistent compounds show a PB and mixed PB and MC type of toxicity associated with liver hyperproliferation, lowered levels of thyroxin and vitamin A, and a dioxin type of toxicity (MC) resulting in thymic atrophy, dermal disorders, and liver necrosis. Metabolization of parent compounds can result in at least two contrasting effects: a decreased level of dioxin type of toxicity and an increased metabolic-specific toxicity such as immunotoxicity. The resultant effect of a lower metabolization capacity therefore depends on the relative contribution of the mitigating influence of a decreased dioxin type of toxicity vs a continued PB and mixed PB/MC induction and the effect of reactive intermediates.

In this respect, attention should be drawn to the often misused concept of toxic equivalency. This concept is based on structure–activity relationships of contaminants with receptors. Tetrachlorodibenzo-dioxins (TCDD) and PCBs have a structure that fits the Ah receptor. The degree of induction by TCDD has been correlated with their toxic effects observed in laboratory animals. Given the similarity in structure of PCB congeners, the ability of these latter compounds to induce the Ah receptor-mediated response is expressed as a ratio to the induction by TCDD. This is called the toxic equivalency factor (TEF), which has been used extensively to assess the toxicity of PCB congeners and their mixtures with DDT and PCDD. That toxicity is calculated by multiplying the TEF of each compound by its concentration, and the sum of the resulting values is considered to be the total toxic equivalent (TEQ) for the mixture of compounds found in the sample.

However, it needs to be stressed that TEFs are based on laboratory animal models. Therefore, the TEQ for a given marine mammal sample only means the effect that the mixture of compounds found in that sample would have on a laboratory animal. Because (1) large differences between species exist in the induction of P450-based enzymes, (2) the toxicities of PCB metabolites are not incorporated into the calculations, and (3) the toxicity of modes other than that of a dioxin type are disregarded, the application of TEQ to assess the toxicological risk to which a particular species is subject to is not necessarily reliable. The same holds for extrapolating TEQ between species. We would therefore emphasize that the frequently used practice of assessing whether the toxic significance of a certain value of TEQ found in a marine mammal is lower/higher than a TEQ value found in a species where effects were observed is unfounded and scientifically unsound.

Another issue that remains to be clarified is the apparent ability of marine mammals, particularly observed in species in the northern arctic regions, to tolerate high levels of some heavy metals, such as mercury, lead, and cadmium. It is known that marine mammals are able to detoxify these metallic compounds by, for example, demethylating the highly toxic form of organic mercury into the less dangerous inorganic mercury, by the binding of metals to metallothioneins, or by the binding of selenium to mercury where inactive salts are produced. It is tempting to speculate whether the animals in those areas have evolved responses to mitigate the effects caused by the naturally occurring contaminants.

VI. Developments in Spatial and Temporal Trends of Pollutants

Data on levels of pollutants in marine mammals are more numerous for western Europe, North America, Canada, and Japan. Limited data are available for many other countries and regions (e.g., Africa, New Zealand, India), and very little information is available for the Southern Hemisphere. As mentioned in an earlier section, the fish-eating marine mammals from the midlatitudes (industrialized and intense agricultural use) of Europe, North America, and Japan have the highest loads (see Fig. 1). Residue tissue concentrations are lowest in the upper north polar region and the Antarctic. Nearly all of the OC contamination in marine mammals in the Arctic and Antarctic has reached these areas via atmospheric transport. Levels of the more volatile OCs are higher compared to PCBs and DDT and are distributed more homogeneously. This pattern of distribution of residue concentrations in marine mammals, however, is gradually changing. Levels of OCs are declining in the midlatitude areas, whereas they are increasing in regions distant from pollution sources. The transfer of OCs released in (sub)tropical countries to the atmosphere causes global redistribution. It is predicted that in the near to midterm future the Arctic and, to some extent, the Antarctic will become the major sinks for OCs.

Temporal trends of contaminants in marine mammals have been relatively little investigated because of the lack of long time series of samples and lack of comparability of the analytical results.

For PCBs, DDT, mercury, lead, and cadmium, some data on tissue concentrations in marine mammals from certain areas are available. In most heavily industrialized and agricultural regions, the production and use of DDT and PCBs was halted in the early 1970s. From the mid-1970s onward, levels of DDT

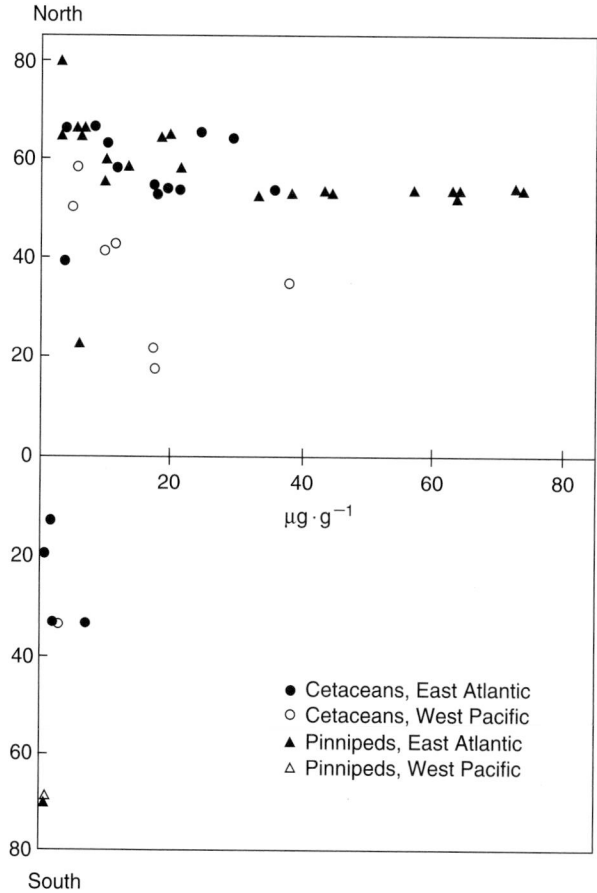

North

μg·g⁻¹

South

● Cetaceans, East Atlantic
○ Cetaceans, West Pacific
▲ Pinnipeds, East Atlantic
△ Pinnipeds, West Pacific

Figure 1 *Latitudinal ΣPCB concentrations (μg·g⁻¹ wet weight) in blubber tissue of various marine mammals from the western Pacific and the eastern Atlantic.*

and PCBs in marine mammal tissues decreased. The decline in DDT levels was stronger than that of PCBs. In pinnipeds the decline was 80–90% for DDTs and 60% for PCBs. The difference is most likely due to less stringent control measures for PCBs; large quantities of these compounds have remained in use in many applications. The overall time trend for PCB and DDT levels in marine mammals is that concentrations have decreased since the mid-1970s. The decrease in DDT levels has continued thereafter. However, PCB levels in some areas leveled off at the end of the 1970s/early 1980s. Figure 2 shows the compartmentation of the global budget of all produced PCBs by industry. Given the fact that only 1% of all the PCBs produced has reached the oceans, that 35% are still in use, 30% are accumulated in dump sites, and that the fate of the other 34% is unknown, it is expected that the observed leveling off of the decrease in marine mammals will not be followed by a strong reduction in the near future. Trends for heavy metal pollution are less apparent. In general, it is accepted that between the mid-1970s and mid-1990s there was no clear trend for mercury and cadmium in pinnipeds from the Canadian Arctic and Greenland. On the contrary, levels of mercury and lead in pinnipeds from the Wadden Sea have decreased considerably.

VII. A Fundamental Approach to Address Pollution Impact on Marine Mammals

It is clear that a considerable amount of fundamental research is needed before it will be possible to adequately address the impact of pollutants on marine mammals. Realizing this situation, the International Whaling Commission (IWC), through its scientific committee, has developed a comprehensive program to investigate pollutant cause–effect relationships in cetaceans: "Pollution 2000+." At a later stage, the International Council for the Exploration of the Sea, through its

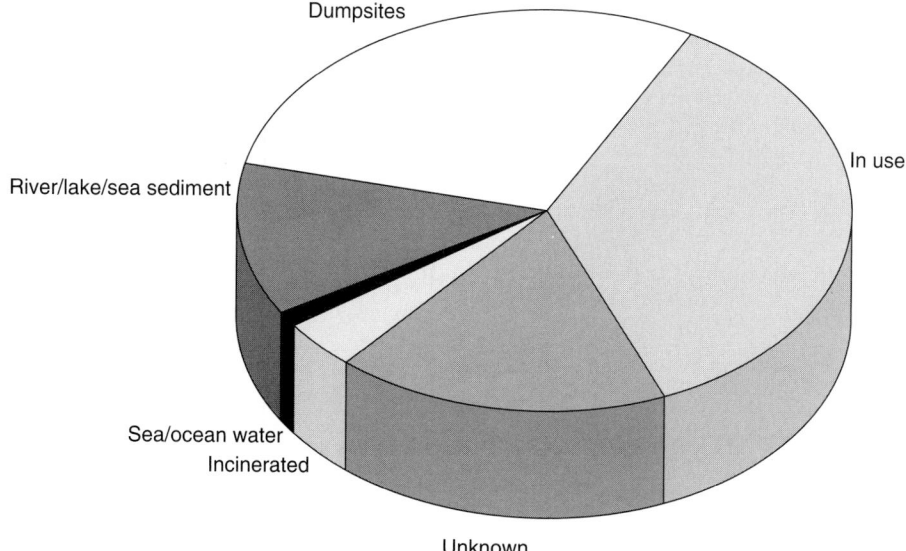

Dumpsites

In use

River/lake/sea sediment

Sea/ocean water
Incinerated

Unknown

Figure 2 *Compartmentation of global budget of produced PCBs (kilotons).*

Marine Mammal Habitat Working Group, developed a similar research program, this time focusing on pinnipeds.

The ultimate objective of pollution studies as related to marine mammal management is to determine a predictive model to link tissue pollutant levels with effects at the population level. This is obviously a longer term goal. It is realized that if any progress is to be made within a reasonable time frame, a multidisciplinary, multinational focused program of research is required that concentrates on species and areas where there is most chance of success.

The mentioned programs earlier focused on PCBs because these chemicals can be used as model compounds of OC pollution. Moreover, PCBs are found at extremely high tissue levels in cetaceans, their effect on mammals are well known, and substantial information is available on their patterns of variation, geographical distribution, and tissue kinetics. The focal species in the IWC program are common bottlenose dolphin and harbor porpoise because both species occur in waters extending over a gradient of pollution and are likely to provide reasonable sample sizes.

In this program, two short-term objectives were established: (1) to select and examine a number of biomarkers of exposure to and/or effects of PCBs and try to determine whether a predictive and quantitative relationship with PCB levels in certain tissues exists and (2) to validate/calibrate sampling and analytical techniques to address such questions for cetaceans, specifically: determine changes in concentrations of variables with postmortem times and examine relationships between concentrations of variables obtained from biopsy sampling with those of concentrations in other tissues that can only be obtained from fresh carcasses.

Because the ultimate aim of the program is to look at potential effects of pollutants at the population level, it was considered necessary to test and develop techniques to feasibly collect data from large numbers of free-living animals. Because biopsy techniques allow such a type of sampling, an initial step in the project has been to calibrate information obtained from biopsy sampling with that collected from dead animals. It was also considered similarly important to ascertain the influence of postmortem time on levels of contaminants and on indicators of exposure and effect. This calibration is needed to ensure that collected samples are representative of actual pollutant loads.

It is clear that Pollution 2000+ is a core program to address some fundamental questions. It does not imply that other research on pollutants and marine mammals is not important. On the contrary, its value is enhanced by cooperation with existing studies and as a context for the development of new programs.

Acknowledgments

The nature of this article in the context of this encyclopedia implies the use at length of work of colleagues in the field of marine mammalogy and ecotoxicology. The views presented here have been shaped partly by our own research, but definitely also by the discussions held with many colleagues within and outside this field of expertise, as well as the reading of published references. We acknowledge these colleagues and sources for their contribution and have tried to stay as close as possible to their original findings and cited directly from their work where relevant.

See Also the Following Articles

Endocrine Systems ■ Health ■ Mass Die-Offs ■ Noise, Effects of ■ Pathology ■ Population Dynamics

References

AMBIO (1992). Seals and seal protection: Population studies, pathological studies, chemical analysis, experimental studies, comparative studies. *Ambio* **21,** 494–606.

Aguilar, A., and Borrell, A. (1994). Abnormally high polychlorinated biphenyl levels in striped dolphins (*Stenella coeruleoalba*) affected by the 1990–1992 Mediterranean epizootic. *Sci. Tot. Environm.* **154,** 237–247.

Aguilar, A., Borrell, A., and Pastor, T. (1999). Biological factors affecting variability of persistent pollutant levels in cetaceans. *J. Cetacean Res. Manage.* (Special Issue 1), 83–116.

Aguilar, A., Borrell, A., and Reijnders, P. J. H. (2000). Geographical and temporal variation in levels of organochlorine contaminants in marine mammals. *Mar. Environ. Res.* (in press).

Boon, J. P., van Arnhem, E., Jansen, S., Kannan, N., Petrick, G., Schultz, D., Duinker, J. C., Reijnders, P. J. H., and Goksør, A. (1992). The toxicokinetics of PCBs in marine mammals with special reference to possible interactions of individuals congeners with the cytochrome P450-dependent mono-oxygenase system: An Overview. *In* "Persistent Pollutants in Marine Ecosystems" (C. H. Walkker and D. R. Livingstone, eds.), pp. 119–159. Pergamon Press, Oxford.

Delong, R. L., Gilmartin, W. G., and Simpson, J. G. (1973). Premature births in California sea lions: Association with high organochlorine pollutant residue levels. *Science* **181,** 1168–1170.

Jepson, P. D., Bennet, P. M., Allchin, C. R., Law, R. J., Kuiken, T., Baker, J. R., Rogan, E., and Kirkwood, J. K. (1999). Investigating potential associations between chronic exposure to polychlorinated biphenyls and infectious disease mortality in harbour porpoises from England and Wales. *Sci. Tot. Environm.* **243,** 339–348.

Martineau, D., de Guise, S., Fournier, M., Shugart, L., Girard, C., Lagacé, A., and Béland, P. (1994). Pathology and toxicology of beluga whales from the St. Lawrence Estuary, Québec, Canada. Past, present and future. *Sci. Tot. Environm.* **154,** 201–215.

Muir, B., Braune, B., DeMarch, B., Norstrom, R., Wagemann, R., Lockhart, L., Hargrave, B., Bright, D., Addison, R., Payne, J., and Reimer, K. (1999). Spatial and temporal trends and effects of contaminants in the Canadian Arctic marine ecosystem: A review. *Sci. Tot. Environm.* **230,** 83–144.

Olsson, M., Karlsson, B., and Ahnland, E. (1994). Diseases and environmental contaminants in seals from the Baltic and the Swedish west coast. *Sci. Tot. Environm.* **154,** 217–227.

O'Shea, T. J. (1999). Environmental contaminants and marine mammals. *In* "Biology of Marine Mammals" (J. E. Reynolds III and S. A. Rommel, eds.), pp. 485–564. Smithsonian Institution Press, Washington, DC.

O'Shea, T. J., Reeves, R. R., and Kirk Long, A. (eds.) (1999). "Marine Mammals and Persistent Ocean Contaminants: Proceedings of the Marine Mammal Commission Workshop," Keystone, Colorado, 12–15 October 1998. Marine Mammal Commission, Bethesda, MD.

Reijnders, P. J. H. (1986). Reproductive failure in harbour seals feeding on fish from polluted coastal waters. *Nature* **324,** 456–457.

Reijnders, P. J. H. (1994). Toxicokinetics of chlorobiphenyls and associated physiological responses in marine mammals, with particular reference to their potential for ecotoxicological risk assessment. *Sci. Tot. Environm.* **154,** 229–236.

Reijnders, P. J. H., Aguilar, A., and Donovan, G. P. (1999). Chemical Pollutants and Cetaceans. *J. Cetacean Res. Manage.* Special Issue 1, 1–273.

Swart, R. L. de, Ross, P. S., Vedder, E. J., Timmerman, H. H., Heisterkamp, S. H., van Loveren, H., Vos, J. G., Reijnders, P. J. H., and Osterhaus, A. D. M. E. (1994). Impairment of immunological functions in harbour seals. (*Phoca vitulina*) feeding on fish from polluted coastal waters. *Ambio* **23,** 155–159.

Tanabe, S., Iwata, H., and Tatsukawa, R. (1994). Global contamination by persistent organochlorines and their ecotoxicological impact on marine mammals. *Sci. Tot. Environm.* **154,** 263–264.

Figure 1 *Human interest in marine mammals: Part of the solution or part of the problem? Paul H. Forestell.*

Popular Culture and Literature

PAUL H. FORESTELL
*Southampton College of Long Island University,
New York*

A general survey of current human behavior provides a rather paradoxical view of our modern cultural perception of marine mammals. On the one hand, it would seem by almost any indicator that human interest in marine mammals is at a fever pitch. Excursions to view marine mammals in the wild draw countless participants from many cultures and age groups (Fig. 1). Attendance at marine parks and display facilities is at an all time high. The past decade has seen an explosion in books and videos about marine mammals. The sale of marine mammal "themed" merchandise includes thousands of products worldwide. Campaigns to protect marine mammals and their habitat exist on every continent, and the number of local and national governments involved in legislative agendas related to marine mammals is staggering. There has been an exponential increase in the number of marine mammal research projects, many involving direct public participation, and the population of marine mammal researchers is growing steadily.

On the other hand, there are at least five species of marine mammal [Chinese river dolphin (*Lipotes vexillifer*), Gulf of California harbor porpoise (*Phocoena sinus*), North Atlantic right whale (*Eubalaena glacialis*), North Pacific right whale (*E. japonica*), and Mediterranean monk seal (*Monachus monachus*)] in danger of becoming extinct because of human activities. Attempts to interact with dolphins in the wild are claimed to be disturbing the dolphins and endangering humans. Efforts to eliminate the killing of marine mammals in fisheries-related by catch are not succeeding. Sealing continues, and efforts to cull seal populations to protect fisheries are on the rise. Commercial whaling has increased over that of a decade ago. The oceans are being poisoned: 3 million pounds of trash were cleaned from beaches in the United States in 1999, and that amount must be assumed to represent only a fraction of material floating in the world's oceans.

In a recent in-depth analysis, David Lavigne, Victor Scheffer, and Stephen Kellert (1999) concluded that there has been a significant evolution in North American attitudes toward marine mammals since the early 1960s. These authors reviewed a diverse range of indicators, including patterns of consumptive exploitation, tourism, media attention, legislation, nongovernmental organization initiatives, and scientific research. They summarized their findings by noting that "With few exceptions . . . North American attitudes toward marine mammals have shifted from a focus on their killing and material utilization to a more aesthetic interest in observing these creatures in the wild, in captivity, and in various media forms." (Lavigne *et al.,* 1999). The most notable exception to the shift in focus is a general willingness to grant special status to aboriginal groups to hunt marine mammals. The authors further concluded that the change in attitudes has resulted primarily from a rapid urbanization of the human population, increased knowledge of marine mammals through new research and extensive media exposure, and a shift in ethical values.

It seems unarguable that North American attitudes (and those of many nations referred to as "industrialized") have changed from a focus on material utilization. However, a qualitative analysis of available behavioral evidence (i.e., what people actually do, rather than what they express as a value or attitude in surveys) suggests that humans (including North Americans) still basically regard marine mammals as a commodity. A consideration of how marine mammals are represented in current literature and the popular media; the types of marine mammal issues that win public attention; the types of activities that humans engage in around marine mammals; and the role that government plays in managing such activities suggest that rather than a shift from material utilization to aesthetic interest, the shift has simply been from one sort of utilization (or exploitation) to another. In brief, while the general trend has been to an increasing effort to protect and conserve, the motivational basis for the change remains highly anthropocentric.

I. Background: Shaping Cultural Perceptions of the 1950s

To understand current perspectives on marine mammals we need not only review current evidence, but also provide a summary characterization of the past. Although more completely detailed elsewhere, a brief history of human interactions with marine mammals is presented here in order to provide an appropriate context for considering current perspectives. A review of patterns in the literature and behavior of humans since 1950s is then undertaken to consider the hypothesis that present attitudes of humans toward marine mammals are, as they have been since the dawn of time, based primarily on an anthropocentric worldview.

The recorded history of human interaction with marine mammals begins with a perspective based on small-scale, direct exploitation in pursuit of vital resources. From opportunistic scavenging of stranded animals to organized HUNTING of locally available species, subsistence efforts were focused on obtaining food and by-products needed by the hunters themselves for survival. By its very nature, subsistence hunting depends on an extended knowledge of the ecosystem in which the hunter and the hunted live. Both predator and prey occupy the same setting and require the same kinds of knowledge to survive. Successful subsistence hunting presupposes a body of knowledge, a set of skills, a motivational focus, and sensitivity to the psyche of the hunted that form little or no part of a strictly "utilitarian" exploitation of natural resources. The Japanese novelist Natsuki Ikezawa described subsistence whale hunting as requiring "a compassion with Nature, a harmony with surroundings. To eat and to be eaten are the two sides of the same deed." It is common within subsistence cultures to promote rituals of compassion and respect for that which provides sustenance, even while it is being killed. This is reflected in Polynesian chants, Inuit ceremonies, and the myths of aboriginal peoples throughout North America, Australia, and Asia.

Subsistence hunting extended, relatively unchanged, over a considerable period of time in human history (Fig. 2). Norwegian petroglyphs (rock drawings) believed to be nearly 4000 years old show men in boats, along with seals, dolphins, and whales. Harpoons with poison tips, nets, and stranding drives were employed by indigenous hunters in many parts of the world, including Japan, North America, the South Pacific, and Europe. Small-scale trading of products frequently occurred within localized networks of nearby communities (e.g., among the Arctic Inuit). However, subsistence hunting of marine mammals up until the 12th century primarily involved hunting for the immediate use of the community; for food, clothing, and other products of domestic utility. In almost all cases, the traditions of these subsistence hunters incorporate myths, chants, and rituals honoring the quarry and the skills needed to overcome it. This is an important consideration in understanding current cultural perspectives on marine mammals: a representation of the development of current attitudes as a linear evolution from utilitarian to aesthetic (as presented by Lavigne *et*

Figure 2 *Tongan whaler c. 1970. Subsistence hunting has been conducted for centuries with essentially unchanged techniques. Photograph by Bill Dawbin.*

al., 1999) is incomplete on two counts. Ancient subsistence attitudes were far more complex than the utilitarian characterization would imply, and current perspectives are far less aesthetic than use of that term would prescribe. In fact, current perspectives on marine mammals probably represent a relatively recent reoccurrence of the same devolution of attitudes from holistic to exploitative that occurred during the transition from subsistence whaling, to commercial whaling, to industrialized whaling. In the present day, we have been witnessing a transition from "subsistence whale watching," to "commercial whale watching," to "industrialized whale watching," as detailed below.

The change from subsistence to commercial whaling was accompanied by a significant change in human perspective on marine mammals during the 11th and 12th centuries. It was during that period that the Basques of Spain apparently developed and implemented the first truly "commercial" hunt and initiated a fundamental change in attitude that would sow the seeds of "industrialized" whaling some 600 years later. The Basques probably learned whaling techniques from the Normans, Norwegian settlers who populated the Bay of Biscay during the 9th century. The Basques, tenacious settlers on a critical access route to the Iberian Peninsula (and onward to Africa), were not subsistence whalers. Originally pastoralists who tended flocks in the mountains, they eventually founded the kingdom of Navarre, which included a number of tiny fishing villages along the Biscayan coast. There they found the North Atlantic right whale, which migrated into the Bay of Biscay each year from autumn through spring. The Basques were poised between a vast, untapped resource on the ocean side and a steady flow of traders to the world's centers of commerce on the other. They found a ready market for almost the entire carcass of the whales they brought ashore. Oil for lamps, lubricants, and soap; BALEEN for corsets, buggy whips, fishing rods, and brushes; bones for fence posts and portals; skin for footwear; and meat, blubber, and tongue for food. For 300 years Basque whalers pursued the right whale: first across the

Bay of Biscay and then to the far side of the Atlantic. So good were they at their craft that, as the need for whale oil grew in Europe, they became teachers of those who would take commercial whaling to the furthest reaches of the world's oceans.

In its earliest form, and as practiced by the Basques, commercial whaling involved the application of subsistence techniques, but for the realization of profit rather than direct consumables. As the theater of endeavor moved from the Biscayan coast to North America, Spitsbergen, and the open seas of the Atlantic and Pacific, the method of catching and killing whales changed little, although the processing of whales was carried out variously either on shore or onboard the whaling ship itself. The perseverance of hunting techniques may be verified by a survey of first-hand accounts of whaling operations. Examples include Guillame Rondelet's rendition of Basque whaling in Europe in 1568; Francis Olmstead's description of American sperm whaling in the Pacific in 1841; and Robert Cushman Murphy's documentation of a 1913 sperm whaling voyage to the South Atlantic. Without benefit of modern machinery, commercial whalers using sailing ships, whale boats, and hand-held harpoons successfully pursued right whales, bowhead whales (*Balaena mysticetus*), and gray whales (*Eschrichtius robustus*) to near extinction and decimated known sperm whale (*Physeter macrocephalus*) populations. A similar fate was visited upon elephant seals (*Mirounga* spp.), walruses (*Odobenus rosmarus*), and sea otters (*Enhydra lutris*), while Steller's sea cow (*Hydrodamalis gigas*) was completely extirpated.

An important change that accompanied the transition from subsistence to commercial hunting was that animals were hunted and killed by humans who were not otherwise connected to the ecosystem or the animals targeted by the enterprise. Strangers came from faraway places for the sole purpose of obtaining as much of the resource as could be most quickly taken away for sale in distant markets (Fig. 3).

By the middle of the 19th century the cost of whale oil climbed as stocks of whales declined. Alternative products were derived from turpentine, cottonseed, and linseed and replaced

Figure 3 *As commercial hunting replaced subsistence hunting, the public came to view whales as circus curiosities rather than honored quarry. The head of a dead right whale forms the backdrop for a family snapshot. (Painting by Larry Foster, 1974).*

whale oil in a number of uses. The distillation of petroleum was the single greatest factor in the decline of commercial whaling as practiced by the sailing ships of the American whaling fleets at the end of the 19th century.

The transition from commercial whaling to industrialized whaling began in 1863 when Svend Føyn, a Norwegian sealer, made his first successful kill of a blue whale (*Balaenoptera musculus*) using a cannon mounted on the bow of a 90-foot steamship that fired a 100-lb. harpoon with an exploding tip. The combination of cannon, exploding harpoon, and steamship meant that large fast whales could be dispatched with relative speed and efficiency. Readily accessible populations of blue, fin (*B. physalus*), and humpback whales (*Megaptera novaeangliae*), unprofitable targets for sailing ships, gave new life to commercial whaling. For 40 years Norwegian-style whaling spread around the world. At the turn of the 20th century, a German chemist discovered a way to convert animal oil into solid fat. Soon after, additional processing was developed to rid whale oil of its inherently strong taste and odor. The result was that whale oil once again became a commodity of worldwide importance, particularly for making margarine and soap, and whaling once more became a profitable enterprise.

Rapid developments in technology soon made it possible to hunt and kill whales more efficiently and in greater numbers. In addition, changes in the economic policies of industrialized nations widened the gulf between consumer and consumed. As both demand and competition grew the philosophies and practices of the industrial revolution were incorporated into whaling enterprises, significantly altering the way whalers did business. The goal was no longer the realization of profit, but maximization of profit. This led, during the first half of this century, to a phenomenal arms race against whales. As the hand-held harpoon was replaced with Svend Føyn's cannon, the trypots gave way to huge pressure cookers; and gigantic hydraulic tail-grabbers dragged entire blue whales out of the ocean, through stern slipways, and onto the flensing decks of pelagic factory ships. By the 1950s a 65-foot whale, weighing 80 tons, could be hauled up the slipway and turned into processed oil, frozen meat, and ground-up fertilizer in just over an hour. By the 1960s nearly every major nation of the world had participated in the decimation of whale stocks—some 30,000 to 50,000 whales per year were taken in the Antarctic alone between 1930 and 1960 (Fig. 4).

By the 1950s it could well be said that the human attitude toward marine mammals had become a strictly utilitarian one. Although the industrialization of slaughter occurred on a grand scale with regard to whales, commercial hunting of elephant seals for oil and sea otters, fur seals and harp seals (*Pagophilus groenlandicus*) for their coats had disastrous results on pinniped populations throughout the world and sea otter populations throughout their range. The hunt for sirenians has never progressed to anything near an industrialized effort. A combination of subsistence and small-scale commercial hunting has generally been sufficient to reduce known populations to extinction (in the case of Steller's sea cow) or near extinction [in the case of manatees (*Trichechus* spp.) and dugongs (*Dugong dugon*)]. Hunting of small cetaceans throughout the world has also been marked by the shortsighted exploitation of pilot

Figure 4 *By the 1950s, commercial whaling had become a sophisticated industrialized operation. Courtesy of The Kendall Whaling Museum, Sharon, Massachusetts.*

whales (*Globicephala* spp.), killer whales (*Orcinus orca*), bottlenose whales (*Hyperoodon* spp.), spinner dolphins (*Stenella longirostris*), spotted dolphins (*S. attenuata* and *S. frontalis*), and common dolphins. What is unique to the history of human interaction with dolphins, compared with our exploitation of other marine mammals, is the massive destruction of pelagic species through bycatch mortality associated with modern drift nets and purse seining. Industrialized fishing techniques have killed more dolphins in the last 50 years than were killed in the entire prior history of humankind. The late David Gaskin succinctly described the attitude underlying the period of industrialized whaling as an ethical failure. "We could have conserved many other resources than just the stocks of large baleen whales, if the dead-end philosophy of the economic industrial growth ethic could be circumvented, and if the aces in the deck of cards with which we play were not invariably short-term economic gain and political expediency."

II. The 1950s: Out with the Old

Beginning in the 1950s there was growing scientific evidence of the complexity of marine mammal BEHAVIOR. Stimulated by the work of John Lilly in the United States, researchers in European, Japanese, and other American laboratories conducted neuroanatomical and electrophysiological studies of cetacean brains that challenged assumptions about the importance of BRAIN SIZE and identified a number of unique characteristics in cortical architecture, structural morphology, degree of lateralization, and brain-wave patterns. Philosopher, anthropologist, and ecologist Gregory Bateson explored the intricacies of dolphin social behavior by studying a group of spinner and pantropical spotted dolphins at Sea Life Park in Hawaii. William Schevill and colleagues (including his wife Barbara Lawrence) began collecting acoustic recordings that would eventually document the underwater sounds of nearly three dozen species of marine mammals, including odontocetes, mysticetes, and pinnipeds. David and Melba Caldwell explored the acoustic and social behaviors of whales and dolphins in the wild.

These researchers heralded an important shift in scientific focus that accompanied the end of the grand age of whaling during the 50s and the beginnings of "modern" marine mammal studies in the 60s. Earlier studies of marine mammals had been carried out, for the most part, on the flensing decks of whaling ships, on the slipways of coastal whaling stations, or on the rocky shores of pinniped haul-out and ROOKERY sites. Information gleaned from such studies was applied to statistical determinations of the size of harvestable stocks and the development of new commercial applications for whale products. R. B. Robertson's account of life aboard an Antarctic factory whaling ship ("Of Whales and Men," A. A. Knopf, 1954) describes such an approach. In one chapter, the author notes some of the questions entertained by the chemist and biologist on board the ship:

> Why has no female sperm whale ever been sighted or killed in the Southern Ocean, though many thousands of old males are killed there every year. . . . At what rate would whale tendon from the flukes be absorbed by the human body if it were put on the market in huge quantities of twenty-foot strands to replace the rare and expensive kangaroo tendon presently used as surgical sutures to repair hernias. . . . What is the nutritive value of a properly cooked fin-whale steak. . . . How could one make the whale—as it should be—the world's main supply of the valuable new pituitary and adrenal hormones (pp. 240–241)?

A similar drive to collect field data to better inform the management of a commercial harvest prompted a group of American biologists to investigate the life history of gray whales by studying dead ones during the 1960s. From 1959 to 1969 special permits were issued to the Del Monte and Golden Gate Fishing Companies in Richmond, California, by the federally run Marine Mammal Biological Laboratory in Seattle, Washington. The permits allowed the collection of 316 gray whales as they traveled along the west coast of the United States during their annual north–south migration between Mexico and the Bering, Chuckchi, and western Beaufort seas. The description of the whales' life history was based on shipboard and aerial observations of their migration numbers and timing, and documentation of a series of body measurements from the 316 whale carcasses. Data included overall length and weight; width of fluke and length of flippers; thickness of blubber; number and size of baleen plates; degree of vertebral fusion; condition of mammary glands, size of ovaries, and diameter of uterine horns in females; length of penis and weight of testes in males; number of wax ear plug growth layers; number and type of external body scars; number and type of endo- and ectoparasites; and stomach contents. Data were used to determine numbers and migratory patterns of different age/sex classes; prey species; reproductive patterns; and susceptibility to parasites and predators (other than humans). Dale Rice and Allen Wolman reported the results of the study in 1971. The expressed purpose of the work was to provide sound biological knowledge to ensure a wise management program if commercial exploitation, ended by international agreement in 1946, should resume.

Compare the questions of Robertson's chemist and the Marine Mammal Biological Laboratory's biologists with those posed by the legendary Kenneth Norris, who studied both captive and wild dolphins in California and Hawaii during the 1960s and early 1970s. In "The Porpoise Watcher" (W. W. Norton, 1974), Norris considers the intricacies of social organization in Hawaiian spinner dolphins:

> . . . most of the subgroups we see in active schools have no clear function. Are they family groups, play groups, are they divided by sex or age. . . . What is it that regulates porpoise school numbers? Why don't their schools grow and grow until every scrap of natural resource is used, as humans are busy doing? What racial wisdom, somehow lost by humans as they adopted civilization, keeps them in balance (pp. 199–200)?

Norris' approach to the study of live animals represented a significant change in marine mammal science that ultimately reshaped the public's perception of marine mammals in many parts of the world (Fig. 5). The change appears to have been born in the observations of live animals by scientists whose first view of their subjects was not from a flensing deck or slaughtering beach. Many of the scientists who developed an interest in marine mammals during the 60s, unlike those of previous decades, did not come out of the whaling industry or from government agencies responsible for managing various types of "fisheries" (including whaling and sealing). Access to marine mammals is logistically and therefore economically challenging. Until World War II there were generally two ways one could study a marine mammal: stumble across a stranded one on a remote beach or accompany commercial harvesters on their hunts. These were the settings that provided material to shape cultural perspectives on marine mammals right up until the 1950s.

Books available to the public during the first half of this century were based primarily on popularized accounts of whaling voyages, or historical descriptions of commercial whaling. For example, in 1913 the Boy Scouts of America sponsored the publication of Frank Bullen's 1898 whaling account, "The Cruise of the Cachalot," as part of the Every Boy's Library series. In the preface the book is proffered as an antidote to "cheap juvenile literature" and as a means of capturing a young

Figure 5 *By the 1960s, biologists such as Ken Norris were becoming more interested in observing the behavior of marine mammals in the wild. © Flip Nicklin (Minden Pictures).*

man's attention with "tales of daring enterprise and adventurous good times." There were few natural history descriptions produced for public consumption prior to World War II. Approximately 230 works of fiction are listed with dates of publication in an informal bibliography of fiction books for adults produced by Trisha Lamb Furstein and colleagues in 1998 that have cetaceans as major or secondary subjects (www.physics.helsinki.fi/whale/literature/fic_main.html). The list shows only two books printed between 1900 and 1950. One is a play and the other appears to be based on ancient myths about a friendship between a dolphin and a boy.

III. The 1960s and 1970s: In with the New

The landscape changed dramatically after World War II, and by the 1960s a significant transition was underway in human perceptions of marine mammals. It would be simplistic to point to one central mechanism behind the groundswell of interest in marine mammals (more specifically, cetaceans) that occurred during this period. The exotic undersea portraits and exciting escapades of Jacques Cousteau; the new age writings of John Lilly; the increasing exposure to whales and dolphins in zoos, parks, and oceanaria; the "protest" mind-set of the 60s and 70s; and the growing awareness of conservation issues due to books by (among others) Aldo Leopold, Rachel Carson, and Victor B. Scheffer all played pivotal roles. The combined effect of these elements was to generate an entirely new look at marine mammals. Four predominant themes emerged quickly and simultaneously throughout North America, the United Kingdom, Europe, and Australia/New Zealand during the 1960s and 1970s: dolphins (and whales) exemplify cognitive capabilities rivaling human capacity; the opportunity to watch dolphins and whales is exciting and entertaining; the worldwide slaughter of whales is unethical and biologically disastrous; and marine mammals demonstrate a number of adaptations to the marine environment that may have important applications to human technology. These have remained the four corners of the box in which humans appear to have marine mammals contained: intelligent, entertaining, endangered, and valuable to humans.

The popularization of cetaceans as highly intelligent animals has been largely due to the writings of John Lilly, beginning with his book "Man and Dolphin" (Pyramid Publication, 1961), chronicling his work with dolphins during the late 1950s. Lilly was not the first scientist to note the possibility of complex cognitive capability in cetaceans—that honor probably goes to psychobiologist Arthur McBride, the first curator of Marine Studios, a facility built in St. Augustine, Florida, in 1938. In 1948 he and the Canadian psychologist Donald Hebb published observations of bottlenose dolphin social behavior. However, Lilly used his position successfully as an acclaimed neurophysiologist with the National Institutes of Health to foster an enduring and widespread public perception of whales and dolphins as near-mythical species with mental capacities rivaling or surpassing those of *Homo sapiens*. Lilly's succession of books for the popular press (but especially "Man and Dolphin," 1961; "The Mind of the Dolphin," Doubleday, 1967; and "Communication Between Man and Dolphin," Crown Publishers, 1978) led to a hostile reaction from the scientific community, which disputed the validity of his claims. Overall, however, this characterization of cetaceans as socially and cognitively complex animals with rich emotional lives guided by a code of ethics resonated with the public. Paraphrasing one of his lectures, John Lilly wrote in "The Mind of the Dolphin" (1967, reprinted in "Lilly on Dolphins," Doubleday, 1975):

> I wish to tell you something of what we have learned of a group of uninhibited nudists who have never worn clothes. They have never walked on their own two feet. They have no property. They cannot write their own names. They have no commerce or stores. They have no radio, no TV. They have no fireplaces, nor furnaces, or any fire at all. They have no atomic or nuclear bombs, or power plants. They have no written or printed records. They have no libraries or paintings. In spite of all these handicaps, they are successful. They have big brains and have readily available food supplies. They have the sense to go south in the winter and to go north in summer. They have the ability to out-think, outmaneuver and fight successfully against their enemies. Finally, they think enough of us to save each of us when they find us in trouble (p. 291).

Sexually liberated, antimaterialistic, antiwar, self-sufficient, intelligent, and altruistic. What suite of characteristics could have more poignantly caught the attention of an up-and-coming generation of "baby boomers" protesting post-World War II materialism, the Vietnam war, restrictive social mores, and cutthroat international capitalism? Lilly's message is clear—sun, surf, and sex—with big brains, and no guilt. Could it get any better?

Lilly's writings and public lectures generated a backdrop against which a new-age philosophy of dolphin as genius, healer, therapist, and spiritual advisor flourished. Although Lilly's interpretations of his findings were frequently based upon gratuitous explanations, he was instrumental in a number of critical discoveries concerning the peculiarities of dolphin acoustic behavior (Fig. 6).

One can hardly discuss the work of the new brand of marine mammal scientist that emerged in the 60s without noting the importance of marine mammal display facilities, beginning with Marine Studios (later called Marineland, Florida). Records of marine mammals being kept in captivity for public amusement date back to ancient times. By far, pinnipeds comprise the majority of display animals. They are relatively easy to take from the wild, adapt well to captivity, and in many cases are trained easily to amuse human spectators. Cetaceans, however, require a degree of knowledge, resources, and commitment that few display facilities are able to meet. Marine Studios was the first facility to make a dedicated commitment to maintaining bottlenose dolphins (*Tursiops truncatus*) in captivity in its role as an "oceanarium" rather than aquarium or zoo. Once Marine Studios established that dolphins could be placed successfully on display, however, the postwar climate fostered the rapid growth of similar facilities, first in the United States and then throughout the world.

From the beginning those responsible for cetaceans at oceanaria encouraged outside scientists to explore the biology and

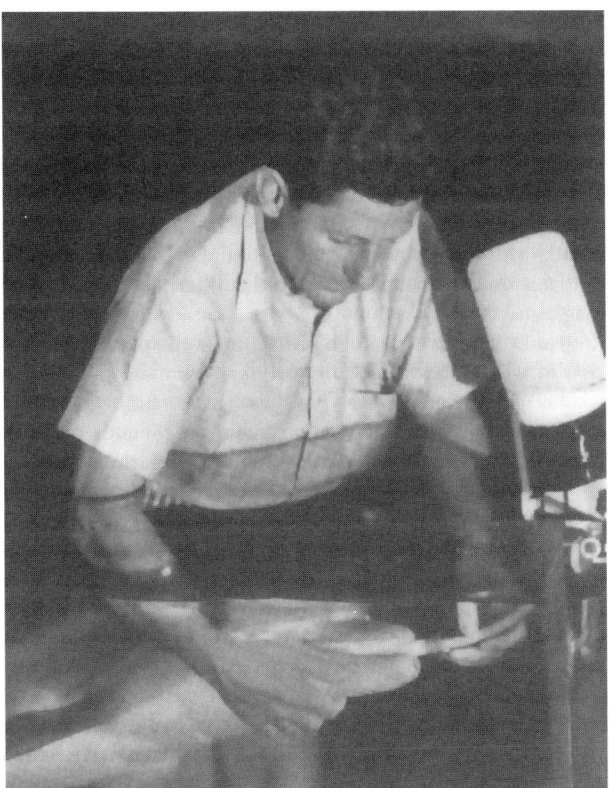

Figure 6 *Dr. John Lilly, neurophysiologist, was a pioneer in the study of dolphin cognition.*

behavior of the animals at their facilities. Arthur McBride, and later Forrest Wood at Marine Studios; Kenneth Norris, the first curator at Marineland of the Pacific (the west coast "sister" institution of Marine Studios); his graduate students and colleagues William Evans at Hubbs-Sea World Research Institute and John Prescott at the New England Aquarium; Tap and Karen Pryor at Sea Life Park; and Murray Newman at the Vancouver Public Aquarium all opened their doors to biologists, neuroscientists, anthropologists, physiologists, psychologists, acoustical engineers, and an ever-widening range of other professionals with an interest in studying live animals. At a time when television and Jacques Cousteau were flooding into American homes, the collaboration between science and the entertainment industry had a tremendous impact on the public perception of marine mammals. Scientists were not used to dealing directly with the public, but those who ran the oceanaria were. As scientists began to discover more and more about the hidden world of whales and dolphins, the oceanaria incorporated much of that knowledge in their decisions about what animals to capture, improvements to their husbandry programs, and, most importantly, in the subject matter of their shows.

One indication of the "coming of age" of dolphins as a significant cultural phenomenon in the United States was the 1960s production of two movies and a television series based on the adventures of "Flipper," a common bottlenose dolphin, and Sandy Ricks, the young son of a fisherman (later changed to a park ranger). The original movie, released in 1963, hinted of the ancient stories of the relationship between a boy and a dolphin, such as described in Iassos, Greece, and the Roman town of Hippo in Africa. In the movie, a dolphin met a young boy and they developed a friendship that led to a bonding that surpassed species distinctions and spawned a sequel. The mythical proportions of the story did not survive long, however, and the subsequent television series seemed to quickly develop into a marine version of "Lassie, Come Home."

While the 1960s initiated a true Renaissance in the portrayal of marine mammals, oceanaria helped emphasize a conceptual "divide" that continues to the present. With limited exceptions, modern public attitudes about marine mammals became (and continue to be) strongly predisposed toward whales and dolphins. In reviewing the status of marine mammals in popular literature and culture, it is important to bear the following caveat in mind. The typical representation of "marine mammal" does not reflect the diversity and extent of actual species. Marine mammals include over 120 SPECIES, divided into four general categories [whales, dolphins, and porpoises; seals, sea lions, and walruses; manatees and dugongs; sea otters and polar bears (*Ursus maritimus*). However, the preponderance of available information on nature excursions, visits to display facilities, and the sales of books, videos, and other merchandise suggest that the prototypical representative of the category "marine mammal" is one of a small set of relatively well-known cetacean species [e.g., common bottlenose dolphin, killer whale, humpback whale, sperm whale].

Although occasional interest in other marine mammals has developed with respect to specific issues [e.g., the killing of neonatal harp seals (*Pagophilus groenlandicus*) for their white fur coats in Canada; the decimation of Caribbean manatee (*Trichechus manatus latirostris*) populations from boat collisions in Florida; competition between sea otters and abalone fishermen in California], it remains clear that cetaceans have drawn public attention in a way that far outdistances the attention generally paid to pinnipeds, sirenians, or sea otters. This difference is despite the fact that about two-thirds of the marine mammals held in captivity since 1973 in North America are either California sea lions (*Zalophus californianus*) or harbor seals (*Phoca vitulina*). The most commonly displayed cetaceans is the common bottlenose dolphin, which generally makes up about 20% of the total number of animals on display, Killer whales, although small in number, have generated a huge marketing success for their keepers. Much of the attraction of cetaceans appears to be related to a fascination with the size, beauty, strength, and perceived INTELLIGENCE of these two most frequently seen examples. While sea lions balance balls on the end of their nose and make hilarious barking sounds on cue in oceanaria and marine parks, captive whales and dolphins cooperate in the development of spectacular water shows that train audiences to see them as powerful, beautiful, fun-loving, and intelligent (Fig. 7).

Marine mammals generated an exponential growth in interest among the public during the 1970s. Expansion in the number of facilities displaying marine mammals, the high-profile efforts of the "Save the Whales" movement, and the development of marine tourism excursions targeting marine mammal viewing

into effect in 1986, both "pirate" whaling and "scientific" whaling continued. Reports by Soviet scientists about illegal and unreported captures of thousands of whales throughout the world between 1950 and 1970 painted an even more destructive picture than determined previously. In addition, tens of thousands of marine mammals have continued to die each year as a result of other forms of human activity (e.g., pollution, bycatch, habitat degradation, ship strikes, overfishing). The "save the whales" movement was, however, one of the most media-genic and culturally diverse platforms for environmental activism during the 20th century. Confrontation between catcher boats and rubber inflatables on the high seas provided a dramatic backdrop to land-based efforts to convince whaling nations to cease the hunt. The story of the persistent efforts to end industrialized whaling is detailed by a number of authors (e.g., Day, 1987). Saving whales constituted a natural complement to the changing perception of marine mammals generated by oceanaria, whale watching excursions, and media attention.

It was also during the 1970s that the United States incorporated changing public opinion in revised federal legislation. Both the Marine Mammal Protection Act of 1972 and the Endangered Species Act of 1973 became international models of marine mammal protection by affording broad powers to federal agencies to prevent not only the killing of marine mammals, but even their disturbance or harassment. These acts were broadly developed to ensure protection of species considered to be of special interest to the public, in part through ecosystem protection, but in general as single species efforts. Both acts provided wide latitude for research and international cooperation in fulfilling conservation agendas. Although revised many times since the original enactment, both these legislative initiatives have received considerable attention around the world as mechanisms for increasing marine mammal protection.

IV. The 1980s and 1990s: Loving Marine Mammals Becomes an Industry

Undoubtedly the most significant demonstration of the cultural importance of marine mammals in modern times is the virtual explosion of interest in excursions to view marine mammals in the wild that has taken place over the past two decades (Fig. 10). The growth of the whale and dolphin watching industry has been detailed in a number of places, but the sheer speed and dimension of the changes that took place during the 1980s and 1990s warrant mention here. A detailed report on worldwide trends in whale watching by Erich Hoyt (2000) documents the dramatic growth of whale and dolphin watching from the 1950s to the present. Since 1980, the number of people, the number of venues, the number of operators, and the number of targeted species have all increased exponentially. The number of countries with marine mammal excursions has tripled since 1990 alone, while the number of passengers has more than doubled to some 9 million people each year. Direct revenues from marine mammal watching excursions now exceed 1 billion U.S. dollars a year. If one considers the "multiplier" effect of the other expenditures associated with whale watching (travel, accommodations, food, and souvenirs), the economic impact must be seen as quite substantial.

Figure 10 *Whale watching grew into a worldwide industry worth hundreds of millions of dollars a year during the 1980s and 1990s.* © *Pacific Whale Foundation.*

The media focus on whale and dolphin research, the oceanariums' "taming" of cetaceans, and public access to marine mammals through commercial whale watching trips during the 1980s were accompanied by a unique phenomenon: the use of laypeople as research assistants in return for financial support of the research project. One of the first instances of such a program took place aboard the *R/V Regina Maris*, a trimasted schooner that operated out of Gloucester, Massachusetts, under the direction of Dr. George Nichols and the nonprofit Ocean Research Expedition Society. This was soon followed by Earthwatch's Center for Field Research in Massachusetts; the School for Field Studies, also in Massachusetts; the Pacific Whale Foundation in Hawaii; Oceanic Society Expeditions in California, and ultimately a host of organizations throughout the world. Such efforts allowed members of the general public, without any prior experience in conducting whale research, the opportunity to participate in field studies of marine mammals. The programs have proven to be immensely popular. Participants are able to spend time in the field with prominent researchers they may have learned about through television documentaries or college courses. Participants generally have the opportunity to observe marine mammals from a perspective that is seldom enjoyed by the layperson on a commercial tour. They also have the satisfaction of some sense of accomplishment in furthering the goals of science and/or conservation. Lavigne *et al.* (1999) pointed out that funding from government agencies in the United States for marine mammal research has shown no real growth in the past three decades. However, the contribution to research efforts from public participation programs has grown tremendously during the same period.

The increase in whale watching has been accompanied by increased concerns that human interest in marine mammals may have undesirable effects on target populations or species. Specific regulations to control human activities around marine mammals have been enacted in a number of countries, including the United States, Australia, New Zealand, Scotland, Argentina, Mexico, and South Africa. Many other countries (Canada, Japan, Norway, and Iceland) are considering such legislation or propose recommendations or voluntary guidelines. The resulting controls vary widely in both philosophy and effectiveness. In the United States, animals may be used in captive swim programs, but it is generally considered to be illegal to swim with them in the wild. In Australia and New Zealand, the display of cetaceans is essentially restricted, but wild swim programs are allowed. In most areas, multiple layers of government controls and restrictions on research activities exist.

The most frequent form of control of human activities around marine mammals throughout the world is to regulate approach distances. However, there is little known empirical basis for the often-used 100 m, and recent studies of distance estimation over water suggest extremely high variability in the accuracy of human performance. A workshop in Italy hosted by the International Fund for Animal Welfare in 1995 focused on developing scientifically based management programs to protect marine mammals from tourism activities around the world. Noting the difficulty in establishing links between human activity and impacts on wild populations, participants in the workshop recommended a precautionary approach based initially on common-sense rules, followed by an iterative process of scientifically monitoring impacts and adjusting rules accordingly. At this time such a procedure seems to have been employed only in New Zealand to develop science-based management of controls. In general, it appears that government regulation is driven by political and socioeconomic considerations, and once implemented, management policies tend not to be amenable to change. When changes do occur, they tend to be based on additional political and socioeconomic considerations rather than sound scientific evidence.

Throughout the 1980s the "discovery" of marine mammals as objects of wonder allowed those with artistic and entrepreneurial abilities the opportunity to develop a wide variety of commercial ventures. Availability of marine mammal books, magazine articles, visual art works, decorated clothing, jewelry, videos, souvenirs and trinkets, marine expeditions, and oceanarium displays increased dramatically during this period. Artistic visual representations of marine mammals benefited enormously from the artists' exposure to live animals (or, perhaps, photographs of live animals). At least two major styles of visual art emerged during the 80s: one incorporated attention to detail based on biological and morphological data and the second attempted to represent the artist's interpretation of the animal in its world. Prominent examples of accuracy in detail include the paintings of Larry Foster (cf. Fig. 2), Pieter Folkens, Janet Biondi (all of California), and Richard Ellis (New York). Realism also characterized the work of sculptors John Perry, Randy Puckett, and Steve Aikenhead. Interpretations of the whale in its world, with less adherence to detailed accuracy, were represented prominently in the works of Robert Lyn Nelson (Fig. 11), Wyland, and Christian Riese Lassen, who reached their pinnacle of commercial success from the island of Maui in Hawaii during the 1990s. There were, of course, a number of other artists who produced work during the 1980s, but those just mentioned comprise a cross section of the most representative and prolific.

Figure 11 *Marine artist Robert Lyn Nelson is a preeminent figure in the marine art world.*

The marketing of marine mammal images grew through the 80s and 90s into a multimillon dollar enterprise, driven largely by art galleries catering to wealthy tourists in places like Lahaina, Monterey, and Provincetown. New techniques were developed to produce prints, posters, lithographs, and other facsimiles of original works that made art available to everyone. Additional products include T-shirts, coffee mugs, note cards, coffee table books, shopping bags, and stickers. Media attention to public relations extravaganzas such as the air brushing of "Whaling Walls" by Wyland in high-visibility settings in major cities throughout the world helped generate a public perception of marine mammals as beautiful, powerful, friendly, and intelligent. Jewelry, sculptures, and a wide variety of trinkets and other memorabilia were produced in mass quantities for the increasing population of whale watchers and marine mammal lovers with money to spend. A major outlet for such products became the retail centers associated with display facilities. Virtually every aquarium or oceanarium in the world that has marine mammals on display has a retail center strategically located so visitors must move past or through it on the way to the exit. Artifacts related to marine mammals constitute a huge source of related income.

As the market for watching marine mammals in the wild and the purchase of memorabilia grew, the increase in publication of books and videos continued. It is possible to identify three major trends in modern published materials for adult readers: new-age, science fiction, and naturalist accounts. The dolphin as a mystic, healer, ancient communicator shows up in books by Horace Dobbs (British), Patricia St. John (American), Olivia de Bergerac (French), Frank Robson (New Zealand), and a host of like-minded writers. The popularity of new-age musings on the significance of marine mammals is based on the use of personal transformation stories that foster an emotional response on the part of the reader. It is important to note that, in most cases, the new-age perspective on marine mammals emphasizes what they (most often cetaceans) can do for humans: as teachers, as healers, as models of a better social order, and as indicators of human mistreatment of the oceans.

A second type of modern adult literature provides science-fiction accounts of marine mammals. These works incorporate extensions of scientific studies of communication and other cognitive capabilities of marine mammals into stories based on themes such as military abuse of power, destruction of the environment by corporate greed, mistreatment of marine mammals by egotistical scientists, or cataclysmic events of nature. "Into the Deep" by Ken Grimwood (William Morrow and Co., New York, 1995) weaves a complex plot that has dolphins communicating with humans to warn of impending natural disaster; representatives of each species eventually combine talents to use telepathic communication to prevent massive destruction of both humans and dolphins. "The Secret Oceans" by Betty Ballentine (Bantam Books, New York, 1994) tells of a group of scientists who become captive subjects in a study of humans conducted deep in the ocean by a form of cetacean more intelligent than humans. "Sounding" by Hank Searls (Ballantine Books, New York, 1982) uses two main characters, an aged bull sperm whale and a sonar officer on a Russian submarine, to explore cetacean behavior and the possibilities of interspecies communication.

The third, and largest, category of books printed during the past two decades includes nonfiction, naturalist accounts of a wide variety of species, including books on all the orders of marine mammals. These books bring together interpretations of scientific findings, stunning art or photography, and personal accounts of fieldwork by accomplished researchers. Two of the earliest are the "Book of Whales" (Alfred A. Knopf, 1980) and "Dolphins and Porpoises" (Alfred A. Knopf, 1982), both by nat-

uralist, painter, and writer Richard Ellis. In "Hawaii's Humpback Whales" by Gregory Kaufman and Paul Forestell (Island Heritage Press, 1986), the authors describe the biology and behavior of humpback whales, based largely on their observations and field studies in Hawaii and Australia. The writing style is meant to model the authors' experiences narrating whale watches and provides a lively, but science-based interpretation of the life of one species. A photographic example of equivalent style is the Japanese photographer Mitsuagi Iwago's "Whales" (Chronicle Books, San Francisco, 1994), which documents his year-long journey to observe humpback whales from Hawaii, to Alaska, to the Ogasawara Islands of Japan.

Attractive, informative, and entertaining treatments of many species of marine mammals are now available. The *National Geographic* production of "Whales, Dolphins and Porpoises" (National Geographic Society, Washington, DC, 1995) featuring photographs by Flip Nicklin (son of pioneering underwater photographer Chuck Nicklin) and edited by James Darling is an unsurpassed synthesis of prominent research findings from around the world. Frank Breummer's "Seals in the Wild" (Laurel Glen Publishing, San Diego, 1998) offers an elegant collection of facts and photographs about pinnipeds. Barbara Sleeper and Jeff Foott team up to present an excellent description of the much overlooked manatee in their book "In the Company of Manatees" (Three Rivers Press, New York, 2000). Ken Norris, Roger Payne, Hal Whitehead, Peter Beamish, Jim Darling, and Carol Howard have written intriguing personal accounts of careers spent investigating marine mammals. A number of field guides for whale watchers in many settings throughout the world are also currently available. These books not only provide a great deal of current information on marine mammals, but serve as important vehicles for conservation efforts.

Certainly not all books published during the 80s and 90s can be categorized according to the three styles just described. Two books that deserve mention in this regard are Heathcote Williams' "Whale Nation" (Harmony Books, New York, 1988) and "The Delicate Art of Whale Watching" (Sierra Club Books, San Francisco, 1982) by Joana Varawa (formerly Joan McIntyre). "Whale Nation" is a celebration of the history of whale/human interactions through a book-length poem and a collection of whale-related excerpts from other sources. Following its publication the author did a world tour of highly acclaimed public readings. Varawa's book is a deceptively small but powerful series of meditations on being in the presence of the sea and its creatures.

More than 300 videos and movies on marine mammals have been produced in the last two decades, more than half of them since the mid-1990s. Most of these are documentaries, generally based on describing the research of one or a few scientists. In some cases, researchers are simply followed in their work while the significance and findings of their efforts are detailed. In other cases, celebrities such as Robin Williams or Chris Reeves accompany scientists to get a first-hand look at their work. A number of television programs have dealt with controversial issues such as dolphins in captivity (*A Whale of a Business*, PBS Frontline, 1997); scientific whaling (*Whale Fever*, BBC News & Current Affairs, 1993); captive-dolphin swim programs (*Dying to Please*, Biosphere Films, 1990); and bycatch of dolphins by the tuna fishery (*If Dolphins Could Talk*, PBS Video, 1990).

The invention of a large-format (70-mm) camera by IMAX in the early 1970s has resulted in three big-screen movies that are based on marine mammals. The first was *Nomads of the Deep*, featuring humpback whales, released in 1979 by IMAX. In 1996, Destination Cinema produced *Whales*, a highly acclaimed and popular big-screen documentary covering a variety of whale species. In 2000, McGillivary Freeman Films released *Dolphins*, a documentary on dolphin communication that focuses on the research being carried out by Kathleen Dudzinski, Alejandro Acevedo-Gutiérrez, Louis Herman, and others.

There have been a handful of feature movies with marine mammals as central characters. A bottlenose dolphin named Snowflake appeared in the comedy *Ace Ventura, Pet Detective* (starring Jim Carrey, 1994). Humpback whales were brought back to the future in the science fiction adventure *Star Trek IV: The Voyage Home* (William Shatner and Leonard Nimoy, 1987). A dog and a dolphin learned to communicate with each other and sparked a romance between the dog's owner and the scientist studying the dolphin in *Zeus and Roxanne* (Steven Guttenberg and Kathleen Quinlan, 1997). *The Secret of Roan Inish* (Jeni Courtney, 1994) was based on a Celtic myth about selkies (seals) and a young girl's search for her lost brother. None of these movies achieved much success on the basis of their representation of marine mammals, however. In 1993 the movie *Free Willy* became a worldwide sensation and was ultimately followed by two sequels. The series chronicled a friendship between a 12-year old boy (Jesse) and an adult male killer whale (Willy). When they met in the first movie, the whale was held captive in a small enclosure in a seaside marine. When Willy refused to perform for audiences and the owner threatened dire consequences, Jesse arranged for the whale's dramatic release. In the second movie, Jesse and Willy ran into each other in the wild and discovered they must work together to avoid the perils of an offshore oil spill. In the third and final movie, the whale was threatened by an illegal whaling operation, but the whaler's son helped Jesse (who had grown into a 17-year old whale researcher) save Willy and his family.

The *Free Willy* movies brought the question of captivity to the forefront of public consciousness and resulted in a long and expensive campaign (still underway) to return the killer whale that was used in the first movie to the ocean. That animal, Keiko, had been kept in rather miserable circumstances in a substandard facility in Mexico. With monies raised through a publicity campaign sponsored by the movies' producers (Richard Donner and Lauren Shuler-Donner), Keiko went through a reorientation program in an Oregon facility and was then moved to Iceland, where he is being reintroduced to the wild.

The public response to *Free Willy* was indicative of an increase in public questioning throughout the 90s about the appropriateness of keeping marine mammals in captivity for public display. With increasing access to information about marine mammals in the wild (through excursions, books, and videos), it became clear that life in a captive facility was a poor substitute for nature. In response, the display industry made significant strides in upgrading facilities, improving husbandry, and refocusing shows to emphasize education and demonstration of the display of animals' natural behaviors. The debate continues on many fronts about the pros and cons of maintaining marine mammals in captivity.

The removal of animals from the wild has decreased significantly in the past two decades. Improved health of captive animals and an increase in the number of animals born in captivity provide an ongoing supply of display animals that have never been in the wild and would be unlikely to survive if released. There seems to be an increasing tendency to move away from keeping cetaceans in captivity, however. A number of facilities (e.g., the Vancouver Public Aquarium) have chosen not to replace whales or dolphins that have died, and others (Maui Ocean Center; Monterey Bay Aquarium) decided from the outset not to hold cetaceans. In a graphic display of the continuing debate about animals in captivity, the producers of the *Free Willy* movies have threatened a Hollywood-sponsored boycott of filmmaking on the Island of Maui if a plan to put the research dolphins from a University of Hawaii laboratory in Honolulu on display at a planned "theme park" on Maui is carried out.

V. International Perspectives

It would be a mistake to assume that North American cultural perspectives are mirrored elsewhere in the world. Clearly, one might expect to find attitudes toward marine mammals to differ across cultures that depend on traditional subsistence hunting practices (such as the Inuit, Aborigines, and some areas of the South Pacific); cultures that opportunistically use marine mammals to prop up collapsing fisheries (as is true throughout much of the Indian subcontinent, parts of Asia, Central and South America, and Africa); cultures that pursue relatively modern programs of commercial exploitation (including Japan and Norway); and cultures that currently forego commercial exploitation (many former whaling nations such as the United States, Canada, Australia, New Zealand, the United Kingdom, and South Africa).

As faulty as simplistic generalizations may be, one can recognize an important dividing line in perspectives on marine mammals that is based more on economic considerations than fundamental differences in cultural attitudes toward nature. Wealthier countries enjoy the luxury of debating whether marine mammals should be harvested. Japan, Norway, the United States, Britain, Australia, and New Zealand are remarkably alike in their common appreciation of marine mammals as intelligent, amusing, and useful. The difference in current behaviors toward marine mammals is one of time, and perhaps degree, for all have participated in the slaughter of marine mammals and each currently reaps the benefits of the worldwide fascination with marine mammals. Less wealthy nations must face immediate considerations of survival before exploring programs devoted to environmental protection or species conservation. Marine mammals are viewed more directly in terms of immediate economic benefit. A major factor in the reduction of a number of small cetacean populations is the readiness with which INCIDENTAL TAKES (accidental ENTANGLEMENT in fishing nets) can be converted into a directed hunt in those areas where there is a market for the meat.

The enthusiasm with which a wide cross section of nations have begun promoting whale and dolphin watching reinforces the notion that marine mammals are regarded neither as sacred nor profane in most parts of the world; they are primarily a commodity whose particular value can change from time to time. Surveys of attitudes and feeding programs in Australia, New Zealand, and the United States make it clear that, regardless of the qualities that participants may attribute to marine mammals, their primary value is in terms of satisfying human needs, and this perspective appears even more entrenched in less affluent areas where marine mammals are harvested directly for food.

One of the most striking demonstrations of the role of marine mammals in modern cultural perspectives in affluent countries is the use of marine mammal images in corporate advertising. The author recently sent a request for information on corporate use of marine mammal images to subscribers of a bulletin board for marine mammal science (http://www.es-cribe.com/science/marmam/). The query resulted in approximately 100 different examples of advertisements. Most of these were print media, but a number of television advertisements were described as well. Almost half the images used were of bottlenose dolphins (Fig. 12). The next most frequent use was of killer whale images. Other species included humpback whale, sperm whale, right whale, beluga whale, false killer whale (*Pseudorca crassidens*), sea lion, walrus, spinner dolphin, humpback dolphin, and polar bear. In general, usage appeared to reflect anticipated familiarity on the part of the public for particular species (i.e., bottlenose dolphin, killer whale, humpback whale, and California sea lion), availability of professional quality images, and the extent to which the image symbolized the message of the advertisement (e.g., speed, power, size, intelligence).

The diversity of products marketed with the help of marine mammal images was extremely broad. In some cases the connection between the product or service and marine mammals was obvious. A number of advertisements used images of marine mammals to promote travel to destinations where marine mammals may be observed. (e.g., Olympic Airways in Europe, Air Nippon in Japan). In some cases there was an identifiable link between the product or service and some perceived feature of the marine mammal. The ruggedness of a Leatherman tool was highlighted with the help of a humpback whale; the warmth of Eddie Bauer clothing was symbolized by a polar bear; and a "school" of dolphins helped promote educational software for children (Fig. 12). More often the relevance of the image to the product was quite obscure. Marine mammals have been used to sell aspirin, automobiles, batteries, beer, chewing gum, chocolates, cigarettes, computers, diapers, electronics, film, gold coins, jet skis, life insurance, orange juice, mobile phones, petrochemicals, potato chips, shower stalls, silverware, televisions, tires, video cameras, and a myriad of investment, real estate, interior decorating, and cosmetic services. One can only wonder about the cultural perspective that associates marine mammals with alcohol, tobacco, and petrochemicals.

Not surprisingly, use of marine mammal images in advertising occurred most frequently in more affluent countries, especially in Asia and Europe. North American and Australian use seemed to be more recent and less frequent, but a comprehensive assessment of corporate use of marine mammal images

has yet to be conducted. It is clear, however, that over the past two decades, increasing public attention to marine mammals has created a kind of "rock-star" status that elicits a classically conditioned positive emotional response to marine mammal images that can be exploited by marketing experts.

VI. Conclusions

The one constant element in human perspectives on marine mammals throughout the 20th century is our recognition of their economic significance. We have never stopped making money from them. They are still a resource. We still regard our needs as more critical than theirs. The proof can be seen in the peculiar and contradictory steps that we take to protect them. Consider, for example, the degree of attention that focused on "Humphrey" the "wrong-way whale" that wandered up the Sacramento River in 1985 and would not come out, or the three gray whales that became caught in ice in Barrow, Alaska, in 1988 and could not get out. Tremendous resources in time, money, and emotions were devoted to "saving" these animals. Compare those situations with the decision to cull seals hanging out at fish ladders in Seattle a few years ago to prevent them from feeding on depleted salmon stocks. On the one hand, no expense was spared to release the whales from natural disasters, while on the other hand, the seals were basically held accountable for the precarious status of salmon stocks that had been decimated by human overfishing and habitat degradation. Japanese scientists have been ridiculed for suggesting that Antarctic minke whale (*Balaenoptera bonaerensis*) numbers need to be culled to promote the recovery of blue whales by reducing competition for prey. At the same time, pinniped populations in North America have been culled to promote the recovery of fisheries by reducing prey competition with humans.

We pass laws to prevent approaching marine mammals too closely, but allow them to be killed if they eat too many fish or get in the way of our fishing equipment. The struggle between nations for military supremacy dictates that we degrade the oceans with the loud sounds of our technical wizardry, reducing its suitability for many marine species. Our inability to control our own population growth forces us to compete against other species for food and then poison them with our waste. Insofar as we can identify interesting features about marine mammals, we subjugate them to serve our needs, whether for amusement, therapy, revelation, or money. While once we killed marine mammals to get their oil, now we kill them by spilling oil on them. While once we chased them and captured them for food, today we chase them and capture them for amusement. Instead of using their baleen for corsets, we use their images to sell diapers.

The development of "whale as icon" during the 1970s seemed to usher in an enlightened period meant to liberate cetaceans from modern commercial whaling. Television shows and oceanaria introduced us to a new understanding of whales and dolphins as entertaining, highly trainable, and utterly fascinating. As a consequence of the discovery of how entertaining and trainable many cetacean species were, scientists began to probe their physiological, behavioral, and even cognitive capabilities. On the surface at least, it looked as if a new age of enlightenment had begun.

Kids learn better in a school of dolphins.

AdventureOnline.com engages your students with educational content based on the subjects they love – from dolphins to dog sledding.

AdventureOnline.com offers the interest-driven, real-time content that kids enjoy – and the reporting and assessment tools that teachers and parents need. Used in more than 2,400 schools during the 1999-2000 school year, AdventureOnline.com helps students learn reading, writing and math.

This fall, our newest product, **AdventureOnline.com 2.0**, will

launch an expanded site of four channels. In addition to the existing channel, Expeditions, the site will offer new daily and weekly educational activities, puzzles and games focused on Sports, Animals and Camping, as well as a Points-for-Performance rewards program.

AdventureOnline.com 1.0 is **FREE** until September 1, 2000, for teachers who register now. Register before September 1 and you'll also receive 50% off the price of membership to AdventureOnline.com 2.0. You'll pay only $4.08/month (regularly $8.16), with no obligation to finish the year! You will not be billed for the first month ($4.08) until September 15, 2000. Visit www.adventureonline.com or call 1-888-609-0095, ext. 119.

LIMITED OFFER
$4.08
PER MONTH

www.adventureonline.com

Figure 12 *A wide range of products and services are marketed with the help of marine mammal images. Not all advertisements seem as relevant or appropriate as those shown here. Courtesy of (A) Learning Outfitters, Inc.; (B) Leatherman Tool Group, Inc.; (C) Eddie Bauer. (continues)*

While it is reasonable to conclude that display facilities and laboratory-based research programs have been extremely important in energizing efforts to limit commercial hunting, they have not led to the liberation of whales and dolphins. Instead they have helped create a different kind of exploitation, a kind of domestication or colonization. Whether we limit the lives of cetaceans by keeping them in tanks, artificial lagoons, open ocean pens, or even let them roam free and control them using standard operant training procedures, one way or the other, we are compelling them to live their lives for our purposes.

That is not liberation. Even when we do not remove animals from their natural environments and even when our only contact is based on "watching" them, rather than directly controlling them, our intrusive presence can alter their natural behaviors. When this is done without their choice, with little or no benefit to them, and only in service to our own needs, we are engaging in a form of colonization.

However, of course all the news is not bad. There does seem to be a tremendous groundswell of respect, hope, and compassion for cetaceans. Thousands of individuals all over the world

Figure 12   *(Continued)*

work tirelessly to assist marine mammals in need. When a humpback whale stranded on a beach near Brisbane, Australia in 1992, more than 5000 people showed up to save it. Our ability to form emotional attachments to individual animals seems boundless (Fig. 13). However, this encourages what right whale biologist Scott Krauss calls "Madison Avenue whale saving"; small-scale, high-profile publicity events that seem noble but do little to save marine mammal populations. Our activities will have little real conservation value as long as we ignore the intrinsic value and importance of marine mammals carrying out

their natural behaviors in their natural habitat, unimpeded by the presence of humans or the POLLUTION we produce.

There are rapidly increasing opportunities for people all over the world to observe marine mammals in the wild. Most of these opportunities involve going out on boats: observing animals at close range for brief periods of time and then returning to shore. Increasingly, however, more varied options are available. In some areas (e.g., Monkey Mia in Western Australia, Tangalooma in Queensland, Australia, and in the Red Sea), dolphins can be approached and fed in shallow water. As we have

Figure 12 *(Continued)*

ment industry that may soon be as out of control as the whaling industry was in earlier times. Economic competition drives some to place more importance in the gain from selling thrills (such as parasailing over herds of dolphins or being towed off the back of a boat through groups of seals, whales, or dolphins) than in the possible harm such activities may cause.

We are faced with a fundamental conflict of interest: our need to exploit marine mammals for our own purpose and their need to be protected from our various forms of interference. Our interest in marine mammals is based on satisfaction of our own needs. In our approaches to whales and dolphins, we make them become what we hope they are and ignore what they may be in reality. We see what we want to see. For example, the sight of a whale or dolphin slapping its tail on the water is often an exciting thing for most people—they want to get closer and see more. The truth is that in most cases, tail slapping is a sign of an angry or otherwise distressed animal. Although the display may not always be related to the presence of a boat or swimmer, it frequently indicates the animal does not want us there. Figure 14 shows two spinner dolphins surfacing. At first it may look like they are having fun. The one on the top seems to be smiling and has its pectoral fin wrapped around the other, like an old friend. On looking closer, however, you can notice that the top dolphin is covering the other dolphin's blowhole, preventing it from breathing as it comes to the surface. That is probably not a friendly gesture. The bottom dolphin has its mouth partly open, which is a common threat gesture. Most people simply cannot interpret dolphin behavior properly, which makes it very likely that they will not know how to avoid disturbing whales and dolphins when they go out on the ocean.

Spinner dolphins got their name because of their habit of jumping above the ocean and spinning in the air. Why do they do it? Well, to most people it seems like something that dolphins do just because it is fun. In fact, the spins may serve a number of purposes. When we look more closely, we notice that many of the spinners have REMORAS, a suckerfish that often attaches itself to the dolphin. One reason for the spinning may be to get the remoras off. The spins may also be a form of social

found out that whales and dolphins are generally friendly and frequently curious, more and more locations around the world are offering chances to swim with dolphins in the wild. For the most part, these types of interactions are based on a fundamental belief in the inherent positive value of whales and dolphins. People who watch marine mammals, feed them, or swim with them seem generally driven by good motives. Unfortunately, people who seek contact with marine mammals are being catered to by an increasingly large and diverse entertain-

Figure 13 *When a marine mammal strands, the public response is typically immediate and highly emotional. Paul H. forestell/Pacific Whale Foundation.*

Figure 14 *Our interpretations of animal behavior are frequently clouded by anthropomorphism. Paul H. Forestell/Pacific Whale Foundation.*

communication. It is just too easy to miss the real explanations. When dolphins come to ride the bow wave of a boat, how often do we assume that the animals are intending to play with us? Generally, only a small subset of individuals attempt to associate with the boat (for reasons that are not clear), while the remainder engage in avoidance behaviors. It is commendable that we so often approach whales and dolphins in friendship. Many would rather idolize marine mammals in the wild than colonize them in oceanaria, but naïve idolization can be as questionable as intentional colonization.

If we are to achieve the status that Lavigne *et al.* (1999) have described (i.e., an aesthetic interest that reflects increased understanding and a change in ethical values), we must undergo a universal transformation in perspective: the ability to value something independently of its potential for meeting our needs. Until that point we may continue to find less cruel ways to put marine mammals in our service, but their value will continue to be defined in anthropocentric terms. Philosophically, that may not seem to be a problem of significant proportion, but insofar as our behaviors put our needs ahead of theirs, they will continue to be at risk.

Acknowledgments

The ideas expressed reflect many years of involvement with the whale and dolphin watching industry in Hawaii and Australia, with occasional visits to Japan, Costa Rica, New Zealand, Alaska, and New England. Many researchers, educators, operators, and members of the public have provided "grist for the mill" during those excursions. I thank Greg Kaufman and the Pacific Whale Foundation for support throughout that time. Japan International Cetacean, Education and Research Center (ICERC) has also provided support and encouragement on a continuing basis. I thank the many individuals who responded to my request for information on the use of marine mammal images in advertising. Trisha Lamb Furstein's web-based bibliographies have been tremendously helpful to me. David Lavigne kindly provided a preprint of his excellent discussion of North American attitudes toward marine mammals. Joana Varawa has given both critical and encouraging insight and much wisdom over the past 5 years. Southampton College of Long Island University provided valuable released time from teaching.

See Also the Following Articles

Captivity ▪ Conservation Efforts ▪ Ethics and Marine Mammals ▪ Folklore and Legends ▪ Marine Parks and Zoos ▪ Museums and Collections ▪ Whale Watching

References

Bullen, F. T. (1898, 1913). "The Cruise of the Cachalot." Grosset and Dunlap, New York.
Carwardine, M. (1994). "On the Trail of the Whale." Thunder Bay Publishing, Guildford, UK.
Darling, J. D., Nicklin, C., Norris, K. S., Whitehead, H., and Würsig, B. (1995). "Whales, Dolphins and Porpoises." National Geographic, Washington, DC.
Ellis, R. (1991). "Men and Whales." Alfred A. Knopf, New York.
Hoyt, E. (2000). "Whale Watching 2000: Worldwide Tourism Numbers, Expenditures, and Expanding Socioeconomic Benefits." International Fund for Animal Welfare, Crowborough, UK.
Kaufman, G. D., and Forestell, P. H. (1986). "Hawaii's Humpback Whales." Island Heritage Press, Honolulu, HI.
Lavigne, D. M., Scheffer, V. B., and Kellert, S. R. (1999). The evolution of North American attitudes toward marine mammals. *In* "Conservation of Marine Mammals" (J. R. Twiss, Jr., and R. R. Reeves, eds.), pp. 10–47. Smithsonian Institute Press, Washington, DC.
Lilly, J. C. (1975). "Lilly on Dolphins." Anchor Press/Doubleday, New York.
McIntyre, J. (1974). "Mind in the Waters." Charles Scribner, New York.
Nolman, J. (1999). "The Charged Border: Where Whales and Humans Meet." Henry Holt, New York.
Norris, K. S. (1974). "The Porpoise Watcher." Norton, New York.
Norris, K. S. (1991). "Dolphin Days." Norton, New York.
Pryor, K. (1975). "Lads Before the Wind." Harper & Row, New York.
Rice, D. W., and Wolman, A. A. (1971). The life history and ecology of the gray whale (*Eschrichtius robustus*). American Society of Mammalogists, Special Publication 3, pp. 1–142.
Robertson, R. B. (1954). "Of Men and Whales." Alfred A. Knopf, New York.
Scheffer, V. B. (1969). "The Year of the Whale." Charles Scribner, New York.
Varawa, J. (1982). "The Delicate Art of Whaling Watching." Sierra Club Books, San Francisco, CA.

Population Dynamics

Paul R. Wade
*National Marine Fisheries Service,
Seattle, Washington*

Population dynamics is the study of changes in population size through time. In other words, it is the study of why populations increase, decrease, or remain the same. One fundamental concept in population dynamics is that the maximum rate at which a population can increase is determined by the intrinsic life history characteristics of the species. Thus, much of the research on marine mammal population dynamics is focused on the better definition of life history parameters. Additionally, many extrinsic factors can potentially influence the dynamics of a population. These include environmental variability, disease, competition, and predation. There is also much interest in the role of density dependence in controlling marine mammal population dynamics. In addition to being scientifically interesting, basic concepts of density dependence have become important to the management and conservation of marine mammals.

I. Rates of Population Increase

A. Population Growth of Long-Lived Animals

Most long-lived animals, such as marine mammals and large terrestrial mammals, have relatively slow intrinsic rates of increase compared to most other kinds of animals. The modest population growth rates are the consequence of their life history characteristics. Characteristics such as the age at which females start reproducing, the number of years between births, and how many years a female will live and reproduce, determine how quickly a population can increase. Most marine mammal species take many years to reach sexual maturity and have long gestation periods that result in the production of, at most, only one young a year. In fact, most species give birth only once every several years. Even when annual rates of survival are very high, these characteristics cause populations to grow slowly. Low rates of population growth make most species of marine mammals vulnerable to overexploitation, as shown by the rapid depletion of many whale populations by commercial harvest.

B. Measuring Population Growth

Population growth can be measured in two ways. In general, the most reliable estimates of population growth come from abundance data collected over many years. Population abundance can be estimated from surveys or counts, and when repeated over several years, the trend (percentage change per year) in a population can be estimated. Specifically, the slope of a linear regression on the natural logarithm of abundance represents the rate of increase (r) of a population experiencing exponential growth. Because population growth is slow and population estimates are imprecise, 10 or more years may be required to directly measure population growth rates.

A less direct way of estimating population growth is from life history data. Estimates of age of sexual maturity, birth rate, survival rate, and maximum age can be compiled in a Leslie matrix or similar model, which can then be used to estimate the rate of increase (usually estimated as $\lambda = e^r$). Although such calculations have been useful for exploring potential population growth rates, relatively few estimates of actual population growth have been made in this way. The main hindrance is the lack of direct data on survival rates of marine mammals. Only a few species have been amenable to survival estimation, usually from mark-recapture studies using individuals recognizable from tags, unique scars, or other markings. Such studies have been undertaken in California sea lions (*Zalophus californianus*), bottlenose dolphins (*Tursiops truncatus*), and killer (*Orcinus orca*), humpback (*Megaptera novaeangliae*), bowhead (*Balaena mysticetus*), and right whales (*Eubalaena* spp.)

C. Taxonomic Differences

Although population growth is comparatively slow for all marine mammal species compared to other vertebrates, there is a considerable range in life history strategies within marine mammals. For example, just within the cetaceans one can contrast a harbor porpoise that becomes sexually mature after a few years, can give birth annually, and rarely lives past 12–15 years with a bowhead whale that becomes sexually mature after perhaps 10–20 years, only gives birth every 3–5 years, and lives for decades, with some individuals apparently living to greater than 100 years. Broad patterns are evident in maximum population growth rates when available information from trend or life history data is examined for different taxonomic groups. Sea otters (*Enhydra lutris*) with maximum observed rates on the order of

20% per year, are perhaps capable of the fastest growth. Next are pinnipeds, many with the potential to reproduce annually, with observed rates ranging from 8 to 13% for otariids [from northern fur seals (*Callorhinus ursinus*) at the low end to Antarctic fur seals (*Arctocephalus gazella*) at the high end] and 6 to 13% for phocids [from Hawaiian monk seals (*Monachus schauinslandi*) at the low end to harbor (*Phoca vitulina*) and gray seals (*Halichoerus grypus*) at the high end]. Maximum rates of increase for mysticetes likely range from about 4 to 10%, or perhaps even higher. Observed rates for gray (*Eschrichtius robustus*) and bowhead whales (3–4%) are at the low end, although the populations that have been studied are likely at high population levels and therefore their growth rate may have already slowed (see Section III). Southern right whales (*Eubalaena australis*) have shown the ability to increase at 7% per year, and humpback whales have been estimated to increase at 7–10% per year. Manatees (*Trichechus* spp.) appear to have maximum population growth rates of at least 7% and perhaps higher, whereas dugongs (*Dugong dugon*) are likely only capable of growing at 5–6% per year. Much uncertainty exists about the maximum rates of increase of odontocetes; what little is known suggests that this group generally has very slow annual population growth rates, as low as 2–3% in some species such as killer whales, and most species are considered unlikely to have a maximum growth rate of more than 4%. There is speculation that a few species, such as harbor porpoises (*Phocoena phocoena*), may have higher rates of increase because of their life history, but such higher rates have not yet been documented.

II. Extrinsic Factors Affecting Population Size

A. Environmental Variance

Another aspect of population dynamics is the study of the effects of extrinsic factors on population growth. The difficulty in precisely estimating population size and life history parameters has made the study of variation in population growth rates of marine mammals difficult. However, at least a few conclusions can be made. Long-lived animals with relatively older ages of sexual maturity cannot respond quickly to favorable environmental conditions. As discussed earlier, many species are limited to relatively slow population growth, even when conditions are good. Consequently, such species cannot decline too often or too rapidly when conditions are bad or they would not have persisted on an evolutionary time scale. Therefore, marine mammals have evolved life history strategies that keep them relatively buffered from interannual variability in environmental conditions, at least compared to other animals such as small terrestrial mammals. From this evolutionary perspective, one can predict that the slower the maximum population growth rate, the lower the variance in growth rate one might expect. Species with these traits are often referred to as "*K*-selected species," meaning they have evolved to maintain relatively stable population sizes at or near the carrying capacity (typically represented by the letter "*K*") of the environment. Therefore, one would expect that cetaceans, in particular, would be less subject to large fluctuations in survival and fecundity from year to year than would sea otters or pinnipeds.

Data sufficient to examine such patterns are rare for marine mammals. Studies of pinnipeds provide the best evidence of the effect of changing oceanographic conditions, particularly because of the ability to closely monitor numbers of pups or adults at rookeries from one year to the next. El Niño oceanographic events, through reductions in prey availability, have led to changes in the life history parameters of several species of otariids in places such as California and the Galapagos. These changes include lower fecundity, lower pup survival, and even lower adult survival during extreme events. In some cases, such as northern fur seals at San Miguel Island, conditions have been bad enough to result in nearly 100% mortality of pups in a given year. In addition, several populations of pinnipeds have experienced long-term (20–30 year) declines that have been mostly unexplained; while some believe that these declines are due to oceanographic regime shifts on that time scale, others believe that direct human-caused mortality or depletion of prey species by commercial fisheries may be at least partially to blame.

Cetaceans probably experience similar responses to environmental change, but these changes are harder to detect. One study did find evidence of nutritional stress in the teeth of dusky dolphins (*Lagenorhynchus obscurus*) off Peru during the 1982–1983 El Niño. Similarly, pregnancy rates of fin whales (*Balaenoptera physalus*) off Iceland have been correlated with changes in food abundance. Within the sirenians, manatees can experience higher mortality during periods of colder water temperatures.

B. Disease and Natural Toxins

Disease or natural toxins also have the potential to cause declines in marine mammal populations. Many species of marine mammal have experienced large mortality events from disease. Although there is often the suspicion by some that such outbreaks of disease have been ultimately triggered from anthropogenic causes, such as degraded habitat or exposure to contaminants, this has proved difficult to confirm from field studies. Disease and toxin-related die-offs may thus represent natural events for marine mammals. Some of these events have been severe enough to be considered a significant influence on the dynamics of these populations. For example, more than 18,000 dead harbor seals were reported from European coasts during a morbillivirus infection in 1988. In the western North Atlantic Ocean, large numbers of common bottlenose dolphins died during a disease-related event in 1987–1988 that may have been triggered by a toxin, and the mortality was thought to have caused a significant decline in the population. Humpback whales in the Atlantic have also experienced mortality events on at least two occasions; in one case at least 14 whales died apparently from consuming toxic dinoflagellates concentrated in fish prey. However, it is not known if such mortality is frequent enough to exert a strong influence at the population level. Such die-offs are particularly dangerous for small populations. The endangered Mediterranean monk seal (*Monachus monachus*) recently experienced a loss of 60–70% of its population off Africa due to a toxin or disease, which has seriously compromised the long-term survival of this species.

C. Competition

Competition from other species may influence the population dynamics of marine mammals, although there is little evidence for this. Whether this is due to competition being unimportant or whether it is simply too difficult to demonstrate is an open question. An increase in crabeater seals (*Lobodon carcinophaga*) was directly attributed to a release from competition following the severe depletion of several species of baleen whales in the Antarctic, leading to an increased availability of KRILL. This explanation has been reevaluated in light of evidence of environmental influences on the population dynamics of Antarctic pinnipeds. It should be noted that some general textbooks (such as books on oceanography) state that competition for krill from Antarctic minke whales (*Balaenoptera bonaerensis*) has prevented the recovery of depleted blue whales (*B. musculus*) in the Antarctic. However, the recent information indicates that the lack of recovery of blue whales in the Antarctic can be fully explained by previously unknown ILLEGAL HARVESTS by the former Soviet Union.

D. Predation

Many marine mammals, especially smaller ones, are preyed upon by other animals, but predation has rarely been suggested as a strong controlling factor in their population dynamics. One exception is a recent study that suggested sea otters may have declined in one part of Alaska because of killer whale predation. Pinniped pups are often vulnerable to predation from predators such as leopard seals (*Hydrurga leptonyx*), great white sharks, and killer whales. While such predation has been shown to affect the growth of local ROOKERIES, it is unclear if it exerts a strong influence on the dynamics of an entire population. Similarly, some pelagic dolphin species experience predation from sharks, and killer whales prey on many cetacean species. Again, even though predation of cetaceans occurs, it is difficult to know whether it influences the dynamics of these populations.

III. Density Dependence

A. Compensation

Another area of great interest is the role of density dependence in controlling the population dynamics of marine mammals. It is generally accepted that marine mammal populations experience density dependence. In other words, as populations become relatively large, they tend to have lower population growth, and eventually stop increasing. This form of density dependence is termed compensation. The level at which a population stabilizes is called its carrying capacity. Evidence has been found for density dependence in life history parameters such as the age of sexual maturity. Females from a population at a level well below K become sexually mature and start reproducing at an earlier age than females from a population at a level close to K. Presumably this is because of access to greater resources such as prey. For example, the age of sexual maturity apparently became younger for fin (*Balaenoptera physalus*) and sei whales (*B. borealis*) in the Antarctic as their populations were depleted by commercial harvest.

It has been hypothesized that the mechanism of the regulation of populations of long-lived mammals would follow a sequence as a population increased, with density dependence first affecting the rate of immature survival, then the age of sexual maturity and the birth rate, and finally the adult survival rate. This hypothesis partially follows from the recognition that a long-lived species that reaches sexual maturity slowly and has a low intrinsic rate of increase must maximize adult survival in order to persist. Adult females of long-lived species may be able to forgo reproduction to maximize individual survival when conditions are poor, but it is unclear if there is necessarily a specific sequence in how density dependence affects all life history traits that are common to all marine mammals.

B. Linear vs Nonlinear Density Dependence

It is difficult to assess how these changes in life history translate into changes in population growth, as few direct data are available on changes in population growth at different population sizes. One debate is whether marine mammals experience linear or nonlinear density dependence. Linear density dependence is a constant decline in the per capita population growth rate as a population increases, illustrated by the simple logistic population model (Fig. 1). Nonlinear density dependence is where a population has no decline in the per capita growth rate as it increases until it reaches a level close to K, where it then has a rapid decline, illustrated by the θ-logistic population model where the value of θ is greater than 1 (Fig. 1). Both linear and nonlinear density dependence occurs in single life history parameters of marine mammals and other large mammals. However, population modeling indicates that nonlinear density dependence in a single life history trait (such as fecundity) may not translate into non-linear density dependence at the popula-

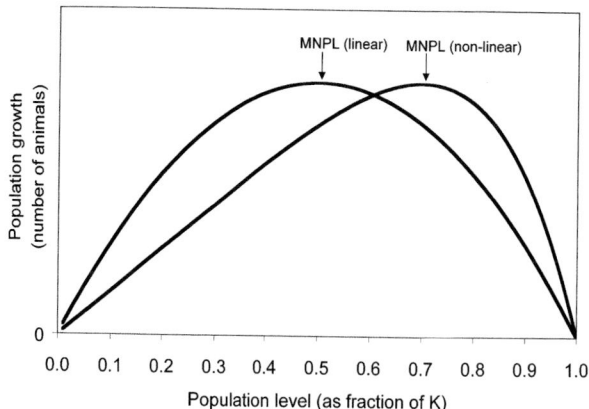

Figure 2 *Population growth (in numbers of animals added to the population each year) as a function of population level, expressed as a fraction of* K. *This quantity is also called the "net productivity," and therefore the peak of this curve is referred to as the maximum net productivity level (MNPL). For linear density dependence, the MNPL is at one-half of* K. *For nonlinear density dependence, the MNPL is closer to* K, *in this case at 0.7K.*

tion level, particularly if density dependence in other life history traits is more linear. The form of the density dependence (linear vs nonlinear) will determine which population size will have the greatest increase in numbers from one year to the next, called the maximum net productivity level (MNPL). For long-lived animals, nonlinear density dependence will generally lead to populations having their MNPL closer to carrying capacity (Fig. 2). A recent review concluded that most marine mammal populations likely have their MNPL between 50 and 85% of their carrying capacity. It has been shown, however, that it is very difficult to estimate MNPL for any marine mammal population given the data that are currently available.

C. Density Dependence and Management

These concepts of density dependence have been incorporated into the management and conservation of marine mammals. Both the International Whaling Commission, in its proposed scheme for the regulation of whale harvests, and the U.S. government, in managing human-caused mortality of marine mammals in U.S. waters, refer to concepts like MNPL, with populations above MNPL considered "healthy."

Many populations of pinnipeds and whales are recovering from unregulated harvests that left them at severely depleted population levels. It will be interesting to observe the recovery of these populations over the next few decades, as many are likely to approach previous population levels and are expected to stop growing at some point. Some populations [such as the eastern North Pacific gray whale (*Eschrichtius robustus*) and some harbor seal populations] are suspected of already having recovered to levels close to their carrying capacity. It is important to understand when density dependence is taking place so that it can be distinguished from human-caused effects on population growth that might require conservation action.

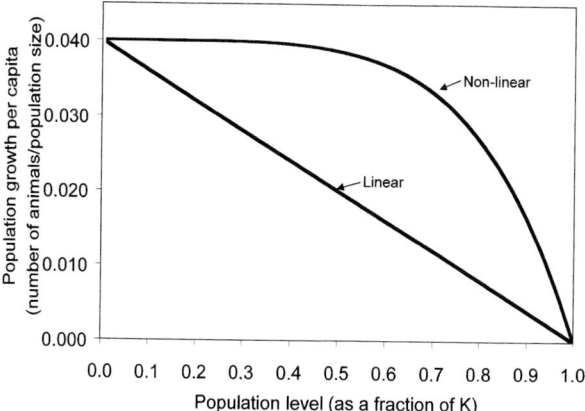

Figure 1 *An illustration of two forms of density dependence. "Linear" density dependence represents a constant decline in population growth per capita as the population level approaches carrying capacity* (K). *Per capita population growth is the number of animals added to the population divided by the total population size, which also represents the population growth rate. "Nonlinear" represents a curved response, where the population growth per capita does not decline until a population level close to* K, *where it declines rapidly. Both curves represent cases where the maximum population growth rate is 0.04 (4% per year).*

D. Allee Effect (Depensation)

Density dependence can also work in the opposite direction, where per capita population growth is slowed at very small population sizes (Fig. 3). This is called the Allee effect by ecologists (and is also called depensation by fisheries biologists). One simple example of the Allee effect would be a reduction in reproduction at very small population sizes due to the inability to find a mate. Allee effects could also occur from inbreeding depression associated with small populations or from behavioral changes that might accompany a reduction to small numbers of animals, such as decreased foraging success or protection from predators. Unfortunately, the difficulties in studying the population dynamics of marine mammals are compounded by the difficulty of studying rare populations so it may take a long time before much is learned about depensation and its role in the dynamics of marine mammals. Several severely depleted baleen whale populations have shown little or no recovery despite decades of apparent relief from human exploitation, raising the possibility that these populations are experiencing Allee effects. However, it has been realized that many of these populations may have continued to experience human-caused mortality from a variety of sources, such as illegal harvest, entanglement in fishery gear, and collisions with ships. This makes it difficult to determine if depensation does play a significant role in the dynamics of small populations of cetaceans.

Similarly, many pinniped species were once harvested to commercial extinction and have since shown recovery, but it is rarely known how small these populations actually were at their low point. One exception is the remarkable story of northern elephant seals (*Mirounga angustirostris*), which were thought to be extinct around 1880 and are now thought to have numbered no more than about 100 individuals around the turn of the century, concentrated in a single colony in Baja California, Mexico. The number of northern elephant seals now exceeds 80,000 animals, and they have recolonized many BREEDING SITES in the United States and Mexico. However, the apparent lack of depensation in one population does not preclude it from being important to other populations or species. Although difficult to study, a decline in growth rate at small population sizes can substantially increase the risk of extinction for a population, so it will continue to be important to consider depensation despite the lack of solid evidence for its occurrence.

See Also the Following Articles

Abundance Estimation ■ Identification Methods ■ Pinniped Life History ■ Predation on Marine Mammals ■ Stock Assessment ■ Sustainability

References

Berta, A., and Sumich, J. L. (1999). "Marine Mammals: Evolutionary Biology." Academic Press, San Diego.

Best, P. B. (1993). Increase rates in severely depleted stocks of baleen whales. *ICES J. Mar. Sci.* **50,** 169–186.

Boyd, I. L., Lockyer, C., and Marsh, H. D. (1999). Reproduction in marine mammals. *In* "Biology of Marine Mammals" (J. E. Reynolds III and S. A. Rommel, eds.), pp. 218–286. Smithsonian Institution Press, Washington, DC.

Brault, S., and H. Caswell. (1993). Pod-specific demography of killer whales (*Orcinus orca*). *Ecology* **7,** 1444–1454.

Cooper, C. F., and Stewart, B. S. (1983). Demography of northern elephant seals, 1911–1982. *Science* **219,** 969–971.

Estes, J. A., Tinker, M. T., Williams, T. M., and Doa, D. F. (1998). Killer whale predation on sea otters linking oceanic and nearshore ecosystems. *Science* **282,** 473–476.

Fowler, C. W. (1987). A review of density dependence in populations of large mammals. *In* "Current Mammalogy" (H. H. Genoways, ed.), Vol. 1, pp. 401–441, Plenum, New York.

Fowler, C. W., and Baker, J. (1991). A review of animal population dynamics at extremely reduced population levels. *Rep. Intl. Whal. Comn.* **41,** 545–554.

Fowler, C. W., and Smith, T. D. (1981). "Dynamics of Large Mammal Populations." Wiley, New York.

Geraci, J. R., Harwood, J., and Lounsbury, V. J. (1999). Marine mammal die-offs: Causes, investigations, and issues. *In* "Conservation and Management of Marine Mammals" (J. R. Twiss and R. R. Reeves, eds.), pp. 367–395. Smithsonian Institution Press, Washington, D.C.

Manzanilla, S. R. (1989). The 1982–1983 El Niño event recorded in dentinal growth layers in teeth of Peruvian dusky dolphins (*Lagenorhynchus obscurus*). *Can. J. Zool.* **67,** 2120–2125.

Olesiuk, P. F., Bigg, M. A., and Ellis, G. M. (1990). Recent trends in the abundance of harbour seals, *Phoca vitulina*, in British Columbia. *Can. J. Fish. Aquat. Sci.* **47,** 992–1003.

O'Shea, T. J., Ackerman, B. B., and Percival, H. F. (eds.) (1995). "Population Biology of the Florida Manatee." Information and Technology Report 1, National Biological Service, U.S. Department of the Interior, Washington, DC.

Ragen, T. J. (1995). Maximum net productivity level estimation for the northern fur seal (*Callorhinus ursinus*) population of St. Paul Island, Alaska. *Mar. Mamm. Sci.* **11,** 275–300.

Read, A. J., and Hohn, A. A. (1995). Life in the fast lane: The life history of harbor porpoises from the Gulf of Maine. *Mar. Mamm. Sci.* **11,** 423–440.

Reilly, S. B., and Barlow, J. (1986). Rates of increase in dolphin population size. *Fish. Bull. U.S.* **84,** 527–533.

Taylor, B. L., and DeMaster, D. P. (1993). Implications of non-linear density dependence. *Mar. Mamm. Sci.* **9,** 360–371.

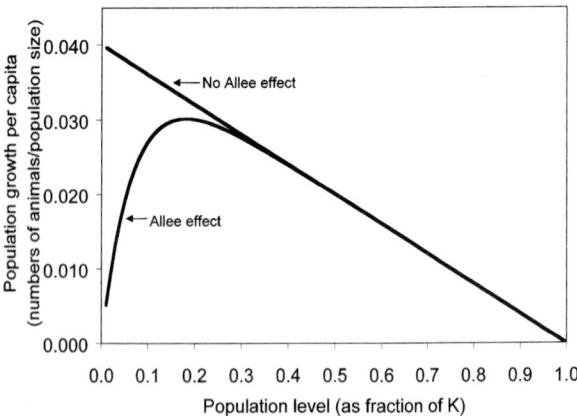

Figure 3 *An illustration of the Allee effect. One curve represents linear density dependence without depensation ("no Allee effect"). The second curve represents linear density dependence with depensation ("Allee effect"). Under depensation, population growth per capita declines at low population levels.*

Trillmich, F., and Ono, K. A. (eds.) (1991). "Pinnipeds and El Niño: Responses to Environmental Stress." Springer Verlag, Berlin.

Wickens, P., and York, A. E. (1997). Comparative population dynamics of fur seals. *Mar. Mamm. Sci.* **13**, 241–292.

Population Status and Trends

JAY BARLOW
*Southwest Fisheries Science Center,
La Jolla, California*

RANDALL R. REEVES
*Okapi Wildlife Associates,
Hudson, Quebec, Canada*

For marine mammals, *status* is a measure of the size or general HEALTH of a population relative to some management standard. A *trend* is a measure of the rate at which a population grows or declines over some (usually long) time period. Taken together, status and trends form the basis for assessing whether management objectives are being met for a given population or management unit.

I. Status

Inherent in the concept of status is the evaluation of populations relative to some standard or metric. Absolute estimates of population size may be included in an assessment of the population status, but an evaluation of status is incomplete without evaluating the significance of this population size relative to some goal. The standards for evaluating the status of populations are typically related either to a harvest or to a conservation objective.

A. Harvest Objectives

Traditionally, to evaluate harvest objectives, population size was evaluated relative to the population level (MSYL) that would give the maximum sustainable yield (MSY). Populations are expected to decline as a result of harvests, but as population size decreases, the population growth rate is expected to increase to compensate for this harvest. The implicit assumption is that populations are resource limited, and as density declines, more per-capita resources are available to support enhanced production, survival, or both. Harvest and growth rates may balance each other over a wide range of possible equilibrium population sizes, but typically harvest will be maximized at one specific equilibrium level (MSYL). Although the MSYL concept has persisted for many years, it was widely recognized by the 1970s that, in practice, it was seriously flawed as a basis for management. Management of populations at their MSYL is a knife-edge balancing act that requires constant conditions and near-perfect data. Usually, when populations slip below that level, the drastic management actions that are required cannot be implemented in time to prevent the collapse of the population as an economic resource. This realization has led to more risk-averse management models that strive to keep populations above their MSYL; this higher target level is sometimes called the optimum sustainable yield level.

Currently, few marine mammal species are managed with explicit harvest objectives. Although a moratorium on commercial whaling has been in place since 1986, the INTERNATIONAL WHALING COMMISSION (IWC) still maintains a harvest-based management framework for large whales [baleen whales, sperm whales (*Physeter macrocephalus*), and bottlenose whales (*Hyperoodon* spp.)]. To determine status, population size is compared to a standard that is based on MSYL. Protected stocks (PS) are less than 0.9 MSYL; sustained management stocks (SMS) are between 0.9 and 1.2 MSYL; and initial management stocks (IMS) are greater than 1.2 MSYL. Using this scheme, the IWC has assessed the current status of many, but not all, of their management stocks (Table I). Acknowledging the risk of managing at the MSYL knife edge, the IWC has, since the mid-1970s, limited harvests to 90% of the estimated MSY. The IWC has been working on a "revised management procedure" (RMP) to replace this MSYL-based management when the moratorium on commercial whaling ends. The performance of the RMP has been tested in simulations where the objective is to maintain populations above 72% of their preexploitation population size.

The Stock Assessment Secretariat of the Canadian Department of Fisheries and Oceans coordinates the production of stock assessment reports for hunted species, notably including harp seals (*Pagophilus groenlandicus*), narwhals (*Monodon monoceros*), and bowhead whales (*Balaena mysticetus*). These reports, prepared by scientists in close consultation with representatives of regional bodies and hunting communities, are intended to provide the basis for managing harvests. However, in the absence of a legal framework equivalent to the U.S. Marine Mammal Protection Act (see later), harvest objectives are ad hoc. A stock assessment report recently completed for the Hudson Bay-Foxe Basin stock of bowhead whales, for example, invokes "sustainability" as an implicit management objective, with no reference to recovery or maximizing yield.

B. Conservation Objectives

The goals of conservation efforts can range from preventing the extinction of species to returning populations to their carrying capacity level. The metrics used to measure the conservation status of populations range over this same spectrum.

When populations become very small or are declining rapidly, their status is often determined by estimating the probability of extinction within a defined time period. For example, the "critically endangered" category of the IUCN Red List includes species whose probability of EXTINCTION is estimated to be at least 50% within 10 years or three generations, whichever is longer. Clearly this is nature's intensive care unit for the nearly hopeless cases. In the IUCN categories of "endangered" and "vulnerable," the values change to 20% chance of extinction within 20 years or five generations and 10% chance within 100 years, respectively. The risk of extinction is estimated using a technique developed in conservation biology known as

TABLE I
Status Classification for Whales Managed under the Authority of the International Whaling Commission[a]

Species	Stock	Status
Blue whale (*Balaenoptera musculus*)	All stocks	Protection stock (PS)
Fin whale (*B. physalus*)	Southern Hemisphere	PS
	North Pacific	PS
	Nova Scotia	PS
	East Greenland–Iceland	Sustained management stock (SMS)
	All other stocks	Not classified
Sei whale (*B. borealis*)	Southern Hemisphere	PS
	North Pacific	PS
	Nova Scotia	PS
	All other stocks	Not classified
Bryde's whales (*B. brydei* and *B. edeni*)	Southern Indian Ocean	Initial management stock (IMS)
	Solomon Islands	IMS
	Western South Pacific	IMS
	Eastern South Pacific	IMS
	Western North Pacific	IMS
	Eastern North Pacific	IMS
	East China Sea	PA
	North Atlantic	IMS
	All other stocks	Not classified
Minke whales (*B. acutorostrata* and *B. bonaerensis*)	Southern hemisphere[b]	Not classified
	Okhotsk Sea–West Pacific	Not classified
	Sea of Japan–Yellow Seal–East China Sea	PS
	North Pacific remainder stocks	Not classified
	West Greenland	PS
	Northeast Atlantic	PS
	Other Atlantic stocks	Not classified
	Indian Ocean	IMS
Right whales (*Eubalaena* spp.)	All stocks	PS
Bowhead whale (*Balaena mysticetus*)	All stocks	PS
Humpback whale (*Megaptera novaeangliae*)	All stocks	PS
Pygmy right whale (*Caperea marginata*)	All stocks	PS
Gray whale (*Eschrichtius robustus*)	Eastern North Pacific	SMS
	Western North Pacific	PS
Sperm whale (*Physeter macrocephalus*)	Western North Pacific	PS
	All other stocks	Not classified
Bottlenose whales (*Hyperoodon* spp.)	North Atlantic	PS
	All other stocks	Not classified

[a]From IWC (1999).

[b]Actually including two species: *Balaenoptera acutorostrata* and *B. bonaerensis*.

population viability analysis (PVA), which evaluates population size, trends in abundance, life history traits, natural variability, trends in habitat loss, and parameter uncertainty. These onerous data requirements have meant that very few marine mammal species or populations have been evaluated using PVA.

Lacking sufficient data to evaluate the risk of extinction in this manner, several surrogate variables may be measured that are highly correlated with extinction risk. Extinction is obviously correlated with declining abundance (see Section II), and the IUCN uses total declines of 80, 50, and 20% over 10 years (or three generations) to classify species as critically endangered, endangered, or vulnerable, respectively. We know that small population size is itself an extinction risk factor, and the IUCN uses population sizes of 50, 250, and 1000 mature individuals to classify species into the same three categories. Other important factors that increase extinction risk include (1) having all individuals in a single location, (2) overdispersal and the resulting loss of mating opportunities and social facilitation, (3) habitat degradation, and (4) extreme fluctuations in population size. The IUCN recognizes the compounding of risk factors and evaluates some of these factors in tandem with population size or trends. Under domestic legislation, the United States maintains a similar endangered species list with categories of "endangered" and "threatened."

The IUCN's development and adoption of quantitative criteria for Red List classifications prompted various national efforts to take a similar approach. For example, the Mammalogical Society of Japan has applied the IUCN criteria to all marine mammal populations in Japanese waters. The status assigned to many of these populations is worse than that indicated by the IUCN's global listing for the species. There are, for example, critically endangered or endangered populations of finless porpoises (*Neophocaena phocaenoides*), striped dolphins (*Stenella coeruleoalba*), short-finned pilot whales (*Globicephala macrohynchus*), and dugongs (*Dugong dugon*) in Japan, whereas these species, overall, are listed by IUCN as either vulnerable, "data deficient" (inadequate information to assess risk of extinction), or "lower risk, conservation dependent" (out of danger but needing continued conservation measures). In Canada, the task of listing species and populations at risk falls on the Committee on the Status of Endangered Wildlife in Canada (COSEWIC). This group consists of technical experts as well as representatives of government agencies and nongovernmental organizations. Its listing decisions are based on status reports prepared by scientists familiar with the species or populations under review. COSEWIC is in the process of developing its own IUCN-style criteria for status determinations.

For species that are above the size at which extinction is a significant risk, status is usually measured relative to historical abundance or environmental carrying capacity (K). Even when the "official" definition of status is based on a percentage of K, historical abundance is often substituted because carrying capacity is so difficult to measure. Historical abundance is, itself, poorly known for many species. If there are no direct measures of historical abundance, it can be estimated by a method called "back-calculation" based on current abundance, a time series of annual human-caused mortality, and a model of population growth.

The U.S. Marine Mammal Protection Act of 1972 is probably the most far-reaching and proactive national legislation for the conservation of marine mammals. It has two main conservation goals: to keep populations at their "optimum sustainable population" (OSP) levels and to keep populations as "functioning elements of their ecosystem." OSP has been interpreted to be a range from a population's maximum net productivity level (typically thought to be between 60 and 80% of K for marine mammals) up to its carrying capacity level. Populations below OSP are considered to be "depleted." The U.S. legislation is significant for having explicit goals to maintain populations near their natural levels instead of protecting them only after they have declined to dangerously low levels.

Similarly, in Europe, the Agreement on the Conservation of Small Cetaceans in the Baltic and North Seas (ASCOBANS) has a conservation goal to restore and/or maintain stocks at a level they would reach when there is the lowest possible anthropogenic influence. ASCOBANS has interpreted this objective to mean restoring and/or maintaining populations at or above 80% of K.

II. Trends

An upward or downward trend in population size is obviously a significant component in evaluating the status of a population; however, it is also one of the more difficult components to determine. Population trends have been directly estimated for only a tiny proportion of all cetacean populations. The primary problems are that population size cannot be estimated very precisely and population growth is typically slow.

Cetacean population size is usually estimated from line-transect SURVEYS. Trends can also be based on an index of relative abundance, such as the number of whales seen per hour on standard transects with consistent survey methods. In either case, the precision of the estimates is measured as a coefficient of variation (CV = standard deviation divided by the mean); CVs of 20% are considered very good and CVs of 30–50% are considered typical. Given their life history constraints, cetacean populations can grow at a maximum rate of about 8% per year, and, for some slow-reproducing species [such as the killer whale (*Orcinus orca*) and sperm whale], maximum growth rates may be as low as 2–3%. There is no similar constraint on the rate at which populations can decline; however, there have been few instances where long-term rates of decline have been found to exceed these values.

Statistical power is a measure of the probability of detecting a significant change in a population if that population is truly growing or declining. Power increases with the actual rate of change in population size and with the sample size (both the amount of survey effort and, more importantly, the length of the monitoring period). As a rule of thumb for cetaceans, at least 10 annual surveys with good precision (CV < 20%) are required to yield a high probability (>80%) of detecting a 50% change in total population size. Thus, even a carefully monitored population can be cut in half before solid evidence of a decline becomes available. The situation is made even more difficult for endangered species [such as the baiji (*Lipotes vexillifer* and vaquita (*Phocena sinus*)], which become almost impossible to census as they become increasingly rare. Similar to cetaceans, the detection of population trends for dugongs and manatees is extremely difficult.

Our ability to detect trends in pinniped populations is considerably better. Although total population size may be difficult to estimate, a segment of the population predictably returns to land or ice to breed, give birth, or molt, and this segment can be counted accurately by ground-based observers or from aerial photographs. Growth or decline in this segment is taken as an index of trend for the entire population so this kind of count is sometimes referred to as an index count. Index counts may be based, for example, on the number of pups born or the number of total seals hauled out at the peak of the molting season. If index counts are used over a wide range of population sizes, ancillary research is typically required to ensure that the fraction of animals counted does not vary in a density-dependent manner (e.g., as per capita resources become scarce, the percentage of time spent foraging away from the beach might increase). Another factor making positive trends in pinniped populations easier to detect is that most are capable of growing at faster rates than cetaceans and sirenians.

See Also the Following Articles

Abundance Estimation ▪ Conservation Biology ▪ Endangered Species and Populations ▪ Management ▪ Population Dynamics ▪ Stock Assessment

References

Gerrodette, T. (1987). A power analysis for detecting trends. *Ecology* **68**(5), 1364–1372.

Gerrodette, T. (1993). Trends: Software for a power analysis of linear regression. *Wildl. Soc. Bull.* **21**, 515–516. Software available at http://mmdshare.ucsd.edu/trends.html.

IUCN (1996). "IUCN Red List of Threatened Animals." IUCN, Gland, Switzerland.

IWC (1999). Schedule of the International Convention for the Regulation of Whaling, as amended by the Commission at the 51st annual meeting 1999, and replacing that dated September 1998. International Whaling Commission.

Klinowska, M. (ed.) (1991). Dolphins, porpoises and whales of the world. *In* "The IUCN Red Data Book." IUCN, Gland, Switzerland.

Marsh, H., and Lefebvre, L. W. (1994). Sirenian status and conservation efforts. *Aquat. Mamm.* **20**, 155–170.

Reijnders, P., Brasseur, S., van der Toorn, J., van der Wolf, P., Boyd, I., Harwood, J., Lavigne, D., and Lowry, L. (1993). Seals, Fur Seals, Sea Lions, and Walrus. Status Survey and Conservation Action Plan. IUCN Species Survival Commission, Seal Specialist Group. IUCN, Gland, Switzerland.

Taylor, B. L., and Gerrodette, T. (1993). The uses of statistical power in conservation biology: The vaquita and northern spotted owl. *Conserv. Biol.* **7**(3), 489–500.

Twiss, J. R., Jr., and Reeves, R. R. (eds.) (1999). "Conservation and Management of Marine Mammals." Smithsonian Institute Press, Washington, DC.

Porpoises, Overview

ANDREW J. READ
*Duke University Marine Laboratory,
Beaufort, North Carolina*

The porpoises are 1 of 10 families that constitute the suborder Odontoceti, or the modern toothed whales. The family Phocoenidae consists of six species, distributed in both hemispheres (Table I). For many years, and in some areas still today, there has been confusion regarding the terms "porpoise" and "dolphin." In part, this is because many early taxonomic accounts included porpoises in the family Delphinidae, although all recent accounts separate the two groups. In addition, many delphinids have been referred to as "porpoises" to distinguish them from the dolphin fish (*Coryphaena hippurus*), also known as mahi-mahi or dorado. Despite their shared vernacular names, porpoises and dolphins are phylogenetically distinct and are as different as horses and cows or dogs and cats. These differences are manifested in their morphology, ecology, and behavior, as outlined here.

TABLE I
Living Species of Porpoises

Common name	Scientific name	Distribution
Harbor porpoise	*Phocoena phocoena*	Coastal Northern Hemisphere
Burmeister's porpoise	*P. spinipinnis*	Coastal South America
Vaquita	*P. sinus*	Gulf of California
Spectacled porpoise	*P. dioptrica*	Southern Ocean
Dall's porpoise	*Phocoenoides dalli*	North Pacific
Finless porpoise	*Neophocaena phocaenoides*	Coastal south and east Asia

I. Origins

Our knowledge regarding the evolution of porpoises comes from inferences drawn from the fossil record and from the morphology and genotypes of living species. The earliest porpoise known is *Salumiphocaena stocktoni*, discovered in late Miocene strata of southern California, from approximately 11 million years ago. These early porpoises appeared at about the same time as the first dolphins. The dates of the origin of the two families are consistent with estimates of divergence between the cytochrome *b* genes of phocoenids and delphinids. Most other early fossil porpoises are known from the Pliocene in North and South America. Porpoises, dolphins, and monodontids, all members of the superfamily Delphinoidea, are likely to have descended from the Kentriodontidae, an extinct family of odontocetes. Like modern porpoises, the kentriodontids were rather small animals, approximately 2 m in length. Kentriodontids occurred in both the Atlantic and Pacific Oceans, but disappeared approximately 10 million years ago.

Analyses of the genome of living porpoises suggest that the finless porpoise is the most primitive, or basal, member of the family. This view is consistent with morphological evidence. Some researchers believe that the earliest phocoenids radiated into temperate waters of both hemispheres from the tropics, where the finless porpoise is still found today. Other researchers maintain that phocoenids originated in temperate waters of the North Pacific Ocean, where the oldest fossils are located. With our current knowledge, it is not possible to resolve which of these two scenarios are correct. Molecular evidence indicates that all modern porpoises have evolved within the past few million years.

The nature of phylogenetic relationships among porpoises has been the subject of recent debate. Researchers studying the morphology of living species have suggested that the spectacled porpoise of the Southern Ocean is closely related to the Dall's porpoise of the North Pacific and that these two species should be classified as a subfamily distinct from the remaining four species. This view has not been supported by analysis of the molecular genome, however, which suggests a close phylogenetic relationship among the vaquita, Burmeister's porpoise, and spectacled porpoise. Most researchers now separate the family into three genera: *Neophocaena*, *Phocoenoides*, and *Pho-*

coena. The first two genera are monotypic (the finless porpoise *Neophocaena phocaenoides* and Dall's porpoise *Phocoenoides dalli*) and the third contains four species (harbor porpoise *Phocoena phocoena*, vaquita *Phocoena sinus,* Burmeister's porpoise *Phocoena spinipinnis,* and spectacled porpoise *Phocoena dioptrica*). More work is required, however, on the exact phylogenetic relationships among these forms.

Periodic cycles of global cooling and warming have had profound effects on the modern biogeography of all cetaceans, and this holds true for the porpoises. For example, the distribution of the vaquita is currently restricted to the northern Gulf of California. In geographical terms, the closest relative of the vaquita is the harbor porpoise, which is found as far south as the coast of central California. Morphological and genetic evidence, however, indicate that the closest living relative of the vaquita is the Burmeister's porpoise, from the coastal waters of temperate South America. It is likely that, during a cool glacial period of the Pleistocene, ancestors of today's Burmeister's porpoises crossed the equator and were subsequently trapped in the Gulf of California by warming equatorial waters. This isolated population eventually gave rise to the vaquita.

The selective factors shaping the evolution of porpoises are poorly understood. It is unclear, for example, why all modern forms exhibit paedomorphosis and small body size (see later). Whatever these selective forces, they appear to have driven similar evolution in the delphinid genus *Cephalorhynchus* of the Southern Hemisphere. The morphology, ecology, and behavior of these dolphins are remarkably similar to porpoises in many ways and they appear to have CONVERGED with porpoises in many aspects of their biology.

II. Morphology

Porpoises share many morphological attributes that distinguish them from other small cetaceans. All six species are small, with no member of the family exceeding 250 cm in body length. Porpoises are stocky, robust animals and lack the rostrum common to most delphinids. The appendages of most species are relatively small and, in the case of the finless porpoise, the dorsal fin is missing altogether. Many of these morphological features may be related to the thermal biology of porpoises and, particularly, to the challenges posed by small body size in a cold and conductive medium.

The skulls of all porpoises exhibit pronounced paedomorphosis, the retention of juvenile characters in the adult form. In particular, the skulls of all porpoises are characterized by short rostra, large and rounded braincases, and delayed fusion of cranial sutures during ontogeny. As a result, the skulls of adult porpoises resemble those of juvenile specimens of other species. Several other cranial features are diagnostic of the family, including raised protuberances on the premaxillae and the spatulate shape of the teeth. The latter feature is an easy way to distinguish porpoises from dolphins, which have conical teeth.

An unusual morphological feature of all phocoenid species, except for the Dall's porpoise, is the presence of epidermal tubercles along the leading edge of the dorsal fin. These small, raised protuberances develop shortly after birth and usually oc-

cur in several rows. Even the finless porpoise, which lacks a dorsal fin, possesses several rows of tubercles on its dorsal ridge. Tubercles are most prominent in the Burmeister's porpoise and their presence provides the basis for the trivial name *spinipinnis.* The function of these unusual structures is unknown.

The pigmentation patterns of phocoenids vary considerably, although there are several features common to the family. Most porpoises possess patches of dark pigmentation surrounding the eye, although these patches are of varying prominence and contrast. The dark eye patch is particularly well developed in the spectacled porpoise, *Phocoena dioptrica,* where it is encircled by a narrow white line, which gives the species both its common and specific names. All porpoises (and most dolphins) exhibit a bridle, a system of stripes extending from the eye and blowhole to the apex of the melon, and most porpoises exhibit the general pattern of countershading, common to many odontocetes, of a dark dorsal cape and a lighter abdomen.

III. Ecology and Behavior

Our general knowledge of the ecology and behavior of porpoises is limited and these areas are fertile subjects for future research. The vaquita was described only as recently as 1958, for example, and few researchers have ever seen the species in the field. Burmeister's and spectacled porpoises are also very poorly known, the latter primarily from a few STRANDINGS in remote areas of the Southern Hemisphere. In contrast, our knowledge of the ecology and behavior of harbor porpoises is much more extensive.

The radiation of phocoenids over the past few million years has allowed porpoises to colonize a variety of habitats in both hemispheres. The harbor porpoise, Burmeister's porpoise, and vaquita are coastal in nature and the range of the finless porpoise extends well upstream into major river systems. The Dall's porpoise and spectacled porpoise are primarily pelagic animals.

There is little evidence of cooperative feeding in the family and it appears that individual porpoises usually forage alone. Most coastal species, such as the harbor and Burmeister's porpoise, feed on small pelagic fish, such as herring, anchovies, and capelin, and supplement this diet with demersal fishes. In contrast, Dall's porpoises feed primarily on the small mesopelagic fish and squid that comprise the deep scattering layer. To date, DIVING BEHAVIOR has been studied for only the harbor porpoise, which can dive to depths of over 200 m. Their small size makes it unlikely that any species of porpoise, even the pelagic Dall's porpoise, are exceptional divers. Dall's porpoises likely take advantage of the daily vertical migration of their prey and feed at night when these mesopelagic fish and squid are near the surface.

Porpoises are among the smallest cetaceans and this aspect of their biology is manifested in their reproductive biology. Compared to many delphinids, porpoises grow rapidly and reach sexual maturity at an early age. For example, many female harbor porpoises attain sexual maturity in their third year of life, at an age where bottlenose dolphins are still accompanying their mothers. Both Dall's and harbor porpoises are capable of annual reproduction, and females of these species are often simultaneously lactating and pregnant. This demanding reproductive schedule is accompanied by a relatively brief life span—very few porpoises live longer than 20 years.

Although our knowledge of the social behavior of phocoenids is limited, we can say that, unlike many pelagic delphinids, porpoises usually occur alone or in small, fluid groups. This is also reflected in patterns of strandings. Porpoises generally strand singly, never in the large groups witnessed in some other odontocete species. Occasionally, hundreds of porpoises have been observed together, but these are likely temporary aggregations, rather than stable groups. In general, it appears that porpoises typically exist in fission–fusion societies in which associations among individuals are extremely dynamic. The only long-term association known to occur is between a lactating female and her dependent calf. The duration of lactation varies among species, and perhaps among populations within species, but is unlikely to surpass 2 years in any phocoenid.

With the exception of the Dall's porpoise, porpoises are generally shy, unobtrusive animals that are difficult to sight and follow at sea. Unlike many delphinids, porpoises seldom, if ever, leap clear of the water. A typical surfacing sequence is characterized by a series of gentle rolls at the surface, followed by a longer submergence. Occasionally on calm days, harbor porpoises will lie quietly at the surface for short periods. Individuals seldom approach boats or ride the bow or stern waves of vessels. In contrast, Dall's porpoises are fast swimmers that often make a characteristic splash or "rooster-tail" when they surface. Dall's porpoises are also the only phocoenid that regularly approaches boats to ride the bow wave, a behavior they may have modified from riding the pressure waves produced in front of swimming baleen whales.

IV. Conservation

Most populations of phocoenids are affected to some degree by human activities. Hunting, bycatches in commercial fisheries, and habitat degradation have had profound impacts on the health and demography of affected populations. Only the spectacled porpoise of the Southern Ocean is largely free of the effects of human influences, although even this species has been taken as bycatch in fisheries off Tierra del Fuego. The nature of human activity posing a conservation threat varies from species to species, as does the conservation status of each species.

Porpoises living in coastal areas are affected by the modification, degradation, and destruction of habitat by humans. Particularly affected in this manner is the finless porpoise, which inhabits the temperate and tropical coasts of Asia and the Indian subcontinent. Human activities, such as dredging, reclamation, pollution, and intense vessel traffic, are felt most keenly in coastal and riverine habitats where the density of humans is high. Thus, particular concern has been expressed for populations of finless porpoises in the South China Sea and adjacent waters. Other coastal species of porpoises are not immune from such effects. Harbor porpoises, for example, may be excluded from portions of their habitat in the North Atlantic and North Pacific by the use of high-intensity acoustic devices designed to keep pinnipeds away from salmon mariculture sites.

Harbor and Dall's porpoises have been hunted for their meat and blubber for many centuries. In the Danish Belt sea, an annual hunt for harbor porpoises occurred from the 14th century until the early 20th century. More than 1000 porpoises were taken annually between 1834 and 1874 in this hunt. Harbor porpoises have also been harvested from other areas, particularly the Bay of Fundy and western Greenland. A large-scale harpoon fishery for Dall's porpoises still exists off the coast of northern Japan. This hunt supplies meat for the domestic Japanese market, partially offsetting the reduction in the availability in whale meat following the moratorium in commercial whaling that began in 1986. The effects of this exploitation on populations of Dall's porpoises are not fully understood, but the magnitude of catches (more than 40,000 in 1988) is cause for concern.

Perhaps the most insidious threat to populations of phocoenids is their bycatch in commercial fisheries. Most porpoises are taken as bycatch in gill net fisheries, which use either floating nets to capture fish near the surface or nets anchored on the bottom to catch demersal species. In both cases, porpoises seldom survive entanglement in this type of fishing gear. The causes of bycatches are unclear; it is uncertain, for example, why animals with such a sophisticated system of ECHOLOCATION do not detect and avoid fishing nets. Nevertheless, all species of porpoises are taken as bycatch, sometimes in staggering numbers. The estimated annual mortality of harbor porpoises in Danish bottom gill net fisheries in the North Sea was almost 7000 between 1994 and 1998, for example. The consequences of such large bycatches for affected populations are unknown. In some areas, such as the coastal waters of Peru, porpoises taken initially as bycatch gained commercial value for human consumption so that the distinction between bycatches and directed hunting is blurred. Bycatches pose by far the most serious threat to the single population of vaquitas in the upper Gulf of California. The size of this unique population is estimated to be little more than 500 individuals and it is still subject to bycatch in several artisanal fisheries. The vaquita is one of the most endangered marine cetaceans; its uncertain future depends on conservation initiatives in Mexico.

See Also the Following Articles

References

Barnes, L. G. (1985). Evolution, taxonomy and antitropical distribution of the porpoises (Phocoenidae, Mammalia). *Mar. Mamm. Sci.* **1**, 149–165.

Bjørge, A., and Donovan, G. P. (eds.) (1995). "Biology of the Phocoenids." Reports of the International Whaling Commission, Special Issue 16. Cambridge, UK.

Gaskin, D. E. (1982). "The Ecology of Whales and Dolphins." Heinemann Educational Books, London.

Jefferson, T. A., and Curry, B. E. (1994). A global review of porpoise (Cetacea: Phocoenidae) mortality in gillnets. *Biol. Conserv.* **67**, 167–183.

Nachtigall, P. E., Lien, J., Au, W. W. L., and Read, A. J. (eds.) (1995). "Harbour Porpoises: Laboratory Studies to Reduce Bycatch." De Spil Publishers, Woerden, The Netherlands.

Northridge, S. P., and Hofman, R. J. (1999). Marine mammal interactions with fisheries. In "Conservation and Management of Marine Mammals" (J. R. Twiss, Jr., and R. R. Reeves, eds.), pp. 99–119. Smithsonian Institution Press, Washington, DC.

Read, A. J. (1999). "Porpoises." Worldlife Library, Colin Baxter Photography, Grantown-on-Spey, Scotland.

Read, A. J., Wiepkema, P. R., and Nachtigall, P. E. (eds.) (1997). "The Biology of the Harbour Porpoise." De Spil Publishers, Woerden, The Netherlands.

Ridgway, S. H., and Harrison, R. (eds.) (1999). "Handbook of Marine Mammals," Vol. 6. Academic Press, San Diego.

Rosel, P. E., Haygood, M. G., and Perrin, W. F. (1995). Molecular relationships among the true porpoises (Cetacea: Phocoenidae). Mol. Phylogenet. Evol. 4, 463–474.

Predation on Marine Mammals

DAVID W. WELLER
*Southwest Fisheries Science Center,
La Jolla, California*

Although marine mammals are regarded as accomplished and sophisticated hunters, they too are preyed upon by a variety of terrestrial, avian, and aquatic predators. Predation is an ecological factor of significant influence on the behavior and organization of animal societies in general, and the need for protection from predation has likely been an important factor in the evolution of most marine mammal social systems. While the risk of predation is of little or no concern for some species, other exist under high levels of predatory pressure. A large portion of all marine mammals, ranging in size from the enormous blue whale (*Balaenoptera musculus*) to the relatively small sea otter (*Enhydra lutris*), are subjected to varying levels of predation. Responses to predators are complex and include detection and avoidance, fleeing, seeking habitat features for cover, and active defense by individuals as well as coordinated groups.

While the topic of predation is expansive and multidimensional, the focus of the following article centers on the hunting and consumption of marine mammals by their predators. The definition of predation used here excludes parasitism, FILTER FEEDING, scavenging (carrion eating), or browsing and is limited to situations in which an animal expends time and energy to locate living prey and exerts additional effort to kill and consume it. Therefore, predation is distinguished from other forms of foraging in that it concludes with the death of an animal that offers some resistance against being discovered and/or being harmed.

I. Predation on Sirenians

The relatively slow moving and rather lethargic behavior of sirenians (manatees, *Trichechus* spp., and dugongs, *Dugong dugon*) makes them seem particularly vulnerable to predation. However, manatees and dugongs actually have few known natural predators and appear to experience only occasional mortality due to predation. Although large sharks, crocodiles (*Crocodylus* spp.), and killer whales (*Orcinus orca*) are all considered to be potential predators, few records exist to confirm these suspicions. Evidence of predation, including tooth scarring indicative of unsuccessful attacks by predators, has been observed only rarely during long-term field studies on manatees (*Trichechus manatus*) in Florida and dugongs off Australia. The limited presence of marine predators in the relatively warm and shallow nearshore waters, rivers, and bays where these animals forage on marine vegetation may partially explain the paucity of observed predatory interactions. Further, the particularly thick skin and exceedingly dense bone characteristic of the sirenians may render them rather unpalatable and serve to deter potential predators.

Predation on sirenians does occur, however. For example, in South America, Amazon manatees (*T. inunguis*) are reported to be preyed on by jaguars (*Panthera onca*) and large sharks, and marine crocodiles may occasionally kill dugongs throughout their distribution. Off Western Australia, predation by killer whales on adult dugongs has been reported, including one occasion when 10 killer whales were observed attacking a group of approximately 40 dugongs in shallow water. During this incident the dugongs were huddled tightly together in an antipredator response, while pieces of flesh and integument floated nearby in blood-stained water. Local residents of Western Australia have also implicated "black porpoises" as predators of dugongs; however, what species these "porpoises" represent is entirely unclear. While some authors suggest that these porpoise attacks were likely to be by killer whales, such records may also refer to one of several other mammal-killing cetaceans such as the false killer whale (*Pseudorca crassidens*), pygmy killer whale (*Feresa attenuata*), or short-finned pilot whale (*Globicephala macrorhynchus*). It is conceivable, of course, that predation on sirenians is considerably higher than has been observed and reported. Predatory attacks on young animals, for example, may be particularly successful, and information regarding the predator-related mortality of species such as the West African (*T. senegalensis*) and Amazon manatees that mainly occur in areas inaccessible to researchers is largely unknown.

II. Predation on Mustelids

Although sharks and killer whales represent the primary predators of sea otters several terrestrial and avian predators have also been documented. Coyotes (*Canis latrans*) are known to prey on recently weaned OTTERS in parts of Alaska, and Russian brown bears (*Ursus arctos*) occasionally kill otters that haul out along the shores of the Kamchatka Peninsula. Near Amchitka Island, Alaska, sea otter pups are hunted by bald eagles (*Haliaeetus leucocephalus*). Pups are particularly vulnerable to avian predation as they float unattended at the surface while their mothers are preoccupied with searching for food. The extraordinary buoyancy of young otter pups prevents them from readily submerging and greatly reduces their chances of escaping attack by diving.

Observations of bald eagles nabbing young otter pups from the surface of nearshore waters confirm that eagles use a hunting strategy similar to that used when capturing large fish. That is, pups are gathered from the water in the talons of an eagle, flown to the nest location, and meticulously devoured. Studies conducted on Amchitka Island between the 1950s and 1970s found that up to 28% of the prey remains in eagle nests were from sea otters. Interestingly, some nests contain high levels of otter remains while other nests have none. This finding suggests that some individual eagles may actually specialize on hunting sea otter pups.

While terrestrial and avian predation account for only a small portion of sea otter mortality, sharks represent a more formidable and common predator. White shark (*Carcharodon carcharias*) attacks on sea otters along the California coast are thought to account for 8–15% of the total otter mortality recorded in this region. Curiously, there is little evidence from examination of white shark stomach contents to suggest that sea otters are actually eaten by the individuals that attack them. Instead, otters are stalked and killed by white sharks off California but are apparently abandoned prior to consumption. The absence of sea otter remains in white shark stomachs cannot be considered conclusive at this time, however, as only a small number of stomachs have been available for examination. Although other shark species are also suspected to occasionally kill sea otters, few specific details are available.

Killer whales are known predators of sea otters, but the small number of observed attacks suggests that otters are not preferred prey. Nonetheless, a substantial increase in the number of killer whale attacks on sea otters was documented between 1992 and 1996 and corresponded with a notable decline in sea otter population levels over a large part of their western Alaska distribution. It is unclear if this increase in observed killer whale attacks was due to a greater observation effort or represents a real increase in sea otter predation. If this change is merely related to increased observation effort, then killer whale predation on sea otters may not be as uncommon as previously suggested. However, if this finding represents a true increase in the rate of sea otter attacks, it may be related to the relatively recent declines of other killer whale prey, such as Steller sea lions (*Eumetopias jubatus*) and harbor seals (*Phoca vitulina*).

III. Predation on Pinnipeds

Of all the marine mammal groups, pinnipeds are probably subjected to the highest level of predation. While some pinniped species experience little or no predation pressure, others are hunted so intensively that important aspects of their natural history, including reproductive strategies, have evolved in response. Not even the largest pinnipeds such as the walrus (*Odobenus rosmarus*), bearded seal (*Erignathus barbatus*), and elephant seals (*Mirounga* spp.) are free from predation. Terrestrial predators of pinnipeds are particularly abundant in the subpolar and polar regions of the Northern Hemisphere, usually appearing in the form of polar bears (*Ursus maritimus*) and Arctic foxes (*Vulpes lagopus*). Southern Hemisphere ice seals are free from land predators, but instead have fierce aquatic predators such as

the leopard seal (*Hydrurga leptonyx*) to contend with. Pinnipeds in temperate and tropical latitudes experience reduced terrestrial predation but are subjected to increased levels of attack by aquatic predators such as sharks and killer whales. When comparing Northern Hemisphere Arctic pinnipeds to Southern Hemisphere Antarctic pinnipeds, clearly divergent predator avoidance tactics are apparent. Arctic pinnipeds escape land predators by fleeing into the water whereas Antarctic pinnipeds escape aquatic predators by retreating onto ice.

All pinnipeds require a land or ice substrate for pupping, and this facet of their natural history makes them particularly vulnerable to attack in regions where terrestrial predators are present. Golden jackals (*Canis aureus*), for example, are common at a Mediterranean monk seal (*Monachus monachus*) colony on the western coast of Mauritania and have been reported to consume freshly dead monk seals and are suspected to also prey on living pups. Freshwater pinnipeds in Russia's Lake Baikal and in the Caspian Sea (*Pusa sibirica* and *P. caspica*, respectively) have no aquatic predators, but instead have an unusually high number of terrestrial adversaries. Wolves (*Canis lupus*) and eagles prey on newborn Caspian seals, and brown bears occasionally hunt Baikal seals. Ringed seal pups (*Pusa hispida*) inhabiting Finland's Lake Saimaa and Russia's Lake Ladoga are preyed upon by red foxes (*Vulpes vulpes*) and are also suspected to suffer some level of mortality due to attacks by ravens (*Corvus corax*), wolves, dogs, and wolverines (*Gulo gulo*). Similarly, brown bears, wolves, and avian predators, including eagles and ravens, sometimes also kill spotted seals (*Phoca largha*) in the Sea of Okhotsk. Glaucous gulls (*Larus hyperboreus*) and ravens may occasionally kill ringed seal pups, and gulls sometimes peck at the eyes of gray seal pups (*Halichoerus grypus*), resulting in some level of mortality.

Additional terrestrial predators also hunt pinnipeds at their haul-out sites. Coyotes, for example, prey on harbor seal pups in the Pacific Northwest and are responsible for at least 16% of the pup mortality within Puget Sound, Washington. Similarly, bears and mountain lions (*Felis concolor*) may have historically preyed on elephant seals (*Mirounga angustirostris*) along the California coast. In the Southern Hemisphere, mountain lions have been reported to prey on southern sea lion pups (*Otaria flavescens*). South African fur seals (*Arctocephalus pusillus*) that breed along the mainland coast of the southern Africa continent are preyed upon by brown hyenas (*Hyaena brunnea*) and black-backed jackals (*Canis mesomelas*), and South American sea lion pups (*Arctocephalus australis*) are probably attacked by mountain lions.

Arctic foxes have been described as hunters of small animals and birds and as a scavenger of marine mammal remains left by polar bears. However, in parts of the eastern and western expanses of the Beaufort Sea, this fox is considered an active predator of newborn ringed seal pups. In early spring, ringed seals birth and rest in "subnivean birth lairs"—ice caves complete with breathing holes constructed beneath the snow. These lairs provide both shelter from cold temperatures and protection from predators by providing a physical barrier that makes it more difficult for surface predators to detect a newborn pup. Nevertheless, foxes and polar bears enter and kill pups concealed within their subnivean homes with relative frequency. Keen olfaction allows foxes to locate lairs that may be buried under as much as 150 cm of snow. In the Beaufort Sea, Arctic foxes enter about 15% of the birth lairs present within an area. Although the annual average predation rate by Arctic foxes on ringed seal pups is about 26%, rates as high as 58% have been recorded. Ringed seals are also preyed upon by polar bears and may occasionally be attacked by red foxes, wolverines, wolves, dogs, and several avian predators. As such, ringed seals are subjected to perhaps the highest level of predation experienced by any of the marine mammals.

A. Polar Bears

Throughout their circumpolar range, the major prey of polar bears consists of pinnipeds. Polar bears are versatile predators and are well adapted for catching Arctic pinnipeds. Predation is particularly heavy on pups, as they represent an easily obtained food resource. Foraging strategies employed by polar bears range from sit-and-wait tactics to active stalking and pursuit of seals on ice and in the water. When stalking seals on ice, bears "creep" along with their heads held low, often momentarily hiding behind snowdrifts and irregularities in the ice. Despite their relative stealth and excellent ability to detect prey by olfaction, bears often have little success sneaking up on seals. Observations of bears hunting, and in at least one instance capturing, free-swimming seals in ice-free waters have also been reported. One of the sit-and-wait strategies employed by polar bears occurs while hunting ringed seals. Ringed seals forage for food under ice-covered waters throughout the winter and must therefore maintain breathing holes in which to surface. Polar bears seek out such breathing holes and often patiently await the arrival of an unsuspecting seal. When a seal surfaces in the hole for a breath of air, the bear quickly grasps it and drags it from the water onto the ice.

The ringed seal is a main staple of the polar bear diet, although in the Canadian Arctic bearded seals and harp seals (*Pagophilus groenlandicus*) are taken to a lesser extent. Harp and hooded seals (*Cystophora cristata*) are particularly vulnerable to predation on the spring pupping grounds, where polar bears may kill more pups than can be consumed. In Alaska, most of the ringed seals attacked by polar bears are over 6 years of age, while in the Canadian Arctic it is mainly 1- to 2-year-old seals that are killed. Polar bears are largely unsuccessful hunting adult ringed seals due to their nearly constant antipredator vigilance. This vigilance behavior is characterized by constant head lifting and scanning of the nearby environment for the presence of predators. In late spring, polar bears enter a period of intense feeding that corresponds with the onset of the ringed seal pupping season. During this time, bears prey heavily on pups by digging into birth lairs; adult female seals attempting to protect their pups are also occasionally killed.

Walruses are occasionally preyed upon by polar bears, but this massive obobenid represents a formidable adversary quite capable of killing predatory bears. The extent of polar bear predation on walruses is not well known and is likely to vary from region to region. Walrus calves, young juveniles, and sick individuals are most vulnerable to polar bear predation. While hunting walruses, bears often cause entire hauled-out herds to "stampede" into the water by rushing toward them. Although most individuals in the stampede easily escape approaching bears, calves, or young animals may be crushed or injured in the ensuing chaos, making subsequent capture substantially easier.

B. Pinnipeds

Several pinniped species are recognized as predators of other pinnipeds and, in some locations, are responsible for a significant portion of the annual mortality incurred by regional populations. The most ferocious pinniped predators include the leopard seal in the Southern Hemisphere and the walrus in the Northern Hemisphere. In addition, several sea lion species are notorious for feeding on pinnipeds. Two types of pinniped–pinniped predation occur, one at the intraspecific level (within species) and another at the interspecific level (between species). In some cases, particular individuals (usually males) specialize in the predation of pinnipeds. For instance, young male Steller sea lions are known to prey on harbor seals off Alaska and have been noted to account for approximately 4–8% of the mortality reported for Northern fur seal pups (*Callorhinus ursinus*) at St. George Island, Alaska. Adult male Steller sea lions (*Eumetopias jubatus*) may also prey on other pinnipeds, as was recorded for one individual at Año Nuevo, California, that was observed feeding on a small California sea lion (*Zalophus californianus*). Similarly, Southern sea lions have been observed preying on South American fur seals, and at Punta San Juan, Peru, over 8% of the fur seal pups are killed by marauding sea lions during the breeding season. Off Macquarie Island in the sub-Antarctic, one young male New Zealand sea lion (*Phocarctos hookeri*) was thought to be responsible for the mortality of 43% of the fur seal pups (*Arctocephalus gazella* and *A. tropicalis*) from a particular year. At the Snares Islands, New Zealand, New Zealand sea lions have also been observed to prey on New Zealand fur seal pups (*Arctocephalus forsteri*). Finally, gray seals have been reported to consume pups of their own species, but it is unclear if this represents actual predation or merely cannibalistic scavenging of beach-cast carcasses.

1. Walruses Walruses are primarily bottom or benthic feeders whose diet consists largely of bivalve mollusks, a variety of invertebrates, and fish. In addition, they also prey on marine mammals and are known to feed on bearded seals, ringed seals, spotted seals, harp seals, and young walruses. Adult and subadult male walruses are typically responsible for pinniped kills, but females in the Chukchi Sea have also been observed eating seals. Some walruses are habitual predators of other marine mammals. Individuals that regularly attack seals develop massive chest and shoulder muscles, have long and slender tusks, and their upper torsos and normally ivory-colored tusks are stained amber from consuming the oil-rich blubber of their prey. In general, walruses kill pups and young individuals, but on occasion mature adult pinnipeds are also taken. Observations of attacks on harp seal pups and bearded seals are characterized by walruses impaling the prey with their tusks. Although very little of the skeletal muscle and bone of their mammalian prey are consumed, walruses methodically devour most, if not all, of the highly caloric hide and blubber.

elephant seals are hunted off Peninsula Valdés, whereas on Possession Island, whales typically take newly weaned southern elephant seal pups. In general, pups and small adult seals are most vulnerable, but adults are also occasionally killed. Once a seal or sea lion has been captured from the beach or nearshore area, it is usually held in the mouth of a killer whale by one of the flippers or taken crossways in the mouth and shaken vigorously. Sometimes, captured pups are exchanged between members of the killer whale pod. Intentional stranding behavior also occurs in the absence of prey, suggesting that adult killer whales may actually teach their youngsters the finer aspects of this foraging strategy.

IV. Predation on Cetaceans

Although killer whales and sharks are responsible for most attacks on whales, dolphins, and porpoises, other cetaceans such as false killer whales, pygmy killer whales, and pilot whales also represent a potential predatory threat. In addition to these aquatic predators, one terrestrial predator, the polar bear, successfully hunts beluga whales (*Delphinapterus leucas*) and narwhals (*Monodon monoceros*) in Arctic areas. River dolphins appear to be the only cetacean group free from natural predation, although it has been suggested that freshwater caiman in South America may occasionally take young dolphins. Finally, killer whales are likely to experience little or no mortality related to predation.

A. Blackfish

Three members of the delphinid family, including the false killer whale, pygmy killer whale, and short-finned pilot whale, are thought to be hunters of other cetaceans. Each of these species has teeth and jaws suitable for killing and handling large mammalian prey, and all have been observed to at least occasionally prey on other dolphins. Of these three "blackfish," the false killer whale is best known for attacks on small pelagic dolphins and also has a record of harassing humpback (*Megaptera novaeangliae*) and sperm whales (*Physeter macrocephalus*). A series of observations, mainly by marine mammal observers onboard purse-seine boats fishing for yellowfin tuna in the eastern tropical Pacific, have detailed false killer whale attacks on pantropical spotted and spinner dolphins (*Stenella attenuata* and *S. longirostris*). Although nearly two dozen attacks were recorded, false killer whale predation on cetaceans outside of the eastern tropical Pacific is rare, suggesting that the high incidence of attack on the yellowfin tuna grounds may be site and circumstance specific. That is to say, false killer whales may be utilizing a prey resource related to tuna fishing operations (i.e., dolphins being released from temporary capture in fishing nets) that is unavailable outside of the eastern tropical Pacific.

Large whales, such as sperm whales and humpback whales, are also subjected to predatory advances by false killer whales. A school of false killer whales has been observed harassing a sperm whale group off the Galápagos Islands. In this event, no sperm whale mortality was recorded, but the false killer whales did inflict at least superficial injury to several individuals and elicited noticeable fear reactions. Similar, albeit uncommon, events have also been suggested for interactions between false killer whales and humpback whales.

Pygmy killer whales have also been observed in predatory attacks on small dolphins during fishery operations in the eastern tropical Pacific, although observations of this nature are less common than those recorded for false killer whales. The predatory habits of pygmy killer whales on other cetaceans are poorly understood. In captivity, this species has been implicated in the death of a young pilot whale and a dusky dolphin (*Lagenorhynchus obscurus*), but it is unclear if these events led to consumption of the victim.

Similarly, few records regarding pilot whale predation on other cetaceans are available. Although pilot whales are not generally known to prey on marine mammals, records from the eastern tropical Pacific suggest that this species does chase, attack, and may occasionally eat dolphins during fishery operations. The incidence at which these predatory events occur, however, is very low. In CAPTIVITY, pilot whales have been noted to eat stillborn and young dolphins. Short-finned pilot whales have been observed harassing sperm whales in the Gulf of Mexico and off the Galápagos Islands, and although such harassment has been observed to be nonlethal, these events nevertheless often elicit a pronounced fear response, called a "marguerite formation" by sperm whale groups. The marguerite is a defensive formation in which group members form a heads-in and tails-out circular arrangement resembling the petals of a flower. By placing the powerful flukes, a source of potential danger for predators, toward the outside and containing particularly vulnerable individuals, such as calves, on the inside of the formation, sperm whales can usually defend themselves from harm. This marguerite response has also been noted for sperm whale groups under lethal attack by killer whales and when being hunted by whalers. Therefore, the formation of a marguerite in response to pilot whale harassment suggests that sperm whales do at times appear to be threatened by this species. It remains unclear, however, if such harassment by pilot whales represents actual predatory intent or if such interactions are merely practice hunting attempts or social play.

As suggested by the accounts presented here, interactions of false killer whales, pygmy killer whales, and pilot whales with other cetaceans are not particularly common. Of the lethal attacks recorded to date for each of these three blackfish species, all have been in relatively unnatural situations. That is, attacks have occurred either in captivity where species that might normally avoid each other are maintained in the same confines or centered around the eastern tropical Pacific tuna fishing operations where smaller dolphins may become available prey due mainly to capture fatigue. Therefore, it is difficult to assess the regularity of marine mammal predation by these several species, and the scarcity of observed predatory events suggests that marine mammal prey is likely to be secondary to an otherwise fish- and squid-based diet.

B. Sharks

Sharks represent a significant predatory threat to some populations of dolphins. Crude estimations of predation rates, as determined by the proportion of dolphins within a study population possessing shark-inflicted scars and injuries, vary greatly. Shark-related scars on odontocetes are particularly notable for some populations, whereas others go seemingly untouched.

Results from several long-term photoidentification studies of bottlenose dolphins (*Tursiops truncatus* and *T. aduncus*) have documented shark bite scars rates as low as 1% off southern California, an intermediate rate of 22% in western Florida, and up to about 37% off eastern Australia. The frequency of scars may also vary for different dolphin species within the same region. For example, off South Africa, where humpbacked dolphins (*Sousa chinensis*) and bottlenose dolphins overlap in distribution and habitat use, the former species has substantially more scarring related to shark attack than the latter.

Interestingly, the proportion of individuals bearing crescent-shaped shark bite wounds is considerably higher for nearshore species than it is for their offshore counterparts. This apparent discrepancy may be attributable to a variety of factors. To date, most long-term studies on dolphin populations have been conducted nearshore, increasing the opportunity to observe shark scarring. Alternative explanations include the idea that predation on oceanic dolphin is less common overall or that shark attacks in the open ocean are generally more successful. One theory that may at least partially explain why nearshore dolphins have higher rates of scarring is related to habitat features. The habitat of nearshore cetaceans offers a variety of "cover" features, such as kelp and surf, which may make escape from a predator more successful, whereas oceanic species have no such cover and depend solely on fleeing or the protection offered by conspecifics within their social group to escape fatal attack.

Tiger sharks, dusky sharks (*Carcharhinus obscurus*), white sharks, and bull sharks (*C. leucas*) are most often implicated in attacks on nearshore dolphins and porpoises. Other sharks, including oceanic white tip and hammerhead sharks, have also been observed to occasionally attack dolphins. Tiger sharks are notorious predators of spinner dolphins off the Hawaiian Islands, whereas white sharks prey on a variety of odontocetes, ranging in size from the small harbor porpoise (*Phocoena phocoena*) to more substantial beaked whales, and perhaps even newborn mysticete whales. Evidence of shark predation on baleen whales is relatively uncommon, but a report of a tiger shark attacking a young humpback whale has been recorded. Similarly, large sharks (and killer whales) were observed circling a group of sperm whales in which one adult female was giving birth, but no direct attack was noted. Although the number of observations regarding shark attack on large whales is few, it is reasonable to assume that some predatory events probably do occur at least occasionally.

While predation by sharks is of particular concern for cetaceans in the tropics and subtropics, attacks in other regions also occur. The remains of a complete southern right whale dolphin (*Lissodelphis peronii*) fetus as well as the genital region of an adult female were found in the stomach of a sleeper shark (*Somniosus pacificus*) off coastal Chile. In addition, Greenland sharks (*S. squamulosus*) have been reported to prey upon narwhals in the eastern Canadian Arctic, and franciscana dolphins (*Ponotoporia blainvillei*) have been found in the stomachs of seven-gilled (*Heptranchias perlo*) and hammerhead sharks off Brazil. While each of these accounts is suggestive of predation, they should be considered with caution, as it is unclear if the aforementioned sharks actually attacked living dolphins or if the remains identified from stomach content analyses were attributable to scavenging.

C. Polar Bears

Although pinnipeds are the principal marine mammal prey of polar bears, they also actively hunt and occasionally consume NARWHALS and BELUGA WHALES. Polar bears off western Alaska, for example, have been observed "fishing" beluga whales and narwhals out of small openings in the ice (Fig. 3), sometimes killing numbers far greater than can possibly be eaten. In one particular event, polar bears killed and dragged onto the ice at least 40 ice-entrapped beluga whales, and in a similar episode, a single male polar bear was seen to successfully capture 13 beluga whales from a small opening in the ice over a short period of time.

Beluga whales regularly swim into extremely shallow estuary and river channel areas. On rising tides, whales penetrate far into rivers and creeks, often moving into waters so shallow that they can rest on the bottom while a considerable portion of their body remains above the surface. This behavior can sometimes result in partial stranding, but the animals are typically able to free themselves. On occasion, however, complete stranding occurs accidentally, during which time individual belugas remain beached until the return of the incoming tide. At least some beluga whale mortality results from polar bears feeding on stranded individuals. In addition to opportunistic foraging on temporarily beached whales, individual bears have been observed wading into shallow waters and chasing whales passing near to shore. Predatory polar bears also actively stalk free-swimming belugas from ice edges. In this situation, bears either roam along the ice edge or remain motionless while awaiting a group of beluga whales to move within striking range. When a whale passes near enough, a polar bear will launch itself from the ice and onto the back of the unsuspecting beluga. In one incident, a single polar bear was observed to use this hunting tactic to capture and kill two beluga whale calves within 24 hr. This hunting technique requires that bears time their jumps accurately and, more amazingly, handle and debilitate their prey in an aquatic medium. Further, once dead, the beluga must be pulled from the water and dragged onto the ice. In cases where

Figure 3 *Polar bear hunting beluga whales. Photo by Sue Flood/BBC Natural History Unit.*

this hunting technique has been observed directly, the captured belugas are generally young, smaller individuals.

Polar bears have also been observed to attempt attacks on belugas while swimming in pursuit of them. Thus far, no successful attacks have been documented for this aquatic hunting tactic, and on at least one occasion, a group of belugas was seen to chase a polar bear out of the water with group-coordinated threat behavior including tail lashing and repeated close approaches toward the swimming bear. Aquatic stalks by polar bears are largely unsuccessful due to the greater mobility and speed of whales in the water. In fact, the willingness of belugas to closely approach bears in the water, either out of curiosity or in a possible attempt to harass them, suggests that they have little fear of this predator when it is waterborne.

In contrast to the inshore habits of beluga whales, narwhals prefer deeper water and are commonly sighted in considerable numbers offshore of beluga groups in the eastern Canadian Arctic. Polar bear predation on narwhals has been observed rarely, with the few attacks reported consisting of narwhals stranded on tidal flats or entrapped by ice. In one incident, three adult female narwhals stranded on a tidal flat were consumed by a single polar bear. All three of the narwhals bore extensive claw marks and their blubber had been stripped dorsally from the head area back to the tail stock.

D. Killer Whales

In addition to pinnipeds, dugongs, and sea otters, mammal-hunting killer whales (termed transients) also prey upon a variety of dolphins and porpoises, and even occasionally attack sperm and baleen whales. Transient killer whales are relentless hunters, spending up to 90% of each daylight period searching for food. In addition to marine mammal prey, terrestrial animals such as deer (*Odocoileus hemionus*) and moose (*Alces alces*) are also taken occasionally. In these cases, killer whales opportunistically intercept individual deer and moose as they swim between coastal islands.

More than most other marine mammals, killer whales are social hunters, often working together to capture prey in a coordinated manner resembling that of pack-hunting social carnivores such as hyenas, wolves, and lions. Transients typically form slightly larger groups while hunting dolphins and porpoises, as compared to group sizes observed during pinniped attacks. Most hunts of small cetaceans have some component of chase, making more individuals necessary to prevent prey escape. Sometimes these high-speed chases result in a killer whale leaping free from the water with a dolphin or porpoise in its mouth. When dolphin prey are assembled in relatively large schools, killer whales often attempt to separate one or a few individuals from the group before commencing active pursuit. Once a prey item becomes exhausted, killer whales then attempt to kill the animal by breaching onto it, ramming it from below, tossing it into the air, or grasping it in their teeth (Fig. 4).

A variety of dolphins and porpoises are hunted by killer whales. Off New Zealand, common dolphins (*Delphinus delphis*) are attacked most commonly, but bottlenose dolphins and dusky dolphins (*Lagenorhynchus obscurus*) are also hunted. Stomach content analysis of a stranded killer whale off southern Brazil found the remains of three franciscana dolphins. In

Figure 4 *Killer whale attacking a Dall's porpoise. Photo by Robin W. Baird.*

the Gulf of Mexico, a pod of killer whales chased and killed a pantropical spotted dolphin, and Pacific white-sided dolphins (*Lagenorhynchus obliquidens*), Dall's porpoise (*Phocoenoides dalli*), and harbor porpoises are some of the more commonly hunted small cetaceans off the west coast of North America. In addition to these relatively small cetaceans, larger prey, including northern bottlenose whales (*Hyperoodon ampullatus*) and long-finned pilot whales (*Globicephala melas*), are also hunted occasionally by killer whales. In Arctic waters, killer whales sometimes herd beluga whales into shallow inlets and creek openings where they then rush into the group to capture young animals. Further, killer whales have been seen feeding on beluga whales and narwhals in open waters and on animals trapped by sea ice.

An interesting study of contrasts exists for transient killer whales off British Columbia and in Prince William Sound, Alaska. Although transients in both regions feed exclusively on marine mammals, harbor seals are the most common prey item of whales off British Columbia, whereas transients in Prince William Sound prey about equally on harbor seals and Dall's porpoises. Low harbor seal abundance in Prince William Sound may account for the apparent preference for porpoise prey in this region.

Although killer whales tend to focus their predatory attentions on pinnipeds and small odontocetes, numerous reports of attacks on sperm whales have also been recorded. In most cases, the sperm whale groups being attacked contained one or more calves. Sperm whales are likely to be difficult for killer whales to kill, as they are excellent deep divers and can escape predation by descending to depth, possess sizable teeth capable of inflicting significant injuries, and actively defend group members when threatened. Regardless of the difficulty in hunting sperm whales, field observations from the Pacific noted killer whales successfully killing at least one adult member of a sperm whale group and fatally injuring at least several others.

Killer whales have been noted to hunt all of the mysticete species except for pygmy right whales (*Caperea marginata*), but observations of attacks on baleen whales are not common. As is true for sperm whales, baleen whales are also difficult to kill, requiring extended effort and coordination between pod members. A typical strategy employed by killer whales during large whale hunts consists of first fatiguing the prey by active pursuit, followed then by delivery of a debilitating attack. It has been suggested that attacking killer whales may grasp large whales by the flukes and pectoral flippers in an attempt to slow or stop their movement or perhaps drown their prey by pulling them underwater.

In the Gulf of California, researchers watched from a small airplane as a group of 15 killer whales attacked and killed a Bryde's whale (*Balaenoptera edeni*). During this event, the killer whales repeatedly swam onto the back and head of the Bryde's whale, a behavior speculated to be useful in hindering life-sustaining respiration of the animal under attack. A similar incident was recorded off British Columbia where killer whales were observed to exhaust and kill a fleeing minke whale (*Balaenoptera acutorostrata*). Humpback whales are also attacked by killer whales, but unlike the more passive escape tactics employed by some of the other mysticetes, humpbacks defend themselves aggressively from killer whales by thrashing at them with their tail flukes and flippers.

Of all the mysticetes, gray whales (*Eschrichtius robustus*) are probably most frequently attacked by killer whales. On an almost predictable basis, gray whales are attacked by "California" killer whales in Monterey Bay each April and May (Fig. 5). Young calves making their first northward migration are particularly vulnerable, even while under the watchful eye of their mothers. Records from beach cast gray whales along the coast of the Chukchi Sea show a similar pattern to that observed off California; whales with the highest incidence of killer whale-induced injuries (i.e., tooth scarring) were generally under 10 m long, suggesting that killer whales in this region also select young gray whales as their primary predatory target.

Figure 5 *Killer whale attack on California gray whales. Photo by Sue Flood/BBC Natural History Unit.*

Figure 6 *Young western gray whale with evident killer whale tooth scarring. Photo by David W. Weller.*

Direct observations of killer whale attacks on large whales are relatively few, but several lines of evidence suggest that predatory interactions may occur more often than suspected. The presence of killer whale tooth rakes on the bodies, flippers, and flukes of many large whales can reach remarkably high proportions (Fig. 6). Photoidentification studies on humpback whales off Newfoundland and Labrador in the north Atlantic found that 33% of the individuals identified had killer whale-inflicted tooth rakes on their bodies. Scars on the flukes of 20–33% of humpback whale calves suggest that predation may be focused on young animals. A similar pattern has also been observed for western gray whales in the Sea of Okhotsk, where nearly 34% of all whales photoidentified possess killer whale tooth rakes. In this case, the western gray whale is highly endangered, making any level of killer whale predation a potentially important source of mortality. Bowhead whales (*Balaena mysticetus*) from the Bering, Chukchi, and Beaufort Sea populations have relatively low rates of killer whale tooth scarring, ranging from about 4 to 8% of the observed individuals. In contrast, 31% of bowhead whales in the Davis Strait population show evidence of scars from killer whales.

The relatively high incidence of killer whale tooth scarring on some regional populations of large whales suggests that predatory attempts are probably more regular than indicated by field observations alone and that many attacks are unsuccessful. Tooth rakes may not be truly indicative of predation attempts by killer whales, however, but may instead represent capture practice or instruction of predatory techniques for younger members of the pod. Finally, rake marks may also result from killer whales testing large whales to assess the presence of particularly vulnerable individuals that may be separated easily from a group and killed.

Although killer whales exert considerable time and energy in pursuit and capture of large whales, they consume relatively little of their victims. Reports from whaling ship logbooks and more recent field observations suggest that killer whales often preferentially consume only the tongue, lips, and portions of the ventrum of large whales before abandoning them. This phenomenon is little understood and stands in stark contrast to the behavior of terrestrial predators that consume all or most of their mammalian prey.

E. Humans

A review of predation on marine mammals would be incomplete without some mention of humans as predators. No other predator has the ability to harvest marine mammals at the same rate or intensity as humans. While killer whales or polar bears may take tens of animals over relatively short periods of time, humans are capable of sometimes killing hundreds of individuals within hours. Although the ecology of the world's oceans is in part maintained by predator–prey interactions, human exploitation of marine mammal populations can have devastating consequences.

See Also the Following Articles

Feeding Strategies and Tactics ■ Hunting of Marine Mammals ■ Killer Whale ■ Leopard Seal ■ Polar Bear ■ Walrus

References

Baird, R. W., and Dill, L. M. (1995). Occurrence and behaviour of transient killer whales: Seasonal and pod-specific variability, foraging behaviour, and prey handling. *Can. J. Zool.* **73,** 1300–1311.

Ford, K. B., and Ellis, G. M. (1999). "Transients: Mammal-Hunting Killer Whales of British Columbia, Washington, and Southeastern Alaska." University of Washington Press, Seattle, WA.

Guinet, C. (1991). Intentional stranding apprenticeship and social play in killer whales (*Orcinus orca*). *Can. J. Zool.* **69,** 2712–2716.

Hammill, M. O., and Smith, T. G. (1991). The role of predation in the ecology of the ringed seal in Barrow Strait, Northwest Territories, Canada. *Mar. Mamm. Sci.* **7,** 123–135.

Jefferson, T. A., Stacey, P. J., and Baird, R. W. (1991). A review of killer whale interactions with other marine mammals: Predation to co-existence. *Mamm. Rev.* **21,** 151–180.

King, J. E. (1983). "Seals of the World," 2nd Ed. Cornell Univ. Press, New York.

Leatherwood, S., and Reeves, R. R. (1983). "The Sierra Club Handbook of Whales and Dolphins." Sierra Club Books, California.

Lopez, J. C., and Lopez, D. (1985). Killer whales (*Orcinus orca*) of Patagonia, and their behavior of intentional stranding while hunting nearshore. *J. Mammol.* **66,** 181–183.

Mann, J., Connor, R. C., Tyack, P. L., and Whitehead, H. (2000). "Cetacean Societies: Field Studies of Dolphins and Whales." University of Chicago Press, Chicago.

McCosker, J. E. (1985). White shark attack behavior: Observations of and speculations about predator and prey strategies. *South. Californ. Acad. Sci. Mem.* **9,** 123–135.

Norris, K. S., Würsig, B., Wells, R. S., and Würsig, M. (1994). "The Hawaiian Spinner Dolphin. University of California Press, California.

Reeves, R. R., Stewart, B. S., and Leatherwood, S. (1992). "The Sierra Club Handbook of Seals and Sirenians." Sierra Club Books, California.

Reidman, M. (1990). "The Pinnipeds: Seals, Seal Lions, and Walruses." University of California Press, California.

Reynolds III, J. E., and Rommel, S. A. (1999). "Biology of Marine Mammals." Smithsonian Institution Press, Washington, DC.

Rice, D. W. (1998). "Marine Mammals of the World: Systematics and Distribution." Special Publication 4, Society for Marine Mammalogy.

Ridgway, S. H., and Harrison, R. (1999). "Handbook of Marine Mammals," Vol. 6. Academic Press, San Diego.

Predator–Prey Relationships

ANDREW W. TRITES
University of British Columbia, Vancouver, Canada

Most marine mammals are predators, but some are also preyed upon by other species. Theoretically, the interaction between marine mammals and their prey influences the structure and dynamics of marine ecosystems. Similarly, predators and prey have shaped each other's behaviors, physiologies, morphologies, and life history strategies. However, there is little empirical evidence of these influences due to the relative scale and complexity of marine ecosystems and the inherent difficulties of observing and documenting marine mammal predator–prey interactions.

I. Evolutionary Time Scales

Predator–prey relationships have been likened to an evolutionary arms race—the prey become more difficult to capture and eat, while the predators perfect their abilities to catch and kill their prey. Just how strong these selective forces are probably depends on the strength of the interactions between the predators and their prey.

As predators, marine mammals feed primarily upon fish, invertebrates, or zooplankton, which in turn feed primarily upon other species of fish, invertebrates, zooplankton, and phytoplankton (Fig. 1). To capture their prey, marine mammals have evolved special sensory abilities (e.g., vision and hearing), morphologies (e.g., dentition), and physiologies (e.g., diving and breath-holding abilities). They have also evolved specialized strategies to capture prey, such as cooperation to corral fish, or the production of curtains of air bubbles used by humpback whales (*Megaptera novaeangliae*) to capture herring. Marine mammals have also evolved specialized feeding behaviors to capture prey that move diurnally up and down the water column or to capture prey that move seasonally across broad geographic ranges. This in turn has likely influenced the life history strategies of marine mammals and their prey. For example, baleen whales feed for about 6 months when plankton are abundant and concentrated in shallow water, and then fast for the remainder of the year when the plankton are too dispersed to make them worth finding.

As prey, marine mammals have had to escape aquatic and terrestrial predators. Some species of pinnipeds for example, are particularly vulnerable to predation by bears and wolves while on land, and to predation by killer whales (*Orcinus orca*) and sharks while in the water. Thus some species of pinnipeds can reduce their risk of being eaten by aquatic predators by hauling out and resting onshore. Similarly, species such as Steller sea lions (*Eumetopias jubatus*) and northern fur seals (*Callorhinus ursinus*) reduce their risk of being eaten by terrestrial predators by breeding and hauling out on offshore rocks and islands where terrestrial predators are absent. Other species, such as ringed seals (*Pusa hispida*), give birth in cav-

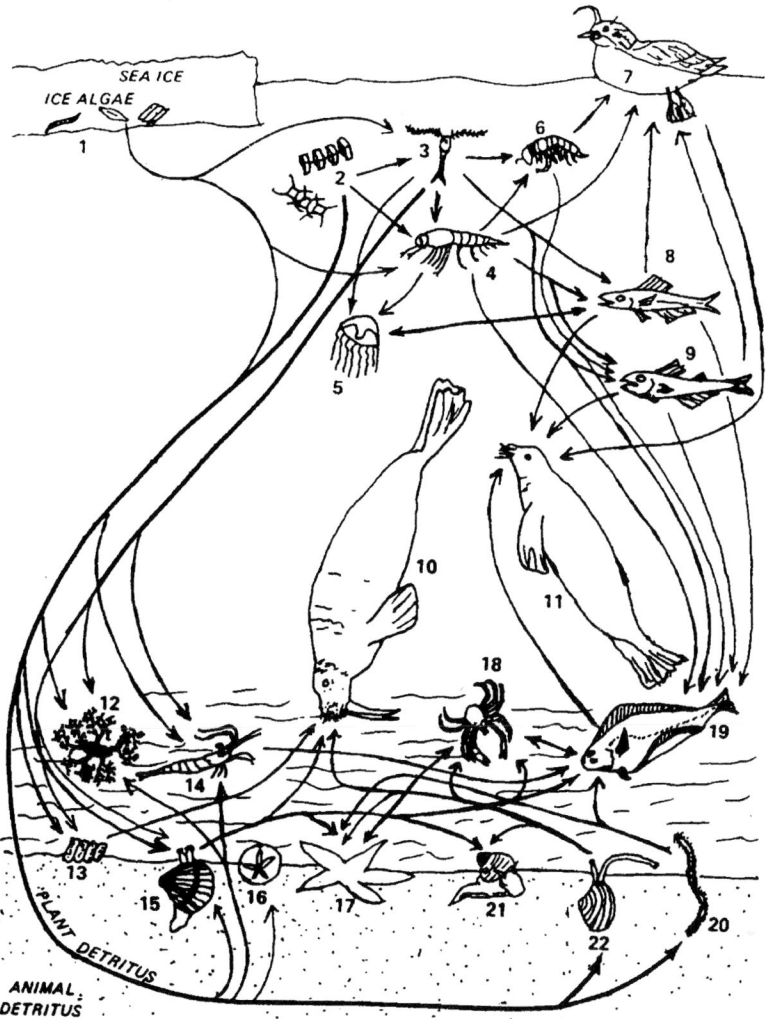

Figure 1 *A simplified depiction of the Bering Sea food web: (1) ice algae; (2) phytoplankton; (3) copepods; (4) mysids and euphausiids; (5) medusae; (6) hyperid amphipods; (7) seabirds; (8,9) pelagic fishes; (10) walrus; (11) seals; (12) basket stars; (13) ascideans; (14) shrimps; (15) filter-feeding bivalves; (16) sand dollars; (17) sea stars; (18) crabs; (19) bottom-feeding fishes; (20) polychaetes; (21) predatory gastropods; and (22) deposit-feeding bivalves. From McConnaughey and McRoy (1976).*

erns formed between ice and snow to avoid predation by polar bears (*Ursus maritimus*).

Fish and other cold-blooded species of prey have evolved a number of strategies to increase their chances of survival. One is cryptic countershading that enables fish to blend in with the bottom when viewed from above, and avoid detection when seen from below against a bright sea surface. Many species of fish, invertebrates, and zooplankton take refuge from predators in the deep, dark waters during the day and move toward the surface to feed under the cover of night. Another strategy evoked by the prey of marine mammals is predator swamping, such as large aggregations of spawning salmon and herring that reduce the numerical effect of predators on their prey populations. Schooling is another antipredator behavior that creates confusion through the sheer volume of stimuli from a fleeing school, making it difficult for a marine mammal to actively select and maintain pursuit of single individuals. Scattering and fleeing is yet another option to reduce predation and is used by some prey when attacked by bulk feeders such as baleen whales (e.g., humpback whales and capelin). The line between feeding and fleeing is undoubtedly fine for species of prey and must be continually evaluated by prey to minimize vulnerability to predation.

Marine mammals may also have indirectly influenced the evolution of nontargeted species in their ecosystems by consuming the predators of these species. The best example of this is the apparent influence of sea otters (*Enhydra lutris*) on kelp and other marine algae. Most species of marine algae use secondary metabolites to defend against herbivores. However, marine algae in the North Pacific have lower levels of chemical defenses where sea otters occur compared to algae species inhabiting the southern oceans where sea otters are not present. Sea otter predation on sea urchins and other herbivores may have removed selective pressure for species of marine algae to defend themselves against herbivores. Because secondary metabolites are expensive to produce, this may have allowed algae, like kelp, to radiate and diversify without the added cost of evolving and producing antigrazer compounds.

II. Ecological Time Scales

On a shorter time scale than the evolutionary one, predators and prey can directly affect the relative abundance of each other, or they can indirectly affect the abundance of other species. Their interaction may also affect the physical complexity of the marine environment.

Predation by sea otters on sea urchins is probably the best example of how marine mammals can alter ecosystem structure and dynamics. Sea otters were hunted to near extinction in the late 1800s throughout their North Pacific range. Without predation, urchin populations grew unchecked and overgrazed the fleshy algae. Kelp did not replace the underwater barrens until reintroduced sea otters once again began preying upon sea urchins.

Primary production has been estimated to be three times higher in areas where sea otters are present compared to those areas where sea otters are absent, allowing those organisms that feed upon primary production to grow faster and attain larger sizes (e.g., mussels and BARNACLES). The increase in primary production may even alter settlement patterns of invertebrates. The kelp also provides habitat for fish and suspension-feeding invertebrates to spawn, grow, and flourish. It can also change water motion and reduce onshore erosion and may even block the shoreward movement of barnacle larvae. Thus a top predator such as the sea otter can change the structure and dynamics of marine ecosystems.

Gray whales (*Eschrichtius robustus*) and walruses (*Odobenus rosmarus*) are other species of marine mammals whose foraging behavior can also affect community structure. For example, gray whales turn over an estimated 9–27% of the bottom substrate each year in the Bering Sea. The feeding pits created by gray whales draw 2–30 times more scavengers and other invertebrates compared to adjacent sediments. The disturbed sediments may also help maintain the high abundance of gray whale prey and other early colonizing species. Similarly, walruses turn over bottom substrate in their search for clams and other bivalves. There is some evidence that they may feed selectively on certain size classes and certain species and that their defecation may result in the redistribution of sediment. Thus, the interaction of benthic feeding marine mammals with their prey can result in food for scavengers and habitat for other species.

Interactions between predators and prey also influence the shapes of their respective life tables (i.e., age-specific survival and pregnancy rates). In Quebec, Canada, for example, there are a number of freshwater lakes that are home to land-locked harbor seals (*Phoca vitulina*). Studies have found that the trout in these lakes are younger, grow faster, attain smaller sizes, and spawn at younger ages compared to adjacent lakes without seals. As for marine mammals, they typically have elevated mortality rates during their first few years of life. This is likely due to a number of factors, including their relative vulnerability to predators and their inexperience at capturing prey and securing optimum nutrition.

In the Gulf of Alaska and Bering Sea, killer whales have been implicated as a contributing factor, but not the main one, in the decline of Steller sea lions and harbor seals through the 1980s. Field observations along the Aleutian Islands indicate that these population declines were followed by a decline of sea otters in the 1990s and that this decline was caused by killer whale predation. Killer whales may have begun supplementing their diet with sea otters because they could not sustain themselves on the low numbers of remaining seals and sea lions. It is not yet clear what ultimately caused the decline of Steller sea lions and began this spiraling change of events. However, it is apparent from mathematical calculations of population sizes and energetic requirements that there are sufficient numbers of killer whales in Alaska to prevent the recovery of pinniped populations. Thus, it is conceivable that populations of pinnipeds and otters may not recover to former levels of abundance until the predation by killer whales is reduced by a reduction in killer whale numbers or by a shift in killer whale diet to other species of mammals such as dolphins and porpoises.

In addition to directly affecting the abundance of their prey, marine mammals can indirectly affect the abundance of other species by outcompeting them or by consuming species that prey upon them. A case in point are harbor seals in British Columbia whose diet was about 4% salmon and 43% hake in the 1980s. Contrary to popular opinion, the harbor seals were likely benefiting salmon because they affected the abundance of hake, a species of fish that is one of the largest predators of salmon smolts. Further north in Alaska's Copper River Delta, harbor seals were culled in the 1960s to reduce the predation on salmon. However, the immediate result of the cull was not an increased number of salmon caught, but a decrease and failure of the razor clam fishery. It turned out that the seals were primarily eating starry flounder, which fed on the razor clams. Without the seals, the predatory flounder population grew unchecked.

In the Antarctic, commercial whaling systematically removed over 84% of the baleen whales and freed an estimated 150 million tons of krill for other predators to consume each year. Species such as crabeater seals (*Lobodon carcinophaga*), Antarctic fur seals (*Arctocephalus gazella*), leopard seals (*Hydrurga leptonyx*) and penguins (chinstrap, Adelie, and macaroni) increased and moved the Antarctic marine ecosystem to new equilibrium levels. Increases were also observed in minke whales (*Balaenoptera bonaerensis*) and squid-eating king penguins due perhaps to reductions in the respective abundance of blue whales (*B. musculus*) and sperm whales (*Physeter macrocephalus*). All of these species appear to have directly benefited from an increase in prey, which was caused by the removal of whales. Penguins and seals may now be hindering the recovery of baleen whale stocks in the Antarctic.

Marine mammals are generally considered to be opportunistic foragers who select from a number of alternative prey according to availability. This is based on the relatively large number of different species that have been reported in the stomachs and feces of marine mammals. Steller sea lions, for example, are known to eat over 50 different species of fish, and even the occasional seabird. However, their diets are typically dominated by 5 or fewer species, suggesting that they may not be truly opportunistic feeders. Little is yet known about the choices that marine mammals make when foraging. Presumably what marine mammals eat is a function of nutritional value, ease of capture, and digestibility, all of which are invariably linked to the abundance of both predators and prey. These are complex biological interactions about which little is known.

Functional response curves represent rates of predation in relation to the density of prey. In most species, the rate of capture rises with the density of prey to some maximum level. These relationships between prey density and predation rates tend to be nonlinear and asymptotic, indicating that there are maximum limits to the rate that predators can capture and process prey, which are independent of prey population size. Establishing these functional relationships for different species of prey is fundamental to fully understanding the foraging ecology of marine mammals. This has not yet been done for marine mammals and will require experimentation in captivity or observational studies in the wild using electronic data collection techniques.

Ecosystem models are another technique for gaining insight into the effects of predator–prey relationships on ecosystem dynamics and structure. Using a series of mathematical equations to account for the flow of energy from one group of species to another, the models can estimate the extent of competition between species and the effect that changes in abundance of one species will have on other species in the ecosystem. One such model was constructed for the Bering Sea to understand whether the declines of Steller sea lions and forage fishes (such as herring) and the increases in pollock and flatfish between the 1970s and the 1980s were related to the commercial removal of whales. Removing historic numbers of whales from the simulated ecosystem resulted in an increase in numbers of pollock. However, the increase was only in the order of 10–20%, not the 400% increase believed to have actually occurred. The ecosystem model suggests that the Bering Sea may exist in two alternative states (consisting of two different complexes of species) and that environmental shifts (from periods of cold to warm water years) may ultimately determine when and for how long these shifts occur. The model also suggests that curtailing fishing on pollock (a major prey of Steller sea lions) may affect the Steller sea lion negatively. The explanation for this counterintuitive prediction is that commercial fisheries primarily remove larger pollock than Steller sea lions consume. Given that pollock are cannibalistic, increasing the size of the adult stock results in the increased predation of younger pollock, leaving fewer fish for Steller sea lions to consume. Thus, ecosystem models are useful tools for exploring the influence of predator–prey interactions on one another and on other components of their ecosystems.

III. Synthesis

Marine mammal predator–prey interactions occur over different spatial and temporal scales, making it difficult to empirically decipher the influences they have on one another and on their ecosystems. However, their coexistence suggests that marine mammal predators and their prey have had profound influences on each other's behaviors, physiologies, morphologies, and life history strategies. The diversity of niches filled by marine mammals makes it difficult to generalize about the evolutionary consequences of their interactions with prey, beyond stating the obvious: marine mammals have adapted to catch food, while their prey have adapted to avoid being caught.

On the shorter ecological time scale, marine mammals can affect the abundance of other species by consuming or outcompeting them. They can also indirectly affect the abundance of nontargeted species by consuming one of their predators, and can have strong impacts on the overall dynamics and structure of their ecosystems. One of the best tools for understanding marine mammal predator–prey interactions is the ecosystem model. However, more work is required through experimental manipulations and observational studies to evaluate the choices made by marine mammals and the costs of obtaining different species of prey.

See Also the Following Articles

Feeding Strategies and Tactics ▪ Hearing ▪ Predation on Marine Mammals ▪ Vision

References

Bowen, W. D. (1997). Role of marine mammals in aquatic ecosystems. *Mar. Ecol. Prog. Ser.* **158,** 267–274.

Estes, J. A. (1996). The influence of large, mobile predators in aquatic food webs: Examples from sea otters and kelp forests. *In* "Aquatic Predators and Their Prey" (S. P. R. Greenstreet and M. L. Tasker, eds.), pp. 65–72. Fishing News Books, Oxford.

Estes, J. A., and Duggins, D. O. (1995). Sea otters and kelp forests in Alaska: Generality and variation in a community ecological paradigm. *Ecol. Monogr.* **65,** 75–100.

Knox, G. A. (1994). "The Biology of the Southern Ocean." Cambridge Univ. Press, Cambridge.

Laws, R. M. (1985). The ecology of the Southern Ocean. *Am. Sci.* **73,** 26–40.

McConnaughey, T., and McRoy, P. (1976). "Food-Web Structure and the Fraction of Carbon Isotopes in the Bering Sea," pp. 296–316. Science in Alaska 1976, Alaska Division of AAAS.

Taylor, R. J. (1984). "Predation" (M. B. Usher and M. L. Rosenzweig, eds.). Chapman and Hall, New York.

Trites, A. W. (1997). The role of pinnipeds in the ecosystem. *In* "Pinniped Populations, Eastern North Pacific: Status, Trends and Issues" (G. Stone, J. Goebel, and S. Webster, eds.), pp. 31–39. New England Aquarium, Conservation Department, Boston.

Trites, A. W., Livingston, P. A., Vasconcellos, M. C., Mackinson, S., Springer, A. M., and Pauly, D. (1999). "Ecosystem Change and the Decline of Marine Mammals in the Eastern Bering Sea: Testing the Ecosystem Shift and Commercial Whaling Hypotheses," Vol. 7(1). Fisheries Centre Research Reports 1999.

Prenatal Development in Cetaceans

JOY S. REIDENBERG AND JEFFREY T. LAITMAN
Mount Sinai School of Medicine,
New York, New York

Very little is known about the specifics of intrauterine growth and development in cetaceans. Indeed, the precise time intervals of such development, the basic genetic determiners, and any distinctive growth trajectories are basically unknown. What is known about cetacean prenatal development is that, as they are mammals, it is to be expected that the same basic stages of early cell division, pattern formation, organogenesis, and growth and differentiation will also be similar. For example, the "embryonic" period is usually defined as the time frame within which an animal's body plan and its organs and organ systems (i.e., integument, skeletal, muscular, nervous, circulatory, respiratory, digestive, urinary and reproductive) are established. Once all organs form, the "fetal" period of growth and distinctive development commences. Cetacean prenatal development will similarly follow this course. It is also to be expected that the absolute time of these periods will differ from terrestrial species, between odontocetes and mysticetes, and among the different species therein.

Many studies that have noted aspects of cetacean prenatal development (most in passing rather than by detailed, systemic analysis) have used terms such as "embryo" or "fetus" in a seemingly imprecise manner. It is, indeed, often difficult to determine from these whether the authors (a) are cognizant of the biological difference between an embryo and a fetus (many reporters are not anatomists or embryologists) or (b) are able to discern differences that afford the distinction to be made with any degree of accuracy. Adding to this complexity is the fact that the precise gestation periods for many cetacean species are not known. In light of these observations, our use of the terms embryo and fetus (or embryonic and fetal periods) should be taken as representing approximate guides to stages of development rather than as a precise descriptor of an absolute time frame.

It is important to remember in discussing cetacean prenatal development that most current knowledge derives from observations on embryonic or fetal specimens discovered in pregnant cetaceans either found stranded or taken aboard whaling ships. In many cases, only a length or weight is recorded (if at all) with an occasional description of external appearance. It is usually impossible to distinguish the age of the specimen, as the date of conception and length of gestation cannot be known with any certainty. As most breeding and calving seasons are known, however, some approximations are available and have been provided (Figs. 1–6).

Figures 1–6 *Fetal long-finned pilot whales (Globicephala melas) obtained postmortem from pregnant, beach-stranded whales on the shores of Cape Cod, Massachusetts. The figures are arranged numerically in order of the specimens' lengths, which should mirror the order of their gestational ages. Each black or white square equals 1 cm on the rulers in Figs. 4 and 6.*

Figure 1 *A fresh specimen of a very small, unpigmented fetus. Based upon its external appearance, this appears to be a very early fetus, probably very close to the transition between embryonic and fetal periods of development. Note the prominent rostrum and the rudimentary development of a dorsal fin, tail flukes, and genital tubercle. The dark spot above the mouth and anterior to the eye appears to be the left nostril.*

Figure 2 *An early fetus preserved in alcohol. Due to dessication, the unpigmented fetal skin (which was pink) is now discolored and shrunken against the skeleton. The fetus is curled in the fetal position, with the tail folded laterally to the left side and the dorsal fin flattened against the body. Note the attached umbilical cord and the prominent rostrum. The dorsum of the skull is depressed (from the dehydration) at the membranous fontanelle. Just anterior to this depression is the blowhole (the paired nostrils appear to have fused into one blowhole and have migrated dorsally and near the midline).*

Figure 3 *A fresh fetus is being held in a pair of gloved hands. The fetus, discovered during an autopsy, is still attached by its umbilical cord to the mother whale and is partially enveloped in its thin layer of fetal membranes (note transparent membrane and vessels at ventrum of tail stock). It has some streaks of gray pigmentation, and the melon has begun to form over the rostrum.*

Figure 4 *A frozen fetus still curled in its fetal position. The skin is pigmented black, but it is covered with whitish-gray deposits of frost from thawing. Note the developing melon over the rostrum, curled tail flukes, and the skin folds (from the folded fetal position) along the concave surface of the right lateral side. The protruding penis can be seen as a conical structure with a hook-like curl at the tip.*

Figure 5 *A thawed, full-term fetus. The lighter patches are areas where the skin has sloughed off postmortem during the freezing. The fetus is curled in the fetal position as it was found in utero, with the tail folded against the left side. The umbilical cord is evident under the tail stock. The left lateral surface of the thorax and abdomen shows a number of dorsoventral stripes, which indicate grooves between puckers where the skin is folded (the skin is particularly crimped at the beginning of the tail stock). The dorsal fin lies flat against the body and the tail flukes are curled ventrally. Note how the well-developed melon now overrides the rostrum.*

Figure 6 *Another full-term fetus (thawed) curled in a slightly different fetal position as it was found in utero. Here the tail bends to the right side and the flukes lie over the dorsum and extend to the left side. The blowhole has migrated to the top of the head, and the bulbous form of the melon can be seen clearly. The dorsal fin, which was flattened along the right side, is being held up by a gloved hand.*

I. Development of Organ Systems

A. Integument and External Characteristics

1. Skin The overall coloring of the embryo appears light pink due to the transparency of the skin (integument) allowing the underlying tissues perfused with blood to be visible. The skin consists of the epidermis (which has four layers), dermis, and hypodermis, which increase in thickness throughout the embryonic period (Meyer *et al.*, 1995). Skin coloration begins during the early fetal period. In mysticetes, dark coloration occurs initially along the rostrum bordering the opening of the oral cavity. As the mysticete fetus grows, dark patches appear along the dorsum of the thorax and abdomen and on the pectoral flippers, tail flukes, and dorsal fin. The separate and irregularly shaped patches fuse and grow into a more uniform pattern (for many species, this is usually a countershaded pattern of dark dorsum and light ventrum that resembles the adult's coloration).

2. Hair Hairs can be found along the surfaces of the upper jaw. In odontocetes, hairs appear on the lateral aspect near the tip of the rostrum, whereas in mysticetes, they are found both laterally and dorsally on the broad rostrum. In some cetacean species, these hairs can also be found on the margins of the lower jaw. These hairs appear to have some tactile properties and may derive from the vibrissae of terrestrial mammals. While most odontocetes will lose these hairs shortly after birth (except perhaps platanistoids), they are retained into adulthood in some species of mysticetes.

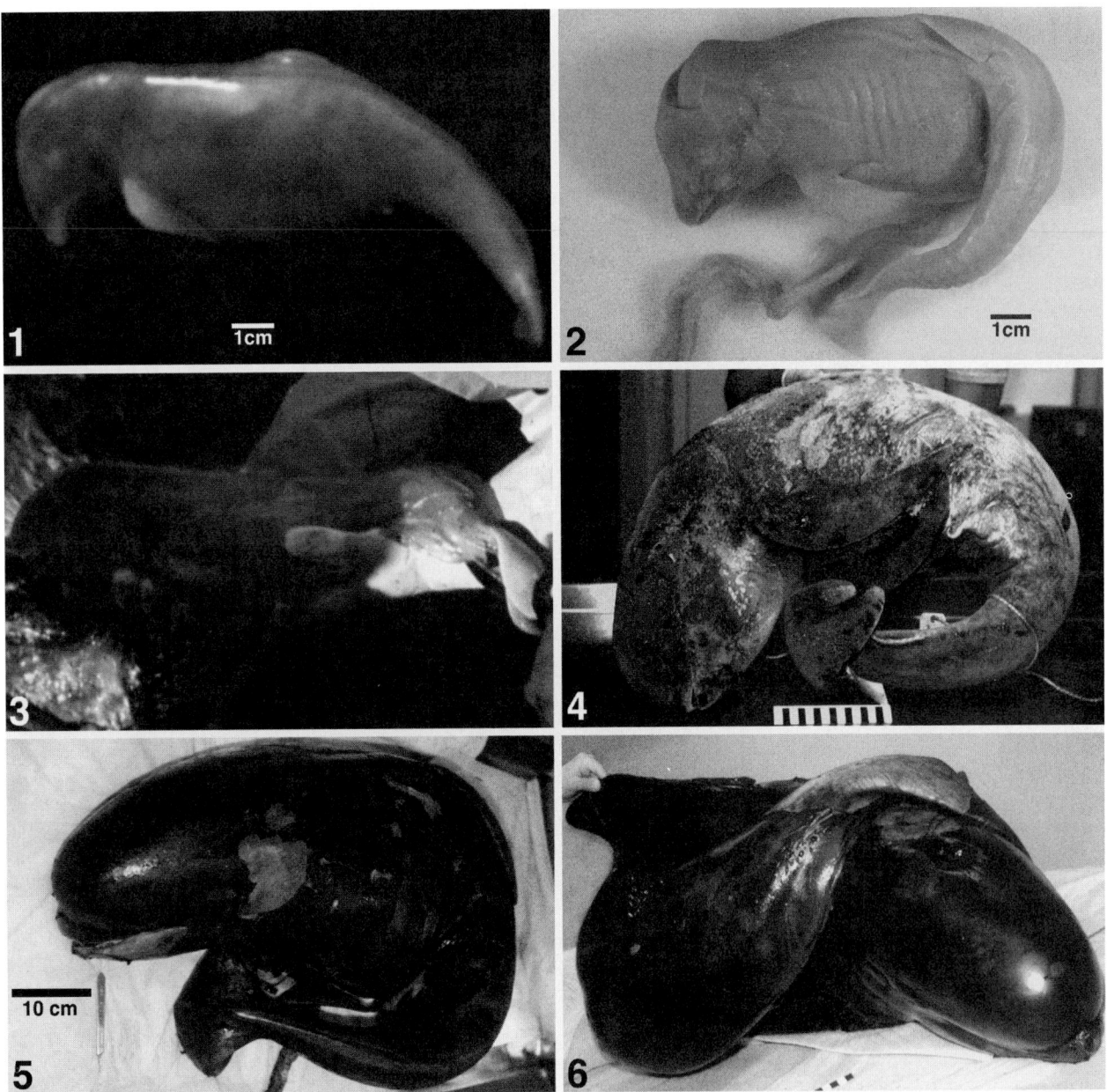

3. External Ears   External ears (pinnae) do not develop, thus maintaining a streamlined surface contour in the ear region of cetaceans. Only a remnant of the external auditory canal is visible as a small hole present in the skin behind the eye.

4. Mammary Glands   Mammary glands (mammae) are epidermal organs derived from modified sweat glands. In terrestrial mammals, and presumably cetaceans, the mammae develop along a mammary ridge (the "milk line"), which extends bilaterally from the axilla (where the forelimb joins the thorax) to the inguinal region (where the hindlimb joins the pelvis). The position of the mammae that eventually develop varies in different species: thoracic (e.g., primates, sirenians), thoracoabdominal (e.g., felids), thoracoinguinal (e.g., canids, suids), and inguinal (e.g., ungulates). Like their ungulate relatives, Cetacea only develop inguinal mammae. In females, the teats (nipples) of the mammae are internalized, being withdrawn into the mammary slits (which are positioned with one on either side of the genital slit). This internal location helps streamline the body contour and thus reduce drag during locomotion.

5. Genitoanal Slit and Contents   It is difficult to sex the cetacean embryo or early fetus. As the genital tubercle develops, however, it is directed cranially in males and caudally in females (Amasaki *et al.*, 1989b). While the penis/clitoris may be totally exposed in an earlier fetus, the external genitalia are usually not completely visible in the full-term fetus as they are withdrawn into the genitoanal slit. (Note that in a postmortem specimen, the tip of the penis usually protrudes through the slit due to relaxation of the retractor penis muscle.) The genitoanal slit opens into a common vestibule occupied caudally by the anus and rostrally by the urogenital openings. In males, the urethra is contained in the penis; in females, the clitoris and urethra are separate, and there is an opening for the vagina. In males, the genitoanal slit is elongated, reaching almost to the navel. In comparison, the genitoanal slit of females is very short, appearing only between the two mammary slits. Both males and females have a streamlined external shape, as the penis or clitoris is withdrawn into the genitoanal slit and there is no scrotum (testes are intraabdominal) or labia, thus further reducing drag during swimming.

B. Musculoskeletal System

1. Extremities   The forelimb extremities of whales are called pectoral flippers. Although cetaceans are derived from a quadrupedal ancestor, adult whales do not possess hind limbs. During the embryonic period, both fore and hind limb buds are present as paddle-shaped projections, with the forelimb developing before the hind limb (Amasaki *et al.*, 1989c). The rudimentary hind limb buds form skeletal element anlagens, vascular plexes, and nerves (Sedmera *et al.*, 1997a) but are completely absorbed by the fetal period. By birth, the only remaining vestige of the hind limb is a skeletal remnant of the femur embedded into the lateral body wall and a rudimentary pelvis that is not attached to the vertebrae. The forelimbs, however, continue developing during the embryonic and fetal periods. Early on, they assume the elongated shape of a typical mammalian arm and forearm, with grooves separating the dig-

its apparent toward the distal edge. The skin overlying the flippers matures faster than the skin over the trunk (Meyer *et al.*, 1995). The stalk-like arm and forearm foreshorten into one functional unit. The skeletal elements (humerus, radius, ulna, and carpal bones) lose their mobility at the elbow and wrist joints, maintaining flexibility only at the shoulder joint.

During the fetal period, the manus of the pectoral flipper fuses into a leaf shape (the distal portion never separates into individual digits, and the interdigital grooves disappear). Odontocete flippers contain five digits—a pattern reminiscent of a terrestrial ancestry. The number of digits within the flipper varies in mysticetes: members of the Balaenidae and Eschrichtiidae families retain all five digits, whereas rorquals (members of the family Balaenopteridae) have reduced that number to four. The tip of the flipper elongates in the caudal direction as differentiation of the phalangeal cartilages progresses proximodistally. Central digits exhibit hyperphalangia (or polyphalangia), i.e., the number of phalangeal elements expands beyond the maximum of three found in most terrestrial mammals (Calzada and Aguilar, 1996). The degree of hyperphalangia varies greatly among species. For example, the second and third digits of *Globicephala melas* have 14–15 and 11 phalanges, respectively, whereas there are seven elements in each of these two digits in *Stenella attenuata* and only five for each of these digits in *Physeter macrocephalus*. Expansion in the number of phalangeal elements, rather than in the lengths of the elements, probably helps support the elongated form of the flipper while retaining some small degree of flexibility that is reminiscent of fin function in fish. Hyperphalangy and elongated pectoral flipper form may also relate functionally to the increasing/decreasing aspect ratio (i.e., relationship between length and width), hydrodynamic form (streamlining effects), or locomotor function (limited to steering, braking, and lift in most species, but can include increased maneuverability or propulsion, e.g., humpback whales).

2. Tail   The tail flukes do not appear until the fetal period, after the hind limbs have regressed. The midline of the tail enlarges dorsally and ventrally in the vertical plane to form the slender and hydrodynamic tail stock. The number of caudal vertebrae may increase above that typically seen in terrestrial mammals (perhaps up to 24 in mysticetes and perhaps up to 48 in odontocetes, compared with up to 21 in ungulates). Note that the actual number of caudal vertebrae is difficult to determine with accuracy, as there are no clear anatomical landmarks to separate the caudal region from the lumbar region. The caudal tip develops two horizontal plates of tissue that do not contain any skeletal elements. These plates form the tail flukes. As the fetus nears full term, the tail flukes curl ventrally at their caudal tips so that they are directed rostromedially. This curling of the flukes makes the tail tip more compact and easier to present through the vagina during birth (see later).

3. Back, Dorsal Fin, and Ribs   At about the same time that the tail flukes appear, a bulge develops along the midline of the back in the region where the dorsal fin will form. The bulge shape is then modified to a species specific shape (e.g., falcate, triangular, rounded, ridge). When sexual dimorphism in fin height is seen (e.g., *Orcinus*), it does not occur prenatally. The vertebrae of the back unfold from the embryonic curvature (ventrally concave) to a horizontally aligned column in the early fetal period. In the late fetal period, however, the growing fetus folds again, only this time the body curves laterally. This flexibility may be possible, in part, due to the lack of a sacrum and lengthening of the vertebral column. There are additional lumbar vertebrae in most cetaceans (perhaps up to 15 in mysticetes and perhaps up to 29 in odontocetes) compared with the usual six of ungulates or five of humans. Again, this number is difficult to determine with accuracy, as there is no sacrum or pelvis, and rib articulations can vary. As the side of the fetal head approximates the tail, the dorsal fin folds flat against the concave side of the body. Dorsal fin folding facilitates vaginal delivery (see later).

The ribs of odontocetes are hinged along the lateral aspect, giving each rib two osseous elements joined by a synovial joint. Postnatally, this will facilitate thoracic cavity collapse during diving (as pressure increases with depth, the volume of air in the lungs will decrease).

4. Head (Position and Shape) and Neck The large embryonic head lies in the typical mammalian pose with the face directed ventrally at 90° to the long axis of the body. The maxillary and mandibular regions form a ventrally projecting, conical rostrum that curves slightly caudally. This projection resembles a parrot's beak, being rather thick at the base. In the early fetal period, the rostrum elongates, particularly in long-beaked species (e.g., *Stenella longirostris, Platanista gangetica*). In the midfetal period, the head and neck junction straightens into the adult position, aligned horizontally with the body. The neck region shortens and stiffens, and in many species (e.g., *Globicephala macrorhynchus*) most of all seven cervical vertebrae become compressed craniocaudally and fuse together (Ogden *et al.*, 1981). This enables a smoother transition in form between the head and the thorax, and a midline head position relative to the body's longitudinal axis. The shortened neck enhances streamlining, and fusion of cervical vertebrae improves head stability during locomotion. Vertebral fusion limits lateral or rotational head motion, leaving only dorsocaudal head movements (which help begin the propulsive body wave) at the large joint between the first cervical vertebra and the skull's enlarged occipital condyles.

5. Hyoid Apparatus The hyoid apparatus is derived from the second and third branchial arches. The single basihyal and paired thyrohyals form the large "U"-shaped plate to which the muscles of the tongue, larynx, and sternum attach, and the paired epihyals, ceratohyals, stylohyals, and tympanohyals form the osseous chains bilaterally connecting the basihyal with the skull (Reidenberg and Laitman, 1994).

6. Skull (Mandible, Ear Ossicles, and Cranium) The mandible (jaw) forms around a cartilaginous precursor (Meckel's cartilage) derived from the first branchial arch. The cetacean mandible is largely composed of a horizontal body, with very little (if any) vertical projection forming the ascending ramus. The condylar process is short, and the condylar

head may appear to rest directly superior to the caudal portion of the mandibular body. In many odontocetes, the condylar head migrates with fetal development to the caudal aspect of the mandible, whereas in some mysticetes, the condylar head occupies the dorsocaudal edge of the mandible. While the ascending ramus develops most of its vertical height postnatally in many terrestrial mammals, the ascending ramus of cetaceans remains practically nonexistent through the adult stage.

The caudal portion of the first branchial arch contributes to the formation of the upper portions of the first two ear ossicles (malleus and incus). The caudal portion of the second branchial arch contributes to the lower portion of these same two ear ossicles as well as the body of the third ossicle (stapes, except for the foot plate, which derives from the otic placode).

In terrestrial mammals, and presumably cetaceans, the SKULL is derived from two types of bone: chondrocranial (that which preforms in cartilage and then ossifies) and desmocranial (that which does not form a cartilaginous stage, but rather ossifies directly in mesenchyme). The portion preformed in cartilage (the skull base) tends to be less plastic in its shape than that which ossifies from membranes (the cranial vault). Cetaceans appears to be no exception to this rule. In fact, they are an excellent example of the plasticity of the cranial vault, as this region is grossly modified compared with terrestrial mammals.

In the fetal period, the elements of the cranial vault begin to shift their relative positions so that the maxilla approximates or meets the occipital dorsally. This process of bony overlapping (called "telescoping") creates a layered appearance to the skull, where portions of bone are buried on the inner surface. In odontocetes, the parietals are depressed laterally and the premaxillary and maxillary bones overlap the frontal bone dorsally, whereas in mysticetes, the premaxilla slides over the frontal and the base of the maxilla moves under the frontal bone. The cranial vault thus changes shape from dolichocephalic (longer than wide) to brachycephalic (wider than long). The ear ossicles begin to rotate into their adult position during the early fetal period. No paranasal air sinuses (i.e., maxillary, ethmoid, sphenoid, or frontal) form within the skull either prenatally or postnatally (a diving adaptation that prevents injury from expanding/contracting the volume of an enclosed air space during depth/pressure changes). The bony nares migrate caudally to the near adult position on the dorsum of the head.

C. Respiratory Tract

1. Nasal Tract Structures of the nasal region form in the early fetal period, but asymmetry is not detected yet. Nasal conchae (bony plates that project from the nasal septum and walls in terrestrial mammals) never form. The nasal plugs (the tissues that close off the airway) are present and may derive from the tissue that forms the upper lip in terrestrial mammals. The odontocete melon, which may also derive from this same tissue, has not yet formed the characteristic bulge in the forehead region. The nasal air sacs, diverticulae of the nasal tract, begin to bud off the soft tissues of the nasal passageways.

The nasal apertures, which appear initially on the dorsum of the rostrum, begin to migrate caudally toward the adult position at the top of the head. They can be found near the junction of the rostrum and the swelling containing the forebrain.

The nasal skull grossly transforms so that the nasal floor projects ahead of the nasal passageways into the rostrum, whereas the lateral parts (which forms the walls of the nasal passages) shift from horizontal to vertical (Klima, 1999). There are two separate bony nasal passageways in all cetaceans. The soft tissues above the skull that surround the nasal passageways are maintained as two separate tubes in mysticetes. In odontocetes, however, the two soft tissue passageways fuse near their exit at the skin into one common blowhole opening. There are further differences within odontocetes in the development of the nasal skull. *Phocoena phocoena* has the most posteriorly positioned nares, whereas *Physeter macrocephalus* has the most anteriorly positioned nares. There are additional specializations in *Physeter* related to the unique forehead containing the spermaceti organ, including asymmetrically sized and positioned narial openings in the skull and soft tissue pathways through the head.

2. Larynx   The larynx (voice box) forms from cartilage elements of the fourth through sixth branchial arches. Its position in cetaceans is similar to that found in terrestrial mammals. The front part (epiglottis) overlaps its ventral surface with the dorsal surface of the soft palate, creating a bridge to channel air from the nasal region into the trachea and lungs. In odontocetes, the larynx undergoes elongation of its rostral portion, forming a "goose beak" shape that inserts into the nasal region. The epiglottis elongates during the midfetal period. The posterior cartilages (corniculates) are still shorter than the epiglottis and will not reach their full proportions (i.e., exceed the epiglottis in height) until the fetus reaches full term. The laryngeal "goose beak" of odontocetes inserts into a muscular sphincter derived from the palatopharyngeal arch of the soft palate. Postnatally, this interlock will keep the rostral opening of the larynx connected with the posterior nasal cavity. This connection imparts circumferential protection from the digestive tract, allowing air to flow between the nasal region and the lungs for sound production while prey is swallowed whole underwater. Internally, the odontocete larynx develops a midline fold (which is bifurcated in *Kogia breviceps*) that appears homologous to the vocal fold of terrestrial mammals (Reidenberg and Laitman, 1988). The mysticete larynx more closely resembles that of terrestrial mammals, except that there is a large and muscular sac attached ventrally in the midline. In *Caperea marginata*, this sac lies on the right side.

3. Trachea and Lungs   Tracheal rings usually develop as "O"-shaped rings, unlike the "C"-shaped rings of most terrestrial mammals. A bronchus leading to the right lung develops above the carina (tracheal bifurcation). As this bronchus emerges directly from the trachea above the primary (main stem) bronchi, it is termed a "tracheal bronchus." A right tracheal bronchus is a feature also found in the closely related artiodactyls.

The lungs mature from the embryonic glandular stage to the fetal canalicular stage (for more information on stages of lung development, see Drabek and Kooyman, 1983). Next, muscular sphincters develop around the terminal bronchioles. Because this feature is not found in terrestrial mammals, it may be an adaptation for diving. The next phase of development is the alveolar stage. By the midfetal period, cartilaginous rings

develop in the terminal bronchioles. This is another feature not found in terrestrial mammals that may also be an adaptation for diving, as cartilage rings may keep the terminal airways patent under high pressures and during lung collapse at depth.

D. Digestive Tract

1. Teeth and Baleen   Late in the fetal period, both odontocetes and mysticetes form tooth buds. Odontocete teeth are single cusped and usually conical in shape with species-specific variations, e.g., narwhal, *Monodon monoceros* (long spiral tusk), *Platanista gangetica* (needle-shaped), beaked whales (flattened and broad), and porpoises (flattened, spade-shaped). Postnatally, teeth are used primarily for grasping and aggression. As there are no incisors, canines, or molars (odontocetes are "homodonts"—all teeth have the same shape), the task of breaking up food is passed onto the stomach (see later). The tooth buds of fetal mysticetes are sometimes multicusped, resembling the teeth of related terrestrial mammals (Slijper, 1979). The mysticete tooth buds are more numerous in the upper jaw than in the lower jaw, but all are usually resorbed before birth. The formation of rudimentary baleen plates, which occurs concurrently with tooth bud degradation, may be induced by the process of tooth bud resorption (Ishikawa and Amasaki, 1995). As the mysticete fetus grows, longitudinal baleen ridges form in the gums of the upper jaw. These longitudinal ridges develop transverse divisions and rows of papillae composed of epidermal folds that become cornified. The cornified papillae are tubular in shape and elongate and coalesce with their neighbors to form baleen plates (Slijper, 1979).

2. Throat Grooves   Throat grooves are a series of parallel, longitudinal folds found on the external, ventral surfaces of the head in rorqual mysticetes that enable expansion of the oral cavity. In other mysticetes and some odontocetes, a single pair of throat grooves can be found at the base of the jaw and may indicate expansion of this region during tongue and hyoid depression (see later). In rorqual mysticetes, throat grooves begin forming in the fetal period, appearing initially between the umbilicus and the pectoral flipper. A second set of ventral throat grooves appears next near the tip of the mandible. Toward the end of gestation, the two sets of throat grooves join to form one complete set running from the mandibular tip to the umbilicus. These throat grooves enable expansion of the floor of the mouth to engulf prey during feeding.

3. Tongue   In both mysticetes and odontocetes, lingual papillae develop along the lateral border of the tongue during the fetal period. Because newborn cetaceans lack lips, these papillae probably play an important role postnatally during nursing in grasping the teat, creating a seal for suction, and forming a channel for milk to flow into the oral cavity. These papillae attain maximal size in the early postnatal period of odontocetes, but can sometimes be found persisting in adults.

The tongues of mysticetes and odontocetes differ greatly in their construction, and this difference is evident in the fetal stage. Odontocete tongues are related more closely to the tongue of terrestrial mammals, being very muscular. Their tongues have large insertions on the broad bones of the hyoid apparatus (Reidenberg and Laitman, 1994). This arrangement helps depress the tongue into the throat like a piston, thereby creating enough negative pressure to draw in prey—a mechanism referred to as "suction feeding." The mysticete tongue (particularly in rorqual whales) is unusual in its structure because it can be flattened against the floor of the mouth and expanded laterally along with the throat pleats during prey engulfing. In addition, there is a fibrocartilage structure in the ventral throat region of rorqual whales that may be related to attachment of the mylohyoid muscle. This structure may aid jaw mechanics and support the tongue and floor of the mouth during expansion/contraction of the throat grooves.

4. Gastrointestinal Tract In mammals, gut development begins with a single, relatively straight gastrointestinal tube that is suspended in the midline of the coelom. As the embryo develops, the gut tube differentiates into the foregut and hindgut, and as each section further develops its specific shape, individual regions of the gut tube begin to rotate into different positions within the coelomic cavity. Toward the end of the embryonic period, the thoracoabdominal wall is distended. This is probably because the stomach is developing its multiple chambers and intestinal rotation is occurring. Cetaceans develop a multichambered stomach [e.g., see Tarpley *et al.* (1997) on the stomach of *Balaena mysticetus*], much like that found in closely related ruminant artiodactyls, the closest group of living land mammals to the cetaceans. The divisions of the cetacean stomach include, from proximal to distal: forestomach, main stomach, and pyloric stomach. As it does in ruminants, the cetacean forestomach arises from the stomach bud rather than the esophagus (Amasaki *et al.,* 1989a), but is not divided into the three small chambers (rumen, reticulum, psalterium) found in, for example, the cow. The size of the forestomach may be dependent on the consistency of the prey. In odontocetes, a large and muscular compartment may signify a function in breaking down whole fish or crustaceans, whereas a smaller compartment possibly relates to a diet of soft prey such as cephalopods. In mysticetes, the forestomach is smaller than the main stomach, perhaps relating to the relatively small size of their prey. The cetacean main stomach and pyloric stomach (which can have up to 12 chambers, e.g., beaked whales) are equivalent to the cow's single rennet stomach (Slijper, 1979).

The process of intestinal rotation probably resembles that of other mammals, involving temporary herniation (protrusion) into the umbilicus, rotation and folding, and then return of the contents back to the abdomen where they lie more compactly. Thus, by the early fetal period the abdomen is no longer distended by the process of intestinal rotation. The cecum and large intestines then differentiate further, developing circular folds that divide the intestines into multiple connected chambers resembling the haustra (sacculations) of terrestrial mammals (Amasaki *et al.,* 1989a).

E. Genitourinary System

1. Urinary Tract In embryos of terrestrial mammals, and presumably cetaceans, the earliest kidney is the mesonephros, composed of ducts and tubules. The embryonic metanephric duct, which buds off the mesonephric duct, becomes the ureter. As the mesonephros regresses, a second kidney structure, the metanephros, develops around the metanephric duct and is retained as the final kidney.

The fetal kidney develops as a cluster of many small and relatively independent kidney units called renicules, which will be retained in the cetacean adult. An adult kidney divided into renicules or lobes is not unusual in mammals (e.g., ox, otter) and may indicate persistence of the fetal condition. The apparent functional advantage of a kidney divided into renicules in large mammals appears to be related to a maximum size for the length of the tubules, which might otherwise be too long for proper function in a large single kidney.

The urogenital sinus (derived from the embryonic cloaca) becomes the urethra. The urinary bladder develops from the proximal portion of the allantois.

2. Reproductive Tract The remnants of the mesonephric duct become the efferent ductules, epididymis, and deferent duct for sperm transport in males. The gonads (ovaries and testes) develop from gonadal ridges, which are paired thickenings of the coelomic epithelium. In females, paramesonephric (Mullerian) ducts develop simultaneously with mesonephric ducts. Paramesonephric ducts become the bicornuate uterus and oviducts in females, but degenerate in males except for the prostatic sinus.

In males, the testes are intraabdominal, i.e., they do not descend as in most terrestrial mammals, and thus there is no scrotal sac. The internalization of these structures helps streamline the body shape, thus reducing drag during swimming. Interestingly, cetaceans develop a gubernaculum (which functions in testicular descent in terrestrial mammals), but do not develop the peritoneal outpocketing that occurs with testicular descent (the vaginal process) during the fetal period (van der Schoot, 1995). A complex vascular plexus supplies the testes (see later), functioning as a countercurrent heat exchanger to keep the testes cool despite their internal location under the insulating blubber.

The genital tubercle gives rise to the penis or clitoris (Amasaki *et al.,* 1989b). External genitalia are not usually visible externally in the full-term fetus as they are withdrawn into the genital slit (see Section I,A,5).

F. Circulatory System

1. Heart During the embryonic period, the heart is visible and has probably undergone a similar differentiation as occurs in other mammals. The heart begins as a straight tube, but during the late embryonic period, it folds and forms septa that eventually divide it into the four chambers found in all mammalian hearts. The cetacean heart, however, shows differences in shape from terrestrial mammals. In both odontocetes and mysticetes, it is laterally (transversely) broad and craniocaudally compressed, with the apex being formed by both ventricles. The cetacean heart has specializations that may be adaptive for diving, such as anastomoses between dorsal and ventral interventricular arteries, and hypertrophy of the right ventricle (Tarpley *et al.,* 1997). Diving adaptations also occur in the great vessels, such as an expandable aortic arch.

2. Blood Vessels The internal carotid artery, which is a major supplier of blood to the brain in terrestrial mammals, tapers dramatically in the neck and terminates under the skull base at the carotid canal before reaching the brain. This reduction of the internal carotid artery probably occurs in all cetaceans that exhibit cervical rete mirabila (see later), as this is the only structure it appears to supply. Interestingly, the internal carotid artery is also reduced or absent in many artiodactyl species. The ductus arteriosus (a fetal vascular connection between the aorta and the pulmonary artery) was thought to remain patent postnatally, but a study in adult pilot whales (*Globicephala* spp.) showed that it does close (Johansen *et al.*, 1988). This is no different from terrestrial mammals and is probably the condition in other cetaceans as well.

The fetus develops complex networks of anastomosing, coiled blood vessels called retia mirabila. These vessel masses are found in regions surrounding the dorsal thoracic cage, the region near the foramen magnum, and the spinal cord. Although the function of retia mirabila is not known, it is thought that they are adaptations to diving and resurfacing. Their vessel structure may compensate for the rapid pressure changes of descent and ascent with a slow and sustained response that moderates blood flow. By dampening oscillations in blood pressure, sensitive tissue such as heart muscle or the brain and spinal cord continue to receive steady perfusion, thus avoiding oxygen debt and lactic acid buildup. As these vessels appear to store blood near vital tissue (e.g., brain, spinal cord, heart), they may thus function as a reservoir, distributing blood to these oxygen-sensitive tissues when normal circulation is affected (e.g., as pressure increases during diving, or metabolism is slowed). A less widely held hypothesis for the function of the retia mirabila is in trapping the nitrogen bubbles (emboli), which may come out of solution in the bloodstream during ascent from a prolonged dive. These bubbles are potentially fatal, as they can block smaller blood vessels and therefore interrupt blood flow in the capillary beds of organs (a condition known in human divers as decompression sickness, or caisson disease).

The fetus also develops a complex network of vessels that supply and drain the testes and uterus. These vessels are arranged in a plexus to effect a thermoregulatory countercurrent exchange. This conserves heat where needed and allows extra heat to be drawn away from these organs. Thus, the male can keep the testes cool and the pregnant female can keep the fetus in the uterus from overheating, despite the internal location under the insulating blubber (Rommel *et al.*, 1993).

G. Nervous System

1. Brain   Brain development in the embryonic period resembles that of other terrestrial mammals. The brain is composed of three main sections: prosencephalon (forebrain), mesencephalon (middle brain), and rhombencephalon (hindbrain). The corticospinal tract does not develop to the same degree as terrestrial mammals, probably due to the loss of the hind limbs and the reduced role of the forelimbs in propulsion. The cochlea is enlarging, while the vestibular system is rudimentary in size—a disparity that will remain in the adult. Olfactory bulbs and nerves are present in both odontocete and mysticete embryonic brains.

The rate of brain growth and degree of encephalization differs for different species (Pirlot and Kamiya, 1975). In the early fetal period, typical cetacean features begin to develop. For example, the olfactory bulbs and nerves disappear in odontocetes. In mysticetes, however, the olfactory bulbs and nerves are retained. Because adult mysticetes retain olfactory mucosa, it is presumed that they use a sense of smell to help locate plankton, particularly swarms of krill. There is some dispute, however, as to the existence of a vomeronasal organ (a chemoreceptive organ that functions mainly in detecting sexual pheromones in terrestrial mammals). Although it had been thought that adult whales had a vomeronasal organ, fetal studies of mysticetes and odontocetes show both the vomeronasal organ and nerve to be absent (Oelschläger *et al.*, 1987). The function of a cetacean vomeronasal organ is purely speculative, but may include detecting the presence and mating status of other whales and perhaps even the odor of food in the mouth. The terminal nerve (a sensory, but not chemosensory, derivative of the olfactory placode sometimes called "cranial nerve 0") persists and may function in the autonomic innervation of intracranial arteries and mucous epithelium of the nasal air sacs.

2. Spinal Cord   In the midfetal period, the head and neck regions align horizontally. During this process, the cervical section of the spinal cord, which was previously flexed ventrally, must now arch under and around the cerebellum to join the thoracic spinal cord (the dorsal aspect is thus concave). As the cervical vertebrae are compressed, much of the cervical spinal cord is contained within the skull.

II. Gestational Length

Because whales are related to terrestrial ungulates, it is not surprising that their gestation is of a similar length (see Table I). Horses, for example, have a gestation of 11 months (compared to 9 months for a human's pregnancy and 22 months for an elephant's gestation). Gestation in mysticetes lasts 10–13 months. The length of gestation in odontocetes, however, is more variable, ranging from 10 to 17 months. The length of the gestation is not correlated with body size (e.g., although the sperm whale is the largest odontocete, its gestation is the same as the smaller pilot whale and less than the Baird's beaked whale, *Berardius bairdii*).

The length of gestation in cetaceans may be related to food supply and MIGRATION. Most mysticetes mate in warm waters, migrate to cold waters to feed, and then migrate back to warm waters to calve. This behavioral cycle, which takes 1 year, thus appears to be related to gestational length. Interestingly, because most feeding occurs in colder waters, a pregnant mysticete whale may well be fasting while simultaneously spending energy in migratory locomotion and nourishing a growing embryo/fetus. In this regard, it is significant that the first half of the pregnancy (largely embryo development) takes place during the migration to the feeding areas, whereas the second half of the pregnancy (largely fetal growth) takes place during the migration back to the calving areas. A whale migrating to feeding areas is not carrying a large load of stored energy compared with a whale returning to calving areas from feeding areas. In some species, pregnant whales may increase food intake by 50–60% above normal during the last 6 months of gestation.

TABLE I
Cetacean Gestations and Newborn Calf vs Adult Measurements

	Gestation (in months)	Newborn calf (length, weight)	Mature adult (length, weight)
Mysticetes			
Right whales (*Eubalaena* spp.)	12	4.5–6 m, 1000 kg	12.5–17.7 m, 30,000–80,000 kg[a]
Bowhead whale (*B. mysticetus*)	12–16	3.6–4.5 m, 1000 kg	11.5–18 m
Pygmy right whale (*Caperea marginata*)	?	2 m?	5.47–6.45 m, 3,100–3,500 kg
Gray whale (*Eschrichtius robustus*)	13–13.5	4.5–5 m, 500–800 kg	13–15.2 m, 14,000–35,000 kg
Humpback whale (*Megaptera novaeangliae*)	11–11.5	4–5 m, 900–1500 kg	11.5–19 m, 25,000–48,000 kg
Minke whales (*Balaenoptera acutorostrata* and *B. bonaerensis*)	10	2.4–3 m, 300–400 kg	6.9–10.7 m, 4000–13,500 kg
Bryde's whales (*B. brydei, B. edeni*)	11–12	3.95–4.3 m, 900 kg	11.6–15.6 m, 16,000–25,000 kg
Sei whale (*B. borealis*)	11.5–12	4.5 m, 780 kg	13–18.3 m, 20,000–25,000 kg
Fin whale (*B. physalus*)	11	6.4–6.5 m, 1750–1800 kg	17.5–27 m, 30,000–90,000 kg
Blue whale (*B. musculus*)	11–12	7–8 m, 2000–3000 kg	19–31 m, 100,000–200,000 kg
Odontocetes			
Sperm whale (*Physeter macrocephalus*)	14–16	3.5–4.5 m, 1000 kg	8.3–20.5 m, 16,000–57,000 kg
Pygmy sperm whale (*Kogia breviceps*)	11	1.2 m	2.7–3.7 m, 400 kg or less
Dwarf sperm whale (*K. sima*)	9	1 m	2.1–2.7 m, 210 kg
Cuvier's beaked whale (*Ziphius cavirostris*)	12	2.5–3 m	5.1–7.5 m, 3,000 kg
Baird's beaked whale (*Berardius bairdii*)	17	4.5–4.8 m	10–12.8 m, 11,000 kg
Northern bottlenosed whale (*Hyperoodon ampullatus*)	12	3–3.5 m	6–9.8 m
Southern bottlenosed whale (*H. planifrons*)	?	2.9 m	5.7–7.45 m
Hector's beaked whale (*Mesoplodon hectori*)	?	2.1 m	4.3–4.43 m
True's beaked whale (*M. mirus*)	?	2.3 m, 136 kg	5.1–5.3 m, 1,394 kg
Gervais' beaked whale (*M. europaeus*)	?	1.6–2.1 m, 49 kg or more	3.7–5.2 m, 1,178 kg or more
Sowerby's beaked whale (*M. bidens*)	12	2.4–2.7 m, 185 kg	5.05–5.5 m
Gray's beaked whale (*M. grayi*)	?	2.42 m	4.74–5.64 m, 1,075–1,100 kg
Hubb's beaked whale (*M. carlhubbsi*)	?	2.5 m	5.3 m, 1,432 kg
Strap-toothed whale (*M. layardii*)	?	0.76–2.8 m	5.8–6.2 m
Blainville's beaked whale (*M. densirostris*)	?	1.9 m, 60 kg	4.56–4.73 m, 1,033 kg
Indian river dolphin (*Platanista gangetica*)	8–12?	0.67–0.9 m	1.7–2.5 m, 85 kg
Amazon river dolphin (boto) (*Inia geoffrensis*)	9–12	0.75–0.8 m, 7–8 kg	2–2.6 m, 100–160 kg
Chinese river dolphin (*Liptotes vexillifer*)	?	0.57–0.95 m, 10 kg or less	2.1–2.5 m, 125–16- kg
Franciscana (*Pontoporia blainvillei*)	10–11	0.75–0.8 m, 7.3–8.5 kg	1.5–1.74 m, 25–53 kg
Beluga whale (*Delphinapterus leucas*)	14–14.5	1.5–1.6 m, 79–80 kg	3–5.5 m, 400–1,500 kg
Narwhal (*Monodon monoceros*)	14–15	1.5 m	3.4–4.7 m
Commerson's dolphin (*Cephalorhynchus commersonii*)	11–12	0.65–0.75 m	1.25–1.75 m, 35–86 kg
Hector's dolphin (*C. hectori*)	?	0.5–0.7 m	1.2–1.8 m, 50–60 kg
Humpbacked dolphin (*Sousa chinensis*)	?	0.9–1 m	2.26–3 m, 195–284 kg
Tucuxi (*Sotalia fluviatilis*)	10	?	1.3–1.9 m, 35–40 kg
Common Bottlenosed dolphin (*Tursiops truncatus*)	12	0.9–1.3 m	1.9–4 m, 90–650 kg
Pantropical spotted dolphin (*Stenella attenuata*)	11–12	0.8–0.89 m	1.82–2.57 m, 119 kg or less
Atlantic spotted dolphin (*S. frontalis*)	?	0.76–1.20 m	2.3 m, 143 kg
Spinner dolphin (*S. longirostris*)	10–11	0.77–0.8 m	1.3–2.16 m, 26.5–75 kg
Striped dolphin (*S. coeruleoalba*)	12–13	1 m	2.16–2.4 m, 156 kg or less
Short-beaked common dolphin (*Delphinus delphis*)[c]	10–11	0.76–0.86 m	1.6–2.6 m, 70–135 kg
Fraser's dolphin (*Lagenodelphis hosei*)	?	0.95 m	2.25–2.65 m, 200 kg or more
White-beaked dolphin (*Lagenorhynchus albirostris*)	?	0.95–1.6 m, 40 kg or more	2–3 m
Atlantic white-sided dolphin (*L. acutus*)	10–12	1.08–1.22 m	2–2.8 m
Pacific white-sided dolphin (*L. obliquidens*)	10–12	0.8–1.24 m	1.7–2.5 m, 75–181 kg
Dusky dolphin (*L. obscurus*)	11	0.55–0.70 m[b]	1.6–2.1 m, 40–80 kg or more
Northern right whale dolphin (*Lissodelphis borealis*)	?	0.8–1 m	2–3.1 m, 115 kg
Risso's dolphin (*Grampus griseus*)	13–14	1.2–1.5 m	2.6–4.3 m, 500 kg
Melon-headed whale (*Peponocephala electra*)	?	1 m	2.2–2.75 m, 160–275 kg
Pygmy killer whale (*Feresa attenuata*)	?	0.8 m	2–2.6 m, 150–225 kg
False killer whale (*Pseudorca crassidens*)	11–15.5	1.6–1.93 m	3.96–6 m, 1,100–2,200 kg
Killer whale (*Orcinus orca*)	16–17	2.06–2.5 m	4.6–9.75 m, 2,600–10,500 kg
Long-finned pilot whale (*Globicephala melas*)	14.5–16	1.75–1.8 m, 70–85 kg	3.8–6.3 m, 280–1750 kg
Short-finned pilot whale (*G. macrorhynchus*)	15	1.4–1.85 m	3.01–7.2 m, 600–3950kg
Irrawaddy dolphin (*Orcaella brevirostris*)	14	0.9–1 m, 12.5 kg	2.15–2.75 m, 90–150 kg
Finless porpoise (*Neophocaena phocaenoides*)	11–12	0.6–0.9 m	1.8–1.9 m
Harbor porpoise (*Phocoena phocoena*)	9–11	0.7–0.9 m, 5–9 kg	1.4–2 m, 40–90 kg
Spectacled porpoise (*P. dioptrica*)	?	0.46 m	1.8–2.4 m
Dall's porpoise (*Phocoenoides dalli*)	11–12	0.95–1 m	1.7–2.2 m, 200 kg

[a]10.09-m male North Atlantic right whale was 9055 kg when weighed intact postmortem.
[b]0.98-m dusky dolphin calf weighed 22 kg; 0.67-m dusky dolphin fetus weighed 3.7 kg.
[c]May include data for *D. capensis*.

Thus, the energy demands on a whale in early pregnancy may be smaller than that of a whale in the later stages of pregnancy, when the fetus is growing at a rapid rate. In odontocetes, these energy constraints appear to have less of a temporal impact on gestational periods. This may be due, in part, to a more constant energy supply (year-round access to a food supply) for those species that migrate, or the lack of migration in other species.

Because the calf must be able to swim, see, hear, and vocalize immediately after birth, the nervous and muscular systems of the calf must be well developed. This also translates to a fairly long gestation, with as much development as possible occurring prenatally (compared with the human baby, which completes much of its neuromuscular development postnatally). The long gestation also enables calves to grow to a large size before birth, reaching approximately one-fourth to one-third of the mother's size. Once the fetus has attained a near adult form, the most dramatic changes appear to be mainly in the overall size of the fetus. As the fetal period progresses, the growth rate increases rapidly. For example, the blue whale (*Balaenoptera musculus*) gains approximately 100 kg/day in the last 2 months alone. Large newborns are also common among the whales' closest land relatives, the ungulates, which also have well-developed neural and muscular systems at birth. An additional advantage of a large calf is a smaller surface-to-volume ratio (which helps the calf conserve heat). Thus, because whales have relatively large calves, it is not surprising that multiple births are a rarity.

III. Maternal Uterus, Placenta, and Umbilical Cord

Cetaceans have a bicornuate uterus (two horns joined in a Y shape). The fetus usually develops in one horn (either horn for mysticetes, but most frequently the left horn for odontocetes), whereas the other horn is generally occupied by the allantois (one of the embryonic membranes) and placenta. The cetacean uterus has a complex vascular plexus that functions in countercurrent heat exchange (Rommel *et al.*, 1993). This keeps the fetus from overheating, despite its insulated location under the maternal blubber and adjacent to the locomotor musculature of the maternal abdomen.

The placenta is epitheliochorial (or cotyledon), which means that the maternal and fetal tissues do not fuse into one tissue, as in humans. Rather, their vascular systems remain separated by two epithelial layers with separate capillary beds. This arrangement ensures that the two layers separate relatively easily at birth, thus minimizing the inevitable loss of blood. Not surprisingly, this type of placenta is also found in the ungulates, the group of terrestrial mammals related most closely to cetaceans.

The umbilical cord is short and thick, with "amnion pearl" knobs on the outer surface that appear to regulate the development (cornification) of fetal skin. It contains two arteries and two veins, as well as an allantoic duct. When the calf is born, the umbilical cord breaks off at the fetal end, allowing the calf to swim unhindered to the surface. Because the mother does not appear to bite off the umbilical cord, it is presumed to break with little force at the moment of birth. The umbilical ring contains invaginations that probably weaken the connection between the fetal epithelium and the umbilical cord. The umbilical arteries and veins are also constricted and weak where they enter the fetal abdomen. The umbilical cord attaches midway along the length of the fetus (unlike the more caudal attachment found in fetuses of other mammals, in which the neck contributes more to fetal length than the tail). Thus, the umbilical cord will be stretched taut to the same degree regardless of whether the head or the tail is delivered first. The stretch from the delivered fetus pulling taut the umbilical cord, which is attached via the placenta to the mother's uterus, may cause its rupture at the umbilicus.

IV. Fetal Position and Birth

Birth takes place underwater. In most observed captive births in odontocetes, the fetal tail emerged first through the vaginal opening. This tail-first presentation may appear unusual, particularly when compared with the usual head-first presentations of most terrestrial mammals. Interestingly, captive manatees have also been observed to deliver their young tail-first underwater. Whereas births in the wild have been documented less frequently, they appear to be more commonly tail-first presentations in odontocetes and may be equally tail first or head first in mysticetes. As the pelvis in whales is rudimentary, it appears to have little, if any, effect on passage of the fetus during birth. In fact, the large size of the cetacean brain at birth may be possible, in part, because of the ease with which the large head of the fetus can be delivered through this rudimentary pelvis. Because there is no significant bony constriction at the pelvic outlet, there does not appear to be a physical need for a head-first delivery as in most large terrestrial mammals.

The higher frequency of tail-first presentations may also be explained by the shape and intrauterine position of the fetus. The cetacean fetus has a fusiform shape, with the rostrum and tail stock both being relatively small in diameter. The tail flukes, dorsal fin, and pectoral flippers are very pliable and are flattened against the body (fin and flippers) or curled back toward the midline to form a small knob (flukes). This folding and curling not only helps keep the fetal body within in the smallest dimensions, but it also enables the fetus to maintain a relatively smooth exterior contour with no protrusions to inhibit delivery through the vagina. In addition, the whole fetus is laterally flexed into a U shape, with the tail recurved toward the head so that the flukes are positioned adjacent to the rostrum. While this fetal folding reduces the intrauterine volume needed for carrying the fetus, it also leaves the fetus with both its rostrum and its tail flukes directed toward the maternal tail. However, because the uterine horn is also folded, only one end of the fetus can thus be directed toward the cervix. In odontocetes, it is most commonly the tail. Because the tail flukes are smaller than the cetacean head, they can therefore slip out of the vagina more easily and thus are more likely to emerge first.

The head being directed away from the cervix before parturition may be a function of either fetal shape or fetal weight. Because both the center of gravity and the largest diameter is closer to the fetal head, its "rest" position may more likely be with the heavier fetal head nearer the center of gravity of the mother. This places the fetal head in the more distensible part of the mother's abdomen and away from the more mobile tail stock (which, due to locomotor constraints, may have less capability for expansion).

Once the fetus is settled into this birth position, continued growth appears to cause it to recurve caudally in order to fit within the mother's abdomen. Unlike terrestrial mammals, the head is not flexed ventrally in the late-term cetacean fetus because there is practically no neck and the cervical vertebrae are largely fused. Rather, the fetus folds in half laterally to conserve space. The curved midsection of the fetus takes up more room in the maternal abdomen than the fetal head and tail. Thus, the fetal abdomen is placed cranially in the mother, where there is more room for expansion, while the fetal head and tail are directed caudally near the less expandable maternal tail stock. Although the fetal head is directed caudally, it is positioned at the tip of the uterine horn (which is thus also folded to face caudally) and not adjacent to the cervix. In this folded position, it is unlikely that the fetus can reposition itself to completely switch from a tail-first to a head-first presentation. As the fetus is delivered, its body must unfold. Thus, midway through parturition, the fetal head will again face toward the maternal head as the fetal body straightens. The newborn calf bears light colored bands and shallow vertical grooves, called "fetal folds," along the skin of the lateral abdomen. These markings indicate the concave side of the fetus as it was folded *in utero*.

Acknowledgment

This work was supported by the Office of Naval Research, Grants N00014-96-1-0764 and N00014-99-1-0815.

See Also the Following Articles

Cetacean Life History ■ Female Reproductive Systems ■ Mammalia ■ Morphology, Functional

References

Amasaki, H., Daigo, M., Taguchi, J., and Nishiyama, S. (1989a). Morphogenesis of the digestive tract in the fetuses of the southern minke whale, *Balaenoptera acutorostrata. Anat Anz.* **169,** 161–168.

Amasaki, H., Ishikawa, H., and Daigo, M. (1989b). Development of the external genitalia in fetuses of the southern minke whale, *Balaenoptera acutorostrata. Acta Anat.* **135,** 142–148.

Amasaki, H., Ishikawa, H., and Daigo, M. (1989c). Developmental changes of the fore- and hind-limbs in the fetuses of the southern minke whale, *Balaenoptera acutorostrata. Anat Anz.* **169,** 145–148.

Calzada, N., and Aguilar, A. (1996). Flipper development in the Mediterranean striped dolphin (*Stenella coeruleoalba*). *Anat Rec.* **245,** 708–714.

Drabek, C. M., and Kooyman, G. L. (1983). Terminal airway embryology of the delphinid porpoises, *Stenella attenuata* and *S. longirostris. J. Morphol.* **175,** 65–72.

Ishikawa, H., and Amasaki, H. (1995). Development and physiological degradation of tooth buds and development of rudiment of baleen plate in southern minke whale, *Balaenoptera acutorostrata. J. Vet. Med. Sci.* **57,** 665–670.

Johansen, K., Elling, F., and Paulev, P. E. (1988). Ductus arteriosus in pilot whales. *Jpn. J. Physiol.* **38,** 387–392.

Klima, M. (1999). Development of the cetacean nasal skull. *Adv. Anat. Embryol. Cell Biol.* **149,** 1–143.

Meyer, W., Neurand, K., and Klima, M. (1995). Prenatal development of the integument in Delphinidae (Cetacea: Odontoceti). *J. Morphol.* **223,** 269–287.

Oelschlager, H. A., Buhl, E. H., and Dann, J. F. (1987). Development of the nervus terminalis in mammals including toothed whales and humans. *Ann. N. Y. Acad. Sci.* **519,** 447–464.

Ogden, J. A., Lee, K. E., Conlogue, G. J., and Barnett, J. S. (1981). Prenatal and postnatal development of the cervical portion of the spine in the short-finned pilot whale *Globicephala macrorhyncha. Anat Rec.* **200,** 83–94.

Pirlot, P., and Kamiya, T. (1975). Comparison of ontogenetic brain growth in marine and coastal dolphins. *Growth* **39,** 507–524.

Reidenberg, J. S., and Laitman, J. T. (1988). Existence of vocal folds in the larynx of odontoceti (toothed whales). *Anat. Rec.* **221,** 884–891.

Reidenberg, J. S., and Laitman, J. T. (1994). Anatomy of the hyoid apparatus in Odontoceti (toothed whales): Specializations of their skeleton and musculature compared with those of terrestrial mammals. *Anat. Rec.* **240,** 598–624.

Rommel, S. A., Pabst, D. A., and McLellan, W. A. (1993). Functional morphology of the vascular plexuses associated with the cetacean uterus. *Anat. Rec.* **237,** 538–546.

Sedmera, D., Misek, I., and Klima, M. (1997a). On the development of cetacean extremities. I. Hind limb rudimentation in the spotted dolphin (*Stenella attenuata*). *Eur. J. Morphol.* **35,** 25–30.

Slijper, E. J. (1979). "Whales," 2nd English Ed. Cornell Univ. Press, Ithaca, NY.

Tarpley, R. J., Hillmann, D. J., Henk, W. G., and George, J. C. (1997). Observations on the external morphology and vasculature of a fetal heart of the bowhead whale, *Balaena mysticetus. Anat Rec.* **247,** 556–581.

van der Schoot, P. (1995). Studies on the fetal development of the gubernaculum in Cetacea. *Anat. Rec.* **243,** 449–460.

Prey, Prey Determination
SEE *Diet*

Pygmy and Dwarf Sperm Whales
Kogia breviceps and *K. sima*

DONALD F. MCALPINE
New Brunswick Museum,
Saint John, Canada

Found throughout temperate and tropical seas, pygmy and dwarf sperm whales are small, toothed cetaceans generally assigned to the family Kogiidae. Although rarely sighted at sea, these whales commonly strand in some regions. The relatively little that is known of their biology has been gleaned from such stranded animals.

I. Diagnosis

In form, *Kogia* spp. are porpoise-like and robust with a distinctive underslung lower jaw. This latter feature has been described as giving these whales a shark-like appearance (Fig. 1). Although height and position of the dorsal fin have been reported as distinguishing the two species, they are probably not separable at sea except under exceptional circumstances. Pygmy sperm whales reach a maximum size of about 3.8 m total length and a weight of 450 kg. Dwarf sperm whales are smaller at 2.7 m and 272 kg. Adults of both species are dark bluish-gray to blackish-brown dorsally with a light venter. On the side of the head between the eye and the flipper there is often a crescent-shaped, light-colored mark referred to as a "false gill." It is only since 1966 that two species of *Kogia* have been recognized, and no subspecies have been described. Some cetologists place *Kogia* in a subfamily Kogiinae within Physeteridae with the sperm whale, *Physeter macrocephalus*. Some anatomical features suggest that the relationship between *Kogia* and other toothed whales remains poorly resolved.

II. Distribution and Status

Dwarf and pygmy sperm whales occur worldwide in temperate and tropical waters of the Atlantic, Pacific, and Indian Oceans. Evidence shows that *K. sima* may prefer warmer seas than *K. breviceps*, but currently there is little indication for seasonality in the distribution or the migration of these whales. Their precise at-sea DISTRIBUTION is unknown, as most records are based on stranded animals. Although many writers have stated that dwarf and pygmy sperm whales are rare, there is insufficient information to classify the world status of *Kogia* species, and their population sizes are unknown. The frequency with which *Kogia* strand on certain coasts, especially in southeastern United States and South Africa, suggests that they may be uncommon rather than rare in some regions.

III. Ecology

Analysis of prey in stranded animals suggests that both species of *Kogia* generally inhabit waters along the continental shelf and slope in the epi- and mesopelagic zones. Differences in the proportion of prey items in stomachs indicate that *K. sima* feeds in shallower water than *K. breviceps*. *Kogia* spp. feed mostly on mid and deep-water cephalopods but also consume fish and occasionally crustaceans, such as shrimp and crabs. Stomach contents that have been analyzed have contained cephalopod beaks from at least 13 families. Most feeding seems to takes place on or near the bottom, probably using ECHOLOCATION to find prey. Kogiid hyoid anatomy suggests powerful suction feeding.

Little is known about disease in *Kogia*. Strandings of unhealthy pygmy and dwarf sperm whales have been attributed to degenerative heart disease, as well as being linked to possible immune system problems associated with the thymus gland. Pneumonia has also been observed in stranded animals. *Kogia* specimens are frequently heavily infected with intestinal nematodes (*Anisakis* sp. and *Terranova cetecola*) and blubber-encysted larval cestodes (*Phyllobothrium delphini*). The parasitic crustacean, *Pennella balaenoptera*, has been observed embedded in the epidermal surface of both species of *Kogia*. Scarring indicates that these whales are attacked by lampreys. A white shark (*Carcharadon carcharias*) attack on a pygmy sperm whale has been documented, and pygmy sperm whale remains have been identified in killer whale (*Orcinus orca*) stomachs. Levels of PREDATION on *Kogia* are otherwise unknown. Heavy infestations with larval cestodes, which probably mature in elasmobranchs, suggest that shark attacks may be more common than the single literature report suggests.

IV. Behavior

There have been no comprehensive behavioral studies of dwarf or pygmy sperm whales. Stranded animals that have been maintained in aquaria have survived no more than several months and usually only live for a few days. At sea, both species occur individually or in small groups of up to 6 (*K. breviceps*) or 10 (*K. sima*) animals of varying age and sex composition. Strandings usually involve single animals. Dwarf and pygmy sperm whales are reported to spend considerable time lying motionless at the surface with the back of the head exposed and the tail hanging down. *K. breviceps* is easily approached, but is timid and slow moving. Normal swimming speed is thought to be about 3 knots. When surfacing, both species rise slowly, produce an inconspicuous blow, and dive without showing the flukes. Neither species is known to be highly vocal.

V. Anatomy and Reproduction

The genus has the shortest rostrum among living cetaceans (Fig. 2) and the skull is markedly asymmetrical. The mandibles are delicate, and the teeth are very sharp, thin, and lack enamel. *K. breviceps* lacks teeth in the upper jaw, but *K. sima* may have up to three pairs of vestigial teeth in this position. Like the Physeteridae, *Kogia* spp. have a spermaceti organ. The function of this organ is unknown, but it has been suggested that it may be involved in maintaining neutral buoyancy or function as an acoustic lens in directing sound beams during ecolocation. *K. sima* reaches sexual maturity at about 2.1 m in length. In *K. breviceps*, males are known to be sexually mature at about 2.7 m and females at a slightly smaller size. Gestation has been cited as 9 or 11 months, with the species about 1 m in length at birth.

Figure 1   (A) Kogia breviceps. *The relatively short and more posteriorly positioned dorsal fin is useful in distinguishing this species from (B) K. sima. Pieter A. Folkens/Higher Porpoise DG.*

Figure 2 *Skull of* Kogia breviceps. *The rostrum is short, the skull asymmetrical, and teeth are lacking in the upper jaw. Photo courtesy of New Brunswick Museum.*

VI. Fossil Records

Fossil forms of Kogiidae have rarely been described from fragments of teeth, cranium, and lower jaws of late Miocene to early Pliocene age. Most of these may be only distantly related to extant *Kogia* spp. However, *Praekogia cedrosensis*, described from the early Pliocene in the Almejas Formation on Isla Cedros Baja California, Mexico, is reported to be clearly ancestral to living *Kogia*.

VII. Human Interactions

The scarcity of pygmy and dwarf sperm whales and the fact that they are rarely encountered at sea mean that direct effects from humans are probably few. However, they have been reported to ingest ocean debris such as plastic bags, possibly resulting in intestinal blockage and death. Their habit of lying quietly at the surface seems to have led to occasional ship strikes. Both species are taken infrequently in commercial harpoon fisheries in the Caribbean and Indian Oceans.

See Also the Following Articles

Skull Anatomy ■ Toothed Whales, Overview

References

Baird, R. W., Nelson, D., Lein, J., and Nagorsen, D. W. (1996). The status of the pygmy sperm whale, *Kogia breviceps*, in Canada. *Can. Field-Nat.* **110**, 525–532.

Caldwell, D. K., and Caldwell, M. C. (1989). Pygmy sperm whale, *Kogia breviceps* (de Blainville, 1838): Dwarf sperm whale, *Kogia simus*, Owen, 1866. *In* "Handbook of Marine Mammals" (S. H. Ridgway and R. Harrison, eds.), Vol. 4, pp. 235–260. Academic Press, San Diego, CA.

McAlpine, D. F., Murison, L. D., and Hoberg, E. P. (1997). New records of the pygmy sperm whale, *Kogia breviceps* (Physeteridae) from Atlantic Canada with notes on diet and parasites. *Mar. Mamm. Sci.* **13**, 701–704.

Willis, P. M., and Baird, R. W. (1998). Status of the dwarf sperm whale, *Kogia simus*, with special reference to Canada. *Can. Field-Nat.* **112**, 114–115.

Pygmy Killer Whale
Feresa attenuata

MEGHAN A. DONAHUE AND WAYNE L. PERRYMAN
*Southwest Fisheries Science Center,
La Jolla, California*

Although called by the common name "whale," the pygmy killer whale, like its close relative, the killer whale, belongs taxonomically to the dolphin family, Delphinidae. Until 1952 this species was only known from two skulls collected in the 19th century. Since that time, a number of specimens have been collected from strandings and fishery catches around the world, yet the pygmy killer whale remains one of the least known of the small cetaceans.

I. Characters and Taxonomic Relationship

The pygmy killer whale has a moderately robust body that narrows posteriorly to the dorsal fin, hence the name *attenuata* from the Latin "to make thin or taper" (Fig. 1). The head is round and blunt and lacks a beak typical of many dolphin species. The head does not narrow or appear triangular when viewed from above as with the melon-headed whale (*Peponocephala electra*). The moderately long flippers are rounded at the tips with convex leading and concave trailing edges.

On the back and portions of the flanks and ventral surface, the pygmy killer whale is dark gray to black. A subtle, dark cape (an area of COLORATION extending from the forehead past the dorsal fin) is present, which extends the greatest distance down the side of the animal below the high, falcate dorsal fin. A paler gray area on each flank is usually present from the tail stock to the eye. On its ventral side, the pygmy killer has an irregularly shaped white patch between the flippers, around the genitals, and occasionally on the tail stock. The lips are also edged with white.

The skull is broad and robust. The upper and lower jaws have less than 15 teeth each, a character that distinguishes the pygmy killer whale from the melon-headed whale, which typically has more than 20 teeth per row.

Length measurements from several specimens average 2.31 m (range 2.14–2.59 m). Differences in lengths between males and females have not been observed in measured specimens.

II. Distribution and Ecology

Pygmy killer whales have been recorded in tropical and subtropical waters worldwide. Sightings have been relatively fre-

Figure 1 *Pygmy killer whale. Photograph by Robert Pitman.*

quent in the eastern tropical Pacific, the Hawaiian Archipelago and off Japan. The migratory status of this species cannot be determined based on available information. However, incidental catches and observations by fishermen suggest that this species is a year-round resident at least in the regions of Sri Lanka and the Lesser Antilles.

Although the feeding habits of pygmy killer whales are not well known, remnants of cephalopods and small fish have been found in specimens from STRANDINGS and incidental fishery catches.

III. Life History and Behavior

Little is known about this species' growth, reproduction, or social organization. An estimated length at sexual maturity of 2 m based on 85% of the mean length of physical maturity (Laws, 1956) is consistent with data collected from three sexually mature males ranging in length from 2.07 to 2.61 m and three pregnant females ranging in length from 2.20 to 2.27 m. Pygmy killer whales are found most commonly in small herds, ranging from 12 to 50 animals, although herds of 100 or more have been encountered. This species has been observed BOW RIDING, performing high LEAPS, and "spyhopping" (raising the head vertically out of the water).

Pygmy killer whales are suspected to be among the small whales that chase, attack, and sometimes eat dolphins (*Stenella* spp. and *Delphinus delphis*) involved in the purse seine fishery for yellowfin tuna in the eastern tropical Pacific (Perryman and Foster, 1980). AGGRESSIVE BEHAVIOR has also been observed by two pygmy killer whales in captivity in Hawaii and South Africa,

but a herd captured off Japan showed no such aggression when placed in an enclosure with other dolphins.

IV. Conservation Status

Although they are not considered rare, the population size and structure of the pygmy killer whale are not known. Between 1986 and 1990, five research vessel surveys were conducted in the eastern tropical Pacific and an abundance of 38,900 (CV = 0.305) pygmy killer whales was estimated for that area (Wade and Gerrodette, 1993).

Pygmy killer whales have been caught directly and incidentally in fisheries. Small cetacean fisheries in St. Vincent and Indonesia have been known to catch pygmy killer whales, but they comprise a small proportion of the catch and these catches are thought to have little effect on the population in those areas. Monitoring of fisheries in which pygmy killer whales are caught incidentally has not been extensive. Mortality in these fisheries, such as those around Sri Lanka, could be greater than that documented and may have a significant impact on pygmy killer whale stocks in those regions. In Sri Lanka, pygmy killer whales have also been harpooned and used as bait in long-line fisheries for sharks, billfish, and other oceanic fishes (Leatherwood and Reeves, 1989).

See Also the Following Articles

False Killer Whale ■ Incidental Catches

References

Laws, R. W. (1956). Growth and sexual maturity in aquatic mammals. *Nature* **178**, 193–194.

Leatherwood, S., and Reeves, R. R. (1989). "Marine Mammal Research and Conservation in Sri Lanka 1985–1986." UNEP, Nairobi, Kenya, Marine Mammal Technical Report Number 1.

Perryman, W. L., and Foster, T. C. (1980). Preliminary report on predation by small whales, mainly the false killer whale, *Pseudorca crassidens*, on dolphins (*Stenella* spp. and *Delphinus delphis*) in the eastern tropical Pacific. La Jolla. SWFSC. Admin. Rep. LJ-80-05, 9 pp.

Wade, P. R., and Gerrodette, T. (1993). Estimates of cetacean abundance and distribution in the eastern tropical Pacific. *Rep. Int. Whal. Comn.* **43**, 477–493.

Pygmy Right Whale
Caperea marginata

CATHERINE M. KEMPER
South Australian Museum
Adelaide

The pygmy right whale *Caperea marginata* (Gray, 1846) is the smallest baleen whale and the only member of the family Neobalaenidae (Fig. 1). Its diagnostic features include long, narrow, creamy-white BALEEN with an outer margin

Figure 1 *Baleen whales include some of the larger whale species. The smallest baleen whale is the pygmy right whale* (Caperea marginata) *which achieves a maximum length not exceeding 6.5 m. Skeletal differences are especially distinctive and warrant the recognition of this species as the sole member of the Neobalaenidae, separate from other baleen whales. Pieter A. Folkens/Higher Porpoise DG.*

of brown or black and very fine bristles; a clearly visible band of white gum at the base of the baleen; a moderately arched rostrum that becomes more pronounced as the animal grows; a small, falcate dorsal fin placed about 25 to 30% of body length from the tail; and shallow throat creases in some animals. The overall body shape of adults is stouter than rorquals but not as broad as right and bowhead whales. From above, the head is broadest at the eyes and narrows sharply into a long and narrow rostrum on which a medial ridge may be present. The flukes are very broad and have a deep medial notch. The body is pale to dark gray above and pale gray to white below. There is a dark eyepatch and an indistinct pale gray chevron across the back behind the blowhole. The oval scars of cookie-cutter sharks *Isistius* spp. are often present. The flippers are small, narrow, and rounded at the tip and are dark gray above (contrasting sharply with the pale color of the sides of the body) and paler below. Mandibular and rostral hairs persist into adulthood but there are no callosities as in true right whales.

At sea, pygmy right whales may be confused with minke whales *Balaenoptera acutorostrata* and *B. bonaerensis*, but close inspection should reveal some of the diagnostic features noted earlier. The blunter rostrum of the pygmy right whale and its habit of swimming with its head "thrown" out of the water at an angle should also help identify it.

The SKULL and SKELETON of the pygmy right whale are unlike any other cetacean. The supraoccipital bone is very long, extending well forward on the skull. The ear bone has a distinctive wrinkle on its outer surface and is squarish in outline.

The mandibles and ribs are very broad and flat. The numerous ribs (18 pairs) extend well along the body. All seven cervical vertebrae are fused and the total number of vertebrae is only 44. The flipper has only four digits.

I. Recent and Fossil Relationships

Although the pygmy right whale has sometimes been included in the family Balaenidae, recent studies of its morphology and genetic makeup have shown that its present position in Neobalaenidae is correct. It is believed to be more closely related to the gray whale (Eschrichtiidae) and rorquals (Balaenopteridae) than to right and bowhead whales (Balaenidae). Geographical variation has not been studied and no subspecies are recognized. No fossil neobalaenids have been described. A reported fossil, *Balaena simpsoni*, from Chile is believed to be related to Balaenidae.

II. Distribution and Habitat

This species is found only in the Southern Hemisphere (Fig. 2). It is circumpolar, between about 30° and 55°S, with records from southern Africa, South America, and Australia and throughout New Zealand. It has also been recorded in the vicinity of the Falkland and Crozet islands and in the open ocean of the South Atlantic and south of Australia.

The pygmy right whale lives in temperate and subantarctic regions where water temperatures are between about 5° and 20°C. It has been seen in oceanic and neritic environments where some individuals have spent up to 2 months very close

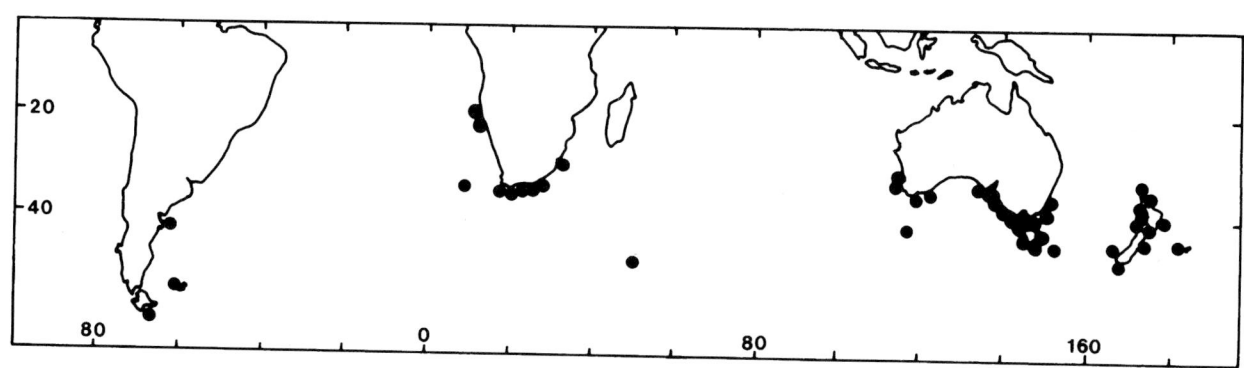

Figure 2 *Distribution of the pygmy right whale. Dots show position only, not number, of records.*

to shore, possibly feeding while there. Seasonal movements in-shore may be related to the availability of food there during spring and summer. Although oceanic feeding has not been observed, animals collected there had full stomachs. The little information available on diet suggests that copepods, euphausiids, and possibly other small PLANKTON are eaten. The predators of pygmy right whales are not known.

III. Behavior

Its behavior is inconspicuous, but because so few observations have been reported, the complete repertoire may not have been recorded. Swimming speeds of 3 to 5 knots have been noted, but it is also capable of very fast acceleration and speed, leaving a conspicuous wake when doing so. One underwater observation of swimming noted that the body action was very flexed. When pygmy right whales dive they remain submerged for up to 4 min and surface briefly before diving again. The blow is inconspicuous and, when visible, is small and oval. The sounds of one solitary juvenile consisted of short, thump-like pulses or tone bursts with a downsweep in frequency and decaying in amplitude. Most energy was between 60 and 120 Hz.

Less than 20 sightings of pygmy right whales "at sea" have been recorded. They have been seen with pilot (*Globicephala melas*), sei (*Balaenoptera borealis*), and minke whales and with dolphins. Mostly these have been of 1 or 2 animals, but groups of up to 10 animals are known and one group of about 80 pygmy right whales was seen in oceanic waters. Strandings throughout the year in Australia and New Zealand suggest that the species does not migrate north–south as do most other baleen whales.

IV. Life History

Relatively little is known of the life history of pygmy right whales. Length at birth is about 2 m and at weaning is about 3.0–3.5 m. Most animals are physically mature at around 6 m, and maximum length and weight are 6.5 m and 3430 kg. Females are slightly longer than males. Sexual maturity may occur at lengths of greater than 5 m. The calving interval, mating season, and gestation period are all not known. The calving season is protracted, possibly year-round. Life expectancy is not known and no age estimates have been made.

V. Interactions with Humans

Pygmy right whales were never targeted by whalers, although they were taken opportunistically. Intentional killing by inshore fisheries and incidental captures in fishing nets are known. No animal has been kept in CAPTIVITY. Toxic contaminants are not believed to be a threat to this species because tissue levels of organochlorines and heavy metals measured in a few animals have been low.

See Also the Following Articles

Baleen Whales ■ Gray Whale ■ Minke Whales

References

Baker, A. N. (1985). Pygmy right whale *Caperea marginata* (Gray, 1846). *In* "Handbook of Marine Mammals" (S. H. Ridgway and R. Harrison, eds.), Vol. 3, pp. 345–354. Academic Press, London.

Bannister, J. L., Kemper, C. M., and Warneke, R. M. (1996). "The Action Plan for Australian Cetaceans." Australian Nature Conservation Agency, Canberra.

Dawbin, W. H., and Cato, D. H. (1992). Sounds of a pygmy right whale (*Caperea marginata*). *Mar. Mamm. Sci.* **8**, 213–219.

Hale, H. M. (1964). The pygmy right whale (*Caperea marginata*) in South Australian waters, part 2. *Rec. South Aust. Mus.* **14**, 679–694.

Ivashin, M. V., Shevchenko, V. I., and Yuchov, V. L. (1972). The pygmy right whale *Caperea marginata* (Cetacea). *Zool. Zhurnal* **51**, 1715–1723.

Jefferson, T. A., Leatherwood, S., and Webber, M. (1993). "Marine Mammals of the World." United Nations Environment Programme Food and Agriculture Organization of the United Nations, Rome.

Leatherwood, S., and Reeves, R. R. (1983). "The Sierra Club Handbook of Whales and Dolphins." Sierra Club Books, San Francisco.

Matsuoka, K., Yoshihiro, F., and Pastene, L. A. (1996). A sighting of a large school of the pygmy right whale, *Caperea marginata*, in the southeast Indian Ocean. *Mar. Mamm. Sci.* **12**, 594–597.

Pavey, C. R. (1992). The occurrence of the pygmy right whale, *Caperea marginata* (Cetacea: Neobalaenidae), along the Australian coast. *Aust. Mammal.* **15**, 1–6.

Ross, G. J. B., Best, P. B., and Donnelly, B. G. (1975). New records of the pygmy right whale (*Caperea marginata*) from South Africa, with comments on distribution, migration, appearance, and behavior. *J. Fish. Res. Board Canada* **32**, 1005–1017.

Sekiguchi, K., Best, P. B., and Kaczmaruk, B. Z. (1992). New information on the feeding habits and baleen morphology of the pygmy right whale *Caperea marginata*. *Mar. Mamm. Sci.* **8**, 288–293.

R

Remoras

DAGMAR FERTL
Minerals Management Service,
U.S. Department of the Interior,
New Orleans, Louisiana

ANDRÉ M. LANDRY, JR.
Texas A&M University, Galveston

Remora, suckerfish, diskfish, and sucker are some of the names describing eight species of marine fishes in the family Echenedidae (=Echeneidae) (Lachner, 1966; Lachner in Fischer, 1978). Remoras inhabit tropical and subtropical waters worldwide, except for the whitefin remora (*Echeneis neucratoides*), which is believed to be restricted to the western Atlantic Ocean.

I. Remora Biology

Remoras use a suction disc to attach to sharks, rays, bony fishes, sea turtles, cetaceans, sirenians, and ships and other floating objects. When attached to these hosts, remoras appear to swim upside down, but the disc is really on top of the head. The oval-shaped disc is a modified dorsal fin that has become split and flattened, forming a series of transverse, plate-like fin rays (disc lamellae) that resemble slats of a venetian blind (Fig. 1). When these fin rays are lifted, a strong vacuum is created between the remora's disc and its host.

The tenacity with which remoras attach to their hosts is best illustrated by the practice of sea turtle fishing by fishermen in the Caribbean and off China and northern Australia, and in Yemen, where it continues to this day. A fisherman ties a line around the tail of a remora and throws the fish into the water. The remora tightly attaches itself to a turtle, and the remora and its "catch" are then hauled ashore.

Suspected benefits of a remora's association with these hosts include transportation, protection from predators, increased courtship/reproduction potential, enhanced respiration, and expanded feeding opportunities. Remoras opportunistically feed on parasitic copepods (which constitute the bulk of their diet), zooplankton and smaller nekton, food scraps from meals of their hosts, and sloughing epidermal tissue and feces of the host.

II. Marine Mammal Hosts

Adult remoras typically attach to the body of a marine mammal (Fig. 2). At least three remora species utilize marine mammals as hosts: whalesucker (*Remora australis* = *Remilegia*

Figure 1 *Lateral and dorsal view of the head of a remora, with suction disc visible. Photographs by Bill Dailey.*

Figure 2   *Short-beaked common dolphin* (Delphinus delphis) *with remora attached.* *Photograph by Bernd Würsig.*

australis), sharksucker (*Echeneis naucrates*), and whitefin remora (=whitefin sharksucker, *Echeneis neucratoides*). Remoras associate with at least 20 cetaceans and 2 sirenian species (dugong and West Indian manatee). The whalesucker has been most often collected and identified from cetaceans (e.g., Rice and Caldwell, 1961; Fertl and Landry, 1999), hence its common name. A sharksucker was recently collected from a common bottlenose dolphin (Fertl and Landry, 1999). Two species of remora have been collected from West Indian manatees; these were positively identified as the whitefin remora (Mignucci-Giannoni *et al.*, 1999) and the sharksucker (Williams and Bunkley-Williams, 1996).

The remora's suction mode of attachment does not hurt the host, or leave scars, as has been suggested. However, a temporary mark resembling the disc imprint can be seen. Wounds attributed to remoras are most likely caused by cookie-cutter sharks (*Isistius brasiliensis*) or lampreys (*Entosphenus tridentatus*), which actually bite their prey or host.

Whether the remora irritates its host is uncertain. A remora may slide all over its host's body, possibly tickling the animal. Dolphins and manatees observed with remoras sliding over their bellies sometimes will jerk, and even roll over. Dolphins of various species leap with remoras attached to them, perhaps to dislodge the "hitchhiker." There also are reports of dolphins dislodging remoras from themselves or their calves and then biting them. Large-sized remoras or multiple remoras on the same host may produce a hydrodynamic drag.

III. Problems with Remora Identification

The whalesucker's preference for cetaceans leads many observers to assume that any remora spotted on a cetacean is of this species. Most remora–marine mammal associations described in the literature are based on visual or photographic observations of a remote, free-swimming host and its passenger(s)

rather than specimens collected from strandings or whaling victims. Host records determined from remote observations should be considered suspect, as the identification of remoras to species is difficult without the actual specimen in hand.

The whalesucker and other host-specific remoras are typically pelagic forms with a specialized morphology consisting of large discs, short stout bodies, and reduced fin size (when compared with inshore counterparts). More commonly reported remoras are slender-bodied, inshore forms, such as the sharksucker, that are least particular about their hosts. The possibility that small, slender remoras, as well as more stocky remoras photographed on cetaceans, may represent different life history stages of one species further complicates positive identification from afar.

See Also the Following Articles

Cetacean Ecology ■ Parasites

References

Fertl, D., and Landry, A. M., Jr. (1999). Sharksucker (*Echeneis naucrates*) on a bottlenose dolphin (*Tursiops truncatus*) and a review of other cetacean—remora associations. *Mar. Mamm. Sci.* **15,** 859–863.

Fischer, W. (1978). "FAO Species Identification Sheets for Fishery Purposes: Western Central Atlantic (Fishing Area 31)," Volume II. Food and Agricultural Organization of the United Nations, Rome.

Helfman, G. S., Collette, B. B., and Facey, D. E. (1997). "The Diversity of Fishes." Blackwell Science, Malden, MA.

Lachner, E. A. (1966). Order Echeneida, family Echeneidae: Diskfishes. *In* "Fishes of the Marshall and Marianas Islands" (L. P. Schultz, ed.), pp. 74–80. United States National Museum Bulletin 202.

Mignucci-Giannoni, A. A., Beck, C. A., Montoya-Ospina, R. A., and Williams, E. H., Jr. (1999). Parasites and commensals of the West Indian manatee from Puerto Rico. *J. Helminthol. Soc. Wash.* **66,** 67–69.

Rice, D. W., and Caldwell, D. K. (1961). Observations on the habits of the whalesucker (*Remilegia australis*). *Norsk Hvalfangst-tidende* **5**, 181–189.

Williams, E. H., Jr., and Bunkley-Williams, L. (1996). "Parasites of Offshore Big Game Fishes of Puerto Rico and the Western Atlantic." Puerto Rico Department of Natural and Environmental Resources and the University of Puerto Rico, Puerto Rico.

Reproductive Behavior

MICHAEL A. FEDAK, BEN WILSON, AND PADDY P. POMEROY
University of St. Andrews, Scotland, United Kingdom

Taking a very broad view, the "function" of marine mammals is to convert prey into offspring. Reproductive behavior is an important part of the process by which this is brought about and must serve to create a situation in which the young can safely be born and nurtured and one which facilitates mating with suitable partners. In long-lived animals, however, reproduction has to be linked to the process of gathering the resources for both reproductive effort and continued survival. Because many marine mammals do not feed where they reproduce, they must also locate breeding areas where reproduction and parental care can take place without compromising nutritional requirements. This article considers the basic problems that the animals must solve to reproduce and gives some illustrative examples of their behavior. We will take just such a broad, strategic view and look at reproductive behavior in a life history context and consider how animals balance their needs for resources and reproduction.

I. Basic Problems to Be Overcome

Although they spend most of their time in the water, seals give birth on land and most newborn pups require a period ashore before being able to cope with life at sea. The vulnerability of pinnipeds on land means that suitable breeding sites need to be isolated from potential predators, limiting the choice of suitable ones. Pinnipeds do not feed while ashore. The widely separated, patchy distribution of resources that typifies most marine ecosystems means that animals are often widely separated from one another while foraging and suitable breeding sites are often few and far between. This necessarily requires the use of stored reserves for periods of days to months. The geographical separation of feeding and breeding sites and the reliance on stored reserves are arguably the most important determinants of seal reproductive strategies and life history patterns.

Whales can give birth, nurse, and mate at sea, but conditions suitable for the birth of young may not be suitable for foraging so these two phases of their annual cycle can take place in widely separated geographical locations. Long migrations between breeding and foraging locations may still be necessary. While foraging, individuals might be widely separated from potential mates, creating difficulties for locating suitable partners. Little food may be available during the birthing and mating period, which therefore can require stored energy and materials for its success. Therefore, even though whales are not constrained to spend time ashore for breeding, in some cases, they face some of the same problems as pinnipeds.

On the whole, smaller cetaceans, including most odontocetes, opt for a different strategy. Foraging, parturition, and calf rearing overlap both spatially and temporally. As a result, annual breeding MIGRATIONS are absent and instead nursing females and calves appear to be aided by associating with conspecifics.

Both seals and the larger whales must move to breeding areas and choose a suitable breeding site where they can safely give birth and protect and feed their young. They must choose a mate, copulate, and produce fertilized eggs. They must protect and feed their young and provide the resources and guidance needed for them to become nutritionally independent and give them a good chance of reaching maturity and recruiting into the breeding population. Then the adults must reestablish successful foraging patterns to provide resources for their own survival and reproductive success in the following year(s).

The marine habitat and the geographic and energetic constraints acting on marine mammals have shaped their life histories and reproductive behaviors to create some of the most dramatic and extreme (some might even say bizarre) reproductive patterns among mammals.

II. Importance of Size

Marine mammal groups contain some of the largest mammals in existence as well as possibly the largest animal to have ever existed. The size adopted by the various species is such an obvious characteristic that we often look past it to other features of the animals without considering its fundamental importance to BEHAVIOR. However, size stands out as being of fundamental importance as to how these animals organize their reproductive behaviors. Because of the scaled relationship between body volume or mass (M) and metabolic rate (MR), where $MR \propto M^{0.75}$, size has obvious implications for diving and foraging behavior. Larger species and individuals will require more prey each year, but they may be able to dive for longer and go longer without food and thus be able to contend with less predictable or widely distributed food distribution. It has equally fundamental implications for variations in reproductive behavior within and between species. Size in large part determines how long animals can fast during reproduction and how often they must leave their pups or the vicinity of potential mates for food. In general, bigger animals can maintain their presence on beaches for longer and can breed farther from food sources. Size also sets the relationship between the duration and the efficiency of lactation (energy used in the process divided by energy stored in the pup). It sets the weaning mass of offspring and the relative cost to the mother of achieving offspring of that mass; larger mothers can produce larger pups without putting themselves at risk. The metabolic overheads (i.e., the amount of energy required to support the metabolism of mother and pup) are relatively lower for larger animals in relation to the stored resources available and delivered to offspring. Size can affect the capability of animals (particularly males) to secure mates and, because of its influence on attendance patterns, can determine the sort of strategies used to gain access to females; larger males can maintain residence for longer on breeding sites. In the variety of strategies and behaviors used to accomplish reproduction and the factors that determine them, size matters.

Although seals and whales face common problems, the fact that whales do not come ashore to carry out any aspects of reproduction means that we have learned about their behavior in very different ways. Behavioral observations are largely confined to activities visible from the surface, and "hands-on" techniques are much more difficult to apply. Cetaceans also have greater opportunity for complex social interactions throughout

the periods of mating and PARENTAL CARE because of the extended times occupied by these activities. The different methodologies used have also led to separation in the approaches used in the study of the two groups, resulting in emphasis placed on different aspects of behavior. It is therefore expedient to treat the two groups separately for much of the remainder of this discussion, even though we will be considering the same basic strategic goals. The walrus, manatees, dugongs, and the sea otter are dealt with briefly.

III. Otariids and Phocids

We consider the strategies of reproductive behavior within the simple life history model (Fig. 1) in which the animal's mass or condition is viewed as the fundamental state variable that determines the constraints on reproductive success. It consid-

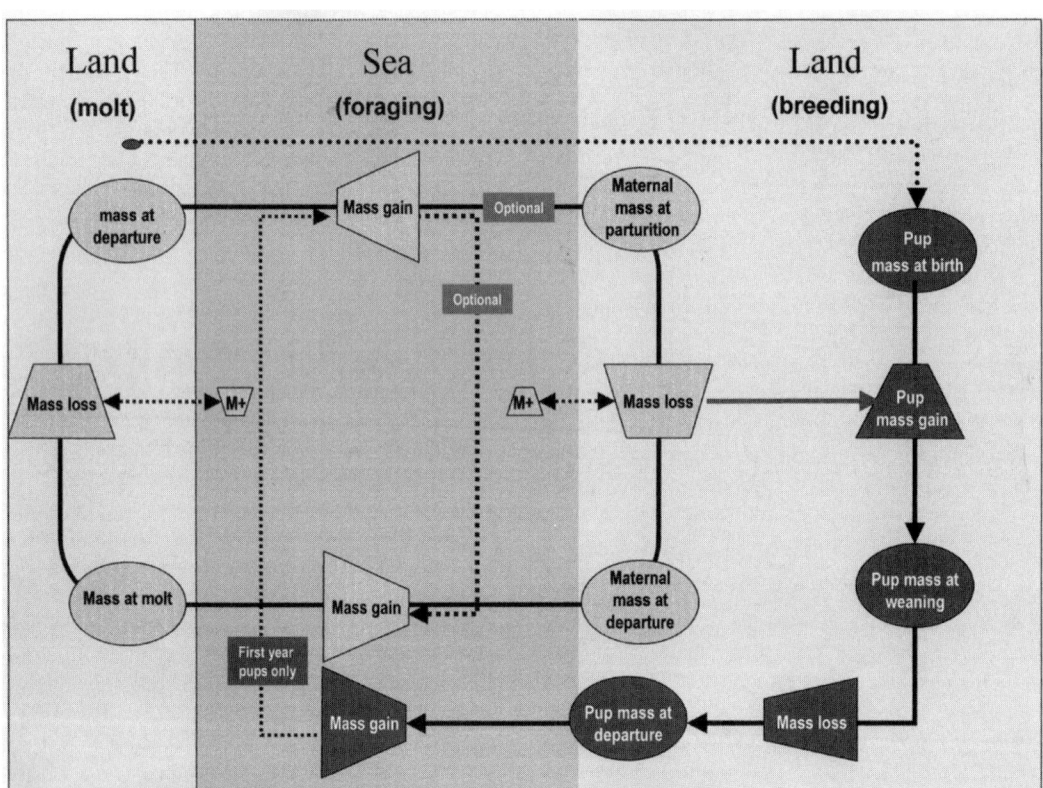

Figure 1 *A diagram of a model of female pinniped life history with mass (as a surrogate for condition) as a state variable determining reproductive success through fecundity and pup quality. Trapezoids represent mass gain and loss. Ovals denote mass at the start and end of lactation and molt. Dotted lines represent optional paths. Some species and/or individuals (particularly smaller phocids and otariid species and smaller individuals of some phocids) may top up body mass during breeding and molt by returning to the sea to feed. Pups usually bypass molt during their first year and do not breed until they have reached a certain critical mass or condition. Females may opt not to produce a pup in years when they are below a critical mass or condition. Pups that are larger or in better condition at weaning are likely to both be able to remain on the beach longer and depart in better condition. They have a better chance of surviving and to breed earlier. In this model, mass and condition determine the path taken and the resulting reproductive success rather than age per se (see Arnbom et al., 1997). In the context of this model, reproductive behavior acts to provide access to suitable mates, influence the transfer of resources to pups and its efficiency, and protect the investment in offspring.*

ers life history as an annual cycle of terrestrial and aquatic phases split among foraging, breeding, and molt (see Fig. 1 legend for details). Virtually all species can be fit into this conceptual framework, and most aspects of reproductive behavior and their links to foraging and molting can be incorporated within it, in terms of how they affect fecundity and offspring quality. As such, it provides a useful framework within which to describe the requirements of behavior.

A. Transition from Foraging to Breeding: Locating a Suitable Place to Breed

This involves behavior that, at one extreme, may occur within a few meters, to the other extreme, movements that traverse the globe. Animals must choose a geographical area, a suitable site within that area such as a particular beach or ice flow, particular conditions within that site, and a position relative to other animals in the colony. BREEDING SITE selection must be accomplished so that parturition can occur on time and suitable mates are also available.

This transition can be considered to occur as soon as animals switch from a period of net gain of resources to net expenditure of body stores to support travel to breeding sites. It is not likely to be a sharp boundary, as animals may encounter food during their trip, but a reduction in resource acquisition is likely because animals are likely to leave prime foraging areas to make their way back to breeding sites. The critical issue is that animals arrive at the breeding site in sufficiently good condition to support the onset of breeding. In the case of many otariids, this means that females must have sufficient reserves to sustain themselves and their pups until the mother's first successful foraging bout. Condition in males will in part determine how long they can remain defending their access to females. In the case of the larger phocids, animals must have sufficient body condition to support the entire breeding effort. Smaller phocids or those with easy access to food may supplement stored reserves with foraging. It could be argued that the expenditure during this phase should be added to those of reproduction, but we know of no studies that have attempted to do this. Navigational skills and previous experience of suitable sites allow the minimum time and energy to be expended.

1. Large-Scale Movements: Choosing a Geographical Location Many species have been shown to have the navigational skills to return to previously used breeding sites from great distances, but the methods they use to accomplish this remain largely unknown. Both southern (*Mirounga leonina*) and northern (*M. angustirostris*) elephant seals have been tracked making directed trips of 1000–3000 km between breeding and foraging locations, arriving on the same beaches they used for breeding the previous season. Their great size (males may weigh over 3000 kg and females average 500 kg at parturition) is important in making such trips energetically feasible. Many other larger species are similarly capable. Smaller species may need a supply of food on route or feed closer to where they breed as the reproductive season approaches. It seems likely that the evolution of large body size in pinnipeds may have had much to do with enabling the uncoupling of geographical locations of feeding from those of breeding (for a contrary view, see Boyd, 1998).

2. The Local Scale: Choosing a Breeding Site within a Locality At the local scale, animals tend to breed where there are other seals present. This aggregative sociality is a key feature of pinniped behavior, although it is modified by the animals' state. The first animals to breed in a season tend to be the older, more experienced animals and their presence ashore encourages others to use the colony. This reinforcement means that the same rookeries tend to be used over a long term. It is possible to envisage at least two ways that new colonies may form. In a growing population, when space at an established colony has become limiting, pregnant females arriving to breed may be forced to move elsewhere if the available habitat is being used. In this scenario, the new colony should be reasonably close to the original one. Younger, primiparous females that breed later in the season may be forced to use an otherwise unoccupied location, and once there, others join them. Main criteria listed for breeding habitats are isolation (protection from land-based predators) and access to resources nearby. Few seals stray far from the sea. Otariids require a plentiful supply of food within reach of the breeding location, otherwise the breeding attempt may fail, but most phocids are functionally divorced from foraging requirements at breeding time by their ability to store resources, principally as blubber. Thus phocids may use breeding locations far removed from their foraging grounds [e.g., gray seals (*Halichoerus grypus*), northern and southern elephant seals]. Harbor seals represent an intermediate breeding and feeding strategy where mothers supplement their stored reserves with food acquired during foraging trips late in the lactation period.

3. Choosing a Position within a Site: The Individual Scale The local topography of the breeding location plays an important role in determining the particular location where pups are born. An almost flat and featureless surface with unlimited access to the sea, such as a sandbank, offers the simplest case in which animals have little to choose except their distance from the sea. Where such a location also has additional resources such as pools of water, these may act as foci for breeding animals. However, seals breeding on many islands or beaches typically arrive at the breeding area through specific access routes. Restricted access produces a radiated pattern of colonization, but also creates thoroughfares where there is continual traffic as animals arrive and leave the colony. Pup mortality in these locations can be substantial.

The degree of topographic variation on a breeding colony at a scale relevant to seals is a primary determinant of their distribution within a site. In fact, the degree of topographical variation on the breeding colony probably also defines the scale of site fidelity shown by gray seals at two Scottish colonies and may explain why fidelity is less apparent at relatively flat, open locations such as Sable Island (see later; Pomeroy *et al.*, 2000).

CONSERVATION OR MANAGEMENT considerations often require information on how animals may use available habitat, particularly where multiple use or potential conflicts occur. Habitat classification within a fine-scale geographic information system has been used to identify suitable breeding areas for gray seals, which was used to make a successful prediction of expansion of breeding areas at the Isle of May colony during the 1990s

(Twiss *et al.*, 2000, 2001). One of the main benefits of having such a detailed digital terrain model is that animals can be mapped accurately and frequently in the field, allowing examination of temporal changes in the distribution of animals. One of the products of such an approach has been the identification of the importance of pools of water on the colony for seal distribution and behavior (Twiss *et al.*, 2000).

4. Assessment of Breeding Locations and Site Choice   Seals that breed on land come ashore for a variable period before parturition occurs. For southern elephant seals, this averages 4.5 days, whereas for gray seals the average time ashore before parturition is 2–3 days, although some animals are present at breeding locations for at least a month prior to parturition. This prepartum period appears to involve some assessment of the breeding location. Female gray seals emerge from the water, looking around intently and sniffing continuously, before making tentative moves inshore. Pregnant females collect together in groups near access points where they remain inactive, and any disturbance is likely to make them return to the sea. In some cases, final selection of the pupping site occurs immediately prior to parturition when females move inland; during this movement they can be seen sniffing the ground. Pomeroy *et al.* (1994) found many cases of females returning to the sea before they came ashore again to pup, sometimes in a different location to that chosen originally. Options for changing the pupping site are often limited to the prepartum period, as in most species, once the pup is born, it is not easy for mother and pup to change location together.

Once a seal has chosen a breeding site, it tends to be used again and again. This breeding site fidelity is shown by Weddell seals (*Leptonychotes weddellii*), gray seals (see later), northern elephant seals, and Antarctic fur seals (*Arctocephalus gazella*) and is probably widespread. Scottish gray seals are faithful to their previous pupping sites. Females return to pup within an average distance of 55 m on North Rona and 24 m on the Isle of May from their previous breeding sites (Pomeroy *et al.*, 1994, 2000). Males that return also show very similar spatial fidelity at both colonies (Twiss *et al.*, 1994; Pomeroy *et al.*, 2000).

Some seal species display philopatry, i.e., they return to breed at the location where they were born, e.g., fur seals (Gentry, 1997) and southern elephant seals. Gray seals also display philopatry, sometimes with remarkable accuracy (Pomeroy *et al.*, 2000).

B. Investing in Young after Birth

Parental care in pinnipeds is the exclusive domain of the mother in all but one species (see later for single exception). Males take no part in the rearing process; the only part they play in breeding is to contribute sperm during mating. In fact, the process of mating is not without risk to current offspring. Pups may become separated from their mothers or be the subject of aggressive behavior from males or females at this time, as well as running the risk of being crushed by males. Therefore, maternal care does not consist simply of feeding but includes all behaviors associated with the pup's welfare, such as mainte-

nance of contact, vigilance, and defense against potential aggressors. A mother has finite resources available to service each breeding attempt and, once the pup is born, she has ultimate control over the feeding schedule and its duration. Mothers must gauge their reproductive effort according to the resources they have available, to do enough for the pup to have a good chance of survival without prejudicing the mother's survival or future breeding chances. The costs for mothers that expend too heavily in 1 year are reduced fecundity and lower breeding success in the next year (Pomeroy *et al.*, 1999; Trillmich, 1996). Consequently, the fundamental maternal trade-off is one of efficiency: supplying resources to the offspring at a low or acceptable expense. The single most important influence of the efficiency of the process is the "physiological time" it takes to accomplish it (Anderson and Fedak, 1987). For most phocids at least, maternal maintenance during the lactation fast must occur in parallel with the demands of feeding the pup. The shorter this time, the smaller the fraction of maternal resources that are lost as heat (generated by the combined metabolism of mother and pup) and the greater the fraction that can appear as pup growth or remain as maternal condition.

The conflict between pups' demands for resources and the requirement of mothers to limit expenditure to that which does not incur a threat to themselves is exemplified by a study of southern elephant seal pups fed as twins. In this case, mothers did not expend resources beyond the level expected for a single pup, so that the cutoff point in this case was fixed by the mother (Arnbom *et al.*, 1997).

1. Maintaining Contact with Pups   Seal breeding colonies are typically sensory-rich environments. Many animals are crowded into a restricted area, with the associated sights, smells, sounds, and actions associated with such a situation. A mother must maintain contact with her pup because, in general, neighbors react aggressively to foreign pups and pups that move away from their mothers may be injured or lose contact with their mothers and starve. At birth, a mother immediately sniffs and interacts with the neonate. By the time the first feed has been completed, mothers have established a bond with their pup that becomes progressively stronger as lactation proceeds. In most species, pups vocalize almost as soon as they are born, with mothers displaying a varying degree of competence in discriminating their own pup's call (fur seals, reviewed in Gentry, 1997). In gray seals which commute from the breeding beach to the water during lactation, a returning mother looks, sniffs, and (presumably) listens when she approaches her pupping site. Often several pups are inspected before one is fed. Reunions involve the approach by the mother, sniffing and flippering of the pup, and finally presenting the nipple to feed. Other pups trying to feed at this time are often dissuaded by aggression, but may also be excluded by the mother turning away. Some gray seal mothers (particularly at the expanding colony on the Isle of May, Scotland) are poor discriminators and feed any pups that approach them. Not surprisingly, these mothers rarely wean large offspring (see earlier discussion). Otariid mothers leave their pups unattended for several days while they forage for food and must recognize their offspring on their return. The primary mechanisms allowing this to occur successfully are smell and sound recognition.

2. Providing Protection Until the pup is born, many species of seals are tolerant of other conspecifics so that, for example, large groups of pregnant female gray seals may lie very close to each other, often touching. As soon as the pup is born this tolerance disappears and the mother becomes fiercely protective of the pup, defending a radius (typically 1.5 body lengths) around it. Any intruders into this space experience an escalating aggressive response, beginning with threats, approaches, then vocalizations, flippering, lunges, and, finally at the most extreme, contact involving biting and flippering. Tolerance of conspecifics varies between mothers, but it is not yet known whether this reflects some form of kinship recognition, familiarity based on nonrelated associations, or simply individual variation in response.

There is a single instance of possible paternal care in pinnipeds, where male Galapagos sea lions mob sharks around colonies (reviewed in Trillmich, 1996).

3. Lactation and Weaning The process of lactation is demanding for mothers. Most phocid species that fast during lactation lose 30–40% of their postpartum mass, much of it blubber, producing the highest-fat milk known in the animal world (up to 60% lipid) in the process. In most phocids the lactation period is short but intense (hooded seals, 4 days, pup growth rate 6.0 kg/day; gray seals, 18 days, pup growth rate 2.0 kg/day; southern elephant seals, 23 days, pup growth rate 4.2 kg/day). Such growth rates can only be achieved by having energy-dense milk, frequent feeds (every 4–5 hr in gray seals), and efficient conversion of maternal resources by the offspring. The concentrated milk also conserves water, which may be of short supply. Otariids have fewer absolute reserves available, although these may be relatively similar to those of phocids, and sustain their energy requirements by foraging throughout their extended but less intense lactation periods. This means that otariid pups receive feeds at intervals several days apart.

Weaning is abrupt in most phocids, as females depart from the rookery to return to the sea, leaving the pups on the beach. In most cases, mating has already occurred, and indeed observation of a successful mating is a good indication of a female's imminent departure. However, there is considerable individual variation in the time that mothers spend with pups after mating; some may remain for several days before returning to the sea. In many otariid species, a long lactation period allows offspring to develop swimming, diving, and foraging while having the option of maternal milk as a food source. As a result, otariid mothers may have a much more prolonged weaning process, as offspring may still be with their mothers in the second year after birth.

C. Locating and Selecting a Mate

Reproduction is the single most important action that individuals of any species carry out in their lifetimes. As such, mate choice is an important consideration. Circumstances dictate the degree of choice likely in that the distribution of females at breeding time controls the mating patterns seen. For example,

solitary hooded seal (*Cystophora cristata*) mothers on fast ice are unlikely to have many options in available mates and may simply mate with the male that has waited persistently beside her until she entered estrus. However, it seems likely that the successful male may have had to defend this position in encounters with others attempting to gain this opportunity. In this situation, males give the appearance of being monogamous. In contrast, large aggregations of female southern elephant seals make it possible for males to attempt to control access for mating, with the result that extreme polygyny occurs. Males in this situation compete vigorously among themselves, as the potential breeding rewards for successful males can be substantial. However, the priorities of each sex are rarely symmetrical. Female elephant seals may be considered to have exercised mate choice just after they arrive at the breeding beach. The 4.5 days spent ashore prior to parturition offer females an opportunity to assess the stability and safety of the harems they enter and the qualities of the guarding males. If pregnant females are disturbed during this preparturition period, they often change locations.

1. Female Mating Behavior Female seals are not receptive to males until they enter estrus. In gray seals, this occurs around day 15 of the average 18-day lactation period. Males that attempt to mate before the female is receptive receive a robust and clear message from the female indicating her unwillingness. Initially, a female will threaten males that approach and her subsequent vocalizations at a persistent male can alert surrounding females to his presence. Neighboring females may join in this threat display to dissuade the male, although in a very few cases, fights between male and female may develop. Because of the sexual dimorphism common to most polygynous mammals, males tend to be favored in such encounters. Females dissuade males using the same repertoire of aggressive behaviors as described earlier, but with the additional consideration that males attempting copulation are likely to have tried to mount the female. In this situation, the female's mobility and lack of cooperation, together with the aggressive display, are usually enough to make the male withdraw. Experienced males rarely attempt more than a preliminary investigation into the female status and seem particularly adept at gauging a female's receptivity.

The ability of females to resist premature advances is perhaps at its most dramatic in elephant seals, where males can be more than 10 times the mass of the females with which they mate. Even here, in a species in which males are not known for their gentility, females can repel unwanted advances. A male holding a harem frequently accesses the receptivity of females by "heading" them, i.e., he approaches and rests his head on the neck of candidate females. If the females are not receptive, they move their hind flippers rapidly from side to side in a swimming motion, slapping the side of the testing male. Most males take heed and move quickly on to test other females.

It is not clear exactly how estrus is signaled in most species. While the general behavioral indications are simply that a female's initial aggressive response to a mating attempt declines to acceptance and passivity, it is not clear what cues a male uses to judge the situation. Olfaction is probably important as males can be seen sniffing during their approach around females

(Gentry, 1997). A successful mating may also signal to other males that a female is receptive. Some females approach males and apparently solicit their attention.

Mate choice may range from having a single candidate, and therefore a passive default, or be an active process involving the assessment of, or competition between, a number of candidates. Competition may even be among sperm, in cases where multiple matings occur (e.g., gray seals, elephant seals, some fur seals). The most comprehensive studies so far come from gray seals and elephant seals. Females that occupy prime sites on gray seal breeding colonies tend to have dominant males nearby (Pomeroy *et al.*, 1994, 2000; Twiss *et al.*, 1994) and most mate with the dominant male. However, the number of pups they produce that are fathered by that male does not reflect the male's behavioral dominance or his mating success. In addition, evidence for genetically dissimilar partners has been put forward (Worthington Wilmer *et al.*, 2000). The reasons for this are not yet clear, but may lie in different attendance patterns of individual females at breeding colonies. There is some anecdotal and circumstantial evidence of mate choice in gray seals. At North Rona, where approximately 1200 pups are born each year, the father of a pup born to female J8 in 1986 was seen next to her in 1993, but both were at the other end of the island from where they had been in 1986. The pup born to J8 the following year was indeed fathered by the male seen with her in 1993. A more intriguing occurrence was first observed in 1997, when a known female left her peripheral pupping site to move about 80 m to the center of the colony where she was mated by a dominant male. She then returned to her pup and the attentions of a peripheral male at her pupping site. Females have been seen initiating copulations, but males initiate most.

2. Male Mating Behavior   A male's reproductive success is dependent on the number of offspring he manages to sire and how many of those eventually manage to reproduce as adults. The first part of this requires successful matings, and to achieve these, males must be able to take up a place among breeding females, avoid or outcompete other males, and gain a successful copulation. Males employ a variety of strategies to achieve success (see later). The second part of his reproductive success is less straightforward, as it is possible to achieve many matings without producing any surviving offspring, let alone grandoffspring.

The first prerequisite is simply to be around breeding females. Males must coordinate their efforts with the availability of receptive females. One of the most effective ways of gaining success for males is to spend a long time on the breeding colony, but this is costly, both in energetic terms, because males usually fast, and in potential injuries inflicted by competing males (Twiss *et al.*, 1998). For these reasons, a large size tends to correlate with male success so that the largest males tend to have advantages of increased energy reserves and of greater competitive abilities.

As discussed previously, the potential for polygamy in these animals depends on the distribution of females. Although the terms monogamy and polygamy usually apply to mating patterns of species, they may be applied to the tactics that individuals employ either throughout or during phases of their lifetimes. However, without complete knowledge of the reproductive histories of individual animals, it is difficult to make generalizations. Evidence from genetic studies can provide useful insights in these areas. In general, the evidence to date from genetics supports the general observational conclusions on mating patterns, e.g., in southern elephant seals and gray seals, although some queries have been raised. One such is the failure of apparently dominant males to account for as many paternities as predicted. Male reproductive longevities are as important as their within-season success. Long-lived, reproductively active males may accrue a greater success than live-fast, die-young males whose activities are conspicuous.

As with many mammals, the risks inherent in engaging in fights over breeding have led to a formalized ritual of aggressive displays in many pinniped species. Dominance hierarchies are common so that disputes lead to fewer actual fights than might be expected. Fights occur, but usually between closely matched opponents where the preliminary assessments could not determine a clear outcome (Arnbom *et al.*, 1997; Twiss *et al.*, 1998). In gray seals, males attempt to control access to groups of females by threatening intruders with open mouth displays, hisses, and vocalizations. Intruders are chased away, but serious challengers may produce fights, which can last up to an hour and leave either or both combatants seriously injured. It is common for losers of such fights to disappear from the breeding colony.

Given the high cost of engaging in the mainstream competition for mating opportunities, it is not surprising that alternatives strategies exist. Younger, less experienced males are seen around the periphery of breeding colonies and may acquire experience gradually. Some males employ a cryptic tactic, using their similarity to females to gain a position among females, making the most of their opportunities when the dominant male is engaged elsewhere. It is becoming evident for some species that aquatic mating may occur to a greater extent than had been suspected and that the phenotypic qualities that are successful on land may not necessarily be the same for aquatically mating males.

D. Mating

In most observable species, males usually initiate copulations. Males often act to immobilize the female in some way by holding or biting the back of the neck. On land, the male's weight applied via his body or flippers can help position the female. In the water, because animals are near neutrally buoyant, the male's weight is less important in restraint. In gray seals breeding on land, a male attempts to mount the female by maneuvering alongside and then throwing his head and shoulders over the female's back. Her response is almost always aggressive, but a female in estrus will accept the male's advances if he persists and manages to grasp the skin of her neck in his jaws. This act is the single best predictor of a female's acquiescence. At the same time, the male tries to achieve intromission by repeated pelvic thrusts, while the female either cooperates by lying still or resists by moving her rear as much as possible. Gray seals also mate underwater. There too, males grasp the females by the back of the neck in their jaws. Because the male cannot restrain the female as easily, she has greater opportunity to avoid the mating. Obviously, both must breathe and both move together to the surface when necessary. It is not clear how the need to breathe is communicated to the other partner, but cooperation is evident. In both situations, when a successful mating is achieved, the pair remains relatively motionless, for anything as brief as 5 min or as long as 40 min (gray seals; Twiss *et al.*, 1998). The function of such long copulations is not known, as males indulging in long copulations are leaving other females unguarded. It is thought that ejaculation occurs toward the end of the copulatory period; certainly females have been observed to have rhythmic contractions of the lower abdomen in the later stages of copulations. Remating of a female may take place soon after a copulation, either by the same or a different male.

E. The Transition to Foraging

1. Postweaning Behavior of Mothers   The mothers of most otariids leave their pups repeatedly to feed and gather the resources to support continued lactation. For them, weaning does not therefore represent a dramatic change in behavior. They simply fail to return. Some species, such as the Galapagos fur seal (*Arctocephalus galapogoensis*) may give birth to the subsequent pup before the prior pup is weaned, but this pattern is unusual. For most, animals may shift to nonbreeding haul-out locations and engage in longer and more distant foraging trips. For phocid mothers, however, weaning occasions an abrupt change in behavior. Typically, soon after mating occurs one or a few times, mothers abandon their pups, leave their position in the colony, and quickly enter the sea. In some species (particularly among otariids), animals may be seen traveling away from breeding sites in groups, but in many phocids, departure appears solitary. Pups are normally left behind. The "decision" to leave is a critical one in relation to the state of body energy reserves. Good foraging areas may be distant from breeding locations and have changed in position and value while animals were breeding. Mothers must have sufficient stored reserves to enable them to reach these without putting themselves at unacceptable risk.

2. Postweaning Behavior of Pups   Weaning prompts dramatic changes in behavior for pups as well. Once pups are weaned, they no longer have the protection of their mothers. They no longer nurse and begin to fast for a time before the transition to nutritional independence. Otariid pups undergo what may be thought of as temporary weaning, when their mothers depart to sea to feed in between bouts of lactation. They have the experience of being left unattended prior to true weaning and show some of the behaviors of weaned pups early in development. In either case, pups typically move to areas where they can avoid contact with adults and may often congregate in large groups. In many cases, this movement is stimulated by aggressive encounters with other mothers and adult males within the colony. In elephant seals, mothers leave the weaned pup behind in the harem, possibly at a central location within it. Other mothers will act aggressively to the approach of pups other than their own, which tends to move unattended pups around, with a net movement to the periphery of the harem. Within a day of weaning, pups are usually out of the harems and then move around the beaches in an apparently

undirected way. When they encounter other pups, they tend to remain with them. The end result of this mobility is that large numbers of pups end up in "crèches" at places on beaches where no harems are present.

Phocid pups often remain in such groups, associated with breeding sites, for periods from days to months after weaning. The function of the time spent in these "postweaning fasts" is not understood, but is thought to involve a period of physiological, behavioral, and/or social development. Pups may interact with one another, exhibiting some variants of adult behavior. For example, male elephant seal pups often engage in mock fights that involve the rearing up and head strikes seen in battles between breeding males. Pups of both sexes tend to move into shallow inshore water or freshwater ponds during the night and swim and dive. Dive depths as great as 271 m have been observed in pups from Macquarie Island during this time (Hindell *et al.*, 1999), but in general dives are short and shallow. An increasing fraction of the day is spent in the water until departure on the first foraging trip, typically after about 30–45 days and after about 35% of the mass attained by weaning has been lost.

For pups too, the decision on when to leave is potentially difficult and critical. Phocid pups have no prior, independent experience of foraging locations. They tend to leave as individuals, not forming into groups to avoid predators, and in any case, any other pups departing after weaning are similarly naïve. The way they choose to locate foraging sites and the cues they use to help them remain largely unknown. Even otariids and phocids with unusually precocious young do not appear to use the opportunity for mothers to lead pups to food. Walruses (*Odobenus rosmarus*) seem to be the only exception to this. Pups travel with mothers prior to weaning and nurse at sea, giving them the opportunity to get geographical information on where to feed. It seems likely that if mothers could direct pups to food, they could obviate the need for a fraction of the material resources given to pups with this information. It seems surprising that this occurs rarely.

IV. Walruses, Sirenians, Sea Otters, and Polar Bears

While facing the same fundamental problems as seals in accomplishing successful reproduction, these groups show some unique variations in reproductive behavior and other life history features. Detailed accounts can be found in Fay (1982) for walruses and Riedman and Estes (1990) for sea otters (*Enhydra lutris*). This section discusses only those features peculiar to these groups, emphasizing the unusual features of reproductive behavior.

A. Walruses

Walruses share many features of their breeding behavior with phocid and otariid seals but differ in that much of the important behavior takes place underwater. Like many species of seals, they are polygynous. Males display and interact aggressively to gain access to females, but unlike some phocid and otariid species, this activity tends to take place exclusively in the water, where mating also occurs. Both sexes are often highly gregarious when hauled out on ice or land, and animals are often seen in closely packed groups. While the activities performed out of water may have important physiological and social functions, with the important exception of parturition, they do not seem to relate directly to reproduction. Pups are born on ice or land but unlike seals, nurse at sea. Nursing is often seen when a mother orients herself vertically with her head above water while the pup nurses upside down, hindflippers at the surface and head down at the nipples. Walruses are the only pinniped group where offspring are known to accompany mothers during maternal foraging and this is made possibly by the ability to feed pups at sea. Because walrus offspring move about with their mothers when they feed, young have the opportunity to learn about foraging locations and techniques from their mothers, which probably has important implications for the demands placed on mothers by lactation.

B. Sirenians

The manatees (*Trichechus* spp.) and dugong (*Dugong dugon*), are unique in being the only group of mammalian marine herbivores. This lifestyle has led to unusual distribution patterns as well as unusual breeding and social behavior. They are not colonial breeders. Individuals of both sexes move in response to the availability of resources such as food and fresh water but they do not to move in herds while foraging. The only apparent long-term social link seems to be between mothers and calves (see later). Florida manatees (*T. manatus latirostris*) living at the northern edge of their range congregate in large groups around warm water sources, such as power plant effluents and warm springs. Although they seem dependent on warm water sites at times of exceptionally cold weather, this opportunistic proximity is not utilized to bring the sexes together for mating. Males and females can range widely at other times. Locating mates seems to be the result of chance encounters between males and estrous females. Little is known about how males locate estrous females, but the increased mobility of estrous females may increase chances of encounters. When a female comes into estrus (lasting up to 3–4 weeks), "mating herds" of hopeful males surround her. The generally quiet and gentle appearance of the species is belied at this time with aggressive behavior between males trying to secure mating opportunities. During estrus, females may mate with several males. Calves are born underwater and nurse there, accompanying their mothers for 1–4 years. Contact between mothers and young is maintained in part acoustically. The extended period of contact between mother and calves while animals forage offers ample opportunity for information to be passed from a mother to her young. This is in marked contrast to the situation described earlier for most pinnipeds and probably has played a role in setting up what has been termed "learned traditional patterns of distribution." The seasonal aggregations of Florida manatees at cold weather refuges such as the springs and power plants mentioned earlier often contain individuals from several generations. Small groups often also congregate briefly at so-called "rendezvous sites" (often places where watercourses intersect) where animals may interact, rubbing and swimming around one another. The social role of these interactions is not understood fully.

C. Sea Otters

Sea otters probably arose from a different evolutionary lineage than the groups discussed earlier from whom they show some important differences in reproductive behavior. Like sirenians, they are not colonial breeders. Although they have a polygynous breeding organization, males occupy and defend a territory that is used for foraging and reproduction. Even though they seem less specialized for a marine existence than seals, sea otters spend virtually all their time at sea, foraging, mating, giving birth, and rearing their young in the shallow coastal zone. However, they are limited to relatively short, shallow dives and spend a greater proportion of time at the water surface than most seals and whales. Most of their reproductive behaviors occur there. Females tend to have partially overlapping home ranges along the coastal zone. Adult males have ranges that overlap with those of one or more females but not of other males. Higher quality male territories (e.g., those rich in food or other resources) tend to overlap with those of more females. Males defend their ranges, but this is often accomplished without obvious aggression, possibly via cues that allow individual recognition and the use of knowledge of past encounters to settle disputes. Males may leave their territories from time to time and may compete elsewhere with other males for access to females.

Mating takes place at the water surface and can involve precopulatory touching and playing. During copulation, males may grasp females by the muzzle while the pair tumble about vigorously. The grip of the males often causes injury and permanent scarring. Pups are born and nurse at sea and may spend 5 months or more with their mothers. Information exchange between mothers and pups is likely to play a role in determining the survival of the young. Pups are often provided with prey by their mothers and may show preference for the same prey types. Sea otters are well known as "tool users" (using hard objects to break up prey) and these skills may also be passed from mothers to pups. Mothers and pups can be quite vocal, calling to each other at the surface using a variety of sounds. Again, males play no role in the care of young.

We are left with the intriguing question: if the ability to give birth at sea, nurse in the water, and lead pups to food is possible in the walrus, sirenians, and sea otter, why do these patterns not occur more often in the "mainstream" pinnipeds? It is true that most extant pinnipeds produce young that are not equipped with the insulation to enter cold water directly from the womb, but in a few species (e.g., harbor seals), pups do enter the water within hours of birth. Aquatic parturition and nursing therefore seem to be possible evolutionary options for the pinnipeds as they are for these others groups and, as discussed later, cetaceans.

D. Polar Bears

Polar bears (*Ursus maritimus*) are considered marine mammals because they range widely over sea ice foraging on seals and often enter the sea to swim relatively short distances to and from ice floes and land. They tend to fast during the summer months when sea ice is absent. Their reproductive patterns are similar to those of brown bears (*U. arctos*). After a period of delayed implantation and a relatively short 3- to 4-month gestation period, the small (approximately 0.5 kg), altricial young are born in dens on land near shore or occasionally on stable sea ice. At this stage both mother and young exist on the mother's stored reserves, acquired from the previous winter's foraging out on the frozen sea. Cubs remain in the den with the mother until they grow to about 10 kg and then may spend some additional time in the den area before mother and cub move away onto the sea ice to look for food. Cubs spend around 2 years with their mother, presumably learning and gaining hunting experience, so there is ample opportunity for information to be transferred from mother to cub.

Male bears will kill and eat cubs if their mother does not defend them, and mothers will not associate with males or mate until cubs are independent. It is not clear how males find and recognize receptive females, although olfaction probably plays an important role.

V. Cetaceans

Unlike pinnipeds, cetaceans have evolved a behavioral and anatomical suite of adaptations allowing them to mate, give birth, suckle, and nurture their young entirely in water. Freed of the spatial and temporal constraints imposed by reliance on land or ice to breed, cetaceans have developed a wide diversity of social systems and life history strategies quite unlike those of the pinnipeds. While some cetaceans, principally the mysticetes, compartmentalize breeding to a temporally and spatially discrete component in their lives, the majority breed and acquire resources simultaneously. Further, consecutive breeding attempts themselves may be superimposed upon each other, with females concurrently rearing calves from different breeding attempts and even contributing directly to the survival of their offspring's own offspring.

Having no need of land to reproduce has assuredly led to the success and ubiquitous nature of the cetaceans, but for the same reason has also severely hampered our abilities to understand them. At the most basic level, discrete acts, such as copulation and birth in most species, have never been observed, let alone quantified, while comparing the success of different tactics employed by individuals within populations is simply impossible as yet. What is known is pieced together from anatomical studies, whaling operations, live captures, individual identification, genetic analyses, and interspecies comparisons. From these fragments, it is clear that the cetaceans have much to teach us about the ecological determinants of reproductive and social behavior and even offer the potential to broaden our understanding of mammalian reproductive behavior as a whole.

A. Seasonality of Reproduction

For the majority of cetaceans, reproduction has a seasonal component. For mysticetes (with the possible exception of Bryde's whales, *Balaenoptera edeni/brydei*), breeding occurs as a discrete phase of each year with other aspects, principally feeding, often being either reduced or halted entirely. The best-known species shuttle on an annual basis between productive feeding regions and areas associated with parturition, early nursing, courtship, and mating. The reproductive behavior of gray whales (*Eschrichtius robustus*) is a prime example. Existing populations survive in the northern Pacific and migrate from high-latitude temperate or polar waters after a summer of feeding, southward along North American and Asian coasts to breed in sheltered coastal waters. Although almost all gray whales migrate, whether reproducing or not, pregnant females move south earlier than males and then 80 or so days later return north again following behind the males and newly mated. Humpback and right whales follow similar patterns, but the behavior of rorquals, such as the blue (*Balaenoptera musculus*), fin (*B. physalus*), and minke (*B. acutorostrata* and *B. bonaerensis*) whales, is more poorly known and, though seasonal, it is as yet unclear for many populations when and where breeding actually occurs.

The lives of odontocete cetaceans are less obviously compartmentalized and breeding takes place simultaneously with other activities. Detecting breeding seasons is consequently harder and is usually estimated from parameters such as the first appearance of neonates at sea, fetal maturity in stranded or captured animals, and seasonal changes in testes. From such studies, it appears that the majority of odontocetes extend their breeding activities over protracted seasons. Interestingly, those that remain in high-latitude areas tend to reproduce at the opposite time of year to neighboring mysticetes. Harbor porpoises (*Phocoena phocoena*) in the North Atlantic, for example, ovulate, mate, and give birth in spring and early summer, whereas seasonally sympatric humpback whales migrate south to breed in winter. Furthermore, within species, the specific timing of reproduction may vary by region or population. Common bottlenose dolphins (*Tursiops truncatus*), for example, show diffuse seasonal peaks in reproduction but these vary in their timing with location.

The reasons why mysticetes and odontocetes adopt such differing behavioral and physiological strategies toward the seasonality of reproduction remain poorly understood. Body size clearly allows the larger mysticetes and sperm whales to store sufficient reserves to forego feeding and dedicate time to breeding. Because most odontocetes are smaller, it is tempting to assume that they have less capacity to fast during a discrete breeding season; however, they are of similar body size or larger than the highly seasonally breeding pinnipeds. It therefore remains a possibility that odontocetes, and females in particular, have protracted breeding seasons, simply because other aspects of their lives allow it. For all cetaceans, it is likely that food availability, risk of predation, water temperature, and sea or river conditions are important in dictating which season is actually selected to breed.

B. Gathering Resources to Invest in Reproduction

At times outside of specific breeding seasons, mysticetes gravitate toward areas that maximize their potential for prey consumption. Migrations into productive, often high-latitude, areas are therefore common. The duration and rate of energy acquisition appear to be important in determining subsequent reproductive interval, ovulation rate, and fecundity. For odontocetes, particularly the smaller species, such migrations are less evident and suggest that their reproductive capabilities allow them to remain in their foraging areas year-round. A different strategy is found in the much larger sperm whales (*Physeter macrocephalus*). Females remain in tropical or subtropical waters year-round, whereas the sexually dimorphic males migrate from productive high-latitude feeding areas toward the equator and their mates to breed. The extreme sexual dimorphism (with males weighing up to three times as much as females) may necessitate such migrations to regions especially rich in prey.

C. Locating a Suitable Place to Breed

Breeding in cetaceans can be broken down into three phases: giving birth, suckling young, and mating. Because gestation in most cetaceans is close to 12 months, these three activities generally occur at a similar time of year and are therefore often considered as if they were one event. However, the factors that influence each differ and thus the choice of breeding habitats may well represent a compromise for the individuals concerned. The processes of giving birth and suckling young may benefit from waters with low predator abundance while these characteristics will be of less importance for mating. Examples of differences in locations for these activities are rare, especially in odontocetes, where little is known about breeding site selection as a whole. One study of harbor porpoises in the North Sea found significantly higher proportions of calves in a specific coastal area relative to neighboring waters, although the reasons why this area was favored are as yet unknown. Mysticetes offer more concrete examples of breeding areas. Those most studied include humpback (*Megaptera novaeangliae*), gray, and right whales (*Eubalaena* spp.), which typically breed near coasts, with the latter species favoring sheltered shallow waters. As might be expected, they also offer some evidence of the differing requirements of raising young and mating, with females with newborn calves favoring slightly different areas to the other breeding individuals. It is unknown if breeding site availability is a significant factor limiting the size of behavior of cetacean populations.

Underlying all of the issues associated with mysticete migrations to breeding sites is the controversy over why the mysticetes migrate at all. Sheltered shallow waters are not unique to the tropics, and some mysticetes, such as the bowhead whale (*Balaena mysticetus*), appear capable of breeding in the same polar waters in which they feed. Factors that pose direct benefits to adults do not appear to withstand scrutiny whereas the thermal constraints on calves do not seem likely when neonate mysticetes are larger than most adult odontocetes and are probably already thermoneutral in colder waters. Instead, the possibility of calf PREDATION by killer whales (*Orcinus orca*) might lie as the root cause of such enormous migrations for at least some of those species large enough to be capable of making them.

Gray whales appear to navigate to and from their breeding areas by following the coastal margins of their respective continents. In contrast, the north–south migrations of humpback whales seem to be deflected by coastlines, currents, and underwater topography rather than guided by them. How these whales find locations such as the Hawaiian Islands each year in waters as large as the Pacific is still unknown. Use of celestial, acoustic, or magnetic markers are distinct possibilities.

D. Giving Birth

Few cetaceans births have been observed in the wild, but in captive odontocetes, most births are accomplished rapidly without direct assistance from conspecifics. However, there are several reports of animals seemingly helping in the birth process, either pulling the fetus or placenta clear of the birth canal. The frequency of such activities, if they occur at all in wild populations, is unknown. Because wild births have been observed so rarely, little is known about how females might reduce the risks of predation and separation in the moments after birth. The proficient swimming abilities of newborn calves permit mother and neonate to vacate an area rapidly where the birth occurred and so minimize the attentions of predators, while the social nature of cetaceans may permit increased predator detection and defense. Newborn calves adopt a swimming posture alongside their mothers permitting tactile communication, camouflage, and slipstreaming.

E. Investing in Young after Birth

Parental care in cetaceans is predominantly the responsibility of the mother, although male and female kin as well as unrelated females may provide additional care. The levels of non-maternal care vary considerably among species, being generally rare in mysticetes and more common in odontocetes. Pilot (*Globicephala* spp.) and killer whales offer some remarkable examples of care and nutritional investment that are notable not just within marine mammals but for all mammals.

1. Maintaining Contact and Providing Protection All cetaceans are born nutritionally dependent on their mothers, but with births occurring without the spatial certainties of land or ice, there is potential for calves to become separated from their mothers and therefore starve rapidly or become prey. For those species that suspend foraging during breeding, females can devote almost continuous attention to their calves. Species that continue to forage face a problem, as the diving abilities of calves may be insufficient to follow their foraging mothers. Babysitting among this latter group appears to be a common solution, with an apparent continuum among odontocetes. This can range from females of similar breeding status schooling with one another and presumably taking turns guarding calves or at least acting as a spatial point of reference, through to related females and their adolescent young accompanying mothers and their neonates in sperm whales. It may even extend to a system demonstrated by killer and pilot whales (apparently unique among mammals) of stable kin groups with neither male nor female dispersal but instead investment in raising their own (females only) and related offspring (females and males). Despite such behavioral safeguards, however, cetacean mobility makes separations between mother and calf inevitable. Individually specific calls are thought to be important in reuniting individuals in species such as the common bottlenose dolphin. However, such mechanisms take time to develop, and neonate mortality, although low compared with other mammals, is substantial.

Our understanding of cetacean reproductive behavior is undoubtedly hampered by the potential ambiguity of the behavior that can be observed. This is particularly prevalent in aspects involving the spatial proximity of individuals and apparently altruistic or cooperative behavior. The social complexities of cetacean societies and the considerable component that appears to be learned and practiced make context an essential component of any behavioral observation. Babysitting is an attractive and logical concept that has been described frequently, but the appearance of a calf with an adult animal other than its mother may result from other motives. Young inexperienced female Indian Ocean bottlenose dolphins (*T. aduncus*), for example, may temporarily kidnap calves and thereby improve their own maternal skills (Mann *et al.*, 2000), whereas males and females of social mammals in general may benefit from capturing and killing another's young.

2. Lactation and Weaning Cetacean calves do not suffer the constraints experienced by phocid seals, which need to rapidly transfer resources to the pup in order to resume feeding, nor do they endure the periods of maternal absence experienced by otariid pups. Instead female cetaceans take their mobile calves with them and are generally only separated for the length of a foraging dive. Thus calves have the opportunity to suckle frequently and match milk intake with energy expenditure and growth. Milk is transferred via two mammary teats, which are located in slits to either side of the genital opening about two-thirds of the way down the mother's body on her ventral side. The presence of bristles on the rostra of neonate cetaceans is thought to help calves orient during suckling while a frilled margin to the tongue and muscular control of milk ejection likely aid efficient milk transfer.

Mysticete calves are generally weaned within a year of birth and, in migratory species, coincides with the pair reaching high-latitude feeding grounds. Weaning may precipitate separation of cow and calf but in its timing offers opportunities for the mother to train a calf in migration routes and the location of feeding areas, and potentially facilitate membership of feeding assemblages. Lactation in the majority of odontocetes is longer, and in some cases far longer, with weaning appearing to be gradual and occurring over a period of months or years. Lactose, for example, has been detected in the stomachs of sperm whales up to 13 years of age. Such long-term maternal investment suggests that many components of odontocete development require a considerable period of learning and training. Foraging tactics in odontocetes are often performed in groups and, while it is unknown what proportion are cooperative or exploitative, their complexity is clear, as is the need for a high degree of interindividual coordination and practice. Calves may learn through observations or dedicated tutoring. The prolonged lactation and consequent investment in young allow calves to develop before facing nutritional independence. Bottlenose dolphins probably wean around 18 months after birth but remain closely associated with their mothers for at least 4 years. Young killer and pilot whales may never separate from their mothers, drawing comparison with elephant matriarchal societies where the eldest animals may function as long-term information stores and guardians for their offspring and offspring's offspring.

F. Locating and Selecting a Mate

1. Female Mating Behavior The number of offspring that female cetaceans produce in a lifetime varies. Groups such as the porpoises and baleen whales may give birth on an annual or biennial cycle and have the potential to produce between 12 and 20 offspring in a lifetime. Others produce much fewer. Some female killer whales, for example, may produce as few as 5 to 6 young in their entire lives. Whether 20 or 5, these numbers are small given the huge investment in time and resources that each calf receives, and therefore the choice of an appropriate mate to father them is a major component of the reproductive fitness for individual females.

In short, we have little solid information on how female cetaceans choose mates. In any species or population there are many potential junctures where a female may be making behavioral or physiological decisions, both before conception and afterward. In some instances, females may have the opportunity to simply select a particular male with which to copulate from a range of alternatives. Such a case has been proposed for humpback whales, where males may engage in communal display behaviors on the breeding grounds without showing any defense of resources. Females would have opportunities to approach males based on the quality of their displays. Female bottlenose dolphins are frequently seen being attended by single or alliances of males. While males may have opportunities to herd a female against her will, the females may also have the opportunity to reject or maintain that contact. Females have been observed rebutting the copulation attempts of males by fleeing or rolling upside down at the surface so that males do not have access to the female's genital opening.

After copulation, females may have a range of behavioral and physiological options to influence the probability of conception. The number of subsequent males with which she mates would influence the probability of a particular male being the father. The large volumes of sperm produced by males of several cetacean species (see later) suggest that females mate with several males and that competition between the sperm themselves may be a frequent occurrence in such species. Mating repeatedly with a particular male would also bias the odds significantly. Whether an egg is available for fertilization is also critical and it appears that ovulation itself may be related to mating and therefore has the potential to be under the female's control.

Even after fertilization has occurred there are opportunities to select whether to continue investment in a particular partner's offspring. These may range from selection abortion, energetic investment in the fetus, and the subsequent level of parental care expended in the calf. At present we have little information to determine whether such behavioral and physiological decisions are made, but because evidence for such has been found in birds and terrestrial mammals, it is possible that such options are open to female cetaceans.

2. Male Mating Behavior The reproductive behavior displayed by males is a function of the social and physical environments in which they live and compete. As with the diversity of habitats and lifestyles exhibited by cetaceans, males of different species and populations show a huge range of tactics to maximize their reproductive potential. At a basic level, males should behave to optimize the number of their own sperm competing to fertilize a female's egg and limit the number of those of competitors. Thus males may increase the probability of obtaining copulations by signaling their quality to females and competing males through physical or acoustic displays (e.g., postural displays in bottlenose dolphins, songs of humpback whales, sperm whale vocalizations), ornamentation or body scarring (teeth and scars in beaked whales), intermale combat (humpback whales), extreme body size (sperm whales), or simply tracking the long-distance migrations of females (humpback, gray, and right whales). Males may attempt to guard receptive or potentially receptive mates to reduce the probability of competitors mating and increase the number of copulations they can obtain themselves. The alliances formed between male bottlenose dolphins may be an example of such behavior where pairs or trios of males may trade off their exclusive access to a female in order to ensure that other males cannot gain mating opportunities. The potential absence of such alliances in some other populations of common bottlenose dolphins (Wilson *et al.*, 1993) suggests that such tactics are context specific and influenced by factors such as the relative abundance of receptive females and potential for males to monopolize them.

Even once copulation has occurred, competition between males need not be over. Gray, bowhead, and right whales and harbor porpoises all have testes substantially larger than their body size would predict. Large quantities of sperm and the ability to copulate frequently may allow males to flush away or dilute the sperm and consequent reproductive chances of other males.

Males may also be able to increase the effective pool of receptive females by influencing the fate of other males' calves. Infanticide is common among terrestrial mammals and is often carried out by males in order to force females to switch from investing resources into a previous calf not sired by themselves and to become reproductively receptive again. It is unknown if such behavior occurs in cetacean societies, but the violent deaths of young bottlenose dolphins in some populations suggest that males may exploit such options.

The long lives and intricate social organization of cetaceans, particularly odontocetes, also offer males the opportunity to increase their fitness, not by maximizing their potential to father offspring, but by investing in their kin. The lack of male dispersal in killer and pilot whales and the absence of interbreeding within pod members suggest that males may remain with their maternally derived relatives to provide care or protection and thus to increase their own inclusive fitness.

G. Mating

Cetaceans live in a three-dimensional environment that facilitates copulation from a variety of orientations. Common positions include ventrum to ventrum with the pair oriented in the same directions, or the male may mount the female from a nonparallel position. Intromission may last only a few seconds or far longer and involve vigorous thrusting or a more passive attitude. Mating may be preceded and followed by prolonged periods of courtship and petting. Mating episodes may be repeated over periods of minutes, hours, or days.

H. The Transition to the Nonbreeding Season

While copulation and parturition are generally seasonal in cetaceans, the investment in reproduction for females is an almost continuous process after reaching sexual maturity. Baleen whales that migrate from breeding grounds with neonate calves wean them on the feeding grounds. They may have several months of intensive feeding before returning to the breeding grounds to mate or be already pregnant following copulation the previous year. Female odontocetes frequently superimpose reproductive events by being both pregnant and lactating or suckling more than one generation of calf at the same time. Perhaps the most intriguing situation is demonstrated by female pilot whales, which appear to cease ovulating after age 40 and yet continue to lactate for well over another decade. In doing so, they have the potential to not only extend long-term care to their own offspring, but also influence the fate of their offspring's own offspring.

See Also the Following Articles

Aggressive Behavior ■ Captive Breeding ■ Courtship Behavior ■ Estrus and Estrous Behavior ■ Female Reproductive Systems ■ Male Reproductive Systems ■ Mating Systems ■ Territorial Behavior

References

Anderson, S. S., and Fedak, M. A. (1987). The energetics of sexual success of gray seals and comparison with the costs of reproduction in other pinnipeds. *Symp. Zool. Soc. Lond.* **57**, 319–341.

Arnbom, T., Fedak, M. A., and Boyd, I. L. (1997). Factors affecting maternal expenditure in southern elephant seals during lactation. *Ecology* **78**(2), 471–483.

Boyd, I. L. (1998). Time and energy constraints in pinniped lactation. *Am. Nat.* **152**(5), 717–728.

Fay, F. H. (1982). Ecology and biology of the Pacific walrus, *Odobenus rosmarus divergens* Illiger. *U.S. Dept. Inter. Fish. Wildl. Ser.* **74**, 279.

Gentry, R. (1997). "Behavior and Ecology of the Northern Fur Seal." Princeton Univ. Press, Princeton, NJ.

Hindell, M. A., McConnell, B. J., Fedak, M. S., Slip, D. J., Burton, H. R., Reijnders, P. J. H., and McMahon, C. R. (1999). Environmental and physiological determinants of successful foraging by naive southern elephant seal pups during their first trip to sea. *Can. J. Zool.* **77**(11, Nov.), 1807–1821.

Le Boeuf, B. J., and Laws, R. M. (eds.) (1994). "Elephant Seals: Population Ecology, Behavior, and Physiology. University of California Press.

Mann, J., Connor, R. C., Tyack, P. L., and Whitehead, H. (eds.) (2000). "Cetacean Societies: Field Studies of Dolphins and Whales." The University of Chicago Press.

McConnell, B. J., and Fedak, M. A. (1996). Movements of southern elephant seals. *Can. J. Zool.* **74**, 1485–1496.

Pomeroy, P. P., Anderson, S. S., Twiss, S. D., and McConnell, B. J. (1994). Dispersion and site fidelity of breeding female gray seals (*Halichoerus grypus*) on North Rona, Scotland. *J. Zool. Lond.* **233**(3), 429–448.

Pomeroy, P. P., Fedak, M. A., Anderson, S. S., and Rothery, P. (1999). Consequences of maternal size for reproductive expenditure and pupping success of gray seals at North Rona, Scotland. *J. Anim. Ecol.* **68**, 235–253.

Pomeroy, P. P., Twiss, S. D., and Duck, C. D. (2000). Expansion of a gray seal (*Halichoerus grypus*) breeding colony: Changes in pupping site use at the Isle of May, Scotland. *J. Zool. Lond.* **250**, 1–12.

Pomeroy, P. P., Twiss, S. D., and Redman, P. (2000). Philopatry, site fidelity and local kin associations within gray seal breeding colonies. *Ethology* **106**, 899–919.

Reynolds, J. E., and Rommel, S. A. (eds.) (1999). "Biology of Marine Mammals." Smithsonian Institution Press, Washington, DC.

Riedman, M. L., and Estes, J. A. (1990). The sea otter (*Enhydra lutris*): Behavior, ecology and natural history. U.S. Fish and Wildlife Service, Biological Report 90(14), Washington, DC.

Trillmich, F. (1996). Parental investment of pinnipeds. *Adv. Study Behav.* **25**, 533–577.

Twiss, S. D., Anderson, S. S., and Monaghan, P. (1998). Limited intraspecific variation in male gray seal (*Halichoerus grypus*) dominance relationships in relation to variation in male mating success and female availability. *J. Zool. Lond.* **246**, 259–267.

Twiss, S. D., Caudron, A., Pomeroy, P. P., Thomas, C. J., and Mills, J. P. (2000). Fine scale topography influences the breeding behavior of female gray seals. *Anim. Behav.* **59**(2), 327–338.

Twiss, S. D., Pomeroy, P. P., and Anderson, S. S. (1994). Dispersion and site fidelity of breeding male gray seals (*Halichoerus grypus*) on North Rona, Scotland. *J. Zool. Lond.* **233**, 683–693.

Twiss, S. D., Thomas, C. J., and Pomeroy, P. P. (2001). Topographic spatial characteristics of grey seal breeding habitats at a sub-seal size spatial grain. *Ecography* **24**, 257–266.

Wilson, B., Thompson, P. M., and Hammond, P. S. (1993). An examination of the social structure of a resident group of bottle-nosed dolphins (*Tursiops truncatus*) in the Moray Firth, NE Scotland. *Eur. Res. Cetaceans* **7**, 54–56.

Worthington Wilmer, J., Overall, A. J., Pomeroy, P. P., Twiss, S. D., and Amos, W. (2000). Patterns of paternal relatedness in British gray seal colonies. *Mol. Ecol.* **9**, 283–292.

Respiration
SEE *Breathing*

Ribbon Seal
Histriophoca fasciata

GENNADIY FEDOSEEV
Marine Mammals Council
Moscow, Russia

I. Diagnostic Characteristics and Taxonomy

The ribbon seal is an endemic species of the North Pacific. Although it used to be grouped together with the Greenland seal into the same genus, Pagophora (Naumov and Smirnov, 1936), it is now considered to belong to the monotypic genus *Histriophoca* (Scheffer, 1958; Geptner *et al.*, 1976).

The ribbon seal's coloring is bright and unusual (Fig. 1). There are four light stripes against a black or brown background: one stripe goes round the neck, another encircles the body at the bottom of the flippers and two more symmetrically underline the base of the pectoral flippers. During the first year the color is fairly dull and becomes brighter only when the seal is 2 years old. The skull of the ribbon seal is short, the cranium braincase and cheekbones are wide, and the face is short and narrow. The teeth are small; their number varies from 32 to 36 (Geptner *et al.*, 1976).

II. Anatomical and Physiological Features

The ribbon seal differs from the rest of the seals by its respiratory anatomy. There are no lobes in the lungs; the trachea in its middle part consists of semirings at the top and a membrane at the bottom. Through a narrow fissure the membrane is connected with a thin-walled air sac that is quite well developed in a mature male. Females have smaller air sacs, whereas immature animals have nondeveloped ones. The air sac also functions as a hydrostatic organ used in phonation (Sokolov *et al.*, 1968).

The ribbon seal has several features that show the degree of its adaptation to deep diving and fast SWIMMING, with well-developed internal organs among them. Their weight as a proportion of body weight is high: heart on average, 8‰, lungs 17‰, diaphragm 7‰, liver 22‰, and kidneys 2.7‰. These indices are higher than those of other seals (Fedoseev, 2000).

The ribbon seal is physiologically well adapted to a pelagic lifestyle and deep diving (up to 600 m): it has the highest number (3.9–4.7 million) and volume (50–72%) of erythrocytes and the highest blood hemoglobin contents (18–26 g%) among all seals (Sokolov, 1966).

III. Distribution and Population Size

The ribbon seal inhabits that area between the southern part of the Chukchi Sea and Japan and Korea. There are three populations of this particular species: two in the Okhotsk Sea and one in the Bering Sea (Fig. 2). A part of the latter in spring and summer migrates to the southern region of the Chuckchi Sea. Single ribbon seals have been seen close to the California coastline (Morro Bay). The range of the migration is little known.

In 1961 the population size of the ribbon seal in the Bering Sea was estimated at 115,000 to 120,000. By 1969, because of sealing, the population was reduced to 60,000 to 70,000. After sealing had been limited to 2000 to 3000 animals a year (vs 9500), the population size by 1987 reached 120,000 to 140,000. In the Okhotsk Sea, according to data collected for many years, the population size fluctuated between 200,000 (1968–1974) and 630,000 (1988–1990). For many years the average number was 370,000 with the central northwestern population being 320,000 and the southern 50,000 (Fedoseev, 2000).

IV. Ecology

The ribbon seal belongs to the ice seals, whose life is closely connected with ice. However, if the ice is thicker than 10–15 cm, the ribbon seal is unable to make holes in it. This is why it dwells in areas with stable white ice broken into huge chunks and avoids areas of solid ice fields. Broken ice usually appears in regions of an abrupt continental shelf, where due to high water circulation this form of ice dominates; moreover, it constantly gets moved around. In such areas there are a lot of fish and squid, which are the main foods for ribbon seals. In the Bering Sea, these areas coincide with

Figure 1 *Male ribbon seal.*

the sea ice edge. In the Okhotsk Sea, where the climate is more severe and ice covers up to 80–90% of the sea surface, reproductive rookeries are located deep in the ice, whereas in the southern part of the sea they can also be found where there is a deep sea bed with cyclic water turnover that facilitates spacing the ice out.

The period of mating and pup rearing is adjusted to spring ice breakup. When it shifts together with the beginning of ice breakup, this results in the reproductive isolation of different populations. It is in the southern part of the Okhotsk Sea where ribbon seals start delivering the pups the earliest, i.e., in the second part of March. In the northwestern part of the Okhotsk Sea and in the Bering Sea it takes place in April, with the peak in the middle of the month.

There are four reproductive ecotypes related to ice formation (Fig. 2). What is special about the reproductive ecotypes is that the variety of breeding conditions defined by different types of ice, its thickness, shape, the hummocks, the amount of snow on it, its location, and the speed of its decomposition determine the time of rearing and pup growth, which is reinforced by natural selection. Due to the existence of ecotypes, the populations of the ribbon seal are adapted to both short-term (within 1 year) and long-term (several years long) modifications of ice conditions: it has a wide range of features adapted to breeding in various conditions and can keep population size relatively stable.

The spatial structure of the populations, their morphological characteristics, and the maturation of animals in different populations are determined by the proportion of the ecotypes in different regions. Ribbon seals that breed on the ice above the deep sea bed in the southern part of the Okhotsk Sea are most specific—both the conditions of reproduction and the foraging situation are different from those in shelf zones and abrupt continental shelf zones in the northwest and in the Bering Sea.

V. Forage

Thirty-five species inhabiting the Okhotsk and the Bering Sea constitute the ribbon seal's food. During the first year it feeds mostly on euphausiids. At the age of 1–2 it feeds mostly on shrimp, whereas adult animals eat cephalopods and fish. In the Okhotsk Sea, adult animals eat mostly Alaskan pollack (65%), whereas in the Bering Sea they mostly feed on squid and octopus (67%). The ribbon seal's daily consumption is 8–10 kg, including invertebrates and fish; the annual consumption of the whole ribbon seal population is more than 300,000 tons in the Bering Sea and about 1,160,000 tons in the Okhotsk Sea (Fedoseev, 2000).

VI. Life Cycle and Behavior

The length of a newborn seal's body is from 73 to 98 cm and it weighs 6–10 kg; specific features of different populations are taken into account. After the first year of life, the length of animals belonging to different populations becomes 128–134 cm and the weight 40–50 kg. During the period of maturation when the animals are 2–3 years old, the average body size is 145–155 cm and the weight is 55–70 kg, whereas mature animals' (who are more than 10 years old) body length is 165–175 cm and weight 72–90 kg. Ribbon seals inhabiting the southern part of the Okhotsk Sea are bigger and heavier than those from the northwestern part and from the Bering Sea.

During the breeding season (March–April), animals do not segregate by age and sex; adult animals dwell on the ice to give birth to pups and immature seals are there for molting. In May and June when adult animals start molting, gathering of ribbon seals on the ice increases because of melting and decomposition of the ice. In June, when there is little ice left, one can see a mass of ribbon seals consisting of different populations. The latter should be taken into consideration when

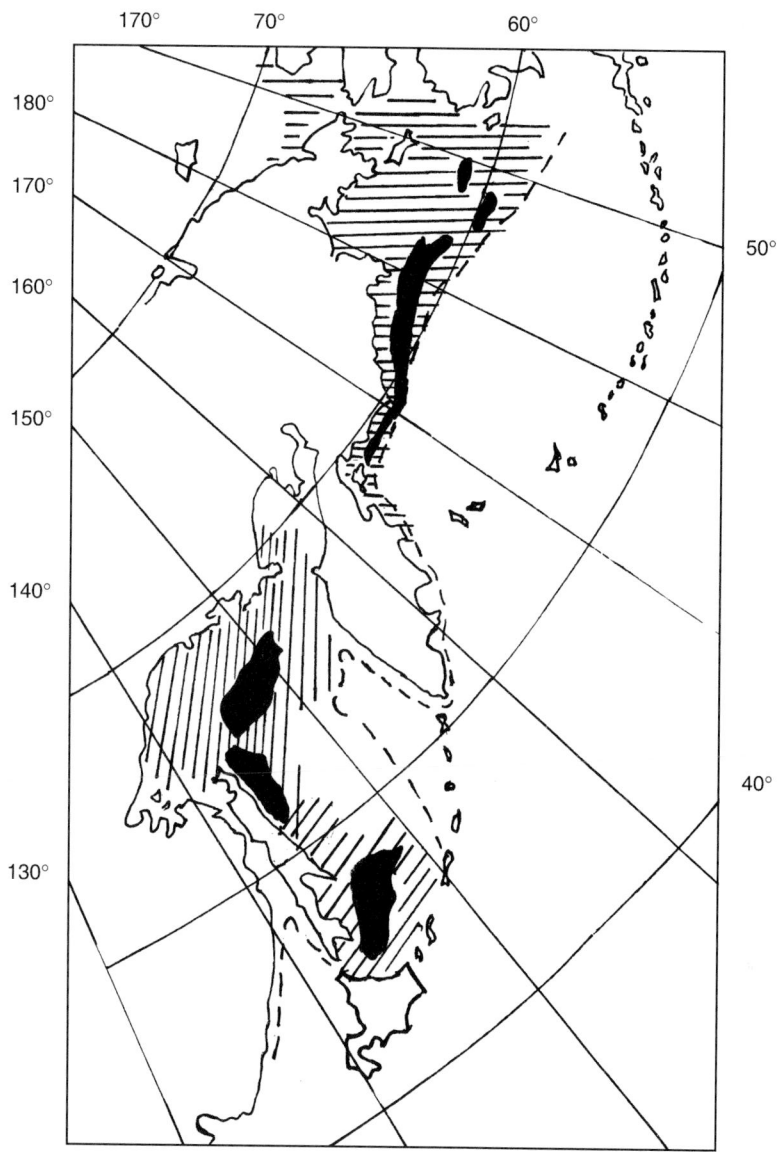

Figure 2 *Different populations and breeding sites of the ribbon seal in North Pacific seas. Broken line, ice edge; solid areas, breeding sites; horizontal stripes, the Bering Sea population; vertical stripes, the northwestern part of the Okhotsk Sea population; and diagonal stripes, the southern part of the Okhotsk Sea population.*

a population approach to data analysis is used. Where the ribbon seal dwells when there is no ice left has been little studied.

Due to the pelagic way of life and dwelling far from the sea coast, the ribbon seal has not developed a fear of humans. This seal is very trusting and hardly reacts at all when a ship approaches and people come out on the ice. It makes the ribbon seal a perfect object for ecological tourists who can take pictures of the animals on the ice. In the open water the ribbon seal is hardly noticed at all as it disappears from the sea surface noiselessly: it sticks its black head out vertically and takes it back like a periscope.

VII. Demographic Parameters of Populations

The ribbon seal's age is defined by horn covers on the claws and the layers of cement on the tusk apexes; analysis shows that it can reach the age of 30 and beyond.

Analysis of age data and the female contingent in the breeding process showed that the proportion of newborns in populations can be 24–31%, and the proportion of females having pups for the first time (recruits) 4–6% (Fedoseev, 2000). During the period of study (1982–1985) the potential of population growth in the Bering Sea was lower than in the Okhotsk Sea because the range of reproductive ecotypes and the distribution in the Okhotsk Sea were wider than in the Bering Sea (Fedoseev and Volokhov, 1991; Fedoseev, 2000).

VIII. Relations with Humans

Between 1956 and 1992, seals living on ice, ribbon seals among them, were the subject of commercial hunting. Up until 1969 the hunt was not limited, with the average annual yield being 11,000 animals in the Okhotsk Sea and 9000 in the Bering Sea. Between 1969 and 1992 the annual yield was reduced to 5000 to 6000 animals in the Okhotsk Sea and 3000 to 4000 in the Bering Sea.

See Also the Following Articles

Breathing ■ Coloration ■ Diving Physiology ■ Pinniped Life History

References

Fedoseev, G. A. (2000). "Population Biology of Ice-Associated Forms of Seals and Their Roles in the Northern Pacific Ecosystems." Center for Russian Environmental Policy, Moscow.

Fedoseev, G. A., and Volokhov, V. I. (1991). "Comparative Demographic Analysis of the Ribbon Seal Population." VNIRO Report Research in Marine Mammals in the North Pacific in 1989–1990. Project 02.0561, "Marine Mammals," pp. 119–130. US–USSR Agreement on Ecology, Moscow.

Geptner, V. G., Chaptsky, K. K., Arseniev, V. A., and Sokolov, V. E. (1976). "The Marine Mammals of the Soviet Union," Vol. 2, pp. 1–718. High School Publishers, Moscow.

Naumov, S. P., and Smirnov, N. A. (1936). "Materials on Classification and Geographic Distribution of *Phocidea* in the Northern Part of the Pacific Ocean," Vol. 3, pp. 161–187. VNIRO Publishing.

Scheffer, V. B. (1958). Seals, sea lions and walruses. *In* "A Review of the Pennipedia," pp. 1–179. Stanford Univ. Press, London.

Sokolov, A. S. (1966). Ecological, functional and age characteristics of North Pacific pinnipeds' red blood. *Acad. Sci. Present.* **169**, 161–187.

Sokolov, A. S., Kosigin, G. M., and Shustov, A. P. (1968). Lung and trachea structure of the Bering Sea pinnipeds. *News TINRO* **62**, 252–263.

Right Whale Dolphins
Lissodelphis borealis and *L. peronii*

JESSICA D. LIPSKY
Southwest Fisheries Science Center,
La Jolla, California

Right whale dolphins are known for their distinctive black and white color patterns and lack of a dorsal fin. These characteristics make these species easy to identify at sea. Although these species were first described in the first half of the 19th century, their overall biology, life history, taxonomy, and behavior are poorly known.

I. Characters and Taxonomic Relationships

While both species of right whale dolphins lack a dorsal fin, they have very different pigmentation patterns. The northern right whale dolphin (*Lissodelphis borealis* Peale, 1848) is mainly black with a white ventral patch that runs from the fluke notch to the throat region (see Fig. 1). This band widens slightly at the urogenital area in males and to a greater extent in females (Leatherwood and Walker, 1979). There is another small white patch on the ventral tip of the rostrum and on the underside of the flippers. The southern right whale dolphin [*L. borealis* (Lacépède, 1804)] has a similar white ventral patch; however, it extends higher on the posterior flanks (see Fig. 2). The back of the dolphin is black, and the white area reaches a high point midway along the body, dipping down at the flipper insertion and covering most of the head and rostrum. Newborn calves are usually dark gray or brown, attaining adult coloration after the first year of life.

Right whale dolphins can grow to lengths of 3 m; males tend to grow larger than females (Leatherwood and Walker, 1979). Weights have been recorded up to 116 kg (Jefferson *et al.*, 1994). The flippers are slender and pointed at the tips. Flukes have a median notch that is moderately deep with concave trailing edges. The teeth are small and sharp, ranging in numbers from 37 to 54 per row in the northern species and 39 to 50 in the southern species, with more teeth in the lower jaw (Jefferson *et al.*, 1994).

Recent classifications have placed the right whale dolphins in a monogeneric delphinid subfamily Lissodelphinae or in the subfamily Delphininae. However, based on an analysis of cytochrome *b* (mtDNA) sequences, LeDuc *et al.* (1999) tentatively placed them in the subfamily Lissodelphinae with *Lagenorhynchus* spp. and *Cephalorhynchus* spp.

II. Distribution and Ecology

Right whale dolphins are found in cool-temperate and sub-Arctic waters in the North Pacific and circumpolar sub-Antarctic and cool-temperate waters in the Southern Ocean (see Fig. 3). In the North Pacific, northern right whale dolphins range from Kuril Islands, Russia, south to Sanriku, Honshu, Japan, extending eastward to the Gulf of Alaska and south to southern California (Rice,

Figure 1 Lissodelphis borealis *off the coast of California. Photo courtesy of Robert L. Pitman.*

Figure 2 Lissodelphis peronii *off the coast of Chile. Photo courtesy of Robert L. Pitman.*

1998). They are distributed approximately from 34°N to 55°N and 145°W to 118°E. Occasional movements south of 30°N are associated with anomalous cold-water temperatures. In the Southern Hemisphere, southern right whale dolphins are found most commonly between 25°S to 55°S in the eastern South Pacific and about 30°S to 65°S. They are found most often between the subtropical and Antarctic convergences, with distributions reflecting the variability in these oceanographic features (Gaskin, 1968). It has also been observed that the range of the southern right whale dolphins often extend northward along eastern cold-water boundary currents.

Migration of both *Lissodelphis* species is not entirely known or understood. In northern right whale dolphins, there appears to be an inshore shift in winter and spring off California, which coincides with peak abundance of their primary food source (squid) (Leatherwood and Walker, 1979). Forney and Barlow (1998) found that northern right whale dolphin abundance was greatest off the Southern California Bight in winter, while in the summer there were no sightings made in this area. In addition, they observed a greater abundance of *L. borealis* offshore in summer and a greater abundance inshore on the Southern California Bight continental shelf in winter (Forney and Barlow, 1998). Southern right whale dolphins tend to occur year-round in a localized area off Namibia, Africa, where high-productivity waters prevail (Newcomer *et al.*, 1996). Off the coast of Chile, southern right whale dolphins are present year-round and have been discovered to migrate northward during winter and spring. Food sources, which are affected by changing water temperatures, appear to be a factor in this species' migration. In addition, these two species are commonly found in oceanic, deep waters, on highly productive continental shelves, or sometimes where deep waters approach the coast.

Northern right whale dolphins have been observed to associate with 14 other species of marine mammals in the North Pacific Ocean. They are mainly observed with Pacific white-sided dolphins, *Lagenorhynchus obliquidens*, which share a similar distribution and habitat (Klumov, 1959; Leatherwood and Walker, 1979). They are also commonly found with pilot whales (*Globicephala macrorhynchus*) and Risso's dolphins, *Grampus griseus* (Leatherwood and Walker, 1979). Southern

right whale dolphins are associated most often with pilot whales (*Globicephala melas*) and dolphins of the genus *Lagenorhynchus* (Jefferson *et al.*, 1994).

PREDATION on right whale dolphins is poorly known; however, killer whales and large shark species are occasional predators. There are two records of predation on southern right whale dolphins: a 0.87-m southern right whale dolphin fetus was found in a 3.6-m sleeper shark (*Somniosus* cf. *pacificus*) off the coast of Valdivia, Chile in 1990 (Crovetto *et al.*, 1992) and a 1.7-m Patagonian toothfish taken off central Chile in 1983 had a 0.86-m southern right whale dolphin neonate in its stomach (Van Waerebeek *et al.*, 1991).

The northern right whale dolphin has been observed to feed primarily on squid and lanternfish; however, other prey species include Pacific hake, saury, and mesopelagic fishes (Leatherwood and Walker, 1979). The southern right whale dolphin feeds primarily on various squid and fish species.

Strandings of northern and southern right whale dolphins are uncommon. An apparent increase in strandings of *L. peronii* is possibly the result of discarded animals from a rapidly developing swordfish gillnet fishery off northern Chile (Van Waerebeek *et al.*, 1991).

III. Behavior and Life History

Right whale dolphins tend to be gregarious animals, often traveling in groups of up to 2000–3000 in the North Pacific (Leatherwood and Walker, 1979) and up to 1000 in the southern species (Gaskin, 1968). Herds are characterized by four main configurations, including V-shaped herds, "chorus line" formation herds, tightly packed herds with no identifiable subgroups, and herds with subgroups within the main group (Leatherwood and Walker, 1979). Both forms have been observed to travel slowly or quickly; this is associated with surfacing modes, BREATHING intervals, and travel speeds. Right whale dolphins can travel up to 40 km per hour (Leatherwood and Reeves, 1983). In some instances, right whale dolphins will bow ride on vessels, especially in the presence of other species, although sometimes they will actively avoid approaching vessels. Aerial behavior such as BREACHING, belly flops, and side and fluke slaps are not uncommon, especially in the fast swimming mode.

Information on growth and reproduction for right whale dolphins is limited. Twenty-three specimens have been examined in the eastern North Pacific, and their data suggest that males attain sexual maturity between 212 and 220 cm and females at about 200 cm (Leatherwood and Walker, 1979). In November 1990 and 1991, 229 northern right whale dolphins were obtained from the Japanese squid drift net fishery and examined for total length, age, and sex (Ferrero and Walker, 1993). Ferrero and Walker (1993) found that the average length for sexually mature males is 214.7 cm and 199.8 cm for females in the northern species. In addition, they calculated that the age at the onset of sexual maturity for males is approximately 9.9 years and for females it is approximately 9.7 years (Ferrero and Walker, 1993). In the western North Pacific, other reports indicate that females attain sexual maturity between 206 and 212 cm (Jefferson *et al.*, 1994). Northern right whale dolphin neonates range between 80 and 100 cm at birth (Jefferson *et al.*, 1994). Ferrero and Walker (1993) found that for northern right whale dolphins length at birth ranged be-

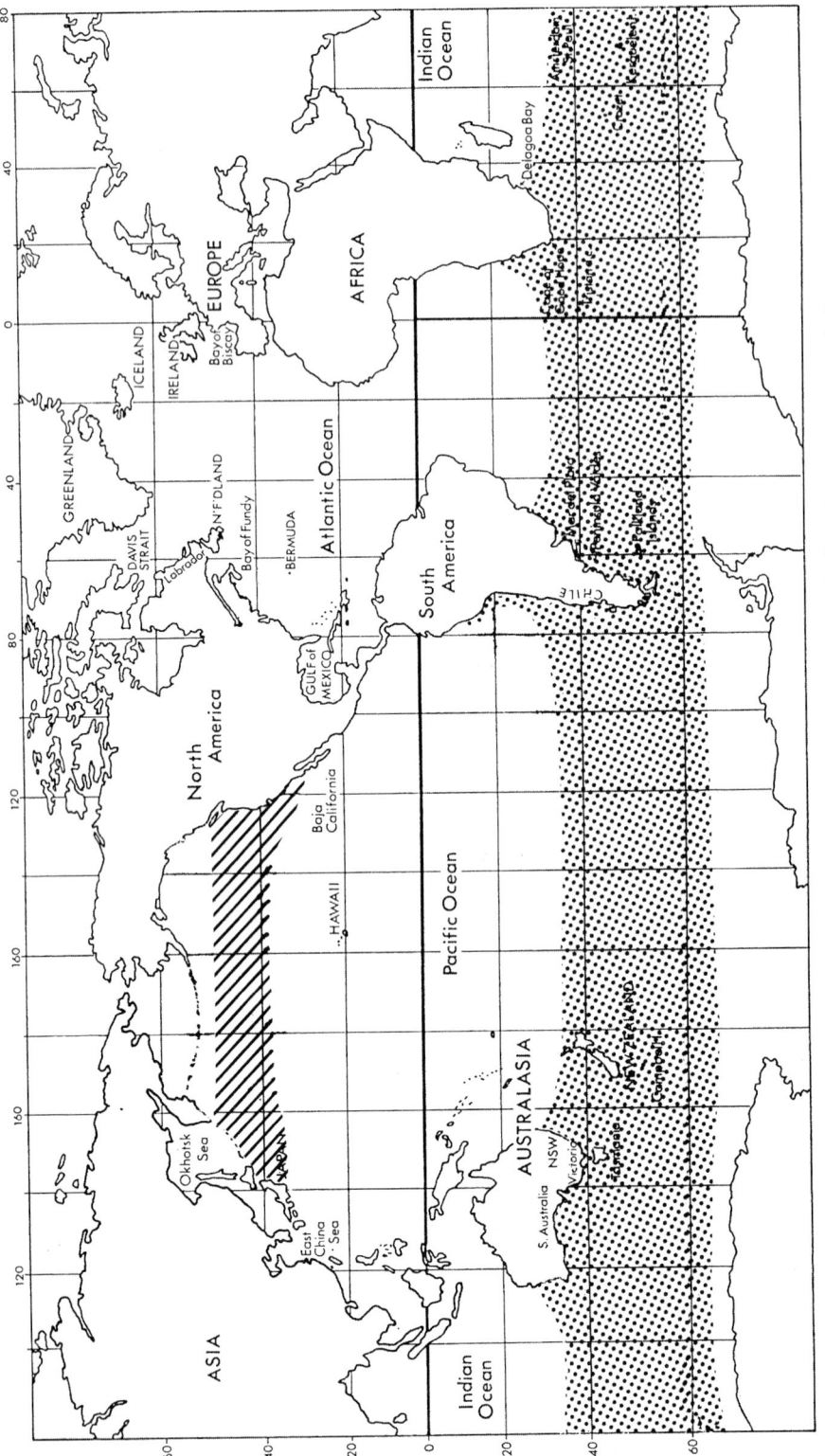

Figure 3 *Approximate ranges of Lissodelphis borealis (hatched) and L. peronii (stippled).*

tween an average of 99.7 and 103.8 cm using three different methods of length estimation. The calving season is unknown; however, small calves are often seen in winter or early spring. In the Southern Ocean, right whale dolphin reproductive biology is largely unknown. Two females measuring 218 and 229 cm and one male measuring 251 cm have been examined and were all sexually mature (Jefferson *et al.*, 1994).

Sound production in northern right whale dolphins has been recorded (Fish and Turl, 1976). Clicks with high repetition rates were recorded, with few whistles. Sound production in southern right whale dolphins has not been described.

There have been few attempts to capture live animals due to the difficulty in maintaining these oceanic species. A northern right whale dolphin was captured live and held for 15 months; however, most live captures have not survived more than 3 weeks (Walker, 1975). There have been no reported attempts to capture live southern right whale dolphins.

IV. Conservation Status

In the 19th century, whalers occasionally took northern right whale dolphins. Although there is currently no direct fishery for right whale dolphins, the northern species is occasionally taken in Japan's harpoon fishery and in the Japanese and Russian purse-seine fisheries (Klumov, 1959). In addition, a few individuals were taken in Japan's salmon gillnet fishery (International Whaling Commission, 1983) and in California's shark and swordfish driftnet fishery. The majority of right whale dolphin bycatches in recent years occurred in the North Pacific squid driftnet fisheries operated by Japan, Korea, and Taiwan (Jefferson *et al.*, 1994). The squid fisheries began in 1978 with small incidental takes of 300–400 dolphins until the mid-1980s when incidental takes were on the order of 15,000–24,000 dolphins per year (Mangel, 1993). It is thought that the stock in this area has been depleted to 24–73% of its pre-exploitation level (Mangel, 1993). In the past few years, southern right whale dolphins have been taken off Chile and Peru for use of their meat and blubber for human consumption or use as crab bait (Newcomer *et al.*, 1996). Since 1989 there has been an increase in southern right whale dolphin bycatches in the developing swordfish gillnet fishery off Chile (Van Waerebeek *et al.*, 1991).

The effects of POLLUTION and contaminants on right whale dolphins are currently unknown. Given the nature of the pelagic habitat of the northern species, the effects of pollution are probably minimal. However, seasonal shifts in migration and distribution could possibly have a negative impact on these species.

See Also the Following Articles

Delphinids, Overview ■ North Pacific Marine Mammals

References

Crovetto, A., Lamilla, J., and Pequeno, G. (1992). *Lissodelphis peronii* Lacépède 1804 (Delphinidae, Cetacea) within the stomach contents of a sleeping shark, *Somniosus* cf. *pacificus*, Bigelow and Schroeder 1944, in Chilean waters. *Mar. Mamm. Sci.* **8**, 312–314.
Ferrero, R. C., and Walker, W. A. (1993). Growth and reproduction of the northern right whale dolphin, *Lissodelphis borealis*, in the offshore waters of the North Pacific Ocean. *Can. J. Zool.* **71**(12), 2335–2344.

Fish, J. F., and Turl, C. W. (1976). Acoustic source levels of four species of small whales. *Naval Undersea Center Techn. Rep.* TP **547**, 1–14.
Forney, K. A., and Barlow, J. (1998). Seasonal patterns in the abundance and distribution of California cetaceans, 1991–1992. *Mar. Mamm. Sci.* **14**(3), 460–489.
Gaskin, D. E. (1968). The New Zealand Cetacea. *Fish. Res. Bull. New Zealand* **1**, 1–92.
International Whaling Commission (1983). Report of the subcommittee on small cetaceans, Annex H. *Rep. Int. Whal. Comm.* **33**, 152–170.
Jefferson, T. A., and Newcomer, M. W. (1993). *Lissodelphis borealis*. *Mamm. Spec.* **425**, 1–6.
Jefferson, T. A., Newcomer, M. W., Leatherwood, S., and Van Waerebeek, K. (1994). Right whale dolphins *Lissodelphis borealis* (Peale, 1848) and *Lissodelphis peronii* (Lacépède, 1804). *In* "Handbook of Marine Mammals" (S. H. Ridgway and R. Harrison, eds.), Vol. 5, pp. 335–362. Academic Press, London.
Klumov, S. K. (1959). Commercial dolphins of the far east. *Pacific Sci. Res. Instit. Fish. Econ. Oceanogr. Rep.* **47**, 154–160. [Translated from Russian.]
Leatherwood, S., and Reeves, R. R. (1983). "The Sierra Club Handbook of Whales and Dolphins." Sierra Club Books, San Francisco.
Leatherwood, S., and Walker, W. A. (1979). The northern right whale dolphin *Lissodelphis borealis* Peale in the eastern North Pacific. *In* "Behavior of Marine Animals" (H. E. Winn and B. L. Olla, eds.), Vol. 3, pp. 85–141. Plenum Press, New York.
Le Duc, R. G., Perrin, W. F., and Dizon, A. E. (1999). Phylogenetic relationships among the delphinid cetaceans based on full cytochrome *b* sequences. *Mar. Mamm. Sci.* **15**(3), 619–648.
Mangel, M. (1993). Effects of high-seas driftnet fisheries on the northern right whale dolphin *Lissodelphis borealis*. *Ecol. Appl.* **3**, 221–229.
Newcomer, M. W., Jefferson, T. A., and Brownell, R. L., Jr. (1996). *Lissodelphis peronii*. *Mamm. Spec.* **531**, 1–5.
Rice, D. W. (1998). "Marine Mammals of the World: Systematics and Distribution." Soc. of Mar. Mamm., Spec. Pub. No. 4. Allen Press.
Van Waerebeek, K., Canto, J., Gonzalez, J., Oporto, J., and Brito, L. (1991). Southern right whale dolphins, *Lissodelphis peronii*, off the Pacific coast of South America. *Zeitsch. Säugetierk.* **56**, 284–295.
Walker, W. A. (1975). Review of the live-capture fishery for smaller cetaceans taken in southern California waters for public display, 1966–73. *J. Fish. Res. Board Can.* **32**, 1197–1211.

Right Whales

SEE *North Atlantic, North Pacific, and Southern Right Whales*

Ringed, Caspian, and Baikal Seals

Pusa hispida, P. caspica, and *P. sibirica*

NOBUYUKI MIYAZAKI
Otsuchi Marine Research Center,
University of Tokyo, Japan

I. Distribution and Taxonomy

The ringed seal (*Pusa hispida*), Baikal seal (*P. sibirica*), and Caspian seal (*P. caspica*) have features in common, such as small size, delicate skull, and affinity for ice.

Baikal seals and Caspian seals are found in Lake Baikal and the Caspian Sea, respectively. The ringed seal is a species of circumpolar Arctic coasts with a broad geographic distribution and is found wherever there is open water in fast ice, even as far as the North Pole, and in fjords and bays, but rarely in the open sea or on floating pack ices (King, 1964). Although numerous populations and/or subspecies have been reported for the ringed seal, at present five distinct subspecies are usually recognized: *P. h. hispida* from the Arctic Ocean and the confluent Bering Sea, *P. h. ochotensis* from the Sea of Okhotsk, *P. h. saimensis* from Lake Saimaa, *P. h. ladogensis* from Lake Ladoga, and *P. h. botnica* from the Baltic Sea (Fig. 1).

II. Morphology

A. External Appearance

Baikal seals, which do not have distinct spots, are uniform dark silver gray dorsally and light yellowish gray ventrally (Fig. 2). Caspian seals are irregularly spotted with brown or black against a light grayish yellow background (Fig. 3). The spots are light-colored rings. In the ringed seal, gray–white rings are found on the generally gray backs, and the belly is usually silver and lacking dark spots (Fig. 4). The rings are separate or somewhat fused together. The pups of these three species are born with a white woolly natal lanugo. This lanugo is considerably finer and longer than that of the two other northern phocids, the spotted seal (*Phoca largha*) and ribbon seal (*Histriophoca fasciata*) (Frost and Lowry, 1981).

B. Skull Morphology

Comparison of skull morphology indicates that Baikal seals have a greater length of jugal, and narrower least interorbital width and width of nasals at the maxillo-frontal suture than Caspian seals. However, in the ringed seal, characters relating to skull width, width of bulla, and greatest length of bulla show larger values, whereas characters relating to condylobasal length and width of snout at canines show smaller ones compared to the other two species. A canonical discriminant analysis was performed using 14 characters lacking sexual differences. Subspecies of the ringed seal are clearly distinguished from Baikal and Caspian seals. Cluster analyses by the neighbor-joining method and the UPGMA (unweighted pair-group method using arithmetic averages) method based on Mahalanobis dis-

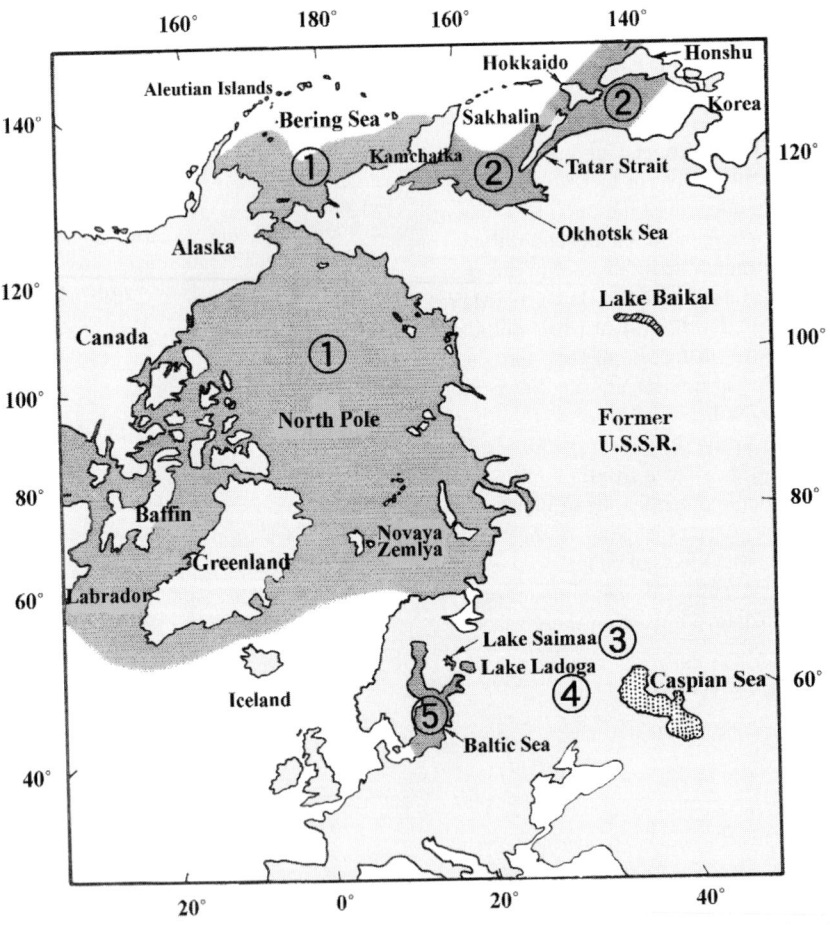

Figure 1 *Distribution of the ringed seal (dark), Caspian seal (dots), and Baikal seal (stripes) (redrawn from King, 1964). Numbers indicate subspecies of the ringed seal; 1, P. h. hispida; 2, P. h. ochotensis; 3, P. h. saimensis; 4, P. h. ladogensis; and 5, P. h. botnica.*

Figure 2 *Adult Baikal seal on ice, Lake Baikal. Photo by S. Tanabe.*

Figure 4 *Ringed seal at Dikson, the Arctic Ocean.*

tance suggest that Baikal seals have closer affinity with the ringed seal than the Caspian. This relationship coincides well with sequence analysis using mitochondria DNA.

III. Age and Growth

The maximum known age in both sexes in Baikal seals is 56 years for females and 52 for males (Pastukhov, 1993). According to Amano *et al.* (2000), the oldest age of Baikal seals in samples (N = 73) collected in 1992 was 24.5 years for females and 35.5 years for males. In Caspian seals collected from Pearl Island in the western North Caspian Sea (N = 118), the oldest age was 43.5 years for females and 33.5 years for males. The maximum known age for the ringed seal is 43 (McLaren, 1958).

Growth in body length of Baikal seals appears to cease around the age of 15 years (Amano *et al.*, 2000). The seals may continue to grow for 8–9 years after the age of sexual maturity (6 years for females and 7 years for males). Asymptotic body length is 140 cm in males and 130 cm in females. In Caspian seals, body length growth appears to cease around the age of 10 years, which is the age of sexual maturity in both sexes. Asymptotic body length is 118 cm in males and 111 cm in females. McLaren (1958) reported that the growth of ringed seals continues throughout the first 8–10 years of life. About 86% of final body

Figure 3 *Caspian seal on Pearl Island, northwestern North Caspian Sea.*

length is attained by sexual maturity of 6–8 years. Average adult lengths for the ringed seal vary from 121 cm in the Chukchi Sea to 128.5 cm in the Bering Sea (Fedoseev, 1975) and 135 cm in the Canadian Arctic (McLaren, 1958).

IV. Behavior

Movements and dive patterns of Baikal seals appear to be associated primarily with seasonal movements of their primary prey, golomyanka and sculpins, and correlated secondarily with patterns of ice formation and thaw. Most dives are to depths of 10–50 m, although a few exceed 300 m (Stewart *et al.*, 1996). Dives may last between 2 and 6 min but a few dives exceed 40 min.

For two adult male Caspian seals, most dives were less than 50 m in depth while a few exceeded 200 m. Dives were mostly less than 50 sec long but some exceeded 200 sec.

V. Reproduction

Most Baikal seals breed by 6 years for females and 7 years for males (Thomas, 1982). Pups are 65 cm in body length and 4.1 kg in body weight on average. A rather high rate of twinning (4% of annual births) is exhibited compared to other seals (Pastukhov, 1968). Mating may occur underwater in March at about the time mothers wean their pups. Mothers nurse the pups in a birthing lair. The lactation period is estimated at 2–3 months. The MATING SYSTEM is assumed to be polygamous with little or no pair bonding. In winter, when Lake Baikal is covered with ice averaging 80–90 cm in thickness with a maximum of 1.5 m, seals are sighted throughout the lake and adjacent to breathing holes in the ice. In Baikal seals of 7 years or more, 84% of females gave birth to pup yearly (Patstukhov, 1993).

Caspian seal pups are born on the ice from the middle of January to the end of February and are about 60 cm in body length. Mating takes place between the end of February and the middle of March. Sexual maturity is attained at 4–6 years in females and 6 years in males (Ognev, 1935; Fedoseev, 1975). The pregnancy rate of Caspian seals over 9 years was 31.3% (N = 16) in 1993 and 20% (N = 30) in 1997 and 1998.

Ringed seal pups, which are born between the middle of March and the middle of April, are on average 65 cm in body length and about 4.5 kg in body weight. They are always born on shore-fast ice, either in a lair under the snow, excavated by the mother, or in a natural hollow in the ice. Ringed seals attain sexual maturity at 6–7 years old in both sexes with wide geographic variation from 3–5 years for *P. hispida botnica* to 6–10 years for *P. hispida hispida* (Frost and Lowry, 1981). The peak of mating activity probably occurs in mid-April, shortly after parturition and while the female is still lactating (King, 1964). The lactation period is nearly 2 months. In ringed seals, pregnancy rates of sexually mature females vary geographically: 91–92% in the Baffin Island area (McLaren, 1958; Smith, 1973), 86% in the southern Chukchi Sea (Johnson *et al.*, 1966), and 53% in Alaska waters in 1975–1977 (Frost and Lowry, 1981).

VI. Food

Baikal seals feed mainly on four fish species: the greater golomyanka (*Comephorus baicalensis*), the lesser golomyanka (*C. dybowskii*), the Baikal yellowfin sculpin (*Cottocomephorus grewingki*), and the longfin sculpin (*C. inermis*), all of which are not of commercial value. In captivity, an adult Baikal seal consumed up to 5.6 kg of fish per day (Pastukhov, 1969).

Caspian seals in the northern Caspian Sea feed on *Clupeonella engrauliformis*, *C. grimmi*, *C. delicata caspia*, Gobiidae, *Rutilus rutilus caspicus*, *Atherina mochon pontica*, *Lucioperca lucioperca*, other fish species, and crustaceans (Khuraskin and Pochtoyeva, 1997). It is estimated that an adult Caspian seal appears to take 2–3 kg fish per day, or approximately 1 ton of fish per year.

Ringed seals feed on small fish and also on a wide variety of small pelagic amphipods, euphausiids, and other crustaceans. Seventy-two food species were identified in the stomachs of seals from the eastern Canadian Arctic. In shallow, inshore waters, the seals feed near the bottom, chiefly on polar cod (*Boreogadus saida*) and on the small crustacean *Mysis*, whereas those in the deeper offshore waters catch the planktonic amphipod *Themisto libellula* (King, 1964).

VII. Population

The population of Baikal seals from 1971 to 1978 was estimated to be between 68,000 and 70,000 animals (Pastukhov, 1978a). The number of seals taken annually has varied: before 1917 about 2000 to 9000; in 1930 about 6000; and currently between 5000 and 6000 (Pastukhov, 1978b). In 1987–1988, an outbreak of morbillivirus infection resulted in a large mass mortality of Baikal seals (Grachev *et al.*, 1989). A mass death of Baikal seals also occurred in 1998.

The Caspian seal population declined from about 1 million animals early in the 20th century to 360,000–400,000 by the end of the 1980s (Krylov, 1990). According to Khuraskin and Pochtoyeva (1997), 115,000–174,000 animals have been hunted annually since the early 19th century. A total of 86,000 animals were killed in 1966. From 1970, seal hunting on the northern ice was limited to a catch of 20,000–25,000 pups. In the spring of 1997, a mass death of several thousand seals occurred.

A minimum population estimate for *P. hispida hispida* is 2.5 million (Bychkov, 1971), although there are many uncertainties in the estimation method, survey season, survey design, data analysis, and so on. Population size was estimated at 800,000–1,000,000 for *P. hispida ochotensis* (Bychkov, 1971), 10,000–50,000 for *P. hispida botnica* (Scheffer, 1958),

2000–5000 for *P. hispida saimensis* (Scheffer, 1958), and 5000–10,000 for *P. hispida ladogensis* (Scheffer, 1958). Many thousands of ringed seals are caught annually from all areas where they occur, mostly for their skins, which are used for leather or for their decorative fur, and also for blubber (King, 1964). A mass death of ringed seals was recorded in 1960 in the Baltic Sea and the North Sea.

See Also the Following Articles

Eared Seals ■ Pinniped Life History

References

Amano, M., Miyazaki, N., and Petrov, E. A. (2000). *In* "Age Determination and Growth of Baikal Seals (*Phoca sibirica*) (A. Rossiter and H. Kawanabe, eds.), pp. 449–462. Academic Press, London.

Bychkov, V. A. (1971). A review of the conditions of the pinniped fauna of the USSR. *In* "Scientific Principles for the Conservation of Nature" (L. K. Shaposhnikov, ed.), pp. 59–74. [Transl. Can. Dept. Foreign Language No. 0929.]

Fedoseev, G. A. (1975). Principal population indicators of dynamics of numbers of seals of the family Phocidae. *Ekologiya* **5**, 62–70. [Transl. Consultants Bureau, New York, 439–446.]

Frost, K. J., and Lowry, L. F. (1981). Ringed, Baikal and Caspian seals— *Phoca hispida* Schreber, 1775; *Phoca sibirica* Gmelin, 1788 and *Phoca caspica* Gmelin, 1788. *In* "Handbook of Marine Mammals" (S. Ridgway and R. J. Harrison, eds.), pp. 29–53. Academic Press, London.

Grachev, M. A., Kumarev, V. P., Mamaev, L. V., Zorin, V. L., Baranova, L. V., Denikina, N. N., Belikov, S. I., Petrov, E. A., Kolesnik, V. S., Kolesnik, R. S., Dorofeev, V. M., Beim, A. M., Kudelin, V. N., Nagieva, F. G., and Sidorov, V. N. (1989). Distemper virus in Baikal seals. *Nature* **338**, 209.

Johnson, M. L., Fiscus, C. H., Ostenson, B. T., and Barbour, M. L. (1966). Marine mammals. *In* "Environment of the Cape Thompson Region, Alaska" (N. J. Wilimovsky and J. N. Wolfe, eds.), pp. 897–924. U. S. Atomic Energy Commission. Oak Ridge, TN.

Khuraskin, L. S., and Pochtoyeva, N. A. (1997). Status of the Caspian seal population. *In* "Caspian Environment Program" (H. Dumont, S. Wilson, and B. Wazniewicz, eds.), pp. 86–94. Proceeding from the first bio-network workshop, Bordeaux, November 1997, World Bank, Washington, DC.

King, J. E. (1964). "Seals of the World." British Museum (Natural History), London.

Krylov, V. I. (1990). Ecology of the Caspian seal. *Finnsih Game Res.* **47**, 32–36.

McLaren, I. A. (1958). The biology of the ringed seal, *Phoca hispida*, in the eastern Canadian Arctic. Fish. Res. Bd. Can. Bull. No. 118.

Ognev, S. I. (1935). Mammals of the USSR and adjacent countries. Vol. III. Carnivora (Fissipedia and Pinnipedia) Moscow: Acad. Sci. USSR. [In Russian; English transl. by A. Birron and Z. S. Coles for Israel Program for Scientific Translations, 1962.]

Pastukhov, V. D. (1968). On twins in *Pusa sibirica* Gmel. *Zool. Zhurnal* **47**, 479–482. [English summary.]

Pastukhov, V. D. (1969). Some results of observations on the Baikal seals under experimental conditions. *In* "IV oye Vsesoyuznaya Konferentsiya po Izucheniyu Morskikh Mlekopitayushchikh," Tezisy Dokladov. Can. Fish. Res. Bd. Translation Series, No. 3544.

Pastukhov, V. D. (1978a). Scientific-production experiment on the Baikal seal. *In* "Morskiye Mlekopitayushchiye," pp. 257–258. Moscow, USSR.

Pastukhov, V. D. (1978b). Baikal seal. *In* "Problemy Baikala" (G. I. Galaziy and K. K. Votintsev, eds.), pp. 251–259. Nauka, Sibirskoye, Otdeleniye, Novosibirsk.

Pastukhov, V. D. (1993). "Baikal Seals." Nauka, Moscow, USSR.

Scheffer, V. D. (1958). "Seals, Sea Lions and Walruses." Stanford Univ. Press, Stanford, CA.

Smith, T. G. (1973). Population dynamics of the ringed seal in the Canadian eastern Arctic. *Fish. Res. Bd. Can. Bull.* **181.**

Stewart, B. S., Petrov, E. A., Baranov, E. A., Timonin, A., and Ivanov, M. (1996). Seasonal movements and dive patterns of juvenile Baikal seals, *Phoca sibirica. Mar. Mamm. Sci.* **12**(4), 528–542.

Thomas, J. (1982). Mammalian species. *Am. Soc. Mammal.* **188,** 1–6.

Risso's Dolphin
Grampus griseus

ROBIN W. BAIRD
*Dalhousie University,
Halifax, Nova Scotia, Canada*

The Risso's dolphin (*Grampus griseus*) is the fifth largest member of the family Delphinidae, with adults of both sexes reaching up to about 4 m in length. The common name comes from the person (Risso), who described the type specimen to G. Cuvier in 1812. Risso's dolphins are unusual looking for a variety of reasons. Their anterior body is extremely robust, tapering to a relatively narrow tail stock, and they have one of the tallest dorsal fins in proportion to body length of any cetacean (Fig. 1). The bulbous head has a distinct vertical crease or cleft along the anterior surface of the melon. Color patterns change dramatically with age. Infants are gray to brown dorsally and creamy-white ventrally, with a white anchor-shaped patch between the pectoral flippers, and white around the mouth. Calves then darken to nearly black, while retaining the ventral white patch. As they mature they lighten (except for the dorsal fin, which remains dark even in adults), and the major-

Figure 2 *Color changes in Risso's dolphin. As adults they become lighter and lighter, as illustrated in this relatively light animal, especially as compared to the animal illustrated in Fig. 1. Photo by R. L. Pitman.*

ity of the dorsal and lateral surfaces of the body become covered with distinctive linear scars, most of which are presumably caused by intraspecific interactions. Older animals can appear almost completely white on the dorsal surface (Fig. 2). No evidence of SEXUAL DIMORPHISM exists. Dentition is unusual, with most individuals having no teeth in the upper jaw and only a small number (two to seven pairs) in the lower jaw. Some researchers have suggested that population division exists both between and within ocean basins (based on morphological data), though sample sizes are small, and no subspecies are currently recognized.

Risso's dolphins are distributed worldwide in temperate and tropical oceans, with an apparent preference for steep shelf-edge habitats between about 400 and 1000 m deep (Fig. 3). Mass strandings of this species are very rare, and because they have not been taken in any numbers in whaling operations, relatively few specimens are available. Because of their typically offshore habits, relatively little is known about the biology or behavior of this species. The inshore presence of Risso's dolphins has been documented in several areas, and such movements have facilitated behavioral and population research. The range of Risso's dolphins seems to be limited by water temperature, with animals most common in waters between 15 and 20°C and rarely found in waters below 10°C. No worldwide population estimates exist, although a number of regional estimates are available.

Risso's dolphins are relatively gregarious in nature, typically traveling in groups of 10–50 individuals, with the largest group observed estimated to contain over 4000 individuals. They frequently travel with other cetaceans. Off southern California they have been documented to "bow ride" on and apparently harass gray whales (*Eschrichtius robustus*), and aggressive behavior directed toward short-finned pilot whales (*Globicephala macrorhynchus*) has also been observed. No evidence of predation by either killer whales (*Orcinus orca*) or large sharks is

Figure 1 *The Risso's dolphin is one of the largest dolphins, characterized by a large dorsal fin, a bulbous head which possesses a distinct cleft on the anterior melon, and a robust body that nevertheless possesses a narrow tail stock. Photo by R. L. Pitman.*

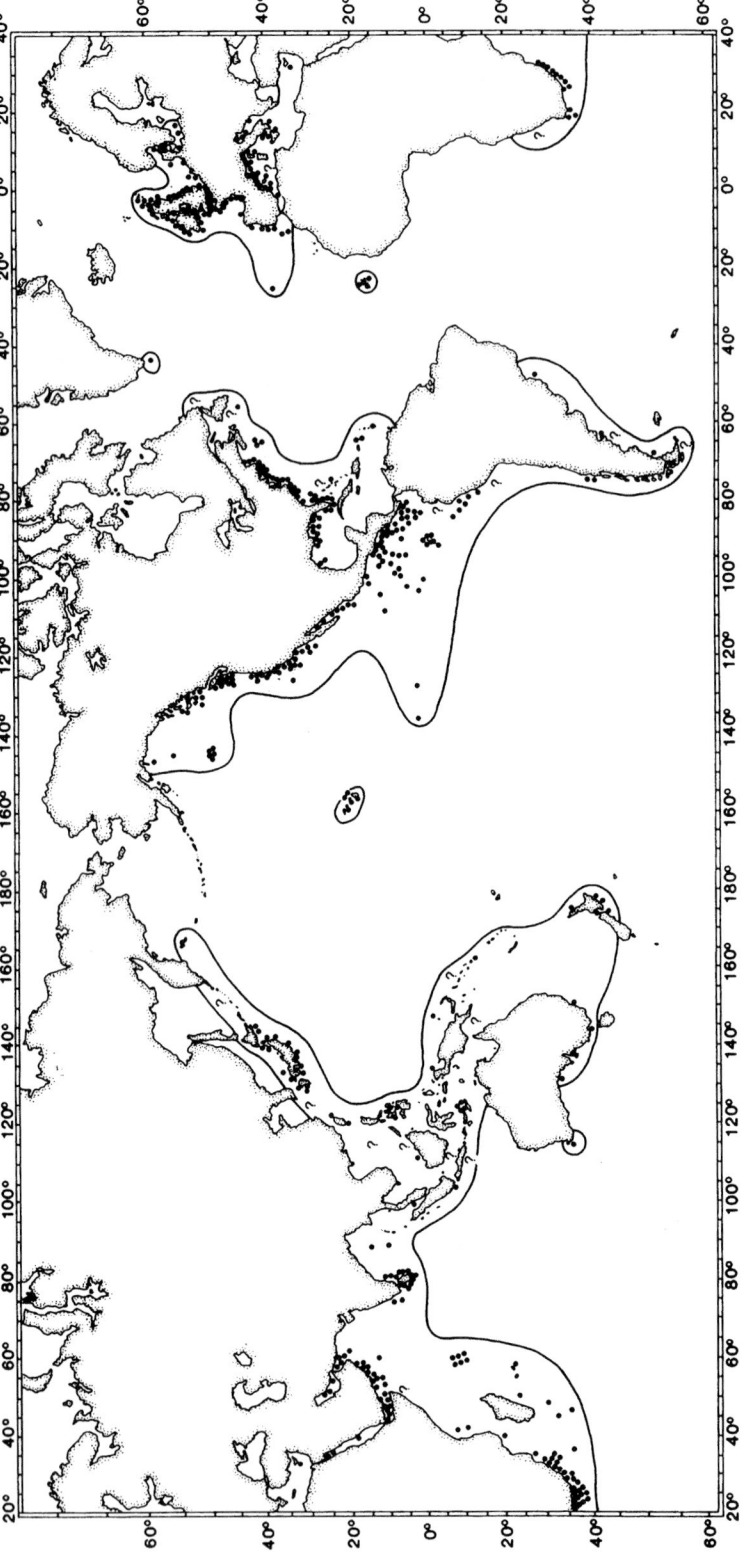

Figure 3 *The known distribution of Risso's dolphins. Dots represent actual sighting and stranding records. Question marks indicate records where the accurate geographical location was not given. From Kruse, Caldwell, and Caldwell (1999).*

available, although both likely prey on Risso's dolphins at least occasionally. Risso's dolphins are thought to feed almost entirely on squid (both neritic and oceanic species), and limited behavioral research suggests that they feed primarily at night. No studies on diving behavior have been undertaken.

Life history information for this species is relatively sparse. The maximum longevity of Risso's dolphins is likely over 30 years, although few aging studies have been undertaken, and growth layer deposition rates in teeth have not been calibrated. Age at sexual maturity and the calving interval are not known. There appears to be a peak in calving seasonality during the winter months.

Interactions with humans are diverse. While they occasionally BOW RIDE on vessels, in most cases Risso's seem to avoid boats. At least one strong exception exists: one Risso's dolphin, nicknamed "Pelorus Jack," escorted boats into Admiralty Bay in New Zealand for over 20 years. Risso's dolphins have been recorded stealing fish from long lines in a number of areas, and these interactions have often resulted in deliberate killing. Small numbers of Risso's dolphins have been killed in small-scale WHALING operations around the world, and off Sri Lanka, these takes may seriously jeopardize the local population. Risso's dolphins are also killed accidentally in fishing gear around the world. Risso's dolphins have been held in aquaria in both Japan and the United States, although they are relatively uncommon in CAPTIVITY compared to other species of cetaceans. A number of hybrids with bottlenose dolphins (*Tursiops truncatus*) have been documented in Japanese aquaria.

See Also the Following Articles

Hybridism ▪ Incidental Catches ▪ Teeth

References

Baird, R. W., and Stacey, P. J. (1991). Status of the Risso's dolphin, *Grampus griseus,* in Canada. *Can. Field-Nat.* **105,** 233–242.

Baumgartner, M. F. (1997). The distribution of Risso's dolphin (*Grampus griseus*) with respect to the physiography of the northern Gulf of Mexico. *Mar. Mamm. Sci.* **13,** 614–638.

Kruse, S., Caldwell, D. K., and Caldwell, M. C. (1999). Risso's dolphin *Grampus griseus* (G. Cuvier, 1812). *In* "Handbook of Marine Mammals" (S. Ridgway, ed.), Vol. 6, pp. 183–212. Academic Press, New York.

Leatherwood, S., Perrin, W. F., Kirby, V. L., Hubbs, C. L., and Dahlheim, M. (1980). Distribution and movements of Risso's dolphin, *Grampus griseus,* in the eastern North Pacific. *Fish. Bull. U.S.* **77,** 951–963.

Shane, S. H. (1994). Occurrence and habitat use of marine mammals at Santa Catalina Island, California from 1983–91. *Bull. South. Calif. Acad. Sci.* **93,** 13–29.

Shane, S. H. (1995). Behavior patterns of pilot whales and Risso's dolphins off Santa Catalina Island, California. *Aqu. Mamm.* **21,** 195–197.

Shane, S. H. (1995). Relationship between pilot whales and Risso's dolphins at Santa Catalina Island, California, USA. *Mar. Ecol. Prog. Ser.* **123,** 5–11.

Shelden, K. E. W., Baldridge, A., and Withrow, D. E. (1995). Observations of Risso's dolphins, *Grampus griseus* with gray whales, *Eschrichtius robustus. Mar. Mamm. Sci.* **11,** 231–240.

River Dolphins

RANDALL R. REEVES
*Okapi Wildlife Associates,
Hudson, Quebec, Canada*

Most people are surprised to learn that some species of dolphins, and one porpoise species, live either entirely or partly in freshwater rivers and lakes. These animals are obviously exceptional, and they are the result of geologic processes that allowed (or forced) marine-adapted species to become established in inland waters. River dolphins exhibit some extreme characteristics in their morphology and sensory systems. They are also among the most seriously threatened cetaceans because their habitat and resources must be shared with many millions of people.

I. Definition and Distribution

The term "river dolphin" is not unambiguous. In Rice's (1998) evaluation of marine mammal SYSTEMATICS, for example, he assigned the term to the "peculiar long-snouted" dolphins in four single-species genera: *Platanista* (the Indian river dolphin), *Lipotes* (the Chinese or Yangtze river dolphin), *Inia* (the Amazon river dolphin), and *Pontoporia* (the La Plata dolphin). He also contends that each of these genera belongs to a separate family and that *Platanista* is the only living representative of the primitive superfamily Platanistoidea. The previous convention had been to consider the four morphologically similar species, or species groups, as all falling within the Platanistoidea. Although the genera *Lipotes* and *Pontoporia* are clearly monospecific, it has been customary to recognize two species of *Platanista*—the Indus dolphin (*P. minor*) and the Ganges dolphin (*P. gangetica*). Rice (1998) found no solid morphological evidence to distinguish them and thus proposed that they be considered subspecies: *P. gangetica minor* and *P. g. gangetica*. There is no question that the two populations have been totally isolated for a considerable time (at least hundreds of years). *P. g. minor* is confined to the Indus drainage in Pakistan, whereas *P. g. gangetica* occurs in the Ganges, Brahmaputra, Megna, Karnaphuli, and Sangu drainage systems of India, Bangladesh, and Nepal. There are three separate populations of the boto (*Inia geoffrensis*): the Bolivian subspecies *I. geoffrensis boliviensis* in the Madeira River drainage above the Teotonio Rapids at Porto Velho, the Amazonian subspecies *I. g. geoffrensis* distributed throughout the Amazon drainage basin except the upper Madeira system, and the Orinoco subspecies *I. g. humboldtiana* distributed throughout the Orinoco drainage basin (Fig. 1). The Yangtze River dolphin or baiji (*Lipotes vexillifer*) is endemic to China's Yangtze River system. In the past, it also occurred at least seasonally in the two large lakes, Dongting and Poyang, appended to the middle reaches of the Yangtze.

However tortuous the arguments may be with regard to the number of species or subspecies, and their systematic relationships, a more immediately practical way to define "river dolphins"

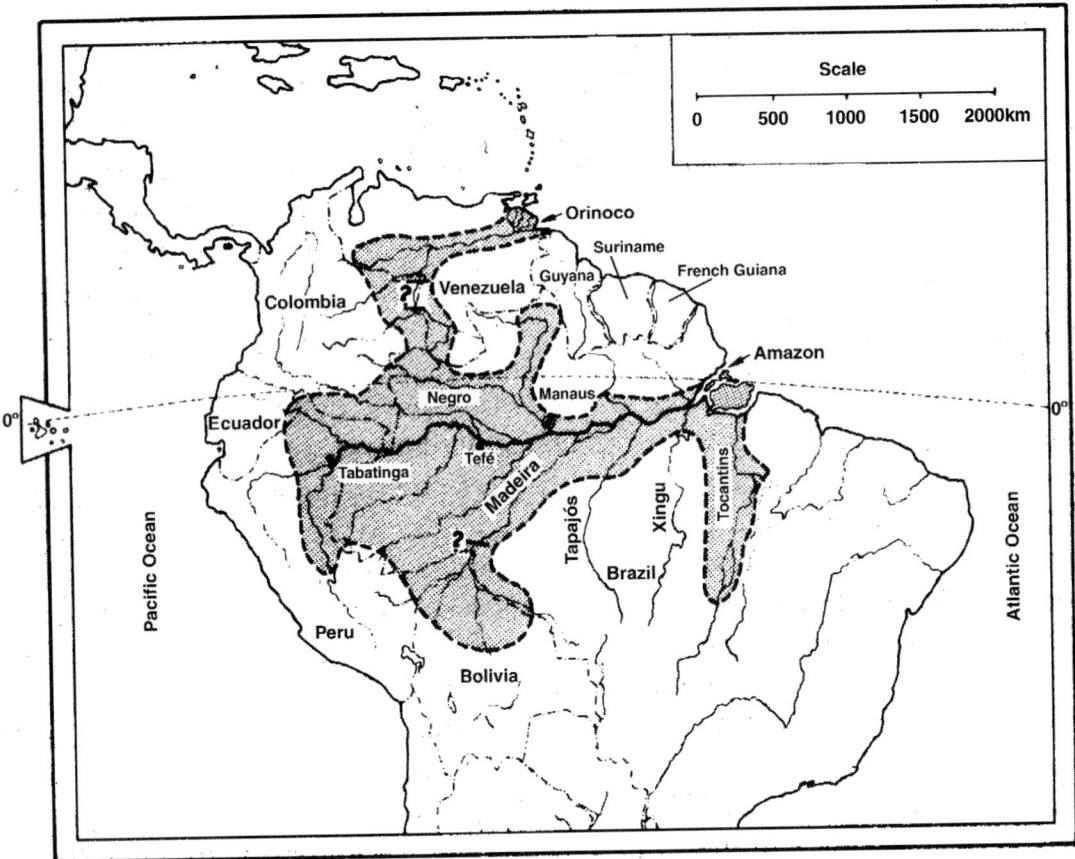

Figure 1 *Distribution of the boto, showing three main river systems inhabited. Upper "?" indicates likely barrier between Orinoco and Amazon subspecies; lower "?" indicates likely barrier between Amazon and Bolivian subspecies.*

is according to their present-day distribution and ecological position in nature. If river dolphins are defined only as those cetaceans that live solely or primarily in rivers, then the La Plata dolphin or franciscana (*Pontoporia blainvillei*) must be excluded because it lives in coastal marine waters of eastern South America, including the estuary of the Rio de la Plata (River Plate) between Argentina and Uruguay. At the same time, several species can be added to the list. The tucuxi (*Sotalia fluviatilis*) inhabits tropical coastal marine waters of eastern South and Central America but also lives far up the Amazon to the Andean foothills, frequently entering lakes and side channels as well as being abundant in the large, fast-flowing main rivers. The species also occurs in the lower Orinoco River and in the lower reaches of rivers in Guyana and Surinam. Rice (1998) recognizes two subspecies: the freshwater *S. f. fluviatilis* and the marine *S. f. guianensis*. The Irrawaddy dolphin (*Orcaella brevirostris*) similarly occurs in nearshore marine and estuarine waters of Southeast Asia, portions of Indonesia, and northern Australia, but it is also present far up several large rivers, including the Irrawaddy, Mekong, and Mahakam. With further study, there will almost certainly be grounds for recognizing subspecies and geographically separate populations of *Orcaella*. The finless porpoise (*Neophocaena phocaenoides*) also fits the category of a facultative freshwater cetacean. Although it occurs primarily in shallow marine and

deltaic waters from the northern Arabian Sea, coastwise, to Japan, one population inhabits the Yangtze River and its adjoining lake systems to as far as 1670 km upriver from Shanghai. This population is classified as a separate subspecies (*N. p. asiaeorientalis*). Finless porpoises also have been known to occur at least several tens of kilometers up the Indus and Yalu rivers.

Regardless of how one defines them, the modern river cetaceans occur in only two continents: South America and Asia. Most questions regarding their origins and how they evolved remain unresolved. In the case of *Inia*, for example, one hypothesis is that their ancestors entered the Amazon basin from the Pacific Ocean approximately 15 million years ago, while another is that they entered from the Atlantic Ocean only 1.8–5 million years ago.

II. Behavior and Ecology

Little, in fact almost nothing, is known about river dolphin societies: how they are structured, how individuals coalesce and disperse to form associations, or whether bonds between individuals are long-lasting or transient. In general, these animals seem not to be highly social, with observed group sizes rarely exceeding 10 or 15 individuals. Yet the densities at which they exist, expressed in terms of individuals per unit area of water surface, sometimes far

Figure 2 *Despite the impression in this photo from Marineland of Florida that botos are sociable, a major problem with captive groups has been the extremely aggressive behavior of mature males. (Photograph from Caldwell, Caldwell, and Brill, 1989.)*

exceed those of marine cetaceans. For example, botos and tucuxis in portions of the upper Amazon system typically occur in densities of 1 to 10 individuals/km² (Vidal *et al.*, 1997).

The distribution of river cetaceans is far from random. They tend to congregate at particular points in a river, especially at confluences (where rivers or streams converge), sharp bends, and sandbanks, and near the downstream ends of islands. In a detailed study of the distribution of Ganges dolphins in Nepal's Karnali River, Smith (1993) found the animals primarily in eddy countercurrent systems of the main river channel. Such areas of interrupted flow occur when fine sand or silt is deposited as a result of stream convergence. It is not entirely clear why the dolphins are attracted to these sites, but it likely has some relation to prey availability and energy saving. As Smith (1993) points out, positions within eddies "require minimal energy to maintain but are near high-velocity currents where the dolphins can take advantage of passing fish." Large confluences may contain tens of dolphins at a given time, but such concentrations appear to be adventitious rather than formed for social reasons. In other words, noninteracting individuals are found in close proximity due to the clumped nature of resources and refugia in the river systems where they are found. There are, of course, differences in degrees of sociability among the species. The author has seen as many as 12–15 tucuxis actively herding prey fish against a riverbank in concert, whereas botos appear to be solitary hunters most of the time, even when they are chasing the same school of fish. This applies equally to Indus and Ganges dolphins, which always seem to be acting individually or in very small groups.

In addition to their freshwater habitat, river dolphins have a number of characteristics that set them apart from other cetaceans. The eyes of Indus and Ganges dolphins lack a crystalline lens, rendering the animals functionally blind. At most, they may be able to perceive gross differences between light and dark. Because most of their habitat is highly turbid, un-

derwater VISION would be of little use. These dolphins usually swim on their side, with one flipper (most often the right one) trailing near the river bottom and the body oriented so that the tail end is somewhat higher in the water column than the head end. Their head nods constantly as they scan acoustically for prey and obstacles, Indus and Ganges dolphins remain active day and night. All river dolphins are endowed with a sophisticated biosonar system, but those other than the Indus and Ganges dolphins also have good vision.

All river dolphins have adapted to living in a highly dynamic environment. Although much of their habitat is silty, they also occur in areas where the water is clear, as in the upper reaches of the Ganges, or "black" (stained by tannic acid), as in many Amazon and Orinoco tributaries. Water levels in the lower Amazon can vary seasonally by as much as 10–12 m. During the dry season, the dolphins (and other fauna) are restricted to the deep channels of lakes and rivers, while during the flood season they can range widely. Amazon dolphins penetrate into rain forests and venture onto grasslands during the floods. Their DIET seems diverse, with at least 45 fish species from 18 families, plus crabs and river turtles, represented in examined stomach contents (Best, 1984). Both schooling and nonschooling fish species are eaten. Amazon dolphins are the only modern cetaceans with a differentiated dentition. The teeth in the front half of the jaw are conical, whereas those in the latter half have a flange on the inside portion of the crown, more reminiscent of molars (for crushing) than canines or incisors (for biting and holding). Presumably, this feature is related to the hard-bodied or spiny character of some of their prey (e.g., armored catfishes, even turtles); large catfish are often torn into smaller pieces before being eaten.

Irrawaddy dolphins engage in "cooperative fishing" with throw-net fishermen in Burma's Irrawaddy River (Smith *et al.*, 1997). The fishermen call the dolphins by repeatedly striking the sides of their canoes with a wooden pin. Then they slap the water surface with a paddle, utter a turkey-like call, and make several practice throws of the net. When conditions are favorable, the dolphins slap the surface with their flukes and begin herding the fish school toward the fishermen. With a signal from one of the dolphins (its partially submerged flukes waving laterally toward the fishermen), the net is thrown. According to the fishermen, catches made with the help of dolphins are consistently better that those made without their assistance. Not surprisingly, the animals are revered and protected by the residents of local fishing communities along the Irrawaddy.

III. Threats and Conservation Concerns

Any description of the river cetaceans must include a section on their conservation status. They are among the most endangered marine mammals (see Smith and Smith, 1998; Reeves *et al.*, 2000). The Yangtze River dolphin is probably the most critically endangered cetacean species. Discovered by Western science as recently as 1918, it apparently was still common and widely distributed along the entire Yangtze River, from Three Gorges to Shanghai, when China's Great Leap Forward began in the autumn of 1958. From that time, the baiji was hunted intensively for meat, oil, and leather. Although legally protected, Yangtze dolphins continue to die accidentally in fishing gear, from collisions with powered vessels, and from exposure to underwater blasting during harbor construction. This mortality, combined with the effects of overfishing, pollution, industrial and vessel noise, and the damming of Yangtze tributaries, has driven the baiji population to the brink of EXTINCTION. Only a few tens of individuals are thought to survive. The finless porpoise that share much of this river dolphin's historical range have also been declining rapidly in recent years, presumably for the same reasons. Efforts to protect both species have been far from adequate. As the controversy surrounding construction of the Three Gorges Dam in the upper Yangtze River has eloquently demonstrated, China is committed to a course that places further industrial development of the Yangtze basin far ahead of preserving the natural environment (Zhou *et al.*, 1998).

The Indus and Ganges dolphins are also classified as endangered, with the former numbering about a thousand and the latter possibly in the low thousands. Indus dolphins occur today only in the main channel of the river, although historically they inhabited several large tributaries as well (Sutlej, Ravi, Chenab, and Jhelum). Their population has been fragmented by irrigation dams, and the subpopulations trapped upriver of these dams have progressively gone extinct. Now, only three subpopulations of Indus dolphins are large enough to be considered potentially viable. The Ganges dolphin has also lost large segments of upstream habitat as a result of dam construction, but its generally broader distribution has meant that it is less immediately threatened with extinction. Like the baiji and Yangtze finless porpoise, the Indus and Ganges dolphins are subjected to incidental capture in fishing gear, especially gill nets. An additional concern for the Ganges dolphin is that fishermen in some parts of India and Bangladesh use dolphin oil as an attractant while fishing for a highly esteemed species of catfish. This means that there is a demand for carcasses and a disincentive for releasing live dolphins found in nets. Some tribal people in remote reaches of the Ganges and Brahmaputra still hunt dolphins for food.

Ultimately, all river cetaceans are threatened by the transformation of their habitat to serve human needs. In addition to impeding the natural movements of dolphins and other aquatic organisms, dams in southern Asia divert water to irrigate farm fields and supply homes and businesses in an arid landscape, reducing directly the amount of habitat available to the dolphins. As water becomes an increasingly strategic resource in a warming world with burgeoning human populations, the prospects for river cetaceans are certain to deteriorate even further.

See Also the Following Articles

Amazon River Dolphin ■ Baiji ■ Endangered Species and Populations ■ Finless Porpoise ■ Franciscana ■ Irrawaddy Dolphin ■ Susu and Bhulan ■ Tucuxi

References

Best, R. C. (1984). The aquatic mammals and reptiles of the Amazon. *In* "The Amazon: Limnology and Landscape Ecology of a Mighty Tropical River and Its Basin" (H. Sioli, ed.), pp. 371–412. Dr W. Junk, Dordrecht, The Netherlands.

Best, R. C., and da Silva, V. M. F. (1989). Amazon river dolphin, boto *Inia geoffrensis* (de Blainville, 1817). *In* "Handbook of Marine Mammals" (S. H. Ridgway and R. Harrison, eds.), Vol. 4, pp. 1–23. Academic Press, London.

Brownell, R. L., Jr. (1989). Franciscana *Pontporia blainvillei* (Gervais and d'Orbigny, 1844). *In* "Handbook of Marine Mammals" (S. H. Ridgway and R. Harrison, eds.), Vol. 4, pp. 45–67. Academic Press, London.

Caldwell, M. C., Caldwell, D. K., and Brill, R. L. (1989). *Inia geoffrensis* in captivity in the United States. *In* "Biology and Conservation of the River Dolphins" Occasional Papers of the IUCN Species Survival Commission No. 3 (W. F. Perrin, R. L. Brownell, Jr., K. Zhou, and J. Liu, eds.), pp. 35–41. IUCN, Gland, Switzerland.

Chen, P. (1989). Baiji *Lipotes vexillifer* Miller, 1918. *In* "Handbook of Marine Mammals" (S. H. Ridgway and R. Harrison, eds.), Vol. 4, pp. 25–43. Academic Press, London.

da Silva, V. M. F., and Best, R. C. (1994). Tucuxi *Sotalia fluviatilis* (Gervais, 1853). *In* "Handbook of Marine Mammals" (S. H. Ridgway and R. Harrison, eds.), Vol. 5, pp. 43–69. Academic Press, London.

Herald, E. S., Brownell, R. L., Jr., Frye, F. L., Morris, E. J., Evans, W. E., and Scott, A. B. (1969). Blind river dolphins: First aide-swimming cetacean. *Science* **166**, 1408–1410.

Kasuya, T. (1999). Finless porpoise *Neophocaena phocaenoides* (G. Cuvier, 1829). *In* "Handbook of Marine Mammals" (S. H. Ridgway and R. Harrison, eds.), Vol. 6, pp. 411–442. Academic Press, San Diego.

Marsh, H., Lloze, R., Heinsohn, G. E., and Kasuya, T. (1989). Irrawaddy dolphin *Orcaella brevirostris* (Gray, 1866). *In* "Handbook of Marine Mammals" (S. H. Ridgway and R. Harrison, eds.), Vol. 4, pp. 101–118. Academic Press, London.

Perrin, W. F., Brownell, R. L., Jr., Zhou, K., and Liu, J. (eds.) (1989). "Biology and Conservation of the River Dolphins." Occasional Papers of the IUCN Species Survival Commission No. 3. IUCN–The World Conservation Union, Gland, Switzerland.

Reeves, R. R., and Brownell, R. L., Jr. (1989). Susu *Platanista gangetica* (Roxburgh, 1801) and *Platanista minor* Owen, 1853. *In* "Handbook of Marine Mammals" (S. H. Ridgway and R. Harrison, eds.), Vol. 4, pp. 69–99. Academic Press, London.

Reeves, R. R., Smith, B. D., and Kasuya, T. (eds.) (2000). "Biology and Conservation of Freshwater Cetaceans in Asia." Occasional Papers of the IUCN Species Survival Commission No. 23. IUCN–The World Conservation Union, Gland, Switzerland.

Rice, D. W. (1998). "Marine Mammals of the World: Systematics and Distribution." Special Publication No. 3, Society for Marine Mammalogy, Lawrence, KS.

Smith, A. M., and Smith, B. D. (1998). Review of status and threats to river cetaceans and recommendations for their conservation. *Environ. Rev.* **6**, 189–206.

Smith, B. D. (1993). 1990 status and conservation of the Ganges river dolphin *Platanista gangetica* in the Karnali River, Nepal. *Biol. Conserv.* **66**, 159–169.

Smith, B. D., Thant, U H., Lwin, J. M., and Shaw, C. D. (1997). Investigation of cetaceans in the Ayeyarwady River and northern coastal waters of Myanmar. *Asian Mar. Biol.* **14**, 173–194.

Vidal, O., Barlow, J., Hurtado, L. A., Torre, J., Cendón, P., and Ojeda, Z. (1997). Distribution and abundance of the Amazon river dolphin (*Inia geoffrensis*) and the tucuxi (*Sotalia fluviatilis*) in the upper Amazon River. *Mar. Mamm. Sci.* **12**, 427–445.

Zhou, K., Sun, J., Gao, A., and Würsig, B. (1998). Baiji (*Lipotes vexillifer*) in the lower Yangtze River: Movements, numbers, threats and conservation needs. *Aqu. Mamm.* **24**, 123–132.

Zhou, K., and Zhang, X. (1991). "Baiji: The Yangtze River Dolphin and Other Endangered Animals of China." Stone Wall Press, Washington, DC.

River Dolphins, Evolutionary History

CHRISTIAN DE MUIZON
National Museum of Natural History,
Paris, France

The term "river dolphins" has been used traditionally to include the Recent odontocetes living in fresh waters and of which affinities to other groups of odontocetes were unclear. They have been generally included in the superfamily Platanistoidea, mainly because of their freshwater habitat and because they present many plesiomorphic characters relative to other groups such as delphinoids, physeteroids, or ziphiids (e.g., Slijper, 1936; Simpson, 1945). The four genera of living "river dolphins" (*Platanista, Lipotes, Inia,* and *Pontoporia*) were, therefore, regarded as belonging to a monophyletic group of primitive odontocetes. Other freshwater (but also partly marine) odontocetes, such as *Orcaella* (Irrawaddy River dolphin) and *Sotalia fluviatilis* (tucuxi), were not included in Platanistoidea because of their unanimously accepted close affinities to Delphinoidea. While there was a widespread assumption of their monophyly, the affinities of recent Platanistoids (i.e., "river dolphins") and referred fossil genera to other living and fossil groups of odontocetes have long been very confused and interpretations diverse. The "platanistoids" or some of their included taxa have been regarded as closely related to several groups of fossil odontocetes, e.g., Squalodontidae, Eurhinodelphinidae, and "Acrodelphinidae."

A possible para- or polyphyly of the traditional concept of Platanistoidea was first considered by Muizon (1984) and was confirmed by further studies (e.g., Muizon, 1987, 1988, 1991, 1994; Heyning, 1989; Fordyce, 1994; Messenger and McGuire, 1998). As expressed in these studies, there now seems to be a consensus on the nonmonophyly of the "river dolphins" and their fossil relatives (Fig. 1). The genus *Platanista* appears to belong to an early diverging group of odontocetes, and the three other genera (*Lipotes, Pontoporia,* and *Inia*) are regarded as closely related to Delphinoidea. *Platanista* is the Recent representative of a monophyletic group of odontocetes, the Platanistoidea, which was well diversified and widely distributed during the Oligocene and the Miocene. This group, in addition to Platanistidae, includes the fossil families Prosqualodontidae, Squalodontidae, Waipatiidae, Squalodelphinidae, and possibly Dalpiazinidae. The other "river dolphins" are included with Delphinoidea within the monophyletic infraorder Delphinida (Muizon, 1988). There is no consensus on their position within Delphinida, although they are generally regarded as basal taxa.

I. Platanistoidea

This monophyletic superfamily of odontocetes includes one Recent genus (*Platanista*) and approximately 15 fossil taxa (according to interpretations) (see Fordyce and Muizon, 2001). As recommended by Muizon (1987) and Fordyce and Muizon (2001), only the genera based, at least, on reasonably complete cranial remains, including ear bones are considered. The monophyly of Platanistoidea is supported by several synapomorphies such as the reduction or loss of the coracoid process of the scapula, the development of an articular ridge or peg on the periotic, and the ventral deflection of the anterior process of the periotic

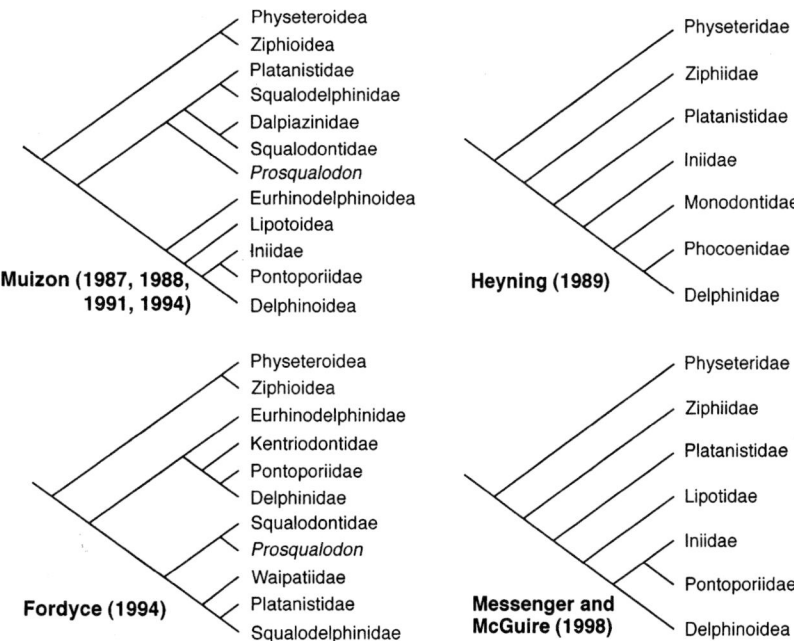

Figure 1 *Cladograms of recent hypotheses on the affinities of "river dolphins."*

(Muizon, 1994; Fordyce, 1994). This superfamily includes five (possibly six) families: Squalodontidae, Prosqualodontidae, Dalpi-azinidae, Waipatiidae, Squalodelphinidae, and Platanistidae. In contrast to their Recent representative, all the fossil platanistoids are marine, which would indicate that adaptation to a freshwater environment is probably a derived condition.

A. Squalodontidae

Squalodonts are the most common fossil platanistoids. They are also called "shark-toothed" dolphins because of their heterodont dentition, with the posterior teeth being triangular with roughly or finely serrated edges resembling shark teeth. This plesiomorphic condition has often been one of the major arguments to refer many fossil taxa to this family. As a consequence, until recently, the Squalodontidae represented a waste basket of fossil odontocetes with heterodont dentition. Squalodontid genera based on partial or complete skulls are *Squalodon, Kelloggia* (a possible synonym of *Squalodon*), *Eosqualodon,* and *Phoberodon.* Synapomorphies of Squalodontidae as defined by Fordyce (1994) are essentially based on the morphology of the periotic, a bone unknown in *Phoberodon, Eosqualodon,* and *Kelloggia.* Muizon (1991, 1994) has proposed some cranial synapomorphies of the Squalodontidae, some of which, as mentioned by Fordyce (1994), are probably not securely established. Therefore, it is clear that the monophyly of Squalodontidae still has to be confirmed by a more thorough study of their auditory region. Furthermore, a better knowledge of the anatomy of the enigmatic genus *Patriocetus* could confirm its introduction into the Squalodontidae, as proposed by Rothausen (1968).

Squalodontidae present the scapula synapomorphies of Platanistoidea (however, this bone is unknown in *Eosqualodon* and incomplete in *Phoberodon*). Some other synapomorphies, present in all the Platanistoids except *Prosqualodon,* are incipiently developed in Squalodontidae; such as the overlap of the palatine by the maxilla and the presence of a shallow subcircular fossa in the squamosal dorsolateral to the periotic (see Muizon, 1987, 1991, 1994).

Squalodontidae are cosmopolitan basal platanistoids. All their remains were found in marine coastal environments. *Squalodon* is present in the Miocene of Europe, Asia, and North America; *Eosqualodon* is present in the Miocene of Europe; *Kelloggia* is present in the late Oligocene of Asia; and *Phoberodon* is from the early Miocene of South America. Undescribed squalodontids have also been found in Australia and New Zealand (Fordyce and Muizon, 2001). Squalodontids are relatively large odontocetes approaching the size of the living *Mesoplodon* spp. (beaked whales). They had a long rostrum with strongly procumbent anterior teeth (Fig. 2). In fact, the medial incisors were almost horizontal. The teeth were strongly heterodont. The vertex was low and the skull was symmetrical. As in all platanistoids, Squalodontidae have enlarged and slightly concave premaxillary fossae anterolateral to the nares. These fossae received premaxillary sacs of the nasal tract. Premaxillary sacs are tightly related to the presence of nasal plugs and melon, and their presence in Squalodontidae is an indication of efficient echolocation ability.

B. Prosqualodontidae

The single genus *Prosqualodon* is included in this family. Initially placed in Squalodontidae (e.g., Simpson, 1945; Rothausen,

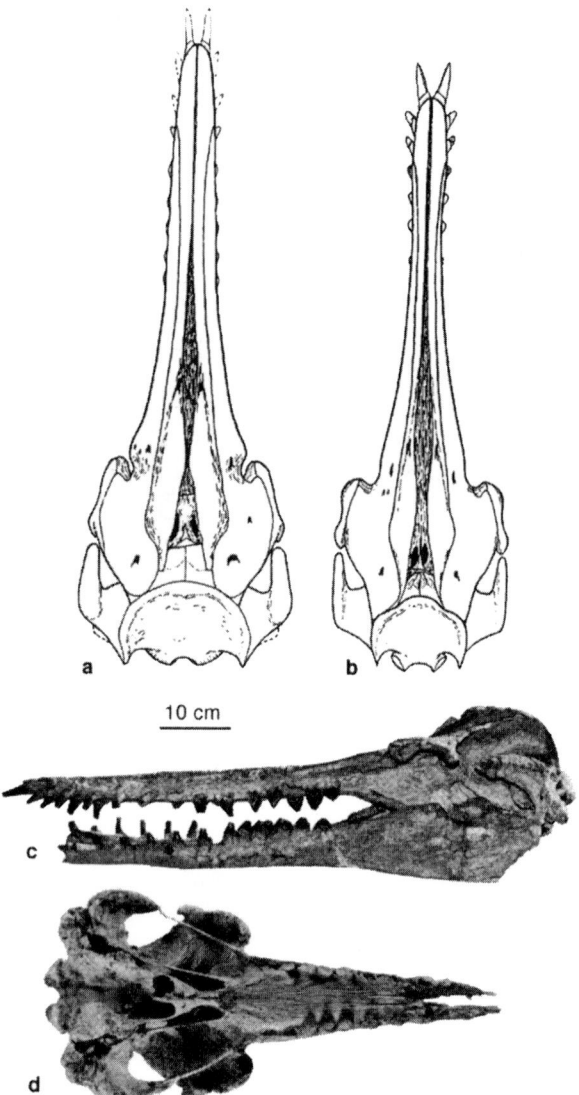

Figure 2 *Skulls of Squalodontidae: (a)* Eosqualodon langewi-eschei *(late Oligocene, Germany), reconstruction of the skull in dorsal view [from Rothausen (1968), modified]; (b)* Squalodon bellunensis *(early Miocene, Italy), reconstruction of the skull in dorsal view [from Rothausen (1968), modified; (c)* S. bellunen-sis *(early Miocene, Italy), skull and mandible in lateral view; (d)* Squalodon bariensis *(early Miocene, France), skull (apex of the rostrum missing) in ventral view. a and b are reproduced with permission of Paläontologische Zeitschrift.*

1968), *Prosqualodon* has been removed from this family by Muizon (1991) because it does not possess the synapomorphies of the auditory region observed in other platanistoids; however, it was maintained in the superfamily because it bears the scapula synapomorphies of the group. Therefore, *Prosqualodon* would be the sister group of all the other platanistoids. However, *Prosqualodon* has also been regarded as member of the the infraorder Delphinida on the basis of the presence of the same apomorphic character of the palatine [presence of a lateral lamina as defined by Muizon (1988)] in the two groups. In fact, as stated by Muizon (1994), the palatine-derived character in Delphinida. Fordyce (1994) regarded *Prosqualodon* as the sister taxon of *Squalodon*, whereas Fordyce and Muizon (2000) retained the taxon Prosqualodontidae, which they regard as a platanistoid family, although they state that *Prosqualodon* could also be a squalodontid.

Prosqualodon is an austral genus that has been found so far in the early Miocene of Argentina, Australia, and New Zealand. It is a medium-sized odontocete, and its size ranges from a small *Globicephala* (pilot whale) to a large *Tursiops* (bottlenose dolphin). As squalodontids, *Prosqualodon* had heterodont teeth. The anterior teeth are elongated conically and project anteroventally; the posterior teeth are triangular, low, transversely compressed with a rugose enamel, and bear several denticles on their anterior and posterior crests. The rostrum is short, the vertex is symmetrical, and the braincase is lower than in *Squalodon*. Premaxillary fossae are clearly present but they are less developed and shallower than in *Squalodon* (Fig. 3).

C. Dalpiazinidae

The single known genus of the family, *Dalpiazina,* has been related to the Platanistoidea on the presence of several simi-

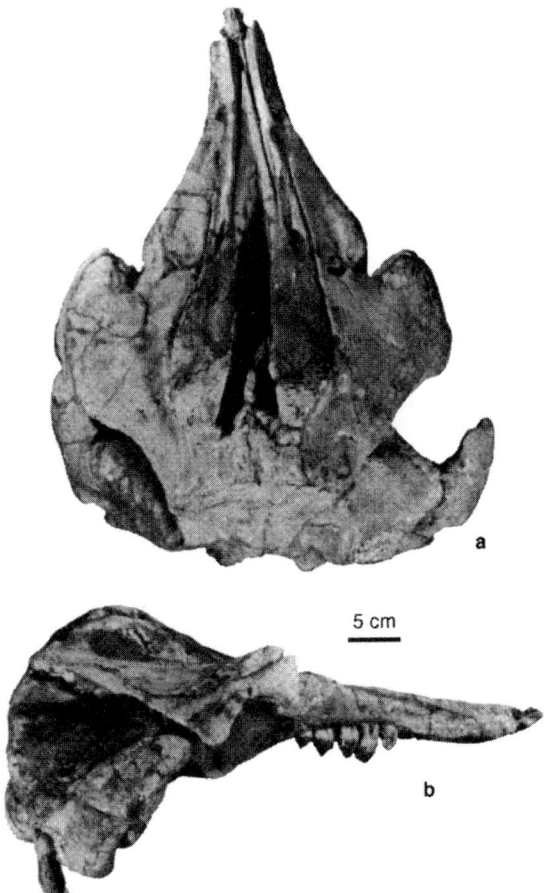

5 cm

Figure 3 *Skull of* Prosqualodon australis *(early Miocene, Argentina) in dorsal (a) and lateral (b) views.*

larities that it shares with *Squalodon* (see Muizon, 1991, 1994). However, it is noteworthy that none of the platanistoid synapomorphies are observable on the specimens available and, therefore, the affinities of this family still have to be confirmed. *Dalpiazina* is a medium-sized odontocete (like a small *Tursiops*). The rostrum is relatively long and bears homodont dentition. It is known from the early Miocene of Italy, and some possible dalpiazinids have been discovered in New Zealand (Fordyce *et al.,* 1994).

D. Waipatiidae

Waipatia is known from a nearly complete skull with ear bones and partial skeleton. This genus presents the auditory region synapomorphies of the platanistoids and, although its scapula is unknown, is better placed in this superfamily than in any other group of odontocete (Fordyce, 1994; Fordyce and Muizon, 2001). It is a medium-sized platanistoid similar in size to *Tursiops.* The rostrum is long and slender (Fig. 4). It bears heterodont teeth but the posterior triangular and double-rooted teeth are smaller than in Squalodontidae. The incisors are conical and strongly procumbent. The skull roof is very low as in squalodontids. The skull of *Waipatia* shows clear directional asymmetry of the bones. The premaxillary fossae are well developed, and the premaxillae extend posteriorly to the nasals and contact the frontals on the vertex as in the other platanistoids. *Waipatia maerewhenua,* the single species clearly referred to this family, is from the late Oligocene of New Zealand.

Sulakocetus is a primitive odontocete from the late Oligocene of Asia (Caucasus). It bears heterodont dentition but its posterior double-rooted teeth are small as in *Waipatia.* Apparently the periotic of the single known skull is unknown (or unprepared) but its tympanic is squalodont-like. The scapula bears a small coracoid process. Because of this character, *Sulakocetus* should be excluded from Platanistoidea. However, it is probable that the small (reduced) size of the process represents an incipient development of the platanistoid condition. This genus has been classified by Fordyce and Muizon (2001) as a possible waipatiid. However, it is clear that more information on its auditory region is needed to clarify the systematic position of *Sulakocetus.*

E. Squalodelphinidae

This family includes the genera *Notocetus, Medocinia, Phocageneus,* and *Squalodelphis.* The four taxa are based on reasonably well-preserved skulls and/or ear bones. Squalodelphinidae present the platanistoid synapomorphies of the scapula (loss of the coracoid process, anterior position of the acromion) and of the ear region (e.g., subcircular fossa, articular ridge of the periotic, morphology of the apex of the tympanic) (see Muizon, 1987, 1994). Squalodelphinidae are cosmopolitan and marine. *Notocetus* is from the early Miocene of South America and New Zealand; *Squalodelphis* and *Medocinia* are from the early to middle Miocene of Europe; and *Phocageneus* is from the early Miocene of North America. Squalodelphinidae are medium-sized odontocetes similar in size to (to slightly smaller than) the living *Tursiops.* The rostrum is of moderate length and slender (Fig. 5). The teeth are more or less homodont: the posterior teeth are single rooted but they are clearly lower and

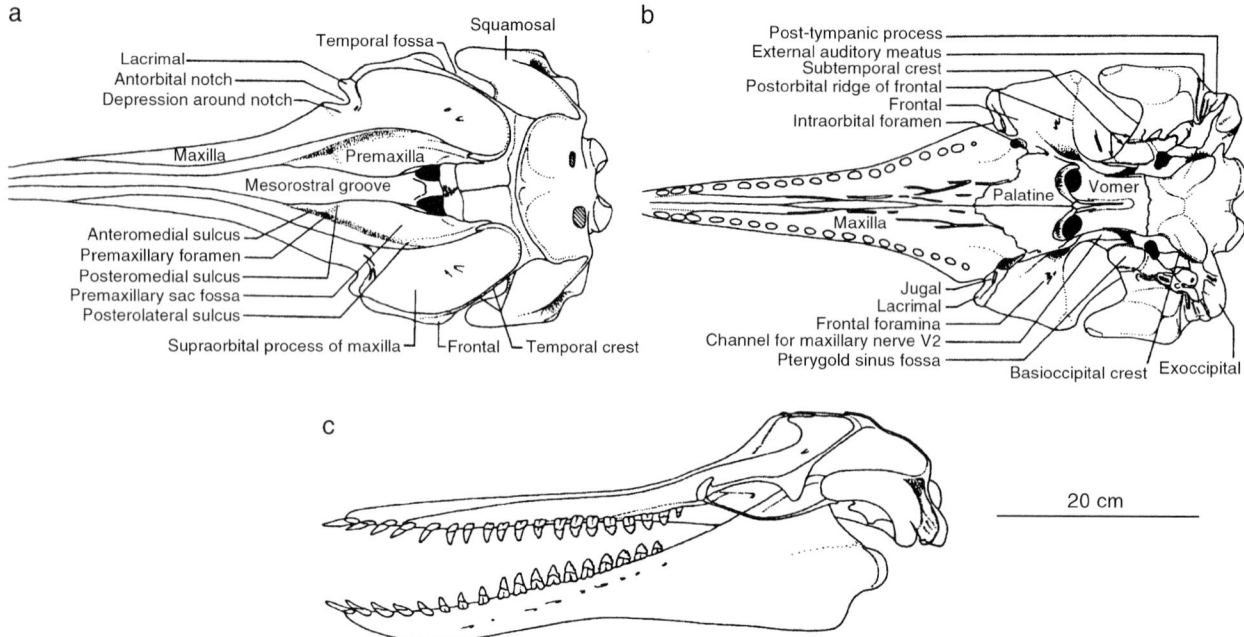

Figure 4 *Reconstruction of the skull of* Waipatia maerewhenua *(late Oligocene, New Zealand): (a) dorsal, (b) ventral, (c) lateral views [from Fordyce (1994), modified]. Reproduced with permission of the San Diego Society of Natural History.*

more triangular than the anterior. An interesting characteristic of Squalodelphinidae is the thickening of the supraorbital region of the skull (maxilla and/or frontal) (Figs. 5b and 5d). This condition seems to foreshadow the extreme morphology observed in Platanistidae (see later and Fig. 6).

F. Platanistidae

Platanistidae are represented in the fossil record by two genera: *Zarhachis* and *Pomatodelphis* (Fig. 6). They both show

all the Platanistoid synapomorphies of the ear region, palatine, and scapula. The main characteristic of Platanistidae is the development of large maxillary (*Platanista*) or maxillofrontal (*Zarhachis*, *Pomatodelphis*) crests, which are already incipiently developed in Squalodelphinidae (see earlier discussion). A peculiarity of *Platanista* is the fact that the palatine is entirely covered by the maxilla and the pterygoid. In *Zarhachis* and *Pomatodelphis* this condition is incipiently developed, as the palatine is partially covered (Muizon, 1987, 1994) and the visible

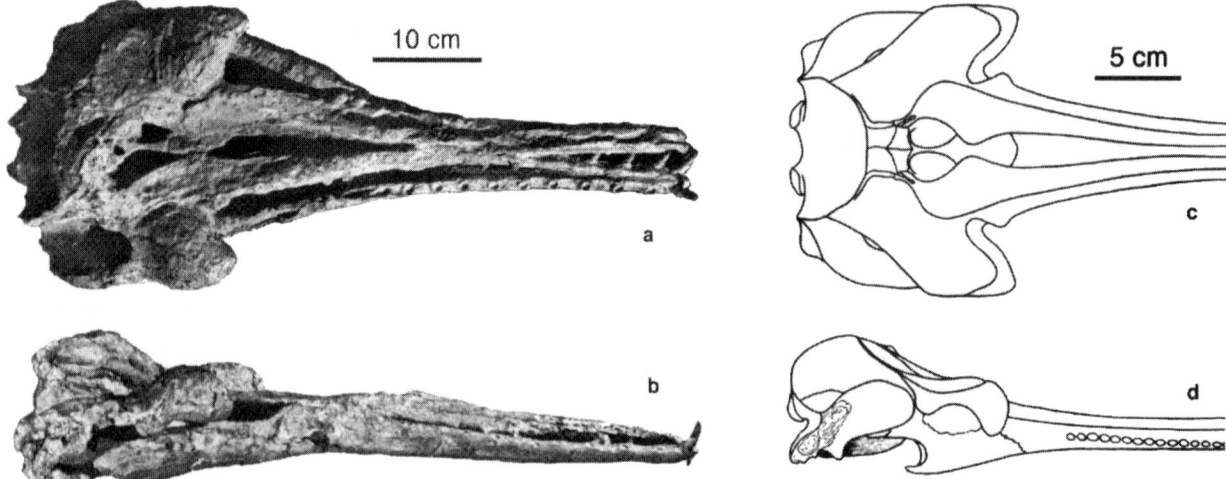

Figure 5 *Skulls of squalodelphinids:* Squalodelphis fabianii *(early Miocene, Italy) skull and mandible in dorsal view (a) and lateral view (c) (note the thickness of the supraorbital region), and reconstruction of the skull of* Medocinia tetragorhina *(early Miocene of France) in dorsal view (b) and lateral view (d) (from Muizon, 1988). Reproduced with permission of the Bulletin du Muséum national d'Histoire naturelle.*

Fordyce and Muizon (2001) recognized three families because the three groups probably do not represent a monophyletic unit. They are included in the infraorder Delphinida on the basis of several synapomorphies (Muizon, 1988), such as the development of a lateral lamina of the palatine, the sigmoid morphology of the involucrum of the tympanic and its posterior excavation, the development of a ventral rim on the ventromedial face of the anterior process of the periotic, and the increase in size of the processus muscularis of the malleus. Lipotidae are the earliest divergent Delphinida. The other Delphinida [Inioidea (Iniidae+Pontoporiidae) and Delphinoidea] differ from Lipotidae in the following synapomorphies: reduction of the anterior process of the periotic, which loses the bullar facet; increase in size of the processus muscularis of the malleus, which is distinctly more developed than the manubrium; and presence of a pair of ventral processes on the anterior region of the sternal manubrium (Muizon, 1988). The second diverging clade is Inioidea. The third clade, Delphinoidea, presents an apomorphic thickening of the apex of the anterior process of the periotic and a great reduction of the dorsal portion of the transverse process of the atlas. Therefore, Delphinida include three superfamilies of odontocetes: Lipotoidea, Inoidea, and Delphinoidea (Muizon, 1988). These phylogenetic hypotheses on the nonplatanistoid "river dolphin" have been partly confirmed by Messenger and McGuire (1998).

A. Lipotoidea

This monofamilial superfamily includes two genera: *Lipotes* (Recent, China) and *Parapontoporia* (Neogene, west coast of North America). *Prolipotes* from the Miocene of China is based on a nondiagnostic mandible fragment and is regarded as *incertae sedis* (Fordyce and Muizon, 2001).

Parapontoporia (Fig. 7) is regarded here as a lipotid, although classified by its author (Barnes, 1985) in Pontoporiidae. However, as noted by Barnes (1985) and Muizon (1988), the periotic of *Parapontoporia* is extremely similar to that of *Lipotes vexillifer*, the baiji, and bears no resemblance to that of pontoporiids. In fact, Barnes (1985) noted that *Parapontoporia* was more similar in many aspects of its skull to *Lipotes* than to *Pontoporia*. The skull of *Parapontoporia* displays a distinct narrowing at the base of the rostrum, which is always present in *Lipotes* and generally absent in pontoporiids (when present in *Pontoporia* it is weak); *Parapontoporia* does not bear the premaxillary eminences that are present in all pontoporiids and iniids; the nasals of *Parapontoporia* tend to be approximately vertical and not horizontal as in pontoporiids.

Parapontoporia is the only fossil lipotid based on good cranial material. Whereas its braincase is only slightly larger than that of *Lipotes*, its rostrum is almost twice as long. The asymmetry is less pronounced than in *Lipotes* but distinct. The teeth are small, single cusped, and numerous (ca. 80 on each side). In *Lipotes* the number of teeth varies from 30 to 36. Lipotids are known in the Northern Hemisphere only (China and California) and it is possible that the evolutionary history of the family took place in the North Pacific.

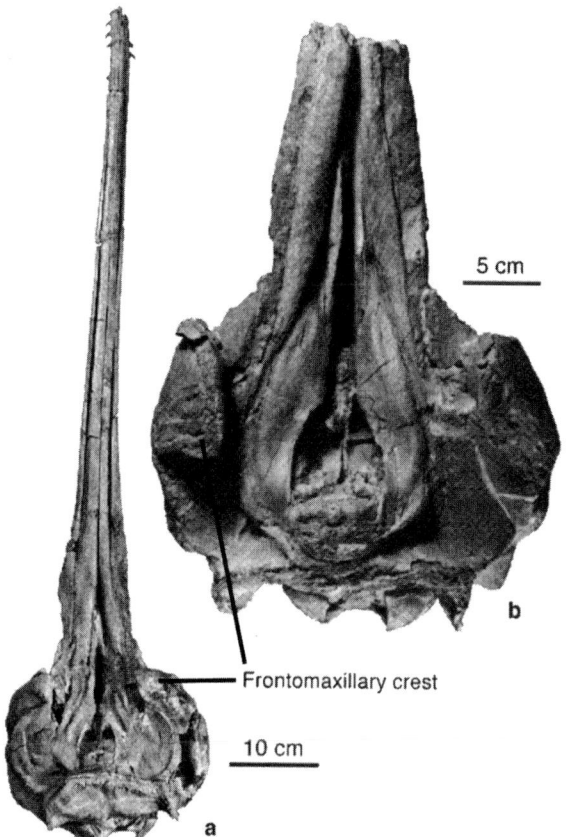

Figure 6 *Skulls of Platanistidae: (a)* Pomatodelphis *cf.* inaequalis *(middle Miocene, Maryland) skull in dorsal view and (b)* Zarhachis flagellator *(middle Miocene, Maryland) skull (most of the rostrum missing) in dorsal view.*

portion of the bone is displaced laterally. Both genera have a very long and slender rostrum bearing homodont teeth. *Zarhachis* is slightly larger than *Pomatodelphis* and approaches the size of a small *Mesoplodon*. *Allodelphis* has been regarded as a platanistid; however, this genus is still too poorly known to be clearly referred to this family. It is provisionally considered to be a platanistoid, pending discovery of more complete specimens. The two undoubted fossil palatanistids, *Zarhachis* and *Pomatodelphis*, were found in a marine environment. They are from the middle Miocene of North America and Europe (*Pomatodelphis* only). No fossil platanistids have been found so far, neither in the Southern Hemisphere nor in Asia, and it is possible that the family had a Tethyan distribution.

II. Nonplatanistoid "River Dolphins"

These are represented by the Recent families Lipotidae (*Lipotes*), Iniidae (*Inia*), and Pontoporiidae (*Pontoporia blainvillei*, the franciscana). Most authors recognize the three families, however, some authors place the three genera in the same family, Iniidae (Heyning, 1989), whereas others (Fordyce *et al.*, 1994) recognize only two families: Pontoporiidae (*Pontoporia* and *Lipotes*) and Iniidae (*Inia*).

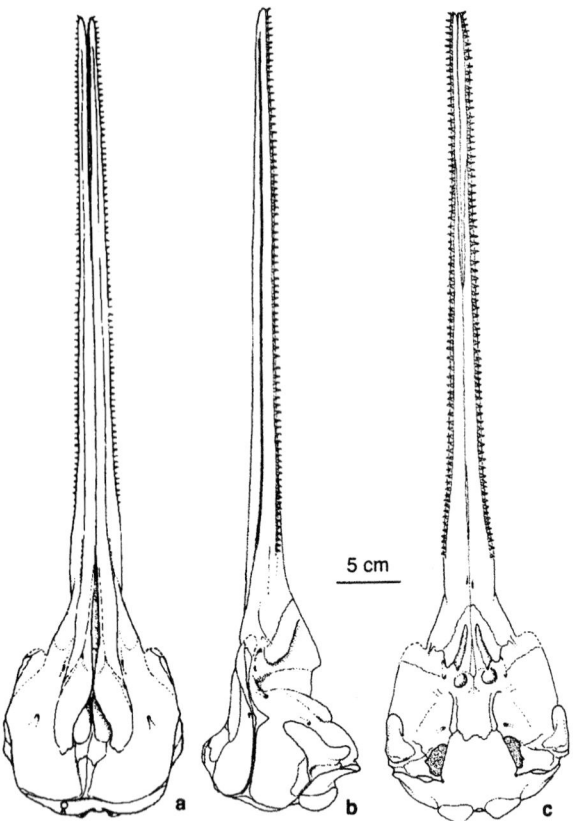

5 cm

Figure 7 *Reconstruction of the skull of* Parapontoporia stern-bergi *(early Pliocene, California): dorsal (a), ventral (b), and lateral (c) views [from Barnes (1985), modified]. Reproduced with permission of the Contributions to Science, Los Angeles County Museum of Natural History.*

B. Inioidea

This superfamily includes Iniidae and Pontoporiidae. The two families share synapomorphies of the periotic (great reduction of the anterior and posterior processes), malleus (unique extreme development of the processus muscularis), premaxilla (presence of premaxillary eminences), and maxilla (frontomaxillary crests: dorsal inflexion of the postorbital edges of the maxilla and frontal). The superfamily is documented by a few well-established fossil genera mostly from South America.

Iniidae are represented in the fossil record by a single genus known by relatively well-preserved cranial remains: *Ischyrorhynchus*. Genera *Saurocetes*, *Plicodontinia*, *Hesperocetus*, *Hesperoinia*, and *Lonchodelphis*, which have been related to Iniidae, are based on nondiagnostic rostra, mandible fragments, or isolated teeth. Although they could be referrable to this family, these genera should be regarded as incertae sedis pending further cranial discoveries. *Goniodelphis* is based on a partial skull, which probably belongs to an iniid. However, because of its incompleteness, this specimen has been only tentatively referred to the family by Fordyce and Muizon (2001).

One of the major characteristics of Iniidae (*Ischyrorhynchus* and *Inia geoffrensis*, the Amazon river dolphin or boto) is the development of a frontal hump on the vertex, which is expanded at its apex. Iniidae also present an extreme reduction of the posterior process of the periotic.

Ischyrorhynchus is approximately 30% larger than *Inia* and its rostrum is proportionally longer. In addition to these features, *Ischyrorhynchus* is very similar to the Recent iniid. It is from the late Miocene of the Paraná Basin (Argentina) and, therefore, was living in a freshwater environment.

Pontoporiidae are known by two fossil genera based on well-preserved cranial remains (with associated ear bones): *Pliopontos* and *Brachydelphis* (Fig. 8). *Pontistes* is another pontoporiid and is based on a single partial skull with a well-preserved vertex. *Pontivaga* has been referred to the pontoporiids; however, this genus, which is based on a partial mandible, is regarded as incertae sedis.

Pontoporiidae share synapomorphies such as a low vertex with flat subhorizontal nasals and the posterior extension of the posterior process of the periotic, which becomes blade-like.

Pliopontos is 50% larger than *Pontoporia*. As in the recent taxon, the rostrum is long and slender with sharp small teeth. Except for its size, *Pliopontos* is very similar to *Pontoporia*. It is from the early Pliocene of Peru and was marine. *Brachydelphis* is a much less classical pontoporiid. It has a very short rostrum, which is as long as the braincase. The latter is much larger than in *Pontoporia* and is similar in size to that of *Pliopontos*. Because of these unique features for a pontoporiid, *Brachydelphis* has been placed in its own subfamily, the Brachydelphinae. It is from the middle Miocene of Peru and lived in a marine environment.

III. Conclusions

There is now a consensus that odontocete taxa which were traditionally placed in the "river dolphins" (the "Platanistoidea" of Simpson 1945) belong to two different groups of dolphins: Platanistoidea, which are an early divergent superfamily of odontocetes, and Inioidea and Lipotoidea, which are included in Delphinida (Lipotoidea + Inioidea + Delphinoidea). Platanistoidea represent the sister group of a clade which includes Delphinida and the fossil superfamily Eurhinodelphinoidea (Fig. 1). Therefore, the "river dolphins" represent a polyphyletic group. They are paraphyletic if only Recent taxa are taken into account. Even the nonplatanistoid "river dolphins" do not represent a monophyletic grouping. Fossil platanistoids are diverse and distributed into several families. Fossil lipotids and inioids are still relatively scarce but can be easily related to one of the three non-platanistoid families of "river dolphins."

As stated earlier, most fossil "river dolphins" are from a marine environment and adaptation to fresh water is a convergence in at least in three groups: Platanistidae, Lipotidae, and Inioidea. Adaptation to this environment also appeared independently in two delphinoids (*Orcaella* and *Sotalia*). It therefore appears necessary to avoid the use of the term "river dolphins," an ecological grouping which is not monophyletic, especially when fossil taxa are considered.

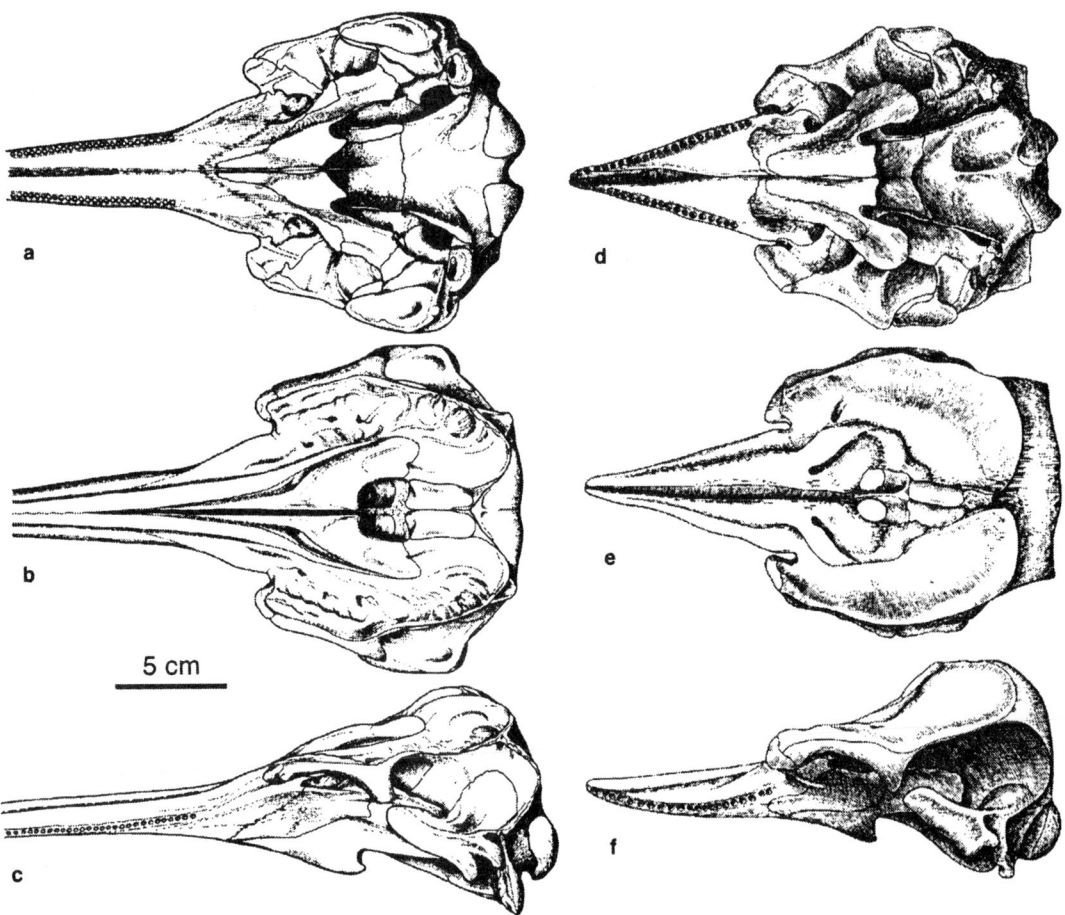

Figure 8 *Skulls of Pontoporiidae:* Pliopontos littoralis *(early Pliocene, Peru) reconstruction of the skull in ventral (a), dorsal (b), and lateral (c) views (from Muizon, 1984); and* Brachydelphys mazeasi *(middle Miocene, Peru) reconstruction of the skull in ventral (d), dorsal (e), and lateral (f) views (from Muizon, 1988). Reproduced with permission of the Institut Français d'Études Andines.*

See Also the Following Articles

Convergent Evolution ▪ Dental Morphology, Evolution of ▪ Fossil Record

References

Barnes, L. G. (1985). Fossil pontoporiid dolphins (Mammalia: Cetacea) from the Pacific coast of North America. *Contrib. Sci. Nat. Hist. Mus. Los Angeles County* **363**, 1–34.

Fordyce, R. E. (1994). *Waipatia maerewhenua*, new genus and new species (Waipatiidae, new family), an archaic Late Oligocene dolphin (Cetacea: Odontoceti: Platanistoidea) from New Zealand. *Proc. San Diego Mus. Nat. Hist.* **29**, 147–176.

Fordyce, R. E., and Muizon, C. de (2001). Evolutionary history of Cetaceans: A review. *In* "Secondary Adaptation to Life in the Water" (V. de Buffrenil and J. M. Mazin, eds.) pp. 169–233. Pfeil Verlag, Munich.

Heyning, J. E. (1989). Comparative facial anatomy of beaked whales (Ziphiidae) and a systematic revision among the families of extant Odontoceti. *Contrib. Sci. Nat. Hist. Mus. Los Angeles County* **405**, 1–64.

Messenger, S. L., and McGuire, J. A. (1998). Morphology, molecules, and the phylogenetics of Cetaceans. *System. Biol.* **47**, 90–124.

Muizon, C. de (1984). Les Vertébrés fossiles de la Formation Pisco (Pérou). Deuxième partie: Les Odontocètes (Cetacea, Mammalia) du Pliocène inférieur de Sud-Sacaco. *Travaux Instit. Français d'Études Andines* **25**, 1–188.

Muizon, C. de (1987). The affinities of *Notocetus vanbenedeni*, an Early Miocene platanistoid (Cetacea, Mammalia) from Patagonia, southern Argentina. *Am. Mus. Novitates* **2904**, 1–27.

Muizon, C. de (1988). Les relations phylogénétiques des Delphinida (Cetacea, Mammalia). *Ann. Paléontol.* **74**, 159–227.

Muizon, C. de (1991). A new Ziphiidae (Cetacea) from the Early Miocene of Washington State (USA) and phylogenetic analysis of the major groups of odontocetes. *Bull. Mus. natl. Hist. nat. Paris* **4**, 279–326.

Muizon, C. de (1994). Are the squalodonts related to the platanistoids? *Proc. Ser. San Diego Soc. Nat. Hist.* **29**, 135–146.

Rothausen, K. (1968). Die systematische Stellung der europaischen Squalodontidae (Odontoceti: Mamm.). *Paläontol. Zeitsch.* **42**, 83–104.

Simpson, G. G. (1945). The principles of classification and a classification of mammals. *Bull. Am. Mus. Nat. Hist.* **85**, 1–350.

Slijper, E. J. (1936). Die Cetaceen vergleichend-anatomisch und systematisch. *Capita Zool.* **7**, 1–590.

River Dolphins, Relationships

JOHN HEYNING
*Natural History Museum of Los Angeles County,
California*

The term "river dolphin" classically includes among the living species the boto (*Inia geoffrensis*), the franciscana (*Pontoporia blainvillei*), the baiji (*Lipotes vexillifer*), and the Indian river dolphin (*Platanista gangetica*). The boto resides in the freshwater rivers of the Amazon and Orinoco drainage, the Indian river dolphin in the Indian subcontinent rivers of the Ganges and Indus River systems, and the baiji in the Yangtze River. However, the franciscana is a "river dolphin" that lives in the coastal marine environments of eastern South America. To add confusion to this already perplexing terminology, several species of distantly related "oceanic dolphins" (family Delphinidae) such as the Irrawaddy River dolphin (*Orcaella brevirostris*) live in freshwater as well as in the ocean.

In former times, scientists classified these four living genera of "river dolphins" together into the family Platanistidae nestled within the superfamily Platanistoidea based more or less on the following logic: they all have long, narrow jaws with primitive architecture of the skull relating to their jaw closing muscles (Fig. 1) and they all inhabit freshwater (the franciscana was described from a specimen found in the mouth of the La Plata River).

The evolutionary status of the superfamily Platanistoidea is problematic in this classical usage. However, virtually all modern phylogenetic analyses indicate that *Platanista* is not closely related to the other extant river dolphins. The only character that appears to unite the river dolphins is the presence of a long mandibular symphysis, a feature that seems to have evolved in parallel. Consequently, the river dolphins have been split into two clades: the Platanistidae and the Iniidae.

I. Family Platanistidae

The modern members of this taxon include the river dolphins of the Indian subcontinent of the genus *Platanista*. These riverine cetaceans have an extremely long snout with teeth of the greatest length at the end of the snout. In this regard, *Platanista* resembles the crocodilian gavials that inhabit the same waters. *Platanista* is also called the blind river dolphin for it lacks a functional eye lens and therefore cannot discern detailed visual images, only degrees of light and dark. As this dolphin inhabits waters that are consistently turbid, sight has limited value. Two synapomorphies unite this family, including the distinctive maxillary crests (Fig. 1). These impressive crests that dominate the top of the skull can be seen less formed in the Miocene form *Zarhachis*.

In one recent cladistic analysis, the family Platanistidae fell into a clade with the extinct families Waipatiidae, Squalodontidae, and Squalodelphidae. The Squalodontidae, or the shark-toothed dolphins, are typified by relatively larger shark-like, sectorial cheek teeth with stoutly built skulls to accommodate large jaw muscles. Squalodonts were a rather common and widely distributed group that flourished during the late Oligocene through the Miocene.

II. Family Iniidae

The Iniidae, as defined herein, are united by three distinct morphological synapomorphies. In the classification used here, there are three extant species of river dolphins in the family Iniidae: *Inia geoffrensis*, *Pontoporia blainvillei*, and *Lipotes vexillifer*. Some taxonomists have separated each of these genera variously into its own family. Such splitting does not negate the conclusion that all these species form a clade, but merely reflects one's interpretation of "how different" the species are from each other.

The Late Miocene or Pliocene *Ischyrorhynchus* and *Saurodelphis* from fluvial deposits of South America are considered related to *Inia*. The only fossil taxon related to *Lipotes* is *Prolipotes* based on a fragment of mandible recovered from freshwater Neogene deposits of China. The extinct long-snouted genus *Parapontoporia* is considered to be intermediate between *Pontoporia* and *Lipotes* by some workers and similar to *Lipotes* by others. *Parapontoporia* is found in Late

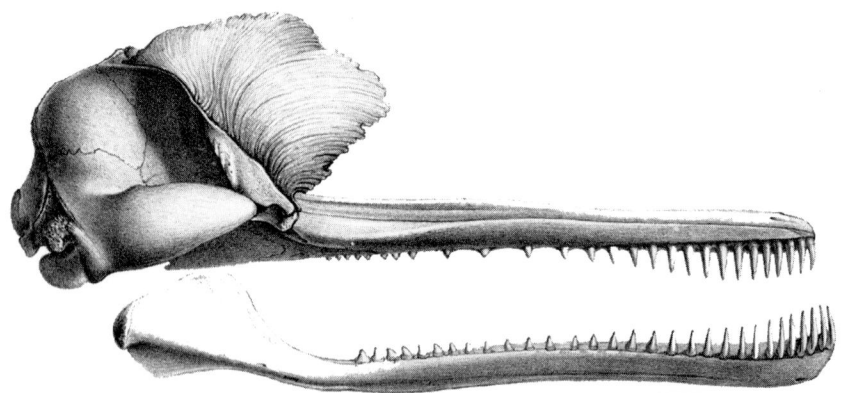

Figure 1   *Skull of the Susu (Platanista gangetica).*

Miocene to Pliocene marine deposits along the eastern North Pacific where it apparently was quite common.

The family Iniidae is currently considered most closely related to the Delphinoidea (the modern dolphins, porpoises, and their relatives) by two synapomorphies: the presence of a vestibular sac along the nasal passages and the reduction of the attachment of the ear bones to the skull. These characters provide evidence that the two groups share a common ancestry. In fact, some of the most primitive delphinoids have been classified as iniids in the past.

See Also the Following Articles

Morphology, Functional ■ River Dolphins, Evolutionary History ■ Skull Anatomy ■ Vision

References

Heyning, J. E. (1989). Comparative facial anatomy of beaked whales (Ziphiidae) and a systematic revision among the families of extant Odontoceti. Contributions in Science. Natural History Museum of Los Angeles County. No. 405.

Messenger, S. L. (1994). Phylogenetic relationships of Platanistoid river dolphins: Assessing the significance of fossil taxa. *In* "Contributions in Marine Mammal Paleontology Honoring Frank C. Whitmore, Jr." (A. Berta and T. Deméré, eds.), pp. 125–134. Proceedings of the San Diego Society of Natural History.

Rookeries

George Anthony Antonelis
*National Marine Fisheries Service,
Honolulu, Hawaii*

Pinnipeds reproduce in a wide range of marine habitats, including various forms of ice, tidal flats, rock outcroppings, and coastal beaches (Scheffer, 1958). Some species form annual breeding aggregations at traditional locations known as rookeries. These reproductive sites are an integral component of the animals' life history patterns, resulting from a complex suite of adaptive factors involving physiology, morphology, ECOLOGY, and DISTRIBUTION. Rookery-breeding pinnipeds exhibit varying forms of polygyny (Boness, 1991); this mode of social organization is believed to have evolved as a consequence of two key traits: parturition on solid substrate and offshore marine foraging (Bartholomew, 1970). The influence of these traits in conjunction with phylogenetic and ecological constraints has likely influenced the development of the polygynous mating systems observed on rookeries today (Emlin and Oring, 1977; Stirling, 1983; Boness, 1991; Boyd, 1991).

Rookery-breeding pinnipeds are subdivided into two families: Otariidae and Phocidae. Rookeries are formed by all otariids (15 species) and three phocids [2 species of elephant seal,

Mirounga angustirostris and *M. leonina*, and gray seals, *Halichoerus grypus* (Reeves *et al.*, 1992; Rice, 1998)]. This article describes the salient social–biological, physical–geographical, and environmental characteristics of pinniped rookeries and provides information on the ecological context in which they occur.

I. Social–Biological Characteristics

Rookeries are formed at specific times and locations, which optimize the reproductive success and survival of offspring (Bartholomew, 1970; Stirling, 1983; Boyd, 1991). After foraging at sea during the nonbreeding season, adult males return to rookeries and begin establishing territories shortly before or about the same time as the arrival of parturient females. Overt aggression, frequent threat vocalizations, and ritualized boundary displays are common among males when establishing and defending territories (Fig. 1). Males also attempt to herd females in an effort to keep them within their areas of influence (Fig. 2). Adult females come ashore to find suitable parturition sites and tend to be highly gregarious. Parturient females frequently threaten one another either vocally or visually and are often aggressive toward offspring of other females. Otariid females suckle their pups for about 4–12 months, although longer periods have been documented for some species (Oftedal *et al.*, 1987; Bowen, 1991). Lactation of rookery-breeding phocid females lasts about 0.5 (gray seals) to 1.0 (elephant seals) month. Estrus occurs early in lactation for otariids and late in lactation for phocids (Oftedal *et al.*, 1987). Most copulations occur on land at or near the parturition site, but a few species commonly breed

Figure 1 *Adult male California sea lions* (Zalophus californianus) *compete for territories at San Miguel Island, California (NMFS, George Antonelis).*

Figure 2 *A much larger and darker adult male California sea lion attempts to block an adult female from leaving his territory at San Miguel Island, California (NMFS, George Antonelis).*

Figure 3 *California sea lions are highly polygynous and form dense aggregations on rookeries commonly found along the shoreline at breeding sites on the California Channel Islands (NMFS, George Antonelis).*

aquatically in the intertidal zone where males maintain aquatic territories. Otariid females intermittently leave the rookery to forage between suckling periods, and phocid females fast on land during the entire lactation period. Thus, some pinnipeds rookeries may be occupied continuously, but most breeding is completed within a relatively short time period of about 2 months.

SEXUAL DIMORPHISM is apparent on pinniped rookeries, as adult males usually have distinctly different characteristics and are larger than females. Each sex and species emits stereotypic vocalizations for long- and short-distance communication (Stirling and Warneke, 1971; Miller, 1991). Males emit loud long-distance threat calls toward other males. Lactating otariid females also emit loud pup attraction calls on rookeries to locate their offspring among hundreds of pups. Short-distance threat vocalizations are common on all pinniped rookeries and have less amplitude than long-distance calls. Noise from rookeries initially may be perceived as a cacophony of sounds, but what seems to be chaos is really a well-organized social structure that has evolved over millions of years.

II. Physical–Geographical Characteristics

Pinniped rookeries are typically found on remote offshore islands, although some occur on mainland beaches. Rookeries are formed near shoreline just above the tidal zone in a variety of substraits, including sand, cobble or boulder beaches, rock shelves, and caves. Breeding aggregations usually occur within several hundred meters of the shoreline and also may occur on hillsides or cliffs overlooking the ocean (Fig. 3). Low-growing vegetation such as low grasses or shrubs is common on some rookeries.

The formation of rookeries on substrate near but above the tidal zone provides several functions that reinforce continued use of each site. To ensure survival to weaning, neonates are usually born in locations where high tides do not wash them away from their mothers or compromise their ability to thermoregulate (Fig. 4). The gregariousness of females at these nearshore locations facilitates the ability of territorial males to monopolize estrous females, a key component of their complex polygynous

mating system (Emlin and Oring, 1977). During anomalous conditions, however, storm surf associated with El Niño events has flooded pinniped rookeries, resulting in significant neonatal mortality and disruption of their polygynous social structure (Trillmich and Ono, 1991). Such events demonstrate the need for rookeries to occur above normal fluctuations in tide height.

Although pinniped rookeries must be located above the tidal zone, they must also be close enough to the water to facilitate access for THERMOREGULATION, foraging trips by lactating females, or escape from terrestrial predators. During high air temperatures in tropical and temperate environments, many rookery-breeding otariids are known to move regularly to the intertidal zone for cooling.

Figure 4 *A female California sea lion suckles her pup in a location protected from the surf on San Miguel Island, California (NMFS, George Antonelis).*

III. Environmental Characteristics

Environmental characteristics related to the formation, timing, and use of pinniped rookeries vary among species and are likely stimulated by proximate factors such as photoperiod, nutrition, and climate, which ultimately relate to survival and reproductive success (Boyd, 1991). The relative importance of these factors is believed to differ according to species on spatial and temporal scales. Pinniped rookeries occur most commonly during the spring and summer months when climatic conditions are relatively warm and the frequency of inclement weather diminishes. Such conditions increase the probability of offspring survival, especially in subpolar climates. Rookery-breeding phocids are the exceptions and form aggregations on rookeries in the fall and winter.

Most rookeries occur in locations where oceanographic conditions result in high productivity. High productivity increases the availability of potential prey resources vital for the foraging success of otariid females, which must feed intermittently during lactation. Rookery-breeding phocid females do not forage during lactation and rely completely on the energy stores obtained before parturition. The availability of prey near rookeries is therefore much more important to otariids than to phocids. However, the availability of prey for pups either before or after weaning may be an essential component for successful transition to foraging self-sufficiency for the young of most pinnipeds born on rookeries.

See Also the Following Articles

Breeding Sites ∎ Mating Systems ∎ Parental Behavior ∎ Pinniped Ecology

References

Bartholomew, G. A. (1970). A model for the evolution of pinniped polygyny. *Evolution* **24**, 546–559.

Boness, D. J. (1991). Determinants of mating systems in the Otariidae (Pinnipedia). *In* "The Behavior of Pinnipeds" (D. Renouf, ed.), pp. 1–44. Chapman and Hall, London.

Bowen, W. D. (1991). Behavioral ecology of pinniped neonates. *In* "The Behavior of Pinnipeds" (D. Renouf, ed.), pp. 66–117. Chapman and Hall, London.

Boyd, I. L. (1991). Environmental and physiological factors controlling the reproductive cycles of pinnipeds. *Can. J. Zool.* **69**, 1135–1148.

Emlen, S. T., and Oring, L. W. (1977). Ecology, sexual selection, and the evolution of mating systems. *Science* **197**, 215–223.

Miller, E. H. (1991). Communication in pinnipeds, with special reference to non-acoustic signalling. *In* "The Behavior of Pinnipeds" (D. Renouf, ed.), pp. 128–215. Chapman and Hall, London.

Oftedal, O. T., Boness, D. J., and Tedman, R. A. (1987). The behavior, physiology, and anatomy of lactation in Pinnipedia. *Curr. Mammal.* **1**, 175–245.

Reeves, R. R., Stewart, B. S., and Leatherwood, S. (1992). "The Sierra Club Handbook of Seals and Sirenians." Sierra Club Books, San Francisco.

Rice, D. W. (1998). "Marine Mammal of the World, Systematics and Distribution," Special Publication No. 4, Society for Marine Mammalogy, Allen Press, Lawrence, KS.

Scheffer, V. B. (1958). "Seals, Sea Lions, and Walruses." Stanford Univ. Press, Palo Alto, CA.

Stirling, I. (1983). The evolution of mating systems in pinnipeds. *In* "Recent Advances in the Study of Mammalian Behavior" (J. F. Eisenberg and D. G. Kleiman, eds.), pp. 487–527. Special Publication, American Society of Mammalogists 7.

Stirling, I., and Warneke, R. M. (1971). Implications of a comparison of the airborne vocalizations as some aspects of the behaviour of two Australian fur seals, *Arctocephalus* spp., on the evolution and present taxonomy of the genus. *Aust. J. Zool.* **19**, 227–241.

Trillmich, F., and Ono, K. (1991). "Pinnipeds and El Niño: Responses to Environmental Stress." Springer-Verlag Press, Berlin.

Ross Seal
Ommatophoca rossii

JEANETTE A. THOMAS
Western Illinois University, Moline

Less is known about the Ross seal (*Ommatophoca rossii*) than any other pinniped. This solitary seal has a circumpolar distribution around Antarctica, occurs in tight pack ice, and does not use traditional hauling out sites. No breeding concentrations of Ross seals have been observed. A variety of Antarctic investigators from Britain, France, America, Soviet Union, Australia, and New Zealand reported sightings of Ross seals from ships. However, there have been no ice-camp or land-based studies of this species. There has been no systematic, continent-wide census of Ross seals, so population estimates are variable.

I. Characters and Taxonomic Relationships

The Ross seal belongs to the family Phocidae and subfamily Monacinae. There is a single species in the genus *Ommatophoca*. Common names include the big-eyed seal and the singing seal. The closest relatives are the other Antarctic seals (crabeater, *Lobodon carcinophaga;* leopard, *Hydrurga leptonyx;* and Weddell, *Leptonychotes weddellii,* seals) and the monk seals (Caribbean, *Monachus tropicalis;* Mediterranean, *M. monachus;* and Hawaiian, *M. schauinslandi*).

As with other phocids, Ross seals crawl on their belly and are not capable of an upright stance or moving the hindlimbs forward. The head is proportionally smaller compared to the body than in other Monacinae and the snout is exceptionally short. The neck is thick and short. The Ross seal often assumes a posture with the head raised and mouth open, pointing upward. Because of this posture they have been called the "singing seal" (Fig. 1). However, this is a misnomer, as the seal rarely emits sound in this posture. More likely the open mouth displays teeth, and thrusting out of the striped chest serves as an aggressive posture (Fig. 2).

The skull has an exceptionally large orbit (hence the Greek name *omma* or eye). The zygomatic arch extends well below the palate, supporting part of the skull weight when placed on

Figure 1 *Microphone positioned to record airborne vocalizations of the "singing" Ross seal.*

a table. As with other Antarctic pinnipeds, it is assumed they have a tapetum, which assists in seeing in low-light levels during the austral winter and during deep dives, and a nictitating membrane that protects their eyes from blowing snow and allows opening their eyes in salt water. Condylobasal length has been measured at 244 mm in males and 242 mm in females. The mastoid width of the skull is at 172 and 170 mm in males and females, respectively. The soft palate is very long, extending posterior to the level of the occipital condyles. The trachea is expanded, and powerful muscles of the tongue and pharynx assist in swallowing large prey. The external ear is absent. The nostrils are normally closed and opened under voluntary control when seals need to respire. They have 15–17 short mystacial whiskers on each lip, only 10–42 mm in length, and superciliary vibrissae. All vibrissae are unbeaded.

Figure 2 *Typical head-up, mouth-open display given by a Ross seal to an intruder. Display emphasizes the distinct striped throat of this species.*

The mouth is small. The incisors and canines are small and recurved, an adaptation for holding slippery cephalopods. The front teeth are not procumbent and the seals do not maintain breathing holes by "ice sawing," as the Antarctic Weddell seal does. The cheek teeth are reduced, often barely breaking the gum line.

Black claws are reduced and probably useful for gripping the ice or scratching. The phalanges are greatly elongated. The fore and hind flippers are proportionally longer than in other phocids, the latter being 22% as long as the body.

The seals stay warm through a thick fur coat and a layer of subcutaneous fat. Ross seals have short, black fur on their back, with grayish silver streaks along their sides that transition into a solid silvery white belly. They have a unique color pattern of vivid strips running from the lower jaw to the chest and alongside the neck. A lanugo is present in the newborn, which is long, thick, and soft black on the back, fading into a bright yellow underbelly. This yellow color may change into silver as the pup grows.

There is no marked sexual dimorphism in body size, but females tend to be slightly larger. Adult males reach 1.68–2.08 m and adult females are 1.96–2.5 m in length. Weight ranges from 129 to 216 kg in adult males and 159 to 201 kg in adult females, with pregnant females obtaining the greatest weights. The exact age of sexual maturity is uncertain, but based on analysis of reproductive tracts is estimated at 3–4 years of age in females and 2–7 years in males. At birth, pups are 105–120 cm in length and weigh 17–27 kg.

II. Distribution and Ecology

Ross seals (estimated at 1–2% of Antarctic pinnipeds) are the least abundant of all the Antarctic pinniped species. This species is uncommon, numbering an estimated 20,000–50,000 according to Scheffer (1958), but as high as 220,000 according to Erickson *et al.* (1971). Nowak (1991) reported Ross seal populations between 100,000 and 650,000, but considers these estimates unreliable. Ross seals are restricted to the heavy, consolidated pack ice of the Southern Ocean, with the farthest northern sighting at Heard Island. Before 1945, there were fewer than 45 sightings of the species. Only with the use of icebreakers into the dense pack ice were more recent sightings obtained, but most often sightings are opportunistic. Sightings usually are of single animals, but sometimes a pocket of small numbers occurs in a local area. The highest recorded density is 2.9/km². Some investigators believe the nonuniform DISTRIBUTION is due to the distribution of their prey, ice type, or both. Occasionally, single seals are seen in southern Australia and some subantarctic islands, such as the South Sandwich, Falkland, Scott, and South Orkney Islands.

III. Behavior and Life History

Mother Ross seals haul out in dense pack ice to give birth during the austral spring (November and December), with peak pupping between 3 and 18 November. The species is solitary and seals do no congregate in large breeding colonies. Although mating has not been observed, it is assumed to be the same as in other Antarctic seals, i.e., takes place in water.

Typically, a single pup is born. When disturbed, mothers vocalize to pups and pups respond with a bawl sound. An observation of a newborn pup reported it swimming in icy water from one floe to another. Pups wean after about 1 month of nursing.

One adult call is an explosive noise emitted with the mouth open, and two adult calls are made with the mouth closed: a pulsed chug and tonal siren call. Calls of adults are used both on the ice surface and under water. Mating occurs after the pup weans in December and molt soon follows. Because of the need to synchronize the time of pupping to annual ice conditions Ross seals exhibit a delayed implantation of 2 months. This delay in pregnancy allows the mother to molt, feed, and recover from the dramatic weight loss associated with lactation before another fetus develops.

Ross seals exhibit a distinct diel pattern of haulout, with most seals hauled out during the midday. This difference in sightability due to time of day makes accurate population estimates difficult. Stewart (cited in Reeves *et al.*, 1992) documented the diving behavior of one Ross seal with a microprocessor-based dive recorder glued to its back. Most dives were deeper than 100 m and lasted around 6 min. The deepest dive was 212 m and for 10 min.

Subadult seals are rarely seen. This seal has small scars around the neck and shoulders that are attributed to intraspecific interactions. The dense pack ice habitat must provide some protection from predators. The Ross seal does not exhibit the large, distinct scars from leopard seal predation seen on crabeater seals. However, in a few areas, Ross seals are sympatric with leopard seals, so the lack of scarring could indicate total hunting success by leopard seals on this smaller seal. Killer whales (*Orcinus orca*) are known to take other Antarctic seals, so likely would feed on Ross seals if not for habitat segregation from this predatory, ice-edge whale. Longevity is unknown, but is at least 21 years.

The DIET of a Ross seal is primarily cephalopods, even larger species than other Antarctic seals eat. Midwater fish and krill are also eaten. Based on stomach contents analysis, the diet of the Ross seal is 47% squid, 34% fish, and 19% invertebrates. Their specialized diet reduces competition with other Antarctic seals or whales.

IV. Interactions with Humans

Ross seals have little fear of humans because there are no natural land predators (such as polar bears) in the Antarctic. James Clark Ross, commander of the *HMS Erebus and Terror*, secured two specimens of this species during his British Antarctic expedition from 1839 to 1843. Gray of the British Museum described and named the species after the explorer. Barrett-Hamilton published one of the first accounts on the Ross seal in 1901, resulting from the voyage of the *S. Y. Belgica* conducted from 1897 to 1899. There is no record of extensive harvest of this species, except for scientific collection. The Ross seal is totally protected under the Antarctic Treaty and the Convention for the Conservation of Antarctic Seals.

See Also the Following Articles

Crabeater Seal ■ Diving Behavior ■ Earless Seals ■ Locomotion, Terrestrial ■ Pinnipedia, Overview ■ Weddell Seal

References

Barrett-Hamilton, G. E. H. Zoologie: seals. Result. Voyage *S. Y. Belgica* 1897–1899. Exped. Antarctica Belge, 1901, 1–20.

Erickson, A. W., Gilbert, J. R., and Ortis, J. (1971). Distributional ecology of Antarctic seals. *In* "Proceedings of the Symposium on Antarctic Ice and Water Masses" (G. Deacon, ed.), pp. 55–75. SCAR Cambridge.

Gray, J. E. (1844). The zoology of the voyage of the *HMS Erebus and Terror, 1839–1843*. 1. Mammalia. 1. The seals of the Southern Hemisphere. E. W. Jansen, London.

Hofman, R. J., Erickson, A., and Siniff, D. B. (1973). The Ross seal (*Ommatophoca rossi*). Publ. New Ser., Suppl. Paper.

King, J. E. (1969). Some aspects of the anatomy of the Ross seal (*Ommatophoca rossi*). British Antarctic Survey. Scientific Rep. No. 63, 1–54.

Nowak R. M. (1991). "Walker's Mammals of the World," Fifth Ed., Vol. II, p. 1256. The Johns Hopkins University Press, Baltimore.

Øristland, T. (1970). Sealing and seal research in the southwest Atlantic pack ice, September–October 1964. *In* "Antarctic Ecology" (M. W. Holdgate, ed.), Vol. 1, pp. 367–3765. Academic Press, London.

Reeves, R. R., Stewart, B. S., and Leatherwood, S. (1992). The Sierra Club Handbook of Seals and Sirenians. Sierra Club Books, San Francisco.

Scheffer, V. (1958). "Seals, Sea Lions and Walruses: A Review of the Pinnipedia." Stanford Univ. Press, Stanford, CA.

Scheffer, V. (1976). Standard measurements of seals. *J. Mammal.* **48**, 459–462.

Solyanick, G. A. (1964). Experiment in marking seals from small ships. *Sov. Antarctic Exped. Bull.* **5**, 212.

Stirling, I. (1966). A technique for handling live seals. *J. Mammal.* **47**, 543–544.

Thomas, J. A., DeMaster, D. P., Stone, S., and Andriashek, D. (1980). Observations of a newborn Ross seal pup (*Ommatophoca rossi*). *Can. J. Zool.* **58**, 2156–2158.

Tikhimirov, E. A. (1975). Biology of the ice forms of seals in the Pacific section of the Antarctic. *Rapp. P. V. Reun. Cons. Int. Explor. Mer.* **169**, 409–412.

Rough-Toothed Dolphin
Steno bredanensis

THOMAS A. JEFFERSON
*Southwest Fisheries Science Center,
La Jolla, California*

Although widely distributed in all three major oceans, the rough-toothed dolphin has not been well studied. It is a tropical to warm-temperate species and is generally found in deep, offshore waters. In most areas where it is known, it does not appear to be among the most abundant of the small cetaceans. This means that it is not commonly encountered.

I. Characteristics and Taxonomic Relationships

The rough-toothed dolphin is very distinctive, when seen at close quarters. It is the only long-beaked dolphin with a

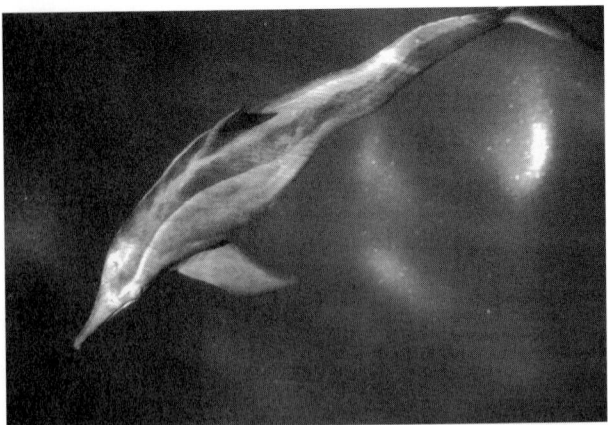

Figure 1 *A rough-toothed dolphin swimming just below the surface in clear waters off Hawaii showing the species' distinctive characteristics. Photo by B. Würsig.*

smoothly sloping melon that does not contain any hint of a crease as it blends into the upper beak (Fig. 1). These dolphins are not particularly slender, and the anterior part of the body may be stocky. The large flippers are set farther back on the body than in most other small cetaceans. They are equal to about 17–19% of the body length. The dorsal fin is tall and generally only slightly recurved. Some large males have a hump of connective tissue posterior to the anus, which gives the appearance of a pronounced keel. These are medium-sized dolphins that can weight up to 155 kg. Males grow to larger sizes than females (known maximums of 265 and 255 cm, respectively), and females may have proportionately longer beaks (Miyazaki and Perrin, 1994).

The color pattern is moderately complex, but consists generally of shades of black, white, and gray (Fig. 2). Rough-toothed dolphins are countershaded, with white bellies and black to dark gray backs. The sides are a medium shade of gray and are separated from the darker back by a dorsal cape that is narrow between the blowhole and the dorsal fin and wider

Figure 2 *Two rough-toothed dolphins socializing at the bow of a research vessel in the Gulf of Mexico. Photo by D. Weller.*

behind the fin. The lower sides and mouth area are often dotted with white patches, splotches, and spots. Many of the white spots are thought to be scars from bites inflicted by cookie-cutter sharks. Young animals have a muted color pattern and generally lack the white spots (Miyazaki and Perrin, 1994).

The skulls of rough-toothed dolphins (adult CBL = 472–555 mm) can be distinguished from those of all other dolphins (except humpback dolphins, *Sousa* spp.) by their combination of long beak, concave rostral and maxillary margins, long mandibular symphysis, and large temporal fossae (Fig. 3). Tooth counts can be used to distinguish them from humpback dolphins: rough-toothed dolphins have 19–28 teeth in each row and humpback dolphins usually have >30. The teeth of *Steno* often have longitudinal ridges, which is the reason for the species' common name (Neuville, 1928). Other differences from the skulls of humpback dolphins are the relatively large orbits and the prominent and long cylindrical ridge on the ventral part of the frontal bones in rough-toothed dolphins (Van Waerebeek *et al.*, 1999).

Traditionally, morphological characters have been used to infer a close relationship between the rough-toothed dolphin and two other genera of dolphins (*Sotalia fluriatilis*, the tucuxi, and *Sousa* spp., the humpback dolphins). Recent genetic analyses (LeDuc *et al.*, 1999) have supported the relationship with *Sotalia* (in the subfamily Stenoninae), but not with *Sousa*, which groups phylogenetically with the Delphininae. In captivity, hybrids between *Steno* and *Tursiops truncatus* (bottlenose dolphin) have occurred (Dohl *et al.*, 1974).

II. Distribution and Ecology

The rough-toothed dolphin is a tropical to warm temperate species and is usually found in oceanic waters (Fig. 4). Records from the Atlantic Ocean are mostly between the southeastern United States and southern Brazil across to the Iberian Peninsula and tropical West Africa, with some (probably extralimital) records from the English Channel and North Sea. The normal range includes the Gulf of Mexico, Caribbean Sea, and Mediterranean Sea.

In the Pacific, they occur from central Japan and northern Australia across to southern Baja California, Mexico, and southern Peru. In the eastern tropical Pacific, they are generally associated with warm, tropical waters lacking major upwellings. The range includes the southern Gulf of California and South China Sea. Records from the west coast of the United States and New Zealand are probably extralimital. In the poorly studied Indian Ocean, there are only a few scattered records, but the species probably has an extensive distribution there north of about 20°S.

Essentially nothing is known about population or stock structure in this species. In fact, the ecology and biology of the species are very poorly studied. There have been only a few reports of feeding habits for this species. In the wild, they feed on a variety of fish and cephalopod species, some coastal and some oceanic. Some large fish may be taken, as suggested by the robust dentition of the species. Algae has been found in the stomachs of stranded specimens, but this may have been ingested incidentally.

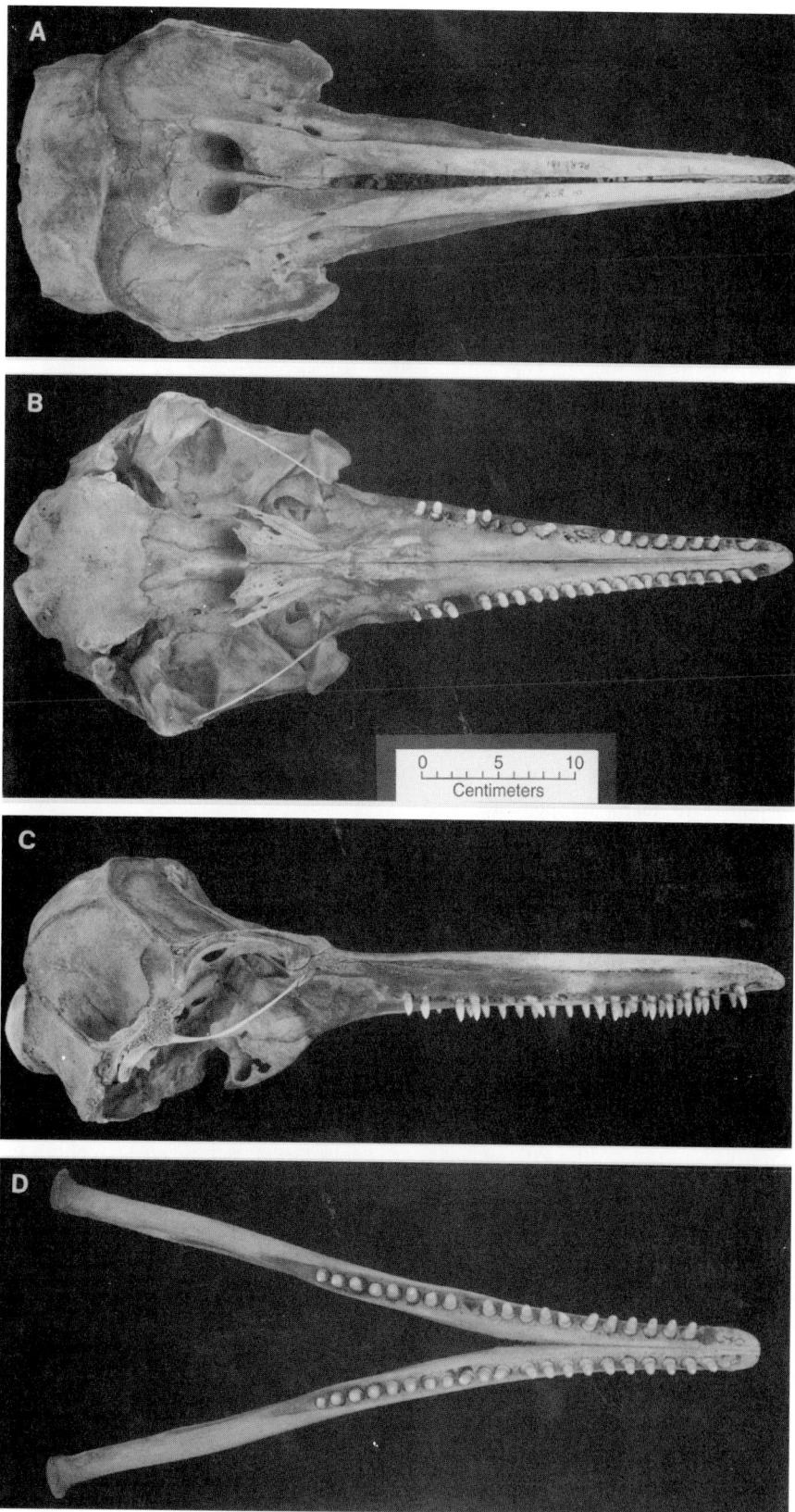

Figure 3 *Dorsal, ventral, and lateral views of the cranium and dorsal view of the mandible of a rough-toothed dolphin. Photos courtesy of W. F. Perrin.*

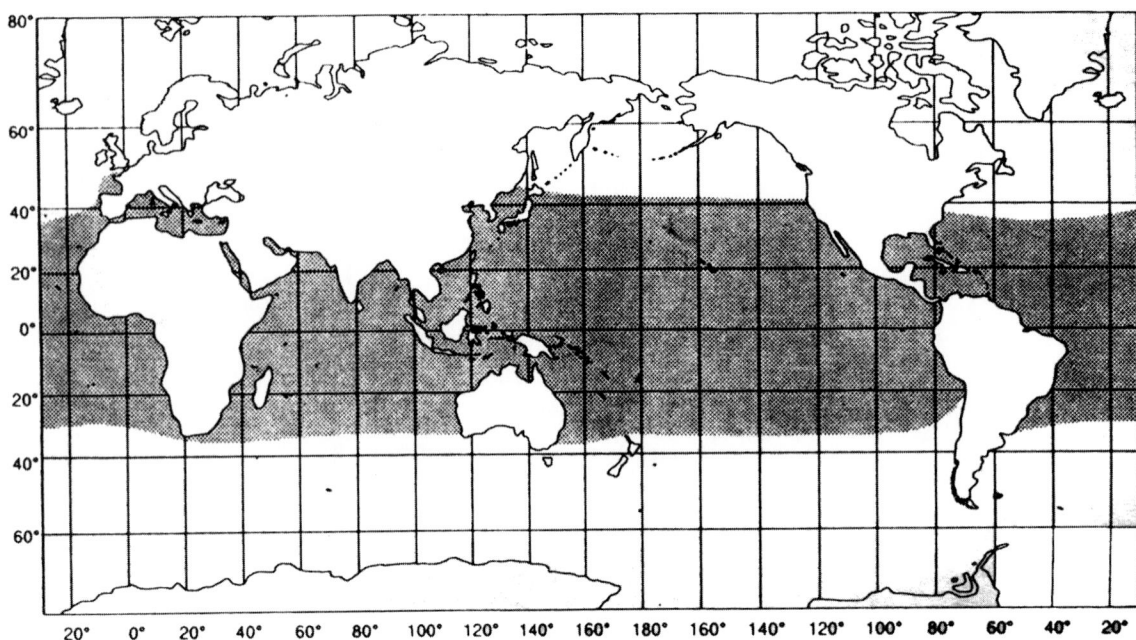

Figure 4 *The approximate distribution of the rough-toothed dolphin. It should be noted that few surveys for marine mammals have been conducted in offshore areas or in the Indian Ocean, so distribution in these regions is especially poorly known.*

III. Behavior and Life History

Rough-toothed dolphins are found in moderate-sized groups, most commonly of 10–20 dolphins, although larger groups have been seen in some areas—over 50 in the eastern tropical Pacific, 300 in Hawaii, and 160 in the Mediterranean (Miyazaki and Perrin, 1994). Mass STRANDINGS have been recorded in several areas of the species' range. They frequently associate with other species of cetaceans, especially other delphinids in the eastern tropical Pacific, where they also often associate with flotsam. Lone animals have been seen with short-finned pilot whales (*Globicephala macrorhynchus*) and Fraser's dolphins (*Lagenodelphis hosei*) in the Sulu Sea.

These animals are not generally fast swimmers and they often appear rather sluggish in the wild. They do ride bow waves and are known for their habit of skimming along the surface at moderate speed with a distinctive splash. Although not highly acrobatic, various leaps and other aerial behaviors have been seen. Rough-toothed dolphins can be found in some shallow, coastal waters in Brazil and elsewhere; in such areas, photo-identification of individual dolphins is considered feasible (Flores and Ximinez, 1997).

Although the maximum recorded dive was only to 70 m, rough-toothed dolphins can probably dive much deeper than this. Behavioral and morphological evidence suggests that they are well adapted for long, deep dives. Submergences of up to 15 min have been recorded. A variety of clicks and whistles have been recorded from these dolphins. Highly directional ECHOLO-CATION clicks, with some pulses as high as 200 kHz, are known.

Detailed studies of life history have only been conducted in Japanese waters. There, males reach sexual maturity at about 14 years and 225 cm, and females at 10 years and 210–220 cm. The maximum age is 32–36 years, although some animals may live significantly longer (Miyazaki, 1980: Miyazaki and Perrin, 1994).

IV. Interactions with Humans

Although not generally the major target, rough-toothed dolphins have been taken in directed dolphin fisheries in Japan, Sri Lanka, Indonesia, the Solomon Islands, Papua New Guinea, St. Vincent, West Africa, and possibly St. Helena in the South Atlantic (Miyazaki and Perrin, 1994). Probably much more significant is the incidental kill of dolphins in fishing nets. Takes in tuna purse seine nets are known for the eastern tropical Pacific, and gill-net catches have been documented at least in Sri Lanka, Brazil, and the offshore North Pacific. Undocumented catches probably occur in most other areas of the range.

Habitat degradation impacts and effects of pollutants are probably much less severe for this species than for other, more coastal small cetaceans. Organochlorine levels have generally been much lower than for other species. However, conservation-oriented studies are almost nonexistent, and therefore the uncertainty that exists about population status for this species should be acknowledged.

Rough-toothed dolphins have been held captive in a number of oceanaria, and some success has been encountered in keeping them alive in the captive environment, especially in Hawaii. One specimen lived for over 12 years in captivity. They have been found to be bold and inventive, and one "creative porpoise" at Sea Life Park in Hawaii astounded its trainers by grasping the concept of inventing novel behaviors (Pryor *et al.*, 1969).

See Also the Following Articles

Captivity ▪ Delphinids ▪ Skull Anatomy ▪ Teeth

References

Dohl, T. P., Norris, K. S., and Kang, I. (1974). A porpoise hybrid: *Tursiops* X *Steno*. *J. Mammal.* **55**, 217–221.

Flores, P. A. D. C., and Ximinez, A. (1997). Observations on the rough-toothed dolphin *Steno bredanensis* off Santa Catarina Island, southern Brazilian coast. *Biotemas* **10**, 71–79.

LeDuc, R. G., Perrin, W. F., and Dizon, A. E. (1999). Phylogenetic relationships among the delphinid cetaceans based on full cytochrome B sequences. *Mar. Mamm. Sci.* **15**, 619–638.

Miyazaki, N. (1980). Preliminary note on age determination and growth of the rough-toothed dolphin *Steno bredanensis*, off the Pacific coast of Japan. *Rep. Int. Whal. Comm. Spec. Issue* **3**, 171–179.

Miyazaki, N., and Perrin, W. F. (1994). Rough-toothed dolphin *Steno bredanensis* (Lesson, 1828). *In* "Handbook of Marine Mammals" (S. H. Ridgway and R. Harrison, eds.), Vol. 5, pp. 1–21. Academic Press, San Diego, CA.

Neuville, H. (1928). Recherches sur le genre "*Steno*" et remarques sur quelques autres Cétacés. *Arch. Mus. Hist. Nat.* **3**, 69–240.

Pryor, K. W., Haag, R., and O'Reilly, J. (1969). The creative porpoise: Training for novel behavior. *J. Exp. Anal. Behav.* **12**, 653–661.

Van Waerebeek, K., Gallagher, M., Baldwin, R., Papastavrou, V., and Al-Lawati-Samira, M. (1999). Morphology and distribution of the spinner dolphin, *Stenella longirostris*, rough-toothed dolphin, *Steno bredanensis*, and melon-headed whale, *Peponocephala electra*, from waters off the Sultanate of Oman. *J. Cet. Res. Manage.* **1**, 167–178.

Scrimshaw

STUART M. FRANK
*Kendall Whaling Museum,
Sharon, Massachusetts*

Scrimshaw is an occupational handicraft of mariners employing by-products of the whale fishery, often in combination with other found materials. Indigenous to the whaling industry, where it was typically a pursuit of leisure time at sea, it was also adopted in other trades and was occasionally practiced ashore. It arose among Pacific Ocean whalers circa 1817–1824, persisted throughout the classic "hand-whaling" era of sailing-ship days into the 20th century, and persisted in degraded form among "modern" whalers on factory ships and shore stations until the industry shut down in the third quarter of the 20th century. Since the early 20th century, similar materials and techniques have simultaneously been employed by nonmariner artisans for both commercial and hobbyist purposes.

There is no consensus regarding etymology. Plausible and eccentric theories alike have been advanced without any creditable evidentiary basis, whereas academic lexicography (notoriously inconclusive respecting nautical terms) fails to present any convincing hypothesis. The term—also rendered *skrimshank, skimshander, skirmshander,* and *skrimshonting*—first appeared in American shipboard usage circa 1826, when the recreational practice of scrimshaw was less than a decade progressed. It originally referred not to whalers' private diversions, but to the fairly common practice whereby crewmen were required to make articles for ship's work (such as tools, tool handles, thole pins, belaying pins, and tackle falls). Sperm whale BONE is ideally suited to such uses: on any "greasy luck" voyage it was in plentiful supply at no cost, its workability is equivalent to the best cabinetmaking hardwoods, its tensile strength is greater than oak, and for many applications its self-lubricating properties were highly desirable. Such was analogously the case regarding the adaptability of cetacean bone and ivory to whales' recreational handicrafts, to which the term *scrimshaw* (and its many variants) came to refer by the 1830s.

I. Materials and Species

Materials associated most commonly with scrimshaw are the ivory teeth and skeletal bone of the SPERM WHALE, the ivory tusks of the WALRUS, and the BALEEN of various mysticete species (the toothless, baleen-bearing whales). In the 19th century the principal prey species were, roughly in descending order of importance, sperm whale (*Physeter macrocephalus*), right whales (*Eubalaena* spp.), Arctic bowhead (*Balaena mysticetus*), gray whale (*Eschrichtius robustus*), and humpback (*Megaptera novaeangliae*). These and the long-finned pilot whale or so-called "blackfish" (*Globicephala melas*), which was hunted primarily from shore, were taken primarily for oil, the mysticetes secondarily for baleen. [The fast-swimming blue whale (*Balaenoptera musculus*) and finback (*B. physalus*) could not be hunted effectively prior to the introduction of steam propulsion and heavy-caliber harpoon cannons in the late 19th century.] From the late 16th century, by reason of geographical proximity of Arctic habitats and similar uses of their meat and oil, the hunt for walruses (*Odobenus rosmarus*) was intimately conjoined with commercial whaling. Later, even when whalers were no longer taking walruses themselves, they characteristically obtained walrus tusks through barter with indigenous Northern peoples.

Commercial uses of walrus ivory were few; there was no significant commercial application for cetacean skeletal bone until the 20th century (when it was ground and desiccated into industrial-grade meal and fertilizer); the utility and market value of baleen ("whalebone") were subject to mercurial fluctuations of fashion; and sperm whale teeth had little or no commodity value. They thus became available for whalers' recreational use, as did teeth of the Antarctic elephant seal (*Mirounga leonina*), the lower mandibles of various dolphins and porpoises, and tusks of the elusive NARWHAL (*Monodon monoceros*). (Narwhal ivory proved too difficult and brittle for anything much beyond canes and analogous shafts, such as hatracks or bedposts.)

The characteristic pigment for highlighting engraved scrimshaw was *lampblack*, which is essentially a viscous suspension of carbon particles in oil. (The notion that sailors used tobacco juice for this is a colorful fabrication with no basis in fact.) Lampblack, collected easily from lamps, stoves, and *tryworks* (shipboard oil-rendering apparatus), was in abundant supply on a whale ship. Colors were introduced almost at the outset: Edward Burdett was using sealing wax and other pigments by 1827 (Fig. 1); full polychrome scrimshaw debuted within the next decade. Sealing wax had the advantages of being universally available, relatively inexpensive, brilliantly colored, and color-fast. Applied properly, it has proven resilient and tenacious, the color as vivid today as when the scrimshaw was new. Improper application—if the cuts are too smooth or insufficiently contoured to grab and hold the wax—results in significant losses from handling and natural desiccation. Sealing wax had the disadvantage of offering only a limited spectrum of colors, all strong. Ambient pigments, however, could be mixed and blended, affording greater subtlety. From the characteristic leeching of pigment into the substrata of some polychrome scrimshaw, a phenomenon that occurs with water- and alcohol-soluble colors but not with waxes or heavy oil-based pigments, it is clear that ambient colors were also favored. Store-bought inks, homemade dyes extracted from berries, and greens from common verdigris seem to predominate; however, their composition has not been investigated comprehensively.

Inlay and other secondary materials—rare on engraved scrimshaw but often encountered on "built" or "architectonic" scrimshaw—were typically obtained at little or no cost, such as other marine byproducts (tortoise shell, mother-of-pearl, sea shells), various woods brought from home or obtained in various ports of call (including exotic tropical species from Africa and Polynesia), and miscellaneous bits of metal (fastenings and finials were often crafted from silver- or copper-alloy coins, typically coins minted in Mexico and South America).

II. Scrimshaw Precursors

Medieval European artistic productions in walrus ivory and cetacean bone were many, but the whalers themselves had no part in them beyond gathering the raw materials. Cetacean bone panels and stilettos survive from the Viking era, some incised with rope patterns and animal figures, and cetacean bones served as beams in vernacular buildings in Norway and the Friesian Islands, but even these do not appear to have been made by whalers and are not known to have been part of their occupational culture. Monastic artisans in Denmark and East Anglia carved walrus ivory and cetacean bone into votive art, primarily altar pieces, friezes, and crosses, whereas craftsmen at Cologne and elsewhere produced secular game pieces and chessmen from the same materials. So important was the "Royal Fish" to the Viking economy that a highly sophisticated body of law evolved to regulate whaling itself and the ownership, taxation, distribution, and export of whale products, whether acquired fortuitously (from stranded carcasses) or by hunting. Pliny the Elder (first century C.E.), Olaus Magnus (1555), Conrad von Gessner (1558), and Ambroise Paré (1582) listed the uses of baleen for whips, springs, garment stays, and umbrella ribs; and the emergence of pelagic Arctic whaling in the 17th century encouraged a search for new applications, especially in Holland. The search, however, proved fruitless and was abandoned by circa 1630, occasioning the appearance of sailor-made baleen objects: there was simply no longer any reason to restrain whalers from using baleen for their private diversions (two centuries later the same principle would make sperm whale teeth available for scrimshaw).

Ditty boxes were the first manifestation of whalers' work. Typically, these have polished baleen sides bent to the oval shape of a wooden bottom 30–35 cm long, to which the baleen is fastened with copper nails and fitted with a wooden top. Two made in 1631 by an anonymous Rotterdam whaling commandeur have baleen sides incised with portraits of whaleships, the wooden tops relief carved with the Dutch lion rampant

Figure 1 *Panbone plaque by Edward Burdett (1805–1833) of Nantucket, circa 1828. The earliest known scrimshaw artist, Burdett was active from 1824 until he was killed by a whale in 1833. His work is characterized by a bold, confident style, with deep blacks and red sealing-wax highlights. He was serving as a mate in the* William Tell *when he engraved this section of sperm whale bone, inscribed "William Tell. of New York. homeward bound. in the latitude of. 50 13. S. long[itude] 80. W. got shipwrecked"; "lost her rudder & c."; "by. E. Burdett." 15.7 × 31.8 cm. From the Kendall Whaling Museum, Sharon, Massachusetts.*

surrounded by nautical symbols, human figures, and decorative borders. A few North Friesian whaleman–artists of the next generation are known by name. Jacob Floer of Amrum engraved buildings, trees, and geometrical borders on the baleen sides of an oval box, signed and dated 1661. Peter Lorenzen of Sylt signed and dated another in 1687. The form continued for the duration of Arctic whaling and was perpetuated with myriad variations by American and British scrimshaw artists in the 19th century.

Another early form was the *mangle* (paddle for folding cloth). An Amsterdam whaling commandeur decorated one with carved geometric ornaments, signed, dated, and inscribed, "Cornelis Floerensen Bettelem. Niet sonder Godt [Not without God]. Anno 1641." A century later a North Friesian whaling master, Lødde Rachtsen of Hooge, made one for his daughter's dowry: it has a pierced-work handle and carved geometric and floral decorations, the broadside portrait of a spouting bowhead whale, and a carved inscription dated 1745. Respecting technical aspects of execution and the iconography of their decoration, this kind of piece is the direct ancestor of the sperm whalemen's decorated baleen corset-busks of the 19th century.

III. Origins and Practice

Pictorial engraving on sperm whale teeth—the quintessential manifestation of scrimshaw—resulted from changing circumstances in the fishery in the aftermath of the Napoleonic wars (of which the American theater was the so-called War of 1812). It arose collectively among British and American whalers in the 1820s in the Hawaiian Islands, where (beginning in 1819) the fleets customarily laid over for weeks on end between seasonal cruises, providing ample opportunities for fraternization and foment.

In the late 17th century, British colonists on the Atlantic seacoast of New York and Massachusetts hunted right whales along shore—an ancillary day fishery, prosecuted by fishers and farmers in rowboats launched from sandy beaches. In the 18th century, expanding markets occasioned offshore cruises in small sailing vessels. The discovery of sperm whales in proximity to New England is ascribed by tradition to Captain Christopher Hussey of Nantucket when he was blown off course while right whaling in 1712. The colonists tuned their technology to accommodate sperm whaling, pioneered the refining of sperm whale oil and the manufacture of spermaceti candles (America's first industry), and developed thriving export networks. Whaling evolved into a full-time occupation, and a distinctive caste of whaler–mariners emerged with its own occupational culture. Scrimshaw would become an integral component of this culture, but it took a whole century for the right circumstances to gel.

Colonial whaling cruises were seasonal, following the Atlantic trade winds on comparatively short passages to and from the grounds. Only a few whales were required to fill the hold before heading home, usually after only a few weeks. The opening of the Pacific grounds in the 1790s changed shipboard dynamics dramatically. Voyages necessarily became longer, as much as 3 or 4 years by the 1840s. The larger catch required to make long voyages profitable mandated larger vessels with larger crews so that three to five whaleboats could be launched for the hunt. The effective result was overmanning and an unprecedented abundance of shipboard leisure—long outward and homeward passages, and idle weeks, even months between whales, when there was little to do but maintain the ship and wait. Most sailors worked "watch-on-watch" in 4-hr shifts, day and night, whenever the ship was underway, but whaleship crews had most nights off: the hunt could not be prosecuted effectively in darkness, and cutting in (flensing) a carcass with sharp blubber spades was dangerous enough even in daylight. Apart from rendering blubber already on hand, there was little work to do evenings. More than in any other seafaring trade, 19th-century whalers had time to spare. They filled it with reading, journal-keeping, drawing, singing, dancing, gamming (visiting among ships at sea), and a host of other diversions.

At the critical juncture, just when things were ripe for scrimshaw, teeth were in short supply. For in the meanwhile, the China Trade, pioneered in the 1780s, had established a network of Far East destinations and products that involved barter with Pacific islanders to obtain goods for Canton. China traders soon realized that many island cultures placed great value on whale teeth, from which they crafted various totemic and decorative objects. Teeth could be obtained cheaply from whalers (there being no other market), so the China merchants bought them up for barter in the Pacific. Such scrimshaw as there was in the 18th century was therefore limited primarily to implements made of skeletal bone—straightedges, hand tools, a few early swifts (yarn-winders), and corset busks; of these, comparatively few were made prior to the florescence of scrimshaw commencing in the 1820s.

Captain David Porter of the U.S. Naval frigate *Essex* was the inadvertent catalyst for the emergence of scrimshaw. Porter's wartime purpose had been to inflict depredations on British shipping and to disrupt British whaling in the Pacific. His narrative (published in 1815, reissued in an expanded edition in 1821) was valued by mariners for its explicit accounts of conditions in the Marquesas and Galapagos Islands and on the coast of Chile and Peru. It also incidentally revealed the barter value of whales teeth in Polynesia and disclosed particulars of how they could be gathered cheaply—this at just around the time the vanguard of the whaling fleet reached Hawaii (1819). There was soon a surplus of whales teeth on the Pacific market; as the teeth were no longer valuable as a commodity, they could be relegated to sailors for private use.

Accordingly, the earliest authentic date on any pictorial sperm whale scrimshaw is 1817—a tooth commemorating a whale taken by the ship *Adam* of London off the Galapagos Islands (Fig. 2); and the earliest provisionally identifiable whaleman–engraver of sperm whale ivory is J. S. King, whaling master of London and Liverpool, to whom two teeth are attributed, one perhaps as early as 1821. These suggest a possible British genesis of pictorial scrimshaw; however, the earliest definitively attributable work is by an American, Edward Burdett of Nantucket, who first went whaling from his native port in 1821 and was scrimshandering by 1824. Fellow Nantucketer Frederick Myrick was the first to sign and date his scrimshaw—three dozen teeth produced during 1828–1829 as a seaman aboard the Nantucket ship *Susan*. Two teeth by Burdett and two "Su-

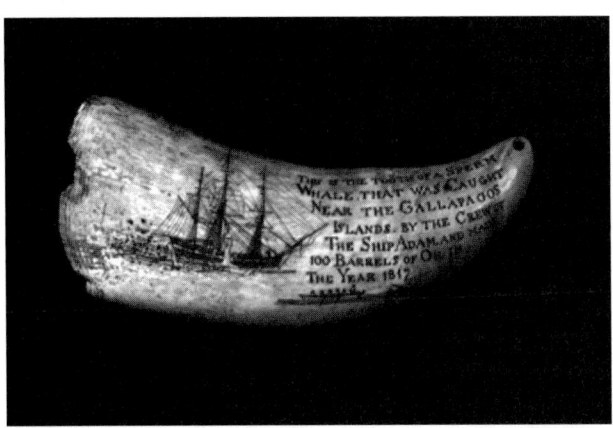

Figure 2 *Genesis of scrimshaw, circa 1817. The oversize tooth is inscribed, "This is the tooth of a sperm whale that was caught near the Galapagos Islands by the crew of the ship Adam [of London], and made 100 barrels of oil in the year 1817." Produced in the wake of the Napoleonic Wars, when the British and American whaling fleets were endeavoring to recover their former prowess in the Pacific, it is believed to be the earliest full-scale work of engraved pictorial scrimshaw. Length 23.5 cm. From the Kendall Whaling Museum, Sharon, Massachusetts.*

san's Teeth" by Myrick were accessioned by the East India Marine Society of Salem, Massachusetts, prior to 1831, while both artists were still living—the first scrimshaw to enter a museum collection. That Myrick's work was listed generically as "Tooth of a Sperm Whale, curiously carved" and "Another, carved by the same hand," with no mention of the exquisitely engraved pictures on them, nor of the artist's name (both are signed), nor of the term *scrimshaw*, testifies to the newness of the genre, perhaps also to the low esteem in which sailors' hobby work was held by the great merchants of Salem at the time.

In the 1830s scrimshaw became widely generalized. On some whaling vessels virtually the entire ship's company participated. In his journal of the New Bedford bark *Abigail* during 1836–1838, Captain William Hathaway Reynard remarked, "The cooper is going ahead making tools for scrimsham. We had a fracas betwixt the cook and the stewart [sic] . . . All hands employed in scrimsa." In other ships the best ivory and bone may have been relinquished to some particularly talented member of the crew, such as Joseph Bogart Hersey of Cape Cod on the Provincetown schooner *Esquimaux* in 1843: "This afternoon we commenced sawing up the large whale's jaws . . . the bone proved to be pretty good and yielded several canes, fids, and busks. I employed a part of my time in engrav[ing] or flowering two busks. Being slightly skilled in the art of flowering; that is drawing and painting upon bone; steam boats, flower pots, monuments, balloons, landscapes &c &c &c; I have many demands made upon my generosity, and I do not wish to slight any; I of course work for all."

The whaleship labor force was very young on average, with green hands often in their early teens; common seamen and even seasoned harpooners were rarely over thirty. Among the greatest scrimshaw artists, Frederick Myrick retired from whal-

ing and from scrimshaw at age 21, Edward Burdett was barely 28 when he was killed by a whale, and Welsh ship's surgeon W. L. Roderick left the fishery at 29. Nevertheless, although in the minority, older hands contributed mightily. Seaman Silas Davenport may have been in his forties when he constructed a fine swift of bone and ebony. Former whaleman N. S. Finney was still engraving walrus ivory on commission in San Francisco in his sixties. Ship's carpenters and coopers—trained craftsmen with skills well adapted to scrimshaw, especially architectural pieces—were normally older than the average crewman. So, too, whaling captains, many of whom were devoted scrimshaw artists: Manuel Enos cut brilliant polychrome portraits into whale ivory right up to the time he was lost at sea at age 55; Frederick Howland Smith was scrimshandering from age 14 until he retired at 61; and the grand old man, Captain Ben Cleveland, was still making napkin rings, mantle ornaments, and scrimshaw gadgets in the 1920s, at age eighty.

Scrimshaw was quintessentially a diversion of the whalemen's ample leisure hours, to fill time, and produce mementos as gifts for loved ones at home. It was occupationally rooted in and wholly indigenous to the deepwater whaling trade, but was eventually also adopted by merchant sailors, navy tars, and occasionally the seafaring wives and children of whaling masters.

Unfortunately, practitioners in whatever trade rarely signed or dated their work, and family provenance has seldom preserved details of the origins of legacy pieces. Thus, scrimshaw has hitherto been mostly anonymous, the names of only a handful of practitioners known. However, systematic forensic studies commencing in the 1980s have made stylistic and iconographical attributions increasingly possible, and the names and works of a few hundred individual artists are now documented with varying degrees of specificity.

IV. Taxonomy

Scrimshaw took many forms. Henry Cheever mentions whalers "working up sperm whales' jaws and teeth and right whale [baleen] into boxes, swifts, reels, canes, whips, folders, stamps, and all sorts of things, according to their ingenuity" (*The Whale and His Captors*, 1850); and Herman Melville alludes to "lively sketches of whales and whaling-scenes, graven by the fishermen themselves on Sperm Whale-teeth, or ladies' busks wrought out of the Right Whale-bone, and other like skrimshander articles, as the whalemen call the numerous little ingenious contrivances they elaborately carve out of the rough material in the hours of ocean leisure" (*Moby-Dick*, 1851). Various tools were used for cutting and polishing, but forensic scrutiny corroborates Melville's observation that the ordinary knife predominated: "Some of them [the whaleman-artisans] have little boxes of dentistical-looking implements, specially intended for the skrimshandering business. But in general, they toil with their jack-knives alone; and, with that almost omnipotent tool of the sailor, they will turn you out anything you please, in the way of a mariner's fancy."

Scrimshaw objects intended for practical use included hand tools, kitchen gadgets, sewing implements, toys, and even full-sized furniture. Some, such as fids, straightedges, tool handles, seam-rubbers, napkin rings, and even some canes, could be carved or turned from a single piece of ivory or bone. Although they had a specific function, corset busks (made of bone, walrus ivory, or baleen) were often elaborately engraved; even apart from being products of painstaking labor, as intimate undergarments they were not bestowed casually. Other implements were constructed from two or more pieces joined or hinged together—pie crimpers with rotating jagging wheels and fold-out forks (Fig. 3); canes with shafts of one material, pummels of another, and inlay of a third. The most elaborate forms were truly "built" and may properly be called *architectural* or *architectonic*. Swifts (yarn winders) have numerous moving parts, with metal pinions and ribbon fastenings (Fig. 4). Bird cages, a labor-intensive technical challenge, could run the gamut of Victorian complexity. Sewing boxes, ditty boxes, chests of drawers, lap secretaries, pocketwatch stands, mantle ornaments, and other composite constructions typically employed combinations of wood, ivory, and bone and may have hinged lids, internal compartments, legs, finials, handles, drawers, drawer pulls, inlay, and all manner of ornamentation.

The quintessential form of purely decorative scrimshaw is engraved ivory and bone, usually rendered in a single medium—a tooth or pair of teeth; a tusk or pair; or a plaque, strip, or section of sperm whale panbone (jawbone). Finished teeth were sometimes set into wooden, silver, or coin-silver mounts; plaques might be framed by the artist; engraved strips of baleen could become oval boxes. Alternately, teeth and tusks could be carved into stand-alone sculptural forms, such as human or animal figures, or could become the components of complex ship models.

Figure 4 *Swift of sperm whale ivory and skeletal bone by Captain James M. Clark, circa 1835–1850. Made by a Yankee whaling captain, this exquisite piece typifies the best of the scrimshaw genre. It is inlaid with abalone shell and baleen, fastened with copper, tied with silk ribbons, fitted with two turnscrews in the form of clenched fists, and has a silver presentation plaque inscribed "R W Vose from Ja Clark." Height 40.7 cm. Swifts were a distinctly American form, used for winding the yarn employed in knitting and occasionally other domestic handicrafts and cottage industries. From the Kendall Whaling Museum, Sharon, Massachusetts.*

Figure 3 *Pie crimper in the form of a mermaid, New Bedford, circa 1875. Practical in origin, these classic kitchen implements inspired some of the scrimshaw's most creative forms and elaborate ornamentation. The jagging wheel was used for crimping pie crusts; they often also had ivory forks to puncture the top of the crust. This one was made aboard the New Bedford ship* Europa, *Captain James H. McKenzie, during 1871–1876. Length 18 cm. From the Kendall Whaling Museum, Sharon, Massachusetts.*

There were no rules and few precedents governing the choice of subject matter for pictures on scrimshaw. The earliest work—by the anonymous *Adam* engraver, J. S. King, Edward Burdett, and Frederick Myrick—was almost exclusively devoted to ship portraiture and whaling scenes. Figures of Columbia, Liberty, and Britannia appeared by around 1830. The ensuing generation enlarged the vocabulary to include patriotic portraiture (notably of Washington and Lafayette), inanimate patriotic devices, female portraiture, landscape, naval engagements, sentimental family scenes, and mortuary motifs (Fig. 5). Gradually, these were canonized as standard genre conventions. Some individual artists developed distinctive styles and themes. George Hilliott's polychrome teeth dialectically juxtapose a Polynesian *wahinee* in a grass skirt on one side and a New England lady in

Figure 5 *Family album wall hanging, New England, circa 1850. This unusual, elaborate construction features 12 tintype photographic portraits mounted in a triangular framework of walrus ivory and bone. The polychrome engraving on the walrus tusks is particularly interesting, as the woman-and-child portrait pair on the right is no doubt copied from a magazine fashion plate (in typical whalers' fashion), whereas the woman-with-telescope on the left appears to be an original image. Height 50 cm. From the Kendall Whaling Museum, Sharon, Massachusetts.*

a fashionable gown on the other. The anonymous "Banknote engraver" did meticulous portraits with banknote-like borders (hence the pseudonym). The "Eagle Artisan" engraved red-and-black American eagles and bold portraits. The "Lambeth Busk Engraver" made busks with London vignettes; a prime example features Lambeth Palace. Much naval scrimshaw is adorned with patriotic devices and naval engagements. Like whalemen's scrimshaw, some examples refer to specific vessels and events. A notable British example is credited to HMS *Beagle* on the same voyage on which Darwin evolved his theory of natural selection. Edward Yorke McCauley—later an admiral and noted Egyptologist—when he was a young midshipman aboard the U.S. Frigate *Powhattan* on Perry's historic Japan expedition in the 1850s, engraved two walrus tusks with portraits of the *Powhattan* and *Susquehanna,* exotic Oriental watercraft, and glimpses of Japan, Hong Kong, and Brunei. Even a Confederate infantryman tried his hand: Hampton Wilson, Irish immigrant, North Carolina sharecropper, Confederate draftee, and Union prisoner of war, while recuperating in a military hospital in Kentucky successfully "flowered" a pair of walrus tusks with military and naval vignettes, using materials and methods presumably supplied by a Yankee whaling veteran among his fellow patients.

Most scrimshaw pictures were inspired by or adapted from illustrations in contemporaneous books and periodicals; copy-

ing and even direct tracing were standard scrimshaw conventions. Because of their specific functional objectives, scrimshaw implements and architectonic forms were also mostly derivative. However, some of the best pieces—and many of the worst—were truly original creations, drawn from the maker's experience or imagination. A few have authentic stature as significant art, whereas others are little more than mere valentines. In the aggregate, anonymity and quality aside, as an indigenous occupational genre scrimshaw comprises some of the most characteristic and revealing documents of any occupational group, capable of providing profound insights into the life, work, and intentionality of the mariners who made them.

V. Museum Collections

Hull Maritime Museum, Kingston on Hull, England. Municipal museum in England's most historic Arctic whaling port; the most significant scrimshaw collection outside the United States.

Kendall Whaling Museum, Sharon, Massachusetts. World's largest and most comprehensive scrimshaw collection; world's only Scrimshaw Forensics Laboratory; annual Scrimshaw Collectors' Weekend; many scrimshaw publications.

Mystic Seaport Museum, Mystic, Connecticut. Comprehensive collection includes important loan deposits from other private and institutional collections; notable for substantive exhibition and informative catalogue.

Nantucket Whaling Museum (Nantucket Historical Association), Nantucket, Massachusetts. Eminent collection in the birthplace of sperm whaling and hometown of scrimshaw pioneers Edward Burdett and Frederick Myrick.

New Bedford Whaling Museum (Old Dartmouth Historical Society), New Bedford, Massachusetts. Large museum; extensive scrimshaw collection in the world's greatest whaling port.

Peabody Essex Museum of Salem, Massachusetts. Outstanding collection, brilliantly exhibited; museum founded 1799 as East India Marine Society; world's oldest collection of scrimshaw (1831).

In addition, there are modest but worthwhile collections at the Christensen Whaling Museum (Sandefjord, Norway), the Providence (Rhode Island) Public Library, the Scott Polar Research Insitute (University of Cambridge, England), South Street Seaport Museum (New York City), Whaler's Village (Lahaina, Maui, Hawaii), the Whaling Museum at Cold Spring Harbor (New York), and several state and municipal museums and libraries in Australia (Sydney; Melbourne; Hobart; and Launceston, Tasmania).

See Also the Following Articles

Folklore and Legends ∎ Museums and Collections ∎ Popular Culture and Literature ∎ Whaling, Traditional

References

Basseches, J., *et al.* (1991). "Edward Burdett (1805–1833): America's First Master Scrimshaw Artist." Kendall Whaling Museum Monograph No. 5, Sharon, MA.

Carpenter, C. H., Jr., and Carpenter, M. G. (1987). "The Decorative Arts and Crafts of Nantucket." Dodd Mead & Co., New York.

Flayderman, E. N. (1972). "Scrimshaw and Scrimshanders, Whales and Whaleman" (R. L. Wilson, ed.), N. Flayderman & Co., New Milford, CT.

Frank, S. M. (1991). "Dictionary of Scrimshaw Artists." Mystic Seaport Museum, Mystic, CT.

Frank, S. M. (1992). The origins of engraved pictorial scrimshaw. *The Magazine Antiques* **142**, 4(October 1992), 510–521.

Frank, S. M. (1998). "More Scrimshaw Artists." Mystic Seaport Museum, Mystic, CT.

Frank, S. M. (2000). Scrimshaw: Occupational art of the whale-hunters. *Marit. Life Trad.* **7**(March 2000), 42–57.

Hellman, N., and Brouwer, N. (1992). "A Mariner's Fancy: The Whaleman's Art of Scrimshaw." South Street Seaport in association with Balsam Press, New York, and University of Washington Press, Seattle.

Malley, R. C. (1983). "Graven by the fishermen themselves: Scrimshaw in Mystic Seaport Museum." Mystic Seaport Museum, Mystic, CT.

McManus, M. (1997). "A Treasury of American Scrimshaw: A Collection of the Useful and Decorative." Penguin Books, New York.

Penniman, T. K. [1952] (1984). "Pictures of Ivory and other Animal Teeth, Bone and Antler; With a brief commentary on their use in identification." Pitt Rivers Museum, Occasional Paper on Technology No. 5, Univ. of Oxford.

Ridley, D. E., *et al.* (2000). "The Scrimshaw of Frederick Myrick (1808–1862): A Catalogue Raisonné and Forensic Survey." The Kendall Whaling Museum, Sharon, MA.

West, J., and Credland, A. G. (1995). "Scrimshaw: The Art of the Whalers." Hull City Museums and Art Galleries and Hutton Press.

Sea Lions, Overview

DARYL J. BONESS

National Zoological Park, Smithsonian Institution, Washington, DC

Sea lions, like the fur seals, are members of the family Otariidae. There are presently seven sea lion species in five genera, with one genus exclusive to the Northern Hemisphere (Steller sea lion, *Eumetopias jubatus*), one that occurs in both hemispheres [in the north, the California (*Zalophus californianus*) and Japanese (*Z. japonicus*) sea lions, and in the south, the Galapagos sea lion, (*Z. wollebaeki*)], and three that are solely in the Southern Hemisphere (southern sea lion, *Otaria flavescens*; Australian sea lion, *Neophoca cinerea*; New Zealand sea lion, *Phocartos hookeri*).

I. Origins, Classification, and Size

Sea lions originated in the North Pacific region, sharing a common ancestor with fur seals. Although the fossil record for sea lions is poor, it appears they crossed into the Southern Hemisphere about three million years ago. Generally the sea lions have been thought of as a separate subfamily (Otariinae) within the family Otariidae. However, as more genetic analyses are done, this view is being questioned. For example, in one analysis the northern fur seal appears to be more closely related to the sea lions than to the other fur seals, which are all

in the genus *Arctocephalus*. In another analysis, two sea lion species clustered with three *Arctocephalus* species to the exclusion of the northern fur seal. The only substantial diagnostic morphological distinction between sea lions and fur seals is the presence of an underhair in fur seals but not in sea lions. Sea lions do tend to be larger than fur seals, with both groups exhibiting substantial differences in body mass, and smaller differences in body length, between males and females, a phenomenon known as sexual dimorphism (Fig. 1). Male sea lions are between two and four times heavier than females and up to one and a half times the length. The body mass of males in the different sea lion species ranges from about 250 to 1000 kg and in females from about 75 to 325 kg. In contrast, the heaviest fur seal male is about 300 kg and the heaviest female is about 75 kg. Lengths of male and female sea lions range from 205 to 330 and 180 to 270 cm, respectively.

II. Morphology and Physiology

Sea lions, like fur seals and walruses, differ anatomically from the true seals (phocids) in several ways. Probably most notably is their ability to walk or run rather than crawl on land. Underlying this capability is the ability to rotate the pelvis to a position that allows bringing the hind flippers under the body. As a result, sea lions have more efficient terrestrial locomotion than phocids. Another obvious anatomical feature of sea lions is the extended and flattened fore flippers. Again, this is a feature they have in common with their otariid fur seal cousins, although the walrus and phocids have relatively short fore flippers. These differences reflect the different swimming modes of these two groups of seals. The sea lions and fur seals use fore flippers to provide thrusting power and the walrus and phocids use their rear flippers.

Consistent with being shallow divers, sea lions have a relatively small lung capacity compared to many other marine mammals. They also have lower oxygen stores (40 ml O$_2$/kg) than true seals (60 ml O$_2$/kg), which are generally deep divers, but still much higher stores than humans (20 ml O$_2$/kg), for example. Additionally, the relative distribution of oxygen stores is different for sea lions. Sea lions have about 47% of their oxygen in blood, 35% in muscle, and 19% in their lungs. Phocids, however, have 64% of their oxygen in blood, 31% in muscle, and only 5% in their lungs. This larger percentage of oxygen in the lungs of sea lions correlates with the smaller degree to which the lungs collapse from water pressure. In humans, 51% of the oxygen is stored in the lungs.

III. Life History and Reproduction

Sea lions follow a life style typical of that of all the otariids, with some characteristics common to all seals. They are long-lived (probably 15–20 years), have delayed sexual maturation, and have physical and social sexual bimaturation with males maturing more slowly than females. For the three sea lion species adequately studied, females normally give birth for the first time at 4–5 years of age. For six of the seven sea lion species there is an annual breeding cycle, but one species, the Australian sea lion, has a unique cycle of just under 18 months. The net result of this cycle is that there is a gradual shift in the time of year and season when the breeding period occurs. For

Figure 1 *A male and several female California sea lions illustrating the size differences between males and females (sexual dimorphism). The male is the largest individual. Note the female on the right that is carrying a newborn pup using her teeth.*

example, over a 19-year period, between 1973 and 1991, the median date of pupping occurred in every month of the year. No other species of seal exhibits such a pattern. Why this pattern exists is unclear, but it does link to a lactation length of about 17 months in this species.

The reproductive behavior of female sea lions follows the typical maternal foraging cycle seen in all other otariids. Females give birth at traditional sites, usually on sandy beaches. Almost without exception a single pup is born to each female. During what has been termed a perinatal period, the female remains with her pup continuously, nursing it frequently. This period ranges from about 7 to 10 days. At the end of the perinatal period, females will have depleted stores of body fat because they have been fasting and begin to make foraging trips to sea, leaving their pups behind on the beach. In some species, most females will come into ESTRUS before they begin foraging trips, whereas in others, estrus will occur after foraging trips have begun. The duration of foraging trips is variable both within and between species (ranging from about 0.5 to 3 days), although they tend to be shorter among the sea lions than among the fur seals (1–12 days). Between foraging trips, females return to their pups on land, nursing them over a period of 0.5 to 1.5 days. This cycle is continued throughout lactation, which lasts about a year for all sea lions, except the Australian sea lion mentioned earlier, which has a 17-month lactation.

A physiological component of this maternal strategy of sea lions is relatively high-fat milk, which provides the energy needed by the pups as they try to grow during the "feast and famine" situation produced by maternal foraging. We do not have measures of milk fat for all sea lion species, but for those that have been studied the fat content of maternal milk ranges from about 15 to 45%, and most likely the level of fat in a milk relates to the typical length of foraging trips. The best example of this is seen in

Zalophus spp. The California species, which has maternal foraging trips of about 2.5 days, produces milk with 43% fat, whereas the Galapagos species, which forages for about a half day before returning to pups, has milk containing only 21% fat. Interestingly, the daily growth rates of sea lion pups, after taking the body size of adults into account, are very similar, suggesting that the maternal strategies are finely tuned to ecological conditions.

The reproductive behavior of male sea lions has been investigated unevenly among the various species. We know almost nothing about the New Zealand and Japanese sea lions but a considerable amount about the California, Galapagos, Steller and southern sea lions. Because females gather on land to give birth and care for their young and estrus is temporally linked to parturition, the conditions are ideal for strong sexual selection through male–male competition. In brief, the tendency for female sea lions to be highly clustered, indeed lying in contact with one another, provides the potential for males to compete for and maintain control over multiple females. The ability to control and mate with multiple females in a given reproductive period is known as polygyny.

As is typical of virtually all otariids, male sea lions return to traditional breeding grounds and vie for positions in areas where females have previously given birth or spent time cooling off during the hottest part of the day. In some species or populations, males may actually defend sites or territories, whereas in others they may be more flexible, defending females directly. Factors that are most important in determining which behavior is typical at a colony are the extent to which females move before they become receptive and the level of competition that exists among males. Female movement is most often associated with the need to cool off because of high ambient temperature. One species for which all studies have shown males only to defend territories is the Steller sea lion. This is likely a result of the high latitude at which this species breeds and the fact that females exhibit little, if any, thermoregulatory movement.

In contrast, the southern sea lion has been shown to behave variably depending on the breeding habitat. At sites where there tend to be large numbers of tidal pools, around which females cluster, males defend territories. However, where there are long narrow sand or pebble beaches, females shift up and down the beach with changes in the state of the tide and air temperatures. Under these conditions, males do not defend territories but shift as females do and defend the females directly.

The level of reproductive success, or number of females mated by the most successful males, may be similar regardless of which pattern of behavior is typical. What seems to constrain the maximum success is the degree to which females are clustered in space and time. If receptive females are too dispersed in time, an individual male may not have enough energy stores to remain competitive throughout the entire season. As food sources are usually not close to the breeding grounds, males must fast during the breeding season; rarely are individual males seen leaving their positions on land. This is true even during the hot part of the day. Minimum estimates of the maximum mating success for the most successful male among the sea lion species are highly variable. The estimate for the most successful Australian sea lion male, a species in which females tend to be quite dispersed, is 7 females mated compared to between 30 and 50 females for California and New Zealand sea lions, species in which females are much more clustered.

This intense competition among males is what produces the extreme sexual dimorphism we see in sea lions and many other seals. At this point it is unclear as to whether the large size of male sea lions is most important in direct competition, i.e., fights and threats with one another, or in the ability to remain ashore for longer periods of time because larger males can store more body fat. In some energetic studies of phocid seals, evidence suggests that it is the amount of energy stores that is more important.

IV. Feeding Habits

Our understanding of sea lion foraging ecology is much poorer than that of fur seals. Diet studies based mainly on analysis of food remains in scats (i.e., feces) do provide a picture of the feeding habits of sea lions, although this is with some bias. In all seven species, evidence suggests that they are primarily fish eaters and secondarily eat cephalopods (e.g., squid and octopus). The southern sea lion and the Australian sea lion, which live in proximity to penguin populations, have been found to prey on penguins occasionally. Penguin predation has not been reported in Galapagos sea lions, however, despite being sympatric with the Galapagos penguin. The two larger sea lions, Steller and southern, also prey periodically on northern and South American fur seal pups, respectively.

V. Population Status

The status of sea lion populations is variable. Two species, the California sea lion and the southern sea lion, are not currently listed as being in trouble. The Japanese sea lion has not been sighted since the 1970s and is now considered extinct.

The World Conservation Union (IUCN) considers the Galapagos sea lion vulnerable because there has been no reliable population estimate since the 1970s, at which time the Galapagos sea lion was thought to number about 30,000. California sea lions are probably in excess of 300,000, and southern sea lion populations probably exceed 200,000. What is not clear is how many southern sea lions in Peru died during the severe El Niño of 1997/1998. The sea lion species for which there is greatest concern at present is the Steller sea lion. Although it is not the smallest population by far (estimated at about 96,000), it has declined by about 80% since the 1970s. It is currently considered endangered by the IUCN and as endangered in the western and threatened in the eastern U.S. stock under the U.S. Endangered Species Act. While the precise cause of the decline is unclear, there is some evidence to implicate a decline in food supply, perhaps resulting from a mixture of commercial fishing and environmental changes known as climatic regime shifts. The Australian and New Zealand sea lions are known for their small populations historically and are at less than 15,000 animals each. The Australian species, which has been classed as rare, has been removed from the IUCN list because the population has been increasing and now appears to have leveled off. The New Zealand species remains listed as threatened, having undergone a major die-off in 1998, with 53% of the pups and perhaps as many as 20% of the adults perishing.

See Also the Following Articles

Breathing ■ Eared Seals ■ Locomotion, Terrestrial ■ Pinniped Life History ■ Pinniped Evolution ■ Sexual Dimorphism

References

Berta, A., and Sumich, J. L. (1999). "Marine Mammals." Academic Press, San Diego, CA.

Bininda-Edmonds, O. R. P., Gittleman, J. L., and Purvis, A. (1999). Building large trees by combining phylogenetic information: A complete phylogeny of the extant Carnivora (Mammalia). *Biol. Rev.* **74,** 143–175.

Boness, D. J., and Bowen, W. D. (1996). The evolution of maternal care in pinnipeds. *BioScience* **46,** 646–654.

Boness, D. J., Bowen, W. D., and Francis, J. M. (1993). Implications of DNA fingerprinting for mating systems and reproductive strategies of pinnipeds. *Symp. Zool. Soc. Lond.* **66,** 61–93.

Boness, D. J., Clapham, P. J., and Mesnick, S. L. (2001). Life history and reproductive strategies. *In* "Marine Mammal Biology: An Evolutionary Approach" (R. Hoelzel, ed.). Blackwell Science, Oxford.

Costa, D. P. (1991). Reproductive and foraging energetics of pinnipeds: Implications for life history patterns. *In* "The Behavior of Pinnipeds" (D. Renouf, ed.), pp. 300–344. Chapman and Hall, London.

Gales, N. J., Shaughnessy, P. D., and Dennis, T. E. (1994). Distribution, abundance and breeding cycle of the Australian sea lion, *Neophoca cinerea* (Mammalia, Pinnipedia). *J. Zool. Lond.* **234,** 353–370.

Lento, G. M., Hickson, R. E., Chambers, G. K., and Penny, D. (1995). Use of spectral analysis to test hypotheses on the origins of pinnipeds. *Mol. Biol. Evol.* **12,** 28–52.

Pabst, D. A., Rommel, S. A., and McLellan, W. A. (1999). The functional morphology of marine mammals. *In* "Biology of Marine Mammals" (J. E. Reynolds III and S. A. Rommel, eds.), pp. 15–72. Smithsonian Institution Press, Washington, DC.

Reijnders, P., Brasseur, S., van der Toorn, J., van der Wolf, P., Boyd, I., Harwood, J., Lavigne, D., and Lowery, L. (1993). "Seals, Fur Seals, Sea Lions and Walrus." International Union for the Conservation of Nature, Gland, Switzerland.

Riedman, M. (1990). "The Pinnipeds." University of California Press, Berkeley.

Wells, R. S., Boness, D. J., and Rathbun, G. B. (1999). Behavior. *In* "Biology of Marine Mammals" (J. E. Reynolds III and S. A. Rommel, eds.), pp. 324–422. Smithsonian Institution Press, Washington, DC.

Sea Otter

SEE *Otters*

Sei Whale

Balaenoptera borealis

JOSEPH HORWOOD

Centre for Environment, Fisheries and Aquaculture Science, Lowestoft, United Kingdom

I. Identification and Nomenclature

The diversity of thought about our great whales is characterized by quotes from the biologist R. C. Haldane on the sei whale. He described the species as the "most graceful of all the whales, as its proportions are so perfect, and wanting the clumsy strength of the two larger *Balaenoptera,* sperms and *Megaptera.*" He added, "it is also far the best to eat, the flesh tasting of something between pork and veal and quite tender."

The sei whale (*Balaenoptera borealis*, Lesson, 1828) is a typical, sleek rorqual, illustrated in Fig. 1. It is the third largest whale, reaching a maximum length of almost 20 m. More typically it is 15 m, weighing 20 metric tons.

Identification of the sei whale at sea can be difficult. By size alone, adult blue and fin whales are obviously larger and minke whales smaller. The dorsal fin is a useful cue, being relatively taller than that of blue and fin whales. It is also strongly concave on its dorsal edge, similar to a minke whale. For a long time it was not distinguished from its close relative, the warm-

water Bryde's whale (*B. edeni*). The Bryde's whale has three distinct ridges, running the length of the head, whereas the sei whale has only one (Fig. 1).

The color helps in identification. It is dark gray dorsally and on the ventral surfaces of the flukes and flippers, and there is no whitening of the lower lip as found in fin whales. However, in a few individuals some white baleen plates occur. Often the body can be heavily scarred with healed lamprey bites. Sei whales dive more by sinking than an arched dive, but the other rorquals can also dive in this quiet manner.

A more detailed external inspection allows a more definite identification. In sei and minke whales the ventral grooves end well before the umbilicus. In other *Balaenoptera* spp., including Bryde's whale, they end at, or posterior to, the umbilicus. The number of ventral grooves varies considerably from whale to whale. In sei whales they vary between 40 and 65 with a mean number of about 50. This is less than in blue (*B. musculus*), fin (*B. physalus*), and minke whales (*B. acutorostrata* and *B. bonaerensis*) but about the same as in Bryde's whales.

The baleen of sei whales is a dark gray, but often with a yellowing-brown hue, and often with some anterior white plates. The plates number about 350 on each side of the jaw, and the largest is less than 80 cm long. The width of the plate is relatively narrow compared to blue, fin, and Bryde's whales. In the sei whale the length-to-breadth ratio is typically over 2.2, whereas in the Bryde's whale it is always less than 2.2. The bristles of the sei whales' baleen are particularly fine. At their base they are about 0.1 mm in diameter compared with 0.3 mm for the other rorquals.

The sei whale is closely related to the Bryde's whale. Wada and Numachi (1991) showed that genetic differentiation of the rorquals took place over 4 million years ago, but the sei and Bryde's whales separated less than 300,000 years ago.

The sei whale is derived from the Norwegian "sejhval," as it would arrive off Norway at the same time as the "seje" or saithe (*Pollachius virens*). There are a variety of other common names, but English forms have disappeared from usage in favor of the sei whale.

II. Distribution

The sei whale can be found in all ocean basins. It undertakes extensive, seasonal, latitudinal MIGRATIONS, spending the summer months feeding in the subpolar higher latitudes and returning to the lower latitudes to calve in winter. Figure 2

Figure 1 *Smaller than blue or fin whales and larger than minke whales, sei whales* (Balaenoptera borealis) *can be distinguished by a relatively larger dorsal fin and by only one distinct ridge extending the length of the head. Sei whales occur in all oceans. Pieter A. Folkens/Higher Porpoise DG.*

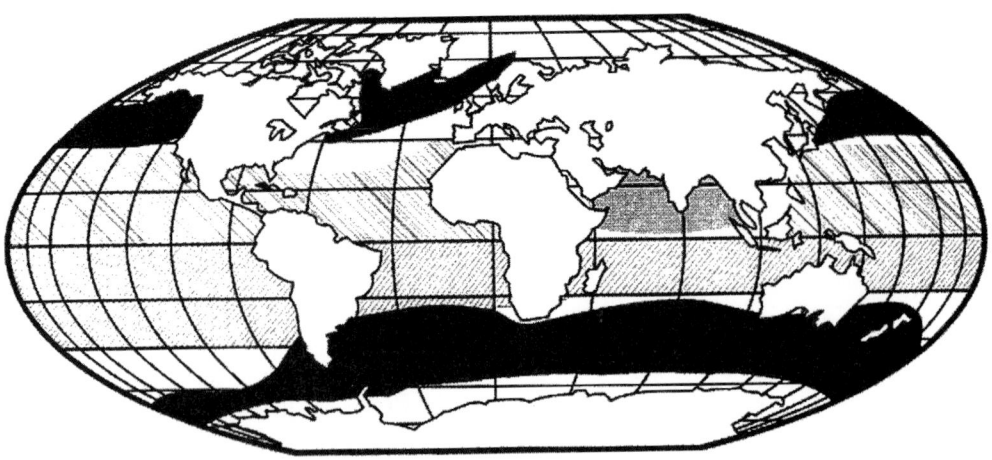

Figure 2 *Global distribution of sei whales. Filled areas are the summer feeding distributions, and hatched areas represent breeding areas.*

shows the global distribution of sei whales in summer and winter. In the Southern Hemisphere, they rarely penetrate as far south as blue, fin, and minke whales, with summer concentrations mainly between the sub-Tropical and Antarctic convergences. Sei whales are seen, and have stranded, in the northern Indian Ocean, but their distribution and migrations are undetermined.

III. Populations and Biology

Genetic studies show the existence of different populations in the Northern and Southern Hemispheres but have failed to identify separate populations within ocean basins. However, biologists have separated populations, for MANAGEMENT purposes, on the basis of different migrations and biological characteristics.

In the Southern Hemisphere, six populations are assumed, as for the other rorquals. In the North Pacific, two or three populations have been proposed. In the North Atlantic, as many as eight populations have been suggested, but only three are considered for management purposes.

The migratory behavior, spatial distribution, and fine baleen structure of the sei whales are associated with the nature of their preferred copepod prey. They feed on patches of copepods, near the surface, usually by skimming, and predominantly at dawn. The other rorquals would probably find such food too scarce, and the sei whale's feeding habits are nearer to those of the right whales than the other rorquals. Sei whales also feed on shoals of fish and squid if they are encountered. Sei whales have a specific feeding niche different from, but sometimes overlapping, that of the other baleen whales.

Maturity is at about 10 years for males and females. Females are larger than males. Sei whales are bigger in the Southern Hemisphere, and here males mature at about 13–14 m and females at 14 m. In most seas, the age of maturity declined by 2 to 3 years, after the populations were depleted by whaling. In the Southern Hemisphere, young are conceived in June and births are in December in northern waters. The young are carried for almost a year and are born at a size of 4.5 m. Most

calves are weaned in 7 months, after they have migrated to colder waters with their mothers.

As for most mammals, sei whales can be expected to have increased rates of mortality when very young or old, but actual rates are poorly known. From observations of age compositions, the rate of natural mortality is typically about 5–10% per year. They die naturally from predators, such as killer whales, and weakening from disease and parasites.

IV. Status of Populations

A fast rorqual, the sei whale was not exploited until the era of modern whaling at the end of the 1800s. Off north Norway, 4000 sei whales were killed between 1885 and 1900. Since then, sei whales were caught in the North Atlantic from land stations in Canada, Faeroes, Iceland, Ireland, Iberia, Norway, and Scotland. In the North Pacific, they were caught from California, Canada, Japan, Kamchatka, and Kuril and by pelagic fleets. In the Southern Hemisphere, they were caught from Brazil, Chile, Peru, South Africa, and South Georgia. The largest catches were made by the Antarctic pelagic fleets, after the numbers of blue and fin whales had been reduced, and between 1960 and 1970 over 110,000 sei whales were killed. Whaling is regulated by the IWC, and whaling for sei whales ceased in the Southern Hemisphere in 1979 and in the North Pacific in 1975. In the North Atlantic, whaling was prohibited from 1986, but limited catches continued for a few years, at Iceland, under a scientific special permit and through subsistence whaling from Greenland.

The size of populations is poorly determined, but whaling significantly depleted populations in all areas. In the Southern Hemisphere the original population was about 100,000, and in 1980 was thought to be 24,000. In the North Pacific, a population of over 60,000 was reduced to about 15,000. By now there may be 70,000 in both areas. The status of the North Atlantic sei whale is more uncertain, but recent sightings surveys indicate about 10,000 sei whales in the central and northeastern North Atlantic.

See Also the Following Articles

Blue Whale ■ Bryde's Whales ■ Fin Whales ■ Minke Whales ■ Whaling, Modern

References

Anonymous (1997). "Report of the Special Meeting of the Scientific Committee of the Sei and Bryde's Whales." p. 150. Rep. Int. Whal. Commn. (Special issue 1).

Carwardine, M., Hoyt, E., Fordyce, R. E., and Gill, P. (1998). "Whales and Dolphins: The Ultimate Guide to Marine Mammals," p. 288. HarperCollins, London.

Gambell, R. (1985). Sei whale: *Balaenoptera borealis*, Lesson 1828. *In* "Handbook of Marine Mammals" (S. Ridgway and R. Harrison, eds.), Vol. 3, pp. 155–170. Academic Press, London.

Horwood, J. (1987). "The Sei Whale: Population Biology, Ecology and Management," p. 375. Croom Helm, London.

Kawamura, A. (1980). A review of food of balaenopterid whales. *Sci. Rep. Whales Res. Inst.* **32**, 155–158.

Masaki, Y. (1976). Biological studies on the North Pacific sei whale. *Bull. Far Sea Fish. Res. Lab.* **14**, 1–104.

Wada, S., and Numachi, K. (1991). Allozyme analyses of genetic differentiation among the populations and species of the Balaenoptera. Rep. Int. Whal. Commn. (Special Issue 13), pp. 125–154.

Sexual Dimorphism

KATHERINE RALLS
*National Zoological Park,
Smithsonian Institution, Washington, DC*

SARAH L. MESNICK
*Southwest Fisheries Science Center,
La Jolla, California*

Dimorphism means two forms. "Sexual dimorphism" means that the two sexes of a species differ in external appearance or other features. Males and females may differ in size, color, shape, the development of appendages (such as horns, teeth, feathers, or fins), and also in scent or sound production. Species in which males and females are identical in appearance or other features are said to be "monomorphic." This article describes the types of dimorphic traits found in marine mammals and explains some of the reasons why these traits might have evolved and what can be inferred about the lives of males and females in a particular species from the pattern of sexual dimorphism. The quality of the information available on sexual dimorphism varies widely across marine mammal species. We know quite a lot about a few species, which are used repeatedly as examples, and virtually nothing about others. Despite the technical difficulties of studying marine mammals, our understanding of the evolution of sexual dimorphism is increasing steadily as observations of rarely encountered species accumulate and new techniques are developed.

I. Evolution of Sexual Dimorphism

Sexual dimorphism has fascinated biologists since before the time of Darwin. Darwin considered that most sexual dimorphism was due to sexual selection, in which evolutionary forces acted separately on the sexes (Darwin, 1871). For example, females might choose to mate with highly ornamented males (e.g., the peacock's tail) or males might develop characters useful for fighting with other males to win in contests for access to females (e.g., large body size and antlers in deer). Today, these two processes are often referred to as female choice and contest competition, respectively. More recently, scientists have learned that males compete not only by physical fighting and display but also, in species where females mate with more than one male, by sperm competition within the female reproductive tract.

Although Darwin's ideas about sexual selection have stood the test of time, some cases of sexual dimorphism seem to be best explained by natural selection. For example, males and females in some species of birds [e.g., Galapagos finches (genus *Geospiza*) and the extinct New Zealand huia (*Neomorpha acutirostris*)] have radically different bill morphologies that are best explained by sex differences in foraging habits (Anderson, 1994). In some species, females appear to be larger than males primarily because big mothers are better mothers (more eggs, better at defending their brood; Ralls, 1976). The emerging view is that the degree of sexual dimorphism in a species is the result of the difference between the sum of all the selective pressures (natural selection and sexual selection) affecting the male and the sum of those affecting the female.

II. Types of Sexual Dimorphism

The adult males and females of a species may differ in size, color, shape, the development of appendages (such as horns, teeth, feathers, or fins), scent, or vocalizations (Fig. 1). In marine mammals, one of the most striking sexually dimorphic characters is size. In some species, males are dramatically larger than females. For example, in southern elephant seals (*Mirounga leonina*), adult males (maximally at 3700 kg) weigh 4–10 times as much as the adult females (which weigh 350–800 kg). Males in some species also possess greatly enlarged TEETH that are lacking in females and are used in fights between males. The best known example is the unicorn-like tusk of the NARWHAL (*Monodon monoceros*). The tusk, which is actually a greatly enlarged left upper tooth, usually erupts only in males and can grow to an extraordinary size, exceeding 3 m in length and 10 kg in weight. In some odontocete species (e.g., bottlenose whales, genus *Hyperoodon*), males have greatly enlarged and densely ossified heads, which they use to ram other males during fights. In otariids, males have thick necks and massive chests that tend to be covered by a dense mane of hair. The noses of males are sometimes bizarrely modified. For example, the most distinctive feature of the male hooded seal (*Cystophora cristata*) is an inflatable hood and bright red nasal sac that may function in agonistic and courtship displays. The appendages of males (flippers, flukes, caudal peduncles, and dorsal fins) are sometimes greatly enlarged. The best known example of dorsal fin enlargement is seen in male killer whales (*Orcinus orca;* Fig. 1b).

Figure 1 *Types of sexual dimorphism in marine mammals. (a) Size. Adult male South American sea lions (O. flavescens) are two to three times heavier than females; males grow to 2.8 m and weights of 340 kg; females to 2.2 m and 144 kg. There is extreme sexual dimorphism in body shape and pelage as well as size; males have massive necks, a broad head with a characteristically upturned muzzle, and a thick mane of long guard hairs. Photo by William Conway. (b) Dorsal fins. A pod of killer whales (O. orca), Alaska. In adult males, the dorsal fin is erect and may grow to 1.8 m in height whereas the dorsal fins of females are less than 0.7 m and distinctly falcate. Sexual dimorphism also occurs in body size, flipper size, and genital pigmentation pattern. Photo by Flip Nicklin (Minden Pictures). (c) Teeth and tusks. Dueling male narwhals (M. monoceros), Canada. The unicorn-like tusk of the narwhal is actually a greatly enlarged left upper tooth. The tusk generally erupts only in males and may exceed 3 m in length and 10 kg in weight. Sexual dimorphism also occurs in body size, pigmentation pattern, and the shape of the flukes and pectoral fins. Photo by Flip Nicklin (Minden Pictures). (d) Noses. Threat vocalizations resonate in the greatly enlarged proboscis of adult male northern elephant seals (M. angustirostris), Año Nuevo, California. There is extreme dimorphism in body size and shape; males grow maximally to 4 m and 2300 kg and females to 3 m and 360–710 kg. The skin on the neck of the adult males is thick, rugose, and scarred by fights, and canine teeth are sexually dimorphic in size and shape. Males are darker brown than females. Photo by Sarah L. Mesnick. (e) Postanal hump. The postanal hump of adult male eastern spinner dolphins (S. longirostris orientalis) is exaggerated tremendously. The dorsal fin of adult males is also forward canted and the tips of the flukes curl up. Photo by Robert L. Pitman.*

Although sexual dimorphism traditionally referred to differences in morphological traits, the sexes can also produce different vocalizations or odors or be differently colored or patterned. Among marine mammals, differences in color are usually limited to fairly minor differences in pattern or density of pigmentation. For example, in ribbon seals (*Histriophoca fasciata*), the banding pattern is similar in both sexes but paler and less distinct in females. There are numerous examples of sexually dimorphic vocalizations in marine mammals, such as the roars and bellows of male sea lions and fur seals (Otariidae), the songs of male humpback whales (*Megaptera novaeangliae*), and the loud clicks of male sperm whales (*Physeter macrocephalus*). In terrestrial mammals, males and females often produce different scents via urine, feces, or specialized scent glands. This has not been observed much in marine mammals but may well occur. It is known, for example, that male ringed seals (*Pusa hispida*) produce a strong odor in the breeding season. Male sea otters (*Enhydra lutris*) frequently investigate the anogenital areas of other otters, and male common bottlenose dolphins (*Tursiops truncatus*) investigate the urogenital region of possibly estrous females with their rostrums.

III. Taxonomic Distribution

A. Baleen Whales

Sexual size dimorphism is "reversed" among the 13 species of baleen whales with females attaining asymptotic lengths that are generally 5% longer than males. Baleen whales typically undertake long-distance migrations between their northern or southern feeding areas and their tropical breeding areas and may not feed while migrating or on the breeding grounds. Females have the added stress of pregnancy and lactation during the nonfeeding periods; a larger size may enable them to store more energy resources in the form of BLUBBER to meet their greater reproductive demands.

Sexually dimorphic vocalizations are well known in humpback whales. Male humpbacks sing lengthy, elaborate songs, the function of which has been the subject of much speculation. Songs might function to attract females, signal status to other males, space males on the breeding grounds, synchronize estrus in females, or some combination of these. The humpback song is particularly intriguing because songs change over time, yet all members of a population sing similar songs at any one time. Male fin whales (*Balaeoptera physalus*) have a patterned call, which has been termed a breeding display, but observations of COURTSHIP or competitive interactions are sparse. Sexually dimorphic vocalizations may also exist in blue whales (*Balaenoptera musculus*). There is dimorphism in the shape of the upper jaw of fin whales and, to a lesser extent, Bryde's whales (*Balaenoptera edeni*), but the function of this dimorphism is unknown.

Observations of clear AGGRESSION between males are known only in humpback and southern right whales (*Eubalaena australis*). Thus, it is not surprising that there are few accounts of sexually dimorphic structures that might be used in contest competition. Male right whales, however, have larger and more numerous callosities (the raised thickened patches of skin on the head) than females, which may function as weapons in contests between males. Male right whales are also scarred more heavily than females.

B. Toothed Whales

The relative size of the sexes varies widely among the 70+ species of toothed whales. Males are larger than females in many species, with the most pronounced dimorphism in sperm whales, killer whales, bottlenose whales, narwhals, belugas (*Delphinapterus leucas*), and pilot whales (genus *Globicephala*). In sperm whales, for example, females reach about 11 m in length and 15 tons, whereas physically mature males are approximately 16 m long and weigh 45 tons. Females are slightly larger than males in Baird's beaked whales (*Berardius bairdii*), the franciscana (*Pontoporia blainvillei*), the Indian river dolphin (*Platanista gangetica*), harbor porpoise (*Phocoena phocoena*), and dolphins in the genus *Cephalorhynchus*. Some species are monomorphic in size, including the Clymene dolphin (*Stenella clymene*), Atlantic spotted dolphins (*Stenella frontalis*), dwarf and pygmy sperm whales (genus *Kogia*), tucuxi (*Sotalia fluviatilis*), and some dolphins in the genus *Lagenorhynchus*.

Differences between the sexes may occur in the size and shape of the head, teeth, thoracic girth, flukes, flippers, dorsal fin, caudal peduncle, postanal hump, and length of the beak. In general, males tend to have larger appendages than females, the exception being the few species in which females have longer rostra than males [both species of south Asian river dolphin, the franciscana, and the rough-toothed dolphin (*Steno bredanensis*)]. Dimorphism in the size and shape of the head may be a result of enlargement of the nose (in sperm whales) or the forehead [in bottlenose whales (genus *Hyperoodon*), pilot whales, and to a lesser extent in bottlenose dolphins] of adult males. The massive nasal complex of adult male sperm whales may be one-quarter to one-third the length of the animal and is probably used in the generation of sound. In bottlenose whales, the forehead is extremely steep and the surface becomes flattened in mature males. Dimorphism in the density of ossification of the bones in the head occurs in bottlenose whales (the cranium) and beaked whales of the genus *Mesoplodon* (the rostrum, which has one of the highest reported densities of any mammalian tissue). Differences between the sexes in the ossification of the head in these species are not surprising given observations of head butting between adult male bottlenose whales and heavy scarring on adult males of several beaked whale species of the genus *Mesoplodon*. The sexually dimorphic pattern of scarring in *Mesoplodon* species is consistent with the idea that adult males use their rostrum, and the exposed teeth on the lower jaw, in fights with other males.

Dimorphism in the size, shape, and/or number of teeth is known in the narwhal, sperm whale, Cuvier's beaked whale (*Ziphius cavirostris*), bottlenose whale, and in beaked whales of the genus *Mesoplodon*. In most of these species (exceptions being sperm whales and narwhals) the teeth erupt only in males and only at sexual maturity.

Differences between the sexes are known in flipper length (killer whales and melonheaded whales, *Peponocephala electra*), serration (Heaviside's dolphins, *Cephalorhynchus heavisidii*), and shape of the trailing edge (belugas). In some species, including sperm whales and Dall's porpoises (*Phocoenoides dalli*), males have deeper caudal peduncles than females, which may function to give more power to the flukes. Postanal humps (thought to be composed of muscle and connective tissue) are exhibited in mature males of several species, although they have been properly described and correlated with age and sex in only a few. The postanal hump of the male eastern spinner dolphin (*Stenella longirostris orientalis*) is exaggerated tremendously (Fig. 1e). While the function of the postanal hump remains unknown, it has been suggested to be an anchor for external genitalia and may serve to enhance sexual performance. It may also serve as a visual signal that makes adult males easily recognizable. Dorsal fins may be larger and more erect in males than females, more hooked, or more forward-canted (Figs. 1b and 1e). The most exaggerated examples of dorsal fin enlargement are seen in male killer whales and pilot whales. The significance of these differences in dorsal fin size and shape is unknown but they may serve a thermoregulatory function and/or as a visual signal. Differences between the sexes also occur in the flukes, which may be longer and broader in males, or differently shaped. In Dall's porpoises and sperm whales, for example, the trailing edge of the flukes of males are convex, and in male eastern spinner dolphins, the tips of the flukes curl up. As is true for mammals in general, the distance between anal and genital openings of odontocetes tends to be greater in males than in females.

Sexual differences in pigmentation patterns are most common in the genital area but are also known to occur on the face, head, and body. Sexual dimorphism in genital pigment patterns is known in several species [killer whales, dolphins in the genus *Cephalorhynchus* and *Lissodelphis*, shortbeaked common dolphins (*Delphinus delphis*), Burmeister's porpoises (*Phocoena spinipinnis*), and Dall's porpoises]. Pigmentation differences may be related to sexual recognition, advertisement (for either males or females), or may help suckling young locate the teats. In most species of beaked whales, the body gets lighter in color with age. The lightening is especially noticeable in adult males and is primarily due to an accumulation of body scars, but may also be due to changes in pigmentation and, in several species, both. In Risso's dolphins (*Grampus griseus*), ontogenetic lightening and an accumulation of body scars make older animals of both sexes appear almost pure white, and the pattern may be more prevalent in males. Adult male spotted dolphins (*Stenella attenuata*) bear conspicuous white rostrum tips, visible at a great distance. In Fraser's dolphins (*Lagenodelphia hosei*), the intensity and thickness of the eye-to-anus stripe becomes more exaggerated (darker and thicker) in adult males. Another type of pigment dimorphism is the occurrence of visible (white or nonpigmented) linear scarring, suggested to result from a lack of repigmentation of damaged tissue from the tooth rake wounds of conspecifics. In some odontocete species, both sexes exhibit heavy scarring [e.g., Baird's beaked whale (genus *Berardius*) and Risso's dolphins]. However, in others (*Mesoplodon* spp., the narwhal, and the sperm whale), males are scarred more heavily than females. In these species, scarring is likely the result of wounds inflicted during male fights.

At present, acoustically dimorphic calls are known only in sperm whales. However, because odontocetes produce a wide range of sounds, dimorphic acoustic signals are likely to occur in several other species as well. Because larger animals make larger sounds, it is also reasonable to expect that other sexually dimorphic species, such as pilot whales, will produce acoustically dimorphic calls.

C. Pinnipeds

The 36 species of pinnipeds show the greatest range in sexual size dimorphism of any higher vertebrate group (Fig. 2). Adult males are up to 10 times as heavy as adult females in some species, whereas females are slightly larger than males in others. For virtually all pinnipeds studied to date, data support, or are highly suggestive of, a polygynous mating system. However, the potential for polygyny varies greatly among species and there is a strong correlation between the degree of polygyny and the degree of dimorphism. The mating system in turn, depends to a large extent on whether breeding takes place on land or at sea. In terrestrially mating pinnipeds [this includes sea lions and fur seals and three species of phocid, the northern and southern elephant seal (genus *Mirounga*) and the gray seal (*Halichoerus grypus*)], extreme polygyny is possible because females gather in dense groups on islands to give birth and mate. Under these conditions, a successful male can defend many females. In these species, males exhibit not only large size but also other characteristics useful in fights over females, such as large canines, massive necks and chests, and dense pelage. Large size may also help males of these species achieve greater reproductive success by enabling them to remain longer on the breeding rookery because larger males have bigger energy reserves in the form of blubber.

In the remaining pinnipeds, the WALRUS (*Odobenus rosmarus*) and nearly all of the phocids, mating takes place in the water. Females of many species give birth on ice (and therefore are not as spatially clumped as terrestrially breeding species in part because they have larger expanses of suitable habitat available) and the mating season is short. Thus, males have less opportunity to defend and mate with multiple females. These characteristics inhibit the development of the extreme polygyny and sexual dimorphism found in terrestrially mating otariids and phocids. In general, males of aquatically mating species are only slightly larger than females or females may be slightly larger than males [bearded (*Erignathus barbatus*), Weddell (*Leptonychotes weddellii*), Ross (*Ommatophoca rossii*), crabeater (*Lobodon carcinophaga*), and leopard (*Hydrurga leptonyx*) seals]. The hooded seal is a notable exception, with males considerably larger than females. In ice-breeding species, large female size may help a mother provide large quantities of fat-rich milk for her pup, and because mating takes place in the water in these species, agility, rather than size or strength, may be important in male contests for females.

In addition to the sexual size dimorphism mentioned earlier, the sexes may also differ in pelage length and color, shape of the head and chest, canine development, and the pattern of scarring on the neck and chest. Adult male otariids tend to be bulkier than females and are distinguished readily by their thicker and more powerful necks and their massive chests. The head, neck, and chest of males tend to be covered by longer, rougher hairs, which gives the impression of a mane [e.g., the South American sea lion (*Otaria flavescens*); Fig. 1a]. In older males, the guard hairs are lighter in color and tinged with white, silver, or tan. Adult male California sea lions (*Zalophus californianus*) also develop a pronounced forehead, or sagittal crest, and adult male southern sea lions have a distinctive upturned muzzle. The skin on the necks of adult male elephant seals and gray seals is thickened and wrinkled and marked by scars from fights. In general, adult male otariids, as well as adult males in some species of phocids, tend to be more darkly pigmented than females.

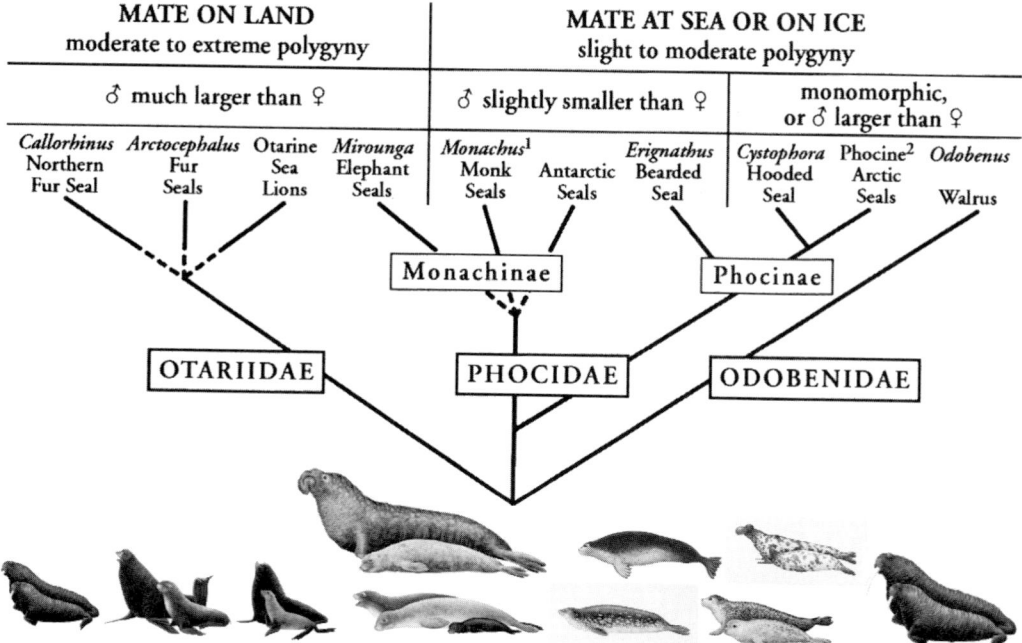

Figure 2 *Sexual size dimorphism in pinnipeds. A composite phylogenetic tree of the Pinnipedia on which sexual dimorphism, mating location, and degree of polygyny have been overlaid. Sexual size dimorphism varies greatly across pinniped species, and there is a strong correlation between the degree of dimorphism and the mating system. In otariids and three species of phocid (both elephant seals and the gray seal), mating takes place on land and extreme polygyny is possible because a successful male can defend many females. Males are much larger than females (2 to 10 times larger) and also exhibit other characteristics useful in fights over females, such as large canines. In the remaining pinnipeds, the walrus and nearly all the phocids, mating takes place in the water or on ice. There is less opportunity for males to mate with multiple females and agility, rather than size or strength, may be important in male contests for females. Males are equal, slightly, or moderately (up to 1.5 times) larger than females or females may be slightly (up to 1.1 times) larger than males in these species.* [1]*Females are slightly larger than males in the Hawaiian monk seal; males are slightly larger than females in the Mediterranean monk seal.* [2]*Among phocines, the gray seal represents a notable case because it can exhibit both terrestrial and aquatic mating and males are maximally 3 times larger than females. Pieter A. Folkens/Higher Porpoise DG.*

Many pinnipeds have sexually dimorphic vocalizations that may function to establish and maintain dominance relationships or to attract females. In some species, the sounds are amplified or resonated in the proboscis (as in hooded and elephant seals; Fig. 1d) or an internal air sac (as in ribbon seals, bearded seals, and walruses). Hooded seals produce numerous sounds as they inflate and deflate their hood and bright red nasal sac in response to disturbances and as part of the courtship display. Male elephant seals also have greatly enlarged noses, and snouts of male gray seals are broader and more elongated than those of females. Males of these species establish dominance hierarchies through stereotyped visual and airborne acoustic threats and, less often, physical aggression. Male harbor, bearded, ribbon, Weddell, ringed, and harp (*Pagophilus groenlandicus*) seals are known for their acoustic courtship displays. Male harbor seals engage in complex hierarchical acoustical mating displays in which several subordinate males passively muzzle singing dominant males underwater. Much of the roaring and bellowing of adult male otariids is thought to in-

timidate rivals but acoustic displays may also be used to advertise to females. The walrus has the most elaborate courtship display of all pinnipeds involving intricate acoustic and visual components. Vocalizing adult male walruses apparently compete for females in lek-like groups in the water near ice floes on which females gather to pup and rest. Their surface vocal repertoire includes barks, whistles, and growls, and underwater vocalizations sound bell like. It is also thought that the massive tusks may play a role as a visual symbol of rank and as a display to females. Male walruses are larger than females in both body and tusk size. In marine mammals, the only well-documented sexually dimorphic scent of which we are aware occurs in the ringed seal. Male ringed seals give off a strong scent during the breeding season.

D. Sirenians, Sea Otters, and Polar Bears

Manatees (genus *Trichechus*) are generally monomorphic in size and appearance. Dugongs (*Dugong dugon*) exhibit no obvious sexual dimorphism apart from the short tusks (upper incisors), which usually erupt in adult males, although females may grow to a slightly larger size than males. Male dugongs compete for access to females by patrolling exclusive areas and engaging in threats, fights, and song. Adult male sea otters are larger than adult females. Second only to pinnipeds, polar bears (*Ursus maritimus*) exhibit the greatest sexual size dimorphism among mammals. Male polar bears may be over twice as heavy as females.

IV. What Can We Infer from Sexual Dimorphism?

The variation in sexual dimorphism among marine mammal taxa is striking. The sexes are visually indistinguishable in some species, whereas in others the differences between the sexes are so extreme that males and females live essentially separate lives except when they meet to mate. This rich variation in sexual dimorphism has prompted scientists to offer a variety of explanations. The various mechanisms of sexual selection—female choice, contest competition, and sperm competition— probably account for a large proportion of sexual dimorphism in marine mammals. However, some dimorphic traits may reflect ecological differences between the sexes (e.g., differences in beak length between the sexes in south Asian river dolphins). Others may be important for females and their young (e.g., larger females make better mothers or urogenital pigment patterns may highlight the mammary glands and help young to find them).

The functional significance of most sexually dimorphic traits in marine mammals remains untested, which is not surprising given the difficulty of observing, let alone experimenting on, most species. In general, the behavior of pinnipeds (which often breed where they can be observed) is better known than the behavior of cetaceans that breed at sea. However, extended observations of behavior have also been possible in a few cetacean species (e.g., bottlenose dolphins, humpbacks, and right whales) that breed close to shore. Nevertheless, we can often infer the functional significance of sexual dimorphism in species whose behavior is poorly known by analogy to well-studied species. The type and degree of sexual dimorphism and its association with other characteristics such as relative testis size and pattern of bodily scarring provide clues to the intensity of sexual selection (the skew in male mating success) in a species and the probable underlying mechanisms of sexual selection.

Based on studies of terrestrial mammals, a positive correlation is generally assumed between the amount of sexual dimorphism in a species and the deviation of the breeding system from monogamy. Thus, in polygynous species, male competition for access to females is severe and males are expected to exhibit traits, such as large size and big canines, favored in fights with other males over access to females. The correlation between sexual size dimorphism and the degree of polygyny has been shown across pinniped taxa (Alexander *et al.*, 1979). For example, among otariids, the northern fur seal (*Callorhinus ursinus*) and Steller sea lion (*Eumetopias jubatus*) show the greatest relative dimorphism in body weight and defend the greatest number of females in their territories as compared to less dimorphic species. Within a species, a large body size has also been shown to be correlated with greater mating success (via dominance rank, endurance, and tenure on rookeries; e.g., elephant seals, gray seals). It is important to note, however, that species that lack sexual size dimorphism do not necessarily lack male–male competition for mates. In these species, sexual selection may be intense, but due to different forms of competition among males for access to mates, and the consequences may not be reflected in size but in other characters, such as song, visual display, or agility.

Sexual dimorphism in size and weaponry (big teeth, enlarged heads, and strong flukes) suggests that contest competition for access to mates plays an important role in the mating strategies of males in many marine mammal species. Contest competition may take the form of ritualized displays (visual or acoustic), by which potential rivals assess relative size or strength, or physical aggression. Among odontocetes, dimorphism in weaponry is correlated to patterns of body scarring and observations of head butting among males. In certain species, such as sperm whales, beaked whales, and narwhals, teeth erupt or are enlarged only in adult males, a pattern that suggests their function has shifted from feeding to use in social interactions. Adult males of many of these same species are heavily scarred, another trait that suggests males use their teeth in physical battles with other males.

Among terrestrial mammals, the relationship between relative testis size and mating system is so strong that relative testes size can be used as a good indicator of the mating system (Gomendio *et al.*, 1998). In general, where copulatory frequency is high, the testes are large, and where copulation is infrequent, the testes are small. In right whales, observations of multiple males mating with single females, together with huge (1 ton) testes, strongly suggest that sperm competition is a principal mating strategy in this species, and probably also in bowhead (*Balaena mysticetus*) and gray (*Eschrichtius robustus*) whales (Brownell and Ralls, 1986). Odontocete species such as sperm whales and beaked whales that exhibit dimorphic traits associated with intense physical combat (e.g., large size, enlarged teeth) tend to have small testes. The testes of sperm and beaked whales represent less than 0.5% of the body weight, weigh only a few kilograms, and can be held in one hand. At the other extreme, species having the largest testes, suggesting the likelihood of significant sperm competition, do not exhibit the extreme dimorphic traits associated with physical combat. These species tend to be sexually monomorphic or have dimorphic traits that may be associated with agility or visual display. For example, harbor porpoises, finless porpoises (*Neophocaena phocaenoides*), and dusky dolphins (*Lagenorhynchus obscurus*) have testes that represent greater than 5% of their body weight. Humans, for comparison, at about the same body mass as these dephinid species, have testes of only 0.08% of body mass (Kenagy and Trombulak, 1986).

In three-dimensional habitats, agility, rather than size or strength, may sometimes determine the outcome of male contests. Agility may be useful in scramble competition for access to mates and it may function as a visual display for female choice. This may be the case in some species, such as the aquatically mating phocids, in which males compete underwater and are smaller than females. Among odontocetes, larger body size typically means that the male's propulsion structures are also proportionally larger than those of the female. The importance of speed and maneuverability is suggested by sexual dimorphism in the flippers, flukes, caudal peduncle, and dorsal fin. Tolley *et al.* (1995) suggested that the larger body size, caudal peduncle, flukes, and dorsal fin of male bottlenose dolphins, and the pattern of dorsal fin scarring, are consistent with males competing for access to dispersed females. Features such as flukes and dorsal fins are used for propulsion, maneuvering, and thermoregulation and in offensive or defensive encounters with other males. More power to the flukes could increase the strength of blows and greater speed could aid in the herding of females.

Traditionally, behavioral ecologists have tended to emphasize the importance of male–male competition in the evolution of exaggerated male traits. More recently, based primarily on bird data, they have found that female choice often plays a critical role. Recordings of male song and the existence of exaggerated morphological traits that make adult males easily recognizable suggest the importance of female choice in marine mammals. The same features that appear to provide advantages in contests between males, such as large size, big canines, or deep roars, may also be used by females to select mates and/or may function to control or intimidate females (Wells *et al.*, 1999). Whether females actually use these traits to assess males or what these traits might signal (e.g., status, fitness, or readiness) is unknown. The enlarged postanal hump of males in some dolphins and porpoises may serve an important biomechanical function for males by facilitating copulation. It may also be important as a visual signal that makes adult males easily recognizable within schools, by both females and other males. Similarly, enlarged dorsal fins, which may have a thermoregulatory function, may also serve as a visual signal. The calls of male pinnipeds may function as male displays to females, in species recognition, and in contests between males. Evidence supporting the idea of lekking in walrus and dugongs suggests an increasing role for female choice in the evolution of vocal mating displays.

Caution is warranted when making inferences about the evolution of sexually dimorphic traits. First, our knowledge of sexual dimorphism across marine mammal taxa is incomplete. There are rarely encountered species for which we have very little information about sexual dimorphism. While our understanding of morphological differences between the sexes is growing, our knowledge of acoustic and pheromonal differences is in its infancy. As we fill in these gaps in our knowledge, our ability to understand the underlying evolutionary patterns and processes will increase. Second, a sexually dimorphic trait may have evolved for different reasons in different species. For example, among odontocetes, males are much larger than females in sperm whales, "resident" killer whales, and long-finned pilot whales (*Globicephala melas*), but it is unlikely that a single explanation fits all three cases. In sperm whales, adult males are solitary and roam great distances searching for females. Males possess large teeth, have massive heads, are scarred, and have been observed ramming each other head on. It is likely that large size serves male sperm whales well in contest competition over access to females. In contrast, adult male "resident" killer whales and long-finned pilot whales live in stable social groups with their maternal relatives, are not scarred, and we know of no accounts of aggressive interactions between

males. In contrast to sperm whales, "resident" male killer whales and long-finned pilot whales may increase their reproductive success, not only by mating with females in other pods, but by assisting kin in their natal pods. At this point, we can only speculate about the advantages that large size confers on males of these species, but assistance in a communal foraging strategy ("resident" killer whales) and protection of the pod (long-finned pilot whales) are possibilities. Females may prefer large males as mates in all these species, but large size may confer different advantages to individuals in each of the three cases.

Despite the technical difficulties of studying marine mammals, our understanding of the evolution of sexual dimorphism is increasing steadily. New techniques, such as scoring molecular genetic markers from tissue samples, are providing insight into social structure and variance in male reproductive success. Video, acoustic recordings, and "critter cams" (small television cameras that can be mounted on individual animals) are providing exciting new data on the behavior and interactions of animals underwater. Clearly, research opportunities abound, and the prospects for increased future understanding of the abundant sexually dimorphic traits in marine mammals are bright.

See Also the Following Articles

Coloration ■ Evolutionary Biology ■ Mating Systems ■ Song ■ Teeth

References

Alexander, R. D., Hoogland, J. L., Howard, R. D., Noonan, K. M., and Sherman, P. W. (1979). Sexual dimorphism and breeding systems in pinnipeds, ungulates, primates, and humans, *In* "Evolutionary Biology and Human Social Behaviour" (N. A. Chagnon and W. D. Irons, eds.), pp. 402–435. Duxbury Press, North Scituate, MA.

Andersson, M., (1994). "Sexual Selection." Princeton Univ. Press, Princeton, NJ.

Boness, D. J., Clapham, P. J., and Mesnick, S. L. (2000). Life history and reproductive behaviour. *In* "Marine Mammal Biology: An Evolutionary Approach" (A. R. Hoelzel, ed.), Blackwell Science, Oxford.

Brownell, R. L., and Ralls, K. (1986). Potential for sperm competition in baleen whales. Rep. Int. Whal. Comn. (Special issue 8). pp. 97–112.

Clapham, P. J. (1996). The social and reproductive biology of humpback whales: An ecological perspective. *Mamm. Rev.* **26**, 27–49.

Cranford, T. W. (1999). The sperm whale nose: Sexual selection on a grand scale? *Mar. Mamm. Sci.* **15**(4), 1133–1157.

Darwin, C. (1871). "The Descent of Man, and Selection in Relation to Sex." Murray, London.

Gomiendo, M., Harcourt, A. H., and Roldan, E. R. S. (1998). Sperm competition in mammals. *In* "Sperm Competition and Sexual Selection" (T. H. Birkhead and A. P. Møller, eds.), pp. 467–755. Academic Press, San Diego, CA.

Jefferson, T. A. (1990). Sexual dimorphism and development of external features in Dall's porpoise *Phocoenoides dalli. Fish. Bull.* **88**, 119–132.

Kenagy, G. J., and Trombulak, S. C. (1986). Size and function of mammalian testes in relation to body size. *J. Mammal.* **67**(1), 1–22.

Leatherwood, S., and Reeves, R. R. (1983). "The Sierra Club Handbook of Whales and Dolphins." Sierra Club Books, San Francisco.

MacLeod, C. D. (1998). Intraspecific scarring in odontocete cetaceans: An indicator of male "quality" in aggressive social interactions? *J. Zool. Lond.* **244**, 71–77.

Morejohn, G. V., Loeb, V., and Baltz, D. M. (1973). Coloration and sexual dimorphism in the Dall porpoise. *J. Mammal.* **54**(4), 977–982.

Ralls, K. (1976). Mammals in which females are larger than males. *Q. Rev. Biol.* **51**, 245–276.

Reeves, R. R., Stewart, B. S., and Leatherwood, S. (1992). "The Sierra Club Handbook of Seals and Sirenians." Sierra Club Books, San Francisco.

Riedman, M. (1990). "The Pinnipeds: Seals, Sea Lions and Walruses." University of California Press, Berkeley.

Tolley, K. A., Read, A. J., Wells, R. S., Urian, K. W., Scott, M. D., Irvine, A. B., and Hohn, A. A. (1995). Sexual dimorphism in wild bottlenose dolphins (*Tursiops truncatus*) from Sarasota, Florida. *J. Mammal.* **74**(4), 1190–1198.

Wells, R. S., Boness, D. J., and Rathbun, G. B. (1999). Behavior. *In* "Biology of Marine Mammals" (J. E. Reynolds III and S. A. Rommel, eds.). Smithsonian Institution Press, Washington, DC.

Shepherd's Beaked Whale
Tasmacetus shepherdi

JAMES G. MEAD
National Museum of Natural History,
Smithsonian Institution, Washington, DC

Shepherd's beaked whale is a rare animal, being known only from 21 strandings in the Southern Hemisphere. It is unusual in that it is the only beaked whale with a full set of functional teeth in both jaws.

I. Diagnostic Characters

Shepherd's beaked whale (*Tasmacetus shepherdi*, Oliver, 1937) is the only beaked whale to have a full set of functional teeth (Fig. 1). It has between 17 and 21 upper teeth on each side and between 18 and 28 lower teeth. The first tooth in the mandible is enlarged, and it is thought that size difference makes it homologous to the single teeth of other beaked whales. It was initially thought to be a primitive member of the beaked whale family (Ziphiidae) on the basis of retention of teeth. There has not been a good study of the evolutionary relationships of the beaked whales, but *Tasmacetus* seems in ways other than the teeth to be just as specialized as the remainder of the family. Adults are between 6 and 7 m in length. The only known calf was 340 cm long. This calf had plastic debris in its stomach, but it is not known whether it had begun to take normal solid food. The pigmentation pattern (Fig. 2) is dark gray dorsally, with a white field ventrally extending dorsally on both the anterior and posterior sides of the flipper. The dark dorsal field extends onto the flipper. Another extension of the white ventral field lies dorsal to the anus.

II. Distribution

The geographic records of *Tasmacetus* are illustrated in Fig. 3. It is known from strandings in New Zealand (12), Australia (1), Argentina (3), Juan Fernandez Islands (2), and the south Sandwich Islands. There are two published sightings attributed to this species: one from New Zealand and one from the Seychelles Islands. It is limited to temperate waters and may not ordinarily come as far north as the tip of Africa.

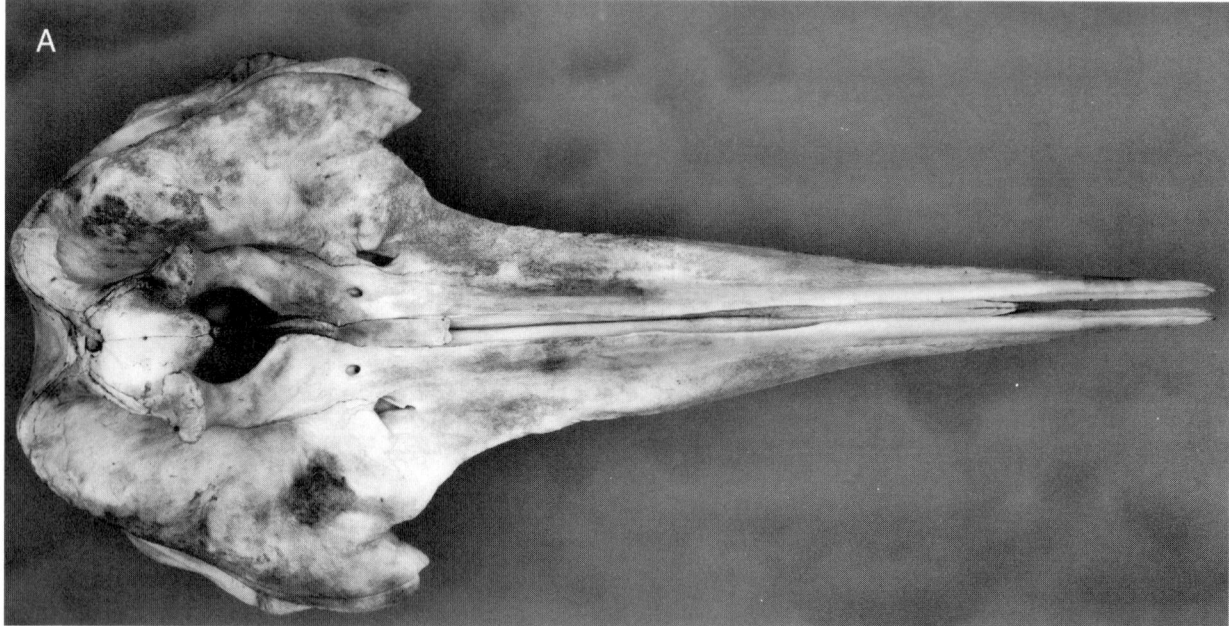

Figure 1 *(A) Dorsal view of skull, (B) ventral view of skull, (C) right lateral view of skull, and (D) right lateral view of mandible of a* Tasmacetus shepherdi *660-cm female. All figures are shot at the same scale; 10-cm scale bar is in D.*

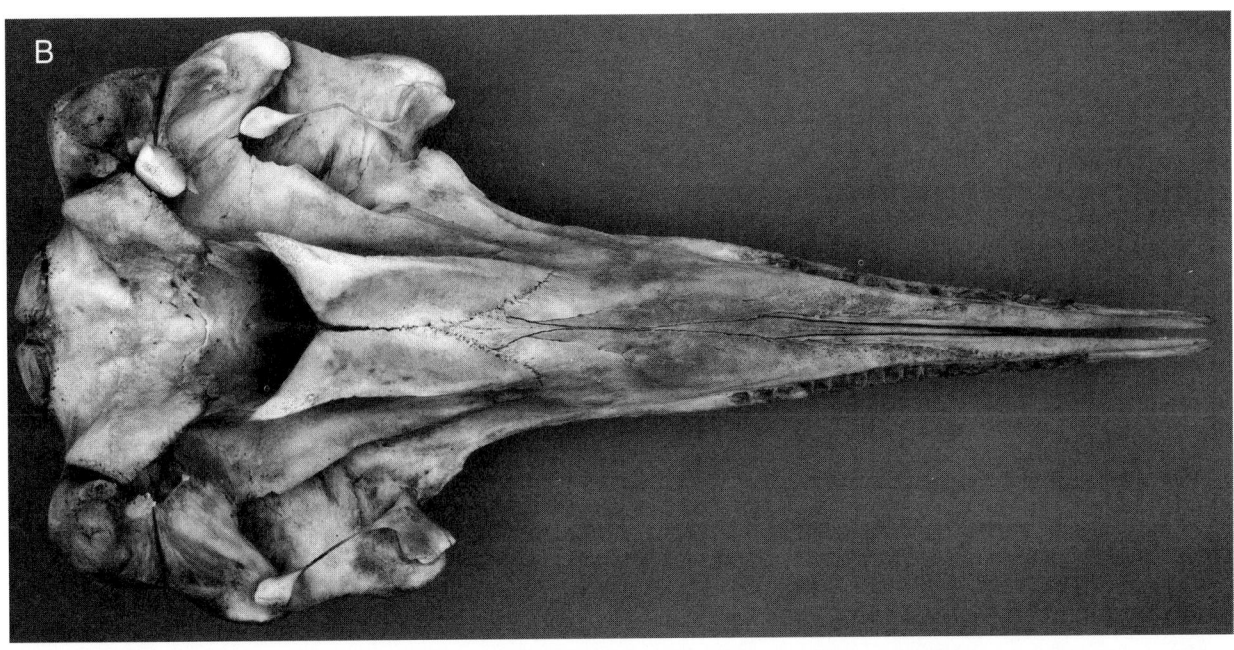

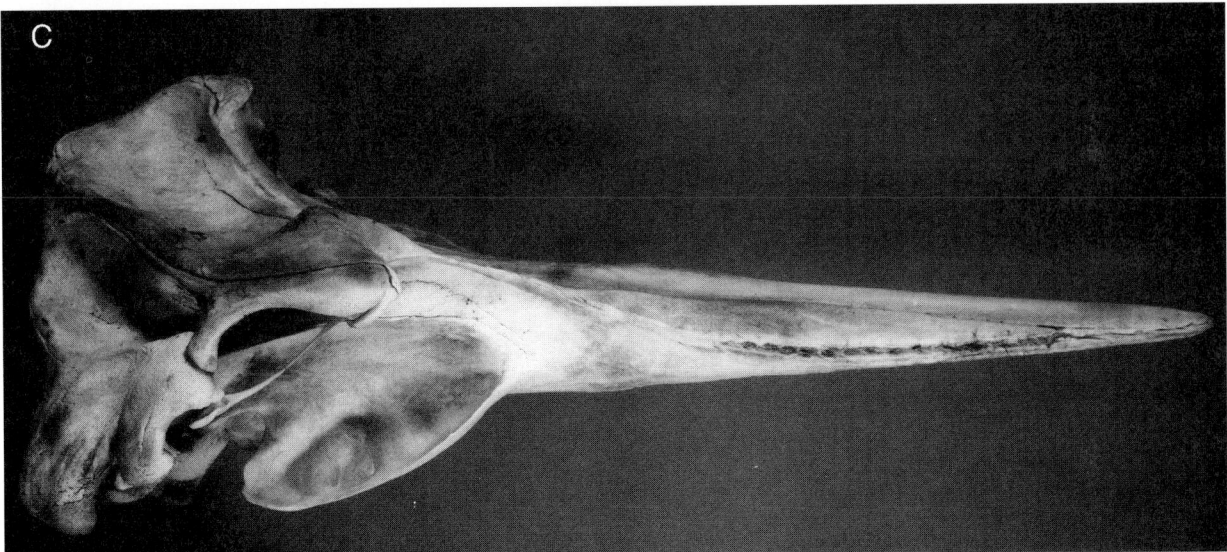

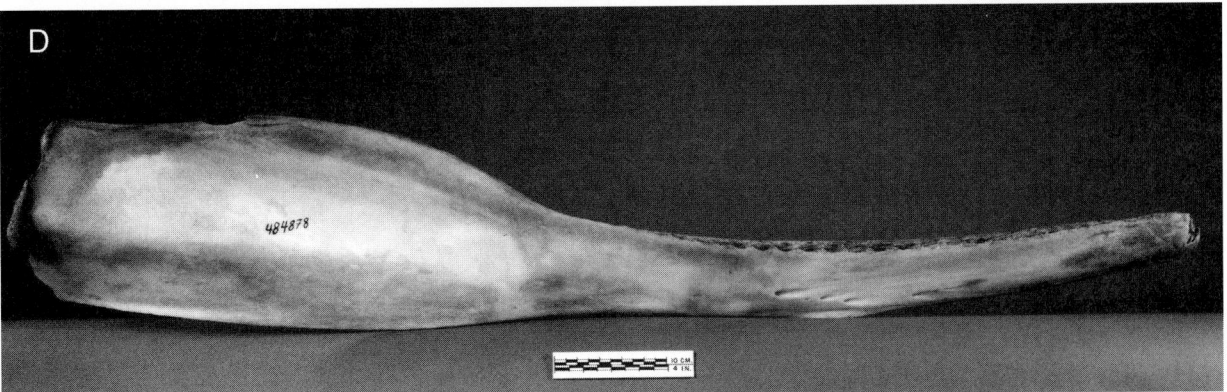

Figure 1   *(Continued)*

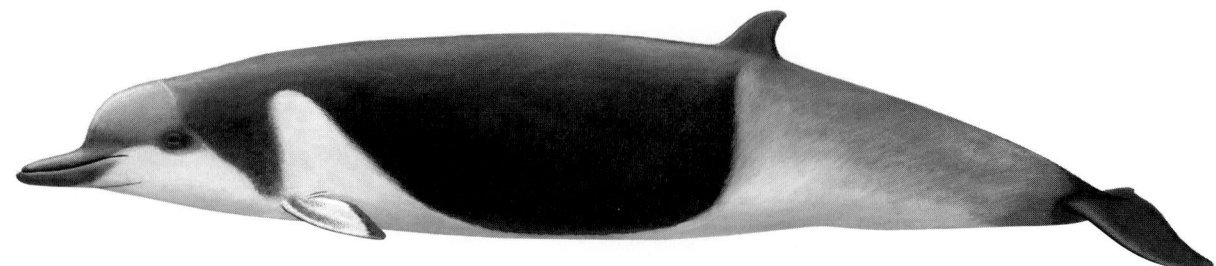

Figure 2   *External view of an adult female* T. shepherdi *(reconstruction by Pieter Folkens).*

III. Ecology (Habitat, Prey, and Predators)

This species presumably feeds primarily on squid and fish. One stranded adult female had traces of bottom fish, squid, and one small crab. All of the beaked whales that are moderately well known feed offshore in deep water.

IV. Interactions with Humans

Shepherd's beaked whale has not been known to be involved in any fisheries and certainly has not been kept in captivity. It was first known to science by a publication on a stranded adult female, which Shepherd found and collected in 1933.

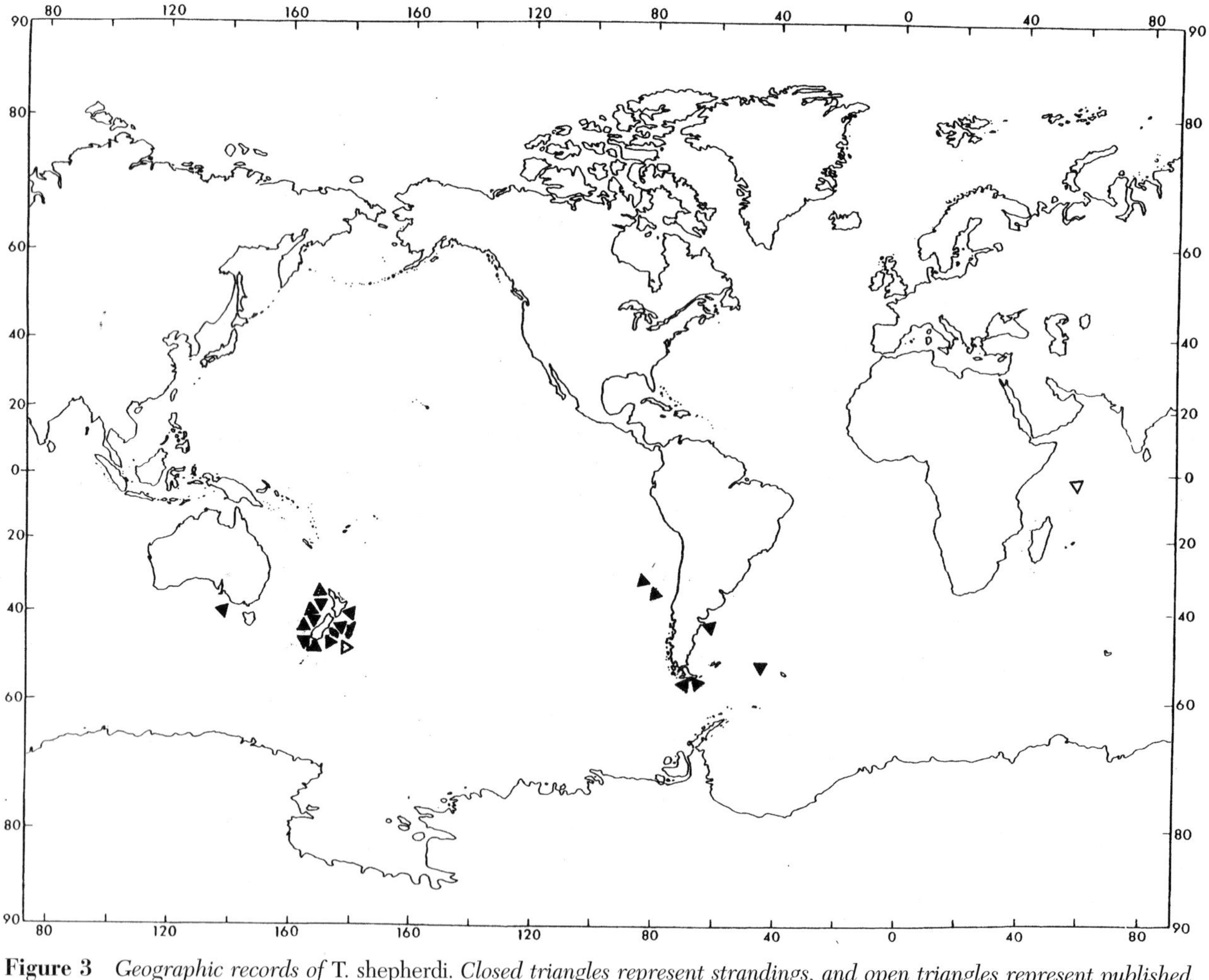

Figure 3   *Geographic records of* T. shepherdi. *Closed triangles represent strandings, and open triangles represent published sightings that are attributed to* T. shepherdi.

See Also the Following Articles

Beaked Whales, Overview ■ Teeth

References

Mead, J. G. (1989). Shepherd's beaked whale, *Tasmacetus shepherdi* Oliver, 1937. *In* "Handbook of Marine Mammals" (S. H. Ridgway and R. Harrison, eds.), Vol. 4, pp. 309–320. Academic Press, London.

Mead, J. G., and Payne, R. S. (1975). A specimen of the Tasman beaked whale, *Tasmacetus shepherdi*, from Argentina. *J. Mammal.* **56**, 213–218.

Signature Whistles

LAELA S. SAYIGH
University of North Carolina, Wilmington

Signature whistles were first described for several species of captive delphinids by David and Melba Caldwell in the 1960s. They found that individual dolphins, while isolated for medical attention, produced primarily one stereotyped individually distinctive whistle contour (Fig. 1). Other researchers had speculated that dolphin whistles might be context specific, i.e., with certain contours being produced in times of distress or to indicate the presence of food. However, it was difficult to support or refute these claims as dolphins do not make any visible movement associated with vocalization; thus it is not possible to identify who is vocalizing within a group. By recording individual dolphins in captivity, the Caldwells were able to get around this problem. Signature whistles have now been documented in more than 300 individual dolphins in a variety of locations, both captive and wild.

Since the Caldwells' pioneering work, most work on signature whistles has focused on the bottlenose dolphin, *Tursiops truncatus,* and it has focused on four major questions. First, do sounds produced during isolation occur in the typical vocal repertoire of dolphins? Second, how do signature whistles function in the natural communication system of dolphins? Third, what is the role of whistle copying in the natural communication system of dolphins? Fourth, how do signature whistles develop in young animals?

The first two of these questions were raised in a study of 10 captive dolphins by McCowan and Reiss (1995), in which no evidence was found for stereotyped signature whistles. However, Janik and Slater (1998) found that captive dolphins tended to produce signature whistles primarily when one group member was voluntarily isolated from the rest of the group. Their study provided strong support for the use of signature whistles as cohesion calls, a finding that was also supported by a study of wild dolphins by Smolker and colleagues (1993). Thus, dolphins that are in close proximity and in visual contact, such as in the McCowan and Reiss study, may not need to produce sig-

nature whistles. Playback experiments with temporarily captured dolphins in Sarasota, Florida, also showed that dolphins can discriminate among different signature whistles and that they are thus likely to function in individual recognition (Sayigh *et al.,* 1999). Again, for captive dolphins that are housed in close proximity to one another, there may be little need for such a function.

Additional evidence in support of the idea that whistles produced during isolation also occur in the typical vocal repertoire of dolphins came from a study of vocal production in free-ranging dolphins in Sarasota, Florida (Hill and Sayigh, 1999). Approximately 50% of whistles produced during normal activities matched signature whistles that had been produced by the same individuals during temporary capture. The other 50% of whistles were comprised by various contours, many of which took on the general form of an upsweep. The functions of these whistles are unknown.

The first study to address copying, or imitation, of signature whistles was conducted by Peter Tyack (1986). He used a telemetry device called a "vocalight," which dolphins wore on their heads and which lit up when a sound was produced, and found that socially interactive dolphins not only produce signature whistles, but that they also produce imitations of each other's signature whistles. Such copying of signature whistles has now been documented in several other studies. Two of these studies (Smolker and Pepper, 1999; Watwood *et al.,* 1999) have found that close male associates tend to converge on a similar whistle type. New technologies that enable localization of vocalizing animals are necessary to study how both signature and nonsignature whistles (including whistle copies) are used in the natural communication system of dolphins.

In addition to discovering signature whistles, David and Melba Caldwell were the first researchers to investigate vocal development in bottlenose dolphins. They found that young dolphins produce tremulous, quavery whistles and then gradually converge on a stereotyped whistle contour during their first year of life. This contour then usually remains stable throughout the life of the animal. In a study of signature whistles produced by dolphins in Sarasota, Florida, Sayigh *et al.* (1990, 1995) found that female calves tended to produce whistles that were highly distinct from those of their mothers. Males were more likely than females to produce whistles similar to those of their mothers. These findings raised the possibility that dolphins may be learning their signature whistles from conspecifics.

Much evidence now supports the idea that dolphins learn their signature whistles, which sets them apart from many other nonhuman mammals, in which vocal production is largely innate. However, little is known about what factors govern the "choice" of whistle contours by calves. Again, studies that utilize acoustic localization with concurrent behavioral observations promise to shed light on this interesting question.

Signature whistles are an important class of vocalizations produced by bottlenose dolphins. They function both in individual recognition and in maintaining group cohesion, and they appear to be learned. Much work remains to be done in order to learn how these whistles develop and how they are used in the natural COMMUNICATION system of dolphins.

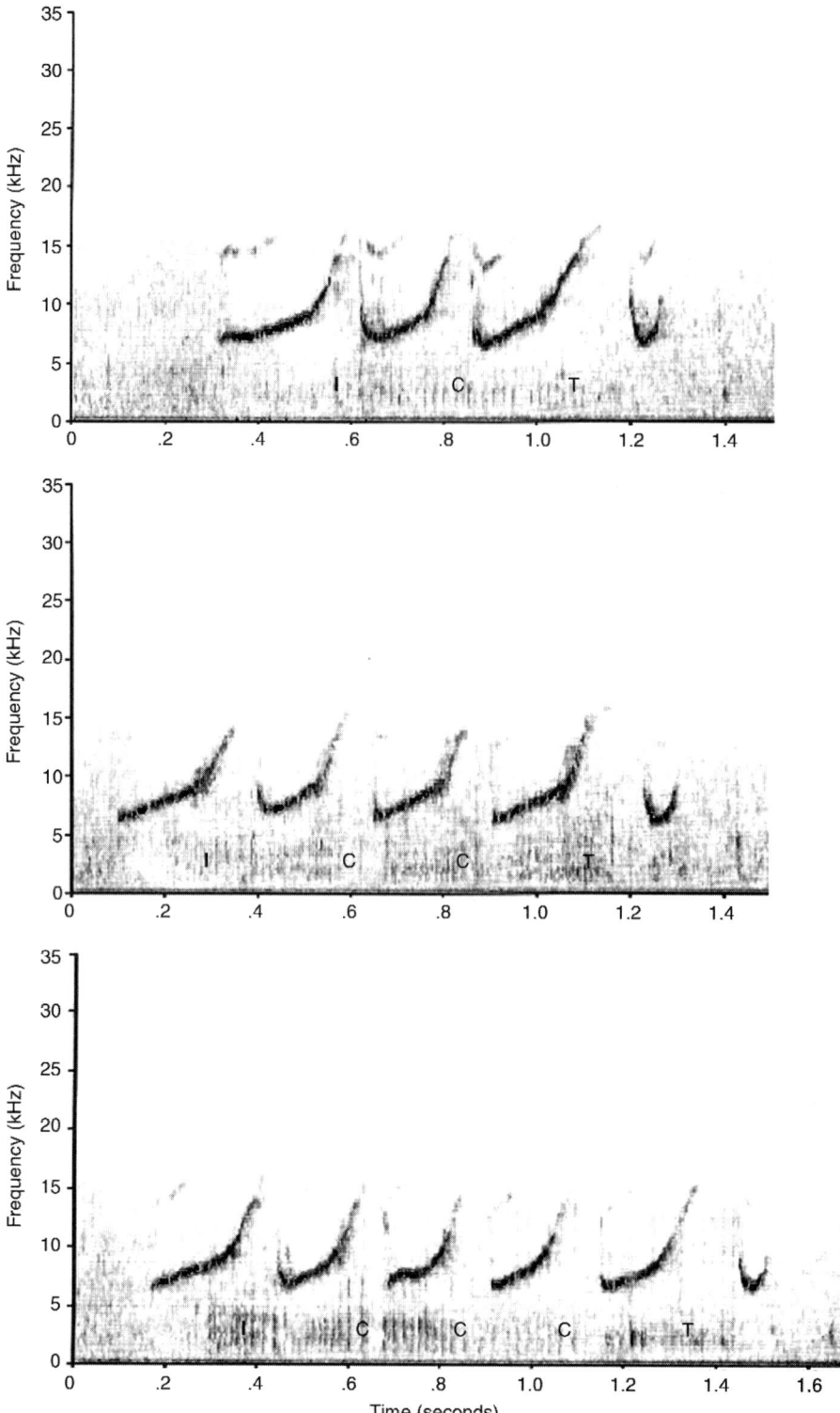

Figure 1 *Examples of a "multilooped" signature whistle of an adult female bottlenose dolphin recorded during temporary capture in Sarasota, Florida, in 1984. Her whistle typically consists of an introductory loop (labeled "I"), a terminal loop (labeled "T"), and a varying number of central loops (labeled "C"). Shown are whistles with one, two, and three central loops. Spectrograms were made with SIGNAL (Engineering Design, Belmont, MA) at a sample rate of 80,000 Hz and a 512 point FFT.*

See Also the Following Articles

Bottlenose Dolphins ▪ Dialects ▪ Mimicry ▪ Song

References

Caldwell, M. C., and Caldwell, D. K. (1965). Individualized whistle contours in bottlenosed dolphins (*Tursiops truncatus*). *Nature* **207,** 434–435.

Caldwell, M. C., Caldwell, D. K., and Tyack, P. L. (1990). Review of the signature whistle hypothesis for the Atlantic bottlenose dolphin, *Tursiops truncatus. In* "The Bottlenose Dolphins" (S. Leatherwood and R. Reeves, eds.), pp. 199–234. Academic Press, New York.

Hill, M., and Sayigh, L. (1999). Quantification of signature whistle production by free-ranging bottlenose dolphins (*Tursiops truncatus*). Proceedings of the 13th Biennial Conference on the Biology of Marine Mammals, Maui, Hawaii. [Abstract]

Janik, V. M., and Slater, P. J. B. (1998). Context-specific use suggests that bottlenose dolphin signature whistles are cohesion calls. *Anim. Behav.* **56,** 829–838.

McCowan, B., and Reiss, D. (1995). Quantitative comparison of whistle repertoires from captive adult bottlenose dolphins (Delphinidae, *Tursiops truncatus*): A reevaluation of the signature whistle hypothesis. *Ethology* **100,** 194–209.

Sayigh, L. S., Tyack, P. L., Wells, R. S., and Scott, M. D. (1990). Signature whistles of free-ranging bottlenose dolphins, *Tursiops truncatus:* Stability and mother-offspring comparisons. *Behav. Ecol. Sociobiol.* **26,** 247–260.

Sayigh, L. S., Tyack, P. L., Wells, R. S., Scott, M. D., and Irvine, A. B. (1995). Sex difference in signature whistle production of free-ranging bottlenose dolphins, *Tursiops truncatus. Behav. Ecol. Sociobiol.* **36,** 171–177.

Sayigh, L. S., Tyack, P. L., Wells, R. S., Solow, A., Scott, M. D., and Irvine, A. B. (1999). Individual recognition in wild bottlenose dolphins: A field test using playback experiments. *Anim. Behav.* **57,** 41–50.

Smolker, R., Mann, J., and Smuts, B. (1993). Use of signature whistles during separations and reunions between bottlenose dolphin mothers and infants. *Behav. Ecol. Sociobiol.* **33,** 393–402.

Smolker, R., and Pepper, J. (1999). Whistle convergence among allied male bottlenose dolphins (Delphinidae, *Tursiops* sp.). *Ethology* **105,** 595–617.

Tyack, P. L. (1986). Whistle repertoires of two bottlenosed dolphins, *Tursiops truncatus:* Mimicry of signature whistles? *Behav. Ecol. Sociobiol.* **18,** 251–257.

Watwood, S., Tyack, P., and Wells, R. (1999). Signature whistle sharing between allied male bottlenose dolphins, *Tursiops truncatus.* Proceedings of the 13th Biennial Conference on the Biology of Marine Mammals, Maui, Hawaii. [Abstract]

Sirenian Evolution

DARYL P. DOMNING
Howard University, Washington, DC

Sirenia are the order of placental mammals comprising modern sea cows (manatees and dugongs) and their extinct relatives. They are the only herbivorous marine mammals now living, and the only herbivorous mammals ever to have become totally aquatic. Sirenians have a known fossil record extending over some 50 million years (early Eocene–Recent). They attained a modest diversity in the Oligocene and Miocene, but since then have declined as a result of climatic cooling, other oceanographic changes, and human depredations. Only two genera and four species survive today: the three species of manatees (*Trichechus*) live along the Atlantic coasts and rivers of the Americas and West Africa; one, the Amazonian manatee, is found only in fresh water. The dugong (*Dugong*) lives in the Indian and southwest Pacific oceans. [For comprehensive references to technical as well as popular publications on fossil and living sirenians, see Domning (1996).]

I. Sirenian Origins

The closest living relatives of sirenians are Proboscidea (elephants). The Sirenia, the Proboscidea, the extinct Desmostylia, and probably the extinct Embrithopoda together make up a larger group called Tethytheria, whose members (as the name indicates) appear to have evolved from primitive hoofed mammals (condylarths) in the Old World along the shores of the ancient Tethys Sea. Together with Hyracoidea (hyraces), tethytheres seem to form a more inclusive group long referred to as Paenungulata. The Paenungulata and (especially) Tethytheria are among the least controversial groupings of mammalian orders and are strongly supported by most morphological and molecular studies. Their ancestry is remote from that of cetaceans or pinnipeds; sirenians reevolved an aquatic lifestyle independently of (though simultaneously with) cetaceans, ultimately displaying strong convergence with them in body form.

II. Early History, Anatomy, and Mode of Life

Sirenians first appear in the fossil record in the early Eocene, and the order was already diverse by the middle Eocene (Fig. 1). As inhabitants of rivers, estuaries, and nearshore marine waters, they were able to spread quickly along the coasts of the world's shallow tropical seas; in fact, the most primitive sirenian known to date (*Prorastomus*) was found not in the Old World but in Jamaica.

The earliest sea cows (families Prorastomidae and Protosirenidae, both confined to the Eocene) were pig-sized, four-legged amphibious creatures. By the end of the Eocene, with the appearance of the Dugongidae, sirenians had taken on their modern, completely aquatic, streamlined body form, featuring flipper-like front legs, no hind legs, and a powerful tail with a horizontal caudal fin, whose up-and-down movements propel them through the water, as in whales and dolphins (Fig. 2). The last-appearing of the four sirenian families (Trichechidae) apparently arose from early dugongids in the late Eocene or early Oligocene. The sirenian fossil record now documents all the major stages of hindlimb and pelvic reduction from completely "terrestrial" morphology to the extremely reduced condition of the pelvis seen in modern manatees, thereby providing one of the most dramatic examples of evolutionary change to be seen among fossil vertebrates.

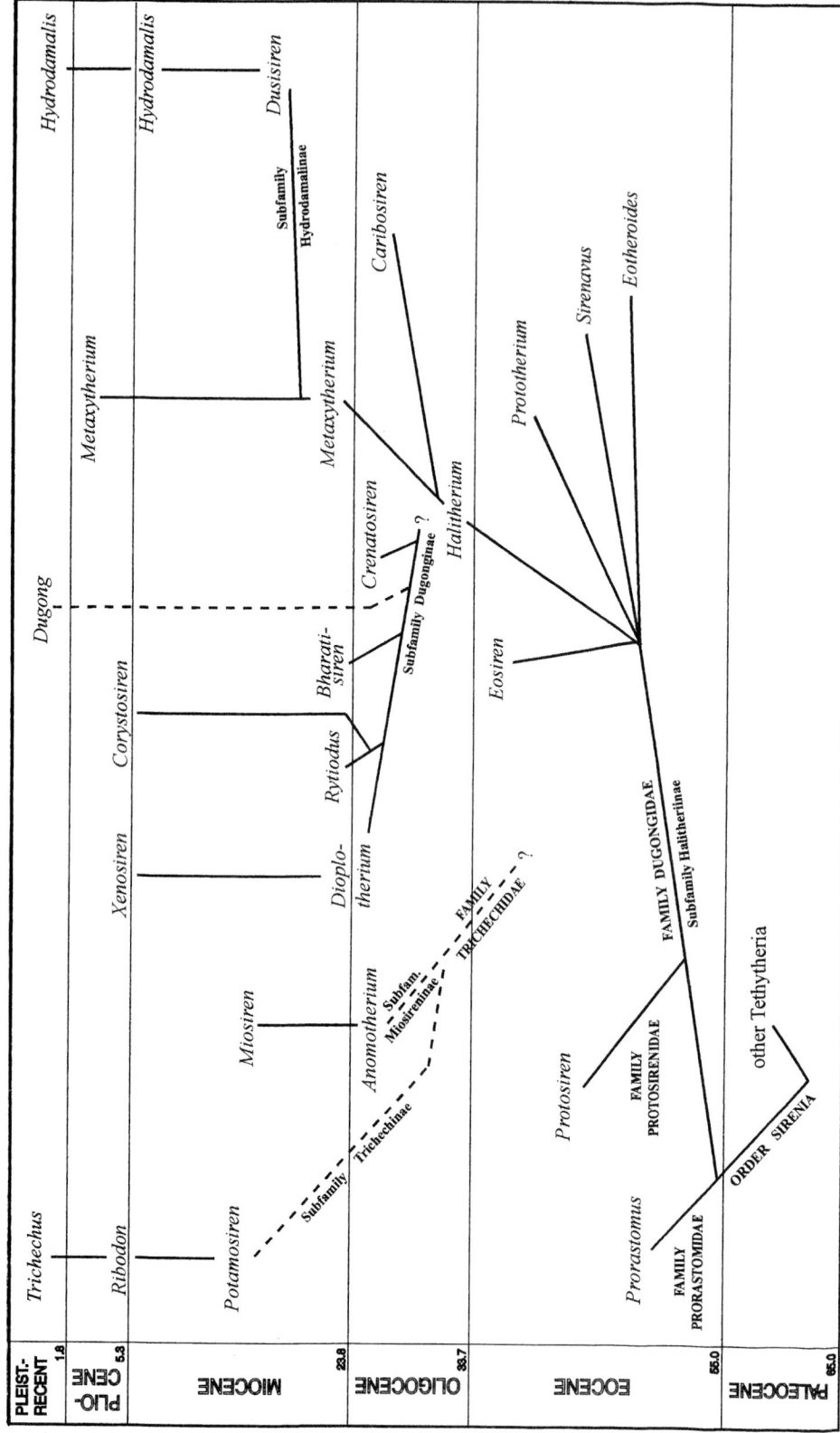

Figure 1 Simplified phylogeny of Sirenia, including only better-known genera. The time scale (at left) is in millions of years. "Ghost lineages" (ancestral groups undocumented by fossils) that span multiple epochal boundaries are shown as dashed lines.

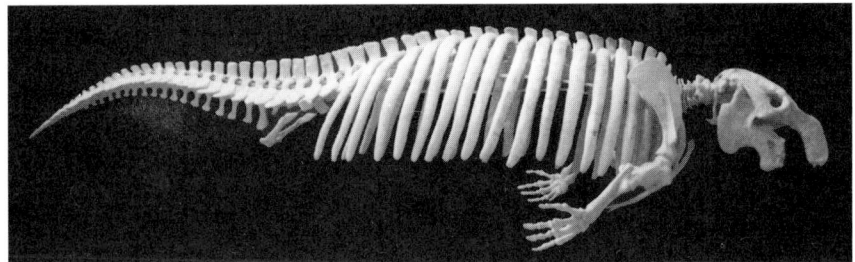

Figure 2 *Skeleton of* Metaxytherium floridanum, *a Miocene halitheriine dugongid. Total length about 3.2 m. After Domning (1988); reproduced with the permission of the* Journal of Vertebrate Paleontology.

From the outset, sirenians were herbivores and probably depended on seagrasses and other aquatic angiosperms (flowering plants) for food. To this day, almost all members of the order have remained tropical, marine, and eaters of angiosperms. No longer capable of LOCOMOTION on land, sirenians are born in the water and spend their entire lives there. Because they are shallow divers with large lungs, they have heavy skeletons, like a diver's weight belt, to help them stay submerged: their bones are both swollen (pachyostotic) and dense (osteosclerotic), especially the ribs, which are often found as fossils.

The sirenian SKULL is characterized by an enlarged and more or less downturned premaxillary rostrum, retracted nasal opening, absence of paranasal air sinuses, laterally salient zygomatic arches, and thick, dense parietals fused into a unit with the supraoccipital. Nasals and lacrimals tend to become reduced or lost, and in most forms the pterygoid processes are large and stout. The periotic is snugly enclosed by a socket in the squamosal and is fused with a ring-shaped tympanic. The mandibular symphysis is long, deep, laterally compressed, and typically fused and downturned; in all but prorastomids the mandibular foramen is enlarged to expose the dental capsule. Incisors, where present, are arranged in parallel, longitudinally aligned rows. In all but the most primitive taxa, the infraorbital and mental foramina are enlarged to accommodate the nerve and blood supply to the large, prehensile, vibrissae-studded lips, which are moved by muscular hydrostats (cf. Marshall *et al.*, 1998).

Eocene sirenians, like Mesozoic mammals but in contrast to other Cenozoic ones, have five instead of four premolars, giving them a 3.1.5.3 dental formula. Whether this condition is truly a primitive retention in the Sirenia is still being debated. The fourth lower deciduous premolar (dp4) is trilobed, like that of many other ungulates; this raises the further unresolved question of whether the three following teeth (dp5, m1, and m2) are actually the homologues of the so-called m1–3 in other mammals.

Although the cheek teeth are relied on for identifying species in many other mammalian groups, they do not vary much in morphology among Sirenia but are almost always low-crowned (brachyodont) with two rows of large, rounded cusps (bunobilophodont). (The most taxonomically informative parts of the sirenian skeleton are the skull and mandible, especially the

frontal and other bones of the skull roof; Fig. 3.) Except for a pair of tusk-like first upper incisors seen in most species, front teeth (incisors and canines) are lacking in all but the earliest fossil sirenians, and cheek teeth in adults are commonly reduced in number to four or five on each side of each jaw: one or two deciduous premolars, which are never replaced, plus three molars. As described later, however, all three of the Recent genera have departed in different ways from this "typical" pattern.

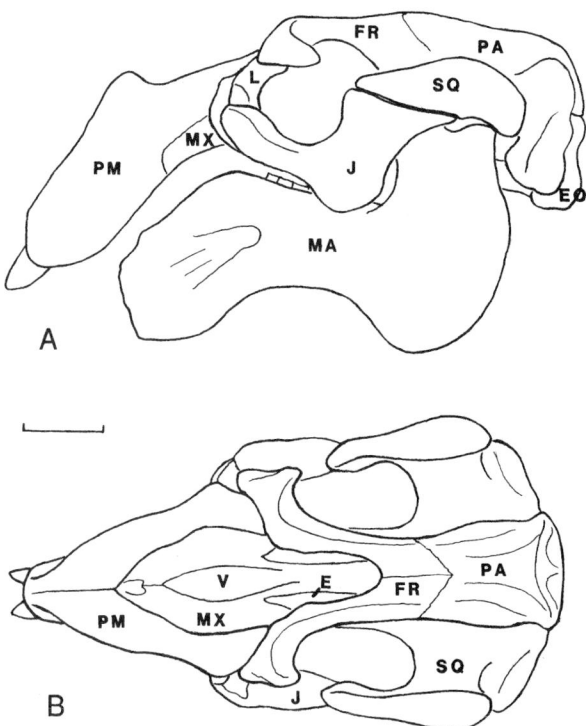

Figure 3 *Skull of* Crenatosiren olseni, *an Oligocene dugongine dugongid, in (A) lateral and (B) dorsal views. Note the large incisor tusks in the premaxillae. E, ethmoid; EO, exoccipital; FR, frontal; J, jugal; L, lacimal; MA, mandible; MX, maxilla; PA, parietal; PM, premaxilla; SQ, squamosal; V, vomer. Scale bar: 5 cm. After Domning (1997); reproduced with the permission of the* Journal of Vertebrate Paleontology.

III. Dugongidae

Dugongids comprise the vast majority of the species and specimens that make up the known fossil record of sirenians. The basal members of this very successful family are placed in the long-lived (Eocene–Pliocene) and cosmopolitan subfamily Halitheriinae (Fig. 1). This paraphyletic group included the well-known fossil genera *Halitherium* and *Metaxytherium*, which were relatively unspecialized seagrass eaters.

Metaxytherium (Fig. 2) gave rise in the Miocene to the Hydrodamalinae, an endemic North Pacific lineage that ended with Steller's sea cow (*Hydrodamalis*)—the largest sirenian that ever lived (up to 9 m or more in length) and the only one to adapt successfully to temperate and cold waters and a diet of marine algae. It was completely toothless, and its truncated, claw-like flippers, used for gathering plants and fending off from rocks, contained no finger bones (phalanges). It was hunted to extinction for its meat, fat, and hide circa A.D. 1768.

Another offshoot of the Halitheriinae, the subfamily Dugonginae, appeared in the Oligocene (Fig. 3). Most dugongines were apparently specialists at digging out and eating the tough, buried rhizomes of seagrasses; for this purpose many of them had large, self-sharpening blade-like tusks (Domning, 2001). The modern *Dugong* is the sole survivor of this group, but it has reduced its dentition (the cheek teeth have only thin enamel crowns, which quickly wear off, leaving simple pegs of dentine) and has (perhaps for that reason) shifted its diet to more delicate seagrasses and ceased to use its tusks for digging.

IV. Trichechidae

Trichechidae have a much less complete FOSSIL RECORD than dugongids. Their definition has been broadened by Domning (1994) to include Miosireninae, a peculiar and little-known pair of genera that inhabited northwestern Europe in the late Oligocene and Miocene (Fig. 1). Miosirenines had massively reinforced palates and dentitions that may have been used to crush shellfish. Such a diet in sirenians living around the North Sea seems less surprising when we consider that modern dugongs and manatees near the climatic extremes of their ranges are known to consume invertebrates in addition to plants.

Manatees in the strict, traditional sense are now placed in the subfamily Trichechinae. They first appeared in the Miocene, represented by *Potamosiren* from freshwater deposits in Colombia. Indeed, much of trichechine history was probably spent in South America, whence they spread to North America and Africa only in the Pliocene or Pleistocene.

During the late Miocene, manatees living in the Amazon basin evidently adapted to a diet of abrasive freshwater grasses by means of an innovation still used by their modern descendants: they continue to add on extra teeth to the molar series as long as they live, and as worn teeth fall out at the front, the whole tooth row slowly shifts forward to make room for new ones erupting at the rear. This type of horizontal tooth replacement has often been likened, incorrectly, to that of elephants, but the latter are limited to only three molars. Only one other mammal, an Australian rock wallaby (*Peradorcas*

concinna), has truly evolved the kind of tooth replacement seen in manatees.

See Also the Following Articles

Dugong ▪ Manatees ▪ Musculature ▪ Steller's Sea Cow

References

Domning, D. (1982). Manatee evolution: A speculative history. *J. Paleontol.* **56,** 599–619.

Domning, D. (1988). Fossil Sirenia of the West Atlantic and Caribbean region. I. *Metaxytherium floridanum* Hay, 1992. *Journal of Vertebrate Paleontology* **8,** 395–426.

Domning, D. (1994). A phylogenetic analysis of the Sirenia. *Proc. San Diego Soc. Nat. Hist.* **29,** 177–189.

Domning, D. (1996). Bibliography and index of the Sirenia and Desmostylia. *Smith. Contrib. Paleobiol.* **80,** 1–611.

Domning, D. (1997). Fossil Sirenia of the West Atlantic and Caribbean region. VI. *Crenatosiren olseni* (Reinhart, 1976). *Journal of Vertebrate Paleontology* **17,** 397–412.

Domning, D. (2001). Sirenians, seagrasses, and Cenozoic ecological change in the Caribbean. *Palaeogeogr. Palaeoclimatol. Palaeoecol.* **166,** 27–50.

Domning, D., and de Buffrénil, V. (1991). Hydrostasis in the Sirenia: Quantitative data and functional interpretations. *Mar. Mamm. Sci.* **7,** 331–368.

Marshall, C., Clark, L., and Reep, R. (1998). The muscular hydrostat of the Florida manatee (*Trichechus manatus latirostris*): A functional morphological model of perioral bristle use. *Mar. Mamm. Sci.* **14,** 290–303.

Reynolds, J., III, and Odell, D. (1991). "Manatees and Dugongs." Facts on File, New York.

Savage, R., Domning, D., and Thewissen, J. (1994). Fossil Sirenia of the West Atlantic and Caribbean region. V. The most primitive known sirenian, *Prorastomus sirenoides* Owen, 1855. *J. Vertebr. Paleontol.* **14,** 427–449.

Sirenian Life History

DANIEL K. ODELL
SeaWorld, Inc., Orlando, Florida

Sirenians are unique among the marine mammals in that they are herbivores. The mammalian order Sirenia has fossil representatives on the order of 50 million years old (Eocene epoch), making them as old as the cetaceans. While there are similarities between the two groups in terms of adaptation to the aquatic environment (e.g., streamlining, loss of hindlimbs, use of the tail for propulsion), the sirenians stayed on the path to herbivory while the cetaceans switched from the herbivorous habits of their terrestrial ancestors to become carnivores. Modern sirenians include three species of MANATEES (family Trichechidae: Amazonian manatee, *Trichechus inunguis;* West African manatee, *T. senegalensis;* West Indian manatee, *T. manatus;* the latter is divided into two subspecies: Antillean manatee, *T. m. manatus,* and Florida manatee, *T. m. latirostris*). The Sirenia also includes the family Dugongidae (dugong, *Dugong dugon*), and STELLER'S SEA COW (family Dugongidae: Steller's sea cow, *Hydrodamalis gigas*). The manatees, as their common names suggest, are distributed in the coastal tropical and subtropical regions of the Atlantic Ocean. The dugong is found in the South Pacific and Indian Ocean regions. The manatees and the dugong are considered ENDANGERED or threatened under various national laws and international conservation schemes. The Steller's sea cow is the exception, not only is it extinct (eliminated by overhunting about 25 years after its discovery in 1765) but it inhabited the islands in the western reaches of the Aleutian archipelago in the North Pacific Ocean. Definitely not a tropical clime!

I. Methodology

Robert Ricklefs (1973) stated that "The life history of an organism can be described in terms of fecundity, growth and development, age at [sexual] maturity, parental care, and longevity." These parameters are closely linked and are acted on by natural selection. While these parameters are easy to state, they are not always easy to measure. It is easy to study a colony of mice in the laboratory but not so with an animal like the Florida manatee that may reach a length of over 4 m and a weight in excess of 1500 kg. How, then, does one get this information for sirenians? Approaches to this question involve both field and "laboratory" (i.e., animals in captivity in marine zoological parks and zoos) studies. In the field, one can learn to identify individual animals using shape, size, and various markings and color patterns or one can mark (using various methods) individual animals. For large, long-lived, aquatic mammals like the sirenians, this becomes an extremely expensive and time-consuming operation. However, in many cases, it is the only way to obtain the desired information. Alternatively, one can collect data from animals that die from both natural and unnatural (i.e., human-related) causes. Animals killed for human consumption or those killed as a result of other human activities provide a sample, albeit potentially biased (nonrandom), of males and females of all age classes. Marine mammal stranding networks collect life history data on animals found dead (i.e., stranded, beach cast, beached). In Florida, the Florida Fish and Wildlife Conservation Commission operates a network dedicated to the collection and examination of all manatees found dead in the state. From these carcasses one can gather information on, among others, reproductive status (immature, mature, pregnant, number of pregnancies) and age. Estimating the age of an animal is critically important in the estimation of life history parameters. Toothed cetaceans and pinnipeds, for example, have permanent teeth with roots that grow continually (but slowly) from birth to death. One can section these teeth and, under the microscope, count growth layer groups (GLGs). If one then knows the frequency with which these GLGs are deposited (e.g., one per year, 2 per year), GLG counts can be directly converted to and estimated age in years. The frequency of GLG deposition can be estimated using known age animals and using chemical markers deposited in the teeth of live animals. There is variability in the reliability of this process from species to species and the only way to know the true age of an animal is to follow it from birth to death.

Growth layers are also often deposited in growing bone. The dugong has tusks that erupt through the gums in the male but not in the female. These tusks grow throughout the life of the animal and GLGs accumulate on an annual (assumed) basis. Manatees, however, do not have permanent teeth. While some growth layers are present in the tooth roots in young animals, these are lost through resorption, as manatees have a continual, horizontal (back to front) tooth replacement throughout life. However, because other growing bones accumulate GLGs, scientists discovered that GLGs in manatee ear bones can be used to estimate age. Studies with chemical markers demonstrated that manatee GLGs accumulate on an annual basis.

Studies with dead animals are very useful and relatively inexpensive, but long-term studies on live animals are necessary to fill in the life history blanks. For example, by following a female from year to year, it is possible to document when she has a calf and how much time elapses between successive calves. While radio and satellite tracking allows scientists to follow individual manatees over long periods of time, it is virtually impossible to watch them from minute to minute. However, studies on manatees in captivity (aka "the laboratory") can fill some of these life history gaps, particularly in the areas of reproductive biology and growth and development. In captivity manatees *can* be observed continually if necessary. Blood and urine samples can be collected to document estrous cycling and gestation in females and testicular activity cycling in males. Parental care of the offspring can be documented in great detail.

II. Longevity

Detailed age estimation studies have only been done for the Florida manatee and for the dugong. Using tetracycline marking, we know that Florida manatees deposit one GLG per year. Based on growth layer groups in the ear bone complex, some Florida manatees may live to be 60 years of age. However, due to bone resorption, 60 years may be a minimum estimate. A manatee born in the Miami Aquarium in 1948 is alive and well today (August 2000) at an aquarium in Bradenton, Florida. This animal provides a good measure of validation for the 60-year estimate. The average longevity for Florida manatees has not been well estimated but is probably on the order of 30 years less. This estimate may be biased downward by manatee deaths resulting from human activities (e.g., watercraft collisions). Average longevity may vary between sexes and among year classes. Longevity data do not exist for Antillean, Amazonian, or West African manatees.

Based on GLGs in their tusks, the oldest dugong examined to date was a female with 73 GLGs, which translates to 73 years if we assume one GLG per year. Tusks in male dugongs erupt through the gums and are worn down, eliminating the early GLGs in older animals. Minimum maximum longevity estimates for male dugongs are about 35 years.

III. Age at Sexual Maturity

Female Florida manatees reach sexual maturity between 2.5 and 6 years of age. Male Florida manatees reach sexual maturity between 2 and 11 years of age based on the presence of active spermatogenesis in the testes. It is important to note that even if a male manatee is sexually mature at an early age, it may not be physically large enough to compete for mating rights in an estrous (mating) herd. However, under circumstances wherein there are not other competing males, small manatees may be able to mate successfully and produce offspring. There are no data on age at sexual maturity for Antillean, Amazonian, or West African manatees.

Even though the estimated maximum life expectancy for the dugong is similar to that of the Florida manatee, dugongs appear to mature at greater ages, and the age at sexual maturity may vary among populations. Female dugongs in Australia and Papua New Guinea mature at 9.5 to 17.5 years. Male dugongs in the same areas mature at 9 to 16 years of age. These differences could reflect distinct genetic differences between populations or the effects of density-dependent factors.

IV. Parental Care

In both manatees and dugongs the male plays no apparent role in the care of the young. After mating, males and females go their separate ways. Florida manatee calves will stay with their mothers for 1–2 years on average. Most calves (about 70%) stay with their mothers through one winter season and the remainder through two winter seasons. Nutritional weaning is a gradual process but there are few data on the length of lactation in the Florida manatee or in the dugong. Lactation may last 1.5 years in dugongs and is probably similar in the manatee. Some wild and captive manatees appear to lactate for several years under certain circumstances. Most Florida manatee and dugong calves start feeding on vegetation at a few weeks of age and may be nutritionally independent before they reach the age of 1 year. However, at least in the case of the Florida manatee, calves probably learn the locations of feeding areas and warm water refugia by following their mothers through at least one winter season. The latter is extremely important for weaned juvenile survival during cold winters.

V. Growth and Development

Florida manatee calves average about 120 cm in length at birth, but viable calves may have a birth length ranging from about 80 to 160 cm. This, along with variable individual growth rates, results in a highly variable length at age distribution. For example, 2-year-old manatees at Blue Spring (Florida) may range from 210 to 260 cm total length. From a sample of carcasses of Florida manatees in the age class >1 and <2 years, total lengths ranged from about 120 to over 260 cm. Florida manatees grow rapidly during the first few years, and body length becomes asymptotic at about 300 cm and an age of 8–10 years. The birth weight for Florida manatees ranges from 30 to 50 kg and the average adult body weight is about 500 kg. The Antillean and West African manatees are probably similar to the Florida manatee in these respects. The Amazonian manatee is smaller and has a birth length of 85–100 cm and a birth weight of 10–15 kg. Large adults may reach lengths of 280 cm and weights of 480 kg. The dugong has a birth length of 100–130 cm and a birth weight of 25–35 kg. Adult dugongs average 270 cm in length and 250–300 kg body weight.

VI. Fecundity

As with other aspects of sirenian life history, data on fecundity are limited and based primarily on studies of the Florida manatee and the dugong. A key factor in assessing fecundity is the gestation period, which, despite numerous conceptions and births of Florida manatees, is not known for any sirenian species. However, scientists generally agree that gestation for the Florida manatee and the dugong is in the range of 12–14 months and that the other species of sirenians are probably similar. The known interbirth interval for wild Florida manatees averages 2.5–2.6 years. The estimated interbirth interval for Florida manatees ranges from 2.5 to 3.0 years when gestation period estimates of 12, 13, and 14 months are applied. This suggests that the true gestation period may be close to 12 months. Estimated interbirth intervals for the dugong range from 2.7 to 5.8 years depending on the length of the gestation period assumed (12, 13, or 14 months) and the population of dugongs used. We do know that Florida manatees resume estrous cycling within 1–2 months after the loss of a calf and become pregnant shortly thereafter. Whereas manatees and dugongs (both males and females) display seasonal reproductive activity, there is scant evidence to suggest that they have reproductive senescence as many mammals, including humans, do.

Other factors important in the calculation of fecundity are the number of offspring per birth and the sex ratio of offspring at birth. Wild Florida manatees have been documented producing twin offspring (but no more) in about 1.4 to 1.8% of births. Estimates of twinning based on Florida manatee carcass studies are as high as 4%. Limited and/or anecdotal data suggest that twinning occurs in the Antillean manatee and in the dugong. The sex ratio of the offspring at birth appears to be 1:1 for both the Florida manatee and the dugong. It is probably reasonable to assume that both Amazonian and West African manatees have similar patterns of twinning and offspring sex ratio.

Applying all of our knowledge of and assumptions about Florida manatee life history, the average fecundity (number of female births per female per year) for age classes 4–29 years was estimated at 0.189, 0.238, and 0.127 using different sets of data obtained from the examination of carcasses. Similarly, estimates of annual pregnancy rate (APR) and gross annual recruitment rate (GARR) ranged from 0.190 to 0.394 and 0.044 to 0.90, respectively, depending on the set of carcass data and gestation period used. Similar calculations for the dugong based on a smaller data set yielded apparent (annual) pregnancy rates of 0.093 to 0.353 over a series of years.

VII. Summary

Even though the Florida manatee and the dugong have been studied intensively since the 1980s, detailed data on many aspects of their life history are only beginning to be elucidated and there is considerably less information on the other species of sirenians. Given the threatened or endangered status of this unique group of marine mammals, every effort should be made to learn about their biology so that we may ensure their survival.

See Also the Following Articles

Age Estimation ■ Captivity ■ Parental Behavior

References

Hartman, D. S. (1979). "Ecology and Behavior of the Manatee (*Trichechus manatus*) in Florida." American Society of Mammalogists, Special Publication No. 5.

O'Shea, T. J., Ackerman, B. B., and Percival, H. F. (eds.) (1995). "Population Biology of the Florida Manatee." Information and Technology Report 1. U.S. Department of the Interior, National Biological Service, Washington, DC.

Reynolds, J. E., III, and Odell, D. K. (1990). "Manatees and Dugongs." Facts On File, New York.

Reynolds, J. E., III, and Rommel, S. A. (eds.) (1999). "Biology of Marine Mammals." Smithsonian Institution Press, Washington, DC.

Ricklefs, R. E. (1973). "Ecology." Chiron Press, Portland, OR.

Ridgway, S. H., and Harrison, R. J. (eds.) (1985). "Handbook of Marine Mammals," Vol. 3. Academic Press, London.

Skeletal Anatomy

SENTIEL A. ROMMEL
*Florida Marine Research Institute,
St. Petersburg*

JOHN E. REYNOLDS III
Eckerd College, St. Petersburg, Florida

The postcranial skeleton includes all the bones and cartilages caudal to the head skeleton. With a few exceptions, our description is limited to the bones of this region. The postcranial skeleton is subdivided into axial components (the vertebral column, ribs, and sternebrae, which are "on" the midline) and appendicular components (the forelimbs, hindlimbs, and pelvic girdle, which are "off" the midline). The skeleton supports and protects soft tissues, controls modes of locomotion, and determines overall body size and shape. The marrow of some bones may generate the precursors to certain blood cells. Skeletal elements may store lipids (particularly in the Cetacea) and calcium (particularly in the Sirenia) and thus influence buoyancy. Because bones are remodeled continuously in response to biochemical and biomechanical demands over the life span of the marine mammal, they offer information that can help interpret life history events after death.

Skeletons of seven different species of marine mammals are discussed: the Florida manatee (*Trichechus manatus latirostris*), the harbor seal (*Phoca vitulina*), the California sea lion (*Zalophus californianus*), the North Atlantic right whale (*Eubalaena glacialis*), the common bottlenose dolphin (*Tursiops truncatus*), the polar bear (*Ursus maritimus*), and the sea otter (*Enhydra lutris*). The domestic dog (*Canis familiaris*) skeleton is used to provide a familiar reference for comparison with marine mammal skeletons. These species were chosen, in part, because much is known about them and they provide a wide range of biomechanical morphologies. The manatee, dolphin, and right whale represent the many species that have lost their hindlimbs and are permanently aquatic; the other species may spend at least some of their life cycles on land. Note that the seven species are representative of generalized types—even within species, sexual dimorphism may produce different features.

Visualize the articulated (assembled) skeleton by first considering individuals in an absolute sense (left and lower parts of Fig. 1) and then in a relative sense (upper right of Fig. 1). Consider what decisions morphologists must make when comparing the "size" of any one feature among individuals that differ substantially in size. Compare the two methods of scaling by using the human swimmer as an example. Because human sensory systems are comparative and relational comparisons (such as proportions and percentages) are intuitive, we use the relative scheme scale comparisons among species for most of this article.

We will refer to the manatee when describing individual morphological structures but refer to the seven selected marine mammal species when discussing marine mammals in general [for details on the same structures in the dolphin, see Rommel (1990); for the true seals, see Pierard (1971) and Pierard and Bisaillon (1978); for the dog, see Evans (1993); for mammals in general, see Flower (1885)].

I. Axial Skeleton

A. Vertebral Structures

A few structural terms are necessary to better grasp the functional morphology of the vertebral column. Each vertebra (plural, vertebrae) has several "standard" parts (Fig. 2A). The body or centrum (plural, centra) of each vertebra is a subcylindrical structure that forms the primary mechanical support of the vertebral (spinal) column (Fig. 2B). Several kinds of projecting parts or *processes* may extend from the centrum and provide attachment sites, protection for soft tissues, or both. The largest and most common lateral processes are transverse processes, which tend to be long and robust in the lumbar vertebrae (and in the first few caudal vertebrae of permanently aquatic species). In some regions of the vertebral column (e.g., cervical), there may be two transverse processes extending from each side of a single vertebra.

Dorsal to the centrum is an arch of bone (the neural arch), which forms the sides of a neural canal. The neural arch protects the spinal cord. Interestingly, in some marine mammal species, the neural arch is considerably enlarged to accommodate relatively large masses of vascular tissue and/or fat that may be juxtaposed to the dura mater of the spinal cord. Because of these enlargements, the neural canal may not reflect the dimensions of the spinal cord as it does in most terrestrial species (Giffen, 1992). Additionally, each neural arch may extend dorsally as a neural spine (spinous process). The neural spines function as levers to increase the mechanical advantage of the epaxial (epiaxial) muscles to dorsiflex the tail.

Relative motion between adjacent vertebrae is controlled, in part, by the size and shape of the intervertebral disks (Fig. 2C). Intervertebral disks resist compression when skeletal

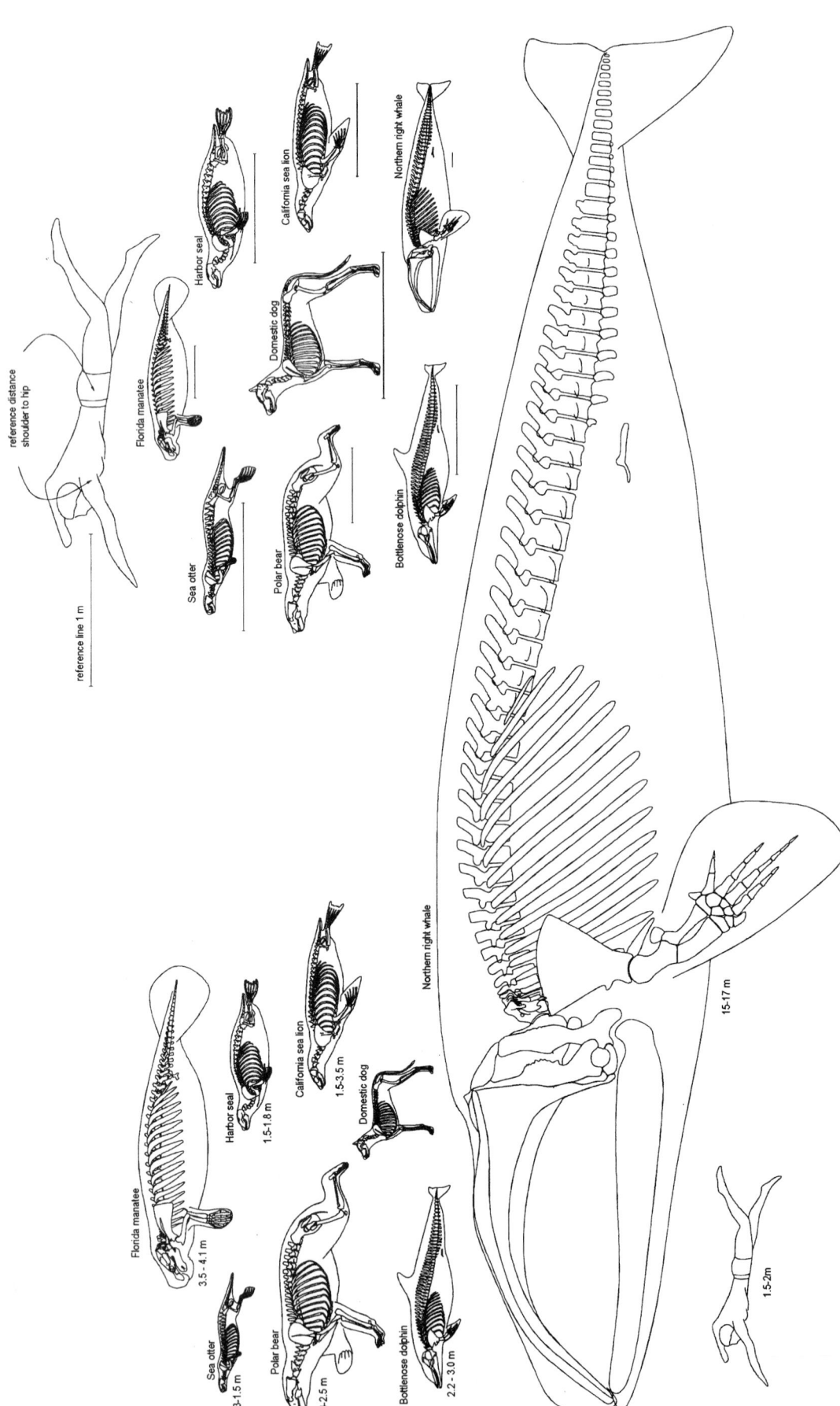

Figure 1 Selected marine mammal skeletons (the Florida manatee, the bottlenose dolphin, the polar bear, and the sea otter) compared with the skeleton of the domestic dog. Drawings on the left and bottom are scaled so that 1 m on any one species equals 1 m on all the others in this group. The range of adult total body lengths (snout tip to tail tip; note that human height is measured differently) is given in meters beneath each drawing. This group is sized to fit the right whale onto the page. Skeletons in the group on the upper right are scaled so that the distance between the shoulder and the hip joints is equal in all seven species. Thus, the body cavities of this group are approximately the same length. A reference line representing 1 m is given below each of these drawings; note that it is different for each skeleton in this scaling scheme. Both groups are sized so that the dolphins on the right and left are equal in length. Copyright S. A. Rommel.

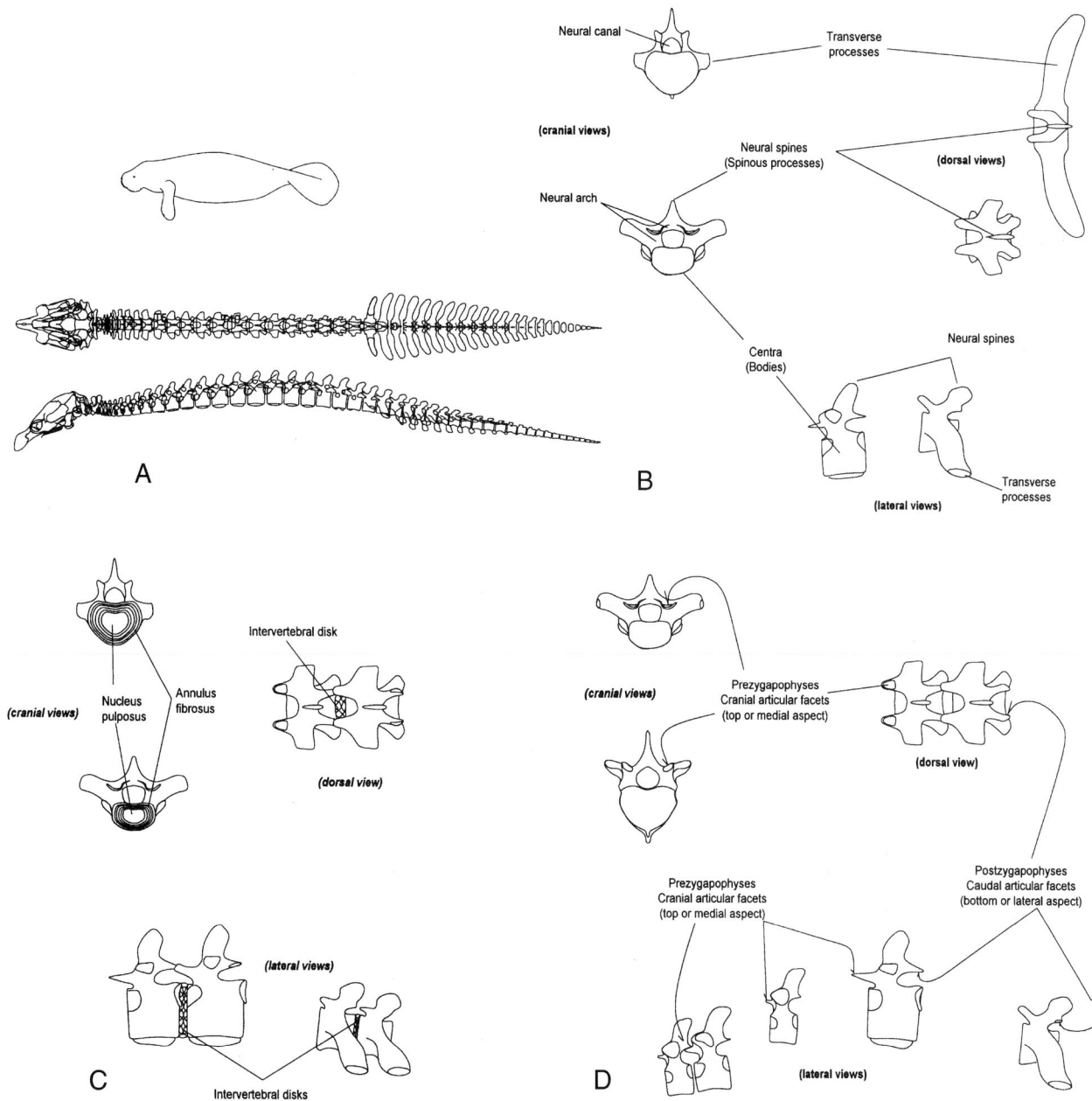

Figure 2 *(A) The manatee and its vertebrae. In this and subsequent illustrations of specific features, we illustrate a particular region of the vertebral column with dorsal and lateral views of that region. (B) Parts of a vertebra. The body, or centrum, of each vertebra is the primary mechanical support of the vertebral column. Vertebral centra help prevent the body of the animal from collapsing when large body muscles contract. Dorsal to the centrum is an arch of bone, the neural arch; within the arch is the neural canal. The neural arch protects the spinal cord and it may extend dorsally as a neural spine or spinous process. (C) The flexible region between adjacent vertebral bodies functions as a tough joint. Intervertebral disks (made of fibrocartilage) have two parts: the inner nucleus pulposus and the outer annulus fibrosus. (D) Zygapophyses are the articular facets that constrain motions between adjacent vertebrae. The cranial pair on each vertebra are prezygapophyses and the caudal pair are postzygapophyses. The region of articulation between prezygapophyses and postzygapophyses of adjacent vertebrae is typically a synovial joint if the facets overlap. In some regions of the vertebral column the facets do not overlap and may be connected by collagen fibers. Copyright S. A. Rommel.*

muscles force vertebrae together. Intervertebral disks are composite structures with a fibrous outer ring, the *annulus fibrosus,* and a semiliquid inner mass, the *nucleus pulposis.* The inner, jelly-like nucleus forms a dynamic joint that can adjust its shape to support complex forces of bending. The outer elastic fibers of the disk are arranged in a resilient ring, or annulus. The same basic design of surrounding a deformable core with elastic fibers is used in baseballs and golf balls.

Flexibility of the vertebral column depends, in part, on the thickness of the disks. Disk thickness varies within and among species, and intervertebral disks represent a substantial proportion (10 to 30%) of the length of the vertebral column. Relaxed (neutral) curvature of the vertebral column is vertebral, *not* intervertebral. That is, the shape of the spinal curve is determined by the shapes of individual vertebrae. Intervertebral disks provide flexibility, whereas curvature is provided by (nonparallel) vertebral body faces (Fig. 3).

In addition to the intervertebral disks, adjacent vertebrae may have other surfaces of articulation (facets). Zygapophyses (singular, zygapophysis) may be located on the neural arch, neural spine, and/or the transverse processes. Zygapophyses are bilaterally paired articular facets, found on the cranial and caudal aspects of each vertebra (Fig. 2D). The cranial pair are termed prezygapophyses, and the caudal pair are postzygapophyses. The region of articulation between prezygapophyses and postzygapophyses is typically a synovial joint if the facets overlap.

Zygapophyses in the neck and cranial thorax tend to be oriented horizontally to allow axial rotation (twisting) and side-to-side motions. Zygapophyses in the caudal thorax and tail tend to be oriented vertically to facilitate up-and-down motions. In some regions of the vertebral column (particularly the tails of most species and under the dorsal fin of dolphins), adjacent vertebrae may have reduced or absent zygapophyses, possibly reflecting the increased flexibility or influence of nonvertebral structures on the mobility of these regions. Test these zygapophyseal constraints on yourself by bending and twisting your body—note how your thorax (rib-bearing region) flexes (most of the axial rotation and some lateral flexion occurs here) when compared with your lumbar (low back) region (where most of the allowable dorsoventral flexion occurs).

B. Vertebral Column

Traditionally, the typical mammalian vertebral column is separated into five regions, each of which is defined by what is or is not attached to the vertebrae. These regions are cervical, thoracic (dorsal in the older literature), lumbar, sacral, and caudal (Fig. 3). The vertebral formula is an alpha-numerical abbreviation for the numbers of vertebrae in each region; note that some regions have similar numbers of vertebrae in most species, whereas other regions have very different numbers of vertebrae in different species. For example, C6:T17:L1:S0:Ca25 describes the 6 cervical, 17 thoracic, 1 lumbar, 0 sacral, and 25 caudal vertebrae in the Florida manatee. Individual vertebrae in each region are also referred to by their position using one or two (regiospecific) letters and a number. For example, the third cervical vertebra is designated as C3, the first thoracic as T1, the first lumbar as L1, the second sacral as S2, and the

tenth caudal as Ca10. In some species the morphological distinctions among vertebrae from each region are unambiguous, whereas in others, the distinctions among vertebrae in adjacent regions are less evident (Fig. 3).

The total number of vertebrae, excluding the caudal vertebrae, is surprisingly close to 30 in most mammals (Flower and Lydekker, 1891). With this in mind, examine the manatee vertebral column and contrast the numbers of different vertebrae it has with the number of vertebrae in the other selected species (Fig. 3).

C. Cervical Region

Cervical vertebrae (Fig. 4A) are located cranial to the rib-bearing vertebrae of the thorax. Some cervical vertebrae have movable lateral processes known as cervical ribs, none of which make contact with the sternum. Most mammals have seven cervical vertebrae, but all sirenians and the two-toed sloth (*Choloepus*) have six and the three-toed sloth (*Bradypus*) has nine. Serial fusion (ankylosis) of two or more cervical vertebrae is common in cetaceans, but all cetaceans have seven cervical vertebrae. Contrary to what has been reported in some published works, cervical vertebrae in some cetaceans (e.g., the narwhal *Monodon monoceros,* beluga *Delphinapterus,* and river dolphins) are unfused and provide considerable neck mobility.

The first two cervical vertebrae have unique names: the atlas (C1) and the axis (C2). Unlike all other vertebrae, the atlas and axis are not separated by an intervertebral disk but rather by a synovial joint, similar to the skull–atlas joint, where the occipital condyles of the skull articulate with the atlas (Fig. 4A). The shape of this joint dictates the range of motion between the skull and the first vertebra. Typical head motions on the atlas are up and down and side to side (Fig. 4A). The axis has an elongated centrum, the odontoid process or dens, that extends into and sometimes beyond the foramen of the atlas. The shape of the dens restricts motions between the atlas and the axis to twists or rotations parallel to the long axis of the body. Test these constraints on yourself by twisting your head and neck. Head nodding, as in expressing "yes," is constrained by the atlas; head shaking, as in expressing "no," is constrained by the axis.

Cervical vertebrae provide support for the head and can aid in FEEDING and LOCOMOTION. As a group, they can, in some cases, flex to provide a wide range of postures. Typically, permanently aquatic marine mammals have short necks, even if they have seven cervical vertebrae. In contrast, pinnipeds have relatively long necks, and the external appearance of a short neck in true seals is misleading. Close comparison of the skeletons of the true seal and sea lion reveals that they have quite similar neck lengths, although the distribution of body mass is different. True seals often hold their heads close to the thorax, which causes the neck to form a deep "S" curve in the neck. This provides true seals with a "slingshot potential" for grasping prey (or careless handlers).

Cervical vertebrae may have distinct transverse foramina that perforate the bases of the transverse processes and run parallel to the long axis of the body. Transverse foramina protect the arteries that supply the head and neck from kinking

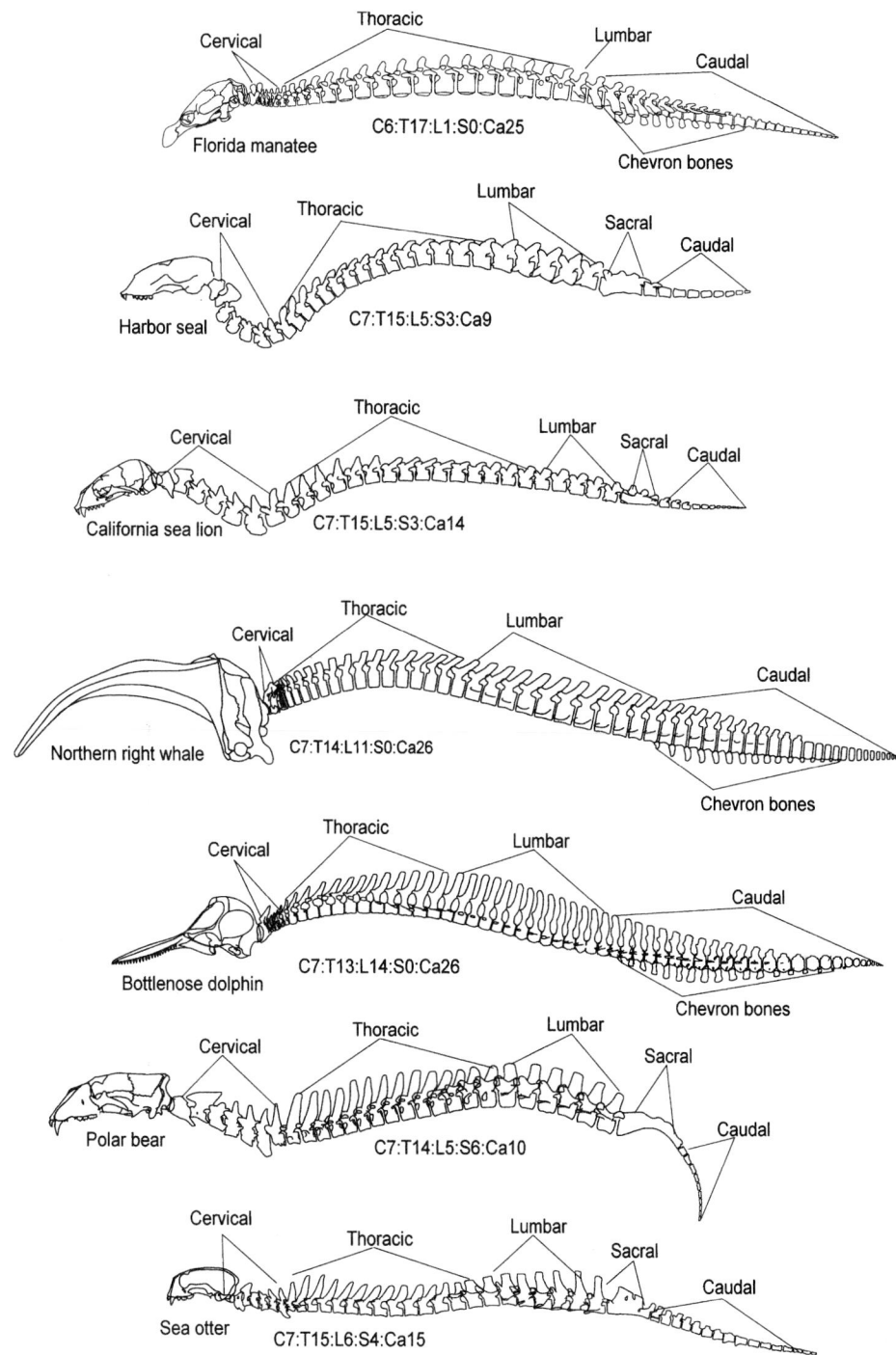

Figure 3 *Vertebral columns of the seven selected marine mammal species; the vertebral column of the dog is similar enough to the vertebral columns of the bear and other carnivorous species to exclude it from this illustration. There are typically five separate regions: cervical, thoracic, lumbar, sacral, and caudal. In this illustration, the first thoracic vertebrae for all the species illustrated are aligned. Each skull and vertebral column is scaled to similar shoulder–"hip" distances. Sacral vertebrae are, by definition, associated with an attached pelvis; therefore, permanently aquatic species with (unattached) pelvic vestiges have no sacral vertebrae. The vertebral formula, an alpha-numerical abbreviation for the numbers of vertebrae in each region, is provided below each vertebral column. The numbers given are for the specimens illustrated; some of these numbers vary between individuals of each species. Copyright S. A. Rommel.*

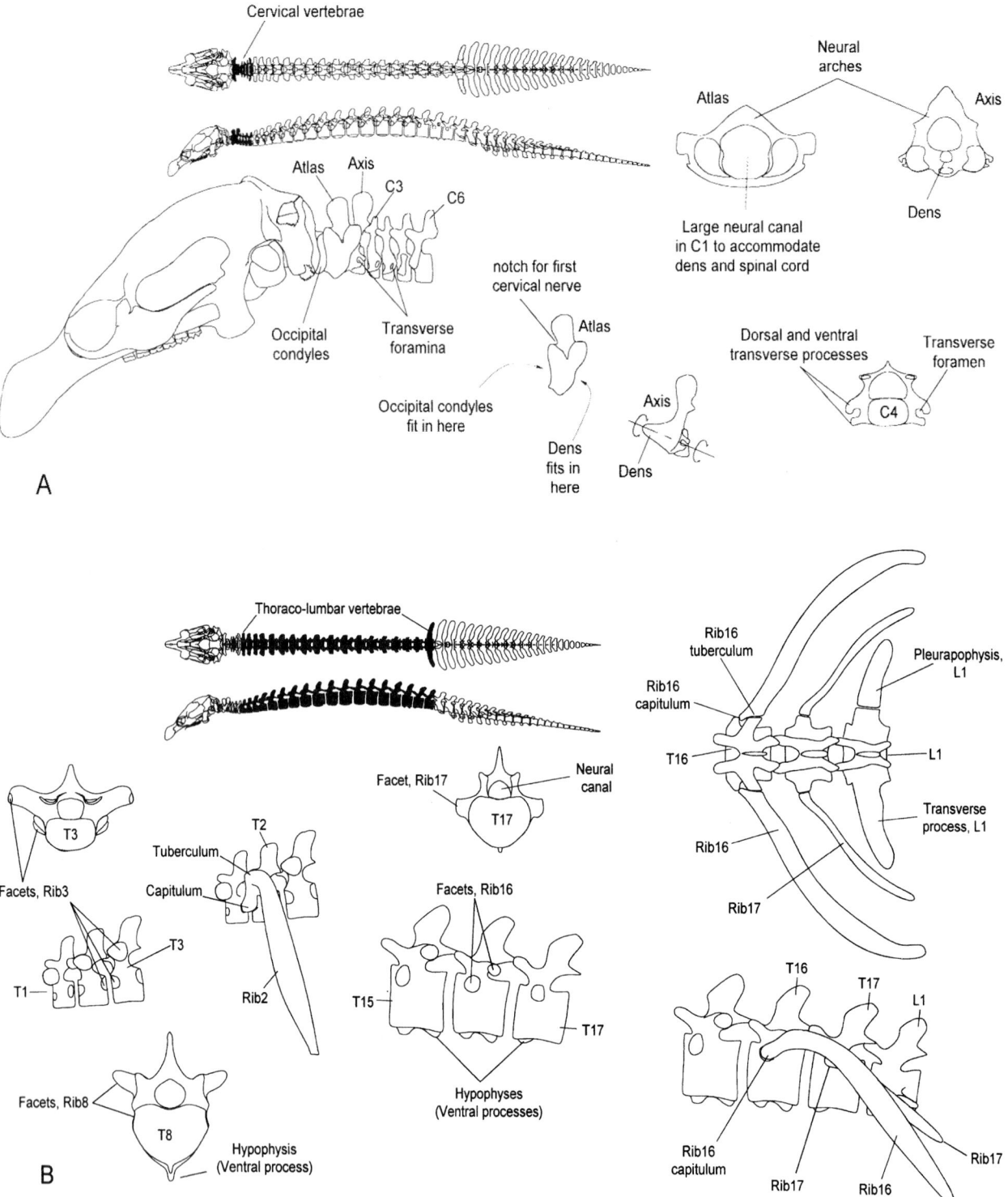

Figure 4 (A) Cervical vertebrae. At the top of this illustration, these vertebrae are filled in on the dorsal and lateral views of the manatee skeleton. The first and second cervical vertebrae are uniquely named the atlas and the axis, respectively. Cervical vertebrae may have transverse foramina, which perforate the bases of the transverse processes. Typical skull motions on the atlas are up and down and side to side. The axis has an elongated centrum, the dens, that extends into the large neural canal of the atlas. The shape of the dens restricts motions between the first two vertebrae to "twists" or rotations parallel to the long axis of the body. (B) Thoraco-lumbar vertebrae. Thoracic vertebrae support the ribs, which anchor the diaphragm muscles. Some ribs are double headed and articulate with their respective vertebrae via capitula and tubercula. (continues)

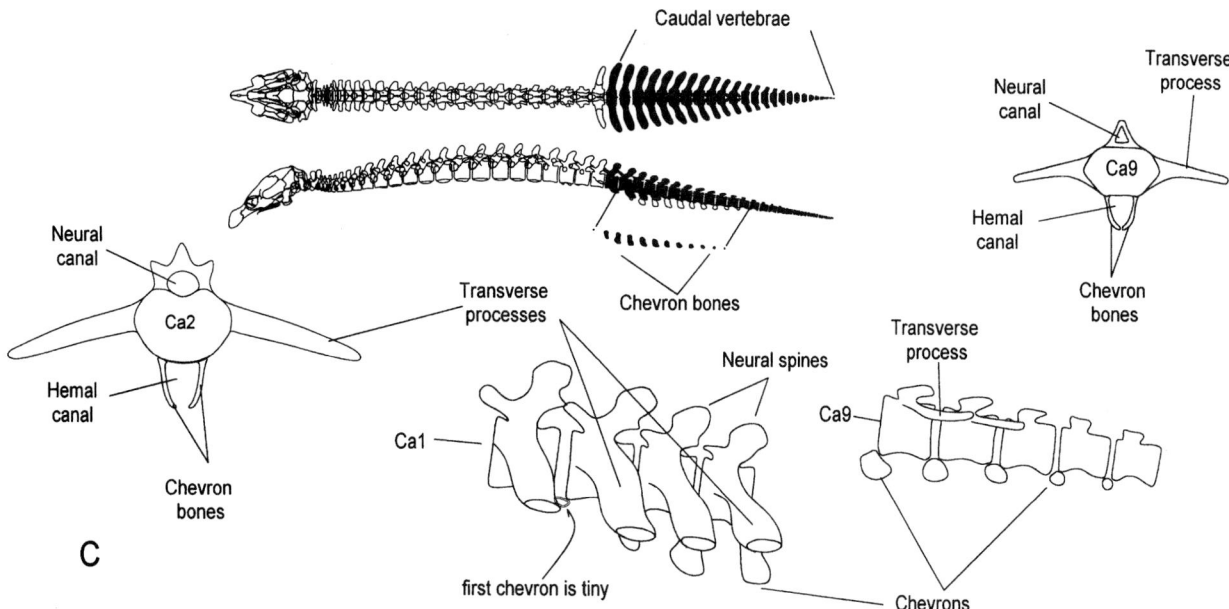

Figure 4 *(Continued)   Each capitulum articulates with one or more costal facets on the vertebral centrum at or near the intervertebral disk. Each tuberculum articulates with a facet on the transverse process of its respective vertebra. Lumbar vertebrae are trunk vertebrae without ribs: they have pronounced transverse processes called pleurapophyses. In manatees, there may be a rib on one side and a transverse process on the other side of the same lumbar vertebra. (C) Caudal vertebrae are found in the tail. At the top of this illustration, these vertebrae are filled in on the dorsal and lateral views of the manatee skeleton. In permanently aquatic marine mammals that have no direct attachment between the pelvic vestiges and the vertebral column, the transition between lumbar and caudal vertebrae is defined by the presence of chevron bones. Chevron bones are ventral intervertebral ossifications. By definition each is associated with the vertebra cranial to it. Copyright S. A. Rommel.*

when the neck is flexed. Transverse foramina are particularly distinct in cervical vertebrae of carnivorous marine mammals (and, of course, the dog). A pair of notches (or foramina) in the neural arch of the atlas may be present (Fig. 4A); this opening is for the first spinal nerve that exits the neural canal between the back of the skull and the bulk of the neural arch of C1. Rarely, one finds cervical ribs, which are movable extensions of the ventral transverse processes. Cervical ribs do not extend to the sternum.

D. Thoracic Region

The thoracic region (commonly referred to as the thorax) is defined by the presence of more or less movable (i.e., not fused to the respective vertebrae) ribs (Figs. 3, 4B, and 5). Thoracic and lumbar vertebrae support the trunk—the region occupied by the thorax (with the lungs and heart) and the abdomen (with the remaining viscera). By definition, the first thoracic vertebra has a pair of ribs that, unlike cervical ribs, extend to the sternum. Most thoracic vertebrae have ribs that possess two segments. The proximal, dorsal segment is attached to the vertebral column and is termed a vertebral rib (Fig. 5). A distal, ventral segment is attached to the sternum and is termed a sternal rib (Romer and Parsons, 1986). A joint between the vertebral rib and the sternal rib allows flexibility of the thorax. The distinction between vertebral and sternal ribs is more common in birds and some reptiles, but it is less common in mam-

mals because most mammals have cartilaginous sternal ribs termed costal cartilages. This distinction is made because, unlike most other mammals, odontocete cetaceans have bony rather than cartilaginous sternal ribs (bony sternal ribs are also found in the armadillo, *Priodontes*).

Thoracic vertebrae are constrained by their rib attachments and, thus, tend to be much less flexible than lumbar and caudal vertebrae. Typically, the thorax is relatively more flexible axially and laterally than it is dorsoventrally because zygapophyses tend to be oriented horizontally.

As suggested earlier, the mechanics of motion of the thoracic vertebrae are different from those of either cervical vertebrae or postthoracic vertebrae. Thoracic vertebrae have dual functions of longitudinal support and support of the diaphragm muscles. This dual function is reflected in the arrangement, size, and shape of the zygapophyses. The last thoracic vertebrae (here, we use first and last to designate cranialmost and caudalmost ribs to avoid complicated sentences confusing caudal as a direction and caudal as a region of the vertebral column) have ribs associated with them, but these last ribs may not be attached to their respective vertebrae (floating ribs), and there is little or no indication on their respective vertebrae to indicate their presence. The last ribs of the cetaceans are closely associated with the superficial aspects of the robust hypaxial muscles and may have little to do with support of the diaphragm, whereas those of manatees are

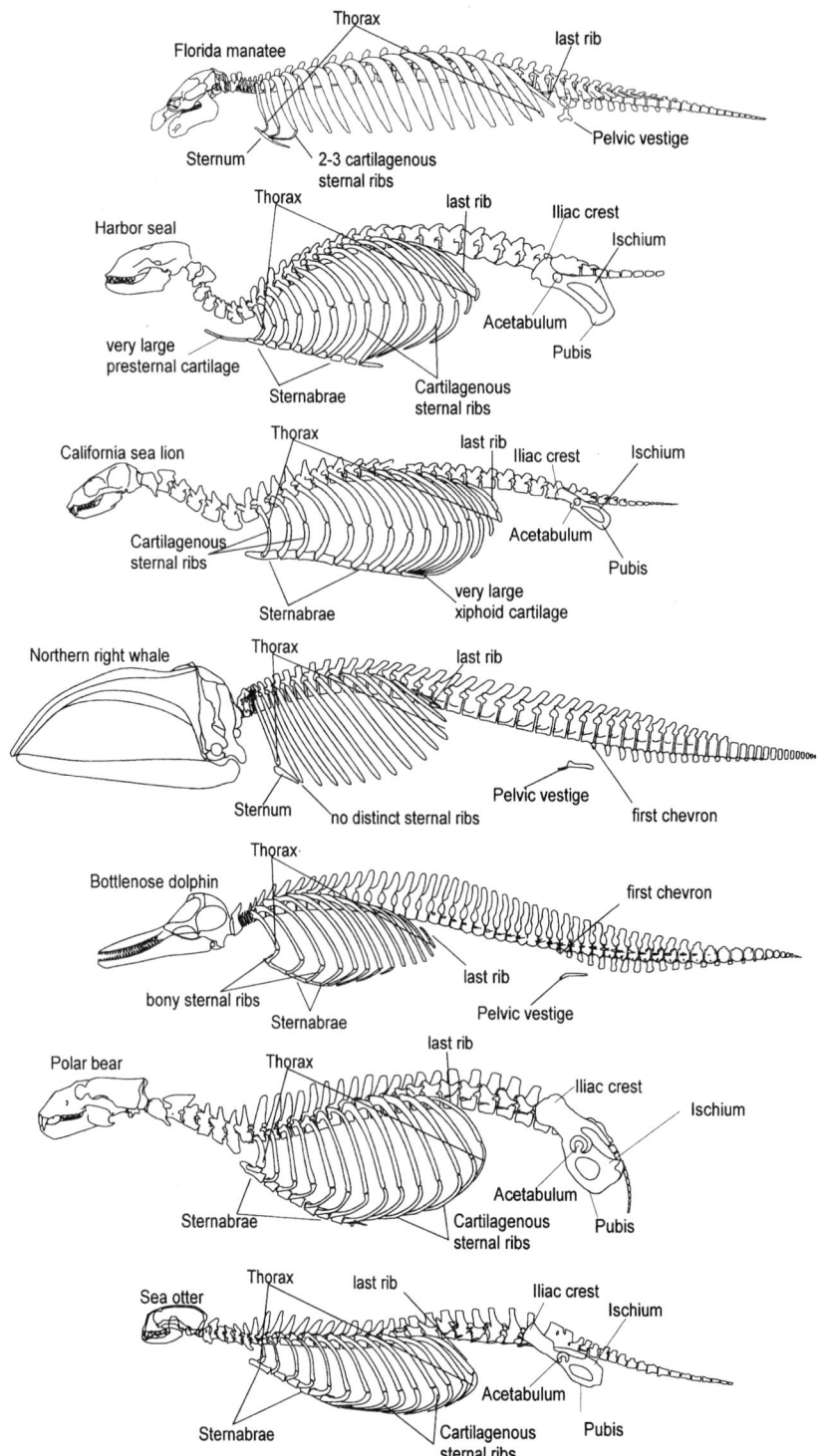

Figure 5 *Axial skeletons plus pelvises of the seven selected marine mammal species; because the axial skeleton and pelvis of the dog are similar to those of the bear and of other carnivorous species, the dog's skeleton was excluded. In this illustration, the first thoracic vertebrae for all the species illustrated are aligned. Vertebral ribs extend from the vertebral column to join the sternal ribs, which extend from the sternum. The sternum develops from a series of sternebrae. In some species (i.e., humans, right whales, manatees), the sternebrae fuse into a single bony unit. Each skull and vertebral column is scaled to similar shoulder–hip distances. Permanently aquatic species (manatee, dolphin, and right whale) have pelvic vestiges. Because the pelvic bones of these three species are not in the same location in relation to the first chevron bones, the first caudal vertebrae (defined by the first chevrons) of these three skeletons are not aligned. Distinct sexual dimorphism has been described for the pelvic vestiges of manatees and dugongs. Copyright S. A. Rommel.*

imbedded in the hypaxial muscles and may provide support for the diaphragm.

The attachment sites of ribs on the vertebrae are called rib facets or costal facets (ribs = costae) if the region of attachment has a distinct articulation (Fig. 4B); rib facets are located on the vertebral centrum, on the transverse processes, or on both. Some rib attachments are made via long connective tissue fibers; the ribs do not actually touch the respective vertebra, and there is no obvious structure on the (cleaned) vertebrae to suggest rib attachment. For this reason, dogmatic adherence to narrow definitions of numbers of thoracic or lumbar vertebrae or both should be avoided. Cetaceans and pinnipeds have very mobile rib cages. Surprisingly, diving mammals have considerable "tilt" to their ribs. In these species, the double-headed ribs (see later) have joints that are aligned to allow for substantial swing of the attached vertebral ribs in the plane of rotation. This extreme mobility of the rib cage can help accommodate the lung volume changes that accompany changes in depth so that ribs will not break. It may also have evolved to accommodate rib cage dynamics caused when abdominal muscle contractions prevent caudal displacement of the diaphragm during inspiration; in this case, ribs could be moved forward to increase lung volume.

Some thoracic vertebrae have ventral projections called hypapophyses—not to be confused with chevron bones, which are intervertebral and are structurally not part of the vertebrae (Figs. 3 and 4C). In the manatee, the diaphragm is firmly attached along the midline of the central tendon to hypapophyses. Hypapophyses also occur in some cetaceans (i.e., *Kogia*, the pigmy and dwarf sperm whales) in the caudal thorax and cranial lumbar regions; it is assumed that these hypapophyses increase the mechanical advantage of the hypaxial muscles.

The neural spines of thoracic vertebrae are often longer than those in any other region of the body (Figs. 3 and 5). Long neural spines provide mechanical advantage to neck muscles that support a head cantilevered in front of the body. Therefore, species with large heads tend to have long neural spines; however, the buoyancy of water negates the need for such long neural spines in permanently aquatic species.

Embryonically, ribs and transverse processes are homologous structures, but the presence of a movable joint ultimately distinguishes a rib from a transverse process in an adult. The unfinished joint may be indicative of developmental age. In some species (i.e., the manatee), there may be a movable "riblet" (pleurapophysis, *pleura* = side or rib) on one side and an attached "transverse process" on the other side of the same (typically the first lumbar) vertebra. Pleurapophyses are more common in some of the lower vertebrates (Romer and Parsons, 1986; Kardong, 1998).

A rib may attach to its respective vertebra (Fig. 4B) at one or more locations (e.g., centrum or transverse process, or both). Typically the cranialmost ribs have two distinct regions of articulation with their juxtaposed vertebrae. The cranialmost articular part of a double-headed rib is the head or capitulum (plural, capitula) and it articulates (via a synovial joint) with the body (or bodies) of one or more vertebrae. The tuberculum (plural, tubercula) is the second point of attachment of a double-headed

rib; it articulates (sometimes with a synovial joint, other times by connective tissue fibers) to a transverse process of the respective vertebra. Since the capitula of some double-headed ribs touch two vertebrae, the vertebra with the transverse process attachment is used to identify the respective vertebra.

The caudalmost ribs have single attachments and are referred to as single headed. In most mammals, the single-headed ribs have lost their tubercula and are attached to their respective vertebrae by the capitulum at the centrum. In contrast, the single-headed ribs of cetaceans lose their capitula and are attached to their respective vertebrae by their tubercula at the transverse processes (as occurs in some reptiles). Thus, single-headed ribs of cetaceans attach to vertebrae farther from the midline than single-headed ribs of other mammals. Additionally, the transverse processes of the thoracic vertebrae in cetaceans can be longer than those of other mammals. The last ribs often "float" free at one or both ends, i.e., they are attached to neither their associated vertebra nor the sternum. These ribs tend to be significantly smaller than the ones cranial to them and are often lost in preparation of the skeleton.

Sternum   The sternum is formed from bilaterally paired serial elements, called sternebrae (the sternal equivalent of vertebrae; Fig. 5). The paired elements fuse on the midline, occasionally imperfectly, leaving foramina in the sternum. The first and last sternebrae are called the manubrium and xiphisternum, respectively. In some species (e.g., some odontocetes, all mysticetes, and all sirenians), sternebrae fuse into a single unit. Sternal ribs extend from the sternum to join the vertebral ribs. True seals have an elongate presternal cartilage extending cranially from the manubrium. Some species, particularly the California sea lion and its close relatives, have enlarged xiphisternal cartilaginous extensions.

E. Lumbar Region

Lumbar vertebrae are trunk vertebrae that typically do not bear movable ribs (Fig. 3). Recall that ribs and transverse processes develop from the same embryonic structures. Occasionally a distinct pleurapophysis may be found in this region. As noted previously, pleurapophysis are commonly found in manatees: there will be a "normal" transverse process on one side and a movable riblet on the other side of the same vertebra. These lumbar ribs are clearly distinct from thoracic ribs (Fig. 4B).

As already noted, the number of lumbar vertebrae is usually closely linked to the number of thoracic vertebrae; an increase in number in one section typically means a reduction in the other. For example, there are 19 thoraco-lumbar vertebrae in all species of Artiodactyla, whereas there are 20 or 21 in Carnivora; compare these numbers with those of the selected marine mammals (Figs. 3 and 4B). Interestingly, the scaling of the skeletons to similar shoulder–hip joint distances helps us avoid some of the "distractions" that are inherent in the variation among species in size, number, and shape of the other regions of the body.

Typically, lumbar vertebrae are more flexible dorsoventrally than they are laterally and they may have no axial flexibility. Some of the mobility of lumbar vertebrae may be constrained by ribs that, although not attached to these vertebrae, are angled backward and extend beyond the tips of some lumbar vertebrae (Figs. 4B and 5).

F. Sacral Region

There are at least two commonly accepted definitions for sacral vertebrae: (a) serial fusion of postlumbar vertebrae, only some of which may attach to the pelvis (the human sacrum), and (b) only those vertebrae that attach to the ileum, whether or not they are fused serially (Fig. 3). Both definitions have merit. In species in which there is contact between the vertebral column and the pelvis, it is relatively easy to define sacral vertebrae. However, it is not easy in cetaceans and sirenians (dugongs have a ligamentous attachment between the vertebral column and the pelvic vestiges) because there is no direct attachment between the pelvic vestige and the vertebral column and there is no vertebral fusion. Thus, permanently aquatic species have (by either definition) no sacral vertebrae. Within a species, the number of serially ankylosed vertebrae may vary, particularly with age. The number of sacral vertebrae is commonly two to five, but there are exceptions.

G. Caudal Region

Caudal (cauda = tail) vertebrae are found in the tail (Fig. 4C). The number of caudal vertebrae varies widely. Long tails usually require numerous caudal vertebrae (up to a maximum of 46 in one African pangolin, *Manis gigantea*). Florida manatees have between 22 and 27 caudal vertebrae, and bottlenose dolphins and right whales both have 26; for cetaceans, the minimum is 13 in *Caperea marginata* (pygmy right whale) and the maximum is 49 in *Phocoenoides dalli* (Dalls porpoise). Note that caudal vertebrae of cetaceans extend to the tip of the tail (fluke notch), whereas manatee vertebrae stop 3–9% of the total body length (as much as 17 cm in a large specimen) from the fluke tip. Caudal vertebrae range from being robust, important locomotory structures in permanently aquatic species to being relatively small, unimportant pieces in pinnipeds and the polar bear.

Chevron Bones The chevron bones or chevrons (Fig. 4C) are ventral intervertebral ossifications found only in the caudal region. They are relatively large in permanently aquatic species but can be found as small ossifications in many other species, including the dog. By definition, each is associated with the vertebra cranial to it [note that there is some controversy over which is the first caudal, see Rommel (1990)]. Pairs of chevron bones form arches, creating a ventral channel called the hemal canal. Within the hemal canal, the blood vessels (caudal arteries and veins) that supply the tail are protected from occlusion when the tail is flexed. In some cetaceans the ventral aspects of each chevron bone pair fuse and may then function to increase the mechanical advantage of the hypaxial muscles to flex the tail ventrally.

II. Appendicular Skeleton
A. Pectoral Limb Complex

The forelimb (Fig. 6) is referred to as a flipper in permanently aquatic species (in true seals and sea lions the hindlimb is also referred to as a flipper). The forelimb includes the scapula, humerus, radius and ulna, and manus.

The scapula is attached to the axial skeleton only by muscles; there is no functional clavicle in marine mammals (Klima *et al.*, 1980; Strickler, 1978). The scapula consists of an essentially flat (slightly concave medially) blade with an elongate scapular spine extending laterally from it. The scapular spine is roughly in the center of the scapular blade in most mammals. However, in cetaceans, the scapular spine is close to the cranial margin of the scapular blade, and both the acromion and the coracoid extend beyond the leading edge of the blade. The distal tip of the spine is the acromion.

The humerus has a ball-and-socket articulation in the glenoid fossa of the scapula; this is a very flexible joint. The humerus articulates distally with the radius and ulna; this is also a flexible joint in most mammals, but it is constrained in cetaceans. The olecranon is a proximal extension of the ulna that increases the mechanical advantage of the triceps muscles that extend the forelimb. In species like the polar bear and the sea lion, the olecranon is robust; however, in the Cetacea it is relatively small. The radius and ulna of manatees fuse at both ends as the animal ages. This fusion prevents axial twists that pronate and supinate the manus. The radius and ulna of Cetacea are also constrained but are not typically fused. The forelimbs of the sea otter are very mobile, allowing for the manipulation food.

The distal radius and ulna articulate with the proximal aspect of the manus. The manus includes the carpals, metacarpals, and phalanges. There are five "columns" of phalanges, each of which is called a digit. The digits are numbered, using Roman numerals, starting from the cranial aspect (the thumb; associated with the radius). Note that the longest digit may be different in different species (Fig. 6).

In many of the marine mammals, the "long" bones of the pectoral limb (humerus, radius, and ulna) are relatively short, and the phalanges are elongated. Cetaceans are unique among mammals in that they have more phalanges than any of the other mammals (Fig. 6); this condition is known as hyperphalangy (Howell, 1930). The number varies between individuals of each species: pilot whales (*Globicephala* spp.) have the most (14), the bottlenose dolphin has a maximum of 9, and the right whale has a maximum of 7.

Notice the "plane" in which each manus is oriented (Fig. 6). The sea otter's manus is much like that of humans and has similar dexterity. The polar bear manus is much like that of the dog and other terrestrial quadrupeds. Permanently aquatic species hold the manus in a plane roughly parallel to the midsagittal plane of the body or parallel to the nearest body surface.

B. Pelvic Limb Complex

The typical mammalian pelvis is made of bilaterally paired bones: ilium, ischium, and pubis (Fig. 7). Each of the halves of the pelvis attaches (via the ilium) to one or more sacral vertebrae. The crest of the ilium, which is a prominent landmark, flares forward and outward beyond the region of attachment between the sacrum and the ilium. The two halves of the pelvis join at the pubic symphysis on the ventral midline. In permanently aquatic marine mammals, there is but a vestige of a

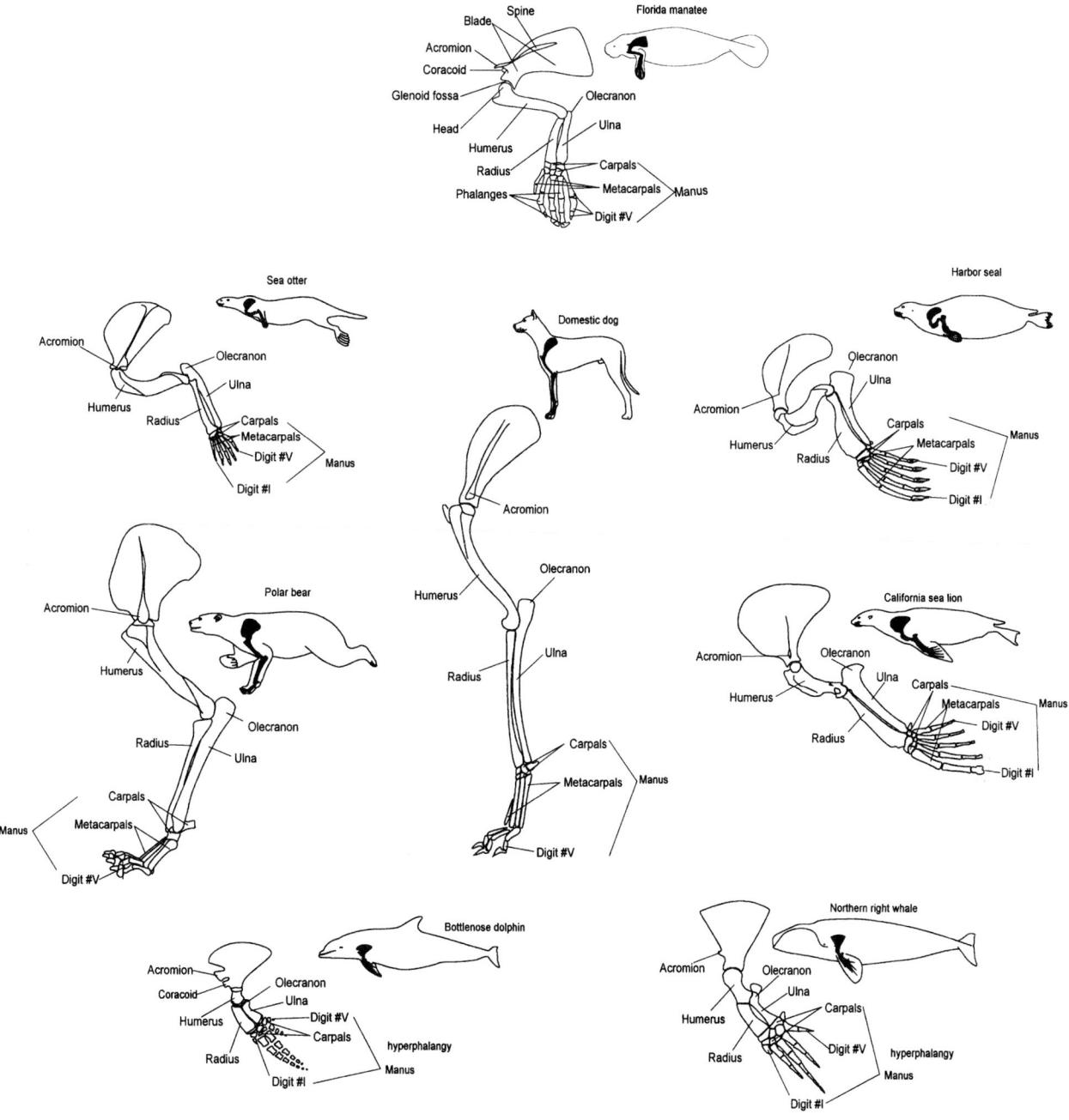

Figure 6 *The forelimb (pectoral appendage) includes the humerus, radius and ulna, and manus. The forelimb attaches to the pectoral girdle, which is made of bilaterally paired scapulae. The scapula is a flat (slightly concave medially) blade with an elongate scapular spine extending laterally from it. The distal tip of the spine is the acromion process. The humerus has a ball-and-socket articulation with the glenoid fossa of the scapula. The humerus articulates with the radius and ulna at the elbow. The radius and ulna articulate with the proximal aspect of the manus at the wrist. The manus includes the carpals, metacarpals, and phalanges. There are five "columns" of phalanges; each "column" is called a digit. The digits are numbered, using Roman numerals, starting from the cranial aspect (associated with the radius, in humans—the thumb). Copyright S. A. Rommel.*

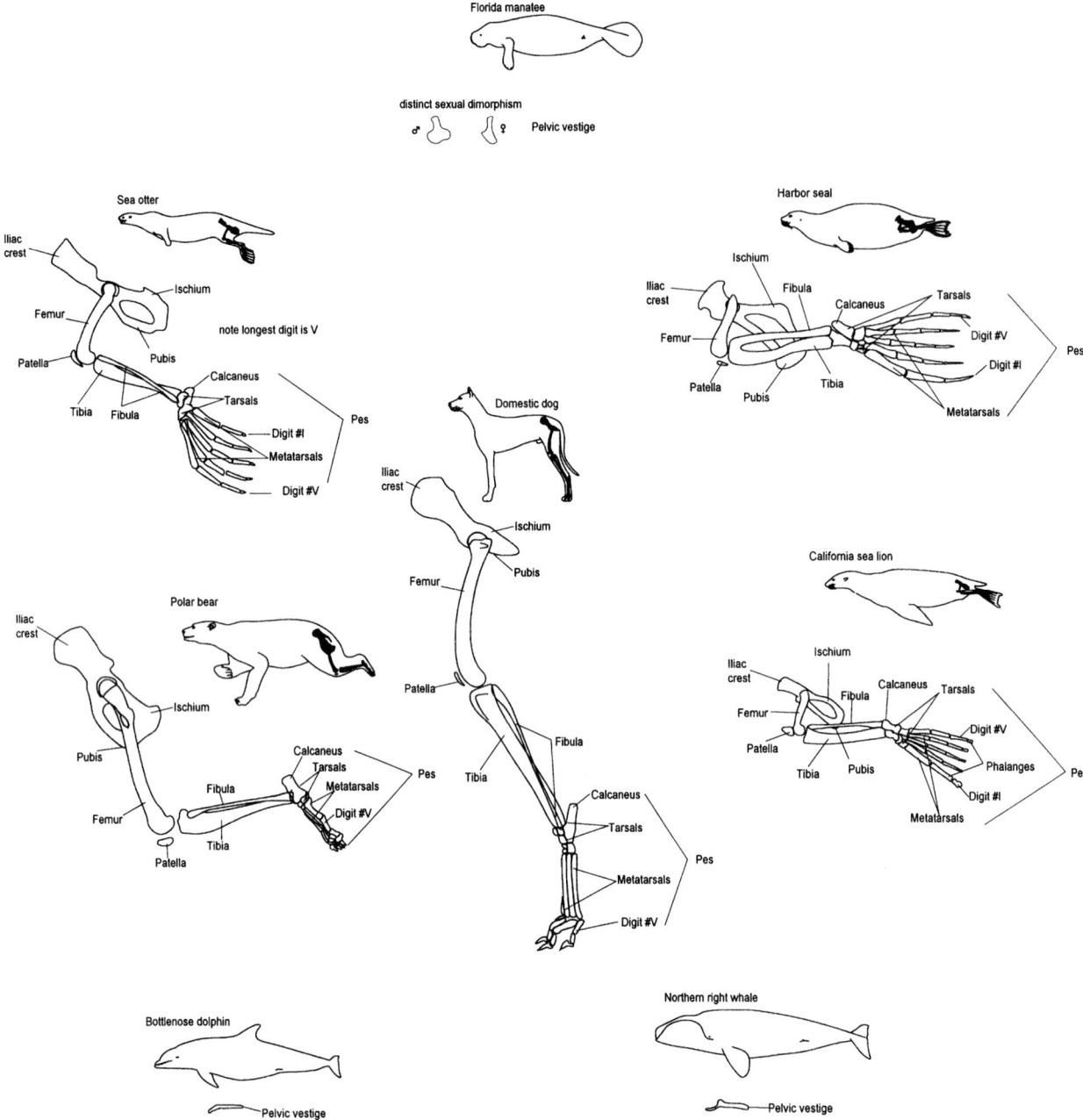

Figure 7 *The pelvic appendage (hindlimb) and pelvic girdle. The typical mammalian pelvis is made of bilaterally paired coxae attached to one or more sacral vertebrae. Each coxa is made of three elements: the ilium, ischium, and pubis. Each half of the pelvis attaches to one or more sacral vertebrae. The crest of the ilium is a prominent landmark; it flares forward and outward beyond the region of attachment between the sacrum and the ilium. The two halves of the pelvis join at the ventral midline pubic symphysis. In permanently aquatic marine mammals there is but a vestige of a pelvis. The hindlimb, if present, is attached to the pelvic girdle via a ball (femoral head)-and-socket (acetabulum) joint at the hip. The proximal limb bone is the femur. Distally, the femur articulates with the tibia and the fibula at the knee joint. The tibia and fibula distally articulate with the pes at the ankle. Note that the pes is oriented in different "planes" in some of the selected species. The pes is made of the tarsals proximally, the metatarsals medially, and the phalanges distally. Copyright S. A. Rommel.*

pelvis to which portions of the rectus muscles of the abdomen may attach. Additionally, the male reproductive organs may be supported by these vestiges. Occasionally, in some of the large whales, a hindlimb articulates with the pelvic vestige.

The hindlimb, if present, is attached to the vertebral column via a ball-and-socket joint at the hip. The proximal limb bone is the femur, whose head articulates with the socket of the pelvis, the acetabulum. Distally the femur articulates with the tibia and the fibula (at the knee joint). The tibia and fibula articulate distally with the pes or foot (Fig. 7). The pes is composed of the tarsals proximally, the metatarsals medially, and the phalanges distally. Note that the digits of the sea lion terminate a significant distance from the tips of the flipper. Of the chosen group of marine mammals, the limbs of the polar bear are closest in gross appearance to those of the terrestrial mammals. In contrast, many of the marine mammals have relatively short "long" bones (femur, tibia, and fibula) and relatively long phalanges. The femurs of some phocids (e.g., the harbor seal) are so short that the knee cannot contact the ground.

Note that the pes is oriented in different "planes" in some of the selected species (Fig. 7). The pes of the polar bear is oriented in a fashion similar to that of the dog. The sea otter, seal, and sea lion all orient the pes parallel to the sagittal plane when swimming. The sea lion can rotate its hind flippers to "walk" on land, and the sea otter can manipulate its hind flippers to hold food. Compare the positions of the first and last digits in the sea otter with those of the seal and sea lion and note how the sea otter's fibula crosses the tibia to achieve this orientation.

Sexual Dimorphisms In many mammals the adult males are larger than the adult females, and among marine mammals, this dimorphism is extremely pronounced in certain pinnipeds and the sperm whales. In contrast, the adult females of the baleen whales and some other species are larger than the adult males. In permanently aquatic marine mammals, there may be sexual dimorphisms in the pelvic vestiges (Fagone *et al.*, 2000). The penises of mammals are supported by tough fibrous structures, the crura, which attach to the pubic bones of the pelvis. The muscles that engorge the penis with blood are also attached to the pelvis, and presumably the mechanical forces associated with these muscles influence pelvic vestige size and shape, particularly in manatees. Males in some groups of mammals, particularly the carnivores, have a bone within the penis (the baculum) that helps support the penis.

III. Sutures and Epiphyses

Many bony features change as the individual develops and matures; these features can be used as milestones in life history studies, and in some cases, they can be fairly accurate estimators of age. Some postcranial elements grow from separate ossifications centers at the margins of the bones (Fig. 8A). Epiphyses (*epi-*, upon, and *physis-*, grow or generate) are regions at the ends of bones (most often at or near a joint) associated with longitudinal growth in mammals. Epiphyses allow the portion of the bone near a joint to grow at a differ-

ent rate than the rest of the bone. Growth occurs at a cartilaginous plate between the epiphysis and the diaphysis (*dia-*, between; body or shaft). At the completion of growth, the epiphysis fuses with the rest of the bone, and the cartilage plate disappears. Eventually the suture between the two parts disappears. Skeletal maturity is defined (for most mammals) when the vertebral epiphyses are all fused to their respective centra. Because most marine mammals are large, their epiphyses are more evident than those of smaller terrestrial mammals.

The best-studied epiphyses are those of the long bones. There are, however, other epiphyses that are less well known. For example, the proximal finger bones may have epiphyses (Fig. 8B). The pattern of epiphysial fusion in the flipper can be used to determine the relative age of an individual when compared to the fusion pattern in other animals of a known age. Occasionally, epiphyses are found on the proximal ends of ribs, on the scapula at or near the coracoid, or at the tips of neural spines and transverse processes.

Among the epiphyses most important to mammalian osteologists are those located at each end of the vertebral centrum (Fig. 8D). Most mammals have indistinct bony plates at each end of their growing vertebrae. These plates are vertebral epiphyses. Manatees (and humans) are among the few species of mammals that have distinct or no vertebral epiphyses. In manatees, the vertebral epiphyses are delicate irregular films of bony tissue found within the cartilaginous surfaces at the ends of the centrum. After these vertebral sutures are completely fused, the skeleton cannot grow in length and the individual is considered skeletally mature. This may or may not coincide with other forms of maturity, such as reproductive maturity.

Nonepiphyseal sutures can also be used to estimate the age of parts of the postcranial skeleton (Fig. 8C). For example, the two halves of the neural arch ankylose at the dorsal midline to form the base of the neural spine. These sutures may persist well into adulthood in cetaceans.

There is much more to learn about the postcranial skeleton. See Pabst *et al.* (1999) for additional functional morphology on marine mammals, Young (1975) for information on mammals in general, and Alexander (1994) for vertebrates. For principles on size and scaling see Calder (1966) and Schmidt-Nielsen (1984).

Acknowledgments

We thank Meghan Bolen, Tim Evans, Llyn French, Judy Leiby, and James Quinn at the Florida Marine Research Institute and Lisa Johnson at Eckerd College for helpful comments on the manuscript. Anatomical illustrations were created with FastCAD (Evolution Computing, Tempe, AZ).

See Also the Following Articles

Bones and Teeth, Histology of ■ Morphology, Functional ■ Musculature ■ Pelvic Anatomy ■ Prenatal Development in Cetaceans

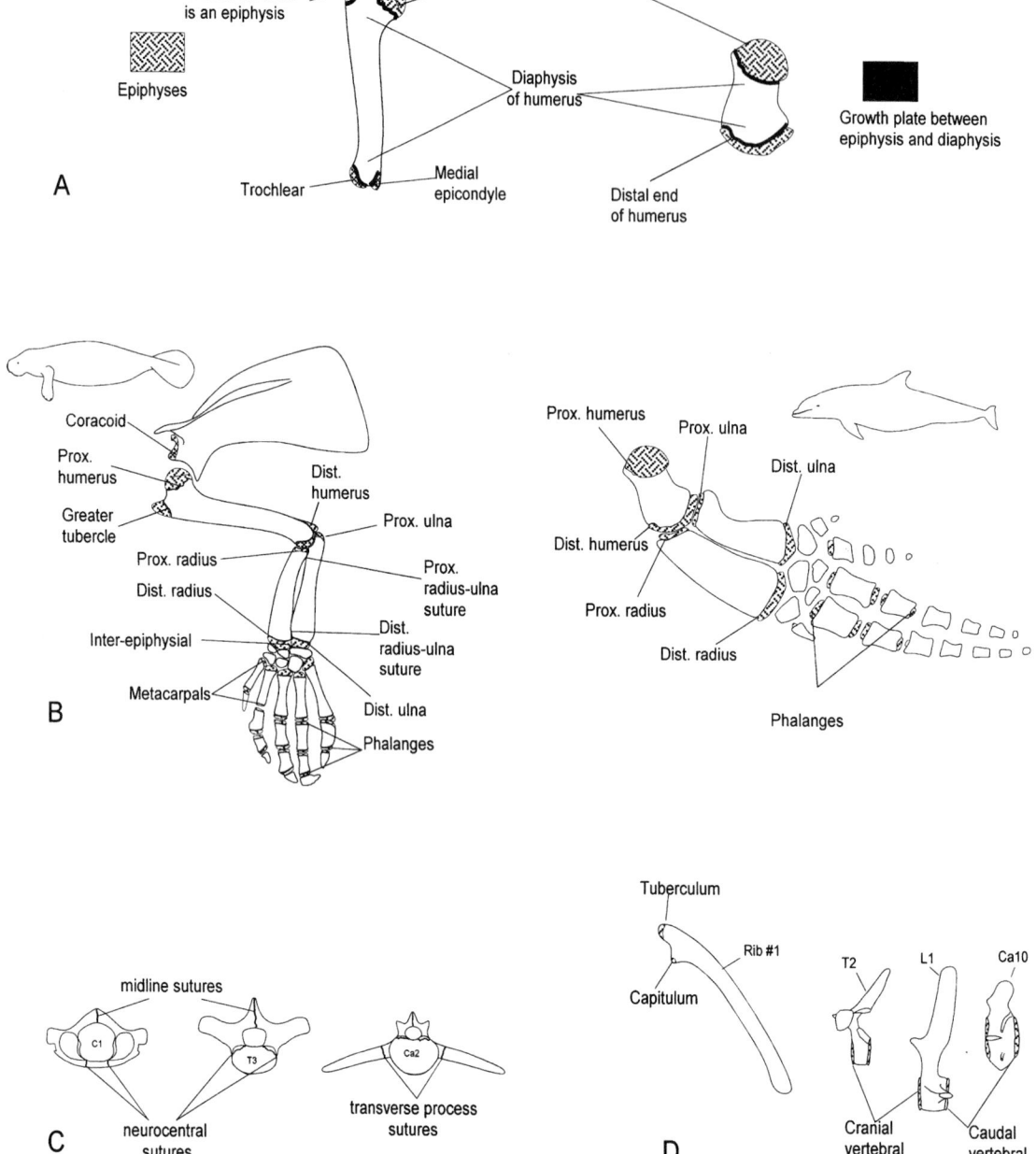

Figure 8 (A) Postcranial features that change with age. Some postcranial elements grow in length by separate ossification centers, the epiphyses, at the margins of the bony element. The humeral head develops as an epiphysis, as does the distal humerus. A cartilaginous growth plate located between the epiphysis and the diaphysis (body or shaft) disappears when growth is completed. (B) Epiphyses are also found at the ends of the phalanges. Patterns of fusion of these epiphyses provide information about aging. The epiphyses illustrated here have been used for relative aging of manatees and dolphins. (C) Nonepiphyseal sutures can also indicate age in the postcranial skeleton. The transverse processes and the neural arches have sutures that disappear early in adolescence. Additionally, the two halves of the neural arch ankylose at the dorsal midline to the base of the neural spine. (D) Epiphyses may occur on the ends of ribs and the tips of transverse processes and neural spines. Most mammals have distinct vertebral epiphyses at each end of their growing vertebrae. Vertebral epiphyses fuse to their respective vertebral centra at physical maturity of the vertebra. Skeletal maturity is achieved when all vertebral epiphyses are fused: after this event, longitudinal growth of the vertebral column ceases. Manatees (and humans) are among the few species that have indistinct or no vertebral epiphyses. Copyright S. A. Rommel.

References

Alexander, R. McN. (1994). "Bones: The Unity of Form and Function." Macmillan, New York.

Calder, W. A., III. (1966). "Size, Function, and Life History." Dover, Mineola, NY.

Evans, H. E. (1993). "Miller's Anatomy of the Dog," 3rd Ed. Saunders, Philadelphia, PA.

Fagone, D. M., Rommel, S. A., and Bolen, M. E. (2000). "Sexual Dimorphism in Vestigial Pelvic Bones of the Florida Manatee. Florida Scientist.

Flower, W. H. (1885). "An Introduction to the Osteology of the Mammalia," 3rd Ed. Macmillan and Co., London. [Reprinted by A. Asher & Co., Amsterdam, 1966.]

Flower, W. H., and Lydekker, R. (1891). "An Introduction to the Study of Mammals Living and Extinct." Adam and Charles Black, London.

Giffen, E. B. (1992). Functional implications of neural canal anatomy in recent and fossil marines carnivores. *J. Morphol.* **214,** 357–374.

Howell, A. B. (1930). "Aquatic Mammals: Their Adaptations to Life in the Water." Charles C. Thomas, Springfield, IL.

Kardong, K. V. (1998). "Vertebrates: Comparative Anatomy, Function, Evolution," 2nd Ed. McGraw-Hill, Boston.

Klima, M., Oeleschlager, H. A., and Wunsch, D. (1980). Morphology of the pectoral girdle in the Amazon dolphin *Inia geoffrensis* with special reference to the shoulder joint and movements of the flippers. *Sonderdeück Zeitschrift Saugetierkunde* **45**(5), 288–309.

Pabst, D. A., Rommel, S. A., and McLellan, W. A. (1999). Functional anatomy of marine mammals. *In* "Marine Mammals" (J. E. Reynolds and S. A. Rommel, eds.), Vol. I, pp. 15–72. Smithsonian Institution Press, Washington, DC.

Pierard, J. (1971). Osteology and myology of the Weddell seal *Leptonychotes weddelli* (Lesson, 1826). *Antarctic Res. Ser.* **18,** 53–108.

Pierard, J., and Bisaillon, A. (1978). Osteology of the Ross seal *Ommatophoca rossi* Gray, 1844. *Antarctic Res. Ser.* **31,** 1–24.

Romer, A. S., and Parsons, T. S. (1986). "The Vertebrate Body." 6th Ed. Saunders, Philadelphia.

Rommel, S. A. (1990). The osteology of the bottlenose dolphin. *In* "The Bottlenose Dolphin" (R. Reeves and S. Leatherwood, eds.), pp. 29–49. Academic Press, San Diego, CA.

Schmidt-Nielsen, K. (1984). "Scaling: Why Is Size so Important? "Cambridge Univ. Press, New York.

Strickler, T. L. (1978). Myology of the shoulder of *Pontoporia blainvillei,* including a review of the literature on shoulder morphology in the Cetacea. *Am. J. Anat.* **152**(3), 419–431.

Young, J. Z. (1975). "The Life of Mammals, Their Anatomy and Physiology." Clarendon Press, Oxford.

Skull Anatomy

SENTIEL A. ROMMEL
Florida Marine Research Institute, St. Petersburg

D. ANN PABST AND WILLIAM A. MCLELLAN
University of North Carolina, Wilmington

To appreciate skull anatomy, take a moment and look at your own face in a mirror. The structures above the neck are designed for the acquisition and initial processing of nutrients, the exchange of respiratory gases, the acquisition of sensory information about light, sound, touch, odor, and taste, and the broadcast of information about your own thoughts and emotions. Sensory and motor information is processed and sent from here to coordinate body functions. Complex signals can be sent to others of our species via vocalizations and/or the contractions of facial muscles. The head is our window for contact, perception, and communication with our world, and the skull provides the framework, the organizing hub, for the head. Thus, the skull is interesting in itself. It is also fundamentally important in our picture of evolutionary biology. This article describes the skull morphology of the evolutionarily diverse group called marine mammals (see Reynolds *et al.,* 1999).

I. Defining the Term "Skull"

The term "skull" is inexact. It has been used to describe the entire skeleton of the head. It has also been used to refer to only the cranium (plural, crania) or housing for the brain and sense organs, and the upper jaw (the part of "poor Yorick" that Hamlet held in his hand). This article uses the word skull to refer to the *entire* head skeleton, including the cranium and the derivatives of the first two visceral arches, i.e., the lower jaw (or mandible, which is composed of right and left dentaries) and the hyoid apparatus. The mandible and hyoid apparatus of marine mammals have received less attention in the literature, but they are particularly important in adaptations for feeding (see later).

The skull acts as a foundation for the fat, muscle, skin, and vascular and sensory structures that form the head. Thus, the skull alone does not dictate the contours of the head. For example, the external profile of the manatee head differs from that of the underlying skull (Fig. 1), and the relationships between the bones and the soft tissues of the head vary among species (Fig. 2). Odontocete cetaceans have a melon, a fatty facial pad, the shape of which is only partly defined by the underlying bones (Mead 1975); the extreme difference in head and skull profiles may be found in the sperm whale (*Physeter macrocephalus*). Contrarily, the dorsal surface of the right whale's head follows closely that of the underlying skull, although the right whale also has huge lower lips that follow the contour of the upper jaw but are not predicted by the outline of the lower jaw.

The shape of the head may also influence the dynamics of LOCOMOTION and balance. The completely aquatic species (cetaceans and sirenians) have shorter necks and less need for "antigravitational" muscles that support the head than do terrestrial mammals (imagine the right whale moving its head around in the air the way a sea lion does).

II. Feeding and Swallowing

The specific characteristics of a skull (including dentition) often reflect the animal's methods of FEEDING (Fig. 3). For example, the "typical" heterodont dentition (Kardong, 1998) of terrestrial mammals such as the dog is also found to various degrees in seals, sea lions, walrus (*Odobenus rosmarus*), sea otter (*Enhydra lutris*), and polar bear (*Ursus maritimus*).

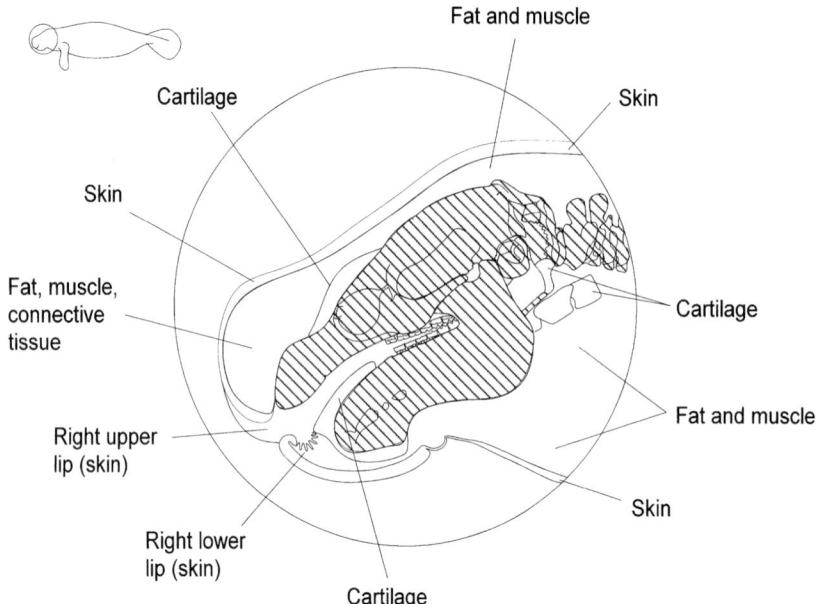

Figure 1 *Positions of skull features within the head vary among species. Here we illustrate how the midline contours of the manatee head differ from a left lateral illustration of the skull. The skeletal elements are cross-hatched to contrast them with soft tissues. Copyright S. A. Rommel.*

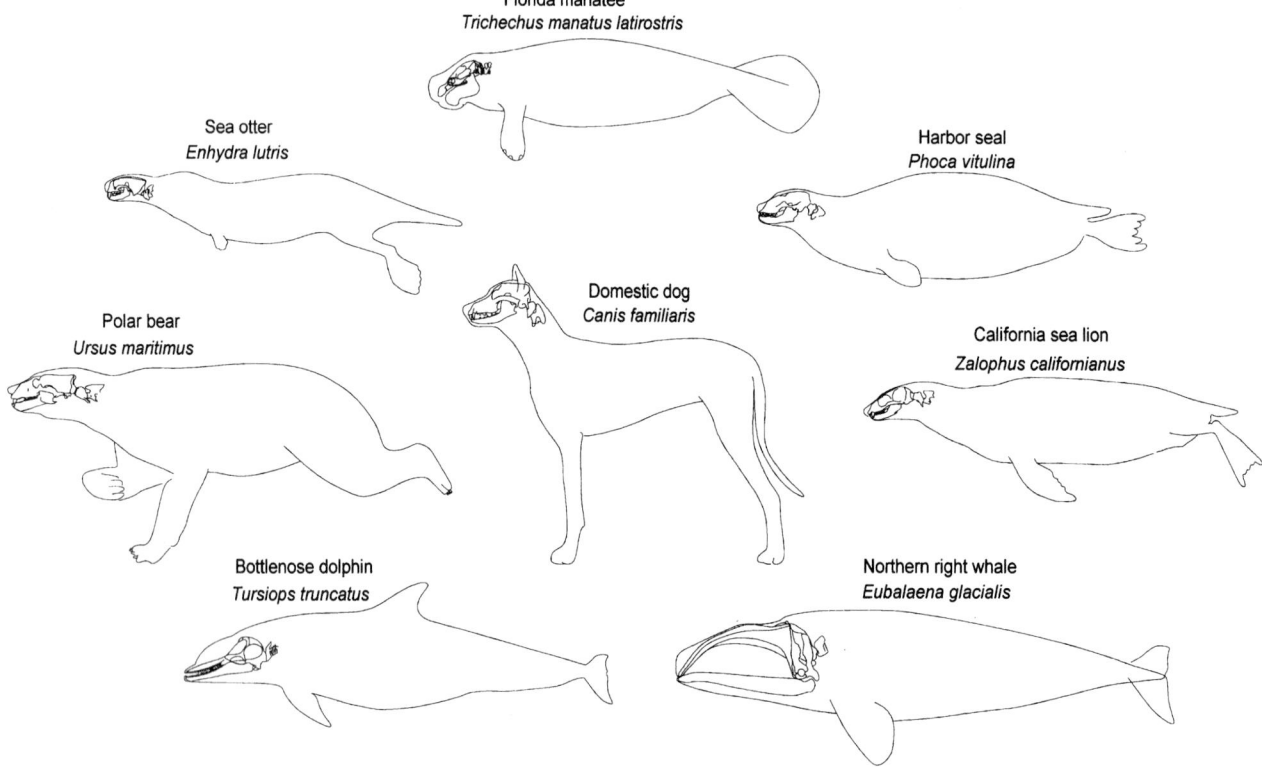

Figure 2 *Skulls and first two cervical vertebrae (unless fused, as it is in cetaceans) of a selection of marine mammals for comparison with those of the dog. Each species is scaled so that the distances between the shoulder and the pelvis are similar; body cavities are therefore roughly similar in length, allowing one to compare head sizes with visceral volumes among species. Copyright S. A. Rommel.*

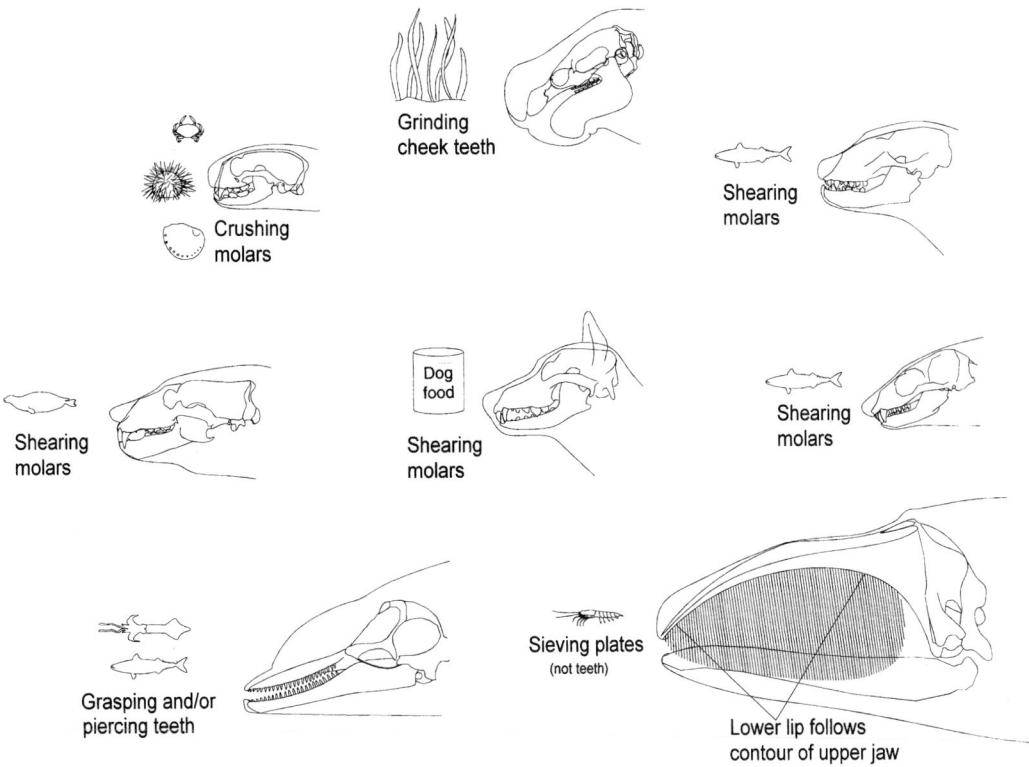

Figure 3 *Feeding apparatus and typical food. Dominant tooth type is also given. Note that the embryonic teeth of the right whale have been replaced with horny plates of baleen in the upper jaw only. Copyright S. A. Rommel.*

Heterodonty is tooth shape differences in different parts of the mouth: incisors and canines rostrally and premolars and molars caudally. Although each of these tooth types may vary in shape, the definitions are specific and are related to tooth position in the bones of the *upper* jaw (Hildebrand, 1995). Incisors are found only in the incisive or premaxillary bone; canines are found in the maxilla, in or very near the suture with the premaxilla; premolars are deciduous cheek teeth (erupting from the maxillary bones) that are found rostral to the molars; and molars are nondeciduous cheek teeth. Typically, deciduous teeth are replaced vertically as they are pushed out by the developing permanent teeth deep to them. Each tooth shape may perform a distinct function, sort of a "Swiss Army mouth" (G. Early, personal communication). Incisors, if chisel like, are for slicing and chipping and, if pointed, for piercing. Long, pointed canines are good for capturing and piercing. Relatively blunt cheek teeth are good for crushing and grinding. Manatees, unlike all other marine mammals (including the dugong, *Dugong dugon*), have continuous, horizontal tooth replacement of their molars (Domning and Hayek, 1984).

In some species, all of the teeth have the same shape—this is the homodont condition. The homodont dentitions of odontocetes and manatees differ in shape and function (they are grasping teeth for dolphins and grinding teeth for manatees).

Baleen whales lose their embryonic teeth and acquire food by sieving PLANKTON, small fish, and (in the case of gray whales, *Eschrichtius robustus*) benthic invertebrates using plates of horny (keratinized) BALEEN suspended from their upper jaws (Pivorunas 1979).

The lower jaw generally mirrors some of the features of the upper jaw, particularly dentition. Articulation of the mandible with the cranium (the temporomandibular joint or TMJ) also reflects feeding style. The lower jaws of the rorquals must support the large, pleated gular sac that fills with water and prey during lunge feeding (Lambertsen, 1983), whereas those of the skimmers, such as the right whales (*Eubalaena* spp.), support massive lower lips that guide an almost continuous stream of water past long baleen plates. Gray whales, which are bottom feeders, have relatively robust lower jaws. In addition to supporting their fish-grasping teeth, the lower jaws of odontocetes are fat filled and perform the acoustic function of receiving and guiding sound energy to the earbones (Norris and Harvey, 1974). The lower jaws of the herbivorous manatees (*Trichechus* spp.) support grinding cheek teeth. The mandible of the manatee also has large mental foramina to transmit the large nerves associated with the vibrissae (see later).

Feeding style may also be reflected in the shape of the rostrum, the zygomatic arch, and the temporal fossa. Relatively large temporal muscles and their fossae are typical of carnivorous mammals that tear or shear flesh without finely dividing that prey in the mouth and/or have teeth for killing and temporarily holding prey (Hildebrand, 1995). Carnivorous mammals have upper and lower tooth rows that do not slide horizontally but rather occlude with a chopping motion, using a hinge joint that is roughly in line with the tooth row (Fig. 3). In contrast, the TMJ in herbivores is typically above the tooth line (Fig. 3). Relatively large masseters and the robust zygomatic arches that support them are more typical of herbivorous mammals that use a crushing and rolling action to chew. This feeding style requires a TMJ that can slide and a large masseter to apply force along the extent of the cheek tooth row. TMJs of the mysticetes accommodate complex axial rotations of the individual dentary bones and absorb shock in lunge feeding of rorquals (Lambertsen, 1983), whereas the two dentaries of the odontocetes tend to fuse with age into a single compound bone.

Feeding includes swallowing. How do the bones of the head accommodate swallowing? Hyoid bones provide the mechanical support of many of the muscles that act upon the tongue and the larynx. Swallowing requires the coordinated action of these muscles as food leaves the oral cavity and moves through the pharynx. The dorsal hyoid bones (epihyoids) are attached by the stylohyoid cartilage to the base of the skull at a position immediately caudal to the tympanic bones (Fig. 4A). They provide the primary hinge to depress and move the tongue caudad. The ventral hyoid bones (basihyoids) form a secondary

hinge that allows sternohyoid (sternum to hyoid) and mylohyoid (mandible to hyoid) muscles to move the basihyoid bones forward and back (Fig. 4B). In suction feeders, such as the squid-eating beaked whales and pilot whales (*Globicephala* spp.), the hyoid apparatus and its associated muscles are massive.

The soft and hard palates that separate the respiratory system from the digestive system are almost unique to mammals and allow extended chewing time or suckling while breathing. The fleshy soft palate extends the bony margin of the nasal cavity (Fig. 5) and makes a very important contribution to the evolutionary forces that shaped mammal skulls. Prolonged chewing, while breathing, means there is additional time for food processing in the mouth. Teeth of mammals are more indicative of food type than is the case for most other vertebrates. The specificity of dentition and the hardness of teeth, which increase preservation in the fossil record, contribute significantly to our current understanding of the ecology and evolution of extinct species.

III. Bony Features vs Bones

One approach to studying the skull is to focus on bony features (Fig. 6). Bony features are morphological characters or landmarks of the skull that are formed by one or more bony elements (individual elements or bones may develop from one or more ossification centers). Bony features reflect evolutionary,

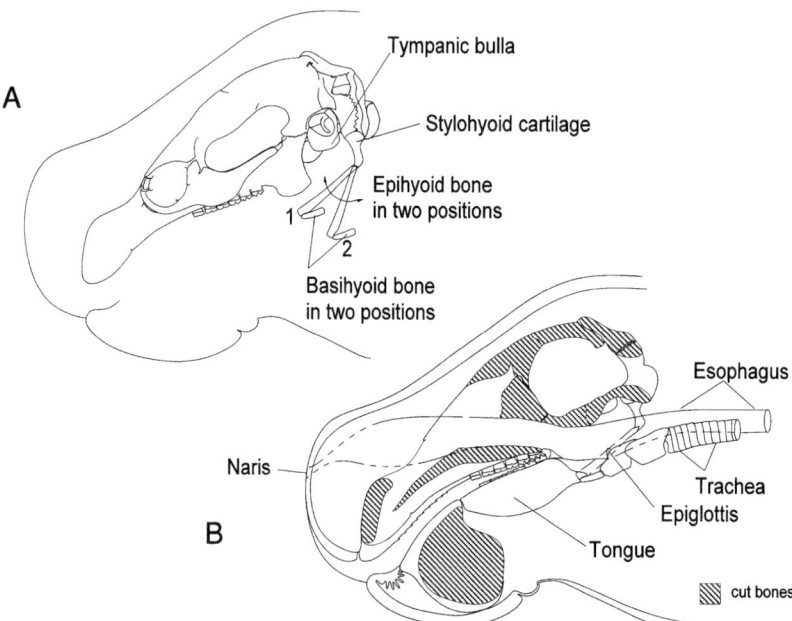

Figure 4   (A) *Lateral view of the manatee cranium with the hyoid apparatus in two positions to illustrate its range of motions. Muscles between the basihyoid and the sternum move the hyoid apparatus down and back. Muscles between the tongue and basihyoid move the hyoid apparatus up and forward. (B) Left midsaggital section of the manatee head. Movements of the hyoid apparatus influence the trajectories of air and food through the head and into the trachea and esophagus, respectively. Copyright S. A. Rommel.*

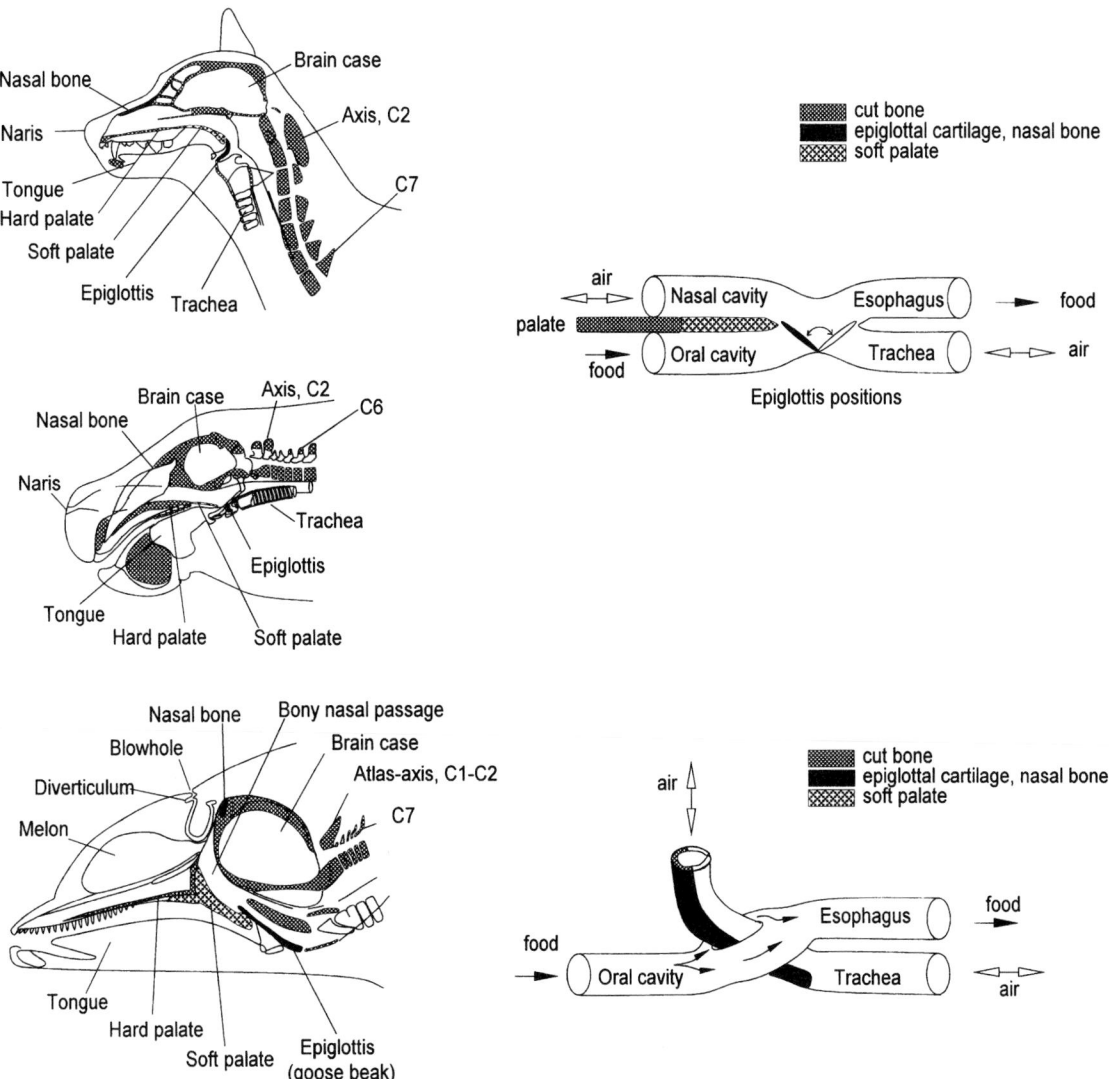

Figure 5 *Comparisons of the morphological adaptations of the mammalian head that allow respiration while food is in the mouth. Separation of oral and nasal cavities accommodates prolonged chewing and allows teeth to be modified accordingly. The dog and manatee (upper two on the left) are represented schematically in the upper right. In the dolphin (lower left and right), further modification is shown to accommodate the migration of the respiratory opening to the top of the head. Copyright S. A. Rommel.*

developmental, and mechanical pressures in a grossly visible manner. In contrast to bony elements or skull bones, bony features are structures that incorporate one or more parts of individual bones. For example, the zygomatic arch, which supports the masseter muscle and helps close the jaws, may be composed of one, two, or three bones. The rostrum or muzzle may be elongate and may include the nasal bones. Thus, to characterize individual skulls without having to identify individual bones, biologists can use the morphology of bony features such as zygomatic arch shape and composition; rostrum length; orbit size, shape, and position; and jaw articulation.

Take, for example, the small zygomatic arch of the right whale and the relatively delicate one of the dolphin. In these species the masseter muscles are relatively small and the temporal muscles are relatively massive. Compare these zygomatic arches with the massive ones in the manatee and polar bear.

In general, large, forward-facing orbits are characteristic of predators that rely on vision as their primary sensory modality, whereas laterally facing orbits are more typical of nonpredatory species. Also note how the orbits of most species in Fig. 6 are open caudally, in contrast to those of the fully aquatic mammals. In species in which the orbit is open, there is a postorbital ligament caudal to the eye that extends between the postorbital process of the frontal bone and zygomatic arch; these postorbital processes, most evident in the polar bear, serve as sites for muscle attachment.

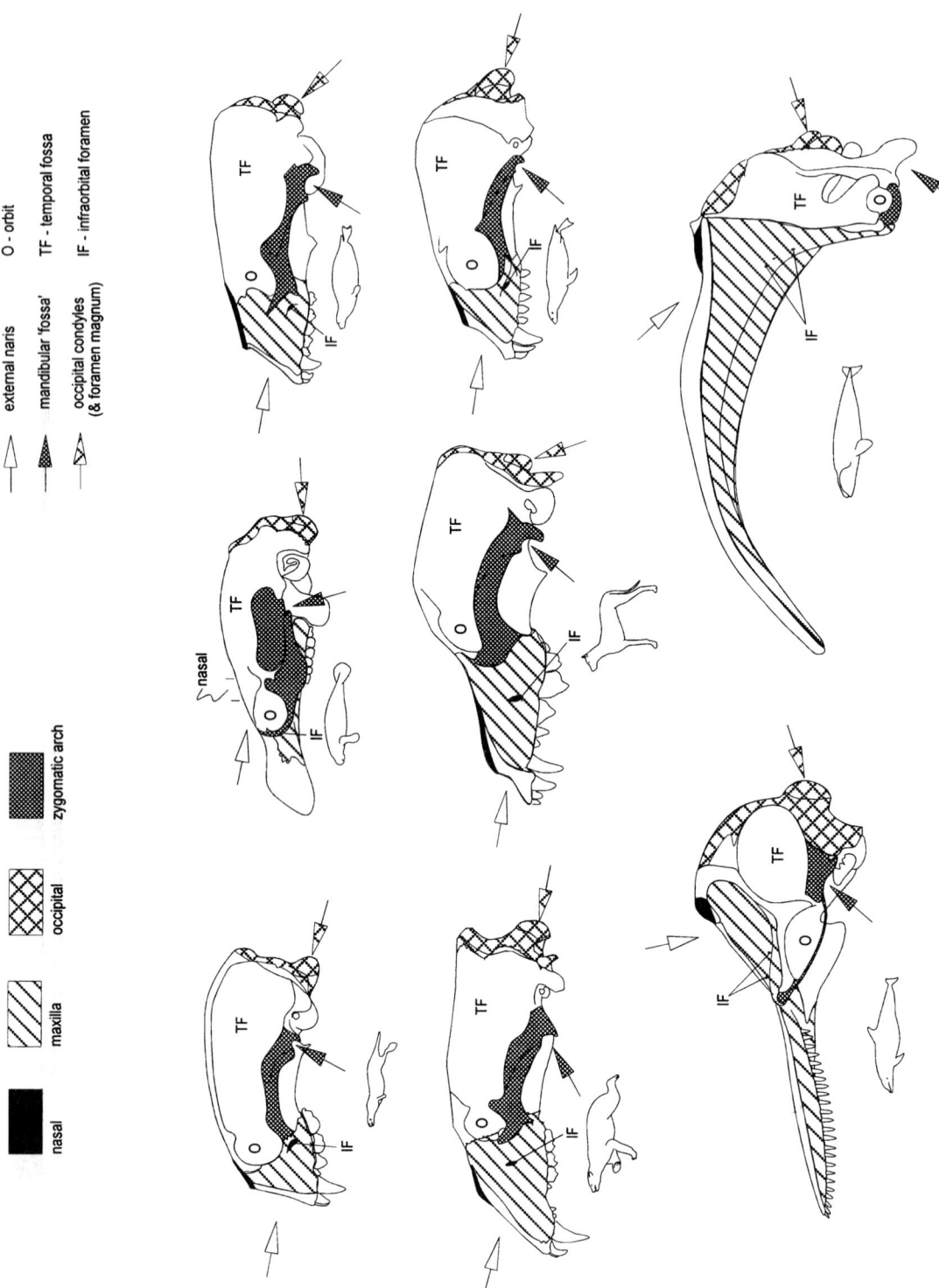

Figure 6 Selected bony features are emphasized. The rostrum, composed of the premaxilla, maxilla, and occasionally the nasal bone, forms the "face" of each species. The zygomatic arch, which supports the masseter muscle, may be composed of a single bone, the jugal, or parts of as many as three bones in some species. Arrows indicate directions of air flow at the external naris, vertebral column articulation at the occipital condyle, and lower jaw articulation at the mandibular fossa. Processes extending from the frontal may be present to help form the dorsocaudal aspect of the orbit, referred to as postorbital processes of the frontal: absent in the seal, small in the otter, and relatively large in the dolphin. Copyright S. A. Rommel.

The positions (on the top of the skull) of the external nares may reflect respiratory adaptations to diving, feeding, and locomotion. Note the positions of the external nares in fully aquatic species. The internal nares or passageways for air through the cranium are bordered by the nasal bones in all species.

The occipital condyles position the head on the neck and influence the flexibility at the terminal end of the neck (Figs. 2 and 6). Some of the species with short necks have two or more fusions of cervical vertebral (Flower, 1885; Slijper, 1979; Rommel, 1990), placing the base of the skull very near the shoulder joint and the thoracic cavity. Species with long necks may have a wide range of neutral head positions and may also have a greater range of movement than fully aquatic species (King, 1983).

IV. Ground Plan of Skull Bones

What other factors shape the skull? In all vertebrates, the skull bones develop from ossification centers in a basic pattern that partially or completely encloses the brain plus the sensory organs of olfaction, vision, HEARING, and balance (Fig. 7). Bones that are preformed in cartilage are referred to as endochondral or replacement bones; those deposited directly as bone within tough connective tissue membranes are referred to as dermal or membrane bones. The distinction between endochondral

and dermal bones is valuable in establishing homologies with skull bones of the lower vertebrates.

How does one determine which elements of the skull, i.e., which of the 40-some bony elements, are homologous between species? A systematic approach that allows one to compare homologous elements (Fig. 8) is a particularly useful way to avoid the "mental indigestion" (Romer and Parsons, 1977) of having to memorize all the individual bones.

In some species, individual bones fuse (ankylose) to form compound bones;[1] these include the occipital, temporal, and sphenoid "bones." Of particular interest is the temporal "bone," which is a compound bone made up of many separate bony elements and/or ossification centers (see Kent and Miller, 1997). In many mammals, the bulk of the temporal is a single unit with no visible sutures between the bony elements once skeletal maturity is reached. Thus, it is common to refer to the temporal as a bone, but this is not the case in several of the marine mammals, particularly cetaceans and sirenians.

The exploded diagram of the cranium of the Florida manatee illustrates the overlapping and/or abutting margins that make up the skull sutures (Fig. 9). Note that the ear bone complex—made up of the periotic and tympanic bones plus the middle ear ossicles (only the malleus is visible in this figure)—is illustrated as a single unit. Another compound bone is the occipital, which in some species is composed of the basioccipital, exoccipitals, supraoccipital, and inter(post) parietal (Jollie, 1973; Kellogg, 1928).

Figure 10 is a left lateral view of the individual cranial bones for our representative marine mammals. Contrast this illustration with Fig. 6 to reinforce the distinction between bones and bony features; see Fig. 2 to compare skull size to total body size. The abbreviations are the same as in the schematic generic cranium (Fig. 8) and the exploded manatee cranium (Fig. 9). Pick any one bone and compare it with the homologous structures in all of the crania.

V. Sutures

The regions between adjacent bones are referred to as sutures. In the exploded illustration of the manatee skull (Fig. 9), the hatched regions represent sutures or regions of overlap; skull bones can meet in several ways and attach by more than one type of material (e.g., cartilage and connective tissue).

In Fig. 11, the sutures of the dog skull are compared with those of the manatee skull. Note that different types of sutures may be found between the same bones in different species. The type of suture generally reflects the mechanical forces necessary to accommodate whatever biological constraints are

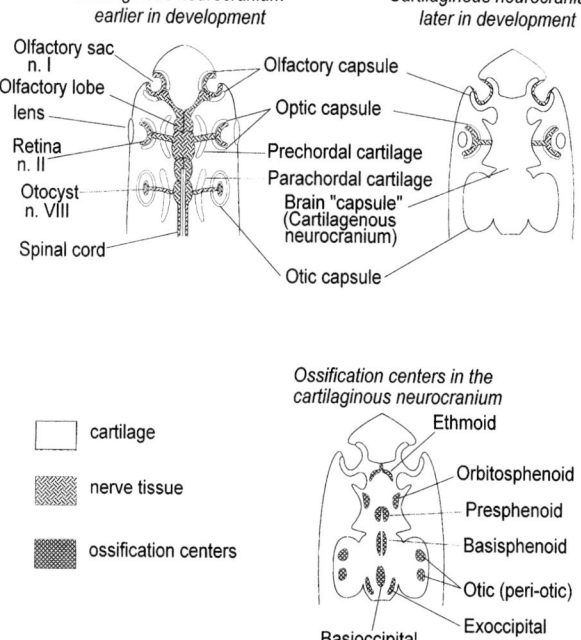

Figure 7 *Schematic illustration of ventral views of the developing vertebrate skull (modified after Kent and Miller, 1997). The basic plan of encapsulation of the senses and brain is illustrated in the top two drawings. The lower drawing illustrates the ossification centers that will eventually become the replacement bones of the cranium. Copyright S. A. Rommel.*

---

[1]We have used three different terms incorporating the word "bone." Bones or bony elements are discrete ossifications that can be traced phylogenetically. Compound bones are structures that appear to be single units in adults because the sutures between the component bones have been resorbed or are not apparent. Bony features are gross features, such as the zygomatic arch, that are made up of discrete and distinct bones or parts of bones (e.g., the zygomatic process of the squamosal bone).

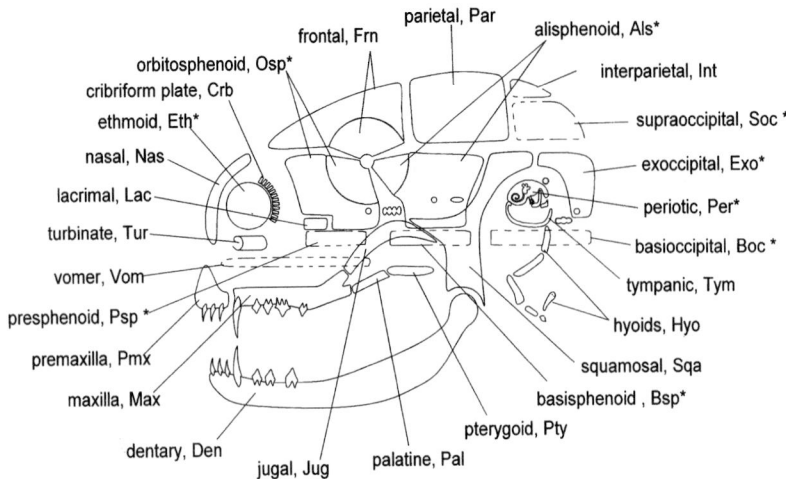

Figure 8 *Left lateral schematic of the mammalian skull illustrating relative bone positions. Most skull bones are bilaterally paired—the unpaired skull bones are positioned along the midline and are dotted in this illustration. This schematic approach has been used for more than 100 years and provides a framework in which to compare a wide variety of mammalian skulls (modified after Flower, 1885; Kent and Miller, 1997; Evans, 1993). Recall that the nose, eyes, and ears are encapsulated early in development; these three sensory areas are represented by circular regions in the schematic. Endochondral bones (first laid down as cartilage) are marked with an asterisk. Copyright S. A. Rommel.*

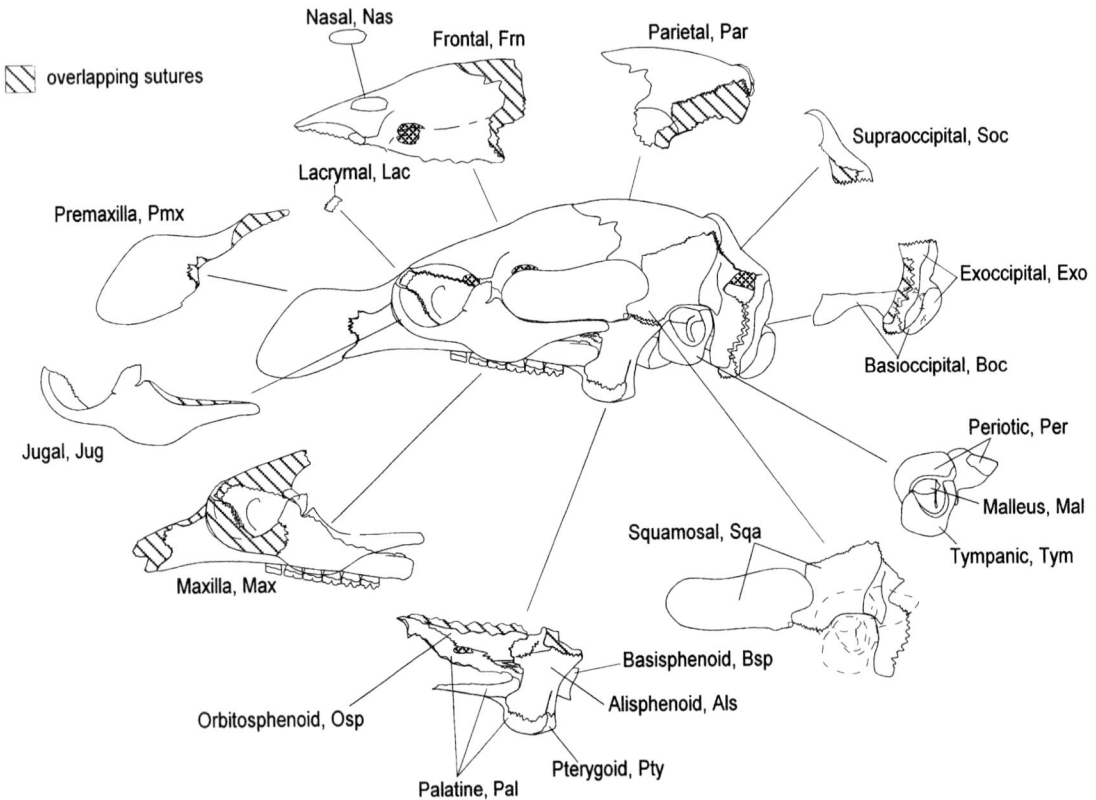

Figure 9 *Exploded cranium of the Florida manatee. The overlapping and/or abutting margins of bones that make up the sutures of the cranium are shown. Copyright S. A. Rommel.*

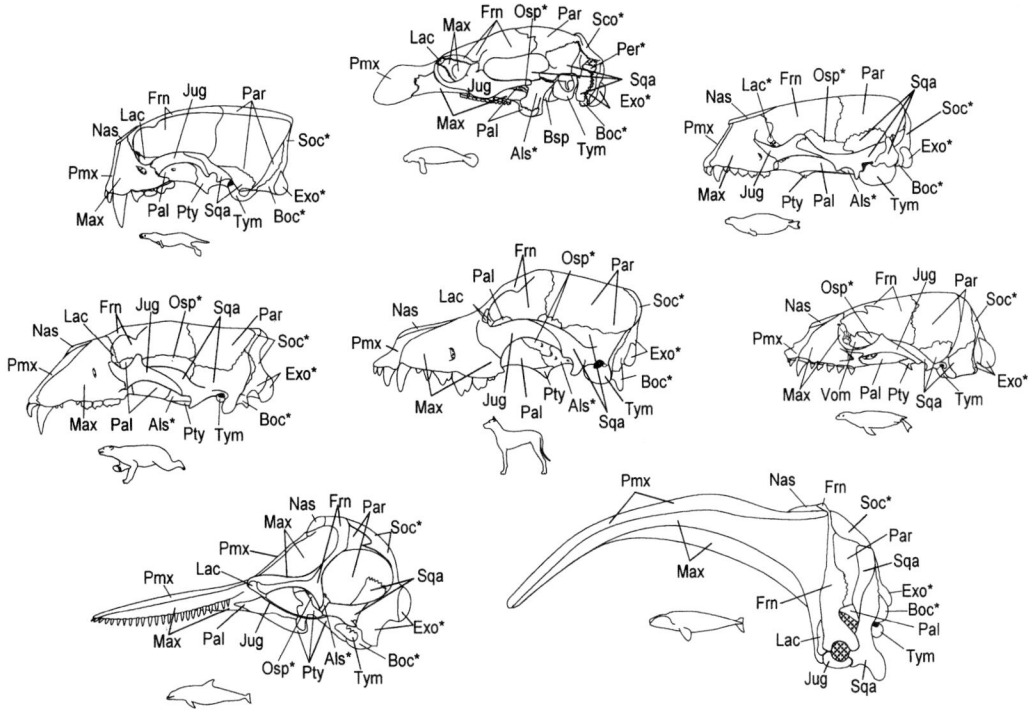

Figure 10 *Left lateral illustrations of individual cranial bones of selected marine mammals and the dog. Abbreviations are the same as in Fig. 8. Use Fig. 8 to help visualize how each species has modified the basic plan of mammalian skull morphology. Copyright S. A. Rommel.*

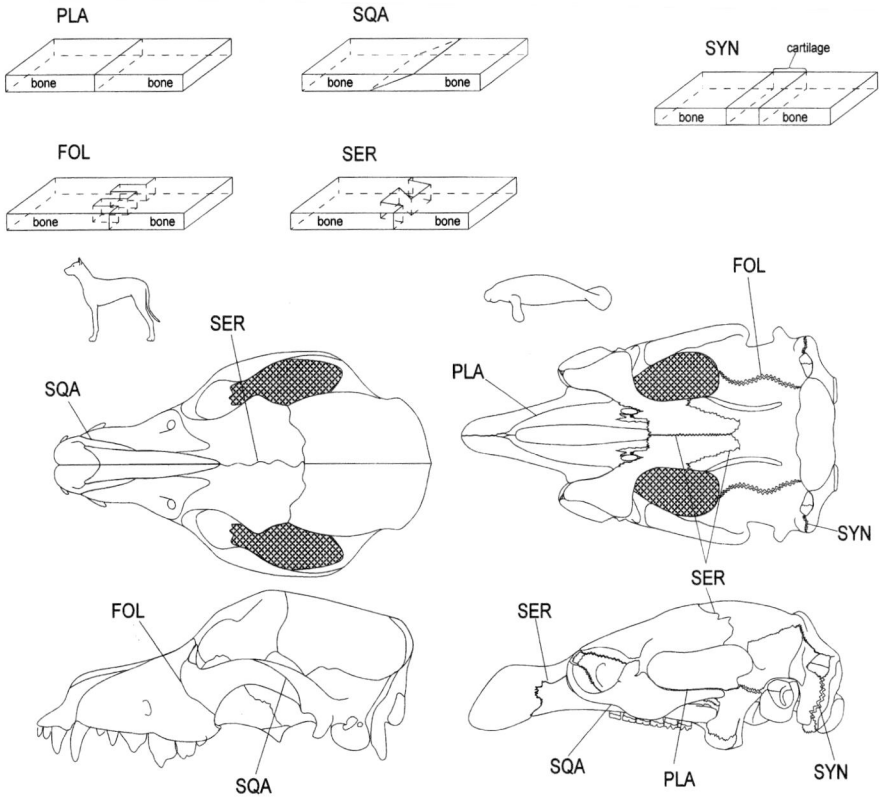

Figure 11 *Suture types in the dog and manatee. Definitions are from a variety of sources, and names in parentheses are from Schaller (1992). Suture types are defined by their shape. PLA, plane or butt joint (harmonious, sutura plana); an approximately straight suture with nearly squared-off margins. SQA, squamous or scarf joint (sutura squamosa); a suture with tapered overlapping margins. FOL, foliate joint (sutura foliata); a regular suture in which adjoining bones interleave. SER, serrate joint (sutura serrata); an irregular suture in which adjoining bones interlock. SYN, synchondrosis; a joint that has persistent cartilage between bones (synchondroses cranii). Copyright S. A. Rommel.*

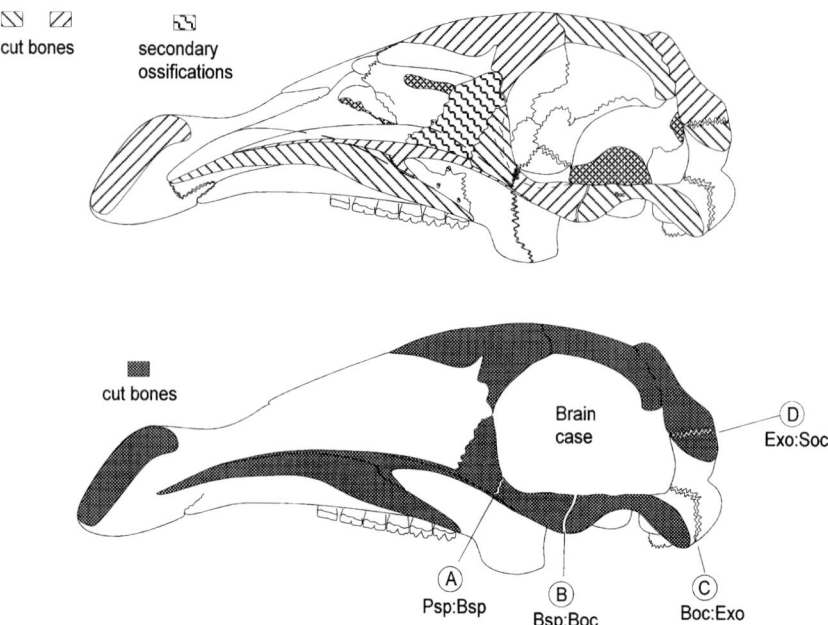

Figure 12 *Sagittally sectioned crania of the manatee. The upper illustration shows the number and complexity of individual bones. The lower drawing is simplified to illustrate four sutures that give some indication of relative age in the manatee. (A) Between the presphenoid bone and the basisphenoid bone. (B) Between the basisphenoid bone and basioccipital bone. Sutures labeled A and B are the first to close and may reflect growth of the braincase. (C) Between the basioccipital bone and the two exoccipital bones (bilaterally paired sutures). (D) Between the two exoccipital bones and the supraoccipital bone. Sutures labeled C and D ankylose later in life and probably reflect changes in the mechanics of head and neck morphology as the animal matures. Copyright S. A. Rommel.*

imposed upon the bones—these constraints include muscle attachments, size and shape of nerves and sensory capsules; and mechanical attachments for feeding, display, locomotion, and social interaction. Interlocking joints can absorb energy. Movable joints eliminate shearing forces. Butt joints can support little shear but are strong in compression. Squamosal or scarf joints allow more surface contact between adjacent bones and are stronger than simple butt joints. Some configurations affect the complexity of joints, and this complexity, in turn, affects the action and strength of the joints.

Relative aging in cetaceans was pioneered by Perrin (1975). He established principles that allow one to use changes in the skeleton that occur during an animal's life to estimate the age of a specimen by comparing it to others. The sutures of a skull can be used for relative aging in some species, e.g., the sutures that allow brain growth ankylose after the brain stops growing. In the manatee, five skull sutures (two associated with the occipital are bilaterally paired) are used to estimate relative age (Fig. 12). The first two (A and B) appear to reflect the changes in brain size and ankylose within the first few years. The three associated with the occipital bone appear to be more related to musculoskeletal changes that occur during juvenile and adult life.

Development of the skull bones proceeds at a pace different from the development of the soft tissues of the head. Bone is constantly being remodeled and thus it can indicate aspects of the evolutionary forces that work to separate species. This plasticity is reflected in the way individual skull bones form around vessels and nerves. The resulting openings, or foramina (singular, foramen), are often phylogenetically conserved and so can be used to establish homologies of the same bones in very different species. An individual nerve or vessel may be completely surrounded by a bone or bones of the skull, resulting in a specific foramen. Because this process occurs early in the development of the individual and appears to have the same essentials in all vertebrates, we use these cranial nerve foramina to help identify the cranial bones (Fig. 13).

VI. Telescoping

Telescoping is a process often cited describing the skulls of cetaceans. The term, coined by Miller (1923), refers to the elongation of the rostral elements and the dorsorostral movement of caudal elements (Kellogg, 1928; Miller, 1923; Rommel, 1990). The relative movement of skull bones in cetaceans

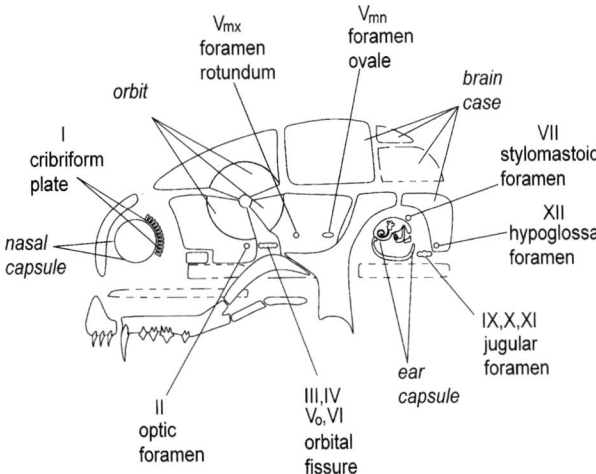

Figure 13 *Openings, or foramina, of the skull. Foramina can be used to establish homologies of the same bones in different species; each foramen (or foramina) associated with 1 or more of the 12 cranial nerves (labeled I through XII) has a name that is used (fairly) consistently by vertebrate morphologists. Thus the cribriform plate, found at the anterior margin of the braincase, is associated with the olfactory nerves (this will be referred to parenthetically as I—olfactory n.) in all of the species that have a sense of smell [even odontocetes, which do not have olfactory nerves as adults, have these perforations (Rommel, 1990)]. The second cranial nerve, associated with the optic nerve, passes through the optic foramen and usually perforates the orbitosphenoid bone (II—optic n.). The orbital fissure is usually at the orbitosphenoid bone–alisphenoid bone suture (also anterior lacerate foramen; III—oculomotor n., IV—trochlear n., V_o—ophthalmic branch of trigeminal n., VI—abducens n.). The foramen rotundum (V_{mx}—maxillary branch of the trigeminal n.) and the foramen ovale (V_{mn}—mandibular branch of the trigeminal n.) perforate the alisphenoid bone. The stylomastoid foramen is located at the tympanic bone–basioccipital bone suture (VII—facial n.). (Nerve VIII—vestibulocochlear n. is not shown; it perforates the periotic bone through its internal auditory meatus.) The jugular foramen (also posterior lacerate foramen) is at the exoccipital bone–basioccipital bone suture (IX—glossopharyngeal n., X—vagus n., XI—accessory n.). The hypoglossal foramen usually perforates the exoccipital (XII—hypoglossal n.). An additional cranial nerve (O—terminal n.) was discovered after the numbering system was developed. This nerve is found rostral to the olfactory nerve of the species illustrated; it has been described only in odontocetes and is not illustrated here. Copyright S. A. Rommel.*

creates sutures with considerable overlap of adjacent bones. If the skull were sectioned, one could observe as many as four different bones overlapping each other; this overlap resembles old-fashioned collapsible telescopes. In dolphins, the external nares have been displaced to the dorsal apex of the skull so the nasal bones are located just caudal to the exter

nal nares but dorsal to the braincase. The premaxillary and maxillary bones have been extended at the rostral tip and also pulled up and back over the frontal bones to maintain their relative positions with the nasal bones. The narial passages are essentially vertical in cetaceans, which eliminates the nasal bones as roofing bones of the nasal passages. The nasal bones are, instead, relatively small vestiges that lie in depressions of the frontal bones; the roof of the cetacean mouth is not the floor of the nasal passages as it is in most other mammals. Telescoping is different in odontocete and mysticete cetaceans (Fig. 14). Changes in the mysticetes are dominated by a ventrocaudal extension of the maxillary bones, whereas in the odontocetes, the premaxillary and maxillary bones are shifted more dorsocaudally (Kellogg, 1928). Interestingly, whereas the odontocete facial muscles have moved up and back over the eye, the temporal muscles of the mysticetes have moved up and forward over the eye. Thus, the temporal fossae of mysticetes are very different from those of odontocetes. Contrast the morphological differences associated with telescoping in cetaceans (Fig. 14) with the homologous structures illustrated in Fig. 6.

Additionally, the remodeling associated with telescoping is reflected in the number and positions of the cranial nerve foramina in the maxilla. Consider the nerves that are associated with the muscles of the face: the (sensory) trigeminal nerve (V) and the (motor) facial nerve (VII). The right and left facial nerves control such muscle activity as facial expression in the dog, feeding in the manatee, and focusing of sonar pulses in the dolphin. The trigeminal nerves signal the brain to coordinate the muscular activities in the same region. The relatively large sizes of these two nerves in the dolphin and manatee reflect the importance of these neuromuscular actions. The nerve diameters are reflected by the size of the foramina (the infraorbital foramina) through which the trigeminal nerve pierces the maxilla (Fig. 15). As the bones and muscles of the face are reshaped to accommodate the evolution of a melon and its need for complex mechanical manipulation, the sensory and motor nerves are moved up and over the orbit. The homologous (and thus same-named) opening, which is *infra*orbital in most species, is now actually above (or *supra*orbital) the orbit in cetaceans!

VII. Additional Influence of Central Nervous System Morphology on the Skull

The shape and dimensions of the central nervous system strongly influence skull morphology. Consider the shape of the brain (Fig. 16) and how it might influence the shape of the braincase (region enclosing the brain). Note the arrangement and sizes of the cranial nerves and envision how the skull foramina would reflect these and their accompanying vessels. In some species, additional vascular adaptations for diving have influenced the development of foramina to excessive proportions (e.g., the *foramen magnum* is a relatively close fit to the spinal cord; but enlarged arteries and veins surrounding the central nervous system in cetaceans, phocids, and manatees have necessitated enlargement of the openings).

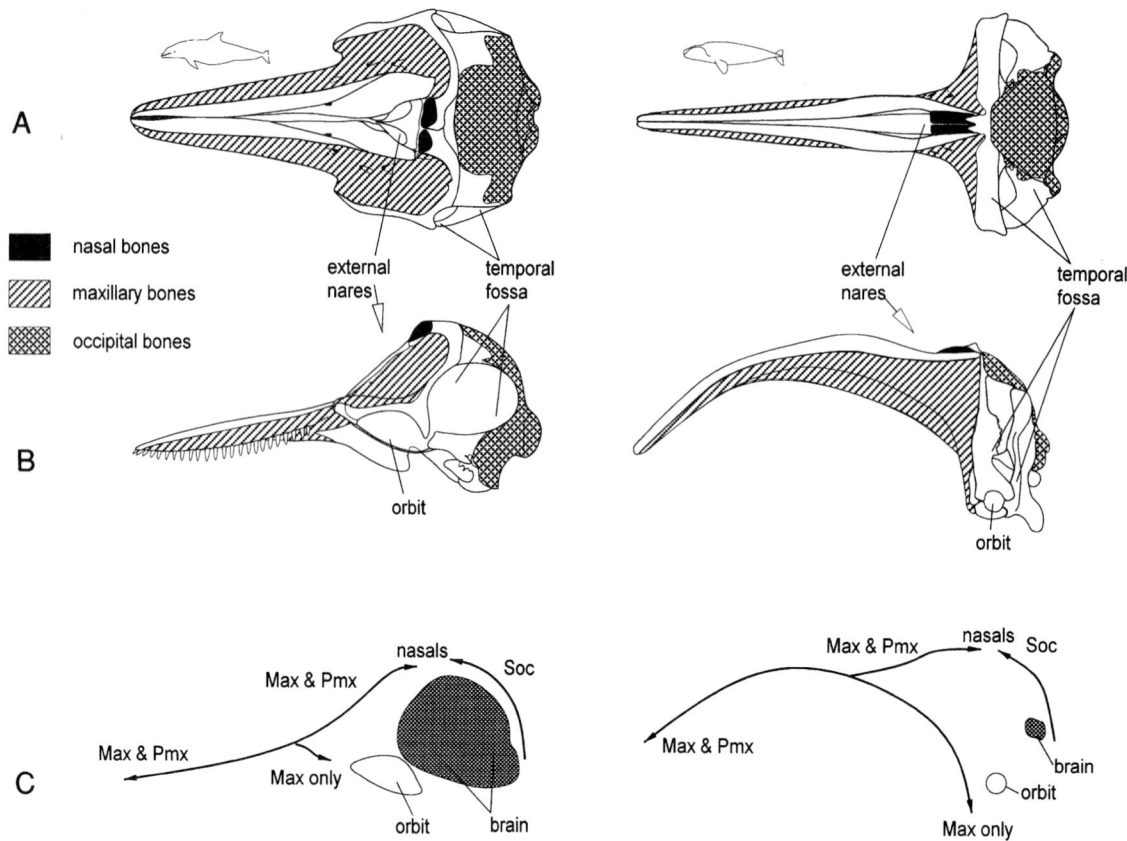

Figure 14 *Dorsal (A) and lateral (B) views illustrating telescoping in odontocetes (left, e.g., Tursiops) and mysticetes (right, e.g., Eubalaena). Telescoping refers to the elongation of the rostral elements [both fore and aft in the case of the premaxillary and maxillary bones (Pmx and Max), the vomer, and mesorostral cartilage], the dorsorostral movement of the caudal elements [particularly the supraoccipital bone (Soc)], and the overlapping of the margins of several bones. This overlap or sliding over each other of these elements resembles old-fashioned telescopes. One result of telescoping is the displacement of the external nares (and the associated nasal bones) toward the dorsal apex of the skull—up and over the rostral margin of the brain! Telescoping is actually quite different in odontocete and mysticete cetaceans; in most odontocetes the rostrum is dorsally concave, whereas in mysticetes the rostrum is ventrally concave. The temporal fossae of the mysticetes have moved up and forward over the eye; the temporal fossae in odontocetes are in a more typical mammalian position. Relatively more bone mass is moved up and over the orbit in odontocetes, whereas relatively more bone mass is moved down and under the orbit in mysticetes. In the lower schematic (C), arrows indicate the directions of relative movement as each skull is remodeled to accommodate the brain and the respiratory, feeding, and acoustic apparatus of the two types of cetaceans. Note that the odontocete brain makes up a larger percentage of the cranial volume than the brain of the mysticete; in C the brains are scaled to fit in the lateral views of the crania in B above them. Copyright S. A. Rommel.*

In conclusion, we can state that a glance at a skull tells a lot about an organism: how it senses its environment, how it feeds, how big its brain is, and so on. Instead of becoming bogged down in trying to memorize names, study a skull with an open mind (pun intended) about adaptations and function.

Acknowledgments

We thank Allen Foley, Llyn French, Lisa Johnson, Judy Leiby, Tom Pitchford, and James Quinn at the Florida Marine Research Institute and John E. Reynolds at Eckerd College for reviewing the manuscript. Anatomical illustrations were created with FastCAD (Evolution Computing, Tempe, AZ).

See Also the Following Articles

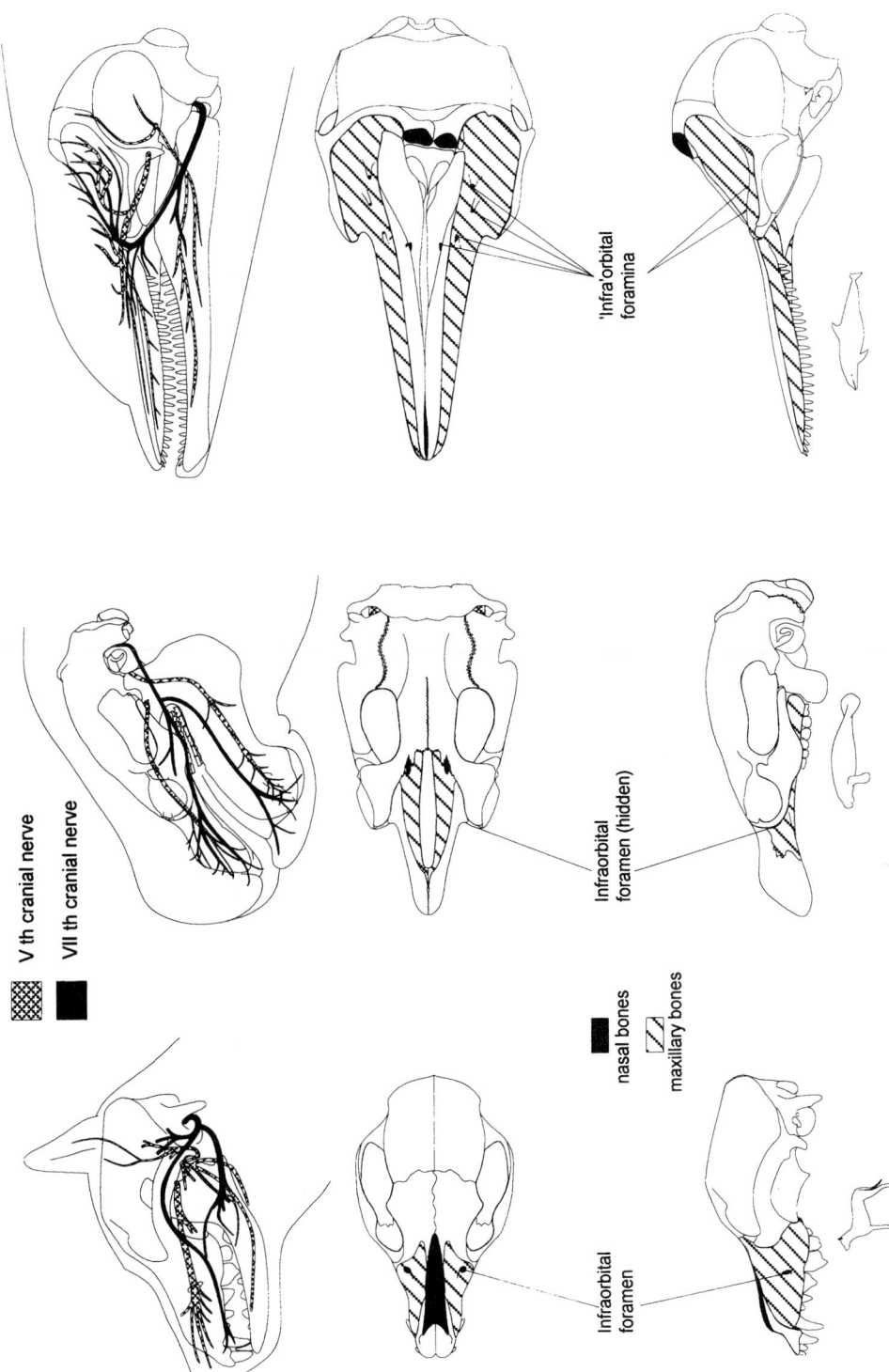

Figure 15 *Nerves associated with the facial muscles also reflect telescoping. These are the trigeminal, or Vth, cranial nerve (a sensory nerve) and the facial, or VIIth, cranial nerve (a motor nerve). Here the dog is compared with the manatee and the dolphin. The relatively large sizes of the Vth nerves are reflected in the relatively large foramina through the maxillae (infraorbital foramina); the larger the nerve, the more information it can carry. The process of telescoping has remodeled the bones of the rostrum, and included in this process are the reshaping of the muscles of the face and the nerves that enervate them. Also note that some odontocete cetaceans have notable bilateral asymmetry in the dorsal elements of the skull. Copyright S. A. Rommel.*

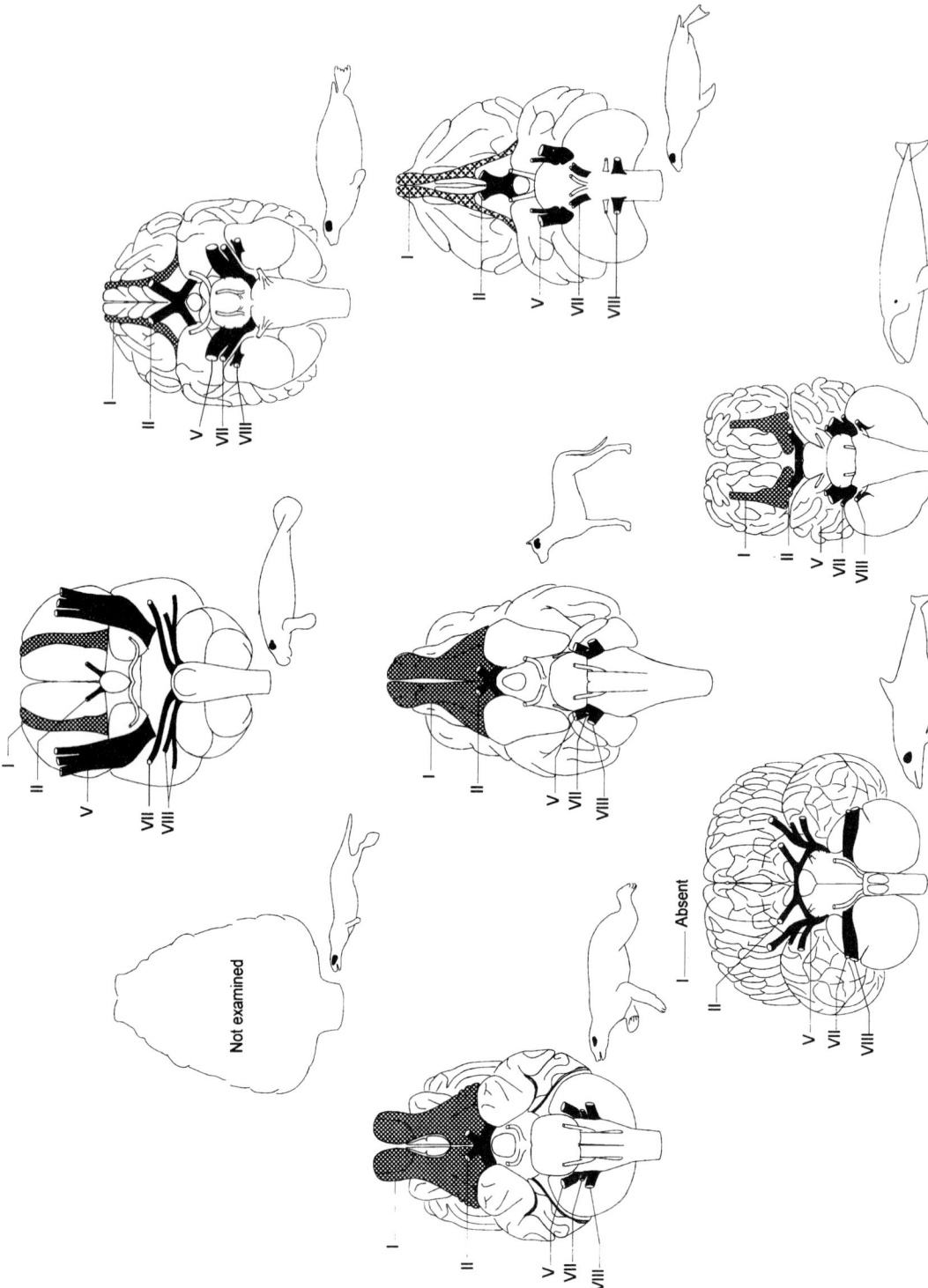

Figure 16 *Ventral views of selected marine mammal brains and of the dog brain. Inserts show lateral views of the brain positioned within the body to illustrate brain size relative to body size; note how relatively tiny the brain of the right whale is. Selected cranial nerves, indicated by Roman numerals, are labeled. Compare the trigeminal (V) and facial (VII) nerves illustrated in Fig. 15 to see the relative importance of the sensory and motor nerves from and to the facial muscles, respectively. Also note the relative sizes of the olfactory (I), optic (II), and acoustic (VIII) nerves; sizes reflect the relative importance of smell, vision, and hearing in the different species. Copyright S. A. Rommel.*

Not examined

Absent

References

Domning, D. P., and Hayek, L.-A. C. (1984). Horizontal tooth replacement in the Amazonian manatee (*Trichechus inunguis*). *Mammalia* **48**(1), 105–127.

Evans, H. E. (1993). "Miller's Anatomy of the Dog," 3rd Ed. Saunders, Philadelphia.

Flower, W. H. (1885). "An Introduction to the Osteology of the Mammalia," 3rd Ed. Macmillan, London. [Reprinted by A. Asher & Co., Amsterdam, 1966.]

Hildebrand, M. (1995). "Analysis of Vertebrate Structure," 4th Ed. Wiley, New York.

Jollie, M. (1973). "Chordate Morphology." R. E. Krieger Pub. Co., Huntington, NY.

Kardong, K. (1998). "Vertebrates, Comparative Anatomy, Function, Evolution." 2nd Ed. McGraw-Hill, New York.

Kellogg, R. (1928). The history of whales: Their adaptations to life in the water. *Q. Rev. Biol.* **3**(1), 29–76 and **3**(1), 174–208.

Kent, G. C., and Miller, L. (1997). "Comparative Anatomy of the Vertebrates," 8th Ed. Wm. C. Brown, Boston.

King, J. E. (1983). "Seals of the World," 2nd Ed. Comstock Publishing Associates, Ithaca, NY.

Lambertsen, R. H. (1983). Internal mechanism of rorqual feeding. *J. Mammal.* **64**(1), 76–88.

Mead, J. G. (1975). Anatomy of the external nasal passages and facial complex in the Delphinidae (Mammalia: Cetacea). *Smith. Contrib. Zoo.* **207**, 1–72.

Miller, G. S., Jr. (1923). The telescoping of the cetacean skull. *Smith. Miscell. Collect.* **76**(5), 1–71.

Norris, K. S., and Harvey, G. W. (1974). Sound transmission in the porpoise head. *J. Acoust. Soci. Am.* **56**(2), 659–664.

Pabst, D. A., Rommel, S. A., and McLellan, W. A. (1999). Functional morphology of marine mammals. *In* "Biology of Marine Mammals" (J. E. Reynolds and S. A. Rommel, eds.), pp. 15–72. Smithsonian Institution Press, Washington, DC.

Perrin, W. F. (1975). "Variation of Spotted and Spinner Porpoise (genus *Stenella*) in the Eastern Tropical Pacific and Hawaii." Univ. Calif. Press, Berkeley, CA.

Pivorunas, A. (1979). The feeding mechanisms of baleen whales. *Am. Sci.* **67**, 432–440.

Reynolds, J. H., Odell, D. K., and Rommel, S. A. (1999). Marine mammals of the world. *In* "Biology of Marine Mammals" (J. E. Reynolds and S. A. Rommel, eds.), pp. 1–14. Smithsonian Institution Press, Washington, DC.

Romer, A. S., and Parsons, T. S. (1977). "The Vertebrate Body." Saunders College, Philadelphia.

Rommel, S. A. (1990). Osteology of the bottlenose dolphin. *In* "The Bottlenose Dolphin" (S. Leatherwood and R. R. Reeves, eds.), pp. 29–49. Academic Press, San Diego, CA.

Schaller, O. (ed.) (1992). "Illustrated Veterinary Anatomical Nomenclature." Ferdinand Enke Verlag, Stuttgart.

Slijper, E. J. (1979). "Whales." Cornell Univ. Press, Ithaca, NY.

Sociobiology

FRITZ TRILLMICH
University of Bielefeld, Germany

I. The Nature of Sociobiological Inquiry

Sociobiology is the study of social behavior and social evolution based on the theory of adaptation through selection. As such it is primarily concerned with the adaptiveness of social behavior and the selective processes producing and maintaining adaptiveness. Understanding the selective processes involved includes studying the ecology, physiology, and behavior, as well as the population genetics of the species in question. Sociobiological investigation increasingly attempts to characterize the genetic breeding system, as well as the population structure, both of which importantly influence the effectiveness of evolutionary processes in molding the characteristics of species.

The sociobiological approach assumes that selection at the individual level is the predominant force producing adaptation. A proper understanding of social phenomena, therefore, needs an understanding of the benefits and costs that the individual derives from its interaction with the social environment. Explicitly, group selection is relegated to a secondary position, as in most circumstances, selection operates more strongly at the individual than at the group level because fertility, dispersal, and mortality events are more frequent and act much more forceful on individuals than on groups. Explanation at the level of the individual of social phenomena such as group formation, parental care, and mating systems form the core of sociobiological inquiry.

As the majority of sociobiological inquiry in the field of marine mammals has been done on whales and pinnipeds, these two groups form the focus of the following sections. Relevant information on sea otters (*Enhydra lutris*) and manatees (*Trichechus* spp.) is mentioned briefly in Section IV.

II. Group Formation

The most obvious phenomenon of social life is group formation. Suitable feeding or breeding habitat may initially lead to an aggregation of individuals, thus setting the stage for selective processes moulding the evolution of elaborate social interactions. In contrast to aggregations, groups imply that individuals of a species come together to derive benefits from interactions that follow from this proximity. Such a grouping may serve social foraging, predator avoidance, or defense against predators. Groups may also be established for mating and to share parental care. These kinds of advantages constitute the selective processes that promote group formation in a wide variety of animals. Sociobiology tries to explain groupings from the advantages and disadvantages incurred by the individuals.

A. Whales

The open ocean habitat offers few options for hiding from predators. Consequently, predation by large sharks and killer whales (*Orcinus orca*), particularly on newborns, is one important factor selecting for group formation in whales and dolphins. Direct observational evidence for this hypothesis is scarce, but signs of scarring provide ample evidence of frequent encounters with predators. For example, about one-third of all humpback whale (*Megaptera novaeangliae*) calves carry tooth marks on their flukes when arriving in the foraging areas, presumably from encounters with killer whales or sharks during migration to the feeding grounds. The most spectacular groupings are found in open ocean species such as spotted dolphins (*Stenella* spp.), which benefit the most from the advantages of grouping as protection against predators, but these species may also benefit from group foraging, as may many others.

Several effects play a role in the protection offered an individual by a group. The "dilution effect" acts by reducing the chances of an individual to be attacked by a predator who has noticed the group (to 1/group size, if a predator takes only one individual out of the group). This effect thus dilutes the chances of an attack on a given individual dramatically. The "confusion effect," many individuals rushing back and forth through the visual field of an attacking predator, makes it more complicated for the predator to concentrate on one prey to catch. Finally, animals in the middle of a group can in effect use other individuals around them as shields against predatory attack, as there will always be other individuals geometrically closer to the predator attacking from the edge of a group. This is the so-called "geometry of the selfish herd," which allows an individual to use other group members as protection. These effects may also partly protect individuals in groups against PARASITES, such as the cookie-cutter shark, which takes little bites out of the side of its victim. The advantages of grouping using dilution, confusion, or cover against predation do not depend on individual recognition or social bonds among the animals joining such a group, but also function perfectly well in a totally anonymous group.

Eventually, the advantage of gaining protection in the group will be counteracted by increasing food competition among individuals in a large group, which will determine the upper size of groups. Because food competition depends on abundance, distribution, and behavior of food organisms, which are little known for whale prey, the importance of food competition is also largely unknown. However, signs of territorial behavior between killer whale groups provide at least some direct evidence for the role of food competition in limiting group size.

For more powerful animals such as killer and sperm whales (*Physeter macrocephalus*), grouping offers the additional option to defend smaller, more vulnerable individuals against predators by taking them into the middle of the group. This kind of group action has a completely different quality than the examples given earlier because it involves bonds among the individuals in the group, which are presumably well cemented by individual recognition. Individuals in these groups actively take some risk to defend others (mostly calves) against predators. This can be understood by the fact that the individuals involved in such cooperative behaviors are kin to each other or it may represent mutualistic cooperation. The evolution of such mutualism would be eased in kin groups through kin selection.

However, grouping can also be of advantage for a predator. Foraging groups have perhaps been best analyzed for killer whales. For the mammal-hunting so-called *transient* killer whales in British Columbia, groups of three individuals proved most efficient in terms of energy intake per unit time when hunting harbor seals (*Phoca vitulina*). Optimal efficiency of this group size results because a group of three cooperatively hunting whales seems to be better detecting and catching prey than single animals or duos. However, competition in such a small group is less than in larger group once it comes to sharing a harbor seal carcass. For more evasive prey such as Dall's porpoise (*Phocoenoides dalli*), larger groups may be more successful because more animals are better at intercepting fleeing prey.

Dusky dolphins (*Lagenorhynchus obscurus*) herd schools of fish cooperatively toward the surface and presumably thereby increase food intake. Such activities have also been observed in other dolphins that herded fish schools into bays or against fishing nets and thus may increase food intake for all individuals in the group.

Social systems in whales differ widely between mysticetes and odontocetes. Mysticetes often live solitarily and, except for the mother–calf association during the rearing period, show little evidence of long-term social bonds. This is most likely caused by the nature of their food resources and the fact that due to their body size they have largely escaped PREDATION. They may, however, aggregate during feeding in particularly productive areas and during the mating season (see later).

In contrast, almost all odontocetes are quite social. The social groupings of sperm whales, killer whales, pilot whales (*Globicephala* spp.), and bottlenose dolphins (*Tursiops* spp.) are best documented. In whales, groupings represent matrilines in which male offspring may (in killer and pilot whales) or may not (sperm whales) remain for life. In sperm whales, the grouping of kin serves protection of offspring. During foraging, these animals routinely dive to 400 m depth and stay at depth for 40–60 min. During this period, young would be unprotected, waiting at the surface. This can be and is avoided by adults in a group staggering their diving in such a way that one or more adults almost always attend the calves at the surface. Clearly, such behavior provides indirect fitness gains if the individuals in a group are relatives (such as mother, daughter, granddaughter). Kin selection effects may also explain the lack of dispersal of young in killer whales (in this case of both sexes). In resident, fish-eating groups, male offspring in a matrilineal group may offer protection to relatives as well as help in the defense of foraging areas. Females that stay with their mother may likewise gain fitness from cooperation with close relatives. The size of the group will be limited by food competition, and indeed large groups split up when they have grown too much into smaller units along matrilineal kinship lines. Limitation of group size by the potential for food competition is particularly evident, as the so-called *transient*, mammal-eating groups are much smaller (no more than three to five animals) than those of fish-eating *residents*, presumably because otherwise the disadvantage of food competition would offset the advantage of close cooperation with kin. Pilot whales show a similar social structure where male and female offspring stay with mothers. Advantages and disadvantages of this social structure are less understood for the pilot whale than for the other species mentioned previously.

Bottlenose dolphins and some other inshore odontocetes live in quite open fission–fusion societies in which females associate frequently with many different partners. These associations may be useful in foraging or vigilance and defense against predators, but this hypothesis seems less likely because females with calves tend to be less gregarious. Whether this is due to food competition is unclear. Females may also sometimes group to avoid harassment by males (see later). Male associations vary between study sites and seem to serve mating purposes.

B. Pinnipeds

As a group, pinnipeds are characterized by an amphibious lifestyle. They forage at sea but females need to return to a firm substrate, land or ice, for parturition. A strong selective force during this period of birth and subsequent pup rearing is to avoid predation on mother and pup. In land-breeding pinnipeds, mostly otariids but also elephant seals (*Mirounga* spp.), monk seals (*Monachus* spp.), and the largest populations of gray seals (*Halichoerus grypus*), predator avoidance has led to the choice of oceanic islands for breeding. In comparison to the ocean used for foraging, these islands are limited in area, automatically leading to high densities of females on land. Otariid females come into estrus shortly after parturition, which sets the stage for sexual selection on males trying to station themselves in these female aggregations to breed (see later).

Primarily, the concentration of females on land has all the characteristics of an aggregation, but females of most land-breeding species stay closer together than necessitated by available habitat. One reason for such active groupings may be to avoid harassment by ardent males trying to mate with females. In addition, subadult males of some species of otariid abduct and herd pups, molesting them to the extent that they may die. In this dangerous situation, grouped pups profit from the dilution effect, which might select for tighter grouping of females.

Thermoregulation may also further grouping: Sea lions and the South African and Australian fur seal (*Arctocephaus pusillus*) tend to clump together as soon as they leave the water, which may serve to keep the body shell warm at minimal metabolic cost to the individual. This is not necessary for a fur seal with its highly insulating, air-trapping fur, and the function of clumping in the South African and Australian fur seal is therefore doubtful.

In contrast, ice-breeding seals tend to be much wider dispersed, be it on pack ice or fast ice. In these species, different factors select for some gregariousness. Most important for many species is predation: in the Arctic by polar bears (*Ursus maritimus*) and in the Antarctic by leopard seals (*Hydrurga leptonyx*) and killer whales. Some kind of a grouping advantage is perhaps operating that might equally benefit animals resting between foraging excursions and animals breeding on the ice. Indeed, harp seals (*Pagophilus groenlandicus*) are well known for breeding—even if dispersed—in well-defined local concentrations. It is not certain whether these concentrations are true social groupings or could also be induced by the characteristics of the ice in combination with the food resources below it, i.e., are aggregations due to resource distribution. It may actually be a combination of both. Weddell seals (*Leptonychotes weddellii*) group around tide cracks in fast ice, which offer holes for entry into the water where the seals forage under the ice. Because suitable holes are limited, this leads to a concentration of animals around entry holes and importantly influences the mating system.

Principally, pinnipeds are solitary foragers. There is little evidence for social foraging except occasional observations of sea lions herding fish into bays and communally preying on the trapped schools of fish. Aggregations of foraging sea lions and fur seals occur quite frequently near large fish schools where pinnipeds, birds, and whales may gather in so-called feeding frenzies. True cooperativity in these aggregations has not been demonstrated. This does not exclude that loose groups of pinnipeds may be more likely to locate food and through signs of foraging activity attract other animals to the site, an advantage frequently exploited by group-foraging birds.

III. Parental Investment Strategies

There is no convincing evidence of paternal investment in the rearing of young in any whale or pinniped. Lack of paternal care is typical for the majority of mammals and reflects that females alone provide for offspring during pregnancy and lactation, which frees males of parental duties. Only in a few whale societies, e.g., the killer whale matrilineal groups, do males act as helpers in defense and perhaps feeding of young. Maternal strategies in whales and pinnipeds are characterized by massive energetic investment in young through the production of large precocial young and extremely lipid-rich milk. However, parental investment is measured not by energy expenditure, but rather by a reduction in future fitness, a cost incurred by the mother through a loss in subsequent fertility or an increase in mortality due to a expenditure that benefits the offspring by increasing its fitness. Evidence of parental investment, so defined, is rare in whales and pinnipeds.

A. Whales

The patterning of parental effort differs widely between mysticetes whales and odontocetes. Mysticete females gather the material and energy for pregnancy (lasting about 1 year) and lactation during a feeding season of about half a year when food is plentiful and starve for the other half-year. Extraordinary fast fetal growth rates (in blue whales, *Balaenoptera musculus*, 27 mm/day) permit even the largest mysticete whale—with the exception of the bowhead (*Balaena mysticetus*), sei (*Balaenoptera borealis*), and gray (*Eschrichtius robustus*) whale—to produce a calf within 1 year. By migrating to warmer oceans in the nonfeeding period, mysticete whale females seem to minimize the metabolic overhead for themselves and newborn and sucking calves. Females suckle their young for a short period of 6–8 months on very lipid-rich milk, which is produced from maternal body stores, and wean abruptly. After lactation, mothers need to replenish body reserves, which usually takes a year. Females, therefore, generally breed every other or every third year. Females of the tropical Bryde's whale (*Balaenaptera edeni*) have similar gestation and lactation lengths but show much less of a seasonal breeding pattern. Despite the impressive energy flux involved in pregnancy and lactation in mysticetes, there is no evidence that this reproductive effort incurs clear fitness costs, i.e., really constitutes parental investment, for example, by systematically reducing the probability of a successful pregnancy in the year following lactation.

Large odontocetes have pregnancy periods lasting longer than 1 year and take 8–36 months to wean their young (but reports on 13-year-old sperm whales with milk in their stomachs also exist). All larger odontocetes need more than a year to wean. Females consequently need much longer than 2 years to complete one reproductive cycle. The long period of nursing, even in dolphins, allows young to profit from the milk supply while gradually developing independent hunting skills.

It is presently speculated that the difference in rearing strategy between the two groups depends largely on the difference in hunting strategy. Mysticetes prey on small schooling prey, which are supposedly easy to catch for recently weaned young, particularly since weaning seems to coincide with the annual peak abundance of prey. In contrast, odontocetes eat large prey singly, which forces them to use more complex prey-hunting tactics. They may need to learn a lot more about prey distribution and behavior before young can feed successfully themselves. This may even involve teaching by mothers as suspected for the technique of beaching themselves used by southern killer whales hunting pinnipeds, e.g., on the breeding beaches of Argentina.

Another peculiar feature of some odontocetes is the occurrence of menopause. This phenomenon was documented for short-finned pilot whales (*Globicephala macrorhynchus*) and killer whales. One idea about the functional significance of menopause, which finds some support in studies on humans, is that menopausal females may help their last born and previous daughters in the group through prolonged maternal and allomaternal care, respectively, and by representing a living memory of how to deal with scarce resources and rare ecological disturbances. In matrilineal societies, menopause could be selected through indirect effects on the fitness of kin if the benefit to kin was larger than the benefit an old female could gain through further reproduction.

Sperm whales, long-finned pilot whales (*Globicephala melas*), false killer whales (*Pseudorca crassidens*), and bottlenose dolphins show evidence of reduced fertility with age, perhaps caused by extended periods of lactation. This change in maternal strategy might be expected from the life history theory because old females have a lower reproductive value and may therefore invest more in current offspring than young females. Alternatively, prolonged lactation in older mothers might be caused by a reduction in condition through previous parental care episodes and could then be considered a cost of reproduction.

Despite marked sexual size dimorphism in many species, there is little evidence for a sex-biased investment in sons vs daughters. Data on short-finned pilot whales suggest that sons are suckled for longer than daughters based on milk in the stomachs of caught individuals. Males had milk in their stomachs up to an age of 15 years, but females only up to an age of 7 years. This fits with the observation that males grow faster than females and are consistently larger at weaning than female juveniles. Similar observations have been reported for sperm whales. If such a difference in effort expended on the two sexes were consistent in the population, one would predict from Fisher's model of sex allocation that the sex ratio at birth or at weaning should be strongly biased toward females. There is as yet no evidence for such a sex ratio bias.

B. Pinnipeds

Pinniped females produce one pup per year (instances of twinning are exceptional). They have a postpartum estrus (otariids) or copulate around the time of or after the weaning of young (phocids). Implantation is delayed and followed, after 3–4 months, by an 8- to 9-month pregnancy. The young are reared on lipid-rich milk and are usually weaned at an age less than 1 year, thus allowing an annual cycle of reproduction. Pinnipeds follow one of two alternative pup-rearing strategies. Fe-

males may nurse pups for a short time (a few weeks) from body reserves and then wean abruptly. This is termed the *fasting strategy* and is typical for larger phocids. Smaller phocids and otariids combine nursing from body reserves for a short (in otariids, 1 week) phase after parturition with regular cycles of alternate foraging close to the colony and nursing ashore (the *foraging cycle strategy* of pup rearing). Phocids wean pups after a shorter duration of lactation (4–65 days) at a smaller mass relative to maternal mass (about 25–35%) the otariids (after 120 days to 3 years, at 35–55% of maternal mass).

Contrary to earlier beliefs, the rearing strategy depends more on female body mass than on phylogenetic grouping. The larger a female, the richer a food resource she needs to support pup rearing by foraging during lactation. This is because the energetic costs of traveling to and from the food resource and maintaining the metabolism of mother and pup during the foraging sojourn become too high for large pinnipeds to make foraging during lactation a feasible strategy. Therefore, large species separate foraging completely from lactation, store massive energy reserves during a long foraging period in rich feeding areas often far away from breeding sites, and then fuel lactation out of body reserves. Because phocid females are on average larger (median maternal mass for all species 140 kg) than otariid females (median 55 kg), this can explain why phocids usually follow a fasting strategy and otariids a foraging cycle strategy. The largest otariid, the Steller sea lion (*Eumetopias jubatus*), with a female mass around 250 kg, needs to take its unweaned young close to the foraging areas to maintain lactation by alternate foraging and suckling. Similarly, walrus (*Odobenus rosmarus*) females take their young to foraging areas where they are suckled while the mother can forage nearby. Ecological constraints thus play an important role in shaping maternal strategy.

Evidence is mixed that the energetic effort expended on pup rearing induces fitness costs of reproduction. In the Northern elephant seal (*Mirounga angustirostris*), primiparous (giving birth for the first time) young females are less likely to bear a pup in the year following birth than older females, thus suffering reduced fertility as a consequence of early onset of reproduction. Also, survival seems to be reduced when females first reproduce at 3 rather than 4 years of age, implying a mortality cost. However, these results were obtained at one but not another site on the Californian islands and the interpretation is not entirely clear. In otariids, Galápagos fur seal (*Arctocephalus galapagoensis*) females incur fitness costs of reproduction in terms of reduced fertility because they frequently lose a newborn pup when still accompanied by their previous young, be it a yearling or a 2 year old, by the time the next pup is born. This arises because young Galápagos fur seals grow extremely slowly and therefore take unusually long (up to 3 years) to become independent of their mothers. They may suckle for a second or third year if environmental conditions are poor, and thus preclude rearing of another pup by their mother. Clear evidence for a fertility cost of reproduction was also found for Antarctic fur seals (*A. gazella*). Parturient females of all ages were significantly less likely to reproduce in the subsequent year than nonparturient females. In addition, females that reproduce in a given year are less likely to survive to the following year than nonreproducing females, a clear mortality cost of reproduction. In this species and the northern fur seal (*Callorhinus ursinus*), females older than about 13 years appear to show reproductive senescence. These old females are less fertile and produce smaller newborns than prime-age females. Particularly for otariids, there is thus evidence that the high energetic effort expended by females on pup rearing indeed constitutes maternal investment because it produces fitness costs to the mother.

It has been claimed repeatedly that female pinnipeds of highly polygynous species invest differentially in male and female offspring. Following sociobiological arguments, this would be expected if an increased investment in males (the larger sex showing greater variance in reproductive success) would lead to a greater expected reproductive success of such males. In many polygynous pinnipeds, males are born heavier and grow faster than females. This was taken as evidence for greater maternal investment in sons. However, this is no proof of greater investment in male offspring because male pups of some otariids store less fat and produce more lean body mass than female pups. Fat has a higher energy density than lean body mass, and consequently smaller female pups may have taken as much energy from their mothers as the larger, leaner male pups. Also, the most important growth spurt determining later male adult size and probably reproductive success occurs generally after weaning, thus making it less likely that male offspring derive direct benefits for their future reproductive success from increased maternal investment.

IV. Mating Systems

Females are producers that are limited, in mammalian reproduction, by the time and resources needed for pregnancy and lactation, as well as by the recovery of condition after a reproductive cycle. This constitutes strong selection to optimize acquisition and efficiency of use of resources for reproduction. Because the maximal reproductive rate of female mammals is necessarily much lower than that of males, which in the extreme need only the time of one copulation to produce offspring, females become a limiting resource for the reproduction of males. This leads to selection on males for an increased ability to get access to and breed with as many females as possible, which is the essence of sexual selection. In mammals such as whales and pinnipeds, where males do not contribute to the care of offspring, males are expected to conform to the description of "ardent" males, eagerly searching for females and even harassing them for copulation. Females distribute themselves in relation to the distribution of resources (food and adequate habitat for reproduction), and males map onto this distribution of females. This difference in the selection on reproductive strategies of males and females leads to the phenomenon that quite often the sexes behave and morphologically look as if they belonged to different, competing species.

Sociobiological reasoning therefore leads to the expectation that observed mating systems represent the compromise emanating from the conflict of the sexes. Competition among males for access to females can take the form of aggressive competition, but can also occur via sperm competition when several

males compulate with one female, as demonstrated for northern elephant seals. Such sperm competition, if occurring regularly, is expected to lead to the evolution of large testis mass, as larger testes produce more sperm and thus increase a male's chances to fertilize the ova of females in competition with sperm of other males. Such an increase in relative testis mass was documented in other mammalian orders where species in which multiple copulation is frequent have larger testes than species where only one male copulates with a given female, whether the social system is monogamous or polygynous.

A. Whales

Whale mating systems are still largely unknown partly due to the problem that copulations are hard to observe. Except for a few particularly observable species, this leaves only indirect methods of investigation, such as genetic analyses, to determine the mating pattern in the more elusive species.

Among mysticete whales, much is known about the humpback whale, so well known for its spectacular songs. During the mating season, males station themselves well spaced out and advertise their position. This is very similar to the lek structures observed in many birds. This song may attract females and keep competing males away, but there is presently no firm evidence for this inference. Alternatively, males follow females, and several males may be doing this simultaneously, leading to competition for proximity to the female. Apparently these males compete over mating access to a female. Because humpback whales have small testes for their size, it is unlikely that females will copulate with several males, thus inducing sperm competition. Such sperm competition is likely to occur regularly in right whales (*Eubalaena* spp.), which—weighing about 50 tons—have testes weighing 1 ton, in strong contrast to blue whales, which weigh about 100 tons but have a testis mass of only about 70 kg. Copulation is observed frequently in right whales but has never been described convincingly in humpbacks, despite much more study of the latter.

Mating patterns in odontocetes are somewhat better known from a few species. Male strategies vary from singly roving males in the sperm whale via mating groups in killer and pilot whales to males that cooperate to herd and perhaps force females into copulation in the bottlenose dolphins. Sperm whales show a pattern similar to elephant mating patterns in which single fully adult, highly aggressive males rove among female groups in search of receptive females. They stay only briefly with each one of the female groups and then go on. Presumably, the long intervals between estrus in females make it unprofitable for these males to stay with any one group of females waiting for one of the females to get into estrous condition. Only fully adult males appear able to compete in this system, and subadult males as well as nonroving males stay at higher latitudes, often in small schools, feeding and maximizing energy intake in this way to grow to a competitive size. It is unclear why these males might stay in small groups as bachelors because they are certainly not endangered by predation and it is unknown which foraging advantages they might derive from grouping.

The killer and pilot whale mating system is the most surprising, least understood of the whale mating systems and has no parallel in terrestrial mammals. Genetic evidence shows that males who remain philopatrically in their maternal group never father offspring in their own group but apparently in other groups. Some evidence suggests that several related males of one group may mate with several females in another group, presumably during repeated encounters of these pods. This was concluded from the genetic observation that offspring in a group seem to be paternally related.

B. Pinnipeds

The pattern of mating interactions among individuals depends greatly on the dispersion of females during the breeding season, which in turn reflects the availability of a suitable habitat for pupping and foraging. In pinnipeds, pack ice, fast ice, and land-breeding species differ widely in this respect.

Phocids breeding on ice floes seem to have a mating system best described as serial monogamy in which a male stays with a female that has recently pupped until she comes in estrus. He then leaves after copulation to search for another female. The reproductive success for males in such a mating system depends more on their aptness to locate females ready to mate than on fighting abilities. In such species, sexual dimorphism tends to be small, slightly reversed (males smaller than females), or nonexistent.

Some fast-ice breeding species also show reversed sexual dimorphism, which is best analyzed in the mating system of the Weddell seal. Females gather around cracks in the fast ice where they dive for food from holes in the ice. During the reproductive season, they pup near these holes and males claim territories under the ice and defend the holes against other males. Under these conditions, maneuverability is considered more important than sheer size in male–male competition. This may be the reason for the observed reverse sexual size dimorphism. Alternatively, females may be selected for larger size, enabling the production of larger young or the storage of more massive fat reserves for lactation, and males may have remained smaller for lack of such selection. Copulation is underwater and consequently little is known about the reproductive success of males in this mating system.

Phocid seals, such as the harbor seal, which breed in the water close to land areas where females loosely aggregate frequently, seem to engage in fights for the best stations where females have to pass by, and such males are often wounded. Fighting males seem to have the highest reproductive success and, in some places, the mating system of this species may be similar in structure to a lek.

When female otariids or land-breeding phocids come together on predator-free islands, the resulting high female density permits males to station themselves among females and attempt to monopolize access to females. This competitive situation sets up sexual selection for an ever increasing male size, leading to impressive sexual size dimorphism in a few phocid and many otariid species, such as the northern fur seal in which males weigh six times as much as females. In addition, a larger size also enables males of these species to remain fasting on territory for long periods, thus increasing their chances to mate with females.

Otariid mating systems have been described as resource defense, female defense, or leks. The presence of resources important to females with pups, such as shade or access to tide pools, on male territories was demonstrated experimentally for a few species. However, resource distribution is not sufficient to predict the exact location of female aggregations nor does it explain female gregariousness, i.e., an active tendency of females to approach each other. Because high female density correlates with increased pup mortality in breeding colonies, there is a marked cost to female gregariousness, which must be compensated by comparable benefits. Bartholomew (1970) suggested that female choice of genetically superior males was responsible for female gregariousness, but little evidence supporting this view has come forward. New studies suggest a strong selection of female gregariousness through avoidance of interaction with large males, whether territorial or not. In many otariid species and in elephant seals, interactions with a much larger male can be dangerous or even deadly to a female. Females can minimize the probability of interaction with territorial males by aggregating into a "selfish herd." Through this effect and by avoidance of dangerous and sometimes directly fatal harassment of females and pups by marginal males, females are selected to group much more closely than can be explained by resource distribution. Thus, the stationing of large adult males on clustered territories among parturient and estrous females creates a resource "peace from marginal male harassment." Except for this socially created resource, the system could also be described as one in which males are lekking on areas where females are forced to stay for a while because they spend the period between parturition and postpartum estrus near-stationary on land. Males may benefit from clustered territories by the reduced chances of losing females when disturbed by marginal subadult or adult males. Within this system, the male defense of access to females varies intra- and interspecifically with habitat and female density from female defense to territorial site defense with larger or smaller territories reminiscent, respectively, of a resource defense or a lekking system.

Dominating males in these land-based breeding systems gain most copulations and can reach quite extreme reproductive success, sometimes up to 100 copulations in one breeding season. However, genetic studies have shown that observed number of copulations does not always correlate well with actual paternity, suggesting that peripheral, apparently "excluded" males may also gain some reproductive success by keeping close to female groups. There is no evidence of female choice beyond the mechanism that females in dense groups attract the strongest males in the best condition because only these are competitive enough to station themselves in the middle of female groups. Female elephant seals protest when they are attacked and forced into copulation by subadult or peripheral males, thereby attracting the attention of the dominant male who will often chase off the smaller male and copulate with the female. In this way females indirectly choose dominant males as copulation partners.

Gray seal mating systems on land strongly resemble the otariid system. Whether female choice plays a more important role in gray seal colonies remains to be seen. Genetic analysis of subsequent offspring of individual females suggests that females copulate year after year with the same male, even though they may stay in the territories of different males. This would suggest that they actively choose a particular male or, in case of multiple copulations, have mechanisms to choose among sperm of several males. In more dispersed breeding species, such as harbor seals, males have no chance to defend access to females as these are too mobile and not available continually in the same areas, depending on tide level and sea conditions. In situations in which females breed on sandbanks, harbor seal males were observed to station themselves in areas where females are likely to pass by and make themselves obvious through vocal display. Whether this and similar observations on walrus males that station themselves near females and produce bell-like sounds can be considered a lek display need further investigation.

V. Sirenians and Sea Otters

Much less is known about the social life of sirenians and sea otters and, therefore, these two groups are only treated briefly here.

The only clearly recognizable social structure in sirenians is the mother–offspring bond, which may last for up to 3 or 4 years. Other than that, it appears that the dispersion of most sirenians relates directly to the distribution pattern of food, aquatic macrophytes, fresh water for drinking (in the Florida manatee, *Trichechus manatus*), and, particularly in winter, warm water areas. Animals may migrate for large distances between such resources. However, it seems possible that underlying the apparent asocial pattern may be a subtle pattern of individualized relationships. This might be hypothesized from "greeting" displays exchanged between individuals that meet only occasionally at widely distant sites.

Cows in estrus seem to induce male scramble competition. In Florida, manatee herds of up to 20 males may follow an estrous female and compete by pushing to get into a favorable position for mating. In dugong males, competition may take a more aggressive form in which males may wound each other with their tusks. For West Australian dugongs, mating competition may lead to a form of lekking. However, the evidence is largely circumstantial.

Sea otter spatial dispersion also related to the need to live close to the coast where they forage relatively shallowly for macroinvertebrates. Females claim often overlapping foraging territories year-round along the coastline. They sometimes aggregate in small groups, so-called "rafts." Young males, and fully adult males outside the reproductive season, also frequently form rafts close to areas of rich feeding resources. Such rafting is presumably related to the reduction of predation risk.

During the reproductive period, fully adult males establish territories that may overlap more than one female territory. This presumably provides males with a chance for a mild form of polygyny, but hard evidence for paternity of such males is at present missing.

VI. Concluding Remarks

Much of the sociobiological interpretation of observations on marine mammals is still in a speculative stage. This situation reflects our lack of detailed knowledge about the marine environment and in particular the macro- and microdistribution of resources vital to marine mammals. Clearly, more observation,

more comparative studies, and especially more experimental work are urgently needed to understand the sociobiology of these magnificent animals. Obviously, experimental work will be particularly challenging and can only be successful if built on the thorough knowledge of marine mammal natural history. However, a well-founded functional understanding of the social behavior of marine mammals cannot be achieved without experimental tests of our many assumptions. Ingenious instrumentation and molecular genetic tools, developed during the last decade, should prove most helpful in making this summary of marine mammal sociobiology soon outdated.

See Also the Following Articles

Energetics ▪ Feeding Strategies and Tactics ▪ Group Behavior ▪ Mating Systems ▪ Parental Behavior ▪ Predator–Prey Relationships ▪ Thermoregulation

References

Amos, B., Schlötterer, C., and Tautz, D. (1993). Social structure of pilot whales revealed by analytical DNA profiling. *Science* **260**, 670–672.

Baird, R. W., and Dill, L. M. (1996). Ecological and social determinants of group size in *transient* killer whales. *Behav. Ecol.* **7**, 408–416.

Bartholomew, G. A. (1970). A model for the evolution of pinniped polygyny. *Evolution* **24**, 546–559.

Boness, D. J. (1991). Determinants of mating systems in the Otariidae (Pinnipedia). *In* "The Behaviour of Pinnipeds" (D. Renouf, ed.), pp. 1–44. Chapman & Hall, London.

Boness, D. J., and Bowen, W. D. (1996). The evolution of maternal care in pinnipeds. *BioScience* **46**, 645–654.

Bowen, W. D. (1991). Behavioral ecology of pinniped neonates. *In* "The Behaviour of Pinnipeds" (D. Renouf, ed.), pp. 66–127. Chapman & Hall, London.

Boyd, I. L. (1998). Time and energy constraints in pinniped lactation. *Am. Nat.* **152**, 717–728.

Clutton-Brock, T. H. (1991). "The Evolution of Parental Care." Princeton Univ. Press, Princeton, NJ.

Costa, D. P. (1993). The relationship between reproductive and foraging energetics and the evolution of the pinnipedia. *Symp. Zool. Soc. Lond* **66**, 293–314.

Hamilton, W. D. (1964). The genetical theory of social behaviour. *J. Theoret. Biol.* **7**, 1–25.

Le Boeuf, B. J. (1991). Pinniped mating systems on land, ice and in the water: Emphasis on the Phocidae. *In* "The Behaviour of Pinnipeds" (D. Renouf, ed.), pp. 45–65. Chapman & Hall, London.

Mace, R. (2000). Evolutionary ecology of human life history. *Anim. Behav.* **59**, 1–10.

Mann, J., Connor, R. C., Tyack, P. L., and Whitehead, H. (eds.) (1999). "Cetacean Societies: Field Studies of Dolphins and Whales," p. 432. Chicago Univ. Press, Chicago.

Stirling, I. (1983). The evolution of mating systems in pinnipeds. *In* "Advances in the Study of Mammalian Behavior" (J. A. Eisenberg and D. G. Kleimann, eds.), pp. 489–527. American Society of Mammalogy, Special Publication No. 7.

Trillmich, F. (1996). Parental investment in pinnipeds. *Adv. Study Behav.* **25**, 533–577.

Trillmich, F., and Trillmich, K. G. K. (1984). The mating systems of pinnipeds and marine iguanas: Convergent evolution of polygyny. *Biol. J. Linn. Soc.* **21**, 209–216.

Song

Jim Darling

West Coast Whale Research Foundation,
Tofino, British Columbia, Canada

Humpback whales (*Megaptera novaeangliae*) produce a complex sequence of sounds ranging from 5 to 30 min in duration and then repeat it with precision. The non-stop repetition of this sequence is called the "song." The song is produced by males during the breeding season: in late fall on the feeding grounds, during MIGRATIONS, and on the winter breeding grounds. The song has a standard structure: several sounds termed *units* are associated to make a *phrase*. A repetition of the phrase composes a *theme,* and several different themes make up the *song*. The composition of the song changes or progresses slowly over time, and the majority of singers in a population sing fundamentally the same version in any season. The songs of humpbacks may be similar across entire ocean basins such as the Pacific, yet different in separate oceans. On the winter breeding grounds singers are usually lone adult males, which may be stationary just beneath the surface or slowly traveling. The singing usually ends with an interaction with other whales. The specific function of the song is not known, but it is clearly a key communication of humpback whales during the breeding season.

Our understanding of the song is in early stages, with aspects of it based on just one or a few studies. The following summarizes the working knowledge of the song, but new findings may change the emphasis substantially.

I. Characteristics

A. First Descriptions

Although likely heard by sailors for centuries, the first recordings of humpback songs were made via U.S. Navy hydrophones in the 1950s off Hawaii and Bermuda. Scientists first recognized these sounds as coming from humpback whales in the 1960s. The first technical description was published in *Science* in 1971 by Roger Payne and Scott McVay with the revealing subheading, "Humpbacks emit sounds in long, predictable patterns ranging over frequencies audible to humans." Their analysis led to a structural context in which to view the sounds introducing the terms units, phrases, themes, and song. With the observation that humpback whales "produce a series of varied sounds then repeat the same series with considerably precision" these authors called the performance "singing" and the repeated series of sounds the "song."

B. Song Structure

The song is composed of a sequence of highly varied sounds ranging from high-pitched squeaks to midrange trumpeting and screeches to lower frequency ratchets and roars, and combinations of all these. This sequence is typically about 10–15 min in duration, although it may range from 5 to 30 min. It is then repeated without a break.

The song has a predictable structure or framework. Discrete sounds are termed *units*. Several different sounds or units in a sequence compose a *phrase*. A phrase is repeated some variable number of times (e.g., 10 times), and this series of the same phrase is called a *theme,* say, "theme 1." After several minutes of singing theme 1 the singer changes to a different set of phrases (composed of different units or sounds) and repeats it a number of times. This might be called "theme 2." This pattern repeats until the whale cycles back to its theme 1. A typical song may contain six themes. A singer may sing in order themes 1-2-3-4-5-6 and then start at 1 again. The number of themes in a song varies from population to population and from year to year. A song session may continue without a break for hours.

Several characteristics of this song display are notable. First is the precision by which a singer repeats its complex song in any song session. An undisturbed singer may repeat the same themes, phrases, and units faithfully for hours as if on a continuous tape. The second characteristic is that all the singers in a humpback assembly, and there may be hundreds, follow the same structural rules and broadcast essentially the same song. At the same time, however, there is some variation in song presentation and some songs are exceptional to the point that they have been termed "aberrant" by investigators. The significance of this variability is not yet understood.

C. Song Progression

One of the more unique characteristics of the humpback song is that it gradually changes over time. Different sounds, and arrangements of sounds, form to create new phrases and themes. These are incorporated into the song gradually, while older patterns are lost. After a period of several years the song my bear little resemblance to the original version. This is a rare, although not unique, characteristic of biological songs. Several bird species are known to vary their song over time.

The song apparently changes as it is being sung; i.e., the progressive changes occur during the singing or breeding season, not during the off-season. For example, as winter progresses, one unit in a particular phrase may be heard less and less while another becomes more common, or two units in one phrase that had been separate may become joined as one sound. These small changes may eventually lead to new phrases and themes. During the nonsinging or summer season, the song appears to remain relatively stable, to begin to progress again once singing begins the following season. The biological forces or mechanism behind the progression of the song is not known.

The change in the song display occurs in a collective or common way throughout a population of humpback whales; i.e., at any one time all the singers in a population produce fundamentally the same version of the gradually changing song. This is a generalization in that there is certainly some range in when individuals incorporate new elements into their song. However, the majority of singers produce many if not all the same clearly recognizable units, phrases, and themes in the same general order in any one season. Changes in the song are apparently incorporated throughout the population. The mechanism by which this occurs is not known. Some researchers speculate that the changes are learned, with some whales making changes and others learning them; others suspect a genetic component to this phenomenon.

The similarity and collective change of the song may occur not only within one population of humpbacks, but between populations separated by thousands of kilometers. For example, humpback populations that winter in Japan, Hawaii, and Mexico sing similar versions of the song at any one time. In contrast, the Atlantic humpback song would be substantially different. The means by which this occurs is not known. Some degree of mixing occurs among all populations in the North Pacific in that individual whales may visit the different winter assembly areas over several years or even in 1 year. Perhaps this mixing is reflected in the similarity of songs. It is important to emphasize that whereas the generalities described here are striking, much further research is required to conclude on the relationship of songs, singers, and populations. This is emphasized by a discovery that over a 2-year period the humpbacks off eastern Australia essentially adopted the western Australian humpback song, apparently introduced to that region by the mixing of whales.

II. Singing Behavior

A. Seasonality

Singing is a seasonal behavior pattern. It occurs primarily during the winter half of the year, throughout the peak of the humpback breeding season. Although a few songs have been recorded during midsummer, they are rare. Songs become increasingly common on the feeding grounds as fall progresses into winter; they are heard regularly during migrations and predominate on the warm water winter breeding grounds. It is not clear when the singing activity declines in the spring, but the decline has occurred by the time the migrant whales reach their summer feeding grounds.

B. A Male Display

The evidence to date indicates that only male humpbacks sing. The gender of a sample of singers has been determined both by photographs of the genital region and by genetic analysis of skin samples, and so far all were males. Also, singing is a behavior pattern common to individual whales that may escort a cow with calf or compete in mating groups—both distinct male behavior patterns. Because there are no confirmed observations of female humpback whales singing, the song may be a male secondary sexual characteristic or display that functions during the breeding season.

A singer on the breeding ground is typically a lone adult male. Also, it is not unusual for an "escort," the adult male accompanying a cow with calf, to be singing. Occasionally a singer has a companion adult or juvenile in the close vicinity. Often the singer is stationary, remaining in one geographic location for many hours while singing. In these situations the singer generally adopts a motionless, head-down, tail-up posture approximately 50–100 feet beneath the surface. It maintains this posture until it surfaces to breathe and then immediately after diving resumes it. At other times a singer travels steadily as it is singing and may move tens if not hundreds of kilometers during a song session.

C. Interactions with Singers

Humpback songs are loud and can be heard underwater for at least several kilometers and, in some circumstances in deep ocean basins, possibly hundreds or even thousands of kilometers. A collection of dozens of singing humpbacks produce a substantial noise, clearly designating the location—on an oceanic scale—of an assembly of humpback whales during the breeding season. It is likely that individual humpbacks interact with the collective singing herd over extensive oceanic distances.

On a smaller scale, several studies have focused on the interaction of individual singers with other whales. Generally, singers sing until one of two things happen: (1) They are joined by other single adults, the singing stops, and the two interact in some way, ranging from a single pass to rolling, tail lobs, or breaches by one or both animals. Often the interactions are very short, with the pair splitting after a few minutes. One or the other may start singing again shortly after the interaction. There is increasing evidence that the lone adults that join male singers are also males. (2) The singer stops singing without any close approach by another whale and then rushes to approach or join a passing group of whales—often a surface active or competitive group; at times a cow with calf. In these cases, the singer joints a group that includes a potentially breeding female.

III. Function

The function of the song has been the subject of much speculation. Much of this has revolved around proposals of the song as a male display that functions to (1) attract females, (2) signal status to other males, or (3) a combination of these. However, other suggestions include that the songs serve as a means to synchronize ESTRUS, a spacing mechanism between males, a migratory beacon, and as sonar by males to find females. Recent research has widened the basis for the speculation but is far from resolving it.

Conservatively, the song is a communication from male humpbacks during the breeding season. It almost certainly serves to provide the location of the singer, and by association the entire herd, and signals that breeding activity is underway. The song also likely broadcasts information about the individual singer, but what information and who the recipient(s) is remain speculative. The collective, gradual change of the song seems to confound many simple explanations as to its function.

Several early investigations proposed that the song by males serves to attract females and went further to suggest that changes in the song are driven by female choice. This notion has persisted and predominates in the popular literature; however, there no data to support it. There is little question that singers join and accompany females, but no evidence that females are attracted by, and voluntarily join, individual singers.

In a related idea, several researchers have proposed that the humpback song functions as part of an aquatic lekking behavior. A lek, as generally defined for land animals, is a seasonal assembly where males display their attributes to attract females for the purpose of mating. This seems to describe the overall breeding assembly of humpbacks, including the male song display; however, if, when, or how females choose a mate is unclear.

Another proposal suggests that the song is a display of males aimed primarily at other males—perhaps signaling social status or position. In this view the song may enable the male–male interactions that are integral to the establishment and maintenance of the dominance-ordered, polygynous mating system that has been hypothesized for humpback whales. Alternatively, it may facilitate temporary male associations on the breeding grounds. This proposal accounts for the common interactions between male singers and male joiners, and the lack of evidence that females approach, choose, or join singers.

Regardless of whether the song display is directed at females, males, or both, these potential explanations suggest that song displays should vary with some attribute of the singer. That is, the song should reflect dominance class, fighting ability, or some other measure of the "fitness" of the male. At this stage, research has not revealed any evidence that this is the case.

In summary, at this time the function(s) of the song is not clearly understood. The notion that it is a sexual display is attractive as it accounts for many of the observations and characteristics of the song, and does so within the larger mammalian context. However, this is far from established, leaving other speculative ideas that do not presume the song to be a sexual display as open possibilities.

See Also the Following Articles

Communication ■ Courtship Behavior ■ Culture in Whales and Dolphins ■ Humpback Whale

References

Chadwick, D. H. (1999). Listening to humpbacks. *Nat. Geogr. Mag.* **196-1,** 111–128.

Darling, J. D. (2001). Interactions of singing humpback whales with other males. *Mar. Mamm. Sci.* **17**(3) (in press).

Darling, J. D., and Bérubé, M. (2001). Interactions of singing humpback whales with other males. *Mar. Mamm. Sci.* **17**(3), 570–584.

Noad, M. J., Cato, D. H., Bryden, M. M., Jenner, M. N., and Jenner, K. C. S. (2000). Cultural revolution in whale songs. *Nature* **408,** 537.

Payne, R. S., and McVay, S. (1971). Songs of humpback whales. *Science* **173,** 587–597.

Payne, K., Tyack, P., and Payne, R. (1983). Progressive changes in the songs of humpback whales (*Megaptera novaeangliae*): A detailed analysis of two seasons in Hawaii. *In* "Communication and Behavior of Whales" (R. Payne, ed.), pp. 9–57. AAAS Selected Symposia Series, Westview Press, Boulder, CO.

Payne, R. S., and Guinee, L. N. (1983). Humpback whale songs as an indicator of "stocks." *In* "Communication and Behavior of Whales" (R. Payne, ed.), pp. 333–358. AAAS Selected Symposia Series, Westview Press, Boulder, CO.

Tyack, P. L. (1981). Interactions between singing Hawaiian humpback whales and conspecifics nearby. *Behav. Ecol. Sociobiol.* **13,** 49–55.

Winn, H. E., Thompson, T. J., Cummings, W. C., Hain, J., Hudnall, J., Hays, H., and Steiner, W. W. (1981). Song of the humpback whale: Population comparisons. *Behav. Ecol. Sociobiol.* **8,** 41–46.

Winn, H. E., and Winn, L. K. (1978). The song of the humpback whale *Megaptera novaeangliae* in the West Indies. *Mar. Biol.* **47,** 97–114.

Sound Production

Adam S. Frankel
Marine Acoustics, Inc., Arlington, Virginia

Most terrestrial mammals rely heavily on vision and smell. These senses are limited in water by the absorption of light and the slow physical movement of water, respectively. As a result, marine mammals have evolved to use sound and hearing as their primary means of communication and sensing their world.

This article briefly reviews basics of sound and the physical ways in which sounds are produced by marine mammals. The main focus is on characteristics of sounds made internally by marine mammals. Certain species stand out as particularly unusual or particularly well studied, and these get special attention. Other species are treated in taxonomic groups, when their acoustic characteristics are similar or less well known. Especially whales and dolphins often leap clear of the water or slap tail or flippers onto the water surface, creating percussive sounds. These sounds are variably thought to be expressions of playing, socializing, or aggressiveness. They are not treated in detail here.

I. Fundamentals of Sound

Imagine throwing a rock into a calm lake. You can easily picture the ripples, or water waves, that move out in an expanding circle. In these ripples, the water's surface moves up and down, in a smooth progression from crest to trough. Sound waves are similar to these water waves. The "crests" of a sound wave are areas of high pressure, and the "troughs" are areas of low pressure. Like the water wave, there is a smooth progression between these areas of high and low pressure. A sound wave is a propagating (moving forward) alternation of these areas of high and low pressure. The wave is formed by a structure that vibrates, such as a radio loudspeaker or our larynx.

Several major terms are used to describe the characteristics of sound. The amount of time it takes for a complete cycle between the highest pressure to the lowest, and back to the highest, is referred to as the period of the wave and is measured in seconds. The reciprocal of that time measurement, the period, is called the frequency of the wave, measured in seconds^{-1}, or more commonly hertz (Hz). The longer the period, the fewer number of cycles will occur in a fixed amount of time, and the lower the frequency of the sound. In music, this low frequency is called one with low pitch, or a bass sound. High-pitched sounds of a flute or violin are high in frequency. Frequency is a physical characteristic of sound, whereas pitch is our perception of sound. Humans hear from about 18 to 20,000 Hz (or 20 kHz). Therefore, sounds below 18 Hz are termed "infrasound," whereas those above 20 kHz are termed "ultrasound."

Marine mammalogists often want to know the loudness of a sound, whether produced by a whale, dolphin, ship, or oil rig. The loudness, or amplitude, of a sound is described in decibels (dB), but "in-air" decibels and "in-water" decibels are different from each other. In water, sound amplitude is measured and referred to as a certain number of dB references to one micropascal. This is usually written XX dB re 1 μPa. Because sound amplitude decreases with distance, the standard of describing amplitude is usually referred to as 1 m from the sound source, or XX dB re 1 μPa at 1 m. Sounds that are not very loud tend to be below 100 dB, and very loud sounds are above 160 to 180 dB.

A signal is said to be frequency modulated (FM) when its frequency changes over time. Dolphin whistles are usually frequency modulated. Amplitude-modulated (AM) signals are those where the amplitude changes over time. Many mysticete calls are amplitude modulated and sound like "growls."

When one examines a display of sound waves per time, or a spectrogram, one often sees "extra lines" above the main "line" of the spectrogram. The "main line" of the spectrogram is the lowest frequency contour, termed the fundamental frequency. Additional "lines" about the fundamental frequency represent the harmonics. Harmonics are integer multiples (e.g., 2×, 3×, 4×) of the fundamental frequency. They are the result of the physical characteristics of the sound-producing structure, and they alter the sound. Harmonics occur with frequency-modulated calls and are strictly integer multiples of the fundamental frequency. Amplitude-modulated calls frequently have a stack of closely spaced bands that look like harmonics, but are side bands that are the product of the rate of amplitude (loudness) modulation (Fig. 1).

II. Sound Production Mechanisms

A. Humans

Unlike most marine mammals, humans exhale to vocalize. The energy for vocalization comes from exhaling. The lungs force air through the larynx. The vocal cords in the larynx open and close as the air rushes past them, breaking the exhalation up into a series of air puffs, or an undifferentiated sound. These air pulses then pass through the vocal tract of the throat, tongue, mouth, and lips. All of these structures move during speech to change the shape of vocal tract, literally shaping the buzzing sound from the larynx into speech.

B. Dolphins

There has been considerable debate over the sound production mechanism in toothed whales, or odontocetes. Only recent work has begun to clarify the issue. As in humans, the energy for vocalizations comes from pushing air. The similarity

ends there, as odontocetes lack vocal cords. Odontocetes have a structure in the upper portion of their heads called dorsal bursae that have structures protruding into the nasal passage called phonic lips. Air appears to be pushed past these phonic lips, and the result is that they slap together and send the neighboring tissue into vibration. Once past the phonic lips, the air enters the vestibular sac, just below the blowhole. The air can either be pushed back into the lower nasal passage or it may be released into the water. Whistling dolphins sometimes emit bubble streams. All odontocetes, except the sperm whale (*Physeter macrocephalus*), have two of these dorsal bursae/phonic lip structures. It appears that they can operate the two structures independently, creating two different sounds simultaneously. Toothed whales produce clicks, click trains (often so rapid as to be heard as a "buzz"), and—especially many species of dolphins—whistles. Some but not all click types are used for echolocation to scan the environment, as also in bats.

Once the sound is produced, it is directed out into the water through a structure called the melon. The melon sits on top of the skull, behind the rostrum of dolphins. The melon is a complex structure that couples the vibrations produced by the dorsal bursae/phonic lips complex to the water. It also functions as an acoustic lens to "focus" clicks into a beam. This is similar in function to the focusing of light into a beam by the lens and reflector of a flashlight (Fig. 2).

C. Baleen Whales

There is no equivalent in mysticetes to the dorsal bursae/phonic lip structure found in odontocetes. Mysticetes have a larynx, but it lacks the vocal cords found in humans. It is likely that the larynx does have a sound production function, but it remains undetermined. Cranial sinuses are also suspected to play a part in vocal production, but the details remain uncertain. Visual observations of singing humpbacks reveal that they do not exhale, even while singing for 20 min or longer.

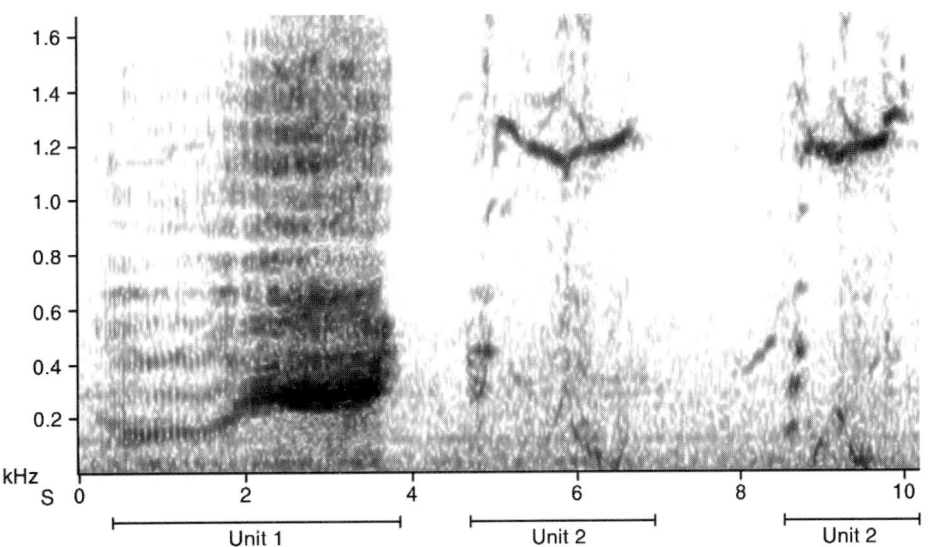

Figure 1 *Spectrogram of part of a humpback whale* (Megaptera novaeangliae) *song. The first sound is shown as a broad, fuzzy structure, typical of amplitude modulated sounds. The second sound, the "V"-shaped structure, is a frequency modulated sound.*

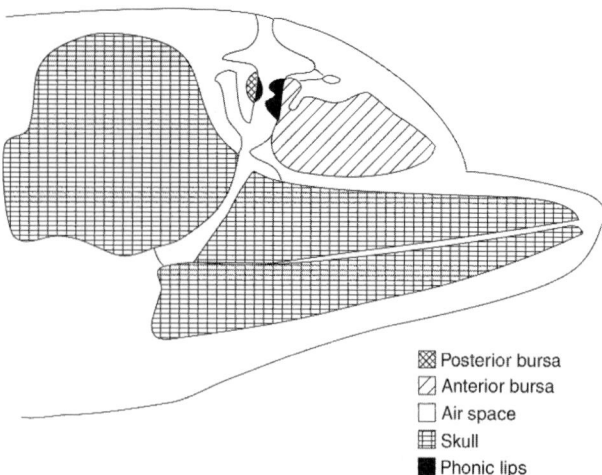

☒ Posterior bursa
⧄ Anterior bursa
☐ Air space
▦ Skull
■ Phonic lips

Figure 2 *Diagram of sound production structures in the head of a dolphin. Redrawn from Cranford (2000).*

This indicates that they are recycling air within their body to vocalize.

D. Sirenians and Carnivores (Pinnipeds, Polar Bear, and Sea/Marine Otters)

Noncetacean marine mammals appear to make sounds similar to those of their terrestrial relatives, generally by vibrations in the throat. Sea lions (Otariidae) bark underwater, and manatees (*Trichechus* spp.) and dugongs (*Dugong dugon*) have a rich repertoire of sounds. Underwater, these sounds are generally made without air being released from the mouth, and it is surmised that air is shunted between the lungs and mouth/nasal structures.

III. Characteristics of Vocalizations by Group and Selected Species

A. Pinnipeds

1. Phocid Seals   Phocid, or true, seals live in a wide variety of environments, from tropical islands to the ice packs of the Arctic and Antarctic. While the underwater sounds of all species have not been described in detail, it is likely that all make underwater sounds. In-air sounds tend to consist of barks or growls, sometimes produced at the surface but more usually produced while mating and pupping on land or ice. Such sounds are often aggressive or warning signals, as when, for example, a human approaches an animal too closely. Pups tend to have higher frequency in-air "mew" sounds, apparently useful for mother/pup interactions and potential pup recognition. Underwater sounds are described in more detail later.

A. WEDDELL SEALS (*LEPTONYCHOTES WEDDELLII*)   Weddell seals are a very vocal species. Their repertoire has been categorized into 12 calls, labeled by letters, and subdivided into 34 types. These calls range between 100 Hz and 12.8 kHz. The function of some of these calls has been suggested. T calls are long downsweeps and are thought to function in territorial advertisement. C calls are pulses of intermediate frequency, which

appear to be used as an aggressive display. M calls are a long low-frequency tone, and H calls are low-frequency pulses. Both M and H calls are thought to be used as a high-intensity threat. G and L calls are both low-frequency-modulated sweeps, thought to be used as a low-intensity threat. Functions have not been proposed in similar detail for other phocid species.

Calling rates are lowest in winter and highest in spring. It has been suggested that calls produced from the pupping season through the mating period function as male advertisement "song." Weddell seals also make numerous in-air vocalizations during this time period that sound similar to those produced underwater, a richer in-air repertoire than produced by most (and possibly all) other phocid seals. Recordings of Weddell seals at different locations indicate that there are geographic differences in calls between breeding populations.

B. LEOPARD SEALS (*HYDRURGA LEPTONYX*)   Recordings of a captive leopard seal included vocalizations reaching as high as 164 kHz. These included frequency-modulated chirps, buzzes, and a click train. The presence of ultrasonic clicks suggested a possible echolocation function, but no data support echolocation in leopard or any other seals.

As in Weddell seals, recordings of leopard seals made in distant locations showed differences in the calls between the two areas. The repertoire of leopard seals shows considerable variability. Their calls range from low-frequency tonal calls, both narrowband and wideband high-frequency pulses, and frequency-modulated calls ranging from 500 Hz to 8 kHz. The most frequent calls of leopard seals are surprisingly soft, given its role as a top-level predator. Many other seals, including the Weddell seal, produce harsh calls during aggressive interactions. It has been suggested that the leopard seal's solitary existence reduced the need for territorial calls. However, when hauled out on ice or land, leopard seals that were approached by a researcher made explosive, broadband vocalizations termed "blasts" and "roars."

C. BEARDED SEALS (*ERIGNATHUS BARBATUS*)   Like Weddell seals, bearded seals are well known for their long calls and songs. Bearded seals produce distinctive trills, which typically consist of a series of discrete frequency sweeps. Trills can begin at 2.5 kHz and gradually sweep down in frequency to a few hundred hertz. Most trills are characterized by decreasing frequency, but some variations can have near constant frequency or even alternations of frequency increases and decreases. Trills last from 4 to 16 sec. In some populations, the long trill is followed by a rapid upsweep to 3 kHz and lasts 1–2 sec. The vocalizations of bearded seals are thought to function as male advertisement displays during the mating season (Fig. 3).

D. RINGED SEALS (*PUSA HISPIDA*)   Ringed seals, like many seals, were long thought to be silent. However, they are now known to produce at least four types of calls. These are low- and high-frequency barks, frequency-modulated tones between 2 and 4 kHz, broadband growls that range up to 6 kHz, and frequency-modulated chirps. It is likely that ringed seals produce higher frequency calls that have not been recorded, as their best hearing sensitivity is at 45 kHz. Animals usually have their best hearing in the same frequency range in which they vocalize. Source levels for all of the recorded ringed seals are relatively low, between 95 and 130 dB re 1 μPa. None of

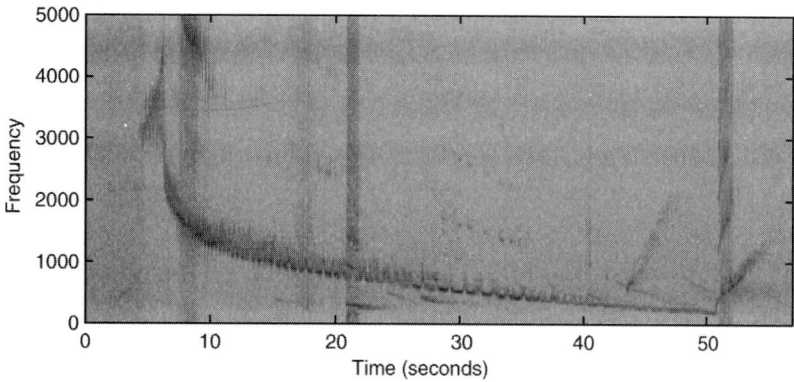

Figure 3 *The long downward trill and rapid upsweep of a bearded seal from Alaska is shown in this spectrogram.*

the recorded calls occurred exclusively within the breeding season. The proportion of calls produced did vary with season, however.

E. HOODED SEALS (*CYSTOPHORA CRISTATA*) The acoustic repertoire of hooded seals can be grouped into three major classes. Class A and B signals are produced by normal vocal mechanisms, whereas class C signals are produced by the hood and septum, a set of specialized anatomical structures in this species. Class A calls tend to be the most common vocalization and are produced both in air and underwater. All of these calls are pulsed and rarely frequency modulated, with energy ranging from 500 Hz to 6 kHz. Class A calls are used in a variety of circumstances, including female responses to displaying males, who also make a variety of A calls. Female and pup interactions also use A calls. Class B calls are described as growls or roars. These growls tend to be made by females fighting with males and by males fighting with other males. Other variants of B calls are used as low-level threats. Class C signals are produced by the inflation and deflation of the hood and septum. They are short-duration, broadband with rapid onsets, and little or no frequency modulation. These calls have been described as "bloops," "wooshes," and metallic "pings," clicks, and "knocking" sounds. It is possible that only males make class C sounds.

F. HARP SEALS (*PAGOPHILUS GROENLANDICUS*) Harp seals aggregate at the ice edge in March in the northwest Atlantic. At least 19 different call types have been described for the species. The call types range from a nearly pure sine wave to pulsed sounds, high-frequency chirps, broadband "warbles," trills, squeaks, and grunts. The maximum source level of these calls is between 135 and 140 dB re 1 μPa. Harp seals also produce clicks that are about 25 dB louder than their other calls. A long-term observation has shown that within one breeding area these call types are stable over a period of tens of years. The calls may help individuals locate the herd; harp seal aggregations can be heard from 30 to 60 km away. Once in the aggregation, calls may be used to find a mate. Comparisons of different breeding aggregations found many shared calls and some that were unique to the breeding areas, again suggesting that there may be distinct breeding populations. In-air record-

ings revealed that harp seal pups have an unusually wide variety of vocalizations. These tend to be longer than calls from other phocids and are very complex, showing a wide variety of frequency and amplitude modulation.

G. ELEPHANT SEALS (*MIROUNGA SPP.*) Elephant seals mate on land and males compete for dominance, and thereby access to females. Males use threat displays and actual fights to establish dominance. Male elephant seals make three main types of calls during AGGRESSION: "snoring," "snorting," and "clap threatening." "Snoring" is used as a low-intensity threat, and dominant males "snort" more aggressively when approached by a challenging male. Snorts range between 200 and 600 Hz. The "clap threat" ranges up to 2.5 kHz. It is thought that in-air threat calls produced by males are also transmitted through the ground, and these seismic signals can produce responses from other elephant seals. Females produce a low-frequency "belch roar" in aggressive situations and a 500 Hz to 1 kHz bark to attract the pup. Pups produced a higher frequency, up to 1.4 kHz, call that is used to maintain contact with the mother or to elicit attention. As in other seal species, both the mother's and the pup's call have individual characteristics that are used to recognize each other acoustically.

Underwater sounds have been recorded, but only described cursorily as sounding like a bell, cymbal, or guitar.

H. WALRUSES (*ODOBENUS ROSMARUS*) Social interactions between walruses hauled out on the ice often include vocalizations. Adults use roars, grunts, and guttural sounds as threats. Roars are long, loud calls that last a second or more and have most of their energy at low frequency. Grunts are brief calls lasting between 100 and 400 msec and vary greatly in amplitude. Grunts range from 50 to 250 Hz. Guttural sounds are the most common threat vocalization, consisting of a series of low-frequency, wideband pulses. These can range from 13 Hz to 4 kHz. Barks are short, frequency-modulated calls that range from 90 to 260 msec in length and between 300 and 500 Hz. They are frequently given as a series. Barks are used by adults to indicate submission, and louder barking may indicate more submission. Calves bark in a wider variety of situations. Calves separated from their mother bark loudly and then may continue to bark softly once they are rejoined. The structure of the bark changes as the calf matures, gradually becoming a longer single call, with both frequency and amplitude modulation. Females produce a soft short call between 0.3 and 1.0 sec long. The call is usually frequency modulated, either as a downsweep or as an alternation of up-and-down sweeps. This call serves as a female contact call, produced when female and pup are in close proximity to each other. Females do not have a loud contact call to attract the calf, as in some other phocid species.

Males on the mating grounds produce a gong-like sound as a courting display, both in air and underwater. This sound is produced by inflated throat pouches, at times augmented by flippers striking the throat. Male walruses also make aggressive clacking sounds with their teeth.

2. Otariid Seals: Sea Lions and Fur Seals   While a considerable detail of sounds for several select phocids has been presented, relatively little time has been spent on eared seals. Most of them are known to make bark-like sounds or other "groans and grunts" in air, and the barks of a group of hauled-out California sea lions (*Zalophus californianus*) are loud and far-reaching (to at least 3 km on a still night). Barks tend to have most energy below 2 kHz. These sounds and other interactions result in a physical structuring of the society on a beach or headland. Sea lions also bark underwater, in similar fashion as in air, but otariids tend not to have the complex social signals of many phocids. This is probably because those phocids tend to mate in the water, and sounds are often, although not always, related to social/sexual interaction. Otariids, however, tend to mate on land, and they are in large part relatively quiet in the sea.

Nevertheless, there are sets of clicks, snorts, bleats, and growls produced underwater by Steller sea lions (*Eumetopias jubatus*), and clicks and high-frequency "sheep-like" bleating sounds by northern fur seals (*Callorhinus ursinus*). The clicks of some species have led some researchers to postulate that otariid seals might echolocate, but repeated investigations in captivity have not shed light on this possibility. If they do use echoes for environmental information (and they are likely to do so at a very basic level, at least), this is not nearly as sophisticated as the echolocation capability of many dolphins and bats.

B. Carnivores

1. Sea Otters (Enhydra lutris)   Sea otters produce airborne sounds that have been described as whines, whistles, growls, cooing sounds, chuckles, and snarls. They may produce harsh screams when stressed. Airborne vocalizations serve to maintain the mother–pup bond. When a mother dives, the pup at the surface often vocalizes continuously until the mother resurfaces. If the mother does not find the pup upon surfacing, she vocalizes and the pup responds. Pups also vocalize to elicit nursing or grooming. The frequencies of these calls lie between 3 and 5 kHz. No underwater vocalizations have been recorded, although they may exist.

2. Polar Bears (Ursus maritimus)   Polar bears may not be quite as vocal as many other carnivores. Males chuff and snort, with powerful rapid exhalations, especially in competitive interactions with other males. Females produce low mew-like calls that may be used for mother/pup recognition.

C. Cetaceans

1. Baleen Whales

A. GRAY WHALES (*ESCHRICHTIUS ROBUSTUS*)   Gray whales most frequently produce sounds referred to as "knocks" and pulses. These range in frequency from < 100 Hz to 2 kHz. A series 2–30 pulses last an average of 1.8 sec. Knocks are most common when gray whales are feeding. They are fairly vocal while feeding, relatively quiet while migrating, and most vocal during mating activities in Mexican waters. The source levels of gray whale vocalizations range between 167 and 188 dB re 1 μPa.

Four major signal types have been recorded from migrating gray whales. The first is composed of clicks and metallic-sounding "boings." Boings consist of 8–14 pulses in bursts of up to 2 sec. The frequencies range from below 100 Hz to above 10 kHz. The second sound is a continuous low-frequency moan, ranging between 100 and 200 Hz. The third is another moan with a pulsed structure. The fourth is a pulsed "grunt," a broadband noise, produced by underwater exhalations. The first two types have been recorded predominantly from whales migrating offshore, in deeper water. Whales migrating nearer to shore in shallower water predominantly make the latter two sounds.

Click trains have been recorded from gray whales in Mexico, off Vancouver Island, and from the captive gray whale "Gigi." The click trains range from 2 to 6 kHz and last between 1 and 2 msec. Between 9.5 and 36 clicks/sec are produced.

B. FIN AND BLUE WHALES (*BALAENOPTERA PHYSALUS* AND *B. MUSCULUS*)   Fin and blue whales are the two largest species of extant cetaceans. These two species share the common characteristic of producing very low-frequency sounds. Both of these species are well known as sources of "20-Hz" sounds. In fact, both of these species do make calls in this extremely low-frequency region, but their range is wider than initially thought.

While fin and blue whales share their low-frequency characteristics, their calls differ in length. Fin whales produce primarily low-frequency downsweeps that are about 1 sec in length, whereas 20-Hz blue whale calls are typically 20 sec in length. Calls from both of these species generally occur in long, regularly spaced patterns. Sometimes the pattern includes "doublets," i.e., a double pulse with a longer interval between successive doublets.

The frequency range of these calls varies in different ocean basins. Most fin whale calls from the Atlantic are 1-sec downsweeps from ~23 to ~18 Hz. These "20-Hz" pulses regularly occur as a long, regular patterned series of calls. The interval between pulses typically ranges from 6 to 37 sec.

These calls are so low in frequency that they are detected on seismic equipment (geophones). It has been suggested that low-frequency sounds occur predominantly in winter and spring, during the mating time. However, fin whales are known to produce 20-Hz signals year-around, and if their calls are used as a reproductive display, then fin whales may be advertising constantly.

Atlantic fin whales also produce higher frequency calls, usually downsweeps, from 100 to 30 Hz. None of these appear to be produced in a regular manner of the 20-Hz pulses. It has been suggested that these calls are used for social communication. Fin whales also produce low-frequency broadband rumbles, centered around 30 Hz. It has been suggested that these calls indicate that the whale has been "surprised," as it was recorded when whales were near drifting ships or other objects. The 30-Hz rumble has also been recorded during social interactions between fin whales. Impulsive sounds have been associated with feeding and may be the result of biomechanical action rather than vocalization.

Blue whale calls are much longer than fin whale calls. Blue whale calls share two common characteristics: a fundamental frequency between about 10 and 40 Hz and a long duration between 10 and 30 sec. Variations on this theme are found. Calls recorded off Chile had fundamentals as low as 12.5 Hz and the harmonics ranged up to 200 Hz. These three-part calls lasted an average of 36.5 sec. All three parts of the call were amplitude modulated. These low-frequency moans were accompanied by short 390-Hz pulses.

Blue whales in the Indian Ocean are reported to produce a song, or repeated sets of sounds. These songs consist of four distinct notes. The first, second, and fourth notes are pulsive while the third is a pure tone. The total length of the four notes is approximately 2 min. Song has not been reported for any other population, and it is possible that song may have been recorded from pygmy blue whales (*B. m. brevicauda*).

Blue and fin whale calls are loud and of generally low frequency. These characteristics allow the calls to travel over great distances, and the sounds of these whales have been recorded to be heard at least several hundred kilometers away. It has been hypothesized that whales may hear each other over extents of ocean basins, especially if sounds are channeled into depths of water where characteristics of temperature and pressure interact to "carry" the sounds efficiently. Even if this occurs infrequently, the normally long distances of communication capability indicate that these whales may be considering themselves as parts of an "acoustic herd" over at least 100 or 200 km and may be exchanging basic information about the whereabouts of food or mating opportunities from afar. Research is only now delving into this possibility, although it was presented as a hypothesis by Roger Payne and Douglas Webb in the 1970s.

C. MINKE WHALES (*B. ACUTOROSTRATA* AND *B. BONAERENSIS*) Minke whales are known to produce a variety of different calls and show significant geographic variation in their call structure. Minke whales are difficult to observe in the wild and therefore it has taken many years to even associate some call types with minke whales. Others have only been described in recent years and others almost certainly remain undiscovered. The one seemingly consistent feature of their vocalizations is a low-frequency, short-duration downsweep. This downsweep generally descends from about 250 to 50 Hz and lasts between 0.2 and 0.3 sec. Australian minke whales (*B. acutorostrata* unnamed subsp.) make a vocalization so unusual that it has been termed the "star wars" call. This call consists of a series of three 100-msec pulses ranging between 50 Hz and 9.4 kHz. These pulses are produced simultaneously with harmonically unrelated low-frequency, amplitude-modulated pulses between 50 and 750 Hz. Following the pulses, the whale produces a pulsed tone at 1.8 kHz along with a tonal call at 80 Hz. The tonal call

shifts up to 140 Hz as the final component of the complex set of vocalizations. The communicative purpose is unknown.

D. HUMPBACK WHALES (*MEGAPTERA NOVAEANGLIAE*) Two baleen whales are known to repeat sets of notes in predictable fashion, or sing. There may be others. The four-note song of Indian Ocean blue whales was mentioned earlier. However, humpback and bowhead whales (discussed later) are well known for their wide vocal repertoires.

Roger Payne and Scott McVay first described the songs of humpback whales in 1971. Only males sing, almost exclusively on the mating grounds in winter, and it is surmised that song is a mating display; intersexual, for males to attract females; intrasexual, as a male dominance display; or both. Humpback whale songs have a hierarchical structure, from the shortest utterance to long bouts of singing that can last for days. Individual calls (somewhat analogous to musical notes) are referred to as song units. These units are repeated and combined to form phrases. These phrases are repeated to form longer themes. A song is typically composed of between 4 and 12 themes. Songs can last from 5 to 30 min in length before beginning again. Individual humpbacks are known to sing for as short as a few minutes and for 48 hr or longer. There is a tremendous variety of song units, including upsweeps, downsweeps, and complex FM sweeps. Numerous types of amplitude-modulated signals are produced, sometimes described as moans, grunts, rumbles, and ratchets (Fig. 4).

Song units range widely in frequency, from about 20 Hz all the way up to 10 kHz. Individual units can range from fractions of a second to several seconds in length.

Individual whales slowly change the structure of their songs over time. To illustrate this, consider a theme with two upsweeps and a growl. One type of change would be the addition of a third and then a fourth upsweep. Alternatively, the FM upsweeps might exhibit less and less change in frequency until they become pure (constant frequency) tones. These could then actually reverse their frequency modulation and gradually become downsweeps. These changes typically occur gradually, over a period of about 1 month. However, the pace of change in song structure is variable as well. In some years the song changes slowly and in other years it evolves rapidly. It is not known what is responsible for this variation in rate. Whales of an area pay attention to and copy each other so that all whales of a population sing essentially the same song, with only minor variations.

Once they have evolved, individual themes do not appear to be reused. All of the themes are created *de novo*. A comparison of 19 years of songs recorded from the west Atlantic found that while general types of song units do reoccur, none of the same combinations ever reappeared.

Humpback whales also produce a rich repertoire of non-song vocalizations, generally termed social sounds. Social sounds can be thought of as a subset of the song units, uttered in a nonpatterned fashion. On the mating grounds, these sounds appear to be used as acoustic threat displays in conjunction with visual displays and direct physical contact.

Finally, humpbacks produce a third class of calls, known as feeding calls. Humpbacks feed through a variety of lunges through the water at different orientations. Their mouth is held

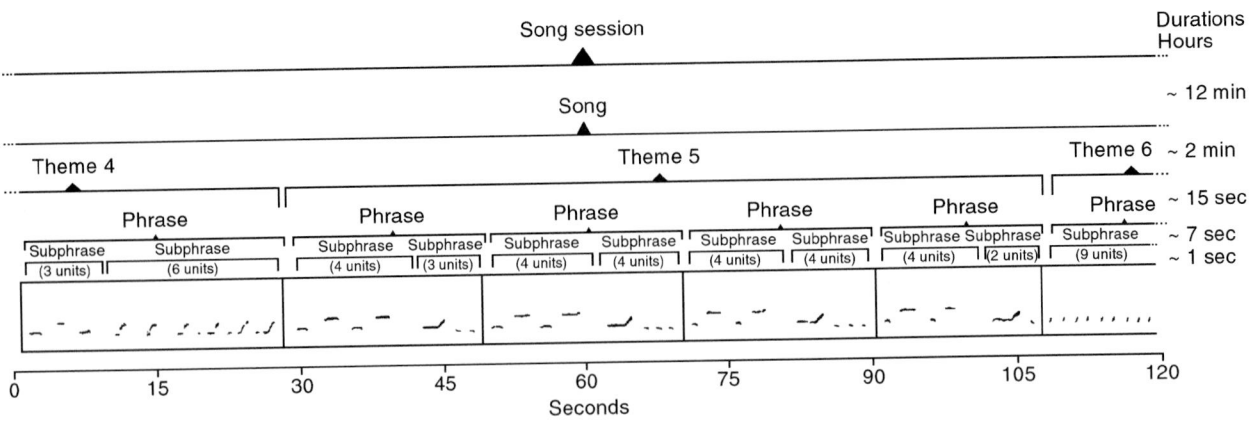

Figure 4 *The hierarchical structure of humpback whale song. Redrawn from Payne* et al. *(1983).*

open during these lunges, and the ventral pleats expand and fill with water and prey. When humpbacks feed on small fish, they may use a cooperative feeding strategy. This entails a group of 6–12 or more humpbacks all vertically lunging through the surface of the water in a coordinated fashion. Humpbacks produce the feeding call while they are maneuvering underwater. The call has been suggested to either coordinate the movements of the whales or manipulate prey. Recent experiments played back feeding calls to herring, which responded by fleeing from the call. These observations suggest that prey manipulation is the most likely function of this call.

The feeding call is a nearly constant frequency tone lasting between 5 and 10 sec. It has been compared to the sound of a train whistle. It is often repeated in a series of calls, and there is sometimes a shift of frequency at the end of the series of calls. There is variation in the frequency of the call, but most are between 500 and 550 Hz.

E. BOWHEAD WHALES (*BALAENA MYSTICETUS*) Humpbacks were the first whale to be recognized as a singer. However, extensive acoustic recording during the population census of bowhead whales off the North Slope of Alaska revealed that bowheads sing as well. Songs have been recorded during the spring migration from the Bering Sea to the Chukchi Sea. Songs are usually heard at the beginning of the migration and less at the end, suggesting that most of the singing occurs in the Bering Sea in the winter.

Song notes are usually longer than nonsong moan and grunt-like calls. Bowheads sing between one and three themes, most often two. Unlike humpback whales, bowhead songs regularly show substantial change in structure in successive years. Within a year, all whales sing the same basic version of song, but there is considerable inter- and intraindividual variation. Most of the sound energy of bowhead calls and song sequences is below 1000 Hz. The songs are frequently composed of both AM and FM components. Bowhead whales produce fewer units and in a narrower frequency range than humpbacks, yet bowheads have an unusually large variation in the tone of their songs, producing a wide variety of different sounding song notes.

In addition to songs, bowheads produce a wide range of calls. There are two main groups, the simple, low-frequency FM calls and complex calls. The FM calls can be categorized by their FM contours, i.e., upsweeps, downsweeps, constant tones, and inflected (change in contour). FM calls are almost always under 400 Hz in frequency. The complex calls have been described as pulsive, pulsed tonal, and high. High calls have frequencies above 400 Hz and sound like a whine. The pulsed tonal is a combination of both frequency and amplitude modulation. Pulsive calls are a mixture of pulses, with both frequency and amplitude modulations. Pulsed tonal calls are often below 400 Hz, but pulsive calls can exceed 1000 Hz.

Because observation conditions are often limited in the Arctic, it has been difficult to associate behaviors with these vocalizations. However, some observations have provided information on some calls. A mother and calf were separated and as they approached each other, loud FM calls were heard. Once they were rejoined, the calling stopped. Migrating bowheads will sometimes produce "signature calls" for a period of minutes up to at least 5 hr (or longer). These signature calls are often made in the context of whales countercalling with each other. Thus it has been suggested that these calls are used to maintain group cohesion. Unlike signature whistles of dolphins, the signature calls of bowheads do not appear to be specific to individuals, and individuals will switch signature call types. These signature calls may also be used to help orient in the ice field. A group of whales has been observed approaching a large block of ice. The early arriving whales only swam around the ice when they were very close. The following whales deflected much earlier, suggesting that they were listening to the echoes of the early whales and using the acoustic information to avoid the ice. Certainly one can imagine that swimming toward quiet areas (where there is no ice to reflect sound) is preferable to swimming toward an area with a strong echo (which could be caused by a large vertical piece of ice).

F. THE REST OF THE BALAENOPTERIDS: SEI AND BRYDE'S WHALES (*B. BOREALIS* AND *B. EDENI/BRYDEI*) Sei whales have been recorded only several times in the northwestern Atlantic. Sei whales produce two phrases each 0.5 to 0.8 sec long. The phrases are composed of 10–20 FM sweeps between 1.5 and 3.5 kHz, each 30–40 msec in duration. There was an interval of 0.4 and 1 sec between the two phrases.

Recordings off California have revealed that Bryde's whales make short low-frequency moans. Moans are between 70 and

245 Hz and last between 0.2 and 1.5 sec. Source levels range between 152 and 174 dB re 1 μPa. Bryde's whales also make a pulsed moan, which ranges between 100 and 900 Hz and between 0.5 and 51 sec in duration. The pulse rate varies between adults and calves. Finally, calves have been recorded making a series of discrete pulses between 700 and 900 Hz. These were recorded from calves when the adult was diving and from a captive juvenile.

Low-frequency moans have been recorded from Bryde's whales in the Caribbean, the eastern tropical Pacific, and off New Zealand. Caribbean whales produced 45-Hz calls that had harmonics and were between 1 and 4 sec long. Whales in the tropical Pacific produced calls of 35 and 42 Hz without harmonics and a call of 52 Hz that did have harmonics. Calls from New Zealand were centered on 22–28 Hz and 70 Hz; these calls did have harmonics.

G. RIGHT WHALES (*CAPEREA MARGINATA* AND *EUBALAENA* SPP.) Calls from pygmy right whales have only recently been recorded. A juvenile was recorded in a harbor and produced only one sound type. It was a short tonal downsweep that began between 90 and 135 Hz and swept down to 60 Hz. Pulses lasted between 140 and 225 msec in duration and were separated by intervals of 430 to 510 msec. Source levels were estimated between 153 and 167 dB re 1 μPa. These calls were very simple and their function is unknown.

The northern (*E. glacialis* and *E. japonica*) and southern (*E. australis*) species of right whales have generally similar calls, but only those of the South Atlantic have been described in some detail. Southern right whale calls have been described in terms of their frequency sweep or structure. They produce an "up" call, an upsweep from 50 to 200 Hz that lasts for about a second. This call appears to be used to bring individuals together, because calling stops once the whales join. "Down" calls are downsweeps from 200 to 100 Hz that are also about a second in length. They may serve to maintain acoustic contact, if not physical proximity. "Constant" calls range between 50 and 500 Hz and are 0.5 to 6 sec in duration. The frequency of these calls remains nearly constant.

Southern right whales make a variety of other sounds, such as high-frequency FM sweeps, amplitude-modulated pulses, and mixed pulses with both amplitude and frequency modulation. Blows (breaths) and slapping body parts on the surface also make sounds. The group type and behavioral context affect the mix of these sounds that are produced. Simple calls were usually used at long ranges and more complex vocalizations were used at shorter ranges.

Baleen rattle is a sound thought to be produced when right whales engage in surface feeding and the upper jaw and the upper portion of the baleen plates are out of the water and the lower portions of the baleen are in the water. As water flows through the baleen plates, they apparently rattle together, producing a series of short broadband pulses between 1 and 9 kHz, with most of the energy between 2 and 4 kHz. This sound is audible both in air and underwater, and may be simply a byproduct of feeding.

2. Toothed Whales The toothed whales, or odontocetes, are all vocal animals *par excellence*. They probably all produce clicks for echolocation, and many produce complicated sets of rapid click trains and whistles, the latter two for communication. This article summarizes sperm whale clicks and the known or believed communication sounds of other odontocetes. Even though toothed whales may be thought of as the most vocally active of all marine mammals, relatively little is known about the details of most species in sound production variability, especially in potential use of meaning of sound. This lack of detail of understanding may, of course, be because of the richness of the repertoires. A school of 500 socializing or feeding dusky dolphins, for example, produces a cacophony of sounds out of which it may be impossible for us to distinguish a particular individual or subgroup. It is likely that the dolphins themselves find meaning in this richness, but then they have evolved into this society, both genetically and behaviorally.

A. SPERM WHALES (*PHYSETER MACROCEPHALUS*) Sperm whales are the largest toothed mammals on earth and have a disproportionately huge head. It is likely that the evolution of that head has in large part been driven by the loud and complicated structure of their clicks, used certainly for communication and probably (although there is some argument on this point) for echolocation as well.

Sperm whales produce a variety of clicks in a variety of contexts. Clicks can occur singly at various intervals, in a short pattern of distinct clicks called a coda, or in a long sequence of tightly spaced clicks known as a creak. The frequency content of clicks differs between sexes. Large males have lower frequency content in their clicks than females and young males.

"Usual" clicks are produced in a regular sequence of clicks at intervals of 0.025 to 1.25 sec and have a duration between 2 and 24 msec. The click interval varies greatly between individuals, but appears to be stable within the click trains of an individual whale (Fig. 5).

Codas are stereotyped, repetitive patterns of clicks. They were originally suggested to serve the function of individual identification, analogous to signature whistles in bottlenose dolphins. Later work, however, has shown that a population of hundreds of sperm whales shared only 13 coda types. This contradicted the individual identification hypothesis, but they do appear to play a role in social communication.

"Creaks" are produced when clicks are produced at a high repetition rate. Creaks sound more like a continuous buzzing sound than individual clicks.

Mature males produce another type of click, called a "slow click" for its low repetition rate. these clicks also have a longer duration, with a mean of 72 msec compared to a 24-msec mean for "usual clicks." Slow clicks have consistent energy concentrations at 1.8 and 2.8 kHz, whereas the energy distribution in the spectra of usual clicks is much more variable.

There are competing hypotheses to explain the unique head of the sperm whale. Malcolm Clarke has pointed out that the spermaceti tissue that occurs in the head is a special lipid that changes density, and therefore buoyancy, based on its temperature. He holds that a sperm whale cools the spermaceti organ to make it denser than water, thus aiding diving. When the whale wants to surface, it warms the spermaceti, thereby decreasing its density and making it buoyant.

The more traditional theory holds that the spermaceti organ is the analogue of the melon in dolphins, and it is used to

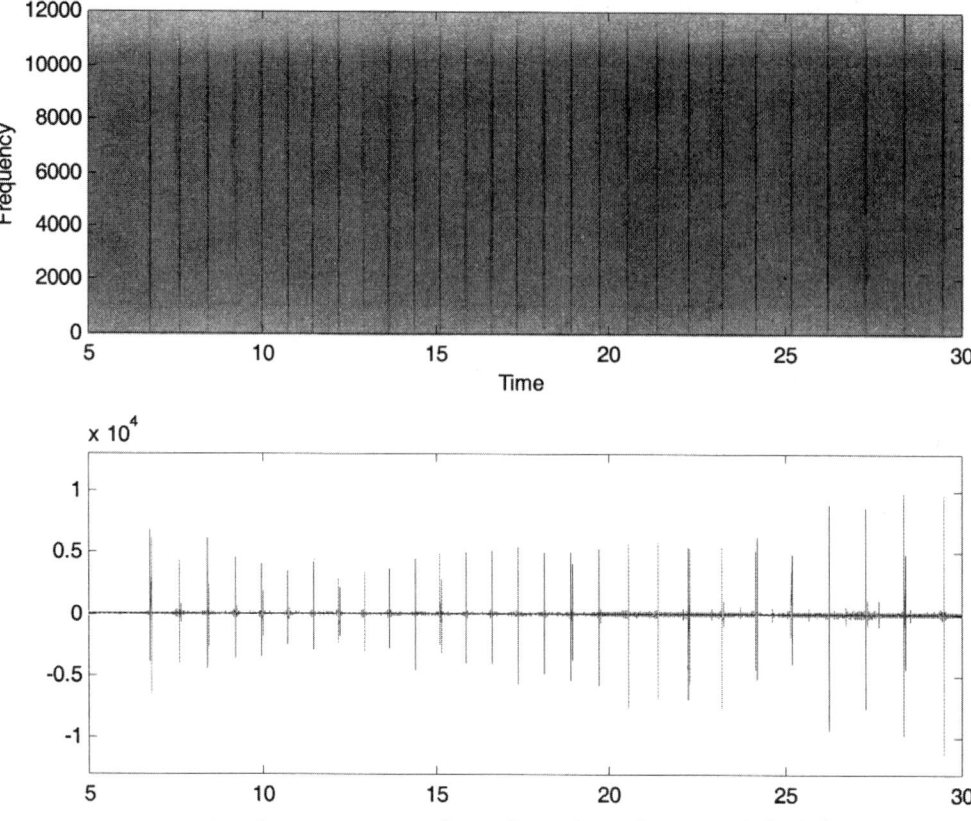

Figure 5 *The spectrogram and waveform of "usual" sperm whale clicks.*

transmit and focus the sound energy of the sperm whale. While there is no doubt that the sperm whale does produce clicks, their functions are still being determined. It is most probable that some sperm whale vocalizations are used for social communication. Differences in structure reveal some information about the sender. However, whether or not clicks are used for echolocation remains unclear. The issue is based on the dual questions of how loud sperm whales are and how much sound is reflected by squid. Squid make up the majority of the diet of the majority of sperm whales. However, squid do not have the bones and air spaces found in bony fish and therefore reflect much less sound than fish. If a sperm whale is going to obtain an echo from a target that reflects little sound, it will have to produce a tremendous amount of sound to get even a weak echo. The question of whether sperm whales can echolocate squid remains unresolved. It is possible that they may be visual predators on squid and acoustic predators on fish. It appears even more likely with relatively recent research findings that sperm whales can at least echolocate onto the surface of the water from below and to the bottom. They therefore can probably place themselves accurately at depth and may even be able to make out details of bottom topography by listening to their click echoes.

Although a detailed description of probable click production is beyond the scope of this article, it is likely that clicks are made near the front of the huge head. A part of a click's energy moves toward the rear of the head, bounces off the up-

ward curving skull of the whale, and is reflected forward. Part of that bounced sound can be reflected backward again because of the placement of an air sac near the front of the head, and thereby one click package can be composed of the main click and a series of tightly spaced intrahead echoes. It is unknown to what degree this complex click package may help in possible echolocation or provide certain information, perhaps size due to differences in click structure by size of head. At any rate, the major sounds of sperm whales are sets of clicks, and these are likely to be critically important in communication among group members. They may also be important for males to gauge each other on the mating grounds and perhaps for females to gauge males. Sperm whale sounds provide a rich set of unanswered questions and potential for further research.

B. KILLER WHALES (*ORCINUS ORCA*) Killer whales that feed on fish in the northeast Pacific tend to live in stable societies of matriarchies that change only by birth, death, and rare splitting, generally when the group or pod becomes very large. Even males born into a pod tend to stay with the pod for life, a condition unusual for any mammalian species. It is not known whether this great stability is a general characteristic for killer whale pods worldwide simply because other groups have not been studied in as much detail (Fig. 6).

On the early 1980s, researcher John Ford discovered that killer whales, who make echolocation clicks, other clicks likely used for communication, pulsed calls of a very rapid click-like structure, and whistles, have different dialects per pod. The

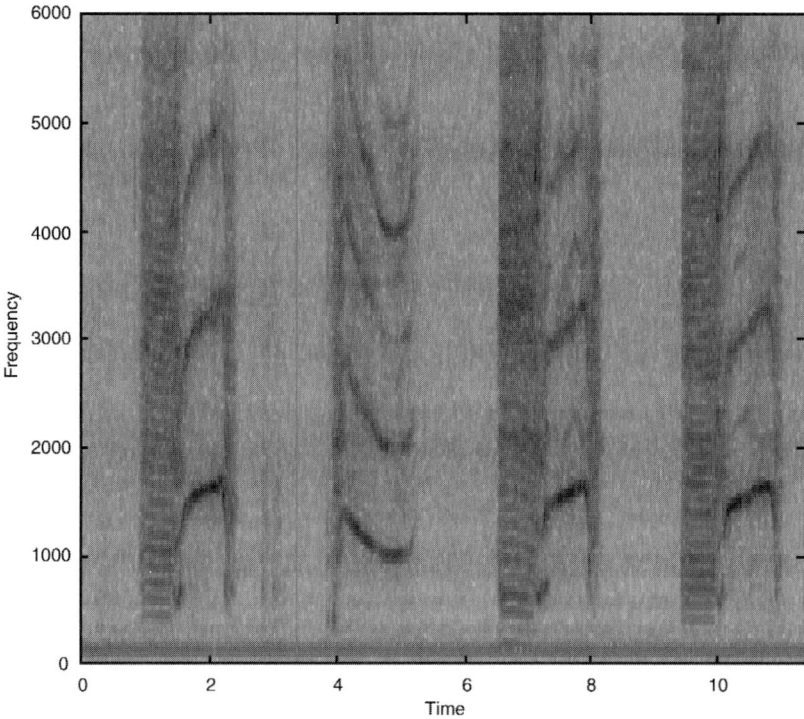

Figure 6 *A typical spectrogram of a killer whale call. This spectrogram also shows amplitude-modulated calls (banded) followed by frequency-modulated upsweeps.*

sounds composing the dialects tend to be within the range of human hearing and predominate around 500 Hz to 10 kHz. Two pods that feed on fish in generally the same area, and who come into acoustic range of one another now and then, can rapidly and at distance distinguish which pod is approaching. Presumably, they can then decide, based on past experience, whether it is prudent to interact at closer range with the other pod or to change course. Researchers who study the pods can distinguish them as well and have placed hydrophones in inlets and bays that announce the arrival or passing by of certain well-known pods.

Dialects may have simply evolved because of behavioral drift as animals that travel together hear each other more than those that do not. However, it is likely that the different dialects now serve important social functions for group cohesion and intergroup recognition. Killer whales in other areas of the world "all" have slightly different sounds as well, but it is presently unknown whether adjacent and at-times interacting pods have a dialect system similar to the system in the Pacific northeast.

C. BELUGAS (*DELPHINAPTERUS LEUCAS*) Sometimes known as the "canaries of the sea," belugas produce a wide variety of whistles, amplitude-modulated pulses, and echolocation clicks. While previous attempts have been made to categorize beluga calls, their nonclick sounds defy categorization. A thorough analysis of the calls of belugas shows that each and every sound can be a point on a continuum between other sounds. Belugas also have the ability to alter the frequency of their echolocation clicks. Individual belugas that were moved

from one area to another shifted the frequency of their echolocation clicks from ~60 kHz to ~100 kHz, apparently to avoid the increased background noise in their new habitat in the lower frequency range. Unlike other odontocetes, belugas are able to alter the physical shape of their melon, perhaps to adapt to sound transmission differences due to their movements between the open ocean and less saline estuarine waters.

D. NARWHALS (*MONODON MONOCEROS*) Narwhal clicks are some of the most intense sounds ever measured, up to 218 dB re 1 μPa. These clicks range up to 100 kHz, with peaks at 20 and 40 kHz. Clicks tend to be produced in two types of series: click trains have between 3 and 10 clicks per second, whereas click bursts contain between 110 and 150 clicks/sec. It has been hypothesized that click trains are used for searching for prey and that the much faster click bursts are used as the animal closes in on its prey, analogous to the click repertoires of some bat species.

E. BEAKED WHALES (*ZIPHIIDAE*) Very little is known about beaked whale sounds. This group is difficult to locate and frequently difficult to approach. There are a few recordings from the wild and from captivity. Whistles are known to range from 500 Hz to at least 10.7 kHz, while clicks range up to 125 kHz. It is believed that these ultrasonic clicks are used for echolocation, and indeed this is a plausible assumption. However, echolocation has not been investigated in detail in the beaked whale group. Whistles, as in dolphins, are most likely used for social interactions.

F. DOLPHINS (*DELPHINIDAE*) Dolphins also have echolocation clicks, burst-pulse sounds that consist of rapid sequences of clicks, and (for many dolphins but not all) pure tone whistles. While echolocation and other clicks can have frequencies above 100 kHz, most whistles of dolphins center around 7 to 15 kHz (and are therefore within the human hearing range). Each single whistle tends to last about 1 to 2 sec., rarely up to 3 sec. While the repertoires of many species and populations of delphinid cetaceans [including the killer whales mentioned previously, as well as pilot whales (*Globicephala* spp.) and other open ocean animals] have been described, we know by far most about the common bottlenose dolphin (*Tursiops truncatus*), mainly due to studies in captivity. Researcher Randy Wells and colleagues, working in the Sarasota–Bradenton area of west Florida, have also carried out excellent sound analyses of vocalizing bottlenose dolphins in nature.

There are no universally accepted classifications of the specifics of communication sounds, although begin-frequency, end-frequency, and kind of frequency modulation within the structures of whistles are usually used for descriptions. In bottlenose dolphins, researchers tend to come up with about 20 to 30 distinct whistle sounds, with many intergrading variations among them.

Bottlenose dolphins (and probably quite a few other species as well) have individualized whistle contours that are called signature whistles. With practice, humans can distinguish up to about seven vocalizing dolphins in an aquarium tank, and the dolphins can probably do even better. Signature whistles are certainly used by dolphins to recognize each other, but they also appear to be used in other social contexts. For example, one dolphin will often mimic the signature whistle of another animal in order to begin social contact. Perhaps the mimicry itself is a form of saying "hello."

Interestingly, signature whistles in Florida bottlenose dolphins appear to be more alike in mothers and their male offspring than in mothers and their daughters. The Sarasota population of dolphins is generally matriarchal, with daughters being closely affiliated for many years or for life, but sons leaving the natal group as subadults. It has been hypothesized that sons and mothers thereby recognize each other easily after prolonged times (perhaps extending to years) apart. This recognition could be important in avoiding inbreeding and in facilitating other kin-related social behaviors, such as lowered aggression. While signature whistles have been studied intensively in recent years, the true functions are not yet known.

There is some debate on which dolphin species and populations exhibit individualized sounds such as signature whistles. In common dolphins (*Delphinus* spp.), there appears to be a lack of sound "signature" per individual, and it has been guessed, but with little data support, that there may well be regional dialects per population or subpopulation, as in killer whales.

The complexity of sounds may well have to do with the complexity of behavior or some aspect of level of individual and therefore group "excitement." In long-finned pilot whales (*Globicephala melas*), for example, whales that are resting make very simple nonwavering whistles. Whistle complexity increases during feeding and bouts of socializing, and variability of whistles and other sounds increases greatly when two groups approach each other. Similar situations appear to exist for other delphinids, although detailed analyses of sounds per patterns of behavior have not been carried out, mainly because behavior below the waves is usually poorly known.

In at least Hawaiian spinner dolphins (*Stenella longirostris*) (and possibly in all other social dolphin and other toothed whale groups), there is a general relationship in amount of whistling, burst pulsing, and echolocation clicking with apparent alertness of the group. A resting group produces few sounds, whereas a feeding or socially active one produces many. This means that it is likely that each individual vocalizes more, but it is unknown whether this is merely an average for the group with much individual variation (perhaps different by age, social status, and gender) or whether each animal indeed indicates its state of alertness by the number (and complexity) of sounds produced.

Whistles are generally regarded as social communication signals, and clicks are typically thought to be used for echolocation. However, certain species, including some phocoenid porpoises, are specialized for very high-frequency hearing. These do not produce whistles, and they may use their high-frequency clicks for communication as well as echolocation.

Much has been written about the supposed "intelligence" of dolphins, and they are indeed large-brained highly behaviorally flexible social mammals. However, the supposition by many lay people that dolphins "have language," which implies sentence structure of some sort, with nouns, verbs, and modifiers, is unlikely to be the case. Dolphin sounds are complex, but appear to serve mainly signaling, emotive, and recognition functions. It is likely that as sounds of toothed cetaceans are studied in more detail in the near future, we will discover further complexities and marvelous adaptations besides that anthropomorphic wish for language.

G. PORPOISES (PHOCOENIDAE) Porpoises show a great range in vocalizations. The most commonly studied species is the harbor porpoise (*Phocoena phocoena*). Harbor porpoise vocalizations cover an extreme frequency range from 40 Hz to at least 150 kHz. Their vocalizations are composed of five major components. The first is of low-frequency calls ranging from 1.4 to 2.5 kHz, produced at high amplitudes. There are probably used for long-range detection. Midfrequency calls, between 30 and 60 kHz, are produced at low amplitude. Broadband midfrequency calls are made between 10 and 100 kHz. High-frequency components range between 110 and 150 kHz and are used for detection and classification of objects. All of these calls are variations of click trains. The repetition rate of clicks appears to be fairly constant within a click train when animals are "at ease" and varies between 15 and 36 clicks per second. When alerted by a prey item or other object at short ranges, harbor porpoises can produce between 500 and 600 clicks per second. The last type of vocalization are whistles, which range between 40 and 600 Hz.

D. Sirenians

1. Manatees The Caribbean manatee (*Trichechus manatus*) tends to produce sounds between 0.15 and 0.5 sec in duration. They show a relatively complex structure and range between 600 Hz and 12 kHz, but are typically 2.5 to 5 kHz. The fundamental frequency is at times less intense than the first harmonic. Calls consist mainly of chirp-squeaks, squeals, and screams. Amazonian manatee (*T. inunguis*) calls are similar in structure, although they may be higher in frequency (6–8 kHz) than those of the Caribbean manatee. The frequency content of vocalizations differs by sex, with female calls lower in frequency than those of males. Neither species is thought to echolocate. Manatees do not vocalize often, but do so under conditions of fear, aggravation, and male sexual arousal. Mothers and calves appear to use acoustic signals to facilitate rejoining each other.

2. Dugongs (Dugong dugon)   Dugongs appear to vocalize more often than the normally quiet manatees, with one study recording vocalizations in over half of recording attempts. Dugongs appear to produce three major types of sounds: chirp-squeaks, barks, and trills. They also produce intermediate sounds with multiple components of the three main types. Chirp-squeaks are short frequency-modulated signals that extend upward to 18 kHz. They are about 60 sec in duration, typically have a slight downward trend in frequency, and have between two and five harmonics. Barks are loud broadband signals that range between 500 Hz and 2.2 kHz. No frequency modulation is seen, and the barks last between 0.03 and 0.12 sec. Trills are a series of individual notes, lasting between 100 and 2200 msec. The notes typically begin at about 3.1 kHz and rise to 3.9 kHz. The frequency sweep is not linear, but rather an oscillating up and down. The functions of these sounds are not yet totally clear, but there are good indications that they are used for social communication.

IV. Conclusions

Marine mammals have a very rich behavioral tapestry of sounds. The carnivores that come to land or ice to breed produce some in-air sounds but generally are not nearly as vociferous as their totally terrestrial taxonomic relatives. Almost all marine mammals become loquacious underwater, however. The basic description here merely hints at this richness in an environment where sight and smell are not transmitted as efficiently as sound. Sound is used for communication and for wresting information from the environment. While only toothed whales are thought to have sophisticated echolocation, it is likely that many sounds give information on depth of water, obstruction ahead, or even silent conspecifics simply by the alteration of sound reflections in different environments.

Our acceptance that sound is critically important to marine mammals also gives us cause for worry. Since the advent of motorized shipping and now ever more with industrial seismic, intense military sonar and other human sources of sound, major parts of the oceans are becoming extremely noisy from nonbiological sources. We do not yet know the details of how noises can affect communication, masking, passive listening, and behavior and nervousness of mammals exposed to them.

See Also the Following Articles

Communication ■ Echolocation ■ Intelligence and Cognition ■ Mimicry ■ Noise, Effects of ■ Signature Whistles ■ Song

References

Anderson, P. K., and Barclay, M. R. (1995). Acoustic signals of solitary dugongs: Physical characteristics and behavioral correlates. *J. Mamm.* **76**, 1226–1237.

Asselin, S., Hammill, M. O., and Barrette, C. (1993). Underwater vocalizations of ice breeding grey seals. *Can. J. Zool.* **71**, 2211–2219.

Clark, C. W. (1982). The acoustic repertoire of the Southern right whale, a quantitative analysis. *Anim. Behav.* **30**, 1060–1071.

Crane, N. L., and Lashkari, K. (1996). Sound production of gray whales, *Eschrichtius robustus,* along their migration route: A new approach to signal analysis. *J. Acoust. Soc. Am.* **100**, 1878–1886.

Cranford, T. (2000). In search of impulse sound sources in odontocetes. *In* "Hearing by Whales and Dolphins" (W. W. L. Au, A. N. Popper, and R. R. Fay, eds.), pp. 109–155. Springer-Verlag, New York.

Cummings, W. C., Thompson, P. O., and Ha, S. J. (1986). Sounds from bryde, *Balaenoptera edeni,* and finback, *B. physalus,* whales in the Gulf of California. *Fish. Bull.* **84**, 359–370.

Edds, P. L., Odell, D. K., and Tershy, B. R. (1993). Vocalizations of a captive juvenile and free-ranging adult-calf pair of Bryde's whales, *Balaenoptera edeni.* *Mar. Mamm. Sci.* **9**, 269–284.

Frankel, A. S., Clark, C. W., Herman, L. M., and Gabriele, C. M. (1995). Spatial distribution, habitat utilization, and social interactions of humpback whales, *Megaptera novaeangliae,* off Hawaii, determined using acoustic and visual techniques. *Can. J. Zool.* **73**, 1134–1146.

Gedamke, J., Costa, D. P., and Dunstan, A. (2000). Localization and visual verification of a complex minke whale vocalization. *J. Acoust. Soc. Am.* **109**, 3038–3047.

Hanggi, E. B., and Schusterman, R. J. (1994). Underwater acoustic displays and individual variation in male harbour seals *Phoca vitulina.* *Anim. Behav.* **48**, 1275–1283.

Knowlton, A., Clark, C. W., and Kraus, S. D. (1991). Sounds recorded in the presence of sei whales, *Balaenoptera borealis. In* "Ninth Biennial Conference on the Biology of Marine Mammals," The Society for Marine Mammalogy, Chicago.

Miller, E. H., and Murray, A. V. (1995). Structure, complexity and organization of vocalizations in harp seal (*Phoca groenlandica*) pups. *In* "Sensory Systems of Aquatic Mammals" (R. A. Kastelein, J. A. Thomas, and P. E. Nachtigall, eds.), pp. 223–236. De Spil, Woerden, The Netherlands.

Miller, L. A., Pristed, J., Mohl, B., and Surlykke, A. (1995). The click-sounds of narwhals (*Monodon monoceros*) in Inglefield Bay, Northwest Greenland. *Mar. Mamm. Sci.* **11**, 491–502.

Payne, R. S., and McVay, S. (1971). Songs of humpback whales. *Science.* **173**, 585–597.

Rogers, T., Cato, D. H., and Bryden, M. M. (1995). Underwater vocal repertoire of the leopard seal (*Hydrurga leptonyx*) in Prydz Bay, Antarctica. *In* "Sensory Systems of Aquatic Mammals" (R. A. Kastelein, J. A. Thomas, and P. E. Nachtigall, eds.), pp. 223–236. De Spil, Woerden, The Netherlands.

Sjare, B. L., and Smith, T. G. (1986). The vocal repertoire of white whales (*Delphinapterus leucas*), summering in Cunningham Inlet, Northwest Territories. *Can. J. Zool.* **64**, 407–415.

Stirling, I. (1973). Vocalization in the ringed seal (*Phoca hispida*). *J. Fish. Res. Board Can.* **30**, 1592–1593.

Stirling, I., and Siniff, D. B. (1979). Underwater vocalizations of leopard seals (*Hydrurga leptonyx*) and crabeater seals (*Lobodon carcinophagus*) near the South Shetland Islands, Antarctica. *Can. J. Zool.* **57**, 1244–1248.

Thomas, J. A., Puddicombe, R. A., George, M., and Lewis, D. (1988). Variations in underwater vocalizations of Weddell seals (*Leptonychotes weddelli*) at the Vestfold Hills as a measure of breeding population discreetness. *Hydrobiologia* **165**, 279–284.

Tyack, P. L. (1999). Communication and cognition. *In* "Biology of Marine Mammals" (J. E. Reynolds III and S. A. Rommel, eds.). Smithsonian Institution Press, Washington, DC.

Van Parijs, S. M., Hastie, G. D., and Thompson, P. M. (2000). Individual and geographic variation in display behavior of male harbour seals in Scotland. *Anim. Behav.* **59,** 559–568.

Verboom, W. C., and Kastelein, R. A. (1997). Structure of harbour porpoise (*Phocoena phocoena*) click train signals. *In* "The Biology of the Harbour Porpoise" (A. J. Read, P. R. Wiepkema, and P. E. Nachtigall, eds.), pp. 343–363. DeSpil Publishers, Woerden, The Netherlands.

Watkins, W. A., and Schevill, W. E. (1977). Sperm whale codas. *J. Acoust. Soc. Am.* **62,** 1485–1490.

Watkins, W. A., Tyack, P., and Moore, K. E. (1987). The 20-Hz signals of finback whales (*Balaenoptera physalus*). *J. Acoust. Soc. Am.* **82,** 1901–1912.

Würsig, B., and Clark, C. (1993). Behavior. *In* "The Bowhead Whale" (J. J. Burns, J. J. Montague, and C. J. Cowles, eds.), pp. 157–199. The Society for Marine Mammalogy, Lawrence, KS.

Sound Reception

SEE *Hearing*

South American Aquatic Mammals

ENRIQUE A. CRESPO

Centro Nacional Patagónico,
Puerto Madryn, Argentina

I. South American Marine and Freshwater Ecosystems

The marine and freshwater ecosystems of South America are very rich in aquatic mammals. Seventy species have been reported to occur within these ecosystems (Table I); most breed locally, and only five species that appear occasionally belong to Antarctic or sub-Antarctic ecosystems. A number of species are found in South America that occur in other parts of the world or the Southern Hemisphere, such as rorquals, ziphiids, and some delphinids. However, 19 species can be considered endemic to the coastal waters or the river systems of South America.

The distribution of marine mammals at sea is related to the distribution pattern of ocean currents that is defined by the characteristics of the major water masses, mainly temperature and salinity. The marine mammal assemblages of South America can be explained in part by the water masses that move around the continent (Fig. 1). However, depth and ocean productivity may also play an important role in the presence, absence, or high concentration of individuals of a given species. Four different water masses each have their own marine mammal assemblage. These are (a) the Humboldt Current,

(b) Equatorial Front of the eastern tropical Pacific, (c) Malvinas Current, and (d) Brazil and South Equatorial Atlantic Currents. In addition, a fifth assemblage is found in continental waters; it is heterogeneous due to the isolation between some of the river basins. Finally, a sixth assemblage that could be defined as "erratic circumpolar" can also be found. However, it is composed of isolated individuals from Antarctic or sub-Antarctic populations that breed or live southward of the Polar Front but move erratically into northern water masses.

II. Marine Mammals of Cold Water Marine Ecosystems

In the extreme south of the continent, the Antarctic circumpolar current moves from west to east and splits into two branches: the Malvinas Current in the Atlantic and the Humboldt Current in the Pacific. The cold marine ecosystem in the Pacific almost reaches the equator with waters between 8 and 15°C, but in the Atlantic the cold-temperate system reaches 40°S. Off Peru an upwelling system gives rise to high levels of primary and secondary productivity.

Several cold-water marine mammals are found in both the Humboldt and the Malvinas Currents. Among those species the most common in coastal waters are two otariids (the South American sea lion *Otaria flavescens* and the South American fur seal *Arctocephalus australis*) and two small cetaceans (the dusky dolphin *Lagenorhynchus obscurus* and Burmeister's porpoise *Phocoena spinipinnis*). Other small cetaceans, such as the dolphins of the genus *Lagenorhynchus* (*L. australis* and *L. cruciger;* the latter more pelagic and less known) and the southern right whale dolphin (*Lissodelphis peronii*), can be included. Two related species, the Chilean dolphin (*Cephalorhynchus eutropia*) in the Pacific and Commerson's dolphin (*C. commersonii*) in the Atlantic, are endemic to the southern parts of the ecosystems.

One of the most conspicuous species in the southwestern Atlantic is the southern right whale (*Eubalaena australis*). With a geographic distribution between 20° and 55°S, one of the highest breeding concentrations is at Península Valdés (42°S). It has been recovering in the present century after a long depletion of its population size, at rates over 7%, like other stocks in South Africa, Australia, and New Zealand. The whales can be observed in several places in the Atlantic: Uruguay, southern Brazil, and Buenos Aires Province in Argentina. At Santa Catarina, Brazil, it seems that a new breeding area is starting to be established. On the Pacific side there are signs of recovery and a possible northward extension of the distribution range.

The spectacled porpoise (*Phocoena dioptrica*) is known from the eastern coast of South America and several sub-Antarctic islands, and the South American marine otter (*Lontra felina*) is known from Perú to Staten Island in the southern South Atlantic. Two fur seals are endemic to the Galápagos and the Juan Fernández Archipelagos (respectively, *Arctocephalus galapagoensis* and *A. phillippii*). The latter is also found in a few other places in Peru and Chile. The Galápagos are also home to an endemic sea lion, *Zalophus wollebaeki*.

The pilot whale (*Globicephala melas*), Risso's dolphin (*Grampus griseus*), and killer whale (*Orcinus orca*) can be in-

TABLE I
Recorded Presence of Species in South American Marine and Freshwater Ecosystems

Taxa	Cold-temperate Pacific	Tropical Pacific	Tropical Atlantic Caribean	Cold-temperate Atlantic	Freshwater	Erratic circumpolar
Sirenidae						
Trichechidae						
1. *Trichechus manatus*			XXX			
2. *T. inunguis*[a]					XXX	
Carnivora						
Pinnipedia						
Otariidae						
3. *Otaria flavescens*[a]	XXX			XXX		
4. *Zalophus wollebaeki*[a]						
5. *Arctocephalus australis*[a]	XXX			XXX		
6. *A. philippii*[a]	XXX					
7. *A. gazella*						XXX
8. *A. galapagoensis*[a]	XXX					
9. *A. tropicalis*						XXX
Phocidae						
10. *Mirounga leonina*	XXX			XXX		XXX
11. *Lobodon carcinophaga*						XXX
12. *Hydrurga leptonyx*						XXX
13. *Leptonychotes weddellii*						XXX
14. *Monachus tropicalis*			Extinct			
Fissipedia						
Mustelidae						
15. *Lontra felina*[a]	XXX			XXX		
16. *L. provocax*[a]					XXX	
17. *L. longicaudis*[a]					XXX	
18. *Pteronura brasiliensis*[a]					XXX	
Cetacea						
Mysticeti						
Balaenidae						
19. *Eubalaena australis*	XXX			XXX		
Neobalaenidae						
20. *Caperea marginata*	XXX			XXX		
Balaenopteridae						
21. *Balaenoptera musculus*	XXX	XXX	XXX	XXX		
22. *B. physalus*	XXX	XXX	XXX	XXX		
23. *B. borealis*	XXX	XXX	XXX	XXX		
24. *B. edeni*	XXX	XXX	XXX	XXX		
25. *B. acutorostrata*	XXX	XXX	XXX	XXX		
26. *B. bonaerensis*	XXX			XXX		
27. *Megaptera novaeangliae*	XXX	XXX	XXX	XXX		
Odontoceti						
Physeteridae						
28. *Physeter macrocephalus*	XXX	XXX	XXX	XXX		
Kogiidae						
29. *Kogia breviceps*	XXX	XXX	XXX	XXX		
30. *K. sima*	XXX	XXX	XXX	XXX		
Ziphiidae						
31. *Berardius arnuxii*	XXX			XXX		
32. *Ziphius cavirostris*	XXX	XXX	XXX	XXX		
33. *Hyperoodon planifrons*	XXX			XXX		
34. *Tasmacetus shepherdi*	XXX			XXX		
35. *Mesoplodon densirostris*	XXX	XXX	XXX	XXX		
36. *M. grayi*	XXX			XXX		
37. *M. hectori*	XXX			XXX		
38. *M. peruvianus*	XXX					

(continues)

TABLE I (Continued)

Taxa	Cold-temperate Pacific	Tropical Pacific	Tropical Atlantic Caribean	Cold-temperate Atlantic	Freshwater	Erratic circumpolar
39. M. europaeus			XXX			
40. M. layardii	XXX			XXX		
41. M. bahamondi[a]	XXX					
42. M. gingkodens	XXX					
Delphinidae						
43. Orcinus orca	XXX	XXX	XXX	XXX		
44. Globicephala melas	XXX			XXX		
45. G. macrorhynchus	XXX	XXX	XXX	XXX		
46. Pseudorca crassidens	XXX	XXX	XXX	XXX		
47. Feresa attenuata	XXX	XXX	XXX	XXX		
48. Peponocephala electra		XXX	XXX	XXX		
49. Sotalia fluviatilis[a]			XXX		XXX	
50. Steno bredanensis		XXX	XXX			
51. Lagenorhynchus obscurus	XXX			XXX		
52. L. cruciger	XXX			XXX		
53. L. australis[a]	XXX			XXX		
54. Grampus griseus	XXX	XXX	XXX	XXX		
55. Tursiops truncatus	XXX	XXX	XXX	XXX		
56. Stenella attenuata		XXX	XXX			
57. S. frontalis			XXX			
58. S. longirostris		XXX	XXX			
59. S. clymene			XXX			
60. S. coeruleoalba		XXX	XXX			
61. Delphinus delphis	XXX	XXX	XXX	XXX		
62. D. capensis	XXX	XXX	XXX	XXX		
63. Lagenodelphis hosei		XXX	XXX	XXX		
64. Lissodelphis peronii	XXX			XXX		
65. Cephalorhynchus commersonii[a]				XXX		
66. C. eutropia[a]	XXX					
Phocoenidae						
67. Phocoena dioptrica[a]	XXX			XXX		
68. P. spinipinnis[a]	XXX			XXX		
Iniidae						
69. Inia geoffrensis[a]					XXX	
Pontoporiidae						
70. Pontoporia blainvillei[a]			XXX	XXX		

[a]Endemic species.

cluded in the cosmopolitan species with locally abundant populations. Eight species of Balaenopteridae and eight ziphiids are common to cold waters of both sides of South America. However, the pygmy beaked whale *Mesoplodon peruvianus* has been recorded only from Peruvian waters (also recorded from México and California) and a recently described species, Bahamonde's beaked whale *M. bahamondi*, from the Juan Fernández Archipelago, Chile.

The dynamics of oceanographic and biologic processes that sustain the high productivity of the Peruvian ecosystem can be disturbed by what has been called the El Niño southern oscillation (ENSO event), whose main characteristic is the inflow of tropical waters into the upwelling region close to Christmas. The nature of ENSO is irregular and unpredictable, and the impact on the intermediate levels of the food chain (e.g., abundance of anchovies) affects finally the seabirds and marine mammals. Demonstrated effects of ENSO events on fur seals, sea lions, dusky dolphins, and seabirds have included those on survival, recruitment, and the general condition of the individuals as a consequence of reduced food availability. An ENSO event is part of a more general pattern of oceanographic change affecting not only the Peruvian ecosystem, but also the entire Southern Ocean.

III. Marine Mammals in Tropical Water Ecosystems

On the Atlantic side, the South Equatorial Current moves from east to west and turns southward becoming the tropical Brazil Current with exclusive influence up to 28°S. The northern Equatorial Current moves northward and turns clockwise in the Caribbean zone. Between 33° and 40°S the Atlantic marine ecosystem shows mixed characteristics of the opposing flows of the Malvinas and Brazil Currents. On the Pacific side of South America the Humboldt Current turns to the west after forming the Tropical Front at around 5°S.

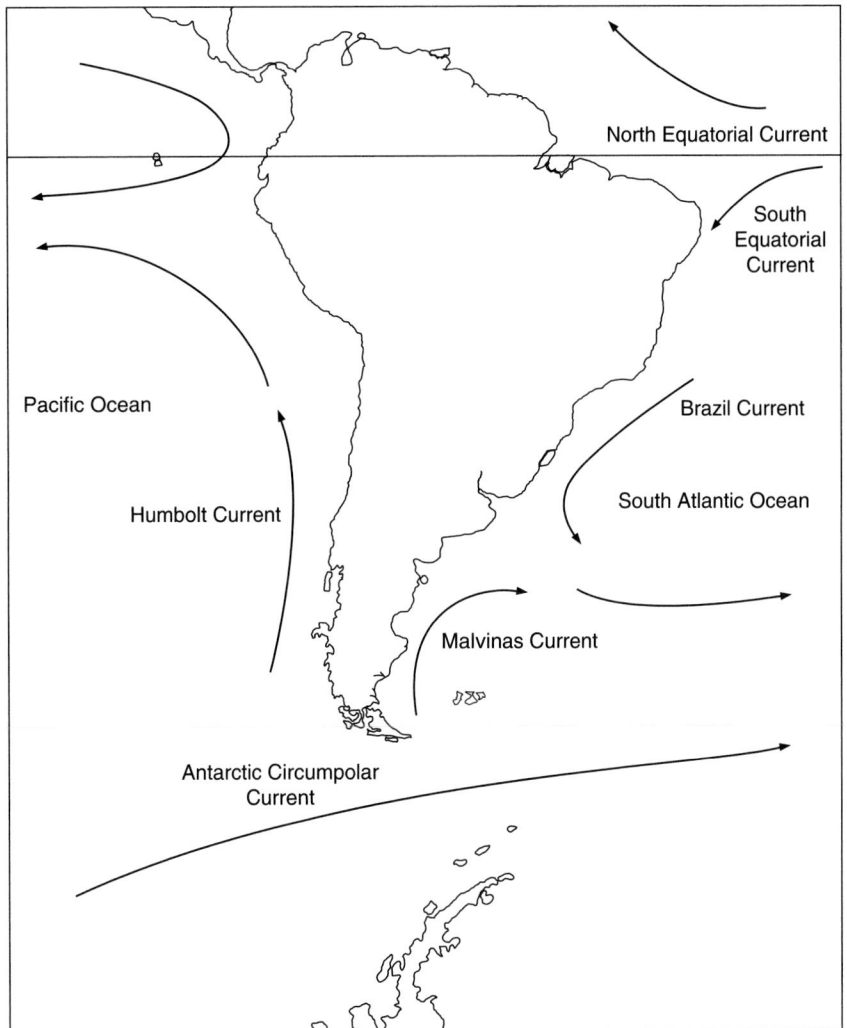

Figure 1 *Oceanic currents around the South American continent.*

Both the Atlantic and Pacific tropical systems show species assemblages typical of warmer waters globally. They include dolphins of the genus *Stenella,* (*S. attenuata, S. longirostris,* and *S. coeruleoalba*), common dolphins (*Delphinus delphis* and *D. capensis*), the melon-headed whale (*Peponocephala electra*), the rough-toothed dolphin (*Steno bredanensis*), Fraser's dolphin (*Lagenodelphis hosei*), Bryde's whale (*Balaenoptera edeni*), both species of Kogiidae, the short-finned pilot whale (*G. macrorhynchus*), the pygmy killer whale (*Feresa attenuata*), and the false killer whale (*Pseudorca crassidens*) among others. The Clymene dolphin (*Stenella clymene*), the Atlantic spotted dolphin (*S. frontalis*), the marine coastal population of the tucuxi (*Sotalia fluviatilis*), and the Caribbean manatee (*Trichechus manatus*) are found only in the Atlantic. Some species, such as the common dolphins, the false killer whale, and the bottle-nose dolphin (*Tursiops truncatus*) range from the tropics far south in the Atlantic, in the area of mixed waters of the Brazil and Malvinas Currents. Gervais' beaked whale *Mesoplodon europaeus* is known from the east coasts of North America and Caribbean islands and the eastern Atlantic Ocean.

IV. Aquatic Mammals in River Basins

South America is very rich in river basins, and all the aquatic mammals in continental waters are endemic. The most important hydrographic systems are the Amazon, the Orinoco, and the Paraná-Uruguay rivers. However, other small river systems are located in cold and high-latitude parts of the continent. Most of those systems are isolated from each other. The Amazon and Orinoco basins hold important populations of aquatic mammals. The most conspicuous include the boto (*Inia geoffrensis*), a freshwater population of the tucuxi (*Sotalia fluviatilis*), and the Amazon manatee (*Trichechus inunguis*). Two otters are also found in those basins and also in the Paraná–Uruguay River systems: the giant otter (*Pteronura brasiliensis*) and the long-tailed otter (*Lontra longicauda*). The cold river systems in southern Argentina and Chile sustain populations of the river otter (*Lontra provocax*), which is also found on Staten Island at the extreme southern end of the continent.

V. Occasional Visitors from the Antarctic

Five species of pinnipeds that breed on sub-Antarctic islands or the Antarctic ice (the Artarctic fur seal *Arctocephalus gazella*, the subantarctic fur seal *A. tropicalis*, the leopard seal *Hydrurga leptonyx*, the crabeater seal *Lobodon carcinophaga*, and the Weddell seal *Leptonychotes weddellii*) move erratically to more northerly waters of both sides of South America. Probably as a consequence of population increases over the last few decades, it has become more frequent to sight individuals moving as far north as the Equator. The southern elephant seal (*Mirounga leonina*) can be included in this list; while most breeding groups are circumpolar, there is an important and increasing breeding stock at Península Valdés, Patagonia, around 42°S in the Atlantic Ocean.

VI. Problems Faced by Aquatic Mammals in South America

Problems faced by marine mammals in South America are much like those they face in other parts of the world. They include incidental catch in fisheries, direct exploitation, competition for fishing resources, and habitat loss and degradation. Most of the species in South America are insufficiently known. For most of them there is no information about abundance and population trends. Exceptions include the pinnipeds, the southern right whale, and few species of small cetaceans such as the dusky and Commerson's dolphins in Patagonia and river dolphins in the Amazon. However, abundance has been estimated for these for only small parts of their ranges.

While large whales are fully protected, giving them the opportunity to recover, a long list of species of small cetaceans are taken incidentally in coastal and high-seas fisheries both in the tropics and in cold-water ecosystems. The most critical situation is probably that of the franciscana, *Pontoporia blainvillei.* However, there are other species or local stocks which interact heavily with different types of fisheries in direct and incidental takes. Examples include the coastal population of the tucuxi in some parts of Brazil, local populations of dusky dolphins in Peru and Patagonia, both species of common dolphins and Burmeister's porpoise in Peru, Commerson's dolphins in Patagonia, and Chilean dolphins in southern Chile. The problem of incidental catch of aquatic mammals in fishing gear has not been addressed by the authorities as part of fishery management to date in the countries of the region.

Direct takes mostly for food, oil, or bait also include the use of genital organs for aboriginal beliefs in the case of small cetaceans and illegal small-scale commerce between South America and Asian countries in a black market in the case of otariids. Cases are known for the Pacific and the Atlantic Oceans and the river basins and include several species. Economic reasons were involved in turning an incidental into a direct catch in Peru during the 1980s. These conditions can be amplified with poverty and hyperinflation. Direct takes are also a matter of concern with other aquatic groups, such as the hunting of river otters for furs in the river systems.

Regarding competition for fishing resources, there is a general perception by fishermen all around the world that some marine mammals, mostly pinnipeds, are currently depleting the target species of fisheries. The perception is sustained by the relative increase of marine mammal populations in the last few decades and the decrease of fishing stocks in the same period. In South America this is thought to be the case with the South American sea lion, *Otaria flavescens.* Even though there has been no formal proposal for culling in the countries of the region, with the exception of Peru where the problem has been discussed by the government, private sector and nongovernmental organizations (NGOs), sea lions are shot illegally by fishermen in many places throughout the distribution range. There is also a general belief that reducing the populations of competitors will increase the stocks of the target species of the fishery, a supposition that to date has not been supported by data; culling carried out around the world has not increased levels of target species or catches.

Loss of habitat is the most important problem that faces freshwater and coastal species. Indiscriminate clearing of rainforest for the use of land in agriculture and cattle growing is currently going on, and many species are declining sharply. Dams and other barriers in large rivers have been or are being planned for hydroelectric power, irrigation, or flood control. These projects have significant impact on regional development and positive benefits for society. Nevertheless, negative consequences in the river ecosystem for river dolphins, manatees, river otters, and other wildlife include, at least, the isolation of populations of aquatic mammals and their prey and unnatural water flows and interruption of migratory paths. In addition to dam construction, other threats include pollution, mining, the use of dynamite for fishing, and incidental and direct catches in fisheries. In coastal areas the most important problems include intensive fishing and POLLUTION by hydrocarbons, agrochemical and heavy metal products, and intensive boat traffic. The importance of each of these varies with the area considered.

VII. Need for International Agreements in Conservation Policies

With the exception of endemic species, most South American aquatic mammals have a wide distribution and occur or breed passing through political boundaries, e.g., those of Brazil, Peru, Uruguay, Argentina, and Chile. Legal protection or status is not the same in each country and sometimes there is little or no enforcement. As an example, the South American sea lion is shot by fishermen in southern Brazil, is one of the tourist targets in Argentine Patagonia, is allowed to be killed if necessary in Chile, and is a potential culling target in Peru. The dusky dolphin is incidentally and directly taken in Peru and northern Chile and incidentally taken in Patagonia. South American fur seals and Commerson's and Peale's dolphins were used for crab bait for many years at the southern tip of South America. The franciscana crosses the boundaries of Brazil, Uruguay, and Argentina and while it is protected by law in the three countries, incidental mortality is still a cause for concern.

At the national level, incidental mortality or bycatch should be considered in fishery management models and decision making. At the international level, agreements should be promoted among the countries of the region in order to give general or particular status of protection for a given species. As an example, in 1991, Colombia, Chile, Ecuador, Panama, and Peru approved the Action Plan for the Conservation of Marine Mammals in the Southeast Pacific in order to help the governments of the region to agree convenient policies for marine mammal conservation and management. The Atlantic region still lacks such an agreement.

See Also the Following Articles

Competition with Fisheries ■ Distribution ■ Habitat Pressures ■ River Dolphins

References

Cappozzo, H. L., and Junin, M. (eds.) (1991). Estado de conservación de los mamíferos marinos del Atlántico Sudoccidental. Informes y estudios del Programa de Mares Regionales del PNUMA No. 138, PNUMA.

Cardenas, J. C., Stutzin, M. E., Oporto, J. A., Cabello, C., and Torres, D. (1986). Manual de Identificación de los cetáceos chilenos. WWF/CODEF, Santiago, Chile.

IUCN (1991). "Plan de acción para las nutrias de latinoamérica" (P. Foster-Turtley, S. Macdonald, and C. Mason, eds.). IUCn/SSC Otter Specialist Group.

Jefferson T. A., Leatherwood, S., and Weber, M. A. (1993). "FAO Species Identification Guide: Marine Mammals of the World." FAO, Rome.

Olson, D., Dinerstein, E., Canevari, P., Davidson, I., Castro, G., Moricet, V., Abell, R., and Toledo, E. (eds.) (1998). "Freshwater Biodiversity of Latin America and the Caribbean: A Conservation Assessment." Biodiversity Support Program, Washington, DC.

Pinedo, M. C., Rosas, F. C. W., and Marmontel, M. (1992). Cetáceos e pinnípedes do Brasil. UNEP/UEFA Manaus, Brasil.

Reeves, R. R., and Leatherwood, S. (1994). "Dolphins, Porpoises, and Whales: 1994–1998 Action Plan for the Conservation of Cetaceans." IUCN, Gland, Switzerland.

Reyes, J. C. (1992). Informe nacional sobre la situación de los mamíferos marinos en Perú. Informes y Estudios del Programa de Mares Regionales del PNUMA No. 145, PNUMA.

Vidal, O. (1992). Los mamíferos marinos del Océano Pacífico Sudeste (Panamá, Colombia, Ecuador, Perú y Chile): diagnostico region. Informes y Estudios del Programa de Mares Regionales del PNUMA No. 142, PNUMA.

South American Sea Lion
Otaria flavescens

HUMBERTO LUIS CAPPOZZO
Museo Argentino de Ciencias Naturales,
Buenos Aires, Argentina

The scientific name of the South American sea lion has been under discussion for many years. Two names were in use until a few years ago: *Otaria flavescens* (Shaw, 1800) and *Otaria byronia* (de Blainville, 1820). More recently, use of *O. flavescens* has been recommended; this is the name used throughout the distribution area in South America. The common name changes with location: *lobo marino de un pelo, león marino del sur, lobo marino del sur, lobo común, lobo chusco, leâo marinho,* or *lobo marinho de um pelo.* "South American sea lion" is preferable to "Southern sea lion," as it prevents confusion with Australian and New Zealand sea lions, species also distributed in the Southern Hemisphere.

This species is one of seven that make up the subfamily Otariinae, part of the family Otariidae, usually called otariids, or pinnipeds with ears. It is one of the largest and most dimorphic otariids. Adult males are much heavier than females. Differences in size between males and females have also been documented among juveniles, and even newborns. The sex ratio at birth is 1:1. Adult males reach a maximum length of around 3 m and weigh 300 to 350 kg; adult females are about 2 m long and weigh up to 150 kg. The newborn sea lion usually weighs between 12 and 15 kg and is 0.75 to 0.86 cm long. Adult males have a characteristic hairy, thick neck. The color is generally brownish, from the very dark brown of adult males to almost yellow in females. Pups are black at birth. After the molt of the embryonic coat (*lanugo*) at 2 months, the first juvenile HAIR is reddish brown, changing in color with age and sex (Fig. 1). The South American sea lion eats mainly fish and squids, but its diet is very variable and it adapts easily to locally abundant prey, including crustaceans and even penguins.

I. Distribution and Population Status

South American sea lions are widely distributed along the Atlantic and Pacific coasts of South America: from Torres in southern Brazil to Cape Horn in the extreme south of the Atlantic coastline, and from Cape Horn to Zorritos in northern Peru, on the Pacific. Total population has been estimated at 110,000 for the southwestern Atlantic coast, concentrated mainly on the Patagonian coast and southern islands. There is no reliable information concerning the Pacific population, but it is considered smaller.

Variation of the substrate and weather phenomena affect DISTRIBUTION. For example, the occurrence of El Niño on the Peruvian Pacific coast influences the whole marine ecosystem by a drastic superficial thermal inversion that affects the entire trophic web from the plankton up to the top predators. Every time El Niño occurs, there is a drastic population decrease due to higher death rates and migration. Also, on the coasts of southern Brazil and Buenos Aires, Argentina, there has been a great reduction of habitat because of human use of the zone.

South American sea lions do not spend long periods away from the coast, as many pinnipeds do, and they gather in groups or colonies. There are both permanent and nonpermanent colonies; the latter are mainly reproductive settlements. At the permanent colonies, during the breeding season, the individuals that do not participate in the reproductive activity stay in the colony segregated in nonreproductive groups.

A. Population Flow

There is little information regarding migrations and seasonal movements. To the north of the Uruguayan breeding

Figure 1 *Adult female, adult male, and offspring of South American sea lion.*

grounds, in southern Brazil, there are only two nonbreeding haul-out rookeries where subadult males predominate, with seasonal movements. Aberrant records for the species have been reported for Rio de Janeiro, Brazil 23°S, and even to 13°S but these are always solitary individuals.

In Argentina, the species was widely distributed from the La Plata river to Cape Horn, but at present only two subadult male rookeries remain in the Buenos Aires Province: at Mar del Plata and Quequén harbors, as well as a breeding rookery at Isla Trinidad, 39°S. To the south, they breed along the Patagonian coast, from Punta Bermeja to Tierra del Fuego. There are 18 breeding and nonbreeding colonies between Punta Bermeja and Punta Leon, in Northern Patagonia, for which a flow of individuals showing a seasonal pattern has been reported. There is a gap of about 200–300 km between the northern and the central Patagonian stocks. The main concentration of this species occurs in central and southern Patagonia, where there are more than 53 breeding and nonbreeding rookeries.

A recent study analyzed the genetic variability between two rookeries: Isla de Lobos, Uruguay, and Punta Norte, Península Valdés, Argentina. The results suggest that both rookeries belong to the same population. The population of the southwestern Atlantic is therefore apparently homogeneous, with movements between rookeries.

II. Reproductive Behavior

Males become sexually mature in their sixth year of age, whereas females produce their first offspring about the fifth year or before. The reproductive behavior of marked individuals was studied for 10 years at Punta Norte rookeries in Península Valdés, Argentina.

A. Mating Strategies

Adult males and females arrive at the breeding rookeries during the first half of December. The males defend a position on the central breeding area, and during the peak of the breeding season they also defend females in ESTRUS. In other rookeries such as Puerto Pirámide (also located at Península Valdés), males defend the territories where females go to mate. In a female-defense polygynous mating system, with males capable of forcing copulation, the female's first priority would be to survive the breeding season and then to mate with high-quality males. Males attempt to mate with as many females as possible. Adult males may maximize their reproductive success through the selective defense of those females that are close to estrus. Adult females develop choice of the male with which they copulate by changing the associated male before giving birth or by mating with more than one male during their maximum sexual receptivity. Thermoregulatory requirements interact with rookery topography to shape mating strategies, variation in mating success, and the mating system type. At Punta Norte, the pebble substrate is homogeneous regarding thermoregulatory advantages. Thus, sites advantageous for THERMOREGULATION are not a limited resource that can be used to attract females. Consequently, males acquire mates by selective female defense or abduction. However, at the Puerto Piramide rookery, a variation in the quality of the substrate with respect to reducing thermal stress favors the development of a territorial system where the best territories contain water or are close to the water. Here, the abduction of females or direct defense of females by males is not required, as females preferentially gather in wet territories. Thus, the topography and substrate of the breeding area, along with thermoregulatory requirements, are driving forces that generate adaptive changes in male mating behavior. Individuals that are sexually mature but cannot compete gather in "bachelor clubs" close to the main breeding area. These subadult or nonterritorial males may develop alternative mating strategies: group raids, solitary breeding (single male with a single female or with a small isolated harem), or female interception (keeping females that leave the main breeding area on the way to and from the water).

B. Breeding

The breeding and pupping season begins in mid-December and extends to early February. Most of the pups are born during January, usually 2–3 days after the mother's arrival at the rookery. Copulations occur on land 6 days after parturition, and females spend 2–3 days more with their pups fasting. Mothers begin leaving their pups temporarily and foraging offshore at between 1 and 4 days; each foraging trip is followed by 2 days of nursing bouts on land. Lactation continues for 8–10 months, although it is not unusual to find females with unweaned yearlings. Lactating females spend around 53% of the time at sea diving, with median and maximum depths ranging from 19 to 62 and from 97 to 175 m, respectively. The pups wander about and tend to gather in groups; they spend most of the time sleeping or playing (Fig. 2). The number of pups in these groups increases as the season goes on. When the mother returns to the colony, she calls her pup and they recognize each other by sound and smell. Each female nurses only its own pup. In 10 years of study at Punta Norte, Península Valdés, only three cases of true adoption were documented, although it is not infrequent to see pups suckling from females other than their mothers, "stealing" milk.

The death rate of pups is usually high, with main reasons being the separation from their mothers, starvation, being crushed by large males where the ground is hard, abuse from juveniles, or predation by other species.

C. Dimorphism

The degree of dimorphism of South American sea lion pups is similar to that of other otariids. Newborn males are 9–18% heavier than females at Península Valdés. Assuming that there is no sex difference in energy expenditure by fetuses, sexual dimorphism in the size of newborn pups suggests that South American sea lion mothers invest more energy in sons than in daughters during gestation. Contrary to other otariids, in which males increase in mass faster than females, no sex difference was found in growth rates among South American sea lions. The size dimorphism present at birth in this species persists during the nursing period, suggesting that sons continue to be more costly to their mothers than daughters. It is not known if differences in size found at birth in the South American sea lions of Península Valdés persist until weaning, but there is some evidence suggesting that this may be true.

Six-month-old male pups at the Malvinas (Falkland) Islands are longer than females; the difference in length remains in 18 month-old individuals and is even more marked in older juveniles. Data for other otariids show that sexual dimorphism at birth continues through weaning.

III. Interactions with Other Species

In some areas of their distribution along the South American coasts, South American sea lions live sympatrically with South American fur seals, but they do not compete with them for breeding space because their respective breeding seasons are out of synchrony. Killer whales prey on sea lions that are at sea and have been seen at Península Valdés preying on groups at the shore by surging out of the water onto the beach and returning with a sea lion in their jaws. Sharks have also been seen eating sea lions in Uruguay.

Human exploitation has been hard on pinnipeds. Some species were hunted down to extinction (Caribbean monk seal) or barely survive (Mediterranean monk seal), whereas others were spared thanks to the remoteness of their location, as the Antarctic phocids and the fur seals at Patagonian islands. Pinnipeds in general were killed to obtain oil, fur, and meat either for subsistence or for commercial purposes. The South American sea lion was hunted mainly for oil, as its fur is not as valuable as that of the South American fur seal (*Arctocephalus australis*). In Southern Chile it is still captured by fishermen as bait for the southern king crab. It is also hunted by fishermen who regard it suspiciously as sea lions are supposed to compete for the same fishes they seek and damage fishing gear. The South American sea lion endured long periods of exploitation with a severe reduction in ABUNDANCE. Although it is estimated that only 20% of their historical number remains, recent signs of recovery have been reported.

See Also the Following Articles

Mating Systems ■ Rookeries ■ Sexual Dimorphism ■ Southern Fur Seals

Figure 2 *Breeding colony of South American sea lions with a group of pups.*

References

Campagna, C. (1985). The breeding cycle of the southern sea lion, *Otaria byronia. Mar. Mamm. Sci.* **1**(3), 210–218.

Campagna, C., and Le Boeuf, B. J. (1988). Reproductive behaviour of southern sea lions. *Behaviour* **104**(3–4), 233–261.

Campagna, C., and Le Boeuf, B. J. (1988). Thermoregulatory behaviour of southern sea lions and its effects on mating strategies. *Behaviour* **107**,(1–2), 72–90.

Campagna, C., Le Boeuf, B. J., and Cappozzo, H. L. (1988). Group raids: A mating strategy of male southern sea lions. *Behaviour* **105**(3–4), 224–249.

Cappozzo, H. L., Campagna, C., and Monserrat, J. (1991). Sexual dimorphism in newborn Southern sea lions. *Mar. Mamm. Sci.* **7**(4), 385–394.

Cappozzo, H. L., and Rosas, F. C. W. (1991). León Marino Sudamericano, *Otaria flavescens* (Shaw, 1800). *In* "Estado de Conservación de los Mamíferos Marinos del Atlántico Sudoccidental" (H. L. Cappozzo and M. Junin, eds.), Vol. 138, pp. 166–170. Publicación científica PNUMA.

Crespo, E. A., and Pedraza, S. N. (1991). Estado actual de la población de lobos marinos de un pelo (*Otaria flavescens*) en el litoral norpatagónico. *Ecol. Austral* **1**(2), 87–96.

Reyes, L. M., Crespo, E. A., and Szapkievich, V. (1999). Distribution and population size of the southern sea lion (*Otaria flavescens*) in central and southern Chubut, Patagonia, Argentina. *Mar. Mamm. Sci.* **15**(2), 478–493.

Riedman, M. L. (1990). "The Pinnipeds: Seals, Sea Lions and Walruses." Univ. of California Press, Berkeley.

Rodriguez, D. H., and Bastida, R. O. (1993). The southern sea lion, *Otaria byronia* or *Otaria flavescens*? *Mar. Mamm. Sci.* **9**(4), 372–381.

Szapkievich, V., Cappozzo, H. L., Crespo, E. A., Bernabeu, R. O., Comas, C., and Mudry, M. (1999). Genetic relatedness in two southern sea lion (*Otaria flavescens*) rookeries in the south-western Atlantic. *Int. J. Mamm. Biol. Z. Saugetierkunde* **64**, 1–5.

Werner, R., and Campagna, C. (1995). Diving behaviour of lactating southern sea lions (*Otaria flavescens*) in Patagonia. *Can. J. Zool.* **73**, 1975–1982.

Southern Fur Seals

Arctocephalus spp.

JOHN P. Y. ARNOULD
Macquarie University, Sydney, Australia

Southern fur seals (genus *Arctocephalus*) are generally recognized as comprising eight species and four subspecies. As the name implies, southern fur seals occur almost exclusively in the Southern Hemisphere with only one species being found north of the equator. They are circumpolar in distribution, occurring in all the Southern Hemisphere oceans. The generic name, *Arctocephalus*, comes from the Greek words *arktos* and *kephale*, meaning "bear headed," and many of their facial characteristics reflect their terrestrial carnivore ancestry.

The majority of Southern fur seals were overexploited during large-scale commercial hunting in the 18th and 19th centuries, and many species were so depleted in numbers that they were considered extinct. Fortunately, because of the isolated nature of the islands on which many southern fur seal species occur, remnant populations persisted and all known species survived. Populations are recovering at various rates, and several species (in particular the Antarctic fur seal, *A. gazella*) have been the focus of extensive research since the 1980s which has improved our understanding of pinniped biology greatly.

I. Diagnostic Characters

The most obvious diagnostic feature separating fur seals from sea lions is the presence of an underfur layer in their pelage. The density of hair follicles in the underfur of fur seals is approximately 50 times greater than that in terrestrial mammals and this layer plays an important role in their THERMOREGULATION. Fur seals are generally smaller than their sea lion conterparts. Another difference between the two groups can also be found in the baculum (penis bone), the tip of this bone being narrow in fur seals whereas in sea lions it is broad. In addition, fur seals have six pairs of upper postcanine teeth, compared to five in sea lions, and the third upper incisor is less circular in horizontal cross section.

The southern fur seals can be distinguished from the northern fur seal, *Callorhinus ursinus*, by the extent to which the fur line extends on the fore flippers. In the northern fur seal the fur stops at the base of the flippers in a sharp line, whereas in southern fur seals it extends across part of the flipper, ending in a line over the metacarpals. A more prominent difference between these species is the shape of the SKULL, with the angle of the slope from the top of the nasal bone to the tip of the premaxilla being much greater in the northern fur seal. This gives the head of the northern fur seal a distinctive shortened-snout appearance in comparison to southern fur seals.

Within the southern fur seals there is a relative uniformity in appearance. This led to some confusion among early researchers about the classification and nomenclature of the various species in the genus. The history surrounding this has been discussed by Bonner (1981, 1994), Gardner and Robbins (1998), and Rice (1998). The classification used here is taken from Rice (1998), and the scientific and common names for each species are given in Table I. The discreteness of their distributions aids in the identification of the different species and in the areas where sympatry occurs morphometric, behavioral, vocalization, and habitat choice differences can be used to separate the species.

Pelage color among southern fur seals is generally uniform dark brown to dark gray on the dorsal surface with a grizzled appearance caused by the tips of the guard hairs (outer fur layer) being white or pale in color. The fur is a lighter color on the ventral surface, especially around the abdomen. There can be considerable variation between individuals of a species in the shading of the pelage and the degree to which it appears grizzled depending on age and sex. For example, older Antarctic fur seal and Australian fur seal (*A. pusillus doriferus*) females often have a lighter, more grizzled appearance than younger individuals. The time elapsed since the last molt, which

TABLE I
Scientific and Common Names of Southern Fur Seal Species (*Arctocephalus* spp.) with Their Mean Adult Body Mass and Most Recent Estimates of Population Size

Scientific name	Common name	Mean adult mass[a] (kg) Female	Male	Population size (× 1000)
A. townsendi (Merriam, 1897)	Guadalupe fur seal	49	124[b]	>7
A. galapagoensis (Heller, 1904)	Galapagos fur seal	27	64	40
A. philippii (Peters, 1866)	Juan Fernandez fur seal	48	140	18
A. australis australis (Zimmerman, 1783)	Falkland Islands fur seal	—	—	20
A. australis gracilis (Nehring, 1887)	South American fur seal	35–58[c]	75–107[c]	215–265
A. forsteri (Lesson, 1828)	New Zealand fur seal	39	127	135
A. pusillus pusillus (Schreber, 1775)	Cape or South African fur seal	57	247	1700
A. pusillus doriferus (Wood Jones, 1925)	Australian fur seal	76	279	35–60
A. tropicalis (Gray, 1872)	Sub-Antarctic or Amsterdam Island fur seal	34–36[c]	88–131[c]	>310
A. gazella (Peters, 1875)	Antarctic or Kerguelen fur seal	45	188	1600

[a]Means of measurements for each species taken during various seasons and stages of the breeding cycle.
[b]$n = 1$.
[c]Range of means from different populations.

occurs annually in late summer–early autumn, also affects the appearance of the pelage due to wear and soiling.

There are two notable exceptions to the general pelage color within southern fur seals. Sub-Antarctic fur seals (*A. tropicalis*) have a distinctive COLORATION: the chest and face (muzzle and around the eyes to just below the ears) are pale yellow or creamy in color while the top of the head and dorsal surface are dark brown-gray (Fig. 1). The demarcation in coloration is more pronounced in males, which also have a conspicuous tuft of dark hair on the forehead that becomes more erect when the animal is excited. In the Antarctic fur seal population on South Georgia, 0.1–0.2% of individuals have a white pelage. These animals are not albinos and have normal skin and eye pigmen-

Figure 1 *Male sub-Antarctic fur seal* (A. tropicalis). *Photo ©* J. Arnould.

tation. This white-phase pelage has not been reported in any of the other Antarctic fur seal populations.

Pups of all the southern fur seal species are born with a black natal coat (lanugo). In the Antarctic fur seal and the New Zealand fur seal (*A. forsteri*), pups develop a grizzled appearance soon after birth, whereas in Cape and Australian fur seals (*A. pusillus* subspp.), the ventral surface can vary from gray to pale yellow. In all species, the natal pelage molts 3–4 months after birth to reveal a silky-smooth gray or brown fur.

The Galapagos fur seal (*A. galapagoensis*) is the smallest of the southern fur seals (and the Otariidae) with mean female and male adult mass at 27 and 64 kg, respectively (see Table I). Most of the other species are slightly larger, with mean masses ranging from 34 to 58 kg for females and 75 to 188 kg for males. The exceptions to this pattern are the Cape fur seal (*A. pursillus pusillus*) and the Australian fur seal (*A. p. doriferus*), which have mean masses of 58 and 76 kg for females and 247 and 279 kg for males, respectively (Fig. 2). These two subspecies also differ from the remainder of the genus in having some behavioral traits reminiscent of the sea lions. In particular, *A. pusillus* subspp. display very thigmotactic tendencies (tolerance of physical contact between individuals), behavior not seen in other fur seals but common in sea lions. In addition, the mode of terrestrial locomotion and aspects of their vocal repertoire resemble those of sea lions more than those of other fur seals do.

Differences in the shape of the flippers can be used as diagnostic characters between several Southern fur seal species, especially when their distributions overlap. For example, the fore flippers of New Zealand seals have a more triangular shape than those of Australian fur seals, which are more paddle shaped and curved (Fig. 3). Similarly, Antarctic fur seals have proportionally

Figure 2 *Female Australian fur seals* (A. pusillus doriferus). *Photo © J. Arnould.*

longer hind flippers than sub-Antarctic fur seals. Snout lengths in southern fur seals also vary, being longest in the Juan Fernandez fur seal (*A. philippii*) and shortest in the Galapagos fur seal. In addition, the rhinarium (soft tissue of the nostrils) is smooth and inconspicuous in Antarctic and Galapagos fur seals, whereas it is inflated and bulbous in Juan Fernandez and New Zealand fur seals.

II. Distribution and Population Sizes

The large-scale HUNTING during the commercial sealing era severely depleted the populations of southern fur seals and it is known to have reduced the breeding distribution of many species. For several species, the presealing distribution and population size are not accurately known, as sealing ships did not always keep detailed records of the number or species

taken. This is particularly so for species that have overlapping ranges such as the New Zealand and Australian fur seals in southern Australia and the Antarctic and sub-Antarctic fur seals on several sub-Antarctic island groups.

The current breeding distributions of southern fur seals are shown in Fig. 4 and their population sizes are given in Table I. The Guadalupe fur seal (*A. townsendi*) is the only southern fur seal species found in the Northern Hemisphere. Its breeding colonies are currently restricted to Isla Guadalupe, situated off the Pacific coast of Mexico, but it once had a wider distribution, including the Channel Islands (of California) and the San Benito Islands (off Baja California). The most recent (1992) estimate of the Guadalupe fur seal population size is >7000 individuals, making this by far the rarest of the southern fur seals.

The distribution of the Galapagos fur seal is limited to the equatorial Galapagos Islands. The population increased following cessation of the extensive sealing in the 19th century and, currently estimated at 40,000 individuals, may have recovered to presealing levels. However, pup production and reproductive success in this species have been shown to be heavily reduced by the periodic El Niño events, which undoubtedly affects the population size (Trillmich and Ono, 1991).

Further to the south, off the coast of Chile, the Juan Fernandez fur seal is confined to the islands of the same name. This species was once very abundant and is estimated to have numbered over four million prior to exploitation. It was thought to have been hunted to extinction until it was rediscovered in 1966. Currently, the Juan Fernandez fur seal occupies four main breeding colonies and has a population size of approximately 18,000, making it the second rarest of the Southern fur seals.

The South American fur seal (*A. australis gracilis*) is found at many sites along the Pacific coast of South America, from Peru down to the islands west of Tierra del Fuego and in the

Figure 3 *Female New Zealand fur seal* (A. forsteri). *Photo © J. Arnould.*

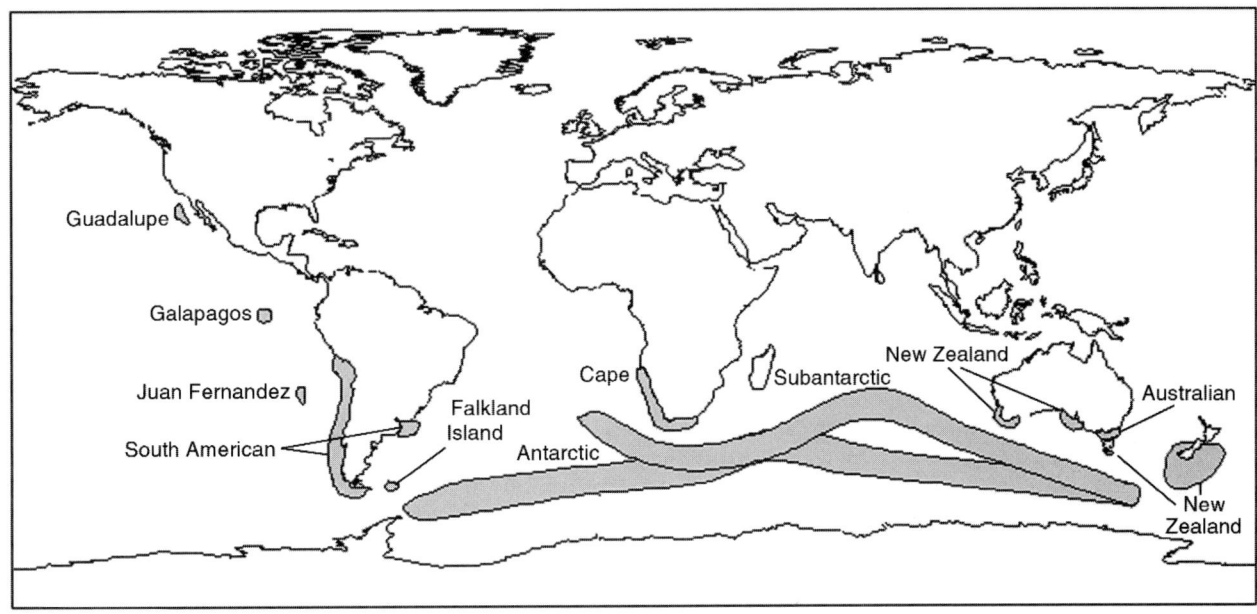

Figure 4 *Distribution of southern fur seals.*

Straits of Magellan, and on the Atlantic coast in southern Brazil and Uruguay. The colonies in Uruguay were exploited regularly in a small, controlled harvest until as recently as the early 1990s. The majority of colonies occur on offshore islands except in Peru where the dramatic increase in population since the 1960s (from 40 in the 1950s to over 20,000) has occurred mostly on three mainland colonies at Punta San Juan, Paracas Peninsula, and San Fernando. The Falkland Islands subspecies (*A. a. australis*) is found on the Falkland Islands where its numbers are still greatly reduced as a result of commercial sealing, which continued into the early part of this century. The total population size for both subspecies is presently estimated at 235,000–285,000 individuals.

In addition to colonies along the coast of the South Island of New Zealand, the breeding distribution of the New Zealand fur seal extends eastward as far as the Bounty, Antipodes, and Chatham Islands and southward to the Snares, Auckland, and Campbell Islands. There are also substantial colonies of New Zealand fur seals along the southern coast of Australia (see Fig. 4). Colonies in the Recherché Archipelago (Western Australia) and Kangaroo Island (South Australia) have been increasing rapidly in size and there is the potential for competition to develop between them and the sympatrically breeding Australian sea lion (*Neophoca cinerea*). The most easterly located colony of New Zealand fur seals in Australia is situated on Maatsuyker Island (<200 individuals; southern coast of Tasmania), which is also an important haul-out site for Australian fur seals. These two species do not presently have sympatric breeding colonies, but historical evidence suggests that the New Zealand fur seal was once found in large numbers alongside the Australian fur seal throughout Bass Strait (southeastern Australia). The total new Zealand fur seal population is estimated at 135,000, with approximately 35,000 being found in Australia. Interestingly, molecular studies have shown significant genetic variation, not only between Australian and New Zealand subpopulations of this species, but also within the latter, suggesting deeply divergent lineages "bordering on species-level distinction" (Lento *et al.*, 1997).

Although separated widely in their distributions, the Cape fur seal and the Australian fur seal show morphological and genetic similarities consistent with their status as subspecies. The Cape fur seal breeds at 25 colonies around the coasts of South Africa and Namibia. It is unique among fur seals in that the four largest colonies are located on mainland sites. The fact that the Namib Desert, which is devoid of large predators, borders these colonies may explain their location. The Cape fur seal has been subject to controlled harvests for over a century, yet the most recent (1993) estimate of population size is 1.7 million individuals, increasing at 3% per annum. In contrast, the Australian fur seal has not been subject to commercial harvesting since 1923, but the population is currently estimated at only 35,000–60,000 individuals and breeding colonies are restricted to nine islands in Bass Strait. While the largest colony (Seal Rocks) has recently experienced increases of 6.2% per annum (Shaughnessy *et al.*, 2000), the population size is still well below the estimated 175,000–225,000 individuals present prior to the commercial sealing era. The slower recovery, and smaller preexploitation size, of the Australian fur seal population in comparison to the Cape fur seal is thought to be attributed to differences in food availability. Australian fur seal colonies are situated in the nutrient-poor waters of Bass Strait, whereas Cape fur seals forage in the highly productive waters of the Benguela Current.

The sub-Antarctic fur seal has a total population of >300,000 and a widely dispersed breeding distribution with colonies on several sub-Antarctic and subtemperate island groups located north of the Antarctic polar front. The main focus for the species is on Gough Island in the South Atlantic Ocean where colonies total over 200,000 seals. A small population (250) is located to the west at nearby Tristan da Cunha. To the east, in the southern Indian Ocean, substantial colonies exist on the Prince Edward Islands (75,000), the Iles Crozet (1000), and Amsterdam Islands (50,000). A small population also exists to the southwest of Australia on Macquarie Island (<200) where the species breeds sympatrically with the Antarctic fur seal and some hybridization has occurred (Goldsworthy et al., 1999). Hybridization between sub-Antarctic fur seals and Antarctic fur seals has also been recorded at Marion Island (Price Edward Islands) where the latter is greatly outnumbered (only 1200). These two species also breed sympatrically on Iles Crozet.

The Antarctic fur seal also has a very large breeding range and is the most southerly breeding of the *Arctocephalus* species. With the total population in 1990 estimated at over 1.6 million and growing at an annual rate of 9.8%, the Antarctic fur seal is now likely to be the most numerous of the Southern fur seals. Over 95% of the population breeds on South Georgia in the South Atlantic Ocean, but colonies are found on numerous island groups spreading eastward as far as Macquarie Island. All colonies are located south of, or on, the Antarctic polar front with the exception of those at Marion Island, Iles Crozet, and Macquarie Island. The species was considered extinct until a small remnant population of 1000–3000 individuals was discovered in 1950 on Bird Island (northwest tip of South Georgia). From this colony emanated the phenomenal recovery of the species with the recolonization of many sites being attributed to emigration from the expanding South Georgia population. The rapid recovery of the South Georgia population is thought to have been facilitated by its predominant prey, Antarctic krill (*Euphausia superba*), being found in superabundance in the surrounding waters.

The nonbreeding range of most southern fur seal species is often much greater than the breeding distribution and is determined by the movements of males and juveniles. These segments of the population are not restricted in the same way as adult females, which must return regularly to the natal colony in order to suckle their pup. Only in Antarctic fur seals, which nurse their pup for just 4 months, do females depart after weaning from the natal colony (to unknown destinations) until the next breeding period. Except for the Guadalupe, Juan Fernandez, and Galapagos fur seals, males of all species undertake some form of seasonal movement. This usually occurs after the breeding season or during the winter months when local food availability is reduced. The evidence for such movements comes from resighting tagged individuals and/or the predictable seasonal appearance of large numbers of individuals far from their breeding colonies. For example, there is a large influx of subadult and juvenile New Zealand fur seal males at Macquarie Island between February and April. Tagged individuals from colonies in New Zealand and Australia, the only potential source for these animals, have been resighted at Macquarie Island (Goldsworthy et al., 1998). Whether such movements constitute structured migrations or reflect individual dispersal patterns may vary between species and different age classes. Satellite telemetry has been used to show that adult male

Antarctic fur seals from South Georgia migrate toward the Antarctic Peninsula following the breeding season (Boyd et al., 1998).

III. Foraging Ecology

Southern fur seals feed on a variety of prey species, including fish, cephalopods (octopus, squid, and cuttlefish), crustaceans (krill, shrimp, rock lobster), and even penguins and other seabirds. Studies of several species have shown great temporal variation in diet composition, with seals exploiting seasonally abundant prey items such as spawning squid or schools of migrating fish. In addition, substantial differences in diet can be found between populations of the same species. For example, the Antarctic fur seal population at South Georgia feeds almost exclusively on Antarctic krill, whereas at Heard and Kerguelen Islands (southern Indian Ocean), seals of this species feed primarily on myctophid fishes. The majority of detailed studies, however, have concentrated on the diet of adult females because natal colonies have been the easiest to sample for feces (the most commonly used technique for determining diet). Consequently, for most species, there is little information on the diet of adult males or how it varies seasonally.

The advent of electronic time-depth recorders in the mid-1970s enabled the foraging behavior of seals to be investigated in great detail for the first time, and numerous studies have subsequently been conducted on most species. As with diet studies, the majority of research on southern fur seal species has involved lactating females due to their relative ease of handling and recapture in comparison to adult males. Although they are known to be able to dive to much greater depths, it has been shown in all species that foraging activity occurs mainly in the surface mixed layer (<50–60 m) at night. The pattern of diving depths throughout the night generally reflects the diel vertical migration of their main prey. The larger body size of males would be expected to enable them to dive deeper than females, and studies of Antarctic fur seals have indeed shown that the mean forage depths of males (100–200 m) are greater than in females (<50 m). Interestingly, these studies have shown that in some areas males forage mostly during daylight hours. Studies on males of additional species are required to determine whether sex differences in foraging behavior are common in southern fur seals.

In recent years, information on the foraging zones of several southern fur seal species has also been obtained through the use of satellite telemetry and geolocation. Studies of lactating female Antarctic, sub-Antarctic, and New Zealand fur seals have found that most foraging activity occurs in up-welling zones, oceanic frontal systems, or continental shelf-edge regions, areas generally high in primary productivity. There is little information about the foraging locations for other age and sex classes in southern fur seals.

IV. Reproduction

The breeding biology of the southern fur seals is similar to that of other otariid seals (sea lions). Females give birth to a single pup each year, and the reproductive cycle consists of a

Figure 5 *Antarctic fur seal* (A. gazella) *females and pups. Photo © J. Arnould.*

perinatal attendance period of 6–10 days, culminating in estrus and copulation; an embryonic diapause lasting 3–4 months; and an active gestation of 8–12 months. Females have two pairs of mammary glands and lactate for 8–12 months, but several species may suckle their pup into part of the second year. The exceptions to this are the Antarctic fur seal (Fig. 5), which has a lactation period lasting only 4 months, and the Galapagos fur seal in which weaning may occur at >3 years of age and females can suckle successive offspring at the same time. Differences in food availability and seasonal predictability may explain these divergent strategies.

Timing of the pupping season is fairly consistent, occurring between October and December in the Southern Hemisphere species and in the corresponding late spring–early summer period of June–July in the Northern Hemisphere for the Guadalupe fur seal (Boyd, 1991). Estrus synchrony, measured as the period over which 90% of births occur, varies from 21 days in the Antarctic fur seal to 70 days in the Galapagos fur seal and is negatively related to latitude (Boness, 1991). High estrus synchrony is considered a prerequisite for polygyny, and resource defense polygyny has been demonstrated clearly in New Zealand, Antarctic, and Australian fur seals. As is common in polygynous species, southern fur seals are sexually dimorphic and have some of the most extreme male:female mass ratios of any mammal. In addition, the operational sex ratio, defined as the average ratio of fertilizable females to sexually active males at any given time, has been shown to be correlated positively to the degree of SEXUAL DIMORPHISM in Southern fur seals.

See Also the Following Articles

Hair and Fur ■ Migration and Movement Patterns ■ Northern Fur Seal

References

Boness, D. J. (1991). Determinants of mating systems in the Otariidae (Pinnipedia). *In* "Behaviour of Pinnipeds" (D. Renouf, eds.), pp. 1–44. Chapman and Hall, New York.

Bonner, W. N. (1981). Southern fur seals *Arctocephalus* (Geoffroy Saint-Hilaire and Cuvier, 1826). *In* "Handbook of Marine Mammals" (S. H. Ridgway and R. J. Harrison, eds.), pp. 161–208. Academic Press, New York.

Bonner, W. N. (1994). "Seals and Sea Lions of the World." Blandford, London.

Boyd, I. L. (1991). Environmental and physiological factors controlling the reproductive cycles of pinnipeds. *Can. J. Zool.* **69,** 1135–1148.

Boyd, I. L., McCaferty, D. J., Reid, K., Taylor, R., and Walker, T. R. (1998). Dispersal of male and female Antarctic fur seals (*Arctocephalus gazella*). *Can. J. Fish. Aquat. Sci.* **55,** 845–852.

Gardner, A. L., and Robbins, C. B. (1998). Generic names of northern and southern fur seals (Mammalia: Otariide). *Mar. Mamm. Sci.* **14,** 544–551.

Goldsworthy, S. D., Boness, D. J., and Fleischer, R. C. (1999). Mate choice among sympatric fur seals: Female preference for conphenotypic males. *Behav. Ecol. Sociobiol.* **45,** 253–267.

Goldsworthy, S. D., Wynen, L., Robinson, S., and Shaughnessy, P. D. (1998). The population status and hybridisation of three sympatric fur seals (*Arctocephalus* spp.) at Macquarie Island. *New Zeal. Nat. Sci.* **23**(Suppl.), 68.

Lento, G. M., Haddon, M., Chambers, G. K., and Baker, C. S. (1997). Genetic variation of southern hemisphere fur seals (*Arctocephalus* spp.): Investigation of population structure and species identity. *J. Hered.* **88,** 202–208.

Rice, D. W. (1998). Marine mammals of the world. Systematics and distribution. *Soc. Mar. Mamm. Spec. Pub.* 4.

Shaughnessy, P. D., Troy, S. K., Kirkwood, R., and Nicholls, A. O. (2000). Australian fur seals at Seal Rocks, Victoria: A second estimate of pup abundance by mark-recapture estimation. *Wildl. Res.* **27,** 629–633.

Trillmich, F., and Ono, K. A. (1991). "Pinnipeds and El Niño: Responses to Environmental Stress." Springer-Verlag, Berlin.

Speciation

RICK LeDUC
Southwest Fisheries Science Center,
La Jolla, California

Speciation, in its most basic sense, is the formation of new species. As such, it is an integral part of the evolution and radiation of life on earth. Typically, one SPECIES (the parent or ancestral species) splits into two or more sister species, although there are also descriptions of one species evolving into another over evolutionary time along a single lineage (allochronic speciation). Speciation is also usually considered non-reversible once it is completed; a single species may split into two, but two species do not combine into one. However, in some taxa, particularly plants, hybrid origins of species are relatively common, although in these cases the parent species continue to persist. A more explicit definition of speciation by divergence (and other rarer processes) is partly dependent on the species concept applied. One may think of it as the splitting of an evolutionary lineage (phylogenetic species) or as the formation of reproductive isolating mechanisms between two populations (biological species). Other species concepts (and their related definitions of speciation) appear in the literature; however, the biological and phylogenetic species concepts are the two most often employed. The perspectives on the speciation process can also affect how the taxa are described. For example, in a cladistic interpretation, the ancestor is considered extinct after a speciation event; the sister species that remain are considered distinct from it as well as from each other. However, one may also see descriptions of one species splitting from another, whereby one of the resulting sister species is equated with the parent species.

I. Elements of the Speciation Process

The origins of taxa at all systematic levels, even kingdoms and phyla, lie in speciation events. Despite its importance, and the abundance of evidence for its occurrence, the process is not easily described or understood. One reason for this difficulty is that speciation takes many generations to complete, which for organisms larger than microbes is usually much longer than the life span of any single research project. For large mammals, it may take thousands of years. Data collected on any given taxon therefore represent a "snapshot" of the process, and it may be difficult to tell from this snapshot whether speciation has already occurred, is occurring, or is merely a future possibility. It is not surprising then that it is even more difficult to understand the forces driving the process.

Despite these difficulties, evolutionary biologists have been able to identify some of the elements that play a role in speciation, either as causes or as effects. One common thread of speciation models for sexually reproducing organisms is the evolution of barriers to gene flow between species, i.e., reproductive isolation. This aspect of speciation is explicit in the biological species concept and implicit in other species concepts. These barriers to gene flow, or reproductive isolating mechanisms, are genetically determined biological mechanisms, in contrast to extrinsic barriers such as geographic isolation. Prezygotic isolating mechanisms prevent the formation of HYBRIDS and include social or behavioral segregation, genital or gametic mismatch, and chromosomal or genetic incompatibility. Postzygotic mechanisms include hybrid inviability or sterility, and selection against hybrids. Although they need to be efficient to maintain the integrity of the species, these isolating mechanisms need not be absolute. Wild and captive hybrids of different marine mammal species have been recorded many times, including a fertile hybrid between the blue whale (*Balaenoptera musculus*) and the fin whale (*B. physalus*). The fact that the parent species in these cases are still quite distinct indicates that breakdowns of isolating mechanisms are relatively rare and possibly inconsequential (in an evolutionary sense) events.

Speciation also typically involves the development of some degree of differentiation between two species, be it morphological, behavioral, physiological, or ecological. One may see discussions of so-called cryptic or sibling species, those that are virtually indistinguishable morphologically, yet are reproductively isolated. However, cryptic species is more of an operational designation than a biological one, as our ability to detect differences is limited by the types of data collected or by our limited perceptions. What looks the same to an investigator may be very different indeed to the potential mates. No two valid species per force can be identical in all respects, and characterizing the differences between groups is integral to understanding if speciation has occurred as well as the mechanisms that underlie it.

Some of the differences between species may have arisen during or subsequent to speciation, i.e., during or after their establishment as reproductively isolated taxa. However, many other differences between species have their origins in the geographic variation that was inherent within the parental species prior to the speciation event. Generally, whether originating before, during, or after speciation, most differences are caused and maintained by differential selection and adaptation to the environment, but they may be subsequently enhanced or even new differences added by the mechanisms of species recognition and mate choice (sexual selection) among individuals. In fact, differences involving species recognition and mate choice may be important factors in prezygotic reproductive isolation. Furthermore, in addition to the differences brought about by natural and sexual selection, many other genetic differences arise that are selectively neutral and often invisible above the molecular level. These are by-products of reproductive isolation, mutation, and genetic drift and are not integral to the speciation process, although they may provide good evidence of its occurrence. In the absence of extensive breeding data, distinct molecular or morphological differences usually provide indirect evidence for reproductive isolation, in much the same way that some differences also provide indirect evidence for differential selection.

In the classic speciation scenario, the geographic range of an ancestral species is split by a vicariance event, creating two allopatric populations. In time, the two populations become differentiated in both genotype and phenotype by adapting to selective differences in their respective environments, by the accumulation of selectively neutral mutations, or both. At this stage, their biological reproductive isolation is largely theoretical, as their geographic isolation precludes any opportunities for interbreeding. If range expansion later leads to the populations becoming partially sympatric, and the gene pools still remain distinct, either by pre- or postzygotic isolating mechanisms, then speciation can be considered complete. If only postzygotic mechanisms have developed, then reinforcement would occur quickly via selection for prezygotic isolating mechanisms, such as social segregation or refinements of mate choice. In other words, if hybrids were inviable or sterile, the selective pressure for positive assortative mating would be strong.

II. Variations of the Process

There have been a number of challenges to and complications of the basic sketch of the speciation process depicted above. For example, is geographic isolation, or allopatry, necessary for the genetic differentiation and formation of the reproductive isolating mechanisms that are needed for speciation, or can such mechanisms arise in sympatry or parapatry? Closely related sympatric species are very common in nature and their geographic origins are not always evident. Due to the swamping effect of even limited interbreeding between incipient species (populations on the verge of speciation), it has been difficult to formulate broadly applicable and plausible evolutionary models to explain speciation in sympatry. Therefore, the observations of sympatric sister species have usually been interpreted as the result of secondary contact between species that originally formed in allopatry. However, not all cases of sympatric sister species can be explained so easily. One well-known example is the monophyletic swarm of closely related African cichlid species that inhabit geologically young Lake Victoria. To use allopatric speciation as a causal mechanism of this swarm, which certainly arose within the confines of the lake, one must invoke allopatry on a small scale, such as between separate reefs along the shoreline, followed by secondary contact. It is debatable that this mechanism could account for the large number of species present. However, recent speciation models that combine ecological variation, mate selection, competition, and differences in morphological and/or behavioral characteristics hold promise as realistic mechanisms for sympatric speciation. In marine mammals, sister species that are at least partially sympatric are common (e.g., the common dolphins, *Delphinus* spp.; the harbor and larga seals, *Phoca vitulina/P. largha*), as well as others who are parapatric (e.g., the two South American species of *Trichechus*, the Caribbean and Amazonian manatees). Generally, the mobility of marine mammals makes microallopatric speciation models implausible, but in some cases, the biogeographic patterns of the species do not lend themselves easily to an explanation involving separation and secondary contact on a larger scale. In addition, the complex behaviors of many marine mammal species could facilitate rapid evolution of the prezygotic isolating mechanisms that are important components of sympatric speciation models. Although it has not been investigated, such models may be applicable to some marine mammal species pairs.

Even the most ardent advocates of sympatric speciation concede that allopatric speciation has played the major role in evolution. However, even within the basic framework of allopatric speciation, questions still need to be addressed. One is the role played by ecological and selective differences in the speciation process. When two populations diverge so that speciation results, are the differences generally due to differing selective pressures or just to simple geographic isolation persisting over sufficiently long periods of time? The incorporation of selection into the speciation process, first argued by Darwin, is integral to the idea of ecological speciation and adaptation. Furthermore, if relatively few mutations can lead to key morphological adaptations, divergence and speciation can be very rapid, whether in sympatry or in allopatry. Such a scenario is plausible in the case of closely related species that demonstrate simple yet dramatic differences, such as long- and short-beaked species of *Delphinus* (*D. capensis* and *D. delphis*, respectively). With behaviorally complex animals such as marine mammals, one must also consider whether morphological adaptations allowed exploitation of a new niche or whether there were behavioral shifts that led to changes in the selective pressures on the morphological characters. In the case of the resident and transient killer whales (*Orcinus orca*) in the northeastern Pacific, the behavioral divergence in foraging strategies appears to be greater than the morphological differentiation between the two groups.

In contrast to the ecological speciation discussed previously, studies on terrestrial animals have found evidence that long periods of geographic isolation, even in nearly identical habitats, can be sufficient to bring about speciation. In these cases, it is thought that ecological adaptation has played a lesser role in speciation than geographic separation. This model may also have relevance to marine mammals. For the 13 or so species of the beaked whale genus *Mesoplodon*, which are clearly distinct species but morphologically and behaviorally very similar, their critical feeding habitats have been hypothesized to be sufficiently uniform over large geographic scales to imply that speciation was more likely driven by differences in sexually selected characters that arose in allopatry rather than by adaptations to differing environments. In other words, ecological adaptation has played a relatively minor role in the radiation of this genus. Under this type of scenario, the differentiation of male secondary sex characteristics, for example (which are presumably selectively neutral relative to the environment), requires concerted evolution of both the genes coding for the male sex traits and for those coding for female mate choice.

Besides the underlying mechanisms for speciation in allopatry, one must also consider the extrinsic mechanisms that lead to populations becoming allopatric. In the vicariance model of biogeography, the range of a parent species becomes split when geographic barriers to gene flow arise within areas previously occupied. The resulting allopatric populations then differentiate via selection or drift into separate species. For example, many terrestrial organisms of the north temperate regions had their ranges bisected during the Pleistocene by glaciers advancing down the middle of the continents. In many cases, the daughter populations on either side of the ice diverged into different species, not interbreeding even when secondary contact occurred with the retreat of the glaciers. For marine organisms, a widely recognized vicariance event is the rise of the Isthmus of Panama approximately 3 million years ago, blocking dispersal between the eastern tropical Pacific and the Caribbean. Many sister species of fish and invertebrates occur on opposite sides of the isthmus and show levels of molecular divergence compatible with 3 million years of separation. However, there are no recognized sister species of marine mammals that are separated by this barrier and thought to have arisen as a result of its formation. This may be due to the extreme mobility of cetaceans (the only marine mammal group with significant diversity in tropical latitudes), which facilitates gene flow on a global scale, at least on the time scales necessary for speciation. After all, most tropical cetaceans, especially the pelagic species, have pantropical distributions. Some other plausible vicariance events in the marine environment that involve physical barriers are a drop in sea level causing the isolation of peripheral seas (e.g., Mediterranean Sea) or the spread of arctic ice sheets isolating the North Atlantic from the North Pacific. The influence of events such as these on marine mammal speciation has yet to be examined.

In many cases, the barriers that arise may be more ecological than physical, such as wide expanses of unsuitable habitat. Examples include warm equatorial waters acting as barriers for cold temperate and polar species or deep pelagic waters for coastal and neritic species. In some cases, these barriers are intermittent, in that certain climatic conditions periodically weaken them as obstacles to dispersal. Dispersing animals may be able to cross such barriers and found new populations, with no regular occupation of the intervening geographic region. This mechanism is comparable to an island model of biogeography for terrestrial organisms, where small groups of founders disperse to islands from a source region and speciate allopatrically. STRANDING records of many marine mammal species contain extralimital records, or stragglers, of individuals or groups that wander far outside their normal range. If such wanderers are able to survive and reproduce in their new haunts but subsequent dispersal is infrequent enough and/or selective pressures are different enough, the new population may differentiate into a new species. This is a plausible mechanism for the rise of species groups characterized by (mostly) widely allopatric species, each with fairly restricted ranges, such as species in the dolphin genus *Cephalorhynchus* and fur seal genus *Arctocephalus*. In one scenario, during glacial maxima, cold boundary currents may have extended far enough into low latitudes for cold temperate species of the Northern Hemisphere to invade the Southern Hemisphere (or *vice versa*), giving rise to the so-called antitropical species pairs, examples of which are found in the genera *Lissodelphis* (right whale dolphins), *Hyperoodon* (bottlenose whales), and *Mirounga* (elephant seals), or to antitropical distributions of conspecific populations, as in the long-finned pilot whale (*Globicephala melas*)

or in some mysticetes. Conversely, in a warm period, the poleward extension of warmer waters may have allowed tropical species to disperse around the Cape of Good Hope, invading the Atlantic Ocean from the Indian Ocean and speciating in the respective ocean basins (e.g., perhaps giving rise to the Atlantic endemic dolphin species of *Stenella, S. frontalis* and *S. clymene*). In some cases, however, it may be difficult to differentiate between a vicariance event and a dispersal event. For example, during the glacial maxima mentioned earlier, it is difficult to tell if a small founding population dispersed across the equator or if the ancestral species was able to extend its normal range to and across the low latitudes, only to be divided anew by a warm water habitat during the interglacial period. Some clues may be found by examining unrelated taxa. During vicariance events, whole biotas may be divided at the same time, giving rise to numerous species pairs of different taxa, but all with similar distribution patterns. Because the different species in a community can be assumed to have widely varying dispersal abilities, dispersal events would be less likely to lead to such a congruence of distribution patterns across unrelated taxa.

When secondary contact occurs between two previously allopatric populations, the reproductive isolation of two new sympatric species is only one possible outcome. Alternatively, one species may outcompete and eliminate the other, or the two might interbreed and a zone of intergradation may arise. Only in the first two cases would speciation have been completed, and the outcome may depend on the degree of differentiation between the two populations or the stringency of mate choice by individuals. However, only the first result, that of two species occurring in sympatry, will be evident to an investigator after the fact. From the other results, it would be difficult or impossible to determine that separation and secondary contact had ever occurred, let alone speciation. This is important to remember when making inferences about the degree of differentiation or duration of separation that is required for speciation; sympatric species represent only those cases where differentiation was sufficient to cause reproductive isolation *and* to allow the coexistence of the two species.

III. Other Factors

In addition to the aforementioned problems in describing the speciation process in general, other problems can arise when interpreting it with real animals. At the onset, inferences about speciation patterns and mechanisms require a good understanding of the systematic relationships among species. This understanding cannot be taken for granted; the systematic relationships within many marine mammal taxa are but poorly understood. For instance, the delphinid genus *Lagenorhynchus* has long been thought to contain three species in the temperate Northern Hemisphere and three in the Southern Hemisphere, their origins representing perhaps three dispersal events across the tropical latitudes. However, it has been suggested that the two species in the North Atlantic are unrelated to the other four, which leaves only one antitropical species pair in the group and limits hypothesized transequatorial speciation to the Pacific Ocean basin. In another example, one hypothesized origin for the species of seals in the genus *Pusa* is that the landlocked species that occur in Lake Baikal and the Caspian Sea (*P. sibirica* and *P. caspica*, respectively) are off-shoots of the mostly marine Arctic Ocean species *P. hispida*, resulting from southward dispersal during a glacial period. Alternatively, they could be descendants of the extinct *P. pontica*, which occupied the vast inland Parathethys Sea during the Miocene and Pliocene. Under the latter scenario, the disappearance of the Parathethys Sea would represent a vicariance event that isolated the ancestors of *P. sibirica* and *P. caspica*. This hypothesis also illustrates the importance of incorporating fossil taxa into inferences about speciation patterns and mechanisms.

The rate of speciation is dependent on many factors, including the population sizes of the incipient species, the differences in selective pressures that they are under, and the level of genetic variation contained in the populations. If a few individuals colonize a new area, or if a vicariance event isolates a small population at the periphery of a species range, the smaller population is likely to be under different selective pressures than the rest and is also likely to represent a biased sample of the source gene pool. These factors together can make their divergence and speciation proceed fairly rapidly. Some genetic macromutations that involve large sections of chromosomes can accelerate the process even more, providing essentially instant reproductive isolation, provided that enough individuals with the new arrangement arise for reproduction to occur and a new gene pool to be established.

Ever since Darwin considered how species arose and proposed a mechanism for evolution, our understanding of the speciation process has been refined. There are many elements that come into play, such as adaptation, selection, genetic drift, competition, dispersal, mate choice, social organization, and mutation. By developing speciation models and collecting appropriate data on various taxa, evolutionary biologists continue to investigate how important these different factors are and how they interact, both in general and for specific speciation events.

See Also the Following Articles

Biogeography ▪ Evolutionary Biology, Overview ▪ Geographic Variation ▪ Habitat Pressures ▪ Systematics

References

Barigozzi, C. (ed.) (1982). "Mechanisms of Speciation." A. R. Liss, New York.

Davies, J. L. (1963). The antitropical factor in cetacean speciation. *Evolution* **17**, 107–116.

Galis, F., and Metz, J. A. J. (1998). Why are there so many cichlid species? *Trends Ecol. Evol.* **13**, 1–2.

Mayr, E. (1963). "Animal Species and Evolution." Harvard Univ. Press, Cambridge, MA.

Orr, M. R., and Smith, T. B. (1998). Ecology and speciation. *Trends Ecol. Evol.* **13**, 502–506.

Otte, D., and Endler, J. A. (eds.) (1989). "Speciation and Its Consequences." Sinauer, Sunderland, MA.

Tregenza, T., and Butlin, R. K. (1999). Speciation without isolation. *Nature* **400**, 311–312.

Species

WILLIAM F. PERRIN
*Southwest Fisheries Science Center,
La Jolla, California*

I. What Is a Species?

Although the term "species" is widely used, there is no agreement on exactly what it means beyond the general conception of a basic taxonomic or evolutionary unit. It is very important to make progress in our understanding of species and how to define them, not least because much of the legal machinery of international conventions, national legislations, and regulations dealing with wildlife and the conservation of biodiversity uses the term in specifying what they seek to protect. If we cannot agree on what constitutes a wild species, we may not be able to agree on how to conserve and protect it.

There are a number of species concepts in use for vertebrates (Claridge *et al.*, 1997). In the *biological species concept* (BSC; Mayr, 1963), a species is a group of interbreeding or potentially interbreeding individuals separated from other such groups by intrinsic (genetically fixed) barriers to gene flow. In cases of sympatry or parapatry, direct evidence of noninterbreeding may be available, but more commonly morphological separation has been used as a proxy to indicate lack of gene flow; in this case the definition works fairly well. In the case of allopatry, however, there is difficulty. While the same operational criterion has been used, in nearly all such cases we do not know whether the groups would interbreed if the chance arose. The barrier to interbreeding may be a purely physical geographic one.

In the more recently developed *phylogenetic species concept* (PSC), a species is "the smallest population or group of populations within which there is a parental pattern of ancestry and descent and which is diagnosable by unique combinations of character states" (Cracraft, 1997), with no inference about potential interbreeding. Indeed, interbreeding is considered a shared primitive character and, as such, is not able to diagnose a lineage, which the PSC requires. In the case of sympatry, this concept does not differ materially from the BSC. Both have difficulties with allopatric populations (biological subspecies may equate to phylogenetic species) and with paraphyly. Phylogenetic taxonomists insist on reciprocal monophyly (having diagnostic unique characteristics, a single common ancestor, and no reticulation) between most closely related (sister) species, but this may be violated by the real phenomenon of lineage sorting in the case of some newly evolved but widely recognized species; e.g., because polar bears evolved from brown bears relatively recently, some brown bears are still related more closely genetically to polar bears than they are to other brown bears. In practice, adherents to the biological species concept use morphological characters in exactly the same way as phylogenetic taxonomists, albeit as proxy indicators of reproductive isolation. One difficulty faced by strict phylogenetic species-level taxonomy is that different genes or sequences may have different phylogenies due to differential modes of inheritance or introgression; a gene tree may not equal a species tree.

The two species concepts are actually complementary perspectives on evolution and can reinforce each other. To quote Avise and Wollenberg (1997), "Historical descent and reproductive ties are related aspects of phylogeny and jointly illuminate discontinuity." Furthermore, more important is not the process of specifying units to fit into the simplifying Linnaean binomial system, but rather describing and understanding in detail the coalescent genealogy and demography of discontinuous populations. Where did they come from, how are they different, and where are they likely to go? Whether they are called species (always, under the phylogenetic species concept) or subspecies (sometimes, under the biological species concept) is little more than semantics rooted in perspective. This is generally true also in the conservation of diversity. For example, named subspecies are accorded the same legal protection as species in the U.S. Endangered Species Act and Marine Mammal Protection Act. However, in the IUCN (International Union for the Conservation of Nature, or World Conservation Union) scheme of categorization of threatened taxa, species are given higher priority for protection than subspecific taxa or populations (IUCN, 1996).

There are additional species concepts, e.g., the *cohesion species concept* (Templeton, 1989), under which a species is "the most inclusive group of organisms having the potential for genetic and/or demographic exchangeability." Some pairs or groups of sympatric or parapatric species, such as wolves (*Canis lupus*) and coyotes (*Canis latrans*), may make up *syngameons* (Templeton's term) of species that are fully interfertile and may at times interbreed extensively but nonetheless have persisted separately for GEOLOGICAL PERIODS of time. Some of the species of southern fur seals (*Arctocephalus* spp.) may fall into this category (discussed later). Despite sporadic or scattered interbreeding, they would qualify as full species under the cohesion species concept.

While the various species concepts are necessary (because of the legal need to define units for conservation) and laudable attempts to model the results of actual evolutionary processes, they are human constructs, and as for all such constructs, in certain circumstances even the most commonly accepted species concepts fray at the edges. One problem of the biological species concept in its application (mentioned earlier) is that of closely related allopatric species, whether they be defined by morphological or genetic characters. If at some point they became sympatric or parapatric because of some geological or climatic change, would they then freely and effectively interbreed (be one species)? Are they a little different from each other only because of inconsequential genetic drift due to physical separation or are their differences the result of important differential ecological selection that has led to a permanent parting of their evolutionary paths? Even if their morphological differences are great, the question remains of whether this is due to large genetic differences, or of pleiotrophic effects of a small difference unassociated with genetic differences that would prevent interbreeding? For the most part, of course, answers to these questions will remain unobtainable. However, we can deduce what might happen from looking at the results of some "natural experiments." In one possible example, a zone of hybridization/intergradation between two subspecies of spinner dolphins, *Stenella longirostris*, in the eastern tropical Pacific is thought to perhaps be the result of recontact after separation during an earlier cooler climatic period (Perrin *et al.*, 1991). HYBRIDIZATION in the putative zone of recontact is evidenced by greater variance in body size and color pattern. The two subspecies, the eastern spinner *S. l. orientalis* and Gray's spinner *S. l. longirostris*, are morphologically so dissimilar that they were formerly thought to belong to different species (eastern spinners were identified as *S. microps* and Gray's spinners in Hawaii as *S. roseiventris*).

Thus, some of our present allopatric species may be transient taxa that will be absorbed or blended back into a pooled line of descent; i.e., they were never really "good" species at all.

Antitropical species seem especially good candidates for potential recontact and subsequent reticulation; sporadic movement of cool-water forms across the equator during cooler geological periods is thought to have played an important role in speciation, e.g., in the porpoises, including the formation of antitropical pairs of putative species. Species pairs that potentially represent such transient situations include the northern and southern right whale dolphins *Lissodelphis borealis* and *L. peronii*, the Pacific white-sided dolphin *Lagenorhynchus obliquidens* and the dusky dolphin *L. obscurus*, the northern and southern right whales *Eubalaena glacialis* and *E. australis*, Baird's and Arnoux's beaked whales *Berardius bairdii* and *B. arnuxii.*, North Atlantic and southern bottlenose whales *Hyperoodon ampullatus* and *H. planifrons*, and the northern and southern elephant seals *Mirounga angustirostris* and *M. leonina*.

Examples of nonantitropical allopatric species complexes include the four *Cephalorhynchus* species *C. hectori*, *C. commersonii*, *C. eutropia*, and *C. heavisidii*; they are sharply distinct morphologically but undoubtedly closely related. The Ganges and Indus susus (formerly *Platanista gangetica* and *P. minor*) are presently separated geographically but are thought to have been in contact in Recent, perhaps historical, time, and Rice (1998) has recognized the likely inconsequential nature of any differences by downgrading them to subspecies in *P. gangetica*. Among pinnipeds, the eight currently recognized species of southern fur seals, *Arctocephalus* spp., are almost entirely allopatric, but there are at least four breeding sites where two species (*A. gazella* and *A. tropicalis*) have begun to co-occur as populations have recovered from severe depletion by sealing in the last century (Rice, 1998). At two of these (Marion Island and Heard Island), limited hybridization has occurred, but at the others (Îles Crozet and Bass Strait) it has not been observed. In the North Pacific, there are two allopatric species of sea lions in the genus *Zalophus* (formerly three; the Japanese sea lion *Z. japonicus* is almost certainly extinct). The three seals in the genus *Pusa* are allopatric, albeit strongly differentiated. The allopatric Mediterranean monk seal *Monachus monachus* and Carribean monk seal *M. tropicalis* are very similar morphologically (the latter is extinct). The three manatee species *Trichechus manatus*, *T. senegalensis*, and *T. inunguis* are allopatric.

All of these are potential cases of geographical forms within a species being perceived as evolutionary species because of allopatry. In some cases, of course, the question is entirely moot. For example, while the Orinoco and Amazon drainages could merge in the future (bringing *T. manatus* and *T. inunguis* into contact and possible full interbreeding), the same could not happen for Amazon and Congo drainages (inhabited by *T. inunguis* and *T. senegalensis*) under any conceivable geological scenario short of reversal of continental drift; thus the latter two forms are "good" species *perforce* under any species concept.

All of the marine mammal species recognized today were described on morphological grounds. For the cryptic pair *Delphinus delphis* and *D. capensis*, genetic data provided part of the basis for recognition of more than one species where previously there had been thought to be only one. [A third species, *D. tropicalis*, is of uncertain status pending both further morphologic and genetic investigation (LeDuc *et al.*, 1999).]

The operational criterion of at least one absolutely differentiating character that is commonly used to indicate species-level difference in morphologically based taxonomy has been extended by some to genetic characters. This is done by considering one fixed difference in DNA sequence between populations (in sympatry or allopatry) to connote species-level difference. However, the potential number of genetic characters (total number of nucleotides in the genome, several billion) is so much greater than the potential number of morphological characters, probably by several orders of magnitude, that this argument by analogy seems injudicious. In fact, genetic investigations have found such fixed differences between certain isolated human populations. Few would argue then that these populations must comprise separate species, as they are undoubtedly fully interfertile and satisfy the criteria of a single species under the biological species concept or any concept of evolutionary species. In another example, Hector's dolphins, *Cephalorhynchus hectori*, from east and west sides of the South Island of New Zealand were found to have no mtDNA control-region sequence haplotypes in common (Pichler *et al.*, 1998). They are thus fully diagnosable and reciprocally monophyletic terminal taxa, qualifying as species under the phylogenetic species concept (Cracraft, 1997), if such a decision were to be based on this single genetic character. Very few would agree with such a decision, although most would probably accept designation of the two populations as *evolutionarily significant units* (something worth saving because of unique genetic or morphological diversity; Ryder, 1986) and agree that they should be managed separately. More recently, the trend has been toward use of a suite of fixed genetic differences (preferably unlinked) as a criterion for species-hood; this parallels the practice of most morphological taxonomists of not recognizing a species based on a single point of difference but rather on the basis of multiple concordant characters (Avise and Wollenberg, 1997).

II. The Future

While Rice's (1998) classification recognized 127 species of marine mammals, the status of many of these is open to question. For example, Rice "lumped" the northern right whale *Balaena glacialis* and the southern right whale *B. australis* into a single species *B. glacialis*, but some authors have preferred

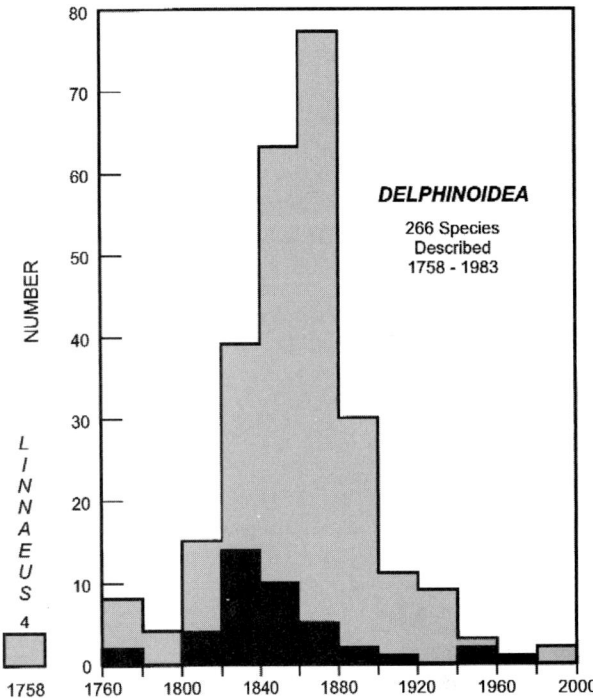

Figure 1 The "Great Victorian Radiation" of described (nominal) species in the cetacean superfamily Delphinoidea (Monodontidae, Delphinidae, and Phocoenidae); all but 44 (shaded) are now placed in synonymy. Data from Hershkovitz (1966) and Rice (1998).

to continue recognition of three species (under the genus *Eubalena*), one reason being the existence of fixed genetic differences between the two forms (but see discussion earlier of allopatric species). [Most recent usage is to recognize three species, including the North Pacific *E. japonica*; see Species List]. In another example, Rice (1998) provisionally listed three species of humpbacked dolphins, *Sousa teuszi*, *S. plumbea*, and *S. chinenis*, but he discussed the results of taxonomic work in progress that indicate that there likely is only one species, with pronounced geographical variation (*S. chinensis* has priority and would be the name borne by a single species). In still another example, the number of recognized species of southern fur seals (*Arctocephalus*) may decrease from eight. These and many other examples demonstrate that the *alpha* (species-level) *taxonomy* of marine mammals is still very much in flux: new species are almost certainly yet to be discovered, some species will be lumped, and others may be split (Fig. 1). There is still work for the taxonomists.

See Also the Following Articles

Biogeography ■ Classification ■ Geographic Variation ■ History of Marine Mammal Research ■ Systematics, Overview

References

Avise, J. C., and Wollenberg, K. (1997). Phylogenetics and the origin of species. *Proc. Natl. Acad. Sci. USA* **94**, 7748–7755.

Claridge, M. F., Dawah, H. A., and Wilson, M. R. (1997). "Species: The Units of Biodiversity." Chapman and Hall, London.

Cracraft, J. (1997). Species concepts in systematics and conservation biology: An ornithological viewpoint. *In* "Species: The Units of Biodiversity" (M. F. Claridge, H. A. Dawah, and M. R. Wilson, eds.), pp. 325–339. Chapman and Hall, London.

Hershkovitz, P. (1966). Catalog of living whales. *U.S. Nat. Mus. Bull.* **246,** 1–259.

IUCN (1996). "1996 IUCN Red List of Threatened Animals." IUCN, Gland, Switzerland.

LeDuc, R. G., Perrin, W. F., and Dizon, A. E. (1999). Phylogenetic relationships among the delphinid cetaceans based on full cytochrome *b* sequences. *Mar. Mamm. Sci.* **15,** 619–648.

Linnaeus, C. (1758). "Systema Naturae," Vol. 1. Laurentii Salvii, Stockholm.

Mayr, E. (1963). "Animal Species and Evolution." Harvard Univ. Press, Cambridge, MA.

Perrin, W. F., Akin, P. A., and Kashiwada, J. V. (1991). Geographic variation in external morphology of the spinner dolphin *Stenella longirostris* in the eastern Pacific and implications for conservation. *Fish. Bull. U.S.* **89,** 411–428.

Pichler, F. B., Dawson, S. M., Slooten, E., and Baker, C. S. (1998). Geographic isolation of Hector's dolphin populations described by mitochondrial DNA sequences. *Conserv. Biol.* **12,** 676–682.

Rice, D. W. (1998). Marine mammals of the world. *Soc. Mar. Mamm. Spec. Publ.* **4,** 231.

Ryder, O. A. (1986). Species conservation and systematics: The dilemma of subspecies. *Trends Ecol. Evol.* **1,** 9–10.

Templeton, A. R. (1989). The meaning of species and speciation. *In* "Speciation and Its Consequences" (D. Otte and J. A. Endler, eds.), pp. 3–27. Sinauer, Sutherland, MA.

Spectacled Porpoise
Phocoena dioptrica

R. NATALIE P. GOODALL
*Centro Austral de Investigaciones Científicas,
Tierra del Fuego, Argentina*

The spectacled porpoise is known from a large number of skeletons, mostly from Tierra del Fuego, but from few fresh specimens. As late as 1976, the species was known from only 10 occurrences (nine specimens), all off southeastern South America. Recent surveys have located isolated crania and made a few at-sea observations near widely scattered offshore islands in the Southern Ocean.

I. Taxonomy

Phocoena dioptrica was first described by Lahille in 1912 from a female collected near Buenos Aires. The only synonym is *P. storni,* based on a cranium from Tierra del Fuego. *Phocaena obtusata* Philippi, 1893, sometimes thought to be of this species, has been shown to be a synonym of *Cephalorhynchus eutropia.* In a revision of the taxonomy of the Phocoenidae, *P.*

dioptrica was put in a new monotypic genus, *Australophocaena,* but recent studies of phocoenid mitochondrial DNA have returned it to the genus *Phocoena.* The specific name, *dioptrica,* refers to the double eye patch seen in most specimens. The common name in Spanish is *marsopa de anteojos.*

II. Diagnostic Characters

The spectacled porpoise is highly distinctive with its unusual pigmentation, small head and facial features, and the large male dorsal fin.

A. Pigmentation

Color patterns have been depicted for few animals. Young animals ($n = 4$) are dark gray dorsally and light gray ventrally with darker gray or brownish streaks, including a well-defined mouth-to-flipper stripe. This pigmentation changes in the adult to black on the dorsal surface, sharply separated on the sides from the pure white ventral region, although there is feathering on some specimens (Fig. 1). Animals seen in the water may appear lighter. The flipper stripe may lighten or disappear with growth. Both young and adults have dark lips surrounded by white; the eye patch is a dark circle, variably outlined by a lighter or white ring. Most animals had a dark upper tail stock and flukes, which are lighter below. The dorsal surface of the flipper varies from white or mottled gray to black, often ending in a sharp line at the base of the flipper.

B. Size and Shape

The spectacled porpoise is a robust animal with a rounded head and no beak. The gape is short, and the flippers are small and situated well forward on the body. The dorsal fin is broadly triangular and shows strong sexual dimorphism, being much larger and more rounded in males. The rather small flukes have rounded tips with a fairly straight posterior border. External measurements are available for 27 specimens. Seven females ranged from 124.6 (a calf) to 203.5 cm in length. Six animals of unknown sex were from 94 cm, the smallest animal found, to 198+ cm. Nine males measured from 109+ (a neonate) to 224 cm, the largest animal known. This probably does not represent the maximum size for the species. Weights are available for five animals, from a fetus at 1.6 kg, to an 85-kg female, and a 115-kg male.

III. Distribution

The spectacled porpoise is circumpolar in cool temperate, sub-Antarctic, and low Antarctic waters. Specimens are known from Brazil (32°S) to Tierra del Fuego, the Falklands (Malvinas), and South Georgia in the southwestern South Atlantic, Auckland and Macquarie Islands in the southwestern South Pacific; Heard and Kerguelen in the southern Indian Ocean, Burney Island, Tasmania and South Australia (Fig. 2). It is associated with the Falkland (Malvinas) current and the West Wind Drift on both sides of the Antarctic Convergence.

Sightings are rare but widely distributed: off Uruguay, South Georgia, Îles Kerguelen, the Auckland Islands, south and southeast of New Zealand, Tasmania, Patagonia, and, recently, Heard

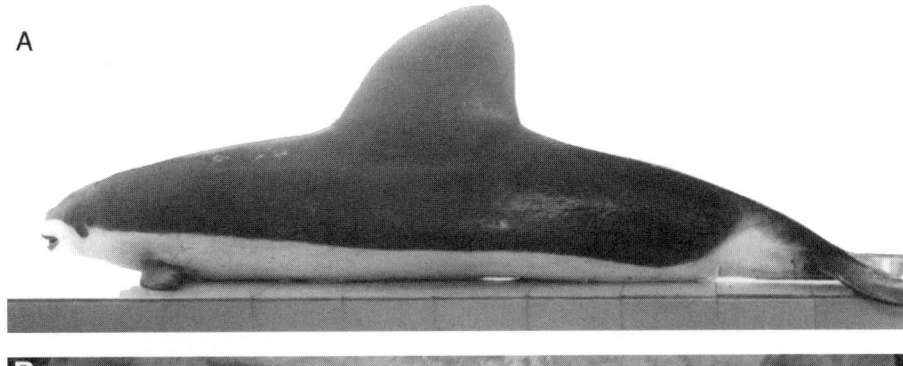

Figure 1 *(A) A male spectacled porpoise found dead in Chubut, Argentina, in June 1985. This animal, with a length of 205 cm, weight of 115 kg, and 6 GLGs, was sexually but not physically mature. Photograph by A. Purgue. (B) A 180-cm lactating female spectacled porpoise found May 27, 2000 at Bahía San Sebastian, Tierra del Fuego. This is the first animal we have seen with white flippers; they have a dark trailing edge. Photograph by R. N. P. Goodall.*

Island. The southernmost sighting was at 58°02′S, 62°19′W in the Drake Passage. The degree of contact among populations in the different oceans is unknown.

IV. Ecology

Once thought to be an inshore animal, recent records from offshore islands show this species to be oceanic in temperate to subantarctic waters. Although sighted in deep waters far from land, spectacled porpoises sometimes enter river estuaries and channels. We do not know why so many specimens are found on the beaches of Tierra del Fuego (some 160 to 1999) and so few elsewhere. Surface water temperature records to date range from 5.5 to 9.3°C.

Nothing is known about the abundance of this species. It is the second-most frequent species found in beach SURVEYS in eastern Tierra del Fuego but is uncommon elsewhere.

There are no observations of feeding behavior. Only four stomachs have been examined. One contained a small piece of algae and another meager amounts of anchovy (*Eugraulis* sp.) and stomatopods. Nothing else is known of the diet of this porpoise.

Killer whales (*Orcinus orca*), leopard seals (*Hydrurga leptonyx*), and sharks are possible predators of this species. The

stomach of a killer whale stranded in Brazil (33°45′S) contained remains of two porpoises, possibly this species.

V. Behavior

Nothing is known of migrations or seasonal movements of this species. Most sightings have been pelagic, but some animals must move inshore off Tierra del Fuego, where they strand or die in nets.

Swimming behavior is described as "unobtrusive." Two animals circling in a small bay at South Georgia in March 1995 reacted to boat noise by STRANDING and were pushed back to deeper water. The swimming before stranding was reported as "slow and steady, with dorsal fins breaking the surface together." A refloated calf had a respiration rate of four to seven breaths a minute. It swam with a slow, rolling motion, arching its back strongly on each dive. A prominent dorsal fin in diving adults was reported for New Zealand.

All sightings to date have been of one to three animals, with one sighting of two large animals with a calf. Mother–calf pairs were stranded twice, but there have been no mass strandings. Five of the seven fresh specimens came ashore alive. Nothing is known of the social behavior, sounds, or associations of this species.

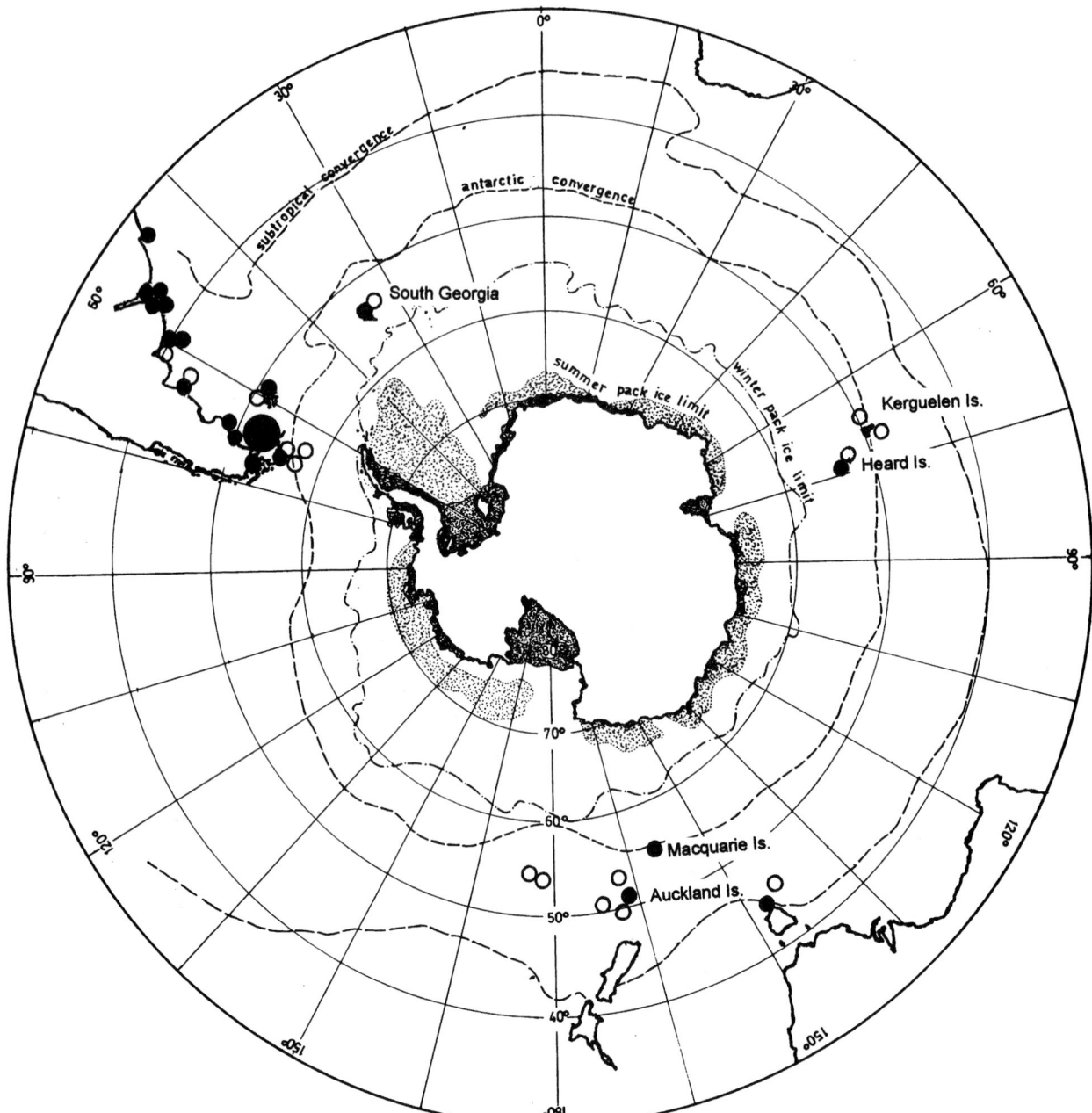

Figure 2 *Map showing the presently known distribution of the spectacled porpoise. Open circles represent sightings, closed circles are single strandings, and the large closed circle represents over 160 specimens.*

VI. Internal Anatomy

The skull has been illustrated and described in several works. Condylobasal lengths ranged from 279 to 324 mm ($n = 46$). The teeth vary from spade to peg shaped and are often hidden in the gums. Counts range from 19 to 26 upper and 19 to 22 lower teeth per side; posterior teeth are in a shallow groove without alveoli. The vertebral count is C7, T14, L14–16, and Ca 32–33 (65–69 total). From three to six cervical vertebrae are fused. There are usually 14 ribs: the first 9 bicipital and 8

sternal ribs. The phalangeal count of one specimen was I = 2, II = 7, III = 4, IV = 3, and V = 4. The facial complex system has been studied, but little else is known of the internal anatomy.

VII. Life History

Little reproductive information is available. Gonads were examined for one male, which at 205 cm and 6 GLGs was sexually mature. Males of 189–224 cm and 4-5 GLGs were physically mature. No ovaries have been examined. Three females of 183–186 cm had calves or fetuses. Female teeth show a change in layering after the second or third GLG, suggesting an early sexual maturation, as found in harbor porpoises. Only one female of 203.5 cm and 6 GLGs was physically mature. The youngest age at complete physical maturity was three animals of unknown sex with 3 GLGs. The maximum age to date ($n = 27$) was an animal of unknown sex with 8 GLGs. The smallest neonates examined measured 94 and 190 cm. Calves of 125–163 cm with unerupted teeth stranded in February to May; birth is probably in late spring or summer (November to February). A calf was seen in December off Heard Island. There is no information on pregnancy rates, gestation, or lactation periods.

No ectoparasites have been reported. Nematodes were found in the blowhole and nasal sacs of the Lahille specimen and in the stomachs of three animals from Tierra del Fuego. Nothing else is known of parasites or disease.

VIII. Interactions with Humans

No spectacled porpoises have been kept in captivity.

Remains of spectacled porpoises were found in 6000- and 1400-year-old kitchen middens of the canoe people in Tierra del Fuego, who harpooned cetaceans for food. In this century, whaling captains off South Georgia and fishermen off Uruguay also hunted spectacled porpoises for food. The first four specimens known came ashore alive and were collected by fishermen. There is no known deliberate exploitation of this species at present. At least some of the specimens found in Tierra del Fuego were taken incidentally in coastal gill nets; the location of the dead animals coincides with that of fishing areas. At least one animal died in offshore bottom or midwater trawls off Patagonia.

None of the biological information needed for conservation and management is available.

See Also the Following Articles

Burmeister's Porpoise ▪ Harbor Porpoise ▪ Vaquita

References

Brownell, R. L., Jr. (1975). *Phocoena dioptrica. Mammal. Spec.* **66**, 1–3.

Brownell, R. L., Jr. (1999). Spectacled porpoise *Australophocaena dioptrica* Lahille, 1912. *In* "Handbook of Marine Mammals: The Second Book of Dolphins and Porpoises" (S. H. Ridgway and R. Harrison, eds.). Vol. 6, pp. 379–391. Academic Press, London.

Crespo, E. A., Corcuera, J. F., and López Cazorla, A. (1994). Interactions between marine mammals and fisheries in some coastal fishing areas of Argentina. *Rep. Int. Whal. Commn.* (Spec. Issue **15**), 269–281.

Fordyce, R. E., Mattlin, R. H., and Dixon, J. M. (1984). Second record of spectacled porpoise from subantarctic southwest Pacific. *Sci. Rep. Whales Res. Int.* **34**, 159–164.

Goodall, R. N. P., and Schiavini, A. C. M. (1995). On the biology of the spectacled porpoise, *Australophocaena dioptrica. Rep. Int. Whal. Commn.* Spec. Issue No. **16**, 411–453.

Goodall, R. N. P., Schiavini, A. C. M., and Fermani, C. (1994). Net fisheries and net mortality of small cetaceans off Tierra del Fuego, Argentina. *Rep. Int. Whal. Commn.* Special Issue No. **15**, 195–304.

Kasamatsu, F., and Joyce, J. (1995). Current status of odontocetes in the Antarctic. *Antarc. Sci.* **7**(4), 365–379.

Rosel, P. E., Haygood, M. G., and Perrin, W. F. (1995). Phylogenetic relationships among the true porpoises (Cetacea: Phocoenidae). *Mol. Phylogen. Evol.* **4**, 463–474.

Speed

FRANK E. FISH
West Chester University, Pennsylvania

The simplest locomotor parameter that has the greatest survival value for marine mammals is swimming speed. High speeds allow for active pursuit of prey, escape from predators, and increased maneuverability. Low swimming speeds are used for foraging and migrating to minimize transport costs and maximize the distance traveled for the work performed. Swimming speed is a function of the power expended by the muscles and transmitted by propulsive movements of the appendages. High swimming speeds require large energy expenditures because the power to generate the propulsive thrust is directly related to the cube of speed. Fast swimming animals use oscillating hydrofoils (i.e., whale flukes, phocid hind flippers, otariid fore flippers) to generate large thrust forces. Paddling, used by the sea otter, is associated with slow swimming due to the high inefficiencies of thrust generation.

Research on swimming speed was motivated by accounts of extraordinary speeds, particularly for dolphins, associated with unresolved mechanisms to reduce hydrodynamic resistance (drag). Accurate swimming speed measurements of free-ranging marine mammals have been rare, and few data are available concerning the duration of swimming efforts. Reports on swimming speed of marine mammals have, in many instances, been anecdotal and often unreliable. The reason for questioning these reports is that estimates of swimming speeds based on observations from ships, airplanes, or shorelines have often been made without fixed reference points, information on currents, or accurate timing instruments. Originally, speeds of cetaceans were reported from whaling operations that were based on distressed and wounded animals. More recently, swimming speeds of animals performing routine behaviors were collected by means including boat measurements, timing of captive animals, theodolite tracking, observations correlated with map locations, sonar tracking, and radio-tagged animals. These methods still have limitations regarding accurate speed measurements. Even with the use of satellite-tracking systems, the resolution and sampling rate are too low to accurately measure short-duration swimming efforts, such as occurs for sprints. The diving pattern and high probability that animals do not swim in straight lines between consecutive locations measured by satellites tend to skew the data toward minimum transit speeds, which are well below routine and maximum speeds.

I. Cetaceans

Swimming speeds are categorized as (1) sustained and (2) sprint swimming. Sustained swimming is characterized by an activity level maintained for long-time durations. These speeds are steady low-level activities. Sustained speeds are the preferred swimming speeds and are probably minimum speeds for the most efficient generation of thrust. Sustained speeds are associated with routine swimming and cruising associated with migration, transit to foraging sites, and DIVING. Sprints are high-speed, maximum effort movements. These are short duration activities. Sprints are used during chases and rapid accelerations.

A. Sustained Swimming

Both mysticete and odontocete whales display low sustained speeds. For mysticetes, the low sustained speeds (1.8–13.0 km/hr) encountered during feeding would be associated with high drag. Mysticetes obtain food by FILTER FEEDING, which involves straining water directly through the whale's baleen plates or gulping a large quantity of water (at least 60 m³ or 70 tons) by depressing the lower jaw and filling the throat pouch. Filter feeding would induce additional drag on the animal, depressing speed.

Odontocetes display a similar range of sustained swimming speeds with mysticete whales. One record for sustained swimming by a killer whale (*Orcinus orca*), although probably incorrect, is unfortunately widespread in the literature. The killer whale was reported to swim for 20 min at 45.0–55.8 km/hr. Data were obtained from questionnaires placed aboard a ship. Because the killer whale was "playing" around the ship, the animal was probably BOW RIDING and thereby swimming at a higher speed with less effort.

When direct observation of swimming speed has not been obtainable, physiological characteristics have been used as an indicator of sustained swimming ability. Species with greater adaptation for deep, long-duration dives are precluded from fast sustainable swimming. Deep divers possess increased oxygen stores via increased hematocrit (i.e., percentage of blood occupied by cellular components) and hemoglobin concentration, but at the expense of oxygen transport capacity for rapid swimming. Belugas (*Delphinapterus leucas*) are regarded as slow swimmers that can dive to great depths and remain submerged for extended periods. The hematocrit and hemoglobin concentration of these whales is greater than for the faster swimming *Cephalorhynchus commersonii*, *Lagenorhynchus obliquidens*, *Orcinus orca*, and *Tursiops truncatus*.

B. Maximum Speeds

Maximum speeds are associated with duration times in seconds and levels of performance indicative of maximum effort. For free-ranging cetaceans, this behavior may be associated with flight or pursuit, where presumably motivation is high. The range of reported maximum speeds varies considerably for mysticetes (7.6–48.2 km/hr) and odontocetes (5.4–55.1 km/hr). Balaenopteridae are the fastest of the mysticetes. The comparative higher speeds of Balaenopteridae may be due to the highly streamlined body contour characteristic of the family. Within Odontoceti, the highest speeds are attained by members of Delphinidae, whereas Platanistidae are the slowest.

Systematic studies of maximum swimming speeds are few. In one study a top speed of 27.9 km/hr for a *Lagenorhynchus obliquidens* swimming in a towing tank was measured. Maximum speeds for trained animals over a 0.7- to 2.8-sec swim were 27.9 km/hr for *Tursiops truncatus*, 28.8 km/hr for *Delphinus delphis*, and 28.8 km/hr for *Pseudorca crassidens*. In open ocean tests, maximum speeds of 39.8 km/hr for *Stenella attenuata* and 29.9 km/hr for *Tursiops* were recorded. The latter study indicated that swimming speed is related to duration of activity. *Tursiops* can swim at 11.1 km/hr indefinitely, 21.9 km/hr for 50 sec, 25.2 km/hr for 10 sec, and 29.9 km/hr for 7.5 sec. Speeds as high as 54.0 km/hr were reported for burst swims by *Tursiops*. The maximum swimming speed of a school of free-ranging *Delphinus* sp. responding to multiple passes by an airplane was 24.1 km/hr, although the average speed for the school was 17.3 km/hr. The lower average speed may have resulted from the presence of newborn dolphins within the school.

C. Size Dependence of Speed

Although there is a considerable range of overlap for maximum swimming speeds regardless of taxonomic status, there is a marked dependence of specific speed on size. Large whales have low-length-specific swimming speeds compared to smaller dolphins and porpoises (Table I). Experiments on trained delphinids showed that smaller dolphins, such as *Stenella attenuata*, can sprint at 6.0 body lengths/sec, whereas a 38-times more massive *Orcinus orca* can sprint at only 1.5 body lengths/sec. Indeed, a 27.4-m blue whale (*Balaenoptera musculus*) sprinting at 36.7 km/hr would have a length-specific speed of 0.37 body lengths/sec. This trend is explained as a matter of scaling whereby animals maintain the same proportion of muscle mass in the body independent of size, but smaller animals have relatively greater muscular power outputs than larger animals.

II. Sirenians

Dugongs (*Dugong dugon*) and manatees (*Trichechus* spp.) are considered to swim in a leisurely manner (2.2–2.9 km/hr). As herbivores, sirenians do not require speed and rapid acceleration to catch prey. Low-speed swimming allows for precise maneuverability, which is necessary to navigate in restricted waterways and feed on rooted plants. Sirenians will migrate and cruise at speeds of 2.9 to 6.8 km/hr. Sprints of over 21.6 km/hr have been reported for sirenians and would aid in escape from predators such as crocodilians and sharks.

III. Pinnipeds

Pinnipeds exhibit sustained swimming speeds for horizontal swimming and diving of 2.2 to 9.4 km/hr. Dive velocities for small pinnipeds are associated with the minimum cost of transport to prolong dive duration and distance covered. However, the additional energy cost of a slightly increased swimming speed for deeper and longer dives by southern elephant seals (*Mirounga leonina*) is offset by pronounced bradycardia to manage available oxygen stores. Maximum swimming speeds

TABLE I
Maximum Length-Specific Velocities for Trained Delphinids

Species	Mass (kg)	Body length (m)	Maximum velocity (km/hr)	Length-specific velocity (body length/sec)
Stenella attenuata	52.7	1.86	39.8	6.0
Tursiops truncatus gilli	89.0	1.91	29.9	4.3
Lagenorhynchus obliquidens	91.0	2.09	27.9	3.7
Delphinus delphis	104.8	1.83	28.8	4.4
Tursiops truncatus	149.2	2.06	27.9	3.8
Pseudorca crassidens	461.8	3.66	28.8	2.2
Orcinus orca	1995.8	5.16	28.5	1.5

are higher for otariids than phocids and the walrus. Otariids can maintain a speed of 18.5 km/hr for 3 to 5 min, although bursts of 24.1 to 28.8 km/hr have been reported. Maximum speed for phocids is 13.0 to 17.3 km/hr.

In addition to swimming, pinnipeds can move over land using terrestrial gaits. Walking by otariids requires alternate movement of the four limbs so that at least three of the limbs are in contact with the ground. As a result, walking speeds are low (2.6 km/hr) for fur seals and sea lions, which must coordinate movements of the enlarged flippers. For faster terrestrial speeds of 7.9 to 13.0 km/hr, a gallop or a bounding gait can be used.

IV. Sea Otter

Surface swimming by the sea otter (*Enhydra lutris*) is generally slower than submerged swimming. This difference in performance is related to the increased resistance to swimming at the surface. The interference pattern of waves generated by the animal at the surface produces a physical limitation to swimming at a higher speed. In addition, at the surface otters use inefficient paddle propulsion. Sea otters swim at speeds ranging from 1.6 to 2.5 km/hr at the surface. Speeds above 2.9 km/hr are performed during bouts of submerged swimming. The maximum escape speed for submerged swimming otters is 9.0 km/hr. During submerged swimming, the otter employs undulation of the body in concert with simultaneous paddling by the hind feet.

See Also the Following Articles

Locomotion, Terrestrial ▪ Musculature ▪ Swimming

References

Fish, F. E. (1996). Transitions from drag-based to lift-based propulsion in mammalian aquatic swimming. *Am. Zool.* **36**, 628–641.
Fish, F. E., and Hui, C. A. (1991). Dolphin swimming: A review. *Mamm. Rev.* **21**, 181–195.
Johannessen, C. L., and Harder, J. A. (1960). Sustained swimming speeds of dolphins. *Science* **132**, 1550–1551.
Kenyon, K. W. (1969). The sea otter in the eastern Pacific Ocean. *N. Am. Fauna* **68**, 1–352.
Kooyman, G. L. (1989). "Diverse Divers: Physiology and Behavior." Springer-Verlag, Berlin.
Lang, T. G. (1975). Speed, power, and drag measurements of dolphins and parameters. *In* "Swimming and Flying in Nature" (T. Y. Wu, C. J. Brokaw, and C. Brennen, eds.), pp. 553–571. Plenum Press, New York.
Norris, K. S., and Prescott, J. H. (1961). Observations on Pacific cetaceans of California and Mexican waters. *Univ. Calif. Publ. Zool.* **63**, 291–402.
Ray, C. (1963). Locomotion in pinnipeds. *Nat. Hist.* **72**, 10–21.
Tomilin, A. G. (1957). "Mammals of the U.S.S.R. and Adjacent Countries," Vol. IX. Izdatel'stvo Akademi Nauk SSSR, Moskva. [Translated from Russian.]
Webb, P. W. (1975). Hydrodynamics and energetics of fish propulsion. *Bull. Fish. Res. Bd. Can.* **190**, 1–158.
Würsig, B., and Würsig, M. (1979). Behavior and ecology of the bottlenose dolphin, *Tursiops truncatus,* in the South Atlantic. *Fish. Bull.* **77**, 399–412.

Spermaceti

DALE W. RICE
National Marine Mammal Laboratory
Seattle, Washington

Spermaceti is the term given to the liquid waxes present in the head of the sperm whale (*Physeter macrocephalus*). The word comes from the late Latin *sperma*, "sperm" or "semen," and the classical Latin *ceti*, genitive singular of *cetus*, "sea monster" or "whale"—literally "semen of the whale." It was bestowed because of the superficial resemblance of these waxes to semen. The vernacular name "sperm whale" is a contraction of "spermaceti whale."

Spermaceti is present in greatest abundance in the spermaceti organ, or "case," and in the "junk." The case is an elongated barrel-shaped organ, which makes up much of the bulk of the sperm whale's huge head. It consists of soft, white, spongy tissue that is saturated with spermaceti. The junk lies below the case and above the rostrum of the skull; it is also white, but it is more solid than the case and is faintly divided into segments by a series of transverse septa composed of denser tissue. It is also heavily saturated with spermaceti (Clarke, 1978; Rice, 1989). At body temperature, the spermaceti will flow freely as a clear, almost watery, fluid from the case and junk when the latter are slashed with a knife. As soon as it cools to about room temperature, however, it solidifies to a whitish wax similar to paraffin but not as hard. The different fractions may be separated by pressing at various temperatures between 10 and 35°C.

The oils of sperm whales and the other toothed whales (suborder Odontoceti) differ in composition from those of the baleen whales (suborder Mysticeti) (Gilmore, 1951). Those of the baleen whales consist exclusively of triglycerides. Triglycerides are esters compounded from one molecule of the tribasic alcohol glycerol ($CH_2OH \cdot CHOH \cdot CH_2OH$) and three molecules of various fatty acids. Lipids of the toothed whales, however, consist of a mixture of triglyceridic oils and wax esters, together with a small proportion ($\geq 2\%$) of diacyl glyceryl ethers in some species. These waxes are compounds formed from one molecule of a higher monobasic alcohol, mostly cetyl ($C_{16}H_{33}OH$) or oleyl ($C_{18}H_{35}OH$), and one molecule of a fatty acid.

Although wax esters are present in all odontocetes, only in the giant sperm whale and the Amazon river dolphin (family Iniidae) do they comprise a majority of the lipids present. In sperm whales the ratio of waxes to triglycerides varies with position in the body. An early study (Hilditch and Lovern, 1928) found that wax esters comprised 74% of the spermaceti in the head, but only 66% of the BLUBBER oil. More detailed studies revealed that even within the head, the proportion of wax esters varies with sex and age as well as position. In the spermaceti case this proportion ranged from 38 to 51% in a calf, 58 to 87% in an adult female, and 71 to 94% in an adult male. Proportions in the junk of the same three whales were 41–62, 65–94, and 91–98%, respectively (Morris, 1975).

Both the triglycerides and the waxes of the sperm whale are characterized by the virtual absence (<2%) of isovaleric acid, a branched, short chain (C_5), unsaturated fatty acid. This trait is shared with pygmy sperm whales (family Kogiidae), beaked whales (family Ziphiidae), and some river dolphins (families Platanistidae and Iniidae). This contrasts with the high proportion of isovalerate lipids in delphinoid cetaceans (families Monodontidae, Delphinidae, and Phocoenidae) (Litchfield et al., 1975).

Wax esters of the sperm whale differ from those of all other toothed whales except the Amazon dolphin (Inia geoffrensis; family Iniidae) in that they consist predominantly of relatively long chain (C_{10}–C_{22}) fatty acids (Litchfield et al., 1975). The fatty acid composition of wax esters in the case is mainly shorter chained, saturated acids at the center, with longer chained, more unsaturated acids at the periphery. In the junk, the anterior portion is mainly shorter chained saturated acids, whereas the posterior portion consists of longer chained, more unsaturated acids. The specific proportions of 25 fatty acid moieties in wax esters from the head of an adult female sperm whale were tabulated by Morris (1975).

Because of its heterogeneous lipid composition, and possible internal temperature gradients, the velocity of SOUND through the spermaceti organ varies in a way that could collimate or focus sound waves (Flewellen and Morris, 1978). This accords with the prevailing hypothesis that this organ functions as an acoustic lens to channel acoustic emissions.

In the 19th century when the old-style open-boat whale fishery was flourishing, the oil in the head of the sperm whale (the spermaceti proper) was simply ladled from the case and junk, and was kept separate from the oil that was rendered from the blubber (the remainder of each carcass was discarded). Spermaceti was favored for making candles, and the liquid blubber oil was used as a lamp fuel. In the years from 1804 to 1925, a total of 164,073,918 gallons of sperm oil (including spermaceti) was landed at United States ports. This quantity was the product of an estimated 262,134 sperm whales killed around the globe. Lesser quantities were taken by vessels out of British, French, and other foreign ports. In the peak year of 1837, American landings totaled 5,349,138 gallons, representing an estimated kill of 7472 sperm whales, and were worth $4,413,039 at the average going price of 82.5 cents per gallon. Discovery of the first commercially recoverable reserves of petroleum in 1859 triggered the phenomenal growth of the oil industry, so the demand for sperm oil plummeted. The last old-style American sperm-whaling voyage was made in 1925.

In the 20th century the modern harpoon-cannon whaling industry was carried out from factory ships and shore stations around the world. Oil was rendered from the entire carcass (minus the meat) of each whale, and that from sperm whales ("sperm oil") was kept separate from the oil of baleen whales ("whale oil"). In the early decades of this fishery, whalers took few sperm whales because the edible oil of baleen whales brought higher returns. By the end of World War II, however, baleen whale populations had been severely depleted, and new uses were being discovered for sperm oil. Because of its unique properties, it was in great demand as a high-pressure lubricant, but it also had a diverse array of other specialized industrial uses, such as an ingredient in hydraulic fluids, inks, detergents, and cosmetics, as a plasticiser, and as an agent in the tanning of leathers and the degreasing of wool (Anonymous, 1957). The rising demand led to rapid increases in sperm whale catches, which reached a peak in 1964, when 29,255 animals were killed, yielding 898,257 barrels (=152,703,690 kg) of sperm oil. The International Whaling Commission shut down the entire whale fishery after the 1984 season. Since the cessation of whaling, substitutes for sperm oil have been found—most notably the oil from the seeds of jojoba (Simmondsia chinensis of the family Buxaceae), a shrub indigenous to the Sonoran Desert. Jojoba oil has chemical and physical properties similar to those of sperm oil (Harris et al., 1975).

See Also the Following Articles

Sperm Whale ■ Toothed Whales, Overview ■ Whaling, Traditional

References

Anonymous (1957). Leviathan in the lab. Laboratory **26**, 34–38.
Clarke, M. R. (1978). Structure and proportions of the spermaceti organ in the sperm whale. J. Mar. Biol. Assoc. UK **58**, 1–17.

Flewellen, C. G., and Morris, R. J. (1978). Sound velocity measurements on samples from the spermaceti organ of the sperm whale (*Physeter catodon*). *Deep-Sea Res.* **A25,** 269–277.

Gilmore, R. M. (1951). The whaling industry: Whales, dolphins, and porpoises. *In* "Marine Products of Commerce" (D. K. Tressler and J. M. Lemon, eds.), pp. 680–715. Reinhold, New York.

Harris, M., Goebel, C. G., Schwartz, A. M., Von Wettberg, E. F., Jr., and Yermanos, D. M. (1975). "Products from Jojoba: A Promising New Crop for Arid Lands," p. 30. National Academy of Sciences, Washington, DC.

Hilditch, T. P., and Lovern, J. A. (1928). The head and blubber oils of the sperm whale. I. Quantitative determinations of the mixed fatty acids present. *J. Soc. Chem. Industr.* **47,** 105T–111T.

Litchfield, C., Greenberg, A. J., Caldwell, D. K., Caldwell, M. C., Sipos, J. C., and Ackman, R. G. (1975). Comparative lipid patterns in acoustical and nonacoustical fatty tissues of dolphins, porpoises and toothed whales. *Comp. Biochem. Physiol. B* **59,** 591–597.

Morris, R. J. (1975). Further studies into the lipid structure of the spermaceti organ of the sperm whale (*Physeter catodon*). *Deep-Sea Res.* **22,** 483–489.

Rice, D. W. (1989). Sperm whale *Physeter macrocephalus* Linnaeus, 1758. *In* "Handbook of Marine Mammals" (S. H. Ridgway and R. Harrison, eds.), Vol. 4, pp. 177–233. Academic Press, London.

Sperm Whale
Physeter macrocephalus

HAL WHITEHEAD
Dalhousie University
Halifax, Nova Scotia, Canada

Sperm whales are animals of extremes. They have unusually large body sizes, sexual dimorphism, brain sizes, home ranges, dive depths, and dive times; they have an ecological role that may be unrivaled in the ocean; their mitochondrial genetic diversity is unexpectedly low; and their vocalizations, social structure, and historical relationship with humans are all remarkable. The likelihood of evolutionary or ecological links between these extreme attributes form one of the scientific attractions of this animal.

I. Names

In 1758, Linnaeus described four sperm whales in the genus *Physeter.* It soon became clear that all refer to the same species, but there has been a long, and sometimes contentious, debate as to whether *P. catodon* or *P. macrocephalus* has precedence. Currently, most, but not all, authorities prefer *P. macrocephalus.*

The common name, "sperm whale," seems to have resulted from whalers misinterpreting the function of the spermaceti oil found in the massive forehead of the whale.

II. Taxonomy and Evolution

The closest living relatives of the sperm whale are the much smaller dwarf and pygmy sperm whales (*Kogia breviceps* and *K. sima*). There has been debate as to the next closest living relative of the three sperm whale species. In the early 1990s, Michel Milinkovitch and colleagues interpreted DNA data as indicating that the sperm whales are related more closely to baleen whales than other toothed whales. However, more recent genetic and morphological analyses support the traditional and intuitive (sperm whales have substantial teeth) inclusion of sperm whales among a monophyletic Odontoceti, or toothed whales. Sperm whales seem to have separated from other odontocetes early in modern cetacean evolution, about 20 million years ago.

III. Description

Sperm whales are the largest of the odontocetes and the most SEXUALLY DIMORPHIC cetaceans in body length and weight (Fig. 1). While adult females reach about 11 m in length and 15 tons, a physically mature male is approximately 16 m and 45 tons.

Figure 1 *Mature male with female sperm whale on the breeding grounds off the Galápagos Islands showing the considerable sexual dimorphism in adults (female in foreground). Photograph courtesy of H. Whitehead laboratory.*

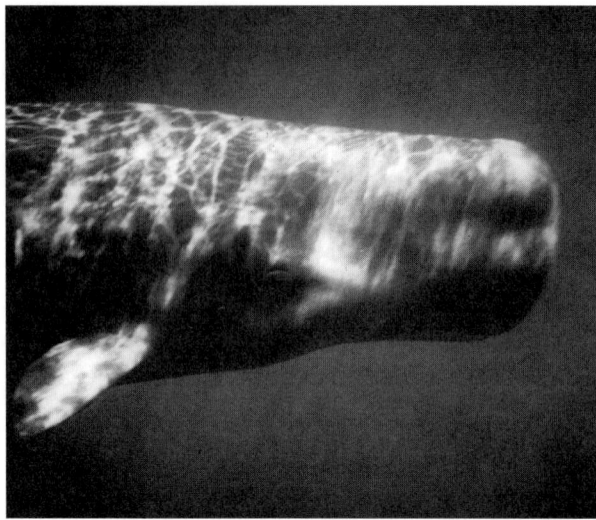

Figure 2 *The head of the sperm whale. The volume above the narrow lower jaw and in front of the eye is largely filled with an oil-filled sac, the spermaceti organ. Photograph courtesy of H. Whitehead laboratory.*

The most distinctive feature of the sperm whale is a massive nasal complex (Fig. 2), one-quarter to one-third of the length of the animal, situated above the lower jaw and in front of the skull. It principally contains the spermaceti organ, which is enclosed in a "case" (Fig. 3). This is a roughly ellipsoidal shaped structure made of spongy tissue filled with spermaceti oil and bounded at both ends by air sacs. Between the spermaceti organ and the upper jaws is the "junk," a complex arrangement of spermaceti oil and connective tissue. Spermaceti oil, which has the properties of a wax, differs chemically from the oils found in the "melons" of most other odontocetes.

There is considerable asymmetry in the parts of the skull and air passages that surround the spermaceti organ. This is externally manifested most clearly by a BLOW, which is pointed forward and to the left from the tip of the snout. Compared with the blows of similarly sized baleen whales, the blow of a sperm whale is comparatively weak, low, and hard to see.

Behind the sperm whale's skull lies the largest BRAIN of any animal (mean of 7.8 kg in mature males). However, as a proportion of body size, the sperm whale's brain is not remarkable, and we have no direct information on the sperm whale's cognitive abilities, although its complex social system is consonant with those found in other cognitively advanced mammals.

The sperm whale has 20–26 large conical teeth in each rodlike lower jaw. These teeth do not seem to be necessary for feeding, as they do not erupt until near puberty, and well-nourished sperm whales have been caught that lack teeth, or even lower jaws. The teeth in the upper jaw seem to be vestigial and rarely erupt.

Large corrugations cover most of the body behind the eye, but the surface of the head and the flukes are smooth. The majority of the body is dark gray in most sperm whales, but there is often a bright white lining to the mouth and sometimes white patches on the belly. The occasional sperm whale has larger white patches, especially in mature males, around the head. The flippers are relatively small and paddle shaped, and the flukes are fairly flat and triangular shaped (Fig. 4). The dorsal fin is low, thick, and usually rounded. Especially in mature females, it may be topped by a white or yellowish rough callus. The dorsal ridge, behind the dorsal fin, consists of a series of large crenulations.

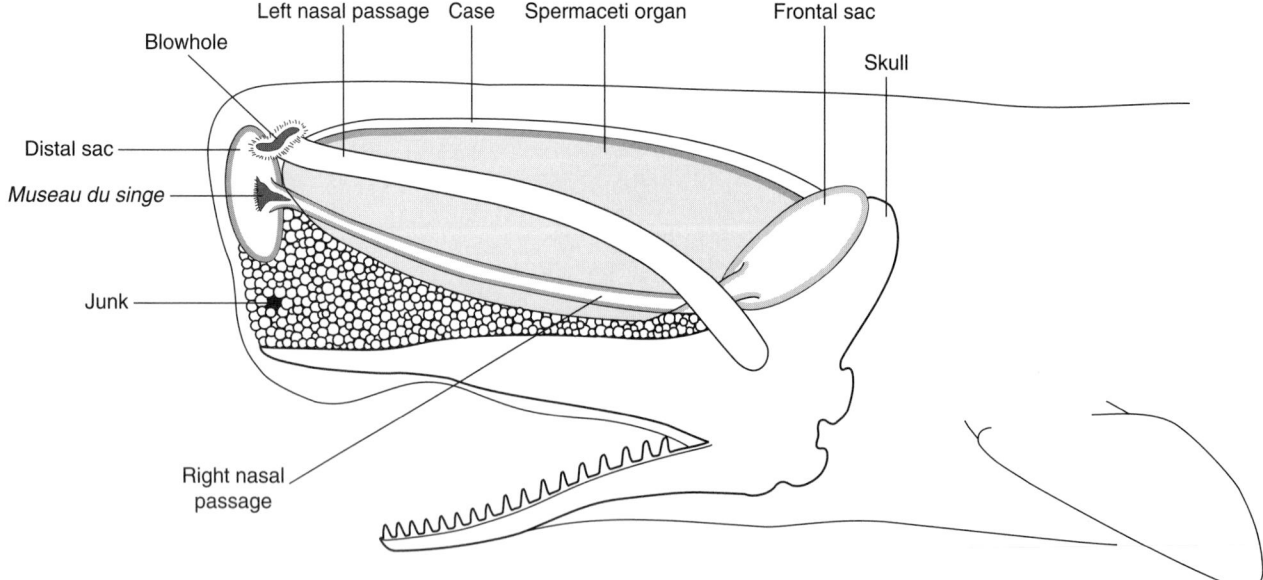

Figure 3 *Diagram of the head of the sperm whale. From Ellis (1980).*

Figure 4 *Flukes, or tail, of diving sperm whale. Photograph courtesy of H. Whitehead laboratory.*

IV. Life History

The sperm whale is the epitome of the "K-selected" mammal, one presumed to have evolved in an environment of competition for resources with members of its own species. It has a very low birth rate, slow growth, slow maturation, and high survival.

Young are born, almost always singly and with an equal sex ratio, at about 4 months following an approximately 14- to 16-month gestation. Although sperm whales eat solid food before their first birthday, they continue suckling for several years. The females reach sexual maturity at about age 9 when roughly 9 m long, at which age growth starts to slow. They give birth roughly once every 5 years, although pregnancy rates have been found to vary between areas and in the same area at different stages of exploitation by humans. Female reproductive rates decline with age. Females reach physical maturity when growth ceases, at about 30 years old and 10.6 m long.

Males, which are slightly larger than females during the first 10 years of life, continue to grow at a substantial rate until well into their 30s, finally reaching physical maturity at about 16 m long when roughly 50 years old. In males, puberty is prolonged, lasting approximately between ages 10 to 20 years old. However, males seem not to generally take much of an active role in breeding before their late twenties.

V. Distribution and Migrations

With the exception of human and killer whales (*Orcinus orca*), few animals on Earth are as widely distributed as the sperm whale. They can be seen close to the edge of pack ice in both hemispheres and are also common along the equator, especially in the Pacific. As with so many aspects of sperm whale biology, the sexes have very different distributions.

Although sperm whales have been sighted in most regions of deep water, there are some areas which the whalers called "grounds," where they are more abundant. Many of the grounds coincide with areas of generally higher primary productivity, usually resulting from upwelling.

Concentrations of a few hundred to a few thousand sperm whales can be found in areas a few hundred kilometers across

characterized by a relatively high deep water biomass and usually situated within grounds. Sometimes aggregations of 50 or more sperm whales can be found within a few kilometers, presumably the result of concentrations of food.

Female sperm whales almost always inhabit water deeper than 1000 m and at latitudes less than 40° (except 50°N in the North Pacific), corresponding roughly to sea surface temperatures greater than 15°C (Fig. 5). Although sometimes seen close to oceanic islands rising from deep ocean floors, female sperm whales are usually far from land.

Young male sperm whales accompany the females in tropical and subtropical waters. On leaving their female relatives, sometime between 4 and 21 years of age, the males gradually move to higher latitudes: the larger and older the male, the higher the average latitude (Fig. 5). Large males may be found near the edge of pack ice in both hemispheres, although from time to time they return to the warm water breeding grounds.

The large males of high latitudes can be found over almost any ice-free deep water, but, like the low-latitude females, they are more likely to be sighted in productive waters, such as those along the edges of continental shelves. However, in some areas, such as off New York and Nova Scotia, the large males are sighted regularly in waters less than 300 m deep.

Migrations of the sperm whale are not as regular or as well understood as those of most baleen whales. In some midlatitudes there appears to be a general seasonal north–south migration, with whales moving poleward in summer, but, in equatorial and some temperate areas, there is no clear seasonal migration.

Scientists have studied the movements of sperm whales using natural and artificial markings, as well as by following groups over periods of a few days. Sperm whales usually move through the water at about 4 km/hr. When feeding conditions are poor, their tracks are straight, resulting in daily displacements of about 90 km per day. However, when food is plentiful, the animals stay in much smaller areas 10–20 km across. Female home ranges seem to be generally of the order of 1000 km across. Some males roam more widely, but others are found quite consistently over several years in restricted coastal waters.

VI. Ecology

The sperm whale has a most catholic DIET consisting of many of the larger organisms that inhabit the deeper regions of the oceans. Females appear to principally eat squids weighing between 0.1 and 10 kg. Favored taxa include Ommastrephidae, Onychoteuthidae, Gonatidae, Pholidoteuthidae, Octopoteuthidae, Histioteuthidae, and Cranchiidae. Of these, the histioteuthids, mesopelagic gelatinous pelagic cephalopods weighing about 0.1–1 kg, have featured at or near the top of the list of preferred food items in several studies of sperm whales. Females will eat much larger prey, such as the giant squid (*Architeuthis*) and the jumbo squid (*Dosidicus*), as well as noncephalopod prey, especially demersal and mesopelagic fish.

Males use the same squid taxa as females, but tend to eat larger individuals. Additionally, they eat species that are largely

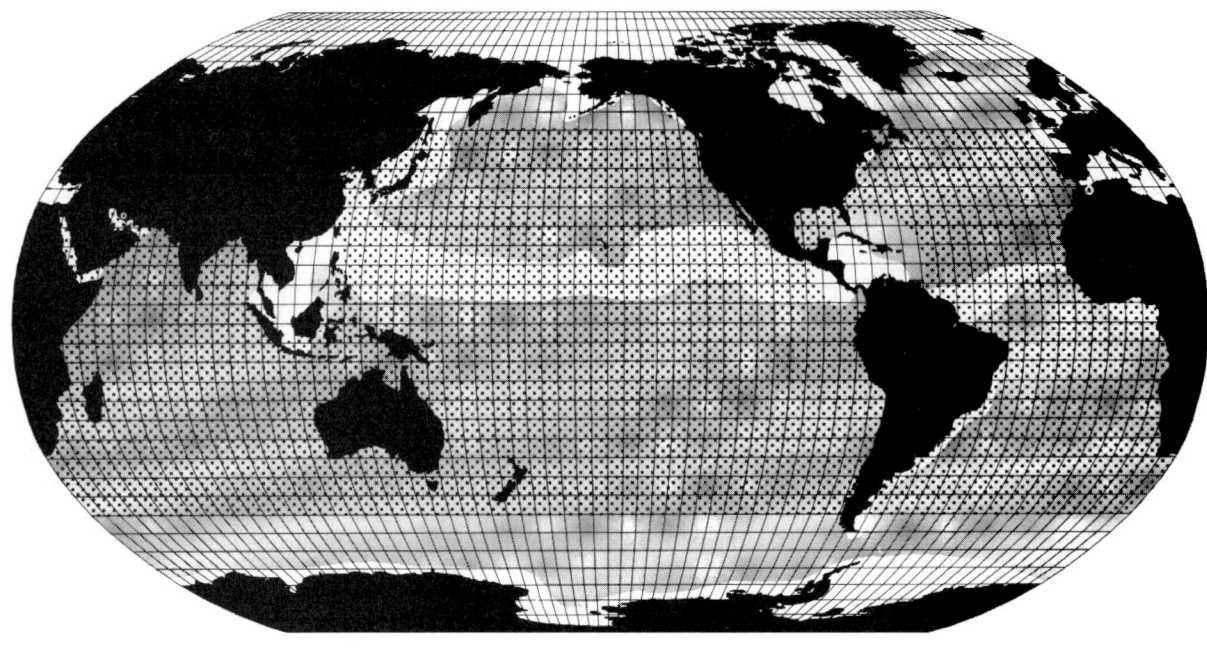

:::::: Approximate female distribution
▓▓▓ Approximate male distribution

Figure 5 *Worldwide distribution of sperm whales: females, dark shading; males, light and dark shading.*

restricted to higher latitudes and are also more likely to eat demersal fish (including sharks, rays, and gadoids) than females. Off Iceland and in the northern Gulf of Alaska, their primary food is fish.

Sperm whales surely have competitors for many of these food items—beaked whale (Ziphiidae) and elephant seals (*Mirounga* spp.) also eat mesopelagic squid—and there may be important nonmammal predators of the species that the sperm whales use. For instance, the jumbo squid eats the smaller histioteuthid squids. Although we know virtually nothing of the quantitative ecology of the deep ocean, sperm whales, because of their size and numbers, seem to dominate this trophic level in terms of biomass removed. Rough estimates of the worldwide consumption of prey by the world's sperm whale population, about 100 million metric tons per year, are comparable with the current annual catch by all human marine fisheries.

VII. Populations and Stocks

There is no valid estimate for the number of sperm whales currently alive in any ocean or globally. Crude extrapolations from the few areas where there are useful estimates suggest that the total population may lie between 200,000 and 1,500,000 animals.

Information on stock structure in sperm whales is confused, but rather few differences have been found between animals in different ocean areas. Modern genetic studies have been unable to find any distinct stocks at geographical scales of less than an ocean basin. In the mitochondrial, maternally transmitted genome, sperm whales worldwide are remarkably homogeneous. Explanations for this unexpected result (also found in pilot, *Globicephala* spp., and killer whales) include a historical population bottleneck, a demographic consequence of the result of these whales' matrilineal social system, increased selection on the mitochondrial genome, or the indirect effects of selection of matrilineally transmitted cultural traits.

VIII. Sound Production

Unusually, the sounds of sperm whales consist almost entirely of clicks (although a few relatively quiet "squeals" and "trumpets" are made). However, these clicks are used in various ways. When sperm whales are at depth they generally produce regular trains of clicks, at rates of about one to two clicks per second, which most scientists believe are used to find food by echolocation. These "usual" trains are occasionally interrupted by periods of silence or accelerating series of clicks, called "creaks," which are thought to indicate that the sperm whale is homing in on potential prey.

While usual clicks are indicative of foraging at depth, "codas," stereotyped series of 3– ~20 clicks lasting 0.2–2 sec, are heard in social situations when whales are interacting. Groups of females have distinctive repertoires of codas, and coda repertoire also shows GEOGRAPHICAL VARIATION. Coda repertoire is probably acquired culturally from within family units.

Whereas codas are heard principally from social females, another form of click, the "slow click" or "clang," in which distinctively ringing clicks are repeated every 6–8 sec, is largely or entirely produced by large males. The function of slow clicks is not clear, but they may attract females, repel other males, or be used to ECHOLOCATE off other whales.

The principal function of the spermaceti organ and its associated structures (the junk, air sacs and passages, and the *museau de singe* clapper system at the front of the organ, see Fig. 3) is thought to be to form, and perhaps focus, the sperm whale's click.

IX. Social Structure

The life of a female sperm whale is overwhelmingly social. She is always in the company of other females, some of whom are her relatives (Fig. 6). The basic element of sperm whale society is a family unit, consisting of about 12 females and their young. Most females probably spend their lives in the same unit, with their close female relatives. However, there are recorded instances of females switching units and nonrelatives showing long-term relationships. Two or more units may travel together for a few days as a cohesive group of about 20 animals.

Members of groups defend themselves and their offspring communally against predators and have been observed risking themselves to assist unit members in peril during a killer whale attack. Female groups occasionally mass strand on shorelines, usually with fatal results for all members.

Young males leave their natal unit when between 4 and 21 years old and then are found in loose aggregations, sometimes called "bachelor schools," with other males of approximately the same size and age. As the males age and grow they move to generally higher latitudes and the aggregation sizes become smaller, until the largest males are usually alone. Although repeat association between males on more than 1 day are rarely observed, mature and maturing males do strand on beaches

together, suggesting significant social relationships on a scale not readily apparent to human boat-based observers.

X. Foraging Behavior

Sperm whales possess two quite distinct behavioral modes: foraging and social/resting. When foraging, the animals make repeated deep dives. Modal dives are to about 400 m and for about 35 min, but dives can be much deeper (to over 1000 m), shallower (e.g., when in shelf waters 200 m deep), or longer (to over 1 hr). Between dives the whales come to the surface to breathe for about 8 min. The dive is usually signaled by the raising of flukes out of the water (Fig. 4). The descent to depth, as well as the return to the surface, can be nearly vertical.

It is not known how sperm whales find and catch their food, but one plausible hypothesis is that they generally locate their prey using the usual click trains. Small gelatinous squid, such as histioteuthids, are usually both relatively unmaneuverable and BIOLUMINESCENT, so these can be captured using visual or acoustic cues. Larger, more muscular animals may need active chasing, including accelerating click trains ("creaks"). It has been suggested that sperm whales may use their clicks to stun squid, making them more easily catchable. However, there is currently only a little circumstantial evidence for stunning.

Groups of females and immatures spread out over 1 km or more of ocean when foraging, often forming a rank aligned perpendicular to the direction of travel. In contrast, males seem to forage independently.

XI. Social/Resting Behavior

Female and young sperm whales spend approximately 75% of their time foraging. However, during periods of several hours, often in the afternoon, they gather at or near the sur-

Figure 6 *A group of female sperm whales. Photograph by François Gohier.*

Figure 7 *Female sperm whales in resting/social mode off the Galápagos Islands. Photograph courtesy of H. Whitehead laboratory.*

face. At these times their behavior is highly variable. Sometimes the animals may lie still and quiet, closely clustered, for hours at a time, apparently resting (Fig. 7). Particularly at the start or end of nonforaging periods their behavior may be much more active, with breaches, lobtails, animals rolling, maneuvering and touching one another, and codas and creaks being emitted. Large males also lie quietly at the surface for long periods, but they are usually alone unless accompanying a group of females.

XII. Calf Care

Within the social units there is communal care for the young, with good, although not definitive, evidence for females suckling calves who are not their own offspring. Young sperm whale calves do not seem to be able to dive to foraging depths for as long as their mothers. Instead they remain at or near the surface, moving between those adult and immature members of their group while they are breathing. This babysitting begins soon after birth. Members of groups with young calves seem to stagger their dives, thus providing better babysitting for the young.

XIII. Defense against Predators

Killer whales have been observed attacking sperm whales, usually unsuccessfully, although they killed at least one sperm whale during a well-documented encounter off California, and sperm whale remains have, very occasionally, been found in the stomachs of killer whales. Pilot whales have been seen harassing sperm whales on a number of occasions, but whether they are any real threat to the much larger sperms is uncertain. Large sharks are also potential predators, especially of young animals.

When faced with predators, particularly killer whales, female sperm whales quickly cluster. Two defensive patterns have been described. In the "marguerite" or "wagon wheel" formation, the members of the group place their heads together at the hub, with the bodies radiating out like spokes. In contrast, when adopting the "heads-out" formation, the sperm whales face their attackers, tightly aligned in a rank, and seem to principally use their jaws for defense. Young calves stay toward the center of whichever defensive formation is adopted.

XIV. Breeding Behavior

The large mature males, in their late 20s and older, return to the tropical breeding grounds to mate, although the timing of such visits is largely unknown. When on the breeding grounds, the large males roam between groups of females, usually spending just a few minutes or hours with each. They are presumably searching for receptive females. There are no clear descriptions of mating itself. The breeding males seem to roam independently and generally avoid one another, although they are sometimes observed within the same group of females, and occasionally fight. These fights are rarely observed, but many large males have deep scars made by the teeth of other males.

XV. Sperm Whales and Humans

In the early 18th century, New Englanders began to hunt sperm whales for commercial purposes off their own shores. Over the next two centuries, sperm whaling grew to be a major worldwide industry. At its peak in the 1830s, about 5000 sperm whales were being killed each year by whalers from several countries, especially the United States, and the oil produced was a vital element of the burgeoning industrial revolution. The whalers sailed all oceans of the world in their stubby square-rigged ships. On sighting sperms, open whaleboats were lowered and rowed or sailed to the whales. The whalers threw harpoons into the animals, and then killed them using lances

(Fig. 8). Dead animals were towed to the whale ship, where the oil was baled from the spermaceti organ, the BLUBBER stripped from the body and boiled to render the oil, and virtually all the remainder of the carcass discarded. In the 19th century, sperm whaling had become such a significant enterprise that it had important effects on fields as diverse as literature (Herman Melville's great novel "Moby Dick") and ocean exploration. The open-boat hunt declined during the latter half of the 19th century because of the development of petroleum products as alternatives for sperm oil, the decline in the sperm whale population, and for other reasons.

After about 50 years of relative peace, the sperm whale populations were again hit hard following World War II. The whalers of the 20th century chased the sperm whales using mechanized catcher vessels equipped with sonar and killed them using explosive harpoons shot from harpoon guns. This whaling was also widely distributed and took a large number of sperms (up to 30,000 per year). However, unlike their earlier counterparts, the modern whalers used almost all of the whale, and preferentially targeted males. Much, but not all, of this whaling was carried out under the auspices of the INTERNATIONAL WHALING COMMISSION (IWC). The effects of 20th century whaling on sperm whale stocks are unclear, but it is likely that many stocks, particularly the male portions, were reduced substantially. Commercial sperm whaling declined in the 1970s and 1980s and virtually ceased with the IWC moratorium in 1988. Currently there is a small catch of sperm whales in Lamalera, Indonesia, using primitive methods, a "scientific" hunt for sperm whales off Japan, as well as some evidence for pirate sperm whaling.

Nowadays most interactions between sperm whales and humans are more benign. Sperm whales are the principal subjects of WHALE-WATCHING operations off Kaikoura (New Zealand),

Andenes (Norway), and Dominica (West Indies). Scientists study their behavior from small vessels in these and a few other locations.

XVI. Conservation

It may seem that the sperm whale has survived the onslaught of the whalers better than most other large whales. There are probably still a few hundred thousand sperm whales left in the ocean, sperm whale food is of little interest to human fishers, and their deep water home is further from most sources of pollution than the preferred habitat of most other marine mammals. However, the effects of whaling seem to be lingering. In the southeast Pacific, where modern whaling on males was particularly severe, large breeding males are still scarce and calving seems depressed below the replacement rate. In other parts of the world, the picture appears brighter. For instance, the population of maturing males in the Antarctic seems to be increasing. However, sperm whales have very low reproductive rates even in the best of times—they do not recover quickly when depleted. They are also killed inadvertently in a range of ways, including ENTRAPMENT in fishing gear, choking on plastic bags, and collision with ships. The chemical POLLUTION levels in their blubber are generally higher than that of baleen whales, but lower than in inshore odontocetes.

XVII. The Colossal Convergence

A potential key to the sperm whale's simultaneous possession of a wide range of extreme biological attributes lies in a series of remarkable parallels with the African elephant (*Loxodonta africana*). These include large body sizes, brain sizes, substantial sexual dimorphism, similar life history variables, large ranges, remarkably congruent matrilineally based social

Figure 8 *Open-boat sperm whaling. From Beale (1839).*

systems, and breeding systems in which males roam between groups of females and generally only mate successfully when in their late twenties or older. The highly evolved spermaceti organ may have paralleled the trunk (another extreme nose) in allowing the animals efficient access to a wide range of resources. Meanwhile, in these animals, large sizes and cooperative societies gave efficient defense against predators, allowing long lives. Long, safe lives, in turn, promote the formation of significant long-term relationships among animals. Thus elephants and sperm whales evolved in highly social populations, near carrying capacity, and become dominant members of their ecological guilds. However, such animals are not well adapted to recovering from sudden depletion. Both the elephant and the sperm whale have been heavily hit, and, because their social structures are so important, exploitation has had consequences beyond animals killed directly.

See Also the Following Articles

Cetacean Ecology ▪ Spermaceti ▪ Toothed Whales, Overview ▪ Whaling, Traditional

References

Beale, T. (1839). "The Natural History of the Sperm Whale." John van Voorst, London.

Berzin, A. A. (1971). "The Sperm Whale." Pacific Scientific Research Institute of Fisheries and Oceanography, Moscow.

Best, P. B. (1979). Social organization in sperm whales, *Physeter macrocephalus*. In "Behavior of Marine Animals" (H. E. Winn and B. L. Olla, eds.), Vol. 3, pp. 227–289. Plenum Press, New York.

Best, P. B., Canham, P. A. S., and Macleod, N. (1984). Patterns of reproduction in sperm whales, *Physeter macrocephalus*. *Rep. Intl. Whal. Comm.* (Special Issue) **6**, 51–79.

Caldwell, D. K., Caldwell, M. C., and Rice, D. W. (1966). Behavior of the sperm whale *Physeter catodon* L. In "Whales, Dolphins and Porpoises" (K. S. Norris, ed.), pp. 677–717. University of California Press, Berkeley.

Clarke, M. R. (1980). Cephalopoda in the diet of sperm whales of the southern hemisphere and their bearing on sperm whale biology. *Dis. Rep.* **37**, 1–324.

Cranford, T. W. (1999). The sperm whale's nose: Sexual selection on a grand scale. *Mar. Mamm. Sci.* **15**, 1133–1157.

Dufault, S., Whitehead, H., and Dillon, M. (1999). An examination of the current knowledge on the stock structure of sperm whales (*Physeter macrocephalus*) worldwide. *J. Cetacean Res. Manag.* **1**, 1–10.

Ellis, R. (1980). "The Book of Whales." Knopf, New York.

Goold, J. C., and Jones, S. E. (1995). Time and frequency domain characteristics of sperm whale clicks. *J. Acoust. Soc. Am.* **98**, 1279–1291.

Gordon, J. C. D. (1987). Sperm whale groups and social behaviour observed off Sri Lanka. *Rep. Intl. Whal. Comm.* **37**, 205–217.

Heyning, J. E. (1997). Sperm whale phylogeny revisited: Analysis of the morphological evidence. *Mar. Mamm. Sci.* **13**, 596–613.

Jaquet, N. (1996). How spatial and temporal scales influence understanding of sperm whale distribution: A review. *Mamm. Rev.* **26**, 51–65.

Kawakami, T. (1980). A review of sperm whale food. *Sci. Rep. Whales Res. Instit. Tokyo* **32**, 199–218.

Mchl, B., Wahlberg, M., Madsen, P. T., Miller, L. A., and Surlykke, A. (2000). Sperm whales clicks: Directionality and source level revisited. *J. Acoust. Soc. Am.* **107**, 638–648.

Melville, H. (1851). "Moby Dick or the Whale." Harper and Brothers, New York.

Milinkovitch, M. C., Guillermo, O., and Meyer, A. (1993). Revised phylogeny of whales suggested by mitochondrial ribosomal DNA sequences. *Nature* **361**, 346–348.

Pitman, R. L., and Chivers, S. J. (1999). Terror in black and white. *Nat. Hist.* **107**, 26–29.

Reeves, R. R., and Whitehead, H. (1997). Status of the sperm whale (*Physeter macrocephalus*) in Canada. *Can. Field Nat.* **111**, 293–307.

Rice, D. W. (1989). Sperm whale. *Physeter macrocephalus* Linnaeus, 1758. *In* "Handbook of Marine Mammals" (S. H. Ridgway, and R. Harrison, eds.), Vol. 4, pp. 177–233. Academic Press, London.

Watkins, W. A. (1980). Acoustics and the behavior of sperm whales. *In* "Animal Sonar Systems" (R. Busnel and J. F. Fish, eds.), pp. 291–297. Plenum Press, New York.

Weilgart, L., Whitehead, H., and Payne, K. (1996). A colossal convergence. *Am. Sci.* **84**, 278–287.

Whitehead, H., Dillon, M., Dufault, S., Weilgart, L., and Wright, J. (1998). Non-geographically based population structure of South Pacific sperm whales: Dialects, fluke-markings and genetics. *J. Anim. Ecol.* **67**, 253–262.

Whitehead, H., and Weilgart, L. (2000). The sperm whale: Social females and roving males. *In* "Cetacean Societies" (J. Mann, R. C. Connor, P. Tyack, and H. Whitehead, eds.), pp. 154–172. University of Chicago Press, Chicago.

Sperm Whales, Evolution

GURAM A. MCHEDLIDZE
Georgian Academy of Sciences, Tbilisi

Sperm whales include the cetacean families Physeteridae (the modern sperm whale) and Kogiidae (the modern pygmy and dwarf sperm whales). Modern sperm whales are represented by two genera only, but sperm whales were much more diverse in past times. Phylogenetically, sperm whales are usually thought to be close to the root of the odontocetes, and they retain many characters that are primitive for odontocetes. They do, however, also have some highly derived features and are not very similar to the primitive Eocene cetaceans (archaeocetes) that they are descended from. Physeterids and kogiids may be subsequent branches of the basal odontocete phylogenetic tree, or both may be derived from a common sperm whale ancestor, such as *Ferecetotherium* (Table I).

I. Physeteridae

Physeterids (including the modern sperm whale *Physeter macrocephalus*) are highly specialized for teutophagy at great depths, and many parts of sperm whale anatomy show adaptations for this behavior. Some specialized morphologies are already present in the oldest known physeterid, *Ferecetotherium kelloggi*, from the late Oligocene of Azerbajan, but many fossil sperm whales were probably fisheaters.

The earliest physeterid, *Ferecetotherium*, was small (approximately 5 m long) and had a small head. An increase in size and head size occurred in the Miocene, and modern *Physeter* has a body length of 16 m in females and 21 m in males. The

TABLE I
Better Known Genera of Physeteridae and Kogiidae

Physeteridae

Ferecetotherium	Late Oligocene	Azerbajan
Diaphorocetus	Early Miocene	Argentina
Aulophyseter	Middle Miocene	United States, west coast
Idiorophus	Middle Miocene	United States, west coast
Orycterocetus	Middle Miocene	United States, east coast
Placoziphius	Upper Miocene	Europe
Scaldicetus	Middle Miocene	Europe, Japan
Physeter	Recent	All oceans, except polar

Kogiidae

Kogiopsis	Late Miocene	United States
Scaphokogia	Miocene	Peru
Praekogia	Pliocene	Mexico
Kogia	Recent	All temporal to tropical oceans

head of *P. macrocephalus* is approximately one-third the size of the body. An increase in body size happened throughout physeterid evolution (Fig. 1).

Ferecetotherium had a relatively narrow rostrum. Miocene sperm whales are characterized by the widening of the rostrum. This widening took different paths in different species. In some clades, widening occurred in the maxillae and premaxillae, whereas in other clades only one of these elements enlarged. In the Miocene *Diaphorocetus* and *Orycterocetus,* the premaxilla and maxilla are nearly equal in width near the base of the rostrum, but the tip of the rostrum consists entirely of the premaxillae. In contrast, in *Physeter* the maxillae make up nearly all of the rostrum, and the premaxillae are only exposed near the tip of the rostrum. Widening of the rostrum was probably the result of the enlargement of the SPERMACETI organ, a large structure housed in a depression (the supracranial basin) on the forehead. The supracranial basin is characteristic of sperm whales, although it is relatively small in older forms. Posteriorly, this basin is bounded by the supraoccipital and the posterior plate of the right premaxilla, and laterally vertical plates of the frontals bound the basin. In modern *Physeter,* the lateral walls of the supraoccipital basin are formed by the maxillae. All physeterids, including *Ferecetotherium,* are also characterized by strong asymmetry of the rostrum, particularly the premaxillae and nasals.

Large evolutionary changes also occurred in the mandible. In modern sperm whales, the maxilla (upper jaw) is much wider than the mandible as a result of the widening of the rostrum and a narrowing of the mandible. These give the lower jaw a peculiar undersized image. This mismatch evolved gradually in physeterids. *Ferecetotherium* has a primitive mandible, not unlike that of archaeocetes. In late Eocene archaeocetes, the lower margin of the mandible is horizontal, but the mandibular depth increases posteriorly and TEETH are present on the ascending ramus. Physeterid specializations in the mandible of *Ferecetotherium* include the parallel edges of the ramus, the superior displacement of the mandibular condyle, and the reduced coronoid process.

As a rule, modern *Physeter* only has teeth in the lower jaw, although occasionally an individual is found with upper teeth. Upper teeth are still present in physeterids from Oligocene through middle Miocene. Teeth in *Physeter* are positioned in the upper edge of the mandible. In *Ferecetotherium,* posterior teeth are rooted in the upper edge of the mandible, but anterior teeth (except the first tooth) are implanted more laterally. The latter condition occurs in all teeth of *Kogia.* All physeterids have homodont and polydont teeth, as do most extant odontocetes. The tooth crowns of physeterids are small, and their roots are more shallow than those of archaeocetes.

In physeterid evolution, the length of mandibular symphysis increased from one-twelfth of the length of the mandible in the Oligocene to one-fourth in the Miocene, and one-half in *Physeter.* The mandible of *Ferecetotherium* has 27 teeth, whereas that of *Diaphorocetus* has 14. In the middle Miocene *Aulophyseter morricei,* upper teeth, if present at all, were only lodged in the gums, and all upper teeth were lost in the upper Miocene *Placoziphius duboisii.*

The great differences in the shape of the face and jaws between early physeterids and modern ones suggest that the teutophagic specializations of *Physeter* may not have been present in extinct relatives. It is possible that Oligocene and Miocene physeterids were ichthyophagous, as were most cetaceans. Modern sperm whales feed on deep-sea squid, and it is likely that early physeterids were not deep-sea animals.

The forelimb of physeterids is very different from that of Eocene cetaceans, but similar to that of other odontocetes. The humerus is shortened and the elbow is immobile. The hand forms a flat, smooth surface with no differentiation of individual fingers, causing the entire limb to be an effective rudder. In early physeterids, the head of the humerus changed considerably. The humeral head of *Ferecetotherium* is deviated externally and faces caudally, but the tubercles are situated anteriorly, as in archaic cetaceans. Whereas *Ferecetotherium* retains a greater and lesser tublercle of the humerus, the lesser tubercle of Miocene physe-

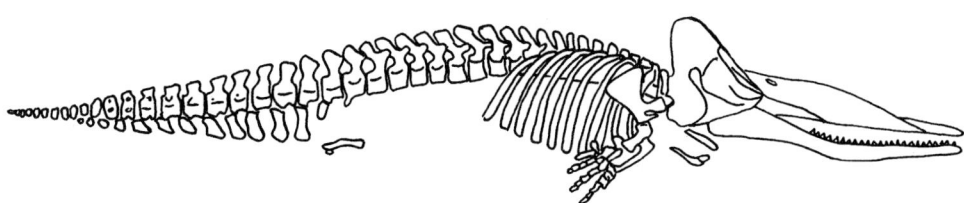

Figure 1 *Skeleton of* Physeter macrocephalus, *the modern sperm whale. Note the disproportionally large head and the supracranial basin.*

terids is enlarged greatly and the greater tublercle reduced. *Physeter* has a weak lesser tubercle, set medially on the humerus.

II. Kogiidae

One genus with two modern species constitutes the Kogiidae (*Kogia breviceps* and *K. sima*). *Kogia* is similar to *Physeter*, but the body of *Kogia* is much smaller (body length approximately 4 m) and the head is smaller (one-sixth to one-eighth of the body). The spermaceti organ is smaller than in *Physeter*, the blowhole more posterior, and the rostrum shorter. Proportions of the head in *Physeter* embryos are similar to those of adult *Kogia*. This suggests that these cetaceans are closely related and that *Physeter* has a more derived facial morphology.

Kogiids are poorly represented in the FOSSIL RECORD, and most specimens are incomplete. The genus *Kogiopsis* is known from a single mandible, and fragmentary SKULLS are known for Miocene *Scaphokogia* and Pliocene *Praekogia*. Another trend in the evolution of kogiids is the reduction of dental enamel. This trend started in the Miocene. In modern *Kogia* some enamel covers the tips of the teeth, whereas Miocene kogiids lack all enamel.

Phylogenetically, *Scaphokogia* is a basal branch of kogiids, retaining primitive morphologies of rostrum, premaxillae, and intermaxillary groove. It is, however, more derived than other kogiids in having a well-developed supracranial basin. *Scaphokogia* may represent an early, specialized branch of kogiids, the subfamily Scaphokogiidae. These went extinct near the end of the Miocene. *Praekogia* is closely related to *Kogia* in that both the nasal passages are anterior due to poorly developed telescoping.

See Also the Following Articles

Cetacean Evolution ■ Dental Morphology, Evolution of ■ Fossil Record ■ Pygmy and Dwarf Sperm Whales

References

Mchedlidze, G. A. (1984). "General Features of the Paleobiological Evolution of Cetacea." Amerind Publ. Corp. (New Delhi) and Smithsonian Institution Libraries.

Spinner Dolphin
Stenella longirostris

WILLIAM F. PERRIN

*Southwest Fisheries Science Center,
La Jolla, California*

The spinner dolphin, *Stenella longirostris* (Gray, 1828), is the most common small cetacean in most tropical pelagic waters. It can be seen at a great distance as it spins high in the air and lands in the water with a great splash.

It also commonly rides the bow wave of boats and thus can be observed closely.

I. Characters and Taxonomic Relationships

The spinner dolphin can be identified externally by its relatively long slender beak, color pattern, and dorsal fin (Fig. 1). The color pattern in most regions is three-part, consisting of a dark-gray cape, light-gray lateral field, and white ventral field. A dark band of even width runs from the eye to the flipper, bordered above by a thin light line. The rostrum is tipped with black or dark gray. In the eastern and Central American subspecies of the eastern tropical Pacific, *S. l. orientalis* and *S. l. centroamericana* (Fig. 2), contrast between the cape and the lateral field is very faint to absent, and the ventral field may be restricted to discontinuous axillary and genital-region patches. Various intermediate color patterns are exhibited by "whitebelly" spinners in a broad zone of HYBRIDIZATION between the eastern form and the more typically patterned Gray's spinner dolphin, *S. l. longirostris*, in the Central and South Pacific. The dorsal fin in adults in all regions is basically triangular, varying from a slightly falcate right triangle to an erect isoceles triangle. In the adult male of the eastern and Central American subspecies the dorsal fin may lean slightly forward, appearing to "be on backwards." This is correlated with the presence of a large postanal ventral hump. Both the dorsal fin and the hump appear to be sexual displays, probably important in male–male or male–female communication or both. In calves and juveniles in all regions, the dorsal fin is on average more falcate than in adults.

Sexually mature adults examined ranged from 129 to 235 cm ($n = 1824$) and weighed 23–78 kg ($n = 33$) (Perrin, 1998). Males are on average slightly larger than females in body size and most skull characters.

The spinner dolphin may be confused in the tropical Atlantic with the endemic Clymene dolphin, *S. clymene*, which is very similar in appearance and has been observed to spin, although not as acrobatically as the spinner dolphin. In the Clymene dolphin, the beak is relatively shorter, the flipper band narrows anteriorly, the lower margin of the cape dips lower toward the ventral field (in the spinner the cape and ventral-field margins are parallel), and there is usually a dark "moustache" mark on the upper side of the beak that has only rarely been observed in spinner dolphins.

The skull (Fig. 3) can be confused with those of *S. coeruleoalba*, *S. clymene*, and *Delphinus* spp.; all have a relatively long and narrow dorsoventrally flattened rostrum, a large number of small slender teeth (about 40–60 in each row), and medially convergent premaxillae and sigmoid ramus (Perrin, 1998). It differs from *Delphinus* in lacking strongly defined palatal grooves. The rostrum is narrower at the base than in *S. coeruleoalba* (57–84 mm vs 93–120 mm). It overlaps *S. clymene* in all skull measurements and tooth counts, but the skull (335–464 mm long in 112 adults) is proportionately longer and narrower. The vertebral count is 69–77 ($n = 90$).

The spinner dolphin is a member of the subfamily Delphininae *sensu stricto* (LeDuc *et al.*, 1999). Morphologically and behaviorally it is most similar to the Clymene dolphin, but in a cladistic phylogenetic analysis based on cytochrome *b* mtDNA sequences it was not closely linked to that species,

Figure 1 *A Gray's spinner dolphin* (Stenella longirostris longirostris) *in the Gulf of Mexico. Photograph by Robert L. Pitman.*

which was the sister species of the striped dolphin, *Stenella coeruleoalba* (LeDuc *et al.*, 1999). The morphological similarly between spinner and Clymene dolphins could be an instance of plesiomorphy (similarity through retention of primitive characters), but the correlated spinning behavior would seem to be highly derived, a probable synapomorphy. This puzzling discordance between molecular and morphological/behavioral characters shows the need for further molecular and morphological investigation; a hybrid origin of *S. clymene* has been suggested as one possible explanation (LeDuc *et al.*, 1999).

II. Distribution and Ecology

The spinner dolphin is pantropical, occurring in all tropical and most subtropical waters around the world between roughly 30–40°N and 20–30°S (Jefferson *et al.*, 1993). It is typically thought of as a high-seas species, but coastal populations and races/subspecies exist in the eastern Pacific, Indian Ocean, Southeast Asia, and likely elsewhere (Perrin, 1998; Perrin *et al.*, 1999; Perryman and Westlake, 1998; Van Waerebeek *et al.*, 1999). In the eastern and western Pacific, the pelagic form has been shown to prey mainly on small mesopelagic fishes and squids, diving to 600 m or deeper (Perrin and Gilpatrick, 1994; Dolar, 1999), but a dwarf subspecies in inner Southeast Asia, *S. l. roseiventris*, inhabits shallow waters in the Gulf of Thailand, Timor Sea, and Arafura Sea and consumes mainly benthic and reef fishes and invertebrates (Perrin *et al.*, 1999).

Predators include sharks, probably killer whales (*Orcinus orca*) and possibly false killer whales (*Pseudorca crassidens*), pygmy killer whales (*Feresa attenuata*), and short-finned pilot

Figure 2 *The Central American spinner* (Stenella longirostris centroamericana) *lacks a bold pattern and has an extremely long beak. Photograph by Robert L. Pitman.*

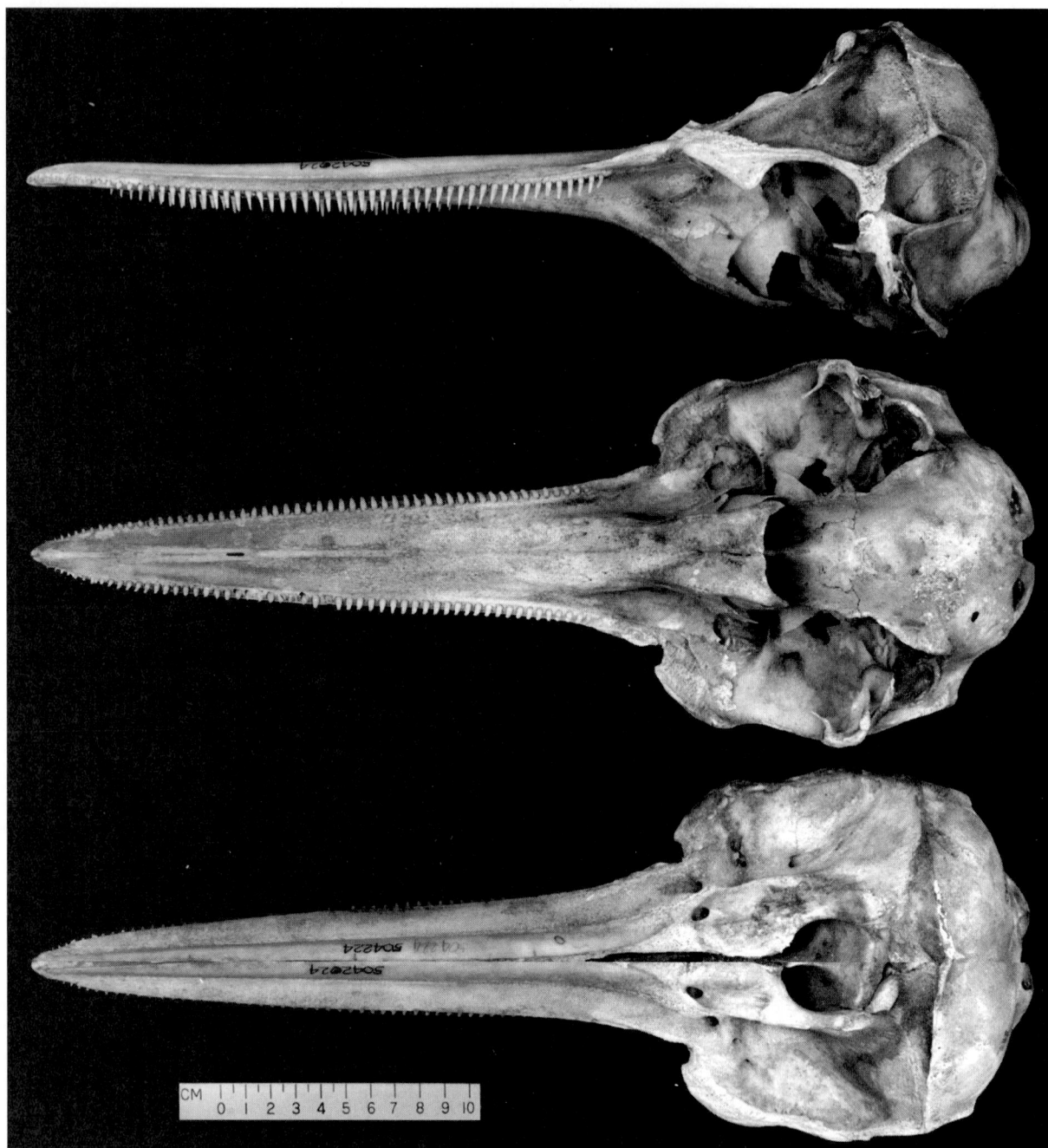

Figure 3 *Skull of* Stenella longirostris longirostris *adult male from Florida. From Perrin and Gilpatrick (1994).*

whales (*Globicephala macrorhynchus*). Parasites may cause direct or indirect mortality.

III. Behavior and Life History

Why the spinner spins is unknown. One animal may spin as many as 14 times in quick succession. It has been suggested that the large underwater bubble plume created by the violent spin and reentry may serve as an ECHOLOCATION target for communication across a widely dispersed school (Norris *et al.*, 1994). It is also probable that spinning is an outgrowth of an alert state and as such has a social facilitation function. It could also—at least at times—represent PLAY.

School size varies greatly, from just a few dolphins to a thousand or more. Spinner dolphins commonly school together with pantropical spotted dolphins, *S. attenuata*, and dolphins

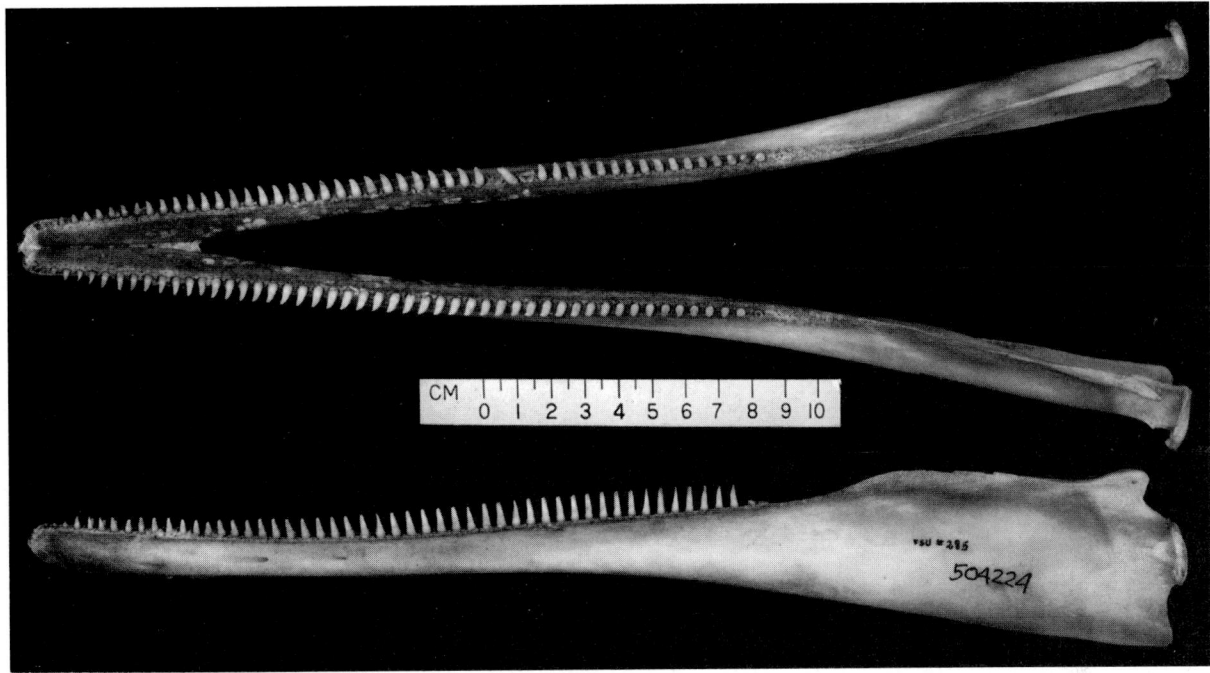

Figure 3   *(Continued)*

and small toothed whales of other species. Social organization in Hawaiian waters is fluid, with schools composed of more or less temporary associations of family units; the associations may vary over days or weeks (Norris *et al.,* 1994). Mating appears to be promiscuous. Adult males form coalitions of up to about a dozen individuals; the function of these is unknown. Maximum recorded movements of individuals are 113 km (over 1220 days) in Hawaii and 275 nmi (over 395 hr) in the eastern Pacific (Perrin, 1998).

Gestation is about 10 months. Average length at birth is about 75–80 cm. Length of nursing is 1 to 2 years. Calving interval is about 3 years. Females attain sexual maturity at 4–7 years and males at 7–10 years. Ovulation may be spontaneous. Breeding is seasonal, more sharply so in some regions than in others.

IV. Interactions with Humans

Spinner dolphins have been kept in CAPTIVITY only in Hawaii, where some have lived for several years. Large numbers have been killed incidentally since the early 1960s by tuna purse seiners in the eastern tropical Pacific; the population of *S. l. orientalis* is estimated to have been reduced to less than one-third of its original size (Perrin, 1998). Likely unsustainably high bycatches also occur in drift nets and purse seines in the Philippines (Dolar, 1994, 1999). As is the case for other small cetaceans caught in fishing nets, local human consumption of by-caught animals in several regions has led to the development of markets and large directed catches from unassessed populations (Perrin, 1999), boding ill for CONSERVATION of the species in these regions.

See Also the Following Articles

Clymene Dolphin ■ Delphinids, Overview ■ Leaping Behavior ■ Tuna–Dolphin Issue

References

Dolar, M. L. L. (1994). Incidental takes of small cetaceans in fisheries in Palawan, Central Visayas and northern Mindanao in the Philippines. *Rep. Int. Whal. Commn. Spec. Iss.* **15,** 355–363.

Dolar, M. L. L. (1999). "Abundance, Distribution and Feeding Ecology of Small Cetaceans in the Eastern Sulu Sea and Tañon Strait, Philippines," p. 241. Ph.D. dissertation, Univ. of California, San Diego.

Jefferson, T. A., Leatherwood, S., and Webber, M. A. (1993). "FAO Species Identification Guide: Marine Mammals of the World," p. 320. FAO, Rome.

LeDuc, R. G., Perrin, W. F., and Dizon, A. E. (1999). Phylogenetic relationships among the delphinid cetaceans based on full cytochrome *b* sequences. *Mar. Mamm. Sci.* **15,** 619–648.

Norris, K. S., Würsig, B., Wells, R. S., and Würsig, M. (1994). "The Hawaiian Spinner Dolphin." Univ. of California Press, Berkeley.

Perrin, W. F. (1998). *Stenella longirostris. Mammal. Species* **599,** 1–7.

Perrin, W. F. (1999). Selected examples of small cetaceans at risk. *In* "Conservation and Management of Marine Mammals" (J. R. Twiss, Jr. and J. Reynolds, eds.). Smithsonian Press, Washington, DC.

Perrin, W. F., Dolar, M. L. L., and Robineau, D. (1999). Spinner dolphins (*Stenella longirostris*) of the western Pacific and southeast Asia: Pelagic and shallow-water forms. *Mar. Mamm. Sci.* **15.**

Perrin, W. F., and Gilpatrick, J. W., Jr. (1994). Spinner dolphin *Stenella longirostris* (Gray, 1828). *In* "Handbook of Marine Mammals" (S. H. Ridgway and R. Harrison, eds.), Vol. 5, pp. 99–128. Academic Press, London.

Perryman, W. L., and Westlake, R. L. (1998). A new geographic form of the spinner dolphin, *Stenella longirostris*, detected with aerial photogrammetry. *Mar. Mamm. Sci.* **14,** 38–50.

Van Waerebeek, K., Baldwin, R., Gallagher, M., and Papastavrou, V. (1999). Morphology and distribution of the spinner dolphin, *Stenella longirostris*, rough-toothed dolphin, *Steno bredanensis,* and melon-headed dolphin, *Peponcephala electra,* from waters off the Sultanate of Oman. *J. Cet. Res. Mgt.* **1,** 167–177.

Spouting

SEE *Blowing*

Spyhopping

SEE *Leaping Behavior*

Steller's Sea Cow

Hydrodamalis gigas

PAUL K. ANDERSON

*University of Calgary,
Alberta, Canada*

DARYL P. DOMNING

Howard University, Washington, D.C.

A very large sirenian once grazed on algae along the shores of the cold North Pacific Ocean. Shipwrecked with Vitus Bering's expedition on Bering Island, the larger of the Kommandorskiye (Commander) Islands, George Wilhelm Steller (b. 1709–d. 1746), the only biologist to observe the species alive before its extinction ca. 1768, observed and described the sea cow and recorded his observations on external and internal anatomy, natural history, and hunting as seen during his enforced stay on the island between November 1741 and August 1742.

I. Characteristics and Systematic Relationships

Two other common names for Steller's sea cow are the German "Borkentier," referring to its rough bark-like hide, and the Russian "kapustnik" (cabbage eater), which Steller attributed to inhabitants of Kamchatka, where dead sea cows reportedly washed ashore after storms. Because Steller's posthumously published observations (1751) predated Linnean taxonomy, the sea cow was given its formal scientific name by later authors on the basis of Steller's description (no type specimen has ever been designated). Among numerous synonyms, the generic and species names now recognized as correct are *Hydrodamalis* (Retzius, 1794) and *Hydrodamalis gigas* (Zimmerman, 1780). The latter combination was authored by Palmer (1895). The generic synonym used most commonly in the 19th century was *Rytina* (Illiger, 1811) (improperly emended to *Rhytina*). For full synonymy and comprehensive references, see Domning (1978, 1996).

Hydrodamalis gigas was a very large toothless sirenian. Sirenians are wholly aquatic, primarily vegetarian, mammals. *H. gigas* was the only sirenian adapted to cool-temperate or cold climates and an algal diet. It differed from all other mammals in feeding predominantly or exclusively on algae and in having a manus composed only of carpals and metacarpals, with the phalanges vestigial or absent. Its pectoral limb had a densely bristled blunt termination that Steller described as claw-like or hoof-like (Fig. 1). It differed from its predecessor (*Hydrodamalis cuestae*) in lacking teeth and from most fossil and living sirenians other than manatees in having a reduced deflection of the rostrum from the occlusal plane (Fig. 2). This characteristic correlated with feeding in the water column above the bottom. Its uniquely large size may have been a factor in its ability to tolerate cold. Surface water temperatures around the Kommandorskiyes range from 0°C in winter to 10°C in summer.

Like other sirenians, sea cows had paired nostrils, abdominal testes, a pair of axillary mammae, no hindlimbs, horny plates on the occlusal surfaces of the rostrum and mandibular symphysis for macerating vegetation, and ribs and other bones that were swollen (pachyostotic) and completely compact (osteosclerotic). It swam by dorsoventral undulation of the body and its horizontally expanded triangular caudal fluke, resembling those of cetaceans. See Domning (1978, 1994) for more detailed morphological diagnoses.

The order Sirenia is agreed to be monophyletic and has been placed in the superorder Tethytheria, which includes the Proboscidea (elephants) as well as an extinct group of marine herbivores, the Desmostylia. The cladistic analysis by Domning (1994), based on morphological characters of skull, mandible, and dentition, supports the conclusion that *H. gigas* was the last of two species in the genus *Hydrodamalis*, sharing the subfamily Hydrodamalinae with its ancestral genus *Dusisiren*. The Hydrodamalinae, Dugonginae, and Halitheriinae constitute the family Dugongidae. The Sirenia also include three other families: the extant Trichechidae (manatees) and the extinct Prorastomidae and Protosirenidae.

II. Distribution

Historically, the sea cow was known only from Bering and Medney (Copper) Islands at the western terminus of the Aleutians, 200 miles to the west of Attu Island and 150 miles east of the Kamchatka Peninsula. Pliocene and Pleistocene fossils demonstrate that the distribution originally extended around the North Pacific from Baja California, Mexico (30°N), along the coast of North America to the Aleutian chain and south in the western Pacific to Honshu, Japan (37°N). No evidence has been adduced for GEOGRAPHIC VARIATION. The northernmost Pacific was likely marginal habitat as members of the Kommandorskiye population may have been stunted. Adults at Bering Island had a body length ca. 750 cm (25 ft.) and a weight of 4500–5900 kg (10,000–13,000 lbs.). Fossil material

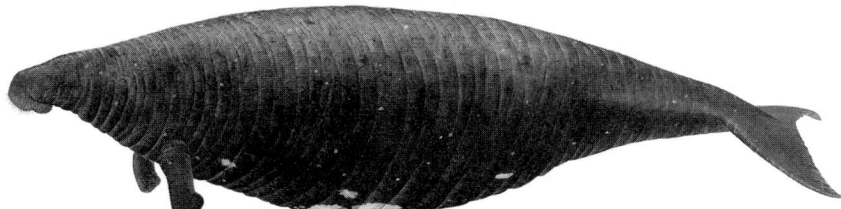

Figure 1 *A reconstruction of Steller's sea cow by Pieter Arend Folkens. Outstanding features are the relatively small head, bulky body, rough skin, whale-like flukes, and blunt unflipper-like pectoral appendages.*

from farther south suggests maximum body lengths may have reached 9–10 m.

III. Natural History and Behavior

Bering Island is approximately 50 miles in length and Medney Island 35 miles long. Both are narrow, high, and rocky. Sea cows foraged on softer parts of large marine algae ("kelp") on open coasts in rocky subtidal and intertidal waters. Steller referred to four types of macroalgae (all still undescribed in his day) as preferred foods; these have been variously identified but probably include such forms as *Agarum* spp., *Alaria praelonga*, *Halosaccion glandiforme*, *Laminaria saccharina*, *Nereocyctis luetkeana*, and *Thalassiophyllum clathrus*. Tougher stems and holdfasts were not eaten, washing up in heaps on the shore where sea cows fed. Seagrasses have also been suggested as elements of the diet, but this is doubtful. Of the two available forms, *Phyllospadix* spp. are among the toughest seagrasses and *Zostera marina* grows on soft bottoms in protected waters. In any case, *H. gigas* was probably dependent on a small number of forage species. Feeding along the shoreline, with backs above water, the sea cows were not observed to submerge; they moved about using their flukes and maintaining position on the rocks with their bristle-tipped pectoral limbs. Algae growing on rock faces were bitten off or pawed free with the forelimbs, and the softer parts were separated by the bristly lips as if "cut off with a dull knife." If the animals were indeed unable to sub-

merge, they could access food only to depths of about a meter below low tide level. Steller wrote that they fed "in herds," keeping the young in the center of the group. He did not specify herd size but referred to family groups as consisting of a pair with their offspring. In addition to rocky shorelines, Steller stated that sea cows were "fond of shallow sandy places along the seashore, but they like especially to live around the mouths of river and creeks, for they love fresh running water."

Sea cows had no fear of humans. Feeding with the head submerged and half of the back above water they were easily approached and could be touched or speared from the rocks. One method of capture was for a swimmer or a boat crew to approach the intended victim with a large hook attached to a line to the shore. When the hook had been driven into the animal's body, a shore crew would attempt to drag it toward the beach. Steller reported that sea cows made no sounds even when wounded by hunters. Of particular interest are the descriptions of both helping behavior and evident pair bonding in this context. Steller reported that while an animal was being dragged ashore other sea cows would attempt to dislodge the hook and/or break the rope. When one female had been dragged ashore, the accompanying male kept station offshore at the site for at least 2 days.

On the basis of his observation of "family groups" and the evidence of strong pair bonds, Steller suspected that the animals were monogamous. Mating apparently involved only a pair, with the male following or herding the female during prolonged

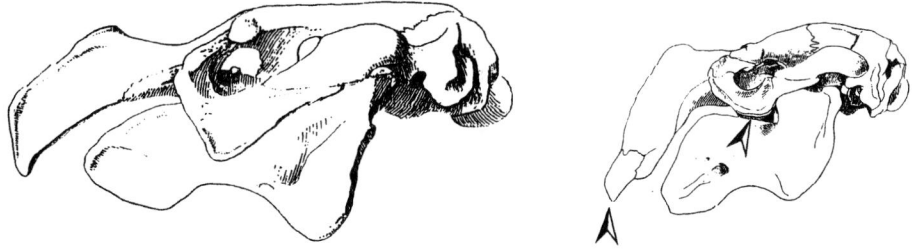

Figure 2 *Skulls of the two recent dugongids drawn to the same scale. The* Hydrodamalis gigas *skull (left) shows the lack of teeth and the relatively undeflected rostrum compared with the downwardly deflected rostrum of* Dugong dugon *(right). The dugong's tusks and molar teeth are indicated by arrows. The sea cow's head was relatively small compared with its body and that of the dugong relatively large so that the size of the two skulls does not reflect relative body size.*

"amorous preludes." Copulation involved mutual clasping with the pectorals, the female in an inverted position and the male above.

Steller's views as to the annual reproductive cycle present some enigmas. He observed mating behavior and copulation in the early spring, and deduced that gestation was more than 12 months, but stated that young were born "at any time of the year, but most frequently in autumn." Fall births would seem likely to put the young at a disadvantage, especially in view of his observations that "during the winter they are often suffocated by the ice that floats about the shore" and "in the winter the animals become so thin that, besides the bones of the spine, all the ribs show." As Steller did not arrive on the island until late November and left the following August, he could not have had direct knowledge of a concentration of births in the fall. He never saw more than one very small calf with a female, implying that births were single.

As an algivore, *Hydrodamalis* may have been dependent on an ecosystem in which the carnivorous sea otter (*Enhydra lutris*) suppressed invertebrate herbivory in shallow coastal waters. Like the sea cow, the sea otter is confined to the shallows within 1 km of shore where it forages for urchins and other invertebrate herbivores, usually to a maximum depth of about 130 ft. (40 m). Within this zone, otters keep sea urchins and other invertebrate algivores in check. *Hydrodamalis* was dependent on fast-growing kelps on rock faces and on the kelp canopy in deeper waters. Where sea otters are extirpated, grazing by urchins results in "kelp barrens." Thus sea cows depended on sea otter predation that maintained the shallow-water kelp community.

Steller reported both external and internal PARASITES. The crustaceans that he described as infesting the sea cows' skin have been interpreted as cyamids or caprellids, but identification will probably remain uncertain. The only specimens purported to exist are ones attached to alleged sea cow skin fragments, but these latter are likely to pertain instead to whales. "White worms half a foot long," observed by Steller in the stomach and intestine of the sea cow, were probably ascarid nematodes; both these and the ectoparasites probably went extinct with the sea cow itself. *Hydrodamalis*, at least the young, may occasionally have been prey for killer whales (*Orcinus orca*) and large sharks.

IV. Evolutionary Origins

Hydrodamalis evolved from tropical and subtropical, seagrass-eating dugongid ancestors that shifted from bottom feeding to surface feeding and from a seagrass in warm protected waters to an algal diet and a high-energy, low-temperature environment. Seagrasses, which have a large part of their biomass below ground in the form of roots and rhizomatous storage organs, are the preferred forage of the extant dugong (*Dugong dugon*) and apparently of the ancestral dugongid line. Kelps have no below-ground biomass. Their softer, more edible, growing portions (the bulk of nutritious biomass of the taller kelps) are at or near the surface, suspended by floats. Fossils record this transition from bottom rooting to surface foraging. *Hydrodamalis* descended from the widely distributed tropical genus *Metaxytherium*, a small-tusked,

seagrass-eating Miocene halitherine dugongid. At the time a diversity of herbivorous marine mammals foraging along the east-central Pacific shorelines was presumably supported by a diverse flora of tropical and subtropical seagrasses, as well as algae (perhaps including kelps in cooler, more exposed microhabitats). The sister group and apparent immediate ancestor of hydrodamalines was the species *M. arctodites*, living in southern California and Baja California by about 14–15 Ma. Probably already present were the earliest members of its descendant genus *Dusisiren*, as well as a dugongine, *Dioplotherium allisoni*. These three sympatric sirenians, representing the three dugongid subfamilies, probably occupied distinct feeding niches. The large-tusked *Dioplotherium* with its strongly deflected rostrum was likely a bottom feeder, rooting for rhizomes of the larger seagrasses. The small-tusked *Metaxytherium* probably depended on the smaller rhizomes and leaves of seagrasses. *Dusisiren* and later hydrodamalines, in contrast, progressively gave up bottom and rhizome feeding, specializing on kelps growing higher in the water column, reducing their rostral deflections, and losing their tusks. Also present with these Miocene sirenians were three or more genera of desmostylians, hippopotamus-like bottom feeders, presumably also herbivorous and feeding on intertidal and subtidal seagrasses and kelps.

As the climate cooled after the Middle Miocene, and tectonic uplift rendered the western North American coastline more emergent (replacing protected embayments with more exposed, higher-energy habitats), the cold-water plants increased their dominance in the marine flora, and the tropical seagrasses finally disappeared altogether. With them went *Dioplotherium*, *Metaxytherium*, and all the desmostylians. By the Late Miocene, *Dusisiren*, having earlier thrown in its lot with the kelps, had increased its cold tolerance and its morphological specialization for kelp eating (passing through the successive evolutionary stages *D. jordani* and *D. dewana*) and had extended its range to the northwestern Pacific. By the end of the Miocene it had evolved into *Hydrodamalis cuestae*, which gave rise in the Pleistocene to *H. gigas*.

The prehistoric distribution of sea cows and sea otters (*E. lutris*) was coterminous. Anderson (1995) speculated on the evolution of a sea cow–sea otter relationship and sea cow morphology as follows: "As the hydrodamaline range shifted northwards, the adoption of a kelp diet may have contributed (along with the thermoregulatory demands of a cooling environment) to selection for large adult body size, high birth weight, prolonged parental care, and low reproductive rate. At the same time, the more effective chemical defenses of kelps growing below the sea otter foraging zone might have favored the sea cow's specialization for surface feeding and its loss of diving ability."

V. Extinction

The vulnerability of the sea cows in their last stronghold made it possible for the weak and scurvy-ridden castaways of Bering's expedition to secure an abundant supply of food and to escape to the mainland in the summer of 1742. Having found ways to capture the sea cows, the survivors were able to divert manpower to salvage materials from the wreck of their ship and build a smaller vessel in which to reach Kamchatka. That voyage, made possible by sea cow vulnerability, carried with it the news of fortunes to be made in hunting sea otters and fur seals. A fur rush followed, fueled by sea cow meat. The first hunting expedition wintered on Bering Island in 1743–1744. By 1763 several parties had spent up to 9 months on the islands, living almost exclusively on sea cows and salting down barrels of meat to provision the 2- to 3-year expeditions to the Aleutian chain and the north Asian and North American coasts in search of furs. Sea cow hides were also used to make large skin boats (baidarkas or umiaks). In 1754–1755 a fur-hunting expedition was forced to winter on Bering Island because the sea cows had been extirpated on Medney Island. The last specimen on Bering Island was reported killed in 1766. Steller's biographer, Leonard Stejneger, summarized accounts of the unrestrained killing by the fur hunters and attributed the extinction of the sea cows to ruthless slaughter.

These magnificent animals would have been at least as vulnerable to healthy and skilled Pleistocene hunters along mainland shores as they were to the weakened Russians on the Kommandorskiyes. Domning (1978) has suggested that human hunting may have extirpated sea cows from areas within reach of aboriginal hunters, leaving only the population around the remote Kommandorskiyes. Over most of its range the sea cow may have been the only marine mammal to succumb to "Pleistocene overkill." The final drama played out on Bering and Medney Islands suggests a complementary and more complex and instructive story. Anderson (1995), following a suggestion by Delphine Haley, proposed that the final extinction of the sea cow resulted not solely from ruthless harvesting, but from a cascade of events beginning with extirpation of the local sea otter population around the islands in the first rush for furs. Decimation of the otters in all probability triggered a sea urchin population explosion and the disappearance of chemically undefended shallow water kelps that were the sea cows' main food supply. Invasion of the shallow waters by chemically defended deep-water kelps left the hunted remnant of the sea cow population with kelp that was likely toxic. Anderson proposed that this may have been a reenactment of events when North Pacific coastlines were first colonized by humans in the waning Pleistocene.

See Also the Following Articles

Extinctions, Specific ■ Otters ■ Sirenian Evolution ■ Tethytheria

References

Anderson, P. K. (1995). Competition, predation, and the evolution and extinction of Steller's sea cow, *Hydrodamalis gigas*. *Mar. Mamm. Sci.* **11**, 391–394.

Domning, D. P. (1978). Sirenian evolution in the North Pacific Ocean. University of California Publ. Geol. Sci. **118**, xi + 176.

Domning, D. P. (1994). A phylogenetic analysis of the Sirenia. *Proc. San Diego Soc. Nat. Hist.* **29**, 177–189.

Domning, D. P. (1996). Bibliography and index of the Sirenia and Desmostylia. *Smithsonian Cont. Paleobiol.* **80**, iii + 611.

Golder, F. A. (1922–25). Bering's voyages: An account of the efforts of the Russians to determine the relation of Asia and America. *Am. Geog. Soc. Res. Ser.* Nos. 1 and No. 2.

Haley, D. (1980). The great northern sea cow. *Oceans* **13**, 7–9.

Scheffer, V. B. (1972). The last days of the sea cow. *Smithsonian* **3**, 64–67.

Stejneger, L. (1887). How the great northern sea-cow (*Rytina*) became exterminated. *Am. Nat.* **21**, 1047–1054.

Stejneger, L. (1936). "Georg Wilhelm Steller: The Pioneer of Alaskan Natural History." Harvard Univ. Press, Cambridge.

Steller, G. W. (1751). The beasts of the sea. *In* "The Fur Seals and Fur Seal Islands of the North Pacific Ocean" (D. S. Jordan, ed.), pp. 179–218. U.S. Government Printing Office, Washington, DC [Translated by W. Miller and J. E. Miller].

Steller's Sea Lion
Eumetopias jubatus

Thomas R. Loughlin
National Marine Mammal Laboratory, Seattle, Washington

The Steller's [or Steller] sea lion (or northern sea lion; *Eumetopias jubatus*) is the largest otariid pinniped and one of the more aesthetically appealing sea lions. In Greek, *Eumetopias* means having a well-developed, broad forehead; in Latin, *jubatus* means having a mane, as in the male. Called *qawax* (pronounced ka-wa; *qawan*, plural) by Aleut natives and *sivuch* (*sivuchi*, plural) in Russian, Steller's sea lions exhibit significant sexual dimorphism with males being larger (Fig. 1). The average standard length of males is 282 cm and of females is 228 cm (maximum of about 325 and 290 cm, respectively). Estimated average weights of males and females are 566 and 263 kg, respectively (maximum of about 1120 and 350 kg, respectively). Pup weight at birth is 16–23 kg and may be slightly larger in the western part of their range. Taxonomically they belong to the family Otariidae, of the order Carnivora, and can be distinguished by a conspicuous diastema between the upper fourth and fifth postcanine teeth. Their range overlaps with the California sea lion (*Zalophus californianus*), which is darker color and smaller in size.

It is an attractive animal. Pups are born with a wavy, chocolate brown fur that molts after about 3 months of age. Adult fur color varies between a light buff to reddish brown with most of the underparts and flippers a dark brown to black. Both sexes become blonder with age. Adult males have long, coarse HAIR on the chest, neck, and shoulders, which are massive.

I. Distribution

The Steller's sea lion was first described by the German physician/theologian George Wilhelm Steller based on a specimen that he obtained from the Russian Commander Islands while serving as naturalist on Vitus Bering's fateful voyage to Alaska in 1741–1742. It was also during this voyage that Steller described the northern fur seal (*Callorhinus ursinus*) and Steller's sea cow (*Hydrodamalis gigas*, now extinct).

Steller's sea lions occur through the north Pacific Ocean rim from Japan to southern California (Fig. 2). They abound on nu-

Figure 1 *An* Eumetopias jubatus *rookery showing pups (foreground) and adult females and males. Note the sexual dimorphism between the larger male and the smaller female. (NMFS photograph).*

merous breeding sites (rookeries) in the Russian Far East, Alaska, and British Columbia with lower numbers in Oregon and California. Washington is the only western coastal state that does not contain a Steller sea lion rookery, although pups were observed at one haul-out site in 1997 and 1998. Seal Rocks in Prince William Sound, Alaska, is the northernmost (60°09′N) rookery and Año Nuevo Island, California, the southernmost (37°06′N). Unlike their more gregarious cousin the California sea lion, Steller's sea lions tend to avoid people and prefer isolated offshore rocks and islands to breed and rest. Although rookeries and rest sites occur in many areas, the locations used are specific and change little from year to year. Steller's sea lions tend to return to their birth island as adults to breed, but they range widely during their first few years and during the nonbreeding season.

Steller's sea lion population numbers have declined since the 1980s by over 90% in most of Alaska and southern California, whereas populations in Oregon, British Columbia, and southeastern Alaska have remained stable or increased slightly (Fig. 3). The worse declines occurred in the Aleutian Islands and Gulf of Alaska, areas that historically were centers of abundance. Causes for these declines have not been identified but may be related to disease, reduced food availability through natural changes in the ocean or by fishing pressure, and the synergistic effects of commercial fishing and climate change. Studies of blood chemistry and body condition of females and pups during the breeding season are within normal ranges for pinnipeds, but similar studies of juveniles, the age class most likely suffering higher mortality rates, have not been accomplished. The large population declines resulted in the species being listed as "threatened" by the United States in 1992. A survey in 1989 provided a total estimate of about 116,000 Steller's sea lions range wide, and another in 1994 resulted in about 100,000 of which about 9% of the total were in Russia, 70% in Alaska, 12%

in British Columbia, 8% in Oregon and California, and 1% in Washington. The decline continued throughout the 1990s.

II. Geographical Variation

Studies of mitochondrial DNA from Steller's sea lions throughout the range suggest that at least two stocks exist: an eastern stock (California through southeastern Alaska) and a western stock (Prince William Sound, Alaska, and westward). Continuing population declines resulted in the western population being listed as "endangered" in 1997; the eastern population remains listed as threatened. The Steller's sea lion has 30 metacentric or submetacentric chromosomes and 4 acrocentric chromosomes; the X chromosome is submetacentric and the Y is acrocentric ($2n = 36$).

III. Ecology

Steller's sea lions eat a variety of fishes and invertebrates. In Alaska, principal prey are walleye pollock (*Theragra chalcogramma*), Pacific cod (*Gadus macrocephalus*), Atka mackerel (*Pleurogrammus monopterygius*), octopus, squid, herring (*Clupea harengus*), flatfishes, sculpins, and a wide variety of other fishes and invertebrates. At specific times of the year other prey may be eaten when plentiful (e.g., Pacific salmon, *Onchorynchus* spp.). During the breeding season, females with pups generally feed at night; territorial males eat very little or not at all while on territory. Feeding occurs during all hours of the day once the breeding season ends.

The variety of the sea lion diet has been correlated with POPULATION DYNAMICS. In some Alaskan areas where the diet is diverse, sea lion numbers have been stable or increasing slightly. In areas where sea lions depend primarily on one prey item, the sea lion population is declining. Whether population trends

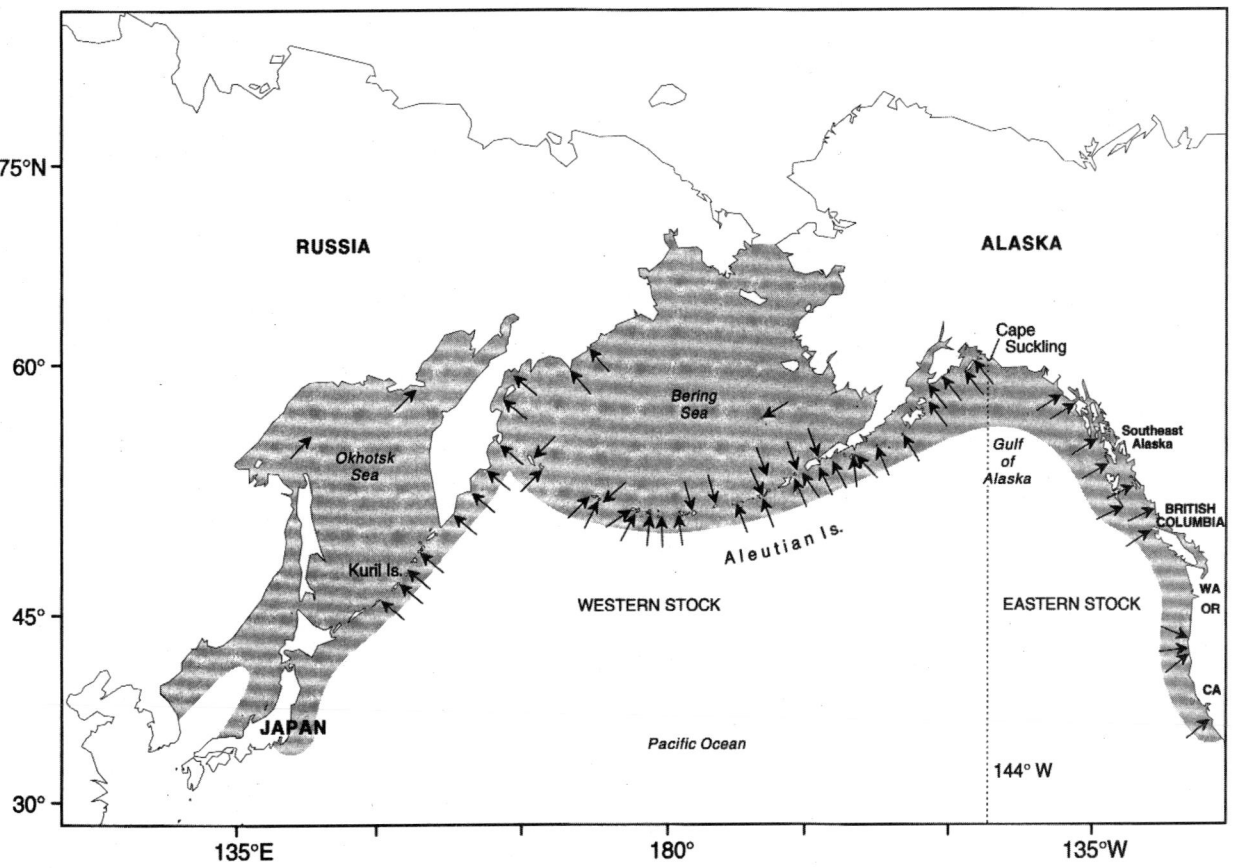

Figure 2 *Map depicting approximate world distribution (shaded area) and rookeries (arrows) of* E. jubatus. *Modified from Loughlin* et al. *(1987).*

are more closely associated with diet diversity is still equivocal, but the availability of a variety of prey is optimal for pinnipeds. Steller's sea lions are eaten by killer whales and sharks, but the possible impact of these predators is unknown.

A cacophony of NOISE engulfs rookeries and haul-out sites with animals of both sexes and all ages vocalizing throughout the day and night. Territorial males use low-frequency roars to signal threats to other males and to court females. Females vocalize frequently, calling to their pups and squabbling with other sea lions of all ages. Pups have a bleating, sheep-like cry and their voices deepen with age.

Grooming is performed by bending the head and neck backward and scratching with the claws of the hind flipper. Sea lions also rub themselves on rocks or other animals. While swimming, the fore flippers are used primarily for movement and the rear flippers for braking and turning.

IV. Notable Behavior

Observations of Steller's sea lions at sea suggest that large groups usually consists of females of all ages and subadult males; adult males sometimes occur in those groups but are usually found individually. On land, all ages and both sexes occur in large aggregations during the nonbreeding season. Breed-

ing season aggregations are segregated by sexual/territorial status. Steller's sea lions are not known to migrate, but they do disperse widely at times of the year other than the breeding season. For example, sea lions marked as pups in the Kuril Islands (Russia) have been sighted near Yokohama, Japan (more than 350 km away), and in China's Yellow Sea (over 750 km away), and pups marked near Kodiak, Alaska, have been sighted in British Columbia, Canada (about 1700 km distant). Generally, animals up to about 4 years of age tend to disperse farther than adults. As they approach breeding age, they have a propensity to stay in the general vicinity of the breeding islands and, as a general rule, return to their island of birth to breed as adults.

The foraging pattern of adult females varies seasonally. Trip duration for females with young pups in summer is approximately 18 to 25 hr, trip length averages 17 km, and they dive approximately 4.7 hr per day. In winter, females may still have a dependent pup, but a mean trip duration is about 200 hr, with a mean trip length of about 130 km, and they dive about 5.3 hr per day. Yearling sea lions in winter exhibit foraging patterns intermediate between summer and winter females in trip distance (mean of 30 km), but shorter in duration (mean of 15 hr) and with less effort devoted to diving (mean of 1.9 hr per day). Estimated home ranges are 320 km^2 for adult females in sum-

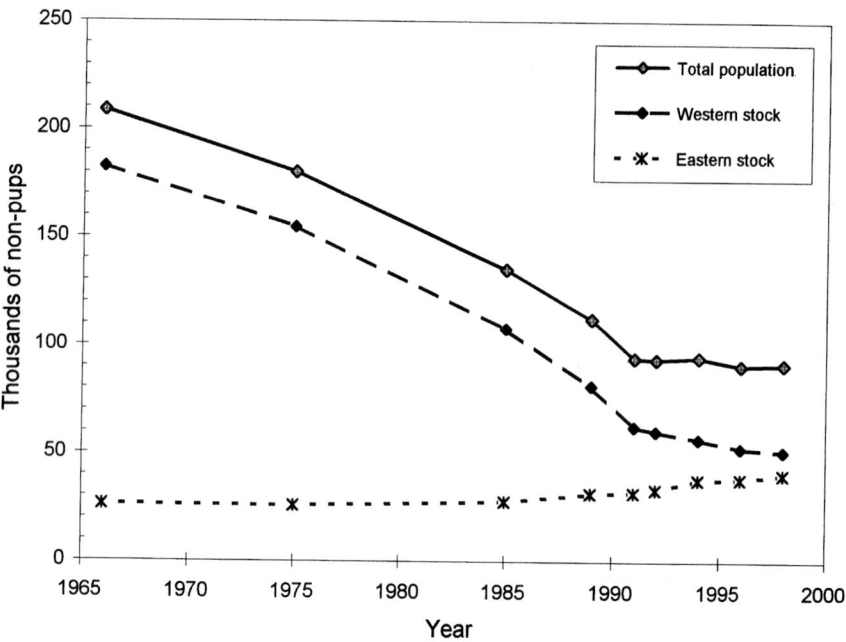

Figure 3 *The population status of Steller's sea lions throughout their range (solid line), the western stock (dashed line), and eastern stock (dotted line) in 1966–1998.*

mer, about 47,600 km² (with large variation) for adult females, and 9200 km² for yearlings in winter.

Compared to some other pinnipeds, the Steller's sea lion tends to make relatively shallow dives, with few dives recorded to depths greater than 250 m. Maximum depths recorded for individual adult females in summer are in the range of 100 to 250 m; maximum depth in winter is greater than 250 m. The maximum depth measured for yearlings in winter was 328 m.

V. Notable Anatomy, Physiology, and Life History

The upper postcanine number 5 is double rooted with the crown directed backward and does not occlude with lower postcanine 5; all other postcanines are singled rooted, slant somewhat forward, and have irregular conical pointed crowns. The diagnostic diastema between upper postcanines 4 and 5 may be caused by rapid growth and extension of the skull rather than by suppression of the fifth and sixth postcanines. Annual growth layers exist in the dentin and cementum with dark layers corresponding to winter and light layers to summer. Dental formula for permanent teeth is i 3/2, c 1/1, pc 5/5, total 34. *E. jubatus* has a double alveolar capillary supply, as in the Cetacea, unique among the pinnipeds.

Like most pinnipeds, Steller's sea lions give birth to a single pup each year; twinning is rare. Males establish territories in May in anticipation of the arrival of females. Viable births begin in late May and continue through early July. The sex ratio at birth is slightly in favor of males. Females breed about 2 weeks after giving birth. Copulations may occur in the water but most are on land. The mother nurses the pup during the day, and after staying with her pup for the first week, she goes

to sea on nightly feeding trips. Pups generally are weaned before the next breeding season, but it is not unusual for a female to nurse her offspring for a year or more.

Females reach sexual maturity between 3 and 8 years of age and may breed into their early 20s. Females can have a pup every year but may skip years as they get older or when stressed nutritionally. Males also reach sexual maturity at about the same age but do not have the physical size or skill to obtain and keep a breeding territory until they are 9 years of age or older. Males may return to the same territory from 1 to 7 years, but rarely more than 3 years. While on their territory during the breeding season, males may not eat for 1–2 months. The rigors of fighting to obtain and hold a territory and the physiological stress over time during the mating season reduce the life expectancy of these animals. They rarely live beyond their mid-teens, while females may live as long as 30 years.

VI. Fossil Record

Otariids probably arose in North Pacific temperate waters from the Enaliarctidae; the earliest known otariid, *Pithanotaria starri*, is between 10 and 12 million years old. The earliest large otariid with most cheek teeth single rooted is from 2 mya Pliocene deposits in Japan and is assigned to the extant genus *Eumetopias*, although it is probably a different species from *E. jubatus*. A skull, teeth, vertebrae, and other parts of a postcranial skeleton of *E. jubatus* were recovered in California from Pleistocene deposits.

VII. Interactions with Humans

Steller's sea lions are rarely seen in aquaria because of their large size and pugnacious nature. Those in captivity in 1999 included about 8 in the United States, 5 in Canada, about 4 in Denmark, and about 15 in Japan. Five in British Columbia at the Vancouver Aquarium and 3 at the Alaska Sealife Center in Alaska were held for both research and exhibit.

Steller's sea lions are an important subsistence resource for Alaskan natives who hunt them for food and other uses. Three hundred or more may be taken a year in Alaska.

Steller's sea lions may be affected by commercial fishing directly through INCIDENTAL CATCH in nets, by entanglement in derelict debris, by shooting, or indirectly through competition for prey, disturbance, or disruption of prey schools. The number of sea lions caught in trawl nets was high during the 1960s and 1970s but has declined since and is presently at very low levels. Incidental entanglement probably contributed to population declines in the Aleutian Islands and western Gulf of Alaska in the 1970s and 1980s, but is not presently considered an important component in the decline. ENTANGLEMENT in derelict gear is rare and unlikely to have contributed to the decline. In some areas, Steller's sea lions were killed deliberately by fishermen, but it is unclear how such killing affected the world population, especially since declines have occurred in areas used uncommonly by commercial fleets (central and western Aleutian Islands) or where fishermen rarely have guns (Russia). Commercial fisheries target on several of the most important prey eaten by Steller's sea lions. In combination, these fisheries remove millions of metric tons of fish. However, the complexity of ecosystem interactions, and limitations of data and models, make it difficult to determine whether fishery removals, directly or indirectly, have negatively impacted Steller's sea lion populations.

See Also the Following Articles

Competition with Fisheries ■ North Pacific Marine Mammals ■ Pinniped Ecology ■ Population Status and Trends ■ Rookeries

References

Alverson, D. L. (1992). A review of commercial fisheries and the Steller sea lion (*Eumetopias jubatus*): The conflict arena. *Rev. Aquat. Sci.* **6,** 203–256.

Calkins, D. G. (1998). Prey of Steller sea lions in the Bering Sea. *Biosph. Conserv.* **1,** 33–44.

Castellini, M. A., Davis, R. W., Loughlin, T. R., and Williams, T. (1993). Blood chemistries and body condition of Steller sea lion pups at Marmot Island, Alaska. *Mar. Mamm. Sci.* **9,** 202–208.

Gentry, R. L. (1970). "Social Behavior of the Steller Sea Lion." Ph.D. dissertation, University of California, Santa Cruz.

Loughlin, T. R. (1997). Using the phylogeographic method to identify Steller sea lion stocks. *In* "Molecular Genetics of Marine Mammals" (A. Dizon, S. J. Chivers, and W. F. Perrin, eds.), pp. 159–171. Special Publication #3 of the Society for Marine Mammalogy.

Loughlin, T. R., Perez, M. A., and Merrick, R. L. (1987). *Eumetopias jubatus.* Mammalian Species Account No. 283, American Society of Mammalogists.

Loughlin, T. R., Perlov, A. S., and Vladimirov, V. A. (1992). Range-wide survey and estimation of total abundance of Steller sea lions in 1989. *Mar. Mamm. Sci.* **8,** 220–239.

Merrick, R. L. (1995). "The Relationship of the Foraging Ecology of Steller Sea Lions (*Eumetopias jubatus*) to Their Population Decline. Ph.D. dissertation, University of Washington, Seattle.

Merrick, R. L., Chumbley, M. K., and Byrd, G. V. (1997). Diet diversity of Steller sea lions (*Eumetopias jubatus*) and their population decline in Alaska: A potential relationship. *Can. J. Fish. Aquat. Sci.* **54,** 1342–1348.

Mitchell, E. D. (1968). The Mio-Pliocene population *Imagotaria. J. Fish. Res. Board Can.* **25,** 1843–1900.

Pitcher, K. W. (1981). Prey of the Steller sea lion, *Eumetopias jubatus,* in the Gulf of Alaska. *Fish. Bull. U.S.* **79,** 467–472.

Pitcher, K. W., and Calkins, D. G. (1981). Reproductive biology of Steller sea lions in the Gulf of Alaska. *J. Mamm.* **62,** 599–605.

Rea, L. D., Castellini, M. A., Fadely, B. S., and Loughlin, T. R. (1998). Health status of young of the year Steller sea lion pups (*Eumetopias jubatus*) as indicated by blood chemistry and hematology. *Comp. Biochem. Physiol. A* **120,** 617–623.

Trites, A. W., and Larkin, P. A. (1996). Changes in the abundance of Steller sea lions (*Eumetopias jubatus*) in Alaska from 1956 to 1992: How many were there? *Aquat. Mamm.* **22,** 153–166.

York, A. (1994). The population dynamics of northern sea lions, 1975–1985. *Mar. Mamm. Sci.* **10,** 38–51.

Stock Assessment

JEFFREY M. BREIWICK AND ANNE E. YORK
*National Marine Mammal Laboratory,
Seattle, Washington*

A stock assessment is an attempt to estimate the productivity or potential of a stock and to predict the future growth of the stock in the face of removals such as incidental catches, directed harvests, or natural causes. It also seeks to measure the capacity of the stock to recover from these removals. The status of the stock with respect to some reference level, such as the unexploited population size, is an element of a stock assessment, as well as the evaluation of the consequences of various management actions. For stocks that are subject to harvest or experience mortality incidental to fishing operations, a goal is to determine allowable removal levels (e.g., harvests that will allow the population to recover to some desired level during some time frame).

The components of a stock assessment will vary with the species considered, its stock identification, the quantity and quality of data available, and the methods and mathematical models employed. It is a process whose steps typically include (1) the definition of the geographic and biological extent of the stock, (2) collection of appropriate data, (3) choice of an assessment model and parameters, (4) specification of performance criteria and evaluation of alternative actions, and (5) presentation of results. While these steps were originally formulated for fishery stock assessments, they are equally applicable to marine mammal stock assessments.

Marine mammal stock assessments are often carried out to determine what level of mortality a stock can sustain. A stock assessment usually requires several pieces of information: current as well as historical abundance, trends in abundance and estimates of biological parameters (such as age at sexual maturity, natural mortality rate, sex ratio, and pregnancy rate), historical harvests, age distribution, maximum sustainable yield level (MSYL), age-specific harvesting mortality, sustainable or replacement yield, spatial distribution of the stock in question, and other relevant factors that may vary by species and population.

I. Productivity

The productivity of a stock is determined by a number of factors, including abundance, rate of increase, population age structure, sex ratio, and the manner in which density dependence operates. Productivity is the amount by which the stock increases over a time interval (usually a year) and is the difference between the number of animals added (by reproduction and immigration) and the number that are lost (due to emigration and from natural causes—all causes not due to harvest). Ideally, immigration and emigration are zero or equal so their effects cancel each other. The amount by which the population increases each year (in the absence of a harvest) is the net production or the replacement yield, and this amount, if harvested, results in the population size at the end of the time period (usually a year) remaining the same size as at the beginning of the year. A related quantity, the sustainable yield, is the productivity when the stock is stable. This occurs when the various population rates (such as natural mortality and reproduction) have remained constant for sufficient time and the environment does not change. Fishery stock assessments generally determine productivity in terms of biomass whereas marine mammal stock assessments most often determine productivity in terms of number of animals. This is not only because marine mammals are difficult to weigh but also because they stop growing after reaching physical maturity, whereas most fish grow throughout their life. It thus becomes increasingly difficult to associate age and size for older animals, especially for cetaceans.

II. Models

Most stock assessments employ a mathematical model of the population to predict the historical trends in abundance as well as future trends under various removal scenarios and choice of model parameters. This approach assumes a constant environment and model parameters. These model parameters are estimated from available data or are fixed at various plausible values. The more reliable the basic data, the more reliable will be the assessment. A simple population model often used for modeling the dynamics of a population is the discrete, generalized logistic model:

$$N_{t+1} = N_t + R_{max}N_t[1 - (N_t/K)^z] - h_t, \qquad (1)$$

where N_t is the population size (in numbers) at the start of year t, R_{max} is the maximum per capita growth rate, K is the carrying capacity or preexploitation abundance of the population, h_t is the harvest in year t, and z is a density-dependent exponent that determines at what population level (between 0 and K) the productivity is maximum. Equation [1] is a difference equation with the population size at time $t + 1$ being a function of the population size at time t.

The population level at which the productivity curve is maximum, the maximum net productivity level (MNPL), is considered to be greater than 50% of K for marine mammals. When the harvest is random with respect to age and sex, MSYL and MNPL are often used interchangeably. The International Whaling Commission (IWC) has usually adopted a value of 60% for the MSYL, which corresponds to a z of 2.39. A value of $z = 1$ results in a symmetric productivity curve with MSYL at 50% of K. This also corresponds to a linear density-dependent relationship between per capita rate of growth and density.

This model is simple in that it combines males and females, it ignores age structure, and it assumes constant environmental conditions and model parameters. It does, however, capture the basic dynamics of the population, including harvesting or other known removals. It has the desirable feature that the recruitment rate is density dependent: it is greatest at small population sizes and decreases as the population increases toward K. The sustainable yield as a function of N for the model shown earlier is given by

$$SY(N) = R_{max}N[1 - (N/K)^z]. \qquad (2)$$

Productivity, therefore, increases as the population size increases, reaches a maximum at some level intermediate between 0 and K (the MSYL), and then declines to 0 at K. This can be seen in Fig. 1, which shows the sustainable yield curves for $z = 1$ (linear density dependence with MSYL = 50% of K) and $z = 2.39$ (nonlinear density dependence with MSYL = 60% of K). By solving for the population size when the productivity is maximum, the MSYL can be shown to be equal to $K(1/1 + z)^{1/z}$. Thus, if $z = 1$, MSYL = $K/2$ or 50% of K.

Equation [1] and similar models, often modified to include sex and age structure, have been used frequently to model cetacean, pinniped, and other marine mammal populations. If an estimate of R_{max} is available along with a time series of removals, then the model can be used to project an initial abundance, N_0, forward to any particular year. This procedure can be programmed to find

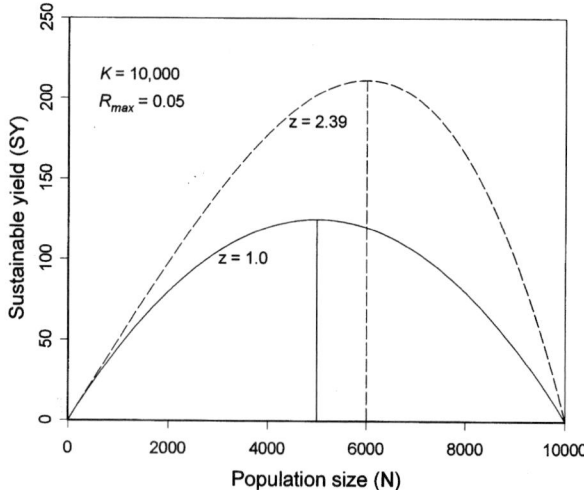

Figure 1 *Plot showing sustainable yield as a function of population size (see Eq. [2]) for a hypothetical population with $R_{max} = 0.05$, $K = 10,000$, and two values for the density dependent exponent, z.*

a N_0 that makes the population trajectory "hit" a current abundance estimate. This is a simplification of a technique employed to assess many marine mammal populations. Parameters of the model can be estimated by minimizing a measure of discrepancy between observed and model-predicted abundance.

III. Population Status

The IWC and the U.S. government (U.S. Marine Mammal Protection Act of 1972; MMPA) have based assessments on classifying stocks by their depletion level with respect to pre-exploitation population size and maximum net productivity level. The IWC new management procedure (NMP) was based on classifying stocks as initial management (IM), sustained management (SM), and protected (P), based on the MSYL and the current depletion level with respect to K. However, for stocks classified as IM or SM, the quota required knowledge of the MSY. The MSY, in turn, depends on, among other things, the level of density dependence (see Fig. 1). The revised management procedure (RMP) addresses shortcomings in the NMP such as the difficulty in determining MSY for a stock and relies primarily on estimates of abundance and their uncertainty and a simple population model such as Eq. [1] requiring few biological parameters.

The U.S. MMPA called for marine mammal populations to be maintained at an "optimum sustainable population" (OSP) level. The U.S. National Marine Fisheries Service defined OSP as a population level between the MNPL and the carrying capacity. The application of this requires the determination of what the current population status is with respect to the MNPL. In some cases, a range of MNPL was used, whereas in other cases an estimate was made whether the abundance was either less than or greater than the MNPL. The 1994 amendments to the MMPA required that a potential biological removal (PBR) be determined for marine mammal stocks. The PBR is equal to the maximum number of animals that can be removed safely from the population annually. It is calculated as the product of the minimum population estimate of the stock, one-half the maximum theoretical or estimated net productivity rate, and a recovery factor between 0.1 and 1. The assessment of allowable removals therefore hinges on an estimation of abundance and the productivity rate. The PBR is a conservative approach whose goal is to allow stocks to reach or maintain their OSP without having to estimate complicated quantities such as MNPL or K.

IV. Uncertainty and Other Considerations

Uncertainty and how to deal with it is a feature of all stock assessments. The environment can change as can the genetic structure of stocks. There is also uncertainty in the underlying population dynamics (termed process error), in the measurement of abundance or indices of abundance (termed measurement error), in the model structure, and in the model parameters. Common methods for dealing with some forms of uncertainty include bootstrapping (a method of resampling the data to estimate variability), maximum likelihood (a method for obtaining parameter estimates and their associated variability), and Bayesian statistical methods.

Bayesian statistical methods are being used increasingly to deal with uncertainty in stock assessments. Bayesian estimation involves integrating the product of the likelihood of the observed data and the prior probability distribution for parameters of interest to obtain what is termed the posterior distribution for the quantity of interest. Due to the complexity involved in integrating this product, it must often be estimated numerically by Monte Carlo methods (based on computer simulations using random numbers). An advantage of Bayesian methodology is that various sources of information on parameters, including observations from other stocks or species, can be incorporated into the assessment. The end result is not just a simple estimate of the growth rate, for example, but a probability distribution, showing the probability of different values of the parameter. A disadvantage of the Bayesian method is that it is often difficult to construct and obtain agreement on a prior distribution to use in an assessment.

Most stock assessments deal with uncertainty in model parameter estimates and management-related quantities but for the most part are still based on single-species population models. Some progress has been made in considering marine mammal stocks as part of a larger ecosystem that includes food webs and interacting (e.g., competing) species. While it would be preferable to consider a host of interacting factors that influence marine mammal stocks, the paucity of available data often precludes the estimation of the parameters necessary to model the various population interactions. It is likely, however, that future stock assessments will increasingly take into account ecosystem considerations.

Another issue that is becoming more important in fisheries and marine mammal stock assessment is the consideration of environmental factors. In the North Pacific, for example, decadal oscillations in climate features have been identified that can affect the distribution and abundance of fish and marine mammals (by affecting the distribution and abundance of prey species). Large amounts of data are required to be able to assess the effects of climate on marine mammals, but these data are beginning to be collected. Without these data it is very difficult to determine whether changes in stock abundance are due to climate, harvesting, or a combination of the two.

V. Examples

A. Cetaceans

Recent stock assessments carried out for both the eastern North Pacific gray whale (*Eschrichtius robustus*) stock and the Western Arctic bowhead whale (*Balaena mysticetus*) have used Bayesian assessment methods. Stock definition is not considered an issue for these two stocks, as there is little evidence for substock structure.

Both of these stocks migrate along the coast where they are counted by shore-based observers. Data collected include counts, which are analyzed to give abundance estimates along with standard errors (measure of uncertainty) and numbers of young and immature animals. Two different assessment models have been used, but both are age- and sex-structured models that incorporate density dependence in the reproductive rate. A method often used to estimate the posterior probability distribution of management-related parameters (such as

MSYL and RY) involves projecting the population model forward using as inputs parameter values that are chosen randomly from their prior probability distributions. A measure of the discrepancy between the observed abundance estimates and the abundances predicted by the population model, the likelihood, is then computed. This is repeated many times and management-related parameters are calculated for each case. These include historic abundance, current abundance, growth rate, MSYL, and RY. From this sample, a smaller sample is taken with probability proportional to the total likelihood computed for each trajectory. This second sample is an estimate of the posterior probability distribution of the management-related parameters. From these distributions the median or mean value can be obtained as well as other statistics of interest, including the answer to such statements such as "What is the probability that the RY is less than 100?" or "What is the probability that the current abundance is greater than the MSYL?" In practice, a conservative approach has often been adopted by computing such quantities as the lower 5th or 10th percentile of the RY and other distribution in determining allowable quotas.

Both the U.S. National Marine Fisheries Service and the IWC have assessed these two stocks. A Bayesian analysis using an age- and sex-structured model for gray whales resulted in a lower 10th percentile for a R_{max} of 4.7%. This, coupled with a minimum population estimate of 24,477 animals and a recovery factor of 1.0, resulted in a PBR of 575 animals, well above the average current annual take, which is less than 180. The IWC assessment was based on a Bayesian analysis using an age structure model. The results included estimates of K (30,000–34,000), N_{1997}/K (0.72–0.80), and RY (550–580). The harvest is much lower than the estimated RY and the population size is estimated to be greater than 70% of K, indicating that current harvests should allow the population to increase, based on this single-species approach to assessment.

The PBR for the western Arctic bowhead whale population was based on a minimum population estimate of 7738, a rate of increase of 3.2% (with a harvest, so an R_{max} of 4% was used), and a recovery factor of 0.5 (because the population is increasing in the presence of a known take). These result in a PBR of 77 animals (7738 × 0.02 × 0.5). The development of a PBR for this stock is required by the MMPA even though the Alaska Eskimo subsistence harvest is managed under the authority of the IWC. Thus, the IWC quota takes precedence over the PBR estimate. Assessments by the IWC were based on Bayesian analyses using age- and sex-structured population models. A reference set of prior distributions was developed for the parameters MSYL, MSYR (a measure of productivity), transition age from juvenile to adult mortality, K, age at sexual maturity, juvenile and adult survival rate, and maximum fecundity rate. Bayesian posterior distributions gave estimates of K from 11,000 to 17,000 and RY from 123 to 276. Thus the smallest RY value is larger than the current IWC maximum strike limit of 82, which should allow the population to increase.

B. Pinnipeds

Commercial harvests of subadult male northern fur seals (*Callorhinus ursinus*) took place on the Pribilof Islands from the time of their discovery in 1786 until 1984. After 1918, the harvest was conducted under the auspices of the Treaty on the Conservation of the Northern Fur Seal. The renegotiated treaty of 1957 provided a vehicle for cooperative research among scientists of the party nations and specified that the population was to be managed to obtain "maximum sustained productivity." Attempts have been made to fit spawner-recruit models to fur seal data and to use them to set the harvest, but these methods largely failed probably due to high variability in year class survival. Using data of numbers of young of the year (pups), counted earlier in the century and the pattern of harvests, it was estimated that the harvests, on average, took about 30% of the number of male seals born or about 15% of the total seals born. When managers of the Pribilof herd learned that age at first reproduction of the Russian herd was, on average, 1 year younger than the Pribilof herd, they justified a large reduction of females in the Pribilof population with the idea that with a lower herd density, age at first reproduction would decrease to the level of the Western Pacific population, and a sustainable harvest of the same size would be obtained from a reduced population. That idea was tried and failed, perhaps because the harvesting regime preferentially killed those females that tended to reproduce at a younger age. At present, there is no commercial harvest, but a subsistence take for food is permitted. The maximum size of the subsistence take is set by the PBR approach at about 18,000 animals, well above the current harvest levels of 800–1000 animals.

The Northwest Atlantic harp seal (*Pagophilus groenlandicus*) population in Canada is currently subject to a harvest. Approximately 450,000 animals (mostly young of the year) were removed from the population each year since 1995. The total population of harp seals (currently estimated at 4–6 million animals) is estimated from a population model based on data of pup production, age-specific reproductive rates, size and age of the commercial harvest, and estimates of size and age of bycatch taken in nearshore fisheries. The model also estimates annual survival rates (assumed to be constant) of adults and juveniles. Confidence intervals for the total population are obtained using a parametric bootstrap procedure that attempts to take account of the uncertainties in all the inputs to the population model.

The replacement harvest for the next year is estimated by projecting current population estimates forward for 1 year, varying the harvest until a constant total population is obtained. The uncertainty in this estimate is determined using a similar bootstrap procedure as in the population model. Using this method, Canadian scientists estimate the replacement yield to be 540,000 (SE = 87,000) animals, or at about 10% of the total population estimate.

See Also the Following Articles

Abundance Estimation ■ Management ■ Population Status and Trends ■ Sustainability

References

Anonymous (1955) United States statement on estimates of maximal sustainable productivity for the Pribilof seal herd. Document 48, presented by the United States during negotiations in Washington,

DC, Dec. 19, 1955, preceding ratification of the 1957 Interim Convention on the Conservation of the North Pacific Fur Seal.

Butterworth, D. S., David, J. H. M., McQuaid, L. H., and Xulu S. S. (1987). Modeling the population dynamics of the South African fur seal, *Arctocephalus pusillus pusillus*. In "Status, Biology, and Ecology of Fur Seals, Proceedings of an International Symposium and Workshop." NOAA Tech. Rep. NMFS 51, U.S. Department of Commerce, Washington, DC.

Chapman, D. G. (1973). Spawner recruit models and the estimation of the level of maximum sustainable catch. *Rapp. P.-V. Reun. Cons. Int. Explor. Mer.* **164**, 325–332.

Gerrodette, T., and DeMaster, D. P. (1990). Quantitative determination of optimum sustainable population level. *Mar. Mamm. Sci.* **6**(1), 1–16.

Goodman, D. (1988). Dynamic response analysis. I. Qualitative estimation of stock status relative to maximum net productivity level from observed dynamics. *Mar. Mamm. Sci.* **4**(3), 183–195.

Hilborn, R. (1997). Uncertainty, risk, and the precautionary principle. In "Global Trends: Fisheries Management" (E. K. Pikitch, D. L. Huppert, and M. P. Sissenwne, eds.), pp. 100–106. American Fisheries Society 20, Bethesda, Maryland.

International Whaling Commission (1977). Report of the Scientific Committee. *Rep. Int. Whal. Commn.* **27**, 36–51.

Kesteven, G. L. (1999). Stock assessments and the management of fishing activities. *Fish. Res.* **44**, 105–112.

National Research Council (1998). "Improving Fish Stock Assessments." National Academy Press, Washington, DC.

Polacheck, T., Hilborn, R., and Punt, A. E. (1993). Fitting surplus production models: Comparing methods and measuring uncertainty. *Can. J. Fish. Aquat. Sci.* **50**(12), 2597–2607.

Punt, A. E., and Hilborn, R. (1997). Fisheries stock assessment and decision analysis: The Bayesian approach. *Rev. Fish Biol. Fish.* **7**(1), 35–63.

Shelton, P. A., Stenson, G. B., Sjare, B., and Warren, W. G. (1996). Model estimates of harp seal numbers-at-age for the Northwest Atlantic. *NAFO Sci. Coun. Stud.* **26**, 1–14.

Smith, T. D. (1983). Changes in size of three dolphin (*Stenella* spp.) populations in the eastern tropical Pacific. *Fish. Bull.* (*U.S.*). **81**(1), 1–13.

Wade, P. R. (1998). Calculating limits to the allowable human-caused mortality of cetaceans and pinnipeds. *Mar. Mamm. Sci.* **14**(1), 1–37.

York, A. E., and Hartley, J. R. (1981). Pup production following harvest of female northern fur seals. *Can. J. Fish. Aqua. Sci.* **38**(1), 84–90.

Stock Identity

JOHN Y. WANG

FormosaCetus Research and Conservation Group, Thornhill, Ontario, Canada

I. Importance of Stock Identity

Determining how a species is divided into stocks (the term stocks is used to refer to biological stocks rather than management stocks; see later) is fundamental to the conservation of marine mammals. Because evolutionary processes act at the intraspecific level, genetic differences and locally adaptive characters will accumulate in stocks over time. This reservoir of genetic and phenotypic diversity increases a

species' ability to persist through environmental changes. Thus, one of the main goals in conservation is to preserve the evolutionary potential of species by maintaining the diversity found in stocks. Another important goal is to maintain species as functioning elements in their ecosystem by preventing regional overexploitation and depletion. Consequently, knowledge of stock structure of species is integral for developing effective management programs to achieve these goals.

The greatest threats to the survival of marine mammals are human activities. Most species experience various levels and kinds of anthropogenic threats in different regions and all exhibit life history characteristics (i.e., long-lived, low fecundity, late age of maturity) that make them especially susceptible to these threats. In order to assess the impact of human activities on marine mammals, it is crucial to identify stocks accurately, establish where the stock boundaries exist, and determine the permeability of the boundaries to genetic exchange with other stocks. This information will influence how the biological data needed for assessments are collected and interpreted and how management plans are designed. Inaccurate stock designations can lead to either unnecessary regulation(s) of fisheries or fallacious management that result in the depletion of a stock and its accompanying genetic material. For example, managing two distinct stocks as one because of a failure to distinguish the two stocks might lead to the depletion of one of the stocks.

Understanding stock structure can also help in streamlining the design of other studies, providing insights into evolution and monitoring ILLEGAL ACTIVITIES [e.g., DNA analysis of cetacean meat products from Japanese markets found species that were prohibited from sale (see Baker and Palumbi, 1994)]. Therefore, it is not surprising that much effort has been directed toward identifying stocks of marine mammals. However, the task remains problematic with two major difficulties: (1) semantic uncertainty and disagreement in the definition of "stock" and (2) studying stock identity with incomplete biological knowledge.

II. Definition of Stock

The term "stock" has been used to refer to both biological and management entities (although in many cases, they are combined or inseparable). A management stock is a group of conspecific individuals that are managed separately. It is very much dependent on the goals of managers and may not be based on biological discontinuities (e.g., International Whaling Commission management stock designations for baleen whales). With the exception of the definition by Moritz (1994), who described a "management unit" (MU) (which he synonymized with "stock" and appears to be equivalent to management stock) as having significant differences in allele frequencies at nuclear or mitochondrial DNA loci, the criteria for determining management stocks may have little to no biological rationale or consistency and are influenced greatly by political interests. Nevertheless, management stocks have been used widely due to the paucity of biological information and will likely continue to play an important role in conservation. Developments in management strategies for situations with incomplete biological information should improve the success of conservation programs (e.g., Taylor, 1997). Although management stocks offer more flexibility in the sense that they can still be the focus of management programs without evidence of biological distinctiveness, conservation goals (e.g., maintaining genetic diversity) are more likely to be achieved if stocks are based on biological data. Therefore, this article focuses mainly on biological stocks.

Biological stocks are characterized by no or low levels of genetic exchange (which means that members of a biological stock tend to interbreed with each other more often than with other individuals). An entity with this property has also been called a population, subpopulation, evolutionary significant unit (ESU), deme, and subspecies (the only intraspecific taxon recognized by the International Commission on Zoological Nomenclature). When gene flow between two groups is absent, there is usually no disagreement that they represent separate biological stocks. However, it is more typical that some level of genetic exchange exists. Even low levels of genetic exchange can obscure stock boundaries and complicate the task of discriminating biological stocks. Although there is no consensus on the threshold level of gene flow above which stock status is no longer recognized, several approaches have been developed to make the identification of biological stocks more objective and explicit.

III. Stock Identification Approaches

Defining stocks is linked inextricably with defining species. There are many concepts that propose species definitions but those advocated most commonly today include biological, evolutionary, and phylogenetic species concepts (for a detailed overview of these and other concepts, see Sites and Crandall, 1997; King, 1993). However, because these concepts all have major limitations, agreement on the best species definition still eludes biologists. Like the species concepts, each approach to stock identification has limitations and weaknesses. In addition, defining stocks can be influenced, and therefore complicated further, by the goals of conservation and legislation. For example, one of the goals of the U.S. Endangered Species Act (ESA) is to decrease the loss of genetic variation. Thus, for this purpose, defining stocks using genetic criteria [e.g., the ESU of Moritz (1994)] is a reasonable proposal [however, see Pennock and Dimmick (1997) and Dimmick *et al.* (1999)]. Unlike the ESA, the U.S. Marine Mammal Protection Act (MMPA) endeavors to keep biological stocks at or beyond their optimum sustainable levels and functioning in their ecological roles. To accomplish the intent of this legislation, defining conservation units will also require demographic information.

There are several operational approaches to stock identification. Whereas some approaches are clear extensions of certain species concepts, the theoretical basis of others may be less explicit or embedded within the methodology. Brief descriptions of the approaches used most commonly are presented.

Morphological characters have been the main evidence for delineating stocks. Because differences between stocks are generally less obvious than between species, examination of a large series of specimens is recommended for the identification of stocks under this approach. However, for most marine mammal species, it is difficult, if not impossible, to obtain a large number of specimens for analysis.

The "phylogeographic" approach proposed by Dizon *et al.* (1992) determines the likelihood that a group of organisms is an ESU. The determination is based on distribution, population response (including demography, behavior, vocalizations), and phenotypic and genotypic information, all of which serve as proxies for reproductive isolation (the essence of the biological species concept). Groups most likely to be ESUs have clear geographic and genetic separation and are assigned to "category I." "Category II" units are characterized by clear genetic separation but little to no geographic partitioning. Units with little genetic differentiation but isolated geographically from other conspecifics define "category III." "Category IV" units are the least likely to be ESUs because they are separated neither geographically nor genetically from other units. This approach has been described as being unwieldy, but it is explicit, transparent, and has performed well. It also seems to provide the most flexibility in stock delineation because several kinds of evidence are used and it offers more than a simple dichotomy for the mosaic of variation present. In addition, by considering information on distribution and population responses, it is much better than the other approaches at detecting recently diverged stocks.

Moritz (1994) proposed definitions for ESU and MU that differ from those of the phylogeographic approach. His definition of ESUs requires these entities to be reciprocally monophyletic (i.e., to have diagnostic differences) in mitochondrial DNA (mtDNA) and significant divergence in nuclear DNA allele frequencies, whereas MUs are defined by significant divergence in either mtDNA or nuclear DNA allele frequencies (irrespective of the distinctiveness of the alleles). Because these definitions are based solely on DNA patterns, they cannot be realized with nonmolecular characters and therefore have limited application. Although DNA information may be more direct for determining whether genetic differences exist (some phenotypic characters can be plastic and influenced by environmental factors), it is not always available and stock differences in phenotypic characters may be established more rapidly after divergence (e.g., demographic response).

The "phylogenetic" approach for defining ESU was advocated by Vogler and DeSalle (1994). Their procedure for recognizing ESUs is similar to how species are delimited under the phylogenetic species concept. Only heritable genetic, morphological, ecological, or behavioral characters are analyzed. An entity is deemed an ESU if it differs from all other entities in having a unique character or a diagnostic combination of characters. However, it is unclear how the definitions of ESU and species differ with this approach, and the process of determining useful characters may require expert knowledge and can be operationally complex.

The "character concordance" approach (Avise and Ball, 1990; Grady and Quattro, 1999) suggests that a group of individuals sharing a common evolutionary history should share characters that are unique to the group and that the level of concordance among independent, shared characters should increase with increasing divergence time. Therefore, high concordance would be strong support for distinctiveness. When concordance is incomplete, the weight of the evidence governs the decision on stock status. Because there are no clear procedural guidelines for interpreting discordant evidence, decisions may be complicated and subjective. Furthermore, many independent characters evolve at different rates so a lack of concordance may be expected for groups that diverged recently. Therefore, this approach may not be effective in identifying recently separated stocks.

Regardless of the approach one decides to use, it is important that clear hypotheses are stated so that interpretation of the results can be objective and divorced from philosophical or conceptual issues. It is also important that the interpretation of data is within the limitations of the hypotheses being tested. For example, if the results of a study do not support distinct units, then the statistical ability (or power) of the study to detect separate units should be examined. With adequate power, the appropriate conclusion would be that differences in the characters examined were not detected rather than that differences do not exist between the units being studied. Without sufficient power, conclusions regarding stock structure would be premature and should not be made. Finally, it may be tempting to combine units when evidence for separating the units is not found. However, this conclusion could be erroneous, is not risk averse, and should be avoided.

In situations where essentially no biological stock information exists, the participants of a workshop on the genetics of marine mammals recommended that the smallest area where exploitation occurs be recognized as a stock (management stock). However, they also cautioned that in certain circumstances (e.g., migratory stocks that experience exploitation in several fisheries in different areas or seasons), this strategy may not be precautionary (Dizon *et al.,* 1997). Therefore, the suitability of this approach should be assessed for each case and only used temporarily while immediate attention is directed at studying biological stock structure.

IV. Analytical Techniques

Several types of information have contributed to our understanding of marine mammal stocks. Which analytical techniques are adopted depends on which stock definition and identification approach are followed.

Analyses of phenotypic characters have dominated this task. Comparisons of osteology, morphology, and pigmentation have contributed the most to stock identification because these characters provide tangible evidence of distinctness. Also, increased computing capabilities have made multivariate analyses of large data sets simple and quick. However, there are few species (and even fewer stocks) for which data from a large series of specimens can be examined because specimens are difficult and expensive to obtain, prepare, and store and some characters can be affected greatly by the condition of the specimen (e.g., decomposition and external morphology; postmortem changes in pigmentation).

Increasingly, attention has been shifting toward molecular characters. Protein analyses were important for stock identification but have become obsolete with the development of efficient DNA technology. Presently, most conclusions regarding stock status are not accepted fully until DNA has been analyzed as well.

Because the properties of mammalian mtDNA are fairly well understood, analysis of mtDNA has dominated molecular studies of marine mammal stocks. For many marine mammal conservation goals, mtDNA evidence is sufficient for designating management stocks [for more details, see Dizon *et al.* (1997)], whereas for designating biological stocks, evidence from characters that are heritable from both parents would be required. Direct analyses of nuclear DNA, which is inherited from both parents, are becoming more common.

Most marine mammal species do not have uniform distributions. There are usually high concentration areas that are separated by areas of low to no concentration of conspecifics. Thus, distribution can provide a first approximation of where stock boundaries may exist. Based on heterogeneous distributions, seasonality of occurrence, oceanographic features (e.g., barriers, water currents, temperature), and geographic distance between areas of high abundance, provisional stocks can be proposed for further studies to test. However, distributional data should always be interpreted in conjunction with additional biological knowledge (e.g., daily and seasonal movement patterns, philopatric behavior).

Most distinct stocks are separated geographically or temporally. Therefore, each stock experiences unique ambient conditions (e.g., differential environmental stresses, food quality or availability, exploitation). Adaptive responses to different conditions may be expressed demographically or ecologically. Different demographic profiles in two groups would be strong evidence of noninterbreeding stocks. Also, demographic differences can reveal recently isolated stocks that have yet to develop genetic or phenotypic distinctiveness. However, to obtain accurate demographic information, a large data set must be analyzed. Because other techniques can address stock identity more directly and efficiently, few studies employ demographic analysis for delineating stocks. If available, demographic information should also be analyzed, especially if stock status based on molecular and phenotypic evidence is uncertain.

Many studies have proposed stocks using analyses of ecological differences. Prey preference, parasitology, pollutant loads, stable isotope ratios, and fatty acid signatures are some of the ecological information used most commonly. Although ecological studies provide another line of evidence for understanding stocks, they act only as proxies for genetic and demographic separation.

V. Complications

Even if there was agreement on a single stock definition and multiple techniques were used, defining stocks would still not be a trivial task. Many situations can obscure and complicate our attempts to delineate stocks, including taxonomic uncertainty; genetic exchange; clinal variation; dispersal and differences between sexes in philopatric behavior; diversity in mating strategies; habitat shifts (e.g., occasional environmental fluctuations may bring stocks that are usually separated geographically into contact and allow interbreeding); fragmentation and genetic bottlenecks resulting from exploitation; mixed stocks; social structure; and short-term and seasonal movements, sometimes across international boundaries. Without knowledge of and consideration for these (and other) attributes, conclusions about stock structure can be compromised.

With so many complications, it is not surprising that the biological stock structure of most marine mammal species (even those that were exploited heavily by commercial harvesting) remains uncertain. However, multidisciplinary techniques, technological advancements, and continued attention should allow us to make rapid progress in identifying biological stocks of marine mammals and to design more effective management programs in the absence of essential biological information.

See Also the Following Articles

Conservation Biology ■ Genetics for Management ■ Geographic Variation ■ Molecular Ecology ■ Species

Acknowledgments

I am grateful to T. Adams, A. E. Dizon, W. F. Perrin, P. E. Rosel, B. N. White, and an anonymous reviewer who provided many helpful suggestions and comments, which improved this article greatly.

References

Avise, J. C., and Ball, R. M., Jr. (1990). Principles of genealogical concordance in species concepts and biological taxonomy. *Oxf. Surv. Evol. Biol.* **7**, 45–67.

Baker, C. S., and Palumbi, S. R. (1994). Which whales are hunted? A molecular genetic approach to monitoring whaling. *Science* **265**, 1538–1539.

Dimmick, W. W., Ghedotti, M. J., Grose, M. J., Maglia, A. M., Meinhardt, D. J., and Pennock, D. S. (1999). The importance of systematic biology in defining units of conservation. *Conserv. Biol.* **13**, 653–660.

Dizon, A. E., Lockyer, C., Perrin, W. F., DeMaster, D. P., and Sisson, J. (1992). Rethinking the stock concept: A phylogeographic approach. *Conserv. Biol.* **6**, 24–36.

Dizon, A. E., Perrin, W. F., Amos, W., Baker, C. S., Chivers, S. J., Costa, A. S., Curry, B. E., Gaggioti, O., Hoelzel, A. R., Hofman, R., LeDuc, R., Loughlin, T. R., Lux, C. A., O'Corry-Crowe, G. M., Rosel, P. E., Rosenberg, A., Scribner, K. T., and Taylor, B. L. (eds.) (1997). Report of the Workshop. *In* "Molecular Genetics of Marine Mammals" (A. E. Dizon, S. J. Chivers, and W. F. Perrin, eds.), Special Publication 3, 3–48. The Society for Marine Mammalogy, Lawrence, KS.

Grady, J. M., and Quattro, J. M. (1999). Using character concordance to define taxonomic and conservation units. *Conserv. Biol.* **13**, 1004–1007.

King, M. (1993). "Species Evolution: The Role of Chromosome Change." Cambridge Univ. Press, Cambridge.

Moritz, C. (1994). Defining "evolutionary significant units" for conservation. *Trends Evol. Ecol.* **9**, 373–375.

Pennock, D. S., and Dimmick, W. W. (1997). Critique of the evolutionary significant unit as a definition for "distinct population segments" under the U.S. Endangered Species Act. *Conserv. Biol.* **11**, 611–619.

Sites, J. W., Jr., and Crandall, K. A. (1997). Testing species boundaries in biodiversity studies. *Conserv. Biol.* **11**, 1289–1297.

Taylor, B. L. (1997). Defining "population" to meet management objectives for marine mammals. *In* "Molecular Genetics of Marine Mammals" (A. E. Dizon, S. J. Chivers, and W. F. Perrin, eds.), Special Publication 3, 49–65. The Society for Marine Mammalogy, Lawrence, KS.

Vogler, A. P., and DeSalle, B. (1994). Diagnosing units of conservation management. *Conserv. Biol.* **8**, 354–363.

Stranding

WILLIAM F. PERRIN
Southwest Fisheries Science Center, La Jolla, California

JOSEPH R. GERACI
National Aquarium in Baltimore, Maryland

Stranded whales have fascinated us through history (Fig. 1). Why do marine mammals strand, what can we learn from their misfortune, and what can we do about it?

I. Why Do Marine Mammals Strand?

Animals that die or become enfeebled at sea of course may be brought passively to shore by wind and wave action. More intriguing are those cases where marine mammals in distress purposely come ashore. A stranded animal when returned to the water may deliberately strand again. This is very frustrating to those who are trying to "rescue" it. It must be understood that an animal may have stranded because it has decided that it cannot keep itself afloat and survive at sea. Thus, deliberate stranding may represent an effort to keep breathing, whatever the ultimate cost. While this may not be adaptive behavior in evolutionary terms because nearly all stranded animals die if unassisted, given the alternative of equally certain but earlier death, the consideration may be moot. A will to survive is adaptive in general, even if not effective in this circumstance.

The natural and unnatural causes of death and disablement leading to single strandings are many: environmental conditions such as anomalously low sea temperature or ice entrapment, parasites, disease, biotoxins, ENTANGLEMENT associated with fisheries, starvation due to decreased food supply, collisions with vessels, contaminants, oil spills, and death or direct injury inflicted by predators, other marine mammals, or at the hands of humans (Geraci and Lounsbury, 1993). Determining the cause of a stranding can be very difficult, even when one is not dealing with a decomposed carcass; symptoms and pathology may be obscure, and two or more factors may be operating simultaneously. Animals that strand in a cluster over a period of a few days may be victims of poisoning, infectious diseases, intensive local fisheries operations, or unusual environmental events. By the time the strandings are investigated, the ultimate cause may no longer be evident.

Certain patterns are exhibited in the strandings of particular species related to their distribution, migrations, and reproduction (Geraci and Lounsbury, 1993). For example, newborn gray whales, *Eschrichtius robustus*, are likely to come ashore in the lagoons of Baja California only during the winter calving season. For all cetaceans, the mother/calf bond is very strong and may continue after lactation ceases. If they come ashore together, it may be impossible to determine who led the way. Young juvenile males of highly social pelagic species may strand after being lost or displaced from bachelor schools; this is

Figure 1 *A sei whale* (Balaenoptera borealis) *stranded in Argentina in 1940. From* Leatherwood *et al. (1982).*

thought to happen to young Atlantic white-sided dolphins, *Lagenorhynchus acutus,* along the U.S. northeast coast during the fall. Some species follow the migrations of prey. Long-finned pilot whales, *Globicephala melas,* for example, pursue squid into shallow waters of Cape Cod Bay during the autumn and early winter and can be expected to strand at these times.

Mass strandings have always been the most puzzling. A large number of whales or dolphins of pelagic species such as sperm whales (*Physeter macrocephalus*), pilot whales (*Globicephala* spp.), false killer whales (*Pseudorca crassidens*), or Fraser's dolphins (*Lagenodelphis hosei*) may come ashore together for no apparent reason and in seeming good health (Fig. 2). While there undoubtedly is more than one cause of mass stranding in these animals, evidence is accumulating that care-giving behavior engendered by tight social bonds may be involved in at least some cases. For example, a herd of 30 false killer whales that semistranded in very shallow water in the Dry Tortugas in 1977 included a large male that was moribund due to illness or injury (Connor, 2000). The other whales clustered around this male and did not return to deep water until he died 3 days later. The whales became agitated when would-be rescuers tried to separate them and insisted on remaining in a tight group around the large male. Similar behavior has been reported for other mass strandings, with rescued individuals deliberately rejoining a group on the beach containing one or more severely ill or injured animals (Geraci *et al.,* 1999). In most cases, in the absence of human intervention, the entire group perishes, victims of a social cohesion that must be highly adaptive in other circumstances. On a population basis, mass stranding must be a rare event, or it could not persist evolutionarily. Suicidal mass stranding has been suggested as a possible mechanism for population regulation (Sergeant, 1982), but this smacks of the group selection hypothesis of Wynne-Edwards (1962), now largely discredited.

In some other instances of mass stranding, careful investigation has uncovered pathological evidence of widespread disease or parasitism that may have been causal or contributory. For ex-

ample, about half of the mature females in a large stranded group of Atlantic white-sided dolphins were severely infected with a nematode to the extent that reproductive success was likely affected (Geraci *et al.,* 1978). In a stranding of 33 or more short-finned pilot whales, *Globicephala macrorhynchus,* in Florida, all live individuals examined were clinically ill, exhibiting an increased respiratory rate, difficulty in breathing, an elevated heart rate, and a wide range of metabolic and hematological abnormalities (Walsh *et al.,* 1991). As Walsh and colleagues noted: "In such a case, the actual etiology of the stranding event may be unknown because the original inciting factor, such as a virus, may have occurred days or even weeks before."

Figure 2 *A mass stranding of sperm whales in Oregon in 1979. Photo by Robert L. Pitman.*

Several additional causes for mass stranding have been posited (Geraci and Lounsbury, 1999). Areas with broad tidal flats, strong or unusual currents, or extreme tidal volume may lead to errors of navigation or judgment that result in stranding. It has been suggested that pelagic animals may run aground in shallow water because their echolocation is impaired. Others believe that cetaceans use the earth's magnetic field for navigation and are led astray by magnetic anomalies or disturbances. However, the only apparent common factor is strong social cohesion, strong enough that when a single animal comes ashore, for whatever reason, others in the group are likely to follow.

Mass strandings may be increasing in frequency, due possibly to anthropogenic causes. For example, during the period 1981 and 1991, there were 20 mass strandings of long-finned pilot whales, on 32 miles of beach in Cape Cod, Massachusetts, whereas only 1 had been reported in the previous 20 years, and epidemiological evidence suggests a possible link with recent morbillivirus outbreaks affecting a number of marine mammal species in the North Atlantic and Mediterranean (Geraci and Lounsbury, 1999; Duignan *et al.*, 1995). The impact of a viral outbreak may be potentiated by organochlorine pollutants that weaken the immune system.

II. What Can Be Learned from Strandings?

All we know about many species of cetaceans is only what we have learned from strandings. This is true for many of the BEAKED WHALES, the *Mesoplodon* species, for example. They are not kept in captivity for exhibit, are hunted only rarely and in remote parts of the world, and are relatively rare, elusive, and notoriously difficult to observe at sea. A rotting carcass on the beach can yield invaluable information on such things as anatomy, life history, genetics, disease, parasites, predators, contaminants, and feeding ecology. A live stranding transported to a holding facility can inform us about physiology, behavior, and cognition. A mass stranding offers a population sample (albeit potentially biased), opening to view parameters such as sex ratio, age structure, pregnancy rate, lactation rate, and relatedness within a group. Every stranding event should be considered a potentially unique opportunity to learn something that cannot be learned any other way.

III. Stranding Programs

In order to take full advantage of the scientific opportunities offered by stranded marine mammals, formal stranding response programs have started up in many parts of the world. One of the first was begun by Frederick True, the noted cetologist and one of the first curators of the U.S. National Museum of Natural History (Smithsonian Institution) in Washington, DC, and resulted in the beginnings of the largest collection of marine mammal specimens in the world. The stranding program has continued to the present (Fig. 3) and is a world model for stranding-response procedures and data and specimen collection and curation (Geraci and Lounsbury, 1993). Another early stranding program began in Britain when

Figure 3 *Stranding poster issued by the Smithsonian Institution in the 1970s.*

the board of trade instructed receivers of wrecks to send telegraphic reports of the stranding of whales to the British Museum (Harmer, 1914). As "Royal Fishe," stranded cetaceans are property of the Crown and thus receive special attention and care. This has resulted in a long series of detailed data reports on Cetacea stranded on British coasts and basic knowledge of many North Atlantic species, as well as an immense and irreplaceable collection of specimens (Fraser, 1974; Sheldrick *et al.*, 1994). Perhaps the oldest stranding program, although not scientifically based or motivated, is in Vietnam, where cetaceans are revered as sea-going friends and souls, and those washing ashore have been collected and their skeletons preserved in Buddhist temples for centuries; these accumulations are now yielding information on the cetacean fauna of the region (Smith *et al.*, 1997).

In recent years, formal stranding programs have been established in many countries, including Australia, Japan, New Zealand, Canada, France, Italy, Argentina, Brazil, Thailand, the Philippines, and others. In the United States, a national stranding alert network and officially mandated regional stranding programs arose after passage of the Marine Mammal Protection Act in 1972. These new programs around the world are motivated not only by scientific considerations, but by the desire to achieve humane treatment of live stranded animals, rescuing them if possible. The goals of these programs are well established (Geraci and Lounsbury, 1993): provide for the welfare of live animals, minimize risk to public health and safety, support scientific investigation, and advance public education.

IV. What Should Be Done with Live-Stranded Marine Mammals?

First, it must be recognized that marine mammals are under legal protection in many countries, and anyone who interferes with them, even though well-meaning, may be breaking the law. The first response should be to notify the relevant authorities so that if there is a formal stranding program in the area it can go into action. A telephone call to the nearest natural history MUSEUM, oceanarium, fisheries laboratory, or marine wildlife agency will help the information get to the right place.

Not every live animal on the beach needs help. Pinnipeds and sea otters spend time out of the water in the course of their normal affairs. Even some cetaceans may come into very shallow water or ashore for brief periods, e.g., a killer whale to snatch a seal or a bottlenose dolphin to ride a wave. However, certain conditions are unambiguous and do demand attention: a live dolphin obviously in distress on a beach, a sea otter coated with oil, a fur seal too feeble to move, or a porpoise trapped in a fishing weir. Although it is not always possible to judge the health of a cetacean by its outward appearance, coastal animals such as bottlenose dolphins, *Tursiops truncatus*, usually strand singly only when ill and likely will need rehabilitation to survive. Many pelagic cetaceans come ashore in apparent good health, or at least free of recognizable disease, and have a reasonable chance of withstanding the rigors of being returned to sea, although their long-term survival is undocumented.

Once a decision is made to do something about a live stranded animal, there are three options: return it to the sea, euthanize it, or transport it to a care facility (oceanarium or marine mammal rescue center) for rehabilitation (Fig. 4). The basic consideration should be to take no action that will only prolong suffering. The basic criteria for making a decision are (1) whether logistical support is available (e.g., a large dolphin or whale cannot be transported without a truck and means to put the animal on it), (2) the number of animals involved (a mass stranding may be a logistical nightmare), (3) the environmental conditions (rough seas, harsh terrain, darkness, or simply a rising tide can increase the risk to the animal and the team, or extremes of heat or cold may affect the animal's ability to thermoregulate), (4) condition of the animal(s) (a healthy animal is resilient, whereas one that is ailing may not survive the ordeal associated with a rescue), and (5) ease of handling (a very large or struggling animal may be impossible to rescue).

Immediate return to the sea is an option when the animal is manageable, healthy, and able to function normally; logistical and environmental conditions are favorable; social obligations (e.g., maternal care for the young) can be met; and the area of release is within the normal range of the animal, suit-

Figure 4 *A stranded pygmy killer whale,* Feresa attenuata, *brought into the Miami Seaquarium. From Leatherwood* et al. *(1982).*

able and navigable. Single-stranded odontocetes, and sea otters or pinnipeds unable to leave the shore, are usually poor candidates for immediate release. Before an animal is released, a plan should be made for monitoring it after release.

Rehabilitation is an option when there is a good chance the animal can be restored to health, facilities are available and equipped for the species and number of animals involved, arrangements can be made for safe and quick transport, the animal is manageable, and, very importantly, there are sufficient funds and staff to provide care for a reasonable period. It should be noted that even where care facilities are increasing in number, more animals come ashore than can be taken into the existing facilities.

Euthanasia is an option when it is necessary to end the suffering of an animal in irreversibly poor condition, no rehabilitation facility is available for orphaned dependent young, rescue is impossible and no care facility is available, animals persistently restrand, or a distressed cetacean ashore is likely to attract others milling nearby to mass strand. The procedure should be carried out humanely by an experienced qualified person and only if essential equipment and materials are available. A clumsy attempt to euthanize an animal without adequate equipment or expertise can cause more suffering than a natural death.

The time between discovery of a stranded cetacean and arrival of a stranding response team can be used by volunteers to relieve stress and improve the animal's chance of recovery. The key is to prevent further injury and keep the animal comfortable while minimizing handling and disturbance. The animal should be protected from blowing sand and kept moist with clean fresh or salt water. Care should be taken to keep water and sand out of the blowhole. In the summer, shade should be provided against the sun. If small, the animal can be positioned on its belly and holes dug to accommodate its flippers. The animal should be out of the surf and protected against lacerations by sharp rocks and seashells.

Moving, release, tagging, transport, rehabilitation, or euthanasia should only be done by qualified, experienced personnel. Detailed guidelines are available in various manuals and government publications (e.g., Geraci and Lounsbury, 1993).

Mass strandings present a special challenge and can only be coped with effectively by an organized stranding response team backed up with adequate resources. One person, the stranding coordinator, must be in charge of all on-site activities. Volunteers are often indispensable, but their activities must be closely overseen by the coordinator. The goal should be the swift release of the largest manageable number of animals that have the best chance of surviving. Live animals should be dealt with first. The animals judged to have the greatest prospect of survival should be given priority, not those near death. The integrity of the group may be important to survival of the released animals. A proven approach is to relocate as many animals as possible to a safe place in shallow water where they can rest and become reoriented among their fellows. After tagging and monitoring of condition, the animals can be released together into open water. Because mass-stranded animals returned to sea may restrand, sometimes immediately but often days or even weeks later, released whales or dolphins should be monitored on a long-term basis, through direct observation (by cooperating fishermen, Coast Guard, sailing clubs, etc.) or telemetry.

V. What Should Be Collected from a Stranded Carcass?

The condition of a carcass determines much about what can be collected from it and should be specified in field notes. Standard condition codes are (1) alive, (2) freshly dead (i.e., edible), (3) decomposed, but organs basically intact, (4) advanced decomposition (i.e., organs not recognizable, carcass intact), and (5) mummified or skeletal remains only. The quality of information that can be obtained depends on a number of additional factors, including location; size, skills, organization, interests, and morale of the team; adherence to clear, detailed protocols; availability of equipment and supplies; number of animals to be examined; amount of time available; and care maintained in packaging, labeling, shipping, and storing samples. It is well recognized that it is not possible or practical to collect maximal samples and data in all cases; the effort must be tailored to the conditions. As a rough guide, three levels of collection have been described (modified slightly from Geraci and Lounsbury, 1993):

Level A Data: Basic Minimum Data

1. Name and institutional address of investigator.
2. Reporting source.
3. Species (including preliminary identification and voucher material in the form of photographs in several views, teeth, skulls, and other specimens).
4. Field number.
5. Number of animals, including total and subgroups.

6. Location (preliminary description, plus longitude and latitude and closest named cartographic feature).
7. Date and time of discovery and of specimen recovery.
8. Length (and girth and weight if possible).
9. Sex.

Level B Data: Supplementary On-Site Information and Samples

1. Weather and tide conditions.
2. Offshore human/predator activity.
3. Presence of prey species.
4. Behavior before and during stranding.
5. Samples collected for life history studies (teeth, earplugs, or bone for age determination, reproductive tracts, stomachs).
6. Samples collected for blood studies.
7. Samples collected for genetic studies.
8. External measurements.
9. Disposition of carcass.

Level C Data: Necropsy Examination and Parasite Collection

1. Collection of tissues for toxicology, microbiology, and gross histopathology.
2. Collection of parasites.

Detailed protocols for collection of data and specimens are contained in stranding and dissection manuals (e.g., Geraci and Lounsbury, 1993; Jefferson *et al.*, 1994). It must be stressed that information has scientific value only when documented carefully. In the case of a mass stranding, it is better, after collecting the minimal basic data for all, to obtain good samples and perform thorough examinations with accurate documentation on a small number of animals than to do a hasty job on many.

See Also the Following Articles

Fishing, Industry ▪ Mass Die-Offs ▪ Pathology ▪ Pollution and Marine Mammals ▪ Telemetry

References

Connor, R. C. (2000). Group living in whales and dolphins. *In* "Cetacean Soicieities: Field Studies of Dolphins and Whales" (J. Mann, R. C. Connor, P. L. Tyack, and H. Whitehead, eds.), pp. 199–218.

Duignan, P. J., House, C., Geraci, J. R., Early, G., Copland, H. G., Walsh, M. T., Bossart, G. D., Cray, C., Sadove, S., St. Aubin, D. J., and Moore, M. (1995). Morbillivirus infection in two species of pilot whales (*Globicephala* sp.) from the western Atlantic. *Mar. Mamm. Sci.* **11**, 150–162.

Fraser, F.C. (1974). "Report on Cetacea Stranded on the British Coasts from 1948 to 1966." Trustees of the British Museum, London.

Gerarci, J. R., Dailey, M. D., and St. Aubin, D. J. (1978). Parasitic mastitis in the Atlantic white-sided dolphin, *Lagenorhynchus acutus,* as a probable factor in herd productivity. *J. Fish. Res. Bd. Can.* **35,** 1350–1355.

Geraci, J. R., Harwood, J., and Lounsbury, V. J. (1999). Marine mammal die-offs. *In* "Conservation and Management of Marine Mammals" (J. R. Twiss, Jr., and R. R. Reeves, eds.), pp. 367–395. Smithsonian Institution Press, Washington, DC.

Geraci, J. R., and Lounsbury, V. J. (1993). "Marine Mammals Ashore: A Field Guide for Strandings." Texas A&M University Sea Grant College Program, Galveston, TX.

Harmer, S. F. (1914). "Report on Cetacea Stranded on the British Coasts during 1913." Trustees of the British Museum, London.

Jefferson, T. A., Myrick, A. C., Jr., and Chivers, S. J. (1994). Small cetacean dissection and sampling: A field guide. NOAA Tech. Memorand. NMFS NOAA-TM-NMFS-SWFSC-198, 1-54 (Available from NTIS).

Leatherwood, S., Reeves, R. R., Perrin, W. F., and Evans, W. E. (1982). Whales, dolphins, and porpoises of the eastern North Pacific and adjacent Arctic waters: A guide to their identification. *NOAA Tech. Rep. NMFS Circ.* **444,** 1–245.

Sergeant, D. E. (1982). Mass strandings of toothed whales (Odontoceti) as a population phenomenon. *Sci. Rep. Whales Res. Inst. Tokyo* **34,** 1–47.

Sheldrick, M. C., Chimonides, P. J., Muir, A. I., George, J. D., Reid, R. J., Kuiken, T., Iskjaer-Ackley, C., and Kitchener, A. (1994). Stranded cetacean records for England, Scotland and Wales, 1987–1992. *Invest. Cetacea* **25,** 259–283.

Smith, B. D., Jefferson, T. A., Leatherwood, S., Dao Tan Ho, Chu Van Thuoc, and Le Hai Quang (1997). Investigations of marine mammals in Vietnam. *Asian Mar. Biol.* **14,** 145–172.

Walsh, M. T., Deusse, D. O., Young, W. G., Lynch, J. D., Asper, E. D., and Odell, D. K. (1991). Medical findings in a mass stranding of pilot whales (*Globicephala macrorhynchus*) in Florida. *In* "Marine Mammal Strandings in the United States: Proceedings of the Second Marine Mammal Stranding Workshop, Miami, Florida, December 3–5, 1987" (J. E. Reynolds III and D. K. Odell, eds.). *NOAA Tech. Rep. NMFS* **98,** 75–83.

Wilkinson, D., and Worthy, G. A. J. (1999). Marine mammal stranding networks. *In* "Conservation and Management of Marine Mammals" (J. R. Twiss, Jr., and R. R. Reeves, eds.), pp. 396–411. Smithsonian Institution Press, Washington, DC.

Wynne-Edwards, V. C. (1962). "Animal Dispersion in Relation to Social Behaviour." Oliver and Boyd, Edinburgh.

Streamlining

FRANK E. FISH
West Chester University, Pennsylvania

Streamlining has a major impact on the ecological performance of marine mammals. Because swimming is an integral behavior of marine mammals that forage, mate, escape predation, disperse, and migrate in water, constraints on performance promoted adaptations for effective locomotion by aquatic mammals. To propel itself through the water at a constant swimming speed, a marine mammal needs to generate a forward force (thrust) at the expense of metabolic energy that is equal and opposite to the sum of resistive forces (drag).

I. Drag

Two major types of drag are experienced by a marine mammal as it swims submerged. These include the pressure or form drag and the viscous or skin friction drag. The pressure drag results from the pressure distribution around the body. As water flows about a body, a high pressure is generated at the upstream face and a lower pressure is generated at the downstream face. This difference in pressure produces a force, pressure drag, which opposes forward movement. Viscous drag is a function of the viscosity or stickiness of the water around the body. Water particles adhere to the body surface within a thin layer of water adjacent to the body, called the boundary layer. Friction within the boundary layer and between the boundary layer and the body create a force in the drag direction. The magnitude of the viscous drag will depend on the wetted surface area of the body and the flow conditions within the boundary layer. Boundary flow can be laminar, turbulent, or transitional. A boundary layer with turbulent flow produces the highest viscous drag.

Drag is minimized primarily by streamlining the shape of the body and the appendages. The streamlined profile of these structures has a fusiform design resembling an elongate teardrop with a rounded leading edge extending to a maximum thickness and a slowly tapering tail. This shape was first investigated in the dolphin by Sir George Cayley (circa 1800) as a solid of least resistance design (Fig. 1). As shown in Fig. 2, marine mammals display a streamlined, fusiform design. This fusiform shape is sculpted by the distribution of blubber and/or fur covering the body. Modern submarines utilize a fusiform design analogous to the body shape of marine mammals. In addition, the appendages, such as the flukes, flippers, and dorsal fin, have a cross-sectional shape similar to conventional aircraft wings and hydrofoils.

Streamlining minimizes drag by reducing the magnitude of the pressure difference over the body. This reduced pressure difference allows water in the boundary layer to flow without separation from the body surface until near the trailing edge. As separation occurs, a wake is generated downstream. The wake behind the body is small, indicating little distortion to the flow and a small pressure drag. Premature separation of the boundary layer occurs because of instabilities in the flow. A laminar boundary flow is inherently less stable and more prone to premature separation than turbulent flow. An animal may pay a higher energetic cost in frictional drag by allowing the development of a turbulent boundary layer, but the pressure and total drags will be substantially lower than if laminar flow with separation transpires.

II. Body Shape

An indicator of the degree of streamlining is the fineness ratio (*FR* = body length/maximum diameter). The *FR* value of 4.5 gives the least drag and surface area for the maximum volume, although only a 10% increase in drag is realized in the *FR* range of 3–7 (Fig. 3). Marine mammals are well streamlined with body dimensions within the optimal range of *FR*. In general, cetaceans, pinnipeds, and sirenians have body shapes with *FR* between 3.3 and 8.0 (Fig. 3). The notable exception for cetaceans is the northern right whale dolphin (*Lissodelphis borealis*) or "snake porpoise" that can have a *FR* up to 10.9. Despite their bulk and specialization of the head for filter feeding, mysticete whales are well streamlined. *FR* for Balaenopteridae ranges from 4.8 to 8.1 and for Balaenidae ranges from 3.3 to 8.0.

FR is, however, a crude indicator of the streamlining of the body because it does not provide information on changes in

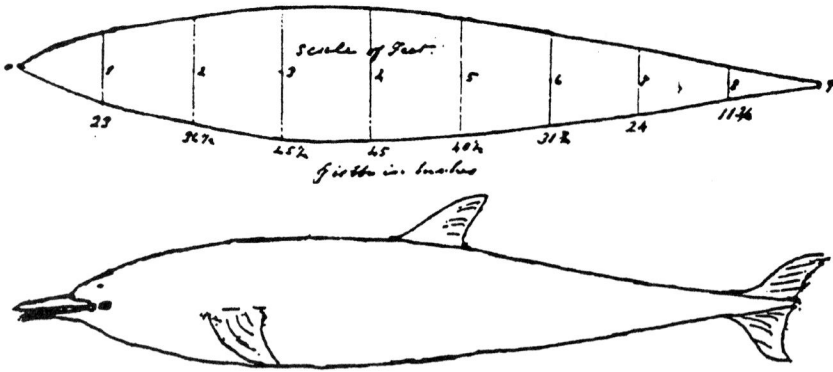

Figure 1 *Sketch of dolphin and body contours by Sir George Cayley (ca. 1800).*

body contour. Another indicator of body streamlining is the position of the maximum thickness, called the shoulder. Shoulder position is important because this is where transition from laminar to turbulent flow and boundary layer separation are likely to occur. Anterior of the shoulder the pressure distribution favors maintenance of a laminar boundary layer. The position of the shoulder in the most rapidly swimming aquatic mammals is displaced posteriorly, which is similar to engineered wings with "laminar" profiles, which reduce drag through maintenance of laminar boundary flow. The shoulder position for dolphins is 34–45% of the body length from the beak.

The shoulder position is located at 40% of body length for otariid seals and 50–60% of body length for phocid seals from the nose. The position can be varied in pinnipeds because the neck is capable of being retracted and extended. Extension of the neck during rapid swimming could modify the flow over the anterior of the seal and reduce drag by extending the region of laminar flow. Such a drag reduction could aid seals in catching fast-swimming, elusive prey.

Experiments on flow visualization using a fluorescent dye applied to a dolphin's melon showed the flow to be laminar over the anterior 32% of the dolphin. Transition began before the dorsal fin with turbulence aft of the fin. Separation of the boundary flow occurred smoothly near the base of the flukes. Flow visualization using BIOLUMINESCENCE within the boundary layer of dolphins and seals similarly indicated a lack of separation from the body surface. Flow separation is restricted to the tips of the flukes, flippers, and dorsal fin. The flow separation has been observed as bioluminescent "contrails" (Fig. 4).

The contrails are vortices generated at the tips of the appendages. A tip vortex is generated from pressure differences along the two surfaces of the appendage. The pressure difference produces a lift force similar to the lift produced by airplane wings. Marine mammals utilize lift generated from the appendages to propel the body, increase stability, regulate depth, and maneuver. A consequence of the tip vortices is the loss of energy from the generation of lift, which is referred to as an induced drag.

The induced drag and interference to the flow over the body due to the addition of the appendages can interfere with streamlining. The dorsal fin, pectoral flippers, and flukes comprised only 2.6, 4.2, and 5.6% of the total surface area of the harbor porpoise (*Phocoena phocoena*), respectively; however,

these appendages are responsible for 35.7% of the total drag (4.3, 18.0, and 13.4%, respectively). Induced drag is reduced by appendages with a long narrow shape and tips that taper sharply to a point (Fig. 5).

Streamlining is also fostered through buoyancy control. Compared to the center of gravity, the center of buoyancy is closer to the head for terrestrial mammals in water. This relationship produces a torque that causes the body to float at an angle to the horizontal. This orientation would decrease streamlining. Internal and external modifications for buoyancy control provide aquatic mammals with longitudinal trim for better streamlining by presenting a smaller body area to the oncoming flow. For sea otters, nonwettable fur provides buoyancy by an entrapped layer of air to maintain a horizontal body orientation when swimming at the surface. The elongate shape of the lungs of marine mammals helps displace the center of buoyancy posteriorly. This arrangement is possible because the diaphragm of marine mammals is oriented obliquely to almost parallel to the spine.

III. Drag Reduction Mechanisms

High swimming speeds attained by many marine mammals have directed some research toward specialized drag reduction mechanisms. In what is known as Gray's paradox, hydrodynamic estimates of dolphin power output at high speeds were inferred to be greater than the power that could be developed for the mass of muscle available for swimming. Resolution of the paradox was believed only possible if the drag was reduced by maintaining laminar flow within the boundary layer, despite a high swimming speed dictating a turbulent boundary flow with increased viscous drag.

While the idea of a special drag reduction mechanism in dolphins has been irresistible, direct evidence of its existence has been elusive. To date, no conclusive evidence has been found of laminar boundary flow over the entire body surface of fast swimming dolphins by mechanisms including compliant skin dampening, secretions, skin cell sloughing, infusion of long-chain polymers into the boundary layer, boundary-layer heating, skin folds, and boundary layer acceleration. Special drag reduction mechanisms are unnecessary to explain Gray's paradox, which is reconciled when one considers that the calculations of power output were based on burst swimming (10

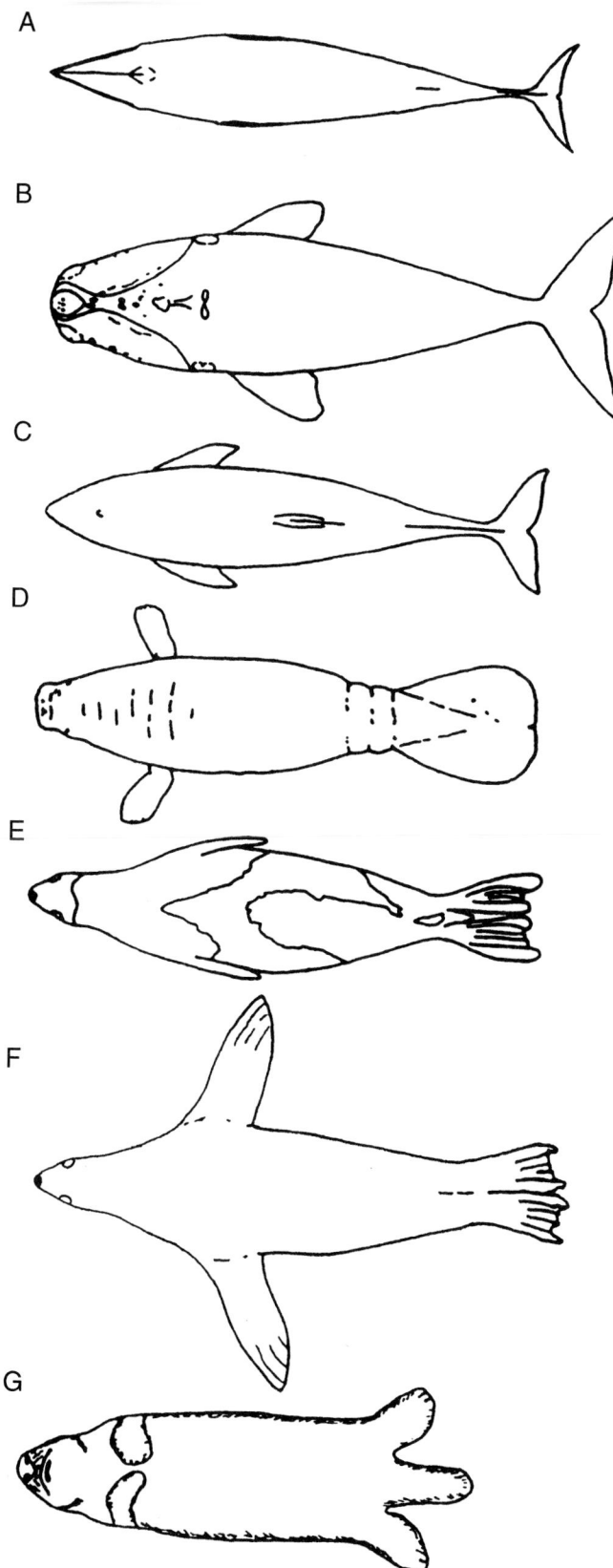

Figure 2 *Fusiform body shape of marine mammals represented by (A) minke whale,* Balaenoptera acutorostrata; *(B) right whale,* Eubalaena glacialis; *(C) harbor porpoise,* Phocoena phocoena; *(D) Florida manatee,* Trichechus manatus; *(E) harp seal,* Phoca groenlandica; *(F) California sea lion,* Zalophus californianus; *and (G) sea otter,* Enhydra lutris.

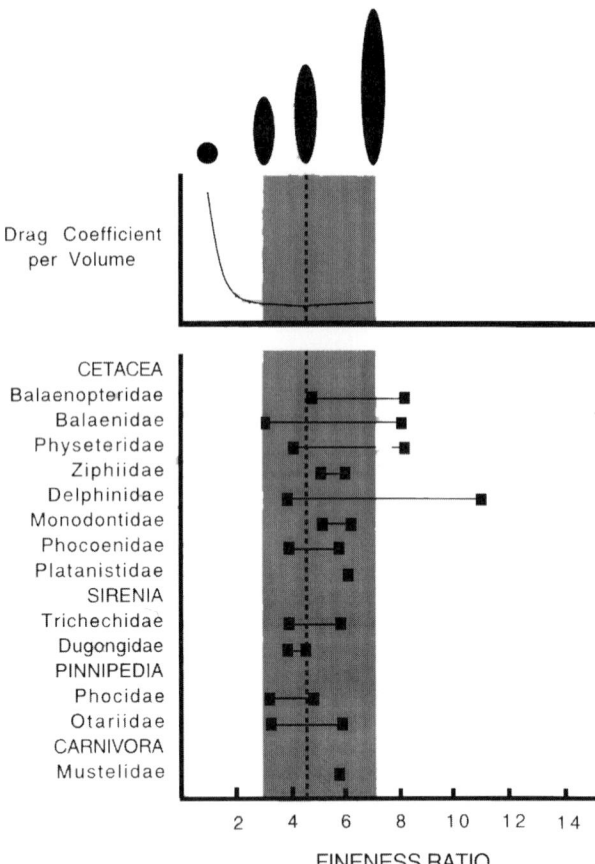

Figure 3 *Comparison of ranges of fineness ratio (FR) for various marine mammal families. The dashed line indicates the optimal FR of 4.5 whereby a body has the least drag for the maximum volume (upper graph). The shaded area represents the range of FR in which drag increases by 10% above the optimum value. Silhouettes show the difference in shape in reference to* FR *from a circular shape (FR = 1) to an elongate form (FR = 7).*

m/sec for 7 sec) and that muscle power output was an underestimate based on sustained performance of dogs and humans.

The naked skin of cetaceans is regarded as a means to maintain a smooth flow with an attached boundary layer over the surface of the body. In addition, cells of the epidermis are produced rapidly, which promotes a high rate of skin sloughing. This increased skin sloughing deters organisms, such as barnacles, from attaching to the skin and thus minimizes drag. The properties of the hair of aquatic mammals are noted to reduce drag by aiding in streamlining of the body. The lack of erector pili muscles in seals and sea otters permits the pelage to lie flat in water, minimizing resistance to swimming.

Behavioral mechanisms are also used to minimize drag. Porpoising, which is performed by the fastest swimmers, is a series of rhythmic leaps. By traveling through the air for a given distance, the animal reduces its drag compared to swimming over an equivalent distance underwater. This reduction in drag is due to the lower density and viscosity of air compared to water. In addition, swimming near the water surface to facilitate breathing incurs a large drag increment due to the energy lost for the animal in the formation of surface waves. Porpoising becomes economical only at high swimming speeds when the cost of LEAPING from the water becomes smaller than the drag on the animal in water. In addition, porpoising is associated with the need to increase ventilation time resulting from greater ENERGETIC demands of rapid swimming.

Many dolphins utilize free-riding behaviors to reduce the energy cost of swimming. In this behavior, the dolphin takes advantage of the pressure field generated by another body and moves along with little or no energetic input. Dolphins have been observed to ride the pressure waves of ships and large whales. By situating itself on the bow wave, the small cetacean can be pushed along or surf down the front slope of the wave. This latter mechanism is analogous to human surfing. Sea lions will surf on breakers to get up onto beaches.

Formation swimming influences the water flow around adjacent individuals. Drafting is the beneficial use of the water flow to

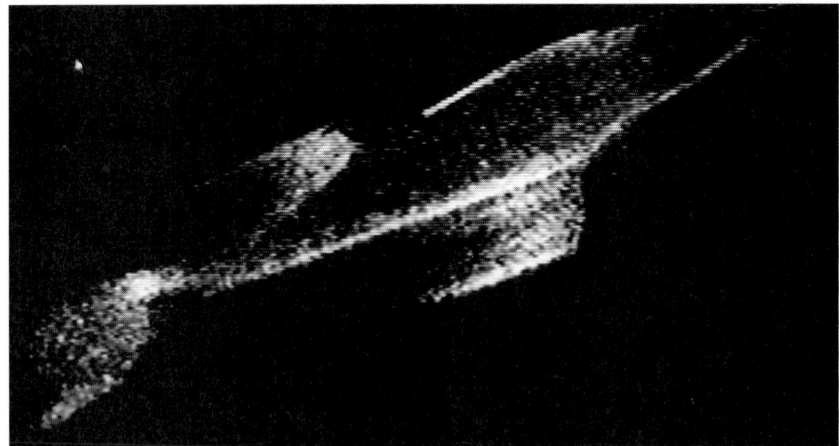

Figure 4 *Bioluminescent image of a gliding dolphin (Tursiops truncatus) showing lack of separation from the body surface. Photograph courtesy of Jim Rohr, Michael Latz, and Mark Stevenson.*

Lang, T. G. (1966). Hydrodynamic analysis of cetacean performance. *In* "Whales, Dolphins and Porpoises," pp. 410–432, K. S. Norris, Ed. Univ. of California Press, Berkeley.

Williams, T. M. (1987). Approaches for the study of exercise physiology and hydrodynamics in marine mammals. *In* "Approaches to Marine Mammal Energetics" (A. C. Huntley, D. P. Costa, G. A. J. Worthy, and M. A. Castellini, eds.), pp. 127–145. Spec. Publ. Soc. Mar. Mamm. No. 1.

Williams, T. M., Friedl, W. A., Fong, M. L., Yamada, R. M., Sedivy, P., and Haun, J. E. (1992). Travel at low energetic cost by swimming and wave-riding bottlenose dolphins. *Nature* **355,** 821–823.

Figure 5 *False killer whale* (Pseudorca crassidens) *displaying highly tapered and elongate appendages.*

reduce drag with a concomitant decrease in the energy cost of locomotion. Large groups of dolphins will organize into side-by-side and echelon formations. Drafting becomes particularly important to young whales to maintain speed with their mothers. A calf swims next to the maximum diameter or near the genital region of the mother. At these locations, the calf is in a hydrodynamically favorable position arising from the interaction of its pressure field with that of the mother's. The calf experiences a thrust component pulling it along with the mother, who experiences increased drag.

See Also the Following Articles

Morphology, Functional ■ Speed ■ Swimming

References

Au, D., and Weihs, D. (1980). At high speeds dolphins save energy by leaping. *Nature* **284,** 548–550.

Fish, F. E. (1996). Transitions from drag-based to lift-based propulsion to mammalian aquatic swimming. *Amer. Zool.* **36,** 628–641.

Fish, F. E., and Hui, C. A. (1991). Dolphin swimming: A review. *Mamm. Rev.* **21,** 181–195.

Gray, J. (1936). Studies in animal locomotion. VI. The propulsive powers of the dolphin. *J. Exp. Biol.* **13,** 192–199.

Hertel, H. (1966). "Structure, Form, and Movement." Rheinhold, New York.

Striped Dolphin
Stenella coeruleoalba

FREDERICK I. ARCHER II
*Southwest Fisheries Science Center,
La Jolla, California*

The striped dolphin, *Stenella coeruleoalba* (Meyen, 1833) is a pelagic small delphinid common in warm-temperate to tropical waters around the world. The trivial name, *coeruleoalba,* refers to the diagnostic pattern of blue and white stripes and blazes along the lateral and dorsal sides of their bodies. These dolphins have been well studied in several regions where they occur, such as the western North Pacific, eastern tropical Pacific, and the Mediterranean. They are known to be social and very acrobatic, but at least in the eastern tropical Pacific, wary of boats and do not bow ride.

I. Characters and Taxonomic Relationships

The most striking characteristic of striped dolphins is their bold blue and white color pattern (Fig. 1). The primary features are an eye-to-anus and an eye-to-flipper stripe that are dark blue or bluish black. The dorsal cape is a muted blue or bluish-gray, usually invaded by a white to light gray spinal blaze that flows from the lateral field. The lateral field is usually darker than the ventral field. A dark stripe bifurcating from the eye-to-anus stripe and continuing into the ventral field is usually present. A faint secondary stripe may also be present in the ventral field.

Striped dolphins are similar to most other small oceanic dolphins, having a long beak (well demarcated from the melon) and falcate dorsal fin. In the field they are most likely to be confused with common dolphins (*Delphinus delphis* and *D. capensis*) Fraser's dolphin (*Lagenodelphis hosei*), spinner dolphins (*S. longirostris*), or Clymene dolphins (*S. clymene*) but can be easily distinguished by their robust body and coloration. The longest recorded specimen was 2.56 m and the heaviest specimen weighed 156 kg. Mean maximum body length in the western Pacific is 2.4 m for males and 2.2 m for females (Archer and Perrin, 1999).

Variation in skull and body size has been documented among several geographical regions and there is evidence that further population subdivision may occur within some of these regions. Cranially, *S. coeruleoalba* most closely resembles *S. clymene,* but

Figure 1   *A striped dolphin,* Stenella coeruleoalba.

is significantly larger (Perrin *et al.*, 1981). Sequence analysis of the cytochrome *b* mitochondrial gene supports a sister species relationship between these two species. The two share an unresolved polytomic clade with *Delphinus* spp., *S. frontalis* and *Tursiops aduncus* (LeDuc *et al.*, 1999). However, as the genus *Stenella* was found to be paraphyletic in this study, the final taxonomic resolution of striped dolphins awaits further work.

II. Distribution and Ecology

The range of striped dolphins is well documented in the western and eastern North and tropical Pacific, where most records are below about 43°N. The species has been recorded from the Atlantic coast of northern South America up to the eastern seaboard of North America, with the northern limit a function of the meanderings of the Gulf Stream. *S. coeruleoalba* is found in the eastern North Atlantic south of the United Kingdom and is the most frequently occurring dolphin in the Mediterranean Sea. It has also been documented from the coast of several countries bordering the Indian Ocean, but its full range in this region is unknown. Striped dolphins can usually be found outside the continental shelf, typically over the continental slope out to oceanic waters, often associated with convergence zones and waters influenced by upwelling. The species has been reported in waters of 10 to 26°C, although most records are from about 18–22°C (Archer and Perrin, 1999).

Striped dolphins feed on a variety of pelagic or benthopelagic fish and squid. Off the coast of Japan and South Africa, fish in the family Myctophidae are the dominant prey items, while they primarily feed on cod in the Northeast Atlantic and prey on squid more in the Mediterranean (Perrin *et al.*, 1994). To reach potential prey, striped dolphins may be diving from 200 to 700 m. Attacks by both sharks and killer whales (*Orcinus orca*) have been reported for this species (Archer and Perrin, 1999).

As a result of a large number of specimens available from the Japanese drive fishery and a morbillivirus epizootic in the western Mediterranean, contaminants and PARASITES have been

studied more intensively in this species than in any other cetacean. High organochlorine loads in the western Mediterranean population have been hypothesized to have caused an immunodepressive state, thus decreasing resistance to infection, which in turn led to the deaths of thousands of dolphins from 1990 to 1992. Blubber PCB levels from animals affected by the epizootic may be among the highest recorded values for any mammal (Archer and Perrin, 1999).

III. Behavior and Life History

Striped dolphins perform a variety of aerial behaviors, such as breaching, chin slaps, and a unique behavior termed "rototailing," in which they make high arcing jumps while rapidly rotating their tail before reentering the water. School sizes vary between regions, ranging from 10–30 to several hundred individuals, but rarely greater than 500 (Archer and Perrin, 1999). In the western Pacific, a complex schooling system has been documented in which individuals move among juvenile, adult, and mixed schools (Miyazaki and Nishiwaki, 1978). The adult and mixed schools are further divided into breeding and nonbreeding schools in what is likely to be a polygynous mating system.

MATING is seasonal, with gestation lasting from 12 to 13 months. Body length at birth has been estimated to be between 93 and 100 cm. Males enter sexual maturity between 7 and 15 years of age, whereas females become sexually mature between 5 and 13 years old. Mean length at sexual maturity is 2.1–2.2 m. Density-dependent changes in several life history parameters have been reported for the western Pacific population, probably as a result of large fishery kills. Maximum estimated age for both males and females is 57.5 years (Archer and Perrin, 1999).

IV. Interactions with Humans

The western North Pacific population of striped dolphins has experienced its heaviest mortalities from Japanese striped dolphin drive and hand-harpoon fisheries. Annual catches in one region during the 1940s and 1950s were ca. 8000–9000 animals, reaching as high as 21,000 in some years. However, self-imposed quotas, lower encounter rates, and the dissolution of a fishery co-operative caused annual catches to drop to below ca. 1000 in the early to mid-1980s (Kishiro and Kasuya, 1993). Other directed takes of the species have occurred in St. Vincent and the Mediterranean for human consumption and protection of fishing gear. INCIDENTAL CATCHES have also been recorded from various gear in fisheries across their range. Striped dolphins have not been successfully maintained in captivity, where most have died within 1–2 weeks after not feeding well (Archer and Perrin, 1999).

See Also the Following Articles

Breaching ■ Common Dolphins ■ Fraser's Dolphin ■ Geographic Variation

References

Archer, F. I., and Perrin, W. F. (1999). *Stenella coeruleoalba*. *Mamm. Spec.* **603**, 1–9.

Kishiro, T., and Kasuya, T. (1993). Review of the Japanese dolphin drive fisheries and their status. *Rep. Intl. Whal. Comm.* **43**, 439–452.

Leduc, R. G., Perrin, W. F., and Dizon, A. E. (1999). Phylogenetic relationships among the delphinid cetaceans based on full cytochrome *b* sequences. *Mar. Mamm. Sci.* **15**(3), 619–648.

Miyazaki, N., and Nishiwaki, M. (1978). School structure of the striped dolphin off the Pacific coast of Japan. *Sci. Rep. Whales Res. Inst.* **30**, 65–115.

Perrin, W. F., Mitchell, E. D., Mead, J. G., Caldwell, D. K., and van Bree, P. J. H. (1981). *Stenella clymene*, a rediscovered tropical dolphin of the Atlantic. *J. Mammal.* **6293**, 583–598.

Perrin, W. F., Wilson, C. E., and Archer, F. I. (1994). Striped dolphin—*Stenella coeruleoalba* (Meyen, 1833). *In* "Handbook of Marine Mammals" (S. H. Ridgway, and R. Harrison, eds.), Vol. 5, pp. 129–160.

Surveys

KARIN A. FORNEY
*Southwest Fisheries Science Center,
La Jolla, California*

Figure 1 *NOAA ship* McArthur, *which has been used for many marine mammal surveys in the eastern Pacific Ocean. Photograph by Karin Forney.*

I. Vessel Surveys

Large oceanographic research vessels (Fig. 1) are the most versatile platform for at-sea surveys of marine mammals. They can carry a dozen or more researchers and remain at sea for weeks at a time, providing the ability to cover extensive marine areas. Search efficiency is greatest aboard these large vessels because observers can search from a greater height above the water (on the flying bridge, bridge, or in a crow's nest) and use high-power, deck-mounted binoculars ("big eyes"; Fig. 2) when searching for and identifying marine mammals. Large research vessels usually also have equipment for collecting oceanographic data for marine mammal habitat studies and may be able to tow hydrophones to detect marine mammals acoustically. A

Surveys are used to address many different marine mammal research questions, including distribution, abundance, trends, and habitat associations. Equipment and methodology vary depending on the species of interest, financial resources, availability of research platforms, and survey objective. Marine mammals at sea are most commonly surveyed aboard ships or large boats or from fixed-wing aircraft. Small boats, helicopters, airships, and land-based viewing stations are also used when appropriate. Line-transect methods are often the most effective for estimating the abundance of marine mammals at sea, although other survey techniques are also used. Marine mammals on or near land, such as pinnipeds, sea otters, or walruses, are more commonly counted from land-based viewing points or using aerial photography.

Figure 2 *"Big eyes" used to search for marine mammals on large survey vessels. Photograph by Jay Barlow.*

Figure 3 *Small survey vessel used in the Philippines. Photograph by Louella Dolar.*

significant disadvantage to large research vessels, however, is their great operating cost: approximately $10,000 per day. Small or medium-sized vessels (Fig. 3), including a variety of fishing boats and sail boats, have been used successfully for surveys at a significantly reduced cost. This is the most feasible option in many parts of the world, particularly in developing countries. On these smaller vessels, searching is usually conducted with hand-held binoculars or by naked eye from the highest stable deck or platform on the ship. In some shallow bays and rivers, even smaller boats (e.g., rigid-hull inflatable boats, Boston Whalers) may be required for safe navigation. Auxiliary studies, including photography, biopsy sampling, diving behavior, and prey sampling, are also often possible during vessel surveys.

II. Aircraft Surveys

The main advantages of aerial surveys are the ability to cover large areas quickly and the lower cost of aircraft compared to large ships. They are most useful for preliminary studies to determine the relative distribution and presence or absence of species in a particular region but can also be used to estimate abundance and monitor trends. Aircraft with high wings and bubble windows (Fig. 4) are best-suited because they allow lateral viewing as well as some downward visibility. An additional downward-viewing ("belly") window enhances sighting efficiency considerably because marine mammals are seen most easily from the air when viewing perpendicular to the water surface. A typical aerial survey observer team consists

Figure 4 *NOAA Twin Otter survey aircraft with bubble window (under propeller) and belly window part (above tire). Photograph by Karin Forney.*

of two observers searching through the two side windows, one data recorder, and, if possible, a belly observer.

III. Land-Based Surveys

A few populations of whales reliably migrate close to shore and have been surveyed successfully from land-based stations. These include bowhead whales at Point Barrow, Alaska, and gray whales along the California coast. During these surveys, visual and acoustic means may be used to record all whales that travel past the observation point during the migratory period. Adjustments are made for unobservable periods, such as nighttime and times of poor weather conditions. Pinnipeds are counted most commonly from land-based viewing stations, such as cliff tops, although aerial photographs can provide an excellent alternative means of surveying these animals along the coastline.

IV. Methodological Considerations

Detectability of marine mammals is a key factor in deciding what type of survey platform to use. Animals that dive for prolonged periods, such as SPERM WHALES and BEAKED WHALES, will be missed much more frequently during aerial surveys than from ships because the aircraft travels much faster. For some species, correction factors have been developed to correct for the proportion of animals missed from airplanes or ships. Vessel attraction or avoidance is another concern when designing shipboard marine mammal surveys. For example, harbor porpoises are known to avoid vessels, and if animals are not detected before they react, resulting abundance estimates may be too low. The opposite problem exists for species that are attracted to vessels to "ride the bow"; in these cases abundance estimates may be too high. Both of these problems can be minimized by using a larger vessel that allows viewing from a greater height and with high-power binoculars; animals can then be detected at a greater distance before they react to the vessel.

See Also the Following Articles

Abundance Estimation ▪ Distribution ▪ Population Status and Trends ▪ Stock Assessment

References

Aragones, L. V., Jefferson, T. A., and Marsh, H. (1997). Marine mammal survey techniques applicable in developing countries. *Asian Mar. Biol.* **14,** 15–39.

Barlow, J. (1995). The abundance of cetaceans in California waters. I. Ship surveys in summer and fall of 1991. *Fish. Bull.* **93,** 1–14.

Buckland, S. T., Anderson, D. R., Burnham, K. P., and Laake, J. L. (1993). "Distance Sampling: Estimating Abundance of Biological Populations." Chapman & Hall, London.

Buckland, S. T., Breiwick, J. M., Cattanach, K. L., and Laake, J. L. (1993). Estimated population size of the California gray whale. *Mar. Mamm. Sci.* **9,** 235–249.

Fiedler, P. C., and S. B. Reilly. (1994). Interannual variability of dolphin habitats in the eastern tropical Pacific. I. Research vessel surveys, 1986–1990. *Fish. Bull.* **92,** 434–450.

Forney, K. A., Barlow, J., and Carretta, J. V. (1995). The abundance of cetaceans in California waters. II. Aerial surveys in winter and spring of 1991 and 1992. *Fish. Bull.* **93,** 15–22.

Garner, G. W., Amstrup, S. C., Laake, J. L., Manly, B. F. J., McDonald, L. L., and Robertson, D. G. (1999). "Marine Mammal Survey and Assessment Methods." A. A. Balkema, Rotterdam.

Heide-Jørgensen, M. P., Teilmann, J., Benke, H., and Wulf, J. (1993). Abundance and distribution of harbor porpoises *Phocoena phocoena* in selected areas of the western Baltic and the North Sea. *Helgoländer Meeresuntersuchungen* **47,** 335–346.

Hiby, A. R., and Hammond, P. S. (1989). Survey techniques for estimating abundance of cetaceans. *Rep. Intl. Whal. Comm. Spec. Issue* **11,** 47–80.

Kraus, S. D., Gilbert, J. R., and Prescott, J. H. (1983). A comparison of aerial, shipboard, and land-based survey methodology for the harbor porpoise, *Phocoena phocoena. Fish. Bull.* **81,** 910–913.

Lowry, M. S. (1999). Counts of California sea lion (*Zalophus californianus*) pups from aerial color photographs and from the ground: A comparison of two methods. *Mar. Mamm. Sci.* **15,** 143–158.

Øien, N. (1991). Abundance of the northeastern Atlantic stock of minke whales based on shipboard surveys conducted in July 1989. *Rep. Intl. Whal. Comm.* **41,** 433–437.

Rathbun, G. (1988). Fixed-wing airplane versus helicopter surveys of manatees (*Trichechus manatus*). *Mar. Mamm. Sci.* **4,** 71–74.

Vidal, O., Barlow, J., Hurtado, L. A., Torre, J., Cendon, P., and Ojeda, Z. (1997). Distribution and abundance of the Amazon river dolphin (*Inia geoffrensis*) and the tucuxi (*Sotalia fluviatilis*) in the upper Amazon River. *Mar. Mamm. Sci.* **13,** 427–445.

Wade, P. R., and Gerrodette, T. (1993). Estimates of cetacean abundance and distribution in the eastern tropical Pacific. *Rep. Intl. Whal. Comm.* **413,** 477–494.

Sustainability

CHARLES W. FOWLER
*National Marine Mammal Laboratory,
Seattle, Washington*

Sustainability has been elusive. Sustainability is important in managing our use of resources (e.g., the harvest of both marine mammals and finfishes). Harvest levels must be established in ways that account for complexity, including the direct effects on harvested populations and such indirect factors as exemplified by the interactions between marine mammals and fish populations. The "sustainable development" of marine mammals is an unlikely option, partly because of their low productivity levels (e.g., Perrin, 1999). However, defining sustainability whether applied to marine mammals, fisheries, or ecosystems remains an important objective.

Historically, the concept of maximum sustainable yield (MSY) has played a major role in the management of our utilization of natural resources. This approach has yet to be assessed in its contribution to worldwide problems such as overharvested fish populations (e.g., Rosenberg *et al.*, 1993; Committee on Ecosystem Management for Sustainable Marine Fisheries, 1999). Commercial whaling and sealing have also involved concepts derived from the MSY approach.

The inadequacies of management based on MSY have been recognized and other methodologies have been developed [e.g., the catch limit algorithm of the INTERNATIONAL WHALING COMMISSION (Slooten, 1996) and the potential biological removal approach (PBR) being used by the National Marine Fisheries Service in the United States (Wade, 1998)]. These alternatives, however, have not escaped the weaknesses of being applicable only to individual species and do not account for complexity. The ecosystem effects of fishing (e.g., Hall, 1999), whaling, or sealing are not adequately considered in current management strategies. The challenges currently facing management are not being met.

It is therefore extremely important to find alternatives that will work. One approach is systemic management (Fowler and Perez, 1999) in which empirical examples are used to define and measure sustainability. Along with other species, marine mammals are sources of information about sustainability that is broadly applicable and meets the demands being made of management.

Marine mammals serve as empirical examples of how sustainable roles, or niches, can be maintained in marine ecosystems because they have persisted as parts of such systems over evolutionary time scales. Resource consumption is a good example of a measure for describing sustainability. Both marine mammals and fisheries consume biomass from various resource species (e.g., fish or squid), making part of their interaction competitive. The various rates at which marine mammals consume biomass exemplify differing levels of sustainability. Importantly, there are limits to variation in such consumption rates. It is now being suggested that conventional approaches to management be replaced by a process that mimics empirical examples of sustainability (systemic management: Fowler *et al.*, 1999; Fowler and Perez, 1999). For example, such management would be carried out to ensure that fishery catches would fall within the normal limits of natural variation for consumption rates among other consumer species. Thus, observed examples of sustainable consumption rates can be used to regulate the catches taken by fisheries while simultaneously conserving resources and habitat for other species. As such, marine mammals, as well as other species, are empirical examples of sustainability that provide information that applies systemically.

I. Management Questions, Empirical Answers

How many tons of whales, seals, fish, cephalopods, or other resources should we harvest each year? What is the appropriate or optimal rate at which to harvest biomass, and when and where should it be harvested? Such questions can be asked with regard to a single resource species or with respect to any area of the oceans, an ecosystem, a group of resource species, a season, or the biosphere. How do we answer all such questions so that the answer for one case will not be in conflict with the answer for another? Empirical information is key to answering such questions consistently, and marine mammals are key elements in providing such information (Fowler, 1999; Fowler *et al.*, 1999).

Figure 1 shows frequency distributions (empirical probability distributions) for consumption rates exhibited by various marine mammals. Also shown, for comparison in each case, is the harvest rate for fisheries: the rate at which humans con-

sume biomass. The top panel illustrates consumption from a population of an individual species, and the bottom panel represents total consumption within the biosphere. Intermediate panels depict consumption from a group of resource species, an ecosystem, and the marine environment. Thus, Fig. 1 represents information for a telescoped series of increasing complexity. In this case, the biosphere contains the marine environment, which, in turn, includes the ecosystem (Bering Sea). Within the ecosystem we find populations of resource species (the finfish), among which is the population of walleye pollock (*Theragra chalcogramma*, one species).

Various requirements are placed on management. It has been made clear that management should promote actions whereby components and processes within systems will fall within their normal range of natural variation (e.g., Christensen *et al.*, 1996; Mangel *et al.*, 1996). These include the components and processes that constitute individuals, species, ecosystems, and the biosphere. Humans (through commercial fishing) are obvious outliers relative to the natural variation illustrated in Fig. 1.

There are things management cannot do. The other species within these distributions are largely beyond our control, especially the collective aggregate of species in each distribution. Individual species can show change opposite to our intentions if we act to influence them directly. In fact, changes we stimulate in these species may result in unwanted reactions in the rest of the system whether we purposely manipulate them individually or as a group.

Management can move forward, however, by focusing on reducing consumption by humans to fall within the normal range of the variation exhibited by other species. Nonhuman species exhibit consumption rates that occur within the context of complexity—all the things that influence these species. That is, all the things that have contributed to the observed rates of consumption, such as those shown in Fig. 1, are taken into account.

Figure 1, then, is a rudimentary guide for the management of our use of fishery resources when achieving sustainability is the ultimate goal. Empirical data exemplify what works in the face of the complete suite of factors that set limits in the consumption of resources. Better options are found among the more numerous examples toward the centers of the distributions compared to the scarcity of examples in the tails (Fig. 1). The risks and constraints that prevent the accumulation of species in such regions are to be avoided. How, then, can we carry out sustainable management?

II. Management to Achieve Sustainability

It has long been recognized that we need a form of management that applies to ecosystems (Christensen *et al.*, 1996). To develop sustainable management strategies, humans must manage by consuming biomass from ecosystems at rates that are sustainable. Sustainable management requires change. For example, reducing consumption by humans to 10% of current harvest rates would place our species squarely within the normal range of natural variation shown in the ecosystem (third)

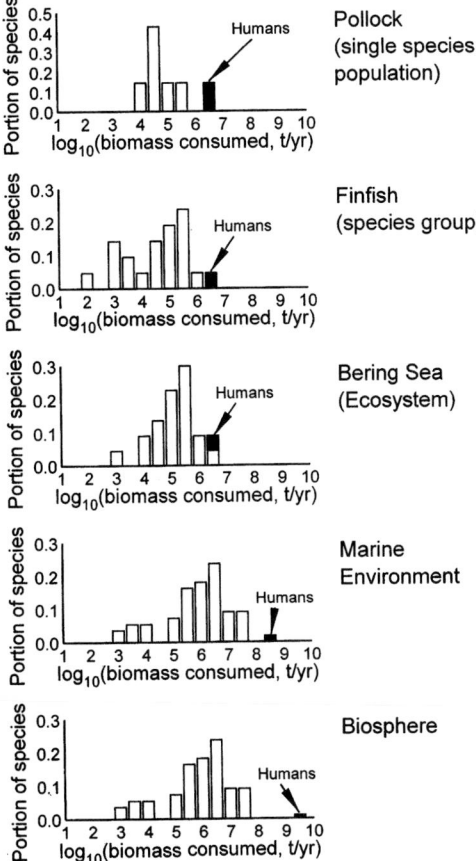

Figure 1 *The frequency distribution of consumption rates (tons of biomass consumed per year, in log$_{10}$ scale) for marine mammals showing optimal consumption rates where most species are concentrated. The rate at which humans harvest biomass is shown for comparison. The top panel shows the natural variation in consumption of walleye pollock (Theragra chalcogramma) as observed for 6 species of marine mammals in the Bering Sea in comparison to recent takes of pollock by commercial fisheries (Livingston, 1993). The second panel shows consumption of finfish in the Bering Sea by 20 species of marine mammals compared to fisheries takes (predominantly pollock; Fowler and Perez, 1999). Total biomass consumption is shown for 20 species of marine mammals in the Bering Sea in the third panel, again compared to the commercial take, which is predominantly pollock (Fowler and Perez, 1999). Total biomass consumption for the entire marine environment (all oceans combined) is shown in the fourth panel for 55 species of marine mammals, here compared to the take of about 110 million metric tons estimated as the harvest of biomass for human use in the late 1990s (Committee on Ecosystem Management for Sustainable Marine Fisheries, 1999). Worldwide consumption of biomass by humans is compared to that of 55 species of marine mammals in the bottom panel. The last two panels are based on indirect estimates (Fowler and Perez, 1999) using population and body size data from the marine mammal series by Ridgway and Harrison (1981–1999) and equations representing ingestion rates as a function of body size in Peters (1983).*

panel of Fig. 1. Such change will require time to accomplish. Maintenance of reduced levels, once achieved, would lead to further challenges for management. These would include responding to changes within ecosystems over seasons, with shifts in climate, or in response to our management.

It has been recognized for some time, in the world of commercial fisheries management, that the historical focus on managing harvests from the individual-species point of view has been insufficient, partly because of the need to consider ecosystems. However, management restricted to ecosystems would be similarly insufficient because of the need to account for both the broader marine environment and the biosphere, as well as individual species.

Management based on empirical examples is an integrated approach that simultaneously helps define sustainable harvests at the level of individual species and the biosphere. For example, reducing harvests to between 1 and 10% of recent harvest levels would be required for the species represented in the top panel of Fig. 1. We can similarly account for multispecies groups and the entire marine environment (second and fourth panels, respectively, Fig. 1). The biosphere can be involved by reducing human consumption to about 0.1% of current consumption levels (Fig. 1, panel 5).

In all cases shown in Fig. 1, management would involve change measured in orders of magnitude to fall within the normal ranges of natural variation. If the distributions themselves change in response to the reduction of harvested biomass by humans, then sustainable management would need to respond to the new information. This requires continuous monitoring through concerted scientific effort to observe such changes.

III. Accounting for Complexity

In management, the list of things to be considered seems endless. For example, we need to account for the effects of evolution, we need to account for endangered species, and we need to deal with multiple complex processes, such as nutrient flow within ecosystems. It is often said that management needs to be interdisciplinary, or an integrated accounting of everything science can study. However, we also need to account for the things we do not, or cannot, study or know about. This is accomplished in two ways when we take advantage of empirical examples provided by species such as marine mammals.

First, every species reflects the effects of everything that influences it. These factors are integrated in the distributions shown in Fig. 1. For example, evolution is taken into account through its influence on the position of every species within each distribution. Each species represents a composite of balances among various opposing forces (e.g., those involved in predation, population growth, evolution, or extinction). Each species reflects the constraints of the system—constraints such as competition for the limited availability of energy in its path through the various trophic levels. The relative importance of each factor compared to every other factor is taken into account implicitly.

Second, we must expand our management questions. We must address the consumption of biomass in space and time. Distributions similar to those of Fig. 1 can be developed for the allocation of biomass consumption over alternative resource species, seasons, and geographic space. We have to deal with complexity by addressing the other management questions before us. Marine mammals can be used as empirical examples in such an endeavor, and through such species we account for the unknowns that influence the position of each species in distributions similar to those of Fig. 1. For marine mammals themselves, finding the rates at which they can be harvested sustainably would involve information regarding consumption by their predators, including other marine mammals!

IV. Consistency

An important aspect of empirical examples is their representation of a system that is internally consistent. Advice for management at the ecosystem level (Fig. 1, panel 3) will be consistent with advice at the individual species level (panel 1) when applied simultaneously. Collective application at the individual species level must be constrained to the limits set by application at the ecosystem level. Nutrients, energy, biomass, and species involved in this systemic consistency guarantee freedom of conflict because the conflicting forces of nature result in what we see in distributions derived empirically.

Marine mammals can serve as empirical sources of information about how SPECIES fit into ecosystems. Through information about other species, such as marine mammals, we are provided with guidance about sustainable harvest levels (e.g., total biomass consumed per year), allocation of harvests over time and space, allocation across resource species, and other management questions.

See Also the Following Articles

Fishing Industry, Effects of ■ Management ■ Population Status and Trends ■ Whaling, Modern

References

Christensen, N. L., Bartuska, A. M., Brown, J. H., Carpenter, S. R., D'Antonio, C., Francis, R., Franklin, J. F., MacMahon, J. A., Noss, R. F., Parsons, D. J., Peterson, C. H., Turner, M. G., and Woodmansee, R. G. (1996). The report of the Ecological Society of America Committee on the Scientific basis for ecosystem management. *Ecol. Appl.* **6**, 665–691.

Committee on Ecosystem Management for Sustainable Marine Fisheries (1999). "Sustaining Marine Fisheries." National Academy Press, Washington, DC.

Fowler, C. W. (1999). Natures's Monte Carlo experiments in sustainability. *In* "Proceedings of the Fifth NMFS Stock Assessment Workshop: Providing Scientific Advice to Implement the Precautionary Approach under the Magnuson-Stevens Fishery Conservation and Management Act" (V. R. Restrepo, ed.), pp. 25–32. U.S. Department of Commerce, NOAA Technical Memorandum NMFS-F/SPO-40.

Fowler, C. W., Baker, J. D., Shelden, K. E. W., Wade, P. R., DeMaster, D. P., and Hobbs, R. C. (1999). Sustainability: Empirical examples and management implications. *In* "Ecosystem Approaches for Fishery Management," pp. 305–314. Univ. of Alaska Sea Grant, Fairbanks, Alaska.

Fowler, C. W., and Perez, M. A. (1999). "Constructing Species Frequency Distributions: A Step toward Systemic Management." U.S.

Department of Commerce, NOAA Technical Memorandum NMFS-AFSC-109.

Hall, S. J. (1999). "The Effects of Fishing on Marine Ecosystems and Communities." Blackwell Science, Oxford.

Livingston, P. A. (1993). Importance of predation by groundfish, marine mammals and birds on walleye pollock *Theragra chalcogramma* and Pacific herring *Clupea pallasi* in the eastern Bering Sea. *Mar. Ecol. Prog. Ser.* **102**, 205–215.

Mangel, M., Talbot, L. M., Meffe, G. K., Agardy, M. T., Alverson, D. L., Barlow, J., Botkin, D. B., Budowski, G., Clark, T., Cooke, J., Crozier, R. H., Dayton, P. K., Elder, D. L., Fowler, C. W., Funtowicz, S., Giske, J., Hofman, R. J., Holt, S. J., Kellert, S. R., Kimball, L. A., Ludwig, D., Magnusson, K., Malayang, B. S., Mann, C., Norse, E. A., Northridge, S. P., Perrin, W. F., Perrings, C., Peterman, R. M., Rabb, G. B., Regier, H. A., Reynolds, J. E., III, Sherman, K., Sissenwine, M. P., Smith, T. D., Starfield, A., Taylor, R. J., Tillman, M. F., Toft, C., Twiss, J. R., Jr., Wilen, J., and Young, T. P. (1996). Principles for the conservation of wild living resources. *Ecol. Appl.* **6**, 338–362.

Perrin, W. F. (1999). Selected examples of small cetaceans at risk. *In* "Conservation and Management of Marine Mammals" (J. Twiss and R. R. Reeves, eds.), pp. 296–310. Smithsonian Press, Washington, DC.

Peters, R. H. (1983). "The Ecological Implications of Body Size." Cambridge Univ. Press, New York.

Ridgway, S. H., and Harrison, R. (eds.) (1981–1999). "Handbook of Marine Mammals," Vols. 1–6. Academic Press, New York.

Rosenberg, A. A., Fogarty, M. J., Sissenwine, M. P., Beddington, J. R., and Shephard, J. G. (1993). Achieving sustainable use of renewable resources. *Science* **262**, 828–829.

Slooten, E. (1998). Risk analysis at the International Whaling Commission. *In* "Statistics in Ecology and Environmental Monitoring 2: decision making and risk assessment in biology." (D. J. Fletcher, L. Kavalieris, and B. F. J. Manly, eds.), pp. 173–180. University of Otago Press, Dunedin, New Zealand.

Wade, P. R. (1998). Calculating limits to the allowable human-caused mortality of cetaceans and pinnipeds. *Mar. Mamm. Sci.* **14**, 1–37.

Susu and Bhulan
Platanista gangetica gangetica and *P. g. minor*

BRIAN D. SMITH
Aquatic Biodiversity Associates, Eureka, California

The susu and bhulan, *Platanista gangetica gangetica* and *P. g. minor,* are two river dolphins of the South Asian subcontinent. Some scientists dispute the existence of one species and consider them to be separate, *P. gangetica* and *P. minor* (Pilleri *et al.,* 1982). A variety of vernacular names are used for the susu and bhulan, mostly connoting the sound made during respiration. For the susu, these include *swongsu* and *sus matsya* (*matsya* = fish) in Nepali, *soonse* and *sunsar* in Hindi, *hiho* and *shihu* in Assamese, and *shushuk* and *sishu-, foo-,* or *hungmaach* (*maach* = fish) in Bengali. For the bhulan, these include *bhoolun* and *sisar.*

I. External Appearance

The subspecies are nearly identical in external appearance (Fig. 1). Their body is supple and robust, attenuating behind

Figure 1 *(A) The susu* (Platanista gangetica gangetica) *and (B) bhulan* (P. g. minor) *are river dolphins with long narrow snouts reminiscent of gharial crocodilians* (Gavialis gangeticus) *that are also found in some of the same rivers. These endangered taxa were once hunted, a practice now prohibited. Entanglement in gillnets, water development, pollution, and occasional poaching are adversely affecting the survival of this species. Pieter A. Folkens/Higher Porpoise DG.*

the dorsal fin in a narrow tail stock. COLORATION is gray overall, becoming blotchy with age. Bellies of young animals are lighter and often have a pinkish hue. The dolphins have a long snout that becomes wider at the tip. Adult females generally have a longer snout, which sometimes curves upward and to one side at the tip when particularly long. Their numerous narrow pointed teeth are curved inward and become longer toward the distal end of the mandibles, becoming visible in larger individuals. Teeth become peg-like in older individuals from wear and accumulation of the cement layer. Their eyes are extremely small and visible as pinhole openings slightly above the upturned mouth. The external auditory meatus is larger than the eye opening and located slightly above it, a unique arrangement among odontocetes. The blowhole is a small slit oriented longitudinally, which is a rare but not unique configuration among cetaceans. A distinct median ridge begins anterior of the blowhole and bisects a convex melon, which becomes less rounded as the dolphin approaches adulthood. The dorsal fin is a low triangular hump with a slightly defined knob at the apex located about two-thirds of the body length posterior of the melon. Broad flippers are squared-off at the end with a crenellated margin, and the arm and hand bones are visible beneath the taut dorsal surface. Flukes are broad with a concave margin and distinct median notch. The opening for male genitals is located much closer to the umbilicus in relation to the anus when compared with most other cetaceans.

II. Taxonomy

Primitive features, such as intestinal cecum, and weak evidence from the nucleotide sequence of the cytochrome *b* gene suggest that *Platanista* may be a sister taxon to all other extant odontocetes (Rice, 1998). Although the superfamily Platanistoidea was previously considered to include three other monotypic genera—*Pontoporia, Lipotes,* and *Inia* (Kasuya, 1973;

Zhou, 1982)—the genus *Platanista* is now recognized as the only extant member of the taxon (Rice, 1998). Pilleri *et al.* (1982) recognized two species of *Platanista* based on differences in the prominence of the nasal crests, caudal height of the maxillary crests, length of the lower transverse processes of the sixth and seventh cervical vertebrae, blood protein composition, and free-esterified cholestrin ratio in the lipids. Their arguments were unconvincing due to the small sample of adult specimens examined and the absence of statistical analyses (see Reeves and Brownell, 1989). Based on differences in tail lengths, Kasuya (1972) proposed that the susu and bhulan be considered subspecies; this was followed by Rice (1998).

III. Habitat

Susus and bhulans generally occur in eddy countercurrents downstream of channel convergences and sharp meanders, and upstream and downstream of mid-channel islands. They also occasionally occur in countercurrents induced by engineering structures such as bridge pilings and groynes. The affinity of *Platanista* for countercurrents is probably greatest in upstream tributaries where productivity is especially clumped and strong downstream currents restrict occupancy to the hydraulic refuge these areas provide (see Smith, 1993; Smith *et al.*, 1998).

IV. Population and Distribution

Although no rigorous range-wide surveys have been conducted, the aggregate populations of susu and bhulan are believed to number in the low thousands and few hundreds, respectively (Smith and Reeves, 2000). A map of historical distribution, charted by the 19th century British naturalist John Anderson (Anderson, 1879), shows the bhulan occurring throughout the Indus River mainstem and in the Sutlej, Ravi, Chenab, and Jhelum tributaries and shows the susu occurring throughout the Ganges–Brahmaputra–Megna and Karnaphuli river systems in Nepal, India, and Bangladesh. Anderson (1879) stated that the range of both species was only limited downstream by increasing salinity in deltas and upstream by rocky barriers or insufficient water. Their distribution has shrunk considerably since then, largely due to water development, which has blocked dolphin movements and degraded their habitat.

Bhulans currently occupies less than 700 km of the Indus mainstem (about one-fifth of the historic range), fragmented into three subpopulations by the Chashma, Taunsa, Guddu, and Sukkur barrages (Fig. 2; Reeves and Chaudhry, 1998). The largest subpopulation is located at the downstream end of the dolphins' range, between the Guddu and Sukkur barrages, with a reported count of 458 individuals (Mirza and Khurshid, 1996). Numbers decline progressively in upstream segments, despite their progressively larger geographical size. Counts of 143 and 39 individuals were reported between the Taunsa and Guddu and the Taunsa and Chasma barrages, respectively (Mirza and Khurshid, 1996). A few scattered individuals may still occur upstream of the Chashma barrage in the Indus and downstream of the Trimmu, Sidhnai, and Panjnad barrages in the Chenab, Ravi, and Sutlej rivers, respectively (Reeves *et al.*, 1991). A few bhulans may also stray below the Sukkur barrage, but these animals are lost to the overall population and have little chance of surviving due to lack of water.

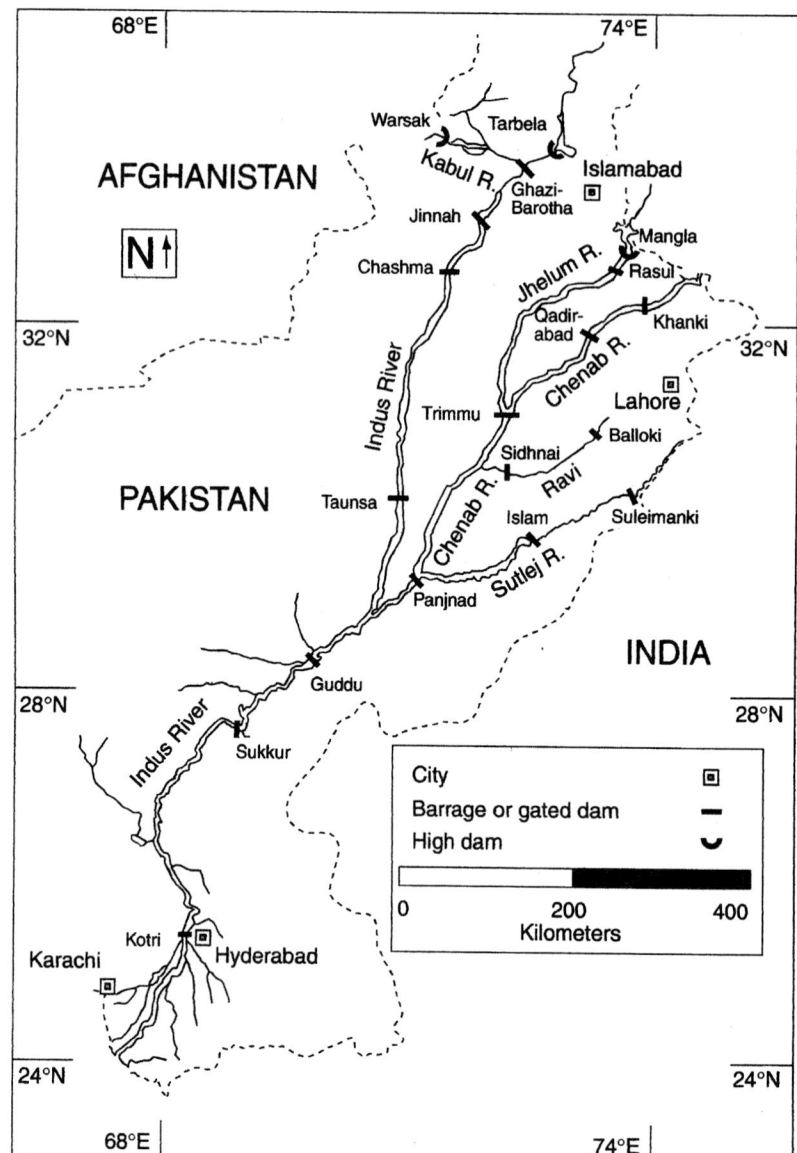

Figure 2 *Map of the Indus river system of Pakistan showing dams and barrages that have fragmented the population of bhulans and degraded their habitat.*

Historically, susus were found in the Ganges River as far upstream as Haridwar, about 100 km above their current range, and in the Yamuna River year round as far upstream as Delhi, probably about 400 km above their current low-water range (Fig. 3; Sinha *et al.*, 2000). In the Ganges mainstem, there are four extant subpopulations isolated between barrages. In the northern tributaries of the Ganges, of the six subpopulations that have been isolated above or between barrages, three have been extirpated (Gandak River above the Gandak barrage and Sarda River above the upper and lower Sarda barrages) (Sinha *et al.*, 2000) and one reduced to insignificant numbers (Kosi River above the Kosi barrage) (Smith *et al.*, 1994). In the Son River, a southern tributary of the Ganges, a small population of susus has been isolated above the Indrapuri barrage, and the

dolphins no longer occur during the dry season for about 100 km below the barrage until the Ganges confluence (Sinha *et al.*, 2000). In southern Bangladesh, susus were documented in the Sangu River (Ahmed, 2000), which is connected to the Karnaphuli by the Sikalbaha–Chandkhali channel. There are occasional reports of dolphins remaining in the reservoir behind Kaptai Dam, but recent surveys produced no sightings.

V. Internal Anatomy

A. Bones

An extraordinary feature of the skull of *Platanista* is the projection of a maxillary crest, upward and forward, covering an air sinus that leads to the tympanic cavity. The crest slants to

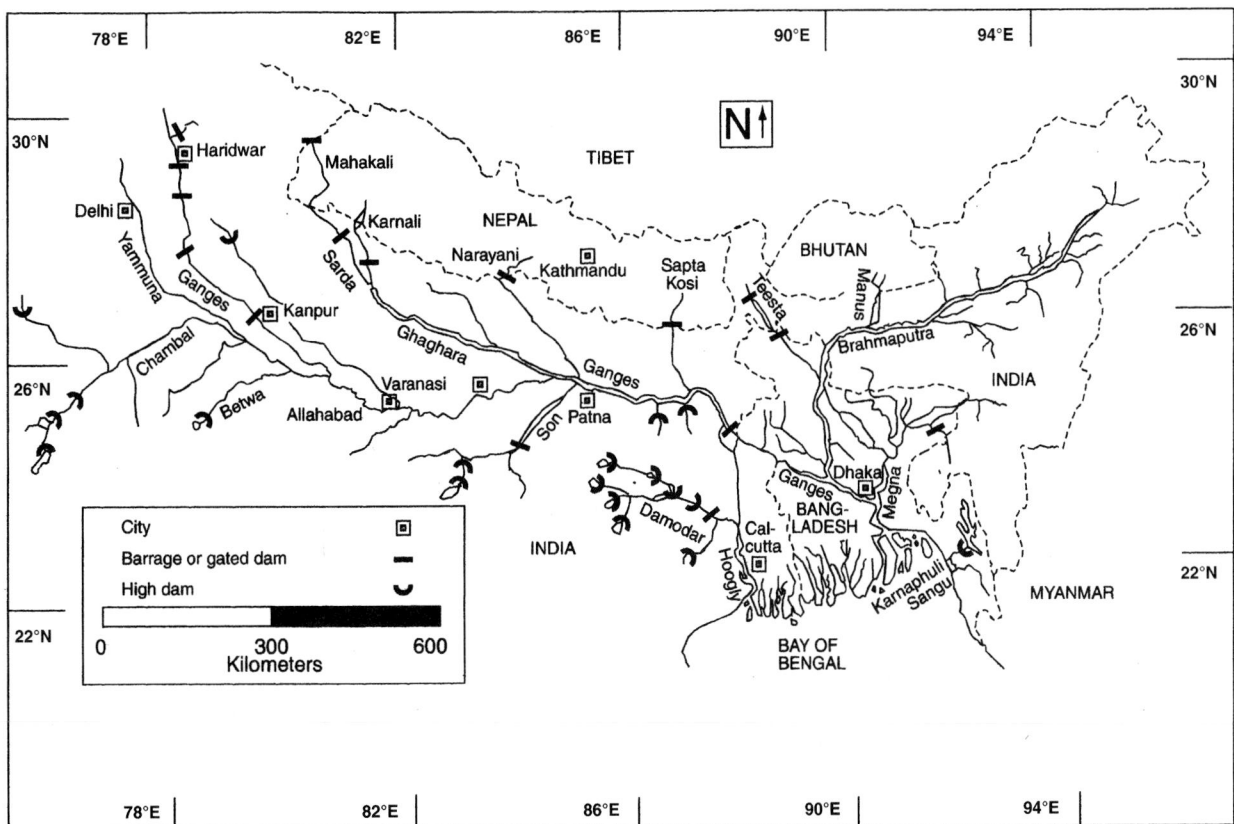

Figure 3 *Map of the Ganges–Brahmaputra–Megna and Karnaphuli–Sangu river systems of Nepal, India, and Bangladesh showing dams and barrages that have fragmented the population of susus and degraded their habitat.*

the left, which makes the SKULL among the most asymmetric of all odontocetes. Other skull features unique to *Platanista* include that the palatine is found in the nasal tube and that the pterygoids are external on the palate and enter the temporal fossa (Reeves and Brownell, 1989). The mandibular symphysis is laterally compressed, slightly upturned toward the end, and it constitutes as much as two-thirds the mandible length of adult females and a little more than half of adult males (Reeves and Brownell, 1989). Cervical vertebrae are unfused, allowing considerable neck movement. Postcranial skeletal features unique to *Platanista* include the location of costal facets of the thoracic vertebrae on the posterior margin of the centrum, a thicker ulna than radius, and the absence of ulnare and pisiform bones in the flippers.

B. Organs

The presence of a cecum between the large and the small intestines, a penis with erectile side lobes, nasolaryngeal air sacs that form a diverticulum of the eustachian tube, and a primitive and relatively unlobulated KIDNEY distinguish *Platanista* from other odontocetes. Compared to other dolphins, their BRAIN is small and neocortical development is low; however, subcortical components associated with acoustical functions are well developed. Their small eyes lack a crystalline lens, giving the dolphins a reputation for being blind, but their retinas do have light-gathering receptors (Herald *et al.*, 1969).

VI. Life History

Males attain sexual maturity at a body length of about 170 cm and physical maturity at 200–210 cm. Females attain sexual maturity at similar or slightly larger body lengths but physical maturity at about 250 cm. The generally larger rostrum of females accounts for this SEXUAL DIMORPHISM, which becomes evident at a body length of about 150 cm. Length at birth is estimated to be about 70 cm. Gestation lasts approximately 1 year, with possible peak birthing seasons in early winter and early summer. Young begin feeding on small prey at about 1 or 2 months and are weaned within a year (Kasuya, 1972).

VII. Behavior and Sensory Abilities

In 1968, three female bhulans were taken to the Steinhart Aquarium in San Francisco, where all died within 7 weeks. During the 1970s, the Swiss neuroanatomist Georgio Pilleri kept a total of seven bhulans in his institute at the University of Berne until they all died within a few years. Studies of these dolphins revealed exceptional aspects of *Platanista* behavior and sensory abilities. The dolphins vocalize almost constantly and swim on their sides in vertical circles. The dolphins continuously emit trains of high-frequency (15–150 kHz) echolocation clicks, interrupted by short pauses of 1–60 sec (Herald *et al.*, 1969). The click trains are focused in two highly directional fields, the dorsal one emitted directly from the melon and the ventral one reflected downward by the maxillary crest, with an acoustic "scotoma" in between (Pilleri *et al.,* 1976). During a dive, the dolphins spin 90° on their lateral axis and position their head down, sweeping it back and forth in a scanning motion, while trailing one flipper along or slightly above the bottom. Shortly before surfacing, the dolphins reverse their spin and surface close to their original position and orientation.

In the wild, susus and bhulans are observed alone or in clusters of 2–3, but occasionally as many as 20 individuals. With the exception of mother–young pairs, the attracting force for these clusters may be more related to the patchy distribution of prey and hydraulic refuge than to survival or reproductive advantages gained by close social affiliations. Susus have been observed surfacing just inside of the upstream end of countercurrent boundaries where the eddies become aligned with main flow. Surfacing at this location may allow the dolphins to minimize energy outputs while monitoring foraging opportunities in the mainstream flow and center pool of the countercurrent (Smith, 1993).

VIII. Threats and Conservation

Both subspecies are classified by the IUCN as endangered. Perhaps the most significant threat to their survival is the existence of numerous dams and barrages that have severely fragmented populations and reduced the amount of suitable habitat. Dams are absolute barriers to dolphin movements. Subpopulations trapped above barrages are believed to lose dolphins when they move downstream during high water while the barrage gates are open. These dolphins probably cannot return due to strong hydraulic forces between the gates while they are open. This apparent involuntary attrition exasperates normal biological problems faced by small isolated populations. Water diverted by barrages, generally for irrigation and flood control, and abstracted by surface pumps and tube wells also results in dolphins competing with humans for the actual substance of their environment: fresh water. During the low-water season, the Indus and Ganges rivers become virtually dry downstream of the Sukkur and Farakka barrages, respectively.

Deliberate killing of bhulans for meat and oil was a traditional practice until at least the early 1970s (Pilleri and Zbinden, 1973–1974). HUNTING is now banned but poaching still occasionally occurs (Reeves and Chaudhry, 1998). Susus are killed by "tribals" in the upper Brahmaputra for their meat and by fishermen in the middle reaches of the Ganges for their oil, which is used as a fish attractant (Smith and Reeves, 2000). Similar to all cetaceans, the susu and bhulan are threatened by entanglement in fishing gear and vessel collisions. Their preferred habitat is often in the same location as primary fishing grounds and ferry crossings, which puts the dolphins at increased risk. The problem of accidental killing in fisheries will undoubtedly worsen as the demand for fish and fishing employment increases. POLLUTION may also be affecting the survival of both species, especially considering the decline in the flushing benefits of abundant water and the aggregate distribution of river dolphins in areas of intensive human use. As top CARNIVORES, the dolphins are particularly vulnerable to persistent contaminants (e.g., PCBs and DDTs), some of which are banned or strictly regulated in more developed countries but widely used in industry and agriculture of south Asia.

In 1974, the government of Sindh declared the Indus River between the Sukkur and Guddu barrages a dolphin reserve and the government of Punjab prohibited deliberate killing (Reeves *et al.,* 1991). Enforcement of these measures seems to have stopped the rapid population decline of the bhulan reported by Pilleri and Zbinden (1973–1974). The susu was perhaps the first cetacean to receive official protection from hunting when it was listed as a protected species in the Moral Edicts of King Asoka in India more than 2000 years ago. Susus currently receive legal protection from deliberate killing in all range states. The Vikramshila Gangetic Dolphin Sanctuary, Bihar, India, located between Sultanganj and Kahalgaon in the mainstem of the Ganges was designated as a protected area for dolphins in August 1991. In a few smaller tributaries, susus receive nominal protection by virtue of small portions of their habitat being included in or adjacent to national parks and sanctuaries.

See Also the Following Articles

Endangered Species and Populations ▪ Habitat Pressures ▪ River Dolphins

References

Ahmed, B. (2000). Water development and the status of the shushuk (*Platanista gangetica*) in southeast Bangladesh. *In* "Biology and Conservation of Freshwater Cetaceans in Asia, IUCN Species Survival Commission Occ. Pap." (R. R. Reeves, B. D. Smith, and T. Kasuya, eds.). pp. 62–66. Gland, Switzerland.

Anderson, J. (1879). "Anatomical and Zoological Researches: Comprising an Account of Zoological Results of the Two Expeditions to Western Yunnan in 1868 and 1875; and a Monograph of the Two Cetacean Genera, Platanista and Orcella." Bernard Quaritch, London.

Herald, E. S., Brownell, R. L., Jr., Frye, F. L., Morris, E. J., Evans, W. E., and Scott, A. B. (1969). Blind river dolphins: First side-swimming cetaceans. *Science* **166,** 1408–1410.

Kasuya, T. (1972). Some informations on the growth of the Ganges dolphin with a comment on the Indus dolphin. *Sci. Rep. Whales Res. Inst.* **24,** 87–108.

Kasuya, T. (1973). Systematic consideration of recent toothed whales based on the morphology of tympano-periotic bone. *Sci. Rep. Whales Res. Inst.* **25,** 1–103.

Mirza, A. H., and Khurshid, S. N. (1996). "Survey of the Indus dolphin, *Platanista minor,* in Sindh." World Wildlife Fund for Nature-Pakistan and Sindh Wildlife Department, Karachi.

Pilleri, G., and Zbinden, K. (1973–1974). Size and ecology of the dolphin population (*Platanista indi*) between the Sukkur and Guddu barrages, Indus River. *Invest. Cetacea* **5,** 59–69.

Pilleri, G., Gihr, M., Purves, P. E., Zbinden, K., and Kraus, C. (1976). On the behaviour, bioacoustics and functional morphology of the Indus River dolphin (*Platanista indi* Blyth, 1859). *Invest. Cetacea* **6,** 11–141.

Pilleri, G., Marcuzzi, G., and Pilleri, O. (1982). Speciation in the Platanistoidea: Systematic, zoogeographical and ecological observations on recent species. *Invest. Cetacea* **14,** 15–46.

Reeves, R. R., and Brownell, R. L., Jr. (1989). Susu *Platanista gangetica* (Roxburgh, 1801) and *Platanista minor* Owen, 1853. *In* "Handbook of Marine Mammals, River Dolphins and Larger Toothed Whales" (S. H. Ridgeway and R. Harrison, eds.), Vol. 4, pp. 69–100. Academic Press, London.

Reeves, R. R., and Chaudhry, A. A. (1998). Status of the Indus river dolphin *Platanista minor. Oryx* **32**(1), 35–44.

Reeves, R. R., Chaudhry, A. A., and Khalid, U. (1991). Competing for water on the Indus Plain: Is there a future for Pakistan's river dolphins? *Environ. Conserv.* **18**, 341–350.

Rice, D. W. (1998). "Marine Mammals of the World, Systematics and Distribution." Special publications of the Society of Marine Mammalology, Lawrence, KS.

Sinha, R. K., Smith, B. D., Sharma, G., Prasad, K., Choudhury, B. C., Sapkota, K., Sharma, R. K., and Behera, S. K. (2000). Status and distribution of the Ganges Susu, *Platanista gangetica*, in the Ganges river system of India and Nepal. *In* "Biology and Conservation of Freshwater Cetaceans in Asia, IUCN Species Survival Commission Occ. Pap." (R. R. Reeves, B. D. Smith, and T. Kasuya, eds.). pp. 42–48. Gland, Switzerland.

Smith, B. D. (1993). 1990 status and conservation of the Ganges River dolphin (*Platanista gangetica*) in the Karnali River Nepal. *Biol. Conserv.* **66**, 159–169.

Smith, B. D., Haque, A. K. M., Hossain, M. S., and Khan, A. (1998). River dolphins in Bangladesh: Conservation and the effects of water development. *Environ. Manage.* **22**(3), 323–335.

Smith, B. D., and Reeves, R. R. (eds.) (2000). Report of the second meeting of the Asian River Dolphin Committee, 22–24 February 1997 Rajendrapur, Bangladesh. *In* "Biology and Conservation of Freshwater Cetaceans in Asia, IUCN Species Survival Commission Occ. Pap." (R. R. Reeves, B. D. Smith, and T. Kasuya, eds.). pp. 1–14. Gland, Switzerland.

Smith, B. D., Sinha, R. K., Regmi, U., and Sapkota, K. (1994). Status of Ganges river dolphins (*Platanista gangetica*) in the Karnali, Mahakali, Narayani and Sapta Kosi Rivers of Nepal and India in 1993. *Mar. Mamm. Sci.* **10**, 368–375.

Zhou, K. (1982). Classification and phylogeny of the superfamily Platanistoidea with notes on evidence for the monophyly of the Cetacea. *Sci. Rep. Whales Res. Inst.* **34**, 93–108.

Figure 1 *A bottlenose dolphin swimming at high speed on the water surface. The generation of waves by the dolphin's movements leads to increases in body drag and elevated energetic costs during surface swimming.*

Swimming

TERRIE M. WILLIAMS
University of California, Santa Cruz

The primary mode of locomotion for marine mammals, with the possible exception of polar bears (*Ursus maritimus*), is swimming. For dolphins, porpoises, and whales it is the only form of locomotion. The duration of swimming among these mammals may be as short as several seconds when moving between prey patches or as long as several months during seasonal migrations across entire ocean basins. Although swimming by marine mammals often appears effortless, it is in reality a delicate balance between precise body streamlining, exceptional thrust production by specialized propulsive surfaces, and locomotor efficiency (Fig. 1).

I. Hydrodynamics and Body Streamlining

One of the most characteristic features of marine mammals is a streamlined body shape. This is not surprising when one considers the forces that the animal has to overcome in order to move through water. When a swimmer moves through water a force, termed drag, acts backward on it resisting its forward motion. The equation describing total body drag is given by

$$\text{drag} = 1/2pV^2ACd, \tag{1}$$

where p is the density of the fluid, V is the velocity of the fluid relative to the body, A is a characteristic area of the body, and Cd is the drag coefficient (a factor that takes into account the shape of the swimmer). Four primary types of drag contribute to total body drag: (1) skin friction drag, which is a tangential force resulting from shear stresses in the water sliding by the body; (2) pressure drag, which is a perpendicular force on the body associated with the pressure of the surrounding fluid; (3) wave drag, which occurs when a swimmer moves on or near the water surface; and (4) induced drag, which is associated with water deflection off of hydrofoil surfaces such as fins, flukes, or flippers. Of these, pressure drag is the component most influenced by body streamlining in marine mammals. The more streamlined a body, the lower the pressure drag and consequently the lower the total body drag of the swimmer.

Mammals whose lifestyles or foraging habits involve prolonged periods of swimming have streamlined body shapes. In contrast to the lanky appearance and appendages of terrestrial mammals, marine mammals tend to have a reduced appendicular skeleton and characteristic teardrop body profile. External features that may disrupt water flow across the body are also reduced or absent in many species of marine mammal. These features include the pinnae (external ears), limbs, and long fur. In highly specialized swimmers such as dolphins the skin contains microscopic ridges that help to direct the flow of water in a controlled manner down the body. All of these adaptations prevent the onset of turbulence in the water surrounding the swimmer, thereby reducing total body drag.

Hydrodynamic theory describes the streamlined body shape as one in which a rounded leading edge slowly tapers to the tail, and total length is three to seven times maximum body diameter. The ratio of these morphological measurements, termed the fineness ratio, can be written

$$\text{fineness ratio} = \frac{\text{maximum body length}}{\text{maximum body diameter}} . \quad (2)$$

The optimum fineness ratio that results in minimum drag with maximum accommodation for volume is 4.5. Calculations of the fineness ratio for a wide variety of marine mammals show that many species have body shapes that conform to the ideal hydrodynamic range (Fig. 2). A review by Fish (1993) showed that many cetaceans, pinnipeds, and sirenians have body shapes with fineness ratios that range from 3.0 to 8.0. The species examined included seals, sea lions, and odontocete whales, which are considered by many to typify a streamlined body profile. However, even the mysticete whales with enlarged heads and jaws specialized for filter feeding maintain a streamlined body profile (Fig. 2).

Despite nearly ideal body streamlining, all marine mammals must contend with drag forces when moving through the water. These forces can be a considerable challenge for the swimmer and will influence how quickly the animal will be able to move. It is apparent from Eq. (1) that the velocity of the swimmer will have a large impact on total body drag. As the swimmer moves faster, body drag increases exponentially. An example of the relationship between total body drag and velocity is presented in Fig. 3 for the sea otter (*Enhydra lutris*). Whether the sea otter swims on the water surface or submerged, body drag increases with velocity. However, body position clearly affects the level of total body drag encountered by the sea otter. At all comparable swimming speeds, body drag is higher for the otter moving on the water surface than when it is swimming submerged.

The same results have been found for other swimmers, including humans and harbor seals (*Phoca vitulina*). In general, body drag for a swimmer moving on or near the water surface is four to five times higher than the level of drag encountered by the submerged swimmer moving at the same speed. Much of this increase in drag at the water surface is due to energy wasted in the formation waves. This can be avoided if the swimmer is able to submerge to a depth equivalent to three body diameters. For a seal or small whale with a maximum body diameter of 1 m, this would mean changing swimming position to at least 3 m in depth to avoid wave drag and the consequent

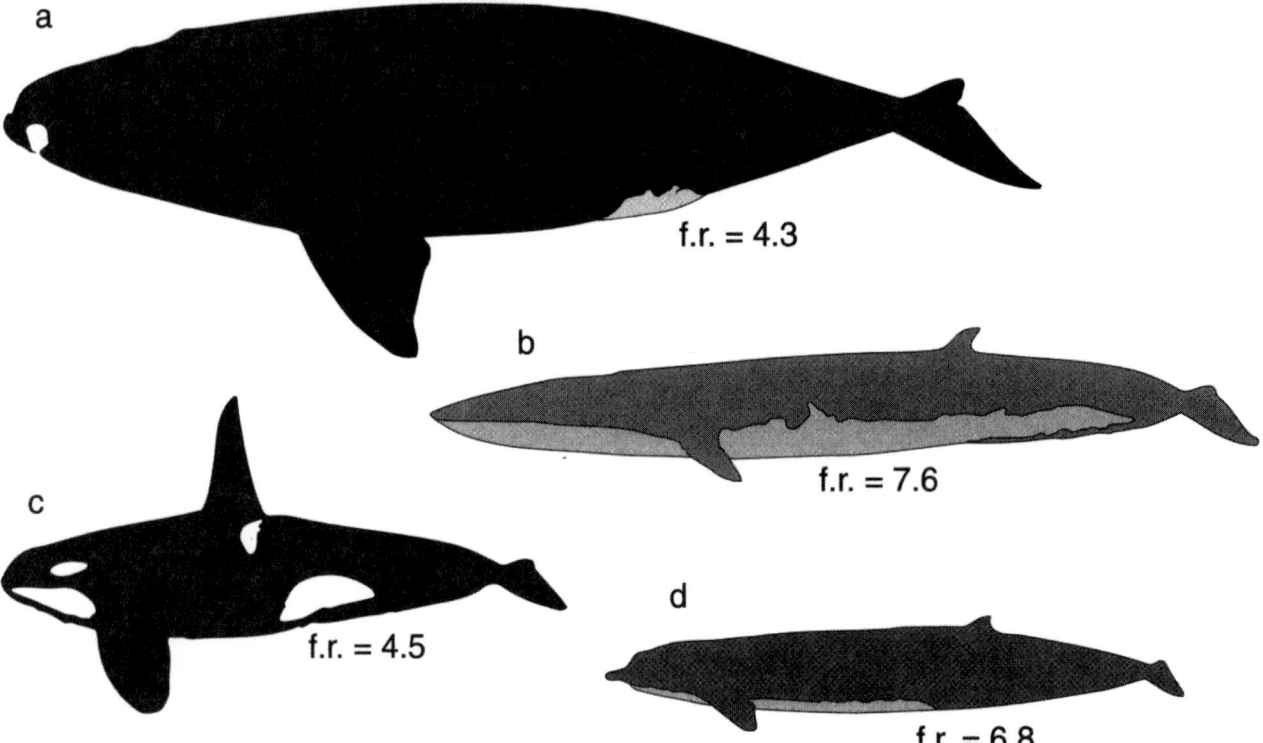

Figure 2 *Body shapes and fineness ratios for cetaceans. Shapes can range from the robust bowhead whale (a) to the long thin tapered body of the rorqual whales (b) and beaked whales (d). The killer whale (c) has the optimum shape in terms of fineness ratio and streamlining. From Berta and Sumich (1999), "Marine Mammals: Evolutionary Biology," Academic Press.*

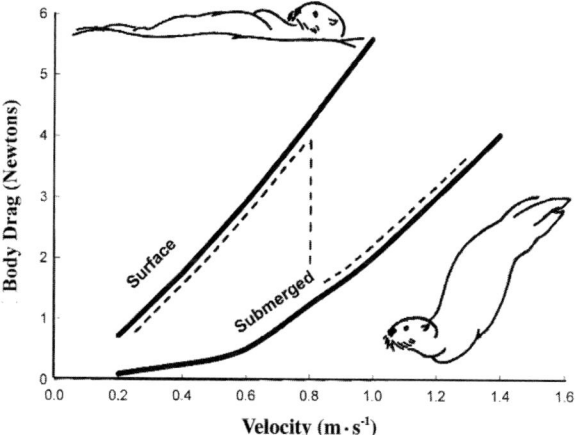

Figure 3 *Comparison of body drag for surface and submerged sea otters in relation to swimming speed. Note that at all comparable speeds, body drag of the sea otter on the water surface is higher than when the otter is submerged. The dashed line denotes the preferred swimming speeds of surface and submerged sea otters.*

elevation in total body drag. This is one of the reasons that swimming is comparatively difficult for humans—all of our performances take place on the water surface where wave drag, and hence total body drag, is the highest.

The ability to swim submerged for prolonged periods is one of the most important adaptations for increasing swimming efficiency and performance in marine mammals. The sea otter provides an excellent example of the advantage provided by this adaptation. Sea otters restrict prolonged periods of surface swimming to speeds less than $0.8 \text{ m} \cdot \text{sec}^{-1}$ and to a maximum body drag of 4.2 N (Fig. 3). For high-speed swimming, sea otters change to a submerged mode of locomotion. In doing so, drag is reduced by 3.5 times and the sea otter is able to reach speeds of $1.4 \text{ m} \cdot \text{sec}^{-1}$ before body drag once again exceeds 4.0 N. Thus, behavioral changes by the sea otter take into account the differences in drag associated with body position in the water and allow the animal to extend its range of swimming speeds. Several other behavioral strategies, such as porpoising and wave riding, are also used by marine mammals to avoid elevated body drag while swimming and are discussed in Section IV.

II. Kinematics

A hallmark of marine mammal swimming is the use of lift-based propulsion that allows thrust to be generated through the entire stroke cycle. This capability is found in highly adapted marine species such as pinnipeds and cetaceans. It contributes to an increase in locomotor efficiency in marine mammals, especially when compared to the inefficient drag-based swimming styles of humans and terrestrial mammals (Fig. 4).

Marine mammals use a wide variety of swimming styles to move through the water (Table I). The most terrestrial species of this group, the POLAR BEAR and sea OTTER, swim by alternate

strokes of the forelimbs or hindlimbs, respectively. Polar bears use a dog style of forelimb paddling with the hindlimbs dragged passively behind or used as an aid to steering. Sea otters are unique among marine mammals in their ability to lie on their backs during surface swimming. Propulsion is provided by either simultaneous or alternate strokes of the hindlimbs. When on the surface, sea otters can also swim ventral surface (belly) down using the hind paws for propulsion. The front paws are held against the submerged chest and do not play a role in propulsion during this mode of swimming. Stroke frequency

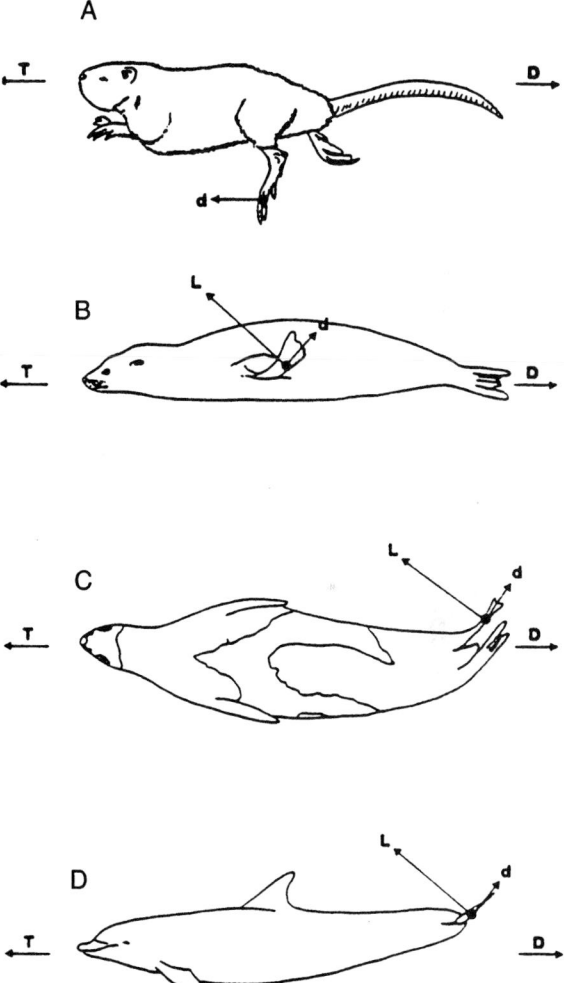

Figure 4 *Swimming modes for semiaquatic and marine mammals. The muskrat (A) is a semiaquatic mammal that uses drag-based propulsion by paddling its hind feet. Otariids (B), phocid seals (C), and cetaceans (D) use life-based propulsion that may involve fore flippers (sea lion), lateral body undulation (seal), or dorsoventral undulation (dolphin). Major forces on the animals and propulsive surfaces are shown. T denotes thrust, and D shows the direction of body drag on the animals. L and d illustrate lift and drag forces on the appendages, respectively. From Fish (1993).*

TABLE I
A Comparison of Swimming Characteristics for Four Major Classes of Marine Mammals

	Sea otter	Otariid	Phocid	Small cetacean
Routine speed (m/sec)	<0.8 (surface) <1.4 (submerged)	2.0–3.0	1.2–2.0, sprints to 4.0	2.0–4.0, sprints to 10.0
Hydrodynamics	Surface/submerged	Submerged	Submerged	Submerged
Kinematics				
Mode	Paddle, row (surface) Undulate (submerged)	Pectoral	Lateral Carangiform	Dorsoventral Thunniform
Energetics[a]				
COT measured	12.0 (surface)			
COT predicted	6.0 (submerged)	2.3–4.0	2.3–4.0	2.1–2.9

[a]The energetic cost of transport (COT) was measured for animals swimming in a flume or swimming freely in open water. The ratio of these values and the predicted values for fish of similar body mass are presented.

has been measured for swimming sea otters and ranges from approximately 30 to 80 strokes per minute while swimming on the water surface.

Polar bears and sea otters are the only marine mammals that rely primarily on drag-based modes of swimming. These modes of swimming have two distinct phases during the stroke cycle: a power phase when thrust is produced and a recovery phase when the foot is repositioned for the next stroke. During the power phase the foot is moved backward relative to the body. Drag created by this motion is subsequently translated into thrust and the animal moves forward through the water. The enlarged hind flippers of sea otters and fore paws of polar bears enable the animals to increase propulsive efficiency by moving a large mass of water during this power phase. The recovery phase of the stroke is only used to bring the limb back to its starting position and occurs without the generation of thrust. Because thrust is produced only during part of the stroke cycle, drag-based modes of swimming are comparatively inefficient.

When sea otters want to move quickly through the water they switch to an undulatory mode of swimming involving dorsoventral body flexion and simultaneous movements of paired hind flippers. The tail and hind flippers are held straight back and trail the undulatory movements of the trunk. The stroke frequency of sea otters remains relatively constant at 55 strokes per minute during submerged undulatory swimming, which suggests that underwater speed is elevated by increasing stroke amplitude.

As observed for submerged swimming sea otters, dolphins and whales use undulatory modes of propulsion. The primary propulsive movements of all cetaceans occur in the vertical plane with the posterior third of the body undulating in a dorsoventral direction. Termed thunniform swimming or carangiform swimming with a semilunate tail, this mode of locomotion is characterized by an undulatory wave that travels with increasing amplitude down the body, caudal peduncle, and finally the flukes (Fig. 5). "Semilunate" refers to the cres-

cent shape of the flukes. This mode of propulsion is shared by other fast-swimming vertebrates, including tuna, hence the name "thunniform." Undulatory propulsion in cetaceans is considered highly efficient and can generate high levels of thrust on both the upstroke and the downstroke. There is no recovery phase and propulsion can be produced throughout the stroke cycle. Stroke frequency using this mode of swimming varies with the speed and size of the cetacean. The range of stroke frequencies for bottlenose dolphins swimming in a pool is 60–180 strokes · min^{-1}. Stroke frequency decreases with increasing body size among the cetaceans. Thus, we find that the largest species of swimming mammal, the 100-ton blue whale (*Balaenoptera musculus*), uses stroke frequencies that are only one-tenth of the range observed for bottlenose dolphins (*Tursiops* spp.). Recent measurements of the stroke frequency of blue whales ascending during a dive were 6–10 strokes · min^{-1}.

Swimming by pinnipeds differs markedly among EARED SEALS (otariids) and true seals (the phocids). Otariids use pectoral appendages to generate propulsive forces during swimming, with the hind flippers trailing passively or occasionally used for steering. In this way, sea lions and fur seals resemble penguins and sea turtles during swimming. Detailed kinematic analyses have been conducted for California sea lions (*Zalophus californianus*) swimming in a flume. These studies revealed three distinct phases to the stroke: (1) the power phase, (2) a paddle phase, and (3) a recovery phase. The majority of thrust is produced during the paddle phase when the fore flippers are moved quickly and forcibly from the water flow to the sides of the animal's body. Stroke frequency for these sea lions increased with swimming speed and ranged from 15 to 50 strokes · min^{-1} as the animals increased speed from 0.5 to 3.0 m · sec^{-1}. In addition to stroke frequency, sea lions increase the amplitude of the fore flipper stroke during high-speed swimming.

When viewed in cross section, the fore flipper of the sea lion resembles a hydrofoil. This specialized shape allows the flipper to produce thrust during the power and recovery phases of the stroke cycle. As found for cetaceans, the specialized flip-

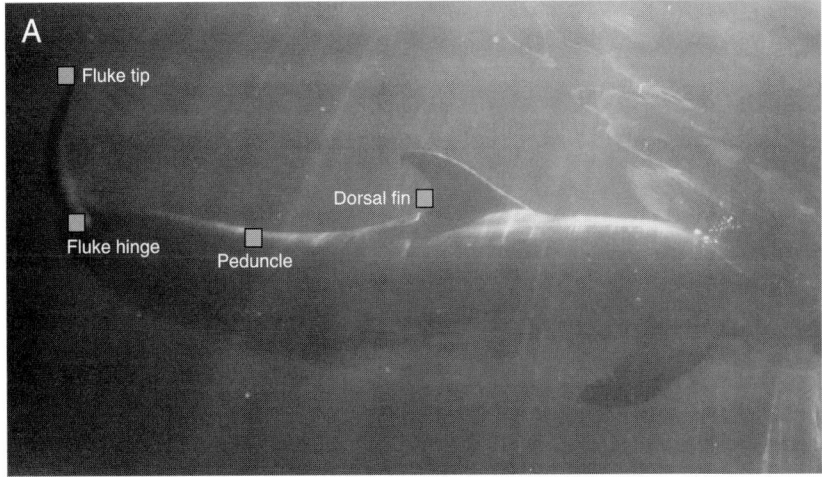

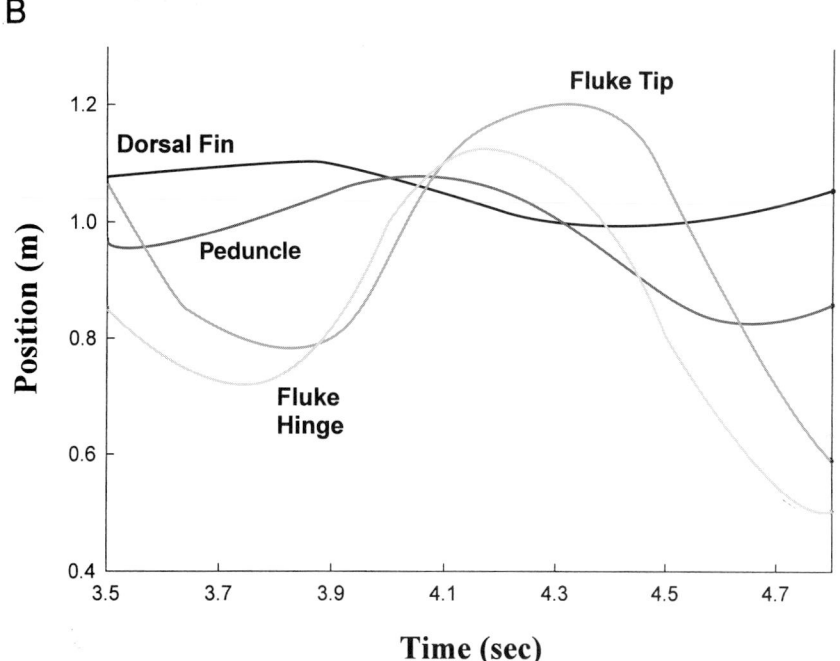

Figure 5 *Video image (A) and range of movement (B) of four anatomical sites during a single stroke for a swimming bottlenose dolphin. Squares correspond to the lines illustrating the movements for each site. Note the traveling wave as it passes from the dorsal fin to the peduncle, fluke hinge, and finally the fluke tip. From Skrovan et al. (1999).*

per movements of otariids result in thrust production throughout the stroke cycle and contribute to overall locomotor efficiency. Several other advantages are provided by fore flipper propulsion. These include stability at slow speeds and maneuverability at high speeds. Consequently, otariids are champion underwater acrobats and are capable of rapid changes in direction and acceleration.

Phocid seals and WALRUSES (*Odobenus rosmarus*) differ from otariids in terms of swimming style and rely on alternate sweeps of the hind flippers for propulsion. In addition to the flippers,

the posterior half of the body flexes during each stroke with the result that body flexion provides nearly 90% of the change in amplitude during the stroke cycle. In phocid seals, both hind flippers are swept in the same direction as the posterior portion of the body during each half of the stroke cycle. The leading flipper remains closed and the trailing flipper maximally expands during the sweep to one side. Once the flippers have moved to the maximum lateral position, the flippers switch their open and closed positions in preparation for the reverse lateral sweep. By reversing the role of each flipper during lateral sweeps, one flipper is able to provide thrust while the other flipper recovers. The result, once again, is the ability to produce propulsive thrust during the entire stroke cycle. Stroke frequency in phocids increases linearly with swimming speed. For harbor seals trained to swim at 1.0 to 1.4 m · sec^{-1} in a water flume, stroke frequency ranged from 60 to 78 strokes · min^{-1}.

III. Energetics

The energetic cost of swimming has been measured for numerous species of semiaquatic and marine mammals using a wide variety of techniques. Smaller swimmers, such as sea otters, seals, and sea lions, have been studied while they swam against a current in water flumes. Similar to placing a human on a treadmill, flume studies have enabled scientists to measure how much energy a swimmer expends while moving at different speeds. Often oxygen consumption is measured during these tests by using a face mask or metabolic hood connected to an oxygen gas analyzer. By training animals to breathe into a metabolic hood, expired respiratory gases can be collected and analyzed for oxygen content. For larger, more powerful swimmers, such as dolphins and whales, most flumes are not adequate in terms of size or challenging water speeds. Instead, investigators have relied on a variety of novel techniques for determining the energetic cost of swimming in cetaceans. Techniques have included using trained dolphins that match their swimming speed to that of a moving boat in open water or having whales swim to metabolic stations where expired gases can be collected for analysis.

To compare swimmers of different size, it is useful to convert the metabolic measurements into a cost of transport. Defined as the amount of fuel it takes to transport one unit of body weight over a unit distance, the cost of transport is analogous to the fuel rating of an automobile. In this case, the cost of transport indicates the "gas per mile" used by the swimmer rather than the "miles per gas" achieved by automobiles. The total cost of transport is calculated from the following equation:

$$\text{total cost of transport} = \frac{\text{oxygen consumption}}{\text{swimming speed}}, \quad (3)$$

where oxygen consumption is in mlO$_2$ · kg^{-1} · sec^{-1} and speed is in m · sec^{-1}, which results in a cost of transport in mlO$_2$ · kg^{-1} · m^{-1}. These values are usually converted to an energetic term and are expressed as Joules expended per kg of body weight per meter traveled (J · kg^{-1} · m^{-1}). The conversion calculation assumes a caloric equivalent of 4.8 kcal per liter of oxygen consumed and a conversion factor of 4.187 × 10^3 J/kcal.

Comparisons of the cost of transport for a wide variety of mammalian swimmers indicate that swimming is energetically expensive for mammals in comparison to fish. The total cost of transport for swimming mammals can also be divided into two distinct groups: semiaquatic mammals and marine mammals (Fig. 6). Swimming costs for semiaquatic mammals, such as minks, muskrats, and humans, are two to five times higher than observed for marine mammals. These high energetic swimming costs are attributed to a wide variety of factors, including elevated body drag associated with a surface swimming position (Fig. 3) and low propulsive efficiency associated with drag-based propulsion.

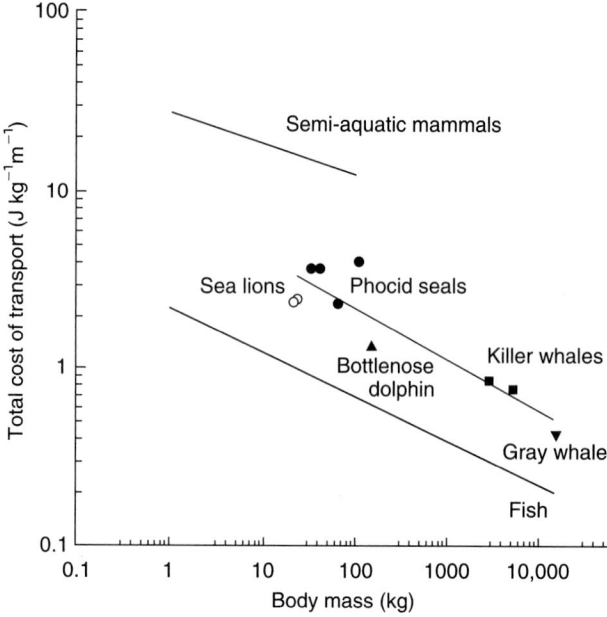

Figure 6 *Total energetic cost of transport in relation to body mass for different classes of swimmers. Marine mammals include gray seals and harbor seals, California sea lions, bottlenose dolphins, killer whales, and a gray whale. The least-squares regression through the data points for marine mammals is presented in the text. This regression is compared to the regressions for swimming semiaquatic mammals (upper solid line) and the predicted regression for salmonid fish (lower solid line). From Williams (1999).*

Mammals specialized for swimming demonstrate comparatively lower energetic costs. The total cost of transport in relation to body mass for swimming marine mammals ranging in size from a 21-kg California sea lion to a 15,000-kg gray whale (*Eschrichtius robustus*) is described by

$$\text{total cost of transport} = 7.79 \, \text{mass}^{-0.29}, \quad (4)$$

where the cost of transport is in J · kg^{-1} · m^{-1} and body mass is in kilograms. Interestingly, the style of swimming used by marine mammals did not affect the cost of transport relationship. Species and swimming styles represented in this equation include sea lions using pectoral fins for propulsion, phocid seals using lateral undulation of paired hind flippers, and odontocete and mysticete whales using dorsoventral undulation of flukes.

As illustrated in Fig. 6, the energetic cost of swimming for marine mammals is greater than predicted for salmonid fish of similar body size. Despite specialization of the body and propulsive surfaces for aquatic locomotion, the cost of transport for swimming by seals and sea lions is 2.3 to 4.0 times higher than predicted for swimming fish. Values for cetaceans are somewhat lower and range from 2.1 to 2.9 times values predicted for fish. Differences in the total cost of transport between marine mammals and fish are due in part to the amount of energy expended for maintenance functions, particularly thermoregulation and the support of a high core body temperature. As endotherms, mammals expend more energy to support the production of endogenous heat than ectothermic fish. In addition, many marine mammals show exceptionally high metabolic rates while resting in water in comparison to terrestrial mammals resting in air. A consequence of these high maintenance costs is an overall increase in the total energy expended during swimming, especially when compared to fish.

IV. Swimming Speeds and Behavior

Although body size varies considerably among marine mammals from the 20-kg sea otter to the 122,000-kg blue whale, routine swimming is limited to a surprisingly narrow range of speeds. Many species of marine mammal routinely swim between approximately 1.0 and 3.6 m · sec^{-1} regardless of body size (Fig. 7). Within this range, pinnipeds generally select slower routine traveling speeds than cetaceans, and mysticete whales swim slower than odontocetes. For example, average swimming speeds for a wide variety of otariids and phocids range from 1.3 to 2.0 m · sec^{-1}. The massive mysticete whales are only slightly faster; routine speeds for this group of marine mammals range from 2.1 to 2.6 m · sec^{-1}. Although they are not the largest marine mammals, odontocetes tend to move the fastest during routine travel. The slowest of the odontocetes represented in Fig. 7 was the beluga whale (*Delphinapterus leucas*) with a routine speed of 1.8 m · sec^{-1}. In comparison, the killer whale

(*Orcinus orca*) demonstrates the fastest routine speed of the marine mammals measured to date and averages 3.6 m · sec^{-1} during casual swimming. These speeds are even more remarkable when compared to the efforts of humans. The routine speed of humans during freestyle swimming is approximately 1.0 m · sec^{-1}, about the same speed as a sea otter swimming underwater.

As would be expected, the sprinting speeds of marine mammals are considerably faster than routine speeds and show much variation among the species measured. Most of the information regarding sprint swimming performance in marine mammals is for cetaceans. However, the speed of adult Weddell seals (*Leptonychotes weddellii*) chasing fish beneath the Antarctic sea ice has been measured and was found to exceed 4.0 m · sec^{-1} during the hunt. Among cetaceans, sprint speeds are even higher. The range of sprinting speeds measured for mysticete whales is 4.1 to 13.3 m · sec^{-1} (Fig. 7); sprint swimming by odontocetes is within the upper end of this range and averages 6.1 to 12.5 m · sec^{-1}. Killer whales remain the fastest of the odontocetes measured and can sprint at 12.5 m · sec^{-1}. This is nearly six times faster than the maximum performance of human swimmers in Olympic sprint competition.

Because marine mammals must surface periodically to breathe, they are subject to high levels of drag associated with the effects of wave formation and splashing, especially during high-speed swimming. To help minimize body drag and energetic costs during these surface intervals, marine mammals have developed a number of unique behavioral strategies to accommodate

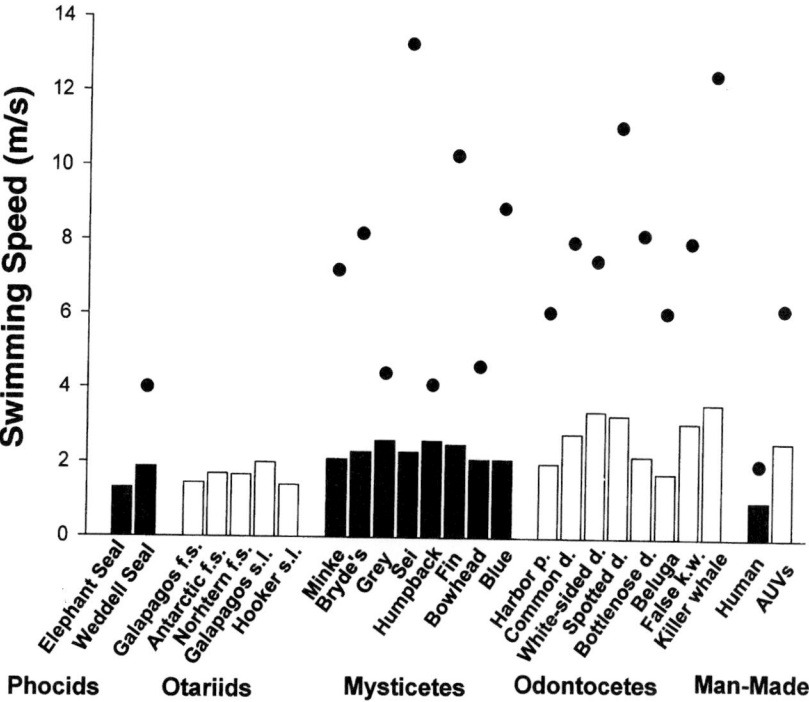

Figure 7 *Swimming speeds for marine mammals. Routine speeds of phocid seals, otariids, mysticetes, odontocetes, humans, and autonomous underwater vehicles (AUVs) are shown. Filled circles above the bars denote the sprinting speeds recorded for each species. Note the similar range of routine speeds for these marine mammals regardless of body size.*

breathing while swimming fast. Porpoising is one such highly visible behavioral strategy used by small cetaceans and some pinnipeds moving at high speed near the water surface. Rather than stroke continuously, the animals leap into the air and simply avoid the elevated wave drag that occurs when swimming near the water surface to breathe. Theoretically, this behavior results in an energetic savings to the animal, although the cost of surface swimming versus leaping has yet to be measured.

Wave riding is another strategy that enables the swimmer to avoid the work of continuous stroking while moving near the water surface. In a study involving bottlenose dolphins trained to swim freely or wave ride next to a moving boat, investigators found that heart rate, respiration rate, and energetic cost were reduced for animals riding the bow wave of the boat. This behavior enabled the dolphins to nearly double their forward traveling speed with only a 13% increase in energetic cost. Consequently, it is not surprising that marine mammals routinely ride waves generated by the wind, surf, the wake of boats, and even large whales. What appears to be an amusing activity also provides an energetic benefit to the swimmer.

Although energetically advantageous when swimming near the water surface, both wave riding and porpoising have been described for only a limited number of marine mammal species moving at high speeds. These locomotor strategies are not possible during slow transit, in large marine mammals such as elephant seals (*Mirounga* spp.) and whales, or in polar regions where ice covers the water surface. Instead, transit swimming is often accomplished by a sawtooth series of sequential dives that allow the animals to remaining submerged except for brief surface intervals to breathe.

V. The Special Case of Swimming at Depth

Most of our information about swimming in marine mammals is from animals moving near the water surface. However, the majority of swimming by these animals occurs at depth in conjunction with diving. When descending or ascending during a dive, marine mammals must contend with buoyant forces and hydrostatic pressure, as well as body drag. As discussed earlier, drag forces resist both forward progression and limb movements of the swimmer. In contrast, buoyant forces act in a vertical direction in the water column and result from the weight, volume, and compressibility of the tissues and air spaces of the animal's body. Hydrostatic pressure results from the weight of the water column above the marine mammal.

The magnitude of buoyant forces and hydrostatic pressure on the swimming marine mammal will depend on where in the water column activity takes place. Hydrostatic pressure increases progressively by 1 ATM for every 10.1 m an animal descends in the water column. This will have a profound effect on compressible spaces or tissues, and hence buoyancy of the animal, especially for marine mammals that may descend and ascend hundreds of meters during the course of a dive. In addition, seasonal changes in blubber content, pregnancy, and lactation will have an effect on the overall buoyancy of the marine mammal.

A consequence of the interrelationships among depth, buoyancy, hydrostatic pressure, and body drag is that the physical forces influencing the animal swimming horizontally near the water surface are very different from those encountered by the diving animal moving vertically through the water column. Detailed studies on diving common bottlenose dolphins and elephant seals have shown that the animals are positively buoyant near the water surface and that buoyancy decreases as the animal descends during the dive. For example, the buoyancy of a bottlenose dolphin changes from positive (24 N) when near the water surface to negative (-26 N) once the animal exceeds 70 m in depth.

These changes in buoyancy are associated with changes in lung compression due to the increase in hydrostatic pressure as marine mammals descend on a dive. Thus, as dolphins, whales, and seals dive, hydrostatic pressure increases and the lungs progressively collapse with the result that overall buoyancy is changed. These marked changes in physical forces with depth affect both locomotor behavior and energetics of the marine mammal as it descends and ascends during a dive (Fig. 8).

Until recently, it was not possible to observe the swimming modes of marine mammals during deep dives. With the development of miniaturized video cameras and instrumentation worn by free-ranging marine mammals, new information about swimming at depth has been obtained (Fig. 9). Videos have revealed that bottlenose dolphins, elephant seals, Weddell seals, and blue whales switch between different modes of swimming during the dive, much like terrestrial mammals switch between gaits. Dive descents usually began with a period of continuous stroking. Once the marine mammals reach 70–80 m in depth they change to a passive glide for the remainder of the descent. For deep divers such as phocid seals, these gliding periods can be quite long. For example, prolonged gliding periods exceed 6 min for Northern elephant seals (*M. angustirostris*) traveling to nearly 400 m and Weddell seals descending to 540 m beneath the Antarctic sea ice. Nearly 80% of the descent of diving seals may be spent gliding passively rather than swimming actively on dives exceeding 200 m in depth.

The ascent portion of a dive requires more effort by marine mammals when compared to the descent. The beginning of the ascent represents the period of greatest swimming effort for mammalian divers. During this period, many species of pinniped and cetacean use sequential, large amplitude strokes to begin moving upward. As the ascent continues, the physical forces impacting the diver are once again altered as they move through the water column. Hydrostatic pressure decreases on ascent. Consequently, the lungs are able to reinflate and the buoyancy of the marine mammal increases. Swimming behavior reflects these changes with the result that the continuous stroking phase is followed by a stroke and glide mode of swimming, and finally a brief glide to the water surface.

By altering the mode of swimming to account for changes in the physical forces that occur during a dive, marine mammals are able to conserve limited oxygen reserves during submergence. One study investigating the metabolic rates of Weddell seals diving from an ice hole found that the incorporation of prolonged glides enabled seals to reduce the energetic cost of individual dives by 9–60%. Such an energetic savings could make the difference between completing the dive aerobically or anaerobically and can increase the time available for hunting or avoiding predators.

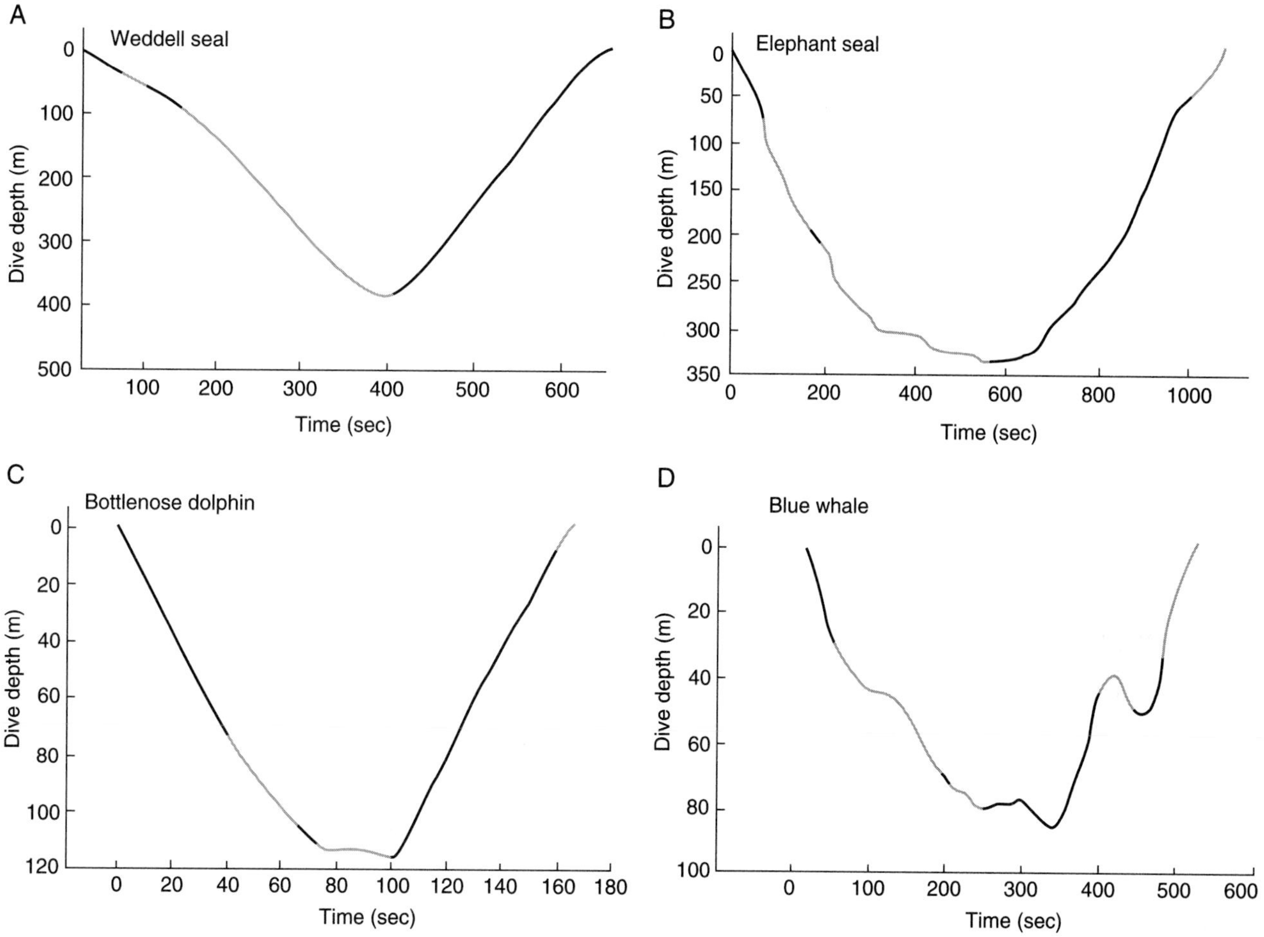

A Weddell seal

B Elephant seal

C Bottlenose dolphin

D Blue whale

Figure 8 *Swimming and gliding activity of four species of diving marine mammal. Representative deep dives are presented for the Weddell seal (A), elephant seal (B), bottlenose dolphin (C), and blue whale (D). Each curve shows dive depth in relation to time elapsed during the dive. The shade of the line corresponds to stroking (black) and gliding (gray) periods. For each species the descent was characterized by prolonged periods of gliding. From Williams et al. (2000).*

Figure 9 *A bottlenose dolphin carries a video camera to record its swimming movements during deep dives. Courtesy of Kevin McDonnell.*

In summary, these studies demonstrate that swimming can be energetically expensive for mammals. Marine adapted species, including sea otters, pinnipeds, and cetaceans, have undergone marked morphological, physiological, and behavioral changes to increase their swimming efficiency. An especially important adaptation that distinguishes marine mammals from semiaquatic mammals is the ability to remains submerged for prolonged periods when swimming. However, prolonged submergence also requires specialized physiological responses associated with oxygen loading and utilization. A major benefit of these adaptations is a capacity for aquatic performance by marine mammals that far exceeds those of semiaquatic mammals and the best Olympic efforts of humans.

See Also the Following Articles

Breathing ▪ Diving Physiology ▪ Energetics ▪ Locomotion, Terrestrial ▪ Musculature ▪ Pelvic Anatomy ▪ Speed ▪ Streamlining

References

Fish, F. E. (1993). Influence of hydrodynamic design and propulsive mode on mammalian swimming energetics. *Aust. J. Zool.* **42**, 79–101.

Skrovan, R. C., Williams, T. M., Berry, P. S., Moore, P. W., and Davis, R. W. (1999). The diving physiology of bottlenose dolphins (*Tursiops truncatus*). II. Biomechanics and changes in buoyancy at depth. *J. Exp. Biol.* **202**, 2749–2761.

Williams, T. M. (1999). The evolution of cost efficient swimming in marine mammals: Limits to energetic optimization. *Phil. Trans. R. Soc. Lond. B* **354**, 193–220.

Williams, T. M., Davis, R. W., Fuiman, L. A., Francis, J., Le Boeuf, B. J., Horning, M., Calambokidis, J., and Croll, D. A. (2000). Sink or swim: Strategies for cost-efficient diving by marine mammals. *Science* **288**, 133–136.

Systematics, Overview

ANNALISA BERTA

San Diego State University, California

Systematics is the study of biological diversity that has as its primary goal the reconstruction of phylogeny, the evolutionary or genealogical history of a particular group of organisms (e.g., species). Because of its emphasis on phylogeny, this discipline is often referred to as phylogenetic systematics or cladistics. Other related goals of systematics include determination of the times at which species originated and became extinct and the origin and rate of change in their characteristics. An important component of systematics is taxonomy, the identification, description, nomenclature, and classification of organisms. Systematics provides a framework for interpreting patterns and processes in evolution using explicit, testable hypotheses.

The rapid pace of research on marine mammals has resulted in renewed interest in their systematics. Phylogenetic systematic methodology as introduced here has gained near universal acceptance. [For a general introduction to the topic readers are referred to texts by Eldredge and Cracraft (1980), Wiley (1980), and Wiley *et al.* (1991).] In addition to their use in elucidating evolutionary relationships, phylogenies are now recognized as powerful tools for unveiling evolutionary patterns of marine mammal diversity in ecological and behavioral settings.

I. Basic Tenets of Phylogenetic Systematics

The recognition of patterns of relationship among SPECIES is founded on the concept of evolution. Patterns of relationship among species are based on changes in the characters of an organism. Characters are diverse, heritable attributes of organisms that include DNA base pairs, anatomical and physiological features, and behavioral traits. Two or more forms of a given character are termed *character states*. For example, among pinnipeds, the character contact between maxillary and frontal bones consists of three character states: (1) V shaped (in bears, extinct desmatophocids, *Enaliarctos,* and phocids), (2) W shaped (in otariids), and (3) transverse (walruses) as described by Berta and Sumich (1999). In the establishment of relationships among groups of organisms, phylogenetic systematics emphasizes evolutionary novelties (derived characters) in contrast to ancestral similarities (primitive characters).

The evolutionary history of a group of organisms can be inferred by sequentially linking species together based on their common possession of derived characters, also known as *synapomorphies.* If derived characters are unique to a particular taxon rather than showing relationships among taxa they are termed *autapomorphies.* An example of an autapomorphy is the transverse contact between maxillary and frontal bones seen only in walruses among pinnipedimorphs (living pinnipeds and their fossil relatives). Derived characters are considered to be *homologous,* a similarity that results from common ancestry. For example, the flipper of a seal and that of a walrus are homologous because their common ancestor had flippers. In contrast to homology, a similarity not due to homology is *homoplasy.* For example, the flipper of a pinniped and that of a whale are homoplasious as flippers because their common ancestor lacked flippers. Homoplasy may arise in one of two ways: convergence (parallelism) or reversal. *Convergence* is the independent evolution of a similar feature in two or more lineages. Thus seal flippers and whale flippers evolved independently as swimming appendages; their similarity is homoplasious by convergent evolution. *Reversal* is the loss of a derived feature coupled with the reestablishment of an ancestral feature. For example, in phocine seals (bearded seal *Erignathus barbatus,* hooded seal *Cystophora cristata,* and the Phocini) the development of strong claws, lengthening of the third digit of the foot, and deemphasis of the first digit of the hand are character reversals because none of them characterize phocids ancestrally but are present in terrestrial arctoid CARNIVORES, common ancestors of PINNIPEDS (Berta and Sumich, 1999).

Relationships among organismal groups are commonly represented in the form of a *cladogram,* a branching diagram that conceptually represents the best estimate of phylogeny (Fig. 1). Derived characters are used to link *monophyletic* groups, groups of taxa that consist of a common ancestor plus all descendants of that ancestor (referred to as a *clade*). For example, a hypothesis of relationships among pinnipedimorphs (inclusive group that includes all pinnipeds and their close fossil relatives) based on morphologic characters proposes that phocid seals (Phocidae) and an extinct lineage (Desmatophocidae) are more closely related to each other than either is to walruses (Odobenidae). Fur seals and sea lions (Otariidae) are positioned as the next closest relative to this clade (walruses + desmatophocids + phocids) with the fossil taxon *Enaliarctos* recognized as the most *basal* lineage (Berta and Sumich, 1999; Fig. 1). According to this hypothesis, relationships among pinnipedimorphs are depicted as sets of nested hierarchies. In this case, four monophyletic groups can be recognized. The most exclusive monophyletic group is that formed by phocid seals and desmatophocids, as this clade shares derived synapomorphies not also exhibited by walruses, otariids, or *Enaliarctos.* At

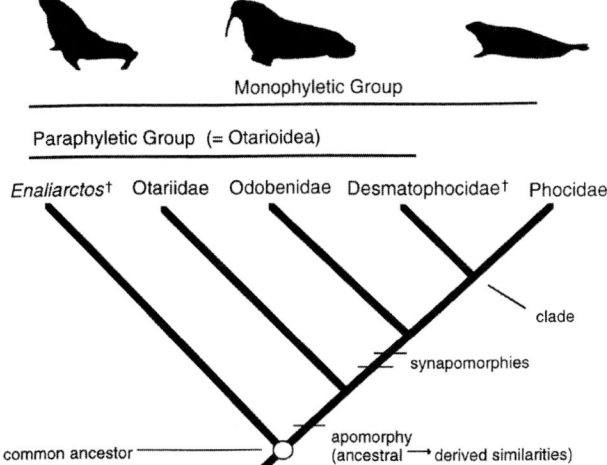

Figure 1 *Hypothesis of pinnipedimorph relationships. †, extinct taxa.*

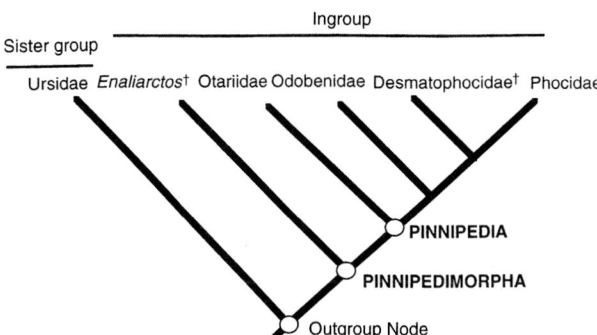

Figure 2 *Pinniped relationships with ingroup and outgroups identified. †, extinct taxa.*

the other extreme, the most inclusive monophyletic group is that formed by *Enaliarctos* + Otariidae + Odobenidae + Desmatophocidae + Phocidae.

The task in inferring a phylogeny for a group of organisms is to determine which characters are derived and which are ancestral. If the ancestral condition of a character or character state is established, then the direction of evolution from ancestral to derived can be inferred, and synapomorphies can be recognized. The methodology for inferring the direction of character evolution is critical to cladistic analysis. *Outgroup comparison* is the most widely used procedure. It relies on the argument that a character state found in close relatives of a group (the outgroup) is likely to be the ancestral or primitive state for the group of organisms in question (the ingroup). Usually more than one outgroup is used in an analysis, the most important being the first or genealogically closest outgroup to the ingroup called the sister group (Maddison *et al.*, 1984). For example, among pinnipedimorphs the ingroup includes the Pho-

cidae (seals), the Desmatophocidae (extinct seal relatives), the Otariidae (fur seals and sea lions), the Odobenidae (walruses) and the fossil taxon *Enaliarctos*. The Phocidae is hypothesized as the sister group of the Desmatophocidae and these taxa together form the sister taxon to the Odobenidae. The Otariidae and *Enaliarctos* are positioned as sequential sister taxa. In the hypothesis presented here, bears (Ursidae) are positioned as the closest pinniped outgroup (Fig. 2), although alternate hypotheses support pinnipeds as either allied with MUSTELIDS or as having an unresolved ancestry with arctoids (ursids, mustelids, and procyonids).

II. Phylogeny Reconstruction

The first step in the reconstruction of phylogeny of a group of organisms is selection and definition of characters and character states for each taxon (e.g., species). Next, the characters and their states are arranged in a data matrix (see Table I). Characters can be further distinguished; those with two states are *binary*, whereas characters with three or more states are *multistate*. For each character, ancestral and derived states are determined. The determination of character state, whether ancestral or derived (also called *polarity* assessment), is done using outgroup comparison. For example, if the distribution of

TABLE I
Data Set for Analysis of Pinnipedimorphs Plus an Outgroup Showing Five Characters and Their Character States

| | *Character/character states* | | | | |
Taxon	1 Lacrimal bone	2 Middle ear bones	3 Orbital/maxilla	4 Maxilla/frontal	5 Squamosal/jugal
Outgroup					
Ursids	Present	Small	No	V shape	Overlapping
Ingroup					
Enaliarctos	Absent	Small	No	V shape	Overlapping
Otariidae	Absent	Small	Yes	W shape	Overlapping
Desmatophocidae	Absent	Large	Yes	V shape	Interlocking
Phocidae	Absent	Large	Yes	V shape	Interlocking
Odobenidae	Absent	Large	Yes	Straight	Overlapping

A

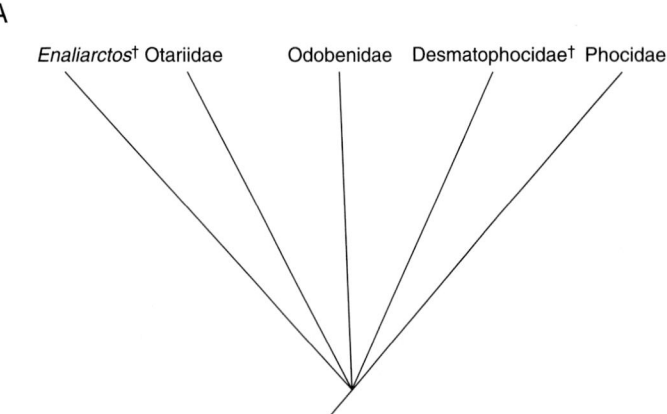

B

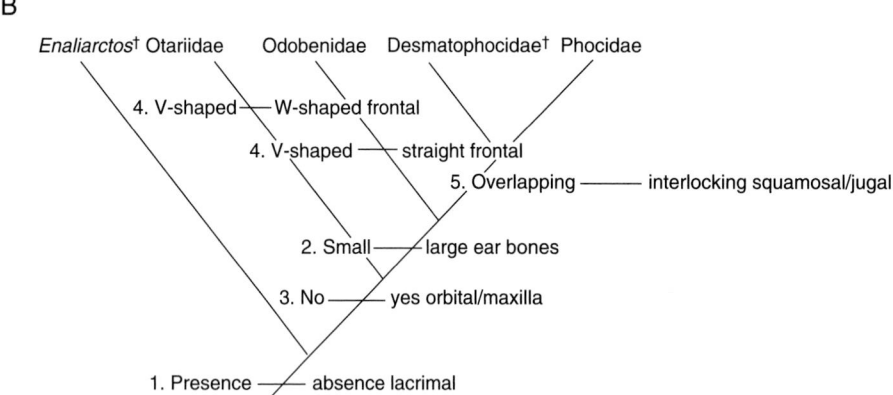

C

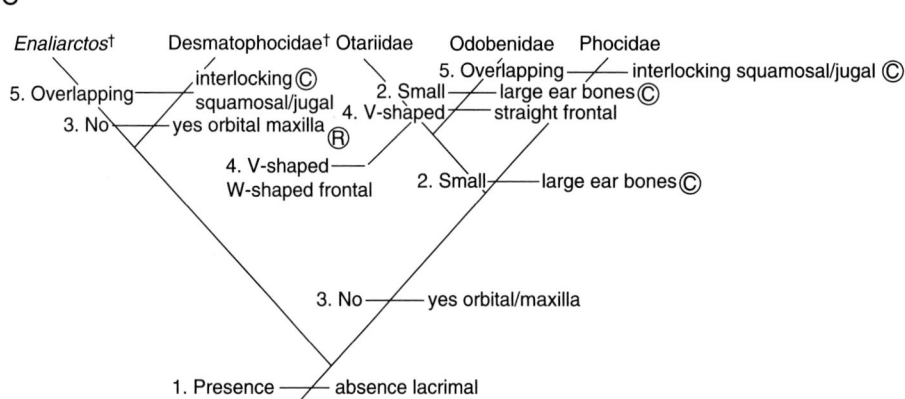

Figure 3 *Three possible cladograms (A–C) of relationships and character-state distributions for the ingroups listed in Table I. Convergences and reversal are denoted by a circled C and R, respectively.*

character No. 1, condition of the lacrimal bone, is considered, two character states are recognized: presence versus absence (or fusion of the lacrimal such that it does not contact the jugal). The outgroup (bears) possess a lacrimal bone condition, which is the ancestral state (Table 1). Ingroup taxa lack the lacrimal bone, which is the derived condition. Because this derived state unites pinnipedimorphs to the exclusion of bears, it is considered a synapomorphy.

A final step in phylogeny reconstruction is the construction of all possible cladograms by sequentially grouping taxa based on the common possession of one or more shared derived character states (Fig. 3). An important aspect of phylogeny reconstruction is the principle of *parsimony*. The basic tenet of parsimony is that the cladogram that contains the fewest number of evolutionary steps, or changes between character states of a given character summed for all characters, is accepted as being the best estimate of phylogeny. In this example, Fig. 3B is the most parsimonious cladogram. Note that an alternative cladogram for the data set (Fig. 3C) showing different relationships among the five taxa requires nine characters state changes, three more than the most parsimonious cladogram (Fig. 3B).

The methods used to search for the most parsimonious tree depend on the size and complexity of the data matrix. These methods are available in several computer programs [e.g., PAUP (Swofford, 1998) and MacClade (Maddison and Maddison, 1992)]. The latter is particularly useful in assessing the evolution of characters. Systematists have become concerned about the relative accuracy of phylogenetic trees (i.e., how much confidence can be placed in a particular phylogenetic reconstruction). Various measures (e.g., tree length) and indices (e.g., consistency index, rescaled consistency index) have been devised to address this concern (reviewed by Swofford, 1998). Another approach that is often used to estimate the value of a particular cladogram with respect to data places confidence limits on the individual branches, a technique known as *bootstrapping*. Another concern stems from the realization that an increase in the size of data sets results in a greater chance of the analysis resulting in more than one equally parsimonious tree. A method of working with multiple trees involves the implementation of *consensus* trees that are useful in identifying the areas of agreement and conflict among competing trees. Perhaps the greatest influence on the accuracy of a cladogram is the number of taxa. When only a few taxa are used (e.g., exemplars of a group), or when only a small number are available, the accuracy of the phylogenetic reconstruction suffers. In summary, many factors should be considered in evaluating phylogenetic hypotheses, among the most important being taxonomic sampling and rigorous analysis, including the underlying assumptions of various methods.

III. Phylogenetic Classification

Taxonomy is the language of biology. One aspect of taxonomy is the classification of organisms that allows us to organize and communicate information about life's diversity. Phylogenetic systematists contend that classification should be based on phylogeny and should include only monophyletic groups. In contrast to monophyletic groups, a *paraphyletic* group (designated by quotation marks) is one that includes a common ancestor and some but not all of the descendants of that ancestor. An example of a paraphyletic taxon is the "Otarioidea," a group that includes walruses, otariid seals, and their extinct relatives to the exclusion of phocid seals (see Fig. 1). The recognition of paraphyletic taxa is to be avoided since by doing so we risk misinterpreting the evolutionary relationships of taxa and their classification. One result of the use of phylogenetic methodology is that traditionally accepted ranks (e.g., phylum, class, order, family, genus, and species) often do not correspond to new information about evolutionary relationships among taxa, thus rendering their classification misleading. One method is the elimination of rank altogether and the indication of relative rank by subordination as shown by indentation. An example, the classification of major lineages of pinnipedimorphs, is shown in Fig. 4 based on the phylogeny in Fig. 2.

IV. Uses of a Phylogeny

Once a phylogenetic framework is produced, one of its most interesting uses is to elucidate questions that integrate evolution, behavior, and ecology. One technique used to facilitate such evolutionary studies is optimization or *mapping*. Once a cladogram is constructed, a feature or condition is selected to be examined in light of the phylogeny of the group. Examples using marine mammals include the evolution of body size in pinnipeds (Wyss, 1994), pinniped host–parasite associations (Hoberg, 1995), and the evolution of locomotor and feeding behavior in pinnipeds (Berta amd Adam, 2001), cetaceans (Thewissen and Fish, 1997), and sirenians (Domning, 2000). One of these studies, the evolution of locomotion among sirenians, is briefly reviewed here. A cladogram of relationships among sirenians was first established based on morphologic data. Next, using this phylogenetic framework, Domning (2000; Fig. 5) mapped locomotor characters onto the tree. Sirenians passed through three locomotor stages from a terrestrial quadrupedal ancestry. In the first stage exemplified by archaic sirenians, the prorastomids, swimming was accomplished by alternate thrusts of the hindlimbs. This was followed by a second stage seen in the extinct taxon *Protosiren* that employed dorsoventral spinal undulation and bilateral thrusts of the hindlimb in SWIMMING. In a final stage, seen in modern sea cows and manatees, sirenians have evolved into completely aquatic animals, swimming with the tail only.

Pinnipedimorpha (including *Enaliarctos* and all other pinnipeds)
　Pinnipedia (fur seals, sea lions, seals, walruses and their extinct relatives)
　　Otariidae (fur seals and sea lions)
　　Phocomorpha (walruses, seals, and their extinct relatives)
　　　Phocoidea (including *Allodesmus*, *Desmatophoca*, seals and walruses)
　　　　Phocidae (seals)
　　　　Odobenidae (walruses)

Figure 4 *Classification of the Pinnipedimorpha.*

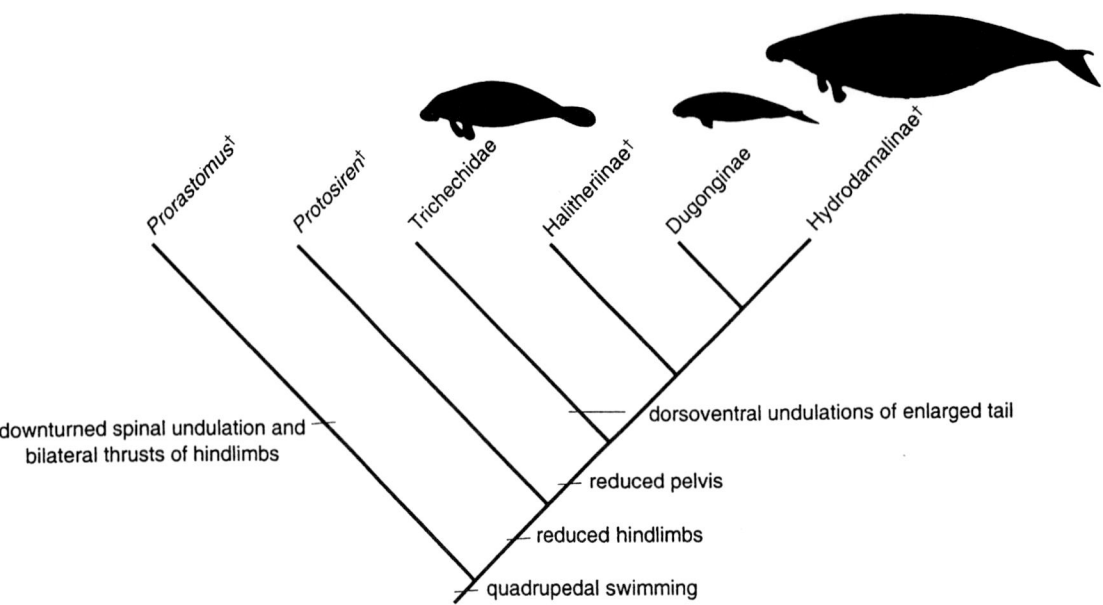

Figure 5 *Evolution of locomotion among sirenians. †, extinct taxa. Based on Domning (2000).*

See Also the Following Articles

Convergent Evolution ▪ Classification ▪ Paleontology

References

Berta, A., and Adam, P. (2001). Evolutionary biology of pinnipeds. *In* "Secondary Adaptation of Tetrapods to Life in Water." (J.-M. Mazin and V. de Buffrénil, eds.), pp. 235–260. Verlag Dr. Frederich Pfeil, Munich.

Berta, A., and Sumich, J. L. (1999). "Marine Mammals: Evolutionary Biology." Academic Press, San Diego.

Domning, D. P. (2000). The readaptation of Eocene sirenians to life in the water. *Historical Biology* **14**, 115–119.

Eldredge, N., and Cracraft, J. (1980). "Phylogenetic Patterns and the Evolutionary Process," Columbia Univ. Press, New York.

Hoberg, E. (1995). Historical biogeography and modes of speciation across high latitude seas of the Holarctic: Concepts for host-parasite co-evolution among the Phocini (Phocidae) and Tetrabothriidae (Eucestoda). *Can. J. Zool.* **73**, 45–57.

Maddison, W. P., Donoghue, M., and Maddison, D. R. (1984). Outgroup analysis and parsimony. *Syst. Zool.* **33**, 83–107.

Maddison, W. P., and Maddison, D. R. (1992). MacClade, version 3. Analysis of phylogeny and character evolution. Sinauer Associates, Sunderland, MA.

Swofford, D. L. (1998). PAUP*: Phylogenetic Analysis Using Parsimony, version 4. Sinauer Associates, Sunderland, MA.

Thewissen, J. G. M., and Fish, F. (1996). Locomotor evolution in the earliest cetaceans: Functional model, modern analogues, and paleontologic evidence. *Paleobiology* **23**(4), 482–490.

Wiley, E. O. (1981). "Phylogenetics: The Theory and Practice of Phylogenetic Systematics," Wiley, New York.

Wiley, E. O., Siegel-Causey, D., Brooks, D. R., and Funk, V. A. (1991). The Complete Cladist: A primer of phylogenetic procedures. Univ. Kans. Mus. Nat. Hist., Spec. Publ., no. 19.

Wyss, A. R. (1994). The evolution of body size in phocids: some ontogenetic and phylogenetic observations. Proc. San Diego Soc. Nat. Hist. **29**, 69–75.

Teeth

NOBUYUKI MIYAZAKI
Otsuchi Marine Research Center,
University of Tokyo, Japan

Teeth in mammals are generally confined to the margins of the upper and lower jaws and are placed in four groups: incisors (I), canines (C), premolars (PM), and molars (M). The dental formula for a typical mammal is written simply as I3/3, C1/1, PM4/4, M3/3 × 2 = 44. All teeth except the molars normally have a "milk" or deciduous set preceding the permanent ones. However, marine mammals, which are well adapted to water life, show various kinds of teeth in number and shape. Among marine mammals, teeth of cetaceans are highly specialized compared with pinnipeds and sirenians. In cetaceans, odontocetes have a wide variety in number and shape of teeth, whereas mysticetes are characterized by a complete absence of teeth. Most of odontocetes exhibit clear differences in tooth numbers and shapes from typical placental mammals. In some odontocetes, such as beaked whales, the pygmy sperm whale (*Kogia breviceps*), and the sperm whale (*Physeter macrocephalus*), there is a reduction in the number of teeth in the jaws. Odontocetes have nearly homodont or peg-like teeth. Thus, in odontocetes the total tooth number for each side of the jaw is simply combined here as a single number (Table I).

I. Cetacea

Mysticetes are characterized by a complete absence of teeth, although teeth may be found in the fetal stage. For feeding on planktonic or micronectonic crustaceans such as krills and/or small pelagic fish, they use a row of baleen plates that project ventrally from the outer edges of the roof of the mouth. Odontocetes differ from mysticetes in feeding principally on fish and/or squid. The diet of odontocetes might reflect on the morphology and number of teeth. In odontocetes, there is no deciduous dentition. The tooth consists of enamel, prenatal dentine, postnatal dentine, cementum, and pulp cavity. Most phocoenids have spatulate teeth and delphinids conical teeth.

TABLE I
Number of Teeth in the Upper/Lower Jaw on Each Side[a]

Cetacea	
Globicephala melas	8–13 / 15–18
G. macrorhynchus	7–8 / 7–8
Pseudorca crassidens	7–12 / 7–12
Lagenorhynchus obliquidens	23–26 / 23–26
L. obscurus	27–36 / 27–36
L. albirostris	22–28 / 22–28
L. acutus	30–40 / 30–40
L. australis	27–33 / 27–33
L. cruciger	26–35 / 27–35
Tursiops truncatus	18–26 / 18–26
Orcinus orca	10–12 / 10–12
Grampus griseus	0 / 2–7
Steno bredanensis	19–26 / 19–28
Sousa chinensis	30–35 / 31–34
S. plumbea	31–38 / 30–37
S. teuszii	27–30 / 27–31
Sotalia fluviatilis	25–36 / 25–36
Stenella attenuata	35–48 / 34–37
S. longirostris	44–64 / 42–62
S. coeruleoalba	39–53 / 39–55
S. clymene	36–49 / 38–48
S. frontalis	32–42 / 30–40
Delphinus delphis	45–60 / 45–60
Lagenodelphis hosei	36–44 / 34–44
Cephalorhynchus commersonii	28–34 / 26–35
C. eutropia	28–34 / 29–33
C. hectori	24–31 / 24–31
C. heavisidii	25–28 / 25–28
Lissodelphis borealis	37–54 / 37–54
L. peronii	39–50 / 39–50
Peponocephala electra	20–26 / 22–25
Feresa attenuata	8–11 / 10–13
Neophocaena phocanoides	13–22 / 13–22
Phocoena spinipinnis	10–23 / 14–23
P. sinus	16–22 / 17–20
P. phocoena	19–28 / 19–28
P. dioptrica	17–23 / 17–20
Phocoenoides dalli	23–28 / 23–28

(continues)

TABLE I (*Continued*)

Monodon monoceros	1 / 0
Delphinapterus leucas	9 / 8
Orcaella brevirostris	17–20 / 15–18
Inia geoffrensis	24–35 / 24–35
Lipotes vexillifer	30–34 / 32–36
Pontoporia blainvillei	53–58 / 51–56
Platanista gangetica	25–39 / 26–35
Physeter macrocephalus	0 / 17–29
Kogia breviceps	0 / 12–16
K. sima	0 / 8–11
Berardius bairdii	0 / 2
B. arnuxii	0 / 2
Ziphius cavirostris	0 / 1
Tasmacetus shepherdi	17–21 / 17–29
Hyperoodon ampullatus	0 / 1
H. planifrons	0 / 1
Indopacetus pacificus	0 / 1
Mesoplodon bidens	0 / 1
M. bowdoini	0 / 1
M. carlhubbsi	0 / 1
M. densirostris	0 / 1
M. europaeus	0 / 1
M. ginkgodens	0 / 1
M. grayi	17–22 / 1
M. hectori	0 / 1
M. layardii	0 / 1
M. mirus	0 / 1
M. stejnegeri	0 / 1
Pinnipedia	
Phoca vitulina	I3/2, C1/1, PC5/5
P. largha	I3/2, C1/1, PC5/5
Pusa hispida	I3/2, C1/1, PC5/5
P. sibirica	I3/2, C1/1, PC5/5
P. caspica	I3/2, C1/1, PC6/5
Pagophilus groenlandicus	I3/2, C1/1, PC5/5
Histriophoca fasciata	I3/2, C1/1, PC5/5
Halichoerus grypus	I3/2, C1/1, PC5–6/5
Erignathus barbatus	I3/2, C1/1, PC5/5
Cystophora cristata	I2/1, C1/1, PC5/5
Monachus monachus	I2/2, C1/1, PC5/5
M. tropicalis	I2/2, C1/1, PC5/5
M. schauinslandi	I2/2, C1/1, PC5/5
Lobodon carcinophaga	I2/1, C1/1, PC5/5
Ommatophoca rossii	I2/2, C1/1, PC5–6/4–6
Hydrurga leptonyx	I2/2, C1/1, PC5/5
Leptonychotes weddellii	I2/2, C1/1, PC5/5
Mirounga leonina	I2/1, C1/1, PC5/5
M. angustirostris	I2/1, C1/1, PC5/5
Odobenus rosmarus	I1/0, C1/1, PC3/3
Phocarctos hookeri	I3/2, C1/1, PC6/5
Otaria flavescens	I3/2, C1/1, PC6/5
Zalophus californianus	I3/2, C1/1, PC5–6/5–6
Neophoca cinerea	I3/2, C1/1, PC5/5
Eumetopias jubatus	I3/2, C1/1, PC5/5
Callorhinus ursinus	I3/2, C1/1, PC6/5
Arctocephalus pusillus	I3/2, C1/1, PC6/5
A. gazella	I3/2, C1/1, PC6/5
A. forsteri	I3/2, C1/1, PC6/5
A. tropicalis	I3/2, C1/1, PC6/5
A. australis	I3/2, C1/1, PC6/5

(*continues*)

TABLE I (*Continued*)

A. galapagoensis	I3/2, C1/1, PC6/5
A. philippii	I3/2, C1/1, PC6/5
A. townsendi	I3/2, C1/1, PC6/5
Sirenia	
Dugong dugon	I2/3, C0/1, PC6/6
Trichechus manatus	I0/0, C0/0, PC5–7/5–7
T. senegalensis	I0/0, C0/0, PC5–7/5–7
T. inunguis	I0/0, C0/0, PC5–7/5–7

[a]From Ridgway and Harrison (1981, 1985, 1989, 1994) and Jefferson *et al.* (1993).

In most odontocetes, there is wide individual variation in tooth number. For example, in the sperm whale (*Physeter macrocephalus*) and the short-beaked common dolphin (*Delphinus delphis*), the total number of teeth may exhibit a 1.5-fold difference among individuals. Among members of the Delphinidae the number varies from 0/2 in Risso's dolphin (*Grampus griseus*) to 44–64/42–62 in the spinner dolphin (*Stenella longirostris*). The spinner dolphin shows a wide geographical difference in number. The average highest tooth count in the lower tooth row is 45–46 in Thailand and northern Australian series, while it is 51–53 in the Central and South Pacific, western Pacific, and Philippines.

The unerupted small teeth of sperm whales remain in the upper jaw during life, as well as the pygmy sperm whale and the dwarf sperm whale (*Kogia sima*), and all of the lower teeth are erupted around sexual maturity. In the sperm whale the number of teeth is 0 for erupted and about 10 for unerupted teeth in each side of the upper jaw and 17–29 in the lower jaw. Pelagic dolphins such as *Stenella* spp. and *Delphinus* spp. are wide ranging in the subtropical and tropical regions of the world and are opportunistic feeders. The pincer-like action of the jaws of these long-snouted forms allows fish to be trapped by the interlocking tips of the teeth when the jaw closes and water is forced out. However, some beaked whales, the narwhal (*Monodon monoceros*), and Risso's dolphin, which capture prey by suction feeding, have a relatively constant and small number of teeth.

Tooth morphology also differs among species (Fig. 1). Teeth of the beaked whales are of various shapes and at various positions in the lower jaw (Fig. 2). In males, one pair of teeth in the lower jaw is erupted at attainment of sexual maturity, whereas in females they remain unerupted during life. It has been suggested that the erupted teeth may function to guide prey in the mouth, but because females and immature males lack these teeth, it is more likely that they function in intraspecific fighting. In the beaked whales of the genus *Mesoplodon*, maxillary teeth are absent with some exceptions. Gray's beaked whale (*Mesoplodon grayi*) and Shepherd's beaked whale (*Tasmacetus shepherdi*) contain functional teeth in the upper jaw. Gray's beaked whales have a series of 17–22 erupted small teeth in the upper jaw; however, they are not rooted in the maxilla but rather in the soft tissue over it. Some specimens of Sowerby's beaked whale (*Mesoplodon bidens*) have vestigial teeth in both upper and lower jaws. Among beaked whales, the position and shape

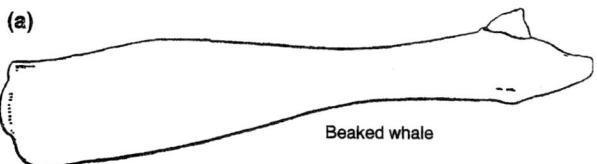

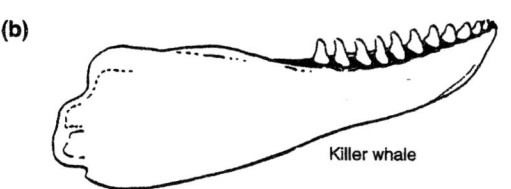

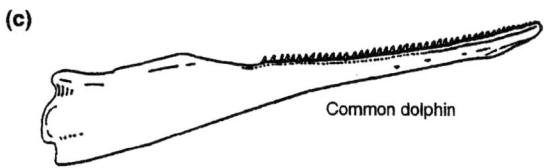

Figure 1 *Comparison of representative lower dentitions of odontocetes: (a) beaked whale (Ziphiidae), (b) killer whale (Orcinus orca), and (c) common dolphin (Delphinus delphis) (Slijper, 1979; Berta and Sumich, 1999).*

of the teeth in the mandibles of adult males are the most important key in identifying species of the genus *Mesoplodon,* although the shape of teeth changes with growth. In Cuvier's beaked whale (*Ziphius cavirostris*), which has one pair of teeth apically, some vestigial teeth are often found in the lower jaw.

The Amazon river dolphin (*Inia geoffrensis*) has 100–140 teeth, which are conical in the anterior half of each jaw but in the posterior have a lingual flange extending from the crown. Wrinkling of the enamel makes the teeth surface rugose. In the franciscana (*Pontoporia blainvillei*), the crown is slender, slightly compressed anteroposteriorly. The enamel is smooth. In the Indian river dolphin (*Platanista gangetica*), the lower teeth are much longer than the upper, and in young animals the anterior upper and lower teeth interlock, overlapping the sides of the jaw. Although narrow and sharp in young, the teeth become worn and flattened with age.

Sperm and beaked whales exhibit a number of adaptations for employing the tongue as a piston for sucking squid into the mouth. This suction-feeding strategy is characterized by a reduced number of teeth and a small gape to allow unobstructed entry of small prey, a ribbed palate to help hold slippery-bodied squid, and throat grooves to allow for expansion of the throat region. The suction-feeding mechanism in BEAKED WHALES involves distension of the floor of the mouth provided by throat grooves and retraction of the tongue by the styloglossus and hyoglossus muscles.

Teeth are used not only for feeding but also as a symbol of social status and/or aggressive display of adult males. Tooth marks of the adult male are often found on the ventral side of

Baird's beaked whale (*Berardius bairdii*). In both sexes of Baird's beaked whale, the lower jaw typically bears four teeth, which are composed of a large triangular pair situated apically and a smaller peg-like pair posterior to a short diastema. The teeth erupt around the onset of sexual maturity in both sexes, with the apical pair projecting obliquely forward well beyond.

Among odontocetes, tusks of NARWHAL, which exceed 3 m in length and have prominent left-handed spiral ridges, are the most specialized teeth. The spiral of the tusk is sinistral (or left-handed) when viewed from the root. The tusk extends anteriorly from the head. The left anterior tooth usually erupts as a tusk only in males. However, male narwhals occasionally develop two tusks or may lack an erupted tusk, and females may develop a left tusk or both tusks. When present, the tusks of females are shorter and less robust than in males. The large disparity in size between male and female narwhals and the

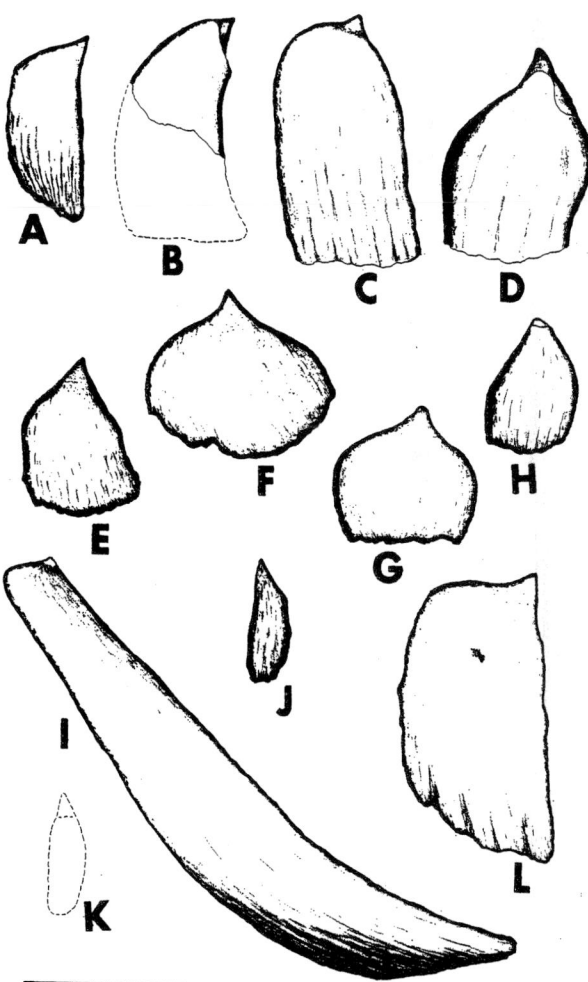

Figure 2 *Lateral views of right teeth of adult male* Mesoplodon: *(A)* M. bidens, *(B)* M. bowdoini, *(C)* M. carlhubbsi, *(D)* M. densirostris, *(E)* M. europaeus, *(F)* M. ginkgodens, *(G)* M. grayi, *(H)* M. hectori, *(I)* M. layardii, *(J)* M. mirus, *(K)* M. Pacificus, and (L) M. stejnegeri (Mead, 1989). Scale bar: 10 cm.

later attainment of sexual maturity in the male strongly suggest that the tusks are used in sexual display among males. It has been hypothesized that the tusk is used for the nonviolent assessment of social status on the basis of relative tusk size during frontal encounter, but evidence from head scarring and broken tusks indicates that adult males may indulge in aggressive displays that involve violent fighting.

II. Pinnipedia

Pinniped species have some reduction of teeth and milk dentition is less developed compared with typical mammals. The teeth posterior to the canines are little differentiated from one another and are referred to as postcanines (PC) or cheek teeth because molars are not clearly distinguished from premolars. Without the need for slicing or chewing while processing food (prey are swallowed whole), the cheek teeth of pinnipeds are typically homodont with a single pointed cusp for gripping slippery prey. Posterior cusps may also be present and are more frequent on posterior cheek teeth. A double-rooted condition of cheek teeth is the ancestral condition, apart from the first premolar, which is single rooted. The permanent incisors and canines normally have milk precursors, but of the postcanines, only the second, third, and fourth are preceded by milk teeth. As the permanent teeth erupt, the small alveoli gradually obliterate the alveoli of the milk teeth in the jaw.

In the WALRUS (*Odobenus rosmarus*), the formula for milk dentition is I3/3, C1/1, PC3/3 × 2 = 28. Milk teeth are shed shortly after birth, but not all of them are succeeded by permanent teeth. Most of the functional teeth erupt in the first and second years after birth. In the adult, the formula for functional dentition is I1/0, C1/1, PC3/3 × 2 = 18. All the teeth are simple conical crowns with a single root. The permanent upper canines erupt at about 4 months of age. They have persistently open pulp cavities and continue to grow throughout the life of the animal, although the tips are abraded constantly during feeding. The upper canine teeth develop into large "tusks" in both sexes. The tusks of males are larger, thicker, spaced farther apart at the base, and more divergent at the tips than those of females, and a single tusk may reach 1 m in length (35 cm is average) and weigh as much as 5.4 kg. Male walruses use their tusks mainly in dominance displays, females use them to defend themselves and their young, and they are occasionally used by both sexes to pull themselves onto ice floes. Tusks are not usually used in feeding, although rarely they are used to stab seal prey. The abrasion to the tips is the result of the tusks being scratched by sediment as an animal is searching for mollusks, its primary prey.

In Otariidae, the retention time of milk teeth depends on when the pups bite floating and other objects, but most of the milk teeth are shed by 4–5 months after birth, although there is some individual variation. Otariids have more teeth than phocids and both have more than the walrus (Table I). The permanent teeth of otariids are less diverse in morphology than those of phocids. The typical otariid dental formula is I3/2, C1/1, PC6/5, with some interspecific differences and individual variation.

Among Phocidae, phocines have incisors 3/2 (except for the hooded seal, *Cystophora cristata*, which has 2/1) and monachines have incisors 2/2 (except for the elephant seals, *Mirounga* spp., which have 2/1). In almost all phocids species the rest of the dental formula is C1/1, PC5/5. There is considerable variation in tooth shape and degree of cusp development (Fig. 3). Cheek teeth of the crabeater seal (*Lobodon carcinophaga*) and, to lesser extent, of the leopard seal (*Hydrurga leptonyx*) are highly modified with complex elongated cusps to trap and strain krill. Both crabeater and leopard seals are thought to suck krill into their open months by retracting their tongues and then forcing excess water out through the sieving cheek teeth. Leopard seals also possess well-developed canines for preying on penguins, other seabirds, and other pinnipeds.

III. Sirenia

The dugong (*Dugong dugon*) has a fixed dental formula of I2/3, C0/1, PM3/3, M3/3 with vertical replacement as in most other mammals. Dugong dentition is distinctly different from that of manatees, which lack incisors and have a different pattern of cheek tooth replacement. Two pairs of upper incisors are small and present in the juvenile dugong. Lower incisors and a canine are vestigial teeth. The deciduous (first) incisors are small, do not erupt, and are resorbed. In males, the deciduous incisors are lost around the time of tusk eruption, and their sockets disappear as the tusks expand. The tusks of males erupt at around 12 to 15 years of age and may be up to 15 cm in length with protrusion of a couple of centimeters. In females, small partially resorbed incisors may persist until the animal is about 30 years old. The permanent incisors (tusks) develop in both sexes but only erupt in males and some old females. In some old females the tusks erupt and wear, indicating that tusk eruption is not always diagnostic of the males. There are six cheek teeth (three premolars and three molars) in each jaw during the life of the dugong. These three premolars are erupted at birth and fall out because the roots of the anterior teeth are progressively resorbed. Their sockets then become occluded with bone. The molars progressively erupt during growth. The second and third molars continue to grow so that the occlusal area of the cheek teeth is maintained and increased even after the loss of anterior cheek teeth.

In the MANATEES (*Trichechus* spp.) there are usually five to seven functional teeth in the upper and the lower jaws. Cheek teeth are replaced from the rear and fall out, as anterior teeth wear from the excessive amount of sand and grit taken during feeding on plants. The total number of teeth per jaw is estimated at 20–30 during the life of an individual. Cheek teeth are brachyodont, or short crowned, and are enameled but lack cementum. There are two vestigial incisors in each jaw at birth, but they are later resorbed. It was thought that tooth replacement in the manatee occurs throughout the life of the individual, but there is evidence that this may not be the case. Teeth are replaced from the rear and, as anterior teeth wear they fall out. The increase in solid food intake after weaning, and hence more chewing, acts as a mechanical stimulus for the initiation and continuation of tooth row movement. Although it is sometimes incorrectly compared to elephants, this combination of horizontal movement with an appar-

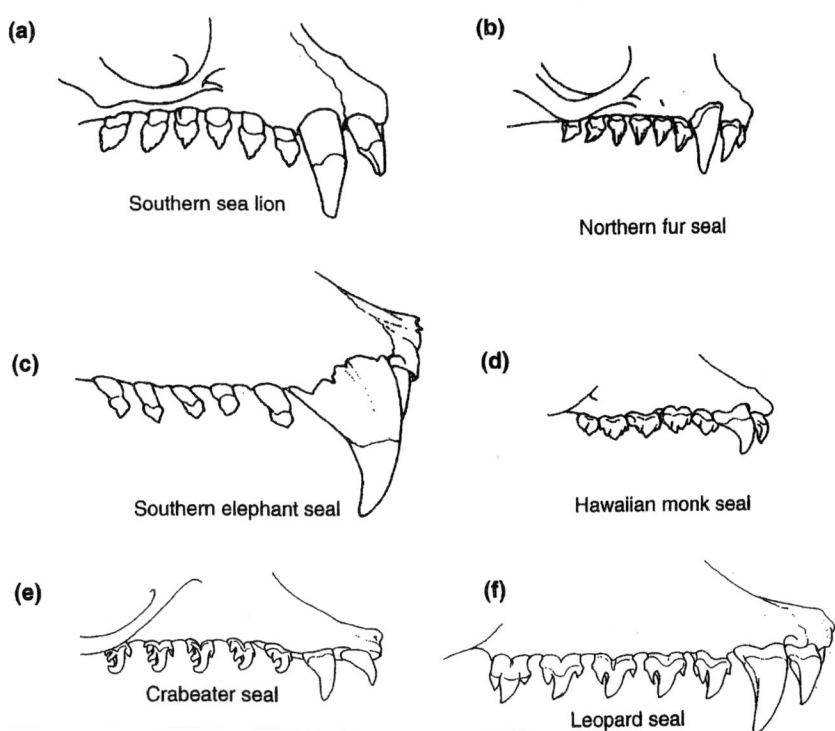

Figure 3 *Comparison of representative otariid (a and b) and phocid (c, d, e, and f) upper dentitions (King, 1964; Berta and Sumich, 1999): (a)* Otaria flavescens, *(b)* Callorhinus ursinus, *(c)* Mirounga leonina, *(d)* Monachus schauinslandi, *(e)* Lobodon carcinophaga, *and (f)* Hydrurga leptonyx.

ent limitless supply of supernumary (extra) molars is found in no other mammal, including the dugong.

IV. Age Determination of Marine Mammals

To understand the life history of aquatic animals, age determination is a fundamental factor. Techniques for age determination of animals vary widely among species and among investigators. One of more reliable methods involves determining an individual's age using teeth, mandibles, and ear bones. As an animal grows, incremental layers accumulate in the above hard tissues. These incremental growth layers, which are analogous to rings in tree trunks, are used for age determination of animals. Thus, counting these layers is now widely used to estimate ages of marine animals. Teeth typically are obtained from dead specimens and from captive or temporarily restrained individuals of all groups of marine mammals except mysticetes, in which growth layers in the large waxy ear plugs, as well as tympanic bullae and skull bones, have been used. Procedures for enhancing the visibility of growth increments usually include thinly slicing and polishing teeth and then etching or staining the polished surface to better resolve the growth layers. Each countable unit of repeating incremental growth layers contains at least one change in tissue density, hardness, or opacity and is referred to as a growth layer group (GLG). In most species examined so far, each GLG is thought to represent an annual increment. Useful methods for age determination by teeth are summarized as (1) preparation of unstained longitudinal thin section, (2) preparation of decalcified and stained longitudinal thin section, (3) etched preparation of the surface of the sanded half-tooth, (4) preparation of scanning electron microscopy, and (5) preparation of microradiography. The latter two methods are not convenient for mass preparation in a short time. The etched preparation of a longitudinally sectioned tooth is a good method for larger dolphins with larger teeth, and thin undecalcified sections are also useful if the longitudinal section is cut accurately or ground to include the largest possible longitudinal section of the pulp cavity. However, such material is suitable for dentinal counts only. Although the number of cemental growth layers is almost the same as the number of dentinal growth layers for young animals, the former exceeds the latter for old ones.

Cemental growth layer groups should be counted in very carefully prepared specimens of the decalcified and stained tooth. Preparation of the decalcified tooth in 5% formic acid for 24 hr and staining with Mayer's hematoxylin solution for about 30 min of a longitudinally thin section (10–20 μm) gives better results for small delphinid teeth.

See Also the Following Articles

Age Estimation ▪ Bones and Teeth, Histology of ▪ Dental Morphology, Evolution of ▪ Skull Anatomy ▪ Toothed Whales, Overview

References

Berta, A., and Sumich, J. L. (1999). "Marine Mammals: Evolutionary Biology." Academic Press, San Diego.

King, J. E. (1964). "Seals of the World." Trustees of the British Museum (Natural History), London.

Jefferson, T. A., Leatherwood, S., and Webber, M. A. (1993). FAO special identification guide. *In* "Marine Mammals of the World." FAO, Rome.

Lambertsen, R. H., Ulrich, N., and Straley, J. (1995). Frontomandibular stay of Balaenopteridae: A mechanism for momentum recapture during feeding. *J. Mammal.* **76**(3), 877–899.

Macdonald, D. W. (1984). "The Encyclopedia of Mammals." Facts On File Publication, New York.

Mead, J. G. (1989). Beaked whales of the genus *Mesoplodon. In* "Handbook of Marine Mammals" (S. H. Ridgway and R. Harrison, eds.), Vol. 4, pp. 349–430. Academic Press, London.

Nishiwaki, M. (1972). General biology. *In* "Mammals of the Sea: Biology and Medicine" (S. H. Ridgway, ed.), pp. 3–204. Charles C. Thomas, Springfield, IL.

Perrin, W. F., and Myrick, A. C., Jr. (1980). Age determination of toothed whales and sirenians. International Whaling Commission, Special Issue 3, Cambridge.

Rice, D. W. (1998). Marine mammals of the world: Systematics and distribution. Special Publication Number 4, The Society for Marine Mammalogy.

Ridgway, S. H., and Harrison, R. (eds.) (1981–1994). "Handbook of Marine Mammals," Vols. 1–5. Academic Press, London.

Slijper, E. J. (1979). "Whales." Hutchinson, London.

Telemetry

ANDREW J. READ

Duke University Marine Laboratory,
Beaufort, North Carolina

Telemetry is the process of obtaining data remotely by transmitting information from a marine mammal or by storing information for later retrieval. Telemetry includes a number of research approaches, from simple radio tags that allow researchers to relocate a tagged animal to complex data loggers that record data from multiple environmental sensors. Recent advances in this field have led to extraordinary insights into the behavior, ECOLOGY, and physiology of marine mammals. It is now possible to obtain data via satellite on the fine-scale behavior of marine mammals from the most remote regions of the world's oceans. It is also possible, for the first time, to record on videotape what a marine mammal sees as it swims through the water column. These advances allow researchers to investigate how marine mammals use their three-dimensional world and to identify and quantify important physical and biological aspects of their environments. To understand the way in which telemetry has changed our understanding of marine mammals, one need only consult some of the older references on the diving capabilities of marine mammals. These early studies relied almost entirely on anecdotal observations, such as observations of sperm whales (*Physeter macrocephalus*) tangled in submarine cables. Today, we have a rich understanding of the diving capabilities of marine mammals from direct observations made using telemetry. The field continues to develop rapidly, fueled by continuing advances in technology and miniaturization.

There are two primary approaches to collecting data with telemetry systems. In the first approach, a data logger is attached to a marine mammal, records data for a predetermined period, and then is recovered, allowing researchers to download the information stored in the package. In the second approach, information is transmitted from a marine mammal via radio or acoustic signals.

I. Recoverable Data Loggers

Data loggers record information from a variety of sensors that provide insight into a marine mammal's behavior, physiology, and environment. The earliest data loggers were simple time-depth recorders that used smoked glass discs rotating past recording needles attached to pressure transducers. These devices were developed in the 1960s by Gerald Kooyman to study the DIVING behavior of Weddell seals (*Leptonychotes weddellii*). To record time in these devices, Kooyman incorporated simple kitchen timers. This ingenuity is typical of the field of marine telemetry, which is too small to support much of its own commercial research and development. Instead, biologists have adopted, modified, and refined technology developed for other purposes.

Modern data loggers are sophisticated digital devices, capable of storing large quantities of information. Data are collected from one or more sensors that measure depth, water temperature, light intensity, or swimming velocity. The sampling interval is set by the researcher and may vary depending on the type of question being asked. The location of a tagged animal can be determined by several methods, but the most common approach is to record light levels and times of dawn and dusk and to back-calculate latitude and longitude after the logger is recovered. It is also possible to record physiological data, such as heart rate and body temperature. Researchers can even record feeding events by transmitting temperature changes in the stomach to external data loggers. To monitor feeding events, a small transmitter, equipped with a temperature sensor, is introduced into the stomach of an animal and transmits data to a data logger mounted on its external surface. Most prey are heterothermic, or cold-blooded, so when they are swallowed, the temperature of the stomach drops abruptly. Eventually the transmitter is passed or regurgitated.

Two recent developments with data loggers have yielded new insights into the BEHAVIOR of marine mammals. The first is the development of a recoverable logger capable of recording sound—in essence, recording the acoustic environment of the tagged animal. These loggers incorporate hydrophones and recording media such as digital audio recorders or miniature hard drives, as well as other sensor systems. In addition to the vocalizations of the animals themselves and the sound of nearby vessels, the loggers have provided unexpectedly rich data on swimming stroke and heart rate. These devices are being used by researchers to study the effects of anthropogenic noise on the behavior of marine mammals.

The second advance in logger technology is the crittercam—a recoverable video recorder system. Developed by researchers from the National Geographic Society, crittercams provide a visual record of everything that a marine mammal sees (if aimed forward) or everything in the path of the animal (if aimed backward). Because humans are visual creatures, these video records provide researchers with powerful insights into the underwater lives of marine mammals.

Data loggers have several advantages over other types of telemetry systems. First, because data storage requires considerably less power than data transmission, fewer and/or smaller batteries are required. In turn, this means that recoverable loggers are generally smaller than transmitting tags. Second, the storage of large quantities of data is possible, particularly with modern digital technology. As noted later, transmitting systems limit the quantity of data that can be relayed from the animal to a receiver.

The primary disadvantage of these systems is the need to recover the data loggers to retrieve stored information. The use of data loggers in studies of pinnipeds is fairly straightforward because these animals haul out at predictable times and locations. Researchers studying elephant seals (*Mirounga* spp.), for example, are able to recover up to 95% of their loggers because of the strong fidelity of these animals to their rookeries. Using data loggers with cetaceans, however, is considerably more challenging, as researchers must first attach the package to a dolphin or whale and then recover the tag after it is jettisoned. One solution to these problems is to attach the loggers with suction cups, fired from a cross bow, and then to recover the buoyant packages after release by homing in on a radio signal emitted by a tag in the package. Several researchers have employed this technique with success, although the logistics of such field work are considerably more complex than the deployment and retrieval of data loggers on pinnipeds.

II. Transmitting Systems

Transmitting systems have also undergone a rapid development over the past several decades. The earliest transmitters were omnidirectional radio or acoustic transmitters that allowed researchers to relocate a tagged animal but did not provide information on its behavior or physiology. These simple systems have evolved into sophisticated systems in which large quantities of data can be recorded, compressed, and transmitted.

Some of the earliest radio transmitters used with marine mammals were developed by Bill Watkins and W. E. Schevill from the Woods Hole Oceanographic Institution. In the 1960s, these researchers developed implantable radio tags fired into the blubber of large whales. Similar tags are still used today to study the movements and behavior of BALEEN WHALES. Radio tags have also been attached to the dorsal fins of dolphins and porpoises or glued to the fur of pinnipeds by other researchers. When successful, these tags have allowed researchers to follow marine mammals at sea and gain insight into their behavior and short-term movements. This labor-intensive field work requires the use of directional receiving antennae to home in on the radio signal produced by the transmitter.

The utility of these simple transmitting systems is limited by several factors. First, because the high-frequency signals emitted by radio tags attenuate rapidly in salt water, it is possible to receive signals only when the transmitter is above the surface. This complicates the tracking of animals because only a few signals are heard at each surfacing. Acoustic signals propagate for much greater distances underwater but often overlap with the hearing range of marine mammals, limiting their applicability. Even under ideal circumstances, radio transmitters have effective ranges of only a few tens of kilometers, so researchers are forced to stay in close proximity to tagged animals. Finally, the large size and cumbersome design of many early radio tags created significant hydrodynamic drag and resulted in the premature detachment of the packages.

Today, researchers have a wide variety of transmitting systems available to them. The most significant advance has been the development of satellite-linked radio transmitters that allow biologists to track the movements and behavior of marine mammals from their offices. The principle underlying these systems is fairly straightforward. Each transmitter emits a stable radio signal to receivers aboard orbiting weather satellites. As a satellite moves across the horizon, the received frequency of the tag changes, due to the Doppler shift, allowing estimation of the position of the transmitter. Each transmission also includes the identity of the transmitter and any associated sensor data. Data are processed and relayed to the user by modem or e-mail. This technological advance obviates the need for researchers to track animals in the field.

Satellite-linked radio transmitters have been coupled with data logging systems to allow the collection of detailed behavioral or environmental data from marine mammals via satellite. This coupling of data logging and transmitting systems has proven to be very successful because it precludes the need to recover the logger package to obtain sensor data. Typical data collected by these systems include depth and swim speed, although in principle any sensor system can be employed.

The advantages of transmitting systems lie primarily in their ability to provide data in real time. Even satellite-linked transmitter systems can provide telemetry data within a few hours, allowing researchers to monitor the movements and behavior of animals in real time from their offices. Simplicity has its virtues, however, and many researchers continue to use conventional radio tags to assist in the relocation of tagged animals in the field. Spending many hours with a particular individual allows for the collection of fine-scale behavioral data. Although it is often possible to use natural features to identify a dolphin or whale, a simple radio tag can greatly facilitate the relocation of a particular individual in the field.

Satellite-linked data loggers are extremely powerful data acquisition systems, but they do have limitations. Their signals can be received only when the transmitter is above the surface and a satellite receiver is overhead. Energy for signal transmission is a significant limitation with current battery technology, although battery life may be conserved by using a salt-water switch, which suppresses transmissions when the tag is submerged. In addition, because the current satellite system limits each transmission to 256 bits, algorithms are required to compress complex data, such as records of individual dive profiles, prior to transmission.

III. Biological Insights

Advances in the field of telemetry have revolutionized our view of marine mammals. As terrestrial observers, we are limited in our ability to study marine mammals and, in the past, have been limited to collecting data from animals at the surface or ashore. While at sea, most marine mammals spend more than 90% of the time submerged, often in remote or harsh environments in which field research is difficult or impossible. Telemetry offers the potential to peer into the lives of whales, dolphins, and seals as they go about their daily activities of FEEDING, finding mates, and avoiding PREDATORS. For the first time, we can ask how a Weddell seal hunts for food under the Antarctic ice, how an elephant seal makes such long dives, or where blue whales go in the winter months. The insights provided by this technology will continue to challenge our thinking about these animals, particularly as new technological developments improve our ability to collect data at sea.

ELEPHANT SEALS have proven to be particularly amenable to study with telemetry. These are large animals that haul out to breed and molt at predictable times and locations but which spend the majority of the year far from shore. Thus it is possible to equip individual elephant seals with fairly large telemetry packages and be confident that most packages will be recovered. Researchers in both hemispheres have equipped a large number of elephant seals with recoverable data loggers and, more recently, satellite-linked data loggers. From this research, we now know that elephant seals in the North Pacific make two long-distance feeding migrations each year, one after breeding and the second after the molt. For example, adult male elephant seals travel from central California to the Gulf of Alaska on each migration, a distance of more than 10,000 km in each round trip. Individuals appear to return to the same feeding area each year and, once on the feeding grounds, forage almost continuously. While feeding, individual elephant seals dive repeatedly, spending more than 90% of their time at sea submerged, and sometimes diving to depths of more than 1500 m. Such behavior is consistent with what we know of the diet of these animals; elephant seals feed primarily on mesopelagic squid found at depths of 200 to 1000 m. These prolonged and continuous dives have raised many physiological questions, particularly with regard to the oxygen storage capacity of these animals (see later). The prodigious diving behavior of elephant seals has led some biologists to refer to them as mesopelagic mammals.

Elephant seals are not alone in possessing an impressive diving capacity. Sperm whales have been tracked using telemetry to depths of more than 2000 m during dives that may last for more than an hour. Beaked whales (Ziphiidae) are also capable of long, deep dives. Northern bottlenose whales (*Hyperoodon ampullatus*), for example, equipped with time-depth recorders attached with suction cups, have made dives to 1500 m and for over an hour in duration. Studies using satellite-linked data loggers attached to smaller whales, such as belugas (*Delphinapterus leucas*) and narwhals (*Monodon monoceros*), indicate that these species are also capable of prolonged, deep dives under the Arctic ice.

Studies with crittercams have provided dramatic findings regarding the behavior of marine mammals. Weddell seals have been videotaped flushing prey from crevices in the ice by blowing bubbles, and researchers have watched Hawaiian monk seals (*Monachus schauinslandi*) sleep and forage on the sea floor. This type of research has particular relevance to conservation because it is believed that monk seals may be endangered, in part, due to conflicts with commercial fisheries. Documenting the availability of prey and the success rate of capture attempts allows us to test such ideas directly for the first time. Backward-mounted crittercams have been used to study the diving behavior of a variety of marine mammals and, in particular, to investigate how whales and seals can make such long dives without exceeding their aerobic capacities. It now appears, for example, that elephant seals and other marine mammals conserve oxygen by gliding extensively during descent. These animals take advantage of the changes in buoyancy brought about by increased pressure at depth and can descend effectively with little extra expenditure of energy or oxygen.

IV. Future Developments

It is difficult to anticipate what surprises the field of telemetry has in store, but it is clear that these techniques will be an integral component of the toolbox of future marine mammal researchers. In particular, it is likely that more sensors will be developed to take advantage of the success of recoverable and satellite-linked data loggers. New sensors may monitor physiological parameters, such as blood oxygen concentration and blubber thickness. Miniaturization and refinement of the crittercam system will allow researchers to ask how deep-diving species, such as elephant seals and sperm whales, find and capture prey at depth. Interest has been expressed in using marine mammals as autonomous oceanographic data collection vehicles. A sample of elephant seals equipped with sensors measuring temperature, salinity, and depth could, for example, provide considerable information on the oceanography of the Southern Ocean. Marine mammals are adept at exploiting fine-scale oceanographic features that concentrate prey, such as frontal systems, and animals instrumented with appropriate sensors could provide considerable information about the location and dynamics of such processes.

New advances in digital technology will undoubtedly result in substantial improvements in our ability to store, transmit, and receive data. Future readers will no doubt find our current suite of data loggers and satellite-based telemetry systems quaint and outdated. Current advances in wireless technology hold great promise for our ability to telemeter data from marine mammals because many current applications in acoustic and video telemetry are limited by bandwidth: the amount of information that can be transmitted from the animal to a receiver. New low-orbit satellite systems could improve our ability to collect more data from a larger sample of animals. So, too, could a network of autonomous underwater data collection systems, similar to those being developed for oceanographic and military purposes, facilitate the collection of data using acoustic telemetry.

Finally, current research efforts are developing attachment techniques that are less invasive and easier to employ, particularly with dolphins, porpoises, and whales. The use of suction cups to attach data loggers to whales for short periods is a good example of this type of innovation. With the advent of improved technology and smaller, more powerful batteries, the size of telemetry packages will also decrease. This will allow the application of this technology to a wider variety of marine mammal species and for longer periods than has been possible to date. The field of telemetry holds great promise to reveal new and exciting insights into the lives of marine mammals.

See Also the Following Articles

Diving Behavior ▪ Identification Methods ▪ Swimming

References

Costa, D. P. (1993). The secret lives of marine mammals: Novel tools for studying their behavior and biology at sea. *Oceanography* **6,** 120–128.

Davis, R. W., Fuiman, L. A., Williams, T. M., Collier, S. O., Hagey, W. P., Kanatous, S. B., Kohin, S., and Horning, M. (1999). Hunting behavior of a marine mammal beneath the Antarctic fast ice. *Science* **283,** 993–996.

Kooyman, G. L. (1989). "Diverse Divers: Physiology and Behavior." Springer-Verlag, Berlin.

Le Boeuf, B. J., and Laws, R. M. (eds.) (1994). "Elephant seals: Population Ecology, Behavior, and Physiology." Univ. California Press, Berkeley.

Stone, G., and Kraus, S. D. (eds.) (1998). Marine animal telemetry tags: What we learn and how we learn it. *Mar. Technol. Soc. J.* **32,** 1–114.

Whitehead, H., Christal, J., and Tyack, P. L. (2000). Studying cetacean social structure in space and time: Innovative techniques. *In* "Cetacean Societies: Field Studies of Dolphins and Whales" (J. Mann, R. C. Connor, P. L. Tyack, and H. Whitehead, eds.), pp. 65–87. Univ. of Chicago Press, Chicago.

Williams, T. M., Davis, R. W., Fuiman, L. A., Francis, J., Le Boeuf, B. J., Horning, M., Calambokidis, J., and Croll, D. A. (2000). Sink or swim: Strategies for cost-efficient diving by marine mammals. *Science* **288,** 133–136.

Territorial Behavior

EDWARD H. MILLER
Memorial University, St. John's,
Newfoundland, Canada

Territory and *territoriality* widely refer to the exclusive use of space by individuals or groups, entailing both spatial defense and advertisement. Such terms have been applied to diverse animal species (e.g., octopuses, insects, primates) and functions (e.g., nesting, mate attraction, feeding) over a broad range of scales in space (millimeters to kilometers) and time (seconds to years). Such broad ranges of biological forms and functions cannot be encompassed by a single definition. Definitions that apply to selected taxonomic groups or functions of interest and over scales of space and time relevant for those groups or functions are best.

In this article, "territoriality" and "territorial behavior" refer to behavior associated with obtaining, defending, or advertising occupancy of space. Of course when two animals interact agonistically, they always do so within a spatial framework, but social interactions by themselves do not constitute territorial behavior. "Territoriality" connotes something other than just behavior relative to the interacting animals themselves or to their arbitrary spatial positions, such as (1) specific spaces relative to other individuals or to the external physical environment or (2) particular spatial features (e.g., a haul-out ledge). Nevertheless it is usually impossible to draw a strict distinction between defense of a space and resources within that space for all situations because species (or individuals, age classes, or social classes) that are usually viewed as "nonterritorial" like walruses (*Odobenus rosmarus*) or elephant seals (*Mirounga* spp.) exhibit territorial-like behavior in some settings, and vice versa for "territorial" species like fur seals and sea lions (Otariidae; "otariid" hereafter). Therefore, in this article the application of the definition given earlier is broad.

Territorial size, clarity and constancy of territorial boundaries, the medium in which territories occur, and functions of territories are interwoven. Cetaceans, sirenians, and polar bears (*Ursus maritimus*) exhibit little behavior suggesting the exclusive use of space defended or advertised by individuals or groups, but territorial behavior is clearly expressed in sea otters (*Enhydra lutris*) and most pinnipeds.

I. Development of Territorial Behavior

Sexual differences in behavior appear early in life in otariids. Male pups are more aggressive than females and engage in play fighting and boundary displays (Section V) within the first few weeks of life. Young otariid males of all ages engage in such behavior commonly in their interactions with one another, both inside and outside the breeding season. Juvenile and subadult walruses (including calves), recently weaned elephant seals, and juvenile and subadult harbor (*Phoca vitulina*) and gray (*Halichoerus grypus*) seals also show sexual differences in social behavior, with males spending more time play fighting with one another.

Territorial males of most otariid species appear to try to keep females on their territories by herding (Section V). Herding behavior appears early in life, and male pups preferentially direct this behavior toward female pups (e.g., Australasian fur seal, *Arctocephalus forsteri*). Herding is also expressed by nonterritorial males during the breeding season, e.g., when they encounter females moving toward or away from breeding aggregations. In all otariid species, numbers of nonterritorial males (including subadults) may rush simultaneously into ("raid") breeding aggregations and herd or interact with females in various ways before they are chased away by territorial males [raiding male

South American sea lions (*Otaria flavescens*) may achieve some direct reproductive benefits]. In species that frequent the colony site year-round (e.g., Australasian fur seal), juvenile and subadult males herd pups and young juveniles occasionally at various times outside the breeding season. An extreme form of this behavior occurs in the South American sea lion, in which nonterritorial males frequenty carry pups away in their mouths following raids on breeding sites. They carry the pups into the ocean (where they may drown) and then to nonbreeding areas where the males herd and mount them, sometimes over several days; pups usually die.

II. Territorial Functions

Nonbreeding pinnipeds engage in many agonistic social interactions in disputes over resting places at haul-out and breeding colony sites. Nonbreeding individuals may also favor and defend particular sites during the breeding season (e.g., otariids, harbor seals). Despite their relatively minor costs and benefits, such airborne interactions constitute territorial behavior in the sense used in this article and employ similar motor patterns to those used in better known and more dramatic examples of territoriality by breeding animals. In otariids, breeding females compete with one another for favored or habitual sites for resting or for nursing the young, and older pups similarly interact agonistically with one another or displace younger pups from favored rest sites.

The best known and most dramatic forms of territoriality in marine mammals are shown by breeding adult males. Male sea otters, otariids, and some seals (Phocidae; "phocid" hereafter) establish territories seasonally where females give birth and enter estrus, or to attract estrous females that have given birth. Sometimes territoriality by male pinnipeds is expressed only after females haul out at a breeding site [e.g., Hooker's sea lion (*Phocarctos hookeri*), gray seal], and males may also change locations facultatively in response to female movements, attend and defend isolated lone females, or even defend female carcasses (Fig. 1). However, the ultimate explanation for male territoriality is to obtain access to estrous females, thereby increasing male reproductive success.

III. Territoriality in Space

Discrete, clearly defined territories are most apparent on small temporal scales, in situations of crowding, in species that have good locomotory abilities so they can efficiently patrol or defend their territories, or where environmental features (e.g., topographical irregularities) occur that can be used by the animals to demarcate territories. Such conditions rarely occur in the lives of cetaceans (especially open-ocean species) so territoriality is rare or nonexistent in that group. RIVER DOLPHINS, with their spatially restricted distributions, or species that feed on concentrated prey that are sedentary or spatially predictable may be exceptions. In Scotland's Moray Firth, year-round resident common BOTTLENOSE DOLPHINS (*Tursiops truncatus*) may be territorial and exclude seasonal (winter) conspecific visitors from deep waters, which are most favorable for feeding.

Most species of otariids breed on crowded colony sites and hold small territories; territories of male Hooker's sea lions are often no more than 3 m in diameter, for example (about 30 m²

Figure 1 *Defense of space and of females is connected intimately in male marine mammals: adult male California sea lion defending carcass of female. From Peterson and Bartholomew (1967).*

in area), and some northern fur seals (*Callorhinus ursinus*) hold territories that are little larger in diameter than a male's body length (Fig. 2). Larger territories occur in related species (e.g., about 200 m² in male Steller sea lions, *Eumetopias jubatus*). Small aquatic territories are held by some male Juan Fernández fur seals (*Arctocephalus philippii*) adjacent to breeding aggregations, and walruses hold small aquatic territories adjacent to mixed herds on ice. In general, however, aquatic territories are large: >100 m in length in some male Weddell seals (*Leptonychotes weddellii*), up to 1 km across in male sea otters, and up to 10 km across in some male harbor seals; aquatic territories are commonly not contiguous (Fig. 3).

Phocids are specialized for aquatic locomotion, so their locomotion on land is slow and energetically costly. Unsurprisingly, territories in land-breeding populations of gray seals exhibit extensive overlap (Fig. 4). The poor locomotory abilities and large size of the two species of elephant seals usually preclude territoriality, although in small confined areas (e.g., coves) the defense of space and of females amounts to the same thing.

Rocks, fracture lines, and other natural features often enable precise delimitation of otariid territories. If such features are present, then parts or all of a territory's boundaries can be discrete. Territories that are partly or completely on featureless terrain (e.g., sandy beaches) cannot be clearly delimited by territory holders, so shifts in size, shape, and even location are common. Aquatic territories may include or be adjacent to prominent geological features or landforms. In species that breed in association with ice, underwater features of ice, or fractures or leads in ice, may be important in determining territorial density, size, or shape (Fig. 3A). In the walrus, male territories are established in the water adjacent to mixed herds on ice (Fig. 5). The ice may be stable land-fast ice (e.g., parts of Canada's arctic archipelago) or unstable drifting pack ice (e.g., Bering Sea). Walrus territories are ill defined and variable due to variability in ice and herd size and dispersion.

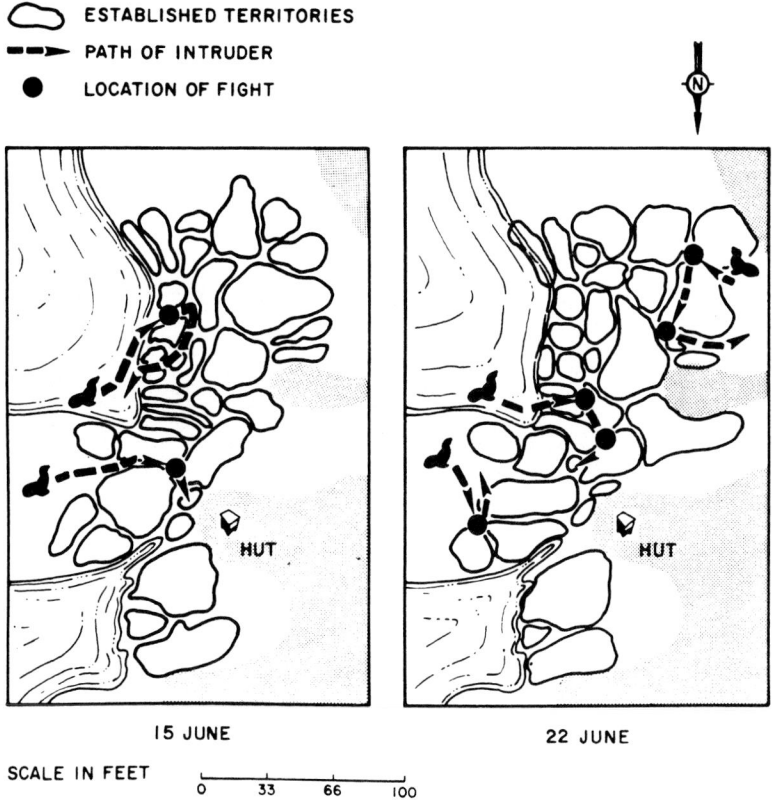

Figure 2　*Small, well-defined breeding territories occur in some species (northern fur seal). Territories are shown as slightly separated for clarity. Routes of entry into the existing territorial structure by males that obtained or failed to obtain territories are shown. From Peterson (1968).*

IV. Territoriality in Time

In otariids, size and shape of territories in shoreline locations are often influenced by the tide. Shoreline territories may alternately be above and partly under water and may be temporarily abandoned at high water levels (e.g., high tides, storms). Inland territories, especially those that lack pools in which males can cool themselves, may be abandoned in the heat of the day [e.g., Afro-Australian fur seal (*Arctocephalus pusillus*), Juan Fernández fur seal]. Walrus territories change in size, shape, and dispersion due to ice movements, amalgamation, and breakup.

Absence of females from their territories sometimes leads to territorial desertion by male otariids, but more commonly males attempt to acquire a new territory where females are present. In Hooker's sea lion, males establish territories several times during the breeding season in response to movement of the female aggregation down the beach. Southern sea lion males defend conventional territories (i.e., defense of space) early in the breeding season but gradually change to defense of females as the season progresses.

Territories of most or all otariids are most clearly defined at the peak of breeding, when territorial density is highest and territorial size is smallest (Figs. 2 and 6). Male otariids habituate to neighboring males and engage in fewer and less aggressive interactions with neighbors over time. A similar effect even

occurs between males that held neighboring territories in the previous year (e.g., Steller sea lion). On Sable Island, Nova Scotia, some male gray seals abandon their territories and then establish new ones on the island some distance away (up to several kilometers) within the same breeding season; smaller scale changes within seasons also occur (Fig. 4).

Long-term fidelity to territorial locations has two components. First, there is a tendency for males to return to breed near the site of their birth (philopatry). Second, males exhibit site fidelity by tending to return in successive years to where they first established a territory. Both forms of site fidelity are known in otariids and terrestrially breeding phocids (elephant seals, gray seals). Site fidelity is especially strong in otariids; for example, one male Antarctic fur seal (*Arctocephalus gazella*) returned to within 1 m^2 of previous territorial locations over seven successive breeding seasons. Males exhibit breeding-site fidelity even in phocids that breed in association with fairly stable land-fast ice [e.g., ring seal (*Pusa hispida*), Weddell seal]. In contrast, some individual male gray seals in eastern Canada have been observed in different breeding seasons at terrestrial breeding sites and at sites where breeding takes place on ice. Male sea OTTERs hold territories in the same location for up to seven successive years. The combination of philopatry with general site fidelity suggests that kin may breed in close proximity to one another in otariids, gray seals,

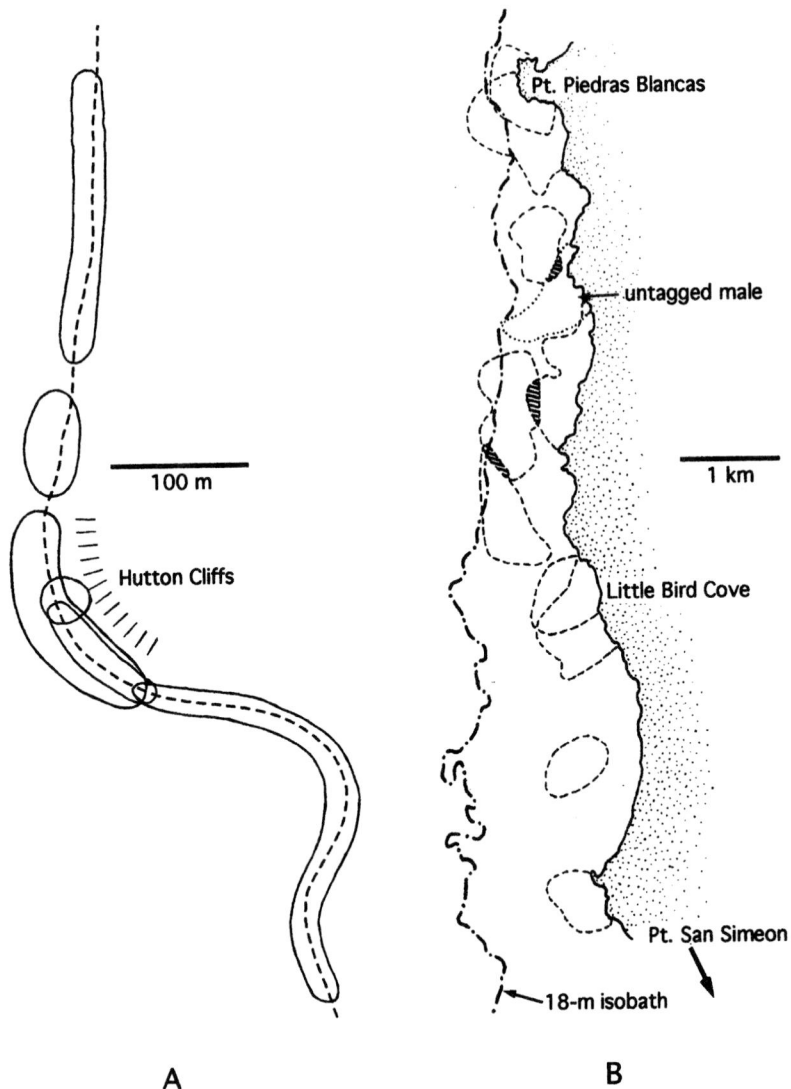

Figure 3 *Aquatic territories are typically large and overlap, but environmental features are important: Weddell seal (A) and sea otter (B). Six territories are outlined in A; the dashed line in A indicates a tidal crack in the ice on which territories were centered (the two small territories by the Hutton Cliffs were less well documented than the others). Eleven territories (10 of tagged males) are shown in B (note placement relative to water depth); overlap of concurrently held territories is indicated by hatching. From Kaufman et al. (1975) and Jameson (1989).*

and some other species. Behavioral effects of this possibility are not documented, but a detailed study of northern fur seals demonstrated that territorial males were careful to not injure pups (by stepping on them, for example)—pups that were sired in the previous year and therefore may have been the offspring of those males.

V. Obtaining, Defending, and Advertising Territories

Dramatic and potentially injurious fights between males occur in all territorial species, particularly when new males attempt to establish themselves. Because severe injuries or (rarely) death can result, fighting is undertaken only in specific circumstances; instead, most interactions between competing or neighboring males involve rich optical, acoustic, tactile, or chemical displays. In the northern fur seal, for example, only about 1% of encounters between territorial males involve actual fights.

The distinctive appearance of adult male otariids is communicatively important in the context of territoriality. For example, in contrast to females or young males, adult male California sea lions (*Zalophus californianus*) are much darker than

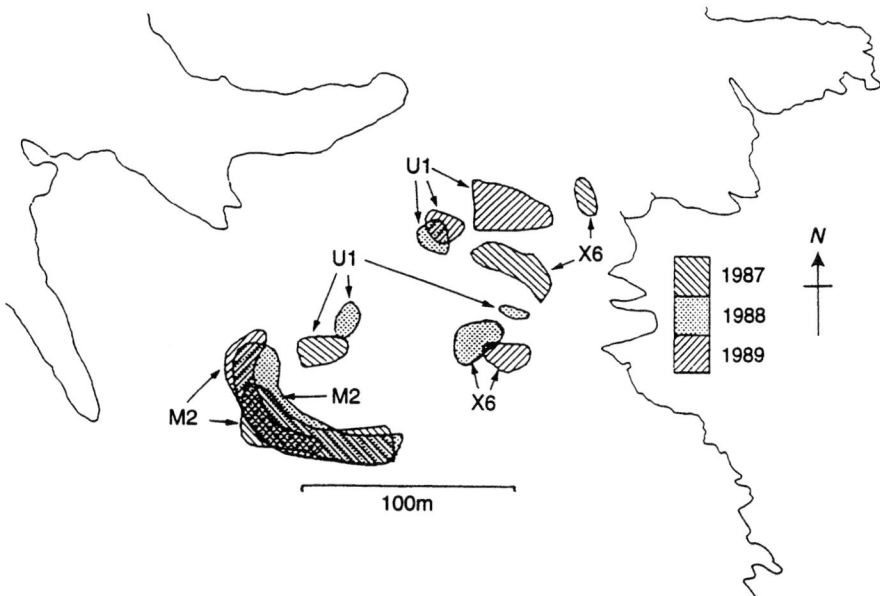

Figure 4 *Movements within and site fidelity between years are important in spacing systems of most or all territorial marine mammals: territorial male gray seals on North Rona, Scotland. From Twiss et al. (1994).*

females or young males, and adult male South American sea lions have distinctive manes. Some of the many optical displays of otariids are relatively passive and undirected (e.g., a distinctive nose-up upright resting posture). Most optical displays are directed toward specific individuals and involve movements, such as distinctive head-and-neck swinging LOCOMOTION, or rapid and complex sequences of motor patterns, including feints, oblique stares, sprawls, and facial expressions, in displays between neighbors across territorial boundaries (Fig. 7). Elephant seals and gray seals have specialized facial features

used in displays over various distances (Fig. 8). In underwater communication, optical displays cannot effectively employ fine features like facial expressions so must rely on general patterns of pelage or on whole body movements or postures.

Acoustic displays can travel long distances and are energetically inexpensive to produce, so unsurprisingly are important in territorial displays of almost all marine mammal species. Not all such displays are vocal. Male sea otters patrol their territory while SWIMMING on their backs and kicking vigorously, thereby generating loud splashing sounds (this display probably represents

Figure 5 *Walruses hold small, dynamic aquatic territories adjacent to mixed herds on ice: territorial male (in water) adjacent to mixed herd in central Canadian arctic. From Sjare and Stirling (1996).*

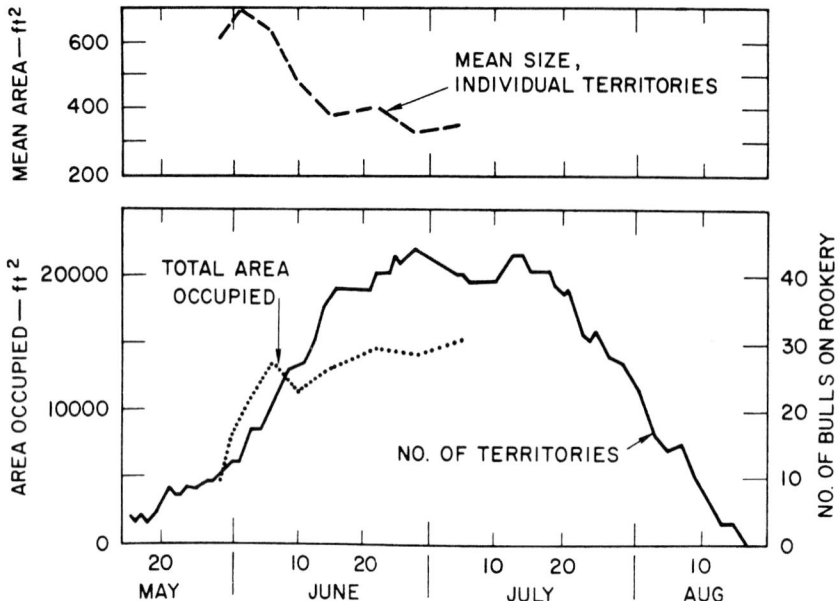

Figure 6   *Territorial size is smallest and defined most clearly in the peak of breeding: seasonal trends in territorial size in the northern fur seal. From Peterson (1968).*

Figure 7   *Complex sequences of motor patterns typify short-range communication between territorial male marine mammals: behavioral sequence (A–H) in representative "boundary display" between male South American sea lions. From Campagna and Le Boeuf (1988).*

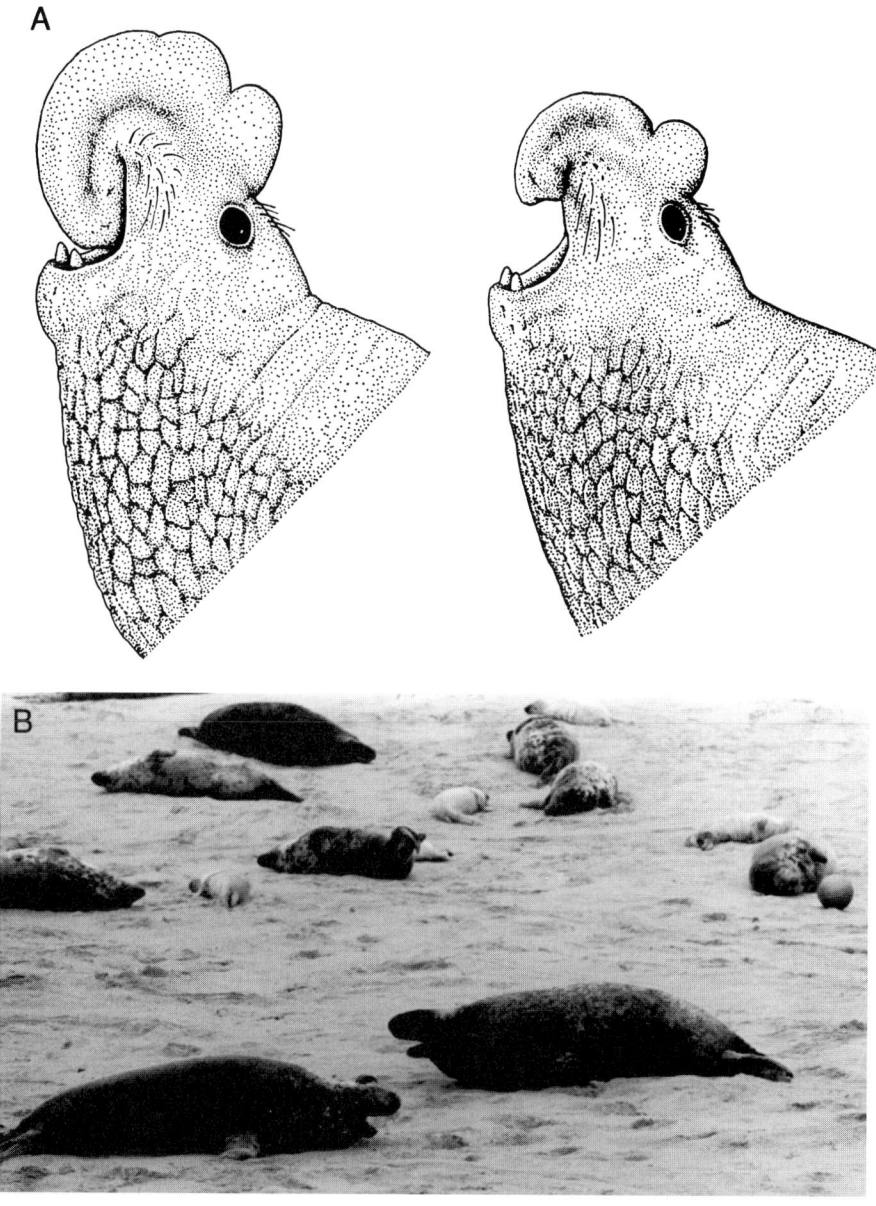

Figure 8 *General appearance and special structures are important aspects of displays between territorial marine mammals: (A) profiles of adult male northern (left) and southern (right) elephant seals* (Mirounga leonina) *with nasal hoods expanded (used in various displays) (from Briggs and Morejohn, 1976) and (B) adult male gray seals (in foreground) in antiparallel display on a breeding colony (Sable Island, Nova Scotia) (from Boness, 1979).*

ritualized swimming). In otariids and elephant seals, loud airborne distinctive vocalizations are given by breeding males. Underwater "songs" are well known in walruses and phocids, including Weddell and bearded seals (*Erignathus barbatus*). Loud sounds are probably designed to signal to a range of listeners (nonterritorial adult or subadult males, other territorial males, females), although in otariids they generally have been interpreted as a communication of "threat" among territorial adult males. Numerous medium- or short-range vocalizations are also given by territorial males; they are often directed toward other males, but are audible hence informative to all nearby animals. There are many unanswered questions about acoustic communication in both territorial and nonterritorial marine mammals; e.g., why do male sea otters and polar bears lack loud vocalizations for long-distance advertisement and why are male Hooker's sea lions, gray seals, and monk seals (*Monachus* spp.) virtually silent?

Chemical COMMUNICATION is probably important in all territorial species of marine mammals although is virtually unstudied. Facial glands are unknown in sea otters but the species has well-developed anatomical (neural) characteristics for olfaction, and individuals often actively smell the air. Glands on the face (many associated with facial vibrissae) occur in otariids, walruses, and phocids and are known to vary seasonally in size and secretory activity in some species. Male otariids emit distinctive odors during boundary displays between neighbors, perhaps from the oral cavity, and have distinctive body odors during the breeding season, but the anatomical source and functions of the smell are unknown. Breeding male ringed seals hold underwater (and underice) territories that are near but do not overlap with areas used by breeding females. During the breeding season, adult males of this species acquire a strong odor, which is the origin of terms like "tiggak" ("stinker") among Inuit and "gasoline seal" among trappers. Male ringed seals may actively deposit secretions from facial glands on entrances to their breathing holes and subnivean ("below snow") resting lairs, and within the latter.

The roles of taste or use of the vomeronasal organ (VNO) are unknown, but otariids commonly exhibit slow, repeated tongue extrusion following agonistic displays (e.g., boundary displays or fights involving males; females, juveniles, and even pups also express the behavior), suggestive of behavior of other mammals that are known to use the VNO in chemical communication (at the very least, tongue extrusion is a conspicuous optical signal and likely has become ritualized as an optical display).

Tactile communication is also important but is essentially unstudied in marine mammals. Breeding males of all species engage physically and contact one another extensively in biting, wrestling, or pushing. Male sea otters often try to bite the opponent's penis, and fractured BACULA (penis bones) of mature males are relatively common in the species. In phocid species that fight aquatically, the rear flippers (necessary for aquatic locomotion) are bitten and injured frequently in fights; walruses use the tusks in physical contests in which males face each other, mainly at or just below the water surface. Physical contests in otariids, walruses, elephant seals, and gray seals on land take place with the animals facing each other, so most pushing, striking, biting, and so on involve anterior body parts.

VI. Costs of Territoriality

Competing for and holding territories are highly competitive and entail risks and costs. In all marine mammal species, physical and behavioral maturation in males is a slow process and differs from females. Males typically attain physiological sexual maturity years before they begin to compete for territorial status because they are neither physically large nor behaviorally experienced enough to effectively obtain and maintain territories.

Costs and risks of competition for territorial status are partly energetic but include dangers of suffering severe physical injury (the relative importance of these factors varies across species). Energetic considerations increase in importance after territorial status is achieved. Energetic costs are borne through advertisement and display, fights, herding, patrolling, alertness, and so on. Males reduce their food intake (e.g., harbor seals) or fast completely (e.g., otariids, elephant seals, gray seals)

when they are territorial. The combined effects of reduced food intake or fasting plus high energetic expenditure are large. On average, territorial male northern fur seals fast and do not drink for about a month (maximum 87 days) and decline by about a third in body mass over that time. This loss in body mass is equivalent to about 0.7% of initial body mass per day, a level exceeded by several other species: Antarctic fur seals (0.8%), gray seals (0.9%), and the aquatically territorial Weddell seal (0.8%). The large northern elephant seal (*Mirounga angustirostris*; ca. 1700 kg) and the harbor seal, in which breeding males do not fast completely, lose less (about 0.4% per day), but overall costs are high in all species and are extreme in some individuals (one territorial male harbor seal lost 30% of his initial body mass).

Territorial male marine mammals must balance the need to economize energetically with the need to be vigilant, defend their territories, and effectively advertise territorial occupancy and their own physical attributes. Studies of time–activity budgets have revealed that territorial males spend much time at rest and little time in the most dangerous or energy-demanding activities such as fights.

Mortality rates of males are similar to those of females until the age of social maturity is attained. Beginning then, mortality rate of territorial males surpass those of females, sometimes by several times.

See Also the Following Articles

Aggressive Behavior, Intraspecific ▪ Breeding Sites ▪ Energetics ▪ Estrus and Estrous Behavior

References

Boness, D. J. (1979). "The Social System of the Grey Seal, *Halichoerus grypus* (Fab.), on Sable Island, Nova Scotia." Ph.D. thesis, Dalhousie University, Halifax, Nova Scotia.

Briggs, K. T., and Morejohn, G. V. (1976). Dentition, cranial morphology and evolution in elephant seals. *Mammalia* **40**, 199–222.

Campagna, C., and Le Boeuf, B. J. (1988). Reproductive behaviour of southern sea lions. *Behaviour* **104**, 233–261.

Campagna, C., Le Boeuf, B. J., and Cappozzo, H. L. (1988). Group raids: A mating strategy of male southern sea lions. *Behaviour* **105**, 224–249.

Connor, R. C. (2000). Group living in whales and dolphins. *In* "Cetacean Societies: Field Studies of Dolphins and Whales" (J. Mann, R. C. Connor, P. L. Tyack, J. Manning, and H. Whitehead, eds.), pp. 199–218. University of Chicago Press, Chicago.

Francis, J. M., and Boness, D. J. (1991). The effect of thermoregulatory behaviour on the mating system of the Juan Fernández fur seal, *Arctocephalus philippii*. *Behaviour* **199**, 104–126.

Gentry, R. L. (1998). "Behavior and Ecology of the Northern Fur Seal." Princeton University Press, Princeton, NJ.

Hardy, M. H., Roff, E., Smith, T. G., and Ryg, M. (1991). Facial skin glands of ringed and grey seals, and their possible function as odoriferous organs. *Can. J. Zool.* **69**, 189–200.

Jameson, R. J. (1989). Movements, home range, and territories of male sea otters off central California. *Mar. Mamm. Sci.* **5**, 159–172.

Miller, E. H. (1991). Communication in pinnipeds, with special reference to non-acoustic signalling. *In* "The Behaviour of Pinnipeds" (D. Renouf, ed.), pp. 128–235. Chapman and Hall, London.

Peterson, R. S. (1968). Social behavior in pinnipeds with particular reference to the northern fur seal. *In* "The Behavior and Physiology of Pinnipeds" (R. J. Harrison, R. C. Hubbard, R. S. Peterson, C. E. Rice, and R. J. Schusterman, eds.), pp. 3–53. Appleton-Century-Crofts, New York.

Peterson, R. S., and Bartholomew, G. A. (1967). "The Natural History and Behavior of the California Sea Lion." American Society of Mammalogists.

Riedman, M. L., and Estes, J. A. (1990). "The Sea Otter (*Enhhydra lutris*): Behavior, Ecology, and Natural History." U.S. Department of the Interior, Fish and Wildlife Service, Washington, DC.

Ryg, M., Solberg, Y., Lydersen, C., and Smith, T. G. (1992). The scent of rutting male ringed seals (*Phoca hispida*). *J. Zool.* **226**, 681–689.

Siniff, D., Reichle, R., Hofman, R., and Kuehn, D. (1975). Movements of Weddell seals in McMurdo Sound, Antarctica, as monitored by telemetry. Rapports et Procès-verbaux des Réunions, Conseil International pour l'Exploration de la Mer, **169**, 387–393.

Sjare, B., and Stirling, I. (1996). The breeding behavior of Atlantic walruses, *Odobenus rosmarus rosmarus*, in the Canadian High Arctic. *Can. J. Zool.* **74**, 897–911.

Smith, T. G., and Hammill, M. O. (1981). Ecology of the ringed seal, *Phoca hispida,* in its fast ice breeding habitat. *Can. J. Zool.* **59**, 966–981.

Twiss, S. D., Pomeroy, P. P., and Anderson, S. S. (1994). Dispersion and site fidelity of breeding male grey seals (*Halichoerus grypus*) on North Rona, Scotland. *J. Zool. Lond.* **233**, 683–693.

Wilson, B., Thompson, P. M., and Hammond, P. S. (1997). Habitat use by bottlenose dolphins: Seasonal distribution and stratified movement patterns in the Moray Firth, Scotland. *J. Appl. Ecol.* **34**, 1365–1374.

Tethytheria

PASCAL TASSY

National Museum of Natural History, Paris, France

The name Tethytheria was coined by the American paleontologist Malcolm C. McKenna in 1975 to group extant sirenians (order Sirenia) and elephants (order Proboscidea) and extinct desmostylians (order Desmostylia). Tethytheria means "beast of the Tethys," with the Tethys Ocean being a kind of a proto-Mediterranean sea, which joined Atlantic and Indian oceans up until the early to mid-Tertiary.

This concept revived an old idea put by the French naturalist Henri Ducrotay de Blainville in the 19th century. Blainville indeed was the first in 1836 to classify together elephants and sirenians only, under the name "les gravigrades," i.e., beasts with a heavy gait, a name first used by him for elephants only. Blainville did not formalize this name scientifically, and Gravigrada was then used for a group of extinct ground sloths. Earlier, Linnaeus in his "Systema Naturae" of 1758 classified elephants and sirenians together under the name Bruta, but also with sloths and anteaters, with a very weak argument (the lack of front teeth). Blainville was impressed by the fact that manatees and elephants shared a distinct dental replace-

ment, called "horizontal tooth displacement," a character that now is considered to be a CONVERGENCE. In elephants and manatees the tooth row is never complete: the anterior molars are lost before posterior ones erupt so that posterior molars take the place of anterior molars, moving anteriorly. In manatees this pattern is associated with polydonty, i.e., the existence of supernumerary molars. In elephants the number of molars is not altered but the three deciduous premolars and the three permanent molars erupt successively during life; in other words, the last molar (M3) is never associated to the first one. Among sirenians this character evolved only in manatees, and it appeared among proboscideans only in Neogene taxa.

I. Characters of Tethytheres

Various characters understood as synapomorphies support a sister group relation between Sirenia and Proboscidea. A few characters based on soft anatomy can only be seen in extant taxa: a bifid apex of the heart, with each ventricle being separated from the other at the tip (a convergence with some large cetaceans), a single pair of pectoral mammae (also seen on other mammalian taxa, but not in those more or less related to tethytheres). Others (osteological and dental characters) can be checked in fossils and define the Tethytheria at their ancestral node. They are a processus zygomaticus of the squamosal expanded laterally, an anterior orbit above the premolars with infraorbital canal under the orbit (Fig. 1), the reduction of the mastoid process, and bilophodont bunolophodont molars. This last character can be explained relatively easily. On the upper molars a posterior crest (loph) is made by three cusps, which are in line (metacone, and metaconule close to hypocone, a condition needed for development of a transverse loph called metaloph); on the lower molars the four main cusps are arranged in two crest-like rows, one anterior and one posterior.

II. Proboscideans

The order Proboscidea is represented by two extant species only, which are the biggest terrestrial mammals today: *Elephas maximus*, the Asian elephant, and *Loxodonta africana,* the African elephant. Approximately 160 fossil species are recognized since the late Paleocene. It is very likely that the earliest proboscideans were amphibious, an adaptation probably inherited from the hypothetical ancestral species of tethytheres.

Eocene proboscidean genera such as *Moeritherium* (discovered in Egypt, Mali, Libya, and Algeria) and *Numidotherium* (discovered in Algeria) were described as having amphibious characters. The low cranium of *Moeritherium* (Fig. 1A), with sirenian proportions, has a tubular shape with no postorbital constriction, high anterior orbits, and high auditory meatus, traits associated with amphibious habits, with the long body and short limbs (Fig. 2). The nasal fossa is not pushed backward, suggesting that no trunk was present. *Moeritherium* was of the size of a pig. *Numidotherium* was slightly taller (tapir sized) with a shorter body. The cranium was higher but still with anterior orbits and high auditory meatus (Fig. 1B). Postcranial bones display distinctive features unknown in

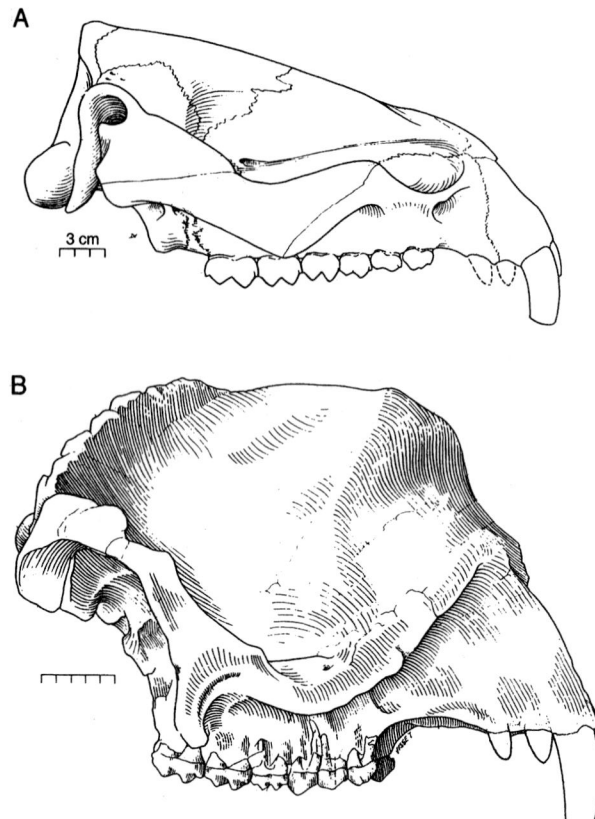

A

3 cm

B

Figure 1 *Crania of Eocene proboscideans showing a tethythere feature despite different overall morphologies: the orbit is located above the anterior premolar. A more derived proboscidean feature is the opening of the orbit in the maxilla and not the jugal. (A)* Moeritherium lyonsi *from the Eocene of Libya. Scale: 3 cm. (B)* Numidotherium koholense *from the Eocene of Algeria. Scale: 5 cm. Drawings by Dominique Visset.*

Figure 2 *Reconstruction of the fossil proboscidean* Moeritherium, *late Eocene and early Oligocene of North Africa (Museum National d'Histoire Naturelle, Paris; photograph by P. Tassy). Proportions of skull, body, and legs are markedly different from those of elephants. Its gait was probably primitive, not the unusual gait of elephants, which lift two feet on one side at the same time. Copyright © Pascal Tassy.*

Moeritherium and in elephant-like proboscideans. The shoulder and hip joints are permanently abducted (or semisprawling), more reminiscent of the semisupinated position of aquatic mammals, such as sirenians, desmostylians, and pinnipeds. This contrasts markedly with the subunguligrade gait of modern proboscideans. Extant sirenians and elephants lack tubular perspiratory glands and this may be a character inherited from an amphibious (but not fully aquatic) common ancestor.

The earliest known proboscidean is *Phosphatherium escuilliei,* a small species (the size of a fox) described in 1996 from the phosphates of the Ouled Abdoun basin, Morocco, an area previously known to yield marine fauna, especially shark teeth. The only fossils known for the genus are two partial maxillae with premolars and molars. Its teeth are relatively generalized, making it difficult to identify these teeth to proboscideans if it were not for the knowledge of Eocene and Oligocene taxa found elsewhere in North Africa: *Moeritherium, Numidotherium,* and *Barytherium.* The teeth of *Phosphatherim* are very similar to those of *Numidotherium koholense* from Alge-

ria, a truly lophodont proboscidean. Hence, during the late Paleocene and Eocene, proboscideans underwent a small radiation, of lophodont and bunolophodont taxa. The largest lophodont taxon is *Barytherium* from Egypt and Libya, which was similar in size to a living elephant. A lophodont lineage, the deinotheres, survived in the Neogene of Africa and Eurasia (up to 1.5 million years in Africa).

Anthracobunids are puzzling tethytheres from the Eocene of Pakistan and India, often closely compared to Proboscidea. Most of the known material consists in portions of maxillae, mandibles, and isolated teeth. Teeth are bunodont and more primitive than those of the bunolophodont contemporaneous African proboscideans. Postcranial remains allocated to anthracobunids (scapula, astragalus) match with those of proboscideans, especially the astragalus, which is the serial type and has a medial tuberculum (however, because no astragalus is known in primitive sirenians the proboscidean characters of the astragalus could well be tethythere characters).

Although early differentiation of elephantoids took place in Africa, elephantoids dispersed into Eurasia in the early Miocene, and probably even earlier, when the Arabo-African plate connected Eurasia closing the eastern Tethys. Elephantiformes such as *Phiomia* are bunolophodont, with an ever growing pair of upper and lower incisors (tusks), a long face with retracted nasal fossa (presence of a trunk), and a long mandibular symphysis. They do not show the horizontal tooth displacement only known in Miocene (and later) taxa. Elephantoids differentiated into several main groups with diverse shapes of crania and upper and lower tusks. Modern elephants are rooted in one of them, usually called "tetralophodont gomphotheres." with extra lophs on molars, shorter crania, and extended lower tusks, which are associated with a shorter mandibular symphysis. Stem elephants (*Stegotetrabelodon*) are known from the late Miocene of Africa and the Arabian Peninsula. The earliest elephants related to the extant genera *Loxodonta* and *Elephas* are contemporaneous in the fossil record, known as soon as the late Miocene (circa 6 millions years) in Uganda and Kenya.

III. Other Fossil Tethytheres

Extinct orders often classified as tethytheres together with Sirenia and Proboscidea are the Desmostylia and Embrithopoda. They share with Sirenia and Proboscidea the anterior orbit and the expanded zygomatic process. They share with the proboscideans the loss of mastoid process ("amastoidy"). However, hands (and feet) are not serial, contrary to both Sirenia and Proboscidea. The horizontal tooth displacement appeared also during the evolution of Miocene desmostylians.

Embrithopods are known in the late Eocene and Oligocene sediments of Egypt, Turkey, and Romania. Despite their very different overall morphology, they share a few derived characters with proboscideans: the loss of hypoglossal foramen (the hypoglossal nerve leaves the cranium by way of the foramen metoticum), the processus ascendens of the palatine bone is reduced greatly in the orbitotemporal fossa, and the astragalus has a prominent medial tuberculum, although the foot is not serial.

Extant tethytheres are also characterized by an unusual ontogenetic character. During the inner ear ontogeny of elephants and sirenians, the perilymphatic foramen fails to subdivide into two foramina so that the perilymphatic duct persists until the adult stage. This trait, probably paedomorphic, is not present in several Paleogene taxa such as the most primitive known sirenian (*Prorastomus* from the Eocene of Jamaica) and the Eocene proboscidean *Numidotherium* from Algeria. As a consequence, it is probably a convergence that appeared among sirenians and proboscideans. Moreover, this character is also known in embrithopods (the genus *Arsinoitherium* from Egypt), which probably is another convergence.

See Also the Following Articles

Classification ■ Dental Morphology, Evolution of ■ Desmostylia ■ Sirenian Evolution

References

Domning, D. P., Ray, C. E., and McKenna, M. C. (1986). Two new Oligocene desmostylians and a discussion of tethytherian systematics. *Smiths. Contr. Paleobiol.* **59,** 1–56.

Fischer, M. S., and Tassy, P. (1993). The interrelation between Proboscidea, Sirenia, Hyracoidea, and Mesaxonia: The morphological evidence. *In* "Mammal Phylogeny: Placentals" (F. S. Szalay, M. Novacek, and M. C. McKenna, eds.), pp. 217–234. Gustav Fischer, New York.

Gheerbrant, E., Sudre, J., and Cappetta, H. (1996). A Palaeocene proboscidean from Morocco. *Nature* **383,** 68–70.

Shoshani, J., and Tassy, P. (eds.) (1996). "The Proboscidea: Evolution and Palaeoecology and Elephants and Their Relatives." Oxford Univ. Press, Oxford.

Tassy, P., and Shoshani, J. (1987). The Tethytheria: Elephants and their relatives. *In* "The Phylogeny and Classification of the Tetrapods" (M. J. Benton, ed.), Vol. 2, pp. 283–315. Clarendon Press, Oxford.

Thermoregulation

Michael Castellini
University of Alaska, Fairbanks

Marine mammals have had to develop methods to retain heat in cold seas (physiologically, biochemically, anatomically, or behaviorally), yet must be able to lose excess heat when they are on land or extremely active in the water. If the problem were one of simply evolving methods to stay warm in a cold ocean, it would be much easier to wrap themselves in deep blubber and fur or to stay very active and not have to deal with the consequences of heat loading. However, the difficulty of that solution is that the animal could get too warm, which would in turn would cause problems with metabolic regulation, reproductive chemistry, neural function, and so on. Thus, the thermoregulatory mechanisms that have evolved in marine mammals function not only to conserve heat, but to dump it when necessary. As poorly insulated humans, we must bring our artificial insulation with us and use exposure suits, wet suits, and a variety of man-made materials if we are to spend any significant time in the sea. For marine mammals, the insulation is already on board and probably serves multiple purposes beyond just thermoregulation.

I. The Physics of Heat and Temperature

The terms "heat" and "temperature" are often incorrectly exchanged for one another, yet they have very different physical aspects. "Heat" is the energy that reflects the molecular motion of atoms and molecules. As energy, heat can flow from an area where the energy is high (something that is "hot") to an area where the energy is low (something that is "cold"). We quantify how hot something is by using a variety of temperature scales (Kelvin, Celsius, Fahrenheit), Thus, the temperature of an object is our definition of the level of heat energy contained by that object. The unit of energy is the calorie, and a single calorie is defined as the amount of heat necessary to raise 1 g of water by 1°C. In common usage in the United States, the calorie associated with food and dieting is actually the kilocalorie (kcal; 1000 calories). In strict scientific terms, a single calorie is defined as 4.184 J.

As with any energy that flows, there is usually some sort of resistance that impedes the flow of that energy. In the field of thermoregulation, that resistance is insulation (the inverse of insulation is thermal conductance). Thus, poor conductors are excellent insulators. Blubber, for example, makes an excellent insulator and conducts heat poorly. The unit of conductance is represented by the term "k" and has the units of calorie $\sec^{-1}$ cm^{-1} $°C^{-1}$. These units define how many calories will flow through an insulator that is 1 cm thick per second for every degree Celsius difference between the hot and cold side. Materials such as silver have an extremely high k value and thus conduct heat very well. Relatively, water is a better insulator than silver by 1000×. However, the point that is relevant to marine mammals is that air is a better insulator than water by

25×. In other words, water conducts heat away from a warm body 25× more effectively than air. This becomes an important point for the discussion of how fur works to help keep some marine mammals warm.

One final physics description is the definition of how heat flows from a warm to a cold object. Heat will flow when there is a temperature gradient between two sides of a conducting material depending on the magnitude of the temperature gradient, the thickness of the material, the inherent thermal conductance of the material, and the area that is exposed to the gradient. In biological terms, this means that heat would flow from the interior of a warm blooded mammal through the fat and skin to the cold outside air or water. Because water conducts heat 25× times more effectively than air, this means that heat flows out of a warm object in the cold water much more efficiently than it does when that same object is in air. Thus, as humans, we can easily stand around outside in 70°F air, but would find being in 70°F water very cold after a while. This principle states that the thicker the insulator, the less heat flow, and the larger the temperature gradient, the greater the driving force. Therefore, to stay warm, an animal would want an effective insulator, a small surface area (reduced appendage size, rolling up into a ball, etc.), a low temperature gradient (seek a warmer area or allow the body temperature to fall, i.e., hibernators), and have a thick insulator. An excellent general discussion of the physics of heat and energy transfer can be found in Schmidt-Nielsen (1997) and in Kooyman (1981) for marine mammals.

II. What Is "Thermoregulation?"

Having discussed the physics of heat, the next step is to define the act of regulating temperature (thermoregulation) in the biological realm. In the broadest sense, animals can be classified as either endotherms or ectotherms, although some animals cross between those two stages. An endotherm is an animal that generates and controls its internal heat so that its body core temperature can be regulated at a level different than ambient. Birds and mammals are the most commonly cited examples of endotherms. However, an ectotherm is an animal that allows its body temperature to mimic and follow the ambient temperature. Most fish and invertebrates are ectotherms. All marine mammals are endotherms and regulate their normal body temperature at about 37°C. Animals that hold their body temperature constant are called homeotherms while those that vary body temperature are called heterotherms. Most mammals are therefore homeothermic endotherms, although there are fascinating new data that suggest marine mammals may let their body temperature vary quite widely.

What is the importance of 37°C to a mammal? Perhaps the most critical point is that the actual value of 37°C is not as important as the constancy of that value. There are no biochemical or physical requirements about 37°C that make it the perfect temperature for a mammal, and most isolated biochemical reactions still work above and below that temperature. However, the mammalian body has evolved to balance its myriad of biochemical reactions at 37°C. If an animal gets much warmer or colder, the system comes out of equilibrium and there can

be significant failures in metabolic regulation. Thus, the study of hibernators is a fascinating example of adaptation to this problem. There is a concept in thermal chemistry called Q_{10}, which is defined as the change in the rate of a reaction for every 10°C change in temperature. In most biological systems, Q_{10} is about two. This means that if the temperature of an animal increased by 10°C, its metabolic rate would probably double. Because the millions of chemical reactions in the body all have slightly different Q_{10} values, it is easy to see how small changes in the core temperature could disrupt the biochemical balance. Some reaction rates would increase more than others and the flow of molecules from one biochemical pathway to another would be compromised.

The goal of thermoregulation then is to maintain a constancy of temperature by adjusting all the properties discussed earlier (the temperature gradient, the conductance, the surface area, etc.). Of course, for an endotherm, there is the additional factor that the animal is generating heat through metabolic processes. To maintain a constant body temperature for an endotherm, the produced heat must equal the heat lost to the environment. Put on too much of an insulator and the core temperature goes up if metabolic heat production stays constant. The perfect example is a human in Alaska wearing a thick, down jacket outside in the winter to stay warm but getting much too hot on cross-country skis wearing that same down jacket. The analogy to marine mammals is straightforward: a whale or a seal may put on large amounts of blubber but, as a consequence, overheat when extremely active. A human can simply take off the down jacket while exercising. A whale, however, does not have the option of taking off its blubber layer; it must be able to dump heat and therefore thermoregulate using other methods.

There are several ways in which heat is transferred to the environment from a warm body. Evaporation is the process of dumping enough heat into a liquid to turn it into a gas. This is the process of cooling down by sweating. Radiation is the movement of heat through the release of electromagnetic energy from the warm body to the cold environment without physical contact (heat energy from the sun traveling through space and warming the earth or a whale radiating in the infrared wavelengths; Cuyler *et al.*, 1992). Conduction refers to the transfer of heat energy (calories) by physical contact between the warm body and the cold environment (putting your hand into cold water). Finally, convection is a specialized case of conduction where the heat that is transferred from the warm body is moved away from the area by a current of air or water. Thus, the environment provides an infinite sink for the heat (wind chill is the example in air or moving through very cold water as opposed to staying still in the water).

On the whole then, in order to maintain a constant body temperature, the heat that is generated in an endothermic mammal must be balanced by the heat lost or gained through radiation, evaporation, conduction, and convection. This is the fundamental equation of thermoregulatory biology.

There is a long and fascinating history to the study of thermoregulation in mammals. In the modern era, the study of arctic mammals and birds under cold conditions and the definition of the "thermoneutral zone (TNZ)" came about largely from the work of Laurence Irving and Per Scholander (1950). The thermoneutral zone is the range of temperature over which an endotherm does not need to regulate its metabolism in order to maintain its body temperature constant. At the lower critical temperature, for example, a mammal would need to increase its heat production in order to stay warm. A great deal of this theory depends on the quality of fur for terrestrial mammals, and detailed studies on the thermal properties of fur were also addressed about the same time by Harold "Ted" Hammel (1955).

III. Thermoregulation in Marine Mammals

So far, all of this discussion could be applied to any biological system, not just marine mammals. Is there anything unique about thermoregulation in this group of marine endotherms? Both King (1983) and Riedman (1990) provide some excellent summaries of the broad field of thermoregulation in this group of mammals.

A unifying characteristic of most marine mammals is that they spend a great portion of their lives, if not their entire lives, in a liquid environment that is significantly colder than their core temperature of 37°C. Based on the discussion earlier on the fundamental aspects of thermoregulation, it should be clear that this aquatic life represents a significant thermal challenge to these mammals. While radiation and evaporation are probably insignificant sources of heat loss, conduction and convection are massive. However, in the Antarctic, seals will move to the relatively warm polar water (at -1.8 C°) when the real or wind chill temperature outside falls below about -40 C°. Clearly, there is a balance where the extreme cold of the ice-covered water, even with its higher thermal conductance, represents less of a thermal challenge than being outside on the surface.

Marine mammals use either fur or blubber for insulation and, like all endotherms, balance their metabolic heat production with various pathways of heat loss. However, the use of blubber or fur has its own biological costs. While blubber is used for thermoregulation, it is also a primary source of metabolic fuel for a marine mammal and plays a role in buoyancy regulation. Blubber is a fairly unique tissue for marine mammals and is not found outside of that group except for a similar tissue in POLAR BEARS (*Ursus maritimus*) and some penguins. Fur, however, is found in both terrestrial and marine mammals. However, the highest quality (density) fur is found in the sea otter. Fur is a very good insulator as long as it is carefully maintained, groomed, and kept dry on the layer next to the skin. Fur seals and sea OTTERS (*Enhydra lutris*) spend up to 12% of the daily energy use just maintaining their fur coats.

A. Heat Conservation and Generation

Marine mammals have no unusual heat-generating mechanisms or tissues that are not seen in any other mammal. For example, while some large warm-bodied fishes have specialized heat-generating tissues behind their eyes, no such organs or tissues exist in marine mammals. Some old data suggest that marine mammals may have an elevated metabolic rate for their mass, but this theory is not generally currently accepted (Lavigne and Kovacs, 1988) with the possible exception of the sea otter. The only heat-generating specialized tissue that has ever been found in marine mammals is brown fat in harp seal (*Pagophilus groenlandicus*) pups (Blix *et al.,* 1979). This tissue is thermogenically active via oxidation of lipid compounds, but only for about the first 3 days after birth. This is an important source of heat for these young pups, but not unique, as brown fat is found in other terrestrial mammals where it serves the same purpose. As noted earlier, marine mammals also have a typical mammalian body core temperature. In fact, upon close examination of the data, there appears to be nothing special about marine mammals that would distinguish them from terrestrial mammals when it comes to heat-generating mechanisms or abilities. Like the hibernators though, they may be able to tolerate more significant alterations in body temperature gradients than most terrestrial mammals.

Given the particularly nondescript aspects of marine mammal heat generation, there must be something that is different about them since they can live in an extremely cold liquid environment that would be fatal to all terrestrial mammals. Again, given the fundamental balance equation of thermoregulation, this suggests that they must have adapted significant ways to alter the heat loss through reduced conduction and convection. They have done this through the use of blubber, fur, and vascular adaptations.

B. Blubber

Blubber is most often incorrectly assumed to be an inert fat layer beneath the skin. However, it actually is a complex, active tissue that consists of a loose, spongy material where the matrix of the sponge is made up of collagen fibers and the volume is made of adipocytes (fat, or lipid cells). As the blubber layer increases or decreases, the collagen matrix remains the same, and it is the movement of lipid in and out of that matrix that accounts for the change in blubber quality and characteristics. However, all blubber is not the same; it varies from species to species in terms of the ratio of collagen to lipid and it can even vary within the same animal from location to location or with depth. Blubber depth can range from just a millimeter or two in newborn pinniped pups to 50 cm in large whales. The key issue here is that blubber, by itself, is a good insulator, as it can be up to 93% lipid with very little water content and has roughly the thermal conductance of asbestos. Because lipid has a conductance of only about one-third that of water, it acts as a relatively good insulator. Blubber acts as an internal insulator for marine mammals because it occurs below the skin layer. Therefore, the skin layer itself will be only marginally warmer than the surrounding water. In polar waters, for example, the skin of a whale or a seal would be just a degree or two above freezing. The thermal gradient exists from the skin surface to the tissues and organs in the core.

In addition to varying between species and with location, blubber can also vary across time in the same animal. This can be seen in the significant seasonal variation in blubber thickness in a seal as it moves between the breeding season (where it is fattest) and the leaner periods associated with molting and mating. For example, northern elephant seals (*Mirounga angustirostris*) can range between 50% to less than 20% body fat depending on the season. Clearly, this temporal change in blubber impacts not only thermoregulation, but also buoyancy and energy reserves during periods of fasting or lactation. Consequently, the role of blubber and its relative thickness as an indicator of nutritional condition has been followed quite closely in recent studies that seek to address the population health of marine mammals. If a population of marine mammals is nutritionally compromised, one would hypothesize that the blubber layer should be reduced due to consumption of the blubber as a fuel source.

Blubber should be thought of as a very dynamic tissue with multiple stressors and pressures on its biology. Because it is a critical tissue for several different processes in marine mammals, it cannot be modeled in a strictly thermal scenario. For example, during a time of fasting, the animal will be utilizing blubber heavily, which would be inconsistent if it were also being challenged with an increasing thermal demand. Hence, fasting periods associated with breeding occur in warmer months or in warmer water for most marine mammals. Rosen and Renouf (1997) have written about the relationships between blubber seasonal distribution and thermal problems in seals.

C. Fur

As with terrestrial mammals, fur in marine mammals functions by trapping dry air next to the skin and keeping water (or cold air for a land mammal) away from the skin surface. Thus, the gradient here is from the skin outward with a warm skin surface and cold outer layers of the fur. The most-cited example of the use of fur by a marine mammal is that of the sea otter and it provides an excellent example of how this animal lives in a cold environment (Williams *et al.*, 1992). The sea otter is faced with a major thermal challenge, as it is a small mammal (large surface area to volume ratio through which to lose heat). It utilizes a dense fur with a series of guard hairs and underfurs to keep its skin warm. However, the cost of this luxurious fur coat is a tremendous amount of maintenance with up to 12% of daily energy expenditure being spent on grooming the coat. This is an absolutely essential cost, however, as without the fur, the animal would lose too much heat to the marine environment.

Many species of seals utilize blubber for thermal protection as adults, but will use a specialized fur, called lanugo, as newborns. Lanugo, or pup fur, is a very effective insulator in the air and is usually both long and very "fluffy." On newborn pups, it functions as protection against the cold air during the time that they are on land or ice for nursing. Lanugo is useless in water and allows the skin to chill to essentially water temperature. A pup must shed its lanugo and develop a significant blubber layer before it can enter the water and be an effective swimmer and diver. Not all species of seal or sea lion pups are born with lanugo, but its purpose is well documented in many cases. Lavigne and Kovacs (1988) provide an excellent description of the first few days of life for harp seals as they adapt from the warm temperature inside the womb to the icy cold of being born on the ice.

It is the reliance on a high-quality fur in the sea otter and fur seals that makes these mammals particularly vulnerable to oil spills. Oil permeates the fur and destroys the air pockets that provide the thermal insulation for the animal. After the *Exxon Valdez* oil spill (EVOS) in Alaska, there was a massive clean-up operation on the hundreds of sea otters that were brought to rescue and rehabilitation centers. The goal was to clean the fur to restore its thermal insulation properties. However, cleaning the fur of man-made oils also cleans the fur of the natural oils (primarily squalene) that help make the fur water resistant. Therefore, small amounts of lipid had to be added and groomed back into the fur of the otters after they were cleaned of the heavy oil. For a general summary of the impact of the EVOS event on marine mammals, see Loughlin (1997); for a detailed discussion on otters, see Williams and Davis (1995).

D. Vascular Adaptations

It is in the area of vascular adaptions for thermoregulation that marine mammals have evolved several unusual adaptations. The first of these is termed the *rete mirabile*, which is Latin for a "wonderful net." This net, which is a countercurrent heat exchanger (Scholander and Schevill, 1955), involves an intertwined network of veins and arteries such that the cold blood returning from the extremities in the veins runs next to the warm blood going out to extremities in the arteries. From the previous discussion on heat flow, it is easy to see how the heat flows from the arteries to the close-by veins thus tending to conserve the heat in the interior and cool the arterial blood going out to the colder regions of the body. Marine mammals have exquisite control of blood flow in their body not only for thermoregulation but also for diving. However, these two demands are themselves interrelated, and the control of one impacts the control of the other. For example, it would do no good for a diving seal to be closely controlling blood flow for oxygen conservation but then to override that control to dump or gain heat. In fact, Elsner and Gooden (1983) discussed some experiments with seals where the diving response inhibited thermoregulatory-driven circulatory adjustments. In another innovative study, divers were able to apply heat flow probes to the skin of dolphins while both divers and dolphins were underwater. The results show that the animals tend to defer heat regulation and favor oxygen conservation vascular adjustments when both must coincide (Noren *et al.*, 1999).

These retes are found in several locations in marine mammals (and in some cold-adapted birds), with the most-cited examples being in the flukes of whales and the flippers of pinnipeds (Tarasoff and Fisher, 1970; Kvadsheim and Folkow, 1997). There has been the fascinating description of another rete, but in this case, the rete is used to cool down the reproductive organs of dolphins and seals by bringing in cold blood from the extremities (Rommel *et al.*, 1995).

The next vascular adjustment seen in marine mammals deals with those mammals that utilize thick blubber as an insulating material. As mentioned several times earlier, this is a good technique for staying warm, but can cause serious problems if trying to cool. In fact, large whales have such a tremendous thermal mass and a low surface area to volume ratio that they may have a much more serious problem dumping heat than conserving it (Hokkanen, 1990). Because marine mammals do not sweat, the answer is that blubber is not just an inert organic blanket surrounding the animal, but is instead vascularized with a series of anastomoses, or blood flow shunts. These shunts can control the amount of blood moving through the blubber and reaching the skin, thereby controlling the amount of heat lost to the environment. If a seal needs to dump heat, the anastomoses open and warm blood can reach the surface of the skin. When Weddell seals (*Leptonychotes weddellii*) in the Antarctic dump excess heat, clouds of steam come off the animal as the blood reaches the surface of their skin. In some cases, the seals get so warm that they partially melt their way into the ice and leave perfect "seal shadows." Conversely, when these shunts are closed, the same seals will be completely covered in snow with no signs of melting at any location except near the eyes and nose.

As mentioned earlier, the balance of blood flow throughout the body of marine mammals can be complex and is controlled by multiple demands: diving, exercise, and heat regulation. Diving requires limited blood circulation, simultaneous underwater exercise requires increased circulation, and thermoregulation can require both. How these animals balance those conflicting demands is an area where much more work needs to be done. This can be seen in even simple manipulations of seals and sea lions. When taking blood samples from the flippers of pinnipeds, the flippers must be warm or there is no blood flow out to the periphery. However, if anesthesia or sedation is required to work with the animal, those procedures may also cause a series of vascular adjustments and can dump great amounts of heat quickly. Then, extra heat needs to be added to the animal to keep the core temperature up and blood flow open to the flippers.

The balance between diving and thermoregulation has another interesting aspect if one looks at it from the point of view of DIVING PHYSIOLOGY. One of the central demands for diving is that oxygen must be conserved in order to extend the dive. This can be done by a variety of means and one of those means is to reduce the demand for oxygen by reducing the metabolic rate. In the discussion earlier, the concept of Q_{10} was mentioned, which describes how a reaction rate changes with temperature. If a marine mammal were to reduce its body temperature while diving, the impact from the Q_{10} relationship would therefore decrease the demand for oxygen, thus extending dive time. There is some evidence from freely diving pinnipeds suggesting that the animals can drop their core temperatures during diving and would thus gain some diving time by reducing the metabolic rate. The exact mechanisms by which this is done are not yet known, but temperature drops have been described in freely diving Weddell seals (Kooyman *et al.*, 1980; Hill *et al.*, 1987) and northern elephant seals (Andrews, 1999).

E. Behavioral Thermoregulation

Most of the mechanisms discussed earlier are biochemical, anatomical, or physiological mechanisms for regulating heat production or loss in a marine mammal. Of course, a marine mammal is not a static system and the animal can alter the demands placed upon it with behavioral modification. For example, sea otters are often seen floating with all four paws out of

the water. The paws are highly vascularized, but not well insulated with fur. Thus, they would be a tremendous source of heat loss if in contact with the water. The otters keep their paws away from the water if they are trying to stay warm. Similarly, it is not unusual to see rafts of California sea lions (*Zalophus californianus*) floating at the surface with their large foreflippers extended out of the water. On the beach, both seals and sea lions will move up or down the tidal zone area to either cool off or warm up. When too hot, sea lions will maximize their surface area by spreading out their flippers, while if too cold, they will lie on top of their flippers. As discussed earlier, Weddell seals will head to the water if the actual or convective temperature drops below about −40°C. However, elephant seals will flip cool sand onto their backs to help keep their body temperature down on sunny days, and Hawaiian monk seals (*Monachus schauinslandi*) will find shade under bushes or in small ravines out on hot, sandy atolls. All of these behavioral mechanisms are not unique to marine mammals, however, except that the animals have the ability to use the sea to cool down as necessary. A good example of both feeding and thermoregulation are the humpback whales (*Megaptera novaeangliae*) that come into cool Alaskan waters during the summer for feeding, but head south to warm, Hawaiian waters for breeding. A review of these behavior patterns for pinnipeds is found in King (1983).

IV. Summary

What are the essential elements of thermoregulation in marine mammals? Like all endotherms, these mammals must obey the physics of heat balance when holding body temperature constant. The methods for producing heat (resting metabolism and exercise) must balance the windows for heat loss (primarily conduction and convection) (Whittow, 1987). Because marine mammals do not appear to have any special adaptations for producing excess heat, most of their ability to thermoregulate comes with their ability to control heat loss. Control of these heat loss mechanisms is via biochemical, anatomical, physiological, and behavioral means. However, as in all levels of adaptation to the environment, systems cannot be considered or modeled in isolation. The problem with balancing blood flow for thermoregulation while also controlling blood flow for diving is an excellent example of this problem.

It is easiest to observe the behavioral means that marine mammals use to stay warm or to cool down: the movement up or down a beach with the tide, the use of shade, flipping of sand, swimming to warmer or colder water, exposing flippers, etc. Behind all of these behavioral patterns are the physiological or anatomical mechanisms that make the behavioral patterns effective. Countercurrent heat exchangers, blood shunts under the blubber, and even the chemistry of the blubber and the microstructure of the fur are all part of the thermoregulatory system. Ultimately, however, we are still left with the paradox of heat balance in marine mammals: they live in a cold, thermally challenging environment that no terrestrial mammal could survive. However, the very means they have utilized to stay warm in cold seas have come at a cost: for many species, they have had to also evolve the means to get rid of excess heat.

The exquisite balance between all these competing demands and systems is what makes the study of thermoregulatory biology in these mammals such a rewarding experience.

See Also the Following Articles

Blubber ∎ Circulatory System ∎ Hair and Fur ∎ Osmoregulation

References

Andrews, R. A. (1999). "Cardio-respiratory Metabolism and Thermoregulatory Physiology of Juvenile Northern Elephant Seals (*Mirounga angustirostris*). Ph.D. thesis, University of British Columbia, Vancouver.

Blix, A. S., Grav, H. J., and Ronald, K. (1979). Some aspects of temperature regulation in newborn harp seal pups. *Am. J. Physiol.* **236**(3), R188–R197.

Cuyler, L. C., Wiulsrod, R., and Oritsland, N. A. (1992). Thermal infrared radiation from free living whales. *Mar. Mamm. Sci.* **8**, 120–134.

Elsner, R., and Gooden, B. (1983). "Diving and Asphyxia." Monographs of the Physiological Society, Number 40. Cambridge Univ. Press, Cambridge.

Hammel, H. T. (1955). Thermal properties of fur. *Am. J. Physiol.* **182**(2), 369–376.

Hill, R. D., Schneider, R. C., Liggins, G. C., Schuette, A. H., Elliot, R. L., Guppy, M., Hochachka, P. W., Qvist, J., Falke, K. J., and Zapol, W. M. (1987). Heart rate and body temperature during free diving of Weddell seals. *Am. J. Physiol.* **253**, R344–R351.

Hind, A. T., and Gurney, W. S. (1997). The metabolic cost of swimming in marine homeotherms. *J. Exp. Biol.* **200**(3), 531–542.

Hokkanen, J. E. (1990). Temperature regulation of marine mammals. *J. Theoret. Biol.* **145**(4), 465–485.

King, J. E. (1983). "Seals of the World." Cornell Univ. Press, Ithaca, NY.

Kooyman, G. L. (1981). "Weddell Seal: Consummate Diver." Cambridge Univ. Press, Cambridge.

Kooyman, G. L., Wahrenbrock, E. A., Castellini, M. A., Davis, R. W., and Sinnett, E. E. (1980). Aerobic and anaerobic metabolism during voluntary diving in Weddell seals: Evidence of preferred pathways from blood biochemistry and behavior. *J. Comp. Physiol.* **138**, 335–346.

Kvadsheim, P. H., and Folkow, L. P. (1997). Blubber and flipper heat transfer in harp seals. *Acta Physiol. Scand.* **161**(3), 385–395.

Lavigne, D. M., and Kovacs, K. M. (1988). "Harps and Hoods: Ice Breeding Seals of the Northwest Atlantic." University of Waterloo Press, Waterloo, Ontario, Canada.

Loughlin, T. R. (ed.) (1994). "Marine Mammals and the *Exxon Valdez*." Academic Press, San Diego.

Noren, D. P., Williams, T. M., Berry, P., and Butler, E. (1999). Thermoregulation during swimming and diving in bottlenose dolphins, *Tursiops truncatus*. *J. Comp. Physiol. B* **169**(2), 93–99.

Reidman, M. (1990). "The Pinnipeds: Seals, Sea Lions and Walruses." University of California Press, Berkeley.

Rommel, S. A., Early, G. A., Matassa, K. A., Pabst, D. A., and McLellan, W. A. (1995). Venous structures associated with thermoregulation of phocid seal reproductive organs. *Anat. Rec.* **243**(3), 390–402.

Rosen, D. A. S., and Renouf, D. (1997). Seasonal changes in blubber distribution in Atlantic harbor seals: Indications of thermodynamic considerations. *Mar. Mamm. Sci.* **13**(2), 229–240.

Schmidt-Nielsen, K. (1997). "Animal Physiology: Adaptation and Environment." Cambridge Univ. Press, Cambridge.

Scholander, P. E., and Schevill, W. E. (1955). Countercurrent vascular heat exchange in the fins of whales. *J. Appl. Physiol.* **8**, 279–282.

Scholander, P. F., Hock, R., Walters, V., and Irving L. (1950). Adaptation to cold in arctic and tropical mammals and birds in relation to body temperature, insulation and basal metabolic rate. *Biol. Bull.* **99**(2), 259–271.

Tarasoff, F. J., and Fisher, H. D. (1970). Anatomy of the hind flippers of two species of seals with reference to thermoregulation. *Can. J. Zool.* **48**(4), 821–829.

Whittow, G. C. (1987). Thermoregulatory adaptations in marine mammals: Interacting effects of exercise and body mass. A review. (1987). *Mar. Mamm. Sci.* **3**(3), 220–241.

Williams, T. D., Allen, D. D., Groff, J. M., and Glass, R. L. (1992). An analysis of California sea otter pelage and integument. *Mar. Mamm. Sci.* **8**(1), 1–18.

Williams, T. M., and Davis, R. W. (eds.) (1995). "Emergency Care and Rehabilitation of Oiled Sea Otters: A Guide for Oil Spills Involving Fur-Bearing Marine Mammals." University of Alaska Press, Fairbanks.

Tool Use

RACHEL SMOLKER
University of Vermont, Burlington

Modern humans make use of a plethora of tools, ranging from the simple hammer and screwdriver to the sublimely complex space probe, guided missile launch, or magnetic resonance imagery device. Even our ancient ancestors crafted fine arrowheads and spears out of stone, bone, and wood. Until quite recently, we considered tool use to be a hallmark of our species, yet another indication of our superior intelligence and manipulative skill. Then in the mid-1960s, Jane Goodall made the startling discovery that chimpanzees used tools—twigs that were carefully selected and modified appropriately—to fish for termites.

Since then, the study of animal behavior has grown tremendously, and more and more wildly diverse creatures have been discovered to use tools in one way or another: Finches probe for grubs under bark using thorns, sea otters (*Enhydra lutris*) pound open abalone shells using rocks, caddisfly larvae encase themselves in suits of armor made up of bits of debris, and spiders weave webs. The need for a more rigorous definition of tool use became necessary, and Benjamin Beck (1980) devised one that is now widely accepted: "The external employment of an unattached environmental object to alter more efficiently the form, position or condition of another object, another organism or the user itself, when the user holds or carries the tool just prior to use and is responsible for the proper and effective orientation of the tool."

I. Sponge Carrying

By this definition, the sponge carrying of Indian Ocean bottlenose dolphins (*Tursiops aduncus*) in Shark Bay, West Australia, is a form of tool use (Smolker *et al.*, 1997). Sponge carrying was first observed in 1984. A dolphin, who was later dubbed "Halffluke" because she lacked half of her tail, surfaced near the author's boat with a large, reddish, knobby "growth" on her face. She took several breaths at the surface and then dove, bringing her damaged tail flukes out of the water as she headed toward the bottom. After watching for some time it became evident that Halffluke was carrying something that she occasionally dropped. It was not a "growth," but rather a sponge. Halffluke was one of several dolphins who regularly carried sponges, usually within a particular deep-water channel within Shark Bay. These "sponge carriers" are all female (most have had offspring, and some of the offspring have subsequently taken up sponge carrying). The sponge carriers have been observed on numerous occasions during every year since 1984, almost invariably carrying sponges. They "specialize" in sponge carrying and have been doing so for years. On a few occasions, the author saw other dolphins carry sponges (see Fig. 1), but these were all single observations—usually just a single surfacing with a sponge—by dolphins that otherwise did not carry sponges.

From the surface, sponge carrying looks no different than other forms of foraging. The dolphin surfaces to BREATHE, traveling slowly along while taking several breaths and then dives, bringing her tail flukes out, an indication that she is diving down toward the bottom. After 2 or 3 min, she resurfaces further along. There is often no predictable travel direction: the dolphin appears to wander around while below the surface. Occasionally the dolphin surfaces without the sponge or trades in one sponge for another. The sponges (*Echinodictyum mesenterinium*) are roughly cone shaped and fit over the dolphin's beak. Some are quite large and flop back over the dolphin's face, even obstructing vision. Others are small and fit like a yarmulke cap, just covering the beak tip. The shape of the sponge and the backward pressure caused as the dolphin swims forward apparently keep it in place. Sometimes they do fall off and are quickly retrieved by the dolphin.

Figure 1 *Sponge carrying was observed first in dolphins of Shark Bay, Western Australia. The practice has been passed from females to female offspring and has an uncertain function.*

After years of observing sponge carrying, we have never seen for sure how the sponges are used, as the dolphins apparently make use of the sponges at or close to the bottom in deep and murky water. Given the amount of time that sponge carriers spend with sponges—similar to the time spent foraging by other dolphins—we suspect that the sponges are used in some way to aid foraging. Also, we sometimes saw sponge carriers surface rapidly without a sponge just prior to surfacing and swallowing some prey item. Thus the sponge is apparently used during search and then dropped during the final pursuit and capture of prey.

All evidence best fits with the hypothesis that the sponges are used as shields, protecting the dolphins' face and beak. They may be protecting themselves from the stingers or spines of noxious organisms encountered while foraging among rocky ledges or from abrasive sand, broken shell, and rocks while poking their beaks into the bottom in search of burrowing fish. We know that there are spiny lionfish in the area as well as stonefish and various other poisonous organisms. It is also known that the dolphins sometimes poke into the sandy bottom in search of prey.

The use of sponges as shields is not the first example of tool use by dolphins: Tayler and Saayman (1973) reported that a captive bottlenose dolphin used pieces of the tile to rub the tank walls, dislodging pieces of seaweed. Brown and Norris (1956) reported observing two captive common bottlenose dolphins (*T. truncatus*) attempting to capture a moray eel. After many attempts, one dolphin sought out and killed a scorpionfish, carried the fish to the crevice where the eel was hiding, and poked at the eel with the spiny fish. The eel abandoned its hiding place and was captured. These examples are indicative of the resourcefulness and intelligence of dolphins, but they are one-time behaviors. Sponge carrying is engaged in by several dolphins and has been ongoing for many years. It has not spread throughout the population as, for example, washing of sweet potatoes among Japanese macaques, but rather is restricted to a few individuals. Thus, like termite fishing by chimpanzees, sponge carrying is a well-established tradition of tool use by an animal living in a vastly different ecosystem and bearing little relatedness to humans.

Dolphins and sea otters appear to be the only marine mammals that engage in tool use.

See Also the Following Articles

Bottlenose Dolphins ■ Feeding Strategies and Tactics ■ Otters

References

Beck, B. (1980). "Animal Tool Behavior." Garland STPM Press, New York.

Brown, D. H., and Norris, K. S. (1956). Observations of wild and captive cetaceans. *Behaviour* **37**, 311–326.

Smolker, R., Richards, A. F., Connor, R. C., Mann, J., and Berggren, P. (1997). Sponge carrying by dolphins (Delphinidae, *Tursiops* sp.): A foraging specialization involving tool use? *Ethology* **103**, 454–465.

Tayler, C. K., and Saayman, G. S. (1973). Imitative behavior by Indian Ocean bottlenose dolphins (*Tursiops aduncus*) in captivity. *Behaviour* **44**, 286–298.

Toothed Whales, Overview

SASCHA K. HOOKER

University of St. Andrews,
Scotland, United Kingdom

Toothed whales comprise the suborder Odontoceti of the order Cetacea. This suborder includes 10 diverse families, 2 of which contain large numbers of species. There are at least 71 species in all, including the true dolphins, monodontids, river dolphins, porpoises, beaked whales, and sperm whales (Table I). These species occur in three primary clades, the superfamilies Delphinoidea (true dolphins, monodontids, and porpoises), Ziphoidea (beaked whales), and Physeteroidea (sperm whales), while the affinities of the river dolphins remain uncertain. With the exception of the sperm whale (males of which reach up to 18 m) and the larger beaked whale species (*Berardius* and *Hyperoodon* spp.), most odontocetes are small to medium-sized cetaceans, ranging in size from the Hector's dolphin (*Cephalorhynchus hectori*) (1.5 m) to the killer whale (*Orcinus orca*) (8.5 m). These species show a range of distributions, with some, such as river dolphins, found only in quite specific areas, whereas others, such as sperm whales or killer whales, show a global distribution.

I. Diagnostic Characters and Taxonomy

Odontocetes and mysticetes differ fundamentally in three major ways: the way that the bones of the SKULL have become telescoped, the specialized ECHOLOCATION system (and associated anatomy) of odontocetes, and the specialized FILTER-FEEDING mechanism of the baleen whales (Table II). The name Odontoceti derives from the Greek *odous* or *odontos* for "tooth" and *ketos* for "sea monster," hence "toothed sea monster," referring to the presence of teeth (Rice, 1998). In contrast, mysticetes do not possess teeth, but instead have baleen plates, which hang from the upper jaw and are used to filter small prey items from the water. However, although all odontocetes possess teeth, in some species (or sexes) these teeth are much reduced and may not erupt.

Other distinctive features include the possession of a single external naris or blowhole, whereas mysticetes have two blowholes. Most odontocetes show some degree of dorsal asymmetry in their skull and facial soft tissue, whereas all mysticetes have a symmetrical skull and facial soft tissue. Odontocetes possess a large ovoid melon in the anterior part of the facial region. This fatty tissue is thought to be an important component of the echolocation system. Although mysticetes possess a fatty structure just anterior to the nasal passages, which may represent a vestigial melon, this is only a fraction of the size of that present in odontocetes.

Odontocetes (with the notable exception of the sperm whale, *Physeter macrocephalus*) tend to be smaller in size than mysticetes, although there is some overlap. Odontocetes also show variation in SEXUAL DIMORPHISM. In some species, males are much larger than females (e.g., sperm whale, killer whale), whereas in others there may be reverse sexual dimorphism, in which females are larger than males [e.g., harbor porpoises (*Phocoena phocoena*), Baird's beaked whale (*Berardius bairdii*)]. Among mysticetes, adult females are always slightly larger than adult males.

The skull and jawbone of odontocetes also contain a number of diagnostic characteristics. The odontocete mandible is symphyseal (the two jawbones lock together in a bony symphysis anteriorly) and each jawbone spreads into a fat-filled hollow pan at the posterior, non tooth-bearing section, whereas mysticete mandibles are nonsymphyseal and are solid. When viewed from above, the odontocete jawbone is relatively straight, whereas the mysticete jawbone curves laterally. The maxilla of both odontocetes and mysticetes has "telescoped," migrating posteriorly to form the long rostrum and dorsal nasal openings. In odontocetes, the maxillae have extended outward over the orbits to form an expanded bony supraorbital process of the frontal bone. This process forms an anchoring point for the facial musculature associated with sound production (see later). In contrast, the maxillae of mysticetes project under the eye orbit and have developed bony protuberances anterior to the eye orbit. Odontocetes lack this antorbital process of the maxilla. The ear bone of odontocetes is also quite different from that of mysticetes. In odontocetes, the tympanic bulla and periotic bone are fused together, are of equal size, and are not fused to the skull. The odontocete tympanic is thin walled and conical, tapering anteriorly. In mysticetes, the tympanic bulla is much larger (nearly twice the volume of the periotic bone) and is thick walled and spherical. Bony flanges of the periotic wedge it against the skull such that only the large bulla can be removed.

The clear morphological differences between odontocetes and mysticetes suggest monophyly of the odontocetes. There has been some disagreement about this issue based on some molecular sequence data that supported a closer relationship between sperm whales and BALEEN WHALES than between sperm whales and other toothed whales. However, reanalysis of these data and more recent work on both morphologic and

TABLE I
Odontocete Families

Family	Common name	No. species
Kogiidae	Pygmy and dwarf sperm whales	2
Physeteridae	Sperm whale	1
Ziphiidae	Beaked whales	20
Delphinidae	True dolphins	36
Monodontidae	Narwhal and beluga	2
Phocoenidae	Porpoises	6
Iniidae	Boto (Amazon river dolphin)	1
Pontoporiidae	Fransiscana	1
Lipotidae	Baiji (Chinese river dolphin)	1
Platanistidae	South Asian river dolphin	1
	Total	71

TABLE II
Major Differences between Odontocete and Mysticete Suborders of Cetaceans

Odontocetes	Mysticetes
Teeth	Baleen plates
Single blowhole	Two blowholes
Dorsally asymmetric skull and facial tissue	Symmetrical skill and facial tissue
Presence of a melon	No melon
Variable size	Always large
Variation in sexual dimorphism	Always reverse sexually dimorphic
Symphyseal mandible	Nonsymphyseal mandible
Hollow pan bone of lower jaw	Lower jawbones solid, no pan bone
Maxillae project outward over expanded suprabital processess	Maxillae project under the eye orbit and possess bony protuberances anterior to the eye orbits
Tympanic and periotic fused and equal sized	Tympanic bulla much larger than periotic from which it separates

molecular data continue to support odontocete monophyly. Relationships among extant odontocetes are somewhat controversial, but there is a general consensus as to the order of branching of the phylogenetic tree, despite controversies over smaller scale relationships (Fig. 1).

Odontocetes first appeared in the fossil record in the late Oligocene, approximately 25 million years ago (Rice, 1998). Odontocete diversity increased during the warm temperatures of the early Miocene, during which time the earliest ziphiids and platanistids appeared in the eastern North Pacific. Middle Miocene diversity was also high and included diversification of these families, together with extensive radiation of the delphinoids, while the platanistoids progressively declined. Modern odontocete families are known from the late Miocene of the eastern North Pacific (monodontids and phocoenids) and western South Atlantic (pontoporiids) (Fig. 2). Evolutionary trends among odontocetes included the expansion and increase in size of the face, shortening of the intertemporal region, elevation of the cranial vertex posterior to the nasals, increased facial asymmetry, and isolation of the ear bones from the skull resulting in the diagnostic features described previously. There has also

been a trend toward either long, slender jaws with an increased number of teeth or short, robust jaws and a reduced number of teeth.

II. Distribution and Range

Different odontocete species can be found across a wide range of habitats in all oceans of the world. The cyclical changes in sea level over the Quaternary (Pleistocene to Recent) period are thought to be responsible for much of the recent speciation within odontocete families. Sea level drops associated with cooling are suggested to have isolated populations, which then speciated allopatrically. The distributions of modern odontocetes

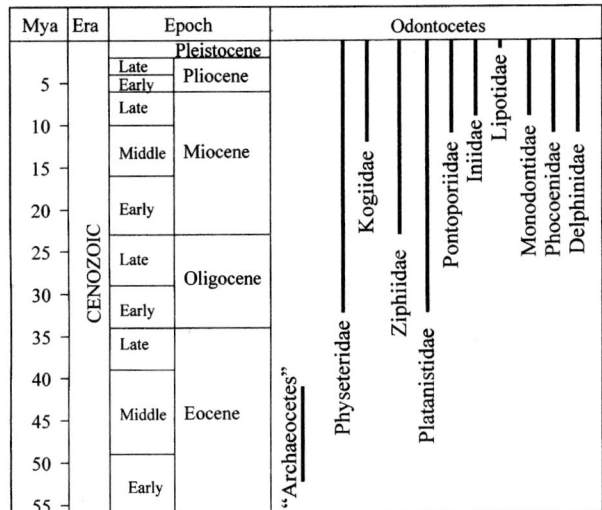

Figure 2　*Chronologic ranges of extant odontocete families. Mya, million years ago. (After Berta and Sumich, 1999; Rice, 1998).*

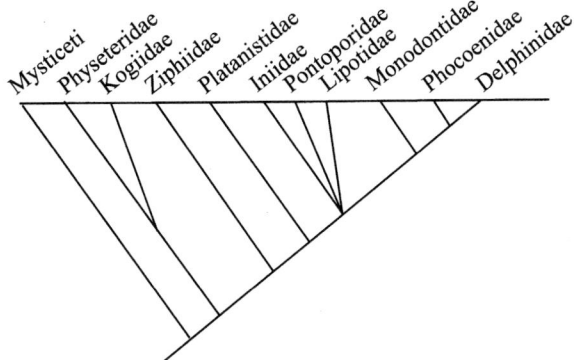

Figure 1　*Generally accepted odontocete phylogeny.*

range from species found globally, such as sperm whales and killer whales, to those with more restricted coastal distributions, such as harbor porpoises. Some species are found only in polar regions, including the narwhal (*Monodon monoceros*) and beluga (*Delphinapterus leucas*) and others only in tropical waters, such as Fraser's dolphin (*Lagenodelphis hosei*) (Jefferson *et al.*, 1993). Segregation by warm tropical waters created antitropical distributions of some species pairs now found only in the Northern and Southern Hemispheres, respectively [e.g., the northern and southern right whale dolphins (*Lissodelphis* spp.), northern and southern bottlenose whales (*Hyperoodon* spp.)].

Few odontocetes show the kind of long-distance seasonal MIGRATION that is found among mysticetes. The only odontocete species known to show a large (ocean basin) scale of movement is the sperm whale. At approximately 10 years of age, male sperm whales leave their natal group in the tropics and subtropics and migrate to cold temperate and subpolar waters where they feed. Once physically mature they return to the tropics to breed. Other odontocetes may show much smaller migrations as they follow their prey movements [e.g., long-finned pilot whales (*Globicephala melas*) in the North Atlantic move from the shelf edge up onto shallower banks in response to changes in squid distribution]. Other species (e.g., killer whales) show negligible seasonal changes in distribution.

Many odontocete species show GEOGRAPHIC VARIATION across their ranges, leading to the definition of several subspecies or races. With increasing scrutiny of both genetic and morphological differences, subspecies and population level differences are continually being identified. In some cases, this is leading to revision of the currently established species (see Rice, 1998).

III. External Appearance

In general, many odontocete species tend to have darker COLORATION on the dorsal surface (back and cape) and lighter pigmentation on the ventral surface (Fig. 3). Such countershading is relatively common in the marine environment and functions essentially as camouflage, such that when viewed

Figure 3 *Common dolphin* (Delphinus *sp.*) *pigmentation shows countershading with dark pigmentation on the dorsal surface and light pigmentation on the ventral surface, common to most odontocetes. Photograph by Sascha K. Hooker.*

from above the dark dorsal surface is seen against the darker depths, and when viewed from below the lighter ventral surface is seen against the brighter downwelling light from the surface (e.g., harbor porpoise). Exceptions to this include the beluga, which has a uniform color pattern, although this is thought to be related to the bright ice-covered habitat that it lives in. The general tendency toward countershading has become quite elaborate in some species, which also have striped and saddle patterns [e.g., striped dolphins (*Stenella coeruleoalba*), short-beaked common dolphins (*Delphinus delphis*)], and has become radically modified in others [e.g., killer whale, Dall's porpoise (*Phocoenoides dalli*)]. It has also been suggested that these color patterns may function in signaling between individuals in a group (Norris *et al.*, 1994), in addition to their function in concealment (from both PREDATORS AND PREY).

Odontocetes tend to show greater body flexibility than mysticetes. This is presumably due to differences in prey capture strategies between odontocetes, which chase fast, mobile, single prey, and mysticetes, which engulf less mobile prey schools. This flexibility is quite pronounced in some odontocete species (e.g., belugas).

IV. Diet and Feeding Strategies

Different odontocete species feed on fish, squid, large crustaceans, birds, and occasionally other marine mammals. They differ from mysticetes in that they generally chase, capture, and swallow single relatively large prey items rather than filtering and swallowing large quantities of small prey items. This more generalized and more adaptable feeding method is thought to account for much of the diversity of odontocete species and their range of habitats. Many odontocetes also feed on much deeper prey than the near-surface plankton diet of many species of mysticetes. Additionally, whereas the diet of baleen whales is highly seasonal due to the seasonal nature of zooplankton biomass and production, that of odontocetes is generally more constant year-round.

Unlike the heterodont condition of most terrestrial mammals, the teeth of odontocetes are uniformly shaped (homodont). There is wide variation between (and some variation within) species in tooth number, size, and shape. The teeth of most odontocete species tend to be cone-shaped with single, open roots. Exceptions to this include the porpoises, which have spade-shaped teeth, and the beaked whales, which show great variation in size, shape, and location of teeth in the jaw.

Narwhals have only two teeth, both in the upper jaw. In females, these teeth usually remain embedded in the upper jawbones, but in males the left tooth grows out through the front of the head into a tusk up to 3 m long. These tusks are thought to be involved in male–male competition. These are used primarily as a display, although males have been seen sparring with their tusks above the water. The teeth of beaked whales have similarly become adapted for use in male–male competition and generally erupt only in males. In some species, such as the dense-beaked whale (*Mesoplodon densirostris*), these teeth form large structures, which protrude above the upper jaw. In the strap-toothed whale (*Mesoplodon layardii*), the teeth, which emerge from the middle of the lower jaw, curl backward and inward, extending over the upper jaw, often preventing it from opening more than a few centimeters.

The DIET of a particular species is generally reflected in the morphology of the jaw and in the type and number of teeth. For example, species that feed primarily on fish tend to have more teeth (e.g., spotted dolphins, *Stenella* spp.) and use these teeth for grasping single prey. Species that feed primarily on squid tend to show reduced dentition [e.g., sperm whales, beaked whales, narwhals, Risso's dolphin (*Grampus griseus*)] and are thought to feed by suction. This suction is achieved by using the tongue as a piston in combination with a small gape. The suction-feeding mechanism of beaked whales was investigated in detail by Heyning and Mead (1996) and involves distension of the floor of the mouth by expanding the throat grooves together with retraction of the tongue by the styloglossus and hyoglossus muscles. Additionally, it has been suggested that some species may debilitate prey by directing high-intensity sounds at them prior to capturing them (Norris and Mohl, 1983). Such feeding methods are thought to explain the occasional observation of healthy animals with severe deformity of the jaws (as has been noted in sperm whales).

Notable specializations associated with odontocete feeding are shown by several bottlenose dolphin populations. In Shark Bay, Australia, a small number of Indian Ocean bottlenose dolphin (*Tursiops aduncus*) females have been observed carrying sponges (Mann *et al.*, 2000). These are thought to function as "tools" to protect the dolphin's rostrum as it roots in bottom coraline sediments in order to flush bottom-dwelling fish out. Another bottlenose dolphin (*T. truncatus*) population in North Carolina has been observed beaching themselves and the fish they were pursuing up onto the surrounding mud banks, thus immobilizing their fish prey, which they can catch easily, and then sliding or wriggling back down into the water (Reynolds and Rommel, 1999). The same behavior has also been observed in killer whales in Patagonia and the Crozet Archipelago in their pursuit of seals up onto steeply shelving beaches.

Foraging specializations of killer whales in British Columbia, Canada, and State of Washington waters have resulted in two separate forms: "transients," which feed on marine mammals, and "residents," which feed on fish. These two forms are thought to potentially represent a case of incipient speciation. The primary difference between these two forms is in their dietary specialization, which has resulted in their social separation and behavioral, morphological, and genetic differences.

Many odontocetes appear to feed throughout the day and night [e.g., sperm whales, northern bottlenose whales (*Hyperoodon ampullatus*)], but many smaller delphinids show marked diurnal differences in feeding [e.g., spinner dolphin (*Stenella longirostris*), pantropical spotted dolphin (*S. attenuata*)]. Because many prey species rise to shallower depths during the darkness of night, it is energetically more efficient for some species to conduct the majority of foraging behavior at night.

As mentioned previously, odontocetes tend to feed at greater depths than mysticetes. In general, however, detailed data on diving and foraging patterns of most species of odontocetes are not yet available. The use of novel technologies, such as the attachment of time-depth recorders to monitor dive profiles of these species, has lagged behind work done on pinnipeds or mysticetes, primarily due to difficulties in deployment, attachment, and recovery (Hooker and Baird, 2001). However, new developments in these technologies and attachment mechanisms are increasingly resulting in the initiation of new studies.

V. Sound Production and Reception

Toothed whales have developed specialized sound production and reception mechanisms for the use of biosonar. All modern odontocetes are thought to use echolocation in the same manner as bats to "image" their environment. Although only a few species of odontocete are unequivocally known to echolocate, all odontocetes known to produce pulse-like sounds in the wild are assumed to be able to echolocate.

The sound production mechanism of odontocetes consists of a sound generator located in the "monkey lips"/dorsal bursae (MLDB complex) associated with the upper nasal passages. In most odontocetes, there are two bilaterally placed MLDB complexes, but sperm whales have only a single complex. The central components of the MLDB complex are fatty dorsal bursae, monkey lips, bursal cartilages, and the blowhole ligament (Cranford *et al.*, 1996). Sounds are generated as air is forced between the monkey lips, setting the MLDB complex into vibration (Fig. 4). Sound is propagated into the water by the melon, a low-density, lipid-filled structure that acts as an acoustic lens to focus a directional sound beam ahead of the animal (see Au, 1993). The short-duration clicks produced as a result of this are used primarily for echolocation, although some species appear to use these clicks in a social context (e.g., Hector's dolphin, *Cephalorhynchus hectori*).

Most odontocete species produce broadband echolocation clicks in the ultrasonic sound range, well above the range of human hearing. The pulse duration, frequency, interclick interval, and source level are adjusted by the animal for optimal performance according to the prevailing conditions of ambient noise, reverberation, target distance, and target characteristics (Au, 1993). With low ambient noise, bottlenose dolphins, belugas, and false killer whales (*Pseudorca crassidens*) often echolocate using frequencies from 20 to 60 kHz, although at higher noise levels they emit stronger pulses at 100–130 kHz. Echolocation clicks of porpoises and many small dolphins are >100 kHz, whereas those of the sperm whale range from <100 Hz to 30 kHz, with most energy from 2 to 4 kHz and 10 to 16 kHz (Richardson *et al.*, 1995).

In addition to echolocation clicks and loud impulse sounds, many toothed whales also produce other sounds, usually described as whistles, squeals, or less distinct pulsed sounds such as cries, grunts, or barks (Richardson *et al.*, 1995). These tend to be narrow band (sometimes pure tone), frequency-modulated sounds, often with a harmonic structure. Most whistles tend to have most of their energy below 20 kHz. These whistles can show a variety of patterns of frequency and amplitude over the duration of the whistle. For any species, the frequency, duration, and level may vary. Observations of captive bottlenose dolphins have shown that individual animals can be identified from the contour of the whistle on a sonogram (a representation of the whistle as sound frequency plotted against

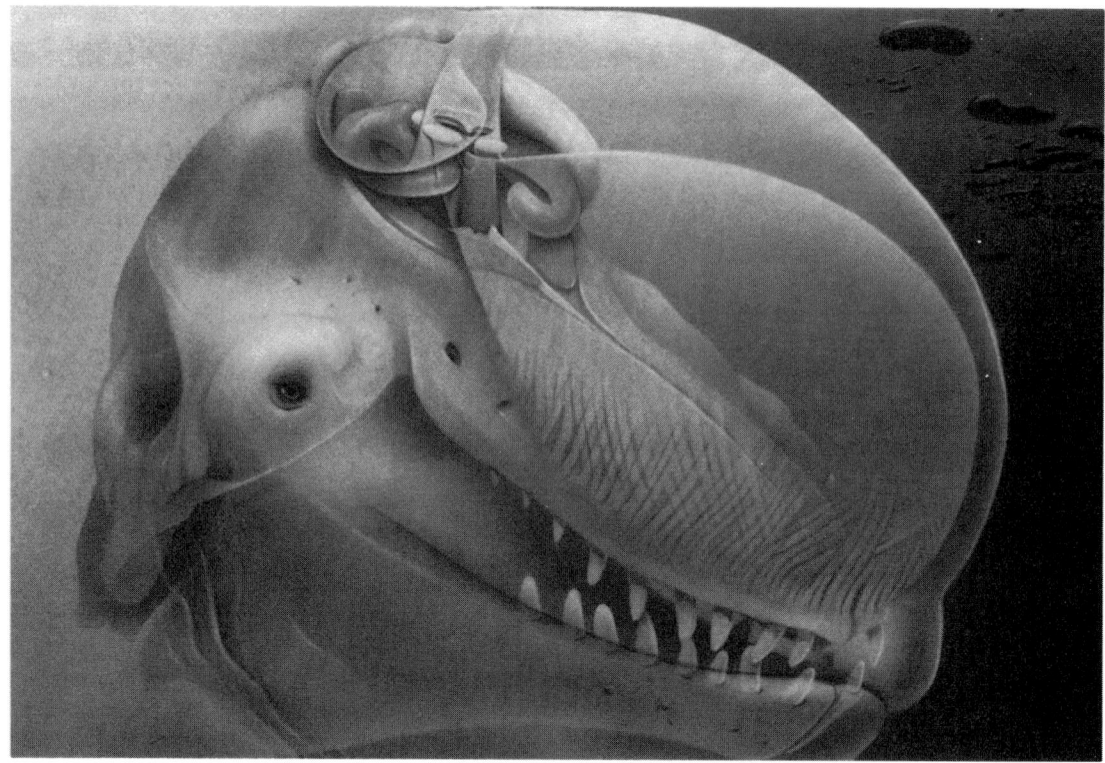

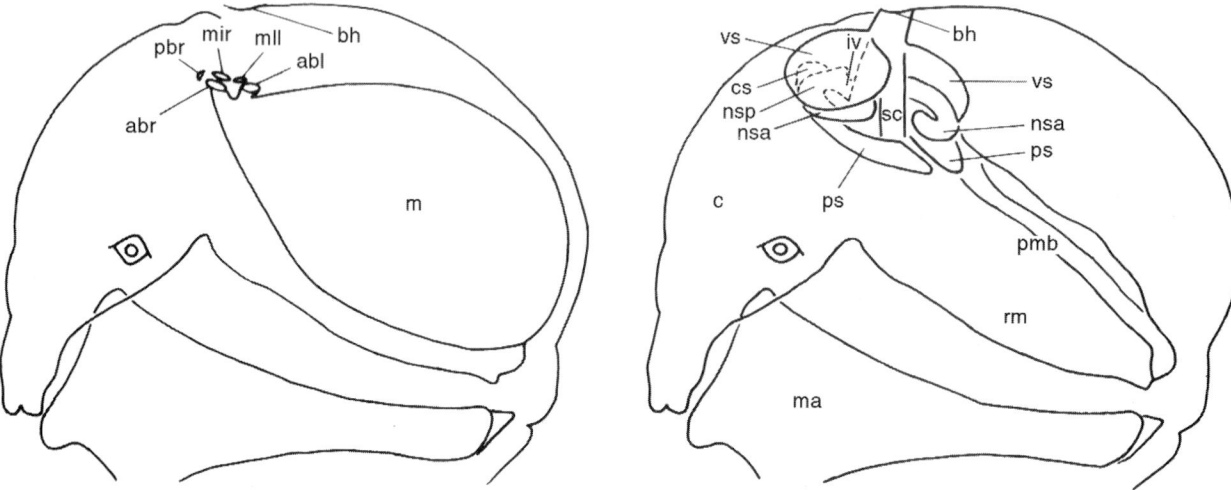

Figure 4 *The sound-producing structures illustrated in a beluga head show anatomical elements of the sound generation and beam formation apparatus within the nasal complex of the forehead. abl, anterior dorsal bursa (left); abr, anterior dorsal bursa (right); bh, blowhole; c, cranium; cs, caudal sac; iv, inferior vestibule; m, melon (main body); ma, mandible; mll, monkey lips (left); mlr, monkey lips (right); nsa, anterior nasofrontal sac; nsp, posterior nasofrontal sac; pbr, posterior dorsal bursa (right); pmb, premaxillary bones; ps, premaxillary sac; rm, rostral muscles; sc, spiracular cavity; vs, vestibular sac. From Cranford* et al. *(1996).* Journal of Morphology *228, 223–285. © John Wiley & Sons, Inc. Reprinted by permission of Wiley-Liss, Inc.*

time). The distinctive character of a whistle is thought to function in identifying the sound producer to other animals. These whistles are therefore known as signature whistles. Evidence also suggests that there may be population or group level distinctions between whistles of bottlenose dolphins.

Group-specific sounds have also been found among KILLER WHALES and sperm whales. Among killer whales, groups appear to have a repertoire of approximately 10 calls. Different groups may share some but not all of the calls within their repertoire, and relationships between groups can be established based on the similarities in their repertoires. In a similar fashion, the codas (rhythmic patterns of clicks) produced by sperm whales also appear to be characteristic of the social group (Mann *et al.*, 2000).

The high-frequency HEARING of odontocetes is reflected in the structures of the jaw and ear. The lower jaw of odontocetes is flared out in a thin hollow pan bone at the rear. This is filled with a fat body that connects directly with the bulla of the middle ear. These fat bodies act as low-density sound channels to conduct sounds to the ears. Within the ear, the tympanic bulla is separated from adjacent bones of the skull by peritympanic sinuses filled with an insulating emulsion of mucus, oil, and air. The bulla is suspended in this emulsion by connective tissue so that the middle ear functions as a sound receiver isolated from the skull to better localize sound signals. The tympanic bulla membrane of odontocetes is also stiffened with bony ligaments, which appear to be associated with ultrasonic hearing (Reynolds and Rommel, 1999).

VI. Social Organization and Culture

Toothed whales are particularly well known for their BRAIN SIZE and rich social lives (Mann *et al.*, 2000). The absolute brain size of odontocetes ranges from 840 g in common dolphins to 7820 g in sperm whales (Berta and Sumich, 1999). However, a more useful way to compare brain sizes is to use the ratio of brain size to body size, the encephalization quotient (EQ). The relative brain sizes of odontocetes (ranging from EQ 0.02% for the sperm whale to EQ 1% for the bottlenose dolphin) are much larger than for most terrestrial mammals, but are similar to those of anthropoid primates (EQ 0.3% gorilla). Captive studies of bottlenose dolphins have shown that dolphins may have a "pecking order" similar to that of chimpanzees, in which both males and females have a social hierarchy, but that generally males are dominant to females.

The social systems of only a few odontocete species are known from long-term studies of wild animals, but these suggest some novel adaptations to standard mammalian patterns (Table III; Connor *et al.*, 1998). In Shark Bay, Western Australia, male Indian Ocean bottlenose dolphins form tight associations with one or two other males with whom they cooperate to form aggressively maintained consortships with individual females. These alliances, or coalitions, further appear to form moderately strong associations with one or two other alliances and will defend each other in competition over females. A completely different strategy is found among the fish-eating "resident" killer whales living along the coast of British Columbia, Canada, and the State of Washington. Here, neither males nor females disperse at matu-

rity, but instead remain with their mothers in stable matrilineal groups. Genetic data suggest that a similar pattern of natal philopatry may exist among long-finned pilot whales. Sperm whales appear to have a different social structure again. Groups of females and immature males are found in the tropical lower latitudes. Juvenile males remain in these natal groups until they reach puberty, at which stage they depart to lead more solitary lives in colder temperate and polar waters. They later return to the tropics when fully mature and search for estrous females with which to mate. Long-term studies of sperm whales have demonstrated strong cooperative bonds between individuals in groups of females and immature males (Fig. 5). These groups were thought to be composed of matrilineally related individuals; however, evidence suggests that they are not purely matrilineal in structure, but are comprised mainly of clusters of related individuals and also of some animals with no close relations. Baird's beaked whales may have one of the most unusual social systems among odontocetes, although this is inferred only from studies of dead animals. It appears that males mature an average of 4 years earlier than females and may live up to 30 years longer (Kasuya *et al.*, 1997). This has led to speculation that males may be providing significant parental care.

An interesting congruence between low levels of mitochondrial diversity and the presence of matrilineal social systems in four species of odontocete may suggest the cultural transmission of matrilineally inherited traits (Whitehead, 1998). Apparent culture in cetaceans includes the learning of particular feeding techniques from the mother or members of the matrilineal group. In fact, many attributes of cetaceans (and odontocetes in particular) favor the evolution of social learning and culture. These include their long lives (~60 to 90 years), advanced cognitive abilities, prolonged parental care, permanent and cohesive groups, and an environment that varies substantially over large spatial and temporal scales (such that individual learning would be costly) (Whitehead, 1998).

A potential offshoot of the advanced sociality observed in some species of odontocetes is the presence of postreproductive care. Short-finned pilot whales (*Globicephala macrorhynchus*) show a decreasing pregnancy rate with increasing maternal age and a parallel age-related decline in the ovulation rate. Up to 25% of adult females may be postreproductive, ceasing to ovulate after age 40 even though the maximum life span exceeds 60 years (Marsh and Kasuya, 1991). However, lactation may continue for up to 15 years after the birth of their last calf. It is unclear whether communal nursing exists.

TABLE III
Social Strategies Employed by Odontocete Species Known from Long-Term Studies of Wild Animals

Species	Social group	Mating system
Shark Bay bottlenose dolphin	Male coalitions Female groups	"Capture" of females by a coalition of males
"Resident" killer whales	Matrilineal groups of females and descendents	Intergroup matings thought to occur when groups meet
Sperm whale	Matrilineal groups of females and juvenile males Solitary adult males	Males rove between groups in search of estrus females

Figure 5 *Group of female and immature sperm whales in the Galápagos Islands. Photograph by Sascha K. Hooker.*

Another apparent consequence of the strong social bonds found among odontocetes is the phenomenon of mass stranding (Sergeant, 1982). This tendency for animals to come ashore in groups to die is found only among the most highly social odontocete families. Several explanations for this behavior have been suggested. These include disorientation due to geomagnetic anomalies, panic flight responses from predators, morbillivirus infections, parasitic infections of the respiratory system, brain, or middle ear, or the strong social bonds of a group causing the entire group to follow one intentional strander. It has also been noted that many locations in which mass strandings occur share certain structural characteristics. These sites are often composed of a sandy peninsula or promontory that extends perpendicular to the coastline and appears to form a "whale trap," potentially due to the loss of ECHOLOCATION abilities in the shallow sandy environment.

VII. Mating System, Reproduction, and Life History

Many odontocete species have promiscuous MATING SYSTEMS in which several adult males may mate with a female. However, other species, such as the bottlenose dolphins in Shark Bay and Dall's porpoises, appear to show a form of mate guarding. Among many beaked whale species, males compete, suggesting that there may be a hierarchical nature to their social organization, probably for access to females. However, detailed comparative data on the mating systems of odontocete species await genetic analyses to determine paternity from offspring within groups.

Gestation periods of odontocetes range between 7 and 17 months and almost all species have interbirth intervals of greater than 1 year (Berta and Sumich, 1999; Reynolds and Rommel, 1999). Length of gestation and fetal growth rate are related to calf size at birth such that larger species tend to require longer periods of gestation. Odontocetes produce a single offspring, which is physically well developed (able to swim and surface to breathe) but socially undeveloped. As a result, odontocete species have characteristically long lactation periods, averaging

between 32 and 100 weeks (Berta and Sumich, 1999). Females continue to feed throughout this long lactation period.

Odontocetes tend to show extended maternal care, resulting in a strong mother–calf bond. For most species, calves will remain with their mothers for a few years, but for some they will remain in close family groups all their lives (e.g., some killer whale populations, possibly long-finned pilot whales). This high level of investment needed to successfully raise calves may have led to the need to practice mothering behavior. Association between infants and nonmothers, termed allomaternal behavior, has been described for a variety of odontocete cetaceans (Mann *et al.,* 2000). Among bottlenose dolphins, such behavior appears to primarily benefit the nonmother, providing experience in parenting. Similar babysitting has also been documented in sperm whale groups, which stagger their deep-diving behavior such that calves are almost always accompanied by an adult at the surface. However, the function of this behavior in these circumstances appears to relate to increasing calf survival and defense against predation rather than to practice mothering (Mann *et al.,* 2000).

Much of the life history data available for odontocetes has come from studies of dead animals (from those taken in WHALING operations, as bycatch to other fisheries, or from STRANDINGS). The age of an odontocete can be estimated from its teeth using much the same technique as counting the growth rings of a tree trunk. As an individual grows, incremental growth layers are deposited in the teeth and bones. In order to determine the age, the tooth is usually thinly sliced and polished and may often be etched or stained to better resolve the growth layer groups. In most species, each growth layer group is thought to represent an annual increment, but this needs to be verified independently in order to use this method for each species. By establishing the ages of animals, researchers are able to investigate the age structure of the population, to look at ages at which animals mature, reproduce, and so on. Long-term studies of odontocete species in the wild (e.g., those listed in Table III) are gradually allowing life history parameters to be recorded from living animals. In some cases, particularly in establishing life span, these are providing records of life span to a much greater age than were estimated from catch records.

Life histories recorded in this manner show large differences between different odontocete species. Harbor porpoises have a maximum longevity of 12 years (although some may live up to 17 years), they reproduce at age 3, and become pregnant every year thereafter. In contrast, killer whales have a maximum longevity of about 80 years and first give birth at approximately 15 years of age, with a 5-year interbirth interval. Many species also show sex-specific differences in life history parameters. For example, sperm whale females become sexually mature at approximately age 9–10, whereas sperm whale males do not appear to become sexually mature until approximately age 26–27.

VIII. Human Interaction and Conservation Status

In the past, the majority of human interaction with odontocetes involved the capture of animals from the wild, either for consumption as part of the whale industry or, more recently, to obtain animals for captive display purposes. Only a few odontocete species were hunted to the same scale as the fisheries for baleen whales in the last two centuries. The main odontocete species taken historically were the sperm whale, some beaked whale species (northern bottlenose whales and Baird's beaked whales), and pilot whales. However, although the moratorium on large whale hunting essentially put an end to the hunting of mysticetes, the only odontocete species included in this moratorium was the sperm whale. Thus today, many odontocete species are still taken in large numbers. Unfortunately, many of these go unmonitored, and so the exact numbers taken are unknown, although they are suspected to be high.

Several odontocete species are or have been maintained in captivity for display, research, or conservation purposes (Twiss and Reeves, 1999). Some species are maintained in captivity for research or public display (e.g., bottlenose dolphin, killer whale, beluga), whereas others have only rarely been kept in captivity following live stranding (e.g., sperm whales, beaked whales). Some species are bred successfully in captivity, including bottlenose dolphins, Commerson's dolphins (*Cephalorhynchus commersonii*), and killer whales, but most other odontocetes species do not appear to fare so well in captive situations.

Humans are also increasingly attempting to interact with odontocetes in the wild. Whale-watching operations are increasing in number worldwide. In the San Juan Island area between British Columbia, Canada, and the state of Washington, the number of whale-watching vessels has increased 10-fold over the last decade. There is currently a good deal of concern about the impact of whale watching on the animals involved. Although there has been little documentation to date of immediate adverse impacts, there has been little investigation of the long-term stresses and consequences for the viability of the populations concerned. Swim-with-dolphin programs are also increasing in frequency. There is also concern about the effect of these, although one of the most pressing concerns is the potential to wrongly educate the public. As a result of these activities, the general public may form the perception that association with wild animals is endorsed. Currently in Florida, many people continue to solicit interactions with wild dolphins and, despite legislation against it, often encourage this by provisioning.

Lone animals of several odontocete species have, at times, been recorded to solicit associations with humans. This behavior has been recorded a number of times all over the world. In all cases, animals have become resident in a small area, where they approach and interact with boats or people in the water. In many cases, the animals involved have been common bottlenose dolphins, but this behavior has also been recorded from belugas and rough-toothed dolphins (*Steno bredanensis*).

Today, CONSERVATION problems exist for nearly all odontocete species. In fact, since the International Whaling Commission moratorium on catching large whale species, many of the current conservation threats faced by cetaceans are likely greater for odontocete species than for mysticetes. Their typically smaller size means they are less likely to be able to free themselves when trapped in nets, leading to a high incidence of bycatch. The squid diet of many species renders them prone to plastic ingestion, and their higher trophic levels magnify their pollutant load. Additionally, the habitat specificity shown by many species means that they have little recourse in the face of habitat destruction.

Other less direct threats to odontocetes include NOISE and disturbance, POLLUTION, and habitat loss and degradation. In general, noise is thought to be less of a problem for many odontocete species than for mysticetes, as much oceanic anthropogenic noise is low frequency. However, higher frequency noise (such as that created by fish finder units) is likely to coincide with the hearing range of many odontocetes (Twiss and Reeves, 1999). Additionally, some odontocete species (such as sperm whales and beaked whales) are thought to be susceptible to disturbance from low-frequency sounds.

See Also the Following Articles

Beaked Whales, Overview ■ Sound Production ■ Sperm Whale Systematics ■ Teeth

References

Au, W. W. L. (1993). "The Sonar of Dolphins." Springer-Verlag, New York.

Berta, A., and Sumich, J. L. (1999). "Marine Mammals: Evolutionary Biology." Academic Press, San Diego.

Connor, R. C., Mann, J., Tyack, P. L., and Whitehead, H. (1998). Social evolution in toothed whales. *Trends Ecol. Evol.* **13**, 228–232.

Cranford, T. W., Amundin, M., and Norris, K. S. (1996). Functional morphology and homology in the odontocete nasal complex: Implications for sound generation. *J. Morphol.* **228**, 223–285.

Heyning, J. E., and Mead, J. G. (1996). Suction feeding in beaked whales: Morphological and observational evidence. *Nat. Hist. Mus. Los Angeles County Contrib. Sci.* **464**, 1–12.

Hooker, S. K., and Baird, R. W. (2001). Diving and ranging behaviour of odontocetes: A methodological review and critique. *Mamm. Rev.* **31**, 81–105.

Jefferson, T. A., Leatherwood, S., and Webber, M. A. (1993). "Marine Mammals of the World." FAO Species Identification Guide, United Nations Environment Programme, Rome.

Kasuya, T., Brownell, R. L., Jr., and Balcomb, K. C. (1997). Life history of Baird's beaked whales off the Pacific coast of Japan. *Rep. Int. Whal. Comm.* **47**, 969–979.

Mann, J., Connor, R. C., Tyack, P., and Whitehead, H. (eds.) (2000). "Cetacean Societies: Field Studies of Dolphins and Whales." University of Chicago Press, Chicago.

Marsh, H., and Kasuya, T. (1991). An overview of the changes in the role of a female pilot whale with age. *In* "Dolphin Societies: Discoveries and Puzzles" (K. Pryor and K. S. Norris, eds.), pp. 281–285. University of California Press, Chicago.

Norris, K. S., and Mohl, B. (1983). Can odontocetes debilitate prey with sound. *Am. Nat.* **122**, 85–104.

Norris, K. S., Würsig, B., Wells, R. S., and Würsig, M. (1994). "The Hawaiian Spinner Dolphin." University of California Press, Berkeley.

Reynolds, J. E., and Rommel, S. A. (1999). "Biology of Marine Mammals." Smithsonian Institution Press, Washington, DC.

Rice, D. W. (1998). "Marine Mammals of the World: Systematics and Distribution." Society for Marine Mammalogy Special Publication No. 4, Allen Press, Lawrence, KS.

Richardson, W. J., Greene, C. R., Jr., Malme, C. I., and Thomson, D. H. (eds.) (1995). "Marine Mammals and Noise." Academic Press, San Diego.

Sergeant, D. E. (1982). Mass strandings of toothed whales (Odontoceti) as a population phenomena. *Sci. Rep. Whales Res. Inst.* **34,** 1–47.

Twiss, J. R., and Reeves, R. R. (eds.) (1999). "Conservation and Management of Marine Mammals." Smithsonian Institution Press, Washington, DC.

Whitehead, H. (1998). Cultural selection and genetic diversity in matrilineal whales. *Science* **282,** 1708–1711.

Training

TED TURNER

Behavior International, Aurora, Ohio

Marine mammal training is a relatively new field that gained formal recognition in the late 1940s. One of the first institutions to successfully maintain a breeding colony of BOTTLENOSE DOLPHINS, Marineland of Florida also became the first institution to begin a formal training program for bottlenose dolphins (*Tursiops truncatus*). As behavioral observations of these animals began to generate questions unanswerable through scientific observation alone, it seemed that further exploration involving dolphin-learning abilities could only be facilitated through training. Early training methods were considered industry secrets, yet those techniques were primitive. Although the methods used at that time were limited, consisting of trial and error, the marine mammal training field now employs proven behavior modification principles and, most predominantly, operant conditioning (behavior that is modified by the consequences) as a means to teach various behaviors to marine mammals. While the use of operant conditioning is not the only mechanism used to train marine mammals, it is arguably the most productive. Therefore, the fundamentals of this science must be assimilated in order to be successful at training marine mammals. Because the field of marine mammal training minimizes the use of punishment techniques, it is highly productive and continues to stand as a model for ethical training. The high exposure and excitement that is created when performing for the general public has focused attention on the cognitive abilities of these animals. Although some disagree with the use of animals in show performances, most are fascinated by the level of complexity, coordination, and consistency in behavior. Still others are amazed at the degree of intimacy and interaction between animal and trainer. Magical to some and practical to others, it is the training process itself that creates the medium for "communication" between marine mammal behaviorists and the animals they closely interact with. In most cases, the shows provide the animals with stimulating enrichment and the general public with the motivation for further education (Fig. 1).

I. Animal Learning and Behavior

Learning is a fundamental biological process necessary for survival. The expression or suppression of certain behaviors enable animals to respond to their environment and adapt to so-

Figure 1 *Marine mammal training and show performances generate public interest in marine mammals and related issues.*

cial and environmental changes. The detailed study of animal learning primarily involves aspects of psychology, such as memory, developmental learning, classical conditioning, operant conditioning, behavior modification, cognition, and other information processing disciplines. When describing the main components of current marine mammal training, the following three primary learning modes become salient.

A. Observational Learning

Also known as modeling, this vicarious process is defined as learning by observing another (model) engage in a behavior. During observational learning, the animal (observer) need not emit the observed behavior nor receive direct consequences in order that learning occurs. Mimicry seems to be a by-product of observational learning. It is a primary learning element for young animals raised in a complex social environment and appears to be of significance in developing socialization skills. The expression of these mimicked behaviors is not relegated to social behavior only. Young marine mammals will attempt to mimic trained behaviors and will often accompany adults in show performances or training sessions. It is common practice for marine mammal behaviorists to "capture" (reinforce) mimicked behaviors in order to rapidly develop and accelerate the learning process for young animals. However, regression often occurs when the model (adult) is no longer present. Training programs that rely on observational learning must often retrain these animals using direct operant conditioning and behavior shaping with individual training sessions. Nonetheless, observational learning plays an important role in the lives of marine mammals.

B. Classical Conditioning

Past understanding of Ivan Pavlov's early work investigating anticipatory salivation in dogs led to a restricted view that classical conditioning [the pairing of a conditioned stimulus (CS) in conjunction with an unconditioned stimulus (UCS)] only produced learning involving reflex systems. In recent years, this view has been replaced with a more contemporary understanding. Associations between conditioned and unconditioned stimuli have been linked to learning phenomena such as sign tracking (movement toward a stimulus that signals the availability of a positive reinforcer such as food), conditioned emotional response (fear), and taste aversion (food preference/avoidance). These seemingly insignificant components have a major impact on subsequent learning during the training process. Accidental pairing of events can create learning via classical conditioning that can interfere with or influence the training of marine mammals. Associating food with a loud noise, for example, can eventually diminish the startle effect the noise might otherwise evoke. However, pairing that same loud noise with a painful medical procedure can exacerbate the negative reaction to that same noise in the future. A minor avoidance response by the animal could intensify into a severe aggressive or panic response, causing injury to the animal or the animal handlers. An experienced marine mammal behaviorist understands this relationship and its effects on the training process. Classical conditioning can be used to help modify behaviors needed for the long-term care of marine mammals, such as introducing new animals into established social groups.

C. Operant Conditioning

Defined as behavior that is modified by its consequences, operant conditioning is currently recognized as the most widely employed training program for marine mammals. The basic components of instrumental learning (learning that occurs as a result of operant conditioning) must occur *in precise order* for learning to be achieved. These three components are the *antecedent* (a stimulus that precedes the behavior), the *behavior* (the resultant observed response emitted by the animal that immediately follows the antecedent), and the *consequence* (a stimulus that immediately follows the response and acts to increase or decrease that behavior). During a training session, a hand signal (antecedent) is usually presented to the animal followed by the animal's response (behavior). If correct, the behavior is usually reinforced (consequence) by the trainer with either primary reinforcement (food) or secondary reinforcement (touch, toys, play, activities). Common in marine mammal training, incorrect behaviors are usually ignored with minimal or no consequence applied. Punishment and negative reinforcement are generally avoided. Other animal training fields use a wide variety of punishment and negative reinforcement techniques to suppress unwanted behavior. In the operant process, each behavior is best described as a learning cycle (Fig. 2).

Principles supporting the successful, technical application of these three elements must be understood by the behaviorist

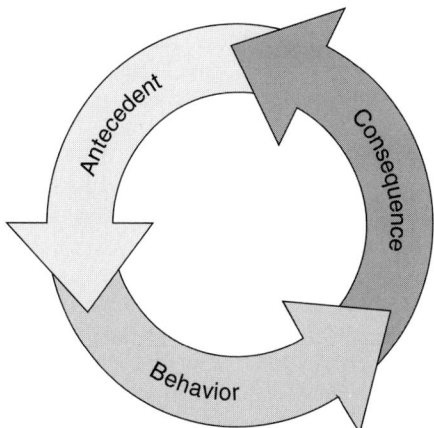

Figure 2 *A typical operant conditioning cycle where the signal (antecedent) precedes the response (behavior), followed by the reinforcement (consequence). A single training session includes many such cycles.*

and include stimulus consistency, stimulus fading, behavioral criterion, behavioral development, delay of reinforcement, schedule of reinforcement, magnitude of reinforcement, and many others. Successful marine mammal trainers understand the implications of precise application. Like other forms of learning, operant conditioning should not be characterized as a "technique" or "system" of training that can be switched on and off conveniently, instead it is a dynamic and ever present environmental learning phenomena that is in continuous action, influencing the acquisition or extinction of specific behaviors. Behaviors are in the constant process of strengthening or diminishing as a result of the consequences that follow. It must be emphasized that not all of the antecedents and consequences experienced by each animal are applied by the marine mammal trainers. Behaviorism views the environment in its entirety with many such antecedent stimuli being processed by the animals in context. Even in the most productive marine mammal training programs, it is rare that more than 3 hr per day, on average, are devoted to actual training sessions, leaving 21 hr per day to other environmental learning influences. For large institutions with formal marine mammal training programs, certain principles and techniques have been refined and operationalized to facilitate staff development and expertise, resulting in efficient animal training. Expertise is often gained through practical hands-on application and is supplemented with course work, seminars, and testing. Other mediums, such as conference attendance and trade groups, provide additional sources of information. The International Marine Animal Trainers Association (IMATA) is an established trade group that offers opportunities to formally exchange specialized marine animal training information. Many marine mammal training programs usually emphasize the following.

1. Establishing "Trust" As an anthropomorphic expression, the word "trust" is used here to describe a more complex set of behavior. Naïve animals (animals without prior training experience or a limited behavior repertoire) often seem cautious and wary of their surroundings. This apprehension can be reduced by minimizing or eliminating those practices that create avoidance or fear response, such as rough handling, aversive procedures, and loud noises. The process of extinguishing avoidance behavior is aided by the cumulative, positive effect of daily feedings and time spent with the animal. Within a few days, marine mammals will readily take food from the familiar hand and allow some calm touching during feeding. This provides the foundation for positive reinforcement (a stimulus that is applied as a consequence and increases the frequency, intensity, and duration of the preceding behavior). Regression can be minimized if aversive events can be delayed or eliminated. As a reinforcement history develops, unavoidable aversive events (such as routine medical treatment) have less impact and cause minimal regression. The opposite appears to be true as well. Repeated aversive procedures early in the process can have a profound negative effect and will delay the positive response to the trainer.

2. Conditioned Reinforcer   The consistent pairing of a stimulus with the applied reinforcer (initially food) establishes the foundation for a conditioned reinforcer (a stimulus that has acquired reinforcing properties through learning). Most marine mammal training facilities call this a "bridging" stimulus. This bridging stimulus acts as a cue that signals the application of reinforcement (a stimulus that increases the frequency, intensity, and duration of a preceding behavior). It is referred to as a "bridge" because it acts to bridge the time delay between the precise behavior targeted for increase and the positive reinforcement. Marine mammal trainers and behaviorists use a variety of tools that act as a bridge. Typically, a high-frequency whistle is used for cetaceans (due in most part to higher hearing ranges), whereas a verbal cue or clicker sound is used for other marine mammals. This bridging stimulus, when paired consistently with the offering of food, quickly becomes recognized by the animal. Soon the sound itself evokes a response and subsequent orientation back to the trainer.

3. Building the Attention Threshold   In order to maintain the animal's attention sufficient to advance learning, an early emphasis on attention threshold (the duration that a stimulus will control behavior) is necessary. This is done by systematically increasing the amount of time the animal will continue to watch the trainer prior to bridging and reinforcement delivery. Increments of a few seconds are literally counted by the trainer when shaping an increased attention span. Eventually, within a few days or weeks, an attention span of several minutes can be conditioned if bridging/reinforcement is precise. Conversely, a lack of precision can delay progress. Inexperienced or impatient marine mammal trainers often overlook this critical element.

4. Following a Target   If the pairing of food with the bridge has been well timed and effective, the animal will quickly learn to touch a hand prior to the bridge and food delivery. By holding the hand close to the animal, the animal will, in all likelihood, touch the hand. The behavior of touching should be reinforced. Eventually, the hand can be moved back and forth in order to strengthen the "following" response. In time, the animal will move to the hand wherever it is. Target poles (a long pole with a small round float attached to the end) are often introduced at this stage to simulate an extended reach. Through the use of this target pole, touching and following can be reinforced. Various behaviors can now be shaped through approximations (Fig. 3).

5. Achieving Stimulus Control   As animals begin to learn the relationship between their emitted response (behavior) and the delivery of reinforcement by the trainer (consequence),

Figure 3   *A California sea lion* (Zalophus californianus) *is trained to follow a target pole. The target enables the trainer to shape behaviors from a distance.*

they also begin to learn which specific behavior will be actually reinforced. Discrimination between learning events occurs when animals learn to respond differently to differing signals (antecedent). The trainer offers reinforcement only for specific responses and thereby facilitates discrimination. This process is aided through the use of specific hand signals (discriminative stimuli) that precede each behavior. When trainers selectively reinforce only desired behaviors, animals learn to recognize the signals and respond appropriately. This is referred to as stimulus control where behaviors are reinforced in the presence of a particular signal or stimulus. Random sampling (performing several behaviors in an attempt to gain access to reinforcement) is a common phenomenon early in the training process. Trainers will ignore all but the requested behavior, thereby aiding the process of stimulus control and discrimination learning.

6. Successive Approximations   Effective trainers break each behavior down into small, easily understood and planned steps. For example, a high jump begins with the lifting of the head in order to touch the target pole and receive reinforcement. With each successive approximation, the target is raised slightly until the desired height is achieved. In effect, each behavior can be visualized by the trainer as a step-by-step process. All behaviors can be broken down in such a manner creating an easier learning process for the animal (Fig. 4).

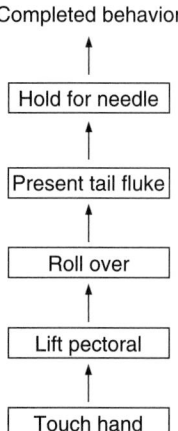

Completed behavior

Hold for needle

Present tail fluke

Roll over

Lift pectoral

Touch hand

Figure 4   *An approximation diagram of a voluntary blood sampling behavior. Approximations can best be described as small "steps" toward a desired goal. Each step represents an advancing behavioral criterion progressing toward the completed behavior. Here, a dolphin is first conditioned to allow a touch of the pectoral flipper. As this behavior is reinforced, the animal learns to follow the trainer's hand with its pectoral flipper into a roll-over behavior. The roll-over behavior is then reinforced, and the animal learns to present itself in a belly-up position after a few trials. At this point, the trainer begins to reinforce the animal for allowing contact systematically closer to the tail flukes. After a few training sessions, the flukes can be held while a needle is slowly inserted for blood removal from a near surface vein. The trainer delivers reinforcement if the animal remains calm and motionless.*

7. Schedules of Reinforcement   This refers to the timing and frequency of reinforcement. As mentioned previously, most all marine mammal training programs emphasize positive reinforcement and generally avoid negative reinforcement (an increase in a response that causes the removal of an aversive stimulus) and punishment (a decrease in a response that is followed by an aversive stimulus). In fact, many marine mammal institutions now consider punishment unethical. Even so, an emphasis on positive reinforcement without regard to the precise schedule of reinforcement can lead to training and subsequent learning problems. Schedules of reinforcement have been derived empirically with many variations currently being studied. Each specific schedule has been shown to manifest certain behavioral characteristics. For example, a simple fixed ratio schedule (where a fixed number of responses is required for reinforcement) can create predictability in the environment and will manifest in episodic motivation with periods of poor response. These behavioral pitfalls are avoided when behavior programs include staff training on the most effective use of reinforcement. It would be an oversimplification to describe the most effective schedule of reinforcement, as each can be equally effective depending on the learning stage of the animal. In general, however, many applied marine mammal training programs are beginning to recognize the importance of variable schedules of reinforcement, including reinforcement variety. Variable schedules of reinforcement provide reinforcement intermittently and unpredictably. This more closely replicates the type of reinforcement schedule most often associated with the wild environment. The challenge of failing and then obtaining primary reinforcement (that which does not depend on learning such as food and water) and secondary reinforcement (an event that becomes reinforcing through learning such as play, games, toys) activates cognition and problem solving. This, in turn, provides enrichment opportunities for each animal while it shapes stronger foraging and hunting skills. (Clearly, it seems logical that an animal that learns to preserve after failure might have a higher survival rate in the wild.) Reinforcement variety (the random delivery of both primary and secondary reinforcement) augments the elements of surprise—an essential component of effective reinforcement.

As mentioned previously, only certain components of operant conditioning have been briefly described for the purpose of identifying those elements that influence the acquisition of trained behaviors most rapidly. Even those institutions that ignore or minimize the use of "training" are most certainly creating positive and negative consequences for behaviors. Formally applied or not, the process of "training" through both operant and classical conditioning continues for the animals. Consider the normal operation of daily cleaning, feeding, and watering. These activities can effectively act as consequences that shape and modify behaviors. The appearance of a cleaning hose, for example, might act as a punisher for certain animals, thereby suppressing the behavior of courtship (depending on the timing of its appearance). However, it could also act as a reinforcer for a different animal, thereby increasing the behavior of vocalizing (again dependent on the timing of appearance). Even stereotypic patterns, such as anticipatory pacing prior to feeding time, create an opportunity for adventitious reinforcement (coincidental cause and effect reinforcement causing an inadvertent increase in the unwanted behavior). Awareness of these learning principles enables caretakers, behaviorists, scientists, and trainers to better understand *all* environmental factors influencing the expression of appropriate or inappropriate behaviors. As we continue to learn about information processing and training, much more needs to be assimilated by the behaviorist. Memory retrieval, short- and long-term memory capacities, can influence behavior acquisition, especially in higher order animals. Learning phenomena such as context-specific learning (learning in one environment that does not necessarily transfer to another environment) can affect the outcome of a reintroduction project, for example. Further investigations into animal emotion and cognition will add enormously to our understanding of animals as will further studies to determine stage-developmental learning.

II. Early Development and Learning

During the first few days of life, the primary challenge for newborns is maternal acceptance. It seems that a mother's instinctive epimeletic (caregiving) behavior patterns should engage automatically after birth, yet newborns are sometimes rejected and thus are unable to survive. Still others have been injured and even killed by their mothers. This disturbing pattern, although rare, has also been observed in most mammal species (both wild and captive animals), including humans. The wide majority of newborns, however, *are* accepted (some immediately after birth and some within a few hours) and their mothers usually allow them to nurse within 24 hr. On rare occasion, offspring have gone 48 hr without taking their first milk. Within a few days, the calf or pup rapidly acquires the motor skills necessary to coordinate successful nursing while its mother is also acquiring and strengthening her own epimeletic repertoire, thereby increasing survivability in subsequent offspring. For the newborn, motor skill development occurs rapidly and, within a few weeks, observations of increasing independence (time and distance away from the mother) begin to emerge as coordination, buoyancy control, visual acuity, ECHOLOCATION, and auditory learning strengthen. Observational learning becomes evident and mimicry increases until the young animal learns to directly manipulate its environment (Fig. 5).

In some cetaceans, between 4 and 7 months of age when teeth begin to erupt and grow, the mother initiates a weaning process through food sharing (either partially digested and regurgitated or with food that has been ripped into small pieces). This seems to coincide with accelerated learning and may signal the beginning of a sensitive learning phase or critical period (a period of rapid acquisition of behaviors). For marine mammals born in captivity, most formal training begins during this stage.

III. An Emphasis on Positive Reinforcement

Marine mammals respond readily to operant conditioning, probably the most widely used training process, and possess a strong aptitude for instrumental learning. The rapid acquisition of behaviors has done little to refute the notion that marine mammals are unusually "intelligent." It must be stated, however, that other animals trained via operant conditioning sometimes show an equal or superior aptitude and learning curve. As stated previously, most marine mammal training programs restrict the use of punishment and applied negative consequences. Instead, they rely heavily on positive reinforcement. Empirical

Figure 5 *Young animals often learn through observation. Here, a killer whale* (Orcinus orca) *calf is observing as her mother interacts with a trainer.*

evidence suggests that positive reinforcement has a much greater influence in motivating animals to voluntarily engage in behaviors, whereas punishment suppresses the expression of behavior. Additionally, punishment serves to increase the development of undesirable behaviors, such as anxiety, escape, active and passive avoidance, frustration, aggression, and learned helplessness (a state of "giving up"). Training programs that use punishment must increase the frequency, intensity, or duration of the punishment in order to maintain its effectiveness. In time, these types of animal training programs develop animals with significant behavioral difficulties, increased stress, and a loss of motivation. The most effective training process develops a foundation of trust through daily care with the elimination of punishment. Most importantly, the desired outcome is *voluntary cooperation* as opposed to compliance under the threat of retaliation. This, at once, seems the most simple and the most difficult concept for the layperson to grasp. However, it may be the marine mammal training industry's most profound contribution to the social sciences.

IV. Animal Acclimation

The assumption that animals will immediately behave toward a new member with acceptance or at best indifference is often erroneous. Animals use AGGRESSION as a means to acquire and establish territories, food sources, breeding rights, sleeping space, and more. Without employment of simple learning principles, the likelihood of any new member being challenged and subject to aggression is highly likely. In a zoologically controlled environment, this is not often desired but can be mediated successfully via training. The use of classical conditioning has shown promise in animal introductions and acclimation. The initial pairing of each new member (CS) with food reinforcement (UCS) helps counter the perception of the new member being a threat. Additionally, after the initial introduction, the use of an operant technique known as differential reinforcement of other behavior (DRO) provides reinforcement to the group for nonaggression. DRO is a decrease procedure that can target antisocial aggression by increasing prosocial behavior. By combining the learning principles of classical conditioning and operant conditioning, animal acclimation can be facilitated.

V. Husbandry Training

Significant achievements in veterinary care have been achieved through husbandry training. Husbandry training generally consists of the development of a behavioral repertoire that facilitates veterinary access without the use of restraint, anesthesia, or isolation. Optimal animal health is achieved by routine examinations. For an untrained marine mammal, simple procedures and checkups can be a risk to both people and animals. Untrained animals usually express fear, apprehension, and a "fight or flight" response (activation syndrome) that can be dangerous for all involved. Since the early 1990s, a variety of procedures have been routinely trained through operant conditioning, approximation training, and positive reinforcement. A sample of these commonly trained and voluntary behaviors include blood sampling, urine collection, body cavity cultures, eye examinations, topical treatments, injections, x rays, tooth drilling,

heart monitoring, body weights, stretcher conditioning, artificial insemination, and gastric intubation.

Once believed impossible to obtain without restraint or anesthesia, routine access to biological samples and simple medical procedures is becoming more and more commonplace in the marine mammal training field which contributes to increased life expectancies. Through training, marine mammal medicine is reaping tremendous benefit from behavioral management. Many zoos are now emulating the success of marine mammal husbandry training and are beginning to experience those benefits with endangered exotic animals as well.

VI. Enrichment and Variety

As we recognize the importance of play in animal development, we have also recognized how important other physical and cognitive challenges are for optimal animal care. An emphasis on behavioral expression pioneered by marine mammal training has gained wide acceptance in zoos, aquariums, and other animal facilities. In the past, this aspect was considered superfluous to animal health and exhibitry. It was believed that "natural" behaviors would be an automatic by-product of a "natural" exhibit. It was soon recognized that these behaviors required more than high-quality food, health care, and housing. A focus on animal enrichment (the application of environmental complexity to stimulate physical activity) has become a necessary component in primary animal care and clearly fits in with the animal training model (i.e., behaviors can be shaped, maintained, and modified using operant techniques). In fact, many successful enrichment programs are managed by experienced animal trainers. They coordinate environmental change, food placement, toys, games, and other behavioral prompts and reinforcers with enormous benefit to the animals. Daily variety and reinforcement of appropriate behaviors help reduce other inappropriate behaviors such as aggression, stereotypic patterns, and stress-related problems.

VII. Long-Term Behavior Management

Applications for marine mammal training principles are expanding into areas traditionally relegated to population management. California sea lions can be extremely aggressive and territorial during the spring and early summer breeding season. Adult males can gain weight rapidly under the influence of increased testosterone production. In zoological settings, the need to replicate this important annual biological cycle serves no purpose for males that are not part of a breeding program and can be problematic. Unless a facility has unlimited space for housing disputing males, intervention must occur to minimize aggression. Current management methods utilize surgical and pharmacological interventions. Progress in behavioral management has proven effective as well. Minimizing body weight fluctuations, paired with direct reinforcement of nonaggression, has been effective. Body weight maintenance requires the training of separation behaviors (separating in and out of a larger group) and daily weight scale readings (the animal is trained to walk onto a platform scale and sit still). The use of DRO decreases inappropriate aggression by increasing behaviors other than aggression.

VIII. Other Uses for Marine Mammal Training

A wide variety of training applications are now being explored in the marine mammal field. In time, these advancements will have as much or more impact than the behavioral advancements preceding these applications. Related to husbandry training, the field of behavior medicine is growing in understanding and in effective use. Behavior medicine (also referred to as health psychology) employs the use of behavior modification to treat chronic and acute medical conditions. Pretreatment baseline data are gathered prior to training (treatment) and subsequent health improvement. A successful treatment model is shown in Fig. 6.

Chronic medical conditions can also be treated by engineering a behavioral training program that targets a specific behavioral goal. All behaviors used for treatment are shaped, trained, and maintained as most other behaviors, although, like traditional medicine, not all treatments are completely successful. A good animal health program utilizes all available expertise, including training (Fig. 7).

Marine mammal training has been employed successfully in open ocean work. The U.S. Navy's marine mammal program has trained a number of marine mammal species a variety of tasks involved in aiding military operations, rescue and recovery, deep diving physiology, echolocation, boat following, and object retrievals to name just a few. These early and ongoing programs continue to generate enormous benefit to scientists in understanding marine mammal learning capabilities, physiology, and motivation. A number of research institutions now use open-ocean training to study certain aspects of marine mammals. Interactive programs that allow in-water interaction are gaining popularity. These high-quality education programs enhance the enrichment that animals receive while giving the general public a chance to get in the water and directly experience the animal's behavioral capabilities. Training techniques have also been employed in the rehabilitation process for sick and injured wild animals. As behavior modification gains

Figure 7 *Advanced application of operant conditioning is used for medical treatment. A Commerson's dolphin (Cephalorhynchus commersonii) is allowing a stomach tube to be inserted for fluid therapy.*

wider application, wildlife managers will continue to increase their knowledge and application of these learning principles. Solutions to human/animal encounters, survival skills acquisition for release of captive-born animals into the wild, and behavior problems associated with reintroduction and relocation efforts will require a specialized understanding of the learning and training processes, pioneered in the learning laboratories and practiced in the field of marine mammal training.

See Also the Following Articles

Behavior, Overview ■ Captivity ■ Intelligence and Cognition ■ Marine Parks and Zoos ■ Mimicry ■ Parental Behavior

References

Chance, (1994). "Learning and Behavior," 3rd Ed. Brooks/Cole Publishing Company, Belmont, CA.

Domjan, M. (1993). "The Principles of Learning and Behavior," 3rd Ed. Brooks/Cole Publishing Company, Belmont, CA.

Honig, W. K., and Staddon, J. E. R. (eds.) (1977). "The Handbook of Operant Behavior." Prentice-Hall, Englewood Cliffs, NJ.

Kazdin, A. E. (1994). "Behavior Modification in Applied Settings," 5th Ed. Brooks/Cole Publishing Company, Belmont, CA.

Pryor, K. (1995). "On Behavior: Essays and Research." Sunshine Books, North Bend, WA.

Ramirez, K. (ed.) (1999). "Animal Training: Successful Animal Management through Positive Reinforcement." Shedd Aquarium Press, Chicago, IL.

Shepherdson, D. J., Mellen, J. D., and Hutchins, M. (eds.) (1998). "Second Nature: Environmental Enrichment for Captive Animals." Smithsonian Institution Press, Washington, DC.

Spear, N. E., and Riccio, D. C. (1994). "Memory: Phenomena and Principles." Allyn & Bacon, Needham Heights, MA.

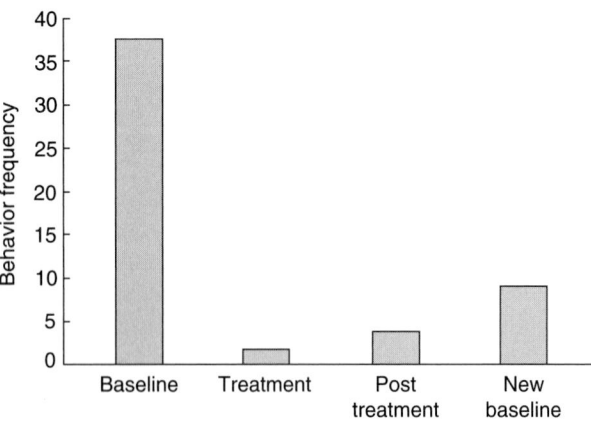

Figure 6 *A model of a typical behavioral medicine application. The effect is to reduce the frequency of the causative health/behavior problem.*

Tucuxi

Sotalia fluviatilis

PAULO A. C. FLORES
PUCRS (Pontifícia Universidade Católico do Rio Grande do Sol), Porto Alegre, Brazil

The tucuxi is listed as insufficiently known by the World Conservation Union (IUCN). Most information on the species comes from specimens collected throughout the Amazon Basin; the scarcity of data on the coastal tucuxi is a result of the lack of long-term longitudinal studies of living animals. The scenario has changed since the early and mid 1990's, mostly through research conducted off Brazil. However, few of these studies' results have been published and so far only presented to scientific conferences and conservation meetings and in academic theses. This article intends to include all available information to provide the best overview of the species.

I. Common Names, Distribution, and General Description

The name tucuxi comes from *tucuchi-una* after the Tupi language of the Mayanas Indians from the Amazon region in Brazil. Other names include boto, boto comum, and golfinho cinza along the Brazilian coast; bufeo gris, bufeo blanco, or bufeo negro in Colombia and Peru; lam in Nicaragua; and gray river dolphin in English. Although up to five species and three subspecies were described, today a single species, *Sotalia fluviatilis,* is recognized, presenting two ecotypes, riverine (or freshwater) and marine, which differ somewhat in length, number of teeth, and color pattern. The riverine tucuxi occurs in the Amazon River basin (drainage) including Brazil and going as far inland as southeastern Colombia, eastern Ecuador, and northeastern Peru. The marine tucuxi is found in the western Atlantic coastal waters of South and Central America from southern Brazil (27°35′S, 48°35′W) to Nicaragua (14°35′N, 83°14′W), including Colombia, Costa Rica, French Guyana, Guyana, Panama, Suriname, Trinidad, and Venezuela, with possible records to Honduras (15°58′N, 79°54′W). In the Orinoco River the species may be seen as far up as Ciudad Bolivar, although whether these specimens belong to the riverine or marine ecotype is not clear. No fossil record is known so far.

The tucuxi looks somewhat like a smaller bottlenose dolphin, *Tursiops truncatus.* It is light gray to bluish gray on the back and pinkish to light gray on the ventral region, with a distinct boundary from the mouth gape to the flipper's leading edge. On the sides the tucuxi has a lighter area between the flippers and the dorsal fin and another one in the midbody at the anus level; marine individuals may have another light gray rounded streak on both sides of the caudal peduncle. The eyes are large, and around the eyelids there are black countershadings. The dorsal fin is triangular and sometimes slightly hooked at the tip (Fig. 1). It has a moderately slender long beak, a rounded melon, and 26 to 36 teeth in each mandibular ramus. The marine tucuxi has more upper teeth and is larger than riverine specimens: total length can reach about 210 cm (with an extreme 220 cm reported) in the marine and about 152 cm in the riverine ecotype.

II. Ecology and Behavior

The marine tucuxi is found mostly in estuaries, bays, and other protected shallow coastal waters. It feeds mainly on demersal and pelagic fishes such as clupeids and scianids, and neritic cephalopods, shrimps, and flounders are occasionally taken. The riverine *S. fluviatilis* occurs in the main channel of rivers, tributaries, and lakes, usually avoiding rapids and not going into the flooded forest as the sympatric boto *Inia geoffrensis.* Its movements may be influenced by seasonal level fluctuation, using channels and lakes during rising and high waters while avoiding them during the low-water season.

Potential predators are the killer whale *Orcinus orca* and various species of coastal sharks for the marine ecotype and the bull shark *Carcharhinus leucas* in the Amazon. However, no record of such predation is known so far.

Figure 1 *Relatively little has been published concerning the Amazon dolphin called the tucuxi, which tends to occupy estuary and other shallow coastal habitats. Pieter A. Folkens/ Higher Porpoise DG.*

The tucuxi is a social dolphin often found in groups of 2 to 6 individuals. Groups of up to 50 or 60 specimens of the marine ecotype can usually be found. Large aggregations of up to 200 are reported in Baía de Sepetiba and around 400 individuals in Baía da Ilha Grande (Lodi and Hetzel 1998), on the southeastern Brazilian coast. Larger aggregations are usually engaged in cooperative feeding. Feeding occurs roughly in two ways: in pairs and cooperatively in larger groups or subgroups when different strategies are employed. During feeding activities marine tucuxis often associate with birds such as the brown booby *Sula leucogaster*, terns (*Sterna* spp.), frigate *Fregata magnificens*, and kelp gull *Larus dominicanus*. In some locations, mixed flocks of up to 100 birds can be seen in such associations. In the Amazon, the tucuxi may feed occasionally in association with terns (*Phaetusa simplex*).

The presence of the tucuxi is recorded throughout the year in many coastal locations such as Baía Norte (South Brazil, coinciding with the species southernmost distribution limit), Cananéia estuary and Baía de Guanabara (both in southeastern Brazil), Baía de Todos os Santos, and around Fortaleza (northeastern Brazil), as well as Bahia Cispatá and Golfo de Morrosquillo (Colombia). Photoidentification studies in few of these areas have demonstrated that at least some animals are resident year-round for up to 7 continuous years (e.g., Flores 1999; Flores *et al.*, 1999). The riverine tucuxi is very common in the Solimões River and its tributaries as well as the Mamirauá area (Brazil), the Tarapoto River as well as Caballococha and El Correo lake systems (Colombia), and near Iquitos (Peru), to give some examples (Fig. 2). From scarce available data, abundance estimations in an area vary with the method employed. In Baía de Guanabara, it was calculated at between 69 and 75 individuals through photoidentification (Pizzorno, 1999) to 418 animals by line transect methods (Geise, 1991). It seems that the marine tucuxi is common but not numerous in such areas. In at least some locations the riverine ecotype is found in one of the highest densities found for cetaceans (Vidal *et al.*, 1997).

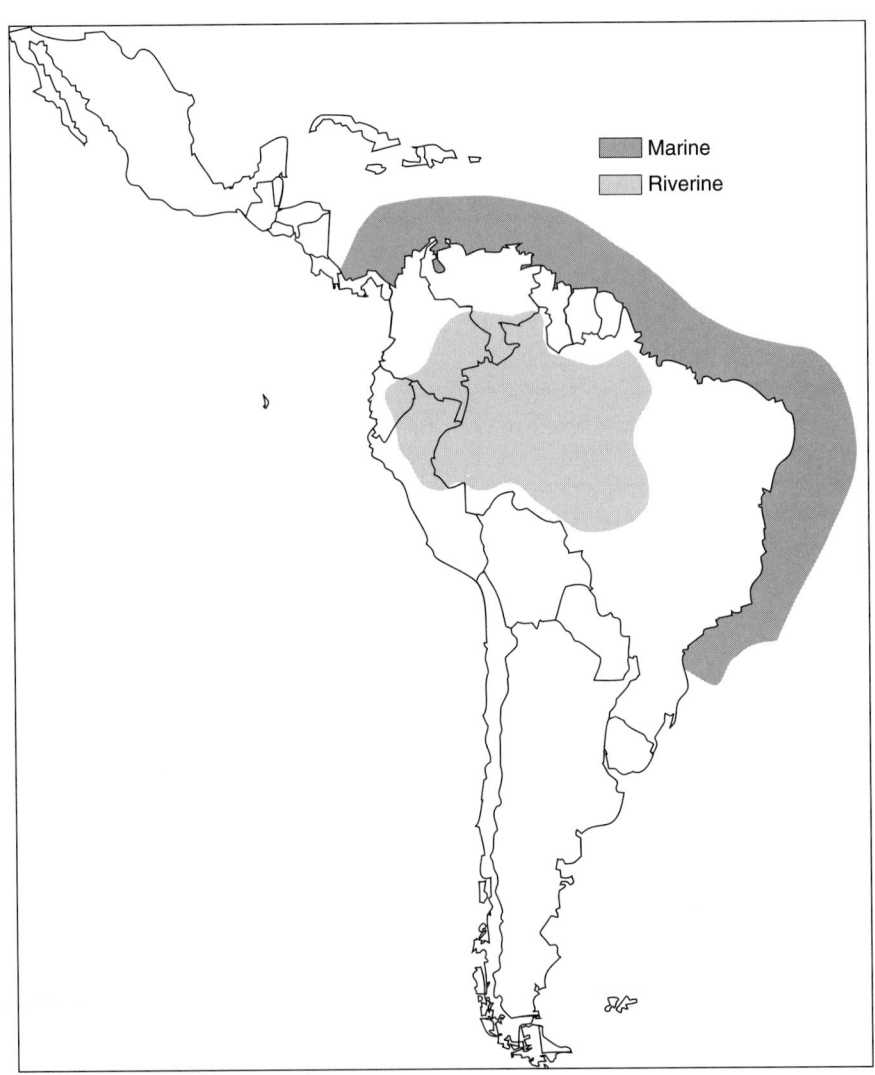

Figure 2 *Map of tucuxi distribution.*

Dive times are about 1.5 to 2 min with shorter surfacings every 5–10 sec in between. A variety of aerial behaviors may be displayed, such as full leaps, somersaults, flukes-up, spy-hopping, surface rolling, and porpoising. They do not bow ride but may surf in waves or wakes produced by passing boats. A lone, wild, and sociable animal, as well as a case of epimeletic behavior, was recorded along the São Paulo coast in southeastern Brazil. Tucuxis may interact with bottlenose (*Tursiops truncatus*) and Amazon river (*Inia geoffrensis*) dolphins. They have been engaged in apparent mating behavior with bottlenose dolphins off Costa Rica.

No mass stranding has been recorded, although dead animals usually wash ashore along its distribution, mainly resulting from INCIDENTAL CAPTURE in fisheries.

III. Natural History

The breeding system of the species is polyandrous, with mating involving tactile contact and copulation occurring belly to belly. Calving is year-round in most of the coastal species' distribution and in October–November during low-water season for the freshwater ecotype. Gestation is estimated to be around 11 to 12 months with calves ranging from 90 to 100 cm (marine ecotype) and around 10 months and 71 to 83 cm (riverine ecotype). The calving interval is believed to be 22–23 months (through photoidentification data). According to tooth growth layer groups, the lifespan can reach 30 and 35 years for marine and riverine ecotypes, respectively. Natural mortality rates are unknown.

IV. Interactions with Humans: Threats and Conservation

The species have not been exploited commercially, although incidental mortality in local and commercial fishing gear such as gill nets and seines are the main direct threat to tucuxis. In some localities they are killed on occasion for shark bait or human consumption, although tucuxis have protection from myths and legends, especially in the Amazon. There, their genital organs and eyes have a small market as love charms. Dams and hydroelectric power facilities may isolate populations and interrupt fish migration as well as reduce fish abundance. Gold mining with mercury, destruction of mangroves and salt marshes, water pollution, and boat traffic are other potential threats. The recent events of hand feeding and the short-term behavioral effects caused by tourism boats also deserve concern.

Tucuxis die easily due to capture stress during transportation or handling. Some animals, though, have been held in captivity, where they exhibit rare aerial behavior and common aggression toward male tucuxis and other species. Today just a few tucuxis are kept in European facilities.

Acknowledgments

I thank Marcos C.O. Santos (Instituto de Biociências/USP, Brasil), Salvatore Siciliano (Museu Nacional/UFRJ, Brasil), Vera M. F. da Silva (Laboratório de Mamíferos Aquáticos/INPA, Brasil), and Fernando Trujillo (Fundación Omacha, Colombia and University of Aberdeen, Scotland) for kindly providing references, information, and useful comments to improve this article.

See Also the Following Articles

Amazon River Dolphin ▪ Bottlenose Dolphins ▪ Identification Methods ▪ Sociobiology

References

Cirilo, P. C. P., Santos, M. C. O., Zampirolli, E., Alvarenga, F., Vicente, A. F., and T. M. A. (1998). Note on a lone and sociable marine tucuxi dolphin *Sotalia fluviatilis* at São Vicente, São Paulo, Brazil. "Abstracts of the 8th Biennial Meeting of Experts on Aquatic Mammals of South America," 25–28 October 1998, Recife, Brasil.

Da Silva, V. M. F., and Best, R. C. (1994). Tucuxi *Sotalia fluviatilis* (Gervais, 1853). *In* "Handbook of Marine Mammals: The First Book of Dolphins" (S. H. Ridgway and R. J. Harrisson, eds.), Vol. 5, pp. 43–49. Academic Press, London.

Da Silva, V. M. F., and Best, R. C. (1996). *Sotalia fluviatilis. Mamm. Species* **527**, 1–7.

Flores, P. A. C. (1999). Preliminary results of a photoidentification study of the marine tucuxi, *Sotalia fluviatilis*, in southern Brazil. *Mar. Mamm. Sci.* **15**(3), 840–847.

Geise, L. (1991). *Sotalia guinensis* (Cetacea, Delphinidae) population in Guanabara Bay, Rio de Janeiro, Brazil. *Mammalia* **55**(3), 371–380.

Lodi, L., and Hetzel, B. (1998). Grandes agregações do boto-cinza (*Sotalia fluviatilis*) na Baía da Ilha Grande, Rio de Janeiro. *Rev. Bioikos, PUC-Campinas* **12**(2), 26–30.

Pizzorno, J. L. A. (1999). "Estimativa populacional do boto-cinza, Sotalia fluviatilis, na Baía de Guanabara, por meio de catálogo de fotoidentificação." M.Sc. thesis, unpublished, Instituto de Florestas, Universidade Federal Rural do Rio de Janeiro, Rio de Janeiro.

Santos, M. C. O., Rosso, S., Siciliano, S., Zerbini, A. N., Zampirolli, E., Vicente, A., and Alvarenga, F. (2000). Behavioral observations of the marine tucuxi dolphin (*Sotalia fluviatilis*) in Sao Paulo estuarine waters, Southeastern Brazil, *Aquatic Mammals* **26**(3), 260–267.

Vidal, O., Barlow, J., Hurtado, L. A., Torre, J., Cendón, P., and Ojeda, Z. (1997). Distribution and abundance of the Amazon River dolphin (*Inia geoffrensis*) and the tucuxi (*Sotalia fluviatiis*) in the upper Amazon River. *Mar. Mamm. Sci.* **13**(3), 427–445.

Tuna–Dolphin Issue

TIMOTHY GERRODETTE
*Southwest Fisheries Science Center,
La Jolla, California*

I. The Problem

In the tropical waters of the Pacific Ocean west of Mexico and Central America, large yellowfin tuna (*Thunnus albacares*) swim together with several species of dolphins (pantropical spotted, *Stenella attenuata;* spinner, *S. longirostris;* and short-beaked common, *Delphinus delphis*). This ecological association of tuna and dolphins is not clearly understood, but it has had two important practical consequences: it has formed the basis of a successful tuna fishery and it has resulted in the deaths of a large number of dolphins. This is the heart of the tuna–dolphin issue.

The bycatch of dolphins in the eastern tropical Pacific (ETP) purse seine tuna fishery stands apart from marine mammal bycatch in other fisheries, not only in scale but in the way the dolphins interact with the fishery. Marine mammals interact with most fishing gear only incidentally, but in the ETP tuna fishery the dolphins are an intrinsic part of the fishing operation. The fishermen intentionally capture both tuna and dolphins together and then release the dolphins from the net. The vast majority (more than 99%) of dolphins captured by the ETP tuna fishery are released alive; thus, an individual dolphin may be chased, captured, and released many times during its lifetime.

The number of dolphins killed since the fishery began some four decades ago is estimated to be over 6 million, the highest known for any fishery. For comparison, the total number of whales of all species killed during commercial whaling in the 20th century is about 2 million. In recent years, the killing of dolphins in the ETP tuna fishery has declined by two orders of magnitude, but even at this level it remains one of the largest documented cetacean kills in the world.

II. Purse Seining for Tuna

Prior to the development of modern purse seines, tropical tuna were caught one at a time using pole-and-line methods. In the late 1950s the twin technological developments of synthetic netting that would not rot in tropical water and a hydraulically driven power block to haul the net made it possible to deploy very large purse seine nets around entire schools of tuna, and thus to catch many tons of fish at a time. Purse seining for tuna in the ETP can be conducted in one of three ways: the net may be set around schools of tuna associated with dolphins ("dolphin sets," which catch large yellowfin tuna), around schools of tuna associated with logs or other floating objects ("log sets," which catch mainly skipjack but also bigeye and small yellowfin tuna), or around unassociated schools of tuna ("school sets," which catch small yellowfin and skipjack tuna). The proportion of each set type has varied considerably, but during the past decade, dolphin and school sets have been roughly equal in number (about 45% of the 15,000–20,000 sets each year), and the remainder (about 10%) have been log sets.

Dolphins are killed almost exclusively in dolphin sets. During "porpoise fishing" (the fishermen's term), schools of tuna are located by first spotting the dolphins or the seabird flocks that are also associated with the fish. Speedboats are used to chase down the dolphins, herd them into a tight group, and set the net around them (Fig. 1). The tuna–dolphin bond is so strong that the tuna stay with the dolphins during this process, and thus tuna and dolphins are captured together in the net (Fig. 2). Dolphins are released from the net during the backdown procedure (Fig. 3). If all goes well, the dolphins are released alive, but the process requires skill by the captain and crew, proper operation of gear, and conducive wind and sea conditions. As with any complicated procedure at sea, things can go wrong, and when they do, dolphins may be killed.

The bycatch in log and school sets is larger than in dolphin sets, but consists primarily of other fish, not dolphins. While the effects of the fishery on dolphin populations have been strong and are relatively well known, the effects on other marine populations of concern, such as sharks and sea turtles, are unknown.

Figure 1 *Purse seine being set on tuna and dolphins in the eastern tropical Pacific Ocean. The net is not yet closed, and four speedboats are driving in tight circles near the opening to keep the dolphins from escaping.*

III. Actions to Reduce Dolphin Bycatch

The magnitude of dolphin mortality in the ETP tuna fishery first came to widespread attention in the mid-1960s. The dolphin kill at that time is not known with precision, but without question was very high (Fig. 4). When the U.S. Marine Mammal Protection Act was passed in 1972, it included provisions for reducing the bycatch to "insignificant levels approaching zero" after a moratorium on regulation during which the tuna industry was expected to solve the problem through the development of improved fishing methods. Under this law, scientific studies were initiated, observers were placed on fishing boats, fishing gear was inspected, and boat captains with high dolphin mortality rates were reviewed. Modifications of fishing gear and procedures were developed to reduce dolphin kill. After much litigation, the first regulations to reduce the dolphin kill on U.S. vessels were promulgated. By the end of the 1970s, the kill had declined from about 500,000 to about 20,000 dolphins per year (Fig. 4).

As the U.S. tuna fleet decreased in size and the fleets of Mexico, Venezuela, Ecuador, and other Latin American countries increased, the dolphin kill began to grow again, and actions to monitor and reduce the dolphin bycatch became international. The Inter-American Tropical Tuna Commission began a dolphin conservation program in 1979 modeled on the U.S. effort. By 1986, an international observer program with all countries participating showed that total dolphin mortality had increased to 133,000/year (Fig. 4). Because U.S. boats operated under restrictions that did not apply to boats of other countries, the United States began requiring that imported tuna be caught at dolphin mortality rates comparable to U.S. boats. The concept of dolphin-safe tuna—tuna caught without

Figure 2 *Spotted dolphins in a purse seine net waiting to be released. Photo by W. High, courtesy of National Marine Fisheries Service.*

setting on dolphins (i.e., log and school sets)—became popular, and by 1994 only dolphin-safe tuna could be sold in the United States. These trade actions were important because the United States is the largest market of the canned tuna product of the fishery.

The dolphin kill declined between 1986 and 1993 due to these various political and economic pressures (Fig. 4). Starting in 1993, the ETP fishing countries voluntarily agreed to increased observer coverage, skipper review panels, and a sched-ule of decreasing dolphin quotas on an individual boat basis (the La Jolla agreement). The Declaration of Panama of 1995 carried these ideas further, proposing observers on every boat over 400 tons and strict per-stock dolphin mortality limits. These features became part of the International Dolphin Conservation Program Agreement, a binding document among the major fishing countries that went into force in 1999. By this time total reported dolphin mortality had fallen to fewer than 3000 dolphins/year (Fig. 4).

Figure 3 *Backdown procedure in progress. As the tuna vessel moves backward, the net is drawn into a long channel. The corkline at the far end is pulled underwater slightly, and the dolphins escape. Speedboats are positioned along the corkline to help keep the net open.*

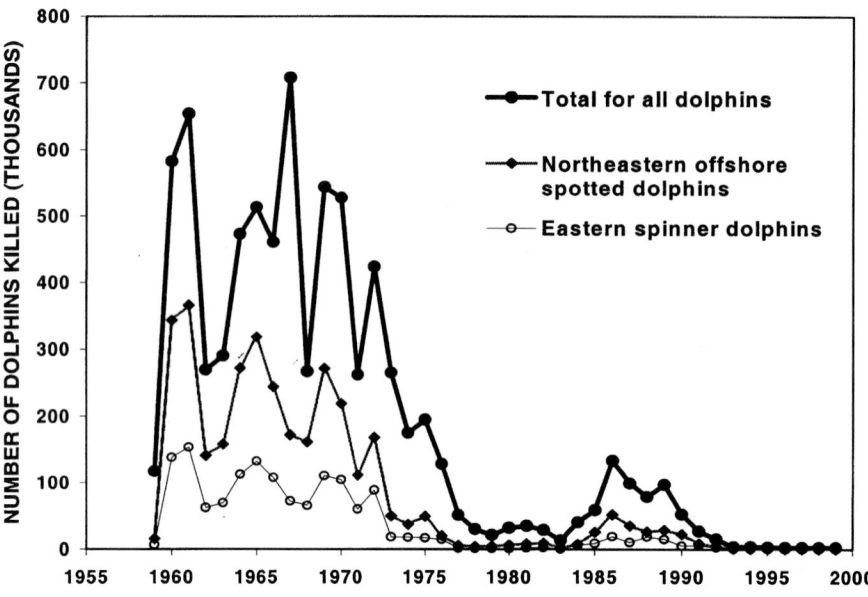

Figure 4 *Estimated annual number of dolphins killed (all dolphins and two dolphin stocks with the highest number killed) in the eastern tropical Pacific purse seine tuna fishery.*

The Declaration of Panama also called for the United States to change its definition of dolphin-safe tuna to include tuna caught by setting on dolphins as long as no dolphins were observed killed or seriously injured *on that set*. Before changing the dolphin-safe label, however, the United States undertook studies to determine if the process of chasing and encircling dolphins was having a significant adverse impact on depleted dolphin populations. At the time of this writing these studies were underway.

IV. Status of Dolphin Populations

The status of ETP dolphin stocks (management units) is based on three time series of data: estimates of the number of dolphins killed, estimates of ABUNDANCE from dedicated research vessel surveys, and estimates of an index of abundance from sightings on tuna vessels. Combining these data in a population model has indicated that the stocks most affected by the tuna fishery are the northeastern stock of the offshore pantropical spotted dolphin (*S. attenuata*) and the ETP

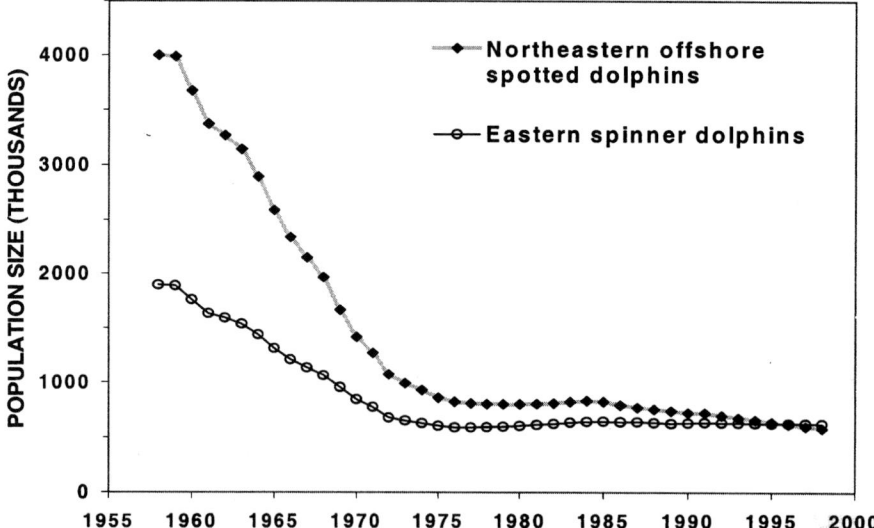

Figure 5 *Estimated population trajectories of northeastern offshore spotted dolphins and eastern spinner dolphins in the eastern tropical Pacific Ocean. The populations declined due to the high number of dolphins killed in the tuna fishery from 1960 to 1975.*

endemic subspecies, the eastern spinner dolphin (*S. longirostris orientalis*). Both populations declined between 1960 and 1975 (Fig. 5), during the period of high mortality on U.S. boats (Fig. 4), but have remained approximately constant since then. As of 1998, northeastern spotted and eastern spinner dolphins were estimated to be at 15 and 34%, respectively, of population sizes when the fishery began. Other stocks have apparently been less affected, although little is known of the small populations of coastal forms of spotted and spinner dolphins.

Since 1993, reported dolphin mortality has been a very small fraction of population size so that recovery of the dolphin populations was expected. By 1999, however, there was no clear indication of a recovery for either northeastern offshore spotted or eastern spinner dolphins (Fig. 5). Several hypotheses could explain these apparent failures to recover: cryptic effects of repeated chase and encirclement on survival and/or reproduction (internal injuries, stress, hyperthermia), separation of nursing calves from their mothers during the fishing process, unobserved or observed but unreported mortality, effects due to breakup of dolphin schools (increased predation, social disruption), ecological effects due to removing tuna from the tuna–dolphin association, and ecosystem or environmental changes. Until there are clear recoveries of the affected dolphin stocks, the tuna–dolphin issue is likely to remain highly controversial.

See Also the Following Articles

Entrapment and Entanglement ■ Fishing Industry, Effects of ■ Incidental Catches ■ Population Status and Trends

References

Gosliner, M. L. (1999). The tuna-dolphin controversy. *In* "Conservation and Management of Marine Mammals" (J. R. Twiss, Jr., R. R. Reeves, eds.), pp. 120–155. Smithsonian Institution Press, Washington, DC.

National Research Council (1992). "Dolphins and the Tuna Industry." National Academy Press, Washington, DC.

Perrin, W. F. (1969). Using porpoise to catch tuna. *World Fish.* **18**, 42–45.

U

Ursidae

RONALD E. HEINRICH
Ohio University, Athens

The eight species of living bears are characterized by having massive skulls, large (25–780 kg) and heavily built bodies, relatively short and powerful limbs, short tails, a shuffling plantigrade gait, and an omnivorous diet aided by a keen sense of smell and large crushing molar teeth. This uniformity of body plan has a relatively recent evolutionary history, however, with the two earlier ursid radiations, the Amphicynodontinae and Hemicyoninae, exhibiting a greater array of dental and postcranial adaptations than the living bears. Throughout their approximately 37 million year history, ursids have maintained a Holarctic distribution, migrating into Africa on at least four separate occasions and into South America at least once, and were the dominant carnivorans of Eurasia during much of the Oligocene and Miocene. Morphology-based analyses indicate that Ursidae is the sister taxon to Pinnipedia, a phylogenetic hypothesis that has also received some support from molecular studies.

I. Ursid Fossil Record

The earliest recognized ursids, members of the subfamily Amphicynodontinae, first appeared in the late Eocene and early Oligocene fossil records of North America and Eurasia. Before going extinct at the end of the early Miocene, this paraphyletic assemblage of CARNIVORANS gave rise to both the hemicyoninine and ursine ursids and possibly to the basal members of Pinnipedia. Amphicynodonts were generally small to medium sized (< 15 kg), plantigrade or possibly semidigitigrade, and tended to retain a primitive dentition possessing a well-developed pair of carnassial teeth for shearing, triangular rather than quadrate upper molars, and relatively small posterior molar teeth (Fig. 1A). Apomorphic characters shared by these earliest ursids with all later members of the family, however, include a posteriorly wide basioccipital bone, a large broad

protocone on the upper carnassial, initial reduction of the parastyle and paraconule on the upper molars (both lost in later ursids), and the presence of a posterior shelf or heel on the lower second molar (Tedford *et al.*, 1994). A more derived amphicynodont was the early Miocene, sea otter-like *Kolponomos*. Collected from coastal deposits along Oregon and Washington. This genus has been reconstructed as semiaquatic with sensitive lips and tactile vibrissae, powerful neck muscles, and broad cheek teeth used to locate, pry off, and then crush, respectively, shelled marine invertebrates found attached to rocky substrates (Tedford *et al.*, 1994). These workers have also argued that among known ursids, *Kolponomos* occupies the position of sister taxon to Pinnipedia.

The second ursid radiation was initiated in the late Oligocene of Eurasia. The Hemicyoninae were the first ursids to attain large body size (upwards of 200 kg) and the only ursids to evolve digitigrady, an adaptation for cursoriality that supports the interpretation of these animals as active PREDATORS comparable in many respects to canids (Hunt, 1998). The dentition of hemicyonines is derived relative to amphicynodonts in that the upper carnassial has a retracted protocone and a shortened metastylar blade, and the upper molars are quadrate due to enlargement and migration of the metaconule (Fig. 1B). Hemicyonines migrated into North America on several occasions and into Africa at least once, but by the latest Miocene they were extinct throughout their geographic range.

The subfamily Ursinae, although primarily a late Miocene and Pliocene radiation, first appeared in the late Oligocene?–early Miocene fossil record of North America. The earliest member of the group, *Ursavus*, was a small to mid-sized Holarctic and plantigrade form that exhibited some of the dental specializations characteristic of later bears, including further reduction (compared to hemicyonines) of the upper carnassial and the development of low-crowned, elongate molars with wrinkled enamel (Hunt, 1998). In modern bears such as *Ursus*, these traits are taken to an extreme with broadening and lengthening of the molar teeth (Fig. 1C), providing an immense grinding surface for processing vegetation. Derived from a species of Eurasian *Ursavus*, *Ursus* first appeared about 5 million years ago (mya) migrating into North America about 4.3 mya and into North Africa about 1 mya. The only bears to have

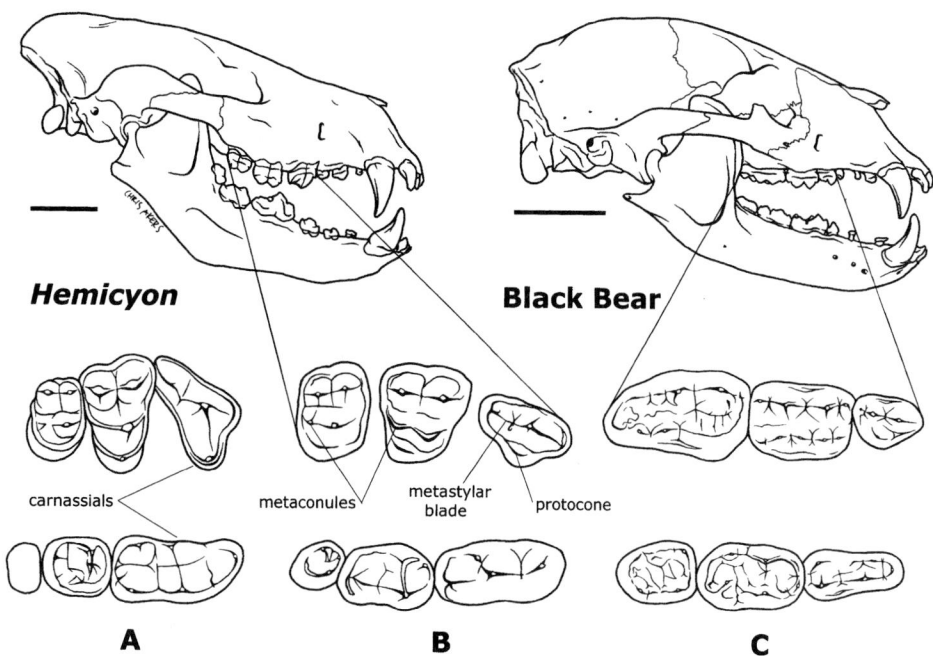

Figure 1 *Comparison of skulls and dentitions of representative members of the three ursid radiations. Lateral views of the skulls of the* Hemicyon *(Miocene) and the living black bear (*Ursus americanus*) and occlusal views of P4-M1 (above) and m1-m3 (below) for* Amphicynodon *(A, Oligocene),* Hemicyon *(B), and* Ursus americanus *(C). Note the relative reduction in size of the carnassial teeth and the increased surface area of the posterior molars in living bears. Figures of* Amphicynodon *and* Hemicyon *are from Cirot and Bonis (1992) and Colbert (1939), respectively. Scale bar: 5 cm.*

reached South America belong to the Ursinae and include *Arctodus*, which arrived less than 2 mya, and the living spectacled bear (*Tremarctos*), which has no South American FOSSIL RECORD.

II. Phylogenetic Relationships of Living Ursids

The eight species of living ursid are currently allocated to as few as three (*Ailuropoda, Tremarctos,* and *Ursus*) and as many as five (additionally *Helarctos* and *Melursus*) genera, all of which are monospecific except *Ursus*. *Tremarctos ornatus* (spectacled bear) and *A. melanoleucus* (giant panda) are successive outgroups to the remaining six species (Fig. 2), and although there has been some contention as to whether the giant panda should be included within the family at all, it is now widely accepted that *Ailuropoda* diverged from other members of Ursinae during the early Miocene. The most recent phylogenetic analysis incorporating both molecular and morphologic data sets (Bininda-Emonds *et al.*, 1999) suggests that the Malayan sun bear (*H. malayanus*) and sloth bear (*M. ursinus*) are sister taxa, that the Asiatic black bear (*U. thibetanus*) is the sister taxon to the clade brown bear (*U. arctos*) + polar bear (*U. maritimus*), and that the position of the American black bear (*U. americanus*) is unresolved with respect to these other two clades (Fig. 2).

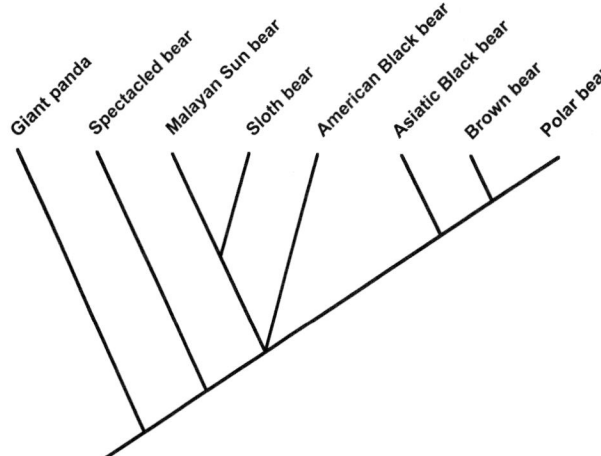

Figure 2 *A cladogram depicting phylogenetic relationships among the eight species of living ursid: the giant panda (*Ailuropoda melanoleucus*), spectacled bear (*Tremarctos ornatus*), Malayan sun bear (*Ursus malayanus*), sloth bear (*U. ursinus*), American black bear (*U. americanus*), Asiatic black bear (*U. thibetanus*), brown bear (*U. arctos*), and polar bear (*U. maritimus*). From Bininda-Emonds et al. (1999).*

The only marine-adapted ursine, the polar bear, is thought to have diverged from the brown bear between 100,000 and 250,000 years ago probably as a result of glacial advances that isolated northern and southern populations. Ancestors of the modern polar bear became adapted for life on the sea ice, evolving a denser, hollow fur for protection against the Arctic cold and morphologic characteristics such as sharper claws and carnassial teeth for grasping and shearing through the skin of captured seals. Despite these and other ecological and morphological differences between the two species, polar bears and brown bears are still capable of producing fertile offspring and some biologists have suggested that the Barren Ground grizzly may be such a HYBRID.

See Also the Following Articles

Carnivora ▪ Dental Morphology, Evolution of ▪ Mustelidae ▪ Pinnipedia ▪ Polar Bear

References

Bininda-Emonds, O. R. P., Gittleman, J. L., and Purvis, A. (1999). Building large trees by combining phylogenetic information: A complete phylogeny of the extant Carnivora (Mammalia). *Biol. Rev.* **74,** 173–175.

Cirot, E., and Bonis L. de (1992). Revision du genre *Amphicynodon,* carnivore de l'Oligocene. *Palaeontograhica* **220,** 103–130.

Colbert, E. H. (1939). Carnivora of the Tung Gur Formation of Mongolia. *Bull. Am. Mus. Nat. Hist.* **76,** 47–81.

Hunt, R. M., Jr. (1998). Ursidae. *In* "Evolution of Tertiary Mammals of North America" (C. M. Janis, K. M. Scott, and L. L. Jacobs, eds.), Vol. 1, pp. 152–173. Cambridge Univ. Press, Cambridge.

Tedford, R. H., Barnes, L. G., and Ray, C. E. (1994). The early Miocene littoral ursoid carnivoran *Kolponomos:* Systematics and mode of life. *In* "Contributions in Marine Mammal Paleontology Honoring Frank C. Whitmore, Jr." (A. Berta and T. A. Demere, eds.), No. 29, pp. 11–32. Proceedings of the San Diego Society of Natural History.

Vaquita
Phocoena sinus

LORENZO ROJAS-BRACHO AND
ARMANDO JARAMILLO-LEGORRETA
*Instituto Nacional de Ecología,
Ensenada, Baja California, Mexico*

The vaquita (*Phocoena sinus* Norris and McFarland, 1958), or Gulf of California harbor porpoise, is endemic to the Gulf of California. It is also one of the two most endangered cetacean species in the world (Jefferson *et al.*, 1993). The vaquita is a relatively new species. It was described in 1958 (Norris and McFarland, 1958). However, it was not until the past decades that important advances were made in our knowledge of this highly endangered porpoise.

I. Description, Characters and Taxonomic Relations

The vaquita is the smallest of all the porpoises. The mean length for females is 140.6 cm, whereas males are slightly smaller (134.9 cm). The vaquita differs from other phocoenids not only in total length, but also in that the flippers of the vaquita are proportionally larger and the dorsal fin is taller and more falcate. Generally, the pigmentation pattern is a dark gray cape, pale gray lateral field, and white ventral field. The most conspicuous features of the pigmentation are the relatively large black eye rings and lip patches (Fig. 1). Skulls of adult vaquitas are smaller and have relatively much broader and shorter rostra than those of other members of the genus. The number of tooth pairs in the upper and lower jaw is 16–22 and 17–20, respectively (Brownell *et al.*, 1987; Vidal *et al.*, 1995). Norris and McFarland (1958) suggested that the vaquita diverged from *P. spinipinnis* rather than from *Phocoena phocoena*, its closest geographic phocoenid neighbor. This hypothesis was corroborated almost 37 years later by Rosel *et al.* (1995) using molecular techniques.

II. Abundance, Distribution, and Ecology

The limited ABUNDANCE, distribution, and narrow habitat specificity of vaquita make it one of the rarest marine mammal species. The most recent published population size estimate by Jaramillo-Legorreta *et al.* (1999) was 567 individuals (CV = 0.51, 95% CI 177–1073). All sightings from systematic ship surveys indicate that its DISTRIBUTION is limited to an area north of 30°45'N (Silber *et al.*, 1994; Gerrodette *et al.*, 1995; Vidal, 1995; Jaramillo-Legorreta *et al.*, 1999) specifically to a portion of the upper Gulf of California (see Fig. 2). This is probably the most restricted distribution of any marine cetacean. Vaquita habitat can be described as a region characterized by strong tidal mixing, convection processes, high turbidity, high nutrient concentration, and primary and secondary productivity. In this environment, most vaquita sightings have been in waters 11–50 m deep, 11 to 25 km from the coast, and over silt and clay bottoms (Silber *et al.*, 1994; Gallo and Torre, 1998). It is known that vaquitas move to the eastern coast of the upper Gulf, mainly from recovered specimens (beached and bycaught) and not so much from sightings (Vidal *et al.*, 1999).

Analysis of stomach contents of 34 vaquitas reveals them to be nonselective feeders (Findley *et al.*, 1995; Pérez-Cortés, 1996). Prey consisted primarily of a wide variety of demersal/benthic fishes (21 species), squid (2 species), and crustaceans (1 species, plus 2 that were fish parasites). Several of the fish prey species (such as croakers, Scianidae) are known to be sound producers so it is possible that vaquitas are frequently using passive sound rather than echolocation to find their prey.

III. Life History and Behavior

The only published paper on the life history of the vaquita is that by Hohn *et al.* (1996). Arguably the most striking aspect of this work is the unusual age distribution of the vaquita. Most individuals (62%) were found to be between 0 and 2 years of age. The remainder of individuals (31%) were between 11 and 16 years of age, with few specimens determined to be between 3 and 10 years. There was a complete absence of individuals between the ages of 3 and 6 years.

The sample of animals may be biased due to spatial segregation of different age classes or the susceptibility of various ages to be captured in nets. Alternatively, if this were the true

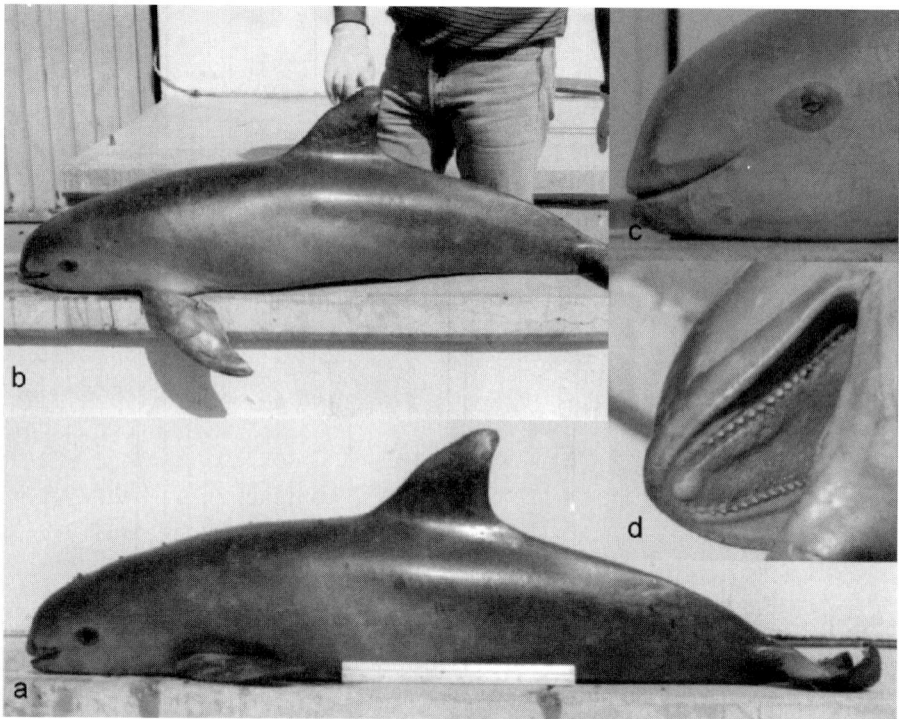

Figure 1   *Morphology of the vaquita. (a and b) Lateral views. Note proportionally large dorsal and pectoral fins. (c) Lateral view of the head showing the dark patches surrounding the eye and lips. (d) Ventral view of the mouth showing the palate and the spade-shaped crown of teeth (courtesy of Armando Jaramillo-Legorreta).*

age distribution of the population, this would represent a complete recruitment failure in recent years.

Like other porpoises, the vaquita is a seasonal reproducer, with most births occurring around March. There are no data, but gestation is probably 10–11 months, as with other porpoises. The maximum observed life span is 21 years, which is consistent with the maximum life span of the harbor porpoise, *Phocoena phocoena*. Age of sexual maturity is difficult to estimate because of the lack of juvenile animals in the sample, but all females less than 3 years were immature and all females older than 6 years were mature. Most female harbor porpoises mature at age 3 and give birth first at age 4, which is consistent with these observations for the vaquita. Reproductive rates could not be estimated accurately, but unlike harbor porpoises in the Gulf of Maine, vaquitas are probably not annual reproducers.

No pathological study of the recovered specimens has been published. However, morphological abnormalities in the vertebrae, unusual number of digits (six), and pathological condition of the ovaries (calcification of corpora albicantia) have been reported (Ortega-Ortiz *et al.*, 1993; Torre-Cosío, 1995; Hohn *et al.*, 1996).

Little is known of the social organization of the vaquita. Mean school size is two, but groups as large as 8 or 10 individuals have been reported. An important aspect is the loose aggregating behavior. The dynamics of these group aggregations are not clear. However, it seems that either their duration is short or they shift locations in short periods of time. For example, Jaramillo-

Legorreta *et al.* (1999) reported 41 vaquita groups loosely aggregated over several hundred square meters. During the next days and couple of weeks later, this group aggregation was not found. There are also indications that suggest that shrimp trawlers are related to this aggregating behavior (Jaramillo-Legorreta *et al.*, 1999).

The vaquita, like other phocoenids, has the capacity to produce narrow band high-frequency clicks (Silber, 1991). The primary use of this sound must be ECHOLOCATION, although its role in COMMUNICATION could be important in the turbid and dynamic environment of the Upper Gulf of California. Evidence indicates that the clicking rate is lower for the vaquita than for the harbor porpoise. The shallow waters inhabited by the vaquita can be a reason for this, as vaquitas can be looking for prey visually at the bottom for longer periods. Further acoustical studies can reveal areas with differential ecological uses by the vaquita.

IV. Conservation Status

The vaquita is classified in the most critical conservation categories by the International Union for the Conservation of Nature (IUCN, 1996), Convention on International Trade in the Endangered Species of Wild Fauna and Flora (1997), and the Mexican government (Norma Oficial Mexicana, DOF 16 de Mayo, 1994). In 1996, the IUCN concluded that the extinction of the vaquita is likely unless CONSERVATION EFFORTS are increased substantially.

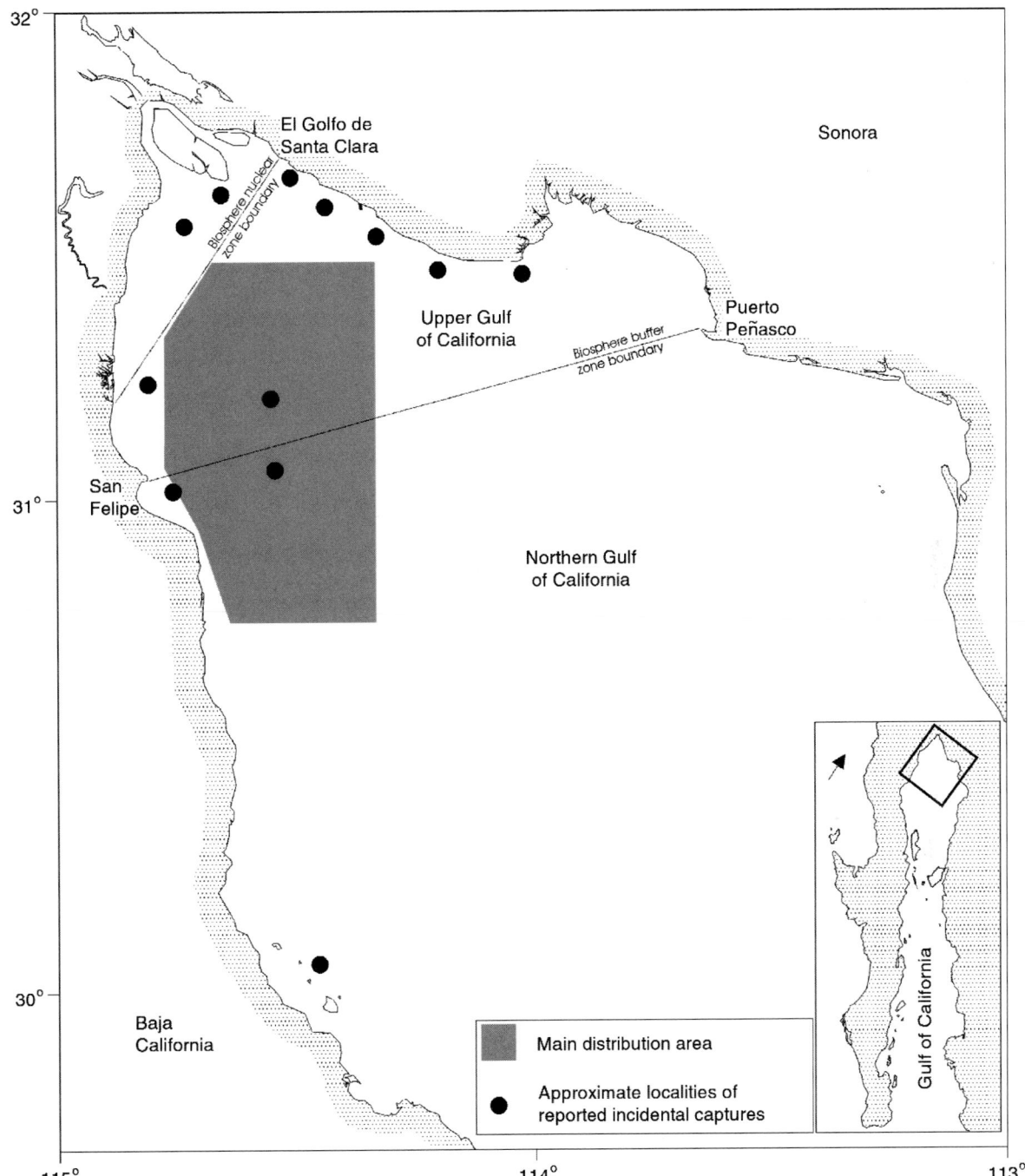

Figure 2 *The Upper Gulf of California. The core area of distribution (gray) was sketched from sightings data. Solid circles represent localities where bycaught vaquitas have been reported.*

To protect the vaquita and other endangered species, the Mexican government created the Upper Gulf of California and Colorado River Delta Biosphere Reserve on June 10, 1993. However, as new knowledge of the vaquita has been gained, it is clear that this measure is insufficient. Results of the surveys in 1993 and 1997 indicate that the boundaries of this biosphere reserve do not correspond well to the distribution of vaquitas. A large percentage of the sightings (40%) lie outside the reserve boundary (Fig. 2). Further, no sightings were within the nuclear zone of the reserve, which is the area where all fishing is prohibited (Gerrodette *et al.*, 1995; Jaramillo-Legorreta *et al.*, 1999).

A more specific action has been the creation of the International Committee for the Recovery of the Vaquita (CIRVA) by the Mexican government. Recognized scientists from the

United Kingdom, Canada, the United States, and Mexico participate in CIRVA. The mandate of this group is to propose a recovery plan based on the best available scientific information. The plan should also contemplate and consider the socioeconomic impacts of any required regulations on the resource users in the affected areas.

During its first meeting, CIRVA concluded that in the short term, gill nets are the greatest risk to the survival of the vaquita. Estimated incidental mortality in gill nets is 39–84 vaquitas per year (D'Agrosa *et al.*, 2000). This represents from 6 to 14% of the current population size estimate and is from only one fishing port. The committee agreed that inbreeding depression, chlorinated pesticide concentrations in the Upper Gulf, and reduced flow from the only freshwater input (the Colorado River) are not risk factors at present (Taylor and Rojas-Bracho, 1999; Rojas-Bracho and Taylor, 1999). It also agreed, however, that, in the long term, changes in vaquita habitat due to reduction of the Colorado River flow are matters of concern and must be investigated.

During the second meeting, CIRVA evaluated the potential mitigation measures: seasonal closures of specific areas, gear restriction, acoustic deterrents, and marine-protected areas.

After analyzing these measures, CIRVA strongly recommended that vaquita bycatch should be reduced to zero as soon as possible, the southern boundary of the biosphere reserve should be expanded to include the entire range of the vaquita, and gill net and trawlers should be banned in the enlarged biosphere reserve. CIRVA recognized that these protective measures would have significant impacts on the resource users of the Upper Gulf of California and, therefore, it was not possible to implement full protection immediately. Therefore, it was recommended that gill net fishing in the areas inhabited by vaquitas be removed in three stages, starting with large-mesh gill nets. Considering this, CIRVA strongly recommended investigating the development of strategies to offset economic hardship imposed by these regulations.

Other recommendations were the effective enforcement of fishing regulations; that acoustic surveys be started immediately to begin monitoring an index of abundance and gather data on seasonal movements of vaquitas; development and testing of alternate gear types to replace gill nets; the design and development of community involvement, education, and public awareness programs; and a description of the critical habitat of vaquita.

Finally, CIRVA recommended inviting the international community and nongovernmental organizations to provide technical and financial assistance to implement the conservation measures to recover this highly endangered porpoise.

See Also the Following Articles

Endangered Species and Populations ■ Habitat Pressures ■ Incidental Catches ■ Porpoises, Overview ■ Surveys

References

Brownell, R. L., Jr., Findley, L. T., Vidal, O., Robles, A., and Manzanilla, S. (1987). External morphology and pigmentation of the vaquita, *Phocoena sinus* (Cetacea: Mammalia). *Mar. Mamm. Sci.* **3**, 22–30.

D'Agrosa, C., Lennert, C. E., and Vidal, O. (2000). Preventing the extinction of a small population: Vaquita (*Phocoena sinus*) fishery mortality and mitigation strategies. *Conserv. Biol.* **14**, 1110–1119.

Findley, L. T., Nava, J. M., and Torre, J. (1995). Food habits of *Phocoena sinus* (Cetacea: Phocoenidae). *In* "Abstracts Eleventh Biennial Conference on the Biology of Marine Mammals," 14–18 December, Orlando, Florida, EUA.

Gallo Reynoso, J. P., and Torre, J. (1998). The critical habitat of the vaquita (*Phocoena sinus*) in the upper Gulf of California. *In* "Abstracts of the World Marine Mammal Science Conference." January 20–24, Monaco.

Gerrodette, T., Fleischer, L. A., Pérez-Cortés, H., and Villa-Ramírez, B. (1995). Distribution of the vaquita, *Phocoena sinus*, based on sightings from systematic surveys. *In* "Biology of the Phocoenids" (A. Bjørge and G. P. Donovan, eds.), Special Issue 16, pp. 273–281. Reports of the International Whaling Commission, Cambridge, UK.

Jaramillo-Legorreta, A. M., Rojas-Bracho, L., and Gerrodette, T. (1999). A new abundance estimate for vaquitas: First step for recovery. *Mar. Mamm. Sci.* **15**, 957–973.

Jefferson, T. A., Leatherwood, S., and Webber, M. A. (1993). "FAO Species Identification Guide. Marine Mammals of the World." FAO, Rome.

Hohn, A. A., Read, A. J., Fernandez, S., Vidal, O., and Findley, L. T. (1996). Life history of the vaquita, *Phocoena sinus* (Phocoenidae, Cetacea). *J. Zool. Lond.* **239**, 235–251.

Norris, K. S., and McFarland, L. T. (1958). A new harbor porpoise of the genus *Phocoena* from the Gulf of California. *J. Mammal.* **39**, 291–340.

Pérez-Cortés Moreno, H. (1996). Contribución al conocimiento de la biología de la vaquita, *Phocoena sinus*. Tesis de Maestría. Instituto de Ciencias del Mar y Limnología. Universidad Nacional Autónoma de México, México, D.F.

Rojas Bracho, L., and Taylor, B. (1999). Risk factors affecting the vaquita (*Phocoena sinus*). *Mar. Mamm. Sci.* **15**, 974–989.

Rosel, P. E., Haygood, M. G., and Perrin, W. F. (1995). Phylogenetic relationship among the true porpoises (Cetacea: Phocoenidae). *Mol. Phylogenet. Evol.* **4**, 463–474.

Silber, G. K. (1991). Acoustic signals of the vaquita (*Phocoena sinus*). *Aquat. Mamm.* **17**, 130–133.

Silber, G. K., Newcomer, M. W., Silber, P. C., Pérez-Cortés, H., and Ellis, G. M. (1994). Cetaceans of the northern Gulf of California: Distribution, occurrence, and relative abundance. *Mar. Mamm. Sci.* **10**, 283–298.

Taylor, B., and Rojas-Bracho, L. (1999). Examining the risk of inbreeding depression in a naturally rare cetacean, the vaquita (*Phocoena sinus*). *Mar. Mamm. Sci.* **15**, 1004–1028.

Vidal, O. (1995). Population biology and exploitation of the vaquita *Phocoena sinus*. *In* "Biology of the Phocoenids" (A. Bjørge and G. P. Donovan, eds.), Special Issue 16, pp. 247–272. Reports of the International Whaling Commission, Cambridge, UK.

Vidal, O., Brownell, R. L., Jr., and Findley, L. T. (1999). Vaquita. *Phocoena sinus*. Norris and McFarland, 1958. *In* "Handbook of Marine Mammals" (S. H. Ridgeway and R. Harrison, eds.), Vol. 6, pp. 357–378. Academic Press, San Digeo.

Vision

ALLA M. MASS AND ALEXANDER YA. SUPIN
Russian Academy of Sciences, Moscow

The vision of marine mammals has a number of specific features associated with its ability to function in both water and air. Although many marine mammals (cetaceans, sirenians) spend their entire life in water, their aerial breathing confines them to a near-surface layer of water. Other marine mammals (pinnipeds, sea otters) spend a significant part of their life on land. As a result, the organization of their visual system fits requirements of both these different media. Although some aspects of organization of the visual system of marine mammals still remain unstudied, many features of their vision are known already.

I. Visual Abilities of Marine Mammals

A. Cetaceans

It was long believed that dolphins—animals with excellent hearing and echolocation—have a poorly developed visual system playing a minor role in their life. However, observations of the visual activity of dolphins have demonstrated the opposite. The ability to catch fish in air, to perform precisely aimed jumps to reach targets above the water, and to recognize their trainers all show that vision in dolphins is well developed. In conditions of captivity, dolphins decrease their use of echolocation and, as their interest in events above the water increases, vision takes on a leading role.

Reviews of Madsen and Herman (1980) and Mobley and Helweg (1990) summarize observations of dolphins in CAPTIVITY and experimental studies that provide a basis for regarding the vision of dolphins as playing an important role in various aspects of their life: in social interactions, in discrimination between individuals and species based on their colors and individual marks, in the search and discrimination of prey, in orientation, in reproductive activity, and in defense. Only vision provides the ability for rapid and precise assessment of distances to objects in air where echolocation does not operate.

Apart from numerous observations, good visual abilities of cetaceans were demonstrated in behavioral experiments for assessing their visual acuity. Precise behavioral measurements performed by Herman and colleagues (see Madsen and Herman, 1980) on the bottlenose dolphin (*Tursiops truncatus*) resulted in an estimate of underwater visual acuity of 8.2 arc min (at the best distance of 1 m) and aerial visual acuity of 12.5 arc min (at distances of 2.5 m and longer). In general, estimates of visual acuity in dolphins varied from 8 to 27 arc min in water and from 12 to 18 arc min in air.

Studies of color vision in cetaceans are very few in number. Only one cone type was found in the bottlenose dolphin, with the best sensitivity at 525 nm; rods are best sensitive to 488 nm. These sensitivity peaks are considerably blue-shifted as compared to those of many terrestrial mammals (Jacobs, 1993). Therefore, the dolphin lacks the common dichromatic vision typical of many terrestrial mammals, which is based on two cone types with different chromatic sensitivity. If color vision is present in dolphins (basing on comparison of signals from rods and cones), it is poorly developed and limited to a blue–green region of the spectrum.

In all cetaceans, the eyes are positioned laterally, thus providing a visual field as wide as 120–130° and panoramic vision. Although positioned laterally, the eyes are directed somewhat forward and downward (ventronasally). On viewing visual objects in air, the dolphin eyes can move forward by 10–15 mm so that the visual fields of the two eyes overlap by 20–30° in the frontal sector, giving a basis for binocular vision. However, un-

crossed optic fibers have not yet been demonstrated in dolphins. Therefore, the existence of true binocular (stereoscopis) vision (based on interaction of crossed and uncrossed optic fibers) in dolphins still remains under question.

Dolphins are equally capable of the perception of complex configurations of objects using both vision and echolocation. Besides, there is also a possibility of intermodal transfer between these two modalities: objects known for a dolphin only by visual appearance can be discriminated and recognized by echolocation, and vice versa. The intermodal transfer is equally successful when visual experience is used for echolocation discrimination and when echolocation experience is used for visual discrimination.

Even in riverine cetaceans inhabiting turbid and low-transparent water (the Amazon river dolphin, *Inia geoffrensis*, the tucuxi, *Sotalia fluviatilis*), the visual system does not exhibit a significant regression. The only exception is the Indian river dolphin, *Platanista gangetica*, in which the visual system is reduced markedly.

B. Pinnipeds

Because pinnipeds spend their life partially in water and partially on land, they use both underwater and aerial vision. On land, vision plays an important role during the reproductive period, during birth and feeding of pups, and for maintaining intrapopulation relationships, as well as for orientation. In water, vision is used for prey detection and recognition, for avoiding predators, and for spatial orientation during MIGRATIONS.

Because of a great diversity of pinniped species in terms of systematic position and ecology, the role of vision diverges widely as well. Walruses (*Odobenus rosmarus*) rely mainly on their vibrissal sensitivity to identify objects during benthic foraging. Other pinnipeds also have a well-developed vibrissal apparatus; however, in aquatic conditions, most seals use both visual and tactile modalities to search for food. Experiments demonstrated that seals are able to distinguish rather small objects visually, to recognize the shape of figures, and to perform a complex analysis of visual images. Data summarized by Fobes and Smock (1981) showed that both otariids and phocids are capable of discriminating objects differing in size from 9 to 24%.

Most pinnipeds (both otariids and phocids) have maximum spectral sensitivity within a range of 496–500 nm. An exception is the southern elephant seal (*Mirounga leonina*), which is sensitive to a shorter wavelength (486 nm).

A possibility of limited color discrimination in a few pinniped species (*Pagophilus groenlandicus, Phoca larga, Arctocephalus pusillus, A. australis*) is indicated by their capability to discriminate blue and green objects from gray ones, although they cannot discriminate red and gray objects. The best rod sensitivity in the harbor seal (*Phoca vitulina*) was found at 496 nm and cone sensitivity at 510 nm; i.e., similar to dolphins, the spectral sensitivity is blue-shifted as compared to terrestrial mammals. No indication was obtained of more than one cone type in pinnipeds.

Measurements of visual acuity based on the use of grids as test stimuli have demonstrated that visual acuity in both water and air is 5–8 arc min in a few otariid species: *Zalophus californianus, Eumetopias jubatus, Arctocephalus pusillus*, and *A. australis*.

C. Other Marine Mammals

1. Sirenians There has been no detailed investigation of visual capabilities in sirenians. A few observations summarized by Piggins *et al.* (1983) showed that the Amazon manatee (*Trichechus inunguis*) is capable of visually driven behavior, particularly visual tracking of underwater objects. A capability of the Caribbean manatee, *Trichechus manatus,* for dichromatic (blue–green) color vision has been shown. It remains unknown whether the manatee has an ability of good aerial vision.

2. Sea Otters Very little is known of the visual abilities of sea otters (*Enhydra lutris*). Inhabiting the coastal zone and feeding underwater, sea otters need to have good vision in both air and water. Observations showed that they actively use vision, and experiments have shown their capability to discriminate objects of different sizes. However, quantitative behavioral measurements of their visual abilities are absent.

II. Eye Anatomy and Optics

A. Cetaceans

Ocular anatomy in cetaceans is markedly different from that in terrestrial mammals by being adjusted to optical properties of water and to a number of other factors: possibility of eye damage because of high density of water and presence of sus-

pended particles, low temperature and low illumination deep in water, significant light scatter, and so on. Characteristic examples of eye structure in cetaceans are shown in Fig. 1. Remarkable features are a thick sclera (especially so in whales, Fig. 1B), a thickened cornea, a highly developed vascular network forming a typical vascular rete mirabilia, which fills a significant part of the orbit behind the eyeball, and massive ocular muscles. All of these structures take part in protecting the eye from underwater cooling and mechanical damage.

While in terrestrial mammals the eyeball is almost spherical, in cetaceans its anterior part is flattened so that the anterior chamber is small and the eyecup is of almost a hemispherical shape. More precisely, the eyecup shape approximates a segment of a sphere of about 150° of arc (Figs. 1A and 1B), and its nasotemporal diameter slightly exceeds the dorsoventral one.

In terrestrial mammals, the convex outer surface of the cornea is the major refractive element of the eye because it separates media with different refractive indices: air with a refractive index of about 1 and the corneal tissue with a refractive index of more than 1.35. However, the refractive index of water is 1.33–1.34, which is very close to that of the cornea and the intraocular media. As a result, the corneal surface plays very little part in underwater light refraction. Therefore, in cetaceans, light refraction and focusing of an image on the retina are almost entirely performed by the lens. This is why

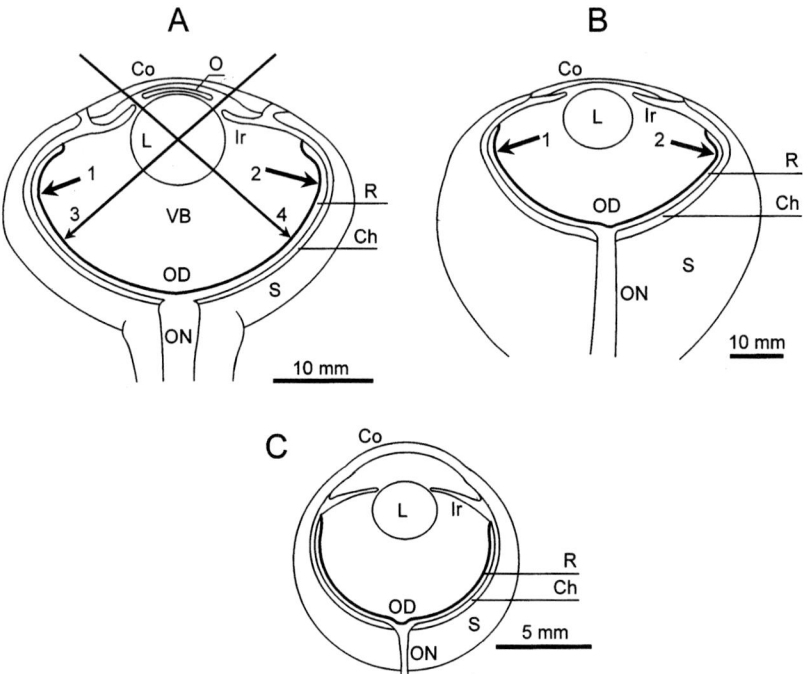

Figure 1 *Schematic presentation of eye anatomy and optics in some cetaceans: (A) the common bottlenose dolphin, (B) the gray whale (Eschrichtius robustus), and (C) the Amazon river dolphin. Co, cornea; L, lens; Ir, iris; O, operculum; S, sclera; Ch, choroid; R, retina; ON, optic nerve; OD, optic disc; VB, vitreous body. Arrows 1 and 2 delimit a part of the eyecup, which can be approximated by a spherical segment of about 150°. Arrows 3 and 4 show directions of light rays passing through the nasal and temporal holes of the pupil and through the lens center to the high-resolution parts of the retina.*

the lens in cetaceans is almost spherical or slightly elliptical. The large curvature of the lens surface provides a sufficiently high refractive power of the lens and well-focused images on the retina, despite very weak refractive power of the corneal surface in water. These optics are similar to those in fish, which is not surprising given that in both cases the eye is adjusted to optical properties of the same medium.

A strongly convex (spherical) lens consisting of homogeneous material has a very strong spherical aberration. The cetacean lens is free of this disadvantage due to a heterogeneous structure: outer layers have a lower refractive index than the inner core.

In the cetacean eye, the spherical lens is located in such a way that its center almost coincides with the center of the spherical segment of the eyecup, so light rays coming from any direction are focused almost identically on the retina. This is significantly different from the case in terrestrial mammals, which provides the best focusing on the eye axis.

In terrestrial mammal eyes, accommodation (refraction adjustment to the distance to the object) is performed by a change in the shape of the lens due to contraction and relaxation of ciliary muscles. In cetaceans, spherical lens shape and center-symmetric optics of the eye led to loss of this accommodatory mechanism. The ciliary muscles are poorly developed in dolphins and are absent in most whales, suggesting that accommodation cannot be achieved by changing the lens shape. It has been suggested that accommodation in cetaceans is performed by another mechanism, namely by axial displacement of the lens due to changes in intraocular pressure. Intraocular pressure can change because of contraction of the retractor/protractor muscles, which produce axial displacements of the eye in the orbit. When the eye is pulled back into the orbit, intraocular pressure increases, thus shifting the lens forward; when the eye is moved forward, the pressure decreases, shifting the lens backward.

The cornea in cetaceans is thicker than in many terrestrial mammals, and this thickness is not uniform: the cornea is thinner in the center and thicker in the periphery. Although major refraction in the cetacean eye is performed by the lens, the refractive role of the cornea is not negligible. Its outer surface is of lower curvature than the inner one; i.e., the cornea has a shape of a divergent lens. Under water, this lens makes a minor contribution to the total refraction power, as the media on both sides of the cornea (water outside and the anterior chamber liquid inside) have refractive indices rather close to that of the cornea. However, some difference between the refractive indices of water (1.33) and the cornea (from 1.37 in the central part to 1.53 in the periphery) does exist. Thus, the cornea acts a weak but nonetheless divergent lens. The total refraction of the cornea and lens makes the cetacean eye well emmetropic within a range of ±1 diopters under water.

Adaptation to underwater vision also affects the cetacean iris and pupil. Cetacean vision functions in conditions of wide and rapid changes of illumination when the animal dives from the well-illuminated water surface into the depths where illumination is very low. This requires the pupil to react in a wide range of illuminations and to have a wide range of sizes. The cetacean pupil is of an unusual shape. The upper part of the iris has a

characteristic protuberance, the operculum. At low illumination, the operculum is contracted (raised), so the pupil, similarly to other mammals, is of a round or slightly oval shape; its horizontal diameter in dolphins is of about 10 mm (Fig. 2A). With illumination increased, the operculum advances downward, turning the pupil into a U-shaped slit (Fig. 2B). At high illumination, the operculum advances so far that the slit becomes closed, leaving only two narrow holes in the temporal and nasal parts of the iris (Fig. 2C). This pupil shape is characteristic for many dolphins, including the bottlenose dolphin, harbor porpoise (*Phocoena phocoena*), common dolphins (*Delphinus* spp.), tucuxi, and also for a number of whales, although in some whales the operculum is small. A known exception is the Amazon river dolphin, which has a round pupil even when it is constricted.

The cetacean eye is well emmetropic in water; however, in air refraction on the outer convex corneal surface adds to the lens refraction. The difference of refractive indices of air and the cornea results in significant refractive power of the central most convex part of the corneal surface: about 20 diopters. The addition of this refraction to the emmetropic lens refraction should make the cetacean eye catastrophically myopic (near-seeing) in air. Nonetheless, dolphins have good visual acuity in both water and air.

The solution of the problem is in the presence of flattened (low-curvature) regions of the cornea. A flat corneal surface does not produced additional refraction in air. Even if the surface is not truly flat but a little convex, its refractive power becomes low enough and may be compensated by some additional mechanisms. Keratoscopis studies in common bottlenose dolphins showed a "spoon" shape of the cornea with lower curvature in its nasal and temporal regions.

Aerial myopia can be partially compensated by accommodatory displacements of the lens. For aerial vision, the dolphin eye moves forward, thus producing a decrease of intraocular pressure; this results in shifting the lens backward and reduced myopia. Additionally, reduction of intraocular pressure decreases the curvature of the cornea. Under water, the eye is retracted into the orbit, which results in increased intraocular pressure and a shift of the lens forward to a position providing underwater emmetropia.

An additional mechanism for the correction of aerial myopia is pupil constriction. Above water, high illumination results in

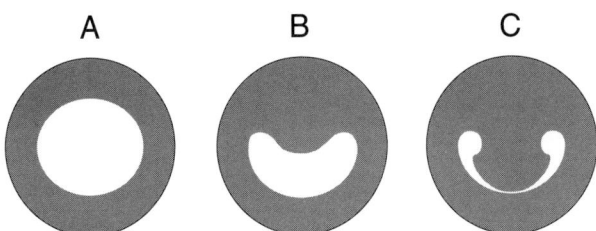

Figure 2 *Shape of the pupil in the common bottlenose dolphin at various levels of illumination: (A) low illumination, nonconstricted oval pupil; (B) moderate illumination, partially constricted U-shaped pupil; and (C) high illumination, strongly constricted pupil reduced to two pinholes.*

strong pupil constriction; the latter corrects all errors of refraction, including aerial myopia, and provides a fairly good
depth of field.

Another adaptation of the cetacean eye to low underwater
illumination is the well-developed reflective layer, the tapetum
lucidum. It lies behind the retinal pigment epithelium within
the choroid. In cetaceans, the tapetum is formed with extra-
cellular collagen fibrils (*tapetum fibrosum*). Multiple reflection
of light from 50 to 70 layers of fibrils results in significant light
reflection back to the retina, thus increasing visual sensitivity in
scotopic conditions.

The tapetum is present in all cetaceans. In most investigated cetaceans, particularly in mysticete whales, it covers all
of the fundus (although it varies in coloration), or at least it covers a large dorsal part of the fundus. Complete coverage of the
fundus by the tapetum is unique among vertebrates: in terrestrial mammals, the tapetum usually does not extend lower than
the horizontal equator of the eyecup.

B. Pinnipeds

In all pinnipeds (except walruses), both the absolute and the
relative sizes of the eyeball are large. The eye structure in pinnipeds (Fig. 3A), despite significant differences from cetaceans,
has some common features arising from adaptation to underwater vision (Jamieson and Fisher, 1972). In particular, a characteristic feature is an almost spherical or slightly elliptical lens.
Although the eyeball does not appear as shortened in the axial
direction, a major part of the eyecup has a shape close to a

hemisphere so a significant part of the retina is almost constantly distant from the lens center. Thus, the eye optics, like
in cetaceans, is almost centrally symmetrical. The difference
between the eyeball shape in cetaceans and pinnipeds (shorter
axial length in cetaceans and longer in pinnipeds) is mainly due
to the larger size of the anterior chamber in pinnipeds.

The iris in pinnipeds is very muscular and heavily vascularized. The dilator is well developed. All pinnipeds have a pupil,
which being constricted becomes pear-shaped, except for a diagonal pupil in the bearded seal, *Erignathus barbatus*. Pupil
size can change over a very wide range; at bright illumination,
it constricts to a very small hole. The ciliary muscle in pinnipeds is well developed, although accommodation is either absent or very weak.

Unlike cetaceans, the central part of the cornea has a clearly
delimited region (6–10 mm in diameter) of almost a flat surface.
It is located near the center of the cornea, slightly shifted to the
nasal direction (FC region in Fig. 3A). Such a flat region of the
cornea was found in a number of both otariids and phocids and
was demonstrated by precise measurements on the Californian
sea lion (*Zalophus californianus*). The flat region of the cornea
serves as an emmetropic "window" in which refraction remains
almost equal in both water and air. In another pinniped, the
hooded seal (*Cystophora cristata*), the flattened part of the
cornea does not look like a delimited region but arises because
of low curvature of the cornea of the extremely large eyeball.

The existence of a flat region in the central part of the cornea
indicates a very specific principle of eye construction in pin-

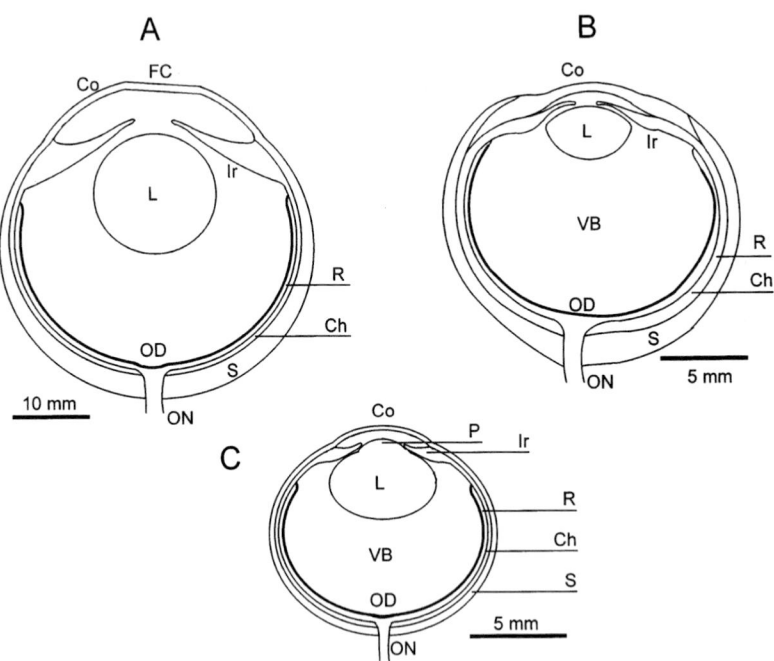

Figure 3 *Schematic presentation of eye anatomy and optics in some representatives of pinnipeds, sirenians, and otters: (A) the northern fur seal, (B) the manatee, and (C) the sea otter. Co, cornea; FC, flattened region of the cornea; L, lens;
Ir, Iris; S, sclera; Ch, choroid; R, retina; ON, optic nerve; OD, optic disc; P, lens
protuberance; VB, vitreous body.*

nipeds. Indeed, the convex shape of the cornea in most animals is a consequence of excessive intraocular pressure, which is necessary for maintaining the shape and size of the eyeball. Direct data on intraocular pressure in pinnipeds are absent, but their flat cornea suggests that this pressure is very low, perhaps about zero. Anatomical observations on the northern fur seal (*Callorhinus ursinus*) showed that its vitreous body is of a rigid rather than a gelatinous consistency, thus taking a part in maintenance of the eyeball shape and dimensions. This way of maintaining the eye shape is evidently used in a number of pinniped species.

The pinniped tapetum is one of the best developed among both terrestrial and aquatic mammals. Contrary to the tapetum fibrosum in cetacean, the tapetum in pinnipeds is formed with intracellular reflective rodlets (tapetum cellulosum). It consists of a large number (20–30 or more) of cell layers and covers all of the fundus.

C. Other Marine Mammals

1. Sirenians Among other marine mammals, the eye anatomy of the manatees (*Trichechus manatus* and *T. inunguis*) is of interest as an example of the order Sirenia, which, apart of cetaceans, is the only group of completely aquatic mammals.

Both in *T. manatus* and *T. inunguis* the eye is rather small (13–19 mm diameter) and is set deeply within the ocular fascia. Its general morphology resembles more that of terrestrial mammals than the cetacean eye (Fig. 3B). The eyeball is almost spherical (the axial length differs little from the equatorial diameter), the anterior chamber is shallow, and the lens is set forward and is not truly spherical: its axial dimension is markedly shorter than the diameter. The sclera is rather thin. Thus, despite the completely aquatic mode of life of the manatee, its eye anatomy exhibits a number of conservative features (Piggins *et al.*, 1983). Underwater, the eye is almost emmetropic or slightly hyperopic, but in air it is strongly myopic. It remains unknown whether the manatee has some mechanism to compensate aerial myopia; thus, its capability to aerial vision remain unknown.

2. Sea Otters To a large extent, the eyeball of the sea otter (*Enhydra lutris*) is similar to those of terrestrial mammals (Fig. 3C): it is almost spherical, the axial length is only a little shorter than the diameter. Contrary to spherical lenses of cetaceans and pinnipeds, the lens of the sea otter is lenticular. However, the front surface of the lens has a protuberance of increased curvature. A characteristic feature of the eye anatomy is that the iris is fastened to the frontal lens surface. Therefore, contraction of iris muscles influences the curvature of the frontal lens surface. This mechanism is capable of providing an accommodation range of up to 60 diopters, thus compensating for the appearance of refraction at the corneal surface in air and its disappearance in water. This accommodation mechanism in the sea otter eye is able to preserve emmetropia in both air and water.

III. Eye Movements

All dolphins and whales have mobile eyes. However, measurements in the bottlenose dolphin indicate that eye mobility is less than in humans, and eye movements are slower.

Oculomotor muscles are well developed in dolphins and whales; an exception is the Indian river dolphin (*Platanista gangetica*), which has reduced eyes and no oculomotor muscles. Other cetaceans have a complete set of muscles known in mammals: four straight and two oblique muscles. These muscles allow eye movements in both the horizontal and the vertical directions. In addition, unlike terrestrial mammals, cetaceans have retractor/protractor muscles (m. retractor bulbi and m. protractor bulbi), which produce axial (in/out) movements of the eye in the orbit. The bottlenose dolphin is capable of moving its eye forward 10–15 mm and pulling it back. As a rule, forward eye movements (protraction) appear when the dolphin examines an object in air visually. These eye movements may be used for the binocular examination of objects. As mentioned earlier, the eye protraction in air can also provide accommodation to avoid aerial myopia.

Another intriguing factor of oculomotor activity in dolphins is the ability to move the left and right eyes independently. Quantitative measurements in dolphins have shown that the correlation of movements of the left and right eyes is very low; i.e., independent eye movements in dolphins are a rule rather than exception.

In addition to independent eye movements, cetaceans have rather independent pupil reflexes of the two eyes. Moreover, eyelids of the left and right eyes can also function independently so one eye can be open while the other is closed. Such observations were made during sleep in dolphins, although similar behavior is also possible in wakefulness: dolphins can swim for long periods with one eye open and the other one closed, with the left and right eye alternating.

As to pinnipeds and sea otters, there is no significant difference from terrestrial mammals in their oculomotor muscle anatomy and the character of eye movements.

IV. The Retina and Optic Nerve
A. Features of the Retina in Cetaceans

The histological structure of the retina has been investigated in a number of cetacean species: the common bottlenose dolphin, short-beaked common dolphin (*Delphinus delphis*), Dall's porpoise (*Phocoenoides dalli*), dwarf sperm whale (*Kogia sima*), Amazon river dolphin, fin whale (*Balaenoptera physalus*), and common minke whale (*B. acutorostrata*). All of these studies have shown that the laminal structure of the cetacean retina is basically similar to that in terrestrial mammals. The retina consists of typical layers as follows (Fig. 4). The receptor layer (the nearest to the pigment epithelium) is composed of densely packed outer segments of photoreceptors. The outer nuclear layer is composed of receptor pericaria arranged in a multilevel manner. The outer plexiform layer contains cell processes establishing connections between receptors and first-order neurons, bipolar cells. The inner nuclear layer is composed mostly of pericaria of bipolar cells; in addition, this layer contains horizontal and amacrine cells, which establish horizontal connections within the outer and the inner plexiform layers. The inner plexiform layer contains processes establishing connections between bipolar and ganglion cells. The ganglion layer contains ganglion cells sending their axons to the optic nerve. Finally, the nerve fiber

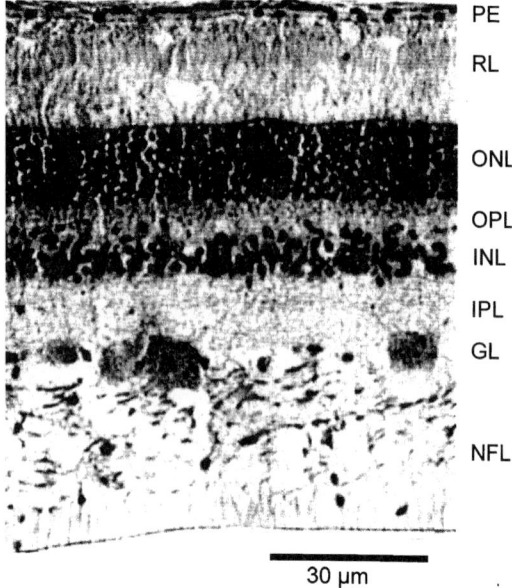

Figure 4 *Microphotographs of a transverse section of the retina of a Dall's porpoise. PE, pigment epithelium; RL, receptor layer; ONL, outer nuclear layer; OPL, outer plexiform layer; INL, inner nuclear layer; IPL, inner plexiform layer; GL, ganglion layer; NFL, nerve fiber layer. From Murayama et al. (1995). Mar. Mamm. Sci. 11(2), 136–149, with permission of the Society for Marine Mammalogy.*

layer (nearest to the vitreous body) contains optic fibers (axons of ganglion cells), which spread along the inner/retinal surface until they reach the optic disk and enter the optic nerve. This laminar structure of the retina is fully developed in all cetaceans. Even in the Indian river dolphin with strongly reduced eyes, the retina contains all the layers. Being basically similar in cetaceans and terrestrial mammals, the retina has a number of specific features in cetaceans. It is markedly thicker than in terrestrial mammals, ranging from 370 to 425 μm (in terrestrial mammals, the retina is 110–240 μm thick).

The most detailed description of the retina is available for the common bottlenose dolphin (Dral, 1977; Dawson, 1980). Its retinal receptor layer consists predominantly of rods (receptors for achromatic vision). The question of the existence of cones (chromatic-vision receptors) in cetaceans is still debatable; if cones exist, they are few in number. This corresponds to behavioral data showing poor color vision in dolphins (see Section I,A). As to amacrine, bipolar, and horizontal cells in the cetacean retina, they are generally similar to those in terrestrial mammals.

A marked difference from terrestrial mammals is in the inner plexiform layer and the ganglion layer of the cetacean retina. The ganglion layer looks like a single row of large, sparsely distributed neurons separated by large intercellular spaces. These neurons have large cell bodies with a clearly defined cell membrane, a large amount of cytoplasm, a well-visible nucleus up to 15 μm in diameter, and a clearly defined nucleolus 4–5 μm in diameter. Cell bodies contain clearly visible, well-stained Nissl granules.

A remarkable feature of the cetacean retina is the large size of ganglion cells, particularly the presence of giant ganglion cells. Bodies of such cells reach 75–80 μm, sometimes more. Giant ganglion cells were described in a number of odontocete species and in a few mysticete whales. In some dolphins, however, retinal ganglion cells do not reach giant sizes: in the Amazon river dolphin and the Indian river dolphin they do not exceed 40–42 and 20 μm, respectively. However, even these cells are large as compared to those in many other mammals. The smallest ganglion cells in cetaceans are as large as 10 μm.

Figures 5A and 5B present ganglion cell size distributions in the retina of the common bottlenose dolphin. The histograms represent samples in different parts of the retina: with high and low concentration of ganglion cells. Despite some difference between the samples (in the area of high cell concentration, cells are a little smaller than in the area of low concentration), both samples demonstrate large cell sizes: the most common size is 20–35 μm, but cells as large as 50–60 μm are also present; there are no cells smaller than 10 μm.

Large cells are not characteristic of all levels of the visual system in cetaceans (the lateral geniculate body, visual cortex); they are typical only in the retina. The largest pyramidal cells in the visual area of the dolphin cerebral cortex are not more than 20–30 μm. In other parts of the dolphin brain, cells do not exceed 20–45 μm either. There is presently no satisfactory explanation why ganglion cells in the cetacean retina are so large. One possible explanation is that large ganglion cells have thick axons with a high velocity of conduction; in a large body, it may be helpful for the fast transmission of signals. However, large terrestrial mammals (e.g., the bull or elephant) have ganglion cells no larger than 25–30 μm.

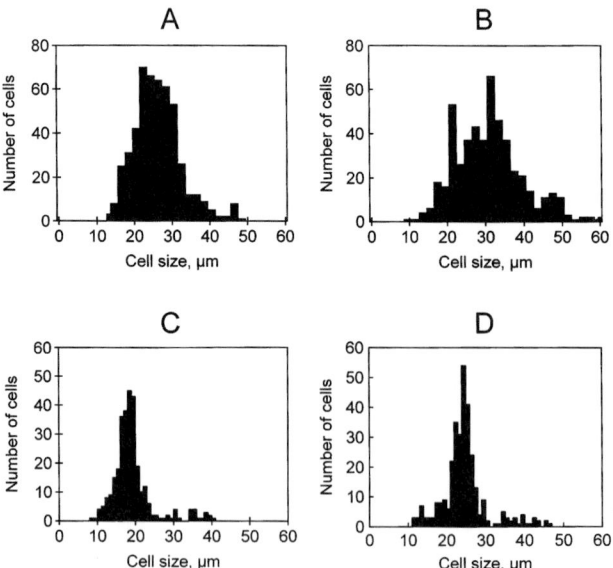

Figure 5 *Histograms showing size distributions of ganglion cells in the retina of cetaceans and pinnipeds. Abscissa axis, cell size; ordinate axis, number of cells of the present size. (A and B) Data from a common bottlenose dolphin; samples from areas of high (A) and low (B) cell densities. (C and D) The same for a northern fur seal.*

Apart from large cell sizes, a characteristic feature of the retinal ganglion layer in cetaceans is low cell density. The large neurons are separated by large intercellular spaces.

The question of separation of retinal ganglion cells into different morphological types has not been solved for cetaceans. Large-size ganglion cells in cetaceans resemble large Y neurons in the visual system of terrestrial mammals, as opposed to smaller X neurons. However, Y neurons in terrestrial mammals constitute no more than 1% of ganglion cells, whereas in cetaceans, large ganglion cells predominate.

B. Optic Nerve Structure in Cetaceans

Retinal ganglion cells send their axons into the optic nerve. Consistent with the large sizes of ganglion cell bodies, the axon diameters in cetaceans are also greater than in terrestrial mammals. In a variety of dolphin species, a significant proportion of optic fibers exceed 15 μm in diameter. For comparison, the maximum fiber diameter in cats and monkeys is no more than 8 μm. The only exception is the Chinese river dolphin, *Lipotes vexillifer*, which has thin optic fibers, although its retina contains ganglion cells as large as 75 μm.

The low density of ganglion cells in the retina of cetaceans corresponds to the low density of fibers in the optic nerve. In cross sections of the optic nerve of dolphins, the density of fibers is less than 50,000/mm^2, whereas in monkeys it exceeds 220,000/mm^2. Thus, although the optic nerve in cetaceans is of a larger diameter, the total number of optic fibers does not exceed that in many terrestrial mammals. More than 50% of the cross section area of the cetacean optic nerve is occupied by intercellular space (as opposed to 12–20% in terrestrial mammals), not by glia.

The total number of optic fibers varies among cetacean species. The smallest number of fibers (14,000–16,000) was found in the Indian river dolphin, *Platanista gangetica*, and the Amazon river dolphin, *Inia geoffrensis*; the number of optic fibers in the Chinese river dolphin, *Lipotes vexillifer*, is a little higher, more than 20,000. In the common bottlenose dolphin, the number of optic fibers is 150,000–180,000. Other odontocetes have an optic fiber number similar to that in the bottlenose dolphin. In mysticetes, the number of optic fibers is within a range of 250,000–420,000.

C. Features of the Retina in Pinnipeds

In general, the retinal structure in pinnipeds is the same as in terrestrial mammals. All layers are present in the pinniped retina, although there are a number of specific features, mainly of the inner nuclear and ganglion layers (Jamison and Fisher, 1972).

The inner nuclear layer in pinnipeds does not have clear margins, in contrast to terrestrial mammals, where this layer is strictly ordered. There are large horizontal cells with very long processes within this layer. The giant horizontal cells are located irregularly among bipolar and amacrine cells, which are also distributed chaotically. Bipolar cells are located mostly in the outer part of the inner nuclear layer while large amacrine cells are located close to the inner plexiform layer.

The ganglion layer in pinnipeds consists of a single row of ganglion cells separated by wide intercellular distances. Ganglion cells have large pericaria, a large amount of Nissl substance in the cytoplasm and long dendrites. Most of these cells are of intermediate size (10–30 μm), although large cells (up to 50 μm) are also encountered (Figs. 5C and 5D). These sizes are smaller than in cetaceans, although larger than in terrestrial mammals.

All pinnipeds have a predominately rod retina. However, light and electron microscopy has shown the presence of cones in the harbor seal and harp seal, although photoreceptors of this type are not numerous. The existence of some amount of cones corresponds to behavioral data showing a limited capability of color discrimination in pinnipeds (see Section I,B).

D. The Retina of Other Marine Mammals

1. Sirenians The retina of the manatees also features the common laminar organization. Receptors are presented mostly by rods; cones are less numerous. Among specific features, the large size of the ganglion cells can be mentioned: up to 60 μm, mostly 15–30 μm, and not less than 10 μm. Thus, the large size of ganglion cells seems to be a common feature of different groups of marine mammals.

2. Sea Otters In the sea otter, the retina has many features similar to those in terrestrial rather than in aquatic mammals. The majority of ganglion cells are not of large size: 7 to 30 μm, mostly 11–15 μm. They can be subdivided into three size groups: large, medium, and small. The retina of the sea otter contains a large number of small amarcine and neuroglial cells.

V. Retinal Topography and Visual Field Organization

A. Cetaceans

Ganglion cells are distributed nonuniformly in the mammalian retina: ganglion cell density (number of cells per area unit) is high in some areas and much lower in the remainder of the retina. Regions of ganglion cell concentration (high density) provide the most detailed analysis of visual images. Characteristics of retinal topography in a variety of mammals are presented in a review by Hughes (1977).

In terrestrial mammals, there are two main types of organization of a region with high cell density. In mammals with frontal vision, highest density is in the fovea or area centralis located in the center of the visual field. This retinal area is little vascularized to avoid its shadowing by blood vessels. In mammals with laterally located eyes, the region of high cell density is shaped as a narrow horizontal strip, the visual streak. All terrestrial mammals studies up until now have only one, if any, region of highest ganglion cell density.

The cetacean retina does not have avascular areas, indicating the presence of fovea or area centralis. Therefore, visual examination of the eye fundus is not capable of revealing such regions. Data on the topography of ganglion cell distribution in the cetacean retina were obtained using retinal whole mounts. Whole mounts are preparation of a total retina flattened on a slide, ganglion layer upward, and stained appropriately. Retinal whole mounts allow one to count ganglion cells systematically across all of the retina surface, thus constructing a topographic map of ganglion cell distribution. Studies of retinal whole mounts have shown that different regions of the cetacean retina have a very different density of ganglion cells (Fig. 6). Beginning from the pioneering studies of Dral (1977), studies of cetacean retinal whole mounts were performed in a number of dolphin species, particularly the common dolphin, common bottlenose dolphin, harbor porpoise, Dall's porpoise, and Pacific white-sided dolphin (*Lagenorhynchus obliquidens*) (see review by Mass, 1997).

The most characteristic feature of these species is that, unlike terrestrial mammals, all of these marine dolphins do not have a single area of high ganglion cell density but two such areas. They are located near the horizontal diameter of the retina, one in the nasal and the other in the temporal sector (Fig. 7A). In the bottlenose dolphin, both of these areas are located at a distance of 15–16 mm from the optic disk, which corresponds to 50–55° of the visual field. Ganglion cell density in each of these areas reaches 700–800 cells/mm^2, which corresponds to 40–50 cells per squared degree of the visual field (cells/deg^2). The two high-density areas are connected by an elongated zone of increased, although somewhat lower, cell density, which runs below the optic disk; this zone looks like a visual streak.

In other dolphin species, the retinal topography is basically similar to that described earlier: there are two areas of high ganglion cell density. Even at low cell density in some cetaceans inhabiting turbid and low-transparent water (e.g., the tucuxi), the retinal topography looks the same (Fig. 7B). However, some quantitative differences do exist. In the bottlenose dolphin, the ganglion cell density is almost equal in the two areas, the nasal and the temporal areas, whereas in the harbor porpoise, the cell density in the temporal area (i.e., the region serving the frontal visual field) is higher than in the nasal region: 28 and 20 cells/deg^2, respectively.

The retinal topography of ganglion cells was studied in two mysticete species: the gray whale, *Eschrichtius robustus,* and common minke whale (*Balaenoptera acutorostrata*). Both of them also have ganglion cell distributions with two area of high cell density in the nasal and temporal sectors (Fig. 7C). Again, the cell density in the temporal area is higher than in the nasal one: 28 and 21 cells/deg^2 in the gray whale.

The significance of the two areas of high ganglion cell density (i.e., of high retinal resolution) is probably associated with the cetacean's capability of good vision both above and under water, in particular, with preventing aerial miopia. Indeed, the high-resolution areas are located just opposite the two small pupil holes formed when the pupil is constricted in air (see Fig. 1A). Because of the centrally symmetric optics of the cetacean eye, light falls onto each of these areas through the opposite hole of the pupil. The areas of the cornea with minimal curvature are located across from these narrow pupil holes. Both the pinhole pupils and the low cornea curvature are devices to prevent aerial miopia. Thus, images are projected onto the high-resolution areas of the retina with minimal distortions.

The two high-resolution retinal areas in cetaceans may be used differently for underwater and aerial vision (see Mass, 1997). When a dolphin looks at an underwater object, it takes a position lateral to the object; i.e., the object is placed into the posterolateral part of the visual field, which projects onto the nasal high-resolution area of the retina. On the contrary, when a dolphin looks at an object above water, it places the object into the ventronasal part of the visual field, which projects onto the temporal high-resolution area of the retina (Fig. 8). Of course, the temporal high-resolution area of the retina also participates in underwater vision. This area serves the frontal part of the visual field, which is very important for forward-moving animals. The existence of two high-resolution areas of the retina can also compensate for the limited head mobility in many cetaceans. At low head mobility, even at high mobility of the eyes, a single high-resolution area allows the animal to inspect only a limited part of the surrounding space, whereas two such areas can provide almost panoramic vision.

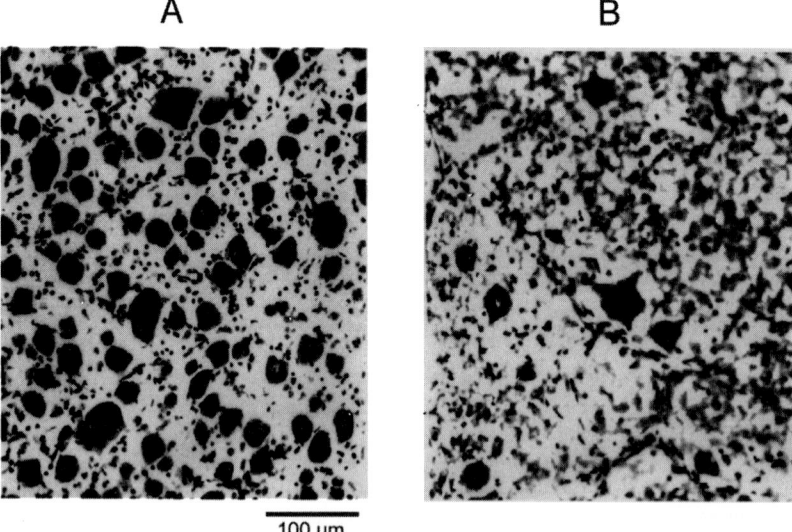

Figure 6 *Microphotographs of the ganglion layer in a retinal whole mount of a bottlenose dolphin: (A) an area of high cell density and (B) an area of low cell density.*

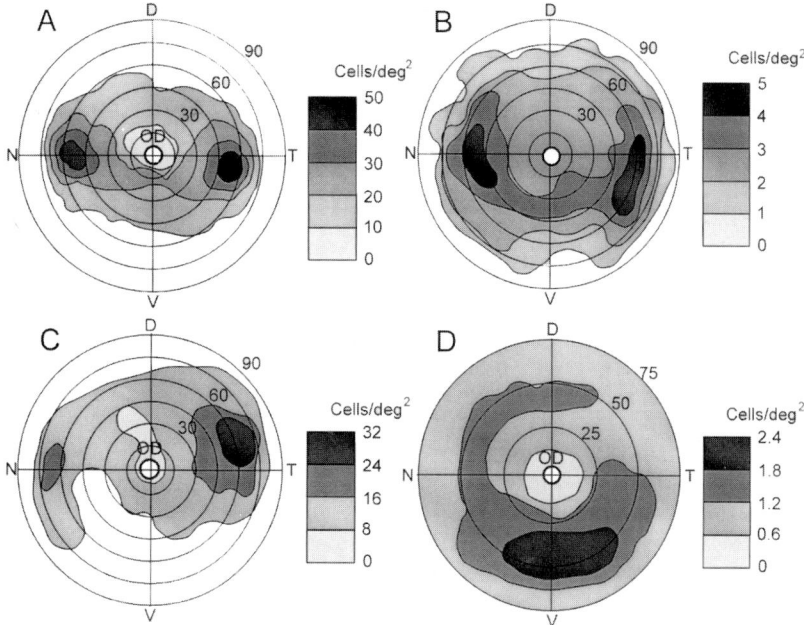

Figure 7 *Topographic distribution of ganglion cell density in the retina of some cetaceans: (A) the common bottlenose dolphin, (B) the tucuxi dolphin, (C) the gray whale, and (D) the Amazon river dolphin. Cell density is expressed as number of cells per squared degree of the visual field and is shown by various shadowing, according to the scales. Concentric circles show angular coordinates on a retinal hemisphere centered on the lens. D, V, N, T, dorsal, ventral, nasal, and temporal poles of the retina, respectively.*

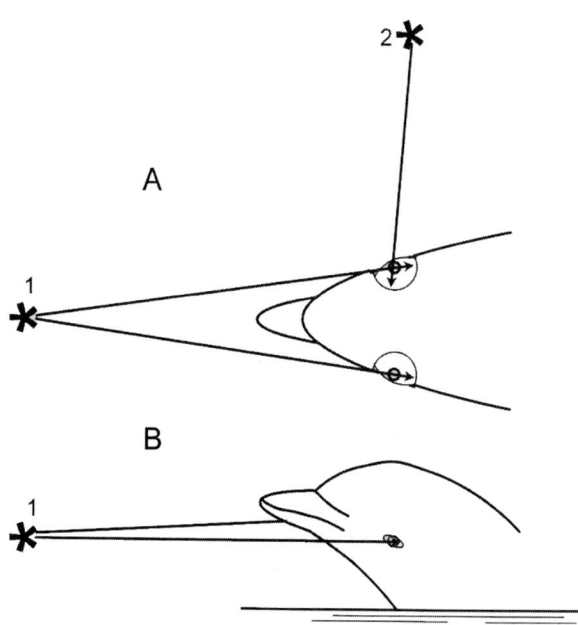

Figure 8 *Characteristic positions of the dolphin body relative to visually inspected objects: (A) Dorsal view (for both underwater and aerial vision) and (B) lateral view (for aerial vision). 1, an above water object; 2, an underwater object. Arrows show directions of light rays from an object to the corresponding high-resolution area of the retina.*

The retina of the Amazon river dolphin is a special case. The visual system of this species is adapted to low-transparent turbid water where vision is possible only at short distances. Contrary to all other investigated cetaceans, the retina of the Amazon river dolphin has only one area of higher ganglion cell density. However, this single area is located not in the center or temporal sector, but in the lower part of the retina, i.e., in the region responsible for the upper part of the visual field (Fig. 7D). In turbid low-transparent water, significant illumination exists only near the water surface, i.e., in the upper part of the visual field of a normally oriented animal. Just this part of the visual field is served by the ventral part of the retina where the Amazon dolphin has higher retinal resolution. The density of ganglion cells in this region reaches 500 cells/mm^2; with the small size of the eyeball, this corresponds to a cell density of about 2 cells/deg^2.

B. Pinnipeds

Few otariids have been subjects of study of the ganglion cell topography in retinal whole mounts (Mass, 1997). The northern fur seal has a typical area centralis, a well-defined area of high concentration of ganglion cells (Fig. 9A). It is located at a distance of 22–23 mm (which corresponds to 35–40° of the visual field) from the optic disk; taking into account the position of the eye in orbit, this place may be at the projection of the vertical meridian of the visual field. Thus, the position of this area is similar to that in terrestrial carnivores. Cell density in this area reaches 1000 cells/mm^2, which corresponds to more than 160 cells/deg^2.

Quite different is the retinal topography in the walrus. The area of increased ganglion cell density is not defined as clearly as in the northern fur seal. It looks like a horizontally extended oval, resembling the visual streak of terrestrial mammals (Fig. 9B). Within this streak, the highest cell density in its temporal part exceeds 1000 cells/mm^2; because of the smaller size of the walrus eye, this cell density corresponds to only about 50 cells/deg^2. No phocid seals have been studied successfully yet with respect to their retinal topography.

C. Other Marine Mammals

1. Sirenians Ganglion cell distribution in the manatee retina represents an example of low specialization. There is no sharply restricted spot of cell concentration. Ganglion cell distribution is not uniform but varies smoothly across the retina: cell density is higher in a large center part of the retina (except in the nearest vicinity of the optic disk) and diminishes toward the edges (Fig. 9C). The highest cell density is about 250–300 cells/mm^2; for a rather small manatee eye, this corresponds to 6–7 cells/deg^2.

2. Sea Otters In the sea otter retina ganglion cells topography (Fig. 9D) has a number of features similar to that of terrestrial mammals. The high-density area resembles nasotemporal streak. Within this streak, in its temporal part, there is a narrow and well-defined spot of the highest cell density, which is similar to the area centralis in terrestrial mammals. The highest ganglion cell density in the sea otter exceeds 4000 cells/mm^2; in the rather small eye of the sea otter, this corresponds to 50–60 cells/deg^2.

D. Estimations of Visual Acuity from Ganglion Cell Density

Visual acuity is determined by two factors: quality of the eye optics and retinal resolution. In normal eyes, these two values are in agreement. Therefore, retinal resolution can be used as a first-order estimate of visual acuity. Retinal resolution depends on the density of ganglion cells (not of other retinal cells, e.g., photoreceptors, as ganglion cells transmit visual information to the brain). Thus, data on ganglion cell topography can be used to estimate the visual acuity of investigated species. When it is possible to compare estimates of visual acuity obtained by behavioral (psychophysical) methods and those based on ganglion cell topography, these estimates are in good agreement.

Retinal resolution (hence, visual acuity) is defined as the mean angular distance between neighboring ganglion cells, i.e., as $s = 1/\sqrt{D}$, where s is the angular distance between cells and D is the cell density per deg^2. The estimation of retinal resolution is different in air and water: if the corneal surface is flat, the retinal image of an object in water is 1.33 times larger than

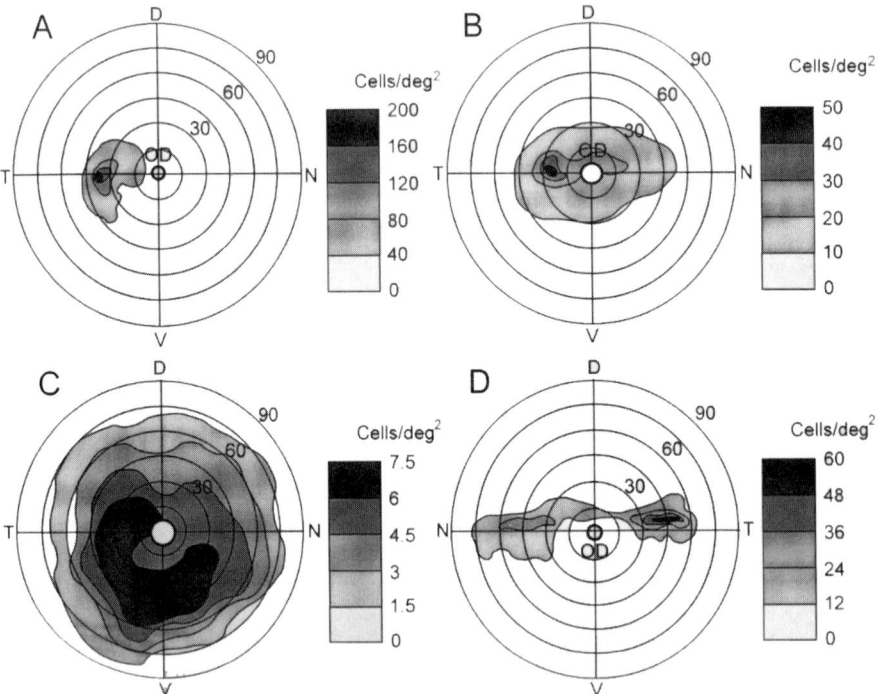

Figure 9 *Topographic distribution of ganglion cell density in the retina of some pinnipeds, sirenians, and sea otters: (A) the northern fur seal, (B) the walrus, (C) the Caribbean manatee, and (D) the sea otter. Cell density is expressed as number of cells per squared degree of the visual field and is shown by various shadowing, according to the scales. Concentric circles show angular coordinates on a retinal hemisphere centered on the lens. D, V, N, T, dorsal, ventral, nasal, and temporal poles of the retina, respectively.*

in air (because the ratio of refraction indices of water and air is 1.33). Therefore, retinal resolution in water is 1.33 times better than in air. If the retinal surface is a little convex, this factor is less than 1.33 but still more than 1.

Retinal resolution in the areas of the highest concentration of ganglion cells (i.e., visual acuity in the best-vision areas of the visual field) was estimated in a number of marine mammal species (Table I). In many cetaceans (except for river dolphins), it varies from 8 to 12 arc min in water, correspondingly from 11 to 15 arc min in air. For the bottlenose dolphin, the estimation of visual acuity obtained from retinal topography almost coincides with that obtained in behavioral experiments. In general the visual acuity of cetaceans is within a range of visual acuities of many terrestrial mammals, except the foveal vision of primates, where the visual acuity is around 1 arc min. The Amazon river dolphin has much worse visual acuity: 40–50 arc min in water; however,

this value is adequate for vision in turbid water where objects are visible at best at a few tens of centimeters.

Among pinnipeds, rather acute vision is characteristic for the northern fur seal: better than 5 arc min in water and better than 7 arc min in air. This is close to estimates obtained in behavioral experiments in a number of both otariids and phocids: 5–8 arc min (Schusterman, 1972). In the walrus, the visual acuity is worse than in seals: around 8 arc min in water and 10 arc min in air.

In the manatee, underwater visual acuity is around 20 arc min (it remains unknown whether the manatee has good aerial vision). In the sea otter, the visual acuity is around 7 arc min in water.

VI. Cerebral Visual Centers

In dolphins, the visual system is well represented in the midbrain (the superior colliculus), thalamus (the lateral geniculate

TABLE I
Visual Acuity of Some Aquatic Mammals[a]

Species	Water	Air	Mode of measurement
Odontoceti			
Common bottlenose dolphin (*Tursiops truncatus*)	8–9	11–12	BR
Short-beaked common dolphin (*Delphinus delphis*)	8		R
Harbor porpoise (*Phocoena phocoena*)	11	15	R
Tucuxi dolphin (*Sotalia fluviatilis*)	25	33	R
Amazon river dolphin (*Inia geoffrensis*)	40	53	R
Dall's porpoise (*Phocoenoides dalli*)	11		R
False killer whale (*Pseudorca crassidens*)	9		R
Pacific white-sided dolphin (*Lagenorhynchus obliquidens*)	11		R
Beluga whale (*Delphinapterus leucas*)	12		R
Mysticeti			
Common minke whale (*Balaenoptera acutorostrata*)	7		R
Gray whale (*Eschrichtius robustus*)	11		R
Phocidae			
Harbor seal (*Phoca vitulina*)	8		B
Otariidae			
Northern fur seal (*Callorhinus ursinus*)	4–5	5–7	R
Steller sea lion (*Eumetopias jubatus*)	6–7		B
California sea lion (*Zalophus californianus*)	5–6	5–7	B
Cape fur seal (*Arctocephalus pusillus*)		6–7	B
Southern fur seal (*A. australis*)		7	B
Walrus (*Odobenus rosmarus*)	8	10	R
Sirenia			
Caribbean manatee (*Trichechus manatus*)	20		R
Lutrinae			
Sea otter (*Enhydra lutris*)	7		R

[a]Visual acuity is presented as the minimal resolvable distance in minutes of arc, rounded to a whole number of minutes. In some cases, a range of variation is indicated (e.g., 11–12 arc min). Estimates of visual acuity are given for underwater (*water*) and aerial (*air*) vision. In the column *air*, data are not presented when none of the authors attempted to interpret their results in terms of aerial visual acuity. When several estimates of visual acuity in different conditions are available (e.g., in the nasal and temporal best-vision areas in cetaceans, at various illumination conditions, etc.), the best estimate (i.e., the minimal resolvable distance) is selected. Mode of measurement: B, behavioral data: R, data on retinal resolution.

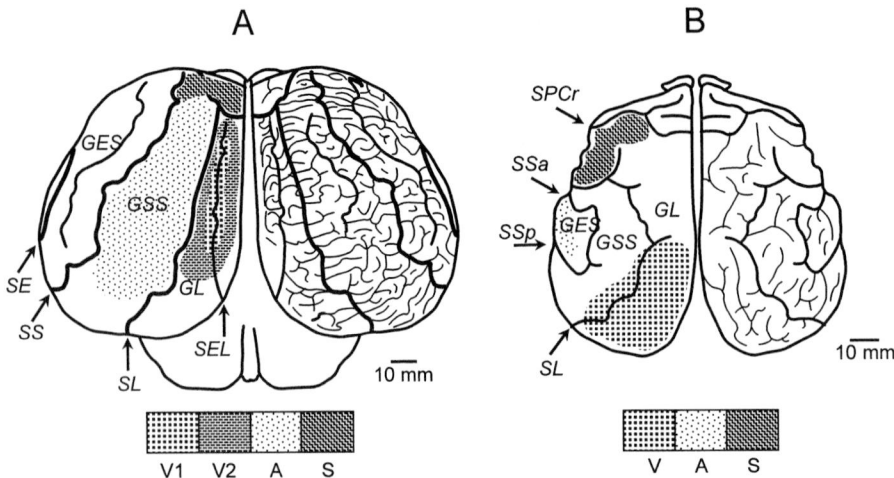

Figure 10 *Position of projection sensory areas (visual, auditory, and somatosensory) in the cerebral cortex of cetaceans and pinnipeds: (A) the common bottlenose dolphin and (B) the northern fur seal. Dorsal view of the cerebral cortex. On the right hemisphere, the pattern of cortical sulci and gyri is shown in more detail. On the left hemisphere, only main cortical sulci are shown and the positions of the visual, auditory, and somatosensory areas are indicated. The main sulci (labeled by arrows at their ends): SE, sulcus ectosylvius; SS, s. suprasylvius; SL, s. lateralis; SEL, s. entolateralis; Ssa, s. suprasylvius anterior; Ssp, s. suprasylvius posterior; SPCr, s. postcruciatus. The main gyri (labeled on their surface): GES, gyrus ectosylvius; GSS, g. suprasylvius; GL, g. lateralis; V, visual area (V1, primary projection zone; V2, nonprimary zone); A, auditory area (only a part of this area is visible in B); S, somatosensory area.*

body), and in the cerebral cortex. However, the visual centers (both the superior colliculus and the lateral geniculate body) are several times less in volume than corresponding parts of the auditory system (the inferior colliculus and medial geniculate body).

In the cerebral cortex of dolphins, visual representation was found by the evoked potential method. This area occupies a part of the cortex named the lateral gyrus (Fig. 10A). The cortical representation of the visual system in dolphins also is not as large as that of the auditory system; nevertheless, it occupies a significant cortical area. There is a differentiation within this area: it contains a zone generating short latency-evoked potentials (i.e., the primary projection zone) and another zone generating evoked potentials of longer latency (a nonprimary zone). The first of them is located in the depth of the entolateral sulcus (a second-order sulcus within the lateral gyrus); the latter occupies the remainder of the lateral gyrus. These two zones differ in cytoarchitectonic features: the primary projection zone contains an incipient layer IV (the layer where visual thalamo-cortical afferent fibers end), whereas this layer is absent in the nonprimary zone.

In mysticetes, which do not have echolocation, the sizes of visual and auditory structures in the midbrain and thalamus are comparable. Their cortical sensory areas were not investigated.

Among pinnipeds, visual representation in the cerebral cortex was found by the evoked potential method in one otariid species—the northern fur seal (Fig. 10B)—and one phocid species—the Caspian seal, *Pusa caspica*. The location of this

area is very similar to that in carnivores: the projection occupies the caudal part of the lateral gyrus.

VII. Conclusions

In general, the visual system of marine mammals demonstrates a rather high degree of development and performance, in particular, good visual acuity, capabilities to precisely aim visually driven BEHAVIOR and intermodal transfer, and well-developed visual brain centers. This system also exhibits a number of specific features associated with adaptation to both aquatic and aerial environment, in particular, specific retinal topography (positions of best-vision areas) along with pupil and cornea structure, which provide emmetropia in both air and water.

See Also the Following Articles

Brain ▪ Echolocation ▪ Hearing

References

Dawson, W. (1980). The cetacean eye. *In* "Cetacean Behavior: Mechanisms and Functions" (L. Herman, ed.), pp. 53–100. Willey Interscience, New York.

Dral, A. (1977). On the retinal anatomy of Cetacea (mainly *Tursiops truncatus*). *In* "Functional Anatomy of Marine Mammals" (R. Harrison, ed.), pp. 81–134. Academic Press, London.

Fobes, J., and Smock, C. (1981). Sensory capacities of marine mammals. *Psychol. Bull.* **89,** 288–307.

Hughes, A. (1977). The topography of vision in mammals of contrasting life style: Comparative optics and retinal organization. *In* "Handbook of Sensory Physiology: The Visual System in Vertebrates" (F. Crescitelli, ed.), Vol. VII/5, pp. 613–756. Springer, Berlin.

Jacobs, G. H. (1993). The distribution and nature of colour vision among the mammals. *Biol. Rev.* **68,** 413–471.

Jamieson, G. S., and Fisher, H. D. (1972). The pinniped eye: A review. *In* "Functional Anatomy of Marine Mammals" (R. J. Harrison, ed.), Vol. 1, pp. 245–261. Academic Press, New York.

Madsen, C., and Herman, L. (1980). Social and ecological correlates of cetacean vision and visual appearance. *In* "Cetacean Behavior: Mechanisms and Functions" (L. Herman, ed.), pp. 101–147. Wiley Interscience, New York.

Mass, A. (1997). Adaptive features of vision in aquatic mammals. *Sensory Sys.* **11,** 181–209.

Mobley, J., and Helweg, D. (1990). Visual ecology and cognition in cetaceans. *In* "Sensory Abilities of Cetaceans" (J. Thomas and R. Kastelein, eds.), pp. 519–536. Plenum, New York.

Piggins, D., Muntz, R., and Best, R. (1983). Physical and morphological aspects of the eye of the manatee *Trichechus inunguis* Natterer 1883 (Sirenia: Mammalia). *Mar. Behav. Physiol.* **9,** 111–130.

Schusterman, R. J. (1972). Visual acuity in pinnipeds. *In* "Behavior of Marine Animals" (H. E. Winn and B. L. Olla, eds.), Vol. 2, pp. 469–492. Plenum, New York.

Walrus

Odobenus rosmarus

RONALD A. KASTELEIN

Harderwijk Marine Mammal Park, The Netherlands

The walrus is the single species of the pinniped family Odobenidae, distinguished by the upper canines in both sexes being prolonged as tusks. Walruses feed mainly on small organisms on the ocean floor, while almost all other pinnipeds feed primarily on highly mobile fish and crustaceans. For an animal of such a large size, this predator consumes organisms that are relatively low in the food chain. The DIET of the walrus influences its biology: compared to other pinnipeds, the walrus has a less streamlined body, swims more slowly, and dives less deeply. Its sensory systems are adapted to its bentic foraging technique.

I. Classification

The Latin name *Odobenus rosmarus* means "tooth walking sea horse." The genus *Odobenus* consists of only one species: *O. rosmarus*. Two subspecies are recognized based on morphological characteristics and on mitochondrial DNA divergence: Pacific walrus, *Odobenus rosmarus divergens* (Illiger, 1815) and Atlantic walrus, *O. rosmarus rosmarus* (Linnaeus, 1758). A potential third subspecies, the Laptev walrus, *O. rosmarus laptevi* (Chapski, 1940), is dubiously distinct from *O. rosmarus divergens*.

II. External Characteristics

The walrus is the largest pinniped except for the male elephant seal (*Mirounga* spp). The body is rotund; the girth at axilla is almost equal to body length. Males are larger than females of the same age. Adult Pacific walruses are on average slightly larger than adult Atlantic walruses of the same gender. Adult male walruses have an average body length of around 320 cm and weight 1200–1500 kg, whereas adult females have a body length of around 270 cm and weigh 600–850 kg. Regional differences in body size per gender and age class exist.

Walruses are easily distinguished from other pinnipeds by their flat noses and enlarged upper canines that form huge tusks. Males have longer and thicker tusks than females. The tusks grow throughout life but growth is usually balanced by tooth wear. The larger whiskers on the upper lips are translucent and yellowish and are directed forward. The eyes are small relative to body size compared to other pinniped species. They are positioned high on the head and can be protruded and retracted. There are no external pinnae. The color of the skin varies from a (lighter) gray when the walrus has just left cold water (Fig. 1) to a (darker) yellowish brown when it is warm and dry (Fig. 2). The fur is very short and in some areas of the body is absent. Although much variation exists, walruses generally molt inconspicuously between May and June and grow new pelage in July and August. The appendages are hairless and the palms and soles are rough. The skin is 2–4 cm thick and very tough. It is thickest around the neck (about 4 cm) and in the area above the whiskers, which is used for plowing through the ocean floor. The skin on the neck of adult males is thicker than that of females and is covered with fibrous tubercles (Fig. 2).

Figure 1 *An adult male Pacific walrus just arriving nearshore from a long foraging journey at sea. Note the light gray color of the skin, which is due to a reduced blood flow to the skin during low temperatures. Photo by Steve Rice, U.S. Fish and Wildlife Service.*

Figure 2 *An adult male Pacific walrus. Note the large diameter of its tusks, the bulging eye, which is turned to look backward, and the pronounced tubercles of the skin on the neck. The latter feature is only seen in adult males. Photo by Steve Rice, U.S. Fish and Wildlife Service.*

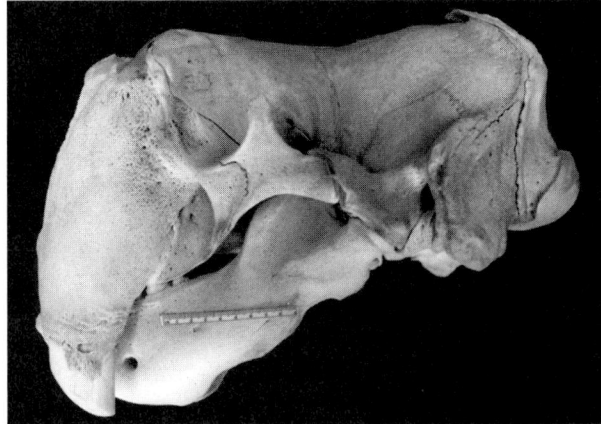

Figure 4 *A side view of a Pacific walrus skull showing the open dorsal side of the orbital cavity and the slit in the zygomatic arch, which in life is filled with cartilage. Photo by Ron Kastelein.*

These tubercles are 1 cm thicker than the surrounding skin, protect the underlying tissues against tusk attacks by other males, and are an important visual sexual characteristic. The blubber layer thickness varies depending on the part of the body and season and can be up to 10 cm.

III. Anatomy

A. Skeleton

The skull has some distinct features that relate to the ecology of the walrus. It is very thick and strong as an adaptation to breaking through ice to make breathing holes. The front of the skull of an adult walrus is much higher and broader than that of other pinnipeds to accommodate the large tusks (Figs. 3 and 4). Males, which have thicker tusks than females, also have broader skulls. The tusks are composed of dentine covered with a thin layer of cementum. The ivory of walruses can be distinguished from that of other mammals by its central globular dentine. The heavy weight of the lower jaw probably serves to increase the impact of the tusks. The walrus is able to use its tusks to haul its body upward onto land or ice because of the strong neck muscles attached to the large mastoid processes and the strong hinge between these processes: the condyle of the occipital bone. The zygomatic arch below the orbital cavity contains a strip of cartilage, probably to dampen shocks from the tusks to the braincase. In contrast to most pinnipeds, the orbital cavity is not closed on the dorsal side of the head, allowing the walrus to look upward during plowing through the substrate.

The spinal column consists of 7 cervical, 14 thoracic (occasionally 15), 6 lumbar (occasionally 5), 4 sacral, and 8 or 9 caudal vertebrae. There are 14–15 pairs of ribs.

The hind flippers of the walrus rotate forward like those of otariids, for LOCOMOTION on ice and land. The females have a 1- to 2-cm-long clitoris bone. The penis bone (BACULUM) of adult males is up to 62 cm long.

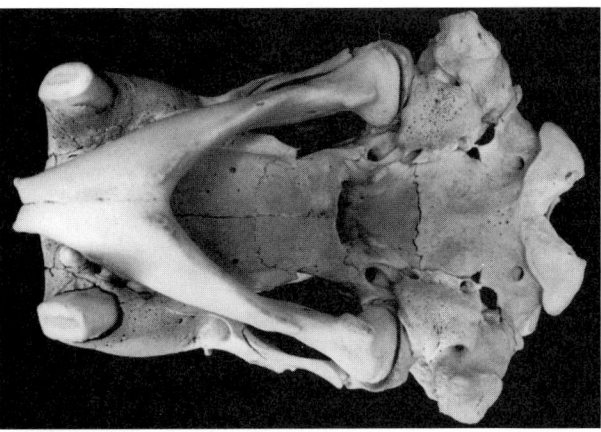

Figure 3 *A ventral view of a Pacific walrus skull showing the large mastoid processes on the sides and well-developed occipital condyle (the hinge to the neck vertebra), the arched roof of the mouth, and the heavy lower jaw. Photo by Ron Kastelein.*

B. Muscles

Many of the muscles in the upper lip of the walrus are used to erect the whiskers in unison, although whiskers can be moved individually, as well.

The tongue muscles are very strong and are used to create a low pressure in the mouth to extract the soft parts from clams. The mastication muscles are not very large, as the walrus usually does not chew its prey but swallows it whole. The cheek muscles are strong so that the walrus can produce powerful water jets from its mouth to wash sediment away from its prey.

The muscles in the hind limbs are similar to those in otariids. The adult walrus is so heavy and rotund that it does not lift its belly off the substrate when it moves on land or ice. Only calves can walk with only their flippers touching the substrate. In the water the hind flippers are used for propulsion and the front flippers mainly for steering.

C. Respiratory System

The trachea is supported by cartilaginous rings throughout its length. It passes between the lungs for a third of their length before bifurcating into bronchi. The lateral walls of the pharynx of subadult and adult males are extremely elastic and, when inflated, form air sacs (pharyncheal pouches). These sacs are used as resonance chambers for the production of the bell-like sounds underwater and also for floatation when resting in the water.

D. Digestive System

The mouth is narrow and bordered by the tusks. The roof of the mouth is very concave, allowing room for the large tongue that acts as a piston when retracted quickly. The force (of up to −119 kPa) generated by this piston is used to extract the soft parts of clams from their shells. Nontusk teeth are worn down to the gums in wild animals, perhaps by sand moving in the mouth during FEEDING. Teeth are pointed in captive walruses that are fed fish. The dental formula of the Pacific walrus' permanent dentition is as follows:

$$\frac{(1)\text{-}(2)\text{-}3}{(1)\text{-}(2)\text{-}(3)} \quad \frac{1}{1} \quad \frac{1\text{-}2\text{-}3\text{-}(4)}{2\text{-}3\text{-}4} \quad \frac{(1)\text{-}(2)}{(1)\text{-}(2)}$$

Incisors ----- Canines -- Premolar ----- Molar ---- ×2 = 18 to 38 teeth

Teeth in parentheses are present in less than 50% of the adult specimens.

The tip of the tongue may be rounded or bifid. The stomach consists of one J-shaped cavity. The intestines are 10–15 times the length of the animal.

E. Cardiovascular System

A conspicuous aortic bulb is present at the base of the aortic arch. The arteries to the fore limbs are larger than those to the hind limbs. This is related to the division of the muscle masses for the limbs; the muscles of the neck and shoulders are more developed than those of the pelvis area.

F. Reproductive System

The testes are situated between the skin and the muscles. The penis is normally retracted into an opening posterior to the umbilicus. The uterus is bicornate and each horn opens separately into the vagina. Walruses usually have four nipples.

IV. Habitat

The walrus is found in the Arctic, where its DISTRIBUTION is limited by the availability of shallow water foraging grounds and thickness of ice. Walruses prefer relatively shallow water over continental shelves because they feed on invertebrates, which occur on the ocean floor up to a depth of about 80 m. Walruses can break through ice up to about 20 cm thick, but when the ice is thicker than this, they retreat to areas with drift ice. Thus, in winter, walruses inhabit those regions of the drifting ice where leads and polynyas (open water) are numerous and where the ice is thick enough to support their weight.

Male Pacific walruses rest in traditional terrestrial haul-out sites (sand, cobble, or boulder beaches), whereas females and calves prefer to haul out on pack ice or ice floes. In summer, males rest and molt while hauled out on land close to their feeding grounds (Fig. 5). Females molt when hauled out on ice. Atlantic walruses also molt in the summer; during that time, both males and females (sometimes in the same groups) haul out on both land and ice.

V. Distribution, Migration, Abundance, and Conservation Status

The Pacific walrus is found principally in the Bering Sea south to Bristol Bay and Kamchatka and in the Chukchi Sea, although in summer it may enter the Beaufort Sea and East Siberian Sea (Fig. 6). Breeding occurs in late winter in the marginal ice zone of the Bering Sea. The location of the main breeding sites varies depending on the state of the ice, but is generally southwest of Nunivak Island and southwest of St. Lawrence Island. Pacific walruses move north in spring with the receding and drifting ice; females give birth on ice floes, which drift north. They move south in autumn, following the movement of the pack ice. Their swimming speed is approximately 10 km/hr. For most of the year, sexes and age classes live separately. The population is estimated at around 200,000 animals.

The Atlantic walrus is found in the western and eastern Atlantic Arctic (Fig. 6). Within their range, Atlantic walruses occur in several (perhaps eight) more or less well-defined subpopulations: five to the west and three to the east of Greenland. In the western Atlantic Arctic, walruses range from the East Canadian Arctic to West Greenland, including Davis Strait, Baffin Bay, the archipelago in the Canadian high Arctic, and Foxe Basin. The western Atlantic population is estimated at >10,000. In the eastern Atlantic Arctic, walruses range from eastern Greenland to Svalbard (Norway), Franz Josef Land (Russia), and the Barents and Kara Seas. The eastern Atlantic population is estimated to be in the low thousands. Movement studies have shown a connection between walruses at Franz Josef Land and Svalbard and between the latter area and eastern Greenland.

The Laptev walrus (if this population is recognized as a subspecies) is found only in the Laptev Sea (Fig. 6). The population has been estimated at 4000–5000 animals.

Figure 5 *A large group of male Pacific walruses hauled out on Round Island, Bristol Bay, Alaska. Note how closely the animals are packed together. Photo by Steve Rice, U.S. Fish and Wildlife Service.*

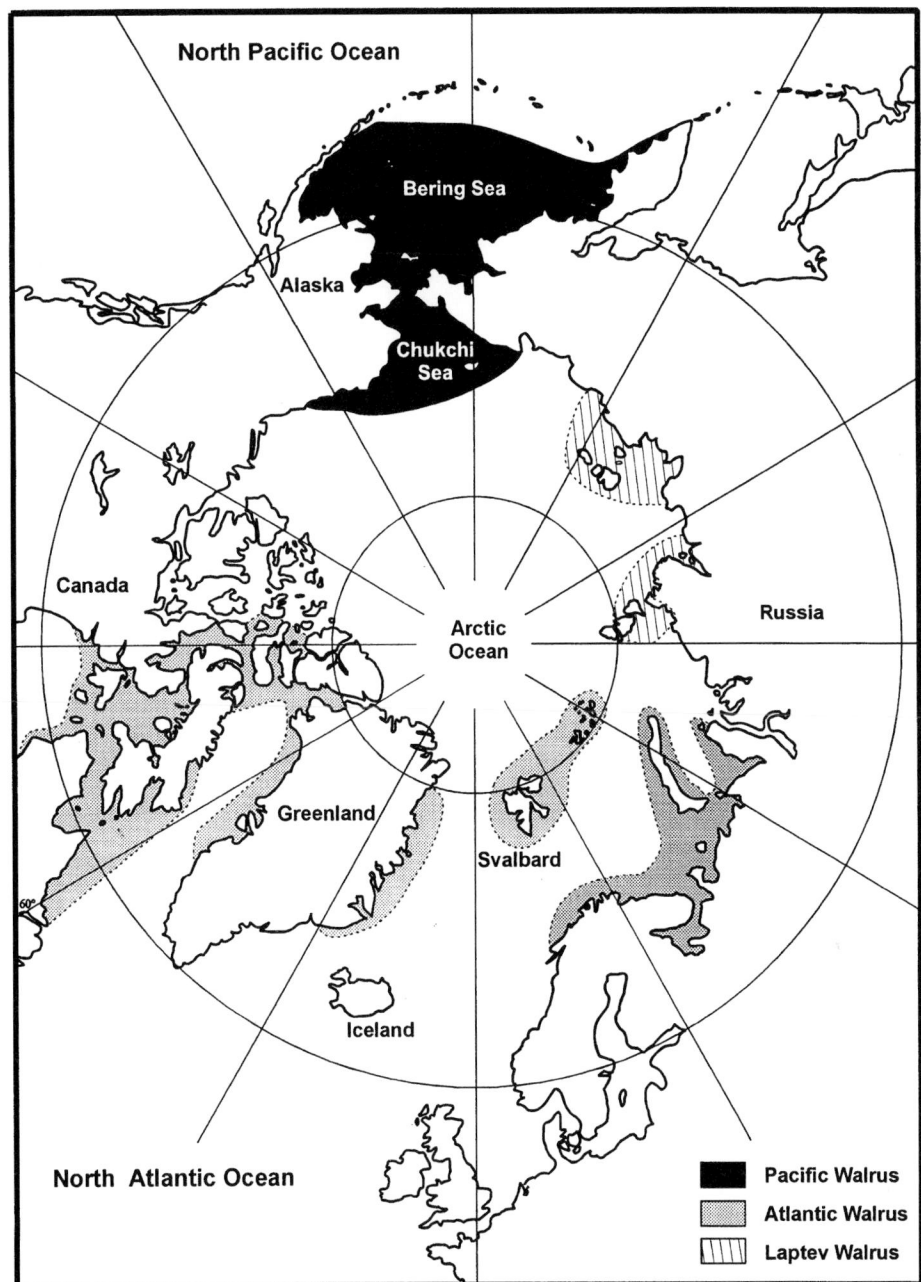

Figure 6 *Distribution of the three walrus subspecies (the recognition of the Laptev walrus as a subspecies is controversial).*

Atlantic walrus stocks were reduced greatly by intense exploitation in the 18th and 19th centuries by European whalers and sealers (for ivory, oil, and hides) and appears not to have recovered fully. The Pacific walrus population recovered from intense hunting, but, for unknown reasons, may have started to decline again.

Atlantic walruses are still hunted for subsistence by native people of Canada and Greenland. Pacific walruses are taken by indigenous peoples of Russia and the United States (Alaska) for the same purpose. Atlantic walruses are fully protected at Svalbard (since 1952) and in the western Russian Arctic (since 1956). Potential threats to walrus populations are overhunting, competition with shellfish fisheries, accidental bycatch by trawlers, POLLUTION (PCBs, heavy metals, nuclear radiation), waterborne NOISE at their foraging areas, aerial acoustic disturbances at their resting places (snowmobiles, aircraft, ships, oil, and gas exploration), and habitat destruction from bottom trawling.

VI. Behavior and Reproduction

Male and female walruses usually do not have contact with each other for most of the year, but animals of each sex congregate in large numbers in both winter and summer. Walruses seem to prefer being in groups. Walruses literally pile on top of each other (probably to conserve heat) in dense aggregations, and spend most of their haul-out time resting (Fig. 5). On land, they are usually in deep sleep and are difficult to wake up. They often lie upside down, perhaps to reduce the pressure on their lungs. In this position their tusks point upward, they can stretch their necks, and surplus ear wax can drain from their ears. Body size and tusk length play important roles in establishing the hierarchy at haul-out sites. The tusks are used for display and as weapons in fights between walruses and against polar bears (*Ursus maritimus*) and killer whales (*Orcinus orca*) (which may prey on walrus calves). They also aid in hauling out on ice floes and shores and serve to enlarge and keep open breathing holes in the ice. In addition they function to hold heads above water by resting on the edges of ice floes.

Several aerial acoustic signals are used in social context: barking, coughing, and roaring when excited, whistling by males during the reproductive period (source levels of around 120 dB re 1 pW have been recorded), and soft calls from the females toward their calves. Alarm calls and other calls of calves have been described. Underwater, bell-like sounds are produced by the air sacs of adult males.

Walruses have well-developed facial muscles and can make many facial expressions, which play a role in short-range COMMUNICATION.

Females generally begin to ovulate at around 7 years of age (but some ovulate already at the age of 5 years) and usually give birth for the first time at the age of 9 years. Males become sexually mature at between 7 and 10 years of age, but become physically and socially mature, and therefore able to mate, at 15 years. Walruses can reach an age of 30 to 40 years.

Adult male and female walruses congregate during the mating season (January–April). Walruses are polygynous. In the mating season, adult males fight intensively in the water, evidently in competition for display sites near females. During courtship, the male walrus emits a stereotyped sequence of underwater sounds consisting of taps, knocks, pulses, and bell-like sounds. This acoustic display probably serves as an advertisement to females and as a warning to other males. Females choose a mate from among the displaying males. Copulation usually occurs in the water, but has been observed to occur on land in captivity, although a pool was available.

After a gestation period of about 15 months (including a period of delayed implantation of 4–5 months), a single calf of around 60 kg and 120 cm in length is born in spring (April–early June). Calves can swim immediately and may sometimes be carried on their mothers' backs for a while. The coat of the calves is slate gray. They are suckled for at least a year (on land, on ice, and in the water) and are usually weaned gradually during their second year, when they begin to forage for invertebrates. Reproductive females can produce about one calf in 3 years. The calves remain near their mothers, in groups of adult females, for several years. Most calves are weaned by the age of 3 years, when males tend to join male herds. The high degree of maternal care and low predation rate result in low natural mortality in walrus calves. This probably allows the walrus to have fewer offspring then other pinnipeds, most of which produce a pup every year.

VII. Foraging Behavior

The diet of walruses consists mainly of benthic invertebrates. By far the most commonly eaten are bivalve mollusks, which are found buried in the sediment in high-density beds. How walruses find these beds is unknown. When they find a mollusk bed, they plough through the sediment with their snouts while swimming with their bodies at a 45° angle to the ocean floor to find prey items (Fig. 7). When foraging by plowing through the sea bed with its snout in search of invertebrates, the walrus uses its tusks and fore flippers as sleds while swimming with its hind flippers. Long furrows in the sediment have been observed in walrus feeding areas.

Once the walrus encounters a potential food item, it is identified quickly by the sensitive whiskers. If it is a bivalve mollusk, the foot or siphon is taken between the mobile lips, and by means of retraction of the large tongue, the soft parts are sucked from the shells and swallowed (Fig. 7). The empty shells are discarded and can be found on the ocean floor near the furrows. The walrus can produce strong water jets with its mouth to excavate its prey but can also remove sediment by producing strong water currents with movements of its fore flippers. The whiskers can be moved individually and are probably used as a tool to manipulate clams so that the foot or siphon is directed toward the walrus' mouth. Once a prey item is in front of the mouth, the tongue probably takes over as a touch and manipulation organ.

Compared to most other pinnipeds, walruses consume organisms that are small and low in the food chain. The large walrus has to consume many small organisms and must have a very

Figure 7 *Walruses foraging. They plow through the sediment on the ocean floor with their noses. When an object is encountered, it is identified with the whiskers. If it is a clam, the foot of the clam is taken between the lips, and the soft parts are sucked from the shell and swallowed. Drawing by Rijkent Vleeshouwer.*

efficient feeding method. In fact, walruses dive for up to 24 min (average around 5 min), usually spend 80% of that time on the bottom, and generally obtain 40–60 clams per dive. Dive times probably vary depending on water depth, prey type, and prey density. Adult walruses require on average about 25 kg of soft clam parts per day, and a walrus has been found with 6000 prey items in its stomach. In captivity, Pacific walruses sometimes eat up to 50 kg of fish per day. Walruses probably store extra fat during summer when they can exploit their inshore foraging grounds and use this during migration when they swim in waters that are too deep to contain their prey.

Uncommonly, individual walruses may take a vertebrate diet. These animals usually kill seals and cetaceans as well as scavenge on their carcasses, from which they suck the blubber and internal organs. Seal-eating walruses thus consume organisms that are higher in the food chain than those consumed by mollusk-eating walruses and therefore carry a relatively high concentration of heavy metals and PCBs. Walruses sometimes also capture and eat seabirds.

VIII. Sensory Systems

A. Vision

The walrus has well-developed extrinsic eye muscles. The orbital cavity is not closed on the dorsal side (Fig. 4). This allows the walrus binocular vision in the frontal and dorsal direction, as it can protrude its eyes, and does so mostly when excited (Fig. 2). The lack of a roof of the orbital cavity suggests that the eyes are vulnerable to mechanical injury. However, the walrus eye has strong retractor muscles. The eyes can be pulled deep into the orbital cavity for protection, and the eye opening can be closed with thick eyelids.

Blood vessels and surrounding fat probably serve to keep the eyes warm and functional under cold conditions. Under high light conditions, the pupil is a vertical slit; during moderate light levels, key-hole shaped; and under low light conditions, circular. Retinal anatomy suggests that the walrus has color vision, but because no psychophysical tests have been carried out, it is unclear which part of the spectrum it can detect. Visual acuity appears to be less than in other pinnipeds investigated so far, and the eyes of the walrus seem to be specialized for short-range VISION.

B. Touch

Because walruses dive up to 130 m and also at night, they often cannot always use vision to detect and process their prey. Instead, they use their sensitive mystacial vibrissae (whiskers). Each walrus has about 450 whiskers, which are highly innervated by sensory and motor nerves. The main sensory nerve (trigeminal nerve) is well developed, and the infraorbital foramen in the skull is proportionally large to accommodate it (Fig. 4). In contrast to most other pinnipeds, which probably use their whiskers to detect vibrations in the water, walruses use their whiskers to examine and manipulate small objects. In a psychophysical test, a captive walrus could distinguish a circle and a triangle that each had a surface area of 0.4 cm^2. The longer and thicker lateral whiskers are used mainly for the detection of objects, whereas the shorter and thinner ones near

the mouth opening are used primarily for identification. The tip of the tongue also contains many mechanoreceptors and can be used to identify or reposition prey.

C. Olfaction

The behavior of the walrus on land and on ice suggests that it probably relies to a high degree on its sense of smell to obtain information about its surroundings. Anatomical evidence also suggests the importance of olfaction to walruses. They have large nares that can be closed during dives, and the nasal passage is highly vascularized so that air that passes through it can be heated. However, no conclusive psychophysical tests have been conducted on the olfactory sensitivity of the walrus.

D. Taste

Compared to many terrestrial mammals, the walrus has relatively few but large taste buds. However, not much is known about the taste abilities and discrimination in walruses. Anecdotal information on captive animals suggests that they are not very sensitive to (bitter) flavors that are disgusting to most terrestrial mammals.

E. Hearing

The walrus has limited ability to locate the source of airborne sounds, as evidenced by its lack of pinnae (Fig. 2). When the walrus is in air, sound reaches the tympanic membrane via a large cartilaginous outer ear tube. The aerial hearing has been tested and is less acute than that of humans in the frequency range tested (125 Hz–8 kHz).

When diving, the walrus closes its auditory meatus and hears by tissue conduction, probably mainly via the vascular lining of the outer ear tube. The underwater hearing range of the walrus has an upper frequency limit of 16 kHz.

IX. Captivity

The first walruses in captivity were in Denmark and Germany. Later, zoological parks in the Netherlands, United States, and Russia began keeping this species.

Captive walruses often ingest foreign objects that they cannot digest. A walrus can even die from consuming too many dead leaves that sometimes drop in pools in autumn. Another problem is the wearing down of tusks on the pool floor and walls, which, in severe cases, can cause root infections.

When cared for properly, walruses are friendly and easily TRAINED. They can perform for the public as well as in psychophysical research projects. Occasionally, males become difficult to handle after they reach maturity. Walruses can be kept in good health on a diet of whole fish, which they swallow without chewing. Reproduction in parks and zoos has improved in the last decade, but only a few calves have been raised by their mothers, as walrus mothers are very protective of their calves toward both conspecifics and humans. This causes the mother to neglect her calf (she spends more time defending the calf than nursing it) or reduces the production of milk. The main problem in the early years of husbandry was the lack of a good formula to raise calves that had been captured from the wild. This problem has been overcome, and calves from the wild or captive-born now can be hand-raised without nutritional deficiencies.

See Also the Following Articles

Blubber ■ Hair and Fur ■ Parental Behavior ■ Skull Anatomy

References

Anderson, L. W., Born, E. W., Gjertz, I., Wiig, Ø., Holm, L.-E., and Bendixen, C. (1998). Population structure and gene flow of the Atlantic walrus (*Odobenus rosmarus rosmarus*) in the eastern Atlantic Arctic based on mitochondrial DNA and microsatellite variation. *Mol. Ecol.* **7,** 1323–1336.

Born, E. W. (1992). (*Odobenus rosmarus* Linnaeus) 1758-Walross. *In,* "Handbuch der Säugetiere Europas" (R. Duguy and D. Robineau, eds.), Band 6, Meeressäuger, Teil II, pp. 269–299. Robben-Pinnipedia, AULA-Verlag, Wiesbaden.

Born, E. W., Dietz, R. Heide-Jørgensen, M. P., and Knutsen, L.Ø. (1997). Historical and present distribution, abundance and exploitation of Atlantic walruses (*Odobenus rosmarus rosmarus* L.) in eastern Greenland. *Meddelser om Grønland Biosci.* **46,** 73.

Born, E. W., Gjertz, I., and Reeves, R. R. (1995). Population assessment of Atlantic walrus (*Odobenus rosmarus rosmarus* L.) Meddelelser no. 138 of Norsk Polar Institutt, Oslo.

Fay, F. H. (1982). Ecology and biology of the Pacific Walrus, *Odobenus rosmarus divergens,* Illiger. North American Fauna No. 74. United States Dept. of the Anterior, Fish and Wildlife Service, Washington, DC.

Fay, F. H. (1985). *Odobenus rosmarus. Mamm. Species.* **238,** 1–7.

Fay, F. H., Eberhardt, L. L., Kelly, B. P., Burns, J. J., and Quakerbush, L. T. (1997). Status of the Pacific walrus population, 1950–1989. *Mar. Mamm. Sci.* **13**(4), 537–565.

Fedoseev, G. A. (1984). Present status of the population of walruses (*Odobenus rosmarus*) in the eastern Arctic and Bering Sea. *In* "Marine Mammals of the Far East" (V. E. Rodin, A. S. Perlov, A. A. Berzin, G. M. Gavrilov, A. I. Shevchenko, N. S. Fadeev, and E. B. Kucheriavenko, eds.). TINRO, Vladivostok. [Translated by F. H. Fay (1985).]

Gjertz, I., and Wiig, Ø. (1993). Status of walrus research in Svalbard and Franz Josef Land in 1992., A review. Appendix IV. *In* "Report of the 2nd Walrus International Technical and Scientific Workshop (WITS), 11–15. January 1993 (R. E. A. Stewart, P. R. Richard, and B. E. Stewart, eds.), Winnipeg, Manitoba, Canada. Can. Tech. Rep. Fish. Aquat. Sci. 1940, 68–84.

Kastelein, R. A., and Gerrits, N. M. (1990). The anatomy of the walrus head (*Odobenus rosmarus*). 1. The skull. *Aquat. Mamm.* **16**(3), 101–119.

Kastelein, R. A., Gerrits, N. M., and Dubbeldam, J. L. (1991). The anatomy of the walrus head (*Odobenus rosmarus*). 2. Description of the muscles and of their role in feeding and haul-out behaviour. *Aquat. Mamm.* **17**(3), 156–180.

Murie, J. (1871). Researches upon the anatomy of the Pinnipedia. I. On the walrus (*Trichechus rosmarus* Linn.). *Trans. Zool. Soc. (Lond.)* **7,** 411–464.

Reijnders, P. J. H., Verriopoulos, G., and Brasseur, S. M. J. M. (1997). "Status of Pinnipeds Relevant to the European Union." IBN Scientific Contributions 8, The Netherlands, 160–182.

Richard, P. R., and Campbell, R. R. (1988). Status of the Atlantic walrus, *Odobenus rosmarus rosmarus* in Canada. *Can. Field Nat.* **102**(2), 337–350.

Wiig, Ø., Gjertz, I., Griffiths, D., and Lydersen, C. (1993). Diving patterns of an Atlantic walrus *Odobenus rosmarus rosmarus* at Svalbard. *Pol. Biol.* **13,** 71–72.

Weddell Seal
Leptonychotes weddellii

JEANETTE A. THOMAS
Western Illinois University, Moline

The Weddell seal has a circumpolar distribution around Antarctica. The seal hauls out through cracks in the fast ice formed from tidal action. This fast ice provides a stable platform for giving birth to pups, hauling out to avoid predators, and resting. Use of these traditional haul-out sites facilitated research on this species since the early 1960s. This species is the best studied of all Antarctic pinnipeds because the seals can be found reliably in breeding colonies at traditional sites. Several investigators maintained research programs near the U.S. bases at McMurdo Sound and Palmer Peninsula, the New Zealand Scott Base, and the Australian base at Davis Station.

I. Characters and Taxonomic Relationships

The Weddell seal belongs to the family Phocidae, subfamily Monacinae, and tribe Lobodontini. There is a single species within the genus *Leptonychotes*. The closest relatives are the other Antarctic seals in the tribe Lobodontini (i.e., crabeater, *Lobodon carcinophaga;* leopard, *Hydrurga leptonyx;* and Ross seals, *Ommatophoca rossii*) and the monk seals in the tribe Monachini (i.e., Caribbean, Mediterranean, and Hawaiian monk seals, *Monachus* spp.).

Weddell seals are black with grayish silver streaks, with individual variations. The adult fur does not have underfur. In pups, the lanugo is longer and thicker than adult fur and is shed by 44 days of age.

There is no dramatic SEXUAL DIMORPHISM in body size; males reach 2.5–2.9 m and females are 2.6–3.3 m in length. Weight ranges from 400 to 500 kg, with pregnant females attaining the greatest weights. At birth, pups weigh 22–29 kg and are about 1.5 m in length.

As with other phocids, the Weddell seal has a fusiform body shape and laboriously crawls on the belly or rolls to move on ice. It is not capable of upright stance or moving the hindlimbs forward. Under water, the Weddell seal propels itself easily by moving the hind flippers in a vertical plane. Manus are flippers and pes are fully webbed. The first metacarpal is noticeably larger than the others. Black claws on the front flippers are large and probably useful for gripping the ice or scratching. Claws on the hind flipper are reduced in size. The tail is distinct and free.

Its large brown eyes often have wet circles around them because of no lacrimal duct. It has a tapetum, which assists in seeing in low-light levels during the austral winter and deep dives. A nictitating membrane protects its eyes from blowing snow and allows its eyes to be open in salt water. The species has excellent VISION under water. Some accounts report that the seal remains in familiar, shallow water areas during the dark austral winter.

The external ear is absent and nothing is known about hearing abilities. The nostrils are oriented vertically and normally are closed, but open when the seal needs to respire. There are seven rows of mystacial vibrissae and superciliary whiskers that are smooth, not beaded as in some pinnipeds. The tip of the tongue is notched. Testes are abdominal, the penis is retractable, and there is a baculum. The uterus is bipartite. Two mammary glands are present and the milk is exceptionally high in fat and protein content.

It has a simple stomach and eats food whole, fish heads down first. It does not drink water, rather obtains water metabolically through diet. It has reniculate KIDNEYS adapted for conserving water and removing high salt loads. There are 34 chromosome pairs in this species.

The dental formula is 2/2, 1/1, 4/4, 1/1. The milk teeth disappear before or soon after birth. Cheek teeth have three points, but are not the exaggerated tricuspid structure of the crabeater and leopard seals. The outer incisors are larger and procumbent or project forward. The canines also are procumbent and, along with the last incisor, are used in "ice-sawing" behavior. The seal maintains circular breathing holes in the fast ice by turning within the hole and raking ice off the rim of the hole with its teeth. Such behavior allows the seals to maintain holes for breathing and hauling out throughout the year. However, ice-sawing behavior has prevented this species from being successfully maintained in captivity.

The seals stay warm through a heavy fur coat and thick layer of subcutaneous fat; their lower critical temperature is −40°C. Weddell seals can vasodilate to dissipate heat or vasoconstrict to conserve body heat. On a warm Antarctic day, steam can be seen rising from their bodies.

II. Distribution and Ecology

The subfamily Monachinae evolved in the North Atlantic during the Miocene. A fossil, *Homiphoca*, found off the South African coast could be the intermediate form between monk seals and Antarctic seals.

No systematic, large-scale population census studies have been conducted, so estimates in the literature are approximate. The Weddell seal is abundant; the estimated range is from 500,000 to 1 million seals. Occasionally, single seals are seen at sub-Antarctic islands, including South Shetlands, the South Orkneys, and South Georgia. Single wandering Weddell seals have been found in remote locations such as Heard, Kerguelen, Macquarie, Auckland, Juan Fernandez, and Falkland Islands, as well as Australia, New Zealand, Patagonia, and Uruguay.

There is no predictable migration in this species, but because there is a trend to move toward the fast ice, seals move northward before the onset of winter as the ice expands and southward during the summer to find pockets of fast ice.

Weddell seals are extremely good divers, commonly diving about 600 m, for up to 82 min (20 min is typical), ranging out to 5 km from a breathing hole and returning on a single dive. A great deal about their DIVING BEHAVIOR has been documented using satellite tags with time-depth recorders and video cameras attached to the seal's back. Before a dive, the seal exhales and the nostrils and mouth are closed. Seals usually make a series of short shallow dives before commencing on a longer dive. Swim speed is 8–12 km/hr and the descent rate for a dive is 35 m/min. During a dive, the Weddell seal collapses the lungs, compresses the trachea, shunts blood to extremities, drops the metabolic rate to 20% of the resting rate, and undergoes bradycardia from about 85 to 16 beats/min. The blood and muscles have a three to five times greater oxygen-carrying capacity than that of humans.

III. Behavior and Life History

Mother Weddell seals return to tidal cracks each austral spring and give birth to pups from late September to early November. Typically, a single pup is born, but twins have been observed and even an albino pup was reported. A few mothers have pups in isolation from other seals, but most give birth in colonies of up to 50 mothers with pups. Mothers and pups maintain individual spacing when hauled out on ice (Fig. 1). When disturbed, mothers vocalize to intruders. Pups nurse for a 7- to 8-week period, gaining almost 2 kg per day. At weaning, pups weigh about 125 kg and mothers have lost nearly that amount of weight because they do not forage during the nursing period. Pups are first enticed into the water by their mother's calling at 10–14 days of age. Pups struggle to swim and stay under the water, popping to the ice surface. Some pups die in the breathing holes, not being able to crawl out of the slippery, steep hole. Mothers and older pups go into the water progressively more often as the pup grows. They exhibit a distinct diel pattern of haul-out, with most seals hauling out in the colony for several hours around midday.

Adult males are not tolerated on the ice in maternal colonies. Only occasionally is a nonbreeding, adult male seen in the periphery of the colony. Soon after females give birth, adult males arrive and compete for underwater territories aligned beneath the tidal cracks. Males patrol using loud (up to 193 dB re 1 μPa) trills to advertise and defend their underwater territories, which cover an area 15–50 m wide by 50–400 m long. Sometimes bloody fights occur between males under water and even continue onto the ice, with the loser evicted from the territory. The underwater repertoire of Weddell seals is elaborate, including some 34 sound types near McMurdo Sound. GEOGRAPHIC VARIATIONS in the

Figure 1 *A mother Weddell seal calls to her pup to maintain spacing within a breeding colony.*

repertoire were documented among McMurdo Sound, Palmer Peninsula, and Davis Station seals. Their sounds are some of the longest among marine mammals, ranging up to 70.0 sec, with 75 calls given per minute by a colony during the height of the breeding season. Mating takes place in water, with about 80% of females becoming pregnant, at least in the McMurdo Sound area. Once the pups wean, the mother enters the water to feed and mating occurs within a male's territory. Adult seals molt after mating. Implantation of the blastocyst is delayed until mid-January. Because of delayed implantation, pups are born the following spring at approximately the same time of year. Delayed implantation allows the mother to molt, feed, and recover from the dramatic weight loss associated with lactation before another fetus starts to develop.

About the time the last pup in the colony weans in mid-December the fast ice breaks up, adults disperse, and newly weaned pups are left to fend for themselves. Some data taken from pups with transmitters indicated pups feed near shore on small fish such as *Pleurograma* spp.

Subadult seals are rarely seen in breeding colonies. They tend to congregate in large groups near the ice edge. Subadult Weddell seals do not exhibit the scars from leopard seal predation that are seen on crabeater seals. However, killer whales are known to take Weddell seals of all ages. When killer whales (*Orcinus orca*) and leopard seals move into areas of Weddell seals, *Leptonychotes* suddenly and dramatically stops calling underwater, perhaps to avoid detection.

An adult seal was seen to surface in a breathing hole with a large, 100-pound, Antarctic cod (*Dissostichus mawsoni*) in its mouth. Video footage from a camera mounted on a seal's back recorded an episode of capture of this large prey. If a fish hides in a crevice, the seal blows bubbles into the water to chase the fish into an open area.

IV. Interactions with Humans

The seals have little fear of humans because there are no natural land predators like the (Arctic) polar bears. James Weddell reported seeing and hearing this seal during his Antarctic expeditions and the seal was named after him. A detailed description of this species was recorded in 1907 by E. A. Wilson while he spent a 2-year period at Hut Point on Captain R. F. Scott's first expedition to Antarctica. The expedition reportedly took seals to feed their sled dogs. In 1934–1935, A. A. Lindsey studied Weddell seals at the Bay of Whales as part of Admiral R. E. Byrd's second Antarctic expedition and specimens now are deposited in the Field Museum of Natural History in Chicago. In 1940, G. C. L. Bertram studied the species as part of the British expedition to the Antarctic Peninsula. J. Sapin-Jaloustre (1952) studied the seal at Terre Adelie Land as part of a French expedition. Some early specimens were collected by the Little America explorers. In 1957, the International Geophysical Year established the importance of conservation and research in the Antarctic. The Antarctic treaty was signed in 1961, establishing the protection of the species and its habitat. During the 1970s and 1980s, inhabitants of the New Zealand station in McMurdo Sound took Weddell seals to feed their sled dogs over the winter. This practice has stopped. Otherwise, there is no record of extensive harvest of this species.

See Also the Following Articles

Diving Physiology ■ Earless Seals ■ Teeth

References

Bertram, G. C. L. (1940). The biology of the Weddell and crabeater seals. British Graham Land Expedition 1934–7. Scientific Reports 1:1–39. British Museum of Natural History, London.

Burt, W. H. (1971). "Antarctic Pinnipedia," Vol. 18. Antarctic Research Series.

Davis, R. W., Fuiman, L. A., Williams, T. M., Collier, S. O., Hagey, W. P., Kanatous, S. G., and Horning, M. (1999). Hunting behavior of a marine mammal beneath the Antarctic fast ice. *Science* **283,** 993–996.

Elsner, R., Kooyman, G. L., and Drabek, C. M. (1970). Diving duration in pregnant Weddell seals. *In* "Antarctic Ecology" (M. W. Holdgate, ed.), pp. 477–482. Academic Press, New York.

Kooyman, G. L. (1981). "Weddell Seal, Consummate Diver." Cambridge Univ. Press, Cambridge.

Lindsey, A. A. (1937). The Weddell seal in the Bay of Whales, Antarctica. *J. Mammal.* **18,** 127–144.

Lugg, D. J. (1966). Annual cycle of the Weddell seal in the Vestfold Hills, Antarctica. *J. Mammal.* **47,** 317–322.

Nowak, R. M. (1991). "Walker's Mammals of the World," 5th Ed., Vol. 2, Johns Hopkins Univ. Press, Baltimore.

Sapin-Jaloustre, J. (1952). Weddell's seal, Mammalia. *In* "National Antarctic Expedition 1901–4." Natural History **2,** 1–66. British Museum, London.

Siniff, D. B., DeMaster, D. P., Hofman, R. J., and Eberhardt, L. L. (1977). An analysis of the dynamics of a Weddell seal population. *Ecol. Monog.* **47,** 319–335.

Smith, M. S. R. (1965). Seasonal movements of the Weddell seal in McMurdo Sound, Antarctica. *J. Wildl. Manage.* **29,** 464–470.

Stirling, I. (1971). Population dynamics of the Weddell seal (*Leptonychotes weddelli*) in McMurdo Sound, Antarctica 1966–1968. *Antarct. Res. Ser.* **18,** 141–161.

Testa, J. W. (1987). Juvenile survival and recruitment in a population of Weddell seals (*Leptonychotes weddellii*) in McMurdo Sound, Antarctica. *Can. J. Zool.* **65,** 2993–2997.

Testa, J. W., Siniff, D. B., Croxall, J. P., and Burton, H. R. (1990). A comparison of reproductive parameters among three populations of Weddell seals (*Leptonychotes weddellii*). *J. Anim. Ecol.* **59,** 1165–1175.

Thomas, J. A., and Kuechle, V. B. (1982). Quantitative analysis of the underwater repertoire of the Weddell seal (*Leptonychotes weddellii*). *J. Acoust. Soc. Am.* **72,** 1730–1738.

Wartzok, D., Elsner, R., Stone, H., and Burns, J. (1992). Under-ice movements and the sensory basis of hold finding in the ringed and Weddell seals. *Can. J. Zool.* **70,** 1712–1722.

Whale Lice

CARL J. PFEIFFER
Virginia Polytechnic Institute and State University, Blacksburg

Whale lice (cyamids) are crustacean ectoparasites living on the skin of some species of cetaceans. Whale lice remain among some of the world's biologically most specialized but least understood crustaceans. The current lack of knowledge about these animals undoubtedly stems from the fact that their natural habitat is limited to the skin surface of primarily slow-moving baleen whales, which themselves are difficult to study because they are generally submerged and constantly moving. Accordingly, it is very difficult for scientists to observe, in their natural setting and over extended periods of time, the behavior and biologic processes of whale lice. Some features have been learned, however, by a small number of scientists who have studied these interesting crustaceans for over a century, and the following brief account reviews these data.

I. Classification

Whale lice are actually not lice but perhaps they acquired this nickname in the 1800s from whalers who noticed that they crawled, presumably as parasites, on the surface of the whale's skin and that their size, in proportion to that of a great whale, was comparable to the size of louse on a human or dog. Whale lice are arthropods of the subphylum Crustaca, class Malacostraca, order Amphipoda, family Cyamidae, and genus *Cyamus*. As such they are closely related to other more commonly observed amphipods such as sand hoppers and caprellids, i.e., common marine peracarids. Altogether there are more than 5000 crustacean species, but only 23 known species of whale lice or cyamids. At least 15 of these comprise the genus Cyamus. Because the practice of WHALING has such a long history, cyamids were historically documented early, including *Cyamus ceti* by Linné in 1758, a genus described further by Latreille in 1796. A number of species were classified in the 1800s, including *C. ovalis* in 1834, *C. delphini* in 1836, *C. boopis* in 1870, and *C. scammoni* in 1872, and these cyamids were found on southern right whales (*Eubalaena australis*), humpback (*Megaptera novaeangliae*), gray (*Eschrichtius robustus*), and other large whales. Species names have occasionally been revised for cyamids and new species were discovered even late in the 20th century, such as the two new cyamids reported on one of the large beaked whales, the Baird's beaked whale *Berardius bairdii* (Waller, 1989). Genus names for cyamids have also been changed over time and the earlier genus *Paracyamus* is no longer considered valid.

II. General Ecology and Morphology of Cyamids

Although information does not exist on the early development of cyamids, their evolution has undoubtedly been closely coupled with the long evolution of cetaceans, which first began around 55 million years ago when terrestrial species returned to the marine environment as precursors of the modern whales. Thus, modern cyamids show little resemblance to other crustaceans and have become greatly specialized for their parasitic and lifelong relationship to whales. They have a high degree of host specificity to whales, and one species, *Cyamus boopis*, is found only on humpback whales and another species, *C. scammoni*, only occurs on the gray whale. Some other species of whale lice overlap their residence on two to four species of whales. The two isolated Arctic odontocetes, beluga whales (*Delphinapterus leucas*) and narwhals (*Monodon monoceros*),

each share the same two cyamids, *C. monodontis* and *C. nodosus*. Table I summarizes the cetacean distribution of cyamids. Cyamids are unable to swim freely in the sea or from whale to whale at any of their developmental stages. Accordingly, they die if they lose their foothold on their host, but can be transferred from cetacean mother to calf or during cetacean mating. Although details on the number of juvenile stages that occur prior to adulthood are not yet established for cyamids, Leung (1976) has estimated that at least seven to eight instar stages exist for *C. scammoni* of the gray whale.

Although the general gross body structure of cyamids has been described elsewhere (Margolis, 1955; Leung, 1967; Berzin and Vlasova, 1982), very few studies have been directed toward the microscopic anatomy of cyamids. Early work on the musculature and very recent work (Levin and Pfeiffer, 1999) on cyamid ocular structure have been reported. Their small, paired eyes appear almost rudimentary. However, ultrastructural analysis of these photoreceptors for *C. ceti* has revealed well-developed sensory organs with each eye containing about 50 visual ommatidial units and an overall organization similar to other amphipod compound eyes. The exoskeleton of cyamids consists of a chitinous cuticle that is similar to that typically observed for other crustaceans. It has an exocuticle with multiple microfibrillar lamellae and an endocuticle traversed by both pore canals and dense fibers, as revealed by electron microscopy (Pfeiffer and Viers, 1998).

Cyamids are SEXUALLY DIMORPHIC. The males are larger and, depending on species, adult cyamids usually range from approximately 6 to 19 mm in length. The most striking features of their appearance are their marked degree of segmentation and prominent gnathopods, or legs, with large dactyli, or hooks, that assure firm attachment to the host. The body is flattened and divided into a small cephalic–cephalon or head with paired, minute eyes, and segmented pereion or body to which are attached two pairs of gills and four pairs of gnathopod-type appendages. Figures 1 and 2 illustrate examples of the general body structure for *C. ovalis* and *C. scammoni*, respectively.

Cyamids are mostly found on those areas of the whale surface most protected from the turbulence of water flow, which on baleen whales include regions around barnacles, skin folds or ventral grooves of the head, protected zones around the blowholes, eyes, and flippers, margins of the lips, on callosities, wounds, and genital slit. In those species of whales that serve as host to several species of cyamids, there may be differences in the spatial distributions of the different species of whale lice, and within one species of cyamid the reproductive status and sex of the cyamid may alter the spatial distribution (Rice and Wolman, 1971; Balbuena and Raga, 1991; Rowntree, 1996). Whale lice do move around on their cetacean hosts; in the case of *C. boopis* of the humpback whale, the larger males may carry their smaller female mates and, in an artificial aquarium setting, were observed to walk at a rate of 4.5 m/hr (Rowntree, 1996).

Whale lice breathe by means of two pairs of external gills, which are much reduced in size in early juvenile stages. It has been reported that they can live for up to 3 days out of an aquatic environment, such as on a stranded whale, suggesting that they can also rely on integumentary respiration (Leung, 1976).

TABLE I
Distribution of Cyamid Species on Cetaceans[a]

Host	Whale lice
Mysticeti	
Bowhead whale, *Balaena mysticetus*	*Cyamus ceti*
Right whales, *Eubalaena* spp.	*C. ceti, C. erraticus, C. gracilis, C. ovalis, C. catadontis*
Gray whale, *Eschrichtius, robustus*	*C. ceti, C. kessleri, C. scammoni*
Blue whale, *Balaenoptera musculus*	*C. balaenopterae, C. bahamondei*
Humpback whale, *Megaptera novaeangliae*	*C. boopis, C. elongatus*
Common minke whale, *B. acutorostrata*	*C. balaenopterae*
Fin whale, *B. physalus*	*C. balaenopterae, C. bahamondei*
Odonticeti	
Sperm whale, *Physeter macrocephalus*	*C. ovalis, C. catadontis, C. bahamondei, Isocyamus delphini, Neocyamus physeteris*
Baird's beaked whale, *Benardius bairdii*	*Platycyamus flaviscutatus, C. orubraedon*
Beluga, *Delphinapteris leucas*	*C. monodontis, C. nodosus*
Narwhal, *Monodon monoceros*	*C. monodontis, C. nodosus*
Northern and southern bottlenose whales, *Hyperoodon* spp.	*C. thompsoni, I. delphini*
Long-finned pilot whale, *Globicephala melas*	*Isocyamus delphini*
Short-finned pilot whale, *G. macrorhynchus*	*I. delphini*
Short-beaked common dolphin, *Delphinus delphis*	*I. delphini, syncyamus pseudorcae*
Risso's dolphin, *Grampus griseus*	*I. delphini*
White-beaked dolphin, *Lagenorhynchus albirostris*	*I. delphini, Scutocyamus parvus*
Harbor porpoise, *Phocoena phocoena*	*I. delphini*
Killer whale, *Orcinus orca*	*I. delphini, C. antarcticensis*
False killer whale, *Pseudorca crassidens*	*I. delphini, S. pseudorcae*
Rough-toothed dolphin, *Steno bredanensis*	*I. delphini*
Gervais' beaked whale, *Mesoplodon densirostris*	*I. delphini*
Pantropical spotted dolphin, *Stenella attenuata*	*Syncyamus* sp.
Striped dolphin, *S. coeruleoalba*	*Syncyamus* sp.
Spinner dolphin, *S. longirostris*	*Syncyamus* sp.
Common bottlenose dolphin, *Tursiops truncatus*	*Syncyamus* sp.

[a]From Margolis (1955), Leung (1967, 1970), Lincoln and Hurley (1974), and Berzin and Vlasova (1982).

III. Feeding Habits of Cyamids

It was speculated for a long time that whale lice fed on whale skin and hence they were deemed ectoparasites. Some workers suggested that they might be ominvorous and ingest algal filaments or suspended materials or plankton in the water near their attachment site. However, their mouth parts are very small compared to their body size and they do not possess claws such as some other crustaceans or food-gathering cirri such as some predatory, sessile crustacean barnacles (Pfeiffer and Lowe, 1989). Cyamids have poorly developed paired mandibles and incisor processes with strong chitinous teeth that appear well suited for piercing and scraping skin. Rowntree (1983) showed that the color of intestinal contents of cyamids from humpback whales matched the skin color (black or white) from which the cyamids were collected. More recent conclusive evidence has proven that whale skin is a principal dietary material of cyamids. Both electron microscopic proof of whale skin keratinocytes within the upper digestive tract of cyamids and stable isotope evidence have shown that the dietary staple of whale lice is whale skin. Analysis of stable carbon and nitrogen isotope ratios from cyamids and skin from six species of whales have shown that the cyamic ratios closely matched those of whale skin, but not those of zooplankton from the sea where the cyamids and whales reside (Schell *et al.*, 1999). Also supporting this conclusion is the evidence of direct damage to the skin by whale lice (Leung, 1976). Thus, cyamids have evolved into the only obligate parasites among the amphipods in distinction to other amphipods, such as caprellids, which are predatory and feed on diatoms, other crustaceans, and so on.

IV. Reproduction in Cyamids

Reproductive and mating behavior has been less studied in cyamids than in other amphipods and, indeed, has not been investigated in most cyamid species. The males practice mate guarding and consorts are formed, but there appears to be less aggressive territorialism than is evident with some other amphipods. Little is known about cyamid copulation. There is morphological evidence of a secretory product being released on the cuticular surface of amphipods (Pfeiffer and Viers, 1998), but it is not known if this serves as a pheromone-type attrac-

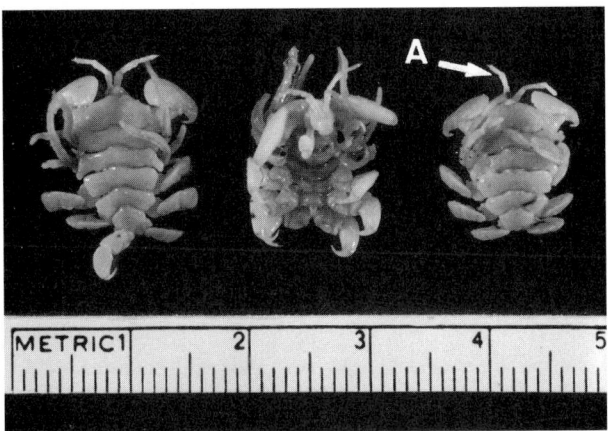

Figure 1 *Three specimens of* Cyamus ovalis *showing ventral surface on center specimen. The head region faces the top. Note segmented body and antennae (A).*

tant for mates or serves some other function. Electron microscopic evidence has shown many tactile sensillae on the antennae and head regions of cyamids, some of which are also likely chemoreceptors. One can question if they always sit on their sole food source, why they have evolved so many sensillae. Female cyamids have a brood pouch (four-plated) or marsupium on their ventral surface, and both unhatched eggs and juvenile whale lice are retained in this cavity. A clutch of 1078 eggs was observed in the marsupium of one female *C. scammoni* (Leung, 1976). The young cyamids measure only about 0.5 mm in length and crawl in and out of the marsupium during development and remain there for at least 2–3 months, when they become about 1.5 mm in length for *C. scammoni*. Several workers have proposed a seasonality for cyamid reproduction, but partly due to the migratory habits of whales, detailed data are not yet available on potential seasonal changes.

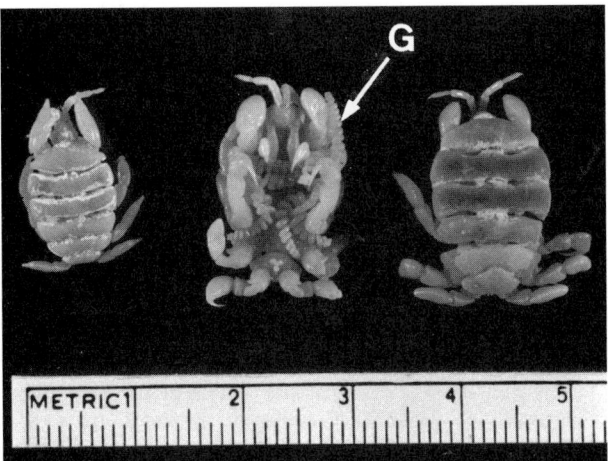

Figure 2 *Three specimens of* Cyamus scammoni *showing ventral surface on center specimen. The gills (G) of this species have a spiral shape.*

See Also the Following Articles

Baleen Whales ■ Callosities ■ Parasites

References

Balbuena, J. A., and Raga, J. A. (1991). Ecology and host relationships of the whale-louse *Isocyamus delphini* (Amphipoda: Cyamidae) parasitizing long-finned pilot whales (*Globecephala melas*) off the Faroe Islands (Northeast Atlantic). *Can. J. Zool.* **69,** 141–145.

Berzin, A. A., and Vlasova, L. P. (1982). Fauna of the Cetacea Cyamidae (Amphipoda) of the world ocean. *Invest. Cetacea* **13,** 149–164.

Leung, Y. M. (1967). An illustrated key to the species of whale-lice (Amphipoda: Cyamidae), ectoparasites of Cetacea, with a guide to the literature. *Crustaceana* **12,** 279–291.

Leung, Y. M. (1970). First record of the whale-louse Genus *Syncyamus* (Cyamidae: Amphipoda) from the western Mediterranean, with notes on the biology of odontocete cyamids. *Invest. Cetacea* **2,** 243–247.

Leung, Y. M. (1976). Life cycle of *Cyamus scammoni* (Amphipoda: Cyamidae), ectoparasite of gray whale, with a remark on the associated species. *Sci. Rep. Whales Res. Inst.* **28,** 153–160.

Levin, M. J., and Pfeiffer, C. J. (1999). Photoreceptor ultrastructure of the amphipod, *Cyamus ceti* (Linné, 1758), an ectoparasite of bowhead, right and gray whales. *J. Submicroscop. Cytol. Pathol.* **31,** 397–405.

Lincoln, R. J., and Hurley, D. E. (1974). *Scutocyamus parvus*, a new genus and species of whale-louse (Amphipoda: Cyamidae) ectoparasitic on the North Atlantic white-beaked dolphin. *Bull. Br. Mus. (Nat. Hist.) Zool.* **27,** 59–64.

Margolis, L. (1955). Notes on the morphology, taxonomy and synonymy of several species of whale-lice (Cyamidae: Amphipoda). *J. Fish. Res. Bd. Canada* **123,** 121–133.

Pfeiffer, C. J., and Viers, V. (1998). Microanatomy of the marsupium, juveniles, eggs and cuticle of cyamid ectoparasites (Crustacea: Amphipoda) of whales. *Aqu. Mamm.* **24,** 83–91.

Pfeiffer, C. J., and Low, K. J. (1989). Cirral structure of the pedunculated marine barnacle *Lepas anatifera* L. (Crustacea: Cirripedia). I. Ultrastructure of the neuromuscular apparatus. *Acta Zool.* **70,** 243–252.

Rice, D. L., and Wolman, A. A. (1971). Parasites and epizooites. *In* "The Life History and Ecology of the Gray Whale (*Eschrichtius robustus*)" (J. N. Lanyne, ed.), Spec. Publ. No. 3, pp. 100–108. Am. Soc. Mammal, Provo, UT.

Rowntree, V. (1983). Cyamids: The louse that moored. *Whalewatcher* **17,** 14–17.

Rowntree, V. (1996). Feeding, distribution, and reproductive behavior of cyamids (Crustacea: Amphipoda) living on humpback and right whales. *Can. J. Zool.* **74,** 103–109.

Schell, D. M., Rowntree, V. J., and Pfeiffer, C. J. (2000). Isotopic evidence that cyamids (Crustacea: Amphipoda) feed on whale skin. *Can. J. Zool.* **78,** 721–727.

Waller, G. N. H. (1989). Two new species of whale lice (Cyamidae) from the ziphioid whale *Berardius bairdii*. *Invest. Cetacea* **22,** 292–297.

Whale Watching

Erich Hoyt
North Berwick, Scotland, United Kingdom

W
hale watching is the human activity of encountering cetaceans in their natural habitat. It can be for scientific, educational, and/or recreational purposes

(sometimes all three). Mostly, whale watching refers to a commercial enterprise, although it is sometimes undertaken privately. The wide variety of whale watching activities includes tours lasting from 1 hr to 2 weeks using platforms ranging from kayaks to cruise ships, from land points including cliffs and beaches, from sea planes and helicopters in the air, as well as SWIMMING and diving activities in which the whale watcher enters the water with cetaceans. Whale watching grew out of the traditions of bird watching and, to a lesser extent, other forms of land-based wildlife watching. To this day, the better whale and dolphin trips include sea birds, seals, turtles, and other marine fauna in order to appeal to more people as well as to give a well-rounded ecological interpretation.

I. The Birth of Whale Watching

The species originally responsible for the development of whale watching was the gray whale (*Eschrichtius robustus*). Beginning in the mid-1940s, students from Scripps Institution of Oceanography, led by Carl L. Hubbs, began participating in annual gray whale counts from university buildings such as Ritter Hall and from coastal headlands and lighthouses. In 1950, the Cabrillo National Monument in San Diego was converted into a public land-based whale watch lookout, attracting 10,000 people the first winter. Year after year, more and more people came to watch whales.

In 1955, the first commercial whale watch operation charged $1 USD to see gray whales on their winter MIGRATION off San Diego. Although the gray whales passed close to shore, the boat tours allowed a closer look. Beginning in 1959, Raymond M. Gilmore, a U.S. Fish and Wildlife Service biologist who had taken over the gray whale counting chores from Carl Hubbs, began serving as the first naturalist on whale watch trips out of San Diego. Through the 1960s, boat tours and land-based whale watching spread slowly up the coast of California to Oregon and Washington, although southern California remained the heart of the industry. In 1972, the first long-range commercial whale watch trip to the Mexican calving lagoons was organized out of San Diego.

In 1971, the Montreal Zoological Society began offering whale watch tours to go down the St. Lawrence River in Canada to see mainly fin (*Balaenoptera physalus*) and minke whales (*Balaenoptera acutorostrata*) and belugas (*Delphinapterus leucas*). This was the first commercial trip on the east coast of North America. These trips became an annual event.

It was the HUMPBACK WHALE (*Megaptera novaeangliae*), however, that really made commercial whale watching into a big industry. Humpback whales tend to be much more active at and above the surface than gray or other whales, frequently breaching clear of the water—ideal for whale watchers wanting photographs. Added to this is the phenomenon of "friendly" behavior—the tendency of certain individual humpback whales to habituate to the presence of whale watch boats and to approach them regularly. This behavior, first observed commonly in humpback whales, has now also been found in certain gray whales, particularly in the mating and calving lagoons of Baja California, Mexico; in certain minke whales; and in orcas (*Orcinus orca*) and bottlenose dolphins (*Tursiops truncatus*), among others.

In New England and Hawaii, tours to see humpbacks began in 1975. For more than a decade before, the Wailupe Whale Watchers, a local club on Oahu, sponsored loosely organized, infrequent tours, but when whale watching began in earnest from Lahaina on Maui, where the humpbacks were more numerous and accessible, it immediately became the center of the humpback whale-watch industry in the Pacific. Most of the Hawaiian tours were strictly commercial.

In New England, however, operators established their own brand of commercial whale watching with strong scientific and educational components—naturalists on every trip who were often working researchers. Educational programs to introduce school children to wild cetaceans—begun in southern California by such groups as the American Cetacean Society—were expanded in New England. Within a decade, the New England industry would attract even more participants than Californian and Hawaiian whale watching. New England was fortunate to have humpback whales on the feeding grounds centered on Stellwagen Bank, 7 miles north of the tip of Cape Cod, as well as North Atlantic right (*Eubalaena glacialis*), fin, minke, and sometimes long-finned pilot whales (*Globicephala melas*), and Atlantic white-sided dolphins (*Lagenorhynchus acutus*). From a commercial point of view, Stellwagen Bank was ideally located close to the large population centers of the U.S. east coast.

II. Scientific Whale Watching

Whale watching for the purposes of research can be traced back to Aristotle, who spent time on boats and with fishermen in the Aegean Sea. In "Historia Animalium," Aristotle writes that the fishermen would nick the tails of the dolphins and that they could tell them apart. This practice foreshadows the studying of animals by watching them, a key feature of the ethology approach for studying birds and land animals pioneered by Konrad Lorenz, Niko Tinbergen, and others. It took longer to attempt such research with cetaceans because of the greater difficulties of approaching close and conducting research at sea. The photographic identification (photo ID) research of cetaceans began in the early 1970s with humpback whales in the North Pacific and North Atlantic, gray whales and killer whales in the eastern North Pacific, and southern right whales (*E. australis*) and bottlenose dolphins off Argentina.

A successful partnership between science and commercial whale watching began in Provincetown, Massachusetts, in 1975, when Al Avellar of the *Dolphin* fleet asked Charles Stormy Mayo to be his naturalist. Mayo soon saw the possibilities for using the boat as a platform for studying whales. He set up the Center for Coastal Studies as a research and educational institution, and the close ties with commercial whale watching have been maintained ever since.

The arrangement works as follows: The Center provides naturalist guides for the *Dolphin* fleet. They are paid a modest amount for helping to direct the boat to the whales, presenting an informal educational lecture, and answering questions. The Center sells T-shirts and other merchandise on board. Most important, Center researchers can conduct their own photo ID research, and often collect other data. Sometimes more than one researcher will come aboard to ensure the maximum use of boat time.

This key partnership between science and commerce has determined the course of whale watching, as well as the practice of whale research, throughout southern New England. As of 1995, 18 of the 21 whale watching operators that mainly go to the Stellwagen Bank area had naturalists guiding boats and lecturing whale watchers, while 10 operations were taking and contributing ID photos. Despite the competitive atmosphere of commercial whale watching in New England, the researchers and their representative institutions have cooperated in setting up the North Atlantic Humpback Whale Catalogue—a photo catalogue and data base covering more than 10,000 individual whales. As a measure of the scientific value of whale watching, at least 30 published papers in refereed journals have come largely from research aboard whale watching boats on Stellwagen Bank.

The New England model of successful whale watching and research, like Yankee WHALING from an earlier century, has had an impact on the development of whale watching in locales as diverse as the Gulf of St. Lawrence in Quebec, northern Norway, and Dominica in the eastern Caribbean. Of course, a large part of scientific research on cetaceans does not lend itself to being conducted from commercial whale watch trips (such as transect surveys, biopsy darting, and collecting skin and fecal samples). In some cases the research and commercial enterprise operate separately, using different boats and personnel, but the commercial operation supports or contributes to the research. In several areas, whale watch operations have discovered new populations of cetaceans, accessible for study. In all, whale watching worldwide has led to at least 50 cetacean photo-ID programs supported in part or conducted aboard commercial whale watch boats. This has contributed to considerable public support for research through much greater familiarity with research programs.

III. The Growth of Commercial Whale Watching

The value of whale watching in 1981 was estimated to be $4.1 million USD in direct revenues and $14 million in total revenues (including travel, accommodation, food, and souvenirs), based on approximately 400,000 boat-based whale watchers. By 1988, these numbers had expanded by more than three times, with the industry still based largely in New England and California with a small amount in Canada, Mexico, and the U.S. northwest (see Table I).

In the late 1980s, whale watching began to spread rapidly to other parts of the world. Between 1987 and 1991, new whale watch industries started up in the Canary Islands, the Azores, Belize, Costa Rica, Dominica, Italy, Madagascar, and New Zealand, while existing industries expanded rapidly in Argentina, Australia, South Africa, and in parts of Canada. The diverse opportunities for whale watching included boat tours to view rare species (Haviside's dolphins, *Cephalorhynchus heavisidii,* in South Africa), observing sperm whales, *Physeter macrocephalus,* from the air (New Zealand), land-based whale watching of southern right whales (South Africa, Australia), and glimpsing various beaked whales in the Azores and the Bahamas. However, by the 1990s whale watching meant for the most part going to sea on large, fast, comfortable purpose-built ships that could take 150 to 400 people to see the whales and get them back to the dock in 2–4 hr.

TABLE I
Estimated Growth of Whale Watching Worldwide

Year[a]	No. of whale watchers	Direct expenditures USD[b]	Total expenditures USD[c]
1981	400,000[d]	$4.1 million	$14 million
1988	1,500,000[d]	$11–16 million	$38.5–56 million
1991	4,046,957[e]	$77.0 million	$317.9 million
1994	5,425,506[e]	$122.4 million	$504.3 million
1998	9,020,196[e]	$299.5 million	$1049.0 million

[a]1981: Kaza, S. (1982). Recreational whale-watching in California: A profile. *Whalewatcher* **16**, 6–8; Kelly, J. E. (1983). The Value of Whale-Watching. Whales Alive Conference, Boston. June 7–11, 1983. Unpublished pp. 1–5, i-vi; David Sergeant, personal communication.
1988: Kraus, S. D. (1989). Whales for profit. *Whalewatcher* **23**, 18–19.
1991: Slightly revised from Hoyt, E. (1992). Whale watching around the world: A report on its value, extent and prospects. *Int. Whale Bull.* **7**, summer, pp. 1–8.
1994: Hoyt, E. (1995). "The Worldwide Value and Extent of Whale Watching," pp. 1–34. Whale and Dolphin Conservation Society, Bath, UK.
1988: Hoyt, E. (2001). "Whale Watching 2001: Worldwide Tourism Numbers, Expenditures, and Expanding Socioeconomic Benefits." pp. 1–158. International Fund for Animal Welfare, Yarmouth Port, MA.
[b]Cost of whale watch tour (ticket price).
[c]Amount spent by tourists going whale watching from point of decision, including transport, food, accommodation, and souvenirs, as well as ticket price, but not including international air fares.
[d]Estimates made by author based on direct revenues; does not include land-based whale watchers.
[e]Includes land-based whale watchers.

During this same period, whale watching became important in Norway and Japan, two countries with strong whaling interests. In both countries, the number of whale watchers increased rapidly year by year until, in 1998, Norway had more than 21,000 whale watchers spending $6.9 million USD, while more than 102,000 whale watchers in 27 Japanese communities spent $32.4 million USD. Norway's whale watching industry has about a dozen operators working from four communities and offering sperm and other whales from May to September or killer whales in October and early November; whale watchers (visitors) primarily come from other European countries. Japan's whale watching industry, however, is a 99% domestic industry with diverse attractions including Bryde's whales (*Balaenoptera edeni*) and sperm whales at several warm-water locations from Shikoku Island and adjacent Honshu; humpback whales in tropical Okinawa and Ogasawara, both island groups off southern Japan; and minke whales, Dall's porpoises (*Phocoenoides dalli*), and Pacific white-sided dolphins (*Lagenorhynchus obliquidens*) from Hokkaido in the north, as well as multiple locations for watching and swimming with bottlenose (*Tursiops truncatus*) and other dolphins off southern and eastern central Japan.

In the late 1990s, Iceland, with accessible populations of blue, fin, humpback, and minke whales, as well as orcas, became the fastest growing whale-watch destination in Europe, with eight communities hosting more than 30,000 whale watchers in 1998 with total expenditures of more than $6.4 million USD. In 1998, three countries attracted more than a million whale watchers per year: the United States, Canada, and Spain's Canary Islands. According to the most recent worldwide figures (1998), more than 9 million people are going whale watching in 87 countries and overseas territories and spending more than $1 billion USD (see Table I).

IV. Whale Watching Conflicts and Regulations

Such explosive whale watching growth has led to management problems. Typical scenarios include too many boats on the water in a limited area, too many close approaches and sometimes collisions with cetaceans, strain on the infrastructure of local communities from too many visitors, and a lack of guidelines or regulations and/or enforcement of them.

Some operators have formed associations to devise self-imposed guidelines, but most have waited for researchers or NGOs to suggest guidelines or for government to try to impose regulations. Yet even where regulations do exist, enforcement tends to be minimal or absent. In the United States, however, cases have been prosecuted with substantial fines levied against boat operators, as well as researchers and photographers, who approach too close or too aggressively to whales or who operate without a permit. In Hawaii, a film maker was fined for harrassing whales when his close-up underwater footage of a pilot whale (*Globicephala macrorhynchus*) mouthing a woman researcher was sold to television.

In 1983, the first whale-watch fatality occurred when a mature gray whale overturned a small boat in Scammons Lagoon, Mexico, killing two tourists. Until 1995, this was the only fatal whale watch accident. Then, all in the space of a year, in the Dominican Republic, the upper deck of a crowded boat collapsed after being hit by a wave, killing one tourist and injuring others, whereas in Kaikoura, New Zealand, a boat overturned, fatally trapping a person underneath. In the same period, on a sightseeing trip near Baffin Island in the Canadian Arctic, a surfacing whale overturned an 18-foot boat and four tourists died of exposure. Only their guide survived. He was wearing a survival suit.

The number of injuries and fatalities is small considering the millions of people who go whale watching every year. Whale watching is by and large safe for both whales and people. Boats have been accidentally overturned by whales, but more accidents have happened due to problems with the boats themselves or with overloading—things that are not specific to whale watching but could happen as part of any marine tourism. Indeed, most if not all of the accidents to date could have been avoided with due care and precaution.

Perhaps the greatest concern for safety is for the tours involving swimming or diving with cetaceans. Even these have a good safety record with thousands of encounters with dolphins in such places as New Zealand, the Bahamas, and Japan. These generally well-regulated enterprises maintain a strict protocol of no touching and approaching the cetaceans. However, some have suggested that swimming with cetaceans should be limited to certain dolphin populations or known individual animals and that experienced researchers should be present as guides to help interpret behavior and ensure safety. Swimming with whales, such as humpback whales on their mating grounds where there is surface active behavior, is potentially more dangerous to humans; despite criticism it continues in several areas of the world, and it remains controversial.

There are other concerns about whale watching—the amount of time boats stay around the same whales, day in and day out, and the presence of boats on critical mating and calving grounds—but little can be determined with certainty until more time passes and more research is done.

V. Whale Watching and Conservation

In 1983, the INTERNATIONAL WHALING COMMISSION (IWC) cosponsored the "Whales Alive" conference in Boston, Massachusetts, which looked at "nonlethal" uses of whales, but it was not until 10 years later, in 1993, that the IWC adopted a whale watching resolution prepared by the Whale and Dolphin Conservation Society and successfully proposed by the United Kingdom at the IWC annual general meeting in Kyoto, Japan. The underlying strength of the argument that the IWC should become involved in whale watching was that, since the IWC moratorium on whaling, the most prevalent "use" of cetaceans among IWC members has been whale watching. However, despite majority agreement on the relevance of whale watching to the IWC, delegates from Japan have repeatedly stated that the IWC has "no competence" concerning whale watching matters.

Beginning in 1995, a series of international workshops, organized by the International Fund for Animal Welfare (IFAW), with assistance from the Whale and Dolphin Conservation Society, Tethys Research Institute, World Wide Fund for Nature, and others, have brought together more than a hundred cetacean experts from some 25 countries to explore the socioeconomic, educational, and legal aspects of whale watching, as well as the scientific aspects of management and the special aspects of watching SPERM WHALES.

The first of these, the seminal Scientific Aspects of Managing Whale Matching workshop, held in Italy, determined that, to date, there were no detrimental long-term effects on cetacean populations attributable to boat-, air-, or land-based whale watching. The participants took the view, however, that the individual welfare of wild cetaceans needs to be considered as well as the health of populations. Even if whale watching is not lessening a whale's ability to survive, it might be diminishing a whale's quality of life. Also, participants acknowledged that long-term effects might not show up for some time. Of course, it is important to recognize human socioeconomic needs when determining whale watching regulations and policy. The workshop recommended a precautionary approach to management with a periodic review of regulations based on continuing research and monitoring into possible effects on cetaceans (sample guidelines for boat operators are provided in Table II).

The overall impact of this and later workshops is still in the early stages, but they have focused the debate on the future of whale watching, pointing out that better regulations were needed as well as enforcement, that whale watching had substantial unrealized potential in terms of education and science, and that, economically, whale watching was worth far more than had previously been determined, although some values were difficult to measure in terms of dollars. A more detailed and complete valuation of whale watching is now underway in a number of communities.

Perhaps the most valuable legacy of whale watching has been the building of a constituency out of the general public that is interested in and sympathetic to marine mammals, the sea, and marine conservation. The designation of Stellwagen

TABLE III
Educational Values of Whale Watching[a]

1. Whales are emblems for promoting awareness of endangered species and habitat protection

2. Whale watching provides the opportunity for people across all ages and cultures to become familiar with environmental issues and to become involved in conservation efforts on a personal, local, regional, national, and international level

3. The development of education programs forges links between the whale watch industry and local communities as well as building bridges between the general public and scientific communities

4. Natural history knowledge gained through whale watching has intrinsic value

5. Whale watching provides an opportunity to observe animals in the wild, transmitting factual information and dispelling myths

6. Whale watching is a model for marine educational programs in adventure travel and ecotourism

7. Whale watching provides the opportunity for appreciation and understanding of local history, culture, and environment

[a]Adapted from IFAW, WWF, and WDCS (1997), used with permission.

Bank as a U.S. National Marine Sanctuary in 1993 was largely the result of public interest in whales in New England and in the northeastern United States through whale watching. Several million people encountered whales in the wild between 1975 and 1992, saw the research being conducted on whale watching boats, and learned about the whales and problems of the sea, which led to overwhelming popular support for the sanctuary.

Since the late 1980s in most areas of the world, however, whale watching has been much less educational (see Table III). A 1998 world SURVEY of whale watch operations found that only 35% of all operators had enlisted naturalists to guide their trips. In terms of the scientific content, about 9% of operators worldwide have researchers or naturalists on board who conduct regular photo ID and other research as part of their trips, whereas 57% never conduct scientific research or even offer information to scientists. Most operations were strictly commercial ventures. Clearly, a great deal more could be done to encourage whale watching tours to offer the maximum benefits to local communities and regions in terms of education, science, and conservation, as well as tourism dollars, while at the same time protecting the whales and ensuring that they will remain in coastal waters and accessible to whale watchers for generations to come.

See Also the Following Articles

Abundance Estimation ■ Ethics and Marine Mammals ■ Popular Culture and Literature

References

Birnie, P., and Moscrop, A. (1999). Report of the Workshop on the Legal Aspects of Whale Watching. Punta Arenas, Chile. Submitted as IWC/51/WW1, International Whaling Commission.

TABLE II
Brief, Useful Guidelines for Boat Operators[a,b]

1. Do not pursue, overtake, head-off, or encircle cetaceans or cause groups to separate

2. Never approach whales/dolphins head on

3. Avoid sudden changes in noise level (gear shifts, avoid reverse unless necessary to back away slowly from a surfacing whale or dolphin group)

4. Reduce speeds in areas where whales may be sighted; approach and leave whales cautiously and slowly

5. Extreme caution is required when the following is present: (1) feeding whales, (2) cow/calf pairs and juveniles, (3) resting, (4) breeding or rowdy and/or (5) socially active groups. Cetaceans behaving in these ways are particularly sensitive to disturbance and may be vulnerable to collisions

[a]Adapted from IFAW, WWF, and WDCS (1997), used with permission.
[b]This is not a complete list covering every situation but is meant to provide some general suggestions and overall direction for the use of operators offering whale watch tours as well as wildlife managers who are establishing guidelines or regulations on whale watching.

Colgan, K., Prasser, S., and Jeffery, A. (eds.) (1996). Encounters with Whales: 1995 Proceedings, Australian Nature Conservation Agency, Canberra, ACT, Australia.

Hoyt, E. (1994). "Whale Watching and the Community: The Way Forward," pp. 1–29. Whale and Dolphin Conservation Society, Bath, UK.

Hoyt, E. (1994). Whale watching worldwide: An overview of the industry and the implications for science and conservation. *In* "European Research on Cetaceans" (P.G.H. Evans, ed.), pp. 24–29. Proc. 8th Ann. Conf. ECS, 2–5 Mar. 1994. Montpellier, France, European Cetacean Society, Cambridge, UK.

Hoyt, E. (1995). "The Worldwide Value and Extent of Whale Watching: 1995." Whale and Dolphin Conservation Society, Bath, UK. Presented as IWC/47/WW2 to the Whale Watching Working Group, International Whaling Commission (IWC), annual meeting, Dublin, Ireland, pp. 1–4 (attached summary by UK government); pp. 1–34.

Hoyt, E. (1995). Whale watching takes off. *Whalewatcher* **29**, 3–7.

Hoyt, E. (1998). Watch a whale; learn from a whale: Enhancing the educational value of whale watching. *In* "Proceedings of the 1998 International Forum on Dolphins and Whales," pp. 5–19. Muroran, Japan.

Hoyt, E. (2001). "Whale Watching 2001: Worldwide Tourism Numbers, Expenditures, and Expanding Socioeconomic Benefits." International Fund for Animal Welfare, Yarmouth Port, MA.

Hoyt, E., and Hvenegaard, G. (1999). "The Development, Value and Study of Whale Watching in the Caribbean." Paper to the Scientific Committee, International Whaling Commission, Grenada, May 1999.

IFAW (1996). Report of the Workshop on the Special Aspects of Watching Sperm Whales, Roseau, Commonwealth of Dominica.

IFAW (1999). Report of the Workshop on the Socioeconomic Aspects of Whale Watching. Kaikoura, New Zealand.

IFAW, Tethys Research Institute and Europe Conservation (1995). Report of the Workshop on the Scientific Aspects of Managing Whale Watching, Montecastello di Vibio, Italy.

IFAW, WWF, and WDCS (1997). Report of the International Workshop on the Educational Values of Whale Watching, Provincetown, MA.

Smith, A., and Hoyt, E. (1995). "Dolphin Watching and Swimming Regulations Worldwide." Paper to Dolphin Workshop at Bienniel Meeting of the Society of Marine Mammalogy, Orlando, FL.

Whaling

Whaling, Early and Aboriginal

RICHARD ELLIS
*American Museum of Natural History,
New York, New York*

Jenkins (1921) writes, "Although the general opinion is that the Basques were the earliest whalers, Noel de la Moriniere says that this is a misapprehension and that the Northmen were really the first in the field." He says that a man called "Ochther" hunted whales and walruses beyond the North Cape, but then he notes that "there is no evidence that it developed into a regular fishery such as that of the Basques."

His "Ochther" was Othere (or Ottar), a Norseman in the service of King Alfred of Wessex around the year 890 A.D. Alfred (called Alfred the Great) is known for his defense of England against the marauding Danes, and also for the initiation of the *Anglo-Saxon Chronicle,* the first history of England. (Our word "whale" comes from the Anglo-Saxon *hwael,* which means "wheel," and probably refers to the shape of the whale's back as it rolls in the water.) Alfred translated many Latin texts, including the one that concerns us here, a description of Europe by one Orosius, who lived four centuries before. To the work of Orosius, Alfred added a description of the northern voyage of Othere, wherein was described the whale and walrus hunting of a northern people known as the Biarmians. From the location (the White Sea in northern Russia), and description of the whales hunted ("50 ells" in length, which by one calculation works out to 187 feet), it would appear that the larger ones—whose size was greatly exaggerated—were bowheads, whereas Othere's "horsewhales" were walruses (*Odobenus rosmarus*). In the history of British *Voyages and Discoveries* compiled in the 16th century, Richard Hakluyt, a diplomat and scholar, wrote that the principal purpose of Ochther's expedition was "to increase the knowledge and discovery of these coasts and countries, for the more commodity of fishing for horse-whales, which have in their teeth bones of great price and excellency: whereof he brought some at his return unto the King. Their skins are also very good to make cables for ships, and so used."

In medieval Scandinavia, whales were very much a part of the lives of the people and were therefore incorporated into their literature. A 13th-century Icelandic account known as *Konungs skuggsjá* (*Speculum Regale* in Latin; *Konegspiel* in German; "King's Mirror" in English) describes the whales that are found off Iceland and includes such mysterious creatures as the horse whale, the red whale, and the pig whale, but also discusses recognizable species, such as the killer whale (*Orcinus orca*), the narwhal (*Monodon monoceros*), and the sperm whale (*Physeter macrocephalus*). The right whale (*Eubalaena glacialis*) is described thus:

> People say it does not eat any food except darkness and the rain which falls on the sea. And when it is caught and its intestines opened, nothing unclean is found in its stomach as would be in other fish that eat food, because its stomach is clean and empty. It cannot open its mouth easily, because the baleen that grows there rise up in the mouth when it is opened, and often causes its death because it cannot shut its mouth. It does no harm to ships: it has no teeth, and is a fat fish and edible.

There is an almost complete lack of information on Norse whaling, but the waters in which they sailed were then (and are still) among the whale-richest in the world. There are right whales, humpbacks (*Megaptera novaeangliae*), fin whales (*Balaenoptera physalus*), sperm whales, belugas (*Delphinapterus leucas*), narwhals, pilot whales (*Globicephata melas*) and various species of dolphins in the cold, productive waters of the North Atlantic. The Norse sagas are silent on the subject of whales and whaling, but it would be hard to imagine these hardy seafarers ignoring a plentiful source of food and oil as they plied the otherwise inhospitable seas around Iceland, Greenland, and Labrador. There are references, however, to battles royal between various "families" as they dispute the ownership of whale carcasses, which indicates the importance of whales—at least of dead whales—in the lives of the early Norsemen. They left no tryworks and their settlements provide no trace of harpoons or lances, but there are tantalizing hints of Norse whaling in some of the more recent discussions. In his 1928 *History of Whaling*, Sydney Harmer says, "The Icelanders seem to have engaged in whaling . . . and the whale known as 'Slettibaka' . . . is believed to have been the Biscay whale." (The modern Icelandic for the right whale is *sletbag*, which means "smooth back.")

I. Early Icelandic Whaling

Iceland's early history is to be found in the sagas, tales of the exploits of the island's early heroes. The Vikings of Norway evidently brought to Iceland knowledge of the techniques employed in driving whales (probably pilot whales) into the fjords for slaughter. There are occasional mentions of disputes over stranded whales in the sagas, but as far as we know, there was no active whale fishery. An Icelandic bestiary from about 1200 describes some of the whales (but not accurately enough for modern cetologists to identify them as to species), and the *Konungs skuggsjá* lists no fewer than 21 sea creatures, some of which can be referred to living whales, dolphins, and pinnipeds, and some of which—mermaids and mermen, for example—are clearly mythological.

In a 17th-century work by an Icelander named Jon Gudmundsson, there is a list of the various whales that might be found in Icelandic waters, including the sperm whale (*Burhvalur*), the narwhal (*Náhvalur*), the right whale (*Slettbakur*), the fin whale (*Geirreydur*), and the blue whale, *B. musculus* (*Steypireydur*). With the exception of the "right whale," which probably refers to the bowhead (*Balaena mysticetus*) and was hunted to extinction in the region after this publication appeared, all these whales can still be seen off Iceland. Also included was something that the author referred to as *Sandloegja*, which has been translated as "sandlier," i.e., one that lies in the sand. Each of the just-mentioned whales is illustrated, so there is little doubt as to its identification. The description of the *Sandloegja* is accompanied by a picture of a whale that has not been seen in the Atlantic since commercial whaling began, and if the interpretation is correct, it depicts the only whale to have become extinct in recent history.

The California gray whale (*Eschrichtius robustus*) is well known from the North Pacific, where it makes the celebrated round-trip migration from Alaska to Baja California. It was the object of an intense fishery in the 19th century, which nearly eliminated the species. Fossil and subfossil remains of a similar—if not identical—species have been found in western Europe (Sweden, England, and the Netherlands) and on the east coast of North America from New Jersey to South Carolina. From the evidence, it appears that there was also an Atlantic gray whale, which probably maintained similar habits to its Pacific cousin: it fed in cold northern waters (perhaps Iceland and Greenland) and then moved south (Spain, France, England?) to breed and calve. With the exception of the fossil evidence, the only clues to the identity of this whale are found in the work by Gudmundsson and in a debatable reference in a New England work of 1725, where Paul Dudley describes the "scrag whale" with characteristics that are not applicable to any other species except the gray whale.

Whether hunted or occasionally appearing on the beach, the gray whale apparently should be listed in the Icelandic cetacean fauna, even though no living Icelander has ever seen one in his own country. (In his analysis of the whales of the *Konungs skuggsjá*, Ian Whitaker writes that "the gray whale was hunted in the Atlantic between 1100 and 1200, although it has not been found there since the 18th Century." He is unable to correlate this species with any of the 19th-century Icelandic names, although he indicates that there are two "unallocated" names, which translate as "hog whale" and "shield whale.")

Whales were caught by the Norwegians off the Tromsø coast as early as the 9th or 10th century. The oil was used for lighting and the baleen for jewelry, coopering, and boatbuilding. "But," as C. B. Hawes wrote in 1924, "with a lamentable lack of foresight, the earliest whaling captains neglected to enlist the services of scholars and historians," so much of the story of early Norse whaling has to be left to conjecture.

One of those who did a lot of conjecturing was Ivan Sanderson. Trained as a zoologist, Sanderson was particularly interested in bizarre phenomena, such as the abdominable snowman and the Loch Ness monster. He wrote several books on zoological and cryptozoological subjects, but he will probably be best remembered for his *Follow the Whale*, which was published in 1956. In this book, along with some rather good accounts of the biology of whales and some excellent maps, he recreates the lives of whalers of the past and the present, "corralling the forgotten and more neglected aspects of whaling history and the new discoveries about the whales themselves, and weaving them into a continuous web of narrative." One of these "neglected aspects" is Norse whaling history, but despite the lack of documentation, Sanderson devotes a whole chapter to the subject, fictionalizing what could not otherwise be ascertained. He has the Norsemen under "Thorvald the Long" trapping sei whales (*B. borealis*) in the fjords of Norway at an unspecified time, along with "Biarni the Yellow standing in the bow holding a trumpet of cow's horn in his hand." He also recounts an Icelandic saga of 1100 A.D., which contains "a delightful passage in which we are told of the stranding of a large rorqual at Rifsker in Iceland and how all the important people who were able went to it." The documentation for this is sparse, but there can be no question that the Norsemen, ranging the North Atlantic from Finnmark to Iceland and from Greenland to North America, had to have encountered whales. Whether they killed the whales in an organized fashion or took them incidentally to their Viking and settlement forays may never be known. They did hunt walruses for their skin and ivory tusks, and narwhals for their spiraled ivory tooth, which was passed off as the horn of the fabled unicorn.

II. Basque Whaling

As far as we can tell, the first western people to hunt large whales in an organized and intentional manner were the Basques. As far back as records go—and even further, perhaps as far back as the Stone Age—these men were hunting whales. In his 1820 *Account of the Arctic Regions*, William Scoresby suggests that "the Biscayans were the first who exercised their courage in waging a war of death with the whales," but he attributes their motivation to the protection of their fishing nets, which "would naturally suggest the necessity of driving these intruding monsters from their coasts." Whatever their reasons, the Basques became the paradigms of the whaling industry, establishing the modus operandi that would characterize the industry for another thousand years. "Historians have only recently begun to realize," wrote Farley Mowat, "that it was the Basques who lit the flame that was eventually to consume the mighty hosts of the whale nations." They discovered the "resource," exploited it, and then pursued it so vigorously that it was uneconomical to continue. They probably took their first whales in the shallows, and then, like the bay whalers who were to follow their lead all around the world, realized that it was considerably more expeditious to go after the whales rather than wait for the whales to come to them. The Basques may also have contributed to the only cetacean extinction in recorded history.

Somewhere around 1000 A.D., these intrepid hunters of the Bay of Biscay began the slow but systematic eradication of the whales that came into the protected bays in the shadow of the Pyrenees. Obviously the Basques did not wait for the first millennium to end before beginning their whaling, but most authors cite this as approximately the time they began. (Ommanney writes: "The industry, founded on the Biscay Right whale, was fully developed by the twelfth century but probably dated from much earlier, possibly from the tenth century when the Basques may have learned the craft from Norse whalers.") The Belgian historian W. M. A. De Smet has searched the literature for references to European whaling *before* the Basques and writes, "Only a few authors are aware of the fact that whaling existed in still earlier days in other European seas, and that it was practiced in the North Sea and the English Channel during the Middle Ages, certainly from the 9th century onward." Although the species of whale in these early instances was rarely recorded, the likelihood is that it was the right whale that was hunted in the North Sea, and perhaps the gray whale, although the precise date of the disappearance of the Atlantic gray is still being debated.

De Smet cites several instances in which whale meat is mentioned in early texts and suggests that "it is clear from the regularity with which whale meat occurred in these markets that it cannot have come from stranded animals alone and there must have been regular landings." After providing for themselves, the enterprising Basques established markets for the meat and blubber, and even had "consulates" in Holland, Denmark, and England to encourage sales. In French, the blubber was known as *lard de carime*, which means "lenten fat," and Europeans were allowed to eat it on designated meatless calendar days. The oil was used for lighting and the manufacture of soap, wool, leather, and paint; the meat was fed to the poor and to the ships' crews; the baleen as put to all sorts of uses (including being shredded into plumes for the decoration of knight's helmets); the vertebrae were used for seats; and the ribs were employed as fence pickets and beams for cheap housing. The tongue was considered a particular delicacy and was reserved for the clergy and royalty.

In the unregulated (and largely undocumented) confusion of the Middle Ages, small pockets of Basques lived along the shores of the Bay of Biscay, speaking their own language, about which a contemporaneous cleric wrote, "The Basques speak among themselves in a tongue they say they understand but I frankly do not believe it." In their strongholds in the crook of the elbow

of the Iberian peninsula, they were isolated from the turmoil of land wars, fiercely intent upon self-preservation, and coincidentally upon the pursuit of the large black whales (which they called *sarda*) that arrived every autumn in their offshore waters.

It is likely that they also hunted the Atlantic gray whale, although there is no evidence to support this supposition. There *is*, however, considerable evidence that the Atlantic gray whale (which was called *otta sotta*) was present in the Atlantic during the days of Basque whaling. Remains have been found on both sides of the ocean: in England, Holland, and Sweden in the east and from New York and New Jersey to North Carolina in the west. An account discovered by Fraser suggests that a gray whale (called *sandloegia* by the Icelanders) existed as recently as 1640 in the waters off Iceland. With nothing more than the absence of gray whales to substantiate his claim, Mowat writes "that by as early as the fourteenth century, the otta sotta had been hunted to virtual extinction in European waters." In their 1984 study of the Atlantic gray whale, Mead and Mitchell recognized only Fraser's *sandloegia*[1]; a 1725 description of the "scrag whale" by Paul Dudley, Esq.[2]; and the 1611 instructions given by the directors of the Muscovy Company to Thomas Edge[3] as "reliable records of gray whales in the North Atlantic." There are no more gray whales in the Atlantic, and while this unfortunate state of affairs might not be directly attributable to the Basques, it is not unreasonable to assign them some part in the disappearance of these whales.

For many years, the most comprehensive study on the subject of Basque whaling was that written by Sir Clements Markham and published in 1881. While writing a study of William Baffin, he learned "that the first English whaling vessels were in the habit of shipping a boat's crew of Basques to harpoon the whales," so he began to investigate and ended up in Spain. He found that King Sancho the Wise of Navarre had granted petitions to the city of San Sebastian in the year 1150 for the warehousing of certain commodities, among which were *boquinas-barbas de ballenas*, plates of whalebone. Markham traced the fishery through the records of various cities and towns (he found the "Casa de Ballenas" in Asturias) and acknowledged that it was the Basques who taught the British how to kill whales. He sums up the Basque contribution as follows: "Of course the English, in due time, learnt to strike the whales themselves; but the Basques were their instructors; and it is therefore to this noble race that we owe the foundations of our whaling trade."

More recently, the Spanish cetologist Alex Aguilar searched the records for written documentation of Basque whaling and has discovered a reference from Bayona, in the Gulf of Biscay, that

---

[1]"Good eating. It has whiter baleen plates, which project from the upper jaw instead of teeth, as in all other baleen whales, which will be discussed later. It is very tenacious of life and can come to land to lie as a seal like to rest the whole day, But in sand it never breaks up."

[2]"The Scrag whale is near a-kin to the Fin-Back, but instead of a Fin on his Back, the Ridge of the After-part of his back is scragged with a half Dozen Knobs or Knuckles; he is nearest the right Whale in Figure and for Quantity of Oil; his Bone is white, but won't split."

[3]"The fourth sort of whale is called Otta Sotta, and it is of the same colour its the Trumpa having finnes in his mouth all white ball white but not above a yard long, being thicker than the Trumpa but not so long. He yeeldes the best oyle but not above 30 hogs' heads."

dates from the year 1059. From the remains of cetaceans examined at some of the settlements on the shore of the Cantabrian Sea (off the northern coast of Spain), it has been suggested that the Basques occasionally hunted sperm whales, but the predominant object of their fishery was the right whale. Ancient whaling bases have been found along the length of this coastline, which encompasses the provinces of Galicia, Asturias, Santander, and the heart of the Spanish Basque country, Guizpuzcoa. From the western tip of northern Spain the sites have Spanish names (Camariñas, Malpica, Antrellusa, Llanes), but as we move eastward, toward the Basque settlements on the Bay of Biscay, the names take on a decidedly Basque flavor: Lequeitio, Ondarroa, Guetaria, and Zarauz. Aguilar quotes several sources (including Markham) for the number of whales killed at Lequeitio from 1517 to 1662, and produces a total of some 62 whales, adults and young, from incomplete records, for a provisional average of a little more than 2 whales per year. Occasional records for Guetaria from 1699 to 1789 provide even lower numbers, suggesting that the Biscayan right whales were on the decline by the 18th century.

Along the French and Spanish Biscayan coasts, there are several towns and villages whose seals and coats-of-arms depict whale-fishers, including Bermeo, Ondarroa, Motrico, and Fuenterrabia in Spain and Biarritz, Hendaye, and Guethary in France. Jenkins writes, "in this fishery the Bayonnais took part, and it is one of the most interesting features in the ancient records of the town of Bayonne." For several centuries, the Basques of Biarritz, St. Jean-de-Luz, Bayonne, San Sebastian, and other towns killed the *sarda* in their inshore and offshore waters. This activity did not go unnoticed by the tax collectors. In 1197, King John of England (acting as the Duke of Guienne) collected a tax on the first two whales taken at Biarritz. In 1261, all whales taken at Bayonne were tithed, a continuation of an earlier, voluntary gift of all whales' tongues to the Church. The kings of Castile and Navarre also extracted taxes from the whalers, often in the form of meat or whalebone. Under a 1324 edict known as *De Praerogativa Regis* (The Royal Prerogative), Edward II of England (1307–1327) collected a duty on every whale captured in British waters, and his successors continued to claim the "royal fish" as Crown property.

To this day, we do not know whence the Basques came, or from whom they were descended. (Their blood type distinguishes them from the French and the Spanish, and biologically as well as linguistically they appear to be distinct from any other people now in existence.) As far as we can ascertain from the scanty records and the ruined stone watchtowers (known as *vigías*) that still stand overlooking the bays, they pursued the right whale. (In 1928, Sydney Harmer wrote, "A watchman who tried to use [the towers] for their original purpose would now have an unprofitable occupation, and he would not be likely to see a single whale of this species during his lifetime.")

Even more significantly, the Basques are said to have invented the on-board try-works, where whales could be processed at sea, avoiding the time-consuming and arduous process of towing the carcass to shore and then winching it up on the beach for rendering. According to Jenkins, this distinction belongs to "a captain of Cibourre named François Sopite," but surprisingly, in a book heavily footnoted with obscure references, this important fact goes undocumented. In Sander-

son's *Follow the Whale*, however, a whole chapter is devoted to a recreation of Sopite's accomplishments, including a description of him standing "silently on the poop with his hands behind his back peering out from under his curious floppy black hat." Sanderson seems to have consulted many of the same references listed by Jenkins, but he does not tell us where the hat comes from or how he knows that Sopite was "smiling wryly" at the success of his experiment. Up to that time, whales were flensed and tried-out on shore, which meant that the whalers could never roam too far from their home ports. As we shall see, however, Sopite's "invention" may have been the invention of some creative authors, as real evidence of the Basque whalers has been uncovered, and there is no indication whatsoever of on-board tryworks.

Even though the hunters never took very many whales in a given season, the right whale (known as the Biscayan whale to distinguish it from the Greenland right whale or bowhead) disappeared from Biscayan waters, and the Basques had to look farther afield for oil and bone. (Clements Markham, in his study of Basque whaling history, has written that each of the whaling villages may have taken no more than a couple of whales per year. This would not be enough to decimate the population, but it is possible that the disturbance caused by the whalers drove the whales to other, less perilous breeding grounds.)

From Iberia, Basque fisherman crossed the North Atlantic seeking new grounds. Some evidence indicates that they may have fished the Labrador–Newfoundland grounds as early as the 14th century but were driven off by the local Eskimos. The vessels that they used were not known until recently, when a Canadian archaeologist named Selma Barkham followed up some vague hints in the historical records of Labrador and, with the help of divers, located the wrecks of several Basque ships in the area known as Red Bay. Found sitting on the bottom of the bay were the remains of a three-masted 90-foot galleon, which is believed to have sunk in a storm in 1565, and the complete hull of one of the frail chalupas.[4] On two of Red Bay's smaller islands, workers found unmistakable evidence of tryworks, where the blubber of the whales was rendered into oil. Because this endeavor took place between the years 1560 and 1570 (ascertained from documents examined in Spanish archives by Barkham), it would appear that Sopite's "invention" of on-board tryworks was either apocryphal or somehow did not extend to the whaling operation at Red Bay.

As the Basques enlarged the scope of their search for whales to the vicinity of Newfoundland and Labrador, they may well have been the first Europeans to fish the Greenland coasts and the Grand Banks, two of the richest cod-fishing grounds in the world. Upon landing, they predated John Cabot and Jacques Cartier, the "official" discoverers of the land known as Terranova. In their pursuit of the sea's bounty, the adventurous Basques visited Ireland, Iceland, Greenland, and evidently sailed as far north as Spitsbergen. They also crossed the Atlantic to find the right whales that inhabited the inshore waters of Newfoundland

---

[4]A *chalupa* (in French a *chaloupe* and in the British fishery a *shallop*) was a 25-foot-long whaleboat, rowed by six oarsmen, from which the whale was harpooned and towed to shore.

and Labrador, but it is unlikely that they made these voyages without island hopping across the perimeter of the North Atlantic, much as the Norse had done before them.

Examination of the bones at Red Bay indicate that bowheads were also processed there by the Basques. This location is considerably south of the known range of the bowhead, which inhabits—or inhabited—eastern Arctic waters and the Bering Sea. (It is likely that the Basques took bowheads farther north and then brought them back for processing, thereby accounting for bowhead bones in a region where bowheads are not known to have lived.) There are no records of Basques hunting humpbacks, but these whales are found off the Canadian Maritime coasts and Greenland.

The rich days of Newfoundland and Labrador whaling ended for the Basques as the 16th century ended. The destruction of the Spanish Armada in 1588 meant that Spanish ships of war could no longer protect fishing fleets so far from home, and the Basque whalers ventured across the Atlantic unprotected. They had established shore stations at Tadoussac and Sept-Iles on the St. Lawrence, where they hunted humpbacks and probably belugas. By 1738 the last Basques had left Canada. Why bother with the transatlantic crossings and hostile North Americans when there were fat Greenland whales for the taking in Spitsbergen? The Basques participated in the early Dutch and British expeditions in Spitsbergen, bringing with them 500 years of whaling experience.

Six Basque harpooners from St.-Jean-de-Luz were part of the crew of the first Muscovy Company expedition to Spitsbergen in 1611. In the early years of the Greenland Fishery (Barendsz had named the Spitsbergen islands "Greenland" when he discovered them in 1596 because he believed they were an extension of the island of that name), the Basques sold their services to whoever was willing to pay, but in addition to their participation in the Dutch/British rivalry, the Spanish Basques also sent their own ships to the northern ice in 1613. No sooner had the Spanish tried to join the fishery on their own than James I of England issued the Muscovy Company an exclusive charter to fish the waters of Spitsbergen, to which the Dutch countered in 1614 by forming their own Noordsche Compagnie with the same objectives.

Although the Spanish Basques had the experience and the expertise, they did not have the naval power to back up their claims, and as the Dutch and the British competed for supremacy in Spitsbergen (the Dutch eventually won the battle because of their more effective management and business practices, but in the end, everybody lost, because they ran out of whales), the Basques faded into whaling oblivion. As time and progress passed them by, their domestic whaling capabilities diminished accordingly. According to J.-P. Proulx, when a whale stranded at St. Jean-de-Luz in 1764, the hunters could only find old and rusty implements with which to cut it up.

In many respects, the Basques were the advance guard of what would eventually become an all-out war on the whales, but in those relatively uncomplicated times, they were only aware of the nutritional needs that could be fulfilled by the taking of these large, inoffensive animals. They would, however, establish a pattern with regard to the right whale fishery that would serve as an example for virtually every nation that fol-

lowed their lead: they took the females and calves, as they were the most accessible, and by so doing guaranteed the catastrophic degeneration of the breeding population. In a review of available data, Aguilar estimates that during the period 1530 to 1610, Basque whalers might have taken as many as 40,000 right whales. Medieval Europeans probably did not have much time to ponder the effects of their actions on future generations, however; certainly not on future generations of whales.

III. The Beginning of Japanese Whaling

Halfway around the world, at approximately the same time that the Basques were winding down their whale fishery, the Japanese were gearing up for what would become one of the predominant whaling enterprises of modern times. As with most early whaling exercises, it began humbly, with people waiting for the occasional whale to wash up, but soon lookouts were being stationed on shore to watch for passing right, sperm, gray, or humpback whales. At first, the villagers drove the whales into a bay and closed the mouth off with a net, but then it occurred to them that they could go after the whales with nets rather than waiting for them in a bay or lagoon. Somewhere around 1675, Yoriharu Wada, now recognized as the "founder" of Japanese net whaling, organized the boatmen and harpooners of Taiji into whaling crews, and when a whale was spotted by lookouts on shore, the whalers went off in pursuit. The whales were herded into the nets by their would-be captors, who then killed them with repeated lance thrusts. The carcass was towed to shore and winched up on to the beach for processing. Unlike their European contemporaries—say in the Greenland bowhead fishery—Japanese whalers wasted no part of the whale. They ate the meat, used the oil for soap and lamps, but they also mixed the oil with vinegar to make an insecticide for rice paddies. The bones were crushed and used for fertilizer; the baleen for fans, fishing rods, lantern handles, and puppet strings. Medicines were made from various internal organs, and predictably, the penis was dried and pulverized into a tonic. The entrails were boiled into soup, and it is said that the membranes of the heart were made into drum heads. The Japanese continued this form of highly efficient, highly ritualized whaling until well into the 19th century, when Yankee sperm whalers, searching the Seven Seas for sperm whales, arrived on the "Japan Grounds" and, by their presence and influence, changed Japanese whaling irrevocably.

IV. Aboriginal Whaling

In its early chapters, the story of whaling was a simple one: man against whale. Very infrequently, the whale won. (In "Moby Dick," despite the rage of Captain Ahab and the skill of the harpooners, the white whale triumphs.) With the passage of time, the hunters changed the nature of the hunt and turned it into an industry. The hunted whales remained unevenly matched with their opponents; all they had was the hope of escape in the depth and expanse of the ocean. As the industry grew more economically important, technological innovations were introduced that altered the odds greatly. The introduction of diesel catcher boats, exploding harpoons, spotter planes, sonar, and asdic changed the nature of the hunt greatly. No longer remotely equitable, it was not even a hunt any more, but a highly mechanized business. The whale had as much of a fighting chance as a tree had against a chain saw.

There are only a few places in the world where people still *hunt* whales. The Caribbean island of Bequia in the Grenadines, for example, has a relic humpback whale fishery that the Bequians learned from Yankee whalers in the 19th century. The Inuit of Greenland hunt minke whales and humpbacks (they are given a quota by the International Whaling Commission), as well as belugas and narwhals, which are considered "small cetaceans" and are not under the jurisdiction of the IWC. Alaskan Inuit "hunt" the bowheads that annually pass their North Slope villages, but because of the complexities of politics and other factors, they have upgraded their weaponry to the point where once again, the whale has hardly any chance of escaping. There are only a couple of places where whaling takes place in a thoroughly primitive manner (and completely unregulated by the IWC): one is in Indonesia and the other is the island of Camiguin in the Philippines, where Bryde's whales are the target of opportunity.

Lomblen, also known as Lembata, is one of a group of islands that make up the Sunda Archipelago (*Nusa Tenggara Timur* to the Indonesians), which includes the large islands of Timor and Flores, as well as the smaller Solor, Adonara, Pantar, and Alor. Lomblen/Lembata is only one of the 13,000 islands that comprise the 3000-mile-long country of Indonesia, but there is something very special about this island. On its southern shore is Lamalarep, one of the few whaling villages in all Indonesia. Lamalarep is the poorest village on the island because it has virtually no industry or agriculture other than whaling, and the success rate of the whalers seems to be rather low. They might capture a whale on 3 trips out of 10; to put it another way, 70% of their trips are unsuccessful. The villagers of Lamalarep do not eat the bulk of the whale meat they take, but dry it in the sun and trade it to other villages for vegetables.

In June of 1979, a research team was sent to Lamalarep by the World Wildlife Fund to investigate the whaling activities there. Unfortunately, on July 17, a giant tsunami inundated Lomblen, causing over 700 casualties and destroying the villages of Wai Teba and Sara Puka. The investigators all survived, however, and remained on Lomblen for 3 weeks. On July 26, on nearby Rote Island, a "giant shark" (species unidentified) was found with the body of what was thought to be a Lomblen fisherman in its stomach. (It is likely that the shark ate one of the victims of the tsunami, rather than taking a swimmer or a fisherman.)

At dawn the Lamalarep fleet sets out for a day's HUNTING. They may roam as far as 17 miles offshore, but the whales are usually found closer to the islands. The boats (known as *peledang*) are about 30 feet long and brightly painted, often with vigilant eyes on the bows. No nails are used in their construction, only wooden pegs; and the sails are patchwork rattan, a single gaff-rigged square sail for each boat. A crew of 10 to 15 men rows (or sails, if the winds are favorable) the boat out to the whaling grounds, south of the islands in the Savu Sea. They look for the forward-angled spouts of the largest of the toothed whales, the sperm, which they call *ikan pails* in Bahasa Indonesia. (In the language of the islands that was employed before the introduc-

tion of this *lingua franca* by President Sukarno, the sperm whale was known as *kotan klema*.) During the 10 weeks that the World Wildlife Fund researchers kept records, the whalers of Lamalarep took sperm whales, killer whales (*Orcinus orca*), pilot whales (*Globicephala macrorhynchus*), and several species of dolphins. Traditionally, the whalers of Lamalarep do not hunt baleen whales. Although the men of Lamalarep are considered whalers, they will also harpoon any large fish, ray, or turtle that they encounter, including sharks, marlin, and ocean sunfish.

The whalers of the island are divided into hereditary "corporations," each of which owns a whaling vessel. The vessels—and their names—are passed down from generation to generation, so when a given boat wears out, the next one built by that clan is given the same name.

When a whale is sighted, the *peledang* crews row stealthily upon it, douse the sail, and, because they are Christians, they whisper a communal *Pater Noster* for their own protection. The harpooner stands on a narrow platform with his bamboo-shafted harpoon poised. At the critical moment, when he is within striking range of the wrinkled, humped back of the whale, the harpooner launches not only the harpoon, but *himself* through the air, using his strength and his weight to drive the iron deep into the flesh of the whale.

As the whale is slowed or stopped by the pain of the harpoon in its back, another harpooner throws himself on the whale and, if necessary, another. The iron must be planted in exactly the right place to kill the whale; otherwise the fragile *peledang* will be towed for miles as the whale pulls the whalers on the Indonesian equivalent of the "Nantucket sleigh ride." There are stories of boats being towed all the way to Timor by a maddened whale. (In fact, there are many tales about maddened whales in the Timor Sea. One of the most notorious of all these was a bull sperm whale named "Timor Jack," who savaged whaleboats for years until he was taken by setting out a barrel on a line which he attacked, allowing whalers to lance him.) Or the whale will dive, pulling the line out rapidly and, unless it is cut, pulling the boat down with it. If the right spot is pierced (the heart or lungs), the whale will spout blood from its blowhole and expire quickly. The dead whale is towed back to the village where it is butchered.

There is a complex system for dividing up the meat of a whale in which the carcass is portioned out according to rank in the clan and the village. The meat is eaten or bartered to other villages; the oil is used for lamps. The men of the village may carve patterns onto the teeth, like scrimshanders everywhere.

The villagers of Lamalarep kill between 30 and 50 sperm whales every year. They do not take the large bulls because the big males do not visit these waters. They cannot eat all the meat, so they barter it in neighboring villages. This is in direct contravention of the regulations of the INTERNATIONAL WHALING COMMISSION, but because Indonesia is not a signatory to the Whaling Convention, the IWC regulations are difficult—if not impossible—to apply.

See Also the Following Articles

Folklore and Legends ▪ Inuit and Marine Mammals ▪ Japanese Whaling

References

Aguilar, A. (1981). The black right whale, *Eubaleana glacialis,* in the Cantabrian Sea. *Rep. Intl. Whal. Commn.* **31,** 457–459.

Aguilar, A. (1985). A review of old Basque whaling and its effect on right whales (*Eubaleana glacialis*) of the North Atlantic. *Rep. Intl. Whal. Commn.* **10,** 191–199.

Barkham, S. (1984). The Basque whaling establishment in Labrador 1536–1632: A summary. *Arctic* **37,** 515–519.

De Smet, W. M. A. (1981). Evidence of whaling in the North Sea and English Channel during the Middle Ages. *In* "Mammals in the Seas." FAO Fisheries Series No. 5, Vol. III, pp. 301–309. Food and Agriculture Organization of the United Nations, Rome.

Dudley, P. (1725). An essay upon the natural history of whales, with particular account of the ambergris found in the sperma ceti whale. *Phil. Trans. Royal Soc. London* **33**(387), 256–259.

Fraser, F. C. (1970). An early 17th century record of the Californian grey whale in Icelandic waters. *Invest. Cetacea* **2,** 13–20.

Harmer, S. F. (1928). The history of whaling. *Proc. Linn. Soc. London* **140,** 51–95.

Hawes, C. B. (1924). "Whaling." Doubleday, Page.

Jenkins, J. T. (1921). "A History of the Whale Fisheries." Reissued 1971, Kennikat Press.

Markham, C. R. (1881). On the whale fisheries of the Basque provinces of Spain. *Proc. Zool. Soc. London* **62,** 969–976.

Morison, S. E. (1971). "The European Discovery of America: The Northern Voyages, A.D. 500–1600." Oxford Univ. Press.

Mowat, F. (1984). "Sea of Slaughter." Atlantic Monthly Press.

Ommanney, F. D. (1971). "Lost Leviathan." Dodd, Mead.

Omura, H. (1986). History of right whale catches in the waters around Japan. *Rep. Int. Whal. Commn.* **10,** 35–41.

Proulx, J.-P. (1986). "Whaling in the North Atlantic from Earliest Times to the Mid-19th Century." Canadian Printing Service.

Sanderson, I. (1956). "Follow the Whale." Little, Brown.

Scoresby, W. (1820). "An Account of the Arctic Regions with a History and a Description of the Northern Whale-Fishery." Archibald Constable, Edinburgh. 1969 edition, David & Charles.

Whitaker, I. (1984). Whaling in classical Iceland. *Polar Rec.* **22**(134), 249–261.

Whitaker, I. (1985). The king's mirror (*Konnungs skuggsjá*). *Polar Rec.* **22**(141), 615–627.

Whitaker, I. (1986). North Atlantic sea creatures in the king's mirror (*Konnungs skuggsjá*). *Polar Rec.* **22**(142), 3–13.

Whaling, Traditional

RICHARD ELLIS
American Museum of Natural History,
New York, New York

Beginning with the Basques around 1000 A.D., there were many people that conducted whale hunts, but until the advent of industrial whaling in the 20th century, there had never been a whale hunt as organized and systematic as the sperm whale fishery out of New England. Founded in the Massachusetts villages of Nantucket and New Bedford, the technology of sperm whale hunting—and the profits to be derived therefrom—spread to many locations in and around New England, including Albany, Long Island, New Jersey, Delaware, and Maine. The whale fishery—immortalized in Herman Melville's 1851 *Moby Dick*—had an enormous effect on the economy of the recently founded United States, and the whale ships spread American culture and customs around the world. The focus here will be on the well-documented American whale fishery, but similar fisheries were operated by many other nations during the same period.

I. The Beginning of Sperm Whaling

The story of Christopher Hussey's accidental encounter with a school of sperm whales (*Physeter macrocephalus*) has been told so often that it probably no longer matters whether it really happened. In 1712, Captain Hussey was cruising the Massachusetts coast (there were still right whales, *Eubaleana glacialis,* to be caught at that time) when an unexpected storm blew him out to sea. When the clouds cleared, he saw the spouts of whales, but they were forward-angled blows, not the vertical, paired plumes of right whales. Hussey managed to capture one of these unusual animals and towed it back to Nantucket. Instead of baleen plates, it had ivory teeth in its underslung lower jaw, and in its head was a great reservoir of clear amber oil, which solidified to wax when exposed to the air.

The first industry practiced by the New England colonists was the export of beaver pelts and furs to England, but these commodities were quickly exhausted, and given the availability of the easily killed right whales close to their shores, they turned their attention from the forests to the sea. The earliest colonial whaling was practiced in the Indian manner; towers were erected along the shore to enable lookouts to watch for whales, and when one was sighted, the whalers took to the boats. As navigation improved, the whalers began to roam farther offshore, occasionally visiting the rich grounds of Georges Bank, and some vessels even ventured south into the vast oceanic river that would become known as the Gulf Stream. The Yankee whalers also headed north, toward the Gulf of St. Lawrence and the Grand Banks of Newfoundland. By the middle of the 18th century, there were some 50 ships bringing oil and bone to England and returning with such things as iron ore, hemp, cloth, and other necessities for the burgeoning new colony. By 1775, Nantucket had a fleet of 150 whalers, which ranged in size from 90 to 180 tons.

The beginning of the sperm whale fishery in 1712 did not automatically spare the remaining right whales. Although sperm oil was enormously desirable for lubrication and candle making, the need for whalebone had not abated. In the middle of the 18th century, European women of fashion still required tight-laced corsets, so the New England whalers captured whatever whales they could find and processed them accordingly. Scammon tells us that "shore-whaling continued for over fifty years, but eventually it was abandoned, for the same reason that the Spitsbergen and Smeerenburg fisheries were—the scarcity of whales near the coast."

Regardless of the species being hunted, the primary product of the whale fishery was oil. (Earlier, however, the Dutch and English whalers of Spitsbergen and Greenland had concentrated on the whalebone to the extent that they sometimes cut the slabs of baleen from the mouth of the whale and discarded the carcass. Much of the commerce of whaling was determined by fashion; by the amount of whalebone that would be required to girdle the ladies.) In America, by contrast, there was no court, no royalty, and, in the mid-18th century Quaker colony of Nantucket, very little fancy dress.

As practiced in the Greenland fishery (and every other whaling operation until that time), the blubber of the whale was cut off in strips (a process known as "making off") and packed directly into casks for transport to the home port. Scoresby, writing in 1820, noted that "in the early ages of the fishery [it was] performed on shore; and even so recently as the middle of the last century, it was customary for ships to proceed into a harbor, and there remain so long as this process was going on." By the middle of the 18th century, an innovation that would change the nature of the entire industry had been introduced: iron caldrons set in a brick furnace enabled the whalers to render the oil from the blubber aboard the ship instead of on shore. This method seems to have evolved around the year 1750, but there is no individual whose name is associated with the invention. Although it is not possible to identify the father of the onboard tryworks, there were many mothers, all of whom had "necessity" as part of their names. Among the reasons for its introduction were the unpleasant odors associated with the onshore boiling of the blubber into oil, and the energetic protests of the people who lived downwind of a noisome blubber works. (Even on board ship, blubber stored in casks tended to spoil quickly, and the stench was overpowering to the whalemen.) On the Spitsbergen and Greenland grounds, the cold climate kept the blubber from turning rancid until the ships could get back to a port, but in New England no such natural refrigeration existed, and the heat often "turned" the oil in the blubber before it could be processed. As long as the whales could be caught within sight—or, at most, a couple of days' sail—of shore, the blubber could be casked and stowed, but when the whales became scarcer in the home waters and longer voyages were required, some other method of processing and stowage was called for. Scoresby (who never employed on-board tryworks, even though the idea existed during his whaling days) wrote that it was less efficient to carry home the blubber, since "blubber in bulk, notwithstanding every precaution . . . generally loses much of its oil."

Now that sperm whales were being processed with some regularity, another change was taking place in the whaling industry. Earlier, all whale oil, casked as blubber or tried-out at sea, was considered usable for lighting and lubrication. (It was often referred to as "train oil," from the Dutch *traan* for "tear" or "drop.") It is a true fat and impregnates every part of the whale, from the bones to the muscles, but most importantly it is found in the blubber. The fat of right whales and bowheads (and the occasional humpback, *Megaptera novaeangliae*) provided the whale oil that was used extensively from the 10th century until the middle of the 19th for heating, lighting, manufacturing of soap and cosmetics, and lubrication of machinery.

Because spermaceti (from the head of the whale) is quite different from train oil, its processing and utilization were also different. Up until the middle of the 19th century, candles were the primary source of indoor light. They were usually made out of wax or tallow and emitted a smelly black smoke as they burned. The head of the sperm whale contained the mysterious fluid that could be used to make a better kind of candle. This wax, which the whale maintains in a liquid form during its lifetime, solidifies when exposed to air, and someone realized that it might be employed in the manufacture of candles. From sometime around 1750, it was used to manufacture smokeless, odorless candles, the best candles known before or since.

In addition to the liquid oil contained in the case, the sperm whale produced a spongy material also impregnated with oil, known to the whalemen as the "junk." The liquid oil in the case and the oil that was squeezed from the junk were collectively known as "head matter" and were used in the finest lamps and candles. The process of manufacture was a fairly complicated one. Upon delivery, the sludge-like substance was heated in a large copper vat and the impurities were drawn off. It was left to congeal in casks and was then bagged in woolen sacks to be pressed later in a large screw press. The oil squeezed from the head matter was the highest quality and was used in lamps. Further processing produced lower qualities of sperm oil, used for candles. Sperm oil candles were particularly popular in Africa and the Caribbean, but as articles of colonial manufacture they could not be imported into England. Both types of oil were considered superior to the train oil of the right whale and were priced accordingly. As the market developed for the finer qualities of oil, colonial entrepreneurs appeared. In 1751, one Benjamin Crabb of Rehoboth, Massachusetts, applied to the state house of representatives for a monopoly of the manufacture and sale of sperm oil candles, which was granted.

As the right whales became scarcer, the tempo picked up for the sperm whale fishery. Starbuck has called the period from 1750 to 1784 "the most eventful era to the whale fishery that it has every passed through." New England whale ships were under constant threat of being captured by privateers (the various wars between France and England for control of the North American colonies were going on at this time), and ships that were not commandeered pursued the fishery as far from home as the Grand Banks and the Bahamas. There were also natural disasters attendant upon the nascent whaling industry: ships were lost to storms and occasionally to whales. For reasons of security and increased profitability, the small sloops that had been the mainstay of the fishery were being replaced by larger ships with tryworks aboard; now the whalers could pursue the sperm whale, "the haughty, elusive aristocrat of the high seas." By this time, the method of lowering boats for whales and fastening to them with harpoons attached to the whaleboats had evolved and would remain the dominant practice for another century. It was this method of whaling—and the great sperm whale—that Melville would immortalize in *Moby Dick*. At this time, the fluctuations in the price of whale oil made for a most uneasy market. Good catches would overload the market and depress the price, while in a bad year, the scarcity of the oil would make it dearer.

The failure of the British whaling industry in the early decades of the 19th century left the field wide open to the New Englanders, and they were quick to capitalize on it. When the French and Indian War ended in 1763 and France conceded her claims to Canada, the New England whalers moved in. They sailed from Massachusetts and New York to the Gulf of St. Lawrence and the Strait of Belle Isle, and by 1776 they had discovered the whaling grounds off western Africa (Angola and

Walvis Bay), the Falkland Islands, and the River Plate grounds of the South Atlantic. In many of these regions, the whalers occasionally encountered a southern right whale (*Eubalaena australis*), but the major object of the fishery by this time had become the sperm whale. These explorers in the name of oil were canvasing the world and perfecting their techniques, but instead of flourishing, they fell deeply into debt. The British government, still trying to support its own collapsing whaling industry, placed a duty on all oil and bone carried to England by colonials. Relations were becoming increasingly strained between the Crown and her rambunctious colony; in the years that followed, the infamous Stamp Act would be enacted and repealed; the Townshend Duties ditto, and finally, the Tea Act was passed in 1773, leading to the Boston Tea Party in the same year. (The East India Company, planning to sell its tea directly to America without having to first sell it to British merchants, shipped 1253 chests of tea from London to Boston on four whale ships: *Beaver, Dartmouth, Eleanor,* and *William and Anne.* It was this tea that the rebels, led by John Hancock and Samuel Adams, dumped into Boston Harbor.) The next 2 years saw the first shots fired at Concord Bridge, and for the ensuing decade, most Americans became preoccupied with things other than whaling. (In April 1775, the same time that the "shot heard round the world" was fired, the ship *Amazon,* Captain Uriah Bunker, was discovering the whaling grounds known as the Brazil Banks, some 500 miles east of that country.)

At approximately the same time that British whalers were depositing their cargoes of convicts and finding themselves in the middle of the rich whaling grounds of Australia and New Zealand, Yankee whalers were cruising almost everywhere in search of sperm whales. It was dangerous work, and sometimes the voyages seemed to last forever, but there were those who saw it as pleasurable and even romantic, the epitome of the wholesome life.

II. Life Aboard a Whaler

In retrospect, however, the voyages were often less than romantic and the weather less than benign. There were indeed fresh breezes, tropical sun, and vast herds of cachalots, but there was also the tedium of years of sailing (the record seems to be the 11-year voyage made by the ship *Nile,* out of New London: 1858 to 1869), as well as gales, blizzards, typhoons, hurricanes, mountainous seas, and howling winds. The crew's quarters were stinking holes; their food was cheap, coarse, and maddeningly monotonous; and the work itself was dirty and dangerous. A voyage aboard a New England whaler was not a luxury cruise.

In the 19th century, the hierarchy of officers and men, so important to the successful operation of a whaling vessel, was rigidly observed, and nowhere was the distinction more evident than in their respective living quarters. The captain lived in relative luxury; the ship's officers had smaller cabins; the boat steerers, the cooper, and the steward occupied the steerage, an irregular compartment fitted with plain bunks. The crew was in the forward section just below the main deck, which followed the shape of the ship: it went from a fairly wide cross section to a narrow, cramped, triangular warren, where the ship's timbers formed the walls, and the pounding of the waves

formed the ambience. The lower portion of the foremast often kept the occupants of the fo'c'sle company, reducing even further their limited space, and the only light that entered this literal and figurative rat hole came from the hatchway cut in the deck for the purpose of giving access to the ladder that allowed the men to climb in and out of their quarters. When the weather turned foul, the hatch was closed, and there was no light but stubby candles, and no ventilation whatever. The number of men that occupied this wretched space often exceeded 20.

No whaleman was ever paid a wage, except in unusual circumstances. If, for instance, a full ship had to take on additional hands on the way home, their share of the profits would be zero (since they had not participated in the whaling) and they were paid a monthly wage. Ordinarily, each man, from the captain to the cabin boy, received a percentage of the profits—called a lay—at the end of the voyage.

The distribution differed from vessel to vessel—larger ships could carry more oil, and therefore the profits to the crew were likely to be proportionately higher—but while a successful voyage could be better for the captain and the officers, it meant precious little indeed to the foremast hands. (On an unsuccessful voyage, where the profits were low or nonexistent, the crew might receive nothing at all.) The captain might earn 1/8 or 1/10 of the net proceeds, whereas a mate could earn 1/15 and a harpooner 1/90. Ordinary seamen could hope at best for 1/150, and there are instances in the records where a green hand signed aboard for 1/350. What did this mean in terms of actual money? On board the *Addison,* first mate Ebenezer Nickerson, whose lay was 1/18, earned $845. Robert Baxter, the second mate with 1/35, earned $554.83, and a boat steerer named Narcisco Manuel, with 1/90, got $376.56. Compare these figures to those of the crew: John Martin, at 1/175, earned a total of $31.95 and Francis Finley got $92.08. During six consecutive voyages totaling 1128 days at sea from 1845 to 1868, the average lay per voyage on the Salem whaler *James Maury* was $321.21, or about 26 cents a day. This compared unfavorably to wages then being paid to unskilled laborers ashore (an average of 90 cents a day), but landlubbers did not get to visit exotic Pacific islands where they might be eaten by cannibals or risk their lives fighting gigantic whales.

Infrequently, the men were paid in the specie of whaling; i.e., they received casks of oil, which they were then able to sell at the prevailing prices in their port of disembarkation. The cooks often received an added benefit: in addition to their lays, they were permitted to save the grease (known as "slush") from their galleys and sell it to soap makers ashore.

The whaleman's food and bunk space were generously provided without charge, but throughout the voyage he was docked for various items that he had to buy from the ship's stores. Additional items of clothing, tobacco, knives, needles, and even thread were charged to each man's account, and if he required spending money in a port of call, this too was deducted from the final reckoning. This was a period where the master's voice was law, and if a man needed a new shirt of a pair of boots, he could "either pay up or go naked." Although most of the whalemen signed aboard voluntarily, they usually did not know of the dangers and hardships that lay ahead of them, and the "profit sharing" that at the outset sounded so attractive often deteriorated into an enforced "risk sharing," which was invariably uncomfortable, inevitably dirty, and frequently dangerous.

Among the more unusual charges assessed to a whaleman was the cost of desertion. If a man jumped ship, his account included the cost of recapturing him, an expense that was obviously nullified if he remained at large. However, there were captains who rewarded the lookouts with bonuses for the sighting of whales. This exercise was glorified in *Moby Dick,* where Ahab nails a gold doubloon to the mainmast and exhorts his crew: "Whosoever of ye raises me a white-headed whale with a wrinkled brow and a crooked jaw; whosoever of ye raises me that white-headed whale, with three holes punctured in his starboard fluke-look ye, whosoever of ye raises me that same white whale, he shall have this gold ounce, my boys!" The "Spanish ounce" that was offered to the crew was a 16-dollar gold piece.

On a 3- or 4-year voyage, a man might earn $100, but the items billed to him often exceeded this amount, so many hands returned to port not only with no spending money, but in debt. The only thing to do to work off this indebtedness was to sign on for another voyage, thus starting the insidious process all over again. If and when they made it back to port, the whalemen were set upon by all sorts of "land sharks," eager to assist them in disposing of their wages by enticing them into taverns, brothels, and other iniquitous dens where they could make up for the pleasures they had been denied for the past several years. In an 1860 issue of *Harper's* magazine, an observer describes the arrival of the whalemen in New Bedford:

> A cart rattles by, loaded with recently discharged whalemen—a motley and a savage-looking crew, unkempt and unshaven, capped with the head-gear of various foreign climes and peoples—under the friendly guidance of a land shark, hastening to the sign of the "Mermaid," the "Whale," or the "Grampus," where, in drunkenness and debauchery, they may soonest get rid of their hard-earned wages, and in the shortest space of time arrive at that condition of poverty and disgust of shore life that must induce them to ship for another four years' cruise.

The system of wages aboard a whaler was obviously not conducive to enthusiasm or hard work. In response to the brutal discipline often administered by the captain, there was bound to be apathy, indifference, and suspicion on the part of the foremast hands. There was also a profound class distinction between the officers and the men. Despite the abuses, hardships, and low earnings that characterized the industry, however, the labor supply was somehow adequate to meet its needs. As Hohman has written, "The steady stream of men pouring into the forecastles proved sufficient to counteract the continuous labor leakage caused by death, illness, incapacity, discharge and desertion." It was possible (although uncommon) for a dedicated seaman to work his way up through the ranks, and there are instances where a green hand, or even a cabin boy, raised his lay from 1/150 to 1/15, and after perhaps 20 years at sea (in 4- or 5-year increments), a man might command a whaling vessel.

The *Benjamin Tucker,* a New Bedford whaler, brought back 73,707 gallons of whale oil, 5348 gallons of sperm oil, and 30,012 pounds of whalebone in a voyage that ended in 1851. At the prevailing prices—43 cents a gallon for whale oil, $1.25 a gallon for sperm oil, and 31 cents a pound for bone—the gross value of this cargo was $47,682.73. From this, $2362.73 was variously deducted, leaving a net of $45,320 to be distributed. Before the profits were divided, however, the owners took a substantial percentage off the top to compensate for their initial outlay and also because these flinty New Englanders were not in the business for the thrill of the chase. In general, the owners took between 60 and 70% of the profits. On the 1805–1807 cruise of the *Lion,* the various oils yielded a total of $37,661.02. Of this, $24,252.74 went directly to the owners, leaving $13,045.53 to be divided among the captain and the crew for 2 years of work.

During its heyday, New Bedford was the richest municipality per capita in America, and Melville described it as "perhaps the dearest place to live in all New England . . . nowhere in America will you find more patrician-like houses, [or] parks and gardens more opulent."

Of course, profits from the whaling industry were not restricted to the owners. They had to repair, refit, and reprovision their ships, which provided work and income for the shipwrights, chandlers, coopers, rope makers, carpenters, and blacksmiths, and ready markets for the farmers and greengrocers. The entire township of New Bedford benefited from the outfitting and victualing of the armada of ships that annually departed her wharves, loaded with food, clothing, and supplies, most of which were bought from local merchants.

The captain had his own cabin, with a proper bunk, a washstand, a table, and perhaps even a sofa and some extra chairs. The captain's quarters of the whale ship *Florida* "opened off the after cabin on the starboard side and extended nearly to the end of the forward cabin. A small room and a toilet room were aft of the stateroom. A large swinging bed was in the captain's cabin instead of the usual fixed berth." The gimballed bed was a special innovation designed by Captain Thomas Williams because he was planning to bring Mrs. Williams along.

Occasionally a captain took his wife and, even more infrequently, he took his entire family. Captain Williams, of the ship *Florida* out of New Bedford, was accompanied by his wife for a voyage that lasted from September 1858 to October 1861. During the voyage, Eliza Azelia Williams gave birth to two children, who spent the first years of their lives at sea. She also kept a detailed journal of her adventures, which allows us a most unusual perspective of life aboard a whale ship. The voyage commenced on September 7, 1858, in New Bedford, and on January 12 of the next year, Mrs. Williams gave birth to a baby boy, whom they named William. (William's arrival might help explain her seasickness early in the voyage, when she wrote, "it remains rugged and I remain Sea sick. I call it a gale, but my Husband laughs at me and tells me I have not seen a gale yet. If this is not one I know I do not want to see one.") On August 5, 1859, off the rugged coasts of Sakhalin in the Okhotsk Sea, the *Florida* spoke the *Eliza F. Mason,* and Mrs. Williams visited another "lady ship," where the captain had brought his wife and child, "a Lady Companion, and a little Girl that they brought from the Bay of Islands, New Zealand." On February 27, 1860, Mrs. Williams wrote, "We have had an addition to the Florida's Crew in the form of our little daughter. . . ."

It was United States maritime law that a logbook be maintained by the mate or the first officer. (The term "log book" originated with the practice of casting a log overboard affixed to the ship by a knotted line. The speed at which the line played out—measured in knots—determined the speed of the ship, and the daily records were originally kept in a book reserved for that purpose. Later, the term "logbook" was used to designate the book used for the keeping of all the ship's records.) For the most part, logbooks and journals were kept by the masters. Although rarely educated in the classical sense, most of these men could read and write passably well, and their records have given us an enduring picture of life aboard a whale ship. Even though the maintenance of a logbook was mandatory, it obviously served the whalers particularly well, as the appearance of whales at a known latitude and longitude in one season might enable the whalers to predict their reappearance at the same location the following year, thereby avoiding aimless wandering.

The more mundane entries consisted of the ship's position, the number of whales caught, and illness and injury aboard ship, but additional dramatic possibilities were vast. Whaling historian Stuart Sherman, in his introduction to the catalogue of the logbook collection of Paul Nicholson, listed "castaways, mutinies, desertions, floggings, women stowaways, drunkenness, illicit shore leave experience, scurvy, fever, collisions, fire at sea, stove boats, drownings, hurricanes, earthquakes, tidal waves, shipwrecks, ships struck by lightning, men falling from the masthead, hostile natives, barratry, brutal skippers, escape from Confederate raiders, hard luck voyages and ships crushed by ice." That is not to say that all logbooks read like *Moby Dick;* dramatic events occurred infrequently, and most of the daily entries—when the ship was not engaged in killing whales—consisted of a remark on the wind direction, the location, and whatever else the keeper of the logbook deemed pertinent.

It is not surprising that few of the foremast hands kept records; their quarters were not conducive to the literary life, and besides, many of them could not write. Francis Olmstead could. Of the literary aspirations of his fo'c'sle companions, he wrote:

> The forecastle of the *North America* is much larger than those of most ships of her tonnage, and is scrubbed out regularly every morning. There is a table and a lamp, so that the men have conveniences for reading and writing if they choose to avail themselves of them; and many of them are practicing writing every day or learning how to write. . . . When not otherwise occupied, they draw books from the library in the cabin and read; or if they do not know how, get someone to teach them. We have a good library on board, consisting of about two hundred volumes. . . .

J. Ross Browne, a journalist who shipped aboard the New Bedford whaler *Bruce* in 1842, kept a journal of his experiences that was published, with major revisions, as *Etchings of a Whaling Cruise* in 1846. Browne wanted to do for whaling what Richard Henry Dana had done for merchant sailing in 1840, i.e., exaggerate the problems so that necessary changes would be implemented. Although his account may contain a certain amount of propaganda in the form of negative commentary, he was aboard a whaler for more than a year, and because he is regarded as a reporter and not a writer of fiction,

much of the material contained in his book can be taken as fact. Here is Browne's description of the place in which he lived:

> The forecastle was black and slimy with filth, very small and hot as an oven. It was filled with a compound of foul air, smoke, sea-chests, soap-kegs, greasy pans, tainted meat, Portuguese ruffians and sea-sick Americans. . . . In wet weather, when most of the hands were below, cursing, smoking, singing and spinning yarns, it was a perfect Bedlam. Think of three or four Portuguese, a couple of Irishmen, and five or six tough Americans, in a hole about sixteen feet wide, and as many perhaps, from the bulkheads to the fore-peak; so low that a full-grown person could not stand upright in it, and so wedged with rubbish as to leave scarcely room for a foothold. It contained twelve small berths, and with fourteen chests in the little area around the ladder, seldom admitted of being cleaned. In warm weather it was insufferably close. It would seem like an exaggeration to say, that I have seen Kentucky pig-sties not half so filthy, and in every respect preferable to this miserable hole; such, however, is the fact.

Rats were more numerous on whale ships than on any other vessels, probably because of the profusion of blood and oil that soaked the decks, despite the regular scrubbings. They were more than any ship's cat could cope with, and then as now, there was nothing that could cope with cockroaches. They were endemic aboard the whalers, and for many seamen, the roaches were a more predominant aspect of a whaling voyage than whales. Francis Olmstead wrote that they made "a noise like a flush of quails among the dry leaves of the forest . . . they are extremely voracious, and destroy almost everything they can find: their teeth are so sharp, the sailors say, that they will eat the edge off a razor."

In *Nimrod of the Sea,* William Davis describes roaches as serving a useful purpose: "His chief recommendation in his insane pursuit of the flea . . .," but then goes on, "it is a horrible experience to awaken at night, in a climate so warm that a finger-ring is the utmost cover you can endure, with the wretched sensation of an army of cockroaches climbing up both legs in search of some Spanish unfortunate! It reminds me of how many times I have placed my tin plate in the overhead nettings of the forecastle, with a liberal lump of duff reserved from dinner, and on taking it down at supper, have found it scraped clean by the same guerrillas. They leave no food alone, and have a nasty odor, which hot water will scarcely remove. But one becomes philosophical at sea in matters of food."

The crew's rations aboard a whale ship ranged from bad to disgusting, but, Browne says, "a good appetite makes almost any kind of food palatable." He describes the usual fare on board the *Bruce* (which he has, for culinary and other reasons, named the *Styx*): "I had seen the time when my fastidious taste revolted at a piece of good wholesome bread without butter, and many a time I had lost a meal by discovering a fly on my plate. I was now glad enough to get a hard biscuit and a piece of greasy pork; and it did not at all affect my appetite to see the mangled bodies of divers well-fed cockroaches in my molasses; indeed, I sometimes thought they gave it a rich flavor." Fresh vegetables were taken on at the outset of a voyage and were often picked up when the vessel put in for provisions, but unless they were used quickly, they rotted. (By Browne's time, the causes of scurvy were known, but if the vegetables were used up and the ship was cruising somewhere off the Aleutian Islands, there was not much anyone could do to prevent the dread disease.) Because of their inability to store much water—and to prevent it from spoiling—the whalers hardly ever drank it. (Scammon tells the story of one captain, who, to preserve the dwindling water supply, had the drinking cup hung from the royal-masthead, requiring any man who wanted a drink to climb all the way up after the cup.) They drank "longlick," a mixture of tea, coffee, and molasses, and if the cook was imaginative, he prepared something known as "lobscouse" (or simply "scouse"), which was a hash made of hard biscuits that had been soaked in the greasy water left over after boiling the salted meat. The mainstay of the whaler's diet was salted meat, which was supposed to be pork or beef, but was occasionally horse. In *Omoo*, Melville described the meat on board a whaleship:

> When opened, the barrels of pork looked as if preserved in iron rust, and diffused an odor like a stale ragout. The beef was worse yet; a mahogany-colored fibrous substance, so tough and tasteless, that I almost believed the cook's story of a horse's hoof with the shoe on having been fished up out of the pickle of one of the casks.

Because the everyday food was so often inedible (Nordhoff describes the duff made by a certain cook as "that potent breeder of heartburns, indigestion, and dyspepsia . . . the very acme of indigestibility," and Ben-Ezra Ely wrote, ". . . no swine that gleans the gutters ever subsisted on viler meat and bread than did our crew"), the opportunity to eat something fresh was a blessing. The cook prepared seabirds, whatever fish they could catch, turtles, dolphins (off the African coast, Nordhoff describes the harpooning and subsequent eating of a hippopotamus), and because they were engaged in the capture of 50- or 60-ton mammals whose carcasses they would otherwise leave for the sharks, they often ate the meat of the whales. On the eating of various parts of the whale, usually during the trying-out, Browne writes:

> About the middle of the watch they get up the bread kid [a kid was a wooden tub] and, after dipping a few biscuits in salt water, heave them into a strainer, and boil them in oil. It is difficult to form any idea of the luxury of this delicious mode of cooking on a long nightwatch. Sometimes, when on friendly terms with the steward, they make fritters of the brains of the whale mixed with flour and cook them in the oil. These are considered a most sumptuous delicacy. Certain portions of the whale's flesh are also eaten with relish, though, to my thinking not a very great luxury being coarse and strong. . . .

It was a different world above decks. On December 28, 1856, the crew of the New Bedford whaler *Addison* caught a porpoise, and Mary Chipman Lawrence (the captain's wife) wrote in her journal, "The meat looks very much like beef. The oil is contained in the skin, which they will boil out tomorrow. Had some of the meat fried for dinner and some made into sausage cakes for supper. They are as nice as pork sausages." If a further demonstration of the disparity between the fare of the men and that of the officers is required, here is Mrs. Lawrence's description of Christmas dinner for that same year: "roast chickens, stuffed potatoes, turnips, onions, stewed cranberries, pickled beets and cucumbers, and a plum duff. For tea I had a tin of preserved grape opened and cut a loaf of fruitcake."

Unlike their British counterparts, American whalers rarely carried any sort of medical man. It commonly fell to the captain to cope with whatever illness or accident befell his crew, and given the master's experience, it was considerably safer to remain healthy. For internal maladies, whale ships were often equipped with medicine chests, which contained various potions and a manual for their dispensation. (Stories were told of masters who, having run out of medicament Number 12, simply administered equal amounts of Numbers 5 and 7.)

Physical injuries were not uncommon, considering the number of sharp-edged tools, whistling whale lines, and hostile natives, not to mention shipboard arguments between men who were almost always armed with knives. Here again, the master served in the role of surgeon, with the same amount of training as he had as apothecary. In *Nimrod of the Sea*, W. M. Davis tells the gory tale of a whaleman who was yanked from his boat by a kinked line and dragged some 125 fathoms from the boat. When he was finally picked up, "it was found that a portion of the hand, including four fingers, had been torn away, and the foot sawed through at the ankle, leaving only the great tendon and the heel suspended to the lacerated stump." Equipped with "his carving knife, carpenter's saw and a fish-hook," the captain "amputated the leg and dressed the hand as best he could."

III. Whale Ships and Whaleboats

As whaling voyages increased in distance and duration, it became expedient to enlarge the ships. In the early days of the fishery (around 1820), the ships averaged around 280 tons burthen, but within two decades, 400-ton vessels were not uncommon. The move toward bigger whale ships contributed to the decline of Nantucket whaling because there was a prominent sandbar across the harbor, and only the smaller, shallower-draft ships could enter. New Bedford, with its excellent harbor facilities, took up the slack.

All whaling vessels were ships—as opposed to *boats*, which were the smaller vessels that the whalers rowed after their quarry. The literature is replete, however, with references to ships, brigs, brigantines, barks, barkentines, and schooners. These differentiations have to do with the rigging of the masts and not with the number of masts, although a three-masted, square-rigged vessel was always known as a ship. If the aftermost mast was rigged fore and aft, with the sail slung between a gaff and a boom, the vessel was known as a *bark*, the commonest plan, because fewer hands were required to handle the sails, and thus there were more men available for the boats. There were further variations, including the *brig*, where the upper courses of the aftermost mast were rigged with square sails, but there was also a fore-and-aft sail known as a "spanker." A *barkentine* was square-rigged only on the foremast, the rest fore and aft, and a *brigantine* had only two masts, the foremast square-rigged and the mainmast fore and aft. A *schooner* had two or more masts, rigged fore and aft.

Whale ships differed from merchantmen of the time in that they usually carried less sail. More canvas meant more men aloft, and the whalers needed as many hands as possible for the boats. One further characteristic of the whaler was the presence of masthead hoops, in which the lookouts stood during the daylight hours to watch for whales.

Square-rigged ships, which gave their name to an era of sailing, ran powerfully before the wind, but were not particularly handy in head- or crosswinds. The whalers did not have to perform any smart sailing maneuvers, nor did they have to sail with great speed. All they had to do was get from one location to another and then lower the boats after the whales. Because of the determined, plodding nature of their craft, the masters rarely sailed at night, preferring instead to furl their sails and wait until dawn before continuing.

It was during the heyday of New England whaling, from 1830 to 1860, that the fabulous clipper ships reached the zenith of sailing-ship design, with their graceful lines, sharply raked bows, and opulence of canvas. In marked contrast to these ocean-going grayhounds, the whalers were sturdy, bluff-bowed, flat-bottomed sailers, designed more for durability and storage than for speed. (The *Lagoda* sailed for 50 years, and the all-time record holder, the *Charles W. Morgan*, sailed for more than 80 years and earned over a million dollars for her owners. The *Lagoda* was copied at half-scale for the New Bedford Whaling Museum, and the *Morgan*, the last of her kind, is now the proud centerpiece of Mystic Seaport in Connecticut.)

A typical whaler was 100 to 150 feet long, and especially broad in the beam to accommodate the fixtures of whaling: heavy brick tryworks on deck, iron caldrons, cooling tanks, davits for the boats, and, of course, the space required to perform the trying-out of the whale. Ordinary seamen, whose voyages did not take 4 or 5 years, belittled the whale ships as "built by the mile and cut off in lengths as you want 'em." They were usually painted black and had mock gun ports painted along the sides, supposedly as a deterrent to pirates or hostile savages.

The naval historian Albert Cook Church wrote: "Whaleships differed materially from any other type of merchant ship or clipper in model and equipment, and in fact, both sides of a whaleship differed from each other above the waterline." The larger ships were equipped with four boats: one on the starboard quarter and three on the port (also known as the "larboard") side. This allowed the cutting stages, which were always on the starboard, to be lowered without interference from davits.

When a whale or a group of whales was sighted, the lookout shouted "She blows!" or "Blows!" and when the captain had ascertained "where away," the boats were lowered and the chase began. All the boats might be lowered, depending on the number of whales sighted. If only a single whale was seen, the captain might designate one boat to chase it. The starboard boat was reserved for the captain (or the fourth mate, if the captain chose to stay aboard ship during the hunt); the larboard, waist, and bow boats were for the first, second, and third mates, respectively. Each boat contained a regular crew, consisting of five oarsmen and a boat steerer/harpooner. Whoever was in command of the whaleboat pulled the steering oar and gave the orders. The boats were double enders; in case they got turned around in the frenzy of the hunt, they would be able to maneuver, and they were among the most graceful and utilitarian boats ever designed.

All the requisite equipment would be stowed carefully aboard the whaleboats, from the line, which was carefully coiled in a tub so it could be let out rapidly, to the knife that might be required to cut it if a man got his leg entangled. In addition to the six adult men who would be required to man the boat, Scammon lists the contents of a fully equipped whaleboat:

> One mast and one yard, one to three sails, five pulling oars, one steering oar, five paddles, three rowlocks, five harpoons, one or two line-tubs, three hand lances, three shortwarps, one boat-spade, three lance-warps, one boat-warp, one boat-hatchet, two boat knives, one boat-waif, one boatcompass, one boat-hook, one drag, one grapnel, one boat-anchor, one sweeping-line, lead, buoy, etc., one boat-keg, one boat-bucket, one piggin, one lantern-keg (containing flint, steel, box of tinder, lantern, candles, bread, tobacco, and pipes), one boat-crotch, one tub-oar crotch, half a dozen chock pins, a roll of canvas, a paper of tacks, two nippers, to which may be added a bomb-gun and four bomb-lances, in all, forty-eight articles, and at least eighty-two pieces.

IV. Killing and Processing the Whale

The lowering of the boats took place as the ship was underway; the captain did not come about for the comfort or convenience of his crews. Often in high seas, the graceful whaleboats took off after the whales with the men facing the stern; the boat steerer was the only man who could see the whales. When they had come within range, the harpooner threw the harpoon. It consisted of a wooden shaft, some 6 feet in length, with a forged iron head. The earliest harpoons had simple fluted arrowhead-shaped heads, but as the fishery developed, more sophisticated designs were introduced. While the two-flued iron pierced the blubber effectively, its razor edges would occasionally pull out as smoothly as they went in. This led to the introduction of the single-flued iron, which held much better. Harpooners and blacksmiths had plenty of time, on board the whalers and in port, to work on harpoon design, and all sorts of elaborate heads with toggles, barbs, and swivels were tried. The most successful of these designs was the double-barbed "Temple" iron, invented in 1848 by a New Bedford blacksmith named Lewis Temple. A graceful, practical device, the Temple iron consisted of a pointed head that was held in the forward position by a wooden shear pin that broke off when withdrawal forces were applied. This rotated the head 90° in the flesh of the whale, forming a T-shaped device that would not pull out because the flattened surfaces were pulling against the meat or blubber. The iron was fastened to the shaft of the harpoon by a line that was bent to the heavy manila line. The line, which Melville calls the "magical, sometimes horrible whale-line," was originally fashioned of hemp, but was later superseded by manila rope, which was stronger and more elastic. "Hemp is a dusky, dark fellow," Melville wrote, "a sort of Indian, but Manilla is as a golden haired Circassian to behold."

Even though tradition demanded that the harpoon and the lance be thrown separately, some creative whalemen tried to design an iron that would fasten to and kill the whale simultaneously. A Scottish toxicologist named Robert Christson invented a poison-headed harpoon, equipped with glass cylinders containing prussic acid, one drop of which is lethal enough to kill a man. There is no evidence that prussic acid harpoons were used in the American fishery, but they were carried on some vessels. The likelihood is that the American harpooners felt that they had enough problems killing the whale without worrying about killing themselves.

If the iron was well placed—the ideal spot was in the flank, forward of the hump—the boat was fast to the whale, and the injured animal took off. Sometimes the whale sounded, taking out the line at such speed that the line smoked as it ran out, and the loggerhead had to be doused with water to keep it from bursting into flames. More often the whale swam at the surface, towing the boat through the waves at a violent clip. Sperm whales are prodigious divers, and no boat could hold enough line for a dive that could be measured in miles. If the whale sounded, another 200-fathom line might be bent to the first and then another. Eventually, the wounded whale had to surface to breathe.

The lance, also known as the "killing iron," was plunged into the "life" of the whale, a vital artery, the lungs, or the heart. The killing iron consisted of a wooden shaft like that of the harpoon, with a scalpel-sharp head. It was not thrown, but rather stabbed repeatedly into the body of the whale. Melville describes the death throes of a whale:

> The red tide now poured from all sides of the monster like brooks down a hill. His tormented body rolled not in brine but in blood, Which bubbled and seethed in furlongs behind in their wake. The slanting sun playing upon this crimson pond in the sea sent back its reflection into every face, so that they all glowed to each other like red men. . . . Stubb slowly churned his long sharp lance into the fish and kept it there, carefully churning and churning, as if cautiously seeking to find some gold watch that the whale might have swallowed, and which he was fearful of breaking ere he could hook it out.

The victory did not always go to the whalers. Sperm whales are immensely powerful creatures and do not take kindly to being stabbed with spears. The most frequent problem occurred when the whale took it into its 20-pound brain to retaliate. A 30-foot whaleboat was no match for an enraged, wounded, 60-ton whale, and the harpooned animal might rise up from the depths and grab the boat in its massive jaws, splintering it into so many matchsticks. Both ends of a wounded whale are lethal; the triangular flukes, which might measure 20 feet across, could function as a formidable weapon, crashing down upon the whaleboat and dumping the men into the sea. Other perils faced the whalemen, where the whistling line might take a turn around a leg or an arm, surgically severing it, or yanking the man into the water. Even if the boat was not destroyed, it might be upended and its occupants dumped into the ocean. Many of them could not swim, so such a plunge often spelled death.

More often than not, however, the world's deadliest predator won the battle and then faced the problem of bringing whale and ship together. If the conquering whaleboat was downwind of the ship, it was a relatively simple matter to sail the ship to the carcass, but if less propitious conditions prevailed, the exhausted whalemen might have to tow the whale back to the ship, often for miles. Then, after an exhausting chase and a laborious haul with a 50-ton deadweight in tow, the real work began. What had been a free-swimming, powerful sea mammal was effectively reduced to a disparate assortment of its parts, the reduction accomplished by literally tearing it apart.

As in virtually every other aspect of New England whaling, the cutting-in process was described better by Melville than anybody else. (In the Yankee whale fishery, the process of removing the whale from his outer integuments was known as "cutting-in," and the rendering of the blubber into oil was known as "trying-out." In the English fishery, these operations were known respectively as "flensing" and "making off.") In *Moby Dick* there is one chapter devoted to the actual process and several more to the by-products, including the "blanket," the "funeral," and the "sphynx"—the last referring to the head of the whale after the body and blubber have been separated from it.

The whale was made fast to the ship by lashing heavy chains through its head and around its flukes. The first part of the whale to be brought aboard was the lower jaw, ripped from the head and laid aside to be dealt with later. Then the whale was decapitated, and if it was a small one, the head was brought aboard. The head of a large whale, often one-third of its 60-foot, 60-ton body, could not be brought on deck (Melville wrote that "even by the immense tackles of the whaler, this were as vain a thing as to attempt weighing a Dutch barn in jeweller's scales") and had to be processed in the water. The "head matter" was saved for last, however, because the carcass of the whale alongside the ship was threatening to the ship by its weight, and the longer it remained unprocessed, the longer the sharks could wreak havoc on the very outer layer of blubber that was of so much interest to the whalers.

By the use of a complicated series of tackles—described by Melville as "ponderous things comprising a cluster of blocks generally painted green, and which no single man can possibly lift—the cutting stages were lowered, and the process of removing the blubber commenced. Sitting or standing on the lowered cutting stages, men with razor-sharp cutting spades began to slice into the whale's rubbery outer covering. A massive iron hook was inserted in the first piece to come off, which was hoisted high into the air while the men on the scaffold sliced the blubber. The whale was rotated in the water, and its blubber "stripped off from the body precisely as an orange is sometimes stripped by spiralizing it." The power for this peeling and dismemberment came from the strong backs of the whalemen, who turned the windlass located forward of the foremast.

As the thick spiral of blubber was peeled from the whale, it was cut into sections approximately 15 feet long and a ton in weight (the "blanket pieces"). These were dropped through a hatch into the blubber room, where they were stored until the carcass of the whale was completely stripped. (With the removal of the blubber and the head, the remainder of the carcass was left for the sharks.) Workers in the dark, bloody blubber room further reduced the blanket pieces to smaller, more manageable "horse pieces," which were then sliced into "Bible leaves," with cuts almost to the skin, making them resemble the splayed pages of a thick-leaved book. (It was believed that the opening of the blubber into "pages" made the oil more accessible.) The Bible leaves were then forked back up through the forehatch to the men who would place them in the trypots.

Although the trypot fires were usually started with wood, the unmelted skin of the whale made a wonderful fuel, and the whale was therefore cooked in a fire of its own kindling. As the oil was separated from the blubber, it was carefully ladled into a copper cooling tank, where it rested before being casked. Aside from the obvious danger of a fire spreading, the process was—like almost every aspect of whaling—hard, messy, and dirty. Oil and blood covered the decks and the people, and the smell was often intolerable. J. Ross Browne called the trying-out process "the most stirring part of the whaling business, and certainly the most disagreeable." He described the nighttime scene aboard the *Styx*:

> Dense clouds of lurid smoke are curling up to the tops, shrouding the rigging from the view. The oil is hissing in the trypots. Half a dozen of the crew are sitting on the windlass, their rough, weather-beaten faces shining in the red glare of the fires, all clothed in greasy duck, and forming about as savage a looking group as ever was sketched by the pencil of Salvator Rosa. The cooper and one of the mates are raking up the fires with long bars of wood or iron. The decks, bulwarks, railing, try-works, and windlass are covered with oil and slime of black-skin, glistering with the red glare of the try-works. Slowly and doggedly the vessel is pitching her way through the rough seas, looking as if enveloped in flames.

At the end of this description, he wrote, "Of the unpleasant effects of the smoke I scarcely know how any idea can be formed, unless the curious inquirer choose to hold his nose over the smoking wick of a sperm oil lamp, and fancy the disagreeable experiment magnified a hundred thousand fold. Such is the romance of life in the whale fishery."

One of the least romantic aspects of the whale fishery was the prospect of fire. Oil-soaked wooden ships upon whose decks fires are being encouraged do not lend themselves to a feeling of security. Care was taken to avoid conflagrations—water was pumped over the decks to keep the planks wet and cool—but occasionally the sails or rigging were ignited by flying sparks, and sometimes the ships burned to the waterline.

When the oil had cooled, it was ladled into the casks that had been assembled by the cooper. Each barrel held 302 gallons, and the figures for the fishery were almost always recorded in barrels. Starbuck's 1878 *History of the Whale Fishery*, which contains the records of every American whaling ship, from every American whaling port, "from its earliest inception to 1876" (insofar as these records were known), lists the result of every whaling voyage in sperm oil (barrels), whale oil (barrels), and whalebone (pounds).

A large female sperm whale might yield 35 barrels of oil, while the largest bulls gave up 75 to 90. As with the sometimes questionable lengths of large bulls, where there were reports of 90 footers (Clifford Ashley writes, "If these whalemen's records are accurate, it would appear that the hundred-foot Sperm Whale is not an impossibility"), the yield of these giants was the subject of occasional exaggeration. Because the reports

were invariably made by men whose reputation would be enhanced by overstating the yield of individual whales, many of the whales in the 100- to 150-barrel range must be questioned.[1]

The amount of oil that could be taken and stored was enormous, but it did not necessarily reflect the success of a voyage. The profits of a voyage could only be calculated when the ship reached port and sold the oil and bone at the prevailing prices. A 31-gallon cask was about 5 feet high and 4 feet in diameter at its bulging middle (Fig. 1). On her maiden whaling voyage, which lasted from October 1841 to September 1843, the *Lagoda* brought home 600 barrels of sperm oil, 2700 barrels of whale oil, and 17,000 pounds of baleen. ("Sperm oil" was the stuff that was ladled out of the whale's "case" and was of a finer quality than "whale oil," which was rendered out of the blubber. Although they were not averse to taking an occasional right whale or humpback, most of the whales hunted by the Yankees were sperm whales.) The *Lagoda* was 108 feet long, with a beam of 27 feet. Hunting concluded when there was no more room for the storage of oil, but the whalers sometimes put into port, off-loaded some of their greasy cargo, and set out again for the whaling grounds. Some of these sweaty, iron-bound vats were probably stored in the blubber room, but most were stored in the hold.

It was the mysterious "head matter" of the sperm whale that made it the primary object of this globe-girdling enterprise. Other whales were encased in blubber, and some of them had the long "finnes" that could be converted into milady's bodices. The spermaceti, however, was the *ne plus ultra* of this business, the pot of liquid gold that attracted the whalers to the Azores and the Galapagos, to Zanzibar and the Japan Grounds, to Kamchatka and the Okhotsk Sea. The stuff is as poorly understood today as it was when some early beachcomber presumed that this vast reservoir in the whale's nose was its seminal fluid. Whatever its purpose to the whale (and it certainly is not its seminal fluid), the amber wax that hardened white as it was exposed to air was worth risking life and limb—and sometimes boat and ship—to the whaler. Kept free from contamination by other oils, sperm oil was worth from three to five times as much as whale oil. In *Nimrod of the Sea*, W. M. Davis records a whale that yielded 27 barrels of spermaceti from the case, and Clifford Ashley's research indicates that the largest bulls gave up something on the order of 30 barrels. At 31.5 gallons per barrel, that works out to 945 gallons of the mysterious liquid wax in the nose of a single whale.

To extract the spermaceti from the head, a much more direct method was employed than the multistep process of turn-

---

[1] If only the whalers' stories remained, we would have no way of verifying the size of the largest whales. There is something that they leave behind, however, and Ashley proposes a novel argument for the existence of gigantic bull sperm whales: he examines a particularly large pair of teeth, over 11 inches long, and suggests that "in the days before the Sperm Whale herds were depleted, there must have been exceptional whales, either larger or older than are found today." Mitchell (1983) finds this argument "well taken, but not conclusive," but a look at these teeth, which are on display in the New Bedford Whaling Museum, certainly gives one cause to wonder.

Figure 1 *Casks of oil line the New Bedford wharf. Photo by Old Dartmouth Historical Society—New Bedford Whaling Museum.*

ing blubber into oil. Because the spermaceti already *was* oil, the whalers only had to remove it from the whale and cask it. A hole was cut in the outer fabric of the whale and a man lowered a bucket into it on a long pole and then turned it over to another man on deck who would empty the bucket into a waiting tub—or as Melville put it, "Tashtego downward guides the bucket into the Tun, till it entirely disappears; then giving the word to the seamen at the whip, up comes the bucket again, all bubbling like a dairy-maid's pail of new milk."

When the oil had all been casked and the casks stowed, the decks were scrubbed down with lye, which had been leached from the cinders and ashes of the tryworks, and the oily, smoky clothes of the whalemen were also scrubbed down, but the pernicious odor of smoked blubber could never really be removed, and until they could exchange their work clothes for new garments, the whalemen usually smelled like disused tryworks.

V. Bowhead Whaling

The gigantic nose of the sperm whale contains an enormous quantity of valuable oil, which was the primary object of the sperm whalers. Bowheads (*Balaena mysticetus*) are fatter than sperm whales, and their blubber thicker, but it was not so much for the blubber that the bowheads were hunted: their huge, arched mouths contained the longest baleen plates of any whale

and it was this that the whalers were after. It was used for the manufacture of skirt hoops, corset stays, horsewhips, and dozens of other necessities of 19th century America.

In July 1848, Captain Thomas Welcome Roys, a sperm whaler out of Sag Harbor (New York), sailed through Bering Strait, a thousand miles farther north than any whale ship had ever gone in the Pacific, and came upon a thriving population of bowheads that had previously been known only to Eskimos. His crews took 11 bowheads in only 35 days during that eventful summer and sailed for home with 1800 barrels of oil, an accomplishment that normally took two or more seasons. According to John Bockstoce, the historian of western Arctic whaling, "Roys's cruise was not only the most important whaling discovery of the nineteenth century, it was also one of the most important events in the history of the Pacific. . . . More than 2,700 whaling voyages were made into Arctic waters at a cost of more than 150 whaleships lost and the near extinction of the bowhead whale, as Roys's whales came to be called." Arctic bowhead whalers also changed the way in which whales were hunted; instead of hand-thrown harpoons, they introduced the efficiency of artillery, shooting at the whales first with shoulder guns, and later killing them with bomb lances.

Many New England whaleships had been captured or burned by the Confederate raider *Shenandoah* in 1865, and to revive the flagging industry, 39 ships sailed to the Arctic in

April 1871. By August, 32 of them were trapped in the ice off Point Barrow, Alaska. The 7 remaining ships took on 1129 men and brought them back to Honolulu, signaling the beginning of the end of Arctic bowhead whaling.

VI. The "Romance" of Whaling

In the 19th century, when so much of the world was still unexplored, whalemen faced even greater hazards than an occasional angry whale. The United States exploring expedition under Wilkes had visited many of the island groups in the Central Pacific and found that some of the stories of hostile savages, often cannibals, were true. The Fiji Islands were known—more or less accurately—as The Cannibal Isles, and whenever possible the whalers avoided them. In 1835, the whaler *Awashonks,* out of Falmouth, was attacked by the natives of Namarik in the Marshall Islands, and the captain, the first and second mates, and four crew members were killed before an enterprising whaleman dynamited the deck where the would-be conquerors were standing, and the ship was retaken. The *Syren* was recaptured from Palauan natives only after a box of tacks was scattered on deck, driving the barefoot raiders howling overboard.

The need for fresh vegetables and water often outweighed the threat of being attacked, and even though many of the captains knew or had heard stories of cannibalism and "cutoffs" (a whaleship captured by natives and its crew massacred), they could not resist the temptations of cheap provisioning. Whale ships hardly ever carried money; the very same "slop chest" that provided the foremast hands with their replacement items of clothing also served as a trading bank. "Recruiting ship" was the term used to describe the acquisition of provisions, and the captain would trade cotton cloth, powder, tobacco, knives and beads for fresh food and water. It was sometimes too much for the parsimonious New Englanders to resist: for a couple of pounds of tobacco or some rusted iron hoops, they could trade for pigs, coconuts, water, wood, and women.

The quality of life aboard a whaler was hardly luxurious, but it was often better than life on the farm. Indeed, many whalemen deserted on the islands, not because they were unduly harassed or flogged, but because life on a lush, green island, with free food and even freer women, was an economic and sociological step upward. By the middle of the 19th century, there may have been as many as 3000 deserters from whale ships scattered throughout the coral archipelagoes of Micronesia and Polynesia.

Hawaii, Tahiti, and the Marquesas are picturebook "South Sea Islands," with tall volcanic mountains, tumbling waterfalls, broad white beaches, and swaying palms. When Herman Melville jumped ship in the Marquesas and subsequently described his experiences in *Typee,* he was responsible for many misinterpretations of life on a tropical island. Not all island groups boasted plentiful paw-paws and willing wahines. Many of these tiny specks in the Pacific were mercilessly unforgiving coral atolls; low rings of sand that had only recently risen out of the Pacific. All along the equator in the Pacific—the grounds known as "On the Line"—sperm whales occurred in some profusion. While the vicinity of the Gilbert Islands (then known as the Kingsmill Group) was a good place to kill sperm whales, it was almost as dangerous for the whalers as it was for the whales. The Gilbertese natives were a particularly aggressive and warlike people, but there were also many beachcombers who had been stranded on these islands, men who were eager to lead the natives in attacks on visiting whalers.

While an occasional whaler was killed by unfriendly or unreceptive islanders, the effect of the ships' landings on the natives was considerably more severe. Unscrupulous captains would often offer to trade for provisions and then sail away without giving anything in return, and it is likely that triggerhappy sailors took the lives of many natives without having to worry about punishment so far from any law but the captain's. In *Nimrod of the Sea,* William Davis recounts the story of three "kanakas" (probably Maoris from New Zealand) who deserted on a tropical island. After demanding their return from the local natives, "the captain doubleshorted his ninepound guns, sent a round into the crowded grass huts of the village, and carried off three natives." From their home ports and pestilent fo'c'sles, the whalemen brought every conceivable communicable disease to the natives, including yaws, influenza, tuberculosis, cholera, syphilis, and the greatest scourge of the unresistant, measles. Prior to exploration, the population of Tahiti was estimated at some 40,000; by 1830, there were only 9000 Tahitians left. A single measles epidemic in 1875 killed nearly 30,000 Fijians.

There was only so much eating, drinking, holystoning of decks, repairing of sails and rigging, and yarn spinning to occupy the sailors on these seemingly endless voyages. To pass the time, some of them created what Clifford Ashley called "the only important indigenous folk art, except for that of the Indians, we have ever had in America; the Art of SCRIMSHAW."

Although there are very few contemporaneous records of scrimshanders at work, probably because the craft was too insignificant to mention, we assume that the whale TEETH were carved during periods of sailing or while waiting in port for provisions or repairs. The baleen of the right and bowhead whales was packed into bundles for commerce at home, but occasionally a piece would be shaped into a busk and decorated with contemporary designs. Baleen was colloquially known to the whalers and merchants of the time as *whalebone,* but it is not bone at all; it is made of keratin, the substance of human hair and fingernails. Whales have bones like any other mammals, but with the exception of the lower jaw—known as the "pan bone"—and the teeth, whale bones are too porous for carving. (Other cultures recognized the attractive nature of whale ivory. Certain Polynesian natives made necklaces of dolphin teeth, and the premissionary Hawaiians crafted the beautiful *le niho palaoa,* a gracefully carved sperm whale tooth that was worn by royalty on a necklace of braided human hair.)

Despite their discomfort, low wages, and even occasional floggings, the crews of whale ships were remarkably docile. The master's word was law, and when the crew became obstreperous, a "taste of the cat" was not unheard of. Only infrequently did they become so desperate that they rebelled. Because harsh treatment, long hours, uncomfortable quarters, and bad food were expected, the whalemen generally endured these indignities in stoic silence. Also, as with any uprising, a leader is required to galvanize men into action, and on the whale ships, these troublemakers were rare. The story of the *Bounty's* mutiny, which had occurred in 1789—and had nothing whatever to do with whales or whaling—was probably known to every seaman and landlubber on either side of the Atlantic. The fate of Fletcher Christian and the mutineers was not known until 1808, when Captain Mayhew Folger of the Nantucket sealer *Topaz* landed at Pitcairn Island and found the survivors.

On Nantucket Island there lived a young man named Samuel Comstock, who may have heard the tale told by Captain Folger. At the age of 19, after three previous cruises, he shipped out aboard the Nantucket whaler *Globe,* departing from Edgartown on the neighboring island of Martha's Vineyard on December 15, 1822. The ship rounded the Horn on March 5 and stopped briefly at Hawaii before heading for the newly discovered Japan grounds. Despite the reports of plentiful whales off Japan, Captain Worth was unable to locate them, and as they sailed in fruitless circles, the crew became increasingly discontented. Rotten meat was an issue, and conditions were so bad that the captain turned back and headed for Hawaii to reprovision. There several members of the crew deserted, and the *Globe's* depleted crew was replenished with beachcombers and drunkards. Repeated conflicts between officers and crew increased the tension, and when the captain had one of the men flogged, Comstock decided to initiate a mutiny.

On January 26, 1824, Samuel Comstock led his followers in one of the bloodiest mutinies in American naval history. They murdered Captain Worth with an axe, slaughtered first mate Beetle with a boarding knife, shot second mate Lumbard in the mouth and then bayonetted him, and shot third mate Fisher in the back of the head. They heaved the bodies overboard and, with Comstock at the helm, looked for a place where they could land. En route, Comstock decided that one of his crew members was plotting against him, held a "trial," and sentenced him to hang. For 2 weeks they wandered around, uncertain of their location or destination, until they decided to land at tiny Mili Atoll, in what was then known as the Mulgrave Islands and is now known as the Ratak chain of the Marshall Islands. It appears that Comstock's original plan was to arrange things so that he was the only survivor, but the natives and his fellow mutineers conspired against his plan for the perfect mutiny. As Comstock began to give the ship's stores to the natives (to ensure their support), the crew members who had signed on in Hawaii realized that they were in for trouble either from their leader or from the natives, and they shot Comstock dead.

Those members of the crew of the *Globe* who had not participated in the mutiny managed to gain control of the ship and sailed away, leaving the mutineers stranded on the island. They would not last long. A bloody conflict between the natives and the whalemen resulted in the death of all the latter but two: William Lay of Saybrook, Connecticut, and Cyrus Hussey of Nantucket. The *Globe* was sailed to Valparaiso, where the news of the mutiny was made known, and then returned to Nantucket. Her crew was cleared of complicity in the mutiny, and the *Dolphin,* under the command of Lieutenant John ("Mad Jack") Percival, was dispatched to the Pacific to find and bring back the mutineers. Hussey and Lay had been with the natives

for almost a year and a half by the time the *Dolphin* arrived, and they looked more like natives than American whalemen. After considerable tension—the Marshallese chiefs were prepared to kill the newly arrived Americans and take their ship—and confusion about who they were, the last of the *Globe's* crew were transported home. Thus ended the story that Starbuck called "the most horrible mutiny that is recounted in the annals of the whale-fishery from any port or nation."

See Also the Following Articles

Blubber ■ Scrimshaw ■ Spermaceti ■ Sperm Whale

References

Allen, E. S. (1973). "Children of the Light: The Rise and Fall of New Bedford Whaling and the Death of the Arctic Fleet." Little, Brown, New York.

Andrews, R. C. (1916). Shore-whaling: A world industry. *Natl. Geogr.* **22**(5), 411–442.

Ashley, C. W. (1938). "The Yankee Whaler." Riverside Press, New York.

Bockstoce, J. R. (1986). "Whales, Ice and Men: The History of Whaling in the Western Arctic." University of Washington Press, Seattle, WA.

Browne, J. R. (1846). "Etchings of a Whaling Cruise, with Notes of a Sojourn on the Island of Zanzibar, to which Is Appended a Brief History of the Whale Fishery, Its Past and Present Condition." Harper & Brothers. Reprinted 1968, Harvard University Press.

Church, A. C. (1938). "Whale Ships and Whaling." Bonanza, New York.

Davis, W. M. (1874). "Nimrod of the Sea, or, The American Whaleman." Christopher, North Quincy, MA.

Ellis, R. (1991). "Men and Whales." Knopf, New York.

Ely, B.-E. (1849). " 'There She Blows:' A Narrative of a Whaling Voyage, in the Indian and South Atlantic Oceans." 1971 edition. Edited by Curtis Dahl. Wesleyan University Press.

Garner, S. (ed.) (1966). "The Captain's Best Mate: The Journal of Mary Shipman Lawrence on the Whaler *Addison* 1856–1860." Brown University Press.

Haley, N. C. (1948). "Whale Hunt: The Narrative of a Voyage by Nelson Cole Haley, Harpooner in the Ship *Charles W. Morgan* 1849–1853." Ives Washburn, New York.

Henderson, D. A. (ed.) (1974). "Journal Aboard the Bark *Ocean Bird*" on a Whaling Voyage to Scammon's Lagoon, Winter of 1858–1859. Dawson's Book Shop, Los Angeles, CA.

Hohman, E. P. (1928). "The American Whaleman." Reissued 1972, Augustus M. Kelley, New Jersey.

Hoyt, E. P. (1975). "Mutiny on the Globe." Random House, New York.

Lay, W., and C. Hussey. (1828). "A Narrative of the Mutiny on Board the Ship Globe of Nantucket in the Pacific Ocean, Jan. 1824 and the Journal of a Residence of Two Years on the Mulgrave Islands: With Observations on the Manners and Customs of the Inhabitants." New London, Connecticut.

Melville, H. (1846). "Typee." New York.

Melville, H. (1847). "Omoo A Narrative of the South Seas." New York.

Melville, H. (1851). "Moby Dick." 1967 Norton Critical Edition, edited by H. Hayford and H. Parker. W. W. Norton, New York.

Miller, P. A. (1979). "And the Whale Is Ours: Creative Writing by American Whalemen." David R. Godine.

Mitchell, E. D. (1983). Potential of logbook data for studying aspects of social structure of the sperm whale *Physeter macrocephalus,* with an example—the ship *Mariner* to the Pacific, 1836–1840. *Rep. Intl. Whal. Commn.* (Special Issue) **5,** 63–80.

Nordhoff, C. (1856). "Whaling and Fishing." Moore, Wilsatch, Keys & Co., London.

Olmstead, F. A. (1841). "Incidents of a Whaling Voyage." Reissued 1936, Charles E. Tuttle, Rutland, VT.

Scammon, C. M. (1874). "The Marine Mammals of the Northwestern Coast of North America; Together with an Account of the American Whale Fishery." Carmany and G. P. Putnam's.

Scoresby, W. (1820). "An Account of the Arctic Regions with a History and Decription of the Northern Whale Fishery." Constable. 1969 edition, David & Charles, Devon.

Sherman, S. C. (1965). "The Voice of the Whaleman." Providence Public Library.

Starbuck, A. (1878). "A History of the American Whale Fishery from Its Earliest Inception to the Year 1876." Part IV, Report to the U.S. Commission on Fish and Fisheries, Washington. Reprinted 1964: Argosy-Antiquarian Ltd., New York.

Whaling, Modern

PHILLIP J. CLAPHAM
Northeast Fisheries Science Center,
Woods Hole, Massachusetts

C. SCOTT BAKER
Auckland University, New Zealand

There is no means known to catch the fin whale, or its fast cousins.

So wrote Herman Melville in the year 1851, thus echoing the common lament of whalers that the fastest (and some of the largest) species of whales, such as the finback (*Balaenoptera physalus*) and the blue (*B. musculus*), lay beyond contemporary means of capture. At the time that Melville wrote *Moby Dick*, the basic technology of whaling had remained essentially unchanged for centuries. Whaling ships plied their trade under sail, and the small boats that they lowered to pursue whales were also powered by wind or by the brute strength of their crew's arms at the oars. The killing of whales required men to bring their frail craft alongside the huge quarry, subduing and fastening to it with hand-thrown harpoons. If this dangerous series of actions succeeded, the whale might ultimately be despatched with a lance thrust deep into some vital organ. Once killed, the carcass would be either towed to shore or brought alongside the whaling vessel for the time-consuming process of flensing and "trying out" (boiling) the blubber for oil.

I. The Emergence of Modern Whaling

These methods had been in use in the 11th century when the Basques began the first sustained commercial whale fishery on right whales (*Eubalaena glacialis*) in the Bay of Biscay. Although many improvements had been made, the technology available to whalers in the middle of the 19th century severely limited the number of animals that could be taken and processed in a working day. Furthermore, as Melville noted, it largely precluded exploitation of the faster species. Blue, fin,

sei (*B. borealis*), and Bryde's whales (*B. brydei/edeni*), all large and desirable targets, were too swift to allow pursuit by oars or sails. Thus it was the slower species such as the humpback, the right, and the sperm whale (*Physeter macrocephalus*) that had borne the brunt of commercial whaling, and they had done so in some cases for almost a thousand years.

By 1860, however, all of this was about to change. Two men, the Norwegian Svend Føyn and an American named Thomas Welcome Roys, were independently experimenting with explosive harpoons. As patented by Roys in 1861, this device was initially fired from a shoulder gun; Føyn developed a different approach using a bow-mounted cannon, which was to become the industry standard. The "bomb lance" was an innovation that was to revolutionize the whaling industry by providing a much more efficient means to dispatch whales, both quickly and from a distance.

About the same time, the use of sail was beginning to give way to steam. This was a key innovation: together with the explosive harpoon, it radically changed the industry and finally allowed the pursuit and capture of any whale. Suddenly, even the fastest rorquals came under the threat of the harpoon as they were chased down by fast motorized catcher boats.

A further innovation, the compressor, solved a long-standing problem with regard to the many species that did not float when dead: by pumping air into the carcass immediately after death, whalers could secure it before it sank, thus reducing the loss rate greatly.

For the industry, this transition into the mechanized age could not have come at a more opportune time. By 1900, many populations of the traditionally hunted species were commercially exhausted. For some populations, such as the bowhead (*Balaena mysticetus*) and the North Atlantic right whale this was the result of exploitation spread over centuries. With some other stocks, decimation had been accomplished in a remarkably short time: the first North Pacific right whale (*E. japonica*) was not killed until 1835, yet 14 years later the population had already been reduced to the point where many whalers switched their focus to the newly established fishery for bowhead whales in the western Arctic. Another quickly depleted stock was that of the eastern North Pacific gray whale (*Eschrichtius robustus*), made vulnerable by its predictable coastal MIGRATION and tendency to concentrate for breeding and calving in the lagoons of Baja California.

The new technology opened up all species to whaling and did so at a time when the industry, spurred on by steam power, was also expanding geographically. By far the most significant development in this regard was the discovery of the vast stocks of whales in the Southern Ocean. In 1904, the Norwegian whaler C. F. Larsen arrived at the South Atlantic island of South Georgia and reported with astonishment, "I see them in hundreds and thousands." Huge pristine populations of rorquals—notably blues, fin whales, and humpbacks (*Megaptera novaeangliae*)—filled the surrounding waters together with southern right whales (*E. australis*) and other species. Modern whaling had found its last and greatest reserve. A slaughter unparalleled in whaling history was about to begin.

Whaling at South Georgia was initially constrained by the need to use land stations for processing of the carcasses. Because of this, the time to spoilage limited the range of the catcher boats; it also left the whaling companies vulnerable to high taxes levied by the British authorities. Despite these difficulties, the industry accomplished the destruction of local stocks of whales with remarkable efficiency. At the height of operations, hundreds of humpback, fin, and blue were taken in a single month. By 1915 the South Georgia population of humpbacks had essentially been extirpated, with a total catch of some 18,557 whales; while occasional catches were made in later years (the largest being 238 humpbacks in 1945/1946), the stock was commercially extinct by the time of the Great War. Blue whales suffered a similar fate: 39,296 were killed at South Georgia between 1904 and 1936, at which point the population had crashed, apparently irretrievably.

The problem of dependence on land stations was solved, at a stroke, with the introduction of the factory ship. The British vessel *Lansing* was the first such floating factory and began operations in Antarctic waters in 1925. It is difficult to overestimate the importance of this innovation to whaling or its contribution to the destruction of whale populations in the Antarctic. Factory ships could operate independently far out to sea for months at a time. They maintained round-the-clock processing operations in the long Antarctic days, their huge flensing decks kept constantly supplied by an attendant fleet of catcher boats. Whale carcasses were hauled up the large stern ramp and dismembered with astonishing mechanical efficiency: an adult fin whale of 70 or 80 feet and 100 tons could be rendered from whole animal down to bone in half an hour. With the factory ship, all of Antarctic waters became open to whalers, their operations limited only by the constant dangers of weather and ice.

Over the six decades following the opening of the Antarctic grounds in 1904, the whaling industry killed approximately 2 million whales in the Southern Hemisphere (Table I). This included 360,000 blue whales, some 200,000 humpbacks, more than 400,000 sperm whales, and a staggering 725,000 fin whales (Table 1). By the 1930s, it was apparent even to the whaling nations that some kind of regulation was required. In 1931, the Convention for the Regulation of Whaling was held and adopted worldwide protection for right whales, an action that came into effect in 1935. The second Convention for the Regulation of Whaling was held in 1937 and provided protection for the much-depleted gray whale. However, neither convention went far enough; among other things, because neither Japan nor the Soviet Union ratified these agreements, both were theoretically free to continue killing the only two species that had been granted protection.

II. Advent of the International Whaling Commission

In 1946, following the virtual cessation of whaling that occurred during World War II, the International Convention for the Regulation of Whaling was developed and signed by all major whaling nations (including Japan and the USSR). Among other things, this landmark convention created the International Whaling Commission (IWC), established to regulate whaling and to oversee research on whale stocks. The latter task had as its principal objective management that would al-

TABLE I
Southern Hemisphere Catch Totals
(1904–2000)[a]

Blue	360,644
Fin	725,116
Sei	203,538
Humpback	208,359
Bryde's	7,757
Minke	116,568
Right	4,338
Sperm	401,670
Other	11,631
Total	2,039,621

[a]Primary sources IWC, 1995; Yablokov *et al.*, 1998.

low the highest viable level of exploitation, a concept widely known as maximum sustainable yield (MSY). This required collection and analysis of information on abundance, population structure, and life history of the great whales to permit setting of quotas.

Unfortunately, the quota system of the IWC was immediately handicapped by an earlier development. In 1932, the whaling nations had developed the "blue whale unit" (bwu). A single bwu was equivalent to one blue whale, two fin whales, two and a half humpbacks, or six sei whales. As such, quotas set in bwu, by permitting whalers to make their own decisions about which whales to take, made no allowance for the conservation status of a particular species, let alone that of a specific population. It was not until 1949 that a species-specific quota (for humpbacks) was established.

The bwu remained in effect until 1972, despite recommendations from IWC scientists as early as 1963 that it be abolished (also in 1963, the same scientists recommended a halt to all humpback and blue whaling; the IWC responded by setting a quota of 10,000 bwu). In some years, the IWC could not agree on a bwu quota, and whaling nations were left to make their own informal agreements on catch levels. Overall, the bwu arguably represents the most ill-conceived and damaging management strategy in IWC history. However, it was far from the only problem in the commission's management of whale populations.

From its inception, the IWC was hampered by the unwillingness of the whaling nations to acknowledge the mounting evidence of decline in whale populations and by the complete lack of any enforcement or independent inspection measures. That the first humpback whale quota in 1949 was immediately exceeded in the three subsequent Antarctic seasons pointedly highlighted the latter issue. Additional examples followed, but it was not until the 1990s that the true extent of this problem became apparent, and it was more egregious than anyone could have predicted.

In 1993, following the end of the Cold War, former Soviet biologists revealed that the USSR had conducted a massive

Figure 1   *These blue whales are among those harvested worldwide during the 20th century. Over 360,000 animals were harvested from the Southern Hemisphere alone. Photo by Whales Research Institute, courtesy of Hideo Omura.*

campaign of illegal whaling beginning shortly after World War II. Soviet factory fleets had killed virtually all whales they encountered, irrespective of size, age, or protected status. The scale of this deception was staggering: in all, the difference between the reported take of the USSR and its actual catches was more than 100,000 animals in the Southern Hemisphere alone (Table II). Of these, the humpback whale was the most heavily impacted: reporting 2710 catches to the IWC, the Soviets had in fact taken more than 48,000. In the Northern Hemisphere, Soviet activities were on a smaller scale, but were nonetheless extremely damaging in some cases. The virtual disappearance of right whales in the eastern North Pacific in the 1960s was recently explained by revelations of Soviet catches of 372 whales from this already depleted stock between 1963 and 1967.

In retrospect, it is possible to see clues to this unfolding catastrophe. Beginning in 1959, at a time when there was increasing discussion of declining populations and the need for diminished quotas, the Soviets began adding a new factory ship each year to their Antarctic fleet. This included the *Sovetskaya Ukraina,* the largest floating factory ever built, with an attendant fleet of 25 catcher vessels. In a single season (1959/1960), *Sovetskaya Ukraina* and a second factory, the *Slava,* killed almost 13,000 humpbacks, mainly from the high-latitude waters south of Australia, New Zealand, and western Oceania. In addition, the intransigence of the USSR in its opposition to a proposed international observer scheme (IOS, to permit independent inspections of catches at sea) is now easy to interpret. The illegal catches continued until adoption and implementation of the IOS was finally accomplished in the early 1970s.

In the latter part of the 1960s the IWC finally began to respond to the increasing evidence that whale populations had been exploited well beyond MSY. Blue whales were protected in 1965, and quotas for fin and sei whales were reduced in the late 1960s in response to declines in catches. Nonetheless, any enforcement remained absent, and the Soviets secretly continued to kill whales irrespective of quota or protection.

III. The Decline of Commercial Whaling

In the following decade, however, a sea change occurred at the IWC. The composition of the commission slowly shifted as nonwhaling nations joined, while others ceased whaling and developed instead into advocates for conservation. A whaling moratorium was proposed by the United States and Mexico as early as 1974, but this and later proposals were rejected by IWC until 1982. In that year, a radically changed commission finally achieved the necessary votes to pass a 10-year moratorium. Predictably, Japan, Norway, and the Soviet Union objected. The moratorium went into effect in 1986, with a zero catch quota for both pelagic and coastal whaling.

TABLE II
Reported versus Actual Catches by the USSR[a]

	Reported	Actual
Blue	3,651	3,642
Pygmy blue	10	8,439
Fin	52,931	41,184
Sei	33,001	50,034
Humpback	2,710	48,477
Bryde's	19	1,418
Minke	17,079	14,002
Right	4	3,212
Sperm	74,834	89,493
Other	1,539	1,745
Total	185,778	261,646

[a]Note that some catches were actually overreported; this was to disguise takes of protected species by overreporting catches of species that were legally huntable at the time. From Yablokov *et al.* (1998).

At this point in time Soviet whaling was coming to an end; with aging capital and the imminent dissolution of the USSR, the nation that had wreaked so much havoc on whale populations (a fact still unknown at this time) slowly removed itself from the business of commercial whaling. Japan, Norway, and Iceland, however, remained active, and in 1987 they effectively circumvented the moratorium by beginning "scientific" whaling. This act exploited a provision in the convention that allows member nations to issue themselves permits to conduct whaling for scientific research; it was originally included at a time when the only way in which any information could be gathered about whales was to kill them. As opponents of scientific whaling pointed out, the emergence in the 1970s of long-term studies of living whales (frequently based on the IDENTIFICATION of individual animals) provided a much better means to study the biology and behavior of cetaceans.

The stated reason for the moratorium was to permit world whale stocks to recover from the overexploitation to which they had been subject. In the meantime, the IWC's scientific committee was charged with developing a new management procedure for future management of stocks and setting of quotas. After considerable debate, the so-called revised management procedure (RMP) was accepted by the scientific committee in 1994. The RMP is a computer model for determining the level of allowable commercial catches for a stock based on its current abundance, history of exploitation, levels of incidental takes, and overlap with adjacent stocks having differing catch histories. However, the scheme by which the RMP would actually be implemented had still not been adopted in 2001, and there remains considerable resistance to the idea of the IWC endorsing a program that would effectively permit the resumption of commercial whaling. In 1994, Norway preempted such an agreement by resuming commercial whaling under "objection" to the moratorium and used the RMP to set its own catch quotas for minke whales in the northeastern North Atlantic.

The reluctance of many nations to implement the RMP stems largely from lingering concerns regarding enforcement and transparency in whaling operations. The whaling nations maintain that adequate measures are now in place to ensure compliance with quotas set under the RMP. Opponents disagree, pointing to the history of deception in modern whaling and noting more recent evidence that such deception continues to exist. In particular, considerable attention has been focused on the use of forensic genetics to test samples of whale meat in Japanese markets; although the only meat that should be found there is that from minke whales (*B. acutorostrata* and *B. bonaerensis*) taken in Japanese scientific catches, numerous other species have been detected. Although some of these animals probably represent bycatch (incidental entrapment in fishing gear), their presence reinforces the fact that there is currently no means to adequately track whale products at every stage from catch to market. A DNA register of all animals taken in Japanese and Norwegian hunts has been proposed, but it has yet to be fully developed and accepted by the IWC; the resolution of this issue is not aided by the whaling nations' insistence that any discussion of trade in whale products lies within the purview not of the IWC but rather of the Convention on International Trade in Endangered Species (CITES) or the World Trade Organization.

IV. Impacts of Whaling on the Stocks of Whales

The impact of modern whaling on the world's stocks of whales has been varied. Many species appear to be recovering despite past exploitation, which in many cases may have reduced numbers by 90% or more from pristine levels. Humpback whales, which were extensively overhunted worldwide and which bore the brunt of illegal Soviet catches in the Southern Hemisphere, are showing strong rates of population growth in the North Atlantic, North Pacific, and some areas of the Southern Ocean. Eastern gray whales number approximately 26,000 animals and have been removed from the U.S. list of endangered species. Although no reliable estimates of abundance exist, populations of fin and sei whales are assumed to be healthy in the Northern Hemisphere; the status of the extensively exploited Antarctic populations is less clear. Similarly, sperm whales are likely to be generally abundant, although in some areas, apparently slow rates of population growth may be attributable to overexploitation of mature males in high latitudes, resulting in insufficient availability of mates that are "acceptable" to adult females. Although there is considerable controversy over Japanese and Norwegian whaling, there is general agreement that some of the targeted stocks of minke whales are abundant. However, some adjacent stocks are considered depleted and could be impacted in areas where the whales mix during migration.

In contrast, other populations of whales appear to be struggling to recover from the indiscriminate exploitation to which they were subject. In some extreme cases, local populations appear to have been extirpated, with no recovery evident in the intervening years. Humpback and blue whales at South Georgia were commercially extinct by 1915 and 1936, respectively, and are rarely observed there today. Blue whales were wiped out from the coastal waters of Japan by about 1948, and no members of this species have been recorded there in recent years despite an often extensive survey effort. Off Gibraltar, a population of fin whales was extirpated with remarkable speed between 1921 and 1927. The population of humpbacks that used the coastal waters of New Zealand as a migratory route crashed in 1960 as a result of shore whaling and the 1959/1960 Soviet catches of almost 13,000 humpbacks in the feeding grounds to the south. A few sightings have been reported off New Zealand in recent years, perhaps suggesting that a slow recovery is underway. In the North Atlantic, right whales were removed from much of their former range largely by historical whaling prior to 1880 (the Labrador stock seems to have been wiped out by Basque whaling as early as 1610), but even here, a remnant population in European waters was extirpated by Norwegians using modern techniques at the beginning of the 20th century. The demise of at least one stock of whales can be attributed exclusively to premodern whaling: the bowhead was commercially extinct from Spitsbergen waters by 1900, and the species is rarely observed there today.

In all of these cases, whaling essentially extirpated a stock of whales. The lack of recovery over a time scale ranging from four decades in the case of New Zealand humpbacks to almost four centuries for right whales off Labrador has important implications for the modern MANAGEMENT of whale populations. Although it is quite likely that the observed lack of recovery was at least partly due to a simultaneous overexploitation of adjacent populations (i.e., those that might otherwise have provided a source for repopulation), these localized extirpations reinforce the belief that management units should be designed carefully on often smaller spatial scales than has been the case in the past.

Of those populations that survived, several are critically endangered. Right whales persist in low numbers in the western North Atlantic and the western North Pacific; the present size of the eastern North Pacific population is unknown, but is clearly precariously small following the immense damage done by the Soviets in the 1960s. In sharp contrast to the eastern ("California") gray whale, the outlook for the western North Pacific population of this species is bleak. Whaling on this small stock continued in Korean waters into the 1960s, and a gray whale was found harpooned in Japan as recently as 1996. Only a hundred or so animals may remain extant today. Furthermore, nothing is known of the location of the breeding grounds for this population; if it is reliant on coastal lagoons for calving (as is a major segment of the eastern stock), the impact of coastal development and other human activities may be severe. Among bowhead populations, that in the Bering/Beaufort/Chukchi Seas is recovering strongly despite continued exploitation by a well-managed Inuit hunt. In the eastern Arctic, a few hundred bowheads remain in Canadian waters (principally Hudson Bay/Foxe Basin and Baffin Bay/Davis Strait), while as noted earlier the Spitsbergen stock appears to be functionally extinct. Finally, blue whales have fared poorly almost everywhere; the only population that appears to be large and healthy is that which feeds off California in summer. Other blue whales, including all of those in the Southern Ocean, remain rare and highly endangered.

It is not clear what the future holds for whaling. Ultimately, it depends on the outcome of developing geopolitics: put simply, on whether the emerging worldview of commercial whaling as an anachronism prevails or, if it does not, on whether whaling can learn the lessons of its grim past. For now, the outcome remains hung in the balance.

See Also the Following Articles

Endangered Species and Populations ■ Illegal and Pirate Whaling ■ International Whaling Commission ■ Japanese Whaling ■ Population Status and Trends ■ Stock Assessment

References

Baker, C. S., and Palumbi, S. R. (1994). Which whales are hunted? A molecular genetic approach to monitoring whaling. *Science* **265**, 1538–1539.

Best, P. B. (1993). Increase rates in severely depleted stocks of baleen whales. *ICES J. Mar. Sci.* **50**, 169–186.

Clapham, P. J., and Hatch, L. (2000). Determining spatial and temporal scales for population management units: Lessons from whaling. Paper SC/52/SD2, available from the International Whaling Commission, Impington, Cambridge, United Kingdom.

Clapham, P. J., Young, S. B., and Brownell, R. L., Jr. (1999). Baleen whales: Conservation issues and the status of the most endangered populations. *Mamm. Rev.* **29**, 35–60.

Ellis, R. (1991). "Men and Whales." Knopf, New York.

International Whaling Commission. (1995). Southern Hemisphere catch data coding: position at 1 July 1994. *Rep. Int. Whal. Commn.* **45**, 129–130.

Tønnessen, J. N., and Johnsen, A. O. (1982). "The History of Modern Whaling." University of California Press, Berkeley.

Yablokov, A. V., Zemsky, V. A., Mikhalev, Y. A., Tormosov, V. V, and Berzin, A. A. (1998). Data on Soviet whaling in the Antarctic in 1947–1972 (population aspects). *Russ. J. Ecol.* **29**, 38–42.

White-Beaked Dolphin
Lagenorhynchus albirostris

CARL CHRISTIAN KINZE
Zoological Museum, University of Copenhagen, Denmark

Recognized only as a separate species in 1846, the white-beaked dolphin was among the last of the commonly occurring North Atlantic dolphins species to enter the cetological theater. Earlier finds remain obscure because they have been confused with bottlenose (*Tursiops truncatus*) or common dolphins (*Delphinus* spp).

The white-beaked dolphin was described twice in 1846, first by the English cetologist Edward Gray based on an animal that was caught off Great Yarmouth, England, and almost simultaneously by the Danish cetologist Daniel F. Eschricht, who described a specimen from the west coast of Jutland, Denmark, under the name *Delphinus ibsenii*.

Figure 1 *A white-beaked dolphin stranded on the Danish west coast. Photograph courtesy of Martin Abrahamsson, Midt-sønderjyllands Museum, Gram, Denmark.*

I. Vernacular Names

Vernacular names of the species in most languages depict the white beak: Weissschnauzen-Delphin (German), dauphin avec bec blanc (French), delfin de hocico blanco (Spanish) Witsnuitdolfijn (Dutch), hvidnæse (Danish), kvitnos or hvitnos (Norwegian), and vitnosdelfin (Swedish).

II. Description

The white-beaked dolphin has a robust appearance. The short beak is usually only between 5 and 8 cm long. There is an erect falcate dorsal fin in the middle of the back. Adults grow between 2.4 and 3.1 m long and may weigh between 180 and 350 kg. Males usually grow larger than females. Newborn animals are 1.2 m long and weigh about 40 kg. There are 25–28 conical teeth in each half of the upper and lower jaws.

The COLORATION is typically black on the back with a white saddle behind the dorsal fin and whitish bands on the flanks that vary in intensity from shining white to ashy gray. The belly and beak are normally white (Figs. 1 and 2). The white beak is sometimes ashy gray or even darker, giving the appearance that the white beak is missing.

III. Distribution

The species is endemic to the temperate and subarctic North Atlantic (Fig. 3). It is frequently sighted in shelf waters and sometimes also dwells in shallow coastal waters. However, it has less oceanic preference than its congener the Atlantic white-sided dolphin (*Lagenorhynchus acutus*). Populations in the eastern and western part of its range have been distinguished by means of skull measurements and osteological features.

In the northeast Atlantic it can be found as far north as the White Sea. It is abundant along the Norwegian coasts and in the northern parts of the North Sea and is not uncommon in the southern North Sea along the seaboard of the United Kingdom, Belgium, the Netherlands, Germany, and Denmark. In certain years small schools or individuals may enter the Kattegat, Danish Straits, and even the Baltic proper: a phenomenon believed to be caused by an increased saline influx to the Baltic Sea. The white-beaked dolphin is also known from the French, Spanish Atlantic coasts and may irregularly reach as far south as the Strait of Gibraltar.

In the northwest Atlantic the species seems to be less abundant. The largest concentrations are found off the Labrador coast and in southwest Greenland, but individuals have been encountered as far south as Cape Cod.

IV. Behavior

Schools of up to 30 animals are common, but much larger schools consisting of several hundred or even thousands of animals have sometimes been recorded. The school structure is not known. Animals can be inquisitive and may approach vessels readily. The aerial behavior is very spectacular and typically

Figure 2 *Close-up of the beak. Photograph courtesy of Martin Abrahamsson, Midtsønderjyllands Museum, Gram, Denmark.*

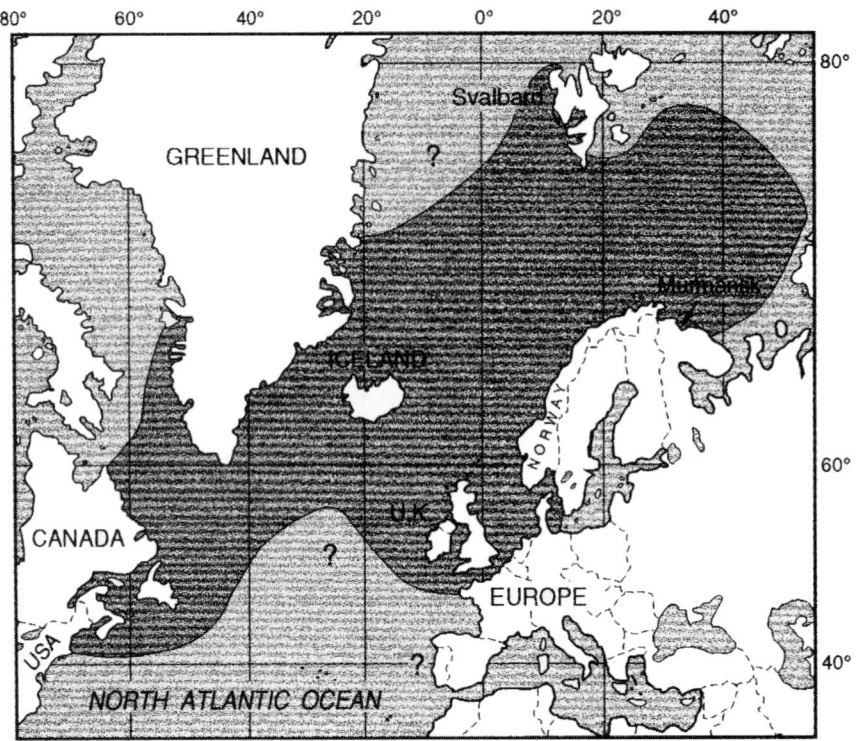

Figure 3 *Distribution of the white-beaked dolphin.*

dolphin-like. Under good sighting conditions the "rooster tail splash" created during speedy swimming is easily visible. The white-beaked dolphin may ride the bow wave of ships and larger whales. The DIVING BEHAVIOR is not known in detail. Mixed schools with white-sided dolphins have been recorded.

V. Reproduction

The available information on reproductive parameters is rather limited. Females reach sexual maturity at about 240 cm in length whereas males mature when around 250 cm long. The age at sexual maturity is not known. Births occur during summer, and gestation lasts between 10 and 11 months. The duration of lactation is not known.

VI. Diet and Natural Enemies

The diet consists of mesopelagic fish species, especially cod, whiting, and other gadids, as well as squids.

Killer whales and also larger shark species probably pose the only natural threat to white-beaked dolphins.

VII. Human Interaction

There is no commercial harvesting of the species. A few animals are shot off southwest Greenland every year. INCIDENTAL CATCHES have occurred in trawls and bottom gill nets (mainly young animals).

See Also the Following Articles

Atlantic White-Sided Dolphin ▪ North Atlantic Marine Mammals

References

Evans, P. G. H., and Smeenk, C. (2000). White-beaked dolphin (*Lagenorhynchus albirostris*). *In* "Handbook of British Mammals" (S. Harris, ed.). Blackwell, London.

Kinze, C. C., Addink, M., Smeenk, C., Garcia-Hartmann, M., Richards, H., Sonntag, R. P., and Benke, H. (1997). The white-beaked dolphin (*Lagenorhynchus albirostris*) and the whitesided dolphin (*Lagenorhynchus acutus*) in the North and Baltic Sea: A Review of available information. *Rep. Intl. Whal. Comm.* **47,** 675–681.

Reeves, R. R., Smeenk, C., Kinze, C. C., Brownell, R. L., Jr., and Lien, J. (1999). White beaked dolphin *Lagenorhynchus albirostris* Gray, 1846. *In* "Handbook of Marine Mammals," Vol 6, pp. 1–30. Academic Press, San Diego.

Zoogeography

SEE *Biogeography*

Zoos

SEE *Marine Parks and Zoos*

MARINE MAMMAL SPECIES[a]

ORDER CARNIVORA

Family Otariidae

Arctocephalus pusillus (Schreber, 1775), Tasmanian and Cape fur seals

Arctocephalus gazella (Peters, 1875), Antarctic fur seal

Arctocephalus tropicalis (Gray, 1872), Subantarctic fur seal

Arctocephalus townsendi Merriam, 1897, Guadalupe fur seal

Arctocephalus philippii (Peters, 1866), Juan Fernández fur seal

Arctocephalus forsteri (Lesson, 1828), South Australian and New Zealand fur seals

Arctocephalus australis (Zimmerman, 1783), South American fur seal

Arctocephalus galapagoensis Heller, 1904, Galapagos fur seal

Callorhinus ursinus (Linnaeus, 1758), Northern fur seal

Zalophus japonicus (Peters, 1866), Japanese sea lion

Zalophus californianus (Lesson, 1828), California sea lion

Zalophus wollebaeki Sivertsen, 1953, Galapagos sea lion

Eumetopias jubatus (Schreber, 1776), Steller sea lion, northern sea lion

Neophoca cinerea (Péron, 1816), Australian sea lion

Phocarctos hookeri (Gray, 1844), New Zealand sea lion, Hooker's sea lion

Otaria flavescens (Shaw, 1800), South American sea lion

Family Odobenidae

Odobenus rosmarus (Linnaeus, 1758), Walrus

Family Phocidae

Erignathus barbatus (Erxleben, 1777), Bearded seal

Phoca vitulina (Linnaeus, 1758), Harbor seal, common seal

Phoca largha (Pallas, 1811), Spotted seal, larga seal

Pusa hispida (Schreber, 1775), Ringed seal

Pusa caspica (Gmelin, 1788), Caspian seal

Pusa sibirica (Gmelin, 1788), Baikal seal

Halichoerus grypus (Fabricius, 1791), Gray seal

Histriophoca fasciata (Zimmerman, 1783), Ribbon seal

Pagophilus groenlandicus (Erxleben, 1777), Harp seal

Cystophora cristata (Erxleben, 1777), Hooded seal

Monachus tropicalis (Gray, 1850), Caribbean monk seal, West Indian monk seal

Monachus monachus (Hermann, 1779), Mediterranean monk seal

Monachus schauinslandi Matschie, 1905, Hawaiian monk seal

Mirounga leonina (Linnaeus, 1758), Southern elephant seal

Mirounga angustirostris (Gill, 1866), Northern elephant seal

Leptonychotes weddellii (Lesson, 1826), Weddell seal

Ommatophoca rossii Gray, 1844, Ross seal

Lobodon carcinophaga (Hombron and Jacquinot, 1842) Crabeater seal

Hydrurga leptonyx (Blainville, 1820), Leopard seal

Family Ursidae

Ursus maritimus Phipps, 1774, Polar bear

---

[a]Living and recently extinct. Author(s) and year of description of the species follow the Latin species name; when these are enclosed in parentheses, species was originally described in a different genus. Classification and scientific names follow Rice (1998), with the exception of the right whales (*Eubalaena* spp.) and the marine otter. Recent genetic evidence strongly supports the recognition of three separate species of right whales (Rosenbaum *et al.*, 2000). In addition, the genus *Eubalaena* (rather than *Balaena* as in Rice, 1998) is retained for the right whales as recommended by the Scientific Committee of the IWC (IWC, 2001). The use of *Lontra* rather than *Lutra* for the marine otter follows Larivière (1998) in recognizing the otters of North and South America as a monophyletic taxon distinct from the otters of Eurasia. For the rest, we do not necessarily endorse Rice's (1998) classification and usage in all their details but have used them as a standard, based on the recent literature, to avoid editorial confusion in the volume.

Family Mustelidae

Enhydra lutris (Linnaeus, 1758), Sea otter
Lontra felina (Molina, 1782), Marine otter, chungungo

ORDER CETACEA

Suborder Mysticeti
Family Balaenidae

Eubalaena glacialis (Müller, 1776), North Atlantic
   right whale
Eubalaena japonica (Lacépède, 1818), North Pacific
   right whale
Eubalaena australis (Desmoulins, 1822), Southern
   right whale
Balaena mysticetus Linnaeus, 1758, Bowhead whale,
   Greenland whale

Family Neobalaenidae

Caperea marginata (Gray, 1846), Pygmy right whale

Family Eschrichtiidae

Eschrichtius robustus (Liljeborg, 1861), Gray whale

Family Balaenopteridae

Megaptera novaeangliae (Borowski, 1781),
   Humpback whale
Balaenoptera acutorostrata Lacépède, 1804,
   Common minke whale
Balaenoptera bonaerensis Burmeister, 1867,
   Antarctic minke whale
Balaenoptera edeni Anderson, 1879, Pygmy Bryde's
   whale, Sittang whale
Balaenoptera brydei Olsen, 1913, Common Bryde's
   whale
Balaenoptera borealis Lesson, 1828, Sei whale
Balaenoptera physalus (Linnaeus, 1758), Fin whale,
   finback whale
Balaenoptera musculus (Linnaeus, 1758), Blue whale

Suborder Odontoceti
Family Physeteridae

Physeter macrocephalus Linnaeus, 1758, Sperm whale

Family Kogiidae

Kogia breviceps (Blainville, 1838), Pygmy sperm whale
Kogia sima (Owen, 1866), Dwarf sperm whale

Family Ziphiidae

Ziphius cavirostris G. Cuvier, 1823, Cuvier's beaked
   whale, goose-beaked whale
Berardius arnuxii Duvernoy, 1851, Arnoux' beaked
   whale
Berardius bairdii Stejneger, 1883, Baird's beaked whale
Tasmacetus shepherdi Oliver, 1937, Shepherd's
   beaked whale, Tasman beaked whale

Indopacetus pacificus (Longman, 1926), Indo-Pacific
   or Longman's beaked whale
Hyperoodon ampullatus (Forster, 1770), North
   Atlantic bottlenose whale
Hyperoodon planifrons (Flower, 1882), Southern
   bottlenose whale
Mesoplodon hectori (Gray, 1871), Hector's beaked
   whale
Mesoplodon mirus True, 1913, True's beaked whale
Mesoplodon europaeus (Gervais, 1855), Gervais'
   beaked whale
Mesoplodon bidens (Sowerby, 1804). Sowerby's
   beaked whale
Mesoplodon grayi von Hast, 1876, Gray's beaked
   whale, scamperdown whale
Mesoplodon peruvianus Reyes, Mead and Van
   Waerebeek, 1991, Pygmy beaked whale, Peruvian
   beaked whale
Mesoplodon bowdoini Andrews, 1908, Andrew's
   beaked whale
Mesoplodon bahamondi Reyes, Van Waerebeek,
   Cárdenas and Yáñez, 1996, Bahamonde's beaked
   whale
Mesoplodon carlhubbsi Moore, 1963, Hubbs' beaked
   whale
Mesoplodon ginkgodens Nishiwaki and Kamiya,
   1958, Ginkgo-toothed beaked whale
Mesoplodon stejnegeri True, 1885, Stejneger's
   beaked whale
Mesoplodon layardii (Gray, 1865), Layard's or strap-
   toothed beaked whale
Mesoplodon densirostris (Blainville, 1817),
   Blainville's or dense-beaked whale

Family Platanistidae

Platanista gangetica (Roxburgh, 1801), Susu, bhulan,
   South Asian river dolphin, Indian river dolphin

Family Iniidae

Inia geoffrensis (Blainville, 1817), Boto, Amazon
   river dolphin

Family Lipotidae

Lipotes vexillifer (Miller, 1918), Baiji, Yangtze river
   dolphin

Family Pontoporiidae

Pontoporia blainvillei (Gervais and d'Orbigny, 1821),
   Franciscana, La Plata river dolphin

Family Monodontidae

Monodon monoceros Linnaeus, 1758, Narwhal
Delphinapterus leucas (Pallas, 1776), Beluga,
   belukha, white whale

Family Delphinidae

Cephalorhynchus commersonii (Lacépède, 1804), Commerson's dolphin

Cephalorhynchus eutropia (Gray, 1846), Chilean dolphin

Cephalorhynchus heavisidii (Gray, 1828), Haviside's dolphin

Cephalorhynchus hectori (van Bénéden, 1881), Hector's dolphin

Steno bredanensis (G. Cuvier *in* Lesson, 1828), Rough-toothed dolphin

Sousa teuszii (Kükenthal, 1892), Atlantic humpbacked dolphin

Sousa plumbea (G. Cuvier, 1829), Indian humpbacked dolphin

Sousa chinensis (Osbeck, 1765), Pacific humpbacked or Chinese white dolphin

Sotalia fluviatilis (Gervais and Deville, 1853), Tucuxi

Tursiops truncatus (Montagu, 1821), Common bottlenose dolphin

Tursiops aduncus (Ehrenberg, 1833), Indian Ocean or Indo-Pacific bottlenose dolphin

Stenella attenuata (Gray, 1846), Pantropical spotted dolphin

Stenella frontalis (G. Cuvier, 1829), Atlantic spotted dolphin

Stenella longirostris (Gray, 1828), Spinner dolphin

Stenella clymene (Gray, 1850), Clymene dolphin

Stenella coeruleoalba (Meyen, 1853), Striped dolphin

Delphinus delphis Linnaeus, 1758, Short-beaked common dolphin

Delphinus capensis Gray, 1828, Long-beaked common dolphin

Delphinus tropicalis van Bree, 1971, Indian Ocean common dolphin

Lagenodelphis hosei Fraser, 1956, Fraser's dolphin

Lagenorhynchus albirostris (Gray, 1846), White-beaked dolphin

Lagenorhynchus acutus (Gray, 1828), Atlantic white-sided dolphin

Lagenorhynchus obliquidens Gill, 1865, Pacific white-sided dolphin

Lagenorhynchus obscurus (Gray, 1828), Dusky dolphin

Lagenorhynchus australis (Peale, 1848), Peale's dolphin

Lagenorhynchus cruciger (Quoy and Gaimard, 1824), Hourglass dolphin

Lissodelphis borealis Peale, 1848, Northern right whale dolphin

Lissodelphis peronii (Lacépède, 1804), Southern right whale dolphin

Grampus griseus (G. Cuvier, 1812), Risso's dolphin, gray grampus

Peponocephala electra (Gray, 1846), Melon-headed whale, Electra dolphin

Feresa attenuata Gray, 1874, Pygmy killer whale

Pseudorca crassidens (Owen, 1846), False killer whale

Orcinus orca (Linnaeus, 1758), Killer whale, orca

Globicephala melas (Traill, 1809), Long-finned pilot whale

Globicephala macrorhynchus Gray, 1846, Short-finned pilot whale

Orcaella brevirostris (Owen *in* Gray, 1866), Irrawaddy dolphin, pesut

Family Phocoenidae

Neophocaena phocaenoides (G. Cuvier, 1829), Finless porpoise

Phocoena phocoena (Linnaeus, 1758), Harbor porpoise, common porpoise

Phocoena sinus (Norris and McFarland, 1958, Vaquita, Gulf of California harbor porpoise

Phocoena spinipinnis Burmeister, 1865, Burmeister's porpoise

Phocoena dioptrica Lahille, 1912, Spectacled porpoise

Phocoenoides dalli (True, 1885), Dall's porpoise, Dall porpoise

ORDER SIRENIA

Family Trichechidae

Trichechus manatus Linnaeus, 1758, Caribbean manatee, West Indian manatee

Trichechus senegalensis Link, 1795, African manatee, West African manatee

Trichechus inunguis (Natterer, 1883), Amazon manatee

Family Dugongidae

Dugong dugon (Müller, 1776), Dugong, sea cow

Hydrodamalis gigas (Zimmerman, 1780), Steller's sea cow, giant sea cow

References

International Whaling Commission (2001). Report of the Scientific Committee. *J. Cet. Res. Management* **3**, *Suppl.* (in press).

Larivière, S. (1998). Lontra felina. *Mammalian Species* **575**, 1–5.

Rice, D. W. (1998). Marine mammals of the world. Systematics and distribution. *Soc. Mar. Mammalogy Spec. Pub.* **4**, 1–231.

Rosenbaum, H., Brownell, R. L., Jr., Brown, M. W., Schaeff, C., Portway, V., White, B. N., Malik, S., Pastene, L. A., Patenaude, N. J., Baker, C. S., Goto, M., Best, P. B., Clapham, P. J., Hamilton, P., Payne, R., Rowntree, V., Tynan, C. T., Bannister, J. L. and DeSalle, R. (2000). World-wide genetic differentiation of *Eubalaena* questioning the number of right whale species. *Mol. Ecol.* **9**, 1793–1802.

BIOGRAPHIES

Science is done by individuals, and the student or practicing scientist often wants to know something about those who came before, who developed the field. The brief biographies listed here are meant to meet that need, however sketchily. The list includes those who, in the opinion of the editors, made an important and lasting contribution to the study of marine mammals. Some of their most important works are also listed, and the student is urged to peruse some of these classics. Living persons are not included, as lifetime contributions cannot be fairly weighed until they are completed.

Abel, Othenio (1875–1946) Austrian paleontologist, the main European researcher on fossil whales in the early 20th century, working on the Belgian Miocene and on *Basilosaurus*. He expanded the perspective of marine mammal paleontologists to include higher-level phylogeny.

Abel, O. (1905). Les Odontocétes du Bolderien (Miocene superieure) d'Anvers. *Mem. Mus. roy. Hist. nat. Belg.* **3**, 1–155.

Abel, O. (1907). The genealogical history of the marine mammals. *Ann. Rep. Smithson. Inst.*, **1907**, 473–496.

Allen, Glover M. (1879–1942) One of the great naturalists of his day; among his enduring works relating to marine mammals are his monograph on the baleen whales of New England and his 1942 book on extinct and endangered mammals, which included the first review of the status of marine mammal species and the conservation problems affecting them.

Allen, G. M. (1916). The whalebone whales of New England. *Mem. Boston Soc. Nat. Hist.* **8**, 109–322.

Allen, G. M. (1942). "Extinct and Vanishing Mammals of the Western Hemisphere with the Marine Species of All the Oceans." American Committee for International Wildlife Protection, Cambridge, Massachusetts. (1972 reprint by Cooper Square, New York).

Allen, Joel A. (1838–1921) Completed the first comprehensive bibliography of cetology and sirenology (1495–1840), which is still highly useful today. He also produced a series of important monographs and papers on the marine mammals of North America.

Allen, J. A. (1880). History of North American pinnipeds: A monograph of the walruses, sea-lions, sea-bears and seals of North America. *U.S. Geol. Surv. Terr. Misc. Pub.* **12**, 1–785.

Allen, J. A. (1881). Preliminary list of works and papers relating to the mammalian orders Cete and Sirenia. *Bull. U.S. Geol. Geogr. Surv. Terr.* **6**, 399–562.

Allen, J. A. (1908). The North Atlantic right whale and its near allies. *Bull. Am. Mus. Nat. Hist.* **24**, 277–329, pl. 19–24.

Andrews, Roy Chapman (1884–1960) Paleontologist, cetologist, educator, and popularizer at the American Museum of Natural History in New York, he traveled the world and authored many important technical papers and monographs as well as best–selling popular books. His cetological work covered porpoises, beaked whales and baleen whales.

Andrews, R. C. (1916). "Whale Hunting with Gun and Camera." Appleton, New York.

Andrews, R. C. (1916). Monographs of the Pacific Cetacea. II. The sei whale (*Balaenoptera borealis* Lesson). 1. History, habits, external anatomy, osteology, and relationships. *Mem. Am. Mus. Nat. Hist.*, N.S. **1**, 293–502.

Arsen'ev, Viktor Alexandrovich (1908–1991) Noted biologist who produced or co–authored several important works on Russian marine mammals; his research focused on the northern fur seal. (See also SOKOLOV.)

Arsen'ev, V. A., and Zemskiy, V. A. (1954). ["In the Land of Whales and Penguins"]. Moskovkiy Gos. Universitet, Moscow. (In Russian).

Zemskiy, V. A., Arsen'ev V. A. and Studenezskaya, I. S. (1973). ["Marine Mammals"]. Pistchevaya Promyshlenost', Moscow.

Bélon du Mans, Pierre (1517–1566) Arguably the first cetologist since Aristotle and Pliny more than a millennium earlier, Bélon was the first to mention cetaceans in the title of a zoological work. The book included an account of the anatomy of a porpoise and listed seven species of whales, dolphins and porpoises. Figures in his second book were much copied by later workers.

Bélon, P. (1541). "L'Histoire Naturelle des Estranges Poissons Marins, avec la vraie Peincture et Description du Dauphin et de plusieurs autres de son espèce." Regnaud Chaudière, Paris.

Belloni, P. (1553). "De aquatilibus, Libri duo." Stephan, Paris.

Berzin, Alfred A. (1930–1996) Latvian whale biologist, born in Rostov-on-Don, Russia; known for his comprehensive research and monograph on the sperm whale. He spent his career at the Pacific Ocean Research Institute for Fisheries and Oceanography (TINRO) in Vladivostock. Most recently, he took part in the recovery and reporting of the true Soviet whaling catch data, which exposed massive unreported kills of protected whales in the 1960s and 1970s that may explain the apparent failure of some stocks to recover.

Berzin, A. A. (1971). "Kashalot" [The sperm whale]. Pischevaya Promyshlennost, Moscow. (Translation by Israel Program for Translation, 1972).

Zemskiy, V. A., Berzin, A. A., Mikhaliev, Y. A., and Tormosov, D. D. (1995). Soviet Antarctic pelagic whaling after WWII: Review of actual catch data. *Rep. Int. Whal. Commn.* **45**, 131–137.

Bigg, Michael A. (1939–1990) Born in London and emigrating to British Columbia as a child, he was the first to use individual identification of killer whales to study long-term social structure, life history and movements. He discovered the existence of sympatric but non-mixing fish-eating and mammal-eating forms/groups ("residents" and "transients") and established the fact that killer whale pods are matrilineal groups, with all offspring remaining with the maternal pod.

Bigg, M. A., Olesiuk, P. F., Ellis, G. M., Ford, J. K. B., and Balcomb, K. C. (1990). Social organization and genealogy of resident killer whales (*Orcinus orca*) in the coastal waters of British Columbia and Washington State. *Rep. Int. Whal. Commn. (Spec. Iss.)* **12**, 383–405.

Bonner, W. Nigel (1928–1994) Biologist at the British Antarctic Survey and a major and primary contributor to the biology, assessment and management of the southern elephant seal, other Antarctic pinnipeds, and British seals. He authored several technical and semi-popular books.

Bonner, W. N. (1968). The fur seal of South Georgia. *Brit. Ant. Surv. Sci. Rep.* **56**, 1–81.

Bonner, W. N. (1982). "Seals and Man: A Study of Interactions." University of Washington Press, Seattle.

Bonner, W. N. (1990). "The Natural History of Seals." Facts on File, New York.

Caldwell, David K. (1928–1990) Cetologist and historian born in Louisville, Kentucky, David Caldwell with his wife and research partner acoustician Melba C. Caldwell discovered that dolphins have individual "signature whistles." He founded and edited the journal *Cetology*.

Caldwell, M. C., and Caldwell, D. K. (1965). Individualized whistle contours in bottlenose dolphins (*Tursiops truncates*). *Nature* **207**, 434–435.

Caldwell, D. K., and Caldwell, M. C. (1972). "The World of the Bottlenose Dolphin." J. B. Lippincott, Philadelphia.

Chapman, Douglas G. (1920–1996) Noted mathematician, wildlife scientist, and teacher, born in Canada and spending his career at the University of Washington in Seattle. He was a member of the famous Committee of Three Scientists established by International Whaling Commission in 1961 as an independent body to advise on status and trends of whale stocks, leading to accurate predictions of collapse of stocks under commercial exploitation. A pioneer in the development and use of quantitative models for marine mammal assessment and management, he worked on pinnipeds as well as cetaceans.

Chapman, D. G. (1961). Population dynamics of the Alaska fur seal herd. *Trans. N. Am. Wildl. Conf.* **26**, 356–369.

Chapman, D. G. (1968). Estimation of fur seal pup populations by ramdomized sampling. *Trans. Am. Fish. Soc.*, **97**, 264–270.

Chapman, D. G. (1971). Status of Antarctic rorqual stocks. In *The Whale Problem. A Status Report* (W. E. Schevill, Ed.), pp. 218–238.

Chapman, D. G. (1981). Evaluation of marine mammal population models. *In* "Dynamics of Large Mammal Populations" (C. W. Fowler and T. D. Smith, Eds), pp. 278–296.

Chapskiy, Konstantin Konstantinovich (1906–1984) The preeminent Russian pinniped biologist of the mid-20th century, he concentrated on the systematics of the phocid seals but also worked on the ecology and morphology of other groups.

Chapskiy, K. K. (1941). ["Marine Animals of the Soviet Arctic"]. Izdatel. Glav., Moscow.

Chapskiy, K. K. (1955). [An attempt at revision of the systematics and diagnostics of seals of the subfamily Phocinae]. *Trud. Zool. Inst. Akad. Nauk SSSR* **17**, 161–199. (*Fish. Res. Bd Can. Transl. Ser.* **114**, 1–57 (1957)).

Chapskiy, K. K. and Sokolov, V. E. (1975). "Morphology and Ecology of Marine Mammals: Seals, Dolphins, Porpoises." Wiley, New York.

Cuvier, Frederic (1773–1838) Comparative anatomist and physiologist, brother of Georges Cuvier; Frederic Cuvier synthesized the available information on cetaceans in his landmark volumes, correcting many errors about cetacean natural history accumulated through the centuries.

Cuvier, F. (1829). Cétacés. *In* "Histoire Naturelle des Mamifères" (E. Geoffroy Saint-Hilaire and F. Cuvier,

Cuvier, F. (1836). "De l'Histoire Naturelle des Cétacés." Roret, Paris.

Cuvier, Georges (1769–1832) Famed paleontologist, zoologist and evolutionary thinker, widely respected in his time as one of the originators of the idea of evolution and the founder of modern biology and comparative anatomy. He spent his career at the Muséum national d'Histoire naturelle, Paris. His enormous treatise on fossils included a volume on cetaceans, and he carried out many basic studies of cetacean anatomy, including the first description of the inner ear, and compiled one of the first stranding reports (1812).

Cuvier, G. (1795). Note sur la découverte de l'oreille interne de Cétacés. *Mag. Encycl.* **6(2)**, 130.

Cuvier, G. (1812). Rapport faità la classe des Sciences mathmathique et physiques, sur divers Cétacés pris sur les côtes de France, principalement sur ceux qui sont échoués près de Paimpol, le 7 janvier 1812. *Ann. Mus. D'Hist. Nat.* **19**, 1–16, pl.1.

Cuvier, G. (1817 and subsequent editions). "Le Règne Animal distribué d'après son organisaton, pour servir de base a l'histoire naturelle des animaux et d'introduction à l'Anatomie Comparée." Deterville, Paris.

Cuvier, G. (1821–1825). "Recherches sur les ossemens fossiles, òu l'on rétablit les caractères des plusiers animaux dont les révolutions du globe ont détruit les espèces." Second ed., 10 vol. (Vol. 5, on Cetacea; 1823). Dufour and d'Ocagne, Paris.

Dawbin, William H. (1921–1998) New Zealand-born biologist, historian and teacher who pioneered much of the early research on distribution and movements of humpback whales and right whales in the Southern Hemisphere. He was launched into marine mammalogy by early years spent in the whaling grounds of the Southern Ocean. His later work focused on catch history of whales in the Southern Hemisphere. He was also a noted herpetologist.

Dawbin, W. H. (1956). The migrations of humpback whales which pass the New Zealand coast. *Trans. Roy. Soc. NZ* **84**, 147–196.

Dawbin, W. H. (1986). Right whales caught in waters around south eastern Australia and New Zealand during the nineteenth and early twentieth centuries. *Rep. Int. Whal. Commn. (Spec. Iss.)* **10**, 261–267.

Dawbin, W. H. (1997). Temporal segregation of humpback whales during migration in southern hemisphere waters. *Mem. Queensland Mus.* **42**, 105–138.

Eschricht, Daniel Frederick (1798–1863) A pioneer in many aspects of modern cetology: the first to consider cetaceans, including harbor porpoises, narwhals and belugas, as migrating animals; the first to describe the embryo of a whale based on dissections; the first to distinguish between the bowhead and right whales; the first (with Owen) to see concentric layers of growth in the teeth of cetaceans; and the first to identify different populations of killer whales in the Arctic. Most of his work was in Danish and largely inaccessible to other marine mammal scientists until published in English by Flower (1866). The gray whale, *Eschrichtius robustus*, bears his name.

Eschricht, D. F. (1840–1842). Om undersögelsen of de nordiske Hvaler. *Skand. Naturf. Förhandl.* **2**, 83–108, 203–227.

Flower, W. H. (Ed.) (1866). "Recent Memoirs on the Cetacea by Professors Eschricht, Reinhardt and Lilljeborg." For Ray Society by Hardwicke, London.

Fay, Francis H. (1927–1994) Born in Massachusetts and spending most of his career at the University of Alaska, "Bud" Fay was the world's leading expert on the walrus. He translated and made accessible to western scientists much of the massive Russian literature on marine mammals of the North Pacific and Arctic.

Fay, F. H. (1982). Ecology and biology of the Pacific walrus, *Odobenus rosmarus divergens* Illiger. *N. Am. Fauna* **74**, 1–279.

Fay, F. H. (1997). Status of the Pacific walrus population, 1950–1989. *Mar. Mamm. Sci.* **13**, 537–565.

Flower, William Henry (1831–1899) Gray's student and his successor as curator of the huge collection of marine mammal specimens at the British Museum in the last quarter of the 18th century, Flower was an extremely able zoologist who contributed major works on the beaked whales and the delphinid cetaceans, as well as publications on manatees and pinnipeds.

Flower, W. H. (1871). On the recent ziphioid whales, with a description of the skeleton of *Berardius arnuxi*. *Trans. Zool. Soc. London* **8**, 203–234, pl. 27–29.

Flower, W. H. (1884). On the characters and divisions of the family Delphinidae. *Proc. Zool. Soc. London* **1883**, 466–513.

Fraas, Eberhardt (1862–1915) German paleontologist, one of the first to study the oldest Eocene whales, e.g., *Protocetus*.

Fraas, E. (1904). Neue Zeuglodonten aus dem untern Mitteleozan van Mkattam bei Cairo. *Geol. Palaeot. Abhandl. Jena* **6** (3), 199–220.

Fraser, Francis C. CBE, FRS (1903–1978) The main practioner of British cetology in the mid-20th century, which was far ahead of its state in North America. A Scot by birth, he was Keeper of Zoology in charge of the whales, dolphins and porpoises at the British Museum for several decades. A UK-wide stranding program had started early in the century (stranded animals are "royal fish" and property of the sovereign), and Fraser kept it going and authored elaborate periodic reports up into the 1970s. His classic work with Peter E. Purves on the comparative functional morphology of the cetacean ear resulted in a classification that has endured and proved to be largely consonant with phylogenies generated by the most modern methods. He described Fraser's dolphin, *Lagenorhynchus hosei*, from a skeleton that had languished in the BM collection for over 50 years as an "abnormal" or hybrid specimen. Much of today's cetology is built on his work.

Fraser, F. C. (1956). A new Sarawak dolphin. *Sarawak Mus. J.* **7**, 478–503. pl. 22–26.

Fraser, F. C. (1960). Hearing in cetaceans. Evolution of the accessory air sacs and the structure and function of the outer and middle ear in Recent cetaceans. *Bull. Brit. Mus. (Nat. Hist.), Zool.* **7**, 1–140, 53 pl.

Fraser, F. C. (1974). "Report on Cetacea stranded on the British Coasts from 1948 to 1966." British Museum (Natural History), London.

Gaskin, David E. (1939–1998) British biologist; started out as a lepidopterist (and remained one), but a stint as a biologist on a UK whaling ship in the Antarctic steered him into whale biology and led him to extensive work on the cetaceans of New Zealand and the porpoises of the North Atlantic. For many years, his book on the ecology of whales and dolphins was the only textbook available to teachers of marine mammalogy.

Gaskin, D. E. (1968). The New Zealand Cetacea. *N.Z. Mar. Dept Fish. Res. Bull. (New Series)* **1**, 1–92.

Gaskin, D. E. (1982). "The Ecology of Whales and Dolphins." Heinemann, London.

Gaskin, D. E. (1992). Status of the harbour porpoise, *Phocoena phocoena*, in Canada. *Can. Field-Nat.* **106**, 36–54.

Gervais, Paul François Louis (1816–1879) French zoologist and paleontologist; he described several cetacean species, explored the anatomy and reproduction of dolphins and whales, and collaborated with Van Beneden to produce the classic atlas of cetacean osteology that is still in use today. (See also VAN BENEDEN).

Gervais, P. (1853). Description de quelques espèces de phoques et de Cétacés. *Ann. Sci. Nat.* **20**, 281–292.

Gervais, P. (1856). Sur trois espèces de dauphins qui vivent dans la région du Haut Amazone. *Comp. Rend. Acad. Sci.* **42**, 806–808.

Gray, John Edward (1800–1875) Had a plan as a young man to "form the largest and most complete Zoological Collection known" and he succeeded, being personally responsible for the building of the huge collections of the British Museum amassed during the Victorian period of wide zoological exploration. He worked and published prolifically on nearly every animal group. To keep track of the current literature, he started the Zoological Record. Among his 1096 publications (1820–1875) were scores on whales, dolphins, porpoises, seals, otters and other aquatic mammals. We still recognize sixteen of the marine mammal species he described (three pinnipeds, one baleen whale, two beaked whales, and ten dolphins and small beaked whales), as well as many of his higher taxa. *Mesoplodon grayi* is named after him.

Gray, J. E. (1866). "Catalog of Seals and Whales in the British Museum." Second Edition. British Museum, London.

Gray, J. E. (1870). The geographical distribution of the Cetacea. *Ann. Mag. Nat. Hist., Ser. 4*, **6**, 387–394.

Gulland, John FRS (1926–1990) Eminent English fisheries mathematician and teacher who was a member of the independent committee of scientists commissioned in the 1960s to assess the status of Antarctic whale stocks; the committee predicted collapse of the stocks due to commercial over-exploitation, and that came to pass. He was the mainstay of the fisheries department of the Food and Agriculture Organization (FAO) of the UN in Rome for nearly 20 years, running highly effective training courses in developing countries around the world, and authored several important texts on fishery assessment and management.

Gulland, J. (1965). The plight of the whales. *Oryx* **8(2)**:74–98.

Gulland, J. (1972). The conservation of Antarctic whales. *Biol. Cons.* **4**, 335–344.

Gulland, J. (1983). "Fish Stock Assessment: A Manual of Basic Methods." J. Wiley and Sons, New York.

Harrison, Sir Richard J. FRS (1920–1999) English anatomist and functional morphologist who first made his mark in human anatomy and medicine. He became deeply interested and intrigued by marine mammals through discussions with Francis Fraser of the British Museum and went on to produce landmark works on the reproduction and general biology of whales, dolphins and seals. He received a knighthood for his contributions to marine biology, education and public service.

Harrison, R., and King, J. E. (1965). "Marine Mammals." Hutchinson Univ. Library, London.

Harrison, R. J. (Ed.). (1972–1977). "Functional Anatomy of Marine Mammals," Vol. 1–3. Academic Press, London.

Ridgway, S. H., and Harrison, R. J. (1981–1999). "Handbook of Marine Mammals," Vol. 1–6. Academic Press, San Diego and London.

Harrison Matthews, Leonard FRS (1901–1986) An early participant in the British *Discovery* investigations in the Antarctic; authored basic monographs on the elephant seal, humpback whale, sperm whale and sei whale. Much of his research focused on growth and reproduction. In later life he wrote several semi-popular books on marine mammals and antarctic research.

Harrison Matthews, L. (1937). The sperm whale, *Physeter catodon*. *Discovery Rep.* **17**, 93–168, pl. 3–11.

Harrison Matthews, L. (1952). "Sea Elephant—The Life and Death of the Elephant Seal." Macgibbon and Kee, London.

Harrison Matthews, L. (1978). "The Natural History of the Whale." Colombia Univ. Press, New York.

Hershkovitz, Phillip (1909–1997) Known mainly for his voluminous works on the systematics and ecology of neotropical terrestrial mammals, Hershkovitz turned to cetaceans at one point and produced his classic *Catalog of Living Whales*, a definitive reference indispensable to any cetacean systematist. While many of his taxonomic "calls" subsequently have been second-guessed, he led the way for cetologists in the last half of the 20th century.

Hershkovitz, P. (1966). Catalog of living whales. *U. S. Nat. Mus. Bull.* **246**, 1–257.

Hubbs, Carl L. (1894–1979) A polymath naturalist born in Arizona whose career and publications spanned over 65 years and included the study of fishes, birds, marine mammals, archeology, geochronology, geology, climatology, evolution, hybridism and the practice of systematics. He carried out the first aerial surveys of gray whales, re-discovered the supposedly extinct Guadaloupe and Juan Fernandez fur seals, and was a major force for the conservation of marine animals and their habitats. He persuaded the Mexican Government to establish the first gray whale sanctuary, in Scammon's Lagoon. Hubb's beaked whale, *Mesoplodon carlhubbsi,* was named in his honor; he and other faculty members at the Scripps Institution of Oceanography ate the type specimen during World War II meat rationing (the osteological portions fortunately were saved).

Hubbs, C. L. (1946). First records of two beaked whales, *Mesoplodon bowdoini* and *Ziphius cavirostris*, from the Pacific coast of the United States. *J. Mamm.* **27**, 242–255.

Hubbs, C. L., and Hubbs, L. C. (1967). Gray whale censuses by airplane in Mexico. *Calif. Fish Game* **53**, 23–27.

Hubbs, C. L., and Norris, K. S. (1971). Original teeming abundance, supposed extinction and survival of the Juan Fernandez fur seal. *Ant. Pinnipedia, Ant. Res. Ser.* **18**, 35–52.

Ichihara, Tadayoshi (1939–1981) Japanese marine mammalogist known for his description of the pygmy blue whale, *Balaenoptera musculus brevicauda,* and for pioneer research on the telemetry of marine animals, including dolphins and seals. This innovative worker did much in his too-brief career to inject rigor and modern technology into Japanese marine mammalogy.

Ichihara, T. (1957). An application of linear discriminant function to external measurements of fin whale. *Sci. Rep. Whales Res. Inst., Tokyo* **12**, 127–189.

Ichihara, T. (1966). Criteria for determining age of fin whale with reference to ear plug and baleen plate. *Sci. Rep. Whales Res. Inst., Tokyo* **20**, 17–82, 8 pl.

Ichihara, T. (1971). [Ultrasonic, radio tags and various problems in fixing them to marine animal body]. *Rep. Fish. Resource Invest. Scient. Fish. Res. Agency Japanese Gov.* **12**, 29–44.

Jonsgård, Åge (1916–1997) Norwegian whale biologist and professor who carried out classical studies on the minke and fin whales and was an influential member of the Scientific Committee of the International Whaling Commission.

Jonsgård, Å. (1951). Studies on the little piked whale or minke whale (*Balaenoptera acutorostrata* Lacépède). Report on Norwegian investigations carried out in the years 1943–1950. *Norsk Hval.-Tid.* **40**, 209–232.

Jonsgård, Å. (1966). Biology of the North Atlantic fin whale *Balaenoptera physalus* (L.). Taxonomy, distribution, migration and food. *Hval. Skr.* **49**, 1–62, fig. 1, map.

Kellogg, Remington (1892–1969) Paleontologist, mammalogist and popularizer of cetaceans at the Smithsonian Institution. He produced dozens of works on fossil whales, dolphins, porpoises and seals, including the classic and massive "Review of the Archeoceti." His doctoral thesis "History of Whales—Their Adaptations to Life in the Water" has served as a primer for cetologists for generations. He took part in the early stages of the organization of international agreements on whaling beginning in the 1930s. The first U.S. Commissioner to the IWC, he battled vigorously but unsuccessfully in the 1960s for limitation of commercial exploitation of whales to sustainable levels. Although he did not win the day, the heat of the controversy led to later independent review of the whale stocks and eventual effective regulation of whaling.

Kellogg, R. (1928). The history of whales—Their adaptation to life in the water. *Quart. Rev. Biol.* **3**, 29–76, 174–208.

Kellogg, R. (1936). A review of the Archeoceti. *Carnegie Inst. Wash. Pub.* **482**, 1–357.

Kellogg, R. (1940). Whales, giants of the seas. *Nat. Geogr.* **77(1)**, 35–90.

Kleinenberg, Sergei Evgenyevich (1909–1968) Russian biologist who studied mammals in the field and laboratory; published extensively on the dolphins and porpoises of the Black Sea and, with three colleagues, produced a monograph on the beluga that has been the basis of all subsequent studies of the species. With his student G. A. Klevezal', he contributed greatly to development of the methodology of determining age in mammals through the examinations of layers in hard tissues.

Kleinenberg, S. (1956). "Mlekopitayushchie Cernogo i Azovskogo morei [Marine mammals of the Black and Azov seas]." Izdatelstvo AN SSR. (In Russian).

Kleinenberg, S., Yablokov, A. V., Bel'kovich, B. M., and Tarasevich, M. N. (1964). ["Beluga *(Delphinapterus leucas)*. Investigation of the Species"]. Izd. Nauk, Moscow. (Translation by Israel Program for Scientific Translation, Jerusalem, 1969).

Klevezal', G. A., and Kleinenberg, S. E. (1967). ["Age Determination of Mammals from Annual Layers in Teeth and Bones"]. Izd. Nauk, Moscow (Translation by Israel Program for Scientific Translation, Jerusalem, 1969).

Lacépède (Bernard Germain Etienne de Laville-sur-Ilon, Compte de Lacépède) (1756–1826) French zoologist who was one of the first modern zoologists to critically review the cetaceans. Colored plates from his book were extensively copied in other authors' works of the 19th century. He listed 34 species in the first edition (more in later editions), many more than any previous worker. Working with the collections in the Muséum National d'Histoire Naturelle, Paris, begin-

ning in 1784, he was highly honored in his time and worked successfully under the monarchy, the republic, the empire and the new monarchy.

La Cepède ["le Citoyen"]. (1804). "Histoire Naturelle des Cétacés." Paris.

Lacépède ["M. le Comte de"]. (1818). Note sur les Cétacés des mers voisines du Japon. *Acad. roy. Sci., Paris* **21 septembre 1818**, 119–121.

Leatherwood, J. Stephen (1944–1997) Marine mammalogist, naturalist, and conservationist, born in Alabama. He authored or coauthored a series of widely used identification guides and guided the establishment of marine mammal research and conservation programs in many of the less-developed nations in Latin America and Asia.

Leatherwood, S. and Reeves, R. R. (1983). "The Sierra Club Handbook of Whales and Dolphins." Sierra Club Books, San Francisco.

Leatherwood, S., Reeves, R. R., Perrin, W. F., and Evans, W. E. (1988). "Whales, dolphins, and porpoises of the eastern North Pacific and Adjacent Arctic Waters. A Guide to their Identification." Dover Press, New York.

Leatherwood, S., and Donovan, G. P. (Eds). (1991). Cetaceans and cetacean research in the Indian Ocean Sanctuary. *UN Env. Pro. Mar. Mam. Tech. Rep.* **3**, 1–287.

Linnaeus, Carolus (1707–1778) Swedish physician and avid naturalist considered the father of modern taxonomy; he worked extensively in the field with birds and plants. Linnaeus named and classified all the then known species of plants and animals in the first edition of his Systema Naturae in 1735, using his binomial system of a species name consisting of genus and trivial name. The official starting point of zoological nomenclature is 1758 (International Commission on Zoological Nomenclature—1985), the year of publication of the 10th edition of his work. In that edition, he gave us 13 species of marine mammals that are still recognized.

International Commission on Zoological Nomenclature. (1985). "International Code of Zoological Nomenclature." Third Ed. Internat. Trust for Zool. Nomen., Brit. Mus. (Nat. Hist.), London.

Linnaeus, C. (1735). "Systema Naturae." Two vol.

Linnaeus, C. (1758). "Systema Naturae." Laurentii Salvii, Stockholm.

Mackintosh, Neil A. (1900–1974) British biological oceanographer and authority on Antarctic whales; his contributions were massive. He was one of the prime organizers of the Discovery investigations that have contributed much to the knowledge of marine mammals and ecology of the Southern Ocean. He authored landmark monographs on whales, plankton and ice, edited and shepherded the Discovery Reports, launched the international whale-marking scheme, was instrumental in the founding of the British Antarctic Survey, and was a leading player in (and first chairman of the Scientific Committee of) the IWC.

Mackintosh, N. A., and Wheeler, J. F. G. (1929). Southern blue and fin whales. *Discovery Rep.* **1**, 257–540, pl. 25–44.

Mackintosh, N. A. (1942). The southern stocks of whalebone whales. *Discovery Rep.* **22**, 197–300.

Mackintosh, N. A. (1965). "The Stocks of Whales." Fishing News (Books), London.

Nemoto, Takahisa (1930–1990) The major authority on the plankton consumed by whales; born in Tokyo and carried out his studies at the Whales Research Institute (1953–1977) and Ocean Research Institute of University of Tokyo (1977–1990). He was instrumental in the organization of the international project BIOMASS to study the ecology of the Southern Ocean.

Nemoto, T. (1959). Food of baleen whales with reference to whale movements. *Sci. Rep. Whales Res. Inst., Tokyo* **14**, 149–290.

Nemoto, T. (1966). Feeding of baleen whales and krill, and the value of krill as a marine resource in the Antarctic. *In* "Symposium on Antarctic Oceanography" (Scientific Committee on Antarctic Research, Sponsor), 240–253. Polar Research Institute, Cambridge, UK.

Nemoto, T., and Kawamura, A. (1977). Characteristics of food habits and distribution of baleen whales with special reference to the abundance of North Pacific sei and Bryde's whales. *Rep. Int. Whal. Commn. (Spec. Iss.)***1**, 80–87.

Nishiwaki, Masaharu (1915–1984) A "force of nature" for the study and conservation of marine mammals in Asia. After serving as a fighter pilot in WWII, he spent several seasons with the Japanese whaling factoryship fleet in the Antarctic; this set the direction of his career. He made basic contributions to knowledge of systematics, life history and reproduction of whales, dolphins and seals and effectively promoted conservation of cetaceans and sirenians in Japan, China, Bangladesh, India, Pakistan, Nepal, Indonesia, Russia and several nations in West Africa. He campaigned against overexploitation of whales and dolphins in Japan; his was a lone voice at the time.

Nishiwaki, M. (1965). ["Whales and Dolphins"]. University of Tokyo Press.

Nishiwaki, M. (1972). General biology. *In* "Mammals of the Sea—Biology and Medicine," S. H. Ridgway (Ed.), 3–204.

Nishiwaki, M., and Marsh, H. (1985). Dugong–*Dugong dugon.* In "Handbook of Marine Mammals, Vol. 3–Sirenians and Baleen Whales," S. H. Ridgway and R. Harrison (Eds.), 1–31.

Norris, Kenneth S. (1924–1998) The virtual founder of modern cetology in the United States; born in Los Angeles. He organized and chaired the first international conference on cetaceans in 1963 in Washington, D.C. Scientist, teacher and natural historian, he co-founded the Society for Marine Mammalogy and served as its first president. With John Prescott, he described the vaquita, *Phocoena sinus*, from the Gulf of California in 1958. He demonstrated echolocation in dolphins, developed theories of hearing and sound production in dolphins and the sperm whale, and carried out extensive innovative research on dolphin behavior and ecology. Also an active "biopolitician," he was influential in development of the landmark U.S. Marine Mammal Protection Act of 1972.

Norris, K. S., Prescott, J. H., Asa-Dorian, P. V., and Perkins, P. (1961). An experimental demonstration of echolocation behavior in the porpoise, *Tursiops truncatus* (Montagu). *Biol. Bull.* **120**, 163–176.

Norris, K. S. (Ed.). (1966). "Whales, Dolphins and Porpoises." University of California Press, Berkeley.

Norris, K. S. (1974). "The Porpoise Watcher." W. W. Norton and Co., New York.

Norris, K. S., Würsig, B., Wells, R. S., and Würsig, M. (Eds). (1994). "The Hawaiian Spinner Dolphin." University of California Press, Berkeley.

Omura, Hideo (1906–1993) Considered the dean of Japanese cetology; published detailed monographs on the osteology, systematics and distribution of baleen whales in Japanese waters and from around the world. He served for decades as Director of the industry-sponsored Whales Research Institute in Tokyo, which fostered and published the bulk of the Japanese research on cetaceans from after World War II until the late 1980s.

Omura, H. (1950). Whales in the adjacent waters of Japan. *Sci. Rep. Whales Res. Inst., Tokyo* **4**, 27–113.

Omura, H., and Ohsumi, S. (1974). Research on whale biology in Japan, with special reference to North Pacific stocks. *In* "The Whale Problem—A Status Report," W. E. Schevill, G. C. Ray and K. S. Norris (Eds.), Harvard Univ. Press, Cambridge, Massachusetts, 196–208.

Omura, H. (1984). History of the gray whale in Japan. *In* "The Gray Whale," M. L. Jones and S. L. Swartz (Eds.), Academic Press, San Diego, 59–77.

Poulter, Thomas C. (1897–1978) U.S. bioacoustician who worked extensively in the Arctic and Antarctic. He long advocated the idea that pinnipeds, like odontocete cetaceans, use echolocation; this was eventually discounted (see Schusterman, 1967). He is most remembered for organizing a series of conferences on biological sonar and diving mammals in the 1960s and 1970s at Stanford Research Institute; these were the forerunners of the current Biennial Conferences on the Biology of Marine Mammals.

Poulter, T. C. (1963). Sonar signals of the sea lion. *Science* **139**, 753–755.

Schusterman, R. J. (1967). Perception and determination of underwater vocalization in the California sea lion. *In* "Les Systèmes Sonar Animaux. Biologie et Bionique," R. G. Busnel (Ed.), 535–617. Laboratoire de Physiologie Acoustique, Jouy-en-josas, France.

Scammon, Charles M. (1825–1911) American whaling captain and naturalist who published only one book and one technical paper, but greatly influenced cetology and the history of whaling. He discovered the breeding grounds of the gray whale in Laguna Ojo de Liebre (also called Scammon's Lagoon) in Baja California in 1855, which led to the near-extinction of the species in the eastern Pacific. His book (rare and valuable in the original edition) is a detailed chronicle of American whaling and description of the marine mammals of the west coast of North America.

Scammon, C. M. (1869). On the cetaceans of the western coast of North America. *Proc. Acad. Nat. Sci. Philadelphia* **1869**, 13–63.

Scammon, C. M. (1874). "The Marine Mammals of the North-western Coast of North America Described and Illustrated together with an Account of the American Whale-fishery." John H. Carmany, San Francisco. (Reprinted 1968, Dover, New York).

Schevill, William E. (1906–1994) Pioneering bioacoustician who was born in Brooklyn and spent most of his career at Harvard. He made the first recordings of underwater cetacean sounds in the wild and recorded and described sounds of many species. He published the first phonograph record of cetacean sounds. With wife and research partner Barbara Lawrence he described the anatomy of the delphinid nasal passages functional in sound production.

Schevill, W. E., and Lawrence, B. (1950). Underwater listening to the white porpoise (*Delphinapterus leucas*). *Science* **109**, 143–144.

Lawrence, B., and Schevill, W. E. (1956). The functional anatomy of the delphinid nose. *Bull. Mus. Comp. Zool.* **165**, 104–152 + 30 fig.

Schevill, W. E., and Watkins, W. A. (1962). *Whale and Porpoise Voices. A Phonograph Record.* Woods Hole Oceanographic Institution, Woods Hole, Massachusetts. 24pp. and record.

Schevill, W. E., Ray, G. C., and Norris, K. S. (1974). "The Whale Problem: A Status Report." Harvard University Press, Cambridge.

Scholander, Per F. ("Pete") (1905–1980) Norwegian-born physiologist who made many basic discoveries and advances in the respiratory physiology and ecology of diving mammals, including observation of the slowing of heart rate (the "diving response") in submerged seals and description of countercurrent heat-exchange structures in the dorsal fin and flukes of cetaceans. He pioneered direct physiological monitoring and experimentation with captive live marine mammals to explore their respiratory and cardiovascular adaptations.

Scholander, P. F. (1940). Experimental investigations on the respiratory function in diving mammals and birds. *Hvalråd. Skr.* **22**, 1–131.

Scholander, P. F., and Schevill, W. E. (1955). Counter-current vascular heat exchange in the fins of whales. *J. Appl. Physiol.* **8**, 279–282.

Scholander, P. F. (1964). Animals in aquatic environments: diving mammals and birds. *In* "Handbook of Physiology, Sect. 4: Adaptations to the Environment," D. B. Dill (Ed.), 729–739. Amer. Physiol. Soc., Washington.

Scoresby, William (1789–1857) Born in Yorkshire, England; accompanied his father on a whaling expedition when he was ten years old, an event that shaped his life; he published his first paper on whales at the age of 19. His most important work "An Account of the Arctic Regions and Description of the Northern Whale Fishery" is the best contemporary account of the Arctic and whaling as it existed in the early nineteenth century. Several equally important books followed.

Scoresby, W. (1808). Account of the *Balaena mysticetus* or great northern or Greenland whale. *Edinb. Mem. Wern. Soc.* **1**, 578–586.

Scoresby, W. (1820). "An Account of the Arctic Regions and Description of the Northern Whale Fishery." Constable, Edinburgh.

Scoresby, W. (1860). "The Whaleman's Adventures." Darton, London.

Stamp, T., and Stamp, C. (1975). "William Scoresby Arctic Scientist." Caedmon of Whidby Press, Cambridge, UK.

Slijper, Everhard Johannes (1907–1968) Like many cetologists of the first half of the 20th century, E. J. Slijper, the Dutch functional anatomist and student of whales and whaling, had his first experience with whale biology on a whaling expedition to the Antarctic. He authored "Walwissen" (1958), which became the famous book "Whales" in English. He analyzed reproduction, locomotion and other functions in cetaceans and other mammals with a comparative anatomical approach that became standard in such studies.

Slijper, E. J. (1962). "Whales." Hutchinson, London.

Slijper, E. J. (1966). Functional morphology of the reproductive system in Cetacea. *In* "Whales, Dolphins and Porpoises," K. S. Norris (Ed.), 277–319. Univ. Calif. Press, Berkeley.

Sokolov, Vladimir Evgenyevich (1928–1998) An important figure in Russian mammalogy and member of the National Academy of Sciences of the USSR, V. E. Sokolov was a specialist on the microanatomy of the skin. He co-authored a three-volume work on the mammals of the Soviet Union, including a volume on seals and the toothed cetaceans.

Sokolov, V. E. (1962). Adaptations of the mammalian skin to the aquatic mode of life. *Nature* **195**, 464–466.

Sokolov, V. E. (1973). Structure of the skin cover in some cetaceans. *In* "Morphology and Ecology of Marine Mammals: Seals, Dolphins and Porpoises," K. K. Chapskiy and V. E. Sokolov (Eds.), 102–118. Wiley, New York.

Geptner, V. G., Chapskiy, K. K., Arseneyev, V. A. and Sokolov, V. E. (1976). ["Mammals of the Soviet Union, Vol 1. Pinnipeds and Toothed Whales"]. Vyshaya Shkola, Moscow.

Steller, Georg W. (1709–1746) German-born botanist who in his relatively short life spent years in the wilds of the Russian Far East with the Great Siberian Expedition and collected specimens and detailed accounts of many new animals, including the Steller sea lion and the extinct Steller's sea cow of Bering Island.

Steller, G. W. (1751). "De Bestiis Marinus." St. Petersburg Acad. Sci. (posthumous; published in translation as *Von Sonderbaren Meerthieren*, Kummel, Halle, Germany, 1753).

Tomilin, Avenir Grigoryevich (1912–2000) Grand old man of Russian cetology who began his 66-year career with a season with the whaling expedition "Aleut" in the Kamchatka, Bering, and Chukchi Seas in 1934. The most famous of his hundreds of technical and semi-popular contributions is the massive volume on Cetacea in the series "Mammals of the U.S.S.R. and Adjacent Countries" (1957), which has been translated into English. Tomilin followed a lifelong commitment to popular scientific education, and his books were read by millions.

Tomilin, A. G. (1937). Kity Dal'nego Vostoka [Whales of the Far East]. Ucheniye Zaliski MGU 8:1119–167.

Tomilin, A. G. (1957). Kitoobraznye. Vol. 9, V. G. Heptner, ed., Zveri SSSR i prelezhashchikh stran. Izdatel'stvo Akademii Nauk SSSR, Moscow. Translation published as "Cetacea, Vol. 9, Mammals of the U.S.S.R.," V. G. Heptner, ed., by Israel Program for Scientific Translations, Jerusalem (1967).

Tomilin, A. G. (1980). "V mirye kitov i delfinov [In the world of whales and dolphins]." Izdatel'stvo "Znaniye," Moscow.

True, Frederick W. (1858–1914) The dominant figure in American cetology at the turn of the century; built the marine mammal collection of the U.S. National Museum into the second-largest in the world, after that of the British Museum. Highly prolific, he documented the American dolphin fisheries in New Jersey and North Carolina and produced scores of reports on the systematics of cetaceans, pinnipeds and other mammals, describing many new species and including the important monographic "Review of the family Delphinidae," which largely delineated the array of dolphin species recognized today, and "Whalebone whales of the North Atlantic."

True, F. W. (1889). Contributions to the natural history of the cetaceans, a review of the family Delphinidae. *Bull. U. S. Nat. Mus.* **36**, 1–191, pl. 1–47.

True, F. W. (1904). The whalebone whales of the North Atlantic. *Smithson. Contrib. Knowledge* **33**, 1–551, pl. 1–50.

True, F. W. (1910). An account of the beaked whales of the family Ziphiidae in the collection of the United States National Museum, with remarks on some specimens in other American museums. *Bull. U. S. Nat. Mus.* **73**, 1–89, pl. 1–42.

Van Beneden, Pierre Joseph (1809–1894) Belgian paleontologist and morphologist of the latter half of the 19th century; he provided some of the first accurate descriptions of fossil whales. With Paul Gervais, he authored a monumental classic work on the osteology of living and fossil cetaceans that is consulted even today by cetologists seeking accurate depictions of dolphin and whale skulls and skeletons.

Van Beneden, P. J. (1864). Recherches sur les ossements provenant du Crag d'Anvers. Les Squalodons. *Acad. roy. Belg.* **2 April 1864**, 1–85, pl. 1–4.

Van Beneden, P. J. and Gervais, P. (1868–1880). "Ostéographie des Cétacés Vivant et Fossiles Comprenant la Description et l'Econographie du Squelette et du Système Dentaire de ces Animaux ainsi que des Documents Relatifs à leur Histoire Naturelle." Bertrand, Paris.

van Utrecht, Willem Lodewijk (1926–1994) Dutch zoologist who was an important early worker in the study of age and growth structures in both toothed cetaceans and baleen whales, working with teeth in the former and both ear plugs and baleen plates in the latter. He turned to the study of aging of fishes in his later years. Much of his research was in collaboration with his wife Clara N. van Utrecht-Cock.

Utrecht, W. L. van (1965). On the growth of the baleen plate of the fin whale and the blue whale. *Bijdragen tot de Dierkunde* **35**, 1–38.

Utrecht, W. L. van, and Utrecht-Cock, C. N. (1969). Comparison of records of baleen plates and of ear plugs in female fin whales, *Balaenoptera physalus* (Linnaeus, 1758). *Bijdragen tot de Dierkunde* **39**, 81–97.

Winn, Howard E. (1926–1995) Oceanographer, whale biologist, and teacher who carried out some of the first acoustic studies of whale songs. He organized and headed the massive Cetacean and Turtle Assessment Program (CETAP), 1978–1982, which provided first estimates of distribution and abundance of whales, dolphins, and porpoises off the eastern U.S., and was active in conservation of the right whale in the North Atlantic.

Winn, H. E., and Winn, L. K. (1978). Song of the humpback whale in the West Indies. *Mar. Biol.* **47**, 97–114.

Winn, H. E., and Olla, B. L. (Eds). (1979). "Behavior of Marine Animals. Vol. 3: Cetaceans." Plenum Press, New York.

Winn, H. E. (1982). "A Characterization of Marine Mammals and Turtles in the Mid- and North Atlantic areas of the U.S. outer continental shelf." Cetacean and Turtle Assessment Program of the University of Rhode Island, Kingston.

GLOSSARY

The following section provides definitions for a list of terms that appear in the text of this Encyclopedia. For the most part these are specialized technical terms used in the field of marine mammalogy that are likely to be unfamiliar to a lay reader. Also included are terms that, though likely to be known in their common sense, have a distinctive or highly specific use in marine mammalogy.

A

abduction   The movement or position of a limb away from the midline of the body.

abuse   In the context of human interaction with marine mammals, a term for mistreatment involving violence that may result in injuries or death.

accessory denticles   Cusps on archaeocete teeth that are not clearly homologous with primitive mammalian tooth cusps.

accidental reinforcement   Another term for ADVENTITIOUS REINFORCEMENT.

acetabulum   A depression in the pelvis in which the head of the femur is secured.

acoustic harassment device   (AHD) A sound-generating device that, because of some combination of intensity, frequency, or other sound characteristic(s), is aversive to marine mammals and keeps or drives them away from an area or structure.

adduction   The movement or position of a limb toward the midline of the body.

adenohypophysis   A glandular structure, also known as the anterior pituitary, located at the base of the hypothalamus in the brain and producing a variety of hormones primarily responsible for stimulating the function of other endocrine glands.

adenosine triphosphate   (ATP) A high-energy compound that, when split by enzymic activity, releases energy in the reaction.

adipocyte   A cell of adipose tissue that stores fat.

advanced   See DERIVED.

adventitious reinforcement   Reinforcement that happens to coincide with a particular response even though that particular response was not responsible for delivery of the reinforcement. In effect, the "wrong" behavior is modified. Also, ACCIDENTAL REINFORCEMENT.

aerial behavior   Any behavior that takes the animal partly or completely out of the water, e.g., lobtails, breaches, spyhops, head, side or back slaps, leaps, flips, and spins.

aerobic   Pertaining to activity or metabolism in which oxygen is the final electron acceptor in the breakdown of glucose. This process produces 39 ATPs from 1 mole of glucose.

aerobic dive limit   The maximum length of a dive accomplished using mainly aerobic metabolic pathways, and with no net production of anaerobic metabolites.

aggregation   A term for a group of individuals that come together because of a common resource, such as a predator-free habitat, rather than by social attraction.

aggressive behavior or **aggression**   A set of social interactions ranging from threats to open fights, reflecting a conflict of interest over limited resources and having the potential to cause injuries and sometimes death to participants. Generally refers to conflict involving members of the same species, but may refer to any interaction of this kind.

agonistic behavior   See AGGRESSION.

agviq   Northern Alaskan Inuit term for a bowhead whale.

air gun   In this context, a device that releases a burst of high-pressure air into the water. Groups ("arrays") of air guns are used by the marine seismic industry to create low-frequency, high-level sound pulses that can characterize rock formations below the seafloor.

aiviq   Northern Alaskan Inuit term for a walrus.

aldosterone   Steroid hormone released by the adrenal gland that increases Na+ retention in the distal convoluted tubule and the collecting duct of the kidney.

algae   A general term for various small aquatic organisms, usually single-celled, which can synthesize organic matter from carbon dioxide using the energy of sunlight and whose reproductive organs are contained in a single cell. Two groups of algae, diatoms and dinoflagellates, form the basis for all other life in the sea.

allantois   Fetal membrane developing from the hindgut or yolk sac that often contributes to the formation of the umbilical cord and placenta in mammals.

Allee effect   A form of density dependence in which the population growth rate slows at smaller population sizes, due to factors such as the difficulty of finding suitable mates. (First described by W. C. *Allee*, 1931)

alleles Alternative forms of a gene, e.g., *W* vs. *w*; Mendel studied pea genes for Wrinkled vs. smooth seeds.

allomaternal care Care provided to offspring by individuals other than the mother.

alloparenting Parental behavior by an animal (male or female) shown toward an offspring that is not its own.

allopatric Describing taxa, species, or populations whose ranges are physically separated.

allopatry The fact of having separate, nonadjacent geographic ranges.

allozygous Describing alleles that are identical by state but not descent.

allozyme A variant form of an enzyme, differing in amino acid sequence.

alternating carpus and tarsus An arrangement of the carpal and tarsal bones in an alternate way, in the manner of bricks in a wall. This is the primitive condition for mammals as seen from various Mesozoic mammalian groups. Cf. SERIAL CARPUS AND TARSUS.

altricial Being born in a helpless state and depending heavily on adult care. In contrast, precocial young are born in an advanced state of maturation and require relatively little adult care.

ambergris A grayish waxy substance formed in the intestines of sperm whales; once widely used in perfumes.

amniote An air-breathing vertebrate with cleidoic eggs having amniotic membranes, which can be laid outside water; i.e., the nonamphibian tetrapods comprising reptiles, birds, and mammals.

amphibious Able to move, feed, and so on, both on land and in water.

amphipod A crustacean of the order Amphipoda, including those on the ocean bottom, fed on by gray whales.

anadromous Describing the migratory behavior of fish that spend most of their lives in the sea, but that then migrate to freshwater to spawn (e.g., salmon).

anaerobic Relating to activity or metabolism in which the breakdown of glucose occurs in the absence of oxygen.

anaerobic dive A dive during which enough of the required metabolic energy comes from anaerobic pathways to result in net production of anaerobic end products.

anaerobic metabolism Chemical processes that temporarily sustain life in the absence of oxygen.

anal canal The terminal specialized portion of the gastrointestinal system.

ancestral species The species from which a present species has evolved.

anestrus The period of quiescence between two periods of sexual receptivity in cyclically breeding mammals.

anlage *plural,* **anlagen** Undifferentiated embryonic cells or tissue from which an organ or part develops; primordium.

annual pregnancy rate The proportion of sexually mature females that are pregnant during any given year, adjusted for the length of the gestation period.

antecedent In operant conditioning, a stimulus (signal or cue) that precedes a behavior emitted by the animal.

anterior Toward the front or head of the body.

anterior pituitary Another term for ADENOHYPOPHYSIS.

anthropogenic Describing a phenomenon or condition created, directly or indirectly, as a result of human activity.

antidiuretic hormone Another term for VASOPRESSIN.

antilipolytic Inhibiting the breakdown of lipids.

antitropical Occupying temperate and/or colder waters in the Northern and Southern Hemispheres, but absent from the intervening tropical latitudes; having distributions north and south of the equator but not in equatorial regions.

apnea A respiratory pause; elapsed time between inspiration and expiration.

apneustic breathing A method of breathing in which the animal rapidly exhales and inhales and then holds its breath for an interval before repeating. This pattern facilitates rapid air exchange while swimming rapidly.

aponeurosis A broad tendon, connecting muscle to muscle or muscle to bone.

appendicular skeleton Bony elements of the limbs.

aquatic Having to do with water; living in or situated in water, which can be either fresh water or the sea.

Archaeoceti (archaeocetes) The suborder of whales that includes the first Cetacea to enter the sea and all their descendants that do not have cranial telescoping.

Arctoidea A phylogenetic grouping that includes the Ursidae (the bear family), Pinnipedia (seals, sea lions, and walruses), Procyonidae (the raccoon family), and Mustelidae (weasels, badgers, otters, and skunks) and that is the sister group to the Cynoidea (the dog family).

arteriovenous anastomoses Vascular shunts between the arterioles and venules in blubber near the body surface.

artificial insemination The introduction of semen into the vagina, cervix, or uterus by artificial means in an attempt to cause conception.

Artiodactyla Even-toed hoofed mammals; the mammalian order that includes hippopotamuses, pigs, peccaries, camels, chevrotains, cattle, antelopes, deer, and giraffes. Among living mammals, probably the closest relatives of whales.

asphyxia A combination of decreased oxygen (hypoxia), increased carbon dioxide (hypercapnia), and the accumulation of acidic metabolic by-products (acidosis).

assemblage A predictable and particular collection of species within a biogeographic unit.

assortative mating Nonrandom mating in which members of a population pair up with individuals who are either more (positive) or less (negative) like themselves than the average for one or more traits.

atavistic Referring to the reappearance of a characteristic or feature in an individual belonging to a lineage whose members have secondarily lost that characteristic.

audiogram A hearing sensitivity curve drawn as a function of frequency.

auditory bulla The ear bone in odontocetes that houses the inner ear structure.

auditory meatus An opening to the ear.

autosomal chromosomes A collective term for all the nuclear chromosomes except the sex chromosomes.

autozygous Describing alleles that are identical by state and descent.

axial skeleton The bony elements of the body, excluding the limbs.

B

bachelor school A term for an apparently loose aggregation of mature or maturing male sperm whales, usually consisting of animals of similar ages.

backdown procedure A procedure used to release dolphins from the purse seine by pulling on the net after it has been pursed.

baleen Plates of dense, hair-like material (keratin) that hang side by side in rows from the roof of the mouth of whales in the order Mysticeti (the baleen whales); forming the "baleen apparatus" for filter feeding on surface plankton. Formerly known as "whalebone" but bearing no actual resemblance to true bone.

barnacles A collective name for various marine crustaceans of the subclass Cirripedia; the adults form a hard outer shell and attach to underwater surfaces such as rocks and ships, as well as to certain whales.

barrage In this context, a low gated dam used to divert water for irrigation, flood control, and/or navigation purposes. Normally the gates remain closed during the low-water season and are opened during the high-water season with differing levels of regulation in between.

basal insectivores Plesiomorphic representatives of Insectivora (hedgehogs, shrews, moles, etc.). They can be used as a theoretical basis for the calculation of encephalization and size indices and for the derivation of more progressive insectivores and other eutherian mammals.

basal metabolic rate The metabolic rate for a adult animal resting within its thermoneutral zone without food in its gut.

basicranial anatomy The base of the skull, particularly the size and shape of bones contributing to the auditory bulla; this has proven to be very important for interpreting phylogenetic relationships within Carnivora.

Bayesian Referring to a method of drawing inference about unknown parameters in which parameters are assigned a prior distribution independent of data and a posterior distribution given the data is calculated. (From Thomas Bayes, 1702-1761, English mathematician.)

beak In cetaceans, a term for the elongated anterior portion of the skull that includes both the upper jaw and the lower jaw.

beaked whales Members of the family Ziphiidae, which includes five current genera: Berardius, Hyperoodon, Mesoplodon, Tasmacetus and Ziphius.

behavior An observable and measurable event performed by an organism. Can also include nonobservable phenomena (such as emotion, cognition, recall) that can be measurable through other assessment strategies.

behavioral adaptation An animal's ability to learn to discern dangerous from nondangerous stimuli, and to react accordingly upon subsequent encounters.

behavioral ecology The influence of the environment on behavior.

behavior medicine Training programs and behavioral procedures intended to specifically condition behaviors that treat or prevent health disorders.

bellyflop A term for a breach in which the animal lands ventrally, on its stomach.

benthic Living in or on the ocean floor.

bicipital groove A groove on the humerus through which a tendon of the biceps muscle runs.

binocular vision The fact of seeing an object with two eyes simultaneously, as in humans. Necessary (but not sufficient) for stereoscopic vision.

binomial name The scientific Latin name of a species, consisting of two parts: the generic name and the trivial name. Note that the trivial name is not the "species names;" two species can have the same trivial name (e.g., the cetaceans *Stenella attenuata* and *Feresa attenuata*), but each has a unique, two-part species name.

biodiversity The diversity of species; a term variously defined but generally construed to mean the relative abundance of different species (and their population size) within a given area, or on Earth as a whole.

bioluminescence Light emitted by certain marine organisms as the result of a chemical reaction; this may be used to avoid or confuse predators, to attract prey, or to attract mates.

biomarker A biological response to a chemical that gives a measure of exposure and sometimes a toxic effect.

biomass A measure of the amount of plant or animal matter in a given context or system, expressed in terms of its aggregate weight.

bipolar Living or found in both polar regions.

blow A term for a cloud of vapor and sea water mixed with air that is exhaled by cetaceans.

blowhole The external opening of a whale's ventilatory system. Two blowholes occur in baleen whales and one in toothed whales.

blue whale unit (BWU) A measure of baleen whale harvest that reflects oil production. One BWU is equivalent to a blue whale, 2 fin whales, 2.5 humpback whales, or 6 sei whales.

blubber A specialized layer of fat that functions as an insulator, found between the skin and underlying muscle of most marine mammals.

bow-riding The action or behavior pattern of riding on the pressure wave in front of the bow of a ship.

bradycardia A condition of decreased heart rate.

brainstem-evoked potential Electrical nervous discharges generated by an external stimulus in a sensory system.

branchial arches Embryonic precursors of mammalian jaw, ear, face, and throat structures that are derived from elements homologous to the gill arches of fish.

breach or breaching A cetacean behavior in which the animal leaps out of the water and reenters on its side or back, often creating a large splash.

breeding sites Traditional land areas (mostly beaches) where adult females give birth and adult males defend territory.

bridging stimulus A conditioned stimulus that signals the imminent delivery of reinforcement.

brine channels Small pockets of brine in seawater ice that have a lower freezing point than the rest of the ice and often remain liquid.

bristles The fine fringe on the inner side of baleen plates that mats to form a sieve for food.

bristle worm A marine worm with short legs.

bunolophodont Referring to a molar having rounded (bunodont) cusps that are joined to form crests (lophs).

buoyancy Upward force exerted on a swimmer's body when it is immersed or floating in water.

burden of proof The duty of proving a disputed assertion. In ecology, traditionally the burden of proof has been placed on scientists to show that harm to resources will result from some human action. The precautionary principle reverses this standard to require proof that a human action will not harm the resource.

bycatch Animals that are caught accidentally in fishing operations; the capture and mortality of all organisms that are not the target species of a fishery.

C

calcareous Composed of calcium carbonate, a chemical compound similar to chalk.

calf A young animal dependent on its mother; applied to the young of cattle and also analogously to other large mammalian species, including some marine mammals.

callosity A patch of thickened, keratinized tissue on the head of a right whale, inhabited by large numbers of whale lice.

cannibalism The consumption by an animal of members of its own species.

carcinoma A cancer arising in an epithelial tissue.

cardiac Relating to or describing the portion of the stomach that lies next to the esophageal opening and contains the cardiac glands.

Carnivora An order of living and extinct mammals that includes such species as dogs, cats, bears, and seals.

carnivoran A term applied to all members of the order Carnivora, distinguishing them from other mammals that may be ecological carnivores.

carnivore An organism that feeds on live prey, in the case of marine mammals including fish and cephalopods as well as mammals, birds, and reptiles.

carnivorous Meat-eating; describes the diet of members of the order Carnivora but many other mammals as well.

carrying capacity The maximum population that can be sustained in a given ecosystem without altering the ecosystem in ways that will ultimately reduce the sustainable population.

catecholamines A class of biologically active compounds, including epinephrine (adrenalin) and norepinephrine (noradrenalin), produced by cells of the adrenal medulla and other neurological structures.

caudal Having to do with or near the tail or fluke; behind.

caudal fin The fin located on the extremity of the tail of fishes, dolphins, and so on.

caudal vertebrae Vertebrae caudal to the sacral vertebrae in most marine mammals; in cetaceans and sirenia, the vertebrae associated with chevron bones.

cecum or **caecum** A large blind pouch forming the beginning of the large intestine.

cementum A calcified tissue that fastens the roots of teeth to the teeth sockets.

centrum The center part of the vertebra.

cephalopod A member of the group of mollusks including squids, cuttlefish, and octopuses.

cervical Relating to or near the neck region.

cervical vertebrae The vertebrae of the neck.

cestode Any flatworm of the class Cestoda, including tapeworms.

Cetacea An order of living and extinct mammals that includes such species as whales, dolphins, and porpoises.

cetacean 1. Any member of the order Cetacea of aquatic, mostly marine mammals that includes whales, dolphins, porpoises, and related forms; among other attributes they have a long tail that ends in two transverse flukes. 2. Belonging to or describing this order.

chevron bone Any of a number of small bones positioned on the ventral aspect of the caudal intervertebral discs. These bones are common in mammalian tails, but they are particularly large in cetaceans and sirenians. In cetaceans, chevron bones are often V-shaped and they embrace the arteries that supply the tail.

chin slap A cetacean behavior pattern in which the animal raises its head out of the water and slaps it back down to make a splash.

chitin A tough insoluble polysaccharide, the main constituent of the arthropod exoskeleton.

Chordata One of the phyla (large groupings) of animals. They are characterized by the presence of a chorda (= notochord), a bar that supports the dorsal side of the animal.

choroid The thin, pigmented, vascular layer between the sclera and the retina of the eye. It consists mainly of blood vessels, which provide nourishment to the retina. The choroid also reduces internal reflection of light.

chromatophore A pigment cell or group of cells, which can be altered in shape or color in response to stimuli from the nervous system or hormones.

circadian Describing a biological rhythm with a period of approximately 24 hours.

circumpolar Distributed around the North or South Pole; in this context, referring to the distribution of species living in the cold water masses of the Southern Ocean.

clade An evolutionary line; a monophyletic group containing all the descendants of the most recent common ancestor of that group.

cladistic Based on a pattern of descent, with the most closely related (sister) taxa having a common immediate ancestor.

cladistics A method of classifying organisms in which species are delimited in time by branching points on an evolutionary tree (speciation events), with taxa being defined solely on the basis of common ancestry.

clan In this context, killer whale pods that share parts of their vocal repertoire (related dialects), reflecting a probable common matrilineal ancestry.

classical conditioning A type of learning in which a conditioned (learned) stimulus is presented in conjunction with an unconditioned stimulus, creating a conditioned (learned) response. During classical conditioning, learning occurs independent of the ongoing activities of the organism.

clean-entry leap A term for a breach in which the animal returns to the water smoothly, head first.

click In phocoenids, a sound of small duration (microseconds) and of narrow band produced at center frequencies between 110 and 150 kHz.

cline Gradual change over a geographic range.

clupeid See CLUPIOID.

clupioid fish A member of an order of schooling marine fishes including sardines, anchovies, and herrings that are often observed in the diets of marine mammals.

coda A patterned series of 3–20 clicks lasting about 0.5–2.5 seconds, used by sperm whales for communication.

codon The sequence of three nucleotides in DNA or messenger RNA that encodes for a particular amino acid.

coefficient of variance A description of the error associated with a statistical estimation.

coelom The body cavity holding the organs.

collagen matrix A protein web that forms the structure of blubber found in marine mammals.

collecting duct The terminal segment of the mammalian nephron.

colonization Occupation or use of an area for breeding by a group of animals.

commensalism A relationship or association of organisms in which one individual benefits at no expense or damage to the other organism (the host).

community An association of interacting groups within a larger population, usually defined by the nature of their interactions or by the place in which they live.

compensation A form of density dependence in which the population growth rate slows at larger population sizes.

competition Interaction between individuals of the same species (intraspecific competition) or between different species (interspecific competition) such that the simultaneous presence of the two competitors is mutually disadvantageous. In terms of competition between humans and marine mammals, the competitive mechanism may be direct, as when a marine mammal eats a fish that could otherwise have been caught by a fisherman, or indirect, as when a marine mammal consumes a fish that is an important prey species of a commercially desirable fish species.

competitive group In this context, a group of three or more humpback whales characterized by a nuclear animal (usually female), a principal escort (male), and one or more other males, who may challenge the principal escort for his position. Competitive groups are assumed to be related to mate choice or mate defense; they are often large and sometimes involve considerable intrasexual aggression.

competitive release A phenomenon of competitive interaction that occurs when predation (natural or from increased fishing) causes a reduction in one of two groups that are in competition. The second group then increases in size as a result of this "competitive release."

computed (axial) tomography A method of using X-rays to produce visual "slices" through an object.

conditioned stimulus A stimulus that attains the ability to "cue" behaviors as a result of learning.

confidence interval In statistics, a range of values, expressed as a minimum and a maximum, in which the real value of an estimate can be placed with a desired confidence.

connecting chamber A stomach compartment in cetaceans that lies between the main and pyloric chambers; lined with pyloric epithelium.

conspecific 1. Of an organism, belonging to the same species as another or others. 2. An individual of the same species as another or others.

contest competition A mechanism of sexual selection in which males compete by fighting with each other to achieve exclusive access to one or more mating females and in which females typically mate with the winners of such battles.

context-specific learning Learning that occurs in an environment in which background stimuli and discrete environmental cues are required for the animal to respond correctly.

continental shelf The part of a continent that is submerged in relatively shallow sea.

continental slope The slope of the sea floor between the continental shelf and the ocean floor, which is steep and 150–4000 m deep.

convergence zone An equatorial region where two north–south currents meet, forming an east–west current.

convergent evolution The evolution by unrelated organisms of structures that are similar in appearance or function.

copepod An organism of the order Copepoda; a small crustacean.

Copepoda An order of very small planktonic crustaceans present in a wide variety and great abundance in marine habitats, forming an important basis of ecosystems.

cortex The outermost layer of tissue of an organ.

cortisol A steroid hormone of the adrenal cortex that is involved in the regulation of protein and carbohydrate metabolism.

cosmopolitan Occurring worldwide or nearly worldwide.

costal Wing-like or lateral.

cotyledon In this context, a unit of the placenta grossly visible as a lobe circumscribed by a deep cleft.

cranial Having to do with the head or nose.

cranial muscle A muscle innervated by one of the cranial nerves.

cranial nerve A nerve that leaves the central nervous system near its anterior end and enters the periphery of the body by traversing one of the foramina of the skull. There are 12 named pairs of cranial nerves in all mammals.

cranial vertex The highest point of the skull, immediately caudal to the bony external nares. The nasal, maxillary, premaxillary, and frontal bones contribute to this region.

crepuscular Having to do with or active at the margins of the day; i.e., at dawn and/or dusk.

Cretaceous A period of geological time, from about 144 million to 65 million years before the present, ending with a mass extinction.

critically endangered A term used to describe a taxon that is facing an extremely high risk of extinction in the wild in the immediate future.

critical period A period of rapid acquisition that is usually associated with early stage development.

crossbow whaling A whaling method using a harpoon that is shot from a rubber-powered crossbow instead of a cannon, originating in Okinawa in 1975 to take advantage of a loophole in whaling regulations.

crown group A clade that includes the most recent common ancestor of all living species, plus all its descendants. Examples in Cetacea: Odontoceti and Mysticeti.

crus The middle region of the hind limb, consisting of the tibia and fibula and surrounding structures.

cryopreservation A process of maintaining the viability of cells, tissues, or organs by storing them at very low subzero temperatures.

cull **1.** To remove certain individuals from a population to control the overall population. **2.** An individual removed in this manner.

culling The process of reducing the population level of a given species. Distinguished from HARVEST, which refers to taking animals for commercial purposes.

cursorial Adapted for moving quickly over land; fast-running.

cusp A point on the grinding surface of a tooth.

cutaneous Referring to the skin surface of an animal.

cyamid A term referring to whale lice, all of which belong to the family Cyamidae.

cyclonic Describing rotation in the same direction as the earth (clockwise in the Northern Hemisphere and counterclockwise in the Southern Hemisphere).

cytochrome b A protein involved in respiration, coded for by an 1100-bp region of the mitochondrial genome.

D

dactyli The fingers or claws on appendages of invertebrates.

deciduous dentition The first generation of teeth of a mammal that erupt early in life and are later replaced by a second generation of permanent teeth. Unlike other mammals, modern cetaceans only have one tooth generation.

deciduous tooth A tooth that will be replaced, a replacement tooth, as opposed to a permanent tooth.

deep scattering layer (DSL) Organisms associated with the edge of the light, or photic, zone in deep water. The DSL tends to migrate vertically on a daily–nightly (diel) basis, being many hundreds of meters below the surface in daytime and closer to the surface at night.

definitive host The host in which a parasite achieves sexual maturity.

delayed implantation **1.** A period of suspended development or growth. **2.** Specifically, in pinnipeds, the suspended development of an embryo between shortly after conception and subsequent attachment (implantation) to the uterine wall.

delphinid A small dolphin species belonging to the family Delphinidae.

demersal A fishing term referring to marine resources captured near the bottom.

demographic Relating to or describing population levels or characteristics.

demographics **1.** The characteristics of a biological population such as population size, proportion of reproductive females, or age at first reproduction. **2.** the study of influences on or changes in such characteristics.

demography **1.** The birth and death rates that determine a population's dynamics. **2.** See DEMOGRAPHICS.

density dependence The dependence of population growth rate on population density or size.

dental formula A numerical summary of the numbers of teeth of each class (incisor, canine, premolar, molar) in the dentition of a mammal. For example, a dog has three incisors, one canine, four premolars and two molars in the upper jaw, and three incisors, one canine, four premolars, and three molars in its lower jaw; thus its dental formula is 3.1.4.2/3.1.4.3.

dentinal growth layer group A layer of dentine, consisting of one translucent and one opaque zone when examined in the longitudinal, thin section of a tooth. The number of these growth layer groups is used to indicate the age of toothed whales.

depleted A term indicating that a species or population stock is below its optimum sustainable population level or that a species or population stock is officially listed as endangered or threatened.

depredation An act of facilitated predation, in which an animal raids or plunders something already caught or otherwise claimed by a fishery.

derived Referring to a character or structure of an organism that has been modified relative to its ancestor; an evolutionary "new" feature.

Desmostylia Extinct "seahorses," a group of herbivorous marine mammals characterized by cheek teeth consisting of clusters of enamel/dentine columns.

diaphragm A musculotendinous sheet between the pleural and peritoneal cavities of mammals; generally considered the most powerful muscle associated with breathing.

diastole The period of ventricular relaxation during the contraction–relaxation cycle of the heart.

diatoms Single-celled algae abundant in marine and freshwater environments.

diel Occurring on a 24-hour cycle.

die-off Mortality on a large scale. This may involve more than one species and can occur over days, weeks, or months.

differential reinforcement A procedure in which reinforcement occurs for any behavior other than the target behavior. The effect is to decrease the target behavior while increasing other (more appropriate) behaviors.

dimethyl sulfoxide (DMSO) A chemical compound that increases cell wall permeability.

dimorphism 1. A difference in form (body size, shape, or color) between two individuals or two groups of individuals. 2. see SEXUAL DIMORPHISM.

dinoflagellates Single-celled algae, which, unlike diatoms, are capable of active movement.

diphyodont Developing two successive sets of teeth.

direct fitness The gene contribution of an individual to the ensuing population through its own offspring.

dispersal 1. The movement of an individual from its birth place to separate feeding grounds, or its integration into a different population. 2. Any outward spreading of organisms from their point of origin.

display behavior Behavior that is evolutionarily specialized for some form of communication, such as bird song. Special structures, colors, or coat patterns commonly evolve as parts of displays.

distinct population segment A population of organisms that is discrete from the other populations of its taxon and that represents a significant (in terms of numbers or range) or ecologically unique portion of a species or subspecies.

diurnal Active during the daylight hours, while inactive or sleeping at night.

diversity 1. See GENETIC DIVERSITY. 2. See BIODIVERSITY.

diving response A suite of physiological and biochemical reactions that are activated when a marine mammal dives, in order to conserve oxygen and extend dive time.

dolphin Any of the small cetaceans in the families Delphinidae (under about 3–4 m in length), Iniidae, Pontoporiidae, Lipotidae and Platanistidae.

dolphin-assisted therapy The use of dolphins to assist in the treatment of such human disorders as depression, autism, cerebral palsy, or mental retardation.

Doppler shift A change in frequency due to the relative motion of a transmitter and observer.

dorsal Toward or relatively near the back and corresponding surface of the head, neck, and tail.

dorsal cape A region of solid color extending along the forward dorsal surface of a dolphin and sweeping up behind the dorsal fin.

dorsal fin The fin on the top midline of the body, as in dolphins.

dorsoventral From top to bottom.

drag The physical force resisting the movement of a body or limb through water or another fluid medium.

drive fishery A style of fishing for dolphins in which speedboats are used to corral a school of dolphins into a bay or shallow water. A net is drawn across the mouth of the bay or around the school, which prevents their escape. The fishermen then wade among the dolphins to kill them. Usually, the entire school is killed.

Drosophila A genus of fruit flies often used as experimental subjects in laboratory studies of chromosome structure and evolution, including *D. melanogaster*, the familiar fly-in-a-bottle of high school and college genetics laboratories.

ductus arteriosus A connection between the pulmonary artery and aorta in mammalian fetal circulation.

dugong A strictly marine, herbivorous mammal foraging at the bottom, primarily on seagrasses, and propelling itself with a triangular caudal fluke.

duodenum The proximal part of the small intestine that is fixed to the dorsal abdominal wall and does not have mesenteries.

dura mater The outermost protective layer covering the brain and spinal cord in vertebrate animals.

E

eared seals Seals of the suborder Otariidae. They propel themselves with their fore flippers underwater and "waddle" with all four limbs on land. So termed because they have external pinnae, or ear flaps, visible to the casual observer.

earless seals Seals of the suborder Phocidae. These "true seals" all use blubber to thermoregulate, do not have external ear pinnae, use their hindlimbs to propel themselves in water, and have a hunching, caterpillar-like mode of locomotion on land.

echolocation The production of high-frequency sound waves and reception of echoes to locate objects and investigate the surrounding environment.

ecophenotypic Resulting from effects of the environment rather than the genome.

ecosystem A biological community and its environment, functioning as a unit in nature.

ecosystem model A mathematical representation of an ecological system that attempts to include all the major components of the system.

ecotype Ecological form or variation of a species.

ectoparasite A type of parasite that resides on the external surface of its host.

ectotympanic One of the bones that make up the skull in mammals. In primitive mammals, it is in the form of a horseshoe, and the tympanic membrane (eardrum) is suspended by it; this shape is retained by sirenians. In most mammals, the ectotympanic consists of additional flanges of bone that may form the wall of the middle ear (as in cetaceans and pinnipeds) and that may fuse with other skull bones.

effective population size The number of individuals in an ideal population with a level of genetic variation equal similar to that observed, i.e., the number of individuals participating in breeding.

electrolytes Charged ions such as sodium, chloride, potassium, calcium, and magnesium that are important components of the internal fluid of an organism.

embryonic diapause Delayed implantation of the blastocyst (embryo) in the uterine wall and a pause in the process of development. Implantation reactivates the blastocyst and allows placental gestation to proceed.

emmetropia The proper eye refraction, as opposed to myopia (near-sight) and hyperopia (far-sight).

empirical Describing phenomena that are directly observable or measurable or that require a minimum of models and assumptions to be characterized.

encephalization An evolutionary increase in brain size relative to body size.

encephalization index or quotient A numerical comparison of brain size to body size.

endangered Describing a species or population that is in danger of extinction in the near future.

endemic Restricted or peculiar to a specific locality or region; describing a species that occurs in only one geographic area (e.g., an island).

endocranium The internal space within a cranium that houses the brain and surrounding tissues.

endocrine Referring to a ductless gland that secretes hormones.

endotherm An animal that generates its own body heat for thermoregulation. Mammals and birds are endotherms.

endothermic Producing its own body heat for temperature regulation; in popular use, warm-blooded.

endothermy The physiological condition wherein the body temperature of an animal is controlled by the generating of heat produced in its own body.

energy reserve The amount of energy usually stored as fat in the body of the animal that is excess to current requirements.

entoconid The cusp (elevated part of a tooth) on the posterolingual side of the posterior heel of a lower cheek tooth.

environmental carrying capacity See CARRYING CAPACITY.

Eocene **1.** The geologic epoch spanning 55 to 34 million years ago during which most, if not all, of the lineages leading to modern carnivorans evolved. **2.** The sediments deposited in this time period.

epimeletic behavior Behavior that supports caregiving, nurturing, and survival in young. This also includes the soliciting of caregiving behavior by other animals.

epipelagic Occurring at or near the surface in pelagic waters.

epithelial cells Cells that form firmly coherent layers of tissue on exposed surfaces.

epizoic Referring to nonparasitic animals that attach themselves to the outer surface of another, normally larger, animal.

epizootic **1.** Referring to a temporarily prevalent and widespread disease in an animal population. **2.** A disease outbreak of this type.

escort A term for an adult male humpback whale that accompanies a female with calf on the breeding ground.

estrous Relating to the stage of the mammalian sexual cycle when females are receptive to copulation.

estrus The time in a female's reproductive cycle when she is "in heat," i.e., fertile and receptive to mating. A characteristic of mammals other than primates.

estrus lordosis See LORDOSIS.

etiology **1.** Generally, the assignment of a cause or reason for some condition or event. **2.** The field of science that studies the causes of diseases.

eumustelids Fossil and extant mustelids that belong to one of the five currently recognized subfamilies; the Mustelinae (weasels), Melinae (badgers), Mellivorinae (honey badger), Lutrinae (otters), and Mephitinae (skunks).

euphausiid Any of an order (Euphausiacea) of small, usually luminescent shrimp-like crustaceans, forming an important part of the marine zooplankton; commonly known as krill.

eustachian tube A slender tube that connects the tympanic cavity with the pharynx and serves to equalize air pressure on either side of the eardrum; the remnant of the embryonic first gill (or pharyngeal) pouch.

Eutheria An infraclass of therian mammals including all living forms except the monotremes and marsupials.

evoked potential An electric response of the brain to sensory stimulation; reflects an integrated simultaneous activity of many neurons evoked by the stimulus.

evolution Descent with modification, with ancestral species giving rise to one or more descendant species.

evolutionary stable strategy Patterns of reproduction or behavior that have equal fitness value.

exon An uninterrupted nucleotide sequence in a locus that is translated into amino acids.

exoskeleton Hard supporting structure made of chitin, secreted by a crustacean and external to the epidermis.

exponential growth Population growth for which a constant percentage is added each year, which means that a greater number of animals are added to the population each year.

ex situ conservation The process of maintaining and breeding animals or plants outside their natural environment.

extant Still existing, not extinct.

extinct No longer living; describing a taxon for which there is no reasonable doubt that the last individual has died. A taxon is considered extinct in the wild when the only surviving individuals exist in cultivation, captivity, or a naturalized population outside the past wild range.

extinction The irreversible disappearance of a population, species, or higher taxonomic group.

extirpation The extinction of an organism in part of its range; loss or removal of part of a population.

extradural Referring to a location superficial to the dura mater, usually within the vertebral canal.

F

facultative "As need dictates;" referring to those animals that can or do live in one particular environment, but are also reasonably well adapted to another environment. For example, facultative river dolphins live in both ocean and river systems. Cf. OBLIGATE.

falcate Recurved and usually pointed; sickle-shaped.

fast ice A term for stable ice that is attached to land.

fasting A limited period of not eating or abstaining from food. Differs from starvation in that it is usually a periodic event for which the animal prepares by laying down extra body fat.

fathometer A sonar that is directed downward to measure water depth.

fatty acid A component of fat molecules. Fats are composed of three fatty acids (of varying carbon chain length; from 2 to 24 or more) attached to a glycerol backbone.

fecundity **1.** The potential level of female reproductive performance in a population; i.e., the average maximum number of offspring that females may bear during their fertile years. **2.** The actual average number of live births (more specifically, of female offspring) per year per female.

feedback loops A process through which substances or conditions occurring in the blood stimulate (positive) or inhibit (negative) the release of hormones responsible for affecting changes that modify the physiological state.

female choice A mechanism of sexual selection in which females prefer to mate with males exhibiting particular qualities, such as bright colors or vigorous displays, and reject other males.

female sexual maturity The stage of female sexual development in which ovulation occurs. Most cetaceans conceive at first ovulation and past ovulations are identifiable on ovaries.

fertilization Union of the sperm and egg.

fibrosis A thickening and scarring of connective tissue, as may result from disease or injury.

filter feeding Method used by some marine mammals to strain small particles of food from the water, using baleen or serrated teeth.

fineness ratio Ratio of body length to maximum diameter.

fisheries The enterprise(s) of catching or harvesting fish for human consumption.

fish ladder A man-made structure that allows fish to pass upstream with no more effort than they would use to swim against natural river rapids.

fissipeds The group of four-legged, mainly carnivorous mammals with toes separate from each other; includes bears, otters, dogs, cats, raccoons, and hyenas.

fitness The ability of an organism to survive and to transmit genetic information to future generations.

flaw zone An area of labile, fractured or broken ice; caused by various factors that result in impingement or interaction among different types of sea ice, e.g., between drifting pack and shore-fast ice.

flensing The process of removing the blubber from a marine mammal carcass.

flipper The flattened forelimb of a marine mammal.

flippering The act of slamming the pectoral fin, or flipper, on the surface of the water.

flotsam The traditional term for bits of debris floating at the surface or in the water column, usually but not necessarily of human origin. Contrasted with **jetsam**, which is human debris that has been purposefully discarded.

flukes The horizontally spread tail of a whale (in contrast to the vertical tails of fish).

fluking The act of raising the tail, or flukes, above the surface of the water during the beginning of a dive.

food web The interconnection of species through their diets.

food web competition A mutually disadvantageous interaction that occurs due to indirect competition for a shared resource, such as when there is overlap of the trophic flows supporting marine mammals and fisheries.

foraging Searching for prey; the process of finding, catching, and eating food.

foramen ovale An opening between the right and left atria of the mammalian fetal heart.

foregut The anterior portion of the digestive tract, which includes the stomach.

forestomach Cetacean stomach chamber lined with stratified squamous epithelium; lies between the esophagus and the main or fundic stomach.

fossil The prehistoric remains of once-living organisms, either extinct or living species.

free hormone Any hormone in the blood not bound to proteins and therefore more readily available to influence cellular functions.

fundic The portion of the stomach distal to the cardiac portion, containing the fundic glands.

G

gadids Fish species of the cod family (Gadidae).

gape The mouth in cetaceans, usually referring to the junction of upper and lower lips.

gastric Relating to or situated in the stomach.

gene The coded information inside cells, which tells the body how to make enzymes, proteins, and other molecules necessary for life.

gene/allele frequency The proportion of gene copies in a population of a particular type; e.g., if 40% of the alleles in a population are type W (as opposed to w), the frequency of W in that population is 0.4.

gene flow The exchange of genetic material within and between populations.

genetic bottleneck A severe reduction in genetic diversity.

genetic diversity Variation in the chromosomes of a community of organisms, due to a large number of slightly dissimilar ancestors.

genetic drift Random changes in the gene frequencies of finite populations.

genetic fingerprinting A method of uniquely identifying an individual animal from a sample of tissue.

genotype The hereditary or genetic makeup of an individual.

geochronology The study of time in relationship to the history of the earth.

geologic range The geologic time span between the first and last appearance of fossil taxa.

gestation The period of time during which the embryo and later the fetus is carried by the mother in the pregnant state.

gilling The process of catching a fish by its gill covers, usually in a mesh of netting, so that the twine, under the gill covers, prevents the fish from moving backwards, while forward movement is prevented by the mesh of the net encircling the fish's head.

gill net A fishing net that is suspended in the water more or less vertically and ensnares fish by the gill covers as they try to swim through its meshes.

glucagon Protein hormone formed in the pancreas that when released to the blood increases the glucose content of the blood by increasing the breakdown of glycogen in the liver.

glucocorticoid   Steroids such as cortisol and corticosterone produced by the adrenal cortex and affecting a broad range of metabolic and immunologic processes.

gluconeogenesis   The synthesis of glucose from other carbon sources, occuring primarily in the liver.

glycogen   The polysaccharide in which carbohydrate is most commonly stored in muscles and liver.

glycolysis   The series of reactions that convert glucose to pyruvic acid.

grade   A cluster of species that have a common pattern of organization but are not necessarily a clade.

granuloma   The product of a type of inflammation, resulting in the formation of distinct nodules of inflammatory tissue.

gravid   Pregnant with young (whether live young or eggs).

gross annual recruitment rate   The product of the proportion of females in a population, the proportion of sexually mature females, and the annual pregnancy rate.

ground   A term for an area of ocean where whalers find abundant sperm whales.

group selection   The principle or phenomenon that differences in the attributes of groups are related to differences in group survival, particularly when the groups are competing with each other.

growth layer group (GLG)   A repeating or semirepeating pattern of adjacent groups of incremental growth layers within the dentine, cement, bone, or other persistent tissue, which is defined as a countable unit. Such a unit must involve at least one change in appearance.

growth layers   Incremental layers deposited in structures that persist and reflect growth or physiological processes at the time of deposition.

groyne   A transverse structure that deflects water flow to prevent erosion of an embankment.

gubernaculum   A fibrous cord connecting the testis to the scrotum that usually plays a significant role in testicular descent.

gyrification   The degree of folding of the cerebral cortex.

H

habituation   A gradual waning of behavioral (or physiological) responsiveness over time, as an animal learns that a repeated or ongoing stimulus is lacking in significant consequences. Cf. SENSITIZATION.

hair cell   A special kind of cell that has tiny hairs projecting from its surface into the intercellular space. Movement of the hairs is registered by neurons that contact the hair cell. Hair cells are found in the ear of mammals.

halogenated organic compounds   Organic compounds, usually anthropogenic or produced by the breakdown of anthropogenic compounds, that contain chlorine or fluorine atoms. Many, including organochlorines such as DDT and the PCBs, break down slowly when released into the environment and are potentially pathogenic.

hand-harpoon whaling   A historic Japanese whaling technique that used hand harpoons and lances.

haplotype   A unique DNA sequence or series of sequences.

harem   A group of females whose breeding is controlled by a single male who seeks to prevent other males from breeding with them.

harpoon   A fishing spear composed of a barbed head, a shaft, and a line that connects the shaft to a speared animal.

harvest   The deliberate taking of wild animals, usually by hunting, netting, or trapping (not including live capture or removals for scientific purposes).

haulout   The act of a seal leaving the ocean and crawling onto land or ice.

heat increment   An increase in metabolism associated with the processing and digestion of food.

hemoglobin   The iron-containing protein that binds to oxygen and is found in red blood cells.

hemopoietic tissue   Tissue in which blood cells are formed (e.g., red blood cells are formed in the spleen of developing mammals and in the bone marrow of adults).

hepatic   Relating to or occurring in the liver.

herbivore   A feeder on plants.

herbivorous   Plant-eating.

hertz   A measure of frequency corresponding to one cycle per second.

heterodont   Having teeth of different types.

heterodont dentition   A type of dentition in which there are different tooth shapes in different parts of the mouth: incisors and canines in front and premolars and molars in back. Cf. HOMODONT DENTITION.

heterodonty   See HETERODONT DENTITION.

heterospecific   Describing an individual of a different species from another.

heteroxenous   Describing a parasite life cycle involving more than one host.

heterozygosity   The proportion of individuals in a population of diploid organisms that has two different alleles at a given locus.

heterozygous   Having one or more pairs of dissimilar alleles (alternative forms of a gene).

hilus   The point where the renal artery enters the kidney and the renal vein and ureter exit.

Hippopotamidae   The mammalian family that includes extant hippopotamuses and their close extinct relatives.

histogram   A graph showing statistical distribution of the probability of events, e.g., how frequently cells of one or another size appear in the overall cell population.

holotype   A single specimen laying at the base of the scientific name of a species or subspecies.

home range   The total area covered or traversed by an individual animal during the course of normal activities.

homodont   Having teeth of the same type.

homodont dentition   A type of dentition in which all of the teeth have similar size and shape. Cf. HETERODONT DENTITION.

homodonty   See HOMODONT DENTITION.

homologous   Pertaining to a relationship between corresponding parts in different organisms due to evolutionary development from the same structure, such as the wing of a bird and the flipper of a dolphin.

homoplastic   Describing a character whose presence in various species is not due to a common origin.

homoplasy Similarity in character state in two or more groups that are not inherited from a common ancestor but evolved independently in each group.

homozygosity The proportion of individuals in a population of diploid organisms that has two identical alleles at a given locus.

homozygous Having one or more pairs of identical alleles (alternative forms of a gene).

hormone An organic compound produced by an endocrine gland that affects other tissues and organs within the body; e.g., the hormone progesterone is secreted by the corpus luteum and stimulates the uterus for implantation.

hormone-sensitive lipase An enzyme inside adipose cells that mobilizes free fatty acids stored in triglycerides and makes them available to other cells.

hydrofoil **1.** An asymmetrical blade whose shape produces lift forces due to differential fluid flow around its upper and lower surfaces. **2.** A wing-like limb flapped underwater to produce lift; e.g., penguin "wings."

hydrophone An underwater microphone.

hyoid apparatus A set of bones and/or cartilages attached to the base of the skull that help move the tongue and larynx during feeding and swallowing.

hyperglycemic Describing a condition in which blood glucose concentration is greater than normal.

hyperoxic Describing a gas mixture having a higher concentration of oxygen than in normal air.

hyperphagia Intense feeding.

hyperventilation Increased total ventilation produced by tidal volume and ventilation rate above the resting rate.

hypoconal crest The crest extending to the cheek side from the hypocone on an upper cheek tooth or to the tongue side from the hypoconid on a lower cheek tooth.

hypocone The cusp on the postero-lingual side of an upper cheek tooth of a derived placental mammal having quadritubercular (four-cusped) teeth.

hypoconid The cusp on the postero-labial side of the posterior heel of a lower cheek tooth.

hypoinsulemic Describing a condition in which the level of insulin in the blood is lower than normal.

hypometabolism A metabolic rate that is less than resting or basal metabolism.

hyponatremia An abnormally low blood concentration of sodium.

hypoxia Reduced availability of oxygen.

Hz Short for HERTZ.

I

ichthyophagous Feeding on fish.

igneous rock Rocks derived from the cooling and solidification of a magma or lava.

ileum The distal part of the small intestine.

immunodepressive Negatively affecting any aspect of the immune response system.

inbreeding Reproduction with a mate more closely related to the individual than the average pair in the population are related to each other.

inbreeding depression Diminution of growth, fertility, or survival as a consequence of inbreeding.

indirect fitness The gene contribution of an individual to the population through the relatives that it successfully aided in rearing and that would not have existed without its help.

infanticide The killing of young by parents or other members of the same species.

infrasonic Describing sound that is lower in frequency than the minimum audible to humans. Some baleen whales produce infrasonic calls.

infrasound Sound below the human hearing threshold, or 20 Hz (cycles per second). These sounds approximate a very low frequency roar or vibration.

insectivorous Feeding mainly on insects, and often also on other small invertebrates such as scorpions, spiders, worms, and snails. Insectivores are members of the Insectivora, a group of small mammals such as shrews and hedgehogs.

***in situ* conservation** The process of conserving animals or plants within their natural environment, in habitats within which evolutionary processes are maintained.

instar A juvenile stage of invertebrate development, which may be a configuration quite distinct from the adult.

insulin A protein hormone released by the pancreas that stimulates the uptake of glucose by peripheral tissues.

intermediate host A host in which a parasite develops to some extent but not to sexual maturity. Where more than one intermediate host is involved in a life cycle, they are ordered in chronological sequence as first, second, etc.

intermodal transfer The transfer of sensory information from one sensory modality to another, e.g., from vision to somatic sense and back, from vision to echolocation and back.

interspecific Involving individuals of different species.

intraspecific Pertaining to a level below or within a species (e.g., subspecies, population, deme).

intrinsic rate of increase The maximum rate of increase a species is capable of; thought to occur at a relatively small population size where a species has its greatest rates of reproduction and survival.

introgression The incorporation of genes of one species into the genetic constitution of another species, as by means of interspecific hybridization.

intron The part of the nucleotide sequence at a protein-coding locus that is transcribed into messenger RNA but not amino acids.

Inuit The native people of Arctic Chukotka, Alaska, Canada, and Greenland, popularly known as Eskimos.

Inupiaq *plural,* **Inupiat** An Inuit person.

invasive sampling The collecting of tissue samples in a manner that requires penetration of the epidermis.

involucrum The internal medial part of the tympanic bulla of cetaceans, which is made of extremely dense and heavy (pachyostotic) bone.

ischemia   A blockage or reduction of blood flow that interrupts oxygen supply.

ischium   The dorsal posterior part of the hip bone.

iteroparous   Describing or referring to animals that reproduce more than once.

J

jejunum   The middle portion of the small intestine (From a Latin word for dry; this part of the intestine was often empty in cadavers because of postmortem peristalsis.)

jetsam   See FLOTSAM.

junk   A complex arrangement of whitish tissue beneath the spermaceti organ, saturated with spermaceti oil.

K

karyotype   A display of the chromosomes.

kelp   Algae belonging to the division Phaeophyta, which includes the large "macroalgae" commonly known as seaweeds.

ketone bodies   Any of three compounds—acetoacetate, acetone, and β-hydroxybutyrate—that can be produced by the liver as a result of lipid oxidization.

ketosis   Accumulation of ketone bodies in the blood, usually associated with prolonged fasting or excessive lipid oxidation.

keystone species   A species whose presence in (or absence from) a given ecosystem has a significant influence on the structure and function of the system, disproportionate to its numerical abundance.

killer whale   The largest species (*Orcinus orca*) of the family Delphinidae, an aggressive cetacean that preys upon other marine mammals.

kinematics   The study of motion. Often used in reference to the movement of limbs relative to the body of active animals.

kin selection   The actions of an individual that lower the individual's own chances of survival and reproduction, but that benefit their relatives and thus indirectly contribute genes to the population.

kleptogyny   Female stealing by males from the territories of other males, for the purpose of mating.

krill   A general term describing shrimp-like crustaceans in the family Euphausiidae. Krill form an important food resource for filter-feeding marine mammals.

k-selection   The reproductive strategy of a species that tends to grow slowly and have relatively few offspring, but that has a longer life span and parental involvement in the rearing of young. Certain large mammals are cited as an example of k-selection. Contrasted with an r-selected species, which has more rapid growth and a large number of offspring, but a shorter life span and negligible parental involvement.

L

lactation   The production of milk by a female mammal for the feeding of young.

lactose   A sugar found in some milks.

Lagrangian study   The study of individual movement.

laminar flow   A smooth and steady flow in which particles travel along a well-defined path.

lance   A spear with a blade on both sides, thrust into a harpooned animal for killing.

land fast ice   Polar ice that forms large sheets continuous with the shore. It is generally more persistent and stable than pack ice, which floats free of the land.

landing sites   Traditional areas used by nonbreeding males for resting, playing, and other activities. In old sealer's jargon, landing sites were called "hauling-out grounds."

lanugo   Soft, dense, downy hair that covers the fetus of some mammals; usually shed before or shortly after birth; this persists for up to several weeks after birth on some pinnipeds.

laparoscopy   The examination of abdominal organs, including the reproductive tract, by means of an illuminated visual instrument (laparascope).

large-type whaling   A historic form of Japanese coastal whaling that took sperm and baleen whales other than minke whales using a vessel and harpoon cannon of large size.

larynx   The group of cartilages and muscles in the neck that house the vocal cords and are used for sound production in most mammals. In odontocetes, the larynx is not the main sound-producing organ, and it projects far rostrally, reaching the posterior side of the palate.

lateral   Away from, or relatively farther from, the middle of the body.

lead   A long linear area of open water or thin, newly formed ice; this occurs along shorelines, at the seaward margin of shore-fast ice, or in the drifting pack; often persistent or recurring.

lead line   A weighted line onto which netting material has been fixed in order to keep part of the netting relatively low in the water column, typically on or near the seabed.

lek   A traditional display site where males gather to defend small individual territories (that lack essential resources) and advertise for mates. (From a Swahili word for a mating ground.)

lekking   The social system in which males gather on a communal display ground, and females choose mates by some criteria of "maleness."

length-specific speed   Speed measured using the body length of the animal as a scale and calculated as the velocity of the animal divided by the body length.

Leydig cells   Testosterone-producing cells between the seminiferous tubules of the testis.

life history   The significant features of the life cycle of an organism, particularly the strategies influencing reproduction and survival.

lighting regime   The ratio of number of hours of light versus number of hours of dark in a 24-hr period (light:dark).

limiting resource   A resource whose relative abundance is a determining factor in the size of a given population.

lineage   Ancestor-descendant populations of organisms through time.

lipoprotein lipase   A tissue-bound enzyme that hydrolyzes circulating triglycerides in the blood, enabling transport of free fatty acids into cells.

lithification   The conversion of a newly deposited sediment into solid rock, involving processes of compaction, cementation, and crystallization.

lithology   A description of a rock unit based on characteristics such as color, mineralogy, and grain size.

lithostratigraphy   The study of rock strata with emphasis on the lithology, succession, and correlation of strata.

lobster krill   Larval crustaceans forming dense, often red, shoals in temperate waters.

lobtailing   A behavior in which a whale slams its flukes down on the water, usually repeatedly.

locus   The position on a chromosome/mtDNA molecule at which the gene for a particular trait or an intron resides.

longevity   The life span of an individual; how long an organism lives. Maximum longevity (the longest any single member of a species or population has lived) is distinguished from average longevity (the average life span of all individuals in a year class or group of year classes). Average longevity cannot be determined until all of the animals in a year class have died.

long line   A buoyed line onto which are attached numerous branch lines each terminating in a baited hook. Long lines may extend for tens of kilometers and are usually left to drift in the surface waters of the ocean to catch large pelagic fish.

lordosis   A reproductive behavior of females in which they arch their back in a manner that raises their rear end upward to make mounting by the male easier.

lower risk   A description of a taxon that has been evaluated for potential population decline, but that does not satisfy appropriate criteria to be classified as vulnerable, endangered, or critically endangered.

low-frequency active sonar (LFA)   A naval sonar system designed to detect and locate quiet submarines at long range. Operates at frequencies below 1 kHz (typically 100–500 Hz) and at nominal source levels up to at least 230 dB re 1 mPa.

luciferase   An enzyme involved in the chemical reaction to create bioluminescence.

luciferin   A substrate involved in the chemical reaction to create bioluminescence.

lumbar   Describing or located in the region near to the portion of the vertebral column between the thorax and the sacrum.

lunge   A term for a thrusting of the forward part of an animal through the water surface, showing less than about 40% of the body (often the result of feeding at the surface).

lutrine   A member of the otter subfamily (the Lutrinae).

M

magnetic declination   The angle between the Earth's surface and magnetic lines of force for the Earth.

magnetite   A magnetic mineral that is composed of iron oxide.

main stomach   The glandular portion of the cetacean stomach, lying between the esophagus or forestomach and the connecting chambers.

maintenance metabolism   The energetic cost associated with the maintenance of all physiological processes necessary to sustain an animal.

maktak   Northern Alaskan Inuit term for the skin and blubber of whales.

male sexual maturity   The stage of male sexual development in which all seminiferous tubules at the testicular center produce spermatozoa. Sperm production starts before this and testicular growth continues after.

management unit   A group of animals that is the target of a particular management measure.

manaq or **manaqtuun**   Northern Alaskan Inuit term for a retrieval hook used during ice edge seal hunting.

manatee   A wholly aquatic herbivorous mammal generalist, foraging throughout the water column in fresh or salt water, and propelling itself with a broad caudal paddle.

maritory   Underwater territories held by males.

mark-recapture   A set of methods for estimating one or more of abundance, survival, and recruitment by recording repeated sightings or captures of animals, some of which are identifiable from marks previously placed on them. Increasingly, natural marks are used, identified from photographs or DNA fingerprinting.

marsupial   Members of the Marsupialia, mammals found mainly in Australasia and South America, which give birth to very small young that then develop while attached to the mother's teats, often in a pouch, e.g., the kangaroo, opossum, wombat.

mass die-off   Mortality on a large scale that may result in many strandings.

mass stranding   The simultaneous stranding of two or more animals other than a female and her offspring.

mastoid process   A component of the periotic bone (which makes the inner ear) that can be seen on the lateral face of the cranium.

mate choice competition   A mechanism of sexual selection in which males compete among themselves by attempting to entice and attract females with visual, acoustic, or pheromonal displays; females typically mate with the male having the most exaggerated display.

maternal care   Any form of maternal behavior that appears likely to increase the fitness of her offspring.

mathematical model   A representation of a system or a synthesis of observations and ideas that is translated into the language of mathematics to facilitate understanding and analysis of the system or process.

matriline   A descent system in which daughters reproduce in their mothers' social unit, leading to a group of maternally related females.

matrilineal Describing a social unit in which group members are descendants of a single female.

matrilineal habitat fidelity The continued return to habitats first learned from the mother during the months before weaning, which may persist for generations.

maxilla or **maxillary bone** One of the two major bones found in the upper jaw of mammals, the other being the premaxilla.

maxillary crest The anterior portion of the maxilla, which is enlarged in cetaceans due to telescoping of the cranial bones.

maximum net productivity level The population size that will result in the maximum growth rate (in number of individuals added to the population per year).

maximum sustainable yield The maximum human harvest from a population that can be sustained continuously and, for a non-selective harvest, that is equivalent to the maximum net productivity level.

medial Toward, or relatively closer to, the middle of the body.

mediastinum In mammals, the region between the lungs.

melanin The pigment responsible for dark coloration in the skin and eyes of many animals.

melon A lump of fatty tissue that forms the "forehead" of toothed whales and that is thought to funtion as a means of focusing sound for echolocation.

Mendelian inheritance Loci that are inherited from both parents.

menopause The termination of reproductive function in a female, as a result of aging.

mesentery The double layer of peritoneum that suspends the abdominal organs from the dorsal wall; it contains blood vessels and nerves that supply the organs.

mesonephros Embryonic kidney that functions only until the metanephros is formed and then regresses. Its duct (Wolffian) is retained in the male as the epididymis and ductus deferens.

mesopelagic Referring to the middle portion of the water column, generally from 200 to 1000 m.

mesosalpinx A special paired mesentery structure, forming part of the broad ligament, and supporting the oviducts.

Mesozoic An era of geological time, from about 285 million to 65 million years before the present, comprising the Triassic, Jurassic, and Cretaceous periods.

metabolic overhead The amount of energy expended by a lactating female for her own maintenance, relative to the total energy expended during lactation.

metabolic water Water that is produced as a result of the oxidation of food or stored body tissue.

metanephros Precursor of the kidney that will function postnatally.

metapopulation An overall population consisting of two or more subpopulations more or less isolated from each other, treated as a single evolutionary or management entity.

metatarsal A bone in the middle portion of the hind foot.

microallopatric Sharing a geographic range but occurring in different habitats or portions of a habitat, so that there is no mixing on a small spatial scale.

microsatellites Non-coding short tandem repeats in the nuclear genome; those having a variable number of repeats (alleles) are used in population analysis.

migration The process of seasonal movement of individuals between different geographic locations.

migration count A sample count of migrating whales carried out from a coastal watch point from which population size is estimated.

mimicry Behavior resembling the behavior of others occurring in the same locale or within some relationship of time and space; an important component of early social development. Various forms of mimicry are exhibited by many animals for separate survival strategies.

mineralocorticoids Steroids such as aldosterone produced primarily in the zona glomerulosa of the adrenal cortex and regulating the metabolism of ions such as sodium and potassium.

Minoan Referring to the Bronze Age civilization centered on the Mediterranean island of Crete (c. 3000–1000 B.C.) prior to the Greek empire. (*Mino* was Crete's legendary king.)

mitochondria Small cytoplasmic organelles found in almost all living cells.

mitochondrial DNA A circular DNA molecule that is found within mitochondria in large numbers of copies. In mammals, mitochondrial DNA is inherited maternally, the gene sequence and contents are very conserved, and the DNA is composed of about 14,000–17,000 nucleotide base pairs.

molar One of the rearmost teeth used mainly for crushing and chewing.

molt or **molting** The relatively abrupt shedding of an old outer covering (in mammals, the skin) to be replaced by a new layer recently formed underneath.

monestrous Having a single estrous cycle once each year.

monogamy A mating system in which one male and one female are together for at least one seasonal breeding cycle.

monomorphism The existence of similar morphology in males and females, usually in both body shape and color.

monophyletic Pertaining to a group that shares a common ancestor and includes all descendants of that ancestor.

monophyly The status of a group that includes an ancestral species and all of its descendants.

monophyodont Developing only a single set of teeth, rather than deciduous and adult sets.

monotypic Being the sole type in its group (genus).

monoxenous Describing a parasite life cycle in which only a single host is used.

monsoon A constant, seasonal wind within the tropics.

morphological Having to do with the appearance or form of an individual or species.

morphology 1. The physical appearance or form of an individual or species. 2. The study of such properties.

motorboating A behavior pattern of very rapid tail slapping (two to three slaps a second) in trains up to 15 sec in length.

motor pattern A pattern of coordinated movements, positions, and postures of the body that is recognizable, repeated, and characteristic of particular classes of movements or functions.

mtDNA See MITOCHONDRIAL DNA.

multiparous Describing a female that has had more than one pregnancy, as contrasted with nulliparous (no pregnancy) and primiparous (first pregnancy).

multireniculate Describing a type of kidney that contains many reniculi.

multispecies model A mathematical representation that includes a minimum of two interacting species in the system.

multivariate analysis Statistical methods that can be used to simultaneously examine multiple variables and their interactions.

muscular hydrostat A three-dimensional array of muscles having constant volume, capable of highly controlled, precise, and varied movements, and not dependent on bony attachments for the support of antagonistic actions of the muscles. Examples include the tentacles of squid, trunks of elephants, and tongues and lips of mammals.

museau de singe In sperm whales, a valve-like clapper system at the end of the right nasal passage, thought to be the producer of the whale's clicks.

mustelids or **mustelidae** Members of the family Mustelidae of carnivorous mammals, which includes the otters, stoats, minks, and fishers.

musteloids All members of the family Mustelidae and its generally accepted sister taxon Procyonidae, as well as those stem taxa that are thought to have given rise to this clade.

mutation An alteration in a gene that causes it to produce a faulty product, or no product at all.

mutualism A relationship of organisms in which individuals engage in actions that are mutually beneficial.

myoglobin A protein in mammalian muscle that carries oxygen. Hemoglobin has a similar function in the blood.

mysid A shrimp-like crustacean of the order Myscidacea.

Mysticeti (mysticetes) The toothless or baleen (whalebone) whales, including the rorquals, gray whale, and right whales; the suborder of whales that includes those that bulk feed and cannot echolocate. Their skulls have an antorbital process of the maxilla, a loose mandibular symphysis, a relatively small pterygoid sinus, and the maxillary bone telescoped beneath the supraorbital process of the frontal.

N

nanuq Northern Alaskan Inuit term for a polar bear.

nares Openings of the nasal cavity.

nasal turbinate Specialized structures in the nasal passage of mammals that allow for the recovery of water.

nasolaryngeal air sacs Air sacs of the larynx that are continuous with the nasal passages.

natal site The group of islands or rookery where animals were born.

natchiq Northern Alaskan Inuit term for a ringed seal.

natriuresis Excretion of sodium ions in the urine.

natural mark(ing) Any distinctive feature of an animal that is useful to researchers for individual identification, without any involvement by the researcher in producing the mark.

natural selection One of several evolutionary mechanisms; it accounts for an attribute becoming more common (selected) because of some benefit that leads to more offspring being produced.

nauplius larvae Developmental series of free-swimming, nonfeeding planktonic euphausiid larval stages. It is the earliest and most basic type of euphausiid larva.

necrosis The death of cells or tissue, caused by disease or injury.

nematodes Cylindrical worm-shaped invertebrates generally known as roundworms; widespread in diverse habitats. Many are parasitic on vertebrates, including marine mammals.

neocortical Referring to the dorsal region of the cerebral cortex, which is the most recently evolved part of the mammalian brain.

neomorph A new structure produced by evolution, in a species or lineage in which no comparable structure existed previously.

neonatal Referring to or occurring in the period immediately after birth.

neonate A newborn individual.

neoplasia The process of formation of a new and abnormal growth of tissue.

nephron The functional unit of the kidney. The nephron is a tubule with regions specialized for the reabsorption and secretion of various solutes, processes which ultimately produce urine from fluid filtered from the blood. Each kidney (or reniculus) can contain millions of nephrons.

neritic Referring to or occurring in waters above the continental shelf.

net sounder An acoustic transmitting device used to try to ensure that fish schools enter a trawl. The device is usually fixed to the trawl head line and provides the ship captain with an indication of the position of the device in relation to the surface and the seabed, and also the position of fish schools underneath the device.

net whaling A historic whaling method in which whales were first driven into nets and then harpooned.

neurohypophysis Downgrowth of the base of the hypothalamus in the brain, containing the secretory extremities of neurons extending from the brain and producing the hormones oxytocin and vasopressin.

neutering Surgical removal of the testicles; often used for permanent contraception.

neutral buoyancy A state in which the upward force of buoyancy is equal to the downward force of gravity.

niche The ecological space or role occupied by a particular species, making it distinct from other species in the same ecological system.

nitrogen narcosis A marked decrease in motor coordination and mental capacities, caused by increased concentrations of nitrogen gas in the brain and other nervous tissue.

nocturnal Active at night, while sleeping or at rest during daylight hours.

nominal species A term for any properly named and described (authored) species, whether or not it is recognized today as a valid species.

nonbasilosaurid archaeocete An archaeocete in the families Pakicetidae, Ambulocetidae, Remingtonocetidae, or Protocetidae. Members of this group were formerly included in Protocetidae.

nonessential amino acids Amino acids that can be synthesized.

nonesterified fatty acids A fatty acid that is not attached to a glycerol molecule.

noninvasive sampling The collection of tissue samples in a manner that does not require penetration of the epidermis.

nonsynonymous mutations Single nucleotide substitutions that result in amino acid replacements.

nontarget species A species that is taken in by a fishery even though it is not one that the fishery is primarily concerned with catching.

Norwegian-type whaling A technique of modern whaling, using a harpoon shot from a cannon mounted on a motor-driven vessel.

nulliparous Describing a female that has had no pregnancy, as contrasted with multiparous (more than one pregnancy) and primiparous (first pregnancy).

O

obligate Those animals consigned to a particular ecological regime by their evolutionary history. Obligate river dolphins can only live in fresh water. Cf. FACULTATIVE.

observational learning A process in which animals can acquire changes in their own behavior by observing as another or others engage in some behavior and experience positive or negative consequences.

oceanic Relating to or occurring in deep water off the continental shelves.

oceanic front The point at which two masses of seawater meet that differ in temperature, salinity, or both.

Odontoceti (odontocetes) The toothed whales, including sperm and killer whales, belugas, narwhals, dolphins, and porpoises; the suborder of whales including those able to echolocate. Their skulls have premaxillary foramina, a relatively large pterygoid sinus that extends anteriorly around the nostril passage, and the maxillary bone telescoped over the supraorbital process of the frontal.

offshore killer whales A provisional name for a genetically distinct population of killer whales recently identified from British Columbia, California, and southeastern Alaska, which appears to range mostly over continental shelf waters rather than nearshore waters. They are not observed to associate with either resident or transient killer whales, and their feeding habits are unknown.

ommatidium A single functional unit of a compound eye in an arthropod.

oocyte A developing egg cell in one of two stages. The primary oocyte begins differentiation near the time of birth; maturation is arrested until after puberty, when recruitment of the egg causes further differentiation toward potential ovulation.

operant conditioning The process of teaching a specific behavior by rewarding or reinforcing the occurrences of that behavior.

optimum sustainable population A term used to describe a population the size of which lies between the carrying capacity of the local environment and the population's maximum net productivity level.

organochlorines Synthetic compounds containing chlorine atoms such as PCBs and pesticides.

organogenesis The formation of organs and organ systems.

osmolality A measure of the osmotic pressure of a solution.

osmolarity Changes in the properties of a solution that result from the number of particles that are dissolved in it. The greater the number of dissolved particles, the greater the osmolarity of the solution.

osmoregulation The process by which an organism maintains its internal salt and water composition constant.

osmotic Referring to properties of solutions that depend on concentrations of dissolved salts and other substances.

osteological Relating to the bony structures.

osteology The study of bones and their structures.

osteosclerosis Increased density of bone due to replacement of spongy or cancellous bone tissue by compact bone tissue. In several groups of marine mammals, this condition is normal and characteristic rather than pathological as it would be in humans.

Otariidae (otariids) The eared seals (sea lions and fur seals), which use their foreflippers for propulsion.

otolith One of three calcareous bodies in the ear of a bony (i.e., teleost) fish, used in orientation.

overexploitation The reduction, as a result of overfishing, of the size of a population to below the level that provides the maximum sustainable yield. Sometimes referred to as "biological overexploitation," because there can be circumstances where it is economically optimal to reduce populations below their maximum sustainable yield levels.

oxidation The process of using oxygen to liberate energy from ingested food or stored tissue, resulting in the production of water and carbon dioxide.

oxygen store The total amount of oxygen available for aerobic metabolism. This oxygen can be a gas in the lungs, bound to hemoglobin in the blood, or bound to myoglobin in muscles.

P

pachyosteosclerosis A combined condition of osteosclerosis and pachyostosis, typically found in the thoracic skeleton of sirenians.

pachyostosis An increased volume of bone, resulting in a swollen appearance of individual skeletal elements.

pack ice Pieces of unstable seawater ice, often called pans, which vary from a few meters to several hundred meters in diameter. These pieces are not attached to land; instead they drift according to local current and wind conditions.

paedomorphosis The retention of juvenile characters in the adult form.

pagophilic Associated with ice.

PAHs polycyclic aromatic hydrocarbons; large groups of naturally occurring aromatic compounds containing two or more benzene rings fused together. Also, oil compounds derived directly (petrogenic) or indirectly via combustion (pyrogenic). Some of these compounds, such as benzopyrene, are known carcinogens.

palatine Either of the two bones that make up the palate.

pan bone A thin area in the posterior section of the mandible of a dolphin.

panmictic Describing the mingling of genetic material throughout a breeding population.

panmixia The free interchange of genes within an interbreeding population.

pantropical Occurring or distributed throughout the tropical regions of the world.

parapatric Referring to populations, species, or taxa whose ranges border each other but do not overlap.

parapatry The fact of occurring in adjacent or slightly overlapping geographical areas.

paraphyletic Describing a taxonomic group that does not contain all of the descendants of the most recent common ancestor of its members.

paraphyly A state in which not all the descendants of an immediate ancestor are included in a taxon; e.g., the taxon Reptilia is paraphyletic because it does not include birds, which are now described as arising from dinosaurs.

paratenic host See TRANSPORT HOST.

paraxonic Describing the hand or foot of a four-footed animal in which the axis of symmetry passes between digits three and four.

parental investment Behavioral and energetic effort expended by a parent on offspring that benefits the offspring at a fitness cost to the parent. Such fitness costs are often measured as a reduction in fertility or as an increase in mortality due to the effort expended on offspring.

parity Number of pregnancies and/or number of times a female has given birth; from the Latin parere "to bear." Nulliparous females have never been pregnant. A primiparous female is pregnant with her first offspring or has had only one offspring. Multiparous females have had more than one pregnancy or offspring.

partial pressure The portion of the total pressure of a gas mixture attributable to one gas component.

parturition The act or fact of giving birth.

pathogen Any organism or chemical that can cause disease as a result of infection or ingestion.

pathogenic Causing disease; harmful to an organism.

PCBs polychlorinated biphenyls; a family of toxic compounds formerly used in industry and manufacturing and frequently discharged into rivers in chemical wastes.

peduncle In cetaceans, the posterior portion of the body bearing the tail, or flukes.

pelage The hair coat of mammals.

pelagic Living or occurring in the open sea; e.g., pelagic dolphins mainly inhabit the deep water of the open ocean.

pelagic trawls Bag-like nets that are towed through the upper parts of the water column rather than near the seabed.

pelagic whaling A whaling operation that does not depend on the use of a land station for processing whales, but uses a floating factory ship (mother ship) or catcher boat with factory facilities.

pelvis A bone complex normally formed from the fusion of three bones (ilium, ischium, and pelvic) that forms the pelvic girdle and attaches to the sacral vertebrae.

perilymphatic duct A duct that carries the lymphatic fluids between the inner ear and the otic space of the cranium.

perinatal Relating to or occurring in the period immediately after giving birth.

permanent ice The core area of ice around both poles that does not melt completely in summer.

periotic The bone housing the inner ear of cetaceans. This complex bone is very dense and is loosely attached to the skull in odontocetes. It is frequently found as an isolated fossil and may be diagnostic as to the level of genus or species.

Perissodactyla (perissodactyls) An order of living and extinct mammals with hooves distinctive for being "odd-toed," a condition in which the weight of the limb is transmitted along digit 3. Examples include horses, tapirs, and rhinos.

peritoneal Relating to the peritoneum, the lining of the abdominal cavity.

pes The foot, consisting of tarsal (ankle) bones, metatarsals, phalanges, and adjacent tissues.

petasma A complex membranous plate on the inner side of a single pair of limbs in the male of several crustaceans. The petasma enlarges during sexual maturation and is used during copulation to aid in the transfer of spermatophores.

P450 enzyme system A group of enzymes induced by and metabolizing a wide range of natural products (e.g., hormones) as well as unnatural chemicals (e.g., dioxins, some PCBs, dibenzofuranes).

phalanges The bones of the toes.

pharynx The anatomical area posterior to the nasal cavity and mouth, where the passage for air crosses that for food. Air passes from the nasal cavity to the nasopharyngeal duct to the pharynx to the larynx to the trachea. Food passes from the oral cavity to the pharynx to the esophagus.

phenotype 1. The visible or observable traits of an organism. 2. More generally, all the collective morphological, physiological, or behavioral aspects of the individual; an expression of the genotype as influenced by environmental factors.

phenotypic Relating to the visible body form that is exhibited by an organism.

pheromone A body secretion that affects or influences the behavior of other individuals, such as a sex attractant.

philopatric Describing individuals that tend to return to breed near the site of their birth.

philopatry The fact of mating as an adult at the site of one's own birth. The importance of philopatry to social behavior and evolution are: (a) adults interacting with one another at a breeding site may be familiar to one another over successive breeding seasons and may even be related; (b) species may show strong genetic differentiation across their geographic ranges. Cf. SITE FIDELITY.

Phocidae (phocids) A family group within the pinnipeds that includes all of the "true" seals (i.e., the "earless" species). Generally used to refer to all recent pinnipeds that are more closely related to *Phoca* than to otariids or the walrus.

Phocoenidae The family of true porpoises.

phoresy The transportation of one species by another so as to benefit the transported species.

photic zone The depth in the ocean into which natural light penetrates.

photoidentification The use of photographs to identify animals individually.

photoperiodism The fact of an animal's being affected by phase changes in a day-night or light-dark cycle.

photophore A light-emitting organ composed of a cluster of light-producing cells, a reflector, and a lens. The luminescence produced by photophores in euphausiids is presumed to be an adaptation for swarming and reproduction.

photopic vision Vision in conditions of high luminance.

phycotoxins Pathogens produced by algae.

phylogenetic Relating to the evolutionary ancestry or lineage of an organism or group of organisms.

phylogenetic analysis A method of codifying and comparing heritable biological data for a group of organisms, including molecules, soft tissues, bones, and behavioral data, to formulate a hypothesis of how the organisms are related to each other (a **phylogenetic tree**).

phylogeny The evolutionary relationships among different taxa, usually at the species level.

phylopatry See PHILOPATRY.

phytoplankton Collectively, planktonic plant life.

pigment A dark compound that imparts color to hair, eyes, and skin.

pinna A flap of skin supported by cartilage, and sometimes moved by small muscles, that projects from the side of the head in vertebrates and functions as a sound funnel. In popular use, it is called the ear, but scientists reserve the term "ear" for the much larger organ of hearing, which includes many parts not visible externally.

pinniped One of a group of aquatic, mostly marine, carnivorous mammals; includes seals, sea lions, and walruses; all their limbs are fin-like, and they spend at least some time on land or ice.

piscivorous Feeding primarily on fish.

pisiform A small bone at the junction of the ulna and the carpus (wrist).

placenta The organ of metabolic interchange between fetus and mother.

placental Describing the members of a major group of mammals whose young develop in the womb supplied with nutriment and oxygen from the mother through a placenta; in contrast to the marsupials, in which significant development takes place outside the womb after birth.

plankton Collectively, the passively floating or weakly swimming plant and animal life of a body of water; usually minute.

plantigrade **1.** Walking on the sole with the heel touching the ground. **2.** An animal that moves in this manner.

plantigrady A foot posture in which the entire palm and sole of the forefoot and hindfoot are in contact with the ground in the standing position.

play An imprecisely defined category of behavior, especially in juvenile animals; generally including activities that, while having no immediate identifiable value, may benefit the animal in some other situation or at some later stage in life.

Pleistocene The first of two epochs in the Quaternary, extending from 2 million years ago until approximately 10,000 years ago.

pleomorphic Having more than one distinct form within a single species, group, or life cycle.

pleopod An anterior abdominal appendage modified for swimming.

plesiomorphic Describing shared primitive characters inherited from a distant ancestor.

plesiomorphy The existence of shared primitive characters (**plesiomorphies**) inherited from a distant ancestor.

plicae circulares (circular folds) Folds in the epithelium of the small intestine that serve to increase the surface area.

plicae semilunares (semilunar folds) Folds of the epithelium of the large intestine produced by the longitudinal muscles.

pneumotachograph A mechanical device to measure the volume and rate of flow of air moved in and out of a mammal's lungs.

pod In resident killer whales, a group of closely related matrilines that regularly associate with each other in preference to other matrilines within the community.

polar front The border between the cold waters of the Southern Ocean and the temperate water masses of the south Atlantic, south Pacific, and Indian Oceans (formerly known as the Antarctic convergence).

polychlorinated biphenyls See PCBS.

polycyclic aromatic hydrocarbons See PAHS.

polydont Having more teeth than the generalized mammalian number.

polydonty For mammals, an increase in the number of teeth above the primitive mammalian formula of 44.

polyestrous Having more than one estrous cycle in a single year.

polygynous Describing a mating system in which one male mates with multiple females during a breeding season.

polygyny A mating system in which one male mates with several females within one breeding season. The term usually refers to a mating system in which relatively few successful males breed with most of the females, while others do not mate. The reverse, one female mating with several males, is rare among mammals.

polynya An irregular-shaped area of open water or thin, recently formed ice completely surrounded by (usually heavier) ice; often persistent or recurring; this varies greatly in size, shape, and causal factors.

polyphyletic Describing or belonging to an artificial taxonomic group containing two or more species not sharing an immediate common ancestor.

polyphyly The status of a group that does not include the most recent common ancestor of all its members.

polyunsaturated fatty acid   A fatty acid that contains two to six double bonds.

population   A group of organisms of the same species occupying a particular area at the same time.

population bottleneck   A rapid reduction in the size of a population, resulting in a loss of genetic diversity.

population stock   See STOCK.

porpoise   Any of several small cetaceans in the family Phocoenidae. The dorsal fin may be absent; teeth are spatulate or vestigial.

porpoising   The behavior of penguins, dolphins, or whales leaping at least partially clear of the water surface during rapid swimming.

postcrania   A collective term for bones that are not part of the skull, e.g., the vertebrae, limbs, hands, or feet.

postpartum   After birth; especially, shortly after birth.

postreproductive   Describing a female of certain toothed whales that ceases ovulating at an advanced age but well before her expected longevity.

practical salinity unit   The measure of salinity of seawater, approximately equal to parts per thousand.

precautionary principle   An ecological management philosophy that seeks to maintain natural resources by managing more conservatively when there is greater ignorance about the status of the resource.

precocial   Being born in an advanced state of maturation and requiring relatively little adult care. In contrast, altricial young are born in a helpless state and depend heavily on adult care.

predation   A situation in which an animal exerts time and energy to locate living prey and expends additional effort to kill and consume it.

predator   An animal that obtains its food primarily or exclusively by hunting and killing other animals.

predatory   Describing an animal that obtains its food by preying on other animals.

preening   In birds, the act of smoothing or cleaning feathers with the beak of the bill.

preformed water   Water that exists as water when it is consumed in the food or drunk.

premaxilla   One of the two major bones found in the upper jaw of mammals. The other is the maxilla.

premaxillary sacs   Premaxillary elements of the air sac system, which is a net of diverticula of the narial passages between the external nares (blowhole) and the bony nares of the skull.

premolar   A tooth behind the large canine, used to slice or chew.

prepartum   Describing or occurring in the time before birth, especially shortly before birth.

prey   Animal that is hunted and killed by another.

primary production (productivity)   Organic matter and/or energy produced or captured by organisms (plants) from inorganic compounds and incident solar radiation.

primiparous   Describing a female that gives birth for the first time, as contrasted with multiparous (more than one pregnancy) and nulliparous (no pregnancy).

primordium   See ANLAGE.

probability of identity   The probability that two different individuals have identical genotypes.

procumbent   Describing a tooth (generally apical) that tends to acquire a subhorizontal to horizontal orientation.

protocetid   A member of the family Protocetidae, a paraphyletic family that is thought to have given rise to the basilosaurids.

proximate explanation   An explanation that describes the immediate or short-term causes of a biological incident or occurrence, in contrast to ultimate explanations, which are historical (specifically, evolutionary) in nature. For example, a threat between neighboring territorial male sea otters can be said to have occurred because the two males encountered one another near their common boundary or had high levels of testosterone (proximate explanation).Or, it can be stated that the threats evolved as a low-risk alternative to fighting (ultimate explanation).

pseudocervix   Folds of tissue, usually vaginal in origin, that are distal to the true cervix, but play a similar role or barrier as the true cervix.

psychophysical test   A test in which an animal is trained to show the experimenter when it detects a physical stimulus, such as light, sound, or a chemical.

psychosphere   Global, nutrient-rich, relatively biologically unexploited deep water.

pteropod   Small molluscans with two "wings," with which they swim through the water.

pterygoid   An anatomical process descending from the sphenoid bone.

pterygoid sinus   An airspace adjacent to the flat pterygoid bones of the skull.

purse seine   A type of fishing gear that surrounds fish schools with a vertical curtain of netting that is then pursed, or closed off, at the bottom by means of ropes and heavy winches, in order to prevent fish from swimming out of the bottom of the net.

pyloric stomach   The most distal portion of the cetacean stomach, adjacent to the pyloric sphincter; contains pyloric glands.

Q

qargi   Northern Alaskan Inuit term for whalers' ceremonial house.

R

race   An interbreeding subgroup of a species that is genetically distinct and usually geographically isolated from other such groups of the same species.

radius   A bone extending from the elbow to the wrist on the outside of the forelimb, opposite the ulna.

range   The maximum extent of geographic area used by a species.

rank   A designated level in a classification scheme.

raptoria   Describing a process of feeding by seizing individual food organisms.

rare   Not common, widespread, or generalized; not generally found.

reciprocity   A behavior pattern in which individuals engage in helpful actions that will be repaid by the recipient.

refractive index   A parameter of a light-transparent medium determining the degree of light refraction at a boundary between media with different indices. The more index difference, the stronger the refraction. The refractive index of a vacuum is 1, that of air is almost the same, and that of water and some eye media (anterior chamber liquid, vitreous body) is 1.33.

regression   The mathematical relationship between two variables, e.g., body mass and brain mass.

reinforcement   An event (reward or punishment) that follows a response and that increases or decreases the likelihood that the response will recur.

reintroduction   The reestablishment of a species within its natural range.

release   The act or technique of moving an animal that has been born in captivity, or that has been in captivity for long enough that it requires retraining in the skills required to survive, back into the wild so that the animal can become free-ranging and no longer under human supervision.

relict species   Persistent remnants of formerly widespread biotas, typically existing in specific isolated areas or habitats.

remora   A type of fish that has a modified dorsal fin to suck onto other fish, marine mammals, or turtles; also called "suckerfish."

renal   Relating to or occurring in the kidney.

renal pelvis   A funnel-like structure into which urine drains from the nephrons.

reniculus   In discrete multireniculate kidneys, the anatomically independent functional units into which the kidney is divided. The reniculus resembles a miniature kidney with a cortex, medulla, vascular supply (renicular artery and vein), and ureteral tubule.

replacement yield   The number of animals, which if removed from a population stock, results in the same number of animals at the end of the year as in the beginning.

reproductive ecotype   An array of species of plants or animals adapted to the conditions of some breeding site and having inheritable features.

reproductive success   **1.** The ability of an organism to reproduce. **2.** Specifically, the number of offspring successfully reared to independence by an individual adopting a particular strategy.

reproductive value   The number of offspring that an individual of a given age can be expected to produce over the rest of its life discounted by the population growth rate.

resident killer whales   A genetically distinct population of killer whales found in coastal waters of the northeastern Pacific Ocean, which feeds preferentially on fish and squid. Cf. OFFSHORE KILLER WHALES, TRANSIENT KILLER WHALES.

resilience   The recuperative power of a population to recover from or absorb sudden environmental stress and deprivation.

resource   Any item, factor, or condition that contributes to the ability of an organism to survive and reproduce; e.g., food, space, and mates.

respiratory acidosis   A decrease in blood pH level, caused by an increase in the partial pressure of carbon dioxide that shifts the normal equilibrium of the bicarbonate-carbonic acid buffer system.

rete mirabile *plural,* **retia mirabilia**   A dense, complex network of numerous small blood vessels.

reticulation   The fact of crossing over or connection between lines of descent.

retroperitoneal   Outside of but projecting into the body cavity; completely or partially covered by peritoneum.

rhizome   An elongate, horizontal subterranean stem of a seagrass or other plant, producing shoots above and roots below and serving for the storage of reserve food material.

ritualization   An evolutionary process through which behavior becomes partly or completely "emancipated" from its original function(s) and acquires new functions for communication. For example, vocalizations presumably represent ritualized breathing.

riverine   Lving in rivers; found in or relating to a river.

rookery   A terrestrial breeding area used by a colony of pinnipeds on a seasonal or permanant basis. (From an earlier use of the term to describe the breeding or nesting area of a large number of rooks, a crowlike bird of Europe.)

rooster tail   A spray of water formed when certain small cetaceans surface at high speed; caused by a cone of water coming off the animal's head. (So named because the form of the spray resembles a rooster's feathers.)

rorqual   Any of six species of baleen whales (the minke, blue, humpback, fin, Bryde's, or sei whale) belonging to the family Balaenopteridae, characterized by a variable number of pleats that run longitudinally from the chin to near the umbilicus. The pleats expand during feeding to increase the capacity of the mouth.

rostral   Having to do with the nose or head; toward the front.

rostral hairs   Small hairs on the outside edge of the rostrum of a cetacean.

rostrum   The beak-like projection found at the front of the skull or head of a cetacean; the term also refers to the skeletal support of the upper jaw, comprising the anterior parts of the maxillary, premaxillary, and vomerine bones of the skull.

r-selection   The reproductive strategy of a species that tends to grow quickly and have a large number of offspring, but that has a relatively short life span and negligible parental involvement in the rearing of young. Certain insects are cited as an example of r-selection. Contrasted with a k-selected species, which has slower growth and a smaller number of offspring, but a longer life span and significant parental involvement.

ruminant   A hoofed mammal that feeds by grazing or browsing; it has a complex four-chambered stomach (**rumen**) and thus is able to regurgitate and then rechew its food (i.e., it "chews its cud"). Ruminants include domestic cattle, sheep, goats, deer, antelopes, and giraffes.

S

sacral vertebrae Vertebrae associated with the pelves; a feature that is not present in cetaceans and manatees.

school An aggregation of fish or other animals that regularly swim together as a unit.

scotoma An area of diminished vision within the visual field.

scotopic vision Vision in conditions of dim luminance.

scramble competition A mechanism of sexual selection in which males compete for access to females by searching and locating females more efficiently than their rivals. Females may mate with one or several males.

sea cow A strictly marine herbivorous mammal foraging near the surface, primarily on large marine algae, and propelling itself with a triangular caudal fluke. The term is also used loosely for any member of the order Sirenia.

seagrass(es) Any of various marine flowering plants belonging to the families Potamogetonaceae and Hydrocharitaceae, consisting of 12 genera and some 50 species.

sea ice Frozen seawater, which forms ice floes (i.e., pack ice).

seal bomb An explosive noise-making device designed to chase marine mammals away from areas where they are regarded as harmful to fisheries, nets, and so on.

seasonality The fact of being affected by or dependent on a change of seasons; e.g., changes in population distribution that are related to changes in seasons.

sebaceous glands Glands located at the base of mammalian hairs that release an oily secretion, sebum.

secondary sexual characteristic A feature of a male or female animal that is produced as sexual maturity occurs.

sedimentology The scientific study of sediments, solid fragmentary material transported and deposited by wind, water, or ice, or precipitated from solution, or secreted by various organisms.

selection The evolutionary process favoring individuals with genotypes that enable them to raise the greatest number of young which survive to breed.

selection coefficient A mathematical term indicating the decrease in disfavored gene types contributed to the next generation; e.g., a selection coefficient of 0.5 for a given genotype indicates a decrease of 50% of that genotype entering the next generation.

semiaquatic Only partly aquatic, e.g., an animal that forages for food in water but otherwise lives and breeds on land.

seminiferous tubules Convoluted loops within the testis where spermatozoa are formed.

sensillum A sensory receptor or a receptor complex in the cuticle of an invertebrate.

sensitization An increased behavioral (or physiological) responsiveness occurring over time, as an animal learns that a repeated or ongoing stimulus has significant consequences. Cf. HABITUATION.

sensor A device that reacts to a certain physical stimulus in a quantifiable manner.

sensory adaptation The lessening of a response to a stimulus due to physiological fatigue.

sentience The ability to sense; in particular, the capacity to experience pleasure and pain.

serial carpus and tarsus An arrangement of the carpal and tarsal bones in a serial way; i.e., the pattern of relations between carpal and tarsal bones is "pillar-like." Each bone of the first row articulates mostly with only one bone just under. This is a derived condition compared to the generalized mammalian condition. Cf. ALTERNATING CARPUS AND TARSUS.

sessile Stationary or fixed, e.g., plants growing on the sea floor or mussels attached to rocks.

sex chromosomes Nuclear chromosomes that are involved in sex determination, e.g., the X and Y chromosomes in mammals.

sexual bimaturation The sexual maturing of one sex significantly earlier than the other.

sexual dimorphism The development of physical and other differences between the sexes of the same species, especially when this difference extends beyond the reproductive organs to overall size, color, and the development of special features; e.g., male elephant seals are vastly larger than the females, with enlarged canine teeth, and thickened skin for fighting other males in competition for mating opportunities.

sexual selection Selection due to the advantage some individuals have over others of the same sex in exclusive relation to reproduction. This may lead to the evolution of costly secondary sexual characters found only in one sex (such as a male peacock's tail or a stag's antlers), which may decrease the probability of survival and which therefore is opposed by natural selection.

shelf A sea zone located close to land and having the same geological structure. A shelf zone is 100–200 m deep.

sibling species A species that is morphologically very similar to another, sometimes to the point of seeming identity, but demonstrated to be a separate species on the grounds of genetic, behavioral, or cryptic morphological evidence.

sister group The closest relative with which a taxon has a common ancestor.

site fidelity The tendency to return to mate at the same site repeatedly in different breeding seasons. This term is used instead of PHILOPATRY if the animal's original birth site is not known.

size index An analog of the ENCEPHALIZATION INDEX, related to single brain components rather than the total brain.

small-type whaling Japanese coastal whaling that uses a vessel and cannon of small size. Operated since the beginning of the 20th century for minke and toothed whales (other than sperm whale); came under the control of the Japanese government in 1947.

soak time A term for the length of time that a piece of fishing gear such as a gill net is left in the water to fish.

social learning The social transfer of information and skill among individuals by processes that include imitation, teaching, exposure, social support, matched dependent learning, stimulus enhancement, observational conditioning, and goal emulation.

somersault A behavior involving an acrobatic stunt in which the body rolls in a complete circle, head over heels, or, in the case of dolphins, head over tail.

sonar   **1.** An acoustic system or device that emits sound into water or another medium, detects echoes received from objects, and processes those signals to characterize and/or visualize the objects. **2.** In marine mammals, a means of detecting, identifying, and tracking prey by the emission of sound pulses and the analysis of the resulting echo patterns.

song   Among marine mammals, a series of sounds repeated in a specific pattern.

sonic lens   An anatomical structure that focuses sound as optical lenses focus light, e.g., the melon of odontocete whales.

sound   A form of energy manifested by small pressure variations in a medium (such as water or air). Depending on the strength and frequency of the sound, it may be detectable by an organ of hearing.

source level   The acoustic pressure that would be measured at a standard distance (usually 1 m) from a point source radiating the same amount of sound as the actual source.

spectral sensitivity   Selective sensitivity of visual receptors to a certain region of the light wavelength spectrum that corresponds to different light colors.

spectrogram   A graphic representation of sound waves per unit of time.

spermaceti   A white, translucent substance, obtained from the distinctive form of oil made of wax esters and triglycerides that is found in the spermaceti organ of sperm whales.

spermaceti organ   An elongated, barrel-shaped structure making up much of the sperm whale's head, made of soft spongy tissue filled with spermaceti oil.

spermatogenesis   The process by which sperm develop and mature from primitive sperm cells (spermatogonia and subsequently spermatids).

spermatophore   A collection of spermatozoa enclosed in a sheath of gelatinous material.

sperm competition   A mechanism of sexual selection in which a female mates with more than one male, and males compete to fertilize her eggs by producing more sperm per ejaculation, or by mating with her more frequently than their rivals.

sphincter   A muscle that forms a ring and contracts upon itself.

splanchnic   Pertaining to the viscera, such as the liver, stomach, and pancreas.

sporta perimedullaris   A sheet-like structure of connective tissue and muscle that lies between the renal cortex and the medulla.

spout   A term for a cloud of vapor and sea water mixed with air that is exhaled by cetaceans.

sprint   A high-speed, maximum effort or movement of short duration.

spyhopping   A behavior that involves raising the head vertically out of the water, then sinking below the surface without much splash.

stable isotope   An alternate form of an element that is more persistent than the common form.

statistical power   The ability of an analysis to correctly reject the null hypothesis, i.e., not to miss a real effect.

stem group   One extinct species, or several extinct species, that form an extinct sister group, or sequence of sister groups, to the crown group.

stereoscopic vision   A capability of vision to discriminate the distance to an object based on a small difference of images (disparity) projected on the retinae of two eyes. This ability requires binocular vision.

stereotyped behavior   A pattern of repetitive movements, such as pacing back and forth or swaying to and fro, that often are performed by captive animals held in impoverished environments, and that are not associated with any behaviors normally performed by these animals in the wild.

stochasticity   A random occurrence of events in space and/or time; random variability.

stock   A group of animals of the same species or smaller taxa, existing in the same locale and interbreeding when mature.

stored reserves   Body components that are supplementary to immediate requirements, most often of fat and blubber.

stranding   A situation in which a marine mammal is helpless. This generally refers to an animal that comes ashore accidentally or because of injury or illness.

strandling   A term for any marine animal that cannot cope with its immediate situation, either ashore or at sea (e.g., a beached whale, an Arctic seal in Florida waters, or a sea otter drenched with oil).

strategy   Genetically distinctive rules for individual behavior. Selection favors the strategy that yields the highest reproductive success in relation to competing strategies.

stratification   A vertical density gradient caused by changes in temperature and salinity with depth.

streamlining   Body shape that causes little distortion to the fluid flow around it and minimizes drag.

striated muscle   The skeletal or locomotor muscles of the body.

strip transect sampling   A sample count method in which counts are conducted within long, narrow strips.

subarachnoid space   A space under the arachnoid membrane, one of the coverings of the brain.

subarctic   Living in an intermediate zone between arctic and temperate waters.

subspecies   A formally described and named geographic entity within a species.

supraoccipital   A bone at the top of the rear of the skull.

surface-active group   A group of right whales, typically consisting of one female and one to many males, engaged in vigorous physical interactions at the surface, which appear to be courtship or other social behavior.

surfing   The behavior of riding on a wave or waves. There are two basic patterns: floating on top of the water and falling down the advancing edge of a wave or riding within a wave being created by water being pushed upward by the movement of a denser medium (such as a whale or vessel). Dolphins engage in both types of surfing or "wave riding."

survival rate   The probability of an animal remaining alive over a given time period; typically expressed as the proportion of individuals in a population surviving from one period (usually defined as a year) to the next.

sustainability   **1.** The combined capacity of a system and its parts to retain their interactions and other natural qualities. **2.** Specifically, the level of human activity in a given ecosystem that can continue without degrading the ecosystem or seriously depleting resources.

sustained swimming Long duration activity generally performed at low speeds.

sympatric Occurring together geographically; coexisting in the same place.

sympatry **1.** The fact of occurring or living in the same geographical area. **2.** Specifically, the existence of two or more populations of closely related species in identical or broadly overlapping geographical areas.

symphysis The merging of two bones.

synapomorphic Sharing a derived character state.

synapomorphy A shared derived character that is the result of common ancestry.

synonymous mutations Single nucleotide substitutions that do not result in amino acid replacements.

systematics **1.** A method of classifying organisms that strives to establish its categories on the basis of evolutionary history and relationships. **2.** The scientific study of such evolutionary relationships.

systemic Broad or universal in application while accounting for everything of importance in proportion to its importance; applying to, and accounting for, complexity.

systole The period of ventricular contraction during the contraction–relaxation cycle of the heart.

T

tachycardia A heart rate above the resting rate.

tag Any artificial mark attached to animals.

tail-slapping Another term for LOBTAILING, sometimes used when referring to dolphins.

taxon *plural,* **taxa** A natural grouping of related organisms, usually named, and of a given rank (e.g., order, family, genus, species, subspecies).

taxonomic(al) Based on or describing the classification of organisms according to species or higher levels.

taxonomy The science of classifying organisms into species, genera, and higher categories.

telemetry The measurement of biological data through transmission or reflection of sound waves or radio waves from an animal to a distant receiver.

telescoping A term for the derived condition in both mysticetes and odontocetes wherein the cranial bones of the rostrum extend posteriorly over the facial region and the bones of the occiput thrust forward toward the cranial vertex.

telson An unpaired terminal abdominal segment of crustaceans.

temporal fossa A cavity or depression in the pair of compound bones that form the sides and base of the skull.

temporary threshold shift A temporary impairment in hearing capability caused by exposure to strong sounds.

tension The partial pressure of a particular gas in solution.

terrestrial On land; occurring or living on land rather than in water or in the air.

territorialism The fact of establishing and defending a territory (see below).

territory A certain roughly delineated area that is aggressively defended by an individual or group against intruders of the same or different species; e.g., a breeding ground that adult males maintain by challenging, threatening, and fighting other males, and in which they mate without harassment from other males.

Tertiary A period of geological time from about 65 million to 2 million years before the present day.

testosterone The primary male hormone, produced in the testis.

teutophagous Feeding on squid.

thelycum A female organ to which spermatophores are attached with a sticky fluid.

thermal insulator A material, such as fur or blubber, that serves to retain heat in an animal.

thermocline In seawater, a steep vertical temperature gradient below a well-mixed surface layer.

thermohaline Relating to the temperature and salinity of seawater.

thermoneutral Falling within the range of temperatures at which thermoregulation can be maintained.

thermoneutral zone The range of environmental temperatures at which an animal does not require an increase in metabolism to maintain its body temperature.

thermoregulation The control of body temperature through the balance of heat production and loss; the ability of an animal to regulate its body temperature.

thoracic Describing the region within the thorax, or the portion of the vertebral column within the thorax.

thoracic vertebrae Rib-bearing vertebrae.

thorax The part of the vertebrate body that contains the heart and the lungs, in mammals divided from the abdomen by the diaphragm.

threatened A designation for a species that is likely to become endangered within the foreseeable future throughout all or a significant portion of its range in the wild.

time-activity budget A quantitative description of the amount of time spent by animals in different behavioral activities. Energetic costs of different behavioral activities are used to construct such budgets.

time-depth (time-data) recorder A mechanical or electronic instrument that measures and stores information on the diving and swimming behavior and environmental variables of aquatic mammals, birds, fish and turtles.

toothed whale See ODONTOCETI.

tooth replacement The shedding of deciduous (or milk) teeth followed by the eruption of permanent teeth.

top-down control The consumer-mediated regulation of the structure and function of an ecological community.

transient killer whales A population of killer whales found in coastal waters of the northeastern Pacific that is sympatric with residents and offshores, but socially and reproductively isolated from them, and genetically distinct. Transients feed preferentially on marine mammals and seabirds.

transport host A host infected by a larval parasite stage in which the parasite survives, and can be transmitted to another host, but does not develop. Transport hosts are not required to complete the life cycle but enhance the chances of transmission to the definitive host.

transverse septum   A membrane that functions as a separator between the heart and the liver.

trawl   A bag-like net that is dragged through the water in order to catch fish.

trematode   Any parasitic flatworm of the class Trematoda, especially a fluke, equipped with hooks or suckers.

triglyceride   An ester of glycerol that is combined with three fatty acids. It is the primary chemical form in which lipids are stored in the tissue.

trophic   Relating to the process of feeding or feeding patterns.

trophic cascade   Interactions among trophic levels through top-down control.

trophic flows   The transfer of food or energy between the different groups of species comprising a food web.

true seals   See EARLESS SEALS.

trying down   The traditional practice of cooking blubber to extract the valuable oil; rendering.

tubercule   **1.** One of a series of small, raised epidermal protuberances found on the dorsal fin of most porpoise species. Their distinctness, density, and covering area vary among species. **2.** A knob-like structure of unknown (but probably sensory) function, found on the head and jaws of humpback whales. Each tubercle contains a single hair (vibrissa).

tumor   A swelling, especially from an abnormal growth of tissue.

tuna purse seine   See PURSE SEINE.

tunica albuginea   A white fibrous layer covering an anatomical structure.

tunicate   A member of a subphylum of chordates containing the classes Ascidiacea (the sessile sea squirt), the free-swimming tadpole-like Larvacea, and the Thaliacea (free-swimming salps). Chordate features (i.e., notochord and nerve chord) are found only in larva and are generally lost in the adult. The adult secretes a tough cellulose sac (tunic) in which the animal is embedded.

turbid   Exhibiting reduced water clarity because of the presence of suspended matter.

turbulent flow   An unsteady and eddying flow of water. Turbulent boundary layers are thicker than laminar boundary layers and likely to separate prematurely.

tympanic cavity   A large, irregularly shaped cavity of the middle ear.

tympanoperiotic earbones   Very dense bones at the base of the skull consisting of the ventral ectotympanic bone or bulla, which encloses the middle ear cavity, and the more dorsal periotic or petrosal bone, which contains the cochlear apparatus of the inner ear and the balance organ (vestibulum plus semicircular canals). The structure of tympanoperiotic bones is used widely to indicate taxonomic relationships of both extant and fossil cetaceans.

U

ugruk   Northern Alaskan Inuit term for a bearded seal.

ultimate explanation   See PROXIMATE EXPLANTION.

ultrasonic   Describing acoustic signals that are above the normal human range of hearing.

ultrasonography   The visualization of deep structures of the body by recording and displaying the echoes of pulse of ultrasonic waves directed into the tissues.

ultrasound   **1.** A sound that is higher in frequency than the maximum audible to humans, usually said to be 20 kHz. Many marine mammals can hear ultrasound. **2.** see ULTRASONOGRAPHY.

umialik   Northern Alaskan Inuit term for a whaling captain.

umiaq *plural,* **umiat**   Northern Alaskan Inuit term for a skin boat used to hunt whales, seals, and walrus.

underhair   A layer of shorter, finer hair that provides better insulation than the longer, thicker hair (called guard hairs) covering it.

ungual   Having claws; claw-bearing.

ungulate   Any of a taxonomic group (Ungulata) comprising the hoofed mammals and their derivatives, including the living Artiodactyla, Cetacea, Perissodactyla, Hyracoidea, Sirenia, and Proboscidea, together with the extinct Desmostylia and numerous other fossil groups.

unguligrade   Referring to limbs that are elongated and slim with an efficient fore-and-aft drive and nearly no capability of rotating. Unguligrade limbs are typical of ungulate mammals.

upwelling   A process by which ocean water rises from a deeper to a shallower level, usually by divergence and offshore currents. This process often occurs along the edges of continental shelves and brings nutrient-rich water to the surface.

ureter   The duct that carries urine from the kidney to the urinary bladder.

urethra   A duct or canal in the penis through which urine and semen are transported.

uropod   Fan-shaped paired appendages on the penultimate segment of euphausiids, used for swimming.

Ursidae   A family of arctoid carnivorans that includes the subfamilies Ursinae, Hemicyoninae, and Amphicynodontinae, the latter two being represented only by fossil taxa.

V

vacuole   A fluid-filled space in a cell, usually small in animal cells; several vacuoles coalesce in the developing follicle to produce the antrum.

vascular   Describing the system of vessels in the body that carry blood. The arteries, veins, and capillaries are part of the vascular system.

vasculitis   The inflammation of a blood vessel.

vasopressin   A peptide hormone (antidiuretic hormone) that is released from the posterior pituitary gland. It increases the permeability of the collecting duct, thereby suppressing urine formation and output.

venae cavae   Major veins returning blood from the periphery to the heart.

ventral Toward the front; in anatomy, toward, or relatively near to, the belly and the corresponding surface of the head, neck, thorax, and tail.

ventral grooves A series of parallel grooves or pleats running longitudinally on the undersurface of the throat and chest region in balaenopterid whales, allowing great expansion of the mouth during feeding.

vertebra *plural* **vertebrae** A bone of the spine; a back bone.

vertebral canal The space within the vertebral column that houses the spinal cord.

vertebral column The row of bones forming the longitudinal axis of the body; the backbone.

vertebrate One of the Vertebrata, the subpyhlum including those animals having a backbone; i.e., fish, amphibians, reptiles, birds, and mammals.

vertex An elevated conjunction of the nasal, frontal, and supraoccipital bones at the top of the skull.

vestigial tooth A tooth that has no function in food processing. Usually these teeth do not erupt out of the jawbone or do not break through the dental gums.

vibrissae Large, tactile, sensory hairs present on the upper lip of mammals.

vicariance event The development of a natural biogeographical barrier causing groups that previously occupied the same area or adjacent areas to become physically separated.

vicariant Describing a major geological or oceanic split in habitat, such as the geological closure of a strait, thus splitting distributions of many organisms.

vicariant speciation Evolution of species triggered by fragmentation of the ancestral species' distribution area.

villus A finger-like anatomical projection that increases the surface area greatly and facilitates exchange between two media (e.g., the surface of the small intestine and its contents) or tissues (e.g., the maternal and fetal parts of the placenta).

visual acuity Sharpness of vision; the capability to discriminate visual pattern details.

vital rate Any of various key indicators describing the dynamics of animal populations, including birth rates, recruitment rates, mortality rates, immigration, and emigration.

vomeronasal organ An organ for chemical reception situated in the anterior part of the roof of the mouth; anatomically and functionally distinct from the senses of smell and taste; well developed in many species of mammals, in which it is especially important in reproduction.

vortex The rotating motion of a fluid around a central core.

vulnerable A designation for a species that is not regarded as critically endangered or endangered at present, but that is facing a high risk of extinction in the wild in the medium-term future.

W

weaning The end of the lactation period; the process of changing from milk to a solid diet in juvenile mammals.

whale A term commonly used for a subset of the mammalian order Cetacea. In general, it includes the larger members of the order, but the term is inconsistently used for animals of vastly different sizes that are not closely related. In some cases, it refers to the Great Whales, i.e., the baleen whales and the (unrelated) sperm whale. Others use whale to include larger members of the dolphin family (such as the killer whale). In that usage, some larger dolphins (such as the bottlenose dolphin) are well above the smallest whales (such as pygmy sperm whales) in size. Paleontologists use the word whale to describe all Eocene forms (some of which are no larger than a dog) and many modern cetaceans. Thus the term whale is scientifically imprecise, and subjects relating to it can be found in this work under more specific headings.

whalebone See BALEEN.

whale louse *plural,* **whale lice** An amphipod crustacean of the family Cyamidae; adapted for living in crevices and other secure places on the skin of cetaceans, on which whale lice largely feed.

whelping The process of giving birth; used to refer to seals.

windkessel A capacitance blood vessel, which due to elastic properties in its wall, first expands and accepts blood volume on blood entry and then passes blood onward due to recoil within its wall.

X

xenobiotic **1.** A chemical compound that is foreign to biological systems, especially one that is toxic. **2.** Relating to or describing a compound of this type.

Z

ziphiid Any of at least 19 extant species of beaked whale, of the family Ziphiidae.

zona glomerulosa The outermost layer of the adrenal cortex principally responsible for the production of the mineralocorticoid aldosterone.

zoonosis *plural,* **zoonoses** **1.** A disease of animals that is transmittable to humans. **2.** The process of disease transfer from animals to humans.

INDEX

E

I

S

Y

Z